Recommended Dietary Allowances (RDA) and Adequate Intakes (AI) for Vitamins

Age (yr)	Thiamin RDA (mg/day)	Riboflavin RDA (mg/day)	Niacin RDA (mg/day)[a]	Biotin AI (µg/day)	Pantothenic Acid AI (mg/day)	Vitamin B_6 RDA (mg/day)	Folate RDA (µg/day)[b]	Vitamin B_{12} RDA (µg/day)	Choline AI (mg/day)	Vitamin C RDA (mg/day)	Vitamin A RDA (µg/day)[c]	Vitamin D RDA (IU/day)[d]	Vitamin E RDA (mg/day)[e]	Vitamin K AI (µg/day)
Infants														
0–0.5	0.2	0.3	2	5	1.7	0.1	65	0.4	125	40	400	400 (10 µg)	4	2.0
0.5–1	0.3	0.4	4	6	1.8	0.3	80	0.5	150	50	500	400 (10 µg)	5	2.5
Children														
1–3	0.5	0.5	6	8	2	0.5	150	0.9	200	15	300	600 (15 µg)	6	30
4–8	0.6	0.6	8	12	3	0.6	200	1.2	250	25	400	600 (15 µg)	7	55
Males														
9–13	0.9	0.9	12	20	4	1.0	300	1.8	375	45	600	600 (15 µg)	11	60
14–18	1.2	1.3	16	25	5	1.3	400	2.4	550	75	900	600 (15 µg)	15	75
19–30	1.2	1.3	16	30	5	1.3	400	2.4	550	90	900	600 (15 µg)	15	120
31–50	1.2	1.3	16	30	5	1.3	400	2.4	550	90	900	600 (15 µg)	15	120
51–70	1.2	1.3	16	30	5	1.7	400	2.4	550	90	900	600 (15 µg)	15	120
>70	1.2	1.3	16	30	5	1.7	400	2.4	550	90	900	800 (20 µg)	15	120
Females														
9–13	0.9	0.9	12	20	4	1.0	300	1.8	375	45	600	600 (15 µg)	11	60
14–18	1.0	1.0	14	25	5	1.2	400	2.4	400	65	700	600 (15 µg)	15	75
19–30	1.1	1.1	14	30	5	1.3	400	2.4	425	75	700	600 (15 µg)	15	90
31–50	1.1	1.1	14	30	5	1.3	400	2.4	425	75	700	600 (15 µg)	15	90
51–70	1.1	1.1	14	30	5	1.5	400	2.4	425	75	700	600 (15 µg)	15	90
>70	1.1	1.1	14	30	5	1.5	400	2.4	425	75	700	800 (20 µg)	15	90
Pregnancy														
≤18	1.4	1.4	18	30	6	1.9	600	2.6	450	80	750	600 (15 µg)	15	75
19–30	1.4	1.4	18	30	6	1.9	600	2.6	450	85	770	600 (15 µg)	15	90
31–50	1.4	1.4	18	30	6	1.9	600	2.6	450	85	770	600 (15 µg)	15	90
Lactation														
≤18	1.4	1.6	17	35	7	2.0	500	2.8	550	115	1200	600 (15 µg)	19	75
19–30	1.4	1.6	17	35	7	2.0	500	2.8	550	120	1300	600 (15 µg)	19	90
31–50	1.4	1.6	17	35	7	2.0	500	2.8	550	120	1300	600 (15 µg)	19	90

NOTE: For all nutrients, values for infants are AI. The table on page Y defines units of nutrient measure.

[a]*Niacin recommendations are expressed as niacin equivalents (NE), except for recommendations for infants younger than 6 months, which are expressed as preformed niacin.*

[b]*Folate recommendations are expressed as dietary folate equivalents (DFE).*

[c]*Vitamin A recommendations are expressed as retinol activity equivalents (RAE).*

[d]*Vitamin D recommendations are expressed as cholecalciferol and assume an absence of adequate exposure to sunlight. Pregnant or lactating girls ages 14-18 also need 15 micrograms vitamin D per day.*

[e]*Vitamin E recommendations are expressed as α-tocopherol.*

Recommended Dietary Allowances (RDA) and Adequate Intakes (AI) for Minerals

Age (yr)	Sodium AI (mg/day)	Chloride AI (mg/day)	Potassium AI (mg/day)	Calcium RDA (mg/day)	Phosphorus RDA (mg/day)	Magnesium RDA (mg/day)	Iron RDA (mg/day)	Zinc RDA (mg/day)	Iodine RDA (µg/day)	Selenium RDA (µg/day)	Copper RDA (µg/day)	Manganese AI (mg/day)	Fluoride AI (mg/day)	Chromium AI (µg/day)	Molybdenum RDA (µg/day)
Infants															
0–0.5	120	180	400	200	100	30	0.27	2	110	15	200	0.003	0.01	0.2	2
0.5–1	370	570	700	260	275	75	11	3	130	20	220	0.6	0.5	5.5	3
Children															
1–3	1000	1500	3000	700	460	80	7	3	90	20	340	1.2	0.7	11	17
4–8	1200	1900	3800	1000	500	130	10	5	90	30	440	1.5	1.0	15	22
Males															
9–13	1500	2300	4500	1300	1250	240	8	8	120	40	700	1.9	2	25	34
14–18	1500	2300	4700	1300	1250	410	11	11	150	55	890	2.2	3	35	43
19–30	1500	2300	4700	1000	700	400	8	11	150	55	900	2.3	4	35	45
31–50	1500	2300	4700	1000	700	420	8	11	150	55	900	2.3	4	35	45
51–70	1300	2000	4700	1000	700	420	8	11	150	55	900	2.3	4	30	45
>70	1200	1800	4700	1200	700	420	8	11	150	55	[illegible]	2.3	4	30	45
Females															
9–13	1500	2300	4500	1300	1250	240	8	8	120	40	[illegible]	[illegible]	[illegible]	[illegible]	[illegible]
14–18	1500	2300	4700	1300	1250	360	15	9	150	55	[illegible]	[illegible]	[illegible]	[illegible]	[illegible]
19–30	1500	2300	4700	1000	700	310	18	8	150	55	[illegible]	[illegible]	[illegible]	[illegible]	[illegible]
31–50	1500	2300	4700	1000	700	320	18	8	150	55	[illegible]	[illegible]	[illegible]	[illegible]	[illegible]
51–70	1300	2000	4700	1200	700	320	8	8	150	55	[illegible]	[illegible]	[illegible]	[illegible]	[illegible]
>70	1200	1800	4700	1200	700	320	8	8	150	55	[illegible]	[illegible]	[illegible]	[illegible]	[illegible]
Pregnancy															
≤18	1500	2300	4700	1300	1250	400	27	12	220	60	1000	2.0	[illegible]	[illegible]	[illegible]
19–30	1500	2300	4700	1000	700	350	27	11	220	60	1000	2.0	3	30	[illegible]
31–50	1500	2300	4700	1000	700	360	27	11	220	60	1000	2.0	3	30	50
Lactation															
≤18	1500	2300	5100	1300	1250	360	10	13	290	70	1300	2.6	3	44	50
19–30	1500	2300	5100	1000	700	310	9	12	290	70	1300	2.6	3	45	50
31–50	1500	2300	5100	1000	700	320	9	12	290	70	1300	2.6	3	45	50

NOTE: For all nutrients, values for infants are AI.

Tolerable Upper Intake Levels (UL) for Vitamins

Age (yr)	Niacin (mg/day)[a]	Vitamin B_6 (mg/day)	Folate (µg/day)[a]	Choline (mg/day)	Vitamin C (mg/day)	Vitamin A (µg/day)[b]	Vitamin D (IU/day)	Vitamin E (mg/day)[c]
Infants								
0–0.5	—	—	—	—	—	600	1000 (25 µg)	—
0.5–1	—	—	—	—	—	600	1500 (38 µg)	—
Children								
1–3	10	30	300	1000	400	600	2500 (63 µg)	200
4–8	15	40	400	1000	650	900	3000 (75 µg)	300
9–13	20	60	600	2000	1200	1700	4000 (100 µg)	600
Adolescents								
14–18	30	80	800	3000	1800	2800	4000 (100 µg)	800
Adults								
19–70	35	100	1000	3500	2000	3000	4000 (100 µg)	1000
>70	35	100	1000	3500	2000	3000	4000 (100 µg)	1000
Pregnancy								
≤18	30	80	800	3000	1800	2800	4000 (100 µg)	800
19–50	35	100	1000	3500	2000	3000	4000 (100 µg)	1000
Lactation								
≤18	30	80	800	3000	1800	2800	4000 (100 µg)	800
19–50	35	100	1000	3500	2000	3000	4000 (100 µg)	1000

[a]The UL for niacin and folate apply to synthetic forms obtained from supplements, fortified foods, or a combination of the two.

[b]The UL for vitamin A applies to the preformed vitamin only.

[c]The UL for vitamin E applies to any form of supplemental α-tocopherol, fortified foods, or a combination of the two.

Tolerable Upper Intake Levels (UL) for Minerals

Age (yr)	Sodium (mg/day)	Chloride (mg/day)	Calcium (mg/day)	Phosphorus (mg/day)	Magnesium (mg/day)[d]	Iron (mg/day)	Zinc (mg/day)	Iodine (µg/day)	Selenium (µg/day)	Copper (µg/day)	Manganese (mg/day)	Fluoride (mg/day)	Molybdenum (µg/day)	Boron (mg/day)	Nickel (mg/day)	Vanadium (mg/day)
Infants																
0–0.5	—	—	1000	—	—	40	4	—	45	—	—	0.7	—	—	—	—
0.5–1	—	—	1500	—	—	40	5	—	60	—	—	0.9	—	—	—	—
Children																
1–3	1500	2300	2500	3000	65	40	7	200	90	1000	2	1.3	300	3	0.2	—
4–8	1900	2900	2500	3000	110	40	12	300	150	3000	3	2.2	600	6	0.3	—
9–13	2200	3400	3000	4000	350	40	23	600	280	5000	6	10	1100	11	0.6	—
Adolescents																
14–18	2300	3600	3000	4000	350	45	34	900	400	8000	9	10	1700	17	1.0	—
Adults																
19–50	2300	3600	2500	4000	350	45	40	1100	400	10,000	11	10	2000	20	1.0	1.8
51–70	2300	3600	2000	4000	350	45	40	1100	400	10,000	11	10	2000	20	1.0	1.8
>70	2300	3600	2000	3000	350	45	40	1100	400	10,000	11	10	2000	20	1.0	1.8
Pregnancy																
≤18	2300	3600	3000	3500	350	45	34	900	400	8000	9	10	1700	17	1.0	—
19–50	2300	3600	2500	3500	350	45	40	1100	400	10,000	11	10	2000	20	1.0	—
Lactation																
≤18	2300	3600	3000	4000	350	45	34	900	400	8000	9	10	1700	17	1.0	—
19–50	2300	3600	2500	4000	350	45	40	1100	400	10,000	11	10	2000	20	1.0	—

[d]The UL for magnesium applies to synthetic forms obtained from supplements or drugs only.

NOTE: An Upper Limit was not established for vitamins and minerals not listed and for those age groups listed with a dash (—) because of a lack of data, not because these nutrients are safe to consume at any level of intake. All nutrients can have adverse effects when intakes are excessive.

SOURCE: Adapted with permission from the Dietary Reference Intakes series, National Academies Press. Copyright 1997, 1998, 2000, 2001, 2002, 2005, 2011 by the National Academies of Sciences.

MindTap Student User Guide

MindTap™

Welcome to MindTap's Student User Guide! This guide will help you get started with MindTap by providing in depth, step-by-step instructions, created specifically for the student user.

Contents

CENGAGE LEARNING
DIGITAL COURSE SUPPORT

Benefits of Using MindTap

MindTap from Cengage Learning represents a new approach to a highly personalized, online learning platform. A cloud-based learning solution, MindTap combines all of your learning tools - readings, multimedia, activities and assessments into a singular Learning Path that guides you through the curriculum. Instructors personalize the experience by customizing the presentation of these learning tools to their students; even seamlessly introducing their own content into the Learning Path via "apps" that integrate into the MindTap platform.

Where to Buy

- Bookstore: **MindTap is available bundled with your textbook for a valuable savings at the bookstore.**

First Time Login

To get started, navigate to: login.cengagebrain.com.
➔If this is your first time using a Cengage Learning Product, you will need to create a new account.

Create Profile

➔ Now you will need to enter the MindTap access code which came with your text.

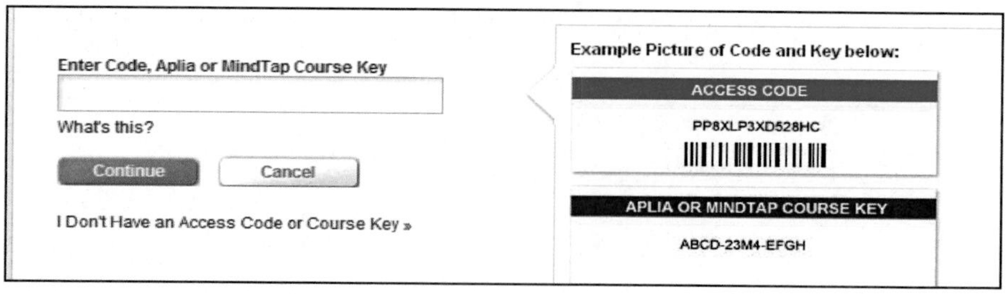

➔ Follow the directions and create a Cengage Learning student account.

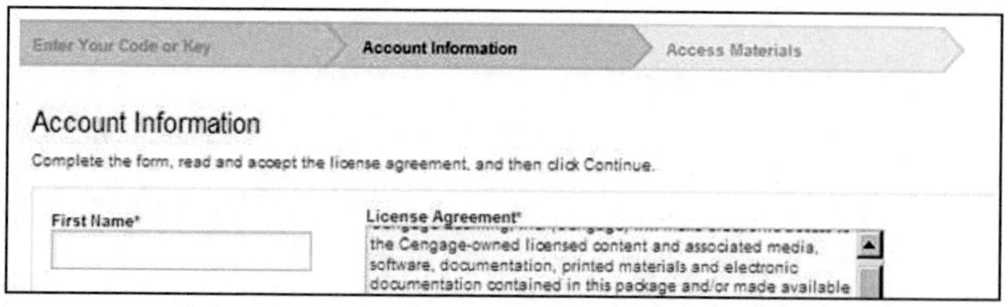

➔ Once completed, you will be directed to your dashboard.

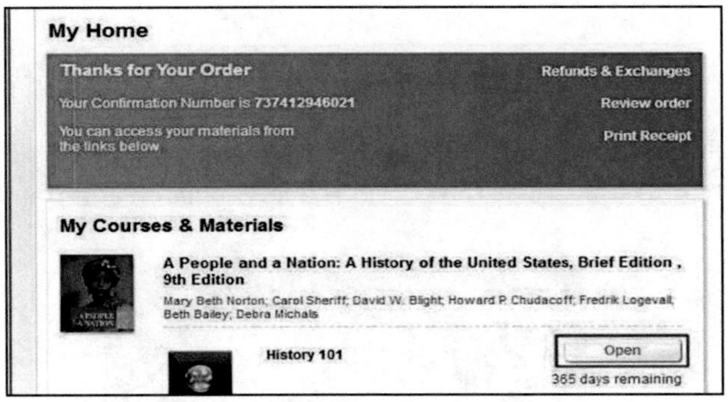

Enter Your Course Key

➔ To access the MindTap product which you just registered, press 

➔ Enter your instructor-provided MindTap CourseKey.

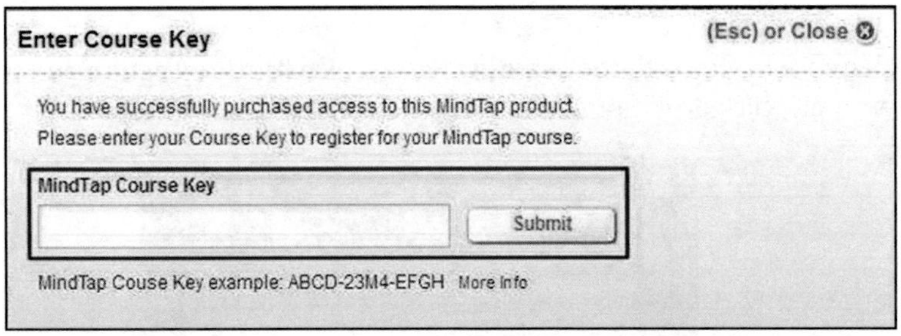

➔ Confirm that the Course Key corresponds to the correct course.

➔ Choose Continue and you are bought directly into your MindTap Course!

Returning User

To get started, navigate to: login.cengagebrain.com. Please log in to your CengageBrain account to activate your product. If you have forgotten your password please select the "Forgot Password" link to retrieve your password.

Walkthrough

Home Page – The Personal Learning Experience

➔ MindTap is a customizable, complete solution that is more than an eBook and different from any learning management system you've ever used. MindTap delivers courses built on authoritative Cengage Learning content.

Learning Path

➔ The MindTap experience begins with the **Learning Path**, which lists Pre-Lecture and Post-Lecture resources for every chapter. For example, click on Chapter 2 for examples of both resources.

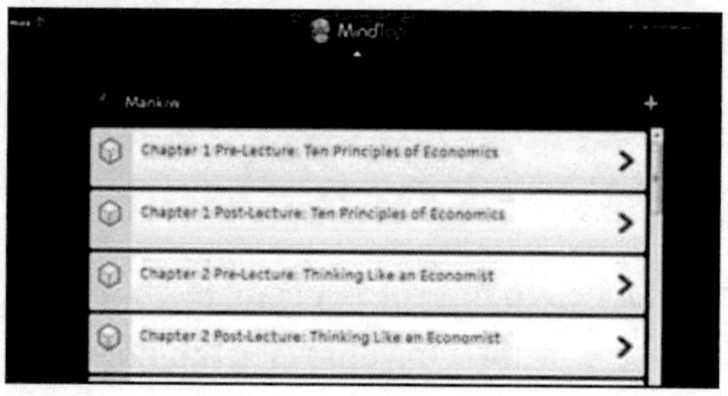

Pre-Lecture

Overview

➔ Each chapter presents the same consistent set of **Pre-Lecture** resources.

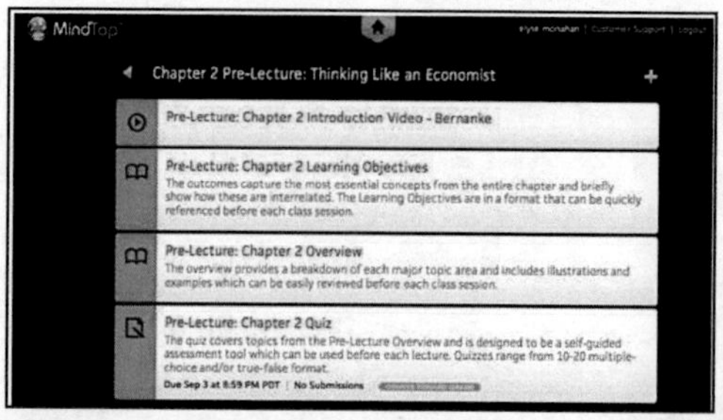

Introduction Video

→ Every MindTap Pre-Lecture section begins with a relevant **Introduction Video** that highlights key topics—and engages interest.

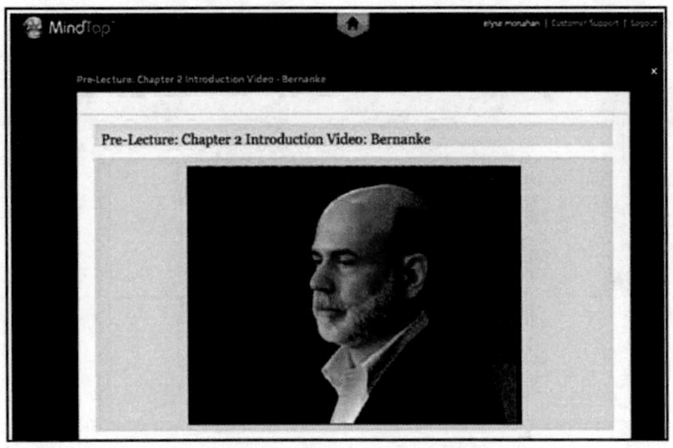

Learning Objectives

→ After watching the video, you can review chapter **Learning Objectives**, which capture essential concepts in an easy-to-read format.

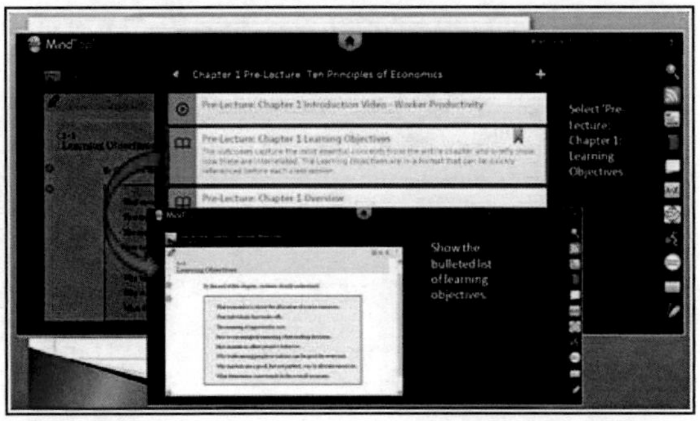

Chapter Overview

→ Next, read the **Overview**. It provides a breakdown of major topic areas and includes illustrations and examples that students can review before class.

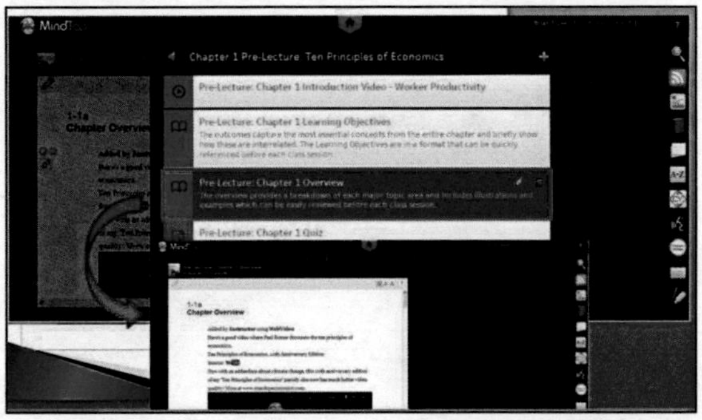

Chapter Quiz

➔ The **Pre-Lecture Quiz** provides you with a clear understanding of their strong and weak areas with respect to the topic, so you can tailor your studying accordingly.

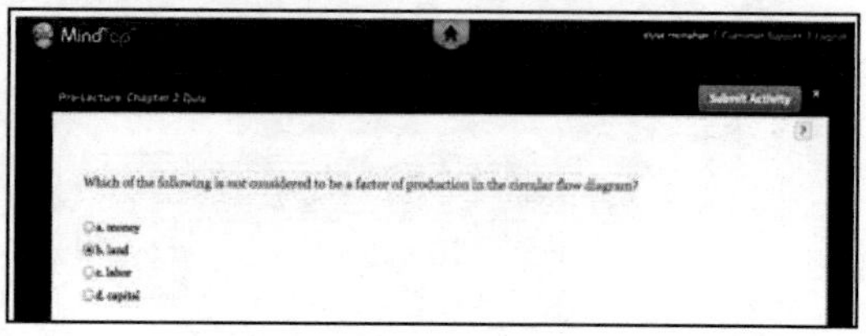

Post-Lecture

Overview

➔ Within the **Post-Lecture**, students easily access their reading and homework assignments.

Reading Activity

➔ An interactive learning resource, the **MindTap Reader** was built from the ground up to create a unique digital reading experience.

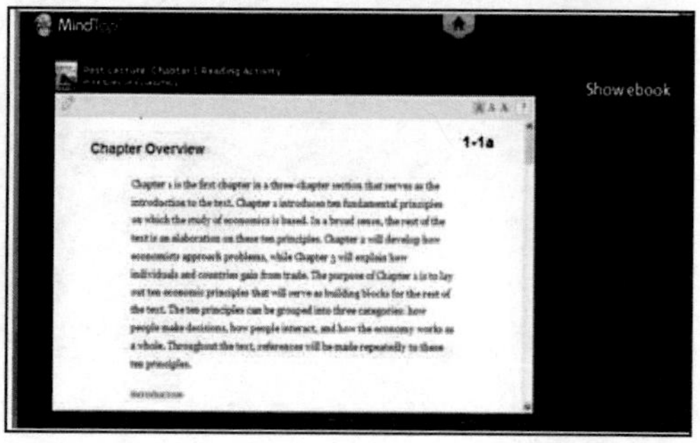

Homework

➔ After completing the Reading Activity, move on to your **Homework**. You can complete and submit homework online—and receive instant feedback.

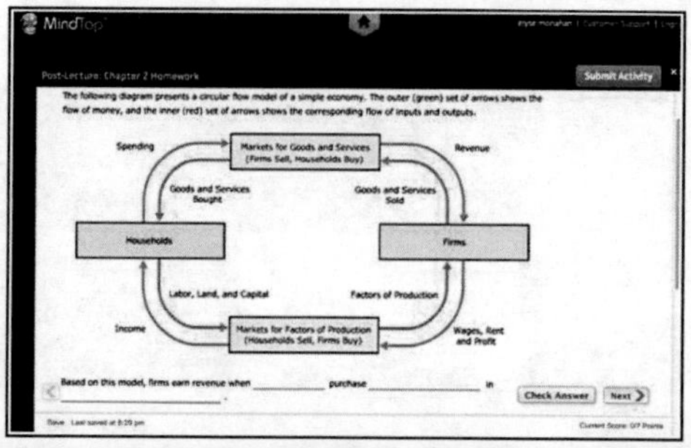

➔ When you answered incorrectly, you can read an explanation that clarifies the correct response.

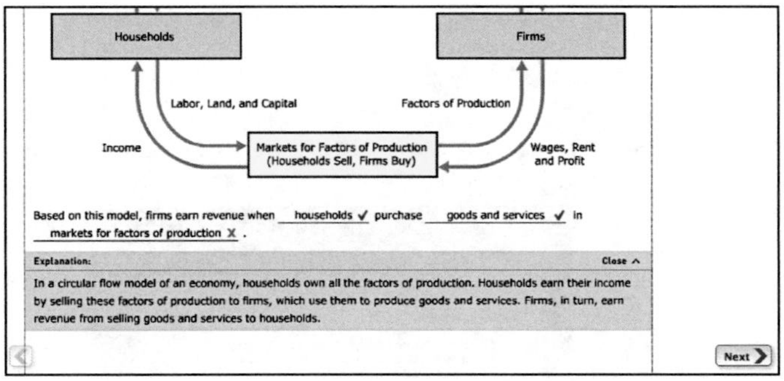

Features

Easy Navigation

➔ Within the reader, users can navigate from section to section or page to page with guidance from page thumbnails and arrows.

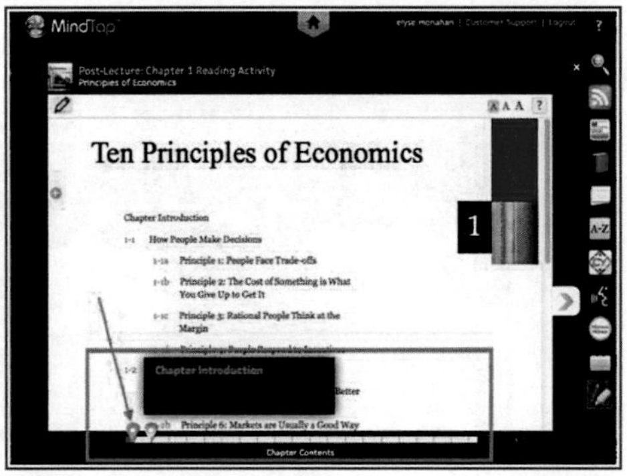

Convenient Tools

➔ The robust functionality of the MindTap Reader allows students to make notes, highlight text, and even find a definition right from the page.

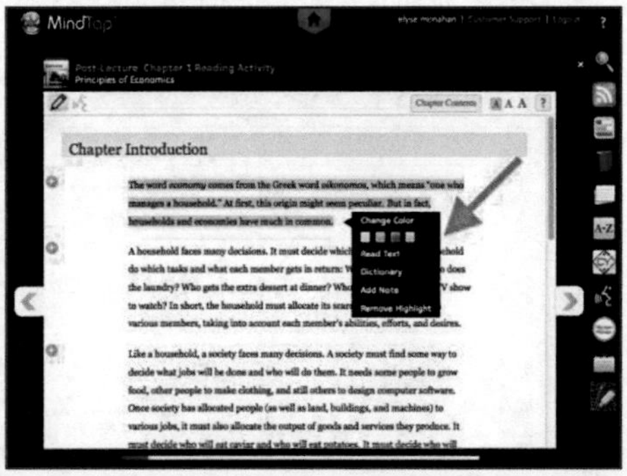

Learning Apps

Overview

➔ A comprehensive library of **Learning Apps** gives students what they need to prepare for a course or exam—all from a single platform.

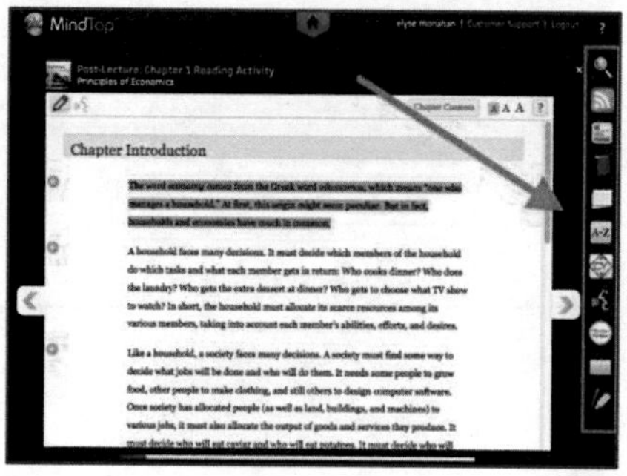

Notebook

➔ By clicking the **My Notes** icon, students can access the **Notebook**, which allows them to write, print, and share their notes. They can also access their notes outside the MindTap environment if they have an EverNote account.

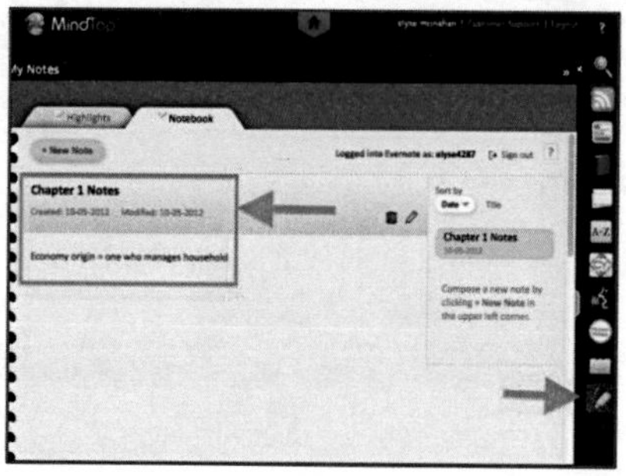

Dictionary

➔ The **Merriam-Webster dictionary app** makes it easy for students to find definitions of new or unfamiliar words. In addition, a click on a word that appears in bold font delivers the definition via the same app.

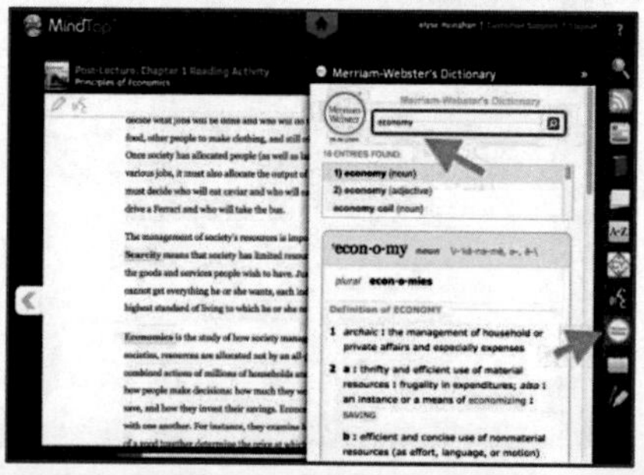

Flashcards

→ Students can either click through MindTap Reader's **Flashcard app** or create their own flashcards to review key terms and concepts.

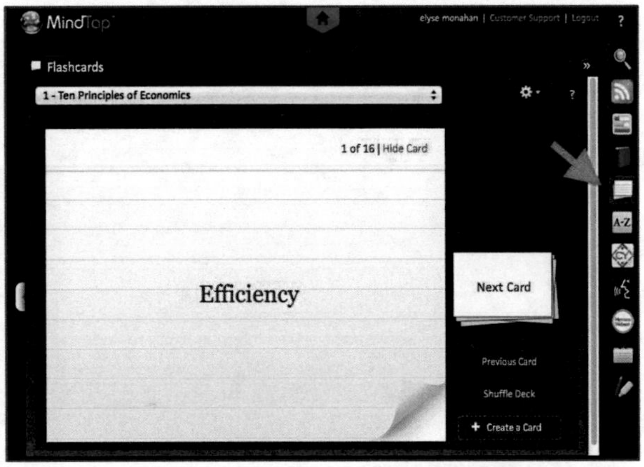

Text-to-Speech ReadSpeaker

→ If students retain more information by hearing instead of reading, they can select a portion of the assignment and listen to it using the **ReadSpeaker text-to-speech app**.

ConnectYard

→ MindTaps boosts engagement by encouraging students to interact with their content, as well as with their peers, instructors, and tutors via the **ConnectYard social app**.

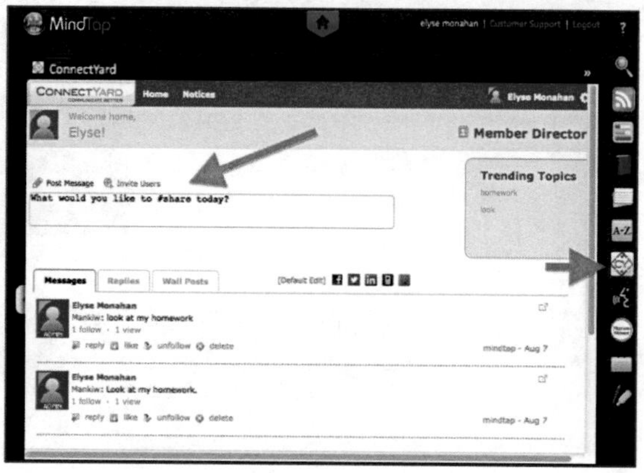

View Progress

→ The **View Progress App** archives grades and makes them easily available to you, keeping you apprised of personal and current class standings.

Product Support

Questions with your CengageBrain account?	Questions regarding MindTap?
• Check the FAQs in the Support area of your CengageBrain home. • Write to cengagebrain.support@cengage.com • Call 866.994.2427 Mon. through Fri. from 8 AM to 6 PM EST	• Go to www.cengage.com/support for 24/7 live chat! • Call 800.354.9706 Mon. through Thurs. 8:30 AM to 9 PM EST and Fri. 8:30 AM to 6 PM EST

NUTRITION: CONCEPTS & CONTROVERSIES

SELECT EDITION

THIRTEENTH EDITION

Frances Sienkiewicz Sizer | Ellie Whitney

CENGAGE Learning

Australia • Brazil • Japan • Korea • Mexico • Singapore • Spain • United Kingdom • United States

CENGAGE Learning

NUTRITION: CONCEPTS & CONTROVERSIES: Special Edition Thirteenth Edition

Senior Project Development Manager:
Linda deStefano

Market Development Manager:
Heather Kramer

Senior Production/Manufacturing Manager:
Donna M. Brown

Production Editorial Manager:
Kim Fry

Sr. Rights Acquisition Account Manager:
Todd Osborne

NUTRITION: CONCEPTS & CONTROVERSIES, THIRTEENTH EDITION
Sizer | Whitney

ISBN-13: 978-1-305-00914-1

ISBN-10: 1-305-00914-2

Cengage Learning
5191 Natorp Boulevard
Mason, Ohio 45040
USA

Cengage Learning is a leading provider of customized learning solutions with office locations around the globe, including Singapore, the United Kingdom, Australia, Mexico, Brazil, and Japan. Locate your local office at:
international.cengage.com/region.
Cengage Learning products are represented in Canada by Nelson Education, Ltd.
For your lifelong learning solutions, visit **www.cengage.com/custom.**
Visit our corporate website at **www.cengage.com.**

Printed in the United States of America

About the Authors

Frances Sienkiewicz Sizer

M.S., R.D., F.A.D.A., attended Florida State University where, in 1980, she received her B.S., and in 1982 her M.S., in nutrition. She is certified as a charter Fellow of the Academy of Nutrition and Dietetics. She is a founding member and vice president of Nutrition and Health Associates, an information and resource center in Tallahassee, Florida, that maintains an ongoing bibliographic database tracking research in more than 1,000 topic areas of nutrition. Her textbooks include *Life Choices: Health Concepts and Strategies; Making Life Choices; The Fitness Triad: Motivation, Training, and Nutrition*; and others. She was a primary author of *Nutrition Interactive*, an instructional college-level nutrition CD-ROM that pioneered the animation of nutrition concepts for use in college classrooms. She has lectured at universities and at national and regional conferences and supports local hunger and homelessness relief organizations in her community.

For the pirates—Nolan, Kayla, Teagan, Kevin, Mackenzie, Lauren, and David.

–Fran

Eleanor Noss Whitney

Ph.D., received her B.A. in biology from Radcliffe College in 1960 and her Ph.D. in biology from Washington University, St. Louis, in 1970. Formerly on the faculty at Florida State University, and a dietitian registered with the Academy of Nutrition and Dietetics, she now devotes full time to research, writing, and consulting in nutrition, health, and environmental issues. Her earlier publications include articles in *Science, Genetics,* and other journals. Her textbooks include *Understanding Nutrition, Understanding Normal and Clinical Nutrition, Nutrition and Diet Therapy,* and *Essential Life Choices* for college students and *Making Life Choices* for high-school students. Her most intense interests presently include energy conservation, solar energy uses, alternatively fueled vehicles, and ecosystem restoration. She is an activist who volunteers full-time for the Citizens Climate Lobby.

To Max, Zoey, Emily, Rebecca, Kalijah, and Duchess with love.

–Ellie

Brief Contents

Contents

Lusoimages/Shutterstock.com

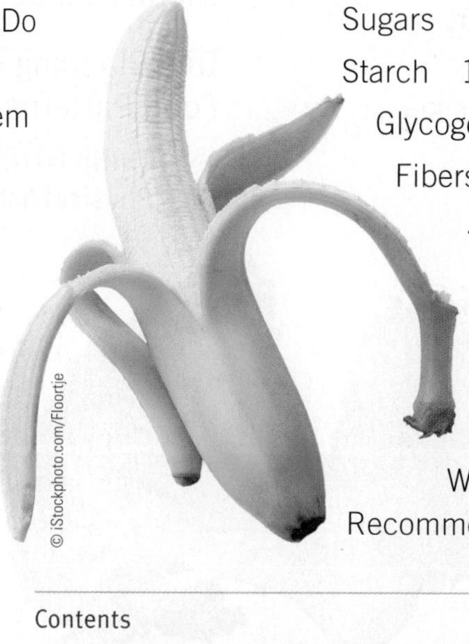

© iStockphoto.com/Floortje

© Sergey Mironov/Shutterstock.com

© ericlefrancais/Shutterstock.com

© Feng Yu/Shutterstock.com

© marilyn barbone/Shutterstock.com

Preface

A billboard in Louisiana reads, "Come as you are. Leave different," meaning that once you've seen, smelled, tasted, and listened to Louisiana, you'll never be the same. This book extends the same invitation to its readers: come to nutrition science as you are, with all of the knowledge and enthusiasm you possess, with all of your unanswered questions and misconceptions, and with the habits and preferences that now dictate what you eat.

But leave different. Take with you from this study a more complete understanding of nutrition science. Take a greater ability to discern between nutrition truth and fiction, to ask sophisticated questions, and to find the answers. Finally, take with you a better sense of how to feed yourself in ways that not only please you and soothe your spirit but nourish your body as well.

For well over a quarter of a century, *Nutrition: Concepts and Controversies* has been a cornerstone of nutrition classes across North America, serving the needs of students and professors in building a healthier future. In keeping with our tradition, in this, our 13th edition, we continue exploring the ever-changing frontier of nutrition science, confronting its mysteries through its scientific roots. We maintain our sense of personal connection with instructors and learners alike, writing for them in the clear, informal style that has become our trademark.

Pedagogical Features

Throughout these chapters, features tickle the reader's interest and inform. For both verbal and visual learners, our logical presentation and our lively figures keep interest high and understanding at a peak. The photos that adorn many of our pages add pleasure to reading.

Many tried-and-true features return in this edition: Each chapter begins with "What Do You Think?" questions to pique interest and set a personal tone for the information that follows. The reader is offered another chance to reflect on these questions with "What Did You Decide?" at the chapter's end. A list of Learning Objectives (LO) offers a sneak peek into the chapter's major goals, and the LO reappear under section headings to make clear the main take-away messages. Streamlined margin entries now bear titles that help readers to grasp their functions at a glance. *My Turn* features invite the reader to hear stories from students in nutrition classes around the nation and to offer evidence-based solutions to real-life situations. *Think Fitness* reminders appear from time to time to alert readers to ways in which physical activity links with nutrition to support health. The *Food Feature* sections that appear in most chapters act as bridges between theory and practice; they are practical applications of the chapter concepts that help readers to choose foods according to sound nutrition principles. The consumer sections, now entitled *A Consumer's Guide To . . .*, have a fresh new contemporary feel. They guide readers through an often bewildering marketplace with scientific clarity, preparing them to move ahead with sound decisions regarding whole-grain foods, safe seafood choices, amino acid supplements, calorie-rich beverages, organic foods, and many others. Each section ends with review questions, new to this edition.

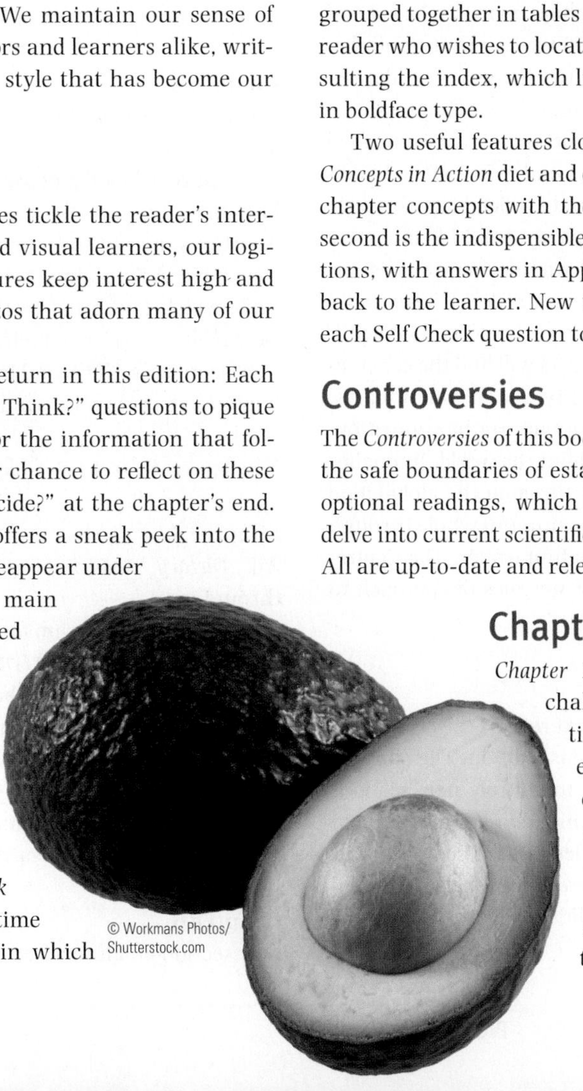
© Workmans Photos/ Shutterstock.com

By popular demand, we have retained our *Snapshots* of vitamins and minerals, this time with a brand-new look. These concentrated capsules of information depict food sources of vitamins and minerals, present the DRI recommended intakes and Tolerable Upper Intake Levels, and offer the chief functions of each nutrient along with deficiency and toxicity symptoms.

New or major terms are defined in the margins of chapter pages or in nearby tables, and they also appear in the Glossary at the end of the book. Definitions in *Controversy* sections are grouped together in tables and also appear in the Glossary. The reader who wishes to locate any term can quickly do so by consulting the index, which lists the page numbers of definitions in boldface type.

Two useful features close each chapter. First, our popular *Concepts in Action* diet and exercise tracking activities integrate chapter concepts with the Diet Analysis Plus program. The second is the indispensible *Self Check* that provides study questions, with answers in Appendix G to provide immediate feedback to the learner. New to this edition, LO numbers anchor each Self Check question to the text for easy reference.

Controversies

The *Controversies* of this book's title invite you to explore beyond the safe boundaries of established nutrition knowledge. These optional readings, which appear at the end of each chapter, delve into current scientific topics and emerging controversies. All are up-to-date and relevant to nutrition science today.

Chapter Contents

Chapter 1 begins the text with a personal challenge to students. It asks the question so many people ask of nutrition educators—"Why should people care about nutrition?" We answer with a lesson in the ways in which nutritious foods affect diseases and present a continuum of diseases from purely genetic in origin to those almost totally preventable by nutrition. After presenting some beginning facts about the genes, nutrients, bioactive food components, and nature of foods, the chapter goes on to present the *Healthy People* goals for the nation. It concludes with a discussion of scientific research in nutrition to lend a perspective on the context in which study results may be rightly viewed.

Chapter 2 brings together the concepts of nutrient allowances, such as the *Dietary Reference Intakes*, and diet planning using the *Dietary Guidelines for Americans* and the USDA *MyPlate* eating patterns. *Chapter 3* presents a thorough, but brief, introduction to the workings of the human body from the genes to the organs, with major emphasis on the digestive system. *Chapters 4–6* are devoted to the energy-yielding nutrients—carbohydrates, lipids, and protein. The concept of inflammation, introduced in Chapter 3, is expanded in discussions of diabetes, colon health, and heart disease. Gene regulation takes its place among major functions of body proteins. *Controversy 4* has renewed its focus on theories and fables surrounding the health effects of dietary carbohydrates. In *Controversy 6* a new emphasis on using MyPlate in diet planning for vegetarians will assist in sound vegetarian meal planning.

Chapters 7 and 8 present the vitamins, minerals, and water. *Chapter 9* relates energy balance to body composition, obesity, and underweight and provides guidance to lifelong weight maintenance. *Chapter 10* presents the relationships between physical activity, athletic performance, and nutrition, with some guidance about products marketed to athletes. *Chapter 11* applies the essence of the first ten chapters to two broad and rapidly changing areas within nutrition: immunity and disease prevention. Readers will revisit the themes of oxidation, inflammation, and disease, introduced in earlier chapters.

Chapter 12 delivers urgently important concepts of food safety. It also addresses the usefulness and safety of food additives, including artificial sweeteners and artificial fats, formerly topics found in Chapters 4 and 5. *Chapters 13 and 14* emphasize the importance of nutrition through the life span and issues surrounding childhood obesity in *Controversy 13*. *Chapter 14* includes nutrition advice for feeding preschoolers, schoolchildren, teens, and the elderly, where readers will find the concluding discussion of inflammation, immunity, and chronic diseases.

Chapter 15 devotes attention to hunger and malnutrition, both in the United States and throughout the world. It touches on the vast network of problems that threaten the global food supply and links each reader to the meaningful whole through sustainable daily choices available to him or her. The *Controversy* introduces some promising new avenues of approach to providing the world's food.

New to This Edition

Every section of each chapter of this text reflects the changes in nutrition science occurring since the last edition. The changes range from subtle shifts of emphasis to entirely new sections that demand our attention. Here, we mention the most salient changes from the last edition. Readers will discover many, many others. Appendix F supplies current references; older references may be viewed in previous editions, available from the publisher.

Chapter 1

2020 Nutrition-Related Health Objectives for the Nation; selected nutrition and weight-related objectives.
Defines the term *eating pattern* (as specified in the *2010 Dietary Guidelines for Americans*).
Introduces new Academy of Nutrition and Dietetics name.
New table of professional responsibilities of dietitians.
New data on diploma mills; new guidance to identify diploma mills.

Chapter 2

2010 Dietary Guidelines for Americans included and applied throughout this edition.
New figure comparing U.S. diet to *Dietary Guidelines.*
USDA Food Guide updated to reflect the *2010 Dietary Guidelines.*
Introduces MyPlate icon and website.
Key nutrients of concern updated.
New figure: Dining Out Trends, United States.
New Consumer's Guide on controlling portion sizes.
New table of antioxidant capacity of selected foods from the USDA ORAC Database, Release 2, with text perspective.

Chapter 3

New organization of Controversy section.
New *Dietary Guidelines 2010* information.
New discussion of strength of evidence for suggested benefits of alcohol.
Binge drinking introduced as *heavy episodic drinking.*

Chapter 4

2010 Dietary Guidelines for Americans for carbohydrates in Table 4–1.
New figure of blood glucose regulation.
Enhanced digestion figure.
New Consumer's Guide on whole grains.
New table: A Sampling of Whole Grains.
Updated label information in figures.
New table of tips for reducing intakes of added sugars.
New section on HFCS and fructose.
Explores emerging links among NAFLD, diabetes, and fructose intake.
New figure: Sources of Added Sugars in the U.S. Diet.

Chapter 5

All *Dietary Guidelines for Americans* material updated throughout.
New Venn-type diagram for choosing fish.
New information on EPA/DHA.
New Consumer's Guide on balancing seafood risks and benefits.
New table of solid fat replacements.
New figure of solid fat sources in the U.S. diet.
Updated material on lipoproteins and heart disease risk.
Expanded discussion of nuts and their potential benefits.
New emphasis on total eating patterns.

Chapter 6

New section on gluten-free diets.

Defines gluten, celiac disease.
Moved world malnutrition discussion to Chapter 15.
Explains vegetarian eating pattern and CVD prevention.
New figure of vegetarian protein foods and milk products in the USDA eating patterns.
Integrates use of USDA eating patterns for lacto-ovo vegetarians and vegans, located in Appendix E.

Chapter 7
Specifies *Dietary Guidelines* vitamins of concern.
New Consumer Guide section on vitamin D sources.
Vitamin D section reflects 2011 DRI scientific background and DRI values.
Includes tocotrienols as forms of vitamin E.
New niacin/CVD information.
New folate/cancer risk information.
All vitamin Snapshots updated with current USDA nutrient data.

Chapter 8
Identifies the *Dietary Guidelines 2010* minerals of national concern.
New Consumer Guide on beverages as calorie sources.
New figure of U.S. calorie intakes from beverages.
Table of water in foods and beverages.
Calcium section reflects 2011 DRI scientific background and DRI values.
Updated figure of current U.S. sodium sources.
New table of promoters and inhibitors of iron absorption.

Chapter 9
Consolidated two figures on body fat analysis techniques.
Narrowed focus and streamlined discussions of theoretical causes of obesity.
Expanded discussions of leptin and ghrelin with new findings.
Simplified table of FDA-approved obesity drugs, including Belviq and Qsymia.
New table of environmental influences on food intake.
New table of food intake for weight gain.
New table of community strategies to combat obesity.
New Consumer's Guide on fad diets.
New table of clues to fad diets and weight-loss scams.
New table summarizing lifestyle strategies used by successful weight losers/maintainers.
Updated and simplified eating disorder diagnostic criteria.

Chapter 10
New table comparing performance-hindering effects of inadequate hydration with symptoms of heat stroke.
Updated sample balanced fitness program.
Increased emphasis on carbohydrate intakes before, during, and after physical activity.
New figure with electron micrographs depicting glycogen stores before and after exercise.
Enhanced discussions of muscle metabolism and the roles of dietary protein in muscle protein synthesis.
New figure of nutritious snacks for athletes.
New table of risk factors and symptoms of hyponatremia.
New table—summary of sports nutrition recommendations.
New fast-food pregame meal option for traveling athletes.
Updated ergogenic aids discussion.

Chapter 11
Reorganized nutrition and immunity section; included inflammation introduction.
New table of micronutrient roles in immune function.
Enhanced the figure on malnutrition and disease interactions.
Enhanced the table on malnutrition and the body's defense systems.
Updated table of recommendations for reducing cancer risks.
New table of strategies for choosing enough fruits, vegetables, and legumes.
Added a brief discussion of acrylamide to the cancer section.
Updated and shortened nutritional genomics section.

Chapter 12
Included the 2010 FDA Food Safety Modernization Act (FSMA).
Introduced the FDA's new Coordinated Outbreak Response and Evaluation Network.
Improved, condensed table of foodborne illness microorganisms.
Reframed *E. coli* O157:H7 to STEC to reflect foodborne illness trends.
Enhanced figure of safe handling and cooking of meats and poultry.
Expanded discussion of imported foods.
Added two tables of foodborne illness myths and truths.
New table weighing estimated health risks from pesticide residues on produce.
Included high-pressure processing and ultrasound technologies for microbial control.
Added luo han guo to nonnutritive sweeteners.
New graphic of increasing production of genetically modified crops.
Addressed genetic engineering advances in food fortification and microbial biofuel research.

Chapter 13
New table of risk factors for gestational diabetes.
New table of warning signs of preeclampsia.
Enhanced discussion of essential fatty acids in breast milk.
Added brief discussion of breastfeeding and reduced risk of SIDS.
New table of tips for successful breastfeeding.
New table of choking prevention.
Updated childhood obesity data and discussion.

Chapter 14
USDA Food Patterns for Young Children, 2011.
New table of tips for feeding picky eaters.
New figure of sugar-sweetened beverage intakes of adolescents.
New table of food skills and developmental milestones of preschool children.
New table of iron needs in adolescence.
Enhanced table of nutrient concerns in aging.

New roles of beverages as nutrient sources for the elderly.
Updated drug–nutrient interactions information.
Included new information on herbs and caffeine interactions.

Chapter 15
New food insecurity data, global and U.S.
Updated world hunger map.
Updated U.S. food security survey.
New section on severe acute malnutrition and chronic malnutrition, including appropriate nutrition therapy.
New table comparing severe acute malnutrition and chronic malnutrition.
New section addressing food waste.
New figure depicting U.S. food waste.
New figure of methods of food waste recovery.

Ancillary Materials

Students and instructors alike will appreciate the innovative teaching and learning materials that accompany this text.

- **MindTap:** A personalized, fully online digital learning platform of authoritative content, assignments, and services that engages your students with interactivity while also offering you choice in the configuration of coursework and enhancement of the curriculum via web-apps known as MindApps. MindApps range from ReadSpeaker (which reads the text out loud to students), to Kaltura (allowing you to insert inline video and audio into your curriculum). MindTap is well beyond an eBook, a homework solution or digital supplement, a resource center website, a course delivery platform, or a Learning Management System. It is the first in a new category—the Personal Learning Experience.
- **Diet Analysis Plus™:** Diet Analysis Plus enables you to track and assess your diet and physical activity online. You can create a personal profile based on height, weight, age, sex, and activity level and use this tool to easily analyze the nutritional value of the food you eat, adjust your diet to meet your personal health goals, and gain a better understanding of how nutrition relates to your life. Diet Analysis Plus includes a 35,000+ food database, 10 reports for analysis, a food recipe feature, the latest Dietary References, and goals and actual percentages of essential nutrients, vitamins, and minerals. Diet Analysis Plus is a valuable tool that you can use in your nutrition course and then continue to use after the course is over.
- **Study Guide**: Provides key-concepts-focused review exercises in a variety of formats, such as practice tests, fill-ins, matching sets, and short-answer questions.
- **Instructor's Manual:** Features ready-to-use assignment materials, including critical thinking questions, food label and diet planning worksheets, and new crossword puzzles. Class preparation tools include ideas for in-class activities—such as quick meal comparisons, new to this edition—lecture presentation outlines, chapter summaries, and text-specific handouts.
- **Test Bank**: Offers a rich assortment of multiple-choice and essay questions, including new food label–based application items.
- **PowerLecture DVD-ROM**: Combines PowerPoint lectures and images, videos, JoinIn quizzes, **ExamView** testing software preloaded with the test bank questions, the instructor's manual, and the test bank into a single resource.

Our Message to You

Our purpose in writing this text, as always, is to enhance our readers' understanding of nutrition science. We also hope the information on this book's pages will reach beyond the classroom into our readers' lives. Take the information you find inside this book home with you. Use it in your life: nourish yourself, educate your loved ones, and nurture others to be healthy. Stay up with the news, too. For despite all the conflicting messages, inflated claims, and even quackery that abound in the marketplace, true nutrition knowledge progresses with a genuine scientific spirit, and important new truths are constantly unfolding.

Acknowledgments

Thank you, Philip, for making everything possible. Our thanks also to our partners Linda Kelly DeBruyne and Sharon Rolfes for 35 years of immeasurable support. Linda, thank you especially, for your TLC in updating Chapter 11 and Chapter 13. To Shannon Gower-Winter, MS, RD, many thanks for your careful and thorough updates of Controversy 13 and Chapter 14. Thank you, Spencer Webb, RD, CSCS, for your design of our unique fitness program and your other assistance in Chapter 10 (and for getting us into shape, too). Rebbecca Skinner, thank you for your skilled attention to the form and content of the Consumer's Guide sections and Controversy 15. Thank you to Carole Sloan for contributing the Critical Thinking questions at the end of each Controversy. To our assistant, Chelsea MacKenzie, many heartfelt thanks for your cheerful, consistent, and careful work.

Our special thanks to our publishing team—Yolanda Cossio, Peggy Williams, Nedah Rose, and Carol Samet—for their dedication to excellence. Thank you to our marketing manager, Tom Ziolkowski, for getting the word out about our new edition.

We would also like to thank the authors of the student and instructor ancillaries for the 13th edition: Alana Cline, who revised and expanded the test bank; Mary Ellen Clark, who contributed materials to the instructor's manual; Jana R. Kicklighter, who authored the study guide; and Jeanne Freeman, who provided content for the PowerLecture and student website.

Reviewers of Recent Editions

As always, we are grateful for the instructors who took the time to comment on this revision. Your suggestions were invaluable in strengthening the book and suggesting new lines of thought. We hope you will continue to provide your comments and suggestions.

Alex Kojo Anderson, *University of Georgia, Athens*
Sharon Antonelli, *San Jose City College*

L. Rao Ayyagari, *Lindenwood University*
James W. Bailey, *University of Tennessee*
Ana Barreras, *Central New Mexico Community College*
Karen Basinger, *Montgomery College*
Leah Carter, *Bakersfield College*
Melissa Chabot, *SUNY at Buffalo*
Priscilla Connors, *University of North Texas*
Monica L. Easterling, *Wayne County Community College District*
Shannon Gower-Winter, MS, RD, *Florida State University*
Jena Nelson Hall, *Butte Community College*
Charlene G. Harkins, *University of Minnesota, Duluth*
Sharon Anne Himmelstein, *Central New Mexico Community College*
David Lightsey, *Bakersfield College*
Craig Meservey, *New Hampshire Technical Institute*
Eimear M. Mullen, *Northern Kentucky University*
Suzanne Linn Nelson, *University of Colorado at Boulder*
Steven Nizielski, *Grand Valley State University*
David J. Pavlat, *Central College*
Begoña Cirera Perez, *Chabot College*
Liz Quintana, *West Virginia University*
Janice M. Rueda, *Wayne State University*
Donal Scheidel, *University of South Dakota*
Carole A. Sloan, *Henry Ford Community College*
Leslie S. Spencer, *Rowan University*
Ilene Sutter, *California State University, Northridge*
Sue Ellen Warren, *El Camino College*
Barbara P. Zabitz, *Wayne County Community College District*
Nancy Zwick, *Northern Kentucky University*

1 Food Choices and Human Health

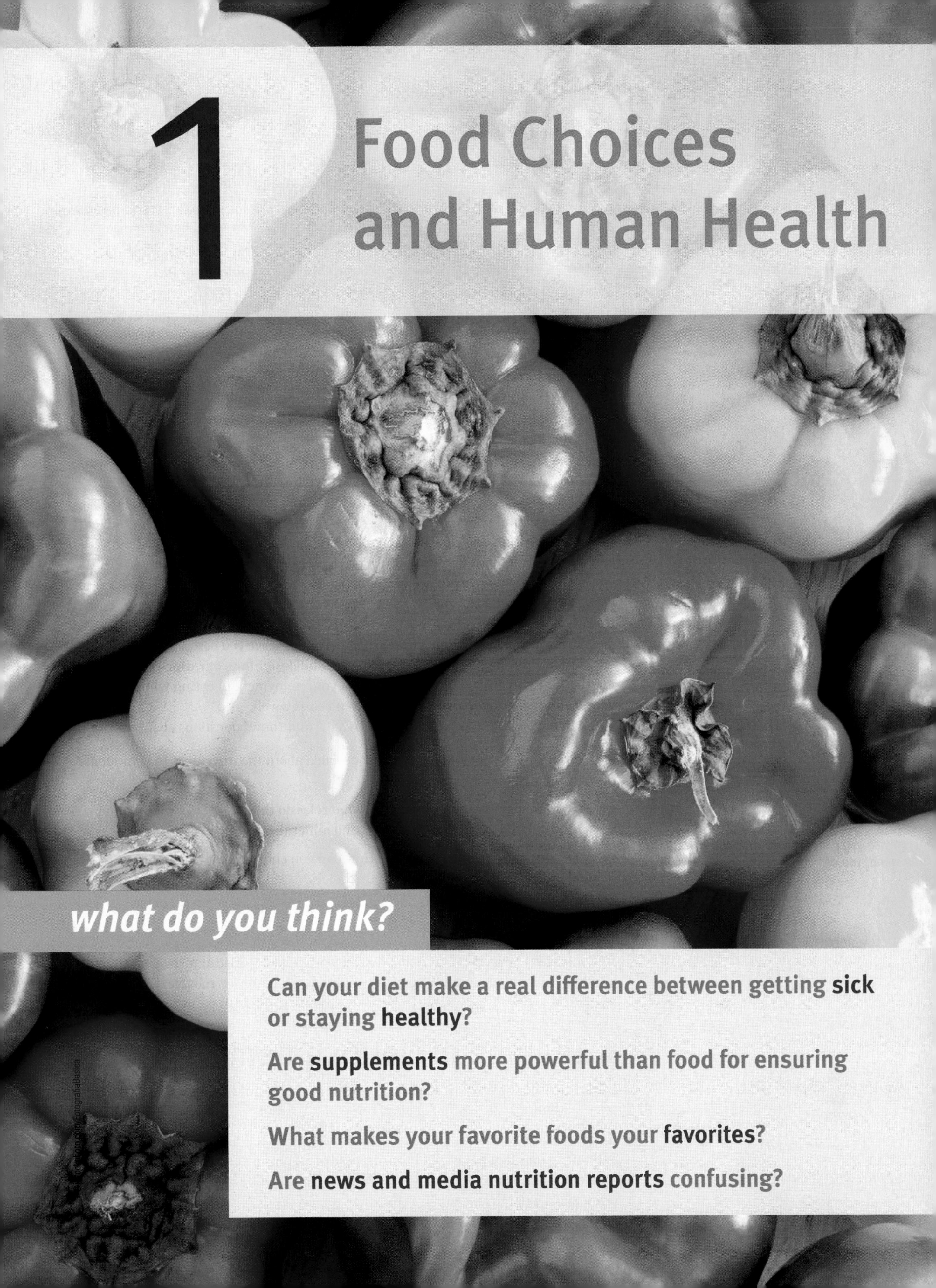

what do you think?

Can your diet make a real difference between getting **sick** or staying **healthy**?

Are **supplements** more powerful than food for ensuring good nutrition?

What makes your favorite foods your **favorites**?

Are **news and media nutrition reports** confusing?

Learning Objectives

After reading this chapter, you should be able to accomplish the following:

LO 1.1 Discuss how daily food choices can help or harm the body's health over time.

LO 1.2 Describe the national *Healthy People* objective for the nation, and identify some nutrition-related objectives.

LO 1.3 Define the term *nutrient* and be able to list the six major nutrients.

LO 1.4 Summarize the five characteristics of a healthy diet and describe cultural or other influences on human food choices.

LO 1.5 Describe the major types of research studies and give reasons why national nutrition research is important for the health of the population.

LO 1.6 List the major steps in behavior change and devise a plan for making successful long-term changes in the diet.

LO 1.7 Define *nutrient density* and explain the advantages of choosing nutrient-dense foods.

LO 1.8 Identify misleading nutrition information in infomercials, advertorials, and other sources in the popular media.

Brand X Pictures/Jupiterimages/Getty Images

When you choose foods with nutrition in mind, you can enhance your own well-being.

If you care about your body, and if you have strong feelings about **food**, then you have much to gain from learning about **nutrition**—the science of how food nourishes the body. Nutrition is a fascinating, much talked about subject. Each day, newspapers, radio, and television present stories of new findings on nutrition and heart health or nutrition and cancer prevention, and at the same time advertisements and commercials bombard us with multicolored pictures of tempting foods—pizza, burgers, cakes, and chips. If you are like most people, when you eat you sometimes wonder, "Is this food good for me?" or you berate yourself, "I probably shouldn't be eating this."

When you study nutrition, you learn which foods serve you best, and you can work out ways of choosing foods, planning meals, and designing your **diet** wisely. Knowing the facts can enhance your health and your enjoyment of eating while relieving your feelings of guilt or worry that you aren't eating well.

This chapter addresses these "why," "what," and "how" questions about nutrition:

- *Why* care about nutrition? Why be concerned about the **nutrients** in your foods? Why not just take supplements?
- *What* are the nutrients in foods, and what roles do they play in the body? What are the differences between vitamins and minerals?
- *What* constitutes a nutritious diet? How can you choose foods wisely, for nutrition's sake? And what motivates your choices?
- *How* do we know what we know about nutrition? How does nutrition science work, and how can a person keep up with changing information?

Controversy 1 concludes the chapter by offering ways to distinguish between trustworthy sources of nutrition information and those that are less reliable.

A Lifetime of Nourishment

LO 1.1 Discuss how daily food choices can help or harm the body's health over time.

If you live for 65 years or longer, you will have consumed more than 70,000 meals, and your remarkable body will have disposed of 50 tons of food. The foods you choose have cumulative effects on your body. As you age, you will see and feel those effects—if you know what to look for.

Your body renews its structures continuously, and each day it builds a little muscle, bone, skin, and blood, replacing old tissues with new. It may also add a little fat if you consume excess food energy (calories) or subtract a little if you consume less than you require. Some of the food you eat today becomes part of "you" tomorrow.

The best food for you, then, is the kind that supports the growth and maintenance of strong muscles, sound bones, healthy skin, and sufficient blood to cleanse and nourish all parts of your body. This means you need food that provides not only the right amount of energy but also sufficient nutrients, that is, enough water, carbohydrates, fats, protein, vitamins, and minerals. If the foods you eat provide too little or too much of any nutrient today, your health may suffer just a little today. If the foods you eat provide too little or too much of one or more nutrients every day for years, then in later life you may suffer severe disease effects.

A well-chosen array of foods supplies enough energy and enough of each nutrient to prevent **malnutrition**. Malnutrition includes deficiencies, imbalances, and excesses of nutrients, alone or in combination, any of which can take a toll on health over time.

KEY POINTS

- The nutrients in food support growth, maintenance, and repair of the body.
- Deficiencies, excesses, and imbalances of energy and nutrients bring on the diseases of malnutrition.

The Diet and Health Connection

Your choice of diet profoundly affects your health, both today and in the future. Only two common lifestyle habits are more influential: smoking and using other forms of tobacco and drinking alcohol in excess. Of the leading causes of death listed in Table 1–1, four are directly related to nutrition, and another—motor vehicle and other accidents—is related to drinking alcohol.

Table 1–1

Leading Causes of Death in the United States

Blue shading indicates that a cause of death is related to nutrition; the light yellow indicates that it is related to alcohol.

	Percentage of Total Deaths
1. Heart disease	24.6%
2. Cancers	23.3%
3. Chronic lung disease	5.6%
4. Strokes	5.3%
5. Accidents	4.8%
6. Alzheimer's disease	3.2%
7. Diabetes mellitus	2.8%
8. Pneumonia and influenza	2.2%
9. Kidney disease	2.0%
10. Suicide	1.5%

Source: Deaths: Preliminary data for 2009, National Vital Statistics Reports, *March 16, 2011, Centers for Disease Control and Prevention, www.cdc.gov/nchs.*

food medically, any substance that the body can take in and assimilate that will enable it to stay alive and to grow; the carrier of nourishment; socially, a more limited number of such substances defined as acceptable by each culture.

nutrition the study of the nutrients in foods and in the body; sometimes also the study of human behaviors related to food.

diet the foods (including beverages) a person usually eats and drinks.

nutrients components of food that are indispensable to the body's functioning. They provide energy, serve as building material, help maintain or repair body parts, and support growth. The nutrients include water, carbohydrate, fat, protein, vitamins, and minerals.

malnutrition any condition caused by excess or deficient food energy or nutrient intake or by an imbalance of nutrients. Nutrient or energy deficiencies are forms of undernutrition; nutrient or energy excesses are forms of overnutrition.

Figure 1–1

Nutrition and Disease

Not all diseases are equally influenced by diet. Some are almost purely genetic, like the anemia of sickle-cell disease. Some may be inherited (or the tendency to develop them may be inherited in the genes) but may be influenced by diet, like some forms of diabetes. Some are purely dietary, like the vitamin and mineral deficiency diseases.

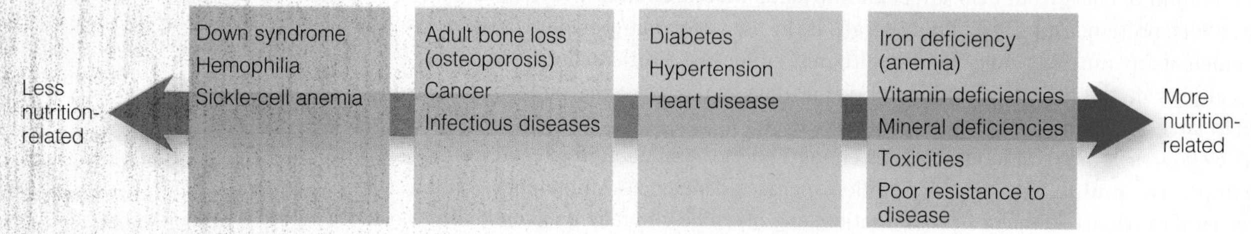

Many older people suffer from debilitating conditions that could have been largely prevented had they known and applied the nutrition principles known today. The **chronic diseases**—heart disease, diabetes, some kinds of cancer, dental disease, and adult bone loss—all have a connection to poor diet.[1]* These diseases cannot be prevented by a good diet alone; they are to some extent determined by a person's genetic constitution, activities, and lifestyle. Within the range set by your genetic inheritance, however, the likelihood of developing these diseases is strongly influenced by your daily choices.

KEY POINT

- Nutrition profoundly affects health.

Genetics and Individuality

Consider the role of genetics. Genetics and nutrition affect different diseases to varying degrees (see Figure 1–1). The anemia caused by sickle-cell disease, for example, is purely hereditary and thus appears at the left of Figure 1–1 as a genetic condition largely unrelated to nutrition. Nothing a person eats affects the person's chances of contracting this anemia, although nutrition therapy may help ease its course. At the other end of the spectrum, iron-deficiency anemia most often results from undernutrition. Diseases and conditions of poor health appear all along this continuum, from almost entirely genetically based to purely nutritional in origin; the more nutrition-related a disease or health condition is, the more successfully sound nutrition can prevent it.

Furthermore, some diseases, such as heart disease and cancer, are not one disease but many. Two people may both have heart disease, but not the same form; one person's cancer may be nutrition-related but another's may not be. Individual people differ genetically from each other in thousands of subtle ways, so no simple statement can be made about the extent to which diet can help any one person avoid such diseases or slow their progress.

The identification of the human **genome** establishes the entire sequence of the **genes** in human **DNA**. This work has, in essence, revealed the body's instructions for making all of the working parts of a human being. A new wealth of information has emerged to explain the workings of the body, and nutrition scientists are working quickly to apply this knowledge to benefit human health. Later chapters expand on the emerging story of nutrition and the genes.

KEY POINTS

- Diet influences long-term health within the range set by genetic inheritance.
- Nutrition has little influence on some diseases but strongly affects others.

Did You Know?

Anemia is a blood condition in which red blood cells, the body's oxygen carriers, are inadequate or impaired and so cannot meet the oxygen demands of the body.

Did You Know?

The human genome is 99.9% the same in all people; all of the normal variations such as differences in hair color, as well as variations that result in diseases such as sickle-cell anemia, lie in the 0.1% of the genome that varies.

[1] Reference notes are found in Appendix F.

Think Fitness — move it! Why Be Physically Active?

Why should people bother to be physically active? While a person's daily food choices can powerfully affect health, the combination of nutrition and physical activity is more powerful still. People who combine regular physical activity with a nutritious diet can expect to receive at least some of these benefits:

Reduced risks of cardiovascular diseases, diabetes, certain cancers, hypertension, others.

Increased endurance, strength, and flexibility.

More cheerful outlook and less likelihood of depression.

Improved mental functioning.

Feeling of vigor.

Feeling of belonging—the companionship of sports.

Stronger self-image.

Reduced body fat, increased lean tissue.

A more youthful appearance, healthy skin, and improved muscle tone.

Greater bone density and lessened risk of adult bone loss in later life.

Increased independence in the elderly.

Sound, beneficial sleep.

Faster wound healing.

Reduced menstrual symptoms.

Improved resistance to infection.

If even half of these benefits were yours for the asking, wouldn't you step up to claim them? In truth, they are yours to claim, at the price of including physical activity in your day. Chapter 10 explores the topics of fitness and physical activity.

start now! Ready to make a change? Go to Diet Analysis Plus online and track your physical activities—all of them—for three days. (The Concepts in Action activity at the end of this chapter will also use this information.) After you have recorded your activities, see how much time you spent exercising at a moderate to vigorous level. Could you increase your level and amount of activity?

Other Lifestyle Choices

Besides food choices, other lifestyle choices also affect people's health. Tobacco use and alcohol and other substance abuse can destroy health. Physical activity, sleep, stress, and other environmental factors can also help prevent or reduce the severity of some diseases. Physical activity is so closely linked with nutrition in supporting health that most chapters of this book offer a feature called Think Fitness, such as the one near here.

KEY POINT

- Life choices, such as being physically active or using tobacco or alcohol, can improve or damage health.

Healthy People: Nutrition Objectives for the Nation

LO 1.2 Describe the national *Healthy People* objective for the nation, and identify some nutrition-related objectives.

In its publication *Healthy People*, the U.S. Department of Health and Human Services sets specific 10-year objectives to guide national health promotion efforts.[2] The vision of *Healthy People 2020* is a society in which all people live long, healthy lives. Table 1–2 provides a quick scan of the nutrition and weight-related objectives set for this decade. The inclusion of nutrition and food-safety objectives shows that public health officials consider these areas to be top national priorities.

In 2010, the nation's health report was mixed: the average blood cholesterol levels had dropped, but most people's diets lacked enough fruits, vegetables, and whole grains; and physical activity levels needed improvement.[3] Positive strides had been made toward reducing harm from certain foodborne infections, heart disease, and several cancers, but on the negative side, the numbers of overweight people and people with diabetes continue to rise. To fully meet the current *Healthy People* goals, our nation must take steps to change its habits.

chronic diseases degenerative conditions or illnesses that progress slowly, are long in duration, and that lack an immediate cure; chronic diseases limit functioning, productivity, and the quality and length of life. Examples include heart disease, cancer, and diabetes.

genome (GEE-nome) the full complement of genetic information in the chromosomes of a cell. In human beings, the genome consists of about 35,000 genes and supporting materials. The study of genomes is *genomics*. Also defined in Controversy 11.

genes units of a cell's inheritance; sections of the larger genetic molecule DNA (deoxyribonucleic acid). Each gene directs the making of one or more of the body's proteins.

DNA an abbreviation for deoxyribonucleic (dee-OX-ee-RYE-bow-nu-CLAY-ick) acid, the thread-like molecule that encodes genetic information in its structure; DNA strands coil up densely to form the chromosomes (Chapter 3 provides more details).

Table 1–2

***Healthy People 2020*, Selected Nutrition and Body Weight Objectives**

Many other Objectives for the Nation are available at www.healthypeople.gov.

Chronic Diseases
■ Reduce the proportion of adults with osteoporosis.
■ Reduce the death rates from cancer, diabetes, heart disease, and stroke.
■ Reduce the annual number of new cases of diabetes.
Food Safety
■ Reduce outbreaks of certain infections transmitted through food.
■ Reduce severe allergic reactions to food among adults with diagnosed food allergy.
Maternal, Infant, and Child Health
■ Reduce the number of low birthweight infants and preterm births.
■ Increase the proportion of infants who are breastfed.
■ Reduce the occurrence of fetal alcohol syndrome (FAS).
■ Reduce iron deficiency among children, adolescents, women of child-bearing age, and pregnant women.
■ Reduce blood lead levels in children.
■ Increase the number of schools offering breakfast.
■ Increase vegetables, fruits, and whole grains in the diets of those aged 2 years and older, and reduce solid fats and added sugars.
Eating Disorders
■ Reduce the proportion of adolescents who engage in disordered eating behaviors in an attempt to control their weight.
Physical Activity and Weight Control
■ Increase the proportion of children, adolescents, and adults who are at a healthy weight.
■ Reduce the proportions of children, adolescents, and adults who are obese.
■ Reduce the proportion of people who engage in no leisure-time physical activity.
■ Increase the proportion of schools that require daily physical education for all students.
Food Security
■ Eliminate very low food security among children in U.S. households.

Source: www.healthypeople.gov.

The next section shifts our focus to the nutrients at the core of nutrition science. As your course of study progresses, the individual nutrients may become like old friends, revealing more and more about themselves as you move through the chapters.

KEY POINT

- Each decade, the U.S. Department of Health and Human Services sets health and nutrition objectives for the nation.

The Human Body and Its Food

LO 1.3 Define the term *nutrient* and be able to list the six major nutrients.

As your body moves and works each day, it must use **energy**. The energy that fuels the body's work comes indirectly from the sun by way of plants. Plants capture and store the sun's energy in their tissues as they grow. When you eat plant-derived foods such as fruits, grains, or vegetables, you obtain and use the solar energy they have stored. Plant-eating animals obtain their energy in the same way, so when you eat animal tissues, you are eating compounds containing energy that came originally from the sun.

Table 1–3

Elements in the Six Classes of Nutrients

The nutrients that contain carbon are organic.

	Carbon	Oxygen	Hydrogen	Nitrogen	Minerals
Water		✓	✓		
Carbohydrate	✓	✓	✓		
Fat	✓	✓	✓		
Protein	✓	✓	✓	✓	[b]
Vitamins	✓	✓	✓	✓[a]	[b]
Minerals					✓

[a]*All of the B vitamins contain nitrogen;* amine *means nitrogen.*

[b]*Protein and some vitamins contain the mineral sulfur; vitamin* B_{12} *contains the mineral cobalt.*

The body requires six kinds of nutrients—families of molecules indispensable to its functioning—and foods deliver these. Table 1–3 lists the six classes of nutrients. Four of these six are **organic**; that is, the nutrients contain the element carbon derived from living things.

Meet the Nutrients

The human body and foods are made of the same materials, arranged in different ways (see Figure 1–2). When considering quantities of foods and nutrients, scientists often measure them in **grams**, units of weight.

The Energy-Yielding Nutrients Foremost among the six classes of nutrients in foods is water, which is constantly lost from the body and must constantly be

Figure 1–2

Components of Food and the Human Body

Foods and the human body are made of the same materials.

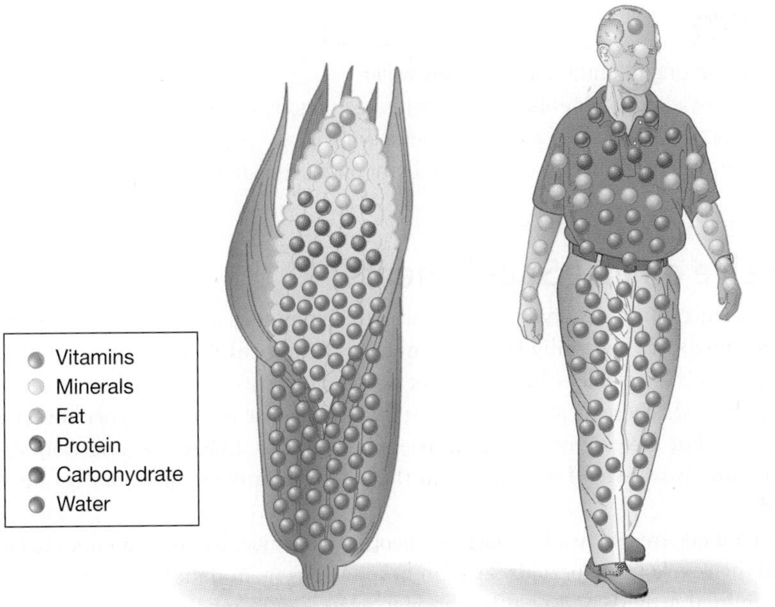

energy the capacity to do work. The energy in food is chemical energy; it can be converted to mechanical, electrical, thermal, or other forms of energy in the body. Food energy is measured in calories, defined on page 8.

organic carbon containing. Four of the six classes of nutrients are organic: carbohydrate, fat, protein, and vitamins. Organic compounds include only those made by living things and do not include compounds such as carbon dioxide, diamonds, and a few carbon salts.

grams units of weight. A gram (g) is the weight of a cubic centimeter (cc) or milliliter (ml) of water under defined conditions of temperature and pressure. About 28 grams equal an ounce.

Table 1–4

Calorie Values of Energy Nutrients

The energy a person consumes in a day's meals comes from these three energy-yielding nutrients; alcohol, if consumed, also contributes energy.

Energy Nutrient	Energy
Carbohydrate	4 cal/g
Fat (lipid)	9 cal/g
Protein	4 cal/g

Note: Alcohol contributes 7 cal/g that the human body can use for energy. Alcohol is not classed as a nutrient, however, because it interferes with growth, maintenance, and repair of body tissues.

© Cengage Learning

Did You Know?

- Energy-yielding nutrients are also called macronutrients because they are needed in relatively large amounts in the diet.
- Vitamins and minerals are also called micronutrients because they are needed in smaller amounts.

© Michael Gray|Dreamstime.com

When you eat foods, you are receiving more than just nutrients.

replaced. Of the four organic nutrients, three are **energy-yielding nutrients**, meaning that the body can use the energy they contain. The carbohydrates and fats (fats are also called lipids) are especially important energy-yielding nutrients. As for protein, it does double duty: it can yield energy, but it also provides materials that form structures and working parts of body tissues. (Alcohol yields energy, too—see the note to Table 1–4).

Vitamins and Minerals The fifth and sixth classes of nutrients are the vitamins and the minerals. These provide no energy to the body. A few minerals serve as parts of body structures (calcium and phosphorus, for example, are major constituents of bone), but all vitamins and minerals act as regulators. As regulators, the vitamins and minerals assist in all body processes: digesting food; moving muscles; disposing of wastes; growing new tissues; healing wounds; obtaining energy from carbohydrate, fat, and protein; and participating in every other process necessary to maintain life. Later chapters are devoted to these six classes of nutrients.

The Concept of Essential Nutrients When you eat food, then, you are providing your body with energy and nutrients. Furthermore, some of the nutrients are **essential nutrients**, meaning that if you do not ingest them, you will develop deficiencies; the body cannot make these nutrients for itself. Essential nutrients are found in all six classes of nutrients. Water is an essential nutrient; so is a form of carbohydrate; so are some lipids, some parts of protein, all of the vitamins, and the minerals important in human nutrition.

Calorie Values Food scientists measure food energy in kilocalories, units of heat. This book uses the common word ***calories*** to mean the same thing. It behooves the person who wishes to control food energy intake and body fatness to learn the calorie values of the energy nutrients, listed in Table 1–4. The most energy-rich of the nutrients is fat, which contains 9 calories in each gram. Carbohydrate and protein each contain only 4 calories in a gram. Weight, measure, and other conversion factors needed for the study of nutrition are found in Appendix C at the back of the book.

Scientists have worked out ways to measure the energy and nutrient contents of foods. They have also calculated the amounts of energy and nutrients various types of people need—by gender, age, life stage, and activity. Thus, after studying human nutrient requirements (in Chapter 2), you will be able to state with some accuracy just what your own body needs—this much water, that much carbohydrate, so much vitamin C, and so forth. So why not simply take pills or **dietary supplements** in place of food? Because, as it turns out, food offers more than just the six basic nutrients.

KEY POINTS

- Foremost among the nutrients in food is water.
- The energy-yielding nutrients are carbohydrates, fats (lipids), and protein.
- The regulator nutrients are vitamins and minerals.
- Food energy is measured in calories; nutrient quantities are often measured in grams.

Can I Live on Just Supplements?

Nutrition science can state what nutrients human beings need to survive—at least for a time. Scientists are becoming skilled at making **elemental diets**—life-saving liquid diets of precise chemical composition for hospital patients and others who cannot eat ordinary food. These formulas, administered for days or weeks, support not only continued life but also recovery from nutrient deficiencies, infections, and wounds. Formulas can also stave off weight loss in the elderly or anyone in whom eating is impaired.

Formula diets are essential to help sick people to survive, but they do not enable people to thrive over long periods. Even in hospitals, elemental diet formulas do not support optimal growth and health, and may even lead to medical complications.[4]

Although serious problems are rare and can be detected and corrected, they show that the composition of these diets is not yet perfect for all people in all settings.

Lately, marketers have taken these liquid supplement formulas out of the medical setting and have advertised them heavily to healthy people of all ages as "meal replacers" or "insurance" against malnutrition. The truth is that real food is superior to such supplements. Most healthy people who eat a nutritious diet need no dietary supplements at all.

Even if a person's basic nutrient needs are perfectly understood and met, concoctions of nutrients still lack something that foods provide. Hospitalized clients who are fed nutrient mixtures through a vein often improve dramatically when they can finally eat food. Something in real food is important to health—but what is it? What does food offer that cannot be provided through a needle or a tube? Science has some partial explanations, some physical and some psychological.

In the digestive tract, the stomach and intestine are dynamic, living organs, changing constantly in response to the foods they receive—even to just the sight, aroma, and taste of food. When a person is fed through a vein, the digestive organs, like unused muscles, weaken and grow smaller. Medical wisdom now dictates that a person should be fed through a vein for as short a time as possible and that real food taken by mouth should be reintroduced as early as possible. The digestive organs also release hormones in response to food, and these send messages to the brain that bring the eater a feeling of satisfaction: "There, that was good. Now I'm full." Eating offers both physical and emotional comfort.

Foods are chemically complex. In addition to their nutrients, foods contain **phytochemicals**, compounds that confer color, taste, and other characteristics to foods. Some may be **bioactive** food components that interact with metabolic processes in the body and may affect disease risks. Even an ordinary baked potato contains hundreds of different compounds. Nutrients and other food components interact with each other in the body and operate best in harmony with one another.[5] In view of all this, it is not surprising that food gives us more than just nutrients. If it were otherwise, *that* would be surprising.

© Brian Chase/Shutterstock.com

Some foods offer phytochemicals in addition to the six classes of nutrients.

KEY POINTS

- Food conveys emotional satisfaction and hormonal stimuli that contribute to health.
- Foods also contain phytochemicals.

The Challenge of Choosing Foods

LO 1.4 Summarize the five characteristics of a healthy diet and describe cultural or other influences on human food choices.

Well-planned meals convey pleasure and are nutritious, too, fitting your tastes, personality, family and cultural traditions, lifestyle, and budget. Given the astounding numbers and varieties available, a consumer can easily lose track of what individual foods contain and how to put them together into a health-promoting diet. A few guidelines can help.

The Abundance of Foods to Choose From

A list of the foods available 100 years ago would be relatively short. It would consist of **whole foods**—foods that have been around for a long time, such as vegetables, fruits, meats, milk, and grains (see Table 1–5 for a glossary of food types). These foods have been called basic, unprocessed, natural, or farm foods. By whatever name, choosing a sufficient variety of these foods each day is an easy way to obtain a nutritious diet. On a given day, however, almost three-quarters of our population consume too few vegetables, and two-thirds of us fail to consume enough fruit.[6] Also, although people generally consume a few servings of vegetables, the vegetable they most often choose is potatoes, usually prepared as French fries. Such dietary patterns make development of chronic diseases more likely.

energy-yielding nutrients the nutrients the body can use for energy—carbohydrate, fat, and protein. These also may supply building blocks for body structures.

essential nutrients the nutrients the body cannot make for itself (or cannot make fast enough) from other raw materials; nutrients that must be obtained from food to prevent deficiencies.

calories units of energy. In nutrition science, the unit used to measure the energy in foods is a kilocalorie (also called *kcalorie* or *Calorie*): it is the amount of heat energy necessary to raise the temperature of a kilogram (a liter) of water 1 degree Celsius. This book follows the common practice of using the lowercase term *calorie* (abbreviated *cal*) to mean the same thing.

dietary supplements pills, liquids, or powders that contain purified nutrients or other ingredients (see Controversy in Chapter 7).

elemental diets diets composed of purified ingredients of known chemical composition; intended to supply all essential nutrients to people who cannot eat foods.

phytochemicals compounds in plant-derived foods (*phyto* means "plant").

bioactive having biological activity in the body. See also the Controversy in Chapter 2.

Table 1–5

Glossary of Food Types

The purpose of this little glossary is to show that good-sounding food names don't necessarily signify that foods are nutritious. Read the comment at the end of each definition.

- **whole foods** milk and milk products; meats and similar foods such as fish and poultry; vegetables, including dried beans and peas; fruits; and grains. These foods are generally considered to form the basis of a nutritious diet. Also called *basic foods*.
- **enriched foods** and **fortified foods** foods to which nutrients have been added. If the starting material is a whole, basic food such as milk or whole grain, the result may be highly nutritious. If the starting material is a concentrated form of sugar or fat, the result may be less nutritious.
- **fast foods** restaurant foods that are available within minutes after customers order them—traditionally, hamburgers, French fries, and milkshakes; more recently, salads and other vegetable dishes as well. These foods may or may not meet people's nutrient needs, depending on the selections made and on the energy allowances and nutrient needs of the eaters.
- **functional foods** whole or modified foods that contain bioactive food components believed to provide health benefits, such as reduced disease risks, beyond the benefits that their nutrients confer. However, all nutritious foods can support health in some ways; Controversy 2 provides details.
- **medical foods** foods specially manufactured for use by people with medical disorders and administered on the advice of a physician.
- **natural foods** a term that has no legal definition but is often used to imply wholesomeness.
- **nutraceutical** a term that has no legal or scientific meaning but is sometimes used to refer to foods, nutrients, or dietary supplements believed to have medicinal effects. Often used to sell unnecessary or unproven supplements.
- **organic foods** understood to mean foods grown without synthetic pesticides or fertilizers. In chemistry, however, all foods are made mostly of organic (carbon-containing) compounds. (See Chapter 12 for details.)
- **processed foods** foods subjected to any process, such as milling, alteration of texture, addition of additives, cooking, or others. Depending on the starting material and the process, a processed food may or may not be nutritious.
- **staple foods** foods used frequently or daily, for example, rice (in East and Southeast Asia) or potatoes (in Ireland). If well chosen, these foods are nutritious.

Did You Know?

In 1900, Americans chose from among 500 or so different foods; today, they choose from among tens of thousands.

The number of foods supplied by the food industry today is astounding. Thousands of foods now line the market shelves—many are processed mixtures of the basic ones, and some are constructed entirely from highly processed ingredients. Ironically, this abundance often makes it more difficult, rather than easier, to plan a nutritious diet.

The food-related terms defined in Table 1–5 reveal that all types of food—including **fast foods** and **processed foods**—offer various constituents to the eater. You may also hear about **functional foods**, a marketing term coined to identify those foods containing substances, natural or added, that might lend protection against chronic diseases. The trouble is, scientists trying to single out the most health-promoting foods find that almost every naturally occurring food—even chocolate—is functional in some way with regard to human health.[7] Controversy 2 in Chapter 2 provides more information about functional foods.

All foods once looked like this . . .

. . . but now many foods look like this.

The extent to which foods support good health depends on the calories, nutrients, and phytochemicals they contain. In short, to select well among foods, you need to know more than their names; you need to know the foods' inner qualities. Even more important, you need to know how to combine foods into nutritious diets. Foods are not nutritious by themselves; each is of value only insofar as it contributes to a nutritious diet. A key to wise diet planning is to make sure that the foods you eat daily, your **staple foods**, are especially nutritious.

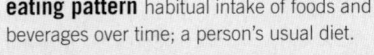

Lusoimages/Shutterstock.com

KEY POINT

- Foods that form the basis of a nutritious diet are whole foods, such as ordinary milk and milk products; meats, fish, and poultry; vegetables and dried peas and beans; fruits; and grains.

How, Exactly, Can I Recognize a Nutritious Diet?

A nutritious diet is really an **eating pattern**, a habitual way of eating, with five characteristics. First is **adequacy**: the foods provide enough of each essential nutrient, fiber, and energy. Second is **balance**: the choices do not overemphasize one nutrient or food type at the expense of another. Third is **calorie control**: the foods provide the amount of energy you need to maintain appropriate weight—not more, not less. Fourth is **moderation**: the foods do not provide excess fat, salt, sugar, or other unwanted constituents. Fifth is **variety**: the foods chosen differ from one day to the next. In addition, to maintain a steady supply of nutrients, meals should occur with regular timing throughout the day. To recap, then, a nutritious diet is an eating pattern that follows the A, B, C, M, V principles: Adequacy, Balance, Calorie control, Moderation, and Variety.

Adequacy Any nutrient could be used to demonstrate the importance of dietary adequacy. Iron provides a familiar example. It is an essential nutrient: you lose some every day, so you have to keep replacing it; and you can get it into your body only by eating foods that contain it.* If you eat too few of the iron-containing foods, you can develop iron-deficiency anemia. With anemia you may feel weak, tired, cold, sad, and unenthusiastic; you may have frequent headaches; and you can do very little muscular work without disabling fatigue. Some foods are rich in iron; others are notoriously poor. If you add iron-rich foods to your diet, you soon feel more energetic. Meat, fish, poultry, and **legumes** are in the iron-rich category, and an easy way to obtain the needed iron is to include these foods in your diet regularly.

Balance To appreciate the importance of dietary balance, consider a second essential nutrient, calcium. A diet lacking calcium causes poor bone development during the growing years and increases a person's susceptibility to disabling bone loss in adult life. Most foods that are rich in iron are poor in calcium. Calcium's richest food sources are milk and milk products, which happen to be extraordinarily poor iron sources. Clearly, to obtain enough of both iron and calcium, people have to balance their food choices among the types of foods that provide specific nutrients. Balancing the whole diet to provide enough but not too much of every one of the 40-odd nutrients the body needs for health requires considerable juggling, however. As you will see in Chapter 2, food group plans that cluster rich sources of nutrients into food groups can help you to achieve dietary adequacy and balance because they recommend specific amounts of foods from each group. An eating pattern with balance among the food groups then becomes the goal.

Calorie Control Energy intakes should not exceed energy needs. Named *calorie control*, this characteristic ensures that energy intakes from food balance energy expenditures required for body functions and physical activity. Eating such a diet helps to control body fat content and weight. The many strategies that promote this goal appear in Chapter 9.

* A person can also take supplements of iron, but as later discussions demonstrate, eating iron-rich foods is preferable.

eating pattern habitual intake of foods and beverages over time; a person's usual diet.

adequacy the dietary characteristic of providing all of the essential nutrients, fiber, and energy in amounts sufficient to maintain health and body weight.

balance the dietary characteristic of providing foods of a number of types in proportion to each other, such that foods rich in some nutrients do not crowd out of the diet foods that are rich in other nutrients. Also called *proportionality*.

calorie control control of energy intake; a feature of a sound diet plan.

moderation the dietary characteristic of providing constituents within set limits, not to excess.

variety the dietary characteristic of providing a wide selection of foods—the opposite of monotony.

legumes (leg-GOOMS, LEG-yooms) beans, peas, and lentils, valued as inexpensive sources of protein, vitamins, minerals, and fiber that contribute little fat to the diet. Also defined in Chapter 6.

Figure 1–3

Components of a Nutritious Diet

All of these factors help to build a nutritious diet.

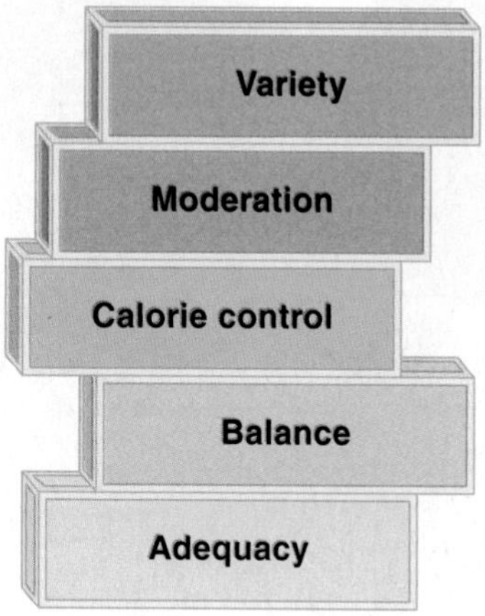

Moderation Intakes of certain food constituents such as saturated fats, cholesterol, added sugars, and salt should be limited for health's sake. Some people take this to mean that they must never indulge in a delicious beefsteak or hot-fudge sundae, but they are misinformed: moderation, not total abstinence, is the key. A steady diet of steak and ice cream might be harmful, but once a week as part of an otherwise healthful eating pattern, these foods may have little impact; as once-a-month treats, these foods would have practically no effect at all. Moderation also means that limits are necessary, even for desirable food constituents. For example, a certain amount of fiber in foods contributes to the health of the digestive system, but too much fiber leads to nutrient losses.

Variety As for variety, nutrition scientists agree that people should not eat the same foods, even highly nutritious ones, day after day. One reason is that a varied diet is more likely to be adequate in nutrients.[8] In addition, some less-well-known nutrients and phytochemicals could be important to health, and some foods may be better sources of these than others. Another reason is that a monotonous diet may deliver large amounts of toxins or contaminants. Such undesirable compounds in one food are diluted by all the other foods eaten with it and are diluted still further if the food is not eaten again for several days. Last, variety adds interest—trying new foods can be a source of pleasure.

Variety applies to nutritious foods consumed within the context of all of the other dietary principles just discussed. Relying solely on the principle of variety to dictate food choices could easily result in a low-nutrient, high-calorie eating pattern with a variety of nutrient-poor snack foods and sweets. If you establish the habit of using all of the principles just described, you will find that choosing a healthful diet becomes as automatic as brushing your teeth or falling asleep. Establishing the A, B, C, M, V habit (summed up in Figure 1–3) may take some effort, but the payoff in terms of improved health is overwhelming. Table 1–6 takes an honest look at some common excuses for *not* eating well.

KEY POINT

- A well-planned diet is adequate, balanced, moderate in energy, and moderate in unwanted constituents, and offers a variety of nutritious foods.

Why People Choose Foods

Eating is an intentional act. Each day, people choose from the available foods, prepare the foods, decide where to eat, which customs to follow, and with whom to dine. Many factors influence food-related choices.

Cultural and Social Meanings Attached to Food

Like wearing traditional clothing or speaking a native language, enjoying traditional **cuisines** and **foodways** can be a celebration of your own or a friend's heritage. Sharing **ethnic foods** can be

Table 1–6

What's Today's Excuse for Not Eating Well?

If you find yourself saying, "I know I should eat well, but I'm too busy" (or too fond of fast food, or have too little money, or a dozen other excuses), take note:

- *No time to cook.* Everyone is busy. Convenience packages of frozen vegetables, jars of pasta sauce, and prepared meats and salads make nutritious meals in little time.
- *Not a high priority.* Priorities change drastically and instantly when illness strikes—better to spend a little effort now nourishing your body's defenses than to spend enormous resources later fighting illnesses.
- *Crave fast food and sweets.* Occasional fast-food meals and sweets in moderation are acceptable in a nutritious diet.
- *Too little money.* Eating right costs no more than eating poorly. Chips, colas, fast food, and premium ice cream cost as much or more per serving as nutritious foods.[a]
- *Take vitamins instead.* Vitamin pills cannot make up for consistently poor food choices.

© Cengage Learning

[a]*For a discussion of this topic, see A. Carlson and E. Frazão, Are healthy foods really more expensive? It depends on how you measure the price, Economic Research Service EIB-96, May 2012, available at www.ers.usda.gov/publications/eib96.*

symbolic: people offering foods are expressing a willingness to share cherished values with others. People accepting those foods are symbolically accepting not only the person doing the offering but also the person's culture. Developing **cultural competence** is particularly important for professionals who help others to achieve a nutritious diet.[9]

Cultural traditions regarding food are not inflexible; they keep evolving as people move about, learn about new foods, and teach each other. Today some people are ceasing to be **omnivores** and are becoming **vegetarians**. Vegetarians often choose this lifestyle because they honor the lives of animals or because they have discovered the health and other advantages associated with eating patterns rich in beans, whole grains, fruits, nuts, and vegetables.[10] The Chapter 6 Controversy explores the pros and the cons of both the vegetarian's and the meat eater's diets.

blue jean images/Getty Images

Sharing ethnic food is a way of sharing culture.

Factors That Drive Food Choices Taste prevails as the number-one factor driving food choices by U.S. consumers, with price a close second.[11] Consumers also value convenience so highly that they are willing to spend almost half of their food budget on meals prepared outside the home.[12] They frequently eat out, bring home ready-to-eat meals, cook meals ahead in commercial kitchens, or have food delivered. In their own kitchens, they want to prepare a meal in 15 to 20 minutes, using only a few ingredients. Such convenience has a cost in terms of nutrition, however: eating away from home reduces intakes of fruit, vegetables, milk, and whole grains and increases intakes of calories, saturated fat, sodium, and added sugars.[13] Convenience doesn't have to mean that nutrition is out the window, however. This chapter's Food Feature explores the trade-offs of time, money, and nutrition that many busy people face today.

Many other factors, psychological, physical, social, and philosophical, all influence how people choose which foods to eat. Some factors include:

- *Advertising.* The media have persuaded you to consume these foods.
- *Availability.* They are present in the environment and accessible to you.[14]
- *Cost.* They are within your financial means.
- *Emotional comfort.* They can make you feel better for a while.
- *Habit.* They are familiar; you always eat them.
- *Personal preference and genetic inheritance.* You like the way these foods taste.
- *Positive or negative associations. Positive:* They are eaten by people you admire, or they indicate status, or they remind you of fun. *Negative*: They were forced on you or you became ill while eating them.
- *Region of the country.* They are foods favored in your area.
- *Social pressure.* They are offered; you feel you can't refuse them.
- *Values or beliefs.* They fit your religious tradition, square with your political views, or honor the environmental ethic.
- *Weight.* You think they will help to control body weight.
- *Nutrition and health benefits.* You think they are good for you.

College students often choose to eat at fast-food and other restaurants to socialize, to get out, to save time, or to date; they are not always conscious of their body's need for nutritious food.

Nutrition understanding depends upon a firm base of scientific knowledge. The next section describes the nature of such knowledge and addresses one of the "how" questions posed earlier in this chapter: how do we know what we know about nutrition?

KEY POINTS

- Cultural traditions and social values often revolve around foodways.
- Many factors other than nutrition drive food choices.

cuisines styles of cooking.

foodways the sum of a culture's habits, customs, beliefs, and preferences concerning food.

ethnic foods foods associated with particular cultural subgroups within a population.

cultural competence having an awareness and acceptance of one's own and others' cultures and abilities leading to effective interactions with all kinds of people.

omnivores people who eat foods of both plant and animal origin, including animal flesh.

vegetarians people who exclude from their diets animal flesh and possibly other animal products such as milk, cheese, and eggs.

The Science of Nutrition

LO 1.5 Describe the major types of research studies and give reasons why national nutrition research is important for the health of the population.

Nutrition is a science—a field of knowledge composed of organized facts. Unlike sciences such as astronomy and physics, nutrition is a relatively young science. Most nutrition research has been conducted since 1900. The first vitamin was identified in 1897, and the first protein structure was not fully described until the mid-1940s. Because nutrition science is an active, changing, growing body of knowledge, scientific findings often seem to contradict one another or are subject to conflicting interpretations. Bewildered consumers complain in frustration, "Those scientists don't know anything. If they don't know what's true, how am I supposed to know?"

Yet, many facts in nutrition are known with great certainty. To understand why apparent contradictions sometimes arise in nutrition science, we need to look first at what scientists do.

Figure 1–4

Animated! The Scientific Method

Research scientists follow the scientific method. Note that most research projects result in new questions, not final answers. Thus, research continues in a somewhat cyclical manner.

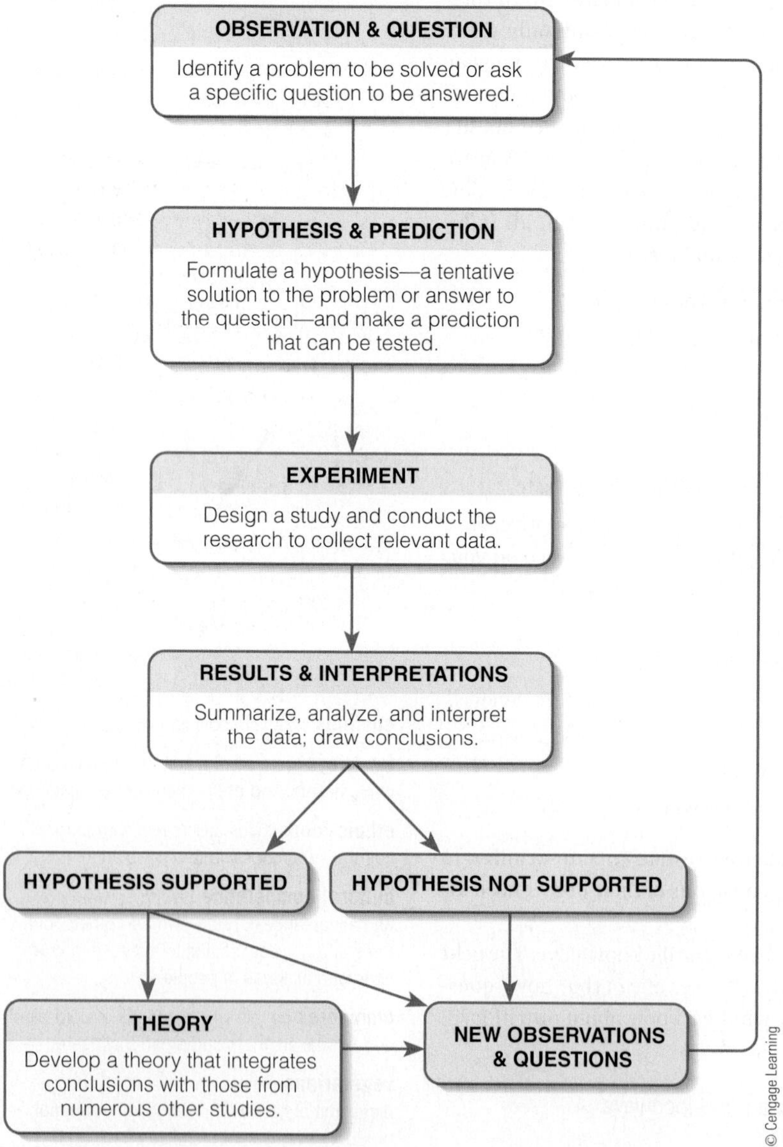

OBSERVATION & QUESTION
Identify a problem to be solved or ask a specific question to be answered.
HYPOTHESIS & PREDICTION
Formulate a hypothesis—a tentative solution to the problem or answer to the question—and make a prediction that can be tested.
EXPERIMENT
Design a study and conduct the research to collect relevant data.
RESULTS & INTERPRETATIONS
Summarize, analyze, and interpret the data; draw conclusions.
HYPOTHESIS SUPPORTED
HYPOTHESIS NOT SUPPORTED
THEORY
Develop a theory that integrates conclusions with those from numerous other studies.
NEW OBSERVATIONS & QUESTIONS

The Scientific Approach

In truth, it is a scientist's business not to know. Scientists obtain facts by systematically asking honest objective questions—that's their job. Following the scientific method (outlined in Figure 1–4), they attempt to answer scientific questions. They design and conduct various experiments to test for possible answers (see Figure 1–5 and Table 1–7, p. 16). When they have ruled out some possibilities and found evidence for others, they submit their findings, not to the news media, but to boards of reviewers composed of other scientists who try to pick the findings apart. Finally, the work is published in scientific journals where still more scientists can read it. Then the news reporters read it and write about it and the public can read about it, too. Table 1–8 (p. 16) explains what you can expect to find in a journal article.

KEY POINT

- Scientists ask questions and then design research experiments to test possible answers.

Scientific Challenge

An important truth in science is that one experiment does not "prove" or "disprove" anything. Even after publication, other scientists try to duplicate the work of the first researchers to support or refute the original finding.

Only when a finding has stood up to rigorous, repeated testing in several kinds of experiments performed by several different researchers is it finally considered confirmed. Even then, strictly speaking, science consists not of facts that are set in stone, but of *theories* that can always be challenged and revised. Some findings, though, like the theory that the earth revolves about the sun, are so well supported by observations and experimental findings that they are generally accepted

Figure 1–5
Examples of Research Design

Case Study

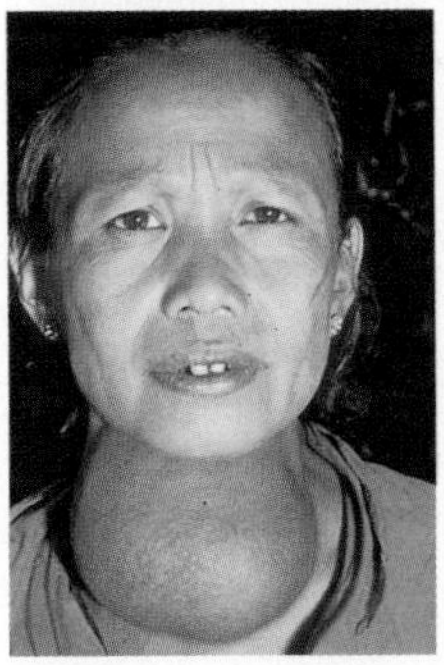

© Lester V. Bergman/CORBIS

"This person eats too little of nutrient X and has illness Y."

Epidemiological Study

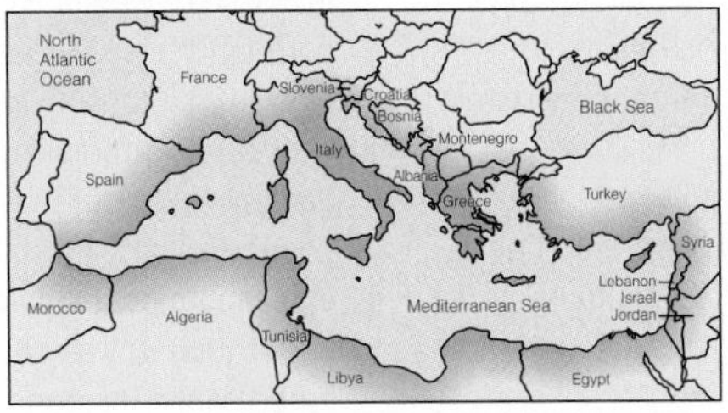

© Cengage Learning

"This country's food supply contains more nutrient X, and these people suffer less illness Y."

Intervention Study

David Buffington/Getty Images

"Let's add foods containing nutrient X to some people's food supply and compare their rates of illness Y with the rates of others who don't receive the nutrient."

Laboratory Study

Leslie Newman & Andrew Flowers/Science Source

"Now let's prove that a nutrient X deficiency causes illness Y by inducing a deficiency in these rats."

The type of study chosen for research depends upon what sort of information the researchers require. Studies of individuals (**case studies**) yield observations that may lead to possible avenues of research. A study of a man who ate gumdrops and became a famous dancer might suggest that an experiment be done to see if gumdrops contain dance-enhancing power.

Studies of whole populations (**epidemiological studies**) provide another sort of information. Such a study can reveal a **correlation**. For example, an epidemiological study might find no worldwide correlation of gumdrop eating with fancy footwork but, unexpectedly, might reveal a correlation with tooth decay.

Studies in which researchers actively intervene to alter people's eating habits (**intervention studies**) go a step further. In such a study, one set of subjects (the **experimental group**) receive a treatment, and another set (the **control group**) go untreated or receive a **placebo** or sham treatment. If the study is a **blind experiment**, the subjects do not know who among the members receives the treatment and who receives the sham. If the two groups experience different effects, then the treatment's effect can be pinpointed. For example, an intervention study might show that withholding gumdrops, together with other candies and confections, reduced the incidence of tooth decay in an experimental population compared to that in a control population.

Finally, **laboratory studies** can pinpoint the mechanisms by which nutrition acts. What is it about gumdrops that contributes to tooth decay: their size, shape, temperature, color, ingredients? Feeding various forms of gumdrops to rats might yield the information that sugar, in a gummy carrier, promotes tooth decay. In the laboratory, using animals or plants or cells, scientists can inoculate with diseases, induce deficiencies, and experiment with variations on treatments to obtain in-depth knowledge of the process under study. Intervention studies and laboratory experiments are among the most powerful tools in nutrition research because they show the effects of treatments.

as facts. What we "know" in nutrition is confirmed in the same way—through years of replicating study findings. This slow path of repeated studies stands in sharp contrast to the media's desire for today's latest news.

To repeat: the only source of valid nutrition information is slow, painstaking, authentic scientific research. We believe a nutrition fact to be true because it has been supported, time and again, in experiments designed to rule out all other possibilities. For example, we know that eyesight depends partly on vitamin A because:

- In case studies, individuals with blindness report having consumed a steady diet devoid of vitamin A, and
- In epidemiological studies, populations with diets lacking in vitamin A are observed to suffer high rates of blindness, and

Table 1–7
Research Design Terms

- **blind experiment** an experiment in which the subjects do not know whether they are members of the experimental group or the control group. In a *double-blind experiment*, neither the subjects nor the researchers know to which group the members belong until the end of the experiment.
- **case studies** studies of individuals. In clinical settings, researchers can observe treatments and their apparent effects. To prove that a treatment has produced an effect requires simultaneous observation of an untreated similar subject (a *case control*).
- **control group** a group of individuals who are similar in all possible respects to the group being treated in an experiment but who receive a sham treatment instead of the real one. Also called *control subjects*. See also *experimental group* and *intervention studies*.
- **controlled clinical trial** a research study design that often reveals effects of a treatment on human beings. Health outcomes are observed in a group of people who receive the treatment and are then compared with outcomes in a control group of similar people who received a placebo (an inert or sham treatment). Ideally, neither subjects nor researchers know who receives the treatment and who gets the placebo (a double-blind study).
- **correlation** the simultaneous change of two factors, such as the increase of weight with increasing height (a *direct* or *positive* correlation) or the decrease of cancer incidence with increasing fiber intake (an *inverse* or *negative* correlation). A correlation between two factors suggests that one may cause the other but does not rule out the possibility that both may be caused by chance or by a third factor.
- **epidemiological studies** studies of populations; often used in nutrition to search for correlations between dietary habits and disease incidence; a first step in seeking nutrition-related causes of diseases.
- **experimental group** the people or animals participating in an experiment who receive the treatment under investigation. Also called *experimental subjects*. See also *control group* and *intervention studies*.
- **intervention studies** studies of populations in which observation is accompanied by experimental manipulation of some population members—for example, a study in which half of the subjects (the *experimental subjects*) follow diet advice to reduce fat intakes while the other half (the *control subjects*) do not, and both groups' heart health is monitored.
- **laboratory studies** studies that are performed under tightly controlled conditions and are designed to pinpoint causes and effects. Such studies often use animals as subjects.
- **placebo** a sham treatment often used in scientific studies; an inert harmless medication. The *placebo effect* is the healing effect that the act of treatment, rather than the treatment itself, often has.

Table 1–8
The Anatomy of a Research Article

Here's what you can expect to find inside a research article:

- *Abstract.* The abstract provides a brief overview of the article.
- *Introduction.* The introduction clearly states the purpose of the current study.
- *Review of literature.* A review of the literature reveals all that science has uncovered on the subject to date.
- *Methodology.* The methodology section defines key terms and describes the procedures used in the study.
- *Results.* The results report the findings and may include summary tables and figures.
- *Conclusions.* The conclusions drawn are those supported by the data and reflect the original purpose as stated in the introduction. Usually, they answer a few questions and raise several more.
- *References.* The references list relevant studies (including key studies several years old as well as current ones).

- In intervention studies (**controlled clinical trials**), vitamin A-rich foods provided to groups of vitamin A-deficient people reduce their blindness rates dramatically, and
- In laboratory studies, animals deprived of vitamin A and only that vitamin begin to go blind; when it is restored soon enough in the diet, their eyesight returns, and
- Further laboratory studies elucidated the molecular mechanisms for vitamin A activity in eye tissues, and
- Replication of these studies provides the same results.

Now we can say with certainty, "eyesight depends upon sufficient vitamin A."

KEY POINT

- Single studies must be replicated before their findings can be considered valid.

Can I Trust the Media to Deliver Nutrition News?

The news media are hungry for new findings, and reporters often latch onto ideas from the scientific laboratories before they have been fully tested. Also, a reporter who lacks a strong understanding of science may misunderstand or misreport complex scientific principles.[15] To tell the truth, sometimes scientists get excited about their findings, too, and leak them to the press before they have been through a rigorous review by the scientists' peers.[16] As a result, the public is often exposed to late-breaking nutrition news stories before the findings are fully confirmed. Then, when the hypothesis being tested fails to hold up to a later challenge, consumers feel betrayed by what is simply the normal course of science at work.

The real scientists are trend watchers. They evaluate the methods used in each study, assess each study in light of the evidence gleaned from other studies, and modify little by little their picture of what may be true. As evidence accumulates, the scientists become more and more confident about their ability to make recommendations that apply to people's health and lives. The Consumer's Guide section near here offers some tips for evaluating news stories about nutrition.

Sometimes media sensationalism overrates the importance of even true, replicated findings. For example, the media eagerly report that oat products lower blood cholesterol, a lipid indicative of heart disease risk. Although the reports are true, they often fail to mention that eating a nutritious diet that is low in certain fats is still the major step toward lowering blood cholesterol. They also may skip over important questions: how much oatmeal must a person eat to produce the desired effect? Do little oat bran pills or powders meet the need? Do oat bran cookies? If so, how many cookies? For oatmeal, it takes a bowl and a half daily to affect blood lipids. A few pills or cookies do not provide nearly so much and certainly cannot undo all the damage from a high-fat meal.

Today, the cholesterol-lowering effect of oats is well-established. The whole process of discovery, challenge, and vindication took almost 10 years of research. Some other lines of research have taken much longer. In science, a single finding almost never makes a crucial difference to our knowledge as a whole, but like each individual frame in a movie, it contributes a little to the big picture. Many such frames are needed to tell the whole story.

KEY POINT

- News media often sensationalize single-study findings, and so may not be trustworthy sources.

My Turn **watch it!**

Lose Weight While You Sleep!

Gabriel

See a student talking about how he learned the truth about nutrition claims made in advertising.

National Nutrition Research

As you study nutrition, you are likely to hear of findings based on ongoing nationwide nutrition and health research projects.[17] A national food and nutrient intake survey, called *What We Eat in America*, reveals what we know about the population's food and supplement intakes. It is conducted as part of a larger research effort, the National Health and Nutrition Examination Surveys (NHANES) that also takes physical examinations and measurements and laboratory tests. Boiled down to its essence, NHANES involves:

- Asking people what they have eaten and
- Recording measures of their health status.

Past NHANES results have provided important data for developing growth charts for children, guiding food fortification efforts, developing national guidelines for reducing chronic diseases, and many other beneficial programs. Some agencies involved with these efforts are listed in the margin.

KEY POINT

- National nutrition research projects, such as NHANES, provide data on U.S. food consumption and nutrient status.

Did You Know?

These agencies are actively engaged in nutrition policy, research, and monitoring:

- Department of Health and Human Services (DHHS)
- U.S. Food and Drug Administration (FDA)
- U.S. Department of Agriculture (USDA)
- Centers for Disease Control and Prevention (CDC)

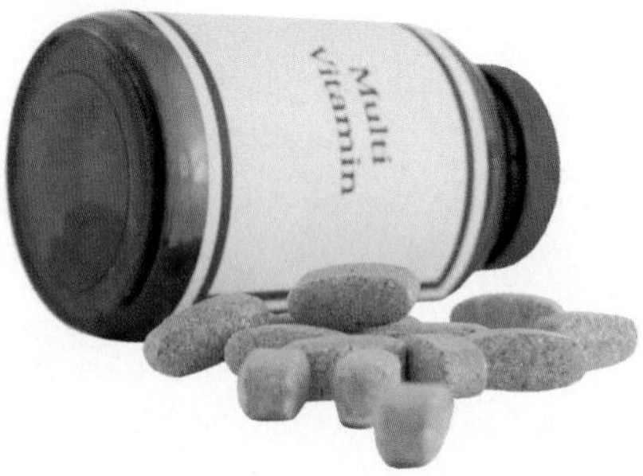

Changing Behaviors

LO 1.6 List the major steps in behavior change and devise a plan for making successful long-term changes in the diet.

Nutrition knowledge is of little value if it only helps people to make A's on tests. The value comes when people use it to improve their diets. To act on knowledge, people must change their behaviors, and while this may sound simple enough, behavior change often takes substantial effort.

The Process of Change

Psychologists often describe the six stages of behavior change, offered in Table 1–9. Knowing where you stand in relation to these stages may help you move along the

Table 1–9
The Stages of Behavior Change

Stage	Characteristics	Actions
Precontemplation	Not considering a change, have no intention of changing; see no problems with current behavior.	Collect information about health effects of current behavior and potential benefits of change.
Contemplation	Admit that change may be needed; weigh pros and cons of changing and not changing.	Commit to making a change and set a date to start.
Preparation	Preparing to change a specific behavior, taking initial steps, and setting some goals.	Write an action plan, spelling out specific parts of the change. Set small-step goals; tell others about the plan.
Action	Committing time and energy to making a change; following a plan set for a specific behavior change.	Perform the new behavior. Manage emotional and physical reactions to the change.
Maintenance	Striving to integrate the new behavior into daily life and striving to make it permanent.	Persevere through lapses. Teach others and help them achieve their own goals. (This stage can last for years.)
Adoption/Moving On	The former behavior is gone and the new behavior is routine.	After months or a year of maintenance without lapses, move on to other goals.

use it! A Consumer's Guide To . . .

Reading Nutrition News

At a coffee shop, Nick, a health-conscious consumer, sets his cup down on the Lifestyle section of the newspaper. He glances at the headline: "Eating Fat OK for Heart Health!" and jumps to a wrong conclusion: "Do you mean to say that I could have been eating burgers and butter all this time? I can't keep up! As soon as I change my diet, the scientists change their story." Nick's frustration is understandable. Like many others, he feels betrayed when, after working for years to make diet changes for his health's sake, headlines seem to turn dietary advice upside down. He shouldn't blame science, however.

Tricks and Traps

The trouble started when Nick was "hooked" by a catchy headline. Media headlines often sensationalize or oversimplify nutrition findings to engage readers' attention and make them want to buy a newspaper or magazine. (By the way, you can read the true story behind changing lipid intake guidelines in the Controversy section of Chapter 5.) Even if Nick had read the entire newspaper article, he could have still been led astray by phrases like "Now we know" or "The truth is." Journalists use such phrases to imply finality, the last word.[1] In contrast, scientists use tentative language, such as "may" or "might," because they know that the conclusions from one study will be challenged, refined, and even refuted by others that follow.

Markers of Authentic Reporting

To approach nutrition news with a trained eye, look for these signs of a scientific approach:

- When an article describes a scientific study, that study should have been published in a peer-reviewed journal, such as the *American Journal of Clinical Nutrition*. An unpublished study or one from a less credible source may or may not be valid; the reader has no way of knowing because the study lacks scrutiny by other experts.
- The news item should describe the researchers' methods; in truth, few popular reports provide these details. It matters whether the study participants numbered 8 or 80,000 or whether researchers personally observed participants' behaviors or relied on self-reports given over the telephone, for example.
- The report should define the study subjects—were they single cells, animals, or human beings? If they were human beings, the more you have in common with them (age and gender, for example), the more applicable the findings may be for you.
- Valid reports also present new findings in the context of previous research. Some reporters in popular media regularly follow developments in a research area and thus acquire the background knowledge needed to report meaningfully. They strive for adequacy, balance, and completeness, and they cover such things as cost of a treatment, potential harms and benefits, strength of evidence, and who might stand to gain from potential sales relating to the finding.*
- For a helpful *scientific* overview of current topics in nutrition, look for review articles written by experts. They regularly appear in scholarly journals such as *Nutrition Reviews.* A relative of the review article, the meta-analysis, uses the power of a computer to combine and reanalyze the results of many previously published studies on a single topic.

* An organization that promotes valid health care reporting is HealthNewsReview.org, available on the Internet.

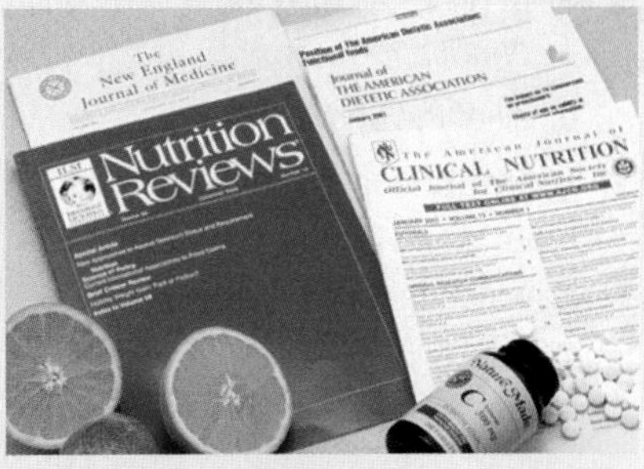

© Cengage Learning

For the whole story on a nutrition topic, read articles from peer-reviewed journals such as these. A review journal examines all available evidence on major topics. Other journals report details of the methods, results, and conclusions of single studies.

The most credible source of scientific nutrition information is the scientific journal. The Controversy section following this chapter addresses other sources of nutrition information and misinformation.

Moving Ahead

Develop a critical eye and let scientific principles guide you as you read nutrition news. When a headline touts a shocking new "answer" to a nutrition question, approach it with caution. It may indeed be a carefully researched report that respects the gradual nature of scientific discovery and refinement, but more often it is a sensational news flash intended to grab your attention and your media dollars.

Review Questions†

1. To keep up with nutrition science, the consumer should ________ .
   a. seek out the health and fitness sections of newspapers and magazines and read them with a trained eye

(continued)

† Answers to Consumer's Guide review questions are found in Appendix G.

b. read studies published in a peer-reviewed journal, such as the *American Journal of Clinical Nutrition*
c. look for review articles published in peer-reviewed journals, such as *Nutrition Reviews*
d. all of the above

2. To answer nutrition questions __________ .
   a. watch for articles that include phrases such as "now we know" or "the answer is" that put nutrition issues to rest
   b. look to science for answers, with the expectation that scientists will continually revise their understandings
   c. realize that problems in nutrition are probably too complex for consumers to understand
   d. a and c

3. Scholarly review journals such as *Nutrition Reviews* __________ .
   a. are behind the times when it comes to nutrition news
   b. discuss all available research findings on a topic in nutrition
   c. are filled with medical jargon
   d. are intended for use by practitioners only, not students

path toward achieving your goals. When offering diet help to others, keep in mind that the other person's stage of change can influence their reaction to your message.[18]

Taking Stock and Setting Goals

To make a change, you must first become aware of a problem. Some problems, such as *never* consuming a vegetable, are easy to spot. More subtle dietary problems, such as failing to meet your need for calcium, may be hidden but can have serious repercussions for health. Tracking food intakes over several days' time and then comparing intakes to standards (see Chapter 2) can reveal all sorts of interesting tidbits about strengths and weaknesses of your eating pattern.

Once a weakness is identified, setting small, achievable goals to correct it becomes the next step to making improvements. The most successful goals are set for specific behaviors, not overall outcomes. For example, if losing 10 pounds is the desired outcome, goals should be set in terms of food intakes and physical activity to help achieve weight loss.[19] After goals are set and changes are underway, a means of tracking progress increases awareness of barriers to changing a behavior. Much more information about achieving goals for weight management is offered in Chapter 9.

© UpperCut Images/Alamy

Many people need to change their daily routines to include physical activity.

Start Now

You may, as you progress through this text, want to change some of your own habits. To help you, little reminders entitled "Start Now" close each chapter's Think Fitness section (on page 5 in this chapter) with an invitation to visit this book's website where you can take inventory of your current behaviors, set goals, track progress, and practice new behaviors until they becomes as comfortable and familiar as the old ones were.

KEY POINTS

- Behavior change follows a predictable pattern.
- Setting goals and monitoring progress facilitate behavior change.

nutrient density a measure of nutrients provided per calorie of food. A *nutrient-dense food* provides vitamins, minerals, and other beneficial substances with relatively few calories.

try it! Food Feature

How Can I Get Enough Nutrients without Consuming Too Many Calories?

LO 1.7 Define *nutrient density* and explain the advantages of choosing nutrient-dense foods.

In the United States, only a tiny percentage of adults manage to choose an eating pattern that achieves both adequacy and moderation. The foods that can help in doing so are foods richly endowed with nutrients relative to their energy contents; that is, they are foods with high **nutrient density**.[20] Figure 1–6 is a simple depiction of this concept. Consider calcium sources, for example. Ice cream and fat-free milk both supply calcium, but a cup of rich ice cream contributes more than 350 calories, whereas a cup of fat-free milk has only 85—and almost double the calcium. Most people cannot, for their health's sake, afford to choose foods without regard to their energy contents. Those who do very often exceed calorie allowances while leaving nutrient needs unmet.

Nutrient density is such a useful concept in diet planning that this book encourages you to think in those terms. Right away, the next chapter asks you to apply your knowledge of nutrient density while developing skills in meal planning. Watch for tables and figures in later chapters that show the best buys among foods, not necessarily in nutrients per dollar, but in nutrients per calorie.

Among foods that often rank high in nutrient density are the vegetables, particularly the nonstarchy vegetables such as broccoli, carrots, mushrooms, peppers, and tomatoes. These inexpensive foods take time to prepare, but time invested in this way pays off in nutritional health. Twenty minutes spent peeling and slicing vegetables for a salad is a better investment in nutrition than 20 minutes spent fixing a fancy, high-fat, high-sugar dessert. Besides, the dessert ingredients often cost more money and strain the calorie budget, too.[21]

Time, however, is another concern. Today's working families, college students, and active people of all ages may have little time to devote to food preparation. Busy chefs should seek out convenience foods that are nutrient-dense, such as bags of ready-to-serve salads, refrigerated prepared low-fat meats and poultry, canned beans, and frozen vegetables. Dried fruit and dry-roasted nuts require only that they be kept on hand and make a tasty, nutritious topper for salads and other foods. To round out the meal, fat-free milk is both nutritious and convenient. Other convenience selections, such as most pot pies, many frozen pizzas, ramen noodles, and "pocket" style sandwiches, are less nutritious overall because they contain too few vegetables and too many calories, making them low in nutrient density. The Food Features of later

Figure 1–6

A Way to Judge Which Foods Are Most Nutritious

Some foods deliver more nutrients for the same number of calories than others do. These two breakfasts provide about 500 calories each, but they differ greatly in the nutrients they provide per calorie. Note that the sausage in the larger breakfast is lower-calorie turkey sausage, not the high-calorie pork variety. Making small choices like this at each meal can add up to large calorie savings, making room in the diet for more servings of nutritious foods and even some treats.

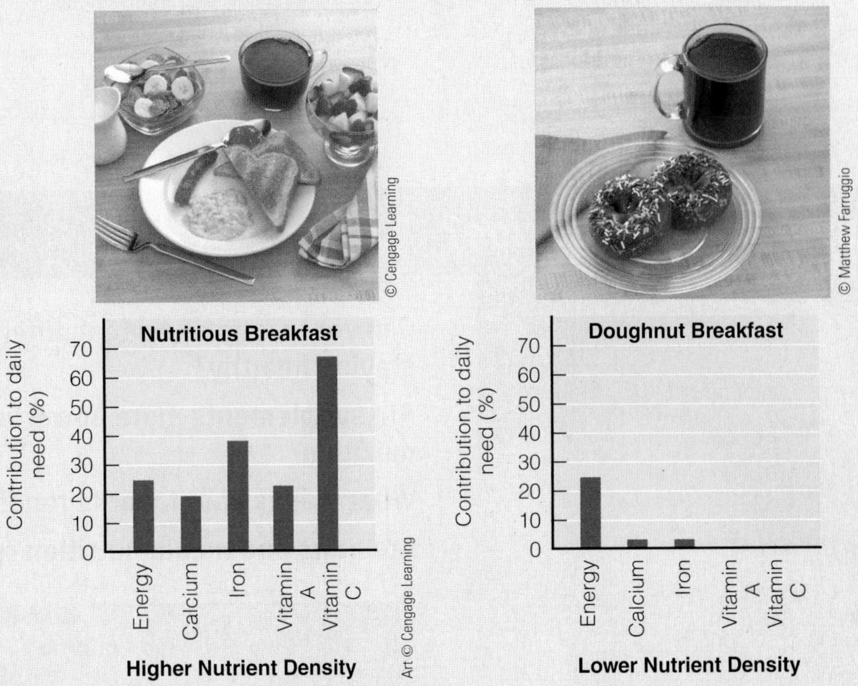

chapters offer many more tips for choosing convenient and nutritious foods.

All of this discussion leads to a principle that is central to achieving nutritional health: It is not the individual foods you choose but your eating pattern—the way you combine foods into meals and the way you arrange meals to follow one another over days and weeks—that determines how well you are nourishing yourself. Nutrition is a science, not an art, but it can be used artfully to create a pleasing, nourishing diet. The remainder of this book is dedicated to helping you make informed choices and combine them artfully to meet all the body's needs.

track it!

Diet Analysis PLUS+ Concepts in Action

Track Your Diet

After each Food Feature section in this text, exercises like this one provide an ongoing diet analysis activity that asks you to apply what you've learned in the chapter to your own diet. To do so, use the Diet Analysis Plus (DA+) program that accompanies this book. Do the following:

1. From the Home page of the DA+ program (after entering your personal data), select the Reports tab from the red navigation bar, then select Profile DRI Goals. Click Create PDF button. You will now have a list of the appropriate DRI values for calories, carbohydrates, and fat for your Profile.
2. For the next three days, with pencil and paper, keep track of everything you eat and drink. Be honest and careful in your record-keeping. Measure or estimate amounts of foods and beverages you consume, as well as margarine or butter, salt, cream sauces, gravies, pasta sauce, ketchup, relish, jams, jellies, and other add-ons. Even a slice of tomato and a lettuce leaf on a sandwich count toward the day's intake. Distribute your data among four meals for each day: breakfast, lunch, dinner, and snacks.
3. Keep track of your physical activity for all three of those days. Record all the minutes spent walking or biking to class, working out, vacuuming rugs, washing cars, playing sports, dancing with friends, or any other nonsedentary behavior. Hold onto this data: you'll need it in chapters to come.
4. From the Home page of DA+ select the Track Diet tab and enter each food item you recorded for Day One, Day Two, and Day Three into the Find Foods area. When finished, select the Reports tab and go to Intake vs. Goals. Click the Generate Report button, and choose all meals. What information on the report most surprised you?
5. From the Reports tab, go to Energy Balance. Using Day Two (from the three-day diet intake) choose all meals and generate a report. Was your calorie intake more or less than the recommended calories (kcal) for your profile? Was it higher or lower than you expected? You will analyze your energy balance in more detail later, in Chapter 9.

what did you decide?

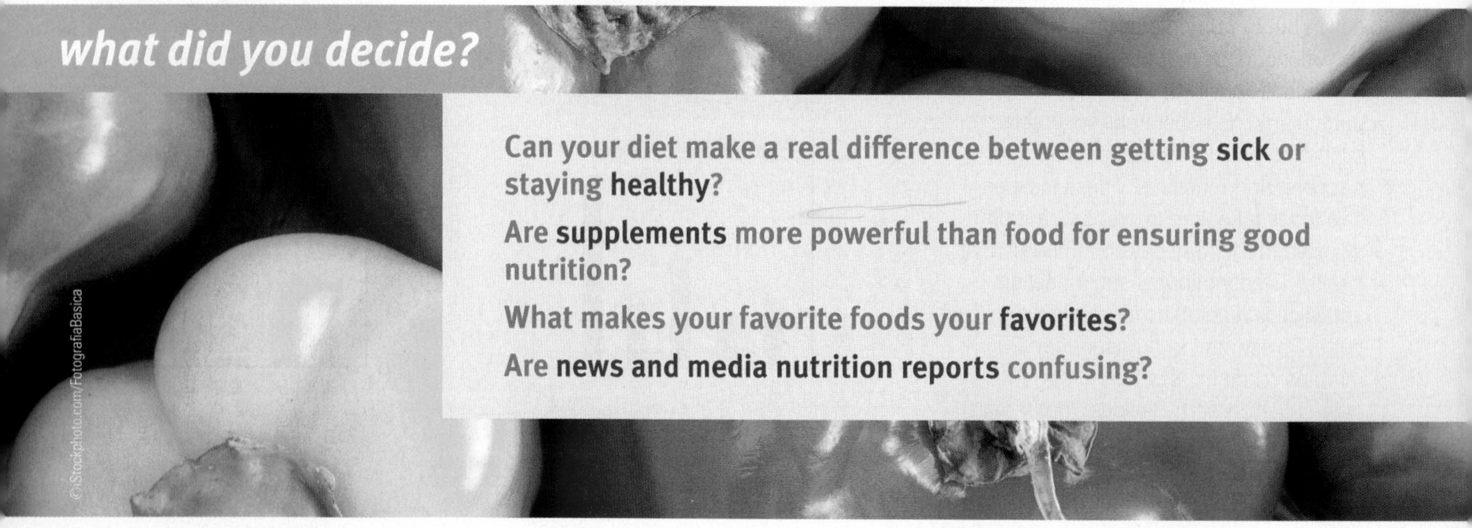

©iStockphoto.com/FotografiaBasica

Can your diet make a real difference between getting sick or staying healthy?

Are supplements more powerful than food for ensuring good nutrition?

What makes your favorite foods your favorites?

Are news and media nutrition reports confusing?

Self Check

1. (LO 1.1) Both heart disease and cancer are due to genetic causes, and diet cannot influence whether they occur.
   T F
2. (LO 1.1) Some conditions, such as ________, are almost entirely nutrition related.
   a. cancer
   b. Down syndrome
   c. iron-deficiency anemia
   d. sickle-cell anemia
3. (LO 1.2) The nutrition objectives for the nation, as part of *Healthy People 2020*,
   a. envision a society in which all people live long, healthy lives.
   b. track and identify cancers as a major killer of people in the United States.
   c. set U.S. nutrition- and weight-related goals, one decade at a time.
   d. a and c.
4. (LO 1.2) According to a national 2010 health report,
   a. most people's diets lacked enough fruits, vegetables, and whole grains.
   b. most people were sufficiently physically active.
   c. the number of overweight people was declining.
   d. the nation had fully met the previous *Healthy People* objectives.
5. (LO 1.3) Energy-yielding nutrients include all of the following except ________ .
   a. vitamins
   b. carbohydrates
   c. fat
   d. protein
6. (LO 1.3) Organic nutrients include all of the following except ________ .
   a. minerals
   b. fat
   c. carbohydrates
   d. protein
7. (LO 1.3) Both carbohydrates and protein have 4 calories per gram.
   T F
8. (LO 1.4) One of the characteristics of a nutritious diet is that the diet provides no constituent in excess. This principle of diet planning is called ________ .
   a. adequacy
   b. balance
   c. moderation
   d. variety
9. (LO 1.4) Which of the following is an example of a processed food?
   a. carrots
   b. bread
   c. nuts
   d. watermelon
10. (LO 1.4) People most often choose foods for the nutrients they provide.
    T F
11. (LO 1.5) Studies of populations in which observation is accompanied by experimental manipulation of some population members are referred to as ________ .
    a. case studies
    b. intervention studies
    c. laboratory studies
    d. epidemiological studies
12. (LO 1.5) An important national food and nutrient intake survey, called *What We Eat in America*, is part of
    a. NHANES
    b. FDA
    c. USDA
    d. none of the above
13. (LO 1.6) Behavior change is a process that takes place in stages.
    T F
14. (LO 1.6) A person who is setting goals in preparation for a behavior change is in a stage called *Precontemplation*.
    T F
15. (LO 1.7) A slice of peach pie supplies 357 calories with 48 units of vitamin A; one large peach provides 42 calories and 53 units of vitamin A. This is an example of ________ .
    a. calorie control
    b. nutrient density
    c. variety
    d. essential nutrients
16. (LO 1.7) A person who wishes to meet nutrient needs while not overconsuming calories is wise to master
    a. the concept of nutrient density
    b. the concept of carbohydrate reduction
    c. the concept of nutrients per dollar
    d. French cooking
17. (LO 1.8) These "red flags" can help to identify nutrition quackery:
    a. enticingly quick and simple answers to complex problems
    b. casts suspicion on the regular food supply
    c. offers solid support and praise from users
    d. all of the above
18. (LO 1.8) In this nation, stringent controls make it difficult to obtain a bogus nutrition credential.
    T F

Answers to these Self Check questions are in Appendix G.

CONTROVERSY 1

Sorting the Imposters from the Real Nutrition Experts

LO 1.8 Identify misleading nutrition information in infomercials, advertorials, and other sources in the popular media.

From the time of salesmen in horse-drawn wagons selling snake oil to today's Internet sales schemes, nutrition **quackery** has been a problem that often escapes government regulation and enforcement. To protect themselves from the quacks, consumers themselves must distinguish between authentic, useful nutrition products or services and a vast array of faulty advice and outright scams.

Each year, consumers spend a deluge of dollars on nutrition-related services and products from both legitimate and fraudulent businesses. Each year, nutrition and other health **fraud** diverts tens of *billions* of consumer dollars from legitimate health care professionals.*[1]

More Than Money at Stake

When scam products are garden tools or stain removers, hoodwinked consumers may lose a few dollars and some pride. When the products are ineffective, untested, or even hazardous "dietary supplements" or "medical devices," consumers stand to lose the very thing they are seeking: good health. When a sick person wastes time with quack treatments, serious problems can easily advance while proper treatment is delayed.[2] And ill-advised "dietary supplements" have inflicted dire outcomes, even liver failure, on previously well people who took them in hopes of *improving* their health.

** Reference notes are found in Appendix F.*

Information Sources

When asked, most people name television as their primary source of nutrition knowledge, with magazine articles a close second, and the Internet gaining quickly from behind.[3] Sometimes, these sources provide sound, scientific, trustworthy information. More often, though, **infomercials**, **advertorials**, and **urban legends** (defined in Table C1–1) pretend to inform but in fact aim primarily to sell products by making fantastic promises for health or weight loss with minimal effort and at bargain prices.

Who speaks on nutrition?

iStockphoto.com/lisegagne

Table C1–1

Quackery and Internet Terms

- **advertorials** lengthy advertisements in newspapers and magazines that read like feature articles but are written for the purpose of touting the virtues of products and may or may not be accurate.
- **anecdotal evidence** information based on interesting and entertaining, but not scientific, personal accounts of events.
- **fraud** or **quackery** the promotion, for financial gain, of devices, treatments, services, plans, or products (including diets and supplements) claimed to improve health, well-being, or appearance without proof of safety or effectiveness. (The word *quackery* comes from the term *quacksalver,* meaning a person who quacks loudly about a miracle product—a lotion or a salve.)
- **infomercials** feature-length television commercials that follow the format of regular programs but are intended to convince viewers to buy products and not to educate or entertain them. The statements made may or may not be accurate.
- **Internet (the Net)** a worldwide network of millions of computers linked together to share information.
- **urban legends** stories, usually false, that may travel rapidly throughout the world via the Internet gaining strength of conviction solely on the basis of repetition.
- **websites** Internet resources composed of text and graphic files, each with a unique URL (Uniform Resource Locator) that names the site (for example, www.usda.gov).
- **World Wide Web** (the Web, commonly abbreviated **www**) a graphical subset of the Internet.

How can people learn to distinguish valid nutrition information from misinformation? Some quackery is easy to identify—like the claims of the salesman in Figure C1–1—whereas other types are more subtle. Between the extremes of accurate scientific data and intentional quackery lies an abundance of nutrition misinformation.† An instructor at a gym, a physician, a health-store clerk, an author of books, or an advocate for juice machines or weight-loss gadgets may all sincerely believe that the nutrition regimens they recommend are beneficial. But what qualifies them to give advice? Would following their advice be helpful or harmful? To sift the meaningful nutrition information from the rubble, you must learn to identify both.

Chapter 1 explained that valid nutrition information arises from scientific research and does not rely on **anecdotal evidence** or testimonials. Scientists who use animals in their research do not apply their findings directly to human beings. And science is first published in peer-reviewed journals. Table C1–2 lists some sources of this authentic nutrition information.

† Quackery-related definitions are available from the National Counsel Against Health Fraud, www.ncahf.org/pp/definitions.html. Consumers with questions or suspicions about fraud can contact the FDA on the Internet at www.FDA.gov or by telephone at (888) INFO-FDA.

Nutrition on the Net

Got a question? The **World Wide Web** on the **Internet** has the answer! In fact, the "net" offers endless access to high-quality information, such as in scientific journals, but it also delivers an

Figure C1–1

Earmarks of Nutrition Quackery

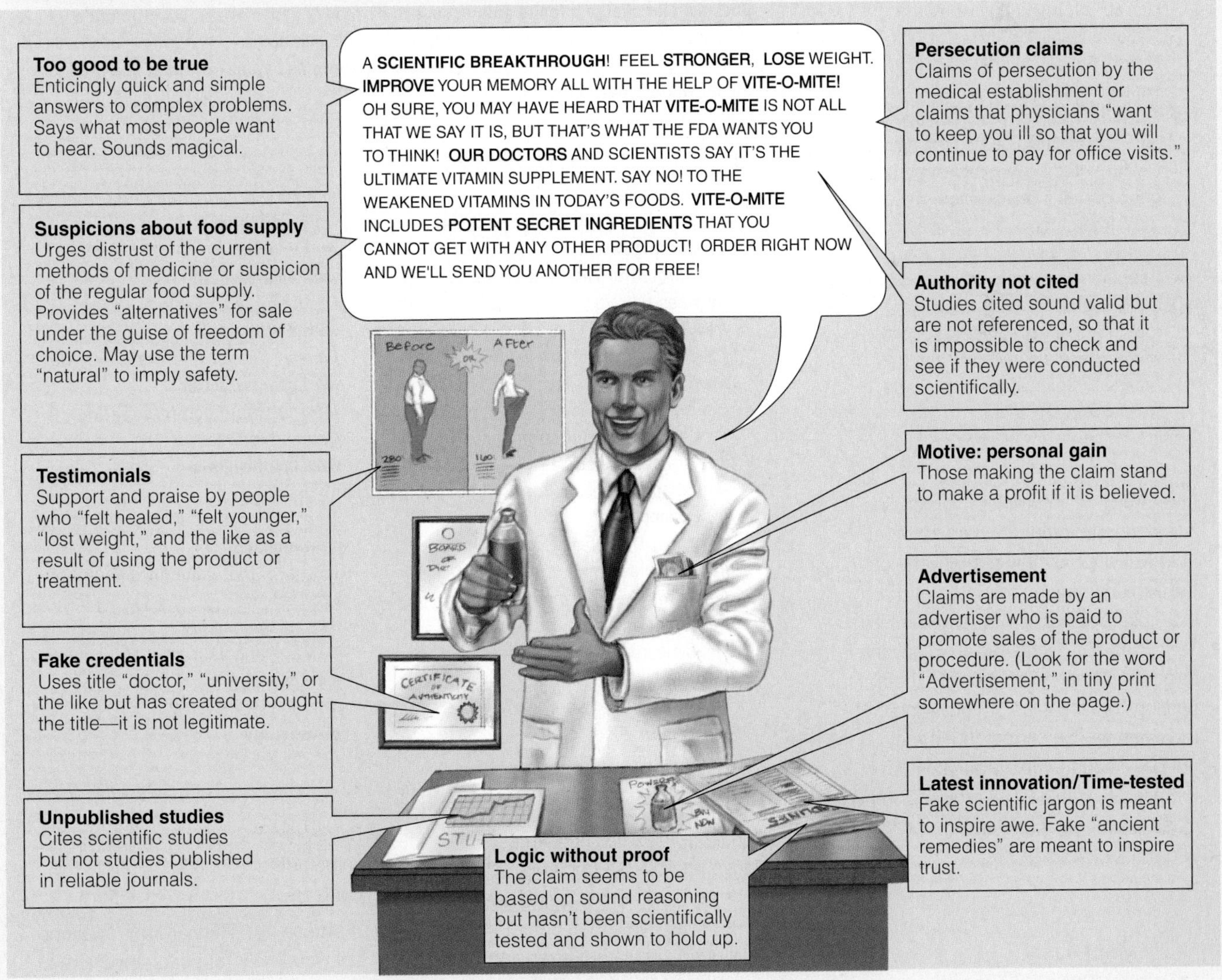

Too good to be true
Enticingly quick and simple answers to complex problems. Says what most people want to hear. Sounds magical.
A SCIENTIFIC BREAKTHROUGH! FEEL STRONGER, LOSE WEIGHT. IMPROVE YOUR MEMORY ALL WITH THE HELP OF VITE-O-MITE! OH SURE, YOU MAY HAVE HEARD THAT VITE-O-MITE IS NOT ALL THAT WE SAY IT IS, BUT THAT'S WHAT THE FDA WANTS YOU TO THINK! OUR DOCTORS AND SCIENTISTS SAY IT'S THE ULTIMATE VITAMIN SUPPLEMENT. SAY NO! TO THE WEAKENED VITAMINS IN TODAY'S FOODS. VITE-O-MITE INCLUDES POTENT SECRET INGREDIENTS THAT YOU CANNOT GET WITH ANY OTHER PRODUCT! ORDER RIGHT NOW AND WE'LL SEND YOU ANOTHER FOR FREE!
Persecution claims
Claims of persecution by the medical establishment or claims that physicians "want to keep you ill so that you will continue to pay for office visits."
Suspicions about food supply
Urges distrust of the current methods of medicine or suspicion of the regular food supply. Provides "alternatives" for sale under the guise of freedom of choice. May use the term "natural" to imply safety.
Authority not cited
Studies cited sound valid but are not referenced, so that it is impossible to check and see if they were conducted scientifically.
Before
After
Testimonials
Support and praise by people who "felt healed," "felt younger," "lost weight," and the like as a result of using the product or treatment.
Motive: personal gain
Those making the claim stand to make a profit if it is believed.
Advertisement
Claims are made by an advertiser who is paid to promote sales of the product or procedure. (Look for the word "Advertisement," in tiny print somewhere on the page.)
Fake credentials
Uses title "doctor," "university," or the like but has created or bought the title—it is not legitimate.
Latest innovation/Time-tested
Fake scientific jargon is meant to inspire awe. Fake "ancient remedies" are meant to inspire trust.
Unpublished studies
Cites scientific studies but not studies published in reliable journals.
Logic without proof
The claim seems to be based on sound reasoning but hasn't been scientifically tested and shown to hold up.

© Cengage Learning

Table C1–2

Credible Sources of Nutrition Information

Professional health organizations, government health agencies, volunteer health agencies, and consumer groups provide consumers with reliable health and nutrition information. Credible sources of nutrition information include:

- Professional health organizations, especially the Academy of Nutrition and Dietetics' National Center for Nutrition and Dietetics (NCND), www.eatright.org/ncnd.html, also the Society for Nutrition Education, www.sne.org and the American Diabetes Association, www.diabetes.org
- Government health agencies such as the Federal Trade Commission (FTC), www.ftc.gov and the National Institutes of Health Office of Dietary Supplements, www.dietary-supplements.info.nih.gov
- Certain consumer watchdog agencies such as the National Council Against Health Fraud, www.ncahf.org, Stephen Barrett's Quackwatch, www.quackwatch.com, and Snopes.com—Rumor Has It, www.snopes.com
- Reputable consumer groups such as the Better Business Bureau, www.bbb.org, the Consumers Union, www.consumersunion.org and the American Council on Science and Health, www.acsh.org

abundance of incomplete, misleading, or inaccurate information on innumerable websites. Simply put: anyone can publish anything on the Internet. For example, popular self-governed Internet "encyclopedia" **websites** allow anyone to post information or change others' postings on all topics.‡ Information on the sites may be correct, but it may not be—readers must evaluate it for themselves. Table C1–3 provides some clues to judging the reliability of nutrition information websites.

‡ *An example is Wikipedia.*

Table C1–3

Is This Site Reliable?

To judge whether an Internet site offers reliable nutrition information, answer the following questions.

- **Who is responsible for the site?** Clues can be found in the three-letter "tag" that follows the dot in the site's name. For example, "gov" and "edu" indicate government and university sites, usually reliable sources of information.
- **Do the names and credentials of information providers appear? Is an editorial board identified?** Many legitimate sources provide e-mail addresses or other ways to obtain more information about the site and the information providers behind it.
- **Are links with other reliable information sites provided?** Reputable organizations almost always provide links with other similar sites because they want you to know of other experts in their area of knowledge. Caution is needed when you evaluate a site by its links, however. Anyone, even a quack, can link a webpage to a reputable site without the organization's permission. Doing so may give the quack's site the appearance of legitimacy, just the effect the quack is hoping for.
- **Is the site updated regularly?** Nutrition information changes rapidly, and sites should be updated often.
- **Is the site selling a product or service?** Commercial sites may provide accurate information, but they also may not, and their profit motive increases the risk of bias.
- **Does the site charge a fee to gain access to it?** Many academic and government sites offer the best information, usually for free. Some legitimate sites do charge fees, but before paying up, check the free sites. Chances are good you'll find what you are looking for without paying.

Some credible websites include:

- Government agencies
  - *Department of Agriculture (USDA)* **www.usda.gov**
  - *Department of Health and Human Services (DHHS)* **www.hhs.gov**
  - *Food and Drug Administration (FDA)* **www.fda.gov**
  - *Health Canada* **www.hc-sc.gc.ca/index-eng.php**
- Volunteer health agencies
  - *American Cancer Society* **www.cancer.org**
  - *American Diabetes Association* **www.diabetes.org**
  - *American Heart Association* **www.heart.org**
- Reputable consumer and professional groups
  - *American Council on Science and Health* **www.acsh.org**
  - *Academy of Nutrition and Dietetics* **www.eatright.org**
  - *American Medical Association* **www.ama-assn.org**
  - *Dietitians of Canada* **www.dietitians.ca**
  - *Federal Citizen Information Center* **www.gsa.gov/portal**
  - *International Food Information Council Foundation* **www.foodinsight.org**
- Journals
  - American Journal of Clinical Nutrition **www.ajcn.org**
  - Journal of the Academy of Nutrition and Dietetics **www.adajournal.org**
  - New England Journal of Medicine **www.nejm.org**
  - Nutrition Reviews **www.ilsi.org**

Personal Internet sites, known as "weblogs" or "blogs," contain the author's personal opinions and are not often reviewed by experts before posting. E-mails often circulate hoaxes and scare stories. Be suspicious when:

- Someone other than the sender or some authority you know wrote the contents.
- A phrase like "Forward this to everyone you know" appears anywhere in the piece.

- The piece states "This is not a hoax"; chances are, it is.
- The information seems shocking or something that you've never heard from legitimate sources.
- The language is overly emphatic or sprinkled with capitalized words or exclamation marks.
- No references are offered or, if present, are of questionable validity when examined.
- Websites such as www.quackwatch.com or www.urbanlegends.com have debunked the message.

In contrast, one of the most trustworthy Internet sites for scientific investigation is the National Library of Medicine's PubMed website, which provides free access to over 10 million abstracts (short descriptions) of research papers published in scientific journals around the world. Many abstracts provide links to full articles posted on other sites. The site is easy to use and offers instructions for beginners. Figure C1–2 introduces this resource.

Who Are the True Nutrition Experts?

Most people turn to their physicians for dietary advice. Physicians are expected to know all about health-related matters. Only about 30 percent of all medical schools in the United States require students to take a comprehensive nutrition course, such as the class taken by students reading this text. Less than half of medical schools require even 25 hours of nutrition instruction. By comparison, your current nutrition class provides an average of 45 hours of instruction.

The exceptional physician has a specialty area in clinical nutrition and is highly qualified to advise on nutrition. Membership in the Academy of Nutrition and Dietetics or the Society for Clinical Nutrition, whose journals are cited many times throughout this text, can be a clue to a physician's nutrition knowledge.

The Academy of Nutrition and Dietetics Proposal

The **Academy of Nutrition and Dietetics**, the professional association of dietitians, proposes that nutrition education should be part of the curriculum for health-care professionals: physicians' assistants, dental hygienists, physical and occupational therapists, social workers, and all others who provide services directly to clients. This plan would bring reliable nutrition information to more people who need it.

Few physicians and other health specialists have the know-how, time, or experience necessary to develop diet plans and provide detailed diet instruction for clients, however. Instead, they refer their clients to nutrition specialists.

Registered Dietitians: The Nutrition Specialists

Fortunately, the credential that indicates a qualified nutrition expert is easy to spot—you can confidently call on a **registered dietitian (RD).** Additionally, some states require that **nutritionists** and **dietitians** obtain a **license to practice**. Meeting state-established criteria in addition to **registration** with the Academy of Nutrition and Dietetics certifies that an expert is the genuine article. Table C1–4 defines nutrition specialists along with some general terms associated with nutrition advice.

RDs are easy to find in most communities because they perform a multitude of duties in a variety of settings (see Table C1–5).[4] They work in foodservice operations, pharmaceutical companies, sports nutrition programs, corporate wellness programs, the food industry, home health agencies, long-term care institutions, private practice, community and public health settings, cooperative extension offices,[§] research centers, universities, hospitals, health maintenance organizations (HMOs), and other

[§] *Cooperative extension agencies are associated with land grant colleges and universities and may be found in the phone book's government listings.*

Figure C1–2

PubMed (www. ncbi.nlm.nih.gov/pubmed): Internet Resource for Scientific Nutrition References

The U.S. National Library of Medicine's PubMed website offers tutorials to help teach the beginner to use the search system effectively. Often, simply visiting the site, typing a query in the *Search for* box, and clicking *Search* will yield satisfactory results.

For example, to find research concerning calcium and bone health, typing in "calcium bone" nets almost 3,000 results. To refine the search, try setting limits on dates, types of articles, languages, and other criteria to obtain a more manageable number of abstracts to peruse.

Type search terms here

Refine the search by setting limits

Use "help" resources to answer questions

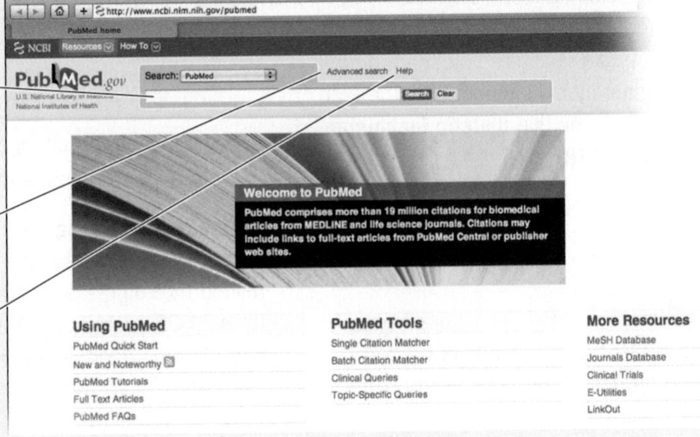

Pub Med/National Center for Biotechnology Information/U.S. National Library of Medicine/NIH

Table C1–4

Terms Associated with Nutrition Advice

- **Academy of Nutrition and Dietetics (AND)** the professional organization of dietitians in the United States (formerly the American Dietetic Association). The Canadian equivalent is the Dietitians of Canada (DC), which operates similarly.
- **certified diabetes educator (CDE)** a health-care professional who specializes in educating people with diabetes to help them manage their disease through medical and lifestyle means. Extensive training, work experience, and an examination are required to achieve CDE status.
- **dietetic technician** a person who has completed a two-year academic degree from an accredited college or university and an approved dietetic technician program. A **dietetic technician, registered** (DTR) has also passed a national examination and maintains registration through continuing professional education.
- **dietitian** a person trained in nutrition, food science, and diet planning. See also *registered dietitian.*
- **license to practice** permission under state or federal law, granted on meeting specified criteria, to use a certain title (such as *dietitian*) and to offer certain services. Licensed dietitians may use the initials LD after their names.
- **medical nutrition therapy** nutrition services used in the treatment of injury, illness, or other conditions; includes assessment of nutrition status and dietary intake and corrective applications of diet, counseling, and other nutrition services.
- **nutritionist** someone who studies nutrition. Some nutritionists are RDs, whereas others are self-described experts whose training is questionable and who are not qualified to give advice. In states with responsible legislation, the term applies only to people who have master of science (MS) or doctor of philosophy (PhD) degrees from properly accredited institutions.
- **public health nutritionist** a dietitian or other person with an advanced degree in nutrition who specializes in public health nutrition.
- **registered dietitian (RD)** food and nutrition experts who have earned at least a bachelor's degree from an accredited college or university with a program approved by the Academy of Nutrition and Dietetics (or the Dietitians of Canada). The dietitian must also serve in an approved internship or coordinated program, pass the registration examination, and maintain professional competency through continuing education.[a] Many states also require licensing of practicing dietitians.
- **registration** listing with a professional organization that requires specific course work, experience, and passing of an examination.

[a] *The five content areas of the registration examination for dietitians are food and nutrition; clinical and community nutrition; education and research; food and nutrition systems; and management. New emphasis is placed on genetics, cultural competency, complementary care, and reimbursement*

facilities. In hospitals, they may offer **medical nutrition therapy** as part of patient care, or they may run the food service operation or they may specialize as **certified diabetes educators (CDE)** to help people with diabetes manage the disease. **Public health nutritionists** play key roles in government agencies as expert consultants and advocates or in direct service delivery.[5] The roles are so diverse that many pages would be required to cover them thoroughly.

In some facilities, a **dietetic technician** assists registered dietitians in both administrative and clinical responsibilities. A dietetic technician has been educated and trained to work under the guidance of a registered dietitian; upon passing a national examination, the technician earns the title **dietetic technician, registered (DTR).**

Detecting Fake Credentials

In contrast to RDs and other credentialed nutrition professionals, thousands of people possess fake nutrition degrees and claim to be nutrition counselors, nutritionists, or "dietists." These and other such titles may sound meaningful, but most of these people lack the established credentials of the Academy of Nutrition and Dietetics–sanctioned dietitian. If you look closely, you can see signs that their expertise is fake.

Educational Background

Take, for example, a nutrition expert's educational background. The minimum standards of education for a dietitian specify a bachelor of science (BS) degree in food science and human nutrition (or related fields) from an **accredited** college or university (Table C1–6 defines this term). Such a degree generally requires four to five years of study.

In contrast, a fake nutrition expert may display a degree from a six-month course of study; such a degree is simply not the same. In some cases, schools posing as legitimate institutions are actually **diploma mills**—fraudulent businesses that sell certificates of competency to anyone who pays the fees, from under a thousand dollars for a bachelor's degree to several thousand for a doctorate. To obtain these "degrees," a candidate need not read any books or pass any examinations, and the only written work is a signature on a check. Here are a few red flags to identify these scams:

- A degree is awarded in a very short time—sometimes just a few days.
- A degree can be based entirely on work or life experience.
- An institution provides only an e-mail address, with vague information on physical location.
- It provides sample views of certificates and diplomas.
- It lets the "student" specify a year of graduation to be printed.[6]

Selling degrees is big business; networks of many bogus institutions are often owned by a single entity. In 2011, more than 2,600 such diploma and accreditation mills were identified and 2,000 more were under investigation.[7]

Table C1–5

Professional Responsibilities of Registered Dietitians

Registered dietitians perform varied and important roles in the workforce. This table lists just a few responsibilities of just a few specialties.

Specialty	Sample Responsibilities
Public Health Nutrition	Influence nutrition policy, regulations, and legislation.
	Plan, coordinate, administer, and evaluate food assistance programs.
	Consult with agencies.
	Plan and manage budgets.
Hospital Health Care/Clinical Care	Design and implement disease prevention services.
	Coordinate patient care with other health care professionals.
	Assess client nutrient status and requirements.
	Provide client care.
	Counsel clients in implementing diet plans.
Food Service Management	Plan and direct an institution's foodservice system, from kitchen to delivery.
	Plan and manage budgets.
	Develop products.
	Market services.
Laboratory Research	Design, execute, and interpret food and nutrition research.
	Write and publish research articles in peer-reviewed journals and lay publications.
	Provide science-based guidance to nutrition practitioners.
	Write and manage grants.
Education	Write curricula to deliver to students appropriate nutrition knowledge for their goals and that meets criteria of accrediting agencies and professional groups.
	Teach and evaluate student progress.
	Often, research and publish.
Health and Wellness	Design and implement research-based programs for individuals or populations to improve nutrition, health, and physical fitness.

Adapted from S. H. Laramee and M. Tate, Dietetics Workforce Demand Study Task Force Supplement: An introduction, Journal of the Academy of Nutrition and Dietetics *112 (2012): S7–S9.*

Table C1–6

Terms Describing Institutions of Higher Learning, Legitimate and Fraudulent

- **accredited** approved; in the case of medical centers or universities, certified by an agency recognized by the U.S. Department of Education.
- **diploma mill** an organization that awards meaningless degrees without requiring its students to meet educational standards. Diploma mills are not the same as diploma forgeries (fake diplomas and certificates bearing the names of real respected institutions). While virtually indistinguishable from authentic diplomas, forgeries can be unveiled by checking directly with the institution.

Accreditation and Licensure

Lack of proper accreditation is the identifying sign of a fake educational institution. To guard educational quality, an accrediting agency recognized by the U.S. Department of Education certifies that certain schools meet the criteria defining a complete and accurate schooling, but in the case of nutrition, quack accrediting agencies cloud the picture. Fake nutrition degrees are available from schools "accredited" by more than 30 phony accrediting agencies.**

State laws do not necessarily help consumers distinguish experts from fakes; some states allow anyone to use the title *dietitian* or *nutritionist.* But other states have responded to the need by allowing

*** To find out whether an online school is accredited, write the Distance Education and Training Council, Accrediting Commission, 1601 Eighteenth Street, NW, Washington, D.C. 20009; call 202-234-5100; or visit their website (www.detc.org).*

To find out whether a school is properly accredited for a dietetics degree, write the Academy of Nutrition and Dietetics, Division of Education and Research, 120 South Riverside Plaza, Suite 2000, Chicago, Illinois 60606–6995, phone: 800-877-1600; or visit their website (www.eatright.org/caade).

The American Council on Education publishes a directory of accredited institutions, professionally accredited programs, and candidates for accreditation in Accredited Institutions of Postsecondary Education Programs *(available at many libraries). For additional information, write the American Council on Education, One Dupont Circle NW, Suite 800, Washington, D.C. 20036; call 202-939-9382; or visit their website (www.acenet.edu)*

only RDs or people with certain graduate degrees and state licenses to call themselves dietitians. Licensing provides a way to identify people who have met minimum standards of education and experience.

A Failed Attempt to Fail

To dramatize the ease with which anyone can obtain a fake nutrition degree, one writer paid $82 to enroll in a nutrition diploma mill that billed itself as a correspondence school. She made every attempt to fail, intentionally giving all wrong answers to the examination questions. Even so, she received a "nutritionist" certificate at the end of the course, together with a letter from the "school" officials explaining that they were sure she must have misread the test.

Would You Trust a Nutritionist Who Eats Dog Food?

In a similar stunt, Mr. Eddie Diekman was named a "professional member" of an association of nutrition "experts." For his efforts, Eddie received a diploma suitable for framing and displaying. Eddie is a cocker spaniel. His owner, Connie B. Diekman, then president of the American Dietetic Association, paid Eddie's tuition to prove that he could be awarded the title "nutritionist" merely by sending in his name.††

Staying Ahead of the Scammers

In summary, to stay one step ahead of the nutrition quacks, check a provider's qualifications. First, look for the degrees and credentials listed after the person's name (such as MD, RD, MS, PhD, or LD). Next find out what you can about the reputations of the institutions that awarded the degrees. Then call your state's health-licensing agency and ask if dietitians are licensed in your state. If they are, find out whether the person giving you dietary advice has a license—and if not, find someone better qualified. Your health is your most precious asset, and protecting it is well worth the time and effort it takes to do so.

Critical Thinking

1. This class will give you the skills to learn how to separate legitimate nutrition claims from those that are questionable. To help practice the skills needed to separate fact from fiction, describe how you would respond to the following situation:

   A friend has started taking Ginseng, a supplement that claims to help her lose weight. You are thinking of trying Ginseng, but you want to learn more about the herb and its effects before deciding. What research would you do, and what questions would you ask your friend to determine if Ginseng is a legitimate weight loss product?

2. Recognizing a nutrition authority that you can consult for reliable nutrition information can be difficult because it is so easy to acquire questionable nutrition credentials. Read the education and experience of the "nutrition experts" described below and put in order beginning with the person with the strongest and most trustworthy nutrition expertise and ending with the person with the least nutrition expertise who should be trusted the least.

   1. Dietetic technician, registered (DTR) working in a clinic
   2. A highly successful athlete/coach who has a small business as a nutrition counselor and sells a line of nutrition supplements
   3. An individual who has completed 30 hours of nutrition training through the American Association of Nutrition Counseling
   4. A registered dietician (RD) associated with a hospital

© Courtesy of eatright.org

Eddie displays his professional credentials.

†† *The stunt described was patterned after that of the late Victor Herbert, whose cat Charlie and poodle Sassafras were also awarded nutritionist credentials by mail.*

2 Nutrition Tools—Standards and Guidelines

what do you think?

How can you tell **how much of each nutrient** you need to consume daily?

Are **government dietary recommendations** too simplistic to be of help?

Are the health claims on food labels **accurate and reliable**?

Can certain **"superfoods"** boost your health with more than just nutrients?

Learning Objectives

After completing this chapter, you should be able to accomplish the following:

LO 2.1 Identify the full names and explain the functions of the RDA, AI, UL, EAR, and AMDR and discuss how the Daily Values differ in nature and use from other sets of nutrient standards.

LO 2.2 List the four major topic areas of the *Dietary Guidelines for Americans* and explain their importance to the population.

LO 2.3 Describe how and why foods are grouped in the USDA Food Patterns, including subgroups.

LO 2.4 Outline the basic steps of diet planning with the USDA Food Patterns, and address limits for solid fats and added sugars.

LO 2.5 Evaluate a food label, delineating the different uses of information found on the Nutrition Facts panel, on the ingredients list, and in any health claims or other claims made for the product.

LO 2.6 State specific nutritional advantages of a carefully planned nutrient-dense diet over a diet chosen without regard for nutrition principles.

LO 2.7 Discuss the positive and negative findings for dietary phytochemicals with regard to health, and make a case for food sources over supplements to provide them.

© Norman Chan/Shutterstock.com

Eating well is easy in theory—just choose foods that supply appropriate amounts of the essential nutrients, fiber, phytochemicals, and energy without excess intakes of fat, sugar, and salt and be sure to get enough physical activity to help balance the foods you eat. In practice, eating well proves harder than it appears. Many people are overweight, or undernourished, or suffer from nutrient excesses or deficiencies that impair their health—that is, they are malnourished. You may not think that this statement applies to you, but you may already have less than optimal nutrient intakes without knowing it. Accumulated over years, the effects of your habits can seriously impair the quality of your life.

Putting it positively, you can enjoy the best possible vim, vigor, and vitality throughout your life if you learn now to nourish yourself optimally. To learn how, you first need some general guidelines and the answers to several basic questions. How much of each nutrient and how many calories should you consume? Which types of foods supply which nutrients? How much of each type of food do you have to eat to get enough? And how can you eat all these foods without gaining weight? This chapter begins by identifying some ideals for nutrient and energy intakes and ends by showing how to achieve them.

Nutrient Recommendations

LO 2.1 Identify the full names and explain the functions of the RDA, AI, UL, EAR, and AMDR and discuss how the Daily Values differ in nature and use from other sets of nutrient standards.

Nutrient recommendations are sets of standards against which people's nutrient and energy intakes can be measured. Nutrition experts use the recommendations to assess intakes and to offer advice on amounts to consume. Individuals may use them to decide how much of a nutrient they need and how much is too much.

Dietary Reference Intakes

The standards in use in the United States and Canada are the **Dietary Reference Intakes (DRI)**. A committee of nutrition experts from the United States and Canada develops, publishes, and updates the DRI.* The DRI committee has set values for all of the vitamins and minerals, as well as for carbohydrates, fiber, lipids, protein, water, and energy.

Dietary Reference Intakes (DRI) a set of four lists of values for measuring the nutrient intakes of healthy people in the United States and Canada. The four lists are Estimated Average Requirements (EAR), Recommended Dietary Allowances (RDA), Adequate Intakes (AI), and Tolerable Upper Intake Levels (UL).

* This is a committee of the Food and Nutrition Board of the National Academy of Sciences' Institute of Medicine, working in association with Health Canada.

Another set of nutrient standards is practical for the person striving to make wise choices among packaged foods. These are the **Daily Values**, familiar to anyone who has read a food label. Nutrient standards—the DRI and Daily Values—are used and referred to so often that they are printed on the inside front and back cover pages of this book: DRI lists—inside front cover pages A, B, and C; Daily Values—inside back cover, page Y.

KEY POINTS

- The Dietary Reference Intakes are U.S. and Canadian nutrient intake standards.
- The Daily Values are U.S. standards used on food labels.

The DRI Lists and Purposes

For each nutrient, the DRI establish a number of values, each serving a different purpose. Most people need to focus on only two kinds of DRI values: those that set goals for nutrient intakes (RDA, AI, and AMDR, described next) and those that describe nutrient safety (UL, addressed later). In total, the DRI include five sets of values:

1. **Recommended Dietary Allowances (RDA)**—adequacy
2. **Adequate Intakes (AI)**—adequacy
3. **Tolerable Upper Intake Levels (UL)**—safety
4. **Estimated Average Requirements (EAR)**—research and policy
5. **Acceptable Macronutrient Distribution Ranges (AMDR)**—healthful ranges for energy-yielding nutrients

RDA and AI—Recommended Nutrient Intakes A great advantage of the DRI values lies in their applicability to the diets of individuals.[1†] People may adopt the Recommended Dietary Allowances and Adequate Intakes values as their own nutrient intake goals.[‡]

The RDA form the indisputable bedrock of the DRI recommended intakes because they derive from solid experimental evidence and reliable observations—they are expected to meet the needs of almost all healthy people. AI values, in contrast, are based as far as possible on the available scientific evidence but also on some educated guesswork. Whenever the DRI committee finds insufficient evidence to generate an RDA, they establish an AI value instead. This book refers to the RDA and AI values collectively as the DRI recommended intakes.

EAR—Nutrition Research and Policy The Estimated Average Requirements, also set by the DRI committee, establishes the average nutrient requirements for given life stages and gender groups that researchers and nutrition policy makers use in their work. Public health officials may also use them to assess nutrient intakes of populations and make recommendations. The EAR values form the scientific basis upon which the RDA values are set (a later section explains how).

UL—Safety Beyond a certain point, it is unwise to consume large amounts of any nutrient, so the DRI committee sets the Tolerable Upper Intake Levels to identify potentially toxic levels of nutrient intake. Usual intakes of a nutrient below this level have a low risk of causing illness. The UL are indispensable to consumers who take supplements or consume foods and beverages to which vitamins or minerals have been added—a group that includes almost everyone. Public health officials also rely on UL values to set safe upper limits for nutrients added to our food and water supplies.

Nutrient needs fall within a range, and a danger zone exists both below and above that range. Figure 2–1 illustrates this point. People's tolerances for high doses of nutrients vary, so caution is in order when nutrient intakes approach the UL values (listed on the inside front cover, page C).

Some nutrients lack UL values. The absence of a UL for a nutrient does not imply that it is safe to consume it in any amount, however. It means only that insufficient data exist to establish a value.

[†] Reference notes are found in Appendix F.
[‡] For simplicity, this book refers to two sets of nutrient goals (AI and RDA) collectively as the *DRI recommended intakes*. The AI values are not the scientific equivalent of the RDA, however.

Did You Know?

The DRI table on the inside front cover distinguishes the RDA from AI values, but both kinds of values are intended as nutrient intake goals for individuals.

Daily Values nutrient standards that are printed on food labels and on grocery store and restaurant signs. Based on nutrient and energy recommendations for a general 2,000-calorie diet, they allow consumers to compare foods with regard to nutrients and calorie contents.

Recommended Dietary Allowances (RDA) nutrient intake goals for individuals; the average daily nutrient intake level that meets the needs of nearly all (97 percent to 98 percent) healthy people in a particular life stage and gender group. Derived from the Estimated Average Requirements (see below).

Adequate Intakes (AI) nutrient intake goals for individuals; the recommended average daily nutrient intake level based on intakes of healthy people (observed or experimentally derived) in a particular life stage and gender group and assumed to be adequate. Set whenever scientific data are insufficient to allow establishment of an RDA value.

Tolerable Upper Intake Levels (UL) the highest average daily nutrient intake level that is likely to pose no risk of toxicity to almost all healthy individuals of a particular life stage and gender group. Usual intake above this level may place an individual at risk of illness from nutrient toxicity.

Estimated Average Requirements (EAR) the average daily nutrient intake estimated to meet the requirement of half of the healthy individuals in a particular life stage and gender group; used in nutrition research and policy making and is the basis upon which RDA values are set.

Acceptable Macronutrient Distribution Ranges (AMDR) values for carbohydrate, fat, and protein expressed as percentages of total daily caloric intake; ranges of intakes set for the energy-yielding nutrients that are sufficient to provide adequate total energy and nutrients while minimizing the risk of chronic diseases.

Figure 2–1

The Naïve View Versus the Accurate View of Optimal Nutrient Intakes

Consuming too much of a nutrient endangers health, just as consuming too little does. The DRI recommended intake values fall within a safety range with the UL marking tolerable upper levels.

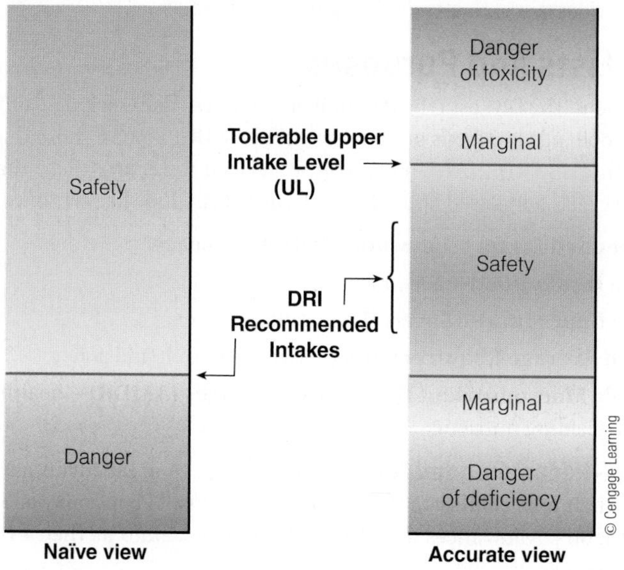

AMDR—Calorie Percentage Ranges The DRI committee also sets healthy ranges of intake for carbohydrate, fat, and protein known as Acceptable Macronutrient Distribution Ranges. Each of these three energy-yielding nutrients contributes to the day's total calorie intake, and their contributions can be expressed as a percentage of the total. According to the committee, a diet that provides adequate energy in the following proportions can provide adequate nutrients while minimizing the risk of chronic diseases:

- 45 to 65 percent of calories from carbohydrate.
- 20 to 35 percent of calories from fat.
- 10 to 35 percent of calories from protein.

The chapters on the energy-yielding nutrients revisit these ranges.

The DRI committee takes chronic disease prevention into account with regard to other nutrients, too. For example, the committee sets intake goals for the mineral calcium at levels known to promote normal bone growth and maintenance, which may in turn help to maintain people's bone strength as they age and reduce their risk of osteoporosis-related bone fractures.[2]

KEY POINTS

- The DRI set nutrient intake goals for individuals, standards for researchers and public policy makers, and tolerable upper limits.
- RDA, AI, UL, and EAR lists are DRI standards, along with AMDR ranges for energy-yielding nutrients.

Don't let the "alphabet soup" of nutrient intake standards confuse you. Their names make sense when you learn their purposes.

Understanding the DRI Recommended Intakes

Nutrient recommendations have been much misunderstood. One young woman posed this question: "Do you mean that some bureaucrat says that I need exactly the same amount of vitamin D as everyone else? Do they really think that 'one size fits all'?" In fact, the opposite is true.

DRI for Population Groups The DRI committee acknowledges differences between individuals and takes them into account when setting nutrient values. It has made separate recommendations for specific groups of people—men, women, pregnant women, lactating women, infants, and children—and for specific age ranges. Children aged 4 to 8 years, for example, have their own DRI recommended intakes. Each individual can look up the recommendations for his or her own age and gender group. Within your own age and gender group, the committee advises adjusting nutrient intakes in special circumstances that may increase or decrease nutrient needs, such as illness, smoking, or vegetarianism. Later chapters provide details about who may need to adjust intakes of which nutrients.

For almost all healthy people, a diet that consistently provides the RDA or AI amount for a specific nutrient is very likely to be adequate in that nutrient. On average, you should try to get 100 percent of the DRI recommended intake for every nutrient over time to ensure an adequate intake.

Other Characteristics of the DRI The following facts will help put the DRI recommended intakes into perspective:

- The values are based on available scientific research to the greatest extent possible and are updated to reflect current scientific knowledge.
- The values are based on the concepts of probability and risk. The DRI recommended intakes are associated with a low probability of deficiency for people of a given life stage and gender group, and they pose almost no risk of toxicity for that group.
- The values are set for optimal intakes, not minimum requirements. They include a generous safety margin and meet the needs of virtually all healthy people in a specific age and gender group.
- The values are set in reference to certain indicators of nutrient adequacy, such as blood nutrient concentrations, normal growth, or reduction of certain chronic diseases or other disorders, rather than prevention of deficiency symptoms alone.
- The values reflect daily intakes to be achieved on average, over time. They assume that intakes will vary from day to day and are set high enough to ensure that the body's nutrient stores will meet nutrient needs during periods of inadequate intakes lasting several days to several months, depending on the nutrient.

The DRI Apply to Healthy People Only The DRI are designed for health maintenance and disease prevention in healthy people, not for the restoration of health or repletion of nutrients in those with deficiencies. Under the stress of serious illness or malnutrition, a person may require a much higher intake of certain nutrients or may not be able to handle even the DRI amount. Therapeutic diets take into account the increased nutrient needs imposed by certain medical conditions, such as recovery from surgery, burns, fractures, illnesses, malnutrition, or addictions.

KEY POINT

- The DRI are up-to-date, optimal, and safe nutrient intakes for healthy people in the United States and Canada.

How the Committee Establishes DRI Values—An RDA Example

A theoretical discussion will help to explain how the DRI committee goes about setting DRI values. Suppose we are the DRI committee members with the task of setting an RDA for nutrient X (an essential nutrient).[§] Ideally, our first step will be to find out how much of that nutrient various healthy individuals need. To do so, we review studies of deficiency states, nutrient stores and their depletion, and the factors influencing

[§] This discussion describes how an RDA value is set; to set an AI value, the committee would use some educated guesswork as well as scientific research results to determine an approximate amount of the nutrient most likely to support health.

Figure 2–2

Individuality of Nutrient Requirements

Each square represents a person. A, B, and C are Mr. A, Mr. B, and Mr. C. Each has a different requirement.

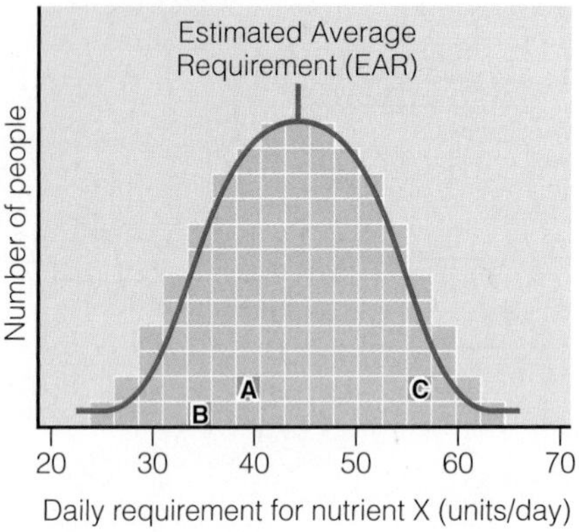

Figure 2–3

Nutrient Recommended Intake: RDA Example

Intake recommendations for most vitamins and minerals are set so that they will meet the requirements of nearly all people (boxes represent people).

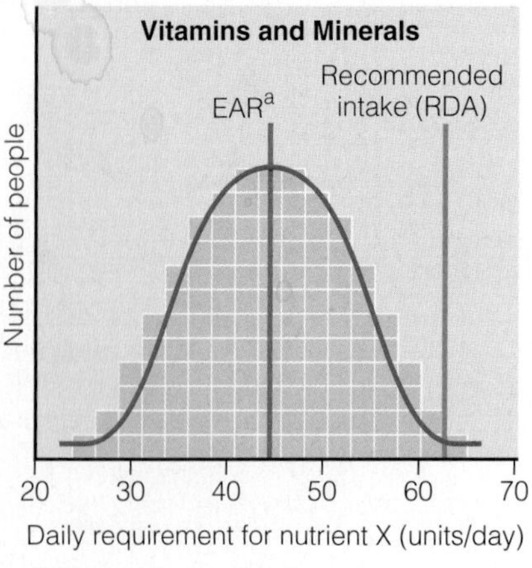

[a]*Estimated Average Requirement*

them. We then select the most valid data for use in our work. Of the DRI family of nutrient standards, the setting of an RDA value demands the most rigorous science and tolerates the least guesswork.

Determining Individual Requirements

One experiment we would review or conduct is a **balance study**. In this type of study, scientists measure the body's intake and excretion of a nutrient to find out how much intake is required to balance excretion. For each individual subject, we can determine a **requirement** to achieve balance for nutrient X. With an intake below the requirement, a person will slip into negative balance or experience declining stores that could, over time, lead to deficiency of the nutrient.

We find that different individuals, even of the same age and gender, have different requirements. Mr. A needs 40 units of the nutrient each day to maintain balance; Mr. B needs 35; Mr. C, 57. If we look at enough individuals, we find that their requirements are distributed as shown in Figure 2–2—with most requirements near the midpoint (here, 45) and only a few at the extremes.

Accounting for the Needs of the Population To set the value, we have to decide what intake to recommend for everybody. Should we set it at the mean (45 units in Figure 2–2)? This is the Estimated Average Requirement for nutrient X, mentioned earlier as valuable to scientists and policy makers but not appropriate as an individual's nutrient goal. The EAR value is probably close to everyone's minimum need, assuming the distribution shown in Figure 2–2. (Actually, the data for most nutrients indicate a distribution that is much less symmetrical.) But if people took us literally and consumed exactly this amount of nutrient X each day, half the population would begin to develop nutrient deficiencies and in time even observable symptoms of deficiency diseases. Mr. C (at 57 units) would be one of those people.

Perhaps we should set the recommendation for nutrient X at or above the extreme, say, at 70 units a day, so that everyone will be covered. (Actually, we didn't study everyone, and some individual we didn't happen to test might have an even higher requirement.) This might be a good idea in theory, but what about a person like Mr. B who requires only 35 units a day? The recommendation would be twice his requirement and to follow it he might spend money needlessly on foods containing nutrient X to the exclusion of foods containing other vital nutrients.

The Decision The decision we finally make is to set the value high enough so that 97 to 98 percent of the population will be covered but not so high as to be excessive (Figure 2–3 illustrates such a value). In this example, a reasonable choice might be 63 units a day. Moving the DRI further toward the extreme would pick up a few additional people, but it would inflate the recommendation for most people, including Mr. A and Mr. B. The committee makes judgments of this kind when setting the DRI recommended intakes for many nutrients. Relatively few healthy people have requirements that are not covered by the DRI recommended intakes.

KEY POINT

- The DRI are based on scientific data and generously cover the needs of virtually all healthy people in the United States and Canada.

Setting Energy Requirements

In contrast to the recommendations for nutrients, the value set for energy, the **Estimated Energy Requirement (EER)**, is not generous; instead, it is set at a level predicted to maintain body weight for an individual of a particular age, gender, height, weight, and physical activity level consistent with good health. The energy DRI values reflect a balancing act: enough food energy is critical to support health and life, but too much energy causes unhealthy weight gain. Because even small amounts of excess energy consumed day after day cause weight gain and associated diseases, the DRI committee did not set a Tolerable Upper Intake Level for energy.

People don't eat energy directly. They derive energy from foods containing carbohydrate, fat, and protein, each in proportion to the others. The Acceptable

Macronutrient Distribution Ranges, described earlier, are designed to achieve a healthy balance among these nutrients and minimize a person's risk of chronic diseases. These ranges resurface in later chapters of this book wherever intakes of the energy-yielding nutrients are discussed with regard to chronic disease risks.

KEY POINT

- Estimated Energy Requirements are predicted to maintain body weight and to discourage unhealthy weight gain.

Why Are Daily Values Used on Labels?

On learning about the Daily Values, many people ask why yet another set of nutrient standards is needed for food labels—why not use the DRI? For one thing, DRI values for a nutrient must vary for different population groups, but food labels must list a single value for each nutrient. The Daily Values, therefore, are based on an "average" person—an individual eating 2,000 to 2,500 calories a day. Also unlike the DRI values, which are updated whenever significant changes occur in nutrition science, the Daily Values remain static.

The Daily Values are set at the highest level of nutrient need among all population groups, from children of age 4 years through aging adults; for example, the Daily Value for iron is 18 milligrams (mg), an amount that far exceeds a man's RDA of 8 mg (but that meets a young woman's high need precisely). Thus, the Daily Values are ideal for allowing comparisons among *foods*, but they cannot serve as nutrient intake goals for individuals. Appropriate use of the Daily Values is demonstrated in a later section.

KEY POINT

- The Daily Values are standards used solely on food labels to enable consumers to compare the nutrient values of foods.

Dietary Guidelines for Americans

LO 2.2 List the four major topic areas of the *Dietary Guidelines for Americans* and explain their importance to the population.

Appendix B presents some World Health Organization guidelines for other countries.

Many countries set dietary guidelines to answer the question, "What should I eat to stay healthy?" The U.S. Department of Agriculture's *Dietary Guidelines for Americans 2010* are part of an overall dietary guidance system and are related to the DRI values. If everyone followed the *Dietary Guidelines*, most people's nutrient and energy intakes would fall into place.

The *Guidelines* Promote Health The *Dietary Guidelines for Americans* (outlined in Table 2–1, p. 38) offer science-based advice to help people age 2 years and older achieve and sustain a healthy weight and to consume a diet of nutrient-dense foods and beverages.[3] People who balance energy intakes with expenditures, eat a nutritious diet, and make physical activity a habit often enjoy the best possible health and reduce their risks of chronic diseases substantially.[4]

Four Major Topic Areas The key recommendations of the *Dietary Guidelines for Americans 2010* fall into four major topic areas:

1. Balance calories to manage a healthy body weight.
2. Increase intakes of certain nutrient-dense foods.
3. Reduce intakes of certain foods and food components.
4. Build a healthy eating pattern.

Notice that the *Dietary Guidelines* do not require that you give up your favorite foods or eat strange, unappealing foods. With a little planning and a few adjustments, almost anyone's diet can approach these ideals. As for physical activity, this chapter's Think Fitness box offers some guidelines, while Chapter 10 provides details.

balance study a laboratory study in which a person is fed a controlled diet and the intake and excretion of a nutrient are measured. Balance studies are valid only for nutrients like calcium (chemical elements) that do not change while they are in the body.

requirement the amount of a nutrient that will just prevent the development of specific deficiency signs; distinguished from the DRI recommended intake value, which is a generous allowance with a margin of safety.

Estimated Energy Requirement (EER) the average dietary energy intake predicted to maintain energy balance in a healthy adult of a certain age, gender, weight, height, and level of physical activity consistent with good health.

Table 2–1

***Dietary Guidelines for Americans 2010*—Key Recommendations**

1. Balancing Calories to Manage Weight	
▪ Prevent and/or reduce overweight and obesity through improved eating and physical activity behaviors. ▪ Control total calorie intake to manage body weight. For people who are overweight or obese, this will mean consuming fewer calories from foods and beverages.	▪ Increase physical activity and reduce time spent in sedentary behaviors. ▪ Maintain appropriate calorie balance during each stage of life—childhood, adolescence, adulthood, pregnancy and breastfeeding, and older age.
2. Foods and Food Components to Reduce	
▪ Reduce daily sodium intake to less than 2,300 milligrams and further reduce intake to 1,500 milligrams among persons who are 51 and older and those of any age who are African American or have hypertension, diabetes, or chronic kidney disease. The 1,500 milligrams recommendation applies to about half of the U.S. population, including children and the majority of adults. ▪ Consume less than 10% of calories from saturated fatty acids by replacing them with monounsaturated and polyunsaturated fatty acids.[a] ▪ Consume less than 300 mg/day of dietary cholesterol.	▪ Keep *trans* fatty acid consumption as low as possible by limiting foods that contain synthetic sources of *trans* fats, such as partially hydrogenated oils, and by limiting other solid fats. ▪ Reduce the intake of calories from solid fats and added sugars. ▪ Limit the consumption of foods that contain refined grains, especially refined grain foods that contain solid fats, added sugars, and sodium. ▪ If alcohol is consumed it should be consumed in moderation—up to one drink per day for women and two drinks per day for men—and only by adults of legal drinking age.
3. Foods and Nutrients to Increase	
▪ Increase vegetable and fruit intake. ▪ Eat a variety of vegetables, especially dark-green and red and orange vegetables, and beans and peas. ▪ Consume at least half of all grains as whole grains. Increase whole-grain intake by replacing refined grains with whole grains. ▪ Increase intake of fat-free or low-fat milk and milk products, such as milk, yogurt, cheese, or fortified soy beverages. ▪ Choose a variety of protein foods, which include seafood, lean meat and poultry, eggs, beans and peas, soy products, and unsalted nuts and seeds.	▪ Increase the amount and variety of seafood consumed by choosing seafood in place of some meat and poultry. ▪ Replace protein foods that are higher in solid fats with choices that are lower in solid fats and calories and/or are sources of oils. ▪ Use oils to replace solid fats where possible. ▪ Choose foods that provide more potassium, dietary fiber, calcium, and vitamin D, which are nutrients of concern in American diets. These foods include vegetables, fruits, whole grains, and milk and milk products.
4. Building Healthy Eating Patterns	
▪ Select an eating pattern that meets nutrient needs over time at an appropriate calorie level. ▪ Account for all foods and beverages consumed and assess how they fit within a total healthy eating pattern.	▪ Follow food safety recommendations when preparing and eating foods to reduce the risk of foodborne illnesses.

[a]*Fatty acids are constituents of fats, as defined in Chapter 5.*

Source: The Dietary Guidelines for Americans 2010, *www.dietaryguidelines.gov*

Jupiterimages

The Dietary Guidelines *recommend physical activity to help balance calorie intakes to achieve and sustain a healthy body weight.*

How Does the U.S. Diet Compare to the Guidelines? The American diet needs improvement.[5] Figure 2–4 shows that people typically take in far too few nutritious foods and far too many less-than-nutritious ones. For most people, meeting the ideals of the *Dietary Guidelines* requires choosing more of these foods:

- Fruits and vegetables (especially dark green vegetables, red and orange vegetables, and legumes)
- Fish and other seafood (to replace some meals of meat and poultry)
- Whole grains
- Fat-free or low-fat milk and milk products

And fewer of these:

- Refined grains
- **Solid fats**: saturated fats, *trans* fats (replace them with unsaturated oils), and cholesterol
- Added sugars
- Salt

Figure 2–4

How Does the Typical U.S. Diet Stack Up?

The bars below reflect the average diet of people in the United States, from toddlers to the elderly. The top part of the figure indicates serious shortages of nutrient-dense foods and nutrients; the bottom part indicates an overabundance of foods and nutrients that should be limited for health's sake.

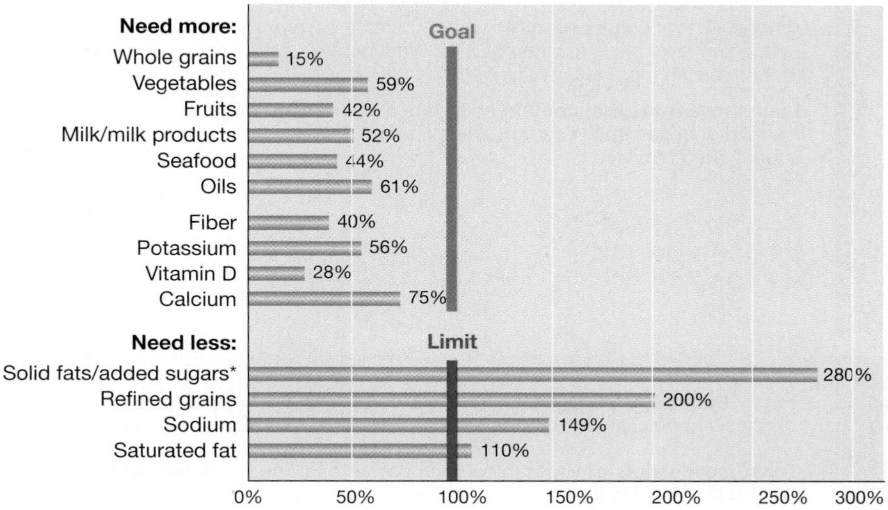

*Measured in calories.

Note: Based on data from U.S. Department of Agriculture, Agricultural Research Service and U.S. Department of Health and Human Services, Centers for Disease Control and Prevention. *What We Eat in America, NHANES 2001–2004 or 2005–2006.*

Source: Dietary Guidelines for Americans, 2010.

In addition, people who drink alcohol should monitor and moderate their alcohol intakes, and many people should reduce their total calorie intakes.

Our Two Cents' Worth If the experts who develop the *Dietary Guidelines* were to ask us, our focus would fall on their last recommendation: enjoy your food but eat less. The joys of eating are physically beneficial to the body because they trigger health-promoting changes in the nervous, hormonal, and immune systems. When the food is nutritious as well as enjoyable, then the eater obtains all the nutrients and phytochemicals needed for healthy body systems, as well as for the healthy skin, glossy hair, and natural attractiveness that accompany robust health.[6] When food is satisfying, eating less doesn't mean feeling deprived. Remember to enjoy your food.

KEY POINTS

- The Dietary Guidelines for Americans address problems of undernutrition and overnutrition.
- They recommend following a healthful eating pattern and being physically active.

Did You Know?

The key nutrients of concern in the U.S. diet are:

- Fiber
- Calcium
- Potassium
- Vitamin D

In addition, these nutrients are of concern to certain groups:

- Iron
- Folate
- Vitamin B_{12}

Diet Planning with the USDA Food Patterns

LO 2.3 Describe how and why foods are grouped in the USDA Food Patterns, including subgroups.

Diet planning connects nutrition theory with the food on the table, and a few minutes invested in meal planning can pay off richly in better nutrition. To help people achieve the goals of the *Dietary Guidelines for Americans*, the USDA employs a **food group plan** known as the USDA Food Patterns.[7] Figure 2–5 (pp. 40–41) displays this plan.

solid fats fats that are high in saturated fat and usually not liquid at room temperature. Some common solid fats include butter, beef fat, chicken fat, pork fat, stick margarine, coconut oil, palm oil, and shortening.

food group plan a diet-planning tool that sorts foods into groups based on their nutrient content and then specifies that people should eat certain minimum numbers of servings of foods from each group.

Figure 2-5

USDA Food Patterns: Food Groups and Subgroups

© Polara Studios, Inc.

1 c fruit =
1 c fresh, frozen, or canned fruit
½ c dried fruit
1 c 100% fruit juice

Fruits contribute folate, vitamin A, vitamin C, potassium, and fiber.

Consume a variety of fruits, and choose whole or cut-up fruits more often than fruit juice.

Apples, apricots, avocados, bananas, blueberries, cantaloupe, cherries, grapefruit, grapes, guava, honeydew, kiwi, mango, nectarines, oranges, papaya, peaches, pears, pineapples, plums, raspberries, strawberries, tangerines, watermelon; dried fruit (dates, figs, prunes, raisins); 100% fruit juices

Limit these fruits that contain solid fats and/or added sugars:
Canned or frozen fruit in syrup; juices, punches, ades, and fruit drinks with added sugars; fried plantains

© Polara Studios, Inc.

1 c vegetables =
1 c cut-up raw or cooked vegetables
1 c cooked legumes
1 c vegetable juice
2 c raw, leafy greens

Vegetables contribute folate, vitamin A, vitamin C, vitamin K, vitamin E, magnesium, potassium, and fiber.

Consume a variety of vegetables each day, and choose from all five subgroups several times a week.

Vegetables subgroups:
Dark-green vegetables: Broccoli and leafy greens such as arugula, beet greens, bok choy, collard greens, kale, mustard greens, romaine lettuce, spinach, turnip greens, watercress

Red and orange vegetables: Carrots, carrot juice, pumpkin, red bell peppers, sweet potatoes, tomatoes, tomato juice, vegetable juice, winter squash (acorn, butternut)

Legumes: Black beans, black-eyed peas, garbanzo beans (chickpeas), kidney beans, lentils, navy beans, pinto beans, soybeans and soy products such as tofu, split peas, white beans

Starchy vegetables: Cassava, corn, green peas, hominy, lima beans, potatoes

Other vegetables: Artichokes, asparagus, bamboo shoots, bean sprouts, beets, brussels sprouts, cabbages, cactus, cauliflower, celery, cucumbers, eggplant, green beans, green bell peppers, iceberg lettuce, mushrooms, okra, onions, seaweed, snow peas, zucchini

Limit these vegetables that contain solid fats and/or added sugars:
Baked beans, candied sweet potatoes, coleslaw, french fries, potato salad, refried beans, scalloped potatoes, tempura vegetables

© Polara Studios, Inc.

1 oz grains =
1 slice bread
½ c cooked rice, pasta, or cereal
1 oz dry pasta or rice
1 c ready-to-eat cereal
3 c popped popcorn

Grains contribute folate, niacin, riboflavin, thiamin, iron, magnesium, selenium, and fiber.

Make most (at least half) of the grain selections whole grains.

Grains subgroups:
Whole grains: amaranth, barley, brown rice, buckwheat, bulgur, cornmeal, millet, oats, quinoa, rye, wheat, wild rice and whole-grain products such as breads, cereals, crackers, and pastas; popcorn

Enriched refined products: bagels, breads, cereals, pastas (couscous, macaroni, spaghetti), pretzels, white rice, rolls, tortillas

Limit these grains that contain solid fats and/or added sugars:
Biscuits, cakes, cookies, cornbread, crackers, croissants, doughnuts, fried rice, granola, muffins, pastries, pies, presweetened cereals, taco shells

Figure 2–5

USDA Food Patterns: Food Groups and Subgroups (continued)

© Polara Studios, Inc.

1 oz protein foods =
- **1 oz cooked lean meat, poultry, or seafood**
- **1 egg**
- **¼ c cooked legumes or tofu**
- **1 tbs peanut butter**
- **½ oz nuts or seeds**

Protein foods contribute protein, essential fatty acids, niacin, thiamin, vitamin B_6, vitamin B_{12}, iron, magnesium, potassium, and zinc.

Choose a variety of protein foods from the three subgroups, including seafood in place of meat or poultry twice a week.

Protein foods subgroups:
Seafood: Fish (catfish, cod, flounder, haddock, halibut, herring, mackerel, pollock, salmon, sardines, sea bass, snapper, trout, tuna), shellfish (clams, crab, lobster, mussels, oysters, scallops, shrimp)

Meats, poultry, eggs: Lean or low-fat meats (fat-trimmed beef, game, ham, lamb, pork, veal), poultry (no skin), eggs

Nuts, seeds, soy products: Unsalted nuts (almonds, cashews, filberts, pecans, pistachios, walnuts), seeds (flaxseeds, pumpkin seeds, sesame seeds, sunflower seeds), legumes, soy products (textured vegetable protein, tofu, tempeh), peanut butter, peanuts

Limit these protein foods that contain solid fats and/or added sugars:
Bacon; baked beans; fried meat, seafood, poultry, eggs, or tofu; refried beans; ground beef; hot dogs; luncheon meats; marbled steaks; poultry with skin; sausages; spare ribs

© Polara Studios, Inc.

1 c milk or milk product =
- **1 c milk, yogurt, or fortified soy milk**
- **1½ oz natural cheese**
- **2 oz processed cheese**

Milk and milk products contribute protein, riboflavin, vitamin B_{12}, calcium, potassium, and, when fortified, vitamin A and vitamin D.

Make fat-free or low-fat choices. Choose other calcium-rich foods if you don't consume milk.

Fat-free or 1% low-fat milk and fat-free or 1% low-fat milk products such as buttermilk, cheeses, cottage cheese, yogurt; fat-free fortified soy milk

Limit these milk products that contain solid fats and/or added sugars:
2% reduced-fat milk and whole milk; 2% reduced-fat and whole-milk products such as cheeses, cottage cheese, and yogurt; flavored milk with added sugars such as chocolate milk, custard, frozen yogurt, ice cream, milk shakes, pudding, sherbet; fortified soy milk

© Matthew Farruggio

1 tsp oil =
- **1 tsp vegetable oil**
- **1 tsp soft margarine**
- **1 tbs low-fat mayonnaise**
- **2 tbs light salad dressing**

Oils are not a food group, but are featured here because they contribute vitamin E and essential fatty acids.

Use oils instead of solid fats, when possible.

Liquid vegetable oils such as canola, corn, flaxseed, nut, olive, peanut, safflower, sesame, soybean, sunflower oils; mayonnaise, oil-based salad dressing, soft *trans*-free margarine; unsaturated oils that occur naturally in foods such as avocados, fatty fish, nuts, olives, seeds (flaxseeds, sesame seeds), shellfish

Limit these solid fats:
Butter, animal fats, stick margarine, shortening

Art © Cengage Learning 2013

Think Fitness move it!

Recommendations for Daily Physical Activity

The USDA's *Physical Activity Guidelines for Americans* suggest that to maintain good health, adults should engage in about 2½ hours of moderate physical activity each week.[8] A brisk walk at a pace of about 100 steps per minute (1,000 steps over 10 minutes) constitutes "moderate" activity.[9] In addition:

- Physical activity can be intermittent, 10 minutes here and there, throughout the week.
- Resistance activity (such as weightlifting) can be a valuable part of the exercise total for the week.

For weight control and additional health benefit more than this minimum amount of physical activity is required. Details can be found in Chapter 10.

start now! Ready to make a change? Set a goal of 30 minutes per day of physical activity (walking, jogging, biking, weight training, etc.) then track your actual activity for 5 days. You can use the Track Activity feature of Diet Analysis Plus.

The A, B, C, M, V principles were explained in **Chapter 1,** pages 11–12.

By using it wisely and by learning about the energy-yielding nutrients, vitamins, and minerals in various foods (as you will in coming chapters), you can achieve the goals of a nutritious diet first mentioned in Chapter 1: adequacy, balance, calorie control, moderation, and variety.

If you design your diet around this plan, it is assumed that you will obtain adequate and balanced amounts of the nutrients of greatest concern along with the two dozen or so other essential nutrients and hundreds of potentially beneficial phytochemicals because all of these compounds are distributed among the same foods. It can also help you to limit calories and potentially harmful food constituents.

Phytochemicals and their potential biological actions are explained in **Controversy 2**.

The Food Groups and Subgroups

Figure 2–5 defines the major food groups and their subgroups. USDA specifies portions of various foods within each group that are nutritional equivalents and thus can be treated interchangeably in diet planning. It also lists the key nutrients provided by foods within each group, information worth noting and remembering. The foods in each group are well-known contributors of the key nutrients listed, but you can count on these foods to supply many other nutrients as well. Note also that the figure sorts foods within each group by nutrient density.

Vegetables Subgroups and Protein Foods Subgroups Not every vegetable supplies every key nutrient attributed to the Vegetables group, so the Food Patterns sort vegetables into subgroups by their nutrient contents. All vegetables provide valuable fiber and the mineral potassium, but many from the "red and orange vegetables" subgroup are known for their vitamin A content; those from the "dark green vegetables" provide a wealth of folate; "starchy vegetables" provide abundant carbohydrate; and "legumes" supply substantial iron and protein.

The Protein Foods group falls into subgroups, too. All Protein Foods dependably supply iron and protein, but their fats vary widely. "Meats" tend to be higher in saturated fats that should be limited. "Seafood" and "nuts, seeds, and soy" foods tend to be low in saturated fat, while providing essential fats that the body requires.

Grains Subgroups and Other Foods Among the grains, the foods of the whole grains subgroup supply fiber and a wide variety of nutrients. Refined grains lack many of these beneficial compounds but provide abundant energy. The Food Patterns suggest that at least half of the grains in a day's meals be whole grains, or that at least three servings of whole-grain foods be included in the diet each day.[10] (Grain serving sizes in 1-ounce equivalents are listed in Figure 2–5.)

Spices, herbs, coffee, and tea provide few if any nutrients but can add flavor and pleasure to meals. Some, such as tea and spices, also provide potentially beneficial phytochemicals—see this chapter's Controversy section.

Variety Among and Within Food Groups Varying food choices, both among the food groups and within each group, helps to ensure adequate nutrient intakes and also protects against consuming large amounts of toxins or contaminants from any one food. Achieving variety may require some effort, but knowing which foods fall into which food groups eases the task.

KEY POINTS

- The USDA Food Patterns divide foods into food groups based on key nutrient contents.
- People who consume the specified amounts of foods from each group and subgroup achieve dietary adequacy, balance, and variety.

Choosing Nutrient-Dense Foods

To help people control calories and achieve and sustain a healthy body weight, the USDA Food Patterns instruct consumers to base their diets on the most nutrient-dense foods from each group. Unprocessed or lightly processed foods are generally best because many processes strip foods of beneficial nutrients and fiber, and others add salt, sugar, or fat. Figure 2–5 identifies many nutrient-dense food choices in each food group and points out some foods of lower **nutrient density** to give you an idea of which are which.

Nutrient density was explained in **Chapter 1**, page 21.

Uncooked (raw) oil is worth notice in this regard. Oil is pure calorie-rich fat and is therefore low in nutrient density, but a small amount of raw oil from sources such as avocado, olives, nuts, and fish, or even raw vegetable oil provides vitamin E and essential lipids that other foods lack. High temperatures used in frying destroy these nutrients, however, so the recommendation specifies *raw* oil.

© Matthew Farruggio

Solid fats, added sugars, and alcohol should be limited.

Solid Fats, Added Sugars, and Alcohol Reduce Nutrient Density Solid fats deliver saturated fat and *trans* fat, terms that will become familiar after reading Chapter 5. Sugars in all their forms (described in Chapter 4) deliver carbohydrate calories. Figure 2–6, p. 44, demonstrates how solid fats and added sugars add "empty" calories to foods, reducing their nutrient density. Solid fats include:

- Naturally occurring fats, such as milk fat and meat fats.
- Added fats, such as butter, cream cheese, hard margarine, lard, sour cream, and shortening.

Added sugars include:

- All caloric sweeteners, such as brown sugar, candy, honey, jelly, molasses, soft drinks, sugar, and syrups.

© iStockphoto.com/koshtu

The heavy syrup of these canned peaches adds 135 calories of extra sugar. A cup of plain peaches provides 60 calories of a nutrient-dense food.

The USDA suggests that combined calories from solid fats and added sugars should not exceed a person's discretionary calorie allowance, as described in the next section.

Alcoholic beverages are a top contributor of calories to the diets of many U.S. adults, but they provide few nutrients.[11] People who drink alcohol should monitor and moderate their intakes, not to exceed one drink a day for women, two for men. People in many circumstances should never drink alcohol (see Controversy 3).

The Concept of Discretionary Calories The concept of a "discretionary calorie allowance" can be useful to those who must limit calorie intakes to prevent excessive weight gain. As Figure 2–7, p. 44, demonstrates, a person needing 2,000 calories of energy in a day to maintain weight may need only 1,700 calories or so of the most

nutrient density a measure of nutrients provided per calorie of food. A *nutrient-dense food* provides vitamins, minerals, and other beneficial substances with relatively few calories. Also defined in Chapter 1.

Figure 2–6

How Solid Fats and Added Sugars Add Calories to Nutrient-Dense Foods

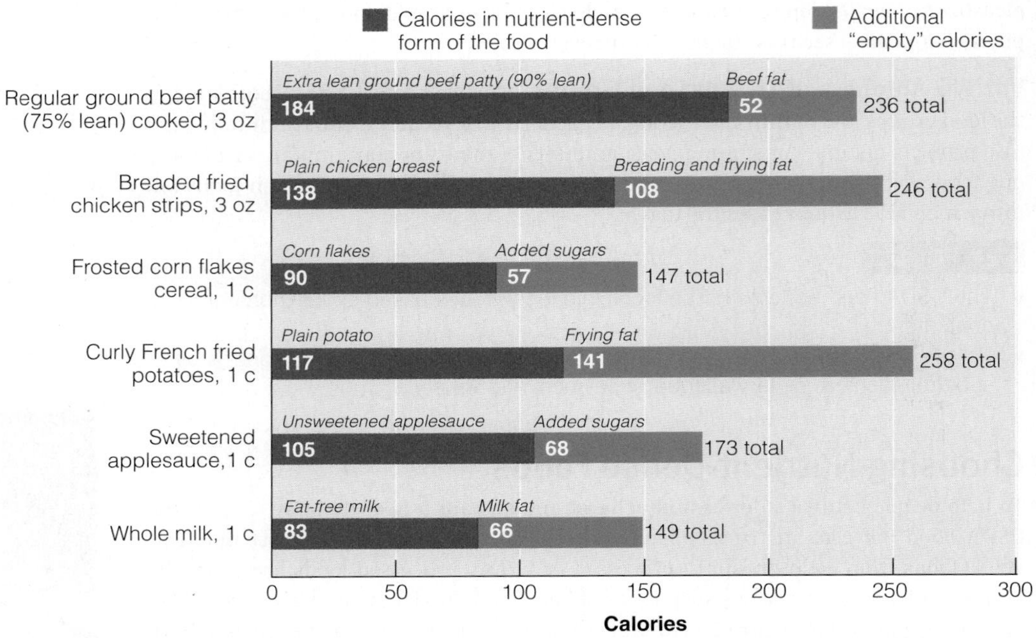

Source: U.S. Department of Agriculture and U.S. Department of Health and Human Services, Dietary Guidelines for Americans 2010, *p. 47.*

Figure 2–7

Discretionary Calorie Concept

The discretionary calorie allowance sets the upper limit for calories from added sugars and solid fats in USDA Food Patterns.

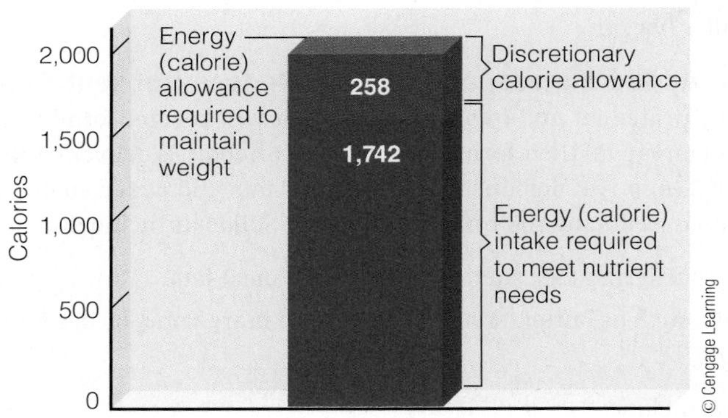

nutrient-dense foods to supply the nutrients required for the day. The difference between the calories needed to maintain weight and those needed to supply nutrients from the most nutrient-dense foods is the person's discretionary calorie allowance (in this case, 258 calories).

In theory, this person may freely choose how to fill the calorie void—with more nutrient-dense foods, foods low in nutrient density, or even some solid fats and added sugars. In practice, few people struggle to meet their calories needs, and few can afford the luxury of daily energy-rich, nutrient-poor treats.

KEY POINTS

- Following the USDA Food Patterns requires choosing nutrient-dense foods most often.
- Solid fats, added sugars, and alcohol should be limited.

Diet Planning Application

LO 2.4 Outline the basic steps of diet planning with the USDA Food Patterns, and address limits for solid fats and added sugars.

© iStockphoto.com

The USDA Food Patterns specify the amounts of food needed from each food group to create a healthful diet for a given number of calories. Look at the top line of Table 2–2 and find yourself among the people described there (for other calorie levels, see Table E–1 of Appendix E). Then look at the column of numbers below for amounts from each food group that meet your calorie need. Note that the more energy spent on physical activity each day, the greater the calorie need.

For vegetables and protein foods, intakes should be divided among all the subgroups over a week's time, as shown in Table 2–3. Look across the top row for your calorie level (obtained from Table 2–2)—a healthful diet includes the listed amounts of each type of vegetable and protein food each *week*. It is not necessary to eat foods from every subgroup each day.

With judicious selections, the diet can supply all the necessary nutrients and provide some luxury items as well. A sample diet plan demonstrates how the theory of the USDA Food Patterns translates to food on the plate. The USDA Food Patterns ensure that a certain amount from each of the five food groups is represented in the diet.

The diet planner begins by assigning each of the food groups to meals and snacks, as shown in Table 2–4. Then the plan can be filled out with real foods to create a menu. For example, this breakfast calls for 1 ounce of grains, 1 cup of milk, and 1/2 cup of fruit. Here's one possibility for this meal:

1 cup ready-to-eat cereal = 1 ounce grains

1 cup fat-free milk = 1 cup milk

1 medium banana = 1/2 cup fruit

Table 2–2

USDA Food Patterns: Daily Amounts from Each Food Group[a]

	Sedentary Women: 51+ yr	Sedentary Women: 26–50 yr	Sedentary Women: 19–25 yr Active Women: 61+ yr Sedentary Men: 61+ yr	Active Women: 31–60 yr Sedentary Men: 41–60 yr	Active Women: 19–30 yr Sedentary Men: 21–40 yr	Active Men: 36–55 yr	Active Men: 19–35 yr
Calories	1,600	1,800	2,000	2,200	2,400	2,800	3,000
Fruits	1 1/2 c	1 1/2 c	2 c	2 c	2 c	2 1/2 c	2 1/2 c
Vegetables[b]	2 c	2 1/2 c	2 1/2 c	3 c	3 c	3 1/2 c	4 c
Grains	5 oz	6 oz	6 oz	7 oz	8 oz	10 oz	10 oz
Protein Foods	5 oz	5 oz	5 1/2 oz	6 oz	6 1/2 oz	7 oz	7 oz
Milk	3 c	3 c	3 c	3 c	3 c	3 c	3 c
Oils	5 tsp	5 tsp	6 tsp	6 tsp	7 tsp	8 tsp	10 tsp
Solid fats/ Added sugars[c]	121 cal	161 cal	258 cal	266 cal	330 cal	395 cal	459 cal

Note: In addition to gender, age, and activity levels, energy needs vary with height and weight (see Chapter 9 and Appendix H).

[a] *Selected calorie levels; see Appendix E for additional calorie and activity levels.*

[b] *Divide these amounts among the vegetable subgroups as specified in Table 2-3.*

[c] *This number defines the calorie limit.*

Table 2–3

USDA Food Patterns: Weekly Amounts from Vegetable and Protein Foods Subgroups

Table 2-2 specifies total intakes per *day*. This table shows those amounts dispersed among five vegetable and three protein subgroups per *week*.

Vegetable Subgroups	1,600 cal	1,800 cal	2,000 cal	2,200 cal	2,400 cal	2,600 cal	2,800 cal	3,000 cal
Dark green	1½ c	1½ c	1½ c	2 c	2 c	2½ c	2½ c	2½ c
Red and orange	4 c	5½ c	5½ c	6 c	6 c	7 c	7 c	7½ c
Legumes	1 c	1½ c	1½ c	2 c	2 c	2½ c	2½ c	3 c
Starchy	4 c	5 c	5 c	6 c	6 c	7 c	7 c	8 c
Other	3½ c	4 c	4 c	5 c	5 c	5½ c	5½ c	7 c
Protein Foods Subgroups								
Seafood	8 oz	8 oz	8 oz	9 oz	10 oz	10 oz	11 oz	11 oz
Meat, poultry, eggs	24 oz	24 oz	26 oz	29 oz	31 oz	31 oz	34 oz	34 oz
Nuts, seeds, soy products	4 oz	4 oz	4 oz	4 oz	5 oz	5 oz	5 oz	5 oz

Table 2–4

A Sample Diet Plan

This diet plan is one of many possibilities for a day's meals. It follows the amounts suggested for a 2,000-calorie diet (with an extra ½ cup of vegetables).

Food Group	Recommended Amounts	Breakfast	Lunch	Snack	Dinner	Snack
Fruits	2 c	½ c		½ c	1 c	
Vegetables	2½ c		1 c		2 c	
Grains	6 oz	1 oz	2 oz	½ oz	2 oz	½ oz
Protein foods	5½ oz		2 oz		3½ oz	
Milk	3 c	1 c		1 c		1 c
Oils	5½ tsp		1½ tsp		4 tsp	
Solid fats/ Added sugars	258 cal					

Our sample diet plan in Table 2–4 has met nutrient needs with about 250 calories remaining—enough for about two extra fruit servings (210 calories), another half-portion of spaghetti (210 calories), one small doughnut (250 calories), or a large soft drink (180 calories). Alternatively, the diet planner can choose to skip such foods and the calories they present when weight loss is a goal.

Then the planner moves on to complete the menu for lunch, supper, and snacks, as shown in Table 2–5, adding only about 100 discretionary calories of solid fats and added sugars. This day's choices are explored further as Monday's Meals in the Food Feature at the end of the chapter.

Table 2–5
A Sample Menu

This sample menu provides about 1,850 calories of the 2,000-calorie plan. About 150 calories remain available to spend on more nutrient-dense foods or luxuries such as added sugars and solid fats.

Amounts	Sample Menu	Energy (Cal)
	BREAKFAST	
1 oz whole grains	1 c whole-grain cereal	108
1 c milk	1 c fat-free milk	100
1/2 c fruit	1 medium banana (sliced)	105
	LUNCH	
2 oz meats, 2 oz whole grains	1 turkey sandwich on whole-wheat roll	272
1 1/2 tsp oils	1 1/2 tbs low-fat mayonnaise	71
1 c vegetables	1 c vegetable juice	50
	SNACK	
1/2 oz whole grains	4 whole-wheat reduced-fat crackers	86
1 c milk	1 1/2 oz low-fat cheddar cheese	74
1/2 c fruit	1 medium apple	72
	DINNER	
1/2 c vegetables	1 c raw spinach leaves	8
1/4 c vegetables	1/4 c shredded carrots	11
1 oz meats	1/4 c garbanzo beans	71
2 tsp oils	2 tbs oil-based salad dressing and olives	76
3/4 c vegetables, 2 1/2 oz meat, 2 oz enriched grains	Spaghetti with meat and tomato sauce	425
1/2 c vegetables	1/2 c green beans	22
2 tsp oils	2 tsp soft margarine	67
1 c fruit	1 c strawberries	49
	SNACK	
1/2 oz whole grains	3 graham crackers	90
1 c milk	1 c fat-free milk	100

Note: This plan meets the recommendations to provide 45–65% of calories from carbohydrate, 20–35% from fat, and 10–35% from protein.

KEY POINT

- The USDA Food Patterns for various calorie levels can guide food choices in diet planning.

MyPlate Educational Tool

For consumers with Internet access, the USDA's MyPlate online suite of educational tools makes applying the USDA Food Patterns easier.[12] Figure 2–8 displays its graphic image. Computer-savvy consumers will find an abundance of MyPlate support materials and diet assessment tools on the Internet (www.choosemyplate.gov). Those without computer access can achieve the same diet-planning goals by following this chapter's principles and working with pencil and paper, as illustrated later.

KEY POINT

- The concepts of the USDA Food Patterns are demonstrated in the MyPlate online educational tools.

Figure 2–8
USDA MyPlate

Note that vegetables and fruits occupy half the plate and that the grains portion is slightly larger than the portion of protein foods. A diet that follows the USDA Food Patterns reflects these ideals.

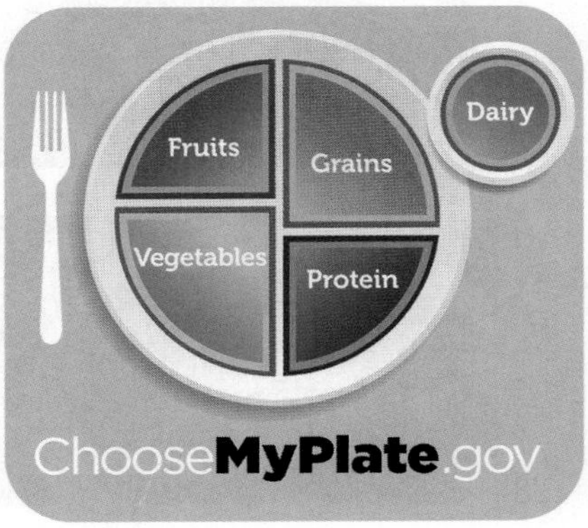

My Turn — watch it!

Right Size—Supersize?

© Cengage Learning

Stephanie

Do you often overeat when you eat out? Listen to a student talk about making healthy choices in restaurants.

Flexibility of the USDA Food Patterns

Although they may appear rigid, the USDA Food Patterns can actually be quite flexible once their intent is understood. For example, the user can substitute fat-free yogurt for fat-free milk because both supply the key nutrients for the Milk and Milk Products group. Legumes, an extraordinarily nutrient-rich food, provide many of the nutrients that characterize the Protein Foods group, but they also constitute a Vegetables subgroup, so legumes in a meal can count either as a serving of meat or of vegetables. Consumers can adapt the plan to mixed dishes such as casseroles and to national and cultural foods as well, as Figure 2–9 illustrates.

Figure 2–9

Ethnic Food Choices

	Grains	Vegetables	Fruits	Protein Foods	Milk
Asian  © Becky Luigart-Stayner/Corbis	Rice, white or rice noodles, millet, wheat or rice wrappers and crepes	Amaranth, baby corn, bamboo shoots, chayote, bok choy, mung bean sprouts, snow peas, mushrooms, water chestnuts, kelp	Carambola, guava, kumquat, lychee, persimmon, melons, mandarin orange	Soybeans and soy products such as miso and tofu, squid, duck eggs, pork, poultry, fish and other seafood, peanuts, cashews	Soy milk
Mediterranean  Photodisc/Getty Images	Pita pocket bread, pastas, rice, couscous, polenta, bulgur, focaccia, Italian bread	Eggplant, tomatoes, peppers, cucumbers, grape leaves	Olives, grapes, figs	Fish and other seafood, gyros, lamb, chicken, beef, pork, sausage, lentils, fava beans	Ricotta, provolone, parmesan, feta, mozzarella, and goat cheeses; yogurt
Mexican  Mitch Hrdlicka/Photodisc/Getty Images	Tortillas (corn or flour), taco shells, rice	Chayote, corn, jicama, tomato salsa, cactus, cassava, tomatoes, yams, chilies	Guava, mango, papaya, avocado, plantain, bananas, oranges	Refried beans, fish, chicken, chorizo, beef, eggs	Cheese, custard

Art © Cengage Learning

Vegetarians can use adaptations of the USDA Food Patterns in making sound food choices, too. The food group that includes the meats also includes nuts, seeds, and products made from soybeans. The vegetable group includes legumes, counted as protein foods for vegetarians.[13] In the food group that includes milk, soy drinks and soy milk (beverages made from soybeans) can fill the same nutrient needs, provided that they are fortified with calcium, riboflavin, vitamin A, vitamin D, and vitamin B_{12}. Appendix E presents two vegetarian adaptations to the USDA Food Patterns, and Controversy 6 provides many diet-planning details. Therefore, for all sorts of careful diet planners, the USDA Food Patterns provide a general road map for designing a healthful diet.

KEY POINT

- The USDA Food Patterns can be used with flexibility by people with different eating styles.

A Note about Exchange Systems

A different kind of diet-planning tool, the **exchange system** (see Appendix D), originally developed for use by people with diabetes, can be useful to anyone wishing to control calories. An exchange system lists estimated grams of carbohydrate, fat, saturated fat, and protein in standardized food portions, as well as their calorie values. These are average gram values for whole groups of foods, and so they often differ from the exacting values given for individual foods in Appendix A. With exchange estimates committed to memory, users of the system can make an informed approximation of the energy-yielding nutrients and calories in almost any food they might encounter. To explore the usefulness of this powerful aid to diet planning, spend some time studying Appendix D.

KEY POINT

- Exchange lists group foods that are similar in carbohydrate, fat, and protein to facilitate control of their consumption.

The Last Word on Diet Planning

All of the dietary changes required to improve nutrition may seem daunting or even insurmountable at first, and taken all at once they may be. However, small steps taken each day can add up to substantial dietary changes over time. If everyone would begin, today, to take such steps, the rewards in terms of less risk of diabetes, obesity, heart disease, and cancer along with a greater quality of life with better health would prove well worth their effort.

Checking Out Food Labels

LO 2.5 Evaluate a food label, delineating the different uses of information found on the Nutrition Facts panel, on the ingredients list, and in any health claims or other claims made for the product.

A potato is a potato and needs no label to tell you so. But what can a package of potato chips tell you about its contents? By law, its label must list the chips' ingredients—potatoes, fat, and salt—and its **Nutrition Facts** panel must also reveal details about their nutrient composition. If the oil is high in saturated fat, the label will tell you so (more about fats in Chapter 5). A label may also warn consumers of a food's potential for causing an allergic reaction (Chapter 14 provides details). In addition to required information, labels may make optional statements about the food being delicious, or good for you in some way, or a great value. Some of these comments, especially some that are regulated by the Food and Drug Administration (FDA), are reliable. Many others are marketing tools, based more on salesmanship than nutrition science.

exchange system a diet-planning tool that organizes foods with respect to their nutrient content and calories. Foods on any single exchange list can be used interchangeably. See Appendix D for details.

Nutrition Facts on a food label, the panel of nutrition information required to appear on almost every packaged food. Grocers may also provide the information for fresh produce, meats, poultry, and seafood.

use it! A Consumer's Guide To . . .

Controlling Portion Sizes at Home and Away

"May I take your order please?" Put on the spot when eating out, a diner must quickly choose from a large, visually exciting menu. No one brings a scale to a restaurant to weigh portions, and physical cues used at home, such as measuring cups are, well, at home. Restaurant portions have no standards. When ordering "a burger," for example, the sandwich may arrive resembling a 2-ounce kids' sandwich or a 3/4-pound behemoth. Even at home, portion sizes can be mystifying—how much spaghetti is enough?

How Big Is Your Bagel?

When college students are asked to bring "medium-sized" foods to class, they reliably bring bagels weighing from 2 to 5 ounces, muffins from 2 to 8 ounces, baked potatoes from 4 to 9 ounces, and so forth. Knowledge of appropriate daily amounts of food is crucial to controlling calorie intakes, but consumers need help to estimate portion sizes, whether preparing a meal at home or choosing from a restaurant menu.

© Matthew Farruggio

How much does your bagel weigh?

Practice with Weights and Measures

At home, practice makes perfect. To estimate the size of food portions, remember these common objects:

- 3 ounces of meat = the size of the palm of a woman's hand or a deck of cards
- 1 medium potato or piece of fruit = the size of a tennis ball
- 1 1/2 ounces cheese = the size of a 9-volt battery
- 1 ounce lunch meat or cheese = 1 slice
- 1 cup cooked pasta = the size of a baseball
- 1 pat (1 tsp) butter or margarine = a slice from a quarter-pound stick of butter about as thick as 280 pages of this book (pressed together).
- Most ice cream scoops hold 1/4 cup = a lump about the size of a golf ball. (Test the size of your scoop—fill it with water and pour the water into a measuring cup. Now you have a handy device to measure portions at home—use the scoop to serve mashed potatoes, pasta, vegetables, rice, and cereals.)

Among volumetric measures, 1 "cup" refers to an 8-ounce measuring cup (not a teacup or drinking glass) filled to level (not heaped up, or shaken, or pressed down). Tablespoons and teaspoons refer to measuring spoons (not flatware), filled to level (not rounded or heaping). Ounces signify weight, not volume. Two ounces of meat, for example, refers to one-eighth of a pound of cooked meat. One ounce (weight) of crispy rice cereal measures a full cup (volume), but take care: 1 ounce of granola cereal measures only 1/4 cup. The Table of Food Composition, Appendix A, can help in determining serving sizes because it lists both weights and volumes for a wide variety of foods.

Colossal Cuisine in Restaurants

Figure 2–10 presents data collected over three decades showing that consumers doubled the amount of food that they typically eat away from home.[1] Two other trends occurred at the same time: food portions grew larger and therefore more caloric (Figure 2–11), and people's body weights increased to new unhealthy levels. Taken together, these trends suggest that restaurant food portions may be affecting public health.

A step in the right direction is a law requiring all chain restaurants, including fast food restaurants, to post calorie information on menus and menu boards

Figure 2–10

Dining Out Trends, United States

The percentage of total calories from foods eaten away from home doubled over the past 30 years; at the same time, the percentage of calories from fast food grew rapidly.

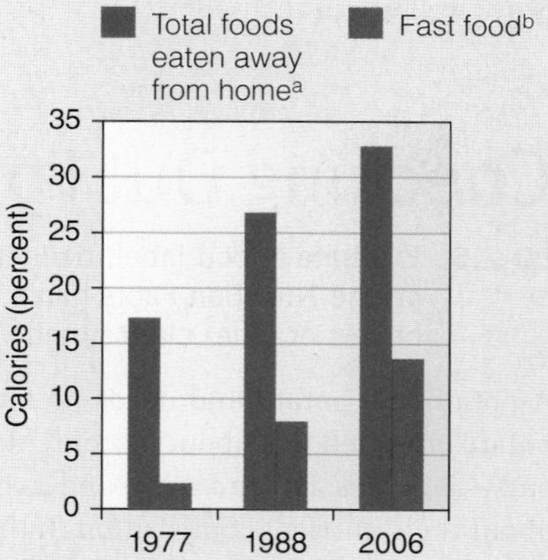

[a] *All foods eaten away from home, including at schools, sports stadiums, restaurants, and other establishments.*

[b] *Includes food from restaurants with counter or drive-through service and cafeterias.*

Source: Data from R. M. Morrison, L. Mancino, and J. N. Variyam, Will calorie labeling in restaurants make a difference? Amber Waves *9 (2011): 10–17.*

Figure 2–11

A Shift toward Colossal Cuisine

The portion sizes of many foods have increased dramatically over past decades.

Food	Typical 1970s	Today's colossal
Cola	10 oz bottle, 120 cal	40–60 oz fountain, 580 cal
French fries	about 30, 475 cal	about 50, 790 cal
Hamburger	3–4 oz meat, 330 cal	6–12 oz meat, 1,000 cal
Bagel	2–3 oz, 230 cal	5–7 oz, 550 cal
Steak	8–12 oz, 690 cal	16–22 oz, 1,260 cal
Pasta	1 c, 200 cal	2–3 c, 600 cal
Baked potato	5–7 oz, 180 cal	1 lb, 420 cal
Candy bar	1½ oz, 220 cal	3–4 oz, 580 cal
Popcorn	1½ c, 80 cal	8–16 c tub, 880 cal

Note: Calories are rounded values for the largest portions in a given range.

© Tony Freeman/PhotoEdit

1970s Today

© Matthew Farruggio (both)

1970s Today

1970s Today

for each standard food item.* Without a gauge readily at hand, consumers most often underestimate both calories and fat in restaurant foods.[2] In local non–chain restaurants where such helpful information may be lacking, people must learn to judge portions on their own.

Knowing not just what to eat, but how much, comes with practice. Try portioning out foods at home until you can easily estimate serving sizes on the go. When you see an enticing menu, look for calorie amounts to use as a gauge.

** The law is the Patient Protection and Affordable Care Act of 2010.*

When portions seem excessively large or calorie-rich, use creative solutions to cut them down to size: order a half portion, ask that half of a regular portion be packaged for a later meal, order a child's portion, or split an entrée with a friend.

Moving Ahead

Portion control is a habit, and a way to defend against overeating. When cooking at home, have measuring tools at the ready. When dining out, your tools are your practiced abilities to judge portion sizes. Then, when the waiter asks "Are you ready to order?" the savvy consumer, armed with portion size know-how, answers confidently, "Yes."

Review Questions†

1. American restaurant portions are stable and consistent; use them as a guide to choosing portion sizes. T F
2. Experimenting with portion sizes at home is a valuable exercise in self-education. T F
3. When consumers guess at the calorie values in restaurant food portions, they generally overestimate. T F

†Answers to Consumer's Guide questions are found in Appendix G.

What Food Labels Must Include

The Nutrition Education and Labeling Act of 1990 set the requirements for certain label information to ensure that food labels truthfully inform consumers about the nutrients and ingredients in the package. In 2012, FDA reviewed the details of this information, but in general, every packaged food must state the following:

- The common or usual name of the product.
- The name and address of the manufacturer, packer, or distributor.
- The net contents in terms of weight, measure, or count.

- The nutrient contents of the product (Nutrition Facts panel).
- The ingredients in descending order of predominance by weight and in ordinary language.
- Essential warnings, such as alerts about ingredients that often cause allergic reactions or other problems.

Not every package need display information about every vitamin and mineral. A large package, such as a box of cereal, must provide all of the information just listed. A smaller label, such as the label on a can of tuna, provides some of the information in abbreviated form. A label on a roll of candy rings provides only a phone number, which is allowed for the tiniest labels.

The Nutrition Facts Panel A little over half of U.S. consumers read food labels.[14] When they do, they often rely on the Nutrition Facts panel, like the one shown in Figure 2–12. Grocers also voluntarily post placards or offer handouts in produce departments to provide consumers with similar sorts of nutrition information for the most popular types of fresh fruits, vegetables, and seafoods. Under a new ruling, packages of meat cuts, ground meats, and poultry also must display a Nutrition Facts panel.[15]

Notice in Figure 2–12 that only the top portion of a food's Nutrition Facts panel conveys information specific to the food inside the package. The bottom portion is identical on every label—it stands as a reminder of the Daily Values.

Figure 2–12

Animated! What's on a Food Label?

This cereal label maps out the locations of information needed to make wise purchases. The text provides details about each label section. Labels may also warn consumers of potential allergy risks (see Chapter 14 for details).

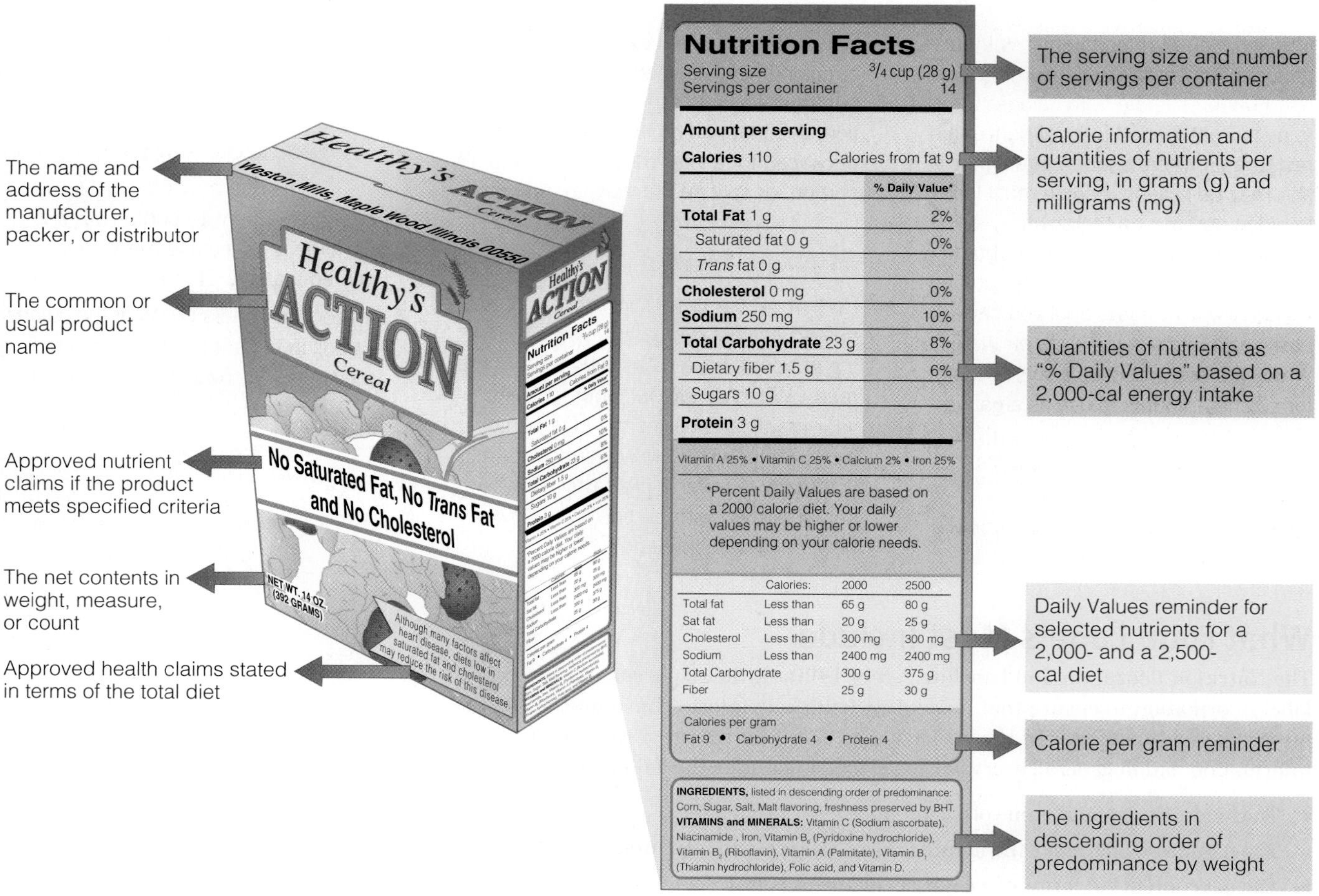

Figure 2–12 points out where the following may be found on a label (from top to bottom):

- *Serving size*. A common household and metric measure of a single serving that provides calorie and the nutrient amounts as listed. A serving of chips may be 10 chips, so if you eat 50 chips, you will have consumed five times the calorie and nutrient amounts listed on the label. When you compare different brands of the same food, check the serving size—it may differ.
- *Servings per container*. Number of servings per box, can, or package.
- *Calories/calories from fat*. Total food energy per serving and energy from fat per serving.
- *Nutrient amounts and percentages of Daily Values*, including:
  - *Total fat*. Grams of fat per serving with a breakdown showing grams of *saturated fat* and *trans fat* per serving.
  - *Cholesterol*. Milligrams of cholesterol per serving.
  - *Sodium*. Milligrams of sodium per serving.
  - *Total carbohydrate*. Grams of carbohydrate per serving, including starch, fiber, and sugars, with a breakdown showing grams of dietary *fiber* and *sugars*. The sugars include those that occur naturally in the food plus any added during processing.
  - *Protein*. Grams of protein per serving.

In addition, the label must state the contents of these nutrients expressed as percentages of the Daily Values: vitamin A, vitamin C, calcium, and iron.

Other nutrients present in significant amounts in the food may also be listed on the label. The percentages of the Daily Values (see the inside back cover, page Y) are given in terms of a 2,000-calorie diet.

- *Daily Values and calories-per-gram reminder*. This portion lists the Daily Values for a person needing 2,000 or 2,500 calories a day and provides a calories-per-gram reminder as a handy reference.

Food labels provide clues for nutrition sleuths.

Ingredients List An often neglected but highly valuable body of information is the list of ingredients. The product's ingredients must be listed in descending order of predominance by weight.

Knowing how to read an ingredients list puts you many steps ahead of the naïve buyer. Anyone diagnosed with a food allergy quickly learns to use them for spotting "off-limits" ingredients in foods. In addition, you can glean clues about the nature of the food. For example, consider the ingredients list on an orange drink powder whose first three entries are "sugar, citric acid, orange flavor." You can tell that sugar is the chief ingredient. Now consider a canned juice whose ingredients list begins with "water, orange juice concentrate, pineapple juice concentrate." This product is clearly made of reconstituted juice. Water is first on the label because it is the main constituent of juice. Sugar is nowhere to be found among the ingredients because no sugar has been added. Sugar occurs naturally in juice, though, so the label does specify sugar grams; details are in Chapter 4.

Now consider a cereal whose entire list contains just one item: "100 percent shredded wheat." No question, this is a whole-grain food with nothing added. Finally, consider a cereal whose first six ingredients are "puffed milled corn, corn syrup, sucrose, honey, dextrose, salt." If you recognize that sugar, corn syrup, honey, and dextrose are all different versions of sugar (and you will after Chapter 4), you might guess that this product contains close to half its weight as added sugar.

More about Percentages of Daily Values The nutrient percentages of Daily Values ("% Daily Value") on labels are for a single serving of food, and they are based on the Daily Values set for a 2,000-calorie diet. For example, if a food contributes 13 milligrams of vitamin C per serving and the Daily Value is 60 milligrams, then a serving of that food provides 22 percent of the Daily Value for vitamin C.

Of course, though the Daily Values are based on a 2,000-calorie diet, people's actual calorie needs vary widely. This makes the Daily Values most useful for comparing one food with another and less useful as nutrient intake targets for individuals. Still, by examining a food's general nutrient profile, you can determine whether the food contributes "a little" or "a lot" of a nutrient, and whether it contributes "more" or "less" than another food.

What Food Labels *May* Include

So far, this section has presented the accurate and reliable food label facts. Another group of reliable statements are the **nutrient claims**.

Nutrient Claims: Reliable Information A food that meets specified criteria may display certain approved nutrient claims on its label. These claims, for example, that a food is "low" in cholesterol or a "good source" of vitamin A, are based on the Daily Values. Table 2–6 provides a list of these regulated, reliable label terms along with their definitions.

Table 2–6

Reliable Nutrient Claims on Food Labels

Energy Terms

- **low calorie** 40 calories or fewer per serving.
- **reduced calorie** at least 25% lower in calories than a "regular," or reference, food.
- **calorie free** fewer than 5 calories per serving.

Fat Terms (Meat and Poultry Products)

- **extra lean**[a]
  less than 5 g of fat *and*
  less than 2 g of saturated fat and *trans* fat combined, *and*
  less than 95 mg of cholesterol per serving.
- **lean**[a]
  less than 10 g of fat *and*
  less than 4.5 g of saturated fat and *trans* fat combined, *and*
  less than 95 mg of cholesterol per serving.

Fat Terms (Main Dishes and Prepared Meals)

- **extra lean**[a]
  less than 5 g total fat *and*
  less than 2 g saturated fat *and*
  less than 95 mg cholesterol per serving.
- **lean**[a]
  less than 8 g total fat *and*
  3.5 g or less saturated fat *and*
  less than 80 mg cholesterol per serving.

Fat and Cholesterol Terms (All Products)

- **cholesterol free**[b]
  less than 2 mg of cholesterol *and*
  2 g or less saturated fat and *trans* fat combined per serving.
- **fat free** less than 0.5 g of fat per serving.
- **less saturated fat** 25% or less saturated fat and *trans* fat combined than the comparison food.

[a] *The word* lean *as part of the brand name (as in "Lean Supreme") indicates that the product contains fewer than 10 g of fat per serving.*

[b] *Foods containing more than 13 g total fat per serving or per 50 g of food must indicate those contents immediately after a cholesterol claim.*

Table 2-6

Reliable Nutrient Claims on Food Labels (continued)

- **low cholesterol**[b]
  20 mg or less of cholesterol *and*
  2 g or less saturated fat per serving.
- **low fat** 3 g or less fat per serving.[a]
- **low saturated fat** 1 g or less saturated fat and less than 0.5 g of *trans* fat per serving.
- **percent fat free** may be used only if the product meets the definition of low fat or fat free. Requires disclosure of grams of fat per 100 g food.
- **reduced** or **less cholesterol**[b]
  at least 25% less cholesterol than a reference food *and*
  2 g or less saturated fat per serving.
- **reduced saturated fat**
  at least 25% less saturated fat *and*
  reduced by more than 1 g saturated fat per serving compared with a reference food.
- **saturated fat free**
  less than 0.5 g of saturated fat *and*
  less than 0.5 g of trans fat.
- ***trans* fat free**
  less than 0.5 g of *trans* fat *and*
  less than 0.5 g of saturated fat per serving.

Fiber Terms

- **high fiber** 5 g or more per serving. (Foods making high-fiber claims must fit the definition of low fat, or the level of total fat must appear next to the high-fiber claim.)
- **good source of fiber** 2.5 g to 4.9 g per serving.
- **more** or **added fiber** at least 2.5 g more per serving than a reference food.

Sodium Terms

- **low sodium** 140 mg or less sodium per serving.
- **reduced sodium** at least 25% lower in sodium than the regular product.
- **sodium free** less than 5 mg per serving.
- **very low sodium** 35 mg or less sodium per serving.

Other Terms

- **free, without, no, zero** none or a trivial amount. **Calorie free** means containing fewer than 5 calories per serving; **sugar free** or **fat free** means containing less than half a gram per serving.
- **fresh** raw, unprocessed, or minimally processed with no added preservatives.
- **good source** 10 to 19% of the Daily Value per serving.
- **healthy** low in fat, saturated fat, *trans* fat, cholesterol, and sodium and containing at least 10% of the Daily Value for vitamin A, vitamin C, iron, calcium, protein, or fiber.
- **high in** 20% or more of the Daily Value for a given nutrient per serving; synonyms include "rich in" or "excellent source."
- **less, fewer, reduced** containing at least 25% less of a nutrient or calories than a reference food. This may occur naturally or as a result of altering the food. For example, pretzels, which are usually low in fat, can claim to provide less fat than potato chips, a comparable food.
- **light** this descriptor has three meanings on labels:
  1. A serving provides one-third fewer calories or half the fat of the regular product.
  2. A serving of a low-calorie, low-fat food provides half the sodium normally present.
  3. The product is light in color and texture, so long as the label makes this intent clear, as in "light brown sugar."
- **more, extra** at least 10% more of the Daily Value than in a reference food. The nutrient may be added or may occur naturally.

[a]*The word* lean *as part of the brand name (as in "Lean Supreme") indicates that the product contains fewer than 10 g of fat per serving.*

[b]*Foods containing more than 13 g total fat per serving or per 50 g of food must indicate those contents immediately after a cholesterol claim.*

nutrient claims claims using approved wording to describe the nutrient values of foods, such as the claim that a food is "high" in a desirable constituent, or "low" in an undesirable one.

Health Claims: Reliable and Not So Reliable In the past, the FDA held manufacturers to the highest standards of scientific evidence before allowing them to place **health claims** (defined in Table 2–7) on food labels. When a label stated "Diets low in sodium may reduce the risk of high blood pressure," for example, consumers could be sure that the FDA had substantial scientific support for the claim. Such reliable health claims are still allowed on food labels, and they have a high degree of scientific validity.

Today, however, the FDA also allows other similar-sounding health claims that are backed by weaker evidence. These are "qualified" claims in the sense that labels bearing them must also state the strength of the scientific evidence backing them up. Unfortunately, most consumers cannot distinguish between scientifically reliable claims and those that are less so.[16]

Structure-Function Claims: Best Ignored Even less reliable are **structure-function claims**. A label-reading consumer is much more likely to encounter this kind of claim on a food or supplement label than the more regulated health claims just described. For the food manufacturer, printing a *health claim* involves acquiring and submitting scientific evidence to the FDA with a request for permission to print the claim, a time-consuming and expensive process. Instead, the manufacturer can easily print a similar-looking structure-function claim that requires only FDA notification and no prior approval. Figure 2–13 compares the three kinds of claims just discussed.

A problem is that, to a reasonable consumer, the two kinds of claims may appear identical:

- "Lowers cholesterol" (health claim)
- "Helps maintain normal cholesterol levels" (structure-function claim)

The first requires advance FDA evaluation and approval. The second can be printed without prior approval.

A label disclaimer (often printed in tiny, easily missed type) must accompany a structure-function claim. It states that the FDA has not evaluated the claim and that the product is not intended to diagnose, treat, cure, or prevent any disease.

Such valid-appearing but unreliable label claims diminish the credibility of all health-related claims on labels. Until laws change to require solid scientific backing, consumers should ignore health-related claims and rely on the Nutrient Facts panels

Table 2–7

Reliable Health Claims on Labels

These claims of potential health benefits are well-supported by research, but other similar-sounding claims may not be.

- Calcium and reduced risk of osteoporosis
- Sodium and reduced risk of hypertension
- Dietary saturated fat and cholesterol and reduced risk of coronary heart disease
- Dietary fat and reduced risk of cancer
- Fiber-containing grain products, fruits, and vegetables and reduced risk of cancer
- Fruits, vegetables, and grain products that contain fiber, particularly soluble fiber, and reduced risk of coronary heart disease
- Fruits and vegetables and reduced risk of cancer
- Folate and reduced risk of neural tube defects
- Sugar alcohols and reduced risk of tooth decay
- Soluble fiber from whole oats and from psyllium seed husk and reduced risk of coronary heart disease
- Soy protein and reduced risk of coronary heart disease
- Whole grains and reduced risk of coronary heart disease and certain cancers
- Plant sterol and plant stanol esters and reduced risk of coronary heart disease
- Potassium and reduced risk of hypertension and stroke

health claims claims linking food constituents with disease states; allowable on labels within the criteria established by the Food and Drug Administration.

structure-function claim a legal but largely unregulated claim permitted on labels of foods and dietary supplements, often mistaken by consumers for a regulated health claim.

Figure 2–13
Label Claims

Nutrient claims characterize the level of a nutrient in the food—for example, "fat free" or "less sodium."

Health claims characterize the relationship of a food or food component to a disease or health-related condition—for example, "soluble fiber from oatmeal daily in a diet low in saturated fat and cholesterol may reduce the risk of heart disease" or "a diet low in total fat may reduce the risk of some cancers."

Structure/function claims describe the effect that a substance has on the structure or function of the body and do not make reference to a disease—for example, "supports immunity and digestive health" or "calcium builds strong bones."

Sam Kolich/Bill Smith Group/Cengage Learning (all)

and ingredients lists for product information. In the world of marketing, current label laws put the consumer on notice: "Let the buyer beware."

Label Short Cuts To some consumers, the information on food labels is daunting—they cannot or will not read it all before making a choice. They want short cuts, such as icons on fronts of packages, to easily and quickly assess a food's nutrient contents. Currently, food manufacturers can pay fees to professional or academic groups to print endorsement symbols or stamps on the fronts of food labels, for example, a "heart healthy" symbol for foods that meet the group's criteria.

According to the Food and Drug Administration, many nutrition-conscious consumers would use such short-cut symbols, so the FDA is working with food industry experts to develop a standardized set. Soon, the front panels of packages of foods that meet the criteria of the *Dietary Guidelines for Americans 2010* may bear informative symbols similar to those depicted in the margin. Their goal is to help consumers of various ages, income brackets, and literacy levels to more easily compare foods and make sound choices based on nutrition.[17]

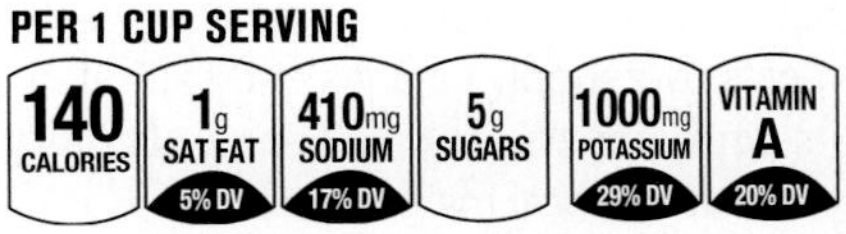

Courtesy Facts Up Front/Grocery Manufacturers Association

Consumers may soon see short-cut icons like these on the fronts of food labels. This set was developed by the Grocery Manufacturers Association and is aligned with FDA regulations.

try it! Food Feature

Getting a Feel for the Nutrients in Foods

LO 2.6 State specific nutritional advantages of a carefully planned nutrient-dense diet over a diet chosen without regard for nutrition principles.

Figures 2–14 and 2–15 illustrate a playful contrast between two days' meals. Monday's meals were selected according to the recommendations of this chapter and follow the sample menu of Table 2–5, shown earlier (page 47). Tuesday's meals were chosen more for convenience and familiarity than out of concern for nutrition.

Comparing the Nutrients

How can a person compare the nutrients that these sets of meals provide? One way is to look up each food in a table of food composition, write down the food's nutrient values, and compare each one to a standard such as the DRI recommended intakes for nutrients, as we've done in Figures 2–14 and 2–15. By this measure, Monday's meals are the clear winners in terms of meeting nutrient needs within a calorie budget. Tuesday's meals oversupply calories and saturated fat while undersupplying fiber and critical vitamins and minerals.

Another useful exercise is to compare the total amounts of foods provided by a day's meals with the recommended amounts from each food group. A tally of the cups and ounces of foods consumed is provided in both Figures 2–14 and 2–15. The totals are then compared with USDA Food Patterns in the tabular portion of the figures. The tables also identify whole grains and Vegetables subgroups and tally calories from solid fats and added sugars to complete the assessment.

Monday's Meals in Detail

Monday's meals provide the necessary servings from each food group along with a small amount of oil needed for health, and the energy provided falls well within the 2,000-calorie allowance. A closer look at Monday's foods reveals that the whole-grain cereal at breakfast, whole-grain sandwich roll at lunch, and whole-grain crackers at snack time meet the recommendation to obtain at least half of the day's grain servings from whole grains.

For the Vegetables subgroups, dark green vegetables, orange vegetables, and legumes are represented in the dinner salad, and "other vegetables" are prominent throughout. To repeat: it isn't necessary to choose vegetables from each subgroup every day, and the person eating this day's meals will need to include vegetables from other subgroups throughout the week. In addition, Monday's eating plan has room to spare in the discretionary calorie allowance for additional servings of favorite foods or for some sweets or fats.

Tuesday's Meals in Detail

Tuesday's meals, though abundant in oils, meats, and enriched grains, completely lack fruit and whole grains and are too low in vegetables and milk to provide adequate nutrients. Tuesday's meals supply too much saturated fat and sugar, as well as excessive meats and refined grains, pushing the calorie total well above the day's allowance. A single day of such fare poses little threat to the eater, but a steady diet of Tuesday's meals presents a high probability of nutrient deficiencies and weight gain and greatly increases the risk of chronic diseases in later life.

Use a Computer—or Not?

If you have access to a computer, it can be a time saver—diet analysis programs perform all of these calculations at lightning speed. This convenience may make working it out for yourself, using paper and a sharp pencil with a big eraser, seem a bit old-fashioned. But there are times when using a laptop or a diet application on a cellular phone may not be practical—such as standing on line at the cafeteria or at a fast-food counter—where real-life food decisions must be made quickly.

People who work out diet analyses for themselves on paper and those who put extra time into studying, changing, and reviewing their computer analysis often learn to "see" the nutrients in foods (a skill you can develop by the time you reach Chapter 10). They can quickly assess their food options and make informed choices at mealtimes. People who fail to develop such skills must wait until they can access their computer programs to find out how well they did after the fact.

Figure 2–14

Monday's Meals—Nutrient-Dense Choices

Breakfast

© Polara Studios, Inc.

Lunch

© Polara Studios, Inc.

Afternoon snack

© Polara Studios, Inc.

Dinner

© Matthew Farruggio

Bedtime snack

© Quest Photographic, Inc.

Foods	Amounts	Energy (cal)	Saturated Fat (g)	Fiber (g)	Vitamin C (mg)	Calcium (mg)
Before heading off to class, a student eats breakfast:						
1 c whole-grain cold cereal	1 oz grains	108	—	3	14	95
1 c fat-free milk	1 c milk	100	—	—	2	306
1 medium banana (sliced)	½ c fruit	105	—	3	10	6
Then goes home for a quick lunch:						
1 roasted turkey sandwich	2 oz meat	50	—	1	60	27
on 2-oz whole-grain roll with	2 oz grains	343	4	2	—	89
1½ tsp low-fat mayonnaise	1½ tsp oils					
1 c low-salt vegetable juice	1 c vegetables					
While studying in the afternoon, the student eats a snack:						
4 whole-wheat reduced-fat crackers	½ oz grains	86	1	2	—	—
1½ oz low-fat cheddar cheese	1 c milk	74	2	—	—	176
1 apple	½ c fruit	72	—	3	6	8
That night, the student makes dinner:						
A salad:						
1 c raw spinach leaves, shredded carrots	1 c vegetables	19	—	2	18	61
¼ c garbanzo beans	1 oz legumes	71	—	3	2	19
5 lg olives and 2 tbs oil-based salad dressing	2 tsp oils	76	1	1	—	2
A main course:						
1 c spaghetti	2 oz grains	425	3	5	15	56
with meat sauce	2½ oz meat	22	—	2	6	29
½ c green beans	1 c vegetables	67	1	—	—	—
2 tsp soft margarine	2 tsp oils					
And for dessert:						
1 c strawberries	1 c fruit	49	—	3	89	24
Later that evening, the student enjoys a bedtime snack:						
3 graham crackers	½ oz grains	90	—	—	—	—
1 c fat-free milk	1 c milk	100	—	—	2	306
Totals:		1,857	12	30	224	1,204
DRI recommended intakes:[a]		2,000	<20[b]	25	75	1,000
Percentage of DRI recommended intakes:		93%	60%	120%	299%	120%

Intakes Compared with Recommended Amounts

Food Group	Breakfast	Lunch	Snack	Dinner	Snack	Monday's Totals	Recommended Amounts
Fruits	½ c		½ c	1 c		2 c	2 c
Vegetables		1 c		2 c		3 c	2½ c
Grains	1 oz	2 oz	½ oz	2 oz	½ oz	6 oz	6 oz
Protein foods		2 oz		3½ oz		5½ oz	5½ oz
Milk	1 c		1 c		1 c	3 c	3 c
Oils		1½ tsp		4 tsp		5½ tsp	5½ tsp
Calorie allowance						1,857 cal	2,000 cal

[a]DRI values for a sedentary woman, age 19–30. Other DRI values are listed on the inside front cover, page B.

[b]The 20-g value listed is the maximum allowable saturated fat for a 2,000-cal diet. The DRI recommends consuming less than 10% of calories from saturated fat.

Figure 2–15

Tuesday's Meals—Less-Nutrient-Dense Choices

Breakfast

© Quest Photographic, Inc.

Lunch

© Quest Photographic, Inc.

Afternoon snack

© Polara Studios, Inc.

Dinner

© Quest Photographic, Inc.

Bedtime snack

© Charles Knowles/Shutterstock.com

Foods	Amounts	Energy (cal)	Saturated Fat (g)	Fiber (g)	Vitamin C (mg)	Calcium (mg)
Today, the student starts the day with a fast-food breakfast:						
1 c coffee	2 oz grains	5	—	—	—	—
1 English muffin with egg, cheese, and bacon	2 oz meat 1 c milk	436	9	2	—	266
Between classes, the student returns home for a quick lunch:						
1 peanut butter and jelly sandwich on white bread	2 oz grains 1 oz legumes	426	4	3	—	93
1 c whole milk	1 c milk	156	6	—	4	290
While studying, the student has:						
12 oz diet cola		—	—	—	—	—
Bag of chips (14 chips)[a]		105	2	—	4	—
That night for dinner, the student eats:						
A salad: 1c lettuce 1 tbs blue cheese dressing	½ c vegetables	84	2	1	2	23
A main course: 6 oz steak	6 oz meat	349	6	—	—	27
½ baked potato	½ c vegetables	161	—	4	17	26
1 tbs butter		102	7	—	—	3
1 tbs sour cream[b]		31	2	—	—	17
12 oz diet cola		—	—	—	—	—
And for dessert: 4 sandwich-type cookies	1 oz grains	158	2	1	—	—
Later on, a bedtime snack:						
2 cream-filled snack cakes	2 oz grains	250	2	2	—	20
1 c herbal tea		—	—	—	—	—
Totals:		2,263	42	13	27	765
DRI recommended intakes:[c]		2,000	<20[d]	25	75	1,000
Percentage of DRI recommended intakes:		113%	210%	52%	36%	77%

Intakes Compared with Recommended Amounts

Food Group	Breakfast	Lunch	Snack	Dinner	Snack	Tuesday's Totals	Recommended Amounts
Fruits						0 c	2 c
Vegetables			[a]	1 c		1 c	2½ c
Grains	2 oz	2 oz		1 oz	2 oz	7 oz	6 oz
Protein foods	2 oz	2 oz		6 oz		9 oz	5½ oz
Milk	1 c	1 c				2 c	3 c
Oils						7½ tsp[b]	5½ tsp
Calorie allowance						2,263 cal	2,000 cal

[a] *The potato in 14 potato chips provides less than ½ c vegetables.*

[b] *The saturated fats of steak, butter, and sour cream are among the solid fats and do not qualify as oils.*

[c] *DRI values for a sedentary woman, age 19–30. Other DRI values are listed on the inside front cover, page B.*

[d] *The 20-g value listed is the maximum allowable saturated fat for a 2,000-cal diet. The DRI recommends consuming less than 10% of calories from saturated fat.*

Art © Cengage Learning

track it!

Diet Analysis PLUS Concepts in Action

Compare Your Intakes with USDA Guidelines

The purpose of this chapter's exercise is to give you a feel for the nutrients in food and to help you consider your sources of solid fats and added sugars. Use the Diet Analysis Plus (DA+) program to help you evaluate your nutritional intake and needs.

1. From the Home page of DA+ select the Reports tab and select MyPlate Analysis. Choose Day Two of your three-day diet intake (from Chapter 1). Choose all meals for that day. Generate a report. Did your intake for that day conform to the MyPlate pattern? Did you consume too few foods from any particular food group(s)? Which, if any, were lacking? Using Table 2–2 (page 45) and Figure 2–5 (pages 42–43) to guide you, suggest ways that you might realistically change your intake to better conform to the USDA Food Patterns.
2. What about fat? Select the Reports tab; then select Macronutrient Ranges. Generate a report. Did your fat intake fall between 20 percent and 35 percent of your total energy? Did you take in enough raw oils to meet your need (see Table 2–2, p. 45)? Which ones? Change your date to include all three days of your record; generate a report to see a fat intake average. How does your single day's fat intake compare with your three-day average?
3. The bottom line of Table 2–2 specifies an upper intake limit calories from solid fats and added sugars. Select the Track Diet tab, and look over your day's food list. Which foods were less-nutrient-dense choices?
4. Breaking this information down further, which foods on your food list contribute added sugars? If you consumed substantial calories of added sugars, suggest realistic ways to reduce your intake.
5. A great feature of the DA+ program is its Source Analysis Report, which allows you to list food sources of calories (kcal) or specific nutrients in order of predominance. From the Reports tab, select Source Analysis, select Day Three, and choose all meals. Generate a report. Which foods provided most to your calorie intake on that day? If you consumed vegetables, where did they fall on the list? In later chapters, you'll use this report again to analyze various nutrients in your diet.

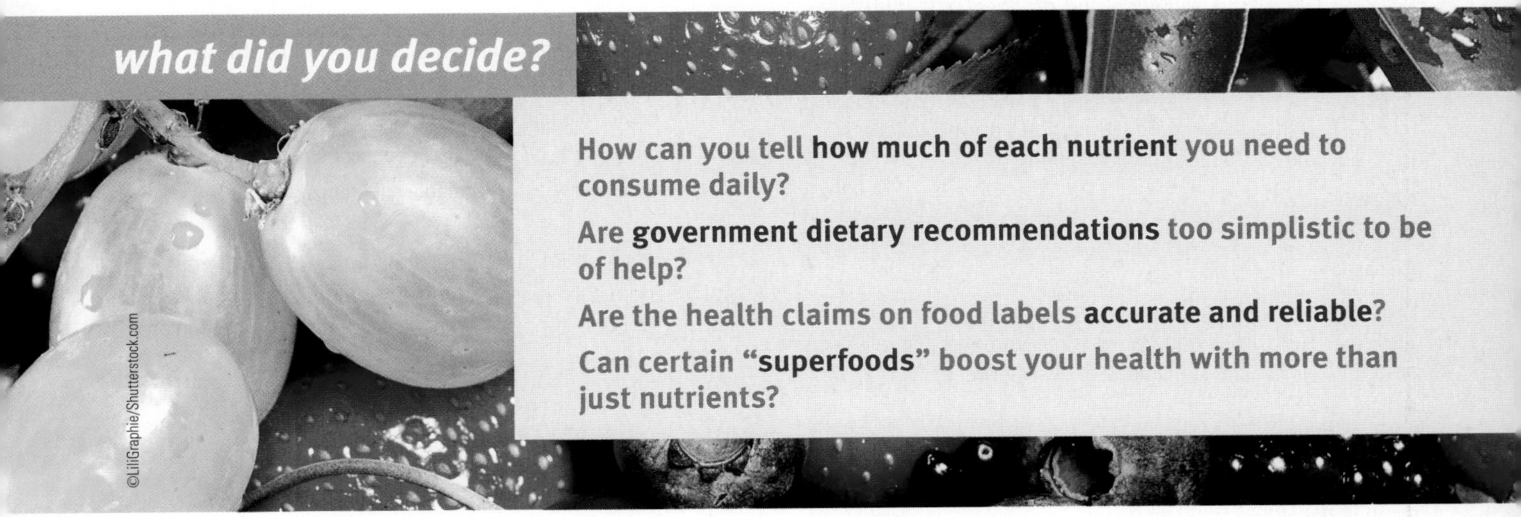

©LiliGraphie/Shutterstock.com

what did you decide?

How can you tell how much of each nutrient you need to consume daily?

Are government dietary recommendations too simplistic to be of help?

Are the health claims on food labels accurate and reliable?

Can certain "superfoods" boost your health with more than just nutrients?

Self Check

1. (LO 2.1) The nutrient standards in use today include all of the following *except* __________ .
   a. Adequate Intakes (AI)
   b. Daily Minimum Requirements (DMR)
   c. Daily Values (DV)
   d. a and c
2. (LO 2.1) The Dietary Reference Intakes were devised for which of the following purposes?
   a. to set nutrient goals for individuals
   b. to suggest upper limits of intakes, above which toxicity is likely
   c. to set average nutrient requirements for use in research
   d. all of the above
3. (LO 2.1) The energy intake recommendation is set at a level predicted to maintain body weight.
   T F
4. (LO 2.1) The Dietary Reference Intakes (DRI) are for all people, regardless of their medical history.
   T F
5. (LO 2.2) The *Dietary Guidelines for Americans* include all of these major topic areas, except:
   a. Balance calories to manage a healthy body weight.
   b. Increase intakes of certain nutrient-dense foods.
   c. Reduce intakes of artificial ingredients.
   d. Reduce intakes of certain foods and food components.
6. (LO 2.2) The *Dietary Guidelines for Americans* recommend physical activity to help balance calorie intakes to achieve and sustain a healthy body weight.
   T F
7. (LO 2.3) According to the USDA Food Patterns, which of the following vegetables should be limited?
   a. carrots
   b. avocados
   c. baked beans
   d. potatoes
8. (LO 2.3) The USDA Food Patterns recommend a small amount of daily oil from which of these sources?
   a. olives
   b. nuts
   c. vegetable oil
   d. all of the above
9. (LO 2.3) People who choose not to eat meat or animal products need to find an alternative to the USDA Food Patterns when planning their diets.
   T F
10. (LO 2.4) To plan a healthy diet that correctly assigns the needed amounts of food from each food group, the diet planner should start by consulting
    a. USDA Food Patterns
    b. Dietary Reference Intakes
    c. sample menus
    d. none of the above
11. (LO 2.4) A properly planned diet controls calories by excluding snacks.
    T F
12. (LO 2.5) Which of the following values is found on food labels?
    a. Recommended Dietary Allowances
    b. Dietary Reference Intakes
    c. Daily Values
    d. Estimated Average Requirements
13. (LO 2.5) By law, food labels must state as a percentage of the Daily Values the amounts of vitamin C, vitamin A, niacin, and thiamin present in food.
    T F
14. (LO 2.5) To be labeled "low fat," a food must contain 3 grams of fat or less per serving.
    T F
15. (LO 2.6) One way to evaluate any eating pattern is to compare the total food amounts that it provides with those recommended by the USDA Food Patterns.
    T F
16. (LO 2.6) A carefully planned diet has which of these characteristics?
    a. It contains sufficient raw oil.
    b. It contains no solid fats or added sugars.
    c. It contains all of the vegetable subgroups.
    d. a and c
17. (LO 2.7) Various whole foods contain so many different phytochemicals that consumers should focus on eating a wide variety of foods instead of seeking out a particular phytochemical.
    T F
18. (LO 2.7) As natural constituents of foods, phytochemicals are safe to consume in large amounts.
    T F

Answers to these Self Check questions are in Appendix G.

CONTROVERSY 2

Are Some Foods Superfoods for Health?

LO 2.7 Discuss the positive and negative findings for dietary phytochemicals with regard to health, and make a case for food sources over supplements to provide them.

Are some foods superfoods for health? Headlines certainly say so: "Forgetful? Blueberries sharpen brain function!" "Too many colds? Supercharge your immune system with soybeans!" "Worried about cancer? Eat tomatoes!" Can the produce aisle double as a medicine chest? Although headlines tend to overstate their talents, **functional foods** do supply **phytochemicals**—nonnutrient components of plants, some of which are under study for their potential to influence human health and disease. (Terms are defined in Table C2–1.)

Table C2–1

Phytochemical and Functional Food Terms

- **antioxidants** (anti-OX-ih-dants) compounds that protect other compounds from damaging reactions involving oxygen by themselves reacting with oxygen (*anti* means "against"; *oxy* means "oxygen"). *Oxidation* is a potentially damaging effect of normal cell chemistry involving oxygen (more in Chapters 5 and 7).
- **bioactive food components** compounds in foods, either nutrients or phytochemicals, that alter physiological processes.
- **broccoli sprouts** the sprouted seed of *Brassica italica*, or the common broccoli plant; believed to be a functional food by virtue of its high phytochemical content.
- **drug** any substance that when taken into a living organism may modify one or more of its functions.
- **edamame** fresh green soybeans, a source of phytoestrogens.
- **flavonoids** (FLAY-von-oyds) a common and widespread group of phytochemicals, with over 6,000 identified members; physiologic effects may include antioxidant, antiviral, anticancer, and other activities. Flavonoids are yellow pigments in foods; *flavus* means "yellow."
- **flaxseed** small brown seed of the flax plant; used in baking, cereals, or other foods. Valued in nutrition as a source of fatty acids, lignans, and fiber.
- **functional foods** whole or modified foods that contain bioactive food components believed to provide health benefits, such as reduced disease risks, beyond the benefits that their nutrients confer. All whole foods are functional in some ways because they provide at least some needed substances, but certain foods stand out as rich sources of bioactive food components. Also defined in Chapter 1.
- **genistein** (GEN-ih-steen) a phytoestrogen found primarily in soybeans that both mimics and blocks the action of estrogen in the body.
- **kefir** (KEE-fur) a liquid form of yogurt, based on milk, probiotic microorganisms, and flavorings.
- **lignans** phytochemicals present in flaxseed, but not in flax oil, that are converted to phytoestrogens by intestinal bacteria and are under study as possible anticancer agents.
- **lutein** (LOO-teen) a plant pigment of yellow hue; a phytochemical believed to play roles in eye functioning and health.
- **lycopene** (LYE-koh-peen) a pigment responsible for the red color of tomatoes and other red-hued vegetables; a phytochemical that may act as an antioxidant in the body.
- **miso** fermented soybean paste used in Japanese cooking. Soy products are considered to be functional foods.
- **organosulfur compounds** a large group of phytochemicals containing the mineral sulfur. Organosulfur phytochemicals are responsible for the pungent flavors and aromas of foods belonging to the onion, leek, chive, shallot, and garlic family and are thought to stimulate cancer defenses in the body.
- **phytochemicals** (FIGH-toe-CHEM-ih-cals) compounds in plants that confer color, taste, and other characteristics. Often, the bioactive food components of functional foods. Also defined in Chapter 1. *Phyto* means "plant."
- **phytoestrogens** (FIGH-toe-ESS-troh-gens) phytochemicals structurally similar to the female sex hormone estrogen. Phytoestrogens weakly mimic estrogen or modulate hormone activity in the human body.
- **plant sterols** phytochemicals that resemble cholesterol in structure but that lower blood cholesterol by interfering with cholesterol absorption in the intestine. Plant sterols include sterol esters and stanol esters, formerly called *phytosterols*.
- **prebiotic** a substance that may not be digestible by the host, such as fiber, but that serves as food for probiotic bacteria and thus promotes their growth.
- **probiotic** a live microorganism which, when administered in adequate amounts, alters the bacterial colonies of the body in ways believed to confer a health benefit on the host.
- **resveratrol** (rez-VER-ah-trol) a phytochemical of grapes under study for potential health benefits.
- **soy milk** a milklike beverage made from soybeans, claimed to be a functional food. Soy drinks should be fortified with vitamin A, vitamin D, riboflavin, and calcium to approach the nutritional equivalency of milk.
- **tofu** a white curd made of soybeans, popular in Asian cuisines, and considered to be a functional food.

A Scientist's View of Phytochemicals

At one time, phytochemicals were known only for their sensory properties in foods, such as taste, aroma, texture, and color. Thank phytochemicals for the burning sensation of hot peppers, the pungent flavors of onions and garlic, the bitter tang of chocolate, the aromatic qualities of herbs, and the beautiful colors of tomatoes, spinach, pink grapefruit, and watermelon.

Today, many phytochemicals are believed to be **bioactive food components**—food constituents that alter body processes beyond the actions of the nutrients (some are listed in Table C2–2). Many phytochemicals are known to have antioxidant activity, and **antioxidants** in the body protect DNA and other cellular compounds from oxidative damage. Some others may interact with genes, affecting protein synthesis; a few others mimic the body's own hormones; and many seem to have no effects or effects awaiting discovery.

Of the tens of thousands of phytochemicals known to exist, few have been studied for health effects, and only a sampling of those are mentioned in this Controversy—enough to illustrate the wide array of foods that supply them and their potential roles in human health. People eat foods, however, not individual phytochemicals, so this section focuses on a few well-known suppliers of these interesting compounds.

Blueberries and the Brain

When researchers fed chow rich in blueberry extracts to a group of rats, they exhibited fewer age-related mental declines than rats on plain chow. This finding set off a flurry of excitement about blueberries as a potential superfood for the brain. To explain their results, the researchers suggest that the **flavonoids** of blueberries, along with grapes and walnuts, may act as antioxidants in the brain, and thus limit damage to brain cells by oxidation.[1]

Population studies hint that a diet high in flavonoids may be associated with fewer cognitive deficits in the elderly, and even in patients with mental illness.[2] However, when researchers evaluated mental decline of older women in relation to a measure of the total antioxidant power of the diet, they found no relationship.[3]

With evidence from animals and populations, and a biological explanation in support, is it safe to say, then, that blueberries constitute a superfood for brain power? Blueberries currently lead the way in flavonoid and brain research, but the definitive test of efficacy is still lacking. Controlled human trials are needed to determine whether adding the berries to a person's diet actually does protect the human brain in aging.

Given the evidence so far, a logical step may be to include blueberries in the daily diet, just in case research proves them to be effective. But how many blueberries might be enough? Can a steady diet of fast food hamburgers, French fries, and colas be offset by a handful of blueberries? Research refutes this idea. People who are observed to benefit from flavonoid-rich diets obtain them from many sources: artichokes, beans, coffee, pomegranates, seeds, spinach, strawberries and other berries, and in fact most fruits and vegetables, whole grains, food mixtures, and even nuts, maple syrup, and seaweed.[4] By focusing only on blueberries, a person could easily miss out on beneficial synergistic interactions among the nutrients and phytochemicals in flavonoid-rich foods.[5]

Rather than gambling on one particular flavonoid or food to sustain optimal brain functioning, the wisest course based on today's understanding is to obtain a variety of phytochemicals from an adequate, balanced diet that is rich in a variety of fruits and vegetables. Blueberries, of course, make a delightful contribution and, as complex whole foods, they probably confer other benefits as well.[6]

Chocolate, Heart, and Mood

Imagine the delight of young research subjects who were paid to eat 3 ounces of dark (bittersweet) chocolate for an experiment. Less appealingly, researchers then drew blood from the subjects to test whether an antioxidant flavonoid in chocolate could be absorbed into the bloodstream. The tests were positive: the flavonoid had indeed accumulated in the subjects' blood. At the same time, the level of potentially harmful oxidizing compounds had dropped by 40 percent.

The heart, arteries, and lungs, organs particularly vulnerable to damage by oxidation, might benefit from such antioxidants. Figure C2–1 (p. 66) compares some antioxidant contributors, but many more foods than this are rich sources, and a food's antioxidants as measured in a test tube may have little relevance to its effects in the human body.*[7] In population studies, chocolate intakes often do correlate with lower cardiovascular disease risks.[8] A recent review supports benefits of chocolate for certain markers of heart health.[9] Still needed are controlled clinical human trials.

Many people also believe that chocolate can lift mood, but evidence does not support the idea. In a study of about 1,000 people, researchers noted that those with greater chocolate intakes had *more* depressive symptoms, not less.[10] Belief in chocolate as a mood lifter may explain the results—it may be that the depressed people in the study were trying to relieve their symptoms by eating chocolate.

For centuries, medicinal use of chocolate has been to promote weight gain. Each 3-ounce piece of chocolate candy offers 400 calories of added sugar and solid fat, calories that most people can little afford to consume. Most people are better off obtaining phytochemicals from nutrient-dense, low-calorie fruits and vegetables—and savoring chocolate as an occasional treat.

**Dark chocolate is rich in flavonoids; milk chocolate or "Dutch" processed chocolate have reduced flavonoid content.*

Table C2–2

Phytochemicals—Potential Health Effects and Food Sources

Chemical Name	Potential Health Effects	Food Sources
Alkylresorcinols (phenolic lipids)	May contribute to the protective effect of grains in reducing the risks of diabetes, heart disease, and some cancers.	Whole-grain wheat and rye
Allicin (organosulfur compound)	Antimicrobial that may reduce ulcers; may lower blood cholesterol.	Chives, garlic, leeks, onions
Capsaicin	Modulates blood clotting, possibly reducing the risk of fatal clots in heart and artery disease.	Hot peppers
Carotenoids (include beta-carotene, lycopene, lutein, and hundreds of related compounds)	Act as antioxidants, possibly reducing risks of cancer and other diseases.	Deeply pigmented fruits and vegetables (apricots, broccoli, cantaloupe, carrots, pumpkin, spinach, sweet potatoes, tomatoes)
Curcumin	Acts as an antioxidant and anti-inflammatory agent; may reduce blood clot formation; may inhibit enzymes that activate carcinogens.	Turmeric, a yellow-colored spice
Flavonoids (include flavones, flavonols, isoflavones, catechins, and others)	Act as antioxidants; scavenge carcinogens; bind to nitrates in the stomach, preventing conversion to nitrosamines; inhibit cell proliferation.	Widespread and common in many foods from plants: berries, black tea, celery, citrus fruits, green tea, olives, onions, oregano, grapes, purple grape juice, soybeans and soy products, vegetables, whole wheat and other whole grains, wine
Genistein and daidzein (isoflavones)	Phytoestrogens that inhibit cell replication in GI tract; may reduce or elevate risk of breast, colon, ovarian, prostate, and other estrogen-sensitive cancers; may reduce cancer cell survival; may reduce risk of osteoporosis.	Soybeans, soy flour, soy milk, tofu, textured vegetable protein, other legume products
Indoles (organosulfur compounds)	May trigger production of enzymes that block DNA damage from carcinogens; may inhibit estrogen action.	Cruciferous vegetables such as broccoli, brussels sprouts, cabbage, cauliflower, horseradish, mustard greens, kale
Isothiocyanates (organosulfur compounds that include sulforaphane)	Act as antioxidants; inhibit enzymes that activate carcinogens; activate enzymes that detoxify carcinogens; may reduce risk of breast cancer, prostate cancer.	Cruciferous vegetables such as broccoli, brussels sprouts, cabbage, cauliflower, horseradish, mustard greens, kale
Lignans	Phytoestrogens that block estrogen activity in cells possibly reducing the risk of cancer of the breast, colon, ovaries, and prostate.	Flaxseed, whole grains
Monoterpenes (including limonene)	May trigger enzyme production to detoxify carcinogens; inhibit cancer promotion and cell proliferation.	Citrus fruit peels and oils
Phenolic acids	May trigger enzyme production to make carcinogens water-soluble, facilitating excretion.	Coffee beans, fruits (apples, blueberries, cherries, grapes, oranges, pears, prunes), oats, potatoes, soybeans
Phytic acid	Binds to minerals, preventing free-radical formation, possibly reducing cancer risk.	Whole grains
Resveratrol	Acts as antioxidant; may inhibit cancer growth; reduce inflammation, LDL oxidation, and blood clot formation.	Red wine, peanuts, grapes, raspberries
Saponins (glucosides)	May interfere with DNA replication, preventing cancer cells from multiplying; stimulate immune response.	Alfalfa sprouts, other sprouts, green vegetables, potatoes, tomatoes
Tannins	Act as antioxidants; may inhibit carcinogen activation and cancer promotion.	Black-eyed peas, grapes, lentils, red and white wine, tea

Figure C2-1

Antioxidant Capacity of Selected Foods[a]

These foods were found to be good sources of antioxidants in one kind of laboratory test, but seasonal differences, variety, storage, testing methods, and other factors can greatly influence a food's antioxidant score.

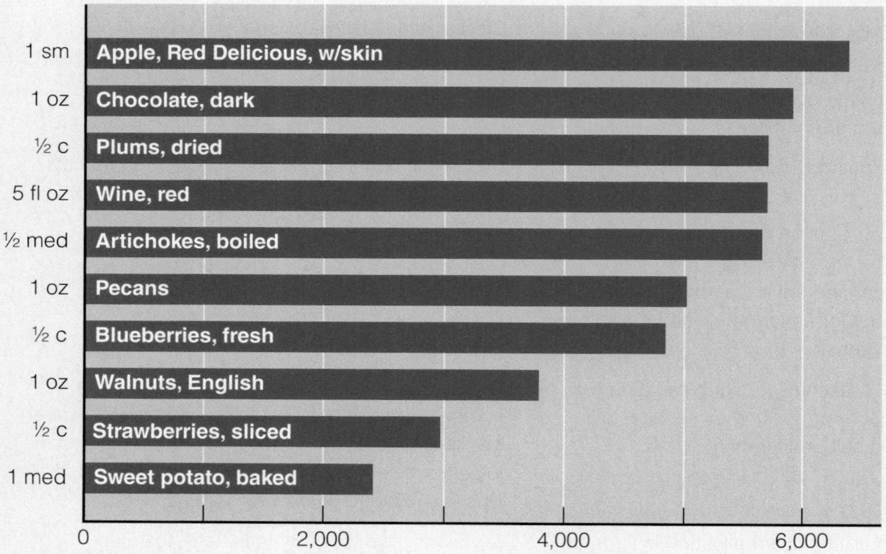

[a]*Measured in micromole TE (Trolox equivalents), a laboratory-derived value used to measure the antioxidant activity of foods. Other laboratory methods yield other results.*

Source: R. M. Bliss, Data on Food Antioxidants Aid Research, November 2007, available at http://www.ars.usda.gov/is/pr/2007/071106.htm.

Flaxseed

Courtesy of Flax Council of Canada

Long valued for relieving constipation and digestive distress, **flaxseed** is showing potential for other health benefits. Flaxseeds are rich in **lignans**, cholesterol-like phytochemicals that can be converted into **phytoestrogens**, compounds that mimic the human hormone estrogen, by bacteria in the digestive tract. In the U.S. diet, lignans are supplied mostly by other seeds, as well as whole wheat and vegetables.

Diets high in lignans may inhibit cholesterol absorption and reduce blood cholesterol, thus reducing heart disease risk.[11] However, population studies linking lignans intake with reduced incidence of heart disease are not convincing, with only two of six studies reporting a positive effect.[12]

Lignans are also under study for potential effects on certain cancers. Some recent research findings include these:

- Rats fed chow high in flaxseed develop fewer cancerous changes and reduced tumor growth in mammary tissue.[13]
- Postmenopausal women with the highest blood levels of a marker for intake of lignans have lower risks of both developing breast cancer and dying of the disease.[14]
- In one study, men with prostate cancer given flaxseed had less cancer cell proliferation than the controls.[15]

Including a little flaxseed in the diet may not be a bad idea, even if research fails to bear out its status as a superfood. Flaxseed richly supplies linolenic acid, an essential fatty acid often lacking in the U.S. diet (see Chapter 5). Moderation is in order, however. Flaxseed also contains compounds that interfere with vitamin or mineral absorption, so high daily flaxseed intakes could cause nutrient deficiencies, and digestive distress is likely.

Garlic

EyeWire, Inc.

For thousands of years, people have consumed garlic for medicinal purposes. Early Egyptian medical writings report its use for headache, heart disease, and tumors.

Today, garlic is studied because its antioxidant **organosulfur compounds** are reported to inhibit cancer development. When oxidizing compounds damage DNA in animal cells, cancerous cell changes can occur. Antioxidants of garlic quench these oxidizing compounds, at least in test tubes. However, how and even whether garlic prevents cancers in people is unknown.[16] Other potential roles for garlic include opposing allergies, heart disease, infections, and ulcers, but these effects also remain uncertain. Evidence to support the use of garlic for disease prevention is limited, and virtually none exists to support taking garlic supplements.[17] If you like garlicky foods, you can consume them with confidence: history and some research is on your side.

Soybeans and Soy Products

Mitch Hrdlicka/Photodisc/Getty Images

Compared with people in the West, Asians living in Asia consume far more soybeans and soy products, such as **edamame**, **miso**, **soy milk**, and **tofu**, and they suffer less frequently from heart disease and certain cancers.[†] Women in Asia also suffer less from problems arising in menopause, the midlife drop in blood estrogen with cessation of menstruation, such as sensations of heat ("hot flashes") and loss of minerals from the bones. When Asians living in the United States adopt

[†] *Among the cancers occurring less often in Asia are breast, colon, and prostate cancers.*

Western diets and habits, however, they experience diseases and symptoms at the same rates as native Westerners.[18]

In research, evidence concerning soy and heart health seems promising.[19] Soy's cholesterol-like **plant sterols** theoretically could inhibit cholesterol absorption in the intestine, and thus lower blood cholesterol.[20] A meta-analysis revealed a significant blood cholesterol–lowering effect from soy foods, attributable partly to soy's metabolic effects on the body and partly to the replacement of saturated fat–rich meats and dairy foods with soy foods in the diet.[21]

With regard to cancer, breast cancer, colon cancer, and prostate cancer can be estrogen-sensitive—cancers that grow when exposed to estrogen.[22] In addition to plant sterols, soy contains phytoestrogens, chemical relatives of human estrogen, and may mimic or oppose its effects.[23] Girls who eat soy foods during childhood and adolescence may have reduced breast cancer risk as young adults.[24] A study of women in China suggests somewhat better survival of breast cancer among soy consumers.[25] Studies that include U.S. women report no effect or mixed results.[26] Clearly, more research is needed before conclusions may be drawn about soy intake and cancer risk.

As for menopause, no consistent findings indicate that soy phytoestrogens can eliminate hot flashes, and in some studies, soy intake accompanied a greater incidence.[27] Some evidence does suggest that soy foods may help to preserve bone density after menopause, but supplements of isolated soy phytoestrogens fail to do so.[28]

Soy's Potential Downsides

Low doses of the soy phytoestrogen **genistein** appear to speed up division of breast cancer cells in laboratory cultures and in mice, whereas high doses seemed to do the opposite.[29] However, it seems unlikely that moderate intakes of soy foods cause harm.[30] If they did, soy-eating peoples would have higher incidences of these cancers. In fact, the opposite is true.[31]

The opposing actions of soy phytoestrogens should raise a red flag against taking supplements, especially by people who have had cancer or have close relatives with cancer. The American Cancer Society recommends that breast cancer survivors and those under treatment for breast cancer should consume only moderate amounts of soy foods as part of a healthy plant-based diet and should not intentionally ingest very high levels of soy or soy phytoestrogens.[32]

Tomatoes

Photodisc/Getty Images

People around the world who eat the most tomatoes, about five tomato-containing meals per week, are less likely to suffer from cancers of the esophagus, prostate, or stomach than those who avoid tomatoes. Among phytochemical candidates for promoting this effect is **lycopene**, a red pigment found in guava, papaya, pink grapefruit, tomatoes (especially cooked tomato products, such as sauce), and watermelon.

Two action of lycopene could, theoretically, inhibit cancer development. First, lycopene and its byproducts are antioxidants that could inhibit the growth of cancer cells, but most research to date does not support this idea.[33] Second, in the skin, lycopene and some of its chemical relatives act as a sort of internal sunscreen, filtering high-energy wavelengths of visible light. This action may protect skin from the damaging sun rays that cause many skin cancers.[‡34]

In contrast to today's media lore, the FDA concludes that no or very little solid evidence links lycopene or tomato consumption with reduced cancer risks. Something else about tomato-eating people may be reducing their risks. Supplements of lycopene seem less harmful than those of its chemical cousins beta-carotene and **lutein**, however, which clearly raise the risk of lung cancer in smokers.[35]

‡ *The other carotenoid relatives of lycopene are lutein and zeaxanthin—more about them in Chapter 7.*

Tea

Everywhere, headlines attribute almost magical benefits to tea—it's a current media darling. People in Asia who drink two cups or more of green tea each day die less often from cardiovascular disease and digestive tract cancers than nondrinkers, possibly by reducing blood pressure or inflammation.[36] For black tea, the type most U.S. consumers drink and a major contributor of flavonoids to the diet, evidence for disease prevention is mixed.[37]

Green tea consumption has the potential to reduce oxidative stress and inflammation and to reduce the levels of harmful blood lipids. Indeed, compiled evidence from 14 short-term controlled human studies supports the idea that green tea may reduce blood lipid concentrations, regardless of whether the subjects took green tea extracts or drank the tea itself.[38] Long-term controlled human studies are needed to confirm these findings, however. As for cancer, a review of 51 studies concluded that the evidence for green tea and cancer was mixed and more research is needed.[39]

In any case, high-dose supplements of green tea extract have been linked with liver toxicity, and high intakes of green tea, with kidney problems.[40] Concentrated flavonoids from tea or soy may also inhibit the release of the thyroid hormone thyroxine, impeding normal energy metabolism.[41] A recent USDA analysis of popular name-brand green tea supplements concluded that while some were of good quality, others lacked any trace of green tea components, and in still others, the flavonoids had decomposed or undeclared additives were present. Supplement quality cannot be judged from label information, not even in leading brand-name pills.[42]

Grapes and Wine

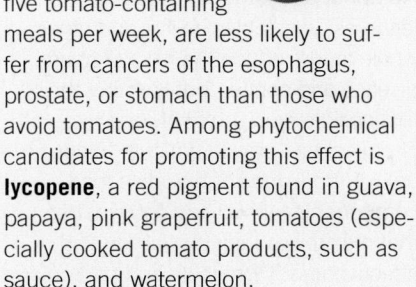

Photodisc/Getty Images

Purple grape juice and red wine contain a number of flavonoids, and among them is a small amount of **resveratrol**.[43] Resveratrol shows promise in research as a disease fighter.[44] In laboratory studies, resveratrol demonstrates the potential to reduce harmful tissue inflammation that often accompanies cancer, diabetes, obesity, and heart disease and to oppose the heart disease development in many other ways.[45] In high doses, resveratrol has also demonstrated some anticancer activities, but such doses are larger than those attainable by diet, and some border on toxic doses.[46] Also to its credit are studies in which resveratrol seemed to extend the life of fish, flies, worms, and yeast cells.[47] To date, no one yet knows if any of these effects are true in human beings.

Hints come from population studies, in which people who regularly consume red wine, grapes and their products, and other fruits and vegetables have a lower incidence of cardiovascular disease than others.[48] As tempting as it may be to jump to the conclusion that grapes and red wine prevent human diseases, the controlled clinical human trials needed to show that people actually benefit from consuming them are still lacking.[49] (Controversy 3 compares the potential risks and benefits of drinking alcohol.)

Yogurt

© Sue Wilson/Alamy

Yogurt is a special case among superfoods. Being a milk product, yogurt lacks typical flavonoids or other phytochemicals from plants. Instead, it contains living *Lactobacillus* or other bacteria that ferment milk into products like yogurt or the liquid yogurt beverage **kefir**. Such microorganisms, called **probiotics**, can set up residence in the digestive tract and alter its functioning in ways that are claimed to reduce colon cancer, ulcers, and other digestive problems; to reduce allergies; or to improve immunity and resistance to infections.[50] *Lactobacillus* and other organisms can help correct the diarrhea that often follows antibiotic drug use.[51] However, reports of both increased mortality among patients with pancreatic diseases and serious infections in those with weakened immunity raise concerns about the safety of probiotic microorganism supplements.[52] Additional research is needed to clarify potential benefits or risks from probiotics.[53]

Other foods provide **prebiotics**, that is, carbohydrates or other constituents upon which the probiotic colony in the digestive tract can feed.[54] A fed colony multiplies rapidly, creating byproducts that are sometimes associated with certain health benefits, such as a decrease in disease-related inflammation of the colon.[55]

Phytochemical Supplements

No doubt exists that diets rich in legumes, vegetables, fruits, and other whole foods reduce the risks of heart disease and cancer, but isolating the responsible food, nutrient, or phytochemical has proved difficult. Foods deliver thousands of bioactive food components, all within a food matrix that maximizes their availability and effectiveness.[56] Broccoli, and particularly **broccoli sprouts**, may contain as many as 10,000 different phytochemicals—each with the potential to influence some action in the body. These foods are under study for their potential to defend against cancers at the DNA level, and Chapter 11 comes back to them.[57]

Even if it were known with certainty which foods protect against which diseases, most isolated supplements, even the most promising ones, fail to actually prevent diseases when they are administered in research.[58] Worse, some such supplements can interfere with and reduce the effectiveness of standard drugs given to people with serious illnesses.[59]

Supporters of Phytochemical Supplements

Users and sellers of phytochemical supplements argue that existing evidence is good enough to recommend that people take supplements of purified phytochemicals. Users, eager for potential benefits, and sellers, hoping for profits, tend to discount the potential for harm from "natural" substances. People have been consuming foods containing phytochemicals for tens of thousands of years, they say, and because the body can handle phytochemicals in foods, it stands to reason that supplements of those phytochemicals are safe as well.

Detractors of Phytochemical Supplements

Such thinking raises concerns among scientists. They point out that the body is equipped to handle the dilute phytochemicals of whole foods but not concentrated supplement doses.[60] Further, the body absorbs only small amounts of these compounds into the bloodstream and quickly destroys most types with its detoxifying equipment.

Physicians and researchers now question the relevance of individual phytochemicals, even antioxidants, to human health, suggesting that the positive outcomes attributed to them may in fact reflect an overall healthy lifestyle that includes eating plenty of fruits, vegetables, seeds, and whole grains.[61] Individual phytochemicals, like actors in a play, are parts of a larger story with intertwining and complementary roles—a fact that reinforces the principle of variety in diet planning.

Consider these facts about phytochemical supplements and health:

- Evidence for the safety of isolated phytochemical supplements in human beings is lacking, and evidence for potential harm is mounting.[62]
- No regulatory body oversees the safety of phytochemicals sold to consumers. No studies are required to prove their safety or effectiveness before they are marketed.
- Phytochemical labels can make structure-function claims that sound good but are generally based on weak or nonexistent evidence.

Phytochemical researchers conclude that the best-known, most effective, and safest sources for bioactive food components are foods, not supplements. Even those in foods, however, can interfere with the activities of certain drugs and undermine the medical treatment of serious diseases. Such food and drug interactions are of critical importance, and the Controversy section of Chapter 14 is devoted to them.

The Concept of Functional Foods

Virtually all whole foods have some special value in supporting health and are therefore functional foods. Modest evidence suggests that cranberries may help to prevent some urinary tract infections, for example.[63] Manufactured functional foods, however, often consist of processed foods that are fortified with nutrients or enhanced with particular bioactive food components (such as herbs) for which little or no supporting evidence exists.

Such novel foods raise questions:

- Is such a food really a food or a **drug**?
- Which is the better choice for the health-conscious diet planner: to eat a food with additives and hope for a benefit or to adjust the diet in ways known to support health?
- Is it a greater benefit to eat fried snack foods and candy bars sprinkled with phytochemicals than to obtain these and other beneficial substances from whole foods?
- What about smoothies packed with medicinal herbs—are these foods safe to consume regularly? Are they safe for children?

The Final Word

In light of all of the evidence for and against phytochemicals and functional foods, a moderate approach is warranted. People who eat abundant and varied fruits and vegetables each day may cut their risk for many diseases by as much as half. Replacing some meat with soy foods may reduce those risks further. Table C2–3 offers some tips for consuming the whole foods known to provide phytochemicals.

A piece of advice: don't try to single out a few superfoods or phytochemicals for their magical health effects, and ignore the hype about packaged products—no evidence exists to support their use. Instead, take a no-nonsense approach and choose a wide variety of whole grains, legumes, nuts, fruits, and vegetables in the context of an adequate, balanced, and varied diet to receive all of the health benefits these foods can offer.[64]

Critical Thinking

1. Divide into two groups. One group will argue in support of using superfoods and one group will argue against the use of superfoods. During the debate be sure to answer the following questions:
   - What is a superfood, and is it appropriate to classify a given food as a superfood?
   - Are there foods that you can reliably say have the characteristics of a superfood? Describe the research you have consulted to support the classification of a food as a superfood.
2. Describe a situation when the intake of a phytochemical supplement or a functional food would be appropriate. Give reasons for using a phytochemical supplement or functional food and also give reasons against its use.

© Craig M. Moore

Functional foods currently on the market promise to "enhance mood," "promote relaxation and good karma," "support alertness," and "benefit memory," among other claims.

Table C2–3

Tips for Consuming Phytochemicals

- Eat more fruit. The average U.S. diet provides little more than 1/2 cup fruit a day. Remember to choose juices and raw, dried, or cooked fruits and vegetables at mealtimes as well as for snacks. Choose dried fruit in place of candy.
- Increase vegetable portions. Double the normal portion of cooked plain, nonstarchy vegetables.
- Use herbs and spices. Cookbooks offer ways to include parsley, basil, garlic, hot peppers, oregano, and other beneficial seasonings.
- Replace some meat. Replace some of the meat in the diet with grains, legumes, and vegetables. Oatmeal, soy meat replacer, or grated carrots mixed with ground meat and seasonings make a luscious, nutritious meat loaf, for example.
- Add grated vegetables. Carrots in chili or meatballs, celery and squash in spaghetti sauce, etc. add phytochemicals without greatly changing the taste of the food.
- Try new foods. Try a new fruit, vegetable, or whole grain each week. Walk through vegetable aisles and visit farmers' markets. Read recipes. Try tofu, fortified soy milk, or soybeans in cooking.

© Cengage Learning

3 The Remarkable Body

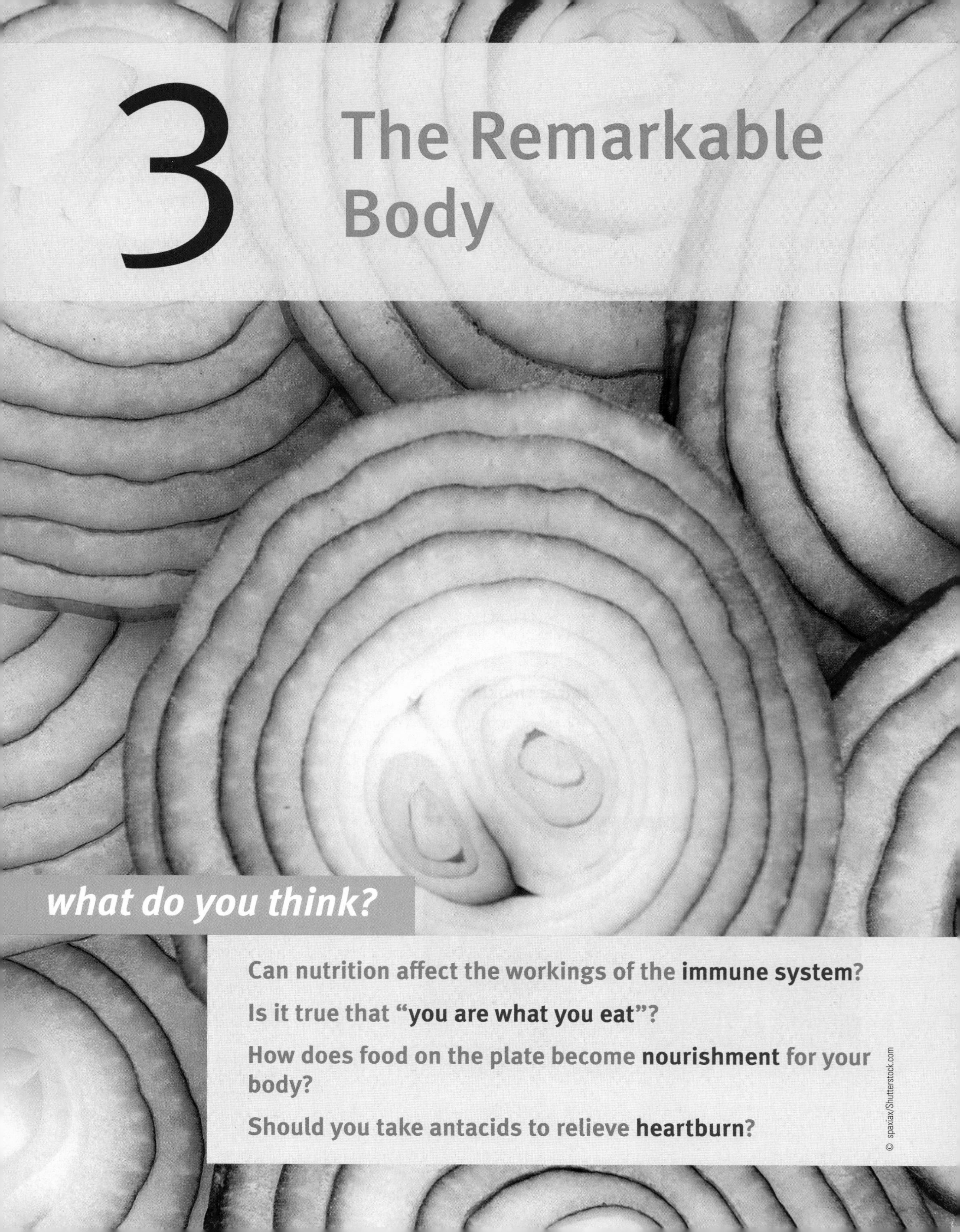

what do you think?

Can nutrition affect the workings of the **immune system**?

Is it true that **"you are what you eat"**?

How does food on the plate become **nourishment** for your body?

Should you take antacids to relieve **heartburn**?

Learning Objectives

After reading this chapter, you should be able to accomplish the following:

LO 3.1 Describe the levels of organization in the body, and identify some basic ways in which nutrition supports them.

LO 3.2 Describe the relationships between the body's fluids and the cardiovascular system and their importance to the nourishment and maintenance of body tissues.

LO 3.3 Summarize the interactions between the nervous and hormonal systems and nutrition.

LO 3.4 State how nutrition and immunity are interrelated, and describe the importance of inflammation to the body's health.

LO 3.5 Compare the terms *mechanical digestion* and *chemical digestion*, and point out where these processes occur along the digestive tract with regard to carbohydrate, fat, and protein.

LO 3.6 Name some common digestive problems and offer suggestions for dietary alterations that may improve them.

LO 3.7 Identify the excretory functions of the lungs, liver, kidneys, and bladder, and state why they are important to maintain normal body functioning.

LO 3.8 Identify glycogen and fat as the two forms of nutrients stored in the body, and identify the liver, muscles, and adipose tissue as the body tissues that store them.

LO 3.9 Define the term *moderate alcohol consumption*, and discuss the potential health effects, both negative and positive, associated with this level of drinking.

At the moment of conception, you received genes in the form of DNA from your mother and father, who, in turn, had inherited them from their parents, and so on into history. Since that moment, your genes have been working behind the scenes, directing your body's development and functioning. Many of your genes are ancient in origin and are little changed from genes of thousands of centuries ago, but here you are—living with the food, the luxuries, the smog, the contaminants, and all the other pleasures and problems of the 21st century. There is no guarantee that a diet haphazardly chosen from today's foods will meet the needs of your "ancient" body. Unlike your ancestors, who nourished themselves from the wild plants and animals surrounding them, you must learn how your body works, what it needs, and how to select foods to meet its needs.

Did You Know?

DNA is the large molecule that encodes all genetic information in its structure; genes are units of a cell's inheritance situated along the DNA strands.

The Body's Cells

LO 3.1 Describe the levels of organization in the body, and identify some basic ways in which nutrition supports them.

The human body is composed of trillions of **cells**, and none of them knows anything about food. *You* may get hungry for fruit, milk, or bread, but each cell of your body needs nutrients—the vital components of foods. The ways in which the body's cells cooperate to obtain and use nutrients are the subjects of this chapter.

Each of the body's cells is a self-contained, living entity (see Figure 3–1, p. 72), but at the same time it depends on the rest of the body's cells to supply its needs. Among the cells' most basic needs are energy and the oxygen with which to burn it. Cells also need water to maintain the environment in which they live. They need building blocks and control systems. They especially need the nutrients they cannot make for themselves—the essential nutrients first described in Chapter 1—which must be supplied from food. The first principle of diet planning is that the foods we choose must provide energy and the essential nutrients, including water.

As living things, cells also die off, although at varying rates. Some skin cells and red blood cells must replenish themselves every 10 to 120 days. Cells lining the digestive

cells the smallest units in which independent life can exist. All living things are single cells or organisms made of cells.

Figure 3–1

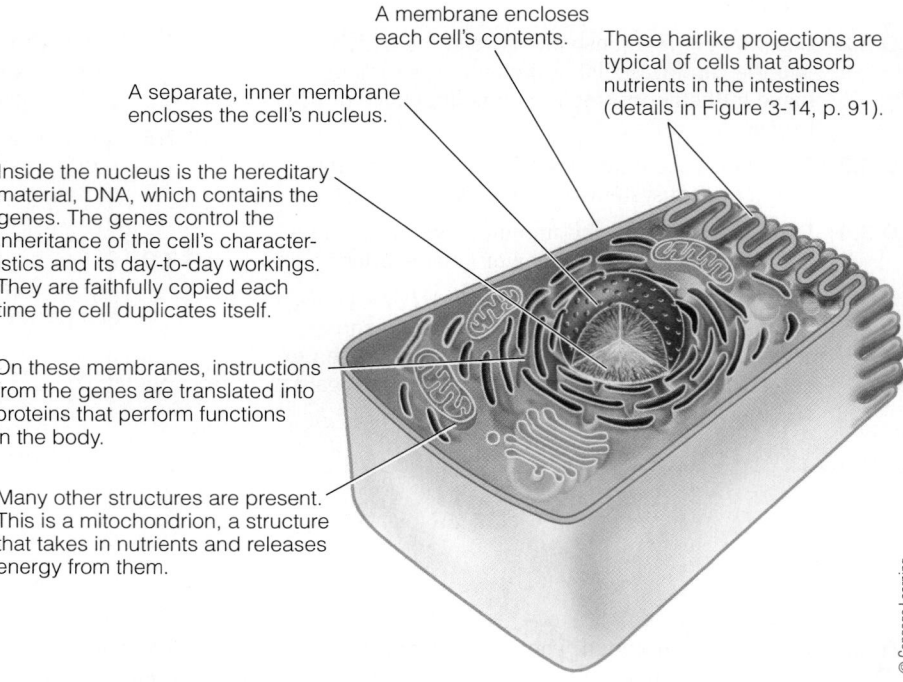

A Cell (Simplified Diagram)

This cell has been greatly enlarged; real cells are so tiny that 10,000 can fit on the head of a pin.

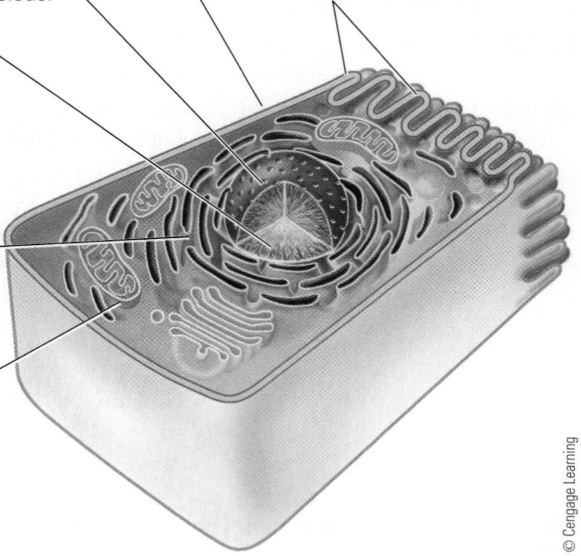

A membrane encloses each cell's contents.
These hairlike projections are typical of cells that absorb nutrients in the intestines (details in Figure 3-14, p. 91).
A separate, inner membrane encloses the cell's nucleus.
Inside the nucleus is the hereditary material, DNA, which contains the genes. The genes control the inheritance of the cell's characteristics and its day-to-day workings. They are faithfully copied each time the cell duplicates itself.
On these membranes, instructions from the genes are translated into proteins that perform functions in the body.
Many other structures are present. This is a mitochondrion, a structure that takes in nutrients and releases energy from them.

tract replace themselves every 3 days. Under ordinary conditions, many muscle cells reproduce themselves only once every few years. Liver cells have the ability to reproduce quickly and do so whenever repairs to the organ are needed. Certain brain cells do not reproduce at all; if damaged by injury or disease, they are lost forever.

The cells work in cooperation with each other to support the whole body. Gene activity within each cell determines the nature of that work.

Genes Control Functions

Each gene is a blueprint that directs the production of one or more proteins, such as an **enzyme** that performs cellular work. Genes also provide the instructions for all of the structural components cells need to survive (see Figure 3–2). Each cell contains a complete set of genes, but different ones are active in different types of cells. For example, in some intestinal cells, the genes for making digestive enzymes are active, but the genes for making keratin of nails and hair are silent; in some of the body's **fat cells**, the genes for making enzymes that metabolize fat are active, but the digestive enzyme genes are silent. Certain nutrients are involved in activating and silencing genes in ways that are just starting to be revealed.

Connections between nutrition and gene activities are emerging in the field of nutritional genomics, described in **Controversy 11**.

Genes affect the way the body handles its nutrients. Certain variations in some of the genes alter the way the body absorbs, metabolizes, or excretes nutrients from the body. Occasionally, a gene variation can cause a lifelong malady—that is, an **inborn error of metabolism**—that may require a special diet to minimize its potential to harm the body. An example is the inborn error **phenylketonuria**, in which a genetic variation compromises the body's ability to handle the amino acid phenylalanine.

enzyme any of a great number of working proteins that speed up a specific chemical reaction, such as breaking the bonds of a nutrient, without undergoing change themselves. Enzymes and their actions are described in Chapter 6.

fat cells cells that specialize in the storage of fat and form the fat tissue. Fat cells also produce fat-metabolizing enzymes; they also produce hormones involved in appetite and energy balance (see Chapter 9).

inborn error of metabolism a genetic variation present from birth that may result in disease.

phenylketonuria (PKU) an inborn error of metabolism that interferes with the body's handling of the amino acid phenylalanine, with potentially serious consequences to the brain and nervous system in infancy and childhood.

Figure 3–2

From DNA to Living Cells

If the human genome were a book of instructions on how to make a human being, then the 23 chromosomes of DNA would be chapters. Each gene would be a word, and the individual molecules that form the DNA would be letters of the alphabet.

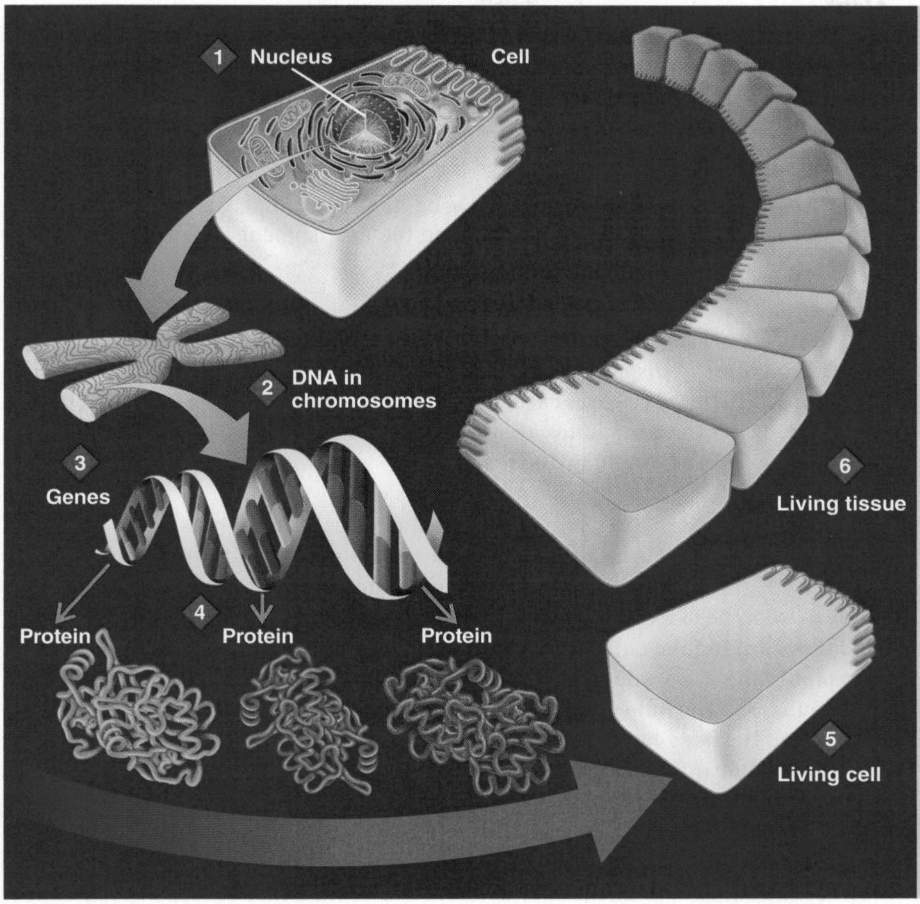

1. Each cell's nucleus contains DNA — the material of heredity in all living things.
2. Long strands of human DNA coil into 23 pairs of chromosomes. If the strands of DNA in all the body's cells were uncoiled and laid end to end, they would stretch to the sun and back four hundred times. Yet DNA strands are so tiny that about 5 million of them could be threaded at once through the eye of a needle.
3. Genes contain instructions for making proteins. Genes are sections along the strands of DNA that serve as templates for the building of proteins. Some genes are involved in building just one protein; others are involved in building more than one.
4. Many other steps are required to make a protein. See Figure 6-6 of Chapter 6.
5. Proteins do the work of living cells. Cells employ proteins to perform essential functions and provide structures.
6. Communities of functioning cells make up the living tissue.

People with this condition must carefully limit their intakes of phenylalanine, so food manufacturers are required to print warning labels on foods, such as certain artificial sweeteners, that contain it.

Nutrients also affect the genes. For example, the concentrations of certain nutrients and phytochemicals in the body fluids and tissues influence the genes to make more or less of certain proteins. These changes, in turn, alter body functions and ultimately hold meaning for health and disease. Controversy 11 presents details.

Cells, Tissues, Organs, Systems

Cells are organized into **tissues** that perform specialized tasks. For example, individual muscle cells are joined together to form muscle tissue, which can contract. Tissues, in turn, are grouped together to form whole **organs**. In the organ we call the heart, for example, muscle tissues, nerve tissues, connective tissues, and others all work together to pump blood. Some body functions are performed by several related organs working together as part of a **body system**. For example, the heart, lungs, and blood vessels cooperate as parts of the cardiovascular system to deliver oxygen to all the body's cells. The next few sections present the body systems with special significance to nutrition.

tissues systems of cells working together to perform specialized tasks. Examples are muscles, nerves, blood, and bone.

organs discrete structural units made of tissues that perform specific jobs. Examples are the heart, liver, and brain.

body system a group of related organs that work together to perform a function. Examples are the circulatory system, respiratory system, and nervous system.

KEY POINTS

- The body's cells need energy, oxygen, and nutrients, including water, to remain healthy and do their work.
- Genes direct the making of each cell's protein machinery, including enzymes.
- Specialized cells are grouped together to form tissues and organs; organs work together in body systems.

The Body Fluids and the Cardiovascular System

LO 3.2 Describe the relationships between the body's fluids and the cardiovascular system and their importance to the nourishment and maintenance of body tissues.

Body fluids supply the tissues continuously with energy, oxygen, and nutrients, including water. The fluids constantly circulate to pick up fresh supplies and deliver wastes to points of disposal. Every cell continuously draws oxygen and nutrients from those fluids and releases carbon dioxide and other waste products into them.

The body's circulating fluids are the **blood** and the **lymph**. Blood travels within the **arteries**, **veins**, and **capillaries**, as well as within the heart's chambers (see Figure 3–3). Lymph travels in separate vessels of its own.

Circulating around the cells are other fluids such as the **plasma** of the blood, which surrounds the white and red blood cells, and the fluid surrounding muscle cells (see Figure 3–4, p. 76). The fluid surrounding cells (**extracellular fluid**) is derived from the blood in the capillaries; it squeezes out through the capillary walls and flows around the outsides of cells, permitting exchange of materials.

Some of the extracellular fluid returns directly to the bloodstream by reentering the capillaries. The fluid remaining outside the capillaries forms lymph, which travels around the body by way of lymph vessels. The lymph eventually returns to the bloodstream near the heart where a large lymph vessel empties into a large vein. In this way, all cells are served by the cardiovascular system.

The fluid inside cells (**intracellular fluid**) provides a medium in which all cell reactions take place. Its pressure also helps the cells to hold their shape. The intracellular fluid is drawn from the extracellular fluid that bathes the cells on the outside.

All the blood circulates to the **lungs**, where it picks up oxygen and releases carbon dioxide wastes from the cells, as Figure 3–5 (p. 77) shows. Then the blood returns to the heart, where the pumping heartbeats push this freshly oxygenated blood from the lungs out to all body tissues. As the blood travels through the rest of the cardiovascular system, it delivers materials cells need and picks up their wastes.

As it passes through the digestive system, the blood delivers oxygen to the cells there and picks up most nutrients other than fats and their relatives from the **intestine** for distribution elsewhere. Lymphatic vessels pick up most fats from the intestine and then transport them to the blood (see Figure 3–6, p. 77). All blood leaving the digestive system is routed directly to the **liver**, which has the special task of chemically altering the absorbed materials to make them better suited for use by other tissues. Later, in passing through the **kidneys**, the blood is cleansed of wastes (look again at Figure 3–3). Note that the blood carries nutrients from the intestine to the liver, which releases them to the heart, which pumps them to the waiting body tissues.

To ensure efficient circulation of fluid to all your cells, you need an ample fluid intake. This means drinking sufficient water to replace the water lost each day. Cardiovascular fitness is essential, too, and constitutes an ongoing project that requires attention to both nutrition and physical activity. Healthy red blood cells also play a role, for they carry oxygen to all the other cells, enabling them to use fuels for energy.

blood the fluid of the cardiovascular system; composed of water, red and white blood cells, other formed particles, nutrients, oxygen, and other constituents.

lymph (LIMF) the fluid that moves from the bloodstream into tissue spaces and then travels in its own vessels, which eventually drain back into the bloodstream (see Figure 3–6).

arteries blood vessels that carry blood containing fresh oxygen supplies from the heart to the tissues (see Figure 3–3).

veins blood vessels that carry blood, with the carbon dioxide it has collected, from the tissues back to the heart (see Figure 3–3).

capillaries minute, weblike blood vessels that connect arteries to veins and permit transfer of materials between blood and tissues (see Figures 3–3 and 3–4).

plasma the cell-free fluid part of blood and lymph.

extracellular fluid fluid residing outside the cells that transports materials to and from the cells.

intracellular fluid fluid residing inside the cells that provides the medium for cellular reactions.

lungs the body's organs of gas exchange. Blood circulating through the lungs releases its carbon dioxide and picks up fresh oxygen to carry to the tissues.

intestine the body's long, tubular organ of digestion and the site of nutrient absorption.

liver a large, lobed organ that lies just under the ribs. It filters the blood, removes and processes nutrients, manufactures materials for export to other parts of the body, and destroys toxins or stores them to keep them out of the circulatory system.

kidneys a pair of organs that filter wastes from the blood, make urine, and release it to the bladder for excretion from the body.

Figure 3–3

Animated! Blood Flow in the Cardiovascular System

The blood is routed through the body as follows:
- Heart to tissues to heart to lungs to heart (repeat).

The portion of the blood that flows through the blood vessels of the intestine travels from:
- Heart to intestine to liver to heart.

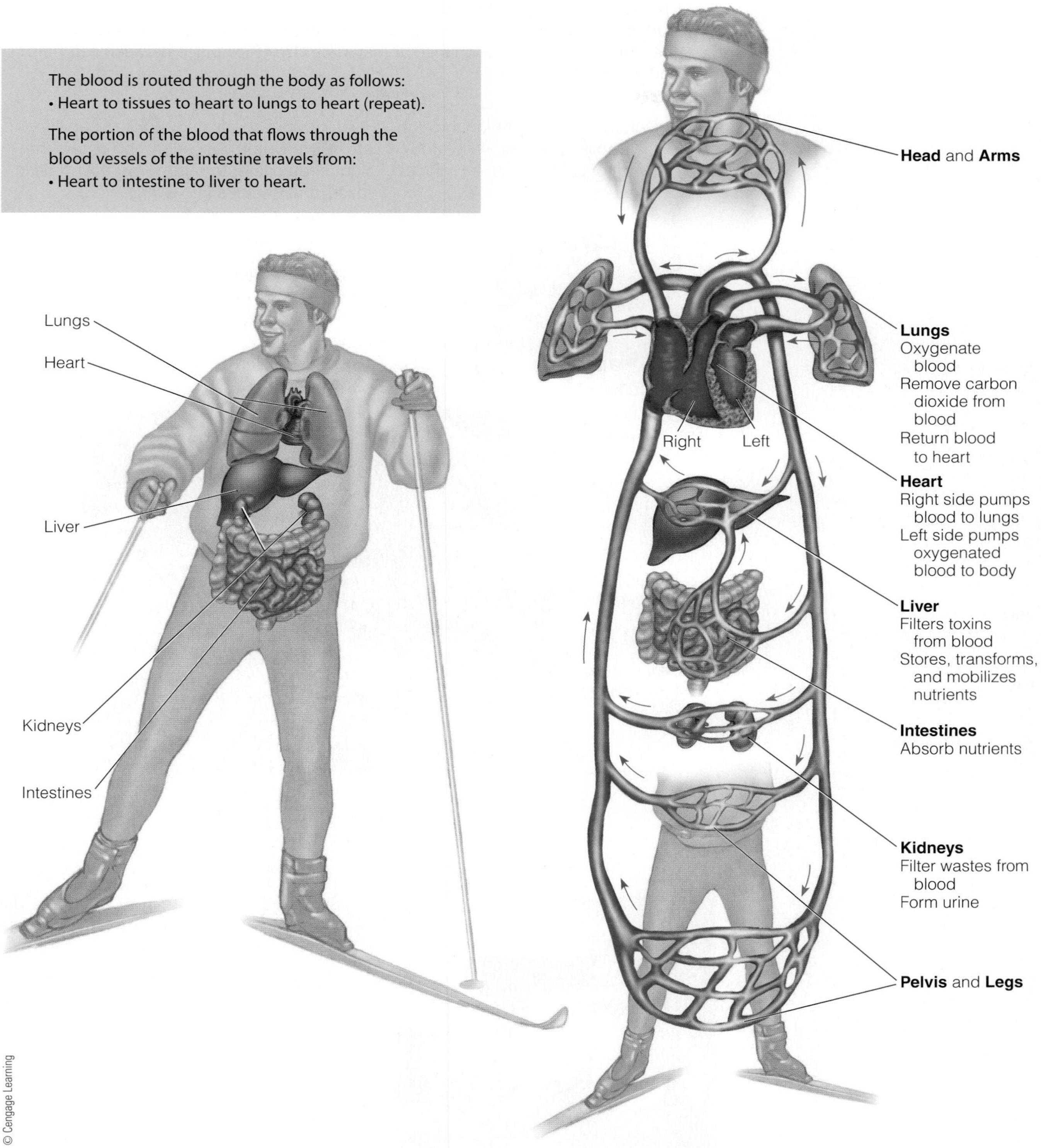

Figure 3–4

Animated! How the Body Fluids Circulate around Cells

The upper box shows a tiny portion of tissue with blood flowing through its network of capillaries (greatly enlarged). The lower box illustrates the movement of the extracellular fluid. Exchange of materials also takes place between cell fluid and extracellular fluid.

1 Fluid filters out of blood through the capillary whose walls are made of cells with small spaces between them.

Fluid may flow back into 2 the capillary or into 3 a lymph vessel. Lymph enters the bloodstream later through a large lymphatic vessel that empties into a large vein.

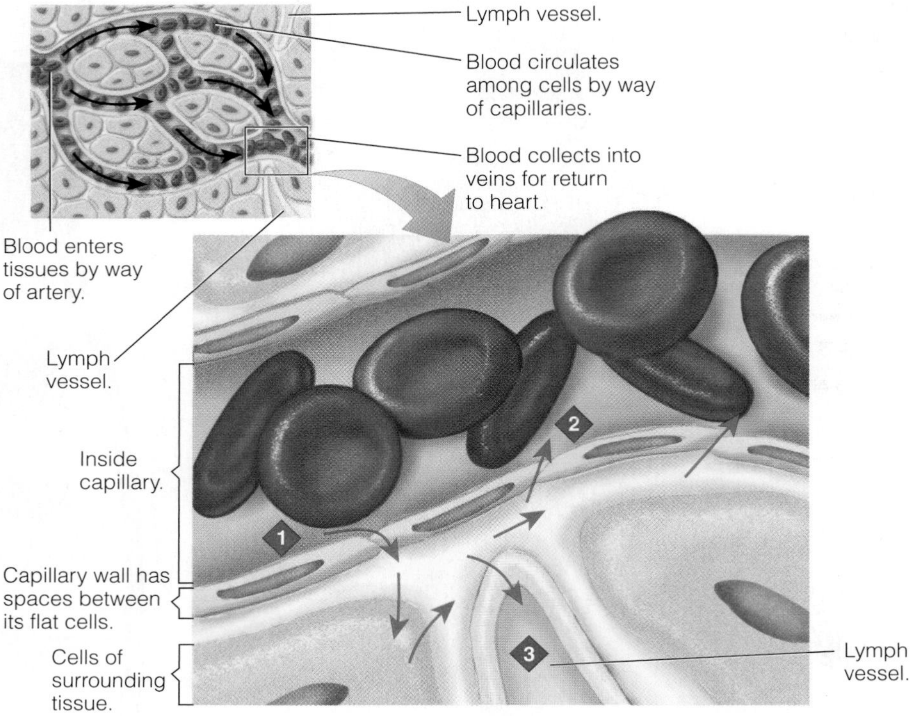

Lymph vessel.
Blood circulates among cells by way of capillaries.
Blood collects into veins for return to heart.
Blood enters tissues by way of artery.
Lymph vessel.
Inside capillary.
Capillary wall has spaces between its flat cells.
Cells of surrounding tissue.
1
2
3
Lymph vessel.

All the body's cells live in water.

Since red blood cells arise, live, and die within about four months, your body replaces them constantly, a manufacturing process that requires many essential nutrients from food. Consequently, the blood is very sensitive to malnutrition and often serves as an indicator of disorders caused by dietary deficiencies or imbalances of vitamins or minerals.

KEY POINTS

- Blood and lymph deliver needed materials to all the body's cells and carry waste materials away from them.
- The cardiovascular system ensures that these fluids circulate properly among all tissues.

Figure 3–5

Oxygen–Carbon Dioxide Exchange in the Lungs

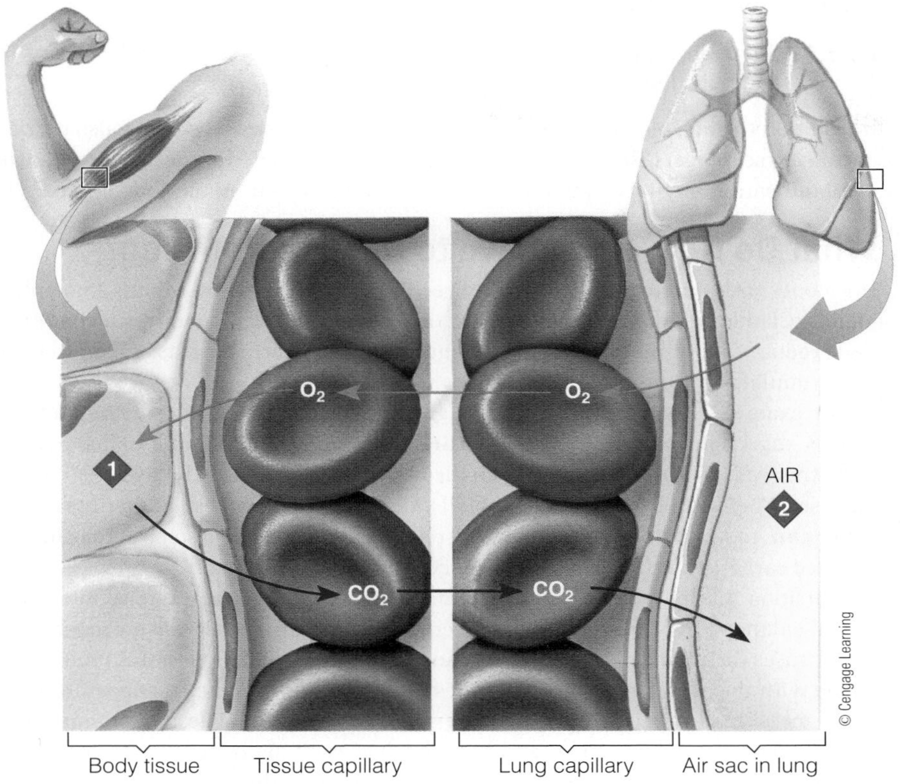

1 In body tissues, red blood cells give up their oxygen (O_2) and absorb carbon dioxide (CO_2).

2 In the air sacs of the lungs, the red blood cells give up their load of carbon dioxide (CO_2) and absorb oxygen (O_2) from air to supply to body tissues.

Figure 3–6

Lymph Vessels and the Bloodstream—Nutrient Flow through the Body

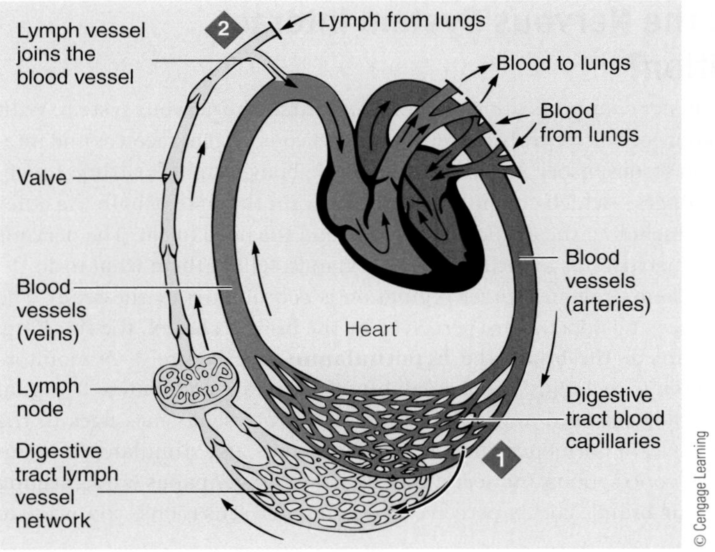

1 Nutrients are absorbed via two kinds of vessels in the intestines: blood capillaries and small lymph vessels. The capillaries lead to larger blood vessels that lead to the liver.

2 The lymph in the lymph vessels carries most of the absorbed dietary fat to the large vein near the heart. Some lymph vessels are depicted in Figure 3-14 (lower right), later on.

The Hormonal and Nervous Systems

LO 3.3 Summarize the interactions between the nervous and hormonal systems and nutrition.

In addition to fluid, blood cells, nutrients, oxygen, and wastes, the blood also carries chemical messengers, **hormones**, from one system of cells to another. Hormones communicate changing conditions that demand responses from the body organs.

What Do Hormones Have to Do with Nutrition?

Hormones are secreted and released directly into the blood by organs known as glands. Glands and hormones abound in the body. Each gland monitors a condition and produces one or more hormones to regulate it. Each hormone acts as a messenger that stimulates various organs to take appropriate actions.

For example, when the **pancreas** (a gland) detects a high concentration of the blood's sugar, glucose, it releases **insulin**, a hormone. Insulin stimulates muscle and other cells to remove glucose from the blood and to store it. The liver also stores glucose. When the blood glucose level falls, the pancreas secretes another hormone, **glucagon**, to which the liver responds by releasing into the blood some of the glucose it stored earlier. Thus, a normal blood glucose level is maintained.

Nutrition affects the hormonal system. Fasting, feeding, and exercise alter hormonal balances. In people who become very thin, for example, altered hormonal balance causes their bones to lose minerals and weaken. Hormones also affect nutrition. Along with the nervous system, hormones regulate hunger and affect appetite. They carry messages to regulate the digestive system, telling the digestive organs what kinds of foods have been eaten and how much of each digestive juice to secrete in response. A hormone produced by the fat tissue informs the brain about the degree of body fatness and helps to regulate appetite. Hormones also regulate the menstrual cycle in women, and they affect the appetite changes many women experience during the cycle and in pregnancy. An altered hormonal state is thought to be at least partially responsible, too, for the loss of appetite that sick people experience. Hormones also regulate the body's reaction to stress, suppressing hunger and the digestion and absorption of nutrients. When there are questions about a person's nutrition or health, the state of that person's hormonal system is often part of the answer.

KEY POINT

- Glands secrete hormones that act as messengers to help regulate body processes.

How Does the Nervous System Interact with Nutrition?

The body's other major communication system is, of course, the nervous system. With the brain and spinal cord as central controllers, the nervous system receives and integrates information from sensory receptors all over the body—sight, hearing, touch, smell, taste, and others—which communicate to the brain the state of both the outer and inner worlds, including the availability of food and the need to eat. The nervous system also sends instructions to the muscles and glands, telling them what to do.

The nervous system's role in hunger regulation is coordinated by the brain. The sensations of hunger and appetite are perceived by the brain's **cortex**, the thinking, outer layer. Deep inside the brain, the **hypothalamus** (see Figure 3–7) monitors many body conditions, including the availability of nutrients and water. To signal hunger, the physiological need for food, the digestive tract sends messages to the hypothalamus by way of hormones and nerves. The signals also stimulate the stomach to intensify its contractions and secretions, causing hunger pangs (and gurgling sounds). When your brain's cortex perceives these hunger sensations, you want to

hormones chemicals that are secreted by glands into the blood in response to conditions in the body that require regulation. These chemicals serve as messengers, acting on other organs to maintain constant conditions.

pancreas an organ with two main functions. One is an endocrine function—the making of hormones such as insulin, which it releases directly into the blood (*endo* means "into" the blood). The other is an exocrine function—the making of digestive enzymes, which it releases through a duct into the small intestine to assist in digestion (*exo* means "out" into a body cavity or onto the skin surface).

insulin a hormone from the pancreas that helps glucose enter cells from the blood (details in Chapter 4).

glucagon a hormone from the pancreas that stimulates the liver to release glucose into the bloodstream.

cortex the outermost layer of something. The brain's cortex is the part of the brain where conscious thought takes place.

hypothalamus (high-poh-THAL-uh-mus) a part of the brain that senses a variety of conditions in the blood, such as temperature, glucose content, salt content, and others. It signals other parts of the brain or body to adjust those conditions when necessary.

Figure 3–7

Cutaway Side View of the Brain Showing the Hypothalamus and Cortex

The hypothalamus monitors the body's conditions and sends signals to the brain's thinking portion, the cortex, which decides on actions. The pituitary gland is called the body's master gland, referring to its roles in regulating the activities of other glands and organs of the body.

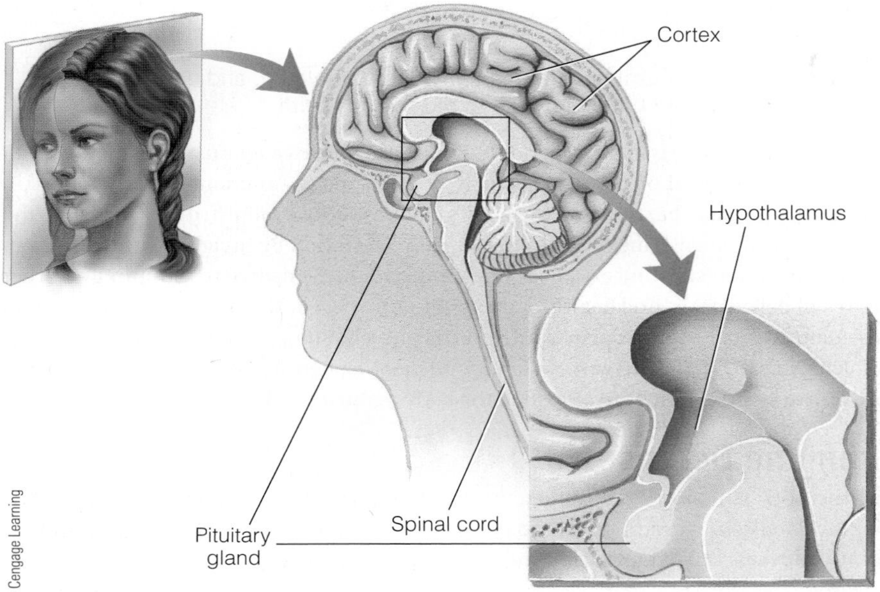

eat. The conscious mind of the cortex, however, can override such signals, and a person can choose to delay eating despite hunger or to eat when hunger is absent.

In a marvelous adaptation of the human body, the hormonal and nervous systems work together to enable a person to respond to physical danger. Known as the **fight-or-flight reaction**, or the *stress response*, this adaptation is present with only minor variations in all animals, showing how universally important it is to survival. When danger is detected, nerves release **neurotransmitters**, and glands supply the compounds **epinephrine** and **norepinephrine**. Every organ of the body responds, and **metabolism** speeds up. The pupils of the eyes widen so that you can see better; the muscles tense up so that you can jump, run, or struggle with maximum strength; breathing quickens and deepens to provide more oxygen. The heart races to rush the oxygen to the muscles, and the blood pressure rises so that the fuel the muscles need for energy can be delivered efficiently. The liver pours forth glucose from its stores, and the fat cells release fat. The digestive system shuts down to permit all the body's systems to serve the muscles and nerves. With all action systems at peak efficiency, the body can respond with amazing speed and strength to whatever threatens it.

In ancient times, stress usually involved physical danger, and the response to it was violent physical exertion. In the modern world, stress is seldom physical, but the body reacts the same way. What stresses you today may be a checkbook out of control or a teacher who suddenly announces a pop quiz. Under these stresses, you are not supposed to fight or run as your ancient ancestor did. You smile at the "enemy" and suppress your fear. But your heart races, you feel it pounding, and hormones still flood your bloodstream with glucose and fat.

Your number-one enemy today is not a saber-toothed tiger prowling outside your cave, but a disease of modern civilization: heart disease. Years of fat and other constituents accumulating in the arteries and stresses that strain the heart often lead to

fight-or-flight reaction the body's instinctive hormone- and nerve-mediated reaction to danger. Also known as the *stress response*.

neurotransmitters chemicals that are released at the end of a nerve cell when a nerve impulse arrives there. They diffuse across the gap to the next cell and alter the membrane of that second cell to either inhibit or excite it.

epinephrine (EP-ih-NEFF-rin) the major hormone that elicits the stress response.

norepinephrine (NOR-EP-ih-NEFF-rin) a compound related to epinephrine that helps to elicit the stress response.

metabolism the sum of all physical and chemical changes taking place in living cells; includes all reactions by which the body obtains and spends the energy from food.

heart attacks, especially when a body accustomed to chronic underexertion experiences sudden high blood pressure. Daily exercise as part of a healthy lifestyle releases pent-up stress and helps to protect the heart.

KEY POINT

- The nervous system and hormonal system regulate body processes, respond to the need for food, govern the act of eating, regulate digestion, and call for the stress response when needed.

The Immune System

LO 3.4 State how nutrition and immunity are interrelated, and describe the importance of inflammation to the body's health.

Many of the body's tissues cooperate to maintain defenses against infection, and all these tissues depend on an ample supply of nutrients to function properly. The skin presents a physical barrier, and the body's cavities (lungs, mouth, digestive tract, and others) are lined with membranes that resist penetration by invading **microbes** and other unwanted substances. These linings are highly sensitive to vitamin and other nutrient deficiencies, and health-care providers inspect both the skin and the inside of the mouth to detect signs of malnutrition. (Later chapters present details of the signs of deficiencies.) If an **antigen**, or foreign invader, penetrates the body's barriers, the **immune system** rushes in to defend the body against harm.

Immune Defenses

Of the 100 trillion cells that make up the human body, one in every hundred is a white blood cell. The actions of two types of white blood cells, the phagocytes and the **lymphocytes**, known as T-cells and B-cells, are of interest:

- **Phagocytes.** These scavenger cells travel throughout the body and are the first to defend body tissues against invaders. When a phagocyte recognizes a foreign particle, such as a bacterium, the phagocyte forms a pocket in its own outer membrane, engulfing the invader. The phagocytes may then attack the invader with oxidizing chemicals in an "oxidative burst" or may otherwise digest or destroy them. Phagocytes also leave a chemical trail that helps other immune cells to find the infection and join the defense.
- **T-cells.** Killer T-cells are lymphocytes that "read" and "remember" the chemical messages put forth by phagocytes to identify invaders. The killer T-cells then seek out and destroy all foreign particles having the same identity. T-cells defend against fungi, viruses, parasites, some bacteria, and some cancer cells. They also pose a formidable obstacle to a successful organ transplant—the physician must prescribe immunosuppressive drugs following surgery to hold down the T-cells' attack against the "foreign" organ. Another group, helper T-cells, does not attack invaders directly but helps other immune cells to do so. People suffering from the disease AIDS (acquired immunodeficiency syndrome) are rendered defenseless against other diseases because the virus that causes AIDS selectively attacks and destroys their helper T-cells.*
- **B-cells.** B-cells respond rapidly to infection by dividing and releasing invader-fighting proteins, **antibodies**, into the bloodstream. Antibodies travel to the site of the infection and stick to the surface of the foreign particles, killing or inactivating them. Like T-cells, B-cells also retain a chemical memory of each invader, and if the encounter recurs, the response is swift. Immunizations work this way: a disabled or harmless form of a disease-causing organism is injected into the body so that the B-cells can learn to recognize it. Later, if the live infectious organism invades, the B-cells quickly release antibodies to destroy it.

* The AIDS virus is the human immunodeficiency virus (HIV).

microbes bacteria, viruses, or other organisms invisible to the naked eye, some of which cause diseases. Also called *microorganisms.*

antigen a microbe or substance that is foreign to the body.

immune system a system of tissues and organs that defend the body against antigens, foreign materials that have penetrated the skin or body linings.

lymphocytes (LIM-foh-sites) white blood cells that participate in the immune response; B-cells and T-cells.

phagocytes (FAG-oh-sites) white blood cells that can ingest and destroy antigens. The process by which phagocytes engulf materials is called *phagocytosis.* The Greek word *phagein* means "to eat."

T-cells lymphocytes that attack antigens. *T* stands for the thymus gland of the neck, where the T-cells are stored and matured.

B-cells lymphocytes that produce antibodies. *B* stands for bursa, an organ in the chicken where B-cells were first identified.

In addition to the phagocytes and lymphocytes, the immune system includes many other categories of white blood cells and many organs and tissues. To function properly, all of these cells and organs depend on a steady flow of nutrients, delivered to the bloodstream from the digestive system.

KEY POINT

- A properly functioning immune system enables the body to resist diseases.

Inflammation

When tissues become injured or irritated they undergo **inflammation**, a condition of increased white blood cells, redness, heat, pain, swelling, and sometimes loss of function of the affected body part. Inflammation is the immune system's normal, healthy response to cell injury.

Many diseases, particularly chronic diseases of later life, such as heart disease, diabetes, and a severe type of arthritis, are associated with chronic tissue inflammation. When chronic, low-grade, unrelieved inflammation exists in chronic diseases, it often foretells an increase in both the severity of the disease and the risk of death from the disease. An important predictor of inflammation is being overweight.[1†] The links among diet, inflammatory processes, and diseases are currently topics of intense research, and later chapters revisit them.

KEY POINTS

- Inflammation is the normal, healthy response of the immune system to cell injury.
- Chronic inflammation is associated with disease development and being overweight.

The Digestive System

LO 3.5 Compare the terms *mechanical digestion* and *chemical digestion*, and point out where these processes occur along the digestive tract with regard to carbohydrate, fat, and protein.

When your body needs food, your brain and hormones alert your conscious mind to the sensation of hunger. Then, when you eat, your taste buds guide you in judging whether foods are acceptable.

Taste buds contain surface structures that detect five basic chemical tastes: sweet, sour, bitter, salty, and umami (ooh-MOM-ee), the Asian name for *savory*.[2] These basic tastes, along with aroma, texture, temperature, and other flavor elements, affect a person's experience of a food's flavor. In fact, the human ability to detect a food's aroma is thousands of times more sensitive than the sense of taste. The nose can detect just a few molecules responsible for the aroma of frying bacon, for example, even when they are diluted in several rooms full of air.

Why Do People Like Sugar, Salt, and Fat?

Sweet, salty, and fatty foods are almost universally desired, but most people have aversions to bitter and sour tastes (see Figure 3–8).[3] The enjoyment of sugars is inborn and encourages people to consume ample energy, especially in the form of foods containing carbohydrates, which provide the energy fuel for the brain.[4] The pleasure of a salty taste prompts eaters to consume sufficient amounts of two very important minerals—sodium and chloride. Likewise, foods containing fats provide concentrated energy and essential nutrients needed by all body tissues. The aversion to bitterness, universally displayed in infants, discourages consumption of foods containing bitter toxins and also affects people's food preferences later on.[5] People with greater aversion to bitter tastes are apt to avoid foods with slightly bitter flavors, such as turnips and broccoli.

† Reference notes are found in Appendix F.

antibodies proteins, made by cells of the immune system, that are expressly designed to combine with and inactivate specific antigens.

inflammation the immune system's response to cellular injury characterized by an increase in white blood cells, redness, heat, pain, and swelling. Inflammation plays a role in many chronic diseases.

Figure 3–8

The Innate Preference for Sweet Taste

This newborn baby is (a) resting; (b) tasting distilled water; (c) tasting sugar; (d) tasting something sour; and (e) tasting something bitter

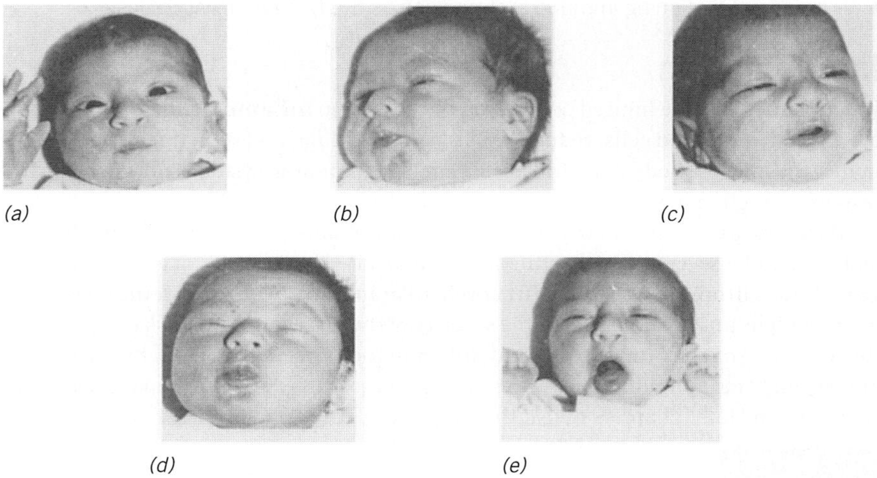

Courtesy of Classic studies of J.E. Steiner, in Taste and Development: The Genesis of Sweet Preference, ed. J.M. Weiffenbach, HHS publication no. NIH 77-1068 (Bethesda, Md: U.S. Department of Health and Human Services, 1977), pp. 173–189, with permission of the author

© Stephen Orsillo/Shutterstock.com

The instinctive liking for sugar, salt, and fat can lead to drastic overeating of these substances. Sugar has become widely available in pure form only in the last hundred years, so it is relatively new to the human diet. Although salt and fat are much older, today all three substances are added liberally to foods by manufacturers to tempt us to eat their products.

KEY POINT

- The preference for sweet, salty, and fatty tastes is inborn and can lead to overconsumption of foods that offer them.

The Digestive Tract

Once you have eaten, your brain and hormones direct the many organs of the **digestive system** to **digest** and **absorb** the complex mixture of chewed and swallowed food. A diagram showing the digestive tract and its associated organs appears in Figure 3–9. The tract itself is a flexible, muscular tube extending from the mouth through the throat, esophagus, stomach, small intestine, large intestine, and rectum to the anus, for a total length of about 26 feet. The human body surrounds this digestive canal. When you swallow something, it still is not inside your body—it is only inside the inner bore of this tube. Only when a nutrient or other substance passes through the wall of the digestive tract does it actually enter the body's tissues. Many things pass into the digestive tract and out again, unabsorbed. A baby playing with beads may swallow one, but the bead will not really enter the body. It will emerge from the digestive tract within a day or two.

The digestive system's job is to digest food to its components and then to absorb the nutrients and some nonnutrients, leaving behind the substances, such as fiber, that are appropriate to excrete. To do this, the system works at two levels: one, mechanical; the other, chemical.

KEY POINTS

- The digestive tract is a flexible, muscular tube that digests food and absorbs its nutrients and some nonnutrients.
- Ancillary digestive organs, such as the pancreas and gallbladder, aid digestion.

digestive system the body system composed of organs that break down complex food particles into smaller, absorbable products. The *digestive tract* and *alimentary canal* are names for the tubular organs that extend from the mouth to the anus. The whole system, including the pancreas, liver, and gallbladder, is sometimes called the *gastrointestinal*, or *GI*, system.

digest to break molecules into smaller molecules; a main function of the digestive tract with respect to food.

absorb to take in, as nutrients are taken into the intestinal cells after digestion; the main function of the digestive tract with respect to nutrients.

Figure 3–9

Animated! The Digestive System

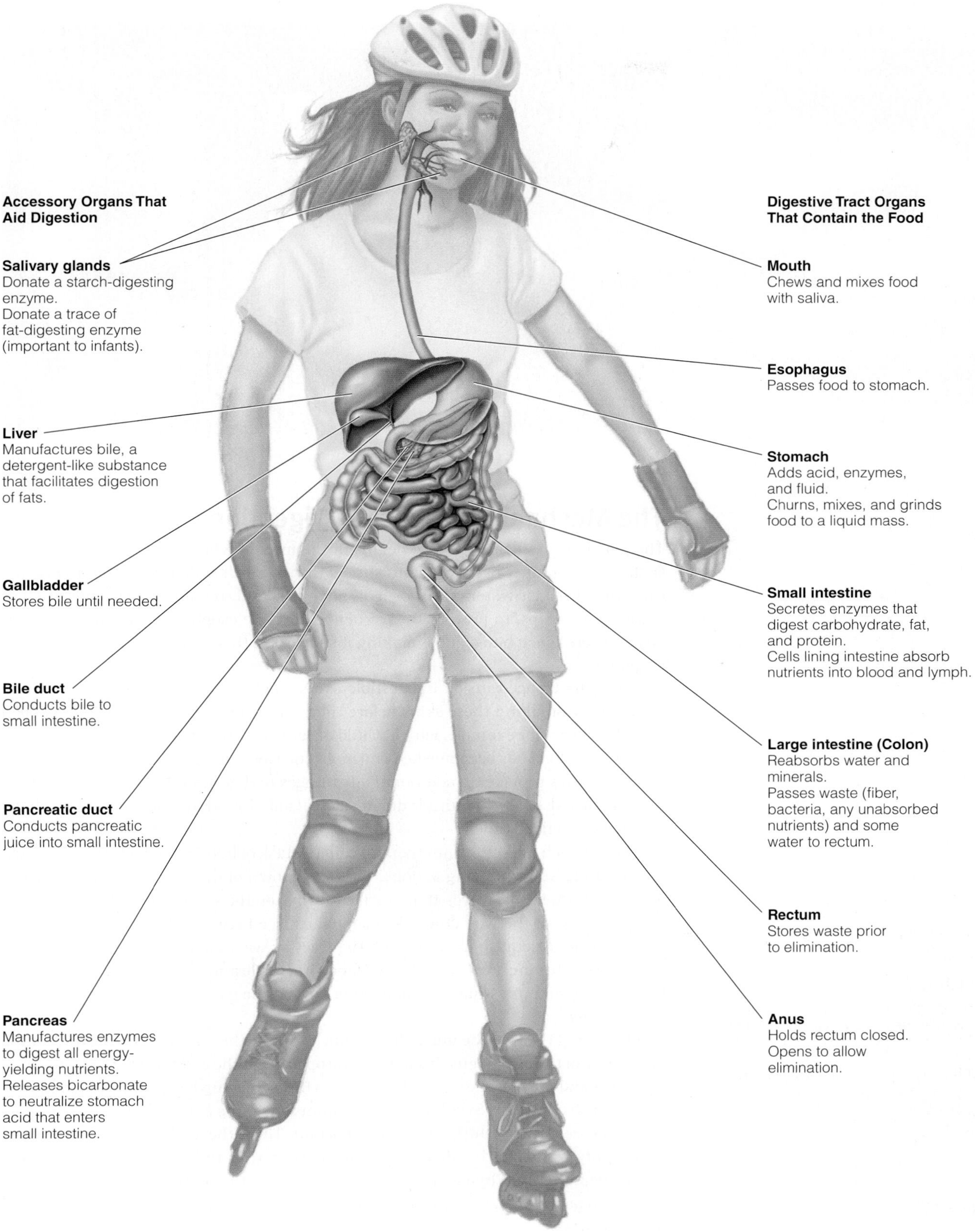

Accessory Organs That Aid Digestion
Salivary glands
Donate a starch-digesting enzyme.
Donate a trace of fat-digesting enzyme (important to infants).
Liver
Manufactures bile, a detergent-like substance that facilitates digestion of fats.
Gallbladder
Stores bile until needed.
Bile duct
Conducts bile to small intestine.
Pancreatic duct
Conducts pancreatic juice into small intestine.
Pancreas
Manufactures enzymes to digest all energy-yielding nutrients.
Releases bicarbonate to neutralize stomach acid that enters small intestine.
Digestive Tract Organs That Contain the Food
Mouth
Chews and mixes food with saliva.
Esophagus
Passes food to stomach.
Stomach
Adds acid, enzymes, and fluid.
Churns, mixes, and grinds food to a liquid mass.
Small intestine
Secretes enzymes that digest carbohydrate, fat, and protein.
Cells lining intestine absorb nutrients into blood and lymph.
Large intestine (Colon)
Reabsorbs water and minerals.
Passes waste (fiber, bacteria, any unabsorbed nutrients) and some water to rectum.
Rectum
Stores waste prior to elimination.
Anus
Holds rectum closed.
Opens to allow elimination.

Figure 3–10

Peristaltic Wave Passing Down the Esophagus and Beyond

Peristalsis moves the digestive tract contents.

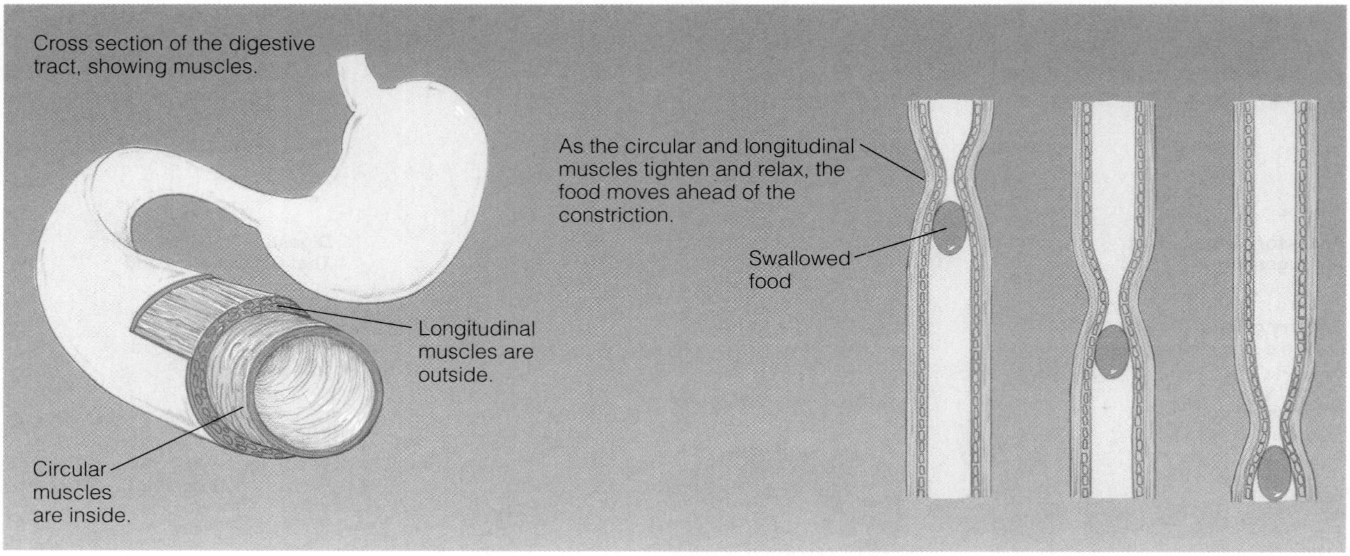

Cross section of the digestive tract, showing muscles.
As the circular and longitudinal muscles tighten and relax, the food moves ahead of the constriction.
Swallowed food
Longitudinal muscles are outside.
Circular muscles are inside.

The Mechanical Aspect of Digestion

The job of mechanical digestion begins in the mouth, where large, solid food pieces such as bites of meat are torn into shreds that can be swallowed without choking. Chewing also adds water in the form of saliva to soften rough or sharp foods, such as fried tortilla chips, to prevent them from tearing the esophagus. Saliva also moistens and coats each bite of food, making it slippery so that it can pass easily down the esophagus.

Nutrients trapped inside indigestible skins, such as the hulls of seeds, must be liberated by breaking these skins before they can be digested. Chewing bursts open kernels of corn, for example, which would otherwise traverse the tract and exit undigested. Once food has been mashed and moistened for comfortable swallowing, longer chewing times provide no additional advantages to digestion. In fact, for digestion's sake, a relaxed, peaceful attitude during a meal aids digestion much more than chewing for an extended time.

The stomach and intestines then take up the task of liquefying foods through various mashing and squeezing actions. The best known of these actions is **peristalsis**, a series of squeezing waves that start with the tongue's movement during a swallow and pass all the way down the esophagus (see Figure 3–10). The stomach and the intestines also push food through the tract by waves of peristalsis. Besides these actions, the **stomach** holds swallowed food for a while and mashes it into a fine paste; the stomach and intestines also add water so that the paste becomes more fluid as it moves along.

Figure 3–11 shows the muscular stomach. Notice the circular **sphincter** muscle at the base of the esophagus. It squeezes the opening at the entrance to the stomach to narrow it and prevent the stomach's contents from creeping back up the esophagus as the stomach contracts. Swallowed food remains in a lump in the stomach's upper portion, squeezed little by little to its lower portion. There the food is ground and mixed thoroughly, ensuring that digestive chemicals mix with the entire thick liquid mass, now called **chyme**. Chyme bears no resemblance to the original food. The starches have been partly split, proteins have been uncoiled and clipped, and fat has separated from the mass.

peristalsis (perri-STALL-sis) the wavelike muscular squeezing of the esophagus, stomach, and small intestine that pushes their contents along.

stomach a muscular, elastic, pouchlike organ of the digestive tract that grinds and churns swallowed food and mixes it with acid and enzymes, forming chyme.

sphincter (SFINK-ter) a circular muscle surrounding, and able to close, a body opening.

chyme (KIME) the fluid resulting from the actions of the stomach upon a meal.

Figure 3–11

The Muscular Stomach

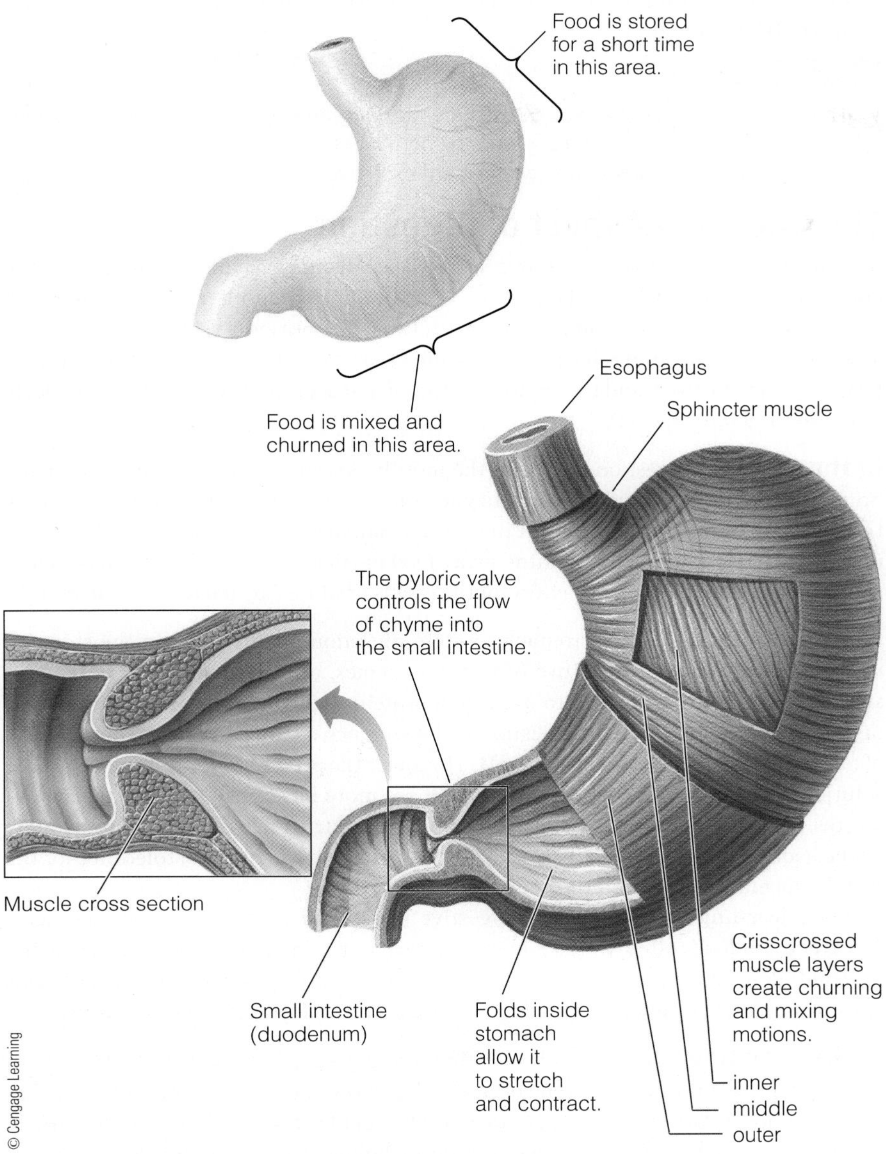
Food is stored for a short time in this area.
Food is mixed and churned in this area.
Esophagus
Sphincter muscle
The pyloric valve controls the flow of chyme into the small intestine.
Muscle cross section
Small intestine (duodenum)
Folds inside stomach allow it to stretch and contract.
Crisscrossed muscle layers create churning and mixing motions.
inner
middle
outer
© Cengage Learning

The stomach also acts as a holding tank. The muscular **pyloric valve** at the stomach's lower end (look again at Figure 3–11) controls the exit of the chyme, allowing only a little at a time to be squirted forcefully into the **small intestine**. Within a few hours after a meal, the stomach empties itself by means of these powerful squirts. The small intestine contracts rhythmically to move the contents along its length.

By the time the intestinal contents have arrived in the **large intestine** (also called the **colon**), digestion and absorption are nearly complete. The colon's task is mostly to reabsorb the water donated earlier by digestive organs and to absorb minerals, leaving a paste of fiber and other undigested materials, the **feces**, suitable for excretion. The fiber provides bulk against which the muscles of the colon can work. The rectum stores this fecal material to be excreted at intervals. From mouth to rectum, the transit of a meal is accomplished in as short a time as a single day or as long as three days.

Some people wonder whether the digestive tract works best at certain hours in the day and whether the timing of meals can affect how a person feels. Timing of meals is important to feeling well, not because the digestive tract is unable to digest food at

pyloric (pye-LORE-ick) **valve** the circular muscle of the lower stomach that regulates the flow of partly digested food into the small intestine. Also called *pyloric sphincter*.

small intestine the 20-foot length of small-diameter intestine, below the stomach and above the large intestine, that is the major site of digestion of food and absorption of nutrients.

large intestine the portion of the intestine that completes the absorption process.

colon the large intestine.

feces waste material remaining after digestion and absorption are complete; eventually discharged from the body.

Table 3–1

Digestive Enzyme Terms

Over 30 digestive enzymes reduce food in the human digestive tract into nutrients that can be absorbed. Naming them all is beyond the scope of this book, but some general enzyme terms may prove useful.

- **-ase** (ACE) a suffix meaning *enzyme.* Categories of digestive and other enzymes and individual enzyme names often contain this suffix.
- **carbohydrase** (car-boh-HIGH-drace) any of a number of enzymes that break the chemical bonds of carbohydrates.
- **lipase** (LYE-pace) any of a number of enzymes that break the chemical bonds of fats (lipids).
- **protease** (PRO-tee-ace) any of a number of enzymes that break the chemical bonds of proteins.

Did You Know?

Alcohol needs no assistance from digestive juices to ready it for absorption; its handling by the body is described in this chapter's Controversy section.

gastric juice the digestive secretion of the stomach.

hydrochloric acid a strong corrosive acid of hydrogen and chloride atoms, produced by the stomach to assist in digestion.

pH a measure of acidity on a point scale. A solution with a pH of 1 is a strong acid; a solution with a pH of 7 is neutral; a solution with a pH of 14 is a strong base.

mucus (MYOO-cus) a slippery coating of the digestive tract lining (and other body linings) that protects the cells from exposure to digestive juices (and other destructive agents). The adjective form is *mucous* (same pronunciation). The digestive tract lining is a *mucous membrane.*

certain times, but because the body requires nutrients to be replenished every few hours. Digestion is virtually continuous, being limited only during sleep and exercise. For some people, eating late may interfere with normal sleep. As for exercise, it is best pursued a few hours after eating because digestion can inhibit physical work (see Chapter 10 for details).

KEY POINTS

- The mechanical digestive actions include chewing, mixing by the stomach, adding fluid, and moving the tract's contents by peristalsis.
- After digestion and absorption, wastes are excreted.

The Chemical Aspect of Digestion

Several organs of the digestive system secrete special digestive juices that perform the complex chemical processes of digestion. Digestive juices contain enzymes that break down nutrients into their component parts (Table 3–1 presents some enzyme terms). The digestive organs that release digestive juices are the salivary glands, the stomach, the pancreas, the liver, and the small intestine. Their secretions were listed previously in Figure 3–9 (on page 83).

In the Mouth Digestion begins in the mouth. An enzyme in saliva starts rapidly breaking down starch, and another enzyme initiates a little digestion of fat, especially the digestion of milk fat (important in infants). Saliva also helps maintain the health of the teeth in two ways: by washing away food particles that would otherwise foster decay and by neutralizing decay-promoting acids produced by bacteria in the mouth.

In the Stomach In the stomach, protein digestion begins. Cells in the stomach release **gastric juice**, a mixture of water, enzymes, and **hydrochloric acid**. This strong acid mixture is needed to activate a protein-digesting enzyme and to initiate digestion of protein—protein digestion is the stomach's main function. The strength of an acid solution is expressed as its **pH**. The lower the pH number, the more acidic the solution; solutions with higher pH numbers are more basic. As Figure 3–12 demonstrates, saliva is only weakly acidic; the stomach's gastric juice is much more strongly acidic. Notice on the right-hand side of Figure 3–12 that the range of tolerance for the blood's normal pH is exceedingly small.

Upon learning of the powerful digestive juices and enzymes within the digestive tract, students often wonder how the tract's own cellular lining escapes being digested along with the food. The answer: specialized cells secrete a thick, viscous substance known as **mucus**, which coats and protects the digestive tract lining.

In the Intestine In the small intestine, the digestive process gets under way in earnest. The small intestine is *the* organ of digestion and absorption, and it finishes what the mouth and stomach have started. The small intestine works with the precision of a laboratory chemist. As the thoroughly liquefied and partially digested nutrient mixture arrives there, hormonal messengers signal the gallbladder to contract and to squirt the right amount of **bile**, an **emulsifier**, into the intestine. Other hormones notify the pancreas to release **pancreatic juice** containing the alkaline compound **bicarbonate** in amounts precisely adjusted to neutralize the stomach acid that has reached the small intestine. All these actions alter the intestinal environment to perfectly support the work of the digestive enzymes.

Meanwhile, as the pancreatic and intestinal enzymes act on the chemical bonds that hold the large nutrients together, smaller and smaller pieces are released into the intestinal fluids. The cells of the intestinal wall also hold some digestive enzymes on their surfaces; these enzymes perform last-minute breakdown reactions required before nutrients can be absorbed. Finally, the digestive process releases pieces small enough for the cells to absorb and use. Digestion by human enzymes and absorption of carbohydrate, fat, and protein are essentially complete by the time the intestinal contents enter the colon. Water, fiber, and some minerals, however, remain in the tract.

Certain kinds of fiber, which cannot be digested by human enzymes, can often be broken down by the billions of living inhabitants of the human colon, the resident

Figure 3–12
pH Values of Digestive Juices and Other Common Fluids

A substance's acidity or alkalinity is measured in pH units. Each step down the scale indicates a tenfold increase in concentration of hydrogen particles, which determine acidity. For example, a pH of 2 is 1,000 times stronger than a pH of 5.

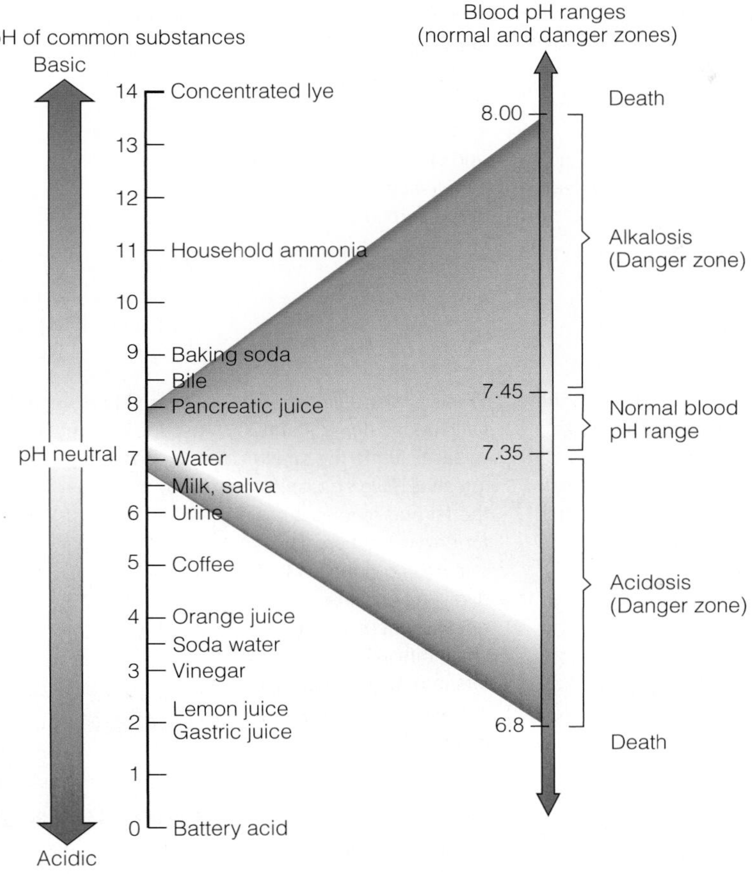

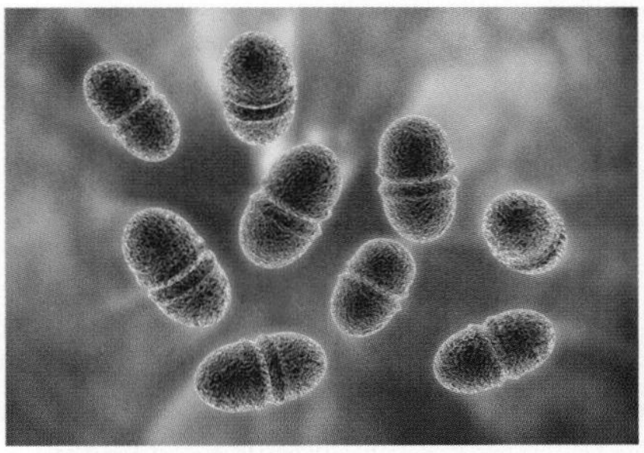

Enterococcus faecalis, *one of the thousands of bacterial species living in the human digestive tract.*

bacteria.[6] So active are these inhabitants in breaking down substances from food that they have been likened to an organ of the body specializing in nutrient salvage. They also affect health in other ways that are just beginning to be understood.[7] The intestinal cells then absorb the small fat fragments released from the fiber to provide a tiny bit of energy. Table 3–2 (p. 88) provides a summary of all the processes involved.

KEY POINTS

- Chemical digestion begins in the mouth, where food is mixed with an enzyme in saliva that acts on carbohydrates.
- Digestion continues in the stomach, where stomach enzymes and acid break down protein.
- Digestion progresses in the small intestine where the liver and gallbladder contribute bile that emulsifies fat, and the pancreas and small intestine donate enzymes that break down food to nutrients.
- Bacteria in the colon break down certain fibers.

Are Some Food Combinations More Easily Digested Than Others?

People sometimes wonder if the digestive tract has trouble digesting certain foods in combination—for example, fruit and meat. Proponents of fad "food-combining"

bile a cholesterol-containing digestive fluid made by the liver, stored in the gallbladder, and released into the small intestine when needed. It emulsifies fats and oils to ready them for enzymatic digestion (described in Chapter 5).

emulsifier (ee-MULL-sih-fire) a compound with both water-soluble and fat-soluble portions that can attract fats and oils into water, combining them.

pancreatic juice fluid secreted by the pancreas that contains both enzymes to digest carbohydrates, fats, and proteins and sodium bicarbonate, a neutralizing agent.

bicarbonate a common alkaline chemical; a secretion of the pancreas; also the active ingredient of baking soda.

Table 3–2

Summary of Chemical Digestion

	Mouth	Stomach	Small Intestine, Pancreas, Liver, and Gallbladder	Large Intestine (Colon)
Sugar and Starch	The salivary glands secrete saliva to moisten and lubricate food; chewing crushes and mixes it with a salivary enzyme that initiates starch digestion.	Digestion of starch continues while food remains in the upper storage area of the stomach. In the lower digesting area of the stomach, hydrochloric acid and an enzyme in the stomach's juices halt starch digestion.	The pancreas produces a starch-digesting enzyme and releases it into the small intestine. Cells in the intestinal lining possess enzymes on their surfaces that break sugars and starch fragments into simple sugars, which then are absorbed.	Undigested carbohydrates reach the large intestine and are partly broken down by intestinal bacteria.
Fiber	The teeth crush fiber and mix it with saliva to moisten it for swallowing.	No action.	Fiber binds cholesterol and some minerals.	Most fiber is excreted with the feces; some fiber is digested by bacteria in the large intestine.
Fat	Fat-rich foods are mixed with saliva. The tongue produces traces of a fat-digesting enzyme that accomplishes some breakdown, especially of milk fats. The enzyme is stable at low pH and is important to digestion in nursing infants.	Fat tends to rise from the watery stomach fluid and foods and float on top of the mixture. Only a small amount of fat is digested. Fat is last to leave the stomach.	The liver secretes bile; the gallbladder stores it and releases it into the small intestine. Bile emulsifies the fat and readies it for enzyme action. The pancreas produces fat-digesting enzymes and releases them into the small intestine to split fats into their component parts (primarily fatty acids), which then are absorbed.	Some fatty materials escape absorption and are carried out of the body with other wastes.
Protein	Chewing crushes and softens protein-rich foods and mixes them with saliva.	Stomach acid (hydrochloric acid) works to uncoil protein strands and to activate the stomach's protein-digesting enzyme. Then the enzyme breaks the protein strands into smaller fragments.	Enzymes of the small intestine and pancreas split protein fragments into smaller fragments or free amino acids. Enzymes on the cells of the intestinal lining break some protein fragments into free amino acids, which then are absorbed. Some protein fragments are also absorbed.	The large intestine carries undigested protein residue out of the body. Normally, almost all food protein is digested and absorbed.
Water	The mouth donates watery, enzyme-containing saliva.	The stomach donates acidic, watery, enzyme-containing gastric juice.	The liver donates a watery juice containing bile. The pancreas and small intestine add watery, enzyme-containing juices; pancreatic juice is also alkaline.	The large intestine reabsorbs water and some minerals.

diets claim that the digestive tract cannot perform certain digestive tasks at the same time, but this is a gross underestimation of the tract's capabilities. The digestive system adjusts to whatever mixture of foods is presented to it. The truth is that all foods, regardless of identity, are broken down by enzymes into the basic molecules that make them up.

Scientists who study digestion suggest that some organs of the digestive tract analyze the diet's nutrient contents and deliver juice and enzymes appropriate for digesting those nutrients. The pancreas is especially sensitive in this regard and has been observed to adjust its output of enzymes to digest carbohydrate, fat, or protein to an amazing degree. The pancreas of a person who suddenly consumes a meal unusually high in carbohydrate, for example, would begin increasing its output of carbohydrate-digesting enzymes within 24 hours, while reducing outputs of other types. This sensitive mechanism ensures that foods of all types are used fully by the body. The next section reviews the major processes of digestion by showing how the nutrients in a mixture of foods are handled.

KEY POINT

- The healthy digestive system can adjust to almost any diet and handle any combination of foods with ease.

If "I Am What I Eat," Then How Does a Peanut Butter Sandwich Become "Me"?

The process of rendering foods into nutrients and absorbing them into the body fluids is remarkably efficient. Within about 24 to 48 hours of eating, a healthy body digests and absorbs about 90 percent of the carbohydrate, fat, and protein in a meal. Figure 3–13 illustrates a typical 24-hour transit time through the digestive system. Next, we follow a peanut butter and banana sandwich on whole wheat, sesame seed bread through the tract.

In the Mouth In each bite, food components are crushed, mashed, and mixed with saliva by the teeth and the tongue. The sesame seeds are crushed and torn open by the teeth, which break through the indigestible fiber coating so that digestive enzymes can reach the nutrients inside the seeds. The peanut butter is the "extra crunchy" type, but the teeth grind the chunks to a paste before the bite is swallowed. The carbohydrate-digesting enzyme of saliva begins to break down the starch of the bread, banana, and peanut butter to sugars. Each swallow triggers a peristaltic wave that travels the length of the esophagus and carries one chewed bite of sandwich to the stomach.

In the Stomach The stomach collects bite after swallowed bite in its upper storage area, where starch continues to be digested until the gastric juice mixes with the salivary enzymes and halts their action. Small portions of the mashed sandwich are pushed into the digesting area of the stomach, where gastric juice mixes with the mass. Acid in gastric juice unwinds proteins from the bread, seeds, and peanut butter; then an enzyme clips the protein strands into pieces. The sandwich has now become chyme. The watery carbohydrate- and protein-rich part of the chyme enters the small intestine first; a layer of fat follows closely behind.

In the Small Intestine Some of the sweet sugars in the banana require so little digesting that they begin to cross the linings of the small intestine immediately on contact. Nearby, the liver donates bile through a duct into the small intestine. The bile blends the fat from the peanut butter and seeds with the watery, enzyme-containing digestive fluids. The nearby pancreas squirts enzymes into the small intestine to break down the fat, protein, and starch in the chemical soup that just an hour ago was a sandwich. The cells of the small intestine itself produce enzymes to complete these processes. As the enzymes do their work, smaller and smaller chemical fragments are liberated from the

Figure 3–13
Typical Digestive System Transit Times

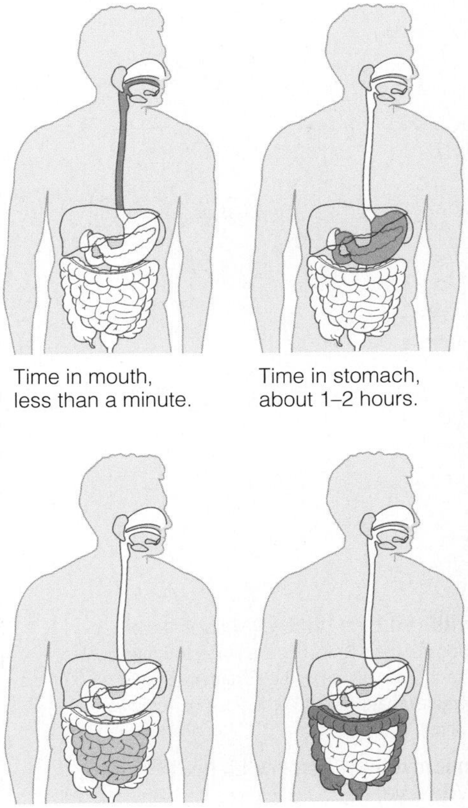

Time in mouth, less than a minute.

Time in stomach, about 1–2 hours.

Time in small intestine, about 7–8 hours.*

Time in colon, about 12–14 hours.*

*Based on a 24-hour transit time. Actual times vary widely.

chemical soup and are absorbed into the blood and lymph through the cells of the small intestine's wall. Vitamins and minerals are absorbed here, too. They all eventually enter the bloodstream to nourish the tissues.

In the Large Intestine (Colon) Only fiber fragments, fluid, and some minerals are absorbed in the large intestine. The fibers from the seeds, whole wheat bread, peanut butter, and banana are partly digested by the bacteria living in the colon, and some of the products are absorbed. Most fiber is not digested, however, and it passes out of the colon along with some other components, excreted as feces.

KEY POINT

- The mechanical and chemical actions of the digestive tract efficiently break down foods to nutrients, and large nutrients to their smaller building blocks.

Absorption and Transport of Nutrients

Once the digestive system has broken down food to its nutrient components, the rest of the body awaits their delivery. First, though, every molecule of nutrient must traverse one of the cells of the intestinal lining. These cells absorb nutrients from the mixture within the intestine and deposit the water-soluble compounds in the blood and the fat-soluble ones in the lymph. The cells are selective: they recognize that some nutrients may be in short supply in the diet. Take the mineral calcium, for example. The less calcium in the diet, the greater the percentage of calcium the intestinal cells absorb from the intestinal contents. The cells are also extraordinarily efficient: they absorb enough nutrients to nourish all the body's other cells.

The Intestine's Absorbing Surface The cells of the intestinal tract lining are arranged in sheets that poke out into millions of finger-shaped projections (**villi**). Every cell on every villus has a brushlike covering of tiny hairlike projections (**microvilli**) that can trap the nutrient particles. Each villus (projection) has its own capillary network and a lymph vessel so that, as nutrients move across the cells, they can immediately mingle with the body fluids. Figure 3–14 provides a close look at these details.

The small intestine's lining, villi and all, is wrinkled into thousands of folds, so its absorbing surface is enormous. If the folds, and the villi that poke out from them, were spread out flat, they would cover a third of a football field. The billions of cells of that surface weigh only 4 to 5 pounds, yet they absorb enough nutrients to nourish the other 150 or so pounds of body tissues.

Nutrient Transport in the Blood and Lymph Vessels After the nutrients pass through the cells of the villi, the blood and lymph vessels transport the nutrients to their ultimate consumers, the body's cells. The lymph vessels initially transport most of the products of fat digestion and the fat-soluble vitamins, ultimately conveying them into a large blood vessel near the heart. The blood vessels directly transport the products of carbohydrate and protein digestion, most vitamins, and the minerals from the digestive tract to the liver. Thanks to these two transportation systems, every nutrient soon arrives at the place where it is needed.

Nourishment of the Digestive Tract The digestive system's millions of specialized cells are themselves exquisitely sensitive to an undersupply of energy, nutrients, or dietary fiber. In cases of severe undernutrition with too little energy and nutrients, the absorptive surface of the small intestine shrinks. The surface may be reduced to a tenth of its normal area, preventing it from absorbing what few nutrients a limited food supply may provide. Without sufficient fiber to provide an undigested bulk for the tract's muscles to push against, the muscles become weak from lack of exercise. Malnutrition that impairs digestion is self-perpetuating because impaired digestion makes malnutrition worse.

The digestive system's needs are few, but important. The body has much to say to the attentive listener, stated in a language of symptoms and feelings that you would

villi (VILL-ee, VILL-eye) fingerlike projections of the sheets of cells lining the intestinal tract. The villi make the surface area much greater than it would otherwise be (*singular*: villus).

microvilli (MY-croh-VILL-ee, MY-croh-VILL-eye) tiny, hairlike projections on each cell of every villus that greatly expand the surface area available to trap nutrient particles and absorb them into the cells (*singular*: microvillus).

Figure 3–14

Details of the Small Intestinal Lining

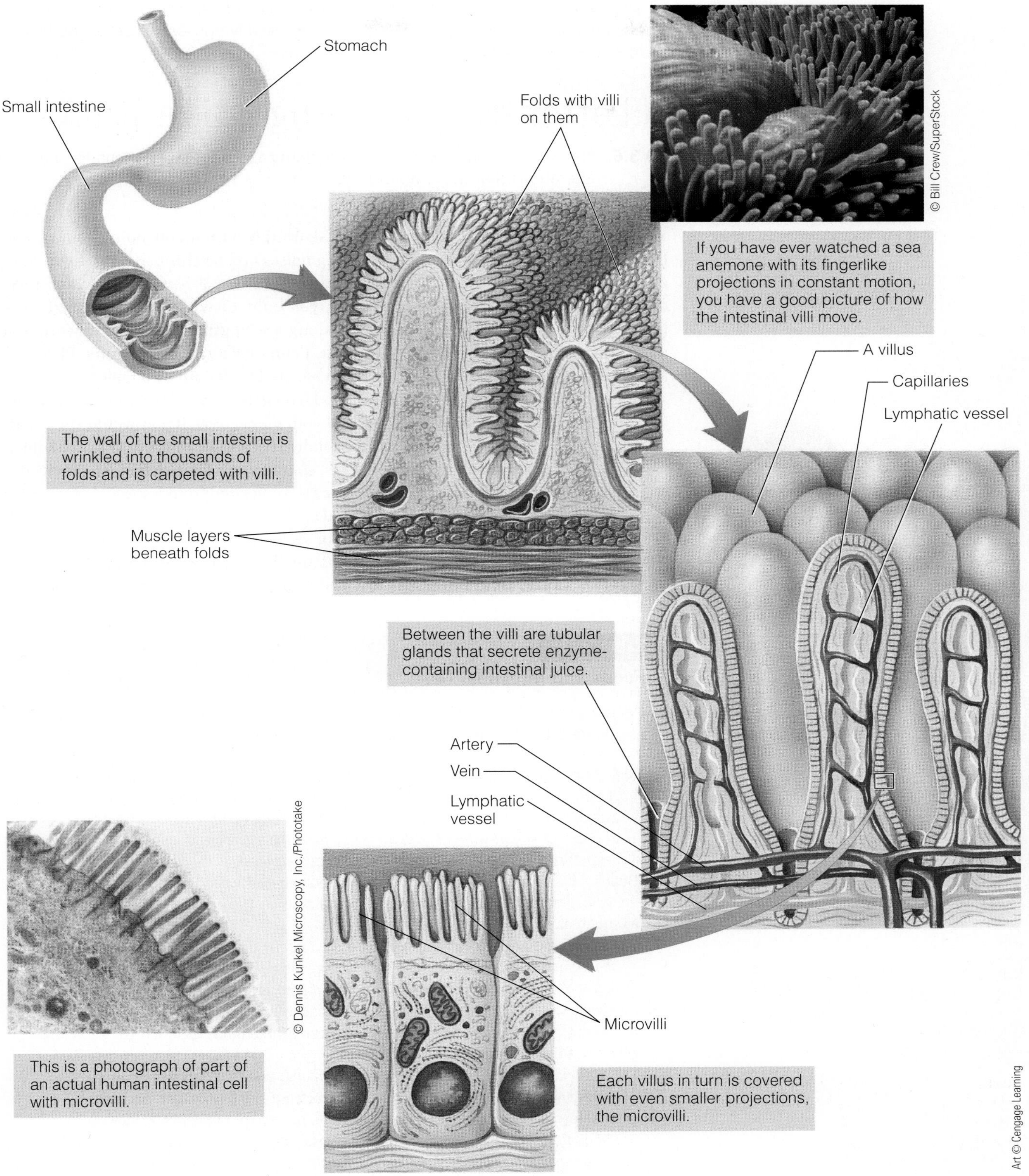
Stomach
Small intestine
Folds with villi on them
© Bill Crew/SuperStock
If you have ever watched a sea anemone with its fingerlike projections in constant motion, you have a good picture of how the intestinal villi move.
A villus
Capillaries
Lymphatic vessel
The wall of the small intestine is wrinkled into thousands of folds and is carpeted with villi.
Muscle layers beneath folds
Between the villi are tubular glands that secrete enzyme-containing intestinal juice.
Artery
Vein
Lymphatic vessel
© Dennis Kunkel Microscopy, Inc./Phototake
Microvilli
This is a photograph of part of an actual human intestinal cell with microvilli.
Each villus in turn is covered with even smaller projections, the microvilli.
Art © Cengage Learning

be wise to study. The next section takes a lighthearted look at what your digestive tract might be trying to tell you.

KEY POINTS

- The digestive system feeds the rest of the body and is itself sensitive to malnutrition.
- The folds and villi of the small intestine enlarge its surface area to facilitate nutrient absorption through countless cells to the blood and lymph, which deliver nutrients to all the body's cells.

A Letter from Your Digestive Tract

LO 3.6 Name some common digestive problems and offer suggestions for dietary alterations that may improve them.

To My Owner,

You and I are so close; I hope that I can speak frankly without offending you. I know that sometimes I *do* offend with my gurgling noises and belching at quiet times and, oh yes, the gas. But, as you can read for yourself in Table 3–3, when you chew gum, drink carbonated beverages, or eat hastily, you gulp air with each swallow. I can't help making some noise as I move the air along my length or release it upward in a noisy belch. And if you eat or drink too fast, I can't help getting **hiccups**. Please sit and relax while you dine. You will ease my task, and we'll both be happier.

Also, when someone offers you a new food, you gobble away, trusting me to do my job. I try. It would make my life easier, and yours less gassy, if you would start with small amounts of new foods, especially those high in fiber. The breakdown of fiber by bacteria produces gas, so introduce fiber-rich foods slowly. But please, if you do notice more gas than normal from a specific food, avoid it. If the gas becomes excessive, check with a physician. The problem could be something simple—or serious.

When you eat or drink too much, it just burns me up. Overeating causes **heartburn** because the acidic juice from my stomach backs up

Table 3–3
Foods and Intestinal Gas

Recent experiments have shed light on the causes and prevention of intestinal gas. Here are some recent findings.

- Milk intake causes gas in those who cannot digest the milk sugar lactose. Most people, however, can consume up to a cup of milk without producing excessive gas.
  Solution: Drink up to 4 ounces of fluid milk at a sitting, or substitute reduced-fat cheeses or yogurt without added milk solids. Use lactose-reduced products, or treat regular products with lactose-reducing enzyme products.
- Beans cause gas because some of their carbohydrates are indigestible by human enzymes, but are broken down by intestinal bacteria. The amount of gas may not be as much as most people fear, however.
  Solution: Use rinsed canned beans or dried beans that are well cooked, because cooked carbohydrates are more readily digestible. Try enzyme drops or pills that can help break down the carbohydrate before it reaches the intestine.
- Air swallowed during eating or drinking can cause gas, as can the gas of carbonated beverages. Each swallow of a beverage can carry three times as much gas as fluid, which some people belch up.
  Solution: Slow down during eating and drinking, and don't chew gum or suck on hard candies that may cause you to swallow air. Limit carbonated beverages.
- Vegetables may or may not cause gas in some people, but research is lacking.
  Solution: If you feel certain vegetables cause gas, try eating small portions of the cooked products. Do try the vegetable again: the gas you experienced may have been a coincidence and unrelated to eating the vegetable.

into my esophagus. Acid poses no problem to my healthy stomach, whose walls are coated with thick mucus to protect them. But when my too-full stomach squeezes some of its contents back up into the esophagus, the acid burns its unprotected surface. Also, those tight jeans you wear constrict my stomach, squeezing the contents up into the esophagus. Just leaning over or lying down after a meal may do the same thing because the muscular sphincter separating the two spaces is much looser than other sphincters. And if we need to lose a few pounds, let's get at it—excess body fat can squeeze my stomach, too. When heartburn is a problem, do me a favor: try to eat smaller meals; drink liquids an hour before or after, but not during, meals; wear reasonably loose clothing; and relax after eating, but sit up (don't lie down). Don't smoke, and go easy on the alcohol and carbonated beverages, too—they all make heartburn likely.

Sometimes your food choices irritate me. Specifically, chemical irritants in foods, such as the "hot" component of chili peppers, chemicals in coffee, fat, chocolate, carbonated soft drinks, and alcohol, may worsen heartburn in some people. Avoid the ones that cause trouble. Above all, do not smoke. Smoking makes my heartburn worse—and you should hear your lungs bellyache about it.

By the way, I can tell you've been taking heartburn medicines again. You must have been watching those misleading TV commercials. You need to know that **antacids** are designed only to temporarily relieve pain caused by heartburn by neutralizing stomach acid for a while. But when the antacids reduce my normal stomach acidity, I respond by producing *more* acid to restore the normal acid condition. Also, the ingredients in antacids can interfere with my ability to absorb nutrients. Please check with our doctor if heartburn occurs more than just occasionally and certainly before you decide that we need to take the heavily advertised **acid reducers**; these restrict my normal ability to produce acid so much that my job of digesting food becomes harder. They may also reduce our defense against serious infections, even pneumonia.[8]

Given a chance, my powerful stomach acid helps to fight off many bacterial infections—most disease-causing bacteria won't survive a bath in my caustic juices. Acid-reducing drugs reduce acid (I'll bet you knew that), and so allow more bacteria to pass through. And, even worse, self-prescribed heartburn medicine can mask the symptoms of **ulcer**, **hernia**, or the destructive form of chronic heartburn known as **gastroesophageal reflux disease (GERD)**.[9] This can be serious because, although the bacterium *H. pylori* that causes most ulcers responds to antibiotic drugs, some ulcers have other causes, such as frequent use of certain painkillers—the *cause* of the ulcer must be treated as well as its symptoms.[‡10] Left untreated, *H. pylori* raises the risk of stomach cancer. A hernia can cause food to back up into the esophagus, so it can feel like heartburn, but many times hernias require corrective treatment by a physician. GERD can feel like heartburn, too, but requires the correct drug therapy to prevent respiratory problems, damage to the esophagus, or even cancer. So please don't wait too long to get medical help for chronic or severe heartburn—it may not be simple indigestion.

When you eat too quickly, I worry about choking (see Figure 3–15, p. 94). Please take time to cut your food into small pieces and chew it until it is crushed and moistened with saliva. Also, refrain from talking or laughing before swallowing, and never attempt to eat when you are breathing hard. Also, for our sake and the sake of others, learn first aid for choking as shown in Figure 3–16, p. 95.

When I'm suffering, you suffer, too, and when **constipation** or **diarrhea** strikes, neither of us is having fun. Slow, hard, dry bowel movements can be painful, and failing to have a movement for too long brings on headaches, backaches, stomachaches, and other ills; if chronic, constipation may cause **hemorrhoids** or other ills.[11] Most people suffer occasional harmless constipation, and laxatives may help, but too frequent use of laxatives and enemas can lead to dependency; upset our fluid, salt, and mineral balances; and, with mineral oil laxatives, can interfere with the absorption of fat-soluble vitamins. (Mineral oil, which is not absorbed, dissolves the vitamins and carries them out of the body with it.)

‡ Anti-inflammatory drugs such as aspirin, ibuprofen, and naproxen sodium.

hiccups spasms of both the vocal cords and the diaphragm, causing periodic, audible, short, inhaled coughs. Can be caused by irritation of the diaphragm, indigestion, or other causes. Hiccups usually resolve in a few minutes but can have serious effects if prolonged. Breathing into a paper bag (inhaling carbon dioxide) or dissolving a teaspoon of sugar in the mouth may stop them.

heartburn a burning sensation in the chest (in the area of the heart) caused by backflow of stomach acid into the esophagus.

antacids medications that react directly and immediately with the acid of the stomach, neutralizing it. Antacids are most suitable for treating occasional heartburn.

acid reducers prescription and over-the-counter drugs that reduce the acid output of the stomach; effective for treating severe, persistent forms of heartburn but not for neutralizing acid already present. Side effects are frequent and include diarrhea, other gastrointestinal complaints, and reduction of the stomach's capacity to destroy alcohol, thereby producing higher-than-expected blood alcohol levels from each drink (see this chapter's Controversy section). Also called *acid controllers*.

ulcer an erosion in the topmost, and sometimes underlying, layers of cells that form a lining. Ulcers of the digestive tract commonly form in the esophagus, stomach, or upper small intestine.

hernia a protrusion of an organ or part of an organ through the wall of the body chamber that normally contains the organ. An example is a *hiatal* (high-AY-tal) *hernia,* in which part of the stomach protrudes up through the diaphragm into the chest cavity, which contains the esophagus, heart, and lungs.

gastroesophageal (GAS-tro-eh-SOFF-ah-jeel) **reflux disease (GERD)** a severe and chronic splashing of stomach acid and enzymes into the esophagus, throat, mouth, or airway that causes injury to those organs. Untreated GERD may increase the risk of esophageal cancer; treatment may require surgery or management with medication.

constipation infrequent, difficult bowel movements often caused by diet, inactivity, dehydration, or medication. Also defined in Chapter 4.

diarrhea frequent, watery bowel movements usually caused by diet, stress, or irritation of the colon. Severe, prolonged diarrhea robs the body of fluid and certain minerals, causing dehydration and imbalances that can be dangerous if left untreated.

hemorrhoids (HEM-or-oids) swollen, hardened (varicose) veins in the rectum, usually caused by the pressure resulting from constipation.

Figure 3–15

Normal Swallowing and Choking

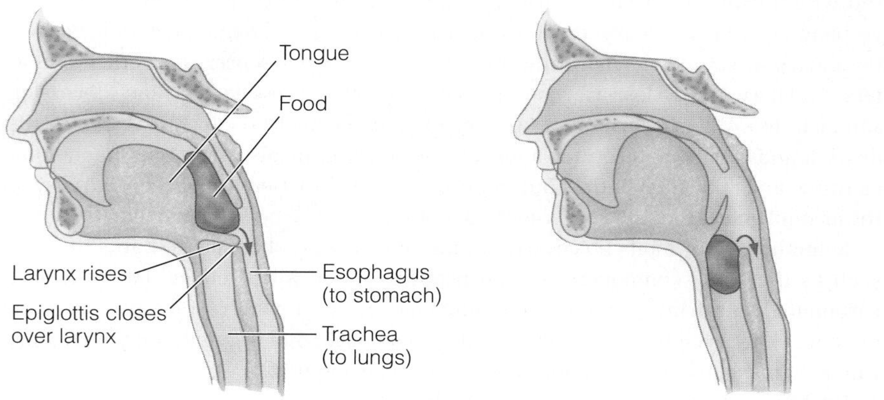

A normal swallow. The epiglottis acts as a flap to seal the entrance to the lungs (trachea) and direct food to the stomach via the esophagus.

Choking. A choking person cannot speak or gasp because food lodged in the trachea blocks the passage of air. The red arrow points to where the food should have gone to prevent choking.

What is your digestive tract trying to tell you?

Instead of relying on laxatives, listen carefully for my signal that it is time to defecate, and make time for it even if you are busy. The longer you ignore my signal, the more time the colon has to extract water from the feces, hardening them. Also, please choose foods that provide enough fiber (some high-fiber foods are listed in Chapter 4, page 121).[§] Fiber attracts water, creating softer, bulkier stools that stimulate my muscles to contract, pushing the contents along. Fiber helps my muscles to stay fit, too, making elimination easier. Be sure to drink enough water because dehydration causes the colon to absorb all the water it can get from the feces. And please make time to be physically active; exercise strengthens not just the muscles of your arms, legs, and torso, but those of the colon, too.

When I have the opposite problem, diarrhea, my system will rob you of water and salts. In diarrhea my intestinal contents have moved too quickly, drawing water and minerals from your tissues into the contents. When this happens, please rest a while and drink fluids (I prefer clear juices and broths). However, if diarrhea is bloody, or if it worsens or persists, call our doctor—severe diarrhea can be life-threatening.

To avoid diarrhea, try not to change my diet too drastically or quickly. I'm willing to work with you and learn to digest new foods, but if you suddenly change your diet, we're both in for it. I hate even to think of it, but one likely cause of diarrhea is foodborne illness. (*Please* read, and use, the tips in Chapter 12 to keep us safe). Also, if diarrhea lasts longer than a day or two, or if it alternates with constipation, it may be **irritable bowel syndrome (IBS)**, and you should see a physician. In IBS, strong contractions speed up the intestinal contents, causing gas, bloating, diarrhea, and frequent or severe abdominal pain.[12] Weakened and slowed contractions may then follow, causing constipation. When you're stressed out, so am I, and stress may contribute to IBS.[13] Try eating smaller meals, avoiding onions or other irritating foods, and using relaxation techniques or exercise to relieve mental stress. If those don't work, by all means, call our doctor—IBS often responds to antispasmodic drugs or even peppermint oil taken under medical supervision.[14]

By the way, I trust you not to believe false claims that health troubles can be solved by washing the colon with a powerful enema machine—in fact, this "colonic

irritable bowel syndrome (IBS) intermittent disturbance of bowel function, especially diarrhea or alternating diarrhea and constipation, often with abdominal cramping or bloating; managed with diet, physical activity, or relief from psychological stress. The cause is uncertain, but IBS does not permanently harm the intestines nor lead to serious diseases.

[§]Rarely, a spastic, constricted bowel causes constipation; this condition requires medical attention, not fiber.

Figure 3–16
First Aid for Choking

First aid for choking relies on abdominal thrusts, sometimes called the Heimlich maneuver. If abdominal thrusts are not successful and the person loses consciousness, lower him to the floor, call 911, remove the object blocking the airway if possible, and begin CPR. Because there is no time for hesitation when called upon to perform this death-defying act, you would do well to take a life-saving course to learn these techniques.

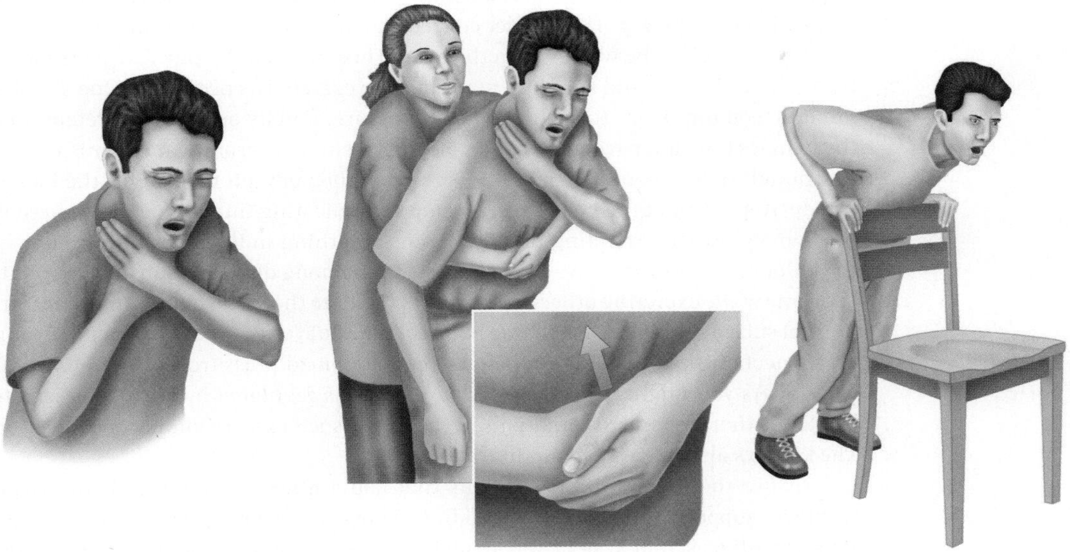

The universal signal for choking alerts others to the need for assistance.

Stand behind the person with your arms wrapped around him. Make a fist with one hand and place the thumb side snugly against the body, slightly above the navel and below the breastbone.

Grasp the fist with your other hand and make a quick upward and inward thrust. Repeat thrusts until the object is dislodged.

To perform abdominal thrusts on yourself, make a fist and place the thumb below your breastbone and above your navel. Grasp your fist with your other hand and press inward with a quick upward thrust. Alternatively, quickly thrust your upper body against a table edge, chair, or railing.

irrigation" is unnecessary and has caused illness and even death from equipment contamination, electrolyte depletion, and intestinal perforation.

Thank you for listening. I know we'll both benefit from communicating like this because you and I are in this together for the long haul.

Affectionately,
Your Digestive Tract

KEY POINT

- Maintenance of a healthy digestive tract requires preventing or responding to symptoms with a carefully chosen diet and sound medical care when problems arise.

The Excretory System

LO 3.7 Identify the excretory functions of the lungs, liver, kidneys, and bladder, and state why they are important to maintain normal body functioning.

Cells generate a number of wastes, and all of them must be eliminated. Many of the body's organs play roles in removing wastes. Carbon dioxide waste from the cells

Did You Know?

The lungs also excrete some small percentage of ingested alcohol—the basis for the "breathalyzer" test given to drivers to determine if they've been drinking.

travels in the blood to the lungs, where it is exchanged for oxygen. Other wastes are pulled out of the bloodstream by the liver. The liver processes these wastes and either tosses them out into the digestive tract with bile, to leave the body with the feces, or prepares them to be sent to the kidneys for disposal in the urine. Organ systems work together to dispose of the body's wastes, but the kidneys are waste- and water-removal specialists.

The kidneys straddle the cardiovascular system and filter the passing blood. Waste materials, dissolved in water, are collected by the kidneys' working units, the **nephrons**. These wastes become concentrated as urine, which travels through tubes to the urinary **bladder**. The bladder collects the urine continuously and empties periodically, removing the wastes from the body. Thus, the blood is purified continuously throughout the day, and dissolved materials are excreted as necessary. One dissolved mineral, sodium, helps to regulate blood pressure, and its excretion or retention by the kidneys is a vital part of the body's blood pressure–controlling mechanism.

Though they account for just 0.5 percent of the body's total weight, the kidneys use up 10 percent of the body's oxygen supply, indicating intense metabolic activity. The kidney's waste-excreting function rivals breathing in its importance to life, but the kidneys act in other ways as well. By sorting among dissolved substances, retaining some while excreting others, the kidneys regulate the fluid volume and concentrations of substances in the blood and extracellular fluid with great precision. Through these mechanisms, the kidneys help to regulate blood pressure (see Chapter 11 for details). As you might expect, the kidneys' work is regulated by hormones secreted by glands that respond to conditions in the blood (such as the sodium concentration). The kidneys also release certain hormones.

Because the kidneys remove toxins that could otherwise damage body tissues, whatever supports the health of the kidneys supports the health of the whole body. A strong cardiovascular system and an abundant supply of water are important to keep blood flushing swiftly through the kidneys. In addition, the kidneys need sufficient energy to do their complex sifting and sorting job, and many vitamins and minerals serve as the cogs of their machinery. Exercise and nutrition are vital to healthy kidney function.

KEY POINT

- The kidneys adjust the blood's composition in response to the body's needs, disposing of everyday wastes and helping remove toxins.

Storage Systems

LO 3.8 Identify glycogen and fat as the two forms of nutrients stored in the body, and identify the liver, muscles, and adipose tissue as the body tissues that store them.

The human body is designed to eat at intervals of about four to six hours, but cells need nutrients around the clock. Providing the cells with a constant flow of the needed nutrients requires the cooperation of many body systems. These systems store and release nutrients to meet the cells' needs between meals. Among the major storage sites are the liver and muscles, which store carbohydrate, and the fat cells, which store fat and other fat-related substances.

When I Eat More Than My Body Needs, What Happens to the Extra Nutrients?

Nutrients collected from the digestive system sooner or later all move through a vast network of capillaries that weave among the liver cells. This arrangement ensures that liver cells have access to the newly arriving nutrients for processing.

Body tissues store excess energy-containing nutrients in two forms (details will follow in later chapters). The liver makes some of the excess into **glycogen** (a carbohydrate), and some is stored as body fat. Liver glycogen can sustain cell activities when the

nephrons (NEFF-rons) the working units in the kidneys, consisting of intermeshed blood vessels and tubules.

bladder the sac that holds urine until time for elimination.

glycogen a storage form of carbohydrate energy (glucose); described more fully in Chapter 4.

intervals between meals become long. Should no food be available, the liver's glycogen supply dwindles; it can be effectively depleted within as few as three to six hours. Muscle cells make and store glycogen, too, but selfishly reserve it for their own use.

Whereas the liver stores glycogen, it ships out fat in packages (see Chapter 5) to be picked up by cells that need it. All body cells may withdraw the fat they need from these packages, and the fat cells of the **adipose tissue** pick up the remainder and store it to meet long-term energy needs. Unlike the liver, fat tissue has virtually infinite storage capacity. It can continue to supply the body's cells with fat for days, weeks, or possibly even months when no food is eaten.

These storage systems for glucose and fat ensure that the body's cells will not go without energy even if the body is hungry for food. Body stores also exist for many other nutrients, each with a characteristic capacity. For example, liver and fat cells store many vitamins, and bones provide reserves of calcium and other minerals. Stores of nutrients are available to keep the blood levels constant and to meet cellular demands.

KEY POINT

- The body stores limited amounts of carbohydrate as glycogen in muscle and liver cells.

Variations in Nutrient Stores

Some nutrients are stored in the body in much larger quantities than others. For example, certain vitamins are stored without limit, even if they reach toxic levels within the body. Other nutrients are stored in only small amounts, regardless of the amount taken in, and these can readily be depleted. As you learn how the body handles various nutrients, pay particular attention to their storage so that you can know your tolerance limits. For example, you needn't eat fat at every meal because fat is stored abundantly. On the other hand, you normally do need to have a source of carbohydrate at intervals throughout the day because the liver stores less than one day's supply of glycogen.

KEY POINT

- The body stores large quantities of fat in fat cells.

Conclusion

In addition to the systems just described, the body has many more: bones, muscles, and reproductive organs, among others. All of these cooperate, enabling each cell to carry on its own life. For example, the skin and body linings defend other tissues against microbial invaders, while being nourished and cleansed by tissues specializing in these tasks. Each system needs a continuous supply of many specific nutrients to maintain itself and carry out its work. Calcium is particularly important for bones, for example; iron for muscles; glucose for the brain. But all systems need all nutrients, and every system is impaired by an undersupply or oversupply of them.

Whereas external events clamor and vie for attention, the body quietly continues its life-sustaining work. Most of the body's work is directed automatically by the unconscious portions of the brain and nervous system, and this work is finely regulated to achieve a state of well-being. But you need to involve your brain's cortex, your conscious thinking brain, to cultivate an understanding and appreciation of your body's needs. In doing so, attend to nutrition first. The rewards are liberating—ample energy to tackle life's tasks, a robust attitude, and the glowing appearance that comes from the best of health. Read on, and learn to let nutrition principles guide your food choices.

KEY POINT

- To nourish a body's systems, nutrients from outside must be supplied through a human being's conscious food choices.

adipose tissue the body's fat tissue, consisting of masses of fat-storing cells and blood vessels to nourish them.

what did you decide?

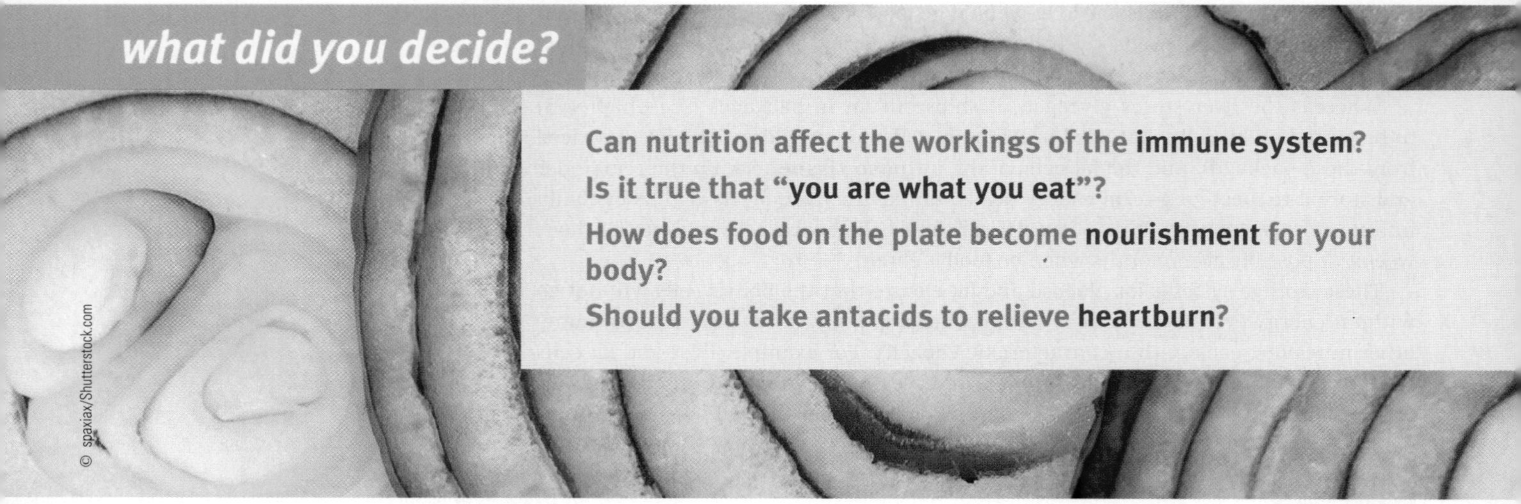

Can nutrition affect the workings of the immune system?

Is it true that "you are what you eat"?

How does food on the plate become nourishment for your body?

Should you take antacids to relieve heartburn?

Self Check

1. (LO 3.1) Cells
   a. are self-contained, living units.
   b. serve the body's needs but have few needs of their own.
   c. remain alive throughout a person's lifetime.
   d. b and c.
2. (LO 3.1) Each gene is a blueprint that directs the production of one or more of the body's organs.
   T F
3. (LO 3.2) After circulating around the cells of the tissues, all extracellular fluid then
   a. evaporates from the body
   b. becomes urine
   c. returns to the bloodstream
   d. a and b
4. (LO 3.2) Blood carries nutrients absorbed from food
   a. from the intestine to the liver
   b. from the lungs to the extremities
   c. from the kidneys to the liver
   d. Nutrients do not travel in blood.
5. (LO 3.3) Hormones
   a. are rarely involved in disease processes
   b. are chemical messengers that travel from one system of cells to affect another
   c. are produced and remain inside single cells for intracellular communications
   d. are unaffected by nutrition status of the body
6. (LO 3.3) The nervous system sends messages to the glands, telling them what to do.
   T F
7. (LO 3.4) T-cells are immune cells that "read" and "remember" chemical messages to identify future invaders.
   T F
8. (LO 3.4) White blood cells include all except the following:
   a. phagocytes
   b. killer T-cells
   c. B-cells
   d. antibodies
9. (LO 3.5) Chemical digestion of all nutrients mainly occurs in which organ?
   a. mouth
   b. stomach
   c. small intestine
   d. large intestine
10. (LO 3.5) Which of the following passes through the large intestine mostly unabsorbed?
    a. starch
    b. vitamins
    c. minerals
    d. fiber
11. (LO 3.5) Absorption of the majority of nutrients takes place across the mucus-coated lining of the stomach.
    T F

12. (LO 3.6) Which of the following increases the production of intestinal gas?
  a. chewing gum
  b. drinking carbonated beverages
  c. eating or drinking hastily
  d. all of the above

13. (LO 3.6) Concerning ulcers, all of the following are correct *except*:
  a. They usually occur in the large intestine.
  b. Some are caused by a bacterium.
  c. If not treated correctly, they can lead to stomach cancer.
  d. Their symptoms can be masked by using antacids regularly.

14. (LO 3.7) The kidneys' working units are
  a. photons
  b. genes
  c. nephrons
  d. villi

15. (LO 3.7) The bladder straddles the cardiovascular system and filters the blood.
  T F

16. (LO 3.8) The body's stores of __________ can sustain cellular activities when the intervals between meals become long.
  a. vitamins
  b. fat
  c. phytochemicals
  d. minerals

17. (LO 3.8) The body's adipose tissue's capacity to store fats is virtually infinite.
  T F

18. (LO 3.9) A drinker may delay intoxication somewhat by
  a. eating plenty of snacks
  b. quickly finishing drinks
  c. drinking on an empty stomach
  d. drinking undiluted drinks

19. (LO 3.9) Alcohol is a natural substance and therefore does no real damage to body tissues.
  T F

Answers to these Self Check questions are in Appendix G.

My Turn **watch it!**

I Am What I Drink

Ashley

Christopher

What do college students think about drinking alcohol? Two students talk about their drinking habits—how much, how often, and where.

CONTROVERSY 3

Alcohol and Nutrition: Do the Benefits Outweigh the Risks?

LO 3.9 Define the term *moderate alcohol consumption*, and discuss the potential health effects, both negative and positive, associated with this level of drinking.

Virtually everyone has heard media reports about positive associations between moderate alcohol consumption and a number of potential health benefits. Equally widely known, however, are alcohol's destructive effects. In the United States, alcohol-related deaths top 79,000 each year, making alcohol a substantial contributor to illness and mortality.[1]

How do researchers weight the significance of these opposing bodies of evidence?[2] Should nondrinkers take up drinking for their health's sake? Or should drinkers stop now to avoid problems? This Controversy presents evidence on both sides of the issue.

U.S. Alcohol Consumption

On a given day, adult drinkers consume about 16% of their total calorie intakes from alcoholic beverages, with men drinking more than women by far.[3] Each individual, however, usually follows a general drinking pattern: some people drink no alcohol at all, many take a glass of wine with meals, many others drink mainly at social functions, and still others take in large quantities of alcohol daily because of a life-shattering addiction.

Both heavy drinking and **heavy episodic drinking** (binge drinking) are common drinking patterns, particularly among college-age people, and pose serious health and social consequences to drinkers and nondrinkers alike.[4] Among U.S. adults, one in six is a binge drinker, a pattern accounting for more than half of the estimated 80,000 annual deaths attributable to alcohol consumption.[5] **Moderate drinkers**, in contrast, limit daily alcohol to one drink each day for women and two for men—no more—and therefore minimize their risks. Table C3–1 provides definitions concerning alcohol and drinking.

Does Moderate Alcohol Use Benefit Health?

Many, but not all, studies report a positive association between moderate drinking among middle-aged people (one drink a day for women, two for men) and reduced risk of heart attacks, strokes, and diabetes.[6] Indirect indicators of heart health, such as improved blood lipids and blood clotting factors, appear to improve with moderate alcohol intake.[7] In fact, moderate drinking often correlates with lower mortality from all causes.[8]

Some research also suggests that light or moderate drinking may be associated with mental acuity in aging, but other findings are inconsistent.[9] Higher alcohol intakes are not associated with benefits of any kind, however.[10]

The Influence of Age

The age of subjects often affects study results. The potential benefits of alcohol are generally observed among middle-aged people, a population segment with a high risk for developing diseases of the heart and arteries. Among teens and young adults, in contrast, the highest mortality rates are not from heart disease but from car crashes, homicides, and other violence, and drinking—even light drinking—raises these risks.[11] For them, any slight potential of alcohol to benefit heart health would be insignificant because the risk of developing heart disease before middle age is low.[12] In people older than middle age, the correlation between alcohol and heart health weakens.[13]

The Influence of Drinking Patterns

Curiously, in some countries, such as France, a light or moderate alcohol intake generally correlates with improved indicators of heart health and lower mortality. In other countries, though, no beneficial correlations can be found in people reporting moderate alcohol intakes.[14]

One explanation may be that drinking *patterns*, not just total intake, influence alcohol's effects on the body.[15] In France, light and moderate daily drinking dispersed throughout the week is the norm. In other countries, people may consume the same amount of alcohol but drink it in heavy episodic patterns. In surveys, drinkers who abstain on many days but then consume four or five drinks or more each weekend night may be counted among light drinkers, but they are, in fact, heavy episodic drinkers who may be increasing their risk of heart disease and mortality.[16]

Is Wine a Special Case?

Anyone you ask will probably tell you that red wine is good for health. U.S. wine labels often sport statements such as, "We encourage you to consult your family doctor about the health effects of wine consumption." Such statements seem to promise that good news about wine and health awaits the information seeker, but the science on wine and health is mixed. For example:

- The good news: the high potassium content and phytochemicals of grape juice may help to maintain normal

Table C3–1

Alcohol and Drinking Terms

- **acetaldehyde** (ass-et-AL-deh-hide) a substance to which ethanol is metabolized on its way to becoming harmless waste products that can be excreted.
- **alcohol dehydrogenase** (dee-high-DRAH-gen-ace) **(ADH)** an enzyme system that breaks down alcohol. The antidiuretic hormone listed below is also abbreviated ADH.
- **alcoholism** a dependency on alcohol marked by compulsive uncontrollable drinking with negative effects on physical health, family relationships, and social health.
- **antidiuretic** (AN-tee-dye-you-RET-ick) **hormone (ADH)** a hormone produced by the pituitary gland in response to dehydration (or a high sodium concentration in the blood). It stimulates the kidneys to reabsorb more water and so to excrete less. (This hormone should not be confused with the enzyme alcohol dehydrogenase, which is also abbreviated ADH.)
- **beer belly** central-body fatness associated with alcohol consumption.
- **cirrhosis** (seer-OH-sis) advanced liver disease, often associated with alcoholism, in which liver cells have died, hardened, turned an orange color, and permanently lost their function.
- **congeners** (CON-jen-ers) chemical substances other than alcohol that account for some of the physiological effects of alcoholic beverages, such as appetite, taste, and aftereffects.
- **drink** a dose of any alcoholic beverage that delivers half an ounce of pure ethanol.
- **ethanol** the alcohol of alcoholic beverages, produced by the action of microorganisms on the carbohydrates of grape juice or other carbohydrate-containing fluids.
- **euphoria** (you-FOR-ee-uh) an inflated sense of well-being and pleasure brought on by a moderate dose of alcohol and by some other drugs.
- **fatty liver** an early stage of liver deterioration seen in several diseases, including kwashiorkor and alcoholic liver disease, in which fat accumulates in the liver cells.
- **fibrosis** (fye-BROH-sis) an intermediate stage of alcoholic liver deterioration. Liver cells lose their function and assume the characteristics of connective tissue cells (become fibrous).
- **formaldehyde** a substance to which methanol is metabolized on the way to being converted to harmless waste products that can be excreted.
- **gout** (GOWT) a painful form of arthritis caused by the abnormal buildup of the waste product uric acid in the blood, with uric acid salt deposited as crystals in the joints.
- **heavy episodic drinking** a drinking pattern that includes occasional or regular consumption of four or more alcoholic beverages in a short time period. Also called *binge drinking*.
- **methanol** an alcohol produced in the body continually by all cells.
- **moderate drinkers** people who do not drink excessively and do not behave inappropriately because of alcohol. A moderate drinker's health may or may not be harmed by alcohol over the long term.
- **nonalcoholic** a term used on beverage labels, such as wine or beer, indicating that the product contains less than 0.5% alcohol. The terms *dealcoholized* and *alcohol removed* mean the same thing. *Alcohol free* means that the product contains no detectable alcohol.
- **problem drinkers** or **alcohol abusers** people who suffer social, emotional, family, job-related, or other problems because of alcohol. A problem drinker is on the way to alcoholism.
- **proof** a statement of the percentage of alcohol in an alcoholic beverage. Liquor that is 100 proof is 50% alcohol, 90 proof is 45%, and so forth.
- **Wernicke-Korsakoff** (VER-nik-ee KOR-sah-koff) **syndrome** a cluster of symptoms involving nerve damage arising from a deficiency of the vitamin thiamin in alcoholism. Characterized by mental confusion, disorientation, memory loss, jerky eye movements, and staggering gait.

blood pressure and reduce inflammation, and both persist when grape juice is made into wine.[17]

- The bad news: alcohol in large amounts raises blood pressure and increases inflammation, so grape juice itself may be better than wine for those with hypertension and heart disease.
- More good news: wine contains phytochemicals that could potentially reduce the risk of heart disease and certain digestive tract cancers.[18]
- More bad news: such phytochemicals are poorly absorbed, and alcohol raises the risk of many digestive tract cancers, sometimes in amounts of less than a drink per day.[19] Emergency room nurses describe a condition called "holiday heart syndrome," irregular heartbeats and other symptoms that follow intakes of alcohol.[20]

And so it goes.

To date, the research in support of health benefits of alcohol is indirect or observational, and other explanations cannot be entirely ruled out. The *Dietary Guidelines for Americans* recommend that no one should begin drinking or drink more frequently in hopes of benefitting their health.[21] Later sections describe increased risks that follow alcohol consumption, and the next section provides some basic facts about alcohol.

What Is Alcohol?

In chemistry, the term *alcohol* refers to a class of chemical compounds whose names end in *-ol*. The glycerol molecule of a triglyceride is an example. Alcohols affect living things profoundly, partly because they act as lipid solvents. Alcohols can easily penetrate a cell's outer lipid membrane and, once inside, denature the cell's protein structures and kill the cell. Because some alcohols kill microbial cells, they make useful disinfectants and antiseptics.

The alcohol of alcoholic beverages, **ethanol**, is somewhat less toxic than others. Sufficiently diluted and taken in moderation, its action in the brain produces **euphoria**, a pleasant sensation that people seek. (More alcohol *impedes* social interactions and diminishes feelings of euphoria, however.[22]) Used in this way, alcohol is a drug, and like many drugs, alcohol presents both benefits and hazards to the taker.

All beverages seem to ease conversation, whether they contain alcohol or not. For example, **nonalcoholic** beers and wines on the market also elevate mood and encourage social interaction, as do tea, coffee, or sodas. The *Dietary Guidelines for Americans* advises that people in many circumstances should not drink alcohol at all (Table C3–2).[23]

What Is a "Drink"?

Alcoholic beverages contain a great deal of water and some other substances, as well as the alcohol ethanol. In beer, wine, and wine coolers, alcohol contributes a relatively low percentage of the beverage's volume—about 5 percent in most beers to about 13 to 15 percent in many wines.* Malt beverages, even those with sugar and fruity flavors added, range from 5 to 10 percent ethanol. In contrast, about 50 percent of the volume of whiskey, vodka, rum, and brandy may be ethanol. The percentage of alcohol is stated as **proof**. Proof equals twice the percentage of alcohol; for example, 100-proof liquor is 50 percent alcohol.

A serving of an alcoholic beverage, commonly called a **drink**, delivers a little over 1/2 ounce of pure ethanol.[†24] Figure C3–1 depicts servings of alcoholic beverages that are considered to be one drink. These standard measures may have little in common with the drinks served by enthusiastic bartenders, however. Many wine glasses easily hold 6 to 8 ounces of wine; wine coolers may come packaged 12 ounces to a bottle; a large beer stein can hold 16, 20, or even more ounces; a strong liquor drink may contain 2 or 3 ounces of various liquors.

* *Nonalcoholic beers and wine may contain a small amount of alcohol, up to 0.5 percent.*

† *One drink contains 0.6 fluid ounces of alcohol.*

Table C3–2

Who Should Not Drink Alcohol?

People in these circumstances should not drink alcoholic beverages at all:

- *People of any age who cannot restrict their drinking to moderate levels.* Examples include people recovering from alcoholism, problem drinkers, and people whose family members have alcohol problems.
- *Anyone younger than the legal drinking age.* Besides being illegal, alcohol consumption increases the risk of drowning, car accidents, and traumatic injury, which are common causes of death in children and adolescents.
- *Women who are pregnant or who may be pregnant.* No safe level of alcohol consumption during pregnancy has been established. (A breastfeeding woman should use caution; if she chooses to drink, she may consume a single alcoholic beverage if she waits at least 4 hours afterward to breastfeed.)
- *People taking medications that can interact with alcohol.* Alcohol alters the effectiveness or toxicity of many medications, and some drugs may increase blood alcohol levels.
- *People with certain specific medical conditions.* Examples are liver disease, high blood lipids, or pancreatitis.
- *People who plan to drive, operate machinery, or take part in other activities that require attention, skill, or coordination or in situations where impaired judgment could cause injury or death.* Examples are swimming, biking, or boating.

Source: Adapted from Dietary Guidelines for Americans 2010, www.dietaryguidelines.gov.

Drinking Patterns

When people congregate to enjoy conversation and companionship, alcoholic beverages may be part of the scene. How alcohol affects the picture depends on how (and whether) it is consumed.

Defining Moderation

Moderation is not easily defined for an individual because tolerance to alcohol differs. In general, women cannot handle as much alcohol as men can, and women should never try to match drinks with men. Genetic makeup also affects tolerance: people of Asian or Native American descent often have lower-than-average tolerance to alcohol, for example.

Figure C3–1

Servings of Alcoholic Beverages That Equal One Drink[a]

[a]A standard drink is equal to 13.7 g (0.6 oz) of pure alcohol.

Health authorities define moderation as:

- No more than two drinks in any one day for the average-sized, healthy man
- No more than one drink in any one day for the average-sized, healthy woman

Doubtless some people can safely consume slightly more than this; others, especially those prone to alcohol addiction, cannot handle nearly so much without significant risk.

These are not average amounts, as noted earlier, but 24-hour maximums. In other words, a person who drinks no alcohol during the week but has seven drinks on Saturday night is not a moderate drinker. Instead, that drinking pattern characterizes heavy episodic drinking—binge drinking.

Problem Drinkers and Alcoholism

In contrast to moderate drinking, the effect of alcohol on problem drinkers or people with **alcoholism** is overwhelmingly negative. For these people, drinking alcohol brings irrational and often

Table C3–3

Symptoms of Problem Drinking and Alcoholism

A health professional can diagnose and evaluate problem drinking or alcohol addiction with the answers to these questions. In the past year, have you:

- Ever ended up drinking more or for longer than you intended?
- Wanted to cut down or stop drinking, or tried to, but couldn't on more than one occasion?
- Endangered yourself more than once while or after drinking (such as driving, swimming, using machinery, walking in a dangerous area, or having unsafe sex) on more than one occasion?
- Found that your usual number of drinks no longer produced the desired effect?
- Continued to drink even though it made you feel depressed or anxious? Or after having had a memory blackout?
- Spent a lot of time drinking, or being sick, or getting over other aftereffects?
- Continued to drink even though it was causing trouble with your family or friends?
- Found that drinking—or being sick from drinking—often interfered with taking care of your home or family? Or caused job troubles? Or school problems?
- Given up or cut back on activities that were important or interesting to you, or gave pleasure, in order to drink?
- Been arrested, been held at a police station, or had other legal problems because of your drinking?
- Found that when the effects of alcohol were wearing off, you had withdrawal symptoms, such as trouble sleeping, shakiness, restlessness, nausea, sweating, racing heartbeat, or seizure? Or sensed things that were not there?

If you have any of these symptoms, or if people close to you are concerned about your drinking, then alcohol may be a cause for concern. The more symptoms you have and the more often you have them, the more urgent the need for change. See a health professional.

Note: These questions are based on symptoms for alcohol use disorders in the American Psychiatric Association's Diagnostic and Statistical Manual (DSM) of Mental Disorders*, Fourth Edition. The DSM is the most commonly used system in the United States for diagnosing mental health disorders.*

Source: Adapted from **Rethinking Drinking: Alcohol and Your Health**, *NIH pub. no. 09-3770, February 2009.*

dangerous behavior, such as driving a car while intoxicated, and regrettable human interactions, such as arguments, violence, or unplanned and risky sexual activity.[25] With continued drinking, such people face psychological depression, physical illness, severe malnutrition, and demoralizing erosion of self-esteem. A tool for self-analysis for alcohol use disorders is found in Table C3–3. If you suspect that your own drinking may not be moderate or if alcohol has caused problems in your life, you should seek a professional evaluation.[‡]

Heavy Episodic Drinking

Young adults enjoy parties, sports events, and other social occasions, but these settings often encourage heavy episodic drinking (defined as at least four drinks in a row for women and five for men, but higher intakes are also reported).[§] Such "binge drinking" skews the statistics, making alcohol use on college campuses appear to be more common that it is in reality. The median number of drinks consumed by all college students is 1.5 per week, but for heavy episodic drinkers, it is 14.5 per week. This destructive drinking pattern accounts for 90 percent of the total alcohol consumed by people 21 years old and younger and is responsible for most of this group's alcohol-related problems.[26]

Harms from Heavy Episodic Drinking

Heavy episodic drinking poses serious health and social consequences for drinkers and nondrinkers alike. Compared with nondrinkers or moderate drinkers, such drinkers are more likely to damage property, to assault other people, to cause fatal automobile accidents, and to engage in risky, unprotected sexual intercourse, resulting in sexually transmitted diseases and unplanned pregnancies.[27] Female heavy episodic drinkers are more likely to be victims of rape.

Heavy episodic drinkers on and off campus may not recognize themselves as **problem drinkers** (refer to Table C3–4, p. 104) until their drinking behavior causes a crisis, such as a car crash, or until they are old enough to have caused substantial damage to their health. The World Health Organization has called for greater worldwide efforts to reduce the millions of annual deaths from heavy alcohol consumption.[28]

Immediate Effects of Alcohol

From the moment an alcoholic beverage is swallowed, the body gives it special attention. As alcohol passes through body tissues, it affects their functioning.

Alcohol Enters the Body

Unlike food, which requires digestion before it can be absorbed, tiny alcohol molecules start diffusing right through the stomach walls, and they reach the brain within a minute. Ethanol is a toxin, and a too-high dose in the stomach triggers one of the body's primary defenses against poison—vomiting. Many times, though, alcohol arrives gradually, diluted in enough fluid or food that the vomiting reflex is suppressed and the alcohol

‡ *The U.S. Center for Facts on Alcohol is the National Clearinghouse for Alcohol and Drug Information: (800) 729-6686.*

§ *This definition of heavy episodic drinking, without specification of time elapsed, is consistent with standard practice in alcohol research.*

Table C3–4

Behaviors Typical of Moderate Drinkers and Problem Drinkers

Moderate Drinkers Typically	Problem Drinkers Typically
▪ Drink slowly, casually.	▪ Gulp or "chug" drinks.
▪ Eat food while drinking or beforehand.	▪ Drink on an empty stomach.
▪ Don't binge drink; know when to stop.	▪ Binge drink; drink to get drunk.
▪ Respect nondrinkers.	▪ Pressure others to drink.
▪ Avoid drinking when solving problems or making decisions.	▪ Turn to alcohol when facing problems or decisions.
▪ Do not admire or encourage drunkenness.	▪ Consider drunks to be funny or admirable.
▪ Remain peaceful, calm, and unchanged by drinking.	▪ Become loud, angry, violent, or silent when drinking.
▪ Cause no problems to others or themselves by drinking.	▪ Physically or emotionally harm themselves, family members, or others when drinking.

passes into the small intestine, which readily absorbs it.

A drinker can soon become intoxicated, especially when drinking on an empty stomach. When the stomach is full of food, molecules of alcohol have less chance of touching the stomach walls and diffusing through, so alcohol reaches the brain more gradually. Also, a full stomach delays alcohol's flow into the small intestine, allowing time for a stomach enzyme to destroy some of it.[29] A person who wants to drink socially and not become intoxicated should eat the snacks provided by the host (but avoid the salty ones; they increase thirst and so may increase alcohol intake).

Alcohol Dehydrates the Tissues

Anyone who has had an alcoholic drink has experienced one of alcohol's physical effects: alcohol increases urine output because alcohol depresses the brain's production of the **antidiuretic hormone**. Loss of body water leads to thirst. The only fluid that relieves dehydration is water, so adding ice to alcoholic drinks to dilute them and alternating alcoholic beverages with nonalcoholic ones will quench thirst. Otherwise, each alcoholic drink may worsen the thirst, leading to more drinking.

The water lost due to hormone depression takes with it important minerals, such as magnesium, potassium, calcium, and zinc, depleting the body's reserves. These minerals are vital to fluid balance and to nerve and muscle coordination. When drinking results in mineral loss, minerals must be made up in subsequent meals to avoid deficiencies.

If a person drinks slowly enough, the liver will collect the alcohol after absorption and process it without much effect on other parts of the body. If a person drinks more rapidly, however, some of the alcohol bypasses the liver and flows for a while through the rest of the body and the brain.

Alcohol Arrives in the Brain

Some people use alcohol as a kind of social lubricant, to help them enjoy social occasions. Many people use alcohol as a sort of medication to help them to relax or become less anxious. However, alcohol can do the opposite and prolong tension and stress.[30] Stress can also reduce feelings of euphoria, an effect that can drive stressed drinkers to drink more to achieve the effect that they seek.

One drink relieves inhibitions, which gives people the impression that alcohol is a stimulant. In reality, alcohol acts as a depressant that sedates the *inhibitory* nerves, allowing excitatory nerves to take over. This effect is temporary, and when blood alcohol rises high enough, it sedates all of the nerve cells (see Figure C3–2).

A Lethal Dose of Alcohol

It is lucky that the brain centers respond to rising blood alcohol in the order shown in Figure C3–2, because a person usually passes out before drinking a lethal dose. If a person drinks fast enough, though, the alcohol continues to be absorbed, and both the blood alcohol level and its effects continue to accelerate after the person has gone to sleep. Every year, deaths attributed to this effect take place during drinking contests. Before passing out, the drinker drinks fast enough to receive a lethal dose. Figure C3–3 shows blood alcohol levels that correspond with progressively greater intoxication.

Alcohol Toxicity, Oxidative Stress, and the Brain

Brain cells are particularly sensitive to exposure to alcohol. The working brain tissue is made largely of lipid (fat) materials, and alcohol is a lipid solvent. With chronic alcohol exposure, brain cells die off and brain tissues shrink, with the extent of the shrinkage in proportion to the amount drunk. Alcohol addicts are prone to brain hemorrhages and strokes; postmortem examinations reveal brain cell loss and diminished functioning of the barrier that protects the brain from toxins.

These conditions may also be related to the oxidative stress that accompanies ethanol metabolism.[31] Free radicals that arise during ethanol metabolism attack brain cell components, causing inflammation. Then, the working brain cells become injured, die off, and disintegrate.

Abstinence from alcohol, together with good nutrition, reverses some of the

Figure C3–2

Effects of Rising Blood Alcohol Levels on the Brain

The higher the blood alcohol, the more severe its effect on brain tissues. This is a typical progression, but individual responses vary to some degree.

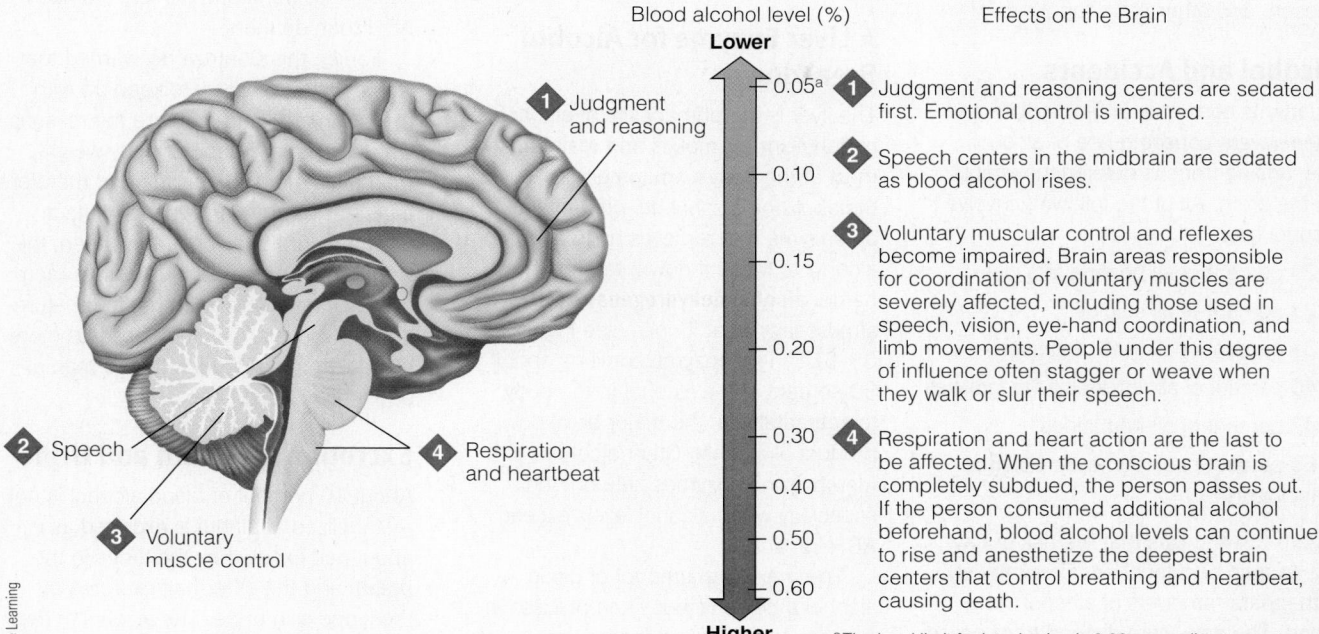

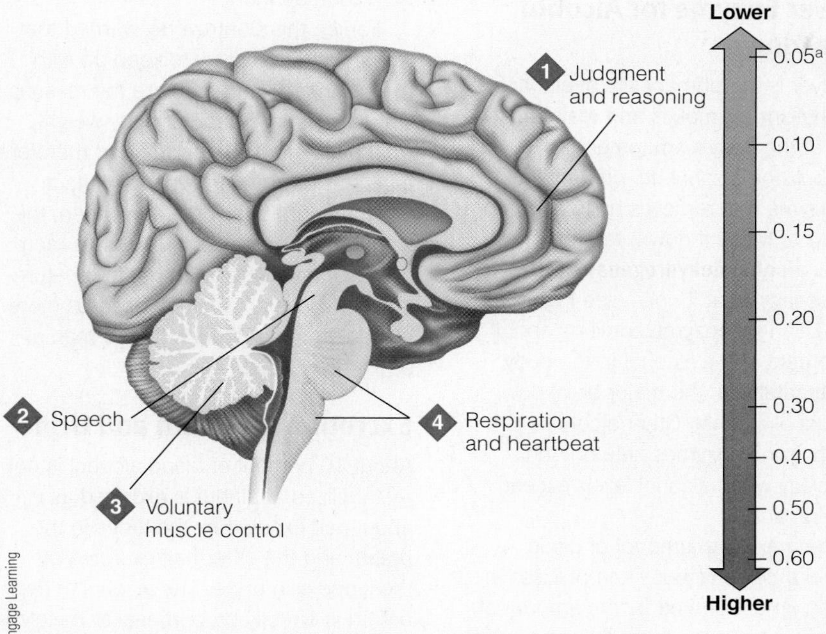

Effects on the Brain

1. Judgment and reasoning centers are sedated first. Emotional control is impaired.
2. Speech centers in the midbrain are sedated as blood alcohol rises.
3. Voluntary muscular control and reflexes become impaired. Brain areas responsible for coordination of voluntary muscles are severely affected, including those used in speech, vision, eye-hand coordination, and limb movements. People under this degree of influence often stagger or weave when they walk or slur their speech.
4. Respiration and heart action are the last to be affected. When the conscious brain is completely subdued, the person passes out. If the person consumed additional alcohol beforehand, blood alcohol levels can continue to rise and anesthetize the deepest brain centers that control breathing and heartbeat, causing death.

[a] *The legal limit for intoxication is 0.08, according to most highway safety ordinances. Driving ability may be impaired at blood alcohol levels below this amount.*

Figure C3–3

Alcohol Doses and Average Blood Level Percentages in Men and Women

Drinks[a]	Body Weight in Pounds—Men: 100	120	140	160	180	200	220	240	
	00	00	00	00	00	00	00	00	ONLY SAFE DRIVING LIMIT
1	.04	.03	.03	.02	.02	.02	.02	.02	IMPAIRMENT BEGINS
2	.08	.06	.05	.05	.04	.04	.03	.03	
3	.11	.09	.08	.07	.06	.06	.05	.05	DRIVING SKILLS SIGNIFICANTLY AFFECTED
4	.15	.12	.11	.09	.08	.08	.07	.06	
5	.19	.16	.13	.12	.11	.09	.09	.08	
6	.23	.19	.16	.14	.13	.11	.10	.09	
7	.26	.22	.19	.16	.15	.13	.12	.11	LEGALLY INTOXICATED
8	.30	.25	.21	.19	.17	.15	.14	.13	
9	.34	.28	.24	.21	.19	.17	.15	.14	
10	.38	.31	.27	.23	.21	.19	.17	.16	

Drinks[a]	Body Weight in Pounds—Women: 90	100	120	140	160	180	200	220	240	
	00	00	00	00	00	00	00	00	00	ONLY SAFE DRIVING LIMIT
1	.05	.05	.04	.03	.03	.03	.02	.02	.02	IMPAIRMENT BEGINS
2	.10	.09	.08	.07	.06	.05	.05	.04	.04	DRIVING SKILLS SIGNIFICANTLY AFFECTED
3	.15	.14	.11	.10	.09	.08	.07	.06	.06	
4	.20	.18	.15	.13	.11	.10	.09	.08	.08	
5	.25	.23	.19	.16	.14	.13	.11	.10	.09	
6	.30	.27	.23	.19	.17	.15	.14	.12	.11	LEGALLY INTOXICATED
7	.35	.32	.27	.23	.20	.18	.16	.14	.13	
8	.40	.36	.30	.26	.23	.20	.18	.17	.15	
9	.45	.41	.34	.29	.26	.23	.20	.19	.17	
10	.51	.45	.38	.32	.28	.25	.23	.21	.19	

NOTE: In some states, driving under the influence is proved when an adult's blood contains 0.08% alcohol, and in others, 0.10. Many states have adopted a "zero-tolerance" policy for drivers under age 21, using 0.02% as the limit.

[a] *Taken within an hour or so: each drink equivalent to 1/2 ounce pure ethanol.*

Source: National Clearinghouse for Alcohol and Drug Information brain damage from heavy drinking if it has not continued for more than a few years. Prolonged drinking beyond an individual's capacity to recover, however, can do severe and irreversible harm to vision, memory, learning, reasoning, speech, and other brain functions.

Alcohol and Accidents

Accidents constitute an immediate and often severe consequence of alcohol use, arising from its deleterious effects on the brain. All of the following involve alcohol use:

- 20 percent of all boating fatalities
- 23 percent of all suicides
- 39 percent of all traffic fatalities
- 40 percent of all residential fire fatalities
- 47 percent of all homicides
- 65 percent of all domestic violence incidents[32]

Figure C3–4 shows that the risk of having an auto accident rises precipitously with greater amounts of alcohol in the blood. The data were derived from police accident reports about people who were driving under the influence of alcohol.

Alcohol Arrives in the Body

The capillaries that surround the digestive tract merge into veins that carry the alcohol-laden blood to the liver. The routing of blood through the liver allows the cells to go right to work detoxifying alcohol and other ingested toxins before they reach other sensitive body organs such as the heart and brain.

A Liver Enzyme for Alcohol Breakdown

The liver is the primary site of alcohol metabolism—it makes and maintains most of the body's equipment for metabolizing alcohol. Its primary tool is an enzyme that removes hydrogens from alcohol to break it down; the enzyme's name, **alcohol dehydrogenase (ADH)**, almost says what it does (see Figure C3–5).** This enzyme converts about 80 percent of the alcohol in the body to **acetaldehyde**, the major breakdown product of alcohol. Other alcohol-metabolizing enzymes help out too, especially when alcohol levels exceed ADH capacity.

The maximum amount of blood alcohol a person's body can process in a given time is limited by the amount of ADH residing in the liver. If more alcohol arrives at the liver than the enzymes can handle, the extra alcohol circulates again and again through the brain, liver, and other organs until enzymes are available to degrade it.

***ADH exists in several variants.*

Alcohol Breakdown in the Stomach

The stomach wall also produces ADH that breaks down some alcohol before it reaches the bloodstream. Research shows that women make less stomach ADH than do men.

Earlier, this Controversy warned that women should not try to keep up with male drinkers, and here are the reasons why: pound for pound of body weight, men have more lean tissue and therefore a greater volume in which to dilute a given amount of alcohol. In women, the same amount of alcohol becomes more concentrated. Also, with her lower stomach ADH levels, a woman absorbs more alcohol from each drink than a man of equal body weight does.

Excretion in Breath and Urine

About 10 percent of blood alcohol is not metabolized at all but is excreted as is, about half exhaled by the lungs in the breath and the other half excreted by the kidneys in urine. The alcohol in the breath is directly proportional to the alcohol in the blood, so the breathalyzer test that law enforcement officers administer to someone suspected of driving while intoxicated accurately reveals the person's degree of intoxication.

Rate of Alcohol Clearance

The liver can process about $^1/_2$ ounce of blood ethanol (one drink's worth) per hour, depending on the person's body size, previous drinking experience, food intake, gender, and general health. Fasting for as little as one day causes degradation of body proteins, including ADH levels, and cuts the rate of alcohol metabolism by half.

The liver's maximum rate of alcohol clearance cannot be accelerated. This explains why only time restores sobriety. Walking doesn't help, because muscles cannot metabolize alcohol. Nor will drinking a cup of coffee be effective. Caffeine is a stimulant, but it won't speed up the metabolism of alcohol. The police say that a cup of coffee only makes a sleepy drunk into a wide-awake drunk. Table C3–5 presents other alcohol myths.

Figure C3–4

Blood Alcohol and Traffic Accidents

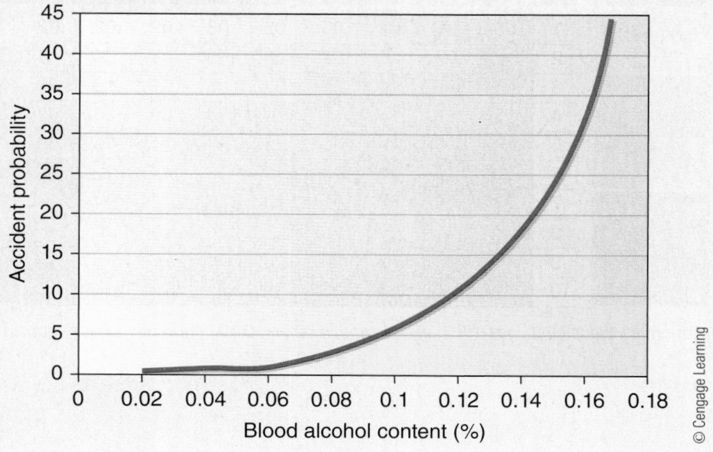

Figure C3–5

Alcohol Breakdown

The major route of alcohol breakdown produces acetaldehyde, creates free radicals, and increases oxidative stress in the tissues.

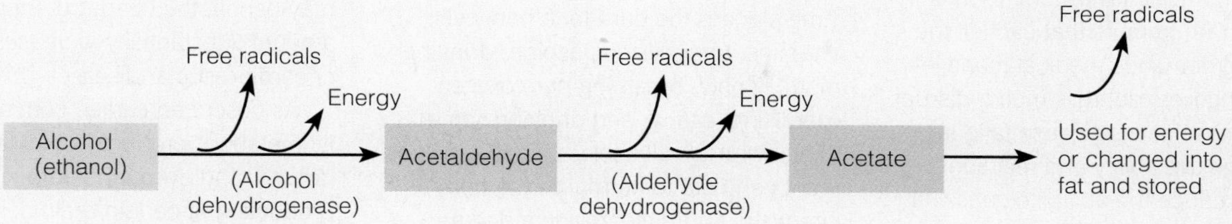

© Cengage Learning

Alcohol Affects the Liver

Among energy sources, ethanol receives the body's highest priority for breakdown. Toxic ethanol cannot be stored in body tissues without first being converted to something safer. Along the way, however, *other* harmful chemicals arise. For example, alcohol's first breakdown product, acetaldehyde, can bind to enzymes and other structures, disrupting their functions. Also, as Figure C3–5 showed, alcohol metabolism generates damaging free radicals and increase oxidative stress (introduced in Controversy 2), a condition linked with inflammation and the development of diabetes, cancer, and other serious diseases. Together, these factors are thought to contribute to the liver damage and other organ damage sustained from drinking ethanol.

Fatty Liver

When presented with alcohol, the liver speeds up its production of fats, which can build up in liver tissues. The first stage of liver deterioration seen in heavy drinkers is therefore known as **fatty liver**; the condition interferes with the distribution of nutrients and oxygen to the liver cells. Fat is known to accumulate in the livers of young men after a single night of heavy episodic drinking and to remain there for more than a day.

Liver Fibrosis and Cirrhosis

If heavy drinking continues for long enough, fibrous scar tissue invades the liver. This is the second stage of liver deterioration, called **fibrosis**. Fibrosis is reversible with good nutrition and abstinence from alcohol, but the next (last) stage, **cirrhosis**, is not. In cirrhosis, the liver cells harden, turn orange, and die, losing function forever. Cirrhosis develops after 10 to 20 years from the cumulative effects of frequent episodes of heavy drinking.

Arthur Glauberman/Science Source

Left, normal liver; center, fatty liver; right, cirrhosis

Table C3–5

Myths and Truths Concerning Alcohol

Myth:	A shot of alcohol warms you up.
Truth:	Alcohol diverts blood flow to the skin making you feel warmer, but it actually cools the body.
Myth:	Wine and beer are mild; they do not lead to addiction.
Truth:	Wine and beer drinkers worldwide have high rates of death from alcohol-related illnesses. It's not what you drink but how much that makes the difference.
Myth:	Mixing drinks is what gives you a hangover.
Truth:	Too much alcohol in any form produces a hangover.
Myth:	Alcohol is a stimulant.
Truth:	Alcohol depresses the brain's activity.
Myth:	Alcohol is legal; therefore, it is not a drug.
Truth:	Alcohol is legal, but it alters body functions and is medically defined as a depressant drug.

© Cengage Learning

The Hangover

The hangover—the awful feeling of headache, pain, unpleasant sensations in the mouth, and nausea the morning after drinking too much—is a mild form of drug withdrawal. (The worst form is a delirium with tremors that can kill the person and demands medical management.) Hangovers depress mood, disrupt sleep, increase anxiety, cause fatigue, reduce cognitive ability and reaction time, and reduce the ability to cope with stress.

Congeners and Dehydration

Hangovers are caused by several factors. One is the toxic effects of **congeners** that accompany the alcohol in alcoholic beverages. Mixing or switching drinks will not prevent hangover because congeners are only one factor. Dehydration of the brain is a second factor: alcohol reduces the water content of the brain cells. When they rehydrate the morning after and swell back to their normal size, nerve pain results.

Formaldehyde and Methanol

Another contributor to the hangover is **formaldehyde**, the smelly chemical laboratories use to preserve dead animals. Formaldehyde arises from **methanol**, another alcohol produced in tiny amounts by cellular metabolic processes. Occupational exposure to formaldehyde is known to raise the risk for certain cancers; no one knows whether formaldehyde generated from alcohol may do the same.[33]

Normally, a set of liver enzymes converts methanol to formaldehyde, with a second set immediately converting the formaldehyde to carbon dioxide and water, harmless waste products that can be excreted. But these same two sets of liver enzymes are the very ones that process ethanol to its own intermediate (and highly toxic) waste product, acetaldehyde, and finally to carbon dioxide and water. The enzymes prefer ethanol 20 times over methanol. Normally, both alcohols are metabolized without delay, but when excess acetaldehyde monopolizes the second set of enzymes, formaldehyde must wait for later detoxification. At that point, formaldehyde starts accumulating, and the hangover begins.

The Sure Cure for Hangover

Time alone is the cure for a hangover. Vitamins, tranquilizers, aspirin, drinking more alcohol, breathing pure oxygen, exercising, eating, and drinking something awful are all useless. Fluid replacement can help to normalize the body's chemistry and may provide a degree of relief. The headache, bad mood, nausea, and other effects of a hangover come simply from drinking too much alcohol. The best cure is prevention: drink less next time.

Alcohol's Long-Term Effects on the Body

A couple of drinks set in motion many destructive processes in the body. The next day's abstinence can reverse them only if the doses taken are moderate, the time between them is ample, and nutrition is adequate. If the doses of alcohol are heavy, however, and the time between them is short, complete recovery cannot take place, and repeated onslaughts of alcohol take a toll on the body.

Effects in Pregnancy

By far the longest-term effects of alcohol are those felt by the child of a woman who drinks during pregnancy. When a pregnant woman takes a drink, her fetus takes the same drink within minutes, and its body is defenseless against the effects. Pregnant women should not drink alcohol—this topic is so important that Chapter 13 devotes a section to it. The rest of this section concerns the effects on drinkers themselves.

Effects on Heart and Brain

Alcohol is directly toxic to skeletal and cardiac muscle, causing weakness and deterioration that is greater, the larger the dose. Alcoholism makes heart disease likely, probably because chronic alcohol use raises blood pressure. At autopsy, the heart of a person with alcoholism appears bloated and weighs twice as much as a normal heart. In middle-aged populations, taking one to two drinks a day (moderate drinking) may benefit the heart, but more than this amount substantially *increases* the risk of cardiovascular diseases.[34]

As described earlier, both alcohol and its metabolic products attack brain cells directly, and even moderate intakes slow production of certain brain cells. Heavy drinking can result in dementia. Mental impairments of alcoholism remain evident even between drinking bouts, but abstinence from alcohol often brings a degree of recovery.

Cancer

Experts include daily ethanol exposure among cancer-causing substances for human beings. Even moderate drinking increases the chances of developing cancers of the breast, colon and rectum, esophagus, liver, mouth, pancreas, prostate gland, stomach, throat, and, in smokers, lung.[††35] Once cancer is established, alcohol seems to speed up its development. Alcohol's byproducts, acetaldehyde and free radicals, may contribute to cancer risk, as well as ethanol itself.

A large body of evidence implicates alcohol in elevating the risk of breast cancer in women. Even one drink per day elevates the risk and, with greater consumption, the risk rises accordingly.[36] In men, moderate drinking increases the risks of cancers at many sites, risks that increase substantially with increasing daily alcohol consumption.[37] A popular myth holds that red wine is safer for women than other types—in reality, all colors of wine present identical breast cancer risks.[38]

Young women who drink may also increase their risk of developing breast cancer.[39] Even a single drink per day, the amount that might provide heart benefits to older people, may raise breast cancer risk in women by as much as 10 percent. More alcohol poses greater risks.

[††] *In 2002, 389,100 cases (3.6 percent) of cancer worldwide were attributable to drinking alcohol.*

Long-Term Effects of Alcohol Abuse

Some of the effects just mentioned may also affect people who drink moderately; however, the long-term effects of alcohol abuse and alcoholism can be devastating. They include the following:

- Bladder, kidney, pancreas, and prostate damage
- Bone deterioration and osteoporosis
- Brain disease, central nervous system damage, and stroke
- Deterioration of the testicles and adrenal glands
- Diabetes (type 2 diabetes)
- Disease of the muscles of the heart
- Feminization and sexual impotence in men
- Impaired immune response
- Impaired memory and balance
- Increased risks of death from all causes
- Major psychological depression, possibly caused by alcohol[40]
- Malnutrition
- Nonviral hepatitis[41]
- Skin rashes and sores
- Ulcers and inflammation of the stomach and intestines

This list is by no means all-inclusive. Alcohol abuse exerts direct toxic effects on all body organs. Monetarily, alcoholism costs our society an estimated $224 *billion* every year in medical services, lost wages, criminal offenses, auto crashes, and other losses.[42]

Alcohol's Effects on Nutrition

Alcohol causes disturbances in nutrition. Its calories are often overlooked by drinkers. Alcohol also causes direct negative effects on nutrients that the body needs to function.

Alcohol and Appetite

Alcoholic beverages affect the appetite. Usually, they reduce it, making people unaware that they are hungry.[43] But in people who are tense and unable to eat or in the elderly who have lost interest in food, a small dose of alcohol, such as a glass of wine, taken 20 minutes before meals may improve the appetite. Although beyond the scope of this discussion, alcohol affects neurotransmitters, hormones, and other signals in ways that modify food intake.

Another example of the beneficial use of alcohol comes from research showing that moderate use of wine in later life improves morale, stimulates social interaction, and promotes restful sleep. In nursing homes, improved patient and staff relations have been attributed to offering a moderate amount of wine or a cocktail to elderly patients who drink.

Alcohol and Body Weight

Alcohol's association with weight gain is complex.[44] Alcohol itself is caloric, and alcoholic beverages can be extremely high calorie. Ethanol yields 7 calories of energy per gram, and drink mixers often present many additional calories (examples are shown in Table C3–6). A small percentage of ethanol's calories escape from the body in breath and urine.

Metabolic interactions occur between fat and alcohol in the body. Presented with both fat and alcohol, the body stores the comparatively harmless fat and rids itself of the toxic alcohol by using it preferentially for energy.[45] Thus, alcohol is reported to promote overweight by increasing fat storage, particularly in the central abdominal area—the **"beer belly"** often seen in drinkers.[46]

Alcohol's Effects on Vitamins

Alcohol has direct toxic effects on body organs, and its abuse damages them indirectly via malnutrition. Like pure sugar and pure fat, alcohol provides empty calories. The more alcohol a person drinks, the less likely that he or she will eat

Table C3–6

Calories in Alcoholic Beverages and Mixers

Beverage	Amount (oz)	Energy (cal)
Malt beverage (sweetened)	16[a]	350
Malt beverage (unsweetened)	16	175
Wine cooler	12	170
Pina colada mix (no alcohol)	4	160
Beer	12	150
Dessert wine	3½	140
Fruit-flavored soda, Tom Collins mix	8	115
Gin, rum, vodka, whiskey (86 proof)	1½	105
Cola, root beer, tonic, ginger ale	8	100
Margarita mix (no alcohol)	4	100
Light beer	12	100
Table wine	3½	85
Tomato juice, Bloody Mary mix (no alcohol)	8	45
Club soda, plain seltzer, diet drinks	8	1

[a] *Typical container size, but up to 32-oz containers are common.*

enough food to obtain the needed nutrients. Simply put, the greater the alcohol intake, the less nutritious the diet.[47]

Alcohol abuse also disrupts every tissue's metabolism of nutrients. In the presence of alcohol, stomach cells oversecrete both acid and histamine, the latter an agent of the immune system that produces inflammation. Intestinal cells fail to absorb thiamin, folate, vitamin B_{12}, and other vitamins. Liver cells lose efficiency in activating vitamin D. Cells of the eye's retina, which normally process the alcohol form of vitamin A (retinol) to the form needed in vision (retinal), must process ethanol instead. Liver cells, too, suffer a reduced capacity to process and use vitamin A. The kidneys excrete needed minerals: magnesium, calcium, potassium, and zinc.

The inadequate food intake and impaired nutrient absorption of alcohol abuse frequently lead to a deficiency of the B vitamin thiamin. In fact, the cluster of thiamin-deficiency symptoms commonly seen in chronic alcoholism has its own name—the **Wernicke-Korsakoff syndrome**. This syndrome is characterized by paralysis of the eye muscles, poor muscle coordination, impaired memory, and damaged nerves. Thiamin supplements may help to repair some of the damage, especially if the person stops drinking.

Most dramatic is alcohol's effect on folate. When an excess of alcohol is present, the body actively expels folate from all of its sites of action and storage. The liver, which normally contains enough folate to meet all needs, leaks its folate into the blood. As blood folate rises, the kidneys excrete it, as if it were in excess. The intestine normally releases and retrieves folate continuously, but it becomes so damaged by folate deficiency and alcohol toxicity that it fails to absorb folate. Alcohol also interferes with the action of what little folate is left. This interference inhibits the production of new cells, especially the rapidly dividing cells of the intestine and the blood.

Nutrient deficiencies are thus an inevitable consequence of alcohol abuse, not only because alcohol displaces food but also because alcohol interferes directly with the body's use of nutrients. People treated for alcohol addiction also need nutrition therapy to reverse deficiencies and to treat deficiency diseases rarely seen in others: night blindness, beriberi, pellagra, scurvy, and acute malnutrition.

The Final Word

This discussion has explored some of the ways alcohol affects health and nutrition. In the end, each person must decide individually whether or not to consume alcohol, a decision that can change at any time.

As for drinking wine or other alcoholic beverages for health's sake, a leading scientist in the field concludes that while his research suggests that middle-aged people who drink moderately may gain some small benefits, far greater benefits come from engaging in regular physical activity and maintaining a healthy body weight.[48] He stated, "I would never recommend that nondrinkers start drinking with the thought of improving their health."‡‡ If you do choose to drink, do so with care and strictly in moderation.

‡‡ *Qi Sun of the Harvard University School of Public Health, as quoted in N. Bakalar, Aging: Health gains from a small drink a day,* New York Times, *September 19, 2011, available at www.nytimes.com*

Critical Thinking

1. Moderate alcohol use has been credited with providing possible health benefits. Construct an argument for why moderate alcohol use to provide protection from heart disease or other health problems may not be a good idea.
2. Your daughter is leaving for college in the fall. Recently, there has been disturbing news about the excessive drinking on college campuses and even a report about the death of one student who had been drinking excessively at the college your daughter is planning to attend. Form a group of four or five people. One group member will play the role of the daughter who is leaving for college. Each remaining member of the group will choose one of the topics listed below and prepare a short (one-minute) speech that attempts to educate your daughter on the dangers of excessive drinking. Be sure to emphasize facts as much as possible with your argument.
   - Explain the physiology of the hangover including congeners, dehydration, formaldehyde, and methanol.
   - Discuss the role of alcohol in weight gain.
   - Describe alcohol's effect on vitamins.
   - Describe the effect of alcohol on the heart and brain.
   - Describe alcohol's effect on the liver and other organs.

4 The Carbohydrates: Sugar, Starch, Glycogen, and Fiber

what do you think?

Do carbohydrates provide only **unneeded calories** to the body?

Why do nutrition authorities unanimously recommend **whole grains**?

Are **low-carbohydrate diets** the best way to lose weight?

Should people with **diabetes** eat sugar?

Learning Objectives

After completing this chapter, you should be able to accomplish the following:

LO 4.1 Compare and contrast the major types of carbohydrates in foods and in the body.

LO 4.2 Explain the important roles of carbohydrates and fiber in the body, and describe the characteristics of whole-grain foods.

LO 4.3 Explain how complex carbohydrates are broken down in the digestive tract and absorbed into the body.

LO 4.4 Describe how hormones control blood glucose concentrations during fasting and feasting, and explain the response of these hormones to various carbohydrates in the diet.

LO 4.5 Describe the scope of the U.S. diabetes problem, and educate someone about the long- and short-term effects of untreated diabetes and prediabetes.

LO 4.6 Identify components of a lifestyle plan to effectively control blood glucose, and describe the characteristics of an eating plan that can help manage type 2 diabetes.

LO 4.7 Describe the symptoms of hypoglycemia, and name some conditions that may cause it.

LO 4.8 Identify the main contributors of various forms of carbohydrates in foods.

LO 4.9 Discuss current research regarding the relationships among added sugars, obesity, diabetes, and other ills.

Carbohydrates are ideal nutrients to meet your body's energy needs, to feed your brain and nervous system, to keep your digestive system fit, and within calorie limits, to help keep your body lean. Digestible carbohydrates, together with fats and protein, add bulk to foods and provide energy and other benefits for the body. Indigestible carbohydrates, which include most of the fibers in foods, yield little or no energy but provide other important benefits.

All carbohydrates are not equal in terms of nutrition. This chapter invites you to learn the differences between foods containing **complex carbohydrates** (starch and fiber) and those made of **simple carbohydrates** (the sugars) and to consider the effects of both on the body. Controversy 4 goes on to explore current theories about how consumption of certain carbohydrates may affect human health.

This chapter on the carbohydrates is the first of three on the energy-yielding nutrients. Chapter 5 deals with the fats and Chapter 6 with protein. Controversy 3 in Chapter 3 already addressed one other contributor of energy to the human diet, alcohol.

A Close Look at Carbohydrates

LO 4.1 Compare and contrast the major types of carbohydrates in foods and in the body.

Carbohydrates contain the sun's radiant energy, captured in a form that living things can use to drive the processes of life. Green plants make carbohydrate through **photosynthesis** in the presence of **chlorophyll** and sunlight. In this process, water (H_2O) absorbed by the plant's roots donates hydrogen and oxygen. Carbon dioxide gas (CO_2) absorbed into its leaves donates carbon and oxygen. Water and carbon dioxide combine to yield the most common of the **sugars**, the single sugar **glucose**. Scientists know the reaction in the minutest detail but have yet to fully reproduce it—green plants are required to make it happen (see Figure 4–1).*[1]

Light energy from the sun drives the photosynthesis reaction. The light energy becomes the chemical energy of the bonds that hold six atoms of carbon together in the sugar glucose. Glucose provides energy for the work of all the cells of the stem, roots, flowers, and fruits of the plant. For example, in the roots, far from the energy-giving

carbohydrates compounds composed of single or multiple sugars. The name means "carbon and water," and a chemical shorthand for carbohydrate is CHO, signifying carbon (C), hydrogen (H), and oxygen (O).

complex carbohydrates long chains of sugar units arranged to form starch or fiber; also called *polysaccharides.*

simple carbohydrates sugars, including both single sugar units and linked pairs of sugar units. The basic sugar unit is a molecule containing six carbon atoms, together with oxygen and hydrogen atoms.

*Reference notes are found in Appendix F.

Figure 4–1

Animated! Carbohydrate—Mainly Glucose—Is Made by Photosynthesis

The sun's energy becomes part of the glucose molecule—its calories, in a sense. In the molecule of glucose on the leaf here, black dots represent the carbon atoms; bars represent the chemical bonds that contain energy.

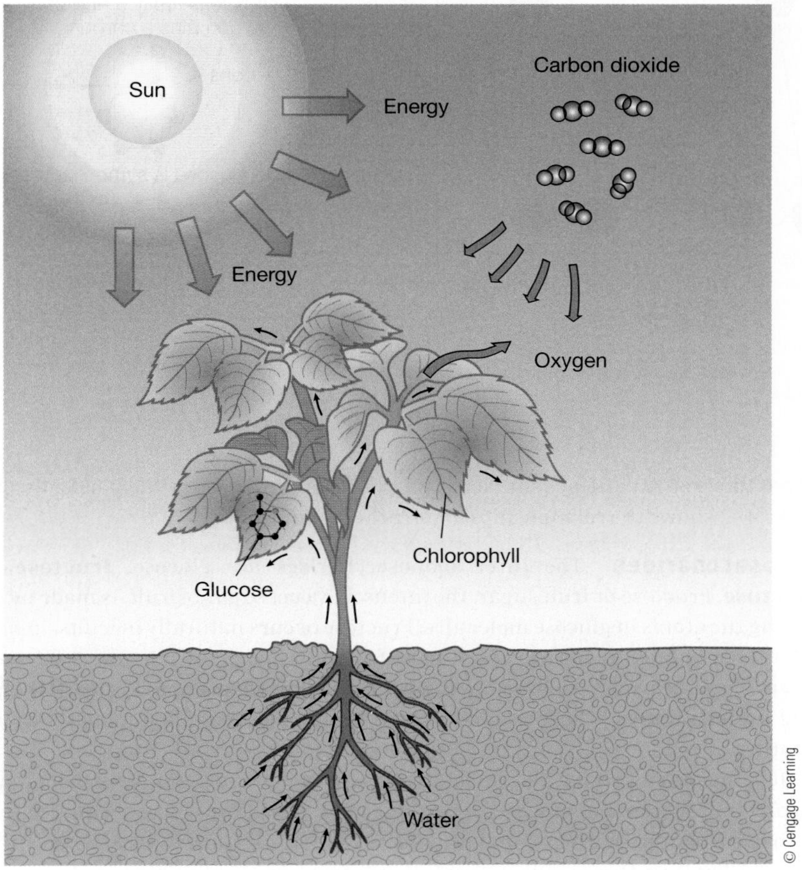

rays of the sun, each cell draws upon some of the glucose made in the leaves, breaks it down (to carbon dioxide and water), and uses the energy thus released to fuel its own growth and water-gathering activities.

Plants do not use all of the energy stored in their sugars, so it remains available for use by the animal or human being that consumes the plant. Thus, carbohydrates form the first link in the food chain that supports all life on earth. Carbohydrate-rich foods come almost exclusively from plants; milk is the only animal-derived food that contains significant amounts of carbohydrate. The next few sections describe the forms assumed by carbohydrates: sugars, starch, glycogen, and fibers.

KEY POINTS

- Through photosynthesis, plants combine carbon dioxide, water, and the sun's energy to form glucose.
- Carbohydrates are made of carbon, hydrogen, and oxygen held together by energy-containing bonds: carbo means "carbon"; hydrate means "water."

Sugars

Six sugar molecules are important in nutrition. Three of these are single sugars, or **monosaccharides**. The other three are double sugars, or **disaccharides**. All of their chemical names end in *ose*, which means "sugar." Although they all sound alike

photosynthesis the process by which green plants make carbohydrates from carbon dioxide and water using the green pigment chlorophyll to capture the sun's energy (*photo* means "light"; *synthesis* means "making").

chlorophyll the green pigment of plants that captures energy from sunlight for use in photosynthesis.

sugars simple carbohydrates; that is, molecules of either single sugar units or pairs of those sugar units bonded together. By common usage, *sugar* most often refers to sucrose.

glucose (GLOO-cose) a single sugar used in both plant and animal tissues for energy; sometimes known as blood sugar or *dextrose*.

monosaccharides (mon-oh-SACK-ah-rides) single sugar units (*mono* means "one"; *saccharide* means "sugar unit").

disaccharides pairs of single sugars linked together (*di* means "two").

Figure 4–2

How Monosaccharides Join to Form Disaccharides

Single sugars are monosaccharides while pairs of sugars are disaccharides.

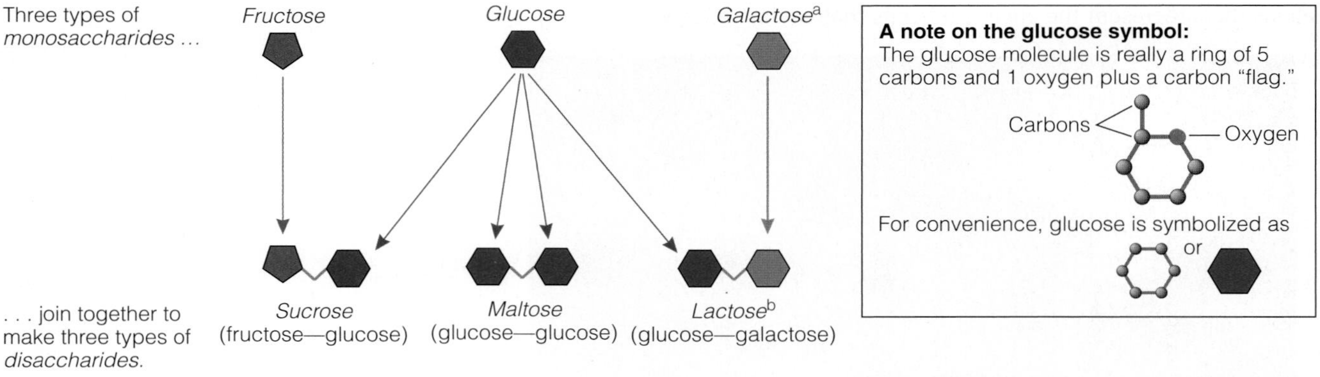

Three types of monosaccharides . . .
Fructose
Glucose
Galactose[a]
. . . join together to make three types of disaccharides.
Sucrose (fructose—glucose)
Maltose (glucose—glucose)
Lactose[b] (glucose—galactose)
A note on the glucose symbol:
The glucose molecule is really a ring of 5 carbons and 1 oxygen plus a carbon "flag."
Carbons
Oxygen
For convenience, glucose is symbolized as
or

[a]Galactose does not occur in foods singly but only as part of lactose.

[b]The chemical bond that joins the monosaccharides of lactose differs from those of other sugars and makes lactose hard for some people to digest—lactose intolerance (see later section, p. 130).

Did You Know?

- Single sugars are monosaccharides.
- Pairs of sugars bonded together are disaccharides.

at first, they exhibit distinct characteristics once you get to know them as individuals. Figure 4–2 shows the relationships among the sugars.

Monosaccharides The three monosaccharides are glucose, **fructose**, and **galactose**. Fructose or fruit sugar, the intensely sweet sugar of fruit, is made by rearranging the atoms in glucose molecules. Fructose occurs naturally in fruits, in honey, and as part of table sugar. However, most fructose is consumed in sweet beverages, desserts, and other foods sweetened with **high-fructose corn syrup (HFCS)** or other **added sugars**.[2] Glucose and fructose are the most common monosaccharides in nature.

The other monosaccharide, galactose, has the same number and kind of atoms as glucose and fructose but in another arrangement. Galactose is one of two single sugars that are bound together to make up the sugar of milk. Galactose rarely occurs free in nature but is tied up in milk sugar until it is freed during digestion.

Disaccharides The three other sugars important in nutrition are disaccharides, which are linked pairs of single sugars. The disaccharides are **lactose**, **maltose**, and **sucrose**. All three contain glucose. In lactose, the milk sugar just mentioned, glucose is linked to galactose. Malt sugar, or maltose, has two glucose units. Maltose appears wherever starch is being broken down. It occurs in germinating seeds and arises during the digestion of starch in the human body.

The last of the six sugars, sucrose, is familiar table sugar, the product most people think of when they refer to *sugar*. In sucrose, fructose and glucose are bonded together. Table sugar is obtained by refining the juice from sugar beets or sugar cane, but sucrose also occurs naturally in many vegetables and fruits. It tastes sweet because it contains the sweetest of the monosaccharides, fructose.

When you eat a food containing monosaccharides, you can absorb them directly into your blood. When you eat disaccharides, though, you must digest them first. Enzymes in your intestinal cells must split the disaccharides into separate monosaccharides so that they can enter the bloodstream. The blood delivers all products of digestion first to the liver, which possesses enzymes to modify nutrients, making them useful to the body. Glucose is the monosaccharide used for energy by all the body's tissues, so the liver releases abundant glucose into the bloodstream for delivery inside the body. Galactose can be converted into glucose by the liver, adding to the body's supply. Fructose, however, is normally used for fuel by the liver or broken down to building blocks for fat or other needed molecules.

fructose (FROOK-tose) a monosaccharide; sometimes known as fruit sugar (*fruct* means "fruit"; *ose* means "sugar").

galactose (ga-LACK-tose) a monosaccharide; part of the disaccharide lactose (milk sugar).

high-fructose corn syrup (HFCS) a widely used commercial caloric sweetener made by adding enzymes to cornstarch to convert a portion of its glucose molecules into sweet-tasting fructose.

added sugars sugars and syrups added to a food for any purpose, such as to add sweetness or bulk or to aid in browning (baked goods). Also called *carbohydrate sweeteners*, they include concentrated fruit juice, glucose, fructose, high-fructose corn syrup, sucrose, and other sweet carbohydrates. Also defined in Chapter 2.

lactose a disaccharide composed of glucose and galactose; sometimes known as milk sugar (*lact* means "milk"; *ose* means "sugar").

Although it is true that the energy of fruits and many vegetables comes from sugars, this doesn't mean that eating them is the same as eating concentrated sweets such as candy or drinking cola beverages. From the body's point of view, fruits are vastly different from purified sugars (as later sections make clear) except that both provide glucose in abundance.

Did You Know?
Strands of many sugars are polysaccharides.

KEY POINTS

- Glucose is the most important monosaccharide in the human body.
- Monosaccharides can be converted by the liver to other needed molecules.

Starch

In addition to occurring in sugars, the glucose in food also occurs in long strands of thousands of glucose units. These are the **polysaccharides** (see Figure 4–3, p. 116). **Starch** is a polysaccharide, as are glycogen and most of the fibers.

Starch is a plant's storage form of glucose. As a plant matures, it not only provides energy for its own needs but also stores energy in its seeds for the next generation. For example, after a corn plant reaches its full growth and has many leaves manufacturing glucose, it links glucose together to form starch, stores packed clusters of starch molecules in **granules**, and packs the granules into its seeds. These giant starch clusters are packed side by side in the kernels of corn. For the plant, starch is useful because it is an insoluble substance that will stay with the seed in the ground and nourish it until it forms shoots with leaves that can catch the sun's rays. Glucose, in contrast, is soluble in water and would be washed away by the rains while the seed lay in the soil. The starch of corn and other plant foods is nutritive for people, too, because they can digest the starch to glucose and extract the sun's energy stored in its chemical bonds. A later section describes starch digestion in detail.

KEY POINT

- Starch is the storage form of glucose in plants and is also nutritive for human beings.

Glycogen

Just as plant tissues store glucose in long chains of starch, animal bodies store glucose in long chains of **glycogen**. Glycogen resembles starch in that it consists of glucose molecules linked together to form chains, but its chains are longer and more highly branched (see Figure 4–3). Unlike starch, which is abundant in grains, potatoes, and other foods from plants, glycogen is nearly undetectable in meats because glycogen breaks down rapidly when the animal is slaughtered. A later section describes how the human body handles its own packages of stored glucose.

KEY POINT

- Glycogen is the storage form of glucose in animals and human beings.

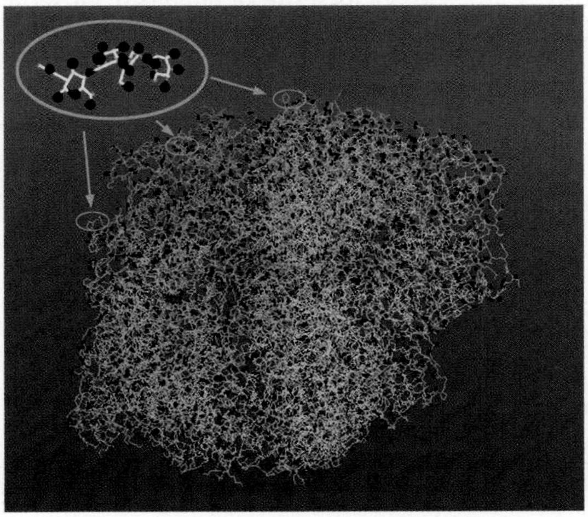

Jon Lomberg/Science Source

A glycogen molecule stores tens of thousands of glucose units nested in an easy-to-retrieve form. In this photo, individual glucose molecules are depicted as black balls linked together with white sticks.

maltose a disaccharide composed of two glucose units; sometimes known as malt sugar.

sucrose (SOO-crose) a disaccharide composed of glucose and fructose; sometimes known as table, beet, or cane sugar and, often, as simply *sugar*.

polysaccharides another term for complex carbohydrates; compounds composed of long strands of glucose units linked together (*poly* means "many"). Also called *complex carbohydrates*.

starch a plant polysaccharide composed of glucose. After cooking, starch is highly digestible by human beings; raw starch often resists digestion.

granules small grains. Starch granules are packages of starch molecules. Various plant species make starch granules of varying shapes.

glycogen (GLY-co-gen) a highly branched polysaccharide that is made and stored by liver and muscle tissues of human beings and animals as a storage form of glucose. Glycogen is not a significant food source of carbohydrate and is not counted as one of the complex carbohydrates in foods.

Figure 4–3

Animated! How Glucose Molecules Join to Form Polysaccharides

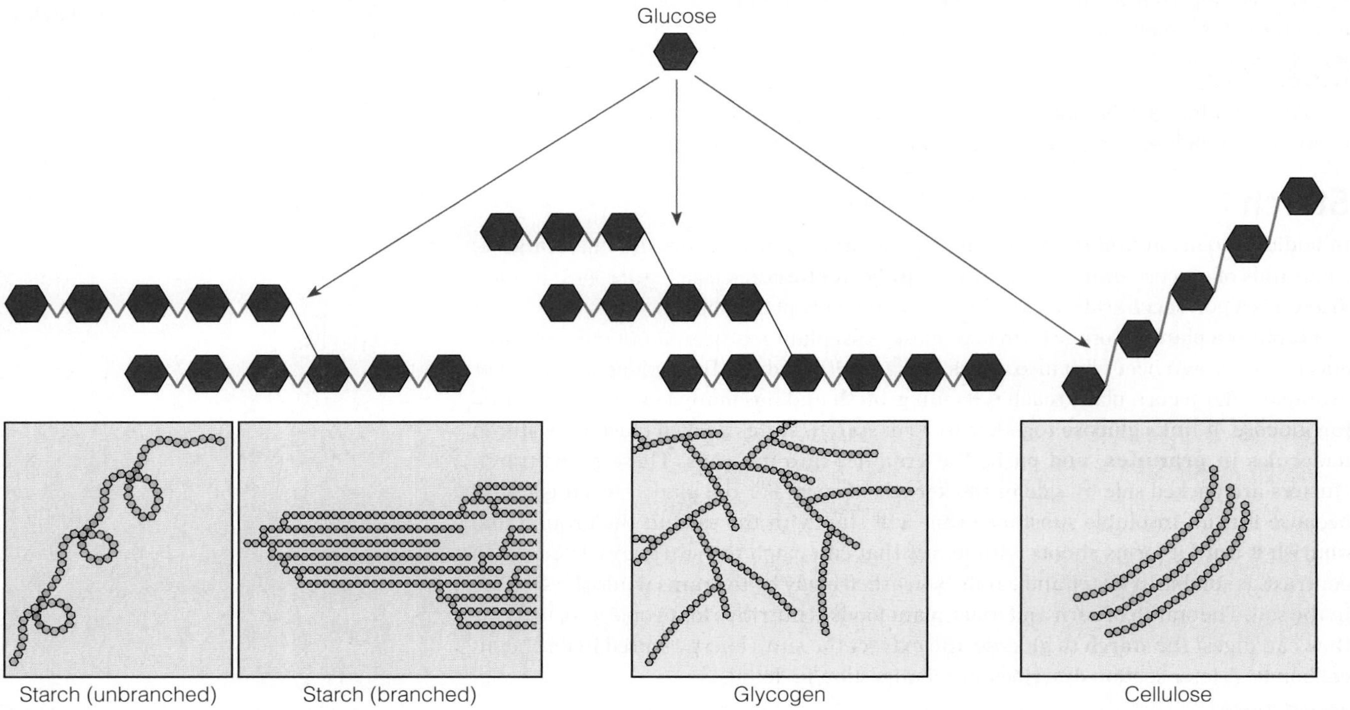

Starch Glucose units are linked in long, occasionally branched chains to make starch. Human digestive enzymes can digest these bonds, retrieving glucose. Real glucose units are so tiny that you can't see them, even with the highest-power light microscope.

Glycogen Glycogen resembles starch in that the bonds between its glucose units can be broken by human enzymes, but the chains of glycogen are more highly branched.

Cellulose (fiber) The bonds that link glucose units together in cellulose are different from the bonds in starch or glycogen. Human enzymes cannot digest them.

Fibers

Some of the **fibers** of a plant form the supporting structures of its leaves, stems, and seeds. Other fibers play other roles, for example, to retain water and thus protect seeds from drying out. Like starch, most fibers are polysaccharides—chains of sugars—but they differ from starch in that the sugar units are held together by bonds that human digestive enzymes cannot break. Most fibers therefore pass through the human body without providing energy for its use. A little energy arises from billions of bacteria residing within the human large intestine (colon) that do possess fiber-digesting enzymes. The bacterial **fermentation** of fiber releases tiny products, mainly fat fragments, which the human colon absorbs. Many animals, such as cattle, depend heavily on their digestive system's bacteria to make the energy of glucose available from the abundant cellulose, a form of fiber, in their fodder. Thus, when we eat beef, we indirectly receive some of the sun's energy that was originally stored in the fiber of the plants. Beef itself contains no fiber, nor do other meats and dairy products.

In summary, plants combine carbon dioxide, water, and the sun's energy to form glucose, which can be stored as the polysaccharide starch. Then animals or people eat the plants and retrieve the glucose. In the body, the liver and muscles may store the glucose as the polysaccharide glycogen, but ultimately it becomes glucose again. The glucose delivers the sun's energy to fuel the body's activities. In the process, glucose breaks down to the waste products carbon dioxide and water, which are excreted. Later, plants use these compounds again as raw materials to make carbohydrate. Fibers are plant constituents that are not digested directly by human enzymes, but intestinal bacteria ferment some fibers, and dietary fiber contributes to the health of the body.

KEY POINTS

- Human digestive enzymes cannot break the chemical bonds of fiber.
- Some fiber is susceptible to fermentation by bacteria in the colon.

The Need for Carbohydrates

LO 4.2 Explain the important roles of carbohydrates and fiber in the body, and describe the characteristics of whole-grain foods.

Glucose from carbohydrate is an important fuel for most body functions. Only two other nutrients provide energy to the body: protein and fats.† Protein-rich foods are usually expensive and, when used to make fuel for the body, provide no advantage over carbohydrates. Moreover, excess dietary protein has disadvantages, as Chapter 6 explains. Fats normally are not used as fuel by the brain and central nervous system. Thus, glucose is a critical energy source, particularly for nerve cells, including those of the brain, and the red cells of the blood. And starchy whole foods that supply complex carbohydrates—and especially the fiber-rich ones—are the preferred source of glucose in the diet.

Sugars also play vital roles in the functioning of body tissues. For example, sugars that dangle from protein molecules, once thought to be mere hitchhikers, are now known to dramatically alter the shape and function of certain proteins. Such a sugar-protein complex is responsible for the slipperiness of mucus, the watery lubricant that coats and protects the body's internal linings and membranes.‡ Sugars also bind to the outsides of cell membranes, affecting cell-to-cell communication, nerve and brain cell function, and certain disease processes. Clearly, the body needs carbohydrates for more than just energy.

If I Want to Lose Weight and Stay Healthy, Should I Avoid Carbohydrates?

Carbohydrates have been wrongly accused of being the "fattening" ingredient of foods, thereby misleading millions of weight-conscious people into eliminating nutritious carbohydrate-rich foods from their diets. In truth, people who wish to lose fat, maintain lean tissue, and stay healthy can do no better than to attend closely to portion sizes and calorie intakes and to design an eating plan around carbohydrate-rich **whole grains**, vegetables, and fruits that supply fiber, other needed nutrients, and beneficial phytochemicals.[3]

Lower in Calories Gram for gram, carbohydrates donate fewer calories than do dietary fats, and converting glucose into fat for storage is metabolically costly. Still, it is possible to consume enough calories of carbohydrate to exceed the need for energy, which reliably leads to weight *gain*. To lose weight, the dieter must plan to consume fewer total calories from all foods and beverages each day.

An Exception: Added Sugars Recommendations to choose carbohydrate-rich foods do not extend to refined added sugars. Purified, refined sugars (mostly sucrose or fructose) contain no other nutrients—no protein, vitamins, minerals, or fiber—and thus are low in nutrient density. A person choosing 400 calories of sugar in place of 400 calories of whole-grain bread loses the nutrients, phytochemicals, and fiber of the bread. You can afford to do this only if you have already met all of your nutrient needs for the day and still have calories to spend.

†Ethanol, the alcohol in alcoholic beverages, also supplies calories, but alcohol is toxic to body tissues.
‡Such combination molecules are known as *glycoproteins*.

Did You Know?

- 1 g carbohydrates = 4 cal
- 1 g fat = 9 cal

fibers the indigestible parts of plant foods, largely nonstarch polysaccharides that are not digested by human digestive enzymes, although some are digested by resident bacteria of the colon. Fibers include cellulose, hemicelluloses, pectins, gums, mucilages, and a few nonpolysaccharides such as lignin.

fermentation the anaerobic (without oxygen) breakdown of carbohydrates by microorganisms that releases small organic compounds along with carbon dioxide and energy.

whole grains grains or foods made from them that contain all the essential parts and naturally occurring nutrients of the entire grain seed (except the inedible husk).

Overuse of added sugars may have other effects as well. Some evidence suggests that, for many obese people, a diet too high in added sugars and other refined carbohydrates may alter blood lipids in ways that may worsen their heart disease risk (Controversy 4 comes back to this topic).[4]

Guidelines For health's sake, then, most people should increase their intakes of fiber-rich whole-food sources of carbohydrates and reduce intakes of foods high in refined white flour and added sugars.[5] Table 4–1 presents carbohydrate recommendations and guidelines from several authorities. This chapter's Consumer's Guide describes various whole-grain foods, and the Food Feature comes back to the sugars in foods. For weight loss, authorities do not recommend omitting carbohydrates. In fact, the opposite is true.

KEY POINTS

- The body tissues use carbohydrates for energy and other functions.
- The brain and nerve tissues prefer carbohydrate as fuel, and red blood cells can use nothing else.
- Intakes of refined carbohydrates should be limited.

Unlike the added sugars in concentrated sweets, the sugars in fruit are diluted with water and naturally packaged with vitamins, minerals, phytochemicals, and fiber.

Table 4–1

Recommendations for Carbohydrate Intakes

1. Total carbohydrate

Dietary Guidelines for Americans 2010
- Consume between 45 and 65% of calories from carbohydrate.

Dietary Reference Intakes (DRI)
- At a minimum, 130 g/day for adults and children to provide glucose to the brain.
- For health, most people should consume between 45 and 65% of total calories from carbohydrate.

USDA Food Patterns
- Grains, fruit, starchy vegetables, and milk contribute to the day's total carbohydrate intake.

2. Added sugars

Dietary Guidelines for Americans 2010
- Reduce intake of calories from added sugars. Limit consumption of foods and beverages that contain added sugars.

USDA Food Patterns
- Added sugars may provide calories within the energy recommendation after meeting all nutrient recommendations with nutritious foods.

The American Heart Association
- A prudent upper limit of not more than 100 cal of added sugars for most women or 150 cal for most men.

3. Whole grains

Dietary Guidelines for Americans 2010
- Reduce intake of refined grains and replace some refined grains with whole grains.
- Increase intake of whole grains.

4. Fiber

USDA Food Patterns
- Increase intakes of whole fruits and vegetables, make at least half the grain choices whole grains, and choose legumes several times per week.

Dietary Reference Intakes (DRI)
- 38 g of total fiber per day for men through age 50; 30 g for men 51 and older.
- 25 g of total fiber per day for women through age 50; 21 g for women 51 and older

Why Do Nutrition Experts Recommend Fiber-Rich Foods?

As mentioned, many carbohydrate-rich foods offer additional benefits. Foods such as whole grains, vegetables, legumes, and fruits supply valuable vitamins, minerals, and phytochemicals, along with fiber and little or no fat. Fiber's best-known health benefits include:

1. Promotion of normal blood cholesterol concentrations and reduced risk of heart and artery disease.
2. Modulation of blood glucose concentrations (reduced risk of diabetes).
3. Maintenance of healthy bowel function (reduced risk of bowel diseases).
4. Promotion of a healthy body weight.[6]

Purified fibers often fail to demonstrate these benefits, however. The obvious choice for anyone placing a value on health is to obtain fibers from a variety of whole foods each day, and not to rely on purified fiber in supplements or added to highly refined foods.

Did You Know?

According to the *Dietary Guidelines for Americans 2010*, fiber is a nutrient of national concern.

Soluble Fibers Researchers often divide fibers into two general groups by their chemical, physical, and functional properties. First are the fibers that dissolve in water, known as **soluble fibers**. These form gels (are **viscous**). Human enzymes cannot digest soluble fibers, but bacteria in the human colon readily ferment them. Commonly found in oats, barley, legumes, okra, and citrus fruits, soluble fibers often lower blood cholesterol and can help to control blood glucose, actions that could improve the odds against heart disease and diabetes.[7] In foods, soluble fibers add pleasing consistency, such as the pectin that puts the gel in jelly and the gums that add thickness to salad dressings.

KEY POINT

- Soluble fibers dissolve in water, form viscous gels, and are easily fermented by colonic bacteria.

Insoluble Fibers Other fibers are **insoluble fibers** that do not dissolve in water, do not form gels (are not viscous), and are less readily fermented. Insoluble fibers, such as cellulose, form structures such as the outer layers of whole grains (bran), the strings of celery, the hulls of seeds, and the skins of corn kernels. These fibers retain their shape and rough texture even after hours of cooking. In the body, they aid the digestive system by easing elimination.

Figure 4–4 (p. 120) shows the diverse effects of different fibers, and Figure 4–5 (p. 121) provides a brief guide to finding these fibers in foods (Appendix A lists the fiber contents of thousands of foods). Most unrefined plant foods contain a mix of fiber types. The following paragraphs describe health benefits associated with daily intakes of fiber-rich foods.

KEY POINT

- Insoluble fibers do not dissolve in water, form structural parts of plants, and are less readily fermented by colonic bacteria.

Lower Cholesterol and Heart Disease Risk Diets rich in legumes, vegetables, and whole grains—and therefore rich in fiber and other complex carbohydrates—may protect against heart disease and stroke.[8] Such diets are also generally low in saturated fat, *trans* fat, and cholesterol and high in vegetable proteins and phytochemicals—all factors associated with a lower risk of heart disease. Oatmeal was first to be identified among cholesterol-lowering foods; apples, barley, carrots, and legumes are also rich in the viscous fibers having a significant cholesterol-lowering effect.[9] In contrast, diets high in refined grains and added sugars may push blood lipids toward elevated heart disease risk.

Foods rich in viscous fibers may lower blood cholesterol by binding with cholesterol-containing compounds in bile. Normally, much of this cholesterol would be

soluble fibers food components that readily dissolve in water and often impart gummy or gel-like characteristics to foods. An example is pectin from fruit, which is used to thicken jellies.

viscous (VISS-cuss) having a sticky, gummy, or gel-like consistency that flows relatively slowly.

insoluble fibers the tough, fibrous structures of fruits, vegetables, and grains; indigestible food components that do not dissolve in water.

Figure 4–4

Characteristics, Sources, and Health Effects of Fibers

People who eat these foods . . .		obtain these types of fibers . . .	with these actions in the body . . .	and these probable health benefits . . .
		Viscous, soluble, more fermentable		
 Stockbyte/Getty Images	• Barley, oats, oat bran, rye, fruits (apples, citrus), legumes (especially young green peas and black-eyed peas), seaweeds, seeds and husks, many vegetables, fibers used as food additives	• Gums • Pectins • Psyllium[a] • Some hemicellulose	• Lower blood cholesterol by binding bile • Slow glucose absorption • Slow transit of food through upper GI tract • Hold moisture in stools, softening them • Yield small fat molecules after fermentation that the colon can use for energy • Increase satiety	• Lower risk of heart disease • Lower risk of diabetes • Lower risk of colon and rectal cancer • Increased satiety, and may help with weight management
		Nonviscous, insoluble, less fermentable		
 Brian Leatart/Getty Images	• Brown rice, fruits, legumes, seeds, vegetables (cabbage, carrots, brussels sprouts), wheat bran, whole grains, extracted fibers used as food additives	• Cellulose • Lignans • Resistant starch • Hemicellulose	• Increase fecal weight and speed fecal passage through colon • Provide bulk and feelings of fullness	• Alleviate constipation • Lower risk of diverticulosis, hemorrhoids, and appendicitis • Lower risk of colon and rectal cancer

[a]Psyllium, a soluble fiber derived from seeds, is used as a laxative and food additive.

Art © Cengage Learning

reabsorbed from the intestine for reuse, but viscous fiber carries some of it out with the feces (Figure 4–6, p. 122).[10] These bile compounds are needed in digestion, so the liver responds to their loss by drawing on the body's cholesterol stocks to synthesize more. Another way in which dietary fiber may reduce cholesterol in the blood is through the actions of one of the small fatty acids released during bacterial fermentation of fiber. This fatty acid is absorbed and travels to the liver, where it may help to reduce cholesterol synthesis. The net result of either mechanism is lowered blood cholesterol.

KEY POINT

- Foods rich in viscous soluble fibers help control blood cholesterol.

Blood Glucose Control High-fiber foods may play a role in reducing the risk of type 2 diabetes.[11] The soluble fibers of foods such as oats and legumes help regulate blood glucose following a carbohydrate-rich meal.[12] Soluble fibers delay the transit of nutrients through the digestive tract, slowing glucose absorption and preventing the glucose surge and rebound often associated with diabetes onset. In people with established diabetes, high-fiber foods can modulate blood glucose and insulin levels, thus helping to prevent medical complications. A later section comes back to insulin in diabetes.

KEY POINT

- Foods rich in viscous fibers help to modulate blood glucose concentrations.

Maintenance of Digestive Tract Health Soluble and insoluble fibers, along with an ample fluid intake, probably play roles in maintaining proper colon function. Soluble fibers help to maintain normal colonic bacteria necessary for intestinal health.[13] Insoluble fibers such as cellulose (as in wheat bran and other cereal brans, fruits, and vegetables) enlarge and soften the stools, easing their passage out of the body and speeding up their transit time through the intestine. Thus, foods rich in these fibers help to alleviate or prevent **constipation**.

Figure 4–5

Fiber Composition of Common Foods

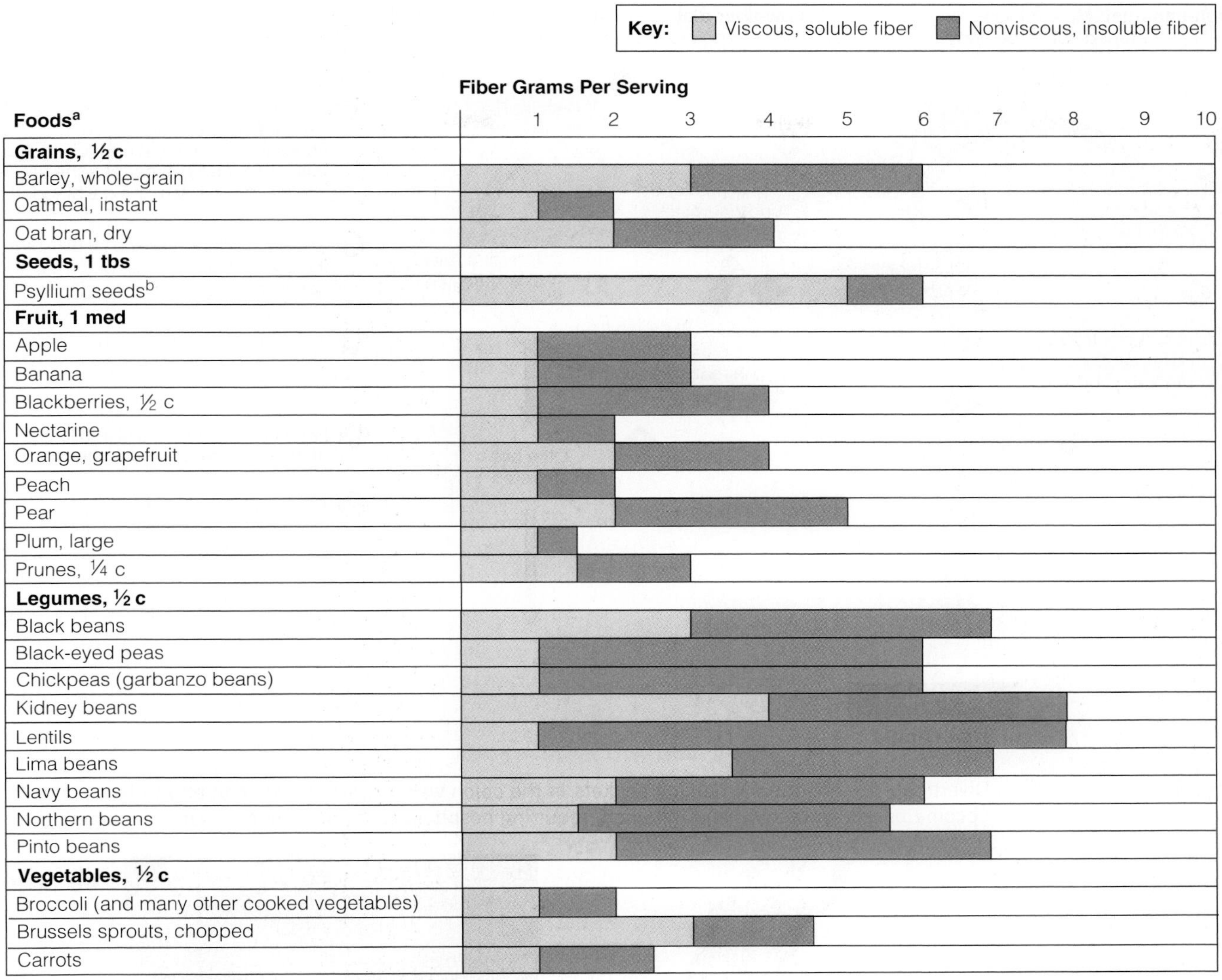

[a]*Values are for cooked or ready-to-serve foods unless specified.*
[b]*Psyllium is used as a fiber laxative and fiber-rich food additive.*

Source: Data from the National Heart, Lung and Blood Institute. Third Report of the National Cholesterol Education Program (NCEP) Expert Panel on Detection, Evaluation and Treatment of High Blood Cholesterol in Adults (Adult Treatment Panel 10, NIH publication no. 02-5215, 2002); V-6; ESHA Research, 2004.

Large, soft stools ease the task of elimination. Pressure is then reduced in the lower bowel (colon), making it less likely that rectal veins will swell (**hemorrhoids**). Fiber prevents compaction of the intestinal contents, which could obstruct the appendix and permit bacteria to invade and infect it (**appendicitis**). In addition, most (but not all) studies show that fiber stimulates the GI tract muscles so that they retain their strength and resist bulging out into pouches known as **diverticula** (illustrated in Figure 4–7, p. 122, bottom).[14]

KEY POINT

- Both kinds of fiber are associated with digestive tract health.

Digestive Tract Cancers Cancers of the colon and rectum claim tens of thousands of lives each year.[15] The risk of these cancers is lower, however, among people with higher dietary fiber intakes.[16] A recent meta-analysis using data from several studies exposed a strong, linear inverse association between dietary fiber and colon

constipation difficult, incomplete, or infrequent bowel movements associated with discomfort in passing dry, hardened feces from the body.

hemorrhoids (HEM-or-oids) swollen, hardened (varicose) veins in the rectum, usually caused by the pressure resulting from constipation.

appendicitis inflammation and/or infection of the appendix, a sac protruding from the intestine.

diverticula (dye-ver-TIC-you-la) sacs or pouches that balloon out of the intestinal wall, caused by weakening of the muscle layers that encase the intestine. The painful inflammation of one or more of the diverticula is known as *diverticulitis.*

Figure 4–6
Animated! One Way Fiber in Food May Lower Cholesterol in the Blood

High-fiber diet: More cholesterol (in bile) is carried out of the body.

Low-fiber diet: More cholesterol (from bile) is reabsorbed and returned to the bloodstream.

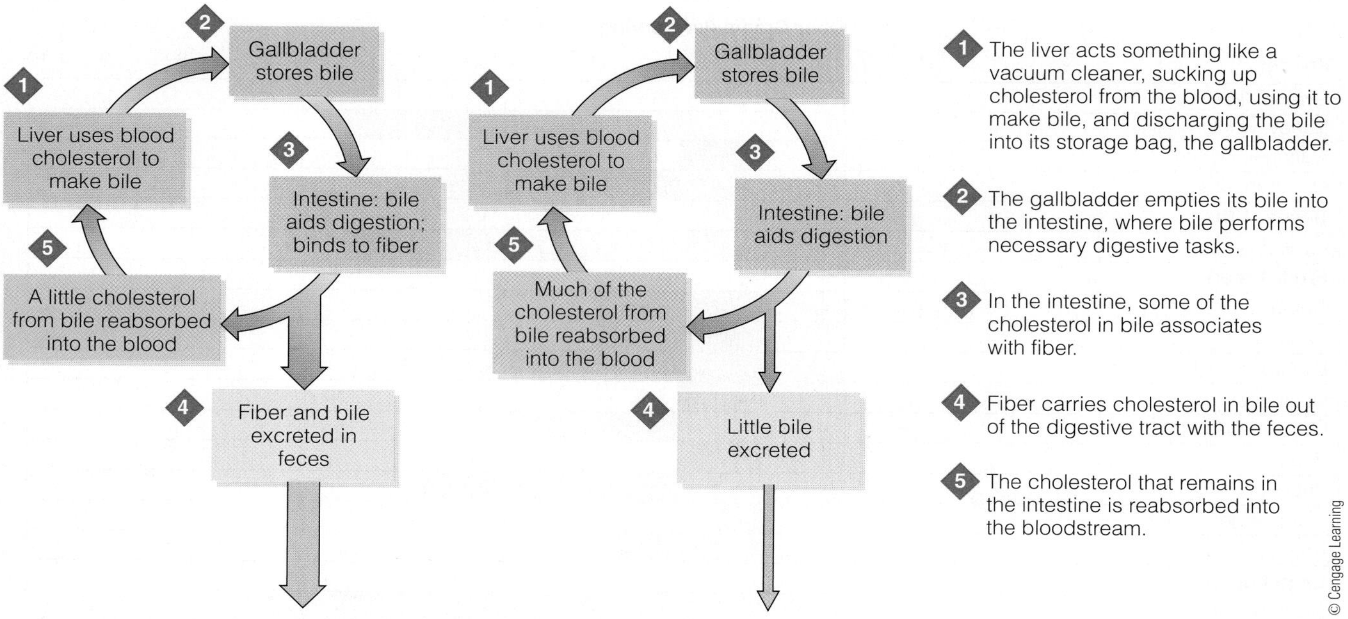
1 Liver uses blood cholesterol to make bile
2 Gallbladder stores bile
3 Intestine: bile aids digestion; binds to fiber
4 Fiber and bile excreted in feces
5 A little cholesterol from bile reabsorbed into the blood
1 Liver uses blood cholesterol to make bile
2 Gallbladder stores bile
3 Intestine: bile aids digestion
4 Little bile excreted
5 Much of the cholesterol from bile reabsorbed into the blood
1 The liver acts something like a vacuum cleaner, sucking up cholesterol from the blood, using it to make bile, and discharging the bile into its storage bag, the gallbladder.
2 The gallbladder empties its bile into the intestine, where bile performs necessary digestive tasks.
3 In the intestine, some of the cholesterol in bile associates with fiber.
4 Fiber carries cholesterol in bile out of the digestive tract with the feces.
5 The cholesterol that remains in the intestine is reabsorbed into the bloodstream.
© Cengage Learning

Figure 4–7
Diverticula

Diverticula are abnormally bulging pockets in the colon wall. These pockets can entrap feces and become painfully infected and inflamed, requiring hospitalization, antibiotic therapy, or surgery.

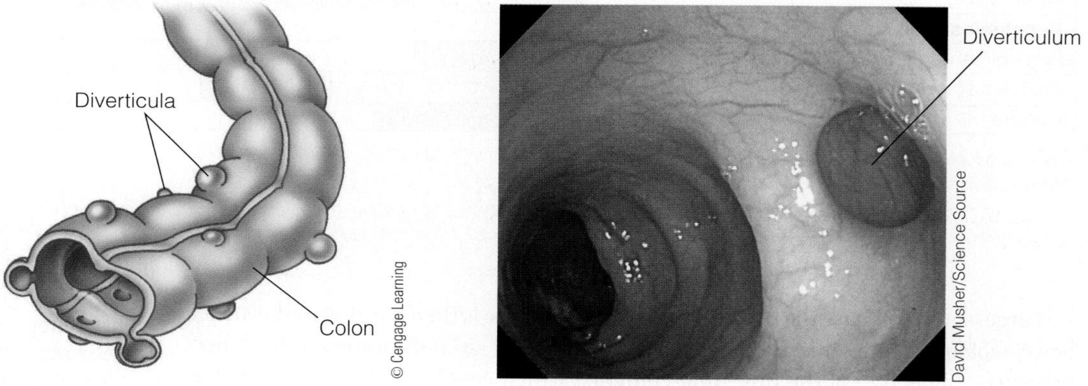
Diverticula
Colon
© Cengage Learning
Diverticulum
David Musher/Science Source

cancer.[17] Subjects who ate the most fiber (24 grams per day) reduced their risk of colon and rectal cancer by almost 30 percent compared with those who ate the least (10 grams per day). Mid-range intakes (18 grams per day) reduced the risk by 20 percent. Importantly, fiber from food but not from supplements demonstrates this association, possibly because fiber supplements lack the nutrients and phytochemicals of whole foods that may also help to protect against cancers.

All plant foods—vegetables, fruits, and whole-grain products—have attributes that may reduce the risks of colon and rectal cancers. Their fiber dilutes, binds, and rapidly removes potential cancer-causing agents from the colon. In addition, bacteria ferment their soluble fiber in the colon, releasing short-chain fatty acids that nourish the colon cells and lower the pH of colon contents. These small fat molecules may also activate cancer-destroying enzymes and inhibit inflammation in the colon.[18]

Other lifestyle factors, including alcohol intake, physical activity, red and processed meat intakes, intake of the vitamin folate and other nutrients, and genetic factors also change colon cancer risks.[19] To lower the risk, experts recommend that dietary fiber come from five to nine 1/2-cup servings of vegetables and fruit, along with generous portions of whole grains and legumes, to obtain all of the benefits that plant foods provide.

KEY POINT

- Fibers in foods help to maintain digestive tract health.

Healthy Weight Management Foods rich in fibers tend to be low in fat and added sugars and can therefore prevent weight gain and promote weight loss by delivering less energy per bite.[20] In addition, fibers absorb water from the digestive juices; as they swell, they create feelings of fullness, delay hunger, and reduce food intake.[21] Soluble fibers may be especially useful for appetite control. In a recent study, soluble fiber from barley shifted the body's mix of appetite-regulating hormones toward reducing food intake.[22] By whatever mechanism, as populations eat more refined low-fiber foods and concentrated sweets, body fat stores creep up.[23] In contrast, people who eat three servings of whole grains each day tend to accumulate less body and abdominal fatness over time.[24]

Commercial weight-loss products often contain bulky fibers such as methylcellulose, but pure fiber compounds are not advised. Instead, consumers should select whole grains, legumes, fruits, and vegetables. High-fiber foods not only add bulk to the diet but are economical, nutritious, and supply health-promoting phytochemicals—benefits that no purified fiber preparation can match.

KEY POINT

- Eating patterns that are adequate in fiber assist the eater in maintaining a healthy body weight.

Table 4–2

A Quick Method for Estimating Fiber Intake

To quickly estimate fiber in a day's meals:

1. Multiply servings (½ c cut up or 1 medium piece) of any fruit or vegetable (excluding juice) by 1.5 g.[a]
   Example: 5 servings of fruits and vegetables × 1.5 = 7.5 g fiber
2. Multiply ½ c servings of refined grains by 1.0 g.
   Example: 4 servings of refined grains × 1.0 = 4.0 g fiber
3. Multiply ½ c servings of whole grains by 2.5 g.
   Example: 3 servings of whole grains × 2.5 = 7.5 g fiber
4. Add fiber values for servings of legumes, nuts, seeds, and high-fiber cereals and breads; look these up in Appendix A.
   Example: ½ c navy beans = 6.0 g fiber
5. Add up the grams of fiber from the previous lines.
   Example: 7.5 + 4.0 + 7.5 + 6.0 = 25 g fiber

Day's total fiber = 25 g fiber

[a]*Most cooked and canned fruits and vegetables contain about this amount, while whole raw fruits and some vegetables contain more.*

Fiber Intakes and Excesses

Few people in the United States or Canada consume sufficient fiber. The DRI intake recommendation for fiber is 14 grams per 1,000 calories, or 25 grams per day for most women, 38 grams for most men—about twice the average current intake of about 14 to 15 grams.[25] DRI recommendations for other age and gender groups are listed on the inside front cover pages. Fiber recommendations are made in terms of total fiber with no distinction between fiber types. This makes sense because most fiber-rich foods supply a mixture of fibers (recall Figure 4–5, page 121).

An effective way to add fiber while lowering fat is to substitute plant sources of protein (legumes) for some of the animal sources of protein (meats and cheeses) in the diet. Another way is to focus on consuming the recommended amounts of fruits, vegetables, legumes, and whole grains each day. You can make a quick approximation of a day's fiber intake by following the instructions in Table 4–2. People choosing high-fiber foods are wise to seek out a variety of fiber sources and to drink extra fluids to help the fiber do its job.

Can My Diet Have Too Much Fiber? No Tolerable Upper Limit has been established for fiber, but consuming purified fibers, such as oat or wheat bran, can be taken to extremes. One overly enthusiastic eater of oat bran muffins required emergency surgery for a blocked intestine; too much oat bran and too little fluid overwhelmed his digestive system. Approach bran and other purified fibers with an attitude of moderation, and be sure to drink an extra beverage with them.

A person who eats only small amounts of food at a time may not be able to eat enough high-fiber food in a day to meet energy or nutrient needs. The malnourished, the elderly, and young children adhering to all-plant (vegan) diets are especially vulnerable to this problem.

Did You Know?

Chelating agents are often sold by supplement vendors to "remove poisons" from the body. There are some valid medical uses, such as treatment of lead poisoning, but most of the chelating agents sold over the counter are based on unproven claims.

The Binders in Fiber Binders in some fibers act as **chelating agents**. This means that they link chemically with important nutrient minerals (iron, zinc, calcium, and others) and then carry them out of the body. The mineral iron is mostly absorbed at the beginning of the intestinal tract, and excess insoluble fibers may limit its absorption by speeding foods through the upper part of the digestive tract.

The next section focuses on the handling of carbohydrates by the digestive system. Table 4–3 sums up the points made so far concerning the functions of carbohydrates in the body and in foods.

KEY POINTS

- Few consume sufficient fiber.
- The best fiber sources are whole foods, and fluid intake should increase along with fiber.
- Very-high-fiber vegetarian diets can pose nutritional risks for some people.

Whole Grains

The USDA food patterns, illustrated in Chapter 2, urge everyone to make at least half of their daily grain choices *whole* grains, an amount equal to at least three 1-ounce equivalents of whole grains a day.[26] To do this, you must distinguish among grain foods that are **refined**, **enriched**, **fortified**, and whole grain (see Table 4–4). This chapter's Consumer's Guide section explains how to find whole-grain foods.

Flour Types The part of a typical grain plant, such as the wheat, that is made into flour (and then into bread, cereals, and pasta) is the seed, or kernel. The kernel has four main parts: the **germ**, the **endosperm**, the **bran**, and the **husk**, as shown in Figure 4–8. The germ is the part that grows into a new plant, in this case wheat, and therefore contains concentrated food to support the new life—it is especially rich in oils, vitamins, and minerals. The endosperm is the soft, white inside portion of the kernel, containing starch and proteins that help nourish the seed as it sprouts. The kernel is encased in the bran, a protective coating that is similar in function to the shell of a nut; the bran is also rich in nutrients and fiber. The husk, commonly called chaff, is the dry outermost layer that is inedible by human beings but can be used in animal feed.

In earlier times, people milled wheat by grinding it between two stones, blowing or sifting out the tough outer chaff, but retaining all the nutrient-rich bran and germ as well as the endosperm. With advances in milling machinery, it became possible to

Table 4–3

Usefulness of Carbohydrates

Carbohydrates in the Body	Carbohydrates in Foods
▪ *Energy source.* Sugars and starch from the diet provide energy for many body functions; they provide glucose, the preferred fuel for the brain and nerves. ▪ *Glucose storage.* Muscle and liver glycogen store glucose. ▪ *Raw material.* Sugars are converted into other compounds, such as amino acids (the building blocks of proteins), as needed. ▪ *Structures and functions.* Sugars interact with protein molecules, affecting their structures and functions. ▪ *Digestive tract health.* Fibers help to maintain healthy bowel function (reduce risk of bowel diseases). ▪ *Blood cholesterol.* Fibers promote normal blood cholesterol concentrations (reduce risk of heart disease). ▪ *Blood glucose.* Fibers modulate blood glucose concentrations (help control diabetes). ▪ *Satiety.* Fibers and sugars contribute to feelings of fullness. ▪ *Body weight.* A fiber-rich diet may promote a healthy body weight.	▪ *Flavor.* Sugars provide sweetness. ▪ *Browning.* When exposed to heat, sugars undergo browning reactions, lending appealing color, aroma, and taste. ▪ *Texture.* Sugars help make foods tender. Cooked starch lends a smooth, pleasing texture. ▪ *Gel formation.* Starch molecules expand when heated and trap water molecules, forming gels. The fiber pectin forms the gel of jellies when cooked with sugar and acid from fruit. ▪ *Bulk and viscosity (thickness).* Carbohydrates lend bulk and increased viscosity to foods. Soluble, viscous fibers lend thickness to foods such as salad dressings. ▪ *Moisture.* Sugars attract water and keep foods moist. ▪ *Preservative.* Sugar in high concentrations dehydrates bacteria and preserves the food. ▪ *Fermentation.* Carbohydrates are fermented by yeast, a process that causes bread dough to rise and beer to brew, among other uses.

Table 4–4

Terms That Describe Grain Foods

- **bran** the protective fibrous coating around a grain; the chief fiber donator of a grain.
- **brown bread** bread containing ingredients such as molasses that lend a brown color; may be made with any kind of flour, including white flour.
- **endosperm** the bulk of the edible part of a grain, the starchy part.
- **enriched, fortified** refers to the addition of nutrients to a refined food product. As defined by U.S. law, these terms mean that specified levels of thiamin, riboflavin, niacin, folate, and iron have been added to refined grains and grain products. The terms *enriched* and *fortified* can refer to the addition of more nutrients than just these five; read the label.[a]
- **germ** the nutrient-rich inner part of a grain.
- **husk** the outer, inedible part of a grain.
- **multi-grain** a term used on food labels to indicate a food made with more than one kind of grain. Not an indicator of a whole-grain food.
- **refined** refers to the process by which the coarse parts of food products are removed. For example, the refining of wheat into white enriched flour involves removing three of the four parts of the kernel—the chaff, the bran, and the germ—leaving only the endosperm, composed mainly of starch and a little protein.
- **refined grains** grains and grain products from which the bran, germ, or other edible parts of whole grains have been removed; not a whole grain. Many refined grains are low in fiber and are enriched with vitamins as required by U.S. regulations.
- **stone ground** refers to a milling process using limestone to grind any grain, including refined grains, into flour.
- **unbleached flour** a beige-colored refined endosperm flour with texture and nutritive qualities that approximate those of regular white flour.
- **wheat bread** bread made with any wheat flour, including refined enriched white flour.
- **wheat flour** any flour made from wheat, including refined white flour.
- **white flour** an endosperm flour that has been refined and bleached for maximum softness and whiteness.
- **white wheat** a wheat variety developed to be paler in color than common red wheat (most familiar flours are made from red wheat). White wheat is similar to red wheat in carbohydrate, protein, and other nutrients, but it lacks the dark and bitter, but potentially beneficial, phytochemicals of red wheat.
- **100% whole grain** a label term for food in which the grain is entirely whole grain, with no added refined grains.
- **whole-wheat flour** flour made from whole-wheat kernels; a whole-grain flour. Also called *graham flour.*

[a]*Formerly,* enriched *and* fortified *carried distinct meanings with regard to the nutrient amounts added to foods, but a change in the law has made these terms virtually synonymous.*

remove the dark, heavy bran and germ, leaving a whiter, smoother-textured flour with a higher starch content and far less fiber. An advantage of this flour, besides producing soft, white baked goods, is its durability—white flour "keeps" much longer than whole-grain flour because the nutrient-rich, oily germ of whole grains turns rancid over time. As food production became more industrialized, suppliers realized that customers also favored this refined soft white flour over the crunchy, dark brown, "old-fashioned" flour.

KEY POINT

- Whole-grain flours retain all edible parts of grain kernels.

Enrichment of Refined Grains In turning to highly refined grains, many people suffered deficiencies of iron, thiamin, riboflavin, and niacin—nutrients formerly obtained from whole grains. To reverse this tragedy, Congress passed the U.S. Enrichment Act of 1942 requiring that iron, niacin, thiamin, and riboflavin be added to all refined grain products before they were sold. In 1996, the vitamin folate (often called *folic acid* on labels) was added to the list. Today, all refined grain products are enriched with at least the nutrients mandated by the Act.

A single serving of enriched grain food is not "rich" in the enrichment nutrients, but people who eat several servings a day obtain significantly more of these nutrients

Figure 4–8

A Wheat Plant and a Single Kernel of Wheat

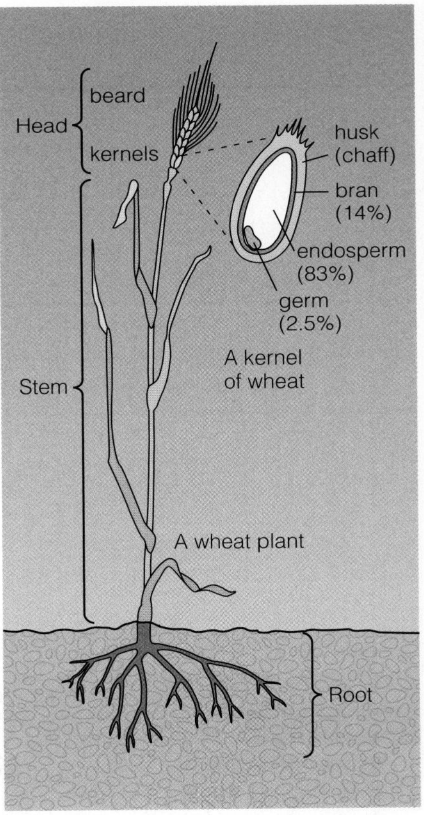

chelating (KEE-late-ing) **agents** molecules that attract or bind with other molecules and are therefore useful in either preventing or promoting movement of substances from place to place.

Figure 4–9

Nutrients in Whole-Grain, Enriched White, and Unenriched White Breads

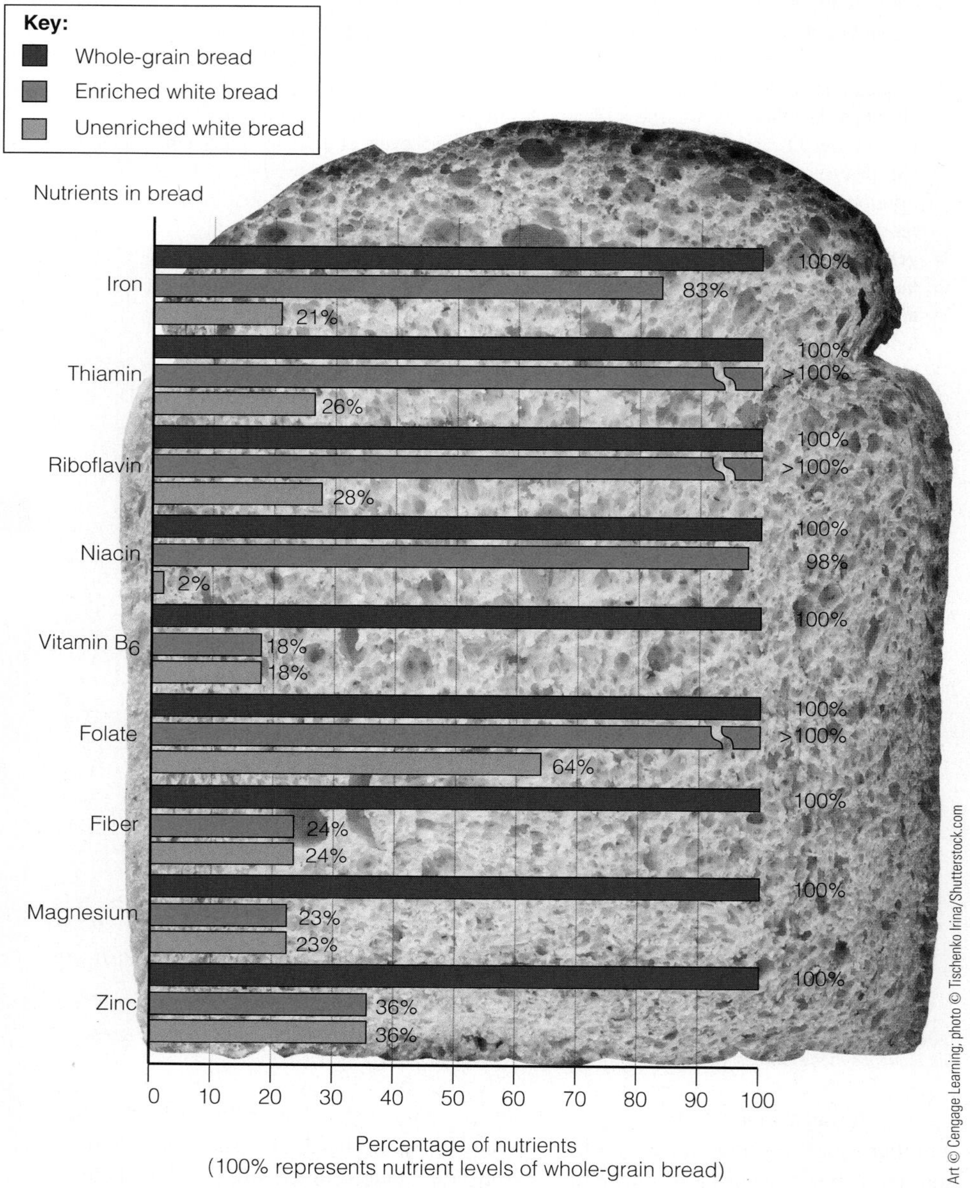

Art © Cengage Learning; photo © Tischenko Irina/Shutterstock.com

than they would from unenriched refined products, as the bread example of Figure 4–9 shows.

Enriched grain foods are nutritionally comparable to whole-grain foods only with respect to their added nutrients; whole grains provide greater amounts of vitamin B_6 and vitamin E, and the minerals magnesium, zinc, and chromium that refined grains lack. Whole grains also provide substantial fiber (see Table 4–5), along with a wide array of potentially beneficial phytochemicals in the bran and the essential oils of the germ.

KEY POINT

- Refined grain products are less nutritious than whole grains.

Health Effects of Whole Grains Whole-grain intakes provide health benefits beyond just nutrients and fiber.[27] People who take in just three daily servings of whole grains often have healthier body weights and less body fatness than other people.[28] It could be that whole grains fill up the stomach, slow down digestion, or promote

Table 4–5

Grams of Fiber in One Cup of Flour

Dark rye, 31 g
Barley flour, 15 g
Whole wheat, 13 g
Buckwheat, 12 g
Whole-grain cornmeal, 9 g
Light rye, 8 g
Enriched white, 3 g

© Cengage Learning

use it! A Consumer's Guide To . . . Finding Whole-Grain Foods

"OK, it's time to take action." A consumer, ready to switch to some whole-grain foods, may find that these good intentions are derailed in the tricky terrain of the grocery store. Even an experienced shopper may feel bewildered in store-length aisles bulging with breads that range from light-as-a-feather, refined enriched white loaves to the heaviest, roughest-textured whole-grain varieties. A baffling array of label claims vies for our shopper's attention, too—and while some are trustworthy, others are not.

Not Every Choice Must Be 100 Percent Whole Grain

If you are just now starting to include whole grains in your diet, keep in mind that various combinations of whole and refined grains can meet the *Dietary Guidelines* recommendation that half of the day's grains be whole grains.[1] Until your taste buds adjust, you may prefer breads, cereals, pastas, and other grain foods made from a half and half blend of whole and refined grains for all of your day's choices. The addition of some refined enriched white flour smoothes the texture of whole grain foods and provides a measure of folate, an important enrichment vitamin in the U.S. diet. Alternatively, you might choose 100 percent whole grains half of the time and refined grains for the other half, or any other combination to meet the need. Research shows that no harm comes from consuming up to half of the day's grains as refined grains.[2]

In addition to whole-grain blends, a new variety of white durum wheat has been developed to mimic the taste and appearance of ordinary enriched refined white flour, while offering nutrients similar to those of whole grains. Such **white wheat** products lack the dark-colored and strong-flavored phytochemicals associated with ordinary whole-wheat products, however, and research has not established whether its effects on the health of the body are equivalent.* (See Table 4–4, page 125, for definitions.)

High Fiber Does Not Equal Whole Grain

An important distinction exists between foods labeled "high-fiber" and those made of whole grains. High-fiber breads or cereals may derive their fiber from the addition of wheat bran or even purified cellulose, and not from whole grains. Label readers can differentiate one kind from the other by scanning the food's ingredients list for words like *bran, cellulose, methylcellulose, gums,* or *psyllium.* Such high-fiber foods may be nutritious and useful in their own way, but they cannot substitute for whole-grain foods in the diet.

Brown Color Does Not Equal Whole Grain

"**Brown bread**" may sound healthy, and white bread less so, but the term *brown* simply refers to color that may derive from brown ingredients, such as molasses. Similarly, whole-grain rice, commonly called brown rice, cannot be judged by color alone. Whole-grain rice comes in red and other colors, too. Also, many rice dishes appear brown because they contain brown-colored ingredients, such as soy sauce, beef broth, or seasonings. Pasta comes in a rainbow of colors, and whole-grain noodles and blends are increasingly available—just read the ingredients list on the label to check that any descriptors on the outside of the package accurately reflect the food inside.

Label Subtleties

A label proclaiming "Multi-Grain Goodness" or "Natural Wheat Bread" may imply healthfulness but can mislead uninformed shoppers, who assume, falsely, that such terms mean "whole grain." Like descriptors such as **multi-grain**, **wheat bread**, and **stone ground**, these terms do not indicate whole grains. To find the real whole grains, look for the words *whole* or *whole grain* preceding the name of a grain in the ingredient list. Learn to recognize individual whole grains by name, too. Many are listed in Table 4–6.

Look at the bread labels in Figure 4–10 (p. 428), and recall from Chapter 2 that ingredients must be listed in descending order of prominence on an ingredients list. It's easy to see from the label of the "Natural Wheat Bread" in the figure that this bread contains no whole grains whatsoever. This loaf is made entirely of refined enriched wheat flour, another name for white flour. The word "Natural" in the name is a market-

* *In 2005, ConAgra began marketing white wheat as* UltraGrain.

Table 4–6

A Sampling of Whole Grains

If some of these sound unfamiliar, why not try them? Other cultures value them, and they could become your favorites, too.

- Amaranth, a grain of the ancient Aztec people.[a]
- Barley (hulled but not pearled).[b]
- Buckwheat.[a]
- Bulgur wheat.
- Corn, including whole cornmeal and popcorn.
- Millet.
- Oats, including oatmeal.
- Quinoa (KEEN-wah), a grain of the ancient Inca people.[a]
- Rice, including brown, red, and others.
- Rye.
- Sorghum (also called milo), a drought-resistant grain.
- Teff, popular in Ethiopia, India, and Australia.
- Triticale, a cross of durum wheat and rye.
- Wheat, in many varieties such as spelt, emmer, farro, einkorn, durum; and forms such as bulgur, cracked wheat, and wheatberries.
- Wild rice.[a]

[a] *Although not botanical grains, these foods are similar to grains in nutrient contents, preparation, and use.*

[b] *Hulling removes only inedible husk; pearling removes beneficial bran.*

ing gimmick, and has no meaning in nutrition.

Now read the label of "Multi-Grain, Honey Fiber Bread." It does contain multiple whole grains, but the major ingredient is still unbleached enriched wheat flour. The key here is the refinement of the wheat berries to yield refined "white" flour that requires enrichment, a flour called "enriched wheat flour" on labels. The bleaching status is irrelevant. Most of the fiber of this bread's name comes from added cellulose and not from its tiny amounts of "multi-grains."

Now focus on the bread labeled "Whole Grain, Whole Wheat." This, at last, is a 100 percent whole-grain food.

After the Salt

Here's a trick: a loaf of bread generally contains about one teaspoon of salt. Therefore, if an ingredient is listed *after* the salt, you'll know that the entire loaf contains less than a teaspoonful of that ingredient, not enough to make a significant contribution to the eater's whole-grain intake. In the "Multi-Grain" bread of the figure, all of the whole grains are listed after the salt.

A Word about Cereals

Ready-to-eat breakfast cereals, from toasted oat rings to granola, are a pleasant way to include whole grains in almost anyone's diet. Like breads, cereals vary widely in their contents of whole grains but, also like breads, they can be evaluated by reading their ingredient lists.

Oatmeal in all its forms, old-fashioned, quick cooking, and even microwavable instant, qualifies as whole grain, but be careful: some instant oatmeal packets contain more sugar than grain. Limit intake of any cereal, hot or cold, with a high sugar, sodium, or saturated fat content, even if it touts "whole grains" on the label.

Moving Ahead

"I've tried buckwheat pancakes, and they're pretty tasty. But what on earth is quinoa?" Admittedly, certain whole grains may be unavailable in mainstream grocery stores. It may take a trip to a "health-food" store to find quinoa, for example. In a welcome trend, larger chain stores are responding to increased consumer demand and stocking more brown rice, wild rice, bulgur, and other whole grain goodies on their shelves.[3]

Once a person begins to enjoy the added taste dimensions of whole grains, he or she may be less drawn to the bland refined foods formerly eaten out of

Figure 4–10
Bread Labels Compared

Natural Wheat Bread

Nutrition Facts	
Serving size 1 slice (30g)	
Servings Per Container 15	
Amount per serving	
Calories 90	Calories from Fat 14
	% Daily Value*
Total Fat 1.5g	2%
***Trans* Fat** 0g	
Sodium 220mg	9%
Total Carbohydrate 15g	5%
Dietary fiber less than 1g	2%
Sugars 2g	
Protein 4g	

INGREDIENTS: UNBLEACHED ENRICHED WHEAT FLOUR [MALTED BARLEY FLOUR, NIACIN, REDUCED IRON, THIAMIN MONONITRATE (VITAMIN B1), RIBOFLAVIN (VITAMIN B2), FOLIC ACID], WATER, HIGH FRUCTOSE CORN SYRUP, MOLASSES, PARTIALLY HYDROGENATED SOYBEAN OIL, YEAST, CORN FLOUR, SALT, GROUND CARAWAY, WHEAT GLUTEN, CALCIUM PROPIONATE (PRESERVATIVE), MONOGLYCERIDES, SOY LECITHIN.

Multi-Grain Honey Fiber

Nutrition Facts	
Serving size 1 slice (43g)	
Servings Per Container 18	
Amount per serving	
Calories 120	Calories from Fat 15
	% Daily Value*
Total Fat 1.5g	2%
***Trans* Fat** 0g	
Sodium 170mg	7%
Total Carbohydrate 9g	3%
Dietary fiber 4g	16%
Sugars 2g	
Protein 5g	

INGREDIENTS: UNBLEACHED ENRICHED WHEAT FLOUR, WATER, WHEAT GLUTEN, CELLULOSE, YEAST, SOYBEAN OIL, HONEY, SALT, BARLEY, NATURAL FLAVOR PRESERVATIVES, MONOCALCIUM PHOSPHATE, MILLET, CORN, OATS, SOYBEAN FLOUR, BROWN RICE, FLAXSEED.

Whole Grain WHOLE WHEAT

Nutrition Facts	
Serving size 1 slice (30g)	
Servings Per Container 18	
Amount per serving	
Calories 90	Calories from Fat 14
	% Daily Value*
Total Fat 1.5g	2%
***Trans* Fat** 0g	
Sodium 135mg	6%
Total Carbohydrate 15g	5%
Dietary fiber 2g	8%
Sugars 2g	
Protein 4g	

MADE FROM: UNBROMATED STONE GROUND 100% WHOLE WHEAT FLOUR, WATER, CRUSHED WHEAT, HIGH FRUCTOSE CORN SYRUP, PARTIALLY HYDROGENATED VEGETABLE SHORTENING (SOYBEAN AND COTTONSEED OILS), RAISIN JUICE CONCENTRATE, WHEAT GLUTEN, YEAST, WHOLE WHEAT FLAKES, UNSULPHURED MOLASSES, SALT, HONEY, VINEGAR, ENZYME MODIFIED SOY LECITHIN, CULTURED WHEY, UNBLEACHED WHEAT FLOUR AND SOY LECITHIN.

habit. More than 90 percent of Americans are stuck in this rut, failing to eat the whole grains they need. Be adventurous with health in mind, and give the hearty flavors of a variety of whole-grain foods a try.

These memory joggers can remind you to choose whole grains during the day:

1. Morning, choose a whole-grain cereal breakfast.
2. Noon, choose whole-grain bread for lunch.
3. Night, choose whole-grain pasta or rice for supper.

Vary your choices, and remember to make at least half of your grain foods whole grains.[4]

Review Questions†

1. When searching for whole-grain bread, a consumer should search the labels __________ .
   a. for words like *multi-grain, wheat bread, brown bread,* or *stone ground*
   b. for the order in which whole grains appear on the ingredients list
   c. for the word "unbleached," which indicates that the food is primarily made from whole grains
   d. b and c

†Answers to Consumer's Guide review questions are found in Appendix G.

2. Whole-grain rice, often called brown rice, __________ .
   a. can be recognized by its characteristic brown color
   b. cannot be recognized by color alone
   c. is often more refined than white rice
   d. b and c
3. A bread labeled "high fiber" __________ .
   a. may not be a whole-grain food
   b. is a good substitute for whole-grain bread
   c. is required by law to contain whole grains
   d. may contain the dangerous chemical cellulose

longer-lasting feelings of fullness than refined grains. The same three daily servings of whole grains also correlate with lower risks of heart disease and type 2 diabetes. Finally, people who make whole grains a habit have lower risks of certain cancers, particularly of the colon. It may be that the fiber, phytochemicals, or nutrients of whole grains improve body tissue health, but these issues need clarification.

Refined grains in amounts of up to one-half of the daily grain intake (without added sugars, fats, or sodium) seem to pose little risk to health.[29] Clearly, however, those who choose to ignore the *Dietary Guidelines for Americans* recommendation to consume sufficient whole grains do so at their peril.

KEY POINT

- A diet rich in whole grains is associated with reduced risks of overweight and certain chronic diseases.

From Carbohydrates to Glucose

LO 4.3 Explain how complex carbohydrates are broken down in the digestive tract and absorbed into the body.

You may eat bread or a baked potato, but the body's cells cannot use foods or even whole molecules of lactose, sucrose, or starch for energy. They need the glucose in those molecules. The various body systems must make glucose available to the cells, not all at once when it is eaten, but at a steady rate all day.

© iStockphoto.com

Digestion and Absorption of Carbohydrate

To obtain glucose from newly eaten food, the digestive system must first render the starch and disaccharides from the food into monosaccharides that can be absorbed through the cells lining the small intestine. The largest of the digestible carbohydrate molecules, starch, requires the most extensive breakdown. Disaccharides, in contrast, need be split only once before they can be absorbed.

Starch Digestion of most starch begins in the mouth, where an enzyme in saliva mixes with food and begins to split starch into shorter units. While chewing a bite of bread, you may notice that a slightly sweet taste develops—the disaccharide maltose is being liberated from starch by the enzyme. The salivary enzyme continues to act on the starch in the bite of bread while it remains tucked in the stomach's upper storage area. As each chewed lump is pushed downward and mixed with the stomach's acid and other juices, the salivary enzyme (made of protein) is deactivated by the stomach's protein-digesting acid. Not all digestive enzymes are susceptible to digestion in the stomach—one enzyme that digests protein works best in the stomach. Its structure protects it from the stomach's acid.

With the breakdown of the salivary enzyme in the stomach, starch digestion ceases, but it resumes at full speed in the small intestine, where another starch-splitting enzyme is delivered by the pancreas. This enzyme breaks starch down into disaccharides and small polysaccharides. Other enzymes liberate monosaccharides for absorption.

Some forms of starch are easily digested. The starch of refined white flour, for example, breaks down rapidly to glucose that is absorbed high up in the small intestine. Other starch, such as that of cooked beans, digests more slowly and releases its glucose later in the digestion process. The least digestible starch, called **resistant starch**, is technically a kind of fiber because much of it passes through the small intestine undigested into the colon.[30] Some resistant starch may be digested, but slowly, and most remains intact until the bacteria of the colon eventually ferment it.[31] Barley, raw or chilled cooked potatoes, cooked dried beans and lentils, oatmeal, and under-ripe bananas all contain resistant starch.

Sugars Sucrose and lactose from food, along with maltose and small polysaccharides freed from starch, undergo one more split to yield free monosaccharides before they are absorbed. This split is accomplished by digestive enzymes attached to the cells of the lining of the small intestine. The conversion of a bite of bread to nutrients for the body is completed when monosaccharides cross these cells and are washed away in a rush of circulating blood that carries them to the waiting liver. Figure 4–11 presents a quick review of carbohydrate digestion.

The absorbed carbohydrates (glucose, galactose, and fructose) travel in the bloodstream to the liver, which can convert fructose and galactose to glucose. The circulatory system transports the glucose and other products to the cells. Liver and muscle cells may store circulating glucose as glycogen; all cells may split glucose for energy.

Fiber As mentioned, although molecules of most fibers are not changed by human digestive enzymes, many of them can be fermented by the bacterial inhabitants of the human colon. A by-product of this fermentation can be any of several odorous gases. Don't give up on high-fiber foods if they cause gas. Instead, start with small servings and gradually increase the serving size over several weeks; chew foods thoroughly to break up hard-to-digest lumps that can ferment in the intestine; and try a variety of fiber-rich foods until you find some that do not cause the problem. Some people also find relief from excessive gas by using commercial enzyme preparations sold for use with beans. Such products contain enzymes that help to break down some of the indigestible fibers in foods before they reach the colon.

KEY POINTS

- The main task of the various body systems is to convert starch and sugars to glucose to fuel the cells' work.
- Fermentable fibers may release gas as they are broken down by bacteria in the large intestine.

Why Do Some People Have Trouble Digesting Milk?

Persistent painful gas may herald a change in digestive tracts' ability to digest the sugar in milk, a condition known as **lactose intolerance**. Its cause is insufficient production of lactase, the enzyme of the small intestine that splits the disaccharide lactose into its component monosaccharides glucose and galactose, which are then absorbed.

Infants produce abundant lactase, which helps them absorb the sugar of breast milk and milk-based formulas; a very few suffer inborn lactose intolerance and must be fed solely on lactose-free formulas. Among adults, the ability to digest the carbohydrate of

resistant starch the fraction of starch in a food that is digested slowly, or not at all, by human enzymes.

lactose intolerance impaired ability to digest lactose due to reduced amounts of the enzyme lactase.

Figure 4–11

Animated! How Carbohydrate in Food Becomes Glucose in the Body

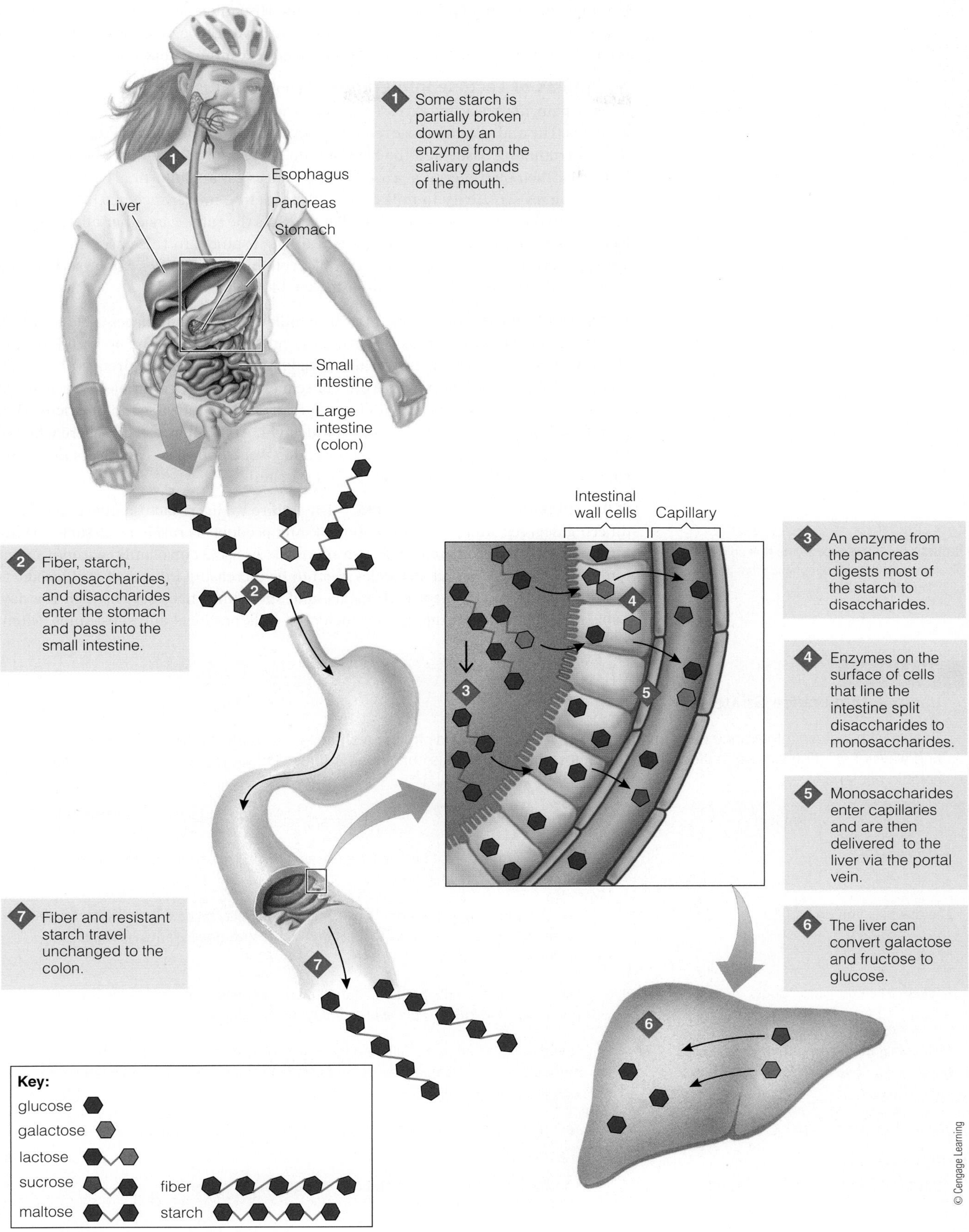

1 Some starch is partially broken down by an enzyme from the salivary glands of the mouth.
Esophagus
Liver
Pancreas
Stomach
Small intestine
Large intestine (colon)
2 Fiber, starch, monosaccharides, and disaccharides enter the stomach and pass into the small intestine.
Intestinal wall cells
Capillary
3 An enzyme from the pancreas digests most of the starch to disaccharides.
4 Enzymes on the surface of cells that line the intestine split disaccharides to monosaccharides.
5 Monosaccharides enter capillaries and are then delivered to the liver via the portal vein.
6 The liver can convert galactose and fructose to glucose.
7 Fiber and resistant starch travel unchanged to the colon.
Key:
glucose
galactose
lactose
sucrose
maltose
fiber
starch

milk varies widely. As they age, upward of 75 percent of the world's people lose much of their ability to produce the enzyme **lactase** to digest the milk sugar lactose.

Only about 12 percent of the entire U.S. population develops lactose intolerance, but up to 80 percent of people of African, Asian, Hispanic, Indian, or Native American descent may develop it.[32] Long ago, all adults may have been lactose intolerant; *tolerance* probably first developed among early herders who used animal milk as food, and thrived.

Symptoms of Lactose Intolerance People with lactose intolerance experience nausea, pain, diarrhea, and excessive gas on drinking milk or eating lactose-containing products. The undigested lactose remaining in the intestine demands dilution with fluid from surrounding tissue and the bloodstream. Intestinal bacteria use the undigested lactose for their own energy, a process that produces gas and intestinal irritants.

Sometimes sensitivity to milk is due not to lactose intolerance but to an allergic reaction to the protein in milk. Milk allergy arises the same way other allergies do—from sensitization of the immune system to a substance. In this case, the immune system overreacts when it encounters milk protein. Food allergies can be serious and should be diagnosed by a specialist—Chapter 14 provides details.

Consequences to Nutrition Because milk is an almost indispensable source of the calcium every child needs for growth, a milk substitute must be found for any child who becomes lactose intolerant. Disadvantaged young children of the developing world sustain the most severe consequences of lactose intolerance when it combines with disease, malnutrition, or parasites to produce a loss of nutrients that greatly reduces the children's chances of survival. And children everywhere who fail to consume enough calcium may later develop weak bones, so caregivers must find substitutes if a child becomes unable to tolerate milk.

Milk Tolerance and Strategies The failure to digest lactose affects people to differing degrees, and total elimination of milk products is rarely necessary.[33] Many affected people can consume up to 6 grams of lactose (1/2 cup of milk) without symptoms. The most successful strategies seem to be increasing intakes of milk products gradually, consuming them with meals, and spreading them out through the day. Table 4–7 offers more strategies for including milk products and substitutes. Often,

lactase the intestinal enzyme that splits the disaccharide lactose to monosaccharides during digestion.

Table 4–7

Lactose Intolerance Strategies

People with lactose intolerance can experiment with milk-based foods to find a strategy that works for them. The trick is to find ways of splitting lactose to glucose and galactose before a food is consumed, rather than providing a lactose feast for colonic bacteria.

Product	Effects/Strategies
Aged cheeses	Bacteria or molds used to create cheeses ferment lactose during the aging process. Use in moderation.
Lactase pills and drops	Lactase added to milk products by consumers or pills taken before milk product consumption split lactose molecules in the digestive tract. Harmless when used as directed by the manufacturer.
Lactase-treated milk products	Lactase added to milk products during manufacturing splits lactose before purchase. Use freely in place of ordinary milk products.
Milk substitutes (soy, nut, or grain-based beverages), cheese and yogurt substitutes	Nonmilk replacments for milk products may or may not be fortified with the nutrients of milk. Compare Nutrition Facts panels for calcium, protein, and vitamin D in particular.
Yogurt (live culture type)	Yogurt-making bacteria can take up residence in the colon where they may reduce lactose intolerance symptoms.
Yogurt (with added milk solids listed on the label)	These contain *extra* lactose and can overwhelm the system.

people overestimate the severity of their lactose intolerance, blaming it for symptoms most probably caused by something else—a mistake that could cost them the health of their bones (details in Chapter 8).[34]

KEY POINTS

- In lactose intolerance, the body fails to produce sufficient amounts of the enzyme needed to digest the sugar of milk, leading to uncomfortable symptoms.
- People with lactose intolerance or milk allergy need alternatives that provide the nutrients of milk.

The Body's Use of Glucose

LO 4.4 Describe how hormones control blood glucose concentrations during fasting and feasting, and explain the response of these hormones to various carbohydrates in the diet.

Glucose is the basic carbohydrate unit used for energy by each of the body's cells. The body handles its glucose judiciously—maintaining an internal store to be used when needed and tightly controlling its blood glucose concentration to ensure a steady supply. Recall that carbohydrates serve functional roles, too, such as forming part of mucus, but they are best known for providing energy.

Splitting Glucose for Energy

Glucose fuels the work of every cell in the body to some extent, but the cells of the brain and nervous system depend almost exclusively on glucose, and the red blood cells use glucose alone. When a cell splits glucose for energy, it performs an intricate sequence of maneuvers that are of great interest to the biochemist—and of no interest at all to most people who eat bread and potatoes. What everybody needs to understand, though, is that there is no good substitute for carbohydrate. Carbohydrate is *essential*, as the following details illustrate.

The Point of No Return At a certain point in the process of splitting glucose for energy, glucose itself is forever lost to the body. First, glucose is broken in half, releasing some energy. Then, two pathways open to these glucose halves. They can be put back together to make glucose again, or they can be broken into smaller molecules. If they are broken further, they cannot be reassembled to form glucose.

The smaller molecules can also take different pathways. They can continue along the breakdown pathway to yield still more energy and eventually break down completely to just carbon dioxide and water. Or, they can be formed into building blocks of protein or be hitched together into units of body fat. Figure 4–12 shows how glucose is broken down to yield energy and carbon dioxide.

Below a Healthy Minimum Although glucose can be converted into body fat, body fat cannot be converted into glucose to feed the brain adequately. When the body faces a severe carbohydrate deficit, it has two problems. Having no glucose, it must turn to protein to make some (the body has this ability), diverting protein from its own critical functions, such as maintaining immune defenses. When body protein is used, it is taken from blood, organ, or muscle proteins; no surplus of protein is stored specifically for such emergencies. Protein is indispensable to body functions, and carbohydrate should be kept available precisely to prevent the use of protein for energy. This is called the **protein-sparing action** of carbohydrate. As for fat, it regenerates a small amount of glucose, but not enough to feed the brain and nerve tissues.

Ketosis The second problem with an inadequate supply of carbohydrate concerns a precarious shift in the body's energy metabolism. Instead of producing energy by following its main metabolic pathway, fat takes another route in which fat fragments combine with each other. This shift causes an accumulation of normally scarce acidic products called **ketone bodies**.

Figure 4–12

Animated! The Breakdown of Glucose Yields Energy and Carbon Dioxide

Cell enzymes split the bonds between the carbon atoms in glucose, liberating the energy stored there for the cell's use. (1) The first split yields two 3-carbon fragments. The two-way arrows mean that these fragments can also be rejoined to make glucose again. (2) Once they are broken down further into 2-carbon fragments, however, they cannot rejoin to make glucose. (3) The carbon atoms liberated when the bonds split are combined with oxygen and released into the air, via the lungs, as carbon dioxide. Although not shown here, water is also produced at each split.

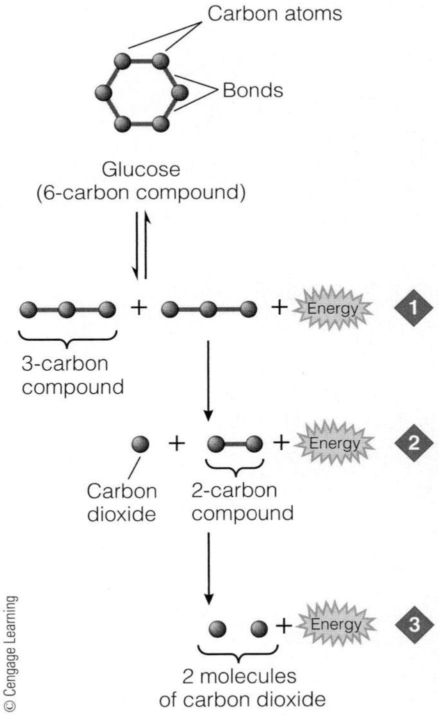

protein-sparing action the action of carbohydrate and fat in providing energy that allows protein to be used for purposes it alone can serve.

ketone (kee-tone) **bodies** acidic, water-soluble compounds that arise during the breakdown of fat when carbohydrate is not available.

Ketone bodies can accumulate in the blood, causing **ketosis**. When they reach high levels, they can disturb the normal acid-base balance, a life-threatening situation. People eating diets that produce ketosis may develop deficiencies of vitamins and minerals, loss of bone minerals, elevated blood cholesterol, impaired mood, and other adverse outcomes.[35] In addition, glycogen stores become too scanty to meet a metabolic emergency or to support vigorous muscular work.

Ketosis isn't all bad, however. Ketone bodies provide a fuel alternative to glucose for brain and nerve cells when glucose is lacking, such as in starvation or very-low-carbohydrate diets. Not all brain tissues can use ketones, however—some rely exclusively on glucose, so the body must still sacrifice some protein to provide it, but at a slower rate. A therapeutic ketogenic diet in addition to medication has substantially reduced seizures in about half of children and adults with hard-to-treat epilepsy, although many find the diet difficult to follow for long periods.[36]

The DRI Minimum Recommendation for Carbohydrate The minimum amount of digestible carbohydrate determined by the DRI committee to adequately feed the brain and reduce ketosis has been set at 130 grams a day for an average-sized person.[37] Several times this minimum is recommended to maintain health and glycogen stores (explained in the next section). The recommended amounts of vegetables, fruits, legumes, grains, and milk presented in Chapter 2 deliver abundant carbohydrates.

KEY POINTS

- Without glucose, the body is forced to alter its uses of protein and fats.
- To help supply the brain with glucose, the body breaks down its protein to make glucose and converts its fats into ketone bodies, incurring ketosis.

How Is Glucose Regulated in the Body?

Should your blood glucose ever climb abnormally high, you might become confused or have difficulty breathing. Should your glucose supplies ever fall too low, you would feel dizzy and weak. The healthy body guards against both conditions with two safeguard activities:

- Siphoning off excess blood glucose into the liver and muscles for storage as glycogen and to the adipose tissue for storage as body fat.
- Replenishing diminished blood glucose from liver glycogen stores.

Two hormones prove critical to these processes. The hormone **insulin** stimulates glucose storage as glycogen, while the hormone **glucagon** helps to release glucose from its glycogen nest.

Insulin After a meal, as blood glucose rises, the pancreas is the first organ to respond. It releases insulin, which signals body tissues to take up glucose from the blood. Muscle tissue responds to insulin by taking up excess blood glucose and using it to build the polysaccharide glycogen. The liver takes up excess blood glucose, too, but it needs no help from insulin to do so. Instead, liver cells respond to insulin by speeding up their glycogen production. Adipose tissue also responds to insulin by taking up excess blood glucose.[38] Simply put, insulin regulates blood glucose by:

- Facilitating blood glucose uptake by the muscles and adipose tissue.
- Stimulating glycogen synthesis in the liver.

Figure 4–13 provides an overview of these relationships.

Tissue Glycogen Stores The muscles hoard two-thirds of the body's total glycogen to ensure that glucose, a critical fuel for physical activity, is available for muscular work. The brain stores a tiny fraction of the total as an emergency reserve to fuel the brain for an hour or two in severe glucose deprivation. The liver stores the remainder and is generous with its glycogen, releasing glucose into the bloodstream

ketosis (kee-TOE-sis) an undesirable high concentration of ketone bodies, such as acetone, in the blood or urine.

insulin a hormone secreted by the pancreas in response to a high blood glucose concentration. It assists cells in drawing glucose from the blood.

glucagon (GLOO-cah-gon) a hormone secreted by the pancreas that stimulates the liver to release glucose into the blood when blood glucose concentration dips.

Figure 4–13

Blood Glucose Regulation—An Overview

The pancreas monitors blood glucose and adjusts its concentration by way of its two opposing hormones, insulin and glucagon. When glucose is high, the pancreas releases insulin; when glucose is low, it releases glucagon. When glucose is restored to the normal range, the pancreas slows its hormone output in an elegant feedback system operating in a healthy body. Many more details about this system are known.

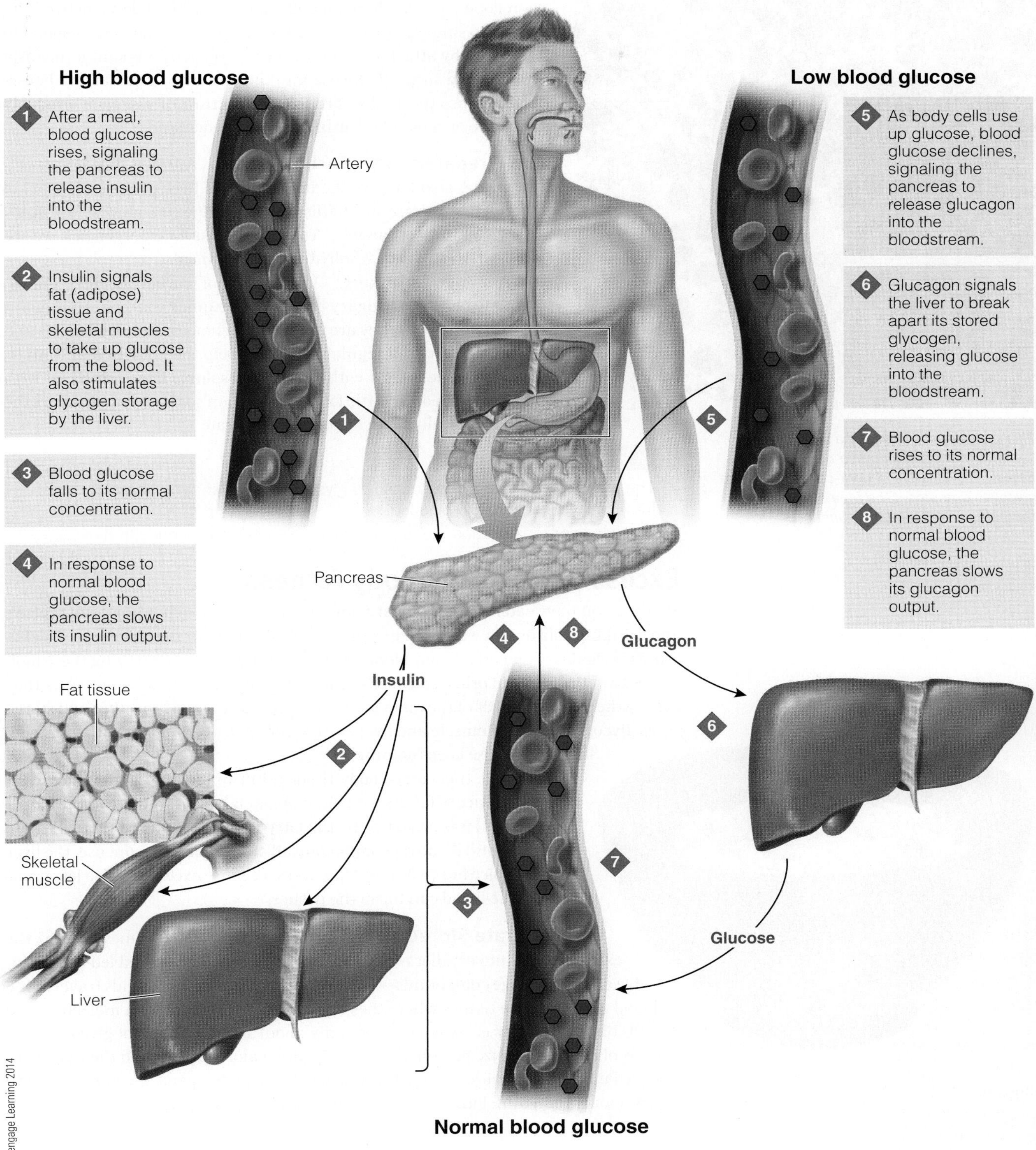

High blood glucose
1 After a meal, blood glucose rises, signaling the pancreas to release insulin into the bloodstream.
2 Insulin signals fat (adipose) tissue and skeletal muscles to take up glucose from the blood. It also stimulates glycogen storage by the liver.
3 Blood glucose falls to its normal concentration.
4 In response to normal blood glucose, the pancreas slows its insulin output.
Artery
Low blood glucose
5 As body cells use up glucose, blood glucose declines, signaling the pancreas to release glucagon into the bloodstream.
6 Glucagon signals the liver to break apart its stored glycogen, releasing glucose into the bloodstream.
7 Blood glucose rises to its normal concentration.
8 In response to normal blood glucose, the pancreas slows its glucagon output.
Pancreas
Insulin
Glucagon
Fat tissue
Skeletal muscle
Liver
Glucose
Normal blood glucose

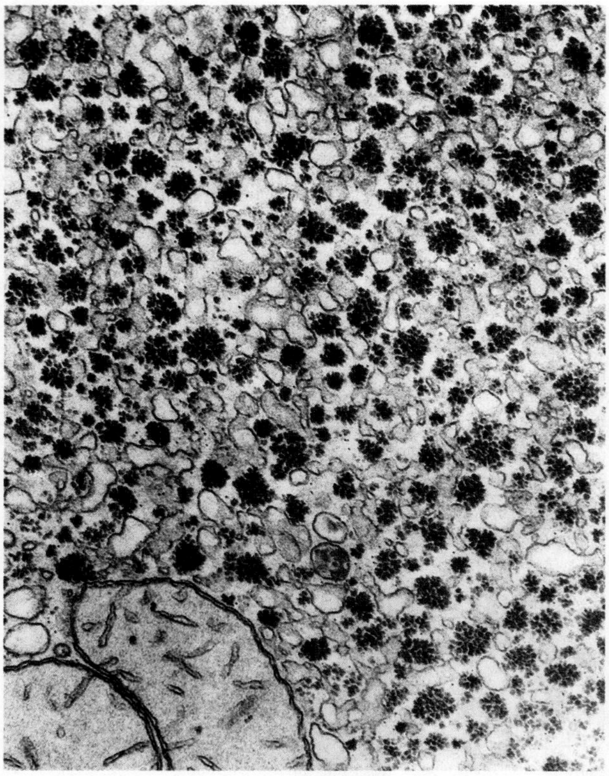

Dr. Donald Fawcett/Visuals Unlimited, Inc.

This photo peeks inside of a single liver cell after a meal (magnified over 100,000 times). The clusters of dark-colored dots are glycogen granules. (The blue structures at the bottom are cellular organelles.)

for the brain or other tissues when the supply runs low. Without carbohydrate from food to replenish it, the liver glycogen stores can be depleted in less than a day.

The Release of Glucose from Glycogen The glycogen molecule is highly branched with hundreds of ends bristling from each molecule's surface (review this structure in Figure 4–3 on page 116). When blood glucose starts to fall too low, the hormone glucagon floods the bloodstream and triggers the breakdown of liver glycogen to single glucose molecules. Enzymes in liver cells respond to glucagon by attacking a multitude of glycogen ends simultaneously to release a surge of glucose into the blood for use by all the body's cells. Thus, the highly branched structure of glycogen uniquely suits the purpose of releasing glucose on demand.

Be Prepared: Eat Carbohydrate Another hormone, epinephrine, also triggers the breakdown of liver glycogen as part of the body's defense mechanism to provide extra glucose for quick action in times of danger.** To store glucose for emergencies, we are well-advised to eat carbohydrate at each meal.

You may be asking, "What kind of carbohydrate?" Candy, "energy bars," and sugary beverages are quick sources of abundant sugar energy, but they are not the best choices. Balanced meals and snacks, eaten on a regular schedule, help the body to maintain its blood glucose. Meals with starch and soluble fiber combined with some protein and a little fat slow digestion so that glucose enters the blood gradually in an ongoing, steady rate.

KEY POINTS

- The muscles and liver store glucose as glycogen; the liver can release glucose from its glycogen into the bloodstream.
- The hormones insulin and glucagon regulate blood glucose concentrations.

Excess Glucose and Body Fatness

Suppose you have eaten dinner and are now sitting on the couch, munching pretzels and drinking cola as you watch a ball game on television. Your digestive tract is delivering molecules of glucose to your bloodstream, and your blood is carrying these molecules to your liver and other body cells. The body cells use as much glucose as they can for their energy needs of the moment. Excess glucose is linked together and stored as glycogen until the muscle and liver stores are full to overflowing with glycogen. Still, the glucose keeps coming.

To handle the excess, body tissues shift to burning more glucose for energy in place of fat. As a result, more fat is left to circulate in the bloodstream until it is picked up by the fatty tissues and stored there. If these measures still do not accommodate all of the incoming glucose, the liver has no choice but to handle the excess, because excess glucose left circulating in the blood can harm the tissues.

© Gene Lee/Shutterstock.com

Carbohydrate Stored as Fat The liver possesses enzymes to break the extra glucose into smaller molecules which can then be assembled into durable energy-storage compounds—fatty acids. Newly made fatty acids travel in the blood to the adipose tissues where they are combined into larger fat molecules and stored. Unlike the liver cells, which store only about 2,000 calories of glycogen, the fat cells of an average-size person store over 70,000 calories of fat, and their capacity to store fat is almost limitless. Moral: You had better play the game if you are going to eat the food. (The Think Fitness feature offers tips to help you play.)

** Epinephrine is also called adrenaline.

Think Fitness move it! What Can I Eat to Make Workouts Easier?

A working body needs carbohydrate fuel to replenish glycogen, and when it runs low, physical activity can seem more difficult. If your workouts seem to drag and never get easier, take a look at your eating pattern. Are your meals regularly timed? Do they provide abundant carbohydrate from nutritious whole foods to fill up glycogen stores so they last through a workout?

Here's a trick: at least an hour before your workout, eat a small snack of about 300 calories of foods rich in complex carbohydrates and drink some extra fluid (see Chapter 10 for ideas). Remember to cut back your intake at other meals by an equivalent amount to prevent unwanted weight gain. The snack provides glucose at a steady rate to spare glycogen, and the fluid helps to maintain hydration.

start now! Choose a one-week period and have a healthy carbohydrate-rich snack of about 300 calories, along with a bottle of water, about an hour before you exercise. Be sure to track your diet in Diet Analysis Plus during this period so that you can accurately track your total calorie intake. Did you have more energy for exercise after you changed your eating plan?

Carbohydrate and Weight Maintenance A balanced eating pattern that provides the recommended complex carbohydrates can help to control body weight and maintain lean tissue. Bite for bite, such carbohydrate-rich foods contribute less to the body's available energy than do fat-rich foods, and they best support physical activity to promote a lean body. Thus, if you want to stay healthy and remain lean, you should make every effort to follow a calorie-appropriate eating pattern providing 45 to 65 percent of its calories from mostly unrefined sources of complex carbohydrates.

This chapter's Food Feature provides the first set of tools required for the job of designing such an eating pattern. Once you have learned to identify the food sources of various carbohydrates, you must then set about learning which fats are which (Chapter 5) and how to obtain adequate protein without overdoing it (Chapter 6). By Chapter 9, you can put it all together with the goal of achieving and maintaining a healthy body weight.

You had better play the game if you are going to eat the food.

KEY POINT

- The liver has the ability to convert glucose into fat, but under normal conditions, most excess glucose is stored as glycogen or used to meet the body's immediate needs for fuel.

The Glycemic Index of Food

Carbohydrate-rich foods vary in the degree to which they elevate both blood glucose and insulin concentrations. A food's average effect in laboratory tests can be ranked on a scale known as the **glycemic index (GI)**. It can then be compared with the score of a standard food, usually glucose, taken by the same person. A food's ranking may surprise you. For example, baked potatoes rank higher than ice cream, partly because ice cream contains sucrose, made of equal parts fructose and glucose. Fructose only slightly elevates blood glucose. In contrast, the starch of the potatoes is all glucose. The milk fat of ice cream also slows digestion and glucose absorption, factors that lower GI ranking. Figure 4–14, p. 138, shows generally where foods have been ranked, but test results often vary widely between laboratories depending upon food ripeness, processing, and seasonal and varietal differences.

In addition to food factors, an individual's own metabolism greatly affects the body's insulin response to carbohydrate.[39] The glycemic response to any one food

glycemic index (GI) a ranking of foods according to their potential for raising blood glucose relative to a standard food such as glucose.

Figure 4–14
Glycemic Index of Selected Foods

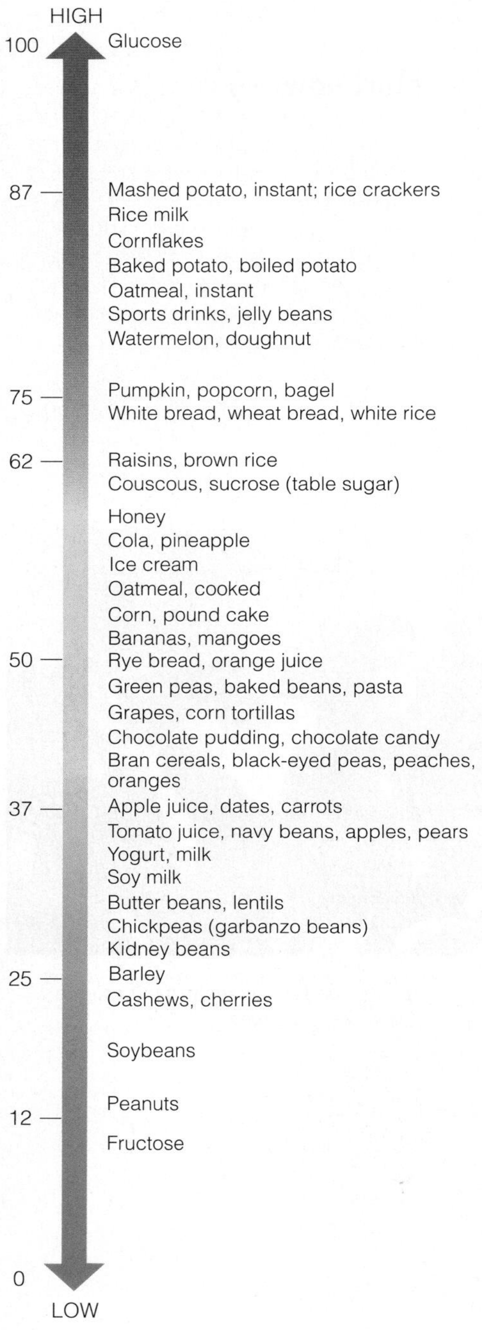

Source: F. S. Atkinson, K. Foster-Powell, and J. C. Brand-Miller, International tables of glycemic index and glycemic load values: 2008, Diabetes Care *31 (2008): 2281–2283.*

varies widely among individual people. Questions have been raised about the validity of calculations of GI values in research studies.[40]

Diabetes and the Glycemic Index The glycemic index, and its mathematical offshoot, **glycemic load (GL)**, may be of interest to people with diabetes who must regulate their blood glucose to protect their health, as the next section describes. Overall, however, little difference in blood glucose control or cardiovascular disease risk is reported between low-GI diets and high-GI diets.[41] Studies are often difficult to interpret because low-GI foods often provide abundant soluble fiber, and soluble fiber slows glucose absorption, sustains feelings of fullness, and improves blood lipids. Therefore, more research is needed to clarify whether the glycemic index, fiber, or some other factor might be responsible for any reported effects.[42]

Nutrition Concerns Choosing foods by GI alone is often not the best choice nutritionally—chocolate candy, for example, has a lower GI than does nutritious brown rice. For people with diabetes, the glycemic index is not of primary concern but may provide a modest benefit when used in addition to using primary strategies for controlling blood glucose. In fact, it may be unnecessary because current guidelines already suggest many low and moderate glycemic index choices: whole grains, legumes, vegetables, fruits, and milk and milk products.

KEY POINTS

- The glycemic index reflects the degree to which a food raises blood glucose.
- The concept of good and bad foods based solely on the glycemic response is an oversimplification.

Diabetes

LO 4.5 Describe the scope of the U.S. diabetes problem, and educate someone about the long- and short-term effects of untreated diabetes and prediabetes.

What happens if the body cannot handle carbohydrates normally? One result is **diabetes**. Diabetes afflicts a rapidly growing number of U.S. adults (see Figure 4–15) and has reached record numbers in children. Almost 26 million people in the United States now have diabetes.[43] Of these, over 7 million are unaware of it and so go untreated. As many as 79 million more have **prediabetes**—their blood glucose is elevated but not yet high enough to be classified as having diabetes.

The Dangers of Diabetes

Diabetes is a leading cause of death in the United States. For people with diabetes, the risk of heart disease, stroke, and dying on any particular day is doubled. Diabetes is also the leading cause of amputations, fatal kidney failure, and permanent blindness.[44] Each year, diabetes costs an estimated $174 billion in U.S. health-care services, disability, lost work, and other costs.[45] The common forms of diabetes are type 1 and type 2, both disorders of blood glucose regulation; their characteristics are summarized in Table 4–8.

Toxicity of Excess Blood Glucose Chronically elevated blood glucose associated with diabetes alters metabolism in virtually every cell of the body. Some cells convert excess glucose to toxic alcohols, causing the cells to swell. Other cells respond by attaching excess glucose to protein molecules in abnormal ways; these altered proteins cannot function, causing many problems.

Chronic inflammation of body tissues accompanies uncontrolled diabetes and may contribute to eye, kidney, heart, and other associated problems.[46] The structures of the blood vessels and nerves become damaged, leading to loss of circulation and nerve function.[47]

Circulation Problems Loss of blood flow to the kidneys damages them, often resulting in the need to cleanse the blood by means of kidney **dialysis**, or, in later stages, to undergo kidney transplant. Poor circulation also increases the likelihood of infections. With loss of both circulation and nerve function, undetected injury and infection may lead to death of tissue (gangrene), necessitating amputation of the limbs (most often the legs or feet).

KEY POINTS

- Diabetes is a major threat to health and life, and its prevalence is increasing.
- Diabetes involves the body's abnormal handling of glucose and the toxic effects of excess glucose.

Figure 4–15

Prevalence of Diabetes among Adults in the United States

The maps below depict regional changes in U.S. diabetes incidence.

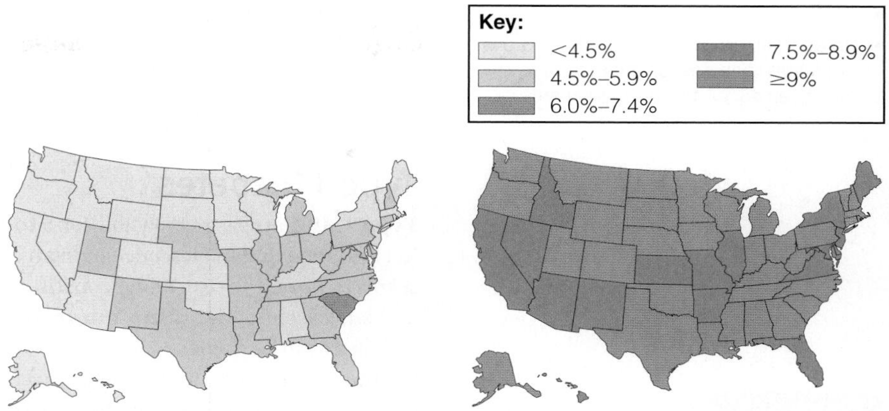

1994: Most states had a prevalence of diabetes of less than 4.5% and only one state had a prevalence of 6% or greater.

2010: No state had a prevalence of diabetes of less than 4.5%; all but three states had a prevalence of 6% or greater.

Source: CDC's Division of Diabetes Translation. National Diabetes Surveillance System available at http://www.cdc.gov/diabetes/statistics

Prediabetes and the Importance of Testing

Prediabetes, a fasting blood glucose level just slightly higher than normal, presents few or none of the warning signs of diabetes (see Table 4–9, p. 140), but tissue damage may progress silently, and type 2 diabetes itself often soon develops.[48] Tens of millions of people in the United States have prediabetes, but few are aware of it.

Diagnosis of diabetes or prediabetes can be made using one of several tests, such as a **fasting plasma glucose test** or an **HbA$_{1c}$ test**.[49††] In a fasting plasma glucose test, a clinician draws blood after a night of fasting and measures whether current

Table 4–8

Type 1 and Type 2 Diabetes Compared

	Type 1	Type 2
Percentage of cases	5–10%	90–95%
Age of onset	<30 years	>40 years[a]
Associated characteristics	Autoimmune diseases, viral infections, inherited factors	Obesity, aging, inherited factors
Primary problems	Destruction of pancreatic beta cells; insulin deficiency	Insulin resistance, insulin deficiency (relative to needs)
Insulin secretion	Little or none	Varies; may be normal, increased, or decreased
Requires insulin	Always	Sometimes
Older names	Juvenile-onset diabetes Insulin-dependent diabetes mellitus (IDDM)	Adult-onset diabetes Noninsulin-dependent diabetes mellitus (NIDDM)

[a]Incidence of type 2 diabetes is increasing in children and adolescence; in more than 90% of these cases, it is associated with overweight or obesity and a family history of type 2 diabetes.

†† Another test for diabetes is the oral glucose tolerance test.

glycemic load (GL) a mathematical expression of both the glycemic index and the carbohydrate content of a food, meal, or diet.

diabetes (dye-uh-BEET-eez) metabolic diseases characterized by elevated blood glucose and inadequate or ineffective insulin, which impair a person's ability to regulate blood glucose. The technical name is *diabetes mellitus* (*mellitus* means "honey-sweet" in Latin, referring to sugar in the urine).

prediabetes condition in which blood glucose levels are higher than normal but not high enough to be diagnosed as diabetes; a major risk factor for diabetes and cardiovascular diseases.

dialysis (die-AL-ih-sis) in kidney disease, treatment of the blood to remove toxic substances or metabolic wastes; more properly, *hemodialysis*, meaning "dialysis of the blood."

fasting plasma glucose test a blood test that measures current blood glucose in a person who has not eaten or consumed caloric beverages for at least 8 hours; the test can detect both diabetes and prediabetes. *Plasma* is the fluid part of whole blood.

HbA$_{1c}$ test a blood test that measures hemoglobin molecules with glucose attached to them (*Hb* stands for *hemoglobin*). The test reflects blood glucose control over the previous few months. Also called *glycosylated hemoglobin test*, or *A1C test*.

Did You Know?

Fasting plasma glucose (milligrams per deciliter):

- Normal: 70–99 mg/dL
- Prediabetes: 100–125 mg/dL
- Diabetes: ≥126 mg/dL
- In addition, an HbA_{1c} value above 6.5% indicates diabetes; a value above 5.7% has been suggested for prediabetes.

blood glucose levels are within the normal range (values are listed in the margin). In an HbA_{1c} test, a nonfasting blood test measures an indicator of how well blood glucose has been controlled over the past few months.[50] A registered dietitian, a Certified Diabetes Educator, or a physician can help those with prediabetes or diabetes learn to manage their condition.

KEY POINTS

- Prediabetes silently threatens the health of tens of millions of people in the United States.
- Medical tests can reveal elevated plasma glucose.

Type 1 Diabetes

Type 1 diabetes is responsible for 5 to 10 percent of diabetes cases. It usually occurs in childhood and adolescence but can occur at any age, even late in life. Its incidence seems to be on the rise, and it constitutes the leading chronic disease among children and adolescents. An **autoimmune disorder** influenced by genetic inheritance, type 1 diabetes arises when the person's own immune system misidentifies the protein insulin as an enemy and attacks the cells of the pancreas that produce it.[51] Soon the damaged pancreas no longer produces enough insulin. Then, after each meal, glucose concentration builds up in the blood, while body tissues are simultaneously starving for glucose, a life-threatening situation. The person must receive insulin from an external source to assist the cells in taking up the glucose they need from the bloodstream that is carrying too much.

Insulin is a protein, and if it were taken orally, the digestive system would digest it. Insulin must therefore be taken as daily shots or pumped from an insulin pump that delivers it through a tiny tube implanted under the skin. Fast-acting and long-lasting forms of insulin allow more flexibility in managing meals and treatments, but users must still plan ahead to balance blood insulin and glucose consumption.

Doing so can make a difference to health—those who control their blood glucose suffer less cardiovascular and other diseases than those who do not.[52] Experimental treatments such as surgical transplants of insulin-producing pancreatic cells and a vaccine to prevent type 1 diabetes are under development.[53]

Table 4–9

Warning Signs of Diabetes

These signs appear reliably in type 1 diabetes and, often, in the later stages of type 2 diabetes.

- Excessive urination and thirst
- Glucose in the urine
- Weight loss with nausea, easy tiring, weakness, or irritability
- Cravings for food, especially for sweets
- Frequent infections of the skin, gums, vagina, or urinary tract
- Vision disturbances; blurred vision
- Pain in the legs, feet, or fingers
- Slow healing of cuts and bruises
- Itching
- Drowsiness
- Abnormally high glucose in the blood

KEY POINTS

- Type 1 diabetes is an autoimmune disease that attacks the pancreas and necessitates an external source of insulin for its management.
- Inadequate insulin leaves blood glucose too high, while cells remain undersupplied with glucose energy.

Type 2 Diabetes

Recent decades have seen a sharp rise in the rate of the predominant type of diabetes mellitus, **type 2 diabetes** (responsible for 90 to 95 percent of cases) in both adults and children. In type 2 diabetes, body tissues lose their sensitivity to insulin. The insulin-resistant muscle and adipose tissues no longer respond to insulin by increasing their uptake of glucose from the blood. As blood glucose climbs higher, the pancreas compensates by producing larger and larger amounts of insulin. Blood insulin may rise abnormally high, but to no avail. Eventually, the overtaxed cells of the pancreas begin to fail and reduce their insulin output, while blood glucose spins further out of control.

Did You Know?

The first use of genetic engineering was to alter the DNA of a bacterium to produce insulin for treatment of diabetes.

Type 2 Diabetes and Obesity Obesity underlies many cases of type 2 diabetes. Middle age and physical inactivity also foreshadow its development. The greater the accumulation of body fat, particularly around the waistline, the more insulin-resistant the cells become, and the higher the blood glucose rises. Even moderate weight gain in adults increases the risk. Aging also increases the risk of developing type 2 diabetes, and so does genetic inheritance.[54] Genetic tests may one day identify susceptible people so that they may take action early to minimize its effects.

One way in which obesity and type 2 diabetes may worsen each other is depicted in Figure 4–16. Many factors may contribute to obesity, but, according to the theory, once obesity sets in, inflammation and other metabolic changes trigger the tissues to resist insulin. Meanwhile, blood lipid levels rise along with blood glucose, resulting in an overabundance of circulating fuels that add lipid to adipose tissue stores. In **insulin resistance**, the brain's complex satiety-signaling system may also become skewed in the direction of increased food intake.[55] As fat mass increases, insulin resistance worsens, blood glucose rises, and both obesity and diabetes are perpetuated. Given this series of events, is it any wonder that obese people with type 2 diabetes have trouble losing weight?

Preventing Type 2 Diabetes Once in the grip of type 2 diabetes, the body struggles to control blood glucose and often fails to stop its damage, even with the best of medical care.[56] Prevention, however, is not only possible but also likely when individuals take action. In research, these five lifestyle factors consistently and dramatically reduce people's risk of developing diabetes:

1. A healthy body weight.
2. A nutritious eating pattern (moderate in calories; low in saturated fat; high in vegetables, legumes, fruit, fish, poultry, and whole grains).
3. Regular physical activity.
4. Moderate alcohol intake among drinkers.
5. Not smoking.[57]

Even people with prediabetes or diabetes can often change their fate by losing weight, exercising, choosing a nutritious diet, and if necessary, faithfully using prescribed medications.[58] It's never too late—even older adults can lower their diabetes risk by changing their lifestyles.

Figure 4–16

An Obesity–Type 2 Diabetes Cycle

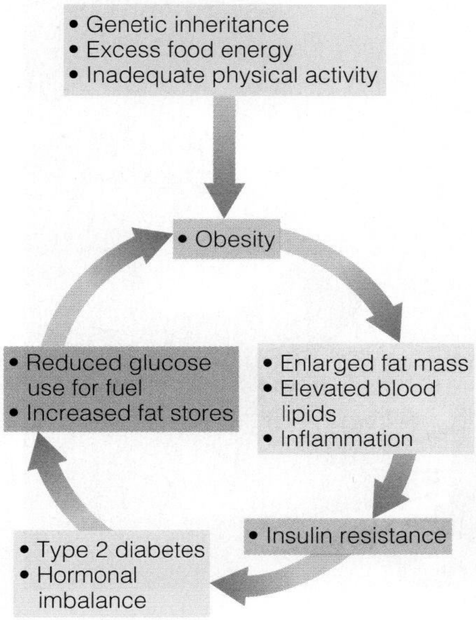

© Cengage Learning

KEY POINT

- Type 2 diabetes risk factors make the disease likely to develop, but prevention is possible.

Management of Diabetes

LO 4.6 Identify components of a lifestyle plan to effectively control blood glucose, and describe the characteristics of an eating plan that can help manage type 2 diabetes.

To reduce harm from diabetes and prediabetes, the primary goal is to keep blood glucose levels within the normal range or as close to normal as is safely and practically possible through a daily routine of proper diet, exercise, glucose monitoring, and medication. In addition, overweight people urgently need to lose weight because overweight worsens type 2 diabetes and its associated conditions—generally, a loss of 5 to 7 percent of body weight is enough to improve well-being.[59] For some obese people,

My Turn **watch it!**

21st Century Epidemic?

Two young people talk about living with diabetes.

© Cengage Learning

Liz

© Cengage Learning

Ariela

type 1 diabetes the type of diabetes in which the pancreas produces no or very little insulin; often diagnosed in childhood, although some cases arise in adulthood. Formerly called *juvenile-onset* or *insulin-dependent diabetes.*

autoimmune disorder a disease in which the body develops antibodies to its own proteins and then proceeds to destroy cells containing these proteins. Examples are type 1 diabetes and lupus.

type 2 diabetes the type of diabetes in which the pancreas makes plenty of insulin but the body's cells resist insulin's action; often diagnosed in adulthood. Formerly called *adult-onset* or *non–insulin-dependent diabetes.*

insulin resistance a condition in which a normal or high level of circulating insulin produces a less-than-normal response in muscle, liver, and adipose tissues; thought to be a metabolic consequence of obesity.

weight-loss surgery may resolve their diabetes, but relapses are common and surgery imposes serious risks of its own (see Chapter 9).[60] Many other diabetes management strategies also aim to maintain the health of the heart and blood vessels because diabetes greatly elevates the risks for developing heart and artery diseases. In making treatment decisions, a person-centered approach that respects the individual's needs, preferences, and values often works best.[61]

© Tony Freeman / Photo Edit

Monitoring is critical to controlling blood glucose.

Nutrition Therapy

Controlling carbohydrate intake plays a central role in controlling blood glucose.[62] A common misconception is that people with diabetes need only to avoid sugary foods, but as far as blood glucose is concerned, the *amount* of carbohydrate often matters more than its *source*.

How Much Carbohydrate Is Best? The total amount of daily carbohydrate recommended for people with diabetes varies with an individual's glucose tolerance, and proper timing of carbohydrate consumption helps to hold blood glucose levels steady. Eating too much carbohydrate at one time can raise blood glucose too high, whereas eating too little can lead to abnormally low glucose levels (**hypoglycemia**). A low-carbohydrate diet (less than 130 grams of carbohydrate per day) is not recommended.[63] Instead, an eating pattern that derives carbohydrates from fruits, vegetables, legumes, whole grains, and low-fat milk, spaced throughout the day, best serves the goals of diabetes treatment.

Sugar Alcohols and Nonnutritive Sweeteners Sugar substitutes are often useful to people wishing to control calorie or sugar intakes. Products sweetened with **sugar alcohols**, such as cookies, sugarless gum, hard candies, and jams and jellies, are safe in moderation. Sugar alcohols provide about half the calories of sugars and trigger a lower glycemic response. One exception, erythritol, cannot be metabolized by human enzymes and so is calorie-free. Table 4–10 names some common sugar alcohols.

Table 4–10
Sugar Alcohols

These common sugar alcohols may be listed on food labels:

- erythritol
- isomalt
- lactitol
- maltitol
- mannitol
- sorbitol
- xylitol

Sugar alcohols are safer for teeth than sugars, making them useful in chewing gums, breath mints, toothpaste, and other products that people keep in their mouths for a while. Mouth bacteria rapidly metabolize regular sugars into acids that cause **dental caries**; sugar alcohols resist such metabolism. Side effects such as gas, abdominal discomfort, and diarrhea arise from ingesting large quantities of sugar alcohols.

In the same vein, **nonnutritive sweeteners**, can sweeten foods without calories, but people have concerns about their use. Their nature and safety are topics of Chapter 12.

Diet Recommendations in Summary The same diet pattern that best controls diabetes can also help to control body weight and support physical activity.[64] This diet pattern is:

- Controlled in total carbohydrate (to regulate blood glucose concentration), delivered in well-timed meals of nutrient dense foods.
- Low in saturated and *trans* fat (these worsen cardiovascular disease risks) and should provide some raw unsaturated oils (to provide essential nutrients).

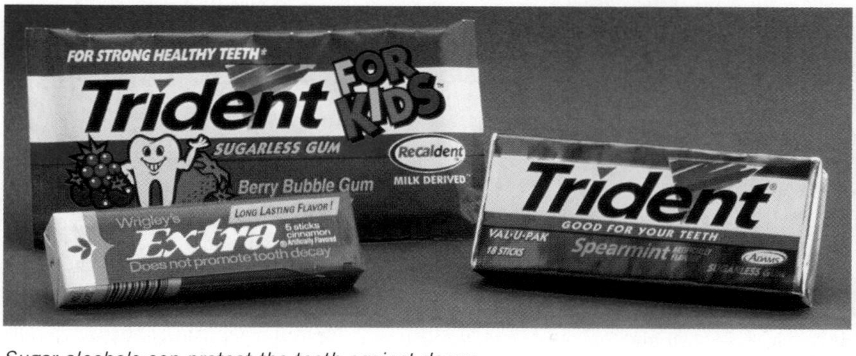

© Craig M. Moore

Sugar alcohols can protect the teeth against decay.

- Adequate in nutrients from food, not supplements (to avoid deficiencies).
- Adequate in fiber (from whole grains, fruits, legumes, and vegetables).
- Limited in added sugars (to help control calories and refined carbohydrate intake).
- Adequate in protein (protein must be limited in kidney disease to reduce risk of harm).

Such a diet also has all the characteristics important to prevention of many chronic diseases and meets the recommendations of the *Dietary Guidelines for Americans 2010*. A person at risk for diabetes can do no better than to adopt such a diet long before symptoms appear.

KEY POINT

- Diet plays a central role in controlling diabetes and the illnesses that accompany it.

Physical Activity

The role of regular physical activity in preventing and controlling diabetes, particularly type 2 diabetes, cannot be overstated.[65] Exercise helps reduce the body's fatness and heightens tissue sensitivity to insulin. Increasing physical activity can help delay onset of type 2 diabetes and help to regulate blood glucose in established cases, sometimes to the degree that medication can be reduced or eliminated. People with type 1 diabetes should check with a physician because physical activity can bring on hypoglycemia. Like a juggler who keeps three balls in motion, the person with diabetes must constantly balance three factors—diet, exercise, and medication—to control the blood glucose level.

KEY POINT

- Regular physical activity, in addition to diet and medication, helps to control blood glucose in diabetes.

Physical activity is a key player in controlling diabetes.

If I Feel Dizzy between Meals, Do I Have Hypoglycemia?

LO 4.7 Describe the symptoms of hypoglycemia, and name some conditions that may cause it.

In healthy people, blood glucose rises after eating and then gradually falls back into the normal range. The transition occurs without notice. Should blood glucose drop below normal, a person would experience the symptoms of hypoglycemia: weakness, rapid heartbeat, sweating, anxiety, hunger, and trembling. Most commonly, hypoglycemia is a consequence of poorly managed diabetes: too much insulin, strenuous physical activity, inadequate food intake, or illness that causes blood glucose levels to plummet.

Hypoglycemia is rare as a true disease, but many people believe they experience its symptoms at times. Most people who experience hypoglycemia need only adjust their diets by replacing refined carbohydrates with fiber-rich whole-food sources of carbohydrate and eating adequate protein at each meal. In addition, smaller meals eaten more frequently may help. Hypoglycemia caused by certain medications, pancreatic tumors, overuse of insulin, alcohol abuse, uncontrolled diabetes, or other illnesses requires medical intervention.

KEY POINT

- In hypoglycemia, blood glucose falls too low; it arises mainly in people with diabetes or other conditions or by medications and is rare among healthy people.

hypoglycemia (HIGH-poh-gly-SEE-mee-ah) an abnormally low blood glucose concentration, often accompanied by symptoms such as anxiety, rapid heartbeat, and sweating.

sugar alcohols sugarlike compounds in the chemical family *alcohol* derived from fruits or manufactured from sugar dextrose or other carbohydrates; sugar alcohols are absorbed more slowly than sugars, are metabolized differently, and do not elevate the risk of dental caries. Also called *polyols*.

dental caries decay of the teeth (*caries* means "rottenness"). Dental caries are a topic of Chapter 14.

nonnutritive sweeteners sugar substitutes that provide negligible, if any, energy. Also defined in Chapter 12.

Conclusion

Part of eating right is choosing wisely among the many foods available. Largely without your awareness, the body responds to the carbohydrates supplied by your diet. Now you take the controls by learning how to integrate carbohydrate-rich foods into an eating pattern that meets your body's needs.

try it! Food Feature

Finding the Carbohydrates in Foods

LO 4.8 Identify the main contributors of various forms of carbohydrates in foods.

To support optimal health, an eating pattern must supply enough of the right kinds of carbohydrate-rich foods. Dietary recommendations for a *health-promoting* 2,000-calorie diet suggest that carbohydrates provide in the range of 45 to 65 percent of calories, or between 225 and 325 grams, each day. This amount more than meets the minimum DRI amount of 130 grams needed to feed the brain and ward off ketosis. People needing more or less energy require proportionately more or less carbohydrate.

If you are curious about your own carbohydrate need, find your DRI estimated energy requirement (see the inside front cover of this text) and multiply by 45 percent to obtain the bottom of your carbohydrate intake range and then by 65 percent for the top; then divide both answers by 4 calories per gram (see the example in the margin).

Do the Math

Example for 45% of calories in a 2,700-calorie diet:

- 2,700 cal × 0.45 = 1,215 cal
- 1,215 cal ÷ 4 cal/g = 304 g

Example for 65% of calories in a 2,700-calorie diet:

- 2,700 cal × 0.65 = 1,755 cal
- 1,775 cal ÷ 4 cal/g = 439 g

The range of carbohydrate intake recommended in a 2,700-calorie eating pattern ranges between about 300 and 440 grams per day.

Breads and cereals, starchy vegetables, fruits, and milk are all good contributors of starch and dilute sugars. Many foods also provide fiber in varying amounts, as Figure 4–17 demonstrates. Concentrated sweets provide sugars but little else, as the last section demonstrates.

Fruits

A fruit portion of 1/2 cup of juice, a small banana, apple, or orange, 1/2 cup of most canned or fresh fruit, or 1/4 cup of dried fruit supplies an average of about 15 grams of carbohydrate, mostly as sugars, including the fruit sugar fructose. Fruits vary greatly in their water and fiber contents and in their sugar concentrations. Juices should contribute no more than one-half of a day's intake of fruit. Except for avocado and olives, which are high in fat, fruits contain insignificant amounts of fat and protein.

Vegetables

Starchy vegetables are major contributors of starch in the diet. Just one small white or sweet potato or 1/2 cup of cooked dry beans, corn, peas, plantain, or winter squash provides 15 grams of carbohydrate, as much as in a slice of bread, though as a mixture of sugars and starch. One-half cup of carrots, okra, onions, tomatoes, cooked greens, or most other nonstarchy vegetables or a cup of salad greens provides about 5 grams as a mixture of starch and sugars.

Grains

Breads and other starchy foods are famous for their carbohydrate. Nutrition authorities encourage people to replace refined grains with whole grains whenever possible, and to make at least half of the grain choices whole grains. A slice of bread, half an English muffin, a 6-inch tortilla, 1/3 cup of rice or pasta, or 1/2 cup of cooked cereal provides about 15 grams of carbohydrate, mostly as starch. Ready-to-eat cereals, particularly those that children prefer, can derive over half their weight in added sugars, so consumers must read labels.

Most grain choices should also be low in solid fats and added sugar. When extra calories are required to meet energy needs, some selections higher in unsaturated fats (see Chapter 5) and added sugar can supply needed calories

Figure 4–17

Fiber in the Food Groups

© Polara Studios, Inc. (all)

Fruits

Food[a]	Fiber (g)	Food	Fiber (g)
Pear, raw, 1 medium	5	Other berries, raw, 1/2 c	2
Blackberries/raspberries, raw, 1/2 c	4	Peach, raw, 1 medium	2
Prunes, cooked, 1/4 c	4	Strawberries, sliced, 1/2 c	2
Figs, dried, 3	3	Cantaloupe, raw, 1/2 c	1
Apple, 1 medium	3	Cherries, raw, 1/2 c	1
Apricots, raw, 4 each	3	Fruit cocktail, canned, 1/2 c	1
Banana, raw, 1	3	Peach half, canned	1
Orange, 1 medium	3	Raisins, dry, 1/4 c	1
		Orange juice, 3/4 c	<1

Vegetables

Food	Fiber (g)	Food	Fiber (g)
Baked potato with skin, 1	4	Mashed potatoes, home recipe, 1/2 c	2
Broccoli, chopped, 1/2 c	3	Bell peppers, 1/2 c	1
Brussels sprouts, 1/2 c	3	Broccoli, raw, chopped, 1/2 c	1
Spinach, 1/2 c	3	Carrot juice, 1/2 c	1
Asparagus, 1/2 c	2	Celery, 1/2 c	1
Baked potato, no skin, 1	2	Dill pickle, 1 whole	1
Cabbage, red, 1/2 c	2	Eggplant, 1/2 c	1
Carrots, 1/2 c	2	Lettuce, romaine, 1 c	1
Cauliflower, 1/2 c	2	Onions, 1/2 c	1
Corn, 1/2 c	2	Tomato, raw, 1 medium	1
Green beans, 1/2 c	2	Tomato juice, canned, 3/4 c	1

Grains

Food	Fiber[a] (g)	Food	Fiber (g)
100% bran cereal, 1 oz	10	Pumpernickel bread, 1 slice	2
Barley, pearled, 1/2 c	3	Shredded wheat, 1 large biscuit	2
Cheerios, 1 oz	3	Cornflakes, 1 oz	1
Whole-wheat bread, 1 slice	3	Muffin, blueberry, 1	1
Whole-wheat pasta,[b] 1/2 c	3	Puffed wheat, 1 1/2 c	1
Wheat flakes, 1 oz	3	White pasta,[b] 1/2 c	1
Brown rice, 1/2 c	2	Cream of wheat, 1/2 c	<1
Light rye bread, 1 slice	2	White bread, 1 slice	<1
Muffin, bran, 1 small	2	White rice, 1/2 c	<1
Oatmeal, 1/2 c	2		
Popcorn, 2 c	2		

Protein Foods

Food	Fiber (g)	Food	Fiber (g)
Lentils, 1/2 c	8	Soybeans, 1/2 c	5
Kidney beans, 1/2 c	8	Almonds or mixed nuts, 1/4 c	4
Pinto beans, 1/2 c	8	Peanuts, 1/4 c	3
Black beans, 1/2 c	7	Peanut butter, 2 tbs	2
Black-eyed peas, 1/2 c	6	Cashew nuts, 1/4 c	1
Lima beans, 1/2 c	5	Meat, poultry, fish, and eggs	0

[a]*All values are for ready-to-eat or cooked foods unless otherwise noted. Fruit values include edible skins. All values are rounded values.*

[b]*Pasta includes spaghetti noodles, lasagna, and other noodles.*

Art © Cengage Learning

and provide pleasure in eating. These choices might include biscuits, cookies, croissants, muffins, ready-to-eat sweetened cereals, and snack crackers.

Protein Foods

With two exceptions, foods of this group provide almost no carbohydrate to the diet. The exceptions are nuts, which provide a little starch and fiber along with their abundant fat, and legumes (dried beans), revered by diet-watchers as high-protein, low-fat sources of both starch and fiber that can reduce feelings of hunger.[66] Just 1/2 cup of cooked beans, peas, or lentils provides 15 grams of carbohydrate, an amount equaling the richest carbohydrate sources. Among sources of fiber, legumes are peerless, providing as much as 8 grams in 1/2 cup.

Milk and Milk Products

A cup of milk or plain yogurt is a generous contributor of carbohydrate, donating about 12 grams. Cottage cheese provides about 6 grams of carbohydrate per cup, but most other cheeses contain little if any carbohydrate. These foods also contribute high-quality protein (a point in their favor), as well as several important vitamins and minerals. Calcium-fortified soy beverages (soy milk) and soy yogurts approximate the nutrients of milk, providing some amount of added calcium and 14 grams of carbohydrate. Milk and soy milk products vary in fat content, an important consideration in choosing among them. Sweetened milk and soy products contain added sugars.

Butter and cream cheese, though dairy products, are not equivalent to milk because they contain little or no carbohydrate and insignificant amounts of the other nutrients important in milk. They are appropriately associated with the solid fats.

Oils, Solid Fats, and Added Sugars

Oils and solid fats are devoid of carbohydrate, but added sugars provide almost pure carbohydrate. Most people enjoy sweets, so it is important to learn something of their nature and to account for them in an eating pattern. First, the definitions of "sugar" come into play (Table 4–11 defines sugar terms).

All sugars originally develop by way of photosynthesis in a plant. A sugar mol-

Table 4–11

Terms That Describe Sugar

Note: The term *sugars* here refers to all of the monosaccharides and disaccharides. On a label's ingredients list, the term *sugar* means sucrose. See Chapter 12 for terms related to noncaloric nonnutritive sweeteners.

- **added sugars** sugars and syrups added to a food for any purpose, such as to add sweetness or bulk or to aid in browning (baked goods). Also called carbohydrate sweeteners, they include glucose, fructose, corn syrup, concentrated fruit juice, and other sweet carbohydrates.
- **agave syrup** a carbohydrate-rich sweetener made from a Mexican plant; a higher fructose content gives some agave syrups a greater sweetening power per calorie than sucrose.
- **brown sugar** white sugar with molasses added, 95% pure sucrose.
- **concentrated fruit juice sweetener** a concentrated sugar syrup made from dehydrated, deflavored fruit juice, commonly grape juice; used to sweeten products that can then claim to be "all fruit."
- **confectioner's sugar** finely powdered sucrose, 99.9% pure.
- **corn sweeteners** corn syrup and sugar solutions derived from corn.
- **corn syrup** a syrup, mostly glucose, partly maltose, produced by the action of enzymes on cornstarch. Includes corn syrup solids.
- **dextrose, anhydrous dextrose** forms of glucose.
- **evaporated cane juice** raw sugar from which impurities have been removed.
- **fructose, galactose, glucose** the monosaccharides.
- **granulated sugar** common table sugar, crystalline sucrose, 99.9% pure.
- **high-fructose corn syrup** a commercial sweetener used in many foods, including soft drinks. Composed almost entirely of the monosaccharides fructose and glucose, its sweetness and caloric value are similar to sucrose.
- **honey** a concentrated solution primarily composed of glucose and fructose, produced by enzymatic digestion of the sucrose in nectar by bees.
- **invert sugar** a mixture of glucose and fructose formed by the splitting of sucrose in an industrial process. Sold only in liquid form and sweeter than sucrose, invert sugar forms during certain cooking procedures and works to prevent crystallization of sucrose in soft candies and sweets.
- **lactose, maltose, sucrose** the disaccharides.
- **levulose** an older name for fructose.
- **malt syrup** a sweetener made from sprouted barley.
- **maple syrup** a concentrated solution of sucrose derived from the sap of the sugar maple tree. This sugar was once common but is now usually replaced by sucrose and artificial maple flavoring.
- **molasses** a thick brown syrup left over from the refining of sucrose from sugar cane. The major nutrient in molasses is iron, a contaminant from the machinery used in processing it.
- **naturally occurring sugars** sugars that are not added to a food but are present as its original constituents, such as the sugars of fruit or milk.
- **nectars** concentrated peach nectar, pear nectar, or others.
- **raw sugar** the first crop of crystals harvested during sugar processing. Raw sugar cannot be sold in the United States because it contains too much filth (dirt, insect fragments, and the like). Sugar sold as "raw sugar" is actually evaporated cane juice.
- **turbinado (ter-bih-NOD-oh) sugar** raw sugar from which the filth has been washed; legal to sell in the United States.
- **white sugar** granulated sucrose, produced by dissolving, concentrating, and recrystallizing raw sugar. Also called *table sugar*.

ecule inside a grape (one of the **naturally occurring sugars**) is chemically indistinguishable from one extracted from sugar cane, grapes, or corn and added at the factory to sweeten strawberry jam. Honey added to food is also an added sugar with similar chemical makeup. All arise naturally and, through processing, are purified of most or all of the original plant material—bees process honey and machines process the other types. The body handles all the sugars in the same way, whatever their source.

Sugar grams listed on the Nutrition Facts panel of a food label reflect sugars from all sources, both added and naturally occurring, listed in one value on the line reading "sugars." Consumers, therefore, cannot estimate their intakes of added sugars from label information.

The *Dietary Guidelines for Americans* offer clear advice on added sugars: reduce intake.[67] Added sugars, when consumed in large amounts, may be linked with health problems (see the Controversy section), and they bring only calories into the diet, with no other significant nutrients.[68] Conversely, the naturally occurring sugars of, say, an orange provide calories but also the vitamins, minerals, fiber, and phytochemicals of oranges. Added sugars can contribute to nutrient deficiencies by displacing nutritious food from the diet. Most people can afford only a little added sugar in their diets if they are to meet nutrient needs within calorie limits. The USDA food patterns suggest about 8 teaspoons of sugar, or almost one soft drink's worth, in a nutrient-dense 2,200 calorie eating pattern (the margin lists other amounts). Table 4–12 provides some tips for taking in less added sugars while still enjoying their sweet taste.

Table 4–12

Tips for Reducing Intakes of Added Sugars

These tricks can help reduce added sugar intake by changing old habits.

- A good use of sugar is to make nutrient-dense but bland or sharp-tasting foods (such as oatmeal or grapefruit) more palatable. Use the least amount possible to do the job.
- Add sweet spices such as cinnamon, nutmeg, allspice, or clove.
- Add a tiny pinch of salt; it will make food taste sweeter.
- Nonnutritive sweeteners add sweetness without calories. Read about them in Chapter 12.
- Choose fruit for dessert most often.
- Choose smaller portions of cake, cookies, ice cream, other desserts, and candy, or skip them.
- Compare sugar contents of similar foods on their Nutrition Facts panels, and choose those with less sugar. For example, a cup of pineapple chunks canned in heavy syrup has 43 g sugar; a cup canned in juice has 26 g. The difference is all added sugars.
- Reduce sugar added to recipes or foods at the table by a third—the difference in taste generally isn't noticeable.
- Replace empty-calorie-rich regular sodas, sports drinks, energy drinks, and fruit drinks with water, fat-free milk, 100% fruit juice, or unsweetened tea or coffee.
- Warm up sweet foods before serving (heat enhances sweet tastes).

The Nature of Sugar

Each teaspoonful of any sweet can be assumed to supply about 16 calories and 4 grams of carbohydrate. An exception is honey, which packs more calories into each teaspoon because its crystals are dissolved in water; the dry crystals of sugar take up more space. If you use ketchup liberally, remember that each tablespoon of it contains a teaspoon of sugar. And for the soft-drink user, a 12-ounce can of sugar-sweetened cola contains at least 8 teaspoons of added sugar.

What about the nutritional value of a product such as molasses or concentrated fruit juice sweetener compared to white sugar? Molasses contains 1 milligram of iron per tablespoon so, if used frequently, it can contribute some of this important nutrient. Molasses is less sweet than the other sweeteners, however, so more molasses is needed to provide the same sweetness as sugar. Also, its iron comes from the machinery in which molasses is made and is in the form of an iron salt not easily absorbed by the body.

As for concentrated juice sweeteners, such as the concentrated grape or pear "juice" used to sweeten foods and beverages, these are highly refined and have lost virtually all of the beneficial nutrients and phytochemicals of the original fruit. A child's fruit punch sweetened with grape juice concentrate, for example, may claim to be "100 percent fruit juice" and sounds nutritious but can contain as much sugar as punches sweetened with sucrose or HFCS. No form of sugar, even honey, is any "more healthy" than white sugar, as Table 4–13, p. 148, shows.

Finally, enjoy whatever sugar you do eat. Sweetness is one of life's great sensations, so enjoy it in moderation.[69]

Did You Know?

The *Dietary Guidelines for Americans 2010* urge most people to reduce their intakes of added sugars to about these levels:

- 4 tsp for 1,600 cal
- 5 tsp for 1,800 cal
- 8 tsp for 2,000 cal
- 8 tsp for 2,200 cal
- 10 tsp for 2,400 cal

Table 4–13

The Empty Calories of Sugar

These data demonstrate the absurdity of trying to rely on any added sugar for nutrient contributions. The 64 calories of honey (1 tablespoon) listed bring 0.1 mg of iron into the diet, but it would take 11,500 calories of honey (180 tablespoons) to provide the needed 18 mg of iron for a young woman. The nutrients of added sugars do not add up as fast as their calories.

Food	Energy (cal)	Protein (g)	Fiber (g)	Calcium (mg)	Iron (mg)	Magnesium (mg)	Potassium (mg)	Zinc (mg)	Vitamin A (µg)	Thiamin (mg)	Riboflavin (mg)	Niacin (mg)	Vitamin B_6 (mg)	Folate (µg)	Vitamin C (mg)
Sugar (1 tbs)	46	0	0	0	0	0	0	0	0	0	0	0	0	0	0
Honey (1 tbs)	64	0	0	1	0.1	0	11	0	0	0	0	0	0	<1	0
Molasses (1 tbs)	55	0	0	42	1.0	50	300	0.1	0	0	0	0.2	0.1	0	0
Concentrated grape or fruit juice sweetener (1 tbs)	30	0	0	0	0	0	0	0	0	0	0	0	0	0	
Jelly (1 tbs)	49	0	0	1	0	1	12	0	0	0	0	0	0	0	<1
Brown sugar (1 tbs)	54	0	0	8	0.2	3	31	0	0	0	0	0	0	0	0
Cola beverage (12 fl oz)	153	0	0	11	0.1	4	4	0	0	0	0	0	0	0	0
Daily Values	2,000	56	25	1,000	18	400	3,500	15	1,000	1.5	1.7	20	2	400	60

track it!

Diet Analysis PLUS

Concepts in Action

Analyze Your Carbohydrate Intake

The purpose of this chapter's exercise is to help you examine the carbohydrate-rich foods in your diet, compare your intakes with recommendations, and help you obtain the recommended daily intake of carbohydrates and soluble and insoluble fiber.

1. In the DA+ program, select the Reports tab, then select the Macronutrient Ranges. Using your 3-day diet records, choose Day Two and choose all meals. Generate a report. Did your intake meet the recommendation to consume between 45 and 65 percent of total calories as carbohydrate?
2. Determine the distribution of carbohydrate among the day's foods. Select Reports, then Source Analysis, and then Carbohydrate from the drop-down box. Generate a separate report for each meal: breakfast, lunch, and dinner. At which meal did you consume the most carbohydrate? Which foods were the greatest contributors?
3. Did your fiber intake fall within the recommended range (25–35 grams per day)? From the Reports tab select Intake vs. Goals. Choose Day One, choose all meals, and generate a report. Did you meet your fiber need?
4. From Reports, select Source Analysis. Using Day Three, choose all meals, and generate a report. Which foods provided the greatest amounts of fiber for the days' intake? If you are short on fiber, take a look at Figure 4–4 (page 120), Figure 4–5 (page 121), and Figure 4–17 (page 145), and suggest fiber-rich foods to increase your intake of both soluble and insoluble fibers.
5. Whole-grain foods add more than just fiber to the diet. From Track Diet, create a new day (do not alter your 3-day record). Enter two food items as a snack: 2.5 cups Froot Loops cereal and 0.5 cup granola (these amounts are equal in calories). Select Reports, Source Analysis, and the mineral magnesium from the drop-down box. Generate a report for the new snack. Which was the better magnesium source?

what did you decide?

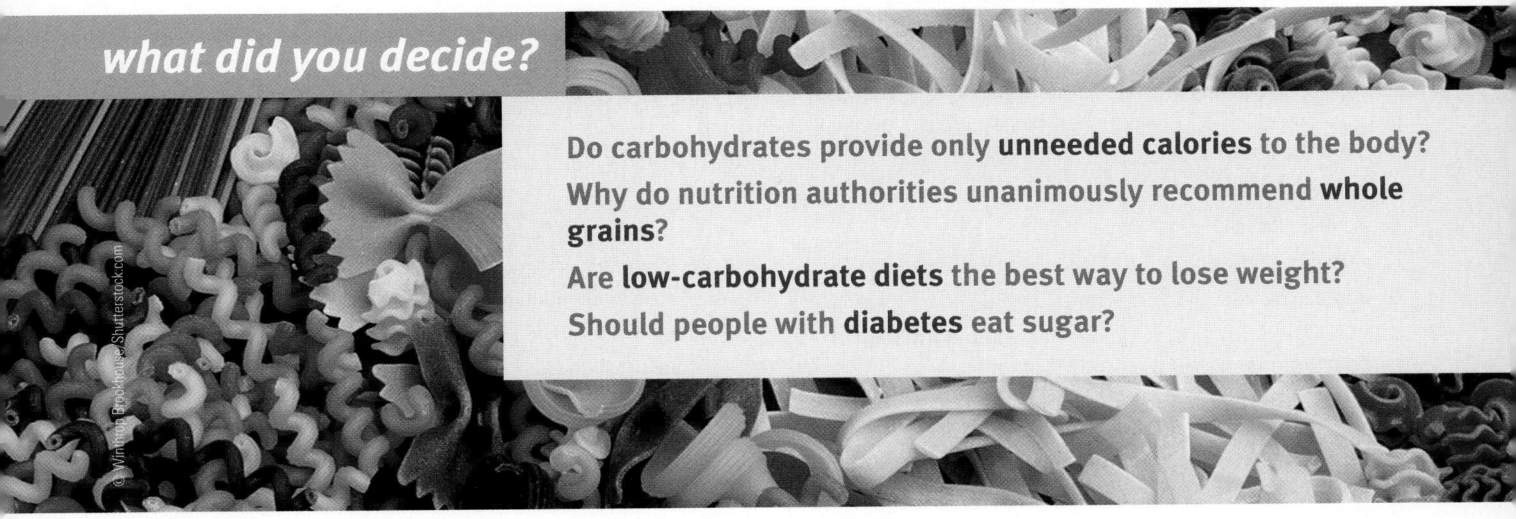

Do carbohydrates provide only unneeded calories to the body?

Why do nutrition authorities unanimously recommend whole grains?

Are low-carbohydrate diets the best way to lose weight?

Should people with diabetes eat sugar?

Self Check

1. (LO 4.1) The dietary monosaccharides include __________.
   a. sucrose, glucose, and lactose
   b. fructose, glucose, and galactose
   c. galactose, maltose, and glucose
   d. glycogen, starch, and fiber
2. (LO 4.1) The polysaccharide that helps form the supporting structures of plants is __________ .
   a. cellulose
   b. maltose
   c. glycogen
   d. sucrose
3. (LO 4.2) Foods rich in soluble fiber lower blood cholesterol.
   T F
4. (LO 4.2) The fiber-rich portion of the wheat kernel is the bran layer.
   T F
5. (LO 4.3) Digestible carbohydrates are absorbed as _________ through the small intestinal wall and are delivered to the liver, which releases ________ into the bloodstream.
   a. disaccharides; sucrose
   b. glucose; glycogen
   c. monosaccharides; glucose
   d. galactose; cellulose
6. (LO 4.3) Around the world, most people are lactose intolerant.
   T F
7. (LO 4.4) When blood glucose concentration rises, the pancreas secretes _______, and when blood glucose levels fall, the pancreas secretes ________.
   a. glycogen; insulin
   b. insulin; glucagon
   c. glucagon; glycogen
   d. insulin; fructose
8. (LO 4.4) The body's use of fat for fuel without the help of carbohydrate results in the production of _______ .
   a. ketone bodies
   b. glucose
   c. starch
   d. galactose
9. (LO 4.5) For people with diabetes, the risk of heart disease, stroke, and dying on any particular day is cut in half.
   T F
10. (LO 4.5) Type 1 diabetes is most often controlled by successful weight-loss management.
    T F
11. (LO 4.6) Type 2 diabetes often improves with a diet that is
    a. low in carbohydrates (less than 130 g per day).
    b. as low in fat as possible.
    c. controlled in carbohydrates and calories.
    d. a and b
12. (LO 4.6) For managing type 2 diabetes, regular physical activity can help by redistributing the body's fluids.
    T F
13. (LO 4.7) Fasting hypoglycemia may be caused by all except
    a. pancreatic tumors
    b. poorly controlled diabetes
    c. overuse of alcohol
    d. lactose intolerance
14. (LO 4.7) Hypoglycemia as a disease is relatively common.
    T F
15. (LO 4.8) Protein foods provide almost no carbohydrate to the U.S. diet, with these two exceptions:
    a. chicken and turkey
    b. beef and pork
    c. fish and eggs
    d. nuts and legumes
16. (LO 4.8) Fruit punch sweetened with grape juice concentrate can contain as much sugar as fruit punch sweetened with high-fructose corn syrup.
    T F
17. (LO 4.9) In the United States, diets high in refined carbohydrate intakes, particularly added sugars from soft drinks, are often associated with increased body fatness.
    T F
18. (LO 4.9) When fructose is consumed in excess of calorie need
    a. it is about as fattening as the same amount of sucrose.
    b. it stimulates a greater insulin response than glucose, and so is more fattening.
    c. it provides more calories per gram than glucose, and so is more fattening.
    d. its metabolism in the body is identical to that of glucose.

Answers to these Self Check questions are in Appendix G.

Are Added Sugars "Bad" for You?

LO 4.9 Discuss current research regarding the relationships among added sugars, obesity, diabetes, and other ills.

© Don Smetzer/Alamy

Recently, the *Dietary Guidelines for Americans* have urged people to strictly limit intakes of added sugars.[1] Does this mean that these food constituents cause harm? This Controversy investigates some of the accusations against these carbohydrates. It also demonstrates the scientific response to questions via research investigation.

Do Added Sugars Cause Obesity?

Over the past several decades, people in the United States have grown dramatically fatter (Figure C4–1). At the same time, their intakes of calories from carbohydrates jumped from 42 percent in the 1970s to 49 percent today, with much of the additional energy coming from refined grains and added sugars.[2] In fact, sugary desserts, such as snack cakes, cookies, and doughnuts, are the number one source of calories for people age 2 years and older.[3] Sugar-sweetened soft drinks follow closely behind for adolescents and young adults.

During the same period, total calorie intakes climbed sharply.[4] The increase, estimated at over 300 calories a day (see Figure C4–2), was more than enough to cause a nation-wide weight gain of two pounds *every month* during this time period.*[5] In addition, as calorie intakes went up, physical activity declined, and most people were not active enough to use up those extra calories.[6] Not surprisingly, then, the average body weight for adults increased by about 20 pounds at the same time.

* *Based on a gain of 1 lb of body weight per 3,500 excess calories; actual amounts vary widely among individuals.*

Figure C4–1
Increases in Adult Body Weight over Time

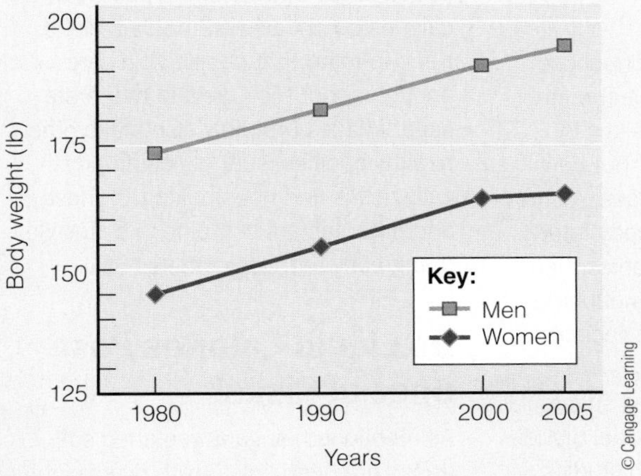

Figure C4–2
Daily Energy Intake over Time

Carbohydrates, and mostly added sugars, account for almost all of the increase in energy intakes during this period.

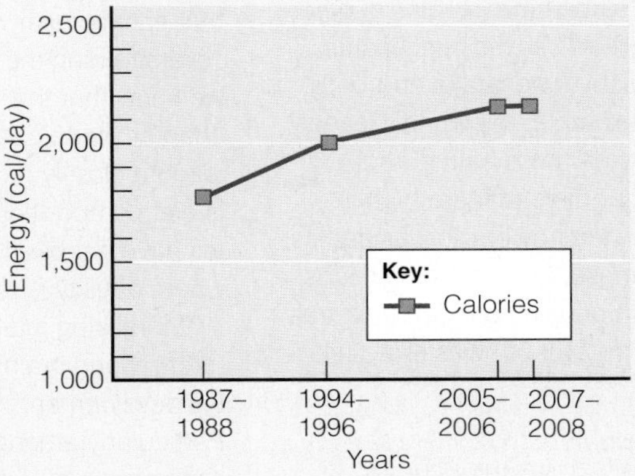

Source: National Center for Health Statistics, 2012.

Intakes of Added Sugars

In past centuries, the only concentrated sweetener was honey, a rare treat. Recent decades have seen a dramatic upward trend in consumption of added sugars. Today, each person in the United States uses on average almost three-quarters of a cup (31 teaspoons) of added sugars in his or her foods and beverages every day.† This amount is enough to provide every man, woman, and child with about 132 pounds of added sugars per year.[7] This number does not account for waste, such as the syrup drained from sweet pickles, nor does it reflect intakes of sweetened imported foods, a fast-growing source of added sugars.

All kinds of sugary foods and beverages taste delicious, cost little money, and are constantly available, making overconsumption extremely likely. More than 95 percent of the sugars in the U.S. diet are now added to foods and beverages by manufacturers (Figure C4–3 depicts sugar sources). In comparison, very little sugar is added from the sugar bowl at home. Because sugar is prepackaged into foods, and because label information does not differentiate between added and naturally occurring sugars, most consumers fail to realize just how much added sugar they take in each day.

Figure C4–3

Sources of Added Sugars in the U.S. Diet

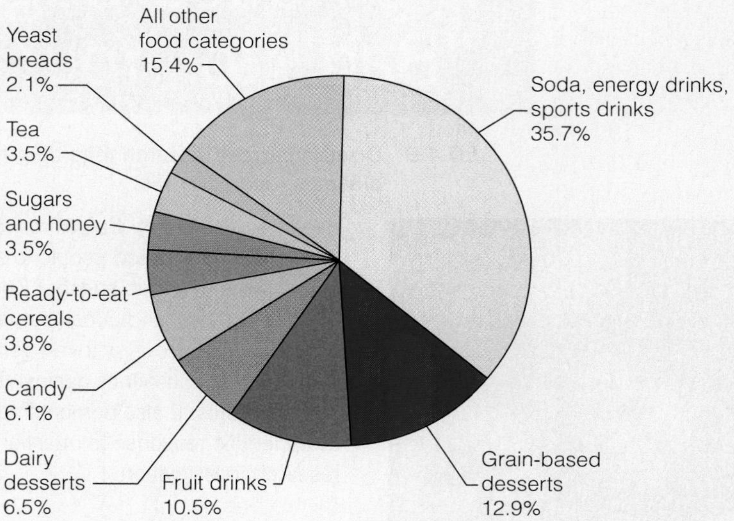

Source: NHANES data, 2005–2006; U.S. Department of Agriculture and U.S. Department of Health and Human Services, Dietary Guidelines for Americans 2010, *available at www.dietaryguidelines.gov, p. 29.*

Carbohydrates or Calories?

In the United States, observational studies often link high refined carbohydrate intakes, particularly added sugars from soft drinks, with increased body fatness.[8] At the same time, studies of other cultures report an *inverse* relationship between carbohydrate intake and body weight.[9] For example, the world's leanest peoples are often those eating traditional low-sugar diets that are high in carbohydrate-rich rice or root vegetables, such as Japanese, Chinese, or Africans.

When such people abandon their traditional diets in favor of "Western" style foods and beverages, they take in far more calories, and their rates of obesity and chronic diseases soar.[10] In addition, as a society gains wealth, its people also consume more meat and cooking fats, so it becomes impossible to tease apart the effects of sugars from the other dietary constituents in causation of obesity and its associated diseases. One such disease, diabetes, is currently advancing worldwide at greatly accelerated rates.

Do Added Sugars Cause Diabetes?

Diabetes impairs blood sugar (glucose) regulation. At one time, people thought that eating sugar *caused* diabetes by "overstraining the pancreas," but now we know that this is not the case. As the chapter made clear, excess body fatness is more closely related to diabetes than is diet composition; high rates of diabetes have not been reported in societies where obesity is rare.

Increasing intakes of added sugars (and refined grains) often parallel diabetes development in population studies. Particularly among certain Native American groups, a profound increase in the prevalence of diabetes is observed when added sugars and refined flour replace the roots, gourds, seeds, and other whole foods of traditional diets. Added sugars in the form of soft drinks also widely correlate with diabetes development.[11] Conversely, eating more whole foods correlates with a lower incidence of type 2 diabetes and cardiovascular disease, a finding that has been repeated many times.[12]

Even strong observational evidence cannot prove causation, however, and other studies report no link between added sugar intakes and diabetes when calories do not exceed the daily need. Keep in mind that people who care enough about their diets to moderate sugar intakes probably also make other healthy choices, such as being physically active and not smoking.[13] These and other factors combine in a lifestyle that greatly reduces diabetes risk.

Do Liquid Calories Pose Special Risks?

As mentioned, sugar-sweetened soft drinks are often linked with overweight in research. It has been suggested that the liquid nature of sugar calories in beverages might elude normal appetite control

† This estimate from the USDA Economic Research Service includes all caloric sweeteners in the U.S. human food supply, including cane and beet sugars, corn sweeteners, honey, and syrups.

mechanisms. To test this idea, subjects were given jelly beans (solid sugar) before a meal. At mealtime, they automatically compensated by eating fewer calories of food. When liquid sugar was substituted for the jelly beans, subjects did not compensate—they ate the full meal. These results seem to indicate that liquid sugars may be particularly fattening, but similar studies have produced only inconsistent results.[14]

It may be that people's expectations modify their intakes: if they do not expect a clear, caloric liquid to make them feel full, that expectation may influence their subsequent eating more than the true energy content of their intakes.[15] In addition, the liquid sugars of fruit punches and soft drinks are easily gulped down—no chewing required. Few people realize that a sugary 16-ounce soft drink can easily deliver 200 calories, and many young people drink several in a day.[16] When overweight people substituted water or diet beverages for caloric beverages, they dropped significant amounts of weight with no other dietary changes.[17]

Hints of Metabolic Mayhem

It may be tempting to close the book on added sugars in foods and beverages as just calorie sources—but before you do, consider some metabolic links among added sugars, obesity, and chronic diseases.[18] Such links have held researchers' attention since the mid-20th century, when a professor called sugar, "pure, white, and deadly."‡

Is It the Insulin?

The hormone insulin has been a target of investigation in this regard. All digestible carbohydrates, including sugars, elevate blood glucose to varying degrees, and blood glucose triggers the release of insulin into the bloodstream. Then, insulin interacts with many tissues to regulate fat metabolism and promote storage of energy nutrients, including storage of body fat in the adipose tissue.[19]

Does sugar cause obesity through insulin's promotion of fat storage, then? In fact, in healthy, normal-weight people who eat a reasonable diet, insulin works in balance with other hormones and mechanisms to dampen the appetite and maintain a normal body weight.[20] In people with insulin resistance, however, cells fail to respond to insulin's effects, upsetting the normal balance. (Insulin resistance was defined earlier in the chapter.) In healthy people, insulin itself is unlikely to trigger obesity. Other metabolic mechanisms involving the monosaccharide fructose, however, may be in play.

Is It the Fructose?

Fructose makes up about half of all sweet sugars (see Figure C4–4). Some people, particularly young people, consume up to 140 grams of fructose a day, the vast majority coming from added sugars.[21] Table C4–1 lists some sources of fructose in the diet. Note that chemically, most added sugars are similar to each other.

Glucose and fructose, despite both being monosaccharides, are handled

‡ *The professor was John Yudkin, as reported in G. A. Bray, Fructose: Pure, white, and deadly? Fructose, by any other name, is a health hazard,* Journal of Diabetes Science and Technology *4 (2010): 1003–1007.*

Table C4–1

Fructose Content of Selected Foods

Food/Beverage (amount)	Fructose (g)[a]
Carbonated soft drinks[b] (12 oz)	16
Fruitcake (small slice)	13
Peaches (1/2 c)	13
Raisins (1/4 c)	12
Punch or lemonade[b] (12 oz)	12
Applesauce, sweetened (1/2 c)	10
Honey (1 tbs)	9
High-fructose corn syrup (1 tbs)	8
Prunes (5 whole)	7
Sucrose	6
Yogurt, sweetened (1 c)	6
Cereal, sweetened (3/4 c)	5
Grapes (10)	4

[a]*Rounded values.*
[b]*Average value.*

Figure C4–4

Glucose and Fructose in Common Added Sugars

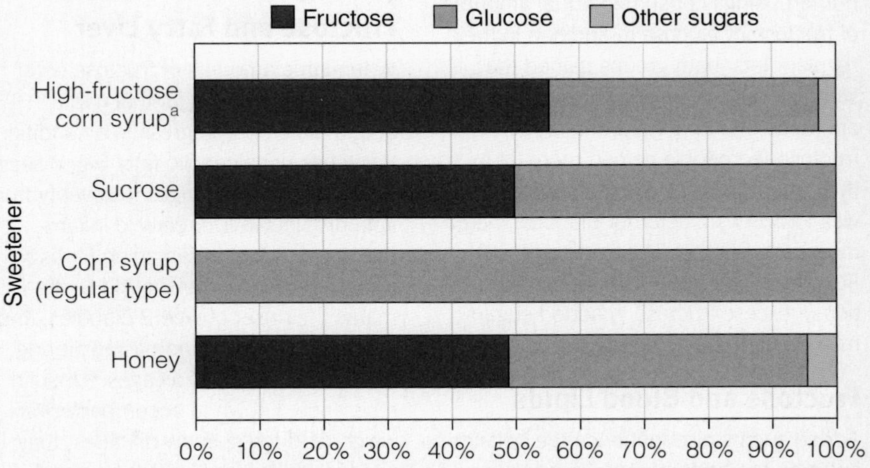

[a]*A typical mixture. Corn syrup purchased for use at home, for example in a pecan pie recipe, is not high-fructose corn syrup; it consists almost entirely of glucose.*

Source: Data from J. S. White, Straight talk about high-fructose corn syrup: What it is and what it ain't, American Journal of Clinical Nutrition *88 (2008): 1716S–1721S.*

differently in the body.[22] All the body's cells pick up glucose from the bloodstream and use it as such. In contrast, the liver soaks up almost all the ingested fructose and quickly converts it into other compounds.[23]

The two sugars affect appetite differently, too. Glucose and the insulin it triggers help to suppress the appetite. Fructose, in contrast, does not raise blood insulin and does not suppress appetite through this mechanism.[24] Also, circulating glucose itself acts on the brain directly in ways that may reduce desire for high-calorie foods.[25] Fructose cannot enter the brain's tissues and so cannot suppress the appetite directly.

Fructose and Body Fatness

In rodents, a diet high in fructose or sucrose often leads to obesity, diabetes, and blood lipid disturbances.[26] For example, when groups of rats are given solutions of glucose, fructose, or sucrose in addition to their chow, all reliably gain body fatness, but the fructose-fed rats gain the most.[27] Fructose-fed rats also reliably develop insulin resistance and other ills. In monkeys, however, no difference in weight gain has been observed between fructose- and glucose-feeding, even after a year's time.

When calories are held constant and fructose is substituted for other carbohydrates, it does not seem to cause weight gain.[28] However, when overweight human beings consumed equal amounts of fructose or glucose in addition to their regular diets, both groups gained the same amount of weight.[29] The researchers noted one important difference—the fructose-fed people gained more fat in their abdomens. Abdominal fatness is a well-known risk factor for diabetes, heart disease, and other ills (see Chapter 9). For overweight people, then, consumption of excess fructose in added sugars may be particularly unwise.[30]

Fructose and Blood Lipids

Added sugars also influence the balance between the body's fat-making and fat-clearing mechanisms, a balance that plays critical roles in the development of heart disease.[31] Fructose stimulates the body's fat-making pathways and impairs its fat-clearing pathways in ways that could lead to an unhealthy buildup of blood lipids (triglycerides; see Chapter 5).[32] This effect was first noted in people given large experimental doses of purified fructose—about a third of their daily calories.[33] Few people eat pure fructose, particularly in such large amounts, however. They eat added sugars that contain fructose.

© Polara Studios, Inc.

Most people are unaware of how much added sugar they consume.

To determine whether added sugars raise blood lipids, researchers studied more than 6,000 healthy adults. They observed that people with higher intakes of added sugars had blood lipid values indicating an increased risk of heart disease.[34] In another study, young adults who consumed more sugar-sweetened beverages had greater abdominal fatness, more harmful blood lipids, and higher blood pressure than those who drank fewer.[35] It may not take an unrealistic amount of added sugars to cause this effect. As little as the equivalent of one or two fructose- or sucrose-sweetened soft drinks a day consumed for only a few weeks significantly changed blood lipids in ways that may pose risks to the heart and arteries.[36]

Fructose and Fatty Liver

When large amounts of fructose enter the liver, it produces fat that can accumulate and progress to a condition known as nonalcoholic fatty liver disease (NAFLD). NAFLD ranges in severity from lipid droplets in liver cells to inflammation, cell death, damaging fibrosis, and even liver cancer.[37] NAFLD often marks the onset of type 2 diabetes, and both conditions are increasing among overweight people of all ages.[38] Insulin resistance, known to accompany overweight and foreshadow diabetes, may cause liver tissues to produce excess fats, resulting in NAFLD.[39] Does fructose *cause* NAFLD and diabetes, as some popular writers claim? Scientists have yet to untangle these relationships.

Moderate amounts of fructose consumed by healthy people in the context of an adequate, nutrient-dense diet, are likely to be safe.[40] In people with obesity, insulin resistance, or diabetes, evidence to date is insufficient to assess the effects of the current average intakes of fructose from added sugars.[41] Given the severity of the problems of obesity and metabolic diseases, more research is urgently needed to clarify these associations.

High-Fructose Corn Syrup (HFCS)

HFCS contributes about half of the added sugars in the U.S. food supply in the form of soft drinks, fruit drinks, candies, salad dressings, jams, breads and baked goods, canned foods, and many, many other foods and beverages.[42] HFCS is as sweet as sugar but cheaper and more convenient for manufacturers to use, making it an attractive option for food businesses, which choose it often.

Is HFCS more harmful to consumers than sucrose? From current research, the answer is uncertain. When tested against each other, some studies observe virtually identical metabolic effects of HFCS and sucrose—an expected result, given their similar chemical makeup. In one recent experiment, however, healthy people were fed either HFCS or sucrose. The HFCS group exhibited a greater rise in blood fructose and glucose, suggesting more efficient monosaccharide absorption, and a small but significant rise in a measure of blood pressure.[43§] No difference was observed in blood triglycerides, blood insulin, or other indicators of chronic disease risk. The authors of the study concluded that, although HFCS may vary somewhat from sucrose in certain metabolic effects, current intakes of both sweeteners are too high and excess intakes of both may play an important role in driving disease risks higher. Until more research proves otherwise, it can be assumed that all common added sugars are similar from the body's point of view, and none should be eaten to excess.

Conclusion

Investigation into the potential health effects of carbohydrates is ongoing. The idea that the nation's obesity problem might be easily solved by removing a single ingredient, such as added sugars, from the food supply is inviting but simplistic.[44]

What is clear is that the *source* of sugars matters to disease risks. Fruits and vegetables package their naturally occurring fructose with fiber, vitamins, minerals, and protective phytochemicals that may modulate its effects. Any advice to eliminate fruits and vegetables from the diet in the belief that their naturally occurring fructose may harm otherwise healthy people should be ignored. In fact, for optimal health, most people need to seek out more fruits and vegetables to meet the recommendations of the *Dietary Guidelines for Americans*.

The pleasure of sweet foods and beverages is part of the enjoyment of life. Just remember to keep them in their place—as occasional treats in the context of a nutritious diet, and not as staple foods or drinks at every meal.

Read about nonnutritive sweeteners and other sugar replacers in **Chapter 12**.

§ *The measure was systolic blood pressure.*

Critical Thinking

1. This controversy addresses five accusations launched against carbohydrate-rich foods as causes of health problems. Break into groups of five. Each person in the group takes one accusation from the list below and presents a one-minute argument in support of the accuracy of that accusation. When each person has completed his or her argument, vote as a group to determine which is most likely to cause health problems.
   - Carbohydrates are making us fat.
   - Carbohydrates cause diabetes.
   - Added sugars cause obesity and illness.
   - High-fructose corn syrup harms health.
   - Blood insulin is to blame.
2. Recommendations about carbohydrate intake can seem to be contradictory. On one hand, it is recommended that the bulk of the diet should be carbohydrates (fruit, vegetables, and whole grains), yet some research indicates that certain carbohydrates may be bad for you. Explain this discrepancy in three paragraphs. Use one paragraph to explain why the bulk of the diet should be carbohydrates, including describing the type of foods that should be eaten. The second paragraph should explain in detail why carbohydrates can be bad for you (give at least three examples). Finally, use the third paragraph to summarize how carbohydrates should be consumed in a way that makes them a healthy part of a good diet.

5 The Lipids: Fats, Oils, Phospholipids, and Sterols

what do you think?

Are **fats** unhealthy food constituents that are best eliminated from the diet?

What are the differences between **“bad”** and **“good”** cholesterol?

Why is choosing **fish** recommended in a healthy diet?

If you trim all **visible fats** from foods, will your diet meet lipid recommendations?

Learning Objectives

After completing this chapter, you should be able to accomplish the following:

LO 5.1 Identify the roles of lipids in both the body and food, and explain why some amount of fat is necessary in the diet.

LO 5.2 Compare and contrast the chemical makeup and physical properties of saturated fats, polyunsaturated fats, monounsaturated fats, and phospholipids.

LO 5.3 Summarize how and where dietary lipids are broken down and absorbed during digestion and how they are transported throughout the body.

LO 5.4 Describe the body's mechanisms for fat storage and use of body fat, including the role of carbohydrate in fat metabolism.

LO 5.5 Summarize the relationships between lipoproteins and disease risks, and explain how various fats and cholesterol in food affect cholesterol in the blood.

LO 5.6 Compare the roles of omega-3 and omega-6 fatty acids in the body, and name important food sources of each.

LO 5.7 Describe the hydrogenation of fat and the formation and structure of a *trans*-fatty acid.

LO 5.8 Outline a diet plan that provides enough of the right kinds of fats within calorie limits.

LO 5.9 Identify at least 10 ways to reduce solid fats in an average diet.

LO 5.10 Discuss evidence for the benefits and drawbacks of specific dietary fats in terms of their effects on human health.

Your bill from a medical laboratory reads "Blood **lipid** profile—$250." A healthcare provider reports, "Your blood **cholesterol** is high." Your physician advises, "You must cut down on the saturated **fats** in your diet and replace them with **oils** to lower your risk of **cardiovascular disease (CVD)**." Blood lipids, cholesterol, saturated fats, and oils—what are they, and how do they relate to health?

No doubt you are expecting to hear that fats have the potential to harm your health, but lipids are also valuable. In fact, lipids are absolutely necessary, and the diet recommended for health is by no means a "no-fat" diet. Luckily, at least traces of fats and oils are present in almost all foods, so you needn't make an effort to eat any extra. The trick is to choose the right ones.

Introducing the Lipids

LO 5.1 Identify the roles of lipids in both the body and food, and explain why some amount of fat is necessary in the diet.

The lipids in foods and in the human body, though many in number and diverse in function, generally fall into three classes.*[1] About 95 percent are **triglycerides**. The other major classes of the lipid family are the **phospholipids** (of which **lecithin** is one) and the **sterols** (cholesterol is the best known of these). Some of these names may sound unfamiliar, but most people will recognize at least a few functions of lipids in the body and in the foods that are listed in Table 5–1, p. 158. More details about each class of lipids follow later.

How Are Fats Useful to the Body?

When people speak of fat, they are usually talking about triglycerides. The term *fat* is more familiar, though, and we will use it in this discussion.

Fuel Stores Fat provides the majority of the energy needed to perform much of the body's muscular work. Fat is also the body's chief storage form for the energy from food eaten in excess of need. The storage of fat is a valuable survival mechanism for

*Reference notes are found in Appendix F.

lipid (LIP-id) a family of organic (carbon-containing) compounds soluble in organic solvents but not in water. Lipids include triglycerides (fats and oils), phospholipids, and sterols.

cholesterol (koh-LESS-ter-all) a member of the group of lipids known as sterols; a soft, waxy substance made in the body for a variety of purposes and also found in animal-derived foods.

fats lipids that are solid at room temperature (70°F or 21°C).

oils lipids that are liquid at room temperature (70°F or 21°C).

cardiovascular disease (CVD) disease of the heart and blood vessels; disease of the arteries of the heart is called *coronary heart disease (CHD)*. Also defined in Chapter 11.

triglycerides (try-GLISS-er-ides) one of the three main classes of dietary lipids and the chief form of fat in foods and in the human body. A triglyceride is made up of three units of fatty acids and one unit of glycerol (*fatty acids* and *glycerol* are defined later). In research, triglycerides are often called *triacylglycerols* (try-ay-seal-GLISS-er-ols).

phospholipids (FOSS-foh-LIP-ids) one of the three main classes of dietary lipids. These lipids are similar to triglycerides, but each has a phosphorus-containing acid in place of one of the fatty acids. Phospholipids are present in all cell membranes.

lecithin (LESS-ih-thin) a phospholipid manufactured by the liver and also found in many foods; a major constituent of cell membranes.

sterols (STEER-alls) one of the three main classes of dietary lipids. Sterols have a structure similar to that of cholesterol.

Table 5–1

The Usefulness of Fats

Fats in the Body	Fats in Food
▪ *Energy fuel.* Fats provide 80 to 90 percent of the resting body's energy, and much of the energy used to fuel muscular work. ▪ *Energy stores.* Fats are the body's chief form of stored energy. ▪ *Emergency reserve.* Fats serve as an emergency fuel supply in times of illness and diminished food intake. ▪ *Padding.* Fats protect the internal organs from shock through fat pads inside the body cavity. ▪ *Insulation.* Fats insulate against temperature extremes by forming a fat layer under the skin. ▪ *Cell membranes.* Fats form the major material of cell membranes. ▪ *Raw materials.* Lipids are converted to other compounds, such as hormones, bile, and vitamin D, as needed.	▪ *Nutrients.* Food fats provide essential fatty acids, fat-soluble vitamins, and other needed compounds. ▪ *Transport.* Fats carry fat-soluble vitamins A, D, E, and K along with some phytochemicals and assist in their absorption. ▪ *Energy.* Food fats provide a concentrated energy source. ▪ *Sensory appeal.* Fats contribute to the taste and smell of foods. ▪ *Appetite.* Fats stimulate the appetite. ▪ *Texture.* Fats make fried foods crisp and other foods tender. ▪ *Satiety.* Fats contribute to feelings of fullness.

people who live a feast-or-famine existence: stored during times of plenty, fat enables them to remain alive during times of famine.

Most body cells can store only limited fat, but some cells are specialized for fat storage. These fat cells seem able to expand almost indefinitely—the more fat they store, the larger they grow. An obese person's fat cells may be many times the size of a thin person's. Far from being a collection of inert sacks of fat, adipose (fat) tissue secretes hormones that help to regulate appetite and influence other body functions in ways critical to health.[2] A fat cell is shown in Figure 5–1.

Efficiency of Fat Stores You may be wondering why the carbohydrate glucose is not the body's major form of stored energy. As mentioned in Chapter 4, glucose is stored in the form of glycogen. Because glycogen holds a great deal of water, it is quite bulky and heavy and the body cannot store enough to provide energy for very long. Fats, however, pack tightly together without water and can store much more energy in a small space. Gram for gram, fats provide more than twice the energy of carbohydrate or protein, making fat the most efficient storage form of energy. The body fat found on a normal-weight person contains more than enough energy to fuel an entire marathon run or to battle disease should the person become ill and stop eating for a while.

Thanks to internal fat pads, vital organs are cushioned from shock.

Cushions, Climate, and Cell Membranes Fat serves many other purposes in the body. Pads of fat surrounding the vital internal organs serve as shock absorbers. Thanks to these fat pads, you can ride a horse or a motorcycle for many hours with no serious internal injuries. A fat blanket under the skin also insulates the body from extremes of temperature, thus assisting with internal climate control. Lipids also play critical roles in all of the body's cells as part of their surrounding envelopes, the cell membranes.

Transport and Raw Material Lipids move around the body in association with other lipids, as described in later sections. Once a lipid arrives at its destination, it may serve as raw material for making any of a large number of needed substances.

KEY POINT

- Lipids provide and store energy, cushion vital organs, insulate against temperature extremes, form cell membranes, transport fat-soluble substances, and serve as raw materials.

Figure 5–1

A Fat Cell

Within the fat cell, lipid is stored in a droplet. This droplet can greatly enlarge, and the fat cell membrane will expand to accommodate its swollen contents. More about fat tissue (also called *adipose tissue*) and body functions in Chapter 9.

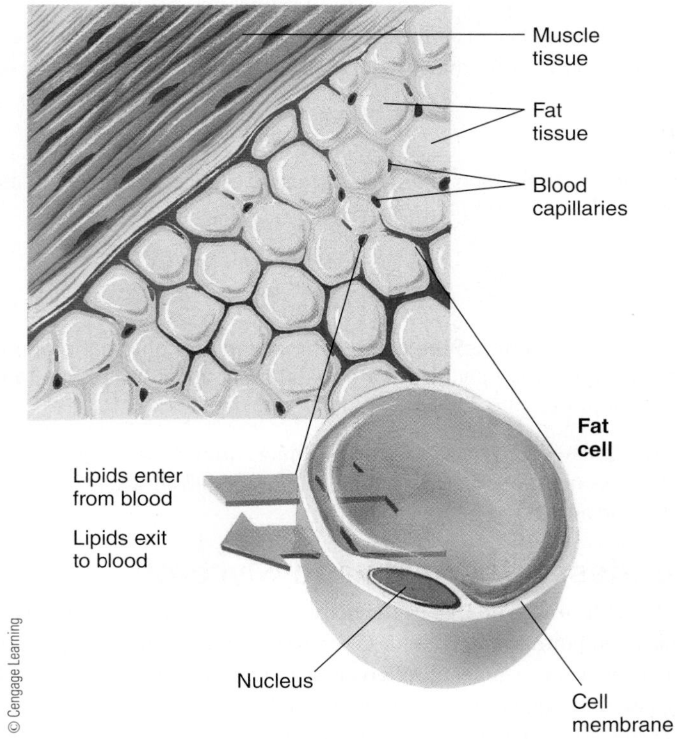

How Are Fats Useful in Food?

Fats in foods are valuable in many ways. They provide concentrated energy and needed substances to the body, and they are notoriously tempting to the palate.

Concentrated Calorie Source Energy-dense fats are uniquely valuable in many situations. A hunter or hiker must consume a large amount of food energy to travel long distances or to survive in intensely cold weather. An athlete must meet often enormous energy needs to avoid weight loss that could impair performance. As Figure 5–2 (p. 160) demonstrates, for such a person fat-rich foods most efficiently provide the needed energy in the smallest package. But for a person who is not expending much energy in physical work, those same high-fat foods may deliver many unneeded calories in only a few bites.

Fat-Soluble Nutrients and Their Absorption Some essential nutrients are lipid in nature and therefore soluble in fat. They often occur in foods that contain fat, and some amount of fat in the diet is necessary for their absorption. These nutrients are the fat-soluble vitamins: A, D, E, and K. Other lipid nutrients are **fatty acids** themselves, including the **essential fatty acids**. Fat also aids in the absorption of some phytochemicals, plant constituents that may be of benefit to health.

Sensory Qualities People naturally like high-fat foods. Fat carries with it many dissolved compounds that give foods enticing aromas and flavors, such as the aroma of frying bacon or French fries. In fact, when a sick person refuses food, dietitians offer

Do the Math

Fats are energy-dense nutrients:

- 1 g fat = 9 cal
- 1 g carbohydrate = 4 cal
- 1 g protein = 4 cal

fatty acids organic acids composed of carbon chains of various lengths. Each fatty acid has an acid end and hydrogens attached to all of the carbon atoms of the chain.

essential fatty acids fatty acids that the body needs but cannot make and so must be obtained from the diet.

Figure 5–2

Two Lunches

Both lunches contain the same number of calories, but the fat-rich lunch takes up less space and weighs less.

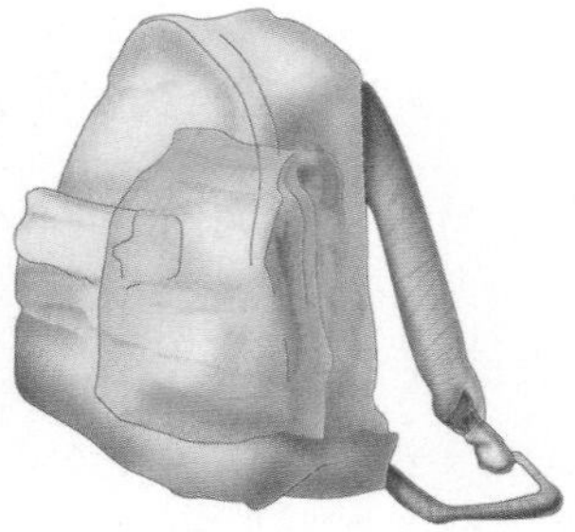

Carbohydrate-rich lunch
1 low-fat muffin
1 banana
2 oz carrot sticks
8 oz fruit yogurt

calories = 550
weight (g) = 500

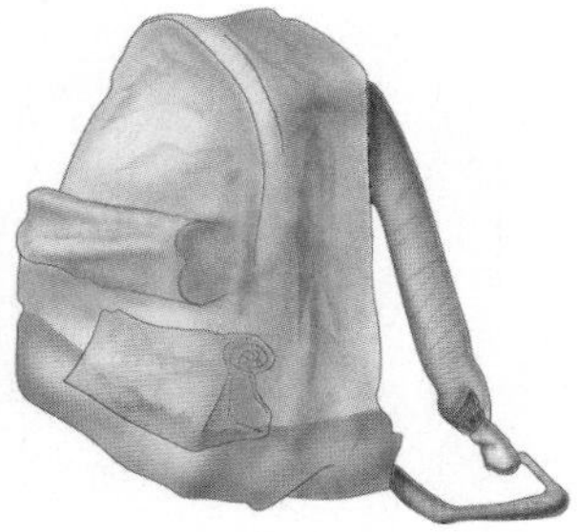

Fat-rich lunch
6 butter-style crackers
$1\frac{1}{2}$ oz American cheese
2 oz trail mix with candy

calories = 550
weight (g) = 115

foods flavored with some fat to spark the appetite and tempt that person to eat again. Fat also lends crispness to fried foods and tenderness to foods such as meats and baked goods. Around the world, as fats becomes less expensive and more available in a given food supply, people consistently choose fatty foods more often.

A Role in Satiety Fat also contributes to **satiety**, the satisfaction of feeling full after a meal. The fat of swallowed food triggers a series of physiological events that slows down the movement of food through the digestive tract and eventually promotes satiety. Even so, before the sensation of fullness stops them, people can easily overeat on fat-rich foods because the delicious taste of fat stimulates eating, and each bite of a fat-rich food delivers many calories. Chapter 9 revisits the topic of appetite and its control.

KEY POINT

- Lipids provide abundant food energy in a small package, enhance aromas and flavors of foods, and contribute to satiety.

A Close Look at Lipids

LO 5.2 Compare and contrast the chemical makeup and physical properties of saturated fats, polyunsaturated fats, monounsaturated fats, and phospholipids.

Each class of lipids—triglycerides, phospholipids, and sterols—possesses unique characteristics. As mentioned, the term *fat* refers to triglycerides, the major form of lipid found in food and in the body.

Triglycerides: Fatty Acids and Glycerol

Very few fatty acids are found free in the body or in foods; most are incorporated into large, complex compounds: triglycerides. The name almost explains itself: three fatty acids *(tri)* are attached to a molecule of **glycerol** to form a triglyceride molecule (Figure 5–3). Tissues all over the body can easily assemble triglycerides or disassemble them as needed. Triglycerides make up most of the lipid present both in the body and in food.

Fatty acids can differ from one another in two ways: in chain length and in degree of saturation (explained next). Triglycerides usually include mixtures of various fatty

Figure 5–3

Triglyceride Formation

Glycerol, a small, water-soluble carbohydrate derivative, plus three fatty acids equals a triglyceride.

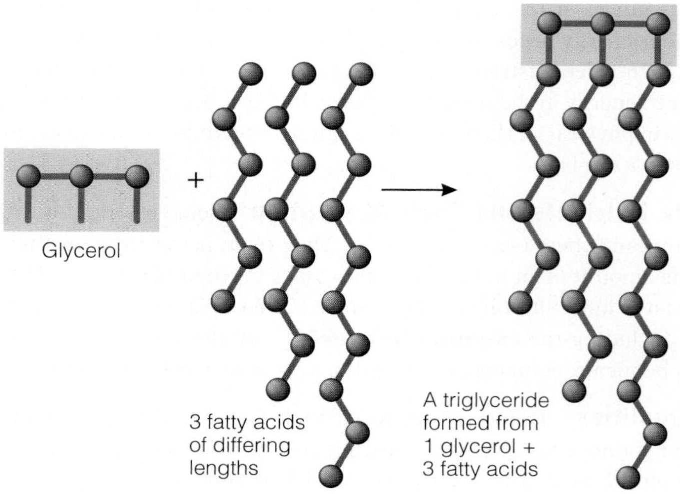

acids. Depending on which fatty acids are incorporated into a triglyceride, the resulting fat will be softer or harder. Triglycerides containing mostly the shorter-chain fatty acids or the more unsaturated ones are softer and melt more readily at lower temperatures.

Each species of animal (including people) makes its own characteristic kinds of triglycerides, a function governed by genetics. Fats in the diet, though, can affect the types of triglycerides made because dietary fatty acids are often incorporated into triglycerides in the body. For example, many animals raised for food can be fed diets containing specific triglycerides to give the meat the types of fats that consumers demand.

KEY POINTS

- The body combines three fatty acids with one glycerol to make a triglyceride, its storage form of fat.
- Fatty acids in food influence the composition of fats in the body.

James Darell/Photodisc/Getty Images

Small amounts of fat offer eaters both pleasure and needed nutrients.

Saturated vs. Unsaturated Fatty Acids

Saturation refers to whether or not a fatty acid chain is holding all of the hydrogen atoms it can hold. If every available bond from the carbons is holding a hydrogen, the chain forms a **saturated fatty acid**; it is filled to capacity with hydrogen. The zigzag structure on the left in Figure 5–4 represents a saturated fatty acid.

Saturation of Fatty Acids Sometimes, especially in the fatty acids of plants and fish, the chain has a place where hydrogens are missing: an "empty spot," or **point of unsaturation**.† A fatty acid carbon chain that possesses one or more points of unsaturation is an **unsaturated fatty acid**. With one point of unsaturation, the fatty acid is a **monounsaturated fatty acid** (see the second structure in Figure 5–4). With two or more points of unsaturation, it is a **polyunsaturated fatty acid**, often abbreviated **PUFA** (see the third structure in Figure 5–4; other examples are given later in the chapter).

Figure 5–4
Three Types of Fatty Acids

The more carbon atoms in a fatty acid, the longer it is. The more hydrogen atoms attached to those carbons, the more saturated the fatty acid is.

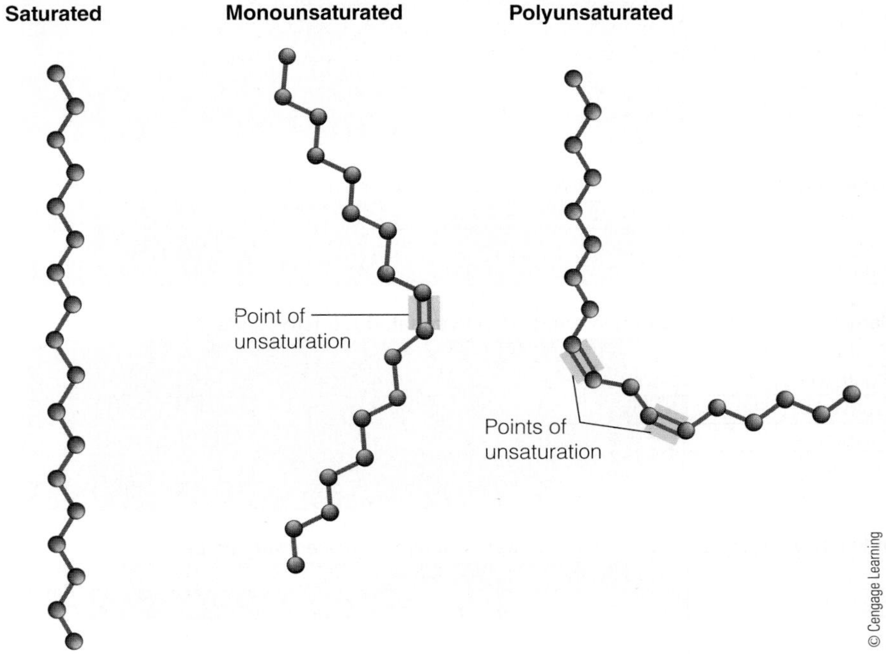

†These points of unsaturation can also be referred to as double bonds.

satiety (sat-EYE-uh-tee) the feeling of fullness or satisfaction that people experience after meals.

glycerol (GLISS-er-all) an organic compound, three carbons long, of interest here because it serves as the backbone for triglycerides.

saturated fatty acid a fatty acid carrying the maximum possible number of hydrogen atoms (having no points of unsaturation). A saturated fat is a triglyceride that contains three saturated fatty acids.

point of unsaturation a site in a molecule where the bonding is such that additional hydrogen atoms can easily be attached.

unsaturated fatty acid a fatty acid that lacks some hydrogen atoms and has one or more points of unsaturation. An unsaturated fat is a triglyceride that contains one or more unsaturated fatty acids.

monounsaturated fatty acid a fatty acid containing one point of unsaturation.

polyunsaturated fatty acid (PUFA) a fatty acid with two or more points of unsaturation.

Did You Know?

There are three types of fatty acids: saturated, monounsaturated, and polyunsaturated.

© Quest Photographic, Inc.

Fats melt at different temperatures. The more unsaturated a fat, the more liquid it is at room temperature. The more saturated a fat, the higher the temperature at which it melts.

Melting Point and Fat Hardness The degree of saturation of the fatty acids in a fat affects the temperature at which the fat melts. Generally, the more unsaturated the fatty acids, the more liquid the fat will be at room temperature. Conversely, the more saturated the fatty acids, the more solid the fat will be at room temperature. Thus, looking at three fats—beef tallow (a type of beef fat), chicken fat, and safflower oil—beef tallow is the most saturated and the hardest; chicken fat is less saturated and somewhat soft; and safflower oil, which is the most unsaturated, is a liquid at room temperature.

If a health-care provider recommends replacing **solid fats**, **saturated fats**, and ***trans* fats** with **monounsaturated fats** and **polyunsaturated fats** to protect your health, you can generally judge by the hardness of the fats which ones to choose. Figure 5–5 compares the percentages of saturated, monounsaturated, and polyunsaturated fatty acids in various fats and oils. To determine the degree of saturation of the fats in the oil you use, place it in a clear container in the refrigerator and watch how solid it becomes. The least saturated oils, such as polyunsaturated vegetable oils, remain clear. Olive oil, mostly monounsaturated fat, may turn cloudy when chilled, but olive oil is still an excellent choice from the standpoint of the health of the heart, as this chapter's Controversy reveals.

Another exception is the solid fat of homogenized milk. Highly saturated milk fat normally collects and floats as a layer of cream (butterfat) on top of the watery milk

Figure 5–5

Fatty Acid Composition of Common Food Fats

Most fats are a mixture of saturated, monounsaturated, and polyunsaturated fatty acids.

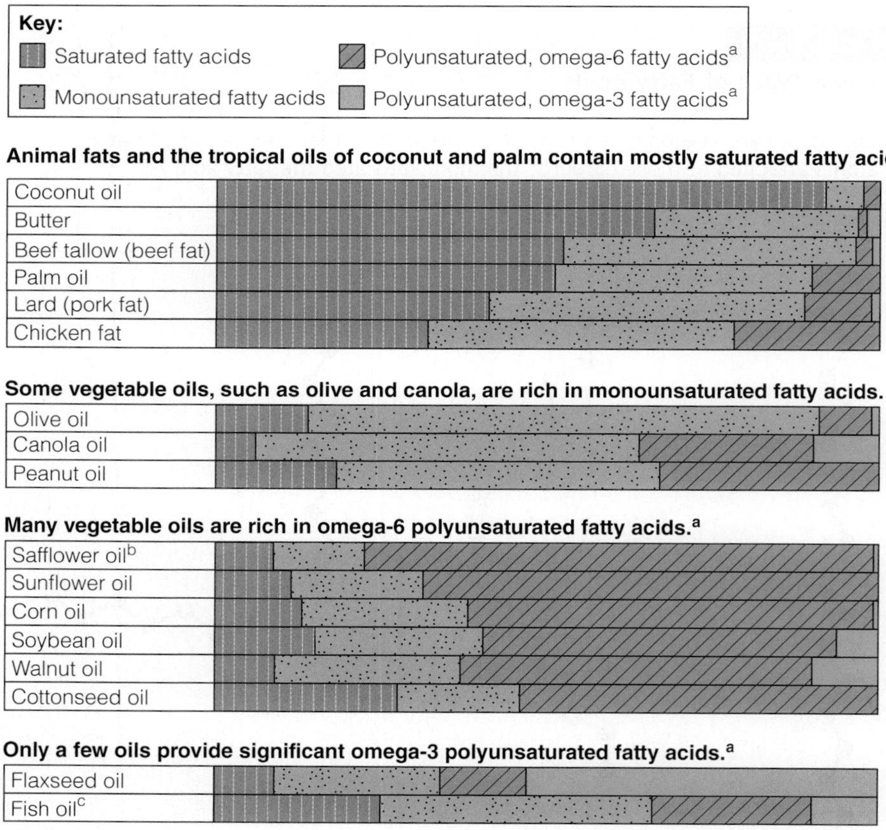

© Cengage Learning

[a]*These families of polyunsaturated fatty acids are explained in a later section.*
[b]*Salad or cooking type over 70% linoleic acid.*
[c]*Fish oil average values derived from USDA data for salmon, sardine, and herring oils.*

fluids. Once churned into butter, the solid fat is revealed and quickly hardens in the refrigerator. During **homogenization**, heated milk and cream are forced under high pressure through tiny nozzle openings to finely divide and disperse the fat droplets evenly throughout the milk. Thus, fluid milk can be a source of solid fat that remains liquid at cold temperatures.

© Workmans Photos, 2011/Shutterstock.com

Where the Fatty Acids Are Found Most vegetable and fish oils are rich in polyunsaturated fatty acids. Some vegetable oils are also rich in monounsaturated fatty acids. Animal fats are generally the most saturated. But you have to know your oils—it is not enough to choose foods with plant oils over those containing animal fats. Some nondairy whipped dessert toppings use coconut oil in place of cream (butterfat). Coconut oil does come from a plant, but it disobeys the rule that plant oils are less saturated than animal fats; the fatty acids of coconut oil—even the heavily advertised "virgin" types—are more saturated than those of cream and may add to heart disease risk.[3] By the way, no solid evidence supports claims made by advertisers for special curative powers of coconut oil. Palm oil, a vegetable oil used in food processing, is also highly saturated and has been shown to elevate blood cholesterol.[4] Likewise, shortenings, stick margarine, and commercially fried or baked products may claim to be "all vegetable fat," but much of their fat may be of the harmful saturated kind, as a later section makes clear.[5]

KEY POINTS

- Fatty acids are energy-rich carbon chains that can be saturated (filled with hydrogens) or monounsaturated (with one point of unsaturation) or polyunsaturated (with more than one point of unsaturation).
- The degree of saturation of the fatty acids in a fat determines the fat's softness or hardness.

Phospholipids and Sterols

Thus far, we have dealt with the largest of the three classes of lipids—the triglycerides and their component fatty acids. The other two classes—phospholipids and sterols—play important structural and regulatory roles in the body.

Phospholipids A phospholipid, like a triglyceride, consists of a molecule of glycerol with fatty acids attached, but it contains two, rather than three, fatty acids. In place of the third is a molecule containing phosphorus, which makes the phospholipid soluble in water, while its fatty acids make it soluble in fat. This versatility permits any phospholipid to play a role in keeping fats dispersed in water—it can serve as an **emulsifier**.

Food manufacturers blend fat with watery ingredients by way of **emulsification**. Some salad dressings separate to form two layers—vinegar on the bottom, oil on top. Other dressings, such as mayonnaise, are also made from vinegar and oil, but they never separate. The difference lies in a special ingredient of mayonnaise, the emulsifier lecithin in egg yolks. Lecithin, a phospholipid, blends the vinegar with the oil to form the stable, spreadable mayonnaise.

Health-promoting properties, such as the ability to lower blood cholesterol, are sometimes attributed to lecithin, but the people making the claims profit from selling supplements. Lecithin supplements have no special ability to promote health—the body makes all of the lecithin it needs.

Phospholipids also play key roles in the body. Phospholipids bind together in a strong double layer that forms the membranes of cells. Because phospholipids have both water-loving and fat-loving characteristics, they help fats travel back and forth across the lipid membranes of cells into the watery fluids on both sides. In addition, some phospholipids generate signals inside the cells in response to hormones, such as insulin, to help modulate body conditions.

Sterols Sterols such as cholesterol are large, complicated molecules consisting of interconnected *rings* of carbon atoms with side chains of carbon, hydrogen, and

solid fats fats that are high in saturated fatty acids and are usually solid at room temperature. Solid fats are found naturally in most animal foods but also can be made from vegetable oils through hydrogenation. Also defined in Chapter 2.

saturated fats triglycerides in which most of the fatty acids are saturated.

***trans* fats** fats that contain any number of unusual fatty acids—*trans*-fatty acids—formed during processing.

monounsaturated fats triglycerides in which most of the fatty acids have one point of unsaturation (are monounsaturated).

polyunsaturated fats triglycerides in which most of the fatty acids have two or more points of unsaturation (are polyunsaturated).

homogenization a process by which milk fat is evenly dispersed within fluid milk; under high pressure, milk is passed through tiny nozzles to reduce the size of fat droplets and reduce their tendency to cluster and float to the top as cream.

emulsifier a substance that mixes with both fat and water and permanently disperses the fat in the water, forming an emulsion.

emulsification the process of mixing lipid with water by adding an emulsifier.

© Matthew Farruggio

Oil and water. Without help from emulsifiers, fats and water separate into layers.

oxygen attached. Cholesterol serves as the raw material for making emulsifiers in **bile** (see the next section for details), important to fat digestion. Cholesterol is also important in the structure of the cell membranes of every cell, making it necessary to the body's proper functioning. Like lecithin, cholesterol can be made by the body, so it is not an essential nutrient. Other sterols include vitamin D, which is made from cholesterol, and the familiar steroid hormones, including the sex hormones.

Cholesterol forms the major part of the plaques that narrow the arteries in atherosclerosis, the underlying cause of heart attacks and strokes. Sterols other than cholesterol exist in plants, and tiny amounts of these sterols can be detected in the human bloodstream.[6] These plant sterols resemble cholesterol in structure and can inhibit cholesterol absorption in the human digestive tract; their effects are currently under study.

KEY POINTS

- Phospholipids play key roles in cell membranes.
- Sterols play roles as part of bile, vitamin D, the sex hormones, and other important compounds.
- Plant sterols in foods inhibit cholesterol absorption.

Lipids in the Body

LO 5.3 Summarize how and where dietary lipids are broken down and absorbed during digestion and how they are transported throughout the body.

From the moment they enter the body, lipids affect the body's functioning and condition. They also demand special handling because fat separates from water, and body fluids consist largely of water.

How Are Fats Digested and Absorbed?

A bite of food in the mouth first encounters the enzymes of saliva. An enzyme produced by the tongue plays a major role in digesting milk fat in infants but is of little importance to lipid digestion in adults.

Fat in the Stomach After being chewed and swallowed, the food travels to the stomach, where droplets of fat separate from the watery components and tend to float as a layer on top. Even the stomach's powerful churning cannot completely disperse the fat, so little fat digestion takes place in the stomach.

Fat in the Small Intestine As the stomach contents empty into the small intestine, the digestive system faces a problem: how to thoroughly mix fats, which are now separated, with its own watery fluids. The solution is an emulsifier: bile. Bile, made by the liver, is stored in the gallbladder and released into the small intestine when it is needed for fat digestion. Bile contains compounds made from cholesterol that work as emulsifiers; one end of each molecule attracts and holds fat, while the other end is attracted to and held by water.

By the time fat enters the small intestine, the gallbladder, which stores the liver's output of bile, has contracted and squirted its bile into the intestine. Bile emulsifies and suspends fat droplets within the watery fluids (see Figure 5–6) until the fat-digesting enzymes contributed by the pancreas can split them into smaller molecules for absorption. These fat-splitting enzymes act on triglycerides to split fatty acids from their glycerol backbones. Free fatty acids, phospholipids, and **monoglycerides** all cling together in balls surrounded by bile emulsifiers.

To review: first, the digestive system mixes fats with bile-containing digestive juices to emulsify the fats. Then, fat-digesting enzymes break the fats down into absorbable pieces. The pieces then assemble themselves into balls that remain emulsified by bile.

People sometimes wonder how a person without a gallbladder can digest food. The gallbladder is just a storage organ. Without it, the liver still produces bile but delivers it into a duct that conducts it into the small intestine instead of into the gallbladder.

Figure 5–6

The Action of Bile in Fat Digestion

Bile and detergents are both emulsifiers and work the same way, which is why detergents are effective in removing grease spots from clothes. Molecule by molecule, the grease is dissolved out of the spot and suspended in the water, where it can be rinsed away.

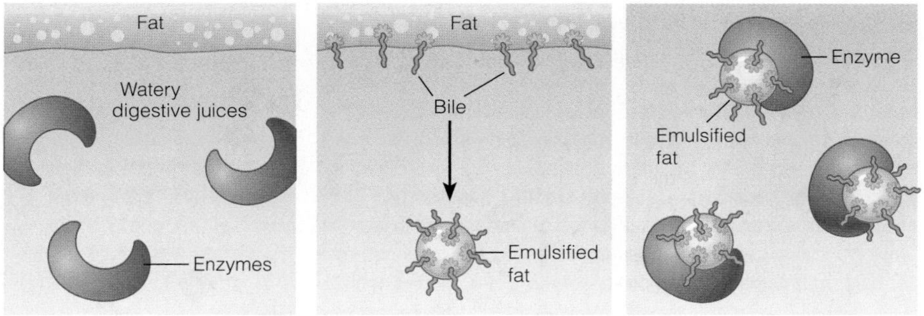

In the stomach, the fat and watery digestive juices tend to separate. Enzymes are in the water and can't get at the fat.

When fat enters the small intestine, the gallbladder secretes bile. Bile compounds have an affinity for both fat and water, so bile can mix the fat into the water.

After emulsification, more fat is exposed to the enzymes, and fat digestion proceeds efficiently.

Fat Absorption Once split and emulsified, the fats face another watery barrier: the watery layer of mucus that coats the absorptive lining of the digestive tract. Fats must traverse this layer to enter the cells of the digestive tract lining. The solution again depends on bile, this time in the balls of digested lipids. The bile shuttles the lipids across the watery mucus layer to the waiting absorptive surfaces on cells of the intestinal villi. The cells then extract the lipids. The bile may be absorbed and reused by the body, or it may flow back into the intestinal contents and exit with the feces, as was shown in Figure 4–6 (p. 122) of Chapter 4.

The digestive tract absorbs triglycerides from a meal with remarkable efficiency: up to 98 percent of fats consumed are absorbed. Very little fat is excreted by a healthy system. The process of fat digestion takes time, though, so the more fat taken in at a meal, the slower the digestive system action becomes. The efficient series of events just described is depicted in Figure 5–7, p. 166.

KEY POINTS

- In the stomach, fats separate from other food components.
- In the small intestine, bile emulsifies the fats, enzymes digest them, and the intestinal cells absorb them.

Transport of Fats

Glycerol and shorter-chain fatty acids pass directly through the cells of the intestinal lining into the bloodstream where they travel unassisted to the liver. The larger lipids, however, present a problem for the body. As mentioned, fat floats in water. Without some mechanism to keep them dispersed, large lipid globules would separate out of the watery blood as it circulates around the body, disrupting the blood's normal functions. The solution to this problem lies in an ingenious use of proteins: many fats travel from place to place in the watery blood as passengers in **lipoproteins**, assembled packages of lipid and protein molecules.

The larger digested lipids, monoglycerides and long-chain fatty acids, must form lipoproteins before they can be released into the lymph in vessels that lead to the bloodstream. Inside the intestinal cells, these lipids re-form into triglycerides and cluster together with proteins and phospholipids to form **chylomicrons** that can safely carry lipids from place to place in the watery blood. Chylomicrons form one

bile an emulsifier made by the liver from cholesterol and stored in the gallbladder. Bile does not digest fat as enzymes do but emulsifies it so that enzymes in the watery fluids may contact it and split the fatty acids from their glycerol for absorption.

monoglycerides (mon-oh-GLISS-er-ides) products of the digestion of lipids; a monoglyceride is a glycerol molecule with one fatty acid attached (*mono* means "one"; *glyceride* means "a compound of glycerol").

lipoproteins (LYE-poh-PRO-teens, LIH-poh-PRO-teens) clusters of lipids associated with protein, which serve as transport vehicles for lipids in blood and lymph. The major lipoproteins include chylomicrons, VLDL, LDL, and HDL.

chylomicrons (KYE-low-MY-krons) lipoproteins formed when lipids from a meal cluster with carrier proteins in the cells of the intestinal lining. Chylomicrons transport food fats through the watery body fluids to the liver and other tissues.

Figure 5–7

Animated! The Process of Lipid Digestion and Absorption

1. In the mouth and stomach:
   Little fat digestion takes place.

2. In the small intestine:
   Digestive enzymes accomplish most fat digestion in the small intestine. There, bile emulsifies fat, making it available for enzyme action. The enzymes cleave triglycerides into free fatty acids, glycerol, and monoglycerides.

3. At the intestinal lining:
   The parts are absorbed by intestinal villi. Glycerol and short-chain fatty acids enter directly into the bloodstream.

4. The cells of the intestinal lining convert large lipid fragments, such as monoglycerides and long-chain fatty acids, back into triglycerides and combine them with protein, forming chylomicrons (a type of lipoprotein) that travel in the lymph vessels to the bloodstream.

5. In the large intestine:
   A small amount of cholesterol trapped in fiber exits with the feces

Note: In this diagram, molecules of fatty acids are shown as large objects, but, in reality, molecules of fatty acids are too small to see even with a powerful microscope, while villi are visible to the naked eye.

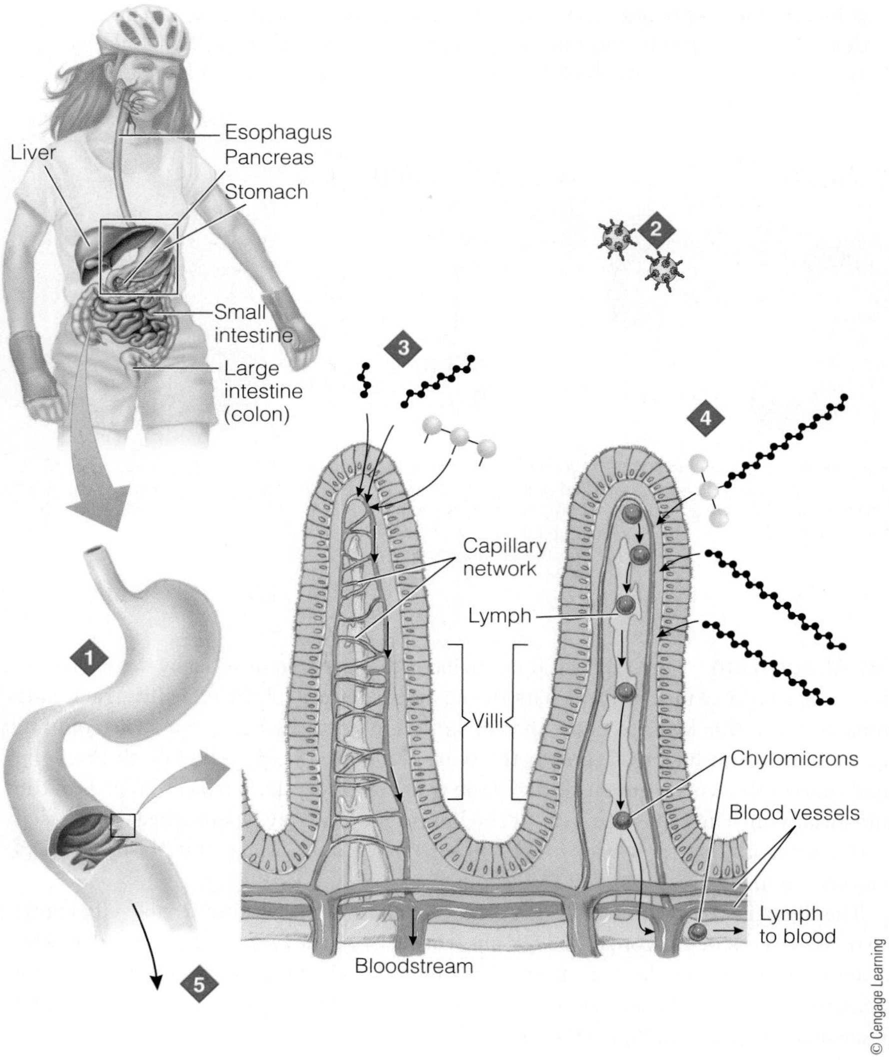

type of lipoprotein (shown in Figure 5–7) and are part of the body's efficient lipid transport system. Other lipoproteins are discussed later with regard to their profound effects on health.

KEY POINTS

- Glycerol and short-chain fatty acids travel in the bloodstream unassisted.
- Other lipids need special transport vehicles—the lipoproteins—to carry them in watery body fluids.

Storing and Using the Body's Fat

LO 5.4 Describe the body's mechanisms for fat storage and use of body fat, including the role of carbohydrate in fat metabolism.

Methodically, the body conserves fat molecules not immediately required for energy. Stored fat serves as a sort of "rainy day" fund to fuel the body's activities at times when food is unavailable, when illness impairs the appetite, or when energy expenditures increase.

The Body's Fat Stores Many triglycerides eaten in foods are transported by the chylomicrons to the fat depots—the external fat layer under the skin, the internal fat pads of the abdomen, the breasts, and others—where they are stored by the body's fat cells for later use. When a person's body starts to run out of available fuel from food, it begins to retrieve this stored fat to use for energy. (It also draws on its stored glycogen, as the last chapter described.)

The body can also store excess carbohydrate as fat, but this conversion is not energy-efficient. Figure 5–8 illustrates a simplified series of conversion steps from carbohydrate to fat. Before excess glucose can be stored as fat, it must first be broken into tiny fragments and then reassembled into fatty acids, steps that require energy to perform. Storing fat itself is more efficient; fat requires fewer chemical steps before storage.

What Happens When the Tissues Need Energy? Fat cells respond to the call for energy by dismantling stored fat molecules (triglycerides) and releasing fatty acids into the blood. Upon receiving these fatty acids, the energy-hungry cells break them down further into small fragments. Finally, each fat fragment is combined with a fragment derived from glucose, and the energy-releasing process continues, liberating energy, carbon dioxide, and water. The way to use more of the energy stored as body fat, then, is to create a greater demand for it in the tissues by decreasing intake of food energy, by increasing the body's expenditure of energy, or both.

Body fat supplies much of the fuel these muscles need to do their work.

Carbohydrate in Fat Breakdown When fat is broken down to provide cellular energy, carbohydrate helps the process run most efficiently. Without carbohydrate, products of incomplete fat breakdown (ketones) build up in the tissues and blood, and they spill out into the urine.

For weight-loss dieters who want to use their body fat for energy, knowing these details of energy metabolism is less important than remembering what research and common sense tell us: successful weight loss depends on taking in less energy than the body needs—not on the proportion of energy nutrients in the diet (Chapter 9 provides details). For the body's health, however, the proportions of certain lipids in the diet matter greatly, as the next section makes clear.

KEY POINTS

- The body draws on its stored fat for energy.
- Carbohydrate is necessary for the complete breakdown of fat.

Figure 5–8

Glucose to Fat

Glucose can be used for energy, or it can be changed into fat and stored.

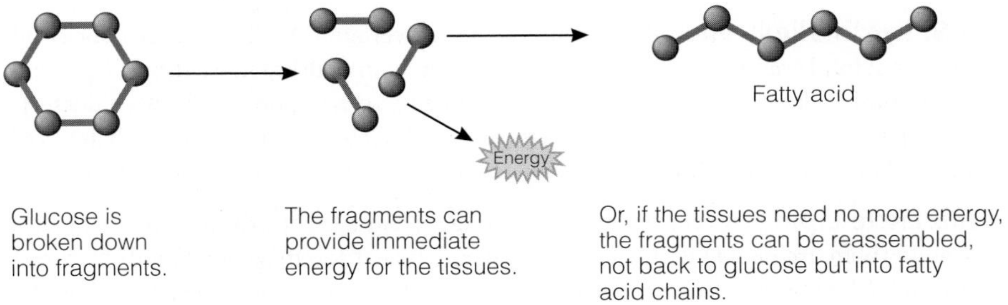

Fatty acid
Energy
Glucose is broken down into fragments.
The fragments can provide immediate energy for the tissues.
Or, if the tissues need no more energy, the fragments can be reassembled, not back to glucose but into fatty acid chains.

Dietary Fat, Cholesterol, and Health

LO 5.5 Summarize the relationships between lipoproteins and disease risks, and explain how various fats and cholesterol in food affect cholesterol in the blood.

High intakes of saturated and *trans* fats are associated with serious diseases, and particularly with heart and artery disease (cardiovascular disease, or CVD), the number-one killer of adults in the United States and Canada. So much research is focused on the links between diet and diseases that an entire chapter, Chapter 11, is devoted to presenting the details of these connections.

Heart and Artery Disease In a person following an eating pattern that provides an abundance of dietary saturated fatty acids and ***trans*-fatty acids**, the blood lipid profile often shifts in ways that increase CVD risks. Saturated fatty acids also contribute to blood clotting associated with heart attacks, whereas polyunsaturated oils from fish oppose this action.

© Colour/Shutterstock.com

When polyunsaturated or monounsaturated fat replaces dietary saturated fat and *trans* fat (discussed later), the blood lipids shift toward a profile associated with good health.[7] Even greater benefits may follow an eating pattern that includes olives and olive oil, nuts, seafood, and soy foods; soluble fiber sources such as legumes, barley, and oatmeal; fruit and vegetables; and other whole foods.[8] Conversely, replacing saturated fats with refined carbohydrates such as added sugars and refined flour may not have a desirable effect on heart disease risk.[9] Reducing saturated fats is important, but what replaces them in the diet matters, too. The Controversy addresses recent debates about saturated fats and their dietary context.

If you are a woman, take note: these observations apply to you. Heart disease kills more women in the United States than any other cause, and the old myth that heart disease is a "man's disease" should be forever put to rest.

Obesity Obesity carries serious risks to health. A diet high in energy-rich fatty foods makes overconsumption of calories likely and encourages unneeded weight gain.[10] An increasing waistline, in turn, often increases blood triglycerides, which can warn of an increasing threat from heart disease and other chronic illnesses.[11] Later chapters provide many more details.

Recommendations for Lipid Intakes

As mentioned, some fat is essential to good health. The *Dietary Guidelines for Americans* recommend that a portion of each day's total fat should come from raw oil (see the margin).[12] A little peanut butter on toast or mayonnaise in tuna salad, for example, can easily meet this need. In addition, the DRI committee sets specific recommended intakes for **linoleic acid** and **linolenic acid**, and they are listed in Table 5–2.

Did You Know?

An adequate intake of the needed fatty acids can be ensured by a small daily intake of raw oil:

- 1,600-cal diet = 22 g (5 tsp)
- 2,000-cal diet = 27 g (6 tsp)
- 2,200-cal diet = 29 g (6 tsp)
- 2,400-cal diet = 31 g (7 tsp)
- 2,800-cal diet = 36 g (8 tsp)

A Healthy Range of Fat Intakes Defining an upper limit—the exact gram amount of fat, saturated fat, *trans* fat, or cholesterol that begins to harm people's health—is difficult, so no Tolerable Upper Intake Levels for the lipids is set.[13] Instead, the DRI committee and the *Dietary Guidelines* suggest an intake range of 20 to 35 percent of daily energy from total fat and less than 10 percent of daily energy intake from saturated fat, as little *trans* fat as possible, and less than 300 milligrams of cholesterol daily. In practical terms, for a 2,000-calorie diet, 20 to 35 percent represents 400 to 700 calories from total fat (roughly 45 to 75 grams, or about 9 to 15 teaspoons).

U.S. Fat Intakes According to surveys, the average U.S. diet provides about 34 percent of total energy from fat, with saturated fat contributing more than 11 percent of the total.[14] The solid fats in foods such as cheese, pizza, and grain-based desserts are top providers of saturated fat (see Figure 5–9), but chicken and chicken dishes, red meats of all kinds, and Mexican-style foods all contribute substantially, as well.

Too Little Lipid A very few people manage to eat too little fat to support health. Among them are people with eating disorders who eat too little of all foods and misguided athletes hoping to improve performance. In the latter case, athletes whose fat

Table 5–2

Lipid Intake Recommendations for Healthy People

1. Total fat[a]

Dietary Reference Intakes

- An acceptable range of fat intake is estimated at 20 to 35% of total calories.

2. Saturated fat

American Heart Association

- Limit saturated fat to less than 7% of total energy.

Dietary Reference Intakes[c]

- Keep saturated fat intake low, less than 10% of calories, within the context of an adequate diet.

Dietary Guidelines for Americans 2010[b]

- Reduce the intake of solid fats.
- Consume less than 10% of calories from saturated fatty acids by replacing them with monounsaturated and polyunsaturated fatty acids.
- Replace protein foods that are high in solid fats with those lower in solid fats and calories and/or are sources of oils.

3. *Trans* fat

American Heart Association

- Limit *trans* fat to less than 1% of total energy.

Dietary Guidelines for Americans 2010[b]

- Keep *trans* fat intake as low as possible by limiting foods that contain synthetic sources of *trans* fats, such as partially hydrogenated oils, and by limiting other solid fats.

4. Polyunsaturated fatty acids

Dietary Reference Intakes[c]

- Linoleic acid (5 to 10% of total calories):
  17 g/day for young men.
  12 g/day for young women.
- Linolenic acid (0.6 to 1.2% of total calories):
  1.6 g/day for men.
  1.1 g/day for women.

5. Cholesterol

American Heart Association, Dietary Guidelines for Americans, and World Health Organization

- Limit cholesterol to less than 300 mg/day.[d]

Dietary Reference Intakes[c]

- Minimize cholesterol intake within the context of a healthy diet.

[a]*Includes monounsaturated fatty acids.*

[b]The Dietary Guidelines for Americans 2010 *use the term* solid fats *to describe sources of saturated and* trans-*fatty acids. Solid fats include milk fat, fats of high-fat meats and cheeses, hard margarines, butter, lard, and shortening.*

[c]*For DRI values set for various life stages, see the inside front cover. Linoleic and linolenic acids are defined later in this chapter.*

[d]*People with heart disease should aim for less than 200 mg/day.*

Figure 5–9

Sources of Solid Fats in the U.S. Diet

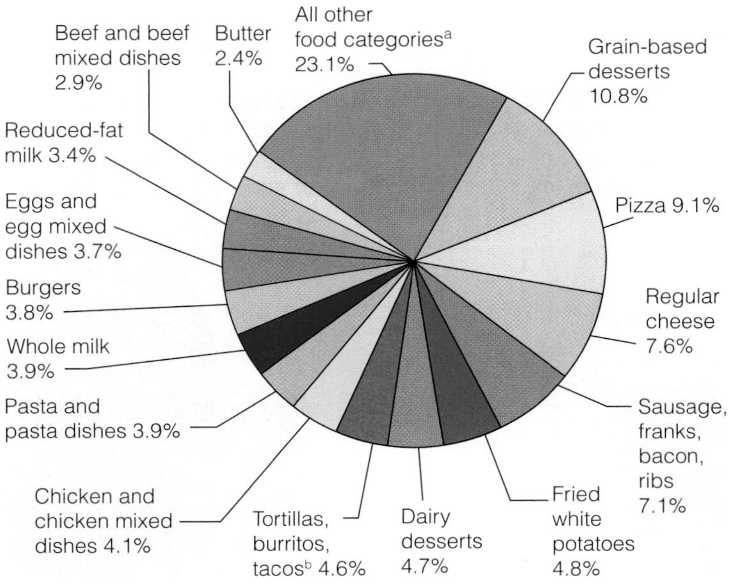

[a]*Food categories that each contribute less than 2% of the total solid fat intake.*
[b]*Includes nachos, quesadillas, and other mixed Mexican dishes.*

Source: U.S. Department of Agriculture and U.S. Department of Health and Human Services, Dietary Guidelines for Americans 2010, *available at www.dietaryguidelines.gov.*

***trans*-fatty acids** fatty acids with unusual shapes that can arise when hydrogens are added to the unsaturated fatty acids of polyunsaturated oils (a process known as *hydrogenation*).

linoleic (lin-oh-LAY-ic) **acid** an essential polyunsaturated fatty acid of the omega-6 family.

linolenic (lin-oh-LEN-ic) **acid** an essential polyunsaturated fatty acid of the omega-3 family. The full name of linolenic acid is *alpha-linolenic acid*.

intake falls short of the 20 percent minimum also short themselves on energy, vitamins, and essential fatty acids, and their performance suffers.

Some points about lipids and heart health are presented next because they form the foundation of lipid intake recommendations. The lipoproteins take center stage because they play important roles concerning the heart.

KEY POINTS

- A small amount of raw oil is recommended each day.
- Energy from fat should provide 20 to 35 percent of the total energy in the diet.

Lipoproteins and Heart Disease Risk

Recall that monoglycerides and long-chain fatty acids from digested food fat depend on chylomicrons, a type of lipoprotein, to transport them around the body. Chylomicrons and other lipoproteins are clusters of protein and phospholipids that act as emulsifiers—they attract both water and fat to enable their large lipid passengers to travel dispersed in the watery body fluids. The tissues of the body can extract whatever fat they need from chylomicrons passing by in the bloodstream. The remnants are then picked up by the liver, which dismantles them and reuses their parts.

Major Lipoproteins: Chylomicrons, VLDL, LDL, HDL The body makes four main types of lipoproteins, distinguished by their size and density.* Each type contains different kinds and amounts of lipids and proteins: the more lipids, the less dense; the more proteins, the more dense. In addition to chylomicrons, the lipoprotein with the least density, the body makes three other types of lipoproteins to carry its fats:

- **Very-low-density lipoproteins (VLDL)**, which carry triglycerides and other lipids made in the liver to the body cells for their use.
- **Low-density lipoproteins (LDL)**, which transport cholesterol and other lipids to the tissues. LDL are made from VLDL after they have donated many of their triglycerides to body cells.
- **High-density lipoproteins (HDL)**, which carry cholesterol away from body cells to the liver for disposal.[15]

Figure 5–10 depicts typical lipoproteins and demonstrates how a lipoprotein's density changes with its lipid and protein contents.

very-low-density lipoproteins (VLDL) lipoproteins that transport triglycerides and other lipids from the liver to various tissues in the body.

low-density lipoproteins (LDL) lipoproteins that transport lipids from the liver to other tissues such as muscle and fat; contain a large proportion of cholesterol.

high-density lipoproteins (HDL) lipoproteins that return cholesterol from the tissues to the liver for dismantling and disposal; contain a large proportion of protein.

Figure 5–10

Lipoproteins

As the graph shows, the density of a lipoprotein is determined by its lipid-to-protein ratio. All lipoproteins contain protein, cholesterol, phospholipids, and triglycerides in varying amounts. An LDL has a high ratio of lipid to protein (about 80 percent lipid to 20 percent protein) and is especially high in cholesterol. An HDL has more protein relative to its lipid content (about equal parts lipid and protein).

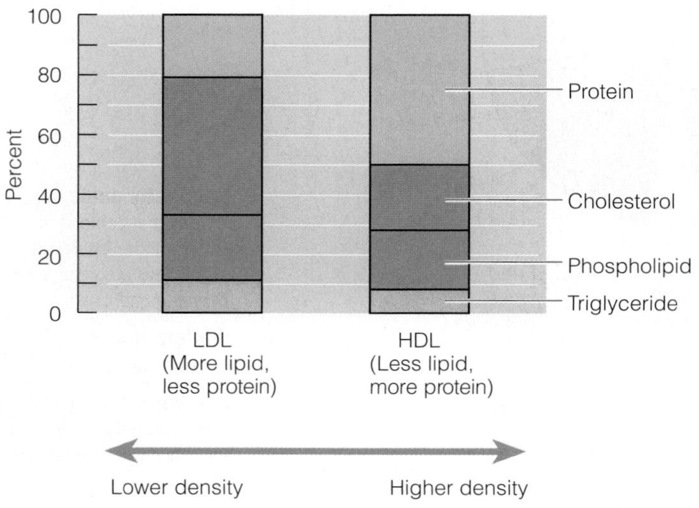

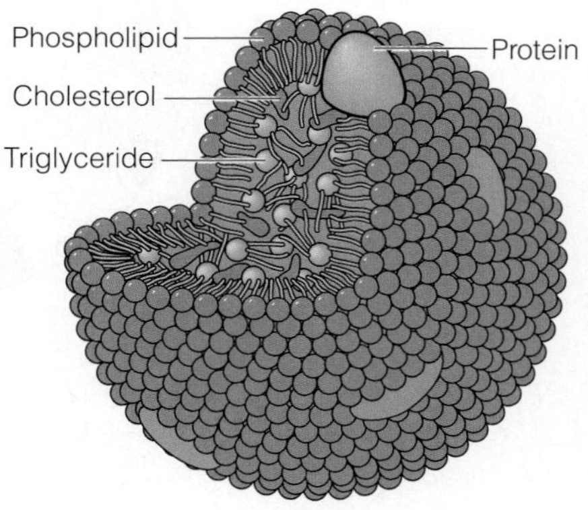

A typical lipoprotein

The LDL and HDL Difference The separate functions and effects of LDL and HDL are worth a moment's attention because they carry important implications for the health of the heart and blood vessels.

- Both LDL and HDL carry lipids in the blood, but LDL are larger, lighter, and richer in cholesterol; HDL are smaller, denser, and packaged with more protein.
- LDL deliver cholesterol to the tissues; HDL scavenge excess cholesterol and other lipids from the tissues and transport them via the bloodstream to the liver for disposal.
- When LDL cholesterol is too high, it contributes to lipid buildup in tissues, particularly in the linings of the arteries, that can trigger **inflammation** and lead to heart disease; HDL cholesterol opposes these effects and when HDL in the blood drops below the recommended level, heart disease risks rise in response.[16]

Both LDL and HDL carry cholesterol but high blood LDL concentrations warn of an increased risk of heart attack, whereas *low* HDL concentrations are associated with a greater risk (Chapter 11 has details). Thus, some people refer to LDL as "bad" cholesterol and HDL as "good" cholesterol—yet they carry the same kind of cholesterol. The difference to health between LDL and HDL lies in the proportions of lipids they contain and the tasks they perform, not in the *type* of cholesterol they carry.

The Importance of Cholesterol Testing The importance of blood cholesterol concentrations to heart health cannot be overstated.‡ The blood lipid profile, a medical test mentioned at the beginning of this chapter, tells much about a person's blood cholesterol and the lipoproteins that carry it. High blood LDL cholesterol and low HDL cholesterol account for two major risk factors for CVD (see Table 5–3). The margin lists the standards for blood lipid profile testing.[17]

Table 5–3

Modifiable Lifestyle Factors in Heart Disease Risk

The more of these factors present in a person's life, the more urgent the need for changes in diet and lifestyle to reduce heart disease risk:

- High blood LDL cholesterol.
- Low blood HDL cholesterol.
- High blood pressure (hypertension).
- Diabetes (insulin resistance).
- Obesity.
- Physical inactivity.
- Cigarette smoking.
- A diet high in saturated fats, including *trans* fats, and low in fish, vegetables, legumes, fruit, and whole grains.

Family history, older age, and male gender are risk factors that cannot be changed.

KEY POINTS

- The chief lipoproteins are chylomicrons, VLDL, LDL, and HDL.
- High blood LDL and low HDL are major heart disease risk factors.

Did You Know?

These numbers (in milligrams per deciliter) represent a desirable blood lipid profile:

- Total cholesterol: <200
- LDL cholesterol: <100
- HDL cholesterol: ≥ 60
- Triglycerides: <150

What Does *Food* Cholesterol Have to Do with *Blood* Cholesterol?

The answer may be "Not as much as most people think." Most saturated food fats (and *trans* fats) raise harmful blood cholesterol more than food *cholesterol* does. When told that dietary cholesterol doesn't matter as much as saturated fat, people may then jump to the wrong conclusion—that blood cholesterol doesn't matter. It does matter. High *blood* LDL cholesterol is a major indicator of CVD risk.[18] The two main food lipids associated with harmful blood LDL cholesterol levels are saturated fat and *trans* fat when intakes exceed recommendations.[19]

Dietary Cholesterol Guidelines and Sources Dietary cholesterol contributes somewhat to elevated blood cholesterol, but its role in heart disease is not clear.[20] The *Dietary Guidelines* recommend holding dietary cholesterol to an intake of below 300 milligrams per day for healthy people (200 for some people with high blood cholesterol or heart disease). On average, women take in about 240 milligrams a day, and men take in 350 milligrams.[21] Foods providing the greatest share of cholesterol to the U.S. diet are eggs and egg dishes, chicken and chicken dishes, beef and beef dishes, and burgers of all kinds.

Genetic Influence Genetic inheritance modifies everyone's ability to handle dietary cholesterol.[22] Most healthy people exhibit little increase in their blood cholesterol when they consume limited amounts of cholesterol-rich foods because the body slows its cholesterol synthesis when the diet provides cholesterol. Cholesterol differs

inflammation (in-flam-MAY-shun) an immune defense against injury, infection, or allergens and marked by heat, fever, and pain. Chronic, low-grade inflammation is associated with many disease states. Also defined in Chapter 3.

‡*Blood, plasma,* and *serum* all refer to about the same thing; this book uses the term *blood* cholesterol. Plasma is blood with the cells removed; in serum, the clotting factors are also removed. The concentration of cholesterol is not much altered by these treatments.

from salt and solid fats and added sugar in this respect: it cannot be omitted from the diet without omitting foods that are nutritious and sometimes low in fat. For example, an average egg contains 200 milligrams of cholesterol, but eggs are a convenient, inexpensive, low-fat, high-protein food that provides many vitamins, minerals, and phytochemicals.[23] One egg a day does not cause heart disease in healthy people, but this amount may raise blood LDL concentrations in people with heart disease or its risk factors.[24]

KEY POINTS

- Saturated fat and *trans* fat intakes raise blood cholesterol.
- Dietary cholesterol raises blood cholesterol to a lesser degree.

Recommendations Applied

In a welcome trend, fewer people in the United States have high blood cholesterol than in past decades.[25] Even so, a large number—more than 13 percent of adults—still tests too high. Heart disease kills more Americans than any other disease, and following the *Dietary Guidelines for Americans* remains an imperative for protecting heart health. To repeat, dietary saturated fat and *trans* fat can trigger a rise in LDL cholesterol in the blood. Conversely, trimming the saturated fat and *trans* fat from foods and replacing them with monounsaturated and polyunsaturated fats, within a reasonable calorie intake, is associated with lower LDL levels.[26]

Importantly, replacing dietary saturated fats with such things as added sugars and refined carbohydrates is often counterproductive.[27] Nutritionists know this: the best diet for health not only replaces saturated fats with polyunsaturated and monounsaturated oils but also is adequate, balanced, calorie-controlled, and based mostly on nutrient-dense whole foods. The context is important, too.

Trimming Saturated Fat to Lower LDL A necessary step toward meeting lipid guidelines is to identify sources of saturated fat and reduce their intakes. Figure 5–11 shows that when food is trimmed of fat, it also loses saturated fat and energy. A pork chop trimmed of its border of fat drops almost 70 percent of its saturated fat, 23 milligrams of cholesterol, and 220 calories. A plain baked potato has no saturated fat or cholesterol and contains about 40 percent of the calories of one with butter and sour cream. Choosing fat-free milk over whole milk provides large savings of saturated fat, calories, and cholesterol. Note that much of the cholesterol of the pork chop remained after trimming because cholesterol forms part of the cell membranes of muscle tissue.

Raising HDL—Recent Revelations As for HDL cholesterol, dietary measures are generally ineffective at raising its concentration. Regular physical activity does effectively raise blood HDL and reduces heart disease risks, as the nearby Think Fitness feature points out.

People with adequate blood HDL concentrations generally enjoy a lower risk of heart disease, but this relationship may be more complex than once thought. When researchers tested medications that increase the blood volume of HDL to levels higher than otherwise achievable, the hoped-for extra reduction in heart disease risk did not materialize.[28] In addition, some people have naturally higher HDL levels, but their risk of having a heart attack is about the same as for other people.[29] Having enough HDL in the blood is still important, but raising HDL to higher than normal levels may not be an effective strategy against heart disease.[30] For now, engaging in regular physical activity remains the best way to raise blood HDL cholesterol.

KEY POINTS

- To lower LDL in the blood, follow a healthy eating pattern that replaces dietary saturated fat and *trans* fat with polyunsaturated and monounsaturated oils.
- To raise HDL and lower heart disease risks, be physically active.

Think Fitness move it!

Why Exercise the Body for the Health of the Heart?

Every leading authority recommends physical activity to promote and maintain the health of the heart. The blood, arteries, heart, and other body tissues respond to exercise in these ways:

- Blood lipids shift toward higher HDL cholesterol.
- The muscles of the heart and arteries strengthen and circulation improves, easing delivery of blood to the lungs and tissues.
- A larger volume of blood is pumped with each heartbeat, reducing the heart's workload.
- The body grows leaner, reducing overall risk of cardiovascular disease.
- Blood glucose regulation is improved, reducing the risk of diabetes.

start now! Ready to make a change? Set a goal of exercising 30 minutes per day at least 5 days per week, then track your activity in Diet Analysis Plus.

Figure 5–11

Cutting Solid Fats Cuts Calories and Saturated Fat

Savings:
110 cal, 10 g solid fat, 4 g saturated fat

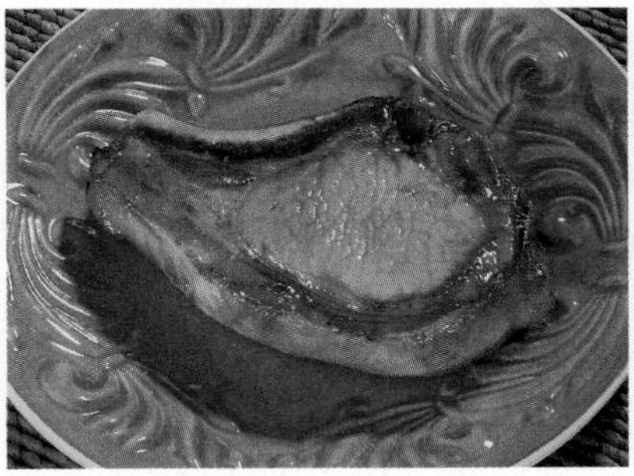

Pork chop with fat (340 cal, 19 g solid fat, 7 g saturated fat)

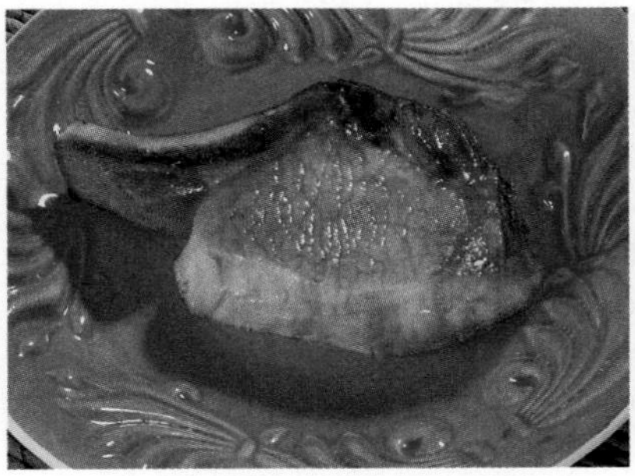

Pork chop trimmed of fat (230 cal, 9 g solid fat, 3 g saturated fat)

Savings:
150 cal, 13 g solid fat, 10 g saturated fat

Potato with 1 tbs butter and 1 tbs sour cream (350 cal, 14 g solid fat, 10 g saturated fat)

Plain potato (200 cal, ≈0 g solid fat, ≈0 g saturated fat)

Savings:
60 cal, 8 g solid fat, 5 g saturated fat

Whole milk, 1 c (150 cal, 8 g solid fat, 5 g saturated fat)

Fat-free milk, 1 c (90 cal, ≈0 g solid fat, ≈0 g saturated fat)

Table 5–4

Functions of the Essential Fatty Acids

These roles for the essential fatty acids are known, but others are under investigation.

- Provide raw material for eicosanoids.
- Serve as structural and functional parts of cell membranes.
- Contribute lipids to the brain and nerves.
- Promote normal growth and vision.
- Assist in gene regulation.
- Maintain outer structures of the skin, thus protecting against water loss.
- Help regulate genetic activities affecting metabolism.
- Support immune cell functions.

Essential Polyunsaturated Fatty Acids

LO 5.6 Compare the roles of omega-3 and omega-6 fatty acids in the body, and name important food sources of each.

The human body needs fatty acids, and it can use carbohydrate, fat, or protein to synthesize nearly all of them. Two are well-known exceptions: linoleic acid and linolenic acid. Body cells cannot make these two polyunsaturated fatty acids from scratch, nor can the cells convert one to the other.

Why Do I Need Essential Fatty Acids?

Linoleic and linolenic acids must be supplied by the diet and are therefore essential nutrients. So important are the essential fatty acids to health that the DRI committee set recommended intake levels for them (see the inside front cover). Table 5–4 summarizes the many established roles of the essential polyunsaturated fatty acids, but new functions continue to emerge.

Deficiencies of Essential Fatty Acids A diet deficient in *all* of the polyunsaturated fatty acids produces symptoms, such as reproductive failure, skin abnormalities, and kidney and liver disorders. In infants, growth is retarded, and vision is impaired. The body stores some essential fatty acids, so extreme deficiency disorders are seldom seen except when intentionally induced in research or on rare occasions when inadequate diets have been provided to infants or hospital patients by mistake. In the United States and Canada, such deficiencies are almost unknown among otherwise healthy adults. The story doesn't end there, however.

Meet the Eicosanoids Essential fatty acids serve as raw materials from which the body makes many needed products. These products include a group of bioactive lipids known as **eicosanoids** that act somewhat like hormones—they send signals to body tissues and elicit responses.[31] Eicosanoids influence a wide range of diverse body functions, such as muscle relaxation and contraction, blood vessel constriction and dilation, blood clot formation, blood lipid regulation, and the immune response to injury and infection such as fever, inflammation, and pain. A familiar drug, aspirin, relieves fever, inflammation, and pain by slowing the synthesis of certain eicosanoids.[32]

KEY POINTS

- Deficiencies of the essential fatty acids are virtually unknown in the United States and Canada.
- Linoleic acid and linolenic acid are converted into eicosanoids, which influence diverse body functions.

Omega-6 and Omega-3 Fatty Acid Families

Linoleic acid is the "parent" member of the **omega-6 fatty acid** family, so named for the chemical structure of these compounds. Given dietary linoleic acid, the body can produce other needed members of the omega-6 family. One of these is **arachidonic acid**, notable for its role as a starting material from which a number of eicosanoids are made. Omega-6 fatty acids are supplied abundantly in vegetable oils.

Linolenic acid is the parent member of the **omega-3 fatty acid** family. Given dietary linolenic acid, the body can make other members of the omega-3 series. Two family members of great interest to researchers are **EPA** and **DHA**. The body makes only limited amounts of these omega-3 fatty acids, but they are found abundantly in the oils of certain fish.

In Heart Health Years ago, someone thought to ask why the native peoples of the extreme north, who eat a diet very high in animal fat, had low rates of heart disease.[33]

The trail led to their abundant intake of fish and marine foods, then to the oils in fish, and finally to EPA and DHA in fish oils. EPA and DHA play important roles in regulating heartbeats, regulating blood pressure, reducing blood clot formation, and reducing inflammation—all factors associated with heart health.[34] Today, as young people of northern tribes trade traditional marine diets for fast foods, added sugars, and saturated fats, their LDL cholesterol, body weights, and blood pressure have soared.[35]

Research often links higher EPA and DHA in the blood and greater intakes of fish in the diet with fewer heart attacks and strokes.[36] Not every study reports lower cardiovascular risks with higher intakes, however, and genetic inheritance may partly determine to what degree a person may benefit from consuming EPA and DHA.[37]

In Cancer Prevention Consuming seafood that provides omega-3 fatty acids is associated with lower rates of some cancers, possibly because they suppress inflammation, a factor in cancer development, or through other mechanisms.[38] After cancer develops, a significant reduction in cancer-related deaths has been associated with higher fish intakes.[39] However, in a surprising twist, men with more DHA in their blood were *more* likely to have aggressive prostate cancer than men with lower levels.[40] So little is known about the relationships between omega-3 fatty acids and cancers that people are wise to eat fish, not take supplements, to provide them.

In Cell Membranes EPA and DHA tend to collect in cell membranes. Unlike straight-backed saturated fatty acids, which physically stack closely together, the kinked shape of unsaturated fatty acids demands more elbow room (look back at Figure 5–4, page 161). When the highly unsaturated EPA and DHA amass in cell membranes, they profoundly change cellular structures and activities in ways that may promote healthy tissue functioning.

In Brain Function and Vision The brain is a fatty organ with a quarter of its dry weight as lipid, and its cell membranes avidly collect EPA and DHA in their structures.[41] Once there, these fatty acids may assist in the brain's communication processes and reduce inflammation associated with brain diseases of aging.[42] Deficiencies are suspected to play roles in age-related vision changes.[43] In infants, breast milk and fortified formula provide abundant DHA, associated with normal growth, visual acuity, immune system functioning, and brain development.[44] Other potential benefits of these remarkable lipids are listed in Table 5–5.

Table 5–5

Potential Health Benefits of Fish Oils

These benefits from fish or fish oil are well-established, but researchers are investigating many others.

Against heart disease

- A shift toward omega-3 eicosanoids by reducing production of omega-6 eicosanoids. This shift may reduce abnormal blood clotting, help sustain more regular heartbeats, and reduce inflammation of many body tissues, including the arteries of the heart.
- Reduced blood triglycerides (in some studies, fish oil supplements elevated blood LDL cholesterol, an opposing, detrimental outcome).
- Retarded hardening of the arteries (atherosclerosis).
- Relaxation of blood vessels, mildly reducing blood pressure.

In infant growth and development

- Normal brain development in infants. DHA concentrates in the brain's cortex, the conscious thinking part.
- Normal vision development in infants. DHA helps to form the eye's retina, the seat of normal vision.

KEY POINTS

- The omega-6 family of polyunsaturated fatty acids includes linoleic acid and arachidonic acid.
- The omega-3 family includes linolenic acid, EPA, and DHA.
- EPA and DHA may play roles in brain communication, disease prevention, and human development.

Where Are the Omega-3 Fatty Acids in Foods?

The *Dietary Guidelines* recommend choosing 8 to 12 ounces of a variety of seafood each week, or about one of every five protein food servings, to provide an average of 250 mg of EPA and DHA per day, along with a beneficial array of nutrients that seafood provides.[45] For example, fish, low in saturated fat and high in protein, contributes not only EPA and DHA but also the mineral selenium, a nutrient of concern for heart health (see Chapter 8).[46]

Special Populations Children and pregnant and lactating women have a critical need for EPA and DHA, but they are also most susceptible to harm from environmental toxins, such as mercury, in fish. For young children, 3 to 6 ounces of seafood each week, including smaller EPA- and DHA-rich species, provides the needed fatty acids and seems safe.[47] For women who are pregnant or breastfeeding, weekly consumption of 8 to 12 ounces of a variety of seafood, including some EPA- and DHA-rich species, is compatible with good health of both mother and child.[48] At the same time, these women must strictly limit their intakes of white albacore tuna to no more than

eicosanoids (eye-COSS-ah-noyds) biologically active compounds that regulate body functions.

omega-6 fatty acid a polyunsaturated fatty acid with its endmost double bond six carbons from the end of the carbon chain. Linoleic acid is an example.

arachidonic (ah-RACK-ih-DON-ik) **acid** an omega-6 fatty acid derived from linoleic acid.

omega-3 fatty acid a polyunsaturated fatty acid with its endmost double bond three carbons from the end of the carbon chain. Linolenic acid is an example.

EPA, DHA eicosapentaenoic (EYE-cossa-PENTA-ee-NO-ick) acid, docosahexaenoic (DOE-cossa-HEXA-ee-NO-ick) acid; omega-3 fatty acids made from linolenic acid in the tissues of fish.

Table 5–6

Food Sources of Omega-6 and Omega-3 Fatty Acids

Omega-6	
Linoleic acid	Seeds, nuts, vegetable oils (corn, cottonseed, safflower, sesame, soybean, sunflower), poultry fat
Omega-3	
Linolenic acid[a]	Oils (canola, flaxseed, soybean, walnut, wheat germ; liquid or soft margarine made from canola or soybean oil) Nuts and seeds (flaxseeds, walnuts, soybeans) Vegetables (soybeans)
EPA and DHA	Human milk Fish and seafood: *>500 mg/3.5 oz serving.* European seabass (bronzini), herring (Atlantic and Pacific), mackerel, oyster (Pacific wild), salmon (wild and farmed), sardines, toothfish (includes Chilian seabass), trout (wild and farmed) *150–500 mg/3.5 oz serving.* black bass, catfish (wild and farmed), clam, cod (Atlantic), crab (Alaskan king), croakers, flounder, haddock, hake, halibut, oyster (eastern and farmed), perch, scallop, shrimp (mixed varieties), sole, swordfish, tilapia (farmed) *<150 mg/3.5 oz serving.* cod (Pacific), grouper, lobster, mahi-mahi, monkfish, red snapper, skate, triggerfish, tuna, wahoo

[a]*Alpha-linolenic acid. Also found in the seed oil of the herb* evening primrose.

6 ounces per week and should not eat tilefish, shark, swordfish, or king mackerel at all because of their high mercury content.[49] The Consumer's Guide near here provides more information about balancing the risks and benefits of seafood.

What about Fish Oil Supplements? Fish, not fish oil supplements, is the preferred source of omega-3 fatty acids. High intakes of omega-3 polyunsaturated fatty acids may increase bleeding time, interfere with wound healing, raise LDL cholesterol, and suppress immune function.[§] Evidence is mixed for people with heart disease—several studies show hopeful results, whereas others reveal no benefits from supplements.[50] Supplements bring risks, such as excessive bleeding; individuals taking daily fish oil supplements need medical supervision.[51] The benefits and risks from EPA and DHA illustrate an important concept in nutrition: too much of a nutrient is often as harmful as too little.

Omega-3 Enriched Foods Manufacturers cannot simply add fish oil to staple foods such as milk, juice, or bread because it adds a fishy taste and quickly undergoes **oxidation** and becomes rancid.[52] Alternatively, food products may be enriched with omega-3 fatty acids indirectly, but rarely do labels identify which fatty acids do the enriching. For example, eggs can be enriched with EPA and DHA by feeding laying hens grains laced with fish oil or algae oil, but most "omega-3 enriched eggs" on the market come from chickens fed on flaxseed and thus are enriched only with linolenic acid, not EPA and DHA. Food scientists are currently working out ways of increasing the EPA and DHA content of enriched eggs without introducing off flavors.[53]

Wild game meat or beef from pasture-fed cattle contains more omega-3 fatty acids (and less saturated fat) than traditional meats. In addition, several forms of marine algae and their oils provide a vegetarian source of DHA, and soybean and yeast sources of EPA are under development.[54]** Common foods that provide essential fatty acids are listed in Table 5–6.

KEY POINTS

- The *Dietary Guidelines* recommend increasing seafood consumption.
- Supplements of omega-3 fatty acids are not recommended.

oxidation interaction of a compound with oxygen; in this case, a damaging effect by a chemically reactive form of oxygen. Chapter 7 provides details.

methylmercury any toxic compound of mercury to which a characteristic chemical structure, a methyl group, has been added, usually by bacteria in aquatic sediments. Methylmercury is readily absorbed from the intestine and causes nerve damage in people.

§ Suppressed immune function is seen with daily intake of 0.9 to 9.4 grams EPA and 0.6 to 6.0 grams DHA for 3 to 24 weeks.
** The fatty acid is SDA, steriodonic acid.

use it! A Consumer's Guide To . . .

Weighing Seafood's Risks and Benefits

Do you ever stand at a seafood counter or sit in a restaurant imagining a healthy fish dinner but wondering what to choose? These days, seafood comes with some questions: which fish provides the needed essential fatty acids? Which fish is lowest in toxins or microorganisms that may pose risks to health? Which is best—farmed or wild?

Finding the EPA and DHA

Fish in many forms—fresh, frozen, and canned—makes a nutritious choice because EPA and DHA along with other key nutrients survive most cooking and processing. However, the *type* of fish is critical—among frozen selections, for example, pre-fried fish sticks and fillets are most often cod, a delicious fish but one that provides little EPA and DHA (look again at Table 5–6).

In fast-food places, fried fish sandwiches are generally cod. These fried fillets derive more of their calories from their oily breading or batter than from the fish itself, and more still from fatty sauces that flavor the bun. Cod, like any fish, provides little solid fat when served grilled, baked, poached, or broiled. And if it displaces fatty meats from the diet, it provides a benefit to the heart—just don't count on cod for EPA and DHA.[1] In sit-down restaurants, diners can almost always find EPA- and DHA-rich species, such as salmon, on menus, but only if they know which is which.

Concerns about Toxins

Analyses of seafood samples have revealed widespread contamination by toxins, raising concerns about seafood safety. Just one among many concerns, the heavy metal mercury escapes from many industries, power plants, and natural sources into the earth's waterways where bacteria in the water convert it into a highly toxic form, **methylmercury**. Methylmercury then concentrates in the flesh of large predatory species of both saltwater and freshwater fish.[2] Cooking and processing do not diminish mercury or other toxins in seafood.

In the U.S. population, blood mercury levels of women are on the rise.[3] Over time mercury accumulates in the liver, immune tissues, brain, and other organs and has the potential to cause harm to a developing fetus. Yet, fish brings health benefits to people of all ages—even a developing fetus, by way of the mother's diet. Currently, for most people, the benefits outweigh the risks.[4]

Pregnant and lactating women should strictly follow recommendations set for them. Rather than give up on fish, it is possible to choose species both rich in omega-3 fatty acids *and* lower in mercury (see Figure 5–12, p. 178), and it's probably wise to do so most often, particularly for women of childbearing age.[5]

Cooked vs. Raw

Many people love sushi, but authorities never recommend eating raw fish and shellfish—it causes many cases of serious or fatal bacterial, viral, and other illness each year (Chapter 12 provides many details). Cooking easily kills off all illness-causing microorganisms, making seafood safe to eat.

Fresh from the Farm

Are farm-raised fish safer? Compared with wild fish, farm-raised fish do tend to collect somewhat less methylmercury in their flesh, and the levels of other harmful pollutants generally test below the maximums set by the U.S. Food and Drug Administration. However, fish "farms" are often giant ocean cages, exposed to whatever contaminants float by in the water. Farmed fish vary in EPA and DHA, too, because they are fed manufactured fish chow that varies in omega-3 fatty acid content.[6] The contamination of fish serves as a reminder that our health is inextricably linked with the health of our planet (details in Chapter 15).

Moving Ahead

Keep these pointers in mind:

- Choose fish and shellfish (prepared without the addition of solid fats) instead of red meat several times a week—people who do generally stay healthier than those who don't.
- Apply the dietary principles of adequacy, moderation, and variety to obtain the benefits of seafood while minimizing risks.
- Avoid eating raw seafood.
- Learn and remember which varieties are the safer EPA- and DHA-rich choices. Stay aware of the latest updates to the lists.

In conclusion, use a variety of seafood to meet your needs—just don't go overboard.

Review Questions*

1. Methylmercury is a toxic industrial pollutant found in highest concentrations in small fish. **T F**
2. Children and pregnant or lactating women should definitely not consume fish because of contamination. **T F**
3. Cod is one of the richest sources of the beneficial fatty acids, EPA, and DHA. **T F**

**Answers to Consumer's Guide review questions are found in Appendix G.*

Figure 5-12

Seafood Species—EPA/DHA and Mercury Contents

The species listed on the left are exceptionally rich in EPA and DHA: two 4-ounce servings a week meet the need. The species on the right are generally lower in mercury, although contamination varies with size and water conditions. Caution: one 4-ounce serving of any of four species—king mackerel, shark, swordfish, and tilefish—often exceeds an entire week's safe mercury intake for pregnant and breastfeeding women. They should choose safer species.

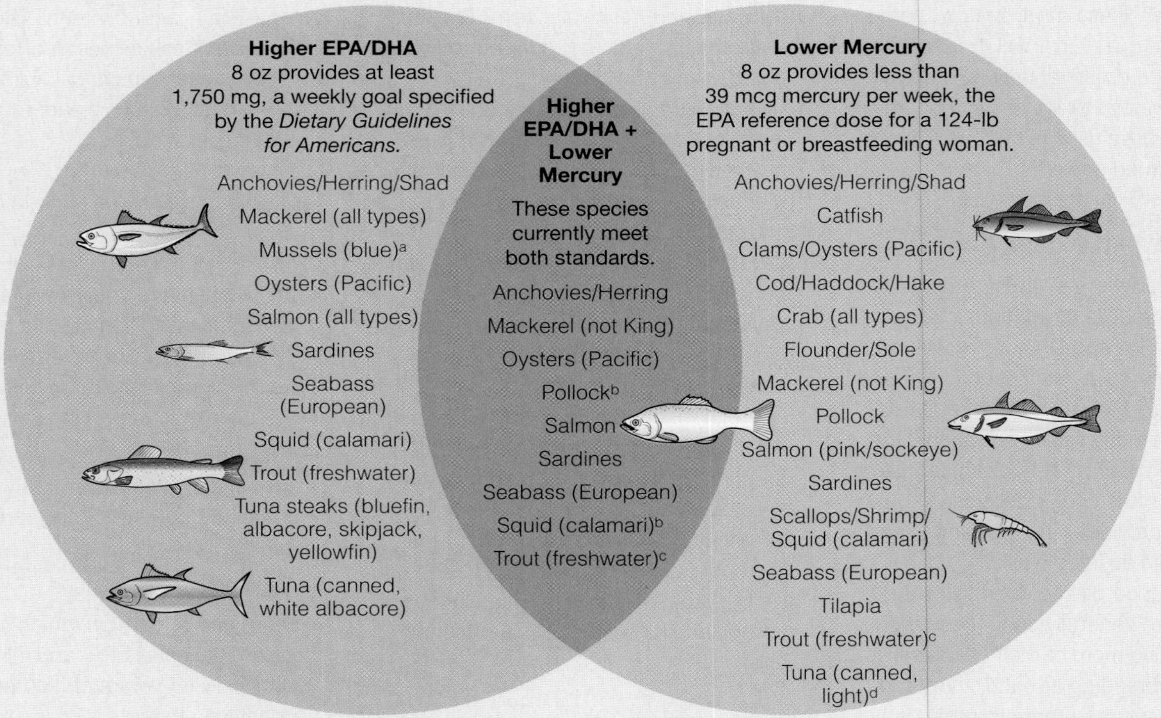

[a] *No mercury data available.*

[b] *Pollock and squid provide somewhat less EPA/DHA per 8 oz but still make substantial contributions.*

[c] *Check local water contamination levels.*

[d] *Light canned tuna provides just 600 mg EPA/DHA per 8 oz but its lower mercury content and popularity make it a notable contributor.*

Source: Data from U.S. Department of Agriculture and U.S. Department of Health and Human Services, Dietary Guidelines for Americans 2010, *Appendix 11, available at www.dietaryguidelines.gov.*

The Effects of Processing on Unsaturated Fats

LO 5.7 Describe the hydrogenation of fat and the formation and structure of a *trans*-fatty acid.

Vegetable oils make up most of the added fat in the U.S. diet because fast-food chains use them for frying, food manufacturers add them to processed foods, and consumers tend to choose margarine over butter. Consumers of vegetable oils may feel safe in choosing them because they are generally less saturated than animal fats. If consumers choose a liquid oil, they may be justified in feeling secure. If the choice is a processed food, however, their security may be questionable, especially if the words *hydrogenated* or *partially hydrogenated* appear on the label's ingredient list.

Figure 5–13

Animated! Hydrogenation Yields Both Saturated and *Trans*-Fatty Acids

Unsaturated fatty acid
Points of unsaturation are places on fatty acid chains where hydrogen is missing. The bonds that would normally be occupied by hydrogen in a saturated fatty acid are shared, reluctantly, as a double bond between two carbons that both carry a slightly negative charge.

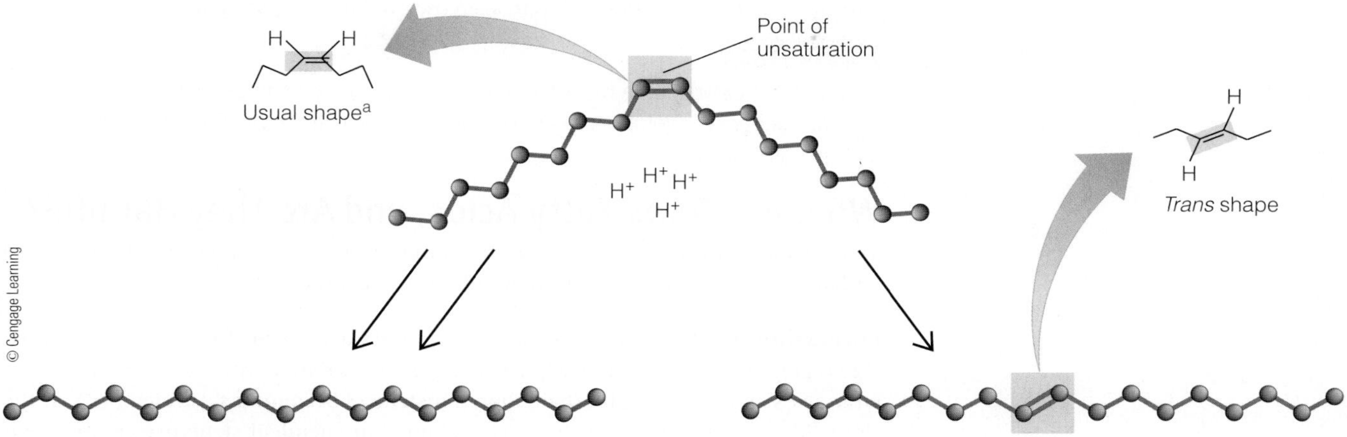

Hydrogenated fatty acid (now fully saturated)
When a positively charged hydrogen is made available to an unsaturated bond, it readily accepts the hydrogen and, in the process, becomes saturated. The fatty acid no longer has a point of unsaturation.

***Trans*-fatty acid**
The hydrogenation process also produces some *trans*-fatty acids. The *trans*-fatty acid retains its double bond but takes a twist instead of becoming fully saturated. It resembles a saturated fatty acid both in shape and in its effects on health.

[a] *The usual shape of the double bond structure is known as a cis (pronounced* sis*) formation.*

What Is "Hydrogenated Vegetable Oil," and What's It Doing in My Chocolate Chip Cookies?

When manufacturers process foods, they often alter the fatty acids in the fat (triglycerides) the foods contain through a process called **hydrogenation**. Hydrogenation of fats makes them stay fresher longer and also changes their physical properties.

Hydrogenation of Oils Points of unsaturation in fatty acids are weak spots that are vulnerable to attack by oxygen damage. When the unsaturated points in the oils of food are oxidized, the oils become rancid and the food tastes "off." This is why cooking oils should be stored in tightly covered containers that exclude air. If stored for long periods, they need refrigeration to retard oxidation.

One way to prevent spoilage of unsaturated fats and also to make them harder and more stable when heated to high temperatures is to change their fatty acids chemically by hydrogenation, as shown on the left side of Figure 5–13. When food producers want to use a polyunsaturated oil such as soybean oil to make a spreadable margarine, for example, they hydrogenate it by forcing hydrogen into the liquid oil. Some of the unsaturated fatty acids become more saturated as they accept the hydrogen, and the oil hardens. The resulting product is more saturated and more spreadable than the original oil. It is also more resistant to damage from oxidation or breakdown from high cooking temperatures. Hydrogenated oil has a high **smoking point**, so it is suitable for frying foods at high temperatures in restaurants.

Hydrogenated oils are thus easy to handle, easy to spread, and store well. Makers of peanut butter often replace a small quantity of the liquid oil from the ground peanuts with hydrogenated vegetable oils to create a creamy paste that does not separate into layers of oil and peanuts as the "old-fashioned" types do. Neither type of peanut butter is high in saturated fat, however.

Baked goods with no trans *fat may still contain a great deal of solid fat as shortening.*

hydrogenation (high-dro-gen-AY-shun) the process of adding hydrogen to unsaturated fatty acids to make fat more solid and resistant to the chemical change of oxidation.

smoking point the temperature at which fat gives off an acrid blue gas.

Nutrient Losses Once fully hydrogenated, oils lose their unsaturated character and the health benefits that go with it. Hydrogenation may affect not only the essential fatty acids in oils but also vitamins, such as vitamin K, decreasing their activity in the body. If you, the consumer, are looking for health benefits from polyunsaturated oils, hydrogenated oils such as those in shortening or stick margarine will not meet your need.

An alternative to hydrogenation for extending a product's shelf life is to add a chemical preservative that will compete for oxygen and thus protect the oil. The additives are antioxidants, and they work by reacting with oxygen before it can do damage. Examples are the additives BHA and BHT[††] listed on snack food labels. Another alternative, already mentioned, is to keep the product refrigerated.

KEY POINTS

- Vegetable oils become more saturated when they are hydrogenated.
- Hydrogenated vegetable oils are useful, but they lose the health benefits of unsaturated oils.

What Are *Trans*-Fatty Acids, and Are They Harmful?

Many consumers identify *trans* fats—that is, fats that contain *trans*-fatty acids–as bad for health. Do *trans* fats warrant their villainous reputation?

Formation of *Trans*-Fatty Acids *Trans*-fatty acids form during hydrogenation. When polyunsaturated oils are hardened by hydrogenation, some of the unsaturated fatty acids end up changing their shapes instead of becoming saturated (look at the right side of Figure 5–13, p. 179). This change in chemical structure creates *trans* unsaturated fatty acids that are similar in shape to saturated fatty acids. The change in shape changes their effects on the health of the body.

Health Effects of *Trans*-Fatty Acids Consuming manufactured *trans* fat poses a risk to the heart and arteries by raising blood LDL cholesterol, lowering HDL cholesterol, and increasing tissue inflammation.[55] In addition, when hydrogenation changes essential fatty acids into their saturated or *trans* counterparts, the consumer loses the health benefits of the original raw oil. The risk to heart health from *trans* fat is similar to or slightly greater than that from saturated fat, so the *Dietary Guidelines for Americans* suggest that people keep *trans* fat intake as low as possible.[56] A small amount of naturally occurring *trans* fat also comes from animal sources, such as milk and lean beef, but these *trans* fats have little effect on blood lipids and are under study for potential health benefits.[‡‡57]

Swapping *Trans* Fats for Saturated Fats? In the past, most commercially fried foods, from doughnuts to chicken, delivered a sizeable load of *trans* fats to

My Turn **watch it!**

Heart to Heart

© Cengage Learning
Jessica

© Cengage Learning
Katy

How often do you think about the consequences of your food choices now on your heart health later in life? Two people talk about planning heart-healthy meals.

[††]BHA and BHT are butylated hydroxyanisole and butylated hydroxytoluene.
[‡‡]The natural *trans* fats of milk are conjugated linoleic acids.

consumers. Today, newly formulated commercial oils and fats perform the same jobs as the old hydrogenated fats but with fewer *trans*-fatty acids.[58]

If a fatty food lacks *trans* fats, is it safe for the heart? It might be, but some new fats merely substitute saturated fat for *trans* fat—and the risk to the heart and arteries from saturated fats is well established.[59] For example, an ounce of crackers or cookies listing 0 grams of *trans* fat on the label can easily contain 6 grams of saturated fat, a substantial contribution to the day's allowance. No health benefits can be expected when saturated fats replace *trans* fats in the diet.

Did You Know?

The Nutrition Facts section of food labels lists the grams of both saturated fat and *trans* fat in foods; a food listing 0 g of either one may actually contain up to 0.5 g in a serving.

KEY POINTS

- The process of hydrogenation creates *trans*-fatty acids.
- *Trans* fats act like saturated fats in the body.

Fat in the Diet

LO 5.8 Outline a diet plan that provides enough of the right kinds of fats within calorie limits.

The remainder of this chapter and its Controversy show you how to choose fats wisely with the goals of providing optimal health and pleasure in eating. A column of Appendix A, entitled *Fat Breakdown (g)*, makes fascinating reading when evaluating the fat contents of foods.

Get to Know the Fats in Foods

As you read about which foods provide what fats, take note of the foods offering unsaturated oils, with their payload of essential nutrients, and those packed with solid fats—the sources of saturated and *trans*-fatty acids. Your ability to choose among them at mealtimes can make a difference to the unseen condition of your arteries.

Essential Fats Everyone needs the essential fatty acids and vitamin E provided by such foods as fish, nuts, and vegetable oils. Infants receive them indirectly via breast milk, but all others must choose the foods that provide them. Luckily, the amount of fat needed to provide these nutrients is small—just a few teaspoons of raw oil a day and two servings of seafood a week are sufficient. Most people consume more than this minimum amount, however. The goal is to choose unsaturated fats in liquid oils instead of saturated solid fats as often as possible.

© iStockphoto.com/dirkr

A serving of ten small olives or a sixth of an avocado each provides about 5 grams of mostly monounsaturated fat, along with essential nutrients and potentially beneficial phytochemicals.

Replace, Don't Add Keep in mind that, whether solid or liquid, essential or nonessential, all fats bring the same abundant calories to the diet and excesses contribute to body fat stores. According to the *Dietary Guidelines for Americans*, no benefits can be expected when oil is added to an already fat-rich diet. Each of these provides about 5 grams of fat, 45 calories, and negligible protein and carbohydrate:

- 1 teaspoon oil or shortening.
- 1½ teaspoon mayonnaise, butter, or margarine.
- 1 tablespoon regular salad dressing, cream cheese, or heavy cream.
- 1½ tablespoon sour cream.

Remember to replace and not add.

Visible vs. Invisible Solid Fats The solid fat of some foods, such as the rim of fat on a steak, is visible (and therefore identifiable and removable). Other solid fats, such as those in candy, cheeses, coconut, hamburger, homogenized milk, and lunchmeats, are invisible (and therefore easily missed or ignored). Equally hidden are the solid fats blended into biscuits, cakes, cookies, chip dips, ice cream, mixed dishes, pastries, sauces, and creamy soups and in fried foods and spreads. Invisible fats supply the majority of the solid fats in the U.S. diet.[60]

Fats in Protein Foods

The marbling of meats and the fat ground into lunchmeat, chicken products, and hamburger conceal a hefty portion of the solid fat that people consume. To help "see"

Definitions of terms relating to the fat contents of meats were provided in **Chapter 2**'s Food Feature.

the fat in meats and poultry, it is useful to think of them in four categories according to their fat contents—very lean, lean, medium-fat, and high-fat—as the exchange lists do in Appendix D. Meats in all four categories contain about equal amounts of protein, but their fat, saturated fat, and calorie amounts vary significantly. Figure 5–14 shows the fat and calorie data on packages of ground meats, and it depicts the amount of solid fat provided by a 3-ounce serving of each kind. New Nutrition Facts panels list the fat contents of many packaged meats.

The USDA eating patterns suggest that most adults limit their intake of protein foods to about 5 to 7 ounces a day. For comparison, the smallest fast-food hamburger weighs about 3 ounces. Steaks served in restaurants often run 8, 12, or 16 ounces, more than a whole day's meat allowance. You may have to weigh a serving or two of meat to see how much you are eating.

Meat: Mostly Protein or Fat? People recognize meat as a protein-rich food, but a close look at some nutrient data reveals a surprising fact. A big (4-ounce) fast-food hamburger sandwich contains 23 grams of protein and 23 grams of fat, more than 8 of them saturated fat.[61] Because protein offers 4 calories per gram and fat offers 9, the meat of the sandwich provides 92 calories from protein but 207 calories from fat. Hot dogs, fried chicken sandwiches, and fried fish sandwiches also provide hundreds of mostly invisible calories of solid fat. Because so much meat fat is hidden from view, meat eaters can easily and unknowingly consume a great many grams of solid fat from this source.

Clues to Lower-Fat Meats When choosing beef or pork, look for lean cuts named *loin* or *round* from which the fat can be trimmed, and eat small portions. Chicken and turkey flesh are naturally lean, but commercial processing and frying add solid fats, especially to "patties," "nuggets," "fingers," and wings. Watch out for ground turkey or

Figure 5–14

Calories, Fat, and Saturated Fat in Cooked Ground Meat Patties[a]

Only the ground round, at 10 percent fat by raw weight, qualifies to bear the word *lean* on its label. To be called "lean," products must contain fewer than 10 grams of fat, 4 grams of saturated fat, and 95 milligrams of cholesterol per 100 grams of food. The red labels on these packages list rules for safe meat handling, explained in Chapter 12.

[a]*Larger servings will, of course, provide more fat, saturated fat, and calories than the values listed here.*

Figure 5–15

Lipids in Milk and Milk Products

Red boxes below indicate foods with higher lipid contents that warrant moderation in their use. Green indicates lower-fat choices.

Nutrition Facts

Amount Per Serving

© Polara Studios, Inc.

Fat-free, skim, zero-fat, no-fat, or nonfat milk, 8 oz (<0.5% fat by weight)

Calories 80	Calories from Fat 0
	% Daily Value*
Total Fat 0g	**0%**
Saturated Fat 0g	**0%**
Cholesterol 5mg	**2%**

Low-fat milk, 8 oz (1% fat by weight)

Calories 105	Calories from Fat 20
	% Daily Value*
Total Fat 2g	**3%**
Saturated Fat 1.5g	**8%**
Cholesterol 10mg	**3%**

Low-fat cheddar cheese, 1.5 oz

Calories 70	Calories from Fat 30
	% Daily Value*
Total Fat 3g	**5%**
Saturated Fat 2g	**10%**
Cholesterol 10mg	**3%**

Strawberry yogurt, 8 oz

Calories 250	Calories from Fat 45
	% Daily Value*
Total Fat 5g	**8%**
Saturated Fat 3g	**15%**
Cholesterol 15mg	**5%**

Whole milk, 8 oz (3.3% fat by weight)

Calories 150	Calories from Fat 70
	% Daily Value*
Total Fat 8g	**12%**
Saturated Fat 5g	**25%**
Cholesterol 24mg	**8%**

Reduced-fat, less-fat milk, 8 oz (2% fat by weight)

Calories 120	Calories from Fat 45
	% Daily Value*
Total Fat 5g	**8%**
Saturated Fat 2g	**10%**
Cholesterol 20mg	**7%**

Cheddar cheese, 1.5 oz

Calories 165	Calories from Fat 130
	% Daily Value*
Total Fat 14g	**22%**
Saturated Fat 9g	**45%**
Cholesterol 40mg	**13%**

Low-fat strawberry yogurt, 8 oz

Calories 240	Calories from Fat 20
	% Daily Value*
Total Fat 2.5g	**4%**
Saturated Fat 2g	**10%**
Cholesterol 15mg	**5%**

Art © Cengage Learning

chicken products. The skin is often ground in to add pleasing moistness, but the food ends up with more solid fat than the amount found in many cuts of lean beef. Also, some people (even famous chefs) misinterpret Figure 5–5 (page 162), reasoning that if poultry or pork fat are less saturated than beef fat, they must be harmless to the heart. Nutrition authorities emphatically state, however, that all sources of saturated fat pose a risk and that even the skin of poultry should be removed before eating the food.

KEY POINT

- Meats account for a large proportion of the hidden solid fat in many people's diets.

Milk and Milk Products

Milk products go by many names that reflect their varying fat contents, as Figure 5–15 shows. A cup of homogenized whole milk contains the protein and carbohydrate of fat-free milk, but in addition it contains about 80 extra calories from butterfat, a solid fat. A cup of reduced-fat (2 percent fat) milk falls between whole and fat-free, with 45 calories of fat. The fat of whole milk occupies only a teaspoon or two of the volume but nearly doubles the calories in the milk.

Milk and yogurt appear together in the Milk and Milk Products group, but cream and butter do not. Milk and yogurt are rich in calcium and protein, but cream and butter are not. Cream and butter are solid fats, as are whipped cream, sour cream, and cream cheese, and they are properly grouped together with other fats. Other cheeses, grouped with milk products, vary in their fat contents and are major contributors of saturated fat in the U.S. diet.[62]

Figure 5–16

Lipids in Grains

Red boxes below indicate foods with higher lipid contents that warrant moderation in their use. Green indicates lower-fat choices.

Nutrition Facts

Amount Per Serving

Low-fat granola, 1/2 c	
Calories 195	Calories from Fat 35
	% Daily Value*
Total Fat 3g	**5%**
Saturated Fat 1g	**5%**
Cholesterol 0mg	**0%**

Crispy oat bran, 1/2 c	
Calories 150	Calories from Fat 45
	% Daily Value*
Total Fat 5g	**8%**
Saturated Fat 1.5g	**8%**
Cholesterol 0mg	**0%**

Buttery crackers, 5 crackers	
Calories 80	Calories from Fat 35
	% Daily Value*
Total Fat 4g	**6%**
Saturated Fat 1g	**5%**
Cholesterol 0mg	**0%**

Fried rice, 1/2 c[a]	
Calories 140	Calories from Fat 65
	% Daily Value*
Total Fat 7g	**11%**
Saturated Fat 1g	**5%**
Cholesterol 20mg	**7%**

A homemade waffle	
Calories 220	Calories from Fat 100
	% Daily Value*
Total Fat 11g	**17%**
Saturated Fat 2g	**10%**
Cholesterol 50mg	**17%**

© Polara Studios, Inc.

A dinner roll	
Calories 80	Calories from Fat 20
	% Daily Value*
Total Fat 2g	**3%**
Saturated Fat 0g	**0%**
Cholesterol 0mg	**0%**

Fettuccine alfredo, 1/2 c	
Calories 250	Calories from Fat 130
	% Daily Value*
Total Fat 14g	**22%**
Saturated Fat 8g	**40%**
Cholesterol 60mg	**20%**

A breakfast bar	
Calories 150	Calories from Fat 55
	% Daily Value*
Total Fat 6g	**9%**
Saturated Fat 2.5g	**13%**
Cholesterol 0mg	**0%**

A muffin	
Calories 160	Calories from Fat 54
	% Daily Value*
Total Fat 6g	**9%**
Saturated Fat 1g	**5%**
Cholesterol 20mg	**7%**

A large biscuit	
Calories 260	Calories from Fat 80
	% Daily Value*
Total Fat 11g	**17%**
Saturated Fat 2.5g	**13%**
Cholesterol 0mg	**0%**

A large croissant	
Calories 270	Calories from Fat 130
	% Daily Value*
Total Fat 14g	**22%**
Saturated Fat 8g	**40%**
Cholesterol 45mg	**15%**

Art © Cengage Learning

KEY POINTS

- Milk products bear names that identify their fat contents.
- Cheeses are major contributors of saturated fat in the U.S. diet.

Grains

Grain foods in their natural state are very low in fat, but fats of all kinds may be added during manufacturing, processing, or cooking (see Figure 5–16). In fact, today's leading single contributor of solid fats to the U.S. diet is grain-based desserts, such as cookies, cakes, and pastries, foods often prepared with butter, margarine, or hydrogenated shortening.[63] Other grain foods made with solid fats include biscuits, cornbread, granola and other ready-to-eat cereals, croissants, doughnuts, fried rice, pasta with creamy or oily sauces, quick breads, snack and party crackers, muffins, pancakes, and homemade waffles. Packaged breakfast bars often resemble vitamin-fortified candy bars in their solid fat and added sugar contents.

KEY POINT

- Solid fat in grain foods can be well hidden.

Now that you know where the fats in foods are found, how can you reduce or eliminate the harmful ones? The Food Feature provides some pointers.

try it! Food Feature

Defensive Dining

LO 5.9 Identify at least 10 ways to reduce solid fats in an average diet.

Meeting today's lipid guidelines can be tricky. To reduce intakes of saturated and *trans*-fatty acids, for example, involves identifying food sources of these fatty acids, that is, foods that contain solid fats. Then, to replace them appropriately involves identifying unsaturated oils and doing something about all of them in the foods you eat. To help simplify these feats, the *Dietary Guidelines* suggest:

1. Selecting the most nutrient-dense foods from all food groups. Solid fats and high-calorie choices lurk in every food group.
2. Consuming fewer and smaller portions of foods and beverages that contain solid fats.
3. Replacing solid fats with liquid oils whenever possible.
4. Checking Nutrition Facts labels and selecting foods with little saturated fat and no *trans* fat.

Such advice is easily dispensed but not easily followed. The first step in doing so is learning which foods contain heavy doses of solid fats. Table 5–7 lists some terms that indicate solid fats on a food label ingredients list.

Table 5–7

Solid Fat Ingredients Listed on Labels

- Beef fat (tallow)
- Butter
- Chicken fat
- Coconut oil
- Cream
- Hydrogenated oil
- Margarine
- Milk fat
- Palm kernel oil; palm oil
- Partially hydrogenated oil
- Pork fat (lard)
- Shortening

In the Grocery Store

The right choices in the grocery store can save you many grams of saturated and *trans* fats. Armed with label information, you can decide whether to use a food often as a staple item or limit it to an occasional treat. For example, plain frozen vegetables without butter or other high-fat sauces are a staple food—they are high in nutrient density and devoid of solid fats.

Make the same distinctions among precooked meats. Avoid those that are coated and fried or prepared in fatty gravies. Try rotisserie chicken from the deli section—rotisserie cooking lets much of the fat and saturated fat drain away. Removing the skin leaves only the chicken—a nutrient-dense food with little solid fat.

Choosing among Margarines

Soft or liquid margarine varieties are made from unhydrogenated oils, which are mostly unsaturated and so are less likely to elevate blood cholesterol than the more saturated solid fats of butter or stick margarine. Some margarines contain olive oil or omega-3 fatty acids, making them preferable to butter and other margarines for the heart. Diet margarines contain fewer calories than regular varieties because water, air, or fillers have been added.

A few margarines advertised to "support heart health" contain added plant sterols, phytochemicals known to lower blood LDL cholesterol somewhat.[64] However, these may not be the best choice because plant sterols seem to accumulate in the tissues of the arteries and brain, and their potential for harm is under investigation.[65] Children should not be given any foods enriched with plant sterol additives until their safety has been established.

With more than 57 types of margarines and spreads on the market in sticks, tubs, sprays, and liquids containing from 0 to 80 percent fat, margarine-buying consumers must read labels and select margarines that they like with the least saturated and *trans* fat. When oils (but not hydrogenated oils) are the first ingredient listed on a margarine label, the margarine is, in all probability, low in saturated and *trans* fats and therefore a good choice for a healthy heart.

Choosing Unsaturated Oils

When choosing oils, trade off among different types to obtain the benefits different oils offer. Peanut and safflower oils are especially rich in vitamin E. Olive oil presents naturally occurring antioxidant phytochemicals with potential heart health benefits (see the Controversy section for details), and canola oil is rich with monounsaturated and essential fatty acids. High temperatures, such as those used in frying, destroy some omega-3 acids and other beneficial constituents, so treat your oils gently. Take care to *substitute* oils for saturated fats in the diet; do not add oils to an already fat-rich diet. No benefits are expected unless oils replace saturated fats in the diet and total calories are controlled.

Fat-Free Products and Artificial Fats

Keep in mind that "fat-free" versions of normally high-fat foods, such as cakes or cookies, do not necessarily provide fewer calories than the original and may not provide a health advantage if added sugars take the place of fats. As Controversy 4 discussed, research indicates that added sugars may pose other risks and so must also be limited in the diet. Products made with unsaturated oils may be very low in solid fats but high in calories. Read labels to evaluate whether such products can fit into your diet.

Some foods contain **fat replacers**—ingredients made from carbohydrate or protein that provide some of the taste and texture of solid fats but with fewer calories and less saturated fat. Others contain **artificial fats**, synthetic compounds offering the sensory properties of fat but none of the calories or fat. For example, "lite" potato chips and other snack foods contain **olestra**, an artificial fat. Chapter 12 comes back to the topic of artificial fats and other food additives.

Revamp Recipes

Once at home, minimize solid fats used as seasonings. This means enjoying the natural flavor of steamed or roasted vegetables, seasoned with lemon pepper, garlic, or herbs; lemon, lime, or other citrus; a touch of olive oil or spray or liquid margarine; or sesame seed oil, nut oils, or some toasted nuts. Seek out recipes that replace solid shortening with liquid vegetable oil such as canola oil and that provide replacements for meat gravies and cheese or cream sauces.

Here are some other tips to revise high-fat recipes that contribute excess fat calories and saturated fats:

- Grill, roast, broil, boil, bake, stir-fry, microwave, or poach foods. Don't fry in solid fats, such as shortening, lard, or butter. Try pan frying in a few teaspoons of olive or vegetable oil instead of deep frying.
- Reduce or eliminate food "add-ons" such as buttery, cheesy, or creamy sauces, sour cream dressings, and bacon bits that drive up the calories and saturated fat. Instead, add a small amount of olives, nuts, or avocado for rich flavor.
- Cut recipe amounts of meat in half; use only lean meats. Fill in the lost bulk with soy meat replacers, shredded vegetables, legumes, pasta, grains, or other low-fat items.
- Replace a thick slice of ham with two or three wafer-thin slices. The serving will be smaller and thus provide less solid fat, but the taste will be satisfying because the ham surface area that imparts flavor to the taste buds is greater.
- Refrigerate meat pan drippings and broth, and lift off the fat when it solidifies. Add the defatted broth to flavor a casserole, cooked vegetables, or soup, or use it as a gravy.
- Make prepared mixes, such as rice or potato mixtures, without the fats called for on the label, or substitute liquid oils for solid fats in preparing them. Avoid heat-and-serve refrigerated potato and other mixtures with high saturated fat contents.

For snacks, make it a habit to choose lower-fat microwave popcorn and then sprinkle on imitation butter or cheese flavored sprinkles that contain no fat and few calories. Keep that flavoring on hand to use in seasoning potatoes, vegetables, and other foods.

Table 5–8 lists some practical ways to cut down on solid fats and replace them with liquid oils in foods. These replacements don't change the taste or appearance much, but they dramatically lower the saturated fat contents of the foods.

Feasting on Fast Foods

All of these suggestions work well when a person plans and prepares each meal at home. But in the real world, people fall behind schedule and don't have time to shop or cook, so they eat fast food. Figure 5–17 (p. 188) compares some fast-food choices and offers tips to reduce the calories and saturated fat to make fast-food meals healthier.

Keep these facts about fast food in mind:

- Salads are a good choice, but beware of toppings such as fried noodles, bacon bits, grease-soaked croutons, sour cream, or shredded cheese that can drive up the calories and solid fat contents.
- If you are really hungry, order a small hamburger, broiled chicken sandwich, or "veggie burger" and a side salad. Hold the cheese (usually full-fat in fast-food restaurants); use mustard or ketchup as condiments.
- A small bowl of chili (hold the cheese and sour cream) poured over a plain baked potato can also satisfy a bigger appetite. Top with chopped raw onions or hot sauce for spice, and pair it with a small salad and fat-free milk for a complete meal.
- Tacos, bean burritos, and other Mexican treats are delicious topped with salsa and onions instead of cheese and sour cream.
- Fast-food fried fish or fried chicken sandwiches can provide as much solid fat as hamburgers. Broiled chicken and fish sandwiches are far less fatty if you order them made without cheese, bacon, or mayonnaise sauces. Even French fries, because they are fried in highly saturated commercial fats, contribute substantial solid fat to the diet.
- Chicken wings are mostly fatty skin, and the tastiest wing snacks are fried in cooking fat (often a saturated hydrogenated type with *trans*-fatty acids), smothered with a buttery, spicy sauce, and then dipped in blue cheese dressing, making wings an extraordinarily high-fat food. If you snack on wings, plan on eating low-fat foods at several other meals to balance them out.

Because fast foods are short on variety, let them be part of a lifestyle in which they complement the other parts. Eat differently, often, elsewhere.

fat replacers ingredients that replace some or all of the functions of fat and may or may not provide energy.

artificial fats zero-energy fat replacers that are chemically synthesized to mimic the sensory and cooking qualities of naturally occurring fats but are totally or partially resistant to digestion.

olestra a noncaloric artificial fat made from sucrose and fatty acids; formerly called *sucrose polyester*. A trade name is *Olean*.

Did You Know?

The foods eaten away from home are higher in fat, saturated fat, *trans* fat, and cholesterol and lower in vitamins and minerals than meals typically eaten at home.

Change Your Habits

The lipid guidelines offered in this chapter do not occur in isolation—they accompany recommendations to achieve and sustain a healthy body weight, to keep calories under control, and to eat a nutrient-dense diet with adequate fruits, vegetables, whole grains, and legumes that provide cholesterol-lowering soluble fiber, as well. By this time you may be wondering if you can realistically make all the changes recommended for your diet.

Be assured that even small changes can yield big dividends in terms of reducing solid fats intake, and most such changes can become habits after a few repetitions. You do not have to give up all high-fat treats, nor should you strive to eliminate all fats. You decide what the treats should be and then choose them in moderation, just for pure pleasure. Meanwhile, make sure that your everyday, ordinary choices are those whole, nutrient-dense foods suggested throughout this book. That way you'll meet all your body's needs for nutrients and never feel deprived.

Table 5–8

Solid Fat Replacements

Select foods that replace solid fats with polyunsaturated or monounsaturated fats. Avoid foods that replace fats with refined white flour or added sugars, as these may present risks of their own. Remember that "light" on a label can refer to color or texture so always compare the Nutrition Facts panel with the regular product.

Instead of these . . .	*. . . try using these*
Solid Fats and Oils	
Regular margarine and butter for spreading, cooking, or baking	Reduced-fat, diet, liquid, or spray margarine; granulated butter replacers; fruit butters, nut butters, or avocado for spreading; olive oil; nut or seed oils for cooking
Regular mayonnaise and salad dressings	Reduced-fat or canola or olive oil mayonnaise and salad dressings
Oils, shortening, or lard in cooking	Nonstick cooking spray for frying; applesauce or oil for baking
Solid fats as seasonings: bacon, bacon fat, butter; fried onion toppers	Herbs, lemons, spices, liquid smoke flavoring, ham-flavored bouillon cubes, broth, wine; toasted nut toppers
Milk Products/Dairy Products	
Whole milk; half and half	Fat-free or reduced-fat milk; fat-free half and half
Regular ricotta cheese; mozzarella cheese; yogurt or sour cream	Part-skim ricotta or fat-free cottage cheese; part-skim mozzarella; fat-free sour cream, "zero" Greek-style yogurt[a]
Regular cheddar, American, or other cheeses; cream cheese	Low fat or fat-free cheeses; fat-free or reduced-fat cream cheese, Neufchatel cheese
Large amounts of mild cheeses	Small amounts of strong-flavored aged cheeses (grated Asiago, Romano, or Parmesan)
Whipped cream	Fat-free whipped cream preparations or fat-free nondairy toppings
Ice cream, mousse, cream custards	"Light" ice cream, frozen yogurt, or other frozen desserts; low-sugar sherbet or sorbet; skim milk low-sugar puddings
Protein Foods	
Bologna, salami, other sliced sandwich meats; hot dogs	Low-fat sandwich meats and hot dogs (95–97% lean, or "light")
Breakfast sausage or bacon	Canadian bacon, lean ham, or soy-based sausage or bacon-like products
High-fat beef, pork, or lamb; ground beef	Leaner cuts trimmed of fat, broiled salmon or other seafood; ground turkey breast (98% lean), soy-based "ground beef" crumbles; legume main dishes
Poultry with skin	Skinless poultry
Commercial fish sticks, breaded fried fish fillets	Plain fish fillets, broiled or rolled in crumbs and pan sautéed in a little oil
Grains and Desserts	
Chips, such as tortilla or potato; appetizer crackers	Baked or "light" chips; reduced fat crackers and cookies, saltine-type crackers; nut, seed, or whole-grain crackers low in saturated fat
Cakes, cookies; doughnuts, pastries, other desserts	Fresh and dried fruit; whole grain muffins, quick breads, or cakes made with oil (not shortening)
Granola, other cereals with coconut oil or hydrogenated fats	Cereals low in saturated fat, with no *trans* fat (compare the Nutrition Facts panel information)
Ramen-type noodles	Soba noodles or other whole-grain noodles cooked in broth, with Asian seasonings added
Other	
Frozen or canned main dishes with more than 2 or 3 g saturated fat per serving	Similar foods with less saturated fat per serving (compare the Nutrition Facts panel information)
Cream-based, cheese, or "loaded" soups	Broth-based, vegetable, or bean soups; or poultry-based, meatless, or other low-fat chili

[a]*If the food must be boiled, stabilize the cottage cheese or yogurt with a small amount of cornstarch or flour.*

Figure 5–17

Compare the Calories and Saturated Fat in Fast-Food Choices

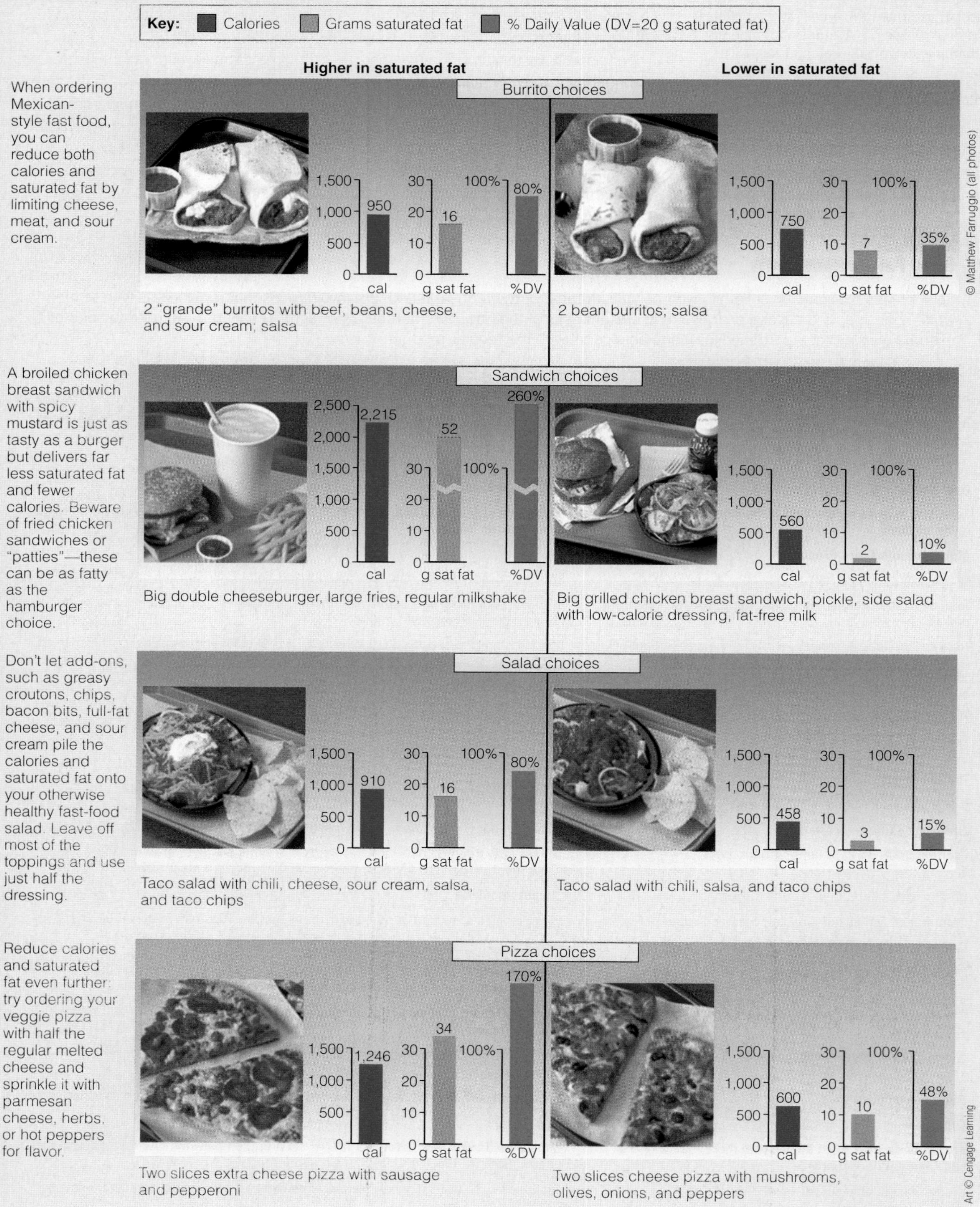
Key: Calories Grams saturated fat % Daily Value (DV=20 g saturated fat)
Higher in saturated fat
Lower in saturated fat
Burrito choices
When ordering Mexican-style fast food, you can reduce both calories and saturated fat by limiting cheese, meat, and sour cream.
950 cal; 16 g sat fat; 80% %DV
2 "grande" burritos with beef, beans, cheese, and sour cream; salsa
750 cal; 7 g sat fat; 35% %DV
2 bean burritos; salsa
Sandwich choices
A broiled chicken breast sandwich with spicy mustard is just as tasty as a burger but delivers far less saturated fat and fewer calories. Beware of fried chicken sandwiches or "patties"—these can be as fatty as the hamburger choice.
2,215 cal; 52 g sat fat; 260% %DV
Big double cheeseburger, large fries, regular milkshake
560 cal; 2 g sat fat; 10% %DV
Big grilled chicken breast sandwich, pickle, side salad with low-calorie dressing, fat-free milk
Salad choices
Don't let add-ons, such as greasy croutons, chips, bacon bits, full-fat cheese, and sour cream pile the calories and saturated fat onto your otherwise healthy fast-food salad. Leave off most of the toppings and use just half the dressing.
910 cal; 16 g sat fat; 80% %DV
Taco salad with chili, cheese, sour cream, salsa, and taco chips
458 cal; 3 g sat fat; 15% %DV
Taco salad with chili, salsa, and taco chips
Pizza choices
Reduce calories and saturated fat even further: try ordering your veggie pizza with half the regular melted cheese and sprinkle it with parmesan cheese, herbs, or hot peppers for flavor.
1,246 cal; 34 g sat fat; 170% %DV
Two slices extra cheese pizza with sausage and pepperoni
600 cal; 10 g sat fat; 48% %DV
Two slices cheese pizza with mushrooms, olives, onions, and peppers
© Matthew Farruggio (all photos)
Art © Cengage Learning

track it!

Diet Analysis PLUS+ Concepts in Action

Analyze Your Lipid Intake

The purpose of this exercise is to help you identify fatty foods in your diet, as well as sources of saturated and *trans* fats. The Diet Analysis Plus (DA+) program will help you learn which foods contain which fats and help you to choose unsaturated fats.

1. No amount of dietary saturated or *trans* fat is required for health. Open DA+ Home page. From the Reports tab, select Fat Breakdown. Choose a date, choose all meals, and then generate a report. Your report will show a breakdown of your fat intake for that day as a percentage of total calories. What are the percentages for saturated, monounsaturated, polyunsaturated, and *trans*-fatty acids in your day's intake?
2. Which foods provide the most fat to your diet? Select Source Analysis from the Report tab. Select a date and choose all meals. Generate a report for saturated fat and select Create PDF. Do the same for monounsaturated, polyunsaturated, and *trans*-fatty acids. What three foods contributed the most saturated and monounsaturated fats? Polyunsaturated and *trans* fats? Which of your foods are listed as saturated fat contributors in Figure 5–11 (page 173)?
3. The Macronutrient Ranges report compares your intakes to the recommended intake ranges. Select a date, click onto the Macronutrient Range report, select the entire day's food intake, and then generate a report. Did your intake for fats fall within the recommended range? What percentage of your calories came from fat?
4. To study your intake of essential fatty acids, select Reports, Intake vs. Goals, and then select Day One, all meals. Generate a report. Look for the essential fatty acid (*efa*) heading. Compared with the DRI goal, how did your intake stack up? Use the Source Analysis Report to find the sources of omega-3 and omega-6 essential fatty acids in your meals. What foods might you change to improve your intake (see Figure 5–5 on page 162)?
5. Toppings and sauces added to nutritious foods drive up calorie and saturated fat intakes. The Food Feature (page 185) gives suggestions for reducing these add-ons. From Track Diet, select the Recipes button, Create New Recipe, to create an appealing heart-healthy salad with little saturated and *trans* fats. Save and close your recipe. Select a new day (not from your 3-day record) and select only your salad. Click on View: Favorites and then click the "i" icon next to the recipe name to display the nutrients in the salad. How did you do?

what did you decide?

© Pakhnyushcha/Shutterstock.com

Are fats unhealthy food constituents that are best eliminated from the diet?

What are the differences between "bad" and "good" cholesterol?

Why is choosing fish recommended in a healthy diet?

If you trim all visible fats from foods, will your diet meet lipid recommendations?

Self Check

1. (LO 5.1) Which of the following is *not* one of the ways fats are useful in foods?
   a. Fats contribute to the taste and smell of foods.
   b. Fats carry fat-soluble vitamins.
   c. Fats provide a low-calorie source of energy compared to carbohydrates.
   d. Fats provide essential fatty acids.
2. (LO 5.1) Fats play few roles in the body, apart from providing abundant fuel in the form of calories.
   T F
3. (LO 5.2) Saturation refers to
   a. the ability of a fat to penetrate a barrier, such as paper.
   b. whether or not a fatty acid chain is holding all of the hydrogen atoms it can hold.
   c. the characteristic of pleasing flavor and aroma.
   d. the fattening power of fat.
4. (LO 5.2) Generally speaking, vegetable and fish oils are rich in saturated fat.
   T F
5. (LO 5.2) A benefit to health is seen when _______ is used in place of _______ in the diet.
   a. saturated fat/monounsaturated fat
   b. saturated fat/polyunsaturated fat
   c. polyunsaturated fat/saturated fat
   d. triglycerides/cholesterol
6. (LO 5.3) Little fat digestion takes place in the stomach.
   T F
7. (LO 5.3) Bile is essential for fat digestion because it
   a. splits triglycerides into fatty acids and glycerol.
   b. emulsifies fats in the small intestine.
   c. works as a hormone to suppress appetite.
   d. emulsifies fat in the stomach.
8. (LO 5.4) When energy from food is in short supply, the body
   a. dismantles its glycogen and releases triglycerides for energy.
   b. dismantles its cholesterol and releases glucose for energy.
   c. converts its glucose to fat for more efficient energy.
   d. dismantles its stored triglycerides and releases fatty acids for energy.
9. (LO 5.4) Fat breakdown without carbohydrate causes ketones to build up in the tissues and blood, and be excreted in the urine.
   T F
10. (LO 5.5) LDL, a class of lipoprotein, delivers triglycerides and cholesterol from the liver to the body's tissues.
    T F
11. (LO 5.5) Chylomicrons, a class of lipoprotein, are produced in the liver.
    T F
12. (LO 5.5) Consuming large amounts of saturated fatty acids lowers LDL cholesterol and thus lowers the risk of heart disease and heart attack.
    T F
13. (LO 5.6) The roles of the essential fatty acids include ________ .
    a. forming parts of cell membranes
    b. supporting infant growth and vision development
    c. supporting immune function
    d. all of the above
14. (LO 5.6) Taking supplements of fish oil is recommended for those who don't like fish.
    T F
15. (LO 5.6) Fried fish from fast-food restaurants and frozen fried fish products are often low in omega-3 and high in solid fats.
    T F
16. (LO 5.7) A way to prevent spoilage of unsaturated fats and make them harder is to change their fatty acids chemically through
    a. acetylation
    b. hydrogenation
    c. oxidation
    d. mastication
17. (LO 5.7) *Trans*-fatty acids arise when unsaturated fats are
    a. used for deep frying
    b. hydrogenated
    c. baked
    d. used as preservatives
18. (LO 5.8) A diet with sufficient essential fatty acids and vitamin E includes
    a. nuts and vegetable oils
    b. ¼ cup of raw oil each day
    c. two servings of seafood a week
    d. a and c
19. (LO 5.8) The majority of solid fats in the U.S. diet are supplied by invisible fats.
    T F
20. (LO 5.9) Solid fats and high-calorie choices lurk in every food group.
    T F
21. (LO 5.10) Which is true of low-fat diets?
    a. They may lack essential fatty acids and certain vitamins.
    b. They may be high in carbohydrate calories.
    c. They are difficult to maintain over time.
    d. All of the above.
22. (LO 5.10) Ways of modifying a diet to obtain the health benefits associated with a traditional Mediterranean diet include
    a. sautéing a steak in olive oil instead of butter.
    b. topping a hamburger with 1/2 cup of sliced olives.
    c. substituting extra virgin olive oil for stick margarine.
    d. choosing a fast-food fried fish sandwich instead of a burger.

Answers to these Self Check questions are in Appendix G.

CONTROVERSY 5

Good Fats, Bad Fats—U.S. Guidelines and the Mediterranean Diet

LO 5.10 Discuss evidence for the benefits and drawbacks of specific dietary fats in terms of their effects on human health.

Jennifer Thermes/Jupiterimages

To consumers, advice about dietary fats appears to change almost daily. "Eat less fat—choose more fatty fish." "Give up butter and margarine—use soft margarine." "Forget soft margarine—replace it with olive oil." To researchers, however, the evolution of advice about fats reflects decades of study to reveal the truth about dietary fats. As scientific understanding has grown, dietary guidelines have become more specific and therefore more meaningful.

This Controversy explores today's guidelines for lipid intakes, focusing on a Mediterranean eating pattern singled out by the *Dietary Guidelines for Americans* as superbly supportive of health despite being higher in fat.*[1] It concludes with the current opinion that, even though specific nutrients influence disease causation or prevention, the repetitive daily food choices that make up a whole eating pattern seem to have the greatest impact on health.

The Objections to "Low-Fat" Guidelines

Over the years, low-fat diets have been a critical centerpiece of treatment plans for people with elevated blood lipids or heart disease and therefore are important in nutrition.[2] Past *Dietary Guidelines* urged all healthy people to cut their fat intakes in everything from hot dogs to salad dressings to preserve their good health. This advice was straightforward: Cut the fat and improve your health.

*Reference notes are found in Appendix F.

In recent years, diet advice for preserving good health has shifted away from reducing total fat to singling out saturated and *trans* fats for restriction, along with a greater emphasis on calorie control to maintain a healthy body weight.[3] Dietary saturated fat was and remains a well-established culprit behind elevated blood cholesterol, but the old guidelines focused on *total* fat, limiting intakes to 30 percent or less of calories. Did this strategy work to cut saturated fat intake? Yes, but only for those few who consistently and correctly applied the advice. Most who tried failed, finding the low-fat diet impossible to maintain over months or years.

Are Low-Fat Diets Helpful—or Not?

In addition to poor compliance, low-fat diets present at least three other problems. For one, a low-fat diet is not necessarily low in calories, and to gain any benefit from the diet, overweight people with heart disease need to reduce calorie consumption. For another, even low-fat diets, when high in refined carbohydrates and added sugars, are linked with high blood triglycerides and low HDL cholesterol, a deleterious combination for heart health.[4] Finally, taken to the extreme, a low-fat diet may exclude nutritious foods, such as fatty fish, nuts, seeds, and vegetable oils that provide the essential fatty acids along with many phytochemicals, vitamins, and minerals.

Research on High-Fat Diets

A classic study leading up to our current understanding was the Seven Countries Study, the first to reveal an association between death rates from heart disease and diets high in saturated fats.[5] Even then, however, evidence for harm from higher intakes of *total* dietary fat was weak. In fact, the two countries with the highest fat intakes (40 percent of total calories) were Finland and the Greek island of Crete; yet Finland had the highest rate of death from heart disease, while Crete had the lowest.

Clearly, total fat was not to blame for a high rate of heart disease—something else had to be responsible. When researchers examined the two diets more closely, they found that the Cretes ate diets high in olive oil but low in saturated fat (less than 10 percent of calories), a pattern they linked with relatively low disease risks. Finns ate diets higher in saturated fats.

Many studies confirmed that peoples consuming a traditional "Mediterranean-type" diet typical of Mediterranean regions in the mid-1900s have low rates of heart and artery diseases, some cancers, and other chronic diseases, and their life expectancies are long.[6] Unfortunately, many young busy Mediterranean people today are swapping labor-intensive traditional diets for fast and convenient Western-style foods. At the same time, their health advantages are rapidly slipping away.

Recent Revelations

Recently, some misleading media stories made it appear that dietary saturated fat is irrelevant to heart disease risk, but this is far from the truth. True, a recent meta-analysis failed to demonstrate an epidemiological link between saturated fat intakes and risks of heart disease, but the data supporting this finding have been challenged.[7] Also, several recent studies suggest that a whole eating pattern approach, rather a single-nutrient focus, may work best for reducing blood LDL cholesterol and heart disease risk.[8] This doesn't mean, however, that high intakes of saturated fats are not harmful. They are.

The Current Guidelines

On reviewing the evidence, the committee writing the *Dietary Guidelines for Americans 2010* concluded that a diet containing up to 35 percent of total calories from total fat (higher than previous recommendations) but reduced in saturated fat and *trans* fat and controlled in calories is compatible with low rates of heart disease, diabetes, obesity, and cancer.[9] To meet these ideals, the guidelines have shifted from a "low-fat" to a "wise-fat" approach and suggest replacing the "bad" saturated and *trans* fatty acids with "good" polyunsaturated fatty acids, and enjoying these fats within calorie limits. However, people do not eat individual fatty acids. They eat foods within overall eating patterns. Therefore, the final recommendations to the public are stated in terms of food choices within beneficial eating patterns observed to support health in real life. The Mediterranean diet is one of those patterns.†

High-Fat Foods of the Mediterranean Diet

Avocados, salami, walnuts, and potato chips are all high-fat foods, yet the fats of these foods differ markedly in their health effects. The following evidence can help to clarify why some high-fat foods rightly belong in a heart-healthy diet and why others are best left on the shelf.

† *The other healthy eating pattern singled out by the* Dietary Guidelines *is the DASH diet, discussed in Chapter 11 and presented in Appendix E.*

Olives and their oil may benefit heart health.

Olive Oil: The Mediterranean Connection

The traditional health-promoting diets of Greece and the Mediterranean region are exemplary in their use of the "good" fats of olives and their oil. Population and laboratory studies reveal lower rates of certain cancers when olive oil takes the place of solid fats in the diet, such as butter, stick margarine, coconut or palm oil, lard, shortening, and meat fats.[10] In addition, when olive oil, and particularly dark green extra virgin olive oil, replaces more saturated fats, it may lower the risks of cardiovascular disease by:

- Providing antioxidant phytochemicals that may reduce LDL cholesterol's vulnerability to oxidation.[11]
- Reducing blood-clotting factors.[12]
- Interfering with the inflammatory process related to disease progression.[13]

The phytochemicals of olives captured in extra virgin olive oil, and not its monounsaturated fatty acids, seem responsible for these potential effects.[14] When processors lighten olive oils to make them more appealing to consumers, they strip away the intensely flavored phytochemicals of the olives, thus diminishing not only the bitter flavor of the oils but also their potential for protecting the health of the heart. Many other liquid unhydrogenated oils, such as avocado oil, grapeseed oil, walnut oil, and other nut oils provide little saturated fat with their abundant unsaturated fats and present their own array of phytochemicals that still await research to reveal potential benefits to health.

Two cautions: First, people who treat olive or nut oil like a magic potion against heart disease are bound to be disappointed. Drizzling olive oil on a saturated fat-laden food, such as a cheese and sausage pizza, does not make the food any healthier. Likewise, cooking a fatty steak in olive oil makes it no better for the heart. Second, like other fats, olive oil delivers 9 calories per gram. Adding oils to foods without cutting down on something else of equal energy can easily add hundreds of calories to a day's intake, making weight gain inevitable (see Chapter 9). Remember: Substitute. Don't add.

The Mediterranean Diet beyond Olive Oil

Olive oil alone cannot fully account for the heart benefits associated with a traditional Mediterranean diet. Such features as low intakes of red meats and higher intakes of nuts, vegetables, and seasonal fruits probably also deserve some credit.[15] Also, a traditional Mediterranean diet often features mostly fresh, whole foods that provide a complex mixture of nutrients and phytochemicals unduplicated in refined foods—even when they are enriched with nutrients.

Each of the countries bordering the Mediterranean Sea has its own culture and dietary mixtures, but researchers have identified some common characteristics. Traditional Mediterranean people focus their diets on crusty breads, nuts, potatoes, and pastas; a variety of vegetables (including wild greens) and legumes; feta and mozzarella cheeses and yogurt; and fruits (particularly lemons, grapes, and figs). They eat fish and other seafood, poultry, a few eggs, and a little meat. Along with olives and olive oil, their principal sources of fat are nuts and fish; they rarely use butter or encounter hydrogenated fats. Consequently, traditional Mediterranean diets are:

- Low in saturated fat.
- Very low or absent in *trans* fat.
- Rich in monounsaturated fats and polyunsaturated fats, including EPA and DHA.
- Rich in carbohydrates and fiber from whole foods.
- Rich in vitamins, minerals, and phytochemicals.

When people with health risks switch to Mediterranean-style diets, lipid profiles often improve, inflammation diminishes, and the risks of heart disease, cancers, and many other conditions decline.[16] Preliminary evidence even suggests that such a diet may help to preserve mental faculties in old age—and the more stringently the diet is followed, the better.[17] Figure C5–1 presents a Mediterranean food pyramid for guidance. The broad base of the pyramid indicates that foods listed there should make up the bulk of every meal, while those at the top should be limited.

Figure C5–1

A Mediterranean Diet Pyramid

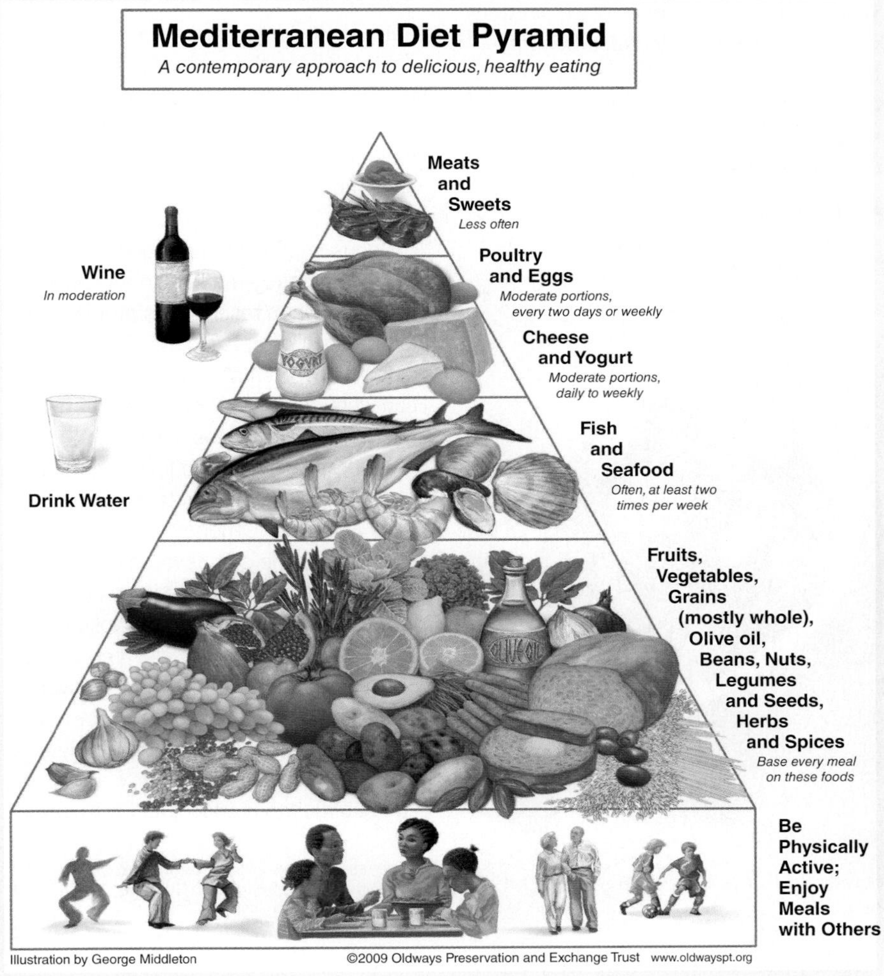

Fatty Fish: A Key Mediterranean Food

The Mediterranean regions are surrounded by the sea, so seafood provides a great deal of the protein in a traditional diet. Fatty fish intake is certainly associated with good health, particularly the

Fish and other seafood contribute key nutrients to the traditional Mediterranean diet.

health of the heart, as the preceding chapter made clear. Increasing omega-3 fatty acids in the diet may lower the risks of developing heart disease and of dying from it.[18]

In traditional Mediterranean diets, omega-3 fatty acids also derive from some atypical foods, such as wild plants and snails, unavailable to U.S. consumers. In addition, because traditionally raised food animals graze in fields, their meat, dairy products, and eggs are richer in omega-3 fatty acids than products from animals fed primarily on grain, as is common elsewhere. Still, even when modern foods comprise a Mediterranean-style diet, benefits accrue.

Nuts: More Than Just a High-Fat Snack

Nuts are extraordinarily popular in traditional Mediterranean cuisines and show up in everything from savory sauces to desserts. In fact, nuts are popular with most people around the world. Not only do nuts taste good, but they may also have favorable effects on heart health.[19] In addition, they may not cause as much weight gain as one might expect judging from their calories alone.

Nuts and Body Weight Nuts and peanuts once had no place in a low-fat or low-calorie diet, with good reason—up to 80 percent of their calories come from fat and an ounce (1/4 cup) of nuts may provide from 130 to more than 200 calories.[20] Despite this, people who regularly eat nuts tend to be leaner, not fatter, and generally have smaller waistlines, better dietary intakes, and lower disease risks than others.[21] No one can yet say, however, whether nut eaters owe their healthier, leaner physiques to an overall health conscious lifestyle or whether nuts themselves provide these benefits.[22] In any case, replacing potato chips or chocolate candy snacks with nuts improves nutrition by reducing intakes of saturated fat, providing lasting satiety, and increasing intakes of vitamin E and other nutrients.[23]

Still, when designing a test diet, researchers must carefully use nuts *instead of*, not in addition to, other fat sources (such as meat, potato chips, or oil) to keep calories constant. If you decide to snack on nuts, you should probably do the same thing and use them to replace other fats in your diet.[24] Remember that nuts, although not associated with obesity or weight gain in research, provide substantially more calories per bite than, say, crunchy raw vegetables.

Nuts for the Heart People who eat an ounce of nuts on two or more days a week appear to have lower heart disease risks and lower risks of sudden death from heart events than those consuming no nuts.[25] In women with diabetes, whose risks are notoriously high, nuts (5 ounces per week) or peanut butter (5 tablespoons per week) are associated with reduced heart disease risk.[26] While as little as 2 ounces of nuts a week were linked with a detectable benefit, higher intakes were associated with greater benefits, with about an 8 percent reduction in cardiovascular risk for each additional weekly serving.[27] Nuts:

- Are rich in monounsaturated fat, moderate in polyunsaturated fat, but low in saturated fat.
- Provide fiber, vegetable protein, essential fatty acids, vitamin E, and other vitamins and minerals.
- Contain plant sterols that can block cholesterol absorption.

In addition, the brown papery skins that surround the nutmeats are rich sources of antioxidant phytochemicals that may help to oppose the inflammation associated with chronic diseases.[28] Such nuts are the easily obtainable, common varieties: almonds, Brazil nuts, cashews, hazelnuts, macadamia nuts, peanuts with skins, pecans, pine nuts, pistachios, and, as mentioned, walnuts and peanut butter.[29] To quote one enthusiastic researcher, " . . . nuts possibly are one of the most cardioprotective foods in the habitual diet."‡

Potential Mechanisms Inflammation is associated with developing heart disease and other chronic diseases, and nuts supply anti-inflammatory phytochemicals and the antioxidant vitamin E that may oppose these harmful processes.[30] Studies of blood levels of certain markers of tissue inflammation demonstrate an inverse relationship with consumption of nuts and seeds—the greater the consumption of these foods, the lower the inflammatory markers in the blood. However, when researchers provided nuts to human subjects as part of a controlled diet, no drop in markers of inflammation were observed.[31] More research is needed to clarify these findings.

With regard to blood lipids, some nuts may be more beneficial than others. Time and again, when walnuts replace other fats in the diet, favorable effects on blood lipids are observed in as little as four weeks, even in people whose total and LDL cholesterol are elevated at the outset.[32] Studies of almonds, however, vary in their results, leading reviewers to conclude that almonds have only a neutral effect on blood lipids.[33] Even so, eating almonds is associated with lower

© Anna Hoychuk/Shutterstock.com

Stay mindful of calories when snacking on nuts.

‡ *The researcher is Emilio Ros, in E. Ros, Health benefits of nut consumption,* Nutrients *2 (2010): 653–682.*

Table C5–1

Food Sources of Fatty Acids

HEALTHFUL FATTY ACID SOURCES		
Monounsaturated	Omega-6 Polyunsaturated	Omega-3 Polyunsaturated
Avocado Nuts (almonds, cashews, filberts, hazelnuts, macadamia nuts, peanuts, pecans, pistachios) Oils (canola, olive, peanut, sesame) Olives Peanut butter (old-fashioned) Seeds (sesame)	Margarine (nonhydrogenated) Mayonnaise Nuts (walnuts) Oils (corn, cottonseed, safflower, soybean) Salad dressing Seeds (pumpkin, sunflower)	Fatty fish (listed in the preceding chapter) Flaxseed Nuts

HARMFUL FATTY ACID SOURCES	
Saturated	*Trans*
Bacon, butter, lard Cheese, whole milk products Chocolate, coconut Cream, half-and-half, cream cheese Meats Oils (coconut, palm, palm kernel) Shortening	Commercial baked goods, including cookies, cakes, pies, or other goodies made with margarine or vegetable shortening Fried foods, particularly restaurant and fast foods Many fried or processed snack foods, including microwave popcorn, chips, and crackers Margarine (hydrogenated or partially hydrogenated) Nondairy creamer Shortening

Note: Keep in mind that foods contain a mixture of fatty acids; see Figure 5-5, p. 162.

heart disease risk, so other mechanisms may be in play.

Fats to Avoid: Saturated Fats and *Trans* Fats

The number-one dietary determinant of LDL cholesterol is saturated fat. Similarly, *trans* fats also raise heart disease risk by elevating LDL cholesterol. A heart-healthy diet limits foods rich in these two types of fat. The previous chapter spelled out the main sources of saturated fats in the contemporary diet.

Designing a diet with zero saturated fat is not possible, even for experts.[34] A nutritionally adequate diet will always provide some saturated fat because foods that provide the essential polyunsaturated fatty acids also supply some amount of saturated fatty acids as well. Diets based on fruits, greens, legumes, nuts, soy products, vegetables, and whole grains can, and often do, deliver less saturated fat than diets based on animal-derived foods, however. Table C5–1 summarizes which foods provide which fats.

Conclusion

Are high-fat diets compatible with good health? Certainly, saturated and *trans* fats seem mostly bad for the health of the heart. Aside from providing energy, which unsaturated fats can do equally well, dietary saturated and *trans* fats bring no indispensable benefits to the body. No harm can come from consuming a diet low in saturated fats and *trans* fats.

In contrast, unsaturated fats can be mostly good or neutral in their effects on the health of the heart. To date, their one proven fault seems to be that they, like all fats, provide abundant energy and so may promote obesity if they drive calorie intakes higher than energy needs.[35] Obesity, in turn, begets many ills (see Chapter 9).

The Synergy of a Whole Foods Diet Pattern and Lifestyle This section has addressed some foods and nutrients in the traditional Mediterranean diet that stand out as beneficial or benign to health. However, adding such foods as nuts to a typical meat- and cheese-rich Western style eating pattern may not improve health risks. Instead, the *Dietary Guidelines for Americans* suggest adopting a whole diet *pattern* based on a variety of nutrient-dense foods, chosen day after day, to best support the health of the body.[36] Apparently, each of the foods in a traditional Mediterranean diet contributes some small benefit that harmonizes with others to produce substantial cumulative or synergistic benefits.[37]

Figure C5–2

Mediterranean-Style Substitutions

These two meals are similar in total fat and calories and are equally delicious, but look what happens to saturated fat when olive oil, fish, and seeds replace butter, meat, and cheese. Note that not every food choice need be a traditional Mediterranean food—avocado is a monounsaturated fat–rich food grown in tropical regions of the world.

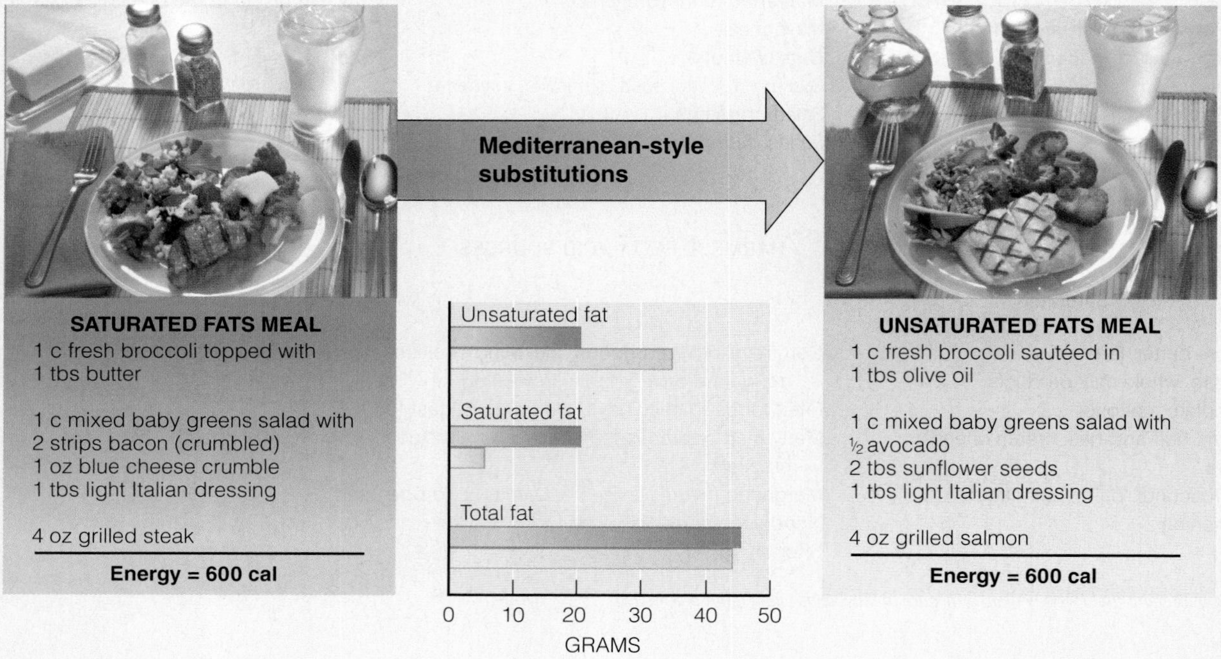

Application To achieve similar results for yourself, try to identify the sources of saturated fat in your diet and design new strategies around traditional Mediterranean principles. Take a look at Figure C5–2 and notice what happened to the saturated and unsaturated fats in this simple meal when foods were chosen with Mediterranean principles in mind. Choosing wisely among fats is paramount to obtaining the benefits of the Mediterranean diet, but attention to fat alone isn't enough—one must also choose abundant vegetables, fruits, whole grains, seafood, and legumes every day. In addition, select portion sizes of all foods that fit within your energy requirement.

Keep in mind that Mediterranean peoples have traditionally led physically active lifestyles, and physical activity reduces disease risks. Therefore, if you love olive oil and generally want to eat like a Greek, you'd better walk, garden, bicycle, and swim like one, too.

Critical Thinking

1. Choose a food item that contains fat or is prepared with fat (e.g., popcorn, cookies, or salad dressing). Go on the Internet, and look up three manufacturers that make the food item you choose. For example, if you choose salad dressing, look up Newman's Own Italian Dressing, Good Season Italian Dressing, and Hidden Valley Italian Dressing. Compare the type of fat in each dressing, and determine the brand of the food item that contains the healthiest fat choices. Share your findings on a discussion board or in class.
2. Choose one restaurant from the following list (or choose popular restaurants near you). Visit the restaurant's website and look at the menu. Choose the menu item at the restaurant that provides the healthiest fat and the food item that contains the least healthy fat. Defend your answer.
   - Cracker Barrel
   - Olive Garden
   - Applebee's
   - Panera Bread

6 The Proteins and Amino Acids

what do you think?

Why does your body need **protein**?

How does heating an **egg** change it from a liquid to a solid?

Do protein or amino acid **supplements** bulk up muscles?

Will your diet lack protein if you don't eat **meat**?

Learning Objectives

After completing this chapter, you should be able to accomplish the following:

LO 6.1 State why some amino acids are essential, nonessential, or conditionally essential to the human body, and outline how the body builds a protein molecule.

LO 6.2 Describe the digestion of protein and the absorption and transport of amino acids in the body.

LO 6.3 List the roles that various proteins and amino acids can play in the body, and describe the influence of carbohydrate on amino acid metabolism.

LO 6.4 Compute the daily protein need for a given individual, and discuss the concepts of nitrogen balance and protein quality.

LO 6.5 Discuss potential physical problems from an eating plan that is too low or too high in protein.

LO 6.6 Identify protein-rich foods, and list some extra advantages associated with legumes.

LO 6.7 Summarize the health advantages and nutrition red flags of vegetarian diets, and develop a lacto-ovo vegetarian eating pattern that meets all nutrient requirements for a given individual.

The proteins are amazing, versatile, and vital cellular working molecules. Without them, life would not exist. First named 150 years ago after the Greek word *proteios* (meaning "of prime importance"), **proteins** have revealed countless secrets of the processes of life and have helped to answer many questions in nutrition: How do we grow? How do our bodies replace the materials they lose? How does blood clot? What gives us immunity? What makes one person different from another? Understanding the nature of the proteins helps to solve these mysteries.

The Structure of Proteins

LO 6.1 State why some amino acids are essential, nonessential, or conditionally essential to the human body, and outline how the body builds a protein molecule.

The structure of proteins enables them to perform many vital functions. One key difference from carbohydrates and fats is that proteins contain nitrogen atoms in addition to the carbon, hydrogen, and oxygen atoms that all three energy-yielding nutrients contain. These nitrogen atoms give the name *amino* (which means "nitrogen containing") to the **amino acids**, the building blocks of proteins. Another key difference is that in contrast to the carbohydrates—whose repeating units, glucose molecules, are identical—the amino acids in a strand of protein are different from one another. A strand of amino acids that makes up a protein may contain 20 *different* kinds of amino acids.

Amino Acids

All amino acids have the same simple chemical backbone consisting of a single carbon atom with both an **amine group** (the nitrogen-containing part) and an acid group attached to it. Each amino acid also has a distinctive chemical **side chain** attached to the center carbon of the backbone (see Figure 6–1). This side chain gives identity and its chemical nature to each amino acid. About 20 amino acids, each with its own different side chain, make up most of the proteins of living tissue.*[1] Other rare amino acids appear in a few proteins.

The side chains make the amino acids differ in size, shape, and electrical charge. Some are negative, some are positive, and some have no charge (are neutral). The first part of

Figure 6–1
An Amino Acid

The "backbone" is the same for all amino acids. The side chain differs from one amino acid to the next. The nitrogen is in the amine group.

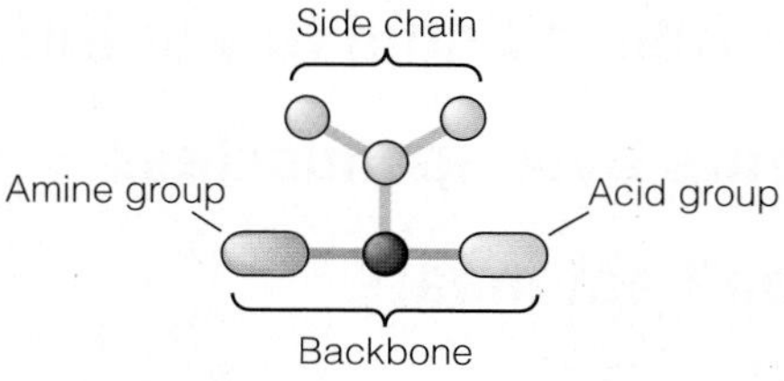

*Reference notes are found in Appendix F.

Figure 6–2
Different Amino Acids Join Together

This is the basic process by which proteins are assembled.

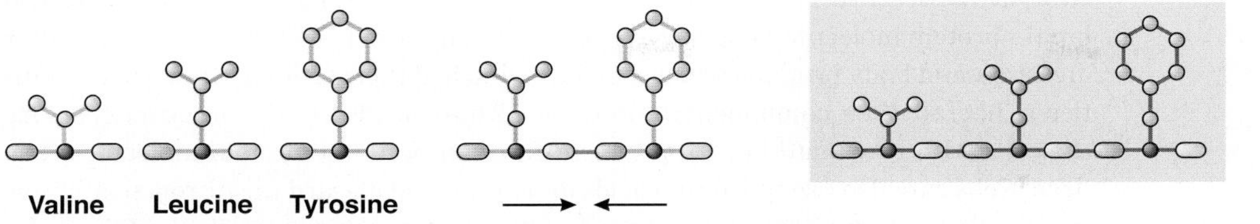

Single amino acids with different side chains . . . can bond to form . . . a strand of amino acids, part of a protein.

Figure 6–2 is a diagram of three amino acids, each with a different side chain attached to its backbone. The rest of the figure shows how amino acids link to form protein strands. Long strands of amino acids form large protein molecules, and the side chains of the amino acids ultimately help to determine the protein's molecular shape and behavior.

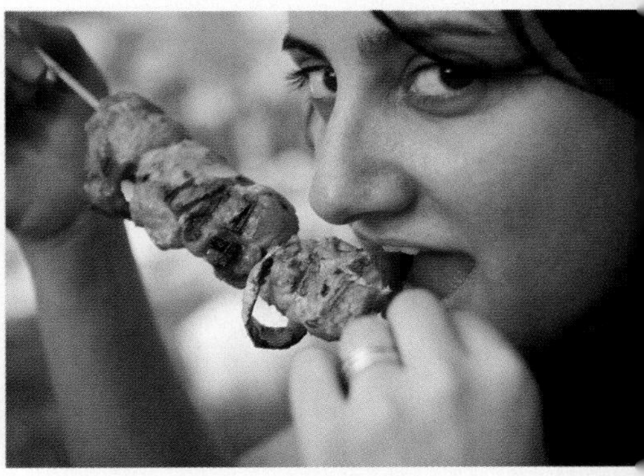

Hair, skin, eyesight, and the health of the whole body depend on proteins from food.

Essential Amino Acids The body can make about half of the 20 amino acids for itself, given the needed parts: fragments derived from carbohydrate or fat to form the backbones and nitrogen from other sources to form the amine groups. The healthy adult body cannot make some amino acids or makes them too slowly to meet its needs. These are the **essential amino acids** (listed in Table 6–1). Without these essential nutrients, the body cannot make the proteins it needs to do its work. Because the essential amino acids can only be replenished from foods, a person must frequently eat the foods that provide them.

Under special circumstances, a nonessential amino acid can become essential. For example, the body normally makes tyrosine (a nonessential amino acid) from the

Table 6–1
Amino Acids Important in Nutrition

The left-hand column lists amino acids that are essential for human beings—the body cannot make them, and they must be provided in the diet. The right-hand column lists other, nonessential amino acids—the body can make these for itself. In special cases, some nonessential amino acids may become conditionally essential (see the text).

Essential Amino Acids (pronunciation)	Nonessential Amino Acids (pronunciation)
Histidine (HISS-tuh-deen)	Alanine (AL-ah-neen)
Isoleucine (eye-so-LOO-seen)	Arginine (ARJ-ih-neen)
Leucine (LOO-seen)	Asparagine (ah-SPAR-ah-geen)
Lysine (LYE-seen)	Aspartic acid (ah-SPAR-tic acid)
Methionine (meh-THIGH-oh-neen)	Cysteine (SIS-the-een)
Phenylalanine (fen-il-AL-ah-neen)	Glutamic acid (GLU-tam-ic acid)
Threonine (THREE-oh-neen)	Glutamine (GLU-tah-meen)
Tryptophan (TRIP-toe-fan, TRIP-toe-fane)	Glycine (GLY-seen)
Valine (VAY-leen)	Proline (PRO-leen)
	Serine (SEER-een)
	Tyrosine (TIE-roe-seen)

proteins compounds composed of carbon, hydrogen, oxygen, and nitrogen and arranged as strands of amino acids. Some amino acids also contain the element sulfur.

amino (a-MEEN-o) **acids** the building blocks of protein. Each has an amine group at one end, an acid group at the other, and a distinctive side chain.

amine (a-MEEN) **group** the nitrogen-containing portion of an amino acid.

side chain the unique chemical structure attached to the backbone of each amino acid that differentiates one amino acid from another.

essential amino acids amino acids that either cannot be synthesized at all by the body or cannot be synthesized in amounts sufficient to meet physiological need. Also called *indispensable amino acids.*

essential amino acid phenylalanine. If the diet fails to supply enough phenylalanine or if the body cannot make the conversion for some reason (as happens in the inherited disease phenylketonuria; see Chapter 3, pp. 72–73), then tyrosine becomes a **conditionally essential amino acid**.

Recycling Amino Acids The body not only makes some amino acids but also breaks protein molecules apart and reuses those amino acids. Both food proteins after digestion and body proteins when they have finished their cellular work are dismantled to liberate their component amino acids. Amino acids from both sources provide the cells with raw materials from which they can build the protein molecules they need. Cells can also use the amino acids for energy and discard the nitrogen atoms as wastes. By reusing intact amino acids to build proteins, however, the body recycles and conserves a valuable commodity while easing its nitrogen disposal burden.

This recycling system also provides access to an emergency fund of amino acids in times of fuel, glucose, or protein deprivation. At such times, tissues can break down their own proteins, sacrificing working molecules before the ends of their normal lifetimes, to supply energy and amino acids to the body's cells.

KEY POINTS

- Proteins are unique among the energy nutrients in that they possess nitrogen-containing amine groups and are composed of 20 different amino acid units.
- Of the 20 amino acids, some are essential and some are essential only in special circumstances.

How Do Amino Acids Build Proteins?

In the first step of making a protein, each amino acid is hooked to the next (as shown in Figure 6–2, p. 199). A chemical bond, called a **peptide bond**, is formed between the amine group end of one amino acid and the acid group end of the next. A string of 10 or more amino acids is known as a **polypeptide**. The side chains bristle out from the backbone of the structure, giving the protein molecule its unique character.

The strand of protein does not remain a straight chain. Figure 6–2 shows only the first step in making all proteins—the linking of amino acid units with peptide bonds until the strand contains from several dozen to as many as 300 amino acids. Amino acids at different places along the strand are chemically attracted to each other, and this attraction can cause some segments of the strand to coil, somewhat like a metal spring. Also, each spot along the strand is attracted to, or repelled from, other spots along its length (demonstrated in Figure 6–3). These interactions often cause the entire coil to fold this way and that to form a globular structure, as shown in Figure 6–4 (p. 202). Other strands may link together in sheets or rods, lending toughness and stability to structures.

The amino acids whose side chains are electrically charged are attracted to water. Therefore, in the body's watery fluids, they orient themselves on the outside of the protein structure. The amino acids whose side chains are neutral are repelled by water and are attracted to one another; these tuck themselves into the center away from the body fluids. All these interactions among the amino acids and the surrounding fluids fold each protein into a unique architecture, a form to suit its function.

One final detail may be needed for the protein to become functional. Several strands may cluster together into a functioning unit, or a metal ion (mineral), a vitamin, or a carbohydrate molecule may join to the unit and activate it.

KEY POINT

- Amino acids link into long strands that make a wide variety of different proteins.

The Variety of Proteins

The particular shapes of proteins enable them to perform different tasks in the body. Those of globular shape, such as some proteins of blood, are water-soluble. Some form hollow balls, which can carry and store materials in their interiors. Some proteins, such as those of tendons, are more than 10 times as long as they are wide, forming

Figure 6–3

Animated! The Coiling and Folding of a Protein Molecule

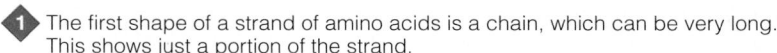

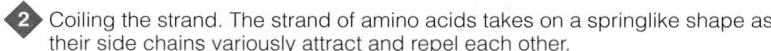

1. The first shape of a strand of amino acids is a chain, which can be very long. This shows just a portion of the strand.
2. Coiling the strand. The strand of amino acids takes on a springlike shape as their side chains variously attract and repel each other.
3. Folding the coil. The coil then folds and flops over on itself to take a functional shape.
4. Once coiled and folded, the protein may be functional as is, or it may need to join with other proteins, or add a carbohydrate molecule or a vitamin or mineral, as the iron of the protein hemoglobin demonstrates in Figure 6-4 (p. 202).

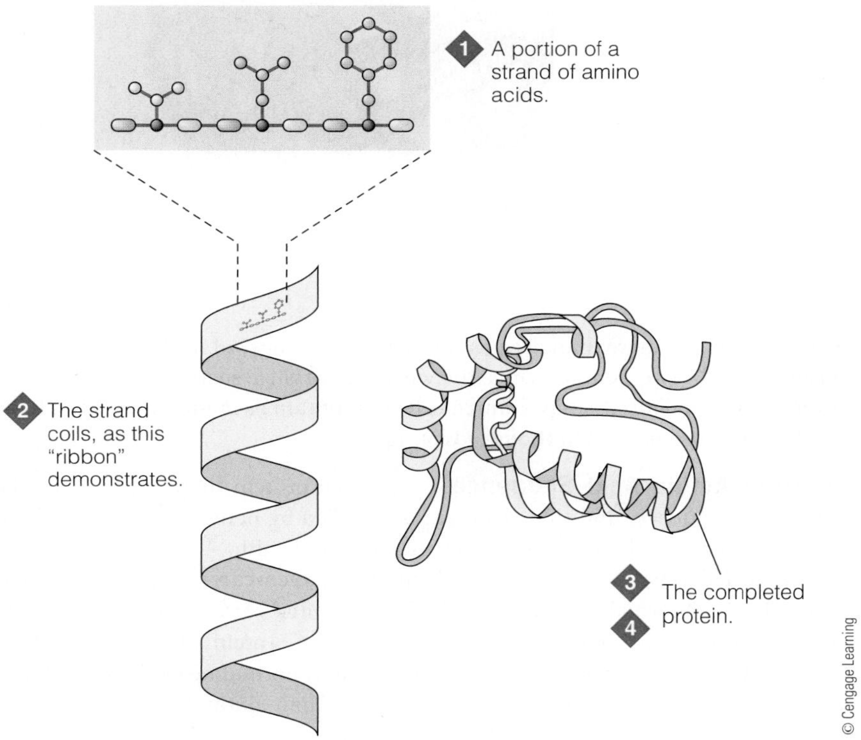
1 A portion of a strand of amino acids.
2 The strand coils, as this "ribbon" demonstrates.
3 4 The completed protein.
© Cengage Learning

stiff, rodlike structures that are somewhat insoluble in water and very strong. A form of the protein **collagen** acts somewhat like glue between cells. The hormone insulin, a protein, helps to regulate blood glucose. Among the most fascinating proteins are the **enzymes**, which act on other substances to change them chemically. More roles of the body's proteins are discussed in a later section.

Some protein strands work alone, while others must associate in groups of strands to become functional. One molecule of **hemoglobin**—the large, globular protein molecule that is packed into the red blood cells by the billions and carries oxygen—is made of four associated protein strands, each holding the mineral iron (see Figure 6–4).

The great variety of proteins in the world is possible because an essentially infinite number of sequences of amino acids can be formed. To understand how so many different proteins can be designed from only 20 or so amino acids, think of how many words are in an unabridged dictionary—all of them constructed from just 26 letters. If you had only the letter "G," all you could write would be a string of Gs: G–G–G–G–G–G–G. But with 26 different letters available, you can create poems, songs, or novels. Similarly, the 20 amino acids can be linked together in a huge variety of

conditionally essential amino acid an amino acid that is normally nonessential but must be supplied by the diet in special circumstances when the need for it exceeds the body's ability to produce it.

peptide bond a bond that connects one amino acid with another, forming a link in a protein chain.

polypeptide (POL-ee-PEP-tide) protein fragments of many (more than 10) amino acids bonded together (*poly* means "many"). A peptide is a strand of amino acids.

collagen (KAHL-ah-jen) a type of body protein from which connective tissues such as scars, tendons, ligaments, and the foundations of bones and teeth are made.

enzymes (EN-zimes) proteins that facilitate chemical reactions without being changed in the process; protein catalysts.

hemoglobin the globular protein of red blood cells, whose iron atoms carry oxygen around the body via the bloodstream (more about hemoglobin in Chapter 8).

Figure 6–4

The Structure of Hemoglobin

Four highly folded protein strands form the globular hemoglobin protein.

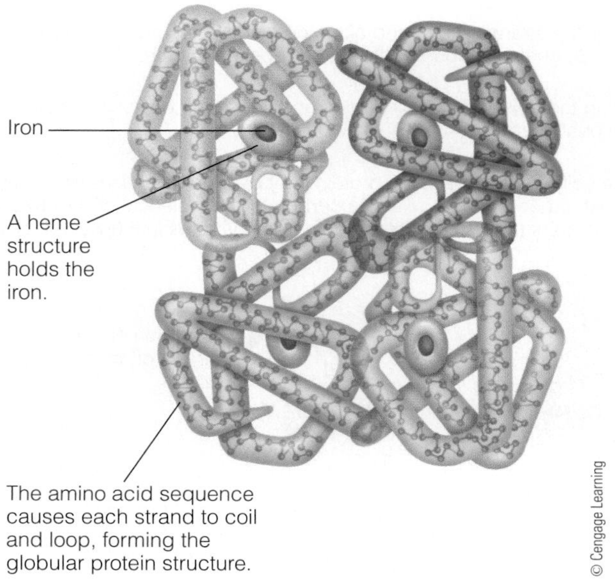

Iron
A heme structure holds the iron.
The amino acid sequence causes each strand to coil and loop, forming the globular protein structure.
© Cengage Learning

sequences—many more than are possible for letters in a word, which must alternate consonant and vowel sounds. Thus, the variety of possible sequences for amino acid strands is tremendous. A single human cell may contain as many as 10,000 different proteins, each one present in thousands of copies.

Inherited Amino Acid Sequences For each protein there exists a standard amino acid sequence, and that sequence is specified by heredity. Often, if a wrong amino acid is inserted, the result can be disastrous to health.

Sickle-cell disease—in which hemoglobin, the oxygen-carrying protein of the red blood cells, is abnormal—is an example of an inherited variation in the amino acid sequence. Normal hemoglobin contains two kinds of protein strands. In sickle-cell disease, one of the strands is an exact copy of that in normal hemoglobin, but in the other strand, the sixth amino acid is valine rather than glutamic acid. This replacement of one amino acid so alters the protein that it is unable to carry and release oxygen. The red blood cells collapse from the normal disk shape into crescent shapes (see Figure 6–5). If too many crescent-shaped cells appear in the blood, the result is abnormal blood clotting, strokes, bouts of severe pain, susceptibility to infection, and early death.[2]

You are unique among human beings because of minute differences in your body proteins that establish everything from eye color and shoe size to susceptibility to certain diseases. These differences are determined by the amino acid sequences of your proteins, which are written into the genetic code you inherited from your parents and they from theirs. Ultimately, the genes determine the sequence of amino acids in each finished protein, and some genes are involved in making more than one protein (how DNA directs protein synthesis is described in Figure 6–6, p. 204). As scientists completed the identification of the human genome, they recognized a still greater task that lies ahead: the identification of every protein made by the human body.[†]

Nutrients and Gene Expression When a cell makes a protein, as shown in Figure 6–6, scientists say that the gene for that protein has been "expressed." Every

[†]The identification of the entire collection of human proteins, the *human proteome* (PRO-tee-ohme), is a work in progress.

Figure 6–5

Normal Red Blood Cells and Sickle Cells

Normal red blood cells are disk shaped. In sickle-cell disease, one amino acid in the protein strands of hemoglobin takes the place of another, causing the red blood cell to change shape and lose function.

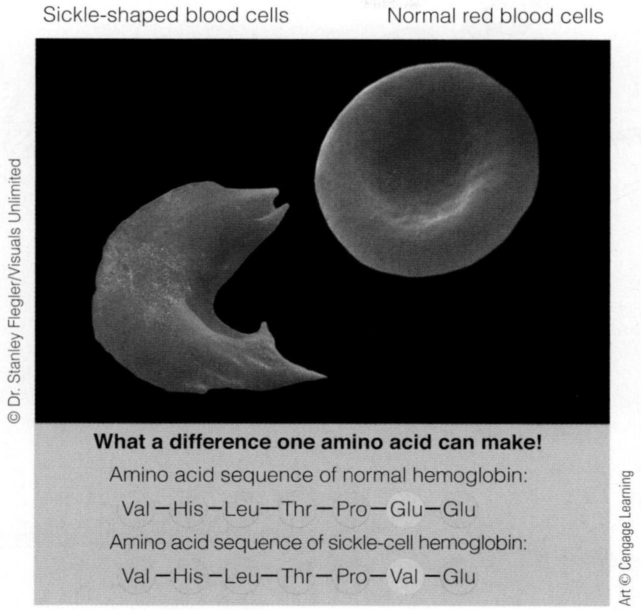

cell nucleus contains the DNA for making every human protein, but cells do not make them all. Some cells specialize in making certain proteins; for example, cells of the pancreas express the gene for the protein hormone insulin. The gene for making insulin exists in all other cells of the body, but it is idle or silenced.

Sickle-shaped blood cells
Normal red blood cells
© Dr. Stanley Flegler/Visuals Unlimited
What a difference one amino acid can make!
Amino acid sequence of normal hemoglobin:
Val—His—Leu—Thr—Pro—Glu—Glu
Amino acid sequence of sickle-cell hemoglobin:
Val—His—Leu—Thr—Pro—Val—Glu
Art © Cengage Learning

Nutrients, including amino acids and proteins, do not change DNA structure, but they greatly influence genetic expression.[3] As research advances, researchers hope to one day use nutrients to influence a person's genes in ways that reduce disease risks, but for now, that day is firmly in the future. The Controversy section of Chapter 11 comes back to this fascinating area of nutrient and gene interactions. The Think Fitness feature near here addresses a related concern of exercisers and athletes about whether extra dietary protein or amino acids can trigger the synthesis of muscle tissue and increase strength.

KEY POINTS

- Each type of protein has a distinctive sequence of amino acids and so has great specificity.
- Often, cells specialize in synthesizing particular types of proteins in addition to the proteins necessary to all cells.
- Nutrients can greatly affect genetic expression.

Denaturation of Proteins

Proteins can be denatured (distorted in shape) by heat, radiation, alcohol, acids, bases, or the salts of heavy metals. The **denaturation** of a protein is the first step in its destruction; thus, these agents are dangerous because they can disrupt a protein's folded structure, making it unable to function in the body. In digestion, however, denaturation is useful to the body.

During the digestion of a food protein, the stomach acid opens up the protein's structure, permitting digestive enzymes to make contact with the peptide bonds and cleave them. Denaturation also occurs during the cooking of foods. Cooking an

denaturation the irreversible change in a protein's folded shape brought about by heat, acids, bases, alcohol, salts of heavy metals, or other agents.

Figure 6–6

Animated! Protein Synthesis

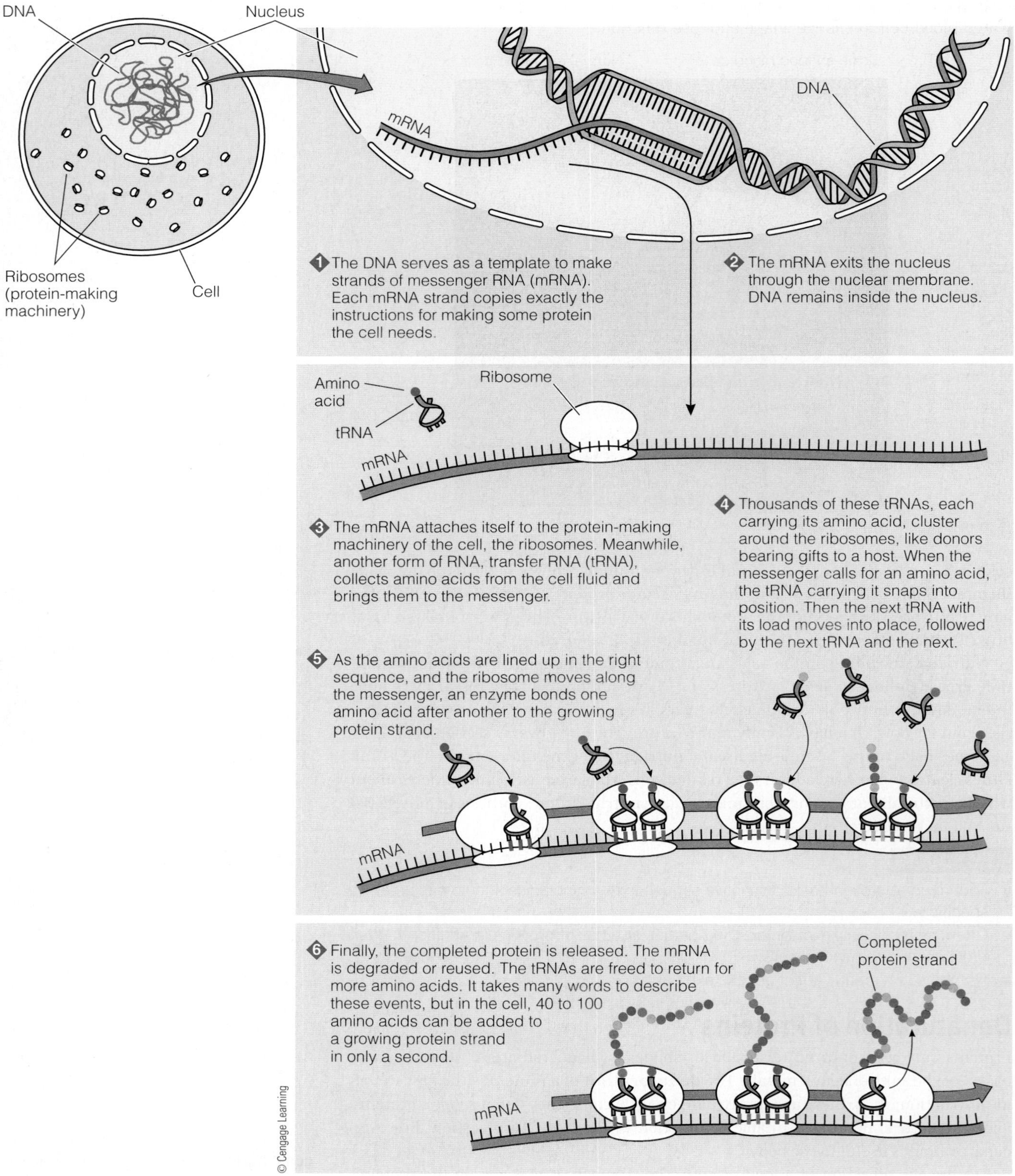
DNA
Nucleus
Ribosomes (protein-making machinery)
Cell
mRNA
DNA
1 The DNA serves as a template to make strands of messenger RNA (mRNA). Each mRNA strand copies exactly the instructions for making some protein the cell needs.
2 The mRNA exits the nucleus through the nuclear membrane. DNA remains inside the nucleus.
Amino acid
Ribosome
tRNA
mRNA
3 The mRNA attaches itself to the protein-making machinery of the cell, the ribosomes. Meanwhile, another form of RNA, transfer RNA (tRNA), collects amino acids from the cell fluid and brings them to the messenger.
4 Thousands of these tRNAs, each carrying its amino acid, cluster around the ribosomes, like donors bearing gifts to a host. When the messenger calls for an amino acid, the tRNA carrying it snaps into position. Then the next tRNA with its load moves into place, followed by the next tRNA and the next.
5 As the amino acids are lined up in the right sequence, and the ribosome moves along the messenger, an enzyme bonds one amino acid after another to the growing protein strand.
mRNA
6 Finally, the completed protein is released. The mRNA is degraded or reused. The tRNAs are freed to return for more amino acids. It takes many words to describe these events, but in the cell, 40 to 100 amino acids can be added to a growing protein strand in only a second.
Completed protein strand
mRNA
© Cengage Learning

Think Fitness move it!

Can Eating Extra Protein Make Muscles Grow Stronger?

The answer is mostly "no" but also a qualified "yes." Athletes and fitness seekers cannot stimulate their muscles to gain size and strength simply by consuming more protein or amino acids. Hard work is necessary to trigger the genes to build more of the muscle tissue needed for sport. The "yes" part of the answer reflects research suggesting that well-timed protein intakes can often further stimulate muscle growth. Protein intake cannot replace exercise in this regard, however, as many supplement sellers would have people believe. Exercise generates cellular messages that stimulate the DNA to begin synthesizing the muscle proteins needed to perform the work. A protein-rich snack—say, a glass of skim milk or soy milk—consumed shortly after strength-building exercise (such as weight lifting) stimulates muscle protein synthesis for a while, but evidence is lacking for a benefit to athletic performance.

Athletes may need somewhat more dietary protein than other people do, and exercise authorities recommend higher protein intakes for athletes pursuing various activities (see Chapter 10 for details).[4] Amino acid or protein supplements, however, offer no advantage over food, and amino acid supplements are more likely to cause problems (as this chapter's Consumer's Guide makes clear). Bottom line: the path to bigger muscles is rigorous physical training with adequate energy and nutrients from balanced, well-timed meals and snacks. Research findings concerning dietary protein and muscles are interesting and important, but this truth remains: extra protein and amino acids without physical work add nothing but excess calories.

start now! Ready to make a change? Go to Diet Analysis Plus and generate an Intake Report for your 3-day dietary tracking. What is your protein intake level? If it is low, create an alternate profile and substitute one 8-oz glass of skim or low-fat milk at two meals (or one meal and one snack). What is the effect on your protein intake?

egg denatures the proteins of the egg and makes it firm, as the margin photo demonstrates. More important for nutrition is that heat denatures two proteins in raw eggs: one binds the B vitamin biotin and the mineral iron, and the other slows protein digestion. Thus, cooking eggs liberates biotin and iron and aids digestion.

Many well-known poisons are salts of heavy metals like mercury and silver; these poisons denature protein strands wherever they touch them. The common first-aid antidote for swallowing a heavy-metal poison is to drink milk. The poison then acts on the protein of the milk rather than on the protein tissues of the mouth, esophagus, and stomach. Later, vomiting can be induced to expel the poison that has combined with the milk.

© iStockphoto.com/Vladimir Glazkov

Heat denatures protein, making it firm.

KEY POINTS

- Proteins can be denatured by heat, acids, bases, alcohol, or the salts of heavy metals.
- Denaturation begins the process of digesting food protein and can also destroy body proteins.

Digestion and Absorption of Dietary Protein

LO 6.2 Describe the digestion of protein and the absorption and transport of amino acids in the body.

Each protein performs a special task in a particular tissue of a specific kind of animal or plant. When a person eats food proteins, whether from cereals, vegetables, beef, fish, or cheese, the body must first alter them by breaking them down into amino acids; only then can it rearrange them into specific human body proteins.

Protein Digestion

Other than being crushed and torn by chewing and moistened with saliva in the mouth, nothing happens to protein until it reaches the stomach. Then, the action begins.

In the Stomach

In the Stomach Strong acid produced by the stomach denatures proteins in food. This acid helps to uncoil the protein's tangled strands so that molecules of the stomach's protein-digesting enzyme can attack the peptide bonds. You might expect that the stomach enzyme, being a protein itself, would be denatured by the stomach's acid. Unlike most enzymes, though, the stomach enzyme functions best in an acid environment. Its job is to break *other* protein strands into smaller pieces. The stomach lining, which is also made partly of protein, is protected against attack by acid and enzymes by its coat of mucus, secreted by its cells.

The whole process of digestion is an ingenious solution to a complex problem. Proteins (enzymes), activated by acid, digest proteins from food, denatured by acid. Digestion and absorption of other nutrients, such as iron, also rely on the stomach's ability to produce strong acid. The normal acid in the stomach is so strong (pH 1.5) that no food is acidic enough to make it stronger; for comparison, the pH of vinegar is about 3.

pH was defined in **Chapter 3** on page 87.

Figure 6–7
A Dipeptide and Tripeptide

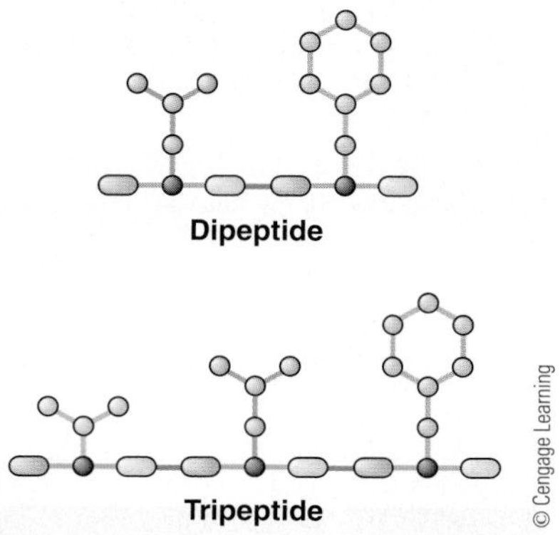

In the Small Intestine By the time most proteins slip from the stomach into the small intestine, they are denatured and cleaved into smaller pieces. A few are single amino acids, but the majority remains in large strands—polypeptides. In the small intestine, alkaline juice from the pancreas neutralizes the acid delivered by the stomach. The pH rises to about 7 (neutral), enabling the next enzyme team to accomplish the final breakdown of the strands. Protein-digesting enzymes from the pancreas and intestine continue working until almost all pieces of protein are broken into single amino acids or into strands of two or three amino acids, **dipeptides** and **tripeptides**, respectively (see Figure 6–7). Figure 6–8 summarizes the whole process of protein digestion.

Common Misconceptions Consumers who fail to understand the basic mechanism of protein digestion are easily misled by advertisers of books and other products who urge, "Take enzyme A to help digest your food" or "Don't eat foods containing enzyme C, which will digest cells in your body." The writers of such statements fail to realize that enzymes (proteins) are digested before they are absorbed, just as all proteins are. Even the stomach's digestive enzymes are denatured and digested when their jobs are through. Similar false claims are made that predigested proteins (amino acid supplements) are "easy to digest" and can therefore protect the digestive system from "overworking." Of course, the healthy digestive system is superbly designed to digest whole proteins with ease. In fact, it handles whole proteins better than predigested ones because it dismantles and absorbs the amino acids at rates that are optimal for the body's use.

KEY POINT

- Digestion of protein involves denaturation by stomach acid and enzymatic digestion in the stomach and small intestine to amino acids, dipeptides, and tripeptides.

What Happens to Amino Acids after Protein Is Digested?

The cells all along the small intestine absorb single amino acids. As for dipeptides and tripeptides, enzymes on the cells' surfaces split most of them into single amino acids, and the cells absorb them, too. Dipeptides and tripeptides are also absorbed as is into the cells, where they are split into amino acids and join with the others to be released into the bloodstream. A few larger peptide molecules can escape the digestive process altogether and enter the bloodstream intact. Scientists believe these larger particles may act as hormones to regulate body functions and provide the body with information about the external environment. The larger molecules may also stimulate an immune response and thus play a role in food allergy.

The cells of the small intestine possess separate sites for absorbing different types of amino acids. Amino acids of the same type compete for the same absorption sites.

dipeptides (dye-PEP-tides) protein fragments that are two amino acids long (*di* means "two").

tripeptides (try-PEP-tides) protein fragments that are three amino acids long (*tri* means "three").

Figure 6–8

Animated! How Protein in Food Becomes Amino Acids in the Body

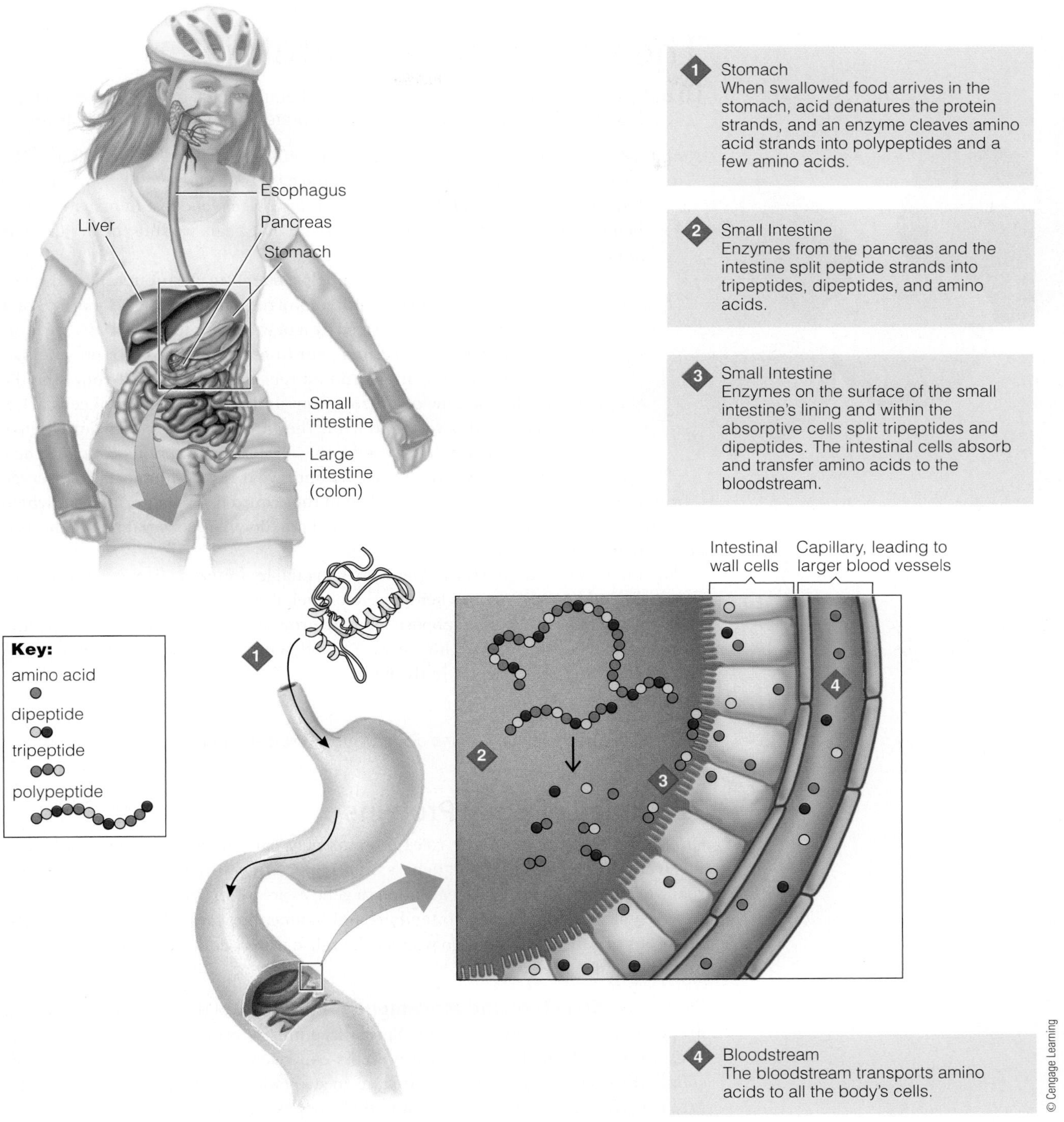

Esophagus
Liver
Pancreas
Stomach
Small intestine
Large intestine (colon)
1 Stomach
When swallowed food arrives in the stomach, acid denatures the protein strands, and an enzyme cleaves amino acid strands into polypeptides and a few amino acids.
2 Small Intestine
Enzymes from the pancreas and the intestine split peptide strands into tripeptides, dipeptides, and amino acids.
3 Small Intestine
Enzymes on the surface of the small intestine's lining and within the absorptive cells split tripeptides and dipeptides. The intestinal cells absorb and transfer amino acids to the bloodstream.
Intestinal wall cells
Capillary, leading to larger blood vessels
Key:
amino acid
dipeptide
tripeptide
polypeptide
1
2
3
4
4 Bloodstream
The bloodstream transports amino acids to all the body's cells.

Consequently, when a person ingests a large dose of any single amino acid, that amino acid may limit absorption of others of its general type. The Consumer's Guide (page 214) cautions against taking single amino acids as supplements partly for this reason.

Once amino acids are circulating in the bloodstream, they are carried to the liver where they may be used or released into the blood to be taken up by other cells of the body. The cells can then link the amino acids together to make proteins that they keep for their own use or liberate into lymph or blood for other uses. When necessary, the body's cells can also use amino acids for energy.

KEY POINT

- The cells of the small intestine complete digestion, absorb amino acids and some larger peptides, and release them into the bloodstream for use by the body's cells.

The Importance of Protein

LO 6.3 List the roles that various proteins and amino acids can play in the body, and describe the influence of carbohydrate on amino acid metabolism.

Amino acids must be continuously available to build the proteins of new tissue. The new tissue may be in an embryo; in the muscles of an athlete in training; in a growing child; in new blood cells needed to replace blood lost in menstruation, hemorrhage, or surgery; in the scar tissue that heals wounds; or in new hair and nails.

Less obvious is the protein that helps to replace worn-out cells and internal cell structures. Each of your millions of red blood cells lives for only 3 or 4 months. Then it must be replaced by a new cell produced by the bone marrow. The millions of cells lining your intestinal tract live for only 3 days; they are constantly being shed and replaced. The cells of your skin die and rub off, and new ones grow from underneath. Nearly all cells arise, live, and die in this way, and while they are living, they constantly make and break down proteins. In addition, cells must continuously replace their own internal working proteins as old ones wear out. Amino acids conserved from these processes provide a great deal of the required raw material from which new structures are built. The entire process of breakdown, recovery, and synthesis is called **protein turnover**.

Each day, about a quarter of the body's available amino acids are irretrievably diverted to other uses, such as being used for fuel. For this reason, amino acids from food are needed each day to support the new growth and maintenance of cells and to make the working parts within them. The following sections spell out some of the critical roles that proteins play in the body.

KEY POINT

- The body needs dietary amino acids to grow new cells and to replace worn-out ones.

The Roles of Body Proteins

Only a sampling of the many roles proteins play can be described here, but these illustrate their versatility, uniqueness, and importance in the body. One important role was already mentioned: regulation of gene expression. Others range from digestive enzymes and antibodies to tendons and ligaments, scars, filaments of hair, the materials of nails, and more. No wonder their discoverers called proteins the primary material of life.

Providing Structure and Movement A great deal of the body's protein (about 40 percent) exists in muscle tissue. Specialized muscle protein structures allow the body to move. In addition, muscle proteins can release some of their amino acids should the need for energy become dire, as in starvation. These amino acids are integral parts of the muscle structure, and their loss exacts a cost of functional protein. Other structural proteins confer shape and strength on bones, teeth, skin, tendons, cartilage, blood vessels, and other tissues. All are important to the workings of a healthy body.

Building Enzymes, Hormones, and Other Compounds Among proteins formed by living cells, enzymes are metabolic workhorses. An enzyme acts as a **catalyst**: it speeds up a reaction that would happen anyway, but much more slowly. Thousands of enzymes reside inside a single cell, and each one facilitates a specific chemical reaction. Figure 6–9 shows how a hypothetical enzyme works—this one synthesizes a compound

Figure 6–9
Enzyme Action

Compounds A and B are attracted to the enzyme's active site and park there for a moment in the exact position that makes the reaction between them most likely to occur. They react by bonding together and leave the enzyme as the new compound, AB.

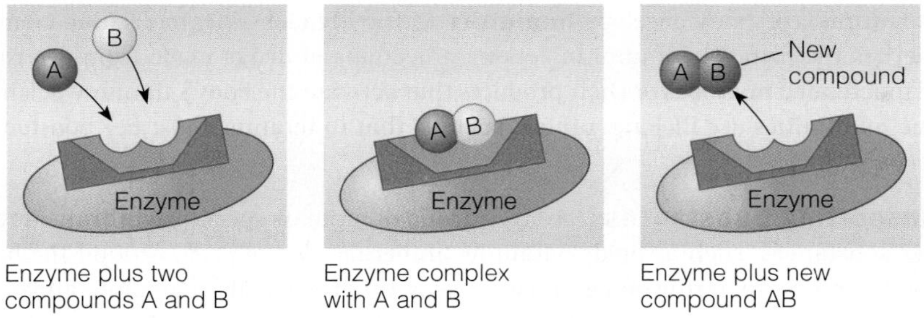

from two chemical components. Other enzymes break compounds apart into two or more products or rearrange the atoms in one kind of compound to make another. A single enzyme can facilitate several hundred reactions in a second.

The body's **hormones** are messenger molecules, and many of them are made from amino acids. Various body glands release hormones when changes occur in the internal environment; the hormones then elicit tissue responses necessary to restore normal conditions. For example, the familiar pair of hormones, insulin and glucagon, oppose each other to maintain blood glucose levels. Both are built of amino acids. For interest, Figure 6–10 shows how many amino acids are linked in sequence to form human insulin. It also shows how certain side groups attract one another to complete the insulin molecule and make it functional.

In addition to serving as building blocks for proteins, amino acids also perform other tasks in the body. For example, the amino acid tyrosine forms parts of the neurotransmitters epinephrine and norepinephrine, which relay messages throughout the nervous system. The body also uses tyrosine to make the brown pigment melanin, which is responsible for skin, hair, and eye color. Tyrosine is also converted into the thyroid hormone **thyroxine**, which regulates the body's metabolism. Another amino acid, tryptophan, serves as starting material for the neurotransmitter **serotonin** and the vitamin niacin.

Building Antibodies Of all the proteins in living organisms, the **antibodies** best demonstrate that proteins are specific to one organism. Antibodies distinguish

Figure 6–10
Amino Acid Sequence of Human Insulin

This picture shows a refinement of protein structure not mentioned in the text. The amino acid cysteine (cys) has a sulfur-containing side group. The sulfur groups on two cysteine molecules can bond together, creating a bridge between two protein strands or two parts of the same strand. Insulin contains three such bridges.

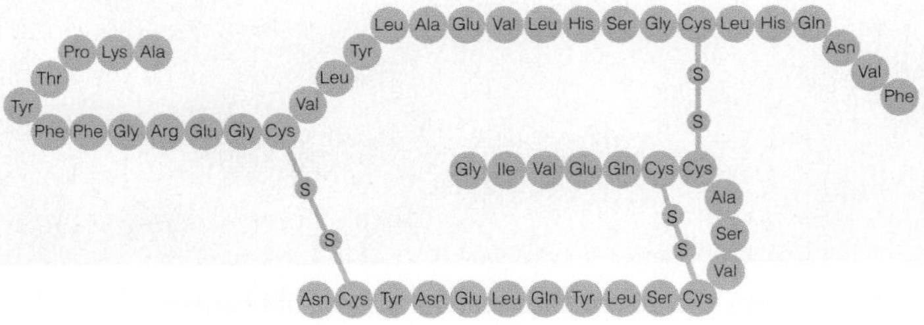

protein turnover the continuous breakdown and synthesis of body proteins involving the recycling of amino acids.

catalyst a substance that speeds the rate of a chemical reaction without itself being permanently altered in the process. All enzymes are catalysts.

hormones chemical messengers secreted by a number of body organs in response to conditions that require regulation. Each hormone affects a specific organ or tissue and elicits a specific response. Also defined in Chapter 3.

thyroxine (thigh-ROX-in) a principal peptide hormone of the thyroid gland that regulates the body's rate of energy use.

serotonin (SARE-oh-TONE-in) a compound related in structure to (and made from) the amino acid tryptophan. It serves as one of the brain's principal neurotransmitters.

antibodies (AN-te-bod-ees) large proteins of the blood, produced by the immune system in response to an invasion of the body by foreign substances (antigens). Antibodies combine with and inactivate the antigens. Also defined in Chapter 3.

foreign particles (usually proteins) from all the proteins that belong in "their" body. When they recognize an intruder, they mark it as a target for attack. The foreign protein may be part of a bacterium, a virus, or a toxin, or it may be present in a food that causes an allergic reaction.

Each antibody is designed to help destroy one specific invader. An antibody active against one strain of influenza is of no help to a person ill with another strain. Once the body has learned how to make a particular antibody, it remembers. The next time the body encounters that same invader, it destroys the invader even more rapidly. In other words, the body develops **immunity** to the invader. This molecular memory underlies the principle of immunizations, injections of drugs made from destroyed and inactivated microbes or their products that activate the body's immune defenses. Some immunities are lifelong; others, such as that to tetanus, must be "boosted" at intervals.

Transporting Substances A large group of proteins specialize in transporting other substances, such as lipids, vitamins, minerals, and oxygen, around the body. To do their jobs, such substances must travel within the bloodstream, into and out of cells, or around the cellular interiors. Two familiar examples are the protein hemoglobin that carries oxygen from the lungs to the cells and the lipoproteins that transport lipids in the watery blood.

Maintaining Fluid and Electrolyte Balance Proteins help to maintain the **fluid and electrolyte balance** by regulating the quantity of fluids in the compartments of the body. To remain alive, cells must contain a constant amount of fluid. Too much can cause them to rupture; too little makes them unable to function. Although water can diffuse freely into and out of cells, proteins cannot, and proteins attract water.

By maintaining stores of internal proteins and also of some minerals, cells retain the fluid they need. By the same mechanism, fluid is kept inside the blood vessels by proteins too large to move freely across the capillary walls. The proteins attract water and hold it within the vessels, preventing it from freely flowing into the spaces between the cells. Should any part of this system begin to fail, too much fluid will soon collect in the spaces between the cells of tissues, causing **edema**.

Not only is the quantity of the body fluids vital to life but so also is their composition. Transport proteins in the membranes of cells also help maintain this composition by continuously transferring substances into and out of cells (see Figure 6–11).

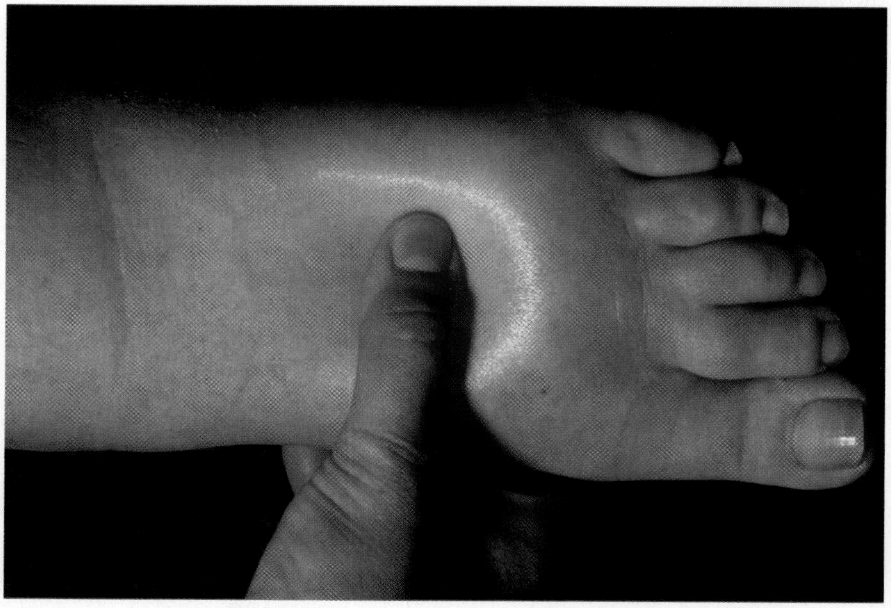

SPL/Science Source

Edema results when body tissues fail to control the movement of water.

immunity protection from or resistance to a disease or infection by development of antibodies and by the actions of cells and tissues in response to a threat.

fluid and electrolyte balance the distribution of fluid and dissolved particles among body compartments (see also Chapter 8).

edema (eh-DEEM-uh) swelling of body tissue caused by leakage of fluid from the blood vessels; seen in protein deficiency (among other conditions).

Figure 6–11

Animated! Proteins Transport Substances into and out of Cells

A transport protein within the cell membrane acts as a sort of two-door passageway—substances enter on one side and are released on the other, but the protein never leaves the membrane. The protein differs from a simple passageway in that it actively escorts the substances in and out of cells; therefore, this form of transport is often called active transport.

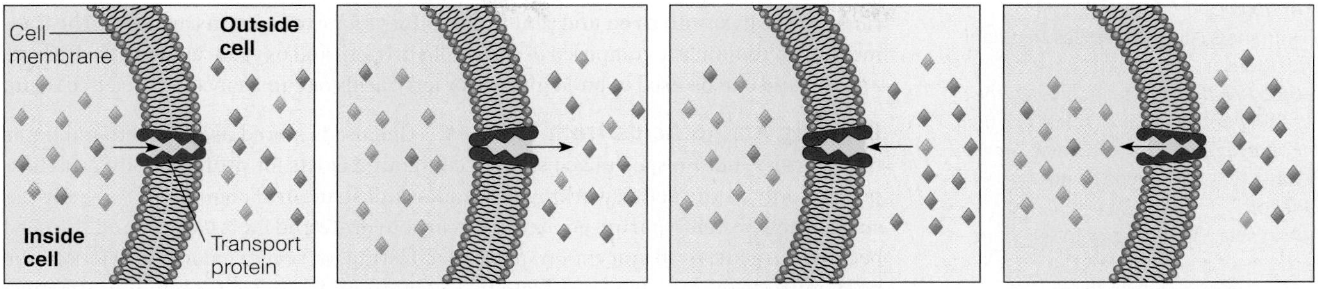

Molecule enters protein from inside cell. | Protein changes shape; molecule exits protein outside the cell. | Molecule enters protein from outside cell. | Molecule exits protein; proper balance restored.

For example, sodium is concentrated outside the cells, and potassium is concentrated inside. A disturbance of this balance can impair the action of the heart, lungs, and brain, triggering a major medical emergency. Cell proteins avert such a disaster by holding fluids and electrolytes in their proper chambers.

Maintaining Acid-Base Balance Normal processes of the body continually produce **acids** and their opposite, **bases**, that must be carried by the blood to the organs of excretion. The blood must do this without allowing its own **acid-base balance** to be affected. This feat is another trick of the blood proteins, which act as **buffers** to maintain the blood's normal pH. The protein buffers pick up hydrogens (acid) when there are too many in the bloodstream and release them again when there are too few. The secret is that negatively charged side chains of amino acids can accommodate additional hydrogens, which are positively charged.

Blood pH is one of the most rigidly controlled conditions in the body. If blood pH changes too much, **acidosis** or the opposite basic condition, **alkalosis**, can cause coma or death. These conditions constitute medical emergencies because of their effect on proteins. When the proteins' buffering capacity is filled—that is, when they have taken on all the acid hydrogens they can accommodate—additional acid pulls them out of shape, denaturing them and disrupting many body processes.

Blood Clotting To prevent dangerous blood loss, special blood proteins respond to an injury by clotting the blood. In an amazing series of chemical events, these proteins form a stringy net that traps blood cells to form a clot. The clot acts as a plug to stem blood flow from the wound. Later, as the wound heals, the protein collagen finishes the job by replacing the clot with scar tissue.

The final function of protein, providing energy, depends upon some metabolic adjustments, as described in the next section. Table 6–2 (p. 212) provides a summary of the functions of proteins in the body.

KEY POINT

- Proteins provide structure and movement; serve as enzymes, hormones, and antibodies; provide molecular transport; help regulate fluid and electrolyte balance; buffer the blood; contribute to blood clotting; and help regulate gene expression.

Providing Energy and Glucose

Only protein can perform all the functions just described, but protein will be surrendered to provide energy if need be. Under conditions of inadequate carbohydrate or energy, protein breakdown speeds up.

acids compounds that release hydrogens in a watery solution.

bases compounds that accept hydrogens from solutions.

acid-base balance equilibrium between acid and base concentrations in the body fluids.

buffers compounds that help keep a solution's acidity or alkalinity constant.

acidosis (acid-DOH-sis) the condition of excess acid in the blood, indicated by a below-normal pH (*osis* means "too much in the blood").

alkalosis (al-kah-LOH-sis) the condition of excess base in the blood, indicated by an above-normal blood pH (*alka* means "base"; *osis* means "too much in the blood").

Table 6–2

Summary of Protein Functions

- *Acid-base balance.* Proteins help maintain the acid-base balance of various body fluids by acting as buffers.
- *Antibodies.* Proteins form the immune system molecules that fight diseases.
- *Blood clotting.* Proteins provide the netting on which blood clots are built.
- *Energy and glucose.* Proteins provide some fuel for the body's energy needs.
- *Enzymes.* Proteins facilitate needed chemical reactions.
- *Fluid and electrolyte balance.* Proteins help to maintain the water and mineral composition of various body fluids.
- *Gene expression.* Proteins associate and interact with DNA, regulating gene expression.
- *Hormones.* Proteins regulate body processes. Some hormones are proteins or are made from amino acids.
- *Structure.* Proteins form integral parts of most body tissues and confer shape and strength on bones, skin, tendons, and other tissues. Structural proteins of muscles allow body movement.
- *Transportation.* Proteins help transport needed substances, such as lipids, minerals, and oxygen, around the body.

Amino Acids to Glucose The body must have energy to live from moment to moment, so obtaining that energy is a top priority. Not only can amino acids supply energy, but many of them can also be converted to glucose, as fatty acids can never be. Thus, if the need arises, protein can help to maintain a steady blood glucose level and serve the glucose need of the brain.

When amino acids are degraded for energy or converted into glucose, their nitrogen-containing amine groups are stripped off and used elsewhere or are incorporated by the liver into **urea** and sent to the kidneys for excretion in the urine. The fragments that remain are composed of carbon, hydrogen, and oxygen, as are carbohydrate and fat, and can be used to build glucose or fatty acids or can be metabolized like them.

Drawing Amino Acids from Tissues Glucose is stored as glycogen and fat as triglycerides, but no specialized storage compound exists for protein. Body protein is present only as the active working molecular and structural components of body tissues. When protein-sparing energy from carbohydrate and fat is lacking and the need becomes urgent, as in starvation, prolonged fasting, or severe calorie restriction, the body must dismantle its tissue proteins to obtain amino acids for building the most essential proteins and for energy. Each protein is taken in its own time: first, small proteins from the blood; then, proteins from the muscles. The body guards the structural proteins of the heart and other organs until forced, by dire need, to relinquish them. Thus, energy deficiency (starvation) always incurs wasting of lean body tissue as well as loss of fat.

Using Excess Amino Acids When amino acids are oversupplied, the body cannot store them. It has no choice but to remove and excrete their amine groups and then use the residues in one of three ways: to meet immediate energy needs, to make glucose for storage as glycogen, or to make fat for energy storage. The body readily converts amino acids to glucose. The body also possesses enzymes to convert amino acids into fat and can produce fatty acids for storage as triglycerides in the fat tissue. An indirect contribution of amino acids to fat stores also exists—the body speeds up its use of amino acids for fuel, burning them instead of fat, making fat more abundantly available for storage in the fat tissue.

The similarities and differences of the three energy-yielding nutrients should now be clear. Carbohydrate offers energy; fat offers concentrated energy; and protein can offer energy plus nitrogen (see Figure 6–12).

Figure 6–12

Three Different Energy Sources

Carbohydrate offers energy; fat offers concentrated energy; and protein, if necessary, can offer energy plus nitrogen. The compounds at the left yield the 2-carbon fragments shown at the right. These fragments oxidize quickly in the presence of oxygen to yield carbon dioxide, water, and energy.

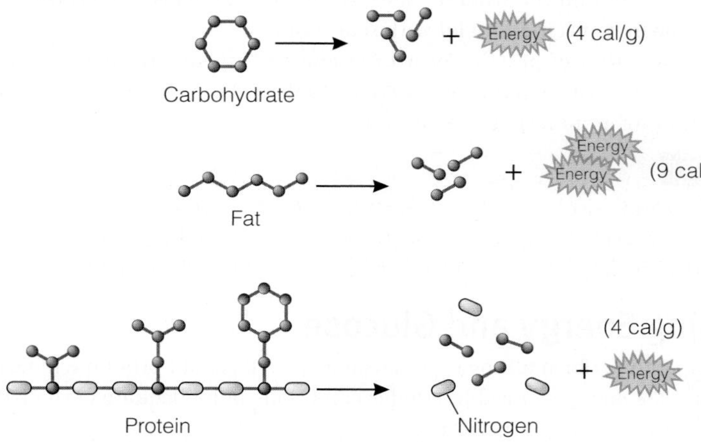

KEY POINTS

- Amino acids can be used as fuel or converted to glucose or fat.
- No storage form of protein exists in the body.

The Fate of an Amino Acid

To review the body's handling of amino acids, let us follow the fate of an amino acid that was originally part of a protein-containing food. When the amino acid arrives in a cell, it can be used in one of several ways, depending on the cell's needs at the time:

- The amino acid can be used as is to build part of a growing protein.
- The amino acid can be altered somewhat to make another needed compound, such as the vitamin niacin.
- The cell can dismantle the amino acid in order to use its amine group to build a different amino acid. The remainder can be used for fuel or, if fuel is abundant, converted to glucose or fat.

In a cell that is starved for energy and has no glucose or fatty acids, the cell strips the amino acid of its amine group (the nitrogen part) and uses the remainder of its structure for energy. The amine group is excreted from the cell and then from the body in the urine. In a cell that has a surplus of energy and amino acids, the cell takes the amino acid apart, excretes the amine group, and uses the rest to meet immediate energy needs or converts it to glucose or fat for storage.

When not used to build protein or make other nitrogen-containing compounds, amino acids are "wasted" in a sense. This wasting occurs under any of four conditions:

1. When the body lacks energy from other sources.
2. When the diet supplies more protein than the body needs.
3. When the body has too much of any single amino acid, such as from a supplement.
4. When the diet supplies protein of low quality, with too few essential amino acids, as described in the next section.

To prevent the wasting of dietary protein and permit the synthesis of needed body protein, the dietary protein must be of adequate quality; it must supply all essential amino acids in the proper amounts; it must be accompanied by enough energy-yielding carbohydrate and fat to permit the dietary protein to be used as such.

Did You Know?

Amino acids in a cell can be:

- Used to build protein.
- Converted to other amino acids or small nitrogen-containing compounds.

Stripped of their nitrogen, amino acids can be:

- Burned as fuel.
- Converted to glucose or fat.

KEY POINTS

- Amino acids can be metabolized to protein, nitrogen plus energy, glucose, or fat.
- They will be metabolized to protein only if sufficient energy is present from other sources.
- When energy is lacking, the nitrogen part is removed from each amino acid, and the resulting fragment is oxidized for energy.

Food Protein: Need and Quality

LO 6.4 Compute the daily protein need for a given individual, and discuss the concepts of nitrogen balance and protein quality.

A person's use of and need for dietary protein depends on many factors. To know whether, say, 60 grams of a particular protein is enough to meet a person's daily needs, one must consider the effects of factors discussed in this section, some pertaining to the body and some to the nature of the protein.

How Much Protein Do People Need?

The DRI recommendation for protein intake is designed to cover the need to replace protein-containing tissue that healthy adults lose and wear out every day. Therefore, it depends on body size: larger people have a higher protein need. For adults of healthy

urea (yoo-REE-uh) the principal nitrogen-excretion product of protein metabolism; generated mostly by removal of amine groups from unneeded amino acids or from amino acids being sacrificed to a need for energy.

use it! A Consumer's Guide To . . .

Evaluating Protein and Amino Acid Supplements

Nature provides protein abundantly in foods, but many people become convinced that they need extra protein and amino acids from supplements. Sorting truth from wishful thinking in advertisements can be tricky: "Take this protein supplement to build muscle," "This one will help you lose weight," "Take an amino acid to get to sleep, grow strong fingernails, cure herpes, build immunity . . ." Can these products really do these things?

Protein Supplements

Athletes often take protein supplements, but most well-fed athletes probably do not need them (Chapter 10 has details). True, protein is necessary for building muscle tissue and, true, consuming protein in conjunction with resistance exercise helps build new muscle proteins. But protein from supplements is not "muscle in a bottle" as advertisers claim. A "couch potato" who takes protein supplements cannot expect gains of muscle tissue or athletic performance. Also, if the supplements create a surplus of nitrogen, it must be metabolized and excreted, placing a burden on the kidneys, particularly if they are already weakened. Much more information about athletes and supplements is found in Chapter 10.

What a boon it would be to people on weight-loss diets if they could take the right protein or amino acids in a milkshake and effortlessly lose weight. Adding extra protein from bars and shakes is unlikely to produce this effect, however, and these products add many calories to the diet. A grain of truth is present in the claims of advertisers, however. Research is ongoing to determine whether sufficient protein content of a meal may help to prolong feelings of fullness or delay the urge to eat.[1] Evidence does not support taking protein supplements to lose weight, however, and common sense opposes it.

Amino Acid Supplements

Enthusiastic popular reports of benefits of amino acid supplements boost their sales. One such amino acid is lysine, touted to prevent or relieve the infections that cause herpes sores on the mouth or genital organs. Lysine does not cure herpes infections. Whether it reduces outbreaks or even whether it is safe is unknown; scientific studies are lacking.

Supplements containing amino acids are often sold to athletes with promises of greater blood flow to muscles or increased muscle protein synthesis. It's true that the essential amino acid leucine is necessary for normal protein synthesis regulation, but all complete protein sources, even a turkey sandwich or a glass of milk, supply plenty of leucine.[2] No clear benefit to muscle tissue has been demonstrated for leucine supplements, and their safety is under review.[3]

Millions of people take amino acid supplements hoping to strengthen soft, dry, weak, easily breakable nails. Nails are sensitive to nutrition. Made largely of protein, nails depend on sulfur bonds between amino acids (this was shown in Figure 6–10) for flexible strength, fatty acids for water resistance, and sufficient water for proper hydration.[4] In addition, nails need many minerals and vitamins to metabolize it all into place. Pills of amino acids or even the protein gelatin, often sold as a nail supplement, cannot provide the complex array of nutrients needed for nail strength. Only a nutritious diet can do so.

Pills of the amino acid tryptophan are sold to relieve pain, depression, and insomnia. Tryptophan provides raw material for making the brain neurotransmitter serotonin, an important regulator of sleep, appetite, mood, and sensory perception. High doses of tryptophan may induce sleepiness, but it may also cause side effects, such as nausea and skin disorders.

What Scientists Say

The body handles whole proteins best. It breaks them into manageable pieces (dipeptides and tripeptides) and then splits these, a few at a time, simultaneously releasing them into the blood. This slow, bit-by-bit assimilation is ideal because groups of chemically similar amino acids compete for the carriers that absorb them into the blood. An excess of one amino acid can tie up a carrier and disturb amino acid absorption, creating a temporary imbalance.[5]

Deep in the cells' nuclei, amino acids play key roles in gene regulation, and amino acid imbalances may affect these processes in unpredictable ways.[6] In mice, for example, excess methionine causes the buildup of an amino acid associated with heart disease (homocysteine) and increases inflammation in the liver. No one knows if the same is true in people. Many supplement takers experience digestive disturbances because a high concentration of amino acids causes excess water to flow into the digestive tract from body tissues, resulting in diarrhea and dehydration. Others have ended up in hospital emergency rooms with racing heartbeats and other serious symptoms.[7]

In cases of disease or malnutrition, particularly among the elderly, a registered dietitian may employ a special protein or amino acid supplement to help treat the condition. Not every patient is a candidate for such therapy, however, because supplemental amino acids can stimulate inflammation, which can worsen some diseases. In addition, protein or amino acid supplements can interfere with the actions of certain medications, allowing disease conditions to advance.[8]

A lack of research prevents the DRI committee from setting Tolerable Upper Intake Levels for amino acids.[9] Therefore, no level of amino acid supplementation can be assumed safe. Some people especially likely to be harmed are listed in Table 6–3. The warning is this: much is still unknown, and the taker of amino acid supplements cannot be certain of their safety or effectiveness, despite convincing marketing materials.[10]

Moving Ahead

Even with all that we've learned from science, it is hard to improve on nature. In almost every case, the complex balance of amino acids and other nutrients found together in whole foods is best for nutrition.[11] Keep it safe and simple: select a variety of protein-rich foods that are low in saturated fat each day, and avoid unnecessary protein and amino acid supplements.

Table 6–3

People Most Likely to Be Harmed by Amino Acid Supplements

Growth or altered metabolism makes these people especially likely to be harmed by self-prescribed amino acid supplements:

- All women of child-bearing age
- Pregnant or lactating women
- Infants, children, and adolescents
- Elderly people
- People with inborn errors of metabolism that affect their bodies' handling of amino acids
- Smokers
- People on low-protein diets
- People with chronic or acute mental or physical illnesses

Review Questions*

1. Commercial shakes and energy bars have proven to be the best protein sources to support weight loss efforts. **T F**
2. Lysine does not cure herpes infections. **T F**
3. In high dose, tryptophan can improve nausea and skin disorders. **T F**

**Answers to Consumer's Guide review questions are found in Appendix G.*

body weight, the DRI recommended intake is set at 0.8 gram for each kilogram (or 2.2 pounds) of body weight (see inside front cover). The minimum amount is set at 10 percent of total calories, although some experts are suggesting that more than this minimum may be needed for optimal health.[5] As mentioned in the Think Fitness box earlier, athletes may need slightly more protein—1.2–1.7 grams per kilogram per day—but even this amount is provided by a well-chosen eating pattern with enough energy for an athlete (see Chapter 10).[6]

For infants and growing children, the protein recommendation, like all nutrient recommendations, is higher per unit of body weight. The DRI committee set an upper limit for protein intake of no more than 35 percent of total calories, an amount significantly higher than average intakes. Table 6–4, p. 216, reviews recommendations for protein intake, and the margin provides a method for determining your own protein need. The following factors also modify protein needs.

Do the Math

The DRI recommended intake for protein (adult) = 0.8 g/kg. To figure out your protein need:

1. Find your body weight in pounds.
2. Convert pounds to kilograms (by dividing pounds by 2.2).
3. Multiply kilograms by 0.8 to find total grams of protein recommended.

For example:

Weight = 130 lb
130 lb ÷ 2.2 = 59 kg
59 kg × 0.8 = 47 g

The Body's Health Malnutrition or infection may greatly increase the need for protein while making it hard to eat even normal amounts of food. In malnutrition, secretion of digestive enzymes slows as the tract's lining degenerates, impairing protein digestion and absorption. When infection is present, extra protein is needed for enhanced immune functions.

Other Nutrients and Energy The need for ample energy, carbohydrate, and fat has already been emphasized. To be used efficiently by the cells, protein must also be accompanied by the full array of vitamins and minerals.

Protein Quality The remaining factor, protein quality, helps determine how well a diet supports the growth of children and the health of adults. Protein quality becomes crucial for people in areas where food is scarce, as described in a later section.

DRI protein intake recommendations assume a normal mixed diet; that is, an eating pattern that provides sufficient nutrients and a combination of animal and plant protein. Because not all proteins are used with 100 percent efficiency, the recommendation is generous. Many healthy people can consume less than the recommended amount and still meet their bodies' protein needs. What this means in terms of food selections is presented in this chapter's Food Feature.

Table 6–4

Protein Intake Recommendations for Healthy Adults

DRI Recommended Intake[a]
■ 0.8 g protein/kg body weight/day. ■ Women: 46 g/day; men: 56 g/day. ■ Acceptable intake range: 10 to 30% of calories from protein.
USDA Food Patterns
■ Every day most adults should eat 5- to 6½-oz equivalents of lean meat, poultry without skin, fish, seafood, legumes, soy products, eggs, nuts, or seeds. ■ Every day most adults need 3 c of fat-free or low-fat milk or yogurt, or the equivalent of fat-free cheese or vitamin- and mineral-fortified soy beverage. ■ Eat a variety of foods to provide small amounts of protein from other sources.

[a]*Protein recommendations for infants, children, and pregnant and lactating women are higher; see inside front cover, page B.*

Growing children end each day with more bone, blood, muscle, and skin cells than they had at the beginning of the day.

KEY POINTS

- The protein intake recommendation depends on size and stage of growth.
- The DRI recommended intake for adults is 0.8 gram of protein per kilogram of body weight.
- Factors concerning both the body and food sources modify an individual's protein need.

Nitrogen Balance

Underlying the protein recommendation are **nitrogen balance** studies, which compare nitrogen lost by excretion with nitrogen eaten in food.[7] In healthy adults, nitrogen-in (consumed) must equal nitrogen-out (excreted). Scientists measure the body's daily nitrogen losses in urine, feces, sweat, and skin under controlled conditions and then estimate the amount of protein needed to replace these losses.[‡8]

Under normal circumstances, healthy adults are in nitrogen equilibrium, or zero balance; that is, they have the same amount of total protein in their bodies at all times. When nitrogen-in exceeds nitrogen-out, people are said to be in positive nitrogen balance; somewhere in their bodies more proteins are being built than are being broken down and lost. When nitrogen-in is less than nitrogen-out, people are said to be in negative nitrogen balance; they are losing protein. Figure 6–13 illustrates these different states.

Positive Nitrogen Balance Growing children add new blood, bone, and muscle cells to their bodies every day, so children must have more protein, and therefore more nitrogen, in their bodies at the end of each day than they had at the beginning. A growing child is therefore in positive nitrogen balance. Similarly, when a woman is pregnant, she must be in positive nitrogen balance until after the birth, when she once again reaches equilibrium.

Negative Nitrogen Balance Negative nitrogen balance occurs when muscle or other protein tissue is broken down and lost. Illness or injury triggers the release of powerful messengers that signal the body to break down some of the less vital proteins, such as those of the skin and muscle.[§] This action floods the blood with amino acids and energy needed to fuel the body's defenses and fight the illness. The result is negative nitrogen balance. Astronauts, too, experience negative nitrogen balance. In the stress of space flight, and with no need to support the body's weight against gravity, the astronauts' muscles waste and weaken. To minimize the inevitable loss of muscle tissue, the astronauts must do special exercises in space.

KEY POINT

- Protein recommendations are based on nitrogen balance studies, which compare nitrogen excreted from the body with nitrogen ingested in food.

My Turn **watch it!**

Veggin' Out

Joshua

Have you ever considered not eating meat? Listen to Joshua discuss how becoming vegetarian affected his social life.

[‡]The average protein is 16 percent nitrogen by weight; that is, each 100 grams of protein contain 16 grams of nitrogen. Scientists can estimate the general amount of protein in a sample of food, body tissue, or other material by multiplying the weight of the nitrogen in it by 6.25.
[§]The messengers are cytokines.

Figure 6–13

Nitrogen Balance

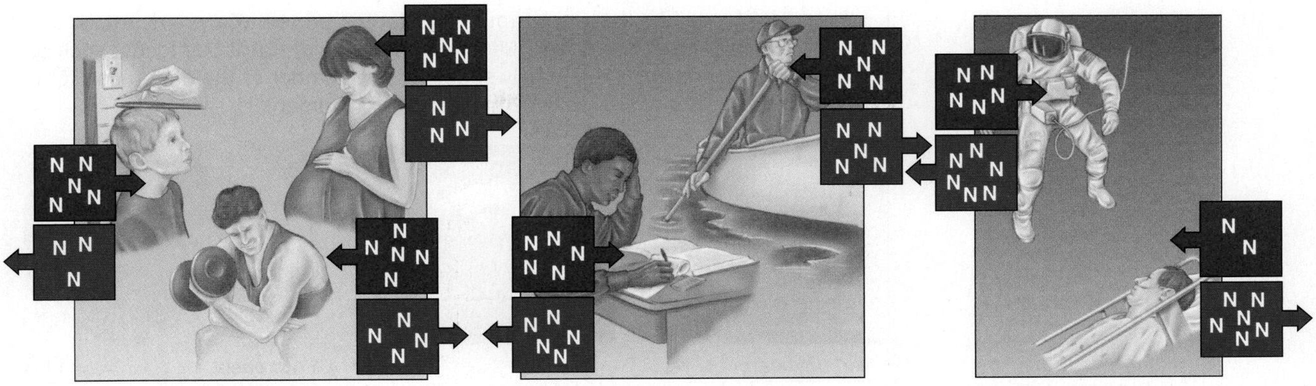

Positive Nitrogen Balance
These people—a growing child, a person building muscle, and a pregnant woman—are all retaining more nitrogen than they are excreting.

Nitrogen Equilibrium
These people—a healthy college student and a young retiree—are in nitrogen equilibrium.

Negative Nitrogen Balance
These people—an astronaut and a surgery patient—are losing more nitrogen than they are taking in.

Protein Quality

Put simply, **high-quality proteins** provide enough of all the essential amino acids needed by the body to create its own working proteins, whereas low-quality proteins don't. Two factors influence a protein's quality: its amino acid composition and its digestibility.

In making their required proteins, the cells need a full array of amino acids. If a nonessential amino acid (that is, one the cell can make) is unavailable from food, the cell synthesizes it and continues attaching amino acids to the protein strands being manufactured. If the diet fails to provide enough of an essential amino acid (one the cell cannot make), the cells begin to adjust their activities. The cells:

- Break down more internal proteins to liberate the needed essential amino acid, and
- Limit their synthesis of proteins to conserve the essential amino acid.[9]

As the deprivation continues, tissues make one adjustment after another in the struggle to survive.

Limiting Amino Acids The measures just described help the cells to channel the available **limiting amino acid** to its wisest use: making new proteins. Even so, the normally fast rate of protein synthesis slows to a crawl as the cells make do with the proteins on hand. When the limiting amino acid once again becomes available in abundance, the cells resume their normal protein-related activities. If the shortage becomes chronic, however, the cells begin to break down their protein-making machinery. Consequently, when protein intakes become adequate again, protein synthesis lags behind until the needed machinery can be rebuilt. Meanwhile, the cells function less and less effectively as their proteins wear out and are only partially replaced.

Thus, a diet that is short in any of the essential amino acids limits protein synthesis. An earlier analogy likened amino acids to letters of the alphabet. To be meaningful, words must contain all the right letters. For example, a print shop that has no letter "N" cannot make personalized stationery for Jana Johnson. No matter how many Js, As, Os, Hs, and Ss are in the printer's possession, the printer cannot use them to replace the missing Ns. Likewise, in building a protein molecule, no amino acid can fill another's spot. If a cell that is building a protein cannot find a needed amino acid, synthesis stops, and the partial protein is released.

Just as each letter of the alphabet is important in forming whole words, each amino acid must be available to build finished proteins.

nitrogen balance the amount of nitrogen consumed compared with the amount excreted in a given time period.

high-quality proteins dietary proteins containing all the essential amino acids in relatively the same amounts that human beings require. They may also contain nonessential amino acids.

limiting amino acid an essential amino acid that is present in dietary protein in an insufficient amount, thereby limiting the body's ability to build protein.

Figure 6–14

Complementary Protein Combinations

Healthful foods like these contribute substantial protein (42 grams total) to this day's meals without meat. Additional servings of nutritious foods, such as milk, bread, and eggs, can easily supply the remainder of the day's need for protein (14 additional grams for men and 4 for women).

3/4 c oatmeal =	5 g
Protein total	5 g

1 c rice =	4 g
1 c beans =	16 g
Protein total	20 g

1½ c pasta =	11 g
1 c vegetables =	2 g
2 tbs Parmesan cheese =	4 g
Protein total	17 g

Figure 6–15

How Complementary Proteins Work Together

Legumes provide plenty of the amino acids isoleucine (Ile) and lysine (Lys), but fall short in methionine (Met) and tryptophan (Trp). Grains have the opposite strengths and weaknesses, making them a perfect match for legumes.

	Ile	Lys	Met	Trp
Legumes	✓	✓		
Grains			✓	✓
Together	✓	✓	✓	✓

Partially completed proteins are not held for completion at a later time when the diet may improve. Rather, they are dismantled and the component amino acids are returned to the circulation to be made available to other cells. If they are not soon inserted into protein, their amine groups are removed and excreted, and the residues are used for other purposes. The need that prompted the call for that particular protein will not be met. Since the other amino acids are wasted, the amine groups are excreted, and the body cannot resynthesize the amino acids later.

Complementary Proteins It follows that if a person does not consume all the essential amino acids in proportion to the body's needs, the body's pools of essential amino acids will dwindle until body organs are compromised. Consuming the essential amino acids presents no problem to people who regularly eat proteins containing ample amounts of all of the essential amino acids, such as those of meat, fish, poultry, cheese, eggs, milk, and most soybean products.

An equally sound choice is to eat a variety of protein foods from plants so that amino acids that are low in some foods will be supplied by the others. The combination of such protein-rich foods yields **complementary proteins** (see Figure 6–14), or proteins containing all the essential amino acids in amounts sufficient to support health.[10] This concept, often employed by vegetarians, is illustrated in Figure 6–15. The figure demonstrates that the amino acids of **legumes** and grains balance each other to provide all of the needed amino acids. The complementary proteins need not be eaten together, so long as the day's meals supply all of them, along with sufficient energy and total protein from a variety of sources.

Protein Digestibility In measuring a protein's quality, digestibility is also important. Simple measures of the total protein in foods are not useful by themselves—even animal hair and hooves would receive a top score by those measures alone. They are made of protein, but not in a form that people can use.

The digestibility of protein varies from food to food and bears profoundly on protein quality. The protein of oats, for example, is less digestible than that of eggs. In general, proteins from animal sources, such as chicken, beef, and pork, are most easily digested and absorbed (over 90 percent). Those from legumes are next (about 80 to 90 percent). Those from grains and other plant foods vary (from 70 to 90 percent). Cooking with moist heat improves protein digestibility, whereas dry heat methods can impair it.

Cooking with moist heat improves protein digestibility, whereas frying makes protein harder to digest.

KEY POINT

- Digestibility of protein varies from food to food, and cooking can improve or impair it.

Perspective on Protein Quality Concern about the quality of individual food proteins is of only theoretical interest in settings where food is abundant. Healthy adults in these places would find it next to impossible *not* to meet their protein needs, even if they were to eat no meat, fish, poultry, eggs, or cheese products at all. They need not pay attention to balancing amino acids, so long as they follow an eating pattern that is varied, nutritious, and adequate in energy and other nutrients—not made up of, say, just cookies, crackers, potato chips, and juices. Protein sufficiency follows effortlessly behind a balanced, nutritious eating pattern.

For people in areas where food sources are less reliable, protein quality can make the difference between health and disease. When food energy intake is limited, where malnutrition is widespread, or when the selection of foods available is severely limited (where a single low-protein food, such as **fufu** made from cassava root,** provides 90 percent of the calories), the primary food source of protein must be checked because its quality is crucial.

KEY POINTS

- A protein's amino acid assortment greatly influences its usefulness to the body.
- Proteins lacking essential amino acids can be used only if those amino acids are present from other sources.

Protein Deficiency and Excess

LO 6.5 Discuss potential physical problems from an eating plan that is too low or too high in protein.

When diets lack sufficient protein from food or sufficient amounts of any of the essential amino acids, symptoms of malnutrition become evident. In contrast, the health effects of protein excess are less well established, but high-protein diets, and particularly high-meat diets, have been implicated in several chronic diseases (see the Controversy section). Evidence is currently insufficient to establish an Upper Level (UL) for protein, but both deficiency and excess are of concern.

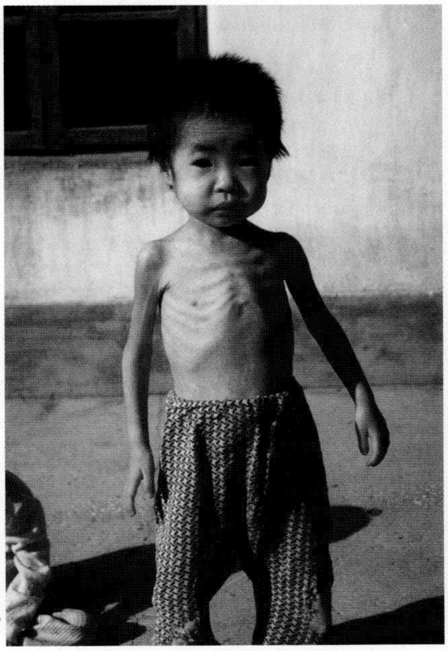

Margaret Prout/Associated Press

Malnutrition: too little food causes deficiencies of protein and many other nutrients. See Chapter 15.

What Happens When People Consume Too Little Protein?

In protein deficiency, when the diet supplies too little protein or lacks a specific essential amino acid relative to the others (a limiting amino acid), the body slows its synthesis of proteins while increasing its breakdown of body tissue protein to liberate the amino acids it needs to build other proteins of critical importance. Without these critical proteins to perform their roles, many of the body's life-sustaining activities would come to a halt. The most recognizable consequences of protein deficiency include slow growth in children, impaired brain and kidney functions, weakened immune defenses, and impaired nutrient absorption from the digestive tract.

In clinical settings, the term *protein-energy malnutrition* is sometimes used to describe the condition that develops when the diet delivers too little protein, too little energy, or both. However, it has become clear that such malnutrition generally reflects insufficient food intake, with not only too little protein and energy but too few vitamins, too few minerals—and in fact, too little of most of the nutrients needed for health and growth. For this reason, the severe malnutrition of starvation and its clinical manifestations are a focus of Chapter 15's discussion of world hunger.

Is It Possible to Consume Too Much Protein?

Overconsumption of protein-rich foods offers no benefits and may pose health risks, particularly for people with weakened kidneys.[11] This section explores current protein intakes and considers the potential for harm from taking in excesses.

complementary proteins two or more proteins whose amino acid assortments complement each other in such a way that the essential amino acids missing from one are supplied by the other.

legumes (leg-GOOMS, LEG-yooms) plants of the bean, pea, and lentil family that have roots with nodules containing special bacteria. These bacteria can trap nitrogen from the air in the soil and make it into compounds that become part of the plant's seeds. The seeds are rich in protein compared with those of most other plant foods. Also defined in Chapter 1.

fufu a low-protein staple food that provides abundant starch energy to many of the world's people; fufu is made by pounding or grinding root vegetables or refined grains and cooking them to a smooth semisolid consistency.

**Cassava is also called *manioc* or *yucca*.

© iStockphoto.com/only_fabrizio

How Much Protein Do People Take In? Most people suspect that Americans eat far too much protein. In fact, the median protein intake for U.S. men is about 16 percent of total calories, with women consuming slightly less.[12] These amounts are well within the DRI suggested range of between 10 and 35 percent of calories.[13] Stated another way, the DRI range for protein intake in a 2,000-calorie diet is 50 to 175 grams; the average U.S. daily intake of protein amounts to about 78 grams.[14]

Protein and Weight Loss Dieting Some popular weight-loss diet advice suggests 65 percent or more of calories from protein as a way to lose weight. True, meeting protein recommendations during weight loss is critical for preserving the body's working lean tissues, such as liver and muscles. Also, much evidence, but not all, suggests a role for protein in helping to control the appetite.[15] However, as Chapter 9 explains, it is calorie reduction alone and not the proportion of energy nutrients in the diet that brings about long-term weight loss. Let it suffice here to say that the best weight-loss plan is reduced in calories from all sources while providing a health-promoting balance of energy nutrients from a wide variety of whole foods.

Did You Know?

An ounce of most protein foods delivers about 7 grams of protein.

Protein Sources in Heart Disease Protein itself is not known to contribute to heart disease and mortality, but some of its food sources may do so.[16] Selecting too many animal-derived protein-rich foods, such as fatty red meats, processed meats, and fat-containing milk products, adds a burden of fat calories and saturated fat to the diet and crowds out fruits, vegetables, legumes, and whole grains.[17] Consequently, it is not surprising that people who habitually take in a great deal of high-fat meat, and particularly processed meat such as lunchmeats and hot dogs, have higher body weights and a greater risk of obesity, heart disease, and diabetes than those who take in less.[18] The Controversy section explores how substituting vegetable protein for at least some of the animal protein in the diet may improve risk factors for heart disease and mortality.[19]

Kidney Disease Animals fed experimentally on high-protein diets often develop enlarged kidneys or livers. In human beings, a high-protein diet increases the kidneys' workload, but this alone does not appear to damage healthy kidneys nor does it cause kidney disease.[20] In people with kidney stones or other kidney diseases, however, a high-protein diet may speed the kidneys' decline. One of the most effective treatments for people with established kidney problems is to limit protein intakes to improve the symptoms of their disease. Taken to an extreme, very-low-protein diets, even when supplemented with essential amino acids, do not delay kidney deterioration further and may increase fatality in such people.[21]

Did You Know?

Red meats include beef, pork, mutton, lamb, veal, goat, venison, and others. White meats include poultry, fish, and other seafood.

Adult Bone Loss When human subjects are given increasing doses of purified protein, they spill larger and larger amounts of calcium from the body into the urine. This fact has raised concerns that high-protein diets may cause or worsen adult bone loss, thereby worsening the crippling bone loss disease, **osteoporosis**. However, research now suggests that, unlike the effect of purified protein, high-protein intakes from whole food such as legumes, meats, or milk may not increase calcium loss from the bones.[22] In fact, the opposite situation—inadequate protein intake—is known to weaken the bones and make fractures likely.[23]

The two groups most likely to have osteoporosis, elderly women and adolescents with anorexia nervosa, are the same ones who typically eat diets too low in protein. For these people, increasing intakes of protein-rich foods may help protect the bones.[24]

Cancer The effects of protein on cancer causation cannot be easily separated from the various foods that provide protein to the diet. When researchers consider all sources of animal protein in the diet, for example, no association with cancers of the colon and rectum are evident.[25] However, when they focus on red meats, such as beef, lamb, and pork, and processed meats, such as frankfurters, sausages, ham, lunchmeats, and bacon, they often report a moderately increased risk for cancers, particularly of the colon and rectum.[26] Limiting processed meats may also be wise for two other reasons: these foods contain large amounts of solid fat and salt, factors associated with heart disease and hypertension. Chapter 11 delves into the links between diet and diseases.

osteoporosis (OSS-tee-oh-pore-OH-sis) a disease of older persons characterized by porous and fragile bones that easily break, leading to pain, infirmity, and death. Also defined in Chapter 8.

gluten (GLOO-ten) a type of protein in certain grain foods that is toxic to the person with celiac disease.

celiac (SEE-lee-ack) **disease** a disorder characterized by intestinal inflammation on exposure to the dietary protein gluten; also called *gluten-sensitive enteropathy* or *celiac sprue*.

Is a Low-Gluten Diet Best for Health? **Gluten**, a protein found in grain foods, has recently gained attention among diet sellers. Gluten is best known for providing a pleasing stretchy texture to yeast breads; it also provides bulk and texture to foods made from wheat, barley, rye, and related grains. In people with a gluten allergy or a condition known as **celiac disease**, gluten consumption causes a range of digestive symptoms and decreases nutrient absorption.[27] Such people often battle unwanted weight loss and severe nutrient deficiencies; eating a gluten-free diet often resolves the worst of these problems.[28] Curiously, some people without celiac disease report relief from mild digestive symptoms when they eat a gluten-free diet, but whether the diet, the placebo effect, or something else is at work is unknown.[29]

Recently, Internet and other media sources have blamed gluten for causing headaches, insomnia, weight gain, and even cancer and Alzheimer's disease, but no evidence supports these accusations. And a low-gluten diet has no special power to spur weight loss in healthy people, despite claims that it does so.[30]

Most people with celiac disease are never diagnosed, and without treatment, they continue to suffer unnecessarily. Ironically, most people following a gluten-free diet in the United States do not have celiac disease or gluten allergy—they buy expensive specialty foods and restrict their intakes of nutritious grain foods needlessly.[31]

A gluten-free diet can bring relief to people with celiac disease.

KEY POINTS

- Most U.S. protein intakes fall within the DRI recommended protein intake range of 10 to 35 percent of calories.
- No Tolerable Upper Intake Level exists for protein, but health risks may follow the overconsumption of protein-rich foods.
- Gluten-free diets often relieve symptoms of celiac disease or gluten allergy, but no evidence supports claims that they cure other ills.

try it! Food Feature

Getting Enough but Not Too Much Protein

LO 6.6 Identify protein-rich foods, and list some extra advantages associated with legumes.

Most foods contribute at least some protein to the diet. The most nutrient-dense selections among them are generally best for nutrition.

Protein-Rich Foods

Foods in the meat, poultry, fish, dry peas and beans, eggs, and nuts group and in the milk, yogurt, and cheese group contribute an abundance of high-quality protein. Two others, the vegetable group and the grains group, contribute smaller amounts of protein, but they can add up to significant quantities. What about the fruit group? Don't rely on fruit for protein; fruit contains only small amounts. Figure 6–16 (p. 222) demonstrates that a wide variety of foods contribute protein to the diet. Figure 6–17 (p. 223) lists the top protein contributors in the U.S. diet.

Protein is critical in nutrition, but too many protein-rich foods can displace other important foods from the diet. Foods richest in protein carry with them a characteristic array of vitamins and minerals, including vitamin B_{12} and iron, but they lack others—vitamin C and folate, for example. In addition, many protein-rich foods such as meat are high in calories, and to overconsume them is to invite obesity.

In Chapter 2, Figures 2–14 and 2–15 (pages 59–60) demonstrated this effect in two diets. What the figure did *not* show was that Monday's meals provided 87 grams of protein from a small amount of meat in harmony with all the other foods needed for the day *and* fell within the calorie budget. The more typical Tuesday's meals provided 106 grams of protein but fell short of meeting many other needs and exceeded the calorie allowance. Meals for both days provided much more than enough protein. Moral: protein-rich meats are not always the best, or even the most desirable, sources of protein in a balanced nutritious diet.

Because American consumption of protein is ample, you can plan meatless or reduced-meat meals with pleasure. Of the many interesting, protein-rich meat equivalents available, one has already been mentioned: the legumes.

The Advantages of Legumes

The protein of some legumes, and soybeans in particular, is of a quality almost comparable to that of meat, an unusual trait in a fiber-rich vegetable. Figure 6–18, p. 223, shows a legume

Figure 6–16

Finding the Protein in Foods[a]

Fruits

Food		Protein g	%DV[b]
Avocado	1/2 c	2	4
Cantaloupe	1/2 c	1	2
Orange sections	1/2 c	1	2
Strawberries	1/2 c	1	2

Vegetables

Food		Protein g	%DV[b]
Corn	1/2 c	3	6
Broccoli	1/2 c	2	4
Collard greens	1/2 c	2	4
Sweet potato	1/2 c	2	4
Baked potato	1/2 c	1	2
Bean sprouts	1/2 c	1	2
Winter squash	1/2 c	1	2

Grains

Food		Protein g	%DV[b]
Pancakes	2 sm	6	12
Bagel	1/2	4	8
Brown rice	1/2 c	3	6
Whole grain bread	1 sl	3	6
Noodles, pasta	1/2 c	3	6
Oatmeal	1/2 c	3	6
Barley	1/2 c	2	4
Cereal flakes	1 oz	2	4

Protein Foods

Food		Protein g	%DV[b]
Roast beef	2 oz	19	33
Turkey leg	2 oz	16	32
Chicken breast	2 oz	15	30
Pork meat	2 oz	15	30
Tuna	2 oz	14	28
Lentils, beans, peas	1/2 c	9	18
Peanut butter	2 tbs	8	16
Almonds	1/4 c	8	16
Hot dog	1 reg	7	14
Lunch meat	2 oz	6	12
Egg	1 lg	6	12
Cashew nuts	1/4 c	5	10

Milk and Milk Products

Food		Protein g	%DV[b]
Cheese, processed	2 oz	13	26
Milk, yogurt	1 c	10	20
Pudding	1 c	5	10

Oils, Solid Fats, and Added Sugars

Not a significant source

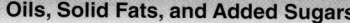

[a]All foods are prepared and ready to eat.

[b]The Daily Value (DV) for protein is 50 g, based on an energy intake of 2,000 cal/day.

Figure 6–17

Top Contributors of Protein to the U.S. Diet[a]

In recent decades, poultry (largely chicken) intakes have been rising steadily while beef intakes have been declining.

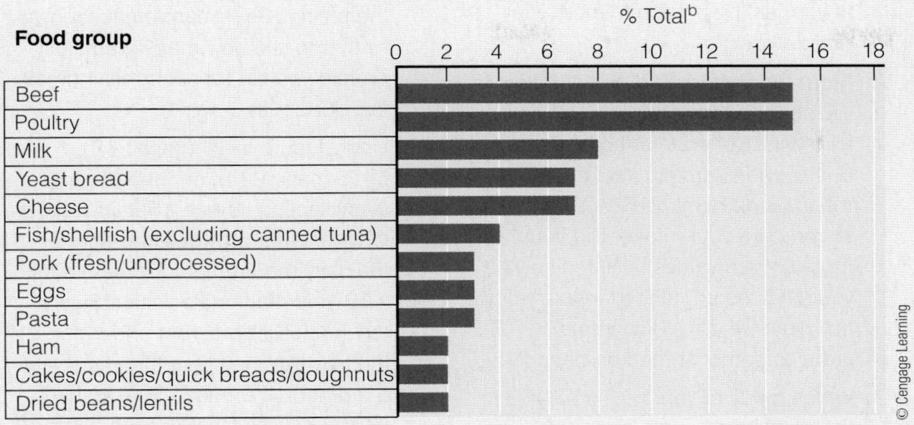

[a] *These foods supply about 70 percent of the protein in the U.S. diet. The remainder comes from foods contributing less than 2 percent of the total.*
[b] *Rounded values.*

Figure 6–18

A Legume

The legumes include such plants as the kidney bean, soybean, green pea, lentil, black-eyed pea, and lima bean. Bacteria in the root nodules can "fix" nitrogen from the air, contributing it to the beans. Ultimately, thanks to these bacteria, the plant accumulates more nitrogen than it can get from the soil and also contributes more nitrogen to the soil than it takes out. The legumes are so efficient at trapping nitrogen that farmers often grow them in rotation with other crops to fertilize fields. Legumes are included with meat in the Protein Foods group in Figure 6–16.

plant's special root system that enables it to make abundant protein by obtaining nitrogen from the soil. Legumes are also excellent sources of many B vitamins, iron, and other minerals, making them exceptionally nutritious foods. On average, a cup of cooked legumes contains about 30 percent of the Daily Values for both protein and iron.[††] Like meats, though, legumes do not offer every nutrient, and they do not make a complete meal by themselves. They contain no vitamin A, vitamin C, or vitamin B_{12}, and their balance of amino acids can be much improved by using grains and other vegetables with them.

Soybeans are versatile legumes, and many nutritious products are made from them. Heavy use of soy products in place of meat, however, inhibits iron absorption. The effect can be alleviated by using small amounts of meat and/or foods rich in vitamin C in the same meal with soy products.

Vegetarians and others sometimes use convenience foods made from **textured vegetable protein** (soy protein) formulated to look and taste like hamburgers or breakfast sausages. Many of these are intended to match the known nutrient contents of animal protein foods, but they often fall short.[‡‡] A wise vegetarian uses such foods in combination with whole foods to supply an entire array of needed nutrients. The nutrients of soybeans are also available as bean curd, or **tofu**, a staple used in many Asian dishes. Thanks to the use of calcium salts when some tofu is made, it can be high in calcium. Check the Nutrition Facts panel on the label.

The Food Features presented so far show that the recommendations for the three energy-yielding nutrients occur in balance with each other. The diets of most people, however, supply too little fiber, too much fat, too many calories, and abundant protein. To bring their diets into line with recommendations, then, requires changing the bulk of intake from calorie-rich fried foods, fatty meats, and sweet treats to lower-calorie complex carbohydrates and fiber-rich choices, such as whole grains, legumes, and vegetables. With these changes, protein totals remain adequate while other constituents automatically fall into place in a healthier eating pattern.

textured vegetable protein processed soybean protein used in products formulated to look and taste like meat, fish, or poultry.

tofu (TOE-foo) a curd made from soybeans that is rich in protein, often enriched with calcium, and variable in fat content; used in many Asian and vegetarian dishes in place of meat. Also defined in Controversy 2.

[††] *Data from the* Food Processor Plus, *ESHA research, version 7.11.*

[‡‡] *In Canada, regulations govern the nutrient contents of such products.*

track it! Diet Analysis PLUS Concepts in Action

Analyze Your Protein Intake

The purpose of this exercise is to make you aware of the effects of choosing protein-rich foods in balancing the three energy-yielding nutrients while planning a nutritious diet.

1. Do you take in adequate protein? From the DA+ Home page, select Reports, then Macronutrient Ranges, choose Day Three, include the entire day's meals, and generate a report. Is your intake within the recommended 10 to 35 percent of total energy intake range, as recommended by the DRI? What percent of your caloric intake consists of protein? If your protein intake is higher than 35 percent, what foods could you choose less often to bring you within range? If your intake is lower than 10 percent of calories, what foods would you add to your eating pattern to meet your protein need?
2. From the Reports tab, select Intake vs. Goals report; choose all Day Two, all meals. Generate a report. Table 6–4 provides protein intake recommendations (page 216) against which to compare your intake. Did your protein gram values fall into line with your DRI recommended intake (multiply your weight in kilograms by 0.8 gram as demonstrated on page 215)?
3. Which foods in your meals provide the greatest amounts of protein? From the Reports tab select Source Analysis, choose all meals for Day Two, select protein in the drop-down box, then generate a report. This report will help you determine your protein sources.
4. Using the same report and date from the previous question, break it down further to see how many grams of protein you eat for each meal: breakfast, lunch, and dinner.
5. Using Figure 6–14 (page 218), Figure 6–16 (page 222), and the Controversy section, create a vegan vegetarian meal that provides one-third of the daily requirement for protein for a 19-year-old female vegan. Select the Track Diet program and input the foods in the meal, then generate a report on the Intake Spreadsheet. How successful were you in achieving protein adequacy? If the meal fell short of the goal, what vegan foods can you change or add to the meal to more closely match this person's protein need?

what did you decide?

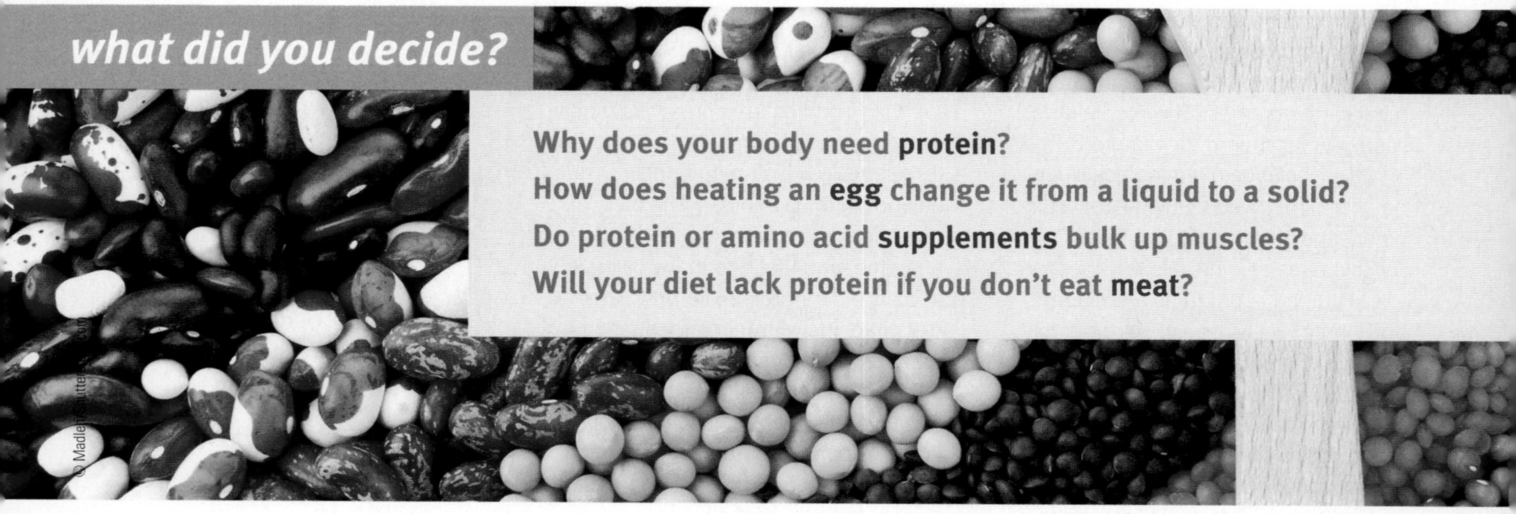

Why does your body need protein?
How does heating an egg change it from a liquid to a solid?
Do protein or amino acid supplements bulk up muscles?
Will your diet lack protein if you don't eat meat?

Self Check

1. (LO 6.1) The basic building blocks for protein are ________.
   a. glucose units
   b. amino acids
   c. side chains
   d. saturated bonds
2. (LO 6.1) The roles of protein in the body include all but
   a. blood clot formation
   b. tissue repair
   c. gas exchange
   d. immunity
3. (LO 6.1) Amino acids are linked together to form a protein strand by
   a. peptide bonds
   b. essential amino acid bonds
   c. side chain attraction
   d. super glue
4. (LO 6.1) Some segments of a protein strand coil, somewhat like a metal spring, because
   a. amino acids at different places along the strand are chemically attracted to each other.
   b. the protein strand has been denatured by acid.
   c. the protein strand is missing one or more essential amino acids.
   d. a coil structure allows access by enzymes for digestion.
5. (LO 6.2) Protein digestion begins in the ________.
   a. mouth
   b. stomach
   c. small intestine
   d. large intestine
6. (LO 6.2) In the intestine, amino acids of the same general type compete for the same absorption sites, so a large dose of any one amino acid can limit absorption of another.
   T F
7. (LO 6.3) Under certain circumstances, amino acids can be converted to glucose and so serve the energy needs of the brain.
   T F
8. (LO 6.3) To prevent wasting of dietary protein, which of the following conditions must be met?
   a. Dietary protein must be adequate in quantity.
   b. Dietary protein must supply all essential amino acids in the proper amounts.
   c. The diet must supply enough carbohydrate and calories.
   d. All of the above.
9. (LO 6.4) For healthy adults, the DRI recommended intake for protein has been set at ________.
   a. 0.8 gram per kilogram of body weight
   b. 2.2 pounds per kilogram of body weight
   c. 12 to 15 percent of total calories
   d. 100 grams per day
10. (LO 6.4) An example of a person in positive nitrogen balance is a pregnant woman.
    T F
11. (LO 6.4) Partially completed proteins are not held for completion at a later time when the diet may improve.
    T F
12. (LO 6.4) The following are complementary proteins:
    a. pasta and tomato sauce
    b. pot roast and carrots
    c. rice and French fries
    d. peanut butter on whole-wheat bread
13. (LO 6.5) Insufficient dietary protein can have severe consequences, but excess dietary protein cannot cause harm.
    T F
14. (LO 6.5) Insufficient dietary protein can cause
    a. slowed protein synthesis
    b. hepatitis
    c. accelerated growth in children
    d. all of the above
15. (LO 6.6) Two tablespoons of peanut butter offer about the same amount of protein as a hot dog.
    T F
16. (LO 6.6) Legumes are a particularly nutritious choice among protein-rich foods because they also provide
    a. vitamin C and vitamin E
    b. fiber
    c. B vitamins, iron, and other minerals
    d. b and c
17. (LO 6.7) Blood LDL values of people eating typical, meat-rich Western diets are generally higher than LDL values of vegetarians.
    T F
18. (LO 6.7) A vegetarian diet planner should be sure to include sources of
    a. carbohydrate
    b. vitamin C
    c. vitamin B_{12}
    d. vitamin E
19. (LO 6.7) Fried banana or vegetable snack chips make a healthy everyday snack choice for vegetarians.
    T F

Answers to these Self Check questions are in Appendix G.

Vegetarian and Meat-Containing Diets: What Are the Benefits and Pitfalls?

LO 6.7 Summarize the health advantages and nutrition red flags of vegetarian diets, and develop a lacto-ovo vegetarian eating pattern that meets all nutrient requirements for a given individual.

In affluent countries, where heart disease and cancer claim many lives, people who eat well-planned **vegetarian** diets have lower risks of chronic diseases, and lower risk of dying from all causes, than people whose diets center on meat.*[1] Should everyone consider using a vegetarian eating pattern, then? If so, is it enough to simply omit meat, or is more demanded of the vegetarian diet planner? What positive contributions do animal products make to the diet that vegetarians must make an effort to replace? This Controversy looks first at the positive health aspects of vegetarian diets and then at the positive aspects of meat eaters' diets. It ends with some practical advice for the vegetarian diet planner.

A vegetarian lifestyle may be immediately associated with a particular culture, religion, political, or other belief system, but there are many reasons why people might choose it, as Table C6–1 makes clear. Vegetarians are not categorized by motivation but by the foods they choose

© Ildi Papp/Shutterstock.com

Can an eating pattern without animal products supply the needed nutrients?

*Reference notes are found in Appendix F.

to eat (see Table C6–2). Such distinctions among vegetarian diets are useful academically; they do not represent uncrossable lines. Some people use meat or broth as a condiment or seasoning for vegetable or grain dishes. Some people eat meat only once a week and use plant protein foods the rest of the time. Others rely mostly on milk products and eggs for protein but will eat fish, too, and so forth. To force people into the categories of "vegetarians" and "meat eaters" leaves out all these in-between eating styles (aptly named *flexitarians* by the press) that have much to recommend them.

Positive Health Aspects of Vegetarian Diets

Today, nutrition authorities state with confidence that a well-chosen vegetarian diet can meet nutrient needs while supporting health superbly.[2] Although much evidence links vegetarian eating patterns with reduced incidence of chronic diseases, particularly heart and artery diseases, such evidence is not easily obtained. It would be easy if vegetarians differed from others only in the absence of meat, but they often have *increased* intakes of whole grains, legumes, nuts, fruits, and vegetables as well. Such eating patterns are rich contributors of carbohydrates, fiber, vitamins, minerals, and phytochemicals that also correlate with low disease risks.

Also, many vegetarians avoid tobacco, use alcohol in moderation if at all, and are more physically active than other adults. When researchers take into account all of the effects of a total health-conscious lifestyle on disease development, the evidence still often weighs in favor of vegetarian eating patterns, as the next sections make clear.

Defense against Obesity

Among both men and women and across many ethnic groups, vegetarians more often maintain a healthier body weight than nonvegetarians.[3] The converse is also true: meat consumption correlates with increased energy intake and increased obesity.[4] The reason for this is not clear but may reflect a lifestyle adopted by many vegetarians that makes health a high priority.

Defense against Heart and Artery Disease

Plant-based eating patterns are associated with low rates of heart and artery disease. Vegetarians die less often from heart disease and related illnesses than do meat-eating people, although not all indicators of heart health are consistently improved in vegetarians.[5]

Vegetarian diets vary, and food choices affect heart disease risk. When vegetarians choose the unsaturated fats of soybeans, seeds, avocados, nuts, olives, and vegetable oils and shun saturated fats from cheese, sour cream, butter, shortening, and other sources, their risks of heart disease are reduced.[6] If the diet also contains nuts and soluble fiber, as most vegetarian diets do, then LDL cholesterol typically falls, and heart

Table C6–1

Reasons for Choosing Eating Styles

Why Some People Are Vegetarians	Why Some People Eat Meat
▪ *Health concerns.* Vegetarian diets are often high in whole grains, fruits, vegetables, and legumes and often low in saturated fats, diet characteristics associated with good health. ▪ *Moral objections.* Some believe that animals should not be killed for food; others object to use of any animal products, such as milk, cheese, eggs, or honey, or use of items made from leather, wool, feathers, or silk. ▪ *Humane treatment of animals.* Many people object to inhumane treatment of livestock and food producing animals. ▪ *Environmental concerns.* Producing meat protein requires a much greater input of resources than does an equal amount of vegetable protein. ▪ *Weight-control efforts.* Some people mistakenly believe that simply eliminating meat will produce weight loss (it doesn't if high-calorie vegetarian foods and treats are consumed in excess of the daily energy need). ▪ *Cover up.* Some adolescents may hide an eating disorder under the guise of being "vegetarian" (Chapter 9 takes up the issues of weight-loss dieting and eating disorders).	▪ *Convenience.* Some people find that a hamburger or chicken salad sandwich makes a convenient lunch. ▪ *Nutrients.* Some people rely on animal products for the energy and key nutrients they supply. ▪ *Taste.* Others enjoy the taste of roasted chicken, barbecued ribs, or a grilled steak. ▪ *Familiarity.* Some people wouldn't know what to eat without meat; they are accustomed to seeing it on the plate. ▪ *Weight-control efforts.* Some people mistakenly believe that eating meat instead of whole grains, fruits and vegetables, and legumes speeds weight loss (it doesn't).

Sources: A. M. Bardone-Cone, The inter-relationships between vegetarianism and eating disorders among females, Journal of the Academy of Nutrition and Dietetics *112 (2012): 12-47-1252; U.S. Department of Agriculture and U.S. Department of Health and Human Services,* Dietary Guidelines for Americans 2010, *available at www .dietaryguidelines.gov; H. J. Marlow and coauthors, Diet and the environment: Does what you eat matter?* American Journal of Clinical Nutrition 89 *(2009): 1699S–1703S; B. M. Popkin, Reducing meat consumption has multiple benefits for the world's health,* Archives of Internal Medicine 169 *(2009): 543–545.*

benefits compound.[7] The lowest blood LDL cholesterol values are found in **vegans**; those of **lacto-ovo vegetarians** are somewhat higher; and LDL values of people eating typical, meat-rich Western diets are the highest of all.

When soy protein replaces saturated fat–rich animal protein in the diet, LDL cholesterol often declines significantly.[8] To achieve this benefit requires consuming about 25 grams of soy protein, or about two and a half average soy "burgers" or a cup and a half of tofu each day. The LDL-lowering effects of soy may arise from the abundant soy protein, polyunsaturated fatty acids, fibers, phytochemicals, vitamins, and minerals in these foods; alternatively, soy foods may simply displace saturated fat sources, such as meats and cheeses, from the diet.[9]

Table C6–2

Terms Used to Describe Vegetarian Diets

Some of the terms below are in common usage, but others are useful only to researchers.

- **fruitarian** includes only raw or dried fruits, seeds, and nuts in the diet.
- **lacto-ovo vegetarian** includes dairy products, eggs, vegetables, grains, legumes, fruits, and nuts; excludes flesh and seafood.
- **lacto-vegetarian** includes dairy products, vegetables, grains, legumes, fruits, and nuts; excludes flesh, seafood, and eggs.
- **macrobiotic diet** a vegan diet composed mostly of whole grains, beans, and certain vegetables; taken to extremes, macrobiotic diets can compromise nutrient status.
- **ovo-vegetarian** includes eggs, vegetables, grains, legumes, fruits, and nuts; excludes flesh, seafood, and milk products.
- **partial vegetarian** a term sometimes used to mean an eating style that includes seafood, poultry, eggs, dairy products, vegetables, grains, legumes, fruits, and nuts; excludes or strictly limits certain meats, such as red meats. Also called *semi-vegetarian.*
- **vegan** includes only food from plant sources: vegetables, grains, legumes, fruits, seeds, and nuts; also called strict vegetarian.
- **vegetarian** includes plant-based foods and eliminates some or all animal-derived foods.

When consumed in sufficient quantity, soy foods, such as this roasted tofu, may improve the health of the heart.

Defense against High Blood Pressure

Vegetarians tend to have lower blood pressure and lower rates of hypertension than average.[10] Often, vegetarians maintain a healthy body weight, and appropriate body weight helps to maintain healthy blood pressure; so does an eating pattern that is high in fiber, fruits and vegetables, the mineral potassium, and soy protein. The mineral sodium promotes high blood pressure, but vegetarian diets are not always low-sodium diets. Other lifestyle factors such as not smoking, moderation of alcohol intake, and being physically active all work together to keep blood pressure normal.

Defense against Cancer

Colon and rectal cancers occur less frequently among people who eat mostly plant-based diets than among those who regularly consume red meat and processed meat.[11] The amount of red or processed meat associated with this effect is surprisingly small and has shown up even with relatively low levels of consumption.[12]

Is the case closed on meat's culpability in colon and rectal cancers, then? In a study of over 60,000 people in the United Kingdom, people who ate fish but not red meats had the lowest overall cancer rates—a finding that agrees with previous studies.[13] However, this evidence is insufficient to allow firm conclusions. Other large studies report no difference in colon cancer rates between vegetarians and meat eaters.[14] The conflict may arise from wide variations in eating patterns within the groups—some meat eaters also love legumes and vegetables and eat them often, while some vegetarians may prefer fried foods, cheeses, and sweets. Other factors, such as body fatness, also affect study results.

Still, evidence is mounting that high intakes of red and processed meat correlate with elevated risks of dying from some types of cancer, along with heart disease and other causes.[15] Such risks appear to increase with:

- Low vegetable and fruit intakes (in any kind of diet).
- Alcohol (moderate-to-high intakes).[16]
- Increased body and abdominal fatness (overweight and obesity).

More details about diet and cancer appear in **Chapter 11.**

Other Health Benefits

In addition to obesity, heart disease, high blood pressure, and cancer, vegetarian eating patterns may help prevent cataracts, diabetes, diverticular disease, gallstones, high blood pressure, and osteoporosis.[17] However, these effects may arise more from what vegetarians include in the diet—abundant fruit, legumes, vegetables, and whole grains—than from omission of meat.

Positive Health Aspects of the Meat Eater's Diet

Unlike vegetarians, for whom suitable replacements exist for meat and milk in a healthy eating pattern, meat eaters find no adequate substitutes for whole grains, fruits, and vegetables (not even antioxidant supplement pills, as the next chapter points out). A meat eater who excludes or minimizes intakes of these foods imperils health. The following sections consider a *balanced, adequate* diet, in which lean meat, poultry, seafood, eggs, and milk play a part.

Both meat eaters and lacto-ovo vegetarians can generally rely on their diets during critical times of life. In contrast, a vegan eating pattern poses challenges. Protein is critical for building new tissues during growth, for fighting illnesses, for building bone during youth, and for maintaining bone and muscle tissue integrity in old age.[18] While protein from plant sources can meet most people's needs, very young children and very elderly people with small appetites may not consume enough legumes, whole grains, or nuts to supply the protein they need.

The chapter made clear that animal protein from meat, fish, milk, and eggs is the clear winner in tests of digestibility and availability to the body, with soy protein a close second. Also, animal-

© iStockphoto.com/Carlos Gawronski

This 5-ounce steak provides almost all of the meat recommended for an entire day's intake in a 2,000-calorie diet.

derived foods provide abundant iron, zinc, vitamin D, and vitamin B_{12} needed by everyone but particularly by pregnant women, infants, children, adolescents, and the elderly (details about these needs appear in later chapters).

In Pregnancy and Infancy

Women who eat seafood, eggs, or milk products can be sure of receiving enough energy, vitamin B_{12}, vitamin D, calcium, iron, and zinc, as well as protein, to support pregnancy and breastfeeding. A woman following a well-planned lacto-ovo vegetarian eating pattern can also relax in the knowledge that she is superbly supplied with energy and all necessary nutrients. And if she also habitually eats abundant vegetables and fruits, she can relax further, knowing that her diet supplies the vitamin folate in amounts needed to protect her developing fetus from certain birth defects. A vegan woman who doesn't meet her nutrient needs, however, may enter pregnancy too thin and with scant nutrient stores to draw on as the nutrient demands of the fetus grow larger.

Of particular interest is vitamin B_{12}, a vitamin abundant in foods of animal origin but absent from vegetables. Obtaining enough vitamin B_{12} poses a challenge to vegans of all ages, who often test low in the vitamin.[19] For pregnant and lactating women, obtaining vitamin B_{12} is critical to prevent serious deficiency-related disorders in infants who do not receive sufficient vitamin

B_{12}.[20] Adults with low vitamin B_{12} may experience vague symptoms—fatigue, indigestion problems, numbness of fingers, and frequent infections.

In Childhood

Children who eat eggs, milk, and fish receive abundant protein, iron, zinc, and vitamin B_{12}; such foods are reliable, convenient sources of nutrients needed for growth.[21] Likewise, children eating well-planned lacto-ovo vegetarian diets also receive adequate nutrients and grow as well as their meat-eating peers.[22] Child-sized servings of vegan foods, however, can fail to provide sufficient energy or several key nutrients needed for normal growth. A child's small stomach can hold only so much food, and the vegan child may feel full before eating enough to meet his or her nutrient needs. Evidence is lacking to evaluate vegan diets for children.[23]

Small, frequent meals of fortified breads, cereals, or pastas with legumes, nuts, nut butters, and sources of unsaturated fats can help to meet protein and energy needs in a smaller volume at each sitting.[24] Because vegan children derive protein only from plant foods, their daily protein requirement may be somewhat higher than the DRI indicates for the general meat-eating population.[25] Other nutrients of concern for vegan children include vitamin B_{12}, vitamin D, calcium, iron, and zinc.

In Adolescence

The healthiest vegetarian adolescents choose balanced diets that are heavy in fruits and vegetables but light on the sweets, fast foods, and salty snacks that tempt the teenage palate. These healthy vegetarian teens often meet national dietary objectives, such as those of *Healthy People 2010* (listed in Chapter 1)—a rare accomplishment in the United States.

Other teens, however, adopt poorly planned vegetarian eating patterns that provide too little energy, protein, vitamin B_{12}, calcium, zinc, and vitamin D. Omissions of protein, calcium, and vitamin D lead to weak bone development at precisely the time when bones must develop strength to protect bone health through later life. Also, teens who fail to obtain vitamin B_{12} risk serious nerve damage from a deficiency. If a vegetarian child or teen refuses sound dietary advice, a registered dietitian can help to identify problems, dispense appropriate guidance, and put unwarranted parental worries to rest.

In Aging and in Illness

For elderly people with diminished appetites or impaired digestion or for people recovering from illnesses, soft or ground meats can provide a well-liked, well-tolerated concentrated source of nutrients. People battling life-threatening diseases may encounter testimonial stories of cures attributed to restrictive eating plans, such as **macrobiotic diets**, but these diets often severely limit food selections and can fail to deliver the energy and nutrients needed for recovery.

Planning a Vegetarian Diet

Eating a nutritious vegetarian diet requires more than just omitting certain foods and food groups—any eating pattern that omits key foods omits essential nutrients. Grains, fruits, and vegetables are naturally abundant in the vegetarian's diet and provide adequate amounts of the nutrients of plant foods: carbohydrate, fiber, thiamin, folate, and vitamins B_6, C, A, and E. Nutrients in animal-derived foods may be of concern, however, including protein, iron, zinc, calcium, vitamin B_{12}, vitamin D, and omega-3 fatty acids. Table C6–3, p. 230, presents good vegetarian sources for these nutrients.

Choosing within the Food Groups

When selecting from the vegetable and fruit groups, vegetarians should emphasize sources of calcium and iron. Green leafy vegetables provide both calcium and iron. Similarly, dried fruits deserve special notice in the fruit group because they can deliver more iron than other fruits. Note that the milk group features fortified soy milk for those who do not use milk, cheese, or yogurt. Soy milk is often fortified with calcium and other nutrients of milk—read the labels to be sure. The Protein Foods group emphasizes legumes, soy products, nuts, and seeds. The oils group encourages the use of vegetable oils, nuts, and seeds rich in unsaturated fats and omega-3 fatty acids. To ensure adequate intakes of vitamin B_{12}, vitamin D, and calcium, vegetarians need to select fortified foods or use supplements daily. Like others, vegetarians need physical activity to round out their plan.

Milk Products and Protein Foods

It takes a little planning to ensure adequate intakes from a variety of vegetarian foods in the Milk and Milk Products group and the Protein Foods Group. Figure C6–1 (p. 231) turns a spotlight on these food groups, and Appendix E provides the USDA Food Patterns in full for both vegans and lacto-ovo vegetarians. The USDA Food Patterns also specify weekly amounts that vegetarians should obtain from the Protein Foods subgroups. For those with Internet access, USDA's MyPlate website provides useful tools for planning vegetarian diets using the USDA food patterns.

Convenience Foods

Prepared frozen or packaged vegetarian foods make food preparation quick and easy—just be sure to scrutinize the label's Nutrition Facts panel when choosing among them. Some products constitute a nutritional bargain, such as vegetarian "hot dogs" or "veggie burgers." Made of soy, these foods look and taste like the original meat product but contain much less fat and saturated fat and no cholesterol. Conversely, banana or vegetable chips, often sold as "healthy" snack food, are no bargain: a quarter cup of banana chips fried in saturated coconut oil provides 150 calories with 7 grams of saturated fat (a big hamburger

Table C6–3

Vegetarian Sources of Key Nutrients

	FOOD GROUPS					
NUTRIENTS	Grains	Vegetables	Fruits	Legumes and Other Protein-Rich Foods	Milk or Soy Milk	Oils
PROTEIN	Whole grains[a]			Legumes, seeds, nuts, soy products (tempeh, tofu, veggie burgers)[a] Eggs (for ovo-vegetarians)	Milk, cheese, yogurt (for lacto-vegetarians); soy milk, soy yogurt, soy cheeses	
IRON	Fortified cereals, enriched and whole grains	Dark green leafy vegetables (spinach, turnip greens)	Dried fruits (apricots, prunes, raisins)	Legumes (black-eyed peas, kidney beans, lentils), soy products		
ZINC	Fortified cereals, whole grains			Legumes (garbanzo beans, kidney beans, navy beans), nuts, seeds (pumpkin seeds)	Milk, cheese, yogurt (for lacto-vegetarians); soy milk, soy yogurt, soy cheeses	
CALCIUM	Fortified cereals	Dark green leafy vegetables (bok choy, broccoli, collard greens, kale, mustard greens, turnip greens, watercress)	Fortified juices, figs	Fortified soy products, nuts (almonds), seeds (sesame seeds)	Milk, cheese, yogurt (for lacto-vegetarians); fortified soy milk, fortified soy yogurt, fortified soy cheese	
VITAMIN B_{12}	Fortified cereals			Eggs (for ovo-vegetarians); fortified soy products	Milk, cheese, yogurt (for lacto-vegetarians); fortified soy milk, fortified soy yogurt, fortified soy cheese	
VITAMIN D	Fortified cereals				Milk, cheese, yogurt (for lacto-vegetarians); fortified soy milk, fortified soy yogurt, fortified soy cheese	
OMEGA-3 FATTY ACIDS		Marine algae and its oils		Flaxseed, walnuts, soybeans, fortified margarine,[b] fortified eggs (for ovo-vegetarians)[b]		Flaxseed oil, walnut oil, soybean oil

[a]Many plant proteins lack certain essential amino acids or contain them in insufficient amounts for human health. A variety of daily plant protein sources, such as grains and legumes, can meet protein needs when energy intake is sufficient.

[b]Fortification sources of EPA and DHA may be fish oil or marine algae oil; read the ingredients list on the label.

Figure C6–1

Filling the Vegetarian MyPlate

Each day, in a 2,000-calorie diet, both vegans and lacto-ovo vegetarians require 3 cups of Milk and Milk Product equivalents and $5^{1}/_{2}$ ounces of Protein Foods. (For details and for other calorie levels, see Appendix E.)

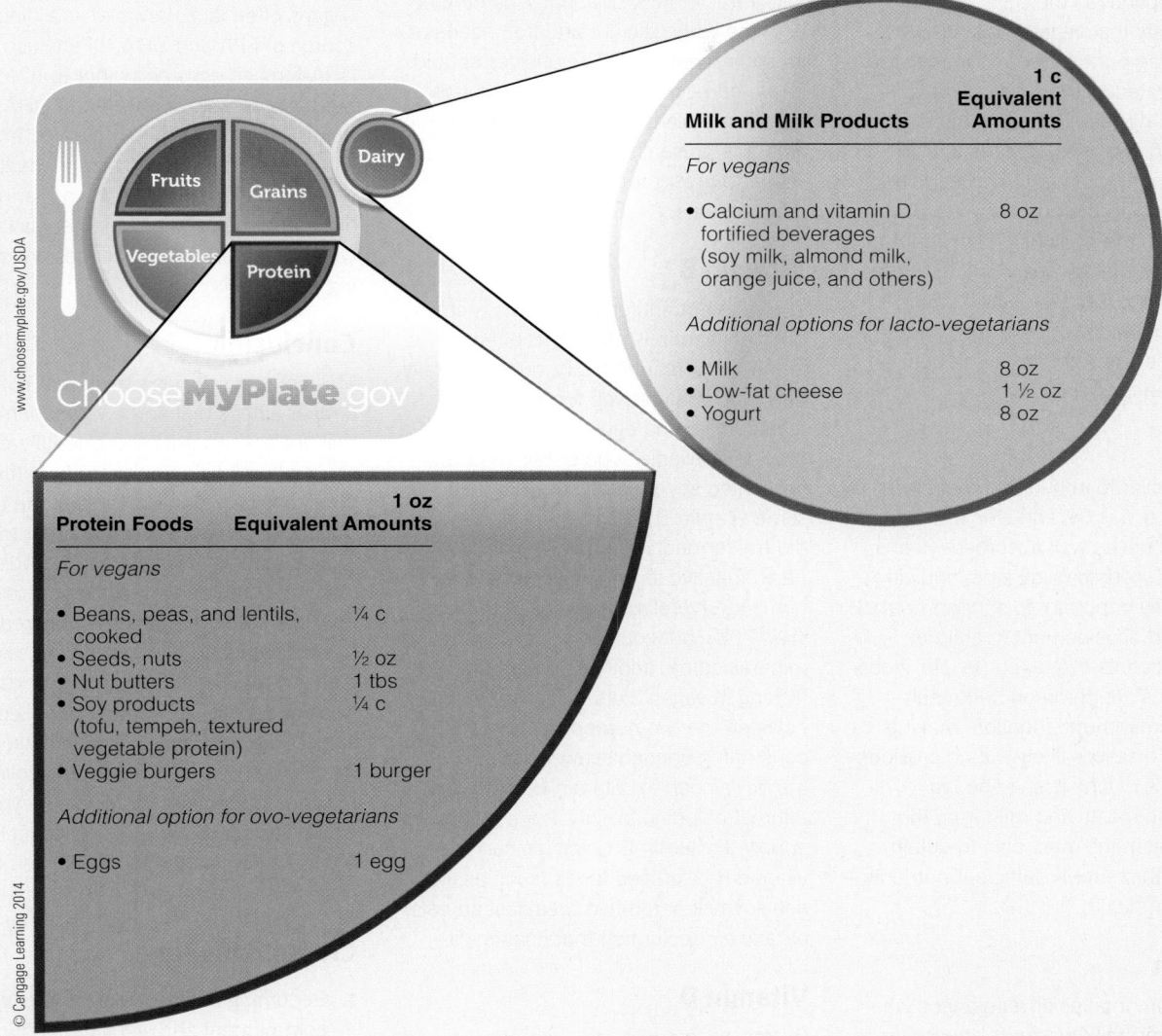

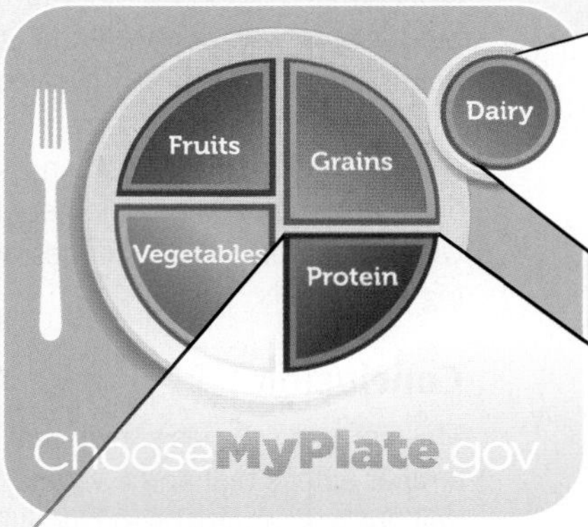

Milk and Milk Products	1 c Equivalent Amounts
For vegans	
• Calcium and vitamin D fortified beverages (soy milk, almond milk, orange juice, and others)	8 oz
Additional options for lacto-vegetarians	
• Milk	8 oz
• Low-fat cheese	1 ½ oz
• Yogurt	8 oz

Protein Foods	1 oz Equivalent Amounts
For vegans	
• Beans, peas, and lentils, cooked	¼ c
• Seeds, nuts	½ oz
• Nut butters	1 tbs
• Soy products (tofu, tempeh, textured vegetable protein)	¼ c
• Veggie burgers	1 burger
Additional option for ovo-vegetarians	
• Eggs	1 egg

has 8 grams). A plain banana has 100 calories and practically no fat. Look for freeze-dried fruit and vegetable "chips"—they have no added fats, but the freeze-drying process creates a pleasing crunch and preserves most nutrients.

A Word about Protein Foods

Vegetarians who consume eggs and low-fat milk products receive high-quality complete protein that provides all the essential amino acids required for health. Even vegans are likely to meet protein needs provided that they meet their energy needs with nutritious whole foods and that their protein sources are varied.[26]

Like the hot dogs just mentioned, many vegetarian protein foods are made of textured vegetable protein (often soy) and formulated to look and taste like meat. While providing protein in abundance, some meat look-alikes may not equal meat in vitamin and mineral content, and they may be higher in salt, sugar, or other additives. Such added items are listed among the product's ingredients on the label, and information about sodium, sugar, vitamins, and minerals is found on the Nutrition Facts panel. Soybeans in other forms, such as plain tofu (bean curd), edamame (cooked green soybeans, pronounced *ed-eh-MAH-may*), or soy flour, also bolster protein intake with fewer unwanted additives.

Iron

Vegetarians, and even certain meat eaters, must be vigilant about obtaining iron. The iron in plant foods such as legumes, dark green leafy vegetables, iron-fortified cereals, and whole-grain breads and cereals is often harder to absorb (see Chapter 8 for more details). Such foods often contain inhibitors of iron absorption, so the DRI committee suggests that vegetarians need 1.8 times the amount of iron recommended for meat eaters.[27] Iron absorption is enhanced by vitamin C consumed with iron-rich foods, and vegetarians typically eat many vitamin C–rich fruits and vegetables with their meals. Also, at some point, the body begins to adapt to a vegetarian diet by absorbing iron more efficiently, but the degree and timing of this adaptation varies. Consequently, vegetarians suffer no more iron deficiency than other people do.

Zinc

Zinc is similar to iron in that meat is its richest food source, and zinc from plant sources is not as well absorbed. Beans, peas, and seeds provide zinc, and zinc is especially important to support normal growth and development in children and adolescents.[28] In aged people, zinc deficiencies are common and result in impaired immune function, making infectious diseases likely.[29] Zinc provides an example of why it isn't enough to simply omit meats and milk from the diet—vegetarians must plan to obtain the foods that supply sufficient nutrients to maintain health.

Calcium

The calcium intakes of vegetarians who use milk and milk products are similar to those of the general population. For vegans, ample quantities of calcium-fortified foods, such as juices, soymilk, and breakfast cereals, are required regularly. Not all such products are well endowed with calcium, however, so label reading must become a passion. This is especially important when feeding children and adolescents whose developing bones demand ample calcium.

Some absorbable calcium is also present in figs, calcium-set tofu, some legumes, some green vegetables (broccoli, kale, and turnip greens, but not spinach—see Chapter 8), some nuts such as almonds, and certain seeds such as sesame seeds.† One cup of cooked kale, for example, provides about 90 milligrams of calcium and 1.2 milligrams of iron, or about 7 percent of both the daily calcium and iron needs of an adolescent girl.[30] The choices should be varied because calcium absorption from some plant foods is limited, and the amounts present are generally insufficient to meet most people's calcium requirements.

Vitamin B_{12}

The requirement for vitamin B_{12} is small, but this vitamin is critical, as already mentioned. Vitamin B_{12} is found naturally only in animal-derived foods, such as meats, milk, and eggs, and these foods provide a reliable source. For vegans, fermented soy products may contain some vitamin B_{12} from the bacteria that did the fermenting, but much of it may be an inactive form. Seaweeds such as nori and chlorella supply just a trace of vitamin B_{12} but copious amounts of the mineral iodine, another nutrient often lacking in vegan diets.[31] The imbalance between the two nutrients means that consuming enough seaweed to supply the day's need for vitamin B_{12} may pose a threat of iodine toxicity. For a reliable supply of vitamin B_{12}, vegans can use vitamin B_{12}–fortified foods (such as fortified soymilk or fortified breakfast cereals) or take a supplement that contains it.

Vitamin D

Overall, vegetarians are similar to nonvegetarians in their vitamin D status; factors such as taking supplements, skin color, and sun exposure have a greater influence on vitamin D than diet.[32] People who do not use vitamin D–fortified foods and do not receive enough exposure to sunlight to synthesize adequate vitamin D may need supplements to fend off bone loss. Of particular concern are infants, children, adolescents, and older adults in northern climates during winter months.

†Calcium salts are often added to tofu during processing to coagulate it.

Omega-3 Fatty Acids

Vegetarian eating patterns typically provide enough flaxseed, walnuts, and their oil, as well as soybeans and canola oil, to supply the essential fatty acids linoleic acid and linolenic acid. What vegans often lack, however, is a dietary source of EPA and DHA.[33] Fatty fish and DHA-fortified eggs and other fortified products can provide EPA and DHA, but as they are ultimately derived from fish, these are unacceptable foods for vegans.[34] Alternatively, certain marine algae and their oils provide beneficial DHA, and more vegan sources are showing promise.[35]

Conclusion

This comparison has shown that both a meat-eater's diet and a vegetarian's diet are best approached scientifically. If you are just beginning to study nutrition, consider adopting the attitude that the choice to make is not whether to be a meat eater or a vegetarian but where along the spectrum to locate yourself. Your preferences should be honored with these caveats: that you plan your own diet and the diets of those in your care to be adequate, balanced, controlled in calories, and varied and that you limit intakes of foods high in sodium, solid fats, and added sugars. Whatever your eating style or reasons for choosing it, choose carefully: the foods that you eat regularly make an impact on your health.

Critical Thinking

1. Becoming a vegan takes a strong commitment and significant education to know how to combine foods and in what quantities to meet nutrient requirements. Most of us will not choose to become vegetarians, but many of us would benefit from a diet of less meat. Identify ways you could alter your diet so that you eat less meat.
2. Outline four nutrients that are most likely to be deficient in the vegetarian's diet. Illustrate foods that the vegetarian should eat to ensure adequate intake of these nutrients.

7 The Vitamins

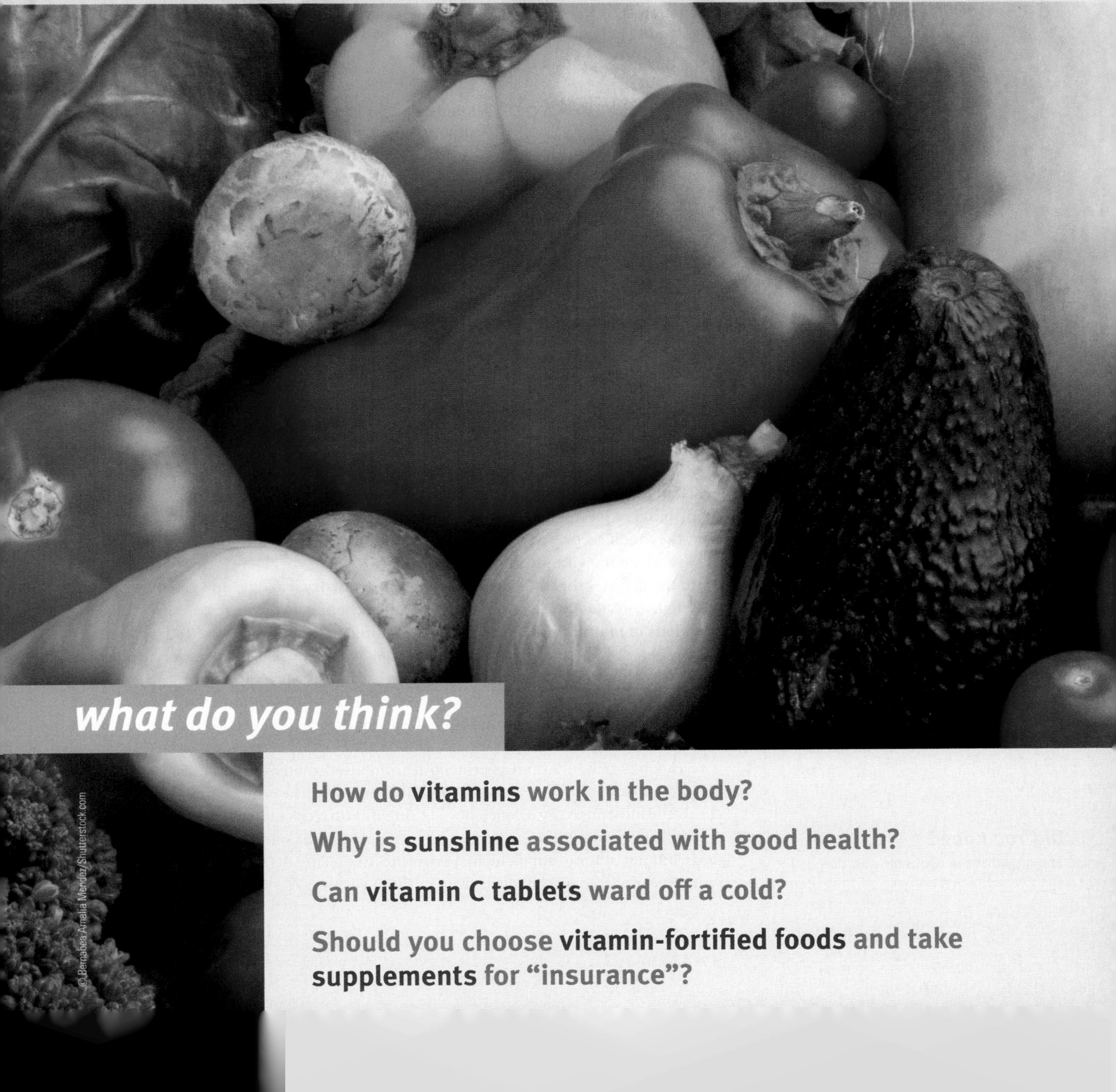

what do you think?

How do **vitamins** work in the body?

Why is **sunshine** associated with good health?

Can **vitamin C tablets** ward off a cold?

Should you choose **vitamin-fortified foods** and take **supplements** for "insurance"?

Learning Objectives

After completing this chapter, you should be able to accomplish the following:

LO 7.1 List the fat-soluble and water-soluble vitamins, and describe how solubility affects the absorption, transport, storage, and excretion of each type.

LO 7.2 Discuss the significance of the fat-soluble nature of some vitamins to human nutrition.

LO 7.3 Summarize the physiological roles of vitamin A and its precursor beta-carotene, name the consequences of deficiencies and toxicities, and list the major food sources of both forms.

LO 7.4 Summarize the physiological roles of vitamin D, name the consequences of deficiencies and toxicities, and list its major food sources.

LO 7.5 Summarize the physiological roles of vitamin E, name the consequences of deficiencies and toxicities, and list its major food sources.

LO 7.6 Summarize the physiological roles of vitamin K, name the consequences of deficiencies, and list its major food sources.

LO 7.7 Describe some characteristics of the water-soluble vitamins.

LO 7.8 Summarize the physiological roles of vitamin C, name the consequences of deficiencies and toxicities, and list its major food sources.

LO 7.9 Describe some of the shared roles of B-vitamins in body systems.

LO 7.10 List and summarize the physiological roles of individual B vitamins in the body, name the consequences of deficiencies, and list their most important food sources.

LO 7.11 Suggest foods that can help to ensure adequate vitamin intakes without providing too many calories.

LO 7.12 Identify both valid and invalid reasons for taking vitamin supplements.

At the beginning of the 20th century, the thrill of the discovery of the first **vitamins** captured the world's imagination as seemingly miraculous cures took place. In the usual scenario, a whole group of people was unable to walk (or were going blind or bleeding profusely) until an alert scientist stumbled onto the substance missing from their diets.*[1] The scientist confirmed the discovery by feeding vitamin-deficient feed to laboratory animals, which responded by becoming unable to walk (or going blind or bleeding profusely). When the missing ingredient was restored to their diet, they soon recovered. People, too, were quickly cured from such conditions when they received the vitamins they lacked.

In the decades that followed, advances in chemistry, biology, and genetics allowed scientists to isolate the vitamins, define their chemical structures, and reveal their functions in maintaining health and preventing deficiency diseases. Today, research hints that certain vitamins may be linked with the development of two major scourges of humankind: cardiovascular disease (CVD) and cancer. Many other conditions, from infections to cracked skin, bear relation to vitamin nutrition, details that unscrupulous sellers of vitamins often use to market their wares (see the Controversy section).

Can foods rich in vitamins protect us from life-threatening diseases? What about vitamin pills? For now, we can say this with certainty: the only disease a vitamin will *cure* is the one caused by a deficiency of that vitamin. As for chronic disease *prevention*, research is ongoing, but evidence so far supports the conclusion that vitamin-rich *foods*, but not vitamin supplements, are protective. (The DRI recommended intakes for vitamins are listed on the inside front cover pages.)

Did You Know?

The only disease a vitamin can cure is the one caused by a deficiency of that vitamin.

*Reference notes are found in Appendix F.

Definition and Classification of Vitamins

LO 7.1 List the fat-soluble and water-soluble vitamins, and describe how solubility affects the absorption, transport, storage, and excretion of each type.

A child once defined a vitamin as "what, if you don't eat, you get sick." Although the grammar left something to be desired, the definition was accurate. Less imaginatively, a vitamin is defined as an essential, noncaloric, organic nutrient needed in tiny amounts in the diet. The role of many vitamins is to help make possible the processes by which other nutrients are digested, absorbed, and metabolized or built into body structures. Although small in size and quantity, the vitamins accomplish mighty tasks.

As each vitamin was discovered, it was given a name, and some were given letters and numbers—vitamin A came before the B vitamins, then came vitamin C, and so forth. This led to the confusing variety of vitamin names that still exists today. This chapter uses the names in Table 7–1; alternative names are given in Tables 7–6 and 7–7 at the end of the chapter.

Table 7–1

Vitamin Names[a]

Fat-Soluble Vitamins
Vitamin A
Vitamin D
Vitamin E
Vitamin K
Water-Soluble Vitamins
B vitamins
Thiamin (B_1)
Riboflavin (B_2)
Niacin (B_3)
Folate
Vitamin B_{12}
Vitamin B_6
Biotin
Pantothenic acid
Vitamin C

[a]Vitamin names established by the International Union of Nutritional Sciences Committee on Nomenclature. Other names are listed in Tables 7-6 and 7-7 (pp. 270–274).

The Concept of Vitamin Precursors

Some of the vitamins occur in foods in a form known as **precursors**. Once inside the body, these are transformed chemically to one or more active vitamin forms. Thus, to measure the amount of a vitamin found in food, we often must count not only the amount of the true vitamin but also the vitamin activity potentially available from its precursors. Tables 7–6 and 7–7 specify which vitamins have precursors.

Two Classes of Vitamins: Fat-Soluble and Water-Soluble

The vitamins fall naturally into two classes: fat-soluble and water-soluble (listed in Table 7–1). Solubility confers on vitamins many of their characteristics. It determines how they are absorbed into and transported around by the bloodstream, whether they can be stored in the body, and how easily they are lost from the body.

Like other lipids, fat-soluble vitamins are mostly absorbed into the lymph, and they travel in the blood in association with protein carriers.[2] Fat-soluble vitamins can be stored in the liver or with other lipids in fatty tissues, and some can build up to toxic concentrations. The water-soluble vitamins are absorbed directly into the bloodstream, where they travel freely. Most are not stored in tissues to any great extent; rather, excesses are excreted in the urine. Thus, the risks of immediate toxicities are not as great as for fat-soluble vitamins.

Vitamins fall into two classes.

Table 7–2, p. 236, outlines the general features of the fat-soluble and water-soluble vitamins. The chapter then goes on to provide important details first about the fat-soluble vitamins and then about the water-soluble ones. At the end of the chapter, two tables sum up the basic facts about all of them.

KEY POINTS

- Vitamins are essential, noncaloric nutrients that are needed in tiny amounts in the diet and help to drive cellular processes.
- Vitamin precursors in foods are transformed into active vitamins by the body.
- The fat-soluble vitamins are vitamins A, D, E, and K.
- The water-soluble vitamins are vitamin C and the B vitamins.

vitamins organic compounds that are vital to life and indispensable to body functions but are needed only in minute amounts; noncaloric essential nutrients.

precursors compounds that can be converted into active vitamins. Also called *provitamins.*

Table 7–2

Characteristics of the Fat-Soluble and Water-Soluble Vitamins

While each vitamin has unique functions and features, a few generalizations about the fat-soluble and water-soluble vitamins can aid understanding.

	Fat-Soluble Vitamins: Vitamins A, D, E, and K	Water-Soluble Vitamins: B Vitamins and Vitamin C
Absorption	Absorbed like fats, first into the lymph, then the blood.	Absorbed directly into the blood.
Transport and Storage	Must travel with protein carriers in watery body fluids; stored in the liver or fatty tissues.	Travel freely in watery fluids; most are not stored in the body.
Excretion	Not readily excreted; tend to build up in the tissues.	Readily excreted in the urine.
Toxicity	Toxicities are likely from supplements, but occur rarely from food.	Toxicities are unlikely but possible with high doses from supplements.
Requirements	Needed in periodic doses (perhaps weeks or even months) because the body can draw on its stores.	Needed in frequent doses (perhaps 1 to 3 days) because the body does not store most of them to any extent.

The Fat-Soluble Vitamins

LO 7.2 Discuss the significance of the fat-soluble nature of some vitamins to human nutrition.

The fat-soluble vitamins—A, D, E, and K—are found in the fats and oils of foods and require bile for absorption. Once absorbed, these vitamins are stored in the liver and fatty tissues until the body needs them. Because they are stored, you need not eat foods containing these vitamins every day. If an eating pattern provides sufficient amounts of the fat-soluble vitamins on average over time, the body can survive for weeks without consuming them. This capacity to be stored also sets the stage for toxic buildup if you take in too much. Excesses of vitamins A and D from supplements and highly fortified foods are especially likely to reach toxic levels.

Deficiencies of the fat-soluble vitamins occur when the diet is consistently low in them. We also know that any disease that produces fat malabsorption (such as liver disease, which prevents bile production) can cause the loss of vitamins dissolved in undigested fat and so bring on deficiencies. In the same way, a person who uses mineral oil (which the body cannot absorb) as a laxative risks losing fat-soluble vitamins because they readily dissolve into the oil and are excreted. Deficiencies are also likely when people follow eating patterns that are extraordinarily low in fat because a little fat is necessary for absorption of these vitamins.

Fat-soluble vitamins play diverse roles in the body. Vitamins A and D act somewhat like hormones, directing cells to convert one substance to another, to store this, or to release that. They also directly influence the genes, thereby regulating protein production. Vitamin E flows throughout the body, guarding the tissues against harm from destructive oxidative reactions. Vitamin K is necessary for blood to clot and for bone health. Each is worth a book in itself.

Did You Know?

U.S. intakes of these three vitamins are low enough to be of concern for public health:

- Vitamin D (for everyone)
- Folate (for some people)
- Vitamin B_{12} (for some people)

Vitamin A

LO 7.3 Summarize the physiological roles of vitamin A and its precursor beta-carotene, name the consequences of deficiencies and toxicities, and list the major food sources of both forms.

Vitamin A has the distinction of being the first fat-soluble vitamin to be recognized. Today, after a century of scientific investigation, vitamin A and its plant-derived precursor, **beta-carotene**, are still very much a focus of research.

beta-carotene an orange pigment with antioxidant activity; a vitamin A precursor made by plants and stored in human fat tissue.

retinol one of the active forms of vitamin A made from beta-carotene in animal and human bodies; an antioxidant nutrient. Other active forms are *retinal* and *retinoic acid*.

Three forms of vitamin A are active in the body. One of the active forms, **retinol**, is stored in specialized cells of the liver. The liver makes retinol available to the bloodstream and thereby to the body's cells. The cells convert retinol to its other two active forms, retinal and retinoic acid, as needed.

Foods derived from animals provide forms of vitamin A that are readily absorbed and put to use by the body. Foods derived from plants provide beta-carotene, which must be converted to active vitamin A before it can be used as such.[3]

Roles of Vitamin A and Consequences of Deficiency

Vitamin A is a versatile vitamin, with roles in gene expression, vision, maintenance of body linings and skin, immune defenses, growth of bones and of the body, and normal development of cells.[4] It is of critical importance for both male and female reproductive functions and for normal development of an embryo and fetus.[5] In short, vitamin A is needed everywhere (its chief functions in the body are listed in the Snapshot on page 242 and in Table 7–6). The following sections provide some details.

Eyesight The most familiar function of vitamin A is to sustain normal eyesight. Vitamin A plays two indispensable roles: in the process of light perception at the **retina** and in the maintenance of a healthy, crystal-clear outer window, the **cornea** (see Figure 7–1).

When light falls on the eye, it passes through the clear cornea and strikes the cells of the retina, bleaching many molecules of the pigment **rhodopsin** that lie within those cells. Vitamin A is a part of the rhodopsin molecule. When bleaching occurs, the vitamin is broken off, initiating the signal that conveys the sensation of sight to the optic center in the brain. The vitamin then reunites with the pigment, but a little vitamin A is destroyed each time this reaction takes place, and fresh vitamin A must replenish the supply.

Night Blindness If the vitamin A supply begins to run low, a lag occurs before the eye can see again after a flash of bright light at night (see Figure 7–2, p. 238). This lag in the recovery of night vision, termed **night blindness**, often indicates a vitamin A deficiency.[6] A bright flash of light can temporarily blind even normal, well-nourished eyes, but if you experience a long recovery period before vision returns, your health-care provider may want to check your vitamin A intake.

Xerophthalmia and Blindness A more profound deficiency of vitamin A is exhibited when the protein **keratin** accumulates and clouds the eye's outer vitamin A–dependent part, the cornea. The condition is known as **keratinization**, and if the deficiency of vitamin A is not corrected, it can worsen to **xerosis** (drying) and then progress to thickening and permanent blindness, **xerophthalmia**.[7] Tragically, a half million of the world's vitamin A–deprived children become blind each year from this often preventable condition; about half die within a year after losing their sight. Vitamin A supplements given early to children developing vitamin A deficiency can reverse the process and save both eyesight and lives.[8] Better still, a child fed a variety of fruits and vegetables regularly is virtually assured protection.

Gene Regulation Vitamin A also exerts considerable influence on an array of other body functions through its interaction with genes—hundreds of genes are regulated by the retinoic acid form of vitamin A.[9] Genes direct the synthesis of proteins, including enzymes, that perform the metabolic work of the tissues. Hence, through its influence on gene expression, vitamin A affects the metabolic activities of the tissues, and, in turn, the health of the body.

Researchers have long known that the presence of genetic equipment needed to make a particular protein does not guarantee that the protein will be made, any more than owning a car guarantees you a ride across town. To get the car rolling, you must also have the right key to start up its engine and to turn it off at the appropriate times. Some dietary components, including the retinoic acid form of vitamin A, are now

Figure 7–1
An Eye (Sectioned)

This eye is sectioned to reveal its inner structures.

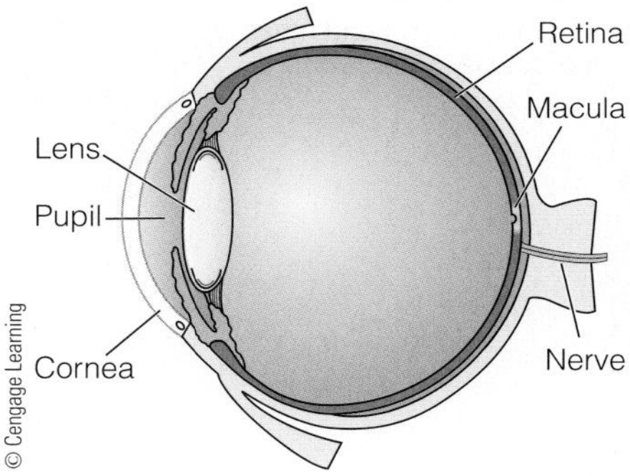

retina (RET-in-uh) the layer of light-sensitive nerve cells lining the back of the inside of the eye.

cornea (KOR-nee-uh) the hard, transparent membrane covering the outside of the eye.

rhodopsin (roh-DOP-sin) the light-sensitive pigment of the cells in the retina; it contains vitamin A (*opsin* means "visual protein").

night blindness slow recovery of vision after exposure to flashes of bright light at night; an early symptom of vitamin A deficiency.

keratin (KERR-uh-tin) the normal protein of hair and nails.

keratinization accumulation of keratin in a tissue; a sign of vitamin A deficiency.

xerosis (zeer-OH-sis) drying of the cornea; a symptom of vitamin A deficiency.

xerophthalmia (ZEER-ahf-THALL-me-uh) progressive hardening of the cornea of the eye in advanced vitamin A deficiency that can lead to blindness (*xero* means "dry"; *ophthalm* means "eye").

Figure 7–2
Night Blindness

This is one of the earliest signs of vitamin A deficiency.

© David Farr/Imagesmythe

In dim light, you can make out the details in this room.

A flash of bright light momentarily blinds you as the pigment in the retina is bleached.

You quickly recover and can see the details again in a few seconds.

With inadequate vitamin A, you do not recover but remain blind for many seconds; this is night blindness.

known to act like such keys—they help to activate or deactivate genes responsible for the production of proteins that perform essential body functions.

Cell Differentiation Vitamin A is needed by all **epithelial tissue** (external skin and internal linings). The cornea of the eye, already mentioned, is such a tissue; so are skin and all of the protective linings of the lungs, intestines, vagina, urinary tract, and bladder. These tissues serve as barriers to infection and other threats.

An example of vitamin A's health-supporting work is the process of **cell differentiation**, in which each type of cell develops to perform a specific function. For example, when goblet cells (cells that populate the linings of internal organs) mature, they specialize in synthesizing and releasing mucus to protect delicate tissues from toxins or bacteria and other harmful elements. In the body's outer layers, vitamin A helps to protect against skin damage from sunlight.

If vitamin A is deficient, cell differentiation is impaired, and goblet cells fail to mature, fail to make protective mucus, and eventually die off. Goblet cells are then displaced by cells that secrete keratin, mentioned earlier with regard to the eye. Keratin is the same protein that provides toughness in hair and fingernails, but in the wrong place, such as skin and body linings, keratin makes the tissue surfaces dry, hard, and cracked. As dead cells accumulate on the surface, the tissue becomes vulnerable to infection (see Figure 7–3). In the cornea, keratinization leads to xerophthalmia; in the lungs, the displacement of mucus-producing cells makes respiratory infections likely; in the urinary tract, the same process leads to urinary tract infections.

Immune Function Vitamin A has gained a reputation as an "anti-infective" vitamin because so many of the body's defenses against infection depend on an adequate supply.[10] Much research supports the need for vitamin A in the regulation of the genes involved in immunity. Without sufficient vitamin A, these genetic interactions produce an altered response to infection that weakens the body's defenses.

When the defenses are weak, especially in vitamin A–deficient children, an illness such as measles can become severe. A downward spiral of malnutrition and infection can set in. The child's body must devote its scanty store of vitamin A to the immune system's fight against the measles virus, but this destroys the vitamin. As vitamin A dwindles further, the infection worsens. Measles takes the lives of more than 450 of the world's children *every day*.[11] Even if the child survives the infection, blindness is likely to occur. The corneas, already damaged by the chronic vitamin A shortage, degenerate rapidly as their meager supply of vitamin A is diverted to the immune system.

Figure 7–3
The Skin in Vitamin A Deficiency

The hard lumps on the skin of this person's arm reflect accumulations of keratin in the epithelial cells.

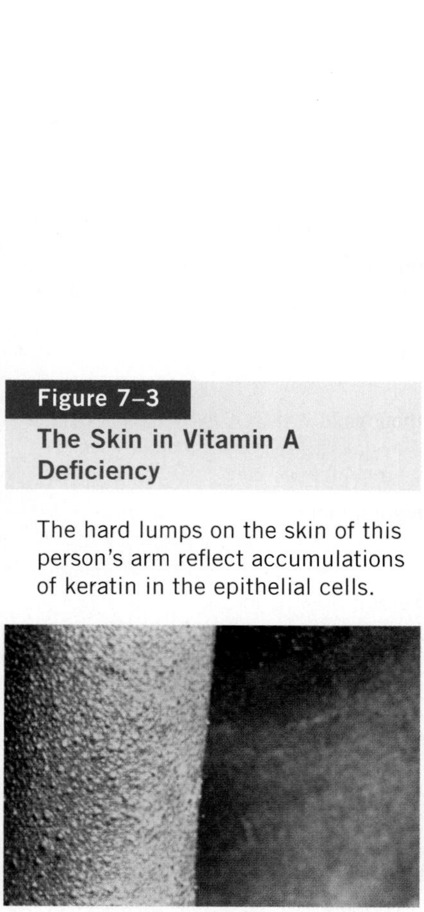

© H. Sanstead, U. of Texas/Galveston

Growth Vitamin A is essential for normal growth of bone (and teeth). Normal children's bones grow longer, and the children grow taller, by remodeling each old bone into a new, bigger version. To do so, the body dismantles the old bone structures and replaces them with new, larger bone parts. Growth cannot take place just by adding on to the original small bone; vitamin A must be present for critical bone dismantling steps.[12] Failure to grow is one of the first signs of poor vitamin A status in a child. Restoring vitamin A to such children is imperative, but correcting dietary deficiencies may be more effective than giving vitamin A supplements alone because many other nutrients from nutritious foods are also needed for children to gain weight and grow taller.

KEY POINTS

- Three active forms of vitamin A and one precursor are important in nutrition.
- Vitamin A plays major roles in gene regulation, eyesight, reproduction, cell differentiation, immunity, and growth.

Vitamin A Deficiency around the World Vitamin A deficiency presents a vast problem worldwide, placing a heavy burden on society. An estimated 5 million of the world's preschool children suffer from signs of vitamin A deficiency—not only night blindness but diarrhea, appetite loss, and reduced food intake that rapidly worsen their condition.[13] A staggering 180 million more children suffer from a milder deficiency that impairs immunity, leaving them open to infections.

In countries where children receive vitamin A supplements, childhood rates of blindness, disease, and death have declined dramatically. Even in the United States, vitamin A supplements are recommended for children with measles.[14] The World Health Organization (WHO) and UNICEF (United Nations International Children's Emergency Fund) are working to eliminate vitamin A deficiency around the world; achieving this goal would greatly improve child survival.

KEY POINT

- Vitamin A deficiency causes blindness, sickness, and death and is a major problem worldwide.

Vitamin A Toxicity

For people who take excess active vitamin A in supplements or fortified foods, toxicity is a real possibility. Figure 7–4, p. 240, shows that toxicity compromises the tissues just as deficiency does and is equally damaging. Symptoms of vitamin A toxicity are many, and they vary depending partly on whether a sudden overdose occurs or too much of the vitamin is taken over time. The Snapshot lists the best-known toxicity symptoms of both kinds. In addition, hair loss, rashes, and a host of uncomfortable general symptoms are possible. Over the years, even relatively small vitamin A excesses may silently weaken the bones and contribute to hip fractures later in life.

Ordinary vitamin supplements taken in the context of today's heavily fortified food supply can add up to small daily excesses of vitamin A.[15] Substantial amounts can be found in fortified cereals, water beverages, energy and candy bars, and even chewing gum (see Table 7–3, p. 240).

Pregnant women, especially, should be wary—excessive vitamin A during pregnancy can injure the spinal cord and other tissues of the developing fetus, causing birth defects.[16] Even a single massive vitamin A dose (100 times the need) can do so. Children, too, can be easily hurt by vitamin A excesses when they mistake chewable vitamin pills and vitamin-laced gum for treats. Even misinformed adolescents put themselves at risk when they take high doses of vitamin A in misguided attempts to cure acne. An effective acne medicine, *Accutane*, and topical prescription acne creams are *derived* from vitamin A but are chemically altered—vitamin A itself has no effect on acne.[17]

KEY POINT

- Vitamin A overdoses and toxicity are possible and cause many serious symptoms.

epithelial (ep-ith-THEE-lee-ull) **tissue** the layers of the body that serve as selective barriers to environmental factors. Examples are the cornea, the skin, the respiratory tract lining, and the lining of the digestive tract.

cell differentiation (dih-fer-en-she-AY-shun) the process by which immature cells are stimulated to mature and gain the ability to perform functions characteristic of their cell type.

Figure 7-4
Vitamin A Deficiency and Toxicity

Danger lies both above and below a normal range of intake of vitamin A.

Vitamin A intake, µg/day	Deficient 0–500		Normal 500–3,000		Toxic 3,000 and over	
	Effects on cells	**Health consequences**	**Effects on cells**	**Health consequences**	**Effects on cells**	**Health consequences**
	Decreased cell division and deficient cell development 	Night blindness Keratinization Xerophthalmia Impaired immunity Reproductive and growth abnormalities Exhaustion Death	Normal cell division and development 	Normal body functioning	Overstimulated cell division 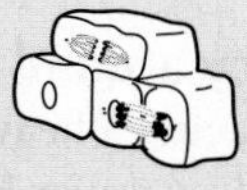	Skin rashes Hair loss Hemorrhages Bone abnormalities Birth defects Fractures Liver failure Death

Vitamin A Recommendations and Sources

The DRI vitamin A intake recommendation is based on body weight. A typical man needs a daily average of about 900 micrograms of active vitamin A; a typical woman, who weighs less, needs about 700 micrograms. During lactation, her need is higher. Children need less. An eating pattern that includes the recommended intakes of fruits and vegetables supplies more than adequate amounts.

The ability of vitamin A to be stored in the tissues means that, although the DRI recommendation is stated as a daily amount, you need not consume vitamin A every day. An intake that meets the daily need when averaged over several months is sufficient.

As for vitamin A supplements, the DRI committee warns against exceeding the Tolerable Upper Intake Level of 3,000 micrograms (for adults older than age 18). The best way to ensure a safe intake of vitamin A is to steer clear of supplements that contain it and to rely on food sources instead.

Table 7-3
Sources of Active Vitamin A

Vitamin A from highly fortified foods and other rich sources can add up. The UL for vitamin A is 3,000 micrograms (µg) per day.

High-potency vitamin pill	3,000 µg
Calf's liver, 1 oz cooked	2,300 µg
Regular multivitamin pill	1,500 µg
Vitamin gumball, 1	1,500 µg
Chicken liver, 1 oz cooked	1,400 µg
"Complete" liquid supplement drink, 1 serving	350–1,500 µg
Instant breakfast drink, 1 serving	600–700 µg
Cereal breakfast bar, 1	350–400 µg
"Energy" candy bar, 1	350 µg
Milk, 1 c	150 µg
Vitamin-fortified cereal, 1 serving	150 µg
Margarine, 1 tsp	55 µg

Food Sources of Vitamin A Active vitamin A is present in foods of animal origin. The richest sources are liver and fish oil but milk and milk products and other vitamin A–fortified foods such as enriched cereals can also be good sources. Even butter and eggs provide some vitamin A. The vitamin A precursor beta-carotene is naturally present in many vegetables and fruit varieties and may be added to cheeses for its yellow color.

Liver: A Lesson in Moderation Foods naturally rich in vitamin A pose little risk of toxicity, with the possible exception of liver. When young laboratory pigs eat daily chow made from salmon parts, including the livers, their growth halts, and they fall ill from vitamin A toxicity. Inuit people and Arctic explorers know that polar bear livers are a dangerous food source because the bears eat whole fish (with the livers) and in turn concentrate large amounts of vitamin A in their own livers.

An *ounce* of ordinary beef or pork liver delivers three times the DRI recommendation for vitamin A intake, and a common portion is 4 to 6 ounces. An occasional serving of liver can provide abundant nutrients and boost nutrient status, but daily use may invite vitamin A toxicity, especially in young children and pregnant women who also routinely take supplements. Snapshot 7–1 is the first of a series of figures that show a sampling of foods that provide more than 10 percent of the Daily Value for a vitamin in a standard-size portion and that therefore qualify as "good" or "rich" sources.

© sarsmis/Shutterstock.com

Colorful foods are often rich in vitamins.

Can Fast Foods Provide Vitamin A? The definitive fast-food meal—a hamburger, fries, and cola—lacks vitamin A. Many fast-food restaurants, however, now offer salads with cheese and carrots, fortified milk, and other vitamin A–rich foods. These selections greatly improve the nutritional quality of fast food.

KEY POINTS

- Vitamin A's active forms are supplied by foods of animal origin.
- Fruits and vegetables provide beta-carotene.

Beta-Carotene

In plants, vitamin A exists only in its precursor forms. Beta-carotene, the most abundant of these **carotenoid** precursors, has the highest vitamin A activity. Even though the other carotenoids are not vitamins, they may play other roles in human health.[18] Eating patterns low in carotenoids, particularly those lacking in dark green, leafy vegetables and orange vegetables, are associated with the most common form of age-related blindness, **macular degeneration**.[†19] The macula, a yellow spot of pigment at the focal center of the retina (identified in Figure 7–1, p. 237), loses integrity, impairing the most important field of vision, the central focus. Supplements of carotenoids and several other nutrients have proven ineffective in preventing this cause of blindness.[20]

Does Eating Carrots Really Promote Good Vision? Bright orange fruits and vegetables derive their color from beta-carotene and are so colorful that they decorate the plate. Carrots, sweet potatoes, pumpkins, mango, cantaloupe, and apricots are all rich sources of beta-carotene—and therefore contribute vitamin A to the eyes and to the rest of the body—so, yes, eating carrots does promote good vision. Another colorful group, *dark* green vegetables, such as spinach, other greens, and broccoli, owes its deep dark green color to the blending of orange beta-carotene with the green leaf pigment chlorophyll. As mentioned, they provide the other carotenoids needed for eye health, as well.

My Turn **watch it!**

Take Your Vitamins?

Claudio talks about vitamins in supplements.

© Cengage Learning

Claudio

carotenoid (CARE-oh-ten-oyd) a member of a group of pigments in foods that range in color from light yellow to reddish orange and are chemical relatives of beta-carotene. Many have a degree of vitamin A activity in the body. Also defined in Controversy 2.

macular degeneration a common, progressive loss of function of the part of the retina that is most crucial to focused vision (the macula is shown on page 237). This degeneration often leads to blindness.

†The carotenoids associated with protection from macular degeneration are lutein (LOO-tee-in) and its close chemical relative zeaxanthin (zee-ZAN-thin).

Snapshot 7-1 Vitamin A and Beta-Carotene

DRI Recommended Intakes

Men: 900 μg/day[a]
Women: 700 μg/day[a]

Tolerable Upper Intake Level

Adults: 3,000 μg vitamin A/day

Chief Functions

Vision; maintenance of cornea, epithelial cells, mucous membranes, skin; bone and tooth growth; regulation of gene expression; reproduction; immunity

Deficiency

Night blindness, corneal drying (xerosis), and blindness (xerophthalmia); impaired bone growth and easily decayed teeth; keratin lumps on the skin; impaired immunity

Toxicity

Vitamin A:
Acute (single dose or short-term): nausea, vomiting, headache, vertigo, blurred vision, uncoordinated muscles, increased pressure inside the skull, birth defects
Chronic: birth defects, liver abnormalities, bone abnormalities, brain and nerve disorders
Beta-carotene: Harmless yellowing of skin

Good Sources*

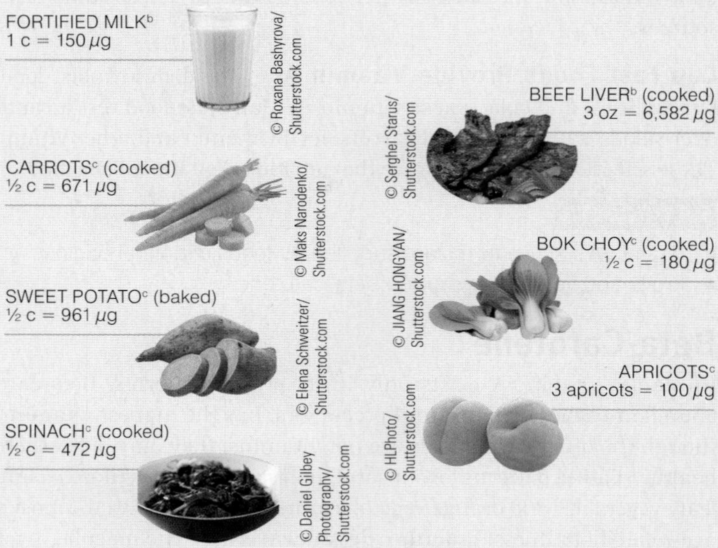

**These foods provide 10% or more of the vitamin A Daily Value in a serving. For a 2,000-cal diet, the DV is 900 μg/day.*
[a]Vitamin A recommendations are expressed in retinol activity equivalents (RAE).
[b]This food contains preformed vitamin A.
[c]This food contains the vitamin A precursor, beta-carotene.

Functional Group of Antioxidants

Key antioxidant vitamins:

- Beta-carotene.
- Vitamin E.
- Vitamin C.

A key antioxidant mineral:

- Selenium.

Antioxidant phytochemicals:

- Carotenoids.
- Flavonoids.
- Others.

Do the Math

1 IU = 0.3 μg retinol

Factors for converting many kinds of units used in nutrition are found in Appendix C.

dietary antioxidants compounds typically found in plant foods that significantly decrease the adverse effects of oxidation on living tissues. The major antioxidant vitamins are vitamin E, vitamin C, and beta-carotene. Many phytochemicals are also antioxidants.

Beta-Carotene, an Antioxidant Beta-carotene is one of many **dietary antioxidants** present in foods—others include vitamin E, vitamin C, the mineral selenium, and many phytochemicals (see Controversy 2 for details).[21] Dietary antioxidants are just one class of a complex array of constituents in whole foods that seem to benefit health synergistically. Foods supply all of these factors and more in ideal amounts and combinations that supplements cannot duplicate.

Antioxidant-rich foods are listed in **Figure C2–1** of **Controversy 2,** page 66.

Measuring Beta-Carotene The conversion of beta-carotene to retinol in the body entails losses, so vitamin A activity for precursors is measured in **retinol activity equivalents (RAE)**. It takes about 12 micrograms of beta-carotene from food to supply the equivalent of 1 microgram of retinol to the body. Some food tables and supplement labels express beta-carotene and vitamin A contents using **IU (international units)**. When comparing vitamin A in foods, be careful to notice whether a food table or supplement label uses micrograms or IU. To convert one to the other, use the factor in the margin.

Toxicity Beta-carotene from food is not converted to retinol efficiently enough to cause vitamin A toxicity. A steady diet of abundant pumpkin, carrots, or carrot juice, however, has been known to turn light-skinned people bright yellow because beta-carotene builds up in the fat just beneath the skin and imparts a harmless yellow cast (see Figure 7–5). Likewise, red-colored carotenoids confer a rosy glow on those who consume the fruits and vegetables that contain them.[22] Food sources of the carotenoids are safe, but concentrated supplements may have adverse effects of their own, as a later section points out.

Food Sources of Beta-Carotene Plants contain no active vitamin A, but many vegetables and fruits provide the vitamin A precursor, beta-carotene. Snapshot 7–1 shows good sources of beta-carotene. Other colorful vegetables, such as red beets, red cabbage, and yellow corn, can fool you into thinking they contain beta-carotene, but these foods derive their colors from other pigments and are poor sources of beta-carotene. As for "white" plant foods such as grains and potatoes, they have none. Some confusion exists concerning the term *yam*. A white-fleshed Mexican root vegetable called "yam" is devoid of beta-carotene, but the orange-fleshed sweet potato called "yam" in the United States is one of the richest beta-carotene sources known. In choosing fruits and vegetables, follow the advice of the *Dietary Guidelines for Americans* of Chapter 2.

KEY POINTS

- The vitamin A precursor in plants, beta-carotene, is an effective antioxidant in the body.
- Many brightly colored plant foods are rich in beta-carotene.

Figure 7–5

Excess Beta-Carotene Symptom: Discoloration of the Skin

The hands on the left show skin discoloration from excess beta-carotene. Another person's normal hand (right) is shown for comparison.

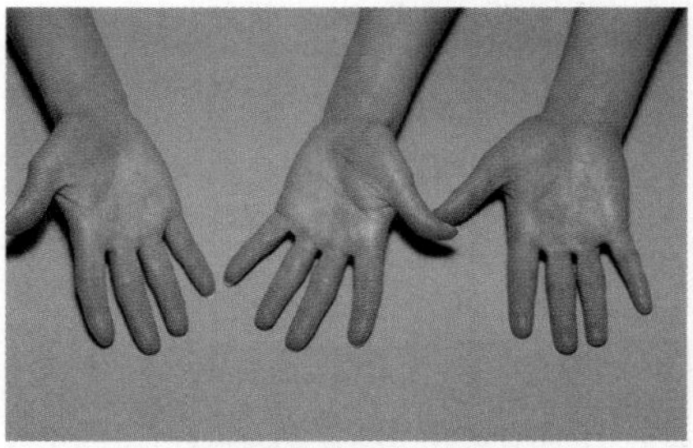

Mediscan/Visuals Unlimited, Inc.

Vitamin D

LO 7.4 Summarize the physiological roles of vitamin D, name the consequences of deficiencies and toxicities, and list its major food sources.

Vitamin D is unique among nutrients in that, with the help of sunlight, the body can synthesize all it needs. In this sense, vitamin D is not an *essential* nutrient—given sufficient sun each day, most people can make enough to meet their need from this source.

As simple as it sounds to obtain vitamin D, some people, particularly adolescents, people with dark skin, women, housebound elderly people, and many overweight and obese people may border on insufficiency.[23] In contrast, even though most people do not take in the recommended amount of vitamin D from food, well over half of the U.S. population has normal blood concentrations, presumably put there by sunshine.[24]

An overall national drop in blood vitamin D levels over the past decades seems apparent.[‡25] Part of the blame may lie with the nation's increasing obesity rates during this period because adipose tissue traps vitamin D, making it less available to the bloodstream. In addition, people began spending less time outdoors.

Do the Math

Vitamin D is measured in micrograms (μg) in research or International Units (IU) on nutrient and supplement fact panels.

- To convert vitamin D amounts from micrograms (μg) to International Units (IU) multiply by 40.
- 1 μg = 40 IU vitamin D.

Roles of Vitamin D

Once in the body, whether made from sunlight or obtained from food, vitamin D must undergo a series of chemical transformations in the liver and kidneys to activate it. Once activated, vitamin D has profound effects on the tissues.

Calcium Regulation Vitamin D is the best-known member of a large cast of nutrients and hormones that interact to regulate blood calcium and phosphorus levels, and thereby maintain bone integrity.[26] Calcium is indispensable to the proper functioning of cells in all body tissues, including muscles, nerves, and glands, which draw calcium from the blood as they need it. To replenish blood calcium, vitamin D acts at three body locations to raise the calcium level. First, the skeleton serves as a vast warehouse of stored calcium that can be tapped when blood calcium begins to fall. Only two other organs can act to increase blood calcium: the digestive tract, which can increase absorption of calcium from food, and the kidneys, which can recycle calcium that would otherwise be lost in urine.

Vitamin D and calcium are inextricably linked in nutrition—no matter how much vitamin D you take in, it cannot make up for a chronic shortfall of calcium. The reverse is also true: excess calcium cannot take the place of sufficient vitamin D for bone health.[27]

retinol activity equivalents (RAE) a new measure of the vitamin A activity of beta-carotene and other vitamin A precursors that reflects the amount of retinol that the body will derive from a food containing vitamin A precursor compounds.

IU (international units) a measure of fat-soluble vitamin activity sometimes used in food composition tables and on supplement labels.

‡ Standards used to determine vitamin D status were < 30 nmol/L = deficiency; 30–49 nmol/L = at risk; 50–125 nmol/L = sufficient.

Functional Group for Bone Health

Key vitamins:

- Vitamin A, vitamin D, vitamin K, vitamin C, other vitamins.

Key minerals:

- Calcium, phosphorus, magnesium, fluoride, other minerals

Key energy nutrient:

- Protein.

Figure 7–6
Rickets

This child has the bowed legs of vitamin D–deficiency disease rickets.

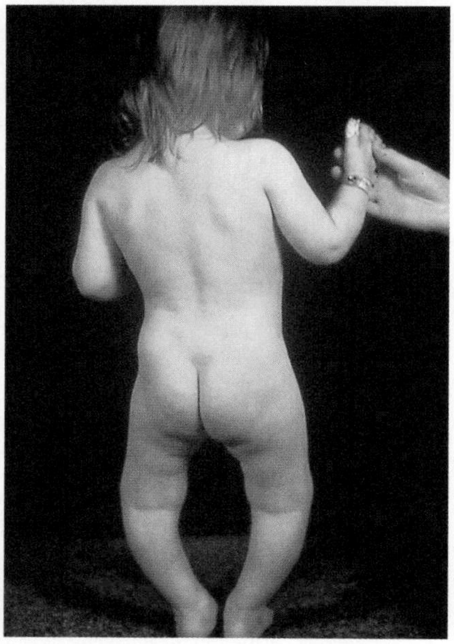

Biophoto Associates/Science Source

This child displays beaded ribs, a symptom of rickets.

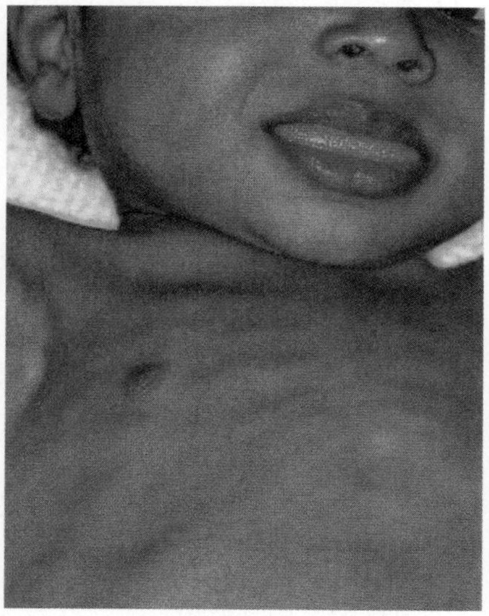

Photo Courtesy of Dr. Norman Carvalho at Childrens Healthcare of Atlanta

Other Vitamin D Roles Activated vitamin D functions as a hormone, that is, a compound manufactured by one organ of the body that acts on other organs, tissues, or cells. Inside cells, for example, vitamin D acts at the genetic level to affect how cells grow, multiply, and specialize. Vitamin D exerts its effects all over the body, from hair follicles, to reproductive system cells, to cells of the immune system.

Research is hinting (sometimes strongly) that to incur a deficit of vitamin D may be to invite problems of many kinds, including macular degeneration, cardiovascular diseases and risk factors, some cancers, respiratory infections such as tuberculosis or flu, inflammatory conditions, multiple sclerosis, and a higher risk of death.[§28] Even so, evidence does not support taking vitamin D supplements to prevent diseases, except those caused by deficiency, and experts recommended against it.[29] The well-established vitamin D roles concern calcium balance and the bones during growth and throughout life, and these form the basis of the DRI intake recommendations.

KEY POINTS

- Low and borderline vitamin D levels are not uncommon in the United States.
- When exposed to sunlight, the skin makes vitamin D from a cholesterol-like compound.
- Vitamin D helps regulate blood calcium and influences other body tissues.

Too Little Vitamin D—A Danger to Bones

Although vitamin D insufficiency is relatively common in the population, overt signs of vitamin D deficiency are rarely reported.[30] The most obvious sign occurs in early life—the abnormality of the bones in the disease **rickets** is shown in Figure 7–6. Children with rickets develop bowed legs because they are unable to mineralize newly forming bone material, a rubbery protein matrix. As gravity pulls their body weight against these weak bones, the legs bow. Many such children also have a protruding belly because of lax abdominal muscles.

Preventing Rickets As early as the 1700s, rickets was known to be curable with cod-liver oil, now recognized as a rich source of vitamin D. More than a hundred years later, a Polish physician linked sunlight exposure to prevention and cure of rickets.

Today, in some areas of the world, such as Mongolia, Tibet, and the Netherlands, more than half of the children suffer the bowed legs, knock-knees, beaded ribs, and protruding (pigeon) chests of rickets. In the United States, rickets is uncommon but not unknown.[31] When it occurs, black children and adolescents—especially females and overweight teens—are most likely to be affected.[32] Many adolescents abandon vitamin D–fortified milk in favor of soft drinks and punches; they may also spend little time outdoors during daylight hours. Soon, their vitamin D values decline, and they may fail to develop the bone mineral density needed to offset bone loss in later life. To prevent rickets and support optimal bone health, the DRI committee recommends that all infants, children, and adolescents consume the recommended amounts of vitamin D each day.[33]

Deficiency in Adults In adults, poor mineralization of bone results in the painful bone disease **osteomalacia**.[34] The bones become increasingly soft, flexible, weak, and deformed. Older people can suffer painful joints if their vitamin D levels are low, a condition easily misdiagnosed as arthritis during examinations. Inadequate vitamin D also sets the stage for a loss of calcium from the bones, which can result in fractures from **osteoporosis**. Vitamin D supplements themselves may not prevent fractures, but they may help to prevent falls. The simple act of taking a vitamin D supplement could easily save the life of an elderly person who might otherwise suffer bone fractures from falls.[35]

Who Should Be Concerned? The DRI committee identified groups of people who may have an increased risk of vitamin D deficiency. People who are overweight

---

[§] Read more about these preliminary findings in the articles listed in this citation (Appendix F).

or obese, for example, may sequester away much of their vitamin D supply in fatty tissue, making it unavailable to the bloodstream and body tissues.[36] People living in northern areas of North America; anyone lacking exposure to sunlight, such as office workers or institutionalized older people; and dark-skinned people, their breastfed infants, and their adolescent children may also lack vitamin D.

In addition, people who restrict intakes of animal and dairy foods may not obtain enough vitamin D from food to meet recommendations. Strict vegetarians and people with milk allergies in particular must seek out other enriched foods or take supplements to be certain of obtaining enough vitamin D. Some medications can also compromise vitamin D status.

KEY POINTS

- A vitamin D deficiency causes rickets in childhood, low bone density in adolescents, or osteomalacia in later life.
- Some groups of people are more likely to develop vitamin D deficiencies.

Too Much Vitamin D—A Danger to Soft Tissues

Vitamin D is the most potentially toxic among the vitamins. Vitamin D intoxication raises the concentration of blood calcium by withdrawing bone calcium, which can then collect in the soft tissues and damage them. With chronic high vitamin D intakes, kidney and heart function decline, blood calcium spins further out of control, and death ensues when the kidneys and heart ultimately fail.

High doses of vitamin D taken for just a few weeks may bring on high blood calcium, nausea, fatigue, back pain, irregular heartbeat, and increased urination and thirst.[37] Moderate supplemental doses, taken over months or years, may be linked with an increased risk of diseases, falls, and fractures, and even an increased risk of death, but more research is needed to clarify these findings.[38]

Bernd Vogel/Getty Images

The sunshine vitamin: vitamin D.

KEY POINTS

- Vitamin D is the most potentially toxic vitamin.
- Overdoses raise blood calcium and damage soft tissues.

Vitamin D from Sunlight

Sunlight supplies the needed vitamin D for most of the world's people. Sunlight presents no risk of vitamin D toxicity because after a certain amount of vitamin D collects in the skin, the sunlight itself begins breaking it down.

Vitamin D Synthesis and Activation When ultraviolet (UV) light rays from the sun reach a cholesterol compound in human skin, the compound is transformed into a vitamin D precursor and is absorbed directly into the blood. Slowly, over the next day and a half, the liver and kidneys finish converting the inactive precursor to the active form of vitamin D. Diseases that affect either the liver or the kidneys can impair this conversion and therefore produce symptoms of vitamin D deficiency.

Like natural sunscreen, the pigments of dark skin protect against UV radiation. To synthesize several days' worth of vitamin D, dark-skinned people require up to 3 hours of direct sun (depending on the climate). Light-skinned people need much less time (an estimated 5 minutes without sunscreen or 10 to 30 minutes with sunscreen). Vitamin D deficiency is especially prevalent when sunlight is weak, such as in the winter months and in the extreme northern regions of the world.[39] The factors listed in Table 7–4, p. 246, can all interfere with vitamin D synthesis.

KEY POINT

- Ultraviolet light from sunshine acts on a cholesterol compound in the skin to make vitamin D.

Vitamin D Intake Recommendations

Vitamin D from the sun, while substantial, varies widely among people of different ages and living in different locations. Measuring the contribution of vitamin D from

rickets the vitamin D–deficiency disease in children; characterized by abnormal growth of bone and manifested in bowed legs or knock-knees, outward-bowed chest, and knobs on the ribs.

osteomalacia (OS-tee-o-mal-AY-shuh) the adult expression of vitamin D–deficiency disease, characterized by an overabundance of unmineralized bone protein (*osteo* means "bone"; *mal* means "bad"). Symptoms include bending of the spine and bowing of the legs.

osteoporosis a weakening of bone mineral structures that occurs commonly with advancing age. Also defined in Chapter 8.

Table 7–4

Factors Affecting Vitamin D Synthesis

The more of these factors present in a person's life, the more critical it becomes to obtain vitamin D from food or supplements.

Factor	Effect on Vitamin D Synthesis
Advanced age	With age, the skin loses some of its capacity to synthesize vitamin D.
Air pollution	Particles in the air screen out the sun's rays.
City living	Tall buildings block sunlight.
Clothing	Most clothing blocks sunlight.
Cloudy skies	Heavy cloud cover reduces sunlight penetration.
Geography	Sunlight exposure is limited: ▪ October through March at latitudes above 43 degrees (most of Canada) ▪ November through February at latitudes between 35 and 43 degrees (many U.S. locations) In locations south of 35 degrees (much of the southern United States), direct sun exposure is sufficient for vitamin D synthesis year-round.
Homebound	Living indoors prevents sun exposure.
Season	Warmer seasons of the year bring more direct sun rays.
Skin pigment	Darker-skinned people synthesize less vitamin D per minute than lighter-skinned people.
Sunscreen	Proper use reduces or prevents skin exposure to sun's rays.
Time of day	Midday hours bring maximum direct sun exposure.

sunlight is difficult, and sun exposure increases skin cancer risks, so the DRI committee made their recommendations in terms of dietary vitamin D alone, with no contribution from the sun.[40] The recommendations do assume an adequate intake of calcium because vitamin D and calcium each alter the body's handling of the other.

The recommendations are set high enough to maintain blood vitamin D levels known to support healthy bones throughout life, but some research suggests that recommendations for some people should be set even higher.[41] The Consumer's Guide provides some details about obtaining vitamin D.

The need for vitamin D remains remarkably steady throughout most of life: people ages 1 year through 70 need 15 micrograms per day; for those 71 and older, the need jumps to 20 micrograms per day because this group faces an increased threat of bone fractures.[42] The DRI committee also set a vitamin D Tolerable Upper Intake Level for adults of all ages: 100 micrograms (4,000 IU), above which the risk of harm increases.

Vitamin D Food Sources

Snapshot 7–2 shows the few significant naturally occurring food sources of vitamin D. In addition, egg yolks provide small amounts, along with butter and cream. Milk, whether fluid, dried, or evaporated, is fortified with vitamin D, so constitutes a major food source for the United States and Canada. Yogurt and cheese products may lack vitamin D, however, whereas orange juice, cereals, margarines, and other foods may be fortified with it, so read the labels.

Snapshot 7-2 Vitamin D

DRI Recommended Intakes

Adults: 15 μg (600 IU)/day (19–70 yr)
20 μg (700 IU)/day (> 70 yr)

Tolerable Upper Intake Level

Adults: 100 μg (4,000 IU)/day

Chief Functions

Mineralization of bones and teeth (raises blood calcium and phosphorus by increasing absorption from digestive tract, withdrawing calcium from bones, stimulating retention by kidneys)

Deficiency

Abnormal bone growth resulting in rickets in children, osteomalacia in adults; malformed teeth; muscle spasms

Toxicity

Elevated blood calcium; calcification of soft tissues (blood vessels, kidneys, heart, lungs, tissues of joints), excessive thirst, headache, nausea, weakness

These foods provide 10% or more of the vitamin D Daily Value in a serving. For a 2,000-cal diet, the DV is 10 μg/day.
[a]Average value.
[b]Avoid prolonged exposure to sun.

Good Sources*

ENRICHED CEREAL (ready-to-eat) ¾ c = 2.5 μg

© stavklem/Shutterstock.com

SARDINES 3 oz = 4.1 μg

© Picsfive/Shutterstock.com

SALMON OR MACKEREL[a] 3 oz = 10.0 μg

© HLPhoto/Shutterstock.com

SUNLIGHT Promotes vitamin D synthesis in the skin.[b]

© Lisa A/Shutterstock.com

COD LIVER OIL 1 tsp = 11 μg

© tab62/Shutterstock.com

FORTIFIED MILK 1 c = 3 μg

© Roxana Bashyrova/Shutterstock.com

TUNA (light, canned) 3 oz = 5.7 μg

© Bizroug/Shutterstock.com

Young adults who drink 3 cups of milk a day receive half of their daily requirement from this source; the other half comes from exposure to sunlight and other foods and supplements. Without adequate sunshine, enriched foods, or supplementation, strict vegetarians cannot meet their vitamin D needs. Certain mushrooms produce significant vitamin D upon brief exposure to ultraviolet light, but these are not widely available as of this writing.[43] For now, vegans must rely on vitamin D–fortified soy products and cereals as their only food sources. Importantly, feeding infants and young children unfortified "health beverages" instead of milk or infant formula can create severe nutrient deficiencies, including rickets.

KEY POINTS

- The DRI committee sets recommended intake levels and a Tolerable Upper Intake Level for vitamin D.
- Food sources of vitamin D include a few naturally rich sources and many fortified foods.

Vitamin E

LO 7.5 Summarize the physiological roles of vitamin E, name the consequences of deficiencies and toxicities, and list its major food sources.

Almost a century ago, researchers discovered a compound in vegetable oils essential for reproduction in rats. This compound was named **tocopherol** from *tokos*, a Greek word meaning "offspring." A few years later, the compound was named vitamin E.

Four tocopherol compounds have long been known to be of importance in nutrition, and each is designated by one of the first four letters of the Greek alphabet: alpha, beta, gamma, and delta. Of these, alpha-tocopherol is the gold standard for vitamin E activity, and the DRI intake recommendations are expressed as alpha-tocopherol.

tocopherol (tuh-KOFF-er-all) a kind of alcohol. The active form of vitamin E is alpha-tocopherol.

use it! A Consumer's Guide To . . .

Sources of Vitamin D

Anyone who complains of feeling tired, or achy, or sleepless may hear this advice: "You're probably low in vitamin D and you need to take a supplement." These days, this bit of counsel has become more and more common from sellers of supplements and some medical professionals, as well. Are supplements really the best way to meet your need for vitamin D? What about foods rich in vitamin D or exposure to the sun? Consumers aiming for optimal vitamin D blood levels while minimizing health risks have a few options.

Finding Vitamin D in Food

Meeting vitamin D needs with foods can be tricky because just these few foods are naturally richly endowed: fatty fish, fish liver oil, beef liver, and eggs yolks. Just 3 ounces of salmon or mackerel provides one-half to three-quarters of a day's vitamin D need for most adults; the next-richest sources, egg yolk and beef liver, provide a little over 10 percent of the day's need per serving.

Today, more and more foods and beverages are fortified with vitamin D largely in response to a barrage of unscientific media hype about "curative powers" of vitamin D. Enriched foods, such as fluid milk, fruit juices, cereals, and breakfast or "energy" bars, are convenient sources of vitamin D in foods, and they effectively contribute to the body's vitamin D supply.[1] A caution is in order, however. A single serving of some beverages or bars can meet an entire day's need; 5 cups of fortified milk will also do so, and when combined with a daily supplement, the day's vitamin D intake can mount up. Read carefully the Nutrition Facts panel of vitamin-D enriched foods, and compare the total provided with your need. Do not exceed the Tolerable Upper Intake Level of 4,000 IU (100 milligrams) per day.

Supplement Speed Bumps

Most daily multivitamin pills provide at least the DRI recommended intake of vitamin D. Going on the false theory that "more is better," some manufacturers have increased the vitamin D in their pills to high levels, and the body readily absorbs a wide range of doses.[2] The serious toxic effects of large overdoses are well-defined but no one knows what consequences may follow mildly elevated vitamin D over time.[3] Women in particular, who often take calcium plus vitamin D pills, may also take daily fish oil, a multivitamin, and some highly enriched foods and beverages that push the day's intake beyond the upper levels of safety.

Sunshine—It's Free, but Is It Safe?

Supplements and enriched foods cost money, but sunshine is freely available to all and, in sunny seasons, can provide much of the vitamin D a person needs. But there's a catch: synthesis of vitamin D requires exposure to the same form of UV radiation that contributes to about a million skin cancers each year in the United States alone.[4] Even small daily exposures to intense sunlight can increase skin cancer risk.* Medical authorities therefore warn against too much sun exposure, but following their advice to use sunscreen or wear sun protective clothing can reduce blood vitamin D concentrations.[5]

Tanning booths are not a safe sun alternative for vitamin D, either. Tanning booths may or may not promote vitamin D synthesis but, like the sun, they deliver UV radiation that promotes skin cancer and prematurely wrinkles the skin.

Moving Ahead

For vitamin D, then, nutrient-dense foods and beverages are the preferred sources because they pose virtually no risk and provide other nutrients that the body needs. Fatty fish, a naturally good source of vitamin D, is also valued for its omega-3 fatty acids, vitamin B_{12}, and high-quality protein. Milk, enriched with vitamin D, is also a key provider of the minerals calcium and phosphorus, the vitamins riboflavin and vitamin B_{12}, among others, and high-quality protein. Enriched whole-grain cereals provide the added vitamin D, along with other vitamins, minerals, fiber, carbohydrates, and the phytochemicals of whole grains.

In contrast, vitamin D–laced sugary refined cereals, high-calorie "complete" beverages, and fatty "energy" bars are no nutritional bargains, and supplements of isolated nutrients are limited to whatever the manufacturer puts into the pills. If you choose to take a supplement, use the principles set forth in this chapter's Controversy section to ensure that the one you choose will meet your needs.

As for the sun exposure, if you live in a southern climate or spend time outdoors in sunny weather, your skin's own vitamin D will contribute to your supply as well. For safety's sake, however, protect your skin from the strong midday sun with sunscreen, wear sun-protective clothing, and take along a big hat.

Review Questions†

1. A daily regimen of supplements and vitamin D–enriched foods increases the risk of vitamin D toxicity. **T F**
2. Just 3 ounces of salmon, a naturally rich source of vitamin D, provides most of an adult's daily need for vitamin D. **T F**
3. All things considered, the best and safest source of vitamin D for people in the United States is exposure to sunshine. **T F**

** For complete information about sun exposure and cancer risk, access the American Cancer Society website: www.cancer.org.*

† Answers to Consumer's Guide review questions are found in Appendix G.

Figure 7–7

Free-Radical Damage and Antioxidant Protection

Free-radical formation occurs during metabolic processes, and it accelerates when diseases or other stresses strike.

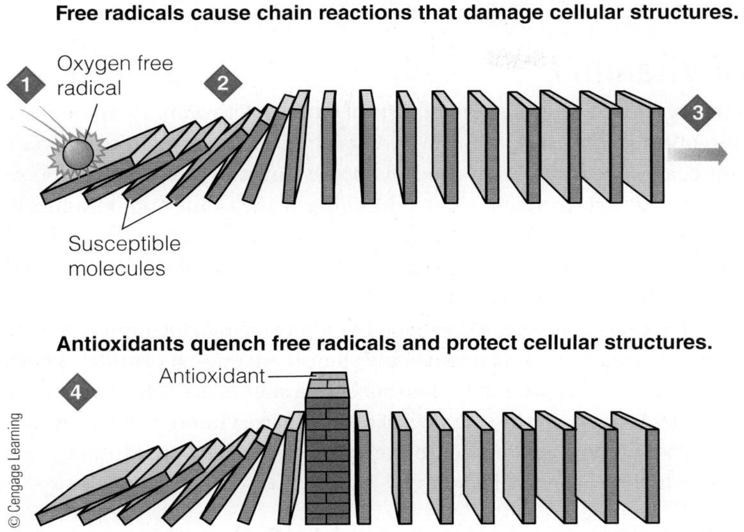
Free radicals cause chain reactions that damage cellular structures.
1
Oxygen free radical
2
3
Susceptible molecules
Antioxidants quench free radicals and protect cellular structures.
Antioxidant
4
© Cengage Learning

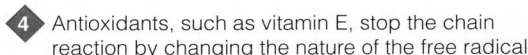

1. A chemically reactive oxygen free radical attacks fatty acid, DNA, protein, or cholesterol molecules, which form other free radicals in turn.
2. This initiates a rapid, destructive chain reaction.
3. The result is:
   - Cell membrane lipid damage.
   - Cellular protein damage.
   - DNA damage.
   - Oxidation of LDL cholesterol.
   - Inflammation.

   These changes may initiate steps leading to diseases such as heart disease, cancer, macular degeneration, and others.
4. Antioxidants, such as vitamin E, stop the chain reaction by changing the nature of the free radical.

Four additional forms of vitamin E have also been identified and are of interest to researchers for potential roles in health.**[44]

Roles of Vitamin E

Vitamin E is an antioxidant and thus acts as a bodyguard against oxidative damage. Such damage occurs when highly unstable molecules known as **free radicals**, formed during normal cell metabolism, run amok. Left unchecked, free radicals create a destructive chain reaction that can damage the polyunsaturated lipids in cell membranes and lipoproteins (LDL), the DNA in genetic material, and the working proteins of cells. This creates inflammation and cell damage associated with aging processes, cancer development, heart disease, and other diseases.[45] Vitamin E, by being oxidized itself, quenches free radicals and reduces inflammation. Figure 7–7 provides an overview of the antioxidant activity of vitamin E and its potential role in disease prevention.[46]

The antioxidant protection of vitamin E is crucial, particularly in the lungs where high oxygen concentrations would otherwise disrupt vulnerable membranes. Red blood cell membranes also need vitamin E's protection as they transport oxygen from the lungs to other tissues. White blood cells that fight diseases equally depend on vitamin E's antioxidant nature, as do blood vessel linings, sensitive brain tissues, and even bones.[47] Tocopherols also perform some nonantioxidant tasks that support the body's health.

Did You Know?

Cooking methods using high heat, such as frying, destroy vitamin E. Only raw or gently cooked oils supply vitamin E.

Vitamin E Deficiency

A deficiency of vitamin E produces a wide variety of symptoms in laboratory animals, but these are almost never seen in healthy human beings. Deficiency of vitamin E, which dissolves in fat, may occur in people with diseases that cause fat malabsorption or in infants born prematurely. Disease or injury of the liver (which makes bile, necessary for digestion of fat), the gallbladder (which delivers bile into the intestine), or the pancreas (which makes fat-digesting enzymes) makes vitamin E deficiency likely. In people without diseases, low blood levels of vitamin E are most likely when diets extremely low in fat are consumed for years.

free radicals atoms or molecules with one or more unpaired electrons that make the atom or molecule unstable and highly reactive

**The other forms of Vitamin E are tocotrienols: alpha, beta, gamma, and delta.

Figure 7–8

Vitamin E Recommendations and Intakes Compared

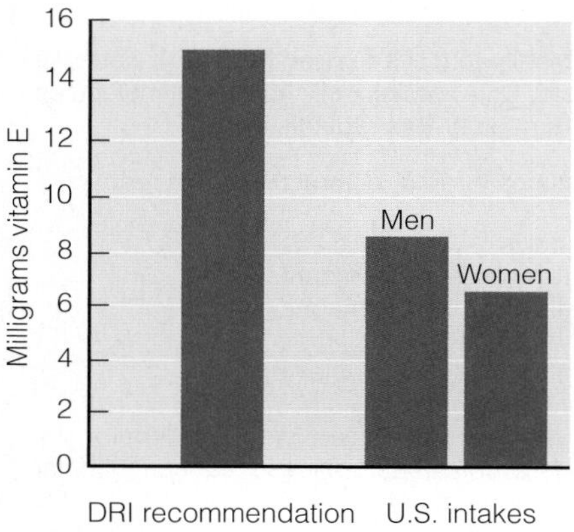

Source: USDA Agricultural Research Service, Table 1, Nutrient Intakes from Food: Mean Amounts Consumed per Individual, One Day, 2005–2006, What We Eat in America: NHANES, *available at www.ars.usda.gov.*

A classic vitamin E deficiency occurs in premature babies born before the transfer of the vitamin from the mother to the infant, which takes place late in pregnancy. Without sufficient vitamin E, the infant's red blood cells rupture (**erythrocyte hemolysis**), and the infant becomes anemic. The few symptoms of vitamin E deficiency observed in adults include loss of muscle coordination and reflexes and impaired vision and speech. Vitamin E corrects all of these symptoms.

Toxicity of Vitamin E

Vitamin E in foods is safe to consume, and reports of vitamin E toxicity symptoms are rare across a broad range of intakes. However, vitamin E in supplements augments the effects of anticoagulant medication used to oppose unwanted blood clotting, so people taking such drugs risk uncontrollable bleeding if they also take vitamin E. Supplemental doses of vitamin E prolong blood clotting times by interfering with the activity of vitamin K.[48] An increase in brain hemorrhages, a form of stroke, has been noted among people taking supplements of vitamin E.[49]

The pooled results from 67 experiments involving almost a quarter-million people suggested that taking vitamin E supplements may slightly increase mortality in both healthy and sick people.[50] Other studies find no effect or a slight decrease in mortality among certain groups.[51] To err on the safe side, people who use vitamin E supplements should probably keep their dosages low, not to exceed the Tolerable Upper Intake Level of 1,000 milligrams alpha-tocopherol per day.

Vitamin E Recommendations and U.S. Intakes

The DRI recommended intake (inside front cover) for vitamin E is 15 milligrams a day for adults. This amount is sufficient to maintain healthy, normal blood values for vitamin E for most people. On average, U.S. intakes of vitamin E fall substantially below the recommendation (see Figure 7–8).[52] The need for vitamin E rises as people consume more polyunsaturated oil because the oil requires antioxidant protection by the vitamin. Luckily, most raw oils also contain vitamin E, so people who eat raw oils also receive the vitamin. Smokers may have higher needs.

Vitamin E Food Sources

Vitamin E is widespread in foods (see Snapshot 7–3). Much of the vitamin E that people consume comes from vegetable oils and products made from them, such as margarine and salad dressings.[53] Wheat germ oil is especially rich in vitamin E. Animal fats have almost none.

Vitamin E is readily destroyed by heat and oxidation—thus, fresh, raw oils and lightly processed vitamin E–rich foods are the best sources. As people choose more highly processed foods, fried fast foods, or "convenience" foods, they lose vitamin E because little vitamin E survives the heating and other processes used to make these foods.

Raw vegetable oils contain substantial vitamin E, but high temperatures destroy it.

KEY POINTS

- Vitamin E acts as an antioxidant in cell membranes.
- Average U.S. intakes fall short of DRI recommendations.
- Vitamin E deficiency disease occurs rarely in newborn premature infants.
- Toxicity is rare but supplements may carry risks.

Vitamin K

LO 7.6 Summarize the physiological roles of vitamin K, name the consequences of deficiencies, and list its major food sources.

Have you ever thought about how remarkable it is that blood can clot? The liquid turns solid in a life-saving series of reactions—if blood did not clot, wounds would just keep bleeding, draining the blood from the body.

Snapshot 7-3 Vitamin E

DRI Recommended Intake
Adults: 15 mg/day

Tolerable Upper Intake Level
Adults: 1,000 mg/day

Chief Functions
Antioxidant (protects cell membranes, regulates oxidation reactions, protects polyunsaturated fatty acids)

Deficiency
Red blood cell breakage, nerve damage

Toxicity
Augments the effects of anticlotting medication

*These foods provide 10% or more of the vitamin E Daily Value in a serving. For a 2,000-cal diet, the DV is 30 IU or 20 mg/day.
[a]Cooking destroys vitamin E.

Good Sources*

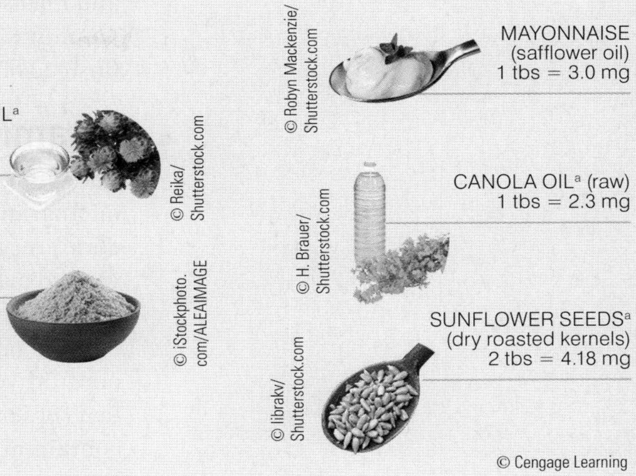

© Reika/Shutterstock.com; © iStockphoto.com/ALEAIMAGE; © Robyn Mackenzie/Shutterstock.com; © H. Brauer/Shutterstock.com; © librakv/Shutterstock.com

© Cengage Learning

Roles of Vitamin K

The main function of vitamin K is to help activate proteins that help clot the blood. Hospitals measure the clotting time of a person's blood before surgery and, if needed, administer vitamin K to reduce bleeding during the operation. Vitamin K is of value only if a vitamin K deficiency exists. Vitamin K does not improve clotting in those with other bleeding disorders, such as the inherited disease hemophilia.

Some people with heart problems need to *prevent* the formation of clots within their circulatory system—this is popularly referred to as "thinning" the blood. One of the best-known medicines for this purpose is warfarin (pronounced WAR-fuh-rin), which interferes with vitamin K's clot-promoting action. Vitamin K therapy may be needed for people on warfarin if uncontrolled bleeding should occur.[54] People taking warfarin who self-prescribe vitamin K supplements risk causing dangerous clotting of their blood; those who suddenly stop taking vitamin K risk causing excess bleeding.[55]

Vitamin K is also necessary for the synthesis of key bone proteins.[56] With low blood vitamin K, the bones produce an abnormal protein that cannot effectively bind the minerals that normally form bones.[57] People who consume abundant vitamin K in the form of green leafy vegetables suffer fewer hip fractures than those with lower intakes.[58] Vitamin K supplements, however, seem ineffective against bone loss, and more research is needed to clarify the links between vitamin K and bone health.[59]

Did You Know?

K stands for the Danish word *koagulation* ("clotting").

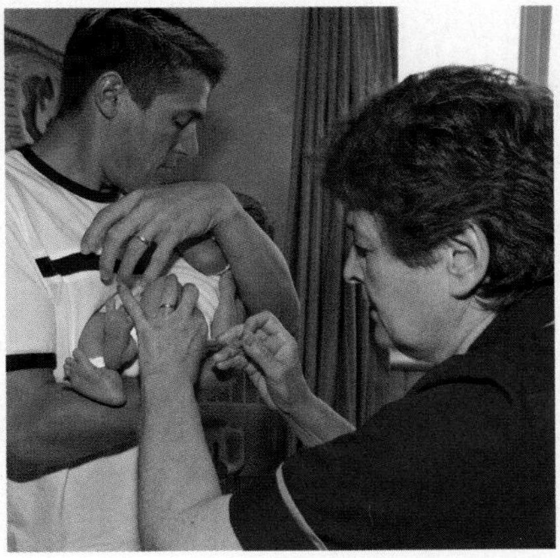

© SHOUT/Alamy

Soon after birth, newborn infants receive a dose of vitamin K.

Vitamin K Deficiency

Few U.S. adults are likely to experience vitamin K deficiency, even if they seldom eat vitamin K–rich foods. This is because, like vitamin D, vitamin K can be obtained from a nonfood source—in this case, the intestinal bacteria. Billions of bacteria normally reside in the intestines, and some of them synthesize vitamin K.

Newborn infants present a unique case with regard to vitamin K because they are born with a sterile intestinal tract, and the vitamin K–producing bacteria take weeks to establish themselves. To prevent hemorrhage, the newborn is given a single dose of vitamin K at birth.[60] People who have taken antibiotics that have killed the bacteria in their intestinal tracts also may develop vitamin K deficiency. In other medical conditions, bile production falters, making lipids, including all of the fat-soluble vitamins, unabsorbable. Supplements of the vitamin are needed in these cases because a vitamin K deficiency can be fatal.

erythrocyte (eh-REETH-ro-sight) **hemolysis** (HEE-moh-LIE-sis, hee-MOLL-ih-sis) rupture of the red blood cells that can be caused by vitamin E deficiency (*erythro* means "red"; *cyte* means "cell"; *hemo* means "blood"; *lysis* means "breaking"). The anemia produced by the condition is *hemolytic* (HEE-moh-LIT-ick) *anemia*.

Vitamin K Toxicity

Reports of vitamin K toxicity among healthy adults are rare, and the DRI committee has not set a Tolerable Upper Intake Level. For infants and pregnant women, however, vitamin K toxicity can result when supplements of a synthetic version of vitamin K are given too enthusiastically.[††] Toxicity induces breakage of the red blood cells and release of their pigment, which colors the skin yellow. A toxic dose of synthetic vitamin K causes the liver to release the blood cell pigment (bilirubin) into the blood (instead of excreting it into the bile) and leads to **jaundice**.

Vitamin K Requirements and Sources

The vitamin K requirement for men is 120 micrograms a day; women require 90 micrograms. As Snapshot 7–4 shows, vitamin K's richest plant food sources include dark green, leafy vegetables such as cooked spinach and other greens, which provide an average of 300 micrograms per half-cup serving. Lettuces, broccoli, brussels sprouts, and other members of the cabbage family are also good sources.

Only one rich animal food source of vitamin K exists: liver. Canola and soybean oils (unhydrogenated liquid oils) provide smaller but still significant amounts; fortified cereals can also be rich sources of added vitamin K. One egg and a cup of milk contain about equal amounts, or 25 micrograms each. Tables of food composition have begun to include the vitamin K contents of foods as better methods of analysis have been developed.

KEY POINTS

- Vitamin K is necessary for blood to clot.
- Vitamin K deficiency causes uncontrolled bleeding.
- Excess vitamin K can cause harm.
- The bacterial inhabitants of the digestive tract produce vitamin K.

Snapshot 7-4 Vitamin K

DRI Recommended Intakes

Men: 120 µg/day

Women: 90 µg/day

Chief Functions

Synthesis of blood-clotting proteins and bone proteins

Deficiency

Hemorrhage; abnormal bone formation

Toxicity

Opposes the effects of anticlotting medication

*These foods provide 10% or more of the vitamin K Daily Value in a serving. For a 2,000-cal diet, the DV is 80 µg/day.
Data from USDA.
ᵃAverage value.

[††] The version of vitamin K responsible for this effect is menadione.

Think Fitness move it! Vitamins for Athletes

Do athletes who strive for top performance need more vitamins than foods can supply? Competitive athletes who choose their diets with reasonable care almost never need nutrient supplements. The reason is elegantly simple. The need for energy to fuel exercise requires that people eat extra calories of food, and if that extra food is of the kind shown in this chapter's Snapshots—fruits, vegetables, milk, eggs, whole or enriched grains, lean meats, and some oils—then the extra vitamins needed to support the activity flow naturally into the body. Chapter 10 comes back to the roles of vitamins in physical activity.

start now! If you haven't already done so, go to Diet Analysis Plus and track your diet for 3 days, including one weekend day. After you have recorded your foods for 3 days, create an Intake Report to see how close you come to meeting the nutrient recommendations for a person of your age, weight, and level of physical activity.

The Water-Soluble Vitamins

LO 7.7 Describe some characteristics of the water-soluble vitamins.

Vitamin C and the B vitamins dissolve in water, which has implications for their handling in food and by the body. In food, water-soluble vitamins easily dissolve and drain away with cooking water, and some are destroyed on exposure to light, heat, or oxygen during processing.[61] Later sections examine vitamin vulnerability and provide tips for retaining vitamins in foods. Recall characteristics of water-soluble vitamins from Table 7–2, earlier.

In the body, water-soluble vitamins are easily absorbed and just as easily excreted in the urine. A few of the water-soluble vitamins can remain in the lean tissues for a month or more, but these tissues actively exchange materials with the body fluids all the time—no real storage tissues exist for any water-soluble vitamins. At any time, the vitamins may be picked up by the extracellular fluids, washed away by the blood, and excreted in the urine.

Advice for meeting the need for these nutrients is straightforward: choose foods rich in water-soluble vitamins to achieve an average intake that meets the recommendation over a few days' time. The snapshots in this section can help to guide your choices. Foods never deliver toxic doses of the water-soluble vitamins, and their easy excretion in the urine protects against toxicity from all but the largest supplemental doses. Normally, though, the most likely hazard to the supplement taker is to the wallet: If you take supplements of the water-soluble vitamins, you may have the most expensive urine in town. The Think Fitness feature asks whether athletes need vitamin supplements.

KEY POINTS

- Water-soluble vitamins are easily absorbed and excreted from the body, and foods that supply them must be consumed frequently.
- Water-soluble vitamins are easily lost or destroyed during food preparation and processing.

Vitamin C

LO 7.8 Summarize the physiological roles of vitamin C, name the consequences of deficiencies and toxicities, and list its major food sources.

More than 200 years ago, any man who joined the crew of a seagoing ship knew he had only half a chance of returning alive—not because he might be slain by pirates

jaundice (JAWN-dis) yellowing of the skin due to spillover of the bile pigment bilirubin (bill-ee-ROO-bin) from the liver into the general circulation.

Long voyages without fresh fruits and vegetables spelled death by scurvy for the crew.

or die in a storm, but because he might contract **scurvy**, a disease that often killed as many as two-thirds of a ship's crew on a long voyage. Ships that sailed on short voyages, especially around the Mediterranean Sea, were safe from this disease. The special hazard of long ocean voyages was that the ship's cook used up the perishable fresh fruits and vegetables early and relied on cereals and live animals for the duration of the voyage.

The first nutrition experiment to be conducted on human beings was devised 250 years ago to find a cure for scurvy. A physician divided some British sailors with scurvy into groups.‡‡ Each group received a different test substance: vinegar, sulfuric acid, seawater, oranges, or lemons. Those receiving the citrus fruits were cured within a short time. Sadly, it took 50 years for the British navy to make use of the information and require all its vessels to provide lime juice to every sailor daily. British sailors were mocked with the term *limey* because of this requirement. The name later given to the vitamin that the fruit provided, **ascorbic acid**, literally means "no-scurvy acid." It is more commonly known today as vitamin C.

The Roles of Vitamin C

Vitamin C performs a variety of functions in the body. It is best known for two of them: its work in maintaining the connective tissues and as an antioxidant.

A Cofactor for Enzymes Vitamin C assists several enzymes in performing their jobs. In particular, the enzymes involved in the formation and maintenance of the protein **collagen** depend on vitamin C for their activity. Collagen forms the base for all of the connective tissues: bones, teeth, skin, and tendons. Collagen forms the scar tissue that heals wounds, the reinforcing structure that mends fractures, and the supporting material of capillaries that prevents bruises. Vitamin C also participates in other synthetic reactions, such as in the production of carnitine, an important compound for transporting fatty acids within the cells and in the creation of certain hormones.

An Antioxidant In addition to assisting enzymes, vitamin C also acts in a more general way as an antioxidant.[62] Vitamin C protects substances found in foods and in the body from oxidation by being oxidized itself. For example, cells of the immune system maintain high levels of vitamin C to protect themselves from free radicals that they generate to use during assaults on bacteria and other invaders. After use, some oxidized vitamin C is degraded irretrievably and must be replaced by the diet. Most of the vitamin, however, is not lost but efficiently recycled back to its active form for reuse. This recycling system plays a key role in maintaining sufficient vitamin C in the cells to allow it to perform its critical work.

In the intestines, vitamin C protects iron from oxidation and so promotes its absorption. Once in the blood, vitamin C protects sensitive blood constituents from oxidation, reduces tissue inflammation, and helps to maintain the body's supply of vitamin E by protecting it and recycling it to its active form. The antioxidant roles of vitamin C are the focus of extensive study, especially in relation to disease prevention. Unfortunately, research has yielded only disappointing results: oral vitamin C supplements are useless against heart disease, cancer, and other diseases unless they are prescribed to treat a deficiency.

In test tubes, a high concentration of vitamin C has the opposite effect from antioxidants; that is, it acts as a **prooxidant** by activating oxidizing elements, such as iron and copper.[63] A few studies suggest that this may happen in people, too, under some conditions. The question of what, if anything, such findings may mean for human health remains unanswered.

Can Vitamin C Supplements Cure a Cold? Many people hold that vitamin C supplements can prevent or cure a common cold, but research most often fails to

‡‡ The physician was James Lind.

support this long-lived belief.[64] In 29 trials of over 11,350 people, no relationship emerged between routine vitamin C supplementation and cold prevention. A few studies do report other modest potential benefits—fewer colds, fewer ill days, and shorter duration of severe symptoms, especially for those exposed to physical and environmental stresses.[65] Fewer upper respiratory tract infections were reported in women consuming *sufficient* vitamin C from food alone; in men, who had lower intakes from food, upper respiratory infections were less frequent among supplement takers.[66] Sufficient vitamin C intake is critical to good health, of course.

Experimentally, supplements of at least 1 gram of vitamin C per day and often closer to 2 grams (the Tolerable Upper Intake Level and not recommended) seem to reduce blood histamine. Anyone who has ever had a cold knows the effects of histamine: sneezing, a runny or stuffy nose, and swollen sinuses. In drug-like doses, then, vitamin C may mimic a weak antihistamine drug, but studies vary in dosing and conditions; drawing conclusions is therefore difficult.[67]

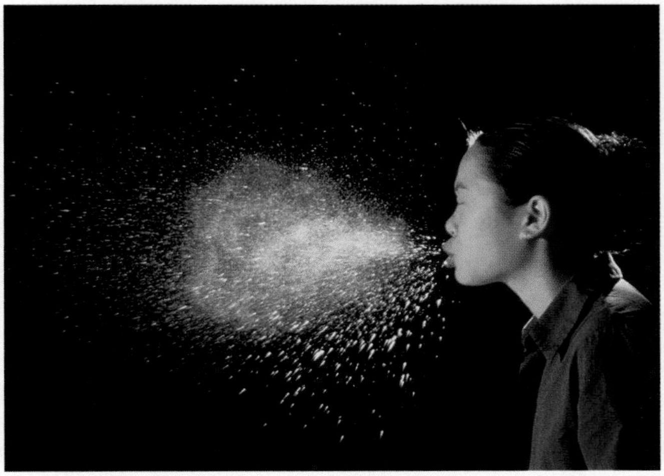

© A. Davidhazy/Custom Medical Stock Photo

Can vitamin C ease the suffering of a person with a cold?

One other effect of taking pills might also provide relief—the placebo effect.[68] In one vitamin C study, some experimental subjects received a placebo but were told they were receiving vitamin C. These subjects reported having fewer colds than the group who had in fact received the vitamin but who thought they were receiving the placebo. At work was the powerful healing effect of faith in a medical treatment.

Deficiency Symptoms

Most of the symptoms of scurvy can be attributed to the breakdown of collagen in the absence of vitamin C: loss of appetite, growth cessation, tenderness to touch, weakness, bleeding gums (shown in Figure 7–9), loose teeth, swollen ankles and wrists, and tiny red spots in the skin where blood has leaked out of capillaries (also shown in the figure). One symptom, anemia, reflects an important role worth repeating—vitamin C helps the body to absorb and use iron. Table 7–7 at the end of the chapter summarizes deficiency symptoms and other information about vitamin C.

In the United States, vitamin C status has been improving, but people who smoke or have low incomes continue to be at risk for deficiency.[69] The disease scurvy is seldom seen today except in a few elderly people, people addicted to alcohol or other drugs, sick people in hospitals, and a few infants who are fed only cow's milk.[70] Breast milk and infant formula supply enough vitamin C, but infants who are fed cow's milk

Figure 7–9

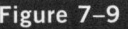

Scurvy Symptoms—Gums and Skin

Vitamin C deficiency causes the breakdown of collagen, which supports the teeth.

Small pinpoint hemorrhages (red spots) appear in the skin indicating that invisible internal bleeding may also be occurring.

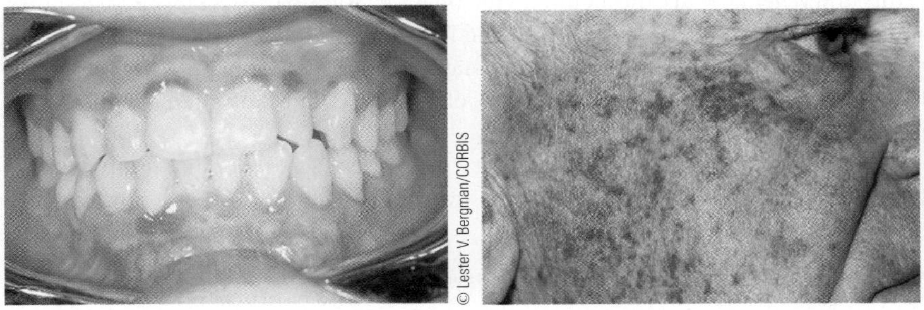

© Lester V. Bergman/CORBIS

Dr. P. Marazzi/Science Source

scurvy the vitamin C–deficiency disease.

ascorbic acid one of the active forms of vitamin C (the other is *dehydroascorbic* acid); an antioxidant nutrient.

collagen (COLL-a-jen) the chief protein of most connective tissues, including scars, ligaments, and tendons, and the underlying matrix on which bones and teeth are built.

prooxidant a compound that triggers reactions involving oxygen.

Figure 7–10

Vitamin C Tower of Recommendations

The DRI Tolerable Upper Intake Level for vitamin C is set at 2,000 mg (2 g)/day. Only 10 mg/day prevents scurvy.

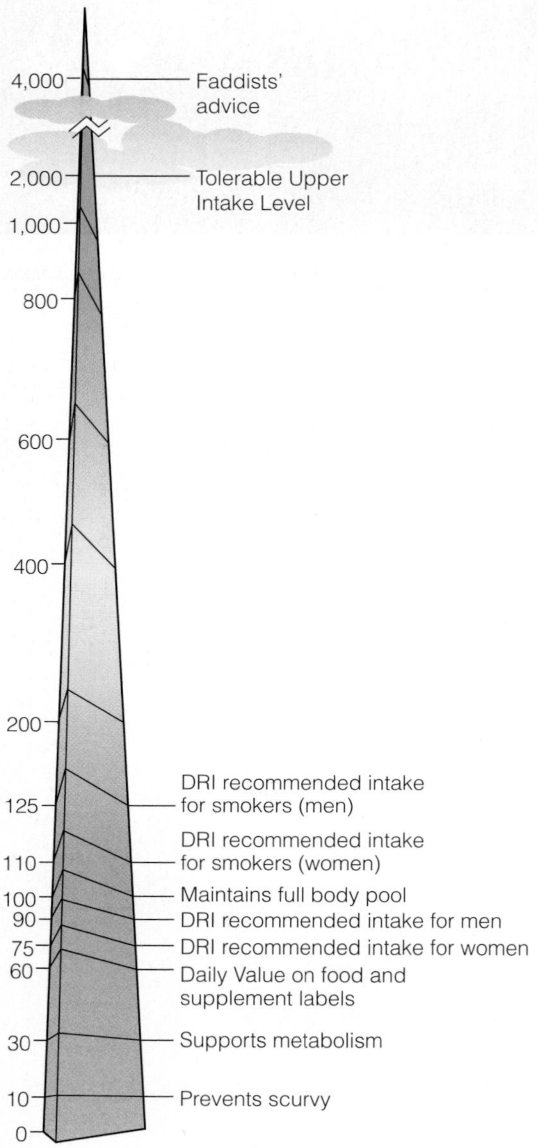

and receive no vitamin C in formula, fruit juice, or other outside sources are at risk. Low intakes of fruits and vegetables and a poor appetite overall lead to low vitamin C intakes and are not uncommon among people aged 65 and older. Vitamin C also supports immune system functions and so protects against infection.

Vitamin C Toxicity

The easy availability of vitamin C in pill form and the publication of books recommending vitamin C to prevent and cure colds and cancer have led thousands of people to take huge doses of vitamin C (see Figure 7–10). These "volunteer" subjects enabled researchers to study potential adverse effects of large vitamin C doses. One effect observed with a 2-gram dose is alteration of the insulin response to carbohydrate in people with otherwise normal glucose tolerances. People taking anticlotting medications may unwittingly counteract the effect if they also take massive doses of vitamin C. Those with kidney disease, a tendency toward gout, or a genetic abnormality that alters vitamin C's breakdown to its excretion products are prone to forming kidney stones if they take large doses of vitamin C.[§§] Vitamin C supplements in any dosage may be unwise for people with an overload of iron in the body because vitamin C increases iron absorption from the intestine and releases iron from storage. Other adverse effects are mild, including digestive upsets, such as nausea, abdominal cramps, excessive gas, and diarrhea.

The safe range of vitamin C intakes seems to be broad, from the absolute minimum of 10 milligrams a day to the Tolerable Upper Intake Level of 2,000 milligrams (2 grams), as Figure 7–10 demonstrates. Doses approaching 10 grams can be expected to be unsafe. Vitamin C from food is always safe.

Vitamin C Recommendations

The adult DRI intake recommendation for vitamin C is 90 milligrams for men and 75 milligrams for women. These amounts are far higher than the 10 or so milligrams per day needed to prevent the symptoms of scurvy. In fact, they are close to the amount at which the body's pool of vitamin C is full to overflowing: about 100 milligrams per day.

Tobacco use introduces oxidants that deplete the body's vitamin C. Thus, smokers generally have lower blood vitamin C levels than nonsmokers.[71] Even "passive smokers" who live and work with smokers and those who regularly chew tobacco need more vitamin C than others. Intake recommendations for smokers are set high, at 125 milligrams for men and 110 milligrams for women, in order to maintain blood levels comparable to those of nonsmokers. Importantly, vitamin C cannot reverse other damage caused by tobacco use. Physical stressors including infections, burns, fever, toxic heavy metals such as lead, and certain medications also increase the body's use of vitamin C.

Vitamin C Food Sources

Fruits and vegetables are the foods to remember for vitamin C, as Snapshot 7–5 shows. A cup of orange juice at breakfast, a salad for lunch, a stalk of broccoli and a potato at dinner easily provide 300 milligrams, making pills unnecessary. People commonly identify citrus fruits and juices as sources of vitamin C, but they often overlook other rich sources that may be lower in calories.

Vitamin C is vulnerable to heat and destroyed by oxygen, so for maximum vitamin C, consumers should treat their fruits and vegetables gently. Losses occurring when

§§ Vitamin C is inactivated and degraded by several routes, and oxalate, which can form kidney stones, is sometimes produced along the way. People may also develop oxalate crystals in their kidneys regardless of vitamin C status.

Snapshot 7-5 Vitamin C

DRI Recommended Intakes

Men:	90 mg/day
Women:	75 mg/day
Smokers:	add 35 mg/day

Tolerable Upper Intake Level

Adults:	2,000 mg/day

Chief Functions

Collagen synthesis (strengthens blood vessel walls, forms scar tissue, provides matrix for bone growth), antioxidant, restores vitamin E to active form, supports immune system, boosts iron absorption

Deficiency

Scurvy, with pinpoint hemorrhages, fatigue, bleeding gums, bruises; bone fragility, joint pain; poor wound healing, frequent infections

Toxicity

Nausea, abdominal cramps, diarrhea; rashes; interference with medical tests and drug therapies; in susceptible people, aggravation of gout or kidney stones

**These foods provide 10% or more of the vitamin C Daily Value in a serving. For a 2,000-cal diet, the DV is 60 mg/day.*

Good Sources*

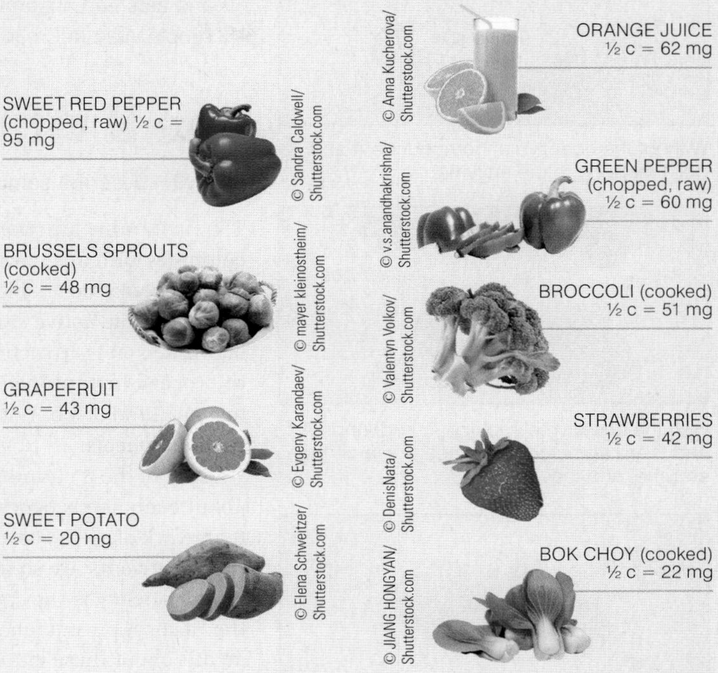

a food is cut, processed, and stored may be large enough to reduce vitamin C's activity in the body. Fresh, raw, and quickly cooked fruits, vegetables, and juices retain the most vitamin C, and they should be stored properly and consumed within a week after purchase. Table 7–5 gives tips for maximizing vitamin retention in foods.

Because of their enormous popularity, white potatoes contribute significantly to vitamin C intakes, despite providing less than 10 milligrams per half-cup serving.

Table 7–5

Minimizing Nutrient Losses

Each of these tactics saves a small percentage of the vitamins in foods but, repeated each day, can add up to significant amounts in a year's time.

Prevent enzymatic destruction:

- Refrigerate most fruits, vegetables, and juices to slow breakdown of vitamins.

Protect from light and air:

- Store milk and enriched grain products in opaque containers to protect riboflavin.
- Store cut fruits and vegetables in the refrigerator in airtight wrappers; reseal opened juice containers before refrigerating.

Prevent heat destruction or losses in water:

- Wash intact fruits and vegetables before cutting or peeling to prevent vitamin losses during washing.
- Cook fruits and vegetables in a microwave oven, or quickly stir fry, or steam them over a small amount of water to preserve heat-sensitive vitamins and to prevent vitamin loss in cooking water. Recapture dissolved vitamins by using cooking water for soups, stews, or gravies.
- Avoid high temperatures and long cooking times.

Figure 7–11

Animated! Coenzyme Action

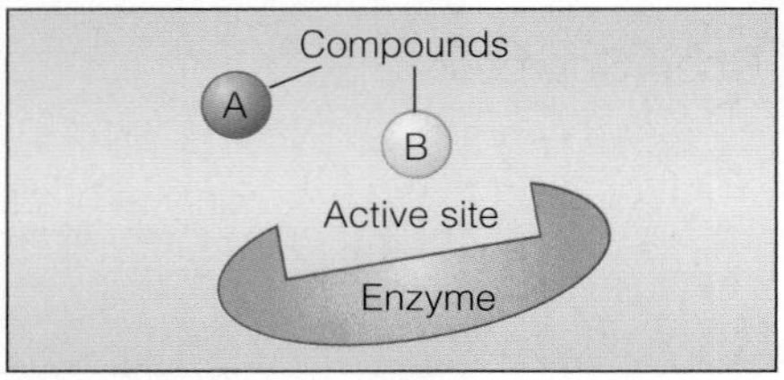

Without the coenzyme, compounds A and B don't respond to the enzyme.

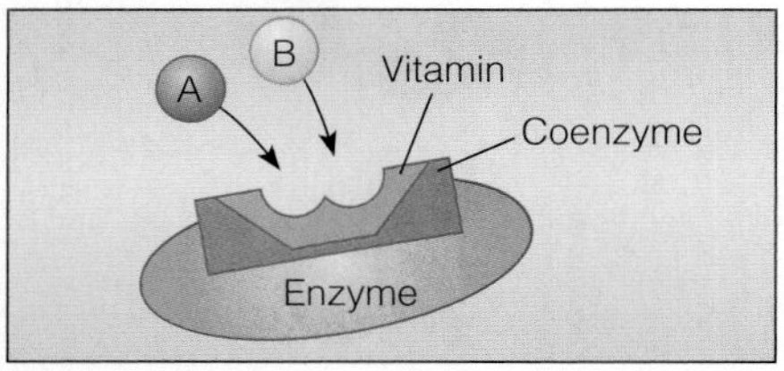

With the coenzyme in place, compounds A and B are attracted to the active site on the enzyme, and they react.

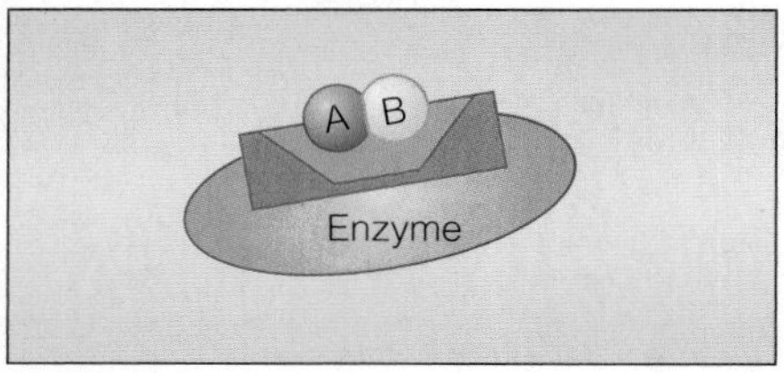

The reaction is completed with the formation of a new product. In this case the product is AB.

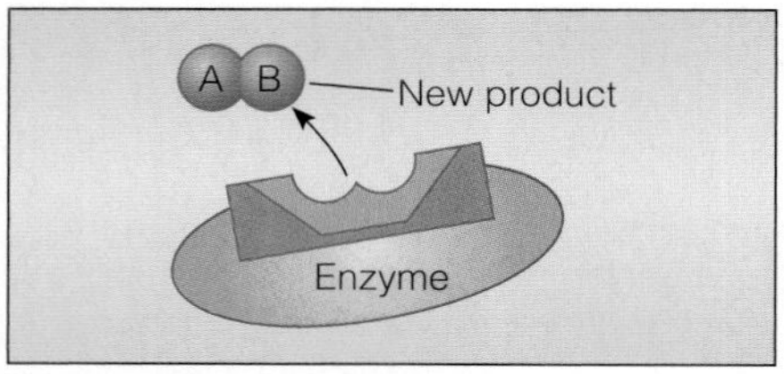

The product AB is released.

Did You Know?

To memorize the names of the eight B vitamins, try remembering this silly sentence or make up one of your own:

Tender	(thiamin)
romance	(riboflavin)
never	(niacin)
fails	(folate)
with 6 or 12	(B_6 and B_{12})
beautiful	(biotin)
pearls	(pantothenic acid)

The sweet potato, often ignored in favor of its paler cousin, is a gold mine of nutrients: a single half-cup serving provides about a third of many people's recommended intake for vitamin C, in addition to its lavish contribution of vitamin A.

KEY POINTS

- Vitamin C maintains collagen, protects against infection, acts as an antioxidant, and aids iron absorption.
- Ample vitamin C can be easily obtained from foods.

The B Vitamins in Unison

LO 7.9 Describe some of the shared roles of B vitamins in body systems.

The B vitamins function as part of coenzymes. A **coenzyme** is a small molecule that combines with an enzyme (described in Chapter 6) and activates it. Figure 7–11 shows how a coenzyme enables an enzyme to do its job. Sometimes the vitamin part of the enzyme is the active site, where the chemical reaction takes place. The substance to be worked on is attracted to the active site and snaps into place, enabling the reaction to proceed instantaneously. The shape of each enzyme predestines it to accomplish just one kind of job. Without its coenzyme, however, the enzyme is as useless as a car without wheels.

Each of the B vitamins has its own special nature, and the amount of detail known about each one is overwhelming. To simplify things, this introduction describes the teamwork of the B vitamins and emphasizes the consequences of deficiencies. Many of these nutrients are so interdependent that it is sometimes difficult to tell which vitamin deficiency is the cause of which symptom; the presence or absence of one affects the absorption, metabolism, and excretion of others. Later sections present a few details about these vitamins as individuals.

B Vitamin Roles in Metabolism

Figure 7–12 shows some body organs and tissues in which the B vitamins help the body metabolize carbohydrates, lipids, and amino acids. The purpose of the figure is not to present a detailed account of metabolism, but to give you an impression of where the B vitamins work together with enzymes in the metabolism of energy nutrients and in the making of new cells.

Many people mistakenly believe that B vitamins supply the body with energy. They do not, at least not directly. The B vitamins are "helpers." The energy-yielding nutrients—carbohydrate, fat, and protein—give the body fuel for energy; the B vitamins *help* the body use that fuel. More specifically, active forms of five of the B vitamins—thiamin, riboflavin, niacin, pantothenic acid, and biotin—participate in the release of energy from carbohydrate, fat, and protein. Vitamin B_6 helps the body use amino acids to synthesize proteins; the body then puts the protein to work in many ways—to build new tissues, to make hormones, to fight infections, or to serve as fuel for energy, to name only a few.

Folate and vitamin B_{12} help cells to multiply, which is especially important to cells with short life spans that must replace themselves rapidly. Such cells include both the red blood cells (which live for about 120 days) and the cells that line the digestive tract (which replace themselves every 3 days). These cells absorb and deliver energy to all the others. In short, each and every B vitamin is involved, directly or indirectly, in energy metabolism.

B Vitamin Deficiencies

As long as B vitamins are present, their presence is not felt. Only when they are missing does their absence manifest itself in a lack of energy and a multitude of other symptoms, as you can imagine after looking at Figure 7–12. The reactions by which B vitamins facilitate energy release take place in every cell, and no cell can do its work without energy. Thus, in a B vitamin deficiency, every cell is affected. Among the symptoms of B vitamin deficiencies are nausea, severe exhaustion, irritability, depression,

Figure 7–12
Some Roles of the B Vitamins in Metabolism: Examples

This figure does not attempt to teach intricate biochemical pathways or names of B vitamin–containing enzymes. Its sole purpose is to show a few of the many tissue functions that depend on a host of B vitamin–containing enzymes working together in harmony. The B vitamins work in every cell, and this figure displays less than a thousandth of what they actually do.

Every B vitamin is part of one or more coenzymes that make possible the body's chemical work. For example, the niacin, thiamin, and riboflavin coenzymes are important in the energy pathways. The folate and vitamin B_{12} coenzymes are necessary for making RNA and DNA and thus new cells. The vitamin B_6 coenzyme is necessary for processing amino acids and, therefore, protein. Many other relationships are also critical to metabolism.

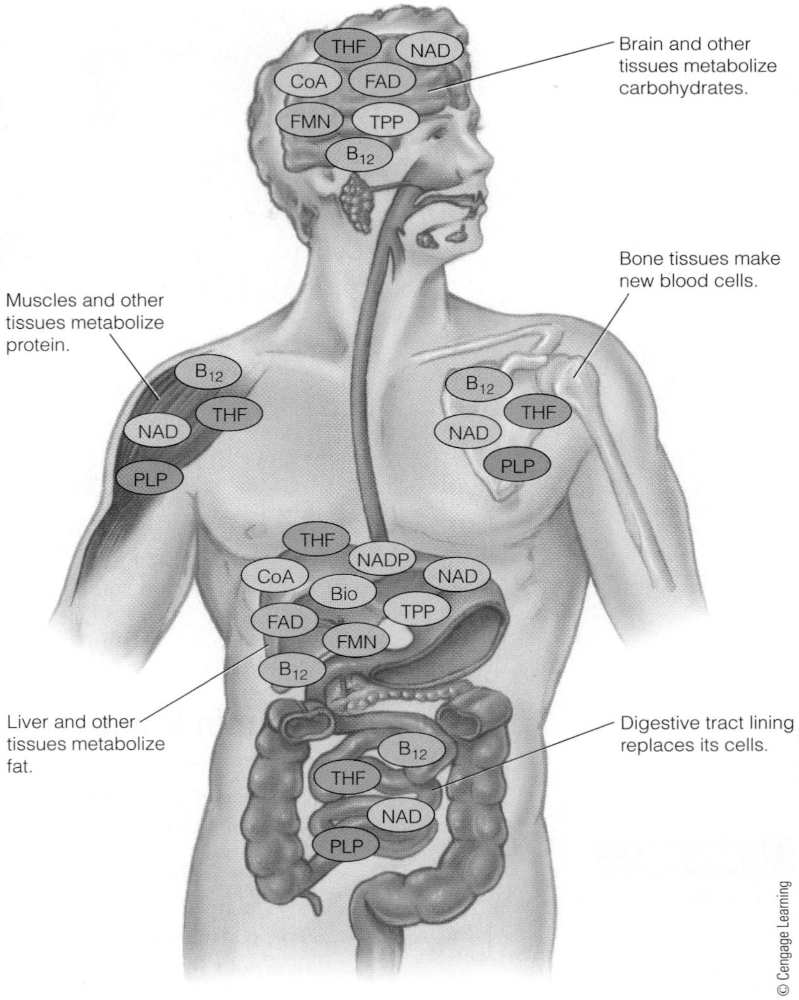

Key:

Coenzyme		Vitamin
TPP	=	thiamin
FAD FMN	=	riboflavin
NAD NADP	=	niacin
PLP	=	vitamin B_6
THF	=	folate
CoA	=	pantothenic acid
Bio	=	biotin
B_{12}	=	vitamin B_{12}

forgetfulness, loss of appetite and weight, pain in muscles, impairment of the immune response, loss of control of the limbs, abnormal heart action, severe skin problems, swollen red tongue, cracked skin at the corners of the mouth, and teary or bloodshot eyes. Figure 7–13, p. 260, shows some of these signs. Because cell renewal depends on energy and protein, which in turn depend on the B vitamins, the digestive tract and the blood are invariably damaged. In children, full recovery may be impossible. In the case of a thiamin deficiency during growth, permanent brain damage can result.

In academic discussions of the B vitamins, different sets of deficiency symptoms are given for each one. Such clear-cut sets of symptoms are found only in laboratory animals that have been fed fabricated diets that lack just one vitamin. In real life, a deficiency of any one B vitamin seldom shows up by itself because people don't eat nutrients singly; they eat foods that contain mixtures of nutrients. A diet low in one B vitamin is likely low in other nutrients, too. If treatment involves giving wholesome food rather than a single supplement, subtler deficiencies and impairments will be corrected along with the major one. The symptoms of B vitamin deficiencies and toxicities are listed in Table 7–7 at the end of the chapter.

KEY POINTS

- As part of coenzymes, the B vitamins help enzymes in every cell do numerous jobs.
- B vitamins help metabolize carbohydrate, fat, and protein.

coenzyme (co-EN-zime) a small molecule that works with an enzyme to promote the enzyme's activity. Many coenzymes have B vitamins as part of their structure (*co* means "with").

Figure 7–13

B Vitamin Deficiency Symptoms of the Tongue and Mouth

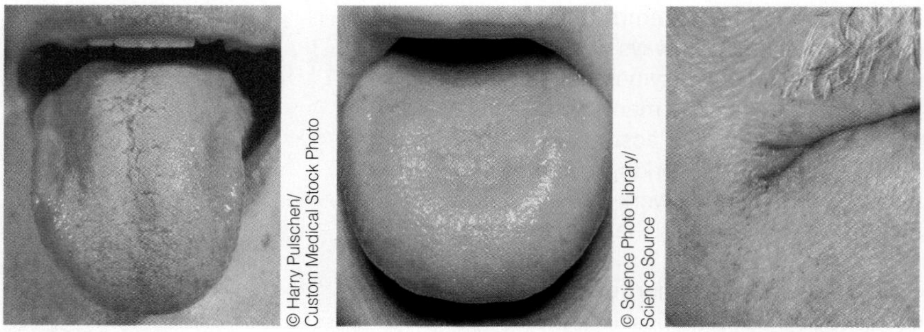

© Harry Pulschen/ Custom Medical Stock Photo

A healthy tongue has a rough and somewhat bumpy surface.

© Science Photo Library/ Science Source

In a B vitamin deficiency, the tongue becomes smooth and swollen.

Science Photo Library/ Custom Medical Stock Photography

In a B vitamin deficiency, the corners of the mouth become inflamed and cracked.

The B Vitamins as Individuals

LO 7.10 List and summarize the physiological roles of individual B vitamins in the body, name the consequences of deficiencies, and list their most important food sources.

Although the B vitamins all work as part of coenzymes and share other characteristics, each B vitamin has special qualities. The next sections provide a few details.

Thiamin Roles

Thiamin plays a critical role in the energy metabolism of all cells. Thiamin also occupies a special site on nerve cell membranes. Consequently, nerve processes and their responding tissues, the muscles, depend heavily on thiamin.

Figure 7–14

Beriberi

Beriberi takes two forms: wet beriberi, characterized by edema (fluid accumulation), and dry beriberi, without edema. This person's ankle retains the imprint of the physician's thumb, showing the edema of wet beriberi.

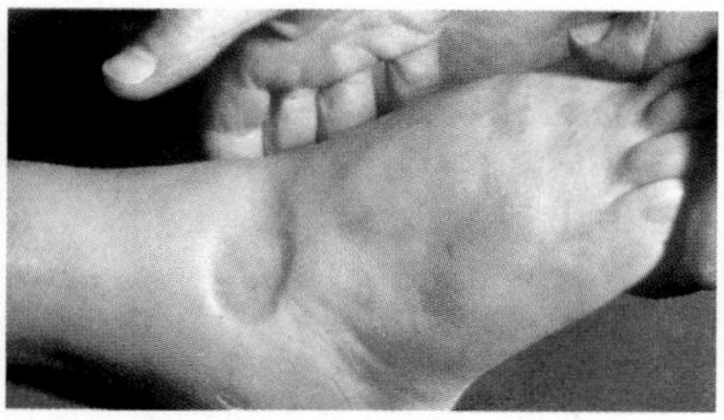

© NMSB/Custom Medical Stock Photo

Thiamin Deficiency The classic thiamin-deficiency disease, **beriberi**, was first observed in East Asia, where rice provided 80 to 90 percent of the total calories most people consumed and was therefore their principal source of thiamin. When the custom of polishing rice (removing its brown coat, which contained the thiamin) became widespread, beriberi swept through the population like an epidemic. Scientists wasted years of effort hunting for a microbial cause of beriberi before they realized that the cause was not something present in the environment, but something absent from it. Figure 7–14 depicts beriberi and describes its two forms.

Just before 1900, an observant physician working in a prison in East Asia discovered that beriberi could be cured with proper diet. The physician noticed that the chickens at the prison had developed a stiffness and weakness similar to that of the prisoners who had beriberi. The chickens were being fed the rice left on prisoners' plates. When the rice bran, which had been discarded in the kitchen, was given to the chickens, their paralysis was cured. The physician met resistance when he tried to feed the rice bran, the "garbage," to the prisoners but it worked—it produced a miracle cure like those described at the beginning of the chapter. Later, extracts of rice bran were used to prevent infantile beriberi; still later, thiamin was synthesized.

In developed countries today, alcohol abuse often leads to a severe form of thiamin deficiency, Wernicke-Korsakoff syndrome, defined in Controversy 3. Alcohol contributes energy but carries almost no nutrients with it and often displaces food from the diet. In addition, alcohol impairs absorption of thiamin from the digestive tract and hastens its excretion in the urine, tripling the risk of deficiency. The syndrome is characterized by symptoms almost indistinguishable from alcohol abuse itself: apathy, irritability, mental confusion, disorientation, memory loss, jerky eye movements, and a staggering gait.[72] Unlike alcohol toxicity, the syndrome responds quickly to an injection of thiamin.

Snapshot 7-6 Thiamin

DRI Recommended Intakes
Men: 1.2 mg/day
Women: 1.1 mg/day

Chief Functions
Part of coenzyme active in energy metabolism

Deficiency[a]
Beriberi with possible edema or muscle wasting; enlarged heart, heart failure, muscular weakness, pain, apathy, poor short-term memory, confusion, irritability, difficulty walking, paralysis, anorexia, weight loss

Toxicity
None reported

**These foods provide 10% or more of the thiamin Daily Value in a serving. For a 2,000-cal diet, the DV is 1.5 mg/day.*
[a]Severe thiamin deficiency is often related to heavy alcohol consumption.

Good Sources*

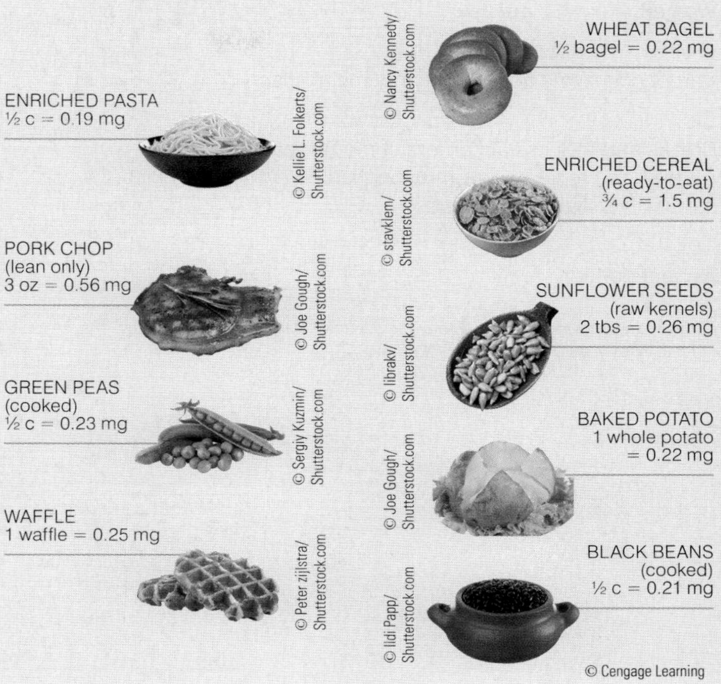

Recommended Intakes and Food Sources The DRI committee set the thiamin intake recommendation at 1.2 milligrams per day for men and at 1.1 milligrams per day for women. Pregnancy and lactation demand somewhat more thiamin (see the DRI, inside front cover, page B). Thiamin occurs in small amounts in many nutritious foods. Ham and other pork products, sunflower seeds, enriched and whole-grain cereals, and legumes are especially rich in thiamin (see Snapshot 7–6). If you keep empty-calorie foods to a minimum and focus your meals on nutritious foods each day, you will easily meet your thiamin needs.

KEY POINTS

- Thiamin works in energy metabolism and in nerve cells.
- Its deficiency disease is beriberi.
- Many foods supply small amounts of thiamin.

Riboflavin Roles and Sources

Like thiamin, **riboflavin** plays a role in the energy metabolism of all cells.[73] When thiamin is deficient, riboflavin may be lacking, too, but its deficiency symptoms, such as cracks at the corners of the mouth, sore throat, or hypersensitivity to light, may go undetected because those of thiamin deficiency are more severe. Worldwide, riboflavin deficiency has been documented among children whose eating patterns lack milk products and meats, and researchers suspect that it occurs among some U.S. elderly as well. An eating pattern that remedies riboflavin deficiency invariably contains some thiamin and so clears up both deficiencies.

Riboflavin recommendations are listed in Snapshot 7–7. People in this country obtain over a quarter of their riboflavin from enriched breads, cereals, pasta, and other grain products, while milk and milk products supply another 20 percent. Certain vegetables, eggs, and meats contribute most of the rest (see Snapshot 7–7).

thiamin (THIGH-uh-min) a B vitamin involved in the body's use of fuels.

beriberi (berry-berry) the thiamin-deficiency disease; characterized by loss of sensation in the hands and feet, muscular weakness, advancing paralysis, and abnormal heart action.

riboflavin (RIBE-o-flay-vin) a B vitamin active in the body's energy-releasing mechanisms.

Snapshot 7-7 Riboflavin

DRI Recommended Intakes
Men: 1.2 mg/day
Women: 1.1 mg/day

Chief Functions
Part of coenzyme active in energy metabolism

Deficiency
Cracks and redness at corners of mouth; painful, smooth, purplish red tongue; sore throat; inflamed eyes and eyelids, sensitivity to light; skin rashes

Toxicity
None reported

**These foods provide 10% or more of the riboflavin Daily Value in a serving. For a 2,000-cal diet, the DV is 1.7 mg/day.*

Good Sources*

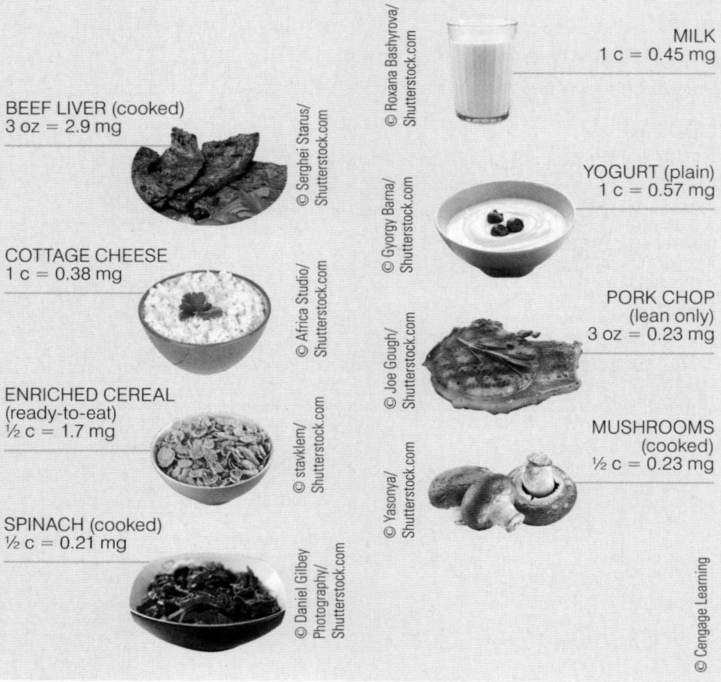

Ultraviolet light and irradiation destroy riboflavin. For these reasons, milk is sold in cardboard or opaque plastic containers and precautions are taken if milk is processed by irradiation. Riboflavin is stable to heat, so cooking does not destroy it.

KEY POINTS

- Riboflavin works in energy metabolism.
- Riboflavin is destroyed by ordinary light.

Niacin

The vitamin **niacin**, like thiamin and riboflavin, participates in the energy metabolism of every cell. Its absence causes serious illness.

Niacin Deficiency The niacin-deficiency disease **pellagra** appeared in Europe in the 1700s when corn from the New World became a staple food. In the early 1900s in the United States, pellagra was devastating lives throughout the South and Midwest. Hundreds of thousands of pellagra victims were thought to be suffering from a contagious disease until this dietary deficiency was identified. The disease still occurs among poorly nourished people living in urban slums and particularly in those with alcohol addiction. Pellagra is also still common in parts of Africa and Asia.[74] Its symptoms are known as the four "Ds": diarrhea, dermatitis, dementia, and, ultimately, death.

Figure 7–15 shows the skin disorder (dermatitis) associated with pellagra. For comparison, Figure 7–3 (p. 238) and Figure 7–18 (p. 268) show skin disorders associated with vitamin A and vitamin B_6 deficiency, respectively, a reminder that any nutrient deficiency affects the skin and all other cells. The skin just happens to be the organ you can see. Table 7–7 at the end of the chapter lists the symptoms of niacin deficiency.

Niacin Toxicity and Pharmacology For over 50 years, large doses of a form of niacin have been prescribed to help improve blood lipids associated with cardiovascular disease.[75***] Its use is limited, however, by the most common side effect of large doses of niacin, the "niacin flush," a dilation of the capillaries of the skin with perceptible tingling that can be painful.[76] Today, effective, well-tolerated drugs are often used instead, and the effectiveness of niacin is in question.[77] Reported risks from large doses of niacin include liver injury, digestive upset, impaired glucose tolerance, or, rarely, vision disturbances. Anyone considering taking large doses of niacin on their own should instead consult a physician who can prescribe safe, effective alternatives.[78]

Niacin Recommendations and Food Sources Niacin recommendations are listed in Snapshot 7–8, p. 264. The key nutrient that prevents pellagra is niacin, but any protein containing sufficient amounts of the amino acid tryptophan will serve in its place. Tryptophan, which is abundant in almost all proteins (but is limited in the protein of corn), is converted to niacin in the body, and it is possible to cure pellagra by administering tryptophan alone. Thus, a person eating adequate protein (as most people in developed nations do) will not be deficient in niacin. The amount of niacin in a diet is stated in terms of **niacin equivalents (NE)**, a measure that takes available tryptophan into account.

Early workers seeking the cause of pellagra observed that well-fed people never got it. From there the researchers defined an eating pattern that reliably produced the disease—one of cornmeal, salted pork fat, and molasses. Corn not only is low in protein but also lacks tryptophan. Salt pork is almost pure fat and contains too little protein to compensate; and molasses is virtually protein-free. Snapshot 7–8 shows some good food sources of niacin.

KEY POINTS

- Niacin deficiency causes the disease pellagra, which can be prevented by adequate niacin intake or adequate dietary protein.
- The amino acid tryptophan can be converted to niacin in the body.

Folate Roles

To make new cells, tissues must have the vitamin **folate**. Each new cell must be equipped with new genetic material—copies of the parent cell's DNA—and folate helps to synthesize DNA. Folate also participates in the metabolism of vitamin B_{12} and several amino acids.[79]

Folate Deficiency Folate deficiencies may result from following an eating pattern too low in folate or from illnesses that impair its absorption, increase its excretion, require medication that interacts with folate, or otherwise increase the body's folate need. However it occurs, folate deficiency has wide-reaching effects.

Immature red and white blood cells and the cells of the digestive tract divide most rapidly and therefore are most vulnerable to folate deficiency. Deficiencies of folate cause anemia, diminished immunity, and abnormal digestive function. The anemia of folate deficiency is related to the anemia of vitamin B_{12} malabsorption because the two vitamins work as teammates in producing red blood cells—see Figure 7–17 (on page 266). Research suggests that a chronic deficiency of folate also elevates the risk of cancer of the cervix (in women infected with a sexually transmitted virus, HPV[†††]), breast cancer (in women who drink alcohol), colon and rectal cancers, and pancreatic cancer.[80]

Of all the vitamins, folate is most likely to interact with medications. Many drugs, including antacids and aspirin and its relatives, have been shown to interfere with the body's use of folate. Occasional use of these drugs to relieve headache or upset stomach presents no concern, but frequent users may need to pay attention to their folate

*** The form of niacin is nicotinic acid.
††† Human papillomavirus.

Figure 7–15
Pellagra

The typical "flaky paint" dermatitis of pellagra develops on skin that is exposed to light. The skin darkens and flakes away.

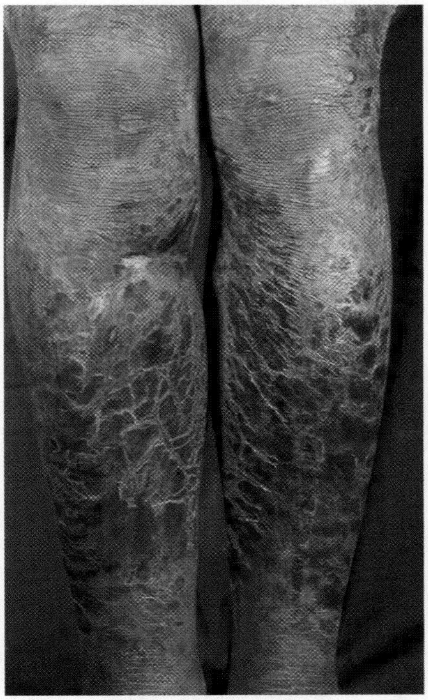

Dr. M.A. Ansary/Science Source

niacin a B vitamin needed in energy metabolism. Niacin can be eaten preformed or made in the body from tryptophan, one of the amino acids. Other forms of niacin are *nicotinic acid*, *niacinamide*, and *nicotinamide*.

pellagra (pell-AY-gra) the niacin-deficiency disease (*pellis* means "skin"; *agra* means "rough"). Symptoms include the "4 Ds": diarrhea, dermatitis, dementia, and, ultimately, death.

niacin equivalents (NE) the amount of niacin present in food, including the niacin that can theoretically be made from its precursor tryptophan that is present in the food.

folate (FOH-late) a B vitamin that acts as part of a coenzyme important in the manufacture of new cells. The form added to foods and supplements is *folic acid*.

Snapshot 7-8 Niacin

DRI Recommended Intakes
Men: 16 mg/day[a]
Women: 14 mg/day

Tolerable Upper Intake Level
Adults: 35 mg/day

Chief Functions
Part of coenzymes needed in energy metabolism

Deficiency
Pellagra, characterized by flaky skin rash (dermatitis) where exposed to sunlight; mental depression, apathy, fatigue, loss of memory, headache; diarrhea, abdominal pain, vomiting; swollen, smooth, bright red or black tongue

Toxicity
Painful flush, hives, and rash ("niacin flush"); excessive sweating; blurred vision; liver damage, impaired glucose tolerance

Good Sources*

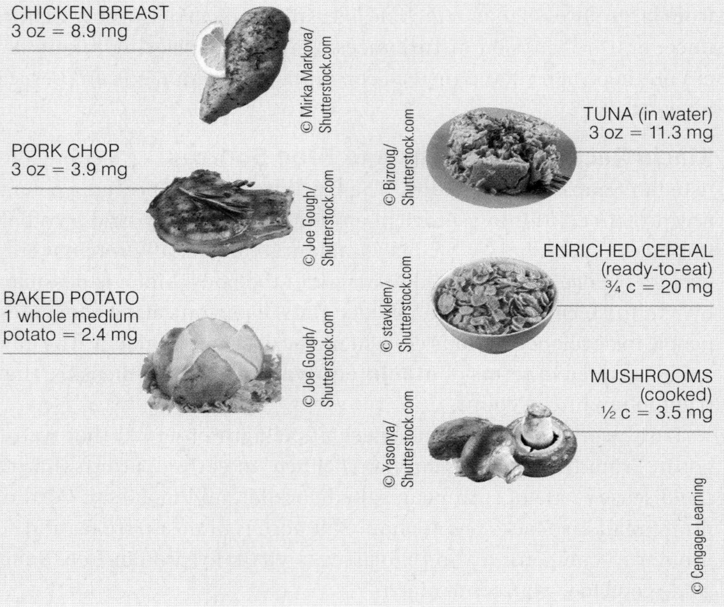

**These foods provide 10% or more of the niacin Daily Value in a serving. For a 2,000-cal diet, the DV is 20 mg/day. The DV values are for preformed niacin, not niacin equivalents.*
[a]Niacin DRI Recommended Intakes are expressed in niacin equivalents (NE); the Tolerable Upper Intake Level refers to preformed niacin.

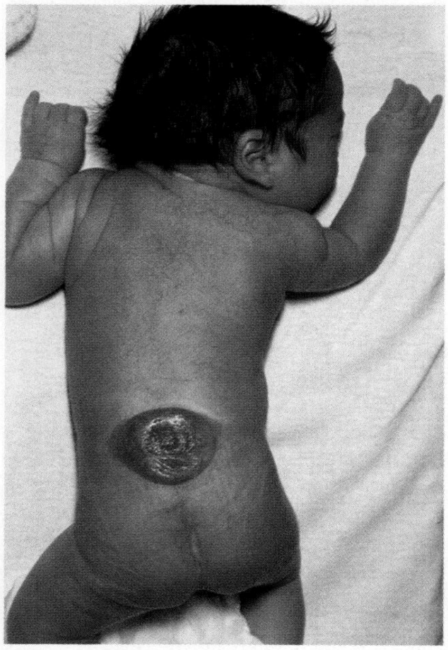
National Medical Slide Bank/Custom Medical Stock Photo

Spina bifida, a neural tube defect characterized by incomplete closure of the bony encasement of the spinal cord. Folate helps to prevent many such defects.

Did You Know?

The B vitamins thiamin, riboflavin, niacin, and folate (as folic acid) are among the enrichment nutrients added to refined grain foods such as breads and cereals.

intakes. These include people with chronic pain or ulcers who rely heavily on aspirin or antacids as well as those who smoke or take oral contraceptives or anticonvulsant medications.

Birth Defects and Folate Enrichment By consuming enough folate during pregnancy, a woman can reduce her child's risk of having one of the devastating birth defects known as **neural tube defects (NTD)**. NTD range from slight problems in the spine to mental retardation, severely diminished brain size, and death shortly after birth. NTD arise in the first days or weeks of pregnancy, long before most women suspect that they are pregnant. Adequate maternal folate may protect against certain other birth defects, as well.[81]

Most young women eat too few fruits and vegetables from day to day to supply even half the folate needed to prevent NTD.[82] In the late 1990s, the FDA ordered all enriched grain products such as bread, cereal, rice, and pasta sold in the United States to be fortified with an absorbable synthetic form of folate, *folic acid*. Since this fortification began, typical folate intakes from fortified foods have increased dramatically, along with average blood folate values.[83] Among women of child-bearing age, for example, prevalence of low serum folate concentrations dropped from 21 percent before folate fortification was introduced to less than 1 percent afterward. During the same period, the U.S. incidence of NTD dropped by a fourth (see Figure 7–16). Miscarriages and certain other birth defects, such as cleft lip, diminished as well.

Folate Toxicity As for folate toxicity, a Tolerable Upper Intake Level for synthetic folic acid from supplements and enriched foods is set at 1,000 micrograms a day for adults. The current level of folate fortification of the food supply appears to be safe for most people but a question remains about the ability of folate to mask a **subclinical deficiency** of vitamin B_{12} (more about this effect later).[84] Also, a lingering suspicion links a high folic acid intake with possibly *increased* risks for developing certain cancers.[85] About 5 percent of the U.S. population exceeds the Tolerable Upper Intake Level for folate, primarily people age 50 and older—the people most at risk for both vitamin B_{12} deficiencies and cancers.[86]

Figure 7–16

Incidence of Spina Bifida Before and After Folate Fortification

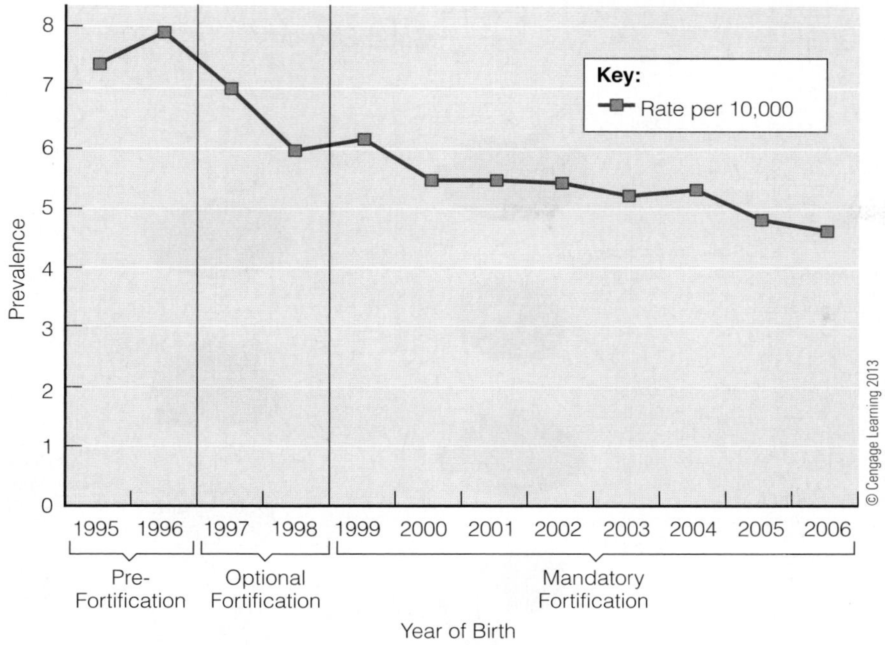

Source: National Center for Health Statistics, Centers for Disease Control and Prevention, available at www.cdc.gov, updated January 2010.

Neural tube defects such as spina bifida have declined since folate fortification began in 1996.

Folate Recommendations The DRI recommended intake for folate for healthy adults is set at 400 micrograms per day. The DRI committee also advises all women of child-bearing age to consume 400 micrograms of *folic acid*, a highly available form of folate, from supplements or enriched foods each day in addition to the folate that occurs naturally in their foods.[87]

Folate Food Sources The name *folate* is derived from the word *foliage*, and sure enough, leafy green vegetables such as spinach and turnip greens provide abundant folate (see Snapshot 7–9, p. 266). Fresh, uncooked vegetables and fruits are often superior sources because the heat of cooking and the oxidation that occurs during storage destroy much of the folate in foods. Eggs also provide some folate.

A difference in absorption between naturally occurring food folate and the synthetic folic acid necessitates compensation when measuring folate. The unit of measure, **dietary folate equivalent**, or **DFE**, converts all forms of folate into micrograms that are equivalent to the folate in foods. Appendix C demonstrates how to use the DFE conversion factor.

KEY POINTS

- Low intakes of folate cause anemia, digestive problems, and birth defects in infants of folate-deficient mothers.
- High intakes can mask the blood symptom of a vitamin B_{12} deficiency.

Vitamin B_{12} Roles

Vitamin B_{12} and folate are closely related: each depends on the other for activation. By itself, vitamin B_{12} also helps to maintain the sheaths that surround and protect nerve fibers.

Vitamin B_{12} Deficiency Symptoms Without sufficient vitamin B_{12}, nerves become damaged and folate fails to do its blood-building work, so vitamin B_{12} deficiency causes an anemia identical to that caused by folate deficiency. The blood symptoms of a deficiency of either folate or vitamin B_{12} include the presence of large, immature red blood

neural tube defects (NTD) abnormalities of the brain and spinal cord apparent at birth and associated with low folate intake in women before and during pregnancy. The neural tube is the earliest brain and spinal cord structure formed during gestation. Also defined in Chapter 13.

subclinical deficiency a nutrient deficiency that has no outward clinical symptoms. Also called *marginal deficiency*.

dietary folate equivalent (DFE) a unit of measure expressing the amount of folate available to the body from naturally occurring sources. The measure mathematically equalizes the difference in absorption between less absorbable food folate and highly absorbable synthetic folate added to enriched foods and found in supplements.

vitamin B_{12} a B vitamin that helps to convert folate to its active form and also helps maintain the sheath around nerve cells. Vitamin B_{12}'s scientific name, not often used, is *cyanocobalamin*.

Snapshot 7-9 Folate

DRI Recommended Intake
Adults: 400 μg DFE/day[a]

Tolerable Upper Intake Level
Adults: 1,000 μg DFE/day

Chief Functions
Part of a coenzyme needed for new cell synthesis

Deficiency
Anemia, smooth, red tongue; depression, mental confusion, weakness, fatigue, irritability, headache; a low intake increases the risk of neural tube birth defects

Toxicity
Masks vitamin B_{12}–deficiency symptoms

**These foods provide 10% or more of the folate Daily Value in a serving. For a 2,000-cal diet, the DV is 400 μg/day.*
[a]Folate recommendations are expressed in dietary folate equivalents (DFE). Note that for natural folate sources, 1 μg = 1 DFE; for enrichment sources, 1 μg = 1.7 DFE.
[b]Some highly enriched cereals may provide 400 μg or more in a serving.

Good Sources*

BEEF LIVER (cooked)
3 oz = 221 μg DFE

PINTO BEANS (cooked)
½ c = 146 μg DFE

ASPARAGUS
½ c = 134 μg DFE

AVOCADO (cubed)
½ c = 61 μg DFE

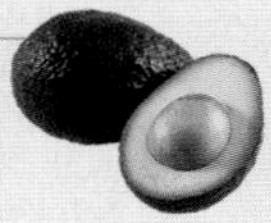

LENTILS (cooked)
½ c = 179 μg DFE

SPINACH (raw)
1 c = 58 μg DFE

ENRICHED CEREAL (ready-to-eat)[b]
¾ c = 400 μg DFE

BEETS
½ c = 68 μg DFE

Figure 7–17

Anemic and Normal Blood Cells

The anemia of folate deficiency is indistinguishable from that of vitamin B_{12} deficiency.

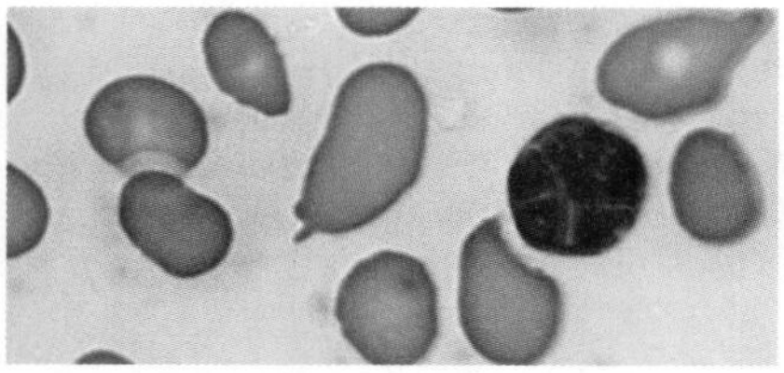

Blood cells of pernicious anemia. *The cells are larger than normal and irregular in shape.*

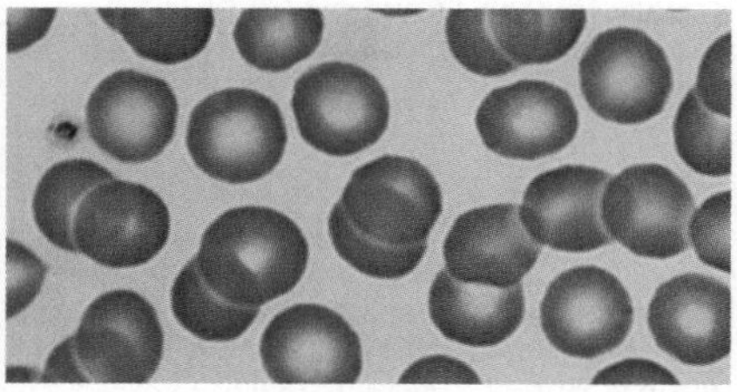

Normal blood cells. *The size, shape, and color of these red blood cells show that they are normal.*

cells. Administering extra folate often clears up this blood condition but allows the deficiency of vitamin B_{12} to continue undetected.[88] Vitamin B_{12}'s other functions then become compromised, and the results can be devastating: damaged nerve sheaths, creeping paralysis, and general malfunctioning of nerves and muscles. Evidence is mounting to suggest that even a marginal vitamin B_{12} deficiency may impair mental functioning in the elderly, worsening dementia.[89]

New NHANES testing efforts will soon yield much-needed information about the nation's vitamin B_{12} status.[90] Meanwhile, in an effort to prevent excessive folate intakes that could mask symptoms of a subclinical vitamin B_{12} deficiency, the FDA specifies the exact amounts of folic acid that can be added to enriched foods.

A Special Case: Vitamin B_{12} Absorption For vitamin B_{12}, deficiencies most often reflect poor absorption that occurs for one of two reasons:

- The stomach produces too little acid to liberate vitamin B_{12} from food.
- **Intrinsic factor**, a compound made by the stomach and needed for absorption, is lacking.

Once the stomach's acid frees vitamin B_{12} from the food proteins that bind it, intrinsic factor attaches to the vitamin, and the complex is absorbed into the bloodstream. The anemia of the vitamin B_{12} deficiency caused by lack of intrinsic factor is known as **pernicious anemia** (see Figure 7–17).

In a few people an inborn defect in the gene for intrinsic factor begins to impair vitamin B_{12} absorption by mid-adulthood. With age, many others lose their ability to produce enough stomach acid and intrinsic factor to allow efficient absorption of vitamin B_{12}.[‡‡‡] Intestinal diseases, surgeries, or stomach infection with an ulcer-causing bacterium can also impair absorption.[91] Taking a common diabetes drug also makes

[‡‡‡] This condition is atrophic gastritis (a-TROH-fik gas-TRY-tis), a chronic inflammation of the stomach accompanied by a diminished size and functioning of the stomach's mucous membrane and glands.

Snapshot 7-10 Vitamin B_{12}

DRI Recommended Intake

Adults: 2.4 μg/day

Chief Functions

Part of coenzymes needed in new cell synthesis; helps to maintain nerve cells

Deficiency

Pernicious anemia;[a] anemia (large-cell type);[b] smooth tongue; tingling or numbness; fatigue, memory loss, disorientation, degeneration of nerves progressing to paralysis

Toxicity

None reported

These foods provide 10% or more of the vitamin B_{12} Daily Value in a serving. For a 2,000-cal diet, the DV is 6 μg/day.

[a]*The name pernicious anemia refers to the vitamin B_{12} deficiency caused by a lack of stomach intrinsic factor, but not to anemia from inadequate dietary intake.*

[b]*Large cell–type anemia is known as either macrocytic or megaloblastic anemia.*

Good Sources*

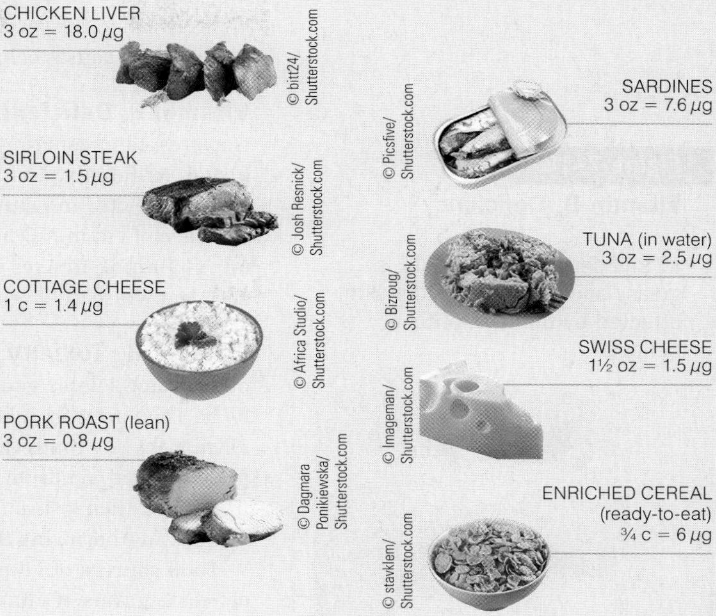

vitamin B_{12} deficiency likely.[§§§92] In cases of malabsorption, vitamin B_{12} must be supplied by injection or via nasal spray to bypass the defective absorptive system.

Vitamin B_{12} Food Sources As Snapshot 7–10 shows, vitamin B_{12} is present only in foods of animal origin, so vitamin B_{12} deficiency poses a threat to strict vegetarians. Controversy 6 discussed vitamin B_{12} sources for vegetarians.

Perspective The way folate masks the anemia of vitamin B_{12} deficiency underscores a point about supplements. It takes a skilled professional to correctly diagnose and treat a nutrient deficiency, and self-diagnosis or acting on advice from self-proclaimed experts poses serious risks. A second point: Since vitamin B_{12} deficiency in the body may be caused by either a lack of the vitamin in the diet or a lack of the intrinsic factor necessary to absorb the vitamin, a dietary change alone may not correct the deficiency; a professional diagnosis can identify such problems.

KEY POINTS

- Vitamin B_{12} occurs only in animal products.
- A deficiency anemia that mimics folate deficiency arises with low intakes or, more often, poor absorption.
- Folate supplements can mask a vitamin B_{12} deficiency.
- Prolonged vitamin B_{12} deficiency causes nerve damage.

Vitamin B_6 Roles

Vitamin B_6 participates in more than 100 reactions in body tissues and is needed to help convert one kind of amino acid, which cells have in abundance, to other nonessential amino acids that the cells lack.[93] In addition, vitamin B_6 functions in these ways:

- Aids in the conversion of tryptophan to niacin.

§§§The diabetes medication is metformin.

intrinsic factor a factor found inside a system. The intrinsic factor necessary to prevent pernicious anemia is now known to be a compound that helps in the absorption of vitamin B_{12}.

pernicious (per-NISH-us) **anemia** a vitamin B_{12}–deficiency disease, caused by lack of intrinsic factor and characterized by large, immature red blood cells and damage to the nervous system (*pernicious* means "highly injurious or destructive").

vitamin B_6 a B vitamin needed in protein metabolism. Its three active forms are *pyridoxine*, *pyridoxal*, and *pyridoxamine*.

- Plays important roles in the synthesis of hemoglobin and neurotransmitters, the communication molecules of the brain. (For example, vitamin B_6 assists the conversion of the amino acid tryptophan to the neurotransmitter **serotonin**.)
- Assists in releasing stored glucose from glycogen and thus contributes to the regulation of blood glucose.
- Has roles in immune function and steroid hormone activity.
- Is critical to the developing brain and nervous system of a fetus; deficiency during this stage causes behavioral problems later.

Figure 7–18
Vitamin B_6 Deficiency

In this dermatitis, the skin is greasy and flaky, unlike the skin affected by the dermatitis of pellagra.

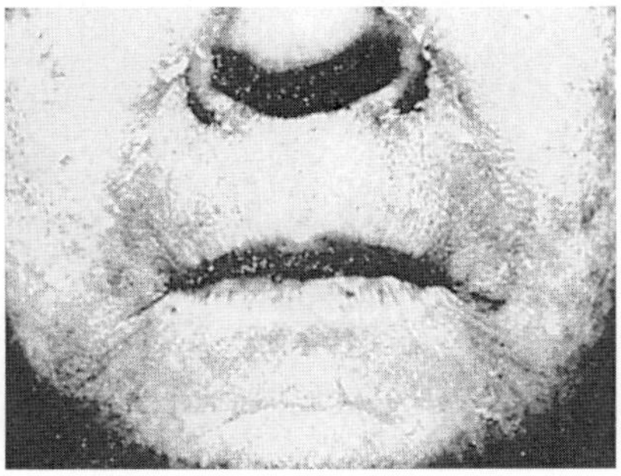

© George L. Blackburn, M.D., Ph. D., Harvard Medical School

Vitamin B_6 Deficiency Because of these diverse functions, vitamin B_6 deficiency is expressed in general symptoms, such as weakness, psychological depression, confusion, irritability, and insomnia. Other symptoms include anemia, the greasy dermatitis depicted in Figure 7–18, and, in advanced cases of deficiency, convulsions. A shortage of vitamin B_6 may also weaken the immune response. Some evidence links low vitamin B_6 intakes with increased risks of some cancers and cardiovascular disease; more research is needed to clarify these associations.[94]

Vitamin B_6 Toxicity Years ago it was generally believed that, like most of the other water-soluble vitamins, vitamin B_6 could not reach toxic concentrations in the body. Then a report told of women who took more than 2 grams of vitamin B_6 daily for months (20 times the current UL of 100 *milligrams* per day), attempting to cure premenstrual syndrome (science doesn't support this use). The women developed numb feet, then lost sensation in their hands, and eventually became unable to walk or work. Withdrawing the supplement reversed the symptoms.

Food sources of vitamin B_6 are safe. Consider that one small capsule can easily deliver 2 grams of vitamin B_6 but it would take almost 3,000 bananas, more than 1,600 servings of liver, or more than 3,800 chicken breasts to supply an equivalent amount. Moral: Stick with food. Table 7–7 (pp. 171–173) lists common deficiency and toxicity symptoms and food sources of vitamin B_6 and other water-soluble vitamins.

Vitamin B_6 Recommendations and Sources Vitamin B_6 plays so many roles in protein metabolism that the body's requirement for vitamin B_6 is roughly proportional to protein intakes. The DRI committee set the vitamin B_6 intake recommendation high enough to cover most people's needs, regardless of differences in protein intakes (see the inside front cover). Meats, fish, and poultry (protein-rich foods); potatoes; leafy green vegetables; and some fruits are good sources of vitamins B_6 (see Snapshot 7–11). Other foods such as legumes and peanut butter provide smaller amounts.

KEY POINT

- Vitamin B_6 works in amino acid metabolism.

Biotin and Pantothenic Acid

Two other B vitamins, **biotin** and **pantothenic acid**, are, like thiamin, riboflavin, and niacin, important in energy metabolism. Biotin is a cofactor for several enzymes in the metabolism of carbohydrate, fat, and protein. Recently, scientists have revealed roles for biotin in gene expression. No adverse effects from high biotin intakes have been reported, but some research indicates that high-dose biotin supplementation may damage DNA. No Tolerable Upper Intake Level has yet been set for biotin.

Pantothenic acid is a component of a key coenzyme that makes possible the release of energy from the energy nutrients. It also participates in more than 100 steps in the synthesis of lipids, neurotransmitters, steroid hormones, and hemoglobin.

Although rare diseases may precipitate deficiencies of biotin and pantothenic acid, both vitamins are readily available in foods. A steady diet of raw egg whites, which contain a protein that binds biotin, can produce biotin deficiency, but you would have to consume more than two dozen raw egg whites daily to produce the effect. Cooking eggs denatures the protein. Healthy people eating ordinary diets are not at risk for deficiencies.

© Alex Staroseltsev/Shutterstock.com

Snapshot 7-11 Vitamin B_6

DRI Recommended Intake
Adults (19–50 yr): 1.3 mg/day

Tolerable Upper Intake Level
Adults: 100 mg/day

Chief Functions
Part of a coenzyme needed in amino acid and fatty acid metabolism; helps to convert tryptophan to niacin and to serotonin; helps to make hemoglobin for red blood cells

Deficiency
Anemia, depression, confusion, abnormal brain wave pattern, convulsions; greasy, scaly dermatitis

Toxicity
Depression, fatigue, impaired memory, irritability, headaches, nerve damage causing numbness and muscle weakness progressing to an inability to walk and convulsions; skin lesions

These foods provide 10% or more of the vitamin B_6 Daily Value in a serving. For a 2,000-cal diet, the DV is 2 mg/day.

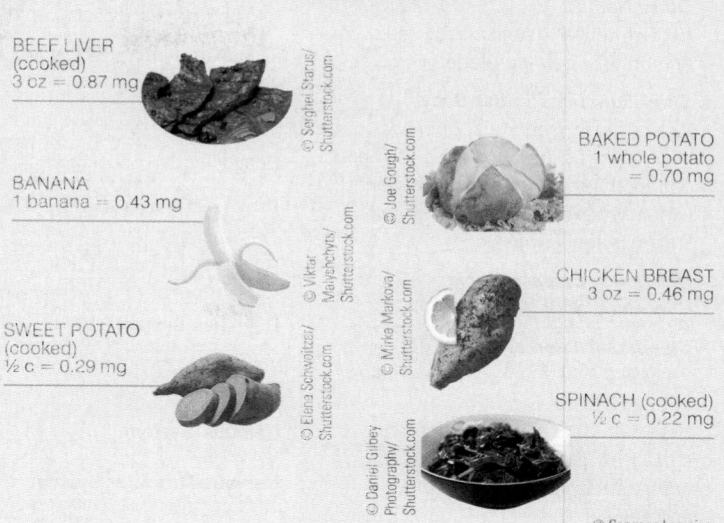

Good Sources*

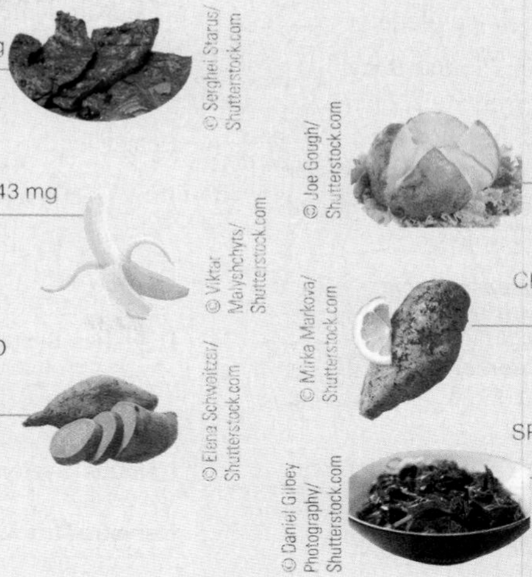

KEY POINT

- Biotin and pantothenic acid are important to the body and are abundant in food.

Non–B Vitamins

Choline, although not defined as a vitamin, might be called a conditionally essential nutrient. When the diet is devoid of choline, the body cannot make enough of the compound to meet its needs, and choline plays important roles in fetal development.[95] Choline is widely supplied by protein foods, and deficiencies are practically unheard of outside the laboratory.[96] DRI recommendations have been set for choline (see inside front cover).

The compounds **carnitine**, **inositol**, and **lipoic acid** might appropriately be called *nonvitamins* because they are not essential nutrients for human beings. Carnitine, sometimes called "vitamin BT," is an important piece of cell machinery, but it is not a vitamin. Although deficiencies can be induced in laboratory animals for experimental purposes, these substances are abundant in ordinary foods. Even if these compounds were essential in human nutrition, supplements would be unnecessary for healthy people eating a balanced diet. Vitamin companies often include these substances to make their formulas appear more "complete," but there is no physiological reason to do so.

Other substances have been mistakenly thought to be essential in human nutrition because they are needed for growth by bacteria or other life-forms. These substances include PABA (para-aminobenzoic acid), bioflavonoids ("vitamin P" or hesperidin), and ubiquinone (coenzyme Q). Other names you may hear are "vitamin B_{15}" and pangamic acid (both hoaxes) or "vitamin B_{17}" (laetrile or amygdalin, not a cancer cure as claimed and not a vitamin by any stretch of the imagination).****

This chapter has addressed all 13 of the vitamins. The basic facts about each one are summed up in Tables 7–6 and 7–7.

KEY POINTS

- Choline is needed in the diet, but it is not a vitamin, and deficiencies are unheard of outside the laboratory.
- Many other substances that people claim are vitamins are not.

serotonin (SER-oh-TONE-in) a neurotransmitter important in sleep regulation, appetite control, and mood regulation, among other roles. Serotonin is synthesized in the body from the amino acid tryptophan with the help of vitamin B_6.

biotin (BY-o-tin) a B vitamin; a coenzyme necessary for fat synthesis and other metabolic reactions.

pantothenic (PAN-to-THEN-ic) **acid** a B vitamin and part of a critical coenzyme needed in energy metabolism, among other roles.

choline (KOH-leen) a nutrient used to make the phospholipid lecithin and other molecules.

carnitine a nonessential nutrient that functions in cellular activities.

inositol (in-OSS-ih-tall) a nonessential nutrient found in cell membranes.

lipoic (lip-OH-ic) **acid** a nonessential nutrient.

**** Read about these and many other claims at the website of the National Council Against Health Fraud, www.ncahf.org.

Table 7–6

The Fat-Soluble Vitamins—Functions, Deficiencies, and Toxicities

VITAMIN A

Other Names
Retinol, retinal, retinoic acid; main precursor is beta-carotene

Chief Functions in the Body
Vision; health of cornea, epithelial cells, mucous membranes, skin; bone and tooth growth; regulation of gene expression; reproduction; immunity
Beta-carotene: antioxidant

Deficiency Disease Name
Hypovitaminosis A

Significant Sources
Retinol: fortified milk, cheese, cream, butter, fortified margarine, eggs, liver
Beta-carotene: spinach and other dark, leafy greens; broccoli; deep orange fruits (apricots, cantaloupe) and vegetables (winter squash, carrots, sweet potatoes, pumpkin)

	Deficiency Symptoms	Toxicity Symptoms
Blood/Circulatory System	Anemia (small-cell type)[a]	Red blood cell breakage, cessation of menstruation, nosebleeds
Bones/Teeth	Cessation of bone growth, painful joints; impaired enamel formation, cracks in teeth, tendency toward tooth decay	Bone pain; growth retardation; increased pressure inside skull; headaches; bone abnormalities
Digestive System	Diarrhea, changes in intestinal and other body linings	Abdominal pain, nausea, vomiting, diarrhea, weight loss
Immune System	Frequent infections	Overreactivity
Nervous/Muscular System	Night blindness (retinal) Mental depression	Blurred vision, uncoordinated muscle, fatigue, irritability, loss of appetite
Skin and Cornea	Keratinization, corneal degeneration leading to blindness,[b] rashes	Dry skin, rashes, loss of hair; cracking and bleeding lips, brittle nails; hair loss
Other	Kidney stones, impaired growth	Liver enlargement and liver damage; birth defects

VITAMIN D

Other Names
Calciferol, cholecalciferol, dihydroxy vitamin D; precursor is cholesterol

Chief Functions in the Body
Mineralization of bones (raises blood calcium and phosphorus via absorption from digestive tract and by withdrawing calcium from bones and stimulating retention by kidneys)

Deficiency Disease Name
Rickets, osteomalacia

Significant Sources
Self-synthesis with sunlight; fortified milk and other fortified foods, liver, sardines, salmon

	Deficiency Symptoms	Toxicity Symptoms
Blood/Circulatory System		Raised blood calcium; calcification of blood vessels and heart tissues
Bones/Teeth	Abnormal growth, misshapen bones (bowing of legs), soft bones, joint pain, malformed teeth	Calcification of tooth soft tissues; thinning of tooth enamel
Nervous/Muscular System	Muscle spasms	Excessive thirst, headaches, irritability, loss of appetite, weakness, nausea
Other		Calcification and harm to soft tissues (kidneys, lungs, joints); heart damage

[a]*Small-cell anemia is termed* microcytic anemia; *large-cell type is* macrocytic *or* megaloblastic anemia.
[b]*Corneal degeneration progresses from* keratinization *(hardening) to* xerosis *(drying) to* xerophthalmia *(thickening, opacity, and irreversible blindness).*

Table 7–6

The Fat-Soluble Vitamins—Functions, Deficiencies, and Toxicities (continued)

VITAMIN E

Other Names
Alpha-tocopherol, tocopherol

Chief Functions in the Body
Antioxidant (quenching of free radicals), stabilization of cell membranes, support of immune function, protection of polyunsaturated fatty acids; normal nerve development

Deficiency Disease Name
(No name)

Significant Sources
Polyunsaturated plant oils (margarine, salad dressings, shortenings), green and leafy vegetables, wheat germ, whole-grain products, nuts, seeds

	Deficiency Symptoms	Toxicity Symptoms
Blood/Circulatory System	Red blood cell breakage, anemia	Augments the effects of anticlotting medication
Digestive System		General discomfort, nausea
Eyes		Blurred vision
Nervous/Muscular System	Nerve degeneration, weakness, difficulty walking, leg cramps	Fatigue

VITAMIN K

Other Names
Phylloquinone, naphthoquinone

Chief Functions in the Body
Synthesis of blood-clotting proteins and proteins important in bone mineralization

Deficiency Disease Name
(No name)

Significant Sources
Bacterial synthesis in the digestive tract; green leafy vegetables, cabbage-type vegetables, soybeans, vegetable oils

	Deficiency Symptoms	Toxicity Symptoms
Blood/Circulatory System	Hemorrhage	Interference with anticlotting medication
Bones	Poor skeletal mineralization	

Table 7–7

The Water-Soluble Vitamins—Functions, Deficiencies, and Toxicities

VITAMIN C

Other Names
Ascorbic acid

Chief Functions in the Body
Collagen synthesis (strengthens blood vessel walls, forms scar tissue, matrix for bone growth), antioxidant, restores vitamin E to active form, hormone synthesis, supports immune cell functions, helps in absorption of iron

Deficiency Disease Name
Scurvy

Significant Sources
Citrus fruits, cabbage-type vegetables, dark green vegetables, cantaloupe, strawberries, peppers, lettuce, tomatoes, potatoes, papayas, mangoes

	Deficiency Symptoms	Toxicity Symptoms
Digestive System		Nausea, abdominal cramps, diarrhea, excessive urination
Immune System	Immune suppression, frequent infections	
Mouth, Gums, Tongue	Bleeding gums, loosened teeth	
Nervous/Muscular System	Muscle degeneration and pain, depression, disorientation	Headache, fatigue, insomnia
Bones	Bone fragility, joint pain	Aggravation of gout
Skin	Pinpoint hemorrhages, rough skin, blotchy bruises	Rashes
Other	Failure of wounds to heal	Interference with medical tests; kidney stones in susceptible people

Table 7–7

The Water-Soluble Vitamins—Functions, Deficiencies, and Toxicities (continued)

THIAMIN

		Deficiency Symptoms	Toxicity Symptoms
Other Names Vitamin B_1 ***Chief Functions in the Body*** Part of a coenzyme needed in energy metabolism, supports normal appetite and nervous system function ***Deficiency Disease Name*** Beriberi (wet and dry) ***Significant Sources*** Occurs in all nutritious foods in moderate amounts; pork, ham, bacon, liver, whole and enriched grains, legumes, seeds	**Blood/Circulatory System**	Edema, enlarged heart, abnormal heart rhythms, heart failure	(No symptoms reported)
	Nervous/Muscular System	Degeneration, wasting, weakness, pain, apathy, irritability, difficulty walking, loss of reflexes, mental confusion, paralysis	
	Other	Anorexia; weight loss	

RIBOFLAVIN

		Deficiency Symptoms	Toxicity Symptoms
Other Names Vitamin B_2 ***Chief Functions in the Body*** Part of a coenzyme needed in energy metabolism, supports normal vision and skin health ***Deficiency Disease Name*** Ariboflavinosis ***Significant Sources*** Milk, yogurt, cottage cheese, meat, liver, leafy green vegetables, whole-grain or enriched breads and cereals	**Mouth, Gums, Tongue**	Cracks at corners of mouth,[a] smooth magenta tongue[b]; sore throat	(No symptoms reported)
	Nervous System and Eyes	Hypersensitivity to light, reddening of cornea	
	Skin	Skin rash	

NIACIN

		Deficiency Symptoms	Toxicity Symptoms
Other Names Nicotinic acid, nicotinamide, niacinamide, vitamin B_3; precursor is dietary tryptophan ***Chief Functions in the Body*** Part of coenzymes needed in energy metabolism ***Deficiency Disease Name*** Pellagra ***Significant Sources*** Synthesized from the amino acid tryptophan; milk, eggs, meat, poultry, fish, whole-grain and enriched breads and cereals, nuts, and all protein-containing foods	**Digestive System**	Diarrhea; vomiting; abdominal pain	Nausea, vomiting
	Mouth, Gums, Tongue	Black or bright red swollen smooth tongue[b]	
	Nervous System	Irritability, loss of appetite, weakness, headache, dizziness, mental confusion progressing to psychosis or delirium	
	Skin	Flaky skin rash on areas exposed to sun	Painful flush and rash, sweating
	Other		Liver damage; impaired glucose tolerance; vision disturbances

[a]*Cracks at the corners of the mouth are termed* cheilosis *(kee-LOH-sis).*

[b]*Smoothness of the tongue is caused by loss of its surface structures and is termed* glossitis *(gloss-EYE-tis).*

Table 7–7

The Water-Soluble Vitamins—Functions, Deficiencies, and Toxicities (continued)

FOLATE

Other Names
Folic acid, folacin, pteroylglutamic acid

Chief Functions in the Body
Part of a coenzyme needed for new cell synthesis

Deficiency Disease Name
(No name)

Significant Sources
Asparagus, avocado, leafy green vegetables, beets, legumes, seeds, liver, enriched breads, cereal, pasta, and grains

	Deficiency Symptoms	Toxicity Symptoms
Blood/Circulatory System	Anemia (large-cell type),[a] elevated homocysteine	Masks vitamin B_{12} deficiency
Digestive System	Heartburn, diarrhea, constipation	
Immune System	Suppression, frequent infections	
Mouth, Gums, Tongue	Smooth red tongue[b]	
Nervous/Muscular System	Increased risk of neural tube birth defects. Depression, mental confusion, fatigue, irritability, headache	Depression, mental confusion, fatigue, irritability, headache

VITAMIN B_{12}

Other Names
Cyanocobalamin

Chief Functions in the Body
Part of coenzymes needed in new cell synthesis, helps maintain nerve cells

Deficiency Disease Name
(No name)[c]

Significant Sources
Animal products (meat, fish, poultry, milk, cheese, eggs)

	Deficiency Symptoms	Toxicity Symptoms
Blood/Circulatory System	Anemia (large-cell type)[a,c]	(No toxicity symptoms known)
Mouth, Gums, Tongue	Smooth tongue[b]	
Nervous/Muscular System	Fatigue, nerve degeneration progressing to paralysis	
Skin	Tingling or numbness	

VITAMIN B_6

Other Names
Pyridoxine, pyridoxal, pyridoxamine

Chief Functions in the Body
Part of a coenzyme needed in amino acid and fatty acid metabolism, helps convert tryptophan to niacin and to serotonin, helps make red blood cells

Deficiency Disease Name
(No name)

Significant Sources
Meats, fish, poultry, liver, legumes, fruits, potatoes, whole grains, soy products

	Deficiency Symptoms	Toxicity Symptoms
Blood/Circulatory System	Anemia (small-cell type)[a]	Bloating
Nervous/Muscular System	Depression, confusion, abnormal brain wave pattern, convulsions	Depression, fatigue, impaired memory, irritability, headaches, numbness, damage to nerves, difficulty walking, loss of reflexes, restlessness, convulsions
Skin	Rashes; greasy, scaly dermatitis	Skin lesions

[a] *Small-cell anemia is termed* microcytic anemia; *large-cell type is* macrocytic *or* megaloblastic anemia.

[b] *Smoothness of the tongue is caused by loss of its surface structures and is termed* glossitis *(gloss-EYE-tis).*

[c] *The name* pernicious anemia *refers to the vitamin B_{12} deficiency caused by lack of intrinsic factor, but not to that caused by inadequate dietary intake.*

Table 7–7

The Water-Soluble Vitamins—Functions, Deficiencies, and Toxicities (continued)

PANTOTHENIC ACID

		Deficiency Symptoms	Toxicity Symptoms
Other Names (None) ***Chief Functions in the Body*** Part of a coenzyme needed in energy metabolism ***Deficiency Disease Name*** (No name) ***Significant Sources*** Widespread in foods	**Digestive System**	Vomiting, intestinal distress	Water retention (infrequent)
	Nervous/Muscular System	Insomnia, fatigue	
	Other	Hypoglycemia, increased sensitivity to insulin	

BIOTIN

		Deficiency Symptoms	Toxicity Symptoms
Other Names (None) ***Chief Functions in the Body*** A cofactor for several enzymes needed in energy metabolism, fat synthesis, amino acid metabolism, and glycogen synthesis ***Deficiency Disease Name*** (No name) ***Significant Sources*** Widespread in foods	**Blood/Circulatory System**	Abnormal heart action	(No toxicity symptoms reported)
	Digestive System	Loss of appetite, nausea	
	Nervous/Muscular System	Depression, muscle pain, weakness, fatigue, numbness of extremities	
	Skin	Dry around eyes, nose, and mouth	

try it! Food Feature

Choosing Foods Rich in Vitamins

LO 7.11 Suggest foods that can help to ensure adequate vitamin intakes without providing too many calories.

On learning how important the vitamins are to their health, most people want to choose foods that are vitamin-rich. How can they tell which are which? Not by food labels—these are required to list only two of the vitamins—vitamin C and vitamin A—of a food's contents. A way to find out more about the vitamin contents of foods is to look down the columns of vitamins and calories in a table of food composition, such as Appendix A at the end of this book, to identify some of the vitamin-rich foods in your own eating pattern. If you are interested in folate, for instance, you can see that cornflakes are an especially good source (folic acid is added to cornflakes), as is orange juice (folate occurs naturally in this food).

Another way of looking at such data appears in Figure 7–19—the long bars show some foods that are rich sources of a particular vitamin and the short or nonexistent bars indicate poor sources. The colors of the bars represent the various food groups.

Which Foods Should I Choose?

After looking at Figure 7–19, don't think that you must memorize the richest sources of each vitamin and eat those foods daily. That false notion would lead you to limit your variety of foods while overemphasizing the components of a few foods. Although it is reassuring to know that your carrot-raisin salad at lunch provided more than your entire day's need for vitamin A, it is a mistake to think that you must then select equally rich sources of all the other vitamins. Such rich sources do not exist for many vitamins—rather, foods work in harmony to provide most nutrients.

For example, a baked potato, not a star performer among vitamin C providers, contributes substantially to a day's need for this nutrient and contributes some thiamin and vitamin B_6 too. By the end of the day, assuming that your food choices were made with reasonable care, the bits of vitamin C, thiamin, and vitamin B_6 from each serving of food have accumulated to more than cover the day's need for them.

A Variety of Foods Works Best

The last two graphs of Figure 7–19 show sources of folate and vitamin C. These nutrients are both richly supplied by fruits and vegetables. The richest source of either may be only a moderate source of the other, but the recommended amounts of fruits and vegetables in the

Figure 7–19

Food Sources of Vitamins Selected to Show a Range of Values

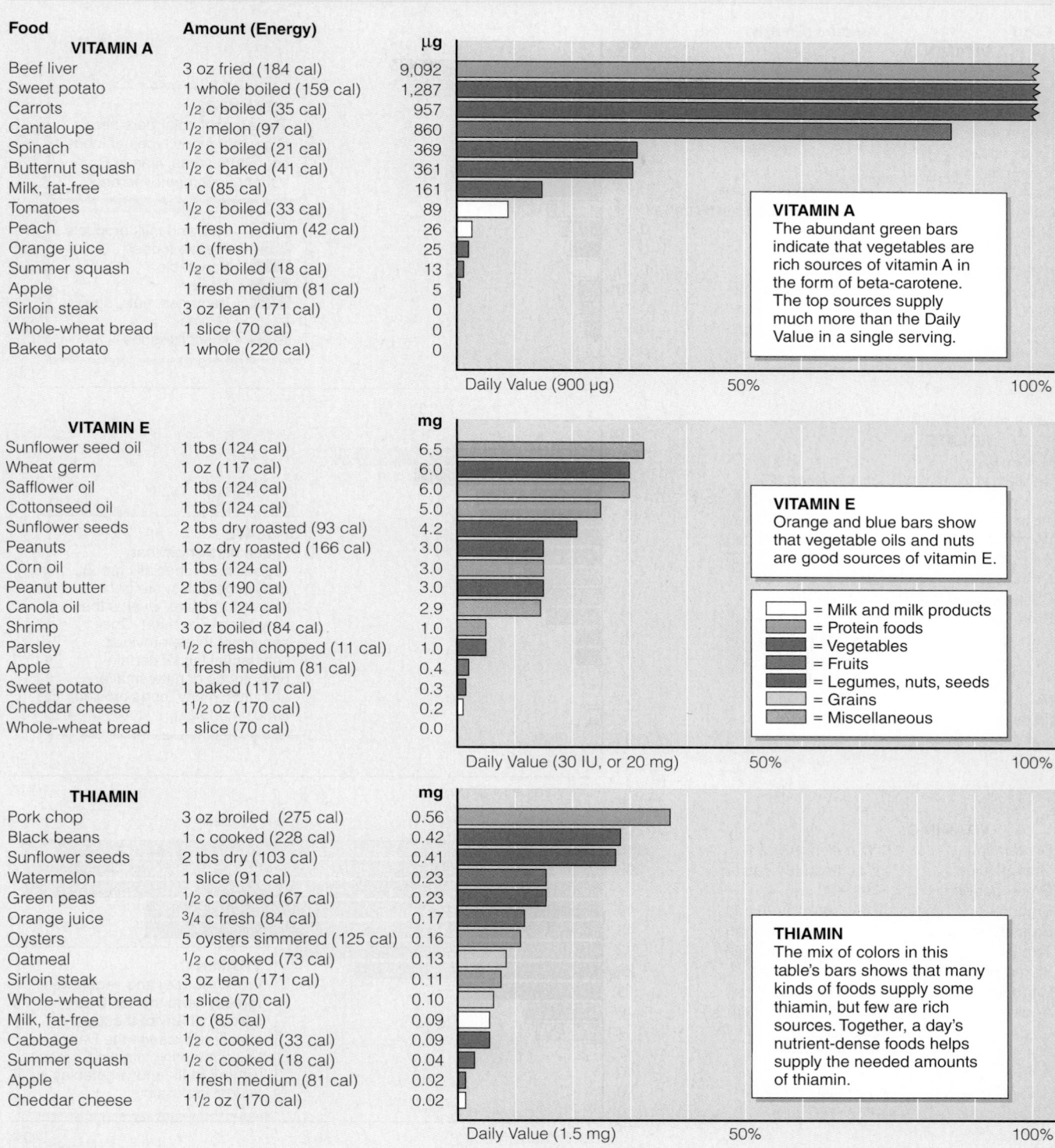

Food	Amount (Energy)	μg
VITAMIN A		
Beef liver	3 oz fried (184 cal)	9,092
Sweet potato	1 whole boiled (159 cal)	1,287
Carrots	1/2 c boiled (35 cal)	957
Cantaloupe	1/2 melon (97 cal)	860
Spinach	1/2 c boiled (21 cal)	369
Butternut squash	1/2 c baked (41 cal)	361
Milk, fat-free	1 c (85 cal)	161
Tomatoes	1/2 c boiled (33 cal)	89
Peach	1 fresh medium (42 cal)	26
Orange juice	1 c (fresh)	25
Summer squash	1/2 c boiled (18 cal)	13
Apple	1 fresh medium (81 cal)	5
Sirloin steak	3 oz lean (171 cal)	0
Whole-wheat bread	1 slice (70 cal)	0
Baked potato	1 whole (220 cal)	0

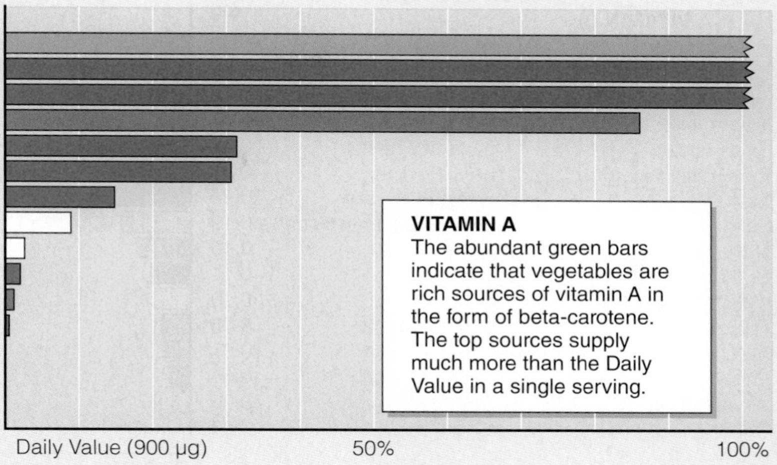

Food	Amount (Energy)	mg
VITAMIN E		
Sunflower seed oil	1 tbs (124 cal)	6.5
Wheat germ	1 oz (117 cal)	6.0
Safflower oil	1 tbs (124 cal)	6.0
Cottonseed oil	1 tbs (124 cal)	5.0
Sunflower seeds	2 tbs dry roasted (93 cal)	4.2
Peanuts	1 oz dry roasted (166 cal)	3.0
Corn oil	1 tbs (124 cal)	3.0
Peanut butter	2 tbs (190 cal)	3.0
Canola oil	1 tbs (124 cal)	2.9
Shrimp	3 oz boiled (84 cal)	1.0
Parsley	1/2 c fresh chopped (11 cal)	1.0
Apple	1 fresh medium (81 cal)	0.4
Sweet potato	1 baked (117 cal)	0.3
Cheddar cheese	1 1/2 oz (170 cal)	0.2
Whole-wheat bread	1 slice (70 cal)	0.0

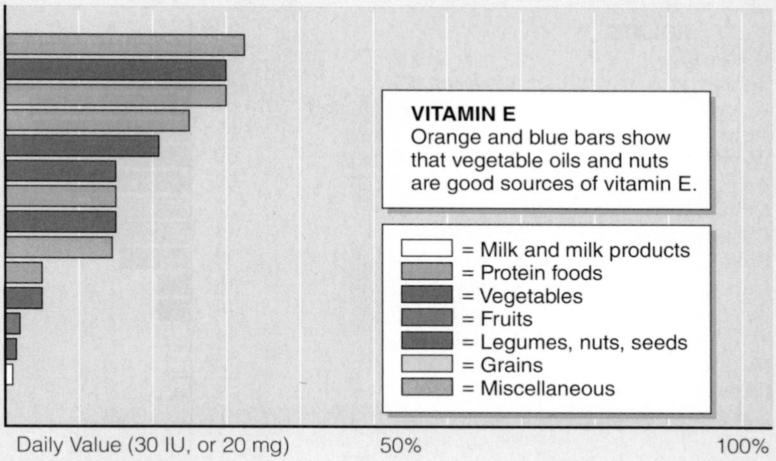

Food	Amount (Energy)	mg
THIAMIN		
Pork chop	3 oz broiled (275 cal)	0.56
Black beans	1 c cooked (228 cal)	0.42
Sunflower seeds	2 tbs dry (103 cal)	0.41
Watermelon	1 slice (91 cal)	0.23
Green peas	1/2 c cooked (67 cal)	0.23
Orange juice	3/4 c fresh (84 cal)	0.17
Oysters	5 oysters simmered (125 cal)	0.16
Oatmeal	1/2 c cooked (73 cal)	0.13
Sirloin steak	3 oz lean (171 cal)	0.11
Whole-wheat bread	1 slice (70 cal)	0.10
Milk, fat-free	1 c (85 cal)	0.09
Cabbage	1/2 c cooked (33 cal)	0.09
Summer squash	1/2 c cooked (18 cal)	0.04
Apple	1 fresh medium (81 cal)	0.02
Cheddar cheese	1 1/2 oz (170 cal)	0.02

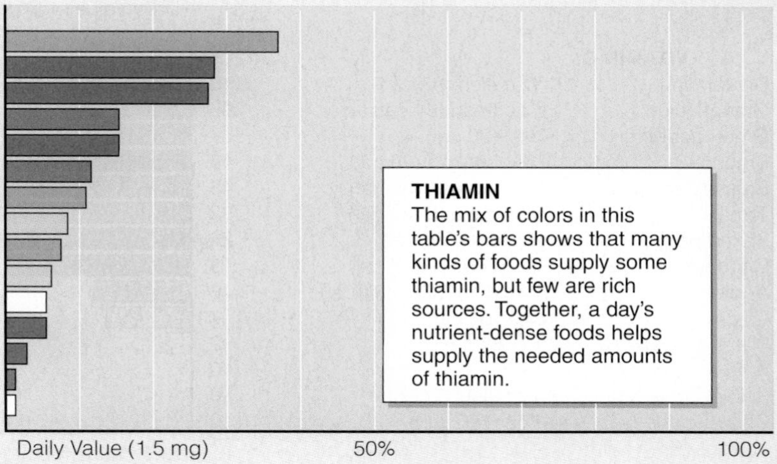

THIAMIN
The mix of colors in this table's bars shows that many kinds of foods supply some thiamin, but few are rich sources. Together, a day's nutrient-dense foods helps supply the needed amounts of thiamin.
Daily Value (1.5 mg)
50%
100%

Figure 7–19

Food Sources of Vitamins Selected to Show a Range of Values (continued)

Food	Amount (Energy)	mg
VITAMIN B_6		
Baked potato	1 whole (220 cal)	0.70
Banana	1 peeled (109 cal)	0.68
Turkey breast	3 oz (133 cal)	0.46
Watermelon	1 slice (91 cal)	0.41
Sirloin steak	3 oz lean (171 cal)	0.38
Pork roast	3 oz lean (175 cal)	0.29
Spinach	1/2 c cooked (21 cal)	0.22
Salmon	3 oz broiled/baked (183 cal)	0.19
Navy beans	1/2 c cooked (129 cal)	0.15
Broccoli	1/2 c cooked (22 cal)	0.11
Milk, fat-free	1 c (85 cal)	0.10
Orange juice	3/4 c fresh (84 cal)	0.08
Apple	1 fresh medium (81 cal)	0.07
Summer squash	1/2 c boiled (18 cal)	0.06
Whole-wheat bread	1 slice (69 cal)	0.05
Cheddar cheese	1 1/2 oz (170 cal)	0.03

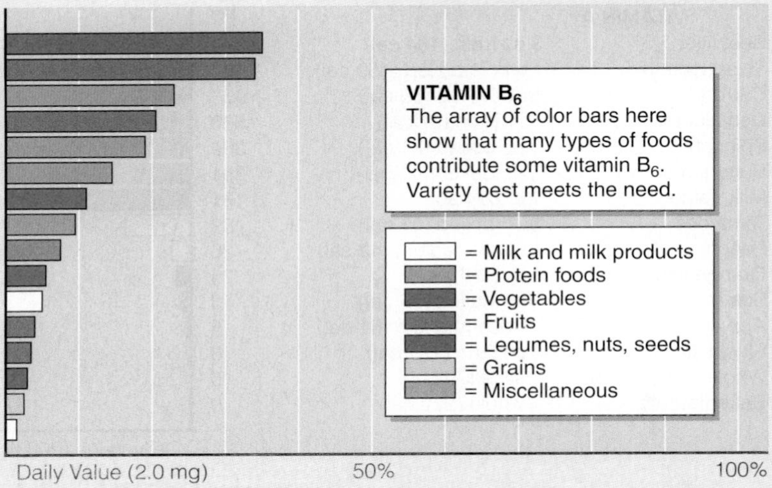
Daily Value (2.0 mg)
50%
100%
VITAMIN B_6
The array of color bars here show that many types of foods contribute some vitamin B_6. Variety best meets the need.
= Milk and milk products
= Protein foods
= Vegetables
= Fruits
= Legumes, nuts, seeds
= Grains
= Miscellaneous

Food	Amount (Energy)	µg
FOLATE		
Beef liver	3 oz fried (184 cal)	185
Spinach	1/2 c cooked (21 cal)	131
Asparagus	4 spears cooked (14 cal)	88
Turnip greens	1/2 c cooked (15 cal)	85
Winter squash	1/2 c cooked (48 cal)	69
Beets	1/2 c cooked (37 cal)	68
Orange juice	3/4 c fresh (84 cal)	57
Cantaloupe	1/2 melon (97 cal)	47
Broccoli	1/2 c cooked (22 cal)	39
Lima beans	1/2 c cooked (85 cal)	18
Summer squash	1/2 c cooked (18 cal)	18
Whole-wheat bread[a]	1 slice (70 cal)	14
Milk, fat-free	1 c (85 cal)	13
Sirloin steak	3 oz lean (171 cal)	8
Cheddar cheese	1 1/2 oz (170 cal)	5
Apple	1 fresh medium (81 cal)	4

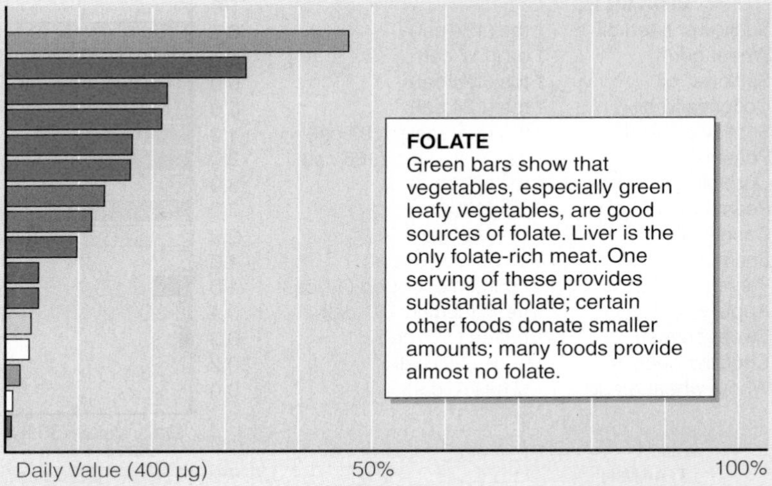
Daily Value (400 µg)
50%
100%
FOLATE
Green bars show that vegetables, especially green leafy vegetables, are good sources of folate. Liver is the only folate-rich meat. One serving of these provides substantial folate; certain other foods donate smaller amounts; many foods provide almost no folate.

Food	Amount (Energy)	mg
VITAMIN C		
Cantaloupe	1/2 melon (97 cal)	116
Orange juice	3/4 c fresh (84 cal)	93
Green peppers	1/2 c (20 cal)	67
Broccoli	1/2 c cooked (26 cal)	48
Brussels sprouts	1/2 c cooked (30 cal)	48
Tomato juice	3/4 c canned (31 cal)	33
Baked potato	1 whole (220 cal)	26
Cabbage	1/2 c cooked (17 cal)	15
Apple	1 fresh medium (81 cal)	8
Oysters	3 oz (69 cal)	7
Milk, fat-free	1 c (85 cal)	2
Whole-wheat bread	1 slice (69 cal)	0
Sirloin steak	3 oz lean (171 cal)	0
Cheddar cheese	1 oz (170 cal)	0

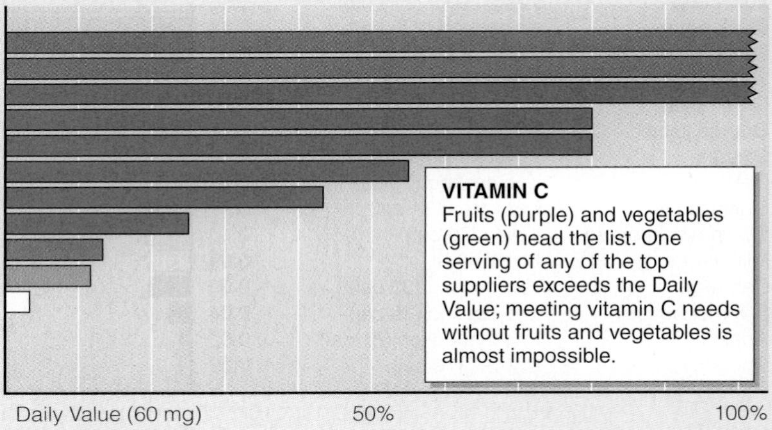
Daily Value (60 mg)
50%
100%
VITAMIN C
Fruits (purple) and vegetables (green) head the list. One serving of any of the top suppliers exceeds the Daily Value; meeting vitamin C needs without fruits and vegetables is almost impossible.

[a]Unenriched.

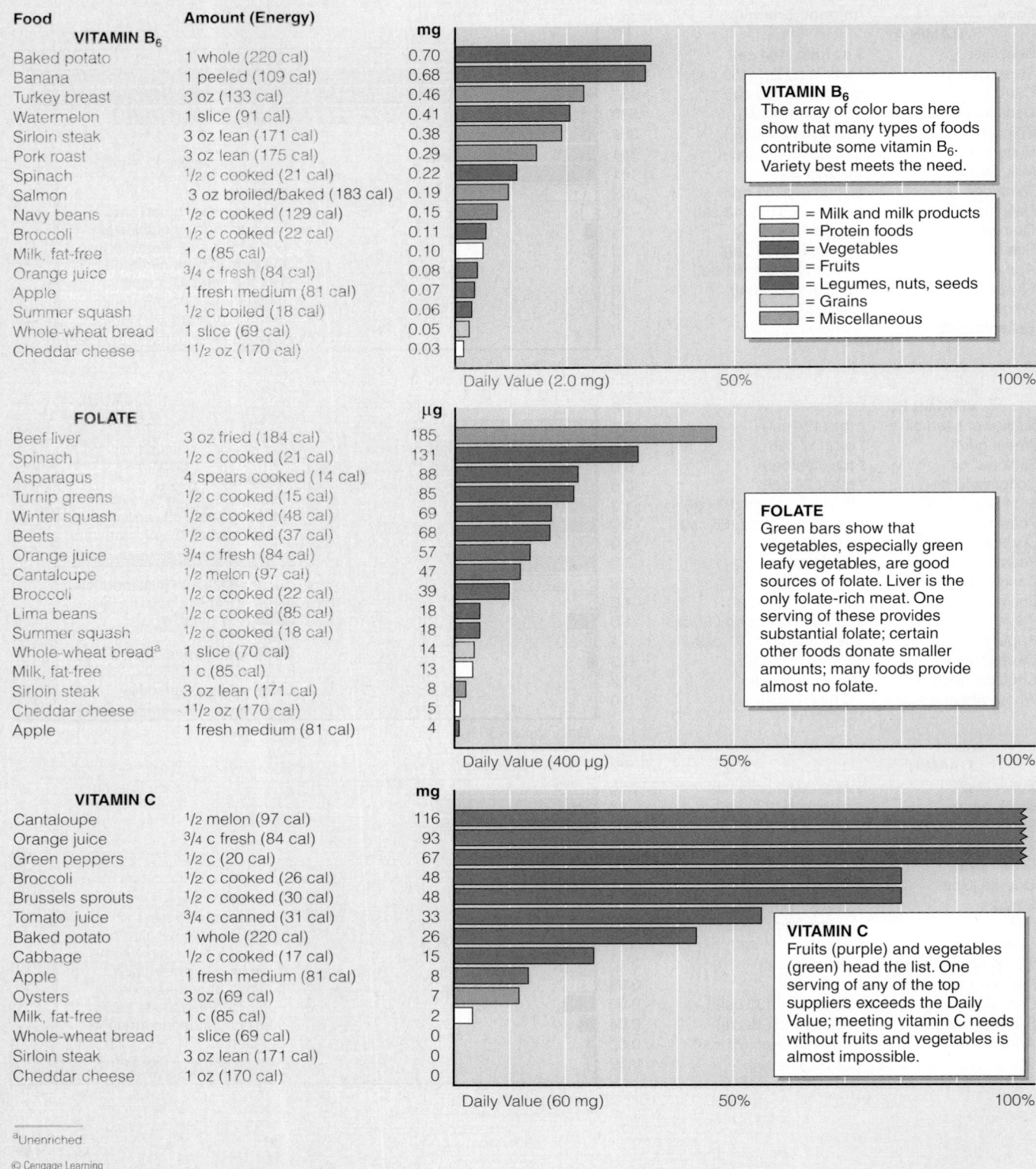

USDA Food Patterns of Chapter 2 cover both needs amply. As for vitamin E, vegetable oils and some seeds and nuts are the richest sources, but vegetables and fruits contribute a little, too.

By now, you should recognize a basic truth in nutrition. The eating pattern that best provides nutrients includes a wide variety of nutrient-dense foods that provide more than just isolated nutrients.[97] Phytochemicals, widespread among whole grains, nuts, fruits, and vegetables, may play roles in human health, as do fiber and other constituents of whole foods. When aiming for adequate intakes of vitamins, therefore, aim for a diet that meets the recommendations of Chapter 2. Even supplements cannot duplicate the benefits of such a diet, a point made in this chapter's Controversy section.

Phytochemicals are the topic of **Controversy 2**.

track it! Diet Analysis PLUS

Concepts in Action

Analyze Your Vitamin Intake

The purpose of this exercise is to help you identify your food sources of water-soluble and fat-soluble vitamins. Many foods rich in vitamins work in harmony to provide a full complement of nutrients, which ultimately contributes to a health-promoting eating pattern.

1. Determine whether your food provides enough vitamins. From the Reports tab select Intake vs. Goals. Choose Day Two, all meals. Generate a report. Did your intakes on that day meet your DRI recommended intake values for vitamins? If not, list those that fall short of the DRI goals. Did any of your intakes exceed DRI values? If so, list those, too.
2. Some fruits and vegetables are good sources of fat-soluble vitamins (see the Snapshots on pages 242, 247, 251, and 252). From the Reports tab, select MyPlate Analysis, and include all meals. Generate a report. Have you met your minimum recommended fruit and vegetable intake? What percentage of your goal have you met for fruits and vegetables? Did you consume any fruits and vegetables listed in the Snapshots for the fat-soluble vitamins? Which ones?
3. From the Reports tab select Source Analysis, choose any day, and include all meals. From the drop-down box, select vitamin C and then folate, and generate a report for each vitamin. What is your best food source for vitamin C? And for folate? Were your best sources shown in the Snapshots on pages 257 and 266?
4. After viewing the Intake vs. Goals report in question 1, if you fell short on any vitamin, what foods could you include that would bring you up to the DRI recommended value? If you exceeded the DRI values, which foods were responsible?
5. The USDA eating patterns suggest that a person who requires 2,000 calories per day should, in a week's time, consume $1\frac{1}{2}$ cups of a variety of dark green, $5\frac{1}{2}$ cups of red and orange, 5 cups of starchy, and 4 cups of other vegetables and $1\frac{1}{2}$ cups of legumes. Create a dish from vegetables or fruits that you enjoy. Get some ideas by using Figure 7–19 (pages 275–276). From the Track Diet tab, enter the ingredients. From the Reports tab select Source Analysis, and select one water-soluble and then one fat-soluble vitamin from the drop-down menu. Generate a report for each. Identify the vitamin-rich foods from the report. What does the bar graph show?

what did you decide?

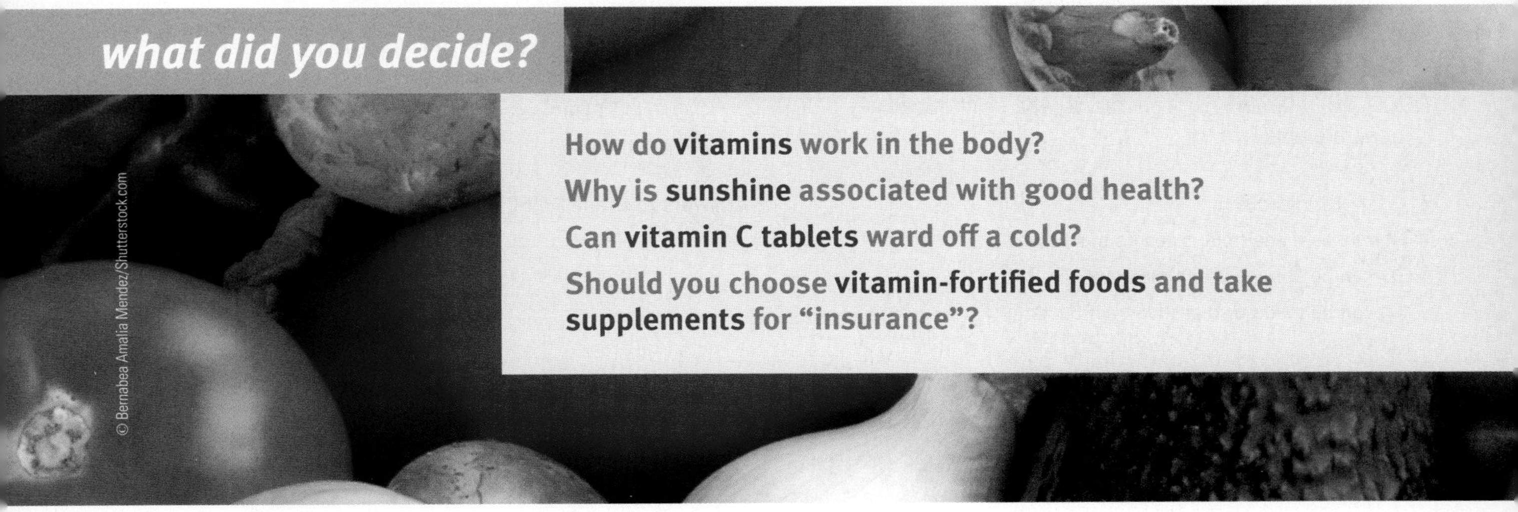

How do vitamins work in the body?

Why is sunshine associated with good health?

Can vitamin C tablets ward off a cold?

Should you choose vitamin-fortified foods and take supplements for "insurance"?

Self Check

1. (LO 7.1) Which of the following vitamins are classified as fat-soluble?
   a. vitamins B and D
   b. vitamins A, D, E, and K
   c. vitamins B, E, D, and C
   d. vitamins B and C
2. (LO 7.1) Which of the following describes the fat-soluble vitamins?
   a. few functions in the body
   b. easily absorbed and excreted
   c. stored extensively in tissues
   d. a and c
3. (LO 7.2) Fat-soluble vitamins are mostly absorbed into the
   a. lymph
   b. blood
   c. extracellular fluid
   d. b and c
4. (LO 7.2) Most water-soluble vitamins are not stored in tissues to any great extent.
   T F
5. (LO 7.3) Which of the following foods is (are) rich in beta-carotene?
   a. sweet potatoes
   b. pumpkin
   c. spinach
   d. all of the above
6. (LO 7.3) Vitamin A supplements can help treat acne.
   T F
7. (LO 7.4) Vitamin D functions as a hormone to help maintain bone integrity.
   T F
8. (LO 7.4) In adults with vitamin D deficiency, poor bone mineralization can lead to
   a. pellagra
   b. pernicious anemia
   c. scurvy
   d. osteomalacia
9. (LO 7.5) Which of the following is (are) rich source(s) of vitamin E?
   a. raw vegetable oil
   b. colorful foods, such as carrots
   c. milk and milk products
   d. raw cabbage
10. (LO 7.5) Vitamin E is famous for its role in
    a. maintaining bone tissue integrity
    b. maintaining connective tissue integrity
    c. protecting tissues from oxidation
    d. as a precursor for vitamin C
11. (LO 7.6) Vitamin K is necessary for the synthesis of key bone proteins.
    T F
12. (LO 7.6) Vitamin K
    a. can be made from exposure to sunlight.
    b. can be obtained from most milk products.
    c. can be made by digestive tract bacteria.
    d. b and c
13. (LO 7.7) Water-soluble vitamins are mostly absorbed into the
    a. lymph
    b. blood
    c. extracellular fluid
    d. b and c
14. (LO 7.7) The water-soluble vitamins are characterized by all of the following, except:
    a. excesses are stored and easily build up to toxic levels.
    b. they travel freely in the blood.
    c. excesses are easily excreted and seldom build up to toxic levels.
    d. b and c
15. (LO 7.8) The theory that vitamin C prevents or cures colds is well supported by research.
    T F
16. (LO 7.8) Vitamin C deficiency symptoms include
    a. red spots
    b. loose teeth
    c. anemia
    d. all of the above
17. (LO 7.9) B vitamins often act as
    a. antioxidants
    b. blood clotting factors
    c. coenzymes
    d. none of the above
18. (LO 7.9) A B vitamin often forms part of an enzyme's active site, where a chemical reaction takes place.
    T F
19. (LO 7.10) A deficiency of niacin may result in which disease?
    a. pellagra
    b. beriberi
    c. scurvy
    d. rickets
20. (LO 7.10) Which of these B vitamins is (are) present only in foods of animal origin?
    a. niacin
    b. vitamin B_{12}
    c. riboflavin
    d. a and c
21. (LO 7.11) The eating pattern that best provides nutrients
    a. singles out a rich source for each nutrient and focuses on these foods.
    b. includes a wide variety of nutrient-dense foods.
    c. is a Western eating style that includes abundant meats and fats.
    d. singles out rich sources of certain phytochemicals and focuses on these foods.
22. (LO 7.12) FDA has extensive regulatory control over supplement sales.
    T F

Answers to these Self Check questions are in Appendix G.

CONTROVERSY 7

Vitamin Supplements: Do the Benefits Outweigh the Risks?

LO 7.12 Identify both valid and invalid reasons for taking vitamin supplements.

At least half of the U.S. population takes dietary supplements, spending $24 *billion* each year to do so.*[1] Most take a daily multivitamin and mineral pill, hoping to make up for dietary shortfalls; others take single nutrient supplements to ward off diseases; and many do both. Do people need all these supplements? If people do need supplements, which ones are best? What about health risks from supplements? This Controversy examines evidence surrounding these questions and concludes with some advice for those choosing a supplement.

Dietary supplements were defined in **Chapter 1**.

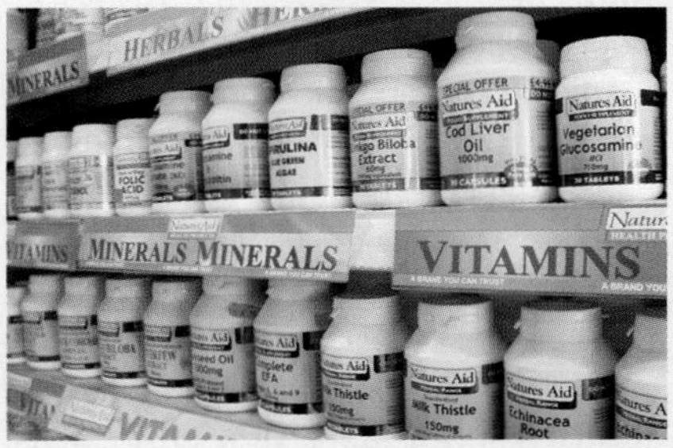

© Alex Segre/Alamy

© RTimages/Shutterstock.com

Which is the best source of vitamins to support good health: supplements or food?

*Reference notes are found in Appendix F.

Arguments in Favor of Taking Supplements

By far, most people can meet their nutrient needs from their diet alone. Indisputably, however, the people listed in Table C7–1 need supplements. For them, nutrient supplements can prevent or reverse illnesses. Because supplements are not risk-free, these people should consult a health-care provider who is alert to potential adverse effects and nutrient-drug interactions.

People with Deficiencies

In the United States and Canada, few adults suffer nutrient deficiency diseases such as scurvy, pellagra, and beriberi. When deficiency diseases do appear, prescribed supplements of the missing nutrients quickly stop or reverse most of the damage (exceptions include vitamin A–deficiency blindness, some vitamin B_{12}–deficiency nerve damage, and birth defects caused by folate deficiency in pregnant women).

Subtle subclinical deficiencies that do not cause classic symptoms are easily overlooked or misdiagnosed—and they often occur. People who diet habitually or elderly people with diminished appetite may eat so little nutritious food that they teeter on the edge of deficiency, with no reserve to handle any

Table C7–1

Some Valid Reasons for Taking Supplements

These people may need supplements:

- People with nutrient deficiencies.
- Women who are capable of becoming pregnant (supplemental or enrichment sources of folic acid are recommended to reduce risk of neural tube defects in infants).
- Pregnant or lactating women (they may need iron and folate).
- Newborns (they are routinely given a vitamin K dose).
- Infants (they may need various supplements, see Chapter 13).
- Those who are lactose intolerant (they need calcium to forestall osteoporosis).
- Habitual dieters (they may eat insufficient food).
- Elderly people often benefit from some of the vitamins and minerals in a balanced supplement (they may choose poor diets, have trouble chewing, or absorb or metabolize nutrients less efficiently; see Chapter 14).
- Victims of AIDS or other wasting illnesses (they lose nutrients faster than foods can supply them).
- Those addicted to drugs or alcohol (they absorb fewer and excrete more nutrients; nutrients cannot undo damage from drugs or alcohol).
- Those recovering from surgery, weight-loss surgery, burns, injury, or illness (they need extra nutrients to help regenerate tissues; weight-loss surgery creates nutrient malabsorption).
- Strict vegetarians (vegans may need vitamin B_{12}, vitamin D, iron, and zinc).
- People taking medications that interfere with the body's use of nutrients.

increase in demand. Similarly, people who omit entire food groups without proper diet planning or are too busy or lack knowledge or lack money are likely to lack nutrients. For them, until they correct their diets, a low-dose, complete vitamin-mineral supplement may help them avoid deficiency diseases.

People with Increased Nutrient Needs

During certain stages of life, many people find it difficult or impossible to meet nutrient needs without supplements. For example, women who lose a lot of blood and therefore a lot of iron during menstruation each month often need an iron supplement. Similarly, pregnant and breastfeeding women have exceptionally high nutrient needs and routinely take special supplements to help meet them. Even newborns require a dose of vitamin K at birth, as the preceding chapter pointed out.[2]

People Coping with Physical Stress

Any interference with a person's appetite, ability to eat, or ability to absorb or use nutrients will impair nutrient status. Prolonged illnesses, extensive injuries, weight-loss or other surgery, and addictions to alcohol or other drugs all have these effects, and such stressors increase nutrient requirements of the tissues. In addition, medications used to treat such conditions often increase nutrient needs. In all these cases, appropriate nutrient supplements can be of benefit.[3]

Arguments against Taking Supplements

In study after study, well-nourished people are the ones found to be taking supplements, adding greater amounts of nutrients to already sufficient intakes.[4] Ironically, people with low nutrient intakes from food generally do not take supplements. Most times, taking supplements is a costly but harmless practice, but occasionally, it is both costly and harmful to health.[5]

Toxicity

Foods rarely cause nutrient imbalances or toxicities, but supplements easily can—and the higher the dose, the greater the risk. Supplement users are more likely to have excessive intakes of certain nutrients—notably iron, zinc, vitamin A, and niacin.

People's tolerances for high doses of nutrients vary, just as their risks of deficiencies do, and amounts tolerable for some may be harmful for others. The DRI Tolerable Upper Intake Levels define the highest intakes that appear safe for *most* healthy people. A few sensitive people may experience toxicities at lower doses, however. Table C7–2 compares Tolerable Upper Intake Levels with typical nutrient doses in supplements.

The true extent of supplement toxicity in this country is unknown, but many adverse events are reported each year from vitamins, minerals, essential oils, herbs, and other supplements.[6] Only an alert health-care professional knowledgeable in nutrition can reliably recognize nutrient toxicity and report it to the U.S. Food and Drug Administration (FDA). Many chronic, subtle toxicities go unrecognized and unreported.

Supplement Contamination and Safety

The FDA recently identified over 140 "dietary supplements" sold on the U.S. market that were contaminated with pharmaceutical drugs, such as steroid hormones and stimulants. Such products are often sold as "natural" alternatives to FDA-approved drugs but their use has caused positive results on tests for banned drugs in athletes.[7] Toxic plant material, toxic heavy metals, bacteria, and other contaminants have also shown up in dietary supplements.[8]

Plain multivitamin and mineral supplements from reputable sources, without herbs or add-ons, generally test free from contamination, although their contents may vary from those stated on the label. Almost twice the label amount of vitamin A was found in a popular multivitamin and more than the Tolerable Upper Intake Level of niacin turned up in a children's chewable tablet.[9]

Many consumers wrongly believe that government scientists, in particular those of the FDA, test each new dietary supplement to ensure its safety and effectiveness before allowing it to be sold. It does not. In fact, under the current Dietary Supplement Health and Education Act, FDA has little control over supplement sales.† The FDA can act to remove *tainted* products from store shelves, however, and does so often.[10]

Most Americans express support for greater regulation of dietary supplements, and most health professionals wholeheartedly agree. Meanwhile, consumers can report adverse reactions to supplements directly to the FDA via its hotline or website.‡

Life-Threatening Misinformation

Another problem arises when people who are ill come to believe that self-prescribed high doses of vitamins or minerals can be therapeutic. On experiencing a warning symptom of a disease, a person might postpone seeking a diagnosis, thinking, "I probably just need a supplement to make this go away." Such self-diagnosis postpones medical care and gives the disease a chance to worsen. Improper dosing can also be a problem. For example, a man who suffered from mental illness arrived at an emergency room with dangerously low blood pressure. He had ingested 11 grams of niacin on the advice of an Internet website that falsely touted niacin as an effective therapy for schizophrenia. The Tolerable Upper Intake Level for niacin is 35 *milligrams*.

Supplements are almost never effective for purposes other than those already listed in Table C7–1. This doesn't stop marketers from making enticing

† The Dietary Supplement Health and Education Act of 1994 regulates supplements, holding them to the same general labeling requirements that apply to foods (labeling terms were defined in Chapter 2).

‡ Consumers should report suspected harm from dietary supplements to their health providers or to the FDA's MedWatch program at (800) FDA-1088 or on the Internet at www.fda.gov/medwatch/.

Table C7–2

Intake Guidelines (Adults) and Supplement Doses

Nutrient	Tolerable Upper Intake Levels[a]	Daily Values	Typical Multivitamin-Mineral Supplement	Average Single-Nutrient Supplement
Vitamins				
Vitamin A	3,000 µg (10,000 IU)	5,000 IU	5,000 IU	8,000 to 10,000 IU
Vitamin D	100 µg (4,000 IU)	400 IU	400 IU	400 to 50,000 IU[b]
Vitamin E	1,000 mg (1,500 to 2,200 IU)[c]	30 IU	30 IU	100 to 1,000 IU
Vitamin K	—[d]	80 µg	40 µg	—[f]
Thiamin	—[d]	1.5 mg	1.5 mg	50 mg
Riboflavin	—[d]	1.7 mg	1.7 mg	25 mg
Niacin (as niacinamide)	35 mg[c]	20 mg	20 mg	100 to 500 mg
Vitamin B_6	100 mg	2 mg	2 mg	100 to 200 mg
Folate	1,000 µg[c]	400 µg	400 µg	400 µg
Vitamin B_{12}	—[d]	6 µg	6 µg	100 to 1,000 µg
Pantothenic acid	—[d]	10 mg	10 mg	100 to 500 mg
Biotin	—[d]	300 µg	30 µg	300 to 600 µg
Vitamin C	2,000 mg	60 mg	10 mg	500 to 2,000 mg
Choline	3,500 mg	—	10 mg	250 mg
Minerals				
Calcium	2,000 to 3,000 mg	1,000 mg	160 mg	250 to 600 mg
Phosphorus	4,000 mg	1,000 mg	110 mg	—[f]
Magnesium	350 mg[e]	400 mg	100 mg	250 mg
Iron	45 mg	18 mg	18 mg	18 to 30 mg
Zinc	40 mg	15 mg	15 mg	10 to 100 mg
Iodine	1,100 µg	150 µg	150 µg	—[f]
Selenium	400 µg	70 µg	10 µg	50 to 200 µg
Fluoride	10 mg	—	—	—[f]
Copper	10 mg	2 mg	0.5 mg	—[f]
Manganese	11 mg	2 mg	5 mg	—[f]
Chromium	—[d]	120 µg	25 µg	200 to 400 µg
Molybdenum	2,000 µg	75 µg	25 µg	—[f]

[a] *Unless otherwise noted, Upper Levels represent total intakes from food, water, and supplements.*

[b] *50,000 IU vitamin D is available by prescription.*

[c] *Upper Levels represent intakes from supplements, fortified foods, or both.*

[d] *These nutrients have been evaluated by the DRI Committee for Tolerable Upper Intake Levels, but none were established because of insufficient data. No adverse effects have been reported with intakes of these nutrients at levels typical of supplements, but caution is still advised, given the potential for harm that accompanies excessive intakes.*

[e] *Upper Levels represent intakes from supplements only.*

[f] *Available as a single supplement by prescription.*

structure-function claims in materials of all kinds—in print, on labels, and on television or the Internet. Such sales pitches often fall far short of the FDA standard that claims should be "truthful and not misleading."

False Sense of Security

Lulled into a false sense of security, a person might eat irresponsibly, thinking, "My supplement will cover my needs." However, no one knows exactly how to formulate the "ideal" supplement, and no standards exist for formulations. What nutrients should be included? How much of each? On whose needs should the choices be based? Which, if any, of the phytochemicals should be added?

Whole Foods Are Best for Nutrients

In general, the body assimilates nutrients best from foods that dilute and disperse them among other substances that facilitate their absorption and use by the body.[11] Taken in pure, concentrated form, nutrients are likely to interfere with one another's absorption or with the absorption of other nutrients from foods eaten at the same time. Such effects are particularly well known among the minerals. For example, zinc hinders copper and calcium absorption, iron hinders zinc absorption, and calcium hinders magnesium and iron absorption. Among vitamins, vitamin C supplements *enhance* iron absorption, making iron overload likely in susceptible people. High doses of vitamin E interfere with vitamin K functions, delaying blood clotting and possibly raising the risk of brain hemorrhage (a form of stroke). These and other interactions present drawbacks to supplement use.

Can Supplements Prevent Chronic Diseases?

Many people take supplements in the belief that they can prevent heart disease and cancer. Can taking a supplement prevent these killers?

Vitamin D and Cancer

Reports that vitamin D supplements might prevent cancers, particularly of the breast, colon, and prostate, have boosted sales. True, low vitamin D intakes have been associated with increased cancer risk in some studies, but overall the connection has proved insignificant. The committee on DRI concludes that insufficient evidence exists to support an association between vitamin D intakes and cancer risk.[12] The U.S. Preventive Services Task Force, a group that offers unbiased advice concerning medical treatments, has recommended against taking vitamin D for cancer prevention.[13]

Oxidative Stress, Subclinical Deficiencies, and Chronic Diseases

Central to the idea that antioxidant nutrients might fight diseases is the theory of **oxidative stress** (terms are defined in Table C7–3). The chapter explained that normal activities of body cells produce free radicals (highly unstable molecules of oxygen) that can damage cell structures. Oxidative stress results when free-radical activity in the body exceeds its antioxidant defenses. When such damage accumulates, it triggers inflammation, which may lead to heart disease and cancer, among other conditions. **Antioxidant nutrients** help to quench these free radicals, rendering them harmless to cellular structures and stopping the chain of events.

Taking antioxidant pills instead of making needed lifestyle changes may sound appealing, but evidence does not support a role for supplements against chronic diseases.[14] In some cases, supplements may even be harmful.[15] For example, taking high doses of vitamin C, an antioxidant nutrient, may lower blood pressure somewhat, a small but potentially beneficial effect, but such doses also *increase* markers of oxidation in the blood and elevate the risk of vision-impairing cataracts in the eyes, potentially serious harms.[16] More research is needed to clarify whether high doses of vitamin C worsen cataract risks.[17]

Vitamin E and Chronic Disease

Hopeful early studies reported that taking vitamin E supplements reduced the rate of death from heart disease.[18] It made sense because in the laboratory, vitamin E opposes blood clotting, tissue inflammation, arterial injury, and lipid oxidation—all factors in heart disease development. After years of follow-up human studies, however, no protective effect is evident.[19] In fact, pooled results revealed a slight but alarming *increased* risk for death among people taking vitamin E supplements. Neither help nor harm is consistently observed with vitamin E supplementation.[20]

Such studies have been criticized for testing too low a dose, testing only the alpha-tocopherol form of vitamin E, failing to establish previous vitamin E status, or failing to account for differences such as illness, radiation exposure, or smoking.[21] For now, the results are disappointing, but research continues.[22]

Currently, much excitement surrounds some preliminary evidence linking some forms of vitamin E with cancer protection.[23] However, much more research is needed to clarify these connections before conclusions can be drawn, and vitamin E supplements, taken for any reason, carry risks.

The Story of Beta-Carotene—A Case in Point

Again and again, population studies confirm that people who eat plenty of fruits and vegetables, particularly those rich in beta-carotene, have low rates of certain cancers. Years ago, researchers focused on beta-carotene, while supplement makers touted it as a powerful anticancer substance. Consumers eagerly bought and took beta-carotene supplements in response.

Then, in a sudden reversal, support for beta-carotene supplements crumbled overnight. Trials around the world were abruptly stopped when scientists noted no benefits but observed a 28 percent *increase* in lung cancer among smokers taking beta-carotene compared with a placebo. Today, beta-carotene supplements are not recommended.[24]

Such reversals might shock and frustrate the unscientific mind, but scientists expect them as research unfolds. In this case, a long-known and basic nutrition principle was reaffirmed: low disease risk accompanies a *diet* of nutritious whole foods, foods that present a balance of nutrients and other beneficial constituents. Whereas a sweet potato and a pill may both contain beta-carotene, the sweet potato presents a balanced array of nutrients, phytochemicals, and fiber

Table C7–3
Antioxidant Terms

- **antioxidant nutrients** vitamins and minerals that oppose the effects of oxidants on human physical functions. The antioxidant vitamins are vitamin E, vitamin C, and beta-carotene. The mineral selenium also participates in antioxidant activities.
- **oxidants** compounds (such as oxygen itself) that oxidize other compounds. Compounds that prevent oxidation are called antioxidants, whereas those that promote it are called prooxidants (*anti* means "against"; *pro* means "for").
- **oxidative stress** damage inflicted on living systems by free radicals.

Table C7–4

Some Invalid Reasons for Taking Supplements

Watch out for plausible-sounding, but false, reasons given by marketers trying to convince you, the consumer, that you need supplements. The invalid reasons listed below have gained strength by repetition among friends, on the Internet, and by the media:

- You fear that foods grown on today's soils lack nutrients (a common false statement made by sellers of supplements).
- You feel tired and falsely believe that supplements can provide energy.
- You hope that supplements can help you cope with stress.
- You wish to build up your muscles faster or without physical activity.
- You want to prevent or cure self-diagnosed illnesses.
- You hope excess nutrients will produce unnamed mysterious beneficial reactions in your body.

People who should never take supplements without a physician's approval include those with kidney or liver ailments (they are susceptible to toxicities), those taking medications (nutrients can interfere with their actions), and smokers (who should avoid products with beta-carotene).

that modulate beta-carotene's effects. The pill provides only beta-carotene, a lone chemical.

For most people, taking an ordinary daily multiple vitamin and mineral supplement is generally safe when they choose an appropriate product and follow dosing directions. And for those who need them, nutrient supplements constitute a modern-day miracle.

SOS: Selection of Supplements

If you fall into one of the categories listed earlier in Table C7–1 and if you absolutely cannot meet your nutrient needs from foods, a supplement containing *nutrients only* can prevent serious problems. In these cases, the benefits outweigh the risks. (Table C7–4 provides some *invalid* reasons for taking supplements in which the risks clearly outweigh the benefits.) Remember, no standard multivitamin and mineral formula exists—the term applies to any combination of nutrients in widely varying doses.

Choosing a Type

Which supplement to choose? The first step is to remain aware that sales of vitamin supplements often approach the realm of quackery because the profits are high and the industry is largely free of oversight. To escape the clutches of the health hustlers, use your imagination, and delete the label pictures of sexy active people and the meaningless, glittering generalities like "Advanced Formula" or "Maximum Power." Also, ignore vague references to functioning of body systems or common complaints, such as cramps (Chapter 2 illustrated such claims)—most of these are overstatements of the truth. Avoid "extras" such as herbs (see Chapter 11). And don't be misled into buying and taking unneeded supplements because none are risk-free.

Reading the Label

Now all you have left is the Supplement Facts Panel (shown in Figure C7–1, p. 284) that lists the nutrients, a list of ingredients, the form of the supplement, and the price—the plain facts. You have two basic questions to answer. The first question: What form do you want—chewable, liquid, or pills? If you'd rather drink your vitamins and minerals than chew them, fine. If you choose a fortified liquid meal replacer, a sugary vitamin drink, or an "energy bar" (a candy bar to which vitamins and other nutrients are added), you must then proportionately reduce the calories you consume in food to avoid gaining unwanted weight. If you choose chewable pills, be aware that vitamin C can erode tooth enamel. Swallow promptly and flush the teeth with a drink of water.

Targeting Your Needs

The second question: Who are you? What vitamins and minerals do you actually need, and in what amounts? Compare the DRI nutrient intake recommendations listed for your age and gender (the tables are on the inside front cover) with the supplement choices. The DRI values meet the needs of all reasonably healthy people.

Choosing Doses

As for doses of nutrients, for most people, an appropriate supplement provides all the vitamins and minerals in amounts smaller than, equal to, or very close to the intake recommendations. Avoid any preparation that in a daily dose provides more than the DRI recommended intake of vitamin A, of vitamin D, or of any mineral, or more than the Tolerable Upper Intake Level of any nutrient. In addition, avoid high doses of iron (more than 10 milligrams per day) except for menstruating women. People who menstruate need more iron, but people who don't, don't. Warning: Expect to reject about 80 percent of available preparations when you choose according to these criteria; be choosy where your health is concerned.

Going for Quality

If you see a USP symbol on the label, it means that a manufacturer has voluntarily paid an independent laboratory to test the product and affirm that it contains the ingredients listed and that it will dissolve or disintegrate in the digestive tract to make the ingredients available for absorption. The symbol does not imply that the supplement has been tested for safety or effectiveness with regard to health, however.

A high price also does not ensure the highest quality; generic brands are often as good as or better than expensive name brand supplements. If they are less expensive, it may mean that their price doesn't have to cover the cost of national advertising. In any case, buy from a well-known retailer who keeps stocks fresh and stores them properly.

Figure C7–1

A Supplement Label

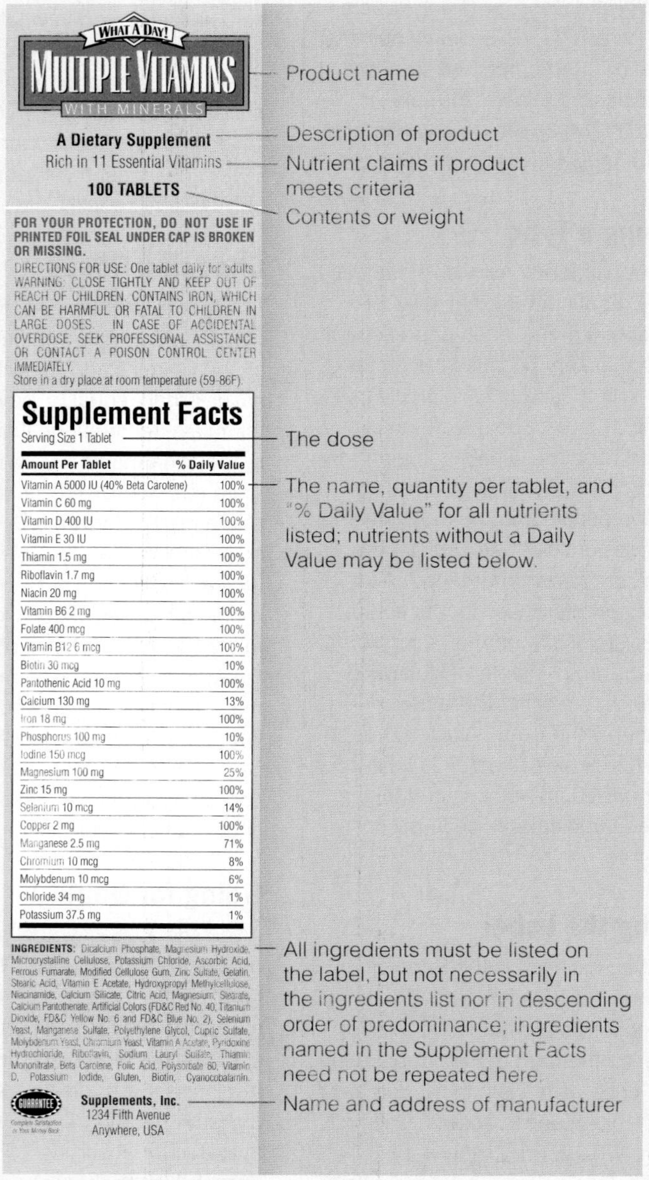
WHAT A DAY!
MULTIPLE VITAMINS
WITH MINERALS
Product name
A Dietary Supplement
Description of product
Rich in 11 Essential Vitamins
Nutrient claims if product meets criteria
100 TABLETS
Contents or weight
FOR YOUR PROTECTION, DO NOT USE IF PRINTED FOIL SEAL UNDER CAP IS BROKEN OR MISSING.
DIRECTIONS FOR USE: One tablet daily for adults.
WARNING: CLOSE TIGHTLY AND KEEP OUT OF REACH OF CHILDREN. CONTAINS IRON, WHICH CAN BE HARMFUL OR FATAL TO CHILDREN IN LARGE DOSES. IN CASE OF ACCIDENTAL OVERDOSE, SEEK PROFESSIONAL ASSISTANCE OR CONTACT A POISON CONTROL CENTER IMMEDIATELY.
Store in a dry place at room temperature (59-86F).
Supplement Facts
Serving Size 1 Tablet
The dose
Amount Per Tablet % Daily Value
Vitamin A 5000 IU (40% Beta Carotene) 100%
Vitamin C 60 mg 100%
Vitamin D 400 IU 100%
Vitamin E 30 IU 100%
Thiamin 1.5 mg 100%
Riboflavin 1.7 mg 100%
Niacin 20 mg 100%
Vitamin B6 2 mg 100%
Folate 400 mcg 100%
Vitamin B12 6 mcg 100%
Biotin 30 mcg 10%
Pantothenic Acid 10 mg 100%
Calcium 130 mg 13%
Iron 18 mg 100%
Phosphorus 100 mg 10%
Iodine 150 mcg 100%
Magnesium 100 mg 25%
Zinc 15 mg 100%
Selenium 10 mcg 14%
Copper 2 mg 100%
Manganese 2.5 mg 71%
Chromium 10 mcg 8%
Molybdenum 10 mcg 6%
Chloride 34 mg 1%
Potassium 37.5 mg 1%
The name, quantity per tablet, and "% Daily Value" for all nutrients listed; nutrients without a Daily Value may be listed below.
INGREDIENTS: Dicalcium Phosphate, Magnesium Hydroxide, Microcrystalline Cellulose, Potassium Chloride, Ascorbic Acid, Ferrous Fumarate, Modified Cellulose Gum, Zinc Sulfate, Gelatin, Stearic Acid, Vitamin E Acetate, Hydroxypropyl Methylcellulose, Niacinamide, Calcium Silicate, Citric Acid, Magnesium Stearate, Calcium Pantothenate, Artificial Colors (FD&C Red No. 40, Titanium Dioxide, FD&C Yellow No. 6 and FD&C Blue No. 2), Selenium Yeast, Manganese Sulfate, Polyethylene Glycol, Cupric Sulfate, Molybdenum Yeast, Chromium Yeast, Vitamin A Acetate, Pyridoxine Hydrochloride, Riboflavin, Sodium Lauryl Sulfate, Thiamin Mononitrate, Beta Carotene, Folic Acid, Polysorbate 80, Vitamin D, Potassium Iodide, Gluten, Biotin, Cyanocobalamin.
All ingredients must be listed on the label, but not necessarily in the ingredients list nor in descending order of predominance; ingredients named in the Supplement Facts need not be repeated here.
GUARANTEE
Supplements, Inc.
1234 Fifth Avenue
Anywhere, USA
Name and address of manufacturer

Avoiding Marketing Traps

In addition, avoid these:

- "For better metabolism." Preparations containing extra biotin may claim to improve metabolism, but no evidence supports this.
- "Organic" or "natural" preparations with added substances. They are no better than standard types, but they cost much more, and the added substances may add risks.
- "High-potency" or "therapeutic dose" supplements. More is not better.
- Items not needed in human nutrition, such as carnitine and inositol. These particular items won't harm you, but they reveal a marketing strategy that makes the whole mix suspect. The manufacturer wants you to believe that its pills contain the latest "new" nutrient that other brands omit, but, in fact, for every valid discovery of this kind, there are 999,999 frauds.
- "Time release." Medications such as some antibiotics or pain relievers often must be sustained at a steady concentration in the blood to be effective; nutrients, in contrast, are incorporated into the tissues where they are needed whenever they arrive.
- "Stress formulas." Although the stress response depends on certain B vitamins and vitamin C, the DRI recommended intake provides all that is needed of these nutrients. If you are under stress (and who isn't?), generous servings of fruits and vegetables will more than cover your need.
- Any supplement sold with claims that today's foods lack sufficient nutrients to support health. Plants make vitamins for their own needs, not ours. A plant lacking a mineral or failing to make a needed vitamin dies before it can bear food for our consumption.

To get the most from a supplement of vitamins and minerals, take it with food. A full stomach retains and dissolves the pill with its churning action.

Conclusion

People in developed nations are far more likely to suffer from *overnutrition* and poor lifestyle choices than from nutrient deficiencies. People wish that swallowing vitamin pills would boost their health. The truth—that they need to improve their eating and exercise habits—is harder to swallow.

Don't waste time and money trying to single out a few nutrients to take as supplements. Invest energy in eating a wide variety of fruits and vegetables in generous quantities, along with the recommended daily amounts of whole grains, protein foods, and milk products every day, and take supplements only when they are truly needed.

Critical Thinking

1. List three reasons why someone might take a multiple vitamin supplement that does not exceed 100 percent of the RDAs. Would you ever take an antioxidant supplement? Why or why not? Suppose you decided that you should take a vitamin supplement because you do not drink milk. How would you determine the best brand of supplement to purchase?
2. Imagine that you are standing in a pharmacy comparing the "Supplement Facts" panels on the labels of two supplement bottles, one a "complete multiple vitamins" product, and the other marked "high potency vitamins." What major differences in terms of nutrient inclusion and doses might you find between these two products? What differences in risk would you anticipate? If you were asked to pick one of these products for an elderly person whose appetite is diminished, which would you choose? Give your justification.

8 Water and Minerals

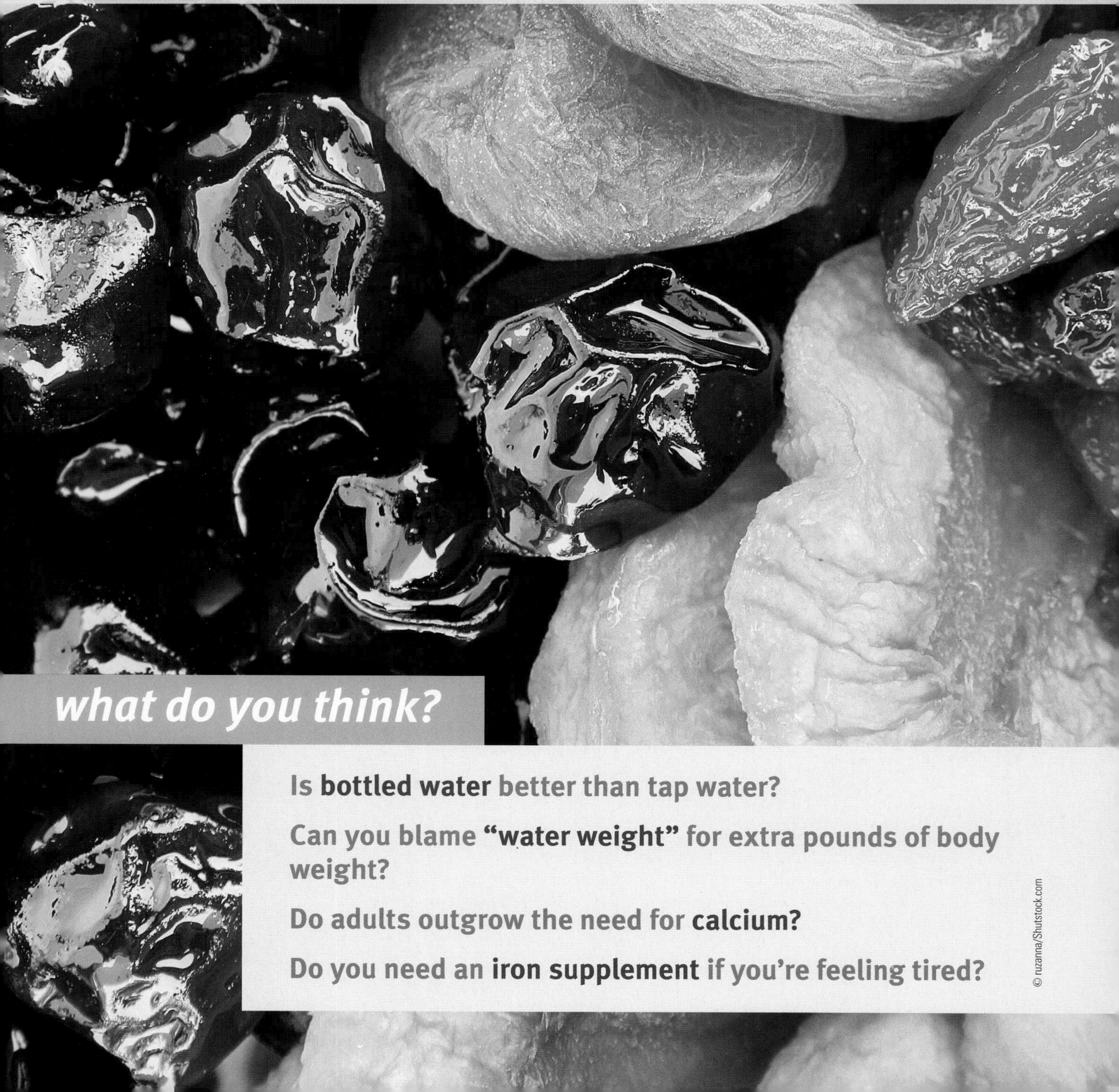

what do you think?

Is **bottled water** better than tap water?

Can you blame **"water weight"** for extra pounds of body weight?

Do adults outgrow the need for **calcium**?

Do you need an **iron supplement** if you're feeling tired?

Learning Objectives

After completing this chapter, you should be able to accomplish the following:

LO 8.1 Describe the body's water sources and routes of water loss, and name factors that influence the need for water.

LO 8.2 Compare and contrast the health effects of various sources of fluid.

LO 8.3 Discuss why electrolyte balance is critical for the health of the body.

LO 8.4 Identify the major minerals important in human nutrition and their physiological roles in the body, the consequences of deficiencies, and their most important food sources.

LO 8.5 Identify the trace minerals important in human nutrition and their physiological roles in the body, the consequences of deficiencies, and their most important food sources.

LO 8.6 Outline a plan for obtaining sufficient calcium from a day's meals.

LO 8.7 Describe the influence of diet during youth on the risk of osteoporosis later in life.

"Ashes to ashes and dust to dust"—it is true that when the life force leaves the body, what is left behind becomes nothing but a small pile of ashes. Carbohydrates, proteins, fats, vitamins, and water are present at first, but they soon disappear. The carbon atoms in all the carbohydrates, fats, proteins, and vitamins combine with oxygen to produce carbon dioxide, which vanishes into the air; the hydrogens and oxygens of those compounds unite to form water; and this water, along with the water that made up a large part of the body weight, evaporates. The ashes left behind are the **minerals**, a small pile that weighs only about 5 pounds. The pile may not be impressive in size, but the work of those minerals is critical to living tissue.

Consider calcium and phosphorus. If you could separate these two minerals from the rest of the pile, you would take away about three-fourths of the total. Crystals made of these two minerals, plus a few others, form the structure of bones and so provide the architecture of the skeleton.

Run a magnet through the pile that remains and you pick up the iron. It doesn't fill a teaspoon, but it consists of billions and billions of iron atoms. As part of hemoglobin, these iron atoms are able to attach to oxygen and make it available at the sites inside the cells where metabolic work is taking place.

If you then extract all the other minerals from the pile of ashes, leaving only copper and iodine, close the windows first. A slight breeze would blow these remaining bits of dust away. Yet the copper in the dust enables iron to hold and to release oxygen, and iodine is the critical mineral in the thyroid hormones. Figure 8–1, p. 288, shows the amounts of the seven **major minerals** and a few of the **trace minerals** in the human body. Other minerals such as gold and aluminum are present in the body but are not known to have nutrient functions.

The distinction between major and trace minerals doesn't mean that one group is more important in the body than the other. A daily deficiency of a few micrograms of iodine is just as serious as a deficiency of several hundred milligrams of calcium. The major minerals are simply present in larger quantities in the body and are needed in greater amounts in the diet.

Water, the first topic of the chapter, is unique among the nutrients and stands alone as the most indispensable of all. The body needs more water each day than any other nutrient—50 times more water than protein and 5,000 times more water than vitamin C. You can survive a deficiency of any of the other nutrients for a long time, in some cases for months or years, but you can survive only a few days without water. In less than a day, a lack of water alters the body's chemistry and metabolism.

Our discussion begins with water's many functions. Next we examine how water and the major minerals mingle to form the body's fluids and how cells regulate the distribution of those fluids. Then we take up the specialized roles of each of the minerals. (Reminder: The DRI lists for water and minerals are listed on the inside front cover pages.)

Did You Know?

U.S. intakes of these three minerals are low enough to be of concern for public health:

- Potassium (for everyone).
- Calcium (for everyone).
- Iron (for some people).

minerals naturally occurring, inorganic, homogeneous substances; chemical elements.

major minerals essential mineral nutrients required in the adult diet in amounts greater than 100 milligrams per day. Also called *macrominerals*.

trace minerals essential mineral nutrients required in the adult diet in amounts less than 100 milligrams per day. Also called *microminerals*.

Figure 8–1

Minerals in a 60-Kilogram (132-Pound) Person, in Grams

The major minerals are needed by the body in larger amounts than the trace minerals, and, as shown in the graph, they are present in larger amounts, too.

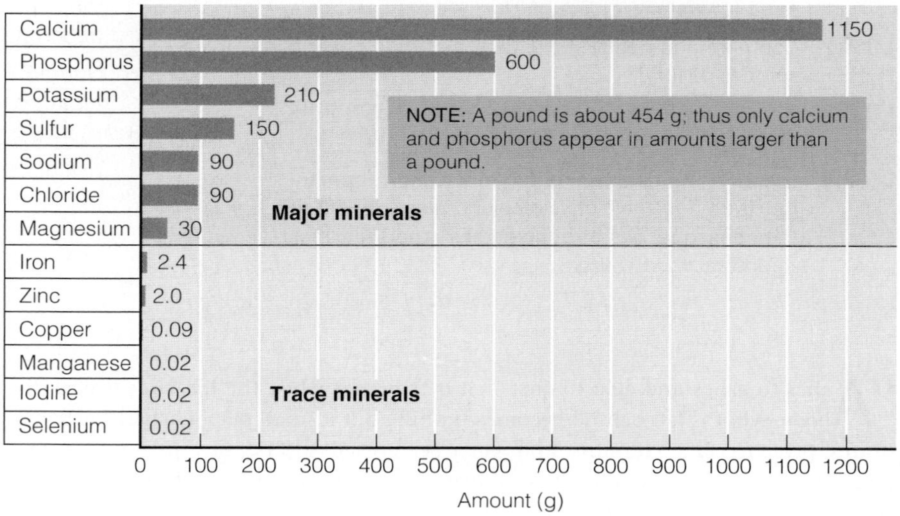

Water

LO 8.1 Describe the body's water sources and routes of water loss, and name factors that influence the need for water.

You began as a single cell bathed in a nourishing fluid. As you became a beautifully organized, air-breathing body of trillions of cells, each of your cells had to remain next to water to stay alive.

Water is the most indispensable nutrient.

Water makes up about 60 percent of an adult person's weight—that's almost 80 pounds of water in a 130-pound person. All this water in the body is not simply a river coursing through the arteries, capillaries, and veins. Soft tissues contain a great deal of water: the brain and muscles are 75 to 80 percent water by weight; even bones contain 25 percent water.*[1] Some of the body's water is incorporated into the chemical structures of compounds that form the cells, tissues, and organs of the body. For example, proteins hold water molecules within them, water that is locked in and not readily available for any other use. Water also participates actively in many chemical reactions.

Why Is Water the Most Indispensable Nutrient?

Water brings to each cell the exact ingredients the cell requires and carries away the end products of its life-sustaining reactions. The water of the body fluids is thus the transport vehicle for all the nutrients and wastes. Without water, cells quickly die.[2]

Solvent Water is a nearly universal **solvent**: it dissolves amino acids, glucose, minerals, and many other substances needed by the cells. Fatty substances, too, can travel freely in the watery blood and lymph because they are specially packaged in water-soluble proteins.

Cleansing Agent Water is also the body's cleansing agent. Small molecules, such as the nitrogen wastes generated during protein metabolism, dissolve in the watery blood and must be removed before they build up to toxic concentrations. The kidneys

*Reference notes are found in Appendix F.

filter these wastes from the blood and excrete them, mixed with water, as urine. When the kidneys become diseased, as can happen in diabetes and other disorders, toxins can build to life-threatening levels. A kidney **dialysis** machine must then take over the task of cleansing the blood by filtering wastes into water contained in the machine.

Lubricant and Cushion Water molecules resist being crowded together. Thanks to this incompressibility, water can act as a lubricant and a cushion for the joints, and it can protect sensitive tissue such as the spinal cord from shock. The fluid that fills the eye serves in a similar way to keep optimal pressure on the retina and lens. From the start of human life, a fetus is cushioned against shock by the bag of amniotic fluid in the mother's uterus. Water also lubricates the digestive tract, the respiratory tract, and all tissues that are moistened with mucus.

Coolant Yet another of water's special features is its ability to help maintain body temperature. The water of sweat is the body's coolant. Heat is produced as a by-product of energy metabolism and can build up dangerously in the body. To rid itself of this excess heat, the body routes its blood supply through the capillaries just under the skin. At the same time, the skin secretes sweat, and its water evaporates. Converting water to vapor takes energy; therefore, as sweat evaporates, heat energy dissipates, cooling the skin and the underlying blood. The cooled blood then flows back to cool the body's core. Sweat evaporates continuously from the skin, usually in slight amounts that go unnoticed; thus, the skin is a major organ through which water is lost from the body. Lesser amounts are lost by way of exhaled breath and the feces.[3]

To sum up, water:

- Carries nutrients throughout the body.
- Serves as the solvent for minerals, vitamins, amino acids, glucose, and other small molecules.
- Cleanses the tissues and blood of wastes.
- Actively participates in many chemical reactions.
- Acts as a lubricant around joints.
- Serves as a shock absorber inside the eyes, spinal cord, joints, and amniotic sac surrounding a fetus in the womb.
- Aids in maintaining the body's temperature.

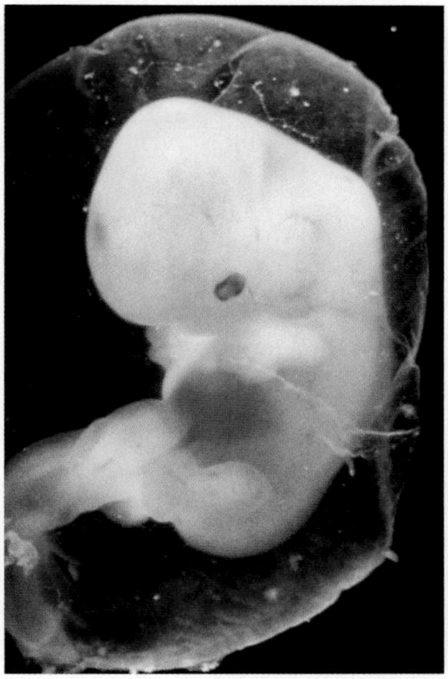
Claude Edelmann/Science Source

Human life begins in water.

KEY POINTS

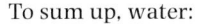

- Water makes up about 60 percent of the body's weight.
- Water provides the medium for transportation, acts as a solvent, participates in chemical reactions, provides lubrication and shock protection, and aids in temperature regulation in the human body.

The Body's Water Balance

Water is such an integral part of us that people seldom are conscious of water's importance, unless they are deprived of it. Since the body loses some water every day, a person must consume at least the same amount to avoid life-threatening losses, that is, to maintain **water balance**. The total amount of fluid in the body is kept balanced by delicate mechanisms. Imbalances such as **dehydration** and **water intoxication** can occur, but the balance is restored as promptly as the body can manage it. The body controls both intake and excretion to maintain water equilibrium.

The amount of the body's water varies by pounds at a time, especially in women who retain water during menstruation. Eating a meal high in salt can temporarily increase the body's water content; the body sheds the excess over the next day or so as the sodium is excreted. These temporary fluctuations in body water show up on the scale, but gaining or losing water weight does not reflect a change in body fat. Fat weight takes days or weeks to change noticeably, whereas water weight can change overnight.

solvent a substance that dissolves another and holds it in solution.

dialysis (dye-AL-ih-sis) a medical treatment for failing kidneys in which a person's blood is circulated through a machine that filters out toxins and wastes and returns cleansed blood to the body. Also called *hemodialysis.*

water balance the balance between water intake and water excretion, which keeps the body's water content constant.

dehydration loss of water. The symptoms progress rapidly, from thirst to weakness to exhaustion and delirium, and end in death.

water intoxication a dangerous dilution of the body's fluids resulting from excessive ingestion of plain water. Symptoms are headache, muscular weakness, lack of concentration, poor memory, and loss of appetite.

An extra drink of water benefits both young and old.

Figure 8–2

Water Balance—A Typical Example

Each day, water enters the body in liquids and foods, and some water is created in the body as a by-product of metabolic processes. Water leaves the body through the evaporation of sweat, in the moisture of exhaled breath, in the urine, and in the feces.

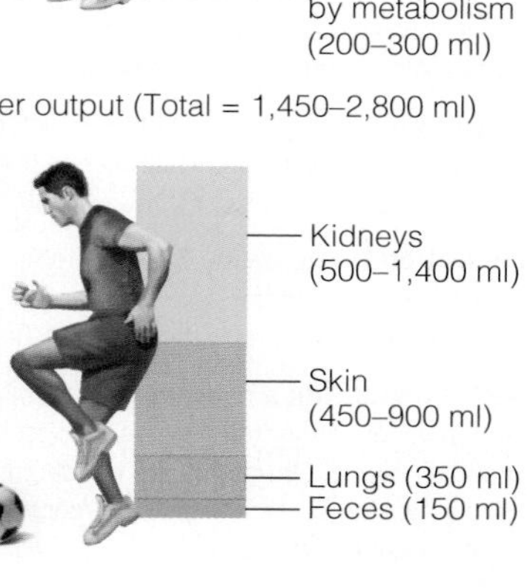

KEY POINT

- A change in the body's water content can bring about a temporary change in body weight.

Quenching Thirst and Balancing Losses

Thirst and satiety govern water intake. When the blood is too concentrated (having lost water but not salt and other dissolved substances), the molecules and particles in the blood attract water out of the salivary glands, and the mouth becomes dry. Water is also drawn from the body's cells, causing them to collapse a little.[4] The brain center known as the hypothalamus (described in Chapter 3) responds to information about the concentrated blood particles, low cell fluid volume, or low blood pressure by initiating nerve impulses to the brain that register as "thirst." The hypothalamus also signals the pituitary gland to release a hormone that directs the kidneys to shift water back into the bloodstream from the fluid destined to become urine. The kidneys themselves respond to the sodium concentration in the blood passing through them and secrete regulatory substances of their own. The net result is that the more water the body needs, the less it excretes. Figure 8–2 shows how water intake and excretion naturally balance out.

Dehydration Thirst lags behind a lack of water. When too much water is lost from the body and is not replaced, dehydration can threaten survival. A first sign of dehydration is thirst, the signal that the body has already lost a cup or two of its total fluid and the need to drink is immediate. But suppose a thirsty person is unable to obtain fluid or, as in many elderly people, fails to perceive the thirst message. Instead of "wasting" precious water in sweat, the dehydrated body diverts most of its water into the blood vessels to maintain the life-supporting blood pressure. Meanwhile, body heat builds up because sweating has ceased, creating the possibility of serious consequences in hot weather (see Table 8–1).

To ignore the thirst signal is to invite dehydration. With a loss of just 1 percent of body weight as fluid, perceptible symptoms appear: headache, fatigue, confusion or forgetfulness, and an elevated heart rate. A loss of 2 percent impairs physical functioning and impedes a wide range of physical activities.[5] People should stay attuned to thirst and drink whenever they feel thirsty to replace fluids lost throughout the day. Older adults in whom thirst is blunted should drink regularly throughout the day, regardless of thirst.

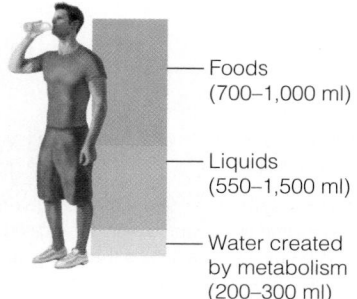

Table 8–1

Effects of Mild Dehydration, Severe Dehydration, and Chronic Lack of Fluid

Mild Dehydration (Loss of <5% Body Weight)	Severe Dehydration (Loss of >5% Body Weight)	Chronic Low Fluid Intake May Increase the Likelihood of:[a]
Thirst	Pale skin	Cardiac arrest (heart attack) and other heart problems
Sudden weight loss	Bluish lips and fingertips	Constipation
Rough, dry skin	Confusion; disorientation	Dental disease
Dry mouth, throat, body linings	Rapid, shallow breathing	Gallstones
Rapid pulse; low blood pressure	Weak, rapid, irregular pulse	Glaucoma (elevated pressure in the eye)
Lack of energy; weakness	Thickening of blood	Hypertension
Impaired kidney function	Shock; seizures	Kidney stones
Reduced quantity of urine; concentrated urine	Coma; death	Pregnancy/childbirth problems
Decreased mental functioning		Stroke
Decreased muscular work and athletic performance		Urinary tract infections
Fever or increased internal temperature		
Fainting and delirium		

[a]*Evidence for bladder and colon cancer is inconsistent.*

Source: B. M. Popkin, K. E. D'Anci, and I. H. Rosenberg, Water, hydration, and health, Nutrition Reviews *68 (2010): 439–458; K. M. Kolasa, C. J. Lacky, and A. C. Grandjean, Hydration and health promotion,* Nutrition Today *44 (2009): 190–201; Standing Committee on the Scientific Evaluation of Dietary Reference Intakes, Food and Nutrition Board, Institute of Medicine,* Dietary Reference Intakes: Water, Potassium, Sodium, Chloride, and Sulfate *(Washington, D.C.: National Academies Press, 2005), pp. 118–127.*

A word about caffeine: People who drink caffeinated beverages lose a little more fluid than when they drink water because caffeine acts as a **diuretic**. The DRI Committee concluded, however, that the mild diuretic effect of moderate caffeine intake does not lead to dehydration, nor does it keep people from meeting their fluid needs. Caffeinated beverages can therefore contribute to daily water intakes. The Controversy section of Chapter 14 discusses other effects of caffeine.

Water Intoxication At the other extreme from dehydration, water intoxication occurs when too much plain water floods the body's fluids and disturbs their normal composition. Most adult victims have consumed several gallons of plain water in a few hours' time. Water intoxication is rare, but when it occurs, immediate action is needed to reverse dangerously diluted blood before death ensues (read more about it in Chapter 10).

KEY POINTS

- Water losses from the body must be balanced by water intakes to maintain hydration.
- The brain regulates water intake; the brain and kidneys regulate water excretion.
- Caloric beverages add to energy intakes.
- Dehydration and water intoxication can have serious consequences.

How Much Water Do I Need to Drink in a Day?

Water needs vary greatly depending on the foods a person eats, the air temperature and humidity, the altitude, the person's activity level, and other factors (see Table 8–2, p. 292). Fluid needs vary widely among individuals and also within the same person in various environmental conditions, so a specific water recommendation is hard to pin down.

Water from Fluids and Foods A wide range of fluid intakes can maintain adequate hydration. As a general guideline, however, the DRI Committee recommends that, given a normal diet and moderate environment, the reference man needs about 13 cups of fluid from beverages including drinking water, and the reference woman needs about 9 cups.[6] This amount of fluid provides about 80 percent of the body's

Did You Know?

Water loss can be expressed as a percentage of body weight. In a 150-lb person,

- A 3-lb loss of body fluid equals 2% of body weight.
  $3 \div 150 \times 100 = 2\%$
- A $4\frac{1}{2}$-lb loss equals 3% of body weight.
  $4.5 \div 150 \times 100 = 3\%$

diuretic (dye-you-RET-ic) a compound, usually a medication, causing increased urinary water excretion; a "water pill."

Table 8–2

Factors That Increase Fluid Needs

These conditions increase a person's need for fluids:

- Alcohol consumption
- Cold weather
- Dietary fiber
- Diseases that disturb water balance, such as diabetes and kidney diseases
- Forced-air environments, such as airplanes and sealed buildings
- Heated environments
- High altitude
- Hot weather, high humidity
- Increased protein, salt, or sugar intakes
- Ketosis
- Medications (diuretics)
- Physical activity
- Pregnancy and breastfeeding (see Chapter 13)
- Prolonged diarrhea, vomiting, or fever
- Surgery, blood loss, or burns
- Very young or old age

Table 8–3

Water in Foods and Beverages

Many solid foods, such as broccoli or steak, are surprisingly high in water.

100%	water, diet soft drinks, seltzer (unflavored), plain tea
95–99%	sugar-free gelatin dessert, clear broth, Chinese cabbage, celery, cucumber, lettuce, summer squash, black coffee
90–94%	sports drinks, grapefruit, fresh strawberries, broccoli, tomatoes
80–89%	sugar-sweetened soft drinks, milk, yogurt, egg white, fruit juices, low-fat cottage cheese, cooked oatmeal, fresh apple, carrot
60–79%	low-calorie mayonnaise, instant pudding, banana, shrimp, lean steak, pork chop, baked potato, cooked rice
40–59%	diet margarine, sausage, chicken, macaroni and cheese
20–39%	bread, cake, cheddar cheese, bagel
10–19%	butter, margarine, regular mayonnaise
5–9%	peanut butter, popcorn
1–4%	ready-to-eat cereals, pretzels
0%	cooking oils, meat fats, shortening, white sugar

Did You Know?

Some beverages are commonly measured in cups and ounces, but others are measured in liters (L). As this list shows, 4 cups (32 ounces or 1 quart) measures a little less than 1 liter:

cups (fluid oz)		***ml (L)***
8 c (64)	=	1,893 (≈ 2)
6½ c (52)	=	1,538 (≈ 1.5)
4 c (32)	=	946 (≈ 1)
2½ c (20)	=	591 (≈ 0.6)
1 c (8)	=	237 (≈ 0.2)

daily water need. On average, most people in the United States consume close to these amounts.[7] The fluids people choose to drink can affect daily calorie intakes, as the Consumer's Guide section makes clear.

Most of the rest of the body's needed daily fluid comes from the water in foods. Nearly all foods contain some water: water constitutes up to 95 percent of the volume of most fruits and vegetables and at least 50 percent of many meats and cheeses (see Table 8–3 and Appendix A). A small percentage of the day's fluid is generated in the tissues themselves as energy-yielding nutrients release **metabolic water** as a product of chemical breakdown.

The Effect of Sweating on Fluid Needs Sweating increases water needs. Especially when performing physical work outdoors in hot weather, people can lose 2 to 4 gallons of fluid in a day. An athlete training in the heat can sweat out more than a half gallon of fluid each hour. The importance of maintaining hydration for athletes exercising in the heat cannot be overemphasized, and Chapter 10 provides detailed instructions for hydrating the exercising body.

KEY POINTS

- Many factors influence a person's need for water.
- Water is provided by beverages and foods and by cellular metabolism.
- Sweating increases fluid needs.
- High-calorie beverages affect daily calorie intakes.

Drinking Water: Types, Safety, and Sources

LO 8.2 Compare and contrast the health effects of various sources of fluid.

In developed countries where clean water is always as close as the tap, people take water for granted, and they often devalue it and waste it. Water, however, could

metabolic water water generated in the tissues during the chemical breakdown of the energy-yielding nutrients in foods.

use it! A Consumer's Guide To . . .

Liquid Calories

Most ordinary beverages help to meet the body's need for fluid. In developed nations such as ours, however, people encounter a constant stream of beverages that contain more than just water.

Mystery Pounds

Derek, an active college student, hasn't thought much about his fluid intake but is lamenting, "... I'm exercising more and I've cut out the junk food but I've still gained five pounds!" What has escaped Derek's attention are the calories that he's been drinking: a big glass of vitamin C–enriched orange punch at breakfast, a soda or two before lunchtime, sometimes a large mocha latte for an afternoon wake-up, and, of course, sports drinks when he works out.

Drinking without Thirst

Like Derek, most people choose beverages for reasons having little to do with thirst. They seek the stimulating effect of caffeine in coffee, tea, or sodas. They choose fluids such as soup, milk, juice, or other beverages at mealtimes. They believe they need the added nutrients in sugar-sweetened "vitamin waters." They think they need the carbohydrate in sports drinks for all physical activities (few exercisers do; read Chapter 10). They drink hot beverages to warm up or cold ones to cool off. Or they drink for pleasure—for the aroma of coffee, the sweet taste of sugar, or the euphoria of alcohol. On each of these drinking occasions, with or without their awareness, people make choices among high-calorie and lower-calorie beverages.

A fancy coffee drink can easily provide 400–700 calories; plain coffee, zero calories.

Weighing In on Extra Fluids

Drinking extra fluid may have some health advantages, such as preventing even minor dehydration and reducing the risk of developing kidney stones.[1] Fluids such as fat-free milk and 100 percent fruit or vegetable juices provide needed nutrients and are thus included in the USDA eating patterns. Other beverages, such as sugary sodas and punches, provide many empty calories of added sugars, and most people must limit their intakes. Just switching from empty-calorie beverages to zero calorie choices could help many people to lose weight.[2]

Young men like Derek top the chart for intakes of calories from beverages, with an average 22 percent of total calories coming from beverages—or 660 calories in a 3,000-calorie diet (see Figure 8–3).[3] Young women drink about 19 percent of their calories each day (about 450 calories of a 2,400-calorie diet). Beverages supply more than half of the added sugars, most of the caffeine, and all of the alcohol in the U.S. diet.[4]

Even among nutritious beverages, daily choices matter. For example, an 8-ounce glass of orange juice provides about 110 calories; tomato juice, a similar choice with regard to vitamins and minerals, provides just 40. A cup of cream-based "bisque" soup may contribute up to 300 calories; a cup of

The pros and cons of sports drinks are discussed in **Chapter 10**. **Controversies 3** and **4** focus on the health effects of alcohol and added sugars.

Figure 8–3

How Many Calories Do We Drink?

The intake of calories from beverages varies widely with age, with 20- to 30-year-old men consuming the greatest amounts by far.

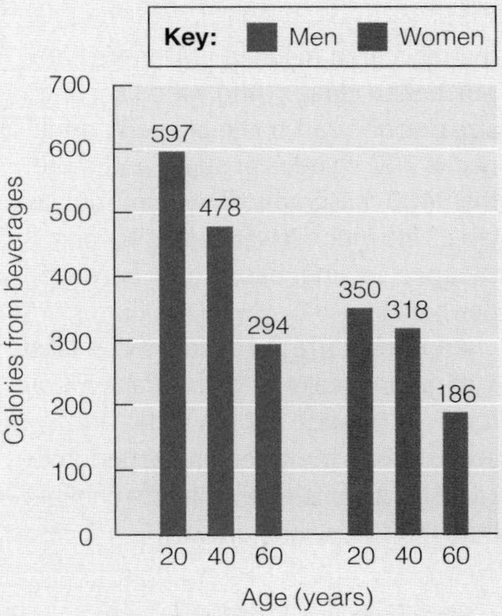

Source: R. P. LaComb and coauthors, Beverage Choices of U.S. Adults: What We Eat in America, NHANES 2007–2008, *Data Brief No. 6, August 2011, available at http://ars.usda.gov.*

broth-based minestrone has just 120 calories and a tiny fraction of the saturated fat. Calories in choices like these add up, and what happened to Derek is the likely outcome—creeping weight gain, even in people who exercise and eliminate "junk food" from the diet.

Seeking an Expert's Advice

"My advice is to track your intake of fluids and add up their calories," says

Nutrition Facts
Serving Size 8 fl oz (240 mL)
Servings Per Container 4
Amount Per Serving
Calories 50
% Daily Value*
Total Fat 0g 0%
Sodium 110mg 5%
Potassium 30mg 1%
Total Carbohydrate 14g 5%
Sugars 14g
Protein 0g
Not a significant source of Calories from Fat, Saturated Fat, Cholesterol, Dietary Fiber, Vitamin A, Vitamin C, Calcium, Iron.
*Percent Daily Values are based on a 2,000 calorie diet.

Compare labels and carefully note the serving size. This Nutrition Facts panel lists calories and sodium for eight ounces—one-fourth of the bottle.

the registered dietitian at Derek's campus health clinic. "And watch serving sizes: your quart bottle of sports drink packs 200 calories of sugar and more than 400 milligrams of sodium, but its label lists much lower values for one 8-ounce serving, based on 4 servings per bottle."

And Derek's reply: "I counted at least 400 random calories that I *drank* every day . . . I'll switch out the sodas and sports drinks for water, and as for coffee, I'll just put some milk in it—it's cheaper than the fancy stuff, anyway."

Looking at Labels

All packaged drinks must carry a Nutrition Facts panel. But what about calories in unlabeled beverages? How many calories are in coffee drinks, iced teas, fountain drinks, or bar drinks? Any beverage that lacks a Nutrition Facts panel requires consumers to look up calorie totals or guess at them, and then remember them to inform their future choices (Appendix A at the end of this text lists the calories in many beverages).

Moving Ahead

All beverages (except alcohol) can readily meet the body's fluid needs, so the question becomes, "What else does this beverage supply?" A 500-calorie smoothie may be the right choice for a person who needs to gain weight, but for most people, the Centers for Disease Control and Prevention recommends these thirst-quenchers instead:

- Plain water.
- Water flavored with cucumber, lemon, lime, mint, or watermelon slices.
- Sparkling water, plain or with a splash of fruit juice.[5]

Other recommendations are plain tea, coffee, nonfat and low-fat milk and soy milk, artificially sweetened beverages, clear soups, 100 percent vegetable juices, and 100 percent fruit juices in moderation (see Chapter 2). If you enjoy regular soft drinks, sweet tea, creamy coffee drinks, punches, and other highly caloric beverages, limit yourself to the smallest size and choose other beverages most of the time to stay hydrated while staying within your calorie budget.

Review Questions*

1. Beverage consumption represents __________ .
   a. 19 percent of a young woman's daily calorie intake
   b. 7 percent of a young woman's daily calorie intake
   c. an insignificant amount of a young woman's daily calorie intake
2. When choosing a beverage, one should __________ .
   a. read the label carefully, especially noting the number of servings in the container and the calories per serving
   b. consider how a beverage's calories fit into the day's calorie needs
   c. consider ingredients in addition to water supplied by the beverage
   d. all of the above
3. The Centers for Disease Control and Prevention recommend __________ .
   a. drinking water, plain or lightly flavored, to quench thirst
   b. staying hydrated with plenty of regular soft drinks, sweet tea, creamy coffee drinks, punches
   c. drinking plain tea, coffee, nonfat and low-fat milk and soy milk, artificially sweetened beverages, clear soups, and 100 percent fruit and vegetable juices, in addition to water
   d. a and c

** Answers to Consumer's Guide review questions are found in Appendix G.*

arguably be the earth's most precious natural resource. Just ask any of the 884 million of the world's people who struggle to stay alive in areas without access to safe drinking water.[8] This section sheds some light on the nature of water and offers perspective on our nation's supply.

Hard Water or Soft Water—Which Is Best?

Water occurs as **hard water** or **soft water**, a distinction that affects your health with regard to three minerals. Hard water has high concentrations of calcium and magnesium. Soft water's principal mineral is sodium. In practical terms, soft water makes more bubbles with less soap; hard water leaves a ring on the tub, a jumble of rocklike crystals in the teakettle, and a gray residue in the wash.

Soft water may seem more desirable, and some homeowners purchase water softeners that remove magnesium and calcium and replace them with sodium. The sodium of soft water, even when it bubbles naturally from the ground, may aggravate **hypertension**, however. Soft water also more easily dissolves certain contaminant metals, such as cadmium and lead, from pipes. Cadmium can harm the body, affecting enzymes by displacing zinc from its normal sites of action. Lead, another toxic metal, is absorbed more readily from soft water than from hard water, possibly because the calcium in hard water protects against its absorption. Old plumbing may contain cadmium or lead, so people living in old buildings should run the cold water tap for a minute to flush out harmful minerals before drawing water for the first use in the morning and whenever no water has been drawn for more than 6 hours.[9]

KEY POINTS

- Hard water is high in calcium and magnesium.
- Soft water is high in sodium, and it dissolves cadmium and lead from pipes.

Did You Know?

Lead is exceptionally harmful to children (details in Chapter 14).

Safety of Public Water

Remember that water is practically a universal solvent: it dissolves almost anything it encounters to some degree. Hundreds of contaminants—including disease-causing bacteria and viruses from human wastes, toxic pollutants from highway fuel runoff, spills and heavy metals from industry, organic chemicals such as pesticides from agriculture, and manure bacteria from farm animals—have been detected, albeit rarely, in public drinking water. Such problems, when they arise, are promptly reported and corrected.

Public water systems remove many hazards. They add disinfectant (usually chlorine) to kill most microorganisms and may expose the water to other treatments to purify it. Private well water is usually not chlorinated, so the 40 million Americans who drink water from private wells should have them tested regularly for harmful microorganisms.

Testing and Reporting All public drinking water must be tested regularly for contamination. The Environmental Protection Agency (EPA) ensures that public water systems meet minimum standards for health. Public utility customers receive a yearly statement, written in plain language, listing the chemicals and bacteria found in local water.

Chlorination and Cancer By-products of water chlorination have been found to cause cancer-related changes in human cells and cancer in laboratory animals.[10] Investigators acknowledge the possibility of a connection between chlorinated drinking water and cancer incidence, but they also passionately defend chlorination as a benefit to public health. In areas of the world without chlorination, an estimated 25,000 people die *each day*—more than are killed by violence, including war—from diseases caused by organisms carried by water and easily killed by chlorine.[11] Substitutes for chlorine exist, but they are too expensive or too slow to be practical for treating a city's water, and some may create by-products of their own.

Water Sources

Meanwhile, what is a consumer to drink? The first option is to drink tap water because municipal water is held to minimum standards for purity, as described. It comes from any of several sources.

Surface Water **Surface water** flowing from lakes, rivers, and reservoirs fills about half of the nation's need for drinking water, mostly in major cities. Surface water is exposed to contamination by acid rain, petroleum products, pesticides, fertilizer, human and animal wastes, and industrial wastes that run directly from pavements, septic tanks, farmlands, and industrial areas into streams that feed surface water bodies. Surface water generally moves faster than **groundwater** and stays above ground where aeration and exposure to sunlight can cleanse it. The plants and

hard water water with high calcium and magnesium concentrations.

soft water water with a high sodium concentration.

hypertension high blood pressure; also defined in Chapter 11.

surface water water that comes from lakes, rivers, and reservoirs.

groundwater water that comes from underground aquifers.

microorganisms that live in surface water also filter it. These processes can remove some contaminants, but others stay in the water.

Groundwater Groundwater comes from protected **aquifers**, deep underground rock formations saturated with water. People in rural areas rely mostly on groundwater pumped from private wells, and some cities tap this resource, too. Groundwater can become contaminated from hazardous waste sites, dumps, oil and gas pipelines, and landfills, as well as downward seepage from surface water bodies. Groundwater moves slowly and is not aerated or exposed to sunlight, so contaminants break down more slowly than in surface water. To mingle with water in the aquifer, surface water must first "percolate," or seep, through soil, sand, or rock, which filters out some contaminants.

Home Water Purification A second option for drinking water is to further purify tap water with home purifying equipment, which ranges in price from about $20 to $5,000. Some home systems do an adequate job of removing lead, chlorine, and other contaminants, but others only improve the water's taste. Each system has advantages and drawbacks, and all require periodic maintenance or filter replacements that vary in price. Not all companies or representatives are legitimate—some perform water tests that yield dramatic-appearing but meaningless results to sell unneeded systems.[12] Verify all claims of contamination before buying a purifying system.

© Rungroj Yongrit/epa/Corbis

Billions of expensive, empty water bottles end up in landfills around the nation each year.

Bottled Water A third option is to use **bottled water** (Table 8–4 provides terms used in marketing of bottled water). About 7 percent of U.S. households turn to bottled water as an alternative to tap water—and they pay 250 to 10,000 times the price of tap water. Are bottled waters worth their price?

Some bottled water may taste fresher than tap because they are disinfected with ozone, which, unlike the chlorine used in most municipal water systems, leaves no flavor or odor. Other bottled waters are simply treated tap water.[13] With regard to safety, the U.S. Food and Drug Administration (FDA) recently tightened bottled water standards after tests revealed contamination with bacteria, arsenic, or synthetic

Table 8–4

Water Terms That May Appear on Labels

- **artesian water** water drawn from a well that taps a confined aquifer in which the water is under pressure.
- **baby water** ordinary bottled water treated with ozone to make it safe but not sterile.
- **caffeine water** bottled water with caffeine added.
- **carbonated water** water that contains carbon dioxide gas, either naturally occurring or added, that causes bubbles to form in it; also called bubbling or sparkling water. Seltzer, soda, and tonic waters are legally soft drinks and are not regulated as water.
- **distilled water** water that has been vaporized and recondensed, leaving it free of dissolved minerals.
- **filtered water** water treated by filtration, usually through activated carbon filters that reduce the lead in tap water, or by reverse osmosis units that force pressurized water across a membrane, removing lead, arsenic, and some microorganisms from tap water.
- **fitness water** lightly flavored bottled water enhanced with vitamins, supposedly to enhance athletic performance.
- **mineral water** water from a spring or well that typically contains at least 250 parts per million (ppm) of naturally occurring minerals. Minerals give water a distinctive flavor. Many mineral waters are high in sodium.
- **natural water** water obtained from a spring or well that is certified to be safe and sanitary. The mineral content may not be changed, but the water may be treated in other ways such as with ozone or by filtration.
- **public water** water from a municipal or county water system that has been treated and disinfected. Also called *tap water.*
- **purified water** water that has been treated by distillation or other physical or chemical processes that remove dissolved solids. Because purified water contains no minerals or contaminants, it is useful for medical and research purposes.
- **spring water** water originating from an underground spring or well. It may be bubbly (carbonated) or "flat" or "still," meaning not carbonated. Brand names such as "Spring Pure" do not necessarily mean that the water comes from a spring.
- **vitamin water** bottled water with a few vitamins added; does not replace vitamins from a balanced diet and may worsen overload in people receiving vitamins from enriched food, supplements, and other enriched products such as "energy" bars.
- **well water** water drawn from groundwater by tapping into an aquifer.

chemicals in about a third of bottled water samples, and the heavy metal lead in about half the samples.[†14]

Under the FDA's rules, water intended for bottling and selling across state lines must be tested yearly for chemical contaminants and weekly for disease-causing bacteria. When contamination shows up in either tap or bottled water, it must be cleared up before the water can be distributed.[15]

Considerable fossil fuels (and many gallons of water) are required to create and transport disposable plastic water bottles. Single serving bottles can be recycled, but 80 percent of the 34.6 *billion* plastic water bottles purchased in the United States each year end up in landfills, incinerators, or as litter. These empties now cost taxpayers hundreds of millions of dollars each year for their disposal and litter cleanup costs.[16]

Whether water comes from the tap or is poured from a bottle, all water comes from the same sources—surface water and groundwater. Given water's importance in the body, the world's supply of clean, wholesome water is a precious resource to be guarded. The remainder of this chapter addresses other important nutrients—the minerals.

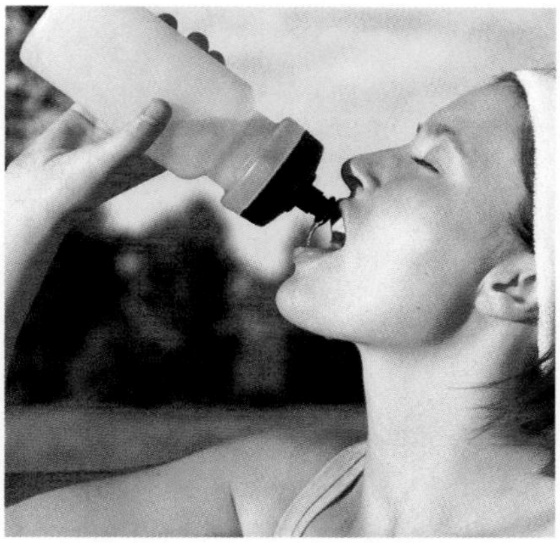

Dougal Waters/The Image Bank/Getty Images

Using refillable bottles saves money and cuts waste.

KEY POINTS

- Public drinking water is tested and treated for safety.
- All drinking water, including bottled water, originates from surface water or groundwater, which are vulnerable to contamination from human activities.

Body Fluids and Minerals

LO 8.3 Discuss why electrolyte balance is critical for the health of the body.

Most of the body's water weight is contained inside the cells, and some water bathes the outsides of the cells. The remainder fills the blood vessels. How do cells keep themselves from collapsing when water leaves them and from swelling up when too much water enters them?

Water Follows Salt

The cells cannot regulate the amount of water directly by pumping it in and out because water slips across membranes freely. The cells can, however, pump minerals across their membranes. The major minerals form **salts** that dissolve in the body fluids; the cells direct where the salts go, and this determines where the fluids flow because water follows salt.

When mineral (or other) salts dissolve in water, they separate into single, electrically charged particles known as **ions**. Unlike pure water, which conducts electricity poorly, ions dissolved in water carry electrical current; for this reason, these electrically charged ions are called **electrolytes**.

As Figure 8–4 (p. 298) shows, when dissolved particles, such as electrolytes, are present in unequal concentrations on either side of a water-permeable membrane, water flows toward the more concentrated side to equalize the concentrations. Cells and their surrounding fluids work in the same way. Think of a cell as a sack made of a water-permeable membrane. The sack is filled with watery fluid and suspended in a dilute solution of salts and other dissolved particles. Water flows freely between the fluids inside and outside the cell but generally moves from the more dilute solution toward the more concentrated one (the photo of salted eggplant slices shows this effect).

© Craig M. Moore

The slices of eggplant on the right were sprinkled with salt. Notice their beads of "sweat," formed as cellular water moves across each cell's membrane (water-permeable divider) toward the higher concentration of salt (dissolved particles) on the surface.

KEY POINT

- Cells regulate water movement by pumping minerals across their membranes; water follows the minerals.

aquifers underground rock formations containing water that can be drawn to the surface for use.

bottled water drinking water sold in bottles.

salts compounds composed of charged particles (ions). An example is potassium chloride (K^+Cl^-).

ions (EYE-ons) electrically charged particles, such as sodium (positively charged) or chloride (negatively charged).

electrolytes compounds that partly dissociate in water to form ions, such as the potassium ion (K^+) and the chloride ion (Cl^-).

† The group was the National Resources Defense Council. Read its report, *Bottled Water: Pure Drink or Pure Hype?*; available at www.nrdc.org/water/drinking/nbw.asp.

Figure 8–4

Animated! How Electrolytes Govern Water Flow

Water flows in the direction of the more highly concentrated solution.

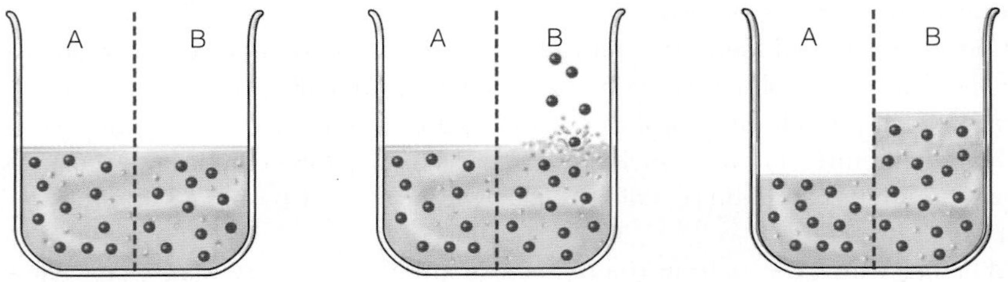

1 With equal numbers of dissolved particles on both sides of a water-permeable divider, water levels remain equal.

2 Now additional particles are added to increase the concentration on side B. Particles cannot flow across the divider. In the case of a cell, the divider (cell membrane) partitions fluids inside and outside the cell.

3 Water can flow both ways across the divider but tends to move from side A to side B, where the concentration of dissolved particles is greater. The *volume* of water increases on side B, and the particle *concentrations* on sides A and B become equal.

Fluid and Electrolyte Balance

To control the flow of water, the body must spend energy moving its electrolytes from one compartment to another (Figure 8–5). Transport proteins form the pumps that move mineral ions across cell membranes, as Chapter 6 described. The result is **fluid and electrolyte balance**, the proper amount and kind of fluid in every body compartment.

If the fluid balance is disturbed, severe illness can develop quickly because fluid can shift rapidly from one compartment to another. For example, in vomiting or diarrhea, the loss of water from the digestive tract pulls fluid from between the cells in every part of the body. Fluid then leaves the cell interiors to restore balance. Meanwhile, the kidneys detect the water loss and attempt to retrieve water from the pool destined for excretion. To do this, they raise the sodium concentration outside the cells, and this pulls still more water out of them. The result is **fluid and electrolyte imbalance**, a medical emergency. Water and minerals lost in vomiting or diarrhea ultimately come from all the body's cells. This loss disrupts the heartbeat and threatens life. It is a cause of death among those with eating disorders.

Figure 8–5

Electrolyte Balance

Transport proteins in cell membranes maintain the proper balance of sodium (mostly outside the cells) and potassium (mostly inside the cells).

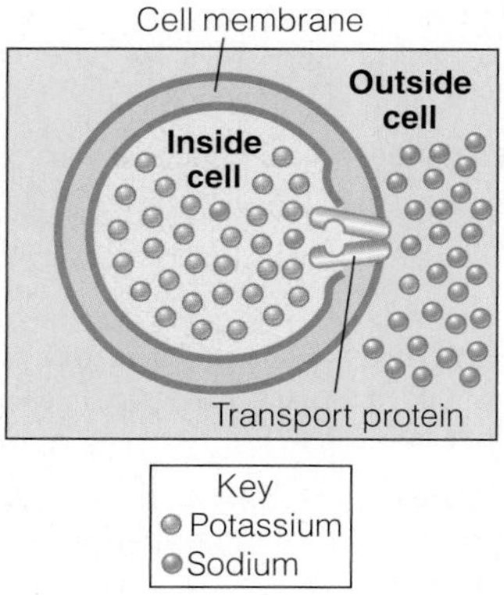

KEY POINT

- Mineral salts form electrolytes that help keep fluids in their proper compartments.

Acid-Base Balance

The minerals help manage still another balancing act, the **acid-base balance**, or pH of the body's fluids. In pure water, a small percentage of water molecules (H_2O) exist as positive (H) and negative (OH) ions, but they exist in equilibrium—the positive charges exactly equal the negatives. When dissolved in watery body fluids, some of the major minerals give rise to acids (H, or hydrogen, ions), and others to bases (OH). Excess H ions in a solution make it an acid; they lower the pH. Excess OH ions in a solution make it a base; they raise the pH.

Maintenance of body fluids at a nearly constant pH is critical to life. Even slight changes in pH drastically change the structure and chemical functions of most biologically important molecules. The body's proteins and some of its mineral salts help prevent changes in the acid-base balance of its fluids by serving as **buffers**—molecules that gather up or release H ions as needed to maintain the correct pH. The kidneys help to control the pH balance by excreting more or less acid (H ions). The lungs also help by excreting more or less carbon dioxide. (Dissolved in the blood, carbon dioxide forms an acid, carbonic acid.) This tight control of the acid-base balance permits all other life processes to continue.

KEY POINT

- Minerals act as buffers to help maintain body fluids at the correct pH to permit life's processes.

Table 8–5

Major Minerals

The major minerals are also called *macrominerals*. The need for each of these is greater than 100 milligrams per day, often far greater.

- Calcium
- Chloride
- Magnesium
- Phosphorus
- Potassium
- Sodium
- Sulfate

The Major Minerals

LO 8.4 Identify the major minerals important in human nutrition and their physiological roles in the body, the consequences of deficiencies, and their most important food sources.

All the major minerals help to maintain the fluid balance, but each one also has some special duties of its own. Table 8–5 lists the major minerals, and Table 8–11 on pages 322–323 summarizes their roles.

Calcium

As Figure 8–1 showed, calcium is by far the most abundant mineral in the body. The roles of calcium are critical to body functioning, but many adults, adolescents, and even some children do not consume enough calcium-rich foods to meet the DRI recommended intake for this mineral.[17] People who do meet their need are likely to be taking calcium supplements.[18]

Nearly all (99 percent) of the body's calcium is stored in the bones and teeth, where it plays two important roles. First, it is an integral part of bone structure. Second, bone calcium serves as a bank that can release calcium to the body fluids if even the slightest drop in blood calcium concentration occurs. Many people think that once deposited in bone, calcium (together with the other minerals of bone) stays there forever—that once a bone is built, it is inert, like a rock. Not so. The minerals of bones are in constant flux, with formation and dissolution taking place every minute of the day and night (see Figure 8–6, p. 300). Almost the entire adult human skeleton is remodeled every 10 years.[19]

Calcium in Bone and Tooth Formation Calcium and phosphorus are both essential to bone formation: calcium phosphate salts crystallize on a foundation material composed of the protein collagen. The resulting **hydroxyapatite** crystals invade the collagen and gradually lend more and more rigidity to a youngster's maturing bones until they are able to support the weight they will have to carry.

Teeth are formed in a similar way: hydroxyapatite crystals form on a collagen matrix to create the dentin that gives strength to the teeth (see Figure 8–7, p. 300). The turnover of minerals in teeth is not as rapid as in bone, but some withdrawal and redepositing do take place throughout life.

Calcium in Body Fluids The fluids that bathe and fill the cells contain the remaining 1 percent of the body's calcium, a tiny amount that is vital to life. It plays these major roles:

- Regulates the transport of ions across cell membranes and is particularly important in nerve transmission.
- Helps maintain normal blood pressure.
- Plays an essential role in the clotting of blood.
- Is essential for muscle contraction and therefore for the heartbeat.
- Allows secretion of hormones, digestive enzymes, and neurotransmitters.
- Activates cellular enzymes that regulate many processes.

Because of its importance, blood calcium is tightly controlled.

Did You Know?

If you could remove all of the minerals from bones, the protein structures that remained (mostly the protein collagen) would be so flexible that you could tie them in a knot.

fluid and electrolyte balance maintenance of the proper amounts and kinds of fluids and minerals in each compartment of the body.

fluid and electrolyte imbalance failure to maintain the proper amounts and kinds of fluids and minerals in every body compartment; a medical emergency.

acid-base balance maintenance of the proper degree of acidity in each of the body's fluids.

buffers molecules that can help to keep the pH of a solution from changing by gathering or releasing H ions.

hydroxyapatite (hi-DROX-ee-APP-uh-tight) the chief crystal of bone, formed from calcium and phosphorus.

Figure 8–6

A Bone

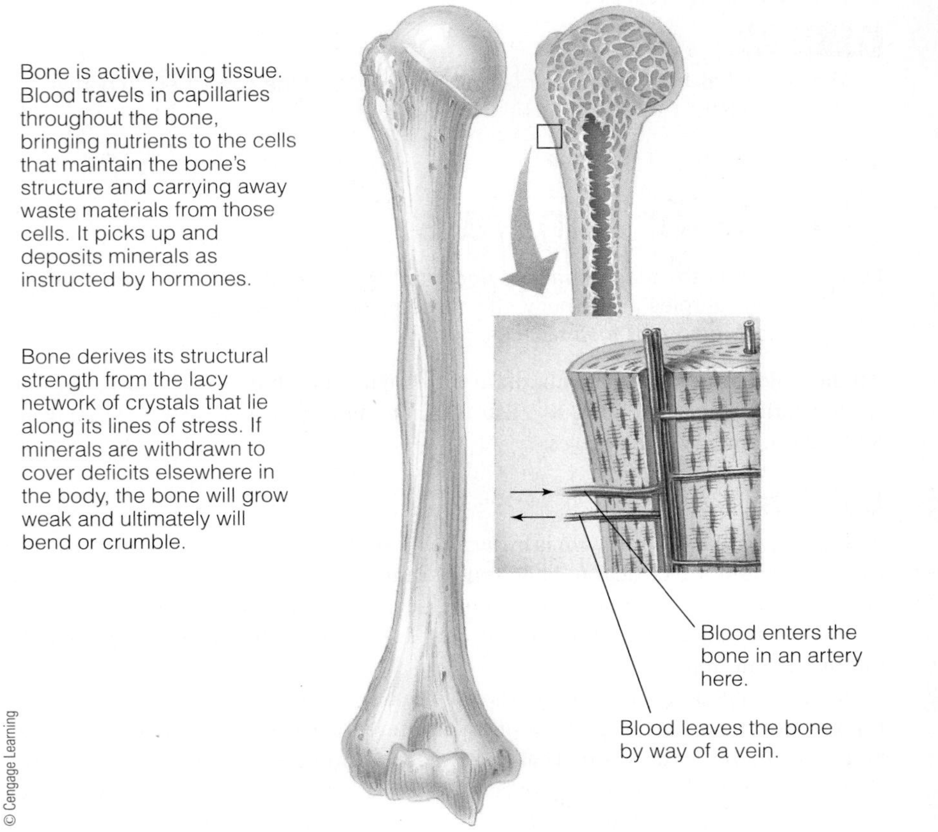

Bone is active, living tissue. Blood travels in capillaries throughout the bone, bringing nutrients to the cells that maintain the bone's structure and carrying away waste materials from those cells. It picks up and deposits minerals as instructed by hormones.
Bone derives its structural strength from the lacy network of crystals that lie along its lines of stress. If minerals are withdrawn to cover deficits elsewhere in the body, the bone will grow weak and ultimately will bend or crumble.
Blood enters the bone in an artery here.
Blood leaves the bone by way of a vein.
© Cengage Learning

Figure 8–7

A Tooth

The inner layer of dentin is bone-like material that forms on a protein (collagen) matrix. The outer layer of enamel is harder than bone. Both dentin and enamel contain hydroxyapatite crystals (made of calcium and phosphorus). The crystals of enamel may become even harder when exposed to the trace mineral fluoride.

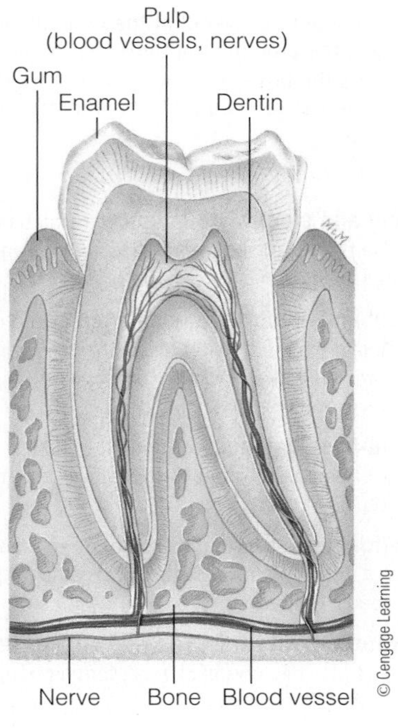

Other roles for calcium are emerging as well. Calcium may help protect against hypertension.[20] Some research also suggests protective relationships between calcium and blood cholesterol, diabetes, and colon and rectal cancers.[21] Several studies link calcium from low-fat milk and milk products (but not from supplements) with having a healthy body weight.[22] Large, well-designed clinical studies are needed to clarify these potential roles of calcium.[23]

Calcium Balance The key to bone health lies in the body's calcium balance, directed by a system of hormones and vitamin D. Cells need continuous access to calcium, so the body maintains a constant calcium concentration in the blood. The body is sensitive to an increased need for calcium but sends no signals to the conscious brain to indicate a calcium need. Instead, three organ systems quietly respond:

1. The intestines increase absorption of calcium from the intestine.
2. The kidneys prevent its loss in the urine.
3. The bones release more calcium into the blood.

The skeleton serves as a bank from which the blood can borrow and return calcium as needed. Thus, a person can go for years with an inadequate calcium intake and still maintain normal blood calcium, but at the expense of **bone density**.

Calcium Absorption Most adults absorb about 25 to 30 percent of the calcium they ingest.[24] When the body needs more calcium, proteins in the intestinal lining increase its absorption.[25] The result is obvious in the case of a pregnant woman, who doubles her absorption. Similarly, breast-fed infants absorb about 60 percent of the calcium in breast milk. Children in puberty absorb almost 35 percent of the calcium they consume.

The body also absorbs a higher percentage of the available calcium when habitual intakes are low.[26] Deprived of the mineral for months or years, an adult may double the calcium absorbed; conversely, when supplied for years with abundant calcium, the same person may absorb only about one-third the normal amount. Despite these adjustments, increased calcium absorption cannot fully compensate for a reduced intake. A person who cuts back on calcium is likely to lose calcium from the bones.

Bone Loss Some bone loss seems an inevitable consequence of aging.[27] Sometime around age 30, the skeleton no longer adds significantly to bone density. After about age 40, regardless of calcium intake, bones begin to lose density. Those who regularly meet calcium, protein, and other nutrient needs and who perform bone-strengthening physical activity may slow down the loss.[28]

A person who reaches adulthood with an insufficient calcium savings account is more likely to develop the fragile bones of **osteoporosis**. Osteoporosis constitutes a major health problem for many older people—its possible causes and prevention are the topics of this chapter's Controversy. To protect against bone loss, attention to calcium intakes during early life is crucial. Too few calcium-rich foods during the growing years may prevent a person from achieving **peak bone mass** (Figure 8–8 illustrates the timing). The margin lists some nutrients that work as a team to support bone health. The role of fluoride, one of the trace minerals, is described later.

How Much Calcium Do I Need and Which Foods Are Good Sources? Setting recommended intakes for calcium is difficult because absorption varies (the Food Feature comes back to calcium absorption). The DRI Committee took such variations into account and set recommendations for calcium at levels that produce maximum calcium retention (see the inside front cover, page B).[29] At lower intakes, the body does not store calcium to capacity; at greater intakes, the excess calcium is excreted and thus is wasted.

Intakes of calcium from supplements may elevate blood calcium or lead to problems such as calcium buildup in soft tissues and kidney stone formation, and findings about their effectiveness in reducing fractures in older women are inconclusive or negative.[30] Because adverse effects are possible with supplemental doses, a Tolerable Upper Intake Level has been established (see inside front cover, page C).

Fiber and the binders phytate (in whole grains) and oxalate (in vegetables) interfere with calcium absorption, but their effects are only minor in typical U.S. eating patterns. Snapshot 8–1, p. 302, provides a look at some foods that are good or excellent sources of calcium, and the Food Feature at the end of the chapter focuses on foods that can help to meet calcium needs.

KEY POINTS

- Calcium makes up bone and tooth structure.
- Calcium plays roles in nerve transmission, muscle contraction, and blood clotting.
- Calcium absorption adjusts somewhat to dietary intakes and altered needs.

Functional Groups for Bones

Key bone vitamins:

- Vitamin A, vitamin D, vitamin K, vitamin C, other vitamins.

Key bone minerals:

- Calcium, phosphorus, magnesium, other minerals.

Key energy nutrient:

- Protein.

Figure 8–8
Bone throughout Life

From birth to about age 20, the bones are actively growing. Between the ages of 12 and 30 years, the bones achieve their maximum mineral density for life—the peak bone mass. Beyond those years, bone resorption exceeds bone formation, and bones lose density.

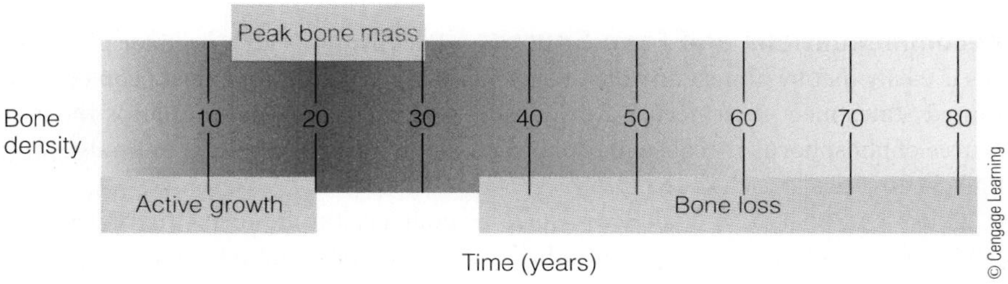

bone density a measure of bone strength, the degree of mineralization of the bone matrix.

osteoporosis (OSS-tee-oh-pore-OH-sis) a reduction of the bone mass of older persons in which the bones become porous and fragile (*osteo* means "bones"; *poros* means "porous"); also known as *adult bone loss*. (Also defined in Chapter 6.)

peak bone mass the highest attainable bone density for an individual; developed during the first three decades of life.

Snapshot 8-1 Calcium

DRI Recommended Intakes
Adults: 1,000 mg/day (19–50 yr men and women; 51–70 yr men)
1,200 mg/day (51–70 yr women; >70 yr men and women)

Tolerable Upper Intake Level
Adults: 2,500 mg/day (19–50 yr)
2,000 mg/day (>50 yr)

Chief Functions
Mineralization of bones and teeth; muscle contraction and relaxation, nerve functioning, blood clotting

Deficiency
Stunted growth and weak bones in children; bone loss (osteoporosis) in adults

Toxicity
Elevated blood calcium; constipation; interference with absorption of other minerals; increased risk of kidney stone formation

Good Sources*

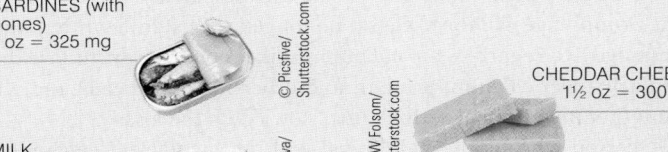

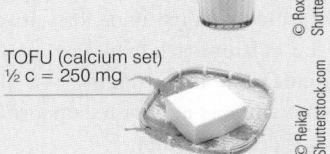

SARDINES (with bones)
3 oz = 325 mg

© Picsfive/Shutterstock.com

MILK
1 c = 300 mg

© Roxana Bashyrova/Shutterstock.com

TOFU (calcium set)
½ c = 250 mg

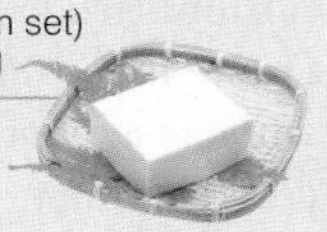

© Reika/Shutterstock.com

BROCCOLI[a] (cooked)
1½ c = 93 mg

© Valentyn Volkov/Shutterstock.com

CHEDDAR CHEESE
1½ oz = 300 mg

© BW Folsom/Shutterstock.com

TURNIP GREENS (cooked)
1 c = 198 mg

© BW Folsom/Shutterstock.com

WAFFLE (whole grain)
1 WAFFLE = 196 mg

© Peter Zijlstra/Shutterstock.com

*These foods provide 10% or more of the calcium Daily Value in a serving. For a 2,000-cal diet, the DV is 1,000 mg/day.
[a]Broccoli, kale, and some other cooked green leafy vegetables are also important sources of bioavailable calcium. Almonds also supply calcium. Spinach and chard contain calcium in an unabsorbable form. Some calcium-rich mineral waters may also be good sources.

Phosphorus

Phosphorus is the second most abundant mineral in the body, next to calcium. More than 80 percent of the body's phosphorus is found combined with calcium in the crystals of the bones and teeth.[31] The rest is everywhere else.

Roles in the Body All body cells must have phosphorus for these functions:

- Phosphorous salts are critical buffers, helping to maintain the acid-base balance of cellular fluids.
- Phosphorus is part of the DNA and RNA of every cell and thus is essential for growth and renewal of tissues.
- Phosphorous compounds carry, store, and release energy in the metabolism of energy nutrients.
- Phosphorous compounds assist many enzymes and vitamins in extracting the energy from nutrients.
- Phosphorus forms part of the molecules of the phospholipids that are principal components of cell membranes (discussed in Chapter 5).
- Phosphorus is present in some proteins.

Did You Know?

The mineral is phosphorus. The adjective form is spelled with an *-ous* (as in phosphorous salts).

Recommendations and Food Sources Luckily, the body's need for phosphorus is easily met by almost any diet, deficiencies are unlikely, and most people in the United States meet their need.[32] As Snapshot 8–2 shows, animal protein is the best source of phosphorus (because phosphorus is abundant in the cells of animals). Milk and cheese are also rich sources.

Phosphorus-based food additives, such as modified starches used in gravies, prepared meals, creamy desserts, and other processed foods, and phosphates added to

My Turn watch it! Drink Your Milk!

Listen to two students talk about how they learned about the importance of calcium.

© Cengage Learning

Kathryn

© Cengage Learning

Cynthia

© Roxana Bashyrova/Shutterstock.com

colas also contribute phosphorus to the diet. Excess phosphorus in the *blood* is associated with indicators of heart disease and osteoporosis, but whether this bears a relationship to phosphorus in the diet is unknown.[33]

KEY POINTS

- Phosphorus is abundant in bones and teeth.
- Phosphorus helps maintain acid-base balance, is part of the genetic material in cells, assists in energy metabolism, and forms part of cell membranes.
- Phosphorus deficiencies are unlikely.

Magnesium

Magnesium qualifies as a major mineral by virtue of its dietary requirement, but only about 1 ounce is present in the body of a 130-pound person, over half of it in the bones. Most of the rest is in the muscles, heart, liver, and other soft tissues, with only 1 percent in the body fluids. The supply of magnesium in the bones can be tapped to maintain a constant blood level whenever dietary intake falls too low. The kidneys can also act to conserve magnesium.

Snapshot 8-2 Phosphorus

DRI Recommended Intake
Adults: 700 mg/day

Tolerable Upper Intake Level
Adults (19–70 yr): 4,000 mg/day

Chief Functions
Mineralization of bones and teeth; part of phospholipids, important in genetic material, energy metabolism, and buffering systems

Deficiency
Muscular weakness, bone pain[a]

Toxicity
Calcification of soft tissues, particularly the kidneys

**These foods provide 10% or more of the phosphorus Daily Value in a serving. For a 2,000-cal diet, the DV is 1,000 mg/day.*
[a]Dietary deficiency rarely occurs, but some drugs can bind with phosphorus, making it unavailable.

Good Sources*

COTTAGE CHEESE
1 c = 358 mg

© Africa Studio/Shutterstock.com

MILK
1 c = 247 mg

© Roxana Bashyrova/Shutterstock.com

NAVY BEANS (cooked)
½ c = 131 mg

© nito/Shutterstock.com

SALMON (canned)
3 oz = 280 mg

© HLPhoto/Shutterstock.com

SIRLOIN STEAK (lean)
3 oz = 209 mg

© Josh Resnick/Shutterstock.com

SUNFLOWER SEEDS
2 tbs = 186 mg

© librakv/Shutterstock.com

© Cengage Learning

Roles in the Body Like phosphorus, magnesium is critical to many cell functions. Magnesium:

- Assists in the operation of hundreds of enzymes and other cellular functions.[34]
- Is needed for the release and use of energy from the energy-yielding nutrients.
- Directly affects the metabolism of potassium, calcium, and vitamin D.
- Is critical to normal heart function.[35]

Magnesium and calcium work together for proper functioning of the muscles: calcium promotes contraction, and magnesium helps relax the muscles afterward. In the teeth, magnesium promotes resistance to tooth decay by holding calcium in tooth enamel.

Magnesium Deficiency A magnesium deficiency may occur as a result of inadequate intake, vomiting, diarrhea, alcoholism, or malnutrition. It may also occur in people who take certain medications, particularly diuretics that cause excess magnesium loss in the urine. Its symptoms include a low blood calcium level, muscle cramps, and seizures. A deficiency also interferes with vitamin D activities and causes hallucinations that can be mistaken for mental illness or drunkenness. In addition, magnesium deficiency may worsen inflammation associated with many chronic diseases and may increase the risk of stroke and sudden death by heart failure, even in otherwise healthy people.[36]

Average U.S. magnesium intakes typically fall below recommendations, which may upset bone metabolism and increase the risk for osteoporosis.[37] Although almost half the U.S. population has intakes below those recommended, deficiency symptoms are rare in healthy people.[38]

Magnesium Toxicity Magnesium toxicity is rare, but it can be fatal. Toxicity occurs only with high intakes from nonfood sources such as supplements. Accidental poisonings may occur in children with access to medicine chests and in older people who take too many magnesium-containing laxatives, antacids, and other medications. The consequences can be severe diarrhea, acid-base imbalance, and dehydration. For safety, be mindful of the Tolerable Upper Intake Level for magnesium when using magnesium-containing medications.

Recommendations and Food Sources Magnesium DRI recommendations vary only slightly among adult age groups; see the inside front cover.[39] Snapshot 8–3 shows magnesium-rich foods. Magnesium is easily washed and peeled away from foods during processing, so lightly processed or unprocessed foods are the best sources. In some parts of the country, water contributes significantly to magnesium intakes, so people living in those regions need less from food.

KEY POINTS

- Magnesium stored in the bones can be drawn out for use by the cells.
- Many people consume less than the recommended amount of magnesium.
- U.S. diets often provide insufficient magnesium.

Sodium

Salt has been known and valued throughout recorded history. "You are the salt of the earth" means that you are valuable. If "you are not worth your salt," you are worthless. Even our word *salary* comes from the Latin word for *salt*. Chemically, sodium is the positive ion in the compound sodium chloride (table salt) and makes up 40 percent of its weight: a gram of salt contains 400 milligrams of sodium.

Roles of Sodium Sodium is a major part of the body's fluid and electrolyte balance system because it is the chief ion used to maintain the volume of fluid outside cells. Sodium also helps maintain acid-base balance and is essential to muscle contraction and nerve transmission. Scientists think that 30 to 40 percent of the body's sodium is stored in association with the bone crystals, where the body can draw on it to replenish the blood concentration.[40]

Did You Know?

To the chemist, a salt results from the reaction between a base and an acid. Sodium chloride, table salt, arises when the base sodium hydroxide reacts with hydrochloric acid.

- Base + acid = salt + water.
- Sodium hydroxide + hydrochloric acid = sodium chloride + water.

Snapshot 8-3 Magnesium

DRI Recommended Intakes
Men (19–30 yr): 400 mg/day
Women (19–30 yr): 310 mg/day

Tolerable Upper Intake Level
Adults: 350 mg/day[a]

Chief Functions
Bone mineralization, protein synthesis, enzyme action, muscle contraction, nerve function, tooth maintenance, and immune function

Deficiency
Weakness, confusion; if extreme, convulsions, uncontrollable muscle contractions, hallucinations, and difficulty in swallowing; in children, growth failure

Toxicity
From nonfood sources only; diarrhea, pH imbalance, dehydration

Good Sources*

SPINACH (cooked) ½ c = 78 mg

© Daniel Gilbey Photography/Shutterstock.com

BLACK BEANS (cooked) ½ c = 60 mg

© Ildi Papp/Shutterstock.com

SOY MILK 1 c = 46 mg

© iStockphoto.com/CraigNeilMcCausland

BRAN CEREAL[b] (ready-to-eat) 1 c = 80 mg

© gcpics/Shutterstock.com

OYSTERS[c] (steamed) 3 oz = 37 mg

© Olga Popova/Shutterstock.com

YOGURT (plain) 1 c = 43 mg

© Gyorgy Barna/Shutterstock.com

*These foods provide 10% or more of the magnesium Daily Value in a serving. For a 2,000-cal diet, the DV is 400 mg/day.
[a]From nonfood sources, in addition to the magnesium provided by food.
[b]Wheat bran provides magnesium, but refined grain products are low in magnesium.
[c]Magnesium in oysters varies.

Sodium Deficiency A deficiency of sodium would be harmful, but no known human diet lacks sodium. Most foods include more salt than is needed, and the body absorbs it freely. The kidneys filter the surplus out of the blood into the urine. They can also sensitively conserve sodium. In the rare event of a deficiency, they can return to the bloodstream the exact amount needed. Small sodium losses occur in sweat, but the amount of sodium excreted in a day equals the amount ingested that day.

Intense activities, such as endurance events performed over several days, can cause sodium losses that reach dangerous levels. Athletes in such events can lose so much sodium and drink so much water that they overwhelm the body's corrective actions and develop **hyponatremia**—the dangerous condition of having too little sodium in the blood. Hyponatremia is caused by excessive sodium losses, not from inadequate sodium intake, a topic taken up again in Chapter 10.[41] To understand why this might happen, you must first understand how sodium and body fluids interact.

Did You Know?
Low blood sodium can pose a danger to endurance athletes performing in hot, humid conditions (see Chapter 10).

How Are Salt and "Water Weight" Related? Blood sodium levels are well controlled. If blood sodium begins to rise, as it will after a person eats salted foods, a series of events trigger thirst and ensure that the person will drink water until the sodium-to-water ratio is restored. Then the kidneys excrete the extra water along with the extra sodium.

Dieters sometimes think that eating too much salt or drinking too much water will make them gain weight, but they do not gain fat, of course. They gain water, but a healthy body excretes this excess water immediately. Excess salt is excreted as soon as enough water is drunk to carry the salt out of the body. From this perspective, then, the way to keep body salt (and "water weight") under control is to control salt intake and drink more, not less, water.

Overly strict use of low-sodium diets in the treatment of hypertension, kidney disease, or heart disease can deplete the body of needed sodium, as can vomiting, diarrhea, or extremely heavy sweating. If blood sodium drops, body water is lost, and both water and sodium must be replenished to avert an emergency.

hyponatremia (high-poh-nah-TREE-mee-ah) a decreased concentration of sodium in the blood.

Figure 8–9

Sodium Intakes of U.S. Adults

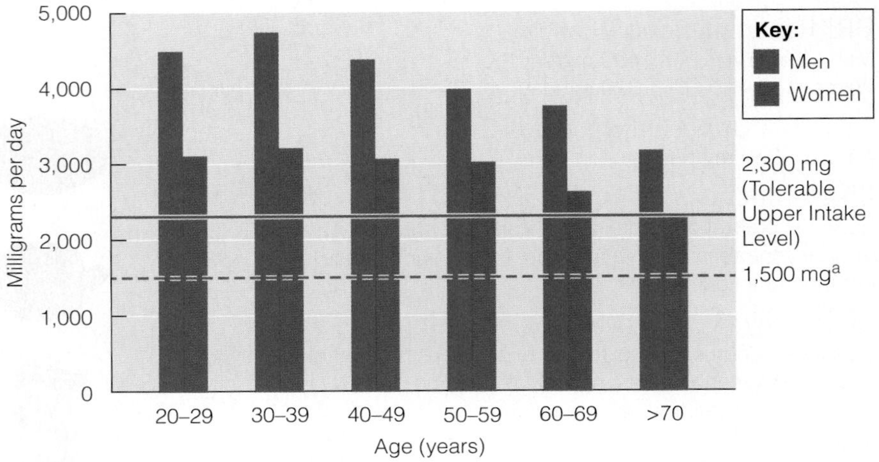

[a]*1,500 mg/day is the Adequate Intake (AI) for young adults; it is the suggested upper limit for about half the population (see text).*

Source: U.S. Department of Agriculture and U.S. Department of Health and Human Services, Dietary Guidelines for Americans 2010, *available at www.dietaryguidelines.gov.*

Sodium Recommendations and Intakes A DRI recommendation for sodium adequacy has been set at 1,500 milligrams for healthy, active young adults; at 1,300 for people ages 51 through 70; and at 1,200 for the elderly.[42] The Tolerable Upper Intake Level (UL) is set at 2,300 milligrams per day (the amount in about 1 tsp of salt). The average U.S. sodium intake tops 3,400 milligrams per day, exceeding the UL by far (see Figure 8–9).[43]

About half of the U.S. population is urged to cut sodium intakes even further. Three groups of people—all those age 51 and older, all African Americans, and everyone with hypertension, diabetes, or chronic kidney disease—should take in no more than 1,500 mg per day (see Table 8–6) because these people often respond most sensitively to the blood pressure–raising effects of sodium.[44]

Table 8–6

Sodium and Salt Intake Guidelines

DRI Recommendations

Recommended intakes for sodium:

Adults: (19–50 years):
1,500 mg/day.
Adults: (51–70 years):
1,300 mg/day.
Adults: (71 years and older):
1,200 mg/day.

Tolerable Upper Intake Level for sodium and salt:

Adults (19 years and older): 2,300 mg sodium, or 5.6 g salt (sodium chloride)/day.

Dietary Guidelines for Americans 2010

Reduce sodium intake to less than 2,300 mg/day.

Further reduced intake to 1,500 mg/day if you:

- are age 51 and older,
- are African American, or
- have hypertension, diabetes, or chronic kidney disease.

Sodium and Blood Pressure High intakes of salt among the world's people correlate with high rates of hypertension, heart disease, and strokes.[45] Over time, a high-salt diet may damage the linings of blood vessels in ways that make hypertension likely to develop.[46] One-third of U.S. adults have hypertension, and among African American adults, the rate is one of the highest in the world—44 percent.[47] An additional 30 percent of U.S. adults have **prehypertension**. Medical standards for both conditions are listed in Chapter 11.

The relationship between salt intake and blood pressure is direct—as chronic sodium intakes increase, blood pressure rises with them in a stepwise fashion.[48] Once hypertension sets in, the risk of death from stroke and heart disease climbs steeply.

Variations exist among people's blood pressure responses to sodium, partly because of their genetic inheritance and possibly because a high-sodium diet itself increases sodium sensitivity.[49] The genetic relationships are complex, but researchers suspect that the genes that affect blood pressure do so by altering the kidneys' handling of sodium.[50]

Can Diet Lower Blood Pressure? A proven eating pattern that can help people to reduce their sodium and increase potassium intakes, and thereby often reduce their blood pressure, is DASH (Dietary Approaches to Stop Hypertension).[51] This pattern calls for greatly increased intakes of potassium-rich fruits and vegetables, with adequate amounts of nuts, fish, whole grains, and low-fat dairy products.

DASH diet details are found in **Chapter 11** and **Appendix E**.

Table 8–7

How to Trim Sodium from a Barbecue Lunch

Lunch #1 exceeds the whole day's Tolerable Upper Intake Level of 2,300 milligrams sodium. With careful substitutions, the sodium drops dramatically in the second lunch, but it is still a high-sodium meal. In lunch #3, three additional changes—omitting the sauce, coleslaw dressing, and salt—cut the sodium by half again.

Lunch #1: Highest	Sodium (mg)	Lunch #2: Lower	Sodium (mg)	Lunch #3: Lowest	Sodium (mg)
▪ Chopped pork sandwich, sauce and meat mixture	950	▪ Sliced pork sandwich, with 1 tbs sauce	400	▪ Sliced pork sandwich (no sauce)	210
▪ Creamed corn, ½ c	460	▪ Corn, 1 cob, soft margarine, salt	190	▪ Corn, 1 cob, soft margarine	50
▪ Potato chips, 2.5 oz	340	▪ Coleslaw, ½ c	180	▪ Green salad, oil and vinegar	10
▪ Dill pickle, ½ medium	420	▪ Watermelon, slice	10	▪ Watermelon, slice	10
▪ Milk, low-fat, 1 c	120	▪ Milk, low-fat, 1 c	120	▪ Milk, low-fat, 1 c	120
▪ Pecan pie, slice	480	▪ Ice cream, low-fat, ½ c	80	▪ Ice cream, low-fat, ½ cup	80
	Total 2,770		**Total 980**		**Total 480**

At the same time, red meat, butter, other high-fat foods, and sweets are held to occasional small portions.

Other Reasons to Cut Salt Intakes Many Americans have much to gain in terms of cardiovascular health and nothing to lose from cutting back on salt as part of an overall lifestyle strategy to reduce blood pressure. Physical activity should also be part of that lifestyle because regular moderate exercise reliably lowers blood pressure.

Excess salt intake also increases calcium excretion, an effect that could potentially compromise the integrity of the bones.[52] Excessive salt may also directly stress a weakened heart or aggravate kidney problems, and high salt intakes have been directly linked with high rates of stomach cancer.[53]

Our Salty Food Supply Cutting down on salt and sodium is easy on paper—just weed out the salt and high-sodium items from the diet. In practice, though, very few people achieve the goal of limiting sodium to 2,300 mg per day, the DRI Tolerable Upper Intake Level.[54] Further, meeting all of a person's nutrient needs from our highly salted food supply and achieving an even lower sodium intake of 1,500 mg per day (many people's recommended limit) can be difficult at best.[55] The lunches of Table 8–7 demonstrate that doing so requires eliminating all salt, sauces, dressings, and salty items such as chips, pickles, and even piecrust.

Many experts today are calling for reductions of sodium in the food supply to give consumers more low-salt options to choose from.[56] Reducing the sodium content in processed foods could prevent an estimated 100,000 deaths and save up to $24 billion in health-care costs in the United States annually.[57]

Reducing Sodium Intakes An obvious step in controlling sodium intake is controlling the saltshaker, but this source may contribute as little as 15 percent of the total salt consumed. As Figure 8–10, p. 308, indicates, a more productive step is to cut down on processed and fast foods, by far our biggest source of sodium. Today, these 10 common foods are the top sodium providers: breads and rolls, cold cuts and cured meats, pizza, fresh and processed poultry, soups, sandwiches (including cheeseburgers), cheese, pasta dishes, meat mixtures (including meatloaf), and salty snacks (including popcorn, chips, and pretzels).[58]

Many people are unaware that foods high in sodium do not always taste salty. Who could guess by taste alone that a single half-cup serving of instant chocolate pudding provides almost one-fifth of the Tolerable Upper Intake Level for sodium? Additives other than salt also increase a food's sodium content: sodium benzoate, monosodium glutamate, sodium nitrite, or sodium ascorbate, to name a few. Moral: Read the Nutrition Facts labels.

Herbs add delicious flavors to foods without adding salt.

Did You Know?

Food labels can help consumers evaluate sodium in foods:

- Foods with a "low sodium" nutrient claim must provide 140 mg or less sodium per serving.
- The Nutrition Facts panel lists the milligrams of sodium in a serving of the food.

prehypertension blood pressure values that predict hypertension. See Chapter 11.

Figure 8–10

Sources of Sodium in the U.S. Diet

Less Processed Foods

Foods that are low in sodium contribute less than 10 percent of the total sodium in the U.S. diet.

Fresh foods higher in sodium
- *Milk*, 120 mg/c
- *Scallops*, 260 mg/3 oz

Fresh meats, about 30 to 70 mg/3 oz
- *Chicken, beef, fish, lamb, pork*

Fresh vegetables, about 30 to 50 mg per $^{1}/_{2}$ c
- *Celery, Chinese cabbage, sweet potatoes*

Fresh vegetables, about 10 to 20 mg per $^{1}/_{2}$ c
- *Broccoli, brussels sprouts, carrots, corn, green beans, legumes, potatoes, salad greens*

Grains (cooked without salt), about 0 to 10 mg per $^{1}/_{2}$ c
- *Barley, oatmeal, pasta, rice*

Salt, Brined Foods, Condiments

Salt added at home, in cooking or at the table, contributes 15 percent of the total sodium in the U.S. diet. Many seasonings and sauces also contribute salt and sodium.

Salts, about 2,000 mg/tsp
- *Salt, sea salt, seasoned salt, onion salt, garlic salt*[a]

Soy sauce, about 300 mg/tsp

Foods prepared in salt or brine, about 300 to 800 mg/serving
- *Anchovies (2 fillets), dill pickles (1), olives (5), sauerkraut ($^{1}/_{2}$ c), chipped beef (1 oz)*

Condiments and sauces, about 100 to 200 mg/tbs
- *Barbecue sauce, ketchup, mustard, salad dressings, sweet pickle relish, taco sauce, Worcestershire sauce*

[a]*Note that herb seasoning blends may or may not contain substantial sodium; read the labels.*

Highly Processed Foods

Processed foods from restaurants or stores contribute 75 percent of the sodium in the U.S. diet.

Dry soup mixes (prepared), about 1,000 to 2,000 mg/c
- *Bouillon cube, noodle soups, onion soup, ramen*

Fast foods and frozen dinners, about 700 to 1,500 mg/serving
- *Breakfast biscuit (cheese, egg, and ham), cheeseburger, chicken wings (10 spicy wings), frozen dinners, pizza, 2 tacos, chili dog, vegetarian soy burger (on bun)*

Canned soups (prepared), about 700 to 1,500 mg/c
- *Bean, beef, or chicken soups, broths, tomato or vegetable soup*

Pasta, frozen or canned, all types, with tomato sauce, 600 to 900 mg/c

Cold cuts/cured meats, about 500 to 700 mg/2 oz
- *Ham products, lunchmeats, hot dogs, smoked sausages*

Cheeses, processed, about 550 mg/oz

Pudding, instant, about 420 mg per $^{1}/_{2}$ c

Canned vegetables, about 200 to 450 mg per $^{1}/_{2}$ c
- *Carrots, corn, green beans, legumes, peas, potatoes*

Snack chips, puffs, crackers, about 200 to 300 mg/oz

Breads and rolls, about 125 mg/1 slice or $^{1}/_{2}$ roll

Remember that the recommendation is to limit sodium, not to eliminate it. Foods eaten without salt may seem less tasty at first but, with repetition, tastes adjust and the delicious natural flavor becomes the preferred taste.

KEY POINTS

- Sodium is the main positively charged ion outside the body's cells.
- Sodium attracts water.
- Too much dietary sodium raises blood pressure; few diets lack sodium.

Potassium

Outside the body's cells, sodium is the principle positively charged ion. *Inside* the cells, potassium takes the role of the principal positively charged ion.

Roles in the Body Potassium plays a major role in maintaining fluid and electrolyte balance and cell integrity. During nerve impulse transmission and muscle contraction, potassium and sodium briefly trade places across the cell membrane. The cell then quickly pumps them back into place. Controlling potassium distribution is a high priority for the body because it affects many aspects of homeostasis, including maintaining a steady heartbeat.

Potassium Deficiency Few people in the United States consume the DRI recommended intake of potassium. Low potassium intakes, especially when combined with high sodium intakes, raise blood pressure and increase the risk of death from heart disease and stroke.[59] Diets with ample potassium, particularly when low in sodium, appear to both prevent and correct hypertension. This effect earns potassium its status as a *Dietary Guidelines* nutrient of national concern.[60]

Severe deficiencies are rare. In healthy people, almost any reasonable diet provides enough potassium to prevent dangerously low blood potassium under ordinary conditions. Dehydration leads to a loss of potassium from inside cells, dangerous partly because potassium is crucial for regular heartbeats. The sudden deaths that occur with fasting, eating disorders, severe diarrhea, or severe malnutrition in children may be due to heart failure caused by potassium loss. Adults are warned not to take diuretics (water pills) that cause potassium loss or to give them to children, except under a physician's supervision. Physicians prescribing diuretics will tell clients to eat potassium-rich foods to compensate for the losses.

Potassium Toxicity Potassium from foods is safe, but potassium injected into a vein can stop the heart. Potassium overdoses from supplements normally are not life-threatening because the kidneys excrete small excesses and large doses trigger vomiting to expel the substance. A person with a weak heart, however, should not go through this trauma, and a baby may not be able to withstand it. Several infants have died when well-meaning parents overdosed them with potassium supplements.

Potassium Intakes and Food Sources A typical U.S. eating pattern, with its low intakes of fruits and vegetables, provides far less potassium than the amount recommended by the DRI Committee.[61] Potassium is found inside all living cells, and cells remain intact until foods are processed; therefore, the richest sources of potassium are fresh, whole foods (see Snapshot 8–4). Most vegetables and fruits are outstanding.

Snapshot 8-4 Potassium

DRI Recommended Intake
Adults: 4,700 mg/day

Chief Functions
Maintains normal fluid and electrolyte balance; facilitates chemical reactions; supports cell integrity; assists in nerve functioning and muscle contractions

Deficiency[a]
Muscle weakness, paralysis, confusion

Toxicity
Muscle weakness; vomiting; for an infant given supplements, or when injected into a vein in an adult, potassium can stop the heart

**These foods provide 10% or more of the potassium Daily Value in a serving. For a 2,000-cal diet, the DV is 3,500 mg/day.*
[a]Deficiency accompanies dehydration.

Good Sources*

ORANGE JUICE 1 c = 496 mg
© Anna Kucherova/Shutterstock.com

BANANA 1 whole banana = 422 mg
© Viktar Malyshchyts/Shutterstock.com

LIMA BEANS (cooked) ½ c = 485 mg
© Lovelle938/Shutterstock.com

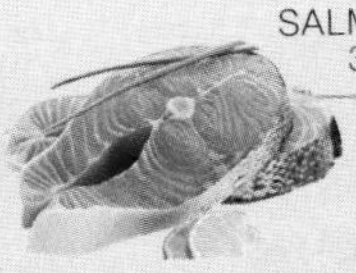
SALMON (cooked) 3 oz = 377 mg
© HLPhoto/Shutterstock.com

BAKED POTATO whole potato = 952 mg
© Joe Gough/Shutterstock.com

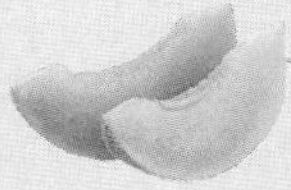
HONEYDEW MELON 1 cup = 388 mg
© HLPhoto/Shutterstock.com

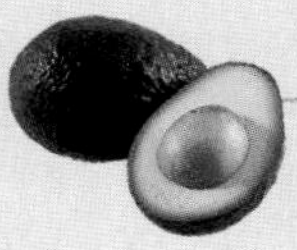
AVOCADO ½ avocado = 534 mg
© Workmans Photos/Shutterstock.com

Bananas, despite their fame as the richest potassium source, are only one of many rich sources, which also include spinach, cantaloupe, and almonds. Nevertheless, bananas are readily available, are easy to chew, and have a likable sweet taste, so health-care professionals often recommend them. Potassium chloride, a salt substitute for people with hypertension, and potassium supplements provide potassium but do not reverse the hypertension associated with a lack of potassium-rich foods.[62]

KEY POINTS

- Potassium, the major positive ion inside cells, plays important metabolic roles and is necessary for a regular heartbeat.
- Americans take in too few potassium-rich fruits and vegetables.
- Potassium excess can be toxic.

Chloride

In its elemental form, chlorine forms a deadly green gas. In the body, the chloride ion plays important roles as the major negative ion. In the fluids outside the cells, it accompanies sodium and so helps to maintain the crucial fluid balances (acid-base and electrolyte balances). The chloride ion also plays a special role as part of hydrochloric acid, which maintains the strong acidity of the stomach necessary to digest protein. The principal food source of chloride is salt, both added and naturally occurring in foods, and no known diet lacks chloride.

KEY POINTS

- Chloride is the body's major negative ion, is responsible for stomach acidity, and assists in maintaining proper body chemistry.
- No known diet lacks chloride.

Sulfate

Sulfate is the oxidized form of sulfur as it exists in food and water. The body requires sulfate for synthesis of many important sulfur-containing compounds. Sulfur-containing amino acids play an important role in helping strands of protein assume their functional shapes. Skin, hair, and nails contain some of the body's more rigid proteins, which have high sulfur contents.

There is no recommended intake for sulfate, and deficiencies are unknown. Too much sulfate in drinking water, either naturally occurring or from contamination, causes diarrhea and may damage the colon. The summary table at the end of this chapter presents the main facts about sulfate and the other major minerals.

KEY POINT

- Sulfate is a necessary nutrient used to synthesize sulfur-containing body compounds.

The Trace Minerals

LO 8.5 Identify the trace minerals important in human nutrition and their physiological roles in the body, the consequences of deficiencies, and their most important food sources.

People require only miniscule amounts of the trace minerals, but these quantities are vital for health and life. Intake recommendations have been established for nine trace minerals—see Table 8–8. Others are recognized as essential nutrients for some animals but have not been proved to be required for human beings.

Table 8–8
Trace Minerals

The trace minerals are also called *microminerals*. They are needed by the body in tiny amounts.

- Iodine
- Iron
- Zinc
- Selenium
- Fluoride
- Chromium
- Copper
- Manganese
- Molybdenum

Iodine

The body needs only traces of iodine, but this amount is indispensable to life. Once absorbed, the form of iodine that does the body's work is the ionic form, iodide.

Iodide Roles Iodide is a part of the hormone thyroxine, made by the thyroid gland. Thyroxine regulates the body's metabolic rate, temperature, reproduction, growth, heart functioning, and more. Iodine must be available for thyroxine to be synthesized.

Iodine Deficiency The ocean is the world's major source of iodine. In coastal areas, kelp, seafood, water, and even iodine-containing sea mist are dependable iodine sources. In many inland areas of the world, however, misery caused by iodine deficiency is all too common. In iodine deficiency, the cells of the thyroid gland enlarge in an attempt to trap as many particles of iodine as possible. Sometimes the gland enlarges making a visible lump in the neck, a **goiter**. People with iodine deficiency this severe may feel cold, become sluggish and forgetful, and may gain weight. Iodine deficiency affects almost 2 billion people globally, including 241 million school-aged children, a huge number but one that represents some improvement over past decades.[63]

Iodine deficiency during pregnancy causes fetal death, reduced infant survival, and extreme and irreversible mental and physical retardation in the infant, known as **cretinism**. It constitutes one of the world's most common and preventable causes of mental retardation.‡ Much of the mental retardation can be averted if the woman's deficiency is detected and treated within the first 6 months of pregnancy, but if treatment comes too late or not at all, the child may have an IQ as low as 20 (100 is average).[64] Children with even a mild iodine deficiency typically have goiters and may perform poorly in school; treatment with iodine relieves the deficiency.[65] Programs to provide iodized salt to the world's iodine-deficient areas now prevent much misery and suffering worldwide.[66]

Iodine Toxicity Excessive intakes of iodine can enlarge the thyroid gland just as a deficiency can. Although average U.S. intakes are generally above the recommended intake of 150 micrograms, they are still below the Tolerable Upper Intake Level of 1,100 micrograms per day for an adult.[67] Harm may begin at only 800 micrograms per day, however.[68] Like chlorine and fluorine, iodine is a deadly poison in large amounts.

Iodine Food Sources and Intakes The iodine in food varies with the amount in the soil in which plants are grown or on which animals graze. Because iodine is plentiful in the ocean, seafood is a dependable source. In the central parts of the United States that were never beneath an ocean, the soil is poor in iodine. In those areas, once widespread iodine deficiencies have been wiped out by the use of iodized salt and the consumption of foods shipped in from iodine-rich areas. Surprisingly, sea salt delivers little iodine because iodine becomes a gas and flies off into the air during the salt-drying process. In the United States, salt labels state whether the salt is iodized; in Canada, all table salt is iodized. Less than a half-teaspoon of iodized salt meets the entire recommendation.

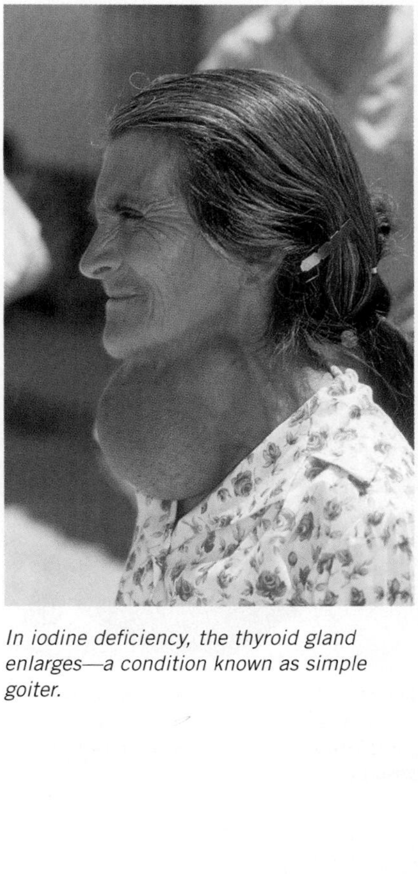

In iodine deficiency, the thyroid gland enlarges—a condition known as simple goiter.

Iodized salt is a source of iodine; plain salt is not.

goiter (GOY-ter) enlargement of the thyroid gland due to iodine deficiency is *simple goiter*; enlargement due to an iodine excess is *toxic goiter*.

cretinism (CREE-tin-ism) severe mental and physical retardation of an infant caused by the mother's iodine deficiency during pregnancy.

‡ Collectively, the problems caused by iodine deficiency are sometimes referred to as *iodine deficiency disorder*.

About 15 percent of the U.S. intake of iodine comes from iodized salt. Much more comes from milk products because most commercial dairies feed iodized grain to dairy cows and sanitize their udders with iodine-rich antiseptics, practices that add iodine to the milk.[69] U.S. consumers rarely need extra iodine—most people meet or exceed the DRI recommended intake, as mentioned, but with one possible exception: iodine intakes of young women barely meet their need.[70]

KEY POINTS

- Iodine is part of the hormone thyroxine, which influences energy metabolism.
- Iodine deficiency diseases are goiter and cretinism.
- Large amounts of iodine are toxic.
- Most people in the United States meet their need for iodine.

Iron

Every living cell, whether plant or animal, contains iron. Most of the iron in the body is a component of two proteins: **hemoglobin** in red blood cells and **myoglobin** in muscle cells.

Roles of Iron Iron-containing hemoglobin in the red blood cells carries oxygen from the lungs to tissues throughout the body. Iron in myoglobin holds and stores oxygen in the muscles for their use.

All the body's cells need oxygen to combine with the carbon and hydrogen atoms released from energy nutrients during their metabolism. This generates carbon dioxide and water waste products that are then removed from the cells; thus, body tissues constantly need fresh oxygen to keep the cells cleansed and functioning. As cells use up their oxygen, iron (in hemoglobin) shuttles fresh oxygen into the tissues from the lungs. In addition to this major task, iron is part of dozens of enzymes, particularly those involved in energy metabolism. Iron is also needed to make new cells, amino acids, hormones, and neurotransmitters.

Iron Stores Iron is clearly the body's gold, a precious mineral to be hoarded. The liver packs iron sent from the bone marrow into new red blood cells and ships them out to the bloodstream. Red blood cells live for about 4 months. When they die, the spleen and liver break them down, salvage their iron for recycling, and send it back to the bone marrow to be kept until it is reused. The body does lose iron from the digestive tract, in nail and hair trimmings, and in shed skin cells, but only in tiny amounts.[71] Bleeding, however, can cause significant iron loss from the body.

The chili dinner provides iron from meat and legumes, and vitamin C from tomatoes. The combination helps to achieve maximum iron absorption.

Special measures are needed to contain iron in the body. Left free, iron is a powerful oxidant that generates free-radical reactions. Free radicals increase oxidative stress and inflammation associated with diseases such as diabetes, heart disease, and cancer.[72] To guard against iron's renegade nature, special proteins transport and store the body's iron supply, and its absorption is tightly regulated.[73]

An Iron-Regulating Hormone—Hepcidin In most well-fed people, only about 10 to 15 percent of iron in the diet is absorbed.[74] However, if the body's iron supply is diminished or if the need for iron increases (say, during pregnancy), absorption can increase several-fold.[75] The reverse is also true: absorption declines when dietary iron is abundant. The hormone **hepcidin**, secreted by the liver, helps to regulate blood iron concentrations by limiting iron absorption from the small intestine and controlling its release from body stores.[76] Many details are known about this process, but, simply described, hepcidin works in an elegant feedback system to control blood iron:

- More abundant iron in the blood (and liver) triggers hepcidin secretion, which reduces iron absorption and inhibits the release of stored iron, thereby reducing the blood iron concentration.
- Less abundant iron in the blood suppresses hepcidin secretion, which permits increased iron absorption and release from stores, raising the blood iron concentration.[77]

Thus, the body adjusts to changing iron needs and iron availability in the diet.[78]

hemoglobin (HEEM-oh-globe-in) the oxygen-carrying protein of the blood; found in the red blood cells (*hemo* means "blood"; *globin* means "spherical protein").

myoglobin (MYE-oh-globe-in) the oxygen-holding protein of the muscles (*myo* means "muscle").

hepcidin (HEP-sid-in) a hormone secreted by the liver in response to elevated blood iron. Hepcidin reduces iron's absorption from the intestine and its release from storage.

Food Factors in Iron Absorption Iron occurs in two forms in foods. Some is bound into **heme**, the iron-containing part of hemoglobin and myoglobin in meat, poultry, and fish. Some is **nonheme iron**, in plants and also in meats. The form affects absorption.[79] Healthy people with adequate iron stores absorb heme iron at a rate of about 23 percent over a wide range of meat intakes. People absorb nonheme iron at rates of 2 to 20 percent, depending on dietary factors and iron stores. (A heme molecule was depicted in Figure 6–4 of Chapter 6.)

Meat, fish, and poultry also contain a peptide factor, sometimes called *MFP factor*, that promotes the absorption of nonheme iron from other foods. Vitamin C also greatly improves absorption of nonheme iron, tripling iron absorption from foods eaten in the same meal. The bit of vitamin C in dried fruit, strawberries, or watermelon helps absorb the nonheme iron in these foods.

Iron Inhibitors Some substances inhibit iron absorption. They include the **tannins** of tea and coffee, the calcium and phosphorus in milk, and the **phytates** that accompany fiber in lightly processed legumes and whole-grain cereals. Ordinary black tea excels at reducing iron absorption—clinical dietitians advise people with **iron overload** to drink it with their meals. For those who need more iron, the opposite advice applies—drink tea between meals, not with food. Thus, the amount of iron absorbed from a regular meal depends partly on the interaction between promoters and inhibitors, listed in Table 8–9.

Table 8–9

Promoters and Inhibitors of Iron Absorption

These dietary factors increase iron absorption:

- Heme form of iron
- Vitamin C
- Meat, fish, poultry (MFP) factor

These dietary factors hinder iron absorption:

- Nonheme form of iron
- Tea and coffee
- Calcium and phosphorus
- Phytates, tannins, and fiber

What Happens in Iron Deficiency? If absorption cannot compensate for losses or low dietary intakes, then iron stores are used up and **iron deficiency** sets in. Iron deficiency and **iron-deficiency anemia** are not one and the same, though they often occur together. Iron deficiency develops in stages, and the distinction between iron deficiency and its **anemia** is a matter of degree. People may be iron deficient, meaning that they have depleted iron stores, without being anemic; with worsening iron deficiency, they may become anemic.

A body severely deprived of iron becomes unable to make enough hemoglobin to fill new blood cells, and anemia results. A sample of iron-deficient blood examined under the microscope shows cells that are smaller and lighter red than normal (see Figure 8–11). These cells contain too little hemoglobin to deliver sufficient oxygen to the tissues. As iron deficiency limits the cells' oxygen and energy metabolism, the person develops fatigue, apathy, and a tendency to feel cold. The blood's lower concentration of its red pigment hemoglobin also explains the pale appearance of fair-skinned iron-deficient people and the paleness of the normally pink tongue and eyelid linings of those with darker skin.

Figure 8–11

Normal and Anemic Blood Cells

Well-nourished red blood cells, shown on the left, are normal in size and color. The cells on the right are typical of iron-deficiency anemia. These cells are small and pale because they contain less hemoglobin.

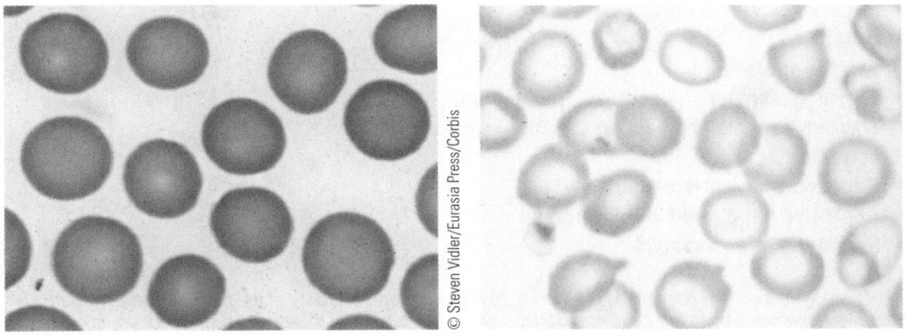

heme (HEEM) the iron-containing portion of the hemoglobin and myoglobin molecules.

nonheme iron dietary iron not associated with hemoglobin; the iron of plants and other sources.

tannins compounds in tea (especially black tea) and coffee that bind iron. Tannins also denature proteins.

phytates (FYE-tates) compounds present in plant foods (particularly whole grains) that bind iron and may prevent its absorption.

iron overload the state of having more iron in the body than it needs or can handle, usually arising from a hereditary defect. Also called *hemochromatosis*.

iron deficiency the condition of having depleted iron stores, which, at the extreme, causes iron-deficiency anemia.

iron-deficiency anemia a form of anemia caused by a lack of iron and characterized by red blood cell shrinkage and color loss. Accompanying symptoms are weakness, apathy, headaches, pallor, intolerance to cold, and inability to pay attention. (For other anemias, see the index.)

anemia the condition of inadequate or impaired red blood cells; a reduced number or volume of red blood cells along with too little hemoglobin in the blood. The red blood cells may be immature and, therefore, too large or too small to function properly. Anemia can result from blood loss, excessive red blood cell destruction, defective red blood cell formation, and many nutrient deficiencies. Anemia is not a disease, but a symptom of another problem; its name literally means "too little blood."

Think Fitness move it! Exercise-Deficiency Fatigue

On hearing about symptoms of iron deficiency, tired people may jump to the conclusion that they need to take iron supplements to restore their pep. More likely, they can obtain help by simply getting their diets in order, getting to bed on time, and getting enough exercise. Few realize that too little exercise over weeks and months is as exhausting as too much—the less you do, the less you're able to do, and the more fatigued you feel. The condition even has a name: "sedentary inertia."

Feeling fatigued, weak, and apathetic does not necessarily mean that you need iron or other supplements. Three actions are called for:

- Get your diet in order.
- Get some exercise.
- If fatigue persists for more than a week or two after making simple changes, consult a physician for a diagnosis.

start now! Using the Track Activity feature in Diet Analysis Plus, track your physical activity for one week, trying to increase your level of activity a little bit each day. See if you can walk briskly, bike, or jog for 30 minutes each day for a week.

Mental Symptoms of Iron Deficiency Long before the red blood cells are affected and anemia is diagnosed, a developing iron deficiency affects behavior.[80] Even slightly lowered iron levels cause fatigue, mental impairments, and impaired physical work capacity and productivity.[81] Symptoms associated with iron deficiency are easily mistaken for behavioral or motivational problems (see Table 8–10). With reduced energy, people work less, play less, and think or learn less eagerly. Lack of energy does not always indicate a need for iron, however—see the Think Fitness feature. Taking supplements for fatigue without a deficiency will not increase energy levels.

Children deprived of iron become restless, irritable, unwilling to work or play, and unable to pay attention, and they may fall behind their peers academically. Some symptoms in children, such as irritability, disappear when iron intake improves. Others, such as academic failure, may linger after iron repletion, although more studies are needed to clarify this association.[82] In iron-deficient adults, mental symptoms clear up reliably when iron is restored.[83]

A poorly understood behavior seen among some iron-deficient people, particularly low-income women and children, is **pica**—the craving and intentional consumption of ice, chalk, starch, clay, soil, and other nonfood substances. Researchers hypothesize that pica may result from hunger, nutrient deficiencies, digestive upsets, or attempts to prevent infections or toxicities.[84] Ingested clay, soil, or raw starch forms a glaze over the intestinal surface that can cause or worsen an iron deficiency by reducing absorption.[85] Soil can also introduce parasites and heavy metals into the body.

Table 8–10

Mental Symptoms of Anemia

- Apathy, listlessness
- Behavior disturbances
- Clumsiness
- Hyperactivity
- Irritability
- Lack of appetite
- Learning disorders (vocabulary, perception)
- Low scores on latency and associative reactions
- Lowered IQ
- Reduced physical work capacity
- Repetitive hand and foot movements
- Shortened attention span

Note: These symptoms are not caused by anemia itself but by iron deficiency in the brain. Children with much more severe anemias from other causes, such as sickle-cell anemia and thalassemia, show no reduction in IQ when compared with children without anemia.

Snapshot 8-5 Iron

DRI Recommended Intakes
Men: 8 mg/day
Women (19–50 yr): 18 mg/day
Women (51+): 8 mg/day

Tolerable Upper Intake Level
Adults: 45 mg/day

Chief Functions
Carries oxygen as part of hemoglobin in blood or myoglobin in muscles; required for cellular energy metabolism

Deficiency
Anemia: weakness, fatigue, headaches; impaired mental and physical work performance; impaired immunity; pale skin, nailbeds, and mucous membranes; concave nails; chills; pica

Toxicity
GI distress; with chronic iron overload, infections, fatigue, joint pain, skin pigmentation, organ damage

Good Sources*

CLAMS[a] (steamed)
3 oz = 23.8 mg

© jreika/Shutterstock.com

BEEF STEAK
3 oz = 1.8 mg

© Josh Resnick/Shutterstock.com

NAVY BEANS[b] (cooked)
½ c = 2.3 mg

© nito/Shutterstock.com

BLACK BEANS (cooked)
½ c = 1.8 mg

© Ildi Papp/Shutterstock.com

ENRICHED CEREAL[c] (ready-to-eat)
¾ c = 18 mg

© stavklem/Shutterstock.com

SPINACH (cooked)
½ c = 3.2 mg

© Daniel Gilbey Photography/Shutterstock.com

SWISS CHARD (cooked)
½ c = 2.0 mg

© Melica/Shutterstock.com

BEEF LIVER (cooked)
3 oz = 5.6 mg

© Serghei Starus/Shutterstock.com

© Cengage Learning

*These foods provide 10% or more of the iron Daily Value in a serving. For a 2,000-cal diet, the DV is 18 mg/day.
Note: Dried figs contain 0.6 mg per ¼ c; raisins contain 0.8 mg per ¼ c.
[a]Some clams may contain less, but most types are iron-rich foods.
[b]Legumes contain phytates that reduce iron absorption.
[c]Enriched cereals vary widely in iron content.

Causes of Iron Deficiency and Anemia Iron deficiency is usually caused by inadequate iron intake, either from sheer lack of food or from a steady diet of iron-poor foods or foods high in iron inhibitors.[86] In developed nations, high-calorie foods that are rich in refined carbohydrates and fats and poor in nutrients often displace nutritious iron-rich foods from an eating pattern.[87] In contrast, Snapshot 8–5 shows some foods that are good or excellent sources of iron.

The number-one nonnutritional factor that can cause anemia is blood loss. Because the majority of the body's iron is in the blood, losing blood means losing iron. Menstrual losses increase women's iron needs to more than double that of men. Digestive tract problems such as ulcers and inflammation can also cause blood loss severe enough to cause anemia.

Who Is Most Susceptible to Iron Deficiency? Women of child-bearing age can easily develop iron deficiency because they not only lose more iron but also eat less food than men, on average. Pregnancy also demands additional iron to support the added blood volume, growth of the fetus, and blood loss during childbirth. Infants and toddlers receive little iron from their high-milk diets, yet they need extra iron to support their rapid growth. The rapid growth of adolescence, especially for males, and the menstrual losses of females also demand extra iron that a typical teen eating pattern may not provide. Finally, iron deficiency is more common among obese people, although the reasons why remain unclear.[88] To summarize, an adequate iron intake is especially important during these stages of life:

- Women in their reproductive years.
- Pregnant women.
- Infants and toddlers.
- Adolescents and teenagers.[89]

In addition, obesity at many stages makes low blood iron more likely to occur.[90]

pica (PIE-ka) a craving and intentional consumption of nonfood substances. Also known as *geophagia* (gee-oh-FAY-gee-uh) when referring to clay eating and *pagophagia* (pag-oh-FAY-gee-uh) when referring to ice craving (*geo* means "earth"; *pago* means "frost"; *phagia* means "to eat").

In the United States, 2.4 million young children suffer from iron deficiency, while almost a half-million are diagnosed with iron-deficiency anemia. Most often, the children are from urban, low-income, and Hispanic families, but children from all groups can develop these conditions. As for women in child-bearing years, the percentage of iron deficiency remains three times higher than the goal of Healthy People 2020 Objectives for the Nation. To combat iron deficiency, the Special Supplemental Feeding Program for Women, Infants, and Children (WIC) provides low-income families with credits redeemable for high-iron foods.

Worldwide, iron deficiency is the most common nutrient deficiency, affecting more than 1.6 billion people, and almost half of preschool children and pregnant women.[91] In developing countries, parasitic infections of the digestive tract cause people to lose blood daily. For their entire lives, they may feel fatigued and listless but never know why. Iron supplements can reverse iron-deficiency anemia from dietary causes in short order, but they may also cause digestive upsets and other problems.

Can a Person Take in Too Much Iron? Iron is toxic in large amounts, largely due to increased oxidative stress in body tissues.[92] Once absorbed inside the body, iron is difficult to excrete. The healthy body defends against iron overload by controlling its entry: the intestinal cells trap some of the iron and hold it within their boundaries. When they are shed, these cells carry out of the intestinal tract the excess iron that they collected during their brief lives.

In healthy people, when iron stores fill up, hepcidin, the iron-suppressing hormone, swings into action, reducing iron absorption and protecting them against iron overload. In people with a genetic failure of this protective system, mostly Caucasian men, excess iron builds up in the tissues.[93] Early symptoms include fatigue, mental depression, or abdominal pain; untreated, the condition can cause liver failure, bone damage, diabetes, and heart failure.[94] Infections are also likely because excess iron can harm the immune system, and bacteria thrive on iron-rich blood.[95] People with the condition must monitor and limit their iron intakes and avoid supplemental iron.

Iron-containing supplements can easily cause accidental poisonings in young children.[96] As few as five ordinary iron tablets have proved fatal in young children. Keep iron-containing supplements out of children's reach.

Do the Math

To calculate the iron RDA for vegetarians, multiply by 1.8:

8 mg × 1.8 = 14 mg/day (vegetarian men)

18 mg × 1.8 = 32 mg/day (vegetarian women, 19 to 50 yr)

Iron Recommendations and Sources The typical eating pattern in the United States provides about 6 to 7 milligrams of iron for every 1,000 calories. Men need 8 milligrams of iron each day, and so do women past age 51, so these people have little trouble meeting their iron needs. For women of child-bearing age, the recommendation is higher—18 milligrams—to replace menstrual losses. During pregnancy, a woman needs even more—27 milligrams a day; pregnant women need a supplement. If a man has a low hemoglobin concentration, his health-care provider should examine him for a blood-loss site. Vegetarians, because vegetable sources of iron are poorly absorbed, should multiply the DRI recommended intake for their age and gender group by 1.8 (see the margin example).

Courtesy of Ray Stanyard

The old-fashioned iron skillet adds supplemental iron to foods.

Cooking foods in an old-fashioned iron pan adds iron salts, somewhat like the iron found in supplements. The iron content of 100 grams of spaghetti sauce simmered in a glass pan is 3 milligrams, but it increases to 87 milligrams when the sauce is cooked in a black iron pan. This iron salt is not as well absorbed as iron from meat, but some does get into the body, especially if the meal also contains meat or vitamin C.

Iron fortification of foods helps some to fend off iron deficiency, but it can be a problem for people who tend toward iron overload. A single ounce of fortified cereal for breakfast, an ordinary ham sandwich at lunch, and a cup of chili with meat for dinner present almost twice the iron a man needs in a day but only about 800 calories. Most men need about 3,000 calories, and more food means still more iron. The U.S. love affair with vitamin C supplements makes matters worse because vitamin C enhances iron absorption. For healthy people, however, fortified foods pose virtually no risk for iron toxicity.

KEY POINTS

- Most iron in the body is in hemoglobin and myoglobin or occurs as part of enzymes in the energy-yielding pathways.
- Iron absorption is affected by the hormone hepcidin, other body factors, and promoters and inhibitors in foods.
- Iron-deficiency anemia is a problem among many groups worldwide.
- Too much iron is toxic.

Zinc

Zinc occurs in a very small quantity in the human body, but it works with proteins in every organ and tissue.[97] Zinc helps more than 50 enzymes to:

- Protect cell structures against damage from oxidation.[98]
- Make parts of the cells' genetic material.
- Make heme in hemoglobin.

Zinc also assists the pancreas with its digestive and insulin functions and helps to metabolize carbohydrate, protein, and fat.

Besides helping enzymes to function, special zinc-containing proteins associate with DNA and help regulate protein synthesis and cell division, functions critical to normal growth before and after birth.[99] Zinc is also needed to produce the active form of vitamin A in visual pigments. Even a mild zinc deficiency can impair night vision. Zinc also:

- Affects behavior, learning, and mood.
- Assists in proper immune functioning.[100]
- Is essential to wound healing, sperm production, taste perception, normal metabolic rate, nerve and brain functioning, bone growth, normal development in children, and many other functions.

When zinc deficiency occurs—even a slight deficiency—it packs a wallop to the body, impairing all of these functions.

© H. Sanstead, University of Texas-Galveston

How old does the boy in the picture appear to be? He is 17 years old but is only 4 feet tall, the height of a 7-year-old in the United States. His reproductive organs are like those of a 6-year-old. The retardation is rightly ascribed to zinc deficiency because it is partially reversible when zinc is restored to the diet. The photo was taken in Egypt.

Problem: Too Little Zinc Zinc deficiency in human beings was first observed a half-century ago in children and adolescent boys in the Middle East who failed to grow and develop normally. Their native diets were typically low in animal protein and high in whole grains and beans; consequently, the diets were high in fiber and phytates, which bind zinc as well as iron. Furthermore, the bread was not **leavened**; in leavened bread, yeast breaks down phytates as the bread rises. Since that time, zinc deficiency has been identified as a substantial contributor to illness throughout the developing world and responsible for almost a half-million deaths each year.[101]

Marginal declines in zinc status also cause widespread problems in pregnancy, infancy, and early childhood. Zinc deficiency alters digestive function profoundly and causes diarrhea, which worsens the malnutrition already present, not only of zinc but of all nutrients. It drastically impairs the immune response, making infections likely.[102] Infections of the intestinal tract then worsen the malnutrition and further increase susceptibility to infections—a classic cycle of malnutrition and disease. Zinc therapy often quickly reduces diarrhea and prevents death in malnourished children, but it can fail to restore normal weight and height if the child returns to a nutrient-poor diet after treatment.[103]

Although zinc deficiencies are not common in developed countries, they do occur among some groups, including pregnant women, young children, the elderly, and the poor. When pediatricians or other health workers note poor growth accompanied by poor appetite in children, they should think zinc.

leavened (LEV-end) literally, "lightened" by yeast cells, which digest some carbohydrate components of the dough and leave behind bubbles of gas that make the bread rise.

Problem: Too Much Zinc Zinc is toxic in large quantities. High doses (over 50 milligrams) of zinc may cause vomiting, diarrhea, headaches, exhaustion, and other

Snapshot 8-6 Zinc

DRI Recommended Intakes

Men: 11 mg/day
Women: 8 mg/day

Tolerable Upper Intake Level

Adults: 40 mg/day

Chief Functions

Activates many enzymes; associated with hormones; synthesis of genetic material and proteins, transport of vitamin A, taste perception, wound healing, reproduction

Deficiency[a]

Growth retardation, delayed sexual maturation, impaired immune function, hair loss, eye and skin lesions, loss of appetite

Toxicity

Loss of appetite, impaired immunity, reduced copper and iron absorption, low HDL cholesterol (a risk factor for heart disease)

Good Sources*

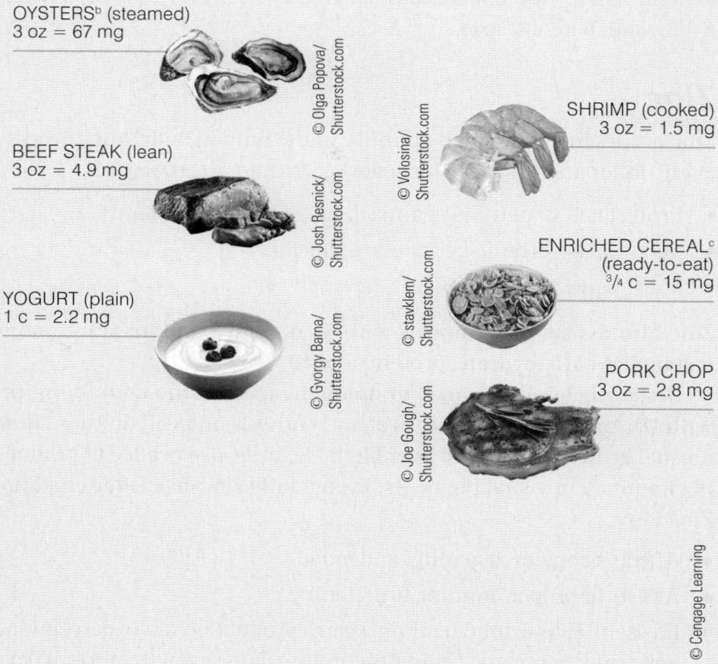

*These foods provide 10% or more of the zinc Daily Value in a serving. For a 2,000-cal diet, the DV is 15 mg/day.
[a]A rare inherited form of zinc malabsorption causes additional and more severe symptoms.
[b]Some oysters contain more or less than this amount, but all types are zinc-rich foods.
[c]Enriched cereals vary widely in zinc content.

symptoms. A UL for adults was set at 40 milligrams—an amount based on degeneration of the heart muscle in animals.

High doses of zinc inhibit iron absorption from the digestive tract. A blood protein that carries iron from the digestive tract to tissues also carries some zinc. If this protein is burdened with excess zinc, little or no room is left for iron to be picked up from the intestine. The opposite is also true: too much iron also inhibits zinc absorption. Zinc from cold-relief lozenges, nasal gels, and throat spray products may sometimes shorten the duration of a cold, but they can upset the stomach and contribute supplemental zinc to the body.[104]

Food Sources of Zinc Meats, shellfish, poultry, and milk products are among the top providers of zinc in the U.S. diet (see Snapshot 8–6). Among plant sources, some legumes and whole grains are rich in zinc but the zinc is not as well absorbed as it is from meat. Most people meet the recommended 11 milligrams per day for men and 8 milligrams per day for women. Vegetarians are advised to plan eating patterns that include zinc-enriched cereals or whole-grain breads well leavened with yeast, which helps make zinc available for absorption.[105] Unlike supplements, food sources of zinc never cause imbalances in the body.

KEY POINTS

- Zinc assists enzymes in all cells with widespread functions.
- Deficiencies cause many diverse maladies.
- Zinc supplements can interfere with iron absorption and can reach toxic doses; zinc in foods is nontoxic.

Selenium

Selenium has attracted the attention of the world's scientists. Hints of its relationships with chronic diseases make fascinating reading.[106]

Roles in the Body Selenium helps to protect vulnerable body molecules against oxidative destruction. Selenium works with a group of enzymes that, in concert with vitamin E, limits the formation of free radicals and prevents oxidative harm to cells and tissues.[107] In addition, selenium-containing enzymes are needed to assist the iodine-containing thyroid hormones that regulate metabolism.[108]

Relationship with Chronic Diseases Evidence is mixed on whether low selenium plays a role in common forms of heart disease, but taking selenium supplements does not reduce the risk.[109] In cancer studies, adequate *blood* selenium seems protective against cancers of the prostate, colon, and other sites.[110] Should everyone take selenium supplements to ward off cancer, then? No. Selenium *deficiency* may increase cancer risk, but U.S. intakes are generally sufficient, and excesses may harm healthy, well-fed people.[111]

Deficiency Without an adequate supply of selenium, the body's ability to make the needed selenium-containing molecules is compromised. Severe deficiencies cause muscle disorders with weakness and pain in people and animals. A specific type of heart disease, prevalent in regions of China where the soil and foods lack selenium, is partly brought on by selenium deficiency.[112] This condition prompted researchers to give selenium its status as an essential nutrient—adequate selenium prevents many cases from occurring.[113]

More subtle deficiencies may also adversely affect body tissues. For example, a shortage of selenium in cells of many tissues may unleash harmful levels of free radicals, increasing inflammation.

Toxicity Toxicity is possible when people take selenium supplements and exceed the Tolerable Upper Intake Level of 400 micrograms per day. Selenium toxicity brings on symptoms such as hair loss and brittle nails; diarrhea and fatigue; and bone, joint, and nerve abnormalities.[114]

Sources Clearly, adequate selenium is important, but research does not support taking selenium supplements. It is widely distributed in meats and shellfish but varies greatly in vegetables, nuts, and grains depending upon whether they are grown on selenium-rich soil.[115] Soils in the United States and Canada vary in selenium, but foods from many regions mingle on supermarket shelves, ensuring that consumers are well supplied with selenium.

KEY POINTS

- Selenium works with an enzyme system to protect body compounds from oxidation.
- Deficiencies are rare in developed countries, but toxicities can occur from overuse of supplements.

Fluoride

Fluoride is not essential to life. It is beneficial in the diet, however, because of its ability to inhibit the development of dental caries in children and adults.

Roles in the Body In developing teeth and bones, fluoride replaces the hydroxy portion of hydroxyapatite, forming **fluorapatite**. During development, fluorapatite enlarges calcium crystals in bones and teeth, decreasing their susceptibility to demineralization. In mature bones, higher intakes of fluoride may also stimulate bone-building cells, but this effect does not seem to prevent spine fractures associated with bone loss in later life.[116]

Fluoride's primary role in health is prevention of **dental caries** throughout life.[117] Once teeth have erupted through the gums, fluoride, particularly when applied to

fluorapatite (floor-APP-uh-tight) a crystal of bones and teeth, formed when fluoride displaces the "hydroxy" portion of hydroxyapatite. Fluorapatite resists being dissolved back into body fluid.

dental caries decay of the teeth, commonly called *cavities*. Also defined in Chapter 14.

tooth surfaces, helps to prevent dental caries by promoting the remineralization of early lesions of the enamel that might otherwise progress to form caries.[118] Fluoride also acts directly on the bacteria of plaque, suppressing their metabolism and reducing the amount of tooth-destroying acid they produce.

© Caroline Fleischer

To prevent fluorosis, young children should not swallow toothpaste.

Deficiency Where fluoride is lacking, dental decay is common, and fluoridation of water is recommended for public dental health. Based on evidence of its benefits, fluoridation has been endorsed by the National Institute of Dental Health, the Academy of Nutrition and Dietetics, the American Medical Association, the National Cancer Institute, and the Centers for Disease Control and Prevention as beneficial and presenting no known risks.

Toxicity In communities where the water contains too much fluoride, discoloration of the teeth, or **fluorosis**, may occur.[119] In bones, skeletal fluorosis causes bone malformations, hardened ligaments, and unusually dense, but weak, fracture-prone bones.[120] Fluorosis in teeth occurs only during tooth development, never after the teeth have formed—and it is irreversible. Skeletal fluorosis has been observed among adults living in high-fluoride areas, who are exposed to industrial sources, or who consume large amounts of fluoridated toothpaste and tea made with fluoridated water.[121] Widespread availability of fluoridated toothpaste and mouthwash, foods made with fluoridated water, and fluoride-containing supplements has led to an increase in the mildest form of dental fluorosis. In this condition, characteristic white spots form in the tooth enamel; a more severe form is shown in Figure 8–12.

To prevent fluorosis, people in areas with fluoridated water should limit other sources, such as fluoride-enriched formula for infants and fluoride supplements for infants or children, unless prescribed by a physician. Children younger than 6 years should use only a pea-sized squeeze of toothpaste and should be told not to swallow their toothpaste when brushing their teeth. The Tolerable Upper Intake Level for fluoride for all people older than 8 years is 10 milligrams per day.

Figure 8–12
Fluorosis

The brown mottled stains on these teeth indicate exposure to high concentrations of fluoride during development.

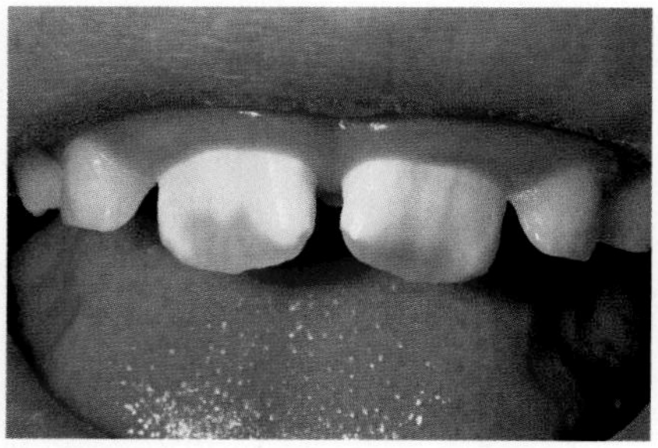

Dr. P. Marazzi/Science Source

Sources of Fluoride Drinking water is the usual source of fluoride. More than 70 percent of the U.S. population has access to public water supplies with an optimal fluoride concentration, which typically delivers about 0.7 milligram per liter.[122] Figure 8–13 shows the percentage of the population in each state with access to fluoridated water. Fluoride is rarely present in bottled waters unless it was added at the source, as in bottled municipal tap water.

KEY POINTS

- Fluoride stabilizes bones and makes teeth resistant to decay.
- Excess fluoride discolors teeth and weakens bones; large doses are toxic.

Chromium

Chromium is an essential mineral that participates in carbohydrate and lipid metabolism. Chromium in foods is safe and essential to health. Industrial chromium is a toxic contaminant, a known carcinogen that damages the DNA.[123]

Roles in the Body Chromium may help maintain glucose homeostasis by enhancing the activity of the hormone insulin, improving cellular uptake of glucose, and other actions.[124] When chromium is lacking, a diabetes-like condition may develop with elevated blood glucose and impaired glucose tolerance, insulin response, and glucagon response. Research is mixed on whether chromium supplements might improve glucose or insulin responses in diabetes.

Chromium Sources Chromium is present in a variety of foods. The best sources are unrefined foods, particularly liver, brewer's yeast, and whole grains. The more refined foods people eat, the less chromium they receive.

Supplement advertisements may convince consumers that they can lose fat and build muscle by taking chromium picolinate. Chromium supplements probably do not reduce body fat or improve muscle strength more than diet and exercise alone, however.

KEY POINTS

- Chromium is needed for normal blood glucose regulation.
- Whole, minimally processed foods are the best chromium sources.

Copper

One of copper's most vital roles is helping to form hemoglobin and collagen. In addition, many enzymes depend on copper for its oxygen-handling ability. Copper plays roles in the body's handling of iron and, like iron, assists in reactions leading to the release of energy. One copper-dependent enzyme helps to control damage from free-radical activity in the tissues.§ Researchers are investigating the possibility that a low-copper diet may contribute to heart disease by suppressing the activity of this enzyme.

Copper deficiency is rare but not unknown: it has been seen in severely malnourished infants fed a copper-poor milk formula. Deficiency can severely disturb growth and metabolism, and in adults, it can impair immunity and blood flow through the arteries. Excess zinc interferes with copper absorption and can cause deficiency. Two rare genetic disorders affect copper status in opposite directions—one causing a functional deficiency and the other toxicity.[125]

Copper toxicity from foods is unlikely, but supplements can cause it. The Tolerable Upper Intake Level for adults is set at 10,000 micrograms (10 milligrams) per day. The best food sources of copper include organ meats, seafood, nuts, and seeds. Water may also supply copper, especially where copper plumbing pipes are used. In the United States, copper intakes are thought to be adequate.[126]

KEY POINTS

- Copper is needed to form hemoglobin and collagen and assists in many other body processes.
- Copper deficiency is rare.

Other Trace Minerals and Some Candidates

DRI intake recommendations have been established for two other trace minerals, molybdenum and manganese. Molybdenum functions as part of several metal-containing enzymes, some of which are giant proteins. Manganese works with dozens of different enzymes that facilitate body processes and is widespread among whole grains, vegetables, fruits, legumes, and nuts.

Several other trace minerals are known to be important to health, but researching their roles in the body is difficult because their quantities are so small and because human deficiencies are unknown. For example, boron influences the activity of many enzymes and may play a key role in bone health, brain activities, and immune response. The richest food sources of boron are noncitrus fruits, leafy vegetables, nuts, and legumes. Cobalt is the mineral in the vitamin B_{12} molecule; the alternative name for vitamin B_{12}, *cobalamin*, reflects cobalt's presence. Nickel may serve as an enzyme cofactor; deficiencies harm the liver and other organs. Future research may reveal key roles played by other trace minerals, including barium, cadmium, lead, lithium, mercury, silver, tin, and vanadium. Even arsenic, a known poison and carcinogen, may turn out to be essential in tiny quantities.

All trace minerals are toxic in excess, and Tolerable Upper Intake Levels exist for boron, nickel, and vanadium (see the inside front cover, page C). Overdoses are most likely to occur in people who take multiple nutrient supplements. Obtaining trace minerals from food is not hard to do—just eat a variety of whole foods in the amounts recommended in Chapter 2. Table 8–11 sums up what this chapter has said about the minerals and fills in some additional information.

KEY POINTS

- Many different trace elements play important roles in the body.
- All of the trace minerals are toxic in excess.

§ The enzyme is superoxide dismutase.

Figure 8–13

U.S. Population with Access to Fluoridated Water through Public Water Systems

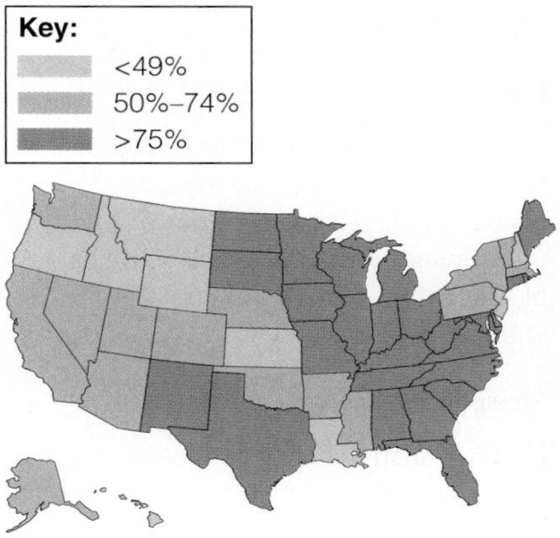

fluorosis (floor-OH-sis) discoloration of the teeth due to ingestion of too much fluoride during tooth development. *Skeletal fluorosis* is characterized by unusually dense but weak, fracture-prone, often malformed bones, caused by excess fluoride in bone crystals.

Table 8–11

The Minerals—A Summary

MINERALS AND CHIEF FUNCTIONS IN THE BODY Major Minerals	Deficiency Symptoms	Toxicity Symptoms	Significant Sources
Calcium The principal mineral of bones and teeth. Also acts in normal muscle contraction and relaxation, nerve functioning, regulation of cell activities, blood clotting, blood pressure, and immune defenses.	Stunted growth in children; adult bone loss (osteoporosis).	High blood calcium; abnormal heart rhythms; soft tissue calcification; kidney stones; kidney dysfunction; interference with absorption of other minerals; constipation.	Milk and milk products, oysters, small fish (with bones), calcium-set tofu (bean curd), certain leafy greens (bok choy, turnip greens, kale), broccoli.
Phosphorus Mineralization of bones and teeth; important in cells' genetic material, in cell membranes as phospholipids, in energy transfer, and in buffering systems.	Appetite loss, bone pain, muscle weakness, impaired growth, and rickets in infants.[a]	Calcification of nonskeletal tissues, particularly the kidney.	Foods from animal sources, some legumes.
Magnesium A factor involved in bone mineralization, the building of protein, enzyme action, normal muscular function, transmission of nerve impulses, proper immune function and maintenance of teeth.	Low blood calcium; muscle cramps; confusion; impaired vitamin D metabolism; if extreme, seizures, bizarre movements; hallucinations, and difficulty in swallowing. In children, growth failure.	Excess magnesium from abuse of laxatives (Epsom salts) causes diarrhea, nausea, and abdominal cramps with fluid and electrolyte and pH imbalances.	Nuts, legumes, whole grains, dark green vegetables, seafoods, chocolate, cocoa.
Sodium Sodium, chloride, and potassium (electrolytes) maintain normal fluid balance and acid-base balance in the body. Sodium is critical to nerve impulse transmission.	Muscle cramps, mental apathy, loss of appetite.	Hypertension.	Salt, soy sauce, seasoning mixes, processed foods, condiments, fast foods.
Potassium Potassium facilitates reactions, including the making of protein; the maintenance of fluid and electrolyte balance; the support of cell integrity; the transmission of nerve impulses; and the contraction of muscles, including the heart.	Deficiency accompanies dehydration; causes muscular weakness, paralysis, and confusion; can cause death.	Causes muscular weakness; triggers vomiting; if given into a vein, can stop the heart.	All whole foods: meats, milk, fruits, vegetables, grains, legumes.
Chloride Chloride is part of the hydrochloric acid found in the stomach, necessary for proper digestion. Helps maintain normal fluid and electrolyte balance.	Growth failure in children; muscle cramps, mental apathy, loss of appetite; can cause death (uncommon).	Normally harmless (the gas chlorine is a poison but evaporates from water); can cause vomiting.	Salt, soy sauce; moderate quantities in whole, unprocessed foods, large amounts in processed foods.

[a]*Seen only rarely in infants fed phosphorus-free formula or in adults taking medications that interact with phosphorus.*

Table 8-11

The Minerals—A Summary (continued)

Major Minerals	Deficiency Symptoms	Toxicity Symptoms	Significant Sources
Sulfate			
A contributor of sulfur to many important compounds, such as certain amino acids, antioxidants, and the vitamins biotin and thiamin; stabilizes protein shape by forming sulfur-sulfur bridges (see Figure 6-10 in Chapter 6, p. 209).	None known; protein deficiency would occur first.	Would occur only if sulfur amino acids were eaten in excess; this (in animals) depresses growth.	All protein-containing foods.
Trace Minerals	**Deficiency Symptoms**	**Toxicity Symptoms**	**Significant Sources**
Iodine			
A component of the thyroid hormone thyroxine, which helps to regulate growth, development, and metabolic rate.	Goiter, cretinism.	Depressed thyroid activity; goiter-like thyroid enlargement.	Iodized salt, seafood, bread, plants grown in most parts of the country and animals fed those plants.
Iron			
Part of the protein hemoglobin, which carries oxygen in the blood; part of the protein myoglobin in muscles, which makes oxygen available for muscle contraction; necessary for the use of energy.	Anemia: weakness, fatigue, pale skin and mucous membranes, pale concave nails, headaches, inability to concentrate, impaired cognitive function (children), lowered cold tolerance.	Iron overload: fatigue, abdominal pain, infections, liver injury, joint pain, skin pigmentation, growth retardation in children, bloody stools, shock.	Red meats, fish, poultry, shellfish, eggs, legumes, green leafy vegetables, dried fruits.
Zinc			
Associated with hormones; needed for many enzymes; involved in making genetic material and proteins, immune cell activation, transport of vitamin A, taste perception, wound healing, the making of sperm, and normal fetal development.	Growth failure in children, dermatitis, sexual retardation, loss of taste, poor wound healing.	Nausea, vomiting, diarrhea, loss of appetite, headache, immune suppression, decreased HDL, reduced iron and copper status.	Protein-containing foods: meats, fish, shellfish, poultry, grains, yogurt.
Selenium			
Assists a group of enzymes that defend against oxidation.	Predisposition to a form of heart disease characterized by fibrous cardiac tissue (uncommon).	Nausea; diarrhea; nail and hair changes; joint pain; nerve, liver, and bone damage.	Seafoods, organ meats, other meats, whole grains, and vegetables depending on soil content.
Fluoride			
Helps form bones and teeth; confers decay resistance on teeth.	Susceptibility to tooth decay.	Fluorosis (discoloration) of teeth, skeletal fluorosis (weak, malformed bones), nausea, vomiting, diarrhea, chest pain, itching.	Drinking water if fluoride-containing or fluoridated, tea, seafood.
Chromium			
Associated with insulin; needed for energy release from glucose.	Abnormal glucose metabolism.	Possibly skin eruptions.	Meat, unrefined grains, vegetable oils.
Copper			
Helps form hemoglobin and collagen; part of several enzymes.	Anemia; bone abnormalities.	Vomiting, diarrhea; liver damage.	Organ meats, seafood, nuts, seeds, whole grains, drinking water.

try it! Food Feature

Meeting the Need for Calcium

LO 8.6 Outline a plan for obtaining sufficient calcium from a day's meals.

Some people behave as though calcium nutrition is of little consequence to their health—they neglect to meet their need.[127] Yet, a low calcium intake is associated with all sorts of major illnesses, including adult bone loss (see the following Controversy), high blood pressure, colon cancer (see Chapter 11), and even lead poisoning (Chapter 14).

Intakes of one of the best sources of calcium—milk—have declined in recent years, while consumption of other beverages, such as sweet soft drinks and fruit drinks, has increased dramatically. This Food Feature focuses on food and beverage sources of calcium and provides guidance about how to include them in an eating pattern that meets nutrient needs.

Milk and Milk Products

Milk and milk products are traditional sources of calcium for people who can tolerate them (see Figure 8–14). People who shun these foods because of lactose intolerance, allergy, a vegan diet, or other reasons can obtain calcium from other sources, but care is needed—*wise* substitutions must be made.[128] This is especially true for children. Children who don't drink milk often have lower calcium intakes and poorer bone health than those who drink milk regularly. Most of milk's many relatives are good choices: yogurt, **kefir**, buttermilk, cheese (especially the low-fat or fat-free varieties), and, for people who can afford the calories, ice milk. Cottage cheese and frozen yogurt desserts contain about half the calcium of milk—2 cups are needed to provide the amount of calcium in 1 cup of milk. Butter, cream, and cream cheese are almost pure fat and contain negligible calcium.

Tinker with milk products to make them more appealing. Add cocoa to milk and fruit to yogurt, make your own fruit smoothies from fat-free milk or yogurt, or add fat-free milk powder to any dish. The cocoa powder added to make chocolate milk does contain a small amount of oxalic acid, which binds with some of milk's calcium and inhibits its absorption, but the effect on calcium balance is insignificant. Sugar lends both sweetness and calories to chocolate milk, so mix your chocolate milk at home where you control the amount of sugary chocolate added to the milk or choose a sugar-free product.

Figure 8–14

Food Sources of Calcium in the U.S. Diet

Milk and milk products contribute over half of the calcium in a typical U.S. diet.

[a]*Other sources include foods contributing at least 1% in descending order: yogurt, ready-to-eat cereal, soft drinks, tortillas, eggs, dried beans and lentils, canned tomatoes, meal replacements and protein supplements, corn bread and corn muffins, hot breakfast cereal, and coffee.*

Source: Data from P. A. Cotton and coauthors, Dietary sources of nutrients among US adults, 1994–1996, Journal of the American Dietetic Association *104 (2004): 921–930, supplemental Table 21 from www.eatright.org.*

Vegetables

Among vegetables, beet greens, bok choy (a Chinese cabbage), broccoli, kale, mustard greens, rutabaga, and turnip greens provide some available calcium. So do collard greens, green cabbage, kohlrabi, parsley, watercress, and probably some seaweeds, such as the **nori** popular in Japanese cookery. Certain other foods, including rhubarb, spinach, and Swiss chard, appear equal to milk in calcium content but provide very little or no calcium to the body because they contain binders that prevent calcium's absorption (see Figure 8–15). The presence of calcium binders does not make spinach an inferior food. Spinach is also rich in iron, beta-carotene, riboflavin, and dozens of other essential nutrients and potentially helpful phytochemicals. Just don't rely on it for calcium.

Figure 8–15

Calcium Absorption from Food Sources

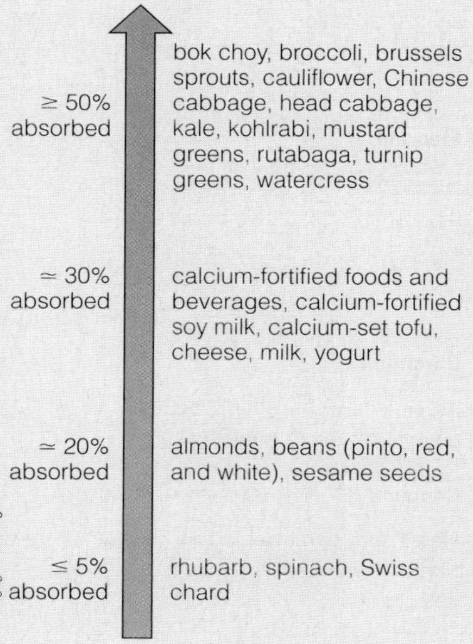

Calcium in Other Foods

For the many people who cannot use milk and milk products, a 3-ounce serving of small fish, such as canned sardines and other canned fishes eaten with their bones, provides as much calcium as a cup of milk. One-third cup of almonds supplies about 100 milligrams of calcium. Calcium-rich mineral water may also be a useful calcium source. The calcium from mineral water, including hard tap water, may be as absorbable as the calcium from milk but with zero calories. Many other foods contribute small but significant amounts of calcium to the diet.

Calcium-Fortified Foods

Some foods contain large amounts of calcium salts by an accident of processing or by intentional fortification. In the processed category are soybean curd, or tofu (calcium salt is often used to coagulate it, so check the label); canned tomatoes (firming agents donate 63 milligrams per cup of tomatoes); **stone-ground flour** and self-rising flour; stone-ground cornmeal and self-rising cornmeal; and blackstrap molasses.

Milk with extra calcium added can be an excellent source; it provides more calcium per cup than any natural milk, 500 milligrams per 8 ounces. Then comes calcium-fortified orange juice, with 300 milligrams per 8 ounces, a good choice because the bioavailability of its calcium is comparable to that of milk. Calcium-fortified soy milk can also be prepared so that it contains more calcium than whole cow's milk.

Finally, calcium supplements are available, sold mostly to people hoping to ward off osteoporosis. The Controversy following this chapter points out that supplements are not magic bullets against bone loss, however.

Figure 8–16

Milk and Milk Products: Average Intakes[a]

On average, people in the United States fall far short of the recommended intake of milk, yogurt, or cheese (or replacements) each day. The picture is worse for the dark green vegetables that supply calcium—only 3 percent of the vegetables consumed each day meet this description.

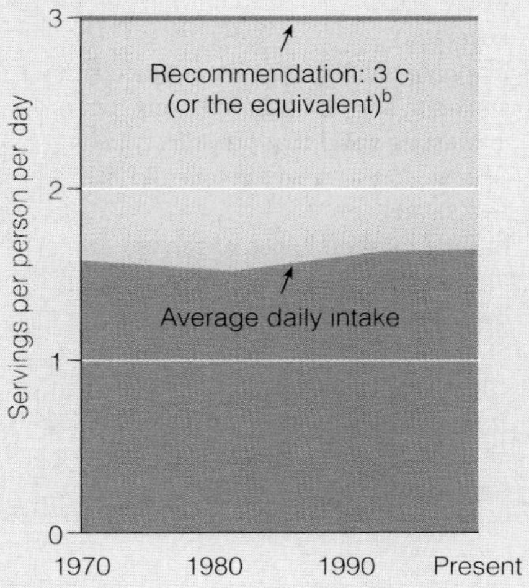

[a]*Includes all forms of milk, yogurt, cheese, and frozen dairy desserts.*

[b]*Recommended amount for adults from the 2010 USDA Food Patterns. Details in Figure 2–5 of Chapter 2.*

Source: Intake Data from U.S. Department of Agriculture, Economic Research Service.

Making Meals Rich in Calcium

For those who tolerate milk, many cooks slip extra calcium into meals by sprinkling a tablespoon or two of fat-free dry milk into almost everything. The added calorie value is small, changes to the taste and texture of the dish are practically nil, but each 2 tablespoons adds about 100 extra milligrams of calcium (see Figure 8–16). Dried buttermilk powder can also add flavor and calcium to baked goods and other dishes and keeps for a year or more when stored in the refrigerator. Table 8–12 provides some more tips for including calcium-rich foods in your meals.

Tracking Calcium

Here is a shortcut for tracking the amount of calcium in a day's meals. To start, memorize these two facts:

1. A cup of milk provides about 300 milligrams of calcium.
2. Adults need 1,000 to 1,200 milligrams each day. Broken down in terms of "cups of milk," the need is $3^1/_3$ to 4 cups each day.

To estimate calcium from an entire day's foods, not just milk, assign "cups of milk" points to various calcium sources. The goal is to achieve $3^1/_2$ to 4 points per day:

- 1 point = 1 cup milk, yogurt, calcium-fortified beverage, or $1^1/_2$ ounces cheese.
- 1 point = 4 ounces canned fish with bones.
- $^1/_2$ point = 1 cup ice cream, cottage cheese, or calcium-rich vegetables (see the text).

Also, because bits of calcium are present in many foods (a bagel has about 50 milligrams, for example):

- 1 point = a well-balanced, adequate, and varied diet.

Example: Say a day's calcium-rich foods include cereal and a cup of milk, a ham and cheese sandwich, and a broccoli and pasta salad.

- 1 point (cup of milk) + 1 point (cheese) + $^1/_2$ point (broccoli) = $2^1/_2$ points

Add 1 point for the other foods eaten that day.

- 1 point + $2^1/_2$ points = $3^1/_2$ points

This day's foods provide a calcium intake that approximates the DRI Committee's recommendation, a worthy goal for everyone's diet.

kefir a yogurt-based beverage.

nori a type of seaweed popular in Asian, particularly Japanese, cooking.

stone-ground flour flour made by grinding kernels of grain between heavy wheels made of limestone, a kind of rock derived from the shells and bones of marine animals. As the stones scrape together, bits of the limestone mix with the flour, enriching it with calcium.

Table 8–12

Calcium in Meals—Breakfast, Lunch, and Supper

Try these techniques for meeting calcium needs.

At Breakfast	At Lunch	At Supper
▪ Choose calcium-fortified orange or vegetable juice. ▪ Lighten tea or coffee, hot or iced, with milk or calcium-fortified replacement, such as soy milk. ▪ Eat cereals, hot or cold, with milk or calcium-rich replacement. ▪ Spread almond butter on toast (2 tbs provides 111 mg calcium, 8 times the amount in peanut butter.) ▪ Cook hot cereals with milk instead of water, then mix in 2 tbs of fat-free dry milk. ▪ Make muffins or quick breads with milk and extra fat-free powdered milk or dried buttermilk powder. ▪ Add milk to scrambled eggs. ▪ Moisten cereals with flavored yogurt.	▪ Add low-fat cheeses to sandwiches, burgers, or salads. ▪ Use a variety of green vegetables, such as watercress or kale, in salads and on sandwiches. ▪ Drink fat-free milk or calcium-fortified soy milk as a beverage or in a smoothie. For tartness and extra calcium, add 2 tbs dried buttermilk powder. ▪ Drink calcium-rich mineral water as a beverage. ▪ Marinate cabbage shreds or broccoli spears in low-fat Italian dressing for an interesting salad that provides calcium. ▪ Choose coleslaw over potato and macaroni salads. ▪ Mix the mashed bones of canned salmon into salmon salad or patties. ▪ Eat sardines with their bones. ▪ Stuff potatoes with broccoli and low-fat cheese. ▪ Try pasta such as ravioli stuffed with low-fat ricotta cheese instead of meat. ▪ Sprinkle parmesan cheese on pasta salads.	▪ Toss a handful of thinly sliced green vegetables, such as kale or young turnip greens, with hot pasta; the greens wilt pleasingly in the steam of the freshly cooked pasta. ▪ Serve a green vegetable every night and try new ones—how about kohlrabi? It tastes delicious when cooked like broccoli. ▪ Remember your dark green leafy vegetables—they can be good, low-calorie calcium sources. ▪ Learn to stir-fry Chinese cabbage and other Asian foods. ▪ Try tofu (the calcium-set kind); this versatile food has inspired whole cookbooks devoted to creative uses. ▪ Add fat-free powdered milk to almost anything—meat loaf, sauces, gravies, soups, stuffings, casseroles, blended beverages, puddings, quick breads, cookies, brownies. Be creative. ▪ Choose frozen yogurt, ice milk, or custards for dessert.

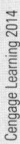

track it! Diet Analysis PLUS

Concepts in Action

Analyze Your Calcium Intakes

The purpose of this exercise is to make you aware of your calcium intake and to give you ideas about how you might meet your DRI recommended intake. Using the Diet Analysis program that accompanies this text, complete the following.

1. From the Reports tab, select Profile DRI Goals. Find your calcium information. What is the DRI Adequate Intake for calcium for your profile?
2. From the Reports tab, select Intake vs. Goals. Choose Day One (from your 3-day diet intake record) and include all meals. What percentage of your calcium DRI did you meet on that day? Was this intake typical?
3. From the Reports tab, select Source Analysis. Choose Day One, include all meals, select calcium from the drop-down box, and generate a report. What were the top three food sources of calcium that day? What were your three lowest sources? Which of your top sources matched those of the calcium Snapshot on page 302?
4. From the Reports tab, select Intake Spreadsheet, choose Day Three, choose breakfast, and then generate a report. Look at the Calcium column. Did the calcium values of any of the foods surprise you? Which ones? How many milligrams of calcium did you consume at breakfast?
5. Using the same Intake Spreadsheet, choose Day Three, and choose lunch and then dinner. At which meal did you consume the most calcium? Which meal had the least calcium: breakfast, lunch, or dinner?
6. Many nondairy foods can provide calcium. Using the tips in the Food Feature of this chapter (pages 324–325), create a calcium-rich side dish without milk or milk products. Select the Track Diet tab and enter the ingredients for your side dish. From the Reports tab, select Source Analysis, select calcium from the drop-down menu, and generate a report. Did you raise your intake of calcium by choosing nondairy calcium sources? How absorbable was the calcium in these foods? (Check Figure 8–16, page 325.)

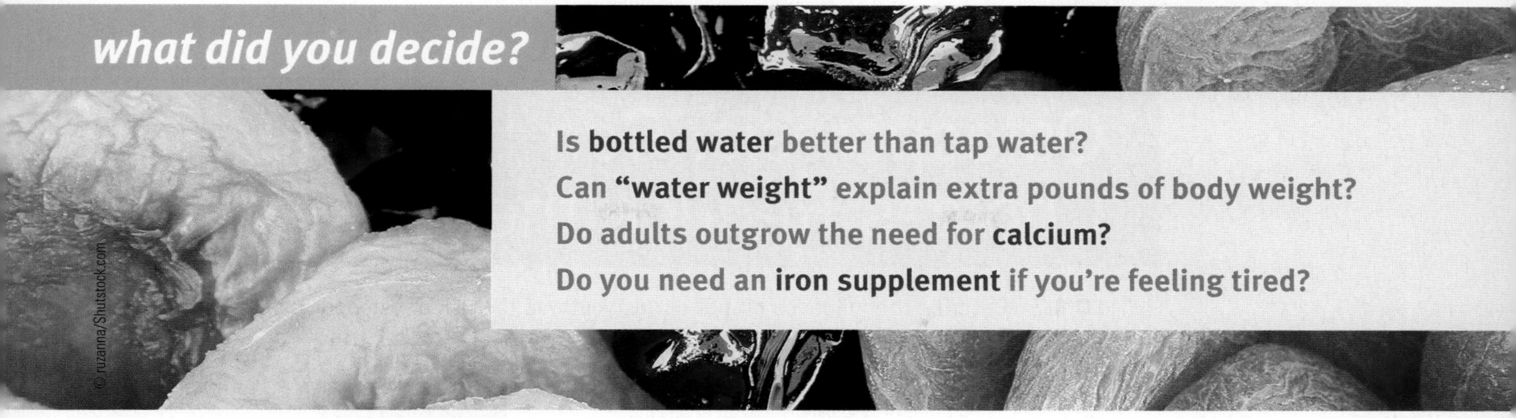

what did you decide?

Is bottled water better than tap water?
Can "water weight" explain extra pounds of body weight?
Do adults outgrow the need for calcium?
Do you need an iron supplement if you're feeling tired?

Self Check

1. (LO 8.1) Water balance is governed by the __________ .
   a. liver b. kidneys c. brain d. b and c
2. (LO 8.1) Water intoxication cannot occur because water is so easily excreted by the body.
   T F
3. (LO 8.2) Water from public water systems
   a. requires frequent home testing for microorganisms.
   b. is less healthful than bottled water.
   c. is disinfected to kill most microorganisms.
   d. is less healthful than private well water.
4. (LO 8.2) On average, young men in the United States obtain __ percent of their calories from beverages.
   a. 12 b. 22 c. 32 d. 42
5. (LO 8.2) Whether from the tap or from a bottle, all water comes from the same sources.
   T F
6. (LO 8.3) To temporarily increase the body's water content, a person need only
   a. consume extra salt c. consume extra sugar
   b. take a diuretic d. consume extra potassium
7. (LO 8.3) Vomiting or diarrhea
   a. causes fluid to be pulled from between the cells in every part of the body.
   b. causes fluid to leave the cell interiors.
   c. causes kidneys to raise the sodium concentration outside the cells.
   d. all of the above.
8. (LO 8.4) Which two minerals are the major constituents of bone?
   a. calcium and zinc c. sodium and magnesium
   b. phosphorus and calcium d. magnesium and calcium
9. (LO 8.4) Magnesium
   a. assists in the operation of enzymes.
   b. is needed for the release and use of energy.
   c. is critical to normal heart function.
   d. all of the above.
10. (LO 8.4) After about 50 years of age, bones begin to lose density.
   T F
11. (LO 8.4) The best way to control salt intake is to cut down on processed and fast foods.
   T F
12. (LO 8.5) The top food sources of zinc include:
   a. grapes c. shellfish
   b. unleavened bread d. potato
13. (LO 8.5) A deficiency of which mineral is a leading cause of mental retardation worldwide?
   a. iron b. iodine c. zinc d. chromium
14. (LO 8.5) Which of these mineral supplements can easily cause accidental poisoning in children?
   a. iron b. sodium c. chloride d. potassium
15. (LO 8.5) The most abundant mineral in the body is iron.
   T F
16. (LO 8.5) The Academy of Nutrition and Dietetics recommends fluoride-free water for the U.S. population.
   T F
17. (LO 8.6) Dairy foods such as butter, cream, and cream cheese are good sources of calcium, whereas vegetables such as broccoli are poor sources.
   T F
18. (LO 8.6) Children who don't drink milk often have lower bone density than milk-drinkers.
   T F
19. (LO 8.7) Trabecular bone readily gives up its minerals whenever blood calcium needs replenishing.
   T F
20. (LO 8.7) Too little ______ in the diet is associated with osteoporosis.
   a. vitamin B_{12} b. protein c. sodium d. niacin

Answers to these Self Check questions are in Appendix G.

CONTROVERSY 8

Osteoporosis: Can Lifestyle Choices Reduce the Risk?

LO 8.7 Describe the influence of diet during youth on the risk of osteoporosis later in life.

An estimated 44 million people in the United States, most of them women older than 50, have or are developing osteoporosis.*[1] Each year, a million and a half people, 30 percent of them men, break a hip, leg, arm, hand, ankle, or other bone as a result of osteoporosis. Of these, hip fractures prove most serious. The break is rarely clean—the bone explodes into fragments that cannot be reassembled. Just removing the pieces is a struggle, and replacing them with an artificial joint requires major surgery. About a third die of complications within a year; many more will never walk or live independently again.[2] Both men and women are urged to do whatever they can to prevent fractures related to osteoporosis.

Development of Osteoporosis

Fractures from osteoporosis occur during the later years, but osteoporosis itself develops silently much earlier. Younger adults are rarely aware of the strength sapping out of their bones until suddenly, 40 years later, a hip gives way. People say, "She fell and broke her hip," but in fact the hip may have been so fragile that it broke *before* she fell.

The causes of osteoporosis are tangled, and many are beyond a person's control. Insufficient dietary calcium, vitamin D, and physical activity certainly play roles, but age, gender, and genetics are also major players. No controversy exists as to the nature of osteoporosis; more controversial, however, are its causes and what people should do about it.

Bone Basics

To understand how the skeleton loses minerals in later years, you must first know a few things about bones. Table C8–1 offers definitions of relevant terms. The photograph on this page shows a human leg bone sliced lengthwise, exposing the lattice of calcium-containing crystals (the **trabecular bone**) inside that are part of the body's calcium bank. Invested as savings during the milk-drinking years of youth, these deposits provide a nearly inexhaustible fund of calcium. **Cortical bone** is the dense, ivorylike bone that forms the exterior shell of a bone and the shaft of a long bone (look closely at the photograph). Both types of bone are crucial to overall bone strength. Cortical bone forms a sturdy outer wall, and trabecular bone provides strength along the lines of stress.

The two types of bone handle calcium in different ways. The lacy crystals of the trabecular bone are tapped to raise blood calcium when the supply from the day's diet runs short; the calcium crystals are redeposited in bone when dietary calcium is plentiful. The calcium of cortical bone fluctuates less.

Bone Loss

Trabecular bone, generously supplied with blood vessels, readily gives up its minerals at the necessary rate whenever blood calcium needs replenishing. Loss of trabecular bone begins to be significant for men and women around age 30. Calcium in cortical bone can also be withdrawn but more slowly.

As bone loss continues (Figure C8–1), bone density declines. Soon, osteoporosis sets in, and bones become so fragile that the body's weight can overburden the spine. Vertebrae may suddenly disintegrate and crush down, painfully pinching major nerves. Or they may compress into wedges, forming what is insensitively called "dowager's hump," the bent posture of many older men and women as they "grow shorter" (see Figure C8–2). Wrists may break as trabecula-rich bone ends weaken, and teeth may loosen or fall out as the trabecular bone of the jaw recedes. As the cortical bone shell weakens as well, breaks often occur in the hip.

Table C8–1

Osteoporosis Terms

- **cortical bone** the ivorylike outer bone layer that forms a shell surrounding trabecular bone and that comprises the shaft of a long bone.
- **trabecular** (tra-BECK-you-lar) **bone** the weblike structure composed of calcium-containing crystals inside a bone's solid outer shell. It provides strength and acts like a calcium storage bank.

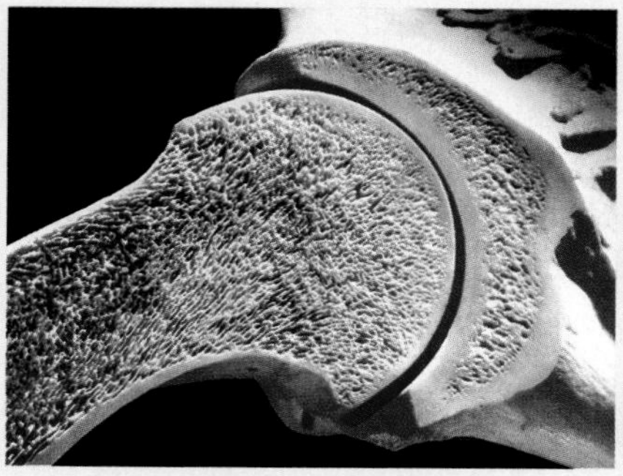

A sectioned bone.

Toward Prevention—Understanding the Causes of Osteoporosis

Scientists are searching for ways to prevent osteoporosis, but they must first establish its causes. Gender and advanced age are clearly associated, but

**Reference notes are found in Appendix F.*

Figure C8–1

Loss of Trabecular Bone

The healthy trabecular bone shown on the left appears thick and dense. The bone on the right is thin and weak, reflecting osteoporosis.

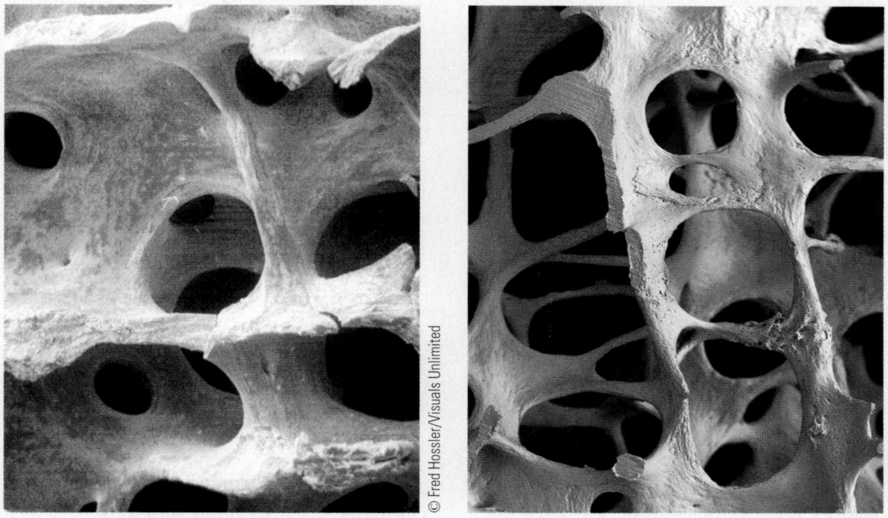

© Fred Hossler/Visuals Unlimited

© Ana Boyde/Visuals Unlimited

genetic inheritance and factors in the environment are also in play. Some risk factors are listed in Table C8–2 (p. 330), and a few details are provided in this section. In addition, a link between inflammation and osteoporosis is under investigation, as well.[3]

Bone Density and the Genes

A strong genetic component contributes to osteoporosis, bone density, and increased risk of fractures.[4] Genes exert influence over:

- The activities of bone-forming cells and bone-dismantling cells.
- The cellular mechanisms that make collagen, a structural bone protein.
- The mechanisms for absorbing and employing vitamin D.
- Many other contributors to bone metabolism.[5]

In addition to genes themselves, nutrients that influence gene activity are under study for their effects on bone density.[6]

Genetic inheritance appears to most strongly influence the maximum bone mass attainable during growth. Risks of osteoporosis differ by race and ethnicity (Caucasians have higher risks than African Americans, for example). Genes set a tendency for strong or weak bones, but diet and other lifestyle choices influence the final outcome, and anyone with risk factors for osteoporosis should take actions to prevent it.[7]

Calcium and Vitamin D

Bone strength later in life depends most on how well the bones were built during childhood and adolescence. Preteen children who consume enough calcium and vitamin D lay more calcium into the structure of their bones than children with less adequate intakes. Unfortunately, most girls in their bone-building years fail to meet their calcium needs.

Milk is not the only food rich in calcium, but milk and milk products supply most of the calcium to the U.S. diet, and they are fortified with vitamin D. Children who do not consume milk do not meet their calcium needs unless they use calcium-fortified foods or supplements.

When people reach the bone-losing years of middle age, those who formed dense bones during youth have more bone tissue to lose before suffering ill effects—see Figure C8–3, p. 330. Building strong bones in youth helps prevent or delay osteoporosis later on.

Dietary calcium and vitamin D in later life cannot make up for earlier deficiencies, but they may help to slow the rate of bone loss. Additionally, calcium absorption declines with age, and older bodies become less efficient at making and activating vitamin D. Many older people take in less calcium and vitamin D than in their earlier years, they absorb less calcium from food and supplements, they often fail to go outdoors for the sunlight necessary to form vitamin D, and their skin becomes less efficient in synthesizing it. For these people, supplements may be of benefit.

Figure C8–2

Loss of Height in a Woman with Osteoporosis

The woman on the left is about 50 years old. On the right, she is 80 years old. Her legs have not grown shorter; only her back has lost length, due to collapse of her spinal bones (vertebrae). When collapsed vertebrae cannot protect the spinal nerves, the pressure of bones pinching the nerves causes excruciating pain.

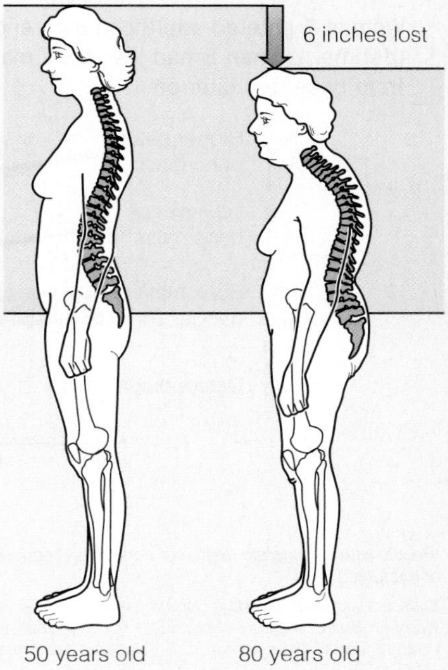

© Cengage Learning

Table C8–2

Risk Factors for Osteoporosis

Nonmodifiable	Modifiable
▪ Female gender	▪ Sedentary lifestyle
▪ Older age	▪ Diet inadequate in calcium and vitamin D
▪ Small frame	▪ Diet excessive in protein, sodium, caffeine
▪ Caucasian, Asian, or Hispanic/Latino heritage	▪ Cigarette smoking
▪ Family history of osteoporosis or fractures	▪ Alcohol abuse
▪ Personal history of fractures	▪ Low body weight
▪ Estrogen deficiency in women (lack of menstruation or menopause, especially early or surgically induced); testosterone deficiency in men	▪ Certain medications, such as glucocorticoids and anticonvulsants

Gender and Hormones

Gender is a powerful predictor of osteoporosis: men have greater bone density than women at maturity, and women often lose more bone, particularly in the 6 to 8 years following menopause when the hormone estrogen diminishes.[8] Thereafter, loss of bone minerals continues throughout the remainder of a woman's lifetime but not at the free-fall pace of the menopause years (refer again to Figure C8–3).

If *young* women fail to produce enough estrogen, they lose bone rapidly, too, and going through menopause early almost doubles a woman's chance of developing osteoporosis.[9] Diseased ovaries are often to blame, but estrogen may be low because the woman suffers from an eating disorder with a dangerously low body weight (see Controversy 9). Even with treatment, the bone loss remains long after the eating disorder has been resolved.

On hearing that bone loss is associated with menopause and estrogen, some men assume that osteoporosis is a "woman's disease." However, each year, millions of men suffer fractures from osteoporosis.[10] Sex hormones, such as testosterone and the small amount of estrogen made by the male body, help to oppose men's osteoporosis, too.[11] Men in whom hormone production has fallen off lose bone and suffer more fractures than others.[†12]

Body Weight

After age and gender, the next risk factor for osteoporosis is being underweight or losing weight. Women who are thin throughout life, or who lose 10 percent or more of their body weight after menopause, face a doubled hip fracture rate. Conversely, researchers are exploring whether excess body fatness with fatty intrusion in bone marrow may have negative effects on bone health.[13]

Physical Activity

Physical activity supports bone growth during adolescence and may protect the bones later on.[14] When combined with adequate calcium intake, the effect is greater still. This relationship may hold in adulthood, too. Researchers have noted that larger, stronger muscles are accompanied by denser, stronger bones in some areas of the body.[15] Conversely, when people lie idle—for example, when they are confined to bed—the bones lose strength just as the muscles do. Astronauts who live without gravity for days or weeks experience rapid and extensive bone loss. The harm to the bones from a sedentary lifestyle equals the harm from nutrient deficiencies or cigarette smoking (Table C8–2).

Preventing falls is a critical focus for fracture prevention in the elderly. The best exercise to keep bones and muscles healthy, and subsequently to prevent falls, is the weight-bearing kind, such as jogging, jumping jacks, jumping rope, vigorous walking, or resistance (weight) training on most days throughout life.[16] In addition to staying physi-

[†]*The condition is known as hypogonadism.*

Figure C8–3

Two Women's Bone Mass History Compared

Woman A entered adulthood with enough calcium in her bones to last a lifetime. Woman B had less bone mass starting out and so suffered ill effects from bone loss later on.

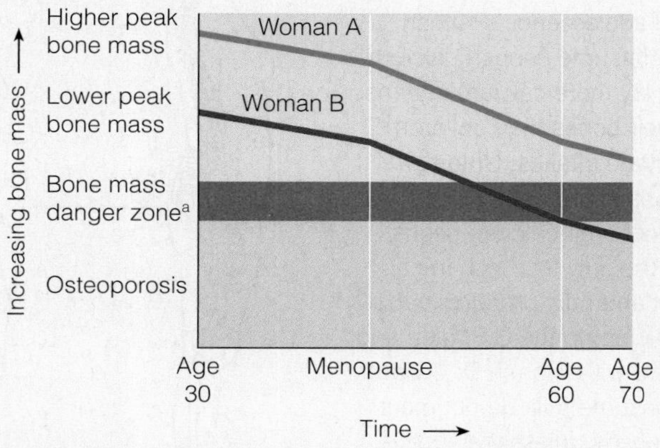

[a]*People with a moderate degree of bone mass reduction are said to have osteopenia and are at increased risk of fractures.*

Source: Data from Standing Committee on the Scientific Evaluation of Dietary Reference Intakes, Food and Nutrition Board, Institute of Medicine, Dietary Reference Intakes for Calcium, Phosphorus, Magnesium, Vitamin D, and Fluoride *(Washington, D.C.: National Academies Press, 2006).*

cally active, evidence supports taking a supplement of vitamin D in the amount of the DRI recommended intake to help prevent falls in independently living adults aged 65 years and older.[17]

Tobacco Smoke and Alcohol

Smoking is hard on the bones. The bones of smokers are less dense than those of nonsmokers. Smoking also increases the risk of fractures and slows fracture healing.[18] Fortunately, quitting can reverse much of the damage. With time, the bone density of former smokers approaches that of nonsmokers.

Heavy drinkers and people who regularly binge drink often have lower bone mineral density and experience more fractures than do nondrinkers and light drinkers.[19] Some evidence suggests that menopausal women who drink moderately may have *higher* bone density than nondrinkers, but more research is needed to determine whether alcohol itself or something else is responsible.[20] Alcoholism is a major cause of osteoporosis in men. Alcohol in high doses may cause calcium excretion in urine and is toxic to body organs that maintain a bone-supporting hormonal balance; it is also directly toxic to the bone-building cells. Finally, drinking contributes to accidents and falls.

Protein

When elderly people take in too little protein, their bones suffer.[21] Recall that the mineral crystals of bone form on a protein matrix—collagen. Restoring protein sources to the diet can often improve bone status and reduce the incidence of hip fractures even in the elderly. However, a diet lacking protein no doubt also lacks energy and other critical bone nutrients, such as vitamin D, vitamin K, and calcium, so restoring a nutritious diet may be of highest importance.[22]

An opposite possibility, that a *high*-protein diet causes bone loss, has also received attention.[23] Excess dietary protein causes urinary calcium losses, but a high protein intake also *increases* calcium absorption, and these opposing effects cancel each other out, producing no net calcium loss from bone.[24]

Milk provides protein along with vitamin A, vitamin D, and calcium, all important nutrients for bone tissue.[25] It follows, then, that vegan vegetarians, who do not consume milk products, would have lower bone mineral density than people who consume milk and milk products—and they generally do.[26] Protein-rich soy foods and beverages may help to oppose bone loss, but whether soy phytochemicals play a role is unknown. [27]

Sodium, Caffeine, Soft Drinks

A high sodium intake is associated with urinary calcium excretion and lowering sodium intakes seems to lessen calcium losses.[28] In study subjects eating the DASH diet, a controlled sodium diet that provides all of the foods in the USDA Food Patterns, urinary calcium losses are reduced.[29] In addition, the DASH diet is higher in calcium than most diets, a critical feature that stands against bone loss. The mechanism behind sodium's effects on the bones is under investigation.[30]

Heavy users of caffeinated beverages, such as coffee, tea, and colas, should be aware that some evidence links caffeine use and bone loss, particularly in rats given high doses.[31] Other findings tend to absolve caffeine use from posing a risk. Cola beverages and processed foods may theoretically speed the dismantling of the bones by way of providing excess phosphorus from phosphoric acid and other food additives.[32] In addition, all soft drinks displace milk from the diet, particularly in children and adolescents.

Other Nutrients Important to Bones

Vitamin K plays roles in the production of at least one bone protein important in bone maintenance. People with hip fractures often have low intakes of vitamin K–rich vegetables, and increasing vegetable intakes may improve both vitamin K status and skeletal health. Giving people vitamin K supplements, however, does not appear to prevent bone loss.[33]

Sufficient vitamin A is needed in the bone-remodeling process, and vitamin C maintains bone collagen. Magnesium may help to maintain bone mineral density.[34] Omega-3 fatty acids may also help preserve bone integrity, and their effects are under study.[35] Clearly, a well-balanced diet that supplies a variety of abundant fruit, vegetables, protein foods, and whole grains along with a full array of nutrients is central to bone health.[36]

The more risk factors of Table C8–2 that apply to you, the greater your chances of developing osteoporosis in the future and the more seriously you should take the advice offered in this Controversy. Treatment, while continuously advancing, remains far from perfect.

Diagnosis and Medical Treatment

Diagnosis of osteoporosis includes measuring bone density using an advanced form of X ray (DEXA, see the nearby photo) or ultrasound.[37] Men with osteoporosis risk factors and all women should have a bone density test after age 50. A thorough examination also includes factors such as race, family history, and physical activity level.

Several drug therapies can reverse bone loss.[38] Some inhibit the activities of the bone-dismantling cells, allowing the bone-building cells to slowly reinforce the bone tissue. Others stimulate the

© oliveromg/Shutterstock.com

These young people are putting bone in the bank.

bone-building cells, resulting in greater bone formation. In some people, such drugs have worked minor miracles in reversing even severe bone loss, but for many others they are ineffective or their side effects prove damaging or intolerable.[39] Estrogen replacement therapy can help nonmenstruating women prevent further bone loss, but questions about safety limits its use. Slow-release forms of fluoride may also increase bone density, but in levels not far above therapeutic thresholds, fluoride poses the serious threat of skeletal fluorosis that weakens the bones.[40]

Calcium Intakes and Recommendations

Adequate calcium nutrition during the growing years is essential for achieving optimal peak bone mass. Yet, at least a quarter of U.S. children older than age 3 fail to take in even 400 milligrams of calcium (the DRI recommended intake is 1,000 milligrams).[41] Only 10 percent of girls and 25 percent of boys fully meet their calcium need during the bone-forming years, and a high percentage of adults also take in too little calcium from food. The DRI committee's recommended intakes for various age groups are found on the inside front cover, and for the most part, they are high—1,300 milligrams for everyone 9 through 18 years of age, for example.

How should you obtain this calcium? Nutritionists strongly recommend the foods and beverages of the USDA eating patterns (see Chapter 2); they reserve supplements for those who cannot meet their needs from foods and beverages. People can do more to support the health of their bones, too, by following the strategies in Table C8–3.

Table C8–3

A Lifetime Plan for Healthy Bones

Ages	Goal	Guidelines
CHILDHOOD		
2 through 12 or 13 years (sexual maturity)	Grow strong bones.	▪ Use milk as the primary beverage to meet the need for calcium within a balanced diet that provides all nutrients. ▪ Play actively in sports or other activities. ▪ Limit television and other sedentary entertainment. ▪ Do not start smoking or drinking alcohol. ▪ Drink fluoridated water.
ADOLESCENCE THROUGH YOUNG ADULTHOOD		
13 or 14 through 30 years	Achieve peak bone mass.	▪ Choose milk as the primary beverage, or if milk causes distress, include other calcium sources. ▪ Commit to a lifelong program of physical activity. ▪ Do not smoke or drink alcohol—if you have started, quit. ▪ Drink fluoridated water.
MATURE ADULT		
31 through 50 years	Maximize bone retention.	▪ Continue as for 13 to 30 year olds. ▪ Adopt bone-strengthening exercises. ▪ Obtain the recommended amount of calcium from food. ▪ Take calcium supplements only if calcium needs cannot be met through foods.
MATURE ADULT		
51 years and above	Minimize bone loss.	▪ Continue as for 13 to 30 year olds. ▪ Continue striving to meet the calcium need from diet. ▪ Continue bone-strengthening exercises. ▪ Obtain a bone density test; follow physician's advice concerning bone-restoring medications and supplements.

Note: The exact ages of cessation of bone accretion and onset of loss vary among people, but in general, data indicate that the skeleton continues to accrete mass for approximately 10 years after adult height is achieved and begins to lose bone around age 35.

A DEXA scan measures bone density to help detect the early stages of bone loss, assess fracture risks, and measure the responses to bone-building treatments. (DEXA stands for dual-energy X-ray absorptiometry.)

Bone loss is not a calcium-deficiency disease comparable to iron-deficiency anemia, in which iron intake reliably reverses the condition. Calcium alone cannot reverse bone loss. For those who are unable to consume enough calcium-rich foods, however, taking calcium supplements—especially in combination with vitamin D—may help to minimize bone loss and reduce the risk of fractures.[42]

Taking self-prescribed calcium supplements entails a few risks (see Table C8–4) and cannot take the place of sound food choices and other healthy habits. One potential threat, a reported link between calcium supplements and heart attacks, has been refuted.[43] Recently, a U.S. Preventive Services Task Force recommended against supplements of calcium with vitamin D as ineffective for preventing fractures in postmenopausal women.[44] Still, millions of people take calcium supplements daily, and the next section provides some details about the variety on the market.

Calcium Supplements

Calcium supplements are often sold as **calcium compounds**, such as calcium carbonate (as in some **antacids**), citrate, gluconate, lactate, malate, or phosphate, and compounds of calcium with amino acids (called **amino acid chelates**). Others are powdered, calcium-rich materials such as **bone meal**, **powdered bone**, **oyster shell**, or **dolomite** (limestone). See Table C8–5 for supplement terms. In choosing a type, consider the answers to the following questions.

Question 1. How much calcium is safe? Although recent evidence suggests that doses of up to 1,000 milligrams present some risks, the DRI committee recommends that habitual calcium intakes from foods and supplements combined should not exceed the Tolerable Upper Intake Level (2,000 to 3,000 milligrams for adults; see the inside front cover, page C).[45] Meeting the need for calcium is important, but more calcium than this provides no additional benefits and may increase risks.[46] Most supplements contain between 250 and 1,000 milligrams of calcium, as stated on the label.

Question 2. How digestible is the supplement? The body cannot use the calcium in a supplement unless the tablet disintegrates in the digestive tract. Manufacturers compress large quantities of calcium into small pills, which the stomach acid must penetrate. To test a supplement, drop a pill into 6 ounces of vinegar and stir occasionally. A digestible pill will dissolve within half an hour.

Question 3. How absorbable is the form of calcium in the supplement? Most healthy people absorb calcium equally well (and as well as from milk) from any of these forms: calcium carbonate, calcium citrate, or calcium phosphate. To improve absorption, divide your dose in half and take it twice a day instead of all at once.

One last pitch: think one more time before you decide to take supplements instead of including calcium-rich foods in your diet. The DRI Committee points out that, particularly among older women, supplements can and do push some people's intakes beyond the Tolerable Upper Intake Level.[47] The Dietary Guidelines for Americans 2010 recommends milk and milk products or calcium and vitamin D fortified soy milk for bone health.[48] The authors of this book are so impressed with the importance of using abundant, calcium-rich foods that we have worked out ways to do so at every meal.

Critical Thinking

1. Osteoporosis occurs during the late years of life; however, it is a disease that develops while one is young. Compose a plan outlining what you can do now to prevent bone loss later in life.
2. Outline the foods you will eat (including quantities) that will provide the recommended RDA for calcium. List lifestyle factors that you can follow to boost your bone density.

Table C8–4

Calcium Supplement Risks

People who take calcium supplements risk:

- *GI distress.* Constipation, intestinal bloating, and excess gas are common.
- *Impaired iron status.* Calcium inhibits iron absorption.
- *Kidney stones or kidney damage.* The risk rises in healthy people taking over 2,000 mg/day. People with a history of kidney stones should be monitored by a physician.
- *Exposure to contaminants.* Some preparations of bone meal and dolomites are contaminated with hazardous amounts of arsenic, cadmium, mercury, and lead.
- *Vitamin D toxicity.* Vitamin D, which is present in many calcium supplements, can be toxic. Users must eliminate other concentrated vitamin D sources.
- *Excess blood calcium.* This complication is seen only with doses of calcium fourfold or greater than customarily prescribed.
- *Other nutrient interactions.* Calcium inhibits absorption of magnesium, phosphorus, and zinc.
- *Drug interactions.* Calcium and tetracycline form an insoluble complex that impairs both mineral and drug absorption.

Table C8–5

Calcium Supplement Terms

- **amino acid chelates** (KEY-lates) compounds of minerals (such as calcium) combined with amino acids in a form that favors their absorption. A chelating agent is a molecule that surrounds another molecule and can then either promote or prevent its movement from place to place (chele means "claw").
- **antacids** acid-buffering agents used to counter excess acidity in the stomach. Calcium-containing preparations (such as Tums) contain available calcium. Antacids with aluminum or magnesium hydroxides (such as Rolaids) can accelerate calcium losses.
- **bone meal** or **powdered bone** crushed or ground bone preparations intended to supply calcium to the diet. Calcium from bone is not well absorbed and is often contaminated with toxic materials such as arsenic, mercury, lead, and cadmium.
- **calcium compounds** the simplest forms of purified calcium. They include calcium carbonate, citrate, gluconate, hydroxide, lactate, malate, and phosphate. These supplements vary in the amount of calcium they contain, so read the labels carefully. A 500-milligram tablet of calcium gluconate may provide only 45 milligrams of calcium, for example.
- **dolomite** a compound of minerals (calcium magnesium carbonate) found in limestone and marble. Dolomite is powdered and is sold as a calcium-magnesium supplement but may be contaminated with toxic minerals, is not well absorbed, and interacts adversely with absorption of other essential minerals.
- **oyster shell** a product made from the powdered shells of oysters that is sold as a calcium supplement but is not well absorbed by the digestive system.

9 Energy Balance and Healthy Body Weight

© andesign101/Shutterstock.com

what do you think?

How can you **control** your body weight, once and for all?

Why are you **tempted** by a favorite treat when you don't feel hungry?

How do extra calories from food become **fat** in your body?

Which popular **diets** are best for managing body weight?

Learning Objectives

After completing this chapter, you should be able to accomplish the following:

LO 9.1 Delineate the health risks of too little and too much body fatness, with emphasis on *central obesity* and its associated health risks.

LO 9.2 Describe the roles of several factors, including BMR, in determining an individual's daily energy needs.

LO 9.3 Calculate the BMI for various people when given their height and weight information, and describe the health implications of any given BMI value.

LO 9.4 Identify several factors, including hormones, that contribute to increased appetite and decreased appetite.

LO 9.5 Describe some inside-the-body factors and theories of obesity development.

LO 9.6 Summarize some outside-the-body factors that may affect weight-control efforts.

LO 9.7 Briefly describe metabolic events occurring in the body during both feasting and fasting.

LO 9.8 Construct a weight-loss plan that includes controlled portions of nutrient-dense foods and sufficient physical activity to produce gradual weight loss while meeting nutrient needs.

LO 9.9 Defend the importance of behavior modification in weight loss and weight maintenance over the long term.

LO 9.10 Compare and contrast the characteristics of anorexia nervosa and bulimia nervosa, and outline strategies for combating eating disorders.

Are you pleased with your body weight? If you answered yes, you are a rare individual. Nearly all people in our society think they should weigh more or less (mostly less) than they do. Their primary concern is usually appearance, but they often perceive, correctly, that physical health is somehow related to weight. Both **overweight** and **underweight** present risks to health and life.*[1]

People also think of their weight as something they should control, once and for all. Three misconceptions in their thinking frustrate their efforts, however—the focus on weight, the focus on *controlling* weight, and the focus on a short-term endeavor. Simply put, it isn't your weight you need to control; it's the fat, or **adipose tissue**, in your body in proportion to the lean—your **body composition**. And controlling body composition directly isn't possible—you can only control your *behaviors*. Sporadic bursts of activity, such as "dieting," are not effective; the behaviors that achieve and maintain a healthy body weight take a lifetime of commitment.[2] Luckily, with time, these behaviors become second nature.

This chapter starts by presenting problems associated with deficient and excessive body fatness and then examines how the body manages its energy budget. The following sections show how to judge body weight on the sound basis of health. The chapter then explores some theories about causes of **obesity** and reveals how the body gains and loses weight. It goes on to present science-based lifestyle strategies for achieving and maintaining a healthy body weight, and it closes with a Controversy section on eating disorders.

The Problems of Too Little or Too Much Body Fat

LO 9.1 Delineate the health risks of too little and too much body fatness, with emphasis on *central obesity* and its associated health risks.

In the United States, too little body fat is not a widespread problem.[3] In contrast, despite a national preoccupation with body image and weight loss, obesity remains at epidemic proportions—see Figure 9–1, p. 336.[4] In 1960, about 13 percent of U.S.

* Reference notes are found in Appendix F.

overweight body weight above a healthy weight; BMI 25 to 29.9 (BMI is defined later).

underweight body weight below a healthy weight; BMI below 18.5.

adipose tissue the body's fat tissue. Adipose tissue performs several functions, including the synthesis and secretion of the hormone leptin involved in appetite regulation.

body composition the proportions of muscle, bone, fat, and other tissue that make up a person's total body weight.

obesity overfatness with adverse health effects, as determined by reliable measures and interpreted with good medical judgment. Obesity is officially defined as a body mass index of 30 or higher.

Figure 9–1

Animated! Increasing Prevalence of Obesity

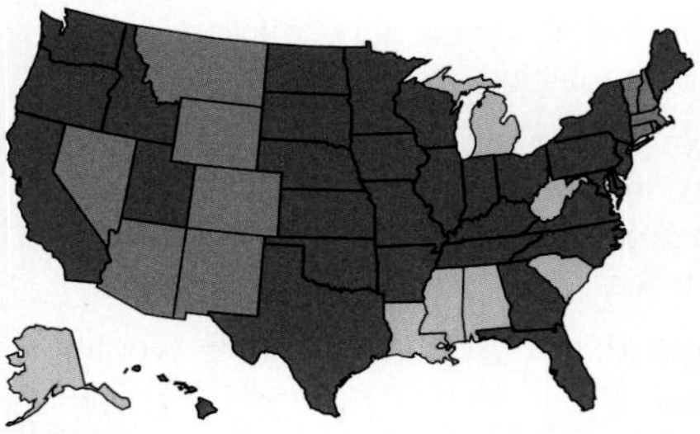

1998: Most states had obesity prevalence rates of less than 20%.

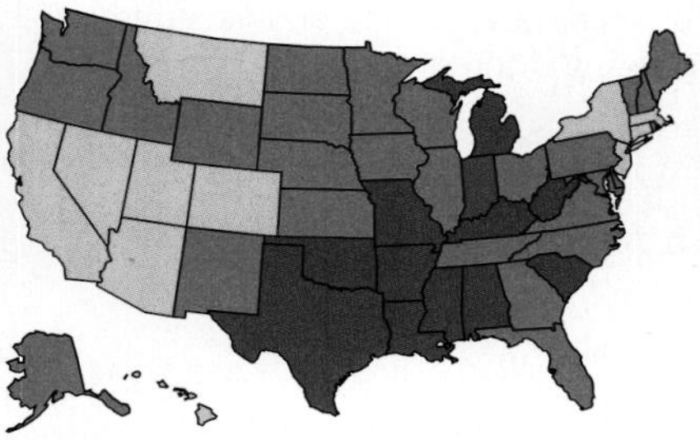

2011: Most states had prevalence rates of greater than 25%, with 12 states reporting prevalence rates of at least 30%.

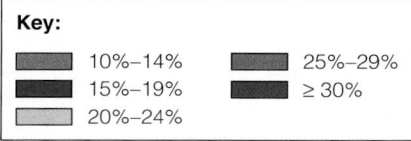

Key:

10%–14%	25%–29%
15%–19%	≥ 30%
20%–24%	

Source: www.cdc.nccdphp/dnpa/obesity/trend/maps/index.htm

adults were obese. Today, an estimated 68 percent of the adults in the United States are now overweight or obese, as illustrated in Figure 9–2, with over 35 percent falling into the obese range.[5] Obesity rates are expected to continue rising into the foreseeable future, although at a somewhat slower rate than in the past.[†6] Even among children and adolescents, 17 percent are obese, and many more are overweight.[7] The United States is caught in a vast global obesity epidemic that is harming the health of people of all ages, in urban and rural areas alike.[8]

Childhood obesity is the topic of **Controversy 13**.

The problem of *underweight*, while affecting fewer than 2 percent of adults in the United States, also poses health threats to those who drop below a healthy minimum.[9] People at either extreme of body weight face increased risks.[10]

What Are the Risks from Underweight?

Thin people die first during a siege or in a famine. Overly thin people are also at a disadvantage in the hospital, where their nutrient status can easily deteriorate if they have to go without food for days at a time while undergoing tests or surgery.[11] Underweight also increases the risk of death for surgical patients and for anyone fighting a **wasting** disease.[12] People with cancer often die not from the cancer itself, but from starvation. Thinner people may also have worse outcomes when they develop heart disease (but heavier people develop it far more often).[13] Thus, excessively underweight people are urged to gain body fat as an energy reserve and to acquire protective amounts of all the nutrients that can be stored.

KEY POINT

- Deficient body fatness threatens survival during a famine or when a person must fight a disease.

What Are the Risks from Too Much Body Fat?

If tomorrow's headlines read, "Obesity Conquered! U.S. Population Loses Excess Fat!" tens of millions of people would be freed from the misery of obesity-related illnesses—heart disease, diabetes, arthritis, certain cancers, and others.[14] In just one year, an estimated 300,000 lives could be saved, along with the estimated $147 billion spent on obesity-related health care.[15] Increased productivity at work would pump an extra $73 billion into the national economy.[16]

Chronic Diseases To underestimate the threat from obesity is to invite personal calamity. Figure 9–3 demonstrates that the risk of dying increases proportionally with increasing body weight.[17] With **extreme obesity**, the risk of dying equals that from smoking.[18] Major obesity-related disease risks include:

- Diabetes.
- Heart disease.[19]
- Hypertension (high blood pressure).
- Gallbladder stones.
- Nonalcoholic fatty liver leading to fibrosis, cirrhosis, and cancer.[20]
- Other cancers.
- Stroke.

Over 70 percent of obese people suffer from at least one other major health problem. Excess body fatness causes up to half of all cases of hypertension, increasing the risk of heart attack and stroke. Obesity triples a person's risk of developing diabetes, and even modest weight gain raises the risk. Most adults with type 2 diabetes are overweight or obese.[21] Chapter 4 presented maps (page 139) depicting increasing U.S. rates of diabetes over the last decade, which parallel increases in obesity (Figure 9–1). Obesity also directly elevates the risk of developing heart disease and certain cancers.

Figure 9–2

Obesity, Overweight, and Underweight in the U.S. Population

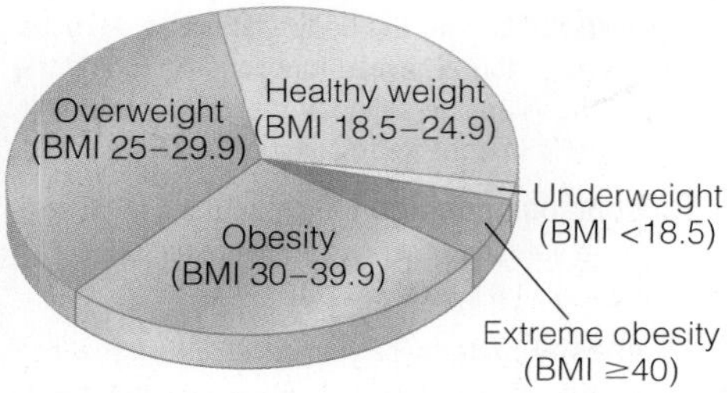

Source: C. D. Fryar and coauthors, Prevalence of overweight, obesity, and extreme obesity among adults: United States, trends 1960–1962 through 2009–2010, September 2012, available at www.cdc.gov.

[†] Defined as more than 100 pounds overweight.

Obesity and Inflammation Why should fat in the body bring extra risk to the heart? Part of the answer may involve **adipokines**, hormones released by adipose tissue. Adipokines help to regulate inflammatory processes and energy metabolism in the tissues.[22]

Metabolic syndrome and chronic diseases are discussed more fully in **Chapter 11.**

In obesity, a shift occurs in the balance of adipokines, among other factors, that favors both tissue inflammation and insulin resistance.[23] The resulting chronic inflammation and insulin resistance may then contribute to diabetes, heart disease, and the other chronic diseases just listed.[24] Calorie-restricted weight-loss diets reduce inflammation.[25]

Other Risks Obese adults also face these threats: abdominal hernias, arthritis, complications in pregnancy and surgery, flat feet, gallbladder disease, gout, high blood lipids, kidney stones, increased risk of medication dosing errors, reproductive disorders, respiratory problems, skin problems, sleep disturbances, sleep apnea (dangerous abnormal breathing during sleep), varicose veins, and even a high accident rate.[26] Some of these maladies start to improve with the loss of just 5 percent of body weight, and risks improve markedly after a 10 percent loss. So great are the harms from obesity that obesity itself is classified as a chronic disease.[27]

KEY POINT

- Obesity raises the risks of developing many chronic diseases and other illnesses.

What Are the Risks from Central Obesity?

A collection of data indicates that fat collected deep within the central abdominal area of the body, called **visceral fat**, poses greater risks of major chronic diseases than does excess fat lying just beneath the skin (**subcutaneous fat**) of the abdomen, thighs, hips, and legs (Figure 9–4).[28] In fact, this **central obesity** elevates the risk of death from *all* causes. Visceral adipose tissue releases more fatty acids into the blood than other types of fat tissue, contributing to a blood lipid profile associated with the **metabolic syndrome** that predicts heart disease.[29] Currently, a measure of central obesity is among the indicators that physicians use to evaluate chronic disease risks.[30]

Men of all ages and women who are past menopause are more prone to develop the "apple" profile that characterizes central obesity, whereas women in their reproductive years typically develop more of a "pear" profile (fat around the hips and thighs that may cling most stubbornly during weight loss).[31] Some women change profiles at menopause, and life-long "pears" may suddenly face the increased disease risks associated with central obesity.

Two other factors also affect body fat distribution. Moderate to high intakes of alcohol associate directly with central obesity, whereas higher levels of physical activity correlate with leanness.[32] A later section explains how to judge whether a person carries too much fat around the middle.

KEY POINTS

- Central obesity is particularly hazardous to health.
- Adipokines are hormones produced by visceral adipose tissue that contribute to inflammation and diseases associated with central obesity.

How Fat Is Too Fat?

People want to know exactly how much body fat is too much. The answer is not the same for everyone, but scientists have developed guidelines.

Evaluating Risks from Body Fatness Obesity experts commonly evaluate the health risks of obesity by way of three indicators (each is described more fully later on). The first is a person's **body mass index (BMI)**. The BMI, which defines average relative weight for height in people older than 20 years, generally (but not always) correlates with body fatness and disease risks. If you are wondering about your own BMI, find it on the inside back cover of the book.

Figure 9–3

Underweight, Overweight, and Mortality

This J-shaped curve associates body mass index (BMI) with mortality. It shows that both underweight and overweight present risks of a premature death. Note that a BMI of 15 generally indicates starvation.

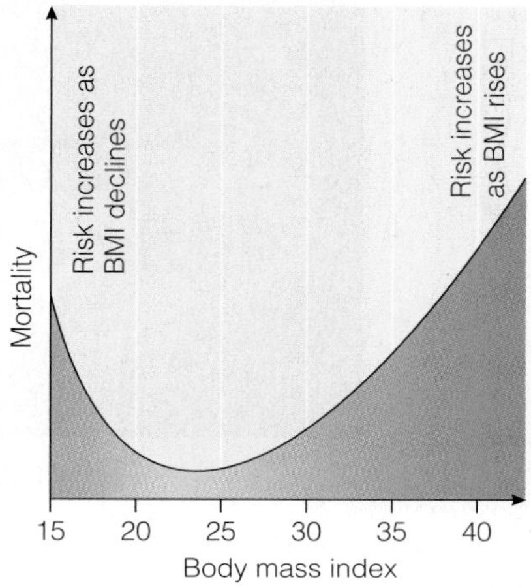

wasting the progressive, relentless loss of the body's tissues that accompanies certain diseases and shortens survival time.

extreme obesity clinically severe overweight, presenting very high risks to health; the condition of having a BMI of 40 or above; also called *morbid obesity.*

adipokines (AD-ih-poh-kynz) protein hormones made and released by adipose tissue (fat) cells.

visceral fat fat stored within the abdominal cavity in association with the internal abdominal organs; also called *intra-abdominal fat.*

subcutaneous fat fat stored directly under the skin (*sub* means "beneath"; *cutaneous* refers to the skin).

central obesity excess fat in the abdomen and around the trunk.

metabolic syndrome a combination of central obesity, high blood glucose (insulin resistance), high blood pressure, and altered blood lipids that greatly increase the risk of heart disease. (Also defined in Chapter 11.)

body mass index (BMI) an indicator of obesity or underweight, calculated by dividing the weight of a person by the square of the person's height.

Figure 9–4

Visceral Fat and Subcutaneous Fat

These abdominal cross sections of an overweight man (left) and woman (right) were produced by CT scans. The people are similar in age and abdominal measurements, but the man's girth is largely from visceral fat; the woman's excess fat is almost all subcutaneous. Excess fat deep within the body's abdominal cavity may pose an especially high risk to health.

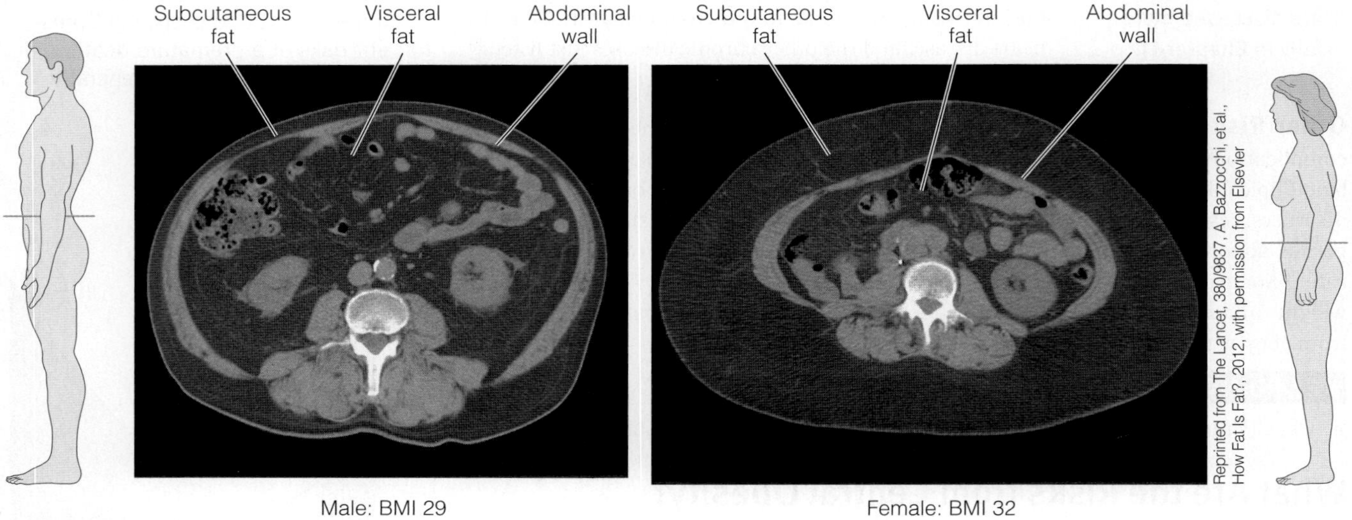

Reprinted from The Lancet, 380/9837, A. Bazzocchi, et al., How Fat Is Fat?, 2012, with permission from Elsevier

The second indicator is **waist circumference**, often reflecting the degree of visceral fatness, or central obesity, in proportion to total body fat (see Table 9–1). The third indicator is the person's disease risk profile, which takes into account other personal factors, such as a diagnosis of hypertension, type 2 diabetes, or elevated blood cholesterol; whether the person smokes; and so forth (see Table 9–2). The more of these factors a person has and the greater the degree of obesity, the greater the urgency to control body fatness.

Table 9–1

Chronic Disease Risks According to BMI Values and Waist Circumference[a]

The degree of risk is heightened by the presence of specific diseases, other risk factors (such as elevated blood LDL cholesterol, as described in Table 9–2), or smoking.

BMI		Waist ≤ 40 in. (Men) or ≤ 35 in. (Women)	Waist ≥ 40 in. (Men) or ≥ 35 in. (Women)
18.5 or less	Underweight	Low risk	—
18.5–24.9	Normal	Low risk	—
25.0–29.9	Overweight	Increased risk	High risk
30.0–34.9	Obese, class I	High risk	Very high risk
35.0–39.9	Obese, class II	Very high risk	Very high risk
40 or greater	Extremely obese, class III	Extremely high risk	Extremely high risk

[a] *Risk for type 2 diabetes, hypertension, and cardiovascular disease.*

Source: National Heart, Lung, and Blood Institute, National Institutes of Health, The Practical Guide: Identification, Evaluation, and Treatment of Overweight and Obesity in Adults, *NIH publication no. 00-4084.*

Table 9–2

Indicators of an Urgent Need for Weight Loss

The National Heart, Lung, and Blood Institute states that aggressive treatment may be needed for extremely obese people who also have any of the following:

- Established cardiovascular disease (CVD)
- Established type 2 diabetes or impaired glucose tolerance
- Sleep apnea, a disturbance of breathing in sleep, including temporary stopping of breathing

The same urgency for treatment exists for an obese person with any three of the following:

- Hypertension
- High LDL
- Smoking
- Low HDL cholesterol
- Sedentary lifestyle
- Age older than 45 years (men) or 55 years (women)
- Heart disease of an immediate family member before age 55 (male) or 65 (female)

Source: National Heart, Lung, and Blood Institute, National Institutes of Health, The Practical Guide: Identification, Evaluation, and Treatment of Overweight and Obesity in Adults, *NIH publication no. 00-4084.*

Why, then, do some obese people remain healthy and live long lives while some average-weight people die young of chronic diseases? Genetic inheritance, smoking habits, and level of physical activity may help to explain why some such individuals stay well while others fall ill. Overweight but fit people have lower risks than normal-weight, unfit ones, for example.[33]

© Joe Sampson, Courtesy of Jennifer Portnick

Being active—even if overweight—is healthier than being sedentary.

Social and Economic Costs of Body Fatness Although a few overfat people escape health problems, no one who is fat in our society quite escapes the social and economic handicaps. Our society places enormous value on thinness, especially for women, and fat people are less sought after for romance, less often hired, and less often admitted to college.[34] They pay higher insurance premiums, they pay more for clothing, and they even pay more in gasoline costs—a car transporting extra weight uses more fuel per mile. In addition, an estimated 30 to 40 percent of all U.S. women (and 20 to 25 percent of U.S. men) are spending $50 billion each year in attempts to lose weight.

Prejudice defines people by their appearance rather than by their ability and character. Obese people suffer emotional pain when others treat them with insensitivity, hostility, and contempt, and they may internalize a sense of guilt and self-deprecation. Health-care professionals, even dietitians, can be among the offenders without realizing it. To free our society of its obsession with body fatness and prejudice against obese people, activists are promoting respect for individuals of all body weights.

KEY POINTS

- Health risks from obesity are reflected in BMI, waist circumference, and a disease risk profile.
- Fit people are healthier than unfit people of the same body fatness.
- Overweight people face social and economic handicaps and prejudice.

The Body's Energy Balance

LO 9.2 Describe the roles of several factors, including BMR, in determining an individual's daily energy needs.

What happens inside the body when you eat too much or too little food? The body ends up with an unbalanced energy budget—you have taken in more or less food energy than you spent over time. The body's energy budget works somewhat like a

waist circumference a measurement of abdominal girth that indicates visceral fatness.

cash budget that grows and dwindles in proportion to the flow of currency. When more food energy is consumed than is needed over days or weeks, excess fat accumulates in the fat cells in the body's adipose tissue where it is stored. When energy supplies run low, stored fat is withdrawn. The daily energy balance can therefore be stated like this:

- Change in energy stores equals food energy taken in minus energy spent on metabolism and muscle activities.

More simply,

- Change in energy stores = energy in − energy out.

Too much or too little fat on the body today does not necessarily reflect today's energy budget.[35] Small imbalances in the energy budget compound over time.

Energy In and Energy Out

The energy in foods and beverages is the only contributor to the "energy in" side of the energy balance equation. Before you can decide how much food will supply the energy you need in a day, you must first become familiar with the amounts of energy in foods and beverages. One way to do so is to look up calorie amounts associated with foods and beverages in the Table of Food Composition (Appendix A) or using a computer program. Such numbers are always fascinating to people concerned with managing body fatness. For example, an apple gives you 70 calories from carbohydrate; a regular-size candy bar gives you about 250 calories mostly from fat and carbohydrate.

You may have heard that for each 3,500 calories you eat in excess of expenditures, you store 1 pound of body fat—a general rule that has previously been used for mathematical estimations. Keep in mind, however, that this number can vary widely with individual metabolic tendencies and efficiencies of nutrient digestion and absorption. Currently, the dynamics of energy storage is a topic of intense scientific investigation.[36]

On the "energy out" side of the equation, no easy method exists for determining the energy an individual spends and therefore needs. Energy expenditures vary so widely among individuals that estimating an individual person's need requires knowing something about the person's lifestyle and metabolism.

KEY POINTS

- The "energy in" side of the body's energy budget is measured in calories taken in each day in the form of foods and beverages.
- No easy method exists for determining the "energy out" side of a person's energy balance equation.

How Many Calories Do I Need Each Day?

Simply put, you need to take in enough calories to cover your energy expenditure each day—your energy budget must balance. One way to estimate your energy need is to monitor your food intake and body weight over a period of time in which your activities are typical and are sufficient to maintain your health. If you keep an accurate record of all the foods and beverages you consume and if your weight is in a healthy range and has not changed during the past few months, you can conclude that your energy budget is balanced. Your average daily calorie intake is sufficient to meet your daily output—your need, therefore, is the same as your current intake.[37] At least 3, and preferably 7, days, including a weekend day, of honest record-keeping are necessary because intakes and activities fluctuate from day to day.

An alternative method of determining energy need is based on energy output. The two major ways in which the body spends energy are (1) to fuel its **basal metabolism** and (2) to fuel its **voluntary activities**. Basal metabolism requires energy

blue jean images/Getty Images

Balancing food energy intake with physical activity can add to life's enjoyment.

to support the body's work that goes on all the time without a person's conscious awareness. A third energy component, the body's metabolic response to food, or the **thermic effect of food**, uses up about 10 percent of a meal's energy value in stepped-up metabolism in the 5 or so hours after finishing a meal.

Basal metabolism consumes a surprisingly large amount of fuel, and the **basal metabolic rate (BMR)** varies from person to person (Figure 9–5). Depending on activity level, a person whose total energy need is 2,000 calories a day may spend as many as 1,000 to 1,600 of them to support basal metabolism. The iodine-dependent hormone thyroxine directly controls basal metabolism—the less secreted, the lower the energy requirements for basal functions. The rate is lowest during sleep.[‡] Many other factors also affect the BMR (see Table 9–3).

People often wonder whether they can speed up their metabolism to spend more daily energy. You cannot increase your BMR very much *today*. You can, however, amplify the second component of your energy expenditure—your voluntary activities. If you do, you will spend more calories today, and if you keep doing so day after day, your BMR will also increase somewhat as you build lean tissue because lean tissue is more metabolically active than fat tissue. Energy spent on voluntary activities depends largely on three factors: weight, time, and intensity. The heavier the weight of the body parts you move, the longer the time you invest in moving them, and the greater the intensity of the work, the more calories you will expend.

Be aware that some ads for weight-loss diets claim that certain substances, such as grapefruit or herbs, can elevate the BMR and thus promote weight loss. This claim is false. Any meal temporarily steps up energy expenditure due to the thermic effect of food, and grapefruit or herbs are not known to accelerate it further.

KEY POINTS

- Two major components of energy expenditure are basal metabolism and voluntary activities.
- A third component of energy expenditure is the thermic effect of food.
- Many factors influence the basal metabolic rate.

Figure 9–5

Components of Energy Expenditure

Typically, basal metabolism represents a person's largest expenditure of energy, followed by physical activity and the thermic effect of food.

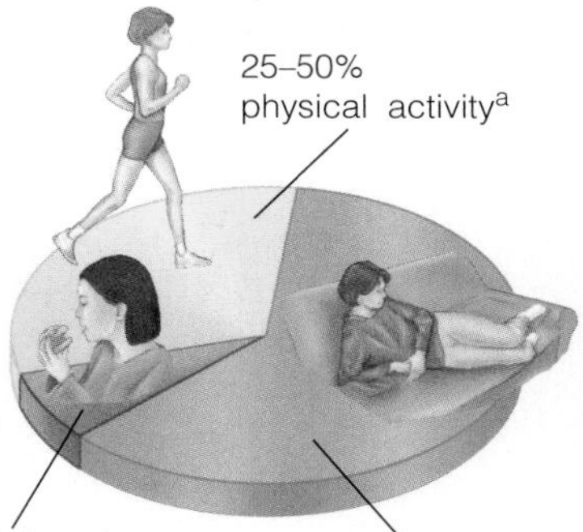

[a]*For a sedentary person, physical activities may account for less than half as much energy as basal metabolism, whereas a very active person's activities may equal the energy cost of basal metabolism.*

Do the Math

To estimate basal energy output:

- Men: kg body weight × 24 = cal/day.
- Women: kg body weight × 23 = cal/day.

(To convert pounds to kilograms [kg], divide pounds by 2.2.)

Table 9–3

Factors That Affect the BMR

Factor	Effect on BMR
Age	The BMR is higher in youth; as lean body mass declines with age, the BMR slows. Physical activity may prevent some of this decline.
Height	Tall people have a larger surface area, so their BMRs are higher.
Growth	Children and pregnant women have higher BMRs.
Body composition	The more lean tissue, the higher the BMR. A typical man has greater lean body mass than a typical woman, making his BMR higher.
Fever	Fever raises the BMR.
Stress	Stress hormones raise the BMR.
Environmental temperature	Adjusting to either heat or cold raises the BMR.
Fasting/starvation	Fasting/starvation hormones lower the BMR.
Malnutrition	Malnutrition lowers the BMR.
Thyroxine	The thyroid hormone thyroxine is a key BMR regulator; the more thyroxine produced, the higher the BMR.

basal metabolism the sum total of all the involuntary activities that are necessary to sustain life, including circulation, respiration, temperature maintenance, hormone secretion, nerve activity, and new tissue synthesis, but excluding digestion and voluntary activities. Basal metabolism is the largest component of the average person's daily energy expenditure.

voluntary activities intentional activities (such as walking, sitting, or running) conducted by voluntary muscles.

thermic effect of food the body's speeded-up metabolism in response to having eaten a meal; also called *diet-induced thermogenesis*.

basal metabolic rate (BMR) the rate at which the body uses energy to support its basal metabolism.

[‡] A measure of energy output taken while the person is awake but relaxed yields a slightly higher number called the *resting metabolic rate*, sometimes used in research.

Did You Know?

The DRI Committee sets an Estimated Energy Requirement (EER) for a reference man and woman:

- Reference man: "Active" physical activity level, 22.5 BMI, 5 ft 10 in. tall, weighing 154 lb.
- Reference woman: "Active" physical activity level, 21.5 BMI, 5 ft 4 in. tall, weighing 126 lb.

Among real people, about 80% fall into the sedentary or moderately active categories.

Estimated Energy Requirements (EER)

A person wishing to know how much energy he or she needs in a day to maintain weight might look up his or her **Estimated Energy Requirement (EER)** value listed on the inside front cover of this book. The numbers listed there seem to imply that for each age and gender group, the number of calories needed to meet the daily requirement is known as precisely as, say, the recommended intake for vitamin A. The printed EER values, however, reflect the needs of only those people who exactly match the characteristics of the "reference man and woman" (see the margin). People who deviate in any way from these characteristics must use other methods for determining their energy needs, and almost everyone deviates.

Taller people need proportionately more energy than shorter people to balance their energy budgets because their greater surface area allows more energy to escape as heat. Older people generally need less than younger people due to slowed metabolism and reduced muscle mass, which occur in part because of reduced physical activity. As Chapter 14 points out, these losses may not be inevitable for people who stay active. On average, though, energy need diminishes by 5 percent per decade beyond the age of 30 years.

In reality, no one is average. In any group of 20 similar people with similar activity levels, one may expend twice as much energy per day as another. A 60-year-old person who bikes, swims, or walks briskly each day may need as many calories as a sedentary person of 30. Clearly, with such a wide range of variation, a necessary step in determining any person's energy need is to study that person.

KEY POINT

- The DRI Committee sets Estimated Energy Requirements for a reference man and woman, but individual energy needs vary greatly.

The DRI Method of Estimating Energy Requirements

The DRI Committee provides a way of estimating EER values for individuals. These calculations take into account the ways in which energy is spent, and by whom. The equation includes:

- *Gender.* Women generally have less lean body mass than men; in addition, women's menstrual hormones influence the BMR, raising it just prior to menstruation.
- *Age.* The BMR declines by an average of 5 percent per decade, as mentioned, so age is a determining factor when calculating EER values.
- *Physical activity.* To help in estimating the energy spent on physical activity each day, activities are grouped according to their typical intensity (Appendix H provides details).
- *Body size and weight.* The higher BMR of taller and heavier people calls for height and weight to be factored in when estimating a person's EER.
- *Growth.* The BMR is high in people who are growing, so pregnant women and children have their own sets of energy equations.

Instructions for estimating EER values are presented in Appendix H. Alternatively, the margin note offers a quick-and-easy way to approximate your own energy need range.

KEY POINT

- The DRI Committee has established a method for determining an individual's approximate energy requirement.

Do the Math

Here's a quick-and-easy method for estimating energy needs:

- First, look up the EER listed for your age and gender group (inside front cover).
- Then, calculate a range of energy needs. For most people, the energy requirement falls within these ranges:

  (Men) EER ± 200 cal. (Women) EER ± 160 cal.
- Virtually everyone's energy requirement falls within these larger ranges:

  (Men) EER ± 400 cal. (Women) EER ± 320 cal.

Body Weight vs. Body Fatness

LO 9.3 Calculate the BMI for various people when given their height and weight information, and describe the health implications of any given BMI value.

For most people, weighing on a scale provides a convenient way to monitor body fatness, but researchers and health-care providers must rely on more accurate assessments. This section describes some of the preferred methods to assess overweight and underweight.

High Body Mass Index (BMI)

Clinicians and researchers use BMI values to help evaluate a person's health risks associated with both underweight and overweight. BMI values correlate significantly with greater body fatness and increased risks of death from diseases, such as heart disease, stroke, diabetes, and nonalcoholic fatty liver disease.[38] The inside back cover of this book provides tables in which to find and evaluate BMI values for adults and adolescents. A formula for determining your BMI is given in the margin.

No one can tell you exactly how much you should weigh, but with health as a value, you have a starting framework in the BMI table (inside back cover). Your weight should fall within the range that best supports your health. As Table 9–1 showed, underweight for adults is defined as BMI of less than 18.5, overweight as BMI of 25.0 through 29.9, and obesity as BMI of 30 or more.

The BMI values have two major drawbacks: they fail to indicate how much of a person's weight is fat and where that fat is located. These drawbacks limit the value of the BMI for use with:

- Athletes (because their highly developed musculature falsely increases their BMI values).
- Pregnant and lactating women (because their increased weight is normal during child-bearing).
- Adults older than age 65 (because BMI values are based on data collected from younger people and because people "grow shorter" with age).
- Women older than age 50 with too little muscle tissue (they may be overly fat for health yet still fall into the normal BMI range).[39]

The bodybuilder in the margin proves this point: with a BMI over 25, he would be classified as overweight by BMI standards alone. However, a clinician would find that his percentage of body fat is well below average and his waist circumference is within a healthy range. For any given BMI value, body fat content can vary widely.[40]

In addition, among some racial and ethnic groups, BMI values may not precisely identify overweight and obesity. African American people of all ages may have more lean tissue per pound of body weight than Asians or Caucasians, for example.[41] In research, BMI may underestimate total obesity in whole populations.[42] Thus, a diagnosis of obesity or overweight requires a BMI value *plus* some measure of body composition and fat distribution. There is no easy way to look inside a living person to measure bones and muscles, but several indirect measures can provide an approximation.

Do the Math

To determine your BMI:

- In pounds and inches

$$\text{BMI} = \frac{\text{weight (lb)} \times 703}{\text{height (in.}^2)}$$

- In kilograms and meters

$$\text{BMI} = \frac{\text{weight (kg)}}{\text{height (m}^2)}$$

At 6'1" tall and 190 lbs., is this athlete too fat for health, as the BMI chart indicates? Further measurements reveal that his body fat content is only 7%, and his health risks are below average.

KEY POINTS

- The BMI values mathematically correlate heights and weights with health risks.
- The BMI concept is flawed for certain groups of people.

Measures of Body Composition and Fat Distribution

A person who stands about 5 feet 10 inches tall and weighs 150 pounds carries about 30 of those pounds as fat. The rest is mostly water and lean tissues: muscles; organs such as the heart, brain, and liver; and the bones of the skeleton (see Figure 9–6, p. 344). This lean tissue is vital to health. The person who seeks to lose weight wants to lose fat, not this precious lean tissue. And for someone who wants to gain weight, it is desirable to gain lean and fat in proportion, not just fat.

Waist circumference measurements indicate visceral fatness (see Figure 9–7, p. 344), and above a certain girth, disease risks rise.[43] Health professionals often use both BMI and waist circumference to assess a person's health risks and monitor changes over time.[44]

Researchers needing more precise measures of body composition may choose any of several techniques to estimate body fatness, including the **skinfold test**. Body fat distribution can be determined by radiographic techniques, such as **dual-energy X-ray absorptiometry**. Mastering these and other sophisticated techniques requires proper

Estimated Energy Requirement (EER) the DRI recommendation for energy intake, accounting for age, gender, weight, height, and physical activity. Also defined in Chapter 2.

skinfold test measurement of the thickness of a fold of skin and subcutaneous fat on the back of the arm (over the triceps muscle), below the shoulder blade (subscapular), or in other places, using a caliper; also called *fatfold test*.

dual-energy X-ray absorptiometry (ab-sorp-tee-OM-eh-tree) a noninvasive method of determining total body fat, fat distribution, and bone density by passing two low-dose X-ray beams through the body. Also used in evaluation of osteoporosis. Abbreviated DEXA.

Figure 9–6

Average Body Composition of Men and Women

The substantially greater fat tissue of women is normal and necessary for reproduction. Normal body fat percentages for people in the healthy BMI range:

- Male: 12 and 20%.
- Female: 20 and 30%.

Figure 9–7

Three Methods Used to Assess Body Fat

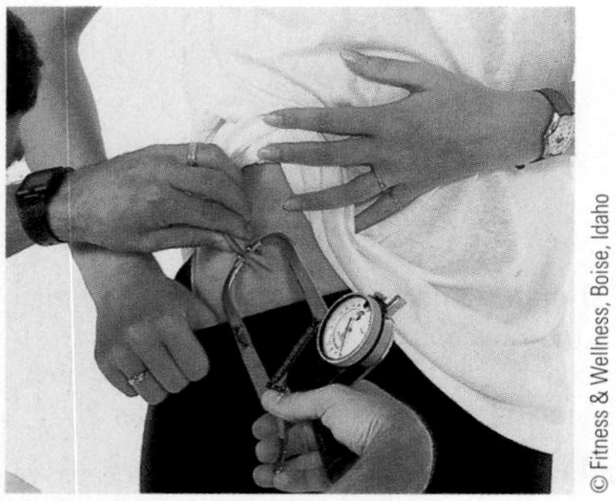

Skinfold measures*. Body fat is measured by using a caliper to gauge the thickness of a fold of skin on the back of the arm (over the triceps), below the shoulder blade (subscapular), and in other places (including lower-body sites), and then comparing these measurements with standards.*

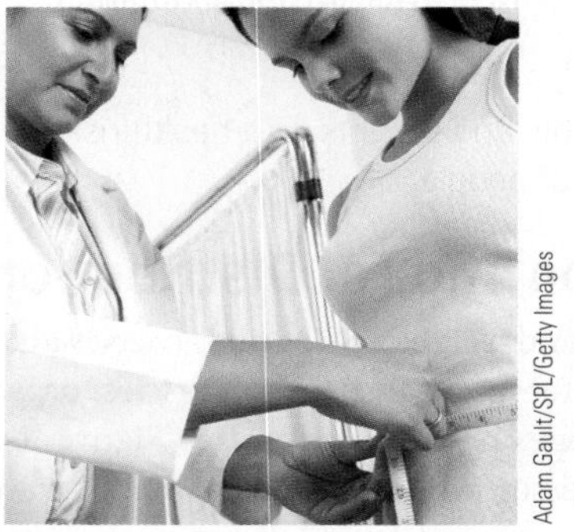

Waist circumference*. Central obesity is measured by placing a nonstretchable measuring tape around the waist just above the bony crest of the hip. The tape is snug but does not compress the skin.*

Art © Cengage Learning

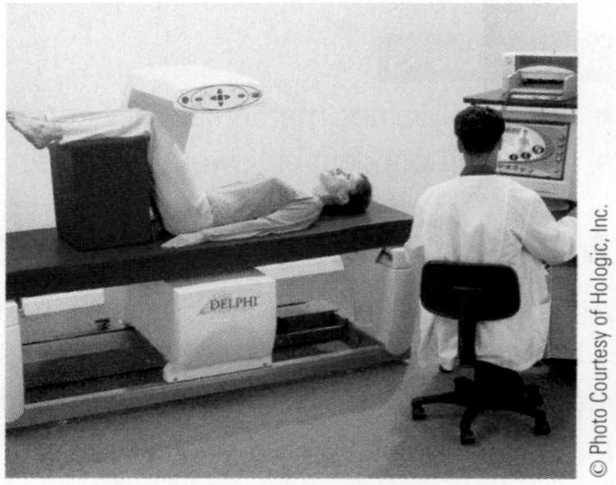

Dual-energy X-ray absorptiometry (DEXA)*. Two low-dose X-rays differentiate among fat-free soft tissue (lean body mass), fat tissue, and bone tissue, providing a precise measurement of total fat and its distribution in all but extremely obese subjects.*

instruction and practice to ensure reliability. Each method has advantages and disadvantages with respect to cost, technical difficulty, and precision of estimating body fat.

KEY POINTS

- Central adiposity can be assessed by measuring waist circumference.
- The percentage of fat in a person's body can be estimated by using skinfold measurements, radiographic techniques, or other methods.
- Body fat distribution can be revealed by radiographic techniques.

How Much Body Fat Is Ideal?

After you have a body fatness estimate, the question arises: What is the "ideal" amount of fat for a body to have? This prompts another question: Ideal for what? If the answer is "society's perfect body shape," be aware that fashion is fickle, and today's popular body shapes are not achievable by most people.

If the answer is "health," then the ideal depends partly on your lifestyle and stage of life. For example, competitive endurance athletes need just enough body fat to provide fuel, insulate the body, and permit normal hormone activity but not so much as to weigh them down. An Alaskan fisherman, in contrast, needs a blanket of extra fat to insulate against the cold. For a woman starting pregnancy, the outcome may be compromised if she begins with too much or too little body fat.

Much remains to be learned about individual requirements for body fat.[45] Questions of how body fat accumulates and how it is controlled are the topics of the next sections.

KEY POINT

- No single body composition or weight suits everyone; needs vary by gender, lifestyle, and stage of life.

The Appetite and Its Control

LO 9.4 Identify several factors, including hormones, that contribute to increased appetite and decreased appetite.

When you grab a snack or eat a meal, you may be aware only of your conscious mind choosing to eat something. However, the choice of when and how much to eat may not be as free as you think—deeper forces of physiology are at work behind the scenes.

Seeking and eating food are matters of life and death, essential for survival. Understandably, therefore, the body's appetite-regulating systems are skewed, tipping in favor of food consumption. **Hunger** demands food. The signals that oppose food consumption, that is, signals for **satiation** and **satiety**, are weaker and more easily overruled. Many signaling molecules, including hormones, help to regulate food intake; the following sections name just a few.

Hunger and Appetite—"Go" Signals

The brain and digestive tract communicate about the need for food and food sufficiency. Their means of communication, hormones and sensory nerve signals, fall roughly into two broad functional categories: "go" mechanisms that stimulate eating and "stop" mechanisms that suppress it. One view of the whole complex process of food intake regulation is summarized in Figure 9–8, p. 346.

Hunger Most people recognize hunger as a strong, unpleasant sensation, the response to a physiological need for food. Hunger makes itself known roughly four to six hours after eating, after the food has left the stomach and much of the nutrient mixture has been absorbed by the intestine. The physical contractions of an empty stomach trigger the hunger signals, as do chemical messengers acting on or originating in the brain's hypothalamus (illustrated in Chapter 3).[46] The hypothalamus has

hunger the physiological need to eat, experienced as a drive for obtaining food; an unpleasant sensation that demands relief.

satiation (SAY-she-AY-shun) the perception of fullness that builds throughout a meal, eventually reaching the degree of fullness and satisfaction that halts eating. Satiation generally determines how much food is consumed at one sitting.

satiety (sah-TIE-eh-tee) the perception of fullness that lingers in the hours after a meal and inhibits eating until the next mealtime. Satiety generally determines the length of time between meals.

Figure 9–8

Hunger, Appetite, Satiation, and Satiety

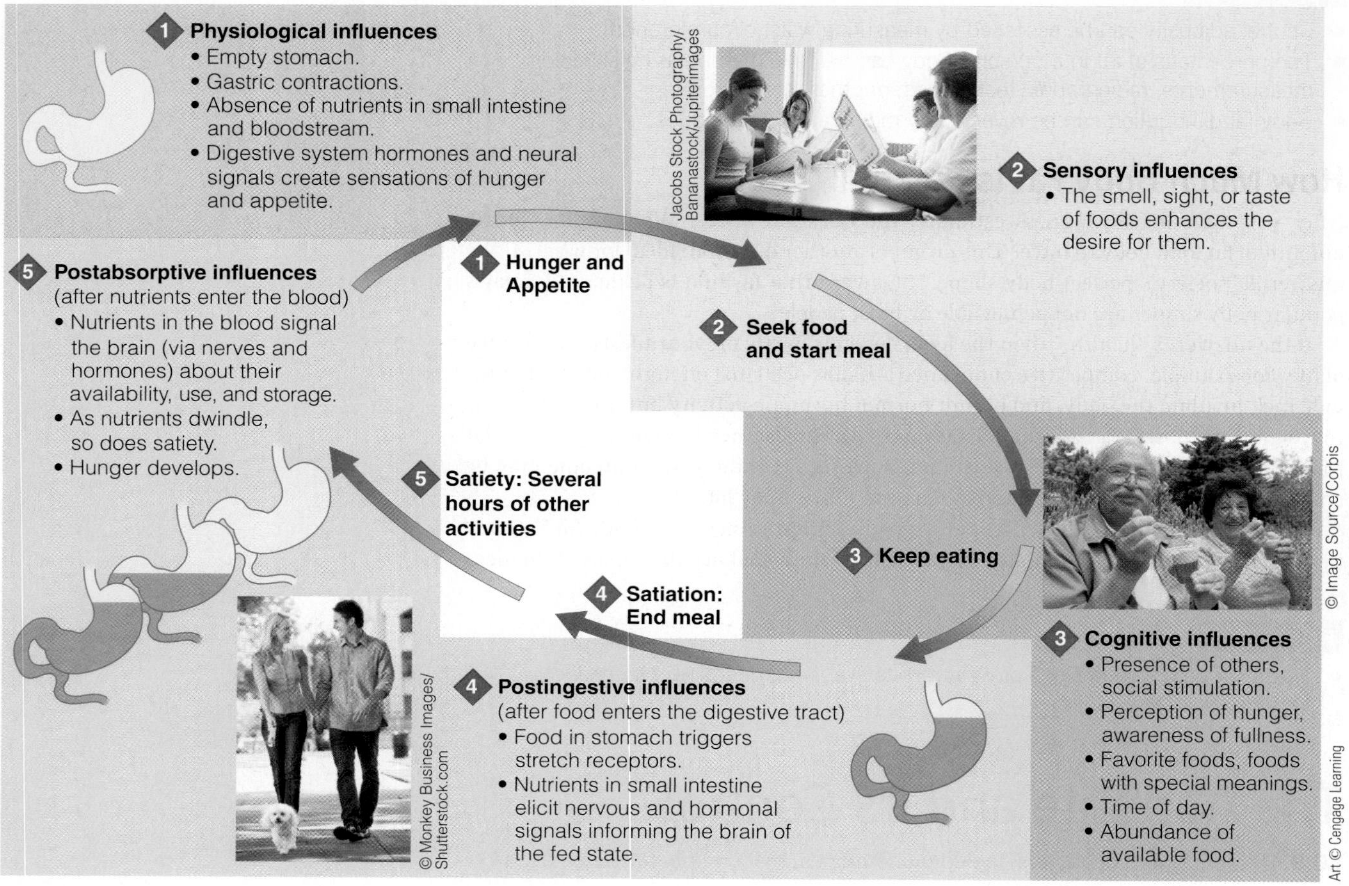

1 Physiological influences
• Empty stomach.
• Gastric contractions.
• Absence of nutrients in small intestine and bloodstream.
• Digestive system hormones and neural signals create sensations of hunger and appetite.
Jacobs Stock Photography/ Bananastock/Jupiterimages
2 Sensory influences
• The smell, sight, or taste of foods enhances the desire for them.
1 Hunger and Appetite
2 Seek food and start meal
5 Postabsorptive influences
(after nutrients enter the blood)
• Nutrients in the blood signal the brain (via nerves and hormones) about their availability, use, and storage.
• As nutrients dwindle, so does satiety.
• Hunger develops.
5 Satiety: Several hours of other activities
3 Keep eating
© Image Source/Corbis
4 Satiation: End meal
3 Cognitive influences
• Presence of others, social stimulation.
• Perception of hunger, awareness of fullness.
• Favorite foods, foods with special meanings.
• Time of day.
• Abundance of available food.
4 Postingestive influences
(after food enters the digestive tract)
• Food in stomach triggers stretch receptors.
• Nutrients in small intestine elicit nervous and hormonal signals informing the brain of the fed state.
© Monkey Business Images/ Shutterstock.com
Art © Cengage Learning

been described as a sort of central hub for energy and body weight regulation, and it can sense molecules representing all three of the energy nutrients.[47]

The polypeptide **ghrelin** is a powerful hunger-stimulating hormone that opposes weight loss. Ghrelin is secreted by stomach cells but works in the hypothalamus and other brain tissues to stimulate **appetite**. Among other things, ghrelin promotes efficient energy storage, contributing to weight gain.[48] Ghrelin may also promote sleep and a lack of sleep may trigger its release, helping to explain why some sleep-deprived people report being extra hungry, eating more food, and ultimately gaining weight.[49]

Ghrelin is just one of many hunger-regulating messengers that informs the brain of the need for food. In fact, the brain itself produces a number of molecular messengers involved in appetite regulation.[50§]

Appetite A person can experience appetite without hunger. For example, the aroma of hot apple pie or the sight of a chocolate butter cream cake after a big meal can trigger a chemical stimulation of the brain's pleasure centers, thereby creating a desire for dessert despite an already full stomach.[51] In contrast, a person who is ill or under stress may physically need food but have no appetite. Other factors affecting appetite include:

- Appetite stimulants or depressants, other medical drugs.
- Cultural habits (cultural or religious acceptability of foods).

[§] One example is neuropeptide Y.

- Environmental conditions (people often prefer hot foods in cold weather and vice versa).
- Hormones (for example, sex hormones).
- Inborn appetites (inborn preferences for fatty, salty, and sweet tastes).[52]
- Learned preferences (cravings for favorite foods, aversion to trying new foods, and eating according to the clock).
- Social interactions (companionship, peer influences).[53]
- Some disease states (obesity may be associated with increased taste sensitivity, whereas colds, flu, and zinc deficiency reduce taste sensitivity).

KEY POINTS

- Hunger outweighs satiety in the appetite control system.
- Hunger is a physiologic response to an absence of food in the digestive tract.
- The stomach hormone ghrelin is one of many contributors to feelings of hunger.

© cloki/Shutterstock.com

Satiation and Satiety—"Stop" Signals

To balance energy in with energy out, eating behaviors must be counterbalanced with ending each meal and allowing periods of fasting between meals. Being able to eat periodically, store fuel, and then use up that fuel between meals is a great advantage. Relieved of the need to constantly seek food, human beings are free to dance, study, converse, wonder, fall in love, and concentrate on endeavors other than eating.

The between-meal interval is normally about 4 to 6 waking hours—about the length of time the body takes to use up most of the readily available fuel—or 12 to 18 hours at night, when body systems slow down and the need is less. As is true for the "go" signals that stimulate food intake, a series of many hormones and sensory nerve messages along with products of nutrient metabolism send "stop" signals to suppress eating. Much more remains to be learned about these mechanisms.[54]

Satiation At some point during a meal, the brain receives signals that enough food has been eaten. The resulting satiation causes continued eating to hold less interest, and limits the size of the meal (consult Figure 9–8 again). Satiation arises from many organs:

- Sensations in the mouth associated with greater food intake trigger increased satiation.[55]
- Nerve stretch receptors in the stomach sense the stomach's distention with a meal and fire, sending a signal to the brain that the stomach is full.
- As nutrients from the meal enter the small intestine, they stimulate other receptor nerves and trigger the release of hormones signaling the hypothalamus about the size and nature of the meal.
- The brain also detects absorbed nutrients delivered by the bloodstream, and it responds by releasing neurotransmitters that suppress food intake.

Together, mouth sensations, stomach distention, and the presence of nutrients trigger nervous and hormonal signals to inform the brain's hypothalamus that a meal has been consumed. Satiation occurs; the eater feels full and stops eating.

Did My Stomach Shrink? Changes in food intake cause rapid adaptations in the body. A person who suddenly eats smaller meals may feel extra hungry for a few days, but then hunger may diminish for a time. During this period, a large meal may make the person feel uncomfortably full, partly because the stomach's capacity has adapted to a smaller quantity of food. A dieter may report "My stomach has shrunk," but the stomach has simply adjusted to smaller meals. At some point in food deprivation, hunger returns with a vengeance and can lead to bouts of extensive overeating.

Just as quickly, the stomach's capacity can adapt to larger meals until moderate portions no longer satisfy. This observation may partly explain the increasing U.S.

Did You Know?

Satiation regulates meal size; satiety regulates meal frequency.

ghrelin (GREL-in) a hormone released by the stomach that signals the brain's hypothalamus and other regions to stimulate eating.

appetite the psychological desire to eat; a learned motivation and a positive sensation that accompanies the sight, smell, or thought of appealing foods.

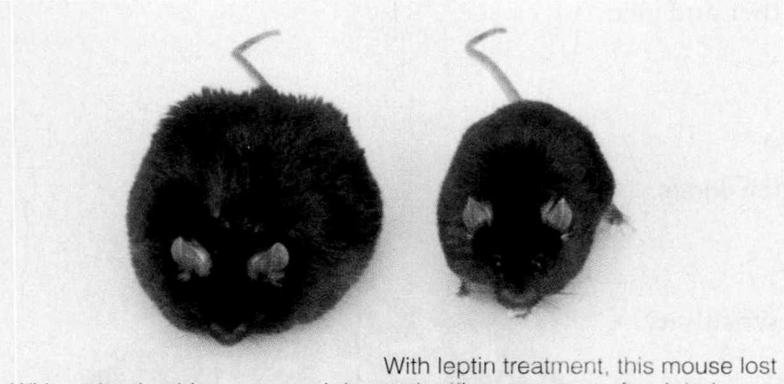

Without leptin, this mouse weighs almost three times as much as a normal mouse.

With leptin treatment, this mouse lost a significant amount of weight but still weighs almost one and a half times as much as a normal mouse.

calorie intakes: popular demand and food industry marketing have led to larger and larger food portions, while stomachs across the nation have adapted to accommodate them.

Satiety After finishing a meal, the feeling of satiety continues to suppress hunger over a period of hours, regulating the frequency of meals. Hormones, nervous signals, and the brain work in harmony to sustain feelings of fullness. At some later point, signals from the digestive tract once again sound the alert that more food is needed.

Leptin, one of the adipokine hormones, is produced in direct proportion to body fatness.** A gain in body fatness stimulates leptin production. Leptin travels from the adipose tissue via the bloodstream to the brain's hypothalamus, where it promotes the release of neurotransmitters that both suppress appetite and increase energy expenditures and, ultimately, body fat loss.[56] A loss of body fatness, in turn, brings the opposite effects—suppression of leptin production, increased appetite, reduced energy expenditure, and accumulation of body fat.[57] Leptin operates on a feedback mechanism—the fat tissue that produces leptin is ultimately controlled by it.

In experiments, obese rats develop both insulin resistance and leptin resistance—the rats fail to respond to leptin's appetite-suppressing effects.[58] In a rare form of human obesity arising from an inherited inability to produce leptin, giving leptin injections quickly reverses both obesity and insulin resistance.[59] More commonly, obese people produce plenty of leptin but are resistant to its effects; giving more leptin does not reverse their obesity.[60]

Energy Nutrients and Satiety The composition of a meal seems to affect satiation and satiety, but the relationships are complex. Of the three energy-yielding nutrients, protein seems to have the greatest satiating effect during a meal.[61] Therefore, including some protein in a meal—even a glass of milk—can improve satiation and possibly even decrease energy intake at the next meal.[62]

Many carbohydrate-rich foods, such as those providing slowly digestible carbohydrate and soluble fiber, also contribute to satiation and satiety.[63] These foods tend to hold blood glucose and insulin steady between meals, minimizing dips in blood glucose that are sensed by the brain, which responds by increasing hunger to restore blood glucose to normal.[64] Added sugars in beverages, in contrast, may not suppress appetite but increase caloric intake.[65] Soluble fibers and digestion-resistant starch also may support colonies of bacteria in the colon that have been associated with leanness in some studies.[66] Finally, fat, famous for triggering a hormone that contributes to long-term satiety, goes almost unnoticed by the appetite control system during consumption of a meal.

Researchers have also reported increased satiety from foods high in water and even from foods that have been puffed up with air. Boredom and sameness of taste and texture also create a sort of sensory-specific satiety—the initial pleasure of eating a particular food diminishes bite after bite, and this decline in liking of a food is a common reason why people stop eating.[67] As dieters await news of dietary tactics against hunger, researchers have not yet identified any one food, nutrient, or attribute—not even protein—that is especially effective for weight loss and its maintenance.[68]

KEY POINTS

- Satiation ends a meal when the nervous and hormonal signals inform the brain that enough food has been eaten.
- Satiety postpones eating until the next meal.
- The adipokine leptin suppresses the appetite and regulates body fatness.
- Protein, carbohydrate, and fat play various roles in satiation and satiety.

** Leptin is also produced in the stomach, where it helps to regulate digestion and contributes to satiation.

Inside-the-Body Theories of Obesity

LO 9.5 Describe some inside-the-body factors and theories of obesity development.

Findings about appetite regulation, the "energy in" side of the body weight equation, do not fully explain why some people gain too much body fat and others stay lean. When given a constant number of excess calories over a period of weeks or months, some people gain many pounds of body fat, but other people gain far fewer.[69] The former seem to use every calorie with great metabolic efficiency, while others may expend calories more freely.

Many theories have emerged to explain these mysteries of obesity in terms of metabolic function and energy expenditure—this section just touches the surface of a few of them. And whenever discussions turn to metabolism, topics in genetics follow closely behind.

Set-Point Theory Like a room's thermostat, the brain and other organs constantly monitor body conditions and respond to slight fluctuations in such essential functions as blood glucose, blood pH levels, and body temperature to maintain them within a narrow range of a physiological set point. The **set-point theory** of obesity holds that, to a degree, this may also be true for body weight.[70] After weight gains or losses, the body adjusts its metabolism somewhat in the direction of restoring the original weight. If weight is gained slowly over time, however, a new set point may be established and defended.[71] Many debates surround the set-point theory of weight regulation.[72]

Thermogenesis Some people tend to expend more energy in metabolism than others. Could that explain obesity? The body's enzymes "waste" a small percentage of energy as heat in a process called **thermogenesis**. Some enzymes expend copious energy in thermogenesis, producing heat but performing no other useful work. As more heat is radiated away from the body, more calories are spent, and fewer calories are available to be stored as body fat.

One tissue extraordinarily gifted in thermogenesis is **brown adipose tissue (BAT)**. BAT, a well-known heat-generating tissue of animals and human infants, has been identified in adult human subjects, too.[73] The subjects with the greatest body fatness in these studies had the least BAT.[74] Intriguingly, a chemical released during muscular work appears to trigger a normally dormant type of adipose cell to act more like BAT metabolically, but the significance of this finding to weight management is unknown.[75]

Is it wise, then, to try to step up thermogenesis to assist in weight loss? Probably not. In rats, the rate of thermogenesis has no effect on overall energy expenditure or body fatness.[76] Also, at a level not far above normal, energy-wasting activity causes cell death. Sham "metabolic" diet products may claim to increase thermogenesis, but no tricks of metabolism can produce effortless fat loss.

Genetics and Obesity If genes carry the instructions for making enzymes, and enzymes control energy metabolism, then genetic variations might reasonably be expected to explain why some people get fat and some stay lean. Indeed, genomic researchers have identified multiple genes likely to play roles in obesity development but have not so far identified a single genetic cause of common obesity.[77] Inherited genes clearly do influence body weight, however. For someone with at least one obese parent, the chance of becoming obese is estimated to fall between 30 and 70 percent.

Complex relationships exist among the many genes related to energy metabolism and obesity, and they each interact with environmental factors, too, even before birth.[78] For example, research suggests that over- or undernutrition of a pregnant female may alter genetic activities of a developing fetus in ways that increase the likelihood of obesity later in life.[79] Even though an individual's genetic inheritance and early influences may make obesity likely, the disease of obesity cannot develop unless the environment—factors that lie outside the body—provides the means of doing so.

leptin an appetite-suppressing hormone produced in the fat cells that conveys information about body fatness to the brain; believed to be involved in the maintenance of body composition (*leptos* means "slender").

set-point theory a theory stating that the body's regulatory controls tend to maintain a particular body weight (the set point) over time, opposing efforts to lose weight by dieting.

thermogenesis the generation and release of body heat associated with the breakdown of body fuels. *Adaptive thermogenesis* describes adjustments in energy expenditure related to changes in environment such as cold and to physiological events such as underfeeding or trauma.

brown adipose tissue (BAT) a type of adipose tissue abundant in hibernating animals and human infants and recently identified in human adults. Abundant pigmented enzymes of energy metabolism give BAT a dark appearance under a microscope; the enzymes release heat from fuels without accomplishing other work. Also called *brown fat*.

My Turn **watch it!**

How Many Calories?

© Cengage Learning

Jackie

© Cengage Learning

Lauren

Two students talk about portion control, physical activity, and dessert.

KEY POINTS

- Metabolic theories attempt to explain obesity on the basis of molecular functioning.
- A person's genetic inheritance greatly influences, but does not ensure, the development of obesity.

Outside-the-Body Theories of Obesity

LO 9.6 Summarize some outside-the-body factors that may affect weight-control efforts.

Food is a source of pleasure, and pleasure drives behavior. Being creatures of free will, people can easily override satiety signals and eat whenever they wish, especially when tempted with delicious treats in large servings. People also value physical ease and seek out labor-savers, such as automobiles and elevators. Over past decades, the abundance of palatable food has increased enormously, while the daily demand for physical activity for survival has all but disappeared.[80]

Environmental Cues to Overeating Here's a common experience: a person walks into a food store feeling not particularly hungry but, after viewing an array of goodies, walks out snacking on a favorite treat. A classic experiment showed that rats, known to precisely maintain body weight when fed standard chow, overeat and rapidly become obese when fed "cafeteria style" on a variety of rich, palatable foods. When offered a delicious smorgasbord, people may do likewise, often without awareness. Like the rats, they respond to external cues. With around-the-clock access to rich palatable foods, we eat more and more often than in decades past—and energy intakes have risen accordingly.[81]

Overeating also accompanies complex human sensations such as loneliness, yearning, craving, addiction, or compulsion. Any kind of stress can also cause overeating and weight gain.[82] ("What do I do when I'm worried? Eat. What do I do when I'm concentrating? Eat!") People who are overweight or obese may be especially responsive to such external cues.[83]

People also overeat in response to large portions of food. In a classic study, moviegoers ate proportionately more popcorn from large buckets than from small bags.[84] In a wry twist, researchers dispensed large and small containers of 14-day-old popcorn to moviegoers who, despite complaining of the staleness, still ate more popcorn from the larger container. Even superior education in nutrition doesn't stop the phenomenon. Nutrition graduate students were invited to a party and offered snack mix from big and small bowls. As predicted, even they ate bigger portions from the bigger bowls. Portion sizes have increased steadily over past decades, and so have calorie intakes and body fatness.[85]

dopamine (DOH-pah-meen) a neurotransmitter with many important roles in the brain, including cognition, pleasure, motivation, mood, sleep, and others.

screen time sedentary time spent using an electronic device, such as a television, computer, or video game player.

built environment The buildings, roads, utilities, homes, fixtures, parks, and all other man-made entities that form the physical characteristics of a community.

food deserts urban and rural low-income areas with limited access to affordable and nutritious foods. Also defined in Chapter 15.

Is Our Food Supply Addictive? People often equate overeating with an addiction.[86] Right away, it should be said that foods, even highly palatable foods, are not comparable to psychoactive drugs in most respects. Yet, emerging evidence supports certain similarities between the brain's chemical responses to both. Pleasure-evoking experiences of all kinds cause brain cells to release the neurotransmitter **dopamine**, which stimulates the reward areas of the brain. The result is feelings of pleasure and desire that create a motivation to repeat the experience. Paradoxically, with repeated exposure to a chemical stimulus (say the drug cocaine) over time, the brain reduces its dopamine response, reducing feelings of pleasure. Soon, larger and larger doses are needed to achieve the desired effect—addiction.

Brain scans reveal reduced brain dopamine activity in people addicted to cocaine or alcohol. In a classic study, brain scans also revealed dopamine reductions in the brains of obese people.[87] This suggests that, similar to an addiction, once these changes are in place, obese people may need more and more delicious food to satisfy their desire for it. Taking the idea one step further, it is plausible that our highly palatable, fat- and sugar-rich food supply could be causing lasting changes in the brain's reward system and making overeating and weight gain likely. It happens reliably in the brains of rats fed cookies, cheese, sugar, and other tasty items, and it may happen in people, too.[88]

Other explanations exist. It may be that consciously restricting intakes of delicious foods increases the desire for them. It may also be that some people are more inclined to "throw caution to the wind" and indulge in treats whenever they present themselves.[89] Future research must untangle these threads before the truth is known.

Physical Inactivity Many people may be obese not because they eat too much, but because they move too little—both in purposeful exercise and in the activities of daily life.[90] Sedentary **screen time** has all but replaced outdoor play for many people. This is a concern because the more time people spend in sedentary activities, the more likely they are to be overweight—and to incur the metabolic risk factors of heart disease (high blood lipids, high blood pressure, and high blood glucose).[91]

Most people also work at sedentary jobs.[92] A hundred years ago, 30 percent of the energy used in farm and factory work came from human muscle power; today, only 1 percent does. The same trend follows at home, at work, at school, and in transportation. The more hours spent sitting still, the higher the risk of dying from heart disease and other causes.[93] The Think Fitness feature offers perspective on the contribution of physical activity to weight management, and Table 9–4 lists the energy costs of some activities.

Can Your Neighborhood Make You Fat? Experts urge people to "take the stairs instead of the elevator" or "walk or bike to work." These are good strategies: climbing stairs provides an impromptu workout, and people who walk or ride a bicycle for transportation most often meet their needs for physical activity. Many people, however, encounter barriers in their **built environment** that prevent such choices.

Few people would choose to walk or bike on roadways that lack safe sidewalks or marked bicycle lanes, where vehicles speed by, or where the air is laden with toxic carbon monoxide gas or other pollutants from gasoline engines.[††] Few would choose to walk up flights of stairs in inconvenient, stuffy, isolated, and unsafe stairwells in modern buildings. In contrast, people living in safe, attractive, affordable neighborhoods with safe biking and walking lanes, public parks, and freely available exercise facilities use them often—their surroundings encourage physical activity.[94] Table 9–5, p. 352, points out some community-wide strategies to help prevent obesity.

In addition, residents of many low-income urban and rural areas lack access to even a single supermarket.[95] Often overweight and lacking transportation, residents of these so-called **food deserts** have limited access to the affordable, fresh, nutrient-dense foods they need.[96] Instead, they shop at local convenience stores and fast-food

[††] Carbon monoxide (CO) avidly binds to hemoglobin in the blood, reducing blood oxygen content; CO in air surrounding roadways can reach levels sufficient to impair driving ability.

Table 9–4

Energy Spent in Activities

To determine the calorie cost of an activity, multiply the number listed by your weight in pounds. Then multiply by the number of minutes spent performing the activity.

Example: Jessica (125 lb) rode a bike at 17 mph for 25 min:

$0.057 \times 125 = 7.125$

$7.125 \times 25 = 178.125$

(about 180 calories)

Activity	Cal/lb Body Weight/min
Aerobic dance (vigorous)	0.062
Basketball (vigorous, full court)	0.097
Bicycling	
13 mph	0.045
15 mph	0.049
17 mph	0.057
19 mph	0.076
21 mph	0.090
23 mph	0.109
25 mph	0.139
Canoeing (flat water, moderate pace)	0.045
Computer sports games[a]	
bowling	0.021
boxing	0.021
tennis	0.022
Cross-country skiing	
8 mph	0.104
Golf (carrying clubs)	0.045
Handball	0.078
Horseback riding (trot)	0.052
Rowing (vigorous)	0.097
Running	
5 mph	0.061
6 mph	0.074
7.5 mph	0.094
9 mph	0.103
10 mph	0.114
11 mph	0.131
Soccer (vigorous)	0.097
Studying	0.011
Swimming	
20 yd/min	0.032
45 yd/min	0.058
50 yd/min	0.070
Table tennis (skilled)	0.045
Tennis (beginner)	0.032
Walking (brisk pace)	
3.5 mph	0.035
4.5 mph	0.048
Weight lifting	
light-to-moderate effort	0.024
vigorous effort	0.048
Wheelchair basketball	0.084
Wheeling self in wheelchair	0.030

[a]*Such as Wii™, by Nintendo.*

Think Fitness move it! Activity for a Healthy Body Weight

Some people believe that physical activity must be long and arduous to obtain benefits, such as improved body composition. Not so. A brisk, 30-minute walk at a pace of about 100 steps per minute on at least 5 days per week can help significantly.[97] To achieve an "active lifestyle" by walking requires an hour a day. Even in increments of 10 minutes throughout the day, exercise can measurably improve fitness.[98]

According to the American College of Sports Medicine,

- 150 to 250 minutes a week of moderate intensity physical activity can prevent initial weight gain; more than 150 minutes a week is associated with modest loss.
- More than 250 minutes per week provides significant weight loss, and may prevent regain after loss.
- Episodes of physical activity lasting at least 10 minutes count toward exercise goals.
- Both aerobic (endurance) and muscle-strengthening (resistance) physical activities are beneficial, but calorie restriction must accompany resistance training to achieve weight loss.[99]

A useful strategy is to augment your planned workouts with bits of physical activity throughout the day. Work in the garden; work your abdominal muscles while you stand in line; stand up straight; walk up stairs; fidget while sitting down; tighten your buttocks while sitting in your chair. Chapter 10 provides many more details.

start now! If you are not currently exercising regularly, try this: go to Diet Analysis Plus and create an alternate profile, adding in 30 minutes of moderate to vigorous physical activity for each day on which you have tracked your food intake. Go to the Reports tab; then choose Energy Balance. What differences do you see when you compare this report with the report you created under your original profile?

places, where they can purchase mostly refined packaged sweets and starches, sugary soft drinks, fatty canned meats, or fast foods, and they often have an eating pattern that predicts nutrient deficiencies and excesses along with the high rates of obesity and type 2 diabetes.[100]

In truth, in most neighborhoods across the United States, the most accessible, affordable, and tempting foods and beverages are high-calorie, good-tasting, inexpensive fare from fast-food restaurants, and it takes a great deal of attention, planning, and time to "go against the flow" to keep calorie intakes reasonable. Today, only about a third of the population succeeds in doing so. The prestigious National Academies' Institute of Medicine has put forth these national goals as most likely to slow or reverse the obesity epidemic and improve the nation's health:

- Make physical activity an integral and routine part of American life.
- Make healthy foods and beverages available everywhere—create food and beverage environments to make healthy food and beverage choices the routine, easy choice.
- Advertise and market what matters for a healthy life.
- Activate employers and health-care professionals.
- Strengthen schools as the heart of health.[101]

Accomplishing any one of these goals on its own might speed up progress in preventing obesity, but if all these goals are realized, their effects will be powerful allies in the nationwide struggle to regain control over weight and health. Such changes require efforts from leaders at all levels and citizenry across all sectors of society working with one goal: improving the health of the nation.

Until these changes occur, the best way for most people to attain a healthy body composition boils down to control in three areas: diet, physical activity, and behavior change. Later sections focus on these areas, while the next section delves into the details of how, exactly, the body loses and gains weight.

Table 9–5

Community Strategies to Combat Obesity

To improve in ways like these, communities must organize and work together for change.

- Create safe communities that support physical activity.
- Promote the availability of affordable healthy food and beverages.
- Support healthy food and beverage choices.
- Encourage physical activity and limit sedentary activity, especially among children and youth.
- Develop legislation, policies, and systems in areas such as zoning, school food service, recreation, and transportation that work to prevent and reduce obesity.

KEY POINTS

- Studies of human behavior identify stimuli that lead to overeating.
- Food environments may trigger brain changes that lead to overeating.
- Too little physical activity, the built environment, and a lack of access to fresh foods are linked with overfatness. National antiobesity efforts are underway.

How the Body Loses and Gains Weight

LO 9.7 Briefly describe metabolic events occurring in the body during both feasting and fasting.

The causes of obesity may be complex, but the body's energy balance is straightforward. To lose or gain body fat requires eating less or more food energy than the body expends. A change in body *weight* of a pound or two may not indicate a change in body fat, however—it can reflect shifts in body fluid content, in bone minerals, in lean tissues such as muscles, or in the contents of the bladder or digestive tract. A weight change often correlates with time of day: people generally weigh the least before breakfast.

The type of tissue lost or gained depends on how you go about losing or gaining it. To lose fluid, for example, you can take a "water pill" (diuretic), causing the kidneys to siphon extra water from the blood into the urine, or you can exercise while wearing heavy clothing in hot weather to cause abundant fluid loss in sweat. (Both practices are dangerous and are not being recommended here.) To gain water weight, you can overconsume salt and water; for a few hours, your body will retain water until it manages to excrete the salt. (This, too, is not recommended.) Most quick weight-change schemes promote large changes in body fluids that register dramatic, but temporary, changes on the scale and accomplish little weight change in the long run.

One other practice is hazardous and not recommended: smoking.[102] Each year, many adolescents, particularly girls, take up smoking as a means to control weight. Nicotine blunts feelings of hunger, and smokers do tend to weigh less than nonsmokers. Fear of weight gain prevents many people from quitting smoking, too. The best advice to smokers trying to quit is to adjust eating and exercise habits to maintain weight during and after cessation. To the person flirting with the idea of taking up smoking for weight control, don't do it—many thousands of people who became addicted as teenagers die from tobacco-related illnesses each year.

Did You Know?

Smoking may keep some people's weight down, but at what cost?

- Cancer.
- Chronic lung diseases.
- Heart disease.
- Low-birthweight babies.
- Miscarriage.
- Osteoporosis.
- Shortened life span.
- Sudden infant death.
- Many others.

Moderate Weight Loss vs. Rapid Weight Loss

When you eat less food energy than you need, your body draws on its stored fuel to keep going. If a person exercises appropriately, moderately restricts calories, and consumes an otherwise balanced diet that meets carbohydrate needs and provides sufficient protein, the body is forced to use up its stored fat for energy.[103] Gradual weight loss will occur.[104] This is preferred to rapid weight loss because lean body mass is spared and fat is lost.

The Body's Response to Fasting If a person doesn't eat for, say, three whole days, then the body makes one adjustment after another. Less than a day into the fast, the liver's glycogen is essentially exhausted. Where, then, can the body obtain glucose to keep its nervous system going? Not from the muscles' glycogen because that is reserved for the muscles' own use. Not from the abundant fat stores most people carry because these are of no use to the nervous system. Fat cannot be converted to glucose—the body lacks enzymes for this conversion.[‡‡] The muscles, heart, and other organs use fat as fuel, but at this stage the nervous system needs glucose. The body does, however, possess enzymes that can convert protein to glucose. Therefore, the

[‡‡] Glycerol, which makes up 5 percent of fat, can yield glucose but is a negligible source.

underfed body sacrifices the proteins in its lean tissue to supply raw materials from which to make glucose.

If the body were to continue to consume its lean tissue unchecked, death would ensue within about 10 days. After all, in addition to skeletal muscle, the blood proteins, liver, digestive tract linings, heart muscle, and lung tissue—all vital tissues—are being burned as fuel. (Fasting or starving people remain alive only until their stores of fat are gone or until half their lean tissue is gone, whichever comes first.) To prevent this, the body puts a key strategy into action: it begins converting fat into **ketone bodies** (introduced in Chapter 4) that some nervous system tissues can use and so forestalls the end. This process is ketosis, an adaptation to prolonged fasting or carbohydrate deprivation.

Ketosis In ketosis, instead of breaking down fat molecules all the way to carbon dioxide and water, the body takes partially broken-down fat fragments and combines them to make ketone bodies, compounds that are normally kept to low levels in the blood. It converts some amino acids—those that cannot be used to make glucose—to ketone bodies, too. These ketone bodies circulate in the bloodstream and help to feed the brain; about half of the brain's cells can make the enzymes needed to use ketone bodies for energy. Under normal conditions, the brain and nervous system devour glucose—about 400 to 600 calories' worth each day. After about 10 days of fasting, the brain and nervous system can meet most, but not all, of their energy needs using ketone bodies.

Thus, indirectly, the nervous system begins to feed on the body's fat stores. Ketosis reduces the nervous system's need for glucose, spares the muscle and other lean tissue from being quickly devoured, and prolongs the starving person's life. Thanks to ketosis, a healthy person starting with average body fat content can live totally deprived of food for as long as six to eight weeks.

In summary,

- The brain and nervous system cannot use fat as fuel and demand glucose.
- Body fat cannot be converted to glucose.
- Body protein can be converted to glucose.
- Ketone bodies made from fat can feed some nervous system tissues and reduce glucose needs, sparing protein from degradation.

Figure 9–9 reviews how energy is used during both feasting and fasting.

Is Fasting Harmful? Respected, wise people in many cultures have practiced fasting as a periodic discipline. The body tolerates short-term fasting, and at least in animals, it may even extend the lifespan.[105] There is no evidence, however, that the body becomes internally "cleansed," as some believe. On the down side, repeated fasting in rats causes greater fat storage without increasing body weight.[106] In people, the fasting body also slows its metabolism to conserve energy—the wrong effect for weight loss. Fasting may also increase the appetite, particularly for starchy foods.[107]

Fasting may become harmful when tissues are deprived of the nutrients they need to assemble new enzymes, red and white blood cells, and other vital components. Also, ketosis imbalances can upset the acid-base balance of the blood and promote excessive mineral losses in the urine. To prevent these effects, the DRI Committee sets a minimum intake for carbohydrate at 130 grams per day.

Food deprivation also leads to a tendency to overeat or even binge when food becomes available.[108] The effect seems to last beyond the point when body weight is regained, sometimes for years. People with eating disorders often report that fasting or severe food restriction heralded the beginning of their loss of control over eating. This indictment applies to extreme dieting and fasting but not to the moderate weight-management strategies described later in this chapter.

KEY POINTS

- When energy balance is negative, glycogen returns glucose to the blood and fat tissue supplies fatty acids for energy.
- When fasting or a low carbohydrate diet causes glycogen to run out, body protein is called upon to make glucose, while fats supply ketone bodies to help feed the brain and nerves.

ketone bodies acidic compounds derived from fat and certain amino acids. Normally rare in the blood, they help to feed the brain during times when too little carbohydrate is available. Also defined in Chapter 4.

Figure 9–9

Feasting and Fasting

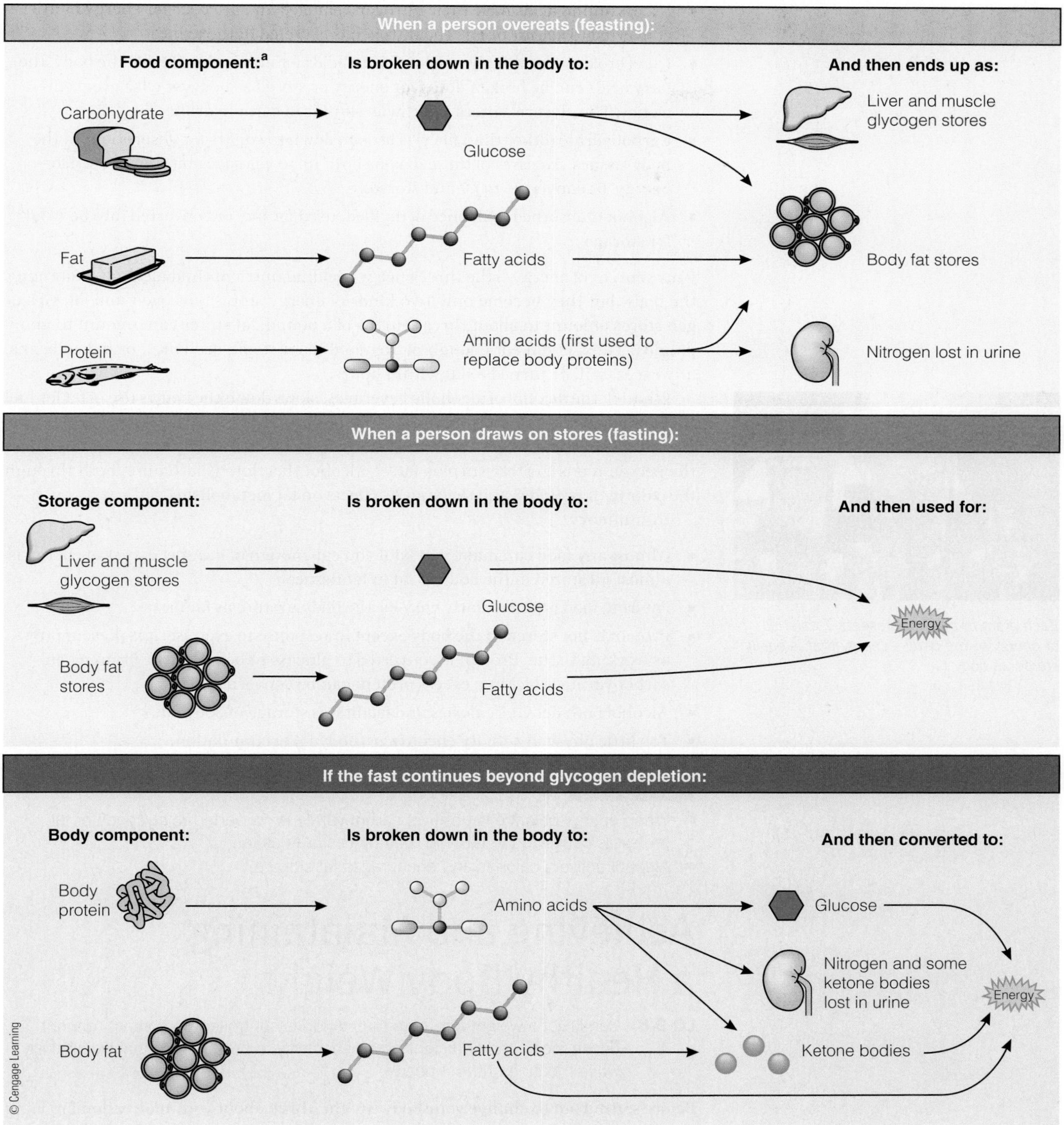

[a]*Alcohol is not included because it is a toxin and not a nutrient, but it does contribute energy to the body. After detoxifying the alcohol, the body uses the remaining two-carbon fragments to build fatty acids and stores them as fat.*

Weight Gain

What happens inside the body when a person does not use up all of the food energy taken in? Previous chapters have already provided the answer—the energy-yielding nutrients contribute the excess to body stores as follows:

- Protein is broken down to amino acids for absorption. Inside the body, these may be used to replace lost body *protein* and, in a person who is exercising or growing, to build new muscle and other lean tissue.
- Excess amino acids have their nitrogen removed and are used for energy or are converted to *glucose* or *fat*. The nitrogen is excreted in the urine.
- Fat is broken down to glycerol and fatty acids for absorption. Inside the body, the fatty acids can be broken down for energy or stored as body *fat* with great efficiency. The glycerol enters a pathway similar to carbohydrate.
- Carbohydrate (other than fiber) is broken down to sugars for absorption. In the body tissues, excesses of these may be built up to *glycogen* and stored, used for energy, or converted to *fat* and stored.
- Alcohol is absorbed and, once detoxified, used for fuel or converted into body fat for storage.

Four sources of energy—the three energy-yielding nutrients and alcohol—may enter the body, but they become only two kinds of energy stores: glycogen and fat. Glycogen stores amount to about three-fourths of a pound; fat stores can amount to many pounds. Thus, if you eat enough of any food, be it steak, brownies, or baked beans, any excess will be turned to fat within hours.

© iStockphoto.com/Dane Wirtzfeld

Each gram of alcohol presents 7 calories of energy to the body—energy that is easily stored as body fat.

Ethanol, the alcohol of alcoholic beverages, slows down the body's use of fat for fuel by as much as a third, causing more fat to be stored. This storage is primarily in the abdominal fat tissue of the "beer drinker's belly" and on the thighs, legs, or anywhere the person tends to store surplus fat.[109] Alcohol therefore is fattening, both through the calories it provides and through its effects on fat metabolism.[§§]

In summary,

- Almost any food can make you fat if you eat enough of it. A net excess of energy is almost all stored in the body as fat in fat tissue.
- Fat from food is particularly easy for the body to store as fat tissue.
- Protein is not stored in the body except in response to exercise; it is present only as working tissue. Protein is converted to glucose to help feed the brain when carbohydrate is lacking; excess protein can be converted to fat.
- Alcohol both delivers calories and facilitates storage of body fat.
- Too little physical activity encourages body fat accumulation.

KEY POINTS

- When energy balance is positive, carbohydrate is converted to glycogen or fat, protein is converted to fat, and food fat is stored as fat.
- Alcohol delivers calories and encourages fat storage.

Achieving and Maintaining a Healthy Body Weight

LO 9.8 Construct a weight-loss plan that includes controlled portions of nutrient-dense foods and sufficient physical activity to produce gradual weight loss while meeting nutrient needs.

Before setting out to change your body weight, think about your motivation for doing so. Many people strive to change their weight, not to improve health, but because their weight fails to meet society's ideals of attractiveness. Unfortunately, this kind of thinking sets people up for disappointment.[110] The human body is not infinitely malleable. Few overweight people will ever become rail-thin, even with the right eating pattern, exercise habits, and behaviors. Likewise, most underweight people will remain on the slim side even after spending much effort to put on some heft.

[§§] People addicted to alcohol are often overly thin because of diseased organs, depressed appetite, and subsequent malnutrition.

Modest weight loss, even 5 or 10 pounds in a person who is still overweight, can quickly produce gains in physical abilities and quality of life, along with improvements in diabetes, blood pressure, and blood lipids.[111] Stair climbing, walking, and other tasks of daily living become noticeably easier. Adopting health or fitness as the ideal rather than some ill-conceived image of beauty can avert much misery.[112] Table 9–6 offers some tips to that end.

© Robyn Mackenzie/Shutterstock.com

The rest of this chapter stresses health and fitness as goals and explains the required actions to achieve them. It uses weight only as a convenient gauge for progress. To repeat, effort in three realms produces results:

- Eating patterns.
- Physical activity.
- Behavior modification.[113]

Eating patterns and physical activity, are explained next. Behavior modification is the topic of this chapter's Food Feature section.

First, a Reality Check Overweight takes years to accumulate. Tackling excess body fatness also takes time, along with patience and perseverance. The person must adopt healthy eating patterns, take on physical activities, create a supportive environment, and seek out behavioral and social support; continue these behaviors for at least 6 months for initial weight loss; and then continue all of it for a lifetime to maintain the losses.[114] Setbacks are a given, and, according to recent evidence, the size of the calorie deficits required to lose a pound of weight initially may be less than those required later on, meaning that weight loss is hard at first, and then it may get harder.[115]

The list of what doesn't work is long: fad diets, skipping meals, "diet foods," special herbs and supplements, and liquid-diet formulas, among others.[116] Many fad diets promise quick and easy weight-loss solutions but, as the Consumer's Guide near here points out, fad diets can interrupt real progress toward life-long weight management. In contrast, people willing to take one step at a time, even if it feels like just a baby step, toward balancing their energy budget are on the right path. An excellent first step is to set realistic goals.

Set Achievable Goals A reasonable first weight goal for an overweight person might be to stop gaining weight. A next goal might be to reduce body weight by 5 to 10 percent over a year's time. This may sound insignificant, but even small losses can improve health and reduce disease risks. Put another way: shoot for a weight that falls two BMI categories lower than a present unhealthy one. For example, a 5-foot-5-inch woman with hypertension weighing 180 pounds (BMI of 30—see the BMI table on the inside back cover) may aim for a BMI of 28, or about 168 pounds. If her health

Table 9–6

Tips for Accepting a Healthy Body Weight

- Value yourself and others for traits other than body weight; focus on your whole self including your intelligence, social grace, and professional and scholastic accomplishments.
- Realize that prejudging people by weight is as harmful as prejudging them by race, religion, or gender.
- Use only positive, nonjudgmental descriptions of your body; never use degrading, negative descriptions.
- Accept positive comments from others.
- Accept that no magic diet exists.
- Stop dieting to lose weight. Adopt a healthy eating and exercise lifestyle permanently.
- Follow the USDA Food Patterns (Chapter 2, pages 42–43). Never restrict food intake below the minimum levels that meet nutrient needs.
- Become physically active, not because it will help you get thin, but because it will enhance your health.
- Seek support from loved ones. Tell them of your plan for a healthy life in the body you have been given.
- Seek professional counseling, not from a weight-loss counselor, but from someone who supports your self-esteem.
- Join with others to fight weight discrimination and stereotypes.

use it! A Consumer's Guide To . . .

Fad Diets

Over the years, Lauren has tried most of the new fad diets, her hopes rising each time, as though she had never been disappointed: "*This* one has the answer. I have *got* to lose 40 pounds. Plus it only costs $30 to start." Who wouldn't pay a few dollars to get trim?

Lauren and tens of millions of people like her have helped to fuel the success of a $33 billion-a-year weight-loss industry. The number of fad-diet books in print could fill a bookstore, and more keep coming out because they continue to make huge profits. Some of them restrict fats or carbohydrates, some disallow certain foods, some advocate certain food combinations, some claim that a person's genetic type or blood type determines the best diet, and others advocate taking unproven weight-loss "dietary supplements."

Unfortunately, most fad diets are more fiction than science. They sound plausible, though, because they are skillfully written. Their authors weave in scientific-sounding words like *eicosanoids* or *adipokines* and bits of authentic nutrition knowledge to set a tone of credibility and convince the skeptical. This makes it hard for people without adequate nutrition knowledge to evaluate them. Table 9–7 presents some clues to identifying scams among fad diets.

Table 9–7

Clues to Fad Diets and Weight-Loss Scams

It may be a fad diet or weight-loss scam if it meets any of the criteria of Figure C1-1 of Controversy 1, or if it:

- Bases evidence for its effectiveness on anecdotal stories and testimonials.
- Blames weight gain on a single nutrient, such as carbohydrate, or constituent, such as gluten.
- Claims to "alter your genetic code" or "reset your metabolism."
- Eliminates an entire food group, such as grains or milk and milk products.
- Fails to include all costs up front.
- Fails to mention potential risks associated with the plan.
- Fails to plan for weight maintenance following loss.
- Guarantees an unrealistic outcome in an unreasonable time period, such as losing 10 pounds in 3 days.
- Promises quick and easy weight loss methods, for example, "Lose weight while you sleep."
- Promotes devices, drugs, products, or procedures not approved by the FDA or scientifically evaluated for safety or effectiveness.
- Sounds too good to be true.
- Specifies a proportion of energy nutrients not in keeping with DRI recommended ranges.
- Recommends using a single food, such as grapefruit, as the key to the program's success.
- Requires you to buy special products not readily available in ordinary supermarkets.

Are Fad Diets All Nonsense?

If fad diets delivered what they promise, the nation's obesity problem would have vanished; if they never worked, people would stop buying into them. In fact, most fad diets do limit calorie intakes and produce weight loss (at least temporarily). Studies demonstrate, however, that fad diets are particularly ineffective for weight-loss maintenance—people may drop some weight, but they quickly gain it back.[1] Straightforward calorie deficit turns out to be the real key to weight loss, and not the elimination of protein, carbohydrate, or fat, or the still unidentified metabolic mechanisms proposed by many fad diets.[2]

For example, diet promoters make much of research showing that high-protein, low-carbohydrate diets produce a little more weight loss than balanced diets over the first few months of dieting. However, in the long run, any low-calorie diet produces about the same degree of loss.[3] In addition, low-carbohydrate, high-protein diets have been reported to increase markers of cardiovascular disease risk and may increase the risk of cardiovascular disease itself (more details are offered in a following section).[4] In the end, most people cannot sustain a fad diet over the long term, and they quickly return to their original or an even higher weight.[5]

Protein nutrition during calorie restriction deserves attention, however. A meal with too little protein may not produce enough satiety to prevent between-meal hunger. Be aware that diets providing more than 35 percent of total calories as protein (the upper end of the recommended DRI intake range) are no more effective for producing weight loss than more balanced calorie-controlled diets.[6] Eating the normal amount of lean protein-rich foods while reducing carbohydrate- and fat-containing foods automatically reduces calories and shifts the balance toward a higher percentage of energy from protein.[7]

Are Fad Diets Nutritious?

Fad diets that severely limit or eliminate one or more food groups cannot meet nutrient needs. To fend off critics, such plans usually recommend nutrient supplements (often conveniently supplied by the diet's originators at greatly inflated prices). Real weight-loss experts know this: no pill, not even the most costly ones, can match the health benefits of whole foods.

Are the Diets Safe?

Although most people can tolerate most diets, exceptions exist. For example, a rare but life-threatening form of blood acid imbalance is associated with a very-low-carbohydrate diet. In addition, such a diet may produce unfavorable changes in blood lipids and artery linings.[8] Recently, a documented case of heart disease and other health problems was reported in a previously healthy man who began following a strict low-carbohydrate diet; his problems resolved after he resumed a normal diet, suggesting that a low-carbohydrate diet may pose a danger to some people.*[9] No one knows the extent to which extreme fad diets might affect people with established diabetes or heart disease—the very people who might try dieting to regain their health.

Moving Ahead

Success for the fad-diet industry is built on failure for the dieter. As one diet shortcut fails, a new version arises to take its place, replenishing industry profits. Success for the dieter takes a longer road: setting realistic goals, eating a nutritious calorie-restricted diet with enough protein, carbohydrate, and fats needed for health. This approach also means adopting a physically active lifestyle that is flexible and comfortable over a lifetime.† Solid plans like this exist; seek them out for serious help with weight loss. Armed with common sense, Lauren and other hopeful dieters can find a sane, stable path and avoid costly detours that sap the will and delay true weight management progress.

Review Questions‡

1. A diet book that addresses eicosanoids and adipokines can be relied upon to reflect current scholarship in nutrition science and provide effective weight loss advice. **T F**
2. Calorie deficit is no longer the primary strategy for weight management. **T F**
3. Diets with sufficient protein may provide more satiety than other diets. **T F**

** The diet was the Atkins diet.*

† An example of a balanced weight loss plan is Weight Watchers®.

‡ Answers to Consumer's Guide review questions are found in Appendix G.

indicators fall into line, she may decide to maintain this weight. If her blood pressure is still high or she has other risks, she may repeat the process to achieve a healthier weight.

Once you have identified your overall target, set specific, achievable, small-step goals for food intake, activity, and behavior changes. One simple and effective first small-step goal might be to eliminate all sugar-sweetened beverages, including sweetened iced tea (fast-food sweet tea provides 280 calories per 32-ounce serving). On average, Americans drink about 350 calories a day in their beverages.

Liquid calories were the topic of **Chapter 8**'s Consumer's Guide section.

Dramatic weight loss overnight is not possible or even desirable; a pound or two of body fat lost each week will safely and effectively bring you to your goal. Losses greater or faster than these are not recommended because they are almost invariably followed by rapid regain. New goals can be built on prior achievements, and a lifetime goal may be to maintain the leaner, healthier body weight.

Keep Records Keeping records is critical to success. Recording your food intake and exercise can help you to spot trends and identify areas needing improvement. The Food Feature, later, demonstrates how to maintain a food and exercise diary. Recording changes in body weight can also provide a rough estimate of changes in body fatness. In addition to weight, measure your waist circumference to track changes in central adiposity.

KEY POINTS

- Setting realistic weight goals provides an important starting point for weight loss.
- Many benefits follow even modest reductions in body fatness among overweight people.
- Successful weight management takes time and effort; fad diets can be counterproductive.

Did You Know?

Children in particular suffer when they learn to dislike their healthy bodies because of unrealistic ideals.

What Food Strategies Are Best for Weight Loss?

Contrary to the claims of faddists, no particular food plan is magical, and no particular food must be either included or excluded. You are the one who will have to live with the plan, so you had better be the one to design it. Remember, you are adopting a healthy eating plan for life, so it must consist of satisfying foods that you like, that are readily available, and that you can afford.

Choose an Appropriate Calorie Intake Nutrition professionals often use an overweight person's BMI to calculate the number of calories to cut from the diet. Dieters with a BMI of 35 or greater are encouraged to reduce their daily calories by about 500 to 1,000 calories from their usual intakes. People with a BMI between 27 and 35 should reduce energy intake by 300 to 500 calories a day. With the assumption that a 3,500-calorie deficit will consistently produce a pound of weight loss, such recommendations predict weight loss of 1 to 2 pounds per week while retaining lean tissue.[117] However, this assumption has been called into question.

Recent research suggests that the number of calories required to lose a pound of weight may increase as weight loss progresses.[118] For some weeks or months, weight loss may proceed rapidly. Eventually, these factors may contribute to a slowdown in the rate of loss:

- Metabolism may slow in response to a lower calorie intake and loss of metabolically active lean tissue.
- Less energy may be expended in physical activity as body weight diminishes.

In addition, the composition of weight loss itself may affect the rate of loss. Body weight lost early in dieting may be composed of a greater percentage of water and lean tissue, which contains fewer calories per pound, as compared with later losses that appear to be composed mostly of fat, which contains far more calories per pound.[119] This may mean that dieters should expect a slowdown in weight loss as they progress past the initial phase.

Newer schemes for determining the calorie deficit required for weight loss reflect changes in energy use by the body over time. These require extensive calculations for accuracy and are most efficiently applied with the power of computer programs.*** A panel of experts recommended this shortcut, however: "every permanent 10-calorie change in energy intake per day will lead to an eventual weight change of 1 pound when the body weight reaches a new steady state. It will take nearly 1 year to achieve 50 percent and 3 years to achieve 95 percent of this weight loss."[120]

In the end, most dieters can lose weight safely on an eating pattern providing approximately 1,000 to 1,200 calories per day for women and 1,200 to 1,600 calories per day for men, while still meeting nutrient needs (as demonstrated in Table 9–8). Very low calorie diets are notoriously unsuccessful at achieving lasting weight loss, lack necessary nutrients, and may set in motion the unhealthy behaviors of eating disorders (see the Controversy) and so are not recommended.[121]

People with a healthy body weight often choose whole grains over refined carbohydrates.

Make Intakes Adequate Healthy eating patterns for weight loss should provide all of the needed nutrients in the form of fresh fruits and vegetables; low-fat milk products or substitutes; legumes, small amounts of lean meats, fish, poultry, and nuts; and whole grains.[122] These foods are necessary for adequate protein, carbohydrate, fiber, vitamins, and minerals and are generally associated with leanness.

Choose fats sensibly by avoiding most solid fats and including enough unsaturated oils (details in Chapter 5) to support health but not so much as to oversupply calories. Nuts provide unsaturated fat and protein, and people who regularly eat nuts often maintain a healthy body weight.[123] Lean meats or other low-fat protein sources also play important roles in weight loss: an ounce of lean ham contains about the same number of calories as an ounce of bread but the ham may produce greater satiety.

*** Two such programs are available at http://bwsimulator.niddk.nih.gov, or at http://www.pbrc.edu/the-research/tools/weight-loss-predictor.

Table 9–8

Eating Patterns for Low-Calorie Diets

These intakes allow most people to lose weight and still meet their nutrient needs with careful selections of nutrient-dense foods. See Chapter 2 for diet-planning details.

Food Group	1,000 Calories	1,200 Calories	1,400 Calories	1,600 Calories
Fruit	1 c	1 c	1½ c	1½ c
Vegetables	1 c	1½ c	1½ c	2 c
Grains	3 oz	4 oz	5 oz	5 oz
Protein Foods	2 oz	3 oz	4 oz	5 oz
Milk	2 c	2½ c	2½ c	3 c
Oils	3 tsp	4 tsp	4 tsp	5 tsp

Sufficient protein foods may also help to preserve lean tissue during weight loss, including muscle tissue.[124] Choose wisely, however—people with high fatty meat intakes are often overweight or obese.[125] Remember to limit alcohol, which lowers inhibitions and can sabotage even the most committed dieter's plans.

A supplement providing vitamins and minerals may be appropriate (Controversy 7 explained how to choose one). If you plan resolutely to include all of the foods from each food group that you need each day, you will be satisfied, well-nourished, and have little appetite left for high-calorie treats.

Avoid Portion Pitfalls Pay careful attention to portion sizes—large portions increase energy intakes, and the monstrous helpings served by restaurants and sold in packages are the enemy of the person striving to control weight.[126] Popular 100-calorie single-serving packages may be useful, but only if the food in the package fits into your calorie budget—100 calories of cookies or fried snacks are still 100 calories that can be safely eliminated. Also, eating a reduced-calorie cookie instead of an ordinary cookie saves calories—but eating half the bag defeats the purpose.

Almost every dieter needs to retrain, using measuring cups for a while to learn to judge portion sizes. Stay focused on calories and portions—don't be distracted by product's claims. Read labels and compare *calories* per serving.

Meal Spacing Three meals a day is standard in our society, but no law says you can't have four or five—just be sure they are smaller, of course. People who eat small, frequent meals can be as successful at weight loss and maintenance as those who eat three.[127] Also, eat regularly, before you become extremely hungry. When you do decide to eat, eat the entire meal you have planned for yourself. Then don't eat again until the next meal or snack.

Pay close attention to snacks. Snacking among Americans has doubled in the past 30 years, and snacks provide almost a third of the empty calories from solid fats and added sugars that most people take in each day.[128] Save calorie-free or favorite foods or beverages for a planned snack at the end of the day if you need insurance against late-evening hunger.

Some people skip breakfast to reduce energy intake, but this is a counterproductive strategy. Breakfast eaters have *lower* BMI values than breakfast skippers, and their overall diet quality is better, too.[129] Additionally, people who skip breakfast are likely to awaken at night to eat, a symptom of **night eating syndrome**.[130]

Identify Calorie Excesses If you doubt that small daily decisions can make a difference to your body weight, try this: turn back to page 51 in Chapter 2 and look at the foods depicted in Figure 2–11. Add up the calories in a 1970s hamburger, cola,

night eating syndrome a disturbance in the daily eating rhythm associated with obesity, characterized by no breakfast, more than half of the daily calories consumed after 7 p.m., frequent nighttime awakenings to eat, and a high calorie intake.

and French fries (similar to today's "small" sizes). Do the same for the calories in the "colossal" size hamburger, cola, and French fries typical of today's meals. Now find the calorie difference between the two meals (subtract the smaller sum from the larger) and multiply the difference by 52 (for weeks in a year):

$$
\begin{array}{r}
\text{Today's calorie total} = 2020 \\
-\ \text{1970s calorie total} = 925 \\
\hline
\text{Calorie difference} = 1095 \times 52 = 56{,}940
\end{array}
$$

In a person who gains a pound of body fat with each excess 3,500 calories (this number varies widely, as mentioned), choosing the largest-sized fast-food meal instead of the smallest *just once per week* will provide enough additional energy in a year's time (56,940 calories) to cause this person to gain over 16 pounds.

Using average U.S. data, each meal eaten away from home increases the daily calorie intake of adults by 134 calories, enough to cause a 2-pound weight gain each year or 20 pounds per decade. New federal labeling requirements will soon demand that chain restaurants list calories for each menu item to let consumers know how many calories they are buying.[131] Whether the labels will change consumer behaviors and calorie intakes is for future research to tell.[132]

Do the Math

The energy density of a food can be calculated mathematically. Find the energy density of carrot sticks and French fries by dividing their calories by their weight in grams.

- A serving of carrot sticks providing 31 cal and weighing 72 g:

  $$\frac{31\ \text{cal}}{72\ \text{g}} = 0.43\ \text{cal/g}$$

- Now do the same for French fries, contributing 167 cal and weighing 50 g:

  $$\frac{167\ \text{cal}}{50\ \text{g}} = 3.34\ \text{cal/g}$$

The higher calories per gram (cal/g), the greater the energy density.

Choose Foods Low in Energy Density People whose eating patterns consist mostly of foods that are high in **energy density** are more often overweight.[133] Turning this around, people who wish to be leaner and to improve their nutrient intakes would be well advised to select mostly foods of low energy density.[134] In general, foods high in fat or low in water, such as cookies or chips, rank high in energy density; foods high in water and fiber, such as fruits and vegetables, rank lower. Lower energy density foods often provide more food and greater satiety for the same number of calories. For example, a snack of grapes with their high water content is lower in energy density than the same weight or volume of their dehydrated counterparts (raisins). Figure 9–10 demonstrates this principle.

Importantly, the *energy* density of a food does not always reflect its *nutrient* density (nutrients per calorie). Beverages provide an example. The energy density of low-fat milk almost equals that of sugary soft drinks (they weigh about the same), but these beverages rank far apart in nutrient density and therefore in their contributions to nutrient needs.

Jim Gathany/Centers for Disease Control and Prevention

Each of these fruit servings provides 100 calories, but the grapes provide more volume.

Consider Nonnutritive Sweeteners Some people who maintain weight loss report using artificially sweetened beverages and fat-modified products liberally.[135] Replacing caloric beverages with water or diet drinks may reduce most people's overall calorie intakes.[136] An idea that nonnutritive sweeteners might trick the brain into craving more calories has been met with debate, and more research will be needed to settle it.[137] In any case, soft drinks of any kind can displace milk from the diet. Milk intake may or may not speed weight loss, but it contributes to the health of the bones.[138]

Demonstration Diet The meals shown in Figure 9–11 demonstrate how a day's meals look before and after trimming 1,100 calories. Full-calorie meals (left side of the figure) were modified in both portions and energy density to produce lower-calorie meals (on the right). Some 350 calories were trimmed by reducing added sugars: less syrup at breakfast and sugar-free gelatin instead of apple pie at lunch. (The diet planner kept the brownie at supper, however—pleasure matters, too.) Try your hand at reducing calories further—the challenge is to keep the diet adequate while doing so.

People who lack the time or ability to make their own low-calorie food selections or control their portion sizes may find it easier to use prepared meal plans. Even though

energy density a measure of the energy provided by a food relative to its weight (calories per gram).

Figure 9–10

Energy Density and Meal Size

Jim Gathany/CDC

Jim Gathany/Centers for Disease Control and Prevention

The larger meal on the right weighs more, provides more fiber, contains more water, and takes far longer to enjoy than the meal on the left. Even foods that are lower in energy density can be overconsumed, so total calories remain important, too.

1/2 c macaroni and cheese
1/2 c baked beans with pork and sauce
1/2 fried chicken breast

$\frac{\text{607 calories}}{\text{343 g total wt}}$ = 1.77 energy density

1 c broccoli
3 large tomato slices
1/2 large sweet potato
1/2 skinless roasted chicken breast

$\frac{\text{293 calories}}{\text{348 g total wt}}$ = 0.84 energy density

Source: Centers for Disease Control and Prevention, Eat More, Weigh Less? How to Manage Your Weight without Being Hungry.

Figure 9–11

Meal Makeover—Reducing the Calories in Meals

Day's meals = about 3,400 cal

Day's meals = about 2,300 cal

2% milk, 1 c, 121 cal
Orange juice, 1 c, 112 cal
Whole-grain waffles, 2 each, 402 cal
Soft margarine, 2 tsp, 68 cal
Syrup, 4 tbs, 210 cal
Banana slices, 1/2 c, 69 cal
Breakfast total: 982

© Michelle Bridwell/PhotoEdit

© Michelle Bridwell/PhotoEdit

Fat-free milk, 1 c, 83 cal
Orange juice, 1 c, 112 cal
Whole-grain waffle, 1 each, 201 cal
Soft margarine, 1 tsp, 34 cal
Syrup, 2 tbs, 105 cal
Banana slices, 1/2 c, 69 cal
Breakfast total: 604

2% milk, 1 c, 121 cal
Hamburger, quarter-pound, 430 cal
French fries, large (about 50), 540 cal
Ketchup, 2 tbs, 32 cal
Apple pie, 1 each, 225 cal
Lunch total: 1,348

© Polara Studios, Inc.

© Polara Studios, Inc.

Fat-free milk, 1 c, 83 cal
Cheeseburger, small, 330 cal
Green salad, 1 c, with light dressing, 1 tbs, 67 cal; croutons, 1/2 c, 50 cal
French fries, regular (about 30), 210 cal
Ketchup, 1 tbs, 16 cal
Gelatin dessert, sugar-free, 20 cal
Lunch total: 776

Italian bread, 2 slices, 162 cal
Soft margarine, 2 tsp, 68 cal
Stewed skinless chicken breast, 4 oz, 202 cal
Tomato sauce, 1/2 c, 40 cal
Brown rice, 1 c, 216 cal
Mixed vegetables, 1/2 c, 59 cal
Regular cheese sauce, 1/4 c, 121 cal
Brownie, 1 each, 224 cal
Supper total: 1,092
Day's total: 3,422

© Polara Studios, Inc.

© Polara Studios, Inc.

Italian bread, 1 slice, 81 cal
Soft margarine, 1 tsp, 34 cal
Stewed skinless chicken breast, 4 oz, 202 cal
Tomato sauce, 1/2 c, 40 cal
Brown rice, 1 c, 216 cal
Mixed vegetables, 1/2 c, 59 cal
Low-fat cheese sauce, 1/4 c, 85 cal
Brownie, 1 each, 224 cal
Supper total: 941
Day's total: 2,321

Day's calorie reduction = 1,100 calories

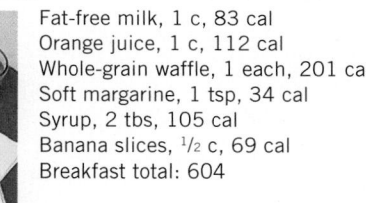

Art © Cengage Learning

they are much more costly than conventional foods, prepared food plans that provide low-calorie, nutritious meals or snacks can support weight loss and ease diet planning.[139] Ideally, the plan should teach users to choose wisely from conventional foods, too, to prevent weight regain from old habits when the plan is ultimately abandoned.

KEY POINTS

- To achieve and maintain a healthy body weight, set realistic goals, keep records, eat regularly (especially at breakfast), and expect to progress slowly.
- Watch energy density, make the diet adequate and balanced, eliminate excess calories, and limit alcohol intakes.

Physical Activity in Weight Loss and Maintenance

The most successful weight losers and maintainers include physical activity in their plans.[140] However, weight loss through physical activity alone is generally not easily achieved. Physical activity guidelines were offered in the Think Fitness feature, earlier.

Advantages of Physical Activity, and a Warning In laboratory studies, among obesity-prone rats made to lose weight and later given unrestricted access to food, the rats made to run regularly on a treadmill stayed leanest.[141] The running rats spontaneously reduced their daily food intakes while the sedentary control rats did not. Many people fear that exercising will increase their hunger. Active people do have healthy appetites, but a workout helps to heighten feelings of satiation during meals, as well.[142]

Muscle-strengthening exercise, performed regularly, adds healthful lean body tissue and provides a trim, attractive appearance. In addition, over the long term, lean muscle tissue burns more calories pound for pound than fat does. Although physical activity slightly raises the BMR in the hours following exercise, this effect requires a sustained high-intensity workout beyond the level achievable by most weight-loss seekers.[143]

Physically active dieters may also avoid some of the bone mineral loss associated with weight-loss dieting, and those who also attend to protein and calcium needs offer their bones even more protection.[144] In addition, plenty of physical activity promotes restful sleep—and getting enough sleep may reduce food consumption and weight gain. Finally, physical activity of all kinds helps to reduce stress, and stress can lead to increased eating.[145]

Here's the warning: nonathletes spend little energy during physical activity. Exercisers who reward themselves with high-calorie treats for "good behavior" can easily negate any calorie deficits incurred.

Which Activities Are Best? A combination of moderate-to-vigorous aerobic exercise along with strength training at a safe level seems best for health. However, some physical activity is better than none.[146] Most important: perform at a comfortable pace within your current abilities. Rushing to improve is practically a guarantee for injury.

Tim Robberts/The Image Bank/Getty Images

Playing an active video sports game burns some calories, but not as many as playing the actual sport.

Active video games and active video fitness programs may help meet their physical activity needs for those who like them, but most people lose interest in just a short while.[147]††† Real sports not only require more energy than their video counterparts but also hold people's interest, year after year.[148]

Fitness also benefits from hundreds of activities required for daily living: washing the car, raking leaves, taking the stairs, and many, many others. However you do it, be active. Walk. Swim. Skate. Dance. Cycle. Skip. Lift weights. Above all, enjoy moving—and move often.

KEY POINT

- Physical activity greatly augments weight-loss efforts.

††† An example is Wii, by Nintendo.

What Strategies Are Best for Weight Gain?

Should a thin person try to gain weight? Not necessarily. If you are healthy, fit, and energetic at your present weight, stay there. However, if your physician has advised you to gain; if you are excessively tired, unable to keep warm, or your BMI is in the "underweight" category of the BMI table (see inside back cover); or if, for women, you have missed at least three consecutive menstrual periods, you may be in danger from being too thin. It can be as hard for a thin person to gain a pound as it is for an overweight person to lose one.

Choose Foods with High Energy Density The weight gainer needs nutritious energy-dense foods. No matter how many sticks of celery you consume, you won't gain weight because celery simply doesn't offer enough calories per bite. Energy-dense foods (the very ones the weight-loss dieter is trying to avoid) are often high in fat, but fat energy is spent in building new tissue; if the fat is mostly unsaturated, such foods will not contribute to heart disease risk. Be sure your choices are nutritious—not just, say, candy bars and potato chips.

Choose an ounce of peanut butter instead of an ounce of lean meat on a sandwich, avocado instead of cucumber on a salad, olives instead of pickles, whole-wheat muffins instead of whole-wheat bread, and flavored milk and milkshakes instead of milk. Make milkshakes of milk, a frozen banana, a tablespoon of vegetable oil, and flavorings for between-meal treats. When you do eat celery, stuff it with tuna salad (use oil-packed tuna); choose flavored coffee drinks over plain coffee, fruit juice over water; use olive oil or mayonnaise-based dressings on salads, whipped toppings on fruit, and soft or liquid margarine on potatoes. Because fat contains more than twice as many calories per teaspoon as sugar, its calories add up quickly without adding much bulk, and its energy is in a form that is easy for the body to store. For those without the skill or ability to create their own high-calorie foods, adding a high-protein, high-calorie liquid or bar-type dietary supplement to regular nutritious meals can sometimes help an underweight person gain or maintain weight.

Portion Sizes and Meal Spacing Increasing portion sizes increases calorie intakes. Choose extra slices of meats and cheeses on sandwiches; use larger plates, bowls, and glasses to disguise the appearance of the larger portions. Expect to feel full, even uncomfortably so. This feeling is normal, and it passes as the stomach gradually adapts to the extra food.

Eat frequently and keep easy-to-eat foods on hand for quick meals. Make three sandwiches in the morning and eat them between classes in addition to the day's three regular meals. Include favorite foods or ethnic dishes often—the more varied and palatable, the better. Drink beverages between meals, not with them, to save space for higher-calorie foods. Always finish with dessert. Other tips for weight gain are listed in Table 9–9.

Physical Activity to Gain Muscle and Fat Food choices alone can cause weight gain, but the gain will be mostly fat. Overly thin people need both muscle and fat, so physical activity is essential in a sound weight-gain plan. Resistance activities are best for building muscles that can help to increase healthy body mass. Start slowly and progress gradually to avoid injury.

Physical activity demands extra calories from food. If you eat just enough to fuel the activity, you can build muscle initially but at the expense of body fat. Hence, gains of both muscle and fat are possible only when the body receives the extra fuel it needs for both. Conventional advice on diet to the person building muscle is to eat about 500 to 700 calories a day above normal energy needs; this range often supports both the added activity and the formation of new muscle. Many more facts about building muscles are provided in Chapter 10.

KEY POINTS

- Weight gain requires an eating pattern of calorie-dense foods, eaten frequently throughout the day.
- Physical activity helps to build lean tissue.

Did You Know?

Grain-based desserts, such as snack cakes and cookies, are the number one contributor of calories to the U.S. diet for all people ages 2 and older.

Table 9–9

Tips for Gaining Weight

In General:

- Eat enough to store more energy than you expend—at least 500 extra calories a day.
- Exercise to build muscle.
- Be patient. Weight gain takes time (1 pound per month would be reasonable).
- Choose energy-dense foods most often.
- Eat at least three meals a day, and add snacks between meals.
- Choose large portions and expect to feel full.
- Drink caloric fluids—juice, chocolate milk, sweet coffee drinks, sweet iced tea.

In Addition:

- Cook and bake often—delicious cooking aromas whet the appetite.
- Invite others to the table—companionship often boosts eating.
- Make meals interesting—try new vegetables and fruit, add crunchy nuts or creamy avocado, and explore the flavors of herbs and spices.
- Keep a supply of favorite snacks, such as trail mix or granola bars, handy for grabbing.
- Control stress and relax. Enjoy your food.

Medical Treatment of Obesity

To someone fatigued from years of battling overweight, the idea of curing obesity by taking pills or undergoing surgery may be attractive. These approaches can cause dramatic weight loss and often save the lives of obese people at critical risk, but they also present serious risks of their own.[149]

Obesity Medications Each year, a million and a half U.S. citizens take prescription weight-loss medications, and many millions more take over-the-counter (OTC) preparations, including many "dietary supplements." Of these people, a fourth are not overweight, but they take the products believing them to be safe.

OTC weight-loss pills, powders, herbs, and other "dietary supplements" are not associated with successful weight loss and maintenance.[150] Further, an FDA investigation revealed an alarmingly widespread adulteration of such products.[151] Strong prescription diuretics, unproven experimental drugs, psychotropic drugs used to treat mental illnesses, and even banned drugs were discovered in OTC weight-loss preparations, posing serious risks to health.

In contrast, for overweight people with a BMI of 30 or above or with elevated disease risks, weight loss achieved with FDA-approved prescription medications (Table 9–10) may be worth the risks.[152] Importantly, weight-loss drugs can help only temporarily while they are being taken; lifestyle changes are still necessary to help manage weight over a lifetime.

Obesity Surgery A person with extreme obesity, that is, someone whose BMI is 40 or above (35 with coexisting disease) may urgently need to reduce body fatness, and surgery is often an option for those healthy enough to withstand it.[153] For obese people without elevated disease risks, however, the expected improvements from surgery must be weighed against its hazards.[154]

Surgical procedures limit a person's food intake by reducing the size of the stomach and delaying the passage of food into the intestine (see Figure 9–12). The results

Table 9–10

FDA Approved Drugs for Weight Loss

Product	Action	Side Effects
Belviq (pronounced BELL-veek) lorcaserin hydrochloride	Stimulates brain serotonin receptors to increase satiety	Headache, dizziness, fatigue, nausea, dry mouth, and constipation; low blood glucose in people with diabetes; serotonin syndrome, including agitation, confusion, fever, loss of coordination, rapid or irregular heart rate, shivering, seizures, and unconsciousness; not for use by pregnant or lactating women, or people with heart valve problems. High doses cause hallucinations.
Orlistat (OR-leh-stat) Trade names: Alli, Xenical	Inhibits pancreatic lipase activity in the GI tract, thus blocking digestion and absorption of dietary fat and limiting energy intake	Cramping, diarrhea, gas, frequent bowel movements, reduced absorption of fat-soluble vitamins; rare cases of liver injury
Phentermine (FEN-ter-mean), diethylpropion (DYE-eth-ill-PRO-pee-on), phendimetrazine (FEN-dye-MEH-tra-zeen)	Enhances the release of the neurotransmitter norepinephrine, which suppresses appetite	Increased blood pressure and heart rate, insomnia, nervousness, dizziness, headache
Qsymia (kyoo-sim-EE-uh)	Combines phentermine (an appetite suppressant) and topirimate (a seizure/ migrane medication) that makes food seem less appealing and increases feelings of fullness	Increased heart rate; can cause birth defects if taken in the first weeks or months of pregnancy; may worsen glaucoma or hyperthyroidism; may interact with other medications

Note: Weight-loss drugs are most effective when taken as directed and used in combination with a reduced-kcalorie diet and increased physical activity.

Figure 9–12

Surgical Obesity Treatments

Both of these surgical procedures limit the amount of food that can be comfortably eaten.

In gastric bypass, the surgeon constructs a small stomach pouch and creates an outlet directly to the lower small intestine. (Dark areas highlight the redirected flow of food.)

In gastric banding, the surgeon uses a gastric band to reduce the opening from the esophagus to the stomach. The size of the opening can be adjusted by inflating or deflating the band.

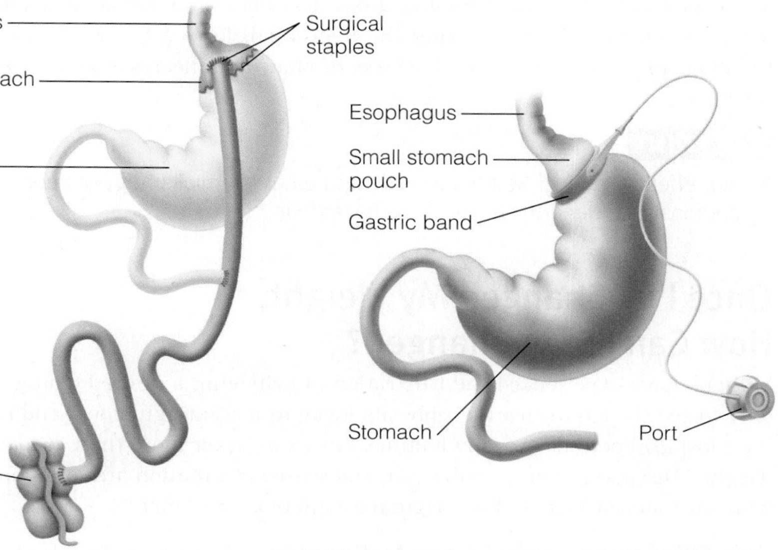

can be dramatic: greater than 90 percent of surgical patients achieve a weight loss of more than 50 percent of their excess body weight, and much of the loss may be maintained over time.[155] More long-term studies are needed, but surgery with weight loss often brings immediate and lasting improvements in such threats as diabetes, insulin resistance, high blood cholesterol, hypertension, heart disease, and sleep apnea.[156]

Surgery is not a sure cure for obesity, despite advertisements claiming so.[157] A few people do not lose the expected pounds, and some who lose weight initially regain much of it in a few years' time. Complications are common, and include infections; nausea, vomiting, and dehydration; hemorrhage; internal hernia; ulcers; and bone abnormalities.[158] In addition, many patients require repeated surgeries. Severe nutrient deficiencies—iron, copper, zinc, vitamin B_{12} and other B vitamins, vitamin A and beta-carotene, and vitamin C—commonly occur and may persist even in patients taking supplements.[159] Calcium metabolism may be disturbed causing bone loss and bone abnormalities.

Life-long nutrition and medical supervision following surgery is a must. The effectiveness of gastric surgery depends, in large part, on compliance with dietary instructions, such as choosing small portions, chewing food completely before swallowing, and drinking beverages separately from meals. For those endangered by their obesity and who cannot achieve meaningful weight loss by other means, however, surgery may prove worth the risks.[160]

KEY POINT

- For people whose obesity threatens their health, medical science offers drugs and surgery.

Herbal Products and Gimmicks

Some herbs or **botanical** products are wildly popular and may be useful for some purposes, but wise consumers avoid products containing substances not proved safe in

botanical pertaining to or made from plants; any drug, medicinal preparation, dietary supplement, or similar substance obtained from a plant.

laboratory studies. The risks are too high. Recently, for example, a previously healthy 28-year-old body builder was hospitalized in a coma after taking a "dietary supplement" containing the known liver toxin usnic acid, sold to her as a "fat burner."[161] Another harmful supplement, ephedra (also called ma huang and banned by the FDA), is sold as a weight-loss "dietary supplement" but has caused cardiac arrest, abnormal heartbeats, hypertension, strokes, seizures, and death. These and many other harmful weight-loss "supplements" remain available on Internet websites.

Also, steam baths and saunas do not melt the fat off the body as claimed, although they may dehydrate you so that you lose water weight. Brushes, sponges, wraps, creams, and massages intended to move, burn, or break up "**cellulite**" are useless for fat loss. Cellulite—the rumpled, dimpled, stubborn fat tissue on the thighs and buttocks—is simply fat, awaiting the body's call for energy. Such nonsense distracts people from the serious business of planning effective weight-management strategies.

KEY POINT

- The effectiveness of herbal products and other gimmicks has not been demonstrated, and they may prove hazardous.

Once I've Changed My Weight, How Can I Stay Changed?

Millions have experienced the frustration of achieving a desired change in weight only to see their hard work visibly slip away in a seemingly never-ending cycle: "I have lost 200 pounds over my lifetime, but I was never more than 20 pounds overweight." Disappointment, frustration, and self-condemnation are common in dieters who have slipped back to their original weight or even higher.[162]

Self-Efficacy and Other Keys to Success Contrary to popular belief, many people who set out to lose weight do so, and many maintain their losses for years. No one can yet say which of their "secrets of success" may be responsible, but the habits of those individuals are of interest to researchers and dieters alike, and they are offered in Table 9–11. In general, such people believe in their ability to control their weight, an attribute known as **self-efficacy**.[163] They also monitor their intakes and body weights, quickly addressing small **lapses** to prevent major ones. Some drink water before meals to ease hunger and fill the stomach.[164] They all use techniques that work for them; people's responses to any one method are highly variable.[165]

Don't forget to drink enough water—it can produce feelings of fullness, and it's calorie-free.

Without a doubt, a key to weight maintenance is accepting the task as a life-long endeavor, not a goal to be achieved and forgotten. Most people who maintain weight loss continue to employ many of the routines that reduced their weight in the first place.[166] They cultivate healthy habits, they remind themselves of the continuing need to manage their weight, they monitor their weight and routine, they renew their commitment to regular physical activity, and they reward themselves for sticking with the plan.[167]

Without a life-long plan, those who try to lose weight may become trapped in endless repeating rounds of weight loss and regain—"yo-yo" dieting. Some evidence suggests that a history of such **weight cycling** may predict a person's susceptibility to weight regain in the future.[168] The Food Feature, next, explores how a person who is ready to change can modify daily eating and exercise behaviors into healthy, life-long habits.

Seek Support Group support can prove helpful when making life changes. Some people find it useful to join a group such as Take Off Pounds Sensibly (TOPS), Weight Watchers (WW), Overeaters Anonymous (OA), or others. Others prefer to form their own self-help groups or find support online.[169] The Internet offers numerous opportunities for weight-loss education, counseling, and virtual group support that may be effective alternatives to face-to-face or telephone counseling programs.[170] Program applications for cellular telephones ("smartphones") and other mobile devices can

Table 9–11

Summary of Lifestyle Strategies for Successful Weight Loss

In addition to calorie control and exercise, people who lost weight and kept it off report using strategies in the four categories listed below. No one strategy is universally useful—responses vary widely, and individualized weight-loss plans work best.

1. General
▪ Make a long-term commitment (greater than 6 months duration). ▪ Target all 3 weight management components (not just eating or exercising alone, for example). ▪ Monitor food intake and body weight (particularly to maintain weight loss). ▪ Follow a commercial weight-loss program (particularly for initial weight loss). ▪ Target weight management specifically rather than other worthy goals, such as disease prevention.
2. Eating Habits
▪ Consume a calorie-reduced diet with adequate protein, controlled in fat and carbohydrate. ▪ Focus on total energy of the diet rather than on reducing specific energy-nutrient components. ▪ Consume low-fat protein sources (particularly for weight maintenance). ▪ Restrict intakes of types of foods (such as high-sugar foods/beverages or restaurant foods). ▪ Eat breakfast every day, and keep dietary routines.
3. Physical Activities
▪ Perform 150–250 min/week of moderate physical activity to prevent weight gain. ▪ Perform more than 250 min/week of moderate physical activity to promote significant weight loss. ▪ Exercise more, on average, an hour a day. ▪ Watch less than 10 hours of television per week.
4. Nutrition Counseling/Behavior Modification
▪ Use behavioral treatment to produce an initial weight loss of 5 to 10% of body weight. ▪ Use cognitive behavior therapy to augment diet and physical activities. ▪ Obtain structured, individualized nutrition counseling to support weight loss efforts. ▪ Use Internet-based education and tracking applications, particularly in the short term. ▪ Weigh on a scale at least once a week. ▪ Recognize and attend to minor lapses.

Sources: S. F. Kirk and coauthors, Effective weight management practice: a review of the lifestyle intervention evidence, International Journal of Obesity *36 (2012): 178–185; J. M. Nicklas and coauthors, Successful weight loss among obese U.S. adults,* American Journal of Preventive Medicine *42 (2012): 481–485; C. N. Sciamanna and coauthors, Practices associated with weight loss versus weight-loss maintenance,* American Journal of Preventive Medicine *41 (2011): 159–166; National Weight Control Registry, available at http://www.nwcr.ws/Research/default.htm; J. P. Moreno and C. A. Johnston, Successful habits of weight losers,* American Journal of Lifestyle Medicine *6 (2012): 113–115.*

effectively track diets and provide calorie information wherever and whenever it's needed.[171] As always, choose wisely and avoid scams.[‡‡‡]

KEY POINT

- People who succeed at maintaining lost weight keep to their eating routines, keep exercising, and keep track of calorie intakes and body weight.

Conclusion

This chapter winds up where it began, considering the U.S. obesity epidemic as a societal problem. Reversing it may depend at least partly on the public will to support healthy lifestyle choices.[172] Meanwhile, individuals can make choices to influence their own behaviors, as the Food Feature points out.

cellulite a term popularly used to describe dimpled fat tissue on the thighs and buttocks; not recognized in science.

self-efficacy a person's belief in his or her ability to succeed in an undertaking.

lapses periods of returning to old habits.

weight cycling repeated rounds of weight loss and subsequent regain, with reduced ability to lose weight with each attempt; also called *yo-yo dieting*.

‡‡‡ A safe, free, user-friendly, and proven program is USDA's Super Tracker, available with smartphone applications at www.choosemyplate.gov.

try it! Food Feature

Behavior Modification for Weight Control

LO 9.9 Defend the importance of behavior modification in weight loss and weight maintenance over the long term.

Supporting changes in both diet and exercise is **behavior modification**. This form of therapy can help the dieter to cement into place all the behaviors that lead to and perpetuate the desired body composition.

How Does Behavior Modification Work?

Behavior modification involves changing both behaviors and thought processes. It is based on the knowledge that habits drive behaviors. Suppose a friend tells you about a shortcut to class. To take it, you must make a left-hand turn at a corner where you now turn right. You decide to try the shortcut the next day, but when you arrive at the familiar corner, you turn right as always. Not until you arrive at class do you realize that you failed to turn left, as you had planned. You can learn to turn left, of course, but at first you will have to make an effort to remember to do so. After a while, the new behavior will become as automatic as the old one was.

A food and activity diary is a powerful ally to help you learn what particular eating stimuli, or cues, affect you. Such self-monitoring is indispensable for learning to control eating and exercising cues, both positive and negative, and for tracking your progress.[173] Figure 9–13 provides a sample of an informal food and activity diary for self-monitoring.

Once you identify the behaviors you need to change, do not attempt to modify all of them at once. No one who attempts too many changes at one time is successful. Set your priorities, and begin with a few behaviors you can handle—practice until they become habitual and automatic, and then select one or two more. For those striving to lose weight, learning to say "No, thank you" might be among the first habits to establish. Learning not to "clean your plate" might follow.

Figure 9–13

A Sample Food and Activity Diary

Record the times and places of meals and snacks, the types and amounts of foods consumed, surroundings and people present, and mood while eating. Describe physical activities, their intensity and duration, and your feelings about them, too. Use this information to structure eating and exercise in ways that serve your physical and emotional needs.

Time	Place	Activity or food eaten	People present	Mood
10:30–10:40	School vending machine	6 peanut butter crackers and 12 oz. cola	by myself	starved
12:15–12:30	Restaurant	Sub sandwich and 12 oz. cola	friends	relaxed & friendly
3:00–3:45	Gym	Weight training	work out partner	tired
4:00–4:10	Snack bar	Small frozen yogurt	by myself	OK

Modifying Behaviors

Behavior researchers have identified six elements useful in replacing old eating and activity habits with new ones:

1. Eliminate inappropriate eating and activity cues.
2. Suppress the cues you cannot eliminate.
3. Strengthen cues to appropriate eating and activities.
4. Repeat the desired eating and physical activity.
5. Arrange or emphasize negative consequences of inappropriate eating or sedentary behaviors.
6. Arrange or emphasize positive consequences of appropriate eating and exercise behaviors.

Table 9–12 provides specific examples of putting these six elements into action.

To begin, set about eliminating or suppressing the cues that prompt you to eat inappropriately. An overeater's life may include many such cues: watching television, talking on the telephone, entering a convenience store, studying late at night. Resolve that you will no longer respond to such cues by eating. If some cues to inappropriate eating behavior cannot be eliminated, suppress them; then strengthen the appropriate cues, and reward yourself for doing so. Some possible activities and rewards to substitute for eating include:

Table 9–12

Behavior Modification Tips for Weight Loss

Use these actions during both weight-loss and maintenance phases of weight management.

1. Eliminate inappropriate eating cues:
   - Don't buy problem foods.
   - Eat only in one room at the designated time.
   - Shop when not hungry.
   - Replace large plates, cups, and utensils with smaller ones.
   - Avoid vending machines, fast-food restaurants, and convenience stores.
   - Turn off the television, video games, and computer or measure out appropriate food portions to eat during entertainment.
2. Suppress the cues you cannot eliminate:
   - Serve individual plates; don't serve "family style."
   - Measure your portions; avoid large servings or packages of food.
   - Remove food from the table after eating a meal to enjoy company and ambience without excess food to trigger overeating.
   - Create obstacles to consuming problem foods—wrap them and freeze them, making them less quickly accessible.
   - Control deprivation; plan and eat regular meals.
   - Plan to spend only one hour per day in sedentary activities, such as watching television or using a computer.
3. Strengthen cues to appropriate eating and exercise:
   - Choose to dine with companions who make appropriate food choices.
   - Store appropriate foods in convenient spots in the refrigerator.
   - Learn appropriate portion sizes.
   - Plan appropriate snacks.
   - Keep sports and play equipment by the door.
4. Repeat the desired eating and exercise behaviors:
   - Slow down eating—put down utensils between bites.
   - Always use utensils.
   - Leave some food on your plate.
   - Move more—shake a leg, pace, stretch often.
   - Join groups of active people and participate.
5. Arrange or emphasize negative consequences for inappropriate eating:
   - Ask that others respond neutrally to your deviations (make no comments—even negative attention is a reward).
   - If you slip, don't punish yourself.
6. Arrange or emphasize positive consequences for appropriate eating and exercise behaviors:
   - Buy tickets to sports events, movies, concerts, or other nonfood amusement.
   - Indulge in a new small purchase.
   - Get a massage; buy some flowers.
   - Take a hot bath; read a good book.
   - Treat yourself to a lesson in a new active pursuit such as horseback riding, handball, or tennis.
   - Praise yourself; visit friends.
   - Nap; relax.

- Attending or participating in sporting events.
- Enjoying leisure activities, card games, or a favorite television show.
- Exercising your muscles at a gym.
- Gardening or indulging in other crafts or hobbies.
- Going to a movie or play.
- Listening to music.
- Napping, reading, relaxing.
- Praising yourself.
- Sprucing up your room or house.
- Taking a bubble bath.
- Telephoning, texting, or e-mailing.
- Vacationing, even for an hour or two in a neighboring town or area.
- Volunteering.
- Window shopping, mall walking.

The list of possibilities is virtually endless.

In addition, be aware that the food marketing industry spends huge sums each year developing cues to modify consumers' behaviors in the opposite direction—toward buying and consuming more snack foods, soft drinks, and other products. These cues work on a subconscious level; they leverage the stronger human hunger and appetite mechanisms to overcome the weaker satiety signals.

Cognitive Skills

Behavior therapists often teach **cognitive skills**, or new ways of thinking, to help dieters solve problems and correct false thinking that can short-circuit healthy eating behaviors. Thinking habits turn out to be as important as eating habits to achieving a healthy body weight, and

behavior modification alteration of behavior using methods based on the theory that actions can be controlled by manipulating the environmental factors that cue, or trigger, the actions.

cognitive skills as taught in behavior therapy, changes to conscious thoughts with the goal of improving adherence to lifestyle modifications; examples are problem-solving skills or the correction of false negative thoughts, termed *cognitive restructuring.*

thinking habits can be changed.[174]* A paradox of making a change is that it takes belief in oneself and honoring of oneself to lay the foundation for changing that self. That is, self-acceptance predicts success, while self-loathing predicts failure. "Positive self-talk" is a concept worth cultivating—many people succeed because their mental dialogue supports, rather than degrades, their efforts. Negative thoughts ("I'm not getting thin anyway, so what's the use of continuing?") should be viewed in the light of empirical evidence ("my starting weight: 174 pounds; today's weight: 163 pounds").

Give yourself credit for your new behaviors; take honest stock of any physical improvements, too, such as lower blood pressure or less painful knees, even without a noticeable change in pant size. Finally, remember to enjoy your emerging fit and healthy self.

* *Psychologists have a term for changing thinking habits:* cognitive restructuring.

track it! Diet Analysis **PLUS** Concepts in Action

Analyze Your Energy Balance

The purpose of this exercise is to help you to use critical thinking to evaluate correlations between nutrition, physical activity, and body weight.

1. Your calorie intake represents the "energy in" part of your energy balance. From the Reports Tab, select Energy Balance Analysis, choose Day One of your three-day diet intake, include all meals, and generate a report to determine your calorie (kCal) intake for the day. How does it compare to the DRI energy recommendation for the reference man or woman of your age, as listed on the inside front cover of the text?
2. Energy balance is affected not just by food eaten but also by energy expended. Compare the effects of two levels of activity on your energy balance. You've already generated an Energy Balance report for Day One. Select the Track Activity tab, and add a new 30-minute activity for Day One.

   Look at the list of activities in Table 9–4 (page 351) for suggestions. Generate a new report. Compare the Energy Balance report with and without the added activity. What changes do you see?
3. If you want to lose body fat, you must expend more energy than you take in. Look over your food diaries. Is there a day that you were in positive energy balance (took in more energy than you used)? If so, develop a revised food record for that day with the goal of reducing calories but still maintaining a wholesome and satisfying diet. Now, select the Track Diet tab to evaluate the revised meal plan. Did you succeed in trimming calories but still consume the recommended nutrients? How many calories did you trim?
4. All three energy-yielding nutrients can contribute excess calories but fat is the least satiating, most highly caloric per gram, and most easily consumed without awareness. From the Reports tab, select Source Analysis and then Total Fat from the drop-down box. Evaluate your daily food records for total fat. Then, choose the day that contained the most fat in grams or the highest percentage of calories from fat. Find the foods that contributed the most fat to your intake. What would you say led to higher intake of fat on that day as compared to another? Which part of the day did you consume the most fat? Were you aware that you were doing so?

what did you decide?

© andesign101/Shutterstock.com

How can you control your body weight, once and for all?

Why are you tempted by a favorite treat when you don't feel hungry?

How do extra calories from food become fat in your body?

Which popular diets are best for managing body weight?

Self Check

1. (LO 9.1) All of the following are health risks associated with excessive body fat except __________ .
   a. respiratory problems
   b. sleep apnea
   c. gallbladder disease
   d. low blood lipids
2. (LO 9.1) Today, an estimated 68 percent of the adults in the United States are overweight or obese.
   T F
3. (LO 9.2) Which of the following statements about basal metabolic rate (BMR) is correct?
   a. The greater a person's age, the higher the BMR.
   b. The more thyroxine produced, the higher the BMR.
   c. Fever lowers the BMR.
   d. Pregnancy lowers the BMR.
4. (LO 9.2) The BMI standard is an excellent tool for evaluating obesity in athletes and the elderly.
   T F
5. (LO 9.2) The thermic effect of food plays a major role in energy expenditure.
   T F
6. (LO 9.3) Body fat can be assessed by which of the following techniques?
   a. a blood lipid test
   b. chest circumference
   c. dual energy X-ray absorptiometry
   d. all of the above
7. (LO 9.3) BMI is of limited value for
   a. athletes
   b. pregnant and lactating women
   c. adults older than age 65
   d. all of the above
8. (LO 9.4) The appetite-stimulating hormone ghrelin is made by the __________ .
   a. brain
   b. fat tissue
   c. pancreas
   d. stomach
9. (LO 9.4) When the brain receives signals that enough food has been eaten, this is called
   a. satiation
   b. ghrelin
   c. adaptation
   d. none of the above
10. (LO 9.5) Brown adipose tissue
    a. develops during starvation
    b. is a well-known heat-generating tissue
    c. develops as fat cells die off
    d. all of the above
11. (LO 9.5) According to genomic researchers, a single inherited gene is the probable cause of common obesity.
    T F
12. (LO 9.6) In many people, any kind of stress can cause overeating and weight gain.
    T F
13. (LO 9.6) A built environment can support physical activity with
    a. safe biking and walking lanes
    b. public parks
    c. free exercise facilities
    d. all of the above
14. (LO 9.7) Which of the following is a physical consequence of fasting?
    a. loss of lean body tissues
    b. lasting weight loss
    c. body cleansing
    d. all of the above
15. (LO 9.7) The nervous system cannot use fat as fuel.
    T F
16. (LO 9.7) A diet too low in carbohydrate brings about responses that are similar to fasting.
    T F
17. (LO 9.8) Efforts in all of the following realms are necessary for weight change:
    a. eating patterns, physical activity, and behavior modification
    b. eating patterns, physical activity, and carbohydrate control
    c. carbohydrate control, physical activity, and behavior modification
    d. eating patterns, physical activity, and psychotherapy
18. (LO 9.8) The number of calories to cut from the diet to produce weight loss should be based on
    a. the amount of weight the person wishes to lose
    b. the person's BMI
    c. the amount of food the person wishes to consume
    d. the RDA for energy for the person's gender and age
19. (LO 9.9) Most people who successfully maintain weight loss do all of the following except
    a. continue to employ many of the routines that reduced their weight in the first place.
    b. obtain at least some guidance from popular diet books.
    c. reward themselves for sticking with their plan.
    d. monitor their weight and routine.
20. (LO 9.10) Adolescents are likely to grow out of early disordered eating behaviors by young adulthood.
    T F

Answers to these Self Check questions are in Appendix G.

CONTROVERSY 9

The Perils of Eating Disorders

LO 9.10 Compare and contrast the characteristics of anorexia nervosa and bulimia nervosa, and outline strategies for combating eating disorders.

Almost 5 million people in the United States, many of them girls and women, suffer from the **eating disorders** of **anorexia nervosa** and **bulimia nervosa**. More than 8 million more, almost 3 million of them men, suffer from **binge eating disorder** or related conditions that imperil physical and mental health. The incidence and prevalence of eating disorders in young people has increased steadily since the 1950s.[1]* Most alarming is the rising prevalence at progressively younger ages. (Table C9–1 defines eating disorder terms.)

An estimated 85 percent of eating disorders start during adolescence. Children of this age often exhibit warnings of disordered eating such as restrained eating, binge eating, purging, fear of fatness, and distorted body image. Many adolescents diet to lose weight and choose unhealthy behaviors associated with disordered eating; by college age the behaviors can be entrenched.[2] Disordered eating behaviors in early life set a pattern that likely continues into young adulthood.[3] Importantly, healthful dieting and physical activity in overweight adolescents do not appear to trigger eating disorders.

Society's Influence

Why do so many people in our society suffer from eating disorders? Most experts agree that eating disorders have many causes: sociocultural, psychological, hereditary, and possibly also genetic and neurochemical.[4] However, excessive pressure to be thin in our society is at least partly to blame. Normal-weight girls as young as 5 years old are placed "on diets" for fear that they are too fat.

When thinness takes on heightened importance, people begin to view the normal, healthy body as too fat—their body images become distorted. People of all shapes, sizes, and ages—including emaciated fashion models with anorexia nervosa—have learned to be unhappy with their "overweight" bodies.[5] Many take serious risks to lose weight. These behaviors and attitudes are almost nonexistent in cultures where body leanness is not central to self-worth.

Barcroft Media/Landov

Table C9–1

Eating Disorder Terms

- **anorexia nervosa** an eating disorder characterized by a refusal to maintain a minimally normal body weight, self-starvation to the extreme, and a disturbed perception of body weight and shape; seen (usually) in teenage girls and young women (anorexia means "without appetite"; nervos means "of nervous origin").
- **binge eating disorder** an eating disorder whose criteria are similar to those of bulimia nervosa, excluding purging or other compensatory behaviors.
- **bulimia** (byoo-LEEM-ee-uh) **nervosa** recurring episodes of binge eating combined with a morbid fear of becoming fat; usually followed by self-induced vomiting or purging.
- **cathartic** a strong laxative.
- **cognitive therapy** psychological therapy aimed at changing undesirable behaviors by changing underlying thought processes contributing to these behaviors; in anorexia, a goal is to replace false beliefs about body weight, eating, and self-worth with health-promoting beliefs.
- **eating disorder** a disturbance in eating behavior that jeopardizes a person's physical or psychological health.
- **emetic** (em-ETT-ic) an agent that causes vomiting.
- **female athlete triad** a potentially fatal triad of medical problems seen in female athletes: disordered eating, amenorrhea, and osteoporosis.

© Cengage Learning

Media Messages

No doubt our society sets unrealistic ideals and devalues those who do not conform to them. The Miss America beauty pageant, for example, puts forth a standard of female desirability—thinner and thinner women over the years have won the crown. Magazines and other media convey a message that to be happy, beautiful, and desirable, one must first be thin. In their quest for identity, adolescent girls are particularly vulnerable to such messages.

Dieting as Risk

Severe food restriction often precedes an eating disorder. Ill-advised "dieting" can create intense stress and extreme hunger that lead to binges.[6] Painful emotions

* Reference notes are found in Appendix F.

such as anger, jealousy, or disappointment may be turned inward by youngsters, some still in kindergarten, who express dissatisfaction with body weight or say they "feel fat." As weight loss and severe food restraint become more and more a focus, psychological problems worsen, and the likelihood of developing full-blown eating disorders intensifies.

Eating Disorders in Athletes

Athletes and dancers are at special risk for eating disorders.[7] They may severely restrict energy intakes in an attempt to enhance performance or appearance or to meet weight guidelines of a sport. In reality, severe energy restriction causes a loss of lean tissue that impairs physical performance and imposes a risk of eating disorders. Risk factors for eating disorders among athletes include:

- Young age (adolescence).
- Pressure to excel in a sport.
- Focus on achieving or maintaining an "ideal" body weight or body fat percentage.
- Participation in sports or competitions that emphasize a lean appearance or judge performance on aesthetic appeal, such as gymnastics, wrestling, figure skating, or dance.
- Unhealthy, unsupervised weight-loss dieting at an early age.

Male athletes—especially dancers, wrestlers, skaters, jockeys, and gymnasts—suffer from eating disorders, too, and their numbers may be increasing.

The Female Athlete Triad

In female athletes, three associated medical problems form the **female athlete triad**: disordered eating (with or without a diagnosed eating disorder), amenorrhea (cessation of menstruation), and osteoporosis.[8] For example, at age 14, Suzanne was a top contender for a spot on the state gymnastics team. Each day her coach reminded team members that they would not qualify to compete if they weighed more than a few ounces above the assigned weights.

Suzanne weighed herself several times a day to ensure that she did not top her 80-pound limit. She dieted and exercised to extremes; unlike many of her friends, she never began to menstruate. A few months before her 15th birthday, Suzanne's coach dropped her back to the second-level team because of a slow-healing stress fracture. Mentally and physically exhausted, she quit gymnastics and began overeating between periods of self-starvation. Suzanne exhibited all the signs of female athlete triad—disordered eating, amenorrhea, and weakened bones—but no one put them together in time to protect her physical and mental health.

An athlete's body must be heavier for a given height than a nonathlete's body because it contains more muscle and dense bone tissue with less fat. However, coaches often use weight standards, such as BMI, that cannot properly assess an athlete's body. For athletes, body composition measures such as skinfold tests yield more useful information.

The prevalence of amenorrhea among premenopausal women in the United States is about 2 to 5 percent overall, but it may be as high as 66 percent among female athletes. Amenorrhea is *not* a normal adaptation to strenuous physical training but a symptom of something going wrong.[9]

For most people, weight-bearing exercise helps to protect bones against the calcium losses of aging. For young women with disordered eating and amenorrhea, strenuous activity can imperil bone health.[10] Bone loss can lead to both stress fractures and osteoporosis (see Figure C9–1).[11]

Male Athletes and Eating Disorders

Male athletes and dancers with eating disorders often deny having them because they mistakenly believe that eating disorders strike only women. Under the same pressures as female athletes, males may also develop eating disorders. They skip meals, restrict fluids, practice in plastic suits, or train in heated rooms to lose a quick 4 to 7 pounds.[12] Many male high school wrestlers, gymnasts, and figure skaters strive for as little as 5 percent body fat (15 percent is average). Wrestlers, especially, must "make weight" to compete in the lowest possible weight class to face smaller opponents. Conversely, male athletes may suffer weight-*gain* problems. When young men with low self-esteem internalize unrealistically bulky male body images, they can become dissatisfied with their own healthy bodies, and this may lead to unhealthy behaviors, even steroid drug abuse.

For young people, unrealistic standards based on appearance, weight, or body type should be replaced with performance-based standards. Table C9–2, p. 376, provides some suggestions to help athletes and dancers protect themselves against eating disorders.

The next sections describe some categories of eating disorders. In many

Figure C9–1
The Female Athlete Triad

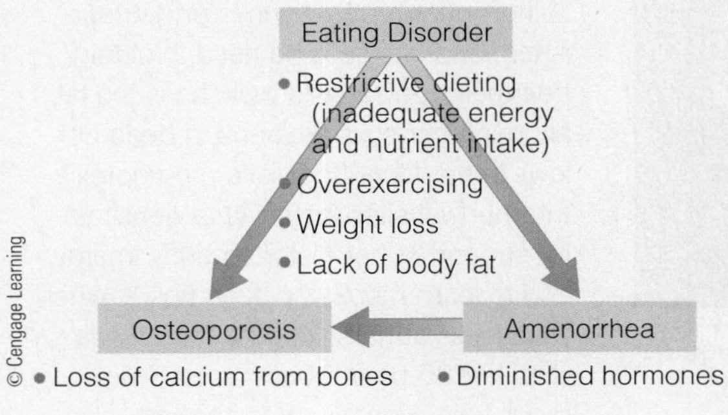

In the female athlete triad, extreme weight loss causes both cessation of menstruation (amenorrhea) and excessive loss of calcium from the bones. The hormone disturbances associated with amenorrhea also contribute to osteoporosis, making the female athlete triad extraordinarily harmful to the bones.

Table C9–2

Tips for Combating Eating Disorders

General Guidelines
▪ Never restrict food intakes to below the amounts suggested for adequacy by the USDA Food Patterns (Chapter 2). ▪ Eat frequently. The person who eats frequently never gets so hungry as to allow hunger to dictate food choices. ▪ If not at a healthy weight, establish a reasonable weight goal based on a healthy body composition. ▪ Allow a reasonable time to achieve the goal. A reasonable rate for losing excess fat is about 1% of body weight per week. ▪ Learn to recognize media image biases, and reject ultrathin standards for beauty. Shift focus to health, competencies, and human interactions; bring behaviors in line with those beliefs.
Specific Guidelines for Athletes and Dancers
▪ Replace weight-based or appearance-based goals with performance-based goals. ▪ Remember that eating disorders impair physical performance. Seek confidential help in obtaining treatment if needed. ▪ Restrict weight-loss activities to the off-season. ▪ Focus on proper nutrition as an important facet of your training, as important as proper technique.

cases, however, these categories overlap; a person may migrate from type to type, or an eating disorder fails to fall into a clear pattern.[13] Three main characteristics of eating disorders have been described:

1. Eating habits or weight-control behaviors have become abnormal.
2. Clinically significant impairment of physical health or psychosocial functioning materializes.
3. The disturbance is not caused by other medical or psychiatric conditions.

Women with anorexia nervosa see themselves as fat, even when they are dangerously underweight.

The problems described in the next section are typical.

Anorexia Nervosa

Julie is 17 years old and a straight-A superachiever in school. She also watches her diet with great care, and she exercises daily, maintaining a heroic schedule of self-discipline. She stands 5 feet 6 inches tall and weighs only 85 pounds, but she is determined to lose weight. She has anorexia nervosa.

Characteristics of Anorexia Nervosa

Julie is unaware that she is undernourished, and she sees no need to obtain treatment. She insists that she is too fat although her eyes are sunk in deep hollows in her face. She visits pro-anorexia Internet websites (or *pro-ana* websites) for support of her distorted body image and to learn more starvation tips.[14] When Julie looks at herself in the mirror, she sees her 85-pound body as fat. The more Julie overestimates her body size, the more resistant she is to treatment and the more unwilling to examine misperceptions.

She stopped menstruating and is moody and chronically depressed but blames external circumstances. She is close to physical exhaustion, but she no longer sleeps easily. Her family is concerned, and although reluctant to push her, they have finally insisted that she see a psychiatrist. Julie's psychiatrist has prescribed group therapy as a start but warns that if Julie does not begin to gain weight soon, she will need to be hospitalized.

No one knows for certain what causes anorexia nervosa, but some influences are associated with its development. Most people with anorexia nervosa come from middle- or upper-class families. Most are female. People with anorexia nervosa are unaware of their condition. They cannot recognize that a distorted body image that overestimates body fatness, a central feature of a diagnosis, is causing the problem. Some general criteria are proposed by the American Psychiatric Association. The entire list is available on the Internet, but, in plain language, they focus on people who:

- Restrict calorie intake to the point of developing a too-low body weight for age, gender, and health.
- Have an intense fear of body fatness or of gaining weight, or strive to prevent weight gain although underweight.
- Hold a false perception of body weight or shape, exaggerate the importance of body weight or shape in their self-evaluation, or deny the danger of being severely underweight.[15]

Many details on diagnostic criteria exist.[16]

The Role of the Family

Certain family attitudes, and especially parental attitudes, stand accused of contributing to or maintaining eating disorders. Families of people with anorexia nervosa are likely to be critical and to overvalue outward appearances. When symptoms appear, unhelpful emotional reactions, such as making excuses or becoming angry, may inadvertently enable and maintain the eating disorder.

Julie is a perfectionist, just as her parents are. She is respectful of authority: polite but controlled, rigid, and unspontaneous. For Julie, rejecting food is a way of gaining control.

Self-Starvation

How can a person as thin as Julie continue to starve herself? Julie uses tremendous discipline to strictly limit her portions of low-calorie foods. She will deny her hunger and, having become accustomed to so little food, she feels full after eating only a few bites. She can recite the calorie contents of dozens of foods and the calorie costs of as many physical activities. If she feels that she has gained an ounce of weight, she runs or jumps rope until she thinks it's gone. She drinks water incessantly to fill her stomach, risking dangerous mineral imbalances and water intoxication. She takes laxatives to hasten the passage of food from her system. She is starving, but she doesn't eat because her need for self-control outweighs her need for food.

Physical Perils

From the body's point of view, anorexia nervosa is starvation and thus brings the same damage as classic severe malnutrition. The person with anorexia depletes the body tissues of needed fat and protein. In young people, growth ceases, normal development falters, and they lose so much lean tissue that basal metabolic rate slows. Bones weaken, too—osteoporosis develops in about a third of those with anorexia nervosa.[17]

Internal organs suffer as nutrient status declines. The heart pumps inefficiently and irregularly, the heart muscle becomes weak and thin, the heart chambers diminish in size, and the blood pressure falls. As iron diminishes, anemia ensues, and the heart struggles to pump oxygen to the tissues. Potassium and other electrolytes that help to regulate the heartbeat go out of balance. Many deaths in people with anorexia nervosa are due to heart failure. Kidneys often fail as well.

Starvation also brings neurological and digestive consequences. The brain loses significant amounts of tissue, nerves cannot function normally, the electrical activity of the brain becomes abnormal, and insomnia is common. Digestive functioning becomes sluggish, the stomach empties slowly, and the lining of the intestinal tract shrinks. The ailing digestive tract fails to adequately digest the small amount of food consumed. The pancreas slows its production of digestive enzymes. Diarrhea sets in, further worsening malnutrition.

In addition to anemia, blood changes include impaired immune response, altered blood lipids, high concentrations of vitamin A and vitamin E, and low blood proteins. Dry skin, low body temperature, and the development of fine body hair (the body's attempt to keep warm) also occur. In adulthood, both women and men lose their sex drives. Mothers with anorexia nervosa may severely underfeed their children, who then fail to thrive and suffer the other harms typical of starvation.

About 1,000 women die of anorexia nervosa each year, mostly from heart abnormalities brought on by malnutrition or from suicide. Anorexia nervosa has the highest mortality rate among psychiatric disorders.[18]

Treatment of Anorexia Nervosa

Treatment of anorexia nervosa requires a multidisciplinary approach that addresses two areas of concern: those relating to food and weight and those involving psychological processes. Teams of physicians, nurses, psychiatrists, family therapists, and dietitians work together to treat people with anorexia nervosa. The expertise of a registered dietitian is essential because an appropriate, individually crafted diet is crucial for normalizing body weight, and nutrition counseling is indispensable.[19]

Professionals classify clients based on the risks posed by the degree of malnutrition present.† Clients with low risks may benefit from family counseling, **cognitive therapy**, behavior modification, and nutrition guidance.[20] Those with greater risks may also need other forms of psychotherapy and supplemental formulas to provide extra nutrients and energy. Antidepressant and other drugs are commonly prescribed but are often ineffective in treating anorexia nervosa.

Clients in later stages are seldom willing to eat, but if they are, chances are they can recover without other interventions. Sometimes, intensive behavior management treatment in a live-in facility helps to normalize food intake and exercise.[21] When starvation leads to severe underweight (less than 75 percent of ideal body weight), high medical risks ensue, and patients require hospitalization. They must be stabilized and fed through a tube to forestall death.[22] Even after recovery, however, energy intakes and eating behaviors may never fully return to normal, and relapses are common.

Before drawing conclusions about someone who is extremely thin, be aware that a diagnosis of anorexia nervosa requires professional assessment. People seeking help for anorexia nervosa for themselves or for others should not delay, but should visit the National Eating Disorders Association website or call them.‡

Bulimia Nervosa

Sophia is a 20-year-old flight attendant, and although her body weight is healthy, she thinks constantly about food. She alternately starves herself and then secretly binges; when she has eaten too much, she vomits. Few people would fail to recognize that these symptoms signify bulimia nervosa.

Characteristics of Bulimia Nervosa

Bulimia nervosa is distinct from anorexia nervosa and is much more prevalent, although the true incidence is difficult to

† *Indicators of malnutrition include a low percentage of body fat, low blood proteins, and impaired immune response.*

‡ *The National Eating Disorders website address is www.nationaleatingdisorders.org; the toll-free referral line is (800) 931–2237.*

establish. People with bulimia nervosa often suffer in secret and, when asked, may deny the existence of a problem. More men suffer from bulimia nervosa than from anorexia nervosa, but bulimia nervosa is still most common in women. Here are some general proposed diagnostic criteria for bulimia nervosa:

- Binge eating behavior, that is, eating a relatively large amount of food in a relatively short period of time.
- An experience of loss of control during binges and compensation behaviors afterwards, such as vomiting or fasting.
- Frequent binges and compensations (at least once a week for three months).
- False perceptions of body weight or shape; exaggerations of the importance of body weight or shape in self-evaluation.[23]

As is typical, Sophia is single, female, and white. She is well educated and close to her ideal body weight, although her weight fluctuates over a range of 10 pounds or so every few weeks. As a young teen, Sophia cycled on and off crash diets.

Sophia seldom lets her bulimia nervosa interfere with her work or other activities. However, she is emotionally insecure, feels anxious at social events, and cannot easily establish close relationships. She is usually depressed and often impulsive. When crisis hits, Sophia responds with an overwhelming urge to eat.[24]

The Role of the Family

Parents and other family members may foster bulimia by example or by direct interactions.[25] Children, particularly daughters, often adopt the dieting behaviors or body dissatisfaction displayed by parents, particularly mothers.[26]

Families may also be controlling but emotionally unsupportive, resulting in a stifling negative self-image believed to perpetuate bulimia (see Figure C9–2). Dieting, arguments, and criticism of body shape or weight commonly arise in families of people with bulimia, and a sensitive child may react with anxiety and self-doubt.[27] The family may also rarely eat meals together, a factor common to those with bulimia nervosa. In the extreme, bulimic women who report having been abused by family members or friends may become emotionally inhibited and continually suffer a sense of being out of control.

Figure C9–2

The Cycle of Bingeing, Purging, and Negative Self-Perception

Each of these factors helps to perpetuate disordered eating.

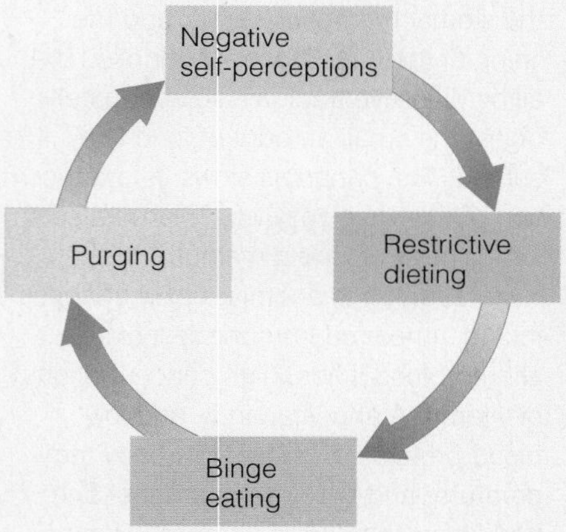

© Cengage Learning

Binge Eating and Purging

A bulimic binge is unlike normal eating. During a binge, Sophia's eating is accelerated by her hunger from previous calorie restriction. She regularly takes in extra food approaching 1,000 calories at each binge, and she may have several binges in a day. Typical binge foods are easy-to-eat, low-fiber, smooth-textured, high-fat, and high-carbohydrate foods, such as cookies, cakes, and ice cream; and she eats the entire bag of cookies, the whole cake, and every spoonful in a carton of ice cream. By the end of the binge, she has vastly overcorrected for her earlier attempts at calorie restriction.

The binge is a compulsion and usually occurs in several stages: anticipation and planning, anxiety, urgency to begin, rapid and uncontrollable consumption of food, relief and relaxation, disappointment, and finally shame or disgust. Then, to purge the food from her body, she may use a **cathartic**—a strong laxative that can injure the lower intestinal tract. Or she may induce vomiting, sometimes with an **emetic**—a drug intended as first aid for poisoning. After the binge she pays the price with hands scraped raw against the teeth during gag-induced vomiting, swollen neck glands and reddened eyes from straining to vomit, and the bloating, fatigue, headache, nausea, and pain that follow.

Physical and Psychological Perils

Purging may seem to offer a quick way to rid the body of unwanted calories,

© Felicia Martinez Photography/Photo Edit

A typical binge consists of easy-to-eat, low-fiber, smooth-textured, high-calorie foods.

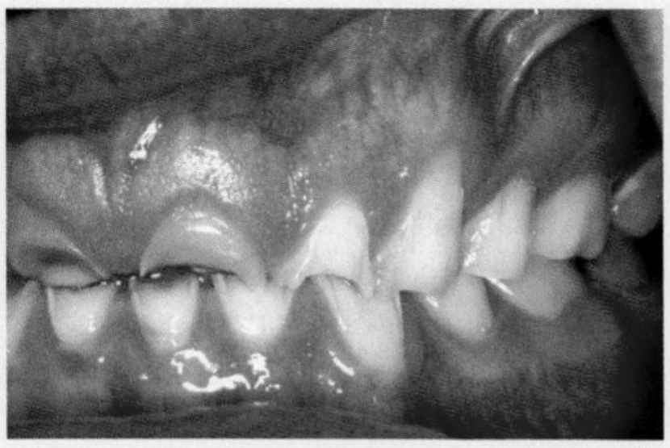

© Edward H. Gill/Custom Medical Stock Photo

Tooth damage from stomach acid in bulimia nervosa.

Table C9–3

Diet Strategies for Combating Bulimia Nervosa

Planning Principles
▪ Plan meals and snacks; record plans in a food diary prior to eating. ▪ Plan meals and snacks that require eating at the table and using utensils. ▪ Refrain from eating "finger foods." ▪ Refrain from "dieting" or skipping meals.
Nutrition Principles
▪ Eat a well-balanced diet and regularly timed meals consisting of a variety of foods. ▪ Include raw vegetables, salad, or raw fruit at meals to prolong eating times. ▪ Choose whole-grain, high-fiber breads, pasta, rice, and cereals to increase bulk. ▪ Consume adequate fluid, particularly water.
Other Tips
▪ Choose meals that provide protein and fat for satiety and bulky, fiber-rich carbohydrates for immediate feelings of fullness. ▪ Try including soups and other water-rich foods for satiety (water contents of foods are listed in Appendix A). ▪ Consume the amounts of food specified in the USDA Food Patterns (Chapter 2). ▪ Select foods that naturally divide into portions. Select one potato, rather than rice or pasta that can be overloaded onto the plate; purchase yogurt and cottage cheese in individual containers; look for small packages of precut steak or chicken; choose frozen dinners with metered portions. ▪ Include 30 minutes or more of physical activity on most days—exercise may be an important tool in controlling bulimia.

but bingeing and purging have serious physical consequences. Fluid and electrolyte imbalances caused by vomiting or diarrhea can lead to abnormal heart rhythms; one common emetic causes heart muscle damage, and its overuse can cause death from heart failure.[§28] Urinary tract infections can lead to kidney failure. Vomiting causes irritation and infection of the pharynx, esophagus, and salivary glands; erosion of the teeth; and dental caries. The esophagus or stomach may rupture or tear. In addition, a correlation is noted between the addictive nature of binge eating and that of drug abuse.[29]

Unlike Julie, Sophia is aware that her behavior is abnormal, and she is deeply ashamed of it. She wants to recover, and this makes recovery more likely for her than for Julie, who clings to denial.

Treatment of Bulimia Nervosa

To gain control over food and establish regular eating patterns requires adherence to a structured eating and exercise plan. Restrictive dieting is forbidden, for it almost always precedes binges. Steady maintenance of weight and prevention of cyclic gains and losses are the goals. Many a former bulimia nervosa sufferer has taken a major step toward recovery by learning to consistently eat enough food to satisfy hunger (at least 1,600 calories a day).

Table C9–3 offers some ways to begin correcting the eating problems of bulimia nervosa. About half of women receiving a diagnosis of bulimia nervosa may recover completely after 5 to 10 years, with or without professional treatment, although treatment or self-help programs probably speed recovery.

Binge Eating Disorder

Charlie is a 40-year-old former baseball outfielder who, after becoming a spectator instead of a player, has gained excess body fat. He believes that he has the willpower to diet until he loses the fat. Periodically he restricts his food intake for several days, only to eventually succumb to cravings for his favorite high-calorie treats. Like Charlie, many overweight people end up bingeing after dieting.[30]

Binge eating behavior responds more readily to treatment than other eating disorders. Intervention, even obtained on an Internet website, improves physical and mental health and may permanently break the cycle of rapid weight losses and gains.

Toward Prevention

Treatments for existing eating disorders have evolved, but prevention of these conditions would be far preferable.[31] One approach may be to provide children and adolescents with defenses against influences that promote eating disorders. A set of suggestions intended to help pediatricians avert eating disorders in their patients might also apply to teachers, coaches, and others who deal with children:

1. Encourage positive eating and physical behaviors that can be maintained over a lifetime; discourage unhealthy dieting.
2. Promote a positive body image; do not use body dissatisfaction as a motivator for behavior change.
3. Encourage frequent and enjoyable family meals consumed at home; discourage hasty meals eaten alone.
4. Help children and teens to nurture their bodies through healthy eating,

§ The heart-damaging emetic is ipecac (IP-eh-kak).

physical activity, and self-talk; divert focus from weight and body shape.

5. Talk to children and teens and their families about mistreatment because of body weight or size; assume it has occurred in overweight children.

Protection against eating disorders in the next generation largely depends upon the actions of adults in authority today. Perhaps a young person's best defense against eating disorders is to learn about normal, expected growth patterns, especially the characteristic weight gain of adolescence (see Chapter 14), and to learn respect for the inherent wisdom of the body. When people discover and honor the body's real needs for nutrition and exercise, they often will not sacrifice health for conformity.

Critical Thinking

1. Eating disorders are common only in cultures where extreme thinness is an ideal. Who in society do you think sets such ideals? How are these ideals conveyed to others? Suggest some steps that schools, parents, and other influential adults might take to help to minimize the impact of idealized body types on children as they develop their own self images.

2. Form a small group. Each member of the group gives an example of a role model that he or she would like to emulate. This can be, for example, a teacher, athlete, movie star, or scientist, among others. State all of the reasons for choosing this person as a role model. Now talk about the body type of each role model. Would you like to achieve that body type? Is it possible to do so? Of all the role models discussed in your group, which role model do you believe is the healthiest and why?

10 Nutrients, Physical Activity, and the Body's Responses

what do you think?

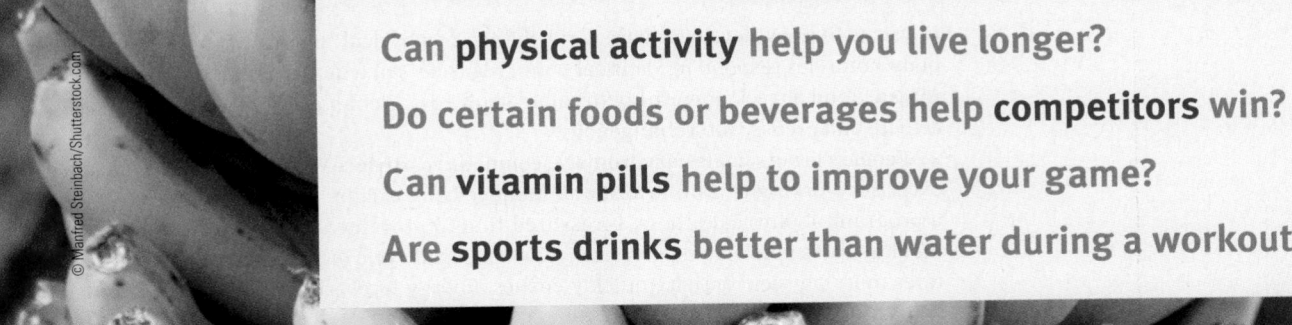

Can **physical activity** help you live longer?

Do certain foods or beverages help **competitors** win?

Can **vitamin pills** help to improve your game?

Are **sports drinks** better than water during a workout?

Learning Objectives

After completing this chapter, you should be able to accomplish the following:

LO 10.1 Provide examples of how regular physical activity benefits the body, and explain how the *Physical Activity Guidelines for Americans* can be incorporated into a healthy person's lifestyle.

LO 10.2 Describe in brief how fitness develops, and explain the beneficial effects of both resistance and cardiorespiratory exercise on the body.

LO 10.3 Explain the importance of glucose, fatty acids, and amino acids to a working athlete before, during, and after vigorous exercise.

LO 10.4 Outline reasons why iron is of special significance for some athletes, and describe the proper roles for nutrient supplements.

LO 10.5 Identify hazards associated with inadequate fluid replacement in the exercising body, and compare the fluid needs of a casual exerciser and an endurance athlete.

LO 10.6 Plan a nourishing and adequate diet for an athlete, including snacks and pregame meals.

LO 10.7 Evaluate whether dietary ergogenic aids are useful for increasing sports performance or obtaining an ideal body composition for sports.

In the body, nutrition and **physical activity** go hand in hand. The working body demands energy-yielding nutrients—carbohydrate, lipid, and protein—to fuel physical activity. It also needs high-quality protein to supply the amino acids necessary to build new muscle tissues. Vitamins and minerals play critical roles in energy metabolism, protein synthesis, and many other functions necessary for physical work.

Physical activity, in turn, also benefits the body's nutrition. Physical activity helps to regulate the use of energy-yielding nutrients, improves body composition, and increases the daily calorie allowance. A person who eats extra calories of nutritious whole foods takes in more beneficial nutrients and phytochemicals, too. Together, a nutritious eating pattern and regular physical activity become a powerful force for human health.

This chapter addresses many of the concerns of physically active people, starting with some basic concepts about health and physical activity. It also provides a basic framework for understanding **performance nutrition**. It describes how foods, fluids, and nutrients help to fuel physical activities, and how the right choices can improve athletic performance, whereas poor choices may hinder it. The Controversy that follows spotlights just a few of the many supplements sold with promises of enhanced athletic performance.

Fitness

LO 10.1 Provide examples of how regular physical activity benefits the body, and explain how the *Physical Activity Guidelines for Americans* can be incorporated into a healthy person's lifestyle.

Physical **fitness** develops with performance of physical activity or **exercise**. The body's muscles respond in identical ways, regardless of whether an individual is running around a track or running to catch a bus, so this chapter uses the terms *physical activity* and *exercise* interchangeably.

People's fitness goals vary from the competitive **athlete** in **training** to the casual exerciser working to gain health and manage body weight. For those just beginning a program of physical fitness, be assured that improvement is not only possible but inevitable. As you become more active, a beneficial cycle of greater fitness that facilitates more physical activity quickly ensues. Energy levels rise, and chronic diseases risks fall. The mechanism also runs in reverse: a sedentary lifestyle robs people of their fitness and fosters the development of several chronic diseases.[1*]

The Nature of Fitness

If you are physically fit, the following describes you: You move with ease and balance. You have endurance that lasts for hours. You are strong and meet daily physical challenges without strain. You are prepared for mental and emotional challenges, too, because physical activity can help to relieve stress, depression, or anxiety. So effective is physical activity that mental health clinicians are encouraged to recommend it as part of their treatment plans. As your fitness improves, you not only begin to feel better and stronger but you look better, too. As you strengthen your muscles, your posture and self-image often improve.

Longevity and Disease Resistance People who regularly engage in moderate physical activity live longer, healthier lives on average than those who are physically inactive.[2] A sedentary lifestyle ranks with smoking and obesity as a powerful predictor of the major killer diseases of our time—cardiovascular disease, some forms of cancer, stroke, diabetes, and hypertension. Sedentary people may even be more likely to catch a cold.[3] Despite the well-known health benefits of physical activity, fewer than half of adults in the United States are sufficiently active to maintain their weight and protect their health.[4]

Some Benefits of Fitness As a person becomes physically active, the health of the entire body improves. Compared with unfit people, physically fit people may enjoy:

- *More restful, beneficial sleep.* Rest and sleep occur naturally after periods of physical activity. During rest, the body repairs injuries, disposes of wastes generated during activity, and builds new physical structures.
- *Improved nutritional health.* Active people expend more energy than sedentary people; those wishing to maintain their weight can meet their increased energy needs with nutrient-rich whole foods.
- *Improved body composition.* A balanced program of physical activity helps limit body fat, particularly abdominal fat, and increases or maintains lean tissue, even during weight loss.[5] Thus, physically active people are often leaner than sedentary people at the same weight.[6]
- *Improved bone density.* Weight-bearing physical activity builds bone strength and protects against the bone loss of osteoporosis.[7]
- *Enhanced resistance to colds and other infectious diseases.* Fitness enhances immunity.[8†]
- *Lower risks of some types of cancers.* Too little physical activity is associated with a 30 to 40 percent increase in cancers of the breast, colon, kidney, pancreas, and uterus, and it adversely affects quality of life and recovery among people with cancer.[9]
- *Stronger circulation and lung function.* Physical activity that challenges the heart and lungs strengthens both the circulatory and the respiratory system.
- *Lower risks of cardiovascular disease.* Physical activity lowers blood pressure, slows resting pulse rate, lowers total blood cholesterol and triglycerides, and raises HDL cholesterol, factors that reduce the risks of heart attacks and strokes.[10] Reduction of abdominal fat through physical activity further reduces disease risks.
- *Lower risk of type 2 diabetes.* Regular physical activity improves glucose tolerance while sedentary behavior worsens it.[11] With regular physical activity, many people's need for diabetes medication is reduced.
- *Reduced risk of gallbladder disease.* Regular physical activity reduces risk of gallbladder disease.[12]

* Reference notes are found in Appendix F.
† Moderate physical activity can stimulate immune function. Intense, vigorous, prolonged activity such as marathon running, however, may compromise immune function.

physical activity bodily movement produced by muscle contractions that substantially increase energy expenditure.

performance nutrition an area of nutrition science that applies its principles to maintaining health and maximizing physical performance in athletes, firefighters, military personnel, and others who must perform at high levels of physical ability.

fitness the characteristics that enable the body to perform physical activity; more broadly, the ability to meet routine physical demands with enough reserve energy to rise to a physical challenge; or the body's ability to withstand stress of all kinds.

exercise planned, structured, and repetitive bodily movement that promotes or maintains physical fitness.

athlete a competitor in any sport, exercise, or game requiring physical skill; for the purpose of this book, anyone who trains at a high level of physical exertion, with or without competition. From the Greek *athlein*, meaning "to contend for a prize."

training regular practice of an activity, which leads to physical adaptations of the body with improvement in flexibility, strength, or endurance.

- *Lower incidence and severity of anxiety and depression.* Physical activity improves mood and enhances the quality of life by reducing depression and anxiety.[13] The sense of achievement that comes from meeting physical challenges also promotes self-confidence.
- *Longer life and higher quality of life in the later years.* Active people live longer, healthier lives and suffer less dementia than sedentary people do.[14] Physical activity supports independence and mobility in later life by reducing falls and, if a fall should occur, by minimizing injuries.

How does physical activity do these wonderful things? At least part of the credit may go to a hormone-like messenger molecule released by working muscles.[‡] This messenger communicates with other tissues, particularly adipose tissue, influencing metabolism in ways thought to protect against a network of often interrelated diseases—obesity, cardiovascular diseases, type 2 diabetes, cancer, dementia, and osteoporosis.[15] This raises the possibility that an "exercise pill" may one day confer the health benefits of exercise without the work but this result is unlikely. The full interplay of muscles, their molecular messengers, and other organs is not yet known. So keep moving.

KEY POINTS

- Physical activity and fitness benefit people's physical and psychological well-being and improve their resistance to disease.
- Physical activity improves survival and quality of life in the later years.
- Hormone-like communication molecules generated by working muscles may trigger healthy changes in body tissues.

Physical Activity Guidelines

What must you do to reap the health rewards of physical activity? You need only meet the *Physical Activity Guidelines for Americans* set forth by the USDA and the Department of Health and Human Services, described next.[16]

Physical Activity Guidelines for Americans The Physical Activity Guidelines for Americans outline how much **aerobic activity** that adults aged 18 to 64 years need to improve or maintain cardiovascular health (see Figure 10–1).[17] The Guidelines also support **resistance training** (strengthening exercises) as beneficial and useful for meeting activity goals. The length of time (exercise duration) required to meet these guidelines varies by **intensity** of the activity—longer duration is required for physical activity of moderate intensity, with shorter duration at more vigorous intensity. Table 10–1 describes activity intensity levels.[18]

Most health benefits occur with the activities listed in Figure 10–1, but additional benefits can result from higher intensity, greater frequency, or longer duration of activity. Older people and the disabled who cannot meet the *Guidelines* should be as active as their conditions allow; those with chronic illnesses should seek advice from a health-care provider. For everyone, some physical activity is better than none.[19]

Note that the guidelines are stated in accumulated weekly totals.[§] This allows individuals to split up their activity into sessions of at least 10 minutes each, performed throughout the days of the week in any combination that suits their lifestyles. Safety is a high priority, and the Think Fitness section provides some tips.

To achieve or maintain a healthy body weight through increasing physical activity demands more than the amount needed for health.[20] Most people with weight-loss goals require at least 4 hours of moderate-intensity physical activity such as brisk walking or bicycling each week.[21]

aerobic activity physical activity that involves the body's large muscles working at light to moderate intensity for a sustained period of time. Brisk walking, running, swimming, and bicycling are examples. Also called *endurance activity.*

resistance training physical activity that develops muscle strength, power, endurance, and mass. Resistance can be provided by free weights, weight machines, other objects, or the person's own body weight. Also called *weight training, resistance exercise,* or *strength exercise.*

intensity in exercise, the degree of effort required to perform a given physical activity.

‡ The molecule is *irisin,* one of the myokines produced by working muscles (from the Greek, *myo* = "muscle," *kino* = "movement").

§ Guidelines from sports medicine experts are stated in metabolic equivalent units (METs) that reflect the ratio of the rate of energy expended during an activity to the rate of energy expended at rest. One MET is equal to the energy expenditure while at rest.

Figure 10–1

Physical Activity Guidelines for Americans[a]

Meeting these guidelines requires physical activity beyond the usual light or sedentary activities required in daily living, such as cooking, cleaning, and walking from an automobile to a store. Table 10–3 provides a sample plan that meets these goals.

Mike Powell/Getty Images

- **Every day—Choose an active lifestyle and engage in flexibility activities.**
  Integrate activity into your day: walk a dog, take the stairs, stand up whenever possible. Stretching exercises lend flexibility for activities such as dance, but minutes spent stretching do not count toward aerobic or strength activity recommendations.

Biddiboo/The Image Bank/Getty Images

- **5 or more days/week—Engage in moderate or vigorous aerobic activities.**
  Perform a minimum of 150 min of moderate-intensity aerobic activity each week by doing activities like brisk walking or ballroom dancing; or, 75 min per week of vigorous aerobic activity, such as bicycling (>10 mph) or jumping rope; or a mix of the two (1 min vigorous activity = 2 min moderate).

© Wesley Hitt/Alamy

- **2 or more days/week—Engage in strength activities.**
  Perform muscle-strengthening activities that are moderate to high intensity and involve all major muscle groups.

© moodboard/Corbis

- **Do seldom—Limit sedentary activities.**
  Limit TV or movie watching, leisure computer time.

[a]*For most men and women, aged 18 to 64 years.*

Source: U.S. Department of Agriculture and U.S. Department of Health and Human Services, 2008 Physical Activity Guidelines for Americans, available at www.health.gov/paguidelines/default.aspx.

Table 10–1

Intensity of Physical Activity

Level of Intensity	Breathing and/or Heart Rate	Perceived Exertion (on a Scale of 0 to 10)	Talk Test	Energy Expenditure	Walking Pace
Light	Little to no increase	<5	Able to sing	<3.5 cal/min	<3 mph
Moderate	Some increase	5 or 6	Able to have a conversation	3.5 to 7 cal/min	3 to 4.5 mph (100 steps per minute or 15 to 20 minutes to walk 1 mile)
Vigorous	Large increase	7 or 8	Conversation is difficult or "broken"	>7 cal/min	>4.5 mph

Source: Centers for Disease Control and Prevention, 2011, available at www.cdc.gov/physicalactivity/everyone.

Guidelines for Sports Performance Athletes who compete in sports require specific types and amounts of physical activity to support their performance, so special guidelines apply to them. However, the "why" behind a person's choice to be physically active doesn't seem to matter much in terms of the resulting gains in fitness. Benefits naturally follow regular physical activity. Table 10–2, p. 386, offers guidelines from the American College of Sports Medicine for sports and fitness that are more specific and also more demanding than the physical activity Guidelines for Americans.[22]

KEY POINTS

- The U.S. *Physical Activity Guidelines for Americans* aim to improve physical fitness and the health of the nation.
- Other guidelines meet other needs.

Table 10–2

American College of Sports Medicine's Guidelines for Physical Fitness

Type of Activity	Aerobic activity that uses large-muscle groups and can be maintained continuously	Resistance activity that is performed at a controlled speed and through a full range of motion	Stretching activity that uses the major muscle groups
Frequency	5 to 7 days per week	2 to 3 nonconsecutive days per week	2 to 7 days per week
Intensity	Moderate (equivalent to walking at a pace of 3 to 4 mph)[a]	Enough to enhance muscle strength and improve body composition	Enough to feel tightness or slight discomfort
Duration	At least 30 minutes per day	2 to 4 sets of 8 to 12 repetitions involving each major muscle group	2 to 4 repetitions of 15 to 30 seconds per muscle group
Examples	Running, cycling, dancing, swimming, inline skating, rowing, power walking, cross-country skiing, kickboxing, water aerobics, jumping rope; sports activities such as basketball, soccer, racquetball, tennis, volleyball	Pull-ups, push-ups, sit-ups, weightlifting, pilates	Yoga

[a]*For those who prefer vigorous-intensity aerobic activity such as walking at a very brisk pace (>4.5 mph) or running (≥5 mph), a minimum of 20 minutes per day, 3 days per week is recommended.*

Source: American College of Sports Medicine position stand, Quantity and quality of exercise for developing and maintaining cardiorespiratory, musculoskeletal, and neuromotor fitness in apparently healthy adults: Guidance for prescribing exercise, Medicine and Science in Sports and Exercise *43 (2011): 1334–1359; W. L. Haskell and coauthors, Physical activity and public health: Updated recommendation for adults from the American College of Sports Medicine and the American Heart Association,* Medicine in Sports & Exercise *39 (2007): 1423–1434.*

The Essentials of Fitness

LO 10.2 Describe in brief how fitness develops, and explain the beneficial effects of both resistance and cardiorespiratory exercise on the body.

To become physically fit, you need to develop enough **flexibility**, **muscle strength**, **muscle endurance**, and **cardiorespiratory endurance** to allow you to meet the everyday demands of life with some to spare. You also need to achieve a reasonable body composition.

So far, the description of fitness applies to anyone interested in improving health. For athletes, however, excelling in sports performance often becomes the primary motivator for working out. Athletes must strive to develop strength and endurance, of course, but they also need **muscle power** to drive their movements, quick **reaction times** to respond with speed, **agility** to instantly change direction, increased resistance to **muscle fatigue**, and mental toughness to carry on when fatigue sets in.

How Do Muscles Adapt to Physical Activity?

A person who engages in physical activity *adapts* by becoming a little more able to perform the activity after each session. People shape their bodies by what they choose to do (and not do). Muscle cells and tissues respond to a physical activity **overload** by building, within genetic limits, the structures and metabolic equipment needed to perform it.[23]

Muscles are under constant renovation. Every day, particularly during the fasting periods between meals, a healthy body degrades a portion of its muscle protein to amino acids and then rebuilds it with available amino acids during fed periods.[24]** A balance between degradation and synthesis maintains the body's lean tissue. To gain muscle tissue protein, however, this balance must more often tip toward synthesis,

** In a healthy, moderately active body, the skeletal muscle protein turnover rate is about 1.2 percent per day, and according to the DRI committee, it represents about 60 to 65 grams of protein.

flexibility the capacity of the joints to move through a full range of motion; the ability to bend and recover without injury.

muscle strength the ability of muscles to overcome physical resistance. This muscle characteristic develops with increasing work load rather than repetition and is associated with muscle size.

muscle endurance the ability of a muscle to contract repeatedly within a given time without becoming exhausted. This muscle characteristic develops with increasing repetition rather than increasing workload and is associated with cardiorespiratory endurance.

cardiorespiratory endurance the ability of the heart, lungs, and metabolism to sustain large-muscle exercise of moderate-to-high intensity for prolonged periods.

muscle power the efficiency of a muscle contraction, measured by force and time.

reaction time the interval between stimulation and response.

agility nimbleness; the ability to quickly change directions.

muscle fatigue diminished force and power of muscle contractions despite consistent or increasing conscious effort to perform a physical activity; muscle fatigue may result from depleted glucose or oxygen supplies or other causes.

Table 10–3

A Sample Balanced Fitness Program

Monday	Tuesday	Wednesday	Thursday	Friday	Saturday or Sunday
5-min warm-up[a]	5-min warm-up[a]	5-min warm-up[a]	5-min warm-up[a]	5-min warm-up[a]	
Resistance training: chest, back, arms, and shoulders 15–45 min[b]	Resistance training: legs, core (abdomen/lower back) 15–45 min		Resistance training: chest, back, arms, and shoulders 15–45 min	Resistance training: legs, core (abdomen/lower back) 15–45 min	Active leisure pursuits: Sports, walking, hiking, biking, swimming
Moderate aerobic activity: 15–20 min	Moderate aerobic activity: 15–20 min	Moderate aerobic activity: 15–20 min	Moderate aerobic activity: 15–20 min	Moderate aerobic activity: 15–20 min	
Stretching: 5 min	Stretching: 5 min	Stretching: 5 min	Stretching: 5 min	Stretching: 5 min	

[a] *The warm-up consists of a slower or less-intense version of the activity ahead and may count toward the week's total activity requirement if it is performed at moderate intensity.*

[b] *Lower-intensity exercise requires more time; higher-intensity exercise requires less time.*

Source: Designed for Nutrition: Concepts and Controversies, *13th ed. (2013) by P. Spencer Webb, RD, CSCS, Exercise/Human Performance Instructor, U.S. Military Special Operations Forces.*

Muscle hypertrophy is an example of positive nitrogen balance, a concept illustrated in **Figure 6–13** of Chapter 6.

a condition called **hypertrophy**, rather than toward muscle breakdown, which results in **atrophy**. Physical activity tips the balance toward muscle hypertrophy. The opposite is also true: unused muscles diminish in size and weaken over time—they atrophy.[25]

The muscles adapt and build only the proteins they need to cope with the work performed. Muscles engaged in activities that require strength develop greater bulk, while those engaged in endurance activities develop more metabolic equipment to combat muscle fatigue.[26] Thus, a tennis player may have one superbly strong arm while the other is just average; cyclists often have well-developed legs that can pedal for many hours but relatively less development of the arms or chest.

Jupiterimages

Bodies are shaped . . .

Courtesy of Angelina Ramos

by the activities they perform.

A Balance of Activities Balanced fitness arises from performing a variety of physical activities that work different muscle groups from day to day. Stretching enhances flexibility, aerobic activity improves cardiorespiratory and muscle endurance, and **resistance training** develops strength, size, and endurance of the worked muscles. Table 10–3 presents one example of a balanced workout program.

Muscles need rest, too, because it takes a day or two to fully replenish muscle fuel supplies and to repair wear and tear. With greater work comes more damage, and muscles require longer rest periods for a full recovery. (A muscle or joint that remains sore after about a week of rest may be injured and in need of medical attention.)

A planned program of training can induce the development of specific muscle tissues and fuel systems. The muscle cells of a trained weight lifter store extra glycogen granules, build up strong connective tissues, and add bulk to the special proteins that contract the muscles, increasing their strength.[††] The muscle cells of a distance swimmer build instead more of the enzymes and structures needed for aerobic metabolism.[27] Therefore, if you wish to become a better jogger, swimmer, or biker, you should

[††] All muscles contain a variety of muscle fibers, but there are two main types—slow-twitch (also called red fibers) and fast-twitch (also called white fibers). Slow-twitch fibers contain extra metabolic equipment to perform aerobic work, which give them a reddish appearance under a microscope; the fast-twitch type store extra glycogen required for anaerobic work, giving them a lighter appearance.

overload an extra physical demand placed on the body; an increase in the frequency, duration, or intensity of an activity. A principle of training is that for a body system to improve, it must be worked at frequencies, durations, or intensities that increase by increments.

hypertrophy (high-PURR-tro-fee) an increase in size (for example, of a muscle) in response to use.

atrophy (AT-tro-fee) a decrease in size (for example, of a muscle) because of disuse.

resistance training physical activity that develops muscle strength, power, endurance, and mass. Resistance can be provided by free weights, weight machines, other objects, or the person's own body weight. Also called *weight training* or *muscular strength exercises.*

train mostly by jogging, swimming, or biking. Your performance will improve as your muscles adapt to a particular activity.

KEY POINTS

- The components of fitness are flexibility, muscle strength, muscle endurance, and cardiorespiratory endurance.
- Muscle protein is built up and broken down every day; muscle tissue is gained when synthesis exceeds degradation.
- Physical activity builds muscle tissues and metabolic equipment needed for the activities they are repeatedly called upon to perform.

Resistance Training for Muscle Strength, Size, Power, and Endurance

Most people know that resistance training helps to build muscle bulk, strength, and endurance. Less well known is that **progressive weight training** may help to prevent and manage several chronic diseases, including cardiovascular disease and osteoporosis, and can enhance psychological well-being, too.[28] By strengthening the muscles of the back and abdomen, resistance training can improve posture, making debilitating back injuries less likely to occur. It also enhances performance in many sports, not just those that demand muscle size. Swimmers can develop a more efficient stroke and tennis players a more powerful serve when they engage in resistance training, for example.

Many factors, including genetic potential and the body's hormones, modulate muscle protein synthesis in response to physical activity. Importantly, the body produces all the hormones it needs—additional hormones are not needed (and they are *not* recommended—see this chapter's Controversy). Some people, particularly many women, fear that resistance training will make their muscles bulky, like those of body builders. In truth, body builders of both genders work to achieve their look by training intensely with heavy weights and by consuming calories by the hundreds or thousands above an average person's intake.[29] In truth, ordinary people who regularly lift weights with sufficient intensity just a day or two each week can slowly develop and maintain a pleasing appearance as muscles strengthen and firm without becoming bulky.[30] If such people also reduce calorie intakes and meet protein needs, they can lose weight without sacrificing their lean body tissue.[31] They may also increase their daily energy expenditure by a little—evidence suggests that just 11 minutes of resistance training on 3 days each week may produce a small but significant rise in 24-hour energy expenditures.[32]

KEY POINTS

- Resistance training can benefit physical and mental health.
- Sports performance, appearance, and body composition improve through resistance training; bulky muscles are the result of intentional body building regimens.

How Does Aerobic Training Benefit the Heart?

Resistance training may provide some heart benefits, but aerobic endurance training reliably and efficiently improves a person's cardiovascular health.[33] Aerobic endurance training enhances the capacity of the heart, lungs, and blood to deliver oxygen to, and remove wastes from, the body's cells.[34] It also reduces many heart disease risk factors, and it encourages leanness and confers a fit, healthy appearance to limbs and torso.

Improvements to Blood, Heart, and Lungs Cardiorespiratory endurance allows the working body to remain active with an elevated heart rate over time. As cardiorespiratory endurance improves, the body delivers oxygen to the tissues more efficiently. In fact, the accepted measure of a person's cardiorespiratory fitness is a measure of the rate at which the tissues consume oxygen—the maximal oxygen uptake (**VO_{2max}**). This measure reflects many facets of oxygen delivery that can improve with regular aerobic exercise.

progressive weight training the gradual increase of a workload placed upon the body with the use of resistance.

VO_{2max} the maximum rate of oxygen consumption by an individual (measured at sea level).

Think Fitness move it! Exercise Safety

Sometimes, physical activity can pose risks.[35] Keeping safe during physical activity involves both common sense and education. The USDA suggests following these guidelines:

- Choose activities appropriate for your current fitness level.
- Gradually increase the amount of physical activity you perform.
- Wear appropriate safety gear, including correct shoes, helmet, pads, and other protection.
- Make sensible choices about when and where to exercise; for example, avoid the hottest hours of the day in southern locales, choose safe bike paths away from heavy traffic, and run with a buddy on isolated trails.
- People with medical problems or increased disease risks should consult a physician before beginning any program of physical activity.[36]

In addition, people can easily injure themselves by using improper techniques during exercise, particularly when it involves equipment. Many people can benefit from consulting with a Certified Personal Trainer (CPT). A CPT can help develop a safe and effective individualized exercise program. Some personal trainers have a more advanced credential, the Certified Strength and Conditioning Specialist (CSCS), that requires completion of a college curriculum that includes human anatomy and exercise physiology; they must also pass a nationally recognized examination. Unless a trainer also possesses a legitimate nutrition credential, he or she is not qualified to dispense diet advice. Fake nutrition credentials were described in Controversy 1; the same kinds of skullduggery occur in the field of physical training, too.

start now! Create a fitness plan that includes gradually increasing both the time and intensity of physical activity. Use a calendar to record your daily plan over at least four weeks, then record your actual activity on each of those days.

The blood's volume and the number of red blood cells partly determine VO_{2max}; the greater the volume and number, the more oxygen the blood can carry. The size and strength of the heart also play roles—as the heart muscle grows stronger and larger, the heart's **cardiac output** increases. Each beat empties the heart's chambers more completely, so the heart pumps more blood per beat—its **stroke volume** increases. The resting heart rate slows because a greater volume of blood is moved with fewer beats. Working muscles also push blood more efficiently through the veins on their return trip to the heart and lungs.

Muscles that inflate and deflate the lungs gain strength and endurance, too, so breathing becomes more efficient. Circulation through the arteries and veins improves, blood moves easily, and blood pressure falls. Figure 10–2, p. 390, shows the major relationships among the heart, lungs, and muscles and Table 10–4, p. 390, describes cardiorespiratory endurance. Anyone who possesses cardiorespiratory endurance can celebrate a lowered risk for cardiovascular diseases.

Cardiorespiratory Training Activities Effective cardiorespiratory training activities have three characteristics:

- They elevate the heart rate for sustained periods of time.
- They use most of the large-muscle groups of the body (for example, legs, buttocks, or upper body muscle groups).

In other words, they are the aerobic activities and they are recommended for heart health. Examples are swimming, cross-country skiing, rowing, fast walking, jogging, running, fast bicycling, soccer, hockey, basketball, in-line skating, lacrosse, and rugby.

The rest of this chapter describes the interactions between nutrients and physical activity. Nutrition alone cannot endow you with fitness or athletic ability but along with consistent physical activity and the right mental attitude, it complements your effort to obtain them. Conversely, unwise food selections can stand in your way.

Did You Know?

An average resting pulse rate is around 70 beats per minute; active people often have a rate of around 50. To take your resting pulse:

1. Sit down and relax for 5 minutes before you begin. Using a watch or clock with a second hand, place your hand over your heart or your finger firmly over an artery at the underside of the wrist or side of the throat under the jawbone.
2. Start counting your pulse at a convenient second, and continue counting for 30 seconds. If a heartbeat occurs exactly on the tenth second, count it as one-half beat. Multiply by 2 to obtain the beats per minute.
3. To ensure a true count, use only fingers, not your thumb, on the pulse point (the thumb has a pulse of its own).
4. Press just firmly enough to feel the pulse. Too much pressure can interfere with the pulse rhythm.

cardiac output the volume of blood discharged by the heart each minute.

stroke volume the volume of oxygenated blood ejected from the heart toward body tissues at each beat.

Figure 10–2

Animated! Delivery of Oxygen by the Heart and Lungs to the Muscles

The cardiorespiratory system responds to increased demand for oxygen by building up its capacity to deliver oxygen. Researchers can measure cardiovascular fitness by measuring the amount of oxygen a person consumes per minute while working out. This measure of fitness, which indicates the person's maximum rate of oxygen consumption, is called VO_{2max}.

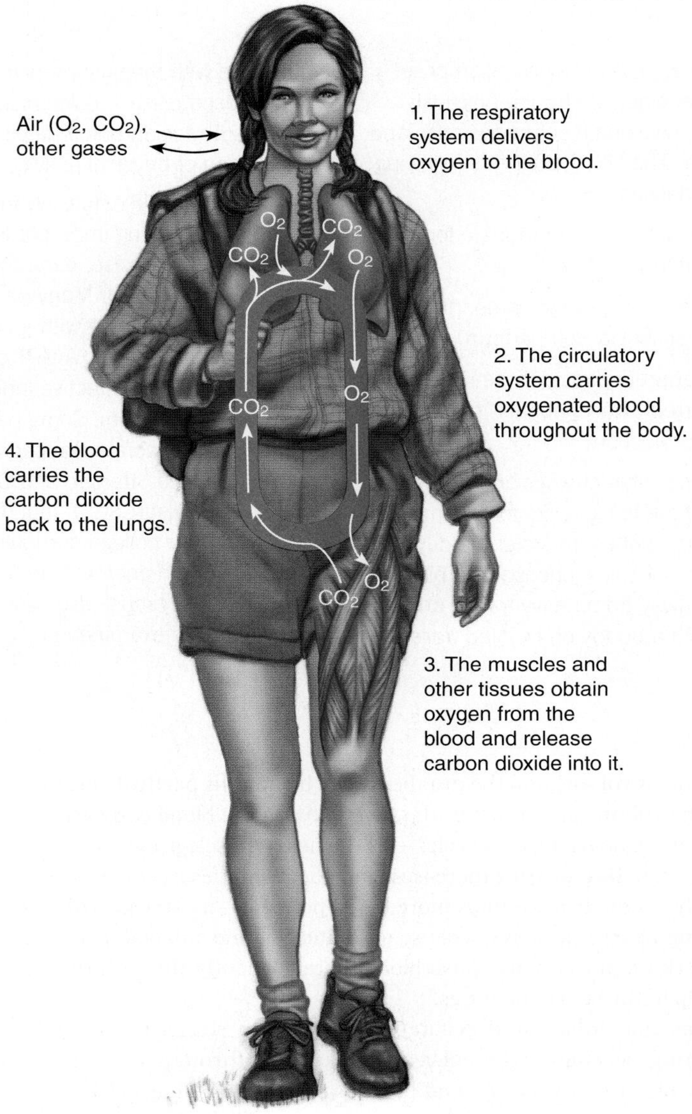

Air (O_2, CO_2), other gases
1. The respiratory system delivers oxygen to the blood.
O_2
CO_2
2. The circulatory system carries oxygenated blood throughout the body.
3. The muscles and other tissues obtain oxygen from the blood and release carbon dioxide into it.
4. The blood carries the carbon dioxide back to the lungs.

Table 10–4

Cardiorespiratory Endurance

Cardiorespiratory endurance is characterized by:

- Increased heart strength and stroke volume
- Slowed resting pulse
- Increased breathing efficiency
- Improved circulation and oxygen delivery
- Reduced blood pressure
- Increased blood HDL cholesterol

Source: E. G. Ciolac and coauthors, Effects of high-intensity aerobic interval training vs. moderate exercise on hemodynamic, metabolic and neuro-humoral abnormalities of young normotensive women at high familial risk for hypertension, Hypertension Research *33 (2010): 836–843; J. J. Whyte and M. H. Laughlin, The effects of acute and chronic exercise on the vasculature,* Acta Physiologica *199 (2010): 441–450; G. Lippi and N. Maffulli, Biological influence of physical exercise on hemostasis,* Seminars in Thrombosis and Hemostasis *35 (2009): 269–276.*

KEY POINTS

- Cardiorespiratory endurance training enhances the ability of the heart and lungs to deliver oxygen to body tissues.
- Cardiorespiratory training activities elevate the heart rate for sustained periods of time and engage the body's large muscle groups.

The Active Body's Use of Fuels

LO 10.3 Explain the importance of glucose, fatty acids, and amino acids to a working athlete before, during, and after vigorous exercise.

Whether belonging to an athlete, a growing child, or an office worker, the human body functions best on a diet of nutrient-dense foods, as explained in Chapter 2.

In addition, however, many people whose lives involve high-level physical demands, such as athletes, firefighters, military personnel, and the like, may benefit from applying the concepts of the next sections.

The Need for Food Energy

Competitive athletes can expend enormous amounts of fuel during training and competition. To prevent unwanted weight loss, such athletes may need to take in an extra 1,000 to 1,500 or even more calories above their ordinary intakes each day. For some minutes or hours following activity, an athlete's metabolism may stay high and continue to expend extra fuels, even during rest. This phenomenon, known as **excess postexercise oxygen consumption (EPOC)**, occurs mainly with activities of high intensity (greater than 70 percent of VO_{2max}).[37]

In contrast, the great majority of physically active people who work out lightly two or three times a week for fitness or weight loss require few or no extra calories. These active people need only consume a nutritious calorie-controlled diet that follows the eating patterns of the *Dietary Guidelines for Americans* (see Chapter 2), along with proper hydration, to perfectly meet their needs. Fitness seekers who dream of a quick and easy workout that "burns fat while they sleep" should be aware that some minimum threshold of intensity and duration must be met to induce even small passive energy expenditures.[38] However, recent research suggests that 30 minutes daily of moderate-intensity aerobic endurance training may be more *efficient* for producing energy deficits than greater amounts of exercise.[39] Study subjects who performed double the work did spend more energy doing so but lost about the same amount of weight and fat as the moderate exercise group.

KEY POINTS

- Food energy needs vary by the goals and activities of the athlete.
- Excess postexercise oxygen consumption (EPOC) can pose weight-loss problems for some athletes but most weight-loss seekers do not achieve significant calorie deficits from EPOC.

Glucose: A Major Fuel for Physical Activity

Glucose is vital to physical activity. In the first few minutes of an activity, muscle glycogen provides the great majority of the extra energy that muscles use for action. This makes sense because glucose quickly yields energy needed for fast action. The muscles store glycogen in their cells, both inside and outside of their contractile fibers, to ensure a quickly accessible supply of glucose energy.[40]

In addition to using their own glycogen, exercising muscles draw available glucose from the bloodstream. You might suspect, then, that exercise would cause a drop in blood glucose concentration, but this is not the case. *Before* a fall in blood glucose can occur, exercise triggers a host of messenger molecules to be released into the bloodstream, including the pancreatic hormone glucagon.[41] Glucagon signals the liver to liberate glucose from its glycogen stores and to make new molecules of glucose for release into the bloodstream that are picked up and used by the working muscles.

Glycogen supplies are not inexhaustible. They are limited to less than 2,000 calories of glucose. An athlete's fat stores, in comparison, can total 70,000 calories or more and can fuel many hours of activity. Fat, however, cannot sustain physical work without glucose, and at some point during sustained physical activity, the body's glycogen begins to run out. The liver simply cannot make glucose fast enough to meet the demand.

Glucagon's effects on the liver are explained in **Chapter 4**.

Glycogen and Endurance The athlete who begins an activity with full glycogen stores has more glucose fuel to last longer during sustained exercise. For most active people, a normal, balanced diet keeps glycogen stores full. For an athlete engaged in heavy training or competition, the more carbohydrate the person eats, the more glycogen the muscles will store (within limits), and the longer the stores will last to support physical activity.

A classic study compared fuel use during physical activity by three groups of runners, each on a different diet.[42] For several days before testing, one of the groups ate a normal mixed diet (55 percent of calories from carbohydrate); a second group ate

excess postexercise oxygen consumption (EPOC) a measure of increased metabolism (energy expenditure) that continues for minutes or hours after cessation of exercise.

Figure 10–3

Animated! The Effect of Diet on Physical Endurance

A high-carbohydrate diet best supports an athlete's endurance. In this classic study, the high-fat diet provided 94 percent of calories from fat and 6 percent from protein; the normal mixed diet provided 55 percent of calories from carbohydrate; and the high-carbohydrate diet provided 83 percent of calories from carbohydrate.

Source: J. Bergstrom and coauthors, Diet, muscle glycogen, and physical performance, Acta Physiologica Scandinavica *71 (1967): 140–150.*

a high-carbohydrate diet (83 percent of calories from carbohydrate); and the third group ate a high-fat diet (94 percent of calories from fat). As Figure 10–3 shows, the high-carbohydrate diet enabled the athletes to work longer before exhaustion. These results, now many times confirmed, established that higher intakes of dietary carbohydrate help to sustain an athlete's endurance by ensuring ample glycogen stores.[43]

Glucose from the Digestive Tract In addition to the body's stored glycogen, glucose in the digestive tract makes its way to the working muscles during activity.[44] Researchers recently gave partial credit to food intake for the success of ultramarathon runners who finished a 100-mile race.[45] The finishers consumed almost twice the calories and carbohydrates per hour during the race as nonfinishers. (They also took in more fluid and a little more fat and sodium.) It may be that the carbohydrate, calories, or other constituents in the food provided what they needed to keep going after others were exhausted. In addition to endurance athletes, those who compete in sports that require repeated bursts of intense activity, such as basketball or soccer may also benefit from taking in extra carbohydrate during an event.[46]

Before concluding that extra glucose during activity might boost your own exercise performance, consider first whether you engage in sustained endurance activity or repeated high-intensity activity. Do you run, swim, bike, or ski nonstop at a rapid pace for more than an hour at a time? Do you compete in high-intensity games lasting for several hours? Does your sport or training demand several bouts of high-intensity activity in one day, or is it repeated on several successive days? If not, you may be better served by eating ample carbohydrate in the context of a regular nutritious diet. Even in athletes, extra carbohydrate during activity is of no help when fatigue is unrelated to glucose supplies, as is true of those who compete in 100-meter sprinting, 100-meter swimming, or power lifting.[47]

KEY POINTS

- During activity, the hormone glucagon helps to prevent a drop in blood glucose.
- Glycogen stores in the liver and muscles affect an athlete's endurance.
- When glucose availability limits performance, carbohydrate taken during training or competition can support physical activity.

The Aerobic and Anaerobic Difference References to "aerobic" and "anaerobic" activities actually refer to two parts of the body's metabolic system for extracting

Figure 10–4

Glucose and Fatty Acids in Their Energy-Releasing Pathways in Muscle Cells

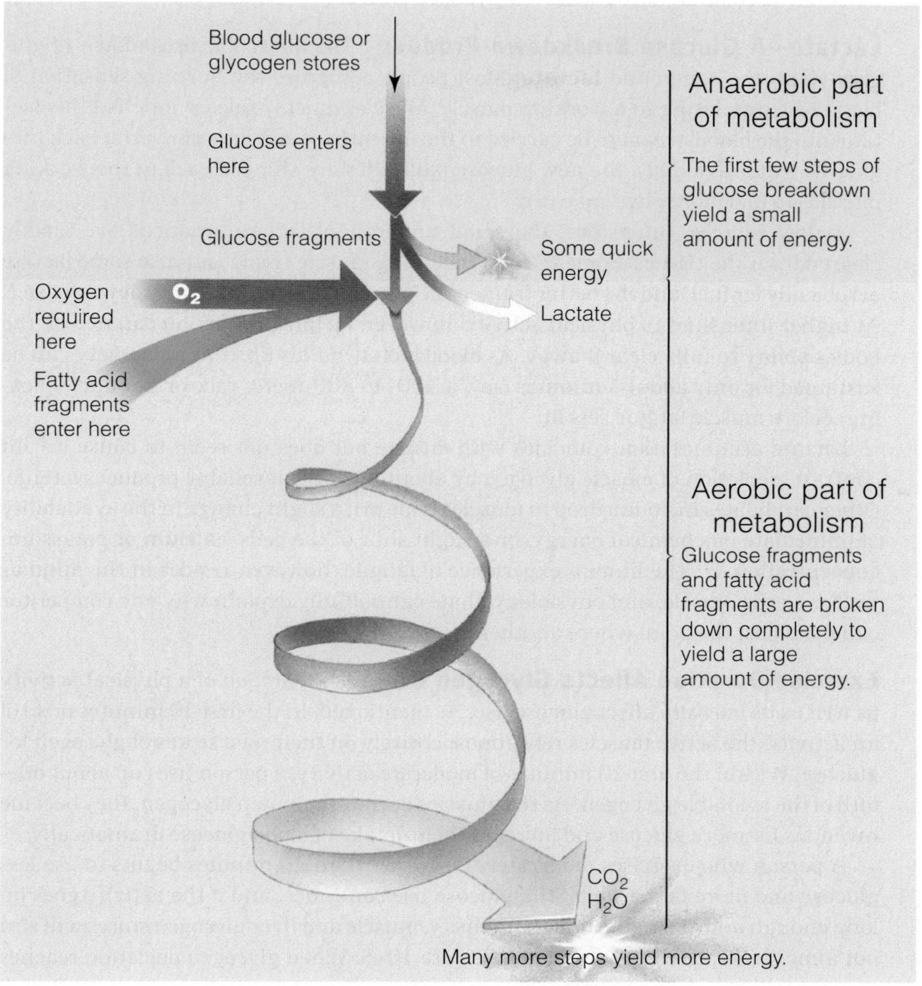

energy from energy-yielding nutrients. One part, the efficient oxygen-dependent **aerobic** part, wrings every last calorie of energy from each glucose molecule, as well as from the body's more abundant fuel, fatty acids. The **anaerobic** part of the system also yields energy but only from glucose, and it extracts less energy per glucose molecule than aerobic metabolism. Its advantage is the ability to provide energy without the input of oxygen, an advantage that gains importance as physical activity intensifies. Figure 10–4 illustrates both parts of the system.

Intense physical activity—the kind that makes it difficult "to catch your breath," such as a quarter-mile race—demands a great deal of energy quickly, so quickly that the demand easily outpaces the body's ability to provide it through its efficient aerobic (oxygen-using) fuel system. The lungs, heart, and blood vessels simply cannot deliver enough oxygen quickly enough to meet the demand. Muscles therefore must rely more on their anaerobic glucose metabolism system to provide the energy for high-intensity work.

Thus, for high-intensity activities, the muscles must draw heavily on their limited glycogen supply to fuel their work. The upper portion of Figure 10–4 illustrates that glucose can yield energy quickly in anaerobic metabolism.

Aerobic Use of Fuel In contrast to high-intensity muscular work, moderate physical activity, such as easy jogging, uses glycogen more slowly. The individual

aerobic (air-ROH-bic) requiring oxygen. Aerobic activity strengthens the heart and lungs by requiring them to work harder than normal to deliver oxygen to the tissues.

anaerobic (AN-air-ROH-bic) not requiring oxygen. Anaerobic activity is of high intensity and short duration.

breathes easily, and the heart beats at a faster pace than at rest but steadily—the activity is aerobic. The bottom half of Figure 10–4 shows that the ample oxygen supplies during aerobic activity allow the muscles to extract their energy from both glucose and fatty acids by way of aerobic metabolism. By depending partly on fatty acids, moderate aerobic activity conserves glycogen stores.

Lactate—A Glucose Breakdown Product The anaerobic breakdown of glucose yields the compound **lactate**. Most people recognize the burning sensation of lactate accumulating in a working muscle. Muscles quickly release much of this lactate into the bloodstream to be carried to the liver where enzymes convert it back into glucose. After assembly, the new glucose molecules are shipped back to the working muscles to fuel more physical work.

At low exercise intensities, the small amounts of lactate produced are readily cleared from the tissues. Some tissues, including muscle tissue, can use some lactate aerobically for fuel, and the better trained the muscles, the more lactate they can use.[48] At higher intensities of physical activity, however, lactate production can exceed the body's ability to fully clear it away. As blood lactate builds up, intense activity can be sustained for only about 3 minutes (say, a 400- to 800-meter race or a round of boxing) before muscle fatigue sets in.

Lactate accumulation coincides with fatigue but does not seem to cause it.[49] In contrast, depletion of muscle glycogen by about 80 percent reliably produces fatigue. Other candidates include a drop in muscle tissue pH, a slight change in the availability of immediate biochemical energy, or a slight shift in the cells' calcium or potassium concentration.[50‡‡] The human experience of fatigue, however, resides in the mind as well as in the muscle, and physiology alone cannot fully explain why one competitor can push past the point where another must stop.[51]

Exercise Duration Affects Glycogen Use The *duration* of a physical activity as well as its *intensity* affect glucose use. As mentioned, in the first 10 minutes or so of an activity, the active muscles rely almost entirely on their own stores of glycogen for glucose. Within the first 20 minutes of moderate activity, a person uses up about one-fifth of the available glycogen. As the muscles devour their own glycogen, they become ravenous for more glucose and increase their uptake of blood glucose dramatically.

A person who exercises moderately for longer than 20 minutes begins to use less glucose and more fat for fuel. Still, glucose use continues, and if the activity goes on long enough and at a high enough intensity, muscle and liver glycogen stores will run out almost completely, as depicted in Figure 10–5. When glycogen depletion reaches a certain point, it brings nervous system function almost to a halt, making continued activity at the same intensity impossible.[52] Marathon runners refer to this point of exhaustion as "hitting the wall" or "bonking."

My Turn watch it! How Much Is Enough?

Listen to two athletes talk about carbohydrate intakes and other eating strategies.

Julian

Adam

‡‡ The most immediate form of biochemical energy is ATP (adenosine triphosphate), which can also be replenished by way of the intermediary compound, phosphocreatine.

Figure 10–5

Glycogen—Before and After Physical Activity

These electron micrographs magnify part of a muscle cell by 20,000 times, revealing the orderly rows of contractile structures within. The dark granulated substance is glycogen. In the photo on the left, the cell's glycogen stores are full; on the right, they have been depleted by exercise.

1. The orderly horizontal rows that appear to be striped at intervals are protein structures that contract the muscles.[a]

2. The elongated black rows between the contractile structures contain much of the muscle's glycogen. More glycogen granules (black dots) are also scattered within the contractile parts (visible at left).

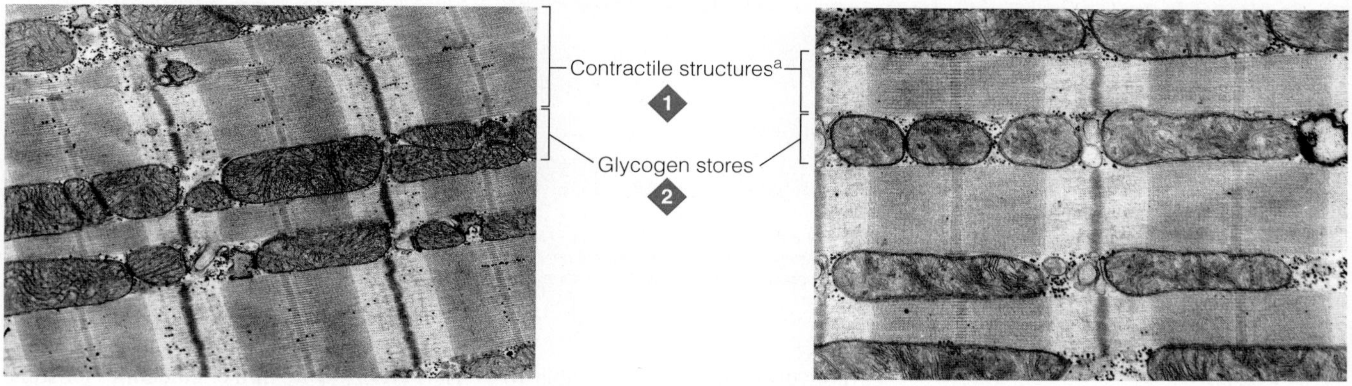

[a]*The contractile structures of the muscle cells are called myofibrils.*

Dr. Donald Fawcett/Visuals Unlimited, Inc. (both photos); Art © Cengage Learning 2014

Degree of Training Affects Glycogen Use Consistent training affects glycogen use during activity in two major ways. First, muscles adapt to their work by storing greater amounts of glycogen needed to support that work. Second, trained muscles burn more fat, and at higher intensities, than untrained muscles so they require less glucose to perform the same work. A person first attempting an activity uses up much more glucose per minute than an athlete trained to perform it.

In summary, these three factors affect glucose use during physical activity:

- Carbohydrate intake.
- Intensity and duration of the activity.
- Degree of training.

KEY POINTS

- Anaerobic breakdown of glucose yields energy and becomes particularly important in high-intensity activity.
- The more intense an activity, the more glucose it demands for anaerobic metabolism.
- Lactate accumulates during anaerobic metabolism but does not appear to cause fatigue.
- Glycogen is used at a rapid rate early in exercise but the rate slows with continued activity.
- When glycogen depletion reaches a certain point, continued activity of the same intensity is impossible.
- Highly trained muscles use less glucose and more fat than do untrained muscles to perform the same work.

Carbohydrate Recommendations for Athletes

To postpone fatigue and maximize performance, athletes must maintain available glucose supplies for as long as they can. To do so, athletes need abundant

lactate a compound produced during the breakdown of glucose in anaerobic metabolism.

Table 10–5

Suggested Daily Carbohydrate Intakes for Athletes

These general research-based guidelines should be adjusted to an athlete's calorie needs, training type, and performance. To calculate daily grams (g) of carbohydrate that may be appropriate for an athlete, first divide pounds of body weight by 2.2 to find kilograms (kg). Then identify the number of carbohydrate grams best suited to the athlete's performance: multiply kg by g.

	RECOMMENDATIONS	CARBOHYDRATE INTAKES (g/day)	
Athletes	Carbohydrate g/kg/day	Reference Male (70 kg)	Reference Female (55 kg)
Casual exercisers (low intensity)	3–5	210–350	165–275
Most athletes (moderate intensity, ≤1 h/day)	5–7	350–490	275–385
Endurance athletes (moderate to high intensity, 1–3 h/day)	6–10	420–840	330–660
Ultraendurance athletes (moderate to high intensity, 4–5 h/day)	8–12	560–840	440–660

Source: Data from C. Rosenbloom and E. J. Coleman, eds., Sports Nutrition: A Practice Manual for Professionals, *5th ed. (Chicago: The Academy of Nutrition and Dietetics, 2012), p. 469; L. M. Burke and coauthors, Carbohydrates for training and competition,* Journal of Sports Sciences *29 (2011): S17–S27; Position of the American Dietetic Association, Dieticians of Canada, and the American College of Sports Medicine,* Journal of the American Dietetic Association *109 (2009): 509–527.*

carbohydrate in their diets. Table 10–5 lists carbohydrate intake guidelines for some athletes. Note that athletes need a minimum number of *grams* of carbohydrate per unit of body weight to achieve full glycogen stores for a given activity, and their recommendations are expressed as grams per kilogram of bodyweight per day (g/kg/d).[53] (The margin note shows how to convert pounds to kilograms.) To maintain adequate glycogen stores for consecutive days of heavy training or competition, some athletes may benefit from large intakes of carbohydrate—perhaps as much as 7 to 12 g/kg/d. The Food Feature of this chapter demonstrates how to design a diet that delivers the carbohydrate an athlete needs.

Do the Math

Find kilograms by dividing pounds by a factor of 2.2. For example, for a 130-lb person, 130 lb ÷ 2.2 = 59 kg (rounded)

Glucose before Activity Most of the athlete's glucose is provided by carbohydrate-rich meals. In addition, however, glucose taken within a few hours before training or competition is thought to "top off" an athlete's glycogen stores to provide the greatest possible glucose supply to support sustained activity. The **pregame meal** can take many forms, and the Food Feature of this chapter describes them in full.

© Isa Ismail/Shutterstock.com

Endurance activities demand fluid and carbohydrate fuel. Don't forget to hydrate.

Glucose during Activity As mentioned, evidence supports the idea that ingestion of carbohydrate during prolonged activity can postpone fatigue.[54] Eating during activity can be tricky, however. The best glucose sources are easily consumed, are smooth-textured, and are low in fiber and fat to facilitate glucose absorption.[55] During long bicycle races, for example, competitors may consume bananas, fruit juices, dried fruit, and energy bars that provide carbohydrate energy and help banish distracting feelings of hunger. (Extreme caution is required to prevent choking.) For athletes who cannot eat solid foods while exercising, commercial **high-carbohydrate energy drinks** or commercial **high-carbohydrate gels** are a portable, easy-to-consume alternative, which most, but not all, athletes can tolerate.[56] Such products are higher in calories and carbohydrate than the fluid-replacement sports drinks discussed later in the Consumer's Guide. Concentrated beverages and gels must be taken with extra water to ensure hydration during activity, however.

Glucose after Activity After an event or training session, the quick recovery of glycogen stores can be important to athletes who compete or train intensely more than once a day or on consecutive days with less than a 24-hour recovery period. A window of opportunity occurs during the hour or two following glycogen-depleting physical activity, when carbohydrate intake appears to speed up the rate of glycogen synthesis.[57] This faster rate of glycogen storage may help to restore glycogen for the next bout of high-intensity training or competition. The concept of recovery meals and its application in an athlete's diet are described in this chapter's Food Feature section.

KEY POINTS

- Carbohydrate recommendations for athletes are stated in grams per kilogram of body weight per day.
- Carbohydrate before, during, and after physical exertion can help to support the performance of certain kinds of activities but not others.

Lipid Fuel for Physical Activity

Unlike the body's limited glycogen stores, fat stores can fuel hours of activity without running out; body fat is (theoretically) an unlimited source of energy for exercise. Even the lean bodies of elite runners carry enough fat to fuel several marathon runs.

Early in activity, muscles begin to draw on fatty acids from two sources—fats stored within the working muscles and fats from fat deposits such as the adipose tissue under the skin.[58] Areas with the most fat to spare donate the greatest amounts. This is why "spot reducing" doesn't work: muscles do not own the fat that surrounds them. Instead, adipose tissue cells release fatty acids into the blood for all the muscles to share. Proof is found in a tennis player's arms: the fatfolds measure the same in both arms, even though the muscles of one arm are more developed than the other.

Activity Intensity and Duration Affect Fat Use Fat can be broken down for energy only by aerobic metabolism. During physical activity of light or moderate intensity, fatty acids from adipose tissue are released into the blood stream and provide the majority of the fuel for muscular work. When the intensity of activity becomes so great that energy demands surpass the ability to provide more energy aerobically, the muscles cannot burn more fat. They burn more glucose instead.[59] Adipose tissue adjusts its delivery of fatty acids to match the needs of the muscles at work, releasing more during moderate activity and restricting lipid release during high-intensity exercise.[60]

The *duration* of activity also affects fat use. At the start of activity, the blood fatty acid concentration falls, but a few minutes into moderate activity, blood flow through the adipose tissue capillaries greatly increases, and hormones, including epinephrine, signal the fat cells to break apart their stored triglycerides. Fatty acids flood into the bloodstream at double or triple the normal rate. After about 20 minutes of sustained, moderate aerobic activity, the fat cells begin to shrink in size as they empty out their lipid stores.[61]

Degree of Training Affects Fat Use Training, performed consistently, stimulates the muscles to develop more fat-burning metabolic enzymes, so trained muscles burn more fat at greater exercise intensities than untrained muscles do. With aerobic training, the heart and lungs also become stronger and better able to deliver oxygen to the muscles during high-intensity activities. This improved oxygen supply, in turn, also helps the muscles to burn more fat.

Fat Recommendations for Athletes For endurance athletes, eating a high-fat, low-carbohydrate diet for even a day or two reduces precious glycogen stores and impairs performance. Eventually, muscles do adapt to such a diet and use more fat to fuel activity, but no performance benefits result from doing so. Athletes on high-fat diets report greater fatigue and perceive physical work as more strenuous than those on high-carbohydrate diets.[62]

Essential fatty acids and fat-soluble nutrients are as important for athletes as they are for everyone else, so experts recommend a diet with 20 to 35 percent of calories from fat.[63] Athletes should make it a point to obtain the needed raw vegetable oils, nuts, olives, fatty fish, and other sources of health-promoting fats each day. Omega-3

pregame meal a meal consumed in the hours before prolonged or repeated athletic training or competition to boost the glycogen stores of endurance athletes.

high-carbohydrate energy drinks flavored commercial beverages used to restore muscle glycogen after exercise or as pregame beverages.

high-carbohydrate gels semi-solid easy-to-swallow supplements of concentrated carbohydrate, commonly with potassium and sodium added; not a fluid source.

fatty acids in particular play roles in reducing inflammation—and tissue inflammation is both the result and the enemy of physical performance. This doesn't mean that athletes need fish oil supplements; they do need to consume the amounts of fatty fish recommended for health (see Chapter 5).

As for saturated and *trans* fats, they pose the same heart disease risk for an athlete as they do for other people. Physical activity offers some protection against cardiovascular disease but athletes still suffer heart attacks and strokes. Controlling saturated and *trans* fat intake is a priority for protecting the health of an athlete's heart and arteries.

To summarize, then, these three factors affect fat use during physical activity:

- Fat intake.
- Intensity and duration of the activity.
- Degree of training.

KEY POINTS

- The intensity and duration of activity, as well as the degree of training, affect fat use.
- A diet high in saturated or trans fat raises an athlete's risk of heart disease.

Protein for Building Muscles and for Fuel

The human body uses amino acids from protein to build and maintain muscle and other lean tissue and, to some extent, to provide fuel. Physical activity provides the primary signal for building muscle proteins to support physical work, but also critical is a sufficient intake of high-quality protein to supply the essential amino acids required to build them.

Repetitive muscle contractions of physical activity send a signal to muscle cells to make more of the specific proteins needed to support the work. For 24 to 48 hours following exercise of sufficient intensity, muscles speed up their rate of protein synthesis.[64] To meet the demand for new proteins, muscle cells need sufficient essential amino acids for building them, some derived from the breakdown of body proteins and more from high-quality protein foods in the diet.[65]

Amino Acids Stimulate Muscle Protein Synthesis When essential amino acids arrive in muscle tissue, protein synthesis speeds up.[66] Experiments have shown that an infusion of essential amino acids, particularly **leucine**, causes the rate of protein synthesis to triple for about an hour or two.[67] After this time, even with a continuing excess of essential amino acids, the rate of muscle protein synthesis quickly drops off. Researchers theorize that after muscles build proteins for a time, they reach a point of being "full" of new proteins, and they stop building them, even though essential amino acids are plentiful.

Courtesy of the Author

Physical activity itself triggers the building of muscle proteins.

This is not to say that people can build bigger muscles by laying on the couch and eating protein—only physical activity can cause a net muscle protein gain. Likewise, consuming more than about 20 grams of protein at a time cannot force the muscles to exceed their protein-building limits. After the "muscle full" state is reached, excess amino acids are dismantled and burned off as fuel.[68] This may explain why taking large doses of protein or amino acids cannot force muscles to gain extra bulk. However, when enough high-quality protein is consumed regularly, and adequate resistance exercise is performed repeatedly, gains in muscle strength and bulk follow reliably behind.

Does Timing of Protein Matter? Research suggests a synergistic effect between physical activity and essential amino acids on the rate of muscle protein synthesis. During the hour or two following intense physical activity, consuming the essential amino acids in, say, a cup-and-a-half to two cups of low-fat milk accelerates muscle protein synthesis beyond the rate expected from either exercise or essential amino acids alone.[69]

> "High quality" means protein with the complete array of essential amino acids needed for protein synthesis, as explained in **Chapter 6**.

Does this mean that, to build bigger, stronger muscles, athletes should take in some protein in the hour or two after exercise? Research has yet to resolve this question, but muscle physiology doesn't support the idea. The speed of muscle protein synthesis is indeed greatest in the two hours following exercise, but synthesis remains elevated to a degree for some 24 to 48 hours longer. During this time, whenever essential amino acids arrive from protein-rich meals, muscle tissues speed up their protein synthesis and build the protein structures they need to perform the activity.[70]

Protein Fuel Use in Physical Activity Studies of nitrogen balance show that the body speeds up its use of amino acids for fuel during physical activity, just as it speeds up its use of glucose and fatty acids. The factors that regulate protein use during activity seem to be the same three that regulate the use of glucose and fat: diet, exercise intensity and duration, and degree of training. As for diet, sufficient carbohydrate spares protein from being used as fuel—too little carbohydrate necessitates the conversion of amino acids to glucose.

Exercise intensity and duration also affect protein fuel use. When endurance athletes train for longer than an hour a day and deplete their glycogen stores, they become more dependent on protein fuel. In contrast, intense anaerobic strength training does not use as much protein fuel but demands protein for building muscle tissue.

Finally, the extent of training also affects the use of protein. Particularly in strength athletes such as body builders, the higher the degree of training, the less protein a person uses during activity at a given intensity. To summarize, the factors that affect protein use during physical activity include:

- Dietary carbohydrate sufficiency.
- Intensity and duration of the activity.
- Degree of training.

Protein Recommendations for Athletes For both endurance and strength, athletes need more protein than others. On learning of their higher protein needs, many athletes go to extremes doubling or tripling the number of protein-rich foods they eat while ignoring other needed foods and nutrients. Also, too much protein generates excess nitrogen, which must then be excreted in urine.

Athletes who take amino acid pills or products in hopes of increasing protein synthesis and inhibiting muscle breakdown should know that the amino acids they seek are delivered in the protein-rich foods that they eat every day, such as milk, chili, or turkey sandwiches.[71] Such foods present plenty of amino acids in the right mix to both stimulate and support protein synthesis.[72] In addition, protein-rich foods present no risk of amino acid imbalances, a known drawback of supplements.

Now you can understand the whole picture of the body's use of fuel during physical activity. At rest and during activities of daily living, fatty acids provide 80 to 90 percent of the body's energy and glucose provides only 5 to 18 percent (amino acids provide a small amount—2 to 5 percent). At the point of exercise intensity (to about 75 percent VO_{2max}) when fat cannot provide sufficient energy to support the work, the muscles shift to using a great deal more glucose to provide the extra energy that is required.[73] The degree of a person's training greatly influences the mix of energy fuels used at any given level of exertion.

Did You Know?

If you are looking for a protein or amino acid supplement with a perfect score of 100 for protein quality, try these: 2½ oz of chicken breast or lean beef, ¾ c low-fat cottage cheese, or 12 ounces of yogurt. Each provides about 20 g (20,000 mg) of protein, with essential amino acids in perfect complement for use by the human body.

KEY POINTS

- Physical activity stimulates muscle cells to both break down and synthesize proteins, resulting in muscle adaptation to activity.
- Athletes use amino acids for building muscle tissue and for energy; dietary carbohydrate spares amino acids.
- Diet, intensity and duration of activity, and degree of training affect protein use during activity.

How Much Protein Should an Athlete Consume?

The DRI does not recommend greater-than-normal protein intakes for athletes, but other authorities do.[74] These recommendations vary by the nature of the activities

leucine one of the essential amino acids; it is of current research interest for its role in stimulating muscle protein synthesis.

Table 10–6

Recommended Protein Intakes for Athletes

	RECOMMENDATIONS (g/kg/day)	PROTEIN INTAKES (g/day)	
		Reference[a] Male (70 kg)	Reference[a] Female (55 kg)
DRI recommended intake	0.8	56	44
Recommended intake for power (strength or speed) athletes	1.2–1.7	84–119	66–94
Recommended intake for endurance athletes	1.2–1.4	84–98	66–77
U.S. average intake		102	70

[a] *Daily protein intakes are based on a 70-kilogram (154-pound) reference man and a 55-kilogram (121-pound) reference woman.Other individuals must calculate their recommended intakes using the numbers of column 1. (For kg divide lb by 2.2.)*

Sources: Position of the American Dietetic Association, Dietitians of Canada, and the American College of Sports Medicine: Nutrition and athletic performance, Journal of the American Dietetic Association *109 (2009): 509–527; U.S. Department of Agriculture, Agricultural Research Service, 2008. Nutrient Intakes from Food: Mean Amounts Consumed per Individual, One Day, 2005–2006. Available: www.ars.usda.gov/ba/bhnrc/fsrg; Committee on Dietary Reference Intakes,* Dietary Reference Intakes for Energy, Carbohydrate, Fiber, Fat, Fatty Acids, Cholesterol, Protein, and Amino Acids *(Washington, D.C.: National Academies Press, 2005), pp. 660–661.*

performed (see Table 10–6).[75] As is true for carbohydrates, the protein recommendations are stated in grams per kilogram of body weight per day (g/kg/d).

You may be wondering whether you eat enough protein for your own activities. In general, a nutritious eating pattern that provides enough total energy and follows the USDA eating patterns provides enough protein for almost everyone—huge meat servings, raw eggs, special foods, or protein supplements do not further improve athletic performance and may even hinder it if protein displaces sufficient carbohydrate from the diet.[76] With too little carbohydrate, athletes will burn off as fuel the very protein they wish to retain in their muscles.

KEY POINT

- Although certain athletes may need some additional protein, a well-chosen diet that provides ample energy and follows the USDA food patterns provides sufficient protein, even for most athletes.

Vitamins and Minerals—Keys to Performance

LO 10.4 Outline reasons why iron is of special significance for some athletes, and describe the proper roles for nutrient supplements.

Vitamins and minerals are crucial for the working body. Many B vitamins participate in releasing energy from fuels. Vitamin C is needed for the formation of the protein collagen, the foundation material of bones, cartilage, and other connective tissues. Folate and vitamin B_{12} help build the red blood cells to carry oxygen to working muscles. Vitamin E helps protect tissues from oxidation. Calcium and magnesium allow muscles to contract, and so on. Do active people need more of these vitamins and minerals to support their work? Do they need supplements?

Do Nutrient Supplements Benefit Athletic Performance?

Many athletes take vitamin and mineral supplements. One of the most common reasons athletes at all levels give for supplement use is "to improve performance."

Contrary to some athletes' beliefs, taking vitamins or minerals just before competition will not help performance. Most vitamins and minerals function as small parts of larger working units. After entering the blood from the digestive tract, they must wait for the cells to combine them with their other parts before they can function. This takes time—hours or days. Nutrients taken right before an event cannot help performance, even if the person is deficient in those nutrients.

More Food Means More Nutrients For the well-nourished athlete, nutrient supplements do not enhance performance. Strenuous physical activity requires abundant energy but athletes and active people who eat enough nutrient-dense foods to meet their greater energy needs automatically consume more of the vitamins and minerals needed to process that energy. Active people eat more food; it stands to reason that with the right choices, they consume more vitamins and minerals, too.

Preventing Deficiencies Deficiencies of vitamins and minerals do impede performance, however. Athletes who starve themselves to meet a sport's weight requirement can easily fail to obtain needed nutrients.[77] Most authorities oppose rigid weight requirements because athletes often risk their health to meet them. For athletes forced to "make weight," or for those who simply cannot eat enough food to maintain body weight during intense periods of training and competition, a balanced multivitamin-mineral tablet providing no more than the DRI recommended intakes may prevent deficiencies.

KEY POINTS

- Vitamins and minerals are essential for releasing the energy trapped in energy-yielding nutrients and for other functions that support physical activity.
- Active people can meet their vitamin and mineral needs if they follow the USDA eating patterns and eat enough nutrient-dense food to meet their energy needs.

Iron—A Mineral of Concern

Iron deficiency impairs performance because iron must be present to deliver oxygen to the working muscles. Iron-containing molecules of aerobic metabolism and the iron-containing muscle protein myoglobin are essential to physical performance. With insufficient iron, aerobic work capacity is compromised and the person tires easily.

Strenuous endurance training promotes destruction of older, more fragile red blood cells: blood cells are squashed when body tissues, such as the soles of the feet, make high-impact contact with an unyielding surface, such as the ground.[78] At the same time, training increases the blood's fluid volume; with fewer red cells distributed in more fluid, the red blood cell count per unit of blood drops. Most researchers view this "sports anemia" as an adaptive, temporary response to endurance training that goes away by itself without treatment.

© James Peragine/Shutterstock.com

Female athletes may be at special risk of iron deficiency.

Physically active young women, particularly those who engage in endurance activities such as distance running, are a special case. Habitually low intakes of iron-rich foods, high iron losses through menstruation, and the high demands of physical performance can contribute to true iron deficiency and anemia in young female athletes. In addition, endurance activities temporarily increase the release of hepcidin, the body's iron-lowering hormone, described in Chapter 8.[79] Further research may reveal whether hepcidin lowers the blood iron of athletes.

Vegetarian athletes may also lack iron because the iron from plant sources is less available than iron from animal sources. To protect against iron deficiency, vegetarian athletes should make it a point to consume fortified cereals, legumes, nuts, and seeds and include some vitamin C–rich foods with each meal—vitamin C enhances iron absorption. A well-chosen vegetarian diet of nutrient-dense foods can meet nutrient needs for health and athletic performance. For anyone found to be iron-deficient through medical testing, prescribed supplements can reverse the condition in short order.

© ZUMA Press, Inc./Alamy

A well-chosen vegetarian diet supports physical activity. Timothy Bradley, a 28-year-old boxing champion, credits his vegan diet for providing an edge over the competition.

© Polara Studios

Foods like these are packed with the nutrients that active people need.

Table 10–7

Symptoms of Heat Stroke

Inadequate hydration hinders performance and elevates the risk of life-threatening heatstroke. Heat stroke is an emergency that demands immediate medical attention. If you suspect heat stroke, don't wait; call 911.

Life-threatening symptoms of heat stroke:

- Clumsiness, stumbling
- Confusion, dizziness, other mental changes, loss of consciousness
- Headache, nausea, vomiting
- Internal (rectal) temperature above 104° Fahrenheit
- Lack of sweating
- Muscle cramping (early warning)
- Racing heart rate
- Rapid breathing
- Skin may feel cool and moist in early stages; hot, dry, and flushed as body temperature rises

KEY POINTS

- Iron-deficiency anemia impairs physical performance because iron is the blood's oxygen handler.
- Sports anemia is a harmless temporary adaptation to physical activity.

Fluids and Temperature Regulation in Physical Activity

LO 10.5 Identify hazards associated with inadequate fluid replacement in the exercising body, and compare the fluid needs of a casual exerciser and an endurance athlete.

The body's need for water, while always greater than the need for any other nutrient, takes on particular urgency during physical activity. If the body loses too much water or the person takes in too much, the body's life-supporting chemistry is compromised.

Water Losses during Physical Activity

The exercising body loses water primarily via sweat; second to that, breathing excretes water, exhaled as vapor. Endurance athletes can lose a quart and a half or more of fluid during *each hour* of activity.

During physical activity, both routes of water loss can be significant, and dehydration is a real threat. The first symptom of dehydration is fatigue. A water loss of greater than 2 percent of body weight can reduce a person's capacity for muscular work.[80] A person with a water loss of about 7 percent is likely to collapse.

Sweat and Temperature Regulation Sweat is the body's coolant. The conversion of water to vapor uses up a great deal of heat, so as sweat evaporates, it cools the skin's surface and the blood flowing beneath it. During exercise, blood flow must divert to the skin to radiate heat away from the body's core. Sufficient water in the bloodstream is therefore crucial to provide sweat, accommodate blood flow to the skin, and still supply muscles with the blood flow they need to perform.

Heat Stroke In hot, humid weather, sweat may fail to evaporate because the surrounding air is already laden with water. Little cooling takes place, and body heat builds up. In such conditions, athletes must take precautions to avoid **heat stroke**—a potentially fatal medical emergency (its symptoms are listed in Table 10–7). To reduce the risk of heat stroke:

heat stroke an acute and life-threatening reaction to heat buildup in the body.

hypothermia a below-normal body temperature.

1. Drink enough fluid before and during the activity.
2. Rest in the shade when tired.
3. Wear lightweight clothing that allows sweat to evaporate.

Never wear rubber or heavy suits sold with promises of weight loss during physical activity. They promote profuse sweating, prevent sweat evaporation, and invite heat stroke.

If you experience any of the symptoms in Table 10–7, stop your activity, sip cold fluids, seek shade, wet your skin and clothing, and ask for help—heat stroke demands immediate medical attention. If someone else is experiencing heat stroke, call for emergency help, and keep them cool and moist by wetting and fanning them; generating a breeze speeds evaporation.

Hypothermia Even in cold weather, the body still sweats and needs fluids. However, the fluids should be warm or at room temperature to help prevent **hypothermia**. Inexperienced runners in long races on cold or wet chilly days may produce too little body heat to keep warm, especially if their clothing is inadequate. Early symptoms of hypothermia include shivers, apathy, and social withdrawal. As body temperature continues to fall, shivering stops; disorientation and slurred speech ensue. People with these symptoms soon become helpless to protect themselves from further body heat losses and need immediate medical attention.

KEY POINTS

- Evaporation of sweat cools the body, regulating body temperature.
- Heat stroke is a threat to physically active people in hot, humid weather, while hypothermia threatens exercisers in the cold.

Active people need extra fluid, even in cold weather.

Fluid and Electrolyte Needs during Physical Activity

The fluid needs of athletes is a topic of scientific debate, but current guidelines urge athletes to hydrate before activity to prepare for fluid losses, and to replace lost fluids both during and after activity.[81] Table 10–8 presents one schedule of hydration for physical activity. No one refutes the dangers of severe dehydration and the threat of heat stroke, but research is mixed on whether athletes who hydrate before and during physical activity perform significantly better than those who do not.[82] Such factors as body weight, genetic tendencies, type of sport, exercise intensity, and variations in ambient temperature and humidity all affect the degree of fluid and sodium losses through sweat.[83]

Table 10–8

Suggested Hydration Schedule for Physical Activity

The amount of fluid required for physical activity varies by the person's weight, genetics, previous hydration level, degree of training, environmental conditions, and other factors.

Timing	Recommendation (ml/kg body weight)	Common Measure	Example: 70-kg Athlete	Example: 55-kg Athlete
≥4 hours before activity	5 to 7 ml/kg	≈1 oz/10 lbs	≈1½ to 2 c	≈1 to 1½ c
2 hours before activity	If heavy sweating is expected, additional 3 to 5 ml/kg	≈0.6 oz/10 lbs	≈1 c (9 ounces)	≈1 c (7 ounces)
During activity	Limit dehydration to <2% body weight	—	Varies[a]	Varies[a]
After activity	—[b]	Drink ≥2 c for each pound of body weight lost[c]	Varies	Varies

[a]*A personal hydration plan, based on prior measures of fluid loss (weight) during the activity, is recommended. Thirst lags behind fluid loss and should be quenched immediately.*

[b]*Research to develop recommendations is ongoing.*

[c]*Hydration is most rapidly achieved with divided doses to provide 2 c every 20 to 30 min after exercise until the total is consumed.*

Source: C. A. Rosenbloom and E. J. Coleman, eds., Sports Nutrition: A Practice Manual for Professionals *(Chicago: Academy of Nutrition and Dietetics, 2012), p. 115; Position of the American Dietetic Association, Dietitians of Canada, and the American College of Sports Medicine: Nutrition and athletic performance,* Journal of the American Dietetic Association *109 (2009): 509–527.*

An athlete's **hourly sweat rate** can be determined by weighing before and after exercise and factoring in the duration of activity. The weight difference is almost all water, and it should be replaced pound for pound (a little more than two cups of water weighs one pound). Even then, in hot weather, the digestive tract may not be able to absorb enough water fast enough to keep up with an athlete's sweat losses, and some degree of dehydration may be inevitable. Athletes should stay mindful of their need for fluid, despite the distractions of competition, and a thirsty athlete shouldn't wait to drink. During activity, thirst is an indicator that some degree of fluid depletion has already taken place.[84] Adequate hydration is imperative for every athlete both in training and in competition. Any coach or athlete who withholds fluids during practice or competition takes a great risk and at the very least, may compromise athletic performance.

Food and Drink/SuperStock

Water What is the best fluid to support physical activity? Most often, just plain cool water, for two reasons: (1) water rapidly leaves the digestive tract to enter the tissues, and (2) it cools the body from the inside out. Endurance athletes are an exception: they may need more from their fluids than water alone. Endurance athletes do need water, but they also may need carbohydrate during prolonged activity to supplement their limited glycogen stores. Sports drinks are designed to provide both fluid and glucose and they are the topic of this chapter's Consumer's Guide.

Electrolyte Losses and Replacement During physical activity, the body loses electrolytes—the minerals sodium, potassium, and chloride—in sweat. Beginners lose these electrolytes to a much greater extent than do trained athletes because the trained body adapts to conserve them.

To replenish lost electrolytes, a person ordinarily needs only to eat a regular diet that meets energy and nutrient needs. During intense activity lasting more than 45 minutes in hot weather, sports drinks provide a convenient way to replace both fluids and electrolytes. Friendly, leisure sporting games almost never require electrolyte replacement. However, even participants in casual events require fluid replacement, particularly in hot weather, and water is the best fluid source under these conditions. Salt tablets can worsen dehydration and impair performance; they increase potassium losses, irritate the stomach, and cause vomiting. They are not recommended.

KEY POINT

- Water is the best drink for most physically active people, but some endurance athletes may need the glucose and electrolytes of sports drinks.

Table 10–9

Hyponatremia: Symptoms and Risk Factors

Hyponatremia, or too little sodium in the blood, can be life threatening.

Symptoms of hyponatremia

- Bloating, puffiness from water retention (shoes tight, rings tight)
- Confusion
- Seizure
- Severe headache
- Vomiting

Risk factors for hyponatremia

- Excessive water consumption during an event (>1.5 L/hr)
- Exercise duration greater than 4 hours
- Low body weight/BMI <20
- Nonsteroidal anti-inflammatory drug use (for example, aspirin or ibuprofen)
- Preexercise overhydration

Sodium Depletion and Water Intoxication

Replenishing electrolytes becomes crucial when competing in endurance sports that last for hours. Athletes who sweat profusely over a long period of time and do not replace lost sodium risk developing dangerous **hyponatremia**. Taking in too much plain water makes hyponatremia a real possibility and can be as life-threatening as taking in too little. The symptoms of hyponatremia differ somewhat from those of dehydration (see Table 10–9).

Athletes who lose more sodium in their sweat may be prone to debilitating **heat cramps**. To prevent both cramps and hyponatremia, endurance athletes who compete and sweat for four or more hours need to replace sodium during the events (not more than one gram of sodium per hour of activity has been recommended). Sports drinks and gels, salty pretzels, and other sodium sources can provide sodium when needed. In the days before the event, especially in hot weather, athletes should not restrict salt intakes.

While hyponatremia can pose a threat to some competitive athletes, most exercising people need not make any special effort to replace sodium. Their regular diets supply all that they need.

KEY POINTS

- During events lasting longer than four hours, athletes need to pay special attention to replacing sodium losses to prevent hyponatremia.
- Most exercising people get enough sodium in their normal foods to replace losses.

hourly sweat rate the amount of weight lost plus fluid consumed during exercise per hour.

hyponatremia (HIGH-poh-nah-TREE-mee-ah) a decreased concentration of sodium in the blood; also defined in Chapter 8.

heat cramps painful cramps of the abdomen, arms, or legs, often occurring hours after exercise; associated with inadequate intake of fluid or electrolytes or heavy sweating.

use it! A Consumer's Guide To . . . Selecting Sports Drinks

Imagine two thirsty people, both in motion:

- Jack, an accountant, striving to shed some pounds, is panting after his 30-minute jog; he wipes the sweat from his eyes and tries to catch his breath.
- Candace, point guard for her college basketball team, powers into her second hour of training, dripping with sweat from her exertion; she's training every muscle fiber for competition.

Both of these physically active people need to replace the fluid they've lost in sweat, but which kind of fluid best meets their needs?

Certainly, **sports drinks**, **flavored waters**, **nutritionally enhanced beverages**, and **recovery drinks** (see Table 10–10 for terms) are popular choices for fluid replacement. Sellers promote these pricey beverages with images of performance excellence, often boosted by celebrity athlete endorsements. Plain, freely available water also meets the fluid needs of most active people, but no celebrities make a case for drinking it. To decide which drink fits what need, consider these three factors: fluid, glucose, and electrolytes.

First: Fluid

Both sports drinks and plain water offer fluid to help offset fluid lost in sweat during physical activity. Some people find sports drinks tasty, and if a drink tastes good, they may drink more of it, ensuring adequate hydration. Bottled flavored waters may also taste good, but a squirt of lemon or other fruit juice added to plain water may do the same thing at a much lower cost. The text pointed out that hot, humid weather, among other factors, also elevates the body's need for fluid.

Second: Glucose

Unlike water, sports drinks offer monosaccharides or **glucose polymers** that can help maintain hydration, contribute to blood glucose, and enhance performance under specific circumstances. Any athlete performing an endurance activity at moderate or vigorous intensity for longer than 1 hour may benefit from some extra carbohydrate. A competitor who participates in a prolonged game that demands repeated intermittent strenuous activity—such as basketball—benefits from extra glucose during activity.[1]

For competitive athletes, not just any sugary beverage will do. To ensure water absorption while providing glucose, most sports drinks contain about 7 percent glucose (half the sugar of ordinary soft drinks). Less than 6 percent glucose may not enhance performance, and more than 8 percent can delay fluid passage from the stomach to the intestine, slowing delivery of the needed water to the tissues. Too much glucose may also cause abdominal cramps, nausea, and diarrhea—uncommon but serious drawbacks to high-carbohydrate gels or other concentrated glucose sources taken during exercise.

Sports drinks provide easy-to-consume glucose, but research shows that for athletes who can eat during activity, such as cyclists, half of a banana taken every 15 minutes during a 2½- to 3-hour bicycle race sustains blood glucose equally well.[2] Bananas better satisfy hunger, and they supply antioxidants, vitamins, minerals, and fiber in a mix of carbohydrates that the body is well-equipped to receive.

Jack, the jogger of our example, needs fluids to replace the fluid he loses in sweat. Advertisers of sports drinks would have him believe that he needs extra glucose in his fluid to hydrate better than water, but far from benefitting his health or performance, the drinks deliver unneeded calories of sugar in a nutrient-poor beverage. In fact, for him, and for anyone who goes for a walk, takes a spin on a bicycle, or exercises to lose weight, sports drinks are most likely counterproductive. Such people don't need the extra calories of these drinks, and extra carbohydrate cannot benefit their performance because their own glycogen is ample for the effort. In addition,

Table 10–10

Sports Drinks and Related Terms

- **flavored waters** lightly flavored beverages with few or no calories, but often containing vitamins, minerals, herbs, or other unneeded substances. Not superior to plain water for athletic competition or training.
- **glucose polymers** compounds that supply glucose, not as single molecules, but linked in chains somewhat like starch. The objective is to attract less water from the body into the digestive tract.
- **nutritionally enhanced beverages** flavored beverages that contain any of a number of nutrients, including some carbohydrate, along with protein, vitamins, minerals, herbs, or other unneeded substances. Such "enhanced waters" may not contain useful amounts of carbohydrate or electrolytes to support athletic competition or training.
- **recovery drinks** flavored beverages that contain protein, carbohydrate, and often other nutrients; intended to support postexercise recovery of energy fuels and muscle tissue. These can be convenient, but are not superior to ordinary foods and beverages, such as chocolate milk or a sandwich, to supply carbohydrate and protein after exercise. Not intended for hydration during athletic competition or training because their high carbohydrate and protein contents may slow water absorption.
- **sports drinks** flavored beverages designed to help athletes replace fluids and electrolytes and to provide carbohydrate before, during, and after physical activity, particularly endurance activities.

they may increase their risk for developing dental caries by often exposing their teeth to the sugar in the drinks. Plain, cool, zero-calorie, almost zero-cost water best meets their fluid needs.

Third: Sodium and Other Electrolytes

Sports drinks offer sodium and other electrolytes to help replace those lost during physical activity, and they may increase fluid retention. The sodium they contain may help to maintain the drive to drink fluid because the sensation of thirst depends partly upon the sodium concentration of the blood. Most athletes do not need to replace the other minerals lost in sweat immediately; a meal eaten within hours of competition replaces these minerals soon enough.

Most sports drinks are relatively low in sodium (55 to 110 milligrams per serving), so they pose little threat of excessive intake in healthy people. Casual exercisers may or may not need to replace electrolytes, depending upon weather conditions and degree of sweating. In Jack's case, the sodium in sports drinks is unnecessary.

Moving Ahead

In the end, most physically active people need fluid but none of the extra ingredients that sports drinks offer. For certain athletes, however, the glucose and sodium in sports drinks may provide advantages over plain water. Remember that regardless of the celebrity sales pitch used to market sports drinks, only Michael Jordan jumps like Michael Jordan—meaning that training and talent do not come in a bottle.

Still, if a sports drink boosts feelings of confidence, little harm can come from using it except to the athlete's finances. A money-saving strategy that reduces plastic bottles in the waste stream is to buy bulk powders to mix with tap water in your own reusable bottle, or fill it with refreshing, cold water.

Review Questions*

1. Many sports drinks offer monosaccharides __________ .
   a. that may help maintain hydration and contribute to blood glucose
   b. also called electrolytes, to help replace those lost during physical activity
   c. that provide a nutrient advantage to most people
   d. a and c
2. Which of these advantages do sports drinks provide over plain water?
   a. They taste good and so may lead people to drink more.
   b. They provide the vitamins and minerals that athletes need to compete.
   c. They improve the body's fitness for sport.
   d. b and c
3. People who take up physical activity for weight loss __________ .
   a. can increase weight loss by using sports drinks
   b. do not need the calories or sodium of sports drinks
   c. receive a performance boost from sports drinks
   d. all of the above

** Answers to Consumer's Guide review questions are found in Appendix G.*

Did You Know?

Beer is calorie-rich, but only one-third of its calories are from carbohydrates. The other two-thirds are from alcohol.

Beer facts:

- Beer is mineral-poor. Beer contains a few minerals, but to replace those lost in sweat, athletes need good sources such as fruit juices.
- Beer is vitamin-poor. Beer contains tiny traces of some B vitamins, but it cannot compete with rich food sources.
- Beer causes fluid losses. Beer contains the diuretic alcohol and causes the body to lose more fluid in urine than is provided by the beer.

Other Beverages

Carbonated beverages are not a good choice for meeting an athlete's fluid needs. Although they are composed largely of water, the air bubbles from the carbonation quickly fill the stomach and so may limit fluid intake and cause uncomfortable gas symptoms. They also provide few nutrients other than carbohydrate. Moderate doses of caffeine in beverages do not seem to hamper athletic performance and may even enhance it (details in the Controversy section).

Like others, athletes sometimes drink alcoholic beverages but these beverages do not serve as fluid replacements. Alcohol is a diuretic—it inhibits a hormone that prevents water loss and so promotes the excretion of water—exactly the wrong effect for fluid balance and athletic performance.[§§] Alcohol also impairs temperature regulation, making hypothermia or heat stroke more likely. It alters perceptions and slows reaction time. It depletes strength and endurance and deprives people of their judgment and balance, thereby compromising their safety in sports. Many sports-related fatalities and injuries each year involve alcohol. Do yourself a favor—choose a nonalcoholic beverage.

KEY POINTS

- Carbonated beverages can limit fluid intake and cause discomfort in exercising people.
- Alcohol use can impair performance in many ways and is not recommended.

[§§] The hormone is antidiuretic hormone (ADH).

Putting It All Together

This chapter opened with the statement that nutrition and physical activity go hand in hand, a relationship that now has been made clear. Training and genetics being equal, who would have the advantage in a competition—the person who arrives at the event with full fluid and nutrient stores and well-met metabolic needs or the one who habitually fails to meet these needs? Of course, the well-fed athlete has the edge. In addition, performance nutrition adaptations may further increase this advantage. Table 10–11 sums up the recommendations for performance nutrition, and the Food Feature, next, demonstrates their application.

Table 10–11

Overview of Performance Nutrition

An individual's personal goals and the intensity, duration, and frequency of his or her physical activity determine which of these recommendations may be of benefit (see the text).

Nutrients	*Dietary Guidelines for Americans/DRI*	Performance Nutrition Recommendations
Energy	Meet but do not exceed calorie needs.	▪ Consume adequate additional calories to support training and performance, and to achieve or maintain optimal body weight. ▪ Calorie deficits for weight loss, when needed, should begin in off season or early in training; calorie deficits can impede performance.
Carbohydrate	Consume between 45% and 65% of calories as carbohydrate; consume at least 130 g carbohydrate per day to prevent ketosis.	▪ Recommendations vary from 3 to 12 g/kg/day (see Table 10–5). ▪ Carbohydrate deficits impede performance. For moderate or vigorous exercise of 1–1.5 hr duration: ▪ *Preexercise:* Consume a high-carbohydrate, low-fiber snack (use proper timing—see text for details). ▪ *Midexercise:* Consume 30–60 g of easy-to-digest carbohydrates (sports drinks, gels, or foods) per hour of exercise. For moderate or vigorous exercise of ≥1.5 hr duration; multiple daily competitive events; or high-intensity weight training, do above plus: ▪ *Postexercise:* Recover lost glycogen with adequate carbohydrate at the next meal (1–3 hr after exercise).
Protein	Consume between 10% and 35% of calories from protein (adults); consume 0.8 g/kg/day of protein.	▪ Recommendations vary from 1.2 to 1.7 g/kg/day (see Table 10–6). ▪ Most U.S. diets supply sufficient protein for muscle growth and maintenance for most athletes. ▪ *Postexercise:* Consume high-quality protein at meals and snacks to facilitate and support muscle protein synthesis. ▪ Food is the preferred protein source.
Fat	Consume between 20% and 35% of calories from fat (adults); hold saturated fat to 10% of calories; keep cholesterol and *trans* fat intake low.	▪ Follow general recommendations.
Vitamins and minerals	Meet the DRI recommended intakes with a well-planned diet of nutrient-dense foods.	▪ Follow general recommendations.
Fluid	A wide range of fluid intakes maintain hydration in individuals, averaging 13 c (men) or 9 c (women).	▪ Balance fluid intake with fluid loss. ▪ *Preexercise; postexercise*: Maintain hydration before and after activity (see Table 10–7). ▪ *Midexercise:* When sweating, drink ½–1 c every 15 min to minimize fluid losses. ▪ *Postexercise:* Drink 2 c for each pound of body weight lost as sweat during activity. ▪ *At all times:* Continue to maintain hydration.

Sources: U.S. Department of Agriculture and U.S. Department of Health and Human Services, Dietary Guidelines for Americans 2010, *available at www.dietaryguidelines.gov; M. Dunford and A. Doyle,* Nutrition for Sport and Exercise *(Belmont: Cengage Learning, 2012), p. 15; C. Rosenbloom, Food and fluid guidelines before, during, and after exercise,* Nutrition Today *47 (2012): 63–69.*

try it! Food Feature

Choosing a Performance Diet

LO 10.6 Plan a nourishing and adequate diet for an athlete, including snacks and pregame meals.

Many different diets can support physical performance—and no one diet works best for everyone, so preferences should be honored. Perhaps most importantly, the diet should comply with standard diet planning principles to protect the person's health while promoting optimal physical performance.

Nutrient Density

Active people need nutrient-dense foods to supply vitamins, minerals, and other nutrients. Athletes must also eat for energy, and, as the chapter mentioned, their energy needs can be immense. Frequent nutritious between-meal snacks can provide the extra calories needed to maintain body weight (Figure 10–6 offers suggestions).

When athletes try to meet their energy needs with mostly empty-calorie, highly refined or highly processed foods, their nutrition suffers. This doesn't mean that athletes can *never* choose a white-bread, bologna, and mayonnaise sandwich with chips, cookies, and a cola for lunch—these foods supply abundant calories but lack nutrients and are rich in solid fats and sugars. Later, they should drink a big glass of fat-free milk, eat a salad with low-fat cheese or chicken, or have a big portion of vegetables, along with whole grains and a serving of lean fish or meat to provide needed nutrients.

Carbohydrate

Full glycogen stores are critical to athletes and other highly active people. Techniques to achieve them vary with intensity and duration of the activity. Those performing at high intensities over a short time, such as sprinters, weight lifters, and hurdlers, require only moderate intakes of carbohydrate—ordinary nutritious balanced diets. Ultraendurance athletes, such as triathletes or bicycle racers who compete in multiday events, need much more.[85] (Refer to Table 10–5, earlier, to review carbohydrate recommendations for athletes.)

A trick used by professional performance nutritionists to maximize an endurance athlete's energy and carbohydrate intakes is to choose vegetable and fruit varieties that are high in both nutrients and energy. A whole cupful of iceberg lettuce supplies few calories or nutrients but a half-cup portion of cooked sweet potatoes is a powerhouse of vitamins, minerals, and carbohydrate energy. Similarly, it takes a whole cup of cubed melon to equal the calories and carbohydrate in a half-cup of fruit canned in juice. Small choices like these, made consistently, can contribute significantly to energy and carbohydrate intakes.

Athletes can have some fun exploring new carbohydrate-rich foods. Try Middle Eastern hummus (chickpea spread) and pita breads, African winter squash or peanut stews, Latin American bean and rice dishes, or Japanese rice or soba noodles. They should avoid "Westernized" ethnic meals, however; these often gain fatty cheeses, meats, sour cream, and the like. Just before a competition is not the time to experiment with new foods—try them early in training or during off season.

Adding carbohydrate-rich foods is a sound and reasonable option for increasing intakes, up to a point. It becomes unreasonable when the athlete cannot eat enough nutrient-dense food to meet the need. At that point, some foods with added sugars may be needed, such as fruit-flavored breakfast bars, "trail mix" or energy bars, sugar-sweetened milk beverages, liquid meal replacers, or commercial products designed to supply carbohydrate.

Figure 10–6

Nutritious Snacks for Active People

© gresei/Shutterstock.com

One ounce of almonds provides protein, fiber, calcium, vitamin E, and unsaturated fats. Similar choices include other nuts or trail mix consisting of dried fruit, nuts, and seeds.

© Apostolos Mastoris/Shutterstock.com

Low-fat Greek yogurt contains more protein per serving than regular yogurt, but a little less calcium. A similar choice is low-fat cottage cheese.

© RoJo Images/Shutterstock.com

Low-fat milk or chocolate milk with fig bars or oatmeal-raisin cookies offer protein and carbohydrate. A similar choice is whole-grain cereal with low-fat milk.

© Polara Studios

Popcorn offers carbohydrate and a fruit smoothie quenches thirst and provides carbohydrate, vitamins, minerals, and other nutrients. A similar choice is pretzels and fruit juice.

Protein

Meats and cheeses often head the list of protein-rich foods, but even highly active people must limit intakes of the fatty varieties of these foods to protect against heart disease. Lean protein foods, such as skinless poultry, fish, eggs, low-fat milk products, low-fat cheeses, legumes with grains, and peanuts and other nuts boost protein intakes while keeping saturated fats within bounds.

Figure 10–7 demonstrates how to meet an athlete's need for extra nutrients by adding nutritious foods to a lower-calorie eating pattern to attain a 3,300-calorie per day intake. These meals supply about 125 grams of protein, equivalent to the highest recommended protein intake for an athlete weighing 160 pounds.

Figure 10–7

Nutritious High-Carbohydrate Meals for Athletes

2,600 Calories	**3,300 Calories**
• 62% cal from carbohydrate (403 g) • 23% cal from fat • 15% cal from protein (96 g)	• 63% cal from carbohydrate (520 g) • 22% cal from fat • 15% cal from protein (125 g)

Additions

Breakfast:
1 c shredded wheat
1 c 1% low-fat milk
1 small banana
1 c orange juice

The regular breakfast *plus*:
2 pieces whole-wheat toast
1/2 c orange juice
4 tsp jelly

Lunch:
1 turkey sandwich on whole-wheat bread
1 c 1% low-fat milk

The regular lunch *plus*:
1 turkey sandwich
1/2 c 1% low-fat milk
Large bunch of grapes

Snack:
2 c plain popcorn
A smoothie made from:
1 1/2 c apple juice
1 1/2 frozen banana

The regular snack *plus*:
1 c popcorn

Dinner:
Salad:
1 c spinach, carrots, and mushrooms
1/2 c garbanzo beans
1 tbs sunflower seeds
1 tbs ranch dressing
1 c spaghetti with meat sauce
1 c green beans
1 slice Italian bread
2 tsp soft margarine
1 1/4 c strawberries
1 c 1% low-fat milk

© Polara Studios, Inc. (all)

The regular dinner *plus*:
1 corn on the cob
1 slice Italian bread
2 tsp soft margarine
1 piece angel food cake
1 tbs whipping cream

Pregame Meals

Athletes who train or compete at moderate or vigorous intensity for longer than one hour may benefit from a small, easily digested high-carbohydrate pregame meal. It should provide enough carbohydrate to "top off" an athlete's glycogen stores but be low enough in fat and fiber to facilitate digestion. It can be moderate in protein and should provide plenty of fluid to maintain hydration in the work ahead (Figure 10–8 provides examples).

Breads, potatoes, pasta, and fruit juices—carbohydrate-rich foods that are low in fat and fiber—form the base of the pregame meal. Although generally desirable, bulky, fiber-rich foods can cause stomach discomfort during activity, and so should be avoided in the hours before exercise. The glycemic index of the food (described in Chapter 4) makes no apparent difference to performance.[86]

Timing of the activity and body weight of the athlete helps to determine the size of the meal. With just an hour remaining before training or competition, an athlete should eat very lightly consuming only about 30 grams of carbohydrate—research suggests that more substantial food eaten within the hour before exercise can inhibit performance. With more time to spare, it is possible to calculate an approximate number of carbohydrate grams that an athlete might need to support performance (but such formulas should not be rigidly applied). Multiply the athlete's body weight in pounds by:

- 0.45 g carbohydrate at 1–2 hours before activity.
- 0.9 g carbohydrate at 2–3 hours before activity.

At three hours or more before activity, a regular mixed meal providing plenty of carbohydrate with a moderate amount of protein and fat is suitable.[87] Here are some suggestions:

- *Try these:* grilled chicken or deli turkey sandwich; hard-boiled egg with toast; oatmeal with yogurt; fruit juices; pasta with red sauce; trail mix, granola bars, or energy bars that contain sufficient carbohydrate.
- *Avoid these:* high-fat meats, cheeses, and milk products; other high-fat foods; high-fiber breads, cereals, and bars; raw vegetables; gas forming foods (broccoli, Brussels sprouts, onions, others).

In addition, because athletes often compete away from home, Figure 10–8 offers a quick restaurant selection. A warning: most fast foods are too high in fat to serve as a pregame meal, so if you must use them, order grilled chicken items and reject add-ons, such as sour cream or full-fat cheese (review the principles of Chapter 5's Food Feature section).

Figure 10–8

Examples of High-Carbohydrate Pregame Meals

Timing modifies the carbohydrate grams in a pregame meal (and so does the weight of the athlete; see text). Any of the choices shown below is suitable for a 150-pound athlete who will work with moderate or vigorous intensity for more than an hour. Athletes often must compete away from home, so the 800-calorie meal uses easy-to-find restaurant foods.

© Matthew Farruggio (left and middle)

(at 1 hour before exercise)
200-calorie meal:
30 g carbohydrate
1 small peeled apple
4 saltine crackers
1 tbs reduced-fat peanut butter

(≈1–2 hours before exercise)
500-calorie meal:
90 g carbohydrate
1 medium bagel
2 tbs jelly
1 c low-fat milk

© Sam Kolich/Bill Smith Group/Cengage Learning

(≈2–3 hours before exercise)[a]
800-calorie meal:
135 g carbohydrate
1 large restaurant-style burrito, with
- 12-inch soft flour tortilla
- Rice
- Chicken
- Black beans[b]
- Pico de gallo (fresh tomato sauce)

14 ounces lemonade

[a]*For an extra 200 calories and about 30 grams of carbohydrate, add an energy bar.*
[b]*If black beans cause gas, replace them with tofu or extra rice.*

Do the Math

Example of a pregame meal carbohydrate calculation for a 130-pound athlete at 2 hours and 3 hours before competition:

- 2 hours before: $130 \times 0.45 = 59$ g carbohydrate
- 3 hours before: $130 \times 0.9 = 117$ g carbohydrate

Most important, athletes should choose what works best for them. One athlete may feel best supported by eating pancakes, eggs, and juice, while another develops nausea and cramps after a hearty meal. During intense physical activity, blood is shunted away from the digestive system to the working muscles, making digestion more difficult. If digestive distress is a problem, finish the pregame meal two to three hours before exercise, or eat less food.

Recovery Meals

An athlete who performs intense practice sessions several times daily or competes for hours on consecutive days needs to quickly replenish both energy and glycogen to be ready for the next effort. Several small recovery meals consumed within several hours after exercise may help to speed the process. A turkey sandwich and a homemade smoothie, taken in divided doses, provides the glucose needed to speed up glycogen recovery. Its protein can speed up protein synthesis, too, and begin the day's work of building new muscle structures.

Athletes who have no appetite for solid food after hard work might try drinking a carbohydrate-rich beverage, such as low-fat or fat-free chocolate milk. A two-cup serving of chocolate milk, taken during the hour or two following exercise has been shown to both maintain muscle glycogen stores and to increase muscle protein synthesis.[88] Table 10–12 makes clear that paying for high-priced brand-name pregame or recovery drinks is needless. Chocolate milk or homemade shakes are inexpensive and easy to prepare, they allow athletes to decide what to add or leave out, and they perform as well as any commercial product. (Don't drop a raw egg in the blender, though, because raw eggs may carry bacteria that can cause illness—see Chapter 12.)

© National Dairy Council

Chocolate milk is a delicious and effective postexercise recovery meal.

In contrast to the athlete just described, most people who work out moderately for fitness or weight loss need only to replace lost fluids and resume a normal, healthy eating pattern after activity. If you meet this description but enjoy a postworkout snack, by all means have one. Just remember to eliminate a similar number of calories from your other meals to allow for it.

Commercial Products

What about drinks, gels, or candy-like sport bars claiming to provide a competitive edge? These mixtures of carbohydrate, protein (usually amino acids), fat, some fiber, and certain vitamins and minerals often taste good, can be convenient to store and carry, and offer extra calories and carbohydrate in a compact package. Read the labels, however: a chocolate candy-based bar may be too high in fat to be useful. Such products tend to be expensive and they have no edge over real food for boosting performance.

Even the most carefully chosen pregame or recovery meals cannot substitute for an overall nutritious diet. Deficits of carbohydrate or fluid, incurred over days or weeks, take a toll on performance that no amount of food or fluid on the day of an event can fully correct.[89] The most vital nutrition choices for athletes are those made day in and day out, in training or off season, with an eating pattern that fully meets nutrient needs.

Table 10–12
Commercial and Homemade Recovery Drinks Compared

	Cost (U.S.)	Energy (cal)	Protein (g)	Carbohydrate (g)	Fat (g)
17-ounce commercial "energy/muscle" drink	about $4.00 per serving	330	32 (39% of calories)	13 (16%)	16 (45%)
12-ounce homemade milkshake[a]	about 80¢ per serving	330	15 (18% of calories)	53 (63%)	7 (19%)
16 ounces low-fat chocolate milk	about $1.00 per serving[b]	330	16 (20% of calories)	53 (64%)	6 (16%)

© Cengage Learning

[a]*Home recipe: 8 oz fat-free milk, 4 oz fat-free or low-fat frozen yogurt, 3 heaping tsp malted milk powder. For even higher carbohydrate and calorie values, blend in ½ mashed banana or ½ c other fruit. For athletes with lactose intolerance, use lactose-reduced milk or soy milk and chocolate or other flavored syrup, with mashed banana or other fruit blended in.*

[b]*Supermarket price; about $2.00 if purchased from a convenience store.*

track it! Diet Analysis PLUS

Concepts in Action

Analyze Your Diet and Activities

The purpose of this exercise is to demonstrate the links between nutrients in the diet and physical activity.

1. The *Physical Activity Guidelines for Americans* (Figure 10–1, page 385) recommend physical activity levels for health. From the Reports tab, select Energy Balance Report. Select Day One of the three-day diet intake; include the entire day's food intake and generate an Energy Balance, choose Day one of the three-day diet intake; include all meals and generate a report at your current activity level. Now, create a fitness program that increases your physical activity by $2\frac{1}{2}$ hours per week. Select moderate physical activities that you enjoy (use Table 10–3, p. 387, as a guide). Include both aerobic and strengthening activities. Enter your activities, and select the Track Activity tab for Day One. Generate a report. Compare the two reports. What changes did you notice?
2. Sweating causes a loss of the minerals sodium and potassium. From the Reports tab, select Intake vs. Goals. Select Day One and generate a report. Is your electrolyte intake deficient, excessive, or within the DRI recommendations? What conditions might change your electrolyte needs? Compare the Intake vs. Goals for each of the two activity levels that you generated in question number 1 above. Discuss how and to what degree the requirements changed when you increased your activity level.
3. The Food Feature in this chapter demonstrates how to choose a performance diet with sufficient carbohydrate. Assume that you need such a diet. Modify your intake for all meals on Day Two with the goal of increasing your carbohydrate intake. For help, use Figure 10–7 (page 409) as a guide. From the Reports tab, select Macronutrient Ranges; generate a report for the modified Day Two. Did you obtain about 400 grams of carbohydrate? This amount would be sufficient for a 150-pound athlete in many activities (consult Table 10–5, page 396).
4. A strategy for maintaining blood glucose during physical activity is to eat a carbohydrate-rich pregame meal a couple of hours beforehand. Calculate the needed carbohydrates for your weight at 2 hours before exercise (use the formula on page 411). Add a high-carbohydrate snack to one of your food records. For ideas, look at the pregame meals in Figure 10–8 (page 410). Submit the new record by selecting Track Diet tab. From the Reports tab, select Program, then go to the Source Analysis. Select Snack, then select Carbohydrate from the drop-down box; generate a report. How did you do? Did your snack contribute an adequate amount of carbohydrate to maintain blood glucose during physical activity?
5. Assume you are an endurance athlete, engaging in vigorous daily training. Calculate the recommended protein intake for an athlete of your weight using Table 10–6 (page 400). Modify Day Two of your diet records to increase the protein to the recommended level. From the Reports tab, select Source Analysis and select Day Two. Select Protein from the drop-down box. Generate a new report. Did your modified diet provide enough protein for an endurance athlete of your size? Were you already consuming enough protein for an endurance athlete without making any changes?
6. Again assume that you are an endurance athlete whose calorie need is 50 calories per kilogram (or 23 calories per pound of body weight). Modify Day Two of your diet to reach the increased calorie goal. From the Reports tab, select Source Analysis, select Day Two, include all meals, and then choose Calories from the drop-down box. Generate a report. Does this diet provide enough calories to support the athlete's increased physical activity level? If not, which foods might you add to obtain adequate calories, and why did you choose them and not others?

what did you decide?

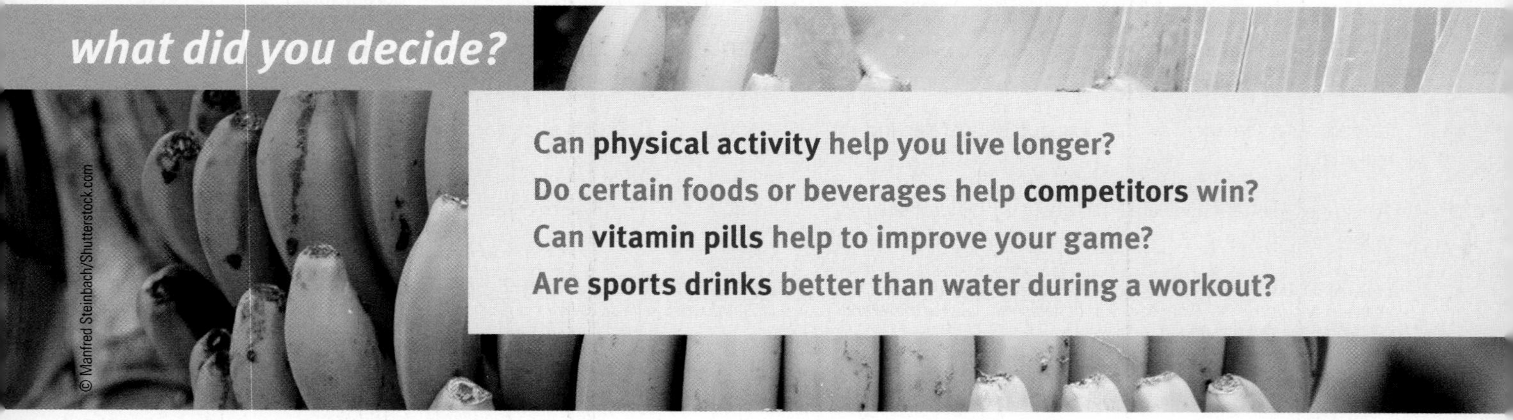

Can physical activity help you live longer?
Do certain foods or beverages help competitors win?
Can vitamin pills help to improve your game?
Are sports drinks better than water during a workout?

Self Check

1. (LO 10.1) All of the following are potential benefits of regular physical activity except
   a. improved body composition
   b. lower risk of sickle-cell anemia
   c. improved bone density
   d. lower risk of type 2 diabetes
2. (LO 10.1) The length of time a person must spend exercising in order to meet the *Physical Activity Guidelines for Americans* varies by
   a. exercise duration
   b. exercise balance
   c. exercise intensity
   d. exercise adequacy
3. (LO 10.2) Weight-bearing activity that improves muscle strength and endurance has no effect on maintaining bone mass.
   T F
4. (LO 10.2) To overload a muscle is never productive.
   T F
5. (LO 10.3) Which of the following provides most of the energy the muscles use in the early minutes of activity?
   a. fat
   b. protein
   c. glycogen
   d. b and c
6. (LO 10.3) Which diet has been shown to increase an athlete's endurance?
   a. high-fat diet
   b. normal mixed diet
   c. high-carbohydrate diet
   d. Diet has not been shown to have any effect.
7. (LO 10.3) A person who exercises moderately for longer than 20 minutes begins to __________ .
   a. use less glucose and more fat for fuel
   b. use less fat and more protein for fuel
   c. use less fat and more glucose for fuel
   d. use less protein and more glucose for fuel
8. (LO 10.3) Aerobically trained muscles burn fat more readily than untrained muscles.
   T F
9. (LO 10.4) Which is required as part of myoglobin?
   a. iron
   b. calcium
   c. vitamin C
   d. potassium
10. (LO 10.4) All of the following statements concerning beer are correct *except* __________ .
    a. beer is poor in minerals
    b. beer is poor in vitamins
    c. beer causes fluid losses
    d. beer gets most of its calories from carbohydrates
11. (LO 10.4) Research does not support the idea that athletes need supplements of vitamins to enhance their performance.
    T F
12. (LO 10.5) An athlete should drink extra fluids in the last few days of training before an event in order to ensure proper hydration.
    T F
13. (LO 10.5) To prevent both muscle cramps and hyponatremia, endurance athletes who compete and sweat heavily for four or more hours need to
    a. replace sodium during the event
    b. avoid salty foods before competition
    c. drink additional plain water during the event
    d. replace glucose during the event
14. (LO 10.6) Athletes should avoid frequent between-meal snacks.
    T F
15. (LO 10.6) Added sugars can be useful in meeting the high carbohydrate needs of some athletes.
    T F
16. (LO 10.6) An athlete's pregame meals should be
    a. low in fat
    b. high in fiber
    c. moderate in protein
    d. a and c
17. (LO 10.6) These foods should form the bulk of the pregame meal:
    a. breads, potatoes, pasta, and fruit juices
    b. meats and cheeses
    c. legumes, vegetables, and whole grains
    d. none of the above
18. (LO 10.7) Before athletic competitions, a moderate caffeine intake
    a. may interfere with concentration
    b. may enhance performance
    c. may increase anxiety
    d. has no effect
19. (LO 10.7) Carnitine supplements
    a. are fat burners that increase cellular energy
    b. raise muscle carnitine concentrations
    c. enhance exercise performance
    d. often produce diarrhea

Answers to these Self Check questions are in Appendix G.

CONTROVERSY 10

Ergogenic Aids: Breakthroughs, Gimmicks, or Dangers?

LO 10.7 Evaluate whether dietary ergogenic aids are useful for increasing sports performance or obtaining an ideal body composition for sports.

Athletes can be sitting ducks for quacks. Many are willing to try almost anything that is sold with promises of producing a winning edge or improved appearance, so long as they perceive it to be safe. Store shelves and the Internet abound with heavily advertised **ergogenic aids**, each striving to appeal to performance-conscious people: protein powders, amino acid supplements, caffeine pills, steroid replacers, "muscle builders," vitamins, and more. Some people spend huge sums of money on these products, often heeding advice from a trusted coach or mentor. Table C10–1 defines some relevant terms in this section and lists many more substances promoted as ergogenic aids. Do these products work as advertised? And most importantly, are they safe?

This Controversy focuses on the scientific evidence for and against a few of the most common dietary supplements for athletes and exercisers. In light of the evidence, it concludes with what many people already know: consistent training and sound nutrition serve an athlete better than any pill, powder, or supplement.*[1]

Paige and DJ

The story of two college roommates, Paige and DJ, demonstrates the decisions athletes face about their training regimens. After enjoying a freshman year when the first things on their minds were tailgate parties and the last thing—the very last thing—was exercise, Paige and DJ have taken up running to shed the "freshman 15" pounds that have crept up on them. Their friendship, once defined by bonding over extra-cheese pizzas and fried chicken wings, now focuses on 5-K races. Both women now compete to win.

Paige and DJ take their nutrition regimens and prerace preparations seriously, but they are as opposite as the sun and moon: DJ takes a traditional approach, sticking to the tried-and-true advice of her older brother, an all-state

* *Reference notes are found in Appendix F.*

Table C10–1

Ergogenic Aid Terms

Additional ergogenic aids are listed in Table C10–2.

- **anabolic steroid hormones** chemical messengers related to the male sex hormone testosterone that stimulate building up of body tissues (*anabolic* means "promoting growth"; *sterol* refers to compounds chemically related to cholesterol).
- **androstenedione** (AN-droh-STEEN-die-own) a precursor of testosterone that elevates both testosterone and estrogen in the blood of both males and females. Often called *andro*, it is sold with claims of producing increased muscle strength, but controlled studies disprove such claims.
- **caffeine** a stimulant that can produce alertness and reduce reaction time when used in small doses but causes headaches, trembling, an abnormally fast heart rate, and other undesirable effects in high doses.
- **carnitine** a nitrogen-containing compound, formed in the body from lysine and methionine, that helps transport fatty acids across the mitochondrial membrane. Carnitine is claimed to "burn" fat and spare glycogen during endurance events, but it does neither.
- **chromium picolinate** a trace element supplement; falsely promoted to increase lean body mass, enhance energy, and burn fat.
- **creatine** a nitrogen-containing compound that combines with phosphate to burn a high-energy compound stored in muscle. Some studies suggest that creatine enhances energy and stimulates muscle growth but long-term studies are lacking; digestive side effects may occur.
- **DHEA (dehydroepiandrosterone)** a hormone made in the adrenal glands that serves as a precursor to the male hormone testosterone; recently banned by the FDA because it poses the risk of life-threatening diseases, including cancer. Falsely promoted to burn fat, build muscle, and slow aging.
- **energy drinks** and **energy shots** sugar-sweetened beverages in various concentrations with supposedly ergogenic ingredients, such as vitamins, amino acids, caffeine, guarana, carnitine, ginseng, and others. The drinks are not regulated by the FDA and are often high in caffeine or other stimulants.
- **ergogenic** (ER-go-JEN-ic) **aids** products that supposedly enhance performance, although few actually do so; the term *ergogenic* implies "energy giving" (*ergo* means "work"; *genic* means "give rise to").
- **whey** (way) the watery part of milk, a by-product of cheese production. Once discarded as waste, whey is now recognized as a high-quality protein source for human consumption.

Training serves an athlete better than any pills or powders.

track and field star. He tells her to train hard, eat a nutritious diet, get enough sleep, drink plenty of fluid on race day, and warm up lightly for 10 minutes before the starting gun. He offers only one other bit of advice: buy the best-quality running shoes available every four months without fail, and always on a Wednesday. Many an athlete admits laughingly to such superstitions as wearing "lucky socks" for a good luck charm.

Paige finds DJ's routine boring and woefully out of date. Paige surfs the Internet for the latest supplements and ergogenic aids advertised in her fitness magazines. She mixes carnitine and protein powders into her beverages, hoping for the promised bonus muscle tissue to help at the weight bench, and she takes a handful of "ergogenic" supplements to get "pumped up" for a race. Her counter is cluttered with bottles of amino acids, caffeine pills, chromium picolinate, and even herbal steroid replacers sold with a promise of speedy recovery from hard runs. No matter what her goal, the Internet stores seem to have a "best selling" product for the job. Sure, it takes money (a *lot* of money) to purchase the products and time to mix the potions and return the occasional wrong shipment—often cutting into her training time. But Paige feels smugly smart in her modern approach. Surely, she will win the most races.

Is Paige correct to expect an athletic edge from taking supplements? Is she safe in taking them?

Ergogenic Aids

Science holds some of the answers to such questions, but finding them requires reading more than just advertising materials. It's easy to see why Paige is misled by fitness magazines—ads often masquerade as informative articles, concealing their true nature. A tangle of valid and invalid ideas in such "advertorials" can appear convincingly scientific, particularly when accompanied by colorful anatomical figures, graphs, and tables. Some ads even cite such venerable sources as the *American Journal of Clinical Nutrition* and the *Journal of the American Medical Association* to create the illusion of credibility. Keep in mind, however, that these "advertorials" are created not to teach, but to *sell*. Supplement companies bring in tens of billions of dollars worldwide—and some unscrupulous sellers will gladly mislead athletes for a share of it.

Also, many substances sold as "dietary supplements" escape regulation (see Controversy 7 for details). This means that athletes are largely on their own in evaluating supplements for effectiveness and safety. So far, the large majority of legitimate research has not supported the claims made for ergogenic aids. Athletes who hear that a product is ergogenic should ask, "Who is making this claim?" and "Who will profit from the sale?"

Antioxidant Supplements

Exercise increases metabolism, and speeded-up metabolism creates extra free radicals that contribute to oxidative stress. It stands to reason, then, that if exercise produces free radicals and oxidative stress, and if antioxidants from foods can quell oxidative stress, then athletes may benefit from taking in more antioxidants. Like many other logical ideas, however, this one falls apart upon scientific examination—research does not support the taking of antioxidant supplements for athletic performance.[2] In fact, free radical production may be a necessary part of a complex signaling system that promotes many of the beneficial responses of the body to physical activity.[3] The possibility exists that flooding the system with excess antioxidants may short-circuit the system and prevent the expected health benefits or improvements in athletic performance from physical activity.[4]

Caffeine

Many athletes (but not all) report that **caffeine** from coffee, tea, **energy drinks** and **energy "shots,"** and other sources provides a physical boost during sports.[5] Caffeine (3–6 mg/kg body weight) may indeed improve performance, particularly during endurance activities, such as cycling and rowing and, to a lesser extent, in high-intensity training.[6] A mild stimulant, caffeine may trigger fatty acid release, enhance alertness and concentration, and reduce the perception of fatigue.

Any potential benefits from caffeine must be weighed against its known adverse effects—high doses cause stomach upset, nervousness, irritability, headaches, dehydration, and diarrhea. Even the caffeine in two to three cups of coffee for a 150-lb person (5 mg/kg body weight) increases the heart rate at a given workload.[7] High doses of caffeine can constrict the arteries and raise blood pressure, making the heart work harder than normal to pump blood to the working muscles. In addition, the acids in such drinks may erode tooth enamel; other constituents can increase water and sodium losses and may impair the ability of cell membranes to heal after being wounded.[8]

Competitors should be aware that college sports authorities prohibit the use of caffeine in amounts greater than 700 milligrams, or the equivalent of eight cups of coffee prior to competition. Controversy 14 lists caffeine doses in common foods and beverages.

Instead of taking caffeine pills before an event, Paige might be better off engaging in some light activity, as DJ does. Pregame activity stimulates the release of fatty acids and warms up the muscles and connective tissues, making them flexible and resistant to injury. Caffeine does not offer these benefits. Instead, caffeine in high doses acts as a diuretic. DJ enjoys a cup or two of coffee before her races, but the amount of caffeine they provide is unlikely to cause problems.[9]

Carnitine

Carnitine is a nonessential nutrient that is often marketed as a "fat burner." In the body, carnitine does help to transfer fatty acids across the membrane that encases the cell's mitochondria. (Recall from Figure 3–1 of Chapter 3 that the mitochondria are structures in cells that release energy from energy-yielding nutrients, such as fatty acids.) Carnitine marketers use this logic: "the more carnitine, the more fat burned, the more energy produced"—but the argument is not valid. Carnitine supplementation neither raises muscle carnitine concentrations nor enhances exercise performance. (Paige found out the hard way that carnitine often produces diarrhea in those taking it.) Vegetarians have less total body carnitine than meat eaters do, but introducing more has no effect on vegetarians' muscle carnitine levels.[10]

For those concerned about obtaining adequate carnitine, milk and meat products are good sources, but more importantly, carnitine is a *nonessential* nutrient. This means that the body makes plenty for itself.

Chromium Picolinate

Advertisements for **chromium picolinate** products bombard consumers with promises of trimming off the most stubborn spare tire and building up rippling muscles. Photos of impossibly fit people, supposedly the "after" shots of those taking chromium picolinate supplements, tempt even people who know that fitness never results from taking a pill.

Chromium is an essential trace mineral involved in carbohydrate and lipid metabolism. One or two initial reports correlated taking supplements with reduced body fatness and increased lean body mass in male weight lifters, but most subsequent studies demonstrate no effect on body fatness, lean body mass, strength, or fatigue.

The safety record of chromium picolinate is not unblemished. One athlete who ingested 1,200 micrograms of chromium picolinate over two days' time developed a dangerous condition of muscle degeneration, with the supplement strongly suspected as the cause.

Creatine

Creatine supplements clearly do not benefit endurance athletes such as runners.[11] For performance of short-term, repetitive, high-intensity activities such as weight lifting or sprinting, however, some studies have reported small but significant increases in muscle strength, power, and size—attributes that support these activities.[12] Other research suggests no benefit, and study authors suggest that resistance training alone during the study period, and not creatine supplements, may account for improvements in studies that report benefits.[13]

Creatine functions in muscles as part of the high-energy storage compound creatine phosphate (or phosphocreatine), and, theoretically, the more creatine phosphate in muscles, the higher the intensity at which an athlete can train.[14] The confirmed effect of creatine, however, is weight gain—a potential boon for some athletes but a bane for others. Unfortunately, the gain may be mostly water because creatine causes muscles to hold water.

Large-scale, long-term safety studies are lacking, but short-term use of creatine in doses suggested by manufacturers has, so far, not proven harmful for healthy adults.[15] However, children as young as 9 years old are taking creatine at the urging of coaches or parents, with unknown consequences. The American Academy of Pediatrics strongly discourages the use of creatine or any other ergogenic supplements in children younger than 18 years old.

Meat, being muscle, is a good supplier of creatine, so anyone who eats meat consumes creatine in abundance. Another safe source is the body's own creatine—human muscles can make all the creatine they need.

Buffers

Sodium bicarbonate (baking soda) acts as a buffer, a compound that neutralizes acids. During high-intensity exercise, acids form in the muscles and may contribute to fatigue. Research indicates that elite male athletes who ingest sodium bicarbonate (0.3 g/kg body weight) before performing high-intensity activities lasting from 1 to 10 minutes may moderately (by 2 percent or so) enhance their performance.[16] Novice exercisers, females, and athletes performing other kinds of longer-duration activities are unlikely to benefit. Unpleasant side effects, such as gas and diarrhea, may make this ergogenic aid impractical, even for elite male athletes.

The buffering effect of the amino acid beta-alanine has recently received attention by exercise researchers.[17] Although beta-alanine may increase the body's buffering capacity to some degree, research has yielded mixed or negative results on exercise performance.[18] A "pins and needles" sensation side effect has been noted.

Amino Acid Supplements

Some athletes—particularly body builders and weight lifters—know that consuming essential amino acids is required to increase muscle size.[19] As mentioned in the chapter, for up to 48 hours following exercise, muscles respond by building up the bulk and strength they need to perform their work. Protein synthesis is held back by a lack of essential amino acids at the critical time. All essential amino acids must be provided for maximum gains, not just a selected few. In addition, the essential amino acids, particularly leucine, signal muscles to speed up their protein synthesis.

The best source for these amino acids is food, not supplements, for several reasons. First, healthy athletes eating a nutritious diet naturally obtain all of the amino acids they need from food, and in an ideal balance not matched by supplements. Second, the amount of amino acids that muscles require is just a few grams—an amount easily provided by any light, protein-containing meal. More than this amount from expensive supplements is unnecessary and ineffective—muscles cannot store excess amino acids, and burn them off as fuel. Third, single amino acid preparations, even leucine, do not improve physical performance or muscle gains.[20]

Fourth, taking amino acid supplements can easily put the body in a too-much–too-little bind. Amino acids compete with each other for carriers in the body, and an overdose of one can limit the availability of another. Finally, supplements can cause digestive disturbances and, for some people in particular, amino acid supplements may pose a hazard (see the Consumer's Guide of Chapter 6).

Whey Protein

Similar to the high-quality protein in lean meat, eggs, milk, and legumes, **whey** supplies all of the essential amino acids, including leucine, needed to initiate and support the building of new muscle tissue—it is a complete protein. Once a discarded by-product of the cheese-making industry, whey is now added to many foods and supplements, including bars, drinks, and powders for athletes. Whey protein is water soluble and stays dissolved in the digestive tract, where it is quickly digested and absorbed. Whey therefore delivers essential amino acids to the bloodstream more rapidly than solid proteins that require greater digestion.[21] Research on whey's effects is ongoing, but despite many claims to the contrary, no clear benefits of whey for athletic performance are evident beyond those of other high-quality protein-rich foods.[22]

Paige believes that by taking a handful of amino acid pills and eating a couple of whey protein bars she can go easier on training and still gain speed on the track, but this is just wishful thinking. Muscles require physically demanding activity, not just protein, to gain in size and performance. Instead of getting faster, Paige will likely get fatter—at 250 calories each, her protein bars contribute 500 calories to her day's intake, an amount far greater than she expends in exercise. Dutifully, her body dismantles the extra protein, removes and excretes the nitrogen from the amino acids, uses what it can for energy, and converts the rest to body fat for storage.

Recently, DJ, who snacks on plain raisins and nuts, placed ahead of Paige in seven of their ten shared competitions. In one of these races, Paige pulled out because of light-headedness—perhaps a consequence of too much caffeine? Still, Paige remains convinced that to win, she must have chemical help, and she is venturing over the danger line by considering hormone-related products. What she doesn't know is very likely to hurt her.

Hormones and Hormone Imitators

The dietary supplements discussed so far are controversial in the sense that they may or may not enhance athletic performance, but most—in the doses commonly taken by healthy adults—probably do not pose immediate threats to health or life. Hormones, however, are clearly damaging and are banned by the World Anti-Doping Agency of the International Olympic committee, and by most professional and amateur sports leagues.

Anabolic Steroids

Among the most dangerous ergogenic practices is the use of **anabolic steroid hormones**. The body's natural steroid hormones stimulate muscle growth in response to physical activity in both men and women. Injections of "fake" hormones produce muscle size and strength far beyond that attainable by training alone, but at great risk to health. These drugs are both illegal in sports and dangerous to the taker, yet athletes often use them without medical supervision, simply taking someone's word for their safety. The list of damaging side effects of steroids is long, and includes:

- Extreme mental hostility; aggression; personality changes; suicidal thoughts.
- Swollen face; severe, scarring acne; yellowing of whites of eyes (jaundice).
- Elevated risk of heart attack, stroke; liver damage, liver tumors, fatal liver failure; kidney damage; bloody diarrhea.
- In females, irreversible deepening of voice, loss of fertility, shrinkage of breasts, permanent enlargement of external genitalia.
- In males, breast enlargement, permanent shrinkage of testes, prostate enlargement, sexual dysfunction, and loss of fertility.
- Many others.

The group of substances discussed in this section is clearly damaging to the body. Don't consider using these products—just steer clear.

Human Growth Hormone

A wide range of athletes, including weight lifters, baseball players, cyclists, and track and field participants, use hGH (human growth hormone) to build lean tissue and improve athletic performance. They inject hGH, believing that, because it isn't a steroid itself, it will provide the muscle gains they seek without the substantial risks of anabolic steroids. However, taken in large quantities, hGH causes the disease acromegaly, in which the body becomes huge and the organs and bones enlarge. Other risks of hGH include diabetes, thyroid disorder, heart disease, menstrual irregularities, diminished sexual desire, and shortened life span.

Steroid Alternatives

Many athletes, and particularly school-age athletes, have tried herbal or insect-derived sterols hawked as "natural" alternatives to steroid drugs. The body cannot readily absorb these, nor does it convert them into human steroids. These products do not enhance muscle size or strength, but some may contain toxins. Remember: "natural" doesn't mean "harmless."

DHEA, a hormone produced in the adrenal glands and liver, is used by the body to make other important hormones, including **androstenedione**, testosterone, and estrogen. Supplements of DHEA or androstenedione produce unpredictable results. In males, such products may have little or no effect because male testes already produce sufficient testosterone. In females, they may disturb hormonal balance, producing a greater proportional blood level of testosterone along with increased estrogens. No evidence supports using DHEA or androstenedione, and even if temporary gains in muscle strength could be achieved, the ill effects from excess steroid hormones in the body can last a lifetime.

Although androstenedione and DHEA are still for sale on the Internet, the National Collegiate Athletic Association, the National Football League, and the International Olympic Committee have banned their use of in competition. The American Academy of Pediatrics and many other medical professional groups have spoken out against the use of these and other "hormone replacement" substances.

Drugs Posing as Supplements

Some ergogenic aids sold as dietary supplements turn out to contain powerful drugs. The FDA recalled a potent thyroid hormone known as TRIAC for interfering with normal thyroid functioning and causing heart attack and stroke. Another is DMAA, a potentially harmful stimulant drug. DMAA is popularly sold as a "natural energy-booster" or "fat-destroyer," but DMAA acts like adrenaline in the body and is suspected of causing fatal strokes or heart attacks, along with panic attacks, seizures, and other adverse effects in users.[23] Although the FDA has banned TRIAC and is acting on DMAA, others are likely to crop up to take their place because the demand is strong and profits are high. Table C10–2 lists a few of them.

Also, a dietary supplement is not always what the label says it is. In one recent study, almost 19 percent of supplements sold to athletes worldwide were found to contain a steroid drug. Another study demonstrated that taking a creatine supplement contaminated with just 0.00005 percent of a steroid drug can produce a positive drug test.[24] An athlete taking such a supplement not only faces the physical risks from unknown substances but also risks being falsely accused of doping and forever banned from competition.

Table C10–2

More Substances Promoted as Ergogenic Aids

Dietary Supplement	Claims	Evidence	Risks
Arginine (an amino acid)	Increased muscle mass	Ineffective	Generally well tolerated; may be harmful to people with heart disease
Boron (trace mineral)	Increased muscle mass	Ineffective	No adverse effects reported with doses up to 10 mg/day; should be avoided by those with kidney disease or women with hormone-sensitive conditions
Casein (milk protein)	Increased muscle mass	As with all dietary protein, contributes to positive nitrogen balance	Well tolerated by many people; triggers allergic reaction in people with milk allergy
DMAA (dimethylamylamine)	Increased energy, concentration, and metabolism	A stimulant, possibly similar to ephedrine or amphetamine	Reports of fatal heart attack, cerebral hemorrhage, liver and kidney failure, seizures, high blood pressure, and rapid heartbeat
Ephedra (ephedrine, ma huang)	Weight loss; muscle enhancement; improved athletic performance	May increase feelings of nervous energy and alertness	Dry mouth, insomnia, nervousness, palpitation, and headache; blood pressure spikes; cardiac arrest; banned by FDA as an unreasonable risk to health
Coenzyme Q10 (carrier in the electron transport chain)	Enhanced exercise performance	Ineffective	Mild indigestion
Gamma-oryzanol (plant sterol)	Increased muscle mass; said to mimic anabolic steroids without side effects	Ineffective	No adverse effects reported with short-term use; no long-term safety studies

Table C10–2

More Substances Promoted as Ergogenic Aids (continued)

Dietary Supplement	Claims	Evidence	Risks
Ginseng (plant)	Enhanced exercise performance	Ineffective	No adverse effects reported with moderate doses; large doses may cause hypertension, nervousness, sleeplessness, acne, edema, headache, and diarrhea; those with diabetes should be aware of hypoglycemic effects; should be avoided by those at risk for estrogen-related cancers, those with blood-clotting issues, and pregnant or lactating women
Glycerol (a 3-carbon molecule that is part of triglycerides and phospholipids)	Improved hydration during exercise; regulation of body temperature during exercise; enhanced exercise performance	Inconsistent findings for improving hydration and regulating body temperature; ineffective for enhancing exercise performance	May cause nausea, headaches, and blurred vision; should be avoided by those with edema, congestive heart failure, kidney disease, hypertension, and other conditions that may be aggravated by fluid retention
Glycine (an amino acid)	Precursor of creatine	Ineffective	Potential amino acid imbalances
Guarana	Enhanced speed and endurance, mental, and sexual functions	Ineffective	High doses may stress the heart and cause panic attacks
HMB (beta-hydroxy-beta-methylbutyrate) (a metabolite of the branched-chain amino acid leucine)	Increased muscle mass and strength	Inconsistent findings	No adverse effects with short-term use and doses up to 76 mg/kg of body weight
Pyruvate (a 3-carbon sugar)	Enhanced endurance	Ineffective	No long-term safety studies; digestive problems with short-term use (< 6 weeks)
Ribose (a 5-carbon sugar)	Increased ATP production and enhanced high-intensity exercise performance	Ineffective	Naturally generated in body; submitted to USDA to become a Generally Recognized As Safe food additive (pending)
Royal jelly (produced by bees)	Enhanced stamina and reduced fatigue	No studies on human beings to date	No adverse effects with doses up to 12 mg/day; should be avoided by those with a history of asthma or allergic reactions
Sodium bicarbonate (baking soda)	Buffers muscle acid; delayed fatigue; enhanced power and strength	May buffer acid and delay muscle fatigue; more research is needed for definitive conclusions	Gastrointestinal distress including diarrhea, cramps, gas, and bloating; should be avoided by those on sodium-restricted diets
Yohimbe	Weight loss; stimulant effects	No evidence available	Kidney failure; seizures

Conclusion

The general scientific response to ergogenic claims is "let the buyer beware." In a survey of advertisements in a dozen popular health and body-building magazines, researchers identified over 300 products containing 235 different ingredients advertised as beneficial, mostly for muscle growth. None had been scientifically shown to be effective.

Athletes like Paige who fall for the promises of better performance through supplements are taking a gamble with their money, their health, or both. They trade one product for another and another when the placebo effect wears thin and the promised miracles fail to materialize. DJ, who takes the scientific approach reflected in this Controversy, faces a problem: how does she tell Paige about the hoaxes and still preserve their friendship?

Explaining to someone that a long-held belief is not true involves a risk: the person often becomes angry with the one telling the truth, rather than with the source of the lie. To avoid this painful outcome, DJ decides to mention only the supplements in Paige's routine that are most likely to cause harm—the chromium picolinate, the overdoses of caffeine, and the hormone replacers. As for the whey protein and other supplements, they are probably just a waste of money, and DJ decides to keep quiet. Perhaps they may serve as harmless superstitions.

When Paige believes her performance is boosted by a new concoction, DJ understands that most likely the power of her mind is at work—the placebo effect. Don't discount that power, by the way, for it is formidable.[25] You don't need to buy unproven supplements for an extra edge because you already have a real one—your mind. And you can use the extra money you save to buy a great pair of running shoes—perhaps on a Wednesday?

Critical Thinking

1. Most of the time, the buyer is wasting his or her money when buying an ergogenic aid to improve performance. Still, even well-educated athletes often take them. What forces do you think might motivate a competitor to "throw caution to the wind" and buy and take unproven supplements sold as ergogenic aids? What role might advertising play?
2. Divide into two groups. One group will argue in favor of ergogenic use by athletes, and one group will argue against the use of ergogenic aids by athletes. Each group will make a list of ergogenic aids that should be allowed for use by athletes and a list of those that should not be allowed.

11 Diet and Health

what do you think?

Can your diet strengthen your **immune system**?

Are your own food choices damaging your **heart**?

Can certain **herbs** improve your health?

Do “natural” foods without **additives** reduce cancer risks?

Learning Objectives

After completing this chapter, you should be able to accomplish the following:

LO 11.1 Describe relationships between immunity and nutrition, and explain how malnutrition and infection worsen each other.

LO 11.2 Identify several risk factors for chronic disease, and explain the relationship between risk factors and chronic diseases.

LO 11.3 Identify dietary factors that increase and decrease CVD risks, and explain the association between high blood LDL and HDL cholesterol concentrations and CVD risk.

LO 11.4 Outline a general eating and exercise plan for a person with prehypertension, and justify the recommendations.

LO 11.5 Describe the associations between diet and cancer observed in research, both positive and negative.

LO 11.6 Make specific recommendations for including sufficient fruits and vegetables in a health-promoting diet.

LO 11.7 Describe some recent advances in nutritional genomics with regard to the health of the body through life.

Can a well-chosen diet protect you from developing a disease? The answer to this question depends on the disease. Two main kinds of diseases afflict people around the world: **infectious diseases** and **chronic diseases**.* Infectious diseases such as tuberculosis, smallpox, influenza, and polio have been major killers of humankind since before the dawn of history. In any society not well defended against them, infectious diseases can cut life so short that the average person dies at 20, 30, or 40 years of age.

With the advent of vaccines and antibiotics, people in developed countries had become complacent about infectious diseases—until recently. Scientists now warn of growing infectious threats: the possibility of the rapid global spread of new human diseases for which no treatments exist and a rising death toll from once-conquered diseases, such as tuberculosis and foodborne infections now resistant to antibiotic drugs.[†1] While scientists work to develop controls for these perils, government health agencies hasten to strengthen emergency response systems and to protect our food and water supplies.

Individuals can take steps to protect themselves, too. Each of us encounters millions of microbes each day, and some of these can cause diseases. Nutrition cannot directly prevent or cure infectious diseases, but good nutrition can strengthen, and malnutrition can weaken, your body's defenses against them.[2] One warning: many so-called "immune-strengthening" foods, dietary supplements, and herbs are hoaxes. For healthy, well-fed people, supplements cannot trigger extra immune power to fend off dangerous infections.

In the United States and Canada, chronic diseases far outrank infections as the leading causes of death and illness.[3] Compare, for example, the death toll attributable to pneumonia and influenza, the only two infectious diseases listed among the top U.S causes of death (Figure 11–1) to deaths from heart disease, cancer, and diabetes. The longer a person dodges life's other perils, the more likely that these diseases will take their toll.

Chronic diseases do not arise from a straightforward cause such as infection, but from a mixture of factors in three areas: genetic inheritance, prior or current diseases such as obesity or hypertension, and lifestyle choices. The first one, inherited susceptibility, people cannot control. The second one, current disease states, people may be able to control or modify to some extent. People control daily life choices directly, however, and these can often prevent or delay the onset of certain diseases.

* The term *disease* is also used to refer to conditions such as birth defects, alcoholism, obesity, and mental disorders.
† Reference notes are found in Appendix F.

Figure 11-1

The Ten Leading Causes of Death in the United States[a]

Many deaths have multiple causes, but diet influences the development of several chronic diseases—notably, heart disease, some types of cancer, stroke, and diabetes.

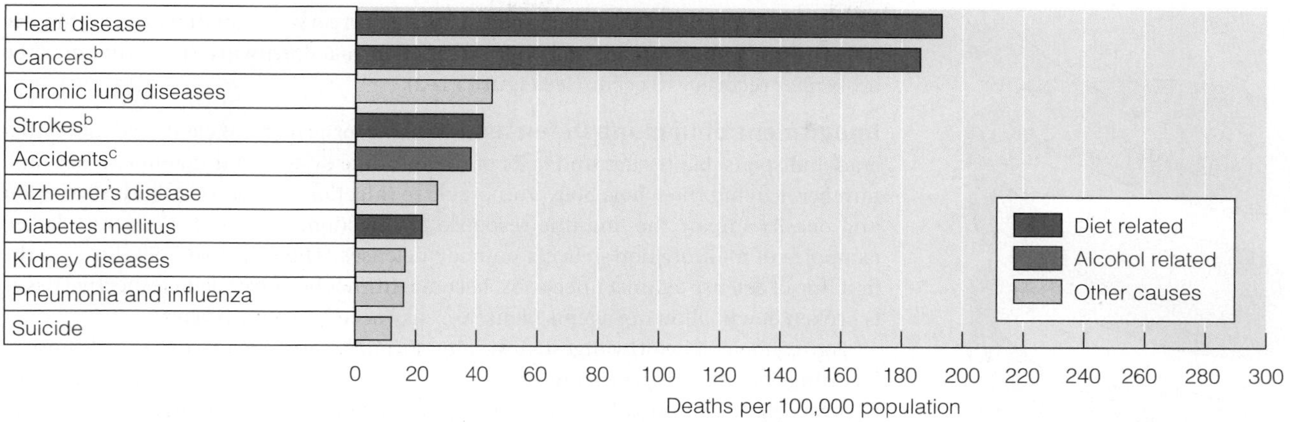

[a]*Rates are age adjusted to allow relative comparisons of mortality among groups and over time.*
[b]*Alcohol increases the risks for some cancers and strokes.*
[c]*Motor vehicle and other accidents are the leading cause of death among people aged 15–24, followed by homicide, suicide, cancer, and heart disease. Alcohol contributes to about half of all accident fatalities.*
Source: Data from National Center for Health Statistics, 2012.

Young people choose whether to nourish their bodies well, to smoke, to exercise, or to abuse alcohol, and these choices are potent determinants of disease risks later on. As people age, their bodies accumulate the effects of a lifetime of choices and, in the later years, these impacts can make the difference between a life of health or one of chronic disability. This discussion begins with the role of nutrition in supporting the body's immune defenses and then reveals the power of the diet to advance or inhibit the development of chronic diseases.

"If there is any deficiency in food or exercise the body will fall sick."

—Hippocrates, a Greek physician, c. 400 B.C.

The Immune System, Nutrition, and Diseases

LO 11.1 Describe relationships between immunity and nutrition, and explain how malnutrition and infection worsen each other.

Without your awareness, your immune system continuously stands guard against thousands of attacks mounted against you by microorganisms and cancer cells. If your immune system falters, you become vulnerable to disease-causing agents, and disease invariably follows.

Immune tissues are among the first to be impaired in the course of a vitamin and mineral deficiency or toxicity.[4] Some nutrient deficiencies are more immediately harmful to immunity than others, determined partly by the physiological roles of the missing nutrient, whether another nutrient can perform some of its tasks, the severity of the deficiency, whether an infection has already taken hold, and being of young or old age.

The Effects of Malnutrition

People who restrict their food intakes, whether because of lack of appetite, illness, an eating disorder, or desire for weight loss, are more likely than others to be caught in the downward spiral of malnutrition and weakened immunity. Also susceptible are those who are one or more of the following: very young or old, poor, hospitalized, or malnourished. Rates of sickness and death increase dramatically when medical tests of a malnourished person indicate that the immune system is compromised.

infectious diseases diseases that are caused by bacteria, viruses, parasites, and other microbes and can be transmitted from one person to another through air, water, or food; by contact; or through vector organisms such as mosquitoes and fleas.

chronic diseases degenerative conditions or illnesses that progress slowly, are long in duration, and that lack an immediate cure; chronic diseases limit functioning, productivity, and quality and length of life. Also defined in Chapter 1.

Figure 11–2

Malnutrition and Disease Worsen Each Other

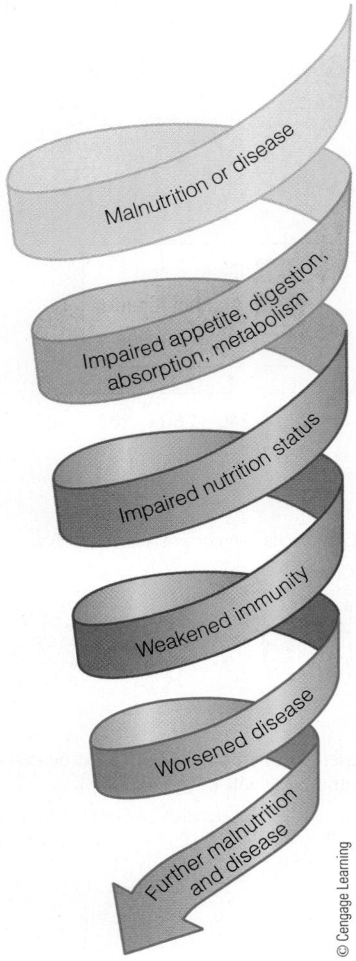

Malnutrition and Disease Worsen Each Other Once a person becomes malnourished, malnutrition often worsens disease, which, in turn, worsens malnutrition. A destructive cycle often begins when impaired immunity opens the way for disease; when disease impairs appetite, interferes with digestion and absorption, increases excretion, or alters metabolism, then nutrition status suffers further. Drugs often become necessary to treat diseases, and many of them impair nutrition status (see Controversy 14). Other treatments, such as surgery or radiation, take a further toll. Thus, together, disease and poor nutrition form a downward spiral that must be broken for recovery to occur (see Figure 11–2).

Impairment of Immune Defenses When deprived of sufficient essential nutrients, indispensable tissues and cells of the immune system can dwindle in size and number, leaving the whole body vulnerable to infection. In addition, both starvation and obesity impair the immune response to infection.[5] Table 11–1 shows selected examples of malnutrition's effects on body defenses. The skin and body linings, the first line of defense against infections, become thinner because their connective tissue is broken down, allowing agents of disease easy access to body tissues.

For example, a healthy digestive system normally musters a formidable defense—its linings not only impose a physical barrier but are also heavily laced with immune tissues, cells, and antibodies that intercept intruders. When malnutrition sets in, all of these defenses diminish and invading infectious agents encounter little resistance. Once inside the warm, moist, nutritious body fluids and tissues, microbes quickly multiply and infection ensues.

A deficiency or a toxicity of just a single nutrient can seriously weaken immune defenses. For example, vitamin A deficiency weakens the body's skin and membranous linings. Vitamin C deficiency robs white blood cells of their killing power. Too little vitamin E may impair immunity in several ways, especially among the aged. Both deficient and excessive zinc intakes impair immunity by reducing the number of effective white blood cells in the first case and impairing the immune response in the second.[6] Table 11–2 lists nutrients known to play key roles in immune function. Clearly, a well-balanced diet is the cornerstone that supports immune system defenses.

Disease Can Worsen Malnutrition Malnutrition can result not only from a lack of available food but also from diseases, such as **AIDS** and cancer, and their treatments. These conditions depress the appetite and speed up metabolism, causing a wasting away of the body's tissues similar to that seen in the last stages of starvation—the body uses its fat and protein reserves for survival. In people with AIDS, wasting or nutrient deficiencies can shorten survival, making medical nutrition therapy a critical need.[7] Nutrients cannot cure AIDS or cancer, of course, but

Table 11–1

Selected Effects of Malnutrition on the Body's Defense Systems

System Component	Effects of Malnutrition
Skin	Thickness, elasticity, and connective tissue are reduced, compromising the skin's ability to serve as a barrier for the protection of underlying tissues; skin sensitivity reaction to antigens is delayed.
Digestive tract membrane and other body linings	Antibody secretions and immune cell numbers are reduced. Barrier functions are also compromised.
Lymph tissues[a]	Immune system organs are reduced in size; cells of immune defense are depleted.
General response	Invader kill time is prolonged; circulating immune cells are reduced; immune response is impaired.

[a] *Thymus gland, lymph nodes, and spleen.*

Table 11–2

Selected Micronutrient Roles in Immune Function

The immune system requires all nutrients for optimal functioning. The vitamins and minerals listed here have well-known, specific roles in immunity.

Nutrient	Key Role(s) in Immune Function
Vitamin A	Maintains healthy skin and other epithelial tissues (barriers to infection); role in cellular replication and specialization that supports immune cell and antibody production and the anti-inflammatory response
Vitamin D	Regulates immune cell (T cells) responses; role in antibody production
Vitamins C and E	Protect against oxidative damage
Vitamin B_6	Helps maintain an effective immune response; role in antibody production
Vitamin B_{12} and folic acid	Assist in cellular replication and specialization that support immune cell and antibody production
Selenium	Protects against oxidative damage
Zinc	Helps maintain an effective immune response; role in antibody production

Source: S. S. Percival, Nutrition and immunity: Balancing diet and immune function, Nutrition Today *46 (2011): 12–17; E. S. Wintergerst, S. Maggini, and D.H. Hornig, Contribution of selected vitamins and trace elements to immune function,* Annals of Nutrition and Metabolism *51 (2007): 301–323; S. Maggini and coauthors, Selected vitamins and trace elements support immune function by strengthening epithelial barriers and cellular and humoral immune responses,* British Journal of Nutrition *98 (2007): S29–S35.*

The immune system was described in **Chapter 3**; starvation and hunger are topics of **Chapter 15**.

an adequate diet may improve responses to drugs, shorten hospital stays, promote independence, and improve the quality of life. Physical activity that strengthens muscles may also help hold wasting to a minimum. In addition, food safety (see Chapter 12) is paramount because common food bacteria and viruses can easily overwhelm a compromised immune system.

To repeat: a *diet* of foods that supplies adequate nutrients ensures the proper functioning of the immune system, but extra daily doses of nutrients, herbs, or other substances do not enhance it. Furthermore, toxic doses clearly diminish it.

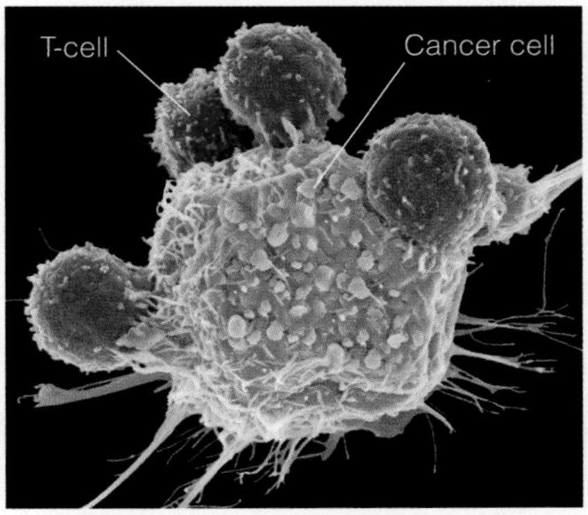

Steve Gschmeissner/Science Source

A killer T-cell (the smaller cell on the top) has recognized a cancer cell and is attacking it with toxic chemicals that punch holes in the cancer cell's surface.

KEY POINTS

- Adequate nutrition is necessary for normal immune system functioning.
- Both deficient and excessive nutrients can harm the immune system.

The Immune System and Chronic Diseases

The immune system's response to infection or injury includes **inflammation**. In inflammation, the blood supply to an infected area increases, and the blood vessels become more permeable, allowing the immune system's white blood cells to rush to the site. Recall from Chapter 3 that white blood cells respond in many ways when injury or infection is present. One part of their response is to release oxidative products, such as deadly hydrogen peroxide (see the photo) to kill microbes or cancer cells, heal an injury, remove damaged tissue, and heal wounds. This acute inflammation response with its oxidative defense is critical to staying healthy.

That same response, however, when it persists, can result in a chronic state of inflammation that harms the tissues. Cells of chronically inflamed tissues produce hormone-like communication molecules, free radicals, blood clotting factors, and

AIDS acquired immune deficiency syndrome; caused by infection with human immunodeficiency virus (HIV), which is transmitted primarily by sexual contact, contact with infected blood, needles shared among drug users, or fluids transferred from an infected mother to her fetus or infant.

inflammation (in-flam-MAY-shun) part of the body's immune defense against injury, infection, or allergens, marked by increased blood flow, release of chemical toxins, and attraction of white blood cells to the affected area (from the Latin *inflammare*, meaning "to flame within"). Also defined in Chapter 5.

other bioactive chemicals that alter functioning and sustain the inflammatory response.[8] Such sustained inflammation threatens health by worsening a number of chronic diseases, as discussed in later sections of this chapter.

KEY POINTS

- Inflammation is part of the body's immune defense system.
- Sustained inflammation can worsen a number of chronic diseases.

The Concept of Risk Factors

LO 11.2 Identify several risk factors for chronic disease, and explain the relationship between risk factors and chronic diseases.

In contrast to the infectious diseases, each of which has a distinct microbial cause such as a bacterium or virus, the chronic diseases have suspected contributors known as **risk factors**. Risk factors show a correlation with a disease—that is, they often occur together with the disease—and although they are candidates for causes, they have not yet been voted in or out. We can say with certainty that a virus causes influenza, but we cannot name the cause of heart disease with such confidence.

An analogy may help clarify the concept of risk factors. A risk factor is like a person who is often seen lurking around the scene of a particular type of crime, say, arson. The police may suspect that person of setting fires, but it may very well be that another, sneakier individual who goes unnoticed is actually pouring the fuel and lighting the match. The evidence against the known suspect is only circumstantial. The police can be sure of guilt only when they observe the criminal in the act. Risk factors have not yet been caught in the act of causing diseases.

Cause versus Increased Risk You may notice a philosophical shift in this chapter from previous chapters. There, we could say "a deficiency of nutrient X causes disease Y." Here, we only cite theories and discuss research that illuminates current thinking. We can say with certainty, for example, that "a diet lacking vitamin C causes scurvy," but to say that a low-fiber diet that lacks vegetables causes cancer would be inaccurate.

Note that in addition to nutrition, disease risk factors are also genetic, environmental, and behavioral, and they tend to occur in clusters and one risk factor may affect several diseases. Food behaviors often underlie many risk factors. Choosing to eat a diet too high in saturated fat, salt, and calories, for example, is choosing to risk becoming obese and contracting atherosclerosis, type 2 diabetes, cancer, **hypertension**, or other diseases.[9] Table 11–3 identifies some diet-related behaviors and other risk factors associated with chronic diseases. In many cases, one chronic disease, such as obesity, contributes to the development or progression of another chronic disease, such as atherosclerosis, as Figure 11–3 shows.

The exact contribution diet makes to each disease is hard to estimate. Many experts believe that diet accounts for about a third of all cases of coronary heart disease. The links between diet and cancer incidence are harder to pin down because each of cancer's many forms associates with different dietary factors.

Estimating Your Risks Some risk factors, such as avoiding tobacco, are important to everyone's health. Others, such as some relating to diet, are more important for people who are genetically predisposed to certain diseases. To pinpoint your own areas of concern, you can search your family's medical history for diseases common to your forebears. Any condition that shows up in several close blood relatives may be a special concern for you (Table 11–4 lists some of these).[‡] For example, a person whose parents, grandparents, or other close blood relatives suffered from diabetes and heart disease is urgently advised to avoid becoming obese and not to smoke. Also,

[‡] The U.S. Surgeon General offers a free online tool, "My Family Health Portrait," to help organize family health information. It is available over the Internet at https://familyhistory.hhs.gov/.

risk factors factors known to be related to (or correlated with) diseases but not proved to be causal.

hypertension higher than normal blood pressure.

Table 11–3
Risk Factors for Chronic Diseases

This table points out that a risk factor associated with one chronic disease often contributes to others, as well. Later tables provide specific risk factors for individual diseases. In addition, Figure 11–3 near here illustrates how chronic diseases themselves can be risk factors for other chronic diseases.

	Cancers	Hypertension	Diabetes (type 2)	Atherosclerosis	Obesity	Stroke
Dietary Risk Factors						
Diets high in added sugars					✓	
Diets high in salty or pickled foods	✓	✓				
Diets high in saturated and/or *trans* fat	✓	✓	✓	✓	✓	✓
Diets low in fruits, vegetables, and other foods rich in fiber and phytochemicals	✓		✓	✓	✓	✓
Diets low in vitamins and/or minerals	✓	✓		✓		
Excessive alcohol intake	✓	✓		✓	✓	✓
Other Risk Factors						
Age	✓	✓	✓	✓		✓
Environmental contaminants	✓					
Genetics	✓	✓	✓	✓	✓	✓
Sedentary lifestyle	✓	✓	✓	✓	✓	✓
Smoking and tobacco use	✓	✓		✓		✓
Stress		✓		✓		✓

Figure 11–3
Interrelationships among Chronic Diseases

The arrows indicate which chronic diseases can lead to others. Note that obesity can lead to diabetes, atherosclerosis, and hypertension. It also increases the risk of some types of cancer.

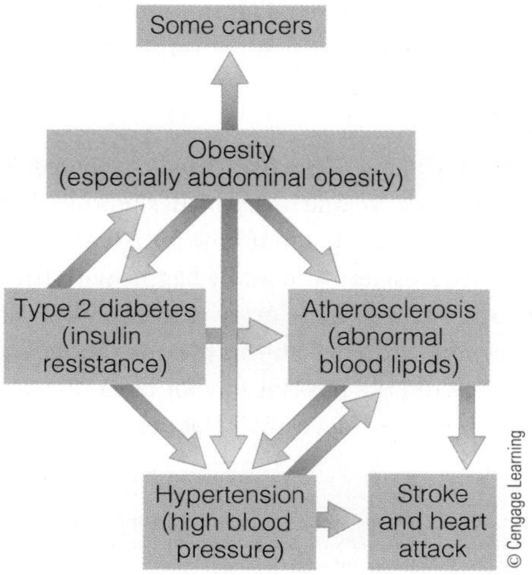

Table 11–4
Family Medical History

These conditions in parents, grandparents, or siblings, especially occurring early in life, may raise a warning flag for you.

- Alcoholism
- Cancer
- Diabetes
- Cardiovascular diseases
- Hypertension
- Liver disease (cirrhosis)
- Osteoporosis

after your next physical examination, find out which test results are out of line. The combination of family medical history and laboratory test results is a powerful predictor of disease. Even people without a family history of diseases can develop them, however; the guidelines presented in this chapter can benefit most people.

KEY POINTS

- The same diet and lifestyle risk factors may contribute to several chronic diseases.
- A person's family history and laboratory test results can reveal elevated disease risks and suggest strategies for disease prevention.

Cardiovascular Diseases

LO 11.3 Identify dietary factors that increase and decrease CVD risks, and explain the association between high blood LDL and HDL cholesterol concentrations and CVD risk.

In the United States today, more than 82 million men and women suffer some form of disease of the heart and blood vessels, collectively known as **cardiovascular disease (CVD)**.[10] Cardiovascular diseases such as heart disease and stroke claim the lives of nearly 1 million people each year in the United States.[§11] These numbers, while unacceptably high, represent a substantial improvement over the numbers of a half-century ago.

One reason heart disease is so deadly is that the heart is one of the least regenerative organs in the body. When cardiac muscle is lost—for example, by way of a heart attack—the heart heals mainly by forming scar tissue. As a result, the contractile function of the heart declines, and heart failure often follows.[12]

The myth that heart disease is a man's disease has been debunked: CVD is the leading cause of death among women. More than 42 million U.S. women have CVD, and the number is increasing despite substantial progress in the awareness, prevention, and treatment of CVD in women.[13] Men still suffer heart attacks more often and earlier in life than women do, but the gap is narrowing. In fact, in all its forms CVD kills more U.S. women, especially those who are past menopause, than any other cause.[14] Learning to recognize the symptoms of a heart attack can be lifesaving because the sooner medical help arrives, the more likely is the person's recovery. In most areas, dial 911 for emergency medical help. Importantly, women may or may not experience classic symptoms such as chest discomfort. Women may instead experience unusual fatigue, dizziness, or weakness.

Did You Know?

Minutes and seconds count when heart attack strikes. Everyone should learn the signs of heart attack listed on the Centers for Disease Control website: www.cdc.gov/heartdisease/signs_symptoms.htm.

How can you minimize your risks of heart attack and stroke? Or, more positively, what steps can you take to help maintain your heart health and vigor throughout life? Many people have done so by quitting smoking or not starting. They have also changed their diets, consuming less saturated fat, less *trans* fat, less cholesterol, more vegetables, more fruits, and more whole grains.[15] In contrast, many people are consuming too many calories, too much sodium, and too few fruits and vegetables; obtaining regular exercise presents a difficult stumbling block for most people.

Atherosclerosis

At the root of most forms of CVD is **atherosclerosis**. Atherosclerosis is the common form of hardening of the arteries. No one is free of all signs of atherosclerosis. The question is not whether you are developing it, but how far advanced it is and what you can do to retard or reverse it. Atherosclerosis usually begins with the accumulation of soft, fatty streaks along the inner walls of the arteries, especially at branch points. These gradually enlarge and become hardened fibrous **plaques** that damage artery walls and make them inelastic, narrowing the passageway for blood to travel through them (see Figure 11–4). Most people have well-developed plaques by the time they reach age 30.[16]

§Deaths from CVD in the United States include 50 percent from coronary heart disease, 16.5 percent from stroke, 7.5 percent from hypertension, 7 percent from congestive heart failure, 3.4 percent from other artery diseases, and the remainder from rheumatic fever/heart disease, congenital cardiovascular defects, and other causes.

Figure 11–4

Animated! **The Formation of Plaques in Atherosclerosis**

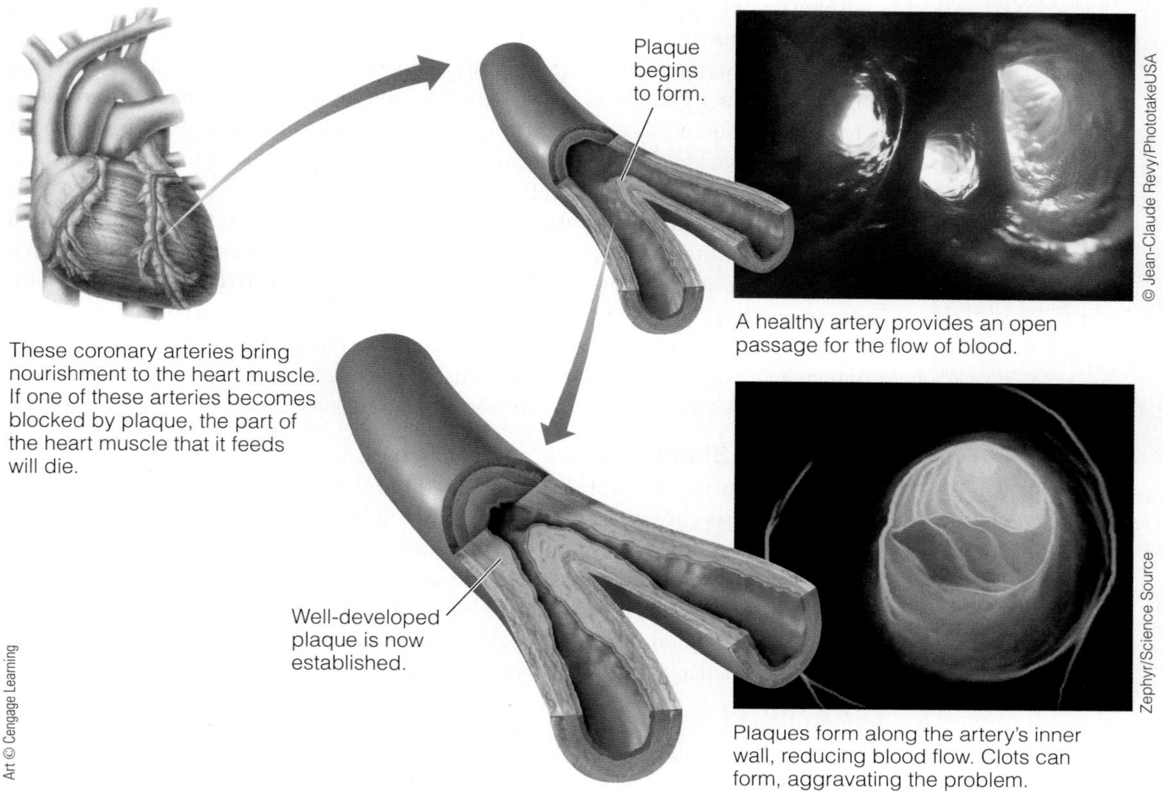

Plaque begins to form.
© Jean-Claude Revy/PhototakeUSA
A healthy artery provides an open passage for the flow of blood.
These coronary arteries bring nourishment to the heart muscle. If one of these arteries becomes blocked by plaque, the part of the heart muscle that it feeds will die.
Well-developed plaque is now established.
Zephyr/Science Source
Art © Cengage Learning
Plaques form along the artery's inner wall, reducing blood flow. Clots can form, aggravating the problem.

How Plaques Form What causes the plaques to form? A diet high in saturated fat is a major contributor to the development of plaques and the progression of atherosclerosis.[17] But atherosclerosis is much more than the simple accumulation of lipids within the artery wall—it is a complex response of the artery to tissue damage and inflammation.[18] Inflammation plays a central role in all stages of atherosclerosis.

The cells lining the arteries may incur damage from a number of factors: high LDL cholesterol, hypertension, diabetes, toxins from cigarette smoking, obesity, or certain viral or bacterial infections.[19] Such damage produces inflammation that triggers the immune system to send white blood cells to the site to try to repair the damage. Soon, particles of LDL cholesterol become trapped in the blood vessel walls, and these become oxidized by the abundant free radicals produced during inflammation. White blood cells—**macrophages**—flood the scene to scavenge and remove the oxidized LDL, but to no avail. As the macrophages become engorged with oxidized LDL, they become known as foam cells, which themselves become triggers of oxidation and inflammation that attract more immune scavengers to the scene. Muscle cells of the arterial wall proliferate in an attempt to heal the damage, but they mix with the foam cells to form hardened areas of plaque. Mineralization increases hardening of the plaques. The process is repeated until many inner artery walls become virtually covered with disfiguring plaques.

Plaque Rupture and Blood Clots Once plaques have formed, a sudden spasm of the artery wall or surge in blood pressure can tear away part of the fibrous coat covering a plaque, causing it to rupture. Unstable plaques with a thin fibrous layer over a large lipid core are most vulnerable to rupture.[20] When a plaque ruptures, the body responds to the damage as an injury—by clotting the blood.

Small, cell-like bodies in the blood, known as **platelets**, cause clots to form when they encounter injuries in blood vessels. Clots form and dissolve in the blood all the

cardiovascular disease (CVD) a general term for all diseases of the heart and blood vessels. Atherosclerosis is the main cause of CVD. When the arteries that carry blood to the heart muscle become blocked, the heart suffers damage known as *coronary heart disease (CHD)*. Also defined in Chapter 5.

atherosclerosis (ath-er-oh-scler-OH-sis) the most common form of cardiovascular disease; characterized by plaques along the inner walls of the arteries (*scleros* means "hard"; *osis* means "too much"). The term *arteriosclerosis* is often used to mean the same thing.

plaques (PLACKS) mounds of lipid material mixed with smooth muscle cells and calcium that develop in the artery walls in atherosclerosis (*placken* means "patch"). The same word is also used to describe the accumulation of a different kind of deposit on teeth, which promotes dental caries.

macrophages (MACK-roh-fah-jez) large scavenger cells of the immune system that engulf debris and remove it (*macro* means "large"; *phagein* means "to eat"). Also defined in Chapter 3.

platelets tiny cell-like fragments in the blood, important in blood clot formation (*platelet* means "little plate").

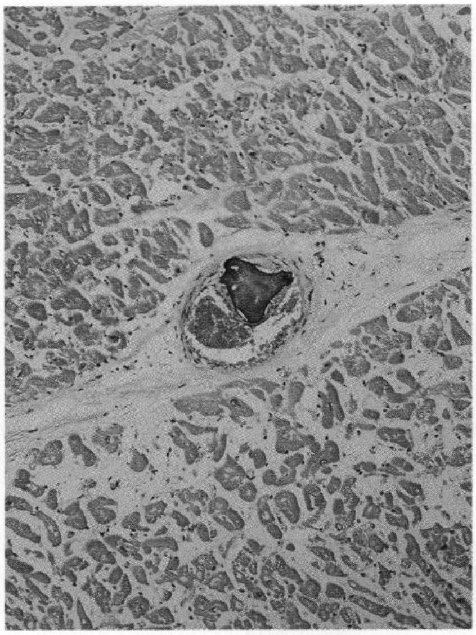

A blood clot in an artery, such as this fatal heart embolism, blocks the blood flow to tissues fed by that artery.

time, and when these processes are balanced, the clots do no harm. That balance is disturbed in atherosclerosis, however. Arterial damage, plaques in the arteries, and inflammation all favor the formation of blood clots.

Abnormal blood clotting can trigger life-threatening events. For example, a clot, once formed, may remain attached to a plaque in an artery and grow until it shuts off the blood supply to the surrounding tissue. The starved tissue slowly dies and is replaced by nonfunctional scar tissue. The stationary clot is called a **thrombus**. When it has grown large enough to close off a blood vessel, it is a **thrombosis**. A clot can also break loose, becoming an **embolus**, and travel along in the bloodstream until it reaches an artery too small to allow its passage. There the clot becomes stuck and is referred to as an **embolism**. The tissues fed by this artery, suddenly robbed of oxygen and nutrients, die rapidly. Such a clot can lodge in an artery of the heart, causing sudden death of part of the heart muscle, a **heart attack**. A clot may also lodge in an artery of the brain, killing a portion of brain tissue, a **stroke**.

Opposing the clot-forming actions of platelets is one of the eicosanoids, an active product of an omega-3 fatty acid in fish oils.[21] A diet lacking in fatty fish may therefore contribute to clot formation and can worsen heart disease risk in other ways as well.[22]

Plaques and Blood Pressure Normally, the arteries expand with each heartbeat to accommodate the pulses of blood that flow through them. Arteries hardened and narrowed by plaques cannot expand, however, so the blood pressure rises. The increased pressure damages the artery walls further and strains the heart. Because plaques are more likely to form at damage sites, the development of atherosclerosis becomes a self-accelerating process. As pressure builds up in an artery, the arterial wall may become weakened and balloon out, forming an **aneurysm**. An aneurysm can burst, and in a major artery such as the **aorta**, this leads to massive bleeding and death.

KEY POINTS

- Atherosclerosis begins with the accumulation of soft, fatty streaks on the inner walls of the arteries.
- The soft, fatty streaks gradually enlarge and become hard plaques.
- Plaques of atherosclerosis trigger hypertension and abnormal blood clotting, leading to heart attacks or strokes.

Risk Factors for CVD

An expert panel of the National Cholesterol Education Program defined major heart disease risk factors already listed briefly in Chapter 5; they are presented in full in Table 11–5. Many of the same factors also predict the occurrence of stroke. All people reaching middle age exhibit at least one of these factors (middle age is a risk factor), and many people have several factors, silently increasing their risk.[23] The more of these risk factors you can control, the lower your risk of CVD and death.[24] In recognition of the urgency to reduce the prevalence of major risk factors for CVD, national initiatives have been developed.[25] The Department of Health and Human Services has launched the Million Hearts campaign with the goal of preventing one million heart attacks and strokes over five years. Million Hearts brings together communities, health systems, nonprofit organizations, federal agencies, and private-sector partners all across the country to prevent heart disease and stroke. The American Heart Association has adopted a 2020 goal of "improving the cardiovascular health of all Americans by 20 percent, while reducing mortality from heart disease and stroke by 20 percent."

It befits a nutrition book to focus on dietary strategies, but Table 11–5 shows that diet is not the only, and perhaps not even the most important, factor in the development of heart disease or stroke. Age, gender, genetic inheritance, cigarette smoking, certain diseases, and physical inactivity predict their development as well. The next few sections address these factors; discussions of diet and physical activity for CVD prevention follow.

Age, Gender, and Genetic Inheritance Three of the major risk factors for CVD cannot be modified by lifestyle choices: age, gender, and genes. The increasing

Table 11–5

Major Risk Factors for Heart Disease

See Figure 11–6 on p. 432 for standards by which to judge blood lipids, obesity, and blood pressure. Risk factors highlighted in color have relationships with diet.

Risk factors that cannot be modified:
- Increasing age.
- Male gender.
- Genetic inheritance.

Risk factors that can be modified:
- High blood LDL cholesterol.
- Low blood HDL cholesterol.
- High blood pressure (hypertension).
- Diabetes.
- Obesity (especially central obesity).
- Physical inactivity.
- Cigarette smoking.
- An "atherogenic" diet (high in saturated fats and *trans* fats and low in vegetables, fruits, and whole grains).

Source: Expert Panel on Detection, Evaluation, and Treatment of High Blood Cholesterol in Adults (Adult Treatment Panel III), Third Report of the National Cholesterol Education Program (NCEP), *NIH publication no. 02-5215 (Bethesda, Md.: National Heart, Lung, and Blood Institute, 2002), pp. II-15–II-20.*

risk associated with growing older reflects the steady progression of atherosclerosis in most people as they age.[26] The speed at which atherosclerosis progresses, however, depends more on the presence or absence of risk factors such as high blood pressure, high blood cholesterol, diabetes, and smoking than on age alone.[27] For example, among men 55 years of age, those with at least two major risk factors were six times as likely to die from CVD by age 80 than those with one or no risk factors. Women of the same age with at least two risk factors were three times as likely to die from CVD by age 80 than those with one or no risk factors.

Gender alters risks at many life stages. In men, aging becomes a significant risk factor for heart disease at age 45 years or older; in women, the risk increases after age 55. Young women can easily become complacent about heart disease because men die earlier of heart attacks than do women. It bears repeating that CVD kills more U.S. women than any other cause.[28]

As for genetic inheritance, early heart disease in immediate family members (siblings or parents) is a major risk factor for developing it. The more family members affected and the earlier the age at which they became ill, the greater the risk to the individual.[29] These relationships suggest a genetic influence on CVD risk but specific genetic links are still under investigation. In the realm of nutritional genomics and CVD risk, scientists are uncovering a vast interrelated network of influences, but the relationships among them are complex and are likely to become more knotty before being untangled.[30] Many of these relationships center on blood lipids.

LDL and HDL Cholesterol Low-density lipoprotein (LDL) cholesterol and high-density lipoprotein (HDL) cholesterol in the blood are strongly linked to a person's risk of developing atherosclerosis and heart disease. The higher the LDL cholesterol, the greater the risk of CVD (see Figure 11–5, p. 432). Conversely, higher HDL is thought to be protective. Some controversy surrounds HDL because drug treatments that elevate its blood concentration do not prevent heart disease.[31] Figure 11–6, p. 432, lists blood lipid values considered to be healthy and those that exceed a healthy level and also presents values for body mass index (BMI) and blood pressure.

LDL cholesterol is made up of the most atherogenic lipoproteins. LDL carry cholesterol to the cells, including the cells that line the arteries, where it can build up as part of the plaques of atherosclerosis described earlier. The lower the LDL cholesterol and blood pressure, the slower the progression of atherosclerosis.[32] In clinical trials,

Did You Know?

The Centers for Disease Control and Prevention has developed WISEWOMAN projects nationwide to provide low-income women with resources needed to reduce their risks of CVD (www.cdc.gov/WISEWOMAN).

thrombus a stationary blood clot.

thrombosis a thrombus that has grown enough to close off a blood vessel. A *coronary thrombosis* closes off a vessel that feeds the heart muscle. A *cerebral thrombosis* closes off a vessel that feeds the brain (*thrombo* means "clot"; the cerebrum is part of the brain).

embolus (EM-boh-luss) a thrombus that breaks loose and travels through the blood vessels (*embol* means "to insert").

embolism an embolus that causes sudden closure of a blood vessel.

heart attack the event in which the vessels that feed the heart muscle become closed off by an embolism, thrombus, or other cause with resulting sudden tissue death. A heart attack is also called a *myocardial infarction* (*myo* means "muscle"; *cardial* means "of the heart"; *infarct* means "tissue death").

stroke the sudden shutting off of the blood flow to the brain by a thrombus, embolism, or the bursting of a vessel (hemorrhage).

aneurysm (AN-you-rism) the ballooning out of an artery wall at a point that is weakened by deterioration.

aorta (ay-OR-tuh) the large, primary artery that conducts blood from the heart to the body's smaller arteries.

Figure 11–5

LDL, HDL, and Risk of Heart Disease

A blood lipid profile with low HDL (<40 mg/dL) and high LDL (≥160 mg/dL) elevates risk.

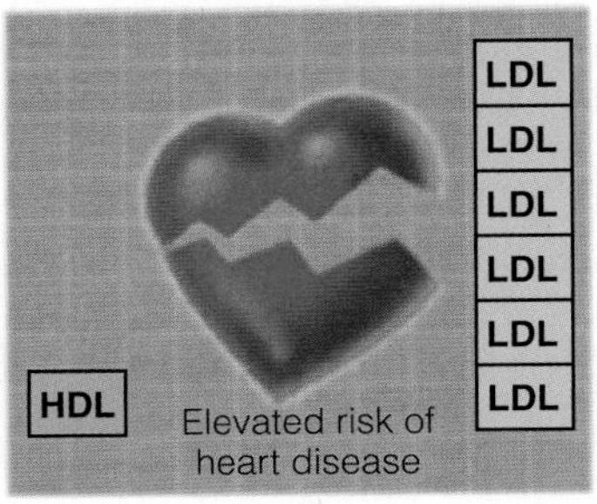

The opposite, high HDL (≥60 mg/dL) and low LDL (<100 mg/dL) lowers risk.

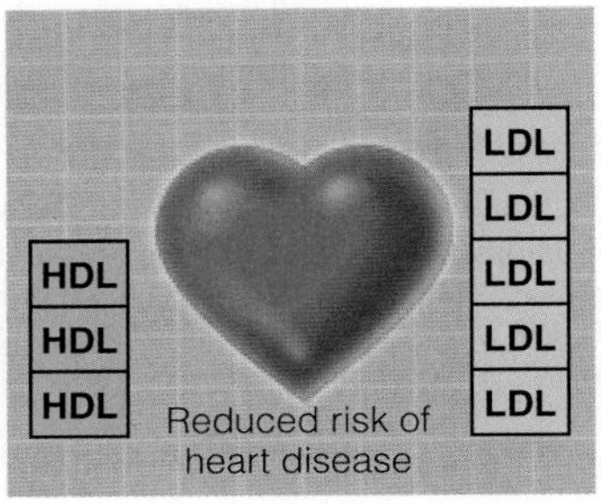

Figure 11–6

Adult Standards for Blood Lipids, Body Mass Index, and Blood Pressure

	Total blood cholesterol (mg/dL)	LDL cholesterol (mg/dL)	HDL cholesterol (mg/dL)	Triglycerides, fasting (mg/dL)	Body mass index (BMI)[a]	Blood pressure systolic / diastolic (mm Hg)
Unhealthy	≥240	160–189[b]	<40	200–499[c]	≥30	≥140 / ≥90
Borderline	200–239	130–159[d]	59–40	150–199	25–29.9	120/80–139/89[e]
Healthy	<200	<100[f]	≥60	<150	18.5–24.9	<120 / <80

[a]Body Mass Index (BMI) was defined in Chapter 9; BMI standards are found on the inside back cover.

[b]>190 mg/dL LDL indicates a very high risk.

[c]>500 mg/dL triglycerides indicates a very high risk.

[d]LDL cholesterol-lowering medication may be needed at 130 mg/dL, depending on other risks.

[e]These values indicate prehypertension.

[f]100–129 mg/dL LDL indicates a near optimal level.

lowering LDL greatly reduces the incidence of heart disease. By one estimate, for every percentage point drop in LDL cholesterol, the risk of heart disease falls proportionately. For this reason, the National Heart, Lung, and Blood Institute urges those at high risk for heart disease to lower blood LDL cholesterol and to use medication if need be to lower it.

High-density lipoproteins (HDL) also carry cholesterol, but they carry it away from the cells to the liver for recycling to other uses or for disposal. Elevated HDL generally indicates a *reduced* risk of atherosclerosis and heart attack. For this reason, heart disease risk assessment inventories, such as the one in Figure 11–7, give extra credit for having a high HDL value.

The different types of lipoproteins are discussed in detail in **Chapter 5**.

High levels of LDL cholesterol in the blood can both initiate and worsen atherosclerosis. When levels are high, LDL enters vulnerable regions of the artery wall where they are oxidized.[33] Oxidized LDL and other factors attract immune cells to the innermost layer of the arterial wall. There, these immune cells undergo transformation into macrophages that form the lipid-rich foam cells characteristic of fatty streaks. This process perpetuates chronic inflammation and may make plaque rupture likely.[34] When plaques rupture, it can cause a heart attack or stroke. In advanced atherosclerosis, a goal of treatment is to lower LDL cholesterol to stabilize existing plaques while slowing the development of new ones.

Not all LDL are the same in regard to heart disease risk—they vary in size and density. The smallest, most dense LDL are considered the most atherogenic, whereas larger, less dense LDL are less atherogenic.

Did You Know?

Cholesterol is carried in several lipoproteins, chief among them LDL and HDL. Remember them this way:

- **L**DL = **L**ow-density lipoproteins = **L**ess healthy.
- **H**DL = **H**igh-density lipoproteins = **H**ealthy.

Hypertension and Atherosclerosis Worsen Each Other

Hypertension and atherosclerosis are twin demons that worsen CVD, and each worsens the other. Hypertension worsens atherosclerosis because a stiffened artery, already strained by each pulse of blood surging through it, is stressed further by high internal pressure. Injuries multiply, more plaques grow, and more weakened vessels become likely to burst and bleed.

Atherosclerosis also worsens hypertension. Since hardened arteries cannot expand, the heart's beats raise the blood pressure. Hardened arteries also fail to

Figure 11–7

How to Assess Your Heart Disease Risk

Do you know your heart disease risk score? This assessment estimates your 10-year risk for heart disease using charts from the Framingham Heart Study.* Be aware that a high score does not mean that you *will* develop heart disease, but it should warn you of the possibility and prompt you to consult a physician about your health. You will need to know your blood cholesterol (ideally, the average of at least two recent measurements) and blood pressure (ideally, the average of several recent measurements). With this information in hand, find yourself in the charts below and add the points for each risk factor.

Age (years)	Points	
	Men	Women
20–34	–9	–7
35–39	–4	–3
40–44	0	0
45–49	3	3
50–54	6	6
55–59	8	8
60–64	10	10
65–69	11	12
70–74	12	14
75–79	13	16

HDL (mg/dL)	Points	
	Men	Women
≥60	–1	–1
50–59	0	0
40–49	1	1
<40	2	2

Systolic Blood Pressure (mm Hg)	Points			
	Untreated		Treated	
	Men	Women	Men	Women
<120	0	0	0	0
120–129	0	1	1	3
130–139	1	2	2	4
140–159	1	3	2	5
≥160	2	4	3	6

Total Cholesterol (mg/dL)	Points									
	Age 20–39		Age 40–49		Age 50–59		Age 60–69		Age 70–79	
	Men	Women	Men	Women	Men	Women	Men	Women	Men	Women
<160	0	0	0	0	0	0	0	0	0	0
160–199	4	4	3	3	2	2	1	1	0	1
200–239	7	8	5	6	3	4	1	2	0	1
240–279	9	11	6	8	4	5	2	3	1	2
≥280	11	13	8	10	5	7	3	4	1	2
Smoking (any cigarette smoking in the past month)										
Smoker	8	9	5	7	3	4	1	2	1	1
Nonsmoker	0	0	0	0	0	0	0	0	0	0

Scoring Your Heart Disease Risk

Add up your total points: ______ . Now find your total in the first column for your gender in the chart at the right and then look to the next column for your approximate risk of developing heart disease within the next 10 years. Depending on your risk category, the following strategies can help reduce your risk:

- *>20% = High risk.* Try to lower LDL using all lifestyle changes and, most likely, lipid-lowering medications as well.
- *10–20% = Moderate risk.* Try to lower LDL using all lifestyle changes and, possibly, lipid-lowering medications.
- *<10% = Low risk.* Maintain or initiate lifestyle choices that help prevent elevation of LDL to prevent future heart disease.

Men		Women	
Total Points	Risk	Total Points	Risk
<0	<1%	<9	<1%
0–4	1%	9–12	1%
5–6	2%	13–14	2%
7	3%	15	3%
8	4%	16	4%
9	5%	17	5%
10	6%	18	6%
11	8%	19	8%
12	10%	20	11%
13	12%	21	14%
14	16%	22	17%
15	20%	23	22%
16	25%	24	27%
≥17	≥30%	≥25	≥30%

**An electronic version of this assessment is available on the ATP III page of the National Heart, Lung, and Blood Institute's website (www.nhlbi.nih.gov/guidelines /cholesterol). Another risk inventory is available from the American Heart Association (www.americanheart.org).*

Source: Adapted from Expert Panel on Detection, Evaluation, and Treatment of High Blood Cholesterol in Adults (Adult Treatment Panel III), Third Report of the National Cholesterol Education Program (NCEP), *NIH publication no. 02-5215 (Bethesda, Md.: National Heart, Lung, and Blood Institute, 2002), section III.*

let blood flow freely through the kidneys, which control blood pressure. The kidneys sense the reduced flow of blood and respond as if the blood pressure were too low; they take steps to raise it further. The higher the blood pressure above normal, the greater the risk of heart attack or stroke. The relationship between hypertension and disease risk holds true for men and women, young and old. A later section gives details about hypertension because it constitutes a major threat to health on its own.

Diabetes Diabetes, a major independent risk factor for all forms of CVD, substantially increases the risk of death from these causes.[35] In diabetes, atherosclerosis progresses rapidly, blocking blood vessels and diminishing circulation. For many people with diabetes, the risk of a future heart attack is roughly equal to that of a person *without* diabetes who has already had a heart attack.[36] When heart disease occurs in conjunction with diabetes, the condition is likely to be severe. In fact, any loss of control of blood glucose, even a transitory one, can be costly in terms of the condition of the arteries. Few people with diabetes recognize that, left uncontrolled, diabetes holds a grave threat of all forms of CVD.

Diabetes is discussed in full in **Chapter 4**.

Physical Inactivity Without routine physical activity, muscles, including the muscles of the heart and arteries, weaken and reduce the heart's ability to meet everyday demands. Regular physical activity expands the heart's capacity to pump blood to the tissues with each beat, thereby reducing the number of heartbeats required and the heart's workload. The slower pulse of fit people reflects the heart's greater pumping capacity.

"Walking is man's best medicine."
—Hippocrates

Physical activity also stimulates development of new arteries to nourish the heart muscle, which may be a factor in the excellent recovery seen in some heart attack victims who exercise. In addition, physical activity favors lean tissue over fat tissue for a healthy body composition, raises HDL cholesterol, improves insulin sensitivity, reduces levels of inflammatory markers, and lowers blood pressure, LDL cholesterol, blood triglyceride levels, and blood glucose.[37] If pursued on at least 5 days each week, 30 minutes or more of brisk walking can improve the odds against heart disease considerably. If you are pressed for time, 15 minutes of more vigorous physical activity, such as jogging, on at least 5 days a week can provide the same benefits. The Think Fitness feature offers suggestions for incorporating physical activity into your daily routine. Chapter 10 provided the *Physical Activity Guidelines for Americans*.

Smoking Cigarette smoking powerfully increases the risk for CVD in men and women. The more a person smokes, the higher the CVD risk. Smoking tobacco in all its forms damages the heart directly with toxins and burdens it by raising the blood pressure. Body tissues starved for oxygen by smoke demand more heartbeats to deliver oxygenated blood, thereby increasing the heart's workload. At the same time, smoking deprives the heart muscle itself of the oxygen it needs to maintain a steady beat. Smoking also damages platelets, making blood clots likely. Damage to the linings of the blood vessels from tobacco smoke toxins makes atherosclerosis likely. When people quit smoking, their risk of heart disease begins to drop within a few months; a year later, their risk has dropped by half, and after 15 years of staying smoke-free, their risks equal those of lifetime nonsmokers.[38]

Atherogenic Diet Diet influences the risk of CVD. An "atherogenic diet"—high in saturated fats, *trans* fats, and cholesterol—increases LDL cholesterol. A high intake of *trans* fatty acids also lowers HDL cholesterol.[39] Fortunately, a well-chosen diet can often lower the risk of CVD and does so to a greater degree than might be expected from its effects on blood lipids alone.[40] A number of beneficial factors in such diets may share the credit, among them the vitamins, minerals, fibers, antioxidant phytochemicals, and omega-3 fatty acids.[41] A diet rich in raw vegetables and fruits may protect against CVD even in people who are genetically susceptible to the disease.[42]

Obesity and Metabolic Syndrome Many of the modifiable risk factors for CVD are directly related to diet (look again at Table 11–5, page 431). Several of these

Think Fitness move it! Ways to Include Physical Activity in a Day

The benefits of physical activity are compelling, so why not tie up your athletic shoes, head out the door, and get going? Here are some ideas to get you started:

- Coach a sport.
- Garden.
- Hike, bike, or walk to nearby stores or to classes.
- Mow, trim, and rake by hand.
- Park a block from your destination and walk.
- Play a sport.
- Play with children.
- Take classes for credit in dancing, sports, conditioning, or swimming.
- Take the stairs, not the elevator.
- Walk a dog.
- Walk 10,000 steps per day. This amounts to about 5 miles, enough to meet the "active" daily activity level. Use an inexpensive pedometer to count your steps.
- Wash your car with extra vigor or bend and stretch to wash your toes in the bath.
- Work out at a fitness club.
- Work out with friends to help one another stay fit.

Also, try these:

- Give two labor-saving devices to someone who needs them.
- Lift small hand weights while talking on the phone, reading e-mail, or watching TV.
- Stretch often during the day.
- Try the President's Challenge for improving fitness: www.presidentschallenge.org.

start now! Using the list above as a guide, make your own list of things you can do today to be physically active. Using the calendar you created in Chapter 10, note on each day for the next month the physical activities you have engaged in for that day.

diet-related risk factors—low blood HDL cholesterol, high blood pressure, elevated fasting blood glucose (insulin resistance), and central obesity—along with high blood triglycerides comprise a cluster of health risks known as the **metabolic syndrome**. The metabolic syndrome underlies several chronic diseases and increases the risk of CVD and type 2 diabetes.[43] The precise cause of the metabolic syndrome is not known, but central obesity and insulin resistance are thought to be primary factors in its development.[44]

Metabolic syndrome, like the chronic diseases associated with it, involves inflammation and elevates the risk for thrombosis.[45] More than one-third of the U.S. adult population meets the criteria for metabolic syndrome shown in Table 11–6, but many are unaware of it and so do not seek treatment.[46] Even obesity alone, especially central obesity, raises LDL cholesterol, lowers HDL cholesterol, raises blood pressure, and promotes insulin resistance.[47] The opposite also holds true: weight loss and physical activity lower LDL, raise HDL, improve insulin sensitivity, and lower blood pressure.[48]

Table 11–6
Metabolic Syndrome

Metabolic syndrome includes any three or more of the following:

- High fasting blood glucose.
- Central obesity.
- Hypertension.
- Low blood HDL.
- High blood triglycerides.

High Blood Triglycerides Some people with heart disease, especially those with diabetes and those who are overweight, have elevated triglycerides. Elevated triglycerides are not directly atherogenic, and therefore do not represent an independent risk factor for CVD.[49] Rather, elevated triglycerides are considered an important marker of CVD risk because of the role they play in lipoprotein metabolism. When triglycerides are moderately high, remnants of very-low-density lipoproteins (VLDL) collect in the blood and are highly atherogenic.

KEY POINTS

- In most people, atherosclerosis progresses steadily as they age.
- Major risk factors for CVD are age, gender, family history, high LDL cholesterol and low HDL cholesterol, hypertension, diabetes, physical inactivity, smoking, an atherogenic diet, and obesity.

Recommendations for Reducing CVD Risk

Recommendations to reduce the risk of CVD focus first on lifestyle changes. To that end, people are encouraged to increase physical activity, lose weight (if necessary),

metabolic syndrome a combination of characteristic factors—high fasting blood glucose or insulin resistance, central obesity, hypertension, low blood HDL cholesterol, and elevated blood triglycerides—that greatly increase a person's risk of developing CVD. Also called *insulin resistance syndrome*.

implement dietary changes, and reduce exposure to tobacco smoke either by quitting smoking or by avoiding secondhand smoke.[50] If such lifestyle changes fail to lower LDL or blood pressure to acceptable levels, then medications are prescribed. Table 11–7 summarizes strategies to reduce the risk of heart disease.[51] The Food Feature, later, offers one example of a heart-healthy diet, the DASH eating plan.

Diet to Reduce CVD Risk What role can diet play in minimizing the risk of developing CVD? The answer focuses primarily on how diet relates to high blood cholesterol. The effects of diet are two sides of the same coin: side one, a diet high in saturated fat and *trans* fatty acids contributes to high blood LDL cholesterol; side two, reducing those fats in the diet lowers blood LDL cholesterol and may reduce the risk of CVD. Table 11–8 demonstrates the power of diet-related factors to reduce LDL cholesterol.

Maximilian Stock Ltd./Photographer's Choice/Getty Images

When diets are rich in whole grains, vegetables, and fruits, life expectancies are long.

Wherever in the world populations consume diets high in saturated fat and low in fish, fruits, vegetables, and whole grains, blood cholesterol is high and heart disease takes a toll on health and life. Conversely, wherever dietary fats are mostly unsaturated and where fish, fruits, vegetables, and whole grains are abundant, blood cholesterol and rates of heart disease are low.

It matters, too, what people choose to eat instead of saturated fats. Relationships between carbohydrate intakes and heart disease are not fully defined, but a diet too high in refined starches and added sugars has the potential to worsen heart disease risk by elevating blood triglycerides and inflammatory markers and reducing HDL cholesterol.[52] People with elevated triglycerides may find that replacing refined starches and sugars with whole grains, legumes, or vegetables helps to improve blood lipids.[53]

Fish oils, rich in omega-3 polyunsaturated fatty acids, may reduce inflammation, lower triglycerides, prevent blood clots, and produce other effects that may reduce the risk of sudden death associated with both heart disease and stroke.[54] For these reasons, the American Heart Association recommends two meals of fish per week. People

Table 11–7

Strategies to Reduce the Risk of CVD

Dietary Strategies	
▪ *Energy:* Balance energy intake and physical activity to prevent weight gain and to achieve or maintain a healthy body weight. ▪ *Saturated fat,* trans *fat, and cholesterol:* Choose lean meats, vegetables, and low-fat milk products; minimize intake of hydrogenated fats. Limit saturated fats to less than 7% of total kcalories, *trans* fat to less than 1% of total kcalories, and cholesterol to less than 300 milligrams a day. ▪ *Soluble fibers:* Choose a diet rich in vegetables, fruits, whole grains, and other foods high in soluble fibers. ▪ *Potassium and sodium:* Choose a diet high in potassium-rich fruits and vegetables, low-fat milk products, nuts, and whole grains. Choose and prepare foods with little or no salt. (Limit sodium intake to less than 2,300 mg/day, and further reduce intake to 1,500 mg/day among those who are 51 and older and those of any age who are African American or have hypertension, diabetes, or chronic kidney disease.)	▪ *Added sugars:* Minimize intake of beverages and foods with added sugars. ▪ *Fish and omega-3 fatty acids:* Consume fatty fish rich in omega-3 fatty acids (salmon, tuna, sardines) at least twice a week. ▪ *Plant sterols and stanols:* Consume food products that contain added plant sterols (defined in Controversy 2, page 63). ▪ *Soy:* Consume soy foods to replace animal and dairy products that contain saturated fat and cholesterol. ▪ *Alcohol:* If alcohol is consumed, limit it to one drink daily for women and two drinks daily for men.
Lifestyle Choices	
▪ *Physical activity:* Participate in at least 30 minutes of moderate-intensity endurance activity on most days of the week. Chapter 10 specifies types and amounts of physical activity required to develop and maintain a healthy body.	▪ *Smoking cessation:* Minimize exposure to any form of tobacco or tobacco smoke.

Sources: American Heart Association, www.heart.org; M. R. Flock and P. M. Kris-Etherton, Dietary Guidelines for Americans 2010: Implications for cardiovascular disease, Current Atherosclerosis Reports *13 (2011): 499–507.*

Table 11–8

How Much Does Changing the Diet Change LDL Cholesterol?

Diet-Related Component	Modification	Possible LDL Reduction
Saturated fat	<7% of calories	8–10%
Dietary cholesterol	<200 mg/day	3–5%
Weight reduction (if overweight)	Lose 10 lb	5–8%
Soluble, viscous fiber	5–10 g/day	3–5%

Source: Expert Panel on Detection, Evaluation, and Treatment of High Blood Cholesterol in Adults (Adult Treatment Panel III), Third Report of the National Cholesterol Education Program (NCEP), *NIH publication no. 02-5215 (Bethesda, Md.: National Heart, Lung, and Blood Institute, 2002), p. V-21.*

diagnosed with heart disease may require more than this amount, preferably from additional servings of fatty fish, but a physician may prescribe fish oil supplements in some cases.[55] Not all studies report benefits, and the supplements may carry their own risks, so they should be taken under the supervision of a physician.[56]

Other Dietary Factors Earlier chapters addressed other dietary factors that may reduce heart disease risks, and many of these are listed in Table 11–7. A potentially helpful innovation is the plant sterols that have been added to certain kinds of margarine, orange juice, and other foods; authorities recommend their use when other lifestyle changes fail to adequately bring blood cholesterol down.

Plant sterols block absorption of cholesterol from the intestine, an effect that can be as powerful as some medications in lowering blood LDL cholesterol. The cost of plant-sterol-enriched foods is high, however, and a caution is in order: plant sterols also reduce absorption of *other* potentially beneficial phytochemicals in the diet.

A small reduction in blood LDL cholesterol may also be realized when soy foods provide the majority of protein in a diet, but the amount of daily soy foods needed to produce the effect is larger than an amount many people may choose.[57] Still, additional benefits may come with soybeans, soy protein products such as imitation meats, and soy milk consumed as part of a heart-healthy diet.

Although diet and physical activity are not the easy route to heart health that everyone hopes for, they form a powerful and safe combination for improving health. The pattern of protection from the recommended diet and physical activity regimen becomes clear—the effects of each small choice add to the beneficial whole. While you are at it, don't smoke. Relax. Meditate or pray. Control stress. Play. Relaxed, happy people who make time to enjoy life have lower blood pressure and fewer heart attacks.[58]

KEY POINTS

- Lifestyle changes to lower the risk of CVD include increasing physical activity, achieving a healthy body weight, reducing exposure to tobacco smoke, and eating a heart-healthy diet.
- Dietary measures to lower LDL cholesterol include reducing intakes of saturated fat, *trans* fat, and cholesterol, along with consuming enough nutrient-dense fruits, vegetables, legumes, fish, and whole grains.

Nutrition and Hypertension

LO 11.4 Outline a general eating and exercise plan for a person with prehypertension, and justify the recommendations.

People with healthy blood pressure generally enjoy a long life and suffer less often from heart disease.[59] Chronic high blood pressure, or hypertension, remains one of the most prevalent forms of CVD, affecting more than 76 million U.S. adults, and its rate has been rising steadily.[60] Hypertension contributes to an estimated one million heart attacks and more than 795,000 strokes each year. The higher above normal the blood pressure goes, the greater the risk.

You cannot tell if you have high blood pressure—it presents no symptoms you can feel. The most effective single step you can take to protect yourself from hypertension is to find out whether you have it. During a checkup, a health-care professional can take an accurate resting blood pressure reading. Self-test machines in drugstores and other public places are often inaccurate. If your resting blood pressure is above normal, the reading should be repeated before confirming the diagnosis of hypertension. Thereafter, blood pressure should be checked at regular intervals.

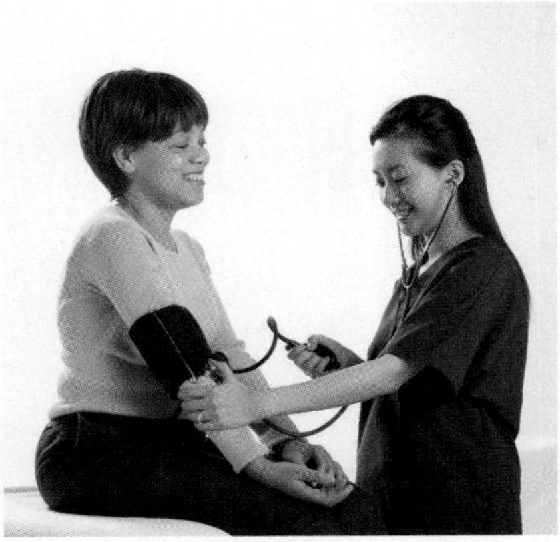

© iofoto/Shutterstock.com

The most effective single step you can take against hypertension is to learn your own blood pressure.

When blood pressure is measured, two numbers are important: the pressure during contraction of the heart's ventricles (large pumping chambers) and the pressure during their relaxation. The numbers are given as a fraction, with the first number representing the **systolic pressure** (ventricular contraction) and the second number the **diastolic pressure** (relaxation). Return to Figure 11–6 (page 432) to see how to interpret your resting blood pressure.

Ideal resting blood pressure is lower than 120 over 80. Just above this value lies **prehypertension**—blood pressure values in the borderline range of up to 139 over 89. Blood pressure in this range means that high blood pressure is likely to develop in the future and that taking steps to keep blood pressure low may avert illness later on.[61] Above this borderline level, though, the risks of heart attacks and strokes rise in direct proportion to increasing blood pressure.

KEY POINTS

- Hypertension silently and progressively worsens atherosclerosis and makes heart attacks and strokes likely.
- All adults should know whether or not their blood pressure falls within the normal range.

How Does Blood Pressure Work in the Body?

Blood pressure is vital to life. It pushes the blood through the major arteries into smaller arteries and finally into tiny capillaries whose thin walls permit exchange of fluids between the blood and the tissues (see Figure 11–8). Blood pressure arises from contractions in the heart muscle that pump blood away from the heart (**cardiac output**) and the resistance blood encounters in the small arteries (**peripheral resistance**).

When either cardiac output or peripheral resistance increases, blood pressure rises. Cardiac output is raised when heart rate or blood volume increases; peripheral resistance is affected mostly by the diameters of the arteries. Blood pressure is therefore influenced by the nervous system, which regulates heart muscle contractions and the arteries' diameters, and hormonal signals, which may cause fluid retention or blood vessel constriction. The kidneys also play a role in regulating blood pressure. If the blood pressure is too low, the kidneys act to increase it—they send hormones to constrict blood vessels and to retain water and salt in the body. When the pressure is right, the cells receive a constant supply of nutrients and oxygen and can release their wastes.

KEY POINTS

- Blood pressure pushes the blood through the major arteries into smaller arteries and capillaries to exchange fluids between the blood and the tissues.
- Blood pressure rises when cardiac output or peripheral resistance increases.

Risk Factors for Hypertension

Several major risk factors predict the development of hypertension:[62]

- *Age.* Hypertension risk increases with age. More than two-thirds of U.S. adults older than 65 have hypertension. Individuals who have normal blood pressure at age 55 still have a 90 percent risk of developing high blood pressure during their lifetime.

Figure 11–8

The Blood Pressure

Three major factors contribute to the pressure inside an artery. First, the heart pushes blood into the artery. Second, the small-diameter arteries and capillaries at the other end resist the blood's flow (peripheral resistance). Third, the volume of fluid in the circulatory system, which depends on the number of dissolved particles in that fluid, adds to the blood pressure.

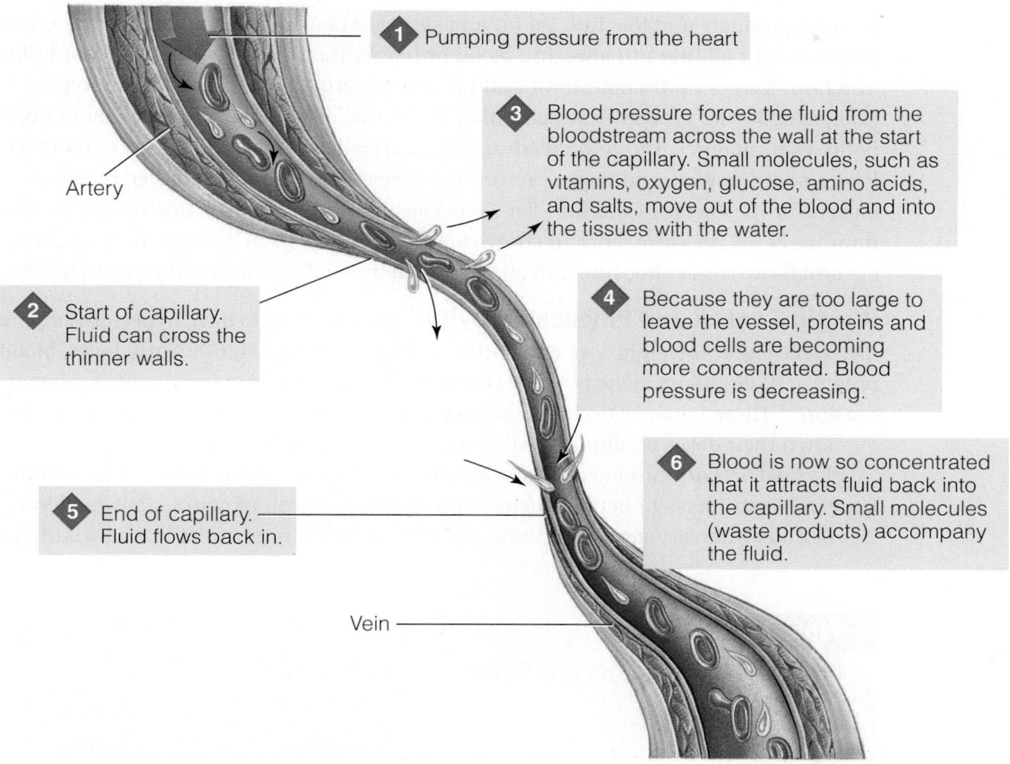

1 Pumping pressure from the heart
3 Blood pressure forces the fluid from the bloodstream across the wall at the start of the capillary. Small molecules, such as vitamins, oxygen, glucose, amino acids, and salts, move out of the blood and into the tissues with the water.
Artery
2 Start of capillary. Fluid can cross the thinner walls.
4 Because they are too large to leave the vessel, proteins and blood cells are becoming more concentrated. Blood pressure is decreasing.
6 Blood is now so concentrated that it attracts fluid back into the capillary. Small molecules (waste products) accompany the fluid.
5 End of capillary. Fluid flows back in.
Vein

- *Genetics.* Hypertension often runs in families and along racial lines: a family history of hypertension raises the risk of developing it, and for African Americans, prevalence of high blood pressure is among the highest in the world.[63] Compared with others, African Americans typically develop high blood pressure earlier in life, and their average blood pressure is much higher.
- *Obesity.* More than half of people with hypertension—an estimated 60 percent—are obese. Obesity raises blood pressure in part by altering kidney function, increasing blood volume, and promoting blood vessel damage through insulin resistance.[64] Excess fat also means miles of extra capillaries through which the blood must be pumped.
- *Salt intake.* As salt intake increases, so does blood pressure.[65] Most people with hypertension can benefit from reducing salt in their diets.
- *Alcohol.* Alcohol, regularly consumed in amounts greater than two drinks per day, is strongly associated with hypertension (details in a later section), and may interfere with drug therapy.[66]
- *Dietary factors.* A person's diet may increase the risk for developing hypertension. As explained in the next section, increasing intakes of three minerals—potassium, calcium, and magnesium—reduces blood pressure.

KEY POINT

- Obesity, age, family background, and race contribute to hypertension risks, as do salt intake and other dietary factors, including alcohol consumption.

systolic (sis-TOL-ik) **pressure** the first figure in a blood pressure reading (the "dupp" sound of the heartbeat's "lubb-dupp" beat is heard), which reflects arterial pressure caused by the contraction of the heart's left ventricle.

diastolic (dye-as-TOL-ik) **pressure** the second figure in a blood pressure reading (the "lubb" of the heartbeat is heard), which reflects the arterial pressure when the heart is between beats.

prehypertension borderline blood pressure between 120 over 80 and 139 over 89 millimeters of mercury, an indication that hypertension is likely to develop in the future.

cardiac output the volume of blood discharged by the heart each minute. Also defined in Chapter 10.

peripheral resistance the resistance to pumped blood in the small arterial branches (arterioles) that carry blood to tissues.

How Does Nutrition Affect Hypertension?

Even mild hypertension can be dangerous, but individuals who adhere to treatment are less likely to suffer illness or early death. Some people need medications to bring their blood pressure down, but diet and physical activity can bring improvements for many people and prevent hypertension in many others. Table 11–9 describes the lifestyle changes that reduce blood pressure, and the following sections address each one.

The DASH Diet The results of the Dietary Approaches to Stop Hypertension (DASH) trial show that a diet rich in fruits, vegetables, nuts, whole grains, and low-fat milk products and low in total fat and saturated fat can significantly lower blood pressure.[67] In addition to lowering blood pressure, the DASH diet improves vascular function, lowers total cholesterol and LDL cholesterol, and reduces inflammation.[68] Compared to the typical American diet, the DASH eating plan provides more fiber, potassium, magnesium, and calcium, emphasizes legumes and fish over red meat, limits added sugars and sugar-containing beverages, and meets other recommendations of the *Dietary Guidelines for Americans*. Eating plans like DASH consistently improve blood pressure, both in study subjects whose diets are provided by researchers and those freely choosing and preparing their own foods according to guidelines.

Weight Control and Physical Activity For people who have hypertension and are overweight, a weight loss of as little as 10 pounds can significantly lower blood pressure. Weight loss alone is one of the most effective nondrug treatments for hypertension.[69] Those who are taking medication to control their blood pressure can often cut down their doses or eliminate their medication if they lose weight.

Physical activity can lower almost everyone's blood pressure, even people without hypertension. Physical activity helps with weight control, of course, but moderate- or vigorous-intensity aerobic activity, such as 30 to 60 minutes of brisk walking or

Table 11–9

Lifestyle Modifications to Reduce Blood Pressure

Modification	Recommendation	Expected Reduction in Systolic Blood Pressure
Weight reduction	Maintain healthy body weight (BMI below 25)	5–20 mm Hg/10 kg lost
DASH eating plan	Adopt a diet rich in fruits, vegetables, and low-fat milk products with reduced saturated fat intake	8–14 mm Hg
Sodium restriction	Reduce dietary sodium intake to less than 2,300 milligrams sodium (less than 6 grams salt) per day[a]	2–8 mm Hg
Physical activity	Perform aerobic physical activity for at least 30 minutes per day, most days of the week	4–9 mm Hg
Moderate alcohol consumption	Men: Limit to 2 drinks per day Women and lighter-weight men: Limit to 1 drink per day	2–4 mm Hg

[a]*According to the Dietary Guidelines and DRI recommendations, sodium intake should be limited to less than 1,500 mg daily for people age 51 and older, people of African American descent, and people with hypertension, diabetes, or chronic kidney disease..*

Source: Adapted from Reference Card from the Seventh Report of the Joint National Committee on Prevention, Detection, Evaluation, and Treatment of High Blood Pressure (JNC 7), *NIH publication no. 03-5231 (Bethesda, Md.: National Institutes of Health, National Heart, Lung, and Blood Institute, and National High Blood Pressure Education Program, May 2003).*

running on most days, also helps to lower blood pressure directly.[70] Even activity performed in manageable 10-minute segments throughout the day that add up to the recommended total provides benefits. Those who engage in regular aerobic activity may not need medication for mild hypertension.

Physical activity also alters the body's hormones in beneficial ways. Physical activity decreases the secretion of stress hormones, reducing stress and lowering blood pressure. Physical activity also redistributes body water and eases transit of the blood through the small arteries that feed the tissues, including those of the heart.

Salt, Sodium, and Blood Pressure As mentioned, high intakes of salt and sodium are associated with hypertension.[71] Lowering sodium intake reduces blood pressure regardless of gender or race, presence or absence of hypertension, or whether people follow the DASH diet or a typical American diet. The combination of the DASH diet with a limited intake of sodium, however, improves blood pressure better than either strategy alone.[72] As salt intakes decrease, blood pressure drops in a stepwise fashion.[73] This direct relationship is reported at all levels of intake, from very low to much higher than average. Research also suggests that reducing salt intake may provide additional protection against heart disease, beyond lowering blood pressure.[74]

The World Health Organization estimates that a significant reduction in sodium intake could reduce by half the number of people requiring medication for hypertension and greatly reduce deaths from CVD. The recommendation of many professionals and agencies is that everyone (even those with normal blood pressure) should moderately restrict salt and sodium intake. No one should consume more than the DRI committee's Tolerable Upper Intake Level—that is, no more than 2,300 milligrams of sodium per day.[75] Certain groups of people, comprising about half of the U. S. population, respond more sensitively than others to sodium intakes: African Americans, people with hypertension, people with kidney problems or diabetes, and older people (those who are 51 or older). These individuals should limit their sodium intakes to no more than 1,500 milligrams of sodium per day.[76]

Alcohol In moderate doses, alcohol initially relaxes the arteries and so reduces blood pressure, but higher doses raise blood pressure.[77] Hypertension is common among people with alcoholism and is apparently caused directly by the alcohol. Hypertension caused by alcohol leads to CVD, the same as hypertension caused by any other factor. Furthermore, alcohol may cause strokes—even *without* hypertension. The *Dietary Guidelines for Americans* urge a sensible, moderate approach for those who drink alcohol.[78] *Moderation* means no more than one drink a day for women or two drinks a day for men, an amount that seems safe relative to blood pressure. The same amount, however, raises women's risk of breast cancer, so other routes to relaxation may prove safer.

Calcium, Potassium, Magnesium, and Vitamin C Other dietary factors may help to regulate blood pressure. A diet providing enough calcium may be one

My Turn **watch it!**

Fast-Food Generation?

How many fast-food meals do you eat each week? Listen to two students tell you about their weekly fast-food fare.

Tami

Alicia

such factor—in both healthy people and those with hypertension, increasing calcium often reduces blood pressure.[79]

Adequate potassium and magnesium may also help in this regard. Hypertension occurs more frequently in areas where diets lack potassium-rich fruits and vegetables. Conversely, diets rich in potassium and low in sodium appear to both prevent and correct hypertension.[80] As for magnesium, deficiency causes the walls of the arteries and capillaries to constrict and so may raise the blood pressure. Similarly, consuming a diet adequate in vitamin C seems to help normalize blood pressure, while vitamin C deficiency may tend to raise it.[81] Other dietary factors may also affect blood pressure; caffeine raises blood pressure temporarily, and the roles of vitamin D, cadmium, selenium, lead, protein, and fat are currently under study.[82]

© Sergey Peterman/Shutterstock.com

How can people be sure of getting all of the nutrients needed to keep blood pressure low? Vitamin and mineral supplements have been disappointing in this regard, showing no promise for lowering blood pressure. Therefore, the best answer is to consume a low-fat diet with abundant fruits, vegetables, and low-fat dairy products that provide the needed nutrients while holding sodium intake within bounds. Should diet and physical activity fail to reduce blood pressure, though, antihypertensive drugs such as diuretics can be lifesaving. Diuretics lower blood pressure by increasing fluid loss and lowering blood volume. Some diuretics may also cause potassium losses. People taking these drugs should make it a point to consume potassium-rich foods daily. Although some diuretics can lead to a potassium deficiency, others spare potassium. A combination of these two types of diuretics may be prescribed to prevent potassium deficiency.

Some people doubt the power of ordinary food to improve health, despite abundant evidence in its favor. In the search for something extra, such people often combine nutrient supplements with herbs and other alternatives to mainstream nutrition and medicine. The Consumer's Guide provides a look at some of these practices.

KEY POINTS

- For most people, maintaining a healthy body weight, engaging in regular physical activity, minimizing salt and sodium intakes, limiting alcohol intake, and eating a diet high in fruits, vegetables, fish, and low-fat dairy products work together to keep blood pressure normal.
- Certain nutrient deficiencies may raise blood pressure.

Nutrition and Cancer

LO 11.5 Describe the associations between diet and cancer observed in research, both positive and negative.

Cancer ranks second only to heart disease as a leading cause of death and disability in the United States. More than 1.6 million new cancer cases and nearly 600,000 deaths from cancer are projected to occur in the United States in 2012.[83] Still, over the past decade, a small but steady trend toward declining cancer deaths has occurred in the United States.[84] Early detection and improved treatment have transformed several common cancers from intractable killers to curable diseases or treatable chronic illnesses. Although the potential for cure is exciting, *prevention* of cancer remains far and away preferable.

Can an individual's chosen behaviors affect the risk of contracting cancer? They can, sometimes powerfully so.[85] Just a few rare cancers are known to be caused primarily by genetic influences and will appear in members of an affected family regardless of lifestyle choices. A few more are linked with microbial infections.** †† For the great majority of cancers, lifestyle factors and environmental exposures become the

** Examples include viral hepatitis and liver cancer, human papilloma virus and cervical cancer, and *H. pylori* bacterium (the ulcer bacterium) and stomach cancer.

†† A free scientific database summarizes evidence concerning breast cancer and environmental exposure to chemical compounds; see the Internet website, www.komen.org/environment.

cancer a disease in which cells multiply out of control and disrupt normal functioning of one or more organs.

use it! A Consumer's Guide To . . .

Deciding about CAM

Have you ever treated a health problem with an herbal remedy or another form of **complementary** and **alternative medicine (CAM)**? (See Table 11–10 for definitions.) If so, you are not alone. U.S. consumers spend more than $34 billion on CAM treatments each year.[1] All of these dollars spur sellers to advertise on thousands of Internet websites, run television infomercials, write innumerable booklets and books, and publish floods of magazine and newspaper advertorials to promote sales.

CAM treatments range from folk medicine to fraud. Some CAM therapies have been used for centuries, but few have been evaluated scientifically for safety or effectiveness.[2] When tested, most prove ineffective or unsafe. Useless remedies continue to sell, however, because an ill person's belief in a treatment, even in a placebo, can sometimes lead to physical healing (*placebo* was defined in Table 1–7 of Chapter 1).[3] Then, undeservedly, the treatment gets the credit.

Table 11–10

Alternative Therapy Terms

- **acupuncture** (ak-you-punk-chur) a technique that involves piercing the skin with long, thin needles at specific anatomical points to relieve pain or illness. Acupuncture sometimes uses heat, pressure, friction, suction, or electromagnetic energy to stimulate the points.
- **complementary** and **alternative medicine (CAM)** a group of diverse medical and health-care systems, practices, and products that are not considered to be a part of conventional medicine. Examples include acupuncture, biofeedback, chiropractic, faith healing, and many others.
- **herbal medicine** a type of CAM that uses herbs and other natural substances to prevent or cure diseases or to relieve symptoms.

CAM Best Bets

This is not to say that all CAM treatments are useless. Dozens of **herbal medicines** contain effective natural drugs. For example, the resin myrrh (pronounced *murr*) contains an analgesic (pain-killing) compound; willow bark contains aspirin; the herb valerian contains a tranquilizing oil; senna leaves produce a powerful laxative. The World Health Organization currently recommends a Chinese herbal medicine, artemisinin derived from a wormwood tree, to fight off malaria in some tropical nations. Herbs, like drugs, can cause side effects.[4] Table 11–11, p. 444, lists the potential actions and risks of selected herbs.*

The National Institutes of Health established its National Center for Complementary and Alternative Medicine (NCCAM) to distinguish alternative therapies that are potentially useful from those are useless or harmful. So far, NCCAM has found that **acupuncture** helps to quell nausea from surgery, chemotherapy, and pregnancy and to ease chronic low-back pain, and possibly migraine headaches, although underlying mechanisms for these effects are not known.[5] After more than a decade of funding studies, the agency has confirmed no effect from most other CAM treatments.[6]

A CAM Worst Case

The toxic drug laetrile, a CAM treatment for cancer, was popularized in the 1970s and remains available with no evidence to support its use, then or now. In fact, its high cyanide (poison) content makes it a hazardous choice.[7] Along with thousands of other sham treatments, laetrile is still sold as a "dietary supplement" to unsuspecting people by way of Internet websites.

Anyone can claim to be an expert in a "new" or "natural" therapy, and many practitioners act knowledgeable but are either misinformed or are frauds (see Controversy 1). Intelligent, clear-minded people can fall for such hoaxes when standard medical therapies fail; loving life and desperate, they fall prey to the worst kind of quackery on the feeblest promise of a cure.

A Curious Case of Anosmia

A popular CAM cold treatment consisting of zinc gel squirted into the nose was widely advertised and sold but lacked FDA approval.† Over the course of a few years, the FDA received over 130 complaints from consumers reporting anosmia, meaning that they lost their sense of smell, sometimes permanently, after using the product. Finally, the FDA took action against the manufacturer, who removed the product from the market. Anosmia may not sound serious, but it dramatically reduces the sense of taste and pleasure in eating, and it poses a danger when it prevents the detection of hazards normally signaled through the sense of smell, such as spoiled food or leaking gas. This case illustrates the trouble with using most CAM products: they are not tested for safety. Without prior testing, the user becomes the tester, and no one can predict the outcome.

Mislabeled Herbs

When common herbal remedies are analyzed, many do not contain the species or the active ingredients stated on

* *A reliable source of information about herbs is V. Tyler,* The Honest Herbal *(New York: Pharmaceutical Products Press). Look for the latest edition.*

† *The products were Zicam Cold Remedy Nasal Gel, Zicam Cold Remedy Gel Swabs, and Zicam Cold Remedy Swabs, Kids Size.*

Table 11–11

Selected Herbs: Claims, Evidence, and Risks

Common Name	Claims	Evidence	Risks[a]
Aloe (gel)	Promote wound healing	May help heal minor burns and abrasions; may cause infections in severe wounds	Generally considered safe
Black cohosh (stems and roots)	Ease menopause symptoms	Conflicting evidence	May cause headaches, stomach discomfort, liver damage
Chamomile (flowers)	Relieve indigestion	Little evidence available	Generally considered safe
Chaparral (leaves and twigs)	Slow aging, "cleanse" blood, heal wounds, cure cancer, treat acne	No evidence available	Acute, toxic hepatitis; liver damage
Cinnamon (bark)	Relieve indigestion, lower blood glucose and blood lipids	May lower blood glucose in type 2 diabetes	May have a "blood-thinning" effect; not safe for pregnant women or those taking diabetes medication
Comfrey (leafy plant)	Soothe nerves	No evidence available	Liver damage
Echinacea (roots)	Alleviate symptoms of colds, flus, and infections; promote wound healing; boost immunity	Ineffective in preventing colds or other infections	Generally considered safe; may cause headache, dizziness, nausea
Ephedra (stems)	Promote weight loss	Little evidence available	Rapid heart rate, tremors, seizures, insomnia, headaches, hypertension; FDA has banned the sale of ephedra-containing supplements
Feverfew (leaves)	Prevent migraine headaches	May prevent migraine headaches	Generally considered safe; may cause mouth irritation, swelling, ulcers, and GI distress; not safe for pregnant women
Garlic (bulbs)	Lower blood lipids and blood pressure	May lower blood cholesterol slightly; conflicting evidence on blood pressure	Generally considered safe; may cause garlic breath, body odor, gas, and GI distress; inhibits blood clotting
Ginger (roots)	Prevent motion sickness, nausea	May relieve pregnancy-induced nausea; conflicting evidence on nausea caused by motion, chemotherapy, or surgery	Generally considered safe
Ginkgo (tree leaves)	Improve memory, relieve vertigo	Little evidence available	Generally considered safe; may cause headache, GI distress, dizziness; may inhibit blood clotting
Ginseng (roots)	Boost immunity, increase endurance	Little evidence available	Generally considered safe; may cause insomnia, headaches, increased oxidative stress, and high blood pressure
Goldenseal (roots)	Relieve indigestion, treat urinary infections	Little evidence available	Generally considered safe; not safe for people with hypertension or heart disease; not safe for pregnant women

[a]*Allergies are always a possible risk; see Controversy 14 for drug interactions. Pregnant women should not use herbal supplements.*

Table 11-11

Selected Herbs: Claims, Evidence, and Risks (continued)

Common Name	Claims	Evidence	Risks[a]
Kava (roots)	Relieves anxiety, promotes relaxation	Little evidence available	Liver failure
Kombucha tea (fermentation product of yeast and bacteria)	Boosts immunity; prevents cancer; improves digestion	No evidence available	Stomach upset; allergic reactions; toxic reactions; metabolic acidosis
Saw palmetto (ripe fruits)	Relieve symptoms of enlarged prostate; diuretic; enhance sexual vigor	Little evidence available	Generally considered safe; may cause nausea, vomiting, diarrhea
St. John's wort (leaves and tops)	Relieve depression and anxiety	May relieve mild depression	Generally considered safe; may cause fatigue, increased sensitivity to sunlight, and GI distress
Tumeric (roots)	Reduces inflammation; relieves heartburn; prevents or treats cancer	No evidence available	Generally considered safe; may cause indigestion; not safe for people with gallbladder disease
Valerian (roots)	Calm nerves, improve sleep	Little evidence available	Long-term use associated with liver damage
Yohimbe (tree bark)	Enhance "male performance"	No evidence available	Kidney failure, seizures

[a]Allergies are always a possible risk; see Controversy 14 for drug interactions. Pregnant women should not use herbal supplements.

their labels.[8] In one analysis, instead of the herbs stated on the label, the CAM products contained unsafe medical drugs. These drugs often interact with prescription medications and can cause a dangerous drop in blood pressure when taken without medical guidance.[9] In another analysis, detectable levels of toxic lead, mercury, or arsenic were found in about a fifth of the samples of a popular herbal remedy.[10] When labels lack veracity and adulteration and contamination are common, consumers cannot make reasonable and safe choices.

If you decide to use an herbal or other CAM product, compare labels and look for the words *U.S. Pharmacopeia* or *Consumer Lab* on the label. These names signify that samples of the product were analyzed and found to contain authentic ingredients in the quantities claimed; these names do *not* indicate safety or effectiveness of the product, however.

Like other drugs, even authentic herbs may interfere with or potentiate the effects of medication (see Controversy 14).[11] For example, because *Ginkgo biloba* impairs blood clotting, it can cause bleeding problems for people on aspirin or other blood-thinning medicines.[12]

Lack of Knowledge

Most people in the market for herbs and other CAM treatments take the advice of herb vendors in stores or online. Few herb sellers, however, possess the training in pharmacology, botany, and human physiology required to appropriately apply herbal remedies; perilous mistakes with herbs are common. The few physicians who are skilled in herbal medicine may integrate the best CAM treatments into their practices. However, most patients using herbs or CAM treatments keep their use a secret, fearing their doctors' disapproval.[13] Such secrecy ups their risk—without knowledge, a physician cannot evaluate the potential for interactions.

Moving Ahead

The consequences of using unproven treatments are unpredictable. If you take prescription drugs, tell your doctor about any herbs you are also taking to rule out incompatibility.

Before taking any herb, find authoritative, scientific sources of information on its potential actions and risks. Don't be led astray by advertisements, rumors, Internet claims, or wishful thinking—investigate and decide for yourself.

Review Questions‡

1. Complementary and alternative medicines (CAM) warrant a cautious approach; these treatments often lack evidence for safety or effectiveness. T F
2. The National Center for Complementary and Alternative Medicine (NCCAM) promotes laetrile therapy. T F
3. For safety, a person seeking medical help should inform their physician about use of herbs or other alternative medicines. T F

‡ *Answers to Consumer's Guide review questions are found in Appendix G.*

major risk factors.[86] For example, if everyone in the United States quit smoking right now, future total cancers would likely drop by almost a third. Obesity and a lack of physical activity almost certainly play a role in the development of colon and breast cancer and probably contribute to pancreatic, esophageal, and renal cancers as well.[87] Alcohol intakes contribute to cancer of the mouth, pharynx, and larynx, cancer of the esophagus, breast cancer, and probably others. Further, incidence of hormone-related breast cancer has dropped significantly at a time when millions of women have ceased taking hormone replacement therapy for symptoms of menopause.[88]

An estimated 30 to 40 percent of cancers are influenced by diet, and these relationships are the focus of this section.[89] Diet patterns that emphasize fat, meat, alcohol, and excess calories and that minimize fruits and vegetables have been the targets of much cancer research. Such constituents of the diet relate to cancer in several ways:

- Foods or their components may cause cancer.
- Foods or their components may promote cancer.
- Foods or their components may protect against cancer.

Also, for the person who has cancer, diet can make a crucial difference in recovery. Some dietary and environmental factors currently believed to be important in cancer causation are listed in Table 11–12.

How Does Cancer Develop?

Cancer arises in the genes. The development of cancer, called **carcinogenesis**, often proceeds slowly and continues for several decades. It often begins when a cell's genetic material (DNA) sustains damage from a **carcinogen**, such as a free-radical compound, radiation, and other factors. Such damage occurs every day, but most is quickly repaired. Sometimes DNA collects bits of damage here and there over time. Usually, if the damage cannot be repaired and the cell becomes unable to faithfully replicate its genome, the cell self-destructs, committing a sort of cellular suicide to prevent its progeny from inheriting faulty genes.

Occasionally, a damaged cell loses its ability to self-destruct and also loses its ability to stop reproducing. In a healthy, well-nourished person, the immune system steps in to destroy these cells.[90] If, however, the immune system falters, the damaged cell replicates uncontrollably, and the result is a mass of abnormal tissue—a tumor. Life-threatening cancer results when the tumor tissue, which cannot perform the critical functions of healthy tissues, overtakes the healthy organ in which it developed or disseminates its cells through the bloodstream to other parts of the body.

Simplified, cancer develops through the following steps (illustrated in Figure 11–9, p. 448):

1. Exposure to a carcinogen.
2. Entry of the carcinogen into a cell.
3. **Initiation** of cancer as the carcinogen damages or changes the cell's genetic material (carcinogenesis).

carcinogenesis the origination or beginning of cancer.

carcinogen (car-SIN-oh-jen) a cancer-causing substance (*carcin* means "cancer"; *gen* means "gives rise to").

initiation an event, probably occurring in a cell's genetic material, caused by radiation or by a chemical carcinogen that can give rise to cancer.

promoters factors such as certain hormones that do not initiate cancer but speed up its development once initiation has taken place.

metastasis (meh-TASS-ta-sis) movement of cancer cells from one body part to another, usually by way of the body fluids.

Table 11-12

Diet-Related Factors and Cancer at Specific Sites

Cancer Sites	Risk Factors	Protective Factors
Breast (postmenopause)	Alcoholic drinks, body fatness, adult attained height,[a] abdominal fatness, adult weight gain	Lactation, physical activity
Breast (premenopause)	Alcoholic drinks, adult attained height,[a] greater birth weight	Lactation, healthy degree of body fatness[b]
Colon and rectum	Red meat, processed meat, alcoholic drinks, body fatness, abdominal fatness, adult attained height[a]	Physical activity, foods containing dietary fiber, garlic, milk, calcium
Endometrium	Body fatness, abdominal fatness	Physical activity
Esophagus	Alcoholic drinks, body fatness	Nonstarchy vegetables, fruits, foods containing beta-carotene, foods containing vitamin C
Gallbladder	Body fatness	
Kidney	Body fatness	
Liver	Aflatoxins,[c] alcoholic drinks	
Lung	Arsenic in drinking water, beta-carotene supplements[d]	Fruits, foods containing carotenoids
Mouth, pharynx, and larynx	Alcoholic drinks	Nonstarchy vegetables, fruits, foods containing carotenoids
Nasopharynx	Cantonese-style salted fish	
Ovary	Adult attained height[a]	
Pancreas	Body fatness, abdominal fatness, adult attained height[a]	Foods containing folate
Prostate	Diets high in calcium	Foods containing lycopene, foods containing selenium, selenium[e]
Skin	Arsenic in drinking water	
Stomach	Salt, salty and salted foods	Nonstarchy vegetables, allium vegetables,[f] fruits

Note: Strength of evidence for all these factors is either "convincing" or "probable."

[a] *Adult attained height is unlikely to directly modify the risk of cancer. It is a marker for genetic, environmental, hormonal, and also nutritional factors affecting growth during the period from preconception to completion of linear growth.*

[b] *Most studies show an increased risk of postmenopausal breast cancer with increased body fatness, but a decreased risk of premenopausal breast cancer with increased body fatness.*

[c] *Aflatoxins are toxins produced by molds or fungi. The main foods that may be contaminated are all types of grains (wheat, rye, rice, corn, barley, oats) and legumes, notably peanuts.*

[d] *The evidence is derived from studies using high-dose supplements (20 mg/day for beta-carotene; 25,000 international units/day for retinol) in smokers.*

[e] *The evidence is derived from studies using supplements at a dose of 200 μg/day. Selenium is toxic at higher doses.*

[f] *This includes vegetables such as garlic, onions, leeks, and shallots.*

Source: World Cancer Research Fund/American Institute for Cancer Research, Food, Nutrition, Physical Activity and the Prevention of Cancer: A Global Perspective *(Washington, D.C.: AICR, 2007).*

4. Acceleration by **promoters** that stimulate cancer cell growth such that the cells multiply out of control—tumor formation.
5. Often, spreading of cancer cells via blood and lymph (**metastasis**).
6. Disruption of normal body functions.

Researchers think that the first four steps, which culminate with tumor formation, are key to cancer prevention. On hearing this, many people mistakenly believe that they should avoid eating all foods that contain carcinogens. Doing so proves

Figure 11–9

Cancer Development

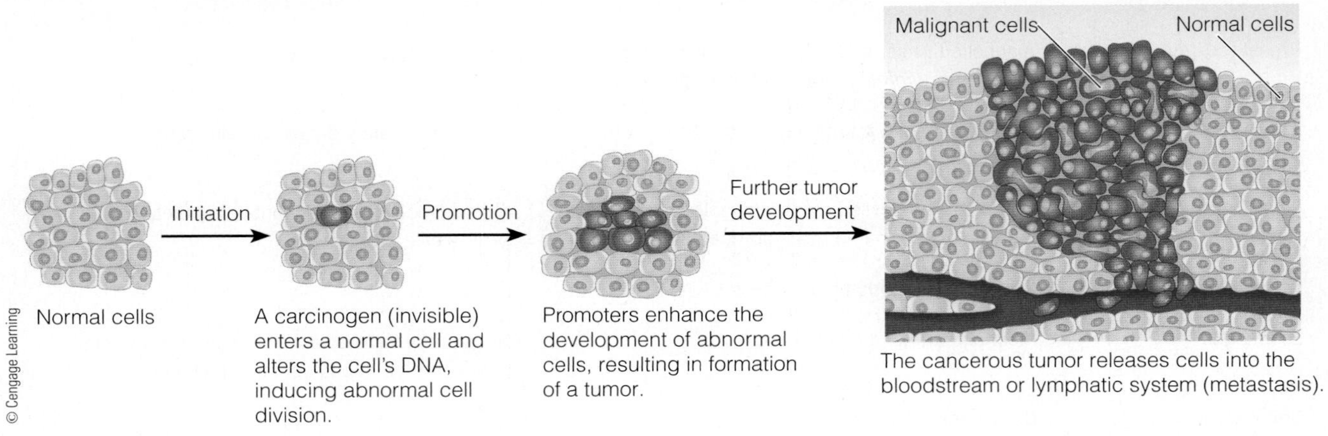
Malignant cells
Normal cells
Initiation
Promotion
Further tumor development
Normal cells
A carcinogen (invisible) enters a normal cell and alters the cell's DNA, inducing abnormal cell division.
Promoters enhance the development of abnormal cells, resulting in formation of a tumor.
The cancerous tumor releases cells into the bloodstream or lymphatic system (metastasis).
© Cengage Learning

Did You Know?

Some chemicals and carcinogens that occur naturally in breakfast foods include the following:

- Coffee: acetaldehyde, acetic acid, acetone, atractylosides, butanol, cafestol palmitate, chlorogenic acid, dimethyl sulfide, ethanol, furan, furfural, guaiacol, hydrogen sulfide, isoprene, methanol, methyl butanol, methyl formate, methyl glyoxal, propionaldehyde, pyridine, 1,3,7,-trimethylxanthine.
- Toast and coffee cake: acetic acid, acetone, butyric acid, caprionic acid, ethyl acetate, ethyl ketone, ethyl lactate, methyl ethyl ketone, propionic acid, valeric acid.

impossible, however, because most carcinogens occur naturally in foods amidst thousands of other chemicals and nutrients needed by the body. The body is well equipped to deal with the minute amounts of carcinogens that occur naturally in common foods like coffee, toast, and coffee cake (see the margin). Of course, consuming coffee, toast, and coffee cake does not elevate a person's risk of developing cancer because the body detoxifies small doses of carcinogens found in foods.

For those who suspect food additives of being carcinogenic, be assured that additives are held to strict standards; no additive approved for use in the United States causes cancer when used appropriately in food. Food *contaminants*, however, that enter foods by accident or toxins that arise through natural processes (for example when a food becomes moldy) may indeed be powerful carcinogens, or they may be converted to carcinogens during the body's attempts to break them down. Most such constituents are monitored in the U.S. food supply and are generally present, if at all, in amounts well below those that could pose risks to consumers.

KEY POINTS

- Cancer arises from genetic damage and develops in steps, including initiation and promotion, which may be influenced by diet.
- Contaminants and naturally occurring toxins can be carcinogenic but they are monitored in the U.S. food supply and the body is equipped to handle tiny doses of most kinds.

Which Diet Factors Affect Cancer Risk?

Certain dietary factors substantially influence cancer development.[91] The degree of risk imposed by food depends partly on the eater's genetic inheritance, but knowledge of these relationships is still unfolding. The following sections explore some of the suspected links between food constituents and prevention or development of certain cancers. The Controversy section delves into areas of scientific advancement that promise to clarify at least some of the relationships between diet and cancer.

Energy Intake When calorie intakes are reduced, cancer rates fall. In animal experiments, this **caloric effect** proves to be one of the most effective dietary interventions to prevent cancer. When researchers establish a cancer-causing condition in laboratory animals and then restrict the energy in their feed, the onset of cancer in the restricted animals is delayed beyond the time when unrestricted animals have died of the disease. Population observations seem to imply that calorie restriction, voluntary or involuntary, may delay cancer in people, too, and clinical experiments to resolve the issue are currently underway.[92] An important note: this effect occurs only in cancer prevention; once started, cancer continues advancing even in a person who is starving. It is also true that when a population's calorie intake rises, cancer rates rise in response; excess calories from carbohydrate, fat, and protein all raise cancer rates. This raises concerns about future U.S. cancer rates among an increasingly overweight population.

Obesity Obesity itself is clearly a risk factor for certain cancers (such as colon, breast in postmenopausal women, endometrial, pancreatic, kidney, and esophageal) and possibly for other types (such as gallbladder) as well.[93] Obesity's influence on cancer development depends on the site as well as other factors. In the case of breast cancer in postmenopausal women, for example, the hormone estrogen is implicated. Obese women have higher levels of circulating estrogen than lean women do because adipose tissue converts other hormones into estrogen and releases it into the blood. In normal-weight women, blood estrogen drops dramatically with onset of menopause. In obese women, extended exposure to estrogen increases their breast cancer risk.[94] A lifestyle that embraces physical activity, no or little alcohol consumption, a healthy body weight, and avoidance of hormone replacement therapy (when medically possible) may substantially reduce a woman's breast cancer risk.

Physical Activity An energy budget that balances calorie intake with physical activity may lower the risk of developing some cancers. People whose lifestyles include regular, vigorous physical activity often have lower risks of colon and breast cancer.[95] Physical activity may protect against cancer by helping to maintain a healthy body weight and by other mechanisms such as changes in hormone levels and immune functions, not related to body weight.[96]

Alcohol Cancers of the head and neck correlate strongly with the combination of alcohol and tobacco use and with low intakes of green and yellow fruits and vegetables. Alcohol intake alone raises the risk of cancers of the mouth, throat, esophagus, and breast, and alcoholism often damages the liver in ways that promote liver cancer.[97]

Fat and Fatty Acids Laboratory studies using animals often reveal that high dietary fat intakes correlate with development of cancer.[98] Simply feeding fat to experimental animals is not enough to get tumors started, however; an experimenter must also expose the animals to a known carcinogen. After that exposure, animals fed the high-fat diet develop more cancers faster than animals fed low-fat diets. Thus, fat appears to be a cancer promoter in animals.

Studies of people, however, have not proved that the effects of fat are independent of the effects of energy intake and physical activity. Overall, evidence associating fats and oils with cancer risk is limited.[99] The type of fat in the diet, however, may influence cancer promotion or prevention.[100] Studies of colon cancer implicate animal fats but not vegetable fat, and a number of studies suggest that omega-3 fatty acids from fish may protect against some cancers.[101] Thus, the same dietary fat advice applies to cancer protection as to heart health: reduce saturated fat intake and increase omega-3 fatty acids.

Red Meats Population studies spanning the globe for over 30 years consistently report that diets high in red meat and processed meat (meat preserved by smoking, curing, or salting, or by the addition of preservatives) increase the risk of colon cancer.[102] Limited evidence suggests that diets high in red and processed meats may also play a role in cancer of the esophagus, lung, breast, stomach, pancreas, kidney, ovary, and prostate.[103] Processed meats may be of special concern. Processed meats contain additives, nitrites or nitrates, that contribute a pink color and deter bacterial growth in meats. In the digestive tract, nitrites and nitrates form other nitrogen-containing compounds that may be carcinogenic.[104]

Cooking meats at high temperatures (frying, broiling) causes amino acids and creatine in the meats to react together and form carcinogens.[105] Grilling meat, fish, or other foods—even vegetables—over a direct flame causes fat and added oils to splash on the fire and then vaporize, creating other carcinogens that rise and stick to the food. Smoking foods has the same effect. Eating these foods, or even well-browned

caloric effect the drop in cancer incidence seen whenever intake of food energy (calories) is restricted.

meats cooked to the crispy well-done stage, introduces carcinogens into the digestive system. These chemicals may or may not cause problems in the digestive tract, but once absorbed, they are detoxified by the liver's competent detoxifying system. A steady diet of foods containing significant amounts of these toxins, however, can overwhelm the system and may increase cancer risk. If you eat broiled, fried, grilled, or smoked foods, choose them in moderation and dilute their effects by varying your choices among foods prepared differently, such as boiled soups, stews, or pastas or baked, steamed, microwaved, or sautéed dishes.

Another reason to limit your intake of fried foods such as French fries and potato chips is the presence of acrylamide, a potential carcinogen. Acrylamide is produced when certain starches such as potatoes are fried or baked at high temperatures. In the body, some acrylamide is metabolized to a substance that may mutate or damage genetic material. As such, acrylamide is classified as "reasonably anticipated to be a human carcinogen."[106] More research is needed to clarify the relationship between acrylamide exposure and health risks.

To summarize, minimize carcinogen formation during cooking:

- Roast or bake meats in the oven.
- When grilling, line the grill with foil, or wrap the food in the foil.
- Take care not to burn foods.
- Marinate meats before cooking.

Fiber-Rich Foods Many studies show that as people increase their dietary fiber intakes, their risk for colon cancer declines.[107] Fiber may protect against cancer by binding, diluting, and rapidly removing potential carcinogens from the GI tract; alternatively, other constituents of fiber-rich foods, such as the phytochemicals of whole grains or the nutrients of fruits and vegetables, may be at work. The mechanisms for a protective role for fiber are not yet known.

© iStockphoto.com/brue

Often, whole foods like these, not individual chemicals, lower people's cancer risks.

As research takes its course, much evidence now weighs in favor of eating a diet rich in high-fiber, low-fat foods. Such a diet helps to regulate blood glucose and blood insulin and is linked with low rates of heart disease as well as some forms of cancer. If a meat-rich, calorie-dense diet is implicated in causation of certain cancers, and if a vegetable-rich, whole-grain-rich diet is associated with prevention, then shouldn't vegetarians have a lower incidence of those cancers? They do, as the many studies cited in Controversy 6 have shown.

Folate and Antioxidant Vitamins Folate may protect against cancer of the esophagus and the colon, although evidence at this time is limited.[108] Folate plays roles in DNA synthesis and repair; thus, inadequate folate intakes may allow DNA damage to accumulate. This reason alone is enough to warrant everyone attending to their folate intake.

Vitamin E, vitamin C, and beta-carotene received attention in Controversy 7. Suffice it to say here that taking supplements has not been proved to prevent or cure cancer. In fact, once cancer is established, such antioxidants may do more harm than good.

Calcium and Vitamin D Sufficient dietary calcium or foods that contain it may be protective against colon cancer.[109] Calcium intakes of about 600 to 1,000 milligrams per day—an amount easily provided by daily calcium-rich foods—appear to trigger the effect.

Another possibility is that the vitamin D added to milk in many areas might also be at work. Some intriguing evidence suggests that an adequate intake of vitamin D may be protective against some cancers.[110] Although no iron-clad case exists for cancer prevention, with all the other points in their favor, prudence dictates that everyone should arrange to meet their calcium and vitamin D needs every day.

Iron Iron, both in the diet and in body stores, is under study for links with promotion of colon cancer. How iron may promote cancer is not known, but iron is a powerful oxidizing agent that can damage DNA and perhaps initiate cancer. A high-meat diet generously supplies iron, and it also correlates with greater risk of colon cancer.[111]

Table 11–13

Recommendations for Reducing Cancer Risk

Overall, follow the USDA food pattern for your appropriate energy level.

Body fatness: Be as lean as possible throughout life without being underweight.

- Maintain body weight within the healthy range.
- Avoid weight gains and increases in waist circumference throughout adulthood.
- For those who are currently overweight or obese, losing even a small amount of weight has health benefits and is a good place to start.

Physical activity: Be physically active as part of everyday life.

- Be moderately physically active for at least 150 min each week, or be vigorously active for at least 75 min each week, or combine an equivalent amount of moderate and vigorous activity throughout the week.
- Limit sedentary habits such as sitting, lying down, watching television, or other forms of screen-based recreation.

Foods and drinks that promote weight gain: Limit consumption of energy-dense foods and avoid sugary drinks.

- Limit intakes of energy-dense foods (225 cal/110 g food).
- Avoid drinks with added sugar.
- Consume "fast foods" sparingly, if at all.

Plant foods: Eat mostly foods of plant origin.

- Eat the daily amounts of nonstarchy vegetables and fruits as recommended by the USDA food patterns.
- Eat relatively unprocessed grains and/or legumes with every meal.
- Limit refined starchy foods.

Animal foods: Limit intake of red meat and avoid processed meat.

- Eat no more than 18 oz of red meat a week, very little if any of which is processed.

Alcoholic drinks: Limit alcoholic drinks.

- If alcoholic drinks are consumed, limit consumption to no more than two drinks a day for men and one drink a day for women.

Preservation, processing, preparation: Limit consumption of salt and avoid moldy grains or legumes.

- Avoid salt-preserved, salted, or salty foods.
- Limit consumption of processed foods with added salt to ensure an intake of less than 6 g of salt (2.4 g of sodium) a day.
- Throw away moldy grains or legumes.

Dietary supplements: Aim to meet nutritional needs through diet.

- Dietary supplements are not recommended for cancer prevention.

Source: L. H. Kushi and coauthors, American Cancer Society Guidelines on Nutrition and Physical Activity for Cancer Prevention, CA: Cancer Journal for Clinicians *62 (2012): 30–67; World Cancer Research Fund/American Institute for Cancer Research,* Food, nutrition, physical activity and the prevention of cancer: A global perspective *(Washington, D.C.: AICR, 2007), pp. 373–390.*

Foods and Phytochemicals In the end, whole foods and whole diets composed of them, not single nutrients, may be most influential in cancer prevention. Fruits and vegetables, for example, contain a wide spectrum of nutrients and phytochemicals that may prevent or reduce oxidative damage to cell structures, including DNA, and a diet rich in fruits and vegetables may provide protection against the development of some cancers.[112] Some phytochemicals in fruits and vegetables are thought to act as **anticarcinogens** that stimulate the buildup of the body's arsenal of carcinogen-destroying enzymes. For example, the **cruciferous vegetables**—broccoli, brussels sprouts, cabbage, cauliflower, collard greens, turnips, and the like—contain a variety of phytochemicals that defend against cancers of the esophagus and endometrium. In addition, fruits and vegetables are rich in fiber. If you are considering making just one change to your diet, here is a place to begin—consume the recommended servings of fruits and vegetables each day. Table 11–13 summarizes dietary and lifestyle recommendations for reducing cancer risk.

Erika Craddock/Science Source

Cruciferous vegetables belong to the cabbage family: arugula, bok choy, broccoli, broccoli sprouts, brussels sprouts, cabbages (all sorts), cauliflower, greens (collard, mustard, turnip), kale, kohlrabi, rutabaga, and turnip root.

KEY POINTS

- Obesity, physical inactivity, alcohol consumption, and diets high in red and processed meats are associated with cancer development.
- Foods containing ample fiber, folate, calcium, many other vitamins and minerals, and phytochemicals may be protective.

Conclusion

Nutrition is often associated with promoting health, and medicine with fighting disease, but no clear line separates nutrition and medicine. Every major agency involved with health promotion or medicine recommends a varied diet of whole foods as part of a lifestyle that provides the best possible chance for a long and healthy life. The Food Feature that follows presents an example of such a diet, the DASH diet.

anticarcinogens compounds in foods that act in any of several ways to oppose the formation of cancer.

cruciferous vegetables vegetables with cross-shaped blossoms—the cabbage family. Their intake is associated with low cancer rates in human populations. Examples are broccoli, brussels sprouts, cabbage, cauliflower, rutabagas, and turnips.

try it! Food Feature

The DASH Diet: Preventive Medicine

LO 11.6 Make specific recommendations for including sufficient fruits and vegetables in a health-promoting diet.

An esteemed former surgeon general once said, "If you do not smoke or drink excessively, your choice of diet can influence your long-term health prospects more than any other action you might take."‡‡ Indeed, healthy young adults today are privileged to be among the first generations with enough nutrition knowledge to lay a foundation of health for today and tomorrow. Figure 11–10 illustrates this point.

Dietary Guidelines and the DASH Diet

The more detailed our knowledge about nutrition science, it seems, the simpler the truth becomes: people who consume the adequate, balanced, calorie-controlled, moderate, and varied diet recommended by the *Dietary Guidelines for Americans* enjoy a longer, healthier life than those who do not. Like the USDA Food Patterns, the DASH eating plan, presented in Table 11–14 at the 2,000-calorie-per-day-level, can help people to meet these goals. Other calorie levels are presented in Table E–2.

‡‡ *C. Everett Koop, 1988.*

"Knowing is not enough; we must apply. Willing is not enough; we must do."
—Goethe

To lower saturated fat and cholesterol intakes, the DASH diet emphasizes fruits, vegetables, whole grains, and fat-free or low-fat milk and milk products. It also features fish, poultry, and nuts instead of some of the red meat so common in U.S. diets. Compared to the typical American diet, the foods of the DASH diet provide greater intakes of fiber, as well as potassium, calcium, and magnesium, minerals shown to lower blood pressure.

Because the DASH diet centers on fresh, unprocessed, or lightly processed foods, it can present less sodium, too. It seems, with regard to sodium, "the lower the better" for reducing blood pressure, with most people benefitting from holding sodium intake to about 1,500 mg per day. Even at higher sodium intakes, however, the DASH diet can still produce a drop in blood pressure, although not as great as with sodium restriction.

Changes in diet are often best attempted a few at a time. A good place to start is by increasing the intake of fruits and vegetables.

Figure 11–10

Proper Nutrition Shields against Diseases

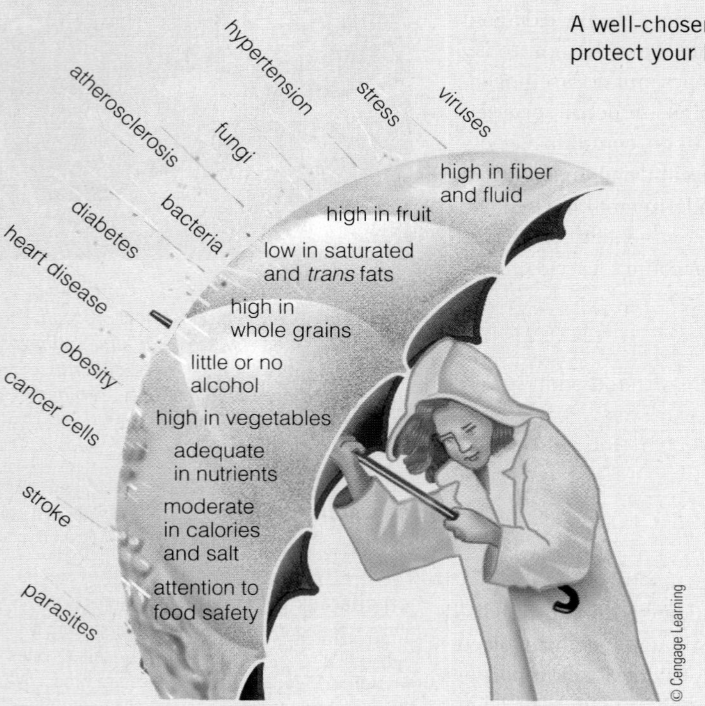

Fruits and Vegetables: More Matters

The National Fruit & Vegetable Program is a confederation composed of the Centers for Disease Control and Prevention, the American Heart Association, the American Diabetes Association, the American Cancer Society, and many other national organizations. These agencies work together to urge people to increase their intakes of a variety of fruits, vegetables, and legumes, not just for nutrients they provide but also for the phytochemicals that combine synergistically to promote health (see Figure 11–11, p. 454). The amount depends upon energy intake and activity level as shown in Table 2–2 (page 45). Alternatively, you can find out how many servings are right for you by visiting the Internet website: www.fruitsandveggiesmatter.org. Table 11–15, p. 454, offers some tips for increasing your intakes of fruits, vegetables, and legumes. Who knows? Foods destined to become your

Table 11–14

The DASH Eating Plan at a 2,000-Calorie Level

Appendix E offers the DASH Eating Plan at the 1,600-, 2,600-, and 3,100-calorie levels.

Food Group	Daily Servings	Serving Sizes	Examples and Notes	Significance of Each Food Group to the DASH Eating Pattern
Grains*	6–8	1 slice bread 1 oz dry cereal[†] ½ cup cooked rice, pasta, or cereal	Whole-wheat bread and rolls, whole-wheat pasta, English muffin, pita bread, bagel, cereals, grits, oatmeal, brown rice, unsalted pretzels and popcorn	Major sources of energy and fiber
Vegetables	4–5	1 cup raw leafy vegetable ½ cup cut-up raw or cooked vegetable ½ cup vegetable juice	Broccoli, carrots, collards, green beans, green peas, kale, lima beans, potatoes, spinach, squash, sweet potatoes, tomatoes	Rich sources of potassium, magnesium, and fiber
Fruits	4–5	1 medium fruit ¼ cup dried fruit ½ cup fresh, frozen, or canned fruit ½ cup fruit juice	Apples, apricots, bananas, dates, grapes, oranges, grapefruit, grapefruit juice, mangoes, melons, peaches, pineapples, raisins, strawberries, tangerines	Important sources of potassium, magnesium, and fiber
Fat-free or low-fat milk and milk products	2–3	1 cup milk or yogurt 1½ oz cheese	Fat-free (skim) or low-fat (1%) milk or buttermilk; fat-free, low-fat, or reduced-fat cheese; fat-free or low-fat regular or frozen yogurt	Major sources of calcium and protein
Lean meats, poultry, and fish	6 or less	1 oz cooked meats, poultry, or fish 1 egg[‡]	Select only lean; trim away visible fats; broil, roast, or poach; remove skin from poultry	Rich sources of protein and magnesium
Nuts, seeds, and legumes	4–5 per week	⅓ cup or 1½ oz nuts 2 tbs peanut butter 2 tbs or ½ oz seeds ½ cup cooked legumes (dry beans and peas)	Almonds, hazelnuts, mixed nuts, peanuts, walnuts, sunflower seeds, peanut butter, kidney beans, lentils, split peas	Rich sources of energy, magnesium, protein, and fiber
Fats and oils[§]	2–3	1 tsp soft margarine 1 tsp vegetable oil 1 tbs mayonnaise 2 tbs salad dressing	Soft margarine, vegetable oil (such as canola, corn, olive, or safflower), low-fat mayonnaise, light salad dressing	The DASH study had 27% of calories as fat, including fat in or added to foods
Sweets and added sugars	5 or less per week	1 tbs sugar 1 tbs jelly or jam ½ cup sorbet, gelatin 1 cup lemonade	Fruit-flavored gelatin, fruit punch, hard candy, jelly, maple syrup, sorbet and ices, sugar	Sweets should be low in fat

*Whole grains are recommended for most grain servings as a good source of fiber and nutrients.

[†]Serving sizes vary between ½ cup and 1¼ cups, depending on cereal type. Check the product's Nutrition Facts label.

[‡]Since eggs are high in cholesterol, limit egg yolk intake to no more than four per week; two egg whites have the same protein content as 1 oz of meat.

[§]Fat content changes serving amount for fats and oils. For example, 1 tbs of regular salad dressing equals one serving; 1 tbs of a low-fat dressing equals one-half serving; 1 tbs of a fat-free dressing equals zero servings.

Figure 11–11

Fruits and Veggies: More Matters

Fill half your plate with fruits and vegetables.

Courtesy of Produce for Better Health Foundatio

favorites may still await you on the produce shelves. An adventurous spirit is a plus in this regard.

Conclusion

In the end, people's choices are their own. Whoever you are, we encourage you to take the time to work out ways of making your diet meet the guidelines you now know will support your health. If you are healthy and of normal weight, if you are physically active, and if your diet on most days follows the *Dietary Guidelines*, then you can indulge occasionally in a cheesy pizza, marbled steak, banana split, or even a greasy fast-food burger and fries without inflicting much damage on your health. (Once a week may be harmless, but less frequently is better.) Especially, take time to enjoy your meals: the sights, smells, and tastes of good foods are among life's greatest pleasures. Joy, even the simple joy of eating, contributes to a healthy life.

David Madison/Photographer's Choice/Getty Images

Don't forget: be physically active all your life.

Table 11–15

Strategies for Consuming Enough Fruits, Vegetables, and Legumes

Many people do not eat the recommended amounts and varieties of fruits, vegetables, and legumes, but these foods are indispensible to a nutritious diet. All nutrient-dense forms count: fresh, frozen, canned, dried, and 100% juice.

Foods	Strategies
All vegetables	▪ Include vegetables of all kinds in meals and snacks; fresh, frozen, and canned vegetables all count, but choose low-fat, low-sodium varieties most often. ▪ Keep cut raw vegetables, such as carrot and celery sticks, in the refrigerator for quick snacks. ▪ Visit a salad bar to buy ready-to-eat vegetables if you are in a hurry. ▪ Try a new vegetable once each month. Read some cookbooks for ideas.
Dark green, red, and orange vegetables	▪ Add chopped dark green leafy vegetables or red and orange vegetables to main dishes, such as stir-fries, soups, casseroles. ▪ Serve side dishes of dark green salad greens or cooked or raw broccoli, spinach, or other dark green vegetables often. Choose cooked or raw red and orange vegetable dishes, too, such as tomato-based dishes, cooked hard squashes, or sliced cooked carrots. ▪ If calories are not a problem for you, try sweet potato fries as an occasional treat. ▪ Order vegetable side dishes when eating out and ask for sauces and dressings to be served on the side.
Legumes (beans, peas, lentils, and soy products)	▪ Keep a variety of low-sodium canned legumes, such as kidney beans, chickpeas (garbanzo beans), black beans, and others on hand. ▪ Use rinsed, drained beans as salad toppers. For interest, marinate them in lemon juice, garlic, and seasonings. ▪ Mash beans with lemon juice, olive oil, and seasonings for topping crackers, celery, or raw zucchini rounds; or use as a dip for vegetable sticks or as a sandwich spread. ▪ Add beans, peas, or lentils to soups and casseroles. ▪ Try new ethnic legume recipes or try new bean dishes in restaurants, such as black beans and rice, white bean chili, lentil veggie burgers, or dal (spicy Indian-style beans, peas, or lentils). ▪ Try using soy products such as soy milk, ground meat and burger replacers, tofu, and soy snacks.
Fruit	▪ Choose whole or cut fruit more often than fruit juice. ▪ Keep a variety of fresh, frozen, low-sugar canned, and dried fruit on hand to choose for snacks or to use in cereal, yogurt, salads, or desserts. ▪ Replace syrup, sugars, and other sweet toppings with berries, cut peaches, applesauce, or fruit mixtures. ▪ Blend smoothies from bananas, fruit juice, and berries with ice or yogurt. ▪ Fruit canned in 100% fruit juice is preferable to fruit canned in sugary syrups.

track it!

Diet Analysis PLUS

Concepts in Action

Analyze Your Diet for Health Promotion

The purpose of this exercise is to increase your awareness of the characteristics of the diet recommended for disease prevention.

1. One way to lower your risk of heart attack is to keep your blood pressure in a normal range. Study Table 11–14 (page 453), which provides an eating plan that supports normal blood pressure. Create a meal that follows the principles of the DASH diet. Select the Track Diet tab from the red navigation bar. Select a date and find the foods that you wish to include in this meal. From the Reports tab, go to Source Analysis Report for that meal, and select Sodium from the drop-down box. Generate a report. How much sodium did your meal contain? Which foods contributed the most sodium to the meal? From the Reports tab, select Intake vs. Goals. Generate a report. Locate sodium, and find the percentage of the DRI intake recommendation provided by your DRI for sodium. If the sodium was higher than 33 percent (one-third) of your allowance, what can you change to bring it into compliance?
2. For people with compromised immune systems, malnutrition demands prompt medical nutrition therapy. Select the Track Diet tab from the red navigation bar. Select a new day, and find foods to create one meal that provides one-third of the day's requirement of high-quality, easily digestible protein for an immune-compromised adult. Take into account a diminished appetite and food safety (you'll learn more about this in Chapter 12), while making the food appealing and easy to eat and digest. From the Reports tab, select Source Analysis and select Protein. Generate a report. How many grams of protein did the meal provide? Now select the Intake vs. Goals Report to see what percentage of the day's protein the meal supplied. Did it supply about a third of the day's need? If not, what adjustments can you make to better meet this person's need?
3. One diet characteristic recommended to reduce many chronic disease risks is reduced saturated fat intake. Create a meal low in saturated fat following the instructions in question number 2 above. Generate a Fat Breakdown Report and an Intake vs. Goals Report for that meal. How much saturated fat did your meal supply? Was it less than 10 percent of total calories for the meal? If not, what can you change to lower it?
4. Together, diet and exercise are a powerful and safe combination for improving heart health. Select the Track Diet tab and select the day you used in question number 3 above. Take a look at Table 11–7 (page 436); then, by adding foods for breakfast, lunch, and snack, create a full day's menu that achieves the diet modifications listed for saturated fat, cholesterol, and soluble fiber (for this activity, ignore the others). Select the Reports tab and the MyPlate Report for that date and generate a report. Did your day's meals meet your goals?
5. Research suggests that vegetarians have a lower incidence of certain cancers as well as a lower rate of heart disease. An ideal vegetarian diet is high in fiber, phytochemicals, whole grains, and vitamins, and it is low in saturated fat. Using Table 11–12 and Table 11–13 (pages 447 and 451) to guide you, create a vegetarian meal that includes foods associated with low cancer risks. Enter the data from the Track Diet tab. Select Intake vs. Goals Report for that date and meal and generate a report. What nutrients would be of interest when evaluating your vegetarian meal for adequacy (consult Table C6–2 on page 227)? Did the meal contain enough protein, vitamins, and minerals? How about fiber? Was it low enough in saturated fat?

what did you decide?

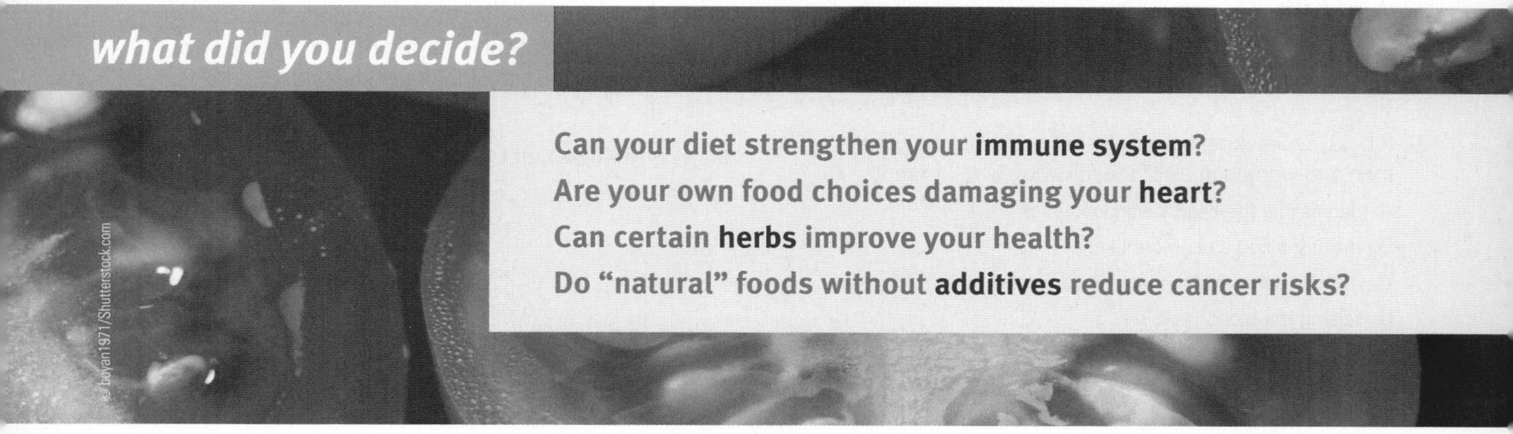

Can your diet strengthen your immune system?
Are your own food choices damaging your heart?
Can certain herbs improve your health?
Do "natural" foods without additives reduce cancer risks?

Self Check

1. (LO 11.1) All of the following are examples of how diseases might worsen malnutrition, except
   a. disease impairs appetite.
   b. disease interferes with digestion and absorption.
   c. disease decreases nutrient excretion.
   d. disease alters metabolism.
2. (LO 11.1) A chronic state of inflammation can be harmful to the tissues.
   T F
3. (LO 11.1) A healthy digestive system defends against invading microbes because
   a. its linings are absorptive.
   b. its linings are heavily laced with immune tissues.
   c. its linings are permeable.
   d. its linings are warm and moist.
4. (LO 11.2) Chronic diseases have distinct causes, known as risk factors.
   T F
5. (LO 11.2) Which of the following is a risk factor for cardiovascular disease?
   a. high blood HDL cholesterol
   b. low blood pressure
   c. low blood LDL cholesterol
   d. diabetes
6. (LO 11.3) By what age do most people have well-developed plaques in their arteries?
   a. 20 years
   b. 30 years
   c. 40 years
   d. 50 years
7. (LO 11.3) Atherosclerosis is simply the accumulation of lipids within the artery wall.
   T F
8. (LO 11.3) An "atherogenic diet" is high in all of the following *except* __________ .
   a. fiber
   b. cholesterol
   c. saturated fats
   d. *trans* fats
9. (LO 11.3) Men suffer more often from heart attacks than women do, making CVD a man's disease.
   T F
10. (LO 11.3) Smoking powerfully raises the risk for CVD in men and women in all of the following ways, except
    a. decreasing the heart's workload.
    b. making blood clots more likely.
    c. directly damaging the heart with toxins.
    d. raising the blood pressure.
11. (LO 11.4) Which of the following minerals may help to regulate blood pressure?
    a. calcium
    b. magnesium
    c. potassium
    d. all of the above
12. (LO 11.4) The most important step that a person can take to protect against hypertension is to be tested for it.
    T F
13. (LO 11.4) Hypertension is more severe and occurs earlier in life among people of European or Asian descent than among African Americans.
    T F
14. (LO 11.5) For the great majority of cancers, lifestyle factors and environmental exposures are the major risk factors.
    T F
15. (LO 11.5) Which of the following have been associated with an increase in cancer risk?
    a. alcohol intake
    b. high intakes of red meat
    c. high intakes of processed meats
    d. all of the above
16. (LO 11.5) When calorie intakes rise, cancer rates also increase.
    T F
17. (LO 11.5) Sufficient intakes of calcium-rich foods may increase the risk of colon cancer.
    T F
18. (LO 11.6) The DASH diet is designed for athletes who compete in sprinting events.
    T F
19. (LO 11.6) The DASH diet is characterized by increased intakes of
    a. fruits and vegetables
    b. whole grains
    c. artificial fats
    d. a and b
20. (LO 11.7) Currently, for the best chance of consuming adequate nutrients and staying healthy, people should obtain an evaluation of their genetic profile.
    T F

Answers to these Self Check questions are in Appendix G.

CONTROVERSY 11

Nutritional Genomics: Can It Deliver on Its Promises?

LO 11.7 Describe some recent advances in nutritional genomics with regard to the health of the body through life.

Health care appears to be standing at the edge of a **genomics** revolution. Today's health-care system emphasizes treatment only after disease symptoms arise. This may one day give way to a system aimed at identifying in healthy people—even children—traits in the **genome** that raise the odds of developing diseases in the future. Once identified, people with those tendencies may be helped to prevent or minimize disease through customized care based on each individual's **genetic profile**.[1]* A registered dietitian, for example, may use the information to individualize medical nutrition therapy for those who already suffer illness, or to provide diets with just the right nutrients and other **bioactive food components** that precisely meet the client's needs.

This Controversy offers just a taste of the exciting research in these areas—there is much more to learn and the science advances daily. To help get started, Table C11–1 distinguishes among the terms *genomics*, ***nutritional genomics***, ***epigenetics***, and others. Then, it offers some evidence that suggests a genomics link between chronic diseases and diet. The closing section brings up some ethical concerns surrounding genetic tests of all kinds and unveils fraud already occurring in the genetic-testing marketplace.

* *Reference notes are found in Appendix F.*

Table C11–1

Nutritional Genomics Terms

- **bioactive food components** nutrients and phytochemicals of foods that alter physiological processes often by interacting, directly or indirectly, with the genes.
- **DNA microarray technology** research tools that analyze the expression of thousands of genes simultaneously and search for particular genes associated with a disease. DNA microarrays are also called *DNA chips*.
- **epigenetics** (ep-ih-gen-EH-tics) the science of heritable changes in gene function that occur without a change in the DNA sequence.
- **epigenome** (ep-ih-GEE-nohm) the proteins and other molecules associated with chromosomes that affect gene expression. The epigenome is modulated by bioactive food components and other factors in ways that can be inherited. *Epi* is a Greek prefix, meaning "above" or "on."
- **genome** (GEE-nohm) the full complement of genetic material in the chromosomes of a cell. Also defined in Chapter 1.
- **genomics** the study of all the genes in an organism and their interactions with environmental factors.
- **genetic profile** the result of an analysis of genetic material that identifies unique characteristics of a person's DNA for forensic or diagnostic purposes.
- **histones** proteins that lend structural support to the chromosome structure and that activate or silence gene expression.
- **methyl groups** small carbon-containing molecules that, among their activities, silence genes when applied to DNA strands by enzymes.
- **mutation** a permanent, heritable change in an organism's DNA.
- **nucleotide** (NU-klee-oh-tied) one of the subunits from which DNA and RNA are composed.
- **nutritional genomics** the science of how food (and its components) interacts with the genome.
- **SNP** a single misplaced nucleotide in a gene that causes formation of an altered protein. The letters SNP stand for *single nucleotide polymorphism*.

Nutritional Genomics Research

With unprecedented speed, new revelations are emerging from laboratories worldwide. Until recently, no one knew *how* identical twins, with their identical DNA, could develop different diseases; or how a pregnant woman's diet might forever affect the health of her grandchildren; or how phytochemicals might alter the course of certain cancers. At least partial answers to these and other mysteries lie in the realm of nutritional genomics.

In general, today's nutritional genomics researchers strive to:

- *Identify the genes*: Which genes can be regulated by diet, and which of these participate in the onset, progression, or severity of chronic disease?

- *Explain the mechanisms*: How exactly do bioactive food components modify the activities of disease-related genes?
- *Develop practical applications*: Which nutrient intake levels best maintain health, and which foods or whole diets might prevent or relieve chronic diseases or other conditions by modifying genetic activity?

With powerful new research tools, such as **DNA microarray technology**, nutrition scientists can now begin to uncover and integrate such information.[2] In DNA microarray technology, robotic arms precisely fasten a single DNA strand of a known sequence onto a slide. Then, a computer compares the pattern of gene expression of the known sequence with that of an unknown DNA sample taken from an individual's cells. This comparison reveals which of the sample genes are expressed (actively making proteins—details in Chapter 6) and which are silenced (inactive), and how they respond under certain conditions. The results allow identification of inherited disease tendencies, unusual nutrient needs, and many other medical concerns.[3] The technique promises major advancements in health care for people worldwide.

DNA Variations, Nutrition, and Disease

Small variations in DNA sequences, called **mutations**, dictate many of the differences among human beings. The most common mutations are **SNPs** (pronounced "snips"), involving the variation of a single tiny molecule (a **nucleotide**) in a strand of DNA. About 10 million possible SNPs are known to exist among people.

SNPs and Diseases

Most individuals carry tens of thousands of SNPs, and most seem to have no functional effect at all.[4] Rarely, however, a single SNP in a high-powered gene can produce a severe disease immediately from birth, such as PKU, described in Chapter 3. More commonly, SNPs do not cause a disease directly but may subtly work with other gene variants and with environmental factors such as diet to increase the risk of developing a chronic disease, such as heart disease, later in life. SNPs set the stage for a chronic disease to occur, but the person's own choices are among the actors that perpetrate it.

As an example, a common SNP in a fat metabolism gene changes the body's response to dietary fats. People with this SNP maintain lower blood LDL cholesterol when they eat a diet rich in polyunsaturated fatty acids (PUFA), and they develop higher LDL when they consume less PUFA. A gene (in this case, a fat metabolism gene with a SNP) interacts with a nutrient from the diet (in this case, PUFA) to influence a risk factor for a disease (LDL cholesterol implicated in heart disease).

Complexity of SNP–Disease Relationships

Genetic risks for chronic diseases may appear to be straightforward—just identify the SNPs to identify an increased disease risk—but these associations are proving difficult to pin down.[5] They often involve SNPs in multiple genes, each of which may interact with many dietary and other environmental factors. Furthermore, another realm of influence on gene behavior exists—the **epigenome**.

Epigenetics

People often think of chromosomes as simple strands of DNA, but chromosomes exist as complex, three-dimensional combinations of DNA, proteins, and other molecules. DNA strands are the primary carrier of inherited information, true, but the epigenome constitutes another parallel bank of inheritable information. The epigenome consists of proteins and other molecules that associate with DNA and interact with it in ways that regulate the expression of genes, turning the genes on or off. In short, like DNA, the epigenome can be inherited from generation to generation but, unlike DNA, it is responsive to environmental influences, including diet, particularly during early development.

To help clarify the concept, the genome and the epigenome have been likened to nature's pen and pencil set. The genome, made of DNA, is written in indelible ink, so to speak, making its sequence more permanent and difficult to change. The epigenome is written in pencil in the margins, allowing for erasures and changes.

A Cell Differentiation Specialist

The special talent of the epigenome is in differentiating one type of cell from another in the body. It does not change the DNA sequence itself but instead controls which genes are turned on or off—that is, which are expressed or silenced. For example, a cone cell of a person's eye and a blood-producing cell of that person's bone marrow contain identical DNA strands. Luckily for the person, the epigenome activates and silences genes on the DNA strands so that each cell type reliably makes only the correct proteins to allow its own specialized functions.

How Epigenetic Regulation Works

Mechanisms for epigenetic gene regulation include, among others, the workings of globular proteins known as **histones** and small organic molecules called **methyl groups**.† Both of these mechanisms can be modified by way of diet and other environmental influences.

Chromosome Structural Proteins in Gene Expression Millions of histones reside within the chromosome (shown in Figure C11–1), supporting its shape. Like thread wound around a spool, sections of DNA "thread" are

† *Other mechanisms include acetylation of DNA (and histones), other chromatin remodeling factors, and noncoding regulatory RNA molecules.*

Figure C11–1

Two Epigenetic Factors and Gene Activity

This figure shows methyl groups attached to a DNA strand and histones, globular protein "spools" wrapped with lengths of DNA. Other epigenetic factors also exist.

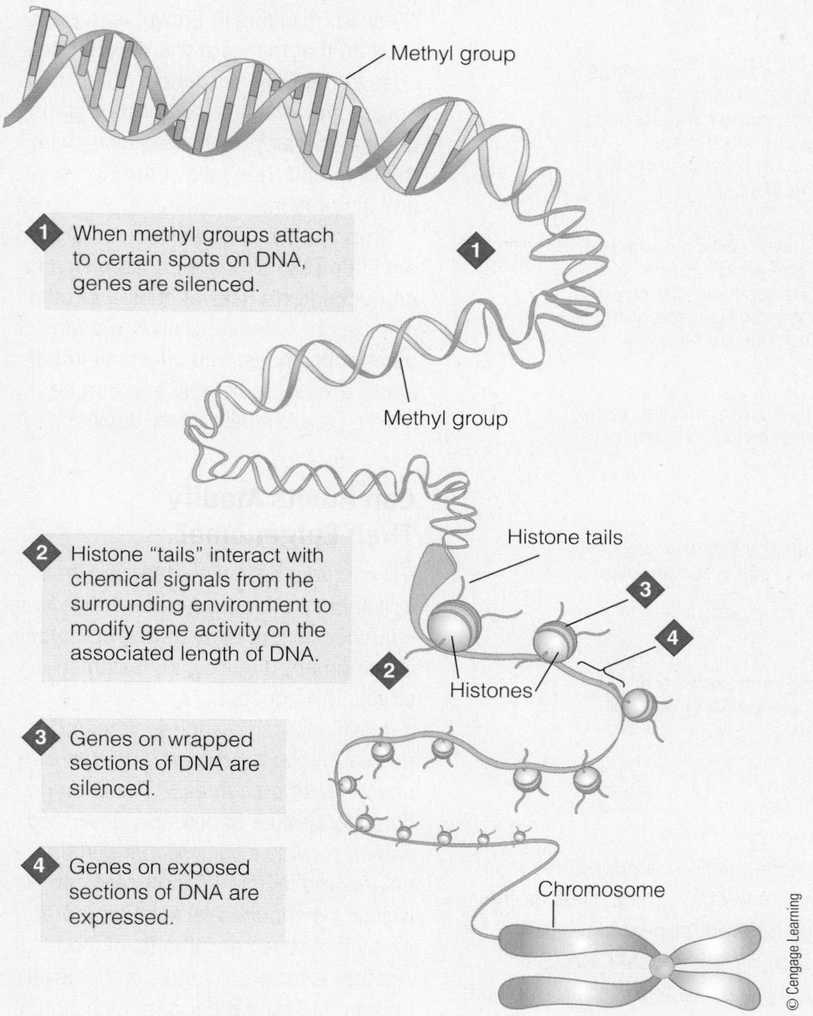

tightly wrapped around protein histones. Thus, the huge DNA molecules are shaped and condensed to fit inside the tiny cell nucleus.

Methyl group
1 When methyl groups attach to certain spots on DNA, genes are silenced.
1
Methyl group
Histone tails
2 Histone "tails" interact with chemical signals from the surrounding environment to modify gene activity on the associated length of DNA.
3
4
2
Histones
3 Genes on wrapped sections of DNA are silenced.
4 Genes on exposed sections of DNA are expressed.
Chromosome
© Cengage Learning

Once believed to confer only structural support to the chromosomes, histones are now known to regulate gene expression, too. Genes on a DNA segment that is wrapped around a histone are silent—they physically lack the room to perform the tasks required for protein synthesis. Histones, though, can change this situation in response to changing environmental conditions.

Histones sport little protein "tails" that stick out from their DNA wrappings. These tails serve as landing sites for many molecules from the environment that signify cellular conditions.

With the right chemical signals, a histone loosens its grip on its wraps of DNA, allowing a bit of the strand to stretch out. Genes on these stretched-out segments can then express their encoded proteins—they are activated. Here's where nutrition comes in: many of the molecular signals to which histones respond arise from the diet—they consist of nutrients and phytochemicals themselves or of compounds generated during their metabolism (Figure C11–2, p. 460).

A Broccoli Phytochemical Example

One phytochemical, sulforafane (see Controversy 2), found in broccoli, broccoli sprouts, and other cabbage-family vegetables, may affect cancer processes by way of histone changes in cancer cells. One characteristic of cancerous tissue is uncontrolled cell division. In cancer cells, histones may inappropriately silence genes that ordinarily prevent cells from multiplying out of control.

In test tubes, sulforaphane reverses those cancer-promoting histone changes and reinstates control of cell division.[6] In mice, sulforaphane inhibits certain cancers. In people, ingestion of one cup of broccoli sprouts alters histone activities in blood cells. Does consumption of broccoli or other cabbage-family food actually prevent cancer in people? People who consume these foods regularly do seem to suffer fewer of some cancers.[7] No one knows whether the foods themselves are protective, however; researchers are still investigating that question.

Many other phytochemicals, including tea flavonoids, curcumin from the spice turmeric, and sulfur compounds from the onion family, along with nutrients, such as folate, vitamin B_{12}, vitamin D, selenium, and zinc add to a growing list of food constituents that affect epigenetic activities in ways that may prevent cancer.[8] Although scientists can duplicate some of these activities with synthetic drugs, the drugs, unlike foods, produce extreme toxicity.[9]

DNA Methyl Groups and Gene Regulation Genes are also regulated by molecules that adhere to the DNA strand itself. Methyl groups, tiny organic compounds, arise from the diet. Methyl groups attach directly onto DNA (look again at Figure C11–1), altering gene

Figure C11–2

Bioactive Food Components and Gene Expression

Nutrients and phytochemicals (bioactive food components) in the diet can affect gene expression directly or indirectly.

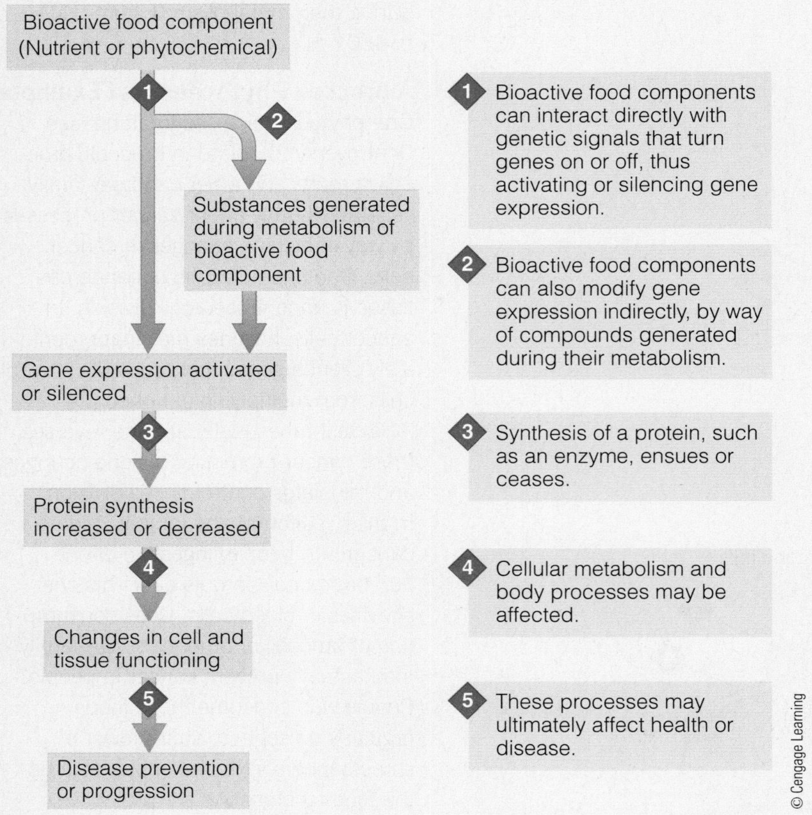
Bioactive food component (Nutrient or phytochemical)
1
2
Substances generated during metabolism of bioactive food component
Gene expression activated or silenced
3
Protein synthesis increased or decreased
4
Changes in cell and tissue functioning
5
Disease prevention or progression
1 Bioactive food components can interact directly with genetic signals that turn genes on or off, thus activating or silencing gene expression.
2 Bioactive food components can also modify gene expression indirectly, by way of compounds generated during their metabolism.
3 Synthesis of a protein, such as an enzyme, ensues or ceases.
4 Cellular metabolism and body processes may be affected.
5 These processes may ultimately affect health or disease.
© Cengage Learning

expression. Attachment of a methyl group to a beginning of a gene on the DNA strand is typically associated with the silencing of that gene; removal of those methyl groups allows gene expression to commence and protein replication to occur.[10]

B Vitamins and Methyl Groups A powerful example of nutritional genomics involves the influence of the B vitamin folate on DNA methylation. Folate (along with other B vitamins) is essential for transferring methyl groups to other molecules, including to DNA. With too little folate, genes may be insufficiently methylated to allow normal suppression of unneeded proteins. The flip side is also true: too much folate can inappropriately silence genes whose proteins are necessary for health.[11]

This effect is illustrated in the accompanying photo of two mice. Despite their strikingly different appearance, both these mice have identical DNA. Both possess a gene that tends to produce fat, yellow pups, but their gene expression

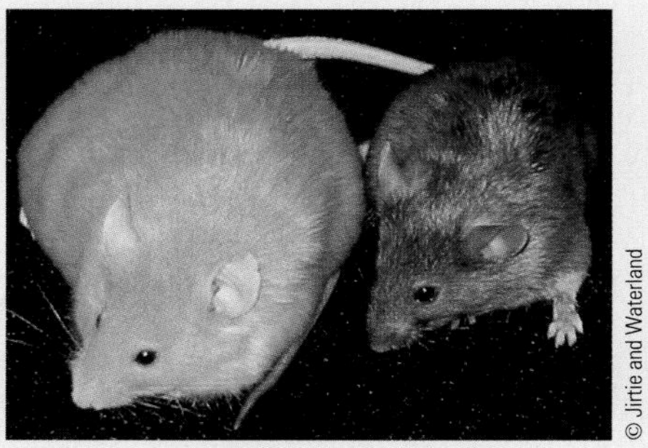

These two mice share an identical gene that tends to produce fat, yellow mice. The mother of the lean, brown mouse received supplemental B vitamins that silenced the gene.

was altered when their mothers were fed different diets during pregnancy. The mother of the lean, brown mouse received doses of the B vitamins folate and vitamin B_{12}. By way of the methyl group transfer activity of these vitamins, the gene for "yellow and fat" was silenced, resulting in brown, lean pups.

Note that the extra vitamins did not change the DNA sequence. Still, such epigenomic changes established during pregnancy can be inherited along with the DNA and thus persist through several generations.

Importantly, pregnant women should not attempt to alter their children's and grandchildren's risks of obesity or other diseases by loading up on B vitamins or other substances. The effects of imbalances are unpredictable and can be severe (see Chapter 13 for details).

Can Adults Modify Their Epigenome?

Researchers believe that the greatest epigenomic changes from environmental influences occur early during embryonic development (Figure C11–3 demonstrates this concept). Some change can still occur into adolescence and even adulthood, however, and they can affect health outcomes.[12] The findings on sulforaphane of broccoli, described earlier, provide evidence that certain epigenome factors in adult cells can indeed be changed, at least temporarily, by bioactive constituents of foods. Another example in adults is the development of liver cancer after ingestion of a mold toxin that can form on corn and other grain (*aflatoxin*, defined in Chapter 12). The toxin is suspected of causing removal of important methyl groups from both histones and the DNA strand, triggering the development of the cancer.[13]

Now a theory emerges to suggest at least a partial solution to the mystery of how identical twins can develop different diseases. Although the twins have identical DNA, they acquire differences in their epigenomes.[14] Each encounters different environmental influences at various times of life that change their genetic expression.

Figure C11–3

An Epigenome Timeline

Environmental influences, including diet, most profoundly alter the epigenome during the earliest stages of development, but some changes are probably still possible later in life.

Arguments Surrounding Genetic Testing

For nutritional genomics to be of practical value, people must undergo genetic testing. Researchers and others debate the merits and demerits of all kinds of genetic testing for currently healthy people.[15] Supporters point out that genetic testing holds enormous promise for improving human health and that application of the technology should keep pace with advances in research.

Critics of this position raise some important points, however.[16] They question whether identifying a genetic marker for disease by way of expensive testing would translate into better health for the nation or waste limited health-care dollars. They also voice fears that certain DNA results, once known, could be misused. Table C11–2 provides the scope of the arguments.

Table C11–2

Genetic Testing: Pros and Cons

Arguments for Genetic Testing	Arguments against Genetic Testing
1. Genetic tests provide additional information for improved understanding of a patient's medical profile.	**1.** More information may not be better. Genetic testing has so far yielded minimally more useful information than less expensive clinical tests.
2. With forewarning, genetically susceptible people could make lifestyle changes to reduce their risks.	**2.** Most people who test positive for risk factors such as diabetes and high LDL cholesterol do not make needed lifestyle changes to reduce their risks. More detailed warnings will probably not help them to do so.
3. Nutritional genomic testing in particular holds the promise of some urgently needed help in fighting today's major killers—heart disease and cancer—despite details in application yet to be worked out. Supporters ask, "Should we let perfectionism stand in the way of progress?"	**3.** Current knowledge does not support application. ▪ Lifestyle changes to match gene profiles have not yet been defined. ▪ Links between specific genetic variations and chronic disease are often not fully defined. ▪ Astronomical numbers of potential interactions between environmental factors with genome and epigenome variations must be pinned down before effective application is possible.
4. Regulations concerning use and ownership of genetic information are under development, and some, such as the Genetic Information Nondiscrimination Act of 2008, are currently in force.	**4.** Regulation loopholes exist, and information may be accessible to other parties. Genetic discrimination can be disguised making it difficult to prove.
5. Ethical concerns are minor and protections and protocol will be established as genetic testing grows more common. Some controls already exist.	**5.** Today's safeguards are incomplete and not universally enforced. Ethical concerns include: ▪ Unintentional disclosure of family relationships (such as adoptions or other parentage issues). ▪ Insurance discrimination or limitations against those with certain genetic traits. ▪ Employer discrimination against certain traits to reduce potential sick time expenses.

Sources: S. Vakili and M. A. Caudill, Personalized nutrition: Nutritional genomics as a potential tool for targeted medical nutrition therapy, Nutrition Reviews *65 (2009): 301–315; M. M. Bergmann, U. Görman, and J. C. Mathers, Bioethical considerations for human nutrigenomics,* Annual Review of Nutrition *28 (2008): 447–467; J. M. Ordovas, Nutrigenetics, plasma lipids, and cardiovascular risk,* Journal of the American Dietetic Association *106 (2006): 1074–1081.*

Nutritional Genomics Fraud

Fraud is a problem in the genetic-testing marketplace today, particularly as it relates to nutrition. Unethical companies may provide bogus assessments of DNA tests or even fail to test the DNA at all, basing their recommendations on simple answers to questionnaires. Some may claim that customers need expensive supplements they sell because they are tailored to meet "personal nutrient requirements" or to "strengthen the body against disease risks" (needs supposedly determined by their fake DNA tests). The supplements can cost up to 30 times the price of comparable products sold in stores.

Certainly, much genetic testing, conducted through legitimate medical providers, is useful and important to proper medical care. As Controversy 1 of this book established, however, consumers must be on guard against all kinds of health and nutrition quackery and should report suspected fraud to the FDA.

Conclusion

No doubt the future of nutrition science will be inextricably linked with the science of genomics, and potential benefits may be enormous. Still, if the authors of this book were to guess the future, based on libraries full of decades of past evidence, most scenarios might go something like this: "Based on our genomics study, Mr. X needs greater amounts of vitamin C from tomato sauces and pink grapefruit, but not from supplements.[17] He needs the fiber, lycopene, carbohydrates, and other bioactive components of a variety of fruits and vegetables, along with less saturated fat, sufficient protein, and a nutritious balanced diet to ward off future problems."

Experience shows that fiber supplements cannot take the place of whole grains for digestive tract health or diabetes control and that calcium pills cannot replace food sources of calcium for bone health, and they may pose risks.[18] Many other examples exist to make the case that eating a well-planned diet of whole foods, as described in Chapter 2, provides the best chance of staying healthy.

The solid and rapidly advancing foundations of nutrition genomics hold great promise for the field of dietetics.[19] Registered dietitians will be key providers of personalized nutrition information and care to minimize disease risks and maximize the health potential of clients.

Critical Thinking

1. Define the status of nutritional genomics research. Provide two examples of where this type of research is leading us.
2. Recall SNPs and explain how they may cause disease.

12 Food Safety and Food Technology

what do you think?

Are most digestive tract symptoms from "**stomach flu**"?

Are most foods from grocery stores **germ-free**?

Should you **refrigerate** leftover party foods after the guests have gone home?

Which poses the greater risk: raw **sushi** from a sushi master or food additives?

Learning Objectives

After completing this chapter, you should be able to accomplish the following:

LO 12.1 Describe two ways in which foodborne microorganisms can cause illness in the body, and describe ways, from purchase to table, in which consumers can reduce their risks of foodborne illnesses.

LO 12.2 Identify foods that often cause foodborne illnesses, and describe ways of increasing their safety.

LO 12.3 Name some recent advances aimed at reducing microbial food contamination, and describe their potential contribution to the safety of the U.S. food supply.

LO 12.4 Describe how pesticides enter the food supply, and suggest possible actions to reduce consumption of residues.

LO 12.5 Discuss potential advantages and disadvantages associated with organic foods.

LO 12.6 Name some functions served by food additives approved for use in the United States, and provide evidence concerning their safety.

LO 12.7 Discuss several ways that food-processing techniques affect nutrients in foods.

LO 12.8 Compare and contrast the advantages and disadvantages of food production by way of genetic modification and conventional farming.

Consumers in the United States and Canada enjoy food supplies ranking among the safest, most pleasing, and most abundant in the world. Along with expanded choices comes a greater consumer responsibility to distinguish between choices leading to food **safety** and those that pose a **hazard**.

As human populations grow and food supplies become more global, new food-safety challenges arise that require new processes, new technologies, and greater cooperation to solve.*[1] Food safety is, therefore, a moving target. The Food and Drug Administration (FDA), the major agency charged with ensuring that the U.S. food supply is safe, wholesome, sanitary, and properly labeled, focuses much effort in these areas of concern:

1. *Microbial foodborne illness.* Recently improved estimation techniques reveal that 48 million U.S. people become ill from foodborne diseases, and about 3,000 of them die.[2]
2. *Natural toxins in foods.* These constitute a hazard mostly when people consume large quantities of single foods either by choice (fad diets) or by necessity (poverty).
3. *Residues in food.*
   a. *Environmental and other contaminants* (other than pesticides). Household and industrial chemicals are increasing yearly in number and concentration, and their impacts are hard to foresee and to forestall.
   b. *Pesticide residues.* A subclass of environmental contaminants, they are listed separately because they are applied intentionally to foods and, in theory, can be controlled.
   c. *Animal drugs.* These include hormones and antibiotics that increase growth or milk production and combat diseases in food animals.
4. *Nutrients in foods.* These require close attention as more and more highly processed and artificially constituted foods appear on the market.
5. *Intentional approved food additives.* These are of little concern because so much is known about them that they pose virtually no risk to consumers and because their use is well regulated.
6. *Genetically modified foods.* Listed last because such foods undergo rigorous scrutiny before going to market.

Other government and world agencies involved in food safety are listed in Table 12–1.

* Reference notes are found in Appendix F.

Table 12–1

Agencies That Monitor the U.S. Food Supply

- **CDC (Centers for Disease Control and Prevention)** a branch of the Department of Health and Human Services that is responsible for monitoring foodborne diseases.
  - Together with the USDA and FDA, CDC operates *FoodNet*, an organization that tracks the prevalence, trends, causes, and interventions of U.S. foodborne illnesses.
  - These groups also conduct molecular DNA tracking of foodborne illness-causing microorganisms through *PulseNet*, a network of scientists in every state who react quickly to identify illness outbreaks.
- **EPA (Environmental Protection Agency)** the federal agency that is responsible for regulating pesticides and establishing water quality standards.
- **FDA (Food and Drug Administration)** the part of the Department of Health and Human Services' Public Health Service that is responsible for ensuring the safety and wholesomeness of all foods sold in interstate commerce except meat, poultry, and eggs (which are under the jurisdiction of the USDA); inspecting food plants and imported foods; and setting standards for food consumption. The FDA also regulates food additives.
- **CORE (Coordinated Outbreak Response and Evaluation Network)** FDA specialist teams that work to prevent and minimize outbreaks of foodborne illness by:
  - continuously monitoring trends and data for signs of outbreak emergence.
  - responding rapidly to stop outbreaks.
  - preventing future outbreaks by improving FDA policies, processes, and guidelines for food industries.
  - informing the media and consumers about outbreaks.
- **USDA (U.S. Department of Agriculture)** the federal agency that is responsible for enforcing standards for the wholesomeness and quality of meat, poultry, and eggs produced in the United States; conducting nutrition research; and educating the public about nutrition.
- **WHO (World Health Organization)** an international agency that develops standards to regulate pesticide use. A related organization is the FAO (Food and Agricultural Organization).

In 2010, the Food Safety and Modernization Act broadened the FDA's focus from reacting after **foodborne illness** strikes to implementing ways to prevent it.[3†] The law stresses safe handling at food processing facilities; gives the FDA more enforcement, inspection, and recall authority; improves foodborne illness surveillance; and improves oversight of imported foods to better safeguard the health of U.S. consumers.

This chapter focuses first on the most immediate food-related threat: foodborne illness. It also addresses concerns about food contamination by both manmade and naturally occurring toxicants. Common food additive questions are then answered, and the Consumer's Guide weighs evidence about organically grown foods. The Controversy addresses the promises and problems of genetically modified foods.

With the privilege of abundance comes the responsibility to choose and handle foods wisely.

Microbes and Food Safety

LO 12.1 Describe two ways in which foodborne microorganisms can cause illness in the body, and describe ways, from purchase to table, in which consumers can reduce their risks of foodborne illnesses.

Some people brush off the threat from foodborne illnesses as less likely and less serious than the threat of flu, but they are misinformed. Foodborne illnesses, caused by **microbes**, can be life-threatening, and some kinds increasingly do not respond to standard antibiotic drug therapy. Even normally mild foodborne illnesses can be lethal

† At the time of publication, FDA rules needed to carry out the law were under review, with implementation anticipated.

safety the practical certainty that injury will not result from the use of a substance.

hazard a state of danger; used to refer to any circumstance in which harm is possible under normal conditions of use.

foodborne illness illness transmitted to human beings through food and water; caused by an infectious agent (*foodborne infection*) or a poisonous substance arising from microbial toxins, poisonous chemicals, or other harmful substances (*food intoxication*). Also commonly called *food poisoning*.

microbes a shortened name for *microorganisms*; minute organisms too small to observe without a microscope, including bacteria, viruses, and others.

for a person who is ill or malnourished; has a compromised immune system; lives in an institution; has liver or stomach illnesses; or is pregnant, very old, or very young.

Celebrated reductions in the number of foodborne illnesses caused by certain organisms each year are countered by unwelcome reports of greater illness caused by other organisms. Achieving the ultimate goal—fewer total foodborne illnesses—will require even more vigilance on the part of regulators, food industries, and consumers.[4]

If digestive tract disturbances are the major or only symptoms of your next bout of what some people erroneously call "stomach flu," chances are that what you really have is a foodborne illness. By learning something about these illnesses and taking a few preventive steps, you can maximize your chances of staying well. Understanding the nature of the microbes responsible is the first step toward defeating them.

How Do Microbes in Food Cause Illness in the Body?

Microorganisms can cause foodborne illness either by infection or by intoxication. Infectious agents, such as *Salmonella* bacteria or hepatitis viruses, infect the tissues of the human body and multiply there. Other microorganisms produce **enterotoxins** or **neurotoxins**, poisonous chemicals that the bacteria release as they multiply. These toxins are absorbed into the tissues and cause various kinds of harm, ranging from mild stomach pain and headache to paralysis and death. The toxins may arise in food during improper preparation or storage or within the digestive tract after a person eats contaminated food. Table 12–2 lists the microbes responsible for 90 percent of U.S. foodborne illnesses, hospitalizations, and deaths, along with their food sources, general symptoms, and prevention methods. Many other illness-causing microbes exist. The steps outlined in this chapter can reduce or eliminate all of them.

Although the most common cause of food intoxication is the *Staphylococcus aureus* bacterium, the most infamous is undoubtedly *Clostridium botulinum*, an organism that produces a toxin so deadly that an amount as tiny as a single grain of salt can kill several people within an hour. *Clostridium botulinum* requires anaerobic conditions such as those found in improperly canned (especially home-canned) foods, home-fermented foods such as tofu, and homemade garlic or herb-flavored oils stored at room temperature.[‡] **Botulism** quickly paralyzes muscles, making seeing, speaking, swallowing, and breathing difficult (symptoms are listed in Table 12–3, p. 468) and demands immediate medical attention.

Some bacterial toxins, such as the botulinum toxin, are heat sensitive and can be destroyed by boiling (but this is not recommended). Others, such as the *Staphylococcus aureus* toxin, are heat-resistant and remain hazardous even after the food is cooked.

KEY POINTS

- Each year in the United States, tens of millions of people suffer mild to life-threatening symptoms caused by foodborne illnesses.
- These people are especially vulnerable to suffer serious harm from foodborne illnesses: pregnant women, infants, toddlers, older adults, and people with weakened immune systems.

Food Safety from Farm to Table

A safe food supply depends upon safe food practices by both domestic and foreign food producers—on the farm or at sea; in processing plants; during transportation; and at supermarkets, institutions, and restaurants (see Figure 12–1, p. 468). Equally critical in the chain of food safety, however, is the final handling of food by people who purchase it and consume it at home. Tens of millions of people needlessly suffer preventable foodborne illnesses each year because they make their own mistakes in purchasing, storing, or preparing their food.

Did You Know?
Anaerobic means "without oxygen."

C Squared Studios/Photodisc/Getty Images

To prevent botulism from homemade flavored oils, wash and dry fresh herbs before use and keep the oil refrigerated. Discard it after a week to 10 days.

enterotoxins poisons that act upon mucous membranes, such as those of the digestive tract.

neurotoxins poisons that act upon the cells of the nervous system.

botulism an often fatal foodborne illness caused by botulinum toxin, a toxin produced by the *Clostridium botulinum* bacterium that grows without oxygen in nonacidic canned foods.

[‡] Complete, up-to-date, home canning instructions are available in the USDA's *Complete Guide to Home Canning*, available from the Superintendent of Documents, Government Printing Office, Washington, DC 20402, or online at www.uga.edu/nchfp/publications/publications_usda.html.

Table 12–2

Major Microbes of Foodborne Illnesses

Organism Name	Most Frequent Food Sources	Onset and General Symptoms	Prevention Methods[a]
Foodborne Infections			
Campylobacter (KAM-pee-loh-BAK-ter) bacterium	Raw and undercooked poultry, unpasteurized milk, contaminated water	Onset: 2 to 5 days. Diarrhea, vomiting, abdominal cramps, fever; sometimes bloody stools; lasts 2 to 10 days.	Cook foods thoroughly; use pasteurized milk; use sanitary food-handling methods.
Clostridium (claw-STRID-ee-um) ***perfringens*** (per-FRINGE-enz) bacterium	Meats and meat products stored at between 120°F and 130°F	Onset: 8 to 16 hours. Abdominal pain, diarrhea, nausea; lasts 1 to 2 days.	Use sanitary food-handling methods; use pasteurized milk; cook foods thoroughly; refrigerate foods promptly and properly.
Escherichia coli; E. coli (esh-eh-REEK-ee-uh-KOH-lye) bacterium (including Shiga toxin-producing strains)[b]	Undercooked ground beef, unpasteurized milk and juices, raw fruits and vegetables, contaminated water, and person-to-person contact	Onset: 1 to 8 days. Severe bloody diarrhea, abdominal cramps, vomiting; lasts 5 to 10 days.	Cook ground beef thoroughly; use pasteurized milk; use sanitary food-handling methods; use treated, boiled, or bottled water.
Norovirus	Person-to-person contact; raw foods, salads, sandwiches	Onset: 1 to 2 days. Vomiting; lasts 1 to 2 days.	Use sanitary food-handling methods.
Listeria (lis-TER-ee-AH) bacterium	Unpasteurized milk; fresh soft cheeses; luncheon meats, hot dogs	Onset: 1 to 21 days. Fever, muscle aches; nausea, vomiting, blood poisoning, complications in pregnancy, and meningitis (stiff neck, severe headache, and fever).	Use sanitary food-handling methods; cook foods thoroughly; use pasteurized milk.
Salmonella (sal-moh-NEL-ah) bacteria (>2,300 types)	Raw or undercooked eggs, meats, poultry, raw milk and other dairy products, shrimp, frog legs, yeast, coconut, pasta, and chocolate	Onset: 1 to 3 days. Fever, vomiting, abdominal cramps, diarrhea; lasts 4 to 7 days; can be fatal.	Use sanitary food-handling methods; use pasteurized milk; cook foods thoroughly; refrigerate foods promptly and properly.
Toxoplasma (TOK-so-PLAZ-ma) ***gondii*** parasite	Raw or undercooked meat; contaminated water; raw goat's milk; ingestion after contact with infected cat feces.	Onset: 7 to 21 days. Swollen glands, fever, headache, muscle pain, stiff neck.	Use sanitary food-handling methods; cook foods thoroughly.
Foodborne Intoxications			
Clostridium (claw-STRID-ee-um) ***botulinum*** (bot-chew-LINE-um) bacterium produces botulin toxin, responsible for causing botulism	Anaerobic environment of low acidity (canned corn, peppers, green beans, soups, beets, asparagus, mushrooms, ripe olives, spinach, tuna, chicken, chicken liver, liver pâté, luncheon meats, ham, sausage, stuffed eggplant, lobster, and smoked and salted fish)	Onset: 4 to 36 hours. Nervous system symptoms, including double vision, inability to swallow, speech difficulty, and progressive paralysis of the respiratory system; often fatal; leaves prolonged symptoms in survivors.	Use proper canning methods for low-acid foods; refrigerate homemade garlic and herb oils; avoid commercially prepared foods with leaky seals or with bent, bulging, or broken cans. Do not give infants honey because it may contain spores of *Clostridium botulinum,* which is a common source of infection for infants.
Staphylococcus (STAF-il-oh-KOK-us) ***aureus*** bacterium produces staphylococcal toxin	Toxin produced in improperly refrigerated meats; egg, tuna, potato, and macaroni salads; cream-filled pastries	Onset: 1 to 6 hours. Diarrhea, nausea, vomiting, abdominal cramps, fever; lasts 1 to 2 days.	Use sanitary food-handling methods; cook food thoroughly; refrigerate foods promptly and properly.

NOTE: Travelers' diarrhea is most commonly caused by E. coli, Campylobacter jejuni, Shigella, *and* Salmonella.

[a]*The "How To" on pp. 628–629 provides more details on the proper handling, cooking, and refrigeration of foods.*

[b]*O157, O145, and other Shiga toxin-producing strains.*

Table 12-3

Dangerous Symptoms of Foodborne Illnesses

Get medical help for these symptoms:

- Bloody stools.
- Dehydration.
- Diarrhea of more than 3 days' duration.
- Fever of longer than 24 hours' duration.
- Headache with muscle stiffness and fever.
- Numbness, muscle weakness, tingling sensations in the skin.
- Rapid heart rate, fainting, dizziness.
- Severe intestinal cramps.

Warning signs of botulism—a medical emergency:

- Difficulty breathing.
- Difficulty swallowing.
- Double vision.
- Weak muscles.

Commercially prepared food is usually safe, but an **outbreak** of illness from this source often makes the headlines because outbreaks can affect many people at once.[5] Dairy farmers, for example, rely on **pasteurization**, a process that heats milk to kill most disease-causing organisms thereby making the milk safe to consume. When a major dairy develops a flaw in its pasteurization system, hundreds of cases of illness can occur as a result.

Other types of farming require other safeguards. Growing food usually involves soil, and soil contains abundant bacterial colonies that can contaminate food. Animal waste deposited onto soil may introduce disease-causing microbes. Additionally, farm workers and other food handlers who are ill can easily pass organisms of illness to consumers through routine handling of foods during and after harvest, a particular concern with regard to foods consumed raw, such as produce.

Attention on *E. coli* Several strains of the *E. coli* bacterium produce a particularly dangerous protein toxin, called *Shiga toxin*, a cause of severe disease. The most notorious strain, *E. coli* O157:H7, caused a large outbreak in 1993, but today most such outbreaks arise from other strains of Shiga toxin-producing *E. coli* (STEC).[§] Outbreaks of

Figure 12-1

Food Safety from Farm to Table

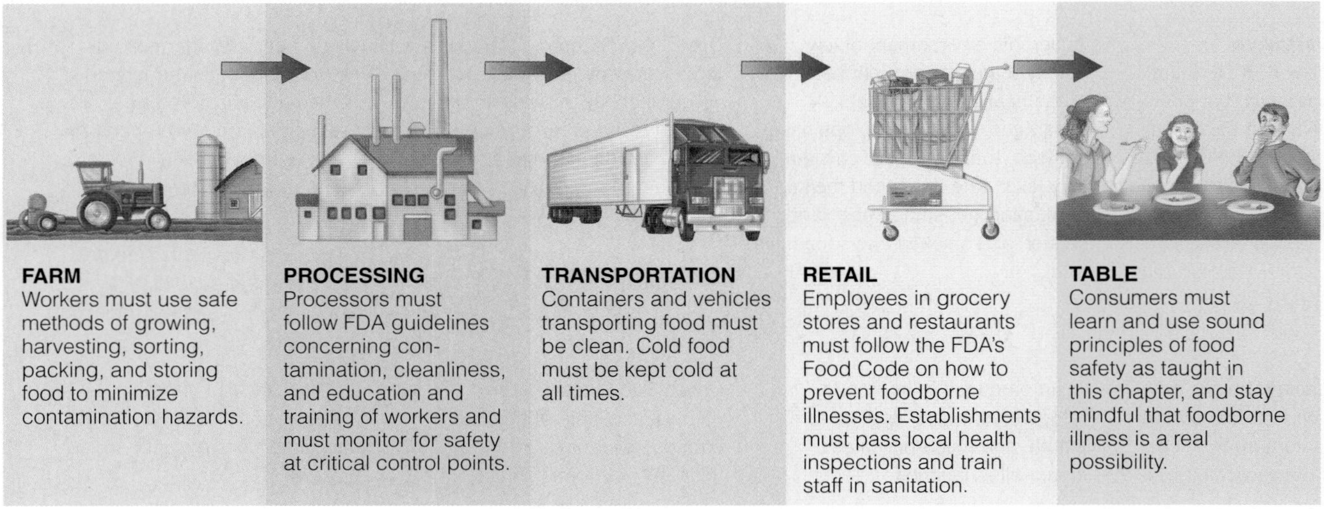

FARM Workers must use safe methods of growing, harvesting, sorting, packing, and storing food to minimize contamination hazards.

PROCESSING Processors must follow FDA guidelines concerning contamination, cleanliness, and education and training of workers and must monitor for safety at critical control points.

TRANSPORTATION Containers and vehicles transporting food must be clean. Cold food must be kept cold at all times.

RETAIL Employees in grocery stores and restaurants must follow the FDA's Food Code on how to prevent foodborne illnesses. Establishments must pass local health inspections and train staff in sanitation.

TABLE Consumers must learn and use sound principles of food safety as taught in this chapter, and stay mindful that foodborne illness is a real possibility.

[§] Shiga toxin was named for the Japanese researcher Kiyoshi Shiga, who discovered the microbial cause of dysentery over 100 years ago.

Table 12-4
Are Your Foods Expiring?

Food manufacturers voluntarily print the following kinds of dates on labels to inform both sellers and consumers of the products' freshness.

- *Sell by:* Specifies the shelf life of the food. After this date, the food may still be safe for consumption if it has been handled and stored properly. Also called *pull date.*
- *Best if used by:* Specifies the last date the food will be of the highest quality. After this date, quality is expected to diminish, although the food may still be safe for consumption if it has been handled and stored properly. Also called *freshness date* or *quality assurance date.*
- *Expiration date:* The last day the food should be consumed. All foods except eggs should be discarded after this date. For eggs, the expiration date refers to the last day the eggs may be sold as "fresh eggs." For safety, purchase eggs before the expiration date, keep them in their original carton in the refrigerator, and use them within 30 days.[a]
- *Open dating:* A general term referring to label dates that are stated in ordinary language that consumers can understand, as opposed to *closed dating,* which refers to dates printed in codes decipherable only by manufacturers. Open dating is used primarily on perishable foods, and closed dating on shelf-stable products such as some canned goods.
- *Pack date:* The day the food was packaged or processed. When used on packages of fresh meats, pack dates can provide a general guide to freshness.

[a]*For best quality, use eggs within 3 weeks of purchase.*

severe or fatal STEC illnesses often grab headlines and focus national attention on two important food-safety issues: raw foods routinely contain live, disease-causing organisms, and strict industry controls are essential to make foods safe. In a new push to control STEC contamination at one of its sources, the USDA has begun routine testing of raw beef used for hamburger meat.[6]

In most cases, STEC disease involves bloody diarrhea, severe intestinal cramps, and dehydration starting a few days after eating tainted meat or raw milk, or contaminated fresh raw produce, such as lettuce, green onions, berries, or even organically grown spinach. In the worst cases, **hemolytic-uremic syndrome** causes a dangerous failure of kidneys and organ systems that very young, very old, or otherwise vulnerable people may not survive. Antibiotics and self-prescribed antidiarrheal medicines can make the condition worse because they increase absorption and retention of the toxin. Severe cases require hospitalization.

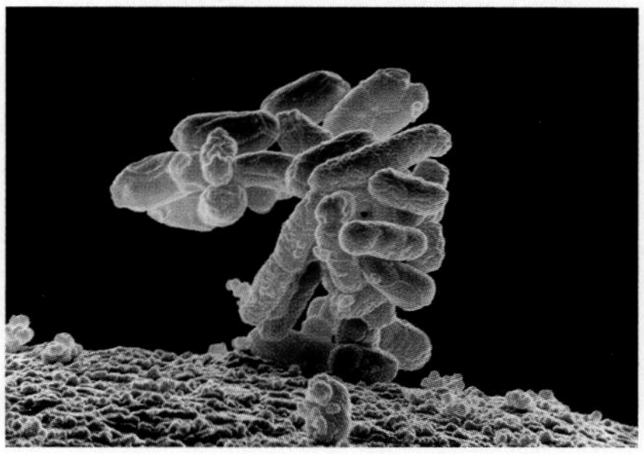

A colony of an E. coli *bacterium (magnified 7,000 times).*

Food Industry Controls Inspections of U.S. meat-processing plants, performed every day by USDA inspectors, help to ensure that these facilities meet government standards. Seafood, egg, produce, and processed food facilities are inspected less often, but all food producers must employ a **Hazard Analysis Critical Control Point (HACCP)** plan to help prevent foodborne illnesses at their source. Each slaughterhouse, producer, packer, distributor, and transporter of susceptible foods must identify "critical control points" in its procedures that pose a risk of food contamination and then devise and implement verifiable ways to eliminate or minimize the risk.

The HACCP system has proved a remarkable success. *Salmonella* contamination of U.S. poultry, eggs, ground beef, and pork has been greatly reduced, and *E. coli* infection from meats has dropped dramatically.

Grocery Safety for Consumers Canned and packaged foods sold in grocery stores are generally safe, but accidents do happen and foods can become contaminated. FDA scientists track outbreaks of illnesses due to large-scale contamination and trace both likely production sources and distribution paths to prevent or minimize consumer exposure. Batch numbering enables the recall of contaminated foods through public announcements in the media.

You can help protect yourself, too. Shop at stores that look and smell clean. Check the freshness dates printed on many food packages and avoid those with expired dates (see Table 12–4). Inspect all seals and wrappers. If a can or package is bulging, leaking, ragged, soiled, or punctured, don't buy it—turn it in to the store manager. A badly dented can or a mangled package is useless in protecting food from microorganisms,

outbreak two or more cases of a disease arising from an identical organism acquired from a common food source within a limited time frame. Government agencies track and investigate outbreaks of foodborne illnesses, but tens of millions of individual cases go unreported each year.

pasteurization the treatment of milk, juices, or eggs with heat sufficient to kill certain pathogens (disease-causing microbes) commonly transmitted through these foods; not a sterilization process. Pasteurized products retain bacteria that cause spoilage.

hemolytic-uremic (HEEM-oh-LIT-ic you-REEM-ick) **syndrome** a severe result of infection with Shiga toxin-producing *E. coli,* characterized by abnormal blood clotting with kidney failure, damage to the central nervous system and other organs, and death, especially among children.

Hazard Analysis Critical Control Point (HACCP) a systematic plan to identify and correct potential microbial hazards in the manufacturing, distribution, and commercial use of food products. *HACCP* may be pronounced "HASS-ip."

insects, or other spoilage. Many jars have safety "buttons" on the lid, designed to pop up once the jar is opened; make sure that they have not "popped." Frozen foods should be solidly frozen, and those in a chest-type freezer case should be stored below the frost line. Check fresh eggs and reject cracked ones. Finally, shop for frozen and refrigerated foods and fresh meats last, just before leaving the store.

KEY POINTS

- Despite government and industry efforts to safeguard the food supply, outbreaks of foodborne illnesses continue to occur.
- Consumers should carefully inspect foods before purchasing them.

Safe Food Practices for Individuals

Some people have come to accept a yearly bout or two of intestinal illness as inevitable, but these illnesses can and should be prevented. Take the safety quiz in Table 12–5 to see how well you follow food-safety rules.

Food can provide ideal conditions for bacteria to multiply and to produce toxins. Disease-causing bacteria require these three conditions to thrive:

- Nutrients.
- Moisture.
- Warmth, 40°F to 140°F (4°C to 60°C).**[7]

To defeat bacteria, you must prevent them from contaminating food or deprive them of one of these conditions. Four core practices illustrated in Figure 12–2 can help to achieve these goals.

Any food with an "off" appearance or odor should be thrown away, of course, and not even tasted. However, you cannot rely on your senses of smell, taste, and sight to warn you because most hazards are not detectable by odor, taste, or appearance. As the old saying goes, "when in doubt, throw it out."

Figure 12–2
Fight Bac!

Four "core" ways to keep food safe. The Fight Bac! website is at www.fightbac.org.

Clean—keep hands, utensils, and surfaces clean.
Separate—keep raw foods separated from ready-to-eat foods.
FIGHT BAC!
CLEAN Wash hands and surfaces often.
CHILL Refrigerate promptly.
COOK Cook to proper temperatures.
BAC
Keep Food Safe From Bacteria
TM
Chill—refrigerate food promptly and keep cold foods cold.
Cook—cook to proper temperatures and keep hot foods hot.

© Cengage Learning

Core Practice #1: Clean Keeping your hands and surfaces clean requires using freshly washed utensils and laundered towels and washing your hands properly, not just rinsing them, particularly before and after handling raw food (see Figure 12–3, p. 472). Normal, healthy skin is covered with bacteria, some of which may cause foodborne illness when deposited on moist, nutrient-rich food and allowed to multiply. Remember to use a nailbrush to clean under fingernails when washing hands and tend to routine nail care—artificial nails, long nails, chipped polish, and even a hangnail harbor more bacteria than do natural, clean, short, healthy nails.

For routine cleansing, washing hands with ordinary soap and warm water is effective; using an alcohol-based hand-sanitizing gel can also provide killing power against many bacteria and most viruses.[8] Following up a good washing with a sanitizer may provide an extra measure of protection useful when someone in the house is ill or when preparing food for an infant, an elderly person, or someone with a compromised immune system.†† If you are ill or have open cuts or sores, stay away from food preparation.

Microbes love to nestle down in small, damp spaces such as the inner cells of sponges or the pores between the fibers of wooden cutting boards. Antibacterial soaps, detergents, and sponges possess a chemical additive intended to deter bacterial growth, but regular products work almost as well and the additive accumulates in the environment.[9] You can ensure the microbial safety of regular sponges by washing them in a dishwasher or by treating them as suggested below.

To eliminate microbes on surfaces, utensils, and cleaning items, you have four choices, each with benefits and drawbacks:

** The FDA suggests these temperatures to consumers at the FDA/CFSAN website; see www.fda.gov. For food industry professionals, the FDA makes other recommendations; see U.S. Public Health Service, *Food Code*, available at www.fda.gov.
†† Effective hand sanitizers contain between 60 and 70 percent isopropyl alcohol.

Table 12–5

Can You Pass the Kitchen Food-Safety Quiz?

How food-safety savvy are you? Give yourself 2 points for each correct answer.

1. The temperature of the refrigerator in my home is
A. 50°F (10°Celsius).
B. 40°F (4°C).
C. I don't know; I don't own a refrigerator thermometer.

2. The last time we had leftover cooked stew or other meaty food, the food was
A. cooled to room temperature, then put in the refrigerator.
B. put in the refrigerator immediately after the food was served.
C. left at room temperature overnight or longer.

3. If I use a cutting board to cut raw meat, poultry, or fish and it will be used to chop another food, the board is
A. reused as is.
B. wiped with a damp cloth or sponge.
C. washed with soap and water.
D. washed with soap and hot water and then sanitized.

4. The last time I had a hamburger, I ate it
A. rare.
B. medium.
C. well-done.

5. The last time there was cookie dough where I live, the dough was
A. made with raw eggs, and I sampled some of it.
B. store-bought, and I sampled some of it.
C. not sampled until baked.

6. I clean my kitchen counters and food preparation areas with
A. a damp sponge that I rinse and reuse.
B. a clean sponge or cloth and water.
C. a clean cloth with hot water and soap.
D. the same as above, then a bleach solution or other sanitizer.

7. When dishes are washed in my home, they are
A. cleaned by an automatic dishwasher and then air-dried.
B. left to soak in the sink for several hours and then washed with soap in the same water.
C. washed right away with hot water and soap in the sink and then air-dried.
D. washed right away with hot water and soap in the sink and immediately towel-dried.

8. The last time I handled raw meat, poultry, or fish, I cleaned my hands afterward by
A. wiping them on a towel.
B. rinsing them under warm tap water.
C. washing with soap and water.

9. Meat, poultry, and fish products are defrosted in my home by
A. setting them on the counter.
B. placing them in the refrigerator.
C. microwaving and cooking promptly when thawed.
D. soaking them in warm water.

10. I realize that eating raw seafood poses special problems for people with
A. diabetes.
B. HIV infection.
C. cancer.
D. liver disease.

ANSWERS

1. Refrigerators should stay at 40°F or less, so if you chose answer B, give yourself 2 points; 0 for other answers.

2. Answer B is the best practice. Give yourself 2 points if you picked it; 0 for other answers.

3. If answer D best describes your household's practice, give yourself 2 points; if C, 1 point.

4. Give yourself 2 points if you picked answer C; 0 for other answers.

5. If you answered A, you may be putting yourself at risk for infection from bacteria in raw shell eggs. Answer C—eating the baked product—will earn you 2 points; answer B, 1 point. Commercial dough is made with pasteurized eggs, but some bacteria may remain.

6. Answer C or D will earn you 2 points each; answer B, 1 point; answer A, 0.

7. Answers A and C are worth 2 points each; other answers, 0.

8. The only correct practice is answer C. Give yourself 2 points if you picked it; 0 for others.

9. Give yourself 2 points if you picked B or C; 0 for others.

10. This is a trick question: all of the answers apply. Give yourself 2 points for knowing one or more of the risky conditions.

RATING YOUR HOME'S FOOD-SAFETY PRACTICES

20 points: Feel confident about the safety of foods served in your home.

12 to 19 points: Reexamine food-safety practices in your home. Some key rules are being violated.

11 points or below: Take steps immediately to correct food-handling, storage, and cooking techniques used in your home. Current practices are putting you and other members of your household in danger of foodborne illness.

Figure 12–3

Proper Hand Washing Prevents Illness

You can avoid many illnesses by following these hand washing procedures before, during, and after food preparation; before eating; after using the bathroom, blowing your nose, or touching your hair; after handling animals or their waste; or when hands are dirty. Wash hands more frequently when someone in the house is sick.

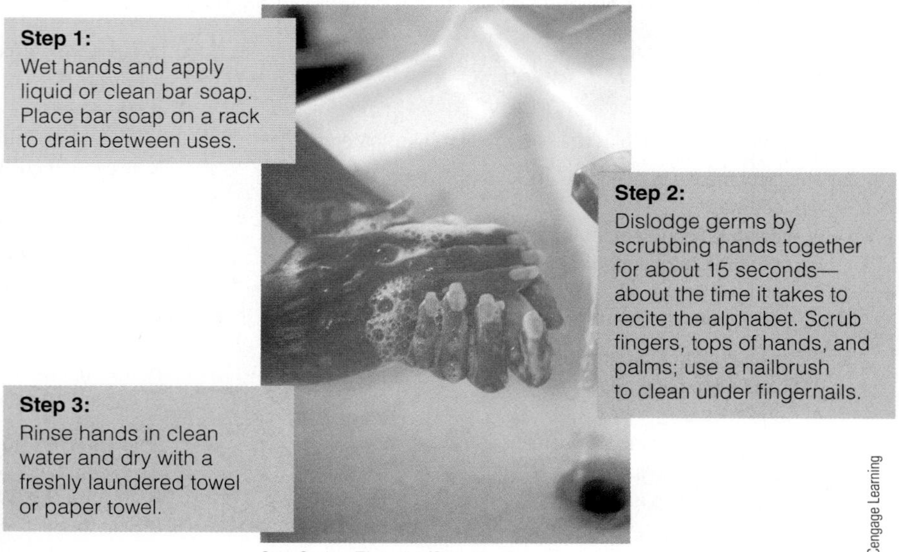
Step 1:
Wet hands and apply liquid or clean bar soap. Place bar soap on a rack to drain between uses.
Step 2:
Dislodge germs by scrubbing hands together for about 15 seconds—about the time it takes to recite the alphabet. Scrub fingers, tops of hands, and palms; use a nailbrush to clean under fingernails.
Step 3:
Rinse hands in clean water and dry with a freshly laundered towel or paper towel.

Suza Scalora/Photodisc/Getty Images

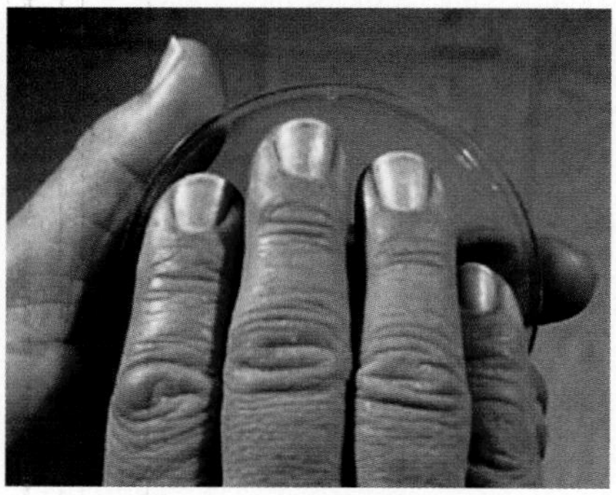

This person's clean-looking but unwashed hand is touching a sterile nutrient-rich gel.

Courtesy of A. Estes Reynolds, George A. Schuler, James A. Christian, and William C. Hurst

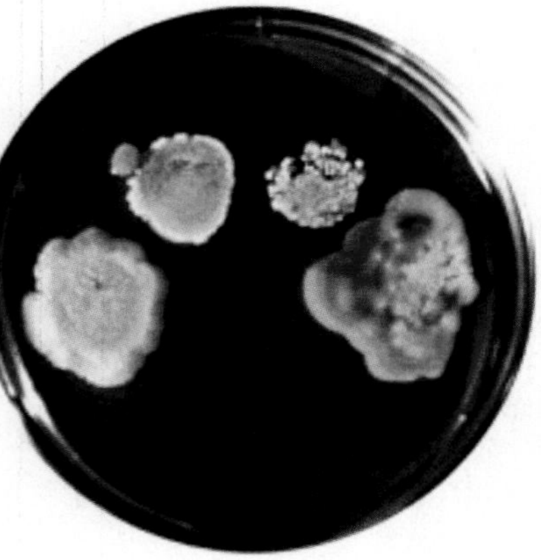

After 24 hours, these large colonies provide visible evidence of the microorganisms that were transferred from the hand to the gel.

Courtesy of A. Estes Reynolds, George A. Schuler, James A. Christian, and William C. Hurst

1. Poison the microbes with highly toxic chemicals such as bleach (one teaspoon per quart of water). Chlorine kills most organisms. However, chlorine is toxic to handle, can ruin clothing, and when washed down household drains into the water supply, it forms chemicals harmful to people and wildlife.
2. Kill the microbes with heat. Soapy water heated to 140°F kills most harmful organisms and washes most others away. This method takes effort, though, since the water must be truly scalding hot, well beyond the temperature of the tap.
3. Use an automatic dishwasher to combine both methods: it washes in water hotter than hands can tolerate, and most dishwasher detergents contain chlorine.
4. Use a microwave to kill microbes on sponges. Place the *wet* sponge in a microwave oven and heat it until steaming hot (times vary). Caution: heat only wet sponges in the microwave oven and watch them carefully; dry sponges or those that contain metal can catch on fire. Also, to prevent scalding your hands, use tongs to remove the steaming-hot sponge.

The third and fourth options turned out to be most effective for sanitizing sponges in an experiment by USDA microbiologists. Washing in a dishwasher and microwaving killed virtually all bacteria trapped in sponges, while soaking in a bleach solution missed over 10 percent. The dishwasher may be preferable, however, for overall safety.

Core Practice #2: Separate Keeping raw food separated means preventing **cross-contamination** of foods. Raw foods, especially meats, eggs, and seafood, are likely to contain illness-causing bacteria. To prevent bacteria from spreading, keep the raw foods and their juices away from ready-to-eat foods. For example, if you take burgers out to the grill on a plate, wash that plate in hot, soapy water before using it to retrieve the cooked burgers. If you use a cutting board to cut raw meat, wash the board, the knife, and your hands thoroughly with soap before handling other foods, and particularly before making a salad or other foods that are eaten raw. Many cooks keep a separate cutting board just for raw meats.

Core Practice #3: Cook Cook foods long enough to reach a safe internal temperature. The USDA urges consumers to use a food thermometer to test the temperatures of cooked foods and not to rely on appearance. Figure 12–4 illustrates various types of thermometers. Table 12–6 provides a glossary of thermometer terms. Figure 12–5 (p. 475) specifies safe internal temperatures for cooked foods.

cross-contamination the contamination of a food through exposure to utensils, hands, or other surfaces that were previously in contact with a contaminated food.

Figure 12–4

Food Safety Temperatures (Fahrenheit) and Household Thermometers

Different thermometers do different jobs. To choose the right one, pay attention to its temperature range: some have high temperature ranges intended to test the doneness of meats and other hot foods (see Figure 12–5). Others have lower ranges for testing temperatures of refrigerators and freezers.

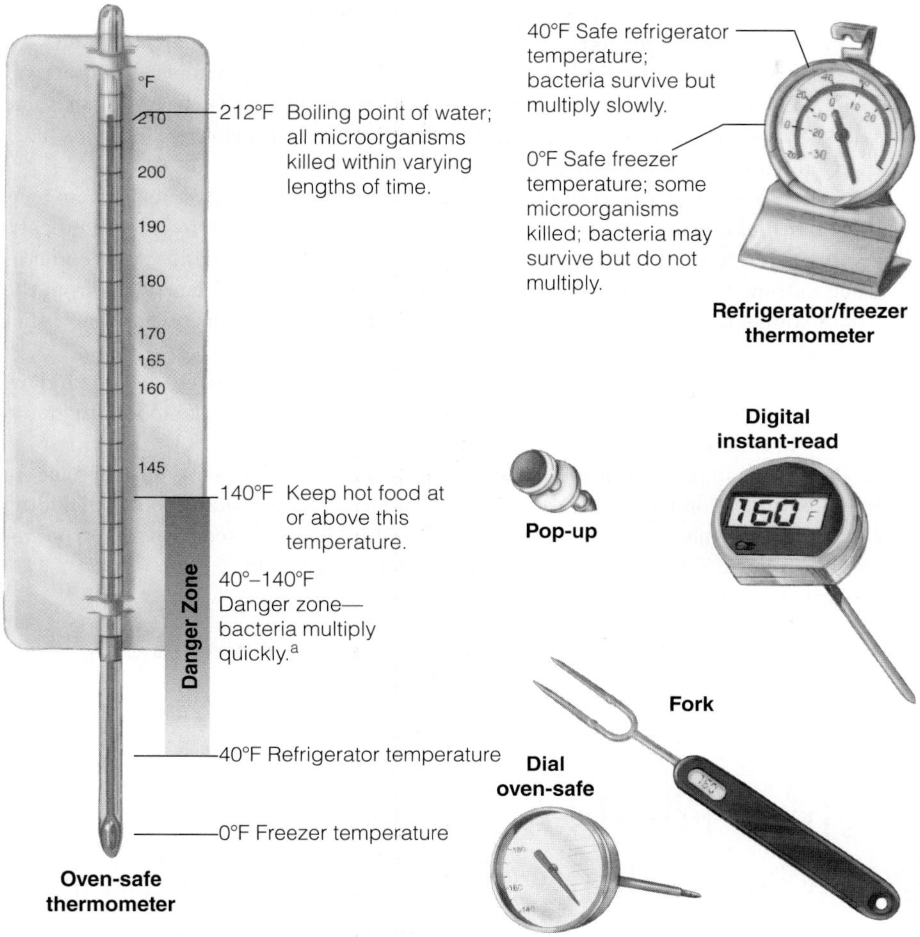

°F
210
200
190
180
170
165
160
145
212°F Boiling point of water; all microorganisms killed within varying lengths of time.
140°F Keep hot food at or above this temperature.
Danger Zone
40°–140°F Danger zone—bacteria multiply quickly.[a]
40°F Refrigerator temperature
0°F Freezer temperature
Oven-safe thermometer
40°F Safe refrigerator temperature; bacteria survive but multiply slowly.
0°F Safe freezer temperature; some microorganisms killed; bacteria may survive but do not multiply.
Refrigerator/freezer thermometer
Digital instant-read
160 F
Pop-up
Fork
Dial oven-safe

© Cengage Learning

[a]*FDA's food storage danger zone for use by consumers. Food professionals adhere to more specific guidelines as put forth in the FDA's Food Code 2009/2011, available at www.cfsan.fda.gov.*

Table 12–6

Glossary of Thermometer Terms

- **appliance thermometer** a thermometer that verifies the temperature of an appliance. An *oven thermometer* verifies that the oven is heating properly; a *refrigerator/freezer thermometer* tests for proper refrigerator (<40°F, or <4°C) or freezer temperature (0°F, or −17°C).
- **fork thermometer** a utensil combining a meat fork and an instant-read food thermometer.
- **instant-read thermometer** a thermometer that, when inserted into food, measures its temperature within seconds; designed to test temperature of food at intervals, and not to be left in food during cooking.
- **oven-safe thermometer** a thermometer designed to remain in the food to give constant readings during cooking.
- **pop-up thermometer** a disposable timing device commonly used in turkeys. The center of the device contains a stainless steel spring that "pops up" when food reaches the right temperature.
- **single-use temperature indicator** a type of instant-read thermometer that changes color to indicate that the food has reached the desired temperature. Discarded after one use, they are often used in commercial food establishments to eliminate cross-contamination.

© Cengage Learning

Table 12–7

Safe Food Storage Times: Refrigerator (≤40°F)

For products with longer shelf lives, rotate them like restaurants do. "First-In-First-Out" means to check dates and use up older products first.

1 to 2 Days
Raw ground meats, breakfast or other raw sausages; raw fish or poultry; gravies
3 to 5 Days
Raw steaks, roasts, or chops; cooked meats, poultry, vegetables, and mixed dishes; lunchmeats (packages opened); mayonnaise salads (chicken, egg, pasta, tuna); fresh vegetables (spinach, green beans, tomatoes)
1 Week
Hard-cooked eggs, bacon, or hot dogs (opened packages); smoked sausages or seafood; milk, cottage cheese
1 to 2 Weeks
Yogurt; carrots, celery, lettuce
2 to 4 Weeks
Fresh eggs (in shells); lunchmeats, bacon, or hot dogs (packages unopened); dry sausages (pepperoni, hard salami); most aged and processed cheeses (Swiss, brick)
2 Months
Mayonnaise (opened jar); most dry cheeses (Parmesan, Romano)

After cooking, keeping hot foods hot requires that they be held at 140°F or higher until served. A temperature of 140°F on a thermometer feels hot, not just warm. Even well-cooked foods, if handled improperly prior to serving, can cause illness. Delicious-looking meatballs on a buffet may harbor bacteria unless they have been kept steaming hot. After the meal, cooked foods should be refrigerated immediately or within two hours at the maximum (one hour if room temperature approaches 90°F, or 32°C). If food has been left out longer than this, toss it out.

Core Practice #4: Chill Chilling and keeping cold food cold starts when you leave the grocery store. If you are running errands, shop last so that the groceries do not stay in the car too long. (If ice cream begins to melt, it has been too long.) An ice chest or insulated bag can help to keep foods cold during transit. Upon arrival home, load foods into the refrigerator or freezer immediately. Table 12–7 lists some safe keeping times for foods stored in the refrigerator at or below 40°F. Foods older than this should be discarded, not ingested.

To ensure safety, thaw frozen meats or poultry in the refrigerator, not at room temperature, and marinate meats in the refrigerator, too. To thaw a food more quickly, submerge it in cold (not hot or warm) water in waterproof packaging or use a microwave to thaw food just before cooking it. Most foods can simply be cooked from the frozen state—just increase the cooking time and use a thermometer to ensure that the food reaches a safe internal temperature.

Chill prepared or cooked foods in shallow containers, not in deep ones. A shallow container allows quick chilling throughout; deeper containers take too many hours to chill through to the center, allowing bacteria time to grow.

Cold foods make a convenient buffet, but keep perishable items safe by placing their containers on ice during serving. This applies to all perishable foods, including custards, cream pies, and whipped-cream or cream-cheese–based treats. Even pumpkin pie, because it contains milk and eggs, should be kept cold.

KEY POINTS

- Foodborne illnesses are common, but the great majority of cases can be prevented.
- To protect themselves, consumers should remember the four cores: clean, separate, cook, chill.

Which Foods Are Most Likely to Cause Illness?

LO 12.2 Identify foods that often cause foodborne illnesses, and describe ways of increasing their safety.

Some foods are more hospitable to microbial growth than others. Foods that are high in moisture and nutrients and those that are chopped or ground are especially favorable hosts. Bacteria in these foods are likely to grow quickly without proper refrigeration.

Protein Foods

Protein-rich foods often require special handling. Packages of raw meats, for example, bear labels to instruct consumers on meat safety (see Figure 12–5).‡‡ Meats in the grocery cooler very often contain bacteria and provide a moist, nutritious environment perfect for microbial growth. Therefore, people who prepare meat should follow these basic meat safety rules:

- Cook all meat and poultry to the suggested temperatures.
- Never defrost meat or poultry at room temperature or in warm water. The warmed outside layer of raw meat fosters bacterial growth.

‡‡ The USDA's Food Information Hotline answers questions about meat, poultry, and seafood safety: 1–888-MPHOTLINE.

prion (PREE-on) a disease agent consisting of an unusually folded protein that disrupts normal cell functioning. Prions cannot be controlled or killed by cooking or disinfecting, nor can the disease they cause be treated; prevention is the only form of control.

bovine spongiform encephalopathy (BOH-vine SPUNJ-ih-form en-SEH-fal-AH-path-ee) **(BSE)** an often fatal illness of the nerves and brain observed in cattle and wild game, and in people who consume affected meats. Also called *mad cow disease.*

- Don't cook large, thick, dense, raw meats or meatloaf in the microwave. Microwaves leave cool spots that can harbor microbes. Never prepare foods that will be eaten raw, such as lettuce or tomatoes, with the same utensils or on the same cutting board as was used to prepare raw meats, such as hamburgers.
- Wash hands thoroughly after handling raw meat.

Unrelated to sanitation, a **prion** disease of cattle and wild game such as deer and elk, **bovine spongiform encephalopathy (BSE)**, causes a rare but fatal brain disorder in human beings who consume meat from afflicted animals.[10§§] U.S. beef industry regulations minimize the risk of contracting BSE from eating beef to almost zero.

Ground Meats Ground meat or poultry is handled more than meats left whole, and grinding exposes much more surface area for bacteria to land on, so experts advise cooking these foods to the well-done stage. Use a thermometer to test the internal temperature of poultry and meats, even hamburgers, before declaring them done. Don't trust appearance alone: burgers often turn brown and appear cooked before their internal temperature is high enough to kill harmful bacteria.

Stuffed Poultry A stuffed turkey or chicken raises concerns because bacteria from the bird's cavity can contaminate the stuffing. During cooking, the center of the stuffing can stay cool enough for long enough for bacteria to multiply.[11] For safe stuffed poultry, follow the Fight Bac core principles—clean, separate, cook, and chill. In addition:

- Cook any raw meat, poultry, or shellfish before adding it to stuffing.
- Mix wet and dry ingredients right before stuffing into the cavity and stuff loosely; cook immediately afterward in a preheated oven set no lower than 325°F (use an oven thermometer to make sure).
- Use a meat thermometer to test the center of the stuffing. It should reach 165°F.

Did You Know?

Meats provide the moisture and nutrients that bacteria require to grow; to thwart them, control temperature.

A safe hamburger is cooked well done (internal temperature of 160°F) and has juices that run clear. Place it on a clean plate when it's done.

Figure 12–5

Safe Handling Instructions and Cooking Temperatures

To avoid foodborne illnesses from meats and poultry, follow the Safe Handling Instructions that appear on all raw meat product packages and cook meats and poultry to proper temperatures (see text for poultry stuffing safety).

Safe handling label for raw meat and poultry

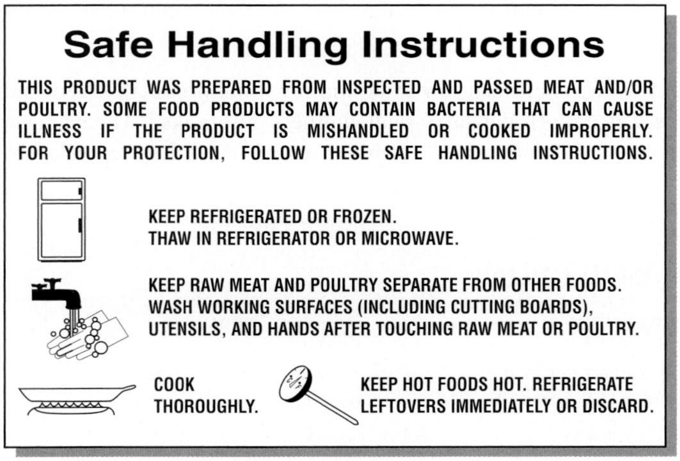

Safe Handling Instructions

THIS PRODUCT WAS PREPARED FROM INSPECTED AND PASSED MEAT AND/OR POULTRY. SOME FOOD PRODUCTS MAY CONTAIN BACTERIA THAT CAN CAUSE ILLNESS IF THE PRODUCT IS MISHANDLED OR COOKED IMPROPERLY. FOR YOUR PROTECTION, FOLLOW THESE SAFE HANDLING INSTRUCTIONS.

KEEP REFRIGERATED OR FROZEN. THAW IN REFRIGERATOR OR MICROWAVE.

KEEP RAW MEAT AND POULTRY SEPARATE FROM OTHER FOODS. WASH WORKING SURFACES (INCLUDING CUTTING BOARDS), UTENSILS, AND HANDS AFTER TOUCHING RAW MEAT OR POULTRY.

COOK THOROUGHLY.

KEEP HOT FOODS HOT. REFRIGERATE LEFTOVERS IMMEDIATELY OR DISCARD.

Recommended cooking temperatures

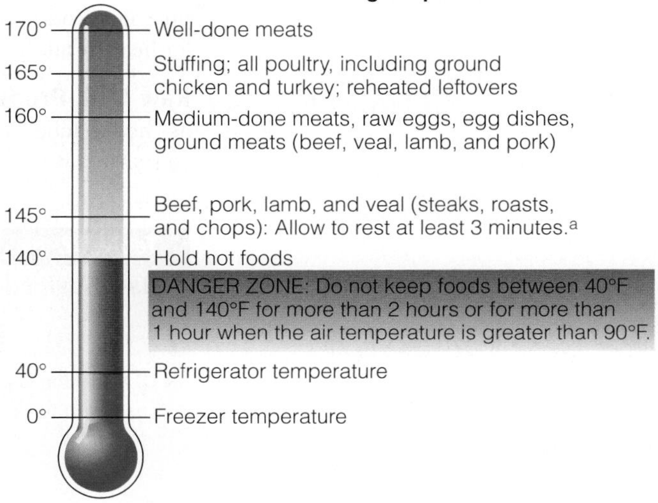

[a]*During the 3 minutes after meat is removed from the heat source, its temperature remains constant or continues to rise, which destroys pathogens.*

§§ The human disease is variant Creutzfeldt-Jakob disease (vCJD).

To repeat: test the stuffing. Even if the poultry meat itself has reached the safe temperature of 165°F, the center of the stuffing may be cool enough to harbor live bacteria.

Eggs Eating undercooked eggs at home accounts for about 30 percent of U.S. *Salmonella* infections.[12] Bacteria from the intestinal tract of hens often contaminate eggs as they are laid, and some bacteria may enter the egg itself. All commercially available eggs are washed and sanitized before packing, and some are pasteurized in the shell to make them safer. To reduce illnesses from *Salmonella*, the FDA requires measures to control bacteria on major egg-producing poultry farms, too.[13] For consumers, egg cartons bear reminders to keep eggs refrigerated, cook eggs until their yolks are firm, and cook egg-containing foods thoroughly before eating them.

What about tempting foods like homemade ice cream, hollandaise sauce, unbaked cake batter, or raw cookie dough that contain raw or undercooked eggs? Healthy adults can enjoy them if they are made safer by choosing pasteurized eggs or liquid egg products. Even these products, because they are made from raw eggs, may contain a few live bacteria that survived pasteurization, making them unsafe for pregnant women, the elderly, young children, or those suffering from immune dysfunction.

Seafood Properly cooked fish and other seafood sold in the United States and Canada is safe from microbial threats. However, even the freshest, most appealing, raw or partly cooked seafood can harbor a variety of microbial dangers from seawater, such as disease-causing viruses; parasites, such as worms and flukes; and bacteria that cause illnesses ranging from self-limiting digestive disturbances to severe, life-threatening illnesses.[14] Table 12–8 lists raw seafood myths that can make people sick.

The dangers posed by seafood are increasing. As burgeoning human populations along the world's shorelines release more contaminants into lakes, rivers, and oceans, the seafood living there becomes less safe to consume. Viruses that cause human diseases have been detected in some 90 percent of the waters off the U.S. coast and easily contaminate filter feeders such as oysters. Government agencies monitor commercial fishing areas and close unsafe waters to harvesters, but illegal harvesting is common. Even reputable seafood shops may unknowingly buy and resell infected fresh seafood.

As for **sushi** or "seared" partially raw fish, even a master chef cannot detect microbial dangers that may lurk within. The marketing term "sushi grade," often applied to seafood to imply wholesomeness, is not legally defined and does not indicate quality, purity, or freshness. Also, freezing does not make raw fish entirely safe to eat. Freezing kills adult parasitic worms, but only cooking can kill all worm eggs and other microorganisms. Safe sushi is made from cooked seafood, seaweed, vegetables, avocados, and other safe delicacies. Experts unanimously agree that today's high levels of microbial contamination makes eating raw or lightly cooked seafood too risky, even for healthy adults.

Raw Milk Products Unpasteurized raw milk and raw milk products (often sold as "health food") cause the majority of dairy-related illness outbreaks.[15] The bacteria counts of raw milk are unpredictable, and sometimes their numbers may be too low to

Table 12–8

Raw Seafood Myths and Truths

Myths	Truths
▪ If a seafood was consumed raw in the past with no ill effect, it is safe to do so today.	▪ Each harvest bears separate risks, and seafood is increasingly contaminated.
▪ Drinking alcohol with raw seafood will "kill the germs."	▪ Alcoholic beverages cannot make contaminated raw seafood safe.
▪ Putting hot sauce on raw oysters and other raw seafood will "kill the germs."	▪ Hot sauce has no effect on microbes in seafood.

cause illness. Other times, even raw milk from a trusted farmer and vendor can cause severe illness. Drinking raw milk presents a real risk without any advantages—the nutrients in pasteurized milk and raw milk are identical.

Even in pasteurized milk, a few bacteria may survive, so milk must be refrigerated to hold bacterial growth to a minimum. Shelf-stable milk, often sold in boxes, is sterilized by an **ultra-high temperature** treatment and so needs no refrigeration until it is opened.

KEY POINTS

- Raw meats and poultry pose special microbial threats and so require special handling.
- Consuming raw eggs, milk, or seafood is too risky.

Raw Produce

Once, meats, eggs, and seafood posed the greatest foodborne illness threats by far, but today the threat from raw produce stands equally among them (see Figure 12–6).[16] Foods such as lettuce, salad spinach, tomatoes, melons, herbs, and scallions grow close to the ground, making bacterial contamination from the soil, animal waste runoff, and manure fertilizers likely. In 2011, for example, one farm recalled 300,000 cases of cantaloupe when illness from *Listeria* killed 30 people and sickened many others across 28 states.[17] Other kinds of produce, and even peanut butter, have been responsible for transmitting dangerous foodborne illnesses to consumers. Such problems often spring from sanitation mistakes made by growers and producers.[18]

Washing produce at home is important, but a quick rinse does not eliminate all microbial contamination. Certain microbes—*E. coli*, among others—exude a **biofilm** that can survive the washing process or even industrial washing.[19] Somewhat more effective is vigorous scrubbing with a vegetable brush to dislodge bacteria, rinsing with vinegar to cut through biofilm, and removing and discarding the outer leaves from heads of leafy vegetables, such as cabbage and lettuce, before washing. Vinegar doesn't sterilize foods, but it can reduce bacterial populations and it's safe for consumption.[20]

Precut Salads and Vegetables Ready-to-eat packaged produce, such as salad greens and cut vegetables, has been triple washed (the label states this) and often rinsed in chlorinated water or treated with oxygen disinfectant (ozone), before packaging. Keep prewashed produce at refrigerator temperatures at all times, and do not rinse it before eating it.

Melons and Berries Rough skins of melons such as cantaloupes provide crevices where bacteria hide and so should be scrubbed with a stiff brush under running water before peeling or cutting. Otherwise, a knife blade or fingers can transfer contaminants from the skin to the edible interior. Raspberries, other berries, and, in fact, all produce should be rinsed thoroughly under running water for at least 10 seconds (see Table 12–9, p. 478). Unpasteurized or raw juices and ciders pose a special problem because microbes on the original fruit may multiply in the juice during storage. Labels of unpasteurized juices must carry the warning shown in Figure 12–7, p. 478. Refrigerated pasteurized juices are generally safe.

Sprouts Sprouts (alfalfa, clover, radish, and others) grow in the warm, moist, nutrient-rich conditions that microbes thrive on.[21] Sprouts are often eaten raw, but the only sure way to make sprouts safe is to cook them. Sprout seeds may harbor *E. coli* or *Salmonella* bacteria that cannot be washed away, so even homegrown, well-rinsed raw sprouts may pose a risk. The elderly, young children, and those with weakened immunity are particularly vulnerable.[22]

KEY POINTS

- Produce causes more foodborne illnesses today than in the past.
- Proper washing and refrigeration can minimize risks.
- Cooking ensures that sprouts are safe to eat.

Figure 12–6

Foodborne Illness from Various Sources

Note that produce is second only to seafood in reported cases.

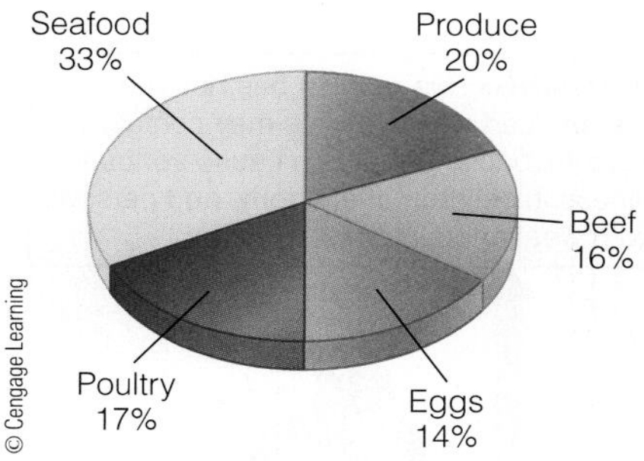

sushi a Japanese dish that consists of vinegar-flavored rice, seafood, and colorful vegetables, typically wrapped in seaweed. Some sushi contains raw fish; other sushi contains only cooked ingredients.

ultra-high temperature a process of sterilizing food by exposing it for a short time to temperatures above those normally used in processing.

biofilm a protective coating of proteins and carbohydrates exuded by certain bacteria; biofilm adheres bacteria to surfaces and can survive rinsing.

Figure 12–7
Warning Label for Unpasteurized Juice

Unpasteurized or untreated juice must bear the following warning on its label:

WARNING: This product has not been pasteurized and therefore may contain harmful bacteria that can cause serious illness in children, the elderly, and persons with weakened immune systems.

Table 12–9
Produce Safety

Cleaning Fresh Fruits and Vegetables
1. Remove and discard the outer leaves from vegetables such as lettuce and cabbage before washing.
2. Wash all fruits and vegetables (including organically grown and homegrown, regardless of place of purchase) just before cooking or eating.
3. Wash fruits and vegetables under clean running water and scrub with clean vegetable brush, or with your hands. Commercial vegetable washing products are safe to use; do not use soap, detergents, or bleach solutions.
4. Dry fruits and vegetables before cutting or eating.
5. Cut away damaged or bruised areas that may contain microbes. Toss out moldy fruit or vegetables.
6. Refrigerate washed or prewashed cut fruits, vegetables, and salads.
Juice Safety
1. Choose chilled pasteurized juices or shelf-stable juices (canned or boxed) that have been treated with high temperature to kill microbes and check their seals to be sure no microbes have entered after processing.
2. Especially infants, children, the elderly, and people with weakened immune systems should never be given raw or unpasteurized juice products.

Other Foods

Careful handling can reduce microbial threats from other foods, too. Foods in this section are common in the food supply, and their safety is worth considering.

Imported Foods and Travel Today, nearly two-thirds of the fruits and vegetables and 80 percent of the seafood consumed in the United States are imported, yet barely 1 percent of imported foods are inspected at U.S. borders.[23] Cooked, frozen, irradiated, or canned imported foods, and foods from developed areas with effective food safety policies, are generally safe. Concerns arise, however, for fresh produce, fish, and other susceptible foods that originate in areas where food safety practices are lax, food handling is unregulated, and where contagious diseases are likely. Fields may be irrigated with contaminated water, crops may be fertilized with untreated manure, and produce may be picked by infected farm workers.

As the food supply becomes more global, illnesses from imported foods appear to be increasing.[24] The FDA is working to close safety gaps in other nations.[25] In addition, to help U.S. consumers to distinguish between imported and domestic foods, regulators now require certain foods, including fish, shellfish, meats, other perishable items, and some nuts to bear a **country of origin label** specifying where they were produced.[26]

People who travel to places where cleanliness standards are lacking have a 50–50 chance of contracting a foodborne illness, commonly called *traveler's diarrhea*. To avoid it, wash hands often; eat only cooked or canned foods or fruits you peel yourself; skip salads; and drink only treated, boiled, canned, or bottled beverages without ice. In general, boil it, cook it, peel it, or forget it.

Honey Honey can contain dormant spores of *Clostridium botulinum* that, when eaten, can germinate and begin to grow and produce their deadly botulinum toxin within the human body. Mature, healthy adults are usually protected against this threat, but infants under one year of age should never be fed honey.

Picnics and Lunch Bags Picnics can be fun and packed lunches a convenience, but to keep them safe, do the following:

© Andrew Holbrooke/Corbis

As many as 400 varieties of fruits and vegetables are imported from other countries.

- Choose foods that are safe without refrigeration, such as whole fruits and vegetables, breads and crackers, shelf-stable foods, and canned spreads and cheeses to open and use on the spot.
- Choose well-aged cheeses, such as cheddar and Swiss; skip fresh cheeses, such as cottage cheese and Hispanic queso fresco. Aged cheese does well without chilling for an hour or two; for longer times, carry it on ice in a cooler or thermal lunch bag.
- Keep meat, egg, cheese, or seafood sandwiches cold until eaten.
- Chill lunch bag foods and use a thermal lunch bag.
- Freeze beverages to pack in with the foods. As the beverages thaw in the hours before lunch, they keep the foods cold.

Note that individual servings of cheese or cold cuts prepackaged with crackers and promoted as lunch foods keep well, but they are high in saturated fat and sodium, they cost triple the price of the foods purchased separately, and their excessive packaging adds to the nation's waste disposal burden.

Mayonnaise, despite its reputation for easy spoilage, is itself somewhat spoilage-resistant because of its acid content. Mayonnaise mixed with chopped ingredients in pasta, meat, or vegetable salads, however, spoils readily. The chopped ingredients have extensive surface areas for bacteria to invade, and cutting boards, hands, and kitchen utensils used in preparation often harbor bacteria. For safe chopped raw foods, start with clean chilled ingredients then chill the finished product in shallow containers; keep it chilled before and during serving; and promptly refrigerate any remainder.

Take-Out Foods and Leftovers Many people rely on take-out foods—rotisserie chicken, pizza, Chinese dishes, and the like—for parties, picnics, or weeknight suppers. When buying these foods, food-safety rules apply: hot foods should be steaming hot and cold foods should be thoroughly chilled.

Leftovers of all kinds make a convenient later lunch or dinner. However, microbes on serving utensils and in the air can quickly contaminate freshly cooked foods; for safety, refrigerate them promptly and reheat them to steaming hot (165°F) before eating. Discard any portion held at room temperature for longer than 2 hours from the time it was served at table until you place it in your refrigerator. Follow the 2, 2, and 4 rules of leftover safety: within 2 hours of cooking, refrigerate the food in clean shallow containers about 2 inches deep, and use it up within 4 days or toss it out.

country of origin label the required label stating the country of origination of many imported meats, chicken, fish and shellfish, other perishable foods, certain nuts, peanuts, and ginseng.

Table 12–10

More Food Safety Myths and Truths

Myths	Truths
▪ "The five-second rule: a food that falls to the floor is safe if it is picked up within five seconds."	▪ A dropped food is immediately contaminated with bacteria.
▪ "If it tastes and smells okay, it's safe to eat."	▪ Most microbial contamination is undetectable by human senses.
▪ "We have always handled our food this way, so it must be safe."	▪ Past generations did not recognize the causes of illness.
▪ "I sampled it a couple of hours ago and didn't get sick, so it is safe to eat."	▪ Illnesses often take half a day or longer to develop.

Exceptions: stuffing and gravy must be used within 2 days, and if room temperature reaches 90°F, all cooked foods must be chilled after 1 hour of exposure. Remember to use shallow containers, not deep ones, for chilling.

Consumers bear a responsibility for food safety, and an essential step is to cultivate awareness that foodborne illness is likely. They must discard old notions that put them at risk (see Table 12–10) and adopt an attitude of self-defense to prevent illness.

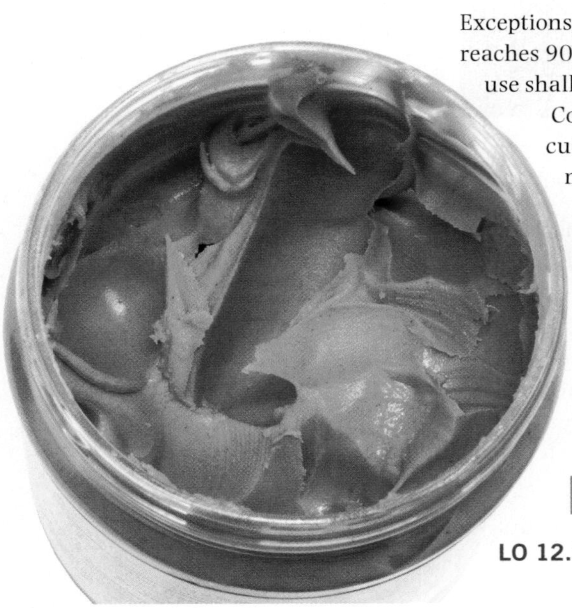

© Stephen VanHorn/Shutterstock.com

KEY POINTS

- The FDA is working to improve the safety of imported foods.
- Honey should never be fed to infants.
- Leftovers can be safely stored when safety rules are followed.

Advances in Microbial Food Safety

LO 12.3 Name some recent advances aimed at reducing microbial food contamination, and describe their potential contribution to the safety of the U.S. food supply.

Advances in technology, such as pasteurization, have dramatically improved the quality and safety of foods over the past century. Today, other technologies promise similar benefits, but some raise concerns among consumers.

Is Irradiation Safe?

Food **irradiation** has been extensively evaluated over the past 50 years. Approved in more than 40 countries, its use is endorsed by numerous health agencies, including the **World Health Organization (WHO)** and the American Medical Association. Food irradiation protects consumers and offers other benefits by:

- Controlling molds, particularly **aflatoxin**.
- Sterilizing spices and teas.
- Controlling insects.
- Extending shelf life in fresh fruits and vegetables (inhibits the growth of sprouts on potatoes and onions and delays ripening in some fruits, such as strawberries and mangoes).
- Destroying disease-causing bacteria in fresh and frozen beef, poultry, lamb, and pork.

Supporters of irradiation say that if more everyday foods were irradiated, the nation's rates of foodborne illnesses would drop dramatically. All irradiated foods except spices must be identified as such on their labels.

How Irradiation Works Irradiation exposes foods to controlled doses of gamma rays from the radioactive compound cobalt 60. As the rays pass through living cells, they disrupt internal DNA, protein, and other structures, killing or deactivating the cells. For example, low radiation doses can kill the growing cells in the "eyes" of potatoes, preventing them from sprouting. Low doses also delay ripening of bananas, avocados, and other fruits. Higher doses easily penetrate tough insect exoskeletons and mold or bacterial cell walls to destroy them. Irradiation works even while food is frozen, making it uniquely useful in protecting foods such as whole frozen turkeys.

Irradiation Effects on Foods Irradiation does not sterilize most foods because doses high enough to kill all microorganisms would also destroy the food. Dried herbs and spices are notable exceptions—they can withstand sterilizing doses. Irradiation does not noticeably change the taste, texture, or appearance of citrus fruits, eggs, certain meats, onions, potatoes, spices, strawberries, and other FDA-approved foods, nor does it make foods radioactive. Some vitamins are destroyed by irradiation, but the losses are comparable to those from other food-processing methods such as canning.

Consumer Concerns about Irradiation Many consumers associate radiation with cancer, birth defects, and mutations, and so they respond negatively to the idea of irradiating foods. Some erroneously fear that food will become contaminated with radioactive particles. More realistic fears concern transport of radioactive materials, training of workers to handle them safely, and safe disposal of spent wastes, which remain radioactive for many years. The food industry echoes these concerns and strives to safeguard both workers and consumers through strict operating standards and compliance with regulations.

Finally, some worry that unscrupulous manufacturers might irradiate old or bacterially tainted foods thereby escaping detection by USDA testers. Instead of being seized or destroyed, the food could be passed off as wholesome to unsuspecting consumers. This objection raises an important point: irradiation is intended to complement, not replace, other traditional food-safety methods. Irradiation cannot entirely protect people from poor sanitation on the farm, in industry, or at home.

Did You Know?

In 2012, the FDA launched a 5-year project called The 100K Genome Project to create a database of the genetic sequences of 100,000 varieties of foodborne illness bacteria. The goal is to offer scientists a resource for discovering how and where the bacteria live, and how they might be controlled.

This "radura" logo is the international symbol for foods treated with irradiation.

KEY POINTS

- Food irradiation kills bacteria, insects, molds, and parasites on foods.
- Consumers have concerns about the effects of irradiation on foods, workers, and the environment.

Other Technologies

The FDA and USDA are improving their monitoring techniques for microbial contamination at all levels of food production. In addition, some food-processing and -packaging technologies are currently helping to reduce microbial threats to consumers, and others show potential for future use.

Improved Testing and Surveillance Microbial testing of foods before they reach consumers is a critical step toward preventing foodborne illnesses. Automated systems have improved testing accuracy from farms to markets. For example, using a mobile laboratory, FDA scientists can now test fresh produce at the growing field and analyze it for many kinds of bacterial contamination.[27] In addition, better detection methods for *E. coli* in water, sediment, and other environmental harbors allow intervention before microbes can contaminate food crops.

Modified Atmosphere Packaging Common packaging methods improve the safety and shelf life of many fresh and prepared foods. Vacuum packaging or **modified atmosphere packaging (MAP)** reduces the oxygen inside a package. This makes it possible for unopened packages of soft pasta noodles, baked goods, prepared foods, fresh and cured meats, seafood, dry beans and other dry products, and

irradiation the application of ionizing radiation to foods to reduce insect infestation or microbial contamination or to slow the ripening or sprouting process. Also called *cold pasteurization*.

World Health Organization (WHO) an agency of the United Nations charged with improving human health and preventing or controlling diseases in the world's people.

aflatoxin (af-lah-TOX-in) a toxin from a mold that grows on corn, grains, peanuts, and tree nuts stored in warm, humid conditions; a cause of liver cancer prevalent in tropical developing nations. (To prevent it, discard shriveled, discolored, or moldy foods.)

modified atmosphere packaging (MAP) a technique used to extend the shelf life of perishable foods; the food is packaged in a gas-impermeable container from which air is removed or to which an oxygen-free gas mixture, such as carbon dioxide and nitrogen, is added.

ground and whole-bean coffee to stay fresh and safe much longer than they would in conventional packaging. Reducing oxygen:

- Reduces growth of oxygen-dependent microbes.
- Prevents discoloration of cut vegetables and fruits.
- Prevents spoilage of fats by rancidity and development of "off" flavors.
- Slows ripening of fruits and vegetables and enzyme-induced breakdown of vitamins.

Perishable foods packaged with MAP must still be chilled properly, however, to keep them safe from microbes that flourish in anaerobic environments, such as the *Clostridium botulinum* bacterium. Chilling of precut salad greens is also a must: temperatures above 50°F cause a dangerous change in *E. coli* bacteria strains present in MAP-bagged lettuces that help them to survive the eater's stomach acid, increasing their ability to cause infection.

High Pressure and Ultrasound High-pressure processing (HPP) technology compresses water to create intense pressure that can kill disease-causing microbes, including norovirus and hepatitis viruses. Previously used to "pasteurize" deli meats, applesauce, and orange juice, HPP may soon make shellfish, such as oysters, clams, and mussels, safer to consume.[28] However, the equipment is expensive, and sufficient pressure to inactivate viruses may alter the taste and texture of the meat.

High-powered ultrasound also holds promise as a sanitizer for organic salad greens. It works by sending high-energy shockwaves through water to dislodge pathogens from the small crevices of leafy greens. It may one day replace chlorine rinses but does not sterilize the food.

© Feng Yu/Shutterstock.com

Antimicrobial Wraps and Films Bacteria-killing food wraps and films hold promise.[29] One biodegradable wrap made from milk whey protein may protect perishable foods from oxidation spoilage and may also receive a dose of antimicrobial materials to prevent bacterial growth. Other films are made from fruit or vegetable purees with a dose of cinnamon or oregano extract, natural antibacterial agents. In addition to protecting the food, the edible wraps may lend a pleasing flavor.

Microbial foodborne illnesses undoubtedly pose the most immediate threat to consumers, but other factors also affect food safety. The next sections address some of these concerns.

KEY POINTS

- Irradiation controls mold, sterilizes spices and teas, controls insects, extends shelf life, and destroys disease-causing bacteria.
- Scientific advances, such as research on high pressure processing, continuously improve food safety.

Toxins, Residues, and Contaminants in Foods

LO 12.4 Describe how pesticides enter the food supply, and suggest possible actions to reduce consumption of residues.

Nutrition-conscious consumers often wonder if our nation's foods are made unsafe by chemical contamination. The FDA, along with the Environmental Protection Agency (EPA), regulates many chemicals in foods that occur as a result of human activities. A later section describes these substances. First, some toxins produced naturally by the foods themselves are worthy of attention.

Natural Toxins in Foods

Some people think they can eliminate all poisons from their diets by eating only "natural" foods. On the contrary, nature has provided many plants with natural poisons to fend off diseases, insects, and other predators. Humans rarely suffer harm from such poisons, but the *potential* for harm does exist.

Herbs and Cabbages The herbs belladonna and hemlock are infamous poisons, but few people know that the herb sassafras contains a carcinogen and liver toxin so potent that it is banned from use in commercially produced foods and beverages.*** Cabbage, turnips, mustard greens, and radishes all contain small quantities of harmful goitrogens, compounds that can enlarge the thyroid gland and aggravate thyroid problems. Ordinarily, cabbages and their relatives are celebrated foods because people who choose them often have low rates of certain cancers. However, in extreme conditions when people have little to eat but cabbages, goitrogens can become a problem; cooking deactivates the compounds, however.[30]

Foods with Cyanogens Other natural poisons are members of a group called cyanogens, precursors to the deadly poison cyanide. Most cassava, a root vegetable staple for many of the world's people, contains just traces of cyanogens; higher amounts in bitter varieties, however, can pose a threat to people with nothing else to eat. Fruit pits also contain cyanogens, but these items are seldom deliberately eaten; an occasional swallowed seed or two presents no danger, but a few dozen pits could be fatal to a small child. One infamous cyanogen extracted from apricot pits is laetrile, a fake cancer cure often passed off as a vitamin.††† True, the poison laetrile kills cancer cells but only at doses that can kill the person, too.

Potatoes Potatoes contain many natural poisons, including solanine, a powerful, bitter, narcotic-like substance. The small amounts of solanine normally found in potatoes are harmless, but solanine can build up to toxic levels when potatoes are exposed to light during storage. Cooking does not destroy solanine, but most of a potato's solanine develops in a thin green layer just beneath the skin so it can be peeled off, making the potato safe to eat. If a potato tastes bitter, however, throw it out.

Seafood Red Tide Toxin At certain times of the year, seafood may become contaminated with the so-called red tide toxin that occurs during algae blooms. Eating seafood contaminated with red tide causes a form of foodborne illness that paralyzes the eater. The FDA monitors fishing waters and closes them to fishing when red tide algae appear.

These examples of naturally occurring toxins serve as a reminder of three principles. First, poisons are poisons, whether made by people or by nature. It's not the source of a compound that makes it hazardous but its chemical structure. Second, because any substance—even pure water—can be toxic when consumed in excess, practice moderation when choosing food portions. Third, by choosing a variety of foods, toxins present in one food are diluted by the volume of the other foods in the diet.

KEY POINTS

- Natural foods contain natural toxins that can be hazardous under some conditions.
- To avoid harm from toxins, eat all foods in moderation, treat toxins from all sources with respect, choose a variety of foods, and respect bans on seafood harvesting.

Pesticides

The use of **pesticides** helps to ensure the survival of food crops, but the damage pesticides do to the environment is considerable and increasing. Moreover, there is some question about whether the widespread use of pesticides has truly increased overall yields of food. Even with extensive pesticide use, the world's farmers lose large quantities of their crops to pests every year.

*** The carcinogen is safrole.

††† Also called *amygdalin* and, erroneously, *vitamin* B_{17}.

Did You Know?

Agricultural pesticides:

- Protect crops from insect damage.
- Increase potential yield per acre.

But they also:

- Accumulate in the food chain.
- Kill pests' natural predators.
- Pollute the water, soil, and air.

pesticides chemicals used to control insects, diseases, weeds, fungi, and other pests on crops and around animals. Used broadly, the term includes *herbicides* (to kill weeds), *insecticides* (to kill insects), and *fungicides* (to kill fungi).

Wash fresh fruits and vegetables to remove pesticide residues.

Did You Know?

Each year, the United States applies billions of pounds of pesticides to:

- Kill pests in and around homes.
- Control pests in flower and vegetable gardens.
- Reduce loss of farm crops to insects.
- Preserve wood products.
- Cure lice, scabies, worms, and other parasites in people and their pets.
- Repel mosquitoes, fleas, and other biting insects from people and pets.
- Many other uses.

The use of pesticides on food crops demonstrates a principle inherent to nutrition decision-making: the expected benefits of an action or inaction must be weighed against its risks. Researchers are working out ways of expressing science-based risks and benefits in context to help decision-makers and consumers determine their best course of action.[31]

Do Pesticides on Foods Pose a Hazard to Consumers? Many pesticides are broad-spectrum poisons that damage all living cells, not just those of pests. Their use can harm the plants and animals in natural systems, and they also present risks to people who produce, transport, and apply them. High doses of pesticides in laboratory animals cause birth defects, sterility, tumors, organ damage, and central nervous system impairment. Equivalent doses are extremely unlikely to occur in human beings, however, except through accidental spills. As Figure 12–8 demonstrates, minute quantities of pesticide **residues** on agricultural products can survive processing, and traces are often present in foods served to people, but these amounts pose negligible risks to most people (see the Consumer's Guide section).[32]

Especially Vulnerable: Infants and Children Infants and children are more susceptible than adults to ill effects of pesticides, for four reasons. First, the immature human detoxifying system cannot effectively cope with poisons, so they tend to stay longer in the body. Second, the developing brain has a more permeable barrier that cannot yet fully exclude pesticides. Many pesticides work by interfering with normal nerve and brain chemistry, and the effects of chronic, low-dose exposure to pesticides on the developing human brain are largely unknown.

Third, children's bodies are small in size, yet their pesticide exposure is often greater than that of adults, meaning that they encounter more pesticide per pound of body weight, which poses a greater risk of harm. Children pick up pesticides through normal child behaviors, such as playing outdoors on treated soil or lawns; handling sticks, rocks, and other contaminated objects; crawling on treated carpets, furniture, and floors; placing fingers and toys in their mouths; seldom washing their hands before eating; and using fingers instead of utensils to grasp foods.

Fourth, children eat proportionally more food per pound of body weight than do adults, and even the trace amounts of pesticides on foods can contribute to total exposure. Fortunately, these traces rarely exceed allowable limits, and most can be further reduced by washing produce thoroughly and following the other guidelines in Table 12–11.[33‡‡‡] Another possibility for reducing pesticide exposure is to choose **organic foods**—read the Consumer's Guide for perspective.

Table 12–11

Ways to Reduce Pesticide Residue Intakes

In addition to these steps, remember to eat a variety of foods to minimize exposure to any one pesticide.

- Trim the fat from meat, and remove the skin from poultry and fish; discard fats and oils in broths and pan drippings. (Pesticide residues concentrate in the animal's fat.)
- Select fruits and vegetables with intact skins.
- Wash fresh produce in warm running water. Use a scrub brush, and rinse thoroughly.
- Use a knife to peel an orange or grapefruit; do not bite into the peel.
- Discard the outer leaves of leafy vegetables such as cabbage and lettuce.
- Peel waxed fruits and vegetables; waxes don't wash off and can seal in pesticide residues.
- Peel vegetables such as carrots and fruits such as apples when appropriate. (Peeling removes pesticides that remain in or on the peel, but also removes fibers, vitamins, and minerals.)
- Choose organically grown foods, which generally contain fewer pesticides.

‡‡‡ For answers to questions about pesticides, call the 24-hour National Pesticide Information Center: 1–800–858-PEST.

Figure 12–8

How Processing Affects Pesticide Residues

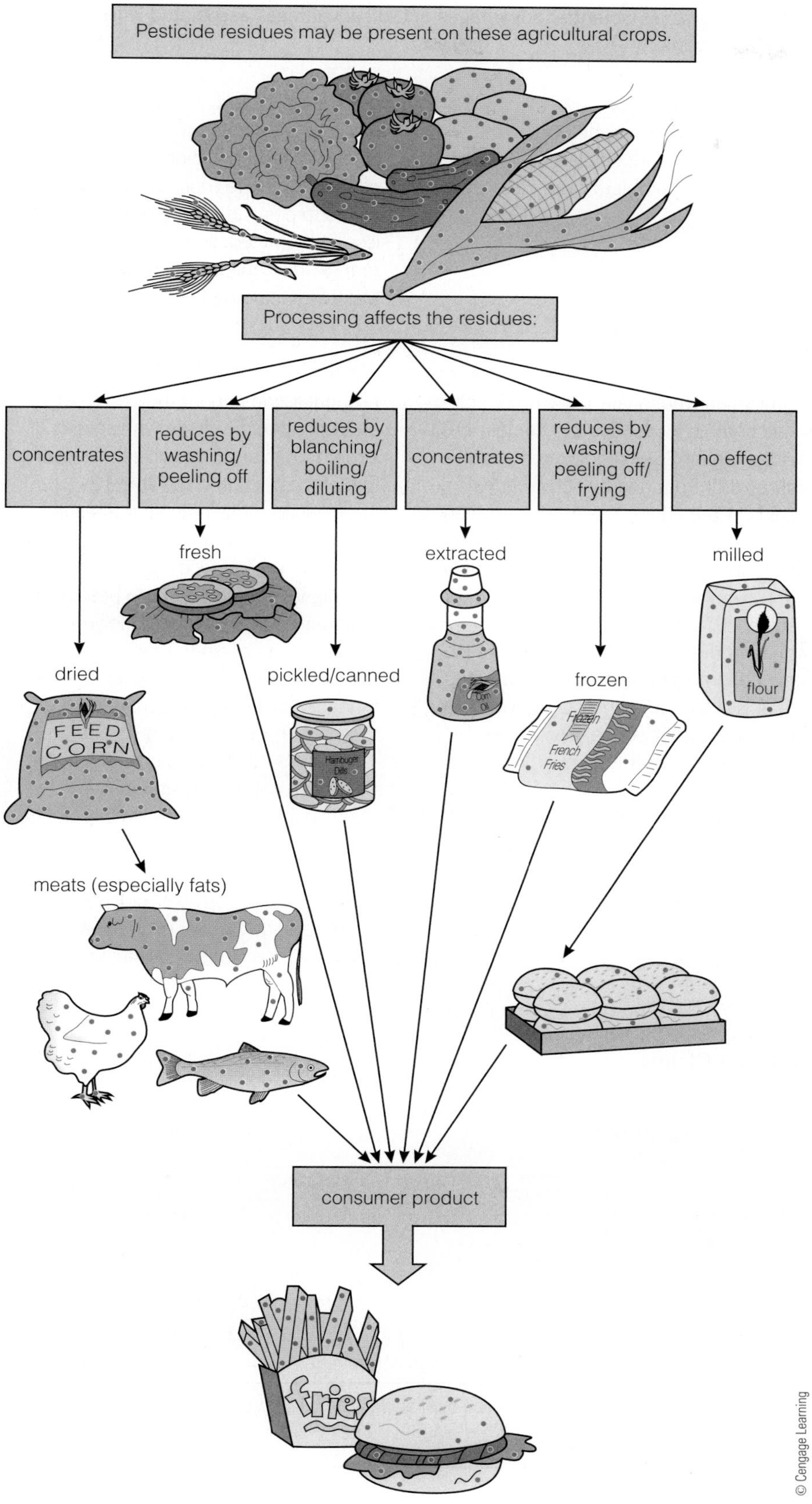

© Cengage Learning

The red dots in the figure represent pesticide residue left on foods from field spraying or postharvest application. Notice that most pesticides follow fats in foods and that some processing methods, such as washing and peeling vegetables, reduce pesticide concentrations, whereas others tend to concentrate them.

residues whatever remains; in the case of pesticides, those amounts that remain on or in foods when people buy and use them.

organic foods foods meeting strict USDA production regulations for *organic*, including prohibition of synthetic pesticides, herbicides, fertilizers, drugs, and preservatives and produced without genetic engineering or irradiation.

use it! A Consumer's Guide To . . .

Understanding Organic Foods

LO 12.5 Discuss potential advantages and disadvantages associated with organic foods.

Sales of certified organic foods have skyrocketed from under $4 billion in 1997 to $28.6 billion in 2011.[1] Even at a 10 to 40 percent higher price, organic foods appeal to consumers who believe that they are buying the freshest, best-tasting, most nutrient-packed, chemical-free, non-genetically modified foods available. Many people are also willing to pay extra for foods produced with little impact on the earth and with respect for animals. Are they getting what they are paying for?

Organic Rules

A U.S. farmer or manufacturer selling *certified organic* food must pass USDA inspections at every step of production, from the seed sown in the ground, through the making of compost for fertilizer, to the manufacturing and labeling of the final product. Figure 12–9 describes the meanings of organic food labels. In contrast, foods labeled with "natural," "free-range," "locally grown," or other wholesome-*sounding* words are not held to any standards to bear out such claims.

The National Organic Program develops, implements, and administers production, handling, and labeling standards for organic agricultural products. Enforcement has proved difficult, however, and compliance problems are common. Program officials are working to solve these problems and close open loopholes.

Pesticide Residues—They're Everywhere

When tested, organic foods generally contain no pesticides or lower levels than similar, conventionally grown products. The organic foods found to contain pesticides (about 25 percent) often acquire spray drift from nearby conventional fields or are contaminated by persistent chemicals in the soil.

Eating a diet of organic foods measurably reduces pesticide exposure. When scientists fed young children an organic diet, markers of pesticide exposure in the children's urine dropped dramatically after just five days and stayed low during the organic diet period. The markers rose again when the children resumed their normal diet.

Does this mean that an organic diet is better for health than a conventional diet? No strong scientific evidence suggests that conventional foods pose excess health risks or that using organic products reduces risks.[2] The typical pesticide exposure in the United States represents an amount 10,000 times below the level at which risks begin to rise. Children are more sensitive than adults to pesticides, and their risks are less well-defined, so parents may wish to reduce their children's exposure from all sources, including foods.

To Bean or Not To Bean

A popular consumer group advocates choosing organically grown varieties of certain fruits and vegetables (listed in the leftmost column of Table 12–12). Their list correctly reflects the results of federal tests for pesticide residues on produce—the foods they name test highest for one or more pesticide residues.[3] So far, so good. However, the group then goes on to urge consumers to choose organic varieties of these foods, implying that they can reduce their health risks by doing so. Table 12–12 tells the whole story—the health risks from eating conventional varieties of those foods are infinitesimally small.[4] The federal testing that generates these data ensures it.

Still, the risk from pesticide residues is not zero, and many people fear harm from unfamiliar chemicals applied to

Figure 12–9
USDA Seal and Organic Food Label Claims

A Food Meeting This Description...	Can Bear This on Its Label.
Made with exclusively 100% organic ingredients	USDA ORGANIC "100% Organic"
Made with at least 95% organic ingredients	USDA ORGANIC "Organic"
Made with at least 70% organic ingredients	"Made with organic ingredients" (May not use seal; may list up to three organic ingredients on front of package)
Made with less than 70% organic ingredients	(May not use seal and must make no claims on the front of the package; may list organic ingredients on side panel.)

Table 12–12

A Perspective on Conventional Produce Risks

The left column lists fruits and vegetables named by a consumer watchdog group as being most highly contaminated with pesticide residues.[a]

The middle column lists a scientific estimate of an average consumer's exposure to any pesticide residue detected in the foods, relative to the Reference Dose (formerly called *tolerance limit*) set by government agencies.[b]

The right column compares estimates of the human health risk from consuming these residues with that of the highest intake that still produces no observable harm in the most sensitive animal species, the No Observable Adverse Effect Level (NOAEL).

Example: a consumer eating an average number of conventionally grown strawberries in a year *may* have a residue exposure of 0.01%, posing a known risk to health one *million* times lower than the NOAEL.

Most Contaminated Produce List	Highest Exposure to Any Pesticide Relative to the Reference Dose	Human Risk Relative to No Observable Adverse Effect Level (NOAEL)
Apples	0.1% to 1%	100,000 to 10,000 times lower
Celery	0.1% to 1%	100,000 to 10,000 times lower
Sweet bell peppers	2.0%	5,000 times lower
Peaches	0.1% to 1%	100,000 to 10,000 times lower
Strawberries	0.01% to 0.1%	1,000,000 to 100,000 times lower
Nectarines (imported)	0.01% to 0.1%	1,000,000 to 100,000 times lower
Grapes	0.01% to 0.1%	1,000,000 to 100,000 times lower
Spinach	0.01% to 0.1%	1,000,000 to 100,000 times lower
Cucumbers	—[c]	
Blueberries (domestic)	0.001% to 0.01%	10,000,000 to 1,000,000 times lower
Kale/greens[d]	0.001% to 0.01%	10,000,000 to 1,000,000 times lower

[a] *The Dirty Dozen, published by the Environmental Working Group.*

[b] *Environmental Protection Agency Reference Dose: the estimated lifetime daily human exposure to a chemical that causes no appreciable harm to health.*

[c] *Data unavailable.*

[d] *May contain very low levels of pesticide residues of special concern.*

food in any amount.[5] Such worries are emotional, however, and not scientific, and they can needlessly put consumers in a bind. If people cannot afford organic foods but fear that conventional foods may harm them, they may limit the amount or variety of fruits and vegetables they take in. This choice increases health risks enormously.[6]

Nutrient Composition

Few nutrient differences exist between conventional and organic foods, and these generally fall within expected variations among food crops.[7] Small nutrient differences occur with varying soil types, soil nutrients, seasonal rainfall, or other factors. In contrast to nutrients, organic foods may be higher in certain phytochemicals.[8] This makes sense because plants, unassisted by pesticides, muster their own phytochemical defenses to ward off insects and other dangers. Even so, organic foods appear to offer no apparent nutrition-related benefits.[9]

The most meaningful nutrient comparisons are not between organic and conventional foods but between whole foods and heavily processed ones. Organic candy bars, soy desserts, and fried vegetable snack chips are no more nutritious (or less fattening) than ordinary treats. Likewise, organic main dishes laden with saturated fats and sodium can throw the health-seeking consumer off course.

Environmental Benefits

Growers of organic foods use *sustainable* agricultural techniques (see Chapter 15 and Controversy 15) that minimize harms to the environment. They add composted animal manure or vegetable matter instead of the synthetic, petroleum-based fertilizers that run off into waterways and pollute them. They battle pests and diseases with a pesticide derived from a bacterial toxin or by rotating crops each season, by introducing predatory insects to kill off pests, or by picking off large insects or diseased plant parts by hand.

Farmers and ranchers who sell organic eggs, dairy products, and meats must keep their animals in surroundings natural to their species with at least some access to the outdoors.[10] Animals raised this way can grow large and stay healthy without growth hormones, daily antibiotics, and the other drugs that become necessary when animals are stressed in overcrowded pens.[11] Without overcrowding, animal waste runoff, a threat to the nation's waterways is greatly reduced, too.

Organics' Potential Pitfalls

Foods contaminated with untreated manure or feces from fertilizer, runoff, or wild animals can harbor dangerous bacteria, but such contamination is equally likely to occur in organic foods and conventional foods. Proper composting of manure-based fertilizers eliminates disease-causing microbes.

Organic ingredients imported from other countries often cost less than domestic ingredients and so make attractive alternatives to dollar-conscious organic food manufacturers. However, overseas farms and producers are difficult to inspect and regulate, and although some adhere to strict standards, others are lax.[12] Also, shipping organic ingredients over long distances violates principles of sustainability.

Moving Ahead

The practical marketplace advice, based on science, is this: Buy safe, affordable conventionally grown fruits and vegetables, wash them well, and consume them with confidence. If you prefer the taste of organic fruits and vegetables, if you appreciate the care of animals and the environment associated with producing them, and if you can afford them, you can choose organics with equal confidence.

If you want organic foods at bargain prices, you might ask for imperfect produce at farmer's markets. Alternatively, try growing some leafy greens, herbs, and tomatoes in pots on a sunny deck—a surprisingly simple and rewarding endeavor. Whatever your choice, choose nutritious fruits and vegetables in abundance. The increased health risks from not doing so are substantial.

Review Questions*

1. To be labeled *100% organic*, a food must ________ .
   a. be inspected before it is sold
   b. contain at least 95% organic ingredients
   c. be labeled "natural" or "free range"
   d. contain only 100% organic ingredients
2. About what percentage of organic foods tested contain pesticide residue?
   a. 40 percent
   b. 25 percent
   c. 15 percent
   d. 10 percent
3. Organic candy bars, soy desserts, and fried vegetable snack chips ________ .
   a. are not more nutritious than ordinary treats
   b. are superior sources of nutrients for children
   c. are a less-fattening alternative to non-organic snack foods
   d. can provide an adequate daily intake of important organic minerals

** Answers to Consumer's Guide review questions are found in Appendix G.*

Regulation of Pesticides The EPA sets a **reference dose** for the maximum residue of an approved pesticide allowable in foods. Over 10,000 regulations set reference doses for the more than 300 pesticide chemicals approved for use on U.S. crops. These limits generally represent between 1/100th to 1/1,000th of the highest dose that still causes *no adverse health effects* in laboratory animals.[34] If a pesticide is misused, growers risk fines, lawsuits, and destruction of their crops.

While the EPA sets limits, both the USDA and FDA occasionally test crop and food product samples for compliance. Over decades of testing, seldom have these agencies found residues above approved limits, so it appears that pesticides are generally applied according to regulations. This makes sense because growers are not anxious to spend extra capital on unneeded chemicals.

Pesticide-Resistant Insects Ironically, some pesticides also promote the survival of the very pests they are intended to wipe out. A pesticide aimed at certain insects may kill almost 100 percent of them, but because of the genetic variability of large populations, a few hardy individuals survive exposure. These resistant insects then multiply free of competition and soon produce offspring with inherited pesticide resistance that attack the crop with enhanced vigor. Controlling resistant insects requires application of different pesticides, which leads to the emergence of a population of insects that survive multiple pesticides. The same biological sequences

My Turn **watch it!**

Organic: Does It Matter?

Two students talk about the pros and cons of choosing organic foods.

© Cengage Learning

Lisa

© Cengage Learning

Jennifer

occur when herbicides and fungicides are repeatedly applied to weeds and fungal pests. One alternative to this destructive series of events is to manage pests using a combination of improved farming techniques and biological controls, as discussed in Controversy 15.

Natural Pesticides Pesticides are not produced only in laboratories; they also occur in nature. The nicotine in tobacco and phytochemicals of celery are examples.§§§ Another is a biodegradable peptide pesticide made by a common soil bacterium that is approved for use in **organic gardens**. (*Peptide* refers to bonds that join amino acids—see Chapter 6.) Peptide pesticides are less damaging to other living things and leave less **persistent** residues in the environment than most others. An ideal pesticide would destroy pests in the field but vanish before consumers ate the food.

© Dario Sabljak/Shutterstock.com

Overcrowding of farm animals makes infection likely.

KEY POINTS

- Pesticides can be part of a safe food production process but can also be hazardous if mishandled.
- Consumers can take steps to minimize their ingestion of pesticide residues in foods.

Animal Drugs—What Are the Risks?

Consumers often worry about consuming meats that may contain hormones, antibiotics, and drugs that contain **arsenic** compounds. However, the most pressing concern to scientists is the emergence and rapid spread of bacterial strains that no longer respond to antibiotic drugs.[35]

Livestock and Antibiotic-Resistant Microbes For a half-century, ranchers and farmers have dosed livestock with antibiotic drugs as part of a daily feeding regimen to ward off infections common in animals living in crowded conditions. These drugs also speed up animal growth and increase feed efficiency. When bacteria too frequently encounter antibiotics, they adapt, losing their sensitivity to the drugs over time.[36] The resulting **antibiotic-resistant bacteria** cause severe infections that do not yield to standard antibiotic therapy, often ending in fatality.

A substantial threat to human health and life arises from antibiotic-resistant bacteria. A limited number of antibiotic drugs exist—the same or related drugs used daily in livestock are also of critical importance for treating illnesses in people. Few treatment options remain for people who become infected with antibiotic-resistant bacteria. So long as antibiotics are overused in animals and people, new resistant strains can be expected to emerge, and once here, they tend to stay.

New federal voluntary guidelines urge farmers to use antibiotics only under veterinary care and only to prevent, control, or treat diseases, but these protections are

§§§ The celery plant produces psoralens that repel insects.

reference dose an estimate of the intake of a substance over a lifetime that is considered to be without appreciable health risk; for pesticides, the maximum amount of a residue permitted in a food. Formerly called *tolerance limit.*

organic gardens gardens grown with techniques of *sustainable agriculture*, such as using fertilizers made from composts and introducing predatory insects to control pests, in ways that have minimal impact on soil, water, and air quality.

persistent of a stubborn or enduring nature; with respect to food contaminants, the quality of remaining unaltered and unexcreted in plant foods or in the bodies of animals and human beings.

arsenic a poisonous metallic element. In trace amounts, arsenic is believed to be an essential nutrient in some animal species. Arsenic is often added to insecticides and weed killers and, in tiny amounts, to certain animal drugs.

antibiotic-resistant bacteria bacterial strains that cause increasingly common and potentially fatal infectious diseases that do not respond to standard antibiotic therapy. An example is MRSA (pronounced MER-suh), a multi drug-resistant *Staphyloccocus aureus* bacterium.

not mandatory so no one can predict their effectiveness.[37] One day, new drugs and vaccines now under development may reduce the need for antibiotics in food animals, but progress is slow.[38] Meanwhile, current food production methods threaten to undo a true medical miracle.

"Trends toward increasing numbers of infection and increasing drug resistance show no sign of abating."

—Joshua Sharfstein, principal deputy commissioner of the FDA, 2009

Growth Hormone in Meat and Milk Cattle producers in the United States commonly inject their herds with a form of **growth hormone—recombinant bovine somatotropin (rbST)**—to increase lean tissue growth, increase milk production, and reduce feed requirements. The hormone, produced by genetically altered bacteria, is identical to growth hormone made in the pituitary gland of the animal's brain. The FDA and WHO deem the use of the drug to be safe, and the FDA does not require testing of food products for traces of it.

Genetic alteration of bacteria and food products is discussed in **Controversy 12.**

Ranchers advocate the use of rbST because more meat and milk on less feed means higher profits. The environment may profit as well. Smaller herds that eat sparingly require less cleared land and fewer resources used to produce and transport feed.

Consumer groups counter with concerns about the safety of another bovine hormone stimulated by rbST, insulin-like growth factor I (IGF-I).[39] However, tests of conventional milk, hormone-free milk, and organic milk reveal no differences in terms of antibiotic, bacteria, hormone, or nutrient contents.

Arsenic in Food Animals The USDA regulates administration of arsenic, a naturally occurring element from the earth's crust and an infamous poison, to food animals. Conventionally raised poultry flocks receive an arsenic-based drug to control parasites that would otherwise stall their growth. Arsenic from the drug accumulates in their meat, wastes, and feathers, which in turn contributes arsenic to the food supply.[40] A surprising source of arsenic is rice: both refined white rice and whole-grain brown rice absorb more arsenic from soil and water than do other grains. The FDA is evaluating ways to limit human arsenic exposure from rice and rice products.[41] Arsenic is also present in foods such as fish, eggs, and milk products, and in drinking water.[42] The FDA and USDA agree that the level of arsenic in foods poses little risk to healthy consumers who consume ordinary portions and vary their food choices.

KEY POINTS

- FDA-approved hormones, antibiotics, and other drugs are used to promote growth or increase milk production in conventionally grown animals.
- Antibiotic resistant bacteria pose a serious and growing threat.

Environmental Contaminants

As world populations increase and become more industrialized, concerns grow about contamination of foods. A **food contaminant** is anything in food that does not belong there.

Harmfulness of Contaminants The potential for harm from a contaminant depends partly on how long it lingers in the environment or in the human body—that is, on how *persistent* it is. Some contaminants are short-lived because microorganisms, sunlight, or oxygen break them down. Some contaminants stay in the body for only a short time because the body rapidly excretes or destroys them. Such contaminants present little cause for concern.

Other contaminants linger and resist environmental breakdown, and they interact with the body's systems without being metabolized or excreted. These contaminants can pass from one species to the next and accumulate at higher concentrations in each level of the food chain, a process called **bioaccumulation**—see Figure 12–10.

The toxic effect of a chemical depends largely upon two factors: the degree of the chemical's **toxicity** and the degree of human exposure.[43] In small enough amounts, even poisonous substances may be tolerable and of no consequence to health; in larger amounts, even innocuous substances may be dangerous. An old saying, "The dose makes the poison," means that with a large enough dose, even normally benign

growth hormone a hormone (somatotropin) that promotes growth and that is produced naturally in the pituitary gland of the brain.

recombinant bovine somatotropin (so-mat-oh-TROPE-in) **(rbST)** growth hormone of cattle, which can be produced for agricultural use by genetic engineering. Also called *bovine growth hormone (bGH).*

food contaminant any substance occurring in food by accident; any food constituent that is not normally present.

bioaccumulation the accumulation of a contaminant in the tissues of living things at higher and higher concentrations along the food chain.

toxicity the ability of a substance to harm living organisms. All substances, even pure water or oxygen, can be toxic in high enough doses.

Figure 12–10

Bioaccumulation of Toxins in the Food Chain

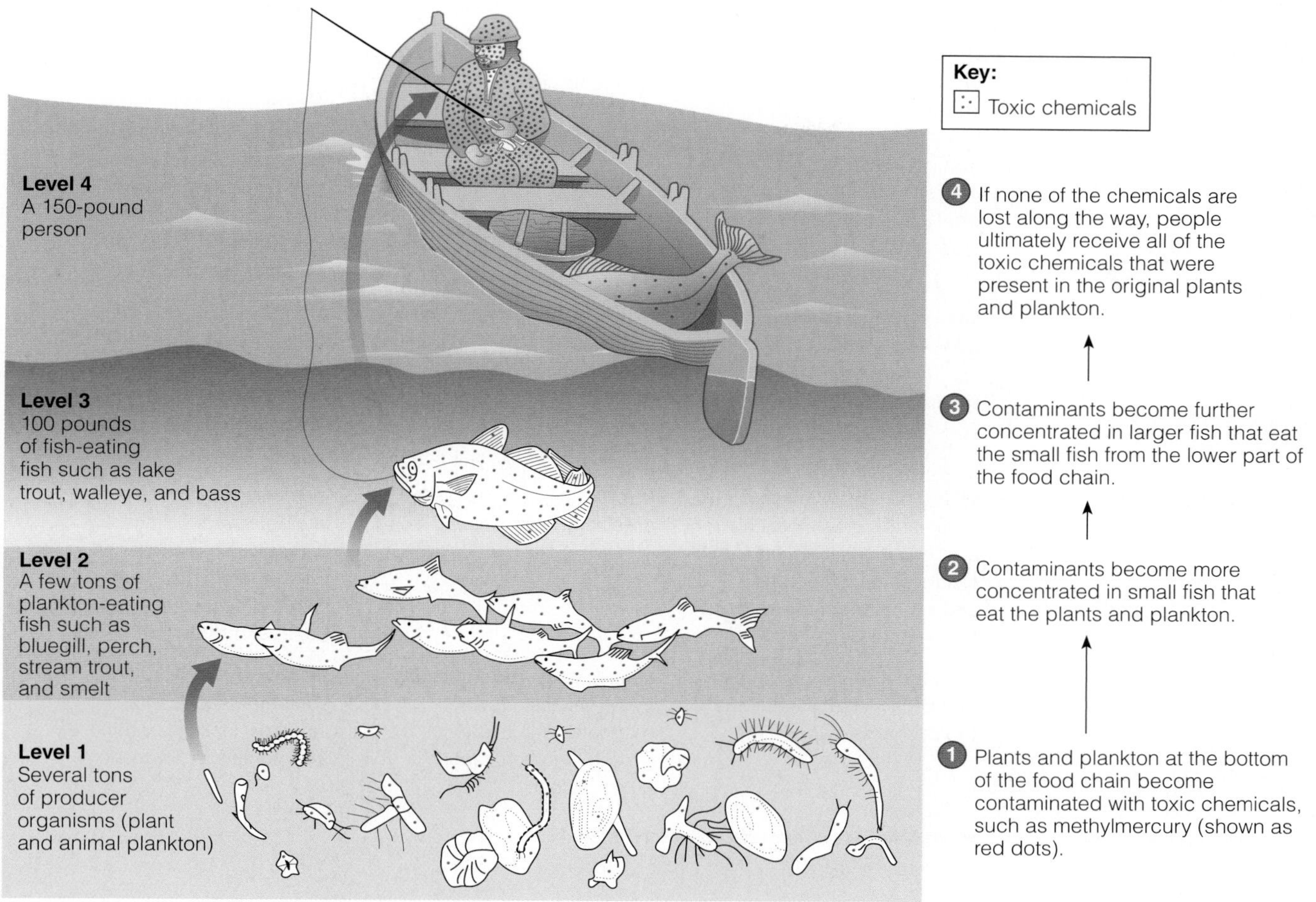
Key:
Toxic chemicals
Level 4
A 150-pound person
Level 3
100 pounds of fish-eating fish such as lake trout, walleye, and bass
Level 2
A few tons of plankton-eating fish such as bluegill, perch, stream trout, and smelt
Level 1
Several tons of producer organisms (plant and animal plankton)
4 If none of the chemicals are lost along the way, people ultimately receive all of the toxic chemicals that were present in the original plants and plankton.
3 Contaminants become further concentrated in larger fish that eat the small fish from the lower part of the food chain.
2 Contaminants become more concentrated in small fish that eat the plants and plankton.
1 Plants and plankton at the bottom of the food chain become contaminated with toxic chemicals, such as methylmercury (shown as red dots).

substances, even sand, can kill a person. The reverse is also true: even poisons can be benign in miniscule doses.

How much of a threat do environmental contaminants pose to the food supply? It depends on the contaminant. In general, the threat remains small because the FDA monitors contaminants in foods and issues warnings when food contamination is evident. Table 12–13, p. 492, describes a few contaminants of greatest concern in foods.

Mercury in Seafood Mercury, **PCBs**, chlordane, dioxins, and DDT are the contaminants most often a problem in food fish species worldwide, but the **heavy metal** mercury leads the list by threefold.[44] Scientists learned of mercury's potential for harm through tragedy. In the mid-20th century, more than 120 people, including 23 infants, in Minamata, Japan, became ill with a strange disease. Mortality was high, and the survivors suffered progressive, irreversible blindness, deafness, loss of coordination, and severe mental and physical retardation.****

Finally, the cause of this misery was discovered: manufacturing plants in the region were discharging mercury into the waters of the bay, where aquatic bacteria metabolized it into the nerve poison methylmercury. The fish in the bay were accumulating the poison in their bodies, and townspeople who regularly ate fish from the bay fell ill. The infants had not eaten any fish, but their mothers had during their pregnancies; the mothers were spared because the poison concentrates in the tissues of the fetus.

**** Minamata disease was named for the location of the disaster.

PCBs (polychlorinated biphenyls) stable oily synthetic chemicals once used in hundreds of U.S. industrial operations that persist today in underwater sediments and contaminate fish and shellfish. Now banned from use in the United States, PCBs circulate globally from areas where they are still in use. PCBs cause cancer, nervous system damage, immune dysfunction, and a number of other serious health effects.

heavy metal any of a number of mineral ions such as mercury and lead, so called because they are of relatively high atomic weight; many heavy metals are poisonous.

Table 12–13

Examples of Contaminants in Foods

Name and Description	Sources	Toxic Effects	Typical Route to Food Chain
Cadmium (heavy metal)	Used in industrial processes including electroplating, plastics, batteries, alloys, pigments, smelters, and burning fuels. Present in cigarette smoke and in smoke and ash from volcanic eruptions.	No immediately detectable symptoms; slowly and irreversibly damages kidneys and liver.	Enters air in smokestack emissions, settles on ground, absorbed into food plants, consumed by farm animals, and eaten in vegetables and meat by people. Sewage sludge and fertilizers leave large amounts in soil; runoff contaminates shellfish.
Lead[a] (heavy metal)	Lead crystal decanters and glassware, painted china, old house paint, batteries, pesticides, old plumbing.	Displaces calcium, iron, zinc, and other minerals from their sites of action in the nervous system, bone marrow, kidneys, and liver, causing failure of function.	Originates from industrial plants and pollutes air, water, and soil. Still present in soil from many years of leaded gasoline use.
Mercury (heavy metal)	Widely dispersed in gases from earth's crust; local high concentrations from industry, electrical equipment, paints, and agriculture; present in most U.S. waterways.	Poisons the nervous system, especially in fetuses.	Inorganic mercury released into waterways by industry and acid rain is converted to methylmercury by bacteria and ingested by food species of fish (tuna, swordfish, and others).
Polychlorinated biphenyls (PCBs) (organic compounds)	No natural source; produced for use in electrical equipment (transformers, capacitors).	Long-lasting skin eruptions, eye irritations, growth retardation in children of exposed mothers, anorexia, fatigue, others.	Discarded electrical equipment; accidental industrial leakage or reuse of PCB containers for food.

[a]For answers to questions concerning lead, call the National Lead Information Center at (800) 424-LEAD.

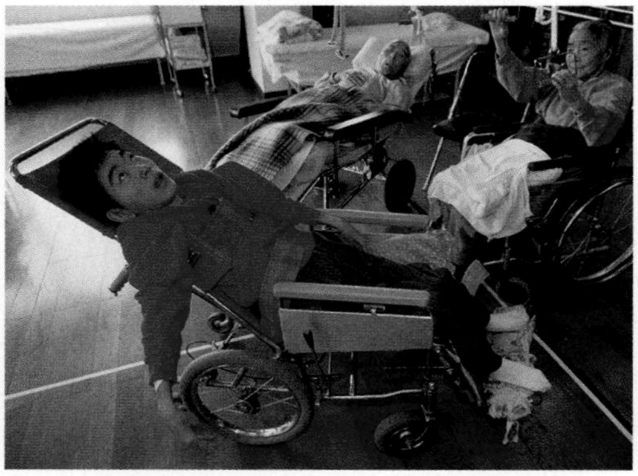

Minimata disease. The effects of mercury contamination can be severe.

Today, in the United States, the FDA and the EPA warn of unacceptably high methylmercury levels in our nation's oceans, freshwater lakes, and streams; they issue advisories urging consumers to limit consumption of many food fish species, particularly large fish that live in the most mercury-contaminated waters.[45] Today, mercury is building up in human tissues, too, posing potential risks to the brain, nerves, and other tissues.[46] The FDA advises all pregnant women, women who may become pregnant, nursing mothers, and young children to avoid certain marine fish species known to be high in methylmercury (Chapter 5 provides details of this warning) and to limit consumption of other mercury-containing species, as well. In addition to mercury, freshwater fish also often contain PCBs, DDT, and other persistent contaminants; check local advisories.

No one expects the tragic results of the 1950s to occur again, but lower doses of methylmercury have been associated with headaches, fatigue, memory loss, impaired ability to concentrate, and muscle or joint pain in adults. In children, the threats may be greater and longer-lasting. Methylmercury is persistent in the environment, so today's efforts to reduce pollution of ocean, lake, and river waters will take years to be effective.

KEY POINTS

- Persistent environmental contaminants present in food pose a small but significant risk to U.S. consumers.
- Mercury and other contaminants pose the greatest threat during pregnancy, lactation, and childhood.

Table 12–14

Selected Food Additives and Their Functions

Agent Type	Function in Foods	Examples
Antimicrobial agents (preservatives)	Prevent food spoilage by mold or bacterial growth.	Acetic acid (vinegar), benzoic acid, nitrates and nitrites, proprionic acid, salt, sugar, sorbic acid.
Antioxidants (preservatives)	Prevent oxidative changes and delay rancidity of fats; prevent browning of fruit and vegetable products.	BHA, BHT, propyl gallate, sulfites, vitamin C, vitamin E.
Artificial colors	Add color to foods.	Certified food colors such as dyes from vegetables (beet juice or beta-carotene) or synthetic dyes (tartrazine and others).
Artificial flavors, flavor enhancers	Add flavors; boost natural flavors of foods.	Amyl acetate (artificial banana flavor), artificial sweeteners, MSG (monosodium glutamate), salt, spices, sugars.
Bleaching agents	Whiten foods such as flour or cheese.	Peroxides.
Chelating (KEE-late-ing) agents (preservatives)	Prevent discoloration, off flavors, and rancidity.	Citric acid, malic acid, tartaric acid (cream of tartar).
Nutrient additives	Improve nutritional value.	Vitamins and minerals.
Stabilizing and thickening agents	Maintain emulsion, foams, or suspensions or lend a desirable thick consistency to foods.	Dextrins (short glucose chains), pectin, starch, or gums such as agar, carrageenan, guar, locust bean, and other gums.

Are Food Additives Safe?

LO 12.6 Name some functions served by food additives approved for use in the United States, and provide evidence concerning their safety.

Many foods contain **additives**, and consumers rightly want to know why they are there and if they are safe to consume. On FDA's list of food worries, food additives rank low. There are 3,000 or so food additives approved for use in the United States, and most are strictly controlled and well studied for safety. In fact, many food additives stop mold and bacterial growth, thereby improving food safety. Often, food additives give foods desirable characteristics: color, flavor, texture, stability, enhanced nutrient composition, or resistance to spoilage and enhanced safety. Some common classes of additives and their functions in foods are listed in Table 12–14.

Without additives, bread would quickly mold and salad dressing would go rancid.

Regulations Governing Additives

Before using a new additive in food products, a manufacturer must test the additive and satisfy the FDA that:

- It is effective (it does what it is supposed to do).
- It can be detected and measured in the final food product.

Then the manufacturer must provide proof that it is safe (causes no birth defects or other injuries) when fed in large doses to experimental animals.[47] Finally, the manufacturer must submit all test results to the FDA for approval. The whole process may take several years. Then, manufacturers must comply with a host of other regulations that ensure the proper use and application of the additive, as well. For example, additives may *not* be used to disguise faulty or inferior products, nor to deceive consumers, nor in any application where they significantly destroy nutrients in foods.

The GRAS List Many additives are exempted from complying with this procedure because they have been used for a long time and their use entails no known hazards. Some 700 substances are on the **generally recognized as safe (GRAS) list**. No additives are permanently approved, however; all are periodically reviewed.

additives substances that are added to foods but are not normally consumed by themselves as foods.

generally recognized as safe (GRAS) list a list, established by the FDA, of food additives long in use and believed to be safe.

The Margin of Safety An important distinction between toxicity and hazard arises during evaluation of an additive's safety. Toxicity is a general property of all substances; hazard is the capacity of a substance to produce injury *under conditions of its use*.[††††] As mentioned, all substances can be toxic at some level of consumption, but they are called hazardous only if they are toxic in the amounts ordinarily consumed. To determine risk, experimenters feed test animals the substance at different concentrations throughout their lifetimes.

An approved food additive has a wide **margin of safety**. Most additives that involve risk are allowed in foods only at concentrations at least 100 times lower than the highest concentration at which the risk is still zero (1/100). Some *natural* toxins produced in food by plants occur at levels that bring their margins of safety close to 1/10. The margin of safety range for vitamins A or D is 1/25 to 1/40; it may be less than 1/10 in infants. For some trace elements, it is about 1/5. People commonly consume table salt in daily amounts only three to five times less than those that cause serious toxicity.

Risks and Benefits of Food Additives Most additives used in foods offer benefits that may outweigh their risks or that make the risks worth taking. In the case of color additives that only enhance the appearance of foods without improving their health value or safety, no amount of risk may be deemed worth taking. Only 10 of an original 80 synthetic color additives are still approved by the FDA for use in foods, and screening of these substances continues. A few food additives receive the most publicity, and consumers have questions about them. The following sections address these additives.

© Gerald Bernard/Shutterstock.com

Sugar and salt: two long-used preservatives.

KEY POINTS

- Food additives must be safe, effective, and measurable in the final product for FDA approval.
- Approved additives have wide margins of safety.

Additives to Improve Safety and Quality

Some additives improve food safety. They restrict bacterial growth or otherwise enhance food quality in ways many people take for granted.[48]

Salt and Sugar Since before the dawn of history, salt has been used to preserve meat and fish; sugar, a relative newcomer to the food supply, serves the same purpose in jams, jellies, and canned and frozen fruits. Both salt and sugar work by withdrawing water from the food; microbes cannot grow without sufficient moisture. Safety questions surrounding these two preservatives center on their overuse as flavoring agents—salt and sugar make foods taste delicious and are often added with a liberal hand. Chapters 4 and 8 provided detailed discussions of these issues.

Nitrites The *nitrites* added to meats and meat products help to preserve their color (especially the pink color of hot dogs and other cured meats) and to inhibit rancidity and thwart bacterial growth. In particular, nitrites prevent the growth of the deadly *Claustridium botulinum* bacterium. Even though nitrites are useful, they raise safety issues. Once in the stomach, nitrites can be converted to nitrosamines, chemicals linked with colon cancer in animals. Other nitrite sources, such as tobacco and beer, may be more significant sources of nitrosamine-related compounds than foods. Still, processed meats are associated with an elevated risk of colon cancer, so the cautious consumer limits his or her intakes.[49]

Did You Know?

Salt and sugar are antimicrobial additives that work by depriving microorganisms of moisture.

Sulfites Sulfites prevent oxidation in many processed foods, in alcoholic beverages (especially wine), and in drugs. Some people experience dangerous allergic reactions to the sulfites, and so their use is strictly controlled. The FDA prohibits sulfite use on food meant to be eaten raw (fresh grapes are an exception), and it requires

---

[††††] The Delaney Clause, a legal requirement of zero cancer risk for additives, is no longer universally applied.

foods and drugs to list on their labels any sulfites that are present. For most people, sulfites do not pose a hazard in the amounts used in products, but they have one other drawback. Because sulfites can destroy a lot of thiamin in foods, you can't count on a food that contains sulfites to contribute to your daily thiamin intake.

KEY POINTS

- Sugar and salt have the longest history of use as additives to prevent food spoilage.
- Nitrites and sulfites have advantages and drawbacks.

Flavoring Agents

Many additives add desirable flavors to foods. One group, the **nonnutritive sweeteners**, may be added by manufacturers or by consumers at home.

Nonnutritive Sweeteners Nonnutritive sweeteners make foods taste sweet without promoting dental decay or providing the empty calories of sugar. The human taste buds perceive many of them as supersweet, so just tiny amounts are added to foods to achieve the desired sweet taste.[50] The FDA endorses the use of nonnutritive sweeteners as safe over a lifetime when used within **acceptable daily intake (ADI)** levels. Table 12–15, p. 496, provides some details about the nonnutritive sweeteners, including ADI levels.

Whether the use of nonnutritive sweeteners promotes weight loss or improves health by reducing total calorie intakes is not known with certainty.[51] Some research even suggests that their use may *promote* weight gain through unknown mechanisms; these are topics of current research.

Through the years, questions have emerged about the safety of nonnutritive sweeteners, particularly saccharine and aspartame. For example, early research indicated that large quantities of saccharin caused bladder tumors in laboratory animals, but these issues have since been resolved. Overloading on huge saccharin doses is probably not safe but consuming moderate amounts poses no known hazard.

Aspartame, a sweetener made from two amino acids (phenylalanine and aspartic acid) is one of the most thoroughly studied food additives ever approved, and scientific evidence linking it with diseases such as cancer is weak or nonexistent.[52] However, aspartame's phenylalanine base poses a threat to those with the inherited disease phenylketonuria (PKU), a disease that, without a low phenylalanine diet, can damage the developing brain in children. Food labels warn people with PKU of the presence of phenylalanine in aspartame-sweetened foods (see Figure 12–11, p. 497). In any case, artificially sweetened foods and drinks have no place in the diets of even healthy infants or toddlers. People who believe a sweetener gives them symptoms should use a different sweetener.

Monosodium Glutamate (MSG) MSG, the sodium salt of the amino acid glutamic acid, is used widely in restaurants, especially Asian restaurants.[‡‡‡‡] In addition to enhancing other flavors, MSG itself presents a basic taste (termed *umami*) independent of the well-known sweet, salty, bitter, and sour tastes.[53]

In a few sensitive individuals, MSG produces adverse reactions known as the **MSG symptom complex**. Plain broth with MSG seems most likely to bring on symptoms in sensitive people, whereas carbohydrate-rich foods, such as rice or noodles, seem to protect against them. MSG, deemed safe for adults, is prohibited in baby foods because very large doses destroy brain cells in developing mice, and the brains of human infants cannot fully exclude such substances. Consumers can easily identify foods that contain MSG because, with the exception of additives used in fresh meats, the FDA requires that food labels disclose each additive by its full name.

KEY POINTS

- People with PKU should avoid the nonnutritive sweetener aspartame.
- The flavor enhancer MSG may cause reactions in people with sensitivities to it.

‡‡‡‡ The MSG trade name is Accent.

margin of safety in reference to food additives, a zone between the concentration normally used and that at which a hazard exists. For common table salt, for example, the margin of safety is 1/5 (five times the amount normally used would be hazardous).

nonnutritive sweeteners sweet-tasting synthetic or natural food additives that offer sweet flavor but with negligible or no calories per serving; also called *artificial sweeteners, intense sweeteners, noncaloric sweeteners,* and *very low-calorie sweeteners*. Also defined in Chapter 4.

acceptable daily intake (ADI) the estimated amount of a sweetener that can be consumed daily over a person's lifetime without any adverse effects.

MSG symptom complex the acute, temporary, and self-limiting reactions, including burning sensations or flushing of the skin with pain and headache, experienced by sensitive people upon ingesting a large dose of MSG. Formerly called *Chinese restaurant syndrome*.

Table 12–15

U.S.-Approved Nonnutritive Sweeteners

Sweetener	Chemical Composition	Digestion/ Absorption	Sweetness Relative to Sucrose[a]	Energy (cal/g)	Acceptable Daily Intake (ADI) and (Estimated Equivalent[b])	Approved Uses
Acesulfame potassium or acesulfame-K (Sunette, Sweet One)	Potassium salt	Not digested or absorbed	200	0	15 mg/kg body weight[c] (30 cans diet soda)	General use, except in meat and poultry Tabletop sweeteners Heat stable
Aspartame (NutraSweet, Equal, others)	Amino acids (phenylalanine and aspartic acid) and a methyl group	Digested and absorbed	180	4[d]	50 mg/kg body weight[e] (18 cans diet soda)	General use in all foods and beverages; warning to population with PKU Degrades when heated
Luo han guo	Curcurbine, glycosides from monk fruit extract	Digested and absorbed	150–300	1	No ADI determined	GRAS[f]; general use as a food ingredient and table top sweetener
Neotame	Aspartame with an additional side group attached	Not digested or absorbed	7,000	0	18 mg/day	General use, except in meat and poultry
Saccharin (SugarTwin, Sweet'N Low, others)	Benzoic sulfimide	Rapidly absorbed and excreted	300	0	5 mg/kg body weight (10 packets of sweetener)	Tabletop sweeteners, wide range of foods, beverages, cosmetics, and pharmaceutical products
Stevia (Sweetleaf, Truvia, PurVia)	Glyosides found in the leaves of the *Stevia rebaudiana* herb	Digested and absorbed	200–300	0	4 mg/kg body weight	GRAS[f]; tabletop sweeteners, a variety of foods and beverages
Sucralose (Splenda)	Sucrose with Cl atoms instead of OH groups	Not digested or absorbed	600	0	5 mg/kg body weight (6 cans diet soda)	Baked goods, carbonated beverages, chewing gum, coffee and tea, dairy products, frozen desserts, fruit spreads, salad dressing, syrups, tabletop sweeteners
Tagatose[g] (Nutralose, Nutrilatose, Tagatesse)	Monosaccharide similar in structure to fructose; naturally occurring or derived from lactose	Not well absorbed	0.9	1.5	7.5 g/day	GRAS[f]; bakery products, beverages, cereals, chewing gum, confections, dairy products, dietary supplements, energy bars, tabletop sweeteners

[a]*Relative sweetness is determined by comparing the approximate sweetness of a sugar substitute with the sweetness of pure sucrose, which has been defined as 1.0. Chemical structure, temperature, acidity, and other flavors of the foods in which the substance occurs all influence relative sweetness.*

[b]*Based on a person weighing 70 kg (154 lb).*

[c]*Recommendations from the World Health Organization limit acesulfame-K intake to 9 mg/kg of body weight per day.*

[d]*Aspartame provides 4 cal/g, as does protein, but because so little is used, its energy contribution is negligible. In powdered form, it is sometimes mixed with lactose, however, so a 1-g packet may provide 4 cal.*

[e]*Recommendations from the World Health Organization and in Europe and Canada limit aspartame intake to 40 mg/kg of body weight per day.*

[f]*Generally Recognized As Safe.*

[g]*Tagatose is a poorly digested sugar and technically not a nonnutritive sweetener.*

Figure 12–11

Nonnutritive Sweeteners on Food Labels

INGREDIENTS: ARTIFICIAL AND NATURAL FLAVORING, TITANIUM DIOXIDE (COLOR), ASPARTAME, ACESULFAME POTASSIUM, STEVIA.
PHENYLKETONURICS: CONTAINS PHENYLALANINE.

Products containing aspartame must carry a warning for people with phenylketonuria.

This partial ingredient list is for a sugar-free food.

Nutrition Facts

Serving Size 8 oz
Servings 6
Calories 0

Amount per serving	% DV*
Total Fat 0g	0%
Sodium 0mg	0%
Total Carb. 0g	0%
Sugars 0g	
Protein 0g	

*Percent Daily Values (DV) are based on a 2,000 calorie diet.

Not a significant source of other nutrients.

Products containing less than 0.5 g of sugar per serving can claim to be "sugarless" or "sugar-free."

Fat Replacers and Artificial Fats

Fat replacers and artificial fats, introduced in Chapter 5, are ingredients that provide some of the taste, texture, and cooking qualities of fats, but with fewer or no calories (see Table 12–16, p. 498). Some fat replacers are derived from carbohydrate, protein, or fat, and these provide a few calories (but fewer than the fats they replace). Carbohydrate-based fat replacers are used primarily as thickeners or stabilizers in foods such as soups and salad dressings. Protein-based fat replacers provide a creamy feeling in the mouth and are often used in foods such as ice creams and yogurts. Fat-based replacers act as emulsifiers and are heat stable, making them most versatile in shortenings used in cake mixes and cookies.

An artificial fat used to make some low-fat snack foods, such as potato chips, is **Olestra**. Digestive enzymes cannot break its chemical bonds, so Olestra cannot be absorbed. Olestra binds fat-soluble vitamins and phytochemicals causing their excretion; to partly prevent these losses, manufacturers saturate Olestra with vitamins A, D, E, and K. Large doses can cause digestive distress, but no serious problems are known to have occurred with normal use.

KEY POINTS

- Fat replacers and artificial fats reduce the fat calories in processed foods, and the FDA deems them safe to consume.
- Olestra in large amounts can cause digestive distress.

Incidental Food Additives

Consumers are often unaware that many substances can migrate into food during production, processing, storage, packaging, or consumer preparation. These substances, although called indirect or **incidental additives**, are really contaminants because no one intentionally adds them to foods. Examples of incidental additives

Olestra a nonnutritive artificial fat made from sucrose and fatty acids; formerly called *sucrose polyester*.

incidental additives substances that can get into food not through intentional introduction, but as a result of contact with the food during growing, processing, packaging, storing, or some other stage before the food is consumed. Also called *accidental* or *indirect additives*.

Table 12–16

A Sampling of Fat Replacers

For comparison, remember that fat has 9 calories per gram.

Fat Replacers	Energy (cal/g)
Carbohydrate-Based Fat Replacers	
■ *Fruit* purees and pastes; add bulk and tenderness to baked goods.	1–4
■ *Maltodextrins* made from corn; powdered and flavored to resemble butter.	1–4
Fiber-Based Fat Replacers	
■ *Gels* derived from cellulose or starch to mimic the texture of fats in fat-free margarine and other products.	0–4[a]
■ *Gums* extracted from beans, sea vegetables, or other sources. Used to thicken salad dressings and desserts.	0–4
■ *Oatrim* derived from oat fiber; has the added advantage of providing satiety; lends creaminess to many foods.	4
■ *Z-trim* a modified form of insoluble fiber; is powdered and feels like fat in the mouth; lends creaminess to many foods.	0
Fat-Based Replacers	
■ *Olestra*[b] a noncaloric artificial fat made from sucrose and fatty acids; formerly called *sucrose polyester*; used for frying and cooking snacks and crackers.	0
■ *Salatrim*[c] derived from fat and contains short- and long-chain fatty acids; can be used in baking but not frying.	5
Protein-Based Fat Replacers	
■ *Microparticulated protein*[d] proteins of milk or egg white processed into mistlike particles that feel and taste like fat. Not suitable for frying.	4

[a] *Energy made available by action of colonic bacteria.*
[b] *Trade name: Olean. Not available in Canada.*
[c] *Trade name: Benefat.*
[d] *Trade names: Simplesse and K-Blazer.*

include compounds released from plastics, tiny bits of glass, paper, metal, and the like from packages, or chemicals from processing, such as the solvent used to decaffeinate some coffees. Incidental additives are well regulated and, once discovered in food, their safety must be confirmed by strict procedures like those governing intentional additives.

BPA The incidental additive BPA migrates into many foods and beverages from plastic-lined food cans, soft drink cans, baby formula containers, and certain clear, hard plastic water bottles.[§§§§] FDA recently banned BPA from baby bottles and toddler "sippy" cups because some studies have raised questions about potential disease risks.[54]

BPA is rapidly broken down by the human body, and exposures are far lower than once feared, so FDA generally supports its safety but is continuing to investigate its

[§§§§] BPA is an abbreviation of bisphenol A, a plastic hardener and component of epoxy resin.

effects on children and adults.[55] Anyone wishing to limit exposure to BPA should avoid hard, clear plastic reusable water bottles stamped with recycle codes 3 or 7, indicating plastics that may contain BPA. In addition:

- Do not use hard clear plastic containers for very hot foods or beverages, do not wash them in the dishwasher, and do not heat them in the microwave because heat releases BPA from plastics.[56]
- Discard scratched hard plastic bottles, which can harbor bacteria and also release BPA.

Microwave Packages Some microwave products are sold in "active packaging" that participates in cooking the food. Pizza, for example, may rest on a cardboard pan coated with a thin film of metal that absorbs microwave energy and may heat up to 500°F (260°C). During the intense heat, some particles of the packaging components migrate into the food. This is expected; the particles have been tested for safety.

In contrast, incidental additives from plastic packages may not be entirely safe for consumption. To avoid them, do not reuse disposable plastic margarine tubs or single-use trays from microwavable meals for microwaving other foods. Use glass or ceramic containers or plastic ones labeled as safe for the microwave. In addition, wrap foods in microwave-safe plastic wraps, waxed paper, cooking bags, parchment paper, or white microwave-safe paper towels instead of ordinary plastic wraps before microwave cooking.

Methylene Chloride Methylene chloride is used to decaffeinate coffee beans and to extract flavor essences from spices or hops, an herb used in making beer. Trace amounts of methylene chloride in the finished products are allowed by FDA at a concentration of less than 30 parts per million. Once diluted in brewed coffee or beer or in spiced foods, its concentration is greatly reduced. Consumers may avoid the small exposure by choosing coffees that are decaffeinated with steam or water, using whole spices instead of extracts, and avoiding commercially brewed beer.

Conclusion

To sum up the messages of this chapter, the ample U.S. food supply is largely safe and hazards are rare. Foodborne microbial illnesses pose the greatest threat by far, and an urgent need exists for new preventive technologies and procedures, along with greater consumer awareness. The Food Feature that follows explores the effects of certain food processing techniques on nutrients in foods, and it offers pointers on the selection, storage, and cooking of foods to preserve their nutrients.

KEY POINTS

- Incidental additives enter food during processing and are regulated; most do not constitute a hazard.
- Consumers should use only microwave-safe containers and wraps for microwaving food.

try it! Food Feature

Processing and the Nutrients in Foods

LO 12.7 Discuss several ways that food-processing techniques affect nutrients in foods.

Nutritionists know that the words *processed foods* are not synonymous with *junk foods*. It is true, however, that in general terms, the more heavily processed a food, the less nutritious it is. Not all processes reduce the nutrient values of foods, however—the effect depends on the food and on the process (Table 12–17 provides examples). As an example, consider the case of orange juice and vitamin C.

The Choice of Orange Juice

Orange juice is available in several forms, each processed a different way. Fresh juice is squeezed from the orange, a process that extracts the fluid juice from the fibrous structures of the whole orange, concentrating its nutrients and calories into a liquid. Each 8-ounce serving of the fresh-squeezed juice contains 124 milligrams of vitamin C. When this juice is condensed by heat, frozen, and then reconstituted, as is the juice from the freezer case of the grocery store, a serving of the reconstituted juice contains just 97 milligrams of vitamin C because vitamin C is destroyed in the condensing process. Canning is even harder on vitamin C: canned orange juice retains only 75 milligrams of vitamin C.

These figures seem to indicate that fresh juice is the superior food, but consider this: most people's recommended intake of vitamin C (75 milligrams for women or 90 milligrams for men) is fully or largely met by one 8-ounce serving of any of the above choices. Thus, for vitamin C, the losses due to processing are not a problem. Besides, processing confers enormous convenience, distribution, and consumer price advantages. Fresh orange juice spoils. Shipping fresh juice to distant places in refrigerated trucks costs much more than shipping frozen juice (which takes up less space) or canned juice (which requires no refrigeration). The fresh product still contains active enzymes that continue to degrade its compounds (including vitamin C) and so cannot be stored indefinitely without compromising nutrient quality. Without canned or frozen juice, people with limited incomes or those with no access to fresh juice would be deprived of this excellent nutrient source. Vitamin C is readily destroyed by oxygen, so whatever the processing methods, orange juice and other vitamin C–rich foods and juices should be stored properly and consumed within a week of opening.

Processing Mischief

Some processing stories are not so rosy. Chapter 8, for instance, explained how processed foods often gain sodium, which people must limit, while needed potassium is leached away. Another misdeed of processors is the addition of sugar and fat—palatable, high-calorie additives that reduce nutrient density. For example, nuts and raisins covered with "natural yogurt" may sound like one healthy food being added to another, but about 75 percent of the weight of the "yogurt" topping is sugar and fat; only 8 percent is yogurt. These sugar- and fat-coated foods taste so good that wishful thinking can take hold, but they are, in reality, candy.

A particularly severe food process is **extrusion**.[57] Extrusion involves grinding and cooking grains, legumes, or other foods, often at high heat and under pressure. The food may then undergo mixing with salt, sugar, flavors, colors, conditioners, and other ingredients; shaping by being pushed through a die; expansion by puffing; and frying. The finished product may be attractive with bright colors, tasty flavors, and pretty shapes, but it has sustained oxidation and heat losses of about 30 percent of its vitamin A, 50 percent of vitamin K, and 90 percent of vitamin C, with similar losses for almost every other vitamin.[58] Beware of processed foods that appear puffy, cute, or colorful—they may lack many of the beneficial constituents that you seek, even if they are made from whole grains and other once-nutritious foods.

Best Nutrient Buys

Here are two good general rules for making food choices:

- Choose whole foods to the greatest extent possible.
- Among processed foods, use those that processing has improved nutritionally or left intact. For example, processing that removes saturated fat, as in fat-free milk, is nutritionally beneficial; commercial prewashing and cutting of fresh vegetables is harmless to nutrients and makes these nutritious foods more accessible to people in a hurry.

Commercially prepared whole-grain breads, frozen cuts of meats, bags of frozen vegetables, and canned or frozen fruit juices do little disservice to nutrition and enable the consumer to eat a wide variety of foods at great savings in time and human energy. The nutrient density of processed foods exists on a continuum:

- Whole-grain bread > refined white bread > sugared doughnuts.
- Milk > fruit-flavored yogurt > canned chocolate pudding.
- Corn on the cob > canned creamed corn > caramel popcorn.
- Oranges > canned orange juice > orange-flavored drink.
- Baked pork loin > ham lunch meat> fried bacon.

Table 12–17

Effects of Food Processing on Nutrients

Process	Method and Purpose	Typical Foods	Effects on Nutrients
Canning	Boil food to sterilize it and seal it in an impervious can or jar to preserve it.	Fruit, fruit preserves, prepared foods such as soups or pasta dishes, vegetables, and meats.	Causes substantial losses of water-soluble vitamins, particularly thiamin and riboflavin; other water-soluble vitamins are dissolved in canning liquid.
Drying	Dehydrate foods to eliminate the water that microbes require for growth.	Fruit, vegetables, meats.	Commercial drying (especially freeze-drying performed at low temperatures) leaves most nutrients intact; home drying may destroy substantial vitamin content; thiamin may be lost in foods treated with sulfur dioxide.
Extruding	Grind, heat, and blend foods with certified colors and flavors and push the resulting paste through screens to form various shapes.	Grains or soybeans, particularly as cereals, baconlike salad toppings, or snack foods in the form of puffs, crisps, or bits.	Considerable nutrient losses occur, notably all vitamins, fiber, and magnesium.
Freezing	Cool a food to its frozen state to stop bacterial reproduction and slow enzymatic reactions.	Fruit, vegetables, ready-to-bake doughs, prepared grain products, meats, soy meat replacers, and mixed dishes.	Negligible effects on nutrients.
Modified atmospheric packaging	Package food in a gas-impermeable container from which air is removed or replaced with other gases to preserve food freshness.	Ready-to-eat salads, cut fruits, soft fresh pasta noodles, baked goods, prepared foods, fresh and preserved meats.	Preserves vitamins by slowing enzymatic breakdown.
Pasteurizing	Expose food to elevated temperature for long enough to reduce bacterial contamination.	Refrigerated foods such as milk, fruit juice, and eggs.	Causes trivial losses of some vitamins.
Ultra-high-temperature processing	Expose food to high temperatures for a short time to eliminate microbial contamination.	Shelf-stable foods such as boxed milk, boxed fruit juice, shelf-stable entrée dishes for microwaving.	Causes trivial losses of some vitamins.

The nutrient continuum is paralleled by another continuum—the nutrition status of the consumer. The closer to the farm the foods you eat, the better nourished you are likely to be, but this doesn't mean that you have to live in the fields. Making wise food choices is half the story of smart nutrition self-care; skillful food preparation is the other half. In general, short cooking times with little water, such as in microwaving, steaming or stir frying, best preserves the nutrients of vegetables, whereas long boiling in copious water that is discarded increases nutrient losses. With reasonable care, if you start with fresh whole foods containing ample amounts of vitamins, you will receive a bounty of the nutrients that they contain.

Purchase mostly whole foods or those that processing has benefited nutritionally.

extrusion processing techniques that transform whole or refined grains, legumes, and other foods into shaped, colored, and flavored snacks, breakfast cereals, and other products.

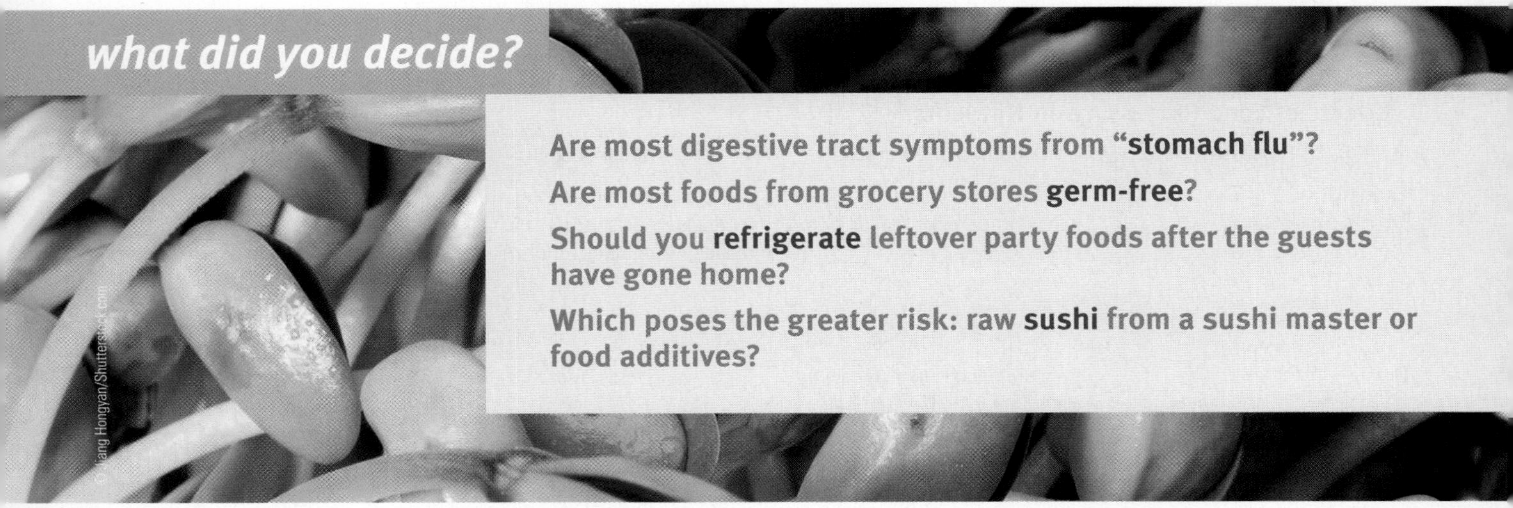

what did you decide?

Are most digestive tract symptoms from "stomach flu"?

Are most foods from grocery stores germ-free?

Should you refrigerate leftover party foods after the guests have gone home?

Which poses the greater risk: raw sushi from a sushi master or food additives?

Self Check

1. (LO 12.1) Microorganisms can cause foodborne illness either by infection or by intoxication.
   T F
2. (LO 12.1) Some microorganisms produce illness-causing __________ .
   a. neurotoxins and enterotoxins
   b. neurotransmitters and aflatoxins
   c. enzymes and hormones
   d. none of the above
3. (LO 12.2) To prevent foodborne illnesses, the refrigerator's temperature should be less than __________ .
   a. 70°F
   b. 65°F
   c. 40°F
   d. 30°F
4. (LO 12.2) Which of the following may be contracted from fresh raw or undercooked seafood?
   a. hepatitis
   b. worms and flukes
   c. viral intestinal disorders
   d. all of the above
5. (LO 12.2) Which of the following organisms can cause hemolytic-uremic syndrome?
   a. *Listeria monocytogenes*
   b. *Campylobacter jejuni*
   c. *Escherichia coli*
   d. *Salmonella*
6. (LO 12.2) The threat of foodborne illness from meats or seafood is far greater than that from produce.
   T F
7. (LO 12.2) Infants under one year of age should never be fed honey because it can contain spores of *Clostridium botulinum*.
   T F
8. (LO 12.3) Which of the following is correct concerning fruits that have been irradiated?
   a. They decay and ripen more slowly.
   b. They lose substantial nutrients.
   c. They lose their sweetness.
   d. They emit gamma radiation.
9. (LO 12.3) Irradiation can
   a. destroy vitamins
   b. sterilize spices
   c. make food radioactive
   d. promote sprouting
10. (LO 12.3) Food packaging can contribute to food safety.
    T F
11. (LO 12.4) It is possible to eliminate all toxins from your diet by eating only "natural" foods.
    T F
12. (LO 12.4) Pregnant women are advised not to eat certain species of fish because the FDA and the EPA have detected unacceptably high lead levels in them.
    T F
13. (LO 12.5) Evidence does not suggest that conventional foods pose health risks or that using organic products reduces risks.
    T F

14. (LO 12.5) Compared with conventionally grown produce, organic produce is often
   a. lower in pesticides.
   b. higher in phytochemicals.
   c. both a and b
   d. none of the above
15. (LO 12.6) Incidental food additives
   a. help to preserve foods.
   b. consist mostly of added sugars and salt.
   c. are really contaminants.
   d. none of the above
16. (LO 12.6) Nitrites added to foods
   a. prevent the growth of the deadly *Claustridium botulinum* bacterium.
   b. preserve the pink color of hot dogs.
   c. are linked with colon cancer in animals.
   d. all of the above
17. (LO 12.7) The words *processed foods* are not synonymous with *junk foods.*
   T F
18. (LO 12.7) Food processing can confer a nutritional advantage by
   a. adding yogurt to the candy coating of raisins.
   b. reducing the costs of nutritious foods.
   c. reducing a food's potassium content.
   d. all of the above
19. (LO 12.8) Selective breeding
   a. involves manipulating an organism's genes in a laboratory.
   b. has been used for thousands of years.
   c. allows scientists to cross species boundaries.
   d. all of the above
20. (LO 12.8) A genetically engineered rice variety in existence today supplies sufficient beta-carotene to fight vitamin A deficiency and childhood blindness worldwide.
   T F

Answers to these Self Check questions are in Appendix G.

CONTROVERSY 12

Genetically Modified Foods: What Are the Pros and Cons?

LO 12.8 Compare and contrast the advantages and disadvantages of food production by way of genetic modification and conventional farming.

With or without their awareness, most people in this country consume foods that contain products of **genetic engineering (GE)**. As Figure C12–1 illustrates, 90 percent of U.S. soybeans and 88 percent of animal feed corn (*not* sweet corn consumed by people) are **genetically engineered organisms (GEOs).** Ubiquitous food additives, such as soy lecithin and high-fructose corn syrup, arise from these GE plant materials and enter the human food supply in a multitude of products. Some consumers recoil from the idea of eating products from GEOs, and whole countries have banned such foods outright. Some objections are based on credible ideas; however, many others arise from emotional fears, distrust of technology, and misinformation.[1] This Controversy sorts the scientific fact from fiction, starting with some definitions of **biotechnology** terms (see Table C12–1).*[2]

Advances in biotechnology have raised hopes of solving some of today's most pressing food and energy problems while boosting profits for farmers and other producers. Although **recombinant DNA (rDNA) technology** may seem futuristic, its roots lie in genetic events that have been occurring unaided for untold millions of years. Human beings have exploited these processes from the advent of agriculture.

Selective Breeding

Season after season, farmers influence the genetic makeup of food plants and animals by selecting only the best farm animals and plants for breeding. Today's lush, hefty, healthy agricultural crops and animals, from cabbage and squash to pigs and cattle, are the result of thousands of years of **selective breeding**. A consumer of today's large cobs of sweet corn, for example, may not recognize the original wild native corn with its sparse four or five kernels to a stalk (shown in the photo).

Today, accelerated selective breeding techniques involve hundreds of thousands of cross-bred seeds planted on vast acreage. To develop crops with desired traits, DNA data from successful seedlings are analyzed by computer. Seedlings with the right genes are grown to maturity and reproduced to yield new breeds in a relatively short time. The unusually colorful carrots in the photo near here, for example, are products of

* *Reference notes are found in Appendix F.*

Figure C12–1

Production of Selected Genetically Engineered Crops, United States 1996–2011

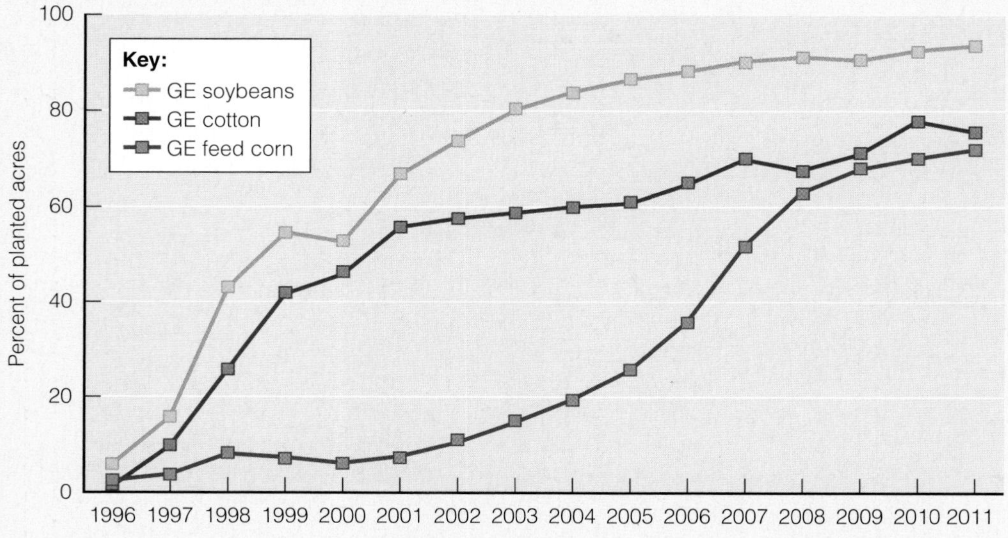

The economic benefits of growing genetically engineered soybeans, cotton, and corn for animal feed has led to widespread replacement of conventional crops on U.S. farms.

Source: Adapted from USDA Economic Research Service, Adoption of genetically engineered crops in the U.S., June 2012, available at www.ers.usda.gov.

Table C12–1
Biotechnology Terms

- **biotechnology** the science of manipulating biological systems or organisms to modify their products or components or create new products; biotechnology includes recombinant DNA technology and traditional and accelerated selective breeding techniques.
- **clone** an individual created asexually from a single ancestor, such as a plant grown from a single stem cell; a group of genetically identical individuals descended from a single common ancestor, such as a colony of bacteria arising from a single bacterial cell; in genetics, a replica of a segment of DNA, such as a gene, produced by genetic engineering.
- **genetic engineering (GE)** the direct, intentional manipulation of the genetic material of living things in order to obtain some desirable inheritable trait not present in the original organism. Also called *recombinant DNA technology*.
- **genetically engineered organism (GEO)** an organism produced by genetic engineering; the term *genetically modified organism (GMO)* is often used to mean the same thing.
- **outcrossing** the unintended breeding of a domestic crop with a related wild species.
- **plant pesticides** substances produced within plant tissues that kill or repel attacking organisms.
- **recombinant DNA (rDNA) technology** a technique of genetic modification whereby scientists directly manipulate the genes of living things; includes methods of removing genes, doubling genes, introducing foreign genes, and changing gene positions to influence the growth and development of organisms.
- **selective breeding** a technique of genetic modification whereby organisms are chosen for reproduction based on their desirability for human purposes, such as high growth rate, high food yield, or disease resistance, with the intention of retaining or enhancing these characteristics in their offspring.
- **stem cell** an undifferentiated cell that can mature into any of a number of specialized cell types. A stem cell of bone marrow may mature into one of many kinds of blood cells, for example.
- **transgenic organism** an organism resulting from the growth of an embryonic, stem, or germ cell into which a new gene has been inserted.

This wild corn, with its sparse kernels, bears little resemblance to today's large, full, sweet ears.

These colorful carrots resulted from intensive selective breeding, not rDNA technology. Researchers bred carrots with high levels of colorful phytochemicals at each generation.

this kind of selective breeding. Selective breeding must stay within the boundaries of a species—a carrot, for example, cannot be crossed with a mosquito. Recombinant DNA technology, however, knows no such limits.

Recombinant DNA Technology

With economy, speed, and precision, rDNA technology can change one or more characteristics of a living thing. The genes for a desirable trait in one organism are transferred directly into another organism's DNA. Figure C12–2, p. 506, compares the genetic results of selective breeding and rDNA technology. Table C12–2, p. 506, presents examples of biotechnology research directions.

Obtaining Desired Traits

Using rDNA technology, scientists can confer useful traits, such as disease resistance, on food crops. To make a disease-resistant potato plant, for example, the process begins with the DNA of an immature cell, known as a **stem cell**, from the "eye" of a potato. Into that stem cell scientists insert a gene snipped from the DNA of a virus that attacks potato plants (enzymes do the snipping). This gene codes for a harmless viral protein, not the infective part.

The newly created stem cell is then stimulated to replicate itself, creating **clone** cells—exact genetic replicas of the modified cell. With time, what was once a single cell grows into a **transgenic organism**, in this case, a potato plant that makes a piece of viral protein in each of

Figure C12-2

Animated! Comparing Selective Breeding and rDNA Technology

Selective Breeding—DNA is a strand of genes, depicted as a strand of pearls. Traditional selective breeding combines many genes from two individuals of the same species.

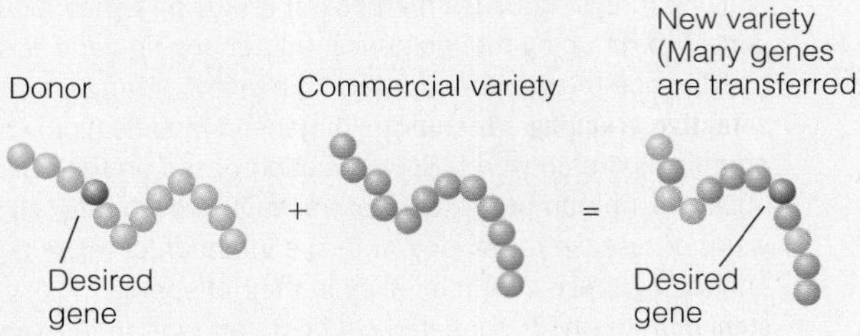

rDNA Technology—Through rDNA technology, a single gene or several may be transferred to the receiving DNA from the same species or others.

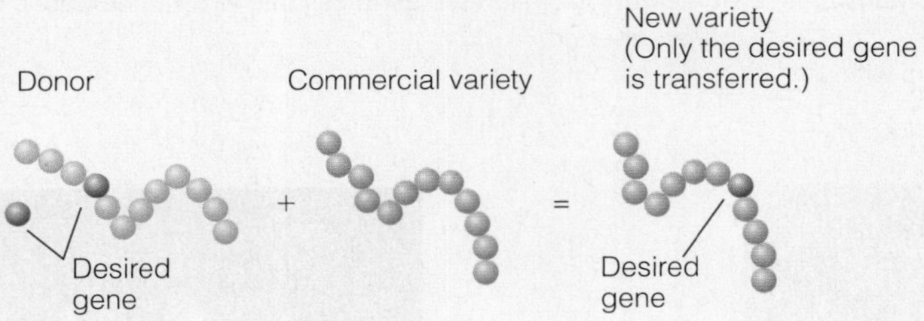

© Cengage Learning

its cells. The presence of the viral protein stimulates the potato plant to develop resistance against an attack from the real wild virus in the potato field.

Plants make likely candidates for genetic engineering because a single plant cell can often be coaxed into producing an entire new plant. Animals can also be modified by rDNA technology, however. Under development is a line of goats that, thanks to a spider's gene, express spider silk protein in their milk (see the photo). Once processed, the stronger-than-steel silk fiber can be used to make artificial ligaments and bulletproof vests.[3]

Courtesy Holly Steinkraus, PhD.

Females of this line of GE goats produce spider silk protein in their milk.

Suppressing Unwanted Traits

rDNA technology can also remove an unwanted protein from a plant by silencing the genes responsible for its creation. For example, scientists are striving to make peanuts safer by silencing the genes for proteins that commonly cause allergic reactions.

The Promises and Problems of rDNA Technology

Supporters hail genetic engineering as nothing short of a revolutionary means of overcoming many of the planet's pressing problems, such as food shortages, nutrient deficiencies, medicine shortages, dwindling farmland, lack of renewable energy sources, and environmental degradation. The Academy of Nutrition and Dietetics takes the position that agricultural and food rDNA technology can enhance the quality, safety, nutritional value, and variety of the food supply while helping to solve problems of production, processing, distribution, and environmental and waste management.[4] A few examples follow.

Table C12-2

Some Examples of Biotechnology Research Directions

Research in genetic engineering is currently directed at creating:

- crops and animals with added desired traits, such as altered nutrient composition, extended shelf life, freedom from allergy-causing constituents, or resistance to diseases or insect pests.
- crops that survive harsh conditions, such as applications of herbicides, heavily polluted or salty soils, or drought conditions.
- microorganisms that produce needed substances, such as pharmaceuticals, hydrocarbon fuels, or other products that do not occur in nature or occur only in limited amounts.

© Cengage Learning

Human Nutrition

Rice leads the way in a genomic revolution of the world's food supply. A GE rice (called *Golden Rice*) provides up to 35 micrograms of beta-carotene per gram of rice (for comparison, carrots have about 80 micrograms), sufficient to fight vitamin A deficiency and childhood blindness worldwide.[5] Other GE rice varieties, some offering 80 percent more iron and zinc than ordinary rice, could relieve much iron-deficiency anemia and zinc deficiency around the world. Still others may resist drought or insects and thus provide more food for hungry populations. Not just rice but worldwide staples like cassava roots or potatoes can be "biofortified" with minerals, vitamins, fatty acids, or promising phytochemicals.[6] In the case of cassava, it can also be made safer by reducing its concentration of naturally occurring toxins.[7]

As for food animals, pigs that develop high levels of the omega-3 fatty acids of fish are under development. Livestock can also be engineered to develop less fat or to produce more milk. Animals themselves also stand to benefit: increased resistance to crippling diseases, such as "mad cow" and udder infections, may soon be attainable.

Molecules from Microbes

The genes of microorganisms have been altered to make pharmaceutical and industrial products. For example, a transgenic bacterial factory now mass-produces the hormone insulin used by people with diabetes. Another bacterium received a bovine gene to make the enzyme rennin, necessary in cheese production. Historically, rennin was harvested from the stomachs of calves, an expensive process. Today, intensive efforts are underway to develop biofuel-producing microbes to supply a more sustainable and price-stable alternative to fossil fuels.[8]

Plants and animals may also play similar roles. Researchers have induced bananas and potatoes to produce a hepatitis vaccine. Animals can be engineered to secrete vaccines in their milk and herds could provide both nourishment and immunization to villages now lacking both food and medicine. Pigs are being engineered for genetic compatibility with human beings to reduce the risk of immune rejection of surgically transplanted pig tissues.[9]

Greater Crop Yields

Most of today's GE crops fall within two categories: herbicide-resistant and insect-resistant, both used to improve yields and protect farmed land. Herbicide-resistant crops, for example, offer weed control with less soil tillage by allowing farmers to spray whole fields, not just weeds, with potent herbicides. The weeds die, their roots hold soil in place between the rows, and GE crops grow normally. After years of such spraying, however, some weeds, such as pigweed, have developed vigorous resistance to today's herbicide. Pigweed grows large and spreads fast despite repeated sprayings, forcing many farmers to return to old tillage methods to control it, thus exposing vast quantities of farm topsoil to wind and water erosion.[†10]

As for insect-resistant crops, these GEOs make what the Environmental Protection Agency (EPA) calls **plant pesticides**—pesticides made by the plant tissues themselves. For example, a type of GE feed corn produces a pesticide that kills a common corn-destroying worm, thereby greatly increasing yields per acre of farmland.

In areas where people cannot afford to lose a single morsel of food, and where plant diseases and insects can claim up to 80 percent of a season's yield, GE plants can save the crop. Such innovations promise at least some relief for the world's chronically hungry people.

Food from Cloned Animals

According to the FDA, milk and meat from cloned adult cattle, pigs, and goats and their offspring are as safe as similar conventional foods, so food labels are not required to distinguish between these sources. However, many people have reservations about consuming food from cloned animals. The high cost of producing animal clones limits their potential in food production.

Issues Surrounding GE Foods

Consumers want to know about any potential risks from rDNA technology. The FDA, in exploring the same issues, asks whether GE foods differ substantially from other foods in their nutrient contents or safety.

Nutrient Composition

In most cases, except for intentional variation created through rDNA technology, the nutrient composition of GE foods is identical to that of comparable traditional foods. From the body's point of view, therefore, eating Golden Rice, mentioned earlier, would be the same as eating plain rice and taking a beta-carotene supplement—and beta-carotene supplements carry risks (see Controversy 7). Thus, while GE foods may contribute to *overdoses* of nutrients or phytochemicals, they pose no unusual threat of deficiencies.

Accidental Ingestion of Drugs from Foods

GE corn, soybeans, rice, and other food crops that make human and animal drugs and industrial proteins must be grown indoors in selected locations. Their containment areas, however, often abut farms where conventional food crops are grown. Critics fear that DNA from drug-producing GE crops might contaminate the regular food supply, despite USDA oversight.[11] Disasters such as tornadoes, floods, or other events could liberate the sequestered plants, and high winds or water could transport their pollen long distances to mingle undetected with food crops.

Pesticide Residues and Resistance

Industry scientists contend that rDNA technology could virtually end problems associated with pesticide use on foods. Human error is eliminated, they say, when genes determine both the nature and the amount of pesticide produced. Critics counter that while GE crops may be protected from one or two common pests that may or may not be present on a particular field, farmers must still spray for other pests devouring their crops. Also, in a worrisome sign, constant exposure is causing crop-destroying insects to become resistant to plant pesticides.

The plant pesticide produced by GE corn, a worm-killing peptide, is the same compound in the "natural pesticide" that organic farmers spray on crops. Originally discovered in a common soil bacterium, this peptide is harvested from industrial bacterial colonies.

Regular pesticides that are sprayed onto crops can be largely removed from food by washing or peeling produce, but consumers cannot remove pesticides that form within the tissues of GE food. Still, plant pesticides are highly unlikely to cause health problems because they are made of peptide chains (small

† *Pigweed is officially known as Palmer amaranth.*

protein strands) that human digestive enzymes readily denature. Plant pesticides, like other pesticide residues, are approved and regulated by the FDA (see the preceding chapter).

Unintended Health Effects

The possibility exists that genetic engineering of food plants or animals may have unintended and therefore unpredictable effects on human consumers. A lesson comes from an unexpected negative effect of selective breeding. Over many years, celery growers had crossed their most attractive celery plants because consumers paid a premium for good-looking celery. Unbeknownst to the growers, however, the most beautiful celery stayed that way because it was especially high in a natural plant pesticide produced by celery. With each breeding cycle, the pesticide concentration increased. Soon, farm and grocery workers who handled the celery began suffering from serious skin rashes until the problem was finally traced to high levels of the natural pesticide in the beautiful plants.

Another example, this time an unintended *benefit* of genetic engineering, involved a cancer-causing fungus that sometimes grows on corn.‡ After several growing seasons, scientists confirmed that GE corn with plant pesticide suffered much less worm damage than conventionally grown corn. Surprisingly, the crops also had far less than the expected growth of the dangerous fungus. It turns out that the worms spread the fungus as they burrow into cobs of ordinary corn, but the plant pesticide in the GE corn killed the worms and stopped the spread of fungal infection. Whether negative or positive, unintended effects are by nature unpredictable.

Environmental Effects

Some striking benefits to the environment have emerged over the past 10 years from the use of GE crops. However, some detrimental environmental issues remain.

Between 1996 and 2006, the planting of GE crops reduced pesticide use by almost 500 million pounds of active ingredients worldwide. Also, herbicide-resistant GE crops require far less plowing to kill weeds and so minimize soil erosion. Traditional farmers must turn over the soil before each planting to reduce weed growth, and exposed soil can easily blow or wash away (more about soil conservation in Controversy 15).

A remaining environmental concern is the likelihood of **outcrossing**, the accidental cross-pollination of plant pesticide crops with related wild weeds. If a weed inherits a pest-resistant trait from a neighboring field of GE crops, it could gain an enormous survival advantage over other important wild species and crowd them out.

Loss of species is a serious threat. Crops on which humankind depends are vulnerable in a changing environment, and valuable genetic traits that could help food crops recover may exist in species teetering on the brink of extinction. One effort to preserve genetic diversity is a global seed bank buried in a frozen mountain in Norway that stores vast numbers of diverse food crop seeds.

GE crops may also directly damage wildlife. In the laboratory, monarch butterfly larvae die when fed pollen from the pesticide-producing corn already described. In real life, wild butterflies do not seem to consume enough toxic corn pollen for populations to be harmed. The new technology may even protect monarchs and other harmless or beneficial insects that now die when they feed on conventionally sprayed fields.

Ethical Arguments about Genetic Engineering

In the end, consumer acceptance determines the applications of genetic engineering.[12] Some fear that by tampering with the basic blueprint of life, rDNA technology in particular will sooner or later unleash mayhem on an unsuspecting world. Any degree of risk is unjustified, they say, because while it raises profits for biotechnology companies and farmers, its products provide no direct benefits to consumers. Others object to rDNA technology on religious grounds, holding that genetic decisions are best left to nature or a higher power. At the very least, these people want food labels to clearly identify food that contain GE ingredients.[13] Table C12–3 summarizes some of these issues.

Proponents of genetic engineering respond that most of the world's people cannot afford the luxury of rejecting the potential benefits of rDNA technology—they lack the abundant foods and fertile lands that protesters take for granted. Delays hurt the poorest of the poor, they say. GE opponents counter that the scope of world hunger far exceeds simple solutions such as increasing food supplies—it involves war, politics, and education. (Chapter 15 explores the tragedy of world hunger.) In 2012, the U.S. Congress denied a petition to require labeling of GE foods nationwide.[14]

The FDA'S Position on GE Foods

The FDA can evaluate only the safety of today's GE fruits, vegetables, and grains for human consumption and takes the position that they are safe, unless they differ substantially from similar foods already in use. Foods from GE animals cannot be marketed without FDA approval. Many scientific organizations agree that rDNA technology can deliver on the promise to improve the food supply if we give it a chance to do so.

The Final Word

For those who would worry themselves into a diet of crackers and water, abundant evidence supports eating sufficient fruits and vegetables regardless of their source. Theoretical future risks from GE foods pale next to the real and immediate perils associated with diets that lack sufficient fruits and vegetables.

Critical Thinking

1. Discuss options and roadblocks to obtaining only non-GEO foods.
2. Outline possible motivations of industry, growers, and consumers for supporting/opposing GE foods.

‡*The fungus (*Aspergillus flavus*) produces the carcinogenic toxin* aflatoxin.

Table C12–3

GE Foods: Point, Counterpoint

Arguments in Opposition to Genetic Engineering	Arguments in Support of Genetic Engineering
1. *Ethical and moral issues.* It's immoral to "play God" by mixing genes from organisms unable to do so naturally. Religious and vegetarian groups object to genes from prohibited species occurring in their allowable foods.	**1.** *Ethical and moral issues.* Scientists throughout history have been persecuted and even put to death by fearful people who accuse them of playing God. Yet today, many of the world's citizens enjoy a long and healthy life of comfort and convenience due to once-feared scientific advances put to practical use.
2. *Imperfect technology.* The technology is young and imperfect; genes rarely function in just one way, their placement is often imprecise, and potential effects are impossible to predict. Toxins are as likely to be produced as the desired trait.	**2.** *Advanced technology.* Recombinant DNA technology is precise and reliable. Many of the most exciting recent advances in medicine, agriculture, and technology were made possible by the application of this technology.
3. *Environmental concerns.* Environmental side effects are unknown. The power of a genetically modified organism to change the world's environments is unknown until such changes actually occur—then the "genie is out of the bottle." Once out, the genie cannot be put back in the bottle because insects, birds, and the wind distribute genetically altered seed and pollen to points unknown.	**3.** *Environmental protection.* Genetic engineering may be the only hope of saving rain forest and other habitats from destruction by impoverished people desperate for arable land. Through genetic engineering, farmers can make use of previously unproductive lands such as salt-rich soils and arid areas.
4. *"Genetic pollution."* Other kinds of pollution can often be cleaned up with money, time, and effort. Once genes are spliced into living things, those genes forever bear the imprint of human tampering.	**4.** *Genetic improvements.* Genetic side effects are more likely to benefit the environment than to harm it.
5. *Crop vulnerability.* Pests and disease can quickly adapt to overtake genetically identical plants or animals around the world. Diversity is key to defense.	**5.** *Improved crop resistance.* Pests and diseases can be specifically fought on a case-by-case basis. Biotechnology is the key to defense.
6. *Loss of gene pool.* Loss of genetic diversity threatens to deplete valuable gene banks from which scientists can develop new agricultural crops.	**6.** *Gene pool preserved.* Thanks to advances in genetics, laboratories around the world are able to stockpile the genetic material of millions of species that, without such advances, would have been lost forever.
7. *Profit motive.* Genetic engineering will profit industry more than the world's poor and hungry.	**7.** *Everyone profits.* Industries benefit from genetic engineering, and a thriving food industry benefits the nation and its people, as witnessed by countries lacking such industries. Genetic engineering promises to provide adequate nutritious food for millions who lack such food today. Developed nations gain cheaper, more attractive, more delicious foods with greater variety and availability year-round.
8. *Unproven safety for people.* Human safety testing of genetically altered products is generally lacking. The population is an unwitting experimental group in a nationwide laboratory study for the benefit of industry.	**8.** *Safe for people.* Human safety testing of genetically altered products is unnecessary because the products are essentially the same as the original foodstuffs.
9. *Increased allergens.* Allergens can unwittingly be transferred into foods.	**9.** *Control of allergens.* Allergens can be transferred into foods, but these are known, and thus can be avoided. Allergen-free peanuts and other foods are under development.
10. *Decreased nutrients.* A fresh-looking tomato or other produce held for several weeks may have lost substantial nutrients.	**10.** *Increased nutrients.* Genetic modifications can easily enhance the nutrients in foods.
11. *No product tracking.* Without labeling, the food industry cannot track problems to the source.	**11.** *Excellent product tracking.* The identity and location of genetically altered foodstuffs are known, and they can be tracked should problems arise.
12. *Overuse of herbicides.* Farmers, knowing that their crops resist herbicide effects, will use them liberally.	**12.** *Conservative use of herbicides.* Farmers will not waste expensive herbicides in second or third applications when the prescribed amount gets the job done the first time.
13. *Increased consumption of pesticides.* When a pesticide is produced by the flesh of produce, consumers cannot wash it off the skin of the produce with running water as they can with most ordinary sprays.	**13.** *Reduced pesticides on foods.* Pesticides produced by plants in tiny amounts known to be safe for consumption are more predictable than applications by agricultural workers who make mistakes. Because other genetic manipulations will eliminate the need for postharvest spraying, fewer pesticides will reach the dinner table.
14. *Lack of oversight.* Government oversight is run by industry people for the benefit of industry—no one is watching out for the consumer.	**14.** *Sufficient regulation, oversight, and rapid response.* The National Academy of Sciences has established protocol for safety testing of GE foods. Government agencies are efficient in identifying and correcting problems as they occur in the industry.

13 Life Cycle Nutrition: Mother and Infant

what do you think?

Can a **man's lifestyle habits** affect a future pregnancy?

How much **alcohol** does it take to harm a developing fetus?

Are **breast milk** and **formula** about the same for an infant?

Can infants **grow and thrive** on only breast milk or formula?

Learning Objectives

After completing this chapter, you should be able to accomplish the following:

LO 13.1 Explain why a nutritionally adequate diet is important before and during a pregnancy, and identify the special nutritional needs of a pregnant teenager.

LO 13.2 Evaluate the statement that "no level of alcoholic beverage intake is safe or advisable during pregnancy."

LO 13.3 Describe the impacts of gestational diabetes and preeclampsia on the health of a pregnant woman and on the fetus.

LO 13.4 Describe maternal nutrition needs for lactation, the impact of malnutrition on breast milk, and contraindications to breastfeeding.

LO 13.5 Identify characteristics of breast milk that make it the ideal food for human infants, and discuss the introduction of solid foods into the diet.

LO 13.6 List some feeding guidelines that encourage normal eating behaviors and autonomy in the child.

LO 13.7 Discuss some relationships between childhood obesity and chronic diseases, and develop a healthy eating and activity plan for an obese child of a given age.

All people need the same nutrients but in differing amounts throughout life. This chapter is the first of two on life's changing nutrient needs. It focuses on the two life stages that might be the most important to an individual's life-long health—pregnancy and infancy.

Pregnancy: The Impact of Nutrition on the Future

LO 13.1 Explain why a nutritionally adequate diet is important before and during a pregnancy, and identify the special nutritional needs of a pregnant teenager.

© iStockphoto.com/Squaredpixels

Both parents can prepare in advance for a healthy pregnancy.

People normally think of nutrition as personal, affecting them alone. For the woman who is pregnant, or who soon will be, however, nutrition choices today profoundly affect the health of her future child and the adult that the child will one day become. The nutrient demands of pregnancy are extraordinary.

Preparing for Pregnancy

Before she becomes pregnant, a woman must establish eating habits that will optimally nourish both the growing **fetus** and herself. She must be well nourished at the outset because early in pregnancy the **embryo** undergoes rapid and significant developmental changes that depend on good nutrition.

Some influences on heritable traits do not directly change DNA, but arise from epigenetic influences during early pregnancy—see **Controversy 11.**

Fathers-to-be are also wise to examine their eating and drinking habits. For example, a sedentary lifestyle and consuming too few fruits and vegetables may affect men's **fertility** (and the fertility of their children), and men who drink too much alcohol or encounter other toxins in the weeks before conception can sustain damage to their sperm's genetic material.*[1] When both partners adopt healthy habits, they will be better prepared to meet the demands of parenting that lie ahead.

Prepregnancy Weight Before pregnancy, all women, but underweight women in particular, should strive for an appropriate body weight. A woman who begins her pregnancy underweight and who fails to gain sufficiently during pregnancy is very likely to bear a baby with a dangerously **low birthweight**.[2] (A later section comes

* Reference notes are found in Appendix F.

fetus (FEET-us) the stage of human gestation from eight weeks after conception until the birth of an infant.

embryo (EM-bree-oh) the stage of human gestation from the third to the eighth week after conception.

fertility the capacity of a woman to produce a normal ovum periodically and of a man to produce normal sperm; the ability to reproduce.

low birthweight a birthweight of less than $5\frac{1}{2}$ pounds (2,500 grams); used as a predictor of probable health problems in the newborn and as a probable indicator of poor nutrition status of the mother before and/or during pregnancy. Low-birthweight infants are of two different types. Some are *premature infants*; they are born early and are the right size for their gestational age. Other low-birthweight infants have suffered growth failure in the uterus; they are small for gestational age (small for date) and may or may not be premature.

Did You Know?

Underweight is defined as BMI <18.5. Obese is defined as BMI ≥30.0 (see Table 13–5 on page 520).

Figure 13–1

Infant Mortality Decline over Time

The graph shows infant deaths per 1,000 live births.

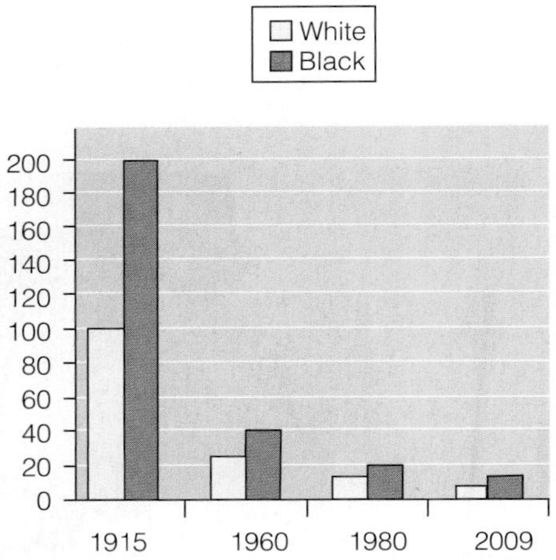

Source: Data from K. D. Kochanek and coauthors, Annual summary of vital statistics: 2009, Pediatrics *129 (2012): 338–348.*

back to the needed gains in pregnancy.) Infant birthweight is the most potent single indicator of an infant's future health. A low-birthweight baby, defined as one who weighs less than 5 1/2 pounds (2,500 grams), is nearly 40 times more likely to die in the first year of life than a normal-weight baby. To prevent low birthweight, underweight women are advised to gain weight before becoming pregnant and to strive to gain adequately thereafter.

When nutrient supplies during pregnancy fail to meet demands, the developing fetus may adapt to the sparse conditions in ways that may make obesity or chronic diseases more likely in later life.[3] Low birthweight is also associated with lower adult IQ and other brain impairments, short stature, and educational disadvantages.[4] Nutrient deficiency coupled with low birthweight is the underlying cause of more than half of all the deaths worldwide of children under 5 years of age. In the United States, the infant mortality rate in 2009 was just under 6.5 deaths per 1,000 live births.[5] This rate, though higher than that of some other developed countries, represents a significant decline over the last two decades and is a tribute to public health efforts aimed at reducing infant deaths (see Figure 13–1).

Low birthweight may also reflect heredity, disease conditions, smoking, and drug (including alcohol) use during pregnancy.[6] Even with optimal nutrition and health during pregnancy, some women give birth to small infants for unknown reasons. Nevertheless, poor nutrition is the major factor in low birthweight—and an avoidable one, as later sections make clear.[7]

Obese women are also urged to strive for healthy weights before pregnancy. Infants born to obese women are more likely to be large for gestational age, weighing more than 9 pounds.[8] Problems associated with a high birthweight include increases in the likelihood of a difficult labor and delivery, birth trauma, and **cesarean section**.[9] Consequently, these babies have a greater risk of poor health and death than infants of normal weight. Infants of obese mothers may be twice as likely to be born with a neural tube defect, too. The vitamin folate may play a role, but a more likely explanation seems to be poor blood glucose control.[10] Obese women themselves are likely to suffer gestational diabetes, hypertension, and complications during and infections after the birth.[11] In addition, both overweight and obese women have a greater risk of giving birth to infants with heart defects and other abnormalities.[12] The obese woman who strives for a healthier prepregnancy body weight helps protect both herself and her future child.

A Healthy Placenta and Other Organs A woman's nutrition before pregnancy is crucial because it determines whether her **uterus** will be able to support the growth of a healthy **placenta** during the first month of **gestation**. The placenta is both a supply depot and a waste-removal system for the fetus. If the placenta works perfectly, the fetus wants for nothing; if it doesn't, no alternative source of sustenance is available, and the fetus will fail to thrive. Figure 13–2 shows the placenta, a mass of tissue in which maternal and fetal blood vessels intertwine and exchange materials. The two bloods never mix, but the barrier between them is notably thin. Nutrients and oxygen move across this thin barrier from the mother's blood into the fetus's blood, and wastes move out of the fetal blood to be excreted by the mother. Thus, by way of the placenta, the mother's digestive tract, respiratory system, and kidneys serve the needs of the fetus whose organs are not yet functional, as well as her own. The **umbilical cord** acts like a pipeline, conducting fetal blood to and from the placenta. The **amniotic sac** surrounds and cradles the fetus, which floats inside its cushioning fluids.

The placenta is a highly metabolic organ that actively gathers up hormones, nutrients, and protein molecules such as antibodies and transfers them into the fetal bloodstream.[13] The placenta also produces a broad range of hormones that act in many ways to maintain pregnancy and prepare the mother's breasts for **lactation**. Is it any wonder that a healthy placenta is essential for the developing fetus?

If the mother's nutrient stores are inadequate during placental development, no amount of nutrients later on in pregnancy can make up for the lack. If the placenta fails to form or function properly, the fetus will not receive optimal nourishment.

cesarean (see-ZAIR-ee-un) **section** surgical childbirth, in which the infant is taken through an incision in the woman's abdomen.

uterus (YOO-ter-us) the womb, the muscular organ within which the infant develops before birth.

placenta (pla-SEN-tuh) the organ of pregnancy in which maternal and fetal blood circulate in close proximity and exchange nutrients and oxygen (flowing into the fetus) and wastes (picked up by the mother's blood).

gestation the period of about 40 weeks (three trimesters) from conception to birth; the term of a pregnancy.

umbilical (um-BIL-ih-cul) **cord** the ropelike structure through which the fetus's veins and arteries reach the placenta; the route of nourishment and oxygen into the fetus and the route of waste disposal from the fetus.

amniotic (AM-nee-OTT-ic) **sac** the "bag of waters" in the uterus in which the fetus floats.

lactation production and secretion of breast milk for the purpose of nourishing an infant.

Figure 13–2

Animated! The Placenta

The placenta is composed of spongy tissue in which fetal blood and maternal blood flow side by side, each in its own vessels. The maternal blood transfers oxygen and nutrients to the fetus's blood and picks up fetal wastes to be excreted by the mother. The placenta performs the nutritive, respiratory, and excretory functions that the fetus's digestive system, lungs, and kidneys will provide after birth.

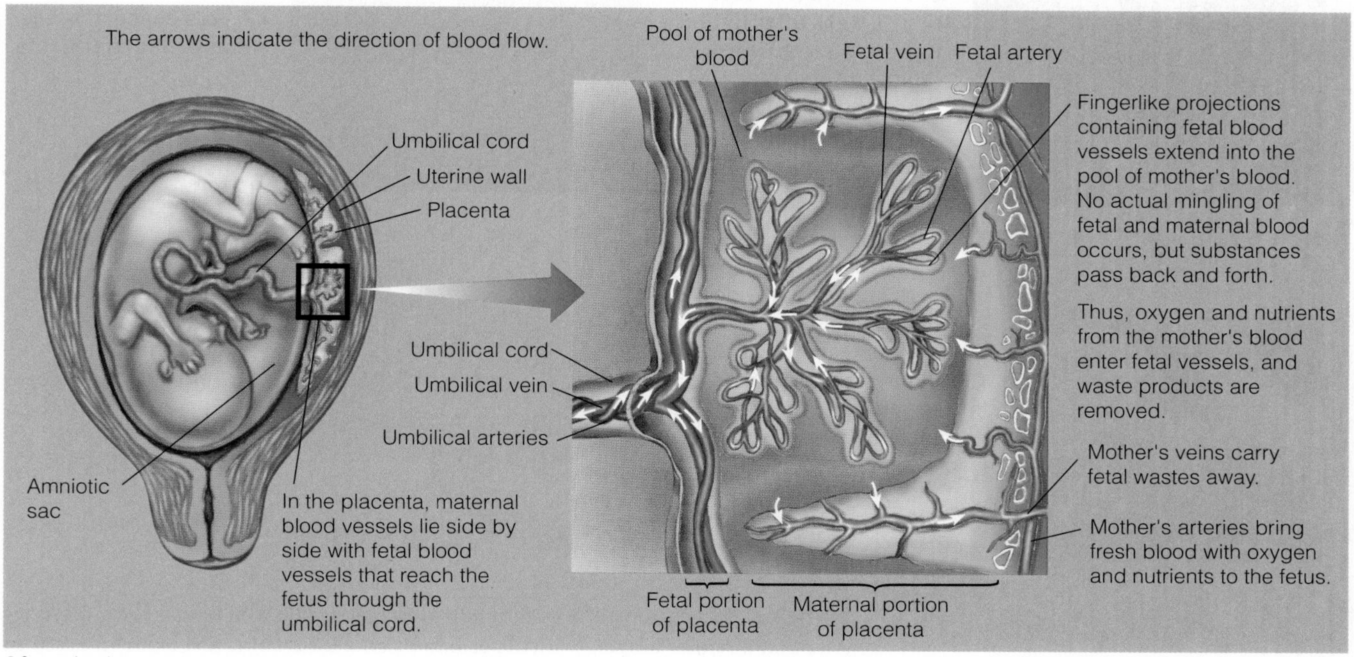

The arrows indicate the direction of blood flow.
Umbilical cord
Uterine wall
Placenta
Amniotic sac
In the placenta, maternal blood vessels lie side by side with fetal blood vessels that reach the fetus through the umbilical cord.
Pool of mother's blood
Fetal vein
Fetal artery
Umbilical cord
Umbilical vein
Umbilical arteries
Fingerlike projections containing fetal blood vessels extend into the pool of mother's blood. No actual mingling of fetal and maternal blood occurs, but substances pass back and forth.
Thus, oxygen and nutrients from the mother's blood enter fetal vessels, and waste products are removed.
Mother's veins carry fetal wastes away.
Mother's arteries bring fresh blood with oxygen and nutrients to the fetus.
Fetal portion of placenta
Maternal portion of placenta

After getting such a poor start on life, the child may be ill equipped, even as an adult, to store sufficient nutrients, and a girl may later be unable to grow an adequate placenta or bear healthy full-term infants. For this and other reasons, a woman's poor nutrition during her early pregnancy could affect her grandchild as well as her child.

KEY POINTS

- Adequate nutrition before pregnancy establishes physical readiness and nutrient stores to support placental and fetal growth.
- Both underweight and overweight women should strive for appropriate body weights before pregnancy.
- Newborns who weigh less than 5½ pounds face greater health risks than normal-weight babies.

The Events of Pregnancy

The newly fertilized **ovum** is called a **zygote**. It begins as a single cell and rapidly divides into many cells during the days after fertilization. Within two weeks, if the cluster of cells embeds itself in the uterine wall in a process known as **implantation**, the placenta begins to grow inside the uterus. Minimal growth in size takes place at this time, but it is a crucial period in development. Adverse influences such as smoking, drug abuse, and malnutrition at this time lead to failure to implant or to abnormalities such as neural tube defects that can cause loss of the developing embryo, often before the woman knows she is pregnant.

The Embryo and Fetus During the next six weeks, the embryo registers astonishing physical changes (see Figure 13–3, p. 514). At eight weeks, the fetus has a complete central nervous system, a beating heart, a fully formed digestive system, well-defined fingers and toes, and the beginnings of facial features.

ovum the egg, produced by the mother, that unites with a sperm from the father to produce a new individual.

zygote (ZYE-goat) the product of the union of ovum and sperm; a fertilized ovum.

implantation the stage of development, during the first two weeks after conception, in which the fertilized egg (fertilized ovum or zygote) embeds itself in the wall of the uterus and begins to develop.

Figure 13–3

Stages of Embryonic and Fetal Development

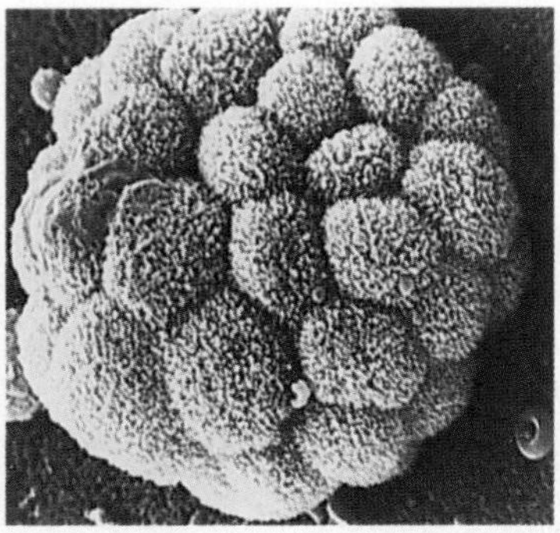

Petit Format/Nestle/Science Source

(1) A newly fertilized ovum, called a zygote, is about the size of the period at the end of this sentence. Less than 1 week after fertilization, the zygote has rapidly divided many times and becomes ready for implantation.

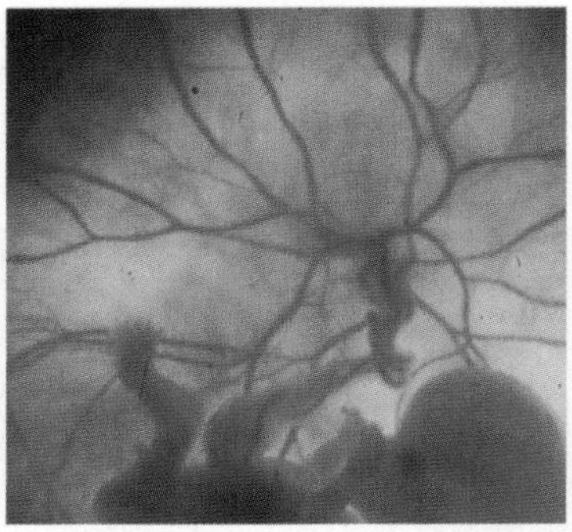

Petit Format/Nestle/Science Source

(3) A fetus after 11 weeks of development is just over an inch long. Notice the umbilical cord and blood vessels connecting the fetus with the placenta.

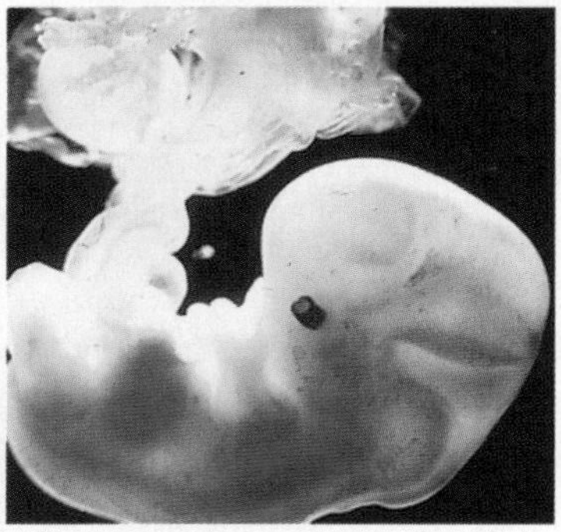

Petit Format/Nestle/Science Source

(2) After implantation, the placenta develops and begins to provide nourishment to the developing embryo. An embryo 5 weeks after fertilization is about $^{1}/_{2}$ inch long.

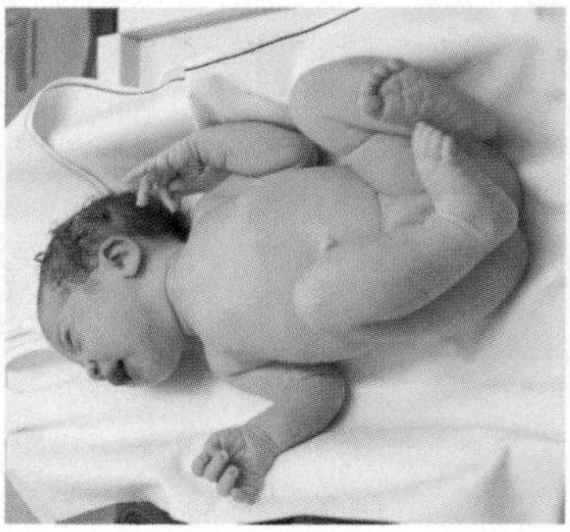

Michael Clement/Radius Images/Masterfile

(4) A newborn infant after 9 months of development measures close to 20 inches in length. The average birthweight is about $7^{1}/_{2}$ pounds. From 8 weeks to term, this infant grew 20 times longer and 50 times heavier.

In the last seven months of pregnancy, the fetal period, the fetus grows 50 times heavier and 20 times longer. Critical periods of cell division and development occur in organ after organ. The amniotic sac fills with fluid, and the mother's body changes. The uterus and its supporting muscles increase in size, the breasts may become tender and full, the nipples may darken in preparation for lactation, and the mother's blood volume increases by half to accommodate the added load of materials it must carry. Gestation lasts approximately 40 weeks and ends with the birth of the infant. The 40 or so weeks of pregnancy are divided into thirds, each of which is called a **trimester**.

A Note about Critical Periods Each organ and tissue type grows with its own characteristic pattern and timing. The development of each takes place only at a certain time—the **critical period**. Whatever nutrients and other environmental conditions are necessary during this period must be supplied on time if the organ is to reach its full potential. If the development of an organ is limited during a critical period, recovery is impossible. For example, the fetus's heart and brain are well developed at 14 weeks; the lungs, 10 weeks later. Therefore, early malnutrition impairs the heart and brain; later malnutrition impairs the lungs.

The effects of malnutrition during critical periods of pregnancy are seen in defects of the nervous system of the embryo (explained later), in the child's poor dental health, and in the adolescent's and adult's vulnerability to infections and possibly higher risks of diabetes, hypertension, stroke, or heart disease.[14] The effects of malnutrition during critical periods are irreversible: abundant and nourishing food, fed after the critical time, cannot remedy harm already done.

Table 13–1 identifies characteristics of a **high-risk pregnancy**. The more factors that apply, the higher the risk. All pregnant women, especially those in high-risk categories, need **prenatal** medical care, including dietary advice.

KEY POINTS

- Implantation, fetal development, and critical period development depend on maternal nutrition status.
- The effects of malnutrition during critical periods are irreversible.

Increased Need for Nutrients

During pregnancy, a woman's nutrient needs increase more for certain nutrients than for others. Figure 13–4 shows the percentage increase in nutrient intakes recommended for pregnant women compared to nonpregnant women. The nutrient demands of pregnancy are high so a woman must make careful food choices, but her body will also do its part by maximizing nutrient absorption and minimizing losses.

Energy, Carbohydrate, Protein, and Fat Energy needs vary with the progression of pregnancy. In the first trimester, the pregnant woman needs no additional energy, but her energy needs rise as pregnancy progresses. She requires an additional 340 daily calories during the second trimester and an extra 450 calories each day during the third trimester.[15] Well-nourished pregnant women meet these demands for more energy in several ways: some eat more food, some reduce their activity, and some store less of their food energy as fat.[16] A woman can easily meet the need for extra calories by selecting more nutrient-dense foods from the five food groups. Table 13–2 (p. 516) offers a sample menu for pregnant and lactating women.

Ample carbohydrate (ideally, 175 grams or more per day and certainly no less than 135 grams) is necessary to fuel the fetal brain and spare the protein needed for fetal growth. Whole grain breads and cereals, dark green and other vegetables, legumes, and citrus and other fruit provide carbohydrates, nutrients, and phytochemicals, along with fiber to help alleviate the constipation that many pregnant women experience.

Table 13–1

High-Risk Pregnancy Factors

- Prepregnancy BMI either <18.5 or ≥25
- Insufficient or excessive pregnancy weight gain
- Nutrient deficiencies or toxicities; eating disorders
- Poverty, lack of family support, low level of education, limited food available
- Smoking, alcohol, or other drug use
- Age, especially 15 years or younger or 35 years or older
- Many previous pregnancies (3 or more to mothers younger than age 20; 4 or more to mothers age 20 or older)
- Short or long intervals between pregnancies (<18 months or >59 months)
- Previous history of problems such as low- or high-birthweight infants
- Twins or triplets
- Pregnancy-related hypertension or gestational diabetes
- Diabetes; heart, respiratory, and kidney disease; certain genetic disorders; special diets and medications

Figure 13–4

Comparison of Selected Nutrient Recommendations for Nonpregnant, Pregnant, and Lactating Women[a]

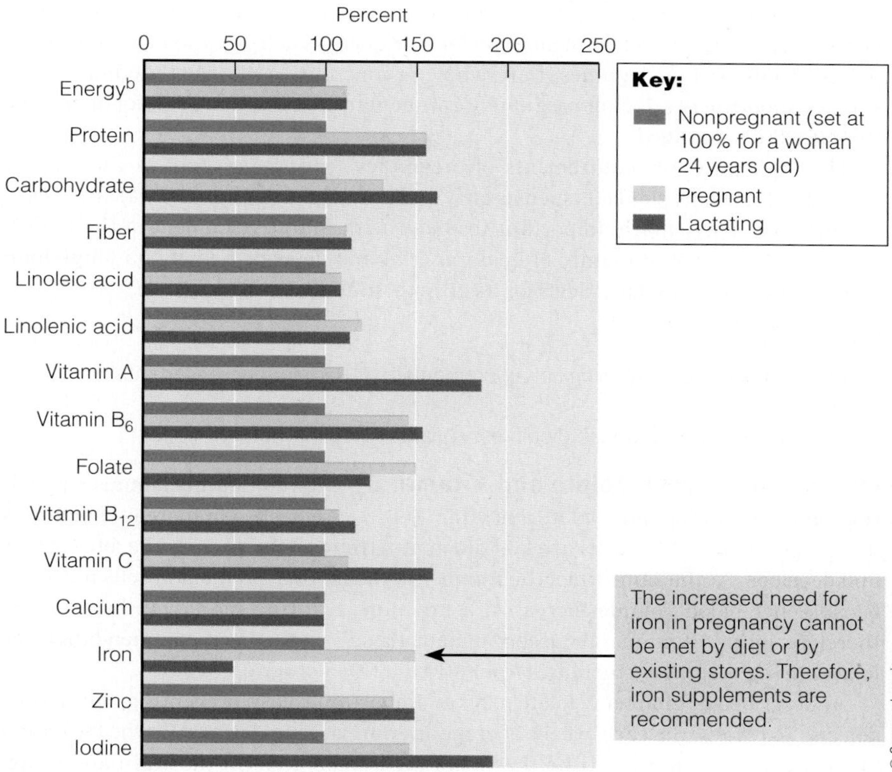

[a]Values for other nutrients are listed on the inside front cover, pages A and B.

[b]Energy allowance during pregnancy is for 2nd trimester; energy allowance during the 3rd trimester is slightly higher; no additional allowance is provided during the 1st trimester. Energy allowance during lactation is for the first 6 months; energy allowance during the second 6 months is slightly higher.

trimester a period representing gestation. A trimester is about 13 to 14 weeks.

critical period a finite period during development in which certain events may occur that will have irreversible effects on later developmental stages. A critical period is usually a period of cell division in a body organ.

high-risk pregnancy a pregnancy characterized by risk factors that make it likely the birth will be surrounded by problems such as premature delivery, difficult birth, retarded growth, birth defects, and early infant death. A *low-risk pregnancy* has none of these factors.

prenatal (pree-NAY-tal) before birth.

Table 13–2

Daily Food Choices for Pregnancy (2nd and 3rd trimesters) and Lactation

Food Group	Amounts	SAMPLE MENU	
Fruits	2 c	**Breakfast** 1 whole-wheat English muffin 2 tbs peanut butter 1 c low-fat vanilla yogurt ½ c fresh strawberries 1 c orange juice	**Dinner** Chicken cacciatore 3 oz chicken ½ c stewed tomatoes 1 c rice ½ c summer squash 1½ c salad (spinach, mushrooms, carrots) 1 tbs salad dressing 1 slice Italian bread 2 tsp soft margarine 1 c low-fat milk
Vegetables	3 c		
Grains	8 oz	**Midmorning snack** ½ c cranberry juice 1 oz pretzels	
Protein Foods	6½ oz	**Lunch** Sandwich (tuna salad on whole-wheat bread) ½ carrot (sticks) 1 c low-fat milk	
Milk	3 c		

Note: This sample meal plan provides about 2,500 calories (55% from carbohydrate, 20% from protein, and 25% from fat) and meets most of the vitamin and mineral needs of pregnant and lactating women.

The protein DRI recommendation for pregnancy is an additional 25 grams per day higher than for nonpregnant women. Most women in the United States, however, need not add protein-rich foods to their diets because they already consume plenty of meats, seafood, poultry and eggs. Low-fat milk and milk products provide protein, calcium, vitamin D, and other nutrients.

Some vegetarian women limit or omit protein-rich meats, eggs, and milk products. For them, meeting the recommendation for food energy each day and including plant-protein foods such as legumes, tofu, whole grains, nuts, and seeds are imperative. Protein supplements during pregnancy can be harmful to infant development, and their use is discouraged.

The high nutrient requirements of pregnancy leave little room in the diet for excess fat, especially solid fats such as fatty meats and butter. The essential fatty acids, however, are particularly important to the growth and development of the fetus.[17] The brain is composed mainly of lipid material and depends heavily on long-chain omega-3 and omega-6 fatty acids for its growth, function, and structure.

KEY POINTS

- Pregnancy brings physiological adjustments that demand increased intakes of energy and nutrients.
- A balanced nutrient-dense diet is essential for meeting nutrient needs.

Of Special Interest: Folate and Vitamin B_{12} Two vitamins famous for their roles in cell reproduction—folate and vitamin B_{12}—are needed in increased amounts during pregnancy. New cells are laid down at a tremendous pace as the fetus grows and develops. At the same time, the number of the mother's red blood cells must rise because her blood volume increases, a function requiring more cell division and therefore more vitamins. To accommodate these needs, the recommendation for folate during pregnancy increases from 400 to 600 micrograms a day.

As described in Chapter 7, folate plays an important role in preventing neural tube defects. To review, the early weeks of pregnancy are a critical period for the formation and closure of the **neural tube** that will later develop to form the brain and spinal cord. By the time a woman suspects she is pregnant, usually around the sixth week of pregnancy, the embryo's neural tube normally has closed. A **neural tube defect (NTD)** occurs when the tube fails to close properly. Each year in the United States, an

neural tube the embryonic tissue that later forms the brain and spinal cord.

neural tube defect (NTD) a group of abnormalities of the brain and spinal cord apparent at birth and caused by interruption of the normal early development of the neural tube.

anencephaly (an-en-SEFF-ah-lee) an uncommon and always fatal neural tube defect in which the brain fails to form.

spina bifida (SPY-na BIFF-ih-duh) one of the most common types of neural tube defects in which gaps occur in the bones of the spine. Often the spinal cord bulges and protrudes through the gaps, resulting in a number of motor and other impairments.

Figure 13–5

Spina Bifida

Spina bifida, a common neural tube defect, occurs when the vertebrae of the spine fail to close around the spinal cord, leaving it unprotected. The B vitamin folate helps prevent spina bifida and other neural tube defects.

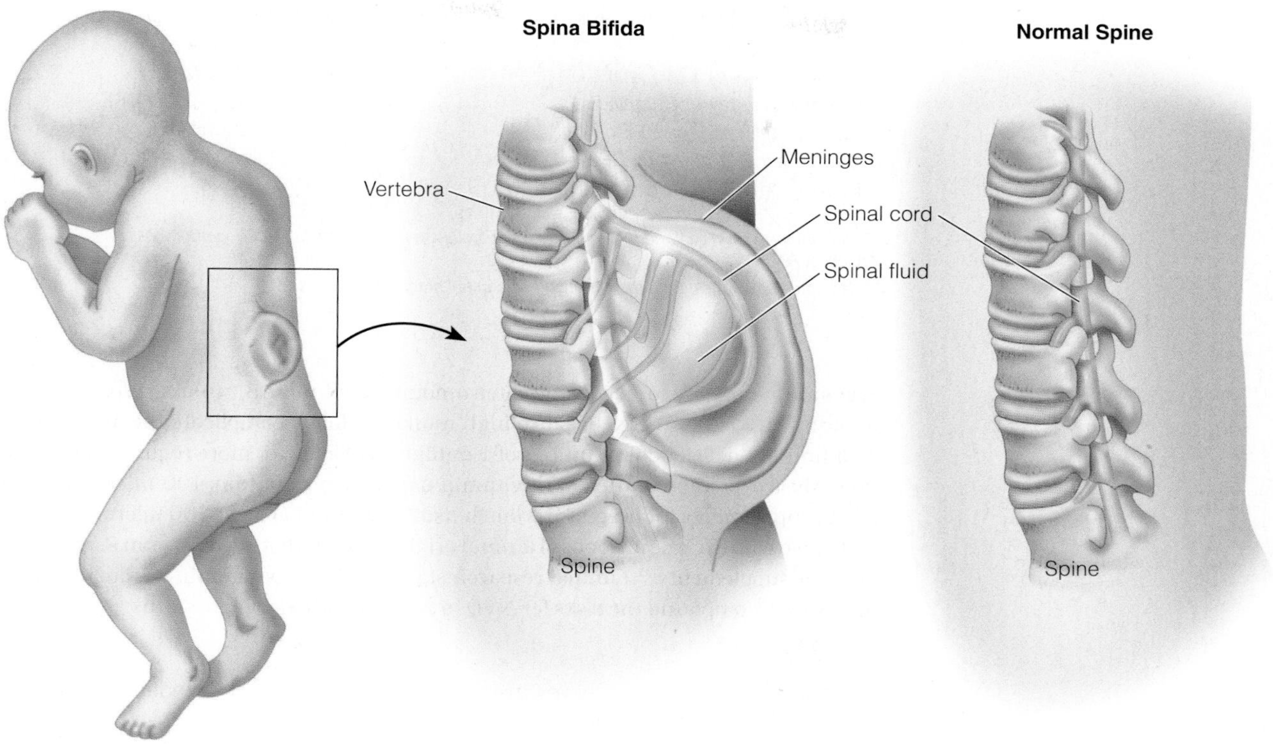

estimated 3,000 pregnancies are affected by a NTD.[18] The two most common types of NTDs are anencephaly (no brain) and spina bifida (split spine).

In **anencephaly**, the upper end of the neural tube fails to close. Consequently, the brain is either missing or fails to develop. Pregnancies affected by anencephaly often end in miscarriage; infants born with anencephaly die shortly after birth.

Spina bifida is characterized by incomplete closure of the spinal cord and its bony encasement (see Figure 13–5). The membranes covering the spinal cord and sometimes the cord itself may protrude from the spine as a sac. Spina bifida often produces paralysis in varying degrees, depending on the extent of spinal cord damage. Mild cases may not be noticed. Moderate cases may involve curvature of the spine, muscle weakness, mental handicaps, and other ills; severe cases can result in death. Table 13–3 lists risk factors for neural tube defects.

To reduce the risk of neural tube defects, women who are capable of becoming pregnant are advised to obtain 400 micrograms of folic acid daily from supplements, fortified foods, or both, *in addition* to eating folate-rich foods (see Table 13–4, p. 518). The DRI committee recommends intake of synthetic folate—folic acid—in supplements and fortified foods because it is better absorbed than the folate naturally present in foods. Foods that naturally contain folate are still important, however, because they contribute to folate intakes while providing other needed vitamins, minerals, fiber, and phytochemicals.

The folic acid enrichment of grain products (cereal, grits, pasta, rice, bread, and the like) sold commercially in the United States has improved the folate status of women of child-bearing age and lowered the number of neural tube defects that occur each year.[19] Researchers expect to see declines in some other birth defects (cleft lip and cleft palate) and miscarriages as well.[20] A safety concern arises, however. The

Table 13–3

Risk Factors for Neural Tube Defects

A pregnancy affected by a neural tube defect can occur in any woman, but these factors make it more likely:

- A personal or family history of a pregnancy affected by a neural tube defect.
- Maternal diabetes.
- Maternal use of certain anti-seizure medications.
- Mutations in folate-related enzymes.
- Maternal obesity.

Table 13–4

Rich Folate Sources[a]

Natural Folate Sources	Fortified Folate Sources
Liver (3 oz) 221 μg DFE	Highly enriched ready-to-eat cereals (¾ c) 680 μg DFE[b]
Lentils (½ c) 179 μg DFE	Pasta, cooked (1 c) 154 (average value) μg DFE
Chickpeas or pinto beans (½ c) 145 μg DFE	Rice, cooked (1 c) 153 μg DFE
Asparagus (½ c) 134 μg DFE	Bagel (1 small whole) 156 μg DFE
Spinach (1 c raw) 58 μg DFE	Waffles, frozen (2) 78 μg DFE
Avocado (½ c) 61 μg DFE	Bread, white (1 slice) 48 μg DFE
Orange juice (1 c) 74 μg DFE	
Beets (½ c) 68 μg DFE	

[a]*Folate amounts for these and thousands of other foods are listed in the Table of Food Composition in Appendix A.*

[b]*Folate in cereals varies; read the Nutrition Facts panel of the label.*

pregnant woman also needs a greater amount of vitamin B_{12} to assist folate in the manufacture of new cells. Because high intakes of folate complicate the diagnosis of a vitamin B_{12} deficiency, quantities of 1 milligram of folate or more require a prescription. Most over-the-counter multivitamin supplements contain 400 micrograms of folate; supplements for pregnant women usually contain at least 800 micrograms.

People who exclude all foods of animal origin from the diet need vitamin B_{12}-fortified foods or supplements.[21] Limited research suggests that low vitamin B_{12} during pregnancy may compound the risks for NTD in women with low folate status.[22]

KEY POINTS

- Folate and vitamin B_{12} play key roles in cell replication and are needed in large amounts during pregnancy.
- Folate plays an important role in preventing neural tube defects.

Vitamin D and Calcium Vitamin D and the minerals involved in building the skeleton—calcium, phosphorus, and magnesium—are in great demand during pregnancy. Insufficient intakes may have adverse effects on fetal bone growth and tooth development.[23]

Vitamin D plays a vital role in calcium absorption and use. Severe maternal vitamin D deficiency interferes with normal calcium metabolism, and, in rare cases, may cause the vitamin D deficiency disease rickets in a newborn.[24] Regular exposure to sunlight and consumption of vitamin D-fortified milk are usually sufficient to provide the recommended amount of vitamin D during pregnancy (15 μg), which is the same as for nonpregnant women.[25] The vitamin D in **prenatal supplements** helps to protect many, but not all, pregnant women from inadequate intakes.[26]

A woman's intestinal absorption of calcium doubles early in pregnancy, and the extra mineral is stored in her bones. Later, as the fetal bones begin to calcify, a dramatic shift of calcium across the placenta occurs. Still unknown is whether the extra calcium added to the mother's bones early in pregnancy is withdrawn later to help meet the fetus's needs.[27] In the final weeks of pregnancy, more than 300 milligrams of calcium a day are transferred to the fetus.[28]

Typically, young women in this country take in too little calcium. Of particular importance, pregnant women younger than age 25, whose own bones are still actively depositing minerals, should strive to meet the DRI recommendation for calcium by increasing their intakes of milk, cheese, yogurt, and other calcium-rich foods. The DRI recommendation for calcium intake is the same for nonpregnant and pregnant women in the same age group. To meet this recommendation, the USDA Food Patterns suggest consuming 3 cups per day of fat-free or low-fat milk or the equivalent in milk products. Women who exclude milk products need calcium-fortified foods such as soy milk, orange juice, and cereals. Less preferred, but still acceptable, is a daily supplement of 600 milligrams of calcium.

Iron A pregnant woman needs iron to help increase her blood volume and to provide for placental and fetal needs. The developing fetus draws heavily on the mother's iron stores to accumulate sufficient stores of its own to last through the first four to six months after birth.[29] The transfer of iron to the fetus is regulated by the placenta, which gives the iron needs of the fetus highest priority.[30] Even a woman with inadequate iron stores transfers a considerable amount of iron to the fetus. In addition, blood losses are inevitable at birth, especially during a delivery by cesarean section, further draining the mother's iron supply. Women who enter pregnancy with iron-deficiency anemia have a greater-than-normal risk of delivering low-birthweight or preterm infants.[31]

During pregnancy, the body makes several adaptations to help meet the exceptionally high need for iron. Menstruation, the major route of iron loss in women, ceases, and absorption of iron increases up to threefold. Even so, to prevent iron supplies from dwindling during pregnancy, the *Dietary Guidelines for Americans 2010* suggest that all women capable of becoming pregnant do these three things:

- Choose foods that supply heme iron (meat, fish, and poultry), which is most readily absorbed.
- Choose additional iron sources, such as iron-rich eggs, vegetables, and legumes.
- Consume foods that enhance iron absorption, such as vitamin C-rich fruits and vegetables.

Restoring maternal iron stores to prepregnancy levels between pregnancies is important. In doing so, women reduce the risk of potential problems related to iron deficiency, which will become greater with subsequent pregnancies.[32] Even so, few women enter pregnancy with adequate iron stores, so a daily 30-milligram iron supplement is recommended early in pregnancy, if not before.[33] A woman with a severe deficiency may need more. To enhance iron absorption, the supplement should be taken between meals and with liquids other than milk, coffee, or tea, which inhibit iron absorption.

Zinc Zinc is vital for protein synthesis and cell development during pregnancy. Typical zinc intakes of pregnant women are lower than recommendations, but fortunately, zinc absorption increases when intakes are low. Large doses of iron can interfere with zinc absorption and metabolism, but most prenatal supplements supply the right balance of these minerals for pregnancy. Zinc is abundant in protein-rich foods such as shellfish, meat, and nuts.

KEY POINTS

- Adequate vitamin D and calcium are indispensable for normal bone development of the fetus.
- Iron supplements are recommended for pregnant women.
- Zinc is needed for protein synthesis and cell development during pregnancy.

Prenatal Supplements A healthy pregnancy and optimal infant development depend heavily on the mother's diet.[34] Pregnant women who make wise food choices can meet most of their nutrient needs, with the possible exception of iron. Even so, physicians routinely recommend daily prenatal multivitamin–mineral supplements for pregnant women.[35] Prenatal supplements typically provide more folate, iron, and calcium than regular supplements (see Figure 13–6, p. 520). Women with poor diets need them urgently, as do women in these high-risk groups: women carrying twins or triplets, and those who smoke cigarettes, drink alcohol, or abuse drugs.[36] For these women in particular, prenatal supplements may reduce the risks of preterm delivery, low infant birthweights, and birth defects.[37] Supplements cannot prevent the vast majority of fetal harm from tobacco, alcohol, and drugs, however, as later sections explain.

prenatal supplements nutrient supplements specifically designed to provide the nutrients needed during pregnancy, particularly folate, iron, and calcium, without excesses or unneeded constituents.

Figure 13–6

Example of a Prenatal Supplement Label

Notice that vitamin A is reduced to guard against birth defects, while extra amounts of folate, iron, and other nutrients are provided to meet the specific needs of pregnant women.

Prenatal Vitamins

Supplement Facts
Serving Size 1 Tablet

Amount Per Tablet	% Daily Value for Pregnant/ Lactating Women
Vitamin A 4000 IU	50%
Vitamin C 100 mg	167%
Vitamin D 400 IU	100%
Vitamin E 11 IU	37%
Thiamin 1.84 mg	108%
Riboflavin 1.7 mg	85%
Niacin 18 mg	90%
Vitamin B6 2.6 mg	104%
Folate 800 mcg	100%
Vitamin B12 4 mcg	50%
Calcium 200 mg	15%
Iron 27 mg	150%
Zinc 25 mg	167%

INGREDIENTS: calcium carbonate, microcrystalline cellulose, dicalcium phosphate, ascorbic acid, ferrous fumarate, zinc oxide, acacia, sucrose ester, niacinamide, modified cellulose gum, di-alpha tocopheryl acetate, hydroxypropyl methylcellulose, hydroxypropyl cellulose, artificial colors (FD&C blue no. 1 lake, FD&C red no. 40 lake, FD&C yellow no. 6 lake, titanium dioxide), polyethylene glycol, starch, pyridoxine hydrochloride, vitamin A acetate, riboflavin, thiamin mononitrate, folic acid, beta carotene, cholecalciferol, maltodextrin, gluten, cyanocobalamin, sodium bisulfite.

Did You Know?

To provide certain key nutrients for pregnancy, lactation, and growth, WIC offers vouchers for:

- Baby foods.
- Eggs, dried and canned beans and peas, tuna fish, peanut butter.
- Fruit, vegetables, and their juices.
- Iron-fortified cereals.
- Milk and cheese.
- Soy-based beverages and tofu.
- Whole-wheat bread, and other whole-grain products.
- Iron-fortified formula for infants who are not breastfed.

KEY POINTS

- Physicians routinely recommend daily prenatal multivitamin–mineral supplements for pregnant women.
- Prenatal supplements are most likely to benefit women who do not eat adequately, are carrying twins or triplets, or who smoke cigarettes, drink alcohol, or abuse drugs.

Food Assistance Programs

The nationwide **Special Supplemental Nutrition Program for Women, Infants, and Children (WIC)** provides vouchers redeemable for nutritious foods, along with nutrition education and referrals to health and social services for low-income pregnant and lactating women and their children.[38] WIC encourages breastfeeding and offers incentives to mothers who do so. For infants given infant formula, WIC also provides iron-fortified formula.

More than 9 million people—most of them infants and young children—receive WIC benefits each month. Proven benefits from WIC participation include improved nutrient status and growth among infants and children, improved iron status among pregnant women, reduced risk of infant mortality and low birthweight, and reduced maternal and newborn medical costs. In addition to WIC, the Supplemental Nutrition Assistance Program (formerly the Food Stamp Program) can also help to stretch the low-income family's grocery dollars. Many communities and organizations such as the Academy of Nutrition and Dietetics and local hospitals also provide educational services and materials, including nutrition, food budgeting, and shopping information.

KEY POINTS

- Food assistance programs such as WIC can provide nutritious food for pregnant women of limited financial means.
- Participation in WIC during pregnancy can reduce iron deficiency, infant mortality, low birthweight, and maternal and newborn medical costs.

How Much Weight Should a Woman Gain during Pregnancy?

Women must gain weight during pregnancy—fetal and maternal well-being depend on it. Ideally, a woman will have begun her pregnancy at a healthy weight, but even more importantly, she will gain within the recommended range for her prepregnancy body mass index (BMI), as shown in Table 13–5. The benefits of proper weight gain include a lower risk of surgical birth, a greater chance of having a healthy birthweight

Table 13–5

Recommended Weight Gains Based on Prepregnancy Weight

	RECOMMENDED WEIGHT GAIN	
Prepregnancy Weight	*For single birth*	*For twin birth*
Underweight (BMI <18.5)	28 to 40 lb (12.5 to 18.0 kg)	Insufficient data to make recommendation
Healthy weight (BMI 18.5 to 24.9)	25 to 35 lb (11.5 to 16.0 kg)	37 to 54 lb (17.0 to 25.0 kg)
Overweight (BMI 25.0 to 29.9)	15 to 25 lb (7.0 to 11.5 kg)	31 to 50 lb (14.0 to 23.0 kg)
Obese (BMI ≥30)	11 to 20 lb (5.0 to 9.0 kg)	25 to 42 lb (11.0 to 19.0 kg)

Source: Institute of Medicine, Weight Gain during Pregnancy: Reexamining the Guidelines *(Washington, D.C.: National Academies Press, 2009).*

Figure 13–7
Components of Weight Gain during Pregnancy

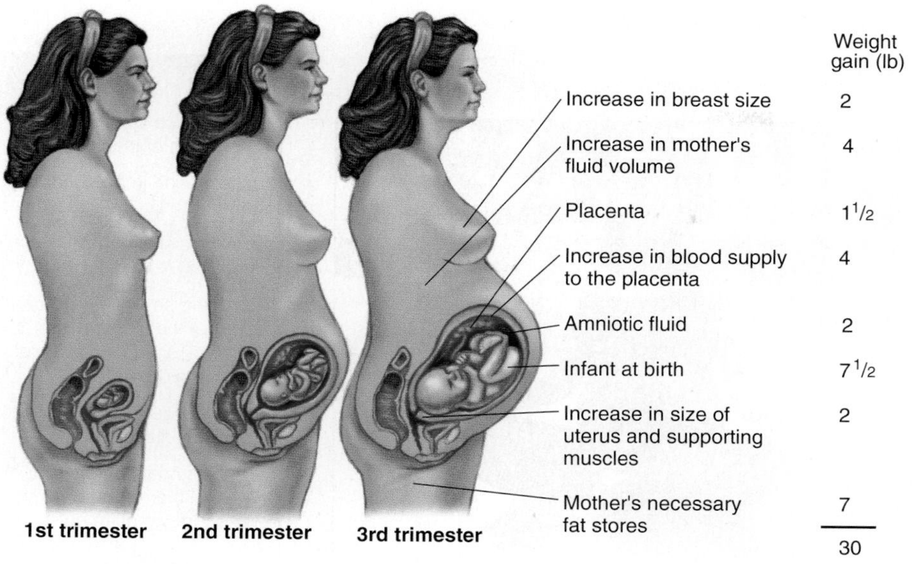

baby, and other positive outcomes for both mothers and infants. Many women exceed the recommended ranges, however, and a few even fall short.[39]

Weight loss during pregnancy is not recommended.[40] Even obese women are advised to gain between 11 and 20 pounds for the best chance of delivering a healthy baby.[41] Overweight women should achieve a healthy body weight before becoming pregnant, avoid excessive weight gain during pregnancy, and postpone weight loss until after childbirth.[42]

The ideal weight-gain pattern for a woman who begins pregnancy at a healthy weight is 3½ pounds during the first trimester and 1 pound per week thereafter. If a woman gains more than is recommended early in pregnancy, she should not restrict her energy intake later on in order to lose weight. Any sudden, large weight gain is a danger signal, however, because it may indicate the onset of preeclampsia (see the section entitled "Troubleshooting").

The weight the pregnant woman gains is nearly all lean tissue: placenta, uterus, blood, milk-producing glands, and the fetus itself (see Figure 13–7). The fat she gains is needed later for lactation. Physical activity can help a pregnant woman cope with the extra weight, as a later section explains.

Weight Loss after Pregnancy

The pregnant woman loses some weight at delivery. In the following weeks, she loses more as her blood volume returns to normal and she loses accumulated fluids. The typical woman does not immediately return to her prepregnancy weight. In general, the more weight a woman gains beyond the needs of pregnancy, the more she retains—mostly as body fat.[43] Even without excessive gain, most women tend to retain a few pounds with each pregnancy. When those few pounds become 7 or more and BMI increases by a unit or more, risks of diabetes and hypertension in future pregnancies, as well as chronic diseases later in life, can increase. Women who achieve a healthy weight prior to the first pregnancy and maintain it between pregnancies best avoid the cumulative weight gain that threatens health later on.

KEY POINTS

- Appropriate weight gain is essential for a healthy pregnancy.
- Appropriate weight gain is influenced by prepregnancy BMI, maternal nutrient needs, and the number of fetuses in the pregnancy.

Special Supplemental Nutrition Program for Women, Infants, and Children (WIC) a USDA program offering low-income pregnant and lactating women and those with infants or preschool children coupons redeemable for specific foods that supply the nutrients deemed most necessary for growth and development. For more information, visit www.usda.gov/FoodandNutrition.

Figure 13–8

Guidelines for Physical Activity during Pregnancy

Pregnant women can enjoy the benefits of physical activity.

DO

- Do exercise regularly (most, if not all, days of the week).
- Do warm up with 5 to 10 minutes of light activity.
- Do 30 minutes or more of moderate physical activity.
- Do cool down with 5 to 10 minutes of slow activity and gentle stretching.
- Do drink water before, after, and during exercise.
- Do eat enough to support the additional needs of pregnancy plus exercise.
- Do rest adequately.

DON'T

- Don't exercise vigorously after long periods of inactivity.
- Don't exercise in hot, humid weather.
- Don't exercise when sick with fever.
- Don't exercise while lying on your back after the first trimester of pregnancy or stand motionless for prolonged periods.
- Don't exercise if you experience any pain or discomfort.
- Don't participate in activities that may harm the abdomen or involve jerky, bouncy movements.
- Don't scuba dive.

Should Pregnant Women Be Physically Active?

An active, physically fit woman experiencing a normal, healthy pregnancy can and should continue to exercise throughout pregnancy, adjusting the intensity and duration as the pregnancy progresses. Staying active improves the fitness of the mother-to-be, facilitates labor, helps to prevent or manage gestational diabetes, and reduces psychological stress. Active women report fewer discomforts throughout their pregnancies and retain habits that help to lose excess weight and regain fitness later.[44]

Pregnant women should choose "low-impact" activities and avoid sports in which they might fall or be hit by other people or objects (for some safe activity suggestions, see the Think Fitness box). Pregnant women with medical conditions or pregnancy complications should seek medical advice before engaging in physical activity. A few more guidelines are offered in Figure 13–8. Several of the guidelines are aimed at preventing excessively high internal body temperature and dehydration, both of which can harm fetal development. To this end, the pregnant woman should also stay out of saunas, steam rooms, and hot whirlpools.

KEY POINTS

- Physically fit women can continue physical activity throughout pregnancy but should choose activities wisely.
- Pregnant women should avoid sports in which they might fall or be hit and avoid becoming overheated or dehydrated.

Teen Pregnancy

The number of infants born to teenage mothers has steadily declined during the last 50 years. Despite the long-term decline, however, the U.S. teen birth rate is still one of highest among industrialized nations.[45] In 2010, more than 367,000 infants were born to teenage mothers.

A pregnant adolescent presents a special case of intense nutrient needs. Young teenage girls have a hard enough time meeting nutrient needs for their own rapid growth and development, let alone those of pregnancy. Many teens enter pregnancy with deficiencies of vitamins B_{12} and D, folate, iron, and calcium that can impair fetal growth.[46] Pregnant adolescents are less likely to receive early prenatal care and are

Think Fitness **move it!** Physical Activities for the Pregnant Woman

Is there an ideal physical activity for the pregnant woman? There might be. Swimming and water aerobics offer advantages over other activities during pregnancy. Water cools and supports the body, provides a natural resistance, and lessens the impact of the body's movement, especially in the later months. Water aerobics can help reduce the intensity of back pain during pregnancy. Other activities considered safe and comfortable for pregnant women include walking, light strength training, rowing, yoga, and climbing stairs.

start now! Ready to make a change? If you weren't exercising regularly before you became pregnant, talk to your doctor before undertaking an activity. Track your activity daily using the Diet Analysis Plus Activity Tracker.

more likely to smoke during pregnancy—two factors that predict low birthweight and infant death.[47] The rates of stillbirths, preterm births, and low-birthweight infants are high when either parent is a teen. Adequate nutrition and appropriate weight gain during pregnancy are indispensable components of prenatal care for teenagers and can substantially improve the outlook for both mother and infant.

A pregnant teenager with a normal BMI is encouraged to gain about 35 pounds. Concerns about obesity raise questions as to whether such recommendations are always appropriate. Compared with older mothers, for example, the adolescent mother's risk of excess weight gain throughout life may be far greater.[48] Researchers agree that optimal weight gain recommendations for pregnant adolescents need focused attention. Meanwhile, pregnant and lactating teenagers can follow the eating pattern presented earlier in Table 2–2 (page 45), making sure to choose a calorie level high enough to support adequate, but not excessive, weight gain.

KEY POINTS

- Pregnant teenage girls have extraordinarily high nutrient needs and an increased likelihood of problem pregnancies.
- Adequate nutrition and appropriate weight gain for pregnant teenagers can substantially improve outcomes for mothers and infants.

Why Do Some Women Crave Pickles and Ice Cream While Others Can't Keep Anything Down?

Does pregnancy give a woman the right to demand pickles and ice cream at 2 am? Perhaps so, but not for nutrition's sake. Food cravings and aversions during pregnancy are common but do not seem to reflect real physiological needs. In other words, a woman who craves pickles is probably not in need of salt. Food cravings and aversions that arise during pregnancy may be due to hormone-induced changes in taste and sensitivities to smells, and they quickly disappear after the birth.

Some pregnant women respond to cravings by eating nonfood items such as laundry starch, clay, soil, or ice—a practice known as pica.[49] Pica may be practiced for cultural reasons that reflect a society's folklore; it is especially common among African American women. Chapter 8 provides more details.

The nausea of "morning" (actually, anytime) sickness seems unavoidable and may even be a welcome sign of a healthy pregnancy because it arises from the hormonal changes of early pregnancy. The problem typically peaks at 9 weeks gestation and resolves within a month or two.[50] Many women complain that odors, especially cooking smells, make them feel nauseated, so minimizing odors may provide some relief. Traditional strategies for quelling nausea are listed in Table 13–6, but little evidence exists to support such advice.[51] Some women do best by simply eating what they desire whenever they feel hungry. Morning sickness can be persistent, however, and if it interferes with normal eating for more than a week or two, the woman should seek medical help to prevent nutrient deficiencies.

Table 13–6

Tips for Relieving Common Discomforts of Pregnancy

To alleviate the nausea of pregnancy:
- On waking, get up slowly.
- Eat dry toast or crackers.
- Chew gum or suck hard candies.
- Eat small, frequent meals whenever hunger strikes.
- Avoid foods with offensive odors.
- When nauseated, do not drink citrus juice, water, milk, coffee, or tea.

To prevent or alleviate constipation:
- Eat foods high in fiber.
- Exercise daily.
- Drink at least 8 glasses of liquids a day.
- Respond promptly to the urge to defecate.
- Use laxatives only as prescribed by a physician; avoid mineral oil—it carries needed fat-soluble vitamins out of the body.

To prevent or relieve heartburn:
- Relax and eat slowly.
- Chew food thoroughly.
- Eat small, frequent meals.
- Drink liquids between meals.
- Avoid spicy or greasy foods.
- Sit up while eating.
- Wait an hour after eating before lying down.
- Wait 2 hours after eating before exercising.

As the hormones of pregnancy alter her muscle tone and the thriving fetus crowds her intestinal organs, an expectant mother may complain of heartburn or constipation. Raising the head of the bed with two or three pillows can help to relieve nighttime heartburn. A high-fiber diet, physical activity, and a plentiful water intake will help relieve constipation. The pregnant woman should use laxatives or heartburn medications only if her physician prescribes them.

KEY POINTS

- Food cravings usually do not reflect physiological needs, and some may interfere with nutrition.
- Nausea arises from normal hormonal changes of pregnancy.

Some Cautions for the Pregnant Woman

Some choices that pregnant women make or substances they encounter can harm the fetus, sometimes severely. Smoking and other threats all deserve consideration, but alcohol constitutes an even greater threat to fetal health and is given a section of its own.

Cigarette Smoking A surgeon general's warning states that parental smoking can kill an otherwise healthy fetus or newborn. Unfortunately, an estimated 10 to 12 percent of pregnant women in the United States smoke, and rates are even higher for unmarried women and those who have not graduated from high school.[52]

Constituents of cigarette smoke, such as nicotine, carbon monoxide, arsenic, and cyanide, are toxic to a fetus.[53] Smoking during pregnancy can damage fetal DNA, which could lead to developmental defects or diseases such as cancer. Smoking restricts the blood supply to the growing fetus and so limits the delivery of oxygen and nutrients and the removal of wastes. It slows fetal growth, can reduce brain size, and may impair the intellectual and behavioral development of the child later in life. Smoking during pregnancy damages fetal blood vessels, an effect that is still apparent at the age of 5 years.[54]

A mother who smokes is more likely to have a complicated birth and a low-birthweight infant. The more a mother smokes, the smaller her baby will be. Of all preventable causes of low birthweight in the United States, smoking has the greatest impact. Smokers tend to have lower intakes of dietary fiber, vitamin A, beta-carotene, folate, and vitamin C—nutrients necessary for a healthy pregnancy. The margin lists complications of smoking during pregnancy.

Did You Know?

Complications associated with smoking during pregnancy:

- Low birthweight.
- Spontaneous abortion.
- Fetal death.
- Sudden infant death syndrome.
- Childhood middle ear infections; cardiac and respiratory diseases.

Smoking during pregnancy interferes with fetal lung development and increases the risks of respiratory infections and childhood asthma.[55] Sudden infant death syndrome (SIDS), the unexplained deaths that sometimes occur in otherwise healthy infants, has been linked to the mother's cigarette smoking during pregnancy.[56] Even in nonsmokers, regular exposure to **environmental tobacco smoke** (or secondhand smoke) during pregnancy increases the risk of low birthweight and the likelihood of SIDS.

Alternatives to smoking—such as using snuff, chewing tobacco, or nicotine-replacement therapy—are not safe during pregnancy.[57] A woman who uses nicotine in any form and is considering pregnancy or who is already pregnant should make every effort to quit.

Medicinal Drugs and Herbal Supplements Medicinal drugs taken during pregnancy can cause serious birth defects. A pregnant woman should not take over-the-counter drugs or any medications not prescribed by a physician; even then, she should read the labels and take warnings seriously.

Some pregnant women mistakenly consider herbal supplements to be safe alternatives to medicinal drugs and take them to relieve nausea, promote water loss, alleviate depression, aid sleep, or for other reasons. Some herbal products may be safe, but almost none have been tested for safety or effectiveness during pregnancy. Pregnant women should stay away from herbal supplements, teas, or other products unless their safety during pregnancy has been ascertained.[58]

Drugs of Abuse Drugs of abuse such as cocaine easily cross the placenta and impair fetal growth and development. Furthermore, such drugs are responsible for preterm births, low-birthweight infants, and sudden infant deaths. If these newborns survive, central nervous system damage is evident: their cries, sleep, and behaviors early in life are abnormal, and their cognitive development later in life is impaired.[59] They may be hypersensitive or underaroused; infants who test positive for drugs suffer the greatest effects of toxicity and withdrawal.[60] Their childhood growth continues, but at a slow rate.

Environmental Contaminants Pregnant women who are exposed to contaminants such as lead often bear low-birthweight infants with delayed mental and psychomotor development who struggle to survive. During pregnancy, the heavy metal lead readily moves across the placenta, inflicting severe damage on the developing fetal nervous system. For pregnant women, choosing a diet free of contamination takes on extra urgency. Adequate dietary calcium can help to defend against lead toxicity by reducing its absorption.

Mercury is a contaminant of concern as well. As discussed in Chapter 5, fatty fish are a good source of omega-3 fatty acids, but some fish contain large amounts of the pollutant mercury, which can harm the developing fetal brain and nervous system. Because the benefits of moderate seafood consumption seem to outweigh the risks, pregnant (and lactating) women need reliable information on which fish are safe to eat.[61] The *Dietary Guidelines for Americans* advise pregnant and lactating women to do the following:

- Avoid eating shark, swordfish, king mackerel, and tilefish (also called golden snapper or golden bass).
- Limit average weekly consumption to 12 ounces (cooked or canned) of seafood *or* to 6 ounces (cooked or canned) of white (albacore) tuna.

Foodborne Illness The vomiting and diarrhea caused by many foodborne illnesses can leave a pregnant woman exhausted and dangerously dehydrated. Particularly threatening, however, is **listeriosis**, which can cause miscarriage, stillbirth, or severe brain or other infections in fetuses and newborns. Pregnant women are about 20 times more likely than other healthy adults to get listeriosis. A woman with listeriosis may develop symptoms such as fever, vomiting, and diarrhea in about 12 hours after eating a contaminated food; serious symptoms may develop a week to six weeks later. A blood test can reliably detect listeriosis, and antibiotics given promptly to the pregnant sufferer can often prevent infection of the fetus or newborn. To protect herself and her fetus from listeriosis, a pregnant woman should follow all of the food safety advice of Chapter 12, and she should observe the following recommendations:

- Use only pasteurized juices and dairy products; do not eat soft cheeses such as feta, brie, Camembert, Panela, "queso blanco," "queso fresco," and blue-veined cheeses such as Roquefort; do not drink raw (unpasteurized) milk or eat foods that contain it.
- Do not eat hot dogs or luncheon or deli meats unless heated until steaming hot.
- Thoroughly cook meat, poultry, eggs, and seafood.
- Wash all fruits and vegetables.
- Avoid refrigerated (not canned) patés or smoked seafood or any fish labeled "nova-style," "lox," or "kippered."

Vitamin–Mineral Overdoses Many vitamins and minerals are toxic when taken in excess. Excessive vitamin A is widely known for its role in fetal malformations of the cranial nervous system. Intakes before the seventh week of pregnancy appear to be the most damaging. For this reason, vitamin A supplements are not given during pregnancy, unless there is specific evidence of deficiency, which is rare.

environmental tobacco smoke the combination of exhaled smoke (mainstream smoke) and smoke from lighted cigarettes, pipes, or cigars (sidestream smoke) that enters the air and may be inhaled by other people.

listeriosis a serious foodborne infection that can cause severe brain infection or death in a fetus or a newborn; caused by the bacterium *Listeria monocytogenes*, which is found in soil and water.

Restrictive Dieting Restrictive dieting, even for short periods, can be hazardous during pregnancy. In particular, low-carbohydrate diets or fasts that cause ketosis deprive the growing fetal brain of needed glucose and may impair cognitive development. Such diets are also likely to lack other nutrients vital to fetal growth. Regardless of prepregnancy weight, pregnant women need an adequate diet to support healthy fetal development.

Sugar Substitutes Artificial sweeteners have been studied extensively and found to be acceptable during pregnancy if used within the FDA's guidelines.[62] Women with the inborn error of metabolism known as phenylketonuria should not use the artificial sweetener aspartame.

Caffeine Caffeine crosses the placenta, and the fetus has only a limited ability to metabolize it. Even so, women can safely consume less than 300 milligrams a day without apparent ill effects on their pregnancy duration or outcome.[63] Limited evidence suggests that heavy use—intake equaling more than three cups of coffee a day—may increase the risk of hypertension and miscarriage.[64] Depending on the quantities consumed and the mother's metabolism, caffeine may also interfere with fetal growth.[65] The most sensible course, therefore, is to limit caffeine consumption to the equivalent of about two cups of coffee or three 12-ounce cola beverages a day. Caffeine amounts in food and beverages are listed in Controversy 14 on pages 590–595.

Did You Know?

One "drink" is the equivalent of:

- 5 oz wine (12% alcohol).
- 10 oz wine cooler.
- 12 oz beer.
- 1½ oz hard liquor (80 proof).

KEY POINTS

- Smoking during pregnancy delivers toxins to the fetus, damages DNA, restricts fetal growth, and limits delivery of oxygen and nutrients and the removal of wastes.
- Smoking and other drugs, contaminants such as mercury, foodborne illnesses, large supplemental doses of nutrients, weight-loss diets, and excessive use of artificial sweeteners and caffeine should be avoided during pregnancy.

Drinking during Pregnancy

LO 13.2 Evaluate the statement that "no level of alcoholic beverage intake is safe or advisable during pregnancy."

Alcohol is arguably the most hazardous drug to future generations because it is legally available, heavily promoted, and widely abused. Society sends mixed messages concerning alcohol. Beverage companies promote an image of drinkers as healthy and active. Opposing this image, health authorities warn that alcohol can injure health, especially during pregnancy (see Figure 13–9). Every container of beer, wine, liquor, or mixed drinks for sale in the United States is required to warn pregnant women of the dangers of drinking during pregnancy.

Figure 13–9

Mixed Messages in Alcohol Advertisements

Labels on alcoholic beverages often display "healthy" images, but their warnings tell the truth.

© Cengage Learning

Alcohol's Effects

Women of childbearing age need to know about alcohol's harmful effects on a fetus. Alcohol crosses the placenta freely and is directly toxic:[66]

- A sudden dose of alcohol can halt the delivery of oxygen through the umbilical cord. The fetal brain and nervous system are extremely vulnerable to a glucose or oxygen deficit, and alcohol causes both by disrupting placental functioning. Alcohol slows cell division, reducing the number of cells produced and inflicting abnormalities on those that are produced and all of their progeny.
- During the first month of pregnancy, the fetal brain is growing at the rate of 100,000 new brain cells a minute. Even a few minutes of alcohol exposure during this critical period can exert a major detrimental effect.
- Alcohol interferes with placental transport of nutrients to the fetus and can cause malnutrition in the mother; then, all of malnutrition's harmful effects compound the effects of the alcohol.
- Before fertilization, alcohol can damage the ovum or sperm in the mother- or father-to-be, leading to abnormalities in the child.

KEY POINTS

- Alcohol crosses the placenta and is directly toxic to the fetus.
- Alcohol limits oxygen delivery to the fetus, slows cell division, and reduces the number of cells organs produce.

Fetal Alcohol Syndrome

Drinking alcohol during pregnancy threatens the fetus with irreversible brain damage, growth restriction, mental retardation, facial abnormalities, vision abnormalities, and many more health problems—a spectrum of symptoms known as **fetal alcohol spectrum disorders**, or **FASD**. Children at the most severe end of the spectrum (those with all of the symptoms) are defined as having **fetal alcohol syndrome**, or **FAS**.[67] The lifelong mental retardation and other tragedies of FAS can be prevented by abstaining from drinking alcohol during pregnancy. Once the damage is done, however, the child remains impaired. Figure 13–10 shows the facial abnormalities of FAS, which are easy to depict. A visual picture of the internal harm is impossible, but that damage seals the fate of the child. An estimated 5 to 20 of every 10,000 children are victims of FAS, making it one of the leading known preventable causes of mental retardation in the world.[68]

Even when a child does not develop full FAS, prenatal exposure to alcohol can lead to less severe, but nonetheless serious, mental and physical problems. The cluster

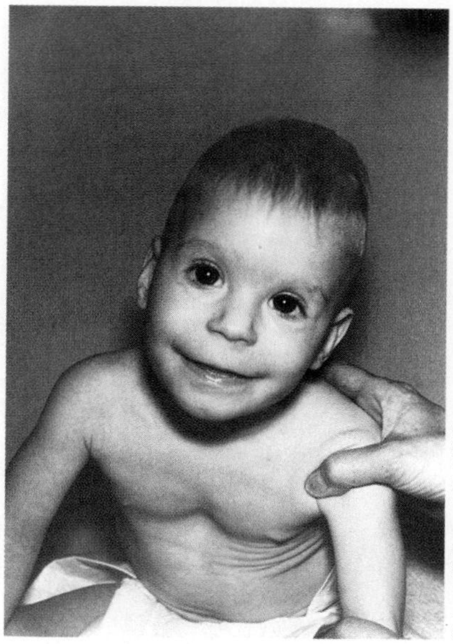

A child with FAS.

Figure 13–10

Typical Facial Characteristics of FAS

The severe facial abnormalities shown here are just outward signs of severe mental impairments and internal organ damage. These defects, though hidden, may create major health problems later.

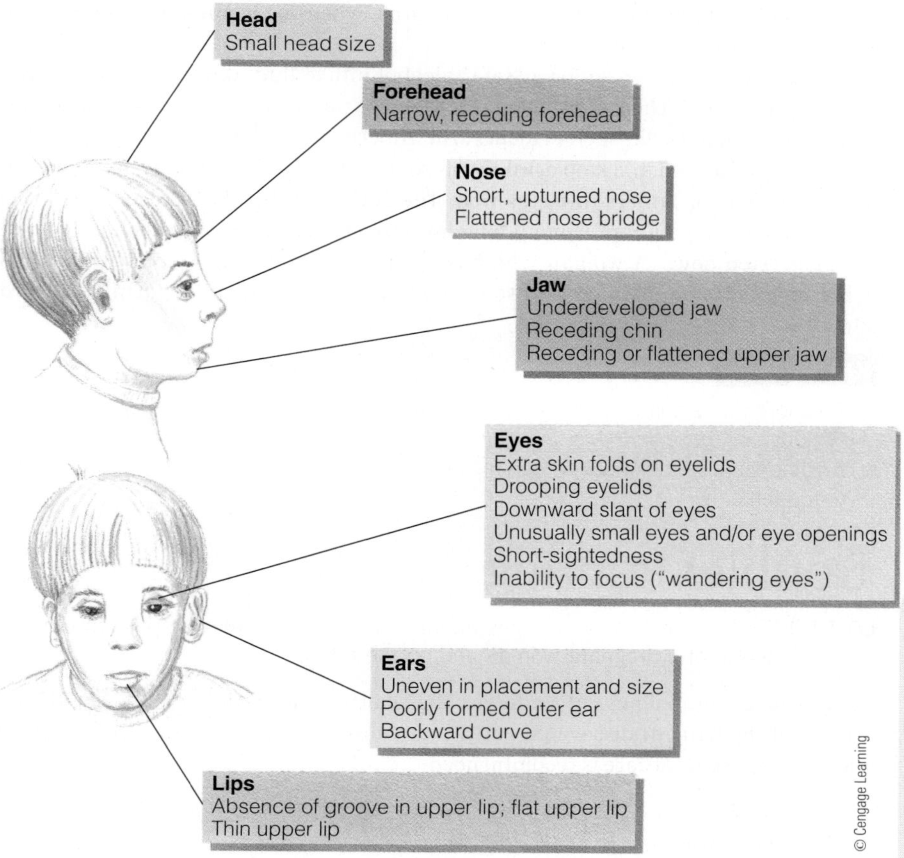

fetal alcohol spectrum disorders (FASD) a spectrum of physical, behavioral, and cognitive disabilities caused by prenatal alcohol exposure.

fetal alcohol syndrome (FAS) the cluster of symptoms including brain damage, growth restriction, mental retardation, and facial abnormalities seen in an infant or child whose mother consumed alcohol during her pregnancy.

of mental problems is known as **alcohol-related neurodevelopmental disorder (ARND)**, and the physical malformations are referred to as **alcohol-related birth defects (ARBD)**.[†] Some of these children show no outward sign of impairment, but others are short in stature or display subtle facial abnormalities. Many perform poorly in school and in social interactions and suffer a subtle form of brain damage. Mood disorders and problem behaviors, such as aggression, are common.

Many children with ARND or ARBD go undiagnosed until problems develop in the preschool years. Upon reaching adulthood, such children are ill equipped for employment, relationships, and the other facets of life most adults take for granted. Alcohol exposure before birth may alter the person's later response to alcohol and other mind-altering drugs, making addictions likely.

KEY POINTS

- The severe birth defects of fetal alcohol syndrome arise from damage to the fetus by alcohol.
- Lesser conditions, ARND and ARBD, also arise from alcohol use in pregnancy.

Experts' Advice

Despite alcohol's potential for harm, 1 out of 8 pregnant women drinks alcohol at some time during pregnancy; 1 out of 75 report "binge" drinking (four or more drinks on one occasion).[69]

Women who know they are pregnant and choose to drink alcohol often ask, "How much alcohol is too much?" The damaging effects are dose dependent, becoming greater as the dose increases.[70] Even one drink a day threatens neurological development and behavior. Low birthweight is reported among infants born to women who drink 1 ounce (two drinks) per day during pregnancy, and FAS is also known to occur with as few as two drinks a day. Birth defects have been reliably observed among the children of women who drink 2 ounces (four drinks) of alcohol daily during pregnancy. Compared to women who do not drink, a sizable and significant increase in stillbirths occurs in women who drink five or more drinks per week. The most severe impact is likely to occur in the first two months, when the woman may not even be aware that she is pregnant.

Researchers have looked for a "safe" alcohol intake limit during pregnancy and have found none.[71] Their conclusion: abstinence from alcohol is the best policy for pregnant women. Given such evidence, the American Academy of Pediatrics (AAP) states that women should stop drinking as soon as they *plan* to become pregnant, an important step for fathers-to-be, as well. The authors of this book recommend this choice, too. For a pregnant woman who has already been drinking alcohol, the best advice is "stop now." A woman who has drunk heavily during the first two-thirds of her pregnancy can still prevent some organ damage by stopping during the third trimester.

KEY POINTS

- Alcohol's damaging effects on the fetus are dose dependent, becoming greater as the dose increases.
- Abstinence from alcohol in pregnancy is critical to preventing irreversible damage to the fetus.

Did You Know?

The *Dietary Guidelines for Americans 2010* lists pregnant women and women who may be pregnant among those who should not drink alcohol at all.

Troubleshooting

LO 13.3 Describe the impacts of gestational diabetes and preeclampsia on the health of a pregnant woman and on the fetus.

Disease during pregnancy can endanger the health of the mother and the health and growth of the fetus. If discovered early, many diseases can be controlled—another reason early prenatal care is recommended.

alcohol-related neurodevelopmental disorder (ARND) behavioral, cognitive, or central nervous system abnormalities associated with prenatal alcohol exposure.

alcohol-related birth defects (ARBD) malformations in the skeletal and organ systems (heart, kidneys, eyes, ears) associated with prenatal alcohol exposure.

† Formerly, ARND and ARBD were grouped together and called fetal alcohol effects (FAE).

Diabetes

Pregnancy presents special challenges for the management of diabetes. Pregnant women with unmanaged type 1 or type 2 diabetes may experience episodes of severe hypoglycemia or hyperglycemia, preterm labor, and pregnancy-related hypertension. Infants may be large or may suffer physical and mental abnormalities or other complications such as respiratory distress. Signs of fetal health problems are apparent even in prediabetes, when maternal glucose is just above normal.

Excellent glycemic control in the first trimester and throughout the pregnancy is associated with the lowest frequency of maternal, fetal, and newborn complications.[72] Ideally, a woman will receive the prenatal care needed to achieve glucose control before conception and continued glucose control throughout pregnancy. Then, continued diabetes management after pregnancy will guard the woman's long-term health.

Some women are prone to develop a pregnancy-related form of diabetes, **gestational diabetes** (see Table 13–7). Gestational diabetes usually resolves after the birth, but some women develop diabetes (usually type 2) later in life, especially if they are overweight.[73] In half of women with gestational diabetes, other forms of diabetes ensue within a few years.[74] When gestational diabetes is identified early and managed properly, the most serious risks, fetal or infant illness or mortality, fall dramatically. Gestational diabetes often leads to surgical birth and high infant birthweight. Physicians screen for the risk factors listed in Table 13–7 and test glucose tolerance in all pregnant women.[75]

Table 13–7

Risk Factors for Gestational Diabetes

- Age 25 or older.
- BMI ≥25 or excessive weight gain.
- Complications in previous pregnancies, including gestational diabetes or high birthweight.
- Prediabetes or symptoms of diabetes.
- Family history of diabetes.
- African American, Asian American, Hispanic American, Native American, Pacific Islander.

Hypertension

Hypertension during pregnancy may be **chronic hypertension** or **gestational hypertension**.[76] In chronic hypertension the condition is generally present before and remains after pregnancy. In women with gestational hypertension, blood pressure usually returns to normal during the first few weeks after childbirth.

Both types of hypertension pose risks to the mother and fetus; the higher the blood pressure, the worse the risk. In addition to heart attack and stroke, high blood pressure may increase the likelihood of growth restriction, preterm birth, and separation of the placenta from the wall of the uterus before the birth.[77]

Table 13–8

Warning Signs of Preeclampsia

- Hypertension.
- Protein in the urine.
- Upper abdominal pain.
- Severe and constant headaches.
- Swelling, especially of the face.
- Dizziness.
- Blurred vision.
- Sudden weight gain (1 lb/day).

Preeclampsia

Preeclampsia involves not only high blood pressure but also protein in the urine.[78] Preeclampsia usually appears in first pregnancies (see Table 13–8 for its warning signs), and starts to disappear within a few days after delivery.[79] Because delivery is the only known cure, preeclampsia is a leading cause of indicated preterm delivery and accounts for about 15 percent of infants who are growth restricted.[80]

Preeclampsia affects almost all of the mother's organs—the circulatory system, liver, kidneys, and brain. If the condition progresses, she may experience seizures; when this occurs, the condition is called **eclampsia**. Maternal mortality during pregnancy is rare in developed countries, but eclampsia is one of the most common causes.[81] Preeclampsia and eclampsia demand prompt medical attention.

KEY POINTS

- If discovered early, many diseases of pregnancy can be controlled—an important reason early prenatal care is recommended.
- Gestational diabetes, hypertension, and preeclampsia are problems of some pregnancies that must be managed to minimize associated risks.

Lactation

LO 13.4 Describe the nutrition needs of lactation, the impact of malnutrition on breast milk, and contraindications to breastfeeding.

As the time of childbirth nears, a woman must decide whether she will feed her baby breast milk, infant formula, or both. These options are the only foods recommended for an infant during the first four to six months of life. A woman who plans

gestational diabetes abnormal glucose tolerance appearing during pregnancy.

chronic hypertension in pregnant women, hypertension that is present and documented before pregnancy; in women whose prepregnancy blood pressure is unknown, the presence of sustained hypertension before 20 weeks of gestation.

gestational hypertension high blood pressure that develops in the second half of pregnancy and usually resolves after childbirth.

preeclampsia (PRE-ee-CLAMP-see-ah) a potentially dangerous condition during pregnancy characterized by hypertension and protein in the urine.

eclampsia (eh-CLAMP-see-ah) a severe complication during pregnancy in which seizures occur.

My Turn watch it!

Bringing Up Baby

Two students talk about some of the choices they make about the care and feeding of their babies.

© Cengage Learning

Cari

© Cengage Learning

Natasha

to breastfeed her baby should begin to prepare toward the end of her pregnancy. No elaborate or expensive preparations are needed, but the expectant mother can read one of the many handbooks available on breastfeeding, or consult a **certified lactation consultant**, employed at many hospitals.[‡] Health-care professionals play an important role in providing encouragement and accurate information on breastfeeding. Part of the preparation involves learning what dietary changes are needed because adequate nutrition is essential to successful lactation.

In rare cases, women produce too little milk to nourish their infants adequately. Severe consequences, including infant dehydration, malnutrition, and brain damage, can occur should the condition go undetected for long. Early warning signs of insufficient milk are dry diapers (a well-fed infant wets about six to eight diapers a day) and infrequent bowel movements.

Nutrition during Lactation

A nursing mother produces about 25 ounces of milk a day, with considerable variation from woman to woman and in the same woman from time to time. The volume produced depends primarily on the infant's demand for milk. The more milk the infant needs, the more the well-nourished mother's body will produce, enough to feed the infant—or even twins—amply.

Energy Cost of Lactation Producing milk costs a woman almost 500 calories per day above her regular need during the first six months of lactation. To meet this energy need, the woman is advised to eat an extra 330 calories of food each day. The other 170 calories can be drawn from the fat stores she accumulated during pregnancy. The food energy consumed by the nursing mother should carry with it abundant nutrients. A lactating woman's nutrient recommendations are listed on the inside front cover; look again at Table 13–2 on page 516 for a sample menu to meet them.

Fluid Need Breast milk contains a lot of water, so the nursing mother is advised to drink plenty of fluid each day (about 13 cups) to protect herself from dehydration.[§] To help themselves remember, many women make a habit of drinking a glass of milk, juice, or water each time the baby nurses as well as at mealtimes.

Variations in Breast Milk A common question is whether a mother's milk may lack a nutrient if she fails to get enough in her diet. The answer differs from one nutrient to the next, but in general, the effect of nutritional deprivation of the mother is to reduce the *quantity*, not the *quality*, of her milk.

Women can produce milk with adequate protein, carbohydrate, fat, folate, and most minerals, even when their own supplies are limited, at the expense of maternal

[‡] La Leche League is an international organization that helps women with breastfeeding concerns: www.lalecheleague.org.
[§] The DRI recommendation for *total* water intake during lactation is 3.8 L/day. This includes 3.1 L, or about 13 cups, as total beverages, including water.

stores. This is most evident in the case of calcium: dietary calcium has no effect on the calcium concentration of breast milk, but maternal bones lose some of their density during lactation if calcium intakes are inadequate.[82] Such losses are generally made up quickly when lactation ends, and breastfeeding has no long-term harmful effects on women's bones. In severe, prolonged malnutrition, breast milk may contain less of certain nutrients such as vitamins B_6, B_{12}, A, and D. Vitamin supplementation of undernourished women appears to help normalize the vitamin concentrations in their milk and may be especially beneficial for these women.

Foods with strong or spicy flavors (such as onions or garlic) may alter the flavor of breast milk. A sudden change in the taste of the milk may annoy some infants. Familiar flavors may enhance enjoyment. Flavors in breast milk from the mother's diet can influence the infant's later food preferences.[83] A mother who is breastfeeding her infant is advised to eat whatever nutritious foods she chooses. If a particular food seems to cause an infant discomfort, the mother can eliminate that food from her diet for a few days and see if the problem goes away.

Generally, infants with a strong family history of food allergies benefit from breastfeeding. Current evidence, however, does not support a major role for maternal dietary restrictions during lactation to prevent or delay the onset of food allergy in infants.[84]

Lactation and Weight Loss Another common question is whether breastfeeding promotes a more rapid loss of the extra body fat accumulated during pregnancy. Studies on this question have not provided a definitive answer. How much weight a woman retains after pregnancy depends on her gestational weight gain and the duration and intensity of breastfeeding. Many women who follow recommendations for gestational weight gain and breastfeeding can readily return to prepregnancy weight by six months after giving birth. Neither the quality nor the quantity of breast milk is adversely affected by moderate weight loss, and infants grow normally. Women often choose to be physically active to lose weight and improve fitness, and this is compatible with breastfeeding and infant growth.[85] A gradual weight loss (1 pound per week) is safe and does not reduce milk output. Too large an energy deficit, especially soon after birth, will inhibit lactation.

KEY POINTS

- The lactating woman needs extra fluid and adequate energy and nutrients for milk production.
- Malnutrition diminishes the quantity of the milk without altering quality.

When Should a Woman Not Breastfeed?

Some substances impair maternal milk production or enter breast milk and interfere with infant development, making breastfeeding an unwise choice. Some medical conditions also prohibit breastfeeding.

Alcohol and Illicit Drugs Alcohol enters breast milk and can adversely affect production, volume, composition, and ejection of breast milk as well as overwhelm an infant's immature alcohol-degrading system.[86] Alcohol concentration peaks within one hour after ingestion of even moderate amounts (equivalent to a can of beer). This amount may alter the taste of the milk to the disapproval of the nursing infant, who may, in protest, drink less milk than normal. Mothers who use illicit drugs should not breastfeed. Breast milk can deliver such high doses of drugs as to cause irritability, tremors, hallucinations, and even cause death in infants.

Tobacco and Caffeine About half the women who quit smoking during pregnancy relapse after delivery.[87] Lactating women who smoke tobacco produce less milk, and milk of lower fat content, than do nonsmokers.[88] Consequently, infants of smokers gain less weight. A lactating woman who smokes not only transfers nicotine and other chemicals to her infant via her breast milk but also exposes the infant to hazardous sidestream smoke.** Babies who are "smoked over" experience a wide

** Also called *environmental tobacco smoke*, or *secondhand smoking*.

certified lactation consultant a health-care provider, often a registered nurse or a registered dietitian, with specialized training and certification in breast and infant anatomy and physiology who teaches the mechanics of breastfeeding to new mothers.

array of health problems—poor growth, hearing impairment, vomiting, breathing difficulties, and even unexplained death.[89]

Excess caffeine can make a breastfed infant jittery and wakeful. As during pregnancy, caffeine consumption should be moderate when breastfeeding.

Medications Many medications pose no danger during breastfeeding, but others may suppress lactation or be secreted into breast milk and may harm the infant.[90] If a nursing mother must take such a medicine, then breastfeeding must be put on hold for the duration of treatment. Meanwhile, the flow of milk can be sustained by pumping the breasts and discarding the milk. A nursing mother should consult with her physician before taking medicines or even herbal supplements—herbs may have unpredictable effects on breastfeeding infants.

Many women wonder about using oral contraceptives during lactation. One type that combines the hormones estrogen and progestin seems to suppress milk output, lower the nitrogen content of the milk, and shorten the duration of breastfeeding. In contrast, progestin-only pills have no effect on breast milk or breastfeeding and are considered appropriate for lactating women.[91]

Environmental Contaminants A woman sometimes hesitates to breastfeed because she has heard warnings that contaminants in fish, water, and other foods (see page 525) may enter breast milk and harm her infant. Although some contaminants do enter breast milk, others may be filtered out. Because formula is made with water, formula-fed infants consume any contaminants that may be in the water supply. Any woman who is concerned about breastfeeding on this basis can consult with a physician or dietitian familiar with the local circumstances. With the exception of rare, massive exposure to a contaminant, the many benefits of breastfeeding far outweigh any small risk that may be associated with environmental hazards in the United States.

Maternal Illness If a woman has an ordinary cold, she can continue nursing without worry. The infant will probably catch it from her anyway, and thanks to immunological protection, a breastfed baby may be less susceptible than a formula-fed baby. A woman who has active, infectious tuberculosis can breastfeed once she has been treated and it is documented that she is no longer infectious.[92] If without treatment, breastfeeding is contraindicated.[93]

The human immunodeficiency virus (HIV), responsible for causing AIDS, can be passed from an infected mother to her infant during pregnancy, at birth, or through breast milk, especially during the early months of breastfeeding. In developed countries such as the United States, where safe alternatives are available, HIV-positive women should not breastfeed their infants.[94] In developing countries, where feeding inappropriate or contaminated formulas causes more than 1 million infant deaths each year, breastfeeding may present the best odds for infant survival.[95] The World Health Organization (WHO) recommends exclusive breastfeeding for infants of HIV-infected women for the first six months of life unless replacement feeding is acceptable, feasible, affordable, sustainable, and safe for mothers and their infants.[96]

Breastfeeding is a natural extension of pregnancy—the mother's body continues to nourish the infant.

KEY POINTS

- Breastfeeding is not advised if the mother's milk is contaminated with alcohol, drugs, or environmental pollutants.
- Most ordinary infections such as colds have no effect on breastfeeding infants, but HIV may be transmitted through milk.

Feeding the Infant

LO 13.5 Identify characteristics of breast milk that make it the ideal food for human infants, and discuss the introduction of solid foods into the diet.

Early nutrition affects later development, and early feedings establish eating habits that influence nutrition throughout life. Trends change, and experts may argue the fine points, but nourishing a baby is relatively simple. Common sense and a nurturing, relaxed environment go far to promote the infant's well-being.

Nutrient Needs

A baby grows faster during the first year of life than ever again, as Figure 13–11 shows. Pediatricians carefully monitor the growth of infants and children because growth directly reflects their nutrition status. An infant's birthweight doubles by about 5 months of age and triples by the age of 1 year. (If a 150-pound adult were to grow like this, the person would weigh 450 pounds after a single year.) The infant's length changes more slowly than weight, increasing about 10 inches from birth to 1 year. By the end of the first year, the growth rate slows considerably; an infant typically gains less than 10 pounds during the second year and grows about 5 inches in height.

Not only do infants grow rapidly, but their basal metabolic rate is also remarkably high—about twice that of an adult's, based on body weight. The rapid growth and metabolism of the infant demand an ample supply of all the nutrients. Of special importance during infancy are the energy nutrients and the vitamins and minerals critical to the growth process, such as vitamin A, vitamin D, and calcium.

Because they are small, babies need smaller *total* amounts of these nutrients than adults do, but as a percentage of body weight, babies need more than twice as much of most nutrients. Infants require about 100 calories per kilogram of body weight per day; most adults require fewer than 40 (see Table 13–9). Figure 13–12 (p. 534) compares a 5-month-old baby's needs (per unit of body weight) with those of an adult man. You can see that differences in vitamin D and iodine, for instance, are extraordinary.

Around 6 months of age, energy needs begin to increase less rapidly as the growth rate begins to slow down, but some of the energy saved by slower growth is spent in increased activity. When their growth slows, infants spontaneously reduce their energy intakes. Parents should expect their babies to adjust their food intakes downward when appropriate and should not force or coax them to eat more.

One of the most important nutrients for infants, as for everyone, is water. The younger a child is, the more of its body weight is water. Breast milk or infant formula normally provides enough water to replace fluid losses in a healthy infant. If the environmental temperature is extremely high, however, infants need supplemental water.[97] Much more of an infant's body water is between the cells and in the vascular space, and this water is easy to lose. Conditions that cause rapid fluid loss, such as vomiting or diarrhea, require an electrolyte solution designed for infants.

KEY POINTS

- An infant's birthweight doubles by about 5 months of age and triples by 1 year.
- Infants' rapid growth and development depend on adequate nutrient supplies, including water from breast milk or formula.

Why Is Breast Milk So Good for Babies?

Many medical and professional organizations advocate breastfeeding for the best infant nutrition, as well as for the many other benefits it provides both infant and

Figure 13–11

Weight Gain of Human Infants and Children in the First Five Years of Life

The colored vertical bars show how the yearly increase in weight gain slows its pace over the years.

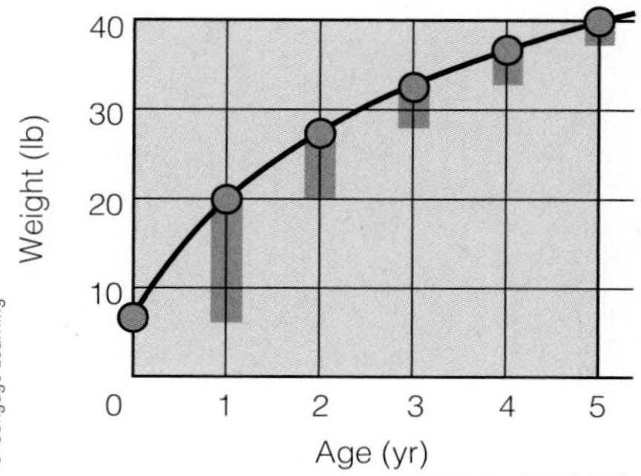

Did You Know?

At the age of 2, healthy children have attained approximately half of their adult height.

After 6 months of age, the energy saved by slower growth is spent on increased activity.

Table 13–9

Infant and Adult Heart Rate, Respiration Rate, and Energy Needs Compared

	Infants	Adults
Heart rate (beats/minute)	120 to 140	70 to 80
Respiration rate (breaths/minute)	20 to 40	15 to 20
Energy needs (cal/body weight)	45/lb (100/kg)	<18/lb (<40/kg)

Figure 13–12

Nutrient Recommendations for a 5-Month-Old Infant and an Adult Male Compared on the Basis of Body Weight

Infants may be relatively small and inactive, but they use large amounts of energy and nutrients in proportion to their body size to keep all their metabolic processes going.

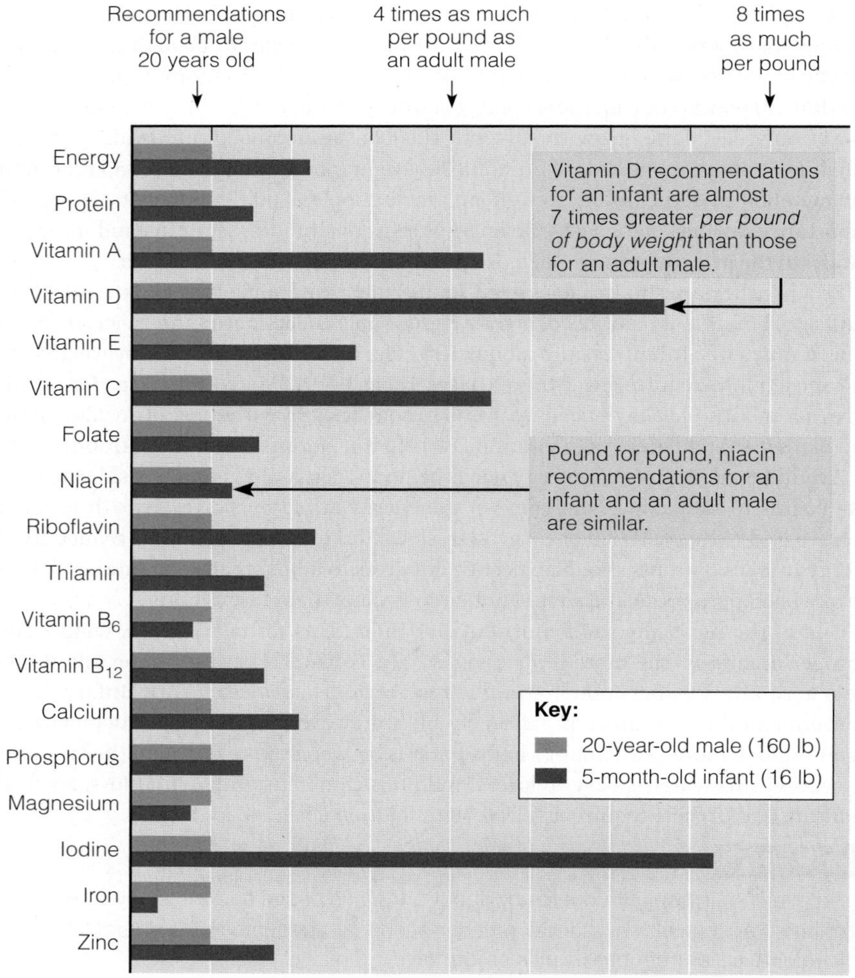

Table 13–10

Benefits of Breastfeeding

For Infants:

- Provides the appropriate composition and balance of nutrients with high bioavailability.
- Provides hormones that promote physiological development.
- Improves cognitive development.
- Protects against a variety of infections.
- May protect against some chronic diseases, such as diabetes and hypertension, later in life.
- Protects against food allergies.
- Reduces the risk of SIDS.

For Mothers:

- Contracts the uterus.
- Delays the return of regular ovulation, thus lengthening birth intervals. (It is not, however, a dependable method of contraception.)
- Conserves iron stores (by prolonging amenorrhea).
- May protect against breast and ovarian cancer.

Other:

- Provides cost savings from not needing medical treatment for childhood illnesses or time off work to care for sick children.
- Provides cost savings from not needing to purchase formula (even after adjusting for added foods in the diet of a lactating mother).
- Provides environmental savings to society from not needing to manufacture, package, and ship formula or dispose of packaging.

mother (see Table 13–10).[98] The AAP and the Academy of Nutrition and Dietetics recommend **exclusive breastfeeding** for 6 months, and breastfeeding with complementary foods for at least 12 months, as an optimal feeding pattern for infants.[99] All legitimate nutrition authorities share this view, but some makers of baby formula try to convince women otherwise—see the Consumer's Guide on page 539.

Breast milk excels as a source of nutrients for the young infant. With the exception of vitamin D (discussed later), breast milk provides all the nutrients a healthy infant needs for the first six months of life.[100] Breast milk also conveys immune factors, which both protect an infant against infection and inform its body about the outside environment.

Breastfeeding Tips Breast milk is more easily and completely digested than infant formula, so breastfed infants usually need to eat more frequently than formula-fed infants do. During the first few weeks, approximately 8 to 12 feedings a day, on demand, as soon as the infant shows early signs of hunger such as increased alertness, activity, or suckling motions, promote optimal milk production and infant growth.

Crying is a late indicator of hunger.[101] An infant who nurses every 2 to 3 hours and sleeps contentedly between feedings is adequately nourished. As the infant gets older, stomach capacity enlarges and the mother's milk production increases, allowing for longer intervals between feedings.

Even though the baby obtains about half the milk from the breast during the first 2 or 3 minutes of suckling, the infant should be encouraged to breastfeed on the first breast for as long as he or she wishes, before being offered the second breast. The infant's suckling, as well as the complete removal of milk from the breast, stimulates lactation. Begin each feeding on the breast offered last.

Energy Nutrients in Breast Milk The energy-nutrient balance of breast milk differs dramatically from that recommended for adults (see Figure 13–13). Yet, for infants, breast milk is the most nearly perfect food, affirming that people at different stages of life have different nutrient needs.

The carbohydrate in breast milk (and standard infant formula) is lactose. In addition to being easily digested, lactose enhances calcium absorption. One of the carbohydrate components of breast milk helps protect the infant from infection by preventing the binding of pathogens to the infant's intestinal cells.[102]

The lipids in breast milk—and infant formula—provide the main source of energy in the infant's diet. Breast milk contains a generous proportion of the essential fatty acids linoleic acid and linolenic acid, as well as their longer-chain derivatives, arachidonic acid and DHA. Most formulas today also contain added arachidonic acid and DHA (read the label). Infants can produce some arachidonic acid and DHA from linoleic and linolenic acid, but some infants may need more than they can make.

DHA is the most abundant fatty acid in the brain and is also present in the retina of the eye. DHA accumulation in the brain is greatest during fetal development and early infancy.[103] Research has focused on the visual and mental development of breastfed infants and infants fed standard formula with and without DHA added.[104] When infants are fed formula with DHA added, about half of the studies show higher visual acuity compared with controls.[105] Factors such as the amount of DHA provided, the sources of the DHA, and varying research methods have led to mixed outcomes. As for mental development, a number of studies suggest that DHA supplementation during development can influence certain measures of cognitive function.[106] Still needed are longer-term studies that follow child development beyond infancy.

The protein in breast milk is largely **alpha-lactalbumin**, a protein the human infant can easily digest. Another breast milk protein, **lactoferrin**, is an iron-gathering compound that helps absorb iron into the infant's bloodstream, keeps intestinal bacteria from getting enough iron to grow out of control, and kills certain bacteria.

Vitamins and Minerals in Breast Milk With one exception—vitamin D—the vitamin content of the breast milk of a well-nourished mother is ample. Even vitamin C, for which cow's milk is a poor source, is supplied generously. The concentration of vitamin D in breast milk is low, however, and vitamin D deficiency impairs bone mineralization.[107] Vitamin D deficiency is most likely in infants who are not exposed to sunlight daily, have darkly pigmented skin, and receive breast milk without vitamin D supplementation.[108] Recently, the vitamin D intake recommendations for infants were changed for two reasons. First, rickets, the vitamin D deficiency disease, has been diagnosed among U.S. infants. Second, the AAP recommends that infants younger than six months be protected from direct sunlight, eliminating this source of vitamin D.

As for minerals, the calcium content of breast milk is ideal for infant bone growth, and the calcium is well absorbed. Breast milk is also low in sodium. The limited amount of iron in breast milk is highly absorbable, and its zinc, too, is absorbed better than from cow's milk, thanks to the presence of a zinc-binding protein.

Supplements for Infants Pediatricians may prescribe supplements containing vitamin D, iron, and fluoride (after 6 months of age) as outlined in Table 13–11, p. 536. Vitamin K nutrition for newborns presents a unique case. A newborn's digestive tract is sterile, and vitamin K–producing bacteria take weeks to establish themselves in the

Figure 13–13

Percentages of Energy-Yielding Nutrients in Breast Milk and Recommended Adult Diets

The proportions of energy-yielding nutrients in human breast milk differ from those recommended for adults.[a]

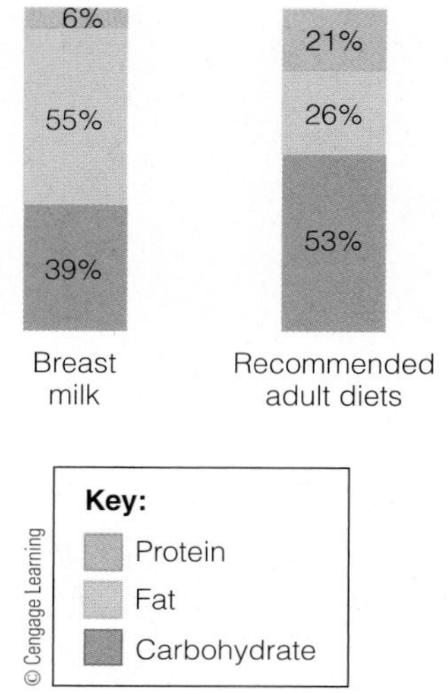

[a]The values listed for adults represent approximate midpoints of the acceptable ranges for protein (10 to 35 percent), fat (20 to 35 percent), and carbohydrate (45 to 65 percent).

exclusive breastfeeding an infant's consumption of human milk with no supplementation of any type (no water, no juice, no nonhuman milk, and no foods) except for vitamins, minerals, and medications.

alpha-lactalbumin (lact-AL-byoo-min) the chief protein in human breast milk. The chief protein in cow's milk is *casein* (CAY-seen).

lactoferrin (lack-toe-FERR-in) a factor in breast milk that binds iron and keeps it from supporting the growth of the infant's intestinal bacteria.

Table 13–11

Supplements for Full-Term Infants

	Vitamin D[a]	Iron[b]	Fluoride[c]
Breastfed infants:			
Birth to 6 months of age	✓		
6 months to 1 year	✓	✓	✓
Formula-fed infants:			
Birth to 6 months of age			
6 months to 1 year		✓	✓

[a]*Vitamin D supplements are recommended for all infants who are exclusively breastfed and for any infants who do not receive at least 1 liter (1,000 milliliters) or 1 quart (32 ounces) of vitamin D–fortified formula per day.*

[b]*All infants 6 months of age need additional iron, preferably in the form of iron-fortified infant cereal and/or infant meats. Formula-fed infants need iron-fortified infant formula.*

[c]*At 6 months of age, breastfed infants and formula-fed infants who receive ready-to-use formulas (these are prepared with water low in fluoride) or formula mixed with water that contains little or no fluoride (less than 0.3 ppm) need supplements.*

Source: Adapted from Committee on Nutrition, American Academy of Pediatrics, Pediatric Nutrition Handbook, *6th ed., ed. R. E. Kleinman (Elk Grove Village, Ill.: American Academy of Pediatrics, 2009).*

baby's intestines. To prevent bleeding in the newborn, the AAP recommends that a single dose of vitamin K be given at birth.

The AAP currently recommends a vitamin D supplement for all infants who are breastfed exclusively and for any infants who do not receive at least 1 liter (1,000 milliliters) or 1 quart (32 ounces) of vitamin D–fortified formula daily.[109] Despite such recommendations, most infants in the United States are consuming inadequate amounts of vitamin D.[110]

Immune Factors in Breast Milk Breast milk offers the infant unsurpassed protection against infection.[111] Its protective factors include antiviral agents, anti-inflammatory agents, antibacterial agents, and infection inhibitors.

During the first two or three days of lactation, the breasts produce **colostrum**, a premilk substance containing antibodies and white cells from the mother's blood. Colostrum (like breast milk) helps protect the newborn infant from infections against which the mother has developed immunity—precisely those in the environment likely to infect the infant. For example, maternal antibodies in colostrum and breast milk inactivate harmful bacteria within the infant's digestive tract before they can start infections.[112] This explains, in part, why breastfed infants have fewer intestinal infections than formula-fed infants.

Breastfeeding also protects against other common illnesses of infancy, such as middle ear infection and respiratory illnesses.[113] In addition, breastfed infants have fewer allergic reactions such as asthma, wheezing, and skin rash.[114] This protection is especially noticeable among infants with a family history of allergies.[115] Even the risk of sudden infant death syndrome (SIDS) is lower among breastfed infants.[116] This protective effect is stronger when breastfeeding is exclusive, but breastfeeding an infant for even the first month has been found to cut the risk of SIDS by half compared to never breastfeeding.[117]

In addition to their protective features, colostrum and breast milk contain hormones and other factors that stimulate the development and maintenance of the infant's digestive tract. Clearly, breast milk is a very special substance.

Other Potential Benefits Breastfeeding may offer some protection against excessive weight gain later, although findings are inconsistent.[118] Many other factors—socioeconomic status, other infant- and child-feeding practices, and especially the mother's weight—strongly predict a child's body weight.[119]

The possibility that breastfeeding may positively affect later intelligence is intriguing. Many studies have suggested such benefits, but when subjected to

strict standards of methodology (for example, large sample size and appropriate intelligence testing), the evidence is less convincing.[120] Most likely, a combination of factors such as DHA in breast milk as well as the feeding process itself are of benefit to the infant's development.[121] More large, well-controlled studies are needed to confirm the effects, if any, of breastfeeding on later intelligence.

KEY POINTS

- With the exception of vitamin D, breast milk provides all the nutrients a healthy infant needs for the first 6 months of life.
- Breast milk offers the infant unsurpassed protection against infection—including antiviral agents, anti-inflammatory agents, antibacterial agents, and infection inhibitors.

Formula Feeding

Formula feeding offers an acceptable alternative to breastfeeding. Nourishment for an infant from formula is adequate, and parents can choose this course with confidence. One advantage is that parents can see how much milk the infant drinks during feedings. Another is that other family members can participate in feeding the infant, giving them a chance to develop the special closeness that feeding fosters.

Mothers who return to work soon after giving birth may choose formula for their infants, but they have another option. Breast milk can be pumped into bottles and given to the baby in day care. At home, mothers may breastfeed as usual. Many mothers use both methods—they breastfeed at first but **wean** their children within 1 to 12 months. If infants are less than a year of age, mothers must wean them onto *infant formula*, not onto plain cow's milk of any kind.

Infant Formula Composition The substitution of formula feeding for breastfeeding involves striving to copy nature as closely as possible. Human milk and cow's milk differ; cow's milk is significantly higher in protein, calcium, and phosphorus, for example, to support the calf's faster growth rate. Thus, to prepare a formula from cow's milk, the formula makers must first dilute the milk and then add carbohydrate and nutrients to make the proportions comparable to those of human milk. Figure 13–14 compares the energy–nutrient balance of breast milk, standard infant formula, and cow's milk. Notice the higher protein concentration of cow's milk, which can stress the infant's kidneys. The AAP recommends that all formula-fed infants receive iron-fortified infant formulas.[122] Use of iron-fortified formulas has increased in recent decades and is credited with the decline of iron-deficiency anemia in U.S. infants.

Special Formulas Standard cow's milk–based formulas are inappropriate for some infants. Special formulas have been designed to meet the dietary needs of infants with specific conditions such as prematurity or inherited diseases. Most infants allergic to milk protein can drink formulas based on soy protein.[123] Soy formulas also use cornstarch and sucrose instead of lactose and so are recommended for infants with lactose intolerance as well. They are also useful as an alternative to milk-based formulas for vegan families. Some infants who are allergic to cow's milk protein may also be allergic to soy protein.[124] For these infants, special formulas based on hydrolyzed protein are available.

The Transition to Cow's Milk For good reasons, the AAP advises that cow's milk is not appropriate for infants younger than one year old.[125] In some infants, particularly those younger than 6 months of age, cow's milk causes intestinal bleeding, which can lead to iron deficiency.[126] Cow's milk is also a poor source of iron, and its higher calcium and lower vitamin C contents inhibit iron absorption. Consequently, plain cow's milk threatens the infant's iron status in three ways: it causes iron loss, it fails to replace iron, and it reduces the bioavailability of iron from infant cereal and other foods. In short, cow's milk is a poor choice during the first year of life; infants need breast milk or iron-fortified formula.

Once the baby is obtaining at least two-thirds of total daily food energy from a balanced mixture of cereals, vegetables, fruits, and other foods (after 12 months of age),

Did You Know?

Options for buying formula include:

- Liquid concentrate (moderately expensive, relatively easy)—mix with equal part water.
- Powdered formula (least expensive, lightest for travel)—follow label directions.
- Ready-to-feed (most expensive, easiest)—pour directly into clean bottles.

Figure 13–14

Percentages of Energy-Yielding Nutrients in Breast Milk, Infant Formula, and Cow's Milk

The average proportions of energy-yielding nutrients in human breast milk and formula differ slightly. In contrast, cow's milk provides too much protein and too little carbohydrate.

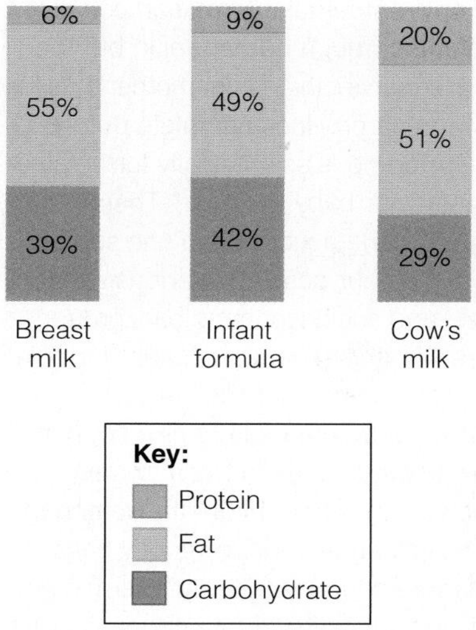

colostrum (co-LAHS-trum) a milklike secretion from the breasts during the first day or so after delivery before milk appears; rich in protective factors.

wean to gradually replace breast milk with infant formula or other foods appropriate to an infant's diet.

use it! A Consumer's Guide To . . .

Formula Advertising versus Breastfeeding Advocacy

Bottle feed or breastfeed? New mothers must answer this question amidst the whirlwind of physical and emotional changes associated with pregnancy and delivery. For a few women, breastfeeding may be proscribed by illness or physical condition; in a few more cases, special needs of the infant may make breastfeeding impossible. The strong scientific consensus holds, however, that breastfeeding is preferable for all other infants, so why do so many women continue to choose formula? For some, the time and logistics required for breastfeeding competes with work or school schedules; for many others, the decision to forgo breastfeeding is influenced by the aggressive advertising of formulas.

Formula Advertising Claims and Tactics

Advertisements of infant formulas often create the illusion that formula is identical to human milk. No formula can match the nutrients, agents of immunity, and environmental information conveyed to infants through human milk, but the ads are convincing: "Like mother's milk, our formula provides complete nutrition" or "Our brand is scientifically formulated to meet your baby's needs." The ads seem to work: according to one survey, one out of four people of various ages, races, and socioeconomic backgrounds agree with the statement "infant formula is as good as breast milk."

Formula manufacturers give coupons and samples of free formula to pregnant women. After childbirth, women in the hospital may receive "goody bags" with more coupons to tempt them to go retrieve their "gifts." More coupons arrive by mail a couple of months later, at a time when many women give up breastfeeding, even though nutrition authorities urge continued breastfeeding for several more months. Aggressive marketing tactics like these can undermine a woman's confidence concerning her choice to breastfeed, and lack of confidence causes many women to quit early.[1]

Table 13–12

Tips for Successful Breastfeeding

- Learn about the benefits of breastfeeding.
- Initiate breastfeeding within 1 hour of birth.
- Ask a health-care professional to explain how to breastfeed and how to maintain lactation.
- Give newborn infants no food or drink other than breast milk, unless medically indicated.
- Breastfeed on demand.
- Give no artificial nipples or pacifiers to breastfeeding infants.[a]
- Find breastfeeding support groups, books, or websites to help troubleshoot breastfeeding problems.

[a] Compared with nonusers, infants who use pacifiers breastfeed less frequently and stop breastfeeding at a younger age.

Breastfeeding Advocacy

National efforts to promote breastfeeding seem to be working, at least to some extent: the percentage of infants who were ever breastfed rose from 60 percent among those born in 1994 to 74 percent among infants born in 2006.[2] This still falls short of national goals, however.[3] Only about 43 percent of infants are still breastfeeding at 6 months of age, and about 23 percent are still doing so at 1 year of age.

Many hospitals employ certified lactation consultants who specialize in helping new mothers establish a healthy breastfeeding relationship with their newborns. Table 13–12 lists tips for successful long-term breastfeeding.

Where Breastfeeding Is Critical

Infant formula is an appropriate substitute when breastfeeding is impossible, but for most infants, the benefits of breast milk beat those of formula. Formula-fed infants in developed nations are healthy and grow normally, but they miss out on the breastfeeding advantages described in the text.

In developing nations, however, the consequence of choosing not to breastfeed can be tragic. Feeding formula is often fatal to the infant when poverty limits access to formula mixes, clean water for safe formula preparation, and medical help when needed. Even in developed nations, a woman may easily lose track of a bottle that spoils in a crib within easy reach of an infant. The World Health Organization (WHO) strongly supports breastfeeding for the world's infants in its "babyfriendly" initiative and opposes the marketing of infant formulas to new mothers.

Moving Ahead

Women are free to choose between breast and bottle. Breast milk is recommended, and is a thrifty choice; infant formula, bottles, and paraphernalia are expensive for anyone's wallet, particularly after the initial coupons run out. During pregnancy, parents-to-be should seek out the facts about each feeding method and be aware that sophisticated formula advertisements are designed to make sales, and not primarily to help potential customers make the best choice.

Review Questions*

1. Commercial infant formula is more reliable than breast milk because it has been scientifically engineered for complete nutrition. **T F**
2. About 60 percent of U.S. infants are still breastfeeding at one year of age. **T F**
3. Lactation consultants are employed by hospitals to help new mothers understand the advantages of feeding their babies with infant formula. **T F**

** Answers to Consumer's Guide review questions are found in Appendix G.*

reduced-fat or low-fat cow's milk (in the context of an overall diet that supplies 30 percent of calories from fat) is an acceptable and recommended beverage.[127] After the age of 2, a transition to fat-free milk can take place, but care should be taken to avoid excessive restriction of dietary fat.

KEY POINTS

- Infant formulas are designed to resemble breast milk in nutrient composition.
- After the baby's first birthday, reduced-fat or low-fat cow's milk can replace formula.

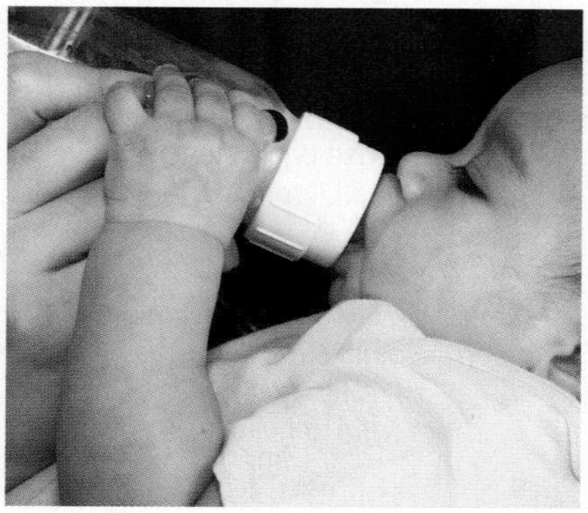

© Jaimie Duplass/Shutterstock.com

The infant thrives on formula offered with affection.

An Infant's First Solid Foods

Foods can be introduced into the diet as the infant becomes physically ready to handle them. This readiness develops in stages. A newborn can swallow only liquids that are well back in the throat. Later (at 4 months or so), the tongue can move against the palate to swallow semisolid food such as cooked cereal. The stomach and intestines are immature at first; they can digest milk sugar (lactose) but not starch. At about 4 months, most infants can begin to digest starchy foods. Still later, the first teeth erupt, but not until sometime during the second year can a baby begin to handle chewy food.

When to Introduce Solid Food The AAP supports exclusive breastfeeding for 6 months but recognizes that infants are often developmentally ready to accept some solid foods between 4 and 6 months of age.[128] Solid foods can provide needed nutrients that are no longer supplied adequately by breast milk or formula alone.[129] The foods chosen must be those that the infant is developmentally capable of handling both physically and metabolically. The exact timing depends on the individual infant's needs, developmental readiness (see Table 13–13, p. 540), and tolerance of the food.

In short, the addition of foods to an infant's diet should be governed by three considerations: the infant's nutrient needs, the infant's physical readiness to handle different forms of foods, and the need to detect and control allergic reactions, as described next. With respect to increased nutrient needs, the nutrient needed first is iron, then vitamin C.

RedChopsticks/Jupiterimages

With the first birthday comes the possibility of tasting cow's milk for the first time.

Foods to Provide Iron and Vitamin C Rapid growth demands iron. At about 4 to 6 months, the infant begins to need more iron than body stores plus breast milk or iron-fortified formula can provide. In addition to breast milk or iron-fortified formula, infants can receive iron from iron-fortified cereals and, once they readily accept solid foods, from meat or meat alternates such as legumes. Iron-fortified cereals contribute a significant amount of iron to an infant's diet, but the iron's bioavailability is poor.[130] Caregivers can enhance iron absorption from iron-fortified cereals by serving vitamin C–rich foods with meals.

The best sources of vitamin C are fruits and vegetables. It has been suggested that infants who are introduced to fruits before vegetables may develop a preference for sweets and find the vegetables less palatable, but there is no evidence to support offering these foods in a particular order.[131] Fruit juice is a source of vitamin C, but excessive juice intake can lead to diarrhea in infants and young children.[132] Furthermore, too much fruit juice contributes excessive calories and displaces other nutrient-rich foods. The AAP recommends limiting juice consumption for infants and young children (1 to 6 years of age) to between 4 and 6 ounces per day.[133] Fruit juices should be diluted and served in a cup, not a bottle, once the infant is 6 months of age or older.

© Polara Studios

Foods such as iron-fortified cereals and formulas, mashed legumes, and strained meats provide iron.

Physical Readiness for Solid Foods Foods introduced at the right times contribute to an infant's physical development. The ability to swallow food develops at around 4 to 6 months, and food offered by spoon helps to develop swallowing ability. At 8 months to a year, a baby can sit up, can handle finger foods, and begins to teethe. At that time, hard crackers and other finger foods may be introduced to promote the development of manual dexterity and control of the jaw muscles. These feedings must occur under the watchful eye of an adult because the baby can also choke on such foods. Babies and young children cannot safely chew and swallow any of the foods listed in

Table 13–13

Infant Development and Recommended Foods

Note: Because each stage of development builds on the previous stage, the foods from an earlier stage continue to be included in all later stages.

Age (mo)	Feeding Skill	Foods Introduced into the Diet
0–4	Turns head toward any object that brushes cheek. Initially swallows using back of tongue; gradually begins to swallow using front of tongue as well. Strong reflex (extrusion) to push food out during first 2 to 3 months.	Feed breast milk or infant formula.
4–6	Extrusion reflex diminishes, and the ability to swallow nonliquid foods develops. Indicates desire for food by opening mouth and leaning forward. Indicates satiety or disinterest by turning away and leaning back. Sits erect with support at 6 months. Begins chewing action. Brings hand to mouth. Grasps objects with palm of hand.	Begin iron-fortified cereal mixed with breast milk, formula, or water. Begin pureed meats, legumes, vegetables, and fruits.
6–8	Able to feed self with fingers. Develops pincher (finger to thumb) grasp. Begins to drink from cup.	Begin textured vegetables and fruits. Begin plain, unsweetened fruit juices from cup.
8–10	Begins to hold own bottle. Reaches for and grabs food and spoon. Sits unsupported.	Begin breads and cereals from table. Begin yogurt. Begin pieces of soft, cooked vegetables and fruit from table. Gradually begin finely cut meats, fish, casseroles, cheese, eggs, and legumes.
10–12	Begins to master spoon, but still spills some.	Add variety. Gradually increase portion sizes.[a]

[a]*Portions of foods for infants and young children are smaller than those for an adult. For example, a grain serving might be ½ slice of bread instead of 1 slice, or ¼ cup rice instead of ½ cup.*

Source: Adapted in part from Committee on Nutrition, American Academy of Pediatrics, Pediatric Nutrition Handbook, *6th ed., ed. R. E. Kleinman (Elk Grove Village, Ill.: American Academy of Pediatrics, 2009), pp. 113–142.*

Courtesy of the Author

Older babies love to eat what their families eat. Let them enjoy their food.

milk anemia iron-deficiency anemia caused by drinking so much milk that iron-rich foods are displaced from the diet.

Table 13–14; they can easily choke on these foods, a risk not worth taking.[134] Nonfood items of small size should always be kept out of the infant's reach to prevent choking.

Some parents want to feed solids as early as possible on the theory that "stuffing the baby" at bedtime will promote sleeping through the night. There is no proof for this theory. Babies start to sleep through the night when they are ready, no matter when solid foods are introduced.

Preventing Food Allergies To prevent allergy or identify one promptly, experts recommend introducing single-ingredient foods, one at a time, in small portions, and waiting three to five days before introducing the next new food.[135] For example, on introducing cereals, try fortified rice cereal first for several days; it causes allergy least often. Try wheat-containing cereal last; it is a common offender. If a food causes an allergic reaction (irritability due to skin rash, digestive upset, or respiratory discomfort), discontinue its use before going on to the next food. If allergies run in your family, use extra caution in introducing new foods. Parents or caregivers who detect allergies early in an infant's life can spare the whole family much grief.

Choice of Infant Foods Infant foods should be selected to provide variety, balance, and moderation. Commercial baby foods in the United States and Canada offer a wide variety of palatable, nutritious foods in a safe and convenient form. Brands vary in their use of starch fillers and sugar—check the ingredients lists (Appendix A lists nutrients in many baby foods). Parents or caregivers should not feed directly from the jar—spoon the needed portion into a dish to prevent contamination of the leftovers that will be stored in the jar.

An alternative to commercial baby food is to process a small portion of the family's table food in a blender, food processor, or baby food grinder. This necessitates cooking without salt or sugar, though, as the best baby food manufacturers do. Adults can season their own food after taking out the baby's portion. Pureed food can be frozen in an ice cube tray to yield a dozen or so servings that can be quickly thawed, heated, and served on a busy day.

Foods to Omit Sweets of any kind (including baby food "desserts") have no place in a baby's diet. The added food energy can promote obesity, and such treats convey few or no nutrients to support growth. Products containing sugar alcohols such as sorbitol should also be limited, as these may cause diarrhea. Salty, canned vegetables are inappropriate for babies, but unsalted varieties provide a convenient source of well-cooked vegetables. Maintaining an awareness of foodborne illness and taking precautions against it are imperative—even a normally mild foodborne illness can cause serious harm to an infant or young child. Honey and corn syrup should never be fed to infants because of the risk of botulism. Infants and young children are vulnerable to foodborne illnesses.

Foodborne illnesses and their prevention are topics of **Chapter 12.**

Beverages and Foods at 1 Year At 1 year of age, reduced-fat or low-fat cow's milk can become a primary source of most of the nutrients an infant needs; 2 to 3 cups a day meet those needs. More milk than this displaces iron-rich foods and can lead to the iron-deficiency anemia known as **milk anemia**. A variety of other foods—meat and meat alternatives, iron-fortified cereal, enriched or whole-grain bread, fruits, and vegetables—should be supplied in amounts sufficient to round out total energy needs. Ideally, the 1-year-old sits at the table, eats many of the same foods everyone else eats, and drinks liquids from a cup, not a bottle. Table 13–15 (p. 542) shows a sample menu that meets the requirements for a 1-year-old.

KEY POINTS

- At 6 months, an infant may be ready to try some solid foods.
- By 1 year, the child should be eating foods from all food groups.

Looking Ahead

The first year of life is the time to lay the foundation for future health. From the nutrition standpoint, the problems most common in later years are obesity and dental disease. Prevention of obesity may also help prevent the obesity-related diseases: atherosclerosis, diabetes, and cancer.

The most important single measure to undertake during the first year is to encourage eating habits that will support continued normal weight as the child grows. This means introducing a variety of nutritious foods in an inviting way (not forcing the baby to finish the bottle or baby food jar), and avoiding concentrated sweets and empty-calorie foods while encouraging physical activity. Parents should not teach babies to seek food as a reward, to expect food as comfort for unhappiness, or to associate food deprivation with punishment. If they cry for companionship, pick them up—don't feed them. If they are hungry, by all means, feed them appropriately. More pointers are offered in this chapter's Food Feature.

Dentists strongly discourage the practice of giving a baby a bottle as a pacifier and recommend limiting treats. Sucking for long periods of time pushes the normal jaw line out of shape and causes a bucktoothed profile: protruding upper and receding lower teeth. Prolonged sucking on a bottle of milk or juice also bathes the upper teeth in a carbohydrate-rich fluid that favors the growth of acid-producing bacteria, which dissolves tooth material. Babies regularly put to bed with a bottle sometimes have teeth decayed all the way to the gum line, a condition known as nursing bottle syndrome, shown in the margin photos.

KEY POINTS

- The early feeding of the infant lays the foundation for lifelong eating habits.
- The most important single measure to undertake during the first year is to encourage eating habits that will support continued normal weight as the child grows.

Table 13–14

Choking Prevention

To prevent choking, do not give infants or young children:
Gum.
Popcorn.
Large raw apple slices.
Whole grapes.
Whole cherries.
Raw celery.
Raw carrots.
Whole beans.
Hot dog slices.
Sausage sticks or slices.
Hard or gel-type candies.
Marshmallows.
Nuts.
Peanut butter.
Keep these nonfood items out of their reach:
Coins.
Balloons.
Small balls.
Pen tops.
Other items of similar size.

Did You Know?

Infants and young children are particularly sensitive to foodborne illnesses and should not receive unpasteurized milk, milk products, or juices; raw or undercooked eggs, meat, poultry, fish, or shellfish; or raw sprouts. Infants also should not have honey or corn syrup.

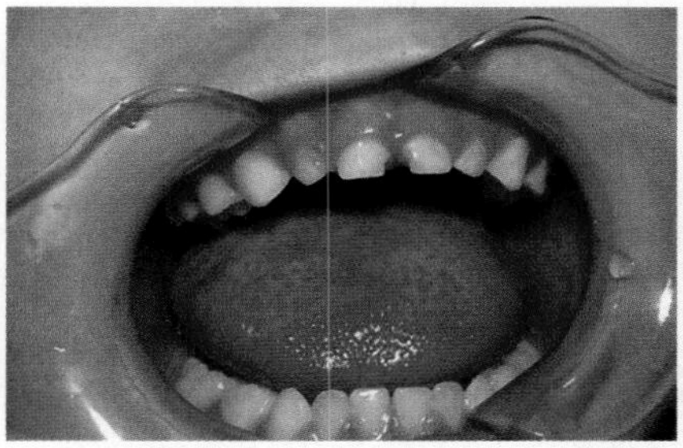

Nursing bottle syndrome in an early stage.

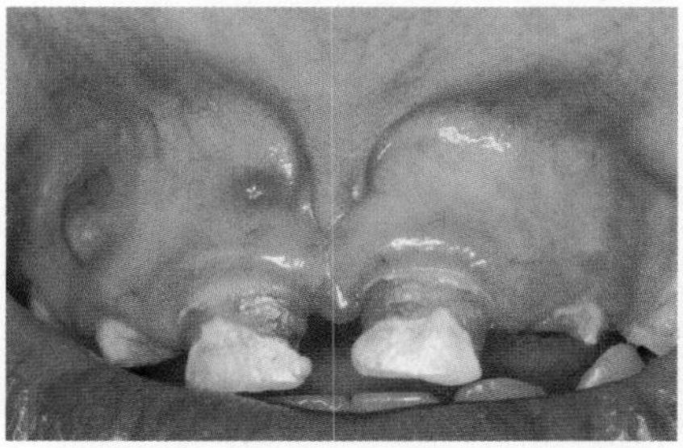

Nursing bottle syndrome, an extreme example. The upper teeth have decayed all the way to the gum line.

try it! Food Feature

Mealtimes with Infants

LO 13.6 List some feeding guidelines that encourage normal eating behaviors and autonomy in the child.

The nurturing of a young child involves more than nutrition. Those who care for young children are responsible for providing not only nutritious foods, milk, and water but also a safe, loving, secure environment in which the children may grow and develop.

Foster a Sense of Autonomy

The person feeding a 1-year-old has to be aware that the child's exploring and experimenting are normal and desirable behaviors. The child is developing a sense of autonomy that, if allowed to develop, will provide the foundation for later assertiveness in choosing when and how much to eat and when to stop eating.

Some Feeding Guidelines

In light of the developmental and nutrient needs of 1-year-olds and in the face of their often contrary and willful behavior, a few feeding guidelines may be helpful:

- *Discourage unacceptable behavior (such as standing at the table or throwing food) by removing the child from the table to wait until later to eat.* Be consistent and firm, not punitive. For example, instead of saying "You make me mad when you don't sit down," say "The fruit salad tastes good—please sit down and eat some with me." The child will soon learn to sit and eat.
- *Let young children explore and enjoy food.* This may mean eating with fingers for a while. Learning to use a spoon will come in time. Children who are allowed to touch, mash, and smell their food while exploring it are more likely to accept it.
- *Don't force food on children.* Rejecting new foods is normal and acceptance is more likely as children become familiar with new foods through repeated opportunities to taste them. Instead of saying "You cannot go outside to play until you taste your carrots," say "You can try the carrots again another time."
- *Provide nutritious foods, and let children choose which ones, and how much, they will eat.* Gradually, they will acquire a taste for different foods.
- *Limit sweets.* Infants and young children have little room for empty-calorie foods in their daily energy allowance. Do not use sweets as a reward for eating meals.
- *Don't turn the dining table into a battleground.* Make mealtimes enjoyable. Teach healthy food choices and eating habits in a pleasant environment. Mealtimes are not the time to fight, argue, or scold.

These recommendations reflect a spirit of tolerance that best serves the emotional and physical interests of the infant. This attitude, carried throughout childhood, helps the child to develop a healthy relationship with food. The next chapter finishes the story of growth and nutrition.

Table 13–15

Sample Meal Plan for a 1-Year-Old

	SAMPLE MENU
BREAKFAST	1 scrambled egg 1 slice whole-wheat toast ½ c whole milk
MORNING SNACK	½ c yogurt ¼ c fruit[a]
LUNCH	½ grilled cheese sandwich: 1 slice whole-wheat bread with 1 slice cheese ½ c vegetables[b] (steamed carrots) ¼ c 100% fruit juice
AFTERNOON SNACK	½ c fruit[a] ½ c toasted oat cereal
DINNER	1 oz chopped meat or ¼ c well-cooked mashed legumes ½ c rice or pasta ½ c vegetables[b] (chopped broccoli) ½ c whole milk

Note: This sample menu provides about 1,000 calories.

[a]*Include citrus fruits, melons, and berries.*

[b]*Include dark green, leafy, and red and orange vegetables.*

track it!

Diet Analysis **PLUS** Concepts in Action

Analyze the Adequacy of a Diet for Pregnancy

The purpose of this exercise is to reinforce the importance of good food choices to provide nutrients to support health during pregnancy, lactation, and growth.

1. To reduce the risk for neural tube defects in infants, women of child-bearing age are urged to obtain 400 micrograms (μg DFE) of folic acid daily in addition to a varied diet. Find folic acid among enriched grains and other fortified foods. Select the Track Diet tab from the red navigation bar. Select a new date, and enter the foods to create a meal that provides folic acid from enriched sources (see Table 13–4, page 518). (*Hint*: A good meal to choose is breakfast.) Select the Reports tab; then select Source Analysis. Select Folate from the drop-down box, and generate a report. How close did your meal come to providing one-third of the needed 400 μg DFE folic acid?
2. A pregnant teenager's need for calcium is 1,300 mg a day. Many teenagers fail to meet their calcium needs, even before pregnancy. Select the Track Diet tab, and select a new day. Add foods to create a high-calcium meal for a pregnant teen. (For tips, see Snapshot 8–1, page 302.) Select the Reports tab: then select Source Analysis. Select Calcium from the drop-down box, and generate a report. How much calcium was provided by the foods in this meal? What did you take into consideration when choosing the foods high in calcium? How can you increase the likelihood that the teenager will consume this meal?
3. During lactation, a woman needs an additional 330 calories per day more than her regular need. Create a new profile from the Profile drop-down box, making it similar to your own but selecting "female" and "pregnant and lactating." To meet this woman's need, choose among nutrient-dense foods (refer to Table 13–2, page 516), and create a one-day diet to meet her increased energy need. Select the Reports tab and then Energy Balance, and generate a report for the day's meals. Did your food choices help this woman to meet her increased energy need? Was "330" listed in the "net kcal" column?
4. Zinc is required for protein synthesis and cell development. Obtaining zinc poses a challenge to vegetarians. Create a vegetarian meal that includes zinc-rich foods. Select the Track Diet tab, and select the profile for the pregnant woman. Select a new date. Choose some zinc-rich foods to create a meal. Select Reports and then Source Analysis. Select Zinc from the drop-down box, and generate a report. What zinc-rich foods would you advise for a pregnant vegetarian?
5. An infant just beginning to eat solid foods needs iron and vitamin C in particular. From the Profile drop down box, create a profile for a 30-inch, 24-pound 1-year-old child. Select the Track Diet tab, and create a breakfast and snack that include food sources of iron and vitamin C. Select the Reports tab, then Source Analysis, and finally Iron from the drop-down box, and generate a report. What were the top sources of iron? Do the same for vitamin C, and name the top sources. Did your food choices supply more than a third of the child's iron and vitamin C requirements? If not, what other foods might you select?

what did you decide?

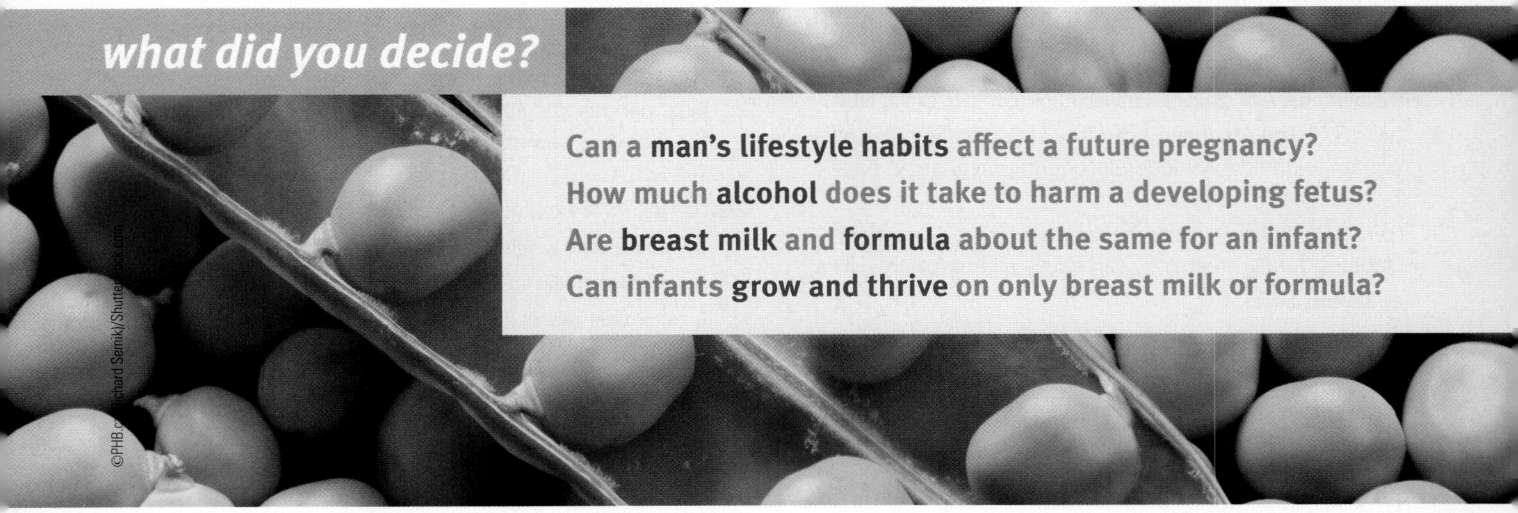

Can a **man's lifestyle habits** affect a future pregnancy?
How much **alcohol** does it take to harm a developing fetus?
Are **breast milk** and **formula** about the same for an infant?
Can infants **grow and thrive** on only breast milk or formula?

Self Check

1. (LO 13.1) A pregnant woman needs an extra 450 calories above the allowance for nonpregnant women during which trimester(s)?
   a. first
   b. second
   c. third
   d. first, second, and third
2. (LO 13.1) A major reason why a woman's nutrition before pregnancy is crucial is that it determines whether her uterus will support the growth of a normal placenta.
   T F
3. (LO 13.1) A deficiency of which nutrient during pregnancy appears to be related to an increased risk of neural tube defects in newborns?
   a. vitamin B_6
   b. folate
   c. calcium
   d. niacin
4. (LO 13.1) The pregnant woman's body helps to conserve iron by
   a. triggering food cravings.
   b. reducing physical activity.
   c. increasing iron excretion.
   d. increasing iron absorption.
5. (LO 13.1) Which of the following preventative measures should a pregnant woman take to avoid contracting listeriosis?
   a. avoid feta cheese
   b. avoid pasteurized milk
   c. thoroughly heat hot dogs
   d. a and c
6. (LO 13.2) Fetal alcohol syndrome (FAS) is one of the leading known preventable causes of mental retardation in the world.
   T F
7. (LO 13.2) All of these characterize the damage done by alcohol during pregnancy, except
   a. halts delivery of oxygen through the umbilical cord.
   b. stimulates maternal appetite and therefore increases fetal nutrition.
   c. slows cell division.
   d. interferes with placental transport of nutrients to the fetus.
8. (LO 13.2) The American Academy of Pediatrics urges all women to drink only moderately during pregnancy.
   T F
9. (LO 13.3) Without proper management, type 1 or type 2 diabetes during pregnancy can cause all except
   a. severe nausea.
   b. severe hypoglycemia or hyperglycemia.
   c. preterm labor.
   d. pregnancy-related hypertension.
10. (LO 13.3) When women in developed countries die of pregnancy complications, the cause is often eclampsia.
    T F
11. (LO 13.4) To support lactation, a breastfeeding woman needs more of the following:
    a. fluid
    b. folate
    c. energy
    d. a and c
12. (LO 13.4) Maternal dietary calcium intake has no effect on the calcium content of breast milk.
    T F
13. (LO 13.4) Lactating women who smoke tobacco
    a. transfer nicotine and other chemicals to their infants through their breast milk.
    b. produce more milk than nonsmokers.
    c. produce milk with a higher fat content, damaging the infant's arteries.
    d. b and c
14. (LO 13.5) Breastfed infants may need supplements of __________ .
    a. fluoride, iron, and vitamin D
    b. zinc, iron, and vitamin C
    c. vitamin E, calcium, and fluoride
    d. vitamin K, magnesium, and potassium
15. (LO 13.5) Protective factors in breast milk include
    a. antiviral agents
    b. anti-inflammatory agents
    c. antibacterial agents
    d. all of the above
16. (LO 13.5) Which of the following foods poses a choking hazard to infants and small children?
    a. pudding
    b. marshmallows
    c. hot dog slices
    d. b and c
17. (LO 13.5) A sure way to get a baby to sleep through the night is to feed solid foods as soon as the baby can swallow them.
    T F
18. (LO 13.6) Fostering a sense of autonomy in a one-year-old includes allowing the child to explore and experiment with her food.
    T F
19. (LO 13.6) In light of the developmental needs of one-year-olds, parents should allow such behaviors as standing at the table or throwing food.
    T F
20. (LO 13.7) To treat obesity in children, a first goal is to
    a. reduce their weight by 10 percent while they grow taller.
    b. quickly achieve their ideal weight.
    c. slow their rate of gain while they grow taller.
    d. a and b

Answers to these Self Check questions are in Appendix G.

CONTROVERSY 13

Childhood Obesity and Early Chronic Diseases

LO 13.7 Discuss some relationships between childhood obesity and chronic diseases, and develop a healthy eating and activity plan for an obese child of a given age.

When most people think of health problems in children and adolescents, they most often think of measles and acne, not type 2 diabetes and hypertension. Today, however, about a third of U.S. children and adolescents 2 to 19 years of age are overweight, and many of these children are obese (shown in Figure C13–1).*[1] Serious risk factors and "adult diseases," such as type 2 diabetes, often accompany obesity, even in a child.[2] U.S. children are not alone in these problems—childhood obesity rates are soaring around the globe.[3]

Although no group has fully escaped the national gain in body weight, obese children tend to have these characteristics:

- Are male.[4]
- Are older.
- Are of African American or Hispanic descent.[5]
- Are sedentary.[6]
- Have parents who are obese.

Additionally, low family income predicts obesity among non-Hispanic white children.[7]

By some measures, childhood obesity rates appear to have steadied or even fallen slightly in a few areas of the country, leading scientists to hope that a turning point may have been reached. In most areas, however, childhood obesity remains an unanswered challenge.[8]

Figure C13–1

Trends in Childhood Obesity

Today, almost 17% of children and adolescents have BMI values at or above the 95th percentile (are obese)[a] as measured on BMI-for-age growth charts (see inside back cover), while almost 12% have BMI values at or above the 97th percentile.

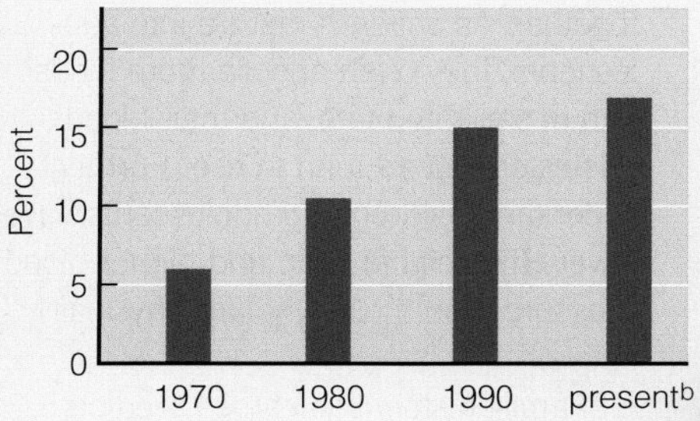

[a]Obesity defined by U.S. Preventive Services Task Force, Screening for obesity in children and adolescents: U.S. Preventive Services Task Force Recommendation Statement, **Pediatrics** *(2010), published online January 18, 2010, doi: 10.1542/peds.2009-2037.*

[b]Childhood obesity rates remained largely unchanged between 1999 and 2008.

Source: Data for 1999–today from C. L. Ogden and coauthors, Prevalence of high body mass index in U.S. children and adolescents, Journal of the American Medical Association *303 (2010): 242–249.*

The Challenge of Childhood Obesity

Obesity takes a heavy toll on the well-being of a child. Education is urgently needed—most overweight children and their parents all but discount the health threats, focusing instead on appearance and the social costs of obesity.

Physical and Emotional Perils

Obese adolescent children often display a risky blood lipid profile that foreshadows development of atherosclerosis—43 percent of obese children test high for total cholesterol, triglycerides, or LDL cholesterol.[9] Overweight children also tend to have high blood pressure; obesity is a leading cause of pediatric hypertension.[10] Without intervention, millions of U.S. children may be destined to develop type 2 diabetes and hypertension in childhood, which together with obesity and high blood cholesterol top the list of factors associated with development of heart disease, also known as cardiovascular disease (CVD), in early adulthood.[11]

Asthma is also much more prevalent among obese children than among their thinner peers.[12] Additionally, obese children suffer more from breathing difficulties while sleeping, a condition known as obstructive sleep apnea.[13] A disease of the liver, nonalcoholic fatty liver disease, also occurs more often, and obese children have a greater risk of complications from anesthesia.[14]

©The Star-Ledger/Saed Hindash/The Image Works

Children with obesity may develop type 2 diabetes, among other ills.

* *Reference notes are found in Appendix F.*

Obese children may also suffer psychologically.[15] Adults may discriminate against them, and peers may make thoughtless comments or reject them. An obese child may develop a poor self-image, a sense of failure, and a passive approach to life. Television shows and movies, two major influences on children's thought processes, often denigrate and stigmatize the fat person as a social misfit.[16] Children have few defenses against these unfair portrayals and quickly internalize negative attitudes toward bulky body sizes.

Overweight or Chubby and Healthy: How Can You Tell?

An accurate assessment of a child's body mass index (BMI) for age is essential—guesswork can lead to unneeded lifestyle changes for a healthy-weight child or to a missed opportunity to help a truly overweight child. Physicians, registered dietitians, and other health-care providers can accurately assess a child's BMI and interpret it using a growth chart (see the inside back cover).[17] Although cutoffs for children generate controversy, children and adolescents are generally considered *overweight* from the 85th to the 94th percentile on the charts and *obese* at the 95th percentile and above.[18]

Unrealistic expectations can undermine good intentions. Most overweight children tend to remain "stocky" long after losing some of their fatness, even into adulthood, and no amount of diet or exercise will make them willowy.[19] Early maturation and the greater bone and muscle mass needed to carry their extra weight contribute to the bulk retained by obese children.[20] In young children, genetic inheritance also plays an important determining role in body size and shape, perhaps even more so than in adults. Still, the child's environment remains a major player in obesity development.[21]

Darla and Gabby

Eight-year-old Gabby and her worried mother Darla tell a typical story of childhood obesity, and they model some appropriate responses. Recently, a note from the school nurse explained that during a routine screening, Gabby's BMI was found to be too high. "The kids next door look skinny to me," says Darla, "like if they got sick they couldn't fight it off." Because Gabby's BMI exceeds the 95th percentile, however, her health may be in peril, and the nurse has suggested further testing for risk factors of chronic diseases.[22] With Gabby's health in danger, Darla's concern grows: "Both my father and his father died of diabetes-related disease, and I'm worried."

Development of Type 2 Diabetes

An estimated 85 percent of the children with type 2 diabetes are obese. Diabetes is most often diagnosed around the age of puberty, but type 2 diabetes is quickly encroaching on younger age groups as children grow fatter. Ethnicity (being Native American or of African, Asian, or Hispanic descent) increases the risk, as does having a family history of type 2 diabetes. Chapter 4 described the risks associated with type 2 diabetes and Chapter 11 revealed its connection with CVD.

Determining exactly how many children suffer from type 2 diabetes is tricky. The child with type 2 diabetes may lack classic telltale symptoms, such as glucose in the urine, ketones in the blood, weight loss, or excessive thirst and urination, so the condition often advances undetected. Undiagnosed diabetes means that children suffering with the condition are left undefended against its ravages.

Development of Heart Disease

Atherosclerosis, first apparent as heart disease in adulthood, begins in youth. By adolescence, most children have formed fatty streaks in their coronary arteries. By early adulthood, the arterial lesions that make heart attacks and strokes likely have formed.

Research is ongoing, but results often indicate that children with the highest risks of developing heart disease in adulthood are sedentary and have central obesity; they may also have diabetes, high blood pressure, and high blood LDL cholesterol.[23] Adolescents who take up smoking greatly compound their risk.

High childhood BMI alone may not always predict increased adulthood heart disease risk, however. Many overweight youngsters appear to grow into adults with average weight and disease risks.[24] Still, authorities recommend that all children aged 6 years and older be screened for obesity, and that obese children be treated with intensive counseling that includes diet, physical activity, and behavior changes.[25]

The note from Gabby's school nurse prompted medical testing, including a family history, a fasting blood glucose test, a blood lipid profile, and a blood pressure test. Luckily, the results for both glucose and blood pressure were normal.

High Blood Cholesterol

Gabby's blood lipid results, however, confirmed her mother's fears: her LDL cholesterol is 135—too high for health. Cholesterol standards for children and adolescents (ages 2 to 18 years) are in Table C13–1.

Obesity, especially central obesity, and high blood cholesterol often occur together. As children mature into adolescents, they often choose more foods rich in saturated fats, and their blood cholesterol levels tend to rise. Further, sedentary children and adolescents have lower HDL, higher LDL, and higher blood pressure than those who are physically active.

Family history sometimes predicts high blood cholesterol. If the parents or grandparents suffered from early heart disease, chances are that a child's blood cholesterol will be higher than average and will remain so through life. For this reason, some experts recommend universal cholesterol screening for all children aged 9–11. Children or adolescents with diabetes, or who are overweight, who smoke, or who consume diets high in saturated fat have a higher risk.

Table C13–1

Cholesterol Values for Children and Adolescents

Disease Risk	Total Cholesterol (mg/dL)	LDL Cholesterol (mg/dL)
Acceptable	<170	<110
Borderline	170–199	110–129
High	≥200	≥130

Note: Adult values appeared in Chapter 11.

High Blood Pressure

High blood pressure in a child or adolescent is a concern—it can signal the early onset of hypertension. Childhood hypertension, left untreated, tends to worsen with time and can accelerate atherosclerosis.[26] Diagnosing hypertension in children must account for age, gender, and height; simple tables like the ones for adults are useless for children.

Dramatic improvements often occur when children with hypertension take up regular aerobic activity and hold their weight down as they grow taller ("grow into their weight"). Restricting sodium intake also causes an immediate drop in most children's and adolescents' blood pressures.[27]

Early Childhood Influences on Obesity

Children begin to learn about behaviors that affect their health from a young age. Parents and other caregivers have a unique opportunity to help children form healthy habits related to the foods they eat, the physical activities they participate in, and their emotional well-being—all of which will pave the way to becoming healthy adults.

Calories—and Cautions

Gabby, who loves sweets, budgets her pocket money (she's saving for a bicycle) to join her friends for a chocolate granola bar (160 calories) every day after school.[28] In addition, she knows how to bake a few peanut butter cookies from a roll of dough kept in the refrigerator to enjoy at bedtime (another 180 calories). Gabby knows that oats and peanut butter are better than candy for health, but she doesn't know that calories from fat and sugar greatly outweigh the healthy ingredients in her granola bars and cookies.

Intuitively, Darla would like to eliminate these treats. However, pediatricians warn parents and caregivers to avoid overly restricting a child's eating; while intentions may be good, excessive restriction of sweets or calories can intensify cravings and spark unnecessary battles about food. Worse, children who feel deprived or hungry may begin to sneak banned foods or hide them and binge on them in secret—behaviors that often predict eating disorders.

Figure C13–2 lists frequent high-calorie snacking as a potential contributing factor in a child's weight gain, but good-tasting snacks and meals are important to all children. A balanced approach may be to include favorite high-calorie treats occasionally in the context of structured, nutritious, and appealing meals and snacks. The next chapter presents more details about designing eating plans for children.

Figure C13–2

Factors Affecting Childhood Weight Gain

The more of these factors in a child's life, the greater the likelihood of unhealthy weight gain.

Food Factors

- Frequent snacks consisting of high-energy foods, such as candies, cookies, crackers, fried foods, and ice cream.
- Irregular or sporadic mealtimes; missed meals.
- Eating when not hungry; eating while watching TV or doing homework.
- Fast-food meals more than once per week.
- Frequent meals of fried or sugary foods and beverages.
- Exposure to advertising that promotes high-calorie foods.

Jose Luis Pelaez Inc./Jupiterimages

Activity Factors

- More than an hour of sedentary activity, such as television, each day.
- Less than 20 minutes of physical activity, such as outdoor play, each day.
- No access to recreational facilities.

Family and Other Factors

- Overweight family members, particularly parents.
- Low-income family.
- Tall for age.

Physical Activity

Children have grown more sedentary, and sedentary children are more often overweight.[29] A child who spends more than an hour or two in "screen time," that is, sitting in front of a television, computer monitor, or other media, often eats fewer family meals and may become obese (Figure C13–3).[30] Children who watch more TV not only move less but they may also snack more, both during television viewing and afterward because of the influence of food advertising.[31] In general, children who are even moderately active have better cardiovascular profiles than sedentary children do.[32]

Darla recalls, "My sisters and I hit the door on Saturday mornings with sandwiches in a bag. We explored, climbed trees, played softball with our friends, jumped in puddles, and played 'tag.' But Gabby and her friends have 252 television channels to choose from, not to mention video games and the Internet—we had only 4 channels when I was little!" The American Academy of Pediatrics (AAP) supports Darla's view and recommends no television time before 2 years of age and a limit of two hours per day of television, computer, and other "screen time" for older children to help prevent obesity.[33]

Figure C13–3

Prevalence of Obesity by Hours of TV per Day, Children Ages 10–15 Years

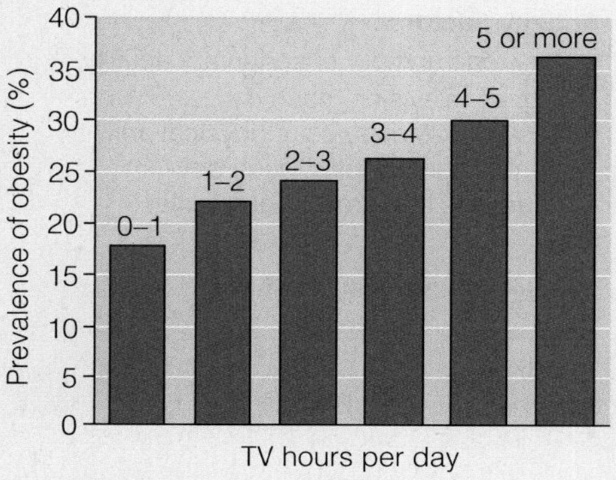

Source: Centers for Disease Control and Prevention, Youth Risk Behavior Survey, available at www.cdc.gov

Food Advertising to Children

Children and youth influence a huge portion of the nation's food spending—up to $200 billion of their own pocket money each year—and influence hundreds of billions more in annual family purchases of foods, beverages, and restaurant meals.[34] As a result, the average child sees an estimated 40,000 TV commercials a year and uncounted Internet commercials—many peddling foods high in sugar, saturated fat, and salt, such as sugar-coated breakfast cereals, candy, "energy" bars, chips, fast foods, and carbonated beverages.[35] The advertisers use developmental psychology to tap children's needs for peer acceptance, fun, love, safety, security, independence, maturity, and identity.[36] Not surprisingly, the more time children spend watching television, the more they ask for the foods and beverages in advertisements—and they get their requests about half of the time.[37]

On the Internet, food marketing agencies develop free, child-attracting "advergames," that is, games built around a manufacturer's foods and beverages, intended to spark brand loyalty in young children. Appealing animated "spokes-characters" speak directly to children, bypassing parents and teachers, to increase children's desire for mostly highly processed, high-fat, high-sugar, low-nutrient treats and fast foods.[38] This approach has proven successful; although children recognize television advertisements with relative ease, they have difficulty separating advertisements from games and other Internet content.[39]

A few major food companies have stopped advertising directly to children; others have agreed to voluntarily promote physical activity and health-promoting products, while reducing the use of beloved animated characters to sell sweets and fats to children, although whether these steps are sufficient is still under debate.[40] This problem is gaining recognition at both national and global levels, as the USDA Food and Nutrition Service has developed a set of consumer-tested messages aimed at helping children develop healthy eating habits, and the World Health Organization recently released a set of recommendations for responsible food marketing to children.[41]

Preventing and Reversing Overweight in Children: A Family Affair

Prevention and treatment of childhood obesity are national priorities.[42] Parents are a starting point: they are encouraged to make major efforts to prevent childhood obesity or to begin treatment early—before adolescence.[43]

For a child who is overweight or obese, an initial goal might be to slow the child's rate of gain while the child grows taller. Weight loss ordinarily is not recommended because diet restriction can easily interfere with normal growth, but this may depend on the severity of the obesity.[44] By including the whole family in an effort to consume balanced meals of appropriate portion sizes and nutritious satisfying snacks and boost physical activity, the goal is often accomplished, and the child does not feel singled out.[45]

Gabby's pediatrician has recommended lifestyle changes to improve both her BMI and blood lipids. Darla is motivated, "I need to take some action!" A warning to Darla: the lifestyle changes may sound easy, but implementing them may prove more difficult than she expects—people's behaviors are notoriously resistant to change. Further, the person, in this case Gabby, must be involved at the planning stage for the changes to be successful.

Parents Set an Example

Parents are among the most influential forces shaping the self-concept, weight concerns, and eating habits of children, and young children learn food behaviors largely from their families.[46] Whole families may be eating too much, dieting inappropriately, and exercising too little.[47] Therefore, successful plans for stabilizing a child's weight center on whole-family lifestyle changes (see

Table C13–2

Family Lifestyle Changes to Help the Overweight Child

The whole family can benefit from health-promoting habits such as these:

- Learn and use appropriate food portions.
- Involve children in shopping for and preparing family meals.
- Set regular mealtimes and dine together frequently.
- For other days, plan and provide a wide variety of nutritious snacks that are low in fat and sugar.
- Provide an appropriate nutritious breakfast every day.
- Provide recommended amounts of fruit juices but no more than this amount.
- Limit high-sugar, high-fat foods, including sugar-sweetened soft drinks and fruit-flavored punches.
- Set a good example and demonstrate positive behaviors for children to imitate.
- Slow down eating and pause to enjoy table companions; stop eating when full.
- Do not use foods to reward or punish behaviors.
- Involve children in daily active outdoor play or structured physical activities, as a family or with friends.
- Limit television time; set a rule to eliminate television-watching during meals.
- Celebrate family special events and holidays with outdoor activities, such as a softball game, a hike, or a summer swim.
- Keep a calendar of scheduled family meals and activity events where everyone can read it.
- Obtain parent and child nutrition and physical activity education and training or family counseling to guide family-based behavioral and other interventions as needed.
- Work with schools to institute school-wide food and activity policies to support a healthy body weight and prevent obesity (see Chapter 14).

Sources: American Medical Association Working Group on Managing Childhood Obesity, Expert Committee recommendations on the assessment, prevention, and treatment of child and adolescent overweight and obesity, June 2007, available at www.ama-assn.org/ama1/pub/upload/mm/433/ped_obesity_recs.pdf; H. Fiore and coauthors, Potentially protective factors associated with healthful body mass index in adolescents with obese and nonobese parents: A secondary data analysis of the Third National Health and Nutrition Examination Survey, 1988–1994, Journal of the American Dietetic Association *106 (2006): 55–64; Position of the American Dietetic Association: Individual-, family-, school-, and community-based interventions for pediatric overweight,* Journal of the American Dietetic Association *106 (2006): 925–945.*

Table C13–2) because when parents set patterns for family behaviors, the children will most often follow their lead.[48]

Lifestyle Changes First, Medications Later

A general rule for treating overweight children is "lifestyle changes first; medications later, if at all." Children with elevated disease risk factors, such as high blood cholesterol or a family history of early heart disease, should still first be treated with diet and physical activity, but if blood cholesterol remains high after 6 to 12 months, then certain drugs may safely be used to lower blood cholesterol without interfering with normal growth or development. Only one obesity drug, orlistat, is approved for limited use in adolescents aged 12 years and older.[49]

Obesity Surgery

Limited research shows that, after surgery, extremely obese adolescents lose significant weight and reduce their risk factors for type 2 diabetes and cardiovascular disease.[50] Surgery may be an option for physically mature adolescents with a BMI of 50 or above or a BMI of 40 or above with significant weight-related health problems who have failed at previous lifestyle modifications and will adhere to the long-term lifestyle changes required after surgery. Whether surgery is a reasonable option for obese teens, however, is the subject of much debate among pediatricians and bariatric surgeons.[51]

Achievable Goals, Loving Support

To preserve the child's healthy sense of self, setting realistic, achievable goals is a first priority. Keeping a positive, upbeat attitude is another. The reverse—impossible goals and a critical, blaming adult—may damage the child's developing self-image and may set the stage for eating disorders later on.

Most of all, Darla must let Gabby know that she is loved, regardless of weight. Blame is a useless concept and can trigger emotional withdrawal of the child just when the opposite—active engagement—is needed most. By being supportive, Darla can help Gabriella grow into a healthy young woman with positive attitudes about food and herself. Meanwhile, she must make some changes to diet and physical activity—but exactly which ones? And how?

Luckily, some government agencies offer help to anyone with computer access. Table C13–3 (p. 550) describes some educational websites that can provide real help for caregivers.

Diet Moderation, Not Deprivation

All children should eat an appropriate amount and variety of foods, regardless of body weight (Chapter 14 provides many details). For the health of the heart, children older than 2 years of age benefit from the same diet recommended for older individuals—that is, a diet limited in fats, especially saturated fat, *trans* fat, and cholesterol; rich in nutrients; and age-appropriate in calories. Such a diet benefits blood lipids without compromising nutrient adequacy, physical growth, or neurological development.

Fruits and vegetables, whole grains, low-fat and nonfat dairy products, beans, fish, and lean meats appropriately make up the bulk of the child's diet. Ice cream, doughnuts, and other high-calorie foods can supply hundreds of excess calories a day, depending upon the child's age and activity level. For perspective, a large (5-inch diameter) glazed doughnut provides 480 calories. Gabby's treats add over 350 calories a day to her intake.

Gabby loves her daily granola bar and the peer interactions it brings.

Table C13–3

Educational Resources for Parents and Children

The USDA Food and Nutrition Service and the White House Task Force on Childhood Obesity have developed a variety of educational materials designed to teach children about healthy food and physical activity choices. The resources below provide more information:

- **MyPlate**

Provides information, tools, games, and other activities to help preschoolers and children make smart choices about eating, being active, and being healthy.
http://www.choosemyplate.gov/preschoolers.html
http://www.choosemyplate.gov/children-over-five.html

- **Eat Smart, Play Hard**

A national campaign aimed at empowering children, parents, and caregivers to make healthy daily choices about eating and exercise. This resource includes fun, interactive materials for children and informative messages and tools for parents.
http://teamnutrition.usda.gov/Resources/eatsmartmaterials.html

- **Let's Move: America's Move to Raise a Healthier Generation of Kids**

A national initiative that provides practical ways to reduce childhood obesity within a generation by giving parents the knowledge they need to foster healthy environments so that children can start out and grow on a path to healthy lifestyles.
http://www.letsmove.gov/

Recognizing that pleasure is important, too, Darla decides to set some goals for providing nutritious, good-tasting lower-calorie foods at regular mealtimes and other snacks to make room for Gabby's favorite treat with her friends. Together, they decide to replace the evening cookies with apple slices spread with a little peanut butter, which cuts the evening snack calories in half without leaving Gabby feeling hungry or deprived. Table C13–4 outlines diet and physical activity recommendations for preventing obesity in children.

Fatty Foods

A steady diet of offerings on most "children's menus" in restaurants, such as fried chicken nuggets, hot dogs, and French fries, easily exceeds a prudent intake of saturated fat, *trans* fat, sodium, and calories and invites both nutrient shortages and gains of body fat. Often, better choices can be found among appetizers, soups, salads, and side selections, and the best establishments offer steamed vegetables, fresh fruit, and broiled or grilled poultry on menus for both children and adults.[52]

Other fatty foods, such as nuts, avocado, vegetable oils, and safer varieties of fish are important for their essential fatty acids. Fatty foods can be calorie-rich, so choosing appropriate, child-sized portions is critical. Low-fat and nonfat milk products or equivalent substitutes deserve a special place in a child's diet for the calcium and other nutrients they supply.

Consumption of Added Sugars

The 2010 Dietary Guidelines recommend limiting intake of added sugars and fats to 5 to 15 percent of daily calories, but children and adolescents consume approximately 16 percent of total caloric intake from added sugars alone.[53] Research has linked sugar-sweetened soft drinks and punches, but not milk or fruit juice, with excess body fatness in children.[54] Although soft drinks amounting to a little more than two cans—the daily consumption of many adolescents—provide an extra 300 calories each day, sugar added to foods has now exceeded sugary beverages as the number one source of added sugar in the American diet.[55] Foods and beverages high in added sugar are best enjoyed in moderation.

Physical Activity

Active children have a better lipid profile and lower blood pressure than more sedentary children. Additionally, the effects of combining a nutritious calorie-controlled diet with exercise can be seen in observable improvements in children's outer measures of health, such as reduced waist circumference and increased muscle strength, along with the inner benefits of greatly improved condition of the heart and arteries.[56] Opportunities to be physically active can include team, individual, and recreational activities (see Figure C13–4).

A new generation of computer games offers some amount of physical activity—they simulate sports games, and participants must move their bodies to play them. These games are better than sedentary screen time, but better still is getting outside and playing the actual sport.[†57] Table C13–3, earlier, listed the *2008 Physical Activity Guidelines for Americans* that apply to children.

Finally, if efforts to stabilize the weight of the youngster fail, the family may benefit from the expertise of an experienced professional. Overweight in children must be sensitively addressed, however. Children are impressionable and can easily come to believe that their worth or lovability is somehow tied to their weight. A registered dietitian or a credentialed childhood weight-loss program may provide assistance.

Darla's Efforts and Gabby's Future

"I'm achieving four of our goals now," says Darla, "and others are planned. First, Gabby and I are getting up a little earlier in the mornings to eat a nutritious breakfast. Gabby's doctor explained that

† *For example, Wii by Nintendo.*

Table C13–4

Recommended Diet and Physical Activity to Prevent Childhood Obesity

American Medical Association Dietary Recommendations for children 2 to 18 years of age

- Limit consumption of sugar-sweetened beverages, such as soft drinks and fruit-flavored punches.
- Eat recommended amounts of fruits and vegetables every day (2 to 4.5 cups per day based on age).
- Learn to eat age-appropriate portions of food.
- Eat foods low in energy density such as those high in fiber and/or water and modest in fat.
- Eat a nutritious breakfast every day.
- Eat a diet rich in calcium.
- Eat a diet balanced in recommended proportions for carbohydrate, fat, and protein.
- Eat a diet high in fiber.
- Eat together as a family as often as possible.
- Limit the frequency of restaurant meals.
- Limit television and other screen time to no more than 2 hours a day.

***Physical Activity Guidelines for Americans* for Children**

- Children and adolescents should do 60 minutes (1 hour) or more physical activity daily.
- Aerobic: Most of the 60 or more minutes a day should be either moderate- or vigorous-intensity aerobic physical activity and should include vigorous-intensity physical activity at least 3 days a week.[a]
- Muscle-strengthening: As part of their 60 minutes of daily physical activity, children and adolescents should include muscle-strengthening physical activity on at least 3 days of the week.
- Bone-strengthening: As part of their 60 or more minutes of daily physical activity, children and adolescents should include bone-strengthening physical activity on at least 3 days of the week.

[a]*Chapter 10 specified activities that characterize various intensity levels.*

breakfast is important because it can help Gabby focus at school and reach a healthier weight.[58] Second, I'm packing Gabby a healthy, tasty, lower-calorie lunch for school. It's easy to make ahead whole-grain sandwiches for the week and freeze them and then toss one into a lunch bag with a low-fat yogurt, or low-fat cheese sticks, and water (not soda!). I'm also including some snacks of good-for-her foods that she loves, like baby carrots and raisins, to tempt her away from the granola bar machine on some days.

"Third, because we both have a sweet tooth, I keep ready-to-eat snacks of fresh fruit, like grapes and strawberries, in clear plastic containers on a refrigerator shelf at eye level. Fourth, although I work days and go to school four nights a week, we have started a new tradition: family meal night each Friday at 6:00 sharp. Gabby and I choose the menu during the week and look forward to making dinner together. We also switched from full-sized dinnerware to pretty new luncheon-sized plates and small dessert-sized bowls. Gabby was charmed with the bright colors, and we both find the smaller portions just as satisfying.

"Although my daughter's idea of a good vegetable has always been a fried potato, she's gradually opening up to trying new foods, which is goal number four. During Friday meal preparation, she's tried bites of broccoli, green beans—even squash! French fries are now just an occasional treat when we eat out. Gabby is doing great, and I'm going to keep offering her healthy new foods to try because her pediatrician said it can take multiple tries for a child to acquire a taste for a new food.[59]

"Goal number five has proved harder: we must start walking together, but when? I need to let her see that I am

Figure C13–4

Age-Appropriate Physical Activity for Kids

Preschoolers (2 to 5 years)

Games in the yard or park
Family walks after dinner
Playing with the dog
Dancing freestyle
Tumbling and gymnastics
T-ball
Playing catch
Family bike rides
Building a snowman
Family swimming at the pool or beach
Playing hide-and-seek

© Anatoly Samara/Shutterstock.com

© Fotokostic/Shutterstock.com

Older children (6 to 12 years)

Throwing a Frisbee
Jumping rope
Bicycling
Playing games and sports such as soccer, softball, baseball, and basketball
Rollerblading
Running
Weight training with light weights
Dancing
Competitive swimming
Snowboarding or skiing
Family kayaking, canoeing, or surfing

Art © Cengage Learning

serious about my personal fitness, but I'm tired after work, and my studies gobble my time. To get Gabby moving after school, I've offered her credits toward her bike in exchange for physical chores, such as raking, planting flowers, and washing the car—and when she gets her bike, she'll be active while riding it, too. Today though, rain or shine, tired or not, I'm going to pull on my running shoes and walk around our neighborhood. And I hope Gabby will join me.

"I love my smart, stubborn, sturdy girl—chubby or lean! But I know her future will be shaped by what we are doing right now. She will grow into her weight if we can hold the line with our new healthy habits. I see her potential to do great things, and what she is learning today about taking care of herself she can pass on to others—to her own children maybe." Darla smiles, "I am so happy we are in this together, and taking charge of our health."

Critical Thinking

1. Who do you believe is responsible for childhood obesity? Organize a chart listing the changes a family and child can make to combat obesity. Include changes in food intake and activity patterns.
2. Draw a picture that represents the concept of energy balance that you could use as a visual aid in explaining this concept to a 10- to 12-year-old.

14 Child, Teen, and Older Adult

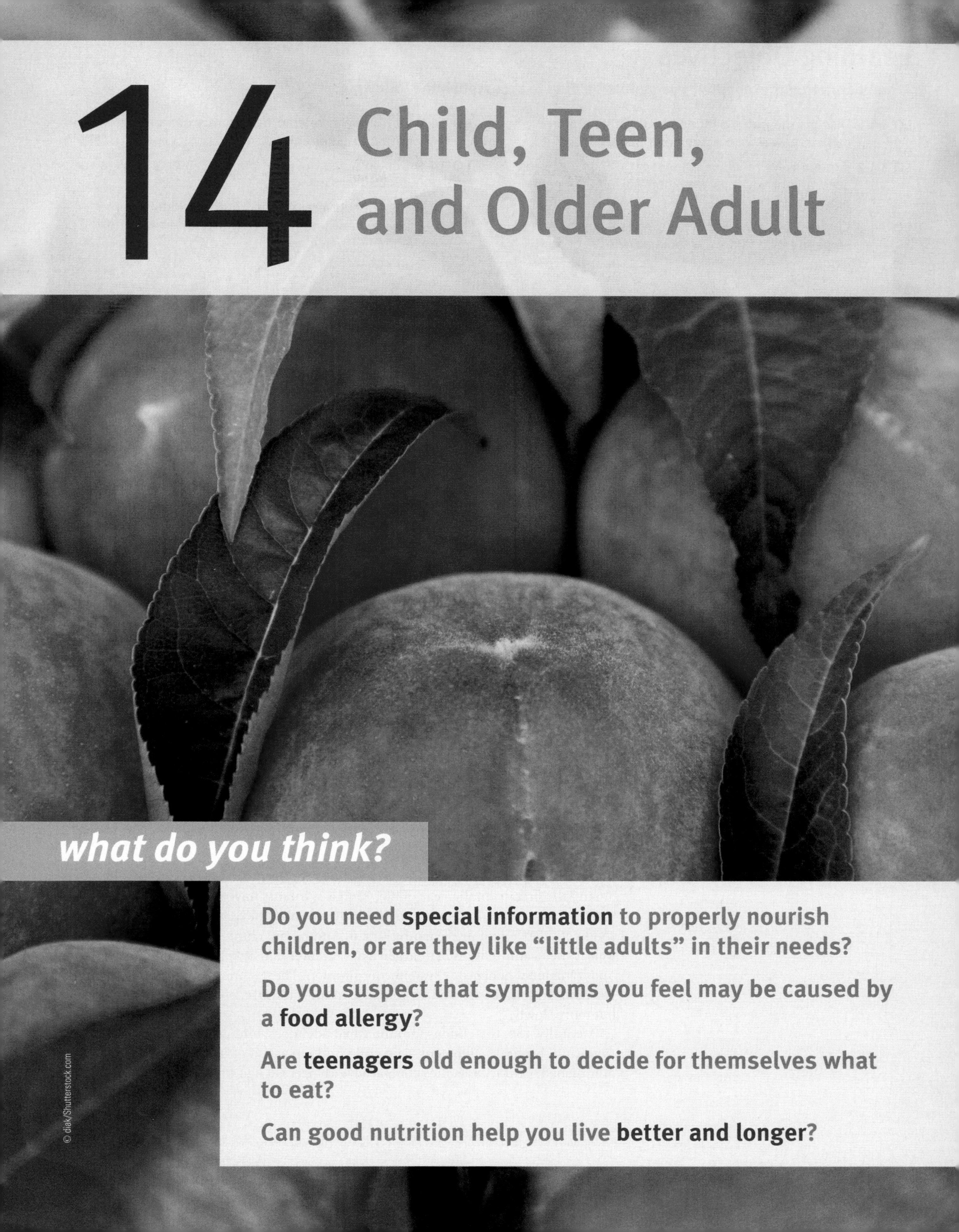

what do you think?

Do you need **special information** to properly nourish children, or are they like "little adults" in their needs?

Do you suspect that symptoms you feel may be caused by a **food allergy**?

Are **teenagers** old enough to decide for themselves what to eat?

Can good nutrition help you live **better and longer**?

Learning Objectives

After completing this chapter, you should be able to accomplish the following:

LO 14.1 Discuss the nutritional needs of young children and explain how a food allergy can impact the diet.

LO 14.2 Explain ways in which a teenager's choice of soda over milk or soy milk may jeopardize nutritional health.

LO 14.3 Contrast life expectancy with life span, and name some lifestyle factors associated with successful aging.

LO 14.4 Outline food-related factors that can predict malnutrition in older adults.

LO 14.5 Design a healthy meal plan for an elderly widower with a fixed income.

LO 14.6 Describe several specific nutrient–drug interactions, and name some herbs that may interfere with the action of medication.

To grow and to function well in the adult world, children need a firm background of sound eating habits, which begin during babyhood with the introduction of solid foods. At that point, the person's nutrition story has just begun; the plot thickens. Nutrient needs change in childhood and throughout life, depending on the rate of growth, gender, activities, and many other factors. Nutrient needs also vary from individual to individual, but generalizations are possible and useful.

In a national assessment of children's diets in the United States, the great majority (81 percent) ranked "poor" or "needing improvement."* The consequences of such diets may not be evident to the casual observer, but nutritionists know that nutrient deficiencies during growth often have far-reaching effects on physical and mental development. Likewise, dietary excesses during childhood often set up a lifelong struggle against obesity and chronic diseases. Anyone who cares about children in their lives, now or in the future, would profit from knowing how to provide the nutrients that children require to reach their potential, while setting patterns that support health throughout life.

Childhood obesity and related chronic diseases are so complex and pervasive that **Controversy 13** was devoted to them.

Early and Middle Childhood

LO 14.1 Discuss the nutritional needs of young children and explain how a food allergy can impact the diet.

Imagine growing 10 inches taller in just one year, as the average healthy infant does during the first dramatic year of life. At age 1, infants have just learned to stand and toddle, and growth has slowed by half; by 2 years, they can take long strides with solid confidence and are learning to run, jump, and climb. These accomplishments reflect the accumulation of a larger mass, greater density of bone and muscle tissue, and refinement of nervous system coordination. These same growth trends, a lengthening of the long bones and an increase in musculature, continue until adolescence but more slowly.

Mentally, too, the child is making rapid advances, and proper nutrition is critical to normal brain development. The child malnourished at age 3 often demonstrates diminished mental capacities compared with peers at age 11.

* As measured by the Healthy Eating Index, a diet assessment tool that measures compliance with the *Dietary Guidelines for Americans* (Chapter 2).

Figure 14–1

Composition of Weight Gain, Infants and Toddlers

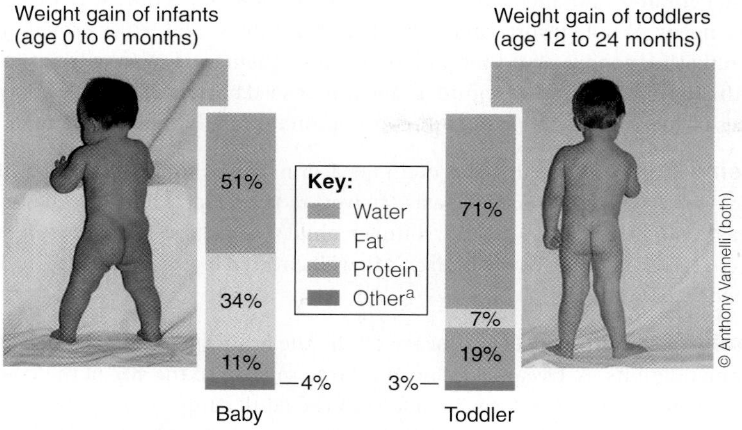

^a "Other" consists of carbohydrate and minerals.

Source: Data from K. L. McCohany and M. F. Picciano, How to grow a healthy toddler—12 to 24 months, Nutrition Today *38 (2003): 156–163.*

These graphs demonstrate that a young infant deposits much more fat than lean tissue, but a toddler deposits more lean than fat. Water follows lean tissue, demonstrated by the water gains in the toddler. You can see that the body shape of a 1-year-old (left photo) changes dramatically by age 2 (right photo). The 2-year-old has lost much baby fat; the muscles (especially in the back, buttocks, and legs) have firmed and strengthened; and the leg bones have lengthened.

Feeding a Healthy Young Child

At no time in life does the human diet change faster than during the second year. From 12 to 24 months, a child's diet changes from infant foods consisting of mostly formula or breast milk to mostly modified adult foods. This doesn't mean, of course, that milk loses its importance in the toddler's diet—it remains a central source of calcium, protein, and other nutrients. Nevertheless, the rapid growth and changing body composition (see Figure 14–1) during this remarkable period demand more nutrients than can be provided by milk alone. Further, the toddling years are marked by bustling activity made possible by new muscle tissue and refined neuromuscular coordination. To support both their activity and growth, toddlers need nutrients and plenty of them.

Appetite Regulation An infant's appetite decreases markedly near the first birthday and fluctuates thereafter. At times children seem insatiable, and at other times they seem to live on air and water. Parents and other caregivers need not worry: given an ample selection of nutritious foods at regular intervals, internal appetite regulation in healthy, normal-weight children guarantees that their overall energy intakes remain remarkably constant and will be right for each stage of growth.[†1]

This ideal situation depends upon the relegation of low-nutrient, high-calorie foods to the status of treats, however. Today's children too often consume a constant stream of tempting foods high in added sugars, saturated fat, refined grains, and calories throughout the day, short-circuiting normal hunger and satiety cues. Children who receive regularly timed snacks and meals of a variety of nutritious foods, with only occasional special treats, often are those who gain weight appropriately and grow normally.[2] The *Dietary Guidelines for Americans* are safe and appropriate goals for the diets of children 2 years of age and older to provide nutrients and energy needed for growth without excesses.

Energy Individual children's energy needs vary widely, depending on their growth and physical activity. On average, though, a 1-year-old child needs about 800 calories a day; at age 6, the child's needs double to about 1,600 daily calories. By age 10, about 1,800 calories a day support normal growth and activity without causing excess storage of body fat. As children age, the total number of calories needed increases, but

† Reference notes are found in Appendix F.

Table 14–1

Estimated Daily Calorie Needs for Children

A child's energy need is also modified by rate of growth, height, weight, and other factors.

Children	Sedentary[a]	Active[b]
2 to 3 yr	1,000	1,400
Females		
4 to 8 yr	1,200	1,800
9 to 13 yr	1,400	2,200
Males		
4 to 8 yr	1,200	2,000
9 to 13 yr	1,600	2,600

[a]Sedentary *describes a lifestyle that includes only the activities typical of day-to-day life.*

[b]Active *describes a lifestyle that includes at least 60 minutes per day of moderate physical activity (equivalent to walking more than 3 miles per day at 3 to 4 miles per hour) in addition to the activities of day-to-day life.*

Did You Know?

The DRI range for total fat intakes:

- 30% to 40% of energy for children 1 to 3 years of age.
- 25% to 35% of energy for children 4 to 18 years of age.

Table 14–2

DRI Recommended Fiber Intakes for Children

Age (yr)	Fiber (g)
1–3	19
4–8	25
9–13	
Boys	31
Girls	26
14–18	
Boys	38
Girls	26

per pound of body weight, the need declines from the extraordinarily high demand of infancy. Table 14–1 shows that both age and activity levels help to determine calorie needs in children.[‡]

Some children, notably those fed a vegan diet, may have difficulty meeting their energy needs. Whole grains, many kinds of vegetables, and fruits provide plenty of fiber and other goodies, but their bulk may make them too low in calories to support growth. Soy products, other legumes, and nut or seed butters offer more concentrated sources of energy and nutrients to support optimal growth and development.[3]

Protein The total amount of protein needed increases somewhat as a child grows larger. On a pound-for-pound basis, however, the older child's need for protein decreases slightly relative to the younger child's need (see the DRI values, inside front cover). Protein needs of children are well covered by typical U.S. diets and well-planned vegetarian diets.

Carbohydrate and Fiber Glucose use by the brain sets the carbohydrate intake recommendations. A 1-year-old's brain is large relative to the size of the body, so the glucose demanded by the 1-year-old falls in the adult range (see inside front cover).[4] Fiber recommendations derive from adult intakes and should be adjusted downward for children who are picky eaters and take in little energy (see Table 14–2).

Fat and Fatty Acids Keeping fat intake within bounds helps to control saturated fat and so may help protect children from developing early signs of adult diseases. Taken to extremes, however, a low-fat diet can lack essential nutrients and energy needed for growth. The essential fatty acids are critical to proper development of nerve, eye, and other tissues.

Children's small stomachs can hold only so much food, and fat provides a concentrated source of food energy needed for growth. The DRI recommended range for fat in a child's diet (see the margin) assumes that energy is sufficient.[5] Specific DRI recommendations are on the inside front cover.

The Need for Vitamins and Minerals As a child grows larger, so does the demand for vitamins and minerals. On a pound-for-pound basis, a 5-year-old's need for, say, vitamin A is about double the need of an adult man. A balanced diet of nutritious foods can meet children's needs for most nutrients, with the exceptions of specific recommendations for fluoride, vitamin D, and iron. Well-nourished children do not need other supplements, and those who receive them typically end up with extra amounts of nutrients already amply provided by their diets.[6] For fluoride, pediatricians may prescribe it for children in areas with fluoride-poor water; vitamin D and iron are discussed next.

Vitamin D According to the DRI committee, children's intakes of vitamin D–fortified milk, ready-to-eat cereals, fortified juices, and other fortified foods should provide 15 micrograms of vitamin D each day to maximize their absorption of calcium and ensure normal, healthy bone growth.[7] Many millions of U.S. children have intakes below this amount.[8] Children who do not consume enough vitamin D from fortified foods should receive a vitamin D supplement to make up the shortfall.

Iron As for iron, iron deficiency is a major problem worldwide and is prevalent in U.S. toddlers 1 to 3 years of age.[9] During the second year of life, toddlers progress from a diet of iron-rich infant foods such as breast milk, iron-fortified formula, and iron-fortified infant cereal to a diet of adult foods and iron-poor cow's milk. Their stores of iron from birth are exhausted, but their rapid growth demands new red blood cells to fill a larger volume of blood. Compounding the problem is the variability in toddlers' appetites: sometimes 2-year-olds are finicky, sometimes they eat voraciously, and they may go through periods of preferring milk and juice while rejecting solid foods for a time. All of these factors—switching to whole milk and unfortified foods, diminished

[‡] DRI estimated energy requirements for infants and children derive from values for weight, age, physical activity, and other parameters.

iron stores, and unreliable food consumption—make iron deficiency likely at a time when iron is critically needed for normal brain growth and development. A later section comes back to iron deficiency and its consequences to the brain.

To prevent iron deficiency, children's foods must deliver 7 to 10 milligrams of iron per day. To achieve this goal, snacks and meals should include iron-rich foods. Milk intake, though critical for the calcium needed for dense, healthy bones, should not exceed daily recommendations to avoid the displacing of lean meats, fish, poultry, eggs, legumes, and whole-grain or enriched grain products from the diet. Table 14–3 lists some iron-rich foods that many children like to eat.

Planning Children's Meals To provide all the needed nutrients, children's meals should include a variety of foods from each food group in amounts suited to their appetites and needs. Table 14–4 provides USDA eating patterns for children who need 1,000 and 1,800 calories per day. MyPlate online resources for preschoolers (2 to 5 years) translates the eating patterns into messages that can help parents ensure that the foods they provide meet their child's needs. For children older than five (6 to 11 years), the site provides an interactive "Blast Off" nutrition teaching game and other resources for teachers, parents, and children themselves (Figure 14–2). These guidelines and resources also stress the importance of balancing calorie intake with calorie expenditure through adequate physical activity to promote growth without increasing the chances of developing obesity. The topic of childhood obesity is so pressing that Controversy 13 was devoted to it.

KEY POINTS

- Other than specific recommendations for fluoride, vitamin D, and iron, well-fed children do not need supplements.
- USDA food patterns provide adequate nourishment for growth without obesity.

Table 14–3

Iron-Rich Foods Kids Like[a]

Breads, Cereals, and Grains
Cream of wheat (½ c)
Enriched ready-to-eat cereals (1 oz)[b]
Noodles, rice, or barley (½ c)
Tortillas (1 flour or whole wheat; 2 corn)
Whole-wheat, enriched, or fortified bread (1 slice)
Vegetables
Cooked snow peas (½ c)
Cooked mushrooms (½ c)
Green peas (½ c)
Mixed vegetable juice (1 c)
Fruits
Canned plums (3 plums)
Cooked dried apricots (¼ c)
Dried peaches (4 halves)
Raisins (1 tbs)
Meats and Legumes
Bean dip (¼ c)
Lean chopped roast beef or cooked ground beef (1 oz)
Liverwurst on crackers (½ oz)
Meat casseroles (½ c)
Mild chili or other bean/meat dishes (¼ c)
Peanut butter and jelly sandwich (½ sandwich)
Sloppy joes (½ sandwich)

[a]Each serving provides at least 1 mg iron, or one-tenth of a child's iron recommendation. Vitamin C–rich foods included with these snacks increase iron absorption. Grains and vegetables are cooked.

[b]Some fortified breakfast cereals contain more than 10 mg iron per half-cup serving (read the labels).

Table 14–4

USDA Eating Patterns for Children (1,000 to 1,800 Calories)

For an estimate of a child's average energy need based on age, gender, and physical activity, access SuperTracker on the MyPlate website (http://www.choosemyplate.gov/); click on Daily Food Plans and follow the prompts. Height, weight, rate of growth, and other factors also alter energy needs.

Food Group	1,000 cal	1,200 cal	1,400 cal	1,600 cal	1,800 cal
Fruits	1 c	1 c	1½ c	1½ c	1½ c
Vegetables	1 c	1½ c	1½ c	2 c	2½ c
Grains	3 oz	4 oz	5 oz	5 oz	6 oz
Protein foods	2 oz	3 oz	4 oz	5 oz	5 oz
Milk	2 c	2½ c	2½ c	3 c	3 c

Figure 14–2

MyPlate Resources for Children

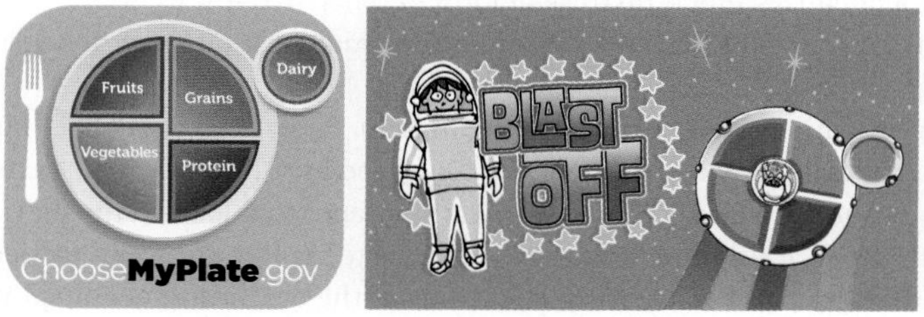

MyPlate resources for preschool children and older children can be found at http://www.choosemyplate.gov/.

Mealtimes and Snacking

The childhood years are the parents' greatest chance to influence their child's food choices.[10] Appropriate eating habits and attitudes toward food, developed in childhood, can help future adults emerge with healthy habits that reduce risks of chronic diseases in later life. The challenge is to deliver nutrients in the form of meals and snacks that are both nutritious and appealing so that children will learn to enjoy a variety of health-promoting nutritious foods.

Current U.S. Children's Food Intakes A comprehensive survey, called the Feeding Infants and Toddlers Study, assessed the food and nutrient intakes of more than 3,000 infants and toddlers.[11] The survey found most infants take in too few fruits and vegetables, and in fact, on a typical day, about 25 percent of those older than 9 months take in none at all.[12] By age 15 months, one vegetable and one fruit stand out as predominant: French fries and bananas, neither a particularly rich source of many needed nutrients. Most children take in too little vitamin E, calcium, magnesium, potassium, and fiber and too much sodium and saturated fat for health.[13] Children's tastes and preferences often lean toward nutrient-poor selections, and providing the nutritious foods they need can prove challenging.

Dealing with Children's Preferences Many children prefer sweet fruits and mild-flavored vegetables served raw or undercooked and crunchy and easy to eat. Cooked foods should be served warm, not hot, because a child's mouth is much more sensitive than an adult's. The flavors should be mild because a child has more taste buds.

Little children prefer small portions of food served at little tables. If offered large portions, children may fill up on favorite foods, ignoring others. Toddlers often go on food jags—consecutive days of eating only one or two favored foods. For food jags lasting a week or so, make no response, because 2-year-olds regard any form of attention as a reward. After two weeks of serving the favored foods, try serving small portions of many foods, including the favored items. Invite the child's friends to occasional meals, and make other foods as attractive as possible.

Bribing a child to eat certain foods by, for example, allowing extra television time as a reward for eating vegetables often fails to produce the desired effect: the child will likely *not* develop a preference for those foods. Likewise, when children are forbidden to eat favorite foods, they yearn for them more—the reverse of the well-meaning caregiver's goal. Include favorites as occasional treats.

Most children can safely enjoy occasional treats of high-calorie foods, but such treats should also be nutritious. From the milk group, ice cream or pudding is good now and then; from the grains group, whole-grain or enriched cakes, oatmeal cookies, snack crackers, or even small doughnuts are an acceptable occasional addition to a nutritious diet. These foods encourage a child to learn that pleasure in eating is important. A steady diet of these treats, however, leads to nutrient deficiencies, obesity, or both.

Fear of New Foods A fear of new foods, **food neophobia**, is almost universal among toddlers and preschoolers.[14] Without so much as a taste, the child rejects the new food on sight, but the reason why this occurs isn't fully known. The child may remember tasting and disliking foods with a similar appearance or aroma. Or, food neophobia may be a protective mechanism that prevented curious ancestral toddlers of ages past from tasting toxic plants in their environment. Regardless of its causes, food neophobia causes much distress among parents striving to nourish their children. Thankfully, the fear diminishes as a child matures, often disappearing completely by adolescence.[15]

In the meanwhile, some practical tips can help. First, keep an upbeat but persistent attitude: a child may ignore or reject a food the first 14 times it is offered but on the 15th may suddenly recognize it as a familiar, accepted food in the diet. Parents' negative attention or attempts to force "just a taste" before the child is ready interrupts this learning process. Offering new foods at the beginning of a meal when the

food neophobia (NEE-oh-FOE-bee-ah) the fear of trying new foods, common among toddlers.

Table 14–5

Tips for Feeding Picky Eaters

If a child fails to eat enough to support healthy growth and development, consult a registered dietitian or physician right away. Otherwise, try these tips.

Get Them Involved	
Children are more likely to try foods when they feel a sense of ownership. Include them in ▪ Meal planning. ▪ Grocery shopping.	▪ Food preparation. ▪ Gardening and harvesting the foods they eat.
Be Creative	
▪ Serve vegetables as finger foods with dips or spreads. ▪ Use cookie cutters to cut fruits and vegetables into fun shapes. ▪ Serve traditional meals out of order (for example, breakfast for dinner).	▪ Encourage (don't force) children's interest and enthusiasm for nutritious foods, such as legumes or whole grains, by using them in craft projects.
Enhance Favorite Recipes	
▪ Blend, slice, or shred vegetables into sauces, casseroles, pancakes, or muffins. ▪ Serve fruit over cereal, yogurt, or ice cream.	▪ Bake brownies with black beans or cookies with lentils as an ingredient (find recipes on the Internet).
Model and Share	
▪ Be a role model to children by eating healthy foods yourself. Offer to share your healthy snack with them. ▪ Children may need multiple exposures to a new food before they accept it, so do continue offering foods that a child initially rejects.	▪ Encourage children to taste at least one bite of each food served at a meal.
Respect and Relax	
▪ Children tend to eat sporadically. They have small stomachs and so tend to fill up fast and become hungry again soon after eating.	▪ Focus on the child's overall weekly intake of food and nutrients rather than on daily consumption.

Source: Adapted from Mayo Clinic Staff, Children's nutrition: 10 tips for picky eaters, 2011, available at http://www.mayoclinic.com/health/childrens-health/HQ01107.

child is hungry often works best, as does serving the child samples of the same foods that adults are enjoying; children follow the examples of adults. The tips offered in Table 14–5 can often make mealtimes go more smoothly.

Child Preferences versus Parental Authority Just as parents are entitled to their likes and dislikes, a child who genuinely and consistently rejects a food should be allowed the same privilege. Also, children should be believed when they say they are full: the "clean-your-plate" dictum should be stamped out for all time.[16] Children who are forced to override their own satiety signals are in training for obesity.

A bright, unhurried atmosphere free of conflict is conducive to good appetite and provides a climate in which a child can learn to enjoy eating. Parents who beg, cajole, and demand that their children eat make power struggles inevitable. A child may find mealtimes unbearable if they are accompanied by a barrage of accusations—"Susie, your hands are filthy . . . your report card . . . and clean your plate!" The child's stomach recoils as both body and mind react to stress of this kind.

© iStockphoto.com/lostinbids

Little children like to eat small portions of food at little tables.

Honoring children's preferences does not mean allowing them to dictate the diet, however, because children naturally prefer fatty, sugary, and salty foods, such as heavily advertised snack chips, cookies, crackers, fast foods, and sugary cereals and

Did You Know?

Here's how to be a role model for a child:

- Set a good example. Eat fruits, vegetables, whole grains, and other nutritious foods when dining with children.
- Shop and cook together. Let the child help pick out and prepare nutritious foods.
- Reward yourself and children with attention, not food. Save treats for dessert.
- Turn off the television. Sit down and eat together.
- Share conversation at mealtimes. Listen to the child.
- Play together. Encourage physical activity.

drinks. When children's tastes are allowed to rule the family's pantry, everyone's nutrition suffers: parents of young children consume much more fat and saturated fat than do adults without children.[17] The responsibility for *what* the child is offered to eat lies squarely with the adult caregiver, but the child should be allowed to decide *how much* and even *whether* to eat.

Many parents overlook perhaps the single most important influence on their child's food habits—their own habits.[18] Parents who don't prepare, serve, and eat carrots shouldn't be surprised when their child refuses to eat carrots.

Snacking As mentioned, parents often find that their children snack so much that they are not hungry at mealtimes. This is not a problem if children are taught how to snack—nutritious snacks are just as health promoting as small meals. Keep snack foods simple and available: milk, cheese, crackers, fruit, vegetable sticks, yogurt, peanut butter sandwiches, and whole-grain cereal.

Restaurant Choices It takes some artful maneuvering to choose nutritious restaurant meals that children can enjoy. Children's menus reliably offer fatty, salty sandwiches, "nuggets," and French fries. For better choices:

- Ask to split a regular meal among several children.
- Choose from appetizers, soups, salads, and side dishes.
- Order vegetable toppings and lean meats on pizza (skip the sausages and hamburger); reduce the saturated fat by requesting half the cheese.
- Request water, fat-free milk, or fruit juice (not punch) for beverages.

Parents who make nutritious restaurant choices for themselves also set good examples for children.

Choking A child who is choking may make no sound, so an adult should keep an eye on children when they are eating. A child who is coughing most often dislodges the food and recovers without help. To prevent choking, encourage the child to sit when eating—choking is more likely when children are running or reclining. Round foods such as grapes, nuts, hard candies, and pieces of hot dog can become lodged in a child's small windpipe. Other potentially dangerous foods include tough meat chunks, popcorn, chips, and peanut butter eaten by the spoonful.

Food Skills Children love to be included in meal preparation, and they like to eat foods they helped to prepare (see Table 14–6). A positive experience is most likely when tasks match developmental abilities and are undertaken in a spirit of enthusiasm and enjoyment, not criticism or drudgery. Praise for a job well done (or at least well attempted) expands a child's sense of pride and helps to develop skills and positive feelings toward healthy foods.

KEY POINTS

- Healthy eating habits are learned in childhood, and parents teach best by example.
- Choking can often be avoided by supervision during meals and avoiding hazardous foods.

How Do Nutrient Deficiencies Affect a Child's Brain?

A child who suffers from nutrient deficiencies exhibits physical and behavioral symptoms: the child feels sick and out of sorts. Such children may be irritable, aggressive, and disagreeable or sad and withdrawn. They may be labeled "hyperactive," "depressed," or "unlikable." Diet–behavior connections are of keen interest to caregivers who both feed children and live with them.

Iron plays key roles in many molecules of the brain and nervous system. An iron deficit in the brain, even before anemia shows up in the blood, has well-known and widespread effects on children's behavior and intellectual performance.[19] The motivation to persist at intellectually challenging tasks is dampened, the attention span is shortened, and the overall intellectual performance is reduced. Administering iron

Table 14–6

Food Skills and Developmental Milestones of Preschool Children[a]

Food Skills	Developmental Milestones
Age 1 to 2 years	
▪ Uses a spoon ▪ Lifts and drinks from a cup ▪ Helps scrub fruits and vegetables, tear lettuce or greens, snap green beans, or dip foods ▪ Can be messy; can be easily distracted	▪ Large muscles develop ▪ Experiences slowed growth and decreased appetite ▪ Develops likes and dislikes ▪ May suddenly refuse certain foods
Age 3 years	
▪ Spears food with a fork ▪ Feeds self independently ▪ Helps wrap, pour, mix, shake, stir, or spread foods ▪ Follows simple instructions	▪ Medium hand muscles develop ▪ May suddenly refuse certain foods ▪ Begins to request favorite foods ▪ Makes simple either/or food choices
Age 4 years	
▪ Uses all utensils and napkin ▪ Helps measure dry ingredients ▪ Learns table manners	▪ Small finger muscles develop ▪ Influenced by TV, media, and peers ▪ May dislike many mixed dishes
Age 5 years	
▪ Measures liquids ▪ Helps grind, grate, and cut (soft foods with dull knife) ▪ Uses hand mixer with supervision	▪ Fine coordination of fingers and hands develops ▪ Usually accepts food that is available ▪ Eats with minor supervision

[a] *These ages are approximate. Healthy, normal children develop at their own pace.*

Source: Adapted from MyPlate for Preschoolers, Behavioral Milestones, available at http://www.choosemyplate.gov/preschoolers/healthy-habits/Milestones.pdf

resolves some of the problems, but others may persist for years after treatment.[20] Both iron deficit and the poverty and poor health often associated with it may contribute to these effects.[21] Despite public health efforts to prevent iron deficiency, such as food fortification, iron deficiency remains a key problem among U.S. children and adolescents.

Only a health-care provider, such as a registered dietitian, should make the decision to give a child a single nutrient iron supplement. Iron is toxic, and overdoses can easily injure or even kill a toddler or child who accidentally ingests iron pills. All supplements should be kept out of children's reach.

KEY POINT

- Iron deficiency and toxicity pose a threat to children.

The Problem of Lead

More than 300,000 children in the United States, most younger than age 6, have blood lead concentrations high enough to cause mental, behavioral, and other health problems.[22] Lead is an indestructible metal element; once inside, the body cannot alter it or easily excrete it.

Sources of Lead Babies love to explore and put everything into their mouths, including chips of old lead paint, pieces of metal that contain lead, and other unlikely substances. Lead may also leach into a home's drinking water supply from old lead pipes and end up in a baby's formula and the family's beverages. In older children, lead dust mixed into outdoor soil can stick to clothing and hands and eventually be consumed. Appreciable amounts of lead have shown up in some chewable vitamins and other children's medications.[23] Once exposed to lead, infants and young children absorb 5 to 10 times as much of the toxin as adults do.

Did You Know?

In 2012, the Centers for Disease Control and Prevention reduced the threshold for lead exposure in a child from 10 micrograms per deciliter of blood to 5 micrograms.

Harm from Lead Lead can build up so silently in a child's body that caregivers may not notice its symptoms until much later, after toxicity has damaged organs. Tragically, once symptoms set in, medical treatments may not reverse all of the functional damage, some of which may linger long beyond childhood.[24]

Impaired thinking, reasoning, perception, and other academic skills, as well as hearing impairment, kidney malfunction, and decreased growth, are associated with even very low blood lead levels in children.[25] Higher levels are associated with scholastic failures, antisocial or hyperactive behavior, and possibly a smaller brain size attained by adulthood.[26] Among adolescents, early lead exposure is linked with lower scores on IQ tests and more arrests for violent crimes.[27] In older women, even low levels of lead in the bones, a measure of lifetime lead exposure, correlates with worsened mental performance.[28]

As lead toxicity slowly injures the kidneys, nerves, brain, bone marrow, and other organs, the child may slip into coma, have convulsions, and may even die if an accurate diagnosis is not made in time. Older children with high blood lead may be suffering physical consequences but be mislabeled as delinquent, aggressive, or learning disabled.

© Tony Freeman/Photo Edit

Old paint is the main source of lead in most children's lives.

Lead and Nutrient Interactions Malnutrition makes lead poisoning especially likely to occur because children absorb more lead from empty stomachs or if they lack calcium, zinc, vitamin C, vitamin D, or iron. A child with iron-deficiency anemia is three times as likely to have elevated blood lead as a child with normal iron status. The chemical properties of lead are similar to those of nutrient minerals like iron, calcium, and zinc, and lead displaces these minerals from their sites of action in body cells but cannot perform their biological functions. Even slight iron and zinc deficiencies may open the door to lead toxicity severe enough to lower a child's scores on tests of verbal ability while increasing feelings of anxiety.[29]

Bans on leaded gasoline, leaded house paint, and lead-soldered food cans have dramatically reduced the amount of lead in the U.S. environment in past decades and have produced a steady decline in children's average blood lead concentrations. However, lead still remains a threat in older communities of homes with lead pipes and layers of old lead paint. Some tips for avoiding lead toxicity are offered in Table 14–7.

KEY POINT

- Blood lead levels have declined in recent times, but even low lead levels can harm children.

Did You Know?

The Environmental Protection Agency (EPA) provides a toll-free hotline for lead information: 1-800-LEAD-FYI (1-800-532-3394).

© William Berry/Shutterstock.com

Table 14–7

Steps to Prevent Lead Poisoning

To protect children:

- If your home was built before 1978, wash floors, windowsills, and other surfaces weekly with warm water and detergent to remove dust released by old lead paint; clean up flaking paint chips immediately.
- Feed children balanced, timely meals with ample iron and calcium.
- Prevent children from chewing on old painted surfaces.
- Wash children's hands, bottles, and toys often.
- Wipe soil off shoes before entering the home.
- Ask a pediatrician whether your child should be tested for lead.

To safeguard yourself:

- Avoid daily use of handmade, imported, or old ceramic mugs or pitchers for hot or acidic beverages, such as juices, coffee, or tea. Commercially made U.S. ceramic, porcelain, and glass dishes or cups are safe. If ceramic dishes or cups become chalky, use them for decorative purposes only.
- Do not use lead crystal decanters for storing alcoholic or other beverages.
- If your home is old and may have lead pipes, run the water for a minute before using, especially before the first use in the morning.

© Cengage Learning

Food Allergy, Intolerance, and Aversion

Parents, when asked, frequently blame food **allergy** for physical and behavioral abnormalities in children, but when children are tested, just 8 percent are diagnosed with true food allergies.[30] Food allergies affect only about 1 or 2 percent of the adult population.[31] The prevalence of food allergy, especially peanut allergy, is on the rise, but no one knows exactly why.[32] About 20 percent of children with peanut allergies "grow out" of them.

Food Allergy A true food allergy occurs when a food protein or other large molecule enters body tissues and triggers an immune response. Most food proteins are dismantled to smaller fragments in the digestive tract before absorption, but some larger fragments enter the bloodstream before being fully digested. The immune system of an allergic person reacts to the foreign molecules as it does to any other **antigen**: it releases **antibodies**, **histamine**, or other defensive agents to attack the invaders. In some people, the result is the life-threatening food allergy reaction of **anaphylactic shock**, which can involve symptoms such as tingling of the tongue, throat, or skin or difficulty breathing. Peanuts, tree nuts, milk, eggs, wheat, soybeans, fish, and shellfish are the foods most likely to trigger this extreme reaction, with peanuts, milk, and shellfish the most prevalent triggers in children.[33]

If a child reacts to allergens with a life-threatening response, two courses of action are required: first, the child's family and school must guard against any ingestion of the allergen. Second, easy-to-administer doses of the life-saving drug **epinephrine** must be kept close at hand.

Allergen Ingestion Parents must teach the child which foods to avoid, but avoiding allergens can be tricky because they often sneak into foods in unexpected ways. For example, a pork chop (an innocent food) may be dipped in egg (egg allergy) and breaded (wheat allergy) before being fried in peanut oil (peanut allergy); marshmallow candies may contain egg whites; lunchmeats may contain milk protein binders; and so forth.

Invisible traces of allergen from, say, peanut butter left on tables, chairs, or other surfaces can easily contaminate the hands of a severely allergic child who may then unknowingly ingest them.[34] Even a trace can cause a life-threatening reaction. Scrupulous cleaning of surfaces with soaps and cleansers and regular hand washing by the allergic child can often prevent such an occurrence. By the way, the protein allergens of peanuts are not volatile—that is, they do not fly off the food into the air under normal conditions, such as when they are being eaten. The distinctive aroma of peanuts can alert people to their presence, but the aroma itself does not cause allergy—only ingestion of peanut protein can do so.[35]

Caregivers of allergic children must pack safe lunches and snacks at home and ask school officials to strictly enforce a "no-swapping" policy in the lunchroom. To prevent nutrient deficiencies, caregivers must also provide adequate substitutes that supply the essential nutrients in the omitted foods.[36] For example, a child allergic to milk must be supplied with calcium from fortified foods, such as calcium-rich orange juice. Nutritional counseling and growth monitoring is recommended for all children with food allergies.[37]

A rumor has it that food allergies can be cured with regular ingestion of small amounts of the offending food. In truth, strictly controlled research studies that provide such oral therapy under medical supervision have noted improvements in immune markers of reactivity.[38] However, significant allergic symptoms may also occur, and anyone attempting such therapy on their own takes a serious risk.

Food Labels Food labels must announce the presence of common allergens in plain language, using common names of the eight foods most likely to cause allergic reactions.[39] For example, a food containing "textured vegetable protein" must say "soy" on its label. Similarly, "casein," a protein in milk, must be identified as "milk." Food producers must also prevent cross-contamination during production and clearly label the foods in which it is likely to occur, as Figure 14–3 (p. 564) demonstrates. Equipment

Ian Boddy/Science Source

An epinephrine "pen" can deliver prompt life-saving treatment to a person suffering from anaphylactic shock.

allergy an immune reaction to a foreign substance, such as a component of food. Also called *hypersensitivity* by researchers.

antigen a substance foreign to the body that elicits the formation of antibodies or an inflammation reaction from immune system cells. Food antigens are usually large proteins. Inflammation consists of local swelling and irritation and attracts white blood cells to the site. Also defined in Chapter 3.

antibodies large protein molecules that are produced in response to the presence of antigens to inactivate them. Also defined in Chapters 3 and 6.

histamine a substance that participates in causing inflammation; produced by cells of the immune system as part of a local immune reaction to an antigen.

anaphylactic (an-ah-feh-LACK-tick) **shock** a life-threatening whole-body allergic reaction to an offending substance.

epinephrine (epp-ih-NEFF-rin) a hormone of the adrenal gland that counteracts anaphylactic shock by opening the airways and maintaining heartbeat and blood pressure.

Figure 14–3
A Food Allergy Warning Label

A food that contains, or could contain, even a trace amount of any of the most common food allergens must clearly say so on its label. For instance, if a product contains the milk protein casein, the label must say "contains milk," or the ingredients list must include "milk." The sunflower seeds below carry a warning about peanut allergy—traces of peanuts may have contaminated the seeds during processing.

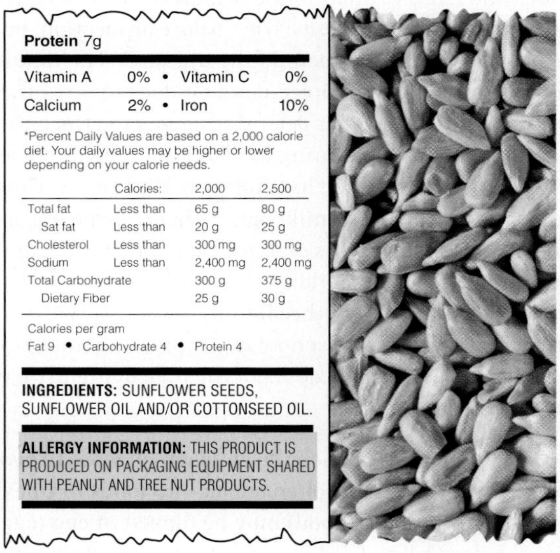
Protein 7g
Vitamin A 0% • Vitamin C 0%
Calcium 2% • Iron 10%
*Percent Daily Values are based on a 2,000 calorie diet. Your daily values may be higher or lower depending on your calorie needs.
Calories: 2,000 2,500
Total fat Less than 65 g 80 g
Sat fat Less than 20 g 25 g
Cholesterol Less than 300 mg 300 mg
Sodium Less than 2,400 mg 2,400 mg
Total Carbohydrate 300 g 375 g
Dietary Fiber 25 g 30 g
Calories per gram
Fat 9 • Carbohydrate 4 • Protein 4
INGREDIENTS: SUNFLOWER SEEDS, SUNFLOWER OIL AND/OR COTTONSEED OIL.
ALLERGY INFORMATION: THIS PRODUCT IS PRODUCED ON PACKAGING EQUIPMENT SHARED WITH PEANUT AND TREE NUT PRODUCTS.

used for making peanut butter must be scrupulously cleaned before being used to pulverize cashew nuts for cashew butter to protect unsuspecting consumers from peanut allergens. Table 14–8 lists symptoms associated with allergic reactions to food.

Detection of Food Allergy Allergies have one or two components. They always involve antibodies; they sometimes involve symptoms. Therefore, allergies cannot be diagnosed from symptoms alone. Anyone who has suffered anaphylaxis or other severe symptoms should not reintroduce suspected foods but should undergo immediate medical testing.

An immediate allergic reaction is easy to identify because symptoms correlate with the time of eating the food. A delayed reaction, taking 24 hours or more, is more difficult to pinpoint. For mild suspected allergy symptoms, a good starting point is to keep a record of food intakes and symptoms. Then, eliminate suspected foods from the diet for a week or two and reintroduce them one at a time to help spot the allergen. An oral challenge test in which the foods are ground up and mixed with other foods or placed in capsules can prevent expectations from skewing the results. If the symptoms correlate with a food, then a blood test for elevated levels of food-specific antibodies and a skin prick test in which a clinician applies droplets of food extracts to

Table 14–8
Symptoms of an Allergic Reaction to Food

Any of these symptoms can occur in minutes or hours after ingesting an allergen:

- *Airway.* Difficulty breathing, wheezing, asthma.
- *Digestive tract.* Vomiting, abdominal cramps, diarrhea.
- *Eyes.* Irritated, reddened eyes.
- *Mouth and throat.* Tingling sensation, swelling of the tongue and throat.
- *Skin.* Hives, swelling, rashes.
- *Other.* Drop in blood pressure, loss of consciousness; in extreme reactions, death.

the skin and then lightly pricks or scratches the skin, or other tests, can confirm the allergy. False positive results can occur, but despite this failing, such tests can support or refute other evidence in making a diagnosis.[40]

Scientific-sounding allergy quackery may deceive people into believing that everything from itchy skin to mental depression is caused by food allergies. Beware of "food sensitivity testing" offered by charlatans on the Internet or elsewhere. It involves fake blood or other tests that supposedly determine which foods to eat and which supplements the "patient" should buy from the quack to relieve the "allergy."

Technology may soon offer new solutions for those with food allergy. Drugs under development may interfere with the immune response that causes allergic reactions, but so far, they remain unavailable.[41] Through genetic engineering, scientists may one day banish allergens from peanuts, soybeans, and other foods to make them safer.[42]

Food Intolerance and Aversion A **food intolerance** is characterized by unpleasant symptoms that reliably occur after consumption of certain foods—lactose intolerance is an example. Unlike allergy, a food intolerance does not involve an immune response. A **food aversion** is an intense dislike of a food that may be a biological response to a food that once caused trouble. Parents are advised to watch for signs of food aversion and to take them seriously. Such a dislike may turn out to be a whim or fancy, but it may turn out to be an allergy or other valid reason to avoid a certain food. Don't prejudge. Test. Then, if an important staple food must be excluded from the diet, find other foods to provide the omitted nutrients.

Foods are often unjustly blamed when behavior problems arise, but children who are sick from any cause are likely to be cranky. The next section singles out one such type of misbehavior.

KEY POINTS

- Food allergy may be diagnosed by the presence of antibodies.
- Food aversions can be related to food allergies or to adverse reactions to food.

Can Diet Make a Child Hyperactive?

Attention-deficit/hyperactivity disorder (ADHD), or **hyperactivity**, is a **learning disability** that occurs in 5 to 10 percent of young, school-aged children—or in 1 to 3 in every classroom of 30 children.[43] ADHD is characterized by the chronic inability to pay attention, along with overly active behavior and poor impulse control. It can delay growth, lead to academic failure, and cause major behavioral problems. Although some children improve with age, many reach the college years or adulthood before they receive a diagnosis and with it the possibility of treatment.

Food Allergies Food allergies have been blamed for ADHD, but research to date has shown no connection. A well-controlled study of close to 300 children suggests that food additives such as artificial colors or sodium benzoate preservative (or both) may worsen hyperactive symptoms such as inattention and impulsivity.[44] More research is needed to confirm or refute these findings; meanwhile, parents who wish to avoid such additives can find them listed with the ingredients on food labels.

Sugar and Behavior Many years ago, sugary foods were accused of causing children to become unruly and adolescents and adults to exhibit antisocial and even criminal behaviors. Science, however, has put the "sugar-behavior" theory to rest. Still, many teachers, parents, grandparents, and others assert that some children react behaviorally to sugar. In a recent study of almost 8,000 European children, those with high intakes of sugary, fatty treats, such as chocolate bars and chips, were often described as more hyperactive than others.[45] Sugary foods clearly displace vegetables and other nutritious choices from the diet, and nutrient deficiencies are known causes of behavioral problems. Sugar itself, however, is unlikely to do so.

Inconsistent Care and Poverty Common sense says that all children get unruly and "hyper" at times. A child who often fills up on caffeinated colas and chocolate, misses lunch, becomes too cranky to nap, misses out on outdoor play, and spends

food intolerance an adverse reaction to a food or food additive not involving an immune response.

food aversion an intense dislike of a food, biological or psychological in nature, resulting from an illness or other negative experience associated with that food.

hyperactivity (in children) a syndrome characterized by inattention, impulsiveness, and excess motor activity; usually diagnosed before age 7, lasts six months or more, and usually does not entail mental illness or mental retardation. Properly called *attention-deficit/hyperactivity disorder (ADHD).*

learning disability a condition resulting in an altered ability to learn basic cognitive skills such as reading, writing, and mathematics.

The **Controversy** section of this chapter lists caffeine amounts in common foods and beverages.

hours in front of a television or other screen media suffers stresses that can trigger chronic patterns of crankiness. Too much television and a poor diet often occur together, and if nutrient deficiencies occur, the child's mood can deteriorate.[46] This detrimental cycle resolves itself when the caregivers begin to limit screen time, and insist on regular hours of sleep, regular mealtimes, a nutritious diet, and regular outdoor physical activity.

Did You Know?

Cranky or rambunctious children may:

- Be chronically hungry or undernourished.
- Be overstimulated.
- Consume too much caffeine from colas or chocolate.
- Desire attention.
- Lack exercise.
- Lack sleep.

Hunger and poverty can cause a child's misbehavior and poor achievement; simply lifting the child from poverty often significantly improves behavior.[47] An estimated 12 million U.S. children are hungry at least some of the time. Such children often lack iron, magnesium, zinc, or omega-3 polyunsaturated fatty acids, and a link between these nutrient deficiencies and ADHD may be emerging.[48] Once the obvious causes of misbehavior are eliminated, a physician can recommend other strategies such as special educational programs or psychological counseling and, in many cases, prescription medication.[49]

KEY POINTS

- ADHD is not caused by food allergies or additives.
- Hunger and poverty may cause behavior problems.

Dental Caries

Dental caries are a serious public health problem afflicting the majority of people in the country, half by the age of 2, with a prevalence rate of 90 percent in some groups.[50] A very lucky few *never* get dental caries because they have an inherited resistance; others have a sealant applied to their teeth during childhood to stop caries before they can begin. Another method used to reduce the incidence of dental decay is fluoridation of community water, considered to be the most effective dental public health measure to date.[51] Perhaps the greatest weapon against caries is simple oral hygiene. But diet has something to do with dental caries, too.

How Caries Develop Caries develop as acids produced by bacterial growth in the mouth eat into tooth enamel (see Figure 14–4). Bacteria form colonies in **plaque**, which sticks more and more firmly to tooth surfaces unless they are brushed, flossed, or scraped away. Eventually, the acid of plaque creates pits that deepen into cavities. The cavities can be treated by a dentist—the decay is removed and replaced with filling material. Scary stories that mercury-containing fillings can harm health are unfounded; the mercury in fillings is present in very small amounts and in a different form from the toxic methylmercury that contaminates many seafoods.[52]

Figure 14–4
Dental Caries

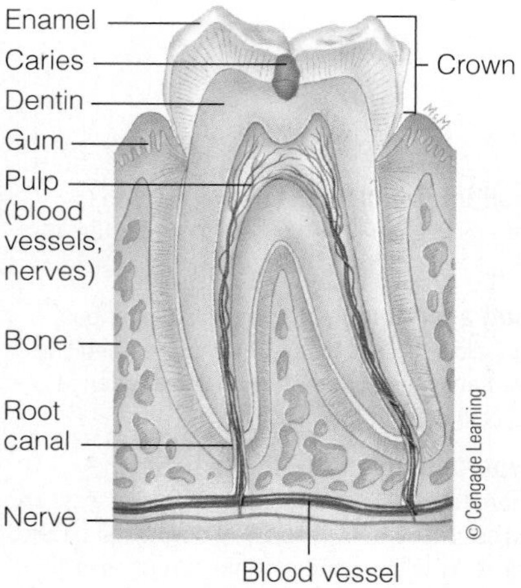

Caries begin when acid dissolves the enamel that covers the tooth. If not repaired, the decay may penetrate the dentin and spread into the pulp of the tooth, causing inflammation and an abscess.

Advanced Dental Disease Left alone, plaque works its way below the gum line until the acid erodes the roots of teeth and the jawbone in which they are embedded, loosening the teeth and leading to infections of the gums. Bacteria from inflamed, infected gums can then migrate by way of the bloodstream to other tissues such as the heart; researchers now suspect a link between these bacteria and heart disease.[53] Gum disease severe enough to threaten tooth loss afflicts the majority of our population by their later years.

Food and Caries Bacteria thrive on carbohydrate, producing acid for 20 to 30 minutes after carbohydrate exposure. Of prime importance is the length of time the teeth are exposed to carbohydrate, and this depends on the food's composition, how sticky it is, how long it lasts in the mouth, frequency of consumption, and especially on whether the teeth are brushed soon afterward. Table 14–9 lists foods of both high and low caries potential. Beverages such as soft drinks, orange juice, and sports drinks not only contain sugar but also have a low pH, and their acidic nature can erode the tooth enamel, weakening it. A growing preference for sugary soft drinks or sports drinks instead of water to quench thirst throughout the day may explain why dental erosion is becoming more common.[54]

Table 14–9

The Caries Potential of Foods

Low Caries Potential	
These foods are less damaging to teeth:	
▪ Eggs, legumes	▪ Pizza
▪ Fresh fruit, fruits packed in water	▪ Popcorn, pretzels
▪ Lean meats, fish, poultry	▪ Sugarless gum and candy,[a] diet soft drinks
▪ Milk, cheese, plain yogurt	▪ Toast, hard rolls, bagels
▪ Most cooked and raw vegetables	
High Caries Potential	
Brush teeth after eating these foods:	
▪ Cakes, muffins, doughnuts, pies	▪ Jams, jellies, preserves
▪ Candied sweet potatoes	▪ Lunchmeats with added sugar
▪ Chocolate milk	▪ Meats or vegetables with sugary glazes
▪ Cookies, granola or "energy" bars, crackers	▪ Oatmeal, oat cereals, oatmeal baked goods[b]
▪ Dried fruits (raisins, figs, dates)	▪ Peanut butter with added sugar
▪ Frozen or flavored yogurt	▪ Potato and other snack chips
▪ Fruit juices or drinks	▪ Ready-to-eat sugared cereals
▪ Fruits in syrup	▪ Sugared gum, soft drinks, candies, honey, sugar, molasses, syrups
▪ Ice cream or ice milk	▪ Toaster pastries

[a]Cariogenic bacteria cannot efficiently metabolize the sugar alcohols in these products, so they do not contribute to dental caries.

[b]The soluble fiber in oats makes this grain particularly sticky and therefore cariogenic.

KEY POINT

- Carbohydrate-rich foods contribute to dental caries.

Is Breakfast Really the Most Important Meal of the Day for Children?

A nutritious breakfast is a central feature of a child's diet that supports healthy growth and development.[55] When a child consistently skips breakfast or is allowed to choose sugary foods (candy or marshmallows) in place of nourishing ones (whole-grain cereals), the child will fail to get enough of several nutrients. Nutrients missed from a skipped breakfast won't be "made up" at lunch and dinner but will be left out completely that day.

Children who eat no breakfast are more likely to be overweight, snack on sweet and fatty foods, perform poorly in tasks requiring concentration, have shorter attention spans, achieve lower test scores, and be tardy or absent more often than their well-fed peers.[56] Common sense tells us that it is unreasonable to expect anyone to study and learn when no fuel has been provided. Even children who have eaten breakfast suffer from distracting hunger by late morning. Chronically underfed children suffer more intensely. Table 14–10 (p. 568) offers some ideas for quick breakfasts.

The U.S. government funds several programs for the purpose of providing nutritious, high-quality meals, including breakfast, to U.S. schoolchildren.[57] Children who eat school breakfast consume less fat and more magnesium during the day and are more replete with vitamin C and folate. When schools participate in federal school meal programs, students often improve not only in terms of nutrition but also in their academic, behavioral, emotional, and social performance. Attendance goes up, while tardiness declines.

KEY POINTS

- Breakfast supports school performance.
- Free or reduced-priced nutritious school meals are available to low-income children.

dental caries decay of the teeth (*caries* means "rottenness"). Also called *cavities.*

plaque (PLACK) a mass of microorganisms and their deposits on the surfaces of the teeth, a forerunner of dental caries and gum disease. The term *plaque* is also used in another connection—arterial plaque in atherosclerosis (see Chapter 11).

Table 14–10

Breakfast Ideas for Rushed Mornings

With some planning, even a rushed morning can include a nutritious breakfast.

- Make ahead sandwiches or tortilla wraps. Freeze, thaw or heat, and serve with juice. Fillings may include peanut butter, low-fat cream cheese or other cheeses, jams, fruit slices, refried beans, or meats.
- Teach school-aged children to help themselves to dry cereals, milk, and juice. Keep unbreakable bowls and cups in low cabinets, and keep milk and juice in small, covered, plastic pitchers on a low refrigerator shelf.
- Keep a bowl of fresh fruit and small containers of shelled nuts, trail mix (the kind without candy), or roasted peanuts for grabbing.
- Mix granola or other grain cereal into 8-oz tubs of yogurt.
- Toast whole-grain frozen waffles—no syrup needed—to grab and go.
- *Nontraditional choices*: Divide carrot sticks among several containers; serve with yogurt or bean dip. Leftover casseroles, stews, or pasta dishes are nutritious choices that children can eat hot or cold.

How Nourishing Are the Meals Served at School?

In the United States today, 50 million children ages 5 to 19 years spend a large portion of each day in school for about nine months of each year. More than 30 million children receive lunches through the National School Lunch Program—more than half of them free or at a reduced price.[58] Ten million children eat breakfast at school through the School Breakfast Program. For many children, particularly those living in poverty, school food programs might constitute their major source of nutrients each day.[59]

The National School Lunch and Breakfast Programs The UDSA-regulated school meals provide age-appropriate servings of needed foods each day (see Table 14–11). The lunches are designed to meet on average at least a third of the recommended intake for energy, total and saturated fat, protein, calcium, iron, vitamin A, and vitamin C, and are often more nutritious than typical lunches brought from home. Students who regularly eat school lunches have higher intakes of many nutrients and fiber than students who do not.[60]

In March 2012, the USDA Food and Nutrition Service issued a final rule updating the meal patterns and nutrition standards for school meals. Changes to school meals include greater availability of fruits, vegetables, whole grains, and fat-free and low-fat milk; decreased levels of sodium, saturated fat, and *trans* fat; and guidelines for meeting nutrient needs within specified calorie ranges based on age/grade groups for school children. Initial implementation of these new standards began on July 1, 2012, while additional changes will be phased in, ending with the final goal for reductions in sodium intake required to be in place by school year 2022–2023.[61]

Competitive Foods at School Recent concern about childhood obesity has focused attention on private vendors in school lunchrooms who offer **competitive foods**.[62] U.S. children develop a taste for these energy-dense, low-nutrient foods early in life and may reject more nutritious meals at school and elsewhere. Policies concerning competitive foods vary widely from state to state.[63] In places that restrict the sale of competitive foods, more students participate in USDA school meal programs. More schools today recognize that promoting fatty foods and sugary beverages to students leads to less nutritious diets and higher body mass index rankings, and more schools now limit these choices.[64]

The recent revisions to the USDA school lunch and breakfast programs promise a generation of healthier children.[65] Realizing this outcome demands the cooperation of everyone—legislators, school district officials, administrators, and parents—to ensure full participation for the benefit of children's nutrition.[66]

competitive foods unregulated meals, including fast foods, that compete side by side with USDA-regulated school lunches.

Table 14–11

School Breakfast and Lunch Patterns for Different Ages

	Breakfast Meal Pattern			Lunch Meal Pattern		
	Grades K–5[a]	Grades 6–8[a]	Grades 9–12[a]	Grades K–5	Grades 6–8	Grades 9–12
Meal Pattern	Amount of food per week (minimum per day)[b]					
Fruits (c)[c,d]	5 (1)[e]	5 (1)[e]	5 (1)[e]	$2\frac{1}{2}$ ($\frac{1}{2}$)	$2\frac{1}{2}$ ($\frac{1}{2}$)	5 (1)
Vegetables (c)[c,d]	0	0	0	$3\frac{3}{4}$ ($\frac{3}{4}$)	$3\frac{3}{4}$ ($\frac{3}{4}$)	5 (1)
Dark green[f]	0	0	0	$\frac{1}{2}$	$\frac{1}{2}$	$\frac{1}{2}$
Red/orange[f]	0	0	0	$\frac{3}{4}$	$\frac{3}{4}$	$1\frac{1}{4}$
Beans/peas (legumes)[f]	0	0	0	$\frac{1}{2}$	$\frac{1}{2}$	$\frac{1}{2}$
Starchy[f]	0	0	0	$\frac{1}{2}$	$\frac{1}{2}$	$\frac{1}{2}$
Other[f,g]	0	0	0	$\frac{1}{2}$	$\frac{1}{2}$	$\frac{3}{4}$
Additional veg to reach total[h]	0	0	0	1	1	$1\frac{1}{2}$
Grains (oz eq)[i]	7–10 (1)[j]	8–10 (1)[j]	9–10 (1)[j]	8–9 (1)	8–10 (1)	10–12 (2)
Meat or meat alternate (oz eq)	0[k]	0[k]	0[k]	8–10 (1)	9–10 (1)	10–12 (2)
Fluid milk (c)[l]	5 (1)	5 (1)	5 (1)	5 (1)	5 (1)	5 (1)
	Other Specifications: Daily amount based on the average for a 5-day week					
Calorie range[m,n,o]	350–500	400–550	450–600	550–650	600–700	750–850
Saturated fat (% of total calories)[n,o]	<10	<10	<10	<10	<10	<10
Sodium (mg)[n,p]						
School year 2014–2015	≤540	≤600	≤640	≤1,230	≤1,360	≤1,420
School year 2017–2018	≤485	≤535	≤570	≤935	≤1,035	≤1,080
School year 2022–2023	≤430	≤470	≤500	≤640	≤710	≤740
Trans fat[n,o]	Nutrition label or manufacturer specifications must indicate zero grams of *trans* fat per serving.					

[a] *In the School Breakfast Program (SBP), the grade groups listed take effect beginning July 1, 2013 (School Year [SY] 2013–2014). In SY 2012–2013 only, schools may continue to use the meal pattern for grades K–12.*

[b] *Minimum creditable serving is $\frac{1}{8}$ c.*

[c] *$\frac{1}{4}$ c of dried fruit counts as $\frac{1}{2}$ c of fruit; 1 c of leafy greens counts as $\frac{1}{2}$ c of vegetables. No more than half of the fruit or vegetable offerings may be in the form of juice. All juice must be 100% full-strength.*

[d] *For breakfast, vegetables substitute for fruits, but the first 2 c per week of any such substitution must be from the dark green, red/orange, beans and peas (legumes), or "Other vegetables" subgroups.*

[e] *The fruit quantity requirement for the SBP (5 c/week and a minimum of 1 c/day) is effective July 1, 2014 (SY 2014–2015).*

[f] *Larger amounts of these vegetables may be served.*

[g] *This category consists of "Other vegetables," that is, any additional amounts from the dark green, red/orange, and beans and peas (legumes) vegetable subgroups.*

[h] *Any vegetable subgroup may be offered to meet the total weekly vegetable requirement.*

[i] *At least half of the grains offered must be whole grain–rich in the NSLP beginning July 1, 2012 (SY 2012–2013), and in the SBP beginning July 1, 2013 (SY 2013–2014). All grains must be whole grain–rich in both the NSLP and the SBP beginning July 1, 2014 (SY 2014–2015).*

[j] *In the SBP, the grain ranges must be offered beginning July 1, 2013 (SY 2013–2014).*

[k] *There is no separate meat/meat alternate component in the SBP. Beginning July 1, 2013 (SY 2013–2014), schools may substitute 1 oz. eq. of meat/meat alternate for 1 oz. eq. of grains after the minimum daily grains requirement is met.*

[l] *Fluid milk must be low-fat (1% milk fat or less, unflavored) or fat-free (unflavored or flavored).*

[m] *The average daily amount of calories for a 5-day school week must be within the range (at least the minimum and no more than the maximum values).*

[n] *Discretionary sources of calories (solid fats and added sugars) may be added to the meal pattern if within the specifications for calories, saturated fat,* trans *fat, and sodium. Foods of minimal nutritional value and fluid milk with fat content greater than 1% milk fat are not allowed.*

[o] *In the SBP, calories and* trans *fat specifications take effect beginning July 1, 2013 (SY 2013–2014).*

[p] *Final sodium specifications are to be reached by SY 2022–2023 or July 1, 2022. Intermediate specifications are established for SY 2014–2015 and 2017–2018.*

Source: U.S. Department of Agriculture, Food and Nutrition Services, Nutrition Standards in the National School Lunch and School Breakfast Programs, Federal Register *77 (2012): 4088–4167.*

KEY POINTS

- School meals are designed to provide at least a third of certain nutrients that children need daily. Recent changes to the school breakfast and lunch menus include increased availability of fruits, vegetables, whole grains, and low-fat and fat-free fluid milk; reduced sodium in meals; and reduced levels of saturated and *trans* fats.
- Competitive foods are often high in calories, saturated fat, salt, and sugar but low in key nutrients.

Nutrition in Adolescence

LO 14.2 Explain ways in which a teenager's choice of soda over milk or soy milk may jeopardize nutritional health.

Teenagers are not fed; they eat. Food choices made during the teen years profoundly affect health, both now and in the future. In the face of new demands on their time, including after-school jobs, social activities, sports, and home responsibilities, these older children easily fall into irregular eating habits, relying on quick snacks or fast foods for meals. Within this setting, **adolescence** brings a transforming physical maturation and a psychological search for identity, acquired largely through trial and error.

Parents, peers, and the media act as primary influences in shaping the adolescent's behaviors and beliefs. Adolescents who frequently eat meals with their families eat more fruits, vegetables, grains, and calcium-rich foods and drink fewer soft drinks than those who seldom eat with their families.[67] They may even be less likely to smoke, drink alcohol, or abuse drugs.[68]

The Adolescent Growth Spurt The adolescent **growth spurt** brings rapid growth and hormonal changes that affect every organ of the body, including the brain. An average girl's growth spurt begins at 10 or 11 years of age and peaks at about 12 years. Boys' growth spurts begin at 12 or 13 years and peak at about 14 years, slowing down at about 19. Two boys of the same age may vary in height by a foot, but if growing steadily, each is fulfilling his genetic destiny according to an inborn schedule of events. A negative assessment of the two may injure the developing identity and open the door to taking such risks as using alcohol, tobacco, and drugs of abuse such as marijuana.

Energy Needs and Physical Activity The energy needs of adolescents vary tremendously depending on growth rate, gender, body composition, and physical activity. Energy balance is often difficult to regulate in this society—an estimated 15 percent of U.S. children and adolescents 6 to 19 years of age are overweight. On the output side, the spontaneous physical activity of childhood diminishes significantly around the age of adolescence, and by age 15 slumps far below the recommended levels.[69]

An active, growing boy of 15 may need 3,500 calories or more a day just to maintain his weight, but an inactive girl of the same age whose growth has slowed may need fewer than 1,800 calories to avoid unneeded weight gain. She may benefit from choosing more low-calorie fruit, vegetables, fat-free milk, whole grains, and other nutritious foods with limited cookies, cakes, soft drinks, fried snacks, and other treats. Extra physical activity throughout the day can help to balance the energy budget, as well.

Nutritious snacks play an important role in an active teen's diet.

Weight Standards and Body Fatness Weight standards meant for adults are useless for adolescents. Physicians use growth charts to track their gains in height and weight, and parents should watch only for smooth progress and guard against comparisons that can diminish the child's self-image.

Girls normally develop a somewhat higher percentage of body fat than boys do, a fact that causes much needless worry about becoming overweight. Teens face tremendous pressures regarding body image, and many readily believe scams that promise slenderness or good-looking muscles through "dietary supplements." Healthy, normal-weight teenagers are often "on diets" and make all sorts of unhealthy weight-loss attempts—even taking up smoking.[70] A few teens without diagnosable eating disorders have been reported to "diet" so severely that they stunted their own growth.

KEY POINTS

- The adolescent growth spurt increases the need for energy and nutrients.
- The normal gain of body fat during adolescence may be mistaken for obesity, particularly in girls.

Nutrient Needs

Needs for vitamins, minerals, the energy-yielding nutrients, and, in fact, all nutrients are greater during adolescence than at any other time of life except pregnancy and lactation (see the DRI table on the inside front cover). The need for iron is particularly high, as all teenagers gain body mass and girls begin menstruation.

The Special Case of Iron The increase in need for iron during adolescence occurs across the genders, but for different reasons. A boy needs more iron at this time to develop extra lean body mass, whereas a girl needs extra not only to gain lean body mass but also to support menstruation. Because menstruation continues throughout a woman's childbearing years, her need stays high until older age. As boys become men, their iron needs drop back to the preadolescent value during early adulthood.

An interesting detail about adolescent iron requirements is that the need increases during the growth spurt, regardless of the age of the adolescent.[71] This shifting requirement makes pinpointing an adolescent's need tricky, as Table 14–12 demonstrates.

Iron intakes often fail to keep pace with increasing needs, especially for girls, who typically consume less iron-rich foods such as meat and fewer total calories than boys. Not surprisingly, iron deficiency is most prevalent among adolescent girls. Adolescent girls and boys who live with food insecurity—that is, they miss meals, eat less expensive, less nutritious foods, or other food-related compromises of poverty—have a threefold greater likelihood of iron deficiency compared with food-secure children.[72]

Calcium and the Bones Adolescence is a crucial time for bone development. The bones are growing longer at a rapid rate (see Figure 14–5) thanks to a special bone structure, the **epiphyseal plate**, which disappears as a teenager reaches adult

Table 14–12
Iron Requirements in Adolescence

Iron DRI intake goals for adolescent boys:

- 9–13 years, 8 mg/day
  - During growth spurt, 10.9 mg/day
- 14–18 years, 11 mg/day
  - During growth spurt, 13.9 mg/day

Iron DRI intake goals for adolescent girls:

- 9–13 years, 8 mg/day
  - If menstruating, 10.5 mg/day
  - If menstruating during growth spurt, 11.6 mg/day
- 14–18 years, 15 mg/day
  - During growth spurt, 16.1 mg/day

Figure 14–5
Growth of Long Bones

Bones grow longer as new cartilage cells accumulate at the top portion of the epiphyseal plate and older cartilage cells at the bottom of the plate are calcified.

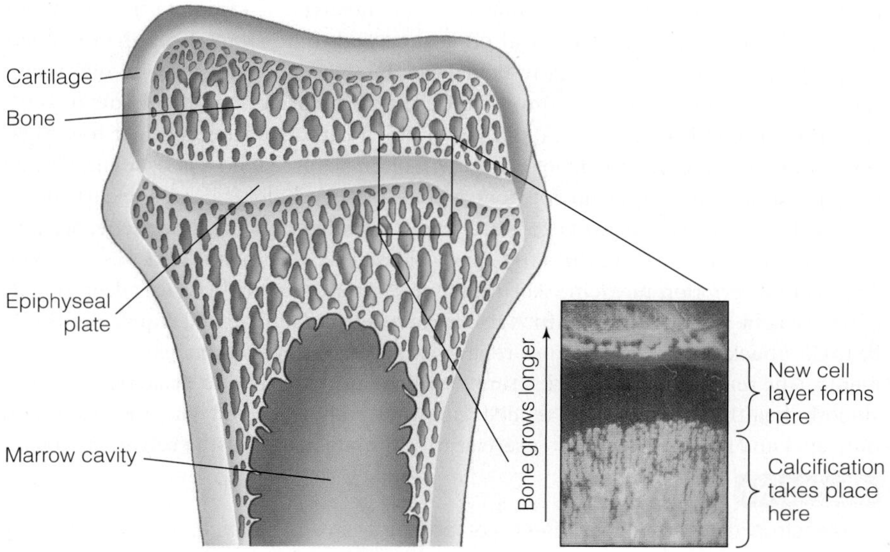

photo, Dr. David Phillips/Visuals Unlimited, Inc.; art © Cengage Learning

adolescence the period from the beginning of puberty until maturity.

growth spurt the marked rapid gain in physical size usually evident around the onset of adolescence.

epiphyseal (eh-PIFF-ih-seal) **plate** a thick, cartilage-like layer that forms new cells that are eventually calcified, lengthening the bone (*epiphysis* means "growing" in Greek).

height. At the same time, the bones are gaining density, laying down the calcium needed later in life. Calcium intakes must be high to support the development of peak bone mass.[73]

Low calcium intakes have reached crisis proportions: 85 percent of females and 70 percent of males ages 9 to 18 years have calcium intakes below recommendations.[74] Paired with a lack of physical activity, low calcium intakes can compromise the development of peak bone mass, greatly increasing the risk of osteoporosis and other bone diseases later on.

Teens often choose soft drinks as their primary beverage (see Figure 14–6). Particularly among girls, this choice displaces calcium-rich milk from the diet (see Figure 14–7) and prevents bones from reaching their full attainable density.[75] Conversely, increasing milk consumption to meet calcium recommendations greatly increases bone density.[76]

Bones also grow stronger with physical activity, but few high schools require students to attend physical activity classes, so most teenagers must make a point to be physically active during leisure hours. Attainment of maximal bone mass during youth and adolescence is the best protection against age-related bone loss and fractures in later life.

Did You Know?

Peak bone mass is the highest attainable bone density for an individual, developed during the first three decades of life.

Vitamin D Vitamin D is also essential for calcium absorption and proper bone growth and development of bone density. Adolescents who do not receive 15 μg of vitamin D from vitamin D–fortified milk (2.9 μg per cup of fat-free milk) and other vitamin D–fortified foods each day should take vitamin D in a supplement.

KEY POINTS

- The need for iron increases during adolescence in both boys and girls.
- Sufficient calcium and vitamin D intakes are also crucial during adolescence.

Figure 14–6

Average Daily Intake of Sugar-Sweetened Beverages by U.S. Children and Adolescents

The *U.S. Dietary Guidelines 2010* suggest limiting consumption of food and beverages containing added sugars. The American Heart Association recommends limiting consumption of sugar-sweetened beverages to 450 calories *per week*, an amount clearly exceeded by many U.S. adolescents.

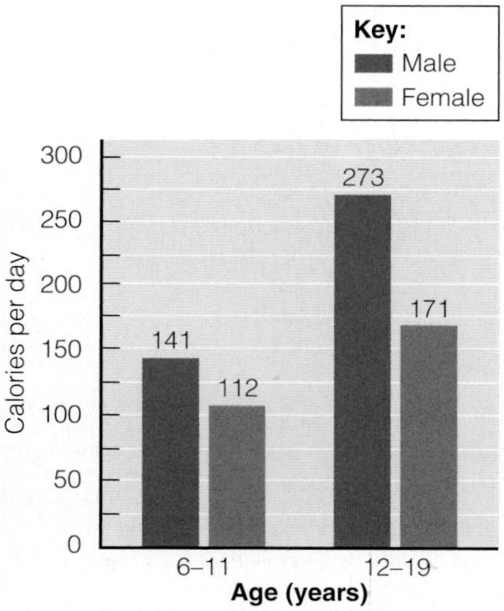

Source: Adapted from C.L. Ogden and coauthors, Consumption of sugar drinks in the United States, 2005–2008, NCHS data brief, National Center for Health Statistics 2011, available at http://www.cdc.gov/nchs/data/databriefs/db71.htm/#Fig2.

Common Concerns

Two other physical changes stand out as important in adolescence. Menstruation and acne pose special concern to many adolescents.

Menstruation Girls face a major change with the onset of menstruation. The hormones that regulate the menstrual cycle affect not just the uterus and the ovaries but the metabolic rate, glucose tolerance, appetite, food intake, and, often, mood and behavior as well. Most women live easily with the cyclic rhythm of the menstrual cycle, but some are afflicted with physical and emotional pain prior to menstruation: **premenstrual syndrome**, or **PMS** (see the Consumer's Guide, pp. 574–575).

Acne Genes clearly play a role in who gets **acne** and who doesn't, but other factors also affect its development.[77] The hormones of adolescence stimulate the oil glands deep in the skin. The skin's natural oil is supposed to flow out through tiny ducts at the skin's surface, but in many teens, the ducts become clogged and oily secretions build up in the ducts causing irritation, inflammation, and breakouts of acne. A "Western" diet high in meat, dairy, and refined carbohydrates is under investigation for an association with acne.[78] Although often accused, chocolate, sugar, French fries, pizza, salt, and iodine do not worsen acne, but psychological stress clearly does.

Vacations from school, sun exposure, and swimming help to relieve acne, perhaps because they are relaxing, the sun's rays kill bacteria, and water cleanses the skin. The oral prescription medicine Accutane, made from vitamin A, cures deep lesions of severe acne. Although vitamin A itself has no effect on acne and supplements can be toxic, quacks market vitamin A–related compounds to young people as acne treatments. One remedy always works: time. While waiting, attend to basic needs. Petal-smooth, healthy skin reflects a tended, cared-for body whose owner provides it with nutrients and fluids to sustain it, exercise to stimulate it, and rest to restore its cells.

KEY POINT

- Menstrual cycle hormones affect metabolism, glucose tolerance, and appetite.
- No single foods have been proved to aggravate acne, but stress can worsen it.

Eating Patterns and Nutrient Intakes

During adolescence, food habits change for the worse, and teenagers often miss out on nutrients they need. Few teens choose sufficient whole grains, for example, which help support adequate nutrient intakes and reduce chronic disease risk.[79] Teens may begin to skip breakfast; choose less milk, fruits, juices, and vegetables; and consume more soft drinks each day. Skipping breakfast and eating fast foods may each bear a relationship to weight gain and higher BMI values.[80]

Roles of Adults Ideally, the adult becomes a **gatekeeper**, controlling the type and availability of food in the teenager's environment.[81] Teenage sons and daughters and their friends should find plenty of nutritious, easy-to-grab food in the refrigerator (meats for sandwiches, raw vegetables, fruit, milk, and fruit juices) and more in the cabinets (breads, peanut butter, nuts, popcorn, cereals). In reality, in many households today, all the adults work outside the home and teens perform many of the gatekeeper's roles, such as shopping for groceries or choosing fast foods or prepared foods.

Snacks On average, about a fourth of a teenager's total daily energy intake comes from snacks, which, if chosen carefully, can contribute needed protein, thiamin, riboflavin, vitamin B_6, magnesium, and zinc. A survey of more than 5,000 adolescents found that those who ate snacks more often were less likely to be overweight or obese and had lower rates of abdominal obesity compared with those who ate snacks less often.[82] Calcium intakes often fall short, but snacks of dairy products can improve this picture. For iron, vitamin A, and other nutrients, a teen could snack on iron-containing meat sandwiches, low-fat bran muffins, or tortillas with spicy bean spread along with a glass of orange juice to help maximize the iron's absorption.

The gatekeeper can help the teenager choose wisely by delivering nutrition information at "teachable moments." Teens prone to weight gain will often open their ears to news about calories in fast foods. Athletic teens may best attend to information about meal timing and sports performance. Still others are fascinated to learn of the skin's need for vitamins. The gatekeeper must set a good example, keep lines of communication open, and stand by with plenty of nourishing food and reliable nutrition information, but the rest is up to the teens themselves. Ultimately, they make the choices.

KEY POINT

- The gatekeeper can encourage teens to meet nutrient requirements by providing nutritious snacks.

Figure 14–7

Percentage of Children and Adolescents Who Drink Milk

Over the past decades, fewer teenagers have been drinking milk. Today, teens rarely take in the recommended 2½ to 3 cups of milk each day, and their calcium intakes suffer.

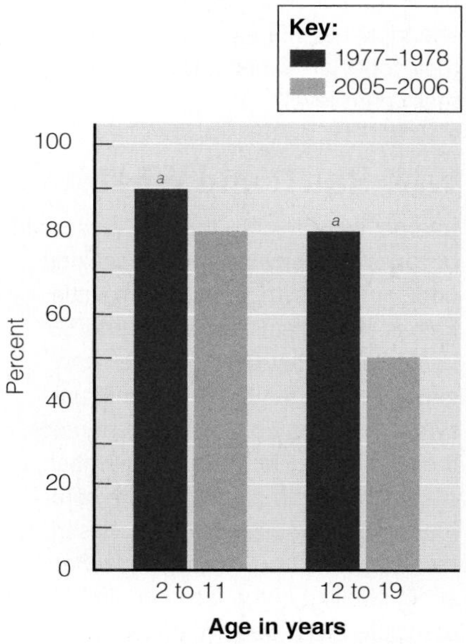

[a]Statistically significant difference between survey years ($p < 0.01$).

Source: Adapted from R. S. Sebastian and coauthors, Fluid milk consumption In the United States, What we eat in America, NHANES 2005–2006, Food Surveys Research Group Dietary Data Brief 3 (2010), available at http://www.ars.usda.gov/SP2UserFiles/Place/12355000/pdf/Dbrief/3_milk_consumption_0506.pdf.

The Later Years

LO 14.3 Contrast life expectancy with life span, and name some lifestyle factors associated with successful aging.

The title of this section may imply it is about older people, but it is relevant even if you are only 20 years old—how you live and think at age 20 affects the quality of your life at 60 or 80. According to an old saying, "as the twig is bent, so grows the tree." Unlike a tree, however, you can bend your own twig.

As the Twig Is Bent . . . Before you will adopt nutrition behaviors to enhance your health in old age, you must accept on a personal level that you, yourself, are aging. To learn what negative and positive views you hold about aging, try answering the questions in the margin of page 575. Your answers reveal not only what you think of older people now but also what will probably become of you.[83] Nutrition has many documented roles that are critical to successful aging.[84] In general, people who reach old age in good mental and physical health most often:

- Are nonsmokers.
- Abstain or drink alcohol only moderately.

premenstrual syndrome (PMS) a cluster of symptoms that some women experience prior to and during menstruation. They include, among others, abdominal cramps, back pain, swelling, headache, painful breasts, and mood changes.

acne chronic inflammation of the skin's follicles and oil-producing glands, which leads to an accumulation of oils inside the ducts that surround hairs; usually associated with the maturation of young adults.

gatekeeper with respect to nutrition, a key person who controls other people's access to foods and thereby affects their nutrition profoundly. Examples are the spouse who buys and cooks the food, the parent who feeds the children, and the caregiver in a day-care center.

use it! A Consumer's Guide To . . .

Nutrition for PMS Relief

Jasmine, seeking relief from premenstrual syndrome (PMS) symptoms, found a promise of a cure on an Internet website. All that is needed to vanquish her PMS, according to the site, is that she triple her vitamin D intake (and buy their "special" variety). Can a vitamin really cure PMS?

Who Has It and What It Is

Internet websites like this are successful because so many women are seeking help for PMS symptoms, and so little help is available. Between 50 and 80 percent of menstruating women and girls report uncomfortable menstrual symptoms, and up to 32 percent meet the criteria for PMS.[1] A great number of possible cyclic symptoms are common, including cramps and aches in the abdomen, back pain, headache, acne, swelling of the face and limbs associated with water retention, food cravings (especially for chocolate and other sweets), abnormal thirst, pain and lumps in the breasts, diarrhea, and mood changes, including both nervousness and depression, sometimes severe enough to interfere with normal daily activities.

In an attempt to standardize diagnostic PMS criteria, researchers have isolated six core symptoms:

- Anxiety and tension.
- Mood swings.
- Aches and pains.
- Increased appetite and food cravings.
- Abdominal cramps.
- Decreased interest in activities.[2]

This smaller group of core symptoms might prove less cumbersome for women tracking bothersome menstrual changes in a diary and in diagnosing premenstrual syndrome.

Causes

PMS symptoms may arise from an altered response to the two major regulatory hormones of the menstrual cycle: estrogen and progesterone. In particular, the hormone estrogen affects mood by altering the brain's neurotransmitter, serotonin. Adequate serotonin in the brain buoys a person's mood, whereas a serotonin deficiency creates psychological depression. In PMS, the natural rise and fall of estrogen affects the actions of serotonin during the last half of each menstrual cycle. Taking oral contraceptives, which supply estrogen, often improves mood by eliminating hormonal peaks and valleys. Antidepressant drugs that amplify serotonin's effects also may provide some relief.[3]

The major connections between PMS and nutrition arise in two research areas. The first, energy metabolism, concerns cyclic food intake and appetite changes. The second, links between PMS and vitamin and mineral intakes, is often misrepresented to promote fraudulent "miracle" PMS cures.

Energy Metabolism

Scientists believe that during the two weeks prior to menstruation:

- The basal metabolic rate during sleep speeds up.
- Appetite, particularly for carbohydrate-rich or fat-rich foods, and calorie intakes increase.[4]
- Alcohol consumption may increase, particularly among women with a long-term history of alcohol use.[5]

For the woman striving to lose weight, then, it may be easier to reduce calorie intakes during the two weeks after menstruation. During the two weeks *before* menstruation, she is fighting a natural, hormone-governed increase in appetite.

Vitamins and Minerals

Links with calcium and vitamin D are intriguing: calcium intakes of 1,000 milligrams per day have significantly improved the fatigue, depression, and changes in appetite associated with PMS.[6] Also, girls and young women taking in plenty of calcium and vitamin D in the form of foods such as low-fat milk are reported as having a lower risk of developing PMS than those taking in less.[7]

As for *therapeutic* doses of vitamin D, one small study of Italian women is suggestive.[8] After a single 300,000 IU dose of vitamin D, women with severe PMS reported less pain and less need for painkillers than those given a placebo. The dose used in the study vastly exceeds the DRI Tolerable Upper Intake Level of 4,000 IU. The dose was so high that if it were taken only once every two months, it would still average about 5,000 IU per day and cannot be recommended.[9] Many more trials are needed to explore vitamin D's safety and effectiveness. Predictably, however, quacks jump the gun and use such preliminary evidence to "prove" that their vitamin D megadoses "cure PMS." Meanwhile, real scientists suggest other possibilities, for example, that the women may have been vitamin D deficient at the start, and that the dose of vitamin D may have both reversed the deficiency and reduced their symptoms.[10]

Years ago, vitamin B_6 was the darling of vitamin sellers after a review of literature concluded that high doses (100 milligrams per day) might alleviate some PMS symptoms.[11] Women who took these doses over time developed numb feet and hands, and they eventually became unable to walk or work. Afterward, vitamin B_6 supplements fell out of favor for treating PMS.

Ongoing Research

Research has *not* shown these to be useful: taking multivitamins, magnesium, or manganese supplements; cutting down on alcohol or sodium; or taking diuretics to relieve water retention. Sodium and water retention just before menstruation may be normal and desirable; diuretic drugs taken to eliminate excess sodium and water also cause a loss of valuable cellular potassium. Caffeine may worsen PMS symptoms, but how much is too much is not clear. And while adequate sleep, physical activity, and stress reduction strategies may help some women, research in these areas is lacking.[12]

Moving Ahead

The good news for Jasmine and others coping with PMS is that, while a cure is elusive, its effects may be lessened with simple, commonsense strategies: be physically active, aim for stress reduction, and eat a diet that meets the DRI recommended intakes for nutrients every day. In particular, eat plenty of low-fat foods rich in calcium and vitamin D each day.

Review Questions*

1. Taking 300,000 IU per day of vitamin D may alleviate some PMS symptoms but this amount exceeds the DRI Tolerable Upper Intake Level. **T F**
2. Taking multivitamins, magnesium, manganese, or diuretics is a scientifically proven strategy to cure PMS. **T F**
3. During the two weeks *before* menstruation, women may experience a natural, hormone-governed lack of appetite. **T F**

** Answers to Consumer's Guide review questions are found in Appendix G.*

- Are physically active (they walk, bike, swim or otherwise spend more than 150 minutes per week in physical activity).
- Are well-nourished, and, particularly, they consume sufficient fruits and vegetables.[85]
- Maintain a healthy body weight.

They also keep a cheerful attitude and are not often depressed.

Life Expectancy The "graying" of America is a continuing trend.[86] Since 1950, the population older than age 65 has almost tripled, and numbers of people older than age 85 have increased sevenfold. People reaching and exceeding age 100 have doubled in number in recent decades, a trend evident among many of the world's populations.

How long a person can expect to live depends on several factors. An estimated 70 to 80 percent of the average person's **life expectancy** depends on individual health-related behaviors, with genes determining the remaining 20 to 30 percent. In the United States, an average person can expect to live almost 78 years.[87] Specifically, life expectancy is almost 81 years for white women and 77 years for black women; for white men, 76 years and for black men, 70 years—all record highs and much higher than the life expectancy of 47 years in 1900.[88] Once a person survives the perils of youth and middle age to reach age 80, women can expect to survive an additional 9 years, on average; men, an additional 7. The racial gap in life expectancy is narrowing, although efforts to reduce cardiovascular diseases, homicide, HIV infections, and infant mortality are needed to reduce it further.[89]

Human Life Span The biological schedule that we call aging cuts off life at a genetically fixed point in time. The **life span** (the maximum length of life possible for a species) of human beings is believed to be 125 years. Even this limit may one day be challenged with advances in medical and genetic technologies.[90] One caution: to date, scientists who study the aging process have found no specific diet or nutrient supplement that will increase **longevity**, despite hundreds of dubious claims to the contrary.

KEY POINTS

- Life expectancy for U.S. adults is increasing, but the human life span is set by genetics.
- Life choices can greatly affect how long a person lives and the quality of life in the later years.

Did You Know?

Your attitudes about aging often predict how you will age:

- In what ways do you expect your appearance to change as you age?
- What physical activities do you see yourself enjoying at age 70?
- What will be your financial status? Will you be independent?
- What will your sex life be like? Will others see you as sexy?
- How many friends will you have? What will you do together?
- Will you be happy? Cheerful? Curious? Depressed? Uninterested in life or new things?

© Shebeko/Shutterstock.com

life expectancy the average number of years lived by people in a given society.

life span the maximum number of years of life attainable by a member of a species.

longevity long duration of life.

Nutrition in the Later Years

LO 14.4 Outline food-related factors that can predict malnutrition in older adults.

Nutrient needs become more individual with age, depending on genetics and individual medical history. For example, one person's stomach acid secretion, which helps in iron absorption, may decline, so that person may need more iron. Another person may excrete more, and therefore need more, folate due to past liver disease. Table 14–13 lists some changes that can affect nutrition.

Courtesy of the Author

Physical activity yields benefits at all stages of life.

Energy and Activity

Energy needs often decrease with advancing age. One reason is that the number of active cells in each organ often decreases and the metabolism-controlling hormone thyroxine diminishes, reducing the body's resting metabolic rate by 3 to 5 percent per decade.[91] Another reason is that as older people reduce their physical activity, their lean tissue diminishes, resulting in **sarcopenia**, an age-related loss of muscle tissue.

Energy Recommendations After about the age of 50, the intake recommendation for energy assumes about a 5 percent reduction in energy output per decade. As in other age groups, obesity increasingly poses a problem. For those who must limit energy intake, there is little leeway in the diet for foods of low nutrient density such as added sugars, fats, and alcohol.

Current findings refute the idea that declining energy needs are unavoidable, however. Staying physically active boosts energy needs and contributes to a healthy immune response and sharp mental functioning, too.[92] Physical activity and an adequate diet also oppose a destructive spiral of sedentary behavior and mental and physical losses in the elderly, sometimes called geriatric failure to thrive or "the dwindles."[93] The set of conditions associated with failure to thrive includes:

- Decreased physical ability to function; inability to shop, cook, or prepare meals.
- Depression or anxiety.
- Malnutrition, which:
  - Impairs immune function.
  - Delays wound healing.
  - Slows recovery time from surgeries.
  - Increases hospitalizations.
- Weight loss and appetite loss with sarcopenia.[94]

Did You Know?

The DRI nutrient intake standards provide separate recommendations for those 51 to 70 years old and for those older than age 70 (see the inside front cover).

Table 14–13

Physical Changes of Aging That Affect Nutrition

DIGESTIVE TRACT	Intestines lose muscle strength resulting in sluggish motility that leads to constipation. Stomach inflammation, abnormal bacterial growth, and greatly reduced acid output impair digestion and absorption. Pain and fear of choking may cause food avoidance or reduced intake.
HORMONES	For example, the pancreas secretes less insulin and cells become less responsive, causing abnormal glucose metabolism.
MOUTH	Tooth loss, gum disease, and reduced salivary output impede chewing and swallowing. Choking may become likely; pain may cause avoidance of hard-to-chew foods.
SENSORY ORGANS	Diminished sight can make food shopping and preparation difficult; diminished senses of smell and taste may reduce appetite, although research is needed to clarify this effect.
BODY COMPOSITION	Weight loss and decline in lean body mass lead to lowered energy requirements. May be preventable or reversible through physical activity.

Involuntary weight loss deserves immediate attention in the older person. It could be the result of some easily treatable condition, such as ulcers. For people older than 70 years, the best health and lowest risk of death have been observed in those who maintain a body mass index (BMI) between 25 and 32, which is higher than the optimal BMI for younger people (18.5 to 25; see the inside back cover). Dealing effectively with weight loss entails finding the causes (physical, psychological, or others) and addressing them. At the same time, offering the person's favorite foods in five or six small, high-calorie meals each day instead of three larger ones often helps stop or reverse weight loss. As an added benefit, frequent snacking or small meal consumption also increases intake of several important vitamins and minerals in older adults.[95]

Physical Activity The Think Fitness feature (p. 578) emphasizes the importance of physical activity to maintaining body tissue integrity throughout life.[96] People spending energy in physical activity can also eat more food, gaining nutrients. Sadly, 80 percent of older adults fail to meet national exercise objectives, and up to 34 percent are simply inactive, missing the opportunity for more robust health and fitness in their later years.[97] Any movement seems better than no movement: even performing daily physical chores seems to help the elderly to expend energy and extend life.[98] Some people in their 90s have improved their balance, added pep to their walking steps, and regained some precious independence after just eight weeks of resistance training.

The photos near here dramatize this point: they compare cross sections of the thigh of a young woman and of an older woman to demonstrate the sarcopenia typical of sedentary aging, which brings with it destructive weakness, poor balance, and deterioration of health and vigor. Resistance training through life helps to prevent at least some of this muscle loss, and consuming sufficient protein may help, too.[99]

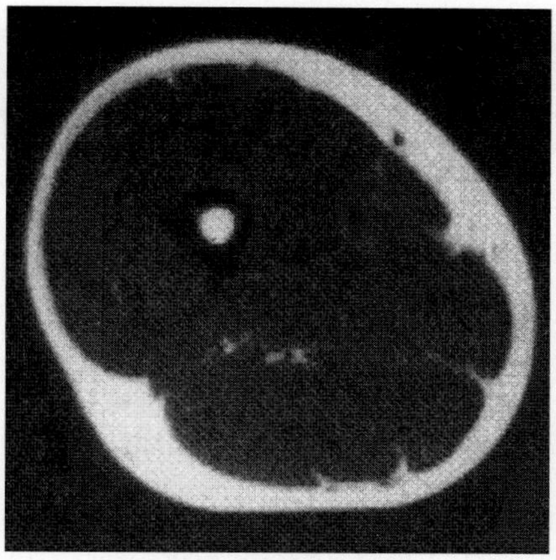

Courtesy of Dr. William Evans

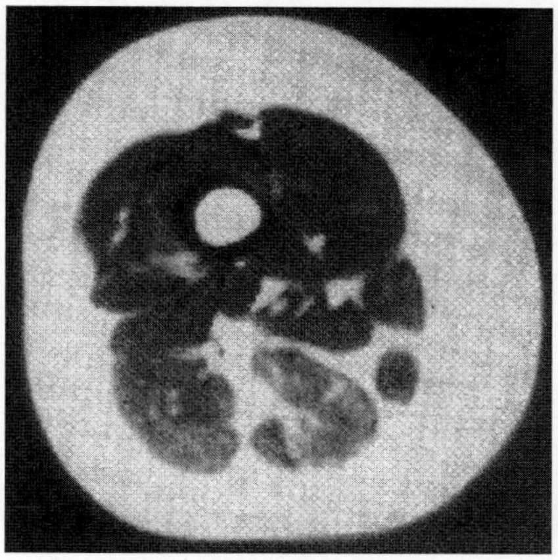

Courtesy of Dr. William Evans

Cross sections of two thighs. These two women's thighs may appear to be about the same size from the outside, but the 20-year-old woman's thigh (top) is dense with muscle tissue (dark areas). The 64-year-old woman's thigh (bottom) has lost muscle and gained fat, changes characteristic of sarcopenia.

KEY POINTS

- Energy needs decrease with age.
- Physical activity maintains lean tissue during aging.

Protein Needs

Protein DRI recommended intakes remain about the same for older people as for young adults (see inside front cover). However, with advancing age, people often take in fewer total calories of food and so may need a greater percentage of calories from protein at each meal in order to prevent losing muscle tissue, bone tissue, and other lean body mass.[100]

For older people who have lost their teeth, chewing tough protein-rich meats sufficiently to allow their proper use by the body becomes next to impossible. They need soft-cooked protein sources, such as well-cooked stewed or chopped meats, milk-based soups, soft cheeses, eggs, or fish. Those with chronic constipation, heart disease, or diabetes may benefit most from fiber-rich, low-fat protein sources, such as legume–whole grain combinations.

As energy needs decrease, lower-calorie protein sources, such as lean tender meats, poultry, fish, boiled eggs, fat-free milk products, and legumes can help hold weight to a healthy level. Underweight or malnourished older adults need the opposite—energy-dense protein sources such as eggs scrambled with margarine, tuna salad with mayonnaise, peanut butter, and milkshakes. Should a flagging appetite reduce food intake, supplemental nutrient-fortified formulas in liquid, pudding, cookie, or other forms between meals can supply needed energy, protein, and other nutrients.

KEY POINT

- Protein needs remain about the same through adult life, but physical conditions dictate appropriate protein sources.

Carbohydrates and Fiber

Ample whole-grain breads, cereals, rice, and pasta provide the steady supply of carbohydrate that the brain demands for optimal functioning. The fiber in these foods takes

sarcopenia (SAR-koh-PEE-nee-ah) age-related loss of skeletal muscle mass, muscle strength, and muscle function.

Think Fitness move it!

Benefits of Physical Activity for the Older Adult

The *Physical Activity Guidelines for Americans* and the American College of Sports Medicine recommend that older adults strive to obtain 150 minutes of physical activity, or whatever amount they can safely and comfortably perform, each week.[101] Older adults who do so have greater flexibility and endurance, more lean body mass, a better sense of balance, greater blood flow to the brain, and stronger immune systems; they suffer fewer falls and broken bones, experience fewer symptoms of arthritis, retain better cognitive skills, enjoy better overall health, and even live longer than their less-fit peers.[102] Those at risk of falling can benefit from exercises that improve balance.

At some point, perhaps around 80 years of age, the body struggles to add new muscle tissue in response to resistance training.[103] Some extra protein at each meal may help in this regard by stimulating protein synthesis in exercised muscles, even at advanced age.[104] Middle-aged and older people who wish to retain vitality should start now and continue to build and defend their muscle mass through life.

Each elderly person faces different degrees and types of physical limitations. Therefore, each should exercise in his or her own way and pace. Even modest exercise, such as a 10-minute walk a day, a workout of upper body flexibility, or resistance training (even while seated) provides progressive benefits.[105] Great achievements are possible and improvements are inevitable.

start now! Ready to make a change? If you are an older person, or if you care for an older person, devise a sensible exercise plan and track your activity in Diet Analysis Plus for one week. At the end of that week, look at your total physical activity, and decide if you can increase your level of activity in the following week.

on extra importance in aging to prevent constipation, a common complaint among older adults and nursing home residents in particular.

Fruits and vegetables supply soluble fibers and other food components to help ward off chronic diseases. With aging, however, come problems of transportation, limited cooking facilities, and chewing disabilities that limit some elderly people's intakes of fresh fruits and vegetables. Even without such problems, most older adults fail to obtain the recommended 25 or so grams of fiber each day (14 grams per 1,000 calories).[106] When low fiber intakes are combined with low fluid intakes, inadequate exercise, and constipating medications, constipation becomes inevitable.

KEY POINT

- Including fiber can help older adults to avoid constipation.

Fats and Arthritis

Older adults must attend to fat intakes for several reasons. Consuming enough of the essential fatty acids supports continued good health, and limiting intakes of saturated and *trans* fats is a priority to minimize the risk of heart disease. Many of the foods lowest in saturated fat are richest in vitamins, minerals, and phytochemicals. In addition, certain fats may affect one type of **arthritis**, a painful deterioration and swelling of the joints. High-fat diets also correlate with obesity, an arthritis risk factor.

Osteoarthritis The common type of arthritis, osteoarthritis, often results from being overweight or from unknown causes as people age.[107] During movement, the ends of healthy bones are protected by small sacs of fluid that act as lubricants. With arthritis, the sacs erode, cartilage and bone ends disintegrate, and joints become malformed and painful to move. Loss of body weight often brings relief, particularly in the knees; physical activities such as walking, bicycling, and swimming can reduce pain and improve physical function, mental health, and quality of life.[108]

Rheumatoid Arthritis Rheumatoid arthritis arises from an immune system malfunction—the immune system mistakenly attacks the bone coverings as if they were foreign tissue. Some individuals report relief from consuming a Mediterranean-style diet, rich in antioxidants of olive oil and vegetables and the omega-3 fatty acid EPA, found in fish oil.[109] Supplemental doses of vitamin C may worsen the condition. Many ineffective or unproven "cures" are sold for arthritis relief, as the margin list shows. Among popular dietary supplements, chondroitin seems to have no effect at all, and glucosamine affects the liver and the immune system, as well as the joints, with uncertain outcomes.[110]

Gout A form of inflammatory arthritis, known as gout, affects millions of U.S. adults, and its prevalence increases with age. An increased incidence of gout has been observed with "triggers" such as insulin resistance, overweight, elevated blood pressure, heart disease, low-level lead exposure, and higher intakes of meat and seafood, beer and spirits, soft drinks, and fructose.[111] Incidences appear to decline with increased consumption of coffee, milk, and vitamin C.

KEY POINT

- Arthritis causes pain and immobility, and older people with arthritis often fall for quack cures.

Vitamin Needs

Vitamin needs change as people age. Changes in absorption, metabolism, and excretion affect the body's needs.

Vitamin A Vitamin A stands alone among the vitamins in that its absorption appears to increase with aging. For this reason, some researchers have proposed lowering the vitamin A requirement for aged populations. Others resist this proposal because foods containing vitamin A and its precursor beta-carotene confer health benefits, and many, such as green leafy vegetables, are frequently lacking in the diet.

Vitamin D For people in their 50s and 60s, maintaining bone mass and minimizing bone loss becomes a critical concern, and the DRI intake recommendation of 15 micrograms a day well covers their needs.[112] For people in their 70s and older, preventing fractures and fracture-related disability and death is primary. Because individual medical concerns and physiological changes with aging can interfere with vitamin D metabolism, a slightly higher intake of 20 micrograms a day is recommended.[113]

As people age, vitamin D synthesis in the skin declines fourfold and the kidneys' ability to activate it diminishes, setting the stage for deficiency. Also, many older adults drink little or no vitamin D–fortified milk and get little or no exposure to sunlight. Recently, however, a U.S. Preventive Services Task Force recommended against supplements of vitamin D and calcium for preventing bone fractures in healthy postmenopausal women because evidence of effectiveness is lacking and their use entails some risk.[114] The Task Force does recommend vitamin D to help prevent falls in elderly people, however, and all women with osteoporosis should follow the advice of a physician regarding supplements and other treatments.[115]

Vitamin B_{12} Adults aged 51 years and older need to obtain 2.4 micrograms of vitamin B_{12} from foods and supplements that supply it. By age 60, however, reduced stomach acid production in about 6 percent of people reduces their ability to absorb vitamin B_{12} from food, making deficiency likely. The number increases with age. Up to 20 percent of elderly people may suffer marginal deficiencies, but most of these cases go unrecognized and untreated.[116] No one yet knows whether dietary insufficiency, malabsorption, or another factor is the primary cause of these deficiencies.[117] Synthetic vitamin B_{12} is reliably absorbed, however, and much misery can be averted by preventing deficiencies of vitamin B_{12} in elderly people.

Other Vitamins and Phytochemicals Antioxidants such as vitamin E may also play roles in conserving immunity, mental functions, and eyesight in the aged. A key aspect of healthy aging is maintaining good vision.[118] Loss of vision in the elderly correlates with loss of life that cannot be explained by other risk factors.[119]

Did You Know?

These are bogus or unproven arthritis treatments:

- Alfalfa tea.
- Aloe vera liquid.
- Any of the amino acids.
- Burdock root.
- Calcium.
- Celery juice.
- Copper or copper complexes.
- Chondroitin.
- Dimethyl sulfoxide (DMSO).
- Fasting.
- Honey.
- Inositol.
- Kelp.
- Lecithin.
- Melatonin.
- Para-aminobenzoic acid (PABA).
- MSM.
- Raw liver.
- Selenium.
- Superoxide dismutase (SOD).
- Vitamin E, other vitamin and mineral supplements.
- Yeast.
- Zinc.
- 100 other substances.

arthritis a usually painful inflammation of joints caused by many conditions, including infections, metabolic disturbances, or injury; usually results in altered joint structure and loss of function.

Dark green leafy vegetables, rich in certain carotenoid phytochemicals, may protect the eyes from one cause of blindness: macular degeneration.[§120] Carotenoid and other nutrient supplements are unproven for eye protection, although some physicians may prescribe them in cases of advanced macular degeneration that threatens vision.[121]

Another problem facing older people is **cataracts**. A cataract is a clouding of the lens that impairs vision and leads to blindness. Only 5 percent of people younger than 50 years have cataracts; afterward, the percentage jumps to between 20 and 30 percent. The lens of the eye is easily oxidized. Some studies suggest that a diet high in *foods* that provide ample antioxidants—carotenoids, vitamin C, and vitamin E—may reduce the risk of early onset and progression of cataracts.[122] Vitamin C supplements, conversely, may increase some people's cataract risk.[123]

KEY POINTS

- Vitamin A absorption increases with aging.
- Elderly people are vulnerable to deficiencies of vitamin D and vitamin B_{12}.

Water and the Minerals

© Yuri Arcurs/Shutterstock.com

Adults of all ages need to attend to daily fluid intakes.

Dehydration is a major risk for older adults. Total body water decreases with age, so even mild stresses, such as a hot day or a fever, can quickly dehydrate the tissues. The thirst mechanism may diminish, and even healthy older people may go for long periods without drinking. The kidneys also become less efficient in recapturing water before it is lost as urine. Dehydration then leads to problems such as constipation, bladder problems, and mental confusion that is easily mistaken for **senile dementia**, an effect that may occur with a water loss of as little as 1 percent of body weight.[124] In a person with asthma, dehydration thickens mucus in the lungs, blocking airways and leading to pneumonia. In a bedridden person, dehydration can lead to **pressure ulcers**. To prevent dehydration, older adults should consume sufficient fluid each day.

Fluid choices, strategically made, can improve the nutrition status of an elderly person. For example, a person with diminished appetite and weight loss may be tempted by a smoothie of bananas, frozen strawberries or other frozen fruit, milk or soy milk, and a touch of chocolate syrup or powdered sugar. Hearty, soft-cooked meat and vegetable soups, milk-based seafood or vegetable bisque, puddings, and commercial liquid meal replacers all provide fluid along with calories, protein, and other nutrients essential for health. Conversely, an overweight elderly person needs tempting low- or no-calorie beverages: plain or sparkling water with lemon or lime, broth-based soups, artificially sweetened tea or coffee, and low-sodium vegetable juices.

Iron Iron status generally improves in later life, especially in women after menstruation ceases and in those who take iron supplements, eat red meat regularly, and include vitamin C–rich fruits in their daily diet. When iron-deficiency anemia does occur, diminished appetite with low food intake is often the cause. Aside from diet, other factors make iron deficiency likely in older people:

- Chronic blood loss from ulcers or hemorrhoids.
- Poor iron absorption due to reduced stomach acid secretion.
- Antacid use, which interferes with iron absorption.
- Use of medicines that cause blood loss, including anticoagulants, aspirin, and arthritis medicines.

Older people take more medicines than others, and drug and nutrient interactions are common.

Zinc Zinc deficiencies, common in older people, are known to impair immune function and may increase the likelihood of infectious diseases, such as pneumonia.[125] Zinc deficiency can also depress the appetite and blunt the sense of taste, thereby reducing

Did You Know?

Beverage recommendations for adults 51+ yr:

- Men: 13 c/day
- Women: 9 c/day

§ The carotenoids are lutein and zeaxanthin, which help to form pigments of the macula of the eye.

Table 14–14

Summary of Nutrient Concerns in Aging

Nutrient	Effects of Aging	Comment
Energy	Need decreases.	Physical activity moderates the decline.
Fiber	Low intakes make constipation likely; beneficial for weight control and reducing the risk of heart disease and type 2 diabetes.	Inadequate water intakes and physical activity, along with some medications, compound risks of constipation.
Protein	Need stays the same; intake often decreases.	Low-fat milk and other high-quality protein foods are appropriate; high-fiber legumes provide protein and other nutrients.
Vitamin A	Absorption increases.	Supplements normally not needed.
Vitamin D	Increased likelihood of inadequate intake; synthesis in skin tissue declines.	Daily moderate exposure to sunlight may be of benefit.
Vitamin B_{12}	Malabsorption of some forms.	Foods fortified with synthetic vitamin B_{12} or a supplement may be of benefit in addition to a balanced diet.
Water	Lack of thirst and increased urine output make dehydration likely.	Mild dehydration is a common cause of confusion.
Iron	In women, status improves after menopause; deficiencies linked to chronic blood losses and low stomach acid output.	Stomach acid is required for absorption; antacid or other medicine use may aggravate iron deficiency; vitamin C and meat enhance absorption.
Zinc	Intakes are often inadequate and absorption may be poor, but needs may also increase.	Medications interfere with absorption; deficiency may depress appetite and sense of taste.
Calcium	Intakes may be low; osteoporosis becomes common.	Lactose intolerance commonly limits milk intake; calcium-rich substitutes or supplements are needed (consider supplements that include vitamin D).
Potassium	Increased intake might decrease the risk of high blood pressure.	Include fruits, vegetables, and low-fat or fat-free milk and yogurt in the diet.
Sodium	Decreasing intake might lower the risk of high blood pressure.	Choose and prepare foods with little to no added salt; consider herbs or salt substitutes to add flavor to foods.
Fat	Increased risk of cardiovascular disease.	Look for foods low in saturated fats, *trans* fats, and cholesterol, and make most fats polyunsaturated or monounsaturated.

food intakes and worsening of zinc status. Many medications interfere with the body's absorption or use of zinc, and an older adult's medicine load can worsen zinc deficiency.

Calcium With aging, calcium absorption declines; at the same time most elderly people fail to consume enough calcium-rich foods. If fresh milk causes stomach discomfort, and the majority of older people report that it does, then lactose-reduced milk or other calcium-rich foods should take its place.

Multinutrient Supplements Overall, elderly people often benefit from a single balanced low-dose vitamin and mineral supplement, and possibly calcium. Older people taking such supplements suffer fewer sicknesses caused by infection, and calcium may slow bone loss. Other single nutrients are tricky: vitamin A has been seen to depress the immunity of elders, while vitamin E may enhance it. A summary of the effects of aging on nutrient needs appears in Table 14–14.

KEY POINT

- Aging alters vitamin and mineral needs; some rise while others decline.

cataracts (CAT-uh-racts) clouding of the lens of the eye that can lead to blindness. Cataracts can be caused by injury, viral infection, toxic substances, genetic disorders, and, possibly, some nutrient deficiencies or imbalances.

senile dementia the loss of brain function beyond the normal loss of physical adeptness and memory that occurs with aging.

pressure ulcers damage to the skin and underlying tissues as a result of unrelieved compression and poor circulation to the area; also called *bed sores*.

Can Nutrition Help People to Live Longer?

The evidence concerning nutrition and longevity is intriguing. Its seems that even though people cannot alter the year of their birth, they can probably alter the length and quality of their lives.

Lifestyle Factors In a classic study, researchers observed that some older adults seemed young for their ages, while others seemed older. To uncover what made the difference, the researchers focused on health habits and identified six factors that affect physiological age. Three of the six factors were related to nutrition:

- Abstinence from, or moderation in, alcohol use.
- Regular nutritious meals.
- Weight control.

The other three were regular adequate sleep, abstinence from smoking, and regular physical activity. The physical health of those who engaged in all six positive health practices was comparable to that of people 30 years younger who engaged in few or none. Numerous studies have confirmed the benefits of these lifestyle factors. Some changes of aging, such as graying of hair and reduced senses of smell, taste, and eyesight are inescapable, whereas others may yield to individual life choices (Table 14–15).

Table 14–15

What to Expect in Aging

Changes with Age You Probably Can Slow or Prevent
By exercising, eating an adequate diet, reducing stress, and planning ahead, you may be able to slow or prevent:
✓ Wrinkling of skin due to sun damage
✓ Some forms of mental confusion
✓ Elevated blood pressure
✓ Accelerated resting heart rate
✓ Reduced lung capacity and oxygen uptake
✓ Increased body fatness
✓ Elevated blood cholesterol
✓ Slowed energy metabolism
✓ Decreased maximum work rate
✓ Loss of sexual functioning
✓ Loss of joint flexibility
✓ Diminished oral health: loss of teeth, gum disease
✓ Bone loss
✓ Digestive problems, constipation
Changes with Age You Probably Must Accept
These changes are probably beyond your control:
✓ Graying of hair
✓ Balding
✓ Some drying and wrinkling of skin
✓ Impairment of near vision
✓ Some loss of hearing
✓ Reduced taste and smell sensitivity
✓ Reduced touch sensitivity
✓ Slowed reactions (reflexes)
✓ Slowed mental function
✓ Diminished visual memory
✓ Menopause (women)
✓ Loss of fertility (men)
✓ Loss of joint elasticity

Energy Restriction Evidence that diet might influence the life span emerged decades ago when researchers fed young rats a diet extremely low in energy. The starved rats stopped growing while a group of control rats ate and grew normally; when the researchers increased food energy in the starved group, growth resumed. Many of the starved rats died young from malnutrition. The few survivors, although permanently deformed from their ordeal, remained alive far beyond the normal life span for such animals and developed diseases of aging much later than normal. Since then, this result has been repeated in many species (see Table 14–16).

With moderate energy restriction, animals also retain youthfulness longer and develop fewer disease risk factors such as high blood pressure, glucose intolerance, and immune system impairments.[126] In monkeys, moderate calorie restriction prolongs life and reduces body weight and the incidence of diabetes, cancer, and cardiovascular disease; it may also reduce the brain shrinkage typical in aging monkeys fed a full diet.[127] In contrast, one ongoing study reports no improvement in survival of monkeys.[128] No one can say conclusively whether any of these findings might also apply to human beings.[129] Monkeys and human beings do share a significant number of genes, making their metabolism more or less similar.

Some questions arise about the safety of severe energy restriction. Energy-restricted mice, for example, die more often from influenza infections, despite evidence that certain immune system functions are maintained.[130] And, although energy restriction may improve chronic disease risks, it stunts growth and may damage some systems to benefit others. Also, without supplements, calorie-restricted diets often lack nutrients.[131] Scientists are hoping to discover drug treatments that mimic the effect of calorie restriction while minimizing risks.[132] For now, however, any supplement or treatment claiming to prolong life is a hoax.

KEY POINT

- In rats and other species, food energy deprivation lengthens the lives of individuals.

Immunity and Inflammation

As people age, the immune system loses function. As they become ill, the immune system becomes overstimulated but less able to cope with the challenge. The combination of an overreactive, inefficient immune response results in a chronic inflammation, with increasing frailty and illness.[133]

Most chronic diseases—such as atherosclerosis, Alzheimer's disease, obesity, and rheumatoid arthritis—involve inflammation.[134] Therefore, inflammation is believed

Table 14–16

Effects of Energy Restriction on Life Span

Differences in maximum life span are observed between animals eating normally and those that are energy restricted.

	Normal Diet	Energy Restricted
Rats	33 months	47 months
Spiders	100 days	139 days
Single-celled animals (protozoa)	13 days	25 days

to be a partly harmful process; yet, at the same time it plays a critical role in destruction of invading organisms and repair of damaged tissues.[135] The trick seems to be to control inflammation enough to prevent harm from its chronic, ineffective processes, but not so much as to prevent its beneficial ones.

Nutrient deficiencies compromise immune function, while a sound diet and regular physical activity can improve it.[136] Typically in old age, people become more sedentary and prone to malnutrition, leaving them particularly vulnerable to infectious diseases. Antibiotics also often lose effectiveness in people with a compromised immune system. Consequently, many older adults die of infectious diseases.

A free-radical hypothesis blames damage from oxidative stress for the physical deterioration associated with aging, a process that appears to be compounded by chronic physical stress.[137] The body's internal antioxidant enzymes diminish with age, and many "age-related" chronic diseases may be linked to free-radical damage. This and related lines of research promote a storm of worthless and sometimes hazardous "life-extending" pills, supplements, and treatments, such as DHEA, testosterone, and growth hormone.[138] Better to spend money on legumes, fresh fruit, and green and yellow vegetables, naturally rich sources of antioxidants linked with many health benefits (see Controversy 2).

KEY POINT

- Claims for life extension through antioxidants or other supplements are common hoaxes.

Can Foods or Supplements Affect the Course of Alzheimer's Disease?

The cause of Alzheimer's disease, the most prevalent form of senile dementia, is unknown, but genetic inheritance clearly contributes.[139] The devastation of Alzheimer's occurs when areas of the brain that coordinate memory and cognition become littered with clumps of abnormal protein fragments and tangles of nerve tissue that damage or kill brain cells.** Soon, memory fails and reasoning powers diminish, followed by loss of communication skills, loss of physical capabilities, and onset of anxiety, delusions, depression, anger, inappropriate behavior, sleep disturbance, and eventually loss of life itself. Once the destruction begins, the outlook for its reversal is bleak. Hope is on the horizon; research is advancing to perfect a drug or vaccine to block the destructive progression of this disease.

Only weak links exist between nutrition and Alzheimer's disease. For example, although the mineral aluminum may build up in the brain with Alzheimer's, a causal connection seems unlikely. Evidence conflicts as to whether supplements of copper, zinc, or other trace minerals worsen Alzheimer's disease, so to err on the safe side, food sources, not concentrated supplements, of trace minerals are advisable for people with the disease.

** The protein fragments are called *beta-amyloid*.

Brain cells of Alzheimer's victims show signs of oxidative damage, so the antioxidant-rich Mediterranean diet is under study for defending against this damage, but evidence so far is mixed.[140] Such a diet also provides fish rich in DHA, which may oppose general cognitive decline in the elderly. While DHA has been used to improve outcomes related to Alzheimer's disease in animal models, its actions in human Alzheimer's patients are not yet established.[141] Suggestions that a moderate alcohol intake may delay cognitive decline appear to be false, as well. [142]To date, no proven benefits are available from vitamin or mineral pills or herbs—even ginkgo biloba—or other remedies, but claims from quacks are all too commonplace.[143]

Preventing weight loss is an important nutrition concern for the person suffering with Alzheimer's disease. Depression and forgetfulness can lead to skipped meals and poor food choices. Caregivers can help by providing well-liked, well-balanced, and well-tolerated meals and snacks served in a cheerful, peaceful atmosphere on brightly colored tableware to spur interest in eating.

KEY POINTS

- Alzheimer's disease causes some degree of brain deterioration in many people older than age 65.
- Nutrition care gains importance as Alzheimer's disease progresses.

Food Choices of Older Adults

Most older people are independent, socially sophisticated, mentally lucid, fully participating members of society who report being happy and healthy. Many older people have heard and heeded nutrition messages: they have cut down on saturated fats in dairy foods and meats and are eating slightly more vegetables and whole-grain breads, although few meet the recommended intakes of these foods. Older people who enjoy a wide variety of foods are better nourished and have a better quality of life than those who subsist on a monotonous diet.[144] Grocers assist the elderly by prominently displaying good-tasting, low-fat, nutritious foods in easy-to-open, single-serving packages with labels that are easy to read.

Obstacles to Adequacy Many factors affect the food choices and eating habits of older people, including whether they live alone or with others, at home or in an institution.[145] Men living alone, for example, are likely to consume poorer-quality diets than those living with spouses. Older people who have difficulty chewing because of tooth loss or loss of taste sensitivity may no longer seek a wide variety of foods. Medical conditions and functional losses can also adversely affect food choices and nutrition. Lack of access to kitchen facilities may be a factor. Many older people become weak when unintentional reductions in food intake result in weight and muscle loss, events often followed by illness or death. It may be that some of these outcomes could have been prevented or delayed if the person had been provided an adequate diet.

Two other factors seem to make older people vulnerable to malnutrition: use of multiple medications and abuse of alcohol. People older than age 65 take about a fourth of all the medications, both prescription and over-the-counter, sold in the United States. Although these medications enable people with health problems to live longer and more comfortably, they also pose a threat to nutrition status because they may interact with nutrients, depress the appetite, or alter the perception of taste (see Controversy 14).

The incidence of alcoholism, alcohol abuse, or problem drinking among the elderly in the United States is estimated at between 2 and 10 percent. Loneliness, isolation, and depression in the elderly accompany overuse of alcohol and detract from nutrient intakes. Table 14–17 provides an easily remembered means of identifying those who might be at risk for malnutrition.

Programs That Help Nutrition professionals are calling for improved nutrition-related services, supported by ongoing research, for all Americans aged 60 years and older regardless of their health status.[146] Currently, several federal programs can provide help for older people. Social Security provides income to retired people older than

Table 14–17

Predictors of Malnutrition in the Elderly

Here is a quick and easy-to-remember list of factors that increase the likelihood of malnutrition in the elderly. The first letters spell the word *DETERMINE.*

To Determine:	Ask:
Disease	■ Do you have an illness or condition that changes the types or amounts of foods you eat?
Eating poorly	■ Do you eat fewer than two meals a day? Do you eat fruits, vegetables, and milk products daily?
Tooth loss or mouth pain	■ Is it difficult or painful to eat?
Economic hardship	■ Do you have enough money to buy the food you need?
Reduced social contact	■ Do you eat alone most of the time?
Multiple medications	■ Do you take three or more different prescribed or over-the-counter medications daily?
Involuntary weight loss or gain	■ Have you lost or gained 10 pounds or more in the last 6 months?
Needs assistance	■ Are you physically able to shop, cook, and feed yourself?
Elderly person	■ Are you older than 80?

age 62 who paid into the system during their working years. The Supplemental Nutrition Assistance Program (SNAP), formerly called the Food Stamp Program, assists the very poor by supplementing their monthly food budgets with a card similar to a credit card, encrypted with benefits and redeemable for food. The Administration on Aging coordinates services governed by the Older Americans Act including providing nutritious meals in a social congregate setting, education and shopping assistance, counseling and referral to other needed services, and transportation to necessary appointments. An estimated 25 percent of the nation's elderly poor benefit from meals provided by the program. For the homebound, Meals on Wheels volunteers deliver meals to the door, a benefit even though the recipients miss out on the social atmosphere of the congregate meals. Nutritionists are wise not to focus solely on nutrient and food intakes of the elderly because enjoyment and social interactions may be as important as food itself.

Many older people, even able-bodied ones with financial resources, find themselves unable to perform cooking, cleaning, and shopping tasks. For anyone living alone, and particularly for those of advanced age, it is important to work through the problems that food preparation presents. This chapter's Food Feature presents some ideas.

KEY POINTS

- Food choices of the elderly are affected by aging, altered health status, and changed life circumstances.
- Federal programs can help to provide nourishment, transportation, and social interactions.

Did You Know?

These and other federal programs help to support the elderly:

- Social Security.
- Supplemental Nutrition Assistance Program (SNAP), formerly called the Food Stamp Program.
- Elderly Nutrition Program (Overseen by the Administration on Aging).
- Meals on Wheels (Overseen by the Administration on Aging).

Shared meals can be the high point of the day.

try it! Food Feature

Single Survival and Nutrition on the Run

LO 14.5 Design a healthy meal plan for an elderly widower with a fixed income.

A single person of any age, whether a busy student in a college dormitory, an elderly person in a retirement apartment, or a professional in an efficiency suite, faces challenges in obtaining nourishing meals. People without access to kitchens and freezers find storing foods problematic, and so often eat out. Following is a collection of ideas gathered from single people who have devised ways to nourish themselves despite obstacles.

Is Eating in Restaurants the Answer?

Restaurant foods are convenient, but can such foods meet nutrient needs or support health as well as homemade foods? The answer is "perhaps," but making it so takes some effort. A few chefs and restaurant owners are concerned with the nutritional health of their patrons, but more often chefs strive to please the palate. Restaurant foods are often overly endowed with calories, fat, saturated fat, sugar, and salt but often lack fiber, iron, or calcium. Vegetables and fruits may be in short supply, but a single meat or pasta portion may exceed a whole day's recommended intake. To improve restaurant meals, follow these suggestions:

- Restrict your portions to sizes that do not exceed your energy needs.
- Ask that excess portions be placed in take-out containers right away.
- Ask for extra vegetables, fruit, or salad.
- Request whole-grain breads and pasta (more restaurants now supply these, and others may do so with repeated requests).
- Make judicious choices of foods that stay within intake guidelines for solid fats, added sugars, and salt.

The Food Feature of Chapter 5 offered specific suggestions for ordering fast food and other foods with an eye to keeping fat intakes within bounds, and Chapter 8 listed foods high in sodium. Table 14–18 provides tips for single survival in the grocery store and at home.

Dealing with Loneliness

For nutrition's sake, among many reasons, it is important to attend to loneliness, and mealtimes provide an opportunity to do so.[147] The person who is living alone must learn to connect food with socializing. Invite guests and make enough food so that you can enjoy the leftovers later on. If you know an older person who eats alone, you can bet that person would love to join you for a meal now and then. Invite that person often.

A Word about Food Safety for Elders

Older adults frequently suffer from foodborne illnesses, and the consequences for them can be more severe. Illnesses that give others an upset stomach can lead to severe diarrhea, vomiting, dehydration, and other serious symptoms in older people and can be fatal. For these reasons, the Dietary Guidelines for Americans 2010 urge older adults to heed food safety rules (see Chapter 12) and offer these additional precautions:

- Older adults should not eat or drink unpasteurized milk, milk products, or juices; raw or undercooked eggs, meat, poultry, fish, or shellfish; or raw sprouts.
- Older adults should eat delicatessen lunchmeats and frankfurters only if they have been reheated to steaming hot.

Shopping for and preparing nutritious foods for one person takes some special know-how.

Table 14–18

Smart Shopping and Creative Cooking

Smart Shopper Tips

- Make a list to reduce impulse buying; buy on sale, and use coupons for needed items.
- Watch sizes: Gallons of milk may be cheaper than pints per ounce, but the savings are lost if the milk sours. Dry milk and small shelf-stable milk boxes often make sense.
- Bulk staple foods, such as dry milk, oatmeal, ready-to-eat cereals, or rice are cheapest, but they must be stored properly (see Chapter 15 for hints to avoid food waste).
- If freezer space allows, buy whole chickens or "family pack" meats at bargain prices. Divide into single servings, wrap well, mark the date, freeze, and use as needed.
- Ask grocers to break open large packages of fresh foods; buy only the amount you can use up. More expensive but convenient small bags of cut and washed fresh vegetables may be an option.
- Frozen vegetables in large resealable bags are more economical than small boxes.
- Freeze a loaf of whole-grain bread; defrost or toast as needed.
- Eggs keep for weeks in the refrigerator; after their sell-by date, hard-boil and refrigerate them for handy protein servings that last weeks longer.
- Buy three pieces of tomatoes, pears, and other fresh fruit in various stages of ripeness: a ripe one to eat right away, a less ripe one to eat soon after, and a green one to ripen in a few days.
- Buy ready-to-heat and -eat foods from the grocery store delicatessen section—these cost less than similar foods from restaurants. Choose nutrient-dense items; skip stuffing, macaroni and cheese, meat loaf and gravy, vegetables in sauce, mayonnaise-dressed mixed salads, and fried foods.
- Buy a ready-roasted chicken; use the main pieces for several dinners; simmer the remainder with herbs and vegetables in a broth for soup.

Creative Kitchen Tricks

- Divide a head of cauliflower or broccoli into thirds. Cook one-third right away; marinate one-third in Italian salad dressing to use later in a salad; toss the remainder into a casserole, soup, or stew, or eat it raw with dip for a crunchy snack.
- Stir-fry ready-to-use blends of cabbage, snow peas, and onions; bags of slaw-cut vegetables; or raw vegetables for a delicious dinner; add Asian seasonings and leftover chicken or seafood. Bonus: one pan to wash.
- Microwavable bags of brown rice cost more but provide a whole-grain food for those less able to cook.
- Treat leftovers with respect: nothing beats a plate of delicious leftovers for speed and convenience—plate, reheat in the microwave, and eat.
- Use nutritious frozen dinners judiciously (caution: these can be very high in solid fats, added sugars, and salt—read Nutrition Facts panels). Round out the meal with a salad, whole-grain roll, and a glass of fat-free milk.

My Turn watch it!

Eating Solo

Allison

Eric

Have you ever felt uninspired by the thought of eating alone? Two students talk about dealing with loneliness, making easy but nutritious meals, and choosing wisely among takeout foods.

track it! Diet Analysis PLUS Concepts in Action

Analyze Three Diets

The purpose of this exercise is to explore food choices and the potential for nutrient deficiencies among young children, teens, and older adults, using three new profiles: a 2-year-old, a 14-year-old, and a 70-year-old.

1. Iron nutrition is required for normal development. Create a new profile for a 2-year-old toddler. Select the Track Diet tab, choose a new day, and then, using Table 14–3 (p. 557), choose foods to create a balanced iron-rich meal for a toddler 2 years of age. Once you've entered the foods, select the Reports tab, select Intake Spreadsheet for that meal, and generate a report. Did the meal supply a third of the iron this toddler needs? Which foods contributed most of the iron?
2. For nutrient adequacy, children's diets should include a variety of foods from each food group (see Figure 14–2, p. 557). Modify your child's meals to provide the needed servings of foods, keeping in mind that children eat small portions and often like colorful, crunchy vegetables and smooth bland foods. Enter the data from the Track Diet tab. Select Intake Spreadsheet and generate reports for both Intake Spreadsheet and MyPlate Analysis for this day's meals. Did your choices provide the right number of servings of a variety of foods from each food group? How can you improve this child's food intake?
3. Nutrients missed at breakfast often cannot be made up at lunch or dinner. Create a nutritious breakfast for your 2-year-old from the above activities, using ideas for a rushed morning. Enter the data from the Track Diet tab. Select Intake Spreadsheet for that date and meal, and generate a report. Did the breakfast meet a significant portion of the child's nutrient needs? If not, what foods or beverages might improve it?
4. Teens' diets often lack calcium, iron, and vitamin A. From the Profile drop-down box, create a new profile for a 14-year-old girl. Select the Track Diet tab, and choose foods that are excellent sources of calcium, iron, and vitamin A for a lunch meal. Recall that vitamin C helps maximize iron absorption from non-heme (non-meat) sources of iron. Select the Reports tab and then Source Analysis. From the drop-down box, generate reports for Calcium, Iron, Vitamin A, and Vitamin C. Did the meal meet the teen's needs for calcium and iron? Did meat or non-heme sources of iron predominate? Which foods also supplied vitamin A and vitamin C?
5. Teens often make snack choices with convenience and taste in mind, but nutritious snacks better suit their nutrient needs. Select the teen's profile (already created), and select the Track Diet tab. Consult Table 14–10 (p. 568) for help in choosing foods to include in a nutritious snack for the morning and afternoon. Make each snack supply about 150 calories of nutritious foods. Select the Intake Spreadsheet, and generate a report for the snacks. Did the snacks provide about 20 percent of nutrients significant for teens? Which? How much saturated fat did the snacks provide?
6. For elderly people, nutritious meals help to retain good health. From the Profile drop-down box, create a new profile for a 70-year-old single adult. Select the Track Diet tab, and choose foods to create a nutritious, convenient, easy-to-eat dinner for this person. Select the Reports tab and then the Intake Spread Sheet for that meal, and generate a report. Did the meal supply enough zinc, protein, vitamin B_{12}, and calcium to meet one-third of the person's need without excessive calories? Did it supply omega-3 fatty acids? If not, suggest ways of improving it.

what did you decide?

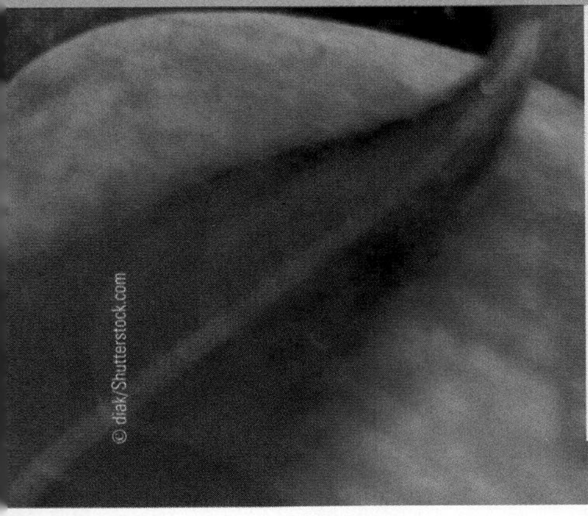

© diak/Shutterstock.com

Do you need special information to properly nourish children, or are they like "little adults" in their needs?

Do you suspect that symptoms you feel may be caused by a food allergy?

Are teenagers old enough to decide for themselves what to eat?

Can good nutrition help you live better and longer?

Self Check

1. (LO 14.1) Children naturally like nutritious foods in all the food groups, *except* __________ .
   a. dairy
   b. meats
   c. vegetables
   d. fruits
2. (LO 14.1) A healthy child's normal appetite control system
   a. cannot be trusted to provide the right level of calories for each stage of growth.
   b. can be short-circuited by a constant stream of foods high in added sugars, saturated fat, and refined grains.
   c. holds the child's appetite constant, without much fluctuation from day to day.
   d. none of the above
3. (LO 14.1) On a pound-for-pound basis, a 5-year-old's need for vitamin A is about double the need of an adult man.
   T F
4. (LO 14.1) Which of the following can contribute to choking in children?
   a. peanut butter eaten by the spoonful
   b. hot dogs and tough meat
   c. grapes and hard candy
   d. all of the above
5. (LO 14.1) Lead poisoning in young children
   a. is no longer a problem in the United States.
   b. is likely because they absorb 5 to 10 times more lead than do adults.
   c. arises primarily from ingesting of foods packed in metal cans.
   d. all of the above
6. (LO 14.1) Research to date supports the idea that food allergies or intolerances are common causes of hyperactivity in children.
   T F
7. (LO 14.1) A child who ate cream of broccoli soup and became ill, now feels ill whenever it is served. The child most likely has a
   a. food allergy.
   b. food intolerance.
   c. food aversion.
   d. food antibody.
8. (LO 14.2) Which of the following is most commonly deficient in adolescents?
   a. folate
   b. zinc
   c. iron
   d. vitamin D
9. (LO 14.2) Which of the following may worsen symptoms of PMS?
   a. vitamin B_6
   b. physical activity
   c. caffeine
   d. calcium
10. (LO 14.2) Which of the following has been shown to improve acne?
    a. avoiding chocolate and fatty foods
    b. vacations
    c. vitamin A supplements
    d. increased stress
11. (LO 14.3) Physical changes of aging that can affect nutrition include __________ .
    a. reduced stomach acid
    b. increased saliva output
    c. tooth loss and gum disease
    d. a and c
12. (LO 14.3) In research, which of the following is associated with a longer lifespan in many species?
    a. energy restriction
    b. superoxide dismutase
    c. omega-3 fatty acids
    d. none of the above
13. (LO 14.4) Nutrition does not seem to play a role in the causation of osteoarthritis.
    T F
14. (LO 14.4) Vitamin A absorption decreases with age.
    T F
15. (LO 14.4) Herbal supplements have been shown to slow down the progression of Alzheimer's disease.
    T F
16. (LO 14.5) A person planning a nutritious diet for an elderly person should pay particular attention to providing enough
    a. vitamin A
    b. vitamin B_{12}
    c. iron
    d. b and c
17. The word DETERMINE is an acronym used in assessing an elderly person's
    a. risk of malnutrition
    b. bone integrity
    c. degree of independence
    d. all of the above
18. (LO 14.6) Nutrient–drug interactions occur only in those taking two or more drugs simultaneously.
    T F
19. (LO 14.6) Nutrients and drugs can interact
    a. before ingestion or absorption.
    b. within the body tissues.
    c. at the level of gene expression.
    d. all of the above

Answers to these Self Check questions are in Appendix G.

CONTROVERSY 14

Nutrient–Drug Interactions: Who Should Be Concerned?

LO 14.6 Describe several specific nutrient–drug interactions, and name some herbs that may interfere with the action of medication.

A 45-year-old Chicago business executive attempts to give up smoking with the help of nicotine gum. She replaces smoking breaks with beverage breaks, drinking frequent servings of tomato juice, coffee, and colas. She is discouraged when her stomach becomes upset and her craving for tobacco continues unabated despite the nicotine gum. Problem: nutrient–drug interaction.

A 14-year-old girl develops frequent and prolonged respiratory infections. Over the past six months, she has suffered constant fatigue despite adequate sleep, has had trouble completing school assignments, and has given up playing volleyball because she runs out of energy on the court. During the same six months, she has been taking antacid pills several times a day because she heard this was a sure way to lose weight. Her pediatrician has diagnosed iron-deficiency anemia. Problem: nutrient–drug interaction.

© iStockphoto.com/AvailableLight

A 30-year-old schoolteacher who benefits from antidepressant medication attends a faculty wine and cheese party. After sampling the cheese with a glass or two of red wine, his face becomes flushed. His behavior prompts others to drive him home. In the early morning hours, he awakens with severe dizziness, a migraine headache, vomiting, and trembling. An ambulance delivers him to an emergency room where a physician takes swift action to save his life. Problem: nutrient–drug interaction.

The Potential for Harm

People sometimes think that medical drugs do only good, not harm. As the opening stories illustrate, however, both prescription and over-the-counter (OTC) medicines can have unintended consequences, among which are significant interactions with nutrition.[1]*

How Drugs and Nutrients Interact

As Figure C14–1 shows, drugs can interact with nutrients in three general realms:

1. *Before ingestion or absorption:*†
   - Chemical incompatibilities may occur in formula mixtures or in the digestive tract, affecting the

* *Reference notes are found in Appendix F.*

† *These three categories of interactions have been called, respectively, pharmaceutical interactions, phamacokinetic interactions, and pharmacodynamic interactions.*

Figure C14–1

How Foods, Drugs, and Herbs Can Interact

The arrows show that foods, drugs, and herbs can interfere with each other's absorption, actions, metabolism, or excretion. Drugs also often change the appetite, affecting food intake.

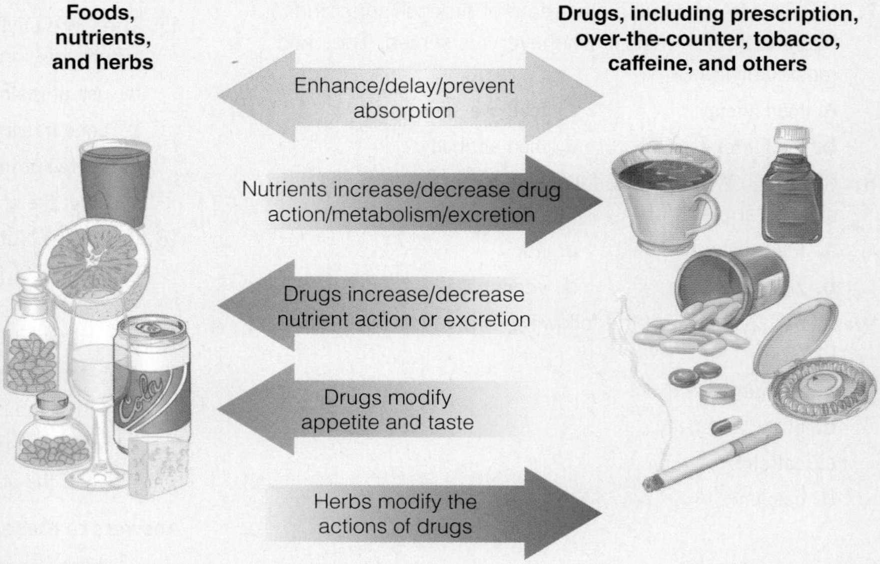

© Cengage Learning

solubility or stability of drugs and nutrients, making them less functional.
   - Food or nutrients in the digestive tract can enhance, delay, or prevent drug absorption.
   - Drugs can enhance, delay, or prevent nutrient absorption.
2. *Within the body tissues*:
   - Nutrients can alter the distribution of a drug among body tissues or interfere with its metabolism, transport, or elimination from the body.
   - Also, the reverse: drugs can alter the distribution of a nutrient among body tissues or interfere with its metabolism, transport, or excretion.
   - Drugs often modify taste, appetite, or food intake, causing weight loss or gain.
   - Herbal supplements can also modify drug effects in any of these ways.[2]
3. *At the level of gene expression*:
   - Drugs and nutrients can alter the production of enzymes, transporters, or receptors for each other, altering their metabolism in the body.

Some drugs are known to interact with specific nutrients (see Table C14–1, p. 592). In addition, alcohol is also infamous for its interactions with nutrients, and the more alcohol ingested, the more likely that a significant nutrient interaction will occur (see Controversy 3).

Factors That Make Interactions Likely

Significant nutrient–drug interactions do not occur every time a person takes a drug. The potential for interactions is greatest in those who take a medicine for a long time, who take multiple drugs, who drink alcohol daily, or who are poorly nourished to begin with. Ninety percent of people over the age of 70 receive at least one prescription medication, and some may take as many as 10 drugs at a time.[3] The risk of an adverse effect rises substantially among older people taking five or more daily medications and compounds further when herbs and other supplements are added to the mix.

People who are overweight or underweight can easily receive too much or too little medication—their nutrition status puts them at higher risk because drug manufacturers often fail to give adequate dosing information for weight extremes.[4] For example, obese people frequently suffer from potentially life-threatening infections after surgery; a dosing error could easily prevent an antibiotic drug from effectively fighting such an infection, with disastrous results.[5] Likewise, a malnourished person whose weight is too low for health, particularly those with kidney or liver impairments, could easily receive a drug overdose.

The details of nutrient–drug interactions are many and far more extensive than can be presented in this section. The following discussions are intended to raise awareness of the most common ones.

Absorption of Drugs and Nutrients

The business executive described earlier felt the effects of the first type of interaction: chemical incompatibility. Acids from the tomato juice, coffee, and colas she drank before chewing the nicotine gum kept the nicotine from being absorbed into the bloodstream through the lining of her mouth and quelling her craving. Instead, it traveled to her stomach and caused nausea.[‡]

Similarly, dairy products or calcium-fortified juices interfere with the absorption of certain antibiotics. Drug label instructions, such as "Take on an empty stomach" or "Do not combine with dairy products," help to avert most such interactions.

Certain drugs can also interfere with the small intestine's absorption of nutrients, particularly minerals. This interaction explains the experience of the tired 14-year-old. Her overuse of antacids eliminated the stomach's normal acidity, on which iron absorption depends. The medicine bound tightly to the iron molecules, forming an insoluble, unabsorbable complex. Her iron stores already bordered on deficiency, as iron stores for young girls typically do, so her misuse of antacids pushed her over the edge into iron-deficiency anemia.

Chronic laxative use can also lead to malnutrition. Laxatives can carry nutrients through the intestines so rapidly that many vitamins have no time to be absorbed. Mineral oil, a laxative the body cannot absorb, can rob a person of important fat-soluble vitamins and potentially beneficial phytochemicals by dissolving them and carrying them out in the feces.

Metabolic Interactions

The teacher who landed in the emergency room was taking an antidepressant medicine, one of the monoamine oxidase inhibitors (MAOI). At the party, he suffered a dangerous chemical interaction between the medicine and the compound tyramine in his cheese and wine. Tyramine is produced during the fermenting process in cheese and wine manufacturing.

The MAOI medication works by depressing the activity of enzymes that destroy the brain neurotransmitter dopamine. With less enzyme activity, more dopamine is left, and depression lifts. As a side effect, the drug also depresses enzymes in the liver that destroy tyramine. Ordinarily, the man's liver would have quickly destroyed the tyramine from the cheese and wine, but due to the MAOI medication, tyramine built up and caused the potentially fatal reaction. Table C14–2 (p. 593) lists some foods high in tyramine.

Other culprits affecting drug metabolism include food phytochemicals and popular herbal supplements. A chemical constituent of grapefruit juice suppresses an enzyme responsible for breaking down many kinds of medical drugs. With less drug breakdown, doses build up in the blood to levels that can have undesirable effects on the body. A person who drinks either

‡ *These items also interfere with the action of nicotine gum: beer; coffee; condiments (ketchup, mustard, and soy sauce); juices (apple, grape, orange, and pineapple); and lemon-lime soda.*

Table C14–1

Selected Nutrient–Drug Interactions

Drug	Effects on Nutrient Absorption	Effects on Nutrient Excretion	Effects on Nutrient Metabolism
Antacids (aluminum containing)	Reduce iron absorption	Increase calcium and phosphorus excretion	May accelerate destruction of thiamin
Antibiotics (long-term usage)	Reduce absorption of fats, amino acids, folate, fat-soluble vitamins, vitamin B_{12}, calcium, copper, iron, magnesium, potassium, phosphate, zinc	Increase excretion of folate, niacin, potassium, riboflavin, vitamin C	Destroy vitamin K–producing bacteria and reduce vitamin K production
Antidepressants (monoamine oxidase inhibitors, MAOI)			Slow breakdown of tyramine, with dangerous blood pressure spike and other symptoms on consuming tyramine-rich foods or drinks: *alcoholic beverages* (sherry, vermouth, red wines, some beers); *cheeses* (aged and processed); *some meats* (caviar, pickled herring, liver, smoked and cured sausages, lunchmeats); *fermented products* (soy sauce, miso, sauerkraut); *others* (brewer's yeast, yeast supplements, yeast paste; baked goods made with baker's yeast are safe; foods past their expiration date)
Aspirin (large doses, long-term usage)	Lowers blood concentration of folate	Increases excretion of thiamin, vitamin C, vitamin K; causes iron and potassium losses through gastric blood loss	
Caffeine		Increases excretion of small amounts of calcium and magnesium	Stimulates release of fatty acids into the blood
Cholesterol-lowering "statin" drugs (Zocor, Lipitor)			Grapefruit juice slows drug metabolism, causing buildup of high drug levels; potentially life-threatening muscle toxicity can result.
Diuretics		Raise blood calcium and zinc; lower blood folate, chloride, magnesium, phosphorus, potassium, vitamin B_{12}; increase excretion of calcium, sodium, thiamin, potassium, chloride, magnesium	Interfere with storage of zinc
Estrogen replacement therapy	May reduce absorption of folate	Causes sodium retention	May raise blood glucose, triglycerides, vitamin A, vitamin E, copper, and iron; may lower blood vitamin C, folate, vitamin B_6, riboflavin, calcium, magnesium, and zinc
Laxatives (effects vary with type)	Reduce absorption of glucose, fat, carotene, vitamin D, other fat-soluble vitamins, calcium, phosphate, potassium	Increase excretion of all unabsorbed nutrients	
Oral contraceptives	Reduce absorption of folate, may improve absorption of calcium	Cause sodium retention	Raise blood vitamin A, vitamin D, copper, iron; may lower blood beta-carotene, riboflavin, vitamin B_6, vitamin B_{12}, vitamin C; may elevate requirements for riboflavin and vitamin B_6; alter blood lipid elevating risk of heart disease in smokers and older women

Table C14–2

Some Foods High in Tyramine

▪ Aged cheeses	▪ Feta cheese
▪ Aged meats	▪ Lima beans
▪ Alcoholic beverages (beer, wine)	▪ Mushrooms
▪ Anchovies	▪ Pickled fish or meat
▪ Caviar	▪ Prepared soy foods (miso, tempeh, tofu)
▪ Fava beans	▪ Smoked fish or meat
▪ Fermented foods (sauerkraut, sausages)	▪ Soy sauce
	▪ Yeast extract (Marmite)

[a]The tyramine content of foods depends on storage conditions and processing; thus the amounts in similar products can vary substantially.

grapefruit or cranberry juice and also takes the blood-thinning drug warfarin may exhibit delayed blood clotting with dangerously prolonged bleeding times.[6] Soy foods may have the opposite effect: an active component in soy seems to oppose the blood thinning effect of warfarin. It makes blood clots more likely by increasing the genetic expression for clotting factors in the blood.[7]

People take the herb ginkgo biloba in hopes of improving memory, but this effect is not consistently supported in research.[8] Takers of ginkgo should know that it opposes blood clotting, increases bleeding time, and poses a risk of hemorrhage when combined with other blood-thinning medications, such as aspirin. The take-home message: many herbs and phytochemicals are known to interact with drugs, sometimes dangerously (see Table C14–3, p. 594).

Drugs often cause nutrient losses, too. Many people take large quantities of aspirin (10 to 12 tablets each day) to relieve the pain of arthritis, backaches, and headaches. This much aspirin can speed up blood loss from the stomach by as much as 10 times, enough to cause iron-deficiency anemia in some people. People who take aspirin regularly should eat iron-rich foods regularly as well.

Two Widely Used Drugs: Caffeine and Tobacco

People in every society use caffeine in some form for its well-known "wake-up" effect. Caffeine's interactions with foods and nutrients are subtle but may be significant because caffeine is ubiquitous in foods and beverages—see Table C14–4 (p. 594). Chocolate bars, colas, and other foods favored by children contain caffeine, and children are more sensitive to its effects. Many popular cold and headache remedies also offer about a cup of coffee's worth of caffeine per dose because, in addition to being a mild pain reliever in its own right, this amount of caffeine remedies the caffeine-withdrawal headache that no other pain reliever can touch.[9]

In many studies, a 200-milligram dose of caffeine, about two cups of coffee's worth, significantly improves the ability to pay attention, especially if subjects are sleepy, but more caffeine is not better in this regard. A single dose of 500 milligrams has been shown to worsen thinking abilities in almost everyone, and more than this may present some risk to health through its actions as a stimulant.

Caffeine is a true stimulant drug. Like all stimulants, it increases the respiratory rate, heart rate, and secretion of stress and other hormones. Caffeine also raises the blood pressure, an effect that lasts for hours after consumption.[10] Because caffeine is a diuretic, high doses promote water loss from the body. Taken in moderation, caffeinated beverages can contribute to daily fluid intakes and do not negatively influence the body's water balance (see Chapter 8 for details).

Some people avoid caffeine for fear it may harm their health. Research on these topics is limited, but it mostly refutes any causative links between caffeine and cancer, cardiovascular disease, or birth defects.[11] Research is mixed on the effects of caffeine or coffee on type 2 diabetes.[12] Much more research is needed to clarify these associations.

As for cigarette and other tobacco use, it is a delivery system for the drug nicotine. Tobacco's dangers are well known, and most are beyond the scope of nutrition. Smoking does depress hunger and, in turn, sometimes reduces body fatness; it also accelerates the breakdown of vitamin C, increasing requirements. Chapters 7 and 9 provided details on these topics.

Illicit Drugs

Like other drugs, illegal drugs modify body functions. Unlike medicines, however, no watchdog agency monitors them for safety, effectiveness, or even purity.

Smoking a marijuana cigarette affects several senses, including the sense of taste. It produces an enhanced enjoyment of eating, especially of sweets. Despite greater food intakes, marijuana abusers often consume fewer nutrients than do nonabusers because the extra foods they choose tend to be high-calorie, low-nutrient snack foods.

These foods and beverages all contain caffeine, but few if any of their labels state how much.

Table C14–3

Herb and Drug Interactions

Herb	Drug	Interaction
Bilberry, dong Quai, feverfew, garlic, ginger, ginkgo biloba, ginseng, meadowsweet, St. John's wort, turmeric, and willow	Warfarin, coumarin (anticlotting drugs, "blood thinners"); aspirin, ibuprofen, and other nonsteroidal anti-inflammatory drugs	Prolonged bleeding time; danger of hemorrhage
Black tea, St. John's wort, saw palmetto	Iron supplement; antianxiety drug	Tannins in herbs inhibit iron absorption; St. John's wort speeds antianxiety drug clearance.
Borage, evening primrose oil	Anticonvulsants	Seizures
Chinese herbs (xaio chai hu tang)	Prednisone (steroid drug)	Decreased blood concentrations of the drug
Echinacea (possible immunostimulant)	Cyclosporine and corticosteroids (immunosuppressants)	May reduce drug effectiveness
Feverfew	Aspirin, ibuprofen, and other nonsteroidal anti-inflammatory drugs	Drugs negate the effect of the herb for headaches.
Garlic supplements	Protease inhibitors (HIV-AIDS[a]) drug	Decreased blood concentrations of the drug
Ginseng	Estrogens, corticosteroids	Enhanced hormonal response
Ginseng, hawthorn, kyushin, licorice, plantain, St. John's wort, uzara root	Digoxin (cardiac antiarrhythmic drug derived from the herb foxglove)	Herbs interfere with drug action and monitoring.
Ginseng, karela	Blood glucose regulators	Herbs affect blood glucose levels.
Kelp (iodine source)	Synthroid or other thyroid hormone replacers	Herb may interfere with drug action.
Licorice	Corticosteroids (oral and topical ointments)	Overreaction to drug (potentiation)
Panax ginseng	Antidepressants	Overexcitability, mania
St. John's wort	Increased enzymatic destruction of many drugs; Cyclosporine (immunosuppressant); antiretroviral drugs (HIV[a] drugs), warfarin (anticoagulant, used to reduce blood clotting) MAOIs (used to treat depression, tuberculosis, or high blood pressure)[b]	Decreased drug effectiveness; increased organ transplant rejection; reduced effectiveness of drugs to treat AIDS[a], reduced anticoagulant effect. Potentiation, with serotonin syndrome (mild): sweating, chills, blood pressure spike, nausea, abnormal heartbeat, muscle tremors, seizures
Valerian	Barbiturates (sedatives)	Enhanced sedation

Note: A valuable free resource for reliable online information about herbs is offered by the Memorial Sloan-Kettering Cancer Center at www.mskcc.org/aboutherbs.

[a]Acquired Immune Deficiency Syndrome, caused by HIV infection (human immunodeficiency virus).

[b]MAOI stands for monoamine oxidase inhibitors.

In contrast to marijuana's appetite-stimulating effect, many other drugs cause loss of appetite, weight loss, and malnutrition among people who abuse them heavily. The stronger the craving for the drug, the less a drug abuser wants nutritious food. Rats given unlimited access to cocaine will choose the drug over food until they die of starvation. Drug abusers face multiple nutrition problems, and an important aspect of addiction recovery is their identification and correction.

Personal Strategy

In conclusion, when you need to take a medicine, do so wisely. Ask your physician, pharmacist, or other health-care provider for specific instructions about

Table C14–4
Caffeine Content of Selected Beverages and Foods

Beverages and Foods	Serving Size	Average (mg)
Coffee		
Brewed	8 oz	95
Decaffeinated	8 oz	2
Instant	8 oz	64
Tea		
Brewed, green	8 oz	30
Brewed, herbal	8 oz	0
Brewed, leaf or bag	8 oz	47
Instant	8 oz	26
Lipton/Nestea bottled iced tea	12 oz	10
Snapple iced tea (all flavors)	16 oz	42
Soft drinks[a]		
A&W Creme Soda	12 oz	29
Barq's Root Beer	12 oz	18
Colas, Dr. Pepper, Mr. Pibb, Sunkist Orange	12 oz	30–35
A&W Root Beer, club soda, Fresca, ginger ale, 7-Up, Sierra Mist, Sprite, Squirt, tonic water, caffeine-free soft drinks	12 oz	0
Mello Yello, Mountain Dew	12 oz	45–50
Energy drinks		
5-hour energy	2 oz	138
AMP, Red Bull	8–8.3 oz	80
Monster	16 oz	160
No Fear, Rock Star	16 oz	174
Wired X344	16 oz	344
Other beverages		
Chocolate milk or hot cocoa	8 oz	5
Starbucks Frappuccino Mocha	9.5 oz	72
Starbucks Frappuccino Vanilla	9.5 oz	64
Yoohoo chocolate drink	9 oz	3
Candies/chewing gum		
Gum, caffeinated	1 pce	95
Dark chocolate covered coffee beans	1 oz	235
Dark chocolate, semisweet	1 oz	18
Milk chocolate	1 oz	6
Java Pops	1 pop	60
White chocolate	1 oz	0
Foods		
Frozen yogurt, Ben & Jerry's coffee fudge	1 c	85
Frozen yogurt, Häagen-Dazs coffee	1 c	40
Ice cream, Starbucks coffee	1 c	50
Ice cream, Starbucks Frappuccino bar	1 bar	15
Yogurt, Dannon coffee flavored	1 c	45

[a]The FDA suggests a maximum of 65 milligrams per 12-ounce cola beverage, but this limit is not mandatory. Because products change, contact the manufacturer for an update on products you use regularly.

the doses, times, and how to take the medication—for example, with meals or on an empty stomach. If you notice new symptoms or if a drug seems not to be working well, consult your physician. The only instruction people need about tobacco and illicit drugs is to avoid them altogether for countless reasons. For drugs with lesser consequences to health, such as caffeine, use moderation.

In general, strive to live life with less chemical assistance. If you are sleepy, try a 15-minute nap or 15 minutes of stretching exercises instead of a 15-minute coffee break. The coffee will stimulate your nerves for an hour, but the alternatives will refresh your attitude for the rest of the day. If you suffer constipation, try getting enough exercise, fiber, and water for a few days. Chances are that a laxative will be unnecessary. Given adequate nutrition, rest, exercise, and hygiene, your body's ability to fine-tune itself may surprise you.

Critical Thinking

1. List all of the foods and drinks that you consume in one day that contain caffeine. Calculate your total caffeine intake. Do you think this is an appropriate caffeine intake for you? Why or why not?
2. Choose three nutrient–drug interactions that are a concern to you. Create a chart that lists the interaction and paraphrases how the interaction affects absorption, excretion and metabolism.

15 Hunger and the Global Environment

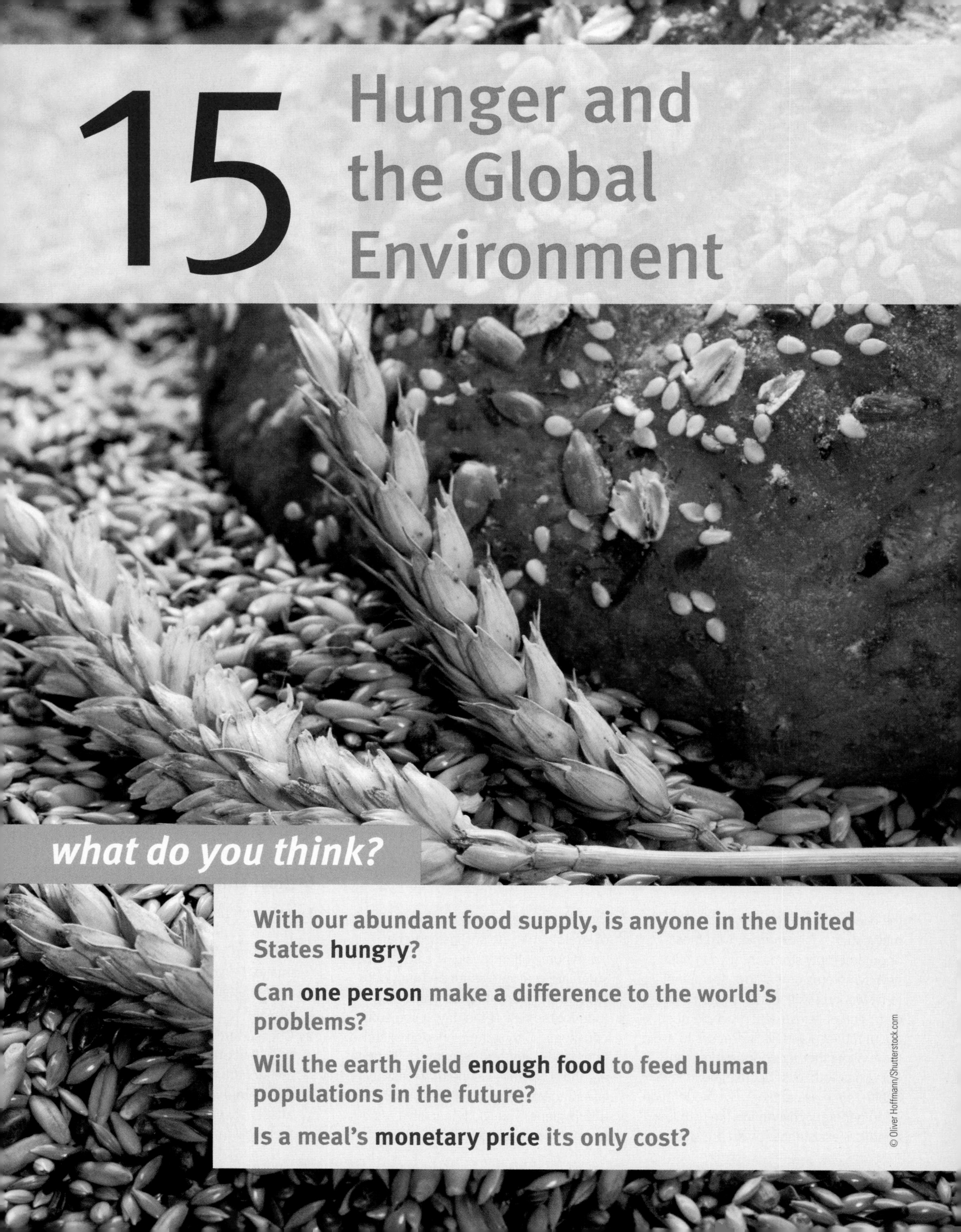

what do you think?

With our abundant food supply, is anyone in the United States **hungry**?

Can **one person** make a difference to the world's problems?

Will the earth yield **enough food** to feed human populations in the future?

Is a meal's **monetary price** its only cost?

Learning Objectives

After completing this chapter, you should be able to accomplish the following:

LO 15.1 Outline the scope and causes of food insecurity in the United States and in the world, and name some U.S. programs aimed at reducing food insecurity.

LO 15.2 Describe the severity of poverty and starvation among the world's poorest peoples.

LO 15.3 Compare and contrast characteristics of acute severe malnutrition with those of chronic malnutrition in children, and describe some basic medical nutrition therapy tools used to restore health.

LO 15.4 Explain how the food supply and environmental conditions are connected, and explain why people in poverty are inclined to have larger families despite food scarcity.

LO 15.5 Identify some steps toward solving the problems that threaten the world's food supply.

LO 15.6 Discuss several steps that governments, businesses, schools, professionals, and individuals can take to help create sustainability and stem waste in the food supply.

LO 15.7 Define the term *ecological footprint*, and list some factors that increase or decrease a person's ecological footprint.

"Never doubt that a small group of thoughtful, committed people can change the world. Indeed, it is the only thing that ever has."

—Margaret Mead

In the United States today, 6.4 million households live with **very low food security**—one or more members of these households, many of them children, repeatedly had little or nothing to eat because of a lack of money.*[1] Another 10.8 million households experienced **low food security** or **marginal food security**, somewhat less dire conditions. Food insecurity often leads to **hunger**—not the healthy appetite triggered by anticipation of a hearty meal but the pain, illness, or weakness caused by a prolonged and involuntary lack of the food.

Worldwide, the problems are much more severe. More than a billion people living in the poorest developing nations suffer chronic food insufficiency, hunger, and severe malnutrition while their neighbors are food secure or even overfed (see Table 15–1).[2] Today, many nations are facing a **food crisis** in which already meager food supplies have dwindled further, and rates of malnutrition have risen sharply.[3]

The tragedy described on these pages may seem at first to be beyond the influence of the ordinary person. What possible difference can one person make? As it turns out, quite a bit. Students in particular can play a powerful role in bringing about change. Students everywhere are helping to change governments, human predicaments, and environmental problems for the better. Student movements persuaded 127 universities and many institutions, corporations, and government agencies to put pressure on South Africa to end the racial divisions of apartheid. Students offer major services to communities through soup kitchens, home repair programs, and childhood education. The young people of today are the world's single best hope for a better tomorrow.

Table 15–1

Global Undernutrition and Overnutrition

Condition	Global Incidence
Hunger, chronic or acute malnutrition	>1 billion people
Vitamin and mineral deficiencies[a]	2.0 billion people
Overnutrition, obesity	≥1.4 billion people

[a]*Vitamin and mineral deficiencies occur in both underfed and overfed populations.*
Source: FAO 2012; World Health Organization 2012.

very low food security a descriptor for households that, at times during the year, experienced disrupted eating patterns or reduced food intake of one or more household members because of a lack of money or other resources for food. Example: a family in which one or more members went to bed hungry, lost weight, or didn't eat for a whole day because they did not have enough food.

low food security a descriptor for households with reduced dietary quality, variety, and desirability but with adequate quantity of food and normal eating patterns. Example: a family whose diet centers on inexpensive, low-nutrient foods such as refined grains, inexpensive meats, sweets, and fats.

marginal food security a descriptor for households with problems or anxiety at times about accessing adequate food, but the quality, variety, or quantity of their food intake were not substantially reduced. Example: a parent worried that the food purchased would not last until the next paycheck.

hunger physical discomfort, illness, weakness, or pain beyond a mild uneasy sensation arising from a prolonged involuntary lack of food; a consequence of food insecurity.

food crisis a steep decline in food availability with a proportional rise in hunger and malnutrition at the local, national, or global level.

* Reference notes are found in Appendix F.

Did You Know?

Nearly 15 million children in the United States—21 percent of all children—live in poverty. About 44 percent live in low-income families that are barely scraping by. In most of these families, parents work full time for low wages.

U.S. Food Insecurity

LO 15.1 Outline the scope and causes of food insecurity in the United States and in the world, and name some U.S. programs aimed at reducing food insecurity.

In the United States, poverty and low food security exist side by side with affluence and the **high food security** enjoyed by most U.S. citizens. Survey questions help to determine the existence and degree of food insecurity in the United States (offered in Table 15–2). Figure 15–1 depicts the latest survey results. In 2011, rates of food insecurity were greatest for single mothers and lowest for elderly people.[4]

Stefanie Keenan/Getty Images

Each person's efforts can help to bring about needed change.

Food Poverty in the United States

In the United States and other developed countries, hunger results primarily from **food poverty**. People go without nourishing meals not because there is no food nearby to purchase, but because they lack sufficient money to pay both for the food they need and other necessities, such as clothing, housing, medicines, and utilities. More than 12 percent of the population of the United States lives in a general state of poverty. The likelihood of food poverty increases with problems, such as abuse of alcohol and other drugs, mental or physical illness, lack of awareness of or access to available food programs, and the reluctance to accept what some perceive as "government handouts" or charity.

Limited Nutritious Food Intakes To stretch meager food supplies, adults may skip meals or cut their portions. When desperate, they may be forced to break social rules—begging from strangers, stealing from markets, consuming pet foods, or even harvesting dead animals from roadsides or scavenging through garbage cans. In the latter cases, such foods may be spoiled or contaminated and inflict dangerous foodborne illnesses that compound harm to health from borderline malnutrition. Children in such families sometimes go hungry for an entire day until the adults can obtain food.

Significant numbers of U.S. children in families of low food security consume enough calories each day but from a steady diet of inexpensive, low-nutrient foods, such as white bread, fats, sugary punches, chips, and snack cakes, with few of the fruits, vegetables, milk products, and other nutritious foods children need to be healthy. The more severe their circumstances, the more likely children are to be in poor or fair health, and the greater their likelihood of hospitalization. Such children are often behind their peers in school, develop behavioral and social problems, are slower to heal from injuries, and are more susceptible to illnesses.[5] These children often misbehave because of malnutrition or in rebellion against their circumstances, and relieving their poverty often improves educational performance and behavior.

Rising Food Costs Dwindling crop supplies and drought have recently pushed global prices of basic foods, such as cereals, sugar, and meats, to record highs.[6] Food prices in the United States rose, too, but not as steeply as in other areas of the world because costs of processed foods favored in this country are more heavily tied to labor, packaging, and transportation than to costs of basic ingredients. Still, for people struggling to make ends meet, any rise in food costs can pose problems.

The Poverty–Obesity Paradox Food insufficiency and obesity often exist side by side—sometimes within the same household or even in the same person.[7] Food insecurity and obesity may logically seem to be mutually exclusive, but research studies consistently show that the highest rates of obesity occur among those living in the greatest poverty and food insecurity.[8] With obesity comes an increased risk of developing chronic diseases, such as diabetes and hypertension. Food insecurity worsens the outlook for controlling those diseases, too.[9]

Low-income urban and rural communities that offer little or no access to affordable nutritious foods, **food deserts** (first mentioned in Chapter 9), lack access to markets that sell fresh produce.[10] Not surprisingly, people living in food deserts often lack

Table 15–2

U.S. Food Security Survey—Short Form

The food security of an individual or household lies along a continuum. The *existence* of food security can be determined by asking the first six questions below. The *degree* of food insecurity is assessed with an additional four questions that follow. Figure 15–1 shows the results of the most recent survey.

1. "The food that (I/we) bought just didn't last, and (I/we) didn't have money to get more." Was that often, sometimes, or never true for (you/your household) in the last 12 months?

[ ] Often true [ ] Sometimes true [ ] Never true

2. "(I/we) couldn't afford to eat balanced meals." Was that often, sometimes, or never true for (you/your household) in the last 12 months?

[ ] Often true [ ] Sometimes true [ ] Never true

3. In the last 12 months, since last (name of current month), did (you/you or other adults in your household) ever cut the size of your meals or skip meals because there wasn't enough money for food?

[ ] Yes [ ] No
If yes, then:

4. How often did this happen—almost every month, some months but not every month, or in only 1 or 2 months?

[ ] Almost every month
[ ] Some months but not every month
[ ] Only 1 or 2 months

5. In the last 12 months, did you ever eat less than you felt you should because there wasn't enough money for food?

[ ] Yes [ ] No

6. In the last 12 months, were you ever hungry but didn't eat because there wasn't enough money for food?

[ ] Yes [ ] No

These additional questions help to determine the severity of food insecurity:

Least severe: Was this statement often, sometimes, or never true for you in the last 12 months? "We worried whether our food would run out before we got money to buy more."
Somewhat more severe: Was this statement often, sometimes, or never true for you in the last 12 months? "We couldn't afford to eat balanced meals."
Midrange severity: In the last 12 months, did you ever cut the size of your meals or skip meals because there wasn't enough money for food?
Most severe: In the last 12 months, did you ever not eat for a whole day because there wasn't enough money for food?
In the last 12 months, did any of the children ever not eat for a whole day because there wasn't enough money for food?

Source: Economic Research Service, USDA, U.S. Household Food Security Survey Module: Six-Item Short Form, 2008. The long survey is also available at http://www.ers.usda.gov/topics/food-nutrition-assistance/food-security-in-the-us/survey-tools.aspx#household.

Figure 15–1

Food Security of U.S. Households

Most U.S. households are food secure.

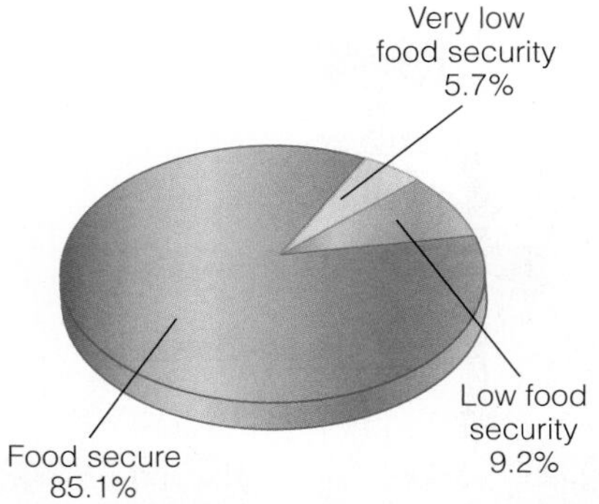

Source: A. Coleman-Jensen and coauthors, Household food security in the United States in 2011 ERS Report Summary, *September 2012, available at www.ers.usda.gov.*

fruits and vegetables in their diets and often fail to meet intake recommendations of the Dietary Guidelines.[11] High-fat, high-sugar, refined, energy-dense foods that are readily available in food deserts infamously lack other needed nutrients. Doughnuts, packaged sweet cakes, sugary punches, hamburgers, and French fries provide a full stomach, are affordable, are easily obtained at any hour, are easily carried, require no preparation, and taste good. Is it any wonder that a steady diet of such high-calorie goodies can result in both obesity and malnutrition?

Economic uncertainty and stress also influence the prevalence of obesity.[12] People who are unsure about their next meal may overeat when food or money become

high food security a descriptor for households with no problems or anxiety about consistently accessing adequate food.

food poverty hunger occurring when enough food exists in an area but some of the people cannot obtain it because they lack money, are being deprived for political reasons, live in a country at war, or suffer from other problems such as lack of transportation.

food deserts a term used to describe urban and rural low-income neighborhoods and communities that have limited access to affordable and nutritious foods.

Figure 15–2

The Poverty–Obesity Paradox

Poverty can lead to both food insecurity and obesity when food choices are limited to calorie-rich, nutrient-poor foods.

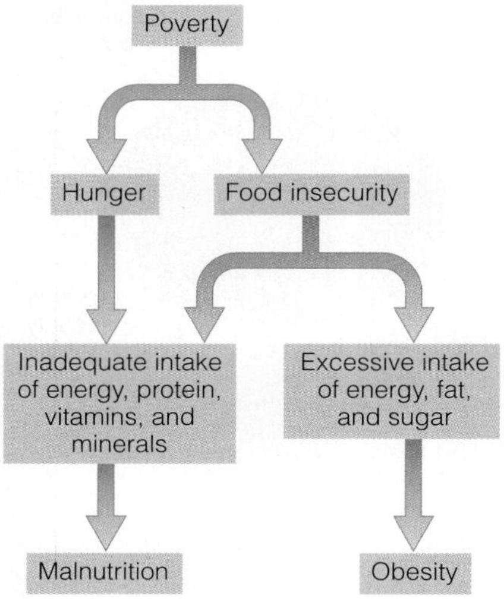

School breakfasts and lunches provide low-income children with nourishment at little or no cost.

available. Interestingly, food insecure people who do *not* participate in food assistance programs have a greater risk of obesity than those who do—it seems that providing a reliable supply of nutritious food may help to prevent obesity among those living with food insecurity.[13] Figure 15–2 shows how poverty and food insecurity can lead to both malnutrition and obesity.

KEY POINTS

- As poverty in the United States increases, food insecurity does, too.
- Children living in food-insecure households often lack the food they need.
- People with low food security may suffer obesity alongside hunger in the same community or family.

What U.S. Food Programs Address Low Food Security?

An extensive network of food assistance programs delivers life-giving food daily to tens of millions of U.S. citizens living in poverty (see Table 15–3). One of every seven Americans receives food assistance of some kind, at a total cost of almost $90 billion per year.[14] Even so, low wages, unemployment, and rising costs of food and other necessities have driven a record number of Americans into poverty or just above it.[15]

Nationwide Efforts The Academy of Nutrition and Dietetics calls for aggressive action to bring an end to U.S. food insecurity and hunger and to achieve food and nutrition security for everyone living here.[16] This premier organization of food and nutrition professionals has recently joined a nationwide initiative designed to help families obtain the food they need and reduce food insecurity in the United States.

The federal government's centerpiece food program for low-income people is the Supplemental Nutrition Assistance Program (SNAP) administered by the U.S. Department of Agriculture (USDA).[17†] It provided assistance to more than 46 million people in 2012; about half of the recipients are children.[18] Eligible households receive electronic debit transfer cards, similar to regular debit cards, through state social services or welfare agencies. Recipients can use the cards like cash to purchase food and food-bearing plants and seeds but not for tobacco, cleaning items, alcohol, or other nonfood items. To help stretch consumer food dollars and SNAP credits, the USDA also provides guidance on planning thrifty meals, complete with daily menus and recipes. Other programs have been described in previous chapters.

Currently, the USDA is studying the spending habits of SNAP recipients to determine whether they are obtaining the nutritious foods that the program intends to provide. A debate surrounds the question of whether foods and beverages of low nutrient

Table 15–3

U.S. Federal and State Food Assistance Programs

This is a sampling of national and state programs aimed at reducing hunger in the United States.

- Commodity Supplemental Food Program.
- Emergency Food Assistance Program.
- Food Distribution Program on Indian Reservations.
- National School Lunch and Breakfast Programs (see Chapter 14).
- Senior Nutrition Program (see Chapter 14).
- Special Supplemental Feeding Program for Women, Infants, and Children (WIC; see Chapter 13).
- Supplemental Nutrition Assistance Program (SNAP), formerly called the Food Stamp Program.

[†] The SNAP program was formerly known as the Food Stamp Program.

density, such as sugary soft drinks and snack cakes, should be restricted from SNAP purchase eligibility.

Community Efforts Recent budget cutbacks have caused some states to scale back or eliminate aid programs, fraying the public safety net and leaving many food-insecure people without needed help.[19] To assist where government programs fall short, concerned citizens in many communities work through local agencies and churches to help deliver food to hungry people. National **food recovery** programs, such as Feeding America, coordinate the efforts of **food banks**, **food pantries**, **emergency kitchens**, and homeless shelters that provide food to tens of millions of people a year.

A combination of various strategies can help to build food security within a community. Table 15–4 presents actions in three stages for developing a hunger-free community. The table also points out that while providing food relief to the hungry is critical, developing **sustainable** long-term solutions that attack the underlying problems of limited food access, low wages, and poverty are equally important. To rephrase a well-known adage: If you give a man a fish, he will eat for a day. If you teach him to fish so that he can buy and maintain his own gear and bait, he will eat for a lifetime and help to feed you, too.

KEY POINT

- Government programs to relieve poverty and hunger are crucial to many people, if not fully successful.

Did You Know?

For information about food pantries, food banks, and other agencies in your community, call the National Hunger Hotline: (800) GLEAN-IT.

Table 15–4

Addressing Community Hunger

Consistently applied, these goals and actions can help to curtail food insecurity within a community.

Stage 1 Short-Term Goals—Fully Use Resources Currently Available

- Make full use of assistance that is available now. Assess available federal, state, local, and private food organizations and make them known to nutrition and other professionals in the community.
- Find out why people in need are not using the services. Especially, assess accessibility of food organizations to people who need them and assess transportation systems.
- Identify food quality and price inequities in low-income neighborhoods.
- Educate consumers and institutions about using local, organic, and seasonal foods.

Stage 2 Medium-Range Goals—Network and Connect Food Relief Agencies and Others to Identify and Address Problems

- Connect emergency food programs with local urban agriculture projects to encourage collection and distribution of available nutritious foods.
- Coordinate food services with parks and recreation programs and other community outlets, such as churches, to which area residents have easy access.
- Integrate public and private hunger-relief agencies, local businesses, and others to create an emergency food delivery network; include low-income participants for input.
- Take a leadership role by serving on community food councils or homeless-relief organizations; organize workshops to educate others.

Stage 3 Long-Range Goals—Redesign Food and Other Systems for Effectiveness and Sustainability

- Mobilize government and community leaders to encourage urban agriculture to foster food self-reliance and improve nutrient intakes.
- Advocate for land-use policies and land grants that allow and encourage urban agriculture, such as community gardens and school gardens.
- Suggest improvements to public transportation to human services agencies and food resources.
- Encourage tax and other financial incentives to attract appropriate food businesses, such as farmer's markets and supermarkets, to low-income neighborhoods.
- Advocate increased minimum wage and more affordable housing.

food recovery collecting wholesome surplus food for distribution to low-income people who are hungry.

food banks facilities that collect and distribute food donations to authorized organizations feeding the hungry.

food pantries community food collection programs that provide groceries to be prepared and eaten at home.

emergency kitchens programs that provide prepared meals to be eaten on-site; often called *soup kitchens.*

sustainable able to continue indefinitely; the use of resources in ways that maintain both natural resources and human life into the future; the use of natural resources at a pace that allows the earth to replace them and does not cause pollution to accumulate.

World Poverty and Hunger

LO 15.2 Describe the severity of poverty and starvation among the world's poorest peoples.

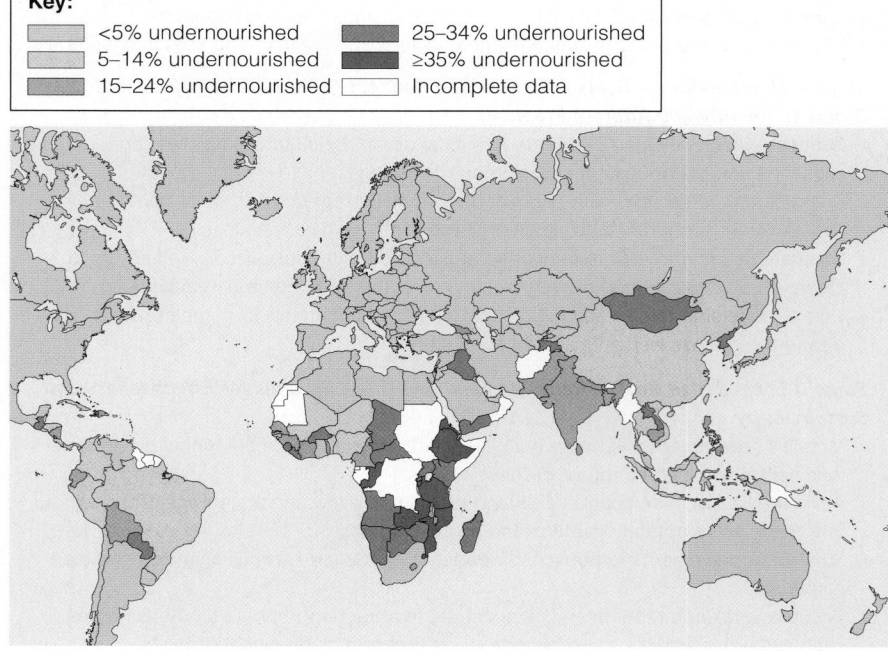

Clean water is precious and lifesaving in many areas of the world.

In the developing world, poverty and hunger are intense and may worsen if global economies slump.[20] Figure 15–3 points out which nations of the world suffer most from insufficiency. The primary problem is still food poverty, but in the hardest hit areas, the poverty is extreme.

The Staggering Statistics Grasping the severity of poverty in the developing world can be difficult, but some statistics may help. One-fifth of the world's 7 billion people have no land and no possessions *at all*. The "poorest poor" survive on less than two dollars a day, they lack water that is safe to drink, and they cannot read or write.[21] Many spend about 80 percent of all they earn on food, but still they are hungry and malnourished. The average U.S. house cat eats twice as much protein every day as one of these people, and the yearly cost of keeping that cat is greater than that person's annual income.

The recent world economic downturn has sent food prices even further out of reach—food prices rose substantially between 2006 and 2011 with a strong likelihood of still higher prices to come.[22] High food prices partly reflect a worldwide economic downturn and a shift to producing **biofuels** from corn and other food crops. Another contributor is drought and flooding that has reduced grain production in many areas.

Women and Children The "poorest poor" are usually women and children. Many societies around the world undervalue females, providing girls with poorer

Figure 15–3

World Hunger Map

Today, more than one billion of the world's people go hungry. Hunger is prevalent in the developing world, with some countries reporting hunger and malnutrition in more than half of their population.

Key:
- <5% undernourished
- 5–14% undernourished
- 15–24% undernourished
- 25–34% undernourished
- ≥35% undernourished
- Incomplete data

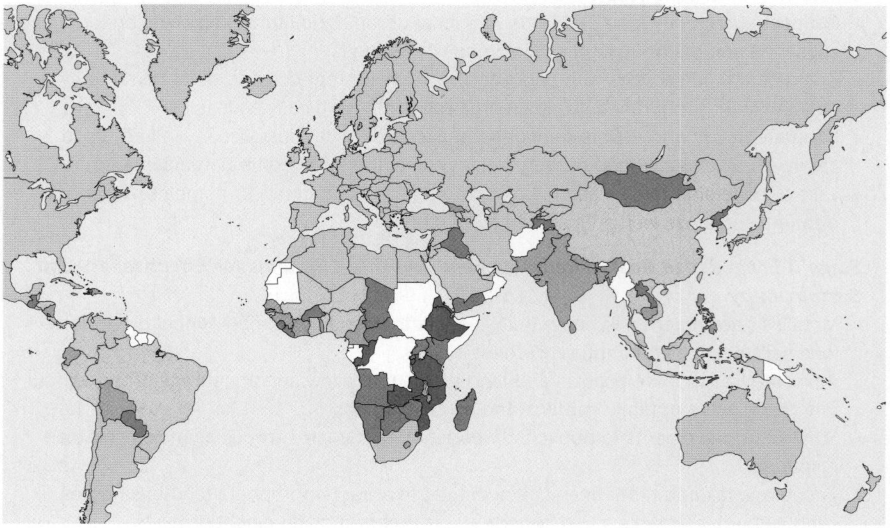

Source: FAO, IFAD and WFP. 2012. The State of Food Insecurity in the World 2012: Economic growth is necessary but not sufficient to accelerate reduction of hunger and malnutrition. *Rome.*

diets and fewer opportunities than boys. Malnourished girls become malnourished mothers who give birth to low-birthweight infants—so the cycle of hunger, malnutrition, and poverty continues. Worldwide, three-fourths of those who die each year from starvation and related illnesses are children.[23] Those who survive simply cannot work hard enough to get themselves out of poverty. Most have no borrowing power, even if credit were available, and lack the money needed to build even small businesses and incomes.[24]

An irony of poverty is that it drives people, even those without sufficient food, to bear more children. An impoverished family depends on its children to farm the land, haul water, and care for the adults in their old age. However, malnutrition and disease cause many young children to die, so parents will have many children to ensure that some will survive to adulthood.[25]

© Jeremy Horner/Corbis

In some countries, every pair of little hands is needed to help feed the family.

Famine The most visible form of hunger is **famine**, a true food crisis in which multitudes of people in an area starve and die. The natural causes of famine—drought, flood, and pests—occur, of course, but they take second place behind the political and social causes.[26] For people of marginal existence, a sudden increase in food prices, a drop in workers' incomes, or a change in government policy can quickly leave millions of people hungry. The World Food Programme of the United Nations responds to food emergencies around the globe.

Intractable hunger and poverty remain enormous challenges to the world. In parts of Africa, killer famines recur whenever human conflict converges with drought in a country that has little food in reserve even in a peaceful year. Racial, ethnic, and religious hatred along with monetary greed often underlie the food deprivation of whole groups of people.[27] Farmers become warriors, and agricultural fields become battlegrounds while citizens starve. Food becomes a weapon when warring factions repel international famine relief in hopes of starving their opponents before they themselves succumb.

KEY POINTS

- Natural causes, along with political and social causes, contribute to hunger and poverty in many developing countries.
- Women and children are generally the world's poorest poor.

The Malnutrition of Extreme Poverty

LO 15.3 Compare and contrast characteristics of acute severe malnutrition with those of chronic malnutrition in children, and describe some basic medical nutrition therapy tools used to restore health.

In the world's most impoverished areas, persistent hunger inevitably leads to malnutrition. A huge number of adults suffer day to day from the effects of malnutrition, but medical personnel often fail to properly diagnose these conditions. Most often, adults with malnutrition feel vaguely ill; they lose fat, muscle, and strength—they are thin and getting thinner.[28] Their energy and enthusiasm is sapped away. With unrelenting food shortages, observable nutrient deficiency diseases develop.

"Hidden Hunger"—Micronutrient Deficiencies

Almost 2 billion people worldwide who consume sufficient calories still lack the variety and quality of foods needed for good nutrition.[29] Nutrient deficiency diseases become apparent as body systems begin to fail. Iron, iodine, vitamin A, and zinc are most commonly lacking, and the results can be severe—learning disabilities, mental retardation, impaired immunity, blindness, incapacity to work, and premature death.[30]

biofuels fuels made mostly of materials derived from recently harvested living organisms. Examples are *biogas*, *ethanol*, and *biodiesel*.

famine widespread and extreme scarcity of food that causes starvation and death in a large portion of the population in an area.

Donated food may temporarily ease hunger for some, but it is usually sporadic and insufficient to prevent nutrient deficiencies or support growth.

The scope of nutrient deficiencies among adults and children is almost impossible to imagine:

- 30 percent or more of the world's population have iron-deficiency anemia, a leading cause of maternal deaths, preterm births, low birthweights, infections, and premature deaths.
- 40 million newborns every year have irreversible mental retardation (cretinism) from iodine deficiency.
- 5 million or more children (younger than age 5) suffer from vitamin A deficiency so severe that they are rendered permanently blind.[31] 180 million more have marginally poor status that reduces their resistance to infections, such as measles.
- 20 percent of the world's population suffers from zinc deficiency that contributes to growth failure, diarrhea, and pneumonia. (The deficiency symptoms of these and other nutrients are presented in Chapters 7 and 8.)

These conditions are devastating, not only to individuals but also to entire nations. When people suffer from mental retardation, blindness, infections, and other consequences of malnutrition, both personal and national economies decline as productivity ceases and health-care costs soar. High infant mortality and short life expectancies plague malnourished nations, as well.

KEY POINTS

- Malnutrition in adults most often appears as general thinness and loss of muscle.
- Individual nutrient deficiencies cause much misery worldwide.

Two Faces of Childhood Malnutrition

In contrast to malnourished adults, young impoverished and malnourished children often exhibit specific, more readily identifiable conditions. The form malnutrition takes in a hungry child depends partly on the nature of the food shortage that caused it. The most perilous condition, **severe acute malnutrition (SAM)**, occurs when food suddenly becomes unavailable, such as in drought or war. Less immediately deadly but still damaging to health is **chronic malnutrition**, the unrelenting chronic food deprivation that occurs in areas where food supplies are usually scanty and food quality is low. Table 15–5 compares key features of SAM with those of chronic malnutrition.

Severe Acute Malnutrition Ten percent of the world's children suffer from SAM, identified by their degree of wasting. In children with SAM, lean and fat tissues have wasted away, burned off to provide energy to stay alive. They weigh too little for their height, and their upper arm circumference measures smaller than normal.[32] In these children, loose skin on the buttocks and thighs often sags down and looks as if the child is wearing baggy pants. Sadly, such children are described as just "skin and bones." The child often feels cold, appears withdrawn or irritable, and is obviously ill.

The starving child faces this threat to life by engaging in as little activity as possible—not even crying for food. The body musters all its forces to meet the crisis, so it cuts down on any expenditure of energy not needed for the functioning of the heart, lungs, and brain. Enzymes are in short supply, and the GI tract lining deteriorates. Consequently, what little food is eaten often cannot be absorbed.

Each year, 7.6 million preschool-aged children, as many as five children *every minute*, die as a result of SAM. Most of them do not starve to death—they die from the diarrhea and dehydration that accompany infections.

Chronic Malnutrition A much greater number, 25 percent of children worldwide, live with chronic malnutrition.[33] They subsist on diluted cereal drinks that supply scant energy and even less protein; such food allows them to survive but not to thrive. Growth ceases because they chronically lack the nutrients required to grow normally.[34] These stunted children may be no larger at age 4 than at age 2, and often suffer the miseries of malnutrition: increased risks of infection and diarrhea, and vitamin and mineral deficiencies.

Table 15–5

Characteristics of Severe Acute Malnutrition and Chronic Malnutrition[39]

	Severe Acute Malnutrition	Chronic Malnutrition
FOOD DEPRIVATION	Current or recent	Long term
PHYSICAL FEATURES	Rapid weight loss Wasting (underweight for height; small upper arm circumference) Edema	Minimal height gains Stunting (short for age)
WORLD PREVALENCE OF CHILDREN YOUNGER THAN AGE 5	5% to 15%	20% to 50%

Note: Vitamin and mineral deficiencies are common in both types of malnutrition.

A fully nourished developing brain normally grows to almost its full adult size within the first 2 years of life. When malnutrition occurs during these years, it can impair brain development and learning ability, effects that may become irreversible.

Kwashiorkor and Marasmus Historically, severe childhood malnutrition was divided between **kwashiorkor**, attributed to protein deficiency, and **marasmus**, attributed to energy deficiency. Traditional thinking has changed, however, because the meager diets of children with these conditions do not differ much—both lack protein and many other nutrients.[35] The following paragraphs describe the symptoms of kwashiorkor and marasmus, two clinical expressions of malnutrition that can appear individually or together in a starving child.[36]

Kwashiorkor is often observed when a mother who has been nursing her first child bears a second child; she weans the first child and puts the second one on the breast. The first child, suddenly switched from nutrient-dense, protein-rich breast milk to a thin gruel of starchy, protein-poor cereal, soon begins to sicken and die. With too little high-quality protein, concentrations of the blood protein albumin fall, causing fluids to shift out of the blood and into the tissues—edema. Some muscle wasting may occur, but it may not be apparent because the child's face, limbs, and abdomen become swollen with edema—a distinguishing feature of kwashiorkor. Loss of hair color is also common because melanin, a dark pigment, is made from the amino acid tyrosine. In addition, telltale patchy and scaly skin develops, often with sores that fail to heal.

Marasmus, appropriately named from the Greek word meaning "dying away," reflects a severe, unrelenting deprivation of food observed in children living in overpopulated and impoverished nations. Children living in severe poverty simply do not have enough food to protect body tissues from being degraded for energy.

KEY POINTS

- Malnutrition in adults is widespread but is often overlooked; severe observable deficiency diseases develop as body systems fail.
- Many of the world's children suffer from wasting, the deadliest form of malnutrition, called severe acute malnutrition.
- In many more children, growth is stunted because they chronically lack the nutrients children need to grow normally.

Rehabilitation

Loss of appetite and impaired food assimilation interfere with any attempts to provide nourishment to a malnourished child, so caring, individualized treatment works best.[37] To restore metabolic balance, physical growth, mental development, and recovery from illnesses, malnourished children need specially formulated fluids and foods.[38] SAM, particularly when complications develop, demands hospitalization, including intensive nursing care, medical nutrition therapy, and medication.[39]

Children dehydrated from diarrhea need immediate rehydration. In severe cases, dramatic fluid and mineral losses cause the blood pressure to drop and heartbeat to

severe acute malnutrition (SAM) malnutrition caused by recent severe food restriction; characterized in children by underweight for height (wasting). *Moderate acute malnutrition* is a somewhat less severe form.

chronic malnutrition malnutrition caused by long-term food deprivation; characterized in children by short height for age (stunting).

kwashiorkor (kwash-ee-OR-core, kwash-ee-or-CORE) severe malnutrition characterized by failure to grow and develop, edema, changes in the pigmentation of hair and skin, fatty liver, anemia, and apathy.

marasmus (ma-RAZ-mus) severe malnutrition characterized by poor growth, dramatic weight loss, loss of body fat and muscle, and apathy. Collectively, kwashiorkor and marasmus may be called *protein-energy malnutrition*.

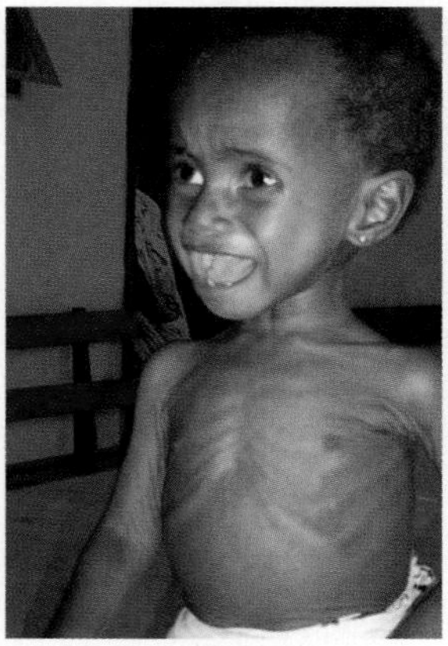

Courtesy Kids Against Hunger

This 2-year-old girl was suffering from severe acute malnutrition.

Courtesy Kids Against Hunger

After a few weeks of nutrition therapy, she gained substantial weight and health along with a new appetite for living.

weaken. The right fluid, given quickly by knowledgeable providers, can help raise the blood pressure and strengthen the heartbeat, thereby averting disaster. Health-care workers around the world save millions of lives each year by effectively reversing dehydration and correcting the diarrhea with **oral rehydration therapy (ORT)**. ORT is a simple, inexpensive, and effective treatment that consists of giving a sugar and salt solution orally. In addition, such children need adequate sanitation and a safe water supply to prevent future infectious diseases.

Once medically stable, malnourished children benefit from **ready-to-use therapeutic food (RUTF)**, commercial products intended to promote rapid reversal of weight loss and nutrient deficiencies.[40] Manufacturers blend smooth pastes of oil and sugars with ground peanuts, powdered milk, or other protein sources and seal pre-measured single doses in sterilized pouches. RUTF need not be mixed with water or prepared in any way, and the pouches resist bacterial contamination. Importantly, RUTF can be safely stored for 3 to 4 months without refrigeration, a rare luxury in many impoverished areas. RUTF may also be used to *prevent* childhood malnutrition in some at-risk populations.[41] RUTF's downside is cost: commercial products are expensive to buy and ship to impoverished areas. Currently, researchers are working on ways of producing RUTF pastes on site from locally available products and at greatly reduced cost.[42] Ideally, optimal breastfeeding for all infants and improved complementary foods for toddlers would prevent malnutrition and save lives before emergencies develop.

KEY POINTS

- Oral rehydration therapy and ready-to-use therapeutic foods, properly applied, can save the life of a starving person.
- Many health and nutrition professionals work to eradicate childhood malnutrition.

The Future Food Supply and the Environment

LO 15.4 Explain how the food supply and environmental conditions are connected, and explain why people in poverty are inclined to have larger families despite food scarcity.

Banishing hunger for all of the world's citizens poses two major challenges. The first is to provide enough food to meet the needs of the earth's expanding population, without destroying natural resources needed to continue producing food. The second challenge is to ensure that all people have access to enough nutritious food to live active, healthy lives.

By all accounts, today's total **world food supply** can abundantly feed the entire current population. Adequate supply alone, however, does not ensure that all people will receive adequate food. The world must develop the intention and momentum to ensure it.

Threats to the Food Supply

Many forces compound to threaten world food production and distribution, both today and in coming decades. The following list names just some of them.

- *Hunger, poverty, and population growth.* Every 60 seconds 105 people die in the world, but in that same 60 seconds 254 are born to replace them.[43] Every year, the earth gains another 78,528,062 new residents to feed (see Figure 15–4), most of them born in impoverished areas.[44] By 2050, 1 billion additional tons of grains will be needed to feed the world's population.
- *Loss of food-producing land.* Food-producing land is becoming saltier, eroding, and being paved over. Each year, the world's farmers try to feed almost 79 million additional people with 24 billion fewer tons of topsoil.
- *Accelerating fossil fuel use.* Fossil fuel use is growing rapidly, with associated pollution of air, soil, and water.[45]
- *Atmosphere and global climate change.* That climate change is occurring is no longer a serious academic debate.[46] In a series of recent reports, the National Academies of Science conclude that "there is a strong, credible body of evidence, based on multiple lines of research, documenting that Earth is warming. Strong evidence also indicates that recent warming is largely caused by human activities, especially the release of greenhouse gases through the burning of fossil fuels."[47] Climbing atmospheric levels of heat-trapping carbon dioxide and other greenhouse gases are a serious concern.[48] The concentration of carbon dioxide is rising, now 26 percent higher than 200 years ago. The associated heat waves, droughts, fires, storms, and floods thwart farmers and destroy crops, particularly in the poorest areas of the world. Arid deserts are projected to expand by 200 million acres in coming years in sub-Sarahan Africa alone. As ocean heat builds up, ocean food chains may fail.

Figure 15–4
World Population Growth

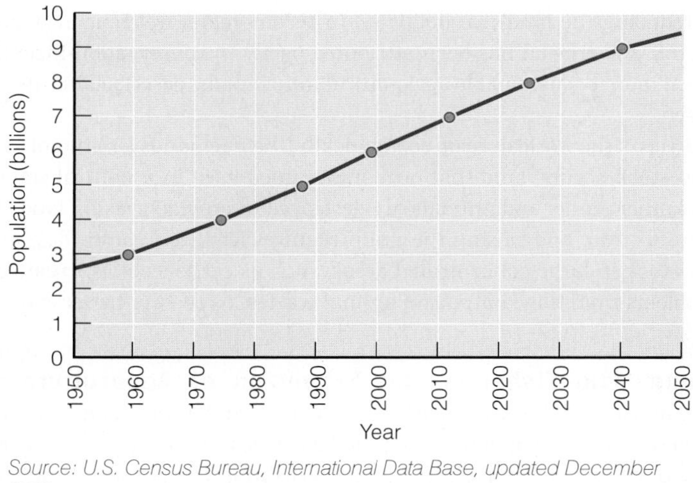

Source: U.S. Census Bureau, International Data Base, updated December 2008.

oral rehydration therapy (ORT) oral fluid replacement for children with severe diarrhea caused by infectious disease. ORT enables parents to mix a simple solution for their child from substances that they have at home. A simple recipe for ORT: ½ L boiled water, 4 tsp sugar, ½ tsp salt.

ready-to-use therapeutic food (RUTF) highly caloric food products offering carbohydrate, lipid, protein, and micronutrients in a soft-textured paste used to promote rapid weight gain in malnourished people, particularly children.

world food supply the quantity of food, including stores from previous harvests, available to the world's people at a given time.

- *Ozone loss from the outer atmosphere.* The outer atmosphere's protective ozone layer is thinning, permitting more harmful radiation from the sun to penetrate. As radiation increases the earth's temperature, polar ice caps are melting, threatening the world's coastlines. Radiation may also directly damage important crops.
- *Fresh water shortages; increased flooding.* The world's supplies of fresh water, critical to human health and life, are dwindling and in many areas becoming more polluted; over a billion people lack access to fresh water today, and over the next 20 years, the average supply of fresh water per person is expected to decline by one-third. At the same time, crop-damaging localized heavy storms with flash floods that run off parched land are expected to gain in frequency.[49]
- *Ocean pollution.* Ocean pollution, particularly agricultural and industrial runoff, is killing fish in large "dead zones" that grow and recede along the world's coasts; overfishing is depleting the fish that remain.[50]

The global problems just described are all related and, often, so are their solutions. To think positively, this means that any initiative a person takes to address one problem will help solve many others.

No part of the world is safely insulated against future food shortages. Developed countries may be the last to feel the effects, but they will ultimately go as the world goes. To limit the threat from climate change will require no less than a major shift in how the world uses and produces energy.[51] This chapter's Controversy highlights one part of that larger effort: new approaches to continued food abundance through sustainable agriculture.

© paul prescott/Shutterstock.com

KEY POINTS

- The world's current food supply is sufficient, but distribution remains a problem.
- Future food security is threatened by many forces.

Environmental Degradation and Hunger

Hunger and poverty interact with a third force—environmental degradation. Poor people often destroy the very resources they need for survival. Desperate to obtain money for food, they sell everything they own—even the seeds that would have produced next year's crops. They cut all available trees for firewood or timber to sell and then lose the topsoil to erosion. Without these resources, they become still poorer. Thus, poverty causes environmental ruin, and the ruin leads to hunger.

Soil Erosion and Grazing Lands Soil erosion affects agriculture in every nation. Deforestation of wild areas dramatically adds to soil loss. Without the forest covering to hold the soil in place, it washes off the rocks or sand beneath, drastically and permanently reducing the land's productivity. In recent years, welcome evidence of the slowing of U.S. soil erosion has been attributed partly to conservation-incentive policies and even more to sustainable agricultural innovations, described in this chapter's Controversy.

As countries of the world develop economically, their demands for animal foods soar. Herds of livestock occupy land that once maintained itself in a natural state. Native grasses and other plants and animals are destroyed to provide grazing land. The animals eat grains, too, and raising the grain requires fertilizers, pesticides, and other inputs. Livestock in large concentrated areas, such as cattle feedlots, create environmental problems from huge outputs of animal wastes, as the Controversy also makes clear.

Diminishing Wild Fisheries and Expansion of Aquaculture Despite intensive expansion of the world's fishing industry, catches of ocean fish are diminishing.[52] Overall, 80 percent of the world food fish stocks are fully exploited or overexploited. The obvious solution is simple: stop overfishing. The problem is how to do so.

International fish species protection guidelines for seasonal species quotas, establishment of "no fishing zones" in oceanic breeding and recovery sanctuaries, and rules against illegal harvesting are in place. What is lacking is the political will to enforce them.

Pollution also limits wild fish populations, and the problem spans the globe. Tiny ocean plants, phytoplankton, that form the base of the food chain diminish as ocean temperature rises. In addition, as a result of climate change, greater storm surges, rising sea levels, more land runoff from flooding, and increasing numbers of parasites and microbes threaten food fish stocks.[53]

Diminished wild fish capture and large potential profits have spurred the rapid growth of **aquaculture** businesses, which now provide almost half of the world's food fish and shellfish.[54] Some aquaculture "fish farms" consist of vast net cages that enclose fish in ocean inlets or freshwater lakes, where natural water flow refreshes the cages. Other types house fish in artificial ponds of various shapes positioned inland close to coast lines. Natural water is diverted through the ponds, bringing in fresh water and washing wastes into streams, lakes, or oceans. Farther inland, pond water is continuously filtered and cleansed. All farmed fish must be fed chow that contains fish, such as sardines, harvested from wild stocks, diverting them from direct human use and from larger wild fish species, such as cod, that depend on them.[55] Adequate regulations and safeguards are a must to prevent overfishing for fish food and to limit the environmental degradation from aquaculture.

Did You Know?

To protect overfished species, consumers can choose these fish options most often:

- Alaskan halibut.
- Alaskan salmon.
- Sardines and other small species.
- Farm-raised tilapia and other farm-raised fish.

Limited Fresh Water As human populations grow, so does the demand for fresh water, but fresh water supplies are limited. In areas of high **water stress**, natural and manmade influences converge to limit access to safe drinking water.[56] Today, over 800 million people must rely on water drawn directly from unprotected and often contaminated ditches, streams, or wells.[57] Poor water management, particularly in agriculture and industry, causes many of the world's water problems, and global fresh water levels are dropping.[58] Each day, people dump 2 million tons of waste into the world's fresh water rivers, lakes, and streams. By 2025, if present patterns continue, two of every three persons on earth will live in water-stressed conditions.

© David Aleksandrowicz/Shutterstock.com

As groundwater is used up, deserts spread.

Overpopulation At the present rate of increase, the human population will exceed the earth's estimated **carrying capacity** by 2033. It took 2 million years for the earth's human population to reach 1 billion; 100 years more to reach 2 billion, and less than 100 additional years to reach its current 7 billion people. Is it any wonder that fresh water and food supplies may one day fall behind demand?

Mark Burnett/Science Source

An open net cage houses fish in the ocean or a lake, where natural flow refreshes the cages.

aquaculture the farming of aquatic organisms for food, generally fish, mollusks, or crustaceans, that involves such activities as feeding immature organisms, providing habitat, protecting them from predators, harvesting, and selling or consuming them.

water stress a measure of the pressure placed on water resources by human activities such as municipal water supplies, industries, power plants, and agricultural irrigation.

carrying capacity the total number of living organisms that a given environment can support without deteriorating in quality.

© Larry Ulrich

Pure rivers, lakes, and streams represent irreplaceable water resources.

Pascal Parrot/Getty Images

Unclean water and poor sanitation cause illnesses that claim the lives of 4,400 preschool-aged children each day.

"Facts are stubborn things; and whatever may be our wishes, our inclinations, or the dictates of our passions, they cannot alter the state of facts and evidence."

—John Adams, 1770

Relieving poverty and hunger may be necessary in curbing population growth. With greater wealth, more children survive, and families are free to choose to limit their numbers. Wealth distribution matters, too. In countries where economic growth has benefited only the rich, birth rates among the rest of the population have remained high.

Food Waste In a hungry world, 1.3 billion tons of nourishing food, one-third of total annual production, is wasted each year, squandering not just the food but the resources spent to produce, package, and transport it.[59] More than 25 percent of all the fresh water used each year is spent producing food that is ultimately wasted. Similarly, about 300 million barrels of oil is spent to fuel the production of that wasted food.

The scope of U.S. food waste is enormous (see Figure 15–5). Discarded food constitutes the single greatest component of municipal waste—even greater than yard wastes or plastics.[60] As food waste decomposes, it generates both methane and carbon dioxide, greenhouse gases that contribute to climate change.

The old proverb, "Waste not, want not," seems to apply: preventing even half of the current food waste would provide food for huge numbers of people without using a single additional acre of farmland, drop of water, or barrel of oil. Better food planning, purchasing, and use by U.S. food service industries and consumers is needed to put food where it belongs: on the plates of hungry people.[61] Figure 15–6 illustrates food recovery methods for food industries, and Table 15–6 provides a guide for individuals in both preventing food waste and saving money.

Figure 15–5

U.S. Food Waste—Calories Per Capita

About 40 percent of the food produced in the United States each year is wasted. For each person, daily food waste amounts to 1,400 calories worth of food, easily enough to cover the energy needs of a hungry child.

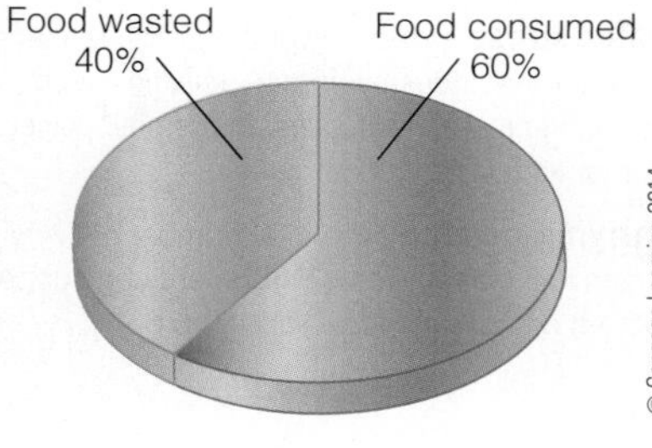

© Cengage Learning 2014

KEY POINTS

- Environmental degradation caused by people is threatening the world's food and water supplies.
- Demand for food and water grows with human population growth.
- Food waste is enormous, and reducing it would increase the food supply without additional production inputs.

Figure 15–6

Food Recovery Hierarchy

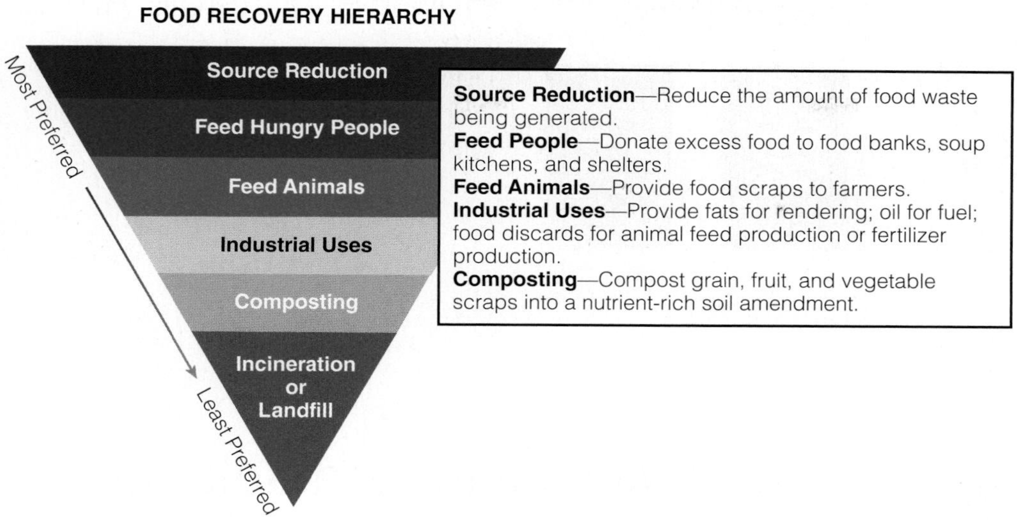
FOOD RECOVERY HIERARCHY
Most Preferred
Least Preferred
Source Reduction
Feed Hungry People
Feed Animals
Industrial Uses
Composting
Incineration or Landfill
Source Reduction—Reduce the amount of food waste being generated.
Feed People—Donate excess food to food banks, soup kitchens, and shelters.
Feed Animals—Provide food scraps to farmers.
Industrial Uses—Provide fats for rendering; oil for fuel; food discards for animal feed production or fertilizer production.
Composting—Compost grain, fruit, and vegetable scraps into a nutrient-rich soil amendment.

Source: Environmental Protection Agency, Generators of food waste, April 26, 2012, available at www.epa.gov.

Table 15–6

How to Stretch Food Dollars and Reduce Waste

Eating well on a budget can pose a challenge, but reducing waste is a good first step. For daily menus and recipes for healthy, thrifty meals, visit the USDA Center for Nutrition Policy and Promotion: www.cnpp.usda.gov.

Plan Ahead	
▪ Plan your menus, write grocery lists, and shop only for foods on your list to avoid expensive "impulse" buying. ▪ Center meals on whole grains, legumes, and vegetables; use smaller quantities of meat, poultry, fish, or eggs. ▪ Use cooked cereals such as oatmeal instead of ready-to-eat breakfast cereals.	▪ Cook large quantities when time and money allow; freeze portions for convenient later meals. ▪ Check for sales, and use coupons for products you need; plan meals to take advantage of sale items.
Shop Smart	
▪ Do not shop when hungry. ▪ Select whole foods instead of convenience foods (raw whole potatoes instead of refrigerated prepared mashed potatoes, for example). ▪ Try store brands. ▪ Buy fresh produce in season; buy canned or frozen items at other times. ▪ Buy large bags of frozen items or dry goods; use as needed and store the remainder.	▪ Buy fat-free dry milk; mix and refrigerate quantities needed for a day or two. Buy fresh milk by the gallon or half gallon only if you can use it up before it spoils. ▪ Buy less expensive cuts of meat, such as beef chuck and pork shoulder roasts; cook with liquid long enough to make the meat tender. ▪ Buy whole chickens instead of pieces; ask a butcher to show you how to cut them up. ▪ Frequent discount stores instead of grocery stores for nonfood items such as toilet paper and detergent.
Reduce Waste	
▪ Change your thinking from "what do I want to eat" to "what do I have available to eat." You paid for the food you have on hand, so use it up. ▪ Buy only the amount of fresh foods that you will eat before it spoils. ▪ Peel away the tough outer layers from stems of asparagus and broccoli; slice and cook the tender stems or add raw to salads. ▪ Scrub, but don't peel, potatoes before cooking—the skins add color, texture, and nutrients to the dish.	▪ Before buying food in bulk, plan how to store it properly. If it spoils before use, you'll throw away your savings. ▪ If your "bargain" bulk food is more than you can use but is still fresh, donate it to your local food bank or homeless shelter. (It won't save you money, but it will provide a wealth of satisfaction.) ▪ If space permits, compost fruit and vegetable scraps to feed shrubs and other outdoor plants.

My Turn watch it!

How Responsible Am I?

Listen to two students talk about what they think about individual responsibility with respect to the environment.

© Cengage Learning

Catie

© Cengage Learning

Jessica

"For every person in the world to reach present U.S. levels of consumption with existing technology would require four more planet Earths."

—E. O. Wilson, 2002

A World Moving toward Solutions

LO 15.5 Identify some steps toward solving the problems that threaten the world's food supply.

Slowly but surely, improvements are becoming evident in developing nations. For example, adult literacy rates have increased by more than 50 percent in some areas since 1970, and the proportion of children being sent to school has risen. Some encouraging progress, particularly in Latin American, Caribbean, Asian, and Pacific nations, has taken place, but much more work is needed to alleviate hunger in other parts of the world. Today, the keys to solving the world's poverty, hunger, and environmental problems are within the reach of both poor and rich nations—if they have the will to employ them.

"We do not inherit the earth from our ancestors, we borrow it from our children."

Ascribed to Chief Seattle, a 19th-century Native American leader

In light of these realities, rich nations need to stem their wasteful and polluting use of resources now, while emerging nations need to quickly develop and adhere to sustainable plans for their energy use, economic development, and industrial growth. Improving all nations' economies is a prerequisite to meeting the world's other urgent needs: population stabilization, arrest of environmental degradation, sustainable use of resources, and relief of hunger.

KEY POINT

- Improvements are slowly taking place in many parts of the world.

How Can People Help?

LO 15.6 Discuss several steps that governments, businesses, schools, professionals, and individuals can take to help create sustainability and stem waste in the food supply.

© marilyn barbone/Shutterstock.com

Every segment of our society can play a role in the fight against poverty, hunger, and environmental degradation. The federal government, the states, local communities, big business and small companies, educators, and all individuals, including dietitians and food service managers, have many opportunities to drive the effort forward.

Government Action

Government policies can change to promote sustainability. For example, the U.S. government is currently devoting record tax dollars and other resources to encouraging development of wind, solar, biofuels, and other sustainable energy sources to reduce reliance on fossil fuels. It also subsidizes conservation programs for agricultural lands. However, more can be done.[62]

Private and Community Enterprises

Businesses can take initiatives to help; some already have—AT&T, Prudential, and Kraft General Foods are major supporters of antihunger programs. Restaurants and other food facilities can plan for less food waste and participate in the nation's gleaning effort by giving their fresh leftover foods to community distribution centers and using other food recovery methods. Food producers are more often choosing to produce their goods sustainably to meet a growing demand for products produced with integrity.

Educators and Students

Educators, including nutrition educators, have a crucial role to play. The nation and world look to scientists to solve problems and innovate for the future, so a solid science curriculum is critical for students at every level of education.[63] While still learning, students can share the knowledge they gain with families, friends, and communities and take action in their communities and beyond.

Food and Nutrition Professionals

Registered dietitians, dietetic technicians, and food service managers can promote sustainable production of food and the saving of resources through choices in procurement, reuse, recycling, energy conservation, water conservation, leadership, and capital improvements both in business and in their personal lives. In addition, the Academy of Nutrition and Dietetics urges its members to work for policy changes in private and government food assistance programs, to intensify education about hunger, and to be advocates on the local, state, and national levels to help end hunger in the United States.

Individuals

All individuals can become involved in these large trends. Many small decisions each day add up to large impacts on the environment. The Consumer's Guide sums up some of these decisions and actions.

KEY POINT

- Government, business, educators, and individuals have opportunities to promote sustainability worldwide and wise resource use at home.

use it! A Consumer's Guide To . . .

Making "Green" Choices

Like the word *natural* on food labels, popular *green* label terms, such as *eco-friendly* and *environmentally neutral*, are not meaningful without solid scientific evidence to back them up. For other terms, such as *biodegradable* and *recyclable*, the Federal Trade Commission issues guidelines for their use, but adherence is voluntary.[1] Consumers of "green" goods on grocery shelves, in home improvement centers, and even in automobile showrooms often believe unquestioningly that buying these products has specific and far-reaching benefits for the environment.

Sometimes, a "green" label reflects a sincere effort by a manufacturer to mitigate environmental harm from the production and use of their goods. These claims are generally specific: "Made with 60% recycled material," for example. Too often, other marketing ploys amount to "greenwashing"—a shallow use of vague, nonspecific terms or catchy symbols that suggest environmental concern, but are intended only to hook unsuspecting consumers and increase profits. Recently, the FTC has put such businesses on alert by bringing several false advertising lawsuits against them.

Less Buying, More Doing

As it turns out, the most beneficial choice for the environment often involves less buying and more doing—a trade many consumers are reluctant to make. Viewed from a broader perspective, the benefits from simple "green" lifestyle actions are not purely altruistic; they can also greatly benefit your health and your budget, as well as your planet.

New Daily Habits

Here are some of these environmentally conscious lifestyle choices—you can probably think of others.

- Ride a bike to work or classes instead of going to the gym to make the best use of time.
- Shop "carless," to save money on gasoline—share rides, use public transportation, bicycle, or walk.
- Reduce food waste (review Table 15–6, p. 611). This single effort saves substantial money and conserves both the food itself and the resources required to make and ship it.
- Carry clean reusable string or cloth grocery sacks when biking, or keep them in the car. Even clean plastic sacks from the store can be reused, and when they wear out, they can be recycled.
- Use fewer electric gadgets. Mix batters, chop vegetables, and open cans by hand to improve hand strength and use less electricity.
- Eat more plant foods and less animal foods. Most people need more of the nutrients that these foods provide; they cost less and require far fewer resources to produce, too.

Choosing Wisely

- Choose small fish most often. Small fish often contain fewer toxins than big fish, and choosing them helps preserve stocks of large species.
- Choose minimally packaged items; buy bulk items or those with reusable or recyclable packaging. Packaging uses resources to produce, is bulky to store and handle, and adds substantially to the cost.
- Choose reusable pans, dishes, cups, napkins, and utensils rather than disposable ones to save cash and reduce trash.
- Buy reusable plastic food storage containers instead of aluminum foil, plastic wraps, or plastic storage bags. The containers quickly pay for themselves in money *not* spent on disposables.
- Use washable kitchen towels and cloths instead of rolls of paper towels. That's one less bulky item to buy and lug home, and fewer trips to the garbage can.
- Choose Fair Trade coffee and other imported food products labeled "Fair Trade" available at many stores.[2] *Fair Trade* indicates that businesses work toward sustainable industries, including food security and wages of workers and conservation of natural resources.
- Plant a vegetable or herb garden, or join a community garden. Gardening provides physical activity and food, too. Even a few pots of herbs, lettuces, and radishes in a sunny spot will soon yield a tasty salad.

Venturing Out

- Shop at farmers' markets and roadside stands for local foods grown close to home. Locally grown foods require less transportation, packaging, and refrigeration than shipped foods.
- Try picking produce at local farms—it's fun, it's exercise, and it saves money, too.*
- Plan bulk food shopping ahead by car. Bulk staple items cost less, and fewer trips to the store mean more free time and fuel saved.

Bigger Ideas

- Join organizations of like-minded people who work to make things better. You'll enjoy meeting new people and making a difference.
- Buy efficient appliances. Energy Star (see Figure 15–7) appliances rank in the top 25 percent for energy efficiency. Soon, a new Super Star ranking may identify products in the top 5 percent. These products save money on utility bills year after year.

** For more on what can you do to support sustainable agriculture, visit the UC Sustainable Agriculture Research and Education Program website at www.sarep.ucdavis.edu.*

Figure 15–7

Energy Star

Money Isn't All You're Saving

www.energystar.gov

Products bearing the U.S. government's Energy Star logo rank highest for energy efficiency. By choosing products with the Energy Star logo when replacing old equipment, the typical household can save almost $400 per year in energy costs.

- Insulate the home to save energy and money.
- Consider using solar power, especially to heat water; check with local utilities for reimbursement grants.
- *Reduce.* The best savings of money, time, and resources come from consuming less. Even recycling has an energy cost.
- *Reuse.* If an item is necessary, go for durable, not disposable.
- *Recycle.* When the last drop of usefulness seems gone, put items into the recycling stream so they can be remade into new useful things.[†]

Moving Ahead

Beyond daily choices, people can make the greatest impact by teaching others and by volunteering with like-minded people in their communities—in local cleanup efforts, in planting trees, and in community gardens. Local food pantries and gleaners also welcome volunteers. If you take action today, you'll soon see the benefits of a "less buying and more doing" lifestyle begin to emerge.

[†] *To help to find out where to recycle common items in your own community, try this website: www.earth911.com.*

Review Questions[‡]

1. A consumer choosing a product that says "green" on the label can be assured that it is safe for the environment. T F
2. Plant foods offer vital nutrients; they require fewer resources to produce and are less expensive to buy than animal foods. T F
3. Adopting some "green" lifestyle habits can __________ .
   a. save money and benefit personal fitness
   b. reduce household trash
   c. help preserve the environment
   d. all of the above

[‡] *Answers to Consumer's Guide review questions are found in Appendix G.*

what did you decide?

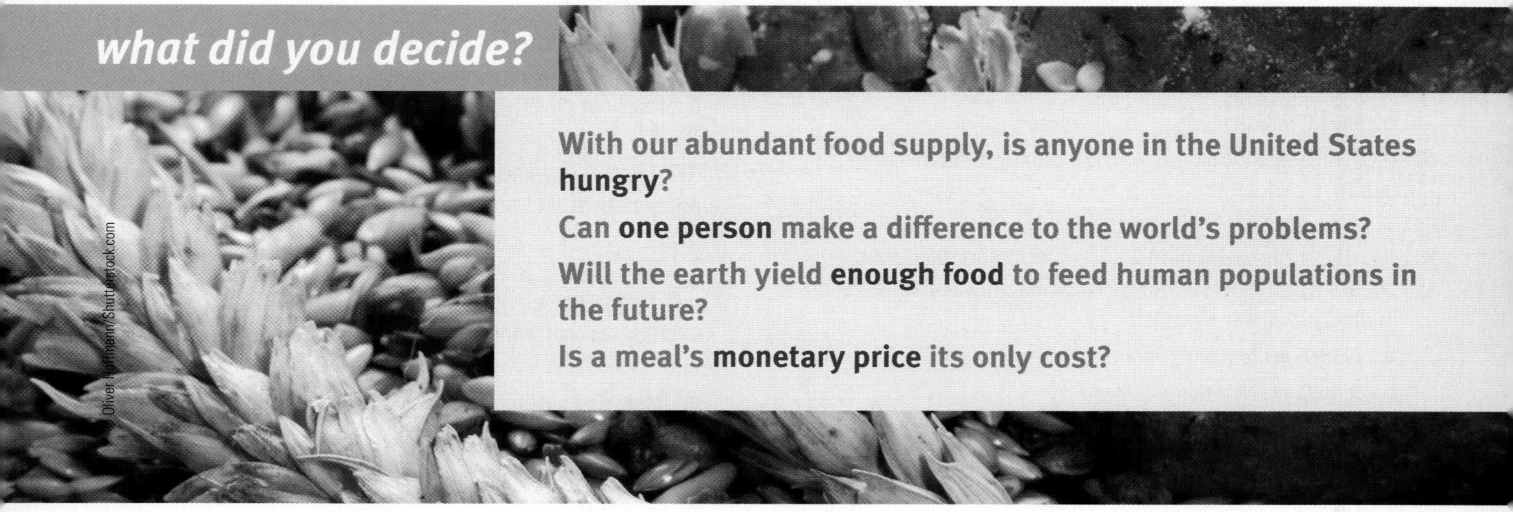

© Oliver Hoffmann/Shutterstock.com

With our abundant food supply, is anyone in the United States hungry?

Can one person make a difference to the world's problems?

Will the earth yield enough food to feed human populations in the future?

Is a meal's monetary price its only cost?

Self Check

1. (LO 15.1) Which of the following is a symptom of food insecurity?
   a. You worry about gaining weight but cannot afford "diet" foods.
   b. You cannot always afford to purchase nutritious foods for balanced meals.
   c. You shop daily to get the best prices and use coupons to stretch your budget.
   d. You buy fresh rather than frozen foods to save money.
2. (LO 15.1) Which of the items can be purchased with electronic debit transfer cards from the Supplemental Nutrition Assistance Program?
   a. hot dogs
   b. cigarettes
   c. dishwashing liquid
   d. red wine
3. (LO 15.2) Today, famine is most often a result of __________.
   a. global food shortage
   b. drought
   c. social causes such as war
   d. flood
4. (LO 15.1) The primary cause of hunger is
   a. a lack of farmable land
   b. a lack of food aid
   c. a lack of nutrition knowledge
   d. poverty
5. (LO 15.2) The world's "poorest poor" spend about 80 percent of their income on food.
   T F
6. (LO 15.2) Worldwide, ____________of those who die each year from starvation and related illnesses are children.
   a. three-fourths
   b. one-third
   c. one-fourth
   d. none of the above
7. (LO 15.3) The malnutrition of poverty inflicts all of the following except
   a. learning disabilities
   b. mental retardation
   c. deafness
   d. blindness
8. (LO 15.3) Most children who die of malnutrition starve to death.
   T F
9. (LO 15.3) The most perilous form of malnutrition that occurs when food suddenly becomes unavailable, such as in drought or war, is called
   a. sudden acute malnutrition
   b. chronic malnutrition
   c. vitamin deficiency malnutrition
   d. b and c
10. (LO 15.4) To save a starving child who has a weak heartbeat and low blood pressure, a necessary first step is to quickly administer
    a. protein supplements
    b. vitamin A supplements
    c. oral rehydration therapy (ORT)
    d. ready-to-use therapeutic food (RUTF)
11. (LO 15.4) Which of the following is an example of environmental degradation?
    a. soil erosion
    b. diminished wild habitat
    c. air pollution
    d. all of the above
12. (LO 15.4) What percentage of its food supply does the United States waste each year?
    a. 20 percent
    b. 30 percent
    c. 40 percent
    d. 50 percent
13. (LO 15.4) Poverty and hunger drive people to bear more children.
    T F
14. (LO 15.4) Relieving poverty and hunger may be necessary in curbing population growth.
    T F
15. (LO 15.5) Today, the keys to solving the world's poverty, hunger, and environmental problems are within the reach of both poor and the rich nations.
    T F
16. (LO 15.5) Necessary steps toward reducing the world's food problems include
    a. reducing energy and food waste
    b. population stabilization
    c. arrest of environmental degradation
    d. all of the above
17. (LO 15.6) Only the federal government and large corporations have the resources necessary to make an impact in the fight against poverty, hunger, and environmental degradation.
    T F
18. (LO 15.7) A vegetarian diet requires just one-third of the energy needed to produce the average meat-containing diet.
    T F
19. (LO 15.7) The scientific discipline that uses ecological theory to study, design, manage, and evaluate productive agricultural systems to conserve critical resources is known as
    a. integrated pest management
    b. sustainability
    c. agroecology
    d. none of the above

Answers to these Self Check questions are in Appendix G.

CONTROVERSY 15

Can We Feed Ourselves Sustainably?

LO 15.7 Define the term *ecological footprint*, and list some factors that increase or decrease a person's ecological footprint.

If predictions hold true, the world's farmers will soon face increased pressure to feed a burgeoning world population.[1]* To produce this food will require vast amounts of land, water, and energy, and it must be accomplished while conserving the resources that make growing crops and animals possible into the future.[2] Sustainability is emerging as a focus in today's agricultural colleges and business schools.[3] Many ideas being generated must be considered and tried to meet the challenges ahead. Not all of them will succeed but, as one business strategist says, "We'll need to try a lot of options, fail fast, and learn quickly" to find what works.[†]

Costs of Current Food Production Methods

Producing food costs the earth dearly. The environmental impacts of agriculture and the food industry take many forms, such as water use and pollution, greenhouse gas emissions, and resource overuse.[4] Table C15–1 offers definitions of terms relevant to these concepts. Important, but beyond the scope of this discussion, are the costs in terms of human health and other problems associated with farm work, such as overexposure to pesticides.[5]

Impacts on Land and Water

To produce food, first, we clear land—prairie, wetland, or forest—replacing native ecosystems with crops or food animals. Crops pull nutrients from the soil. With each harvest, some of those nutrients are removed, so manufactured fertilizers are applied to replace them. Some of the nitrogen in this fertilizer flies off as gas, contributing to greenhouse gas emissions.[6]

With rain or irrigation, fertilizer from fields plus manure from grazing lands and feed lots run off into waterways causing algae overgrowth. The algae dies and decomposes, forming ocean **dead zones** as whole areas are depleted of oxygen.[7] Some plowed soil runs off, too, clouding the water and burying aquatic plants and animals. To protect crops, herbicides and pesticides are applied. These poisons also kill native plants, native insects, and animals that eat those plants and insects. Meanwhile, with continued chemical use, weeds and pests grow resistant to their effects.[8]

Finally, we irrigate, a practice that adds salts to the soil—the water evaporates, but the salts do not. As soils become salty, plant growth fails. Irrigation can also deplete the fresh water supply over time because much of

Table C15–1
Sustainability Terms

- **agroecology** a scientific discipline that combines biological, physical, and social sciences with ecological theory to develop methods for producing food sustainably.
- **alternative (low-input,** or **sustainable) agriculture** agriculture practiced on a small scale using individualized approaches that vary with local conditions so as to minimize technological, fuel, and chemical inputs.
- **farm share** an arrangement in which a farmer offers the public a "subscription" for an allotment of the farm's products throughout the season.
- **dead zones** columns of oxygen-depleted ocean water in which marine life cannot survive; often caused by algae blooms that occur when agricultural fertilizers and waste runoff enter natural waterways.
- **integrated pest management (IPM)** management of pests using a combination of natural and biological controls and minimal or no application of pesticides.

Vast areas under plow are exposed to erosion, and those that must be irrigated can, over time, become salty and unusable.

* *Reference notes are in Appendix F.*

[†] *J. Koomey,* Cold Cash, Cool Climate: Science-based Advice for Ecological Entrepreneurs *(Burlingame, CA: Analytics Press, 2012).*

the water taken from surface or underground supplies evaporates or runs off. This process, carried to an extreme, can dry up whole rivers and lakes and lower the water table of entire regions. The lower the water table, the more farmers must irrigate; and the more they irrigate, the more groundwater they use up.

Soil Depletion

The soil can also be depleted by other agricultural practices, particularly indiscriminate land clearing (deforestation) and overuse by cattle (overgrazing). Traditional farming methods that turn over all topsoil each season expose vast areas to the forces of wind and water. Exposed topsoil blows away on the wind or washes into the sea, leaving an unfertile area behind.

Unsustainable agriculture has already destroyed many once-fertile regions where civilizations formerly flourished. The dry, salty deserts of North Africa were once plowed and irrigated wheat fields, the breadbasket of the Roman Empire. Today's mistreatment of soil and water is causing destruction on an unprecedented scale.

Loss of Species

Agriculture also weakens its own underpinnings when it fails to conserve species diversity. By the year 2050, some 40,000 plant species, existing today, may go extinct. The United Nations Food and Agriculture Organization attributes many of the losses to modern farming practices, as well as to human population growth.

Global eating habits are growing more uniform, a trend that contributes to species loss. As people everywhere eat the same limited array of foods, demand for local, genetically diverse, native plants is insufficient to make them financially worth preserving. Yet, in the future, as the climate warms, those very plants may be needed for food. A wild species of corn that grows in a dry climate, for example, might contain just the genetic information necessary to help make the domestic corn crop resistant to drought. For this and other reasons, protecting biodiversity is a critical human need.

Fuel Use and Energy Sources

Energy and fertilizers from fossil fuels have spurred unprecedented gains in agricultural output, but scientists now recognize that, with limited fossil fuel resources, such gains are not sustainable into the future.[9] Fossil fuel use itself also threatens the future of food production by contributing to pollution and global climate changes.

Biofuels made from renewable corn and soybeans were once hailed as safer alternatives to fossil fuels, but these also carry high environmental costs. Strong world demand for corn or soybean ethanol triggers the conversion of wild native habitats into corn and soybean fields, diverts resources away from growing food crops needed to feed hungry local populations, and increases greenhouse gas emissions.[10] Other materials, such as native grasses, a type of cane, and even genetically engineered algae appear to be more promising materials than food crops for sustainable biofuel production.[11] Other potential energy sources, such as wind and solar energy, remain underdeveloped.[12]

Fossil Energy in Food Production

For the roughly 300 calories of food energy available in a can of corn, more than 6,000 calories of fuel (including the can and transportation) are used to produce it; add 2,000 more calories of fuel to that if the corn comes frozen. Food production represents one-quarter of U.S. fossil energy consumption, including fertilizers, pesticides, and irrigation. Corn and soybeans account for most of the pesticide use, but studies show that these two important crops can be grown successfully using half the current level of pesticide without reducing crop yield.[13] Clearly that can of corn from the supermarket carries an additional cost to the environment—a constellation of inputs not simple to grasp—and not reflected in its price tag. These "hidden" costs must come to account if 21st-century food systems are to be sustainable.

The Problems of Livestock

Raising livestock takes an enormous toll on land and energy resources. Like plant crops, herds of livestock occupy land that once maintained itself in a natural state. The land suffers losses of native plants and animals, soil erosion, water depletion, and desert formation.

U.S. Meat Production

If animals are raised in concentrated areas such as cattle feedlots or giant hog "farms," huge masses of manure produced in these overcrowded, factory-style farms leach into local soils and water supplies, polluting them. In an effort to control this source of pollution, the U.S. Environmental Protection Agency (EPA) offers incentives to livestock farmers who agree to clean up their wastes and allow their operations to be monitored for pollution.

In addition, animals in such feedlots must be fed, and grain is grown for them on other land (Figure C15–1 compares the grain required to produce various foods). That grain may require fertilizers, herbicides, pesticides, and irrigation, too. In the United States, one-fifth of all cropland is used to produce feed for livestock—more land than is used to produce grain for people.

World Trends in Meat Consumption

A worldwide trend toward increased meat and dairy product consumption in places with growing economies is putting pressure on ecological systems.[14] In 1989, for example, less than 40 percent of Chinese citizens derived a significant proportion of their calories from animal-derived foods; that number jumped to 67 percent in 2006. This trend has been underestimated in long-term projections of the world's demand for food and energy.

A Sustainable Future Starts Now

For each of the problems just described, sustainable solutions are being devised, and their use is growing worldwide.[15] Across the nation, ideas for sustainable

Figure C15–1

Pounds of Grain Needed to Produce One Pound of Bread and One Pound of Animal Weight Gain

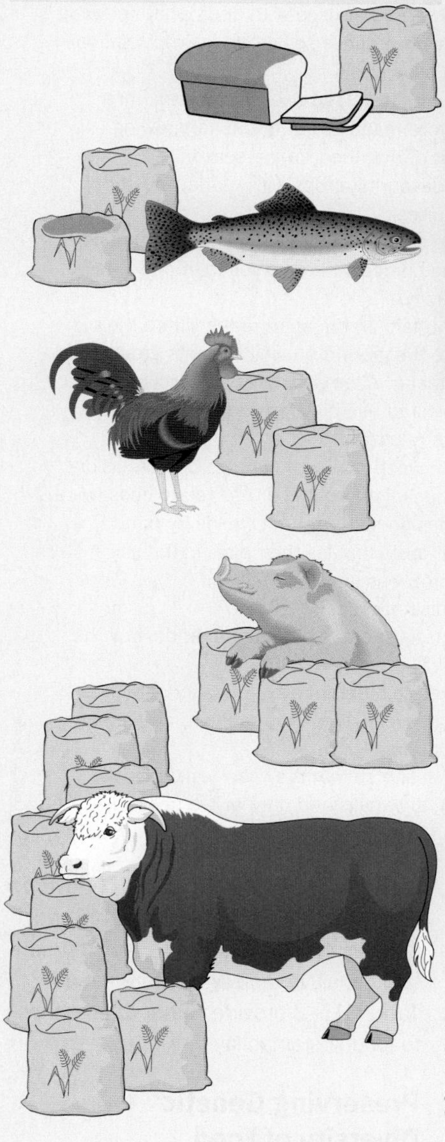

© Cengage Learning

© Stephen Morrison/epa/Corbis

Vertical farms make use of air space instead of acreage to produce food.

food production are emerging from a new field of study: **agroecology**. This scientific discipline applies ecological theory to study, design, manage, and evaluate productive agricultural systems that conserve critical resources. Agroecology seeks to optimize food production and minimize damage to the environment and society. Agricultural production issues cannot be considered separately from environmental or human issues, and this new interdisciplinary framework integrates the biological and physical sciences with ecology and social sciences to generate global solutions that work on many levels.[16]

Sustainable agriculture is not one system but a set of practices that can be matched to particular needs in local areas.[17] The crop yields from farms that employ these practices often compare favorably with those from farms using less sustainable methods, but farmers wishing to employ them must first do some learning and be willing to change. The first of these ideas, sustainable **alternative**, or **low-input**, **agriculture**, emphasizes careful use of natural processes wherever possible, rather than chemically intensive methods.

Low-Input and Precision Agriculture

Farmers may use low-input agriculture, adopting **integrated pest management (IPM)** strategies, such as rotating crops and introducing natural predators to control pests, rather than depending on pesticides alone. Table C15–2 (p. 620) contrasts low-input agriculture methods with unsustainable methods. Many low-input techniques are not really new—they would be familiar to our great-grandparents. Many farmers today are rediscovering the benefits of old techniques while also taking advantage of technological advances that their predecessors could not have imagined.

Low-input agriculture works. Farmers who use it are part of a food production system that contributes substantially to the food supply while restoring soil and water resources and reducing reliance on fossil fuels.

High-tech methods, such as precision agriculture, can also work. The meaning of *precision agriculture* is much the way it sounds: farmers adjust soil and crop management to target the precise needs of various areas of the farm. *Global positioning satellite (GPS)* units in the sky beam data about a field to GPS receiving devices on equipment here on earth. Farmers use the information to target, within a meter's accuracy, land areas that need treatments. The potential dollar and environmental savings in terms of water, fertilizers, and pesticides are enormous. The initial costs of the equipment, however, are high.

Soil Conservation

The U.S. Conservation Reserve Program provides federal assistance to farmers and ranchers who wish to improve their conservation of soil, water, and related natural resources on environmentally sensitive lands.[18] It encourages farmers

Table C15–2

High-Input and Low-Input Agricultural Techniques Compared

Unsustainable Practice	Sustainable Practice
▪ Growing the same crop repeatedly on the same patch of land. This takes more and more nutrients out of the soil, makes fertilizer use necessary; favors soil erosion; and invites weeds and pests to become established, making pesticide use necessary.	▪ Rotate crops. This increases nitrogen in the soil so there is less need to buy fertilizers. If used with appropriate plowing methods, crop rotation reduces soil erosion. Crop rotation also reduces weeds and pests.
▪ Using fertilizers generously. Excess fertilizer pollutes ground and surface water and costs both farmers' household money and consumers' tax money.	▪ Reduce the use of fertilizers and use livestock manure more effectively. Store manure during the nongrowing season and apply it during the growing season. ▪ Alternate nutrient-devouring crops with nutrient-restoring crops, such as legumes. ▪ Compost on a large scale, including all plant residues not harvested. Plow the compost into the soil to improve its water-holding capacity.
▪ Feeding livestock in feedlots where their manure produces major water and soil pollutant problems. Piled in heaps or held in huge ditches, manure also releases methane, a global-warming gas.	▪ Feed livestock or buffalo on the open range where their manure will fertilize the ground on which plants grow and will release no methane. Alternatively, at least collect feedlot animals' manure and use it as fertilizer, or, at the very least, treat it before release.
▪ Spraying herbicides and pesticides over large areas to wipe out weeds and pests.	▪ Apply technology in weed and pest control. Use precision agriculture techniques if affordable or use rotary hoes twice instead of herbicides once. Spot-treat weeds by hand. ▪ Rotate crops to foil pests that lay their eggs in the soil where last year's crop was grown. ▪ Use genetically resistant crops. ▪ Use biological controls such as predators that destroy the pests.
▪ Plowing the same way everywhere, allowing unsustainable water runoff and erosion.	▪ Plow in ways tailored to different areas. Conserve both soil and water by using cover crops, crop rotation, no-till planting, and contour plowing.
▪ Injecting animals with antibiotics to prevent disease in livestock.	▪ Maintain animals' health so that they can resist disease.
▪ Irrigating on a large scale.	▪ Irrigate only during dry spells and only where needed.

to plant native grasses, food plants for wildlife, or trees instead of cash crops on highly erodible cropland, wetlands, or other environmentally sensitive acreage. It also encourages conservative techniques such as shallow tilling and planting grassy strips to facilitate the flow of water off fields. In exchange, farmers receive annual rental payments and other assistance over a 10- to 15-year contract period. The goals of the program are to:

- Reduce soil erosion.
- Protect production of food and fiber.
- Reduce sedimentation in streams and lakes.
- Improve water quality.
- Establish wildlife habitat.
- Enhance forest and wetland resources.

Other programs offer incentives for improving air quality or water quality or for purchasing sensitive lands for conservation. Private foundations or other groups may get help in funding such programs from local, state, or federal agencies.

The Potential of Genetic Engineering

Many farmers worldwide report both financial and conservation benefits from planting genetically engineered crops.[19] Growing herbicide-resistant crops, for example, requires less tilling of the soil for reducing weed growth, reducing soil loss to wind and water erosion. Pesticide-resistant crops demand less use of petroleum-based pesticides and less fuel to run the equipment to apply it. Salt-resistant crops can grow in salty areas where conventional crops wither. If approached carefully (see Controversy 12), genetically engineered plants and animals promise economic, environmentally sensitive, and agricultural lands.[20] Table C15–3 provides other approaches to food sustainability.

Preserving Genetic Diversity of Food

With loss of biodiversity and species extinction, humankind loses some of the raw genetic material on which to draw for improvement of agricultural plants and food animals. The Svalbard Global Seed Vault in Norway houses a collection of 1.1 million genetic samples of the world's 64 most important food crops. Materials submitted by 120 participating nations advance the bank's goal: protecting plant genetic resources and securing genetic biodiversity for food researchers,

Table C15–3

Twelve Steps toward Food Sustainability

1. ***Employ agroforestry.*** Planting trees in and around farms reduces soil erosion by providing a natural barrier against strong winds and rainfall, and roots stabilize and nourish soils.
2. ***Improve soil management.*** Alternating crop species allows soil to rest, restores nutrients, and controls pests. Soil amendments, such as biochar, help soils retain moisture near plant roots.
3. ***Increase crop diversity.*** Growing many crop varieties reduces pests and diseases and decreases reliance on single varieties, increasing domestic food security.
4. ***Increase livestock diversity.*** Genetic diversity in food animals strengthens disease resistance. Lesser-known livestock such as North American Bison are often hardier and produce richer milk.
5. ***Improve food production from existing livestock.*** Feeding grass rather than corn or soybeans to animals lowers the demand for feedstuffs and reduces pressure on global human food supplies.
6. ***Support "Meatless Mondays."*** Forgoing meat on one day a week reduces environmental impacts; the same choice is also widely associated with lower risks of chronic disease in people.
7. ***Use smarter irrigation.*** Installing water sensors or micro-irrigation technology and planning water-efficient gardens or farms using specific crops and locations can conserve crucial water supplies.
8. ***Use integrated farming systems.*** Integrated farming systems, such as permaculture, improve soil fertility and agricultural productivity by using natural resources as sustainably and efficiently as possible. Research and implementation of permaculture techniques, such as recycling wastewater or planting groups of plants that use the same resources in related ways, are expanding rapidly across the United States.
9. ***Use organic and agroecological farming.*** Organic and agroecological farming methods are designed to build soil quality and promote plant and animal health in harmony with local ecosystems.
10. ***Support small-scale farmers.*** Small farms often specialize in growing fruits and vegetables for human consumption; large farms often focus on corn and soybeans for industrial uses.
11. ***Re-evaluate ethanol as fuel.*** Encouraging clean energy alternatives to crop-based biofuels will increase the amount of food available for consumption.
12. ***Support agricultural research.*** Government support for agricultural research and its applications can help address issues such as hunger, malnutrition, and poverty.

Source: Adapted from Worldwatch Institute, 12 Innovations to Combat Drought, Improve Food Security, and Stabilize Food Prices, August 2, 2012, available at http://www.worldwatch.org/12-innovations-combat-drought-improve-food-security-and-stabilize-food-prices.

breeders, and farmers around the world. If a devastating blight attacks a major food crop, then this repository might contain the information needed to counter the threat. Such resource banks are imperative if our food supply is to adapt to a changing climate.[21]

Energy Conservation

Worldwide, the demand for energy increases daily, while fossil-fuel supplies dwindle. To be sustainable, our consumption of energy and our means of producing and using it must change. The food industry, for example, is evaluating and redesigning energy inputs for producing, processing, packaging, transporting, storing, and preparing foods.

Work done on corn and soybean ethanol, the first generation of biofuels, is paving the way for a second generation of fuels, such as:

- Switchgrass, a fast-growing native plant that thrives without fertilizer on land too rocky and nutrient-poor for growing food, offers 540 percent more energy than the amount used to produce it. Other cellulose materials are under study.
- Light-capturing algae, an aquatic plant that can be cultivated to produce fuel. Algae, not typically used for food, can be grown in fresh or saltwater.[22]
- Harnessing wind, sunlight, and the natural heat of the earth, which hold potential as sustainable sources of power.

In the struggle to secure sustainable global energy supplies, no resource can be overlooked, and none wasted.

Energy Recycling

A new paradigm for sustainable energy calls for converting wastes into energy at many levels of food production. For example:

- Vermont's dairy farmers who join the state-sponsored "Cow Power" program convert methane from cow manure into electricity.[23] Wisconsin dairies are producing a type of biogas from cow manure and other wastes.[24]
- Other farmers turn plant waste into charcoal and bury it, trapping carbon underground and fertilizing the soil, thus avoiding applications of fossil-fuel–based fertilizer.[25]
- Some community utilities provide home composters to citizens at reduced cost; the composters turn household garbage into fertilizer for home garden and landscape plants, reducing both trash in the landfill and commercial fertilizer use.[26]
- Other communities capture methane gas, also called "natural gas," produced by decomposing garbage in landfills, and use it to fuel vehicles, such as garbage trucks.[27]

Consumer dollars also represent the energy spent to earn them, and they factor in to the sustainability equation. Consumers can help by spending their dollars on foods that require low-energy inputs from industry, a choice that is described next.

Roles of Consumers

Conscientious consumers can reduce pollution and the use of resources through the choices they make.[28] Some new, fresh ways of thinking about how to obtain foods, and which foods to choose, can enliven the diet and enrich daily life.

Keeping Local Profits Local

Farmers selling their broccoli, carrots, and apples at city farmers' markets and country roadside stands often net a higher profit, especially when consumers perceive, often correctly, that farm-fresh foods taste best. Another benefit: families who buy homegrown produce tend to eat a greater quantity and selection of fruits and vegetables, and the health benefits of this outcome are well known.[29] Through a **farm share**, consumers can buy a weekly share of a local farmer's crops, harvested in season and picked up while fresh. Some consumers also value knowing where and how their foods are grown and handled.

Farmers' markets and farm share arrangements provide fresh foods from local growers.

Eating Lower on the Food Chain

Overall, a vegetarian diet requires just one-third of the energy needed to produce the average meat-containing diet. While 60 percent of the world's crops are eaten by people, animal feed absorbs 35 percent, and 5 percent is used for biofuels and industry. About 30 pounds of grain is required to produce 1 pound of beef, and meat production entails tremendous amounts of water and land, thus reducing the resources available to grow staple foods for human beings.[30] An exception is livestock raised on the open range; these animals eat grass and require low energy inputs. So much of our beef is grain-fed, however, that the average energy requirement to produce it is high. In general, eating more foods derived from plants and fewer foods derived from animals conserves resources.[31]

An often overlooked point: food choices that benefit the environment also benefit human health. Little doubt remains that a diet higher in legumes, whole grains, and vegetables and lower in fatty red meats and highly processed foods is associated with lower risks of developing chronic diseases.

All in all, our choices as a nation add up to a measurable "ecological footprint"—the productive land and water required to supply all of the resources an individual consumes and to absorb all of the wastes generated using prevailing practices.[32] The footprint of each individual is four times larger in an industrialized country than in a developing one (see Figure C15–2). To help size up your own ecological footprint, take the quiz in Table C15–4.

Conclusion

Sustainability problems are global in scope, yet the actions of individual people lie at the heart of their solutions. Do what you can to tread lightly on the earth. Celebrate changes that are possible today by making them a permanent part of your life and reap the benefits of increased health and well-being; do the same with changes that become possible tomorrow, and every day thereafter.

Figure C15–2
Ecological Footprints

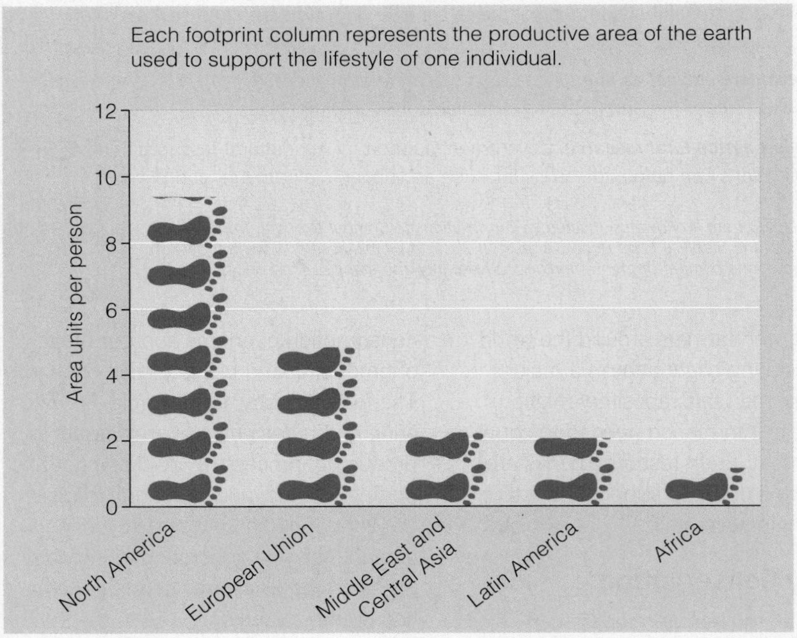

Source: Ecological Footprint and Biocapacity, 2006, available at www.footprintnetwork.org/index.php.

Table C15–4

How Big Is Your Ecological Footprint?

This quiz can help you evaluate your impact on the earth. The higher you score, the smaller your "footprint."

At home, do you

1. Recycle everything you can: newspapers, cans, glass bottles and jars, scrap metal, used oil, etc.?
2. Use cold water in the washer whenever possible?
3. Avoid using appliances (such as electric can openers) to do things you can do by hand?
4. Reuse grocery bags to line your wastebasket? Reuse or recycle bread bags, butter tubs, etc.?
5. Store food in reusable containers rather than plastic wrap, disposable bags and containers, or aluminum foil?

In the yard, do you

6. Pull weeds instead of using herbicides?
7. Fertilize with manure and compost, rather than with chemical fertilizers?
8. Compost your leaves and yard debris, rather than burning them?
9. Take extra plastic and rubber pots back to the plant nursery?

On vacation, do you

10. Turn down the heat and turn off the hot water heater before you leave?
11. Carry reusable cups, dishes, and flatware (and use them)?
12. Dispose of trash appropriately (never litter)?
13. Buy no souvenirs made from wild or endangered animals?
14. Stay on roads and trails, and not trample dunes and fragile undergrowth?

About your car, do you

15. Keep your car tuned up for maximum fuel efficiency?
16. Use public transit whenever possible?
17. Ride your bike or walk whenever possible?
18. Plan to replace your car with a more fuel-efficient model when you can?
19. Recycle your engine oil?

At school or work, do you

20. Recycle paper whenever possible?
21. Use scrap paper for notes to yourself and others?
22. Print or copy on both sides of the paper?
23. Reuse large envelopes and file folders?
24. Use the stairs instead of the elevator whenever you can?

When buying, do you

25. Buy as little plastic and foam packaging as possible?
26. Buy permanent, rather than disposable, products?
27. Buy paper rather than plastic, if you must buy disposable products?
28. Buy fresh produce grown locally?
29. Buy in bulk to avoid unnecessary packaging?

In other areas, do you

30. Volunteer your time to conservation projects?
31. Encourage your family, friends, and neighbors to save resources, too?
32. Write letters to support conservation issues?

Scoring

First, give yourself 4 points for answering this quiz: ___

Then, give yourself 1 point each for all the habits you know people should adopt. This is to give you credit for your awareness, even if you haven't acted on it yet (total possible points = 32): ___

Finally, give yourself 2 more points for each habit you have adopted—or honestly would if you could (total possible points = 64): ___

Total score:

1 to 25: You are a beginner in stewardship of the earth. Try to improve.

26 to 50: You are on your way and doing better than many consumers.

51 to 75: Good. Pat yourself on the back, and keep on improving.

76 to 100: Excellent. You are a shining example for others to follow.

Source: Adapted from Conservation Action Checklist, *produced by the Washington Park Zoo, Portland, Oregon, and available from Conservation International, 1015 18th St. NW, Suite 1000, Washington, DC 20036: 1-800-406-2306 (website: www.conservation.org). Call or write for copies of the original or for more information.*

Appendix Table of Contents

Appendix A

Table of Food Composition

This table of food composition is updated periodically to reflect current nutrient data for foods, to remove outdated foods, and to add foods that are new to the marketplace.* The nutrient database for this appendix is compiled from a variety of sources, including the USDA Nutrient Database and manufacturers' data. The USDA database provides data for a wider variety of foods and nutrients than other sources. Because laboratory analysis for each nutrient can be quite costly, manufacturers tend to provide data only for those nutrients mandated on food labels. Consequently, data for their foods are often incomplete; any missing information on this table is designated as a dash. Keep in mind that a dash means only that the information is unknown and should not be interpreted as a zero. A zero means that the nutrient is not present in the food.

When using nutrient data, remember that many factors influence the nutrient contents of foods. These factors include the mineral content of the soil, the diet fed to the animal or the fertilizer used on the plant, the season of harvest, the method of processing, the length and method of storage, the method of cooking, the method of analysis, and the moisture content of the sample analyzed. With so many influencing factors, users should view nutrient data as a close approximation of the actual amount.

For updates, corrections, and a list of more than 35,000 foods and codes found in the diet analysis software that accompanies this text, visit www.cengagebrain.com and click on Diet Analysis Plus.

- ***Fats*** Total fats, as well as the breakdown of total fats to saturated, monounsaturated, and polyunsaturated, are listed in the table. The fatty acids seldom add up to the total in part due to rounding but also because values may include some non-fatty acids, such as glycerol, phosphate, or sterols. *Trans*-fatty acids are not listed separately in this edition because newer hydrogenated fats generally add less than 0.5 g *trans* fat to a serving of food, an amount often reported as 0.
- ***Vitamin A, Vitamin E, and Folate*** In keeping with the DRI values for vitamin A, this appendix presents data for vitamin A in micrograms (µg) RAE. Similarly, because the DRI intake values for vitamin E are based only on the alpha-tocopherol form of vitamin E, this appendix reports vitamin E data in milligrams (mg) alpha-tocopherol, listed on the table as Vit E (mg α). Folate values are listed in µg DFE, a unit that equalizes the bioavailability of naturally occurring folate and added folic acid in enriched foods.
- ***Bioavailability*** Keep in mind that the availability of nutrients from foods depends not only on the quantity provided by a food as reflected in this table, but also on the amount absorbed and used by the body.
- ***Using the Table*** The foods and beverages in this table are organized into several categories, which are listed at the head of each right-hand page. Page numbers are provided, and each group is color-coded to make it easier to find individual foods.

*This food composition table has been prepared by Cengage Learning. The nutritional data are supplied by Axxya Systems.

TABLE A-1 Table of Food Composition

(Computer code is for Cengage Diet Analysis Plus program)

DA+ Code	Food Description	QTY	Measure	Wt (g)	H_2O (g)	Ener (cal)	Prot (g)	Carb (g)	Fiber (g)	Fat (g)	Fat Breakdown (g) Sat	Mono	Poly
	Breads, Baked Goods, Cakes, Cookies, Crackers, Chips, Pies												
	Bagels												
8534	Cinnamon and raisin	1	item(s)	71	22.7	194	7.0	39.2	1.6	1.2	0.2	0.1	0.5
14395	Multi-grain	1	item(s)	61	—	170	6.0	35.0	1.0	1.5	0.5	0.1	0.4
8538	Oat bran	1	item(s)	71	23.4	181	7.6	37.8	2.6	0.9	0.1	0.2	0.3
4910	Plain, enriched	1	item(s)	71	25.8	182	7.1	35.9	1.6	1.2	0.3	0.4	0.5
4911	Plain, enriched, toasted	1	item(s)	66	18.7	190	7.4	37.7	1.7	1.1	0.2	0.3	0.6
	Biscuits												
25008	Biscuits	1	item(s)	41	15.8	121	2.6	16.4	0.5	4.9	1.4	1.4	1.8
16729	Scone	1	item(s)	42	11.5	148	3.8	19.1	0.6	6.2	2.0	2.5	1.3
25166	Wheat biscuits	1	item(s)	55	21.0	162	3.6	21.9	1.4	6.7	1.9	1.9	2.5
	Bread												
325	Boston brown, canned	1	slice(s)	45	21.2	88	2.3	19.5	2.1	0.7	0.1	0.1	0.3
8716	Bread sticks, plain	4	item(s)	24	1.5	99	2.9	16.4	0.7	2.3	0.3	0.9	0.9
25176	Cornbread	1	piece(s)	55	25.9	141	4.7	18.3	0.9	5.4	2.1	1.4	1.5
327	Cracked wheat	1	slice(s)	25	9.0	65	2.2	12.4	1.4	1.0	0.2	0.5	0.2
9079	Croutons, plain	¼	cup(s)	8	0.4	31	0.9	5.5	0.4	0.5	0.1	0.2	0.1
8582	Egg	1	slice(s)	40	13.9	113	3.8	19.1	0.9	2.4	0.6	0.9	0.4
8585	Egg, toasted	1	slice(s)	37	10.5	117	3.9	19.5	0.9	2.4	0.6	1.1	0.4
329	French	1	slice(s)	32	8.9	92	3.8	18.1	0.8	0.6	0.2	0.1	0.3
8591	French, toasted	1	slice(s)	23	4.7	73	3.0	14.2	0.7	0.5	0.1	0.1	0.2
42096	Indian fry, made with lard (Navajo)	3	ounce(s)	85	26.9	281	5.7	41.0	—	10.4	3.9	3.8	0.9
332	Italian	1	slice(s)	30	10.7	81	2.6	15.0	0.8	1.1	0.3	0.2	0.4
1393	Mixed grain	1	slice(s)	26	9.6	69	3.5	11.3	1.9	1.1	0.2	0.2	0.5
8604	Mixed grain, toasted	1	slice(s)	24	7.6	69	3.5	11.3	1.9	1.1	0.2	0.2	0.5
8605	Oat bran	1	slice(s)	30	13.2	71	3.1	11.9	1.4	1.3	0.2	0.5	0.5
8608	Oat bran, toasted	1	slice(s)	27	10.4	70	3.1	11.8	1.3	1.3	0.2	0.5	0.5
8609	Oatmeal	1	slice(s)	27	9.9	73	2.3	13.1	1.1	1.2	0.2	0.4	0.5
8613	Oatmeal, toasted	1	slice(s)	25	7.8	73	2.3	13.2	1.1	1.2	0.2	0.4	0.5
1409	Pita	1	item(s)	60	19.3	165	5.5	33.4	1.3	0.7	0.1	0.1	0.3
7905	Pita, whole wheat	1	item(s)	64	19.6	170	6.3	35.2	4.7	1.7	0.3	0.2	0.7
338	Pumpernickel	1	slice(s)	32	12.1	80	2.8	15.2	2.1	1.0	0.1	0.3	0.4
334	Raisin, enriched	1	slice(s)	26	8.7	71	2.1	13.6	1.1	1.1	0.3	0.6	0.2
8625	Raisin, toasted	1	slice(s)	24	6.7	71	2.1	13.7	1.1	1.2	0.3	0.6	0.2
10168	Rice, white, gluten free, wheat free	1	slice(s)	38	—	130	1.0	18.0	0.5	6.0	0	—	—
8653	Rye	1	slice(s)	32	11.9	83	2.7	15.5	1.9	1.1	0.2	0.4	0.3
8654	Rye, toasted	1	slice(s)	29	9.0	82	2.7	15.4	1.9	1.0	0.2	0.4	0.3
336	Rye, light	1	slice(s)	25	9.3	65	2.0	12.0	1.6	1.0	0.2	0.3	0.3
8588	Sourdough	1	slice(s)	25	7.0	72	2.9	14.1	0.6	0.5	0.1	0.1	0.2
8592	Sourdough, toasted	1	slice(s)	23	4.7	73	3.0	14.2	0.7	0.5	0.1	0.1	0.2
491	Submarine or hoagie roll	1	item(s)	135	40.6	400	11.0	72.0	3.8	8.0	1.8	3.0	2.2
8596	Vienna, toasted	1	slice(s)	23	4.7	73	3.0	14.2	0.7	0.5	0.1	0.1	0.2
8670	Wheat	1	slice(s)	25	8.9	67	2.7	11.9	0.9	0.9	0.2	0.2	0.4
8671	Wheat, toasted	1	slice(s)	23	5.6	72	3.0	12.8	1.1	1.0	0.2	0.2	0.4
340	White	1	slice(s)	25	9.1	67	1.9	12.7	0.6	0.8	0.2	0.2	0.3
1395	Whole wheat	1	slice(s)	46	15.0	128	3.9	23.6	2.8	2.5	0.4	0.5	1.4
	Cakes												
386	Angel food, prepared from mix	1	piece(s)	50	16.5	129	3.1	29.4	0.1	0.2	0	0	0.1
8772	Butter pound, ready to eat, commercially prepared	1	slice(s)	75	18.5	291	4.1	36.6	0.4	14.9	8.7	4.4	0.8
28517	Carrot	1	slice(s)	131	56.6	339	4.8	56.5	1.9	11.1	1.0	5.7	3.8
4931	Chocolate with chocolate icing, commercially prepared	1	slice(s)	64	14.7	235	2.6	34.9	1.8	10.5	3.1	5.6	1.2
8756	Chocolate, prepared from mix	1	slice(s)	95	23.2	352	5.0	50.7	1.5	14.3	5.2	5.7	2.6
393	Devil's food cupcake with chocolate frosting	1	item(s)	35	8.4	120	2.0	20.0	0.7	4.0	1.8	1.6	0.6
8757	Fruitcake, ready to eat, commercially prepared	1	piece(s)	43	10.9	139	1.2	26.5	1.6	3.9	0.5	1.8	1.4
1397	Pineapple upside down, prepared from mix	1	slice(s)	115	37.1	367	4.0	58.1	0.9	13.9	3.4	6.0	3.8
411	Sponge, prepared from mix	1	slice(s)	63	18.5	187	4.6	36.4	0.3	2.7	0.8	1.0	0.4

Chol (mg)	Calc (mg)	Iron (mg)	Magn (mg)	Pota (mg)	Sodi (mg)	Zinc (mg)	Vit A (µg)	Thia (mg)	Vit E (mg α)	Ribo (mg)	Niac (mg)	Vit B_6 (mg)	Fola (µg DFE)	Vit C (mg)	Vit B_{12} (µg)	Sele (µg)
0	13	2.69	19.9	105.1	228.6	0.80	14.9	0.27	0.22	0.19	2.18	0.04	123.5	0.5	0	22.0
0	60	1.08	—	—	310.0	—	0	—	—	—	—	—	—	0	—	—
0	9	2.18	22.0	81.7	360.0	0.63	0.7	0.23	0.23	0.24	2.10	0.03	95.1	0.1	0	24.3
0	63	4.29	15.6	53.3	318.1	1.34	0	0.42	0.07	0.18	2.82	0.04	160.5	0.7	0	16.2
0	65	2.97	15.8	56.1	316.8	0.85	0	0.39	0.07	0.17	2.88	0.04	134.0	0	0	16.6
0	38	0.94	6.0	47.4	206.0	0.20	—	0.16	0.01	0.12	1.20	0.01	46.8	0.1	0.1	7.1
49	79	1.35	7.1	48.7	277.2	0.29	64.7	0.14	0.42	0.15	1.19	0.02	49.1	0	0.1	10.9
0	57	1.21	16.1	81.0	321.1	0.42	—	0.19	0.01	0.14	1.65	0.03	63.2	0.1	0.1	0
0	32	0.94	28.4	143.1	284.0	0.22	11.3	0.01	0.14	0.05	0.50	0.03	6.3	0	0	9.9
0	5	1.02	7.7	29.8	157.7	0.21	0	0.14	0.24	0.13	1.26	0.01	61.0	0	0	9.0
21	94	0.91	10.5	71.5	209.8	0.48	—	0.14	0.32	0.15	1.03	0.04	78.7	1.7	0.2	6.2
0	11	0.70	13.0	44.3	134.5	0.31	0	0.09	—	0.06	0.92	0.08	19.0	0	0	6.3
0	6	0.30	2.3	9.3	52.4	0.06	0	0.04	—	0.02	0.40	0.00	15.7	0	0	2.8
20	37	1.21	7.6	46.0	196.8	0.31	25.2	0.17	0.10	0.17	1.93	0.02	52.0	0	0	12.0
21	38	1.23	7.8	46.6	199.8	0.31	25.5	0.14	0.10	0.16	1.77	0.02	47.7	0	0	12.2
0	14	1.16	9.0	41.0	208.0	0.29	0	0.13	0.05	0.09	1.52	0.03	73.6	0.1	0	8.7
0	11	0.89	7.1	32.2	165.6	0.24	0	0.10	0.04	0.09	1.24	0.02	49.9	0	0	6.8
6	48	3.43	15.3	65.5	279.8	0.29	0	0.36	0.00	0.18	3.91	0.03	166.7	—	0	15.8
0	23	0.88	8.1	33.0	175.2	0.25	0	0.14	0.08	0.08	1.31	0.01	91.2	0	0	8.2
0	27	0.65	20.3	59.8	109.2	0.44	0	0.07	0.09	0.03	1.05	0.06	19.5	0	0	8.6
0	27	0.65	20.4	60.0	109.7	0.44	0	0.06	0.10	0.03	1.05	0.07	16.8	0	0	8.6
0	20	0.93	10.5	44.1	122.1	0.26	0.6	0.15	0.13	0.10	1.44	0.02	36.0	0	0	9.0
0	19	0.92	9.2	33.2	121.0	0.28	0.5	0.12	0.13	0.09	1.29	0.01	28.1	0	0	8.9
0	18	0.72	10.0	38.3	161.7	0.27	1.4	0.10	0.13	0.06	0.84	0.01	23.5	0	0	6.6
0	18	0.74	10.3	38.5	162.8	0.28	1.3	0.09	0.13	0.06	0.77	0.02	18.7	0.1	0	6.7
0	52	1.57	15.6	72.0	321.6	0.50	0	0.35	0.18	0.19	2.77	0.02	99.0	0	0	16.3
0	10	1.95	44.2	108.8	340.5	0.97	0	0.21	0.39	0.05	1.81	0.17	22.4	0	0	28.2
0	22	0.91	17.3	66.6	214.7	0.47	0	0.10	0.13	0.09	0.98	0.04	40.0	0	0	7.8
0	17	0.75	6.8	59.0	101.4	0.18	0	0.08	0.07	0.10	0.90	0.01	40.6	0	0	5.2
0	17	0.76	6.7	59.0	101.8	0.19	0	0.07	0.07	0.09	0.81	0.02	35.5	0.1	0	5.2
0	100	1.08	—	—	140	—	—	0.15	—	0.10	1.20	—	47.5	0	—	—
0	23	0.90	12.8	53.1	211.2	0.36	0	0.13	0.10	0.10	1.21	0.02	48.3	0.1	0	9.9
0	23	0.89	12.5	53.1	210.3	0.36	0	0.11	0.10	0.09	1.09	0.02	42.9	0.1	0	9.9
0	20	0.70	3.9	51.0	175.0	0.18	0	0.10	—	0.08	0.80	0.01	5.3	0	0	8.0
0	11	0.91	7.0	32.0	162.5	0.23	0	0.11	0.05	0.07	1.19	0.03	57.5	0.1	0	6.8
0	11	0.89	7.1	32.2	165.6	0.24	0	0.10	0.04	0.09	1.24	0.02	49.9	0	0	6.8
0	100	3.80	—	128.0	683.0	—	0	0.54	—	0.33	4.50	0.04	—	0	—	42.0
0	11	0.89	7.1	32.2	165.6	0.24	0	0.10	0.04	0.09	1.24	0.02	49.9	0	0	6.8
0	36	0.87	12.0	46.0	130.3	0.30	0	0.09	0.05	0.08	1.30	0.03	24.8	0.1	0	7.2
0	38	0.94	13.6	51.3	140.5	0.34	0	0.10	0.06	0.09	1.44	0.04	23.0	0	0	7.7
0	38	0.94	5.8	25.0	170.3	0.19	0	0.11	0.06	0.08	1.10	0.02	42.8	0	0	4.3
0	15	1.42	37.3	144.4	159.2	0.69	0	0.13	0.35	0.10	1.83	0.09	35.9	0	0	17.8
0	42	0.11	4.0	67.5	254.5	0.06	0	0.04	0.01	0.10	0.08	0.00	14.5	0	0	7.7
166	26	1.03	8.3	89.3	298.5	0.34	111.8	0.10	—	0.17	0.98	0.03	46.5	0	0.2	6.6
0	65	2.18	23.0	279.6	367.7	0.44	—	0.25	0.01	0.19	1.73	0.10	98.2	4.6	0	14.7
27	28	1.40	21.8	128.0	213.8	0.44	16.6	0.01	0.62	0.08	0.36	0.02	14.7	0.1	0.1	2.1
55	57	1.53	30.4	133.0	299.3	0.65	38.0	0.13	—	0.20	1.08	0.03	37.1	0.2	0.2	11.3
19	21	0.70	—	46.0	92.0	—	—	0.04	—	0.05	0.30	—	8.1	0	—	2.0
2	14	0.89	6.9	65.8	116.1	0.11	3.0	0.02	0.38	0.04	0.34	0.02	13.8	0.2	0	0.9
25	138	1.70	15.0	128.8	366.9	0.35	71.3	0.17	—	0.17	1.36	0.03	44.9	1.4	0.1	10.8
107	26	0.99	5.7	88.8	143.6	0.37	48.5	0.10	—	0.19	0.75	0.03	33.4	0	0.2	11.7

TABLE A-1 Table of Food Composition *(continued)*

(Computer code is for Cengage Diet Analysis Plus program)

DA+ Code	Food Description	QTY	Measure	Wt (g)	H_2O (g)	Ener (cal)	Prot (g)	Carb (g)	Fiber (g)	Fat (g)	Fat Breakdown (g) Sat	Mono	Poly
	Breads, Baked Goods, Cakes, Cookies, Crackers, Chips, Pies—continued												
8817	White with coconut frosting, prepared from mix	1	slice(s)	112	23.2	399	4.9	70.8	1.1	11.5	4.4	4.1	2.4
8819	Yellow with chocolate frosting, ready to eat, commercially prepared	1	slice(s)	64	14.0	243	2.4	35.5	1.2	11.1	3.0	6.1	1.4
8822	Yellow with vanilla frosting, ready to eat, commercially prepared	1	slice(s)	64	14.1	239	2.2	37.6	0.2	9.3	1.5	3.9	3.3
	Snack cakes												
8791	Chocolate snack cake, creme filled, with frosting	1	item(s)	50	9.3	200	1.8	30.2	1.6	8.0	2.4	4.3	0.9
25010	Cinnamon coffee cake	1	piece(s)	72	22.6	231	3.6	35.8	0.7	8.3	2.2	2.6	3.0
16777	Funnel cake	1	item(s)	90	37.6	276	7.3	29.1	0.9	14.4	2.7	4.7	6.1
8794	Sponge snack cake, creme filled	1	item(s)	43	8.6	155	1.3	27.2	0.2	4.8	1.1	1.7	1.4
	Snacks, chips, pretzels												
29428	Bagel chips, plain	3	item(s)	29	—	130	3.0	19.0	1.0	4.5	0.5	—	—
29429	Bagel chips, toasted onion	3	item(s)	29	—	130	4.0	20.0	1.0	4.5	0.5	—	—
38192	Chex traditional snack mix	1	cup(s)	45	—	197	3.0	33.3	1.5	6.1	0.8	—	—
654	Potato chips, salted	1	ounce(s)	28	0.6	155	1.9	14.1	1.2	10.6	3.1	2.8	3.5
8816	Potato chips, unsalted	1	ounce(s)	28	0.5	152	2.0	15.0	1.4	9.8	3.1	2.8	3.5
5096	Pretzels, plain, hard, twists	5	item(s)	30	1.0	114	2.7	23.8	1.0	1.1	0.2	0.4	0.4
4632	Pretzels, whole wheat	1	ounce(s)	28	1.1	103	3.1	23.0	2.2	0.7	0.2	0.3	0.2
4641	Tortilla chips, plain	6	item(s)	11	0.2	53	0.8	7.1	0.6	2.5	0.3	0.8	0.5
	Cookies												
8859	Animal crackers	12	item(s)	30	1.2	134	2.1	22.2	0.3	4.1	1.0	2.3	0.6
8876	Brownie, prepared from mix	1	item(s)	24	3.0	112	1.5	12.0	0.5	7.0	1.8	2.6	2.3
25207	Chocolate chip cookies	1	item(s)	30	3.7	140	2.0	16.2	0.6	7.9	2.1	3.3	2.1
8915	Chocolate sandwich cookie with extra creme filling	1	item(s)	13	0.2	65	0.6	8.9	0.4	3.2	0.7	2.1	0.3
14145	Fig Newtons cookies	1	item(s)	16	—	55	0.5	11.0	0.5	1.3	0	—	—
8920	Fortune cookie	1	item(s)	8	0.6	30	0.3	6.7	0.1	0.2	0.1	0.1	0
25208	Oatmeal cookies	1	item(s)	69	12.3	234	5.7	45.1	3.1	4.2	0.7	1.3	1.8
25213	Peanut butter cookies	1	item(s)	35	4.1	163	4.2	16.9	0.9	9.2	1.7	4.7	2.3
33095	Sugar cookies	1	item(s)	16	4.1	61	1.1	7.4	0.1	3.0	0.6	1.3	0.9
9002	Vanilla sandwich cookie with creme filling	1	item(s)	10	0.2	48	0.5	7.2	0.2	2.0	0.3	0.8	0.8
	Crackers												
9012	Cheese cracker sandwich with peanut butter	4	item(s)	28	0.9	139	3.5	15.9	1.0	7.0	1.2	3.6	1.4
9008	Cheese crackers (mini)	30	item(s)	30	0.9	151	3.0	17.5	0.7	7.6	2.8	3.6	0.7
33362	Cheese crackers, low sodium	1	serving(s)	30	0.9	151	3.0	17.5	0.7	7.6	2.9	3.6	0.7
8928	Honey graham crackers	4	item(s)	28	1.2	118	1.9	21.5	0.8	2.8	0.4	1.1	1.1
9016	Matzo crackers, plain	1	item(s)	28	1.2	112	2.8	23.8	0.9	0.4	0.1	0	0.2
9024	Melba toast	3	item(s)	15	0.8	59	1.8	11.5	0.9	0.5	0.1	0.1	0.2
9028	Melba toast, rye	3	item(s)	15	0.7	58	1.7	11.6	1.2	0.5	0.1	0.1	0.2
14189	Ritz crackers	5	item(s)	16	0.5	80	1.0	10.0	0	4.0	1.0	—	—
9014	Rye crispbread crackers	1	item(s)	10	0.6	37	0.8	8.2	1.7	0.1	0	0	0.1
9040	Rye wafer	1	item(s)	11	0.6	37	1.1	8.8	2.5	0.1	0	0	0
432	Saltine crackers	5	item(s)	15	0.8	64	1.4	10.6	0.5	1.7	0.2	1.1	0.2
9046	Saltine crackers, low salt	5	item(s)	15	0.6	65	1.4	10.7	0.5	1.8	0.4	1.0	0.3
9052	Snack cracker sandwich with cheese filling	4	item(s)	28	1.1	134	2.6	17.3	0.5	5.9	1.7	3.2	0.7
9054	Snack cracker sandwich with peanut butter filling	4	item(s)	28	0.8	138	3.2	16.3	0.6	6.9	1.4	3.9	1.3
9048	Snack crackers, round	10	item(s)	30	1.1	151	2.2	18.3	0.5	7.6	1.1	3.2	2.9
9050	Snack crackers, round, low salt	10	item(s)	30	1.1	151	2.2	18.3	0.5	7.6	1.1	3.2	2.9
9044	Soda crackers	5	tem(s)	15	0.8	64	1.4	10.6	0.5	1.7	0.2	1.1	0.2
9059	Wheat cracker sandwich with cheese filling	4	item(s)	28	0.9	139	2.7	16.3	0.9	7.0	1.2	2.9	2.6
9061	Wheat cracker sandwich with peanut butter filling	4	item(s)	28	1.0	139	3.8	15.1	1.2	7.5	1.3	3.3	2.5
9055	Wheat crackers	10	item(s)	30	0.9	142	2.6	19.5	1.4	6.2	1.6	3.4	0.8

Chol (mg)	Calc (mg)	Iron (mg)	Magn (mg)	Pota (mg)	Sodi (mg)	Zinc (mg)	Vit A (µg)	Thia (mg)	Vit E (mg α)	Ribo (mg)	Niac (mg)	Vit B_6 (mg)	Fola (µg DFE)	Vit C (mg)	Vit B_{12} (µg)	Sele (µg)
1	101	1.29	13.4	110.9	318.1	0.37	13.4	0.14	0.13	0.21	1.19	0.03	57.12	0.1	0.1	12.0
35	24	1.33	19.2	113.9	215.7	0.39	21.1	0.07	—	0.10	0.79	0.02	20.5	0	0.1	2.2
35	40	0.68	3.8	33.9	220.2	0.16	12.2	0.06	—	0.04	0.32	0.01	25.6	0	0.1	3.5
0	58	1.80	18.0	88.0	194.5	0.52	0.5	0.01	0.54	0.03	0.46	0.07	17.5	1.0	0	1.7
26	55	1.36	9.9	91.9	277.6	0.30	—	0.17	0.23	0.16	1.29	0.02	66.1	0.3	0.1	9.6
62	126	1.90	16.2	152.1	269.1	0.65	49.5	0.23	1.54	0.32	1.86	0.04	75.6	0	0.3	17.7
7	19	0.54	3.4	37.0	155.1	0.12	2.1	0.06	0.50	0.05	0.52	0.01	23.0	0	0	1.3
0	0	0.72	—	45.0	70.0	—	0	—	—	—	—	—	—	0	0	—
0	0	0.72	—	50.0	300.0	—	0	—	—	—	—	—	—	0	0	—
0	0	0.55	—	75.8	621.2	—	0	0.09	—	0.05	1.21	—	38.5	0	—	—
0	7	0.45	19.8	465.5	148.8	0.67	0	0.01	1.91	0.06	1.18	0.20	21.3	5.3	0	2.3
0	7	0.46	19.0	361.5	2.3	0.30	0	0.04	2.58	0.05	1.08	0.18	12.8	8.8	0	2.3
0	11	1.29	10.5	43.8	514.5	0.25	0	0.13	0.10	0.18	1.57	0.03	86.0	0	0	1.7
0	8	0.76	8.5	121.9	57.6	0.17	0	0.12	—	0.08	1.85	0.07	15.3	0.3	0	—
0	19	0.25	15.8	23.2	45.5	0.26	0	0.00	0.46	0.01	0.13	0.02	2.2	0	0	0.7
0	13	0.82	5.4	30.0	117.9	0.19	0	0.10	0.03	0.09	1.04	0.01	49.5	0	0	2.1
18	14	0.44	12.7	42.2	82.3	0.23	42.2	0.03	—	0.05	0.24	0.02	9.4	0.1	0	2.8
13	11	0.69	12.4	62.1	108.8	0.24	—	0.08	0.54	0.06	0.87	0.01	14.1	0	0	4.1
0	2	1.01	4.7	17.8	45.6	0.10	0	0.02	0.25	0.02	0.25	0.00	9.0	0	0	1.1
0	10	0.36	—	—	57.5	—	0	—	—	—	—	—	—	0	—	—
0	1	0.12	0.6	3.3	21.9	0.01	0.1	0.01	0.00	0.01	0.15	0.00	8.4	0	0	0.2
0	26	1.93	48.8	176.7	311.1	1.42	—	0.26	0.23	0.13	1.35	0.09	65.6	0.3	0	17.4
13	27	0.65	21.1	112.8	154.1	0.46	—	0.08	0.73	0.09	1.85	0.05	35.0	0.1	0.1	4.8
18	5	0.30	1.7	12.2	49.4	0.08	—	0.04	0.28	0.05	0.31	0.01	13.1	0	0	3.1
0	3	0.22	1.4	9.1	34.9	0.04	0	0.02	0.16	0.02	0.27	0.00	8.2	0	0	0.3
0	14	0.76	15.7	61.0	198.8	0.29	0.3	0.15	0.66	0.08	1.63	0.04	39.8	0	0.1	2.3
4	45	1.43	10.8	43.5	298.5	0.33	8.7	0.17	0.01	0.12	1.40	0.16	72.3	0	0.1	2.6
4	45	1.43	10.8	31.8	137.4	0.33	5.1	0.17	0.09	0.12	1.40	0.16	40.2	0	0.1	2.6
0	7	1.04	8.4	37.8	169.4	0.22	0	0.06	0.09	0.08	1.15	0.01	18.5	0	0	2.9
0	4	0.89	7.1	31.8	0.6	0.19	0	0.11	0.01	0.08	1.10	0.03	4.8	0	0	10.5
0	14	0.55	8.9	30.3	124.4	0.30	0	0.06	0.06	0.04	0.61	0.01	29.0	0	0	5.2
0	12	0.55	5.9	29.0	134.9	0.20	0	0.07	—	0.04	0.70	0.01	19.4	0	0	5.8
0	20	0.72	—	10.0	135.0	—	—	—	—	—	—	—	—	0	—	—
0	3	0.24	7.8	31.9	26.4	0.23	0	0.02	0.08	0.01	0.10	0.02	6.5	0	0	3.7
0	4	0.65	13.3	54.5	87.3	0.30	0	0.04	0.08	0.03	0.17	0.03	5.0	0	0	2.6
0	10	0.84	3.3	23.1	160.8	0.12	0	0.01	0.14	0.06	0.78	0.01	33.2	0	0	1.5
0	18	0.81	4.1	108.6	95.4	0.11	0	0.08	0.01	0.06	0.78	0.01	33.2	0	0	2.9
1	72	0.66	10.1	120.1	392.3	0.17	4.8	0.12	0.06	0.19	1.05	0.01	44.8	0	0	6.0
0	23	0.77	15.4	60.2	201.0	0.31	0.3	0.13	0.57	0.07	1.71	0.04	34.2	0	0	3.0
0	36	1.08	8.1	39.9	254.1	0.20	0	0.12	0.60	0.10	1.21	0.01	55.8	0	0	2.0
0	36	1.08	8.1	106.5	111.9	0.20	0	0.12	0.60	0.10	1.21	0.01	55.8	0	0	2.0
0	10	0.84	3.3	23.1	160.8	0.12	0	0.01	0.14	0.06	0.78	0.01	33.2	0	0	1.5
2	57	0.73	15.1	85.7	255.6	0.24	4.8	0.10	—	0.12	0.89	0.07	26.9	0.4	0	6.8
0	48	0.74	10.6	83.2	226.0	0.23	0	0.10	—	0.08	1.64	0.03	26.0	0	0	6.1
0	15	1.32	18.6	54.9	238.5	0.48	0	0.15	0.15	0.09	1.48	0.04	56.1	0	0	1.9

TABLE A-1 Table of Food Composition *(continued)*

(Computer code is for Cengage Diet Analysis Plus program)

DA+ Code	Food Description	QTY	Measure	Wt (g)	H_2O (g)	Ener (cal)	Prot (g)	Carb (g)	Fiber (g)	Fat (g)	Fat Breakdown (g) Sat	Mono	Poly
	Breads, Baked Goods, Cakes, Cookies, Crackers, Chips, Pies—continued												
9057	Wheat crackers, low salt	10	item(s)	30	0.9	142	2.6	19.5	1.4	6.2	1.6	3.4	0.8
9022	Whole wheat crackers	7	item(s)	28	0.8	124	2.5	19.2	2.9	4.8	1.0	1.6	1.8
	Pastry												
16754	Apple fritter	1	item(s)	17	6.4	61	1.0	5.5	0.2	3.9	0.9	1.7	1.1
41565	Cinnamon rolls with icing, refrigerated dough	1	serving(s)	44	12.3	145	2.0	23.0	0.5	5.0	1.5	—	—
4945	Croissant, butter	1	item(s)	57	13.2	231	4.7	26.1	1.5	12.0	6.6	3.1	0.6
9096	Danish, nut	1	item(s)	65	13.3	280	4.6	29.7	1.3	16.4	3.8	8.9	2.8
9115	Doughnut with creme filling	1	item(s)	85	32.5	307	5.4	25.5	0.7	20.8	4.6	10.3	2.6
9117	Doughnut with jelly filling	1	item(s)	85	30.3	289	5.0	33.2	0.8	15.9	4.1	8.7	2.0
4947	Doughnut, cake	1	item(s)	47	9.8	198	2.4	23.4	0.7	10.8	1.7	4.4	3.7
9105	Doughnut, cake, chocolate glazed	1	item(s)	42	6.8	175	1.9	24.1	0.9	8.4	2.2	4.7	1.0
437	Doughnut, glazed	1	item(s)	60	15.2	242	3.8	26.6	0.7	13.7	3.5	7.7	1.7
10617	Toaster pastry, brown sugar cinnamon	1	item(s)	50	5.3	210	3.0	35.0	1.0	6.0	1.0	4.0	1.0
30928	Toaster pastry, cream cheese	1	item(s)	54	—	200	3.0	23.0	0	11.0	4.5	—	—
	Muffins												
25015	Blueberry	1	item(s)	63	29.7	160	3.4	23.0	0.8	6.0	0.9	1.5	3.3
9189	Corn, ready to eat	1	item(s)	57	18.6	174	3.4	29.0	1.9	4.8	0.8	1.2	1.8
9121	English muffin, plain, enriched	1	item(s)	57	24.0	134	4.4	26.2	1.5	1.0	0.1	0.2	0.5
29582	English muffin, toasted	1	item(s)	50	18.6	128	4.2	25.0	1.5	1.0	0.1	0.2	0.5
9145	English muffin, wheat	1	item(s)	57	24.1	127	5.0	25.5	2.6	1.1	0.2	0.2	0.5
8894	Oat bran	1	item(s)	57	20.0	154	4.0	27.5	2.6	4.2	0.6	1.0	2.4
	Granola bars												
38161	Kudos milk chocolate granola bars w/fruit and nuts	1	item(s)	28	—	90	2.0	15.0	1.0	3.0	1.0	—	—
38196	Nature Valley banana nut crunchy granola bars	2	item(s)	42	—	190	4.0	28.0	2.0	7.0	1.0	—	—
38187	Nature Valley fruit 'n' nut trail mix bar	1	item(s)	35	—	140	3.0	25.0	2.0	4.0	0.5	—	—
1383	Plain, hard	1	item(s)	25	1.0	115	2.5	15.8	1.3	4.9	0.6	1.1	3.0
4606	Plain, soft	1	item(s)	28	1.8	126	2.1	19.1	1.3	4.9	2.1	1.1	1.5
	Pies												
454	Apple pie, prepared from home recipe	1	slice(s)	155	73.3	411	3.7	57.5	2.3	19.4	4.7	8.4	5.2
470	Pecan pie, prepared from home recipe	1	slice(s)	122	23.8	503	6.0	63.7	—	27.1	4.9	13.6	7.0
33356	Pie crust mix, prepared, baked	1	slice(s)	20	2.1	100	1.3	10.1	0.4	6.1	1.5	3.5	0.8
9007	Pie crust, ready to bake, frozen, enriched, baked	1	slice(s)	16	1.8	82	0.7	7.9	0.2	5.2	1.7	2.5	0.6
472	Pumpkin pie, prepared from home recipe	1	slice(s)	155	90.7	316	7.0	40.9	—	14.4	4.9	5.7	2.8
	Rolls												
8555	Crescent dinner roll	1	item(s)	28	9.7	78	2.7	13.8	0.6	1.2	0.3	0.3	0.6
489	Hamburger roll or bun, plain	1	item(s)	43	14.9	120	4.1	21.3	0.9	1.9	0.5	0.5	0.8
490	Hard roll	1	item(s)	57	17.7	167	5.6	30.0	1.3	2.5	0.3	0.6	1.0
5127	Kaiser roll	1	item(s)	57	17.7	167	5.6	30.0	1.3	2.5	0.3	0.6	1.0
5130	Whole wheat roll or bun	1	item(s)	28	9.4	75	2.5	14.5	2.1	1.3	0.2	0.3	0.6
	Sport bars												
37026	Balance original chocolate bar	1	item(s)	50	—	200	14.0	22.0	0.5	6.0	3.5	—	—
37024	Balance original peanut butter bar	1	item(s)	50	—	200	14.0	22.0	1.0	6.0	2.5	—	—
36580	Clif Bar chocolate brownie energy bar	1	item(s)	68	—	240	10.0	45.0	5.0	4.5	1.5	—	—
36583	Clif Bar crunchy peanut butter energy bar	1	item(s)	68	—	250	12.0	40.0	5.0	6.0	1.5	—	—
36589	Clif Luna Nutz over Chocolate energy bar	1	item(s)	48	—	180	10.0	25.0	3.0	4.5	2.5	—	—
12005	PowerBar apple cinnamon	1	item(s)	65	—	230	9.0	45.0	3.0	2.5	0.5	1.5	0.5
16078	PowerBar banana	1	item(s)	65	—	230	9.0	45.0	3.0	2.5	0.5	1.0	0.5
16080	PowerBar chocolate	1	item(s)	65	6.4	230	10.0	45.0	3.0	2.0	0.5	0.5	1.0
29092	PowerBar peanut butter	1	item(s)	65	—	240	10.0	45.0	3.0	3.5	0.5	—	—
	Tortillas												
1391	Corn tortillas, soft	1	item(s)	26	11.9	57	1.5	11.6	1.6	0.7	0.1	0.2	0.4
1669	Flour tortilla	1	item(s)	32	9.7	100	2.7	16.4	1.0	2.5	0.6	1.2	0.5

Chol (mg)	Calc (mg)	Iron (mg)	Magn (mg)	Pota (mg)	Sodi (mg)	Zinc (mg)	Vit A (µg)	Thia (mg)	Vit E (mg α)	Ribo (mg)	Niac (mg)	Vit B_6 (mg)	Fola (µg DFE)	Vit C (mg)	Vit B_{12} (µg)	Sele (µg)
0	15	1.32	18.6	60.9	84.9	0.48	0	0.15	0.15	0.09	1.48	0.04	21.6	0	0	10.1
0	14	0.86	27.7	83.2	184.5	0.60	0	0.05	0.24	0.02	1.26	0.05	7.8	0	0	4.1
14	9	0.26	2.2	22.4	6.8	0.09	7.1	0.03	0.07	0.04	0.23	0.01	9.2	0.2	0.1	2.6
0	—	0.72	—	—	340.1	—	0	—	—	—	—	—	—	—	—	—
38	21	1.15	9.1	67.3	424.1	0.42	117.4	0.22	0.47	0.13	1.24	0.03	74.1	0.1	0.1	12.9
30	61	1.17	20.8	61.8	236.0	0.56	5.9	0.14	0.53	0.15	1.49	0.06	79.3	1.1	0.1	9.2
20	21	1.55	17.0	68.0	262.7	0.68	9.4	0.28	0.24	0.12	1.90	0.05	92.7	0	0.1	9.2
22	21	1.49	17.0	67.2	249.1	0.63	14.5	0.26	0.36	0.12	1.81	0.08	88.4	0	0.2	10.6
17	21	0.91	9.4	59.7	256.6	0.25	17.9	0.10	0.90	0.11	0.87	0.02	32.9	0.1	0.1	4.4
24	89	0.95	14.3	44.5	142.8	0.23	5.0	0.01	0.08	0.02	0.19	0.01	27.3	0	0	1.7
4	26	0.36	13.2	64.8	205.2	0.46	2.4	0.53	—	0.04	0.39	0.03	13.2	0.1	0.1	5.0
0	0	1.80	—	70.0	190.0	—	—	0.15	—	0.17	2.00	0.20	21.0	0	0	—
10	100	1.80	—	—	220.0	—	—	0.15	—	0.17	2.00	—	21.0	0	0.6	—
20	56	1.02	7.8	70.2	289.4	0.28	—	0.17	0.75	0.15	1.25	0.02	62.5	0.4	0.1	8.8
15	42	1.60	18.2	39.3	297.0	0.30	29.6	0.15	0.45	0.18	1.16	0.04	63.8	0	0.1	8.7
0	30	1.42	12.0	74.7	264.5	0.39	0	0.25	—	0.16	2.21	0.02	57.0	0	0	—
0	95	1.36	11.0	71.5	252.0	0.38	0	0.19	0.16	0.14	1.90	0.02	62.5	0.1	0	13.5
0	101	1.63	21.1	106.0	217.7	0.61	0	0.24	0.25	0.16	1.91	0.05	46.7	0	0	16.6
0	36	2.39	89.5	289.0	224.0	1.04	0	0.14	0.37	0.05	0.23	0.09	79.2	0	0	6.3
0	200	0.36	—	—	60.0	—	0	—	—	—	—	—	—	0	0	—
0	20	1.08	—	120.0	160.0	—	0	—	—	—	—	—	—	0	—	—
0	0	0.00	—	—	95.0	—	0	—	—	—	—	—	—	0	—	—
0	15	0.72	23.8	82.3	72.0	0.50	0	0.06	—	0.03	0.39	0.02	5.8	0.2	0	4.0
0	30	0.72	21.0	92.3	79.0	0.42	0	0.08	—	0.04	0.14	0.02	6.8	0	0.1	4.6
0	11	1.73	10.9	122.5	327.1	0.29	17.1	0.22	—	0.16	1.90	0.05	58.9	2.6	0	12.1
106	39	1.80	31.7	162.3	319.6	1.24	100.0	0.22	—	0.22	1.03	0.07	41.5	0.2	0.2	14.6
0	12	0.43	3.0	12.4	145.8	0.07	0	0.06	—	0.03	0.47	0.01	22.2	0	0	4.4
0	3	0.36	2.9	17.6	103.5	0.05	0	0.04	0.42	0.06	0.39	0.01	16.3	0	0	0.5
65	146	1.96	29.5	288.3	348.8	0.71	660.3	0.14	—	0.31	1.21	0.07	43.4	2.6	0.1	11.0
0	39	0.93	5.9	26.3	134.1	0.18	0	0.11	0.02	0.09	1.16	0.02	47.6	0	0.1	5.5
0	59	1.42	9.0	40.4	206.0	0.28	0	0.17	0.03	0.13	1.78	0.03	73.5	0	0.1	8.4
0	54	1.87	15.4	61.6	310.1	0.53	0	0.27	0.23	0.19	2.41	0.02	86.1	0	0	22.3
0	54	1.86	15.4	61.6	310.1	0.53	0	0.27	0.23	0.19	2.41	0.01	86.1	0	0	22.3
0	30	0.69	24.1	77.1	135.5	0.57	0	0.07	0.26	0.04	1.04	0.06	8.5	0	0	14.0
3	100	4.50	40.0	160.0	180.0	3.75	—	0.37	—	0.42	5.00	0.50	102.0	60.0	1.5	17.5
3	100	4.50	40.0	130.0	230.0	3.75	—	0.37	—	0.42	5.00	0.50	102.0	60.0	1.5	17.5
0	250	4.50	100.0	370.0	150.0	3.00	—	0.37	—	0.25	3.00	0.40	80.0	60.0	0.9	14.0
0	250	4.50	100.0	230.0	250.0	3.00	—	0.37	—	0.25	3.00	0.40	80.0	60.0	0.9	14.0
0	350	5.40	80.0	190.0	190.0	5.25	—	1.20	—	1.36	16.00	2.00	400.0	60.0	6.0	24.5
0	300	6.30	140.0	125.0	100.0	5.25	—	1.50	—	1.70	20.00	2.00	400.0	60.0	6.0	—
0	300	6.30	140.0	190.0	100.0	5.25	0	1.50	—	1.70	20.00	2.00	400.0	60.0	6.0	—
0	300	6.30	140.0	200.0	95.0	5.25	0	1.50	—	1.70	20.00	2.00	400.0	60.0	6.0	5.1
0	300	6.30	140.0	130.0	120.0	5.25	0	1.50	—	1.70	20.00	2.00	400.0	60.0	6.0	—
0	21	0.32	18.7	48.4	11.7	0.34	0	0.02	0.07	0.02	0.39	0.06	1.3	0	0	1.6
0	41	1.06	7.0	49.6	203.5	0.17	0	0.17	0.06	0.08	1.14	0.01	64.3	0	0	7.1

TABLE A-1 Table of Food Composition *(continued)*

(Computer code is for Cengage Diet Analysis Plus program)

											Fat Breakdown (g)		
DA+ Code	Food Description	QTY	Measure	Wt (g)	H_2O (g)	Ener (cal)	Prot (g)	Carb (g)	Fiber (g)	Fat (g)	Sat	Mono	Poly
Breads, Baked Goods, Cakes, Cookies, Crackers, Chips, Pies—continued													
	Pancakes, waffles												
8926	Pancakes, blueberry, prepared from recipe	3	item(s)	114	60.6	253	7.0	33.1	0.8	10.5	2.3	2.6	4.7
5037	Pancakes, prepared from mix with egg and milk	3	item(s)	114	60.3	249	8.9	32.9	2.1	8.8	2.3	2.4	3.3
1390	Taco shells, hard	1	item(s)	13	1.0	62	0.9	8.3	0.6	2.8	0.6	1.6	0.5
30311	Waffle, 100% whole grain	1	item(s)	75	32.3	200	6.9	25.0	1.9	8.4	2.3	3.3	2.1
9219	Waffle, plain, frozen, toasted	2	item(s)	66	20.2	206	4.7	32.5	1.6	6.3	1.1	3.2	1.5
500	Waffle, plain, prepared from recipe	1	item(s)	75	31.5	218	5.9	24.7	1.7	10.6	2.1	2.6	5.1
Cereal, Flour, Grain, Pasta, Noodles, Popcorn													
	Grain												
2861	Amaranth, dry	½	cup(s)	98	9.6	365	14.1	64.5	9.1	6.3	1.6	1.4	2.8
1953	Barley, pearled, cooked	½	cup(s)	79	54.0	97	1.8	22.2	3.0	0.3	0.1	0	0.2
1956	Buckwheat groats, cooked, roasted	½	cup(s)	84	63.5	77	2.8	16.8	2.3	0.5	0.1	0.2	0.2
1957	Bulgur, cooked	½	cup(s)	91	70.8	76	2.8	16.9	4.1	0.2	0	0	0.1
1963	Couscous, cooked	½	cup(s)	79	57.0	88	3.0	18.2	1.1	0.1	0	0	0.1
1967	Millet, cooked	½	cup(s)	120	85.7	143	4.2	28.4	1.6	1.2	0.2	0.2	0.6
1969	Oat bran, dry	½	cup(s)	47	3.1	116	8.1	31.1	7.2	3.3	0.6	1.1	1.3
1972	Quinoa, dry	½	cup(s)	85	11.3	313	12.0	54.5	5.9	5.2	0.6	1.4	2.8
	Rice												
129	Brown, long grain, cooked	½	cup(s)	98	71.3	108	2.5	22.4	1.8	0.9	0.2	0.3	0.3
2863	Brown, medium grain, cooked	½	cup(s)	98	71.1	109	2.3	22.9	1.8	0.8	0.2	0.3	0.3
37488	Jasmine, saffroned, cooked	½	cup(s)	280	—	340	8.0	78.0	0	0	0	0	0
30280	Pilaf, cooked	½	cup(s)	103	74.0	129	2.1	22.2	0.6	3.3	0.6	1.5	1.0
28066	Spanish, cooked	½	cup(s)	244	184.2	241	5.7	50.2	3.3	1.9	0.4	0.6	0.7
2867	White glutinous, cooked	½	cup(s)	87	66.7	84	1.8	18.3	0.9	0.2	0	0.1	0.1
484	White, long grain, boiled	½	cup(s)	79	54.1	103	2.1	22.3	0.3	0.2	0.1	0.1	0.1
482	White, long grain, enriched, instant, boiled	½	cup(s)	83	59.4	97	1.8	20.7	0.5	0.4	0	0.1	0
486	White, long grain, enriched, parboiled, cooked	½	cup(s)	79	55.6	97	2.3	20.6	0.7	0.3	0.1	0.1	0.1
1194	Wild brown, cooked	½	cup(s)	82	60.6	83	3.3	17.5	1.5	0.3	0	0	0.2
	Flour and grain fractions												
505	All purpose flour, self-rising, enriched	½	cup(s)	63	6.6	221	6.2	46.4	1.7	0.6	0.1	0	0.2
503	All purpose flour, white, bleached, enriched	½	cup(s)	63	7.4	228	6.4	47.7	1.7	0.6	0.1	0	0.2
1643	Barley flour	½	cup(s)	56	5.5	198	4.2	44.7	2.1	0.8	0.2	0.1	0.4
383	Buckwheat flour, whole groat	½	cup(s)	60	6.7	201	7.6	42.3	6.0	1.9	0.4	0.6	0.6
504	Cake wheat flour, enriched	½	cup(s)	69	8.6	248	5.6	53.5	1.2	0.6	0.1	0.1	0.3
426	Cornmeal, degermed, enriched	½	cup(s)	69	7.8	255	5.0	54.6	2.8	1.2	0.1	0.2	0.5
424	Cornmeal, yellow whole grain	½	cup(s)	61	6.2	221	4.9	46.9	4.4	2.2	0.3	0.6	1.0
1978	Dark rye flour	½	cup(s)	64	7.1	207	9.0	44.0	14.5	1.7	0.2	0.2	0.8
1644	Masa corn flour, enriched	½	cup(s)	57	5.1	208	5.3	43.5	5.5	2.1	0.3	0.6	1.0
1976	Rice flour, brown	½	cup(s)	79	9.4	287	5.7	60.4	3.6	2.2	0.4	0.8	0.8
1645	Rice flour, white	½	cup(s)	79	9.4	289	4.7	63.3	1.9	1.1	0.3	0.3	0.3
1980	Semolina, enriched	½	cup(s)	84	10.6	301	10.6	60.8	3.2	0.9	0.1	0.1	0.4
2827	Soy flour, raw	½	cup(s)	42	2.2	185	14.7	14.9	4.1	8.8	1.3	1.9	4.9
1990	Wheat germ, crude	2	tablespoon(s)	14	1.6	52	3.3	7.4	1.9	1.4	0.2	0.2	0.9
506	Whole wheat flour	½	cup(s)	60	6.2	203	8.2	43.5	7.3	1.1	0.2	0.1	0.5
	Breakfast bars												
39230	Atkins Morning Start apple crisp breakfast bar	1	item(s)	37	—	170	11.0	12.0	6.0	9.0	4.0	—	—
10571	Nutri-Grain apple cinnamon cereal bar	1	item(s)	37	—	140	2.0	27.0	1.0	3.0	0.5	2.0	0.5
10647	Nutri-Grain blueberry cereal bar	1	item(s)	37	5.4	140	2.0	27.0	1.0	3.0	0.5	2.0	0.5
10648	Nutri-Grain raspberry cereal bar	1	item(s)	37	5.4	140	2.0	27.0	1.0	3.0	0.5	2.0	0.5
10649	Nutri-Grain strawberry cereal bar	1	item(s)	37	5.4	140	2.0	27.0	1.0	3.0	0.5	2.0	0.5
	Breakfast cereals, hot												
1260	Cream of Wheat, instant, prepared	½	cup(s)	121	—	388	12.9	73.3	4.3	0	0	0	0
365	Farina, enriched, cooked w/water and salt	½	cup(s)	117	102.4	56	1.7	12.2	0.3	0.1	0	0	0

Chol (mg)	Calc (mg)	Iron (mg)	Magn (mg)	Pota (mg)	Sodi (mg)	Zinc (mg)	Vit A (µg)	Thia (mg)	Vit E (mg α)	Ribo (mg)	Niac (mg)	Vit B_6 (mg)	Fola (µg DFE)	Vit C (mg)	Vit B_{12} (µg)	Sele (µg)
64	235	1.96	18.2	157.3	469.7	0.61	57.0	0.22	—	0.31	1.73	0.05	60.4	2.5	0.2	16.0
81	245	1.48	25.1	226.9	575.7	0.85	82.1	0.22	—	0.35	1.40	0.12	61.6	0.7	0.4	—
0	13	0.25	11.3	29.7	51.7	0.21	0.1	0.03	0.09	0.01	0.25	0.03	11.1	0	0	0.6
71	194	1.60	28.5	171.0	371.3	0.87	48.8	0.15	0.32	0.25	1.47	0.08	38.3	0	0.4	20.0
10	203	4.56	15.8	95.0	481.8	0.35	262.7	0.34	0.64	0.46	5.86	0.68	78.5	0	1.9	8.3
52	191	1.73	14.3	119.3	383.3	0.51	48.8	0.19	—	0.26	1.55	0.04	51.0	0.3	0.2	34.7
0	149	7.40	259.3	356.8	20.5	3.10	0	0.06	—	0.20	1.24	0.20	47.8	4.1	0	—
0	9	1.04	17.3	73.0	2.4	0.64	0	0.06	0.01	0.04	1.61	0.09	12.6	0	0	6.8
0	6	0.67	42.8	73.9	3.4	0.51	0	0.03	0.07	0.03	0.79	0.06	11.8	0	0	1.8
0	9	0.87	29.1	61.9	4.6	0.51	0	0.05	0.01	0.02	0.91	0.07	16.4	0	0	0.5
0	6	0.30	6.3	45.5	3.9	0.20	0	0.05	0.10	0.02	0.77	0.04	11.8	0	0	21.6
0	4	0.75	52.8	74.4	2.4	1.09	0	0.12	0.02	0.09	1.59	0.13	22.8	0	0	1.1
0	27	2.54	110.5	266.0	1.9	1.46	0	0.55	0.47	0.10	0.43	0.07	24.4	0	0	21.2
0	40	3.88	167.4	478.5	4.2	2.62	0.8	0.30	2.06	0.26	1.28	0.40	156.4	0	0	7.2
0	10	0.41	41.9	41.9	4.9	0.61	0	0.09	0.02	0.02	1.49	0.14	3.9	0	0	9.6
0	10	0.51	42.9	77.0	1.0	0.60	0	0.09	—	0.01	1.29	0.14	3.9	0	0	38.0
0	—	2.16	—	—	780.0	—	—	—	—	—	—	—	—	—	—	—
0	11	1.16	9.3	54.6	390.4	0.37	33.0	0.13	0.28	0.02	1.23	0.06	73.1	0.4	0	4.3
0	37	1.52	95.4	330.5	97.1	1.40	—	0.27	0.12	0.05	3.24	0.38	89.9	22.6	0	14.3
0	2	0.12	4.4	8.7	4.4	0.35	0	0.01	0.03	0.01	0.25	0.02	0.9	0	0	4.9
0	8	0.94	9.5	27.7	0.8	0.38	0	0.12	0.03	0.01	1.16	0.07	76.6	0	0	5.9
0	7	1.46	4.1	7.4	3.3	0.40	0	0.06	0.01	0.01	1.43	0.04	97.4	0	0	4.0
0	15	1.43	7.1	44.2	1.6	0.29	0	0.16	0.01	0.01	1.82	0.12	107.4	0	0	7.3
0	2	0.49	26.2	82.8	2.5	1.09	0	0.04	0.19	0.07	1.05	0.11	21.3	0	0	0.7
0	211	2.90	11.9	77.5	793.7	0.38	0	0.42	0.02	0.24	3.64	0.02	193.4	0	0	21.5
0	9	2.90	13.7	66.9	1.2	0.42	0	0.48	0.02	0.30	3.68	0.02	183.3	0	0	21.2
0	16	0.70	45.4	185.9	4.5	1.04	0	0.06	—	0.02	2.56	0.14	4.5	0	0	2.0
0	25	2.42	150.6	346.2	6.6	1.86	0	0.24	0.18	0.10	3.68	0.34	32.4	0	0	3.4
0	10	5.01	11.0	71.9	1.4	0.42	0	0.61	0.01	0.29	4.65	0.02	194.6	0	0	3.4
0	2	2.98	24.1	104.9	4.8	0.48	7.6	0.42	0.10	0.28	3.66	0.12	231.1	0	0	8.0
0	4	2.10	77.5	175.1	21.3	1.10	6.7	0.22	0.24	0.12	2.20	0.18	15.2	0	0	9.4
0	36	4.12	158.7	467.2	0.6	3.58	0.6	0.20	0.90	0.16	2.72	0.28	21.1	0	0	22.8
0	80	4.10	62.7	169.9	2.8	1.00	0	0.80	0.08	0.42	5.60	0.20	190.9	0	0	8.5
0	9	1.56	88.5	228.3	6.3	1.92	0	0.34	0.94	0.06	5.00	0.58	12.6	0	0	—
0	8	0.26	27.6	60.0	0	0.62	0	0.10	0.08	0.02	2.04	0.34	3.2	0	0	11.9
0	14	3.64	39.2	155.3	0.8	0.86	0	0.66	0.20	0.46	5.00	0.08	219.2	0	0	74.6
0	87	2.70	182.0	1067.0	5.5	1.65	2.5	0.24	0.82	0.48	1.83	0.18	146.4	0	0	3.2
0	6	0.90	34.4	128.2	1.7	1.76	0	0.27	—	0.07	0.97	0.18	40.4	0	0	11.4
0	20	2.32	82.8	243.0	3.0	1.74	0	0.26	0.48	0.12	3.82	0.20	26.4	0	0	42.4
0	200	—	—	90.0	70.0	—	—	0.22	—	0.25	3.00	—	—	9.0	—	—
0	200	1.80	8.0	75.0	110.0	1.50	—	0.37	—	0.42	5.00	0.50	40.0	0	—	—
0	200	1.80	8.0	75.0	110.0	1.50	—	0.37	—	0.42	5.00	0.50	40.0	0	0	—
0	200	1.80	8.0	70.0	110.0	1.50	—	0.37	—	0.42	5.00	0.50	40.0	0	0	—
0	200	1.80	8.0	55.0	110.0	1.50	—	0.37	—	0.42	5.00	0.50	40.0	0	0	—
0	862	34.91	21.4	150.8	732.6	0.86	—	1.59	—	1.47	21.55	2.15	122.2	0	0	—
0	5	0.58	2.3	15.1	383.3	0.09	0	0.07	0.01	0.05	0.57	0.01	139.2	0	0	10.6

TABLE A-1 Table of Food Composition *(continued)*

(Computer code is for Cengage Diet Analysis Plus program)

DA+ Code	Food Description	QTY	Measure	Wt (g)	H_2O (g)	Ener (cal)	Prot (g)	Carb (g)	Fiber (g)	Fat (g)	Fat Breakdown (g) Sat	Mono	Poly
	Cereal, Flour, Grain, Pasta, Noodles, Popcorn—continued												
363	Grits, white corn, regular and quick, enriched, cooked w/water and salt	½	cup(s)	121	103.3	71	1.7	15.6	0.4	0.2	0	0.1	0.1
8636	Grits, yellow corn, regular and quick, enriched, cooked w/salt	½	cup(s)	121	103.3	71	1.7	15.6	0.4	0.2	0	0.1	0.1
8657	Oatmeal, cooked w/water	½	cup(s)	117	97.8	83	3.0	14.0	2.0	1.8	0.4	0.5	0.7
5500	Oatmeal, maple and brown sugar, instant, prepared	1	item(s)	198	150.2	200	4.8	40.4	2.4	2.2	0.4	0.7	0.8
5510	Oatmeal, ready to serve, packet, prepared	1	item(s)	186	158.7	112	4.1	19.8	2.7	2.0	0.4	0.7	0.8
	Breakfast cereals, ready to eat												
1197	All-Bran	1	cup(s)	62	1.3	160	8.1	46.0	18.2	2.0	0.4	0.4	1.3
1200	All-Bran Buds	1	cup(s)	91	2.7	212	6.4	72.7	39.1	1.9	0.4	0.5	1.2
1199	Apple Jacks	1	cup(s)	33	0.9	130	1.0	30.0	0.5	0.5	0	—	—
1204	Cap'n Crunch	1	cup(s)	36	0.9	147	1.3	30.7	1.3	2.0	0.5	0.4	0.3
1205	Cap'n Crunch Crunchberries	1	cup(s)	35	0.9	133	1.3	29.3	1.3	2.0	0.5	0.4	0.3
1206	Cheerios	1	cup(s)	30	1.0	110	3.0	22.0	3.0	2.0	0	0.5	0.5
3415	Cocoa Puffs	1	cup(s)	30	0.6	120	1.0	26.0	0.2	1.0	—	—	—
1207	Cocoa Rice Krispies	1	cup(s)	41	1.0	160	1.3	36.0	1.3	1.3	0.7	0	0
5522	Complete wheat bran flakes	1	cup(s)	39	1.4	120	4.0	30.7	6.7	0.7	—	—	—
1211	Corn Flakes	1	cup(s)	28	0.9	100	2.0	24.0	1.0	0	0	0	0
1247	Corn Pops	1	cup(s)	31	0.9	120	1.0	28.0	0.3	0	0	0	0
1937	Cracklin' Oat Bran	1	cup(s)	65	2.3	267	5.3	46.7	8.0	9.3	4.0	4.7	1.3
1220	Froot Loops	1	cup(s)	32	0.8	120	1.0	28.0	1.0	1.0	0.5	0	0
38214	Frosted Cheerios	1	cup(s)	37	—	149	2.5	31.1	1.2	1.2	—	—	—
372	Frosted Flakes	1	cup(s)	41	1.1	160	1.3	37.3	1.3	0	0	0	0
38215	Frosted Mini Chex	1	cup(s)	40	—	147	1.3	36.0	0	0	0	0	0
10268	Frosted Mini-Wheats	1	cup(s)	59	3.1	208	5.8	47.4	5.8	1.2	0	0	0.6
38216	Frosted Wheaties	1	cup(s)	40	—	147	1.3	36.0	0.3	0	0	0	0
1223	Granola, prepared	½	cup(s)	61	3.3	298	9.1	32.5	5.5	14.7	2.5	5.8	5.6
2415	Honey Bunches of Oats honey roasted	1	cup(s)	40	0.9	160	2.7	33.3	1.3	2.0	0.7	1.2	0.1
1227	Honey Nut Cheerios	1	cup(s)	37	0.9	149	3.7	29.9	2.5	1.9	0	0.6	0.6
2424	Honeycomb	1	cup(s)	22	0.3	83	1.5	19.5	0.8	0.4	0	—	—
10286	Kashi whole grain puffs	1	cup(s)	19	—	70	2.0	15.0	1.0	0.5	0	—	—
41142	Kellogg's Mueslix	1	cup(s)	83	7.2	298	7.6	60.8	6.1	4.6	0.7	2.4	1.5
1231	Kix	1	cup(s)	24	0.5	96	1.6	20.8	0.8	0.4	—	—	—
30569	Life	1	cup(s)	43	1.7	160	4.0	33.3	2.7	2.0	0.3	0.6	0.6
1233	Lucky Charms	1	cup(s)	24	0.6	96	1.6	20.0	0.8	0.8	—	—	—
38220	Multi Grain Cheerios	1	cup(s)	30	—	110	3.0	24.0	3.0	1.0	—	—	—
1201	Multi-Bran Chex	1	cup(s)	63	1.3	216	4.3	52.9	8.6	1.6	0	0	0.5
13633	Post Bran Flakes	1	cup(s)	40	1.5	133	4.0	32.0	6.7	0.7	0	—	—
1241	Product 19	1	cup(s)	30	1.0	100	2.0	25.0	1.0	0	0	0	0
32432	Puffed rice, fortified	1	cup(s)	14	0.4	56	0.9	12.6	0.2	0.1	0	—	—
32433	Puffed wheat, fortified	1	cup(s)	12	0.4	44	1.8	9.6	0.5	0.1	0	—	—
13334	Quaker 100% natural granola oats and honey	½	cup(s)	48	—	220	5.0	31.0	3.0	9.0	3.8	4.1	1.2
13335	Quaker 100% natural granola oats, honey, and raisins	½	cup(s)	51	—	230	5.0	34.0	3.0	9.0	3.6	3.8	1.1
2420	Raisin Bran	1	cup(s)	59	5.0	190	4.0	46.0	8.0	1.0	0	0.1	0.4
1244	Rice Chex	1	cup(s)	31	0.8	120	2.0	27.0	0.3	0	0	0	0
1245	Rice Krispies	1	cup(s)	26	0.8	96	1.6	23.2	0	0	0	0	0
5593	Shredded Wheat	1	cup(s)	49	0.4	177	5.8	40.9	6.9	1.1	0.1	0	0.2
1248	Smacks	1	cup(s)	36	1.1	133	2.7	32.0	1.3	0.7	—	—	—
1246	Special K	1	cup(s)	31	0.9	110	7.0	22.0	0.5	0	0	0	0
3428	Total corn flakes	1	cup(s)	23	0.6	83	1.5	18.0	0.6	0	0	0	0
1253	Total whole grain	1	cup(s)	40	1.1	147	2.7	30.7	4.0	1.3	—	—	—
1254	Trix	1	cup(s)	30	0.6	120	1.0	27.0	1.0	1.0	—	—	—
382	Wheat germ, toasted	2	tablespoon(s)	14	0.8	54	4.1	7.0	2.1	1.5	0.3	0.2	0.9
1257	Wheaties	1	cup(s)	36	1.2	132	3.6	28.8	3.6	1.2	—	—	—
	Pasta, noodles												
449	Chinese chow mein noodles, cooked	½	cup(s)	23	0.2	119	1.9	12.9	0.9	6.9	1.0	1.7	3.9
1995	Corn pasta, cooked	½	cup(s)	70	47.8	88	1.8	19.5	3.4	0.5	0.1	0.1	0.2

Chol (mg)	Calc (mg)	Iron (mg)	Magn (mg)	Pota (mg)	Sodi (mg)	Zinc (mg)	Vit A (μg)	Thia (mg)	Vit E (mg α)	Ribo (mg)	Niac (mg)	Vit B_6 (mg)	Fola (μg DFE)	Vit C (mg)	Vit B_{12} (μg)	Sele (μg)
0	4	0.73	6.1	25.4	269.8	0.08	0	0.10	0.02	0.07	0.87	0.03	46.0	0	0	3.8
0	4	0.73	6.1	25.4	269.8	0.08	2.4	0.10	0.02	0.07	0.87	0.03	44.8	0	0	3.3
0	11	1.05	31.6	81.9	4.7	1.17	0	0.09	0.09	0.02	0.26	0.01	7.0	0	0	6.3
0	26	6.83	49.9	126.4	403.5	1.03	0	1.02	—	0.05	1.56	0.30	42.2	0	0	11.1
0	21	3.96	44.7	112.4	240.9	0.92	0	0.60	—	0.04	0.77	0.18	18.7	0	0	3.8
0	241	10.90	224.4	632.4	150.0	3.00	300.1	1.40	—	1.68	9.16	7.44	1362.8	12.4	12.0	5.8
0	57	13.64	186.4	909.1	614.5	4.55	464.5	1.09	1.42	1.27	15.45	6.09	2054.8	18.2	18.2	26.3
0	0	4.50	8.0	30.0	130.0	1.50	150.2	0.37	—	0.42	5.00	0.50	196.0	15.0	1.5	2.4
0	5	6.80	20.0	73.3	266.7	5.00	2.5	0.51	—	0.57	6.68	0.67	946.8	0	0	6.7
0	7	6.53	18.7	73.3	240.0	5.13	2.4	0.51	—	0.57	6.68	0.67	910.7	0	0	6.7
0	100	8.10	40.0	95.0	280.0	3.75	150.3	0.37	—	0.42	5.00	0.50	493.2	6.0	1.5	11.3
0	100	4.50	8.0	50.0	170.0	3.75	0	0.37	—	0.42	5.00	0.50	165.9	6.0	1.5	2.0
0	53	6.00	10.7	66.7	253.3	2.00	200.1	0.49	—	0.56	6.67	0.67	442.8	20.0	2.0	5.8
0	0	24.00	53.3	226.7	280.0	20.00	300.1	2.00	—	2.27	26.67	2.67	909.1	80.0	8.0	4.1
0	0	8.10	3.4	25.0	200.0	0.16	149.8	0.37	—	0.42	5.00	0.50	221.8	6.0	1.5	1.4
0	0	1.80	2.5	25.0	120.0	1.50	150.0	0.37	—	0.42	5.00	0.50	174.7	6.0	1.5	2.0
0	27	2.40	80.0	293.3	200.0	2.00	299.9	0.49	—	0.56	6.67	0.67	217.1	20.0	2.0	14.4
0	0	4.50	8.0	35.0	150.0	1.50	150.1	0.37	—	0.42	5.00	0.50	166.1	15.0	1.5	2.3
0	124	5.60	19.9	68.4	261.3	4.67	—	0.46	—	0.52	6.22	0.62	444.4	7.5	1.9	—
0	0	6.00	3.7	26.7	200.0	0.20	200.1	0.49	—	0.56	6.67	0.67	260.4	8.0	2.0	1.8
0	133	12.00	—	33.3	266.7	4.00	—	0.49	—	0.56	6.67	0.67	266.7	8.0	2.0	—
0	0	16.66	69.4	196.7	5.8	1.74	0	0.43	—	0.49	5.78	0.58	193.5	0	1.7	2.4
0	133	10.80	0	46.7	266.7	10.00	—	1.00	—	1.13	13.33	1.33	901.2	8.0	4.0	—
0	48	2.58	106.8	329.4	15.3	2.45	0.6	0.44	6.77	0.17	1.30	0.17	50.0	0.7	0	17.0
0	0	10.80	21.3	0	253.3	0.40	—	0.49	—	0.56	6.67	0.67	549.6	0	2.0	—
0	124	5.60	39.8	112.0	336.0	4.67	—	0.46	—	0.52	6.22	0.62	444.4	7.5	1.9	8.8
0	0	2.03	6.0	26.3	165.4	1.13	—	0.28	—	0.32	3.74	0.37	126.1	0	1.1	—
0	0	0.36	—	60.0	0	—	0	0.03	—	0.03	0.80	0.00	—	0	—	—
0	48	6.83	74.2	363.3	257.5	5.67	136.7	0.67	6.00	0.67	8.33	3.08	1030.0	0.3	9.2	14.4
0	120	6.48	6.4	28.0	216.0	3.00	120.2	0.30	—	0.34	4.00	0.40	317.8	4.8	1.2	4.8
0	149	11.87	41.3	120.0	213.3	5.33	0.9	0.53	—	0.60	7.12	0.71	607.6	0	0	10.7
0	80	3.60	12.8	48.0	168.0	3.00	—	0.30	—	0.34	4.00	0.40	300.7	4.8	1.2	4.8
0	100	18.00	24.0	85.0	200.0	15.00	—	1.50	—	1.70	20.00	2.00	699.3	15.0	6.0	—
0	108	17.50	64.8	237.7	410.6	4.05	171.1	0.40	—	0.45	5.40	0.54	893.3	6.5	1.6	4.9
0	0	10.80	80.0	266.7	280.0	2.00	—	0.49	—	0.56	6.67	0.67	453.3	0	2.0	—
0	0	18.00	16.0	50.0	210.0	15.00	225.3	1.50	—	1.70	20.00	2.00	675.9	60.0	6.0	3.6
0	1	4.43	3.5	15.8	0.4	0.14	0	0.36	—	0.25	4.94	0.01	2.7	0	0	1.5
0	3	3.80	17.4	41.8	0.5	0.28	0	0.31	—	0.21	4.23	0.02	3.8	0	0	14.8
0	61	1.20	51.0	220.0	20.0	1.05	0.5	0.13	—	0.12	0.82	0.07	16.8	0.2	0.1	8.3
0	59	1.20	49.0	250.0	20.0	0.99	0.5	0.13	—	0.12	0.80	0.08	15.8	0.4	0.1	8.8
0	20	10.80	80.0	360.0	360.0	2.25	—	0.37	—	0.42	5.00	0.50	248.4	0	2.1	—
0	100	9.00	9.3	35.0	290.0	3.75	—	0.37	—	0.42	5.00	0.50	389.4	6.0	1.5	1.2
0	0	1.44	12.8	32.0	256.0	0.48	120.1	0.30	—	0.34	4.80	0.40	237.4	4.8	1.2	4.1
0	18	2.90	60.3	179.3	1.1	1.37	0	0.14	—	0.12	3.47	0.18	21.1	0	0	2.0
0	0	0.48	10.7	53.3	66.7	0.40	200.2	0.49	—	0.56	6.67	0.67	224.6	8.0	2.0	17.5
0	0	8.10	16.0	60.0	220.0	0.90	225.1	0.52	—	0.59	7.00	2.00	675.8	21.0	6.0	7.0
0	752	13.53	0	22.6	157.9	11.28	112.8	1.13	22.56	1.28	15.04	1.50	518.2	45.1	4.5	1.2
0	1333	24.00	32.0	120.0	253.3	20.00	200.4	2.00	31.32	2.27	26.67	2.67	901.2	80.0	8.0	1.9
0	100	4.50	0	15.0	190.0	3.75	150.3	0.37	—	0.42	5.00	0.50	155.4	6.0	1.5	6.0
0	6	1.28	45.2	133.8	0.6	2.35	0.7	0.23	2.25	0.11	0.79	0.13	49.7	0.8	0	9.2
0	24	9.72	38.4	126.0	264.0	9.00	180.4	0.90	—	1.02	12.00	1.20	403.6	7.2	3.6	1.7
0	5	1.06	11.7	27.0	98.8	0.31	0	0.13	0.78	0.09	1.33	0.02	31.1	0	0	9.7
0	1	0.18	25.2	21.7	0	0.44	2.1	0.04	—	0.02	0.39	0.04	4.2	0	0	2.0

TABLE A-1 Table of Food Composition *(continued)*

(Computer code is for Cengage Diet Analysis Plus program)

DA+ Code	Food Description	QTY	Measure	Wt (g)	H2O (g)	Ener (cal)	Prot (g)	Carb (g)	Fiber (g)	Fat (g)	Fat Breakdown (g) Sat	Mono	Poly
Cereal, Flour, Grain, Pasta, Noodles, Popcorn—continued													
448	Egg noodles, enriched, cooked	½	cup(s)	80	54.2	110	3.6	20.1	1.0	1.7	0.3	0.5	0.4
1563	Egg noodles, spinach, enriched, cooked	½	cup(s)	80	54.8	106	4.0	19.4	1.8	1.3	0.3	0.4	0.3
440	Macaroni, enriched, cooked	½	cup(s)	70	43.5	111	4.1	21.6	1.3	0.7	0.1	0.1	0.2
2000	Macaroni, tricolor vegetable, enriched, cooked	½	cup(s)	67	45.8	86	3.0	17.8	2.9	0.1	0	0	0
1996	Plain pasta, fresh-refrigerated, cooked	½	cup(s)	64	43.9	84	3.3	16.0	—	0.7	0.1	0.1	0.3
1725	Ramen noodles, cooked	½	cup(s)	114	94.5	104	3.0	15.4	1.0	4.3	0.2	0.2	0.2
2878	Soba noodles, cooked	½	cup(s)	95	69.4	94	4.8	20.4	—	0.1	0	0	0
2879	Somen noodles, cooked	½	cup(s)	88	59.8	115	3.5	24.2	—	0.2	0	0	0.1
493	Spaghetti, al dente, cooked	½	cup(s)	65	41.6	95	3.5	19.5	1.0	0.5	0.1	0.1	0.2
2884	Spaghetti, whole wheat, cooked	½	cup(s)	70	47.0	87	3.7	18.6	3.2	0.4	0.1	0.1	0.1
	Popcorn												
476	Air popped	1	cup(s)	8	0.3	31	1.0	6.2	1.2	0.4	0	0.1	0.2
4619	Caramel	1	cup(s)	35	1.0	152	1.3	27.8	1.8	4.5	1.3	1.0	1.6
4620	Cheese flavored	1	cup(s)	36	0.9	188	3.3	18.4	3.5	11.8	2.3	3.5	5.5
477	Popped in oil	1	cup(s)	11	0.1	64	0.8	5.0	0.9	4.8	0.8	1.1	2.6
Fruit and Fruit Juices													
	Apples												
952	Juice, prepared from frozen concentrate	½	cup(s)	120	105.0	56	0.2	13.8	0.1	0.1	0	0	0
225	Juice, unsweetened, canned	½	cup(s)	124	109.0	58	0.1	14.5	0.1	0.1	0	0	0
224	Slices	½	cup(s)	55	47.1	29	0.1	7.6	1.3	0.1	0	0	0
946	Slices without skin, boiled	½	cup(s)	86	73.1	45	0.2	11.7	2.1	0.3	0	0	0.1
223	Raw medium, with peel	1	item(s)	138	118.1	72	0.4	19.1	3.3	0.2	0	0	0.1
948	Dried, sulfured	¼	cup(s)	22	6.8	52	0.2	14.2	1.9	0.1	0	0	0
226	Applesauce, sweetened, canned	½	cup(s)	128	101.5	97	0.2	25.4	1.5	0.2	0	0	0.1
227	Applesauce, unsweetened, canned	½	cup(s)	122	107.8	52	0.2	13.8	1.5	0.1	0	0	0
38492	Crabapples	1	item(s)	35	27.6	27	0.1	7.0	0.9	0.1	0	0	0
	Apricot												
228	Fresh without pits	4	item(s)	140	120.9	67	2.0	15.6	2.8	0.5	0	0.2	0.1
229	Halves with skin, canned in heavy syrup	½	cup(s)	129	100.1	107	0.7	27.7	2.1	0.1	0	0	0
230	Halves, dried, sulfured	¼	cup(s)	33	10.1	79	1.1	20.6	2.4	0.2	0	0	0
	Avocado												
233	California, whole, without skin or pit	½	cup(s)	115	83.2	192	2.2	9.9	7.8	17.7	2.4	11.3	2.1
234	Florida, whole, without skin or pit	½	cup(s)	115	90.6	138	2.5	9.0	6.4	11.5	2.2	6.3	1.9
2998	Pureed	⅛	cup(s)	28	20.2	44	0.5	2.4	1.8	4.0	0.6	2.7	0.5
	Banana												
4580	Dried chips	¼	cup(s)	55	2.4	285	1.3	32.1	4.2	18.5	15.9	1.1	0.3
235	Fresh whole, without peel	1	item(s)	118	88.4	105	1.3	27.0	3.1	0.4	0.1	0	0.1
	Blackberries												
237	Raw	½	cup(s)	72	63.5	31	1.0	6.9	3.8	0.4	0	0	0.2
958	Unsweetened, frozen	½	cup(s)	76	62.1	48	0.9	11.8	3.8	0.3	0	0	0.2
	Blueberries												
959	Canned in heavy syrup	½	cup(s)	128	98.3	113	0.8	28.2	2.0	0.4	0	0.1	0.2
238	Raw	½	cup(s)	73	61.1	41	0.5	10.5	1.7	0.2	0	0	0.1
960	Unsweetened, frozen	½	cup(s)	78	67.1	40	0.3	9.4	2.1	0.5	0	0.1	0.2
	Boysenberries												
961	Canned in heavy syrup	½	cup(s)	128	97.6	113	1.3	28.6	3.3	0.2	0	0	0.1
962	Unsweetened, frozen	½	cup(s)	66	56.7	33	0.7	8.0	3.5	0.2	0	0	0.1
35576	**Breadfruit**	1	item(s)	384	271.3	396	4.1	104.1	18.8	0.9	0.2	0.1	0.3
	Cherries												
967	Sour red, canned in water	½	cup(s)	122	109.7	44	0.9	10.9	1.3	0.1	0	0	0
3000	Sour red, raw	½	cup(s)	78	66.8	39	0.8	9.4	1.2	0.2	0.1	0.1	0.1
3004	Sweet, canned in heavy syrup	½	cup(s)	127	98.2	105	0.8	26.9	1.9	0.2	0	0.1	0.1
969	Sweet, canned in water	½	cup(s)	124	107.9	57	1.0	14.6	1.9	0.2	0	0	0
240	Sweet, raw	½	cup(s)	73	59.6	46	0.8	11.6	1.5	0.1	0	0	0

Chol (mg)	Calc (mg)	Iron (mg)	Magn (mg)	Pota (mg)	Sodi (mg)	Zinc (mg)	Vit A (μg)	Thia (mg)	Vit E (mg α)	Ribo (mg)	Niac (mg)	Vit B_6 (mg)	Fola (μg DFE)	Vit C (mg)	Vit B_{12} (μg)	Sele (μg)
23	10	1.17	16.8	30.4	4.0	0.52	4.8	0.23	0.13	0.11	1.66	0.03	110.4	0	0.1	19.1
26	15	0.87	19.2	29.6	9.6	0.50	8.0	0.19	0.46	0.09	1.17	0.09	75.2	0	0.1	17.4
0	5	0.90	12.6	30.8	0.7	0.36	0	0.19	0.04	0.10	1.18	0.03	83.3	0	0	18.5
0	7	0.33	12.7	20.8	4.0	0.30	3.4	0.08	0.14	0.04	0.72	0.02	71.0	0	0	13.3
21	4	0.73	11.5	15.4	3.8	0.36	3.8	0.13	—	0.10	0.64	0.02	66.6	0	0.1	—
18	9	0.89	8.5	34.5	414.5	0.30	—	0.08	—	0.04	0.71	0.03	4.0	0.1	0	—
0	4	0.45	8.5	33.2	57.0	0.11	0	0.09	—	0.02	0.48	0.03	6.6	0	0	—
0	7	0.45	1.8	25.5	141.7	0.19	0	0.01	—	0.03	0.08	0.01	1.8	0	0	—
0	7	1.00	12.4	51.5	0.5	0.35	0	0.12	0.04	0.07	0.90	0.04	77.4	0	0	40.0
0	11	0.74	21.0	30.8	2.1	0.57	0	0.08	0.21	0.03	0.50	0.06	3.5	0	0	18.1
0	1	0.25	11.5	26.3	0.6	0.25	0.8	0.01	0.02	0.01	0.18	0.01	2.5	0	0	0
2	15	0.61	12.3	38.4	72.5	0.20	0.7	0.02	0.42	0.02	0.77	0.01	1.8	0	0	1.3
4	40	0.79	32.5	93.2	317.4	0.71	13.6	0.04	—	0.08	0.52	0.08	3.9	0.2	0.2	4.3
0	0	0.22	8.7	20.0	116.4	0.34	0.9	0.01	0.27	0.00	0.13	0.01	2.8	0	0	0.2
0	7	0.31	6.0	150.6	8.4	0.05	0	0.00	0.01	0.02	0.05	0.04	0	0.7	0	0.1
0	9	0.46	3.7	147.6	3.7	0.04	0	0.03	0.01	0.02	0.12	0.04	0	1.1	0	0.1
0	3	0.06	2.7	58.8	0.5	0.02	1.6	0.01	0.10	0.01	0.05	0.02	1.6	2.5	0	0
0	4	0.16	2.6	75.2	0.9	0.03	1.7	0.01	0.04	0.01	0.08	0.04	0.9	0.2	0	0.3
0	8	0.16	6.9	147.7	1.4	0.05	4.1	0.02	0.24	0.03	0.12	0.05	4.1	6.3	0	0
0	3	0.30	3.4	96.8	18.7	0.04	0	0.00	0.11	0.03	0.20	0.03	0	0.8	0	0.3
0	5	0.44	3.8	77.8	3.8	0.05	1.3	0.01	0.26	0.03	0.24	0.03	1.3	2.2	0	0.4
0	4	0.14	3.7	91.5	2.4	0.03	1.2	0.01	0.25	0.03	0.22	0.03	1.2	1.5	0	0.4
0	6	0.12	2.5	67.9	0.4	—	0.7	0.01	0.20	0.01	0.03	—	2.0	2.8	0	—
0	18	0.54	14.0	362.6	1.4	0.28	134.4	0.04	1.24	0.05	0.84	0.07	12.6	14.0	0	0.1
0	12	0.38	9.0	180.6	5.2	0.14	80.0	0.02	0.77	0.02	0.48	0.07	2.6	4.0	0	0.1
0	18	0.87	10.5	381.5	3.3	0.12	59.1	0.00	1.42	0.02	0.85	0.05	3.3	0.3	0	0.7
0	15	0.66	33.3	583.0	9.2	0.78	8.0	0.08	2.23	0.16	2.19	0.31	102.3	10.1	0	0.4
0	12	0.19	27.6	403.6	2.3	0.45	8.0	0.02	3.03	0.04	0.76	0.08	40.3	20.0	0	—
0	3	0.14	8.0	134.1	1.9	0.17	1.9	0.01	0.57	0.03	0.48	0.07	22.4	2.8	0	0.1
0	10	0.69	41.8	294.8	3.3	0.40	2.2	0.04	0.13	0.01	0.39	0.14	7.7	3.5	0	0.8
0	6	0.30	31.9	422.4	1.2	0.17	3.5	0.03	0.11	0.08	0.78	0.43	23.6	10.3	0	1.2
0	21	0.45	14.4	116.6	0.7	0.38	7.9	0.01	0.84	0.02	0.47	0.02	18.0	15.1	0	0.3
0	22	0.60	16.6	105.7	0.8	0.19	4.5	0.02	0.88	0.03	0.91	0.05	25.7	2.3	0	0.3
0	6	0.42	5.1	51.2	3.8	0.09	2.6	0.04	0.49	0.07	0.14	0.05	2.6	1.4	0	0.1
0	4	0.20	4.4	55.8	0.7	0.12	2.2	0.03	0.41	0.03	0.30	0.04	4.4	7.0	0	0.1
0	6	0.14	3.9	41.9	0.8	0.05	1.6	0.03	0.37	0.03	0.40	0.05	5.4	1.9	0	0.1
0	23	0.55	14.1	115.2	3.8	0.24	2.6	0.03	—	0.04	0.29	0.05	43.5	7.9	0	0.5
0	18	0.56	10.6	91.7	0.7	0.15	2.0	0.04	0.57	0.02	0.51	0.04	41.6	2.0	0	0.1
0	65	2.07	96.0	1881.6	7.7	0.46	0	0.42	0.38	0.11	3.45	0.38	53.8	111.4	0	2.3
0	13	1.67	7.3	119.6	8.5	0.09	46.4	0.02	0.28	0.05	0.22	0.05	9.8	2.6	0	0
0	12	0.25	7.0	134.1	2.3	0.08	49.6	0.02	0.05	0.03	0.31	0.03	6.2	7.8	0	0
0	11	0.44	11.4	183.4	3.8	0.12	10.1	0.02	0.29	0.05	0.50	0.03	5.1	4.6	0	0
0	14	0.45	11.2	162.4	1.2	0.10	9.9	0.03	0.29	0.05	0.51	0.04	5.0	2.7	0	0
0	9	0.26	8.0	161.0	0	0.05	2.2	0.02	0.05	0.02	0.11	0.04	2.9	5.1	0	0

TABLE A-1 Table of Food Composition *(continued)*

(Computer code is for Cengage Diet Analysis Plus program)

											Fat Breakdown (g)		
DA+ Code	Food Description	QTY	Measure	Wt (g)	H_2O (g)	Ener (cal)	Prot (g)	Carb (g)	Fiber (g)	Fat (g)	Sat	Mono	Poly
	Fruit and Fruit Juices—continued												
	Cranberries												
3007	Chopped, raw	½	cup(s)	55	47.9	25	0.2	6.7	2.5	0.1	0	0	0
1717	Cranberry apple juice drink	½	cup(s)	123	102.6	77	0	19.4	0	0.1	0	0	0.1
1638	Cranberry juice cocktail	½	cup(s)	127	109.0	68	0	17.1	0	0.1	0	0	0.1
241	Cranberry juice cocktail, low calorie, with saccharin	½	cup(s)	119	112.8	23	0	5.5	0	0	0	0	0
242	Cranberry sauce, sweetened, canned	¼	cup(s)	69	42.0	105	0.1	26.9	0.7	0.1	0	0	0
	Dates												
244	Domestic, chopped	¼	cup(s)	45	9.1	125	1.1	33.4	3.6	0.2	0	0	0
243	Domestic, whole	¼	cup(s)	45	9.1	125	1.1	33.4	3.6	0.2	0	0	0
	Figs												
975	Canned in heavy syrup	½	cup(s)	130	98.8	114	0.5	29.7	2.8	0.1	0	0	0.1
974	Canned in water	½	cup(s)	124	105.7	66	0.5	17.3	2.7	0.1	0	0	0.1
973	Raw, medium	2	item(s)	100	79.1	74	0.7	19.2	2.9	0.3	0.1	0.1	0.1
	Fruit cocktail and salad												
245	Fruit cocktail, canned in heavy syrup	½	cup(s)	124	99.7	91	0.5	23.4	1.2	0.1	0	0	0
978	Fruit cocktail, canned in juice	½	cup(s)	119	103.6	55	0.5	14.1	1.2	0	0	0	0
977	Fruit cocktail, canned in water	½	cup(s)	119	107.6	38	0.5	10.1	1.2	0.1	0	0	0
979	Fruit salad, canned in water	½	cup(s)	123	112.1	37	0.4	9.6	1.2	0.1	0	0	0
	Gooseberries												
982	Canned in light syrup	½	cup(s)	126	100.9	92	0.8	23.6	3.0	0.3	0	0	0.1
981	Raw	½	cup(s)	75	65.9	33	0.7	7.6	3.2	0.4	0	0	0.2
	Grapefruit												
251	Juice, pink, sweetened, canned	½	cup(s)	125	109.1	57	0.7	13.9	0.1	0.1	0	0	0
249	Juice, white	½	cup(s)	124	111.2	48	0.6	11.4	0.1	0.1	0	0	0
3022	Pink or red, raw	½	cup(s)	114	100.8	48	0.9	12.2	1.8	0.2	0	0	0
248	Sections, canned in light syrup	½	cup(s)	127	106.2	76	0.7	19.6	0.5	0.1	0	0	0
983	Sections, canned in water	½	cup(s)	122	109.6	44	0.7	11.2	0.5	0.1	0	0	0
247	White, raw	½	cup(s)	115	104.0	38	0.8	9.7	1.3	0.1	0	0	0
	Grapes												
255	American, slip skin	½	cup(s)	46	37.4	31	0.3	7.9	0.4	0.2	0.1	0	0
256	European, red or green, adherent skin	½	cup(s)	76	60.8	52	0.5	13.7	0.7	0.1	0	0	0
3159	Grape juice drink, canned	½	cup(s)	125	106.6	71	0	18.2	0.1	0	0	0	0
259	Grape juice, sweetened, with added vitamin C, prepared from frozen concentrate	½	cup(s)	125	108.6	64	0.2	15.9	0.1	0.1	0	0	0
3060	Raisins, seeded, packed	¼	cup(s)	41	6.8	122	1.0	32.4	2.8	0.2	0.1	0	0.1
987	**Guava, raw**	1	item(s)	55	44.4	37	1.4	7.9	3.0	0.5	0.2	0	0.2
35593	**Guavas, strawberry**	1	item(s)	6	4.8	4	0	1.0	0.3	0	0	0	0
3027	**Jackfruit**	½	cup(s)	83	60.4	78	1.2	19.8	1.3	0.2	0	0	0.1
990	**Kiwi fruit or Chinese gooseberries**	1	item(s)	76	63.1	46	0.9	11.1	2.3	0.4	0	0	0.2
	Lemon												
262	Juice	1	tablespoon(s)	15	13.8	4	0.1	1.3	0.1	0	0	0	0
993	Peel	1	teaspoon(s)	2	1.6	1	0	0.3	0.2	0	0	0	0
992	Raw	1	item(s)	108	94.4	22	1.3	11.6	5.1	0.3	0	0	0.1
	Lime												
269	Juice	1	tablespoon(s)	15	14.0	4	0.1	1.3	0.1	0	0	0	0
994	Raw	1	item(s)	67	59.1	20	0.5	7.1	1.9	0.1	0	0	0
995	**Loganberries, frozen**	½	cup(s)	74	62.2	40	1.1	9.6	3.9	0.2	0	0	0.1
	Mandarin orange												
1038	Canned in juice	½	cup(s)	125	111.4	46	0.8	11.9	0.9	0	0	0	0
1039	Canned in light syrup	½	cup(s)	126	104.7	77	0.6	20.4	0.9	0.1	0	0	0
999	**Mango**	½	cup(s)	83	67.4	54	0.4	14.0	1.5	0.2	0.1	0.1	0
1005	**Nectarine, raw, sliced**	½	cup(s)	69	60.4	30	0.7	7.3	1.2	0.2	0	0.1	0.1

PAGE KEY: A-2 = Breads/Baked Goods A-8 = Cereal/Rice/Pasta A-12 = Fruit A-18 = Vegetables/Legumes A-28 = Nuts/Seeds A-30 = Vegetarian A-32 = Dairy A-40 = Eggs A-40 = Seafood A-44 = Meats A-48 = Poultry A-48 = Processed Meats A-50 = Beverages A-54 = Fats/Oils A-56 = Sweets A-58 = Spices/Condiments/Sauces A-62 = Mixed Foods/Soups/Sandwiches A-68 = Fast Food A-88 = Convenience A-90 = Baby Foods

A

Chol (mg)	Calc (mg)	Iron (mg)	Magn (mg)	Pota (mg)	Sodi (mg)	Zinc (mg)	Vit A (µg)	Thia (mg)	Vit E (mg α)	Ribo (mg)	Niac (mg)	Vit B_6 (mg)	Fola (µg DFE)	Vit C (mg)	Vit B_{12} (µg)	Sele (µg)
0	4	0.13	3.3	46.8	1.1	0.05	1.7	0.01	0.66	0.01	0.05	0.03	0.6	7.3	0	0.1
0	4	0.09	1.2	20.8	2.5	0.02	0	0.00	0.15	0.00	0.00	0.00	0	48.4	0	0
0	4	0.13	1.3	17.7	2.5	0.04	0	0.00	0.28	0.00	0.05	0.00	0	53.5	0	0.3
0	11	0.05	2.4	29.6	3.6	0.02	0	0.00	0.06	0.00	0.00	0.00	0	38.2	0	0
0	3	0.15	2.1	18.0	20.1	0.03	1.4	0.01	0.57	0.01	0.06	0.01	0.7	1.4	0	0.2
0	17	0.45	19.1	291.9	0.9	0.12	0	0.02	0.02	0.02	0.56	0.07	8.5	0.2	0	1.3
0	17	0.45	19.1	291.9	0.9	0.12	0	0.02	0.02	0.02	0.56	0.07	8.5	0.2	0	1.3
0	35	0.36	13.0	128.2	1.3	0.14	2.6	0.03	0.16	0.05	0.55	0.09	2.6	1.3	0	0.3
0	35	0.36	12.4	127.7	1.2	0.15	2.5	0.03	0.10	0.05	0.55	0.09	2.5	1.2	0	0.1
0	35	0.36	17.0	232.0	1.0	0.14	7.0	0.06	0.10	0.04	0.40	0.10	6.0	2.0	0	0.2
0	7	0.36	6.2	109.1	7.4	0.09	12.4	0.02	0.49	0.02	0.46	0.06	3.7	2.4	0	0.6
0	9	0.25	8.3	112.6	4.7	0.11	17.8	0.01	0.47	0.02	0.48	0.06	3.6	3.2	0	0.6
0	6	0.30	8.3	111.4	4.7	0.11	15.4	0.02	0.47	0.01	0.43	0.06	3.6	2.5	0	0.6
0	9	0.37	6.1	95.6	3.7	0.10	27.0	0.02	—	0.03	0.46	0.04	3.7	2.3	0	1.0
0	20	0.42	7.6	97.0	2.5	0.14	8.8	0.03	—	0.07	0.19	0.02	3.8	12.6	0	0.5
0	19	0.23	7.5	148.5	0.8	0.09	11.3	0.03	0.28	0.02	0.23	0.06	4.5	20.8	0	0.5
0	10	0.45	12.5	202.2	2.5	0.08	0	0.05	0.05	0.03	0.40	0.03	12.5	33.6	0	0.1
0	11	0.25	14.8	200.1	1.2	0.06	1.2	0.05	0.27	0.02	0.25	0.05	12.4	46.9	0	0.1
0	25	0.09	10.3	154.5	0	0.07	66.4	0.04	0.14	0.03	0.23	0.06	14.9	35.7	0	0.1
0	18	0.50	12.7	163.8	2.5	0.10	0	0.04	0.11	0.02	0.30	0.02	11.4	27.1	0	1.1
0	18	0.50	12.2	161.0	2.4	0.11	0	0.05	0.11	0.03	0.30	0.02	11.0	26.6	0	1.1
0	14	0.07	10.4	170.2	0	0.08	2.3	0.04	0.15	0.02	0.30	0.05	11.5	38.3	0	1.6
0	6	0.13	2.3	87.9	0.9	0.02	2.3	0.04	0.09	0.02	0.14	0.05	1.8	1.8	0	0
0	8	0.27	5.3	144.2	1.5	0.05	2.3	0.05	0.14	0.05	0.14	0.07	1.5	8.2	0	0.1
0	9	0.16	7.5	41.3	11.3	0.04	0	0.28	0.00	0.44	0.18	0.04	1.3	33.1	0	0.1
0	5	0.13	5.0	26.3	2.5	0.05	0	0.02	0.00	0.03	0.16	0.05	1.3	29.9	0	0.1
0	12	1.06	12.4	340.3	11.6	0.07	0	0.04	—	0.07	0.46	0.07	1.2	2.2	0	0.2
0	10	0.14	12.1	229.4	1.1	0.12	17.1	0.03	0.40	0.02	0.59	0.06	27.0	125.6	0	0.3
0	1	0.01	1.0	17.5	2.2	—	0.3	0.00	—	0.00	0.03	0.00	—	2.2	0	
0	28	0.49	30.5	250.0	2.5	0.35	12.4	0.02	—	0.09	0.33	0.09	11.5	5.5	0	0.5
0	26	0.23	12.9	237.1	2.3	0.10	3.0	0.02	1.11	0.01	0.25	0.04	19.0	70.5	0	0.2
0	1	0.00	0.9	18.9	0.2	0.01	0.2	0.00	0.02	0.00	0.02	0.01	2.0	7.0	0	0
0	3	0.01	0.3	3.2	0.1	0.01	0.1	0.00	0.01	0.00	0.01	0.00	0.3	2.6	0	0
0	66	0.75	13.0	156.6	3.2	0.10	2.2	0.05	—	0.04	0.21	0.11	—	83.2	0	1.0
0	2	0.02	1.2	18.0	0.3	0.01	0.3	0.00	0.03	0.00	0.02	0.01	1.5	4.6	0	0
0	22	0.40	4.0	68.3	1.3	0.07	1.3	0.02	0.14	0.01	0.13	0.02	5.4	19.5	0	0.3
0	19	0.47	15.4	106.6	0.7	0.25	1.5	0.04	0.64	0.03	0.62	0.05	19.1	11.2	0	0.1
0	14	0.34	13.7	165.6	6.2	0.64	53.5	0.10	0.12	0.04	0.55	0.05	6.2	42.6	0	0.5
0	9	0.47	10.1	98.3	7.6	0.30	52.9	0.07	0.13	0.06	0.56	0.05	6.3	24.9	0	0.5
0	8	0.10	7.4	128.7	1.7	0.03	31.4	0.05	0.92	0.04	0.48	0.11	11.6	22.8	0	0.5
0	4	0.19	6.2	138.7	0	0.12	11.7	0.02	0.53	0.02	0.78	0.02	3.5	3.7	0	0

TABLE A-1 Table of Food Composition *(continued)*

(Computer code is for Cengage Diet Analysis Plus program)

DA+ Code	Food Description	QTY	Measure	Wt (g)	H2O (g)	Ener (cal)	Prot (g)	Carb (g)	Fiber (g)	Fat (g)	Fat Breakdown (g) Sat	Mono	Poly
	Fruit and Fruit Juices—continued												
	Melons												
271	Cantaloupe	½	cup(s)	80	72.1	27	0.7	6.5	0.7	0.1	0	0	0.1
1000	Casaba melon	½	cup(s)	85	78.1	24	0.9	5.6	0.8	0.1	0	0	0
272	Honeydew	½	cup(s)	89	79.5	32	0.5	8.0	0.7	0.1	0	0	0
318	Watermelon	½	cup(s)	76	69.5	23	0.5	5.7	0.3	0.1	0	0	0
	Orange												
14412	Juice with calcium and vitamin D	½	cup(s)	120	—	55	1.0	13.0	0	0	0	0	0
29630	Juice, fresh squeezed	½	cup(s)	124	109.5	56	0.9	12.9	0.2	0.2	0	0	0
14411	Juice, not from concentrate	½	cup(s)	120	—	55	1.0	13.0	0	0	0	0	0
278	Juice, unsweetened, prepared from frozen concentrate	½	cup(s)	125	109.7	56	0.8	13.4	0.2	0.1	0	0	0
3040	Peel	1	teaspoon(s)	2	1.5	2	0	0.5	0.2	0	0	0	0
273	Raw	1	item(s)	131	113.6	62	1.2	15.4	3.1	0.2	0	0	0
274	Sections	½	cup(s)	90	78.1	42	0.8	10.6	2.2	0.1	0	0	0
	Papaya, raw												
16830	Dried, strips	2	item(s)	46	12.0	119	1.9	29.9	5.5	0.4	0.1	0.1	0.1
282	Papaya	½	cup(s)	70	62.2	27	0.4	6.9	1.3	0.1	0	0	0
35640	**Passion fruit, purple**	1	item(s)	18	13.1	17	0.4	4.2	1.9	0.1	0	0	0.1
	Peach												
285	Halves, canned in heavy syrup	½	cup(s)	131	103.9	97	0.6	26.1	1.7	0.1	0	0	0.1
286	Halves, canned in water	½	cup(s)	122	113.6	29	0.5	7.5	1.6	0.1	0	0	0
290	Slices, sweetened, frozen	½	cup(s)	125	93.4	118	0.8	30.0	2.3	0.2	0	0.1	0.1
283	Raw, medium	1	item(s)	150	133.3	59	1.4	14.3	2.3	0.4	0	0.1	0.1
	Pear												
8672	Asian	1	item(s)	122	107.7	51	0.6	13.0	4.4	0.3	0	0.1	0.1
293	D'Anjou	1	item(s)	200	168.0	120	1.0	30.0	5.2	1.0	0	0.2	0.2
294	Halves, canned in heavy syrup	½	cup(s)	133	106.9	98	0.3	25.5	2.1	0.2	0	0	0
1012	Halves, canned in juice	½	cup(s)	124	107.2	62	0.4	16.0	2.0	0.1	0	0	0
291	Raw	1	item(s)	166	139.0	96	0.6	25.7	5.1	0.2	0	0	0
1017	**Persimmon**	1	item(s)	25	16.1	32	0.2	8.4	—	0.1	0	0	0
	Pineapple												
3053	Canned in extra heavy syrup	½	cup(s)	130	101.0	108	0.4	28.0	1.0	0.1	0	0	0
1019	Canned in juice	½	cup(s)	125	104.0	75	0.5	19.5	1.0	0.1	0	0	0
296	Canned in light syrup	½	cup(s)	126	108.0	66	0.5	16.9	1.0	0.2	0	0	0.1
1018	Canned in water	½	cup(s)	123	111.7	39	0.5	10.2	1.0	0.1	0	0	0
299	Juice, unsweetened, canned	½	cup(s)	125	108.0	66	0.5	16.1	0.3	0.2	0	0	0.1
295	Raw, diced	½	cup(s)	78	66.7	39	0.4	10.2	1.1	0.1	0	0	0
1024	**Plantain, cooked**	½	cup(s)	77	51.8	89	0.6	24.0	1.8	0.1	0.1	0	0
300	**Plum, raw, large**	1	item(s)	66	57.6	30	0.5	7.5	0.9	0.2	0	0.1	0
1027	**Pomegranate**	1	item(s)	154	124.7	105	1.5	26.4	0.9	0.5	0.1	0.1	0.1
	Prunes												
5644	Dried	2	item(s)	17	5.2	40	0.4	10.7	1.2	0.1	0	0	0
305	Dried, stewed	½	cup(s)	124	86.5	133	1.2	34.8	3.8	0.2	0	0.1	0
306	Juice, canned	1	cup(s)	256	208.0	182	1.6	44.7	2.6	0.1	0	0.1	0
	Raspberries												
309	Raw	½	cup(s)	62	52.7	32	0.7	7.3	4.0	0.4	0	0	0.2
310	Red, sweetened, frozen	½	cup(s)	125	90.9	129	0.9	32.7	5.5	0.2	0	0	0.1
311	**Rhubarb, cooked with sugar**	½	cup(s)	120	81.5	140	0.5	37.5	2.7	0.1	0	0	0.1
	Strawberries												
313	Raw	½	cup(s)	72	65.5	23	0.5	5.5	1.4	0.2	0	0	0.1
315	Sweetened, frozen, thawed	½	cup(s)	128	99.5	99	0.7	26.8	2.4	0.2	0	0	0.1
16828	**Tangelo**	1	item(s)	95	82.4	45	0.9	11.2	2.3	0.1	0	0	0
	Tangerine												
1040	Juice	½	cup(s)	124	109.8	53	0.6	12.5	0.2	0.2	0	0	0
316	Raw	1	item(s)	88	74.9	47	0.7	11.7	1.6	0.3	0	0.1	0.1

Chol (mg)	Calc (mg)	Iron (mg)	Magn (mg)	Pota (mg)	Sodi (mg)	Zinc (mg)	Vit A (µg)	Thia (mg)	Vit E (mg α)	Ribo (mg)	Niac (mg)	Vit B_6 (mg)	Fola (µg DFE)	Vit C (mg)	Vit B_{12} (µg)	Sele (µg)
0	7	0.17	9.6	213.6	12.8	0.14	135.2	0.03	0.04	0.01	0.59	0.05	16.8	29.4	0	0.3
0	9	0.29	9.4	154.7	7.7	0.06	0	0.01	0.04	0.03	0.20	0.14	6.8	18.5	0	0.3
0	5	0.15	8.8	201.8	15.9	0.07	2.7	0.03	0.01	0.01	0.37	0.07	16.8	15.9	0	0.6
0	5	0.18	7.6	85.1	0.8	0.07	21.3	0.02	0.04	0.01	0.13	0.03	2.3	6.2	0	0.3
0	175	0.00	12.0	225.0	0	—	0	0.08	—	0.03	0.40	0.06	30.0	36.0	0	—
0	14	0.25	13.6	248.0	1.2	0.06	12.4	0.11	0.05	0.04	0.50	0.05	37.2	62.0	0	0.1
0	10	0.00	12.5	225.0	0	0.06	0	0.08	—	0.03	0.40	0.06	30.0	36.0	0	0.1
0	11	0.12	12.5	236.6	1.2	0.06	6.2	0.10	0.25	0.02	0.25	0.06	54.8	48.4	0	0.1
0	3	0.01	0.4	4.2	0.1	0.01	0.4	0.00	0.01	0.00	0.01	0.00	0.6	2.7	0	0
0	52	0.13	13.1	237.1	0	0.09	14.4	0.11	0.23	0.05	0.36	0.07	39.3	69.7	0	0.7
0	36	0.09	9.0	162.9	0	0.06	9.9	0.07	0.16	0.03	0.25	0.05	27.0	47.9	0	0.4
0	73	0.30	30.4	782.9	9.2	0.21	83.7	0.06	2.22	0.08	0.93	0.05	58.0	37.7	0	1.8
0	17	0.07	7.0	179.9	2.1	0.05	38.5	0.02	0.51	0.02	0.24	0.01	26.6	43.3	0	0.4
0	2	0.28	5.2	62.6	5.0	0.01	11.5	0.00	0.00	0.02	0.27	0.01	2.5	5.4	0	0.1
0	4	0.35	6.6	120.5	7.9	0.11	22.3	0.01	0.64	0.03	0.80	0.02	3.9	3.7	0	0.4
0	2	0.39	6.1	120.8	3.7	0.11	32.9	0.01	0.59	0.02	0.63	0.02	3.7	3.5	0	0.4
0	4	0.46	6.3	162.5	7.5	0.06	17.5	0.01	0.77	0.04	0.81	0.02	3.8	117.8	0	0.5
0	9	0.37	13.5	285.0	0	0.25	24.0	0.03	1.09	0.04	1.20	0.03	6.0	9.9	0	0.2
0	5	0.00	9.8	147.6	0	0.02	0	0.01	0.14	0.01	0.26	0.02	9.8	4.6	0	0.1
0	22	0.50	12.0	250.0	0	0.24	—	0.04	1.00	0.08	0.20	0.03	14.6	8.0	0	1.0
0	7	0.29	5.3	86.5	6.7	0.10	0	0.01	0.10	0.02	0.32	0.01	1.3	1.5	0	0
0	11	0.36	8.7	119.0	5.0	0.11	0	0.01	0.10	0.01	0.25	0.02	1.2	2.0	0	0
0	15	0.28	11.6	197.5	1.7	0.16	1.7	0.02	0.19	0.04	0.26	0.04	11.6	7.0	0	0.2
0	7	0.62	—	77.5	0.3	—	0	—	—	—	—	—	—	16.5	0	—
0	18	0.49	19.5	132.6	1.3	0.14	1.3	0.11	—	0.03	0.36	0.09	6.5	9.5	0	—
0	17	0.35	17.4	151.9	1.2	0.12	2.5	0.12	0.01	0.02	0.35	0.09	6.2	11.8	0	0.5
0	18	0.49	20.2	132.3	1.3	0.15	2.5	0.11	0.01	0.03	0.36	0.09	6.3	9.5	0	0.5
0	18	0.49	22.1	156.2	1.2	0.15	2.5	0.11	0.01	0.03	0.37	0.09	6.2	9.5	0	0.5
0	16	0.39	15.0	162.5	2.5	0.14	0	0.07	0.03	0.03	0.25	0.13	22.5	12.5	0	0.1
0	10	0.22	9.3	84.5	0.8	0.09	2.3	0.06	0.02	0.03	0.39	0.09	14.0	37.0	0	0.1
0	2	0.45	24.6	358.1	3.9	0.10	34.7	0.04	0.10	0.04	0.58	0.19	20.0	8.4	0	1.1
0	4	0.11	4.6	103.6	0	0.06	11.2	0.02	0.17	0.02	0.27	0.02	3.3	6.3	0	0
0	5	0.46	4.6	398.9	4.6	0.18	7.7	0.04	0.92	0.04	0.46	0.16	9.2	9.4	0	0.9
0	7	0.16	6.9	123.0	0.3	0.07	6.6	0.01	0.07	0.03	0.32	0.03	0.7	0.1	0	0
0	24	0.51	22.3	398.0	1.2	0.24	21.1	0.03	0.24	0.12	0.90	0.27	0	3.6	0	0.1
0	31	3.02	35.8	706.6	10.2	0.53	0	0.04	0.30	0.17	2.01	0.55	0	10.5	0	1.5
0	15	0.42	13.5	92.9	0.6	0.26	1.2	0.02	0.54	0.02	0.37	0.03	12.9	16.1	0	0.1
0	19	0.81	16.3	142.5	1.3	0.22	3.8	0.02	0.90	0.05	0.28	0.04	32.5	20.6	0	0.4
0	174	0.25	16.2	115.0	1.0	—	—	0.02	—	0.03	0.25	—	—	4.0	0	—
0	12	0.30	9.4	110.2	0.7	0.10	0.7	0.02	0.21	0.02	0.28	0.03	17.3	42.3	0	0.3
0	14	0.59	7.7	125.0	1.3	0.06	1.3	0.01	0.30	0.09	0.37	0.03	5.1	50.4	0	0.9
0	38	0.09	9.5	172.0	0	0.06	10.5	0.08	0.17	0.03	0.26	0.05	28.5	50.5	0	0.5
0	22	0.25	9.9	219.8	1.2	0.04	16.1	0.07	0.16	0.02	0.12	0.05	6.2	38.3	0	0.1
0	33	0.13	10.6	146.1	1.8	0.06	29.9	0.05	0.18	0.03	0.33	0.07	14.1	23.5	0	0.1

TABLE A-1 Table of Food Composition *(continued)*

(Computer code is for Cengage Diet Analysis Plus program)

											Fat Breakdown (g)		
DA+ Code	Food Description	QTY	Measure	Wt (g)	H_2O (g)	Ener (cal)	Prot (g)	Carb (g)	Fiber (g)	Fat (g)	Sat	Mono	Poly
Vegetables, Legumes													
	Amaranth												
1043	Leaves, boiled, drained	½	cup(s)	66	60.4	14	1.4	2.7	—	0.1	0	0	0.1
1042	Leaves, raw	1	cup(s)	28	25.7	6	0.7	1.1	—	0.1	0	0	0
8683	**Arugula leaves, raw**	1	cup(s)	20	18.3	5	0.5	0.7	0.3	0.1	0	0	0.1
	Artichoke												
1044	Boiled, drained	1	item(s)	120	100.9	64	3.5	14.3	10.3	0.4	0.1	0	0.2
2885	Hearts, boiled, drained	½	cup(s)	84	70.6	45	2.4	10.0	7.2	0.3	0.1	0	0.1
	Asparagus												
566	Boiled, drained	½	cup(s)	90	83.4	20	2.2	3.7	1.8	0.2	0	0	0.1
568	Canned, drained	½	cup(s)	121	113.7	23	2.6	3.0	1.9	0.8	0.2	0	0.3
565	Tips, frozen, boiled, drained	½	cup(s)	90	84.7	16	2.7	1.7	1.4	0.4	0.1	0	0.2
	Bamboo shoots												
1048	Boiled, drained	½	cup(s)	60	57.6	7	0.9	1.2	0.6	0.1	0	0	0.1
1049	Canned, drained	½	cup(s)	66	61.8	12	1.1	2.1	0.9	0.3	0.1	0	0.1
	Beans												
1801	Adzuki beans, boiled	½	cup(s)	115	76.2	147	8.6	28.5	8.4	0.1	0	—	—
511	Baked beans with franks, canned	½	cup(s)	130	89.8	184	8.7	19.9	8.9	8.5	3.0	3.7	1.1
513	Baked beans with pork in sweet sauce, canned	½	cup(s)	127	89.3	142	6.7	26.7	5.3	1.8	0.6	0.6	0.5
512	Baked beans with pork in tomato sauce, canned	½	cup(s)	127	93.0	119	6.5	23.6	5.1	1.2	0.5	0.7	0.3
1805	Black beans, boiled	½	cup(s)	86	56.5	114	7.6	20.4	7.5	0.5	0.1	0	0.2
14597	Chickpeas, garbanzo beans or bengal gram, boiled	½	cup(s)	82	49.4	134	7.3	22.5	6.2	2.1	0.2	0.5	0.9
569	Fordhook lima beans, frozen, boiled, drained	½	cup(s)	85	62.0	88	5.2	16.4	4.9	0.3	0.1	0	0.1
1806	French beans, boiled	½	cup(s)	89	58.9	114	6.2	21.3	8.3	0.7	0.1	0	0.4
2773	Great northern beans, boiled	½	cup(s)	89	61.1	104	7.4	18.7	6.2	0.4	0.1	0	0.2
2736	Hyacinth beans, boiled, drained	½	cup(s)	44	37.8	22	1.3	4.0	—	0.1	0.1	0.1	0
570	Lima beans, baby, frozen, boiled, drained	½	cup(s)	90	65.1	95	6.0	17.5	5.4	0.3	0.1	0	0.1
515	Lima beans, boiled, drained	½	cup(s)	85	57.1	105	5.8	20.1	4.5	0.3	0.1	0	0.1
579	Mung beans, sprouted, boiled, drained	½	cup(s)	62	57.9	13	1.3	2.6	0.5	0.1	0	0	0
510	Navy beans, boiled	½	cup(s)	91	58.1	127	7.5	23.7	9.6	0.6	0.1	0.1	0.4
32816	Pinto beans, boiled, drained, no salt added	½	cup(s)	63	58.8	14	1.2	2.6	—	0.2	0	0	0.1
1052	Pinto beans, frozen, boiled, drained	½	cup(s)	47	27.3	76	4.4	14.5	4.0	0.2	0	0	0.1
514	Red kidney beans, canned	½	cup(s)	128	99.0	108	6.7	19.9	6.9	0.5	0.1	0.2	0.2
1810	Refried beans, canned	½	cup(s)	127	96.1	119	6.9	19.6	6.7	1.6	0.6	0.7	0.2
1053	Shell beans, canned	½	cup(s)	123	111.1	37	2.2	7.6	4.2	0.2	0	0	0.1
1670	Soybeans, boiled	½	cup(s)	86	53.8	149	14.3	8.5	5.2	7.7	1.1	1.7	4.4
1108	Soybeans, green, boiled, drained	½	cup(s)	90	61.7	127	11.1	9.9	3.8	5.8	0.7	1.1	2.7
1807	White beans, small, boiled	½	cup(s)	90	56.6	127	8.0	23.1	9.3	0.6	0.1	0.1	0.2
575	Yellow snap, string or wax beans, boiled, drained	½	cup(s)	63	55.8	22	1.2	4.9	2.1	0.2	0	0	0.1
576	Yellow snap, string or wax beans, frozen, boiled, drained	½	cup(s)	68	61.7	19	1.0	4.4	2.0	0.1	0	0	0.1
	Beets												
584	Beet greens, boiled, drained	½	cup(s)	72	64.2	19	1.9	3.9	2.1	0.1	0	0	0.1
2730	Pickled, canned with liquid	½	cup(s)	114	92.9	74	0.9	18.5	3.0	0.1	0	0	0
581	Sliced, boiled, drained	½	cup(s)	85	74.0	37	1.4	8.5	1.7	0.2	0	0	0.1
583	Sliced, canned, drained	½	cup(s)	85	77.3	26	0.8	6.1	1.5	0.1	0	0	0
580	Whole, boiled, drained	2	item(s)	100	87.1	44	1.7	10.0	2.0	0.2	0	0	0.1
585	**Cowpeas or black-eyed peas, boiled, drained**	½	cup(s)	83	62.3	80	2.6	16.8	4.1	0.3	0.1	0	0.1
	Broccoli												
588	Chopped, boiled, drained	½	cup(s)	78	69.6	27	1.9	5.6	2.6	0.3	0.1	0	0.1
590	Frozen, chopped, boiled, drained	½	cup(s)	92	83.5	26	2.9	4.9	2.8	0.1	0	0	0.1
587	Raw, chopped	½	cup(s)	46	40.6	15	1.3	3.0	1.2	0.2	0	0	0

Chol (mg)	Calc (mg)	Iron (mg)	Magn (mg)	Pota (mg)	Sodi (mg)	Zinc (mg)	Vit A (µg)	Thia (mg)	Vit E (mg α)	Ribo (mg)	Niac (mg)	Vit B_6 (mg)	Fola (µg DFE)	Vit C (mg)	Vit B_{12} (µg)	Sele (µg)
0	138	1.49	36.3	423.1	13.9	0.58	91.7	0.01	—	0.09	0.37	0.12	37.6	27.1	0	0.6
0	60	0.65	15.4	171.1	5.6	0.25	40.9	0.01	—	0.04	0.18	0.05	23.8	12.1	0	0.3
0	32	0.29	9.4	73.8	5.4	0.09	23.8	0.01	0.09	0.02	0.06	0.01	19.4	3.0	0	0.1
0	25	0.73	50.4	343.2	72.0	0.48	1.2	0.06	0.22	0.10	1.33	0.09	106.8	8.9	0	0.2
0	18	0.51	35.3	240.2	50.4	0.33	0.8	0.04	0.16	0.07	0.93	0.06	74.8	6.2	0	0.2
0	21	0.81	12.6	201.6	12.6	0.54	45.0	0.14	1.35	0.12	0.97	0.07	134.1	6.9	0	5.5
0	19	2.21	12.1	208.1	347.3	0.48	49.6	0.07	1.47	0.12	1.15	0.13	116.2	22.3	0	2.1
0	16	0.50	9.0	154.8	2.7	0.36	36.0	0.05	1.08	0.09	0.93	0.01	121.5	22.0	0	3.5
0	7	0.14	1.8	319.8	2.4	0.28	0	0.01	—	0.03	0.18	0.06	1.2	0	0	0.2
0	5	0.21	2.6	52.4	4.6	0.43	0.7	0.02	0.41	0.02	0.09	0.09	2.0	0.7	0	0.3
0	32	2.30	59.8	611.8	9.2	2.03	0	0.13	—	0.07	0.82	0.11	139.2	0	0	1.4
8	62	2.24	36.3	304.3	556.9	2.42	5.2	0.08	0.21	0.07	1.17	0.06	38.9	3.0	0.4	8.4
9	75	2.08	41.7	326.4	422.5	1.73	0	0.05	0.03	0.07	0.44	0.07	10.1	3.5	0	6.3
9	71	4.09	43.0	373.2	552.8	6.93	5.1	0.06	0.12	0.05	0.62	0.08	19.0	3.8	0	5.9
0	23	1.80	60.2	305.3	0.9	0.96	0	0.21	—	0.05	0.43	0.05	128.1	0	0	1.0
0	40	2.36	39.4	238.6	5.7	1.25	0.8	0.09	0.28	0.05	0.43	0.11	141.0	1.1	0	3.0
0	26	1.54	35.7	258.4	58.7	0.62	8.5	0.06	0.24	0.05	0.90	0.10	17.9	10.9	0	0.5
0	56	0.95	49.6	327.5	5.3	0.56	0	0.11	—	0.05	0.48	0.09	66.4	1.1	0	1.1
0	60	1.88	44.3	346.0	1.8	0.77	0	0.14	—	0.05	0.60	0.10	90.3	1.2	0	3.6
0	18	0.33	18.3	114.0	0.9	0.16	3.0	0.02	—	0.03	0.20	0.01	20.4	2.2	0	0.7
0	25	1.76	50.4	369.9	26.1	0.49	7.2	0.06	0.57	0.04	0.69	0.10	14.4	5.2	0	1.5
0	27	2.08	62.9	484.5	14.5	0.67	12.8	0.11	0.11	0.08	0.88	0.16	22.1	8.6	0	1.7
0	7	0.40	8.7	62.6	6.2	0.29	0.6	0.03	0.04	0.06	0.51	0.03	18.0	7.1	0	0.4
—	63	2.14	48.2	354.0	0	0.93	0	0.21	0.01	0.06	0.59	0.12	127.4	0.8	0	2.6
0	9	0.41	11.3	61.7	32.1	0.10	0	0.04	—	0.03	0.45	0.03	18.3	3.8	0	0.4
0	24	1.27	25.4	303.6	39.0	0.32	0	0.12	—	0.05	0.29	0.09	16.0	0.3	0	0.7
0	32	1.62	35.8	327.7	330.2	2.09	0	0.13	0.02	0.11	0.57	0.10	25.6	1.4	0	0.6
10	44	2.10	41.7	337.8	378.2	1.48	0	0.03	0.00	0.02	0.39	0.18	13.9	7.6	0	1.6
0	36	1.21	18.4	133.5	409.2	0.33	13.5	0.04	0.04	0.07	0.25	0.06	22.1	3.8	0	2.6
0	88	4.42	74.0	442.9	0.9	0.98	0	0.13	0.30	0.24	0.34	0.20	46.4	1.5	0	6.3
0	131	2.25	54.0	485.1	12.6	0.82	7.2	0.23	—	0.14	1.13	0.05	99.9	15.3	0	1.3
0	65	2.54	60.9	414.4	1.8	0.97	0	0.21	—	0.05	0.24	0.11	122.6	0	0	1.2
0	29	0.80	15.6	186.9	1.9	0.23	2.5	0.05	0.28	0.06	0.38	0.04	20.6	6.1	0	0.3
0	33	0.59	16.2	85.1	6.1	0.32	4.1	0.02	0.03	0.06	0.26	0.04	15.5	2.8	0	0.3
0	82	1.36	49.0	654.5	173.5	0.36	275.8	0.08	1.30	0.20	0.35	0.09	10.1	17.9	0	0.6
0	12	0.46	17.0	168.0	299.6	0.29	1.1	0.01	—	0.05	0.28	0.05	30.6	2.6	0	1.1
0	14	0.67	19.6	259.3	65.5	0.30	1.7	0.02	0.03	0.03	0.28	0.05	68.0	3.1	0	0.6
0	13	1.54	14.5	125.8	164.9	0.17	0.9	0.01	0.02	0.03	0.13	0.04	25.5	3.5	0	0.4
0	16	0.79	23.0	305.0	77.0	0.35	2.0	0.02	0.04	0.04	0.33	0.06	80.0	3.6	0	0.7
0	106	0.92	42.9	344.9	3.3	0.85	33.0	0.08	0.18	0.12	1.15	0.05	104.8	1.8	0	2.1
0	31	0.52	16.4	228.5	32.0	0.35	60.1	0.04	1.13	0.09	0.43	0.15	84.2	50.6	0	1.2
0	30	0.56	12.0	130.6	10.1	0.25	46.9	0.05	1.21	0.07	0.42	0.12	51.5	36.9	0	0.6
0	21	0.33	9.6	143.8	15.0	0.19	14.1	0.03	0.36	0.05	0.29	0.08	28.7	40.6	0	1.1

TABLE A-1 Table of Food Composition *(continued)*

(Computer code is for Cengage Diet Analysis Plus program)

DA+ Code	Food Description	QTY	Measure	Wt (g)	H_2O (g)	Ener (cal)	Prot (g)	Carb (g)	Fiber (g)	Fat (g)	Fat Breakdown (g) Sat	Mono	Poly
	Vegetables, Legumes—continued												
16848	**Broccoflower, raw, chopped**	½	cup(s)	32	28.7	10	0.9	1.9	1.0	0.1	0	0	0
	Brussels sprouts												
591	Boiled, drained	½	cup(s)	78	69.3	28	2.0	5.5	2.0	0.4	0.1	0	0.2
592	Frozen, boiled, drained	½	cup(s)	78	67.2	33	2.8	6.4	3.2	0.3	0.1	0	0.2
	Cabbage												
595	Boiled, drained, no salt added	1	cup(s)	150	138.8	35	1.9	8.3	2.8	0.1	0	0	0
35611	Chinese (pak choi or bok choy), boiled with salt, drained	1	cup(s)	170	162.4	20	2.6	3.0	1.7	0.3	0	0	0.1
16869	Kim chee	1	cup(s)	150	137.5	32	2.5	6.1	1.8	0.3	0	0	0.2
594	Raw, shredded	1	cup(s)	70	64.5	17	0.9	4.1	1.7	0.1	0	0	0
596	Red, shredded, raw	1	cup(s)	70	63.3	22	1.0	5.2	1.5	0.1	0	0	0.1
597	Savoy, shredded, raw	1	cup(s)	70	63.7	19	1.4	4.3	2.2	0.1	0	0	0
35417	**Capers**	1	teaspoon(s)	4	—	2	0	0	0	0	0	0	0
	Carrots												
8691	Baby, raw	8	item(s)	80	72.3	28	0.5	6.6	2.3	0.1	0	0	0.1
601	Grated	½	cup(s)	55	48.6	23	0.5	5.3	1.5	0.1	0	0	0.1
1055	Juice, canned	½	cup(s)	118	104.9	47	1.1	11.0	0.9	0.2	0	0	0.1
600	Raw	½	cup(s)	61	53.9	25	0.6	5.8	1.7	0.1	0	0	0.1
602	Sliced, boiled, drained	½	cup(s)	78	70.3	27	0.6	6.4	2.3	0.1	0	0	0.1
32725	**Cassava or manioc**	½	cup(s)	103	61.5	165	1.4	39.2	1.9	0.3	0.1	0.1	0
	Cauliflower												
606	Boiled, drained	½	cup(s)	62	57.7	14	1.1	2.5	1.4	0.3	0	0	0.1
607	Frozen, boiled, drained	½	cup(s)	90	84.6	17	1.4	3.4	2.4	0.2	0	0	0.1
605	Raw, chopped	½	cup(s)	50	46.0	13	1.0	2.6	1.2	0	0	0	0
	Celery												
609	Diced	½	cup(s)	51	48.2	8	0.3	1.5	0.8	0.1	0	0	0
608	Stalk	2	item(s)	80	76.3	13	0.6	2.4	1.3	0.1	0	0	0.1
	Chard												
1057	Swiss chard, boiled, drained	½	cup(s)	88	81.1	18	1.6	3.6	1.8	0.1	0	0	0
1056	Swiss chard, raw	1	cup(s)	36	33.4	7	0.6	1.3	0.6	0.1	0	0	0
	Collard greens												
610	Boiled, drained	½	cup(s)	95	87.3	25	2.0	4.7	2.7	0.3	0	0	0.2
611	Frozen, chopped, boiled, drained	½	cup(s)	85	75.2	31	2.5	6.0	2.4	0.3	0.1	0	0.2
	Corn												
29614	Yellow corn, fresh, cooked	1	item(s)	100	69.2	107	3.3	25.0	2.8	1.3	0.2	0.4	0.6
615	Yellow creamed sweet corn, canned	½	cup(s)	128	100.8	92	2.2	23.2	1.5	0.5	0.1	0.2	0.3
612	Yellow sweet corn, boiled, drained	½	cup(s)	82	57.0	89	2.7	20.6	2.3	1.1	0.2	0.3	0.5
614	Yellow sweet corn, frozen, boiled, drained	½	cup(s)	82	63.2	66	2.1	15.8	2.0	0.5	0.1	0.2	0.3
618	**Cucumber**	¼	item(s)	75	71.7	11	0.5	2.7	0.4	0.1	0	0	0
16870	**Cucumber, kim chee**	½	cup(s)	75	68.1	16	0.8	3.6	1.1	0.1	0	0	0
	Dandelion greens												
620	Chopped, boiled, drained	½	cup(s)	53	47.1	17	1.1	3.4	1.5	0.3	0.1	0	0.1
2734	Raw	1	cup(s)	55	47.1	25	1.5	5.1	1.9	0.4	0.1	0	0.2
1066	**Eggplant, boiled, drained**	½	cup(s)	50	44.4	17	0.4	4.3	1.2	0.1	0	0	0
621	**Endive or escarole, chopped, raw**	1	cup(s)	50	46.9	8	0.6	1.7	1.5	0.1	0	0	0
8784	**Jicama or yambean**	½	cup(s)	65	116.5	49	0.9	11.4	6.3	0.1	0	0	0.1
	Kale												
623	Frozen, chopped, boiled, drained	½	cup(s)	65	58.8	20	1.8	3.4	1.3	0.3	0	0	0.2
29313	Raw	1	cup(s)	67	56.6	33	2.2	6.7	1.3	0.5	0.1	0	0.2
	Kohlrabi												
1072	Boiled, drained	½	cup(s)	83	74.5	24	1.5	5.5	0.9	0.1	0	0	0
1071	Raw	1	cup(s)	135	122.9	36	2.3	8.4	4.9	0.1	0	0	0.1

PAGE KEY: A-2 = Breads/Baked Goods A-8 = Cereal/Rice/Pasta A-12 = Fruit A-18 = Vegetables/Legumes A-28 = Nuts/Seeds A-30 = Vegetarian A-32 = Dairy A-40 = Eggs A-40 = Seafood A-44 = Meats A-48 = Poultry A-48 = Processed Meats A-50 = Beverages A-54 = Fats/Oils A-56 = Sweets A-58 = Spices/Condiments/Sauces A-62 = Mixed Foods/Soups/Sandwiches A-68 = Fast Food A-88 = Convenience A-90 = Baby Foods

Chol (mg)	Calc (mg)	Iron (mg)	Magn (mg)	Pota (mg)	Sodi (mg)	Zinc (mg)	Vit A (µg)	Thia (mg)	Vit E (mg α)	Ribo (mg)	Niac (mg)	Vit B_6 (mg)	Fola (µg DFE)	Vit C (mg)	Vit B_{12} (µg)	Sele (µg)
0	11	0.23	6.4	96.0	7.4	0.20	2.6	0.02	0.01	0.03	0.23	0.07	18.2	28.2	0	0.2
0	28	0.93	15.6	247.3	16.4	0.25	30.4	0.08	0.33	0.06	0.47	0.13	46.8	48.4	0	1.2
0	20	0.37	14.0	224.8	11.6	0.18	35.7	0.08	0.39	0.08	0.41	0.22	78.3	35.4	0	0.5
0	72	0.24	22.5	294.0	12.0	0.30	6.0	0.08	0.20	0.04	0.36	0.16	45.0	56.2	0	0.9
0	158	1.76	18.7	630.7	459.0	0.28	360.4	0.04	0.14	0.10	0.72	0.28	69.7	44.2	0	0.7
0	144	1.26	27.0	379.5	996.0	0.36	288.0	0.06	0.36	0.10	0.80	0.32	88.5	79.6	0	1.5
0	28	0.33	8.4	119.0	12.6	0.12	3.5	0.04	0.10	0.02	0.16	0.08	30.1	25.6	0	0.2
0	31	0.56	11.2	170.1	18.9	0.15	39.2	0.04	0.07	0.05	0.29	0.14	12.6	39.9	0	0.4
0	24	0.28	19.6	161.0	19.6	0.18	35.0	0.05	0.11	0.02	0.21	0.13	56.0	21.7	0	0.6
0	0	0.00	—	—	140	—	0	—	—	—	—	—	—	0	—	—
0	26	0.71	8.0	189.6	62.4	0.13	552.0	0.02	—	0.02	0.44	0.08	21.6	2.1	0	0.7
0	18	0.16	6.6	176.0	37.9	0.13	459.2	0.03	0.36	0.03	0.54	0.07	10.4	3.2	0	0.1
0	28	0.54	16.5	344.6	34.2	0.21	1128.1	0.11	1.37	0.07	0.46	0.26	4.7	10.0	0	0.7
0	20	0.18	7.3	195.2	42.1	0.15	509.4	0.04	0.40	0.04	0.60	0.08	11.6	3.6	0	0.1
0	23	0.26	7.8	183.3	45.2	0.15	664.6	0.05	0.80	0.03	0.50	0.11	10.9	2.8	0	0.5
0	16	0.27	21.6	279.1	14.4	0.35	1.0	0.08	0.19	0.04	0.87	0.09	27.8	21.2	0	0.7
0	10	0.19	5.6	88.0	9.3	0.10	0.6	0.02	0.04	0.03	0.25	0.10	27.3	27.5	0	0.4
0	15	0.36	8.1	125.1	16.2	0.11	0	0.03	0.05	0.04	0.27	0.07	36.9	28.2	0	0.5
0	11	0.22	7.5	151.5	15.0	0.14	0.5	0.03	0.04	0.03	0.26	0.11	28.5	23.2	0	0.3
0	20	0.10	5.6	131.3	40.4	0.07	11.1	0.01	0.14	0.03	0.16	0.04	18.2	1.6	0	0.2
0	32	0.16	8.8	208.0	64.0	0.10	17.6	0.01	0.21	0.04	0.25	0.05	28.8	2.5	0	0.3
0	51	1.98	75.3	480.4	156.6	0.29	267.8	0.03	1.65	0.08	0.32	0.07	7.9	15.8	0	0.8
0	18	0.64	29.2	136.4	76.7	0.13	110.2	0.01	0.68	0.03	0.14	0.03	5.0	10.8	0	0.3
0	133	1.10	19.0	110.2	15.2	0.21	385.7	0.03	0.83	0.10	0.54	0.12	88.4	17.3	0	0.5
0	179	0.95	25.5	213.4	42.5	0.22	488.8	0.04	1.06	0.09	0.54	0.09	64.6	22.4	0	1.3
0	2	0.61	32.0	248.0	242.0	0.48	13.0	0.20	0.09	0.07	1.60	0.06	46.0	6.2	0	0.2
0	4	0.48	21.8	171.5	364.8	0.67	5.1	0.03	0.09	0.06	1.22	0.08	57.6	5.9	0	0.5
0	2	0.36	21.3	173.8	0	0.50	10.7	0.17	0.07	0.05	1.32	0.04	37.7	5.1	0	0.2
0	2	0.38	23.0	191.1	0.8	0.51	8.2	0.02	0.05	0.05	1.07	0.08	28.7	2.9	0	0.6
0	12	0.20	9.8	110.6	1.5	0.14	3.8	0.01	0.01	0.01	0.07	0.03	5.3	2.1	0	0.2
0	7	3.61	6.0	87.8	765.8	0.38	—	0.02	—	0.02	0.34	0.08	17.3	2.6	0	—
0	74	0.95	12.6	121.8	23.1	0.15	179.6	0.07	1.28	0.09	0.27	0.08	6.8	9.5	0	0.2
0	103	1.70	19.8	218.3	41.8	0.22	279.4	0.10	1.89	0.14	0.44	0.13	14.8	19.2	0	0.3
0	3	0.12	5.4	60.9	0.5	0.06	1.0	0.04	0.20	0.01	0.30	0.04	6.9	0.6	0	0
0	26	0.41	7.5	157.0	11.0	0.39	54.0	0.04	0.22	0.03	0.20	0.01	71.0	3.2	0	0.1
0	16	0.78	15.5	194.0	5.2	0.20	1.3	0.02	0.59	0.04	0.25	0.05	15.5	26.1	0	0.9
0	90	0.61	11.7	208.7	9.8	0.11	477.8	0.02	0.59	0.07	0.43	0.05	9.1	16.4	0	0.6
0	90	1.14	22.8	299.5	28.8	0.29	515.2	0.07	—	0.08	0.66	0.18	19.4	80.4	0	0.6
0	21	0.33	15.7	280.5	17.3	0.26	1.7	0.03	0.43	0.02	0.32	0.13	9.9	44.6	0	0.7
0	32	0.54	25.7	472.5	27.0	0.04	2.7	0.06	0.64	0.02	0.54	0.20	21.6	83.7	0	0.9

TABLE A-1 Table of Food Composition *(continued)*

(Computer code is for Cengage Diet Analysis Plus program)

											Fat Breakdown (g)		
DA+ Code	Food Description	QTY	Measure	Wt (g)	H_2O (g)	Ener (cal)	Prot (g)	Carb (g)	Fiber (g)	Fat (g)	Sat	Mono	Poly
	Vegetables, Legumes—continued												
	Leeks												
1074	Boiled, drained	½	cup(s)	52	47.2	16	0.4	4.0	0.5	0.1	0	0	0
1073	Raw	1	cup(s)	89	73.9	54	1.3	12.6	1.6	0.3	0	0	0.1
	Lentils												
522	Boiled	¼	cup(s)	50	34.5	57	4.5	10.0	3.9	0.2	0	0	0.1
1075	Sprouted	1	cup(s)	77	51.9	82	6.9	17.0	—	0.4	0	0.1	0.2
	Lettuce												
625	Butterhead leaves	11	piece(s)	83	78.9	11	1.1	1.8	0.9	0.2	0	0	0.1
624	Butterhead, Boston or Bibb	1	cup(s)	55	52.6	7	0.7	1.2	0.6	0.1	0	0	0.1
626	Iceberg	1	cup(s)	55	52.6	8	0.5	1.6	0.7	0.1	0	0	0
628	Iceberg, chopped	1	cup(s)	55	52.6	8	0.5	1.6	0.7	0.1	0	0	0
629	Looseleaf	1	cup(s)	36	34.2	5	0.5	1.0	0.5	0.1	0	0	0
1665	Romaine, shredded	1	cup(s)	56	53.0	10	0.7	1.8	1.2	0.2	0	0	0.1
	Mushrooms												
15585	Crimini (about 6)	3	ounce(s)	85	—	28	3.7	2.8	1.9	0	0	0	0
8700	Enoki	30	item(s)	90	79.7	40	2.3	6.9	2.4	0.3	0	0	0.1
1079	Mushrooms, boiled, drained	½	cup(s)	78	71.0	22	1.7	4.1	1.7	0.4	0	0	0.1
1080	Mushrooms, canned, drained	½	cup(s)	78	71.0	20	1.5	4.0	1.9	0.2	0	0	0.1
630	Mushrooms, raw	½	cup(s)	48	44.4	11	1.5	1.6	0.5	0.2	0	0	0.1
15587	Portabella, raw	1	item(s)	84	—	30	3.0	3.9	3.0	0	0	0	0
2743	Shiitake, cooked	½	cup(s)	73	60.5	41	1.1	10.4	1.5	0.2	0	0.1	0
	Mustard greens												
2744	Frozen, boiled, drained	½	cup(s)	75	70.4	14	1.7	2.3	2.1	0.2	0	0.1	0
29319	Raw	1	cup(s)	56	50.8	15	1.5	2.7	1.8	0.1	0	0	0
	Okra												
16866	Batter coated, fried	11	piece(s)	83	55.6	156	2.1	12.7	2.0	11.2	1.5	3.7	5.5
32742	Frozen, boiled, drained, no salt added	½	cup(s)	92	83.8	26	1.9	5.3	2.6	0.3	0.1	0	0.1
632	Sliced, boiled, drained	½	cup(s)	80	74.1	18	1.5	3.6	2.0	0.2	0	0	0
	Onions												
635	Chopped, boiled, drained	½	cup(s)	105	92.2	46	1.4	10.7	1.5	0.2	0	0	0.1
2748	Frozen, boiled, drained	½	cup(s)	106	97.8	30	0.8	7.0	1.9	0.1	0	0	0
1081	Onion rings, breaded and pan fried, frozen, heated	10	piece(s)	71	20.2	289	3.8	27.1	0.9	19.0	6.1	7.7	3.6
633	Raw, chopped	½	cup(s)	80	71.3	32	0.9	7.5	1.4	0.1	0	0	0
16850	Red onions, sliced, raw	½	cup(s)	57	50.7	24	0.5	5.8	0.8	0	0	0	0
636	Scallions, green or spring onions	2	item(s)	30	26.9	10	0.5	2.2	0.8	0.1	0	0	0
16860	**Palm hearts, cooked**	½	cup(s)	73	50.7	84	2.0	18.7	1.1	0.1	0	0	0.1
637	**Parsley, chopped**	1	tablespoon(s)	4	3.3	1	0.1	0.2	0.1	0	0	0	0
638	**Parsnips, sliced, boiled, drained**	½	cup(s)	78	62.6	55	1.0	13.3	2.8	0.2	0	0.1	0
	Peas												
639	Green peas, canned, drained	½	cup(s)	85	69.4	59	3.8	10.7	3.5	0.3	0.1	0	0.1
641	Green peas, frozen, boiled, drained	½	cup(s)	80	63.6	62	4.1	11.4	4.4	0.2	0	0	0.1
35694	Pea pods, boiled with salt, drained	½	cup(s)	80	71.1	32	2.6	5.2	2.2	0.2	0	0	0.1
1082	Peas and carrots, canned with liquid	½	cup(s)	128	112.4	48	2.8	10.8	2.6	0.3	0.1	0	0.2
1083	Peas and carrots, frozen, boiled, drained	½	cup(s)	80	68.6	38	2.5	8.1	2.5	0.3	0.1	0	0.2
2750	Snow or sugar peas, frozen, boiled, drained	½	cup(s)	80	69.3	42	2.8	7.2	2.5	0.3	0.1	0	0.1
640	Snow or sugar peas, raw	½	cup(s)	32	28.0	13	0.9	2.4	0.8	0.1	0	0	0
29324	Split peas, sprouted	½	cup(s)	60	37.4	77	5.3	16.9	—	0.4	0.1	0	0.2
	Peppers												
644	Green bell or sweet, boiled, drained	½	cup(s)	68	62.5	19	0.6	4.6	0.8	0.1	0	0	0.1
643	Green bell or sweet, raw	½	cup(s)	75	69.9	15	0.6	3.5	1.3	0.1	0	0	0
1664	Green hot chili	1	item(s)	45	39.5	18	0.9	4.3	0.7	0.1	0	0	0
1663	Green hot chili, canned with liquid	½	cup(s)	68	62.9	14	0.6	3.5	0.9	0.1	0	0	0
1086	Jalapeno, canned with liquid	½	cup(s)	68	60.4	18	0.6	3.2	1.8	0.6	0.1	0	0.3
8703	Yellow bell or sweet	1	item(s)	186	171.2	50	1.9	11.8	1.7	0.4	0.1	0	0.2

Chol (mg)	Calc (mg)	Iron (mg)	Magn (mg)	Pota (mg)	Sodi (mg)	Zinc (mg)	Vit A (μg)	Thia (mg)	Vit E (mg α)	Ribo (mg)	Niac (mg)	Vit B_6 (mg)	Fola (μg DFE)	Vit C (mg)	Vit B_{12} (μg)	Sele (μg)
0	16	0.56	7.3	45.2	5.2	0.02	1.0	0.01	—	0.01	0.10	0.04	12.5	2.2	0	0.3
0	53	1.86	24.9	160.2	17.8	0.10	73.9	0.05	0.81	0.02	0.35	0.20	57.0	10.7	0	0.9
0	9	1.65	17.8	182.7	1.0	0.63	0	0.08	0.05	0.04	0.52	0.09	89.6	0.7	0	1.4
0	19	2.47	28.5	247.9	8.5	1.16	1.5	0.17	—	0.09	0.86	0.14	77.0	12.7	0	0.5
0	29	1.02	10.7	196.4	4.1	0.16	137.0	0.04	0.14	0.05	0.29	0.06	60.2	3.1	0	0.5
0	19	0.68	7.1	130.9	2.7	0.11	91.3	0.03	0.09	0.03	0.19	0.04	40.1	2.0	0	0.3
0	10	0.22	3.8	77.5	5.5	0.08	13.7	0.02	0.09	0.01	0.07	0.02	15.9	1.5	0	0.1
0	10	0.22	3.8	77.5	5.5	0.08	13.7	0.02	0.09	0.01	0.07	0.02	15.9	1.5	0	0.1
0	13	0.31	4.7	69.8	10.1	0.06	133.2	0.02	0.10	0.02	0.13	0.03	13.7	6.5	0	0.2
0	18	0.54	7.8	138.3	4.5	0.13	162.4	0.04	0.07	0.03	0.17	0.04	76.2	13.4	0	0.2
0	0	0.67	—	—	32.6	—	0	—	—	—	—	—	—	0	0	—
0	1	0.98	14.4	331.2	2.7	0.54	0	0.16	0.01	0.14	5.31	0.07	46.8	0	0	2.0
0	5	1.35	9.4	277.7	1.6	0.67	0	0.05	0.01	0.23	3.47	0.07	14.0	3.1	0	9.3
0	9	0.61	11.7	100.6	331.5	0.56	0	0.06	0.01	0.01	1.24	0.04	9.4	0	0	3.2
0	1	0.24	4.3	152.6	2.4	0.25	0	0.04	0.01	0.19	1.73	0.05	7.7	1.0	0	4.5
0	39	0.35	—	—	9.9	—	0	—	—	—	—	—	—	0	0	—
0	2	0.31	10.2	84.8	2.9	0.96	0	0.02	0.02	0.12	1.08	0.11	15.2	0.2	0	18.0
0	76	0.84	9.8	104.3	18.8	0.15	265.5	0.03	1.01	0.04	0.19	0.08	52.5	10.4	0	0.5
0	58	0.81	17.9	198.2	14.0	0.11	294.0	0.04	1.12	0.06	0.45	0.10	104.7	39.2	0	0.5
2	54	1.13	32.2	170.8	109.7	0.44	14.0	0.16	1.50	0.12	1.29	0.11	43.7	9.2	0	3.6
0	88	0.61	46.9	215.3	2.8	0.57	15.6	0.09	0.29	0.11	0.72	0.04	134.3	11.2	0	0.6
0	62	0.22	28.8	108.0	4.8	0.34	11.2	0.10	0.21	0.04	0.69	0.15	36.8	13.0	0	0.3
0	23	0.24	11.5	174.3	3.1	0.21	0	0.03	0.02	0.02	0.17	0.12	15.7	5.5	0	0.6
0	17	0.32	6.4	114.5	12.7	0.06	0	0.02	0.01	0.02	0.14	0.06	13.8	2.8	0	0.4
0	22	1.20	13.5	91.6	266.3	0.29	7.8	0.19	—	0.09	2.56	0.05	73.1	1.0	0	2.5
0	18	0.16	8.0	116.8	3.2	0.13	0	0.03	0.01	0.02	0.09	0.09	15.2	5.9	0	0.4
0	13	0.10	5.7	82.4	1.7	0.09	0	0.02	0.01	0.01	0.04	0.08	10.9	3.7	0	0.3
0	22	0.44	6.0	82.8	4.8	0.11	15.0	0.01	0.16	0.02	0.15	0.01	19.2	5.6	0	0.2
0	13	1.23	7.3	1318.4	10.2	2.72	2.2	0.03	0.36	0.12	0.62	0.53	14.6	5.0	0	0.5
0	5	0.23	1.9	21.1	2.1	0.04	16.0	0.00	0.02	0.00	0.05	0.00	5.8	5.1	0	0
0	29	0.45	22.6	286.3	7.8	0.20	0	0.06	0.78	0.04	0.56	0.07	45.2	10.1	0	1.3
0	17	0.80	14.5	147.1	214.2	0.60	23.0	0.10	0.02	0.06	0.62	0.05	37.4	8.2	0	1.4
0	19	1.21	17.6	88.0	57.6	0.53	84.0	0.22	0.02	0.08	1.18	0.09	47.2	7.9	0	0.8
0	34	1.57	20.8	192.0	192.0	0.29	41.6	0.10	0.31	0.06	0.43	0.11	23.2	38.3	0	0.6
0	29	0.96	17.9	127.5	331.5	0.74	368.5	0.09	—	0.07	0.74	0.11	23.0	8.4	0	1.1
0	18	0.75	12.8	126.4	54.4	0.36	380.8	0.18	0.41	0.05	0.92	0.07	20.8	6.5	0	0.9
0	47	1.92	22.4	173.6	4.0	0.39	52.8	0.05	0.37	0.09	0.45	0.13	28.0	17.6	0	0.6
0	14	0.65	7.6	63.0	1.3	0.08	17.0	0.04	0.12	0.02	0.19	0.05	13.2	18.9	0	0.2
0	22	1.34	33.6	228.6	12.0	0.62	4.8	0.12	—	0.08	1.84	0.14	86.4	6.2	0	0.4
0	6	0.31	6.8	112.9	1.4	0.08	15.6	0.04	0.34	0.02	0.32	0.15	10.9	50.6	0	0.2
0	7	0.25	7.5	130.4	2.2	0.09	13.4	0.04	0.27	0.02	0.35	0.16	7.5	59.9	0	0
0	8	0.54	11.3	153.0	3.2	0.13	26.6	0.04	0.31	0.04	0.42	0.12	10.4	109.1	0	0.2
0	5	0.34	9.5	127.2	797.6	0.10	24.5	0.01	0.46	0.02	0.54	0.10	6.8	46.2	0	0.2
0	16	1.28	10.2	131.2	1136.3	0.23	57.8	0.03	0.47	0.03	0.27	0.13	9.5	6.8	0	0.3
0	20	0.85	22.3	394.3	3.7	0.31	18.6	0.05	—	0.04	1.65	0.31	48.4	341.3	0	0.6

TABLE A-1 Table of Food Composition *(continued)*

(Computer code is for Cengage Diet Analysis Plus program)

DA+ Code	Food Description	QTY	Measure	Wt (g)	H₂O (g)	Ener (cal)	Prot (g)	Carb (g)	Fiber (g)	Fat (g)	Fat Breakdown (g) Sat	Mono	Poly
	Vegetables, Legumes—continued												
1087	**Poi**	½	cup(s)	120	86.0	134	0.5	32.7	0.5	0.2	0	0	0.1
	Potatoes												
1090	Au gratin mix, prepared with water, whole milk and butter	½	cup(s)	124	97.7	115	2.8	15.9	1.1	5.1	3.2	1.5	0.2
1089	Au gratin, prepared with butter	½	cup(s)	123	90.7	162	6.2	13.8	2.2	9.3	5.8	2.6	0.3
5791	Baked, flesh and skin	1	item(s)	202	151.3	188	5.1	42.7	4.4	0.3	0.1	0	0.1
645	Baked, flesh only	½	cup(s)	61	46.0	57	1.2	13.1	0.9	0.1	0	0	0
1088	Baked, skin only	1	item(s)	58	27.4	115	2.5	26.7	4.6	0.1	0	0	0
5795	Boiled in skin, flesh only, drained	1	item(s)	136	104.7	118	2.5	27.4	2.1	0.1	0	0	0.1
5794	Boiled, drained, skin and flesh	1	item(s)	150	115.9	129	2.9	29.8	2.5	0.2	0	0	0.1
647	Boiled, flesh only	½	cup(s)	78	60.4	67	1.3	15.6	1.4	0.1	0	0	0
648	French fried, deep fried, prepared from raw	14	item(s)	70	32.8	187	2.7	23.5	2.9	9.5	1.9	4.2	3.0
649	French fried, frozen, heated	14	item(s)	70	43.7	94	1.9	19.4	2.0	3.7	0.7	2.3	0.2
1091	Hashed brown	½	cup(s)	78	36.9	207	2.3	27.4	2.5	9.8	1.5	4.1	3.7
652	Mashed with margarine and whole milk	½	cup(s)	105	79.0	119	2.1	17.7	1.6	4.4	1.0	2.0	1.2
653	Mashed, prepared from dehydrated granules with milk, water, and margarine	½	cup(s)	105	79.8	122	2.3	16.9	1.4	5.0	1.3	2.1	1.4
2759	Microwaved	1	item(s)	202	145.5	212	4.9	49.0	4.6	0.2	0.1	0	0.1
2760	Microwaved in skin, flesh only	½	cup(s)	78	57.1	78	1.6	18.1	1.2	0.1	0	0	0
5804	Microwaved, skin only	1	item(s)	58	36.8	77	2.5	17.2	4.2	0.1	0	0	0
1097	Potato puffs, frozen, heated	½	cup(s)	64	38.2	122	1.3	17.8	1.6	5.5	1.2	3.9	0.3
1094	Scalloped mix, prepared with water, whole milk and butter	½	cup(s)	124	98.4	116	2.6	15.9	1.4	5.3	3.3	1.5	0.2
1093	Scalloped, prepared with butter	½	cup(s)	123	99.2	108	3.5	13.2	2.3	4.5	2.8	1.3	0.2
	Pumpkin												
1773	Boiled, drained	½	cup(s)	123	114.8	25	0.9	6.0	1.3	0.1	0	0	0
656	Canned	½	cup(s)	123	110.2	42	1.3	9.9	3.6	0.3	0.2	0	0
	Radicchio												
8731	Leaves, raw	1	cup(s)	40	37.3	9	0.6	1.8	0.4	0.1	0	0	0
2498	Raw	1	cup(s)	40	37.3	9	0.6	1.8	0.4	0.1	0	0	0
657	**Radishes**	6	item(s)	27	25.7	4	0.2	0.9	0.4	0	0	0	0
1099	**Rutabaga, boiled, drained**	½	cup(s)	85	75.5	33	1.1	7.4	1.5	0.2	0	0	0.1
658	**Sauerkraut, canned**	½	cup(s)	118	109.2	22	1.1	5.1	3.4	0.2	0	0	0.1
	Seaweed												
1102	Kelp	½	cup(s)	40	32.6	17	0.6	3.8	0.5	0.2	0.1	0	0
1104	Spirulina, dried	½	cup(s)	8	0.4	22	4.3	1.8	0.3	0.6	0.2	0.1	0.2
1106	**Shallots**	3	tablespoon(s)	30	23.9	22	0.8	5.0	—	0	0	0	0
	Soybeans												
1670	Boiled	½	cup(s)	86	53.8	149	14.3	8.5	5.2	7.7	1.1	1.7	4.4
2825	Dry roasted	½	cup(s)	86	0.7	388	34.0	28.1	7.0	18.6	2.7	4.1	10.5
2824	Roasted, salted	½	cup(s)	86	1.7	405	30.3	28.9	15.2	21.8	3.2	4.8	12.3
8739	Sprouted, stir fried	½	cup(s)	63	42.3	79	8.2	5.9	0.5	4.5	0.6	1.0	2.5
	Soy products												
1813	Soy milk	1	cup(s)	240	211.3	130	7.8	15.1	1.4	4.2	0.5	1.0	2.3
2838	Tofu, dried, frozen (koyadofu)	3	ounce(s)	85	4.9	408	40.8	12.4	6.1	25.8	3.7	5.7	14.6
13844	Tofu, extra firm	3	ounce(s)	85	—	86	8.6	2.2	1.1	4.3	0.5	0.9	2.8
13843	Tofu, firm	3	ounce(s)	85	—	75	7.5	2.2	0.5	3.2	0	0.9	2.3
1816	Tofu, firm, with calcium sulfate and magnesium chloride (nigari)	3	ounce(s)	85	72.2	60	7.0	1.4	0.8	3.5	0.7	1.0	1.5
1817	Tofu, fried	3	ounce(s)	85	43.0	230	14.6	8.9	3.3	17.2	2.5	3.8	9.7
13841	Tofu, silken	3	ounce(s)	85	—	42	3.7	1.9	0	2.3	0.5	—	—
13842	Tofu, soft	3	ounce(s)	85	—	65	6.5	1.1	0.5	3.2	0.5	1.1	2.2
1671	Tofu, soft, with calcium sulfate and magnesium chloride (nigari)	3	ounce(s)	85	74.2	52	5.6	1.5	0.2	3.1	0.5	0.7	1.8

Chol (mg)	Calc (mg)	Iron (mg)	Magn (mg)	Pota (mg)	Sodi (mg)	Zinc (mg)	Vit A (μg)	Thia (mg)	Vit E (mg α)	Ribo (mg)	Niac (mg)	Vit B_6 (mg)	Fola (μg DFE)	Vit C (mg)	Vit B_{12} (μg)	Sele (μg)
0	19	1.06	28.8	219.6	14.4	0.26	3.6	0.16	2.76	0.05	1.32	0.33	25.2	4.8	0	0.8
19	103	0.39	18.6	271.0	543.3	0.29	64.4	0.02	—	0.10	1.16	0.05	8.7	3.8	0	3.3
28	146	0.78	24.5	485.1	530.4	0.85	78.4	0.08	—	0.14	1.22	0.21	16.0	12.1	0	3.3
0	30	2.18	56.6	1080.7	20.2	0.72	2.0	0.12	0.08	0.09	2.84	0.62	56.6	19.4	0	0.8
0	3	0.21	15.3	238.5	3.1	0.18	0	0.06	0.02	0.01	0.85	0.18	5.5	7.8	0	0.2
0	20	4.08	24.9	332.3	12.2	0.28	0.6	0.07	0.02	0.06	1.77	0.35	12.8	7.8	0	0.4
0	7	0.42	29.9	515.4	5.4	0.40	0	0.14	0.01	0.02	1.95	0.40	13.6	17.7	0	0.4
0	13	1.27	34.1	572.0	7.4	0.46	0	0.14	0.01	0.03	2.13	0.44	15.0	18.4	0	—
0	6	0.24	15.6	255.8	3.9	0.21	0	0.07	0.01	0.01	1.02	0.21	7.0	5.8	0	0.2
0	16	1.05	30.8	567.0	8.4	0.39	0	0.08	0.09	0.03	1.34	0.37	16.1	21.2	0	0.4
0	8	0.51	18.2	315.7	271.6	0.26	0	0.09	0.07	0.02	1.55	0.12	19.6	9.3	0	0.1
0	11	0.43	27.3	449.3	266.8	0.37	0	0.13	0.01	0.03	1.80	0.37	12.5	10.1	0	0.4
1	23	0.27	19.9	344.4	349.6	0.31	43.0	0.09	0.44	0.04	1.23	0.25	9.4	11.0	0.1	0.8
2	36	0.21	21.0	164.8	179.5	0.26	49.3	0.09	0.53	0.09	0.90	0.16	8.4	6.8	0.1	5.9
0	22	2.50	54.5	902.9	16.2	0.72	0	0.24	—	0.06	3.46	0.69	24.2	30.5	0	0.8
0	4	0.31	19.4	319.0	5.4	0.25	0	0.10	—	0.01	1.26	0.25	9.3	11.7	0	0.3
0	27	3.44	21.5	377.0	9.3	0.29	0	0.04	0.01	0.04	1.28	0.28	9.9	8.9	0	0.3
0	9	0.41	10.9	199.7	307.2	0.21	0	0.08	0.15	0.02	0.97	0.08	9.0	4.0	0	0.4
14	45	0.47	17.4	252.2	423.7	0.31	43.5	0.02	—	0.06	1.28	0.05	12.4	4.1	0	2.0
15	70	0.70	23.3	463.1	410.4	0.49	0	0.08	—	0.11	1.29	0.22	16.0	13.0	0	2.0
0	18	0.69	11.0	281.8	1.2	0.28	306.3	0.03	0.98	0.09	0.50	0.05	11.0	5.8	0	0.2
0	32	1.70	28.2	252.4	6.1	0.20	953.1	0.02	1.29	0.06	0.45	0.06	14.7	5.1	0	0.5
0	8	0.23	5.2	120.8	8.8	0.25	0.4	0.01	0.90	0.01	0.10	0.02	24.0	3.2	0	0.4
0	8	0.23	5.2	120.8	8.8	0.25	0.4	0.01	0.90	0.01	0.10	0.02	24.0	3.2	0	0.4
0	7	0.09	2.7	62.9	10.5	0.07	0	0.00	0.00	0.01	0.06	0.01	6.8	4.0	0	0.2
0	41	0.45	19.6	277.1	17.0	0.30	0	0.07	0.27	0.04	0.61	0.09	12.8	16.0	0	0.6
0	35	1.73	15.3	200.6	780.0	0.22	1.2	0.03	0.17	0.03	0.17	0.15	28.3	17.3	0	0.7
0	67	1.12	48.4	35.6	93.2	0.48	2.4	0.02	0.32	0.04	0.16	0.00	72.0	1.2	0	0.3
0	9	2.14	14.6	102.2	78.6	0.15	2.2	0.18	0.38	0.28	0.96	0.03	7.1	0.8	0	0.5
0	11	0.36	6.3	100.2	3.6	0.12	18.0	0.02	—	0.01	0.06	0.09	10.2	2.4	—	0.4
0	88	4.42	74.0	442.9	0.9	0.98	0	0.13	0.30	0.24	0.34	0.20	46.4	1.5	0	6.3
0	120	3.39	196.1	1173.0	1.7	4.10	0	0.36	—	0.64	0.90	0.19	176.3	4.0	0	16.6
0	119	3.35	124.7	1264.2	140.2	2.70	8.6	0.08	0.78	0.12	1.21	0.17	181.5	1.9	0	16.4
0	52	0.25	60.4	356.6	8.8	1.32	0.6	0.26	—	0.12	0.69	0.10	79.9	7.5	0	0.4
0	60	1.53	60.0	283.2	122.4	0.28	0	0.14	0.26	0.16	1.23	0.18	43.2	0	0	11.5
0	310	8.27	50.2	17.0	5.1	4.16	22.1	0.42	—	0.27	1.01	0.24	78.2	0.6	0	46.2
0	65	1.16	84.1	—	0	—	0	—	—	—	—	—	—	0	0	—
0	108	1.16	56.1	—	0	—	0	—	—	—	—	—	—	0	0	—
0	171	1.36	31.5	125.9	10.2	0.70	0	0.05	0.01	0.05	0.08	0.06	16.2	0.2	0	8.4
0	316	4.14	51.0	124.2	13.6	1.69	0.9	0.14	0.03	0.04	0.08	0.08	23.0	0	0	24.2
0	56	0.34	33.1	—	4.7	—	0	—	—	—	—	—	—	0	1.7	—
0	108	1.16	35.5	—	0	—	0	—	—	—	—	—	—	0	1.9	—
0	94	0.94	23.0	102.1	6.8	0.54	0	0.04	0.01	0.03	0.45	0.04	37.4	0.2	0	7.6

TABLE A-1 Table of Food Composition *(continued)*

(Computer code is for Cengage Diet Analysis Plus program)

DA+ Code	Food Description	QTY	Measure	Wt (g)	H2O (g)	Ener (cal)	Prot (g)	Carb (g)	Fiber (g)	Fat (g)	Fat Breakdown (g) Sat	Mono	Poly
Vegetables, Legumes—continued													
	Spinach												
663	Canned, drained	½	cup(s)	107	98.2	25	3.0	3.6	2.6	0.5	0.1	0	0.2
660	Chopped, boiled, drained	½	cup(s)	90	82.1	21	2.7	3.4	2.2	0.2	0	0	0.1
661	Chopped, frozen, boiled, drained	½	cup(s)	95	84.5	32	3.8	4.6	3.5	0.8	0.1	0	0.4
662	Leaf, frozen, boiled, drained	½	cup(s)	95	84.5	32	3.8	4.6	3.5	0.8	0.1	0	0.4
659	Raw, chopped	1	cup(s)	30	27.4	7	0.9	1.1	0.7	0.1	0	0	0
8470	Trimmed leaves	1	cup(s)	32	27.5	3	0.9	0	2.8	0.1	—	—	—
	Squash												
1662	Acorn winter, baked	½	cup(s)	103	85.0	57	1.1	14.9	4.5	0.1	0	0	0.1
29702	Acorn winter, boiled, mashed	½	cup(s)	123	109.9	42	0.8	10.8	3.2	0.1	0	0	0
29451	Butternut, frozen, boiled	½	cup(s)	122	106.9	47	1.5	12.2	1.8	0.1	0	0	0
1661	Butternut winter, baked	½	cup(s)	102	89.5	41	0.9	10.7	3.4	0.1	0	0	0
32773	Butternut winter, frozen, boiled, mashed, no salt added	½	cup(s)	121	106.4	47	1.5	12.2	—	0.1	0	0	0
29700	Crookneck and straightneck summer, boiled, drained	½	cup(s)	65	60.9	12	0.6	2.6	1.2	0.1	0	0	0.1
29703	Hubbard winter, baked	½	cup(s)	102	86.8	51	2.5	11.0	—	0.6	0.1	0	0.3
1660	Hubbard winter, boiled, mashed	½	cup(s)	118	107.5	35	1.7	7.6	3.4	0.4	0.1	0	0.2
29704	Spaghetti winter, boiled, drained, or baked	½	cup(s)	78	71.5	21	0.5	5.0	1.1	0.2	0	0	0.1
664	Summer, all varieties, sliced, boiled, drained	½	cup(s)	90	84.3	18	0.8	3.9	1.3	0.3	0.1	0	0.1
665	Winter, all varieties, baked, mashed	½	cup(s)	103	91.4	38	0.9	9.1	2.9	0.4	0.1	0	0.2
1112	Zucchini summer, boiled, drained	½	cup(s)	90	85.3	14	0.6	3.5	1.3	0	0	0	0
1113	Zucchini summer, frozen, boiled, drained	½	cup(s)	112	105.6	19	1.3	4.0	1.4	0.1	0	0	0.1
	Sweet potatoes												
666	Baked, peeled	½	cup(s)	100	75.8	90	2.0	20.7	3.3	0.2	0	0	0.1
667	Boiled, mashed	½	cup(s)	164	131.4	125	2.2	29.1	4.1	0.2	0.1	0	0.1
668	Candied, home recipe	½	cup(s)	91	61.1	132	0.8	25.4	2.2	3.0	1.2	0.6	0.1
670	Canned, vacuum pack	½	cup(s)	100	76.0	91	1.7	21.1	1.8	0.2	0	0	0.1
2765	Frozen, baked	½	cup(s)	88	64.5	88	1.5	20.5	1.6	0.1	0	0	0
1136	Yams, baked or boiled, drained	½	cup(s)	68	47.7	79	1.0	18.7	2.7	0.1	0	0	0
32785	**Taro shoots, cooked, no salt added**	½	cup(s)	70	66.7	10	0.5	2.2	—	0.1	0	0	0
	Tomatillo												
8774	Raw	2	item(s)	68	62.3	22	0.7	4.0	1.3	0.7	0.1	0.1	0.3
8777	Raw, chopped	½	cup(s)	66	60.5	21	0.6	3.9	1.3	0.7	0.1	0.1	0.3
	Tomato												
16846	Cherry, fresh	5	item(s)	85	80.3	15	0.7	3.3	1.0	0.2	0	0	0.1
671	Fresh, ripe, red	1	item(s)	123	116.2	22	1.1	4.8	1.5	0.2	0	0	0.1
675	Juice, canned	½	cup(s)	122	114.1	21	0.9	5.2	0.5	0.1	0	0	0
75	Juice, no salt added	½	cup(s)	122	114.1	21	0.9	5.2	0.5	0.1	0	0	0
1699	Paste, canned	2	tablespoon(s)	33	24.1	27	1.4	6.2	1.3	0.2	0	0	0.1
1700	Puree, canned	¼	cup(s)	63	54.9	24	1.0	5.6	1.2	0.1	0	0	0.1
1118	Red, boiled	½	cup(s)	120	113.2	22	1.1	4.8	0.8	0.1	0	0	0.1
3952	Red, diced	½	cup(s)	90	85.1	16	0.8	3.5	1.1	0.2	0	0	0.1
1120	Red, stewed, canned	½	cup(s)	128	116.7	33	1.2	7.9	1.3	0.2	0	0	0.1
1125	Sauce, canned	¼	cup(s)	61	55.6	15	0.8	3.3	0.9	0.1	0	0	0
8778	Sun dried	¼	cup(s)	27	3.9	70	3.8	15.1	3.3	0.8	0.1	0.1	0.3
8783	Sun dried in oil, drained	¼	cup(s)	28	14.8	59	1.4	6.4	1.6	3.9	0.5	2.4	0.6
	Turnips												
678	Turnip greens, chopped, boiled, drained	½	cup(s)	72	67.1	14	0.8	3.1	2.5	0.2	0	0	0.1
679	Turnip greens, frozen, chopped, boiled, drained	½	cup(s)	82	74.1	24	2.7	4.1	2.8	0.3	0.1	0	0.1
677	Turnips, cubed, boiled, drained	½	cup(s)	78	73.0	17	0.6	3.9	1.6	0.1	0	0	0
	Vegetables, mixed												
1132	Canned, drained	½	cup(s)	82	70.9	40	2.1	7.5	2.4	0.2	0	0	0.1
680	Frozen, boiled, drained	½	cup(s)	91	75.7	59	2.6	11.9	4.0	0.1	0	0	0.1

Chol (mg)	Calc (mg)	Iron (mg)	Magn (mg)	Pota (mg)	Sodi (mg)	Zinc (mg)	Vit A (µg)	Thia (mg)	Vit E (mg α)	Ribo (mg)	Niac (mg)	Vit B_6 (mg)	Fola (µg DFE)	Vit C (mg)	Vit B_{12} (µg)	Sele (µg)
0	136	2.45	81.3	370.2	28.9	0.48	524.3	0.02	2.08	0.14	0.41	0.11	104.8	15.3	0	1.5
0	122	3.21	78.3	419.4	63.0	0.68	471.6	0.08	1.87	0.21	0.44	0.21	131.4	8.8	0	1.4
0	145	1.86	77.9	286.9	92.2	0.46	572.9	0.07	3.36	0.16	0.41	0.12	115.0	2.1	0	5.2
0	145	1.86	77.9	286.9	92.2	0.46	572.9	0.07	3.36	0.16	0.41	0.12	115.0	2.1	0	5.2
0	30	0.81	23.7	167.4	23.7	0.16	140.7	0.02	0.61	0.06	0.22	0.06	58.2	8.4	0	0.3
0	25	2.13	25.5	134.1	38.0	0.18	—	0.03	—	0.05	0.18	0.07	0	7.5	0	—
0	45	0.95	44.1	447.9	4.1	0.17	21.5	0.17	—	0.01	0.90	0.19	19.5	11.1	0	0.7
0	32	0.68	31.9	322.2	3.7	0.13	50.2	0.12	—	0.01	0.65	0.14	13.5	8.0	0	0.5
0	23	0.70	10.9	161.9	2.4	0.14	203.3	0.06	0.14	0.05	0.56	0.08	19.5	4.3	0	0.6
0	42	0.61	29.6	289.6	4.1	0.13	569.1	0.07	1.31	0.01	0.99	0.12	19.4	15.4	0	0.5
0	23	0.70	10.9	161.2	2.4	0.14	202.4	0.06	—	0.05	0.56	0.08	19.4	4.2	0	0.6
0	14	0.31	13.6	137.1	1.3	0.19	5.2	0.03	—	0.02	0.29	0.07	14.9	5.4	0	0.1
0	17	0.48	22.4	365.1	8.2	0.15	308.0	0.07	—	0.04	0.57	0.17	16.3	9.7	0	0.6
0	12	0.33	15.3	252.5	5.9	0.11	236.0	0.05	0.14	0.03	0.39	0.12	11.8	7.7	0	0.4
0	16	0.26	8.5	90.7	14.0	0.15	4.7	0.02	0.09	0.01	0.62	0.07	6.2	2.7	0	0.2
0	24	0.32	21.6	172.8	0.9	0.35	9.9	0.04	0.12	0.03	0.46	0.05	18.0	5.0	0	0.2
0	23	0.45	13.3	247.0	1.0	0.23	267.5	0.02	0.12	0.07	0.51	0.17	20.5	9.8	0	0.4
0	12	0.32	19.8	227.7	2.7	0.16	50.4	0.04	0.11	0.04	0.39	0.07	15.3	4.1	0	0.2
0	19	0.54	14.5	216.3	2.2	0.22	10	0.05	0.13	0.04	0.43	0.05	8.9	4.1	0	0.2
0	38	0.69	27.0	475.0	36.0	0.32	961.0	0.10	0.71	0.10	1.48	0.28	6.0	19.6	0	0.2
0	44	1.18	29.5	377.2	44.3	0.33	1290.7	0.09	1.54	0.08	0.88	0.27	9.8	21.0	0	0.3
7	24	1.03	10.0	172.6	63.9	0.13	0	0.01	—	0.03	0.36	0.03	10.0	6.1	0	0.7
0	22	0.89	22.0	312.0	53.0	0.18	399.0	0.04	1.00	0.06	0.74	0.19	17.0	26.4	0	0.7
0	31	0.47	18.4	330.1	7.0	0.26	913.3	0.05	0.67	0.04	0.49	0.16	19.3	8.0	0	0.5
0	10	0.35	12.2	455.6	5.4	0.13	4.1	0.06	0.23	0.01	0.37	0.15	10.9	8.2	0	0.5
0	10	0.28	5.6	240.8	1.4	0.37	2.1	0.02	—	0.03	0.56	0.07	2.1	13.2	0	0.7
0	5	0.42	13.6	182.2	0.7	0.15	4.1	0.03	0.25	0.02	1.25	0.03	4.8	8.0	0	0.3
0	5	0.41	13.2	176.9	0.7	0.15	4.0	0.03	0.25	0.02	1.22	0.04	4.6	7.7	0	0.3
0	9	0.22	9.4	201.5	4.3	0.14	35.7	0.03	0.45	0.01	0.50	0.06	12.8	10.8	0	0
0	12	0.33	13.5	291.5	6.2	0.20	51.7	0.04	0.66	0.02	0.73	0.09	18.5	15.6	0	0
0	12	0.52	13.4	278.2	326.8	0.18	27.9	0.06	0.39	0.04	0.82	0.14	24.3	22.2	0	0.4
0	12	0.52	13.4	278.2	12.2	0.18	27.9	0.06	0.39	0.04	0.82	0.14	24.3	22.2	0	0.4
0	12	0.97	13.8	332.6	259.1	0.20	24.9	0.02	1.41	0.05	1.00	0.07	3.9	7.2	0	1.7
0	11	1.11	14.4	274.4	249.4	0.22	16.3	0.01	1.23	0.05	0.91	0.07	6.9	6.6	0	0.4
0	13	0.82	10.8	261.6	13.2	0.17	28.8	0.04	0.67	0.03	0.64	0.10	15.6	27.4	0	0.6
0	9	0.24	9.9	213.3	4.5	0.15	37.8	0.03	0.48	0.01	0.53	0.07	13.5	11.4	0	0
0	43	1.70	15.3	263.9	281.8	0.22	11.5	0.06	1.06	0.04	0.91	0.02	6.4	10.1	0	0.8
0	8	0.62	9.8	201.9	319.6	0.12	10.4	0.01	0.87	0.04	0.59	0.06	6.7	4.3	0	0.1
0	30	2.45	52.4	925.3	565.7	0.53	11.9	0.14	0.00	0.13	2.44	0.09	18.4	10.6	0	1.5
0	13	0.73	22.3	430.4	73.2	0.21	17.6	0.05	—	0.10	0.99	0.08	6.3	28.0	0	0.8
0	99	0.58	15.8	146.2	20.9	0.10	274.3	0.03	1.35	0.05	0.30	0.13	85.0	19.7	0	0.6
0	125	1.59	21.3	183.7	12.3	0.34	441.2	0.04	2.18	0.06	0.38	0.06	32.0	17.9	0	1.0
0	26	0.14	7.0	138.1	12.5	0.09	0	0.02	0.02	0.02	0.23	0.05	7.0	9.0	0	0.2
0	22	0.86	13.0	237.2	121.4	0.33	475.1	0.04	0.24	0.04	0.47	0.06	19.6	4.1	0	0.2
0	23	0.74	20.0	153.8	31.9	0.44	194.7	0.06	0.34	0.10	0.77	0.06	17.3	2.9	0	0.3

TABLE A-1 Table of Food Composition *(continued)*

(Computer code is for Cengage Diet Analysis Plus program)

											Fat Breakdown (g)		
DA+ Code	Food Description	QTY	Measure	Wt (g)	H_2O (g)	Ener (cal)	Prot (g)	Carb (g)	Fiber (g)	Fat (g)	Sat	Mono	Poly
	Vegetables, Legumes—continued												
7489	V8 100% vegetable juice	½	cup(s)	120	—	25	1.0	5.0	1.0	0	0	0	0
7490	V8 low sodium vegetable juice	½	cup(s)	120	—	25	0	6.5	1.0	0	0	0	0
7491	V8 spicy hot vegetable juice	½	cup(s)	120	—	25	1.0	5.0	0.5	0	0	0	0
	Water chestnuts												
31073	Sliced, drained	½	cup(s)	75	70.0	20	0	5.0	1.0	0	0	0	0
31087	Whole	½	cup(s)	75	70.0	20	0	5.0	1.0	0	0	0	0
1135	**Watercress**	1	cup(s)	34	32.3	4	0.8	0.4	0.2	0	0	0	0
	Nuts, Seeds, and Products												
	Almonds												
32940	Almond butter with salt added	1	tablespoon(s)	16	0.2	101	2.4	3.4	0.6	9.5	0.9	6.1	2.0
1137	Almond butter, no salt added	1	tablespoon(s)	16	0.2	101	2.4	3.4	0.6	9.5	0.9	6.1	2.0
32886	Blanched	¼	cup(s)	36	1.6	211	8.0	7.2	3.8	18.3	1.4	11.7	4.4
32887	Dry roasted, no salt added	¼	cup(s)	35	0.9	206	7.6	6.7	4.1	18.2	1.4	11.6	4.4
29724	Dry roasted, salted	¼	cup(s)	35	0.9	206	7.6	6.7	4.1	18.2	1.4	11.6	4.4
29725	Oil roasted, salted	¼	cup(s)	39	1.1	238	8.3	6.9	4.1	21.7	1.7	13.7	5.3
508	Slivered	¼	cup(s)	27	1.3	155	5.7	5.9	3.3	13.3	1.0	8.3	3.3
1138	**Beechnuts, dried**	¼	cup(s)	57	3.8	328	3.5	19.1	5.3	28.5	3.3	12.5	11.4
517	**Brazil nuts, dried, unblanched**	¼	cup(s)	35	1.2	230	5.0	4.3	2.6	23.3	5.3	8.6	7.2
1166	**Breadfruit seeds, roasted**	¼	cup(s)	57	28.3	118	3.5	22.8	3.4	1.5	0.4	0.2	0.8
1139	**Butternuts, dried**	¼	cup(s)	30	1.0	184	7.5	3.6	1.4	17.1	0.4	3.1	12.8
	Cashews												
32931	Cashew butter with salt added	1	tablespoon(s)	16	0.5	94	2.8	4.4	0.3	7.9	1.6	4.7	1.3
32889	Cashew butter, no salt added	1	tablespoon(s)	16	0.5	94	2.8	4.4	0.3	7.9	1.6	4.7	1.3
1140	Dry roasted	¼	cup(s)	34	0.6	197	5.2	11.2	1.0	15.9	3.1	9.4	2.7
518	Oil roasted	¼	cup(s)	32	1.1	187	5.4	9.6	1.1	15.4	2.7	8.4	2.8
	Coconut, shredded												
32896	Dried, not sweetened	¼	cup(s)	23	0.7	152	1.6	5.4	3.8	14.9	13.2	0.6	0.2
1153	Dried, shredded, sweetened	¼	cup(s)	23	2.9	116	0.7	11.1	1.0	8.3	7.3	0.4	0.1
520	Shredded	¼	cup(s)	20	9.4	71	0.7	3.0	1.8	6.7	5.9	0.3	0.1
	Chestnuts												
1152	Chinese, roasted	¼	cup(s)	36	14.6	87	1.6	19.0	—	0.4	0.1	0.2	0.1
32895	European, boiled and steamed	¼	cup(s)	46	31.3	60	0.9	12.8	—	0.6	0.1	0.2	0.2
32911	European, roasted	¼	cup(s)	36	14.5	88	1.1	18.9	1.8	0.8	0.1	0.3	0.3
32922	Japanese, boiled and steamed	¼	cup(s)	36	31.0	20	0.3	4.5	—	0.1	0	0	0
32923	Japanese, roasted	¼	cup(s)	36	18.1	73	1.1	16.4	—	0.3	0	0.1	0.1
4958	**Flax seeds or linseeds**	¼	cup(s)	43	3.3	225	8.4	12.3	11.9	17.7	1.7	3.2	12.6
32904	**Ginkgo nuts, dried**	¼	cup(s)	39	4.8	136	4.0	28.3	—	0.8	0.1	0.3	0.3
	Hazelnuts or filberts												
32901	Blanched	¼	cup(s)	30	1.7	189	4.1	5.1	3.3	18.3	1.4	14.5	1.7
32902	Dry roasted, no salt added	¼	cup(s)	30	0.8	194	4.5	5.3	2.8	18.7	1.3	14.0	2.5
1156	**Hickory nuts, dried**	¼	cup(s)	30	0.8	197	3.8	5.5	1.9	19.3	2.1	9.8	6.6
	Macadamias												
32905	Dry roasted, no salt added	¼	cup(s)	34	0.5	241	2.6	4.5	2.7	25.5	4.0	19.9	0.5
32932	Dry roasted, with salt added	¼	cup(s)	34	0.5	240	2.6	4.3	2.7	25.5	4.0	19.9	0.5
1157	Raw	¼	cup(s)	34	0.5	241	2.6	4.6	2.9	25.4	4.0	19.7	0.5
	Mixed nuts												
1159	With peanuts, dry roasted	¼	cup(s)	34	0.6	203	5.9	8.7	3.1	17.6	2.4	10.8	3.7
32933	With peanuts, dry roasted, with salt added	¼	cup(s)	34	0.6	203	5.9	8.7	3.1	17.6	2.4	10.8	3.7
32906	Without peanuts, oil roasted, no salt added	¼	cup(s)	36	1.1	221	5.6	8.0	2.0	20.2	3.3	11.9	4.1
	Peanuts												
2807	Dry roasted	¼	cup(s)	37	0.6	214	8.6	7.9	2.9	18.1	2.5	9.0	5.7
2806	Dry roasted, salted	¼	cup(s)	37	0.6	214	8.6	7.9	2.9	18.1	2.5	9.0	5.7

Chol (mg)	Calc (mg)	Iron (mg)	Magn (mg)	Pota (mg)	Sodi (mg)	Zinc (mg)	Vit A (µg)	Thia (mg)	Vit E (mg α)	Ribo (mg)	Niac (mg)	Vit B_6 (mg)	Fola (µg DFE)	Vit C (mg)	Vit B_{12} (µg)	Sele (µg)
0	20	0.36	12.9	260.0	310.0	0.24	100.0	0.05	—	0.03	0.87	0.17	—	30.0	0	—
0	20	0.36	—	450.0	70.0	—	100.0	0.02	—	0.02	0.75	—	—	30.0	0	—
0	20	0.36	12.9	240.0	360.0	0.24	50.0	0.05	—	0.03	0.88	0.17	—	15.0	0	—
0	7	0.00	—	—	5.0	—	0	—	—	—	—	—	—	2.0	—	—
0	7	0.00	—	—	5.0	—	0	—	—	—	—	—	—	2.0	—	—
0	41	0.06	7.1	112.2	13.9	0.03	54.4	0.03	0.34	0.04	0.06	0.04	3.1	14.6	0	0.3
0	43	0.59	48.5	121.3	72.0	0.49	0	0.02	4.16	0.10	0.46	0.01	10.4	0.1	0	0.8
0	43	0.59	48.5	121.3	1.8	0.48	0	0.02	—	0.09	0.46	0.01	10.4	0.1	0	—
0	78	1.34	99.7	249.0	10.2	1.13	0	0.07	8.95	0.20	1.32	0.04	10.9	0	0	1.0
0	92	1.55	98.7	257.4	0.3	1.22	0	0.02	8.97	0.29	1.32	0.04	11.4	0	0	1.0
0	92	1.55	98.7	257.4	117.0	1.22	0	0.02	8.97	0.29	1.32	0.04	11.4	0	0	1.0
0	114	1.44	107.5	274.4	133.1	1.20	0	0.03	10.19	0.30	1.43	0.04	10.6	0	0	1.1
0	71	1.00	72.4	190.4	0.3	0.83	0	0.05	7.07	0.27	0.91	0.03	13.5	0	0	0.7
0	1	1.39	0	579.7	21.7	0.20	0	0.16	—	0.20	0.48	0.38	64.4	8.8	0	4.0
0	56	0.85	131.6	230.7	1.1	1.42	0	0.21	2.00	0.01	0.10	0.03	7.7	0.2	0	671.0
0	49	0.50	35.3	616.7	15.9	0.58	8.5	0.22	—	0.12	4.20	0.22	33.6	4.3	0	8.0
0	16	1.21	71.1	126.3	0.3	0.94	1.8	0.12	—	0.04	0.31	0.17	19.8	1.0	0	5.2
0	7	0.81	41.3	87.4	98.2	0.83	0	0.05	0.15	0.03	0.26	0.04	10.9	0	0	1.8
0	7	0.81	41.3	87.4	2.4	0.83	0	0.05	—	0.03	0.26	0.04	10.9	0	0	1.8
0	15	2.06	89.1	193.5	5.5	1.92	0	0.07	0.32	0.07	0.48	0.09	23.6	0	0	4.0
0	14	1.95	88.0	203.8	4.2	1.73	0	0.12	0.30	0.07	0.56	0.10	8.1	0.1	0	6.5
0	6	0.76	20.7	125.2	8.5	0.46	0	0.01	0.10	0.02	0.13	0.07	2.1	0.3	0	4.3
0	3	0.45	11.6	78.4	60.9	0.42	0	0.01	0.09	0.00	0.11	0.06	1.9	0.2	0	3.9
0	3	0.48	6.4	71.2	4.0	0.21	0	0.01	0.04	0.00	0.11	0.01	5.2	0.7	0	2.0
0	7	0.54	32.6	173.0	1.4	0.33	0	0.05	—	0.03	0.54	0.15	26.1	13.9	0	2.6
0	21	0.80	24.8	328.9	12.4	0.11	0.5	0.06	—	0.03	0.32	0.10	17.5	12.3	0	—
0	10	0.32	11.8	211.6	0.7	0.20	0.4	0.08	0.18	0.05	0.48	0.18	25.0	9.3	0	0.4
0	4	0.19	6.5	42.8	1.8	0.14	0.4	0.04	—	0.01	0.19	0.03	6.1	3.4	0	—
0	13	0.75	23.2	154.8	6.9	0.51	1.4	0.15	—	—	0.24	0.14	21.4	10.1	0	—
0	142	2.13	156.1	354.0	11.9	1.83	0	0.06	0.14	0.06	0.59	0.39	118.4	0.5	0	2.3
0	8	0.62	20.7	390.2	5.1	0.26	21.5	0.17	—	0.07	4.58	0.25	41.4	11.4	0	—
0	45	0.98	48.0	197.4	0	0.66	0.6	0.14	5.25	0.03	0.46	0.17	23.4	0.6	0	1.2
0	37	1.31	51.9	226.5	0	0.74	0.9	0.10	4.58	0.03	0.61	0.18	26.4	1.1	0	1.2
0	18	0.64	51.9	130.8	0.3	1.29	2.1	0.26	—	0.04	0.27	0.06	12.0	0.6	0	2.4
0	23	0.88	39.5	121.6	1.3	0.43	0	0.23	0.19	0.02	0.76	0.12	3.4	0.2	0	3.9
0	23	0.88	39.5	121.6	88.8	0.43	0	0.23	0.19	0.02	0.76	0.12	3.4	0.2	0	3.9
0	28	1.24	43.6	123.3	1.7	0.44	0	0.40	0.18	0.05	0.83	0.09	3.7	0.4	0	1.2
0	24	1.27	77.1	204.5	4.1	1.30	0.3	0.07	—	0.07	1.61	0.10	17.1	0.1	0	1.0
0	24	1.26	77.1	204.5	229.1	1.30	0	0.06	3.74	0.06	1.61	0.10	17.1	0.1	0	2.6
0	38	0.92	90.4	195.8	4.0	1.67	0.4	0.18	—	0.17	0.70	0.06	20.2	0.2	0	—
0	20	0.82	64.2	240.2	2.2	1.20	0	0.16	2.52	0.03	4.93	0.09	52.9	0	0	2.7
0	20	0.82	64.2	240.2	296.7	1.20	0	0.16	2.84	0.03	4.93	0.09	52.9	0	0	2.7

(Computer code is for Cengage Diet Analysis Plus program)

DA+ Code	Food Description	QTY	Measure	Wt (g)	H_2O (g)	Ener (cal)	Prot (g)	Carb (g)	Fiber (g)	Fat (g)	Fat Breakdown (g) Sat	Mono	Poly
	Nuts, Seeds, and Products—continued												
1763	Oil roasted, salted	¼	cup(s)	36	0.5	216	10.1	5.5	3.4	18.9	3.1	9.4	5.5
1884	Peanut butter, chunky	1	tablespoon(s)	16	0.2	94	3.8	3.5	1.3	8.0	1.3	3.9	2.4
30303	Peanut butter, low sodium	1	tablespoon(s)	16	0.2	95	4.0	3.1	0.9	8.2	1.8	3.9	2.2
30305	Peanut butter, reduced fat	1	tablespoon(s)	18	0.2	94	4.7	6.4	0.9	6.1	1.3	2.9	1.8
524	Peanut butter, smooth	1	tablespoon(s)	16	0.3	94	4.0	3.1	1.0	8.1	1.7	3.9	2.3
2804	Raw	¼	cup(s)	37	2.4	207	9.4	5.9	3.1	18.0	2.5	8.9	5.7
	Pecans												
32907	Dry roasted, no salt added	¼	cup(s)	28	0.3	198	2.6	3.8	2.6	20.7	1.8	12.3	5.7
32936	Dry roasted, with salt added	¼	cup(s)	27	0.3	192	2.6	3.7	2.5	20.0	1.7	11.9	5.6
1162	Oil roasted	¼	cup(s)	28	0.3	197	2.5	3.6	2.6	20.7	2.0	11.3	6.5
526	Raw	¼	cup(s)	27	1.0	188	2.5	3.8	2.6	19.6	1.7	11.1	5.9
12973	**Pine nuts or pignolia, dried**	1	tablespoon(s)	9	0.2	58	1.2	1.1	0.3	5.9	0.4	1.6	2.9
	Pistachios												
1164	Dry roasted	¼	cup(s)	31	0.6	176	6.6	8.5	3.2	14.1	1.7	7.4	4.3
32938	Dry roasted, with salt added	¼	cup(s)	32	0.6	182	6.8	8.6	3.3	14.7	1.8	7.7	4.4
1167	**Pumpkin or squash seeds, roasted**	¼	cup(s)	57	4.0	296	18.7	7.6	2.2	23.9	4.5	7.4	10.9
	Sesame												
32912	Sesame butter paste	1	tablespoon(s)	16	0.3	94	2.9	3.8	0.9	8.1	1.1	3.1	3.6
32941	Tahini or sesame butter	1	tablespoon(s)	15	0.5	89	2.6	3.2	0.7	8.0	1.1	3.0	3.5
1169	Whole, roasted, toasted	3	tablespoon(s)	10	0.3	54	1.6	2.4	1.3	4.6	0.6	1.7	2.0
	Soy nuts												
34173	Deep sea salted	¼	cup(s)	28	—	119	11.9	8.9	4.9	4.0	1.0	—	—
34174	Unsalted	¼	cup(s)	28	—	119	11.9	8.9	4.9	4.0	0	—	—
	Sunflower seeds												
528	Kernels, dried	1	tablespoon(s)	9	0.4	53	1.9	1.8	0.8	4.6	0.4	1.7	2.1
29721	Kernels, dry roasted, salted	1	tablespoon(s)	8	0.1	47	1.5	1.9	0.7	4.0	0.4	0.8	2.6
29723	Kernels, toasted, salted	1	tablespoon(s)	8	0.1	52	1.4	1.7	1.0	4.8	0.5	0.9	3.1
32928	Sunflower seed butter with salt added	1	tablespoon(s)	16	0.2	93	3.1	4.4	—	7.6	0.8	1.5	5.0
	Trail mix												
4646	Trail mix	¼	cup(s)	38	3.5	173	5.2	16.8	2.0	11.0	2.1	4.7	3.6
4647	Trail mix with chocolate chips	¼	cup(s)	38	2.5	182	5.3	16.8	—	12.0	2.3	5.1	4.2
4648	Tropical trail mix	¼	cup(s)	35	3.2	142	2.2	23.0	—	6.0	3.0	0.9	1.8
	Walnuts												
529	Dried black, chopped	¼	cup(s)	31	1.4	193	7.5	3.1	2.1	18.4	1.1	4.7	11.0
531	English or Persian	¼	cup(s)	29	1.2	191	4.5	4.0	2.0	19.1	1.8	2.6	13.8
Vegetarian Foods													
	Prepared												
34222	Brown rice and tofu stir-fry (vegan)	8	ounce(s)	227	244.4	302	16.5	18.0	3.2	21.0	1.7	4.7	13.4
34368	Cheese enchilada casserole (lacto)	8	ounce(s)	227	80.3	385	16.6	38.4	4.1	17.8	9.5	6.1	1.1
34247	Five bean casserole (vegan)	8	ounce(s)	227	175.8	178	5.9	26.6	6.0	5.8	1.1	2.5	1.9
34261	Lentil stew (vegan)	8	ounce(s)	227	227.9	188	11.5	35.9	11.0	0.7	0.1	0.1	0.3
34397	Macaroni and cheese (lacto)	8	ounce(s)	227	352.1	391	18.1	37.1	1.0	18.7	9.8	6.0	1.8
34238	Steamed rice and vegetables (vegan)	8	ounce(s)	227	222.9	587	11.2	87.9	5.8	23.1	4.1	8.7	9.1
34308	Tofu rice burgers (ovo-lacto)	1	piece(s)	218	77.6	435	22.4	68.6	5.6	8.4	1.7	2.4	3.5
34276	Vegan spinach enchiladas (vegan)	1	piece(s)	82	59.2	93	4.9	14.5	1.8	2.4	0.3	0.6	1.3
34243	Vegetable chow mein (vegan)	8	ounce(s)	227	163.3	166	6.5	22.1	2.0	6.4	0.7	2.7	2.5
34454	Vegetable lasagna (lacto)	8	ounce(s)	227	178.9	208	13.7	29.9	2.6	4.1	2.3	1.1	0.3
34339	Vegetable marinara (vegan)	8	ounce(s)	252	200.7	104	3.0	16.7	1.4	3.1	0.4	1.4	1.0
34356	Vegetable rice casserole (lacto)	8	ounce(s)	227	178.9	238	9.7	24.4	4.0	12.5	4.9	3.5	3.1
34311	Vegetable strudel (ovo-lacto)	8	ounce(s)	227	63.1	478	12.0	32.4	2.5	33.8	11.5	16.7	3.9
34371	Vegetable taco (lacto)	1	item(s)	85	46.5	117	4.2	13.6	2.9	5.6	2.1	1.9	1.3
34282	Vegetarian chili (vegan)	8	ounce(s)	227	191.4	115	5.6	21.4	7.1	1.5	0.2	0.3	0.7
34367	Vegetarian vegetable soup (vegan)	8	ounce(s)	227	257.9	111	3.2	16.0	3.2	5.0	1.0	2.1	1.6
	Boca burger												
32067	All American flamed grilled patty	1	item(s)	71	—	90	14.0	4.0	3.0	3.0	1.0	—	—
32074	Boca chik'n nuggets	4	item(s)	87	—	180	14.0	17.0	3.0	7.0	1.0	—	—

A

Chol (mg)	Calc (mg)	Iron (mg)	Magn (mg)	Pota (mg)	Sodi (mg)	Zinc (mg)	Vit A (µg)	Thia (mg)	Vit E (mg α)	Ribo (mg)	Niac (mg)	Vit B_6 (mg)	Fola (µg DFE)	Vit C (mg)	Vit B_{12} (µg)	Sele (µg)
0	22	0.54	63.4	261.4	115.2	1.18	0	0.03	2.49	0.03	4.97	0.16	43.2	0.3	0	1.2
0	7	0.30	25.6	119.2	77.8	0.45	0	0.02	1.01	0.02	2.19	0.07	14.7	0	0	1.3
0	6	0.29	25.4	107.0	2.7	0.47	0	0.01	1.23	0.02	2.14	0.07	11.8	0	0	1.2
0	6	0.34	30.6	120.4	97.2	0.50	0	0.05	1.20	0.01	2.63	0.06	10.8	0	0	1.4
0	7	0.30	24.6	103.8	73.4	0.47	0	0.01	1.44	0.02	2.14	0.09	11.8	0	0	0.9
0	34	1.67	61.3	257.3	6.6	1.19	0	0.23	3.04	0.04	4.40	0.12	87.6	0	0	2.6
0	20	0.78	36.8	118.3	0.3	1.41	1.9	0.12	0.35	0.03	0.32	0.05	4.5	0.2	0	1.1
0	19	0.75	35.6	114.5	103.4	1.36	1.9	0.11	0.34	0.03	0.31	0.05	4.3	0.2	0	1.1
0	18	0.68	33.3	107.8	0.3	1.23	1.4	0.13	0.70	0.03	0.33	0.05	4.1	0.2	0	1.7
0	19	0.69	33.0	111.7	0	1.23	0.8	0.18	0.38	0.04	0.32	0.06	6.0	0.3	0	1.0
0	1	0.47	21.6	51.3	0.2	0.55	0.1	0.03	0.80	0.02	0.37	0.01	2.9	0.1	0	0.1
0	34	1.29	36.9	320.4	3.1	0.71	4.0	0.26	0.59	0.05	0.44	0.39	15.4	0.7	0	2.9
0	35	1.34	38.4	333.4	129.6	0.73	4.2	0.26	0.61	0.05	0.45	0.40	16.0	0.7	0	3.0
0	24	8.48	303.0	457.4	10.2	4.22	10.8	0.12	0.00	0.18	0.99	0.05	32.3	1.0	0	3.2
0	154	3.07	57.9	93.1	1.9	1.17	0.5	0.04	—	0.03	1.07	0.13	16.0	0	0	0.9
0	21	0.66	14.3	68.9	5.3	0.69	0.5	0.24	—	0.02	0.85	0.02	14.7	0.6	0	0.3
0	94	1.40	33.8	45.1	1.0	0.68	0	0.07	—	0.02	0.43	0.07	9.3	0	0	0.5
0	59	1.07	—	—	148.1	—	0	—	—	—	—	—	—	0	—	—
0	59	1.07	—	—	9.9	—	0	—	—	—	—	—	—	0	—	—
0	7	0.47	29.3	58.1	0.8	0.45	0.3	0.13	2.99	0.03	0.75	0.12	20.4	0.1	0	4.8
0	6	0.30	10.3	68.0	32.8	0.42	0	0.01	2.09	0.02	0.56	0.06	19.0	0.1	0	6.3
0	5	0.57	10.8	41.1	51.3	0.44	0	0.03	—	0.02	0.35	0.07	19.9	0.1	0	5.2
0	20	0.76	59.0	11.5	83.2	0.85	0.5	0.05	—	0.05	0.85	0.13	37.9	0.4	0	—
0	29	1.14	59.3	256.9	85.9	1.20	0.4	0.17	—	0.07	1.76	0.11	26.6	0.5	0	—
2	41	1.27	60.4	243.0	45.4	1.17	0.8	0.15	—	0.08	1.65	0.09	24.4	0.5	0	—
0	20	0.92	33.6	248.2	3.5	0.41	0.7	0.15	—	0.04	0.51	0.11	14.7	2.7	0	—
0	19	0.97	62.8	163.4	0.6	1.05	0.6	0.01	0.56	0.04	0.14	0.18	9.7	0.5	0	5.3
0	29	0.85	46.2	129.0	0.6	0.90	0.3	0.10	0.20	0.04	0.32	0.15	28.7	0.4	0	1.4
0	353	6.34	118.3	501.4	142.2	2.03	—	0.23	0.07	0.14	1.49	0.36	51.8	24.8	0	14.8
39	441	2.44	34.6	191.2	1139.7	1.84	—	0.31	0.05	0.35	2.23	0.11	118.3	20.4	0.4	20.0
0	48	1.78	40.8	364.1	613.6	0.61	—	0.10	0.52	0.07	0.93	0.11	64.5	8.3	0	3.3
0	34	3.23	50.0	548.8	436.5	1.42	—	0.24	0.14	0.16	2.31	0.29	202.7	26.4	0	12.1
43	415	1.71	45.4	267.8	1641.0	2.32	—	0.32	0.27	0.48	2.18	0.13	162.7	0.9	0.8	33.3
0	91	3.31	153.1	810.1	3117.8	2.04	—	0.37	3.03	0.21	6.16	0.64	70.1	35.2	0	18.8
52	467	9.01	89.7	455.6	2449.5	2.06	—	0.27	0.12	0.26	3.43	0.29	167.7	2.0	0.1	43.0
0	117	1.13	40.4	170.5	134.2	0.68	—	0.07	—	0.07	0.53	0.10	20.3	1.8	0	5.1
0	189	3.70	28.0	310.3	372.7	0.76	—	0.13	0.05	0.11	1.43	0.14	76.8	8.0	0	6.5
10	176	1.86	41.9	470.0	759.4	1.14	—	0.26	0.05	0.25	2.49	0.22	124.5	19.0	0.4	21.8
0	17	0.94	19.1	189.9	439.6	0.42	—	0.15	0.55	0.08	1.36	0.12	88.4	23.5	0	10.8
17	190	1.28	29.3	414.2	626.0	1.24	—	0.16	0.35	0.29	2.00	0.19	154.8	56.0	0.2	5.8
29	200	2.15	24.5	181.0	512.1	1.24	—	0.28	0.20	0.31	2.88	0.11	111.4	17.4	0.2	19.7
7	77	0.88	26.3	174.1	280.7	0.59	—	0.08	0.04	0.06	0.49	0.08	38.7	4.6	0	3.0
0	65	1.98	41.0	543.1	390.7	0.74	—	0.14	0.15	0.10	1.31	0.18	47.7	20.3	0	4.4
0	46	1.87	34.9	550.3	729.5	0.56	—	0.13	0.55	0.09	1.99	0.27	49.9	29.9	0	1.4
5	150	1.80	—	—	280.0	—	0	—	—	—	—	—	—	0	—	—
0	40	1.44	—	—	500.0	—	—	—	—	—	—	—	—	0	—	—

TABLE A-1 Table of Food Composition *(continued)*

(Computer code is for Cengage Diet Analysis Plus program)

DA+ Code	Food Description	QTY	Measure	Wt (g)	H_2O (g)	Ener (cal)	Prot (g)	Carb (g)	Fiber (g)	Fat (g)	Fat Breakdown (g) Sat	Mono	Poly
	Vegetarian Foods—continued												
32075	Boca meatless ground burger	½	cup(s)	57	—	60	13.0	6.0	3.0	0.5	0	—	—
32072	Breakfast links	2	item(s)	45	—	70	8.0	5.0	2.0	3.0	0.5	—	—
32071	Breakfast patties	1	item(s)	38	—	60	7.0	5.0	2.0	2.5	0	—	—
35780	Cheeseburger meatless burger patty	1	item(s)	71	—	100	12.0	5.0	3.0	5.0	1.5	—	—
33958	Original meatless chik'n patties	1	item(s)	71	—	160	11.0	15.0	2.0	6.0	1.0	—	—
32066	Original patty	1	item(s)	71	—	70	13.0	6.0	4.0	0.5	0	—	—
32068	Roasted garlic patty	1	item(s)	71	—	70	12.0	6.0	4.0	1.5	0	—	—
37814	Roasted onion meatless burger patty	1	item(s)	71	—	70	11.0	7.0	4.0	1.0	0	—	—
	Gardenburger												
37810	BBQ chik'n with sauce	1	item(s)	142	—	250	14.0	30.0	5.0	8.0	1.0	—	—
39661	Black bean burger	1	item(s)	71	—	80	8.0	11.0	4.0	2.0	0	—	—
39666	Buffalo chik'n wing	3	item(s)	95	—	180	9.0	8.0	5.0	12.0	1.5	—	—
39665	Country fried chicken with creamy pepper gravy	1	item(s)	142	—	190	9.0	16.0	2.0	9.0	1.0	—	—
37808	Flamed grilled chik'n	1	item(s)	71	—	100	13.0	5.0	3.0	2.5	0	—	—
37803	Garden vegan	1	item(s)	71	—	100	10.0	12.0	2.0	1.0	—	—	—
39663	Homestyle classic burger	1	item(s)	71	—	110	12.0	6.0	4.0	5.0	0.5	—	—
37807	Meatless breakfast sausage	1	item(s)	43	—	50	5.0	2.0	2.0	3.5	0	—	—
37809	Meatless meatballs	6	item(s)	85	—	110	12.0	8.0	4.0	4.5	1.0	—	—
37806	Meatless riblets with sauce	1	item(s)	142	—	160	17.0	11.0	4.0	5.0	0	—	—
29913	Original	1	item(s)	71	—	90	10.0	8.0	3.0	2.0	0.5	—	—
39662	Sun-dried tomato basil burger	1	item(s)	71	—	80	10.0	11.0	3.0	1.5	0.5	—	—
29915	Veggie medley	1	item(s)	71	—	90	9.0	11.0	4.0	2.0	0	—	—
	Loma Linda												
9311	Big franks, canned	1	item(s)	51	—	110	11.0	3.0	2.0	6.0	1.0	1.5	3.5
9323	Fried chik'n with gravy	2	piece(s)	80	45.9	150	12.0	5.0	2.0	10	1.5	2.5	5.0
9326	Linketts, canned	1	item(s)	35	21.0	70	7.0	1.0	1.0	4.0	0.5	1.0	2.5
9336	Redi-Burger patties, canned	1	slice(s)	85	50.5	120	18.0	7.0	4.0	2.5	0.5	0.5	1.5
9350	Swiss Stake pattie with gravy, frozen	1	piece(s)	92	65.7	130	9.0	9.0	3.0	6.0	1.0	1.5	3.5
9354	Tender Rounds meatball substitute, canned in gravy	6	piece(s)	80	53.9	120	13.0	6.0	1.0	4.5	0.5	1.5	2.5
	Morningstar Farms												
33707	America's Original Veggie Dog links	1	item(s)	57	—	80	11.0	6.0	1.0	0.5	0	—	—
9362	Better'n Eggs egg substitute	¼	cup(s)	57	50.3	20	5.0	0	0	0	0	0	0
9371	Breakfast bacon strips	2	item(s)	16	6.8	60	2.0	2.0	0.5	4.5	0.5	1.0	3.0
9368	Breakfast sausage links	2	item(s)	45	26.8	80	9.0	3.0	2.0	3.0	0.5	1.5	1.0
33705	Chik'n nuggets	4	piece(s)	86	—	190	12.0	18.0	2.0	7.0	1.0	2.0	4.0
11587	Chik patties	1	item(s)	71	36.3	150	9.0	16.0	2.0	6.0	1.0	1.5	2.5
2531	Garden veggie patties	1	item(s)	67	40.1	100	10.0	9.0	4.0	2.5	0.5	0.5	1.5
33702	Spicy black bean veggie burger	1	item(s)	78	—	140	12.0	15.0	3.0	4.0	0.5	1.0	2.5
9412	Vegetarian chili, canned	1	cup(s)	230	172.6	180	16.0	25.0	10.0	1.5	0.5	0.5	0.5
	Worthington												
9424	Chili, canned	1	cup(s)	230	167.0	280	24.0	25.0	8.0	10.0	1.5	1.5	7.0
9436	Diced chik, canned	¼	cup(s)	55	42.7	50	9.0	2.0	1.0	0	0	0	0
9440	Dinner roast, frozen	1	slice(s)	85	53.2	180	14.0	6.0	3.0	11.0	1.5	4.5	5.0
9420	Meatless chicken slices, frozen	3	slice(s)	57	38.9	90	9.0	2.0	0.5	4.5	1.0	1.0	2.5
36702	Meatless chicken style roll, frozen	1	slice(s)	55	—	90	9.0	2.0	1.0	4.5	1.0	1.0	2.5
9428	Meatless corned beef, sliced, frozen	3	slice(s)	57	31.2	140	10.0	5.0	0	9.0	1.0	2.0	5.0
9470	Meatless salami, sliced, frozen	3	slice(s)	57	32.4	120	12.0	3.0	2.0	7.0	1.0	1.0	5.0
9480	Meatless smoked turkey, sliced	3	slice(s)	57	—	140	10.0	4.0	0	9.0	1.5	2.0	5.0
9462	Prosage links	2	item(s)	45	26.8	80	9.0	3.0	2.0	3.0	0.5	0.5	2.0
9484	Stakelets patty beef steak substitute, frozen	1	piece(s)	71	41.5	150	14.0	7.0	2.0	7.0	1.0	2.5	3.5
9486	Stripples bacon substitute	2	item(s)	16	6.8	60	2.0	2.0	0.5	4.5	0.5	1.0	3.0
9496	Vegetable Skallops meat substitute, canned	½	cup(s)	85	—	90	17.0	4.0	3.0	1.0	0	0	0.5
Dairy													
	Cheese												
1433	Blue, crumbled	1	ounce(s)	28	12.0	100	6.1	0.7	0	8.1	5.3	2.2	0.2
884	Brick	1	ounce(s)	28	11.7	105	6.6	0.8	0	8.4	5.3	2.4	0.2

PAGE KEY: A-2 = Breads/Baked Goods A-8 = Cereal/Rice/Pasta A-12 = Fruit A-18 = Vegetables/Legumes A-28 = Nuts/Seeds A-30 = Vegetarian A-32 = Dairy A-40 = Eggs A-40 = Seafood A-44 = Meats A-48 = Poultry A-48 = Processed Meats A-50 = Beverages A-54 = Fats/Oils A-56 = Sweets A-58 = Spices/Condiments/Sauces A-62 = Mixed Foods/Soups/Sandwiches A-68 = Fast Food A-88 = Convenience A-90 = Baby Foods

A

Chol (mg)	Calc (mg)	Iron (mg)	Magn (mg)	Pota (mg)	Sodi (mg)	Zinc (mg)	Vit A (µg)	Thia (mg)	Vit E (mg α)	Ribo (mg)	Niac (mg)	Vit B_6 (mg)	Fola (µg DFE)	Vit C (mg)	Vit B_{12} (µg)	Sele (µg)
0	60	1.80	—	—	270.0	—	0	—	—	—	—	—	—	0	—	—
0	20	1.44	—	—	330.0	—	0	—	—	—	—	—	—	0	—	—
0	20	1.08	—	—	280.0	—	0	—	—	—	—	—	—	0	—	—
5	80	1.80	—	—	360.0	—	—	—	—	—	—	—	—	0	—	—
0	40	1.80	—	—	430.0	—	—	—	—	—	—	—	—	0	—	—
0	60	1.80	—	—	280.0	—	0	—	—	—	—	—	—	0	—	—
0	60	1.80	—	—	370.0	—	0	—	—	—	—	—	—	0	—	—
0	100	2.70	—	—	300.0	—	—	—	—	—	—	—	—	0	—	—
0	150	1.08	—	—	890.0	—	—	—	—	—	—	—	—	0	—	—
0	40	1.44	—	—	330.0	—	—	—	—	—	—	—	—	0	—	—
0	40	0.72	—	—	1000.0	—	—	—	—	—	—	—	—	0	—	—
5	40	1.44	—	—	550.0	—	—	—	—	—	—	—	—	0	—	—
0	60	3.60	—	—	360.0	—	—	—	—	—	—	—	—	0	—	—
0	40	4.50	—	—	230.0	—	—	—	—	—	—	—	—	0	—	—
0	80	1.44	—	—	380.0	—	—	—	—	—	—	—	—	0	—	—
0	20	0.72	—	—	120.0	—	—	—	—	—	—	—	—	0	—	—
0	60	1.80	—	—	400.0	—	—	—	—	—	—	—	—	0	—	—
0	60	1.80	—	—	720.0	—	—	—	—	—	—	—	—	3.6	—	—
0	80	1.08	30.4	193.4	490.0	0.89	—	0.10	—	0.15	1.08	0.08	10.1	1.2	0.1	7.0
5	60	1.44	—	—	260.0	—	—	—	—	—	—	—	—	3.6	—	—
0	40	1.44	27.0	182.0	290.0	0.46	—	0.07	—	0.08	0.90	0.09	10.6	9.0	0	4.0
0	0	0.77	—	50.0	220.0	—	0	0.22	—	0.10	2.00	0.70	—	0	2.4	—
0	20	1.80	—	70.0	430.0	0.33	0	1.05	—	0.34	4.00	0.30	—	0	2.4	—
0	0	0.36	—	20.0	160.0	0.46	0	0.12	—	0.20	0.80	0.16	—	0	0.9	—
0	0	1.06	—	140.0	450.0	—	0	0.15	—	0.25	4.00	0.40	—	0	1.2	—
0	0	0.72	—	200.0	430.0	—	0	0.45	—	0.25	10.00	1.00	—	0	5.4	—
0	20	1.08	—	80.0	340.0	0.66	0	0.75	—	0.17	2.00	0.16	—	0	1.2	—
0	0	0.72	—	60.0	580.0	—	0	—	—	—	—	—	—	0	—	—
0	20	0.72	—	75.0	90.0	0.60	37.5	0.03	—	0.34	0.00	0.08	24.0	—	0.6	—
0	0	0.36	—	15.0	220.0	0.05	0	0.75	—	0.04	0.40	0.07	—	0	0.2	—
0	0	1.80	—	50.0	300.0	—	0	0.37	—	0.17	7.00	0.50	—	0	3.0	—
0	20	2.70	—	320.0	490.0	—	0	0.52	—	0.25	5.00	0.30	—	0	1.5	—
0	0	1.80	—	210.0	540.0	—	0	1.80	—	0.17	2.00	0.20	—	0	1.2	—
0	40	0.72	—	180.0	350.0	—	—	—	—	—	—	—	—	0	—	—
0	40	1.80	—	320.0	470.0	—	0	—	—	—	0.00	—	—	0	—	—
0	40	3.60	—	660.0	900.0	—	—	—	—	—	—	—	—	0	—	—
0	40	3.60	—	330.0	1130.0	—	0	0.30	—	0.13	2.00	0.70	—	0	1.5	—
0	0	1.08	—	100.0	220.0	0.24	0	0.06	—	0.10	4.00	0.08	—	0	0.2	—
0	20	1.80	—	120.0	580.0	0.64	0	1.80	—	0.25	6.00	0.60	—	0	1.5	—
0	250	1.80	—	250.0	250.0	0.26	0	0.37	—	0.13	4.00	0.30	—	0	1.8	—
0	100	1.08	—	240.0	240.0	—	0	0.37	—	0.13	4.00	0.30	—	0	1.8	—
0	0	1.80	—	130.0	460.0	0.26	0	0.45	—	0.17	5.00	0.30	—	0	1.8	—
0	0	1.08	—	95.0	800.0	0.30	0	0.75	—	0.17	4.00	0.20	—	0	0.6	—
0	60	2.70	—	60.0	450.0	0.23	0	1.80	—	0.17	6.00	0.40	—	0	3.0	—
0	0	1.44	—	50.0	320.0	0.36	0	1.80	—	0.17	2.00	0.30	—	0	3.0	—
0	40	1.08	—	130.0	480.0	0.50	0	1.20	—	0.13	3.00	0.30	—	0	1.5	—
0	0	0.36	—	15.0	220.0	0.05	0	0.75	—	0.03	0.40	0.08	—	0	0.2	—
0	0	0.36	—	10.0	390.0	0.67	0	0.03	—	0.03	0.00	0.01	—	0	0	—
21	150	0.08	6.5	72.6	395.5	0.75	56.1	0.01	0.07	0.10	0.28	0.04	10.2	0	0.3	4.1
27	191	0.12	6.8	38.6	158.8	0.73	82.8	0.00	0.07	0.10	0.03	0.01	5.7	0	0.4	4.1

TABLE A-1 Table of Food Composition *(continued)*

(Computer code is for Cengage Diet Analysis Plus program)

DA+ Code	Food Description	QTY	Measure	Wt (g)	H_2O (g)	Ener (cal)	Prot (g)	Carb (g)	Fiber (g)	Fat (g)	Fat Breakdown (g) Sat	Mono	Poly
	Dairy—continued												
885	Brie	1	ounce(s)	28	13.7	95	5.9	0.1	0	7.8	4.9	2.3	0.2
34821	Camembert	1	ounce(s)	28	14.7	85	5.6	0.1	0	6.9	4.3	2.0	0.2
5	Cheddar, shredded	¼	cup(s)	28	10.4	114	7.0	0.4	0	9.4	6.0	2.7	0.3
888	Cheddar or colby	1	ounce(s)	28	10.8	112	6.7	0.7	0	9.1	5.7	2.6	0.3
32096	Cheddar or colby, low fat	1	ounce(s)	28	17.9	49	6.9	0.5	0	2.0	1.2	0.6	0.1
889	Edam	1	ounce(s)	28	11.8	101	7.1	0.4	0	7.9	5.0	2.3	0.2
890	Feta	1	ounce(s)	28	15.7	75	4.0	1.2	0	6.0	4.2	1.3	0.2
891	Fontina	1	ounce(s)	28	10.8	110	7.3	0.4	0	8.8	5.4	2.5	0.5
8527	Goat cheese, soft	1	ounce(s)	28	17.2	76	5.3	0.3	0	6.0	4.1	1.4	0.1
893	Gouda	1	ounce(s)	28	11.8	101	7.1	0.6	0	7.8	5.0	2.2	0.2
894	Gruyere	1	ounce(s)	28	9.4	117	8.5	0.1	0	9.2	5.4	2.8	0.5
895	Limburger	1	ounce(s)	28	13.7	93	5.7	0.1	0	7.7	4.7	2.4	0.1
896	Monterey jack	1	ounce(s)	28	11.6	106	6.9	0.2	0	8.6	5.4	2.5	0.3
13	Mozzarella, part skim milk	1	ounce(s)	28	15.2	72	6.9	0.8	0	4.5	2.9	1.3	0.1
12	Mozzarella, whole milk	1	ounce(s)	28	14.2	85	6.3	0.6	0	6.3	3.7	1.9	0.2
897	Muenster	1	ounce(s)	28	11.8	104	6.6	0.3	0	8.5	5.4	2.5	0.2
898	Neufchatel	1	ounce(s)	28	17.6	74	2.8	0.8	0	6.6	4.2	1.9	0.2
14	Parmesan, grated	1	tablespoon(s)	5	1.0	22	1.9	0.2	0	1.4	0.9	0.4	0.1
17	Provolone	1	ounce(s)	28	11.6	100	7.3	0.6	0	7.5	4.8	2.1	0.2
19	Ricotta, part skim milk	¼	cup(s)	62	45.8	85	7.0	3.2	0	4.9	3.0	1.4	0.2
18	Ricotta, whole milk	¼	cup(s)	62	44.1	107	6.9	1.9	0	8.0	5.1	2.2	0.2
20	Romano	1	tablespoon(s)	5	1.5	19	1.6	0.2	0	1.3	0.9	0.4	0
900	Roquefort	1	ounce(s)	28	11.2	105	6.1	0.6	0	8.7	5.5	2.4	0.4
21	Swiss	1	ounce(s)	28	10.5	108	7.6	1.5	0	7.9	5.0	2.1	0.3
	Imitation cheese												
42245	Imitation American cheddar cheese	1	ounce(s)	28	15.1	68	4.7	3.3	0	4.0	2.5	1.2	0.1
53914	Imitation cheddar	1	ounce(s)	28	15.1	68	4.7	3.3	0	4.0	2.5	1.2	0.1
	Cottage cheese												
9	Low fat, 1% fat	½	cup(s)	113	93.2	81	14.0	3.1	0	1.2	0.7	0.3	0
8	Low fat, 2% fat	½	cup(s)	113	89.6	102	15.5	4.1	0	2.2	1.4	0.6	0.1
	Cream cheese												
11	Cream cheese	2	tablespoon(s)	29	15.6	101	2.2	0.8	0	10.1	6.4	2.9	0.4
17366	Fat-free cream cheese	2	tablespoon(s)	30	22.7	29	4.3	1.7	0	0.4	0.3	0.1	0
10438	Tofutti Better than Cream Cheese	2	tablespoon(s)	30	—	80	1.0	1.0	0	8.0	2.0	—	6.0
	Processed cheese												
24	American cheese food, processed	1	ounce(s)	28	12.3	94	5.2	2.2	0	7.1	4.2	2.0	0.3
25	American cheese spread, processed	1	ounce(s)	28	13.5	82	4.7	2.5	0	6.0	3.8	1.8	0.2
22	American cheese, processed	1	ounce(s)	28	11.1	106	6.3	0.5	0	8.9	5.6	2.5	0.3
9110	Kraft deluxe singles pasteurized process American cheese	1	ounce(s)	28	—	108	5.4	0	0	9.5	5.4	—	—
23	Swiss cheese, processed	1	ounce(s)	28	12.0	95	7.0	0.6	0	7.1	4.5	2.0	0.2
	Soy cheese												
10437	Galaxy Foods vegan grated parmesan cheese alternative	1	tablespoon(s)	8	—	23	3.0	1.5	0	0	0	0	0
10430	Nu Tofu cheddar flavored cheese alternative	1	ounce(s)	28	—	70	6.0	1.0	0	4.0	0.5	2.5	1.0
	Cream												
26	Half and half cream	1	tablespoon(s)	15	12.1	20	0.4	0.6	0	1.7	1.1	0.5	0.1
32	Heavy whipping cream, liquid	1	tablespoon(s)	15	8.7	52	0.3	0.4	0	5.6	3.5	1.6	0.2
28	Light coffee or table cream, liquid	1	tablespoon(s)	15	11.1	29	0.4	0.5	0	2.9	1.8	0.8	0.1
30	Light whipping cream, liquid	1	tablespoon(s)	15	9.5	44	0.3	0.4	0	4.6	2.9	1.4	0.1
34	Whipped cream topping, pressurized	1	tablespoon(s)	3	1.8	8	0.1	0.4	0	0.7	0.4	0.2	0
	Sour cream												
30556	Fat-free sour cream	2	tablespoon(s)	32	25.8	24	1.0	5.0	0	0	0	0	0
36	Sour cream	2	tablespoon(s)	24	17.0	51	0.8	1.0	0	5.0	3.1	1.5	0.2
	Imitation cream												
3659	Coffeemate nondairy creamer, liquid	1	tablespoon(s)	15	—	20	0	2.0	0	1.0	0	0.5	0
40	Cream substitute, powder	1	teaspoon(s)	2	0	11	0.1	1.1	0	0.7	0.7	0	0
904	Imitation sour cream	2	tablespoon(s)	29	20.5	60	0.7	1.9	0	5.6	5.1	0.2	0

Chol (mg)	Calc (mg)	Iron (mg)	Magn (mg)	Pota (mg)	Sodi (mg)	Zinc (mg)	Vit A (µg)	Thia (mg)	Vit E (mg α)	Ribo (mg)	Niac (mg)	Vit B_6 (mg)	Fola (µg DFE)	Vit C (mg)	Vit B_{12} (µg)	Sele (µg)
28	52	0.14	5.7	43.1	178.3	0.67	49.3	0.02	0.06	0.14	0.10	0.06	18.4	0	0.5	4.1
20	110	0.09	5.7	53.0	238.7	0.67	68.3	0.01	0.06	0.14	0.18	0.06	17.6	0	0.4	4.1
30	204	0.19	7.9	27.7	175.4	0.87	74.9	0.01	0.08	0.10	0.02	0.02	5.1	0	0.2	3.9
27	194	0.21	7.4	36.0	171.2	0.87	74.8	0.00	0.07	0.10	0.02	0.02	5.1	0	0.2	4.1
6	118	0.11	4.5	18.7	173.5	0.51	17.0	0.00	0.01	0.06	0.01	0.01	3.1	0	0.1	4.1
25	207	0.12	8.5	53.3	273.6	1.06	68.9	0.01	0.06	0.11	0.02	0.02	4.5	0	0.4	4.1
25	140	0.18	5.4	17.6	316.4	0.81	35.4	0.04	0.05	0.23	0.28	0.12	9.1	0	0.5	4.3
33	156	0.06	4.0	18.1	226.8	0.99	74.0	0.01	0.07	0.05	0.04	0.02	1.7	0	0.5	4.1
13	40	0.53	4.5	7.4	104.3	0.26	81.6	0.02	0.05	0.10	0.12	0.07	3.4	0	0.1	0.8
32	198	0.06	8.2	34.3	232.2	1.10	46.8	0.01	0.06	0.09	0.01	0.02	6.0	0	0.4	4.1
31	287	0.04	10.2	23.0	95.3	1.10	76.8	0.01	0.07	0.07	0.03	0.02	2.8	0	0.5	4.1
26	141	0.03	6.0	36.3	226.8	0.59	96.4	0.02	0.06	0.14	0.04	0.02	16.4	0	0.3	4.1
25	211	0.20	7.7	23.0	152.0	0.85	56.1	0.00	0.07	0.11	0.02	0.02	5.1	0	0.2	4.1
18	222	0.06	6.5	23.8	175.5	0.78	36.0	0.01	0.04	0.08	0.03	0.02	2.6	0	0.2	4.1
22	143	0.12	5.7	21.5	177.8	0.82	50.7	0.01	0.05	0.08	0.02	0.01	2.0	0	0.6	4.8
27	203	0.11	7.7	38.0	178.0	0.79	84.5	0.00	0.07	0.09	0.02	0.01	3.4	0	0.4	4.1
22	21	0.07	2.3	32.3	113.1	0.14	84.5	0.00	—	0.05	0.03	0.01	3.1	0	0.1	0.9
4	55	0.04	1.9	6.3	76.5	0.19	6.0	0.00	0.01	0.02	0.01	0.00	0.5	0	0.1	0.9
20	214	0.14	7.9	39.1	248.3	0.91	66.9	0.01	0.06	0.09	0.04	0.02	2.8	0	0.4	4.1
19	167	0.27	9.2	76.9	76.9	0.82	65.8	0.01	0.04	0.11	0.04	0.01	8.0	0	0.2	10.3
31	127	0.23	6.8	64.6	51.7	0.71	73.8	0.01	0.06	0.12	0.06	0.02	7.4	0	0.2	8.9
5	53	0.03	2.1	4.3	60	0.12	4.8	0.00	0.01	0.01	0.00	0.00	0.4	0	0.1	0.7
26	188	0.15	8.5	25.8	512.9	0.59	83.3	0.01	—	0.16	0.20	0.03	13.9	0	0.2	4.1
26	224	0.05	10.8	21.8	54.4	1.23	62.4	0.01	0.10	0.08	0.02	0.02	1.7	0	0.9	5.2
10	159	0.08	8.2	68.6	381.3	0.73	32.3	0.01	0.07	0.12	0.03	0.03	2.0	0	0.1	4.3
10	159	0.09	8.2	68.6	381.3	0.73	32.3	0.01	0.07	0.12	0.04	0.03	2.0	0	0.1	4.3
5	69	0.15	5.7	97.2	458.8	0.42	12.4	0.02	0.01	0.18	0.14	0.07	13.6	0	0.7	10.2
9	78	0.18	6.8	108.5	458.8	0.47	23.7	0.02	0.02	0.20	0.16	0.08	14.7	0	0.8	11.5
32	23	0.34	1.7	34.5	85.8	0.15	106.1	0.01	0.08	0.05	0.02	0.01	3.8	0	0.1	0.7
2	56	0.05	4.2	48.9	163.5	0.26	83.7	0.01	0.00	0.05	0.04	0.01	11.1	0	0.2	1.5
0	0	0.00	—	—	135.0	—	0	—	—	—	—	—	—	0	—	—
23	162	0.16	8.8	82.5	358.6	0.90	57.0	0.01	0.06	0.14	0.04	0.02	2.0	0	0.4	4.6
16	159	0.09	8.2	68.6	381.3	0.73	49.0	0.01	0.05	0.12	0.03	0.03	2.0	0	0.1	3.2
27	156	0.05	7.7	47.9	422.1	0.80	72.0	0.01	0.07	0.10	0.02	0.02	2.3	0	0.2	4.1
27	338	0.00	0	33.8	459.0	1.22	114.0	—	—	0.14	—	—	—	0	0.2	—
24	219	0.17	8.2	61.2	388.4	1.02	56.1	0.00	0.09	0.07	0.01	0.01	1.7	0	0.3	4.5
0	60	0.00	—	75.0	97.5	—	—	—	—	—	—	—	—	—	—	—
0	200	0.36	—	—	190.0	—	—	—	—	—	—	—	—	0	—	—
6	16	0.01	1.5	19.5	6.2	0.08	14.6	0.01	0.05	0.02	0.01	0.01	0.5	0.1	0	0.3
21	10	0.00	1.1	11.3	5.7	0.03	61.7	0.00	0.15	0.01	0.01	0.00	0.6	0.1	0	0.1
10	14	0.01	1.4	18.3	6.0	0.04	27.2	0.01	0.08	0.02	0.01	0.01	0.3	0.1	0	0.1
17	10	0.00	1.1	14.6	5.1	0.03	41.9	0.00	0.13	0.01	0.01	0.00	0.6	0.1	0	0.1
2	3	0.00	0.3	4.4	3.9	0.01	5.6	0.00	0.01	0.00	0.00	0.00	0.1	0	0	0
3	40	0.00	3.2	41.3	45.1	0.16	23.4	0.01	0.00	0.04	0.02	0.01	3.5	0	0.1	1.7
11	28	0.01	2.6	34.6	12.7	0.06	42.5	0.01	0.14	0.03	0.01	0.00	2.6	0.2	0.1	0.5
0	0	0.00	—	30.0	0	—	0	0.01	—	0.01	0.20	—	—	0	—	—
0	0	0.02	0.1	16.2	3.6	0.01	0	0.00	0.01	0.00	0.00	0.00	0	0	0	0
0	1	0.11	1.7	46.3	29.3	0.34	0	0.00	0.21	0.00	0.00	0.00	0	0	0	0.7

TABLE A-1 Table of Food Composition *(continued)*

(Computer code is for Cengage Diet Analysis Plus program)

DA+ Code	Food Description	QTY	Measure	Wt (g)	H_2O (g)	Ener (cal)	Prot (g)	Carb (g)	Fiber (g)	Fat (g)	Fat Breakdown (g) Sat	Mono	Poly
	Dairy—continued												
35972	Nondairy coffee whitener, liquid, frozen	1	tablespoon(s)	15	11.7	21	0.2	1.7	0	1.5	0.3	1.1	0
35976	Nondairy dessert topping, frozen	1	tablespoon(s)	5	2.4	15	0.1	1.1	0	1.2	1.0	0.1	0
35975	Nondairy dessert topping, pressurized	1	tablespoon(s)	4	2.7	12	0	0.7	0	1.0	0.8	0.1	0
	Fluid milk												
60	Buttermilk, low fat	1	cup(s)	245	220.8	98	8.1	11.7	0	2.2	1.3	0.6	0.1
54	Low fat, 1%	1	cup(s)	244	219.4	102	8.2	12.2	0	2.4	1.5	0.7	0.1
55	Low fat, 1%, with nonfat milk solids	1	cup(s)	245	220.0	105	8.5	12.2	0	2.4	1.5	0.7	0.1
57	Nonfat, skim or fat free	1	cup(s)	245	222.6	83	8.3	12.2	0	0.2	0.1	0.1	0
58	Nonfat, skim or fat free with nonfat milk solids	1	cup(s)	245	221.4	91	8.7	12.3	0	0.6	0.4	0.2	0
51	Reduced fat, 2%	1	cup(s)	244	218.0	122	8.1	11.4	0	4.8	3.1	1.4	0.2
52	Reduced fat, 2%, with nonfat milk solids	1	cup(s)	245	217.7	125	8.5	12.2	0	4.7	2.9	1.4	0.2
50	Whole, 3.3%	1	cup(s)	244	215.5	146	7.9	11.0	0	7.9	4.6	2.0	0.5
	Canned milk												
62	Nonfat or skim evaporated	2	tablespoon(s)	32	25.3	25	2.4	3.6	0	0.1	0	0	0
63	Sweetened condensed	2	tablespoon(s)	38	10.4	123	3.0	20.8	0	3.3	2.1	0.9	0.1
61	Whole evaporated	2	tablespoon(s)	32	23.3	42	2.1	3.2	0	2.4	1.4	0.7	0.1
	Dried milk												
64	Buttermilk	¼	cup(s)	30	0.9	117	10.4	14.9	0	1.8	1.1	0.5	0.1
65	Instant nonfat with added vitamin A	¼	cup(s)	17	0.7	61	6.0	8.9	0	0.1	0.1	0	0
5234	Skim milk powder	¼	cup(s)	17	0.7	62	6.1	9.1	0	0.1	0.1	0	0
907	Whole dry milk	¼	cup(s)	32	0.8	159	8.4	12.3	0	8.5	5.4	2.5	0.2
909	**Goat milk**	1	cup(s)	244	212.4	168	8.7	10.9	0	10.1	6.5	2.7	0.4
	Chocolate milk												
33155	Chocolate syrup, prepared with milk	1	cup(s)	282	227.0	254	8.7	36.0	0.8	8.3	4.7	2.1	0.5
33184	Cocoa mix with aspartame, added sodium and vitamin A, no added calcium or phosphorus, prepared with water	1	cup(s)	192	177.4	56	2.3	10.8	1.2	0.4	0.3	0.1	0
908	Hot cocoa, prepared with milk	1	cup(s)	250	206.4	193	8.8	26.6	2.5	5.8	3.6	1.7	0.1
69	Low fat	1	cup(s)	250	211.3	158	8.1	26.1	1.3	2.5	1.5	0.8	0.1
68	Reduced fat	1	cup(s)	250	205.4	190	7.5	30.3	1.8	4.8	2.9	1.1	0.2
67	Whole	1	cup(s)	250	205.8	208	7.9	25.9	2.0	8.5	5.3	2.5	0.3
70	**Eggnog**	1	cup(s)	254	188.9	343	9.7	34.4	0	19.0	11.3	5.7	0.9
	Breakfast drinks												
10093	Carnation Instant Breakfast classic chocolate malt, prepared with skim milk, no sugar added	1	cup(s)	243	—	142	11.1	21.3	0.7	1.3	0.7	—	—
10092	Carnation Instant Breakfast classic French vanilla, prepared with skim milk, no sugar added	1	cup(s)	273	—	150	12.9	24.0	0	0.4	0.4	—	—
10094	Carnation Instant Breakfast stawberry sensation, prepared with skim milk, no sugar added	1	cup(s)	243	—	142	11.1	21.3	0	0.4	0.4	—	—
10091	Carnation Instant Breakfast strawberry sensation, prepared with skim milk	1	cup(s)	273	—	220	12.5	38.8	0	0.4	0.4	—	—
1417	Ovaltine rich chocolate flavor, prepared with skim milk	1	cup(s)	258	—	170	8.5	31.0	0	0	0	0	0
8539	**Malted milk, chocolate mix, fortified, prepared with milk**	1	cup(s)	265	215.8	223	8.9	28.9	1.1	8.6	5.0	2.2	0.5
	Milkshakes												
73	Chocolate	1	cup(s)	227	164.0	270	6.9	48.1	0.7	6.1	3.8	1.8	0.2
3163	Strawberry	1	cup(s)	226	167.8	256	7.7	42.8	0.9	6.3	3.9	—	—
74	Vanilla	1	cup(s)	227	169.2	254	8.8	40.3	0	6.9	4.3	2.0	0.3

Chol (mg)	Calc (mg)	Iron (mg)	Magn (mg)	Pota (mg)	Sodi (mg)	Zinc (mg)	Vit A (μg)	Thia (mg)	Vit E (mg α)	Ribo (mg)	Niac (mg)	Vit B_6 (mg)	Fola (μg DFE)	Vit C (mg)	Vit B_{12} (μg)	Sele (μg)
0	1	0.00	0	28.9	12.0	0.00	0.2	0.00	0.12	0.00	0.00	0.00	0	0	0	0.2
0	0	0.00	0.1	0.9	1.2	0.00	0.3	0.00	0.05	0.00	0.00	0.00	0	0	0	0.1
0	0	0.00	0	0.8	2.8	0.00	0.2	0.00	0.04	0.00	0.00	0.00	0	0	0	0.1
10	284	0.12	27.0	370.0	257.3	1.02	17.2	0.08	0.12	0.37	0.14	0.08	12.3	2.5	0.5	4.9
12	290	0.07	26.8	366.0	107.4	1.02	141.5	0.04	0.02	0.45	0.22	0.09	12.2	0	1.1	8.1
10	314	0.12	34.3	396.9	127.4	0.98	144.6	0.09	—	0.42	0.22	0.11	12.3	2.5	0.9	5.6
5	306	0.07	27.0	382.2	102.9	1.02	149.5	0.11	0.02	0.44	0.23	0.09	12.3	0	1.3	7.6
5	316	0.12	36.8	419.0	129.9	1.00	149.5	0.10	0.00	0.42	0.22	0.11	12.3	2.5	1.0	5.4
20	285	0.07	26.8	366.0	100.0	1.04	134.2	0.09	0.07	0.45	0.22	0.09	12.2	0.5	1.1	6.1
20	314	0.12	34.3	396.9	127.4	0.98	137.2	0.09	—	0.42	0.22	0.11	12.3	2.5	0.9	5.6
24	276	0.07	24.4	348.9	97.6	0.97	68.3	0.10	0.14	0.44	0.26	0.08	12.2	0	1.1	9.0
1	93	0.09	8.6	105.9	36.7	0.28	37.6	0.01	0.00	0.09	0.05	0.01	2.9	0.4	0.1	0.8
13	109	0.07	9.9	141.9	48.6	0.36	28.3	0.03	0.06	0.16	0.08	0.02	4.2	1.0	0.2	5.7
9	82	0.06	7.6	95.4	33.4	0.24	20.5	0.01	0.04	0.10	0.06	0.01	2.5	0.6	0.1	0.7
21	359	0.09	33.3	482.5	156.7	1.21	14.9	0.11	0.03	0.48	0.27	0.10	14.2	1.7	1.2	6.2
3	209	0.05	19.9	289.9	93.3	0.75	120.5	0.07	0.00	0.30	0.15	0.06	8.5	1.0	0.7	4.6
3	214	0.05	20.3	296.0	95.3	0.76	123.1	0.07	0.00	0.30	0.15	0.06	8.7	1.0	0.7	4.7
31	292	0.15	27.2	425.6	118.7	1.06	82.2	0.09	0.15	0.38	0.20	0.09	11.8	2.8	1.0	5.2
27	327	0.12	34.2	497.8	122.0	0.73	139.1	0.11	0.17	0.33	0.67	0.11	2.4	3.2	0.2	3.4
25	251	0.90	50.8	408.9	132.5	1.21	70.5	0.11	0.14	0.46	0.38	0.09	14.1	0	1.1	9.6
0	92	0.74	32.6	405.1	138.2	0.51	0	0.04	0.00	0.20	0.16	0.04	1.9	0	0.2	2.5
20	263	1.20	57.5	492.5	110.0	1.57	127.5	0.09	0.07	0.45	0.33	0.10	12.5	0.5	1.1	6.8
8	288	0.60	32.5	425.0	152.5	1.02	145.0	0.09	0.05	0.41	0.31	0.10	12.5	2.3	0.9	4.8
20	273	0.60	35.0	422.5	165.0	0.97	160.0	0.11	0.10	0.45	0.41	0.06	5.0	0	0.8	8.5
30	280	0.60	32.5	417.5	150.0	1.02	65.0	0.09	0.15	0.40	0.31	0.10	12.5	2.3	0.8	4.8
150	330	0.50	48.3	419.1	137.2	1.16	116.8	0.08	0.50	0.48	0.26	0.12	2.5	3.8	1.1	10.7
9	444	4.00	88.9	631.1	195.6	3.38	—	0.33	—	0.45	4.44	0.44	4.0	26.7	1.3	8.0
9	500	4.50	100.0	665.0	192.0	3.75	—	0.37	—	0.51	5.00	0.49	100.0	30.0	1.5	9.0
9	444	4.00	88.9	568.9	186.7	3.38	—	0.33	—	0.45	4.44	0.44	88.9	26.7	1.3	8.0
9	500	4.47	100.0	665.0	288.0	3.75	—	0.37	—	0.51	5.07	0.50	100.0	30.0	1.5	8.8
5	350	3.60	100.0	—	270.0	3.75	—	0.37	—	—	4.00	0.40	—	12.0	1.2	—
27	339	3.76	45.1	577.7	230.6	1.16	903.7	0.75	0.15	1.31	11.08	1.01	13.3	31.8	1.1	12.5
25	300	0.70	36.4	508.9	252.2	1.09	40.9	0.10	0.11	0.50	0.28	0.05	11.4	0	0.7	4.3
25	256	0.24	29.4	412.0	187.9	0.81	58.9	0.10	—	0.44	0.39	0.10	6.8	1.8	0.7	4.8
27	332	0.22	27.3	415.8	215.8	0.88	56.8	0.06	0.11	0.44	0.33	0.09	15.9	0	1.2	5.2

TABLE A-1 Table of Food Composition *(continued)*

(Computer code is for Cengage Diet Analysis Plus program)

DA+ Code	Food Description	QTY	Measure	Wt (g)	H2O (g)	Ener (cal)	Prot (g)	Carb (g)	Fiber (g)	Fat (g)	Fat Breakdown (g) Sat	Mono	Poly
Dairy—continued													
	Ice cream												
4776	Chocolate	½	cup(s)	66	36.8	143	2.5	18.6	0.8	7.3	4.5	2.1	0.3
12137	Chocolate fudge, no sugar added	½	cup(s)	71	—	100	3.0	16.0	2.0	3.0	1.5	—	—
16514	Chocolate, soft serve	½	cup(s)	87	49.9	177	3.2	24.1	0.7	8.4	5.2	2.4	0.3
16523	Sherbet, all flavors	½	cup(s)	97	63.8	139	1.1	29.3	3.2	1.9	1.1	0.5	0.1
4778	Strawberry	½	cup(s)	66	39.6	127	2.1	18.2	0.6	5.5	3.4	—	—
76	Vanilla	½	cup(s)	72	43.9	145	2.5	17.0	0.5	7.9	4.9	2.1	0.3
12146	Vanilla chocolate swirl, fat-free, no sugar added	½	cup(s)	71	—	100	3.0	14.0	2.0	3.0	2.0	—	—
82	Vanilla, light	½	cup(s)	76	48.3	125	3.6	19.6	0.2	3.7	2.2	1.0	0.2
78	Vanilla, light, soft serve	½	cup(s)	88	61.2	111	4.3	19.2	0	2.3	1.4	0.7	0.1
	Soy desserts												
10694	Tofutti low fat vanilla fudge nondairy frozen dessert	½	cup(s)	70	—	140	2.0	24.0	0	4.0	1.0	—	—
15721	Tofutti premium chocolate supreme nondairy frozen dessert	½	cup(s)	70	—	180	3.0	18.0	0	11.0	2.0	—	—
15720	Tofutti premium vanilla nondairy frozen dessert	½	cup(s)	70	—	190	2.0	20.0	0	11.0	2.0	—	—
	Ice milk												
16517	Chocolate	½	cup(s)	66	42.9	94	2.8	16.9	0.3	2.1	1.3	0.6	0.1
16516	Flavored, not chocolate	½	cup(s)	66	41.4	108	3.5	17.5	0.2	2.6	1.7	0.6	0.1
	Pudding												
25032	Chocolate	½	cup(s)	144	109.7	155	5.1	22.7	0.7	5.4	3.1	1.7	0.2
1923	Chocolate, sugar free, prepared with 2% milk	½	cup(s)	133	—	100	5.0	14.0	0.3	3.0	1.5	—	—
1722	Rice	½	cup(s)	113	75.6	151	4.1	29.9	0.5	1.9	1.1	0.5	0.1
4747	Tapioca, ready to eat	1	item(s)	142	102.0	185	2.8	30.8	0	5.5	1.4	3.6	0.1
25031	Vanilla	½	cup(s)	136	109.7	116	4.7	17.6	0	2.8	1.6	0.9	0.2
1924	Vanilla, sugar free, prepared with 2% milk	½	cup(s)	133	—	90	4.0	12.0	0.2	2.0	1.5	—	—
	Frozen yogurt												
4785	Chocolate, soft serve	½	cup(s)	72	45.9	115	2.9	17.9	1.6	4.3	2.6	1.3	0.2
1747	Fruit varieties	½	cup(s)	113	80.5	144	3.4	24.4	0	4.1	2.6	1.1	0.1
4786	Vanilla, soft serve	½	cup(s)	72	47.0	117	2.9	17.4	0	4.0	2.5	1.1	0.2
	Milk substitutes												
	Lactose free												
16081	Fat-free, calcium fortified [milk]	1	cup(s)	240	—	80	8.0	13.0	0	0	0	0	0
36486	Low fat milk	1	cup(s)	240	—	110	8.0	13.0	0	2.5	1.5	—	—
36487	Reduced fat milk	1	cup(s)	240	—	130	8.0	12.0	0	5.0	3.0	—	—
36488	Whole milk	1	cup(s)	240	—	150	8.0	12.0	0	8.0	5.0	—	—
	Rice												
10083	Rice Dream carob rice beverage	1	cup(s)	240	—	150	1.0	32.0	0	2.5	0	—	—
17089	Rice Dream original rice beverage, enriched	1	cup(s)	240	—	120	1.0	25.0	0	2.0	0	—	—
10087	Rice Dream vanilla enriched rice beverage	1	cup(s)	240	—	130	1.0	28.0	0	2.0	0	—	—
	Soy												
34750	Soy Dream chocolate enriched soy beverage	1	cup(s)	240	—	210	7.0	37.0	1.0	3.5	0.5	—	—
34749	Soy Dream vanilla enriched soy beverage	1	cup(s)	240	—	150	7.0	22.0	0	4.0	0.5	—	—
13840	Vitasoy light chocolate soymilk	1	cup(s)	240	—	100	4.0	17.0	0	2.0	0.5	0.5	1.0
13839	Vitasoy light vanilla soymilk	1	cup(s)	240	—	70	4.0	10.0	0	2.0	0.5	0.5	1.0
13836	Vitasoy rich chocolate soymilk	1	cup(s)	240	—	160	7.0	24.0	1.0	4.0	0.5	1.0	2.5
13835	Vitasoy vanilla delite soymilk	1	cup(s)	240	—	120	7.0	13.0	1.0	4.0	0.5	1.0	2.5
	Yogurt												
3615	Custard style, fruit flavors	6	ounce(s)	170	127.1	190	7.0	32.0	0	3.5	2.0	—	—
3617	Custard style, vanilla	6	ounce(s)	170	134.1	190	7.0	32.0	0	3.5	2.0	0.9	0.1
32101	Fruit, low fat	1	cup(s)	245	184.5	243	9.8	45.7	0	2.8	1.8	0.8	0.1

PAGE KEY: A-2 = Breads/Baked Goods A-8 = Cereal/Rice/Pasta A-12 = Fruit A-18 = Vegetables/Legumes A-28 = Nuts/Seeds A-30 = Vegetarian A-32 = Dairy A-40 = Eggs A-40 = Seafood A-44 = Meats A-48 = Poultry A-48 = Processed Meats A-50 = Beverages A-54 = Fats/Oils A-56 = Sweets A-58 = Spices/Condiments/Sauces A-62 = Mixed Foods/Soups/Sandwiches A-68 = Fast Food A-88 = Convenience A-90 = Baby Foods

Chol (mg)	Calc (mg)	Iron (mg)	Magn (mg)	Pota (mg)	Sodi (mg)	Zinc (mg)	Vit A (µg)	Thia (mg)	Vit E (mg α)	Ribo (mg)	Niac (mg)	Vit B_6 (mg)	Fola (µg DFE)	Vit C (mg)	Vit B_{12} (µg)	Sele (µg)
22	72	0.61	19.1	164.3	50.2	0.38	77.9	0.02	0.19	0.12	0.14	0.03	10.6	0.5	0.2	1.7
10	100	0.36	—	—	65.0	—	—	—	—	—	—	—	—	0	—	—
22	103	0.32	19.0	192.0	43.3	0.45	66.6	0.03	0.22	0.13	0.11	0.03	4.3	0.5	0.3	2.5
0	52	0.13	7.7	92.6	44.4	0.46	9.7	0.02	0.02	0.08	0.07	0.02	6.8	5.6	0.1	1.3
19	79	0.13	9.2	124.1	39.6	0.22	63.4	0.03	—	0.16	0.11	0.03	7.9	5.1	0.2	1.3
32	92	0.06	10.1	143.3	57.6	0.49	85.0	0.03	0.21	0.17	0.08	0.03	3.6	0.4	0.3	1.3
10	100	0.00	—	—	65.0	—	—	—	—	—	—	—	—	0	—	—
21	122	0.14	10.6	158.1	56.2	0.55	97.3	0.04	0.09	0.19	0.10	0.03	4.6	0.9	0.4	1.5
11	138	0.05	12.3	194.5	61.6	0.46	25.5	0.04	0.05	0.17	0.10	0.04	4.4	0.8	0.4	3.2
0	0	0.00	—	8.0	90.0	—	0	—	—	—	—	—	—	0	—	—
0	0	0.00	—	7.0	180.0	—	0	—	—	—	—	—	—	0	—	—
0	0	0.00	—	2.0	210.0	—	0	—	—	—	—	—	—	0	—	—
6	94	0.15	13.1	155.2	40.6	0.36	15.7	0.03	0.05	0.11	0.08	0.02	3.9	0.5	0.3	2.2
16	76	0.05	9.2	136.2	48.5	0.47	90.4	0.02	0.05	0.11	0.06	0.01	3.3	0.1	0.2	1.3
35	149	0.46	31.3	226.7	137.0	0.71	—	0.05	0.00	0.22	0.15	0.06	8.3	1.2	0.5	4.9
10	150	0.72	—	330.0	310.0	—	—	0.06	—	0.26	—	—	—	0	—	—
7	113	0.28	15.8	201.4	66.4	0.52	41.6	0.03	0.05	0.17	0.34	0.06	4.5	0.2	0.2	4.8
1	101	0.15	8.5	130.6	205.9	0.31	0	0.03	0.21	0.13	0.09	0.03	4.3	0.4	0.3	0
35	146	0.17	17.2	188.9	136.4	0.52	—	0.04	0.00	0.22	0.10	0.05	8.0	1.2	0.5	4.6
10	150	0.00	—	190.0	380.0	—	—	0.03	—	0.17	—	—	—	0	—	—
4	106	0.90	19.4	187.9	70.6	0.35	31.7	0.02	—	0.15	0.22	0.05	7.9	0.2	0.2	1.7
15	113	0.52	11.3	176.3	71.2	0.31	55.4	0.04	0.10	0.20	0.07	0.04	4.5	0.8	0.1	2.1
1	103	0.21	10.1	151.9	62.6	0.30	42.5	0.02	0.07	0.16	0.20	0.05	4.3	0.6	0.2	2.4
3	500	0.00	—	—	125.0	—	100.0	—	—	—	—	—	—	0	0	—
10	300	0.00	—	—	125.0	—	100.0	—	—	—	—	—	—	0	—	—
20	300	0.00	—	—	125.0	—	98.2	—	—	—	—	—	—	0	—	—
35	300	0.00	—	—	125.0	—	58.1	—	—	—	—	—	—	0	—	—
0	20	0.72	—	82.5	100.0	—	—	—	—	—	—	—	—	1.2	—	—
0	300	0.00	13.3	60.0	90.0	0.24	—	0.06	—	0.00	0.84	0.07	—	0	1.5	—
0	300	0.00	—	53.0	90.0	—	—	—	—	—	—	—	—	0	1.5	—
0	300	1.80	60.0	350.0	160.0	0.60	33.3	0.15	—	0.06	0.80	0.12	60.0	0	3.0	—
0	300	1.80	40.0	260.0	140.0	0.60	33.3	0.15	—	0.06	0.80	0.12	60.0	0	3.0	—
0	300	0.72	24.0	200.0	140.0	0.90	—	0.09	—	0.34	—	—	24.0	0	0.9	—
0	300	0.72	24.0	200.0	120.0	0.90	—	0.09	—	0.34	—	—	24.0	0	0.9	—
0	300	1.08	40.0	320.0	150.0	0.90	—	0.15	—	0.34	—	—	60.0	0	0.9	—
0	40	0.72	—	320.0	115.0	—	0	—	—	—	—	—	—	0	—	—
15	300	0.00	16.0	310.0	100.0	—	—	—	—	0.25	—	—	—	0	—	—
15	300	0.00	16.0	310.0	100.0	—	—	—	—	0.25	—	—	—	0	—	—
12	338	0.14	31.9	433.7	129.9	1.64	27.0	0.08	0.04	0.39	0.21	0.09	22.1	1.5	1.1	6.9

TABLE A-1 Table of Food Composition *(continued)*

(Computer code is for Cengage Diet Analysis Plus program)

DA+ Code	Food Description	QTY	Measure	Wt (g)	H_2O (g)	Ener (cal)	Prot (g)	Carb (g)	Fiber (g)	Fat (g)	Fat Breakdown (g) Sat	Mono	Poly
Dairy—continued													
29638	Fruit, nonfat, sweetened with low-calorie sweetener	1	cup(s)	241	208.3	123	10.6	19.4	1.2	0.4	0.2	0.1	0
93	Plain, low fat	1	cup(s)	245	208.4	154	12.9	17.2	0	3.8	2.5	1.0	0.1
94	Plain, nonfat	1	cup(s)	245	208.8	137	14.0	18.8	0	0.4	0.3	0.1	0
32100	Vanilla, low fat	1	cup(s)	245	193.6	208	12.1	33.8	0	3.1	2.0	0.8	0.1
5242	Yogurt beverage	1	cup(s)	245	199.8	172	6.2	32.8	0	2.2	1.4	0.6	0.1
38202	Yogurt smoothie, nonfat, all flavors	1	item(s)	325	—	290	10.0	60.0	6.0	0	0	0	0
	Soy yogurt												
34617	Stonyfield Farm O'Soy strawberry-peach pack organic cultured soy yogurt	1	item(s)	113	—	100	5.0	16.0	3.0	2.0	0	—	—
34616	Stonyfield Farm O'Soy vanilla organic cultured soy yogurt	1	item(s)	170	—	150	7.0	26.0	4.0	2.0	0	—	—
10453	White Wave plain silk cultured soy yogurt	8	ounce(s)	227	—	140	5.0	22.0	1.0	3.0	0.5	—	—
Eggs													
	Eggs												
99	Fried	1	item(s)	46	31.8	90	6.3	0.4	0	7.0	2.0	2.9	1.2
100	Hard boiled	1	item(s)	50	37.3	78	6.3	0.6	0	5.3	1.6	2.0	0.7
101	Poached	1	item(s)	50	37.8	71	6.3	0.4	0	5.0	1.5	1.9	0.7
97	Raw, white	1	item(s)	33	28.9	16	3.6	0.2	0	0.1	0	0	0
96	Raw, whole	1	item(s)	50	37.9	72	6.3	0.4	0	5.0	1.5	1.9	0.7
98	Raw, yolk	1	item(s)	17	8.9	54	2.7	0.6	0	4.5	1.6	2.0	0.7
102	Scrambled, prepared with milk and butter	2	item(s)	122	89.2	204	13.5	2.7	0	14.9	4.5	5.8	2.6
	Egg substitute												
4028	Egg Beaters	¼	cup(s)	61	—	30	6.0	1.0	0	0	0	0	0
920	Frozen	¼	cup(s)	60	43.9	96	6.8	1.9	0	6.7	1.2	1.5	3.7
918	Liquid	¼	cup(s)	63	51.9	53	7.5	0.4	0	2.1	0.4	0.6	1.0
Seafood													
	Cod												
6040	Atlantic cod or scrod, baked or broiled	3	ounce(s)	85	64.6	89	19.4	0	0	0.7	0.1	0.1	0.2
1573	Atlantic cod, cooked, dry heat	3	ounce(s)	85	64.6	89	19.4	0	0	0.7	0.1	0.1	0.2
2905	**Eel, raw**	3	ounce(s)	85	58.0	156	15.7	0	0	9.9	2.0	6.1	0.8
	Fish fillets												
25079	Baked	3	ounce(s)	84	79.9	99	21.7	0	0	0.7	0.1	0.1	0.3
8615	Batter coated or breaded, fried	3	ounce(s)	85	45.6	197	12.5	14.4	0.4	10.5	2.4	2.2	5.3
25082	Broiled fish steaks	3	ounce(s)	85	68.1	128	24.2	0	0	2.6	0.4	0.9	0.8
25083	Poached fish steaks	3	ounce(s)	85	67.1	111	21.1	0	0	2.3	0.3	0.8	0.7
25084	Steamed	3	ounce(s)	85	72.2	79	17.2	0	0	0.6	0.1	0.1	0.2
25089	**Flounder, baked**	3	ounce(s)	85	64.4	113	14.8	0.4	0.1	5.5	1.1	2.2	1.4
1825	**Grouper, cooked, dry heat**	3	ounce(s)	85	62.4	100	21.1	0	0	1.1	0.3	0.2	0.3
	Haddock												
6049	Baked or broiled	3	ounce(s)	85	63.2	95	20.6	0	0	0.8	0.1	0.1	0.3
1578	Cooked, dry heat	3	ounce(s)	85	63.1	95	20.6	0	0	0.8	0.1	0.1	0.3
1886	**Halibut, Atlantic and Pacific, cooked, dry heat**	3	ounce(s)	85	61.0	119	22.7	0	0	2.5	0.4	0.8	0.8
1582	**Herring, Atlantic, pickled**	4	piece(s)	60	33.1	157	8.5	5.8	0	10.8	1.4	7.2	1.0
1587	**Jack mackerel, solids, canned, drained**	2	ounce(s)	57	39.2	88	13.1	0	0	3.6	1.1	1.3	0.9

Chol (mg)	Calc (mg)	Iron (mg)	Magn (mg)	Pota (mg)	Sodi (mg)	Zinc (mg)	Vit A (µg)	Thia (mg)	Vit E (mg α)	Ribo (mg)	Niac (mg)	Vit B_6 (mg)	Fola (µg DFE)	Vit C (mg)	Vit B_{12} (µg)	Sele (µg)
5	369	0.62	41.0	549.5	139.8	1.83	4.8	0.10	0.16	0.44	0.49	0.10	31.3	26.5	1.1	7.0
15	448	0.19	41.7	573.3	171.5	2.18	34.3	0.10	0.07	0.52	0.27	0.12	27.0	2.0	1.4	8.1
5	488	0.22	46.6	624.8	188.7	2.37	4.9	0.11	0.00	0.57	0.30	0.13	29.4	2.2	1.5	8.8
12	419	0.17	39.2	536.6	161.7	2.03	29.4	0.10	0.04	0.49	0.26	0.11	27.0	2.0	1.3	12.0
13	260	0.22	39.2	399.4	98.0	1.10	14.7	0.11	0.00	0.51	0.30	0.14	29.4	2.1	1.5	—
5	300	2.70	100.0	580.0	290.0	2.25	—	0.37	—	0.42	5.00	0.50	100.0	15.0	1.5	—
0	100	1.08	24.0	5.0	20.0	—	0	0.22	—	0.10	—	0.04	—	0	0	—
0	150	1.44	40.0	15.0	40.0	—	—	0.30	—	0.13	—	0.08	—	0	0	—
0	400	1.44	—	0	30.0	—	0	—	—	—	—	—	—	0	—	—
210	27	0.91	6.0	67.6	93.8	0.55	91.1	0.03	0.56	0.23	0.03	0.07	23.5	0	0.6	15.7
212	25	0.59	5.0	63.0	62.0	0.52	84.5	0.03	0.51	0.25	0.03	0.06	22.0	0	0.6	15.4
211	27	0.91	6.0	66.5	147.0	0.55	69.5	0.02	0.48	0.20	0.03	0.06	17.5	0	0.6	15.8
0	2	0.02	3.6	53.8	54.8	0.01	0	0.00	0.00	0.14	0.03	0.00	1.3	0	0	6.6
212	27	0.91	6.0	67.0	70.0	0.55	70.0	0.03	0.48	0.23	0.03	0.07	23.5	0	0.6	15.9
210	22	0.46	0.9	18.5	8.2	0.39	64.8	0.03	0.43	0.09	0.00	0.06	24.8	0	0.3	9.5
429	87	1.46	14.6	168.4	341.6	1.22	174.5	0.06	1.33	0.53	0.09	0.14	36.6	0.2	0.9	27.5
0	20	1.08	4.0	85.0	115.0	0.60	112.5	0.15	—	0.85	0.20	0.08	60.0	0	1.2	—
1	44	1.18	9.0	127.8	119.4	0.58	6.6	0.07	0.95	0.23	0.08	0.08	9.6	0.3	0.2	24.8
1	33	1.32	5.6	207.1	111.1	0.82	11.3	0.07	0.17	0.19	0.07	0.00	9.4	0	0.2	15.6
47	12	0.41	35.7	207.5	66.3	0.49	11.9	0.07	0.68	0.06	2.13	0.24	6.8	0.8	0.9	32.0
47	12	0.41	35.7	207.5	66.3	0.49	11.9	0.07	0.68	0.06	2.13	0.24	6.8	0.9	0.9	32.0
107	17	0.42	17.0	231.3	43.4	1.37	887.0	0.13	3.40	0.03	2.97	0.05	12.8	1.5	2.6	5.5
44	8	0.31	29.1	489.0	86.1	0.48	—	0.02	—	0.05	2.47	0.46	8.1	3.0	1.0	44.3
29	15	1.79	20.4	272.2	452.5	0.37	9.4	0.09	—	0.09	1.78	0.08	17.0	0	0.9	7.7
37	55	0.97	96.7	524.3	62.9	0.49	—	0.05	—	0.08	6.47	0.36	12.6	0	1.2	42.5
32	48	0.85	84.0	455.6	54.7	0.42	—	0.05	—	0.07	5.92	0.33	11.5	0	1.1	37.0
41	12	0.29	24.7	319.3	41.7	0.34	—	0.06	—	0.06	1.89	0.21	6.1	0.8	0.8	32.0
44	19	0.34	47.3	224.7	280.2	0.20	—	0.06	0.40	0.07	2.02	0.18	7.4	2.8	1.6	33.5
40	18	0.96	31.5	404.0	45.1	0.43	42.5	0.06	—	0.01	0.32	0.29	8.5	0	0.6	39.8
63	36	1.15	42.5	339.4	74.0	0.40	16.2	0.03	0.42	0.03	3.94	0.29	6.8	0	1.2	34.4
63	36	1.14	42.5	339.3	74.0	0.40	16.2	0.03	—	0.03	3.93	0.29	11.1	0	1.2	34.4
35	51	0.91	91.0	489.9	58.7	0.45	45.9	0.05	—	0.07	6.05	0.33	11.9	0	1.2	39.8
8	46	0.73	4.8	41.4	522.0	0.31	154.8	0.02	1.02	0.08	1.98	0.10	1.2	0	2.6	35.1
45	137	1.15	21.0	110.0	214.9	0.57	73.7	0.02	0.58	0.12	3.50	0.11	2.8	0.5	3.9	21.4

TABLE A-1 Table of Food Composition *(continued)*

(Computer code is for Cengage Diet Analysis Plus program)

DA+ Code	Food Description	QTY	Measure	Wt (g)	H_2O (g)	Ener (cal)	Prot (g)	Carb (g)	Fiber (g)	Fat (g)	Fat Breakdown (g) Sat	Mono	Poly
	Seafood—continued												
8580	**Octopus, common, cooked, moist heat**	3	ounce(s)	85	51.5	139	25.4	3.7	0	1.8	0.4	0.3	0.4
1831	**Perch, mixed species, cooked, dry heat**	3	ounce(s)	85	62.3	100	21.1	0	0	1.0	0.2	0.2	0.4
1592	**Pacific rockfish, cooked, dry heat**	3	ounce(s)	85	62.4	103	20.4	0	0	1.7	0.4	0.4	0.5
	Salmon												
2938	Coho, farmed, raw	3	ounce(s)	85	59.9	136	18.1	0	0	6.5	1.5	2.8	1.6
1594	Broiled or baked with butter	3	ounce(s)	85	53.9	155	23.0	0	0	6.3	1.2	2.3	2.3
29727	Smoked chinook (lox)	2	ounce(s)	57	40.8	66	10.4	0	0	2.4	0.5	1.1	0.6
154	**Sardine, Atlantic with bones, canned in oil**	3	ounce(s)	85	50.7	177	20.9	0	0	9.7	1.3	3.3	4.4
	Scallops												
155	Mixed species, breaded, fried	3	item(s)	47	27.2	100	8.4	4.7	—	5.1	1.2	2.1	1.3
1599	Steamed	3	ounce(s)	85	64.8	90	13.8	2.0	0	2.6	0.4	1.0	0.8
1839	**Snapper, mixed species, cooked, dry heat**	3	ounce(s)	85	59.8	109	22.4	0	0	1.5	0.3	0.3	0.5
	Squid												
1868	Mixed species, fried	3	ounce(s)	85	54.9	149	15.3	6.6	0	6.4	1.6	2.3	1.8
16617	Steamed or boiled	3	ounce(s)	85	63.3	89	15.2	3.0	0	1.3	0.4	0.1	0.5
1570	**Striped bass, cooked, dry heat**	3	ounce(s)	85	62.4	105	19.3	0	0	2.5	0.6	0.7	0.9
1601	**Sturgeon, steamed**	3	ounce(s)	85	59.4	111	17.0	0	0	4.3	1.0	2.0	0.7
1840	**Surimi, formed**	3	ounce(s)	85	64.9	84	12.9	5.8	0	0.8	0.2	0.1	0.4
1842	**Swordfish, cooked, dry heat**	3	ounce(s)	85	58.5	132	21.6	0	0	4.4	1.2	1.7	1.0
1846	**Tuna, yellowfin or ahi, raw**	3	ounce(s)	85	60.4	92	19.9	0	0	0.8	0.2	0.1	0.2
	Tuna, canned												
159	Light, canned in oil, drained	2	ounce(s)	57	33.9	112	16.5	0	0	4.6	0.9	1.7	1.6
355	Light, canned in water, drained	2	ounce(s)	57	42.2	66	14.5	0	0	0.5	0.1	0.1	0.2
33211	Light, no salt, canned in oil, drained	2	ounce(s)	57	33.9	112	16.5	0	0	4.7	0.9	1.7	1.6
33212	Light, no salt, canned in water, drained	2	ounce(s)	57	42.6	66	14.5	0	0	0.5	0.1	0.1	0.2
2961	White, canned in oil, drained	2	ounce(s)	57	36.3	105	15.0	0	0	4.6	0.7	1.8	1.7
351	White, canned in water, drained	2	ounce(s)	57	41.5	73	13.4	0	0	1.7	0.4	0.4	0.6
33213	White, no salt, canned in oil, drained	2	ounce(s)	57	36.3	105	15.0	0	0	4.6	0.9	1.4	1.9
33214	White, no salt, canned in water, drained	2	ounce(s)	57	42.0	73	13.4	0	0	1.7	0.4	0.4	0.6
	Yellowtail												
8548	Mixed species, cooked, dry heat	3	ounce(s)	85	57.3	159	25.2	0	0	5.7	1.4	2.2	1.5
2970	Mixed species, raw	2	ounce(s)	57	42.2	83	13.1	0	0	3.0	0.7	1.1	0.8
	Shellfish, meat only												
1857	Abalone, mixed species, fried	3	ounce(s)	85	51.1	161	16.7	9.4	0	5.8	1.4	2.3	1.4
16618	Abalone, steamed or poached	3	ounce(s)	85	40.7	177	28.8	10.1	0	1.3	0.3	0.2	0.2
	Crab												
1851	Blue crab, canned	2	ounce(s)	57	43.2	56	11.6	0	0	0.7	0.1	0.1	0.2
1852	Blue crab, cooked, moist heat	3	ounce(s)	85	65.9	87	17.2	0	0	1.5	0.2	0.2	0.6
8562	Dungeness crab, cooked, moist heat	3	ounce(s)	85	62.3	94	19.0	0.8	0	1.1	0.1	0.2	0.3
1860	**Clams, cooked, moist heat**	3	ounce(s)	85	54.1	126	21.7	4.4	0	1.7	0.2	0.1	0.5
1853	**Crayfish, farmed, cooked, moist heat**	3	ounce(s)	85	68.7	74	14.9	0	0	1.1	0.2	0.2	0.4

PAGE KEY: A-2 = Breads/Baked Goods A-8 = Cereal/Rice/Pasta A-12 = Fruit A-18 = Vegetables/Legumes A-28 = Nuts/Seeds A-30 = Vegetarian A-32 = Dairy A-40 = Eggs A-40 = Seafood A-44 = Meats A-48 = Poultry A-48 = Processed Meats A-50 = Beverages A-54 = Fats/Oils A-56 = Sweets A-58 = Spices/Condiments/Sauces A-62 = Mixed Foods/Soups/Sandwiches A-68 = Fast Food A-88 = Convenience A-90 = Baby Foods

A

Chol (mg)	Calc (mg)	Iron (mg)	Magn (mg)	Pota (mg)	Sodi (mg)	Zinc (mg)	Vit A (µg)	Thia (mg)	Vit E (mg α)	Ribo (mg)	Niac (mg)	Vit B_6 (mg)	Fola (µg DFE)	Vit C (mg)	Vit B_{12} (µg)	Sele (µg)
82	90	8.11	51.0	535.8	391.2	2.85	76.5	0.04	1.02	0.06	3.21	0.55	20.4	6.8	30.6	76.2
98	87	0.98	32.3	292.6	67.2	1.21	8.5	0.06	—	0.10	1.61	0.11	5.1	1.4	1.9	13.7
37	10	0.45	28.9	442.3	65.5	0.45	60.4	0.03	1.32	0.07	3.33	0.22	8.5	0	1.0	39.8
43	10	0.29	26.4	382.7	40.0	0.36	47.6	0.08	—	0.09	5.79	0.56	11.1	0.9	2.3	10.7
40	15	1.02	26.9	376.6	98.6	0.56	—	0.13	1.14	0.05	8.33	0.18	4.2	1.8	2.3	41.0
13	6	0.48	10.2	99.2	1134.0	0.17	14.7	0.01	—	0.05	2.67	0.15	1.1	0	1.8	21.6
121	325	2.48	33.2	337.6	429.5	1.10	27.2	0.04	1.70	0.18	4.43	0.14	10.2	0	7.6	44.8
28	20	0.38	27.4	154.8	215.8	0.49	10.7	0.02	—	0.05	0.70	0.06	23.3	1.1	0.6	12.5
27	20	0.22	45.9	238.0	358.7	0.78	32.3	0.01	0.16	0.05	0.84	0.11	10.2	2.0	1.1	18.2
40	34	0.20	31.5	444.0	48.5	0.37	29.8	0.04	—	0.00	0.29	0.39	5.1	1.4	3.0	41.7
221	33	0.85	32.3	237.3	260.3	1.48	9.4	0.04	—	0.39	2.21	0.04	11.9	3.6	1.0	44.1
227	31	0.62	28.9	192.1	356.2	1.49	8.5	0.01	1.17	0.32	1.69	0.04	3.4	3.2	1.0	43.7
88	16	0.91	43.4	279.0	74.8	0.43	26.4	0.09	—	0.03	2.17	0.29	8.5	0	3.8	39.8
63	11	0.59	29.8	239.7	388.5	0.35	198.9	0.06	0.52	0.07	8.30	0.19	14.5	0	2.2	13.3
26	8	0.22	36.6	95.3	121.6	0.28	17.0	0.01	0.53	0.01	0.18	0.02	1.7	0	1.4	23.9
43	5	0.88	28.9	313.8	97.8	1.25	34.9	0.03	—	0.09	10.02	0.32	1.7	0.9	1.7	52.5
38	14	0.62	42.5	377.6	31.5	0.44	15.3	0.37	0.42	0.04	8.33	0.77	1.7	0.8	0.4	31.0
10	7	0.79	17.6	117.3	200.6	0.51	13.0	0.02	0.49	0.07	7.03	0.06	2.8	0	1.2	43.1
17	6	0.87	15.3	134.3	191.5	0.43	9.6	0.01	0.19	0.04	7.52	0.19	2.3	0	1.7	45.6
10	7	0.78	17.6	117.4	28.3	0.51	0	0.02	—	0.06	7.03	0.06	2.8	0	1.2	43.1
17	6	0.86	15.3	134.4	28.3	0.43	0	0.01	—	0.04	7.52	0.19	2.3	0	1.7	45.6
18	2	0.36	19.3	188.8	224.5	0.26	2.8	0.01	1.30	0.04	6.63	0.24	2.8	0	1.2	34.1
24	8	0.55	18.7	134.3	213.6	0.27	3.4	0.00	0.48	0.02	3.28	0.12	1.1	0	0.7	37.2
18	2	0.36	19.3	188.8	28.3	0.26	0	0.01	—	0.04	6.63	0.24	2.8	0	1.2	34.1
24	8	0.54	18.7	134.4	28.3	0.27	3.4	0.00	—	0.02	3.28	0.12	1.1	0	0.7	37.3
60	25	0.53	32.3	457.6	42.5	0.56	26.4	0.14	—	0.04	7.41	0.15	3.4	2.5	1.1	39.8
31	13	0.28	17.0	238.1	22.1	0.29	16.4	0.08	—	0.02	3.86	0.09	2.3	1.6	0.7	20.7
80	31	3.23	47.6	241.5	502.6	0.80	1.7	0.18	—	0.11	1.61	0.12	11.9	1.5	0.6	44.1
144	50	4.84	68.9	295.0	980.1	1.38	3.4	0.28	6.74	0.12	1.89	0.21	6.0	2.6	0.7	75.6
50	57	0.47	22.1	212.1	188.8	2.27	1.1	0.04	1.04	0.04	0.77	0.08	24.4	1.5	0.3	18.0
85	88	0.77	28.1	275.6	237.3	3.58	1.7	0.08	1.56	0.04	2.80	0.15	43.4	2.8	6.2	34.2
65	50	0.36	49.3	347.0	321.5	4.65	26.4	0.04	—	0.17	3.08	0.14	35.7	3.1	8.8	40.5
57	78	23.78	15.3	534.1	95.3	2.32	145.4	0.12	—	0.36	2.85	0.09	24.7	18.8	84.1	54.4
117	43	0.94	28.1	202.4	82.5	1.25	12.8	0.03	—	0.06	1.41	0.11	9.4	0.4	2.6	29.1

TABLE **A-1** Table of Food Composition *(continued)*

(Computer code is for Cengage Diet Analysis Plus program)

DA+ Code	Food Description	QTY	Measure	Wt (g)	H_2O (g)	Ener (cal)	Prot (g)	Carb (g)	Fiber (g)	Fat (g)	Fat Breakdown (g) Sat	Mono	Poly
Seafood—continued													
	Oysters												
8720	Baked or broiled	3	ounce(s)	85	68.6	89	5.6	3.2	0	5.8	1.3	2.1	1.9
152	Eastern, farmed, raw	3	ounce(s)	85	73.3	50	4.4	4.7	0	1.3	0.4	0.1	0.5
8715	Eastern, wild, cooked, moist heat	3	ounce(s)	85	59.8	117	12.0	6.7	0	4.2	1.3	0.5	1.6
8584	Pacific, cooked, moist heat	3	ounce(s)	85	54.5	139	16.1	8.4	0	3.9	0.9	0.7	1.5
1865	Pacific, raw	3	ounce(s)	85	69.8	69	8.0	4.2	0	2.0	0.4	0.3	0.8
1854	**Lobster, northern, cooked, moist heat**	3	ounce(s)	85	64.7	83	17.4	1.1	0	0.5	0.1	0.1	0.1
1862	**Mussel, blue, cooked, moist heat**	3	ounce(s)	85	52.0	146	20.2	6.3	0	3.8	0.7	0.9	1.0
	Shrimp												
158	Mixed species, breaded, fried	3	ounce(s)	85	44.9	206	18.2	9.8	0.3	10.4	1.8	3.2	4.3
1855	Mixed species, cooked, moist heat	3	ounce(s)	85	65.7	84	17.8	0	0	0.9	0.2	0.2	0.4
Beef, Lamb, Pork													
	Beef												
4450	Breakfast strips, cooked	2	slice(s)	23	5.9	101	7.1	0.3	0	7.8	3.2	3.8	0.4
174	Corned beef, canned	3	ounce(s)	85	49.1	213	23.0	0	0	12.7	5.3	5.1	0.5
33147	Cured, thin siced	2	ounce(s)	57	32.9	100	15.9	3.2	0	2.2	0.9	1.0	0.1
4581	Jerky	1	ounce(s)	28	6.6	116	9.4	3.1	0.5	7.3	3.1	3.2	0.3
	Ground beef												
5898	Lean, broiled, medium	3	ounce(s)	85	50.4	202	21.6	0	0	12.2	4.8	5.3	0.4
5899	Lean, broiled, well done	3	ounce(s)	85	48.4	214	23.8	0	0	12.5	5.0	5.7	0.3
5914	Regular, broiled, medium	3	ounce(s)	85	46.1	246	20.5	0	0	17.6	6.9	7.7	0.6
5915	Regular, broiled, well done	3	ounce(s)	85	43.8	259	21.6	0	0	18.4	7.5	8.5	0.5
	Beef rib												
4241	Rib, small end, separable lean, 0″ fat, broiled	3	ounce(s)	85	53.2	164	25.0	0	0	6.4	2.4	2.6	0.2
4183	Rib, whole, lean and fat, ¼″ fat, roasted	3	ounce(s)	85	39.0	320	18.9	0	0	26.6	10.7	11.4	0.9
	Beef roast												
16981	Bottom round, choice, separable lean and fat, ⅛″ fat, braised	3	ounce(s)	85	46.2	216	27.9	0	0	10.7	4.1	4.6	0.4
16979	Bottom round, separable lean and fat, ⅛″ fat, roasted	3	ounce(s)	85	52.4	185	22.5	0	0	9.9	3.8	4.2	0.4
16924	Chuck, arm pot roast, separable lean and fat, ⅛″ fat, braised	3	ounce(s)	85	42.9	257	25.6	0	0	16.3	6.5	7.0	0.6
16930	Chuck, blade roast, separable lean and fat, ⅛″ fat, braised	3	ounce(s)	85	40.5	290	22.8	0	0	21.4	8.5	9.2	0.8
5853	Chuck, blade roast, separable lean, 0″ trim, pot roasted	3	ounce(s)	85	47.4	202	26.4	0	0	9.9	3.9	4.3	0.3
4296	Eye of round, choice, separable lean, 0″ fat, roasted	3	ounce(s)	85	56.5	138	24.4	0	0	3.7	1.3	1.5	0.1
16989	Eye of round, separable lean and fat, ⅛″ fat, roasted	3	ounce(s)	85	52.2	180	24.2	0	0	8.5	3.2	3.6	0.3
	Beef steak												
4348	Short loin, t-bone steak, lean and fat, ¼″ fat, broiled	3	ounce(s)	85	43.2	274	19.4	0	0	21.2	8.3	9.6	0.8
4349	Short loin, t-bone steak, lean, ¼″ fat, broiled	3	ounce(s)	85	52.3	174	22.8	0	0	8.5	3.1	4.2	0.3
4360	Top loin, prime, lean and fat, ¼″ fat, broiled	3	ounce(s)	85	42.7	275	21.6	0	0	20.3	8.2	8.6	0.7
	Beef variety												
188	Liver, pan fried	3	ounce(s)	85	52.7	149	22.6	4.4	0	4.0	1.3	0.5	0.5
4447	Tongue, simmered	3	ounce(s)	85	49.2	242	16.4	0	0	19.0	6.9	8.6	0.6
	Lamb chop												
3275	Loin, domestic, lean and fat, ¼″ fat, broiled	3	ounce(s)	85	43.9	269	21.4	0	0	19.6	8.4	8.3	1.4

Chol (mg)	Calc (mg)	Iron (mg)	Magn (mg)	Pota (mg)	Sodi (mg)	Zinc (mg)	Vit A (µg)	Thia (mg)	Vit E (mg α)	Ribo (mg)	Niac (mg)	Vit B_6 (mg)	Fola (µg DFE)	Vit C (mg)	Vit B_{12} (µg)	Sele (µg)
43	36	5.30	37.4	125.0	403.8	72.22	60.4	0.07	0.98	0.06	1.04	0.04	7.7	2.8	14.7	50.7
21	37	4.91	28.1	105.4	151.3	32.23	6.8	0.08	—	0.05	1.07	0.05	15.3	4.0	13.8	54.1
89	77	10.19	80.8	239.0	358.9	154.45	45.9	0.16	—	0.15	2.11	0.10	11.9	5.1	29.8	60.9
85	14	7.82	37.4	256.8	180.3	28.27	124.2	0.10	0.72	0.37	3.07	0.07	12.8	10.9	24.5	131.0
43	7	4.34	18.7	142.9	90.1	14.13	68.9	0.05	—	0.20	1.70	0.04	8.5	6.8	13.6	65.5
61	52	0.33	29.8	299.4	323.2	2.48	22.1	0.01	0.85	0.05	0.91	0.06	9.4	0	2.6	36.3
48	28	5.71	31.5	227.9	313.8	2.27	77.4	0.25	—	0.35	2.55	0.08	64.6	11.6	20.4	76.2
150	57	1.07	34.0	191.3	292.4	1.17	0	0.11	—	0.11	2.60	0.08	33.2	1.3	1.6	35.4
166	33	2.62	28.9	154.8	190.5	1.32	57.8	0.02	1.17	0.02	2.20	0.10	3.4	1.9	1.3	33.7
27	2	0.71	6.1	93.1	509.2	1.44	0	0.02	0.06	0.05	1.46	0.07	1.8	0	0.8	6.1
73	10	1.76	11.9	115.7	855.6	3.03	0	0.01	0.12	0.12	2.06	0.11	7.7	0	1.4	36.5
23	6	1.53	10.8	243.2	815.9	2.25	0	0.04	0.00	0.10	2.98	0.19	6.2	0	1.5	16.0
14	6	1.53	14.5	169.2	627.4	2.29	0	0.04	0.13	0.04	0.49	0.05	38.0	0	0.3	3.0
58	6	2.00	17.9	266.2	59.5	4.63	0	0.05	—	0.23	4.21	0.23	7.6	0	1.8	16.0
69	12	2.21	18.4	250.0	62.4	5.86	0	0.08	—	0.23	5.10	0.16	9.4	0	1.7	19.0
62	9	2.07	17.0	248.3	70.6	4.40	0	0.02	—	0.16	4.90	0.23	7.6	0	2.5	16.2
71	12	2.30	18.5	242.4	72.4	5.18	0	0.08	—	0.23	4.93	0.17	8.5	0	1.6	18.0
65	16	1.59	21.3	319.8	51.9	4.64	0	0.06	0.34	0.12	7.15	0.53	8.5	0	1.4	29.2
72	9	1.96	16.2	251.7	53.6	4.45	0	0.06	—	0.14	2.85	0.19	6.0	0	2.1	18.7
68	6	2.29	17.9	223.7	35.7	4.59	0	0.05	0.41	0.15	5.05	0.36	8.5	0	1.7	29.3
64	5	1.83	14.5	182.0	29.8	3.76	0	0.05	0.34	0.12	3.92	0.29	6.8	0	1.3	23.0
67	14	2.15	17.0	205.8	42.5	5.93	0	0.05	0.45	0.15	3.63	0.25	7.7	0	1.9	24.1
88	11	2.66	16.2	198.2	55.3	7.15	0	0.06	0.17	0.20	2.06	0.22	4.3	0	1.9	20.9
73	11	3.12	19.6	223.7	60.4	8.73	0	0.06	—	0.23	2.27	0.24	5.1	0	2.1	22.7
49	5	2.16	16.2	200.7	32.3	4.28	0	0.05	0.30	0.15	4.69	0.34	8.5	0	1.4	28.0
54	5	1.98	15.3	193.1	31.5	3.95	0	0.05	0.34	0.13	4.37	0.31	7.7	0	1.5	25.2
58	7	2.56	17.9	233.9	57.8	3.56	0	0.07	0.18	0.17	3.29	0.27	6.0	0	1.8	10.0
50	5	3.11	22.1	278.1	65.5	4.34	0	0.09	0.11	0.21	3.93	0.33	6.8	0	1.9	8.5
67	8	1.88	19.6	294.3	53.6	3.85	0	0.06	—	0.15	3.96	0.31	6.0	0	1.6	19.5
324	5	5.24	18.7	298.5	65.5	4.44	6586.3	0.15	0.39	2.91	14.86	0.87	221.1	0.6	70.7	27.9
112	4	2.22	12.8	156.5	55.3	3.47	0	0.01	0.25	0.25	2.96	0.13	6.0	1.1	2.7	11.2
85	17	1.53	20.4	278.1	65.5	2.96	0	0.08	0.11	0.21	6.03	0.11	15.3	0	2.1	23.3

TABLE A-1 Table of Food Composition *(continued)*

(Computer code is for Cengage Diet Analysis Plus program)

DA+ Code	Food Description	QTY	Measure	Wt (g)	H_2O (g)	Ener (cal)	Prot (g)	Carb (g)	Fiber (g)	Fat (g)	Fat Breakdown (g) Sat	Mono	Poly
	Beef, Lamb, Pork—continued												
	Lamb leg												
3264	Domestic, lean and fat, ¼″ fat, cooked	3	ounce(s)	85	45.7	250	20.9	0	0	17.8	7.5	7.5	1.3
	Lamb rib												
182	Domestic, lean and fat, ¼″ fat, broiled	3	ounce(s)	85	40.0	307	18.8	0	0	25.2	10.8	10.3	2.0
183	Domestic, lean, ¼″ fat, broiled	3	ounce(s)	85	50.0	200	23.6	0	0	11.0	4.0	4.4	1.0
	Lamb shoulder												
186	Shoulder, arm and blade, domestic, choice, lean and fat, ¼″ fat, roasted	3	ounce(s)	85	47.8	235	19.1	0	0	17.0	7.2	6.9	1.4
187	Shoulder, arm and blade, domestic, choice, lean, ¼″ fat, roasted	3	ounce(s)	85	53.8	173	21.2	0	0	9.2	3.5	3.7	0.8
3287	Shoulder, arm, domestic, lean and fat, ¼″ fat, braised	3	ounce(s)	85	37.6	294	25.8	0	0	20.4	8.4	8.7	1.5
3290	Shoulder, arm, domestic, lean, ¼″ fat, braised	3	ounce(s)	85	41.9	237	30.2	0	0	12.0	4.3	5.2	0.8
	Lamb variety												
3375	Brain, pan fried	3	ounce(s)	85	51.6	232	14.4	0	0	18.9	4.8	3.4	1.9
3406	Tongue, braised	3	ounce(s)	85	49.2	234	18.3	0	0	17.2	6.7	8.5	1.1
	Pork, cured												
29229	Bacon, Canadian style, cured	2	ounce(s)	57	37.9	89	11.7	1.0	0	4.0	1.3	1.8	0.4
161	Bacon, cured, broiled, pan fried or roasted	2	slice(s)	16	2.0	87	5.9	0.2	0	6.7	2.2	3.0	0.7
35422	Breakfast strips, cured, cooked	3	slice(s)	34	9.2	156	9.8	0.4	0	12.5	4.3	5.6	1.9
189	Ham, cured, boneless, 11% fat, roasted	3	ounce(s)	85	54.9	151	19.2	0	0	7.7	2.7	3.8	1.2
29215	Ham, cured, extra lean, 4% fat, canned	2	2 ounce(s)	57	41.7	68	10.5	0	0	2.6	0.9	1.3	0.2
1316	Ham, cured, extra lean, 5% fat, roasted	3	ounce(s)	85	57.6	123	17.8	1.3	0	4.7	1.5	2.2	0.5
16561	Ham, smoked or cured, lean, cooked	1	slice(s)	42	27.6	66	10.5	0	0	2.3	0.8	1.1	0.3
	Pork chop												
32671	Loin, blade, chops, lean and fat, pan fried	3	ounce(s)	85	42.5	291	18.3	0	0	23.6	8.6	10	2.6
32672	Loin, center cut, chops, lean and fat, pan fried	3	ounce(s)	85	45.1	236	25.4	0	0	14.1	5.1	6.0	1.6
32682	Loin, center rib, chops, boneless, lean and fat, braised	3	ounce(s)	85	49.5	217	22.4	0	0	13.4	5.2	6.1	1.1
32603	Loin, center rib, chops, lean, broiled	3	ounce(s)	85	55.4	158	21.9	0	0	7.1	2.4	3.0	0.8
32478	Loin, whole, lean and fat, braised	3	ounce(s)	85	49.6	203	23.2	0	0	11.6	4.3	5.2	1.0
32481	Loin, whole, lean, braised	3	ounce(s)	85	52.2	174	24.3	0	0	7.8	2.9	3.5	0.6
	Pork leg or ham												
32471	Pork leg or ham, rump portion, lean and fat, roasted	3	ounce(s)	85	48.3	214	24.6	0	0	12.1	4.5	5.4	1.2
32468	Pork leg or ham, whole, lean and fat, roasted	3	ounce(s)	85	46.8	232	22.8	0	0	15.0	5.5	6.7	1.4
	Pork ribs												
32693	Loin, country style, lean and fat, roasted	3	ounce(s)	85	43.3	279	19.9	0	0	21.6	7.8	9.4	1.7
32696	Loin, country style, lean, roasted	3	ounce(s)	85	49.5	210	22.6	0	0	12.6	4.5	5.5	0.9
	Pork shoulder												
32626	Shoulder, arm picnic, lean and fat, roasted	3	ounce(s)	85	44.3	270	20.0	0	0	20.4	7.5	9.1	2.0
32629	Shoulder, arm picnic, lean, roasted	3	ounce(s)	85	51.3	194	22.7	0	0	10.7	3.7	5.1	1.0
	Rabbit												
3366	Domesticated, roasted	3	ounce(s)	85	51.5	168	24.7	0	0	6.8	2.0	1.8	1.3
3367	Domesticated, stewed	3	ounce(s)	85	50.0	175	25.8	0	0	7.2	2.1	1.9	1.4

PAGE KEY: A-2 = Breads/Baked Goods A-8 = Cereal/Rice/Pasta A-12 = Fruit A-18 = Vegetables/Legumes A-28 = Nuts/Seeds A-30 = Vegetarian A-32 = Dairy A-40 = Eggs A-40 = Seafood A-44 = Meats A-48 = Poultry A-48 = Processed Meats A-50 = Beverages A-54 = Fats/Oils A-56 = Sweets A-58 = Spices/Condiments/Sauces A-62 = Mixed Foods/Soups/Sandwiches A-68 = Fast Food A-88 = Convenience A-90 = Baby Foods

A

Chol (mg)	Calc (mg)	Iron (mg)	Magn (mg)	Pota (mg)	Sodi (mg)	Zinc (mg)	Vit A (µg)	Thia (mg)	Vit E (mg α)	Ribo (mg)	Niac (mg)	Vit B_6 (mg)	Fola (µg DFE)	Vit C (mg)	Vit B_{12} (µg)	Sele (µg)
82	14	1.59	19.6	263.7	61.2	3.79	0	0.08	0.11	0.21	5.66	0.11	15.3	0	2.2	22.5
84	16	1.59	19.6	229.5	64.6	3.40	0	0.07	0.10	0.18	5.95	0.09	11.9	0	2.2	20.3
77	14	1.87	24.7	266.1	72.3	4.47	0	0.08	0.15	0.21	5.56	0.12	17.9	0	2.2	26.4
78	17	1.67	19.6	213.4	56.1	4.44	0	0.07	0.11	0.20	5.22	0.11	17.9	0	2.2	22.3
74	16	1.81	21.3	225.3	57.8	5.13	0	0.07	0.15	0.22	4.89	0.12	21.3	0	2.3	24.2
102	21	2.03	22.1	260.3	61.2	5.17	0	0.06	0.12	0.21	5.66	0.09	15.3	0	2.2	31.6
103	22	2.29	24.7	287.5	64.6	6.20	0	0.06	0.15	0.23	5.38	0.11	18.7	0	2.3	32.1
2130	18	1.73	18.7	304.5	133.5	1.70	0	0.14	—	0.31	3.87	0.19	6.0	19.6	20.5	10.2
161	9	2.23	13.6	134.4	57.0	2.54	0	0.06	—	0.35	3.13	0.14	2.6	6.0	5.4	23.8
28	5	0.38	9.6	195.0	798.9	0.78	0	0.42	0.11	0.09	3.53	0.22	2.3	0	0.4	14.2
18	2	0.22	5.3	90.4	369.6	0.56	1.8	0.06	0.04	0.04	1.76	0.04	0.3	0	0.2	9.9
36	5	0.67	8.8	158.4	713.7	1.25	0	0.25	0.08	0.12	2.58	0.11	1.4	0	0.6	8.4
50	7	1.13	18.7	347.7	1275.0	2.09	0	0.62	0.26	0.28	5.22	0.26	2.6	0	0.6	16.8
22	3	0.53	9.6	206.4	711.6	1.09	0	0.47	0.09	0.13	3.00	0.25	3.4	0	0.5	8.2
45	7	1.25	11.9	244.1	1023.1	2.44	0	0.64	0.21	0.17	3.42	0.34	2.6	0	0.6	16.6
23	3	0.39	9.2	132.7	557.3	1.07	0	0.28	0.10	0.10	2.10	0.19	1.7	0	0.3	10.7
72	26	0.74	17.9	282.4	57.0	2.71	1.7	0.52	0.17	0.25	3.35	0.28	3.4	0.5	0.7	29.7
78	23	0.77	24.7	361.5	68.0	1.96	1.7	0.96	0.21	0.25	4.76	0.39	5.1	0.9	0.6	33.2
62	4	0.78	14.5	329.1	34.0	1.76	1.7	0.44	—	0.20	3.66	0.26	3.4	0.3	0.4	28.4
56	22	0.57	21.3	291.7	48.5	1.91	0	0.48	0.08	0.18	6.68	0.57	0	0	0.4	38.6
68	18	0.91	16.2	318.1	40.8	2.02	1.7	0.53	0.20	0.21	3.75	0.31	2.6	0.5	0.5	38.5
67	15	0.96	17.0	329.1	42.5	2.10	1.7	0.56	0.17	0.22	3.90	0.32	3.4	0.5	0.5	41.0
82	10	0.89	23.0	318.1	52.7	2.39	2.6	0.63	0.18	0.28	3.95	0.26	2.6	0.2	0.6	39.8
80	12	0.85	18.7	299.4	51.0	2.51	2.6	0.54	0.18	0.26	3.89	0.34	8.5	0.3	0.6	38.5
78	21	0.90	19.6	292.6	44.2	2.00	2.6	0.75	—	0.29	3.67	0.37	4.3	0.3	0.7	31.6
79	25	1.09	20.4	296.8	24.7	3.24	1.7	0.48	—	0.29	3.96	0.37	4.3	0.3	0.7	36.0
80	16	1.00	14.5	276.4	59.5	2.93	1.7	0.44	—	0.25	3.33	0.29	3.4	0.2	0.6	28.6
81	8	1.20	17.0	298.5	68.0	3.46	1.7	0.49	—	0.30	3.66	0.34	4.3	0.3	0.7	32.7
70	16	1.93	17.9	325.7	40.0	1.93	0	0.07	—	0.17	7.17	0.40	9.4	0	7.1	32.7
73	17	2.01	17.0	255.1	31.5	2.01	0	0.05	0.37	0.14	6.09	0.28	7.7	0	5.5	32.7

A

TABLE A-1 Table of Food Composition *(continued)*

(Computer code is for Cengage Diet Analysis Plus program)

DA+ Code	Food Description	QTY	Measure	Wt (g)	H_2O (g)	Ener (cal)	Prot (g)	Carb (g)	Fiber (g)	Fat (g)	Fat Breakdown (g) Sat	Mono	Poly
Beef, Lamb, Pork—continued													
	Veal												
3391	Liver, braised	3	ounce(s)	85	50.9	163	24.2	3.2	0	5.3	1.7	1.0	0.9
3319	Rib, lean only, roasted	3	ounce(s)	85	55.0	151	21.9	0	0	6.3	1.8	2.3	0.6
1732	Deer or venison, roasted	3	ounce(s)	85	55.5	134	25.7	0	0	2.7	1.1	0.7	0.5
Poultry													
	Chicken												
29562	Flaked, canned	2	ounce(s)	57	39.3	97	10.3	0.1	0	5.8	1.6	2.3	1.3
	Chicken, fried												
29632	Breast, meat only, breaded, baked or fried	3	ounce(s)	85	44.3	193	25.3	6.9	0.2	6.6	1.6	2.7	1.7
35327	Broiler breast, meat only, fried	3	ounce(s)	85	51.2	159	28.4	0.4	0	4.0	1.1	1.5	0.9
36413	Broiler breast, meat and skin, flour coated, fried	3	ounce(s)	85	48.1	189	27.1	1.4	0.1	7.5	2.1	3.0	1.7
36414	Broiler drumstick, meat and skin, flour coated, fried	3	ounce(s)	85	48.2	208	22.9	1.4	0.1	11.7	3.1	4.6	2.7
35389	Broiler drumstick, meat only, fried	3	ounce(s)	85	52.9	166	24.3	0	0	6.9	1.8	2.5	1.7
35406	Broiler leg, meat only, fried	3	ounce(s)	85	51.5	177	24.1	0.6	0	7.9	2.1	2.9	1.9
35484	Broiler wing, meat only, fried	3	ounce(s)	85	50.9	179	25.6	0	0	7.8	2.1	2.6	1.8
29580	Patty, fillet or tenders, breaded, cooked	3	ounce(s)	85	40.2	256	14.5	12.2	0	16.5	3.7	8.4	3.7
	Chicken, roasted, meat only												
35409	Broiler leg, meat only, roasted	3	ounce(s)	85	55.0	162	23.0	0	0	7.2	1.9	2.6	1.7
35486	Broiler wing, meat only, roasted	3	ounce(s)	85	53.4	173	25.9	0	0	6.9	1.9	2.2	1.5
35138	Roasting chicken, dark meat, meat only, roasted	3	ounce(s)	85	57.0	151	19.8	0	0	7.4	2.1	2.8	1.7
35136	Roasting chicken, light meat, meat only, roasted	3	ounce(s)	85	57.7	130	23.1	0	0	3.5	0.9	1.3	0.8
35132	Roasting chicken, meat only, roasted	3	ounce(s)	85	57.3	142	21.3	0	0	5.6	1.5	2.1	1.3
	Chicken, stewed												
1268	Gizzard, simmered	3	ounce(s)	85	57.8	124	25.8	0	0	2.3	0.6	0.4	0.3
1270	Liver, simmered	3	ounce(s)	85	56.8	142	20.8	0.7	0	5.5	1.8	1.2	1.7
3174	Meat only, stewed	3	ounce(s)	85	56.8	151	23.2	0	0	5.7	1.6	2.0	1.3
	Duck												
1286	Domesticated, meat and skin, roasted	3	ounce(s)	85	44.1	287	16.2	0	0	24.1	8.2	11.0	3.1
1287	Domesticated, meat only, roasted	3	ounce(s)	85	54.6	171	20.0	0	0	9.5	3.5	3.1	1.2
	Goose												
35507	Domesticated, meat and skin, roasted	3	ounce(s)	85	44.2	259	21.4	0	0	18.6	5.8	8.7	2.1
35524	Domesticated, meat only, roasted	3	ounce(s)	85	48.7	202	24.6	0	0	10.8	3.9	3.7	1.3
1297	Liver pate, smoked, canned	4	tablespoon(s)	52	19.3	240	5.9	2.4	0	22.8	7.5	13.3	0.4
	Turkey												
3256	Ground turkey, cooked	3	ounce(s)	85	50.5	200	23.3	0	0	11.2	2.9	4.2	2.7
3263	Patty, batter coated, breaded, fried	1	item(s)	94	46.7	266	13.2	14.8	0.5	16.9	4.4	7.0	4.4
219	Roasted, dark meat, meat only	3	ounce(s)	85	53.7	159	24.3	0	0	6.1	2.1	1.4	1.8
222	Roasted, fryer roaster breast, meat only	3	ounce(s)	85	58.2	115	25.6	0	0	0.6	0.2	0.1	0.2
220	Roasted, light meat, meat only	3	ounce(s)	85	56.4	134	25.4	0	0	2.7	0.9	0.5	0.7
1303	Turkey roll, light and dark meat	2	slice(s)	57	39.8	84	10.3	1.2	0	4.0	1.2	1.3	1.0
1302	Turkey roll, light meat	2	slice(s)	57	42.5	56	8.4	2.9	0	0.9	0.2	0.2	0.1
Processed Meats													
	Beef												
1331	Corned beef loaf, jellied, sliced	2	slice(s)	57	39.2	87	13.0	0	0	3.5	1.5	1.5	0.2
	Bologna												
13459	Beef	1	slice(s)	28	15.1	90	3.0	1.0	0	8.0	3.5	4.3	0.3
13461	Light, made with pork and chicken	1	slice(s)	28	18.2	60	3.0	2.0	0	4.0	1.0	2.0	0.4
13458	Made with chicken and pork	1	slice(s)	28	15.0	90	3.0	1.0	0	8.0	3.0	4.1	1.1
13565	Turkey bologna	1	slice(s)	28	19.0	50	3.0	1.0	0	4.0	1.0	1.1	1.0
	Chicken												
7125	Breast, smoked	1	slice(s)	10	—	10	1.8	0.3	0	0.2	0	—	—

Chol (mg)	Calc (mg)	Iron (mg)	Magn (mg)	Pota (mg)	Sodi (mg)	Zinc (mg)	Vit A (µg)	Thia (mg)	Vit E (mg α)	Ribo (mg)	Niac (mg)	Vit B_6 (mg)	Fola (µg DFE)	Vit C (mg)	Vit B_{12} (µg)	Sele (µg)
435	5	4.34	17.0	279.8	66.3	9.55	8026	0.15	0.57	2.43	11.18	0.78	281.5	0.9	72.0	16.4
98	10	0.81	20.4	264.5	82.5	3.81	0	0.05	0.30	0.24	6.37	0.23	11.9	0	1.3	9.4
95	6	3.80	20.4	284.9	45.9	2.33	0	0.15	—	0.51	5.70	—	—	0	—	11.0
35	8	0.89	6.8	147.4	408.2	0.79	19.3	0.01	—	0.07	3.58	0.19	2.3	0	0.2	—
67	19	1.05	24.7	222.6	450.2	0.84	—	0.08	—	0.09	10.97	0.46	4.3	0	0.3	—
77	14	0.96	26.4	234.7	67.2	0.91	6.0	0.06	0.35	0.10	12.57	0.54	3.4	0	0.3	22.3
76	14	1.01	25.5	220.3	64.6	0.93	12.8	0.06	0.39	0.11	11.68	0.49	6.0	0	0.3	20.3
77	10	1.13	19.6	194.8	75.7	2.45	21.3	0.06	0.65	0.19	5.13	0.29	9.4	0	0.3	15.6
80	10	1.12	20.4	211.8	81.6	2.73	15.3	0.06	—	0.20	5.22	0.33	7.7	0	0.3	16.7
84	11	1.19	21.3	216.0	81.6	2.53	17.0	0.07	0.38	0.21	5.68	0.33	7.7	0	0.3	16.0
71	13	0.96	17.9	176.9	77.4	1.80	15.3	0.03	0.40	0.10	6.15	0.50	3.4	0	0.3	21.6
49	11	0.75	19.6	244.8	411.4	0.79	4.3	0.09	1.04	0.12	5.99	0.24	35.7	0	0.2	13.9
80	10	1.11	20.4	205.8	77.4	2.43	16.2	0.06	0.22	0.19	5.37	0.31	6.8	0	0.3	18.8
72	14	0.98	17.9	178.6	78.2	1.82	15.3	0.03	0.22	0.10	6.21	0.50	3.4	0	0.3	21.0
64	9	1.13	17.0	190.5	80.8	1.81	13.6	0.05	—	0.16	4.87	0.26	6.0	0	0.2	16.7
64	11	0.91	19.6	200.7	43.4	0.66	6.8	0.05	0.22	0.07	8.90	0.45	2.6	0	0.3	21.9
64	10	1.02	17.9	194.8	63.8	1.29	10.2	0.05	—	0.12	6.70	0.34	4.3	0	0.2	20.9
315	14	2.71	2.6	152.2	47.6	3.75	0	0.02	0.17	0.17	2.65	0.06	4.3	0	0.9	35.0
479	9	9.89	21.3	223.7	64.6	3.38	3385.8	0.24	0.69	1.69	9.39	0.64	491.6	23.7	14.3	70.1
71	12	0.99	17.9	153.1	59.5	1.69	12.8	0.04	0.22	0.13	5.20	0.22	5.1	0	0.2	17.8
71	9	2.29	13.6	173.5	50.2	1.58	53.6	0.14	0.59	0.22	4.10	0.15	5.1	0	0.3	17.0
76	10	2.29	17.0	214.3	55.3	2.21	19.6	0.22	0.59	0.39	4.33	0.21	8.5	0	0.3	19.1
77	11	2.40	18.7	279.8	59.5	2.22	17.9	0.06	1.47	0.27	3.54	0.31	1.7	0	0.3	18.5
82	12	2.44	21.3	330.0	64.6	2.69	10.2	0.07	—	0.33	3.47	0.39	10.2	0	0.4	21.7
78	36	2.86	6.8	71.8	362.4	0.47	520.5	0.04	—	0.15	1.30	0.03	31.2	0	4.9	22.9
87	21	1.64	20.4	229.6	91.0	2.43	0	0.04	0.28	0.14	4.09	0.33	6.0	0	0.3	31.6
71	13	2.06	14.1	258.5	752.0	1.35	9.4	0.09	0.87	0.17	2.16	0.18	57.3	0	0.2	20.8
72	27	1.98	20.4	246.6	67.2	3.79	0	0.05	0.54	0.21	3.10	0.30	7.7	0	0.3	34.8
71	10	1.30	24.7	248.3	44.2	1.48	0	0.03	0.07	0.11	6.37	0.47	5.1	0	0.3	27.3
59	16	1.14	23.8	259.4	54.4	1.73	0	0.05	0.07	0.11	5.81	0.45	5.1	0	0.3	27.3
31	18	0.76	10.2	153.1	332.3	1.13	0	0.05	0.19	0.16	2.72	0.15	2.8	0	0.1	16.6
19	4	0.21	10.8	242.1	590.8	0.50	0	0.01	0.07	0.08	4.05	0.23	2.3	0	0.2	7.4
27	6	1.15	6.2	57.3	540.4	2.31	0	0.00	—	0.06	0.99	0.06	4.5	0	0.7	9.8
20	0	0.36	3.9	47.0	310.0	0.56	0	0.01	—	0.03	0.67	0.04	3.6	0	0.4	—
20	40	0.36	5.6	45.6	300.0	0.45	0	—	—	—	—	—	—	0	—	—
30	20	0.36	5.9	43.1	300.0	0.39	0	—	—	—	—	—	—	0	—	—
20	40	0.36	6.2	42.6	270.0	0.51	0	—	—	—	—	—	—	0	—	—
4	0	0.00	—	—	100.0	—	0	—	—	—	—	—	—	0	—	—

TABLE A-1 Table of Food Composition *(continued)*

(Computer code is for Cengage Diet Analysis Plus program)

DA+ Code	Food Description	QTY	Measure	Wt (g)	H_2O (g)	Ener (cal)	Prot (g)	Carb (g)	Fiber (g)	Fat (g)	Fat Breakdown (g) Sat	Mono	Poly
Processed Meats—continued													
	Ham												
7127	Deli-sliced, honey	1	slice(s)	10	—	10	1.7	0.3	0	0.3	0.1	—	—
7126	Deli-sliced, smoked	1	slice(s)	10	—	10	1.7	0.2	0	0.3	0.1	—	—
8614	**Beef and pork mortadella, sliced**	2	slice(s)	46	24.1	143	7.5	1.4	0	11.7	4.4	5.2	1.4
1323	**Pork olive loaf**	2	slice(s)	57	33.1	133	6.7	5.2	0	9.4	3.3	4.5	1.1
1324	**Pork pickle and pimento loaf**	2	slice(s)	57	34.2	128	6.4	4.8	0.9	9.1	3.0	4.0	1.6
	Sausages and frankfurters												
37296	Beerwurst beef, beer salami (bierwurst)	1	slice(s)	29	16.6	74	4.1	1.2	0	5.7	2.5	2.7	0.2
37257	Beerwurst pork, beer salami	1	slice(s)	21	12.9	50	3.0	0.4	0	4.0	1.3	1.9	0.5
35338	Berliner, pork and beef	1	ounce(s)	28	17.3	65	4.3	0.7	0	4.9	1.7	2.3	0.4
37298	Bratwurst pork, cooked	1	piece(s)	74	42.3	181	10.4	1.9	0	14.3	5.1	6.7	1.5
37299	Braunschweiger pork liver sausage	1	slice(s)	15	8.2	51	2.0	0.3	0	4.5	1.5	2.1	0.5
1329	Cheesefurter or cheese smokie, beef and pork	1	item(s)	43	22.6	141	6.1	0.6	0	12.5	4.5	5.9	1.3
1330	Chorizo, beef and pork	2	ounce(s)	57	18.1	258	13.7	1.1	0	21.7	8.2	10.4	2.0
8600	Frankfurter, beef	1	item(s)	45	23.4	149	5.1	1.8	0	13.3	5.3	6.4	0.5
202	Frankfurter, beef and pork	1	item(s)	45	25.2	137	5.2	0.8	0	12.4	4.8	6.2	1.2
1293	Frankfurter, chicken	1	item(s)	45	28.1	100	7.0	1.2	0.2	7.3	1.7	2.7	1.7
3261	Frankfurter, turkey	1	item(s)	45	28.3	100	5.5	1.7	0	7.8	1.8	2.6	1.8
37275	Italian sausage, pork, cooked	1	item(s)	68	32.0	234	13.0	2.9	0.1	18.6	6.5	8.1	2.2
37307	Kielbasa or kolbassa, pork and beef	1	slice(s)	30	18.5	67	5.0	1.0	0	4.7	1.7	2.2	0.5
1333	Knockwurst or knackwurst, beef and pork	2	ounce(s)	57	31.4	174	6.3	1.8	0	15.7	5.8	7.3	1.7
37285	Pepperoni, beef and pork	1	slice(s)	11	3.4	51	2.2	0.4	0.2	4.4	1.8	2.1	0.3
37313	Polish sausage, pork	1	slice(s)	21	11.4	60	2.8	0.7	0	5.0	1.8	2.3	0.5
206	Salami, beef, cooked, sliced	2	slice(s)	52	31.2	136	6.5	1.0	0	11.5	5.1	5.5	0.5
37272	Salami, pork, dry or hard	1	slice(s)	13	4.6	52	2.9	0.2	0	4.3	1.5	2.0	0.5
40987	Sausage, turkey, cooked	2	ounce(s)	57	36.9	111	13.5	0	0	5.9	1.3	1.7	1.5
8620	Smoked sausage, beef and pork	2	ounce(s)	57	30.6	181	6.8	1.4	0	16.3	5.5	6.9	2.2
8619	Smoked sausage, pork	2	ounce(s)	57	32.0	178	6.8	1.2	0	16.0	5.3	6.4	2.1
37273	Smoked sausage, pork link	1	piece(s)	76	29.8	295	16.8	1.6	0	24.0	8.6	11.1	2.8
1336	Summer sausage, thuringer, or cervelat, beef and pork	2	ounce(s)	57	25.6	205	9.9	1.9	0	17.3	6.5	7.4	0.7
37294	Vienna sausage, cocktail, beef and pork, canned	1	piece(s)	16	10.4	37	1.7	0.4	0	3.1	1.1	1.5	0.2
	Spreads												
1318	Ham salad spread	¼	cup(s)	60	37.6	130	5.2	6.4	0	9.3	3.0	4.3	1.6
32419	Pork and beef sandwich spread	4	tablespoon(s)	60	36.2	141	4.6	7.2	0.1	10.4	3.6	4.6	1.5
	Turkey												
13604	Breast, fat free, oven roasted	1	slice(s)	28	—	25	4.0	1.0	0	0	0	0	0
13606	Breast, hickory smoked fat free	1	slice(s)	28	—	25	4.0	1.0	0	0	0	0	0
16049	Breast, hickory smoked slices	1	slice(s)	56	—	50	11.0	1.0	0	0	0	0	0
16047	Breast, honey roasted slices	1	slice(s)	56	—	60	11.0	3.0	0	0	0	0	0
16048	Breast, oven roasted slices	1	slice(s)	56	—	50	11.0	1.0	0	0	0	0	0
7124	Breast, oven roasted	1	slice(s)	10	—	10	1.8	0.3	0	0.1	0	0	0
13567	Turkey ham, 10% water added	2	slice(s)	56	40.9	70	10.0	2.0	0	3.0	0	0.4	0.6
37270	Turkey pastrami	1	slice(s)	28	20.3	35	4.6	1.0	0	1.2	0.3	0.4	0.3
3262	Turkey salami	2	slice(s)	57	39.1	98	10.9	0.9	0.1	5.2	1.6	1.8	1.4
37318	Turkey salami, cooked	1	slice(s)	28	20.4	43	4.3	0.1	0	2.7	0.8	0.9	0.7
Beverages													
	Beer												
866	Ale, mild	12	fluid ounce(s)	360	332.3	148	1.1	13.3	0.4	0	0	0	0
686	Beer	12	fluid ounce(s)	356	327.7	153	1.6	12.7	0	0	0	0	0
16886	Beer, non alcoholic	12	fluid ounce(s)	360	328.1	133	0.8	29.0	0	0.4	0.1	0	0.2
31609	Bud Light beer	12	fluid ounce(s)	355	335.5	110	0.9	6.6	0	0	0	0	0
31608	Budweiser beer	12	fluid ounce(s)	355	327.7	145	1.3	10.6	0	0	0	0	0

Chol (mg)	Calc (mg)	Iron (mg)	Magn (mg)	Pota (mg)	Sodi (mg)	Zinc (mg)	Vit A (µg)	Thia (mg)	Vit E (mg α)	Ribo (mg)	Niac (mg)	Vit B_6 (mg)	Fola (µg DFE)	Vit C (mg)	Vit B_{12} (µg)	Sele (µg)
4	0	0.12	—	—	100.0	—	0	—	—	—	—	—	—	0.6	—	—
4	0	0.12	—	—	103.3	—	0	—	—	—	—	—	—	0.6	—	—
26	8	0.64	5.1	75.0	573.2	0.96	0	0.05	0.10	0.07	1.23	0.06	1.4	0	0.7	10.4
22	62	0.30	10.8	168.7	842.9	0.78	34.1	0.16	0.14	0.14	1.04	0.13	1.1	0	0.7	9.3
33	62	0.75	19.3	210.7	740.7	0.95	44.3	0.22	0.22	0.06	1.41	0.23	21.0	4.4	0.3	4.5
18	3	0.44	3.5	66.5	264.9	0.71	0	0.02	0.05	0.03	0.98	0.04	0.9	0	0.6	4.7
12	2	0.15	2.7	53.3	261.0	0.36	0	0.11	0.03	0.04	0.68	0.07	0.6	0	0.2	4.4
13	3	0.32	4.3	80.2	367.7	0.70	0	0.10	—	0.06	0.88	0.05	1.4	0	0.8	4.0
44	33	0.95	11.1	156.9	412.2	1.70	0	0.37	0.01	0.13	2.36	0.15	1.5	0.7	0.7	15.7
24	1	1.42	1.7	27.5	131.5	0.42	641.0	0.03	0.05	0.23	1.27	0.05	6.7	0	3.1	8.8
29	25	0.46	5.6	88.6	465.3	0.96	20.2	0.10	0.10	0.06	1.24	0.05	1.3	0	0.7	6.8
50	5	0.90	10.2	225.7	700.2	1.93	0	0.35	0.12	0.17	2.90	0.30	1.1	0	1.1	12.0
24	6	0.67	6.3	70.2	513.0	1.10	0	0.01	0.09	0.06	1.06	0.04	2.3	0	0.8	3.7
23	5	0.51	4.5	75.2	504.0	0.82	8.1	0.09	0.11	0.05	1.18	0.05	1.8	0	0.6	6.2
43	33	0.52	9.0	90.9	379.8	0.50	0	0.02	0.09	0.11	2.10	0.14	3.2	0	0.2	10.4
35	67	0.66	6.3	176.4	485.1	0.82	0	0.01	0.27	0.08	1.65	0.06	4.1	0	0.4	6.8
39	14	0.97	12.2	206.7	820.8	1.62	6.8	0.42	0.17	0.15	2.83	0.22	3.4	0.1	0.9	15.0
20	13	0.44	4.9	84.4	283.0	0.61	0	0.06	0.06	0.06	0.87	0.05	1.5	0	0.5	5.4
34	6	0.37	6.2	112.8	527.3	0.94	0	0.19	0.32	0.07	1.55	0.09	1.1	0	0.7	7.7
13	2	0.15	2.0	34.7	196.7	0.30	0	0.05	0.00	0.02	0.59	0.04	0.7	0.1	0.2	2.4
15	2	0.29	2.9	37.3	199.3	0.40	0	0.10	0.04	0.03	0.71	0.03	0.4	0.2	0.2	3.7
37	3	1.14	6.8	97.8	592.8	0.92	0	0.04	0.08	0.08	1.68	0.08	1.0	0	1.6	7.6
10	2	0.16	2.8	48.4	289.3	0.53	0	0.11	0.02	0.04	0.71	0.07	0.3	0	0.4	3.3
52	12	0.84	11.9	169.0	377.1	2.19	7.4	0.04	0.10	0.14	3.24	0.18	3.4	0.4	0.7	0
33	7	0.42	7.4	101.5	516.5	0.71	7.4	0.10	0.07	0.06	1.66	0.09	1.1	0	0.3	0
35	6	0.33	6.2	273.9	468.9	0.74	0	0.12	0.14	0.10	1.59	0.10	0.6	0	0.4	10.4
52	23	0.87	14.4	254.6	1136.6	2.13	0	0.53	0.18	0.19	3.43	0.26	3.8	1.5	1.2	16.4
42	5	1.15	7.9	147.4	737.1	1.45	0	0.08	0.12	0.18	2.44	0.14	1.1	9.4	3.1	11.5
14	2	0.14	1.1	16.2	155.0	0.25	0	0.01	0.03	0.01	0.25	0.01	0.6	0	0.2	2.7
22	5	0.35	6.0	90.0	547.2	0.66	0	0.26	1.04	0.07	1.25	0.09	0.6	0	0.5	10.7
23	7	0.47	4.8	66.0	607.8	0.61	15.6	0.10	1.04	0.08	1.03	0.07	1.2	0	0.7	5.8
10	0	0.00	—	—	340.0	—	0	—	—	—	—	—	—	0	—	—
10	0	0.00	—	—	300.0	—	0	—	—	—	—	—	—	0	—	—
25	0	0.72	—	—	720.0	—	0	—	—	—	—	—	—	0	—	—
20	0	0.72	—	—	660.0	—	0	—	—	—	—	—	—	0	—	—
20	0	0.72	—	—	660.0	—	0	—	—	—	—	—	—	0	—	—
4	0	0.06	—	—	103.3	—	0	—	—	—	—	—	—	0	—	—
40	0	0.72	12.3	162.4	700.0	1.44	0	—	—	—	—	—	—	0	—	—
19	3	1.19	4.0	97.8	278.1	0.61	1.1	0.01	0.06	0.07	1.00	0.07	1.4	4.6	0.1	4.6
43	23	0.70	12.5	122.5	569.3	1.31	1.1	0.24	0.13	0.17	2.25	0.24	5.7	0	0.6	15.0
22	11	0.35	6.2	61.2	284.6	0.65	0.6	0.12	0.06	0.08	1.12	0.12	2.8	0	0.3	7.5
0	18	0.07	21.6	90.0	14.4	0.03	0	0.03	0.00	0.10	1.62	0.18	21.6	0	0.1	2.5
0	14	0.07	21.4	96.2	14.3	0.03	0	0.01	0.00	0.08	1.82	0.16	21.4	0	0.1	2.1
0	25	0.21	25.2	28.8	46.8	0.07	—	0.07	0.00	0.18	3.99	0.10	50.4	1.8	0.1	4.3
0	18	0.14	17.8	63.9	9.0	0.10	0	0.03	—	0.10	1.39	0.12	14.6	0	0	4.0
0	18	0.10	21.3	88.8	9.0	0.07	0	0.02	—	0.09	1.60	0.17	21.3	0	0.1	4.0

TABLE A-1 Table of Food Composition *(continued)*

(Computer code is for Cengage Diet Analysis Plus program)

											Fat Breakdown (g)		
DA+ Code	Food Description	Qty	Measure	Wt (g)	H_2O (g)	Ener (cal)	Prot (g)	Carb (g)	Fiber (g)	Fat (g)	Sat	Mono	Poly
	Beverages—continued												
869	Light beer	12	fluid ounce(s)	354	335.9	103	0.9	5.8	0	0	0	0	0
31613	Michelob beer	12	fluid ounce(s)	355	323.4	155	1.3	13.3	0	0	0	0	0
31614	Michelob Light beer	12	fluid ounce(s)	355	329.8	134	1.1	11.7	0	0	0	0	0
	Gin, rum, vodka, whiskey												
857	Distilled alcohol, 100 proof	1	fluid ounce(s)	28	16.0	82	0	0	0	0	0	0	0
687	Distilled alcohol, 80 proof	1	fluid ounce(s)	28	18.5	64	0	0	0	0	0	0	0
688	Distilled alcohol, 86 proof	1	fluid ounce(s)	28	17.8	70	0	0	0	0	0	0	0
689	Distilled alcohol, 90 proof	1	fluid ounce(s)	28	17.3	73	0	0	0	0	0	0	0
856	Distilled alcohol, 94 proof	1	fluid ounce(s)	28	16.8	76	0	0	0	0	0	0	0
	Liqueurs												
33187	Coffee liqueur, 53 proof	1	fluid ounce(s)	35	10.8	113	0	16.3	0	0.1	0	0	0
3142	Coffee liqueur, 63 proof	1	fluid ounce(s)	35	14.4	107	0	11.2	0	0.1	0	0	0
736	Cordials, 54 proof	1	fluid ounce(s)	30	8.9	106	0	13.3	0	0.1	0	0	0
	Wine												
861	California red wine	5	fluid ounce(s)	150	133.4	125	0.3	3.7	0	0	0	0	0
858	Domestic champagne	5	fluid ounce(s)	150	—	105	0.3	3.8	0	0	0	0	0
690	Sweet dessert wine	5	fluid ounce(s)	147	103.7	235	0.3	20.1	0	0	0	0	0
1481	White wine	5	fluid ounce(s)	148	128.1	121	0.1	3.8	0	0	0	0	0
1811	Wine cooler	10	fluid ounce(s)	300	267.4	159	0.3	20.2	0	0.1	0	0	0
	Carbonated												
31898	7 Up	12	fluid ounce(s)	360	321.0	140	0	39.0	0	0	0	0	0
692	Club soda	12	fluid ounce(s)	355	354.8	0	0	0	0	0	0	0	0
12010	Coca-Cola Classic cola soda	12	fluid ounce(s)	360	319.4	146	0	40.5	0	0	0	0	0
693	Cola	12	fluid ounce(s)	368	332.7	136	0.3	35.2	0	0.1	0	0	0
2391	Cola or pepper-type soda, low calorie with saccharin	12	fluid ounce(s)	355	354.5	0	0	0.3	0	0	0	0	0
9522	Cola soda, decaffeinated	12	fluid ounce(s)	372	333.4	153	0	39.3	0	0	0	0	0
9524	Cola, decaffeinated, low calorie with aspartame	12	fluid ounce(s)	355	354.3	4	0.4	0.5	0	0	0	0	0
1415	Cola, low calorie with aspartame	12	fluid ounce(s)	355	353.6	7	0.4	1.0	0	0.1	0	0	0
1412	Cream soda	12	fluid ounce(s)	371	321.5	189	0	49.3	0	0	0	0	0
31899	Diet 7 Up	12	fluid ounce(s)	360	—	0	0	0	0	0	0	0	0
12031	Diet Coke cola soda	12	fluid ounce(s)	360	—	2	0	0.2	0	0	0	0	0
29392	Diet Mountain Dew soda	12	fluid ounce(s)	360	—	0	0	0	0	0	0	0	0
29389	Diet Pepsi cola soda	12	fluid ounce(s)	360	—	0	0	0	0	0	0	0	0
12034	Diet Sprite soda	12	fluid ounce(s)	360	—	4	0	0	0	0	0	0	0
695	Ginger ale	12	fluid ounce(s)	366	333.9	124	0	32.1	0	0	0	0	0
694	Grape soda	12	fluid ounce(s)	372	330.3	160	0	41.7	0	0	0	0	0
1876	Lemon lime soda	12	fluid ounce(s)	368	330.8	147	0.2	37.4	0	0.1	0	0	0
29391	Mountain Dew soda	12	fluid ounce(s)	360	314.0	170	0	46.0	0	0	0	0	0
3145	Orange soda	12	fluid ounce(s)	372	325.9	179	0	45.8	0	0	0	0	0
1414	Pepper-type soda	12	fluid ounce(s)	368	329.3	151	0	38.3	0	0.4	0.3	0	0
29388	Pepsi regular cola soda	12	fluid ounce(s)	360	318.9	150	0	41.0	0	0	0	0	0
696	Root beer	12	fluid ounce(s)	370	330.0	152	0	39.2	0	0	0	0	0
12044	Sprite soda	12	fluid ounce(s)	360	321.0	144	0	39.0	0	0	0	0	0
	Coffee												
731	Brewed	8	fluid ounce(s)	237	235.6	2	0.3	0	0	0	0	0	0
9520	Brewed, decaffeinated	8	fluid ounce(s)	237	234.3	5	0.3	1.0	0	0	0	0	0
16882	Cappuccino	8	fluid ounce(s)	240	224.8	79	4.1	5.8	0.2	4.9	2.3	1.0	0.2
16883	Cappuccino, decaffeinated	8	fluid ounce(s)	240	224.8	79	4.1	5.8	0.2	4.9	2.3	1.0	0.2
16880	Espresso	8	fluid ounce(s)	237	231.8	21	0	3.6	0	0.4	0.2	0	0.2
16881	Espresso, decaffeinated	8	fluid ounce(s)	237	231.8	21	0	3.6	0	0.4	0.2	0	0.2
732	Instant, prepared	8	fluid ounce(s)	239	236.5	5	0.2	0.8	0	0	0	0	0
	Fruit drinks												
29357	Crystal Light sugar-free lemonade drink	8	fluid ounce(s)	240	—	5	0	0	0	0	0	0	0
6012	Fruit punch drink with added vitamin C, canned	8	fluid ounce(s)	248	218.2	117	0	29.7	0.5	0	0	0	0
31143	Gatorade Thirst Quencher, all flavors	8	fluid ounce(s)	240	—	50	0	14.0	0	0	0	0	0
260	Grape drink, canned	8	fluid ounce(s)	250	210.5	153	0	39.4	0	0	0	0	0

Chol (mg)	Calc (mg)	Iron (mg)	Magn (mg)	Pota (mg)	Sodi (mg)	Zinc (mg)	Vit A (μg)	Thia (mg)	Vit E (mg α)	Ribo (mg)	Niac (mg)	Vit B_6 (mg)	Fola (μg DFE)	Vit C (mg)	Vit B_{12} (μg)	Sele (μg)
0	14	0.10	17.7	74.3	14.2	0.03	0	0.01	0.00	0.05	1.38	0.12	21.2	0	0.1	1.4
0	18	0.10	21.3	88.8	9.0	0.07	0	0.02	—	0.09	1.60	0.17	21.3	0	0.1	4.0
0	18	0.14	17.8	63.9	9.0	0.10	0	0.03	—	0.10	1.39	0.12	14.6	0	0	4.0
0	0	0.01	0	0.6	0.3	0.01	0	0.00	—	0.00	0.00	0.00	0	0	0	0
0	0	0.01	0	0.6	0.3	0.01	0	0.00	0.00	0.00	0.00	0.00	0	0	0	0
0	0	0.01	0	0.6	0.3	0.01	0	0.00	0.00	0.00	0.00	0.00	0	0	0	0
0	0	0.01	0	0.6	0.3	0.01	0	0.00	0.00	0.00	0.00	0.00	0	0	0	0
0	0	0.01	0	0.6	0.3	0.01	0	0.00	—	0.00	0.00	0.00	0	0	0	0
0	0	0.02	1.0	10.4	2.8	0.01	0	0.00	0.00	0.00	0.05	0.00	0	0	0	0.1
0	0	0.02	1.0	10.4	2.8	0.01	0	0.00	—	0.00	0.05	0.00	0	0	0	0.1
0	0	0.02	0.6	4.5	2.1	0.01	0	0.00	0.00	0.00	0.02	0.00	0	0	0	0.1
0	12	1.43	16.2	170.6	15.0	0.14	0	0.01	0.00	0.04	0.11	0.05	1.5	0	0	—
0	—	—	—	—	—	—	—	—	—	—	—	—	—	—	0	—
0	12	0.34	13.2	135.4	13.2	0.10	0	0.01	0.00	0.01	0.30	0.00	0	0	0	0.7
0	13	0.39	14.8	104.7	7.4	0.18	0	0.01	0.00	0.01	0.15	0.06	1.5	0	0	0.1
0	18	0.75	15.0	129.0	24.0	0.18	—	0.01	0.03	0.03	0.13	0.03	3.0	5.4	0	0.6
0	—	—	—	0.6	75.0	—	—	—	—	—	—	—	—	—	—	—
0	18	0.03	3.5	7.1	74.6	0.35	0	0.00	0.00	0.00	0.00	0.00	0	0	0	0
0	—	—	—	0	49.5	—	0	—	—	—	—	—	—	0	—	—
0	7	0.41	0	7.4	14.7	0.06	0	0.00	0.00	0.00	0.00	0.00	0	0	0	0.4
0	14	0.06	3.5	14.2	56.8	0.11	0	0.00	0.00	0.00	0.00	0.00	0	0	0	0.3
0	7	0.08	0	11.2	14.9	0.03	0	0.00	0.00	0.00	0.00	0.00	0	0	0	0.4
0	11	0.06	0	24.9	14.2	0.03	0	0.02	0.00	0.08	0.00	0.00	0	0	0	0.3
0	11	0.39	3.5	28.4	28.4	0.03	0	0.02	0.00	0.08	0.00	0.00	0	0	0	0
0	19	0.18	3.7	3.7	44.5	0.26	0	0.00	0.00	0.00	0.00	0.00	0	0	0	0
0	—	—	—	77.0	45.0	—	—	—	—	—	—	—	—	—	—	—
0	—	—	—	18.0	42.0	—	0	—	—	—	—	—	—	0	—	—
0	—	—	—	70.0	35.0	—	—	—	—	—	—	—	—	—	—	—
0	—	—	—	30.0	35.0	—	—	—	—	—	—	—	—	—	—	—
0	—	—	—	109.5	36.0	—	0	—	—	—	—	—	—	0	—	—
0	11	0.65	3.7	3.7	25.6	0.18	0	0.00	0.00	0.00	0.00	0.00	0	0	0	0.4
0	11	0.29	3.7	3.7	55.8	0.26	0	0.00	—	0.00	0.00	0.00	0	0	0	0
0	7	0.41	3.7	3.7	33.2	0.14	0	0.00	0.00	0.00	0.05	0.00	0	0	0	0
0	—	—	—	0	70.0	—	—	—	—	—	—	—	—	—	—	—
0	19	0.21	3.7	7.4	44.6	0.36	0	0.00	—	0.00	0.00	0.00	0	0	0	0
0	11	0.14	0	3.7	36.8	0.14	0	0.00	—	0.00	0.00	0.00	0	0	0	0.4
0	—	—	—	0	35.0	—	—	—	—	—	—	—	—	—	—	—
0	18	0.18	3.7	3.7	48.0	0.26	0	0.00	0.00	0.00	0.00	0.00	0	0	0	0.4
0	—	—	—	0	70.5	—	0	—	—	—	—	—	—	0	—	—
0	5	0.02	7.1	116.1	4.7	0.04	0	0.03	0.02	0.18	0.45	0.00	4.7	0	0	0
0	7	0.14	11.8	108.9	4.7	0.00	0	0.00	0.00	0.03	0.66	0.00	0	0	0	0.5
12	144	0.19	14.4	232.8	50.4	0.50	33.6	0.04	0.09	0.27	0.13	0.04	7.2	0	0.4	4.6
12	144	0.19	14.4	232.8	50.4	0.50	33.6	0.04	0.09	0.27	0.13	0.04	7.2	0	0.4	4.6
0	5	0.30	189.6	272.6	33.2	0.11	0	0.00	0.04	0.42	12.34	0.00	2.4	0.5	0	0
0	5	0.30	189.6	272.6	33.2	0.11	0	0.00	0.04	0.42	12.34	0.00	2.4	0.5	0	0
0	10	0.09	9.5	71.6	9.5	0.01	0	0.00	0.00	0.00	0.56	0.00	0	0	0	0.2
0	0	0.00	—	160.0	40.0	—	0	—	—	—	—	—	—	0	—	—
0	20	0.22	7.4	62.0	94.2	0.02	5.0	0.05	0.04	0.05	0.05	0.02	9.9	89.3	0	0.5
0	0	0.00	—	30.0	110.0	—	0	—	—	—	—	—	—	0	—	—
0	130	0.17	2.5	30.0	40.0	0.30	0	0.00	0.00	0.01	0.02	0.01	0	78.5	0	0.3

TABLE A-1 Table of Food Composition *(continued)*

(Computer code is for Cengage Diet Analysis Plus program)

DA+ Code	Food Description	QTY	Measure	Wt (g)	H_2O (g)	Ener (cal)	Prot (g)	Carb (g)	Fiber (g)	Fat (g)	Fat Breakdown (g) Sat	Mono	Poly
Beverages—continued													
17372	Kool-Aid (lemonade/punch/fruit drink)	8	fluid ounce(s)	248	220.0	108	0.1	27.8	0.2	0	0	0	0
17225	Kool-Aid sugar free, low calorie tropical punch drink mix, prepared	8	fluid ounce(s)	240	—	5	0	0	0	0	0	0	0
266	Lemonade, prepared from frozen concentrate	8	fluid ounce(s)	248	221.6	99	0.2	25.8	0	0.1	0	0	0
268	Limeade, prepared from frozen concentrate	8	fluid ounce(s)	247	212.6	128	0	34.1	0	0	0	0	0
14266	Odwalla strawberry C monster smoothie blend	8	fluid ounce(s)	240	—	160	2.0	38.0	0	0	0	0	0
10080	Odwalla strawberry lemonade quencher	8	fluid ounce(s)	240	—	110	0	28.0	0	0	0	0	0
10099	Snapple fruit punch fruit drink	8	fluid ounce(s)	240	—	110	0	29.0	0	0	0	0	0
10096	Snapple kiwi strawberry fruit drink	8	fluid ounce(s)	240	211.2	110	0	28.0	0	0	0	0	0
	Slim Fast ready-to-drink shake												
16054	French vanilla ready to drink shake	11	fluid ounce(s)	325	—	220	10.0	40.0	5.0	2.5	0.5	1.5	0.5
40447	Optima rich chocolate royal ready-to-drink shake	11	fluid ounce(s)	330	—	180	10.0	24.0	5.0	5.0	1.0	3.5	0.5
16055	Strawberries n cream ready to drink shake	11	fluid ounce(s)	325	—	220	10.0	40.0	5.0	2.5	0.5	1.5	0.5
	Tea												
33179	Decaffeinated, prepared	8	fluid ounce(s)	237	236.3	2	0	0.7	0	0	0	0	0
1877	Herbal, prepared	8	fluid ounce(s)	237	236.1	2	0	0.5	0	0	0	0	0
735	Instant tea mix, lemon flavored with sugar, prepared	8	fluid ounce(s)	259	236.2	91	0	22.3	0.3	0.2	0	0	0
734	Instant tea mix, unsweetened, prepared	8	fluid ounce(s)	237	236.1	2	0.1	0.4	0	0	0	0	0
733	Tea, prepared	8	fluid ounce(s)	237	236.3	2	0	0.7	0	0	0	0	0
	Water												
1413	Mineral water, carbonated	8	fluid ounce(s)	237	236.8	0	0	0	0	0	0	0	0
33183	Poland spring water, bottled	8	fluid ounce(s)	237	237.0	0	0	0	0	0	0	0	0
1821	Tap water	8	fluid ounce(s)	237	236.8	0	0	0	0	0	0	0	0
1879	Tonic water	8	fluid ounce(s)	244	222.3	83	0	21.5	0	0	0	0	0
Fats and Oils													
	Butter												
104	Butter	1	tablespoon(s)	14	2.3	102	0.1	0	0	11.5	7.3	3.0	0.4
2522	Butter Buds, dry butter substitute	1	teaspoon(s)	2	—	5	0	2.0	0	0	0	0	0
921	Unsalted	1	tablespoon(s)	14	2.5	102	0.1	0	0	11.5	7.3	3.0	0.4
107	Whipped	1	tablespoon(s)	9	1.5	67	0.1	0	0	7.6	4.7	2.2	0.3
944	Whipped, unsalted	1	tablespoon(s)	11	2.0	82	0.1	0	0	9.2	5.9	2.4	0.3
	Fats, cooking												
2671	Beef tallow, semisolid	1	tablespoon(s)	13	0	115	0	0	0	12.8	6.4	5.4	0.5
922	Chicken fat	1	tablespoon(s)	13	0	115	0	0	0	12.8	3.8	5.7	2.7
5454	Household shortening with vegetable oil	1	tablespoon(s)	13	0	115	0	0	0	13.0	3.4	5.5	2.7
111	Lard	1	tablespoon(s)	13	0	115	0	0	0	12.8	5.0	5.8	1.4
	Margarine												
114	Margarine	1	tablespoon(s)	14	2.3	101	0	0.1	0	11.4	2.1	5.5	3.4
5439	Soft	1	tablespoon(s)	14	2.3	103	0.1	0.1	0	11.6	1.7	4.4	2.1
32329	Soft, unsalted, with hydrogenated soybean and cottonseed oils	1	tablespoon(s)	14	2.5	101	0.1	0.1	0	11.3	2.0	5.4	3.5
928	Unsalted	1	tablespoon(s)	14	2.6	101	0.1	0.1	0	11.3	2.1	5.2	3.5
119	Whipped	1	tablespoon(s)	9	1.5	64	0.1	0.1	0	7.2	1.2	3.2	2.5
	Spreads												
54657	I Can't Believe It's Not Butter!, tub, soya oil (non-hydrogenated)	1	tablespoon(s)	14	2.3	103	0.1	0.1	0	11.6	2.8	2.0	5.1
2708	Mayonnaise with soybean and safflower oils	1	tablespoon(s)	14	2.1	99	0.2	0.4	0	11.0	1.2	1.8	7.6
16157	Promise vegetable oil spread, stick	1	tablespoon(s)	14	4.2	90	0	0	0	10.0	2.5	2.0	4.0

Chol (mg)	Calc (mg)	Iron (mg)	Magn (mg)	Pota (mg)	Sodi (mg)	Zinc (mg)	Vit A (µg)	Thia (mg)	Vit E (mg α)	Ribo (mg)	Niac (mg)	Vit B_6 (mg)	Fola (µg DFE)	Vit C (mg)	Vit B_{12} (µg)	Sele (µg)
0	14	0.45	5.0	49.6	31.0	0.19	—	0.03	—	0.05	0.04	0.01	4.3	41.6	0	1.0
0	0	0.00	—	10.1	10.1	—	0	—	—	—	—	—	—	6.0	—	—
0	10	0.39	5.0	37.2	9.9	0.05	0	0.01	0.02	0.05	0.04	0.01	2.5	9.7	0	0.2
0	5	0.00	4.9	24.7	7.4	0.02	0	0.01	0.00	0.01	0.02	0.01	2.5	7.7	0	0.2
0	20	0.72	—	0	20.0	—	0	—	—	—	—	—	—	600.0	0	—
0	0	0.00	—	70.0	10.0	—	0	—	—	—	—	—	—	54.0	0	—
0	0	0.00	—	20.0	10.0	—	0	—	—	—	—	—	—	0	0	—
0	0	0.00	—	40.0	10.0	—	0	—	—	—	—	—	—	0	0	—
5	400	2.70	140.0	600.0	220.0	2.25	—	0.52	—	0.59	7.00	0.70	120.0	60.0	2.1	17.5
5	1000	2.70	140.0	600.0	220.0	2.25	—	0.52	—	0.59	7.00	0.70	120.0	30.0	2.1	17.5
5	400	2.70	140.0	600.0	220.0	2.25	—	0.52	—	0.59	7.00	0.70	120.0	60.0	2.1	17.5
0	0	0.04	7.1	87.7	7.1	0.04	0	0.00	0.00	0.03	0.00	0.00	11.9	0	0	0
0	5	0.18	2.4	21.3	2.4	0.09	0	0.02	0.00	0.01	0.00	0.00	2.4	0	0	0
0	5	0.05	2.6	38.9	5.2	0.02	0	0.00	0.00	0.00	0.02	0.00	0	0	0	0.3
0	7	0.02	4.7	42.7	9.5	0.02	0	0.00	0.00	0.01	0.07	0.00	0	0	0	0
0	0	0.04	7.1	87.7	7.1	0.04	0	0.00	0.00	0.03	0.00	0.00	11.9	0	0	0
0	33	0.00	0	0	2.4	0.00	0	0.00	—	0.00	0.00	0.00	0	0	0	0
0	2	0.02	2.4	0	2.4	0.00	0	0.00	—	0.00	0.00	0.00	0	0	0	0
0	7	0.00	2.4	2.4	7.1	0.00	0	0.00	0.00	0.00	0.00	0.00	0	0	0	0
0	2	0.02	0	0	29.3	0.24	0	0.00	0.00	0.00	0.00	0.00	0	0	0	0
31	3	0.00	0.3	3.4	81.8	0.01	97.1	0.00	0.32	0.01	0.01	0.00	0.4	0	0	0.1
0	0	0.00	0	1.6	120.0	0.00	0	0.00	0.00	0.00	0.00	0.00	0	0	0	—
31	3	0.00	0.3	3.4	1.6	0.01	97.1	0.00	0.32	0.01	0.01	0.00	0.4	0	0	0.1
21	2	0.01	0.2	2.4	77.7	0.01	64.3	0.00	0.21	0.00	0.00	0.00	0.3	0	0	0.1
25	3	0.00	0.2	2.7	1.3	0.01	78.0	0.00	0.26	0.00	0.00	0.00	0.3	0	0	0.1
14	0	0.00	0	0	0	0.00	0	0.00	0.34	0.00	0.00	0.00	0	0	0	0
11	0	0.00	0	0	0	0.00	0	0.00	0.34	0.00	0.00	0.00	0	0	0	0
0	0	0.00	0	0	0	0.00	0	0.00	—	0.00	0.00	0.00	0	0	0	—
12	0	0.00	0	0	0	0.01	0	0.00	0.07	0.00	0.00	0.00	0	0	0	0
0	4	0.01	0.4	5.9	133.0	0.00	115.5	0.00	1.26	0.01	0.00	0.00	0.1	0	0	0
0	4	0.00	0.3	5.5	155.4	0.00	142.7	0.00	1.00	0.00	0.00	0.00	0.1	0	0	0
0	4	0.00	0.3	5.4	3.9	0.00	103.1	0.00	0.98	0.00	0.00	0.00	0.1	0	0	0
0	2	0.00	0.3	3.5	0.3	0.00	115.5	0.00	1.80	0.00	0.00	0.00	0.1	0	0	0
0	2	0.00	0.2	3.4	97.1	0.00	73.7	0.00	0.45	0.00	0.00	0.00	0.1	0	0	0
0	4	0.00	0.3	5.5	155.3	0.00	142.6	0.00	0.72	0.00	0.00	0.00	0.1	0	0	0
8	2	0.06	0.1	4.7	78.4	0.01	11.6	0.00	3.03	0.00	0.00	0.08	1.1	0	0	0.2
0	10	0.18	—	8.7	90.0	—	—	0.00	—	0.00	0.00	—	—	0.6	—	

A

TABLE A-1 Table of Food Composition *(continued)*

(Computer code is for Cengage Diet Analysis Plus program)

											Fat Breakdown (g)		
DA+ Code	Food Description	QTY	Measure	Wt (g)	H_2O (g)	Ener (cal)	Prot (g)	Carb (g)	Fiber (g)	Fat (g)	Sat	Mono	Poly
Fats and Oils—continued													
	Oils												
2681	Canola	1	tablespoon(s)	14	0	120	0	0	0	13.6	1.0	8.6	3.8
120	Corn	1	tablespoon(s)	14	0	120	0	0	0	13.6	1.8	3.8	7.4
122	Olive	1	tablespoon(s)	14	0	119	0	0	0	13.5	1.9	9.9	1.4
124	Peanut	1	tablespoon(s)	14	0	119	0	0	0	13.5	2.3	6.2	4.3
2693	Safflower	1	tablespoon(s)	14	0	120	0	0	0	13.6	0.8	10.2	2.0
923	Sesame	1	tablespoon(s)	14	0	120	0	0	0	13.6	1.9	5.4	5.7
128	Soybean, hydrogenated	1	tablespoon(s)	14	0	120	0	0	0	13.6	2.0	5.8	5.1
130	Soybean, with soybean and cotton-seed oil	1	tablespoon(s)	14	0	120	0	0	0	13.6	2.4	4.0	6.5
2700	Sunflower	1	tablespoon(s)	14	0	120	0	0	0	13.6	1.8	6.3	5.0
357	**Pam original no stick cooking spray**	1	serving(s)	0	0.2	0	0	0	0	0	0	0	0
	Salad dressing												
132	Blue cheese	2	tablespoon(s)	30	9.7	151	1.4	2.2	0	15.7	3.0	3.7	8.3
133	Blue cheese, low calorie	2	tablespoon(s)	32	25.4	32	1.6	0.9	0	2.3	0.8	0.6	0.8
1764	Caesar	2	tablespoon(s)	30	10.3	158	0.4	0.9	0	17.3	2.6	4.1	9.9
29654	Creamy, reduced calorie, fat-free, cholesterol-free, sour cream and/or buttermilk and oil	2	tablespoon(s)	32	23.9	34	0.4	6.4	0	0.9	0.2	0.2	0.5
29617	Creamy, reduced calorie, sour cream and/or buttermilk and oil	2	tablespoon(s)	30	22.2	48	0.5	2.1	0	4.2	0.6	1.0	2.4
134	French	2	tablespoon(s)	32	11.7	146	0.2	5.0	0	14.3	1.8	2.7	6.7
135	French, low fat	2	tablespoon(s)	32	17.4	74	0.2	9.4	0.4	4.3	0.4	1.9	1.6
136	Italian	2	tablespoon(s)	29	16.6	86	0.1	3.1	0	8.3	1.3	1.9	3.8
137	Italian, diet	2	tablespoon(s)	30	25.4	23	0.1	1.4	0	1.9	0.1	0.7	0.5
139	Mayonnaise-type	2	tablespoon(s)	29	11.7	115	0.3	7.0	0	9.8	1.4	2.6	5.3
942	Oil and vinegar	2	tablespoon(s)	32	15.2	144	0	0.8	0	16.0	2.9	4.7	7.7
1765	Ranch	2	tablespoon(s)	30	11.6	146	0.1	1.6	0	15.8	2.3	5.2	7.6
3666	Ranch, reduced calorie	2	tablespoon(s)	30	20.5	62	0.1	2.2	0	6.1	1.1	1.8	2.9
940	Russian	2	tablespoon(s)	30	11.6	107	0.5	9.3	0.7	7.8	1.2	1.8	4.4
939	Russian, low calorie	2	tablespoon(s)	32	20.8	45	0.2	8.8	0.1	1.3	0.2	0.3	0.7
941	Sesame seed	2	tablespoon(s)	30	11.8	133	0.9	2.6	0.3	13.6	1.9	3.6	7.5
142	Thousand Island	2	tablespoon(s)	32	14.9	118	0.3	4.7	0.3	11.2	1.6	2.5	5.8
143	Thousand Island, low calorie	2	tablespoon(s)	30	18.2	61	0.3	6.7	0.4	3.9	0.2	1.9	0.8
	Sandwich spreads												
138	Mayonnaise with soybean oil	1	tablespoon(s)	14	2.1	99	0.1	0.4	0	11.0	1.6	2.7	5.8
140	Mayonnaise, low calorie	1	tablespoon(s)	16	10.0	37	0	2.6	0	3.1	0.5	0.7	1.7
141	Tartar sauce	2	tablespoon(s)	28	8.7	144	0.3	4.1	0.1	14.4	2.2	3.8	7.7
Sweets													
4799	**Butterscotch or caramel topping**	2	tablespoon(s)	41	13.1	103	0.6	27.0	0.4	0	0	0	0
	Candy												
1786	Almond Joy candy bar	1	item(s)	45	4.3	220	2.0	27.0	2.0	12.0	8.0	3.3	0.7
1785	Bit-O-Honey candy	6	item(s)	40	—	190	1.0	39.0	0	3.5	2.5	—	—
33375	Butterscotch candy	2	piece(s)	12	0.6	47	0	10.8	0	0.4	0.2	0.1	0
1701	Chewing gum, stick	1	item(s)	3	0.1	7	0	2.0	0.1	0	0	0	0
33378	Chocolate fudge with nuts, prepared	2	piece(s)	38	2.9	175	1.7	25.8	1.0	7.2	2.5	1.5	2.9
1787	Jelly beans	15	item(s)	43	2.7	159	0	39.8	0.1	0	0	0	0
1784	Kit Kat wafer bar	1	item(s)	42	0.8	210	3.0	27.0	0.5	11.0	7.0	3.5	0.3
4674	Krackel candy bar	1	item(s)	41	0.6	210	2.0	28.0	0.5	10.0	6.0	3.9	0.4
4934	Licorice	4	piece(s)	44	7.3	154	1.1	35.1	0	1.0	0	0.1	0
1780	Life Savers candy	1	item(s)	2	—	8	0	2.0	0	0	0	0	0
1790	Lollipop	1	item(s)	28	—	108	0	28.0	0	0	0	0	0
4679	M & Ms peanut chocolate candy, small bag	1	item(s)	49	0.9	250	5.0	30.0	2.0	13.0	5.0	5.4	2.1
1781	M & Ms plain chocolate candy, small bag	1	item(s)	48	0.8	240	2.0	34.0	1.0	10.0	6.0	3.3	0.3
4673	Milk chocolate bar, Symphony	1	item(s)	91	0.9	483	7.7	52.8	1.5	27.8	16.7	7.2	0.6

Chol (mg)	Calc (mg)	Iron (mg)	Magn (mg)	Pota (mg)	Sodi (mg)	Zinc (mg)	Vit A (μg)	Thia (mg)	Vit E (mg α)	Ribo (mg)	Niac (mg)	Vit B_6 (mg)	Fola (μg DFE)	Vit C (mg)	Vit B_{12} (μg)	Sele (μg)
0	0	0.00	0	0	0	0.00	0	0.00	2.37	0.00	0.00	0.00	0	0	0	0
0	0	0.00	0	0	0	0.00	0	0.00	1.94	0.00	0.00	0.00	0	0	0	0
0	0	0.07	0	0.1	0.3	0.00	0	0.00	1.93	0.00	0.00	0.00	0	0	0	0
0	0	0.00	0	0	0	0.00	0	0.00	2.11	0.00	0.00	0.00	0	0	0	0
0	0	0.00	0	0	0	0.00	0	0.00	4.63	0.00	0.00	0.00	0	0	0	0
0	0	0.00	0	0	0	0.00	0	0.00	0.19	0.00	0.00	0.00	0	0	0	0
0	0	0.00	0	0	0	0.00	0	0.00	1.10	0.00	0.00	0.00	0	0	0	0
0	0	0.00	0	0	0	0.00	0	0.00	1.64	0.00	0.00	0.00	0	0	0	0
0	0	0.00	0	0	0	0.00	0	0.00	5.58	0.00	0.00	0.00	0	0	0	0
0	0	0.00	0	0.3	1.5	0.01	0.1	0.00	0.00	0.00	0.00	0.00	0	0	0	0
5	24	0.06	0	11.1	328.2	0.08	20.1	0.00	1.80	0.03	0.03	0.01	7.8	0.6	0.1	0.3
0	28	0.16	2.2	1.6	384.0	0.08	—	0.01	0.08	0.03	0.01	0.01	1.0	0.1	0.1	0.5
1	7	0.05	0.6	8.7	323.4	0.03	0.6	0.00	1.56	0.00	0.01	0.00	0.9	0	0	0.5
0	12	0.08	1.6	42.6	320.0	0.05	0.3	0.00	0.21	0.01	0.01	0.01	1.9	0	0	0.5
0	2	0.03	0.6	10.8	306.9	0.01	—	0.00	0.71	0.00	0.01	0.01	0	0.1	0	0.5
0	8	0.25	1.6	21.4	267.5	0.09	7.4	0.01	1.60	0.01	0.06	0.00	0	0	0	0
0	4	0.27	2.6	34.2	257.3	0.06	8.6	0.01	0.09	0.01	0.14	0.01	0.6	0	0	0.5
0	2	0.18	0.9	14.1	486.3	0.03	0.6	0.00	1.47	0.01	0.00	0.01	0	0	0	0.6
2	3	0.19	1.2	25.5	409.8	0.05	0.3	0.00	0.06	0.00	0.00	0.02	0	0	0	2.4
8	4	0.05	0.6	2.6	209.0	0.05	6.2	0.00	0.60	0.01	0.00	0.01	1.8	0	0.1	0.5
0	0	0.00	0	2.6	0.3	0.00	0	0.00	1.46	0.00	0.00	0.00	0	0	0	0.5
1	4	0.03	1.2	8.4	354.0	0.01	5.4	0.00	1.84	0.01	0.00	0.00	0.3	0.1	0	0.1
0	5	0.01	1.5	8.4	413.7	0.01	0.9	0.00	0.72	0.01	0.00	0.00	0.3	0.1	0	0.1
0	6	0.20	3.0	51.9	282.3	0.06	13.2	0.01	0.98	0.01	0.16	0.02	1.5	1.4	0	0.5
2	6	0.18	0	50.2	277.8	0.02	0.6	0.00	0.12	0.00	0.00	0.00	1.0	1.9	0	0.5
0	6	0.18	0	47.1	300.0	0.02	0.6	0.00	1.50	0.00	0.00	0.00	0	0	0	0.5
8	5	0.37	2.6	34.2	276.2	0.08	4.5	0.46	1.28	0.01	0.13	0.00	0	0	0	0.5
0	5	0.27	2.1	60.6	249.3	0.05	4.8	0.01	0.30	0.01	0.13	0.00	0	0	0	0
5	1	0.03	0.1	1.7	78.4	0.02	11.2	0.01	0.72	0.01	0.00	0.08	0.7	0	0	0.2
4	0	0.00	0	1.6	79.5	0.01	0	0.00	0.32	0.00	0.00	0.00	0	0	0	0.3
8	6	0.20	0.8	10.1	191.5	0.05	20.2	0.00	0.97	0.00	0.01	0.07	2.0	0.1	0.1	0.5
0	22	0.08	2.9	34.4	143.1	0.07	11.1	0.01	—	0.03	0.01	0.01	0.8	0.1	0	0
0	18	0.33	30.3	126.5	65.0	0.36	0	0.01	—	0.06	0.21	—	—	0	—	—
0	20	0.00	—	—	150.0	—	0	—	—	—	—	—	—	0	—	—
1	0	0.00	0	0.4	46.9	0.01	3.4	0.00	0.01	0.00	0.00	0.00	0	0	0	0.1
0	0	0.00	0	0.1	0	0.00	0	0.00	0.00	0.00	0.00	0.00	0	0	0	0
5	22	0.74	20.9	69.5	14.8	0.54	14.4	0.02	0.09	0.03	0.12	0.03	6.1	0.1	0	1.1
0	1	0.05	0.9	15.7	21.3	0.02	0	0.00	0.00	0.01	0.00	0.00	0	0	0	0.5
3	60	0.36	16.4	126.0	30.0	0.51	0	0.07	—	0.22	1.07	0.05	59.6	0	0.1	2.0
3	40	0.36	—	168.8	50.0	—	0	—	—	—	—	—	—	0	—	—
0	0	0.22	2.6	28.2	126.3	0.07	0	0.01	0.07	0.01	0.04	0.00	0	0	0	—
0	0	0.00	—	0	0	—	0	0.00	—	0.00	0.00	—	—	0	—	0
0	0	0.00	—	—	10.8	—	0	0.00	—	0.00	0.00	—	—	0	—	1.0
5	40	0.36	36.5	170.6	25.0	1.13	14.8	0.03	—	0.06	1.60	0.04	17.3	0.6	0.1	1.9
5	40	0.36	19.6	127.4	30.0	0.46	14.8	0.02	—	0.06	0.10	0.01	2.9	0.6	0.1	1.4
22	228	0.82	61.0	398.6	91.9	1.00	0	0.06	—	0.25	0.14	0.10	10.9	2.0	0.4	—

TABLE A-1 Table of Food Composition *(continued)*

(Computer code is for Cengage Diet Analysis Plus program)

DA+ Code	Food Description	QTY	Measure	Wt (g)	H_2O (g)	Ener (cal)	Prot (g)	Carb (g)	Fiber (g)	Fat (g)	Fat Breakdown (g) Sat	Mono	Poly
	Sweets—continued												
1783	Milky Way bar	1	item(s)	58	3.7	270	2.0	41.0	1.0	10.0	5.0	3.5	0.3
1788	Peanut brittle	1½	ounce(s)	43	0.3	207	3.2	30.3	1.1	8.1	1.8	3.4	1.9
1789	Reese's peanut butter cups	2	piece(s)	51	0.8	280	6.0	19.0	2.0	15.5	6.0	7.2	2.7
4689	Reese's pieces candy, small bag	1	item(s)	43	1.1	220	5.0	26.0	1.0	11.0	7.0	0.9	0.4
33399	Semisweet chocolate candy, made with butter	½	ounce(s)	14	0.1	68	0.6	9.0	0.8	4.2	2.5	1.4	0.1
1782	Snickers bar	1	item(s)	59	3.2	280	4.0	35.0	1.0	14.0	5.0	6.1	2.9
4694	Special Dark chocolate bar	1	item(s)	41	0.4	220	2.0	25.0	3.0	12.0	8.0	4.6	0.4
4695	Starburst fruit chews, original fruits	1	package(s)	59	3.9	240	0	48.0	0	5.0	1.0	2.1	1.8
4698	Taffy	3	piece(s)	45	2.2	179	0	41.2	0	1.5	0.9	0.4	0.1
4699	Three Musketeers bar	1	item(s)	60	3.5	260	2.0	46.0	1.0	8.0	4.5	2.6	0.3
4702	Twix caramel cookie bars	2	item(s)	58	2.4	280	3.0	37.0	1.0	14.0	5.0	7.7	0.5
4705	York peppermint pattie	1	item(s)	39	3.9	160	0.5	32.0	0.5	3.0	1.5	1.2	0.1
	Frosting, icing												
4760	Chocolate frosting, ready to eat	2	tablespoon(s)	31	5.2	122	0.3	19.4	0.3	5.4	1.7	2.8	0.6
4771	Creamy vanilla frosting, ready to eat	2	tablespoon(s)	28	4.2	117	0	19.0	0	4.5	0.8	1.4	2.2
17291	Dec-A-Cake variety pack candy decoration	1	teaspoon(s)	4	—	15	0	3.0	0	0.5	0	—	—
536	White icing	2	tablespoon(s)	40	3.6	162	0.1	31.8	0	4.2	0.8	2.0	1.2
	Gelatin												
13697	Gelatin snack, all flavors	1	item(s)	99	96.8	70	1.0	17.0	0	0	0	0	0
2616	Sugar free, low calorie mixed fruit gelatin mix, prepared	½	cup(s)	121	—	10	1.0	0	0	0	0	0	0
548	**Honey**	1	tablespoon(s)	21	3.6	64	0.1	17.3	0	0	0	0	0
	Jams, jellies												
550	Jam or preserves	1	tablespoon(s)	20	6.1	56	0.1	13.8	0.2	0	0	0	0
42199	Jams, preserves, dietetic, all flavors, w/sodium saccharin	1	tablespoon(s)	14	6.4	18	0	7.5	0.4	0	0	0	0
552	Jelly	1	tablespoon(s)	21	6.3	56	0	14.7	0.2	0	0	0	0
545	**Marshmallows**	4	item(s)	29	4.7	92	0.5	23.4	0	0.1	0	0	0
4800	**Marshmallow cream topping**	2	tablespoon(s)	40	7.9	129	0.3	31.6	0	0.1	0	0	0
555	**Molasses**	1	tablespoon(s)	20	4.4	58	0	14.9	0	0	0	0	0
4780	**Popsicle or ice pop**	1	item(s)	59	47.5	47	0	11.3	0	0.1	0	0	0
	Sugar												
559	Brown sugar, packed	1	teaspoon(s)	5	0.1	17	0	4.5	0	0	0	0	0
563	Powdered sugar, sifted	⅓	cup(s)	33	0.1	130	0	33.2	0	0	0	0	0
561	White granulated sugar	1	teaspoon(s)	4	0	16	0	4.2	0	0	0	0	0
	Sugar substitute												
1760	Equal sweetener, packet size	1	item(s)	1	—	0	0	0.9	0	0	0	0	0
13029	Splenda granular no calorie sweetener	1	teaspoon(s)	1	—	0	0	0.5	0	0	0	0	0
1759	Sweet N Low sugar substitute, packet	1	item(s)	1	0.1	4	0	0.5	0	0	0	0	0
	Syrup												
3148	Chocolate syrup	2	tablespoon(s)	38	11.6	105	0.8	24.4	1.0	0.4	0.2	0.1	0
29676	Maple syrup	¼	cup(s)	80	25.7	209	0	53.7	0	0.2	0	0.1	0.1
4795	Pancake syrup	¼	cup(s)	80	30.4	187	0	49.2	0	0	0	0	0
Spices, Condiments, Sauces													
	Spices												
807	Allspice, ground	1	teaspoon(s)	2	0.2	5	0.1	1.4	0.4	0.2	0	0	0
1171	Anise seeds	1	teaspoon(s)	2	0.2	7	0.4	1.1	0.3	0.3	0	0.2	0.1
729	Bakers' yeast, active	1	teaspoon(s)	4	0.3	12	1.5	1.5	0.8	0.2	0	0.1	0
683	Baking powder, double acting with phosphate	1	teaspoon(s)	5	0.2	2	0	1.1	0	0	0	0	0
1611	Baking soda	1	teaspoon(s)	5	0	0	0	0	0	0	0	0	0
8552	Basil	1	teaspoon(s)	1	0.8	0	0	0	0	0	0	0	0
34959	Basil, fresh	1	piece(s)	1	0.5	0	0	0	0	0	0	0	0

Chol (mg)	Calc (mg)	Iron (mg)	Magn (mg)	Pota (mg)	Sodi (mg)	Zinc (mg)	Vit A (μg)	Thia (mg)	Vit E (mg α)	Ribo (mg)	Niac (mg)	Vit B_6 (mg)	Fola (μg DFE)	Vit C (mg)	Vit B_{12} (μg)	Sele (μg)
5	60	0.18	19.8	140.1	95.0	0.41	15.1	0.02	—	0.06	0.20	0.02	5.8	0.6	0.2	3.3
5	11	0.51	17.9	71.4	189.2	0.37	16.6	0.05	1.08	0.01	1.12	0.03	19.6	0	0	1.1
3	40	0.72	45.4	217.4	180.0	0.93	0	0.12	—	0.08	2.35	0.07	28.1	0	0.1	2.3
0	20	0.00	18.9	169.9	80.0	0.32	0	0.04	—	0.06	1.22	0.03	12.0	0	0.1	0.8
3	5	0.44	16.3	51.7	1.6	0.23	0.4	0.01	—	0.01	0.06	0.01	0.4	0	0	0.5
5	40	0.36	42.3	—	140.0	1.37	15.3	0.03	—	0.06	1.60	0.05	23.5	0.6	0.1	2.7
0	0	1.80	45.5	136.0	50.0	0.59	0	0.01	—	0.02	0.16	0.01	0.8	0	0	1.2
0	10	0.18	0.6	1.2	0	0.00	—	0.00	—	0.00	0.00	0.00	0	30	0	0.5
4	4	0.00	0	1.4	23.4	0.09	12.2	0.01	0.04	0.01	0.00	0.00	0	0	0	0.3
5	20	0.36	17.5	80.3	110.0	0.33	14.5	0.01	—	0.03	0.20	0.01	0	0.6	0.1	1.5
5	40	0.36	18.5	116.8	115.0	0.45	15.0	0.09	—	0.13	0.69	0.01	13.9	0.6	0.1	1.2
0	0	0.33	23.4	66.1	10.0	0.28	0	0.01	—	0.03	0.31	0.01	1.5	0	0	—
0	2	0.44	6.4	60.0	56.1	0.09	0	0.00	0.48	0.00	0.03	0.00	0.3	0	0	0.2
0	1	0.04	0.3	9.5	51.5	0.01	0	0.00	0.43	0.08	0.06	0.00	2.2	0	0	0
0	0	0.00	—	—	15.0	—	0	—	—	—	—	—	—	0	—	—
0	4	0.01	0.4	5.6	76.4	0.01	44.4	0.00	0.32	0.01	0.00	0.00	0	0	0	0.3
0	0	0.00	—	0	40.0	—	0	—	—	—	—	—	—	0	—	—
0	0	0.00	0	0	50.0	0.00	0	0.00	0.00	0.00	0.00	0.00	0	0	0	—
0	1	0.08	0.4	10.9	0.8	0.04	0	0.00	0.00	0.01	0.02	0.01	0.4	0.1	0	0.2
0	4	0.10	0.8	15.4	6.4	0.01	0	0.00	0.02	0.02	0.01	0.00	2.2	1.8	0	0.4
0	1	0.56	0.7	9.7	0	0.01	0	0.00	0.01	0.00	0.00	0.00	1.3	0	0	0.2
0	1	0.04	1.3	11.3	6.3	0.01	0	0.00	0.00	0.01	0.01	0.00	0.4	0.2	0	0.1
0	1	0.06	0.6	1.4	23.0	0.01	0	0.00	0.00	0.00	0.02	0.00	0.3	0	0	0.5
0	1	0.08	0.8	2.0	32.0	0.01	0	0.00	0.00	0.00	0.03	0.00	0.4	0	0	0.7
0	41	0.94	48.4	292.8	7.4	0.05	0	0.01	0.00	0.00	0.18	0.13	0	0	0	3.6
0	0	0.31	0.6	8.9	4.1	0.08	0	0.00	0.00	0.00	0.00	0.00	0	0.4	0	0.1
0	4	0.03	0.4	6.1	1.3	0.00	0	0.00	0.00	0.00	0.01	0.00	0	0	0	0.1
0	0	0.01	0	0.7	0.3	0.00	0	0.00	0.00	0.00	0.00	0.00	0	0	0	0.2
0	0	0.00	0	0.1	0	0.00	0	0.00	0.00	0.00	0.00	0.00	0	0	0	0
0	0	0.00	0	0	0	0.00	0	0.00	0.00	0.00	0.00	0.00	0	0	0	0
0	0	0.00	—	—	0	—	—	0.00	—	0.00	0.00	—	—	0	0	—
0	0	0.00	—	—	0	—	0	—	0.00	—	—	—	—	0	—	—
0	5	0.79	24.4	84.0	27.0	0.27	0	0.00	0.01	0.01	0.12	0.00	0.8	0.1	0	0.5
0	54	0.96	11.2	163.2	7.2	3.32	0	0.01	0.00	0.01	0.02	0.00	0	0	0	0.5
0	2	0.02	1.6	12.0	65.6	0.06	0	0.01	0.00	0.01	0.00	0.00	0	0	0	0
0	13	0.13	2.6	19.8	1.5	0.01	0.5	0.00	—	0.00	0.05	0.00	0.7	0.7	0	0.1
0	14	0.77	3.6	30.3	0.3	0.11	0.3	0.01	—	0.01	0.06	0.01	0.2	0.4	0	0.1
0	3	0.66	3.9	80.0	2.0	0.25	0	0.09	0.00	0.21	1.59	0.06	93.6	0	0	1.0
0	339	0.51	1.8	0.2	363.1	0.00	0	0.00	0.00	0.00	0.00	0.00	0	0	0	0
0	0	0.00	0	0	1258.6	0.00	0	0.00	0.00	0.00	0.00	0.00	0	0	0	0
0	2	0.02	0.6	2.6	0	0.01	2.3	0.00	0.01	0.00	0.01	0.00	0.6	0.2	0	0
0	1	0.01	0.4	2.3	0	0.00	1.3	0.00	—	0.00	0.00	0.00	0.3	0.1	0	0

TABLE A-1 Table of Food Composition *(continued)*

(Computer code is for Cengage Diet Analysis Plus program)

											Fat Breakdown (g)		
DA+ Code	Food Description	Qty	Measure	Wt (g)	H_2O (g)	Ener (cal)	Prot (g)	Carb (g)	Fiber (g)	Fat (g)	Sat	Mono	Poly
Spices, Condiments, Sauces—continued													
808	Basil, ground	1	teaspoon(s)	1	0.1	4	0.2	0.9	0.6	0.1	0	0	0
809	Bay leaf	1	teaspoon(s)	1	0	2	0	0.5	0.2	0.1	0	0	0
11720	Betel leaves	1	ounce(s)	28	—	17	1.8	2.4	0	0	—	—	—
730	Brewers' yeast	1	teaspoon(s)	3	0.1	8	1.0	1.0	0.8	0	0	0	0
11710	Capers	1	teaspoon(s)	5	—	0	0	0	0	0	0	0	0
1172	Caraway seeds	1	teaspoon(s)	2	0.2	7	0.4	1.0	0.8	0.3	0	0.2	0.1
1173	Celery seeds	1	teaspoon(s)	2	0.1	8	0.4	0.8	0.2	0.5	0	0.3	0.1
1174	Chervil, dried	1	teaspoon(s)	1	0	1	0.1	0.3	0.1	0	0	0	0
810	Chili powder	1	teaspoon(s)	3	0.2	8	0.3	1.4	0.9	0.4	0.1	0.1	0.2
8553	Chives, chopped	1	teaspoon(s)	1	0.9	0	0	0	0	0	0	0	0
51420	Cilantro (coriander)	1	teaspoon(s)	0	0.3	0	0	0	0	0	0	0	0
811	Cinnamon, ground	1	teaspoon(s)	2	0.2	6	0.1	1.9	1.2	0	0	0	0
812	Cloves, ground	1	teaspoon(s)	2	0.1	7	0.1	1.3	0.7	0.4	0.1	0	0.1
1175	Coriander leaf, dried	1	teaspoon(s)	1	0	2	0.1	0.3	0.1	0	0	0	0
1176	Coriander seeds	1	teaspoon(s)	2	0.2	5	0.2	1.0	0.8	0.3	0	0.2	0
1706	Cornstarch	1	tablespoon(s)	8	0.7	30	0	7.3	0.1	0	0	0	0
1177	Cumin seeds	1	teaspoon(s)	2	0.2	8	0.4	0.9	0.2	0.5	0	0.3	0.1
11729	Cumin, ground	1	teaspoon(s)	5	—	11	0.4	0.8	0.8	0.4	—	—	—
1178	Curry powder	1	teaspoon(s)	2	0.2	7	0.3	1.2	0.7	0.3	0	0.1	0.1
1179	Dill seeds	1	teaspoon(s)	2	0.2	6	0.3	1.2	0.4	0.3	0	0.2	0
1180	Dill weed, dried	1	teaspoon(s)	1	0.1	3	0.2	0.6	0.1	0	0	0	0
34949	Dill weed, fresh	5	piece(s)	1	0.9	0	0	0.1	0	0	0	0	0
4949	Fennel leaves, fresh	1	teaspoon(s)	1	0.9	0	0	0.1	0	0	—	—	—
1181	Fennel seeds	1	teaspoon(s)	2	0.2	7	0.3	1.0	0.8	0.3	0	0.2	0
1182	Fenugreek seeds	1	teaspoon(s)	4	0.3	12	0.9	2.2	0.9	0.2	0.1	—	—
11733	Garam masala, powder	1	ounce(s)	28	—	107	4.4	12.8	0	4.3	—	—	—
1067	Garlic clove	1	item(s)	3	1.8	4	0.2	1.0	0.1	0	0	0	0
813	Garlic powder	1	teaspoon(s)	3	0.2	9	0.5	2.0	0.3	0	0	0	0
1068	Ginger root	2	teaspoon(s)	4	3.1	3	0.1	0.7	0.1	0	0	0	0
1183	Ginger, ground	1	teaspoon(s)	2	0.2	6	0.2	1.3	0.2	0.1	0	0	0
35497	Leeks, bulb and lower-leaf, freeze-dried	¼	cup(s)	1	0	3	0.1	0.6	0.1	0	0	0	0
1184	Mace, ground	1	teaspoon(s)	2	0.1	8	0.1	0.9	0.3	0.6	0.2	0.2	0.1
1185	Marjoram, dried	1	teaspoon(s)	1	0	2	0.1	0.4	0.2	0	0	0	0
1186	Mustard seeds, yellow	1	teaspoon(s)	3	0.2	15	0.8	1.2	0.5	0.9	0	0.7	0.2
814	Nutmeg, ground	1	teaspoon(s)	2	0.1	12	0.1	1.1	0.5	0.8	0.6	0.1	0
2747	Onion flakes, dehydrated	1	teaspoon(s)	2	0.1	6	0.1	1.4	0.2	0	0	0	0
1187	Onion powder	1	teaspoon(s)	2	0.1	7	0.2	1.7	0.1	0	0	0	0
815	Oregano, ground	1	teaspoon(s)	2	0.1	5	0.2	1.0	0.6	0.2	0	0	0.1
816	Paprika	1	teaspoon(s)	2	0.2	6	0.3	1.2	0.8	0.3	0	0	0.2
817	Parsley, dried	1	teaspoon(s)	0	0	1	0.1	0.2	0.1	0	0	0	0
818	Pepper, black	1	teaspoon(s)	2	0.2	5	0.2	1.4	0.6	0.1	0	0	0
819	Pepper, cayenne	1	teaspoon(s)	2	0.1	6	0.2	1.0	0.5	0.3	0.1	0	0.2
1188	Pepper, white	1	teaspoon(s)	2	0.3	7	0.3	1.6	0.6	0.1	0	0	0
1189	Poppy seeds	1	teaspoon(s)	3	0.2	15	0.5	0.7	0.3	1.3	0.1	0.2	0.9
1190	Poultry seasoning	1	teaspoon(s)	2	0.1	5	0.1	1.0	0.2	0.1	0	0	0
1191	Pumpkin pie spice, powder	1	teaspoon(s)	2	0.1	6	0.1	1.2	0.3	0.2	0.1	0	0
1192	Rosemary, dried	1	teaspoon(s)	1	0.1	4	0.1	0.8	0.5	0.2	0.1	0	0
11723	Rosemary, fresh	1	teaspoon(s)	1	0.5	1	0	0.1	0.1	0	0	0	0
2722	Saffron powder	1	teaspoon(s)	1	0.1	2	0.1	0.5	0	0	0	0	0
11724	Sage	1	teaspoon(s)	1	—	1	0	0.1	0	0	—	—	—
1193	Sage, ground	1	teaspoon(s)	1	0.1	2	0.1	0.4	0.3	0.1	0	0	0
30189	Salt substitute	¼	teaspoon(s)	1	—	0	0	0	0	0	0	0	0
30190	Salt substitute, seasoned	¼	teaspoon(s)	1	—	1	0	0.1	0	0	0	—	—
822	Salt, table	¼	teaspoon(s)	2	0	0	0	0	0	0	0	0	0
1194	Savory, ground	1	teaspoon(s)	1	0.1	4	0.1	1.0	0.6	0.1	0	—	—
820	Sesame seed kernels, toasted	1	teaspoon(s)	3	0.1	15	0.5	0.7	0.5	1.3	0.2	0.5	0.6
11725	Sorrel	1	teaspoon(s)	3	—	1	0.1	0.1	0	0	0	—	—
11721	Spearmint	1	teaspoon(s)	2	1.6	1	0.1	0.2	0.1	0	0	0	0
35498	Sweet green peppers, freeze-dried	¼	cup(s)	2	0	5	0.3	1.1	0.3	0	0	0	0
11726	Tamarind leaves	1	ounce(s)	28	—	33	1.6	5.2	0	0.6	—	—	—
11727	Tarragon	1	ounce(s)	28	—	14	1.0	1.8	0	0.3	—	—	—

Chol (mg)	Calc (mg)	Iron (mg)	Magn (mg)	Pota (mg)	Sodi (mg)	Zinc (mg)	Vit A (µg)	Thia (mg)	Vit E (mg α)	Ribo (mg)	Niac (mg)	Vit B_6 (mg)	Fola (µg DFE)	Vit C (mg)	Vit B_{12} (µg)	Sele (µg)
0	30	0.58	5.9	48.1	0.5	0.08	6.6	0.00	0.10	0.00	0.09	0.03	3.8	0.9	0	0
0	5	0.25	0.7	3.2	0.1	0.02	1.9	0.00	—	0.00	0.01	0.01	1.1	0.3	0	0
0	110	2.29	—	155.9	2.0	—	—	0.04	—	0.07	0.19	—	—	0.9	0	—
0	6	0.46	6.1	50.7	3.3	0.21	0	0.41	—	0.11	1.00	0.06	104.3	0	0	0
0	—	—	—	—	105.0	—	—	—	—	—	—	—	—	—	0	—
0	14	0.34	5.4	28.4	0.4	0.11	0.4	0.01	0.05	0.01	0.07	0.01	0.2	0.4	0	0.3
0	35	0.89	8.8	28.0	3.2	0.13	0.1	0.01	0.02	0.01	0.06	0.01	0.2	0.3	0	0.2
0	8	0.19	0.8	28.4	0.5	0.05	1.8	0.00	—	0.00	0.03	0.01	1.6	0.3	0	0.2
0	7	0.37	4.4	49.8	26.3	0.07	38.6	0.01	0.75	0.02	0.20	0.09	2.6	1.7	0	0.2
0	1	0.01	0.4	3.0	0	0.01	2.2	0.00	0.00	0.00	0.01	0.00	1.1	0.6	0	0
0	0	0.01	0.1	1.7	0.2	0.00	1.1	0.00	0.01	0.00	0.00	0.00	0.2	0.1	0	0
0	23	0.19	1.4	9.9	0.2	0.04	0.3	0.00	0.05	0.00	0.03	0.00	0.1	0.1	0	0.1
0	14	0.18	5.5	23.1	5.1	0.02	0.6	0.00	0.17	0.01	0.03	0.01	2.0	1.7	0	0.1
0	7	0.25	4.2	26.8	1.3	0.02	1.8	0.01	0.01	0.01	0.06	0.00	1.6	3.4	0	0.2
0	13	0.29	5.9	22.8	0.6	0.08	0	0.00	—	0.01	0.03	—	0	0.4	0	0.5
0	0	0.03	0.2	0.2	0.7	0.01	0	0.00	0.00	0.00	0.00	0.00	0	0	0	0.2
0	20	1.39	7.7	37.5	3.5	0.10	1.3	0.01	0.07	0.01	0.09	0.01	0.2	0.2	0	0.1
0	20	—	—	43.6	4.8	—	—	—	—	—	—	—	—	—	—	—
0	10	0.59	5.1	30.9	1.0	0.08	1.0	0.01	0.44	0.01	0.06	0.02	3.1	0.2	0	0.3
0	32	0.34	5.4	24.9	0.4	0.10	0.1	0.01	—	0.01	0.05	0.01	0.2	0.4	0	0.3
0	18	0.48	4.5	33.1	2.1	0.03	2.9	0.00	—	0.00	0.02	0.01	1.5	0.5	0	—
0	2	0.06	0.6	7.4	0.6	0.01	3.9	0.00	0.01	0.00	0.01	0.00	1.5	0.9	0	—
0	1	0.02	—	4.0	0.1	—	—	0.00	—	0.00	0.01	0.00	—	0.3	0	—
0	24	0.37	7.7	33.9	1.8	0.07	0.1	0.01	—	0.01	0.12	0.01	—	0.4	0	—
0	7	1.24	7.1	28.5	2.5	0.09	0.1	0.01	—	0.01	0.06	0.02	2.1	0.1	0	0.2
0	215	9.24	93.6	411.1	27.5	1.07	—	0.09	—	0.09	0.70	—	0	0	0	—
0	5	0.05	0.8	12.0	0.5	0.03	0	0.01	0.00	0.00	0.02	0.03	0.1	0.9	0	0.4
0	2	0.07	1.6	30.8	0.7	0.07	0	0.01	0.01	0.00	0.01	0.08	0.1	0.5	0	1.1
0	1	0.02	1.7	16.6	0.5	0.01	0	0.00	0.01	0.00	0.02	0.01	0.4	0.2	0	0
0	2	0.20	3.3	24.2	0.6	0.08	0.1	0.00	0.32	0.00	0.09	0.01	0.7	0.1	0	0.7
0	3	0.06	1.3	19.2	0.3	0.01	0.1	0.01	—	0.00	0.02	0.01	2.9	0.9	0	0
0	4	0.23	2.8	7.9	1.4	0.03	0.7	0.01	—	0.01	0.02	0.00	1.3	0.4	0	0
0	12	0.49	2.1	9.1	0.5	0.02	2.4	0.00	0.01	0.00	0.02	0.01	1.6	0.3	0	0
0	17	0.32	9.8	22.5	0.2	0.18	0.1	0.01	0.09	0.01	0.26	0.01	2.5	0.1	0	4.4
0	4	0.06	4.0	7.7	0.4	0.04	0.1	0.01	0.00	0.00	0.02	0.00	1.7	0.1	0	0
0	4	0.02	1.5	27.1	0.4	0.03	0	0.01	0.00	0.00	0.01	0.02	2.8	1.3	0	0.1
0	8	0.05	2.6	19.8	1.1	0.04	0	0.01	0.01	0.00	0.01	0.02	3.5	0.3	0	0
0	24	0.66	4.1	25.0	0.2	0.06	5.2	0.01	0.28	0.01	0.09	0.01	4.1	0.8	0	0.1
0	4	0.49	3.9	49.2	0.7	0.08	55.4	0.01	0.62	0.03	0.32	0.08	2.2	1.5	0	0.1
0	4	0.29	0.7	11.4	1.4	0.01	1.5	0.00	0.02	0.00	0.02	0.00	0.5	0.4	0	0.1
0	9	0.60	4.1	26.4	0.9	0.03	0.3	0.00	0.01	0.01	0.02	0.01	0.2	0.4	0	0.1
0	3	0.14	2.7	36.3	0.5	0.04	37.5	0.01	0.53	0.01	0.15	0.04	1.9	1.4	0	0.2
0	6	0.34	2.2	1.8	0.1	0.02	0	0.00	—	0.00	0.01	0.00	0.2	0.5	0	0.1
0	41	0.26	9.3	19.6	0.6	0.28	0	0.02	0.03	0.01	0.02	0.01	1.6	0.1	0	0
0	15	0.53	3.4	10.3	0.4	0.04	2.0	0.00	0.02	0.00	0.04	0.02	2.1	0.2	0	0.1
0	12	0.33	2.3	11.3	0.9	0.04	0.2	0.00	0.01	0.00	0.03	0.01	0.9	0.4	0	0.2
0	15	0.35	2.6	11.5	0.6	0.03	1.9	0.01	—	0.01	0.01	0.02	3.7	0.7	0	0.1
0	2	0.04	0.6	4.7	0.2	0.01	1.0	0.00	—	0.00	0.01	0.00	0.8	0.2	0	—
0	1	0.07	1.8	12.1	1.0	0.01	0.2	0.00	—	0.00	0.01	0.01	0.7	0.6	0	0
0	4	—	1.1	2.7	0	0.01	—	0.00	—	—	—	—	—	—	0	—
0	12	0.19	3.0	7.5	0.1	0.03	2.1	0.01	0.05	0.00	0.04	0.01	1.9	0.2	0	0
0	7	0.00	0	603.6	0.1	—	0	—	—	—	—	—	—	0	—	—
0	0	0.00	—	476.3	0.1	—	0	—	—	—	—	—	—	0	—	—
0	0	0.01	0	0.1	581.4	0.00	0	0.00	0.00	0.00	0.00	0.00	0	0	0	0
0	30	0.53	5.3	14.7	0.3	0.06	3.6	0.01	—	—	0.05	0.02	—	0.7	0	0.1
0	3	0.21	9.2	10.8	1.0	0.27	0.1	0.03	0.01	0.01	0.15	0.00	2.6	0	0	0
0	—	—	—	—	0.1	—	—	—	—	—	—	—	—	—	—	—
0	4	0.22	1.2	8.7	0.6	0.02	3.9	0.00	—	0.00	0.01	0.00	2.0	0.3	0	—
0	2	0.16	3.0	50.7	3.1	0.03	4.5	0.01	0.06	0.01	0.11	0.03	3.7	30.4	0	0.1
0	85	1.48	20.2	—	—	—	—	0.06	—	0.02	1.16	—	—	0.9	0	—
0	48	—	14.5	128.1	2.6	0.17	—	0.04	—	—	—	—	—	0.6	0	—

TABLE A-1 Table of Food Composition *(continued)*

(Computer code is for Cengage Diet Analysis Plus program)

DA+ Code	Food Description	QTY	Measure	Wt (g)	H₂O (g)	Ener (cal)	Prot (g)	Carb (g)	Fiber (g)	Fat (g)	Fat Breakdown (g) Sat	Mono	Poly
Spices, Condiments, Sauces—continued													
1195	Tarragon, ground	1	teaspoon(s)	2	0.1	5	0.4	0.8	0.1	0.1	0	0	0.1
11728	Thyme, fresh	1	teaspoon(s)	1	0.5	1	0	0.2	0.1	0	0	0	0
821	Thyme, ground	1	teaspoon(s)	1	0.1	4	0.1	0.9	0.5	0.1	0	0	0
1196	Turmeric, ground	1	teaspoon(s)	2	0.3	8	0.2	1.4	0.5	0.2	0.1	0	0
11995	Wasabi	1	tablespoon(s)	14	10.7	10	0.7	2.3	0.2	0	—	—	—
	Condiments												
674	Catsup or ketchup	1	tablespoon(s)	15	10.4	15	0.3	3.8	0	0	0	0	0
703	Dill pickle	1	ounce(s)	28	26.7	3	0.2	0.7	0.3	0	0	0	0
138	Mayonnaise with soybean oil	1	tablespoon(s)	14	2.1	99	0.1	0.4	0	11.0	1.6	2.7	5.8
140	Mayonnaise, low calorie	1	tablespoon(s)	16	10.0	37	0	2.6	0	3.1	0.5	0.7	1.7
1682	Mustard, brown	1	teaspoon(s)	5	4.1	5	0.3	0.3	0	0.3	—	—	—
700	Mustard, yellow	1	teaspoon(s)	5	4.1	3	0.2	0.3	0.2	0.2	0	0.1	0
706	Sweet pickle relish	1	tablespoon(s)	15	9.3	20	0.1	5.3	0.2	0.1	0	0	0
141	Tartar sauce	2	tablespoon(s)	28	8.7	144	0.3	4.1	0.1	14.4	2.2	3.8	7.7
	Sauces												
685	Barbecue sauce	2	tablespoon(s)	31	18.9	47	0	11.3	0.2	0.1	0	0	0.1
834	Cheese sauce	¼	cup(s)	63	44.4	110	4.2	4.3	0.3	8.4	3.8	2.4	1.6
32123	Chili enchilada sauce, green	2	tablespoon(s)	57	53.0	15	0.6	3.1	0.7	0.3	0	0	0.1
32122	Chili enchilada sauce, red	2	tablespoon(s)	32	24.5	27	1.1	5.0	2.1	0.8	0.1	0	0.4
29688	Hoisin sauce	1	tablespoon(s)	16	7.1	35	0.5	7.1	0.4	0.5	0.1	0.2	0.3
1641	Horseradish sauce, prepared	1	teaspoon(s)	5	3.3	10	0.1	0.2	0	1.0	0.6	0.3	0
16670	Mole poblano sauce	½	cup(s)	133	102.7	156	5.3	11.4	2.7	11.3	2.6	5.1	3.0
29689	Oyster sauce	1	tablespoon(s)	16	12.8	8	0.2	1.7	0	0	0	0	0
1655	Pepper sauce or Tabasco	1	teaspoon(s)	5	4.8	1	0.1	0	0	0	0	0	0
347	Salsa	2	tablespoon(s)	32	28.8	9	0.5	2.0	0.5	0.1	0	0	0
52206	Soy sauce, tamari	1	tablespoon(s)	18	12.0	11	1.9	1.0	0.1	0	0	0	0
839	Sweet and sour sauce	2	tablespoon(s)	39	29.8	37	0.1	9.1	0.1	0	0	0	0
1613	Teriyaki sauce	1	tablespoon(s)	18	12.2	16	1.1	2.8	0	0	0	0	0
25294	Tomato sauce	½	cup(s)	150	132.8	63	2.2	11.9	2.6	1.8	0.2	0.4	0.9
728	White sauce, medium	¼	cup(s)	63	46.8	92	2.4	5.7	0.1	6.7	1.8	2.8	1.8
1654	Worcestershire sauce	1	teaspoon(s)	6	4.5	4	0	1.1	0	0	0	0	0
	Vinegar												
30853	Balsamic	1	tablespoon(s)	15	—	10	0	2.0	0	0	0	0	0
727	Cider	1	tablespoon(s)	15	14.0	3	0	0.1	0	0	0	0	0
1673	Distilled	1	tablespoon(s)	15	14.3	2	0	0.8	0	0	0	0	0
12948	Tarragon	1	tablespoon(s)	15	13.8	2	0	0.1	0	0	0	0	0
Mixed Foods, Soups, Sandwiches													
	Mixed dishes												
16652	Almond chicken	1	cup(s)	242	186.8	281	21.8	15.8	3.4	14.7	1.8	6.3	5.6
25224	Barbecued chicken	1	serving(s)	177	99.3	327	27.1	15.7	0.5	17.1	4.8	6.8	3.8
25227	Bean burrito	1	item(s)	149	81.8	326	16.1	33.0	5.6	14.8	8.3	4.7	0.9
9516	Beef and vegetable fajita	1	item(s)	223	143.9	397	22.4	35.3	3.1	18.0	5.9	8.0	2.5
16796	Beef or pork egg roll	2	item(s)	128	85.2	225	9.9	18.4	1.4	12.4	2.9	6.0	2.6
177	Beef stew with vegetables, prepared	1	cup(s)	245	201.0	220	16.0	15.0	3.2	11.0	4.4	4.5	0.5
30233	Beef stroganoff with noodles	1	cup(s)	256	190.1	343	19.7	22.8	1.5	19.1	7.4	5.7	4.4
16651	Cashew chicken	1	cup(s)	242	186.8	281	21.8	15.8	3.4	14.7	1.8	6.3	5.6
30274	Cheese pizza with vegetables, thin crust	2	slice(s)	140	76.6	298	12.7	35.4	2.5	12.0	4.9	4.7	1.6
30330	Cheese quesadilla	1	item(s)	54	18.3	190	7.7	15.3	1.0	10.8	5.2	3.6	1.3
215	Chicken and noodles, prepared	1	cup(s)	240	170.0	365	22.0	26.0	1.3	18.0	5.1	7.1	3.9
30239	Chicken and vegetables with broccoli, onion, bamboo shoots in soy based sauce	1	cup(s)	162	125.5	180	15.8	9.3	1.8	8.6	1.7	3.0	3.1
25093	Chicken cacciatore	1	cup(s)	244	175.7	284	29.9	5.7	1.3	15.3	4.3	6.2	3.3
28020	Chicken fried turkey steak	3	ounce(s)	492	276.2	706	77.1	68.7	3.6	12.0	3.4	2.9	3.9
218	Chicken pot pie	1	cup(s)	252	154.6	542	22.6	41.4	3.5	31.3	9.8	12.5	7.1
30240	Chicken teriyaki	1	cup(s)	244	158.3	364	51.0	15.2	0.7	7.0	1.8	2.0	1.7
25119	Chicken waldorf salad	½	cup(s)	100	67.2	179	14.0	6.8	1.0	10.8	1.8	3.1	5.2
25099	Chili con carne	¾	cup(s)	215	174.4	198	13.7	21.4	7.5	6.9	2.5	2.8	0.5
1062	Coleslaw	¾	cup(s)	90	73.4	70	1.2	11.2	1.4	2.3	0.3	0.6	1.2

Chol (mg)	Calc (mg)	Iron (mg)	Magn (mg)	Pota (mg)	Sodi (mg)	Zinc (mg)	Vit A (μg)	Thia (mg)	Vit E (mg α)	Ribo (mg)	Niac (mg)	Vit B_6 (mg)	Fola (μg DFE)	Vit C (mg)	Vit B_{12} (μg)	Sele (μg)
0	18	0.51	5.6	48.3	1.0	0.06	3.4	0.00	—	0.02	0.14	0.03	4.4	0.8	0	0.1
0	3	0.14	1.3	4.9	0.1	0.01	1.9	0.00	—	0.00	0.01	0.00	0.4	1.3	0	—
0	26	1.73	3.1	11.4	0.8	0.08	2.7	0.01	0.10	0.01	0.06	0.01	3.8	0.7	0	0.1
0	4	0.91	4.2	55.6	0.8	0.09	0	0.00	0.06	0.01	0.11	0.04	0.9	0.6	0	0.1
0	13	0.11	—	—	—	—	—	0.02	—	0.01	0.07	—	—	11.2	0	—
0	3	0.07	2.9	57.3	167.1	0.03	7.1	0.00	0.21	0.02	0.21	0.02	1.5	2.3	0	0
0	12	0.10	2.0	26.1	248.1	0.03	2.6	0.01	0.02	0.01	0.03	0.01	0.3	0.2	0	0
5	1	0.03	0.1	1.7	78.4	0.02	11.2	0.01	0.72	0.01	0.00	0.08	0.7	0	0	0.2
4	0	0.00	0	1.6	79.5	0.01	0	0.00	0.32	0.00	0.00	0.00	0	0	0	0.3
0	6	0.09	1.0	6.8	68.1	0.01	0	0.00	0.09	0.00	0.01	0.00	0.2	0.1	0	—
0	3	0.07	2.5	6.9	56.8	0.03	0.2	0.01	0.01	0.00	0.02	0.00	0.4	0.1	0	1.6
0	0	0.13	0.8	3.8	121.7	0.02	9.2	0.00	0.08	0.01	0.03	0.00	0.2	0.2	0	0
8	6	0.20	0.8	10.1	191.5	0.05	20.2	0.00	0.97	0.00	0.01	0.07	2.0	0.1	0.1	0.5
0	4	0.06	3.8	65.0	349.7	0.04	3.8	0.00	0.20	0.01	0.15	0.01	0.6	0.2	0	0.4
18	116	0.13	5.7	18.9	521.6	0.61	50.4	0.00	—	0.07	0.01	0.01	2.5	0.3	0.1	2.0
0	5	0.36	9.5	125.7	61.9	0.11	—	0.02	0.00	0.02	0.63	0.06	5.7	43.9	0	0
0	7	1.05	11.1	231.3	113.8	0.14	—	0.01	0.00	0.21	0.61	0.34	6.6	0.3	0	0.3
0	5	0.16	3.8	19.0	258.4	0.05	0	0.00	0.04	0.03	0.18	0.01	3.7	0.1	0	0.3
2	5	0.00	0.5	6.7	14.6	0.01	8.0	0.00	0.02	0.01	0.00	0.00	0.5	0.1	0	0.1
1	38	1.81	58.3	280.9	304.8	1.15	13.3	0.06	1.72	0.08	1.84	0.09	15.9	3.4	0.1	1.1
0	5	0.02	0.6	8.6	437.3	0.01	0	0.00	0.00	0.02	0.23	0.00	2.4	0	0.1	0.7
0	1	0.05	0.6	6.4	31.7	0.01	4.1	0.00	0.00	0.00	0.01	0.01	0.1	0.2	0	0
0	9	0.14	4.8	95.0	192.0	0.11	4.8	0.01	0.37	0.01	0.02	0.05	1.3	0.6	0	0.3
0	4	0.43	7.3	38.7	1018.9	0.08	0	0.01	0.00	0.03	0.72	0.04	3.3	0	0	0.1
0	5	0.20	1.2	8.2	97.5	0.01	0	0.00	—	0.01	0.11	0.03	0.2	0	0	—
0	5	0.30	11.0	40.5	689.9	0.01	0	0.01	0.00	0.01	0.22	0.01	1.4	0	0	0.2
0	23	1.24	28.9	536.8	268.6	0.36	—	0.08	0.52	0.08	1.64	0.20	23.2	32.0	0	1.0
4	74	0.20	8.8	97.5	221.3	0.25	—	0.04	—	0.11	0.25	0.02	3.1	0.5	0.2	—
0	6	0.30	0.7	45.4	55.6	0.01	0.3	0.00	0.00	0.01	0.03	0.00	0.5	0.7	0	0
0	0	0.00	—	—	0	—	0	—	—	—	—	—	—	0	—	—
0	1	0.03	0.7	10.9	0.7	0.01	0	0.00	0.00	0.00	0.00	0.00	0	0	0	0
0	1	0.09	0	2.3	0.1	0.00	0	0.00	0.00	0.00	0.00	0.00	0	0	0	5.0
0	0	0.07	—	2.3	0.7	—	—	0.07	—	0.07	0.07	—	—	0.3	0	—
41	68	1.86	58.1	539.7	510.6	1.50	31.5	0.07	4.11	0.22	9.57	0.43	26.6	5.1	0.3	13.6
120	26	1.70	32.4	419.7	500.9	2.67	—	0.09	0.01	0.24	6.87	0.40	15.0	7.9	0.3	19.5
38	333	3.01	52.5	447.6	510.6	1.98	—	0.28	0.01	0.30	1.92	0.19	134.4	8.2	0.3	15.9
45	85	3.65	37.9	475.0	756.0	3.52	17.8	0.38	0.80	0.29	5.33	0.39	102.6	23.4	2.1	28.3
74	31	1.68	20.5	248.3	547.8	0.89	25.6	0.32	1.28	0.24	2.55	0.18	38.4	4.0	0.3	17.5
71	29	2.90	—	613.0	292.0	—	—	0.15	0.51	0.17	4.70	—	—	17.0	0	15.0
74	69	3.25	35.8	391.7	816.6	3.63	69.1	0.21	1.25	0.30	3.80	0.21	69.1	1.3	1.8	27.9
41	68	1.86	58.1	539.7	510.6	1.50	31.5	0.07	4.11	0.22	9.57	0.43	26.6	5.1	0.3	13.6
17	249	2.78	28.0	294.0	739.2	1.42	47.6	0.29	1.05	0.33	2.84	0.14	89.6	15.3	0.4	18.6
23	190	1.04	13.5	75.6	469.3	0.86	58.3	0.11	0.43	0.15	0.89	0.02	32.4	2.4	0.1	9.2
103	26	2.20	—	149.0	600.0	—	—	0.05	—	0.17	4.30	—	75.5	0	—	29.0
42	28	1.19	22.7	299.7	620.5	1.32	81.0	0.07	1.11	0.14	5.28	0.36	16.2	22.5	0.2	12.0
109	47	1.97	40.0	489.3	492.1	2.13	—	0.11	0.00	0.20	9.81	0.57	25.3	14.0	0.3	22.6
156	423	8.79	110.3	1182.9	880.4	6.18	—	0.72	0.00	1.05	20.16	1.20	180.1	2.7	1.3	97.6
68	66	3.32	37.8	390.6	652.7	1.94	259.6	0.39	1.05	0.39	7.25	0.23	113.4	10.3	0.2	27.0
156	51	3.26	68.3	588.0	3208.6	3.75	31.7	0.15	0.58	0.36	16.68	0.88	24.4	2.0	0.5	36.1
42	20	0.82	23.9	202.5	246.5	1.13	—	0.05	0.62	0.09	4.06	0.25	15.8	2.5	0.2	10.7
27	42	2.83	50.6	636.8	864.8	2.36	—	0.15	0.01	0.22	3.18	0.19	58.1	10.3	0.6	7.3
7	41	0.53	9.0	162.9	20.7	0.18	47.7	0.06	—	0.05	0.24	0.11	24.3	29.4	0	0.6

TABLE A-1 Table of Food Composition *(continued)*

(Computer code is for Cengage Diet Analysis Plus program)

											Fat Breakdown (g)		
DA+ Code	Food Description	QTY	Measure	Wt (g)	H_2O (g)	Ener (cal)	Prot (g)	Carb (g)	Fiber (g)	Fat (g)	Sat	Mono	Poly
	Mixed Foods, Soups, Sandwiches—continued												
1574	Crab cakes, from blue crab	1	item(s)	60	42.6	93	12.1	0.3	0	4.5	0.9	1.7	1.4
32144	Enchiladas with green chili sauce (enchiladas verdes)	1	item(s)	144	103.8	207	9.3	17.6	2.6	11.7	6.4	3.6	1.0
2793	Falafel patty	3	item(s)	51	17.7	170	6.8	16.2	—	9.1	1.2	5.2	2.1
28546	Fettuccine alfredo	1	cup(s)	244	88.7	279	13.1	46.1	1.4	4.2	2.2	1.0	0.4
32146	Flautas	3	item(s)	162	78.0	438	24.9	36.3	4.1	21.6	8.2	8.8	2.3
29629	Fried rice with meat or poultry	1	cup(s)	198	128.5	333	12.3	41.8	1.4	12.3	2.2	3.5	5.7
16649	General Tso chicken	1	cup(s)	146	91.0	296	18.7	16.4	0.9	17.0	4.0	6.3	5.3
1826	Green salad	¾	cup(s)	104	98.9	17	1.3	3.3	2.2	0.1	0	0	0
1814	Hummus	½	cup(s)	123	79.8	218	6.0	24.7	4.9	10.6	1.4	6.0	2.6
16650	Kung pao chicken	1	cup(s)	162	87.2	434	28.8	11.7	2.3	30.6	5.2	13.9	9.7
16622	Lamb curry	1	cup(s)	236	187.9	257	28.2	3.7	0.9	13.8	3.9	4.9	3.3
25253	Lasagna with ground beef	1	cup(s)	237	158.4	284	16.9	22.3	2.4	14.5	7.5	4.9	0.8
442	Macaroni and cheese, prepared	1	cup(s)	200	122.3	390	14.9	40.6	1.6	18.6	7.9	6.4	2.9
29637	Meat filled ravioli with tomato or meat sauce, canned	1	cup(s)	251	198.7	208	7.8	36.5	1.3	3.7	1.5	1.4	0.3
25105	Meat loaf	1	slice(s)	115	84.5	245	17.0	6.6	0.4	16.0	6.1	6.9	0.9
16646	Moo shi pork	1	cup(s)	151	76.8	512	18.9	5.3	0.6	46.4	6.9	15.8	21.2
16788	Nachos with beef, beans, cheese, tomatoes and onions	1	serving(s)	551	253.5	1576	59.1	137.5	20.4	90.8	32.6	41.9	9.4
6116	Pepperoni pizza	2	slice(s)	142	66.1	362	20.2	39.7	2.9	13.9	4.5	6.3	2.3
29601	Pizza with meat and vegetables, thin crust	2	slice(s)	158	81.4	386	16.5	36.8	2.7	19.1	7.7	8.1	2.2
655	Potato salad	½	cup(s)	125	95.0	179	3.4	14.0	1.6	10.3	1.8	3.1	4.7
25109	Salisbury steaks with mushroom sauce	1	serving(s)	135	101.8	251	17.1	9.3	0.5	15.5	6.0	6.7	0.8
16637	Shrimp creole with rice	1	cup(s)	243	176.6	309	27.0	27.7	1.2	9.2	1.7	3.6	2.9
497	Spaghetti and meatballs with tomato sauce, prepared	1	cup(s)	248	174.0	330	19.0	39.0	2.7	12.0	3.9	4.4	2.2
28585	Spicy thai noodles (pad thai)	8	ounce(s)	227	73.3	221	8.9	35.7	3.0	6.4	0.8	3.3	1.8
33073	Stir fried pork and vegetables with rice	1	cup(s)	235	173.6	348	15.4	33.5	1.9	16.3	5.6	6.9	2.6
28588	Stuffed shells	2½	item(s)	249	157.5	243	15.0	28.0	2.5	8.1	3.1	3.0	1.3
16821	Sushi with egg in seaweed	6	piece(s)	156	116.5	190	8.9	20.5	0.3	7.9	2.2	3.2	1.5
16819	Sushi with vegetables and fish	6	piece(s)	156	101.6	218	8.4	43.7	1.7	0.6	0.2	0.1	0.2
16820	Sushi with vegetables in seaweed	6	piece(s)	156	110.3	183	3.4	40.6	0.8	0.4	0.1	0.1	0.1
25266	Sweet and sour pork	¾	cup(s)	249	205.9	265	29.2	17.1	1.0	8.1	2.6	3.5	1.5
16824	Tabouli, tabbouleh or tabuli	1	cup(s)	160	123.7	198	2.6	15.9	3.7	14.9	2.0	10.9	1.6
25276	Three bean salad	½	cup(s)	99	82.2	95	1.9	9.7	2.6	5.9	0.8	1.4	3.5
160	Tuna salad	½	cup(s)	103	64.7	192	16.4	9.6	0	9.5	1.6	3.0	4.2
25241	Turkey and noodles	1	cup(s)	319	228.5	270	24.0	21.2	1.0	9.2	2.4	3.5	2.3
16794	Vegetable egg roll	2	item(s)	128	89.8	201	5.1	19.5	1.7	11.6	2.5	5.7	2.6
16818	Vegetable sushi, no fish	6	piece(s)	156	99.0	226	4.8	49.9	2.0	0.4	0.1	0.1	0.1
	Sandwiches												
1744	Bacon, lettuce and tomato with mayonnaise	1	item(s)	164	97.2	341	11.6	34.2	2.3	17.6	3.8	5.5	6.7
30287	Bologna and cheese with margarine	1	item(s)	111	45.6	345	13.4	29.3	1.2	19.3	8.1	7.0	2.4
30286	Bologna with margarine	1	item(s)	83	33.6	251	8.1	27.3	1.2	12.1	3.7	5.0	2.1
16546	Cheese	1	item(s)	83	31.0	261	9.1	27.6	1.2	12.7	5.4	4.2	2.1
8789	Cheeseburger, large, plain	1	item(s)	185	78.9	564	32.0	38.5	2.6	31.5	12.5	10.2	1.0
8624	Cheeseburger, large, with bacon, vegetables, and condiments	1	item(s)	195	91.4	550	30.8	36.8	2.5	30.9	11.9	10.6	1.3
1745	Club with bacon, chicken, tomato, lettuce, and mayonnaise	1	item(s)	246	137.5	546	31.0	48.9	3.0	24.5	5.3	7.5	9.4
1908	Cold cut submarine with cheese and vegetables	1	item(s)	228	131.8	456	21.8	51.0	2.0	18.6	6.8	8.2	2.3
30247	Corned beef	1	item(s)	130	74.9	265	18.2	25.3	1.6	9.6	3.6	3.5	1.0
25283	Egg salad	1	item(s)	126	72.1	278	10.7	28.0	1.4	13.5	2.9	4.2	5.0
16686	Fried egg	1	item(s)	96	49.7	226	10.0	26.2	1.2	8.6	2.3	3.2	1.9
16547	Grilled cheese	1	item(s)	83	27.5	291	9.2	27.9	1.2	15.8	6.0	5.7	3.0
16659	Gyro with onion and tomato	1	item(s)	105	68.3	163	12.0	20.0	1.1	3.5	1.3	1.3	0.5

Chol (mg)	Calc (mg)	Iron (mg)	Magn (mg)	Pota (mg)	Sodi (mg)	Zinc (mg)	Vit A (µg)	Thia (mg)	Vit E (mg α)	Ribo (mg)	Niac (mg)	Vit B_6 (mg)	Fola (µg DFE)	Vit C (mg)	Vit B_{12} (µg)	Sele (µg)
90	63	0.64	19.8	194.4	198.0	2.45	34.2	0.05	—	0.04	1.74	0.10	36.6	1.7	3.6	24.4
27	266	1.07	38.5	251.4	276.3	1.26	—	0.07	0.02	0.16	1.27	0.17	44.6	59.3	0.2	6.0
0	28	1.74	41.8	298.4	149.9	0.76	0.5	0.07	—	0.08	0.53	0.06	47.4	0.8	0	0.5
9	218	1.83	38.4	163.9	472.9	1.24	—	0.41	0.00	0.35	2.85	0.09	225.0	1.6	0.4	38.4
73	146	2.66	61.3	222.9	885.7	3.43	0	0.10	0.10	0.16	3.00	0.26	95.7	0	1.2	36.7
103	38	2.77	33.7	196.0	833.6	1.34	41.6	0.33	1.60	0.18	4.17	0.27	146.5	3.4	0.3	22.0
66	26	1.46	23.4	248.2	849.7	1.40	29.2	0.10	1.62	0.18	6.28	0.28	23.4	12.0	0.2	19.9
0	13	0.65	11.4	178.0	26.9	0.21	59.0	0.03	—	0.05	0.56	0.08	38.3	24.0	0	0.4
0	60	1.91	35.7	212.8	297.7	1.34	0	0.10	0.92	0.06	0.49	0.49	72.6	9.7	0	3.0
65	50	1.96	63.2	427.7	907.2	1.50	38.9	0.15	4.32	0.14	13.22	0.58	42.1	7.5	0.3	23.0
90	38	2.95	40.1	493.2	495.6	6.60	—	0.08	1.29	0.28	8.03	0.21	28.3	1.4	2.9	30.4
68	233	2.22	40.1	420.1	433.6	2.70	—	0.21	0.21	0.29	3.06	0.22	91.1	15.0	0.8	21.4
34	310	2.06	40.0	258.0	784.0	2.06	180.0	0.27	0.72	0.43	2.18	0.08	100.0	0	0.5	30.6
15	35	2.10	20.1	283.6	1352.9	1.28	27.6	0.19	0.70	0.16	2.77	0.14	60.2	21.6	0.4	13.3
85	59	1.87	21.8	300.8	411.7	3.40	—	0.08	0.00	0.27	3.72	0.13	18.7	0.9	1.6	17.9
172	32	1.57	25.7	333.7	1052.5	1.82	49.8	0.49	5.39	0.36	2.88	0.31	21.1	8.0	0.8	30.0
154	948	7.32	242.4	1201.2	1862.4	10.68	259.0	0.29	7.71	0.81	6.39	1.09	148.8	16.0	2.6	44.1
28	129	1.87	17.0	305.3	533.9	1.03	105.1	0.26	—	0.46	6.09	0.11	76.7	3.3	0.4	26.1
36	258	3.14	31.6	352.3	971.7	2.02	49.0	0.37	1.13	0.37	3.77	0.19	91.6	15.6	0.6	22.8
85	24	0.81	18.8	317.5	661.3	0.38	40.0	0.09	—	0.07	1.11	0.17	8.8	12.5	0	5.1
60	74	1.94	23.8	314.5	360.5	3.45	—	0.10	0.00	0.27	3.95	0.13	20.8	0.7	1.6	17.4
180	102	4.68	63.2	413.1	330.5	1.72	94.8	0.29	2.06	0.11	4.75	0.21	121.5	12.9	1.2	49.3
89	124	3.70	—	665.0	1009.0	—	81.5	0.25	—	0.30	4.00	—	—	22.0	—	22.0
37	31	1.56	49.4	181.3	591.6	1.05	—	0.18	0.35	0.13	1.82	0.17	55.8	22.6	0.1	3.2
46	38	2.71	33.0	396.9	569.5	2.08	—	0.51	0.38	0.20	5.07	0.30	162.4	18.8	0.4	22.8
30	188	2.26	49.4	403.0	471.5	1.41	—	0.26	0.00	0.26	3.83	0.24	161.2	17.9	0.2	28.9
214	45	1.84	18.7	135.7	463.3	0.98	106.1	0.13	0.67	0.28	1.35	0.13	82.7	1.9	0.7	20.3
11	23	2.15	25.0	202.8	340.1	0.78	45.2	0.26	0.24	0.07	2.76	0.14	121.7	3.6	0.3	13.9
0	20	1.54	18.7	96.7	152.9	0.68	25.0	0.19	0.12	0.03	1.86	0.13	118.6	2.3	0	9.8
74	40	1.76	35.6	619.9	621.8	2.53	—	0.81	0.20	0.37	6.69	0.66	14.7	11.9	0.7	49.6
0	30	1.21	35.2	249.6	796.8	0.48	54.4	0.07	2.43	0.04	1.11	0.11	30.4	26.1	0	0.5
0	26	0.96	15.5	144.8	224.2	0.30	—	0.02	0.88	0.04	0.26	0.04	32.2	10.0	0	2.7
13	17	1.02	19.5	182.5	412.1	0.57	24.6	0.03	—	0.07	6.86	0.08	8.2	2.3	1.2	42.2
77	69	2.56	33.2	400.8	577.1	2.51	—	0.23	0.28	0.30	6.41	0.29	109.5	1.4	1.1	33.4
60	29	1.65	17.9	193.3	549.1	0.48	25.6	0.15	1.28	0.20	1.59	0.09	46.1	5.5	0.2	11.3
0	23	2.38	21.8	157.6	369.7	0.82	48.4	0.28	0.15	0.05	2.44	0.12	135.7	3.7	0	8.1
21	79	2.36	27.9	351.0	944.6	1.08	44.3	0.32	1.16	0.24	4.36	0.21	105.0	9.7	0.2	27.1
40	258	2.38	24.4	215.3	941.3	1.88	102.1	0.31	0.55	0.35	2.97	0.14	91.0	0.2	0.8	20.4
17	100	2.24	16.6	138.6	579.3	1.02	44.8	0.29	0.49	0.21	2.92	0.12	88.8	0.2	0.5	16.0
22	233	2.04	19.9	127.0	733.7	1.24	97.1	0.25	0.47	0.30	2.25	0.05	88.8	0	0.3	13.1
104	309	4.47	44.4	401.5	986.1	5.75	0	0.35	—	0.77	8.26	0.49	129.48	0	2.8	38.9
98	267	4.03	44.9	464.1	1314.3	5.20	0	0.33	—	0.67	8.25	0.47	122.9	1.4	2.4	6.6
71	157	4.57	46.7	464.9	1087.3	1.82	41.8	0.54	1.52	0.40	12.82	0.61	172.2	6.4	0.4	42.3
36	189	2.50	68.4	394.4	1650.7	2.57	70.7	1.00	—	0.79	5.49	0.13	109.4	12.3	1.1	30.8
46	81	3.04	19.5	127.4	1206.4	2.26	2.6	0.23	0.20	0.24	3.42	0.11	88.4	0.3	0.9	31.2
219	85	2.25	18.8	159.0	423.1	0.87	—	0.27	0.12	0.43	2.06	0.15	112.1	0.9	0.6	29.9
206	104	2.79	17.3	117.1	438.7	0.92	89.3	0.26	0.66	0.40	2.26	0.10	110.4	0	0.6	24.2
22	235	2.05	19.9	128.7	763.6	1.26	129.5	0.19	0.72	0.28	2.05	0.05	58.1	0	0.2	13.2
28	47	1.77	22.1	218.4	235.2	2.33	9.5	0.23	0.26	0.20	3.12	0.13	63.0	3.2	0.9	18.3

TABLE A-1 Table of Food Composition *(continued)*

(Computer code is for Cengage Diet Analysis Plus program)

DA+ Code	Food Description	QTY	Measure	Wt (g)	H_2O (g)	Ener (cal)	Prot (g)	Carb (g)	Fiber (g)	Fat (g)	Fat Breakdown (g) Sat	Mono	Poly
Mixed Foods, Soups, Sandwiches—continued													
1906	Ham and cheese	1	item(s)	146	74.2	352	20.7	33.3	2.0	15.5	6.4	6.7	1.4
31890	Ham with mayonnaise	1	item(s)	112	56.3	271	13.0	27.9	1.9	11.6	2.8	4.0	4.0
756	Hamburger, double patty, large, with condiments and vegetables	1	item(s)	226	121.5	540	34.3	40.3	—	26.6	10.5	10.3	2.8
8793	Hamburger, large, plain	1	item(s)	137	57.7	426	22.6	31.7	1.5	22.9	8.4	9.9	2.1
8795	Hamburger, large, with vegetables and condiments	1	item(s)	218	121.4	512	25.8	40.0	3.1	27.4	10.4	11.4	2.2
25134	Hot chicken salad	1	item(s)	98	48.4	242	15.2	23.8	1.3	9.2	2.9	2.5	3.0
25133	Hot turkey salad	1	item(s)	98	50.1	224	15.6	23.8	1.3	6.9	2.3	1.6	2.5
1411	Hotdog with bun, plain	1	item(s)	98	52.9	242	10.4	18.0	1.6	14.5	5.1	6.9	1.7
30249	Pastrami	1	item(s)	134	71.2	328	13.4	27.8	1.6	17.7	6.1	8.3	1.2
16701	Peanut butter	1	item(s)	93	23.6	345	12.2	37.6	3.3	17.4	3.4	7.7	5.2
30306	Peanut butter and jelly	1	item(s)	93	24.2	330	10.3	41.9	2.9	14.7	2.9	6.5	4.4
1909	Roast beef submarine with mayonnaise and vegetables	1	item(s)	216	127.4	410	28.6	44.3	—	13.0	7.1	1.8	2.6
1910	Roast beef, plain	1	item(s)	139	67.6	346	21.5	33.4	1.2	13.8	3.6	6.8	1.7
1907	Steak with mayonnaise and vegetables	1	item(s)	204	104.2	459	30.3	52.0	2.3	14.1	3.8	5.3	3.3
25288	Tuna salad	1	item(s)	179	102.2	415	24.5	28.4	1.6	22.4	3.5	6.2	11.4
30283	Turkey submarine with cheese, lettuce, tomato, and mayonnaise	1	item(s)	277	168.0	529	30.4	49.4	3.0	22.8	6.8	6.0	8.6
31891	Turkey with mayonnaise	1	item(s)	143	74.5	329	28.7	26.4	1.3	11.2	2.6	2.6	4.8
	Soups												
25296	Bean	1	cup(s)	301	253.1	191	13.8	29.0	6.5	2.3	0.7	0.8	0.5
711	Bean with pork, condensed, prepared with water	1	cup(s)	253	215.9	159	7.3	21.0	7.3	5.5	1.4	2.0	1.7
713	Beef noodle, condensed, prepared with water	1	cup(s)	244	224.9	83	4.7	8.7	0.7	3.0	1.1	1.2	0.5
825	Cheese, condensed, prepared with milk	1	cup(s)	251	206.9	231	9.5	16.2	1.0	14.6	9.1	4.1	0.5
826	Chicken broth, condensed, prepared with water	1	cup(s)	244	234.1	39	4.9	0.9	0	1.4	0.4	0.6	0.3
25297	Chicken noodle soup	1	cup(s)	286	258.4	117	10.8	10.9	0.9	2.9	0.8	1.1	0.7
827	Chicken noodle, condensed, prepared with water	1	cup(s)	241	226.1	60	3.1	7.1	0.5	2.3	0.6	1.0	0.6
724	Chicken noodle, dehydrated, prepared with water	1	cup(s)	252	237.3	58	2.1	9.2	0.3	1.4	0.3	0.5	0.4
823	Cream of asparagus, condensed, prepared with milk	1	cup(s)	248	213.3	161	6.3	16.4	0.7	8.2	3.3	2.1	2.2
824	Cream of celery, condensed, prepared with milk	1	cup(s)	248	214.4	164	5.7	14.5	0.7	9.7	3.9	2.5	2.7
708	Cream of chicken, condensed, prepared with milk	1	cup(s)	248	210.4	191	7.5	15.0	0.2	11.5	4.6	4.5	1.6
715	Cream of chicken, condensed, prepared with water	1	cup(s)	244	221.1	117	3.4	9.3	0.2	7.4	2.1	3.3	1.5
709	Cream of mushroom, condensed, prepared with milk	1	cup(s)	248	215.0	166	6.2	14.0	0	9.6	3.3	2.0	1.8
716	Cream of mushroom, condensed, prepared with water	1	cup(s)	244	224.6	102	1.9	8.0	0	7.0	1.6	1.3	1.7
25298	Cream of vegetable	1	cup(s)	285	250.7	165	7.2	15.2	1.9	8.6	1.6	4.6	1.9
16689	Egg drop	1	cup(s)	244	228.9	73	7.5	1.1	0	3.8	1.1	1.5	0.6
25138	Golden squash	1	cup(s)	258	223.9	145	7.6	20.4	0.4	4.1	0.8	2.2	0.9
16663	Hot and sour	1	cup(s)	244	209.7	161	15.0	5.4	0.5	7.9	2.7	3.4	1.1
28054	Lentil chowder	1	cup(s)	244	202.8	153	11.4	27.7	12.6	0.5	0.1	0.1	0.2
28560	Macaroni and bean	1	cup(s)	246	138.8	146	5.8	22.9	5.1	3.7	0.5	2.2	0.6
714	Manhattan clam chowder, condensed, prepared with water	1	cup(s)	244	225.1	73	2.1	11.6	1.5	2.1	0.4	0.4	1.2
28561	Minestrone	1	cup(s)	241	185.4	103	4.5	16.8	4.8	2.3	0.3	1.4	0.4
717	Minestrone, condensed, prepared with water	1	cup(s)	241	220.1	82	4.3	11.2	1.0	2.5	0.6	0.7	1.1

Chol (mg)	Calc (mg)	Iron (mg)	Magn (mg)	Pota (mg)	Sodi (mg)	Zinc (mg)	Vit A (µg)	Thia (mg)	Vit E (mg α)	Ribo (mg)	Niac (mg)	Vit B_6 (mg)	Fola (µg DFE)	Vit C (mg)	Vit B_{12} (µg)	Sele (µg)
58	130	3.24	16.1	290.5	770.9	1.37	96.4	0.30	0.29	0.48	2.68	0.20	78.8	2.8	0.5	23.1
34	91	2.47	23.5	210.6	1097.6	1.13	5.6	0.57	0.50	0.26	3.80	0.25	90.7	2.2	0.2	20.2
122	102	5.85	49.7	569.5	791.0	5.67	0	0.36	—	0.38	7.57	0.54	110.7	1.1	4.1	25.5
71	74	3.57	27.4	267.2	474.0	4.11	0	0.28	—	0.28	6.24	0.23	80.8	0	2.1	27.1
87	96	4.92	43.6	479.6	824.0	4.88	0	0.41	—	0.37	7.28	0.32	115.5	2.6	2.4	33.6
39	115	1.88	20.0	172.4	505.1	1.19	—	0.23	0.28	0.22	4.84	0.19	79.5	0.5	0.2	19.9
37	114	1.99	21.3	189.0	494.5	1.07	—	0.22	0.28	0.20	4.26	0.22	79.6	0.5	0.2	23.4
44	24	2.31	12.7	143.1	670.3	1.98	0	0.23	—	0.27	3.64	0.04	60.7	0.1	0.5	26.0
51	80	3.02	22.8	182.2	1364.1	2.70	2.7	0.28	0.26	0.26	4.97	0.14	89.8	0.3	1.0	14.3
0	110	2.92	66.0	226.0	580.3	1.32	0	0.31	2.39	0.23	6.72	0.18	130.2	0	0	13.2
0	94	2.50	56.7	198.1	492.9	1.12	—	0.26	2.01	0.20	5.66	0.15	110.7	0.1	0	11.2
73	41	2.80	67.0	330.5	844.6	4.38	30.2	0.41	—	0.41	5.96	0.32	88.6	5.6	1.8	25.7
51	54	4.22	30.6	315.5	792.3	3.39	11.1	0.37	—	0.30	5.86	0.26	68.1	2.1	1.2	29.2
73	92	5.16	49.0	524.3	797.6	4.52	0	0.40	—	0.36	7.30	0.36	128.5	5.5	1.6	42.0
59	78	2.97	35.9	316.3	724.7	1.02	—	0.27	0.34	0.26	12.07	0.46	99.6	1.8	2.4	76.9
64	307	4.59	49.9	534.6	1759.0	2.74	74.8	0.49	1.19	0.65	3.91	0.31	171.7	10.5	0.6	42.1
67	100	3.46	34.3	304.6	564.9	3.00	5.7	0.28	0.74	0.32	6.84	0.46	94.4	0	0.3	40
5	79	3.05	61.8	588.8	689.0	1.41	—	0.27	0.02	0.15	3.63	0.23	140.1	3.6	0.2	7.9
3	78	1.89	43.0	371.9	883.0	0.96	43.0	0.08	1.08	0.03	0.52	0.03	30.4	1.5	0	7.8
5	20	1.07	7.3	97.6	929.6	1.51	12.2	0.06	1.22	0.05	1.03	0.03	29.3	0.5	0.2	7.3
48	289	0.80	20.1	341.4	1019.1	0.67	358.9	0.06	—	0.33	0.50	0.07	10.0	1.3	0.4	7.0
0	10	0.51	2.4	209.8	775.9	0.24	0	0.01	0.04	0.07	3.34	0.02	4.9	0	0.2	0
24	25	1.38	16.4	340.1	774.5	0.77	—	0.15	0.02	0.16	5.57	0.13	37.2	1.8	0.3	10.2
12	14	1.59	9.6	53.0	638.7	0.38	26.5	0.13	0.07	0.10	1.30	0.04	28.9	0	0	11.6
10	5	0.50	7.6	32.8	577.1	0.20	2.5	0.20	0.12	0.07	1.08	0.02	27.7	0	0.1	9.6
22	174	0.86	19.8	359.6	1041.6	0.91	62.0	0.10	—	0.27	0.88	0.06	29.8	4.0	0.5	8.0
32	186	0.69	22.3	310.0	1009.4	0.19	114.1	0.07	—	0.24	0.43	0.06	7.4	1.5	0.5	4.7
27	181	0.67	17.4	272.8	1046.6	0.67	178.6	0.07	—	0.25	0.92	0.06	7.4	1.2	0.5	8.0
10	34	0.61	2.4	87.8	985.8	0.63	163.5	0.02	—	0.06	0.82	0.01	2.4	0.2	0.1	7.0
10	164	1.36	19.8	267.8	823.4	0.79	81.8	0.10	1.01	0.29	0.62	0.05	7.4	0.2	0.6	6.0
0	17	1.31	4.9	73.2	775.9	0.24	9.8	0.05	0.97	0.05	0.50	0.00	2.4	0	0	2.9
1	80	1.20	17.5	340.7	787.9	0.56	—	0.12	1.05	0.18	3.32	0.12	39.5	10.7	0.3	4.3
102	22	0.75	4.9	219.6	729.6	0.48	41.5	0.02	0.29	0.19	3.02	0.05	14.6	0	0.5	7.6
4	262	0.78	42.4	542.2	515.6	0.88	—	0.16	0.52	0.30	1.14	0.16	32.7	12.5	0.7	6.0
34	29	1.24	19.5	373.3	1561.6	1.43	—	0.26	0.12	0.24	4.96	0.20	14.6	0.5	0.4	19.3
0	48	4.38	59.3	626.2	26.7	1.57	—	0.24	0.06	0.12	1.87	0.32	192.7	16.1	0	3.5
0	59	1.90	32.4	275.9	531.0	0.51	—	0.16	0.37	0.12	1.44	0.10	92.3	9.1	0	8.8
2	27	1.56	9.8	180.6	551.4	0.87	48.8	0.02	1.22	0.03	0.77	0.09	9.8	3.9	3.9	9.0
0	62	1.70	29.9	287.5	442.7	0.42	—	0.09	0.23	0.09	0.70	0.07	62.0	13.3	0	3.5
2	34	0.91	7.2	313.3	911.0	0.74	118.1	0.05	—	0.04	0.94	0.09	50.6	1.2	0	8.0

TABLE A-1 Table of Food Composition *(continued)*

(Computer code is for Cengage Diet Analysis Plus program)

DA+ Code	Food Description	QTY	Measure	Wt (g)	H_2O (g)	Ener (cal)	Prot (g)	Carb (g)	Fiber (g)	Fat (g)	Fat Breakdown (g) Sat	Mono	Poly
	Mixed Foods, Soups, Sandwiches—continued												
28038	Mushroom and wild rice	1	cup(s)	244	199.7	86	4.7	13.2	1.7	0.3	0	0	0.2
828	New England clam chowder, condensed, prepared with milk	1	cup(s)	248	212.2	151	8.0	18.4	0.7	5.0	2.1	0.7	0.6
28036	New England style clam chowder	1	cup(s)	244	227.5	61	3.8	8.8	1.8	0.2	0.1	0	0
28566	Old country pasta	1	cup(s)	252	183.3	146	6.5	18.3	3.6	4.5	2.0	2.4	0.9
725	Onion, dehydrated, prepared with water	1	cup(s)	246	235.7	30	0.8	6.8	0.7	0	0	0	0
16667	Shrimp gumbo	1	cup(s)	244	207.2	166	9.5	18.2	2.4	6.7	1.3	2.9	2.0
28037	Southwestern corn chowder	1	cup(s)	244	217.8	98	4.9	17.0	2.4	0.5	0.1	0.1	0.2
30282	Soybean (miso)	1	cup(s)	240	218.6	84	6.0	8.0	1.9	3.4	0.6	1.1	1.4
25140	Split pea	1	cup(s)	165	119.7	72	4.5	16.0	1.6	0.3	0.1	0	0.2
718	Split pea with ham, condensed, prepared with water	1	cup(s)	253	206.9	190	10.3	28.0	2.3	4.4	1.8	1.8	0.6
726	Tomato vegetable, dehydrated, prepared with water	1	cup(s)	253	238.4	56	2.0	10.2	0.8	0.9	0.4	0.3	0.1
710	Tomato, condensed, prepared with milk	1	cup(s)	248	213.4	136	6.2	22.0	1.5	3.2	1.8	0.9	0.3
719	Tomato, condensed, prepared with water	1	cup(s)	244	223.0	73	1.9	16.0	1.5	0.7	0.2	0.2	0.2
28595	Turkey noodle	1	cup(s)	244	216.9	114	8.1	15.1	1.9	2.4	0.3	1.1	0.7
28051	Turkey vegetable	1	cup(s)	244	220.8	96	12.2	6.6	2.0	1.1	0.3	0.2	0.3
25141	Vegetable	1	cup(s)	252	228.1	82	5.2	16.5	4.5	0.3	0	0	0.1
720	Vegetable beef, condensed, prepared with water	1	cup(s)	244	224.0	76	5.4	9.9	2.0	1.9	0.8	0.8	0.1
28598	Vegetable gumbo	1	cup(s)	252	184.5	170	4.4	28.9	3.6	4.7	0.7	3.2	0.5
721	Vegetarian vegetable, condensed, prepared with water	1	cup(s)	241	222.7	67	2.1	11.8	0.7	1.9	0.3	0.8	0.7
Fast Food													
	Arby's												
36094	Au jus sauce	1	serving(s)	85	—	43	1.0	7.0	0	1.3	0.4	—	—
751	Beef 'n cheddar sandwich	1	item(s)	195	—	445	22.0	44.0	2.0	21.0	6.0	—	—
9279	Cheddar curly fries	1	serving(s)	198	—	631	8.0	73.0	7.0	37.4	6.8	—	—
34770	Chicken breast fillet sandwich, grilled	1	item(s)	233	—	414	32.0	36.0	3.0	17.0	3.0	—	—
36131	Chocolate shake, regular	1	serving(s)	397	—	507	13.0	83.0	0	13.0	8.0	—	—
36045	Curly fries, large size	1	serving(s)	198	—	631	8.0	73.0	7.0	37.0	7.0	—	—
36044	Curly fries, medium size	1	serving(s)	128	—	406	5.0	47.0	5.0	24.0	4.0	—	—
752	Ham 'n cheese sandwich	1	item(s)	167	—	304	23.0	35.0	1.0	7.0	2.0	—	—
36048	Homestyle fries, large size	1	serving(s)	213	—	566	6.0	82.0	6.0	37.0	7.0	—	—
36047	Homestyle fries, medium size	1	serving(s)	142	—	377	4.0	55.0	4.0	25.0	4.0	—	—
33465	Homestyle fries, small size	1	serving(s)	113	—	302	3.0	44.0	3.0	20.0	4.0	—	—
9249	Junior roast beef sandwich	1	item(s)	125	—	272	16.0	34.0	2.0	10.0	4.0	—	—
9251	Large roast beef sandwich	1	item(s)	281	—	547	42.0	41.0	3.0	28.0	12.0	—	—
39640	Market Fresh chicken salad with pecans sandwich	1	item(s)	322	—	769	30.0	79.0	9.0	39.0	10.0	—	—
39641	Market Fresh Martha's Vineyard salad, without dressing	1	serving(s)	330	—	277	26.0	24.0	5.0	8.0	4.0	—	—
34769	Market Fresh roast turkey and Swiss sandwich	1	serving(s)	359	—	725	45.0	75.0	5.0	30.0	8.0	—	—
9267	Market Fresh roast turkey ranch and bacon sandwich	1	serving(s)	382	—	834	49.0	75.0	5.0	38.0	11.0	—	—
39642	Market Fresh Santa Fe salad, without dressing	1	serving(s)	372	—	499	30.0	42.0	7.0	23.0	8.0	—	—
39650	Market Fresh Southwest chicken wrap	1	serving(s)	251	—	567	36.0	42.0	4.0	29.0	9.0	—	—
37021	Market Fresh Ultimate BLT sandwich	1	item(s)	294	—	779	23.0	75.0	6.0	45.0	11.0	—	—
750	Roast beef sandwich, regular	1	item(s)	154	—	320	21.0	34.0	2.0	14.0	5.0	—	—
36132	Strawberry shake, regular	1	serving(s)	397	—	498	13.0	81.0	0	13.0	8.0	—	—
2009	Super roast beef sandwich	1	item(s)	198	—	398	21.0	40.0	2.0	19.0	6.0	—	—
36130	Vanilla shake, regular	1	serving(s)	369	—	437	13.0	66.0	0	13.0	8.0	—	—
	Auntie Anne's												
35371	Cheese dipping sauce	1	serving(s)	35	—	100	3.0	4.0	0	8.0	4.0	—	—
35353	Cinnamon sugar soft pretzel	1	item(s)	120	—	350	9.0	74.0	2.0	2.0	0	—	—

Chol (mg)	Calc (mg)	Iron (mg)	Magn (mg)	Pota (mg)	Sodi (mg)	Zinc (mg)	Vit A (μg)	Thia (mg)	Vit E (mg α)	Ribo (mg)	Niac (mg)	Vit B_6 (mg)	Fola (μg DFE)	Vit C (mg)	Vit B_{12} (μg)	Sele (μg)
0	30	1.38	27.3	376.9	283.8	1.00	—	0.06	0.07	0.23	3.21	0.14	27.3	3.7	0.1	4.7
17	169	3.00	29.8	456.3	887.8	0.99	91.8	0.20	0.54	0.43	1.96	0.17	22.3	5.2	11.9	10.9
3	89	1.30	29.2	503.7	256.8	0.52	—	0.06	0.02	0.10	1.30	0.15	24.2	11.8	3.0	3.8
5	57	2.43	51.9	500.4	355.7	0.78	—	0.21	0.01	0.14	2.64	0.20	105.2	22.0	0.1	10.3
0	22	0.12	9.8	76.3	851.2	0.12	0	0.03	0.02	0.03	0.15	0.06	0	0.2	0	0.5
51	105	2.85	48.8	461.2	441.6	0.90	80.5	0.18	1.90	0.12	2.52	0.19	102.5	17.6	0.3	14.9
1	83	1.03	26.3	434.4	217.4	0.57	—	0.08	0.09	0.13	1.81	0.21	32.1	39.6	0.2	1.7
0	65	1.87	36.0	362.4	988.8	0.86	232.8	0.06	0.96	0.16	2.61	0.15	57.6	4.6	0.2	1.0
0	28	1.26	29.8	328.1	602.4	0.52	—	0.10	0.00	0.07	1.50	0.16	49.5	8.1	0	0.7
8	23	2.27	48.1	399.7	1006.9	1.31	22.8	0.14	—	0.07	1.47	0.06	2.5	1.5	0.3	8.0
0	20	0.60	10.1	169.5	334.0	0.20	10.1	0.06	0.43	0.09	1.26	0.06	12.7	3.0	0.1	2.0
10	166	1.36	29.8	466.2	711.8	0.84	94.2	0.09	0.44	0.31	1.35	0.15	5.0	15.6	0.6	9.2
0	20	1.31	17.1	273.3	663.7	0.29	24.4	0.04	0.41	0.07	1.23	0.10	0	15.4	0	6.1
26	28	1.40	24.1	223.8	395.9	0.75	—	0.21	0.01	0.12	2.85	0.15	65.1	7.1	0.1	13.7
21	38	1.48	25.0	423.4	348.7	0.99	—	0.09	0.01	0.09	3.64	0.27	23.9	10.6	0.2	10.3
0	40	2.08	39.1	681.5	670.3	0.67	—	0.16	0.00	0.09	2.70	0.26	36.3	22.4	0	2.2
5	20	1.09	7.3	168.4	773.5	1.51	190.3	0.03	0.58	0.04	1.00	0.07	9.8	2.4	0.3	2.7
0	56	1.85	38.9	360.2	518.4	0.64	—	0.18	0.64	0.08	1.77	0.17	107.5	21.9	0	4.1
0	24	1.06	7.2	207.3	814.6	0.45	171.1	0.05	1.39	0.04	0.90	0.05	9.6	1.4	0	4.3
0	0	—	—	—	1510.0	—	—	—	—	—	—	—	—	—	—	—
51	80	3.96	—	—	1274.0	—	—	—	—	—	—	—	—	1.8	—	—
0	80	3.24	—	—	1476.0	—	—	—	—	—	—	—	—	9.6	—	—
9	90	3.06	—	—	913.0	—	—	—	—	—	—	—	—	10.8	—	—
34	510	0.54	—	—	357.0	—	—	—	—	—	—	—	—	5.4	—	—
0	80	3.24	—	—	1476.0	—	—	—	—	—	—	—	—	9.6	—	—
0	50	1.98	—	—	949.0	—	—	—	—	—	—	—	—	6.0	—	—
35	160	2.70	—	—	1420.0	—	—	—	—	—	—	—	—	1.2	—	—
0	50	1.62	—	—	1029.0	—	—	—	—	—	—	—	—	12.6	—	—
0	30	1.08	—	—	686.0	—	—	—	—	—	—	—	—	8.4	—	—
0	30	0.90	—	—	549.0	—	—	—	—	—	—	—	—	6.6	—	—
29	60	3.06	—	—	740.0	—	0	—	—	—	—	—	—	0	—	—
102	70	6.30	—	—	1869.0	—	0	—	—	—	—	—	—	0.6	—	—
74	180	4.32	—	—	1240.0	—	—	—	—	—	—	—	—	30.0	—	—
72	200	1.62	—	—	454.0	—	—	—	—	—	—	—	—	33.6	—	—
91	360	5.22	—	—	1788.0	—	—	—	—	—	—	—	—	10.2	—	—
109	330	5.40	—	—	2258.0	—	—	—	—	—	—	—	—	11.4	—	—
59	420	3.60	—	—	1231.0	—	—	—	—	—	—	—	—	36.6	—	—
88	240	4.50	—	—	1451.0	—	—	—	—	—	—	—	—	7.8	—	—
51	170	4.68	—	—	1571.0	—	—	—	—	—	—	—	—	16.8	—	—
44	60	3.60	—	—	953.0	—	0	—	—	—	—	—	—	0	—	—
34	510	0.72	—	—	363.0	—	—	—	—	—	—	—	—	6.6	—	—
44	70	3.78	—	—	1060.0	—	—	—	—	—	—	—	—	6.0	—	—
34	510	0.36	—	—	350.0	—	—	—	—	—	—	—	—	5.4	—	—
10	100	0.00	—	—	510.0	—	—	—	—	—	—	—	—	0	—	—
0	20	1.98	—	—	410.0	—	0	—	—	—	—	—	—	0	—	—

TABLE A-1 Table of Food Composition *(continued)*

(Computer code is for Cengage Diet Analysis Plus program)

DA+ Code	Food Description	QTY	Measure	Wt (g)	H_2O (g)	Ener (cal)	Prot (g)	Carb (g)	Fiber (g)	Fat (g)	Fat Breakdown (g) Sat	Mono	Poly
	Fast Food—continued												
35354	Cinnamon sugar soft pretzel with butter	1	item(s)	120	—	450	8.0	83.0	3.0	9.0	5.0	—	—
35372	Marinara dipping sauce	1	serving(s)	35	—	10	0	4.0	0	0	0	0	0
35357	Original soft pretzel	1	serving(s)	120	—	340	10.0	72.0	3.0	1.0	0	—	—
35358	Original soft pretzel with butter	1	item(s)	120	—	370	10.0	72.0	3.0	4.0	2.0	—	—
35359	Parmesan herb soft pretzel	1	item(s)	120	—	390	11.0	74.0	4.0	5.0	2.5	—	—
35360	Parmesan herb soft pretzel with butter	1	item(s)	120	—	440	10.0	72.0	9.0	13.0	7.0	—	—
35361	Sesame soft pretzel	1	item(s)	120	—	350	11.0	63.0	3.0	6.0	1.0	—	—
35362	Sesame soft pretzel with butter	1	item(s)	120	—	410	12.0	64.0	7.0	12.0	4.0	—	—
35364	Sour cream and onion soft pretzel	1	item(s)	120	—	310	9.0	66.0	2.0	1.0	0	—	—
35366	Sour cream and onion soft pretzel with butter	1	item(s)	120	—	340	9.0	66.0	2.0	5.0	3.0	—	—
35373	Sweet mustard dipping sauce	1	serving(s)	35	—	60	0.5	8.0	0	1.5	1.0	—	—
35367	Whole wheat soft pretzel	1	item(s)	120	—	350	11.0	72.0	7.0	1.5	0	—	—
35368	Whole wheat soft pretzel with butter	1	item(s)	120	—	370	11.0	72.0	7.0	4.5	1.5	—	—
	Boston Market												
34978	Butternut squash	¾	cup(s)	143	—	140	2.0	25.0	2.0	4.5	3.0	—	—
35006	Caesar side salad	1	serving(s)	71	—	40	3.0	3.0	1.0	20.0	2.0	—	—
35013	Chicken Carver sandwich with cheese and sauce	1	item(s)	321	—	700	44.0	68.0	3.0	29.0	7.0	—	—
34979	Chicken gravy	4	ounce(s)	113	—	15	1.0	4.0	0	0.5	0	—	—
35053	Chicken noodle soup	¾	cup(s)	283	—	180	13.0	16.0	1.0	7.0	2.0	—	—
34973	Chicken pot pie	1	item(s)	425	—	800	29.0	59.0	4.0	49.0	18.0	—	—
35054	Chicken tortilla soup with toppings	¾	cup(s)	227	—	340	12.0	24.0	1.0	22.0	7.0	—	—
35007	Cole slaw	¾	cup(s)	125	—	170	2.0	21.0	2.0	9.0	2.0	—	—
35057	Cornbread	1	item(s)	45	—	130	1.0	21.0	0	3.5	1.0	—	—
34980	Creamed spinach	¾	cup(s)	191	—	280	9.0	12.0	4.0	23.0	15.0	—	—
34998	Fresh vegetable stuffing	1	cup(s)	136	—	190	3.0	25.0	2.0	8.0	1.0	—	—
34991	Garlic dill new potatoes	¾	cup(s)	156	—	140	3.0	24.0	3.0	3.0	1.0	—	—
34983	Green bean casserole	¾	cup(s)	170	—	60	2.0	9.0	2.0	2.0	1.0	—	—
34982	Green beans	¾	cup(s)	91	—	60	2.0	7.0	3.0	3.5	1.5	—	—
34984	Homestyle mashed potatoes	¾	cup(s)	221	—	210	4.0	29.0	3.0	9.0	6.0	—	—
34985	Homestyle mashed potatoes and gravy	1	cup(s)	334	—	225	5.0	33.0	3.0	9.5	6.0	—	—
34988	Hot cinnamon apples	¾	cup(s)	145	—	210	0	47.0	3.0	3.0	0	—	—
34989	Macaroni and cheese	¾	cup(s)	221	—	330	14.0	39.0	1.0	12.0	7.0	—	—
51193	Market chopped salad with dressing	1	item(s)	563	—	580	11.0	31.0	9.0	48.0	9.0	—	—
34970	Meatloaf	1	serving(s)	218	—	480	29.0	23.0	2.0	33.0	13.0	—	—
39383	Nestle Toll House chocolate chip cookie	1	item(s)	78	—	370	4.0	49.0	2.0	19.0	9.0	—	—
34965	Quarter chicken, dark meat, no skin	1	item(s)	134	—	260	30.0	2.0	0	13.0	4.0	—	—
34966	Quarter chicken, dark meat, with skin	1	item(s)	149	—	280	31.0	3.0	0	15.0	4.5	—	—
34963	Quarter chicken, white meat, no skin or wing	1	item(s)	173	—	250	41.0	4.0	0	8.0	2.5	—	—
34964	Quarter chicken, white meat, with skin and wing	1	item(s)	110	—	330	50.0	3.0	0	12.0	4.0	—	—
34968	Roasted turkey breast	5	ounce(s)	142	—	180	38.0	0	0	3.0	1.0	—	—
35011	Seasonal fresh fruit salad	1	serving(s)	142	—	60	1.0	15.0	1.0	0	0	0	0
51192	Spinach with garlic butter sauce	1	serving(s)	170	—	130	5.0	9.0	5.0	9.0	6.0	—	—
34969	Spiral sliced holiday ham	8	ounce(s)	227	—	450	40.0	13.0	0	26.0	10.0	—	—
35003	Steamed vegetables	1	cup(s)	136	—	50	2.0	8.0	3.0	2.0	0	—	—
35005	Sweet corn	¾	cup(s)	176	—	170	6.0	37.0	2.0	4.0	1.0	—	—
35004	Sweet potato casserole	¾	cup(s)	198	—	460	4.0	77.0	3.0	17.0	6.0	—	—
	Burger King												
29731	Biscuit with sausage, egg, and cheese	1	item(s)	191	—	610	20.0	33.0	1.0	45.0	15.0	—	—
14249	Cheeseburger	1	item(s)	133	—	330	17.0	31.0	1.0	16.0	7.0	—	—
14251	Chicken sandwich	1	item(s)	219	—	660	24.0	52.0	4.0	40.0	8.0	—	—
3808	Chicken Tenders, 8 pieces	1	serving(s)	123	—	340	19.0	21.0	0.5	20.0	5.0	—	—
14259	Chocolate shake, small	1	item(s)	315	—	470	8.0	75.0	1.0	14.0	9.0	—	—
29732	Croissanwich with sausage and cheese	1	item(s)	106	37.2	370	14.0	23.0	0.5	25.0	9.0	12.7	3.3

Chol (mg)	Calc (mg)	Iron (mg)	Magn (mg)	Pota (mg)	Sodi (mg)	Zinc (mg)	Vit A (µg)	Thia (mg)	Vit E (mg α)	Ribo (mg)	Niac (mg)	Vit B_6 (mg)	Fola (µg DFE)	Vit C (mg)	Vit B_{12} (µg)	Sele (µg)
25	30	2.34	—	—	430.0	—	—	—	—	—	—	—	—	0	—	—
0	0	0.00	—	—	180.0	—	0	—	—	—	—	—	—	0	—	—
0	30	2.34	—	—	900.0	—	0	—	—	—	—	—	—	0	—	—
10	30	2.16	—	—	930.0	—	—	—	—	—	—	—	—	0	—	—
10	80	1.80	—	—	780.0	—	—	—	—	—	—	—	—	1.2	—	—
30	60	1.80	—	—	660.0	—	—	—	—	—	—	—	—	1.2	—	—
0	20	2.88	—	—	840.0	—	0	—	—	—	—	—	—	0	—	—
15	20	2.70	—	—	860.0	—	—	—	—	—	—	—	—	0	—	—
0	30	1.98	—	—	920.0	—	—	—	—	—	—	—	—	0	—	—
10	40	2.16	—	—	930.0	—	—	—	—	—	—	—	—	0	—	—
40	0	0.00	—	—	120.0	—	0	—	—	—	—	—	—	0	—	—
0	30	1.98	—	—	1100.0	—	0	—	—	—	—	—	—	0	—	—
10	30	2.34	—	—	1120.0	—	—	—	—	—	—	—	—	0	—	—
10	59	0.80	—	—	35.0	—	—	—	—	—	—	—	—	22.2	—	—
0	60	0.43	—	—	75.0	—	—	—	—	—	—	—	—	5.4	—	—
90	211	2.85	—	—	1560.0	—	—	—	—	—	—	—	—	15.8	—	—
0	0	0.00	—	—	570.0	—	0	—	—	—	—	—	—	0	—	—
55	0	1.07	—	—	220.0	—	—	—	—	—	—	—	—	1.8	—	—
115	40	4.50	—	—	800.0	—	—	—	—	—	—	—	—	1.2	—	—
45	123	1.32	—	—	1310.0	—	—	—	—	—	—	—	—	18.4	—	—
10	41	0.48	—	—	270.0	—	—	—	—	—	—	—	—	24.5	—	—
5	0	0.71	—	—	220.0	—	0	—	—	—	—	—	—	0	—	—
70	264	2.84	—	—	580.0	—	—	—	—	—	—	—	—	9.5	—	—
0	41	1.48	—	—	580.0	—	—	—	—	—	—	—	—	2.5	—	—
0	0	0.85	—	—	120.0	—	0	—	—	—	—	—	—	14.3	—	—
5	20	0.72	—	—	620.0	—	—	—	—	—	—	—	—	2.4	—	—
0	43	0.38	—	—	180.0	—	—	—	—	—	—	—	—	5.1	—	—
25	51	0.46	—	—	660.0	—	—	—	—	—	—	—	—	19.2	—	—
25	100	0.59	—	—	1230.0	—	—	—	—	—	—	—	—	24.9	—	—
0	16	0.28	—	—	15.0	—	—	—	—	—	—	—	—	0	—	—
30	345	1.65	—	—	1290.0	—	—	—	—	—	—	—	—	0	—	—
10	—	—	—	—	2010.0	—	—	—	—	—	—	—	—	—	—	—
125	140	3.77	—	—	970.0	—	—	—	—	—	—	—	—	1.8	—	—
20	0	1.32	—	—	340.0	—	—	—	—	—	—	—	—	0	—	—
155	0	1.52	—	—	260.0	—	0	—	—	—	—	—	—	0	—	—
155	0	2.14	—	—	660.0	—	0	—	—	—	—	—	—	0	—	—
125	0	0.89	—	—	480.0	—	0	—	—	—	—	—	—	0	—	—
165	0	0.78	—	—	960.0	—	0	—	—	—	—	—	—	0	—	—
70	20	1.80	—	—	620.0	—	0	—	—	—	—	—	—	0	—	—
0	16	0.29	—	—	20.0	—	—	—	—	—	—	—	—	29.5	—	—
20	—	—	—	—	200.0	—	—	—	—	—	—	—	—	—	—	—
140	0	1.73	—	—	2230.0	—	0	—	—	—	—	—	—	0	—	—
0	53	0.46	—	—	45.0	—	—	—	—	—	—	—	—	24.0	—	—
0	0	0.43	—	—	95.0	—	—	—	—	—	—	—	—	5.8	—	—
20	44	1.18	—	—	210.0	—	—	—	—	—	—	—	—	9.8	—	—
210	250	2.70	—	—	1620.0	—	89.9	—	—	—	—	—	—	0	—	—
55	150	2.70	—	—	780.0	—	—	0.24	—	0.31	4.17	—	—	1.2	—	—
70	64	2.89	—	—	1440.0	—	—	0.50	—	0.32	10.29	—	—	0	—	—
55	20	0.72	—	—	960.0	—	—	0.14	—	0.11	10.93	—	—	0	—	—
55	333	0.79	—	—	350.0	—	—	0.11	—	0.61	0.26	—	—	2.7	0	—
50	99	1.78	20.1	217.3	810.0	1.51	—	0.34	1.03	0.33	4.33	—	—	0	0.6	22.2

TABLE A-1 Table of Food Composition *(continued)*

(Computer code is for Cengage Diet Analysis Plus program)

DA+ Code	Food Description	QTY	Measure	Wt (g)	H2O (g)	Ener (cal)	Prot (g)	Carb (g)	Fiber (g)	Fat (g)	Fat Breakdown (g) Sat	Mono	Poly
Fast Food—continued													
14261	Croissanwich with sausage, egg, and cheese	1	item(s)	159	71.4	470	19.0	26.0	0.5	32.0	11.0	15.8	6.1
3809	Double cheeseburger	1	item(s)	189	—	500	30.0	31.0	1.0	29.0	14.0	—	—
14244	Double Whopper sandwich	1	item(s)	373	—	900	47.0	51.0	3.0	57.0	19.0	—	—
14245	Double Whopper with cheese sandwich	1	item(s)	398	—	990	52.0	52.0	3.0	64.0	24.0	—	—
14250	Fish Filet sandwich	1	item(s)	250	—	630	24.0	67.0	4.0	30.0	6.0	—	—
14255	French fries, medium, salted	1	serving(s)	116	—	360	4.0	41.0	4.0	20.0	4.5	—	—
14262	French toast sticks, 5 pieces	1	serving(s)	112	37.6	390	6.0	46.0	2.0	20.0	4.5	10.6	2.9
14248	Hamburger	1	item(s)	121	—	290	15.0	30.0	1.0	12.0	4.5	—	—
14263	Hash brown rounds, small	1	serving(s)	75	27.1	230	2.0	23.0	2.0	15.0	4.0	—	—
14256	Onion rings, medium	1	serving(s)	91	—	320	4.0	40.0	3.0	16.0	4.0	—	—
39000	Tendercrisp chicken sandwich	1	item(s)	286	—	780	25.0	73.0	4.0	43.0	8.0	—	—
37514	TenderGrill chicken sandwich	1	item(s)	258	—	450	37.0	53.0	4.0	10.0	2.0	—	—
14258	Vanilla shake, small	1	item(s)	296	—	400	8.0	57.0	0	15.0	9.0	—	—
1736	Whopper sandwich	1	item(s)	290	—	670	28.0	51.0	3.0	39.0	11.0	—	—
14243	Whopper with cheese sandwich	1	item(s)	315	—	760	33.0	52.0	3.0	47.0	16.0	—	—
	Carl's Jr												
33962	Carl's bacon Swiss crispy chicken sandwich	1	item(s)	268	—	750	31.0	91.0	—	28.0	28.0	—	—
10801	Carl's Catch fish sandwich	1	item(s)	215	—	560	19.0	58.0	2.0	27.0	7.0	—	1.9
10862	Carl's Famous Star hamburger	1	item(s)	254	—	590	24.0	50.0	3.0	32.0	9.0	—	—
10785	Charbroiled chicken club sandwich	1	item(s)	270	—	550	42.0	43.0	4.0	23.0	7.0	—	2.9
10866	Charbroiled chicken salad	1	item(s)	437	—	330	34.0	17.0	5.0	7.0	4.0	—	1.0
10855	Charbroiled Santa Fe chicken sandwich	1	item(s)	266	—	610	38.0	43.0	4.0	32.0	8.0	—	—
10790	Chicken stars, 6 pieces	1	serving(s)	85	—	260	13.0	14.0	1.0	16.0	4.0	—	1.6
34864	Chocolate shake, small	1	serving(s)	595	—	540	15.0	98.0	0	11.0	7.0	—	—
10797	Crisscut fries	1	serving(s)	139	—	410	5.0	43.0	4.0	24.0	5.0	—	—
10799	Double Western Bacon cheeseburger	1	item(s)	308	—	920	51.0	65.0	2.0	50	21.0	—	6.6
14238	French fries, small	1	serving(s)	92	—	290	5.0	37.0	3.0	14.0	3.0	—	—
10798	French toast dips without syrup, 5 pieces	1	serving(s)	155	—	370	8.0	49.0	0	17.0	5.0	—	1.4
10802	Onion rings	1	serving(s)	128	—	440	7.0	53.0	3.0	22.0	5.0	—	0.8
34858	Spicy chicken sandwich	1	item(s)	198	—	480	14.0	48.0	2.0	26.0	5.0	—	—
34867	Strawberry shake, small	1	serving(s)	595	—	520	14.0	93.0	0	11.0	7.0	—	—
10865	Super Star hamburger	1	item(s)	348	—	790	41.0	52.0	3.0	47.0	14.0	—	—
38925	The Six Dollar burger	1	item(s)	429	—	1010	40.0	60.0	3.0	66.0	26.0	—	—
10818	Vanilla shake, small	1	item(s)	398	—	314	10.0	51.5	0	7.4	4.7	—	—
10770	Western Bacon cheeseburger	1	item(s)	225	—	660	32.0	64.0	2.0	30.0	12.0	—	4.8
	Chick Fil-A												
38746	Biscuit with bacon, egg, and cheese	1	item(s)	163	—	470	18.0	39.0	1.0	26.0	9.0	—	—
38747	Biscuit with egg	1	item(s)	135	—	350	11.0	38.0	1.0	16.0	4.5	—	—
38748	Biscuit with egg and cheese	1	item(s)	149	—	400	14.0	38.0	1.0	21.0	7.0	—	—
38753	Biscuit with gravy	1	item(s)	192	—	330	5.0	43.0	1.0	15.0	4.0	—	—
38752	Biscuit with sausage, egg, and cheese	1	item(s)	212	—	620	22.0	39.0	2.0	42.0	14.0	—	—
38771	Carrot and raisin salad	1	item(s)	113	—	170	1.0	28.0	2.0	6.0	1.0	—	—
38761	Chargrilled chicken Cool Wrap	1	item(s)	245	—	390	29.0	54.0	3.0	7.0	3.0	—	—
38766	Chargrilled chicken garden salad	1	item(s)	275	—	180	22.0	9.0	3.0	6.0	3.0	—	—
38758	Chargrilled chicken sandwich	1	item(s)	193	—	270	28.0	33.0	3.0	3.5	1.0	—	—
38742	Chicken biscuit	1	item(s)	145	—	420	18.0	44.0	2.0	19.0	4.5	—	—
38743	Chicken biscuit with cheese	1	item(s)	159	—	470	21.0	45.0	2.0	23.0	8.0	—	—
38762	Chicken Caesar Cool Wrap	1	item(s)	227	—	460	36.0	52.0	3.0	10.0	6.0	—	—
38757	Chicken deluxe sandwich	1	item(s)	208	—	420	28.0	39.0	2.0	16.0	3.5	—	—
38764	Chicken salad sandwich on wheat bun	1	item(s)	153	—	350	20.0	32.0	5.0	15.0	3.0	—	—
38756	Chicken sandwich	1	item(s)	170	—	410	28.0	38.0	1.0	16.0	3.5	—	—
38768	Chick-n-Strip salad	1	item(s)	327	—	400	34.0	21.0	4.0	20.0	6.0	—	—
38763	Chick-n-Strips	4	item(s)	127	—	300	28.0	14.0	1.0	15.0	2.5	—	—
38770	Cole slaw	1	item(s)	128	—	260	2.0	17.0	2.0	21.0	3.5	—	—
38776	Diet lemonade, small	1	cup(s)	255	—	25	0	5.0	0	0	0	0	0

Chol (mg)	Calc (mg)	Iron (mg)	Magn (mg)	Pota (mg)	Sodi (mg)	Zinc (mg)	Vit A (µg)	Thia (mg)	Vit E (mg α)	Ribo (mg)	Niac (mg)	Vit B_6 (mg)	Fola (µg DFE)	Vit C (mg)	Vit B_{12} (µg)	Sele (µg)
180	146	2.63	28.6	313.2	1060.0	2.08	—	0.38	1.66	0.51	4.72	0.28	—	0	1.1	38.0
105	250	4.50	—	—	1030.0	—	—	0.26	—	0.44	6.37	—	—	1.2	—	—
175	150	8.07	—	—	1090.0	—	—	0.39	—	0.59	11.05	—	—	9.0	—	—
195	299	8.08	—	—	1520.0	—	—	0.39	—	0.66	11.03	—	—	9.0	—	—
60	101	3.62	—	—	1380.0	—	—	—	—	—	—	—	—	3.6	—	—
0	20	0.71	—	—	590.0	—	0	0.15	—	0.48	2.30	—	—	8.9	—	—
0	60	1.80	21.3	124.3	440.0	0.57	—	0.31	0.98	0.19	2.88	0.05	—	0	0	13.7
40	80	2.70	—	—	560.0	—	—	0.25	—	0.28	4.25	—	—	1.2	—	—
0	0	0.36	—	—	450.0	—	0	0.11	0.83	0.06	1.35	0.17	—	1.2	—	—
0	100	0.00	—	—	460.0	—	0	0.14	—	0.09	2.32	—	—	0	—	—
75	79	4.43	—	—	1730.0	—	—	—	—	—	—	—	—	8.9	—	—
75	57	6.82	—	—	1210.0	—	—	—	—	—	—	—	—	5.7	—	—
60	348	0.00	—	—	240.0	—	—	0.11	—	0.63	0.21	—	—	2.4	0	—
51	100	5.38	—	—	1020.0	—	—	0.38	—	0.43	7.30	—	—	9.0	—	—
115	249	5.38	—	—	1450.0	—	—	0.38	—	0.51	7.28	—	—	9.0	—	—
80	200	5.40	—	—	1900.0	—	—	—	—	—	—	—	—	2.4	—	—
80	150	2.70	—	—	990.0	—	60.0	—	—	—	—	—	—	2.4	—	—
70	100	4.50	—	—	910.0	—	—	—	—	—	—	—	—	6.0	—	—
95	200	3.60	—	—	1330.0	—	—	—	—	—	—	—	—	9.0	—	—
75	200	1.80	—	—	880.0	—	—	—	—	—	—	—	—	30.0	—	—
100	200	3.60	—	—	1440.0	—	—	—	—	—	—	—	—	9.0	—	—
35	19	1.02	—	—	470.0	—	0	—	—	—	—	—	—	0	—	—
45	600	1.08	—	—	360.0	—	0	—	—	—	—	—	—	0	—	—
0	20	1.80	—	—	950.0	—	0	—	—	—	—	—	—	12.0	—	—
155	300	7.20	—	—	1730.0	—	—	—	—	—	—	—	—	1.2	—	—
0	0	1.08	—	—	170.0	—	0	—	—	—	—	—	—	21.0	—	—
3	0	0.00	—	—	470.0	—	0	0.25	—	0.23	2.00	—	—	0	—	—
0	20	0.72	—	—	700.0	—	0	—	—	—	—	—	—	3.6	—	—
40	100	3.60	—	—	1220.0	—	—	—	—	—	—	—	—	6.0	—	—
45	600	0.00	—	—	340.0	—	0	—	—	—	—	—	—	0	—	—
130	100	7.20	—	—	980.0	—	—	—	—	—	—	—	—	9.0	—	—
145	279	4.29	—	—	1960.0	—	—	—	—	—	—	—	—	16.7	—	—
30	401	0.00	—	—	234.0	—	0	—	—	—	—	—	—	0	—	—
85	200	5.40	—	—	1410.0	—	60.0	—	—	—	—	—	—	1.2	—	—
270	150	2.70	—	—	1190.0	—	—	—	—	—	—	—	—	0	—	—
240	80	2.70	—	—	740.0	—	—	—	—	—	—	—	—	0	—	—
255	150	2.70	—	—	970.0	—	—	—	—	—	—	—	—	0	—	—
5	60	1.80	—	—	930.0	—	0	—	—	—	—	—	—	0	—	—
300	200	3.60	—	—	1360.0	—	—	—	—	—	—	—	—	0	—	—
10	40	0.36	—	—	110.0	—	—	—	—	—	—	—	—	4.8	—	—
65	200	3.60	—	—	1020.0	—	—	—	—	—	—	—	—	6.0	—	—
65	150	0.72	—	—	620.0	—	—	—	—	—	—	—	—	30	—	—
65	80	2.70	—	—	940.0	—	—	—	—	—	—	—	—	6.0	—	—
35	60	2.70	—	—	1270.0	—	0	—	—	—	—	—	—	0	—	—
50	150	2.70	—	—	1500.0	—	—	—	—	—	—	—	—	0	—	—
80	500	3.60	—	—	1350.0	—	—	—	—	—	—	—	—	1.2	—	—
60	100	2.70	—	—	1300.0	—	—	—	—	—	—	—	—	2.4	—	—
65	150	1.80	—	—	880.0	—	—	—	—	—	—	—	—	0	—	—
60	100	2.70	—	—	1300.0	—	—	—	—	—	—	—	—	0	—	—
80	150	1.44	—	—	1070.0	—	—	—	—	—	—	—	—	6.0	—	—
65	40	1.44	—	—	940.0	—	—	—	—	—	—	—	—	0	—	—
25	60	0.36	—	—	220.0	—	—	—	—	—	—	—	—	36.0	—	—
0	0	0.36	—	—	5.0	—	0	—	—	—	—	—	—	15.0	—	—

TABLE A-1 Table of Food Composition *(continued)*

(Computer code is for Cengage Diet Analysis Plus program)

DA+ Code	Food Description	QTY	Measure	Wt (g)	H_2O (g)	Ener (cal)	Prot (g)	Carb (g)	Fiber (g)	Fat (g)	Fat Breakdown (g) Sat	Mono	Poly
Fast Food—continued													
38755	Hashbrowns	1	serving(s)	84	—	260	2.0	25.0	3.0	17.0	3.5	—	—
38765	Hearty breast of chicken soup	1	cup(s)	241	—	140	8.0	18.0	1.0	3.5	1.0	—	—
38741	Hot buttered biscuit	1	item(s)	79	—	270	4.0	38.0	1.0	12.0	3.0	—	—
38778	IceDream, small cone	1	item(s)	135	—	160	4.0	28.0	0	4.0	2.0	—	—
38774	IceDream, small cup	1	serving(s)	227	—	240	6.0	41.0	0	6.0	3.5	—	—
38775	Lemonade, small	1	cup(s)	255	—	170	0	41.0	0	0.5	0	—	—
38777	Nuggets	8	item(s)	113	—	260	26.0	12.0	0.5	12.0	2.5	—	—
38769	Side salad	1	item(s)	108	—	60	3.0	4.0	2.0	3.0	1.5	—	—
38767	Southwest chargrilled salad	1	item(s)	303	—	240	25.0	17.0	5.0	8.0	3.5	—	—
40481	Spicy chicken cool wrap	1	serving(s)	230	—	380	30.0	52.0	3.0	6.0	3.0	—	—
38772	Waffle potato fries, small, salted	1	serving(s)	85	—	270	3.0	34.0	4.0	13.0	3.0	—	—
	Cinnabon												
39572	Caramellata Chill w/whipped cream	16	fluid ounce(s)	480	—	406	10.0	61.0	0	14.0	8.0	—	—
39571	Cinnabon Bites	1	serving(s)	149	—	510	8.0	77.0	2.0	19.0	5.0	—	—
39570	Cinnabon Stix	5	item(s)	85	—	379	6.0	41.0	1.0	21.0	6.0	—	—
39567	Classic roll	1	item(s)	221	—	813	15.0	117.0	4.0	32.0	8.0	—	—
39568	Minibon	1	item(s)	92	—	339	6.0	49.0	2.0	13.0	3.0	—	—
39573	Mochalatta Chill w/whipped cream	16	fluid ounce(s)	480	—	362	9.0	55.0	0	13.0	8.0	—	—
39569	Pecanbon	1	item(s)	272	—	1100	16.0	141.0	8.0	56.0	10.0	—	—
	Dairy Queen												
1466	Banana split	1	item(s)	369	—	510	8.0	96.0	3.0	12.0	8.0	—	—
38552	Brownie Earthquake®	1	serving(s)	304	—	740	10.0	112.0	0	27.0	16.0	—	—
38561	Chocolate chip cookie dough blizzard,® small	1	item(s)	319	—	720	12.0	105.0	0	28.0	14.0	—	—
1464	Chocolate malt, small	1	item(s)	418	—	640	15.0	111.0	1.0	16.0	11.0	—	—
38541	Chocolate shake, small	1	item(s)	397	—	560	13.0	93.0	1.0	15.0	10.0	—	—
17257	Chocolate soft serve	½	cup(s)	94	—	150	4.0	22.0	0	5.0	3.5	—	—
1463	Chocolate sundae, small	1	item(s)	163	—	280	5.0	49.0	0	7.0	4.5	—	—
1462	Dipped cone, small	1	item(s)	156	—	340	6.0	42.0	1.0	17.0	9.0	4.0	3.0
38555	Oreo cookies blizzard, small	1	item(s)	283	—	570	11.0	83.0	0.5	21.0	10.0	—	—
38547	Royal Treats Peanut Buster® Parfait	1	item(s)	305	—	730	16.0	99.0	2.0	31.0	17.0	—	—
17256	Vanilla soft serve	½	cup(s)	94	—	140	3.0	22.0	0	4.5	3.0	—	—
	Domino's												
31606	Barbeque buffalo wings	1	item(s)	25	—	50	6.0	2.0	0	2.5	0.5	—	—
31604	Breadsticks	1	item(s)	30	—	115	2.0	12.0	0	6.3	1.1	—	—
37551	Buffalo Chicken Kickers	1	item(s)	24	—	47	4.0	3.0	0	2.0	0.5	—	—
37548	CinnaStix	1	item(s)	30	—	123	2.0	15.0	1.0	6.1	1.1	—	—
37549	Dot, cinnamon	1	item(s)	28	7.6	99	1.9	14.9	0.7	3.7	0.7	—	—
31605	Double cheesy bread	1	item(s)	35	—	123	4.0	13.0	0	6.5	1.9	—	—
31607	Hot buffalo wings	1	item(s)	25	—	45	5.0	1.0	0	2.5	0.5	—	—
	Domino's Classic hand tossed pizza												
31573	America's favorite feast, 12″	1	slice(s)	102	—	257	10.0	29.0	2.0	11.5	4.5	—	—
31574	America's favorite feast, 14″	1	slice(s)	141	—	353	14.0	39.0	2.0	16.0	6.0	—	—
37543	Bacon cheeseburger feast, 12″	1	slice(s)	99	—	273	12.0	28.0	2.0	13.0	5.5	—	—
37545	Bacon cheeseburger feast, 14″	1	slice(s)	137	—	379	17.0	38.0	2.0	18.0	8.0	—	—
37546	Barbeque feast, 12″	1	slice(s)	96	—	252	11.0	31.0	1.0	10.0	4.5	—	—
37547	Barbeque feast, 14″	1	slice(s)	131	—	344	14.0	43.0	2.0	13.5	6.0	—	—
31569	Cheese, 12″	1	slice(s)	55	—	160	6.0	28.0	1.0	3.0	1.0	—	—
31570	Cheese, 14″	1	slice(s)	75	—	220	8.0	38.0	2.0	4.0	1.0	—	—
37538	Deluxe feast, 12″	1	slice(s)	201	101.8	465	19.5	57.4	3.5	18.2	7.7	—	—
37540	Deluxe feast, 14″	1	slice(s)	273	138.4	627	26.4	78.3	4.7	24.1	10.2	—	—
31685	Deluxe, 12″	1	slice(s)	100	—	234	9.0	29.0	2.0	9.5	3.5	—	—
31694	Deluxe, 14″	1	slice(s)	136	—	316	13.0	39.0	2.0	12.5	5.0	—	—
31686	Extravaganzza, 12″	1	slice(s)	122	—	289	13.0	30.0	2.0	14.0	5.5	—	—
31695	Extravaganzza, 14″	1	slice(s)	165	—	388	17.0	40.0	3.0	18.5	7.5	—	—
31575	Hawaiian feast, 12″	1	slice(s)	102	—	223	10.0	30.0	2.0	8.0	3.5	—	—
31576	Hawaiian feast, 14″	1	slice(s)	141	—	309	14.0	41.0	2.0	11.0	4.5	—	—
31687	Meatzza, 12″	1	slice(s)	108	—	281	13.0	29.0	2.0	13.5	5.5	—	—

Chol (mg)	Calc (mg)	Iron (mg)	Magn (mg)	Pota (mg)	Sodi (mg)	Zinc (mg)	Vit A (µg)	Thia (mg)	Vit E (mg α)	Ribo (mg)	Niac (mg)	Vit B_6 (mg)	Fola (µg DFE)	Vit C (mg)	Vit B_{12} (µg)	Sele (µg)
5	20	0.72	—	—	380.0	—	—	—	—	—	—	—	—	0	—	—
25	40	1.08	—	—	900.0	—	—	—	—	—	—	—	—	0	—	—
0	60	1.80	—	—	660.0	—	0	—	—	—	—	—	—	0	—	—
15	100	0.36	—	—	80.0	—	—	—	—	—	—	—	—	0	—	—
25	200	0.36	—	—	105.0	—	—	—	—	—	—	—	—	0	—	—
0	0	0.36	—	—	10.0	—	0	—	—	—	—	—	—	15.0	—	—
70	40	1.08	—	—	1090.0	—	0	—	—	—	—	—	—	0	—	—
10	100	0.00	—	—	75.0	—	—	—	—	—	—	—	—	15.0	—	—
60	200	1.08	—	—	770.0	—	—	—	—	—	—	—	—	24.0	—	—
60	200	3.60	—	—	1090.0	—	—	—	—	—	—	—	—	3.6	—	—
0	20	1.08	—	—	115.0	—	0	—	—	—	—	—	—	1.2	—	—
46	—	—	—	—	187.0	—	—	—	—	—	—	—	—	—	—	—
35	—	—	—	—	530.0	—	—	—	—	—	—	—	—	—	—	—
16	—	—	—	—	413.0	—	—	—	—	—	—	—	—	—	—	—
67	—	—	—	—	801.0	—	—	—	—	—	—	—	—	—	—	—
27	—	—	—	—	337.0	—	—	—	—	—	—	—	—	—	—	—
46	—	—	—	—	252.0	—	—	—	—	—	—	—	—	—	—	—
63	—	—	—	—	600.0	—	—	—	—	—	—	—	—	—	—	—
30	250	1.80	—	—	180.0	—	—	—	—	—	—	—	—	15.0	—	—
50	250	1.80	—	—	350.0	—	—	—	—	—	—	—	—	0	—	—
50	350	2.70	—	—	370.0	—	—	—	—	—	—	—	—	1.2	—	—
55	450	1.80	—	—	340.0	—	—	—	—	—	—	—	—	2.4	—	—
50	450	1.44	—	—	280.0	—	—	—	—	—	—	—	—	2.4	—	—
15	100	0.72	—	—	75.0	—	—	—	—	—	—	—	—	0	—	—
20	200	1.08	—	—	140.0	—	—	—	—	—	—	—	—	0	—	—
20	200	1.08	—	—	130.0	—	—	—	—	—	—	—	—	1.2	—	—
40	350	2.70	—	—	430.0	—	—	—	—	—	—	—	—	1.2	—	—
35	300	1.80	—	—	400.0	—	—	—	—	—	—	—	—	1.2	—	—
15	150	0.72	—	—	70.0	—	—	—	—	—	—	—	—	0	—	—
26	10	0.36	—	—	175.5	—	—	—	—	—	—	—	—	0	—	—
0	0	0.72	—	—	122.1	—	—	—	—	—	—	—	—	0	—	—
9	0	0.00	—	—	162.5	—	—	—	—	—	—	—	—	0	—	—
0	0	0.72	—	—	111.4	—	—	—	—	—	—	—	—	0	—	—
0	6	0.59	—	—	85.7	—	—	—	—	—	—	—	—	0	—	—
6	40	0.72	—	—	162.3	—	—	—	—	—	—	—	—	0	—	—
26	10	0.36	—	—	254.5	—	—	—	—	—	—	—	—	1.2	—	—
																—
22	100	1.80	—	—	625.5	—	—	—	—	—	—	—	—	0.6	—	—
31	140	2.52	—	—	865.5	—	—	—	—	—	—	—	—	0.6	—	—
27	140	1.80	—	—	634.0	—	—	—	—	—	—	—	—	0	—	—
38	190	2.52	—	—	900.0	—	—	—	—	—	—	—	—	0	—	—
20	140	1.62	—	—	600.0	—	—	—	—	—	—	—	—	0.6	—	—
27	190	2.16	—	—	831.5	—	—	—	—	—	—	—	—	0.6	—	—
0	0	1.80	—	—	110.0	—	0	—	—	—	—	—	—	0	—	—
0	0	2.70	—	—	150.0	—	0	—	—	—	—	—	—	0	—	—
40	199	3.56	—	—	1063.1	—	—	—	—	—	—	—	—	1.4	—	—
53	276	4.84	—	—	1432.2	—	—	—	—	—	—	—	—	1.8	—	—
17	100	1.80	—	—	541.5	—	—	—	—	—	—	—	—	0.6	—	—
23	130	2.34	—	—	728.5	—	—	—	—	—	—	—	—	1.2	—	—
28	140	1.98	—	—	764.0	—	—	—	—	—	—	—	—	0.6	—	—
37	190	2.70	—	—	1014.0	—	—	—	—	—	—	—	—	1.2	—	—
16	130	1.62	—	—	546.5	—	—	—	—	—	—	—	—	1.2	—	—
23	180	2.34	—	—	765.0	—	—	—	—	—	—	—	—	1.2	—	—
28	130	1.80	—	—	739.5	—	—	—	—	—	—	—	—	0	—	—

TABLE A-1 Table of Food Composition *(continued)*

(Computer code is for Cengage Diet Analysis Plus program)

DA+ Code	Food Description	QTY	Measure	Wt (g)	H_2O (g)	Ener (cal)	Prot (g)	Carb (g)	Fiber (g)	Fat (g)	Fat Breakdown (g) Sat	Mono	Poly
Fast Food—continued													
31696	Meatzza, 14″	1	slice(s)	146	—	378	17.0	39.0	2.0	18.0	7.5	—	—
31571	Pepperoni feast, extra pepperoni and cheese, 12″	1	slice(s)	98	—	265	11.0	28.0	2.0	12.5	5.0	—	—
31572	Pepperoni feast, extra pepperoni and cheese, 14″	1	slice(s)	135	—	363	16.0	39.0	2.0	17.0	7.0	—	—
31577	Vegi feast, 12″	1	slice(s)	102	—	218	9.0	29.0	2.0	8.0	3.5	—	—
31578	Vegi feast, 14″	1	slice(s)	139	—	300	13.0	40.0	3.0	11.0	4.5	—	—
	Domino's thin crust pizza												
31583	America's favorite, 12″	1	slice(s)	72	—	208	8.0	15.0	1.0	13.5	5.0	—	—
31584	America's favorite, 14″	1	slice(s)	100	—	285	11.0	20.0	2.0	18.5	7.0	—	—
31579	Cheese, 12″	1	slice(s)	49	—	137	5.0	14.0	1.0	7.0	2.5	—	—
31580	Cheese, 14″	1	slice(s)	68	27.0	214	8.8	19.0	1.4	11.4	4.6	2.9	2.5
31688	Deluxe, 12″	1	slice(s)	70	—	185	7.0	15.0	1.0	11.5	4.0	—	—
31697	Deluxe, 14″	1	slice(s)	94	—	248	10.0	20.0	2.0	15.0	5.5	—	—
31689	Extravaganzza, 12″	1	slice(s)	92	—	240	11.0	16.0	1.0	15.5	6.0	—	—
31698	Extravaganzza, 14″	1	slice(s)	123	—	320	14.0	21.0	2.0	20.5	8.0	—	—
31585	Hawaiian, 12″	1	slice(s)	71	—	174	8.0	16.0	1.0	9.5	3.5	—	—
31586	Hawaiian, 14″	1	slice(s)	100	—	240	11.0	21.0	2.0	13.0	5.0	—	—
31690	Meatzza, 12″	1	slice(s)	78	—	232	11.0	15.0	1.0	15.0	6.0	—	—
31699	Meatzza, 14″	1	slice(s)	104	—	310	14.0	20.0	2.0	20	8.0	—	—
31581	Pepperoni, extra pepperoni and cheese, 12″	1	slice(s)	68	—	216	9.0	14.0	1.0	14.0	5.5	—	—
31582	Pepperoni, extra pepperoni and cheese, 14″	1	slice(s)	93	—	295	13.0	20.0	1.0	19.0	7.5	—	—
31587	Vegi, 12″	1	slice(s)	71	—	168	7.0	15.0	1.0	9.5	3.5	—	—
31588	Vegi, 14″	1	slice(s)	97	—	231	10.0	21.0	2.0	13.5	5.0	—	—
	Domino's Ultimate deep dish pizza												
31596	America's favorite, 12″	1	slice(s)	115	—	309	12.0	29.0	2.0	17.0	6.0	—	—
31702	America's favorite, 14″	1	slice(s)	162	—	433	17.0	42.0	3.0	23.5	8.0	—	—
31590	Cheese, 12″	1	slice(s)	90	—	238	9.0	28.0	2.0	11.0	3.5	—	—
31591	Cheese, 14″	1	slice(s)	128	53.9	351	14.5	41.0	2.9	13.2	5.2	3.8	2.5
31589	Cheese, 6″	1	item(s)	215	—	598	22.9	68.4	3.9	27.6	9.9	—	—
31691	Deluxe, 12″	1	slice(s)	122	—	287	11.0	29.0	2.0	15.0	5.0	—	—
31700	Deluxe, 14″	1	slice(s)	156	—	396	15.0	42.0	3.0	20.0	7.0	—	—
31692	Extravaganzza, 12″	1	slice(s)	136	—	341	14.0	30.0	2.0	19.0	7.0	—	—
31701	Extravaganzza, 14″	1	slice(s)	186	—	468	20.0	43.0	3.0	25.5	9.5	—	—
31599	Hawaiian, 12″	1	slice(s)	114	—	275	12.0	30.0	2.0	13.0	5.0	—	—
31600	Hawaiian, 14″	1	slice(s)	162	—	389	17.0	43.0	3.0	18.0	6.5	—	—
31693	Meatzza, 12″	1	slice(s)	121	—	333	14.0	29.0	2.0	19.0	7.0	—	—
31703	Meatzza, 14″	1	slice(s)	167	—	458	19.0	42.0	3.0	25.0	9.5	—	—
31593	Pepperoni, extra pepperoni and cheese, 12″	1	slice(s)	110	—	317	13.0	29.0	2.0	17.5	6.5	—	—
31594	Pepperoni, extra pepperoni and cheese, 14″	1	slice(s)	155	—	443	18.0	42.0	3.0	24.0	9.0	—	—
31602	Vegi, 12″	1	slice(s)	114	—	270	11.0	30.0	2.0	13.5	5.0	—	—
31603	Vegi, 14″	1	slice(s)	159	—	380	15.0	43.0	3.0	18.0	6.5	—	—
31598	With ham and pineapple tidbits, 6″	1	item(s)	430	—	619	25.2	69.9	4.0	28.3	10.2	—	—
31595	With Italian sausage, 6″	1	item(s)	430	—	642	24.8	69.6	4.2	31.1	11.3	—	—
31592	With pepperoni, 6″	1	item(s)	430	—	647	25.1	68.5	3.9	32.0	11.7	—	—
31601	With vegetables, 6″	1	item(s)	430	—	619	23.4	70.8	4.6	28.7	10.1	—	—
	In-n-Out Burger												
34391	Cheeseburger with mustard and ketchup	1	serving(s)	268	—	400	22.0	41.0	3.0	18.0	9.0	—	—
34374	Cheeseburger	1	serving(s)	268	—	480	22.0	39.0	3.0	27.0	10.0	—	—

PAGE KEY: A-2 = Breads/Baked Goods A-8 = Cereal/Rice/Pasta A-12 = Fruit A-18 = Vegetables/Legumes A-28 = Nuts/Seeds A-30 = Vegetarian A-32 = Dairy A-40 = Eggs A-40 = Seafood A-44 = Meats A-48 = Poultry A-48 = Processed Meats A-50 = Beverages A-54 = Fats/Oils A-56 = Sweets A-58 = Spices/Condiments/Sauces A-62 = Mixed Foods/Soups/Sandwiches A-68 = Fast Food A-88 = Convenience A-90 = Baby Foods

A

Chol (mg)	Calc (mg)	Iron (mg)	Magn (mg)	Pota (mg)	Sodi (mg)	Zinc (mg)	Vit A (µg)	Thia (mg)	Vit E (mg α)	Ribo (mg)	Niac (mg)	Vit B_6 (mg)	Fola (µg DFE)	Vit C (mg)	Vit B_{12} (µg)	Sele (µg)
37	190	2.52	—	—	983.5	—	—	—	—	—	—	—	—	0	—	—
24	130	1.62	—	—	670.0	—	70.9	—	—	—	—	—	—	0	—	—
33	180	2.34	—	—	920.0	—	104.7	—	—	—	—	—	—	0	—	—
13	130	1.62	—	—	489.0	—	—	—	—	—	—	—	—	0.6	—	—
18	180	2.34	—	—	678.0	—	—	—	—	—	—	—	—	0.6	—	—
23	100	0.90	—	—	533.0	—	—	—	—	—	—	—	—	2.4	—	—
32	140	1.26	—	—	736.5	—	—	—	—	—	—	—	—	3.0	—	—
10	90	0.54	—	—	292.5	—	60.0	—	—	—	—	—	—	1.8	—	—
14	151	0.48	17.7	125.1	338.0	0.02	64.6	0.05	1.01	0.07	0.69	—	—	2.4	0.5	24.1
19	100	0.90	—	—	449.0	—	—	—	—	—	—	—	—	2.4	—	—
24	130	1.08	—	—	601.0	—	—	—	—	—	—	—	—	3.6	—	—
29	140	1.08	—	—	671.5	—	—	—	—	—	—	—	—	2.4	—	—
38	190	1.44	—	—	886.5	—	—	—	—	—	—	—	—	3.6	—	—
17	130	0.72	—	—	454.0	—	—	—	—	—	—	—	—	3.0	—	—
24	180	0.90	—	—	637.5	—	—	—	—	—	—	—	—	3.6	—	—
29	140	0.90	—	—	647.0	—	—	—	—	—	—	—	—	1.8	—	—
38	190	1.26	—	—	865.5	—	—	—	—	—	—	—	—	2.4	—	—
26	130	0.72	—	—	577.0	—	80.0	—	—	—	—	—	—	1.8	—	—
35	80	1.08	—	—	792.5	—	105.8	—	—	—	—	—	—	2.4	—	—
14	130	0.72	—	—	396.5	—	—	—	—	—	—	—	—	2.4	—	—
19	180	1.08	—	—	550.5	—	—	—	—	—	—	—	—	3.0	—	—
25	120	2.34	—	—	796.5	—	—	—	—	—	—	—	—	0.6	—	—
34	170	3.24	—	—	1110.0	—	—	—	—	—	—	—	—	0.6	—	—
11	110	1.98	—	—	555.5	—	70.0	—	—	—	—	—	—	0	—	—
18	189	3.78	32.0	209.9	718.1	1.75	99.8	0.29	1.13	0.31	5.44	—	—	0	0.6	45.6
36	295	4.67	—	—	1341.4	—	174.0	—	—	—	—	—	—	0.5	—	—
20	120	2.16	—	—	712.0	—	—	—	—	—	—	—	—	1.2	—	—
26	170	3.06	—	—	974.5	—	—	—	—	—	—	—	—	1.2	—	—
31	160	2.52	—	—	934.5	—	—	—	—	—	—	—	—	1.2	—	—
40	220	3.42	—	—	1260.0	—	—	—	—	—	—	—	—	1.2	—	—
19	150	1.98	—	—	717.0	—	—	—	—	—	—	—	—	1.2	—	—
26	210	2.88	—	—	1011.0	—	—	—	—	—	—	—	—	1.8	—	—
31	160	2.34	—	—	910.5	—	—	—	—	—	—	—	—	0	—	—
40	220	3.24	—	—	1230.0	—	—	—	—	—	—	—	—	0.6	—	—
27	150	2.16	—	—	840.5	—	86.5	—	—	—	—	—	—	0	—	—
37	220	3.06	—	—	1166.0	—	115.4	—	—	—	—	—	—	0.6	—	—
15	150	2.16	—	—	659.5	—	—	—	—	—	—	—	—	0.6	—	—
21	220	3.06	—	—	924.0	—	—	—	—	—	—	—	—	1.2	—	—
43	298	4.84	—	—	1497.8	—	—	—	—	—	—	—	—	1.5	—	—
45	302	4.89	—	—	1478.1	—	—	—	—	—	—	—	—	0.6	—	—
47	299	4.81	—	—	1523.7	—	167.9	—	—	—	—	—	—	0.6	—	—
36	307	5.10	—	—	1472.5	—	—	—	—	—	—	—	—	4.7	—	—
60	200	3.60	—	—	1080.0	—	—	—	—	—	—	—	—	12.0	—	—
60	200	3.60	—	—	1000.0	—	—	—	—	—	—	—	—	9.0	—	—

TABLE A-1 Table of Food Composition *(continued)*

(Computer code is for Cengage Diet Analysis Plus program)

DA+ Code	Food Description	QTY	Measure	Wt (g)	H_2O (g)	Ener (cal)	Prot (g)	Carb (g)	Fiber (g)	Fat (g)	Fat Breakdown (g) Sat	Mono	Poly
Fast Food—continued													
34390	Cheeseburger, lettuce leaves instead of buns	1	serving(s)	300	—	330	18.0	11.0	3.0	25.0	9.0	—	—
34377	Chocolate shake	1	serving(s)	425	—	690	9.0	83.0	0	36.0	24.0	—	—
34375	Double-Double cheeseburger	1	serving(s)	330	—	670	37.0	39.0	3.0	41.0	18.0	—	—
34393	Double-Double cheeseburger with mustard and ketchup	1	serving(s)	330	—	590	37.0	41.0	3.0	32.0	17.0	—	—
34392	Double-Double cheeseburger, lettuce leaves instead of buns	1	serving(s)	362	—	520	33.0	11.0	3.0	39.0	17.0	—	—
34376	French fries	1	serving(s)	125	—	400	7.0	54.0	2.0	18.0	5.0	—	—
34373	Hamburger	1	item(s)	243	—	390	16.0	39.0	3.0	19.0	5.0	—	—
34389	Hamburger with mustard and ketchup	1	serving(s)	243	—	310	16.0	41.0	3.0	10.0	4.0	—	—
34388	Hamburger, lettuce leaves instead of buns	1	serving(s)	275	—	240	13.0	11.0	3.0	17.0	4.0	—	—
34379	Strawberry shake	1	serving(s)	425	—	690	9.0	91.0	0	33.0	22.0	—	—
34378	Vanilla shake	1	serving(s)	425	—	680	9.0	78.0	0	37.0	25.0	—	—
	Jack in the Box												
30392	Bacon ultimate cheeseburger	1	item(s)	338	—	1090	46.0	53.0	2.0	77.0	30.0	—	—
1740	Breakfast Jack	1	item(s)	125	—	290	17.0	29.0	1.0	12.0	4.5	—	—
14074	Cheeseburger	1	item(s)	131	—	350	18.0	31.0	1.0	17.0	8.0	—	—
14106	Chicken breast strips, 4 pieces	1	serving(s)	201	—	500	35.0	36.0	3.0	25.0	6.0	—	—
37241	Chicken club salad, plain, without salad dressing	1	serving(s)	431	—	300	27.0	13.0	4.0	15.0	6.0	—	—
14064	Chicken sandwich	1	item(s)	145	—	400	15.0	38.0	2.0	21.0	4.5	—	—
14111	Chocolate ice cream shake, small	1	serving(s)	414	—	880	14.0	107.0	1.0	45.0	31.0	—	—
14073	Hamburger	1	item(s)	118	—	310	16.0	30.0	1.0	14.0	6.0	—	—
14090	Hash browns	1	serving(s)	57	—	150	1.0	13.0	2.0	10.0	2.5	—	—
14072	Jack's Spicy Chicken sandwich	1	item(s)	270	—	620	25.0	61.0	4.0	31.0	6.0	—	—
1468	Jumbo Jack hamburger	1	item(s)	261	—	600	21.0	51.0	3.0	35.0	12.0	—	—
1469	Jumbo Jack hamburger with cheese	1	item(s)	286	—	690	25.0	54.0	3.0	42.0	16.0	—	—
14099	Natural cut french fries, large	1	serving(s)	196	—	530	8.0	69.0	5.0	25.0	6.0	—	—
14098	Natural cut french fries, medium	1	serving(s)	133	—	360	5.0	47.0	4.0	17.0	4.0	—	—
1470	Onion rings	1	serving(s)	119	—	500	6.0	51.0	3.0	30.0	6.0	—	—
33141	Sausage, egg, and cheese biscuit	1	item(s)	234	—	740	27.0	35.0	2.0	55.0	17.0	—	—
14095	Seasoned curly fries, medium	1	serving(s)	125	—	400	6.0	45.0	5.0	23.0	5.0	—	—
14077	Sourdough Jack	1	item(s)	245	—	710	27.0	36.0	3.0	51.0	18.0	—	—
37249	Southwest chicken salad, plain, without salad dressing	1	serving(s)	488	—	300	24.0	29.0	7.0	11.0	5.0	—	—
14112	Strawberry ice cream shake, small	1	serving(s)	417	—	880	13.0	105.0	0	44.0	31.0	—	—
14078	Ultimate cheeseburger	1	item(s)	323	—	1010	40.0	53.0	2.0	71.0	28.0	—	—
14110	Vanilla ice cream shake, small	1	serving(s)	379	—	790	13.0	83.0	0	44.0	31.0	—	—
	Jamba Juice												
31645	Aloha Pineapple smoothie	24	fluid ounce(s)	730	—	500	8.0	117.0	4.0	1.5	1.0	—	—
31646	Banana Berry smoothie	24	fluid ounce(s)	719	—	480	5.0	112.0	4.0	1.0	0	—	—
31656	Berry Lime Sublime smoothie	24	fluid ounce(s)	728	—	460	3.0	106.0	5.0	2.0	1.0	—	—
31647	Carribean Passion smoothie	24	fluid ounce(s)	730	—	440	4.0	102.0	4.0	2.0	1.0	—	—
38422	Carrot juice	16	fluid ounce(s)	472	—	100	3.0	23.0	0	0.5	0	—	—
31648	Chocolate Moo'd smoothie	24	fluid ounce(s)	634	—	720	17.0	148.0	3.0	8.0	5.0	—	—
31649	Citrus Squeeze smoothie	24	fluid ounce(s)	727	—	470	5.0	110.0	4.0	2.0	1.0	—	—
31651	Coldbuster smoothie	24	fluid ounce(s)	724	—	430	5.0	100.0	5.0	2.5	1.0	—	—
31652	Cranberry Craze smoothie	24	fluid ounce(s)	793	—	460	6.0	104.0	4.0	0.5	0	—	—
31654	Jamba Powerboost smoothie	24	fluid ounce(s)	738	—	440	6.0	105.0	6.0	1.0	0	—	—
38423	Lemonade	16	fluid ounce(s)	483	—	300	1.0	75.0	0	0	0	0	0
31657	Mango-a-go-go smoothie	24	fluid ounce(s)	690	—	440	3.0	104.0	4.0	1.5	0.5	—	—
38424	Orange juice, freshly squeezed	16	fluid ounce(s)	496	—	220	3.0	52.0	0.5	1.0	0	—	—
38426	Orange/carrot juice	16	fluid ounce(s)	484	—	160	3.0	37.0	0	1.0	0	—	—
31660	Orange-a-peel smoothie	24	fluid ounce(s)	726	—	440	8.0	102.0	5.0	1.5	0	—	—
31662	Peach Pleasure smoothie	24	fluid ounce(s)	720	—	460	4.0	108.0	4.0	2.0	1.0	—	—
31665	Protein Berry Pizzaz smoothie	24	fluid ounce(s)	710	—	440	20.0	92.0	5.0	1.5	0	—	—
31668	Razzmatazz smoothie	24	fluid ounce(s)	730	—	480	3.0	112.0	4.0	2.0	1.0	—	—
31669	Strawberries Wild smoothie	24	fluid ounce(s)	725	—	450	6.0	105.0	4.0	0.5	0	—	—
38421	Strawberry Tsunami smoothie	24	fluid ounce(s)	740	—	530	4.0	128.0	4.0	2.0	1.0	—	—

Chol (mg)	Calc (mg)	Iron (mg)	Magn (mg)	Pota (mg)	Sodi (mg)	Zinc (mg)	Vit A (µg)	Thia (mg)	Vit E (mg α)	Ribo (mg)	Niac (mg)	Vit B_6 (mg)	Fola (µg DFE)	Vit C (mg)	Vit B_{12} (µg)	Sele (µg)
60	200	2.70	—	—	720.0	—	—	—	—	—	—	—	—	12.0	—	—
95	300	0.72	—	—	350.0	—	—	—	—	—	—	—	—	0	—	—
120	350	5.40	—	—	1440.0	—	—	—	—	—	—	—	—	9.0	—	—
115	350	5.40	—	—	1520.0	—	—	—	—	—	—	—	—	12.0	—	—
120	350	4.50	—	—	1160.0	—	—	—	—	—	—	—	—	12.0	—	—
0	20	1.80	—	—	245.0	—	0	—	—	—	—	—	—	0	—	—
40	40	3.60	—	—	650.0	—	—	—	—	—	—	—	—	9.0	—	—
35	40	3.60	—	—	730.0	—	—	—	—	—	—	—	—	12.0	—	—
40	40	2.70	—	—	370.0	—	—	—	—	—	—	—	—	12.0	—	—
85	300	0.00	—	—	280.0	—	—	—	—	—	—	—	—	0	—	—
90	300	0.00	—	—	390.0	—	—	—	—	—	—	—	—	0	—	—
140	308	7.38	—	540.0	2040.0	—	—	—	—	—	—	—	—	0.6	—	—
220	145	3.48	—	210.0	760.0	—	—	—	—	—	—	—	—	3.5	—	—
50	151	3.61	—	270.0	790.0	—	40.2	—	—	—	—	—	—	0	—	—
80	18	1.60	—	530.0	1260.0	—	—	—	—	—	—	—	—	1.1	—	—
65	280	3.35	—	560.0	880.0	—	—	—	—	—	—	—	—	50.4	—	—
35	100	2.70	—	240.0	730.0	—	—	—	—	—	—	—	—	4.8	—	—
135	460	0.47	—	840.0	330.0	—	—	—	—	—	—	—	—	0	—	—
40	100	3.60	—	250.0	600.0	—	0	—	—	—	—	—	—	0	—	—
0	10	0.18	—	190.0	230.0	—	0	—	—	—	—	—	—	0	—	—
50	150	1.80	—	450.0	1100.0	—	—	—	—	—	—	—	—	9.0	—	—
45	164	4.92	—	380.0	940.0	—	—	—	—	—	—	—	—	9.8	—	—
70	234	4.20	—	410.0	1310.0	—	—	—	—	—	—	—	—	8.4	—	—
0	20	1.42	—	1240.0	870.0	—	0	—	—	—	—	—	—	8.9	—	—
0	19	1.01	—	840.0	590.0	—	0	—	—	—	—	—	—	5.6	—	—
0	40	2.70	—	140.0	420.0	—	40.0	—	—	—	—	—	—	18.0	—	—
280	88	2.36	—	310.0	1430.0	—	—	—	—	—	—	—	—	0	—	—
0	40	1.80	—	580.0	890.0	—	—	—	—	—	—	—	—	0	—	—
75	200	4.50	—	430.0	1230.0	—	—	—	—	—	—	—	—	9.0	—	—
55	274	4.10	—	670.0	860.0	—	—	—	—	—	—	—	—	43.8	—	—
135	466	0.00	—	750.0	290.0	—	—	—	—	—	—	—	—	0	—	—
125	308	7.39	—	480.0	1580.0	—	—	—	—	—	—	—	—	0.6	—	—
135	532	0.00	—	750.0	280.0	—	—	—	—	—	—	—	—	0	—	—
5	200	1.80	60.0	1000.0	30.0	0.30	—	0.37	—	0.34	2.00	0.60	60.0	102.0	0	1.4
0	200	1.44	40.0	1010.0	115.0	0.60	—	0.09	—	0.25	0.80	0.70	24.0	15.0	0.2	1.4
5	200	1.80	16.0	510.0	35.0	0.30	—	0.06	—	0.25	6.00	0.70	140.0	54.0	0	1.4
5	100	1.80	24.0	810.0	60.0	0.30	—	0.09	—	0.25	5.00	0.50	100.0	78.0	0	1.4
0	150	2.70	80.0	1030.0	250.0	0.90	—	0.52	—	0.25	5.00	0.70	80.0	18.0	0	5.6
30	500	1.08	60.0	810.0	380.0	1.50	—	0.22	—	0.76	0.40	0.16	16.0	6.0	1.5	4.2
5	100	1.80	80.0	1170.0	35.0	0.30	—	0.37	—	0.34	1.90	0.60	100.0	180.0	0	1.4
5	100	1.08	60.0	1260.0	35.0	16.50	—	0.37	—	0.34	3.00	0.40	121.5	1302.0	0	1.4
0	250	1.44	16.0	500.0	50.0	0.30	—	0.03	—	0.25	5.00	0.60	120.0	54.0	0	1.4
0	1200	1.80	480.0	1070.0	45.0	16.50	—	5.55	—	6.12	68.00	7.40	640.0	288.0	10.8	77.0
0	20	0.00	8.0	200.0	10.0	0.00	—	0.03	—	0.17	14.00	1.80	320.0	36.0	0	0
5	100	1.08	24.0	780.0	50.0	0.30	—	0.15	—	0.25	5.00	0.70	120.0	72.0	0	1.4
0	60	1.08	60.0	990.0	0	0.30	—	0.45	—	0.13	2.00	0.20	160.0	246.0	0	0
0	100	1.80	60.0	1010.0	125.0	0.60	—	0.45	—	0.25	3.00	0.50	120.0	132.0	0	2.8
0	250	1.80	80.0	1380.0	160.0	0.90	—	0.45	—	0.42	2.00	0.50	140.0	240.0	0.6	1.4
5	100	0.72	32.0	740.0	60.0	0.30	—	0.06	—	0.25	4.00	0.60	80.0	18.0	0	1.4
0	1100	2.62	60.0	650.0	240.0	0.58	—	0.08	—	0.17	1.20	0.70	58.3	60.0	0	5.6
5	150	1.80	32.0	810.0	70.0	0.30	—	0.09	—	0.34	6.00	1.00	160.0	60.0	0	1.4
5	250	1.80	40.0	1050.0	180.0	0.90	—	0.12	—	0.34	0.80	0.40	40.0	60.0	0.6	1.4
5	100	1.08	24.0	480.0	10.0	0.30	—	0.06	—	0.34	14.00	1.80	320.0	90.0	0	1.4

TABLE A-1 **Table of Food Composition** *(continued)*

(Computer code is for Cengage Diet Analysis Plus program)

DA+ Code	Food Description	QTY	Measure	Wt (g)	H_2O (g)	Ener (cal)	Prot (g)	Carb (g)	Fiber (g)	Fat (g)	Fat Breakdown (g) Sat	Mono	Poly
	Fast Food—continued												
38427	Vibrant C juice	16	fluid ounce(s)	448	—	210	2.0	50.0	1.0	0	0	0	0
38428	Wheatgrass juice, freshly squeezed	1	ounce(s)	28	—	5	0.5	1.0	0	0	0	0	0
	Kentucky Fried Chicken (KFC)												
31850	BBQ baked beans	1	serving(s)	136	—	220	8.0	45.0	7.0	1.0	0	—	—
31853	Biscuit	1	item(s)	57	—	220	4.0	24.0	1.0	11.0	2.5	—	—
51223	Boneless Fiery Buffalo Wings	6	item(s)	211	—	530	30.0	44.0	3.0	26.0	5.0	—	—
39386	Boneless Honey BBQ Wings	6	item(s)	213	—	570	30.0	54.0	5.0	26.0	5.0	—	—
51224	Boneless Sweet & Spicy Wings	6	item(s)	203	—	550	30.0	50.0	3.0	26.0	5.0	—	—
31851	Cole slaw	1	serving(s)	130	—	180	1.0	22.0	3.0	10.0	1.5	—	—
31842	Colonel's Crispy Strips	3	item(s)	151	—	370	28.0	17.0	1.0	20.0	4.0	—	—
31849	Corn on the cob	1	item(s)	162	—	150	5.0	26.0	7.0	3.0	1.0	—	—
51221	Double Crunch sandwich	1	item(s)	213	—	520	27.0	39.0	3.0	29.0	5.0	—	—
3761	Extra Crispy chicken, breast	1	item(s)	162	—	370	33.0	10.0	2.0	22.0	5.0	—	—
3762	Extra Crispy chicken, drumstick	1	item(s)	60	—	150	12.0	4.0	0	10.0	2.5	—	—
3763	Extra Crispy chicken, thigh	1	item(s)	114	—	290	17.0	16.0	1.0	18.0	4.0	—	—
3764	Extra Crispy chicken, whole wing	1	item(s)	52	—	150	11.0	11.0	1.0	7.0	1.5	—	—
51218	Famous Bowls mashed potatoes with gravy	1	serving(s)	531	—	720	26.0	79.0	6.0	34.0	9.0	—	—
51219	Famous Bowls rice with gravy	1	serving(s)	384	—	610	25.0	67.0	5.0	27.0	8.0	—	—
31841	Honey BBQ chicken sandwich	1	item(s)	147	—	290	23.0	40.0	2.0	4.0	1.0	—	—
31833	Honey BBQ wing pieces	6	item(s)	157	—	460	27.0	26.0	3.0	27.0	6.0	—	—
10859	Hot wings pieces	6	piece(s)	134	—	450	26.0	19.0	2.0	30.0	7.0	—	—
42382	KFC Snacker sandwich	1	serving(s)	119	—	320	14.0	29.0	2.0	17.0	3.0	—	—
31848	Macaroni and cheese	1	serving(s)	136	—	180	8.0	18.0	0	8.0	3.5	—	—
31847	Mashed potatoes with gravy	1	serving(s)	151	—	140	2.0	20.0	1.0	5.0	1.0	—	—
10825	Original Recipe chicken, breast	1	item(s)	161	—	340	38.0	9.0	2.0	17.0	4.0	—	—
10826	Original Recipe chicken, drumstick	1	item(s)	59	—	140	13.0	3.0	0	8.0	2.0	—	—
10827	Original Recipe chicken, thigh	1	item(s)	126	—	350	19.0	7.0	1.0	27.0	7.0	—	—
10828	Original Recipe chicken, whole wing	1	item(s)	47	—	140	10.0	4.0	0	9.0	2.0	—	—
51222	Oven roasted Twister chicken wrap	1	item(s)	269	—	520	30.0	46.0	4.0	23.0	3.5	—	—
31844	Popcorn chicken, small or individual	1	item(s)	114	—	370	19.0	21.0	2.0	24.0	4.5	—	—
31852	Potato salad	1	serving(s)	128	—	180	2.0	22.0	2.0	9.0	1.5	—	—
10845	Potato wedges, small	1	serving(s)	102	—	250	4.0	32.0	3.0	12.0	2.0	—	—
31839	Tender Roast chicken sandwich with sauce	1	item(s)	236	—	430	37.0	29.0	2.0	18.0	3.5	—	—
	Long John Silver												
39392	Baked cod	1	serving(s)	101	—	120	22.0	1.0	0	4.5	1.0	—	—
3777	Batter dipped fish sandwich	1	item(s)	177	—	470	18.0	48.0	3.0	23.0	5.0	—	—
37568	Battered fish	1	item(s)	92	—	260	12.0	17.0	0.5	16.0	4.0	—	—
37569	Breaded clams	1	serving(s)	85	—	240	8.0	22.0	1.0	13.0	2.0	—	—
37566	Chicken plank	1	item(s)	52	—	140	8.0	9.0	0.5	8.0	2.0	—	—
39404	Clam chowder	1	item(s)	227	—	220	9.0	23.0	0	10.0	4.0	—	—
39398	Cocktail sauce	1	ounce(s)	28	—	25	0	6.0	0	0	0	0	0
3770	Coleslaw	1	serving(s)	113	—	200	1.0	15.0	3.0	15.0	2.5	1.8	4.1
39400	French fries, large	1	item(s)	142	—	390	4.0	56.0	5.0	17.0	4.0	—	—
3774	Fries, regular	1	serving(s)	85	—	230	3.0	34.0	3.0	10.0	2.5	—	—
3779	Hushpuppy	1	piece(s)	23	—	60	1.0	9.0	1.0	2.5	0.5	—	—
3781	Shrimp, batter-dipped, 1 piece	1	piece(s)	14	—	45	2.0	3.0	0	3.0	1.0	—	—
39399	Tartar sauce	1	ounce(s)	28	—	100	0	4.0	0	9.0	1.5	—	—
39395	Ultimate Fish sandwich	1	item(s)	199	—	530	21.0	49.0	3.0	28.0	8.0	—	—
	McDonald's												
50828	Asian salad with grilled chicken	1	item(s)	362	—	290	31.0	23.0	6.0	10.0	1.0	—	—
2247	Barbecue sauce	1	item(s)	28	—	45	0	11.0	0	0	0	0	0
737	Big Mac hamburger	1	item(s)	219	—	560	25.0	47.0	3.0	30.0	10.0	—	—
29777	Caesar salad dressing	1	package(s)	44	—	150	1.0	5.0	0	13.0	2.5	—	—
38391	Caesar salad with grilled chicken, no dressing	1	serving(s)	278	230.6	181	26.4	10.5	3.1	6.0	2.9	1.7	0.8
38393	Caesar salad without chicken, no dressing	1	serving(s)	190	170.4	84	6.0	8.1	3.0	3.9	2.2	0.9	0.3
738	Cheeseburger	1	item(s)	119	—	310	15.0	35.0	1.0	12.0	6.0	—	—
29775	Chicken McGrill sandwich	1	item(s)	213	—	400	27.0	38.0	3.0	16.0	3.0	—	—

Chol (mg)	Calc (mg)	Iron (mg)	Magn (mg)	Pota (mg)	Sodi (mg)	Zinc (mg)	Vit A (µg)	Thia (mg)	Vit E (mg α)	Ribo (mg)	Niac (mg)	Vit B_6 (mg)	Fola (µg DFE)	Vit C (mg)	Vit B_{12} (µg)	Sele (µg)
0	20	1.08	40.0	720.0	0	0.30	—	0.30	—	0.10	1.60	0.40	80.0	678.0	0	0
0	0	1.80	8.0	80.0	0	0.00	0	0.03	—	0.03	0.40	0.04	16.0	3.6	0	2.8
0	100	2.70	—	—	730.0	—	—	—	—	—	—	—	—	1.2	—	—
0	40	1.80	—	—	640.0	—	—	—	—	—	—	—	—	0	—	—
65	40	1.80	—	—	2670.0	—	—	—	—	—	—	—	—	1.2	—	—
65	40	1.80	—	—	2210.0	—	—	—	—	—	—	—	—	1.2	—	—
65	60	1.80	—	—	2000.0	—	—	—	—	—	—	—	—	1.2	—	—
5	40	0.72	—	—	270.0	—	—	—	—	—	—	—	—	12.0	—	—
65	40	1.44	—	—	1220.0	—	0	—	—	—	—	—	—	1.2	—	—
0	60	1.08	—	—	10.0	—	—	—	—	—	—	—	—	6.0	—	—
55	100	2.70	—	—	1220.0	—	—	—	—	—	—	—	—	6.0	—	—
85	20	2.70	—	—	1020.0	—	—	—	—	—	—	—	—	1.2	—	—
55	0	1.44	—	—	300.0	—	0	—	—	—	—	—	—	0	—	—
95	20	2.70	—	—	700.0	—	—	—	—	—	—	—	—	—	—	—
45	20	1.08	—	—	340.0	—	—	—	—	—	—	—	—	0	—	—
35	200	5.40	—	—	2330.0	—	—	—	—	—	—	—	—	6.0	—	—
35	200	4.50	—	—	2130.0	—	—	—	—	—	—	—	—	6.0	—	—
60	80	2.70	—	—	710.0	—	—	—	—	—	—	—	—	2.4	—	—
140	40	1.80	—	—	970.0	—	—	—	—	—	—	—	—	21.0	—	—
115	40	1.44	—	—	990.0	—	—	—	—	—	—	—	—	1.2	—	—
25	60	2.70	—	—	690.0	—	—	—	—	—	—	—	—	2.4	—	—
15	150	0.72	—	—	800.0	—	—	—	—	—	—	—	—	1.2	—	—
0	40	1.44	—	—	560.0	—	—	—	—	—	—	—	—	1.2	—	—
135	20	2.70	—	—	960.0	—	—	—	—	—	—	—	—	6.0	—	—
70	20	1.08	—	—	340.0	—	—	—	—	—	—	—	—	0	—	—
110	20	2.70	—	—	870.0	—	—	—	—	—	—	—	—	1.2	—	—
50	20	1.44	—	—	350.0	—	0	—	—	—	—	—	—	1.2	—	—
60	40	6.30	—	—	1380.0	—	—	—	—	—	—	—	—	15.0	—	—
25	40	1.80	—	—	1110.0	—	0	—	—	—	—	—	—	0	—	—
5	0	0.36	—	—	470.0	—	—	—	—	—	—	—	—	6.0	—	—
0	20	1.08	—	—	700.0	—	0	—	—	—	—	—	—	0	—	—
80	80	2.70	—	—	1180.0	—	—	—	—	—	—	—	—	9.0	—	—
90	20	0.72	—	—	240.0	—	—	—	—	—	—	—	—	0	—	—
45	60	2.70	—	—	1210.0	—	—	—	—	—	—	—	—	2.4	—	—
35	20	0.72	—	—	790.0	—	—	—	—	—	—	—	—	4.8	—	—
10	20	1.08	—	—	1110.0	—	0	—	—	—	—	—	—	0	—	—
20	0	0.72	—	—	480.0	—	0	—	—	—	—	—	—	2.4	—	—
25	150	0.72	—	—	810.0	—	—	—	—	—	—	—	—	0	—	—
0	0	0.00	—	—	250.0	—	—	—	—	—	—	—	—	0	—	—
20	40	0.36	—	222.7	340.0	0.70	—	0.07	—	0.08	2.34	—	—	18.0	—	—
0	0	0.00	—	—	580.0	—	0	—	—	—	—	—	—	24.0	—	—
0	0	0.00	—	370.0	350.0	0.30	0	0.09	—	0.01	1.60	—	—	15.0	—	—
0	20	0.36	—	—	200.0	—	0	—	—	—	—	—	—	0	—	—
15	0	0.00	—	—	160.0	—	0	—	—	—	—	—	—	1.2	—	—
15	0	0.00	—	—	250.0	—	0	—	—	—	—	—	—	0	—	—
60	150	2.70	—	—	1400.0	—	—	—	—	—	—	—	—	4.8	—	—
65	150	3.60	—	—	890.0	—	—	—	—	—	—	—	—	54.0	—	—
0	0	0.00	—	55.0	260.0	—	—	—	—	—	—	—	—	0	—	—
80	250	4.50	—	400.0	1010.0	—	—	—	—	—	—	—	—	1.2	—	—
10	40	0.18	—	30.0	400.0	—	—	—	—	—	—	—	—	0.6	—	—
67	178	1.77	—	708.9	767.3	—	—	0.15	—	0.19	10.62	—	127.9	29.2	0.2	—
10	163	1.15	17.1	410.4	157.7	—	—	0.08	—	0.07	0.40	—	102.6	26.8	0	0.4
40	200	2.70	—	240.0	740.0	—	60.0	—	—	—	—	—	—	1.2	—	—
70	150	2.70	—	510.0	1010.0	—	—	—	—	—	—	—	—	6.0	—	—

TABLE A-1 Table of Food Composition *(continued)*

(Computer code is for Cengage Diet Analysis Plus program)

DA+ Code	Food Description	QTY	Measure	Wt (g)	H2O (g)	Ener (cal)	Prot (g)	Carb (g)	Fiber (g)	Fat (g)	Fat Breakdown (g) Sat	Mono	Poly
Fast Food—continued													
1873	Chicken McNuggets, 6 piece	1	serving(s)	96	—	250	15.0	15.0	0	15.0	3.0	—	—
3792	Chicken McNuggets, 4 piece	1	serving(s)	64	—	170	10.0	10.0	0	10.0	2.0	—	—
29774	Crispy chicken sandwich	1	item(s)	232	121.8	500	27.0	63.0	3.0	16.0	3.0	5.7	7.4
743	Egg McMuffin	1	item(s)	139	76.8	300	17.0	30.0	2.0	12.0	4.5	3.8	2.5
742	Filet-O-Fish sandwich	1	item(s)	141	—	400	14.0	42.0	1.0	18.0	4.0	—	—
2257	French fries, large	1	serving(s)	170	—	570	6.0	70.0	7.0	30.0	6.0	—	—
1872	French fries, small	1	serving(s)	74	—	250	2.0	30.0	3.0	13.0	2.5	—	—
33822	Fruit 'n Yogurt Parfait	1	item(s)	149	111.2	160	4.0	31.0	1.0	2.0	1.0	0.2	0.1
739	Hamburger	1	item(s)	105	—	260	13.0	33.0	1.0	9.0	3.5	—	—
2003	Hash browns	1	item(s)	53	—	140	1.0	15.0	2.0	8.0	1.5	—	—
2249	Honey sauce	1	item(s)	14	—	50	0	12.0	0	0	0	0	0
38397	Newman's Own creamy caesar salad dressing	1	item(s)	59	32.5	190	2.0	4.0	0	18.0	3.5	4.6	9.6
38398	Newman's Own low fat balsamic vinaigrette salad dressing	1	item(s)	44	29.1	40	0	4.0	0	3.0	0	1.0	1.2
38399	Newman's Own ranch salad dressing	1	item(s)	59	30.1	170	1.0	9.0	0	15.0	2.5	9.0	3.7
1874	Plain Hotcakes with syrup and margarine	3	item(s)	221	—	600	9.0	102.0	2.0	17.0	4.0	—	—
740	Quarter Pounder hamburger	1	item(s)	171	—	420	24.0	40.0	3.0	18.0	7.0	—	—
741	Quarter Pounder hamburger with cheese	1	item(s)	199	—	510	29.0	43.0	3.0	25.0	12.0	—	—
2005	Sausage McMuffin with egg	1	item(s)	165	82.4	450	20.0	31.0	2.0	27.0	10.0	10.9	4.6
50831	Side salad	1	item(s)	87	—	20	1.0	4.0	1.0	0	0	0	0
	Pizza Hut												
39009	Hot chicken wings	2	item(s)	57	—	110	11.0	1.0	0	6.0	2.0	—	—
14025	Meat Lovers hand tossed pizza	1	slice(s)	118	—	300	15.0	29.0	2.0	13.0	6.0	—	—
14026	Meat Lovers pan pizza	1	slice(s)	123	—	340	15.0	29.0	2.0	19.0	7.0	—	—
31009	Meat Lovers stuffed crust pizza	1	slice(s)	169	—	450	21.0	43.0	3.0	21.0	10.0	—	—
14024	Meat Lovers thin 'n crispy pizza	1	slice(s)	98	—	270	13.0	21.0	2.0	14.0	6.0	—	—
14031	Pepperoni Lovers hand tossed pizza	1	slice(s)	113	—	300	15.0	30.0	2.0	13.0	7.0	—	—
14032	Pepperoni Lovers pan pizza	1	slice(s)	118	—	340	15.0	29.0	2.0	19.0	7.0	—	—
31011	Pepperoni Lovers stuffed crust pizza	1	slice(s)	163	—	420	21.0	43.0	3.0	19.0	10.0	—	—
14030	Pepperoni Lovers thin 'n crispy pizza	1	slice(s)	92	—	260	13.0	21.0	2.0	14.0	7.0	—	—
10834	Personal Pan pepperoni pizza	1	slice(s)	61	—	170	7.0	18.0	0.5	8.0	3.0	—	—
10842	Personal Pan supreme pizza	1	slice(s)	77	—	190	8.0	19.0	1.0	9.0	3.5	—	—
39013	Personal Pan Veggie Lovers pizza	1	slice(s)	69	—	150	6.0	19.0	1.0	6.0	2.0	—	—
14028	Veggie Lovers hand tossed pizza	1	slice(s)	118	—	220	10.0	31.0	2.0	6.0	3.0	—	—
14029	Veggie Lovers pan pizza	1	slice(s)	119	—	260	10.0	30.0	2.0	12.0	4.0	—	—
31010	Veggie Lovers stuffed crust pizza	1	slice(s)	172	—	360	16.0	45.0	3.0	14.0	7.0	—	—
14027	Veggie Lovers thin 'n crispy pizza	1	slice(s)	101	—	180	8.0	23.0	2.0	7.0	3.0	—	—
39012	Wing blue cheese dipping sauce	1	item(s)	43	—	230	2.0	2.0	0	24.0	5.0	—	—
39011	Wing ranch dipping sauce	1	item(s)	43	—	210	0.5	4.0	0	22.0	3.5	—	—
	Starbucks												
38052	Cappuccino, tall	12	fluid ounce(s)	360	—	120	7.0	10.0	0	6.0	4.0	—	—
38053	Cappuccino, tall nonfat	12	fluid ounce(s)	360	—	80	7.0	11.0	0	0	0	0	0
38054	Cappuccino, tall soymilk	12	fluid ounce(s)	360	—	100	5.0	13.0	0.5	2.5	0	—	—
38059	Cinnamon spice mocha, tall nonfat w/o whipped cream	12	fluid ounce(s)	360	—	170	11.0	32.0	0	0.5	—	—	—
38057	Cinnamon spice mocha, tall w/whipped cream	12	fluid ounce(s)	360	—	320	10.0	31.0	0	17.0	11.0	—	—
38051	Espresso, single shot	1	fluid ounce(s)	30	—	5	0	1.0	0	0	0	0	0
38088	Flavored syrup, 1 pump	1	serving(s)	10	—	20	0	5.0	0	0	0	0	0
32562	Frappuccino bottled coffee drink, mocha	9½	fluid ounce(s)	298	—	190	6.0	39.0	3.0	3.0	2.0	—	—
32561	Frappuccino coffee drink, all bottled flavors	9½	fluid ounce(s)	281	—	190	7.0	35.0	0	3.5	2.5	—	—
38073	Frappuccino, mocha	12	fluid ounce(s)	360	—	220	5.0	44.0	0	3.0	1.5	—	—
38067	Frappuccino, tall caramel w/o whipped cream	12	fluid ounce(s)	360	—	210	4.0	43.0	0	2.5	1.5	—	—
38070	Frappuccino, tall coffee	12	fluid ounce(s)	360	—	190	4.0	38.0	0	2.5	1.5	—	—
39894	Frappuccino, tall coffee, light blend	12	fluid ounce(s)	360	—	110	5.0	22.0	2.0	1.0	0	—	—

Chol (mg)	Calc (mg)	Iron (mg)	Magn (mg)	Pota (mg)	Sodi (mg)	Zinc (mg)	Vit A (μg)	Thia (mg)	Vit E (mg α)	Ribo (mg)	Niac (mg)	Vit B_6 (mg)	Fola (μg DFE)	Vit C (mg)	Vit B_{12} (μg)	Sele (μg)
35	20	0.72	—	240.0	670.0	—	—	—	—	—	—	—	—	1.2	—	—
25	0	0.36	—	160.0	450.0	—	—	—	—	—	—	—	—	1.2	—	—
60	80	3.60	62.6	526.6	1380.0	1.53	41.8	0.46	2.27	0.39	12.85	—	94.2	6.0	0.4	—
230	300	2.70	26.4	218.2	860.0	1.59	—	0.36	0.82	0.51	4.31	0.20	109.8	1.2	0.9	—
40	150	1.80	—	250.0	640.0	—	36.2	—	—	—	—	—	—	0	—	—
0	20	1.80	—	—	330.0	—	0	—	—	—	—	—	—	9.0	—	—
0	20	0.72	—	—	140.0	—	0	—	—	—	—	—	—	3.6	—	—
5	150	0.67	20.9	248.8	85.0	0.53	0	0.06	—	0.17	0.35	—	19.4	9.0	0.3	—
30	150	2.70	—	210.0	530.0	—	5.0	—	—	—	—	—	—	1.2	—	—
0	0	0.36	—	210.0	290.0	—	0	—	—	—	—	—	—	1.2	—	—
0	0	0.00	—	0	0	—	0	—	—	—	—	—	—	0	—	—
20	61	0.00	3.0	16.0	500.0	0.20	—	0.01	15.43	0.02	0.01	0.64	2.4	0	0.1	0.1
0	4	0.00	1.3	8.8	730.0	0.01	—	0.00	0.00	0.00	0.00	0.00	0	2.4	0	0
0	40	0.00	1.8	70.4	530.0	0.03	0	0.01	—	0.08	0.01	0.02	0.6	0	0	0.2
20	150	2.70	—	280.0	620.0	—	—	—	—	—	—	—	—	0	—	—
70	150	4.50	—	390.0	730.0	—	10.0	—	—	—	—	—	—	1.2	—	—
95	300	4.50	—	440.0	1150.0	—	100.0	—	—	—	—	—	—	1.2	—	—
255	300	3.60	29.7	282.2	950.0	2.01	—	0.43	0.82	0.56	4.83	0.24	—	0	1.2	—
0	20	0.72	—	—	10.0	—	—	—	—	—	—	—	—	15.0	—	—
70	0	0.36	—	—	450.0	—	—	—	—	—	—	—	—	0	—	—
35	150	1.80	—	—	760.0	—	—	—	—	—	—	—	—	6.0	—	—
35	150	2.70	—	—	750.0	—	—	—	—	—	—	—	—	6.0	—	—
55	250	2.70	—	—	1250.0	—	—	—	—	—	—	—	—	9.0	—	—
35	150	1.44	—	—	740.0	—	—	—	—	—	—	—	—	6.0	—	—
40	200	1.80	—	—	710.0	—	57.7	—	—	—	—	—	—	2.4	—	—
40	200	2.70	—	—	700.0	—	57.7	—	—	—	—	—	—	2.4	—	—
55	300	2.70	—	—	1120.0	—	—	—	—	—	—	—	—	3.6	—	—
40	200	1.44	—	—	690.0	—	58.0	—	—	—	—	—	—	2.4	—	—
15	80	1.44	—	—	340.0	—	38.5	—	—	—	—	—	—	1.4	—	—
20	80	1.86	—	—	420.0	—	—	—	—	—	—	—	—	3.6	—	—
10	80	1.80	—	—	280.0	—	—	—	—	—	—	—	—	3.6	—	—
15	150	1.80	—	—	490.0	—	—	—	—	—	—	—	—	9.0	—	—
15	150	2.70	—	—	470.0	—	—	—	—	—	—	—	—	9.0	—	—
35	250	2.70	—	—	980.0	—	—	—	—	—	—	—	—	9.0	—	—
15	150	1.44	—	—	480.0	—	—	—	—	—	—	—	—	9.0	—	—
25	20	0.00	—	—	550.0	—	0	—	—	—	—	—	—	0	—	—
10	0	0.00	—	—	340.0	—	0	—	—	—	—	—	—	0	—	—
25	250	0.00	—	—	95.0	—	—	—	—	—	—	—	—	1.2	0	—
3	200	0.00	—	—	100.0	—	—	—	—	—	—	—	—	0	0	—
0	250	0.72	—	—	75.0	—	—	—	—	—	—	—	—	0	0	—
5	300	0.72	—	—	150.0	—	—	—	—	—	—	—	—	0	0	—
70	350	1.08	—	—	140.0	—	—	—	—	—	—	—	—	2.4	0	—
0	0	0.00	—	—	0	—	0	—	—	—	—	—	—	0	0	—
0	0	0.00	—	—	0	—	0	—	—	—	—	—	—	0	0	—
12	219	1.08	—	530.0	110.0	—	—	—	—	—	—	—	—	0	—	—
15	250	0.36	—	510.0	105.0	—	—	—	—	—	—	—	—	0	—	—
10	150	0.72	—	—	180.0	—	—	—	—	—	—	—	—	0	0	—
10	150	0.00	—	—	180.0	—	—	—	—	—	—	—	—	0	0	—
10	150	0.00	—	—	180.0	—	—	—	—	—	—	—	—	0	0	—
0	150	0.00	—	—	220.0	—	—	—	—	—	—	—	—	0	—	—

TABLE A-1 Table of Food Composition *(continued)*

(Computer code is for Cengage Diet Analysis Plus program)

DA+ Code	Food Description	QTY	Measure	Wt (g)	H_2O (g)	Ener (cal)	Prot (g)	Carb (g)	Fiber (g)	Fat (g)	Fat Breakdown (g) Sat	Mono	Poly
	Fast Food—continued												
38071	Frappuccino, tall espresso	12	fluid ounce(s)	360	—	160	4.0	33.0	0	2.0	1.5	—	—
39897	Frappuccino, tall mocha, light blend	12	fluid ounce(s)	360	—	140	5.0	28.0	3.0	1.5	0	—	—
39887	Frappuccino, tall Strawberries and Creme, w/o whipped cream	12	fluid ounce(s)	360	—	330	10.0	65.0	0	3.5	1.0	—	—
38063	Frappuccino, tall Tazo chai creme w/o whipped cream	12	fluid ounce(s)	360	—	280	10.0	52.0	0	3.5	1.0	—	—
38066	Frappuccino, tall Tazoberry	12	fluid ounce(s)	360	—	140	0.5	36.0	0.5	0	0	0	0
38065	Frappuccino, tall Tazoberry Crème	12	fluid ounce(s)	360	—	240	4.0	54.0	0.5	1.0	0	—	—
38080	Frappuccino, tall vanilla w/o whipped cream	12	fluid ounce(s)	360	—	270	10.0	51.0	0	3.5	1.0	—	—
39898	Frappuccino, tall white chocolate mocha, light blend	12	fluid ounce(s)	360	—	160	6.0	32.0	2.0	2.0	1.0	—	—
38074	Frappuccino, tall white chocolate w/o whipped cream	12	fluid ounce(s)	360	—	240	5.0	48.0	0	3.5	2.5	—	—
39883	Java Chip Frappuccino, tall w/o whipped cream	12	fluid ounce(s)	360	—	270	5.0	51.0	1.0	7.0	4.5	—	—
33111	Latte, tall w/nonfat milk	12	fluid ounce(s)	360	335.3	120	12.0	18.0	0	0	0	0	0
33112	Latte, tall w/whole milk	12	fluid ounce(s)	360	—	200	11.0	16.0	0	11.0	7.0	—	—
33109	Macchiato, tall caramel w/nonfat milk	12	fluid ounce(s)	360	—	170	11.0	30.0	0	1.0	0	—	—
33110	Macchiato, tall caramel w/whole milk	12	fluid ounce(s)	360	—	240	10.0	28.0	0	10.0	6.0	—	—
33107	Mocha coffee drink, tall nonfat, w/o whipped cream	12	fluid ounce(s)	360	—	170	11.0	33.0	1.0	1.5	0	—	—
38089	Mocha syrup	1	serving(s)	17	—	25	1.0	6.0	0	0.5	0	—	—
33108	Mocha, tall mocha w/whole milk	12	fluid ounce(s)	360	—	310	10.0	32.0	1.0	17.0	10.0	—	—
38042	Steamed apple cider, tall	12	fluid ounce(s)	360	—	180	0	45.0	0	0	0	0	0
38087	Tazo chai black tea, soymilk, tall	12	fluid ounce(s)	360	—	190	4.0	39.0	0.5	2.0	0	—	—
38084	Tazo chai black tea, tall	12	fluid ounce(s)	360	—	210	6.0	36.0	0	5.0	3.5	—	—
38083	Tazo chai black tea, tall nonfat	12	fluid ounce(s)	360	—	170	6.0	37.0	0	0	0	0	0
38076	Tazo iced tea, tall	12	fluid ounce(s)	360	—	60	0	16.0	0	0	0	0	0
38077	Tazo tea, grande lemonade	16	fluid ounce(s)	480	—	120	0	31.0	0	0	0	0	0
38045	Vanilla crème steamed nonfat milk, tall w/whipped cream	12	fluid ounce(s)	360	—	260	11.0	33.0	0	8.0	5.0	—	—
38046	Vanilla crème steamed soymilk, tall w/whipped cream	12	fluid ounce(s)	360	—	300	8.0	37.0	1.0	12.0	6.0	—	—
38044	Vanilla crème steamed whole milk, tall w/whipped cream	12	fluid ounce(s)	360	—	330	10.0	31.0	0	18.0	11.0	—	—
38090	Whipped cream	1	serving(s)	27	—	100	0	2.0	0	9.0	6.0	—	—
38062	White chocolate mocha, tall nonfat w/o whipped cream	12	fluid ounce(s)	360	—	260	12.0	45.0	0	4.0	3.0	—	—
38061	White chocolate mocha, tall w/ whipped cream	12	fluid ounce(s)	360	—	410	11.0	44.0	0	20.0	13.0	—	—
38048	White hot chocolate, tall nonfat w/o whipped cream	12	fluid ounce(s)	360	—	300	15.0	51.0	0	4.5	3.5	—	—
38050	White hot chocolate, tall soymilk w/ whipped cream	12	fluid ounce(s)	360	—	420	11.0	56.0	1.0	16.0	9.0	—	—
38047	White hot chocolate, tall w/whipped cream	12	fluid ounce(s)	360	—	460	13.0	50.0	0	22.0	15.0	—	—
	Subway												
15842	Cheese steak sandwich, 6″, wheat bread	1	item(s)	250	—	360	24.0	47.0	5.0	10.0	4.5	—	—
40478	Chicken and bacon ranch sandwich, 6″, white or wheat bread	1	serving(s)	297	—	540	36.0	47.0	5.0	25.0	10.0	—	—
38622	Chicken and bacon ranch wrap with cheese	1	item(s)	257	—	440	41.0	18.0	9.0	27.0	10.0	—	—
32045	Chocolate chip cookie	1	item(s)	45	—	210	2.0	30.0	1.0	10.0	6.0	—	—
32048	Chocolate chip M&M cookie	1	item(s)	45	—	210	2.0	32.0	0.5	10.0	5.0	—	—
32049	Chocolate chunk cookie	1	item(s)	45	—	220	2.0	30.0	0.5	10.0	5.0	—	—
4024	Classic Italian B.M.T. sandwich, 6″, white bread	1	item(s)	236	—	440	22.0	45.0	2.0	21.0	8.5	—	—

CHOL (mg)	CALC (mg)	IRON (mg)	MAGN (mg)	POTA (mg)	SODI (mg)	ZINC (mg)	VIT A (µg)	THIA (mg)	VIT E (mg α)	RIBO (mg)	NIAC (mg)	VIT B_6 (mg)	FOLA (µg DFE)	VIT C (mg)	VIT B_{12} (µg)	SELE (µg)
10	100	0.00	—	—	160.0	—	—	—	—	—	—	—	—	0	0	—
0	150	0.72	—	—	220.0	—	—	—	—	—	—	—	—	0	—	—
3	350	0.00	—	—	270.0	—	—	—	—	—	—	—	—	21.0	—	—
3	350	0.00	—	—	270.0	—	—	—	—	—	—	—	—	3.6	0	—
0	0	0.00	—	—	30.0	—	0	—	—	—	—	—	—	0	0	—
0	150	0.00	—	—	125.0	—	0	—	—	—	—	—	—	1.2	0	—
3	350	0.00	—	—	370.0	—	—	—	—	—	—	—	—	3.6	0	—
3	150	0.00	—	—	250.0	—	—	—	—	—	—	—	—	0	—	—
10	150	0.00	—	—	210.0	—	—	—	—	—	—	—	—	0	0	—
10	150	1.44	—	—	220.0	—	—	—	—	—	—	—	—	0	—	—
5	350	0.00	39.8	—	170.0	1.35	—	0.12	—	0.47	0.36	0.13	17.5	0	1.3	—
45	400	0.00	46.6	—	160.0	1.28	—	0.12	—	0.54	0.34	0.14	16.8	2.4	1.2	—
5	300	0.00	—	—	160.0	—	—	—	—	—	—	—	—	1.2	—	—
30	300	0.00	—	—	135.0	—	—	—	—	—	—	—	—	2.4	—	—
5	300	2.70	—	—	135.0	—	—	—	—	—	—	—	—	0	—	—
0	0	0.72	—	—	0	—	0	—	—	—	—	—	—	0	0	—
55	300	2.70	—	—	115.0	—	—	—	—	—	—	—	—	0	—	—
0	0	1.08	—	—	15.0	—	0	—	—	—	—	—	—	0	0	—
0	200	0.72	—	—	70.0	—	—	—	—	—	—	—	—	0	0	—
20	200	0.36	—	—	85.0	—	—	—	—	—	—	—	—	1.2	0	—
5	200	0.36	—	—	95.0	—	—	—	—	—	—	—	—	0	0	—
0	0	0.00	—	—	0	—	0	—	—	—	—	—	—	0	0	—
0	0	0.00	—	—	15.0	—	0	—	—	—	—	—	—	4.8	0	—
35	350	0.00	—	—	170.0	—	—	—	—	—	—	—	—	0	0	—
30	400	1.44	—	—	130.0	—	—	—	—	—	—	—	—	0	0	—
65	350	0.00	—	—	140.0	—	—	—	—	—	—	—	—	0	0	—
40	0	0.00	—	—	10.0	—	—	—	—	—	—	—	—	0	0	—
5	400	0.00	—	—	210.0	—	—	—	—	—	—	—	—	0	0	—
70	400	0.00	—	—	210.0	—	—	—	—	—	—	—	—	2.4	0	—
10	450	0.00	—	—	250.0	—	—	—	—	—	—	—	—	0	0	—
35	500	1.44	—	—	210.0	—	—	—	—	—	—	—	—	0	0	—
75	500	0.00	—	—	250.0	—	—	—	—	—	—	—	—	3.6	0	—
35	150	8.10	—	—	1090.0	—	—	—	—	—	—	—	—	18.0	—	—
90	250	4.50	—	—	1400.0	—	—	—	—	—	—	—	—	21.0	—	—
90	300	2.70	—	—	1680.0	—	—	—	—	—	—	—	—	9.0	—	—
15	0	1.08	—	—	150.0	—	—	—	—	—	—	—	—	0	—	—
10	20	1.00	—	—	100.0	—	—	—	—	—	—	—	—	0	—	—
10	0	1.00	—	—	100.0	—	—	—	—	—	—	—	—	0	—	—
55	150	2.70	—	—	1770.0	—	—	—	—	—	—	—	—	16.8	—	—

TABLE A-1 Table of Food Composition *(continued)*

(Computer code is for Cengage Diet Analysis Plus program)

DA+ Code	Food Description	QTY	Measure	Wt (g)	H_2O (g)	Ener (cal)	Prot (g)	Carb (g)	Fiber (g)	Fat (g)	Fat Breakdown (g) Sat	Mono	Poly
	Fast Food—continued												
15838	Classic tuna sandwich, 6″, wheat bread	1	item(s)	250	—	530	22.0	45.0	4.0	31.0	7.0	—	—
15837	Classic tuna sandwich, 6″, white bread	1	item(s)	243	—	520	21.0	43.0	2.0	31.0	7.5	—	—
16397	Club salad, no dressing and croutons	1	item(s)	412	—	160	18.0	15.0	4.0	4.0	1.5	—	—
3422	Club sandwich, 6″, white bread	1	item(s)	250	—	310	23.0	45.0	2.0	6.0	2.5	—	—
4030	Cold cut combo sandwich, 6″, white bread	1	item(s)	242	—	400	20.0	45.0	2.0	17.0	7.5	—	—
34030	Ham and egg breakfast sandwich	1	item(s)	142	—	310	16.0	35.0	3.0	13.0	3.5	—	—
3885	Ham sandwich, 6″, white bread	1	item(s)	238	—	310	17.0	52.0	2.0	5.0	2.0	—	—
3888	Meatball marinara sandwich, 6″, wheat bread	1	item(s)	377	—	560	24.0	63.0	7.0	24.0	11.0	—	—
4651	Meatball sandwich, 6″, white bread	1	item(s)	370	—	550	23.0	61.0	5.0	24.0	11.5	—	—
15839	Melt sandwich, 6″, white bread	1	item(s)	260	—	410	25.0	47.0	4.0	15.0	5.0	—	—
32046	Oatmeal raisin cookie	1	item(s)	45	—	200	3.0	30.0	1.0	8.0	4.0	—	—
16379	Oven-roasted chicken breast sandwich, 6″, wheat bread	1	item(s)	238	—	330	24.0	48.0	5.0	5.0	1.5	—	—
32047	Peanut butter cookie	1	item(s)	45	—	220	4.0	26.0	1.0	12.0	5.0	—	—
4655	Roast beef sandwich, 6″, wheat bread	1	item(s)	224	—	290	19.0	45.0	4.0	5.0	2.0	—	—
3957	Roast beef sandwich, 6″, white bread	1	item(s)	217	—	280	18.0	43.0	2.0	5.0	2.5	—	—
16378	Roasted chicken breast, 6″, white bread	1	item(s)	231	—	320	23.0	46.0	3.0	5.0	2.0	—	—
34028	Southwest steak and cheese sandwich, 6″, Italian bread	1	item(s)	271	—	450	24.0	48.0	6.0	20.0	6.0	—	—
4032	Spicy Italian sandwich, 6″, white bread	1	item(s)	220	—	470	20.0	43.0	2.0	25.0	9.5	—	—
4031	Steak and cheese sandwich, 6″, white bread	1	item(s)	243	—	350	23.0	45.0	3.0	10.0	5.0	—	—
32050	Sugar cookie	1	item(s)	45	—	220	2.0	28.0	0.5	12.0	6.0	—	—
40477	Sweet onion chicken teriyaki sandwich, 6″, white or wheat bread	1	serving(s)	281	—	370	26.0	59.0	4.0	5.0	1.5	—	—
38623	Turkey breast and bacon melt wrap with chipotle sauce	1	item(s)	228	—	380	31.0	20.0	9.0	24.0	7.0	—	—
15834	Turkey breast and ham sandwich, 6″, white bread	1	item(s)	227	—	280	19.0	45.0	2.0	5.0	2.0	—	—
16376	Turkey breast sandwich, 6″, white bread	1	item(s)	217	—	270	17.0	44.0	2.0	4.5	2.0	—	—
15841	Veggie Delite sandwich, 6″, wheat bread	1	item(s)	167	—	230	9.0	44.0	4.0	3.0	1.0	—	—
16375	Veggie Delite, 6″, white bread	1	item(s)	160	—	220	8.0	42.0	2.0	3.0	1.5	—	—
32051	White chip macadamia nut cookie	1	item(s)	45	—	220	2.0	29.0	0.5	11.0	5.0	—	—
	Taco Bell												
29906	7-Layer burrito	1	item(s)	283	—	490	17.0	65.0	9.0	18.0	7.0	—	—
744	Bean burrito	1	item(s)	198	—	340	13.0	54.0	8.0	9.0	3.5	—	—
749	Beef burrito supreme	1	item(s)	248	—	410	17.0	51.0	7.0	17.0	8.0	—	—
33417	Beef Chalupa Supreme	1	item(s)	153	—	380	14.0	30.0	3.0	23.0	7.0	—	—
34474	Beef Gordita Baja	1	item(s)	153	—	340	13.0	29.0	4.0	19.0	5.0	—	—
29910	Beef Gordita Supreme	1	item(s)	153	—	310	14.0	29.0	3.0	16.0	6.0	—	—
2014	Beef soft taco	1	item(s)	99	—	200	10.0	21.0	3.0	9.0	4.0	—	—
10860	Beef soft taco supreme	1	item(s)	135	—	250	11.0	23.0	3.0	13.0	6.0	—	—
34472	Chicken burrito supreme	1	item(s)	248	—	390	20.0	49.0	6.0	13.0	6.0	—	—
33418	Chicken Chalupa Supreme	1	item(s)	153	—	360	17.0	29.0	2.0	20.0	5.0	—	—
34475	Chicken Gordita Baja	1	item(s)	153	—	320	17.0	28.0	3.0	16.0	3.5	—	—
29909	Chicken quesadilla	1	item(s)	184	—	520	28.0	40.0	3.0	28.0	12.0	—	—
29907	Chili cheese burrito	1	item(s)	156	—	390	16.0	40.0	3.0	18.0	9.0	—	—
10794	Cinnamon twists	1	serving(s)	35	—	170	1.0	26.0	1.0	7.0	0	—	—
29911	Grilled chicken Gordita Supreme	1	item(s)	153	—	290	17.0	28.0	2.0	12.0	5.0	—	—
14463	Grilled chicken soft taco	1	item(s)	99	—	190	14.0	19.0	1.0	6.0	2.5	—	—

Chol (mg)	Calc (mg)	Iron (mg)	Magn (mg)	Pota (mg)	Sodi (mg)	Zinc (mg)	Vit A (µg)	Thia (mg)	Vit E (mg α)	Ribo (mg)	Niac (mg)	Vit B6 (mg)	Fola (µg DFE)	Vit C (mg)	Vit B12 (µg)	Sele (µg)
45	100	5.40	—	—	1030.0	—	—	—	—	—	—	—	—	21.0	—	—
45	100	3.60	—	—	1010.0	—	—	—	—	—	—	—	—	16.8	—	—
35	60	3.60	—	—	880.0	—	—	—	—	—	—	—	—	30.0	—	—
35	60	3.60	—	—	1290.0	—	—	—	—	—	—	—	—	13.8	—	—
60	150	3.60	—	—	1530.0	—	—	—	—	—	—	—	—	16.8	—	—
190	80	4.50	—	—	720.0	—	66.7	—	—	—	—	—	—	3.6	—	—
25	60	2.70	—	—	1375.0	—	—	—	—	—	—	—	—	13.8	—	—
45	200	7.20	—	—	1610.0	—	—	—	—	—	—	—	—	36.0	—	—
45	200	5.40	—	—	1590.0	—	—	—	—	—	—	—	—	31.8	—	—
45	150	5.40	—	—	1720.0	—	—	—	—	—	—	—	—	24.0	—	—
15	20	1.08	—	—	170.0	—	—	—	—	—	—	—	—	0	—	—
45	60	4.50	—	—	1020.0	—	—	—	—	—	—	—	—	18.0	—	—
15	20	0.72	—	—	200.0	—	—	—	—	—	—	—	—	0	—	—
20	60	6.30	—	—	920.0	—	—	—	—	—	—	—	—	18.0	—	—
20	60	4.50	—	—	900.0	—	—	—	—	—	—	—	—	13.8	—	—
45	60	2.70	—	—	1000.0	—	—	—	—	—	—	—	—	13.8	—	—
45	150	8.10	—	—	1310.0	—	—	—	—	—	—	—	—	21.0	—	—
55	60	2.70	—	—	1650.0	—	—	—	—	—	—	—	—	16.8	—	—
35	150	6.30	—	—	1070.0	—	—	—	—	—	—	—	—	13.8	—	—
15	0	0.72	—	—	140.0	—	—	—	—	—	—	—	—	0	—	—
50	80	4.50	—	—	1220.0	—	—	—	—	—	—	—	—	24.0	—	—
50	200	2.70	—	—	1780.0	—	—	—	—	—	—	—	—	6.0	—	—
25	60	2.70	—	—	1210.0	—	—	—	—	—	—	—	—	13.8	—	—
20	60	2.70	—	—	1000.0	—	—	—	—	—	—	—	—	13.8	—	—
0	60	4.50	—	—	520.0	—	—	—	—	—	—	—	—	18.0	—	—
0	60	2.70	—	—	500.0	—	—	—	—	—	—	—	—	13.8	—	—
15	20	0.72	—	—	160.0	—	—	—	—	—	—	—	—	0	—	—
25	250	5.40	—	—	1350.0	—	—	—	—	—	—	—	—	15.0	—	—
5	200	4.50	—	—	1190.0	—	5.9	—	—	—	—	—	—	4.8	—	—
40	200	4.50	—	—	1340.0	—	9.9	—	—	—	—	—	—	6.0	—	—
40	150	2.70	—	—	620.0	—	—	—	—	—	—	—	—	3.6	—	—
35	100	2.70	—	—	780.0	—	—	—	—	—	—	—	—	2.4	—	—
40	150	2.70	—	—	620.0	—	—	—	—	—	—	—	—	3.6	—	—
25	100	1.80	—	—	630.0	—	—	—	—	—	—	—	—	1.2	—	—
40	150	2.70	—	—	650.0	—	—	—	—	—	—	—	—	3.6	—	—
45	200	4.50	—	—	1360.0	—	—	—	—	—	—	—	—	9.0	—	—
45	100	2.70	—	—	650.0	—	—	—	—	—	—	—	—	4.8	—	—
40	100	1.80	—	—	800.0	—	—	—	—	—	—	—	—	3.6	—	—
75	450	3.60	—	—	1420.0	—	—	—	—	—	—	—	—	1.2	—	—
40	300	1.80	—	—	1080.0	—	—	—	—	—	—	—	—	0	—	—
0	0	0.37	—	—	200.0	—	0	—	—	—	—	—	—	0	—	—
45	150	1.80	—	—	650.0	—	—	—	—	—	—	—	—	4.8	—	—
30	100	1.08	—	—	550.0	—	14.6	—	—	—	—	—	—	1.2	—	—

TABLE **A-1** Table of Food Composition *(continued)*

(Computer code is for Cengage Diet Analysis Plus program)

											Fat Breakdown (g)		
DA+ Code	Food Description	QTY	Measure	Wt (g)	H_2O (g)	Ener (cal)	Prot (g)	Carb (g)	Fiber (g)	Fat (g)	Sat	Mono	Poly
Fast Food—continued													
29912	Grilled Steak Gordita Supreme	1	item(s)	153	—	290	15.0	28.0	2.0	13.0	5.0	—	—
29904	Grilled steak soft taco	1	item(s)	128	—	270	12.0	20.0	2.0	16.0	4.5	—	—
29905	Grilled steak soft taco supreme	1	item(s)	135	—	235	13.0	21.0	1.0	11.0	6.0	—	—
2021	Mexican pizza	1	serving(s)	216	—	530	20.0	42.0	7.0	30.0	8.0	—	—
29894	Mexican rice	1	serving(s)	131	—	170	6.0	23.0	1.0	11.0	3.0	—	—
10772	Meximelt	1	serving(s)	128	—	280	15.0	22.0	3.0	14.0	7.0	—	—
2011	Nachos	1	serving(s)	99	—	330	4.0	32.0	2.0	21.0	3.5	—	—
2012	Nachos Bellgrande	1	serving(s)	308	—	770	19.0	77.0	12.0	44.0	9.0	—	—
2023	Pintos 'n cheese	1	serving(s)	128	—	150	9.0	19.0	7.0	6.0	3.0	—	—
34473	Steak burrito supreme	1	item(s)	248	—	380	18.0	49.0	6.0	14.0	7.0	—	—
33419	Steak Chalupa Supreme	1	item(s)	153	—	360	15.0	28.0	2.0	21.0	6.0	—	—
747	Taco	1	item(s)	78	—	170	8.0	13.0	3.0	10.0	3.5	—	—
2015	Taco salad with salsa, with shell	1	serving(s)	548	—	840	30.0	80.0	15.0	45.0	11.0	—	—
14459	Taco supreme	1	item(s)	113	—	210	9.0	15.0	3.0	13.0	6.0	—	—
748	Tostada	1	item(s)	170	—	230	11.0	27.0	7.0	10.0	3.5	—	—
Convenience Meals													
	Banquet												
29961	Barbeque chicken meal	1	item(s)	281	—	330	16.0	37.0	2.0	13.0	3.0	—	—
14788	Boneless white fried chicken meal	1	item(s)	286	—	310	10.0	21.0	4.0	20.0	5.0	—	—
29960	Fish sticks meal	1	item(s)	207	—	470	13.0	58.0	1.0	20.0	3.5	—	—
29957	Lasagna with meat sauce meal	1	item(s)	312	—	320	15.0	46.0	7.0	9.0	4.0	—	—
14777	Macaroni and cheese meal	1	item(s)	340	—	420	15.0	57.0	5.0	14.0	8.0	—	—
1741	Meatloaf meal	1	item(s)	269	—	240	14.0	20.0	4.0	11.0	4.0	—	—
39418	Pepperoni pizza meal	1	item(s)	191	—	480	11.0	56.0	5.0	23.0	8.0	—	—
33759	Roasted white turkey meal	1	item(s)	255	—	230	14.0	30.0	5.0	6.0	2.0	—	—
1743	Salisbury steak meal	1	item(s)	269	196.9	380	12.0	28.0	3.0	24.0	12.0	—	—
	Budget Gourmet												
1914	Cheese manicotti with meat sauce entrée	1	item(s)	284	194.0	420	18.0	38.0	4.0	22.0	11.0	6.0	1.3
1915	Chicken with fettucini entrée	1	item(s)	284	—	380	20.0	33.0	3.0	19.0	10.0	—	—
3986	Light beef stroganoff entrée	1	item(s)	248	177.0	290	20.0	32.0	3.0	7.0	4.0	—	—
3996	Light sirloin of beef in herb sauce entrée	1	item(s)	269	214.0	260	19.0	30.0	5.0	7.0	4.0	2.3	0.3
3987	Light vegetable lasagna entrée	1	item(s)	298	227.0	290	15.0	36.0	4.8	9.0	1.8	0.9	0.6
	Healthy Choice												
9425	Cheese French bread pizza	1	item(s)	170	—	340	22.0	51.0	5.0	5.0	1.5	—	—
9306	Chicken enchilada suprema meal	1	item(s)	320	251.5	360	13.0	59.0	8.0	7.0	3.0	2.0	2.0
3821	Familiar Favorites lasagna bake with meat sauce entrée	1	item(s)	255	—	270	13.0	38.0	4.0	7.0	2.5	—	—
13744	Familiar Favorites sesame chicken with vegetables and rice entrée	1	item(s)	255	—	260	17.0	34.0	4.0	6.0	2.0	2.0	2.0
9316	Lemon pepper fish meal	1	item(s)	303	—	280	11.0	49.0	5.0	5.0	2.0	1.0	2.0
9322	Traditional salisbury steak meal	1	item(s)	354	250.3	360	23.0	45.0	5.0	9.0	3.5	4.0	1.0
9359	Traditional turkey breasts meal	1	item(s)	298	—	330	21.0	50.0	4.0	5.0	2.0	1.5	1.5
	Stouffers												
2313	Cheese French bread pizza	1	serving(s)	294	—	380	15.0	43.0	3.0	16.0	6.0	—	—
11138	Cheese manicotti with tomato sauce entrée	1	item(s)	255	—	360	18.0	41.0	2.0	14.0	6.0	—	—
2366	Chicken pot pie entrée	1	item(s)	284	—	740	23.0	56.0	4.0	47.0	18.0	12.4	10.5
11116	Homestyle baked chicken breast with mashed potatoes and gravy entrée	1	item(s)	252	—	270	21.0	21.0	2.0	11.0	3.5	—	—
11146	Homestyle beef pot roast and potatoes entrée	1	item(s)	252	—	260	16.0	24.0	3.0	11.0	4.0	—	—
11152	Homestyle roast turkey breast with stuffing and mashed potatoes entrée	1	item(s)	273	—	290	16.0	30.0	2.0	12.0	3.5	—	—
11043	Lean Cuisine Comfort Classics baked chicken and whipped potatoes and stuffing entrée	1	item(s)	245	—	240	15.0	34.0	3.0	4.5	1.0	2.0	1.0
11046	Lean Cuisine Comfort Classics honey mustard chicken with rice pilaf entrée	1	item(s)	227	—	250	17.0	37.0	1.0	4.0	1.0	1.0	1.0

Chol (mg)	Calc (mg)	Iron (mg)	Magn (mg)	Pota (mg)	Sodi (mg)	Zinc (mg)	Vit A (µg)	Thia (mg)	Vit E (mg α)	Ribo (mg)	Niac (mg)	Vit B_6 (mg)	Fola (µg DFE)	Vit C (mg)	Vit B_{12} (µg)	Sele (µg)
40	100	2.70	—	—	530.0	—	—	—	—	—	—	—	—	3.6	—	—
35	100	2.70	—	—	660.0	—	—	—	—	—	—	—	—	3.6	—	—
35	120	1.44	—	—	565.0	—	29.2	—	—	—	—	—	—	3.6	—	—
40	350	3.60	—	—	1000.0	—	—	—	—	—	—	—	—	4.8	—	—
15	100	1.44	—	—	790.0	—	—	—	—	—	—	—	—	3.6	—	—
40	250	2.70	—	—	880.0	—	—	—	—	—	—	—	—	2.4	—	—
3	80	0.71	—	—	530.0	—	0	—	—	—	—	—	—	0	—	—
35	200	3.60	—	—	1280.0	—	—	—	—	—	—	—	—	4.8	—	—
15	150	1.44	—	—	670.0	—	—	—	—	—	—	—	—	3.6	—	—
35	200	4.50	—	—	1250.0	—	9.9	—	—	—	—	—	—	9.0	—	—
40	100	2.70	—	—	530.0	—	—	—	—	—	—	—	—	3.6	—	—
25	80	1.08	—	—	350.0	—	—	—	—	—	—	—	—	1.2	—	—
65	450	7.20	—	—	1780.0	—	—	—	—	—	—	—	—	12.0	—	—
40	100	1.08	—	—	370.0	—	—	—	—	—	—	—	—	3.6	—	—
15	200	1.80	—	—	730.0	—	—	—	—	—	—	—	—	4.8	—	—
																—
50	40	1.08	—	—	1210.0	—	0	—	—	—	—	—	—	4.8	—	—
45	80	1.44	—	—	1200.0	—	—	—	—	—	—	—	—	18.0	—	—
55	20	1.44	—	—	710.0	—	—	—	—	—	—	—	—	0	—	—
20	100	2.70	—	—	1170.0	—	—	—	—	—	—	—	—	0	—	—
20	150	1.44	—	—	1330.0	—	0	—	—	—	—	—	—	0	—	—
30	0	1.80	—	—	1040.0	—	0	—	—	—	—	—	—	0	—	—
35	150	1.80	—	—	870.0	—	0	—	—	—	—	—	—	0	—	—
25	60	1.80	—	—	1070.0	—	—	—	—	—	—	—	—	3.6	—	—
60	40	1.44	—	—	1140.0	—	0	—	—	—	—	—	—	0	—	—
85	300	2.70	45.4	484.0	810.0	2.29	—	0.45	—	0.51	4.00	0.22	30.7	0	0.7	—
85	100	2.70	—	—	810.0	—	—	0.15	—	0.42	6.00	—	—	0	—	—
35	40	1.80	38.9	280.0	580.0	4.71	—	0.17	—	0.36	4.28	0.27	18.9	2.4	2.5	—
30	40	1.80	57.7	540.0	850.0	4.81	—	0.15	—	0.29	5.53	0.37	38.4	6.0	1.6	—
15	283	3.03	78.5	420.0	780.0	1.39	—	0.22	—	0.45	3.13	0.32	74.8	59.1	0.2	—
10	350	3.60	—	—	600.0	—	—	—	—	—	—	—	—	0	—	—
30	40	1.44	—	—	580.0	—	—	—	—	—	—	—	—	3.6	—	—
20	100	1.80	—	—	600.0	—	—	—	—	—	—	—	—	0	—	—
35	18	0.72	—	—	580.0	—	—	—	—	—	—	—	—	12.0	—	—
35	20	0.36	—	—	580.0	—	—	—	—	—	—	—	—	30.0	—	—
45	80	2.70	—	—	580.0	—	—	—	—	—	—	—	—	21.0	—	—
35	40	1.80	—	—	600.0	—	—	—	—	—	—	—	—	0	—	—
30	200	1.80	—	230.0	660.0	—	—	—	—	—	—	—	—	2.4	—	—
70	250	1.44	—	550.0	920.0	—	—	—	—	—	—	—	—	6.0	—	—
65	150	2.70	—	—	1170.0	—	—	—	—	—	—	—	—	2.4	—	—
55	20	0.72	—	490.0	770.0	—	0	—	—	—	—	—	—	0	—	—
35	20	1.80	—	800.0	960.0	—	—	—	—	—	—	—	—	6.0	—	—
45	40	1.08	—	490.0	970.0	—	—	—	—	—	—	—	—	3.6	—	—
25	40	1.16	—	500.0	650.0	—	—	—	—	—	—	—	—	3.6	—	—
30	64	0.38	—	370.0	650.0	—	—	—	—	—	—	—	—	0	—	—

A

TABLE A-1 Table of Food Composition *(continued)*

(Computer code is for Cengage Diet Analysis Plus program)

DA+ Code	Food Description	QTY	Measure	Wt (g)	H2O (g)	Ener (cal)	Prot (g)	Carb (g)	Fiber (g)	Fat (g)	Fat Breakdown (g) Sat	Mono	Poly
Convenience Meals—continued													
9479	Lean Cuisine Deluxe French bread pizza	1	item(s)	174	—	310	16.0	44.0	3.0	9.0	3.5	0.5	0.5
360	Lean Cuisine One Dish Favorites chicken chow mein with rice	1	item(s)	255	—	190	13.0	29.0	2.0	2.5	0.5	1.0	0.5
11054	Lean Cuisine One Dish Favorites chicken enchilada Suiza with Mexican-style rice	1	serving(s)	255	—	270	10.0	47.0	3.0	4.5	2.0	1.5	1.0
9467	Lean Cuisine One Dish Favorites fettucini alfredo entrée	1	item(s)	262	—	270	13.0	39.0	2.0	7.0	3.5	2.0	1.0
11055	Lean Cuisine One Dish Favorites lasagna with meat sauce entrée	1	item(s)	298	—	320	19.0	44.0	4.0	7.0	3.0	2.0	0.5
	Weight Watchers												
11164	Smart Ones chicken enchiladas suiza entrée	1	item(s)	255	—	340	12.0	38.0	3.0	10.0	4.5	—	—
39763	Smart Ones chicken oriental entrée	1	item(s)	255	—	230	15.0	34.0	3.0	4.5	1.0	—	—
11187	Smart Ones pepperoni pizza	1	item(s)	198	—	400	22.0	58.0	4.0	9.0	3.0	—	—
39765	Smart Ones spaghetti bolognese entrée	1	item(s)	326	—	280	17.0	43.0	5.0	5.0	2.0	—	—
31512	Smart Ones spicy szechuan style vegetables and chicken	1	item(s)	255	—	220	11.0	34.0	4.0	5.0	1.0	—	—
Baby Foods													
787	Apple juice	4	fluid ounce(s)	127	111.6	60	0	14.8	0.1	0.1	0	0	0
778	Applesauce, strained	4	tablespoon(s)	64	56.7	26	0.1	6.9	1.1	0.1	0	0	0
779	Bananas with tapioca, strained	4	tablespoon(s)	60	50.4	34	0.2	9.2	1.0	0	0	0	0
604	Carrots, strained	4	tablespoon(s)	56	51.7	15	0.4	3.4	1.0	0.1	0	0	0
770	Chicken noodle dinner, strained	4	tablespoon(s)	64	54.8	42	1.7	5.8	1.3	1.3	0.4	0.5	0.3
801	Green beans, strained	4	tablespoon(s)	60	55.1	16	0.7	3.8	1.3	0.1	0	0	0
910	Human milk, mature	2	fluid ounce(s)	62	53.9	43	0.6	4.2	0	2.7	1.2	1.0	0.3
760	Mixed cereal, prepared with whole milk	4	ounce(s)	113	84.6	128	5.4	18.0	1.5	4.0	2.2	1.2	0.4
772	Mixed vegetable dinner, strained	2	ounce(s)	57	50.3	23	0.7	5.4	0.8	0	—	—	0
762	Rice cereal, prepared with whole milk	4	ounce(s)	113	84.6	130	4.4	18.9	0.1	4.1	2.6	1.0	0.2
758	Teething biscuits	1	item(s)	11	0.7	44	1.0	8.6	0.2	0.6	0.2	0.2	0.1

PAGE KEY: A-2 = Breads/Baked Goods A-8 = Cereal/Rice/Pasta A-12 = Fruit A-18 = Vegetables/Legumes A-28 = Nuts/Seeds A-30 = Vegetarian A-32 = Dairy A-40 = Eggs A-40 = Seafood A-44 = Meats A-48 = Poultry A-48 = Processed Meats A-50 = Beverages A-54 = Fats/Oils A-56 = Sweets A-58 = Spices/Condiments/Sauces A-62 = Mixed Foods/Soups/Sandwiches A-68 = Fast Food A-88 = Convenience A-90 = Baby Foods

Chol (mg)	Calc (mg)	Iron (mg)	Magn (mg)	Pota (mg)	Sodi (mg)	Zinc (mg)	Vit A (µg)	Thia (mg)	Vit E (mg α)	Ribo (mg)	Niac (mg)	Vit B6 (mg)	Fola (µg DFE)	Vit C (mg)	Vit B12 (µg)	Sele (µg)
20	150	2.70	—	300.0	700.0	—	—	—	—	—	—	—	—	15.0	—	—
25	40	0.72	—	380.0	650.0	—	—	—	—	—	—	—	—	2.4	—	—
20	150	0.72	—	350.0	510.0	—	—	—	—	—	—	—	—	2.4	—	—
15	200	0.72	—	290.0	690.0	—	0	—	—	—	—	—	—	0	—	—
30	250	1.47	—	610.0	690.0	—	—	—	—	—	—	—	—	2.4	—	—
40	200	0.72	—	—	800.0	—	—	—	—	—	—	—	—	2.4	—	—
35	40	0.72	—	—	790.0	—	—	—	—	—	—	—	—	6.0	—	—
15	200	1.08	—	401.0	700.0	—	69.1	—	—	—	—	—	—	4.8	—	—
15	150	3.60	—	—	670.0	—	—	—	—	—	—	—	—	9.0	—	—
10	40	1.44	—	—	890.0	—	—	—	—	—	—	—	—	0	—	—
0	5	0.72	3.8	115.4	3.8	0.03	1.3	0.01	0.76	0.02	0.10	0.03	0	73.4	0	0.1
0	3	0.12	1.9	45.4	1.3	0.01	0.6	0.01	0.36	0.02	0.04	0.02	1.3	24.5	0	0.2
0	3	0.12	6.0	52.8	5.4	0.04	1.2	0.01	0.36	0.02	0.08	0.04	3.6	10.0	0	0.4
0	12	0.20	5.0	109.8	20.7	0.08	320.9	0.01	0.29	0.02	0.25	0.04	8.4	3.2	0	0.1
10	17	0.40	9.0	89.0	14.7	0.32	70.4	0.03	0.12	0.04	0.44	0.04	8.3	0	0	2.4
0	23	0.40	12.0	87.6	3.0	0.12	10.8	0.02	0.04	0.04	0.20	0.02	14.4	0.2	0	0
9	20	0.02	1.8	31.4	10.5	0.10	37.6	0.01	0.04	0.02	0.10	0.01	3.1	3.1	0	1.1
12	249	11.82	30.6	225.7	53.3	0.80	28.4	0.49	—	0.65	6.54	0.07	10.2	1.4	0.3	
—	12	0.18	6.2	68.6	4.5	0.08	77.1	0.01	—	0.02	0.28	0.04	4.5	1.6	0	0.4
12	271	13.82	51.0	215.5	52.2	0.72	24.9	0.52	—	0.56	5.90	0.12	6.8	1.4	0.3	4.0
0	11	0.39	3.9	35.5	28.4	0.10	3.1	0.02	0.02	0.05	0.47	0.01	7.6	1.0	0	2.6

Appendix B

World Health Organization Nutrition Recommendations and Guidelines

The World Health Organization is the source of nutrition guidance for many of the world's populations. The nutrient intake recommendations set the basis for country-specific dietary guidance, as demonstrated in Table B–1.

TABLE B-1 World Health Organization (WHO) Nutrient Intake Guidelines

WHO has assessed the relationships between diet and the development of chronic diseases. Its recommendations include:

- Energy: sufficient to support growth, physical activity, and a healthy body weight (BMI between 18.5 and 24.9) and to avoid weight gain greater than 11 lb (5 kg) during adult life
- Total fat: 15% to 35% of total energy
- Saturated fatty acids: <10% of total energy
- Polyunsaturated fatty acids: 6% to 11% of total energy
- Omega-6 polyunsaturated fatty acids: 2.5% to 9% of total energy
- Omega-3 polyunsaturated fatty acids: 0.5% to 2% of total energy
- *Trans*-fatty acids: <1% of total energy
- Total carbohydrate: 55% to 75% of total energy
- Sugars: <10% of total energy
- Protein: 10% to 15% of total energy
- Cholesterol: <300 mg/day
- Salt (sodium): <5 g salt/day (<2 g sodium/day), appropriately iodized
- Fruits and vegetables: ≥400 g/day (about 1 lb)
- Total dietary fiber: >25 g/day from foods
- Physical activity: 1 hr of moderate-intensity activity, such as walking, on most days of the week

Source: Compiled from tables found at http://www.who.int/publications/guidelines/nutrition/en/index.html

Appendix C

Aids to Calculations

Mathematical problems have been worked out for you as examples at appropriate places in the text. This appendix aims to help with the use of the metric system and with those problems not fully explained elsewhere.

Conversion Factors

Conversion factors are useful mathematical tools in every day calculations, like the ones encountered in the study of nutrition. A conversion factor is a fraction in which the numerator (top) and the denominator (bottom) express the same quantity in different units. For example, 2.2 pounds (lb) and 1 kilogram (kg) are equivalent; they express the same weight. The conversion factor used to change pounds to kilograms or vice versa is:

$$\frac{2.2 \text{ lb}}{1 \text{ kg}} \quad \text{or} \quad \frac{1 \text{ kg}}{2.2 \text{ lb}}$$

Because both factors equal 1, measurements can be multiplied by the factor without changing the value of the measurement. Thus, the units can be changed.

The correct factor to use in a problem is the one with the unit you are seeking in the numerator (top) of the fraction. Following are some examples of problems commonly encountered in nutrition study; they illustrate the usefulness of conversion factors.

Example 1

Convert the weight of 130 pounds to kilograms.

1. Choose the conversion factor in which the unit you are seeking is on top:

$$\frac{1 \text{ kg}}{2.2 \text{ lb}}$$

2. Multiply 130 pounds by the factor:

$$130 \text{ lb} \times \frac{1 \text{ kg}}{2.2 \text{ lb}} = \frac{130 \text{ kg}}{2.2}$$

$$= 59 \text{ kg (rounded off to the nearest whole number)}$$

Example 2

How many grams (g) of saturated fat are contained in a 3-ounce (oz) hamburger?

1. Appendix A shows that a 4-ounce hamburger contains 7 grams of saturated fat. You are seeking grams of saturated fat; therefore, the conversion factor is:

$$\frac{7 \text{ g saturated fat}}{4 \text{ oz hamburger}}$$

2. Multiply 3 ounces of hamburger by the conversion factor:

$$3 \text{ oz hamburger} \times \frac{7 \text{ g saturated fat}}{4 \text{ oz hamburger}} = \frac{3 \times 7}{4} = \frac{21}{4}$$

$$= 5 \text{ g saturated fat (rounded off to the nearest whole number)}$$

Energy Units

1 calorie* (cal) = 4.2 kilojoules

1 millijoule (MJ) = 240 cal

1 kilojoule (kJ) = 0.24 cal

1 gram (g) carbohydrate = 4 cal = 17 kJ

1 g fat = 9 cal = 37 kJ

1 g protein = 4 cal = 17 kJ

1 g alcohol = 7 cal = 29 kJ

Nutrient Unit Conversions

Sodium

To convert milligrams of sodium to grams of salt:

mg sodium ÷ 400 = g of salt

The reverse is also true:

g salt × 400 = mg sodium

Folate

To convert micrograms (µg) of synthetic folate in supplements and enriched foods to Dietary Folate Equivalents (µg DFE):

µg synthetic folate × 1.7 = µg DFE

For naturally occurring folate, assign each microgram folate a value of 1 µg DFE:

µg folate = µg DFE

Example 3

Consider a pregnant woman who takes a supplement and eats a bowl of fortified cornflakes, 2 slices of fortified bread, and a cup of fortified pasta.

1. From the supplement and fortified foods, she obtains synthetic folate:

Supplement	100 µg folate
Fortified cornflakes	100 µg folate
Fortified bread	40 µg folate
Fortified pasta	60 µg folate
	300 µg folate

2. To calculate the DFE, multiply the amount of synthetic folate by 1.7:

300 µg × 1.7 = 510 µg DFE

3. Now add the naturally occurring folate from the other foods in her diet—in this example, another 90 µg of folate.

510 µg DFE + 90 µg = 600 µg DFE

Notice that if we had not converted synthetic folate from supplements and fortified foods to DFE, then this woman's intake would appear to fall short of the 600 µg recommendation for pregnancy (300 µg + 90 µg = 390 µg). But as this example shows, her intake does meet the recommendation. At this time, supplement and fortified food labels list folate in µg only, not µg DFE, making such calculations necessary.

Throughout this book and in the appendixes, the term calorie is used to mean kilocalorie. Thus, when converting calories to kilojoules, do not enlarge the calorie values—they are kilocalorie values.

Vitamin A

Equivalencies for vitamin A:

1 µg RAE = 1 µg retinol
= 12 µg beta-carotene
= 24 µg other vitamin A carotenoids

1 international unit (IU) = 0.3 µg retinol
= 3.6 µg beta-carotene
= 7.2 µg other vitamin A carotenoids

To convert older RE values to micrograms RAE:

1 µg RE retinol = 1 µg RAE retinol

6 µg RE beta-carotene = 12 µg RAE beta-carotene

12 µg RE other vitamin A carotenoids = 24 µg RAE other vitamin A carotenoids

International Units (IU)

To convert IU to:

- µg vitamin D: divide by 40 or multiply by 0.025.
- 1 IU natural vitamin E = 0.67 mg alpha-tocopherol.
- 1 IU synthetic vitamin E = 0.45 mg alpha-tocopherol.
- vitamin A, see above.

Percentages

A percentage is a comparison between a number of items (perhaps your intake of energy) and a standard number (perhaps the number of calories recommended for your age and gender—your energy DRI). The standard number is the number you divide by. The answer you get after the division must be multiplied by 100 to be stated as a percentage (percent means "per 100").

Example 4

What percentage of the DRI recommendation for energy is your energy intake?

1. Find your energy DRI value on the inside front cover. We'll use 2,368 calories to demonstrate.
2. Total your energy intake for a day—for example, 1,200 calories.
3. Divide your calorie intake by the DRI value:

1,200 cal (your intake) ÷ 2,368 cal (DRI) = 0.507

4. Multiply your answer by 100 to state it as a percentage:

0.507 × 100 = 50.7 = 51% (rounded off to the nearest whole number)

In some problems in nutrition, the percentage may be more than 100. For example, suppose your daily intake of vitamin A is 3,200 and your DRI is 900 μg. Your intake as a percentage of the DRI is more than 100 percent (that is, you consume more than 100 percent of your recommendation for vitamin A). The following calculations show your vitamin A intake as a percentage of the DRI value:

$$3{,}200 \div 900 = 3.6 \text{ (rounded)}$$
$$3.6 \times 100 = 360\% \text{ of DRI}$$

Example 5

Food labels express nutrients and energy contents of foods as percentages of the Daily Values. If a serving of a food contains 200 milligrams of calcium, for example, what percentage of the calcium Daily Value does the food provide?

1. Find the calcium Daily Value on the inside back cover, page Y.
2. Divide the milligrams of calcium in the food by the Daily Value standard:

$$\frac{200}{1{,}000} = 0.2$$

3. Multiply by 100:

$$0.2 \times 100 = 20\% \text{ of the Daily Value}$$

Example 6

This example demonstrates how to calculate the percentage of fat in a day's meals.

1. Recall the general formula for finding percentages of calories from a nutrient:

(one nutrient's calories ÷ total calories) × 100 = the percentage of calories from that nutrient

2. Say a day's meals provide 1,754 calories and 54 grams of fat. First, convert fat grams to fat calories:

54 g × 9 cal per g = 486 cal from fat

3. Then apply the general formula for finding percentage of calories from fat:

(fat calories ÷ total calories) × 100 = percentage of calories from fat

(486 ÷ 1,754) × 100 = 27. 7 (28%, rounded)

Weights and Measures

Length

1 inch (in.) = 2.54 centimeters (cm)
1 foot (ft) = 30.48 cm
1 meter (m) = 39.37 in

Temperature

	Celsius‡	Fahrenheit	
Steam	100°C	212°F	Steam
Body temperature	37°C	98.6°F	Body temperature
Ice	0°C	32°F	Ice

- To find degrees Fahrenheit (°F) when you know degrees Celsius (°C), multiply by 9/5 and then add 32.
- To find degrees Celsius (°C) when you know degrees Fahrenheit (°F), subtract 32 and then multiply by 5/9.

Volume

Used to measure fluids or pourable dry substances such as cereal.

1 milliliter (ml) = $\frac{1}{5}$ teaspoon or 0.034 fluid ounce or $\frac{1}{1000}$ liter
1 deciliter (dL)= $\frac{1}{10}$ liter
1 teaspoon (tsp or t) = 5 ml or about 5 grams (weight) salt
1 tablespoon (tbs or T) = 3 tsp or 15 ml
1 ounce, fluid (fl oz) = 2 tbs or 30 ml
1 cup (c) = 8 fl oz or 16 tbs or 250 ml
1 quart (qt) = 32 fl oz or 4 c or 0.95 liter
1 liter (L) = 1.06 qt or 1,000 ml
1 gallon (gal) = 16 c or 4 qt or 128 fl oz or 3.79 L

Weight

1 microgram (μg or mcg) = $\frac{1}{1000}$ milligram
1 milligram (mg) = 1,000 mcg or $\frac{1}{1{,}000}$ gram
1 gram (g) = 1,000 mg or $\frac{1}{1{,}000}$ kilogram
1 ounce, weight (oz) = about 28 g or $\frac{1}{16}$ pound
1 pound (lb) =16 oz (wt) or about 454 g
1 kilogram (kg) =1,000 g or 2.2 lb

‡Also known as centigrade.

Appendix D

Exchange Lists for Diabetes

Chapter 2 introduces the exchange system, and this appendix provides details from the *2008 Choose Your Foods: Exchange Lists for Diabetes.* Exchange lists can help people with diabetes to manage their blood glucose levels by controlling the amount and kinds of carbohydrates they consume. These lists can also help in planning diets for weight management by controlling calorie and fat intake.

The Exchange System

The exchange system sorts foods into groups by their proportions of carbohydrate, fat, and protein (Table D-1 on p. D-2). These groups may be organized into several exchange lists of foods (Tables D-2 through D-12 on pp. D-3–D-14). For example, the carbohydrate group includes these exchange lists:

- Starch
- Fruits
- Milk (fat-free, reduced-fat, and whole)
- Sweets, Desserts, and Other Carbohydrates
- Nonstarchy Vegetables

Then any food on a list can be "exchanged" for any other on that same list. Another group for alcohol has been included as a reminder that these beverages often deliver substantial carbohydrate and calories, and therefore warrant their own list.

Serving Sizes

The serving sizes have been carefully adjusted and defined so that a serving of any food on a given list provides roughly the same amount of carbohydrate, fat, and protein, and, therefore, total energy. Any food on a list can thus be exchanged, or traded, for any other food on the same list without significantly affecting the diet's energy-nutrient balance or total calories. For example, a person may select 17 small grapes or ½ large grapefruit as one fruit exchange, and either choice would provide roughly 15 grams of carbohydrate and 60 calories. A whole grapefruit, however, would count as 2 fruit exchanges.

To apply the system successfully, users must become familiar with the specified serving sizes. A convenient way to remember the serving sizes and energy values is to keep in mind a typical item from each list (review Table D-1).

The Foods on the Lists

Foods do not always appear on the exchange list where you might first expect to find them. They are grouped according to their energy-nutrient contents rather than by their source (such as milks), their outward appearance, or their vitamin and mineral contents. For example, cheeses are grouped with meats (not milk) because, like meats, cheeses contribute energy from protein and fat but provide negligible carbohydrate.

For similar reasons, starchy vegetables such as corn, green peas, and potatoes are found on the Starch list with breads and cereals, not with the vegetables. Likewise, bacon is grouped with the fats and oils, not with the meats.

D

Diet planners learn to view mixtures of foods, such as casseroles and soups, as combinations of foods from different exchange lists. They also learn to interpret food labels with the exchange system in mind.

Controlling Energy, Fat, and Sodium

The exchange lists help people control their energy intakes by paying close attention to serving sizes. People wanting to lose weight can limit foods from the Sweets, Desserts, and Other Carbohydrates and Fats lists, and they might choose to avoid the Alcohol list altogether. The Free Foods list provide low-calorie choices.

By assigning items like bacon to the Fats list, the exchange lists alert consumers to foods that are unexpectedly high in fat. Even the Starch list specifies which grain products contain added fat (such as biscuits, cornbread, and waffles) by marking them with a symbol to indicate added fat (the symbols are explained in the table keys). In addition, the exchange lists encourage users to think of fat-free milk as milk and of whole milk as milk with added fat, and to think of lean meats as meats and of medium-fat and high-fat meats as meats with added fat. To that end, foods on the milk and meat lists are separated into categories based on their fat contents (review Table D-1). The Milk list is subdivided for fat-free, reduced fat, and whole; the meat list is subdivided for lean, medium fat, and high fat. The meat list also includes plant-based proteins, which tend to be rich in fiber. Notice that many of these foods (p. D-9) bear the symbol for "high fiber."

People wanting to control the sodium in their diets can begin by eliminating any foods bearing the "high sodium" symbol. In most cases, the symbol identifies foods that, in one serving, provide 480 milligrams or more of sodium. Foods on the Combination Foods or Fast Foods lists that bear the symbol provide more than 600 milligrams of sodium. Other foods may also contribute substantially to sodium (consult Chapter 8 for details).

TABLE D-1 The Food Lists

Lists	Typical Item/Portion Size	Carbohydrate (g)	Protein (g)	Fat (g)	Energy[a] (cal)
Carbohydrates					
Starch[b]	1 slice bread	15	0–3	0–1	80
Fruits	1 small apple	15	—	—	60
Milk					
Fat-free, low-fat, 1%	1 c fat-free milk	12	8	0–3	100
Reduced-fat, 2%	1 c reduced-fat milk	12	8	5	120
Whole	1 c whole milk	12	8	8	160
Sweets, desserts, and other carbohydrates[c]	2 small cookies	15	varies	varies	varies
Nonstarchy vegetables	½ c cooked carrots	5	2	—	25
Meat and Meat Substitutes					
Lean	1 oz chicken (no skin)	—	7	0–3	45
Medium-fat	1 oz ground beef	—	7	4–7	75
High-fat	1 oz pork sausage	—	7	8+	100
Plant-based proteins	½ c tofu	varies	7	varies	varies
Fats	1 tsp butter	—	—	5	45
Alcohol	12 oz beer	varies	—	—	100

[a]The energy value for each exchange list represents an approximate average for the group and does not reflect the precise number of grams of carbohydrate, protein, and fat. For example, a slice of bread contains 15 grams of carbohydrate (60 calories), 3 grams protein (12 calories), and a little fat—rounded to 80 calories for ease in calculating. A ½ cup of vegetables (not including starchy vegetables) contains 5 grams carbohydrate (20 calories) and 2 grams protein (8 more), which has been rounded down to 25 calories.

[b]The Starch list includes cereals, grains, breads, crackers, snacks, starchy vegetables (such as corn, peas, and potatoes), and legumes (dried beans, peas, and lentils).

[c]The Sweets, Desserts, and Other Carbohydrates list includes foods that contain added sugars and fats such as sodas, candy, cakes, cookies, doughnuts, ice cream, pudding, syrup, and frozen yogurt.

Planning a Healthy Diet

To obtain a daily variety of foods that provide healthful amounts of carbohydrate, protein, and fat, as well as vitamins, minerals, and fiber, the meal plan for adults and teenagers should include at least:

- two to three servings of nonstarchy vegetables
- two servings of fruits
- six servings of grains (at least three of whole grains), beans, and starchy vegetables
- two servings of low-fat or fat-free milk
- about 6 ounces of meat or meat substitutes
- *small* amounts of fat and sugar

The actual amounts are determined by age, gender, activity levels, and other factors that influence energy needs.

TABLE D-2 Starch

The Starch list includes bread, cereals and grains, starchy vegetables, crackers and snacks, and legumes (dried beans, peas, and lentils). 1 starch choice = 15 grams carbohydrate, 0–3 grams protein, 0–1 grams fat, and 80 calories.

NOTE: In general, one starch exchange is ½ cup cooked cereal, grain, or starchy vegetable; ⅓ cup cooked rice or pasta; 1 ounce of bread product; ¾ ounce to 1 ounce of most snack foods.

Bread

Food	Serving Size
Bagel, large (about 4 oz)	¼ (1 oz)
▽ Biscuit, 2½ inches across	1
Bread	
☺ reduced-calorie	2 slices (1½ oz)
white, whole-grain, pumpernickel, rye, unfrosted raisin	1 slice (1 oz)
Chapatti, small, 6 inches across	1
▽ Cornbread, 1¾ inch cube	1 (1½ oz)
English muffin	½
Hot dog bun or hamburger bun	½ (1 oz)
Naan, 8 inches by 2 inches	¼
Pancake, 4 inches across, ¼ inch thick	1
Pita, 6 inches across	½
Roll, plain, small	1 (1 oz)
▽ Stuffing, bread	⅓ cup
▽ Taco shell, 5 inches across	2
Tortilla, corn, 6 inches across	1
Tortilla, flour, 6 inches across	1
Tortilla, flour, 10 inches across	⅓
▽ Waffle, 4-inch square or 4 inches across	1

Cereals and Grains

Food	Serving Size
Barley, cooked	⅓ cup
Bran, dry	
☺ oat	¼ cup
☺ wheat	½ cup
☺ Bulgur (cooked)	½ cup

Cereals and Grains—continued

Food	Serving Size
Cereals	
☺ bran	½ cup
cooked (oats, oatmeal)	½ cup
puffed	1½ cups
shredded wheat, plain	½ cup
sugar-coated	½ cup
unsweetened, ready-to-eat	¾ cup
Couscous	⅓ cup
Granola	
low-fat	¼ cup
▽ regular	¼ cup
Grits, cooked	½ cup
Kasha	½ cup
Millet, cooked	⅓ cup
Muesli	¼ cup
Pasta, cooked	⅓ cup
Polenta, cooked	⅓ cup
Quinoa, cooked	⅓ cup
Rice, white or brown, cooked	⅓ cup
Tabbouleh (tabouli), prepared	½ cup
Wheat germ, dry	3 Tbsp
Wild rice, cooked	½ cup

Starchy Vegetables

Food	Serving Size
Cassava	⅓ cup
Corn	½ cup
on cob, large	½ cob (5 oz)
☺ Hominy, canned	¾ cup

(continued)

KEY

☺ = More than 3 grams of dietary fiber per serving.

▽ = Extra fat, or prepared with added fat. (Count as 1 starch + 1 fat.)

🧂 = 480 milligrams or more of sodium per serving.

D

TABLE D-2 Starch *(continued)*

Starchy Vegetables—continued	
Food	**Serving Size**
☺ Mixed vegetables with corn, peas, or pasta	1 cup
☺ Parsnips	½ cup
☺ Peas, green	½ cup
Plantain, ripe	⅓ cup
Potato	
baked with skin	¼ large (3 oz)
boiled, all kinds	½ cup or ½ medium (3 oz)
⚠ mashed, with milk and fat	½ cup
french fried (oven-baked)[a]	1 cup (2 oz)
☺ Pumpkin, canned, no sugar added	1 cup
Spaghetti/pasta sauce	½ cup
☺ Squash, winter (acorn, butternut)	1 cup
☺ Succotash	½ cup
Yam, sweet potato, plain	½ cup

Crackers and Snacks[b]	
Food	**Serving Size**
Animal crackers	8
Crackers	
⚠ round-butter type	6
saltine-type	6
⚠ sandwich-style, cheese or peanut butter filling	3
⚠ whole-wheat regular	2–5 (¾ oz)
☺ whole-wheat lower fat or crispbreads	2–5 (¾ oz)
Graham cracker, 2½-inch square	3
Matzoh	¾ oz
Melba toast, about 2-inch by 4-inch piece	4
Oyster crackers	20
Popcorn	3 cups
⚠ ☺ with butter	3 cups
☺ no fat added	3 cups
☺ lower fat	3 cups
Pretzels	¾ oz
Rice cakes, 4 inches across	2
Snack chips	
fat-free or baked (tortilla, potato), baked pita chips	15–20 (¾ oz)
⚠ regular (tortilla, potato)	9–13 (¾ oz)

Beans, Peas, and Lentils[c]

The choices on this list count as 1 starch + 1 lean meat.

Food	Serving Size
☺ Baked beans	⅓ cup
☺ Beans, cooked (black, garbanzo, kidney, lima, navy, pinto, white)	½ cup
☺ Lentils, cooked (brown, green, yellow)	½ cup
☺ Peas, cooked (black-eyed, split)	½ cup
🧂 ☺ Refried beans, canned	½ cup

KEY

☺ = More than 3 grams of dietary fiber per serving.

⚠ = Extra fat, or prepared with added fat. (Count as 1 starch + 1 fat.)

🧂 = 480 milligrams or more of sodium per serving.

[a]*Restaurant-style french fries are on the Fast Foods list.*

[b]*For other snacks, see the Sweets, Desserts, and Other Carbohydrates list. For a quick estimate of serving size, an open handful is equal to about 1 cup or 1 to 2 ounces of snack food.*

[c]*Beans, peas, and lentils are also found on the Meat and Meat Substitutes list.*

TABLE D-3 Fruits

Fruit[a]

The Fruits list includes fresh, frozen, canned, and dried fruits and fruit juices. 1 fruit choice = 15 grams carbohydrate, 0 grams protein, 0 grams fat, and 60 calories.
NOTE: In general, one fruit exchange is ½ cup canned or fresh fruit or unsweetened fruit juice; 1 small fresh fruit (4 ounces); 2 tablespoons dried fruit.

Food	Serving Size
Apple, unpeeled, small	1 (4 oz)
Apples, dried	4 rings
Applesauce, unsweetened	½ cup
Apricots	
canned	½ cup
dried	8 halves
☺ fresh	4 whole (5½ oz)
Banana, extra small	1 (4 oz)
☺ Blackberries	¾ cup
Blueberries	¾ cup
Cantaloupe, small	⅓ melon or 1 cup cubed (11 oz)
Cherries	
sweet, canned	½ cup
sweet fresh	12 (3 oz)
Dates	3
Dried fruits (blueberries, cherries, cranberries, mixed fruit, raisins)	2 Tbsp
Figs	
dried	1½
☺ fresh	1½ large or 2 medium (3½ oz)
Fruit cocktail	½ cup

(continued)

[a]*The weight listed includes skin, core, seeds, and rind.*

TABLE D-3 Fruits *(continued)*

Fruit—continued	
Food	**Serving Size**
Grapefruit	
large	½ (11 oz)
sections, canned	¾ cup
Grapes, small	17 (3 oz)
Honeydew melon	1 slice or 1 cup cubed (10 oz)
☺ Kiwi	1 (3½ oz)
Mandarin oranges, canned	¾ cup
Mango, small	½ (5½ oz) or ½ cup
Nectarine, small	1 (5 oz)
☺ Orange, small	1 (6½ oz)
Papaya	½ or 1 cup cubed (8 oz)
Peaches	
canned	½ cup
fresh, medium	1 (6 oz)
Pears	
canned	½ cup
fresh, large	½ (4 oz)
Pineapple	
canned	½ cup
fresh	¾ cup

Fruit—continued	
Food	**Serving Size**
Plums	
canned	½ cup
dried (prunes)	3
small	2 (5 oz)
☺ Raspberries	1 cup
☺ Strawberries	1¼ cup whole berries
☺ Tangerines, small	2 (8 oz)
Watermelon	1 slice or 1¼ cups cubes (13½ oz)
Fruit Juice	
Food	**Serving Size**
Apple juice/cider	½ cup
Fruit juice blends, 100% juice	⅓ cup
Grape juice	⅓ cup
Grapefruit juice	½ cup
Orange juice	½ cup
Pineapple juice	½ cup
Prune juice	⅓ cup

KEY

☺ = More than 3 grams of dietary fiber per serving.

⚠ = Extra fat, or prepared with added fat. (Count as 1 starch + 1 fat.)

🧂 = 480 milligrams or more of sodium per serving.

TABLE D-4 Milk

The Milk list groups milks and yogurts based on the amount of fat they have (fat-free/low fat, reduced fat, and whole). Cheeses are found on the Meat and Meat Substitutes list and cream and other dairy fats are found on the Fats list.

NOTE: In general, one milk choice is 1 cup (8 fluid ounces or ½ pint) milk or yogurt.

Milk and Yogurts	
Food	**Serving Size**
Fat-free or low-fat (1%)	
1 fat-free/low-fat milk choice = 12 g carbohydrate, 8 g protein, 0–3 g fat, and 100 cal.	
Milk, buttermilk, acidophilus milk, Lactaid	1 cup
Evaporated milk	½ cup
Yogurt, plain or flavored with an artificial sweetener	⅔ cup (6 oz)
Reduced-fat (2%)	
1 reduced-fat milk choice = 12 g carbohydrate, 8 g protein, 5 g fat, and 120 cal.	
Milk, acidophilus milk, kefir, Lactaid	1 cup
Yogurt, plain	⅔ cup (6 oz)
Whole	
1 whole milk choice = 12 g carbohydrate, 8 g protein, 8 g fat, and 160 cal.	
Milk, buttermilk, goat's milk	1 cup
Evaporated milk	½ cup
Yogurt, plain	8 oz

(continued)

TABLE D-4 Milk *(continued)*

Dairy-Like Foods		
Food	**Serving Size**	**Count As**
Chocolate milk		
fat-free	1 cup	1 fat-free milk + 1 carbohydrate
whole	1 cup	1 whole milk + 1 carbohydrate
Eggnog, whole milk	½ cup	1 carbohydrate + 2 fats
Rice drink		
flavored, low fat	1 cup	2 carbohydrates
plain, fat-free	1 cup	1 carbohydrate
Smoothies, flavored, regular	10 oz	1 fat-free milk + 2½ carbohydrates
Soy milk		
light	1 cup	1 carbohydrate + ½ fat
regular, plain	1 cup	1 carbohydrate + 1 fat
Yogurt		
and juice blends	1 cup	1 fat-free milk + 1 carbohydrate
low carbohydrate (less than 6 grams carbohydrate per choice)	⅔ cup (6 oz)	½ fat-free milk
with fruit, low-fat	⅔ cup (6 oz)	1 fat-free milk + 1 carbohydrate

TABLE D-5 Sweets, Desserts, and Other Carbohydrates

1 other carbohydrate choice = 15 grams carbohydrate, variable grams protein, variable grams fat, and variable calories.
NOTE: In general, one choice from this list can substitute for foods on the Starch, Fruits, or Milk lists.

Beverages, Soda, and Energy/Sports Drinks		
Food	**Serving Size**	**Count As**
Cranberry juice cocktail	½ cup	1 carbohydrate
Energy drink	1 can (8.3 oz)	2 carbohydrates
Fruit drink or lemonade	1 cup (8 oz)	2 carbohydrates
Hot chocolate		
regular	1 envelope added to 8 oz water	1 carbohydrate + 1 fat
sugar-free or light	1 envelope added to 8 oz water	1 carbohydrate
Soft drink (soda), regular	1 can (12 oz)	2½ carbohydrates
Sports drink	1 cup (8 oz)	1 carbohydrate
Brownies, Cake, Cookies, Gelatin, Pie, and Pudding		
Food	**Serving Size**	**Count As**
Brownie, small, unfrosted	1¼-inch square, ⅞ inch high (about 1 oz)	1 carbohydrate + 1 fat
Cake		
angel food, unfrosted	1/12 of cake (about 2 oz)	2 carbohydrates
frosted	2-inch square (about 2 oz)	2 carbohydrates + 1 fat
unfrosted	2-inch square (about 2 oz)	1 carbohydrate + 1 fat
Cookies		
chocolate chip	2 cookies (2¼ inches across)	1 carbohydrate + 2 fats
gingersnap	3 cookies	1 carbohydrate
sandwich, with crème filling	2 small (about ⅔ oz)	1 carbohydrate + 1 fat
sugar-free	3 small or 1 large (¾–1 oz)	1 carbohydrate + 1–2 fats
vanilla wafer	5 cookies	1 carbohydrate + 1 fat
Cupcake, frosted	1 small (about 1¾ oz)	2 carbohydrates + 1–1½ fats
Fruit cobbler	½ cup (3½ oz)	3 carbohydrates + 1 fat
Gelatin, regular	½ cup	1 carbohydrate
Pie		
commercially prepared fruit, 2 crusts	⅙ of 8-inch pie	3 carbohydrates + 2 fats
pumpkin or custard	⅛ of 8-inch pie	1½ carbohydrates + 1½ fats
Pudding		
regular (made with reduced-fat milk)	½ cup	2 carbohydrates
sugar-free or sugar- and fat-free (made with fat-free milk)	½ cup	1 carbohydrate

(continued)

TABLE D-5 Sweets, Desserts, and Other Carbohydrates *(continued)*

Candy, Spreads, Sweets, Sweeteners, Syrups, and Toppings		
Food	**Serving Size**	**Count as**
Candy bar, chocolate/peanut	2 "fun size" bars (1 oz)	1½ carbohydrates + 1½ fats
Candy, hard	3 pieces	1 carbohydrate
Chocolate "kisses"	5 pieces	1 carbohydrate + 1 fat
Coffee creamer		
dry, flavored	4 tsp	½ carbohydrate + ½ fat
liquid, flavored	2 Tbsp	1 carbohydrate
Fruit snacks, chewy (pureed fruit concentrate)	1 roll (¾ oz)	1 carbohydrate
Fruit spreads, 100% fruit	1½ Tbsp	1 carbohydrate
Honey	1 Tbsp	1 carbohydrate
Jam or jelly, regular	1 Tbsp	1 carbohydrate
Sugar	1 Tbsp	1 carbohydrate
Syrup		
chocolate	2 Tbsp	2 carbohydrates
light (pancake type)	2 Tbsp	1 carbohydrate
regular (pancake type)	1 Tbsp	1 carbohydrate
Condiments and Sauces[a]		
Food	**Serving Size**	**Count as**
Barbeque sauce	3 Tbsp	1 carbohydrate
Cranberry sauce, jellied	¼ cup	1½ carbohydrates
🧂 Gravy, canned or bottled	½ cup	½ carbohydrate + ½ fat
Salad dressing, fat-free, low-fat, cream-based	3 Tbsp	1 carbohydrate
Sweet and sour sauce	3 Tbsp	1 carbohydrate
Doughnuts, Muffins, Pastries, and Sweet Breads		
Food	**Serving Size**	**Count as**
Banana nut bread	1-inch slice (1 oz)	2 carbohydrates + 1 fat
Doughnut		
cake, plain	1 medium (1½ oz)	1½ carbohydrates + 2 fats
yeast type, glazed	3¾ inches across (2 oz)	2 carbohydrates + 2 fats
Muffin (4 oz)	¼ muffin (1 oz)	1 carbohydrate + ½ fat
Sweet roll or Danish	1 (2½ oz)	2½ carbohydrates + 2 fats
Frozen Bars, Frozen Desserts, Frozen Yogurt, and Ice Cream		
Food	**Serving Size**	**Count as**
Frozen pops	1	½ carbohydrate
Fruit juice bars, frozen, 100% juice	1 bar (3 oz)	1 carbohydrate
Ice cream		
fat-free	½ cup	1½ carbohydrates
light	½ cup	1 carbohydrate + 1 fat
no sugar added	½ cup	1 carbohydrate + 1 fat
regular	½ cup	1 carbohydrate + 2 fats
Sherbet, sorbet	½ cup	2 carbohydrates
Yogurt, frozen		
fat-free	⅓ cup	1 carbohydrate
regular	½ cup	1 carbohydrate + 0–1 fat
Granola Bars, Meal Replacement Bars/Shakes, and Trail Mix		
Food	**Serving Size**	**Count as**
Granola or snack bar, regular or low-fat	1 bar (1 oz)	1½ carbohydrates
Meal replacement bar	1 bar (1⅓ oz)	1½ carbohydrates + 0–1 fat
Meal replacement bar	1 bar (2 oz)	2 carbohydrates + 1 fat
Meal replacement shake, reduced calorie	1 can (10–11 oz)	1½ carbohydrates + 0–1 fat
Trail mix		
candy/nut-based	1 oz	1 carbohydrate + 2 fats
dried fruit-based	1 oz	1 carbohydrate + 1 fat

KEY

🧂 = 480 milligrams or more of sodium per serving

[a] *You can also check the Fats list and Free Foods list for other condiments.*

D

TABLE D-6 Nonstarchy Vegetables

The Nonstarchy Vegetables list includes vegetables that have few grams of carbohydrates or calories; starchy vegetables are found on the Starch list. 1 nonstarchy vegetable choice = 5 grams carbohydrate, 2 grams protein, 0 grams fat, and 25 calories.

NOTE: In general, one nonstarchy vegetable choice is ½ cup cooked vegetables or vegetable juice or 1 cup raw vegetables. Count 3 cups of raw vegetables or 1½ cups of cooked vegetables as one carbohydrate choice.

Nonstarchy Vegetables[a]
Amaranth or Chinese spinach
Artichoke
Artichoke hearts
Asparagus
Baby corn
Bamboo shoots
Beans (green, wax, Italian)
Bean sprouts
Beets
🧂 Borscht
Broccoli
☺ Brussels sprouts
Cabbage (green, bok choy, Chinese)
☺ Carrots
Cauliflower
Celery
☺ Chayote
Coleslaw, packaged, no dressing
Cucumber
Eggplant
Gourds (bitter, bottle, luffa, bitter melon)
Green onions or scallions
Greens (collard, kale, mustard, turnip)
Hearts of palm
Jicama
Kohlrabi
Leeks
Mixed vegetables (without corn, peas, or pasta)
Mung bean sprouts
Mushrooms, all kinds, fresh
Okra
Onions
Oriental radish or daikon
Pea pods
☺ Peppers (all varieties)
Radishes
Rutabaga
🧂 Sauerkraut
Soybean sprouts
Spinach
Squash (summer, crookneck, zucchini)
Sugar pea snaps
☺ Swiss chard
Tomato
Tomatoes, canned
🧂 Tomato sauce
🧂 Tomato/vegetable juice
Turnips
Water chestnuts
Yard-long beans

KEY

☺ = More than 3 grams of dietary fiber per serving.

🧂 = 480 milligrams or more of sodium per serving

[a]*Salad greens (like chicory, endive, escarole, lettuce, romaine, spinach, arugula, radicchio, watercress) are on the Free Foods list.*

TABLE D-7 Meat and Meat Substitutes

The Meat and Meat Substitutes list groups foods based on the amount of fat they have (lean meat, medium-fat meat, high-fat meat, and plant-based proteins).

Lean Meats and Meat Substitutes

1 lean meat choice = 0 grams carbohydrate, 7 grams protein, 0–3 grams fat, and 100 calories.

Food	Amount
Beef: Select or Choice grades trimmed of fat: ground round, roast (chuck, rib, rump), round, sirloin, steak (cubed, flank, porterhouse, T-bone), tenderloin	1 oz
🧂 Beef jerky	1 oz
Cheeses with 3 grams of fat or less per oz	1 oz
Cottage cheese	¼ cup
Egg substitutes, plain	¼ cup
Egg whites	2
Fish, fresh or frozen, plain: catfish, cod, flounder, haddock, halibut, orange roughy, salmon, tilapia, trout, tuna	1 oz
🧂 Fish, smoked: herring or salmon (lox)	1 oz
Game: buffalo, ostrich, rabbit, venison	1 oz
🧂 Hot dog with 3 grams of fat or less per oz (8 dogs per 14 oz package) *Note: May be high in carbohydrate.*	1
Lamb: chop, leg, or roast	1 oz
Organ meats: heart, kidney, liver *Note: May be high in cholesterol.*	1 oz
Oysters, fresh or frozen	6 medium
Pork, lean	
🧂 Canadian bacon	1 oz
rib or loin chop/roast, ham, tenderloin	1 oz
Poultry, without skin: Cornish hen, chicken, domestic duck or goose (well-drained of fat), turkey	1 oz
Processed sandwich meats with 3 grams of fat or less per oz: chipped beef, deli thin-sliced meats, turkey ham, turkey kielbasa, turkey pastrami	1 oz
Salmon, canned	1 oz
Sardines, canned	2 medium
🧂 Sausage with 3 grams of fat or less per oz	1 oz
Shellfish: clams, crab, imitation shellfish, lobster, scallops, shrimp	1 oz
Tuna, canned in water or oil, drained	1 oz
Veal, lean chop, roast	1 oz

(continued)

TABLE D-7 Meat and Meat Substitutes *(continued)*

Medium-Fat Meat and Meat Substitutes

1 medium-fat meat choice = 0 grams carbohydrate, 7 grams protein, 4–7 grams fat, and 130 calories.

Food	Amount
Beef: corned beef, ground beef, meatloaf, Prime grades trimmed of fat (prime rib), short ribs, tongue	1 oz
Cheeses with 4–7 grams of fat per oz:	
feta, mozzarella, pasteurized processed cheese spread, reduced-fat cheeses, string	1 oz
Egg *Note: High in cholesterol, so limit to 3 per week.*	1
Fish, any fried product	1 oz
Lamb: ground, rib roast	1 oz
Pork: cutlet, shoulder roast	1 oz
Poultry: chicken with skin; dove, pheasant, wild duck, or goose; fried chicken; ground turkey	1 oz
Ricotta cheese	2 oz or ¼ cup
🧂 Sausage with 4–7 grams of fat per oz	1 oz
Veal, cutlet (no breading)	1 oz

High-Fat Meat and Meat Substitutes

1 high-fat meat choice = 0 grams carbohydrate, 7 grams protein, 8+ grams fat, and 150 calories. These foods are high in saturated fat, cholesterol, and calories and may raise blood cholesterol levels if eaten on a regular basis. Try to eat 3 or fewer servings from this group per week.

Food	Amount
Bacon	
🧂 pork	2 slices (16 slices per lb or 1 oz each, before cooking)
🧂 turkey	3 slices (½ oz each before cooking)
Cheese, regular: American, bleu, brie, cheddar, hard goat, Monterey jack, queso, and Swiss	1 oz
⚠ 🧂 Hot dog: beef, pork, or combination (10 per lb-sized package)	1
🧂 Hot dog: turkey or chicken (10 per lb-sized package)	1
Pork: ground, sausage, spareribs	1 oz
Processed sandwich meats with 8 grams of fat or more per oz: bologna, pastrami, hard salami	1 oz
🧂 Sausage with 8 grams fat or more per oz: bratwurst, chorizo, Italian, knockwurst, Polish, smoked, summer	1 oz

Plant-Based Proteins

1 plant-based protein choice = variable grams carbohydrate, 7 grams protein, variable grams fat, and variable calories. Because carbohydrate content varies among plant-based proteins, you should read the food label.

Food	Serving Size	Count As
"Bacon" strips, soy-based	3 strips	1 medium-fat meat
☺ Baked beans	⅓ cup	1 starch + 1 lean meat
☺ Beans, cooked: black, garbanzo, kidney, lima, navy, pinto, white[a]	½ cup	1 starch + 1 lean meat
☺ "Beef" or "sausage" crumbles, soy-based	2 oz	½ carbohydrate + 1 lean meat
"Chicken" nuggets, soy-based	2 nuggets (1½ oz)	½ carbohydrate + 1 medium-fat meat
☺ Edamame	½ cup	½ carbohydrate + 1 lean meat
Falafel (spiced chickpea and wheat patties)	3 patties (about 2 inches across)	1 carbohydrate + 1 high-fat meat
Hot dog, soy-based	1 (1½ oz)	½ carbohydrate + 1 lean meat
☺ Hummus	⅓ cup	1 carbohydrate + 1 high-fat meat
☺ Lentils, brown, green, or yellow	½ cup	1 carbohydrate + 1 lean meat
☺ Meatless burger, soy-based	3 oz	½ carbohydrate + 2 lean meats
☺ Meatless burger, vegetable- and starch-based	1 patty (about 2½ oz)	1 carbohydrate + 2 lean meats
Nut spreads: almond butter, cashew butter, peanut butter, soy nut butter	1 Tbsp	1 high-fat meat
☺ Peas, cooked: black-eyed and split peas	½ cup	1 starch + 1 lean meat
🧂 ☺ Refried beans, canned	½ cup	1 starch + 1 lean meat
"Sausage" patties, soy-based	1 (1½ oz)	1 medium-fat meat
Soy nuts, unsalted	¾ oz	½ carbohydrate + 1 medium-fat meat
Tempeh	¼ cup	1 medium-fat meat
Tofu	4 oz (½ cup)	1 medium-fat meat
Tofu, light	4 oz (½ cup)	1 lean meat

KEY

☺ = More than 3 grams of dietary fiber per serving.

⚠ = Extra fat, or prepared with added fat. (Count as 1 starch + 1 fat.)

🧂 = 480 milligrams or more of sodium per serving (based on the sodium content of a typical 3-oz serving of meat, unless 1 or 2 oz is the normal serving size).

[a]*Beans, peas, and lentils are also found on the Starch list; nut butters in smaller amounts are found in the Fats list.*

TABLE D-8 Fats

Fats and oils have mixtures of unsaturated (polyunsaturated and monounsaturated) and saturated fats. Foods on the Fats list are grouped together based on the major type of fat they contain.

1 fat choice = 0 grams carbohydrate, 0 grams protein, 5 grams fat, and 45 calories.

NOTE: In general, one fat exchange is 1 teaspoon of regular margarine, vegetable oil, or butter; 1 tablespoon of regular salad dressing.

When used in large amounts, bacon and peanut butter are counted as high-fat meat choices (see Meat and Meat Substitutes list). Fat-free salad dressings are found on the Sweets, Desserts, and Other Carbohydrates list. Fat-free products such as margarines, salad dressings, mayonnaise, sour cream, and cream cheese are found on the Free Foods list.

Monounsaturated Fats

Food	Serving Size
Avocado, medium	2 Tbsp (1 oz)
Nut butters (*trans* fat-free): almond butter, cashew butter, peanut butter (smooth or crunchy)	1½ tsp
Nuts	
almonds	6 nuts
Brazil	2 nuts
cashews	6 nuts
filberts (hazelnuts)	5 nuts
macadamia	3 nuts
mixed (50% peanuts)	6 nuts
peanuts	10 nuts
pecans	4 halves
pistachios	16 nuts
Oil: canola, olive, peanut	1 tsp
Olives	
black (ripe)	8 large
green, stuffed	10 large

Polyunsaturated Fats

Food	Serving Size
Margarine: lower-fat spread (30%–50% vegetable oil, *trans* fat-free)	1 Tbsp
Margarine: stick, tub (*trans* fat-free) or squeeze (*trans* fat-free)	1 tsp
Mayonnaise	
reduced-fat	1 Tbsp
regular	1 tsp
Mayonnaise-style salad dressing	
reduced-fat	1 Tbsp
regular	2 tsp
Nuts	
Pignolia (pine nuts)	1 Tbsp
walnuts, English	4 halves
Oil: corn, cottonseed, flaxseed, grape seed, safflower, soybean, sunflower	1 tsp
Oil: made from soybean and canola oil—Enova	1 tsp
Plant stanol esters	
light	1 Tbsp
regular	2 tsp
Salad dressing	
reduced-fat *Note: May be high in carbohydrate.*	2 Tbsp
regular	1 Tbsp
Seeds	
flaxseed, whole	1 Tbsp
pumpkin, sunflower	1 Tbsp
sesame seeds	1 Tbsp
Tahini or sesame paste	2 tsp

Saturated Fats

Food	Serving Size
Bacon, cooked, regular or turkey	1 slice
Butter	
reduced-fat	1 Tbsp
stick	1 tsp
whipped	2 tsp
Butter blends made with oil	
reduced-fat or light	1 Tbsp
regular	1½ tsp
Chitterlings, boiled	2 Tbsp (½ oz)
Coconut, sweetened, shredded	2 Tbsp
Coconut milk	
light	⅓ cup
regular	1½ Tbsp
Cream	
half and half	2 Tbsp
heavy	1 Tbsp
light	1½ Tbsp
whipped	2 Tbsp
whipped, pressurized	¼ cup
Cream cheese	
reduced-fat	1½ Tbsp (¾ oz)
regular	1 Tbsp (½ oz)
Lard	1 tsp
Oil: coconut, palm, palm kernel	1 tsp
Salt pork	¼ oz
Shortening, solid	1 tsp
Sour cream	
reduced-fat or light	3 Tbsp
regular	2 Tbsp

KEY

= 480 milligrams or more of sodium per serving

TABLE D-9 Free Foods

A "free" food is any food or drink choice that has less than 20 calories and 5 grams or less of carbohydrate per serving.

- Most foods on this list should be limited to 3 servings (as listed here) per day. Spread out the servings throughout the day. If you eat all 3 servings at once, it could raise your blood glucose level.
- Food and drink choices listed here without a serving size can be eaten whenever you like.

Low Carbohydrate Foods

Food	Serving Size
Cabbage, raw	½ cup
Candy, hard (regular or sugar-free)	1 piece
Carrots, cauliflower, or green beans, cooked	¼ cup
Cranberries, sweetened with sugar substitute	½ cup
Cucumber, sliced	½ cup
Gelatin	
dessert, sugar-free	
unflavored	
Gum	
Jam or jelly, light or no sugar added	2 tsp
Rhubarb, sweetened with sugar substitute	½ cup
Salad greens	
Sugar substitutes (artificial sweeteners)	
Syrup, sugar-free	2 Tbsp

Modified Fat Foods with Carbohydrate

Food	Serving Size
Cream cheese, fat-free	1 Tbsp (½ oz)
Creamers	
nondairy, liquid	1 Tbsp
nondairy, powdered	2 tsp
Margarine spread	
fat-free	1 Tbsp
reduced-fat	1 tsp
Mayonnaise	
fat-free	1 Tbsp
reduced-fat	1 tsp
Mayonnaise-style salad dressing	
fat-free	1 Tbsp
reduced-fat	1 tsp
Salad dressing	
fat-free or low-fat	1 Tbsp
fat-free, Italian	2 Tbsp
Sour cream, fat-free or reduced-fat	1 Tbsp
Whipped topping	
light or fat-free	2 Tbsp
regular	1 Tbsp

Condiments

Food	Serving Size
Barbecue sauce	2 tsp
Catsup (ketchup)	1 Tbsp
Honey mustard	1 Tbsp
Horseradish	

Condiments—continued

Food	Serving Size
Lemon juice	
Miso	1½ tsp
Mustard	
Parmesan cheese, freshly grated	1 Tbsp
Pickle relish	1 Tbsp
Pickles	
dill	1½ medium
sweet, bread and butter	2 slices
sweet, gherkin	¾ oz
Salsa	¼ cup
Soy sauce, light or regular	1 Tbsp
Sweet and sour sauce	2 tsp
Sweet chili sauce	2 tsp
Taco sauce	1 Tbsp
Vinegar	
Yogurt, any type	2 Tbsp

Drinks/Mixes

Any food on the list—without a serving size listed—can be consumed in any moderate amount.

- Bouillon, broth, consommé
- Bouillon or broth, low-sodium
- Carbonated or mineral water
- Club soda
- Cocoa powder, unsweetened (1 Tbsp)
- Coffee, unsweetened or with sugar substitute
- Diet soft drinks, sugar-free
- Drink mixes, sugar-free
- Tea, unsweetened or with sugar substitute
- Tonic water, diet
- Water
- Water, flavored, carbohydrate free

Seasonings

Any food on this list can be consumed in any moderate amount.

- Flavoring extracts (for example, vanilla, almond, peppermint)
- Garlic
- Herbs, fresh or dried
- Nonstick cooking spray
- Pimento
- Spices
- Hot pepper sauce
- Wine, used in cooking
- Worcestershire sauce

KEY

= 480 milligrams or more of sodium per serving

D

TABLE D-10 Combination Foods

Many foods are eaten in various combinations, such as casseroles. Because "combination" foods do not fit into any one choice list, this list of choices provides some typical combination foods.

Entrees		
Food	**Serving Size**	**Count as**
🧂 Casserole type (tuna noodle, lasagna, spaghetti with meatballs, chili with beans, macaroni and cheese)	1 cup (8 oz)	2 carbohydrates + 2 medium-fat meats
🧂 Stews (beef/other meats and vegetables)	1 cup (8 oz)	1 carbohydrate + 1 medium-fat meat + 0–3 fats
Tuna salad or chicken salad	½ cup (3½ oz)	½ carbohydrate + 2 lean meats + 1 fat
Frozen Meals/Entrees		
Food	**Serving Size**	**Count as**
🧂 ☺ Burrito (beef and bean)	1 (5 oz)	3 carbohydrates + 1 lean meat + 2 fats
🧂 Dinner-type meal	generally 14–17 oz	3 carbohydrates + 3 medium-fat meats + 3 fats
🧂 Entrée or meal with less than 340 calories	about 8–11 oz	2–3 carbohydrates + 1–2 lean meats
Pizza		
🧂 cheese/vegetarian, thin crust	¼ of a 12 inch (4½–5 oz)	2 carbohydrates + 2 medium-fat meats
🧂 meat topping, thin crust	¼ of a 12 inch (5 oz)	2 carbohydrates + 2 medium-fat meats + 1½ fats
🧂 Pocket sandwich	1 (4½ oz)	3 carbohydrates + 1 lean meat + 1–2 fats
🧂 Pot pie	1 (7 oz)	2½ carbohydrates + 1 medium-fat meat + 3 fats
Salads (Deli-Style)		
Food	**Serving Size**	**Count as**
Coleslaw	½ cup	1 carbohydrate + 1½ fats
Macaroni/pasta salad	½ cup	2 carbohydrates + 3 fats
🧂 Potato salad	½ cup	1½–2 carbohydrates + 1–2 fats
Soups		
Food	**Serving Size**	**Count as**
🧂 Bean, lentil, or split pea	1 cup	1 carbohydrate + 1 lean meat
🧂 Chowder (made with milk)	1 cup (8 oz)	1 carbohydrate + 1 lean meat + 1½ fats
🧂 Cream (made with water)	1 cup (8 oz)	1 carbohydrate + 1 fat
🧂 Instant	6 oz prepared	1 carbohydrate
🧂 with beans or lentils	8 oz prepared	2½ carbohydrates + 1 lean meat
🧂 Miso soup	1 cup	½ carbohydrate + 1 fat
🧂 Oriental noodle	1 cup	2 carbohydrates + 2 fats
Rice (congee)	1 cup	1 carbohydrate
🧂 Tomato (made with water)	1 cup (8 oz)	1 carbohydrate
🧂 Vegetable beef, chicken noodle, or other broth-type	1 cup (8 oz)	1 carbohydrate

KEY

☺ = More than 3 grams of dietary fiber per serving.

▽! = Extra fat, or prepared with added fat.

🧂 = 600 milligrams or more of sodium per serving (for combination food main dishes/meals).

TABLE D-11 Fast Foods

The choices in the Fast Foods list are not specific fast-food meals or items, but are estimates based on popular foods. Ask the restaurant or check its website for nutrition information about your favorite fast foods.

Breakfast Sandwiches

Food	Serving Size	Count as
Egg, cheese, meat, English muffin	1 sandwich	2 carbohydrates + 2 medium-fat meats
Sausage biscuit sandwich	1 sandwich	2 carbohydrates + 2 high-fat meats + 3½ fats

Main Dishes/Entrees

Food	Serving Size	Count as
Burrito (beef and beans)	1 (about 8 oz)	3 carbohydrates + 3 medium-fat meats + 3 fats
Chicken breast, breaded and fried	1 (about 5 oz)	1 carbohydrate + 4 medium-fat meats
Chicken drumstick, breaded and fried	1 (about 2 oz)	2 medium-fat meats
Chicken nuggets	6 (about 3½ oz)	1 carbohydrate + 2 medium-fat meats + 1 fat
Chicken thigh, breaded and fried	1 (about 4 oz)	½ carbohydrate + 3 medium-fat meats + 1½ fats
Chicken wings, hot	6 (5 oz)	5 medium-fat meats + 1½ fats

Oriental

Food	Serving Size	Count as
Beef/chicken/shrimp with vegetables in sauce	1 cup (about 5 oz)	1 carbohydrate + 1 lean meat + 1 fat
Egg roll, meat	1 (about 3 oz)	1 carbohydrate + 1 lean meat + 1 fat
Fried rice, meatless	½ cup	1½ carbohydrates + 1½ fats
Meat and sweet sauce (orange chicken)	1 cup	3 carbohydrates + 3 medium-fat meats + 2 fats
Noodles and vegetables in sauce (chow mein, lo mein)	1 cup	2 carbohydrates + 1 fat

Pizza

Food	Serving Size	Count as
Pizza		
cheese, pepperoni, regular crust	⅛ of a 14 inch (about 4 oz)	2½ carbohydrates + 1 medium-fat meat + 1½ fats
cheese/vegetarian, thin crust	¼ of a 12 inch (about 6 oz)	2½ carbohydrates + 2 medium-fat meats + 1½ fats

Sandwiches

Food	Serving Size	Count as
Chicken sandwich, grilled	1	3 carbohydrates + 4 lean meats
Chicken sandwich, crispy	1	3½ carbohydrates + 3 medium-fat meats + 1 fat
Fish sandwich with tartar sauce	1	2½ carbohydrates + 2 medium-fat meats + 2 fats
Hamburger		
large with cheese	1	2½ carbohydrates + 4 medium-fat meats + 1 fat
regular	1	2 carbohydrates + 1 medium-fat meat + 1 fat
Hot dog with bun	1	1 carbohydrate + 1 high-fat meat + 1 fat
Submarine sandwich		
less than 6 grams fat	6-inch sub	3 carbohydrates + 2 lean meats
regular	6-inch sub	3½ carbohydrates + 2 medium-fat meats + 1 fat
Taco, hard or soft shell (meat and cheese)	1 small	1 carbohydrate + 1 medium-fat meat + 1½ fats

KEY

= More than 3 grams of dietary fiber per serving.

= Extra fat, or prepared with added fat.

= 600 milligrams or more of sodium per serving (for fast-food main dishes/meals).

TABLE D-11 Fast Foods *(continued)*

SALADS		
FOOD	**SERVING SIZE**	**COUNT AS**
Salad, main dish (grilled chicken type, no dressing or croutons)		1 carbohydrate + 4 lean meats
Salad, side, no dressing or cheese	Small (about 5 oz)	1 vegetable
SIDES/APPETIZERS		
FOOD	**SERVING SIZE**	**COUNT AS**
French fries, restaurant style	small	3 carbohydrates + 3 fats
	medium	4 carbohydrates + 4 fats
	large	5 carbohydrates + 6 fats
Nachos with cheese	small (about 4½ oz)	2½ carbohydrates + 4 fats
Onion rings	1 serving (about 3 oz)	2½ carbohydrates + 3 fats
DESSERTS		
FOOD	**SERVING SIZE**	**COUNT AS**
Milkshake, any flavor	12 oz	6 carbohydrates + 2 fats
Soft-serve ice cream cone	1 small	2½ carbohydrates + 1 fat

KEY

= More than 3 grams of dietary fiber per serving.

= Extra fat, or prepared with added fat.

= 600 milligrams or more of sodium per serving (for fast-food main dishes/meals).

TABLE D-12 Alcohol

1 alcohol equivalent = variable grams carbohydrate, 0 grams protein, 0 grams fat, and 100 calories.

NOTE: In general, one alcohol choice (½ ounce absolute alcohol) has about 100 calories. For those who choose to drink alcohol, guidelines suggest limiting alcohol intake to 1 drink or less per day for women, and 2 drinks or less per day for men. To reduce your risk of low blood glucose (hypoglycemia), especially if you take insulin or a diabetes pill that increases insulin, always drink alcohol with food. While alcohol, by itself, does not directly affect blood glucose, be aware of the carbohydrate (for example, in mixed drinks, beer, and wine) that may raise your blood glucose.

ALCOHOLIC BEVERAGE	SERVING SIZE	COUNT AS
Beer		
light (4.2%)	12 fl oz	1 alcohol equivalent + ½ carbohydrate
regular (4.9%)	12 fl oz	1 alcohol equivalent + 1 carbohydrate
Distilled spirits: vodka, rum, gin, whiskey, 80 or 86 proof	1½ fl oz	1 alcohol equivalent
Liqueur, coffee (53 proof)	1 fl oz	1 alcohol equivalent + 1 carbohydrate
Sake	1 fl oz	½ alcohol equivalent
Wine		
dessert (sherry)	3½ fl oz	1 alcohol equivalent + 1 carbohydrate
dry, red or white (10%)	5 fl oz	1 alcohol equivalent

Appendix E

Food Patterns to Meet the *Dietary Guidelines for Americans 2010*

This appendix presents several eating patterns that meet the ideals of the *Dietary Guidelines for Americans 2010*. First, Table E–1 lists the USDA eating patterns in full. Next, Table E–2 presents the Dietary Approaches to Stop Hypertension, or DASH, Eating Plan. Although it was originally developed to fight high blood pressure, the DASH plan has proved useful for cutting people's risks of many diseases while meeting nutrient needs superbly.

Two adaptations of the USDA eating patterns, offered in Table E–3 and Table E–4, demonstrate the flexibility of the patterns. These tables provide guidance for vegetarians and show how to meet nutrient needs without eating meat, and in the case of vegans, without eating any animal products.

TABLE E-1 USDA Eating Patterns

For each food group or subgroup,[a] recommended average daily intake amounts[b] at all calorie levels. Recommended intakes from vegetable and protein foods subgroups are per week. For more information and tools for application, go to www.choosemyplate.gov.

Calorie Level of pattern[c]	1,000	1,200	1,400	1,600	1,800	2,000	2,200	2,400	2,600	2,800	3,000	3,200
Food Group	Food group amounts shown in cup (c) or ounce-equivalents (oz-eq), with number of servings (srv) in parentheses when it differs from the other units. Oils are shown in grams (g).											
Fruits	1 c	1 c	1½ c	1½ c	1½ c	2 c	2 c	2 c	2 c	2½ c	2½ c	2½ c
Vegetables[d]	1 c	1½ c	1½ c	2 c	2½ c	2½ c	3 c	3 c	3½ c	3½ c	4 c	4 c
Dark green vegetables	½ c/wk	1 c/wk	1 c/wk	1 ½ c/wk	1 ½ c/wk	1 ½ c/wk	2 c/wk	2 c/wk	2 ½ c/wk	2 ½ c/wk	2 ½ c/wk	2 ½ c/wk
Red and orange vegetables	2½ c/wk	3 c/wk	1 c/wk	4 c/wk	5 ½ c/wk	5 ½ c/wk	6 c/wk	6 c/wk	7 c/wk	7 c/wk	7½ c/wk	7½ c/wk
Beans and peas (legumes)	½ c/wk	½ c/wk	½ c/wk	1 c/wk	1 ½ c/wk	1 ½ c/wk	2 c/wk	2 c/wk	2½ c/wk	2½ c/wk	3 c/wk	3 c/wk
Starchy vegetables	2 c/wk	3½ c/wk	3½ c/wk	4 c/wk	5 c/wk	5 c/wk	6 c/wk	6 c/wk	7 c/wk	7 c/wk	8 c/wk	8 c/wk
Other vegetables	1 ½ c/wk	2½ c/wk	2½ c/wk	3½ c/wk	4 c/wk	44 c/wk	5 c/wk	5 c/wk	5½ c/wk	5½ c/wk	7 c/wk	7 c/wk
Grains[e]	3 oz-eq	4 oz-eq	5 oz-eq	5 oz-eq	6 oz-eq	6 oz-eq	7 oz-eq	8 oz-eq	9 oz-eq	10 oz-eq	10 oz-eq	10 oz-eq
Whole grains	1½	2	2½	3	3	3	3½	4	4½	5	5	5
Enriched grains	1½	2	2½	2	3	3	3½	4	4½	5	5	5
Protein foods[d]	2 oz-eq	3 oz-eq	4 oz-eq	5 oz-eq	5 oz-eq	5½ oz-eq	6 oz-eq	6½ oz-eq	6½ oz-eq	7 oz-eq	7 oz-eq	7 oz-eq
Seafood	3 oz/wk	5 oz/wk	6 oz/wk	8 oz/wk	8 oz/wk	8 oz/wk	9 oz/wk	10 oz/wk	10 oz/wk	11 oz/wk	11 oz/wk	11 oz/wk
Meat, poultry, eggs	10 oz/wk	14 oz/wk	19 oz/wk	24 oz/wk	24 oz/wk	26 oz/wk	29 oz/wk	31 oz/wk	31 oz/wk	34 oz/wk	34 oz/wk	34 oz/wk
Nuts, seeds, soy products	1 oz/wk	2 oz/wk	3 oz/wk	4 oz/wk	4 oz/wk	4 oz/wk	4 oz/wk	5 oz/wk	5 oz/wk	5 oz/wk	5 oz/wk	5 oz/wk
Dairy (milk and milk products)[f]	2 c	2 ½ c	2 ½ c	3 c	3 c	3 c	3 c	3 c	3 c	3 c	3 c	3 c
Oils[g]	15 g	17 g	17 g	22 g	24 g	27 g	29 g	31 g	34 g	36 g	44 g	51 g
Maximum SoFAS[h] limit, calories (% of calories)	137 (14%)	121 (10%)	121 (9%)	121 (8%)	161 (9%)	258 (13%)	266 (12%)	330 (14%)	362 (14%)	395 (14%)	459 (15%)	596 (19%)

Notes on page E-2

TABLE E-1 USDA Eating Patterns *(continued)*

Notes for Table E-1

[a]All foods are assumed to be in nutrient-dense forms, lean or low-fat and prepared without added fats, sugars, or salt. Solid fats and added sugars may be included up to the daily maximum limit identified in the table. Food items in each group and subgroup are:

Fruits	All fresh, frozen, canned, and dried fruits and fruit juices: for example, oranges and orange juice, apples and apple juice, bananas, grapes, melons, berries, raisins.
Vegetables	
Dark green vegetables	All fresh, frozen, and canned dark-green leafy vegetables and broccoli, cooked or raw: for example, broccoli; spinach; romaine; collard, turnip, and mustard greens.
Red and orange vegetables	All fresh, frozen, and canned red and orange vegetables, cooked or raw: for example, tomatoes, red peppers, carrots, sweet potatoes, winter squash, and pumpkin.
Beans and peas (legumes)	All cooked beans and peas: for example, kidney beans, lentils, chickpeas, and pinto beans. Does not include green beans or green peas. (See additional comment under protein foods group.)
Starchy vegetables	All fresh, frozen, and canned starchy vegetables: for example, white potatoes, corn, green peas.
Other vegetables	All fresh, frozen, and canned other vegetables, cooked or raw: for example, iceberg lettuce, green beans, and onions.
Grains[e]	
Whole grains	All whole-grain products and whole grains used as ingredients: for example, whole-wheat bread, whole-grain cereals and crackers, oatmeal, and brown rice.
Enriched grains	All enriched refined-grain products and enriched refined grains used as ingredients: for example, white breads, enriched grain cereals and crackers, enriched pasta, white rice.
Proteins foods[d]	All meat, poultry, seafood, eggs, nuts, seeds, and processed soy products. Meat and poultry should be lean or low-fat and nuts should be unsalted. Beans and peas are considered part of this group as well as the vegetable group, but should be counted in one group only.
Dairy (milk and milk products)[f]	All milks, including lactose-free and lactose-reduced products and fortified soy beverages, yogurts, frozen yogurts, dairy desserts, and cheeses. Most choices should be fat-free or low-fat. Cream, sour cream, and cream cheese are not included due to their low calcium content.

[b]Food group amounts are shown in cup (c) or ounce-equivalents (oz-eq). Oils are shown in grams (g). Quantity equivalents for each food group are:

- *Grains, 1 ounce-equivalent is: 1 one-ounce slice bread; 1 ounce uncooked pasta or rice; ½ cup cooked rice, pasta, or cereal; 1 tortilla (6" diameter); 1 pancake (5" diameter); 1 ounce ready-to-eat cereal (about 1 cup cereal flakes).*
- *Vegetables and fruits, 1 cup equivalent is: 1 cup raw or cooked vegetable or fruit; ½ cup dried vegetable or fruit; 1 cup vegetable or fruit juice; 2 cups leafy salad greens.*
- *Protein foods, 1 ounce-equivalent is: 1 ounce lean meat, poultry, seafood; 1 egg; 1 Tbsp peanut butter; ½ ounce nuts or seeds. Also, ¼ cup cooked beans or peas may also be counted as 1 ounce-equivalent.*
- *Dairy, 1 cup equivalent is: 1 cup milk, fortified soy beverage, or yogurt; 1½ ounces natural cheese (e.g., cheddar); 2 ounces of processed cheese (e.g., American).*

[c]Food intake patterns at 1,000, 1,200, and 1,400 calories meet the nutritional needs of children ages 2 to 8 years. Patterns from 1,600 to 3,200 calories meet the nutritional needs of children ages 9 years and older and adults. If a child ages 4 to 8 years needs more calories and, therefore, is following a pattern at 1,600 calories or more, the recommended amount from the dairy group can be 2½ cups per day. Children ages 9 years and older and adults should not use the 1,000, 1,200, or 1,400 calorie patterns.

[d]Vegetable and protein foods subgroup amounts are shown in this table as weekly amounts, because it would be difficult for consumers to select foods from all subgroups daily.

[e]Whole-grain subgroup amounts shown in this table are minimums. More whole grains up to all of the grains recommended may be selected, with offsetting decreases in the amounts of enriched refined grains.

[f]The amount of dairy foods in the 1,200 and 1,400 calorie patterns have increased to reflect new RDAs for calcium that are higher than previous recommendations for children ages 4 to 8 years.

[g]Oils and soft margarines include vegetable, nut, and fish oils and soft vegetable oil table spreads that have no trans fats.

[h]SoFAS are calories from solid fats and added sugars. The limit for SoFAS is the remaining amount of calories in each food pattern after selecting the specified amounts in each food group in nutrient-dense forms (forms that are fat-free or low-fat and with no added sugars). The number of SoFAS is lower in the 1,200, 1,400, and 1,600 calorie patterns than in the 1,000 calorie pattern. The nutrient goals for the 1,200 to 1,600 calorie patterns are higher and require that more calories be used for nutrient-dense foods from the food groups.

Source: USDA Guidelines. Found at: http://www.cnpp.usda.gov/Publications/USDAFoodPatterns/USDAFoodPatternsSummaryTable.pdf.

TABLE E-2 The DASH Eating Plan at 1,600-, 2,000-, 2,600-, and 3,100-Calorie Levels[a]

The number of daily servings to choose from each food group depends on a person's energy requirement (see Chapter 9).

Food Groups	1,600 Calories	2,000 Calories	2,600 Calories	3,100 Calories	Serving Sizes	Examples and Notes	Significance of Each Food Group to the DASH Eating Plan
Grains[b]	6 servings	7–8 servings	10–11 servings	12–13 servings	1 slice bread, 1 oz dry cereal,[c] ½ cup cooked rice, pasta, or cereal	Whole wheat bread, English muffin, pita bread, bagel, cereals, grits, oatmeal, crackers, unsalted pretzels, and popcorn	Major sources of energy and fiber
Vegetables	3–4 servings	4–5 servings	5–6 servings	6 servings	1 cup raw leafy vegetable, ½ cup cooked vegetable, 6 oz vegetable juice	Tomatoes, potatoes, carrots, green peas, squash, broccoli, turnip greens, collards, kale, spinach, artichokes, green beans, lima beans, sweet potatoes	Rich sources of potassium, magnesium, and fiber
Fruits	4 servings	4–5 servings	5–6 servings	6 servings	6 oz fruit juice 1 medium fruit ¼ cup dried fruit ½ cup fresh, frozen, or canned fruit	Apricots, bananas, dates, grapes, oranges, orange juice, grapefruit, grapefruit juice, mangoes, melons, peaches, pineapples, prunes, raisins, strawberries, tangerines	Important sources of potassium, magnesium, and fiber
Low-fat or fat-free dairy foods	2–3 servings	2–3 servings	3 servings	3–4 servings	8 oz milk 1 cup yogurt 1½ oz cheese	Fat-free or low-fat milk, fat-free or low-fat buttermilk, fat-free or low-fat regular or frozen yogurt, low-fat and fat-free cheese	Major sources of calcium and protein
Meat, poultry, fish	1–2 servings	2 or less servings	2 servings	2–3 servings	3 oz cooked meats, poultry, or fish	Select only lean; trim away visible fats; broil, roast, or boil instead of frying; remove skin from poultry	Rich sources of protein and magnesium
Nuts, seeds, legumes	3–4 servings/ week	4–5 servings/ week	1 serving	1 serving	⅓ cup or 1½ oz nuts, 2 Tbsp or ½ oz seeds ½ cup cooked dry beans or peas	Almonds, filberts, mixed nuts, peanuts, walnuts, sunflower seeds, kidney beans, lentils	Rich sources of energy, magnesium, potassium, protein, and fiber
Fat and oils[d]	2 servings	2–3 servings	3 servings	4 servings	1 tsp soft margarine 1 Tbsp low-fat mayonnaise 2 Tbsp light salad dressing, 1 tsp vegetable oil	Soft margarine, low-fat mayonnaise, light salad dressing, vegetable oil (such as olive, corn, canola, or safflower)	DASH has 27 percent of calories as fat (low in saturated fat), including fat in or added to foods
Sweets	0 servings	5 servings/week	2 servings	2 servings	1 Tbsp sugar, 1 Tbsp jelly or jam, ½ oz jelly beans, 8 oz lemonade	Maple syrup, sugar, jelly, jam, fruit-flavored gelatin, jelly beans, hard candy, fruit punch, sorbet, ices	Sweets should be low in fat

[a] *NIH publication No. 03–4082; Karanja NM et al. JADA 8:S19–27, 1999.*

[b] *Whole grains are recommended for most servings to meet fiber recommendations.*

[c] *Equals ½–1¼ cups, depending on cereal type. Check the product's Nutrition Facts Label.*

[d] *Fat content changes serving counts for fats and oils: For example, 1 Tbsp of regular salad dressing equals 1 serving; 1 Tbsp of a low-fat dressing equals ½ serving; 1 Tbsp of a fat-free dressing equals 0 servings.*

Source: U.S. Department of Agriculture and U.S. Department of Health and Human Services, Dietary Guidelines for Americans 2010, available at www.dietaryguidelines.gov

TABLE E-3 Lacto-ovo Vegetarian Adaptation of the USDA Food Patterns

For each food group or subgroup,[a] recommended average daily intake amounts[b] at all calorie levels. Recommended intakes from vegetable and protein foods subgroups are per week. For more information and tools for application, go to MyPyramid.gov.

Calorie Level of pattern[c]	1,000	1,200	1,400	1,600	1,800	2,000	2,200	2,400	2,600	2,800	3,000	3,200
Fruits	1 c	1 c	1½ c	1½ c	1½ c	2 c	2 c	2 c	2 c	2½ c	2½ c	2½ c
Vegetables[d]	1 c	1½ c	1½ c	2 c	2½ c	2½ c	3 c	3 c	3½ c	3½ c	4 c	4 c
Dark-green vegetables	½ c/wk	1 c/wk	1 c/wk	1½ c/wk	1½ c/wk	1½ c/wk	2 c/wk	2 c/wk	2½ c/wk	2½ c/wk	2½ c/wk	2½ c/wk
Red and orange vegetables	2½ c/wk	3 c/wk	3 c/wk	4 c/wk	5½ c/wk	5½ c/wk	6 c/wk	6 c/wk	7 c/wk	7 c/wk	7½ c/wk	7½ c/wk
Beans and peas (legumes)	½ c/wk	½ c/wk	½ c/wk	1 c/wk	1½ c/wk	1½ c/wk	2 c/wk	2 c/wk	2½ c/wk	2½ c/wk	3 c/wk	3 c/wk
Starchy vegetables	2 c/wk	3½ c/wk	3½ c/wk	4 c/wk	5 c/wk	5 c/wk	6 c/wk	6 c/wk	7 c/wk	7 c/wk	8 c/wk	8 c/wk
Other vegetables	1½ c/wk	2½ c/wk	2½ c/wk	3½ c/wk	4 c/wk	4 c/wk	5 c/wk	5 c/wk	5½ c/wk	5½ c/wk	7 c/wk	7 c/wk
Grains[e]	3 oz-eq	4 oz-eq	5 oz-eq	5 oz-eq	6 oz-eq	6 oz-eq	7 oz-eq	8 oz-eq	9 oz-eq	10 oz-eq	10 oz-eq	10 oz-eq
Whole grains	1½ oz-eq	2 oz-eq	2½ oz-eq	3 oz-eq	3 oz-eq	3 oz-eq	3½ oz-eq	4 oz-eq	4½ oz-eq	5 oz-eq	5 oz-eq	5 oz-eq
Refined grains	1½ oz-eq	2 oz-eq	2½ oz-eq	2 oz-eq	3 oz-eq	3 oz-eq	3½ oz-eq	4 oz-eq	4½ oz-eq	5 oz-eq	5 oz-eq	5 oz-eq
Protein foods[d]	2 oz-eq	3 oz-eq	4 oz-eq	5 oz-eq	5 oz-eq	5½ oz-eq	6 oz-eq	6½ oz-eq	6½ oz-eq	7 oz-eq	7 oz-eq	7 oz-eq
Eggs	1 oz-eq/wk	2 oz-eq/wk	3 oz-eq/wk	4 oz-eq/wk	4 oz-eq/wk	4 oz-eq/wk	4 oz-eq/wk	5 oz-eq/wk	5 oz-eq/wk	5 oz-eq/wk	5 oz-eq/wk	5 oz-eq/wk
Beans and peas[f]	3½ oz-eq/wk	5 oz-eq/wk	7 oz-eq/wk	9 oz-eq/wk	9 oz-eq/wk	10 oz-eq/wk	10 oz-eq/wk	11 oz-eq/wk	11 oz-eq/wk	12 oz-eq/wk	12 oz-eq/wk	12 oz-eq/wk
Soy products	4 oz-eq/wk	6 oz-eq/wk	8 oz-eq/wk	11 oz-eq/wk	11 oz-eq/wk	12 oz-eq/wk	13 oz-eq/wk	14 oz-eq/wk	14 oz-eq/wk	15 oz-eq/wk	15 oz-eq/wk	15 oz-eq/wk
Nuts and seeds	5 oz-eq/wk	7 oz-eq/wk	10 oz-eq/wk	12 oz-eq/wk	12 oz-eq/wk	13 oz-eq/wk	15 oz-eq/wk	16 oz-eq/wk	16 oz-eq/wk	17 oz-eq/wk	17 oz-eq/wk	17 oz-eq/wk
Dairy[g]	2 c	2½ c	2½ c	3 c	3 c	3 c	3 c	3 c	3 c	3 c	3 c	3 c
Oils[h]	12 g	13 g	12 g	15 g	17 g	19 g	21 g	22 g	25 g	26 g	34 g	41 g
Maximum SoFAS[i] limit, calories (% total calories)	137 (14%)	121 (10%)	121 (9%)	121 (8%)	161 (9%)	258 (13%)	266 (12%)	330 (14%)	362 (14%)	395 (14%)	459 (15%)	596 (19%)

a,b,c,d,e. See Table E–1, notes a through e.

f. Total recommended beans and peas amounts would be the sum of amounts recommended in the vegetable and the protein foods groups. An ounce-equivalent of beans and peas in the protein foods group is ¼ cup, cooked. For example, in the 2,000 calorie pattern, total weekly beans and peas recommendation is (10 oz-eq/4) + 1½ cups = about 4 cups, cooked.

g,h,i. See Table E–1, notes f, g, and h.

Source: Dietary Guidelines for Americans 2010

TABLE E-4 Vegan Adaptation of the USDA Food Patterns

For each food group or subgroup,[a] recommended average daily intake amounts[b] at all calorie levels. Recommended intakes from vegetable and protein foods subgroups are per week. For more information and tools for application, go to MyPyramid.gov.

Calorie Level of pattern[c]	1,000	1,200	1,400	1,600	1,800	2,000	2,200	2,400	2,600	2,800	3,000	3,200
Fruits	1 c	1 c	1½ c	1½ c	1½ c	2 c	2 c	2 c	2 c	2½ c	2½ c	2½ c
Vegetables[d]	1 c	1½ c	1½ c	2 c	2½ c	2½ c	3 c	3 c	3½ c	3½ c	4 c	4 c
Dark-green vegetables	½ c/wk	1 c/wk	1 c/wk	1½ c/wk	1½ c/wk	1½ c/wk	2 c/wk	2 c/wk	2½ c/wk	2½ c/wk	2½ c/wk	2½ c/wk
Red and orange vegetables	2½ c/wk	3 c/wk	3 c/wk	4 c/wk	5½ c/wk	5½ c/wk	6 c/wk	6 c/wk	7 c/wk	7 c/wk	7½ c/wk	7½ c/wk
Beans and peas (legumes)	½ c/wk	½ c/wk	½ c/wk	1 c/wk	1½ c/wk	1½ c/wk	2 c/wk	2 c/wk	2½ c/wk	2½ c/wk	3 c/wk	3 c/wk
Starchy vegetables	2 c/wk	3½ c/wk	3½ c/wk	4 c/wk	5 c/wk	5 c/wk	6 c/wk	6 c/wk	7 c/wk	7 c/wk	8 c/wk	8 c/wk
Other vegetables	1½ c/wk	2½ c/wk	2½ c/wk	3½ c/wk	4 c/wk	4 c/wk	5 c/wk	5 c/wk	5½ c/wk	5½ c/wk	7 c/wk	7 c/wk
Grains[e]	3 oz-eq	4 oz-eq	5 oz-eq	5 oz-eq	6 oz-eq	6 oz-eq	7 oz-eq	8 oz-eq	9 oz-eq	10 oz-eq	10 oz-eq	10 oz-eq
Whole grains	1½ oz-eq	2 oz-eq	2½ oz-eq	3 oz-eq	3 oz-eq	3 oz-eq	3½ oz-eq	4 oz-eq	4½ oz-eq	5 oz-eq	5 oz-eq	5 oz-eq
Refined grains	1½ oz-eq	2 oz-eq	2½ oz-eq	2 oz-eq	3 oz-eq	3 oz-eq	3½ oz-eq	4 oz-eq	4½ oz-eq	5 oz-eq	5 oz-eq	5 oz-eq
Protein foods[d]	2 oz-eq	3 oz-eq	4 oz-eq	5 oz-eq	5 oz-eq	5½ oz-eq	6 oz-eq	6½ oz-eq	6½ oz-eq	7 oz-eq	7 oz-eq	7 oz-eq
Beans and peas[f]	5 oz-eq/wk	7 oz-eq/wk	10 oz-eq/wk	12 oz-eq/wk	12 oz-eq/wk	13 oz-eq/wk	15 oz-eq/wk	16 oz-eq/wk	16 oz-eq/wk	17 oz-eq/wk	17 oz-eq/wk	17 oz-eq/wk
Soy products	4 oz-eq/wk	5 oz-eq/wk	7 oz-eq/wk	9 oz-eq/wk	9 oz-eq/wk	10 oz-eq/wk	11 oz-eq/wk	11 oz-eq/wk	11 oz-eq/wk	12 oz-eq/wk	12 oz-eq/wk	12 oz-eq/wk
Nuts and seeds	6 oz-eq/wk	8 oz-eq/wk	11 oz-eq/wk	14 oz-eq/wk	14 oz-eq/wk	15 oz-eq/wk	17 oz-eq/wk	18 oz-eq/wk	18 oz-eq/wk	20 oz-eq/wk	20 oz-eq/wk	20 oz-eq/wk
Dairy (vegan)[g]	2 c	2½ c	2½ c	3 c	3 c	3 c	3 c	3 c	3 c	3 c	3 c	3 c
Oils[h]	12 g	12 g	11 g	14 g	16 g	18 g	20 g	21 g	24 g	25 g	33 g	40 g
Maximum SoFAS[i] limit, calories (% total calories)	137 (14%)	121 (10%)	121 (9%)	121 (8%)	161 (9%)	258 (13%)	266 (12%)	330 (14%)	362 (14%)	395 (14%)	459 (15%)	596 (19%)

a,b,c,d,e. See Table E–1, notes a through e.

f. Total recommended beans and peas amounts would be the sum of amounts recommended in the vegetable and the protein foods groups. An ounce-equivalent of beans and peas in the protein foods group is ¼ cup, cooked. For example, in the 2,000 calorie pattern, total weekly beans and peas recommendation is (13 oz-eq/4) + 1½ cups = about 5 cups, cooked.

g. The vegan "dairy group" is composed of calcium-fortified beverages and foods from plant sources. For analysis purposes the following products were included: calcium-fortified soy beverage, calcium-fortified rice milk, tofu made with calcium-sulfate, and calcium-fortified soy yogurt. The amounts in the 1,200 and 1,400 calorie patterns have increased to reflect new RDAs for calcium that are higher than previous recommendations for children ages 4 to 8 years.

h,i. See Table E–1, notes g and h.

Source: Dietary Guidelines for Americans 2010

Appendix F

Notes

Chapter 1

1. S. S. Gidding and coauthors, Implementing American Heart Association pediatric and adult nutrition guidelines, *Circulation* 119 (2009): 1161–1175; Preventability of cancer, in World Cancer Research Fund/American Institute for Cancer Research, Policy and Action for Cancer Prevention, *Food, Nutrition, and Physical Activity: A Global Perspective* (Washington, D.C.: AICR, 2009).
2. U.S. Department of Health and Human Services, *Healthy People 2020* (Washington, D.C.: U.S. Government Printing Office, 2010), available online at www.healthypeople.gov; Centers for Disease Control and Prevention, QuickStats: Average total cholesterol level among men and women aged 20–74 Years—National Health and Nutrition Examination Survey, United States, 1959–1962 to 2007–2008, *Morbidity and Mortality Weekly Report* 58 (2009): 1045.
3. E. J. Sondik and coauthors, Progress toward the Healthy People 2010 goals and objectives, *Annual Review of Public Health* 31 (2010): 271–281.
4. B. A. Mizock, Immunonutrition and critical illness: an update, *Nutrition* 26 (2010): 701–707.
5. D. R. Jacobs, M. D. Gross, and L. C. Tapsell, Food synergy: An operational concept for understanding nutrition, *American Journal of Clinical Nutrition* 89 (2009): 1543S–1548S.
6. Centers for Disease Control and Prevention, State indicator report on fruits and vegetables, 2009, available at www.fruitsandveggiesmatter.gov/downloads/StateIndicatorReport2009.pdf.
7. Position of the American Dietetic Association, Functional foods, *Journal of the American Dietetic Association* 109 (2009): 735–746.
8. U.S. Department of Agriculture and U.S. Department of Health and Human Services, *Dietary Guidelines for Americans 2010*, available at www.dietaryguidelines.gov.
9. K. Stein, Navigating cultural competency: In preparation for an expected standard in 2010, *Journal of the American Dietetic Association* 109 (2009): 1676–1688.
10. Position of the American Dietetic Association and Dietitians of Canada, Vegetarian diets, *Journal of the American Dietetic Association* 109 (2009): 1266–1283.
11. International Food Information Council Foundation, *2011 Food and Health Survey: Consumer Attitudes toward Food Safety, Nutrition, and Health*, May 2011, available at www.foodinsight.org.
12. R. M. Morrison, L. Mancino, and J. N. Variyam, Will calorie labeling in restaurants make a difference? *Amber Waves* 9 (2011): 10–17.
13. J. E. Todd, L. Mancino, and B. Lin, The Impact of Food Away From Home on Adult Diet Quality: ERR-90, U.S. Department of Agriculture, Econ. Res. Serv., February 2010, *Advances in Nutrition* 2 (2011): 442–443.
14. M. Franco and coauthors, Availability of healthy foods and dietary patterns: The multi-ethnic study of atherosclerosis, *American Journal of Clinical Nutrition* 89 (2009): 897–904.
15. D. Quagliani and M. Hermann, Communicating accurate food and nutrition information: practice paper of the Academy of Nutrition and Dietetics (abstract), *Journal of the Academy of Nutrition and Dietetics* 112 (2012): 759, epub ahead of print, doi:10.1016/j.jand.2012.03.006.
16. J. P. A. Loannidis, An Epidemic of False Claims, *Scientific American*, June 2011, available at http://www.scientificamerican.com/article.cfm?id=an-epidemic-of-false-claims.
17. USDA, Monitoring America's nutritional health, *Agricultural Research,* March 2012, pp. 4–23.
18. H. Mochari-Greenberger, M. B. Terry, and L. Mosca, Does stage of change modify the effectiveness of an educational intervention to improve diet among family members of hospitalized cardiovascular disease patients?, *Journal of the American Dietetic Association* 110 (2010):1027–1035.
19. N. T. Artinian and coauthors, Interventions to promote physical activity and dietary lifestyle changes for cardiovascular risk factor reduction in adults: a scientific statement from the American Heart Association, *Circulation* 122 (2010): 406–441.
20. G. D. Miller and coauthors, It is time for a positive approach to dietary guidance using nutrient density as a basic principle, *Journal of Nutrition* 139 (2009): 1198–1202.
21. L. M. Lipsky, Are energy-dense foods really cheaper? Reexamining the relation between food price and energy density, *American Journal of Clinical Nutrition* 90 (2009): 1397–1401.

Consumer's Guide 1

1. D. Quagliani and M. Hermann, Communicating accurate food and nutrition information: Practice paper of the Academy of Nutrition and Dietetics (abstract), *Journal of the Academy of Nutrition and Dietetics* 112 (2012): 759, epub ahead of print, doi:10.1016/j.jand.2012.03.006.

Controversy 1

1. Food and Drug Administration, FDA 101: Health Fraud Awareness, FDA Consumer Health Information, July 2009, available at www.fda.gov.
2. E. B. Cohen and R. Winch, *Diploma and Accreditation Mills: New Trends in Credential Abuse* (Bedford, Great Britain: Verifile Accredibase, 2011) available at www.acredibase.com.
3. D. Quagliani and M. Hermann, Communicating accurate food and nutrition information: practice paper of the Academy of Nutrition and Dietetics (abstract), *Journal of the Academy of Nutrition and Dietetics* 112 (2012): 759, epub ahead of print, doi:10.1016/j.jand.2012.03.006.
4. S. H. Laramee and M. Tate, Dietetics Workforce Demand Study Task Force supplement: An introduction, *Journal of the Academy of Nutrition and Dietetics* 112 (2012): S7–S9.
5. Laramee and Tate, Dietetics Workforce Demand Study Task Force supplement, 2012.
6. E. B. Cohen and R. Winch, *Diploma and Accreditation Mills: Exposing Academic Credential Abuse* (Bedford, Great Britain: Verifile Accredibase, 2011), available at www.acredibase.com.
7. Cohen and Winch, *Diploma and Accreditation Mills: Exposing Academic Credential Abuse*, 2011.

Chapter 2

1. Standing Committee on the Scientific Evaluation of Dietary Reference Intakes, Food and Nutrition Board, Institute of Medicine, *Dietary Reference Intakes: Applications in Dietary Assessment* (Washington, D.C.: National Academy Press, 2000), pp. 5–7.
2. Committee on Dietary Reference Intakes, *Dietary Reference Intakes for Calcium and Vitamin D* (Washington, D.C.: National Academies Press, 2011), pp. 359–360.
3. U.S. Department of Agriculture and U.S. Department of Health and Human Services, *Dietary Guidelines for Americans 2010*, available at www.dietaryguidelines.gov.
4. M. R. Flock and P. M. Kris-Etherton, Dietary Guidelines for Americans 2010: Implications for cardiovascular disease, *Current Atherosclerosis Reports* (2011), epub, doi:10.1007/s11883-011-0205-0.
5. U.S. Department of Agriculture and U.S. Department of Health and Human Services, *Dietary Guidelines for Americans 2010*, available at www.dietaryguidelines.gov.
6. R. D. Whitehead and coauthors, You are what you eat: Within-subject increases in fruit and veg-

etable consumption confer beneficial skin-color changes, *PLoS ONE* 7 (2012), epub, doi:10.1371/journal.pone.0032988.
7. U.S. Department of Agriculture and U.S. Department of Health and Human Services, *Dietary Guidelines for Americans 2010.*
8. S. J. Marshall and coauthors, Translating physical activity recommendations into a pedometer-based step goal, *American Journal of Preventive Medicine* 36 (2009): 410–415.
9. U.S. Department of Health and Human Services, 2008 Physical Activity Guidelines for Americans, available at www.health.gov/paguidelines/default.aspx.
10. Food and Drug Administration, *The Scoop on Whole Grains*, May 6, 2009, available at www.fda.gov/consumer.
11. *Dietary Guidelines for Americans 2010*, p. 15.
12. R. C. Post and coauthors, What's new on MyPlate? A new message, redesigned web site, and SuperTracker debut, *Journal of the Academy of Nutrition and Dietetics* 112 (2012): 18–22.
13. *Dietary Guidelines for Americans 2010*, p. 80.
14. U.S. Food and Drug Administration, Food Label Use, in *2008 Health and Diet Survey*, available at www.fda.gov.
15. R. M. Bliss, Nutrient data in time for the new year, *Agricultural Research*, January 2012, pp. 20–21.
16. Position of the American Dietetic Association, Functional foods, *Journal of the American Dietetic Association* 109 (2009): 735–746.
17. E. A. Wartella and coauthors, *Front-of-Package Nutrition Rating Systems and Symbols: Promoting Healthier Choices* (Washington, D.C.: National Academies Press, 2011), available at www.iom.edu/Reports/2011/Front-of-Package-Nutrition-Rating-Systems-and-Symbols-Promoting-Healthier-Choices.aspx.

Consumer's Guide 2

1. R. M. Morrison, L. Mancino, and J. N. Variyam, Will calorie labeling in restaurants make a difference? *Amber Waves*, March 2011 available at www.ers.usda.gov/AmberWaves.
2. Morrison, Mancino, and Variyam, Will calorie labeling in restaurants make a difference? 2011.

Controversy 2

1. J. A. Joseph, B. Shukitt-Hale, and L. M. Willis, Grape juice, berries, and walnuts affect brain aging and behavior, *Journal of Nutrition* 139 (2009): 1813S–1817S.
2. S. Jang and R. W. Johnson, Can consuming flavonoids restore old microglia to their youthful state? *Nutrition Reviews* 68 (2010): 719–728.
3. E. Devore and coauthors, Total antioxidant capacity of diet in relation to cognitive function and decline, *American Journal of Clinical Nutrition* 92 (2010): 1157–1164.
4. A. S. Chang, B. Yeong, and W. Koh, Symposium on plant polyphenols: Nutrition, health and innovations, June 2009, *Nutrition Reviews* 68 (2010): 246–252.
5. D. O. Kennedy and E. L. Wightman, Herbal extracts and phytochemicals: plant secondary metabolites and the enhancement of human brain function, *Advances in Nutrition* 2 (2011): 32–50.
6. N. M. Wedick and coauthors, Dietary flavonoid intakes and risk of type 2 diabetes in US men and women, *American Journal of Clinical Nutrition* 95 (2012): 925–933.
7. Anon, ORAC: over-rated antioxidant claims, *Berkeley Wellness Newsletter*, October 2012, available at www.wellnessletter.com; O. Khawaja, J. M. Gaziano, and L. Djoussé, Chocolate and coronary heart disease: A systematic review, *Current Atherosclerosis Reports* 13 (2011): 447–452.
8. A. Cassidy and coauthors, Habitual intake of flavonoid subclasses and incident hypertension in adults, *American Journal of Clinical Nutrition* 93 (2011): 338–347; A. Buitrago-Lopez and coauthors, Chocolate consumption and cardiometabolic disorders: systematic review and meta-analysis, *British Medical Journal* 343 (2011), epub, doi:10.1136/BMJ.D4488.
9. L. Hooper and coauthors, Effects of chocolate, cocoa, and flavan-3-ols on cardiovascular health: A systematic review and meta-analysis of randomized trials, *American Journal of Clinical Nutrition* 95 (2012): 740–751.
10. N. Rose, S. Koperski, and B. A. Golomb, Chocolate and depressive symptoms in a cross-sectional analysis, *Archives of Internal Medicine* 170 (2010): 699–703.
11. S. B. Racette and coauthors, Dose effects of dietary plant sterols on cholesterol metabolism: A controlled feeding study, *American Journal of Clinical Nutrition* 91 (2010): 32–38; J. H. van Ee, Soy constituents: Modes of action in low-density lipoprotein management, *Nutrition Reviews* 67 (2009): 222–234; A. Pan and coauthors, Meta-analysis of the effects of flaxseed interventions on blood lipids, *American Journal of Clinical Nutrition* 90 (2009): 288–297.
12. P. C. H. Hollman and coauthors, The biological relevance of direct antioxidant effects of polyphenols for cardiovascular health in humans is not established, *Journal of Nutrition* 141 (2011): 989S–1009S.
13. J. Chen and coauthors, Flaxseed and pure secoisolariciresinol diglucoside, but not flaxseed hull, reduce human breast tumor growth (MCF7) in athymic mice, *Journal of Nutrition* 139 (2009): 2061–2066.
14. K. Buck and coauthors, Serum enterolactone and prognosis of postmenopausal breast cancer, *Journal of Clinical Oncology* 29 (2011): 3730–3738.
15. W. Demark-Wahnefried and coauthors, Flaxseed supplementation (not dietary fat restriction) reduces prostate cancer proliferation rates in men presurgery, *Cancer Epidemiology, Biomarkers, and Prevention* 17 (2008): 3577–3587.
16. J. Y. Kim and O. Kwon, Garlic intake and cancer risk: An analysis using the Food and Drug Administration's evidence-based review system for the scientific evaluation of health claims, *American Journal of Clinical Nutrition* 89 (2009): 257–264.
17. S. Mukherjee and coauthors, Freshly crushed garlic is a superior cardioprotective agent than processed garlic, *Journal of Agricultural and Food Chemistry* 57 (2009): 7137–7144; J. Y. Kim and O. Kwon, Garlic intake and cancer risk: Analysis using the Food and Drug Administration's evidence-based review system for the scientific evaluation of health claims, *American Journal of Clinical Nutrition* 89 (2009): 257–264; R. S. Rivlin, Can garlic reduce risk of cancer? (editorial), *American Journal of Clinical Nutrition* 89 (2009): 17–18.
18. N. Mehrotra, S. Gaur, and A. Petrova, Health care practices of the foreign born Asian Indians in the United States. A community based survey, *Journal of Community Health* 37 (2012): 328–334.
19. D. J. Jenkins and coauthors, Soy protein reduces serum cholesterol by both intrinsic and food displacement mechanisms, *Journal of Nutrition* 140 (2010): 2302S-2311S.
20. S. B. Racette and coauthors, Dose effects of dietary plant sterols on cholesterol metabolism: A controlled feeding study, *American Journal of Clinical Nutrition* 91 (2010): 32–38.
21. Jenkins and coauthors, Soy protein reduces serum cholesterol by both intrinsic and food displacement mechanisms, 2010.
22. P. L. de Souza and coauthors, Clinical pharmacology of isoflavones and its relevance for potential prevention of prostate cancer, *Nutrition Reviews* 68 (2010): 542–555; X. O. Shu and coauthors, Soy food intake and breast cancer survival, *Journal of the American Medical Association* 302 (2009): 2437–2443.
23. J. H. van Ee, Soy constituents: Modes of action in low-density lipoprotein management, *Nutrition Reviews* 67 (2009): 222–234.
24. S. A. Lee and coauthors, Adolescent and adult soy food intake and breast cancer risk: Results from the Shanghai Women's Health Study, *American Journal of Clinical Nutrition* 89 (2009): 1920–1926; M. Messina and A. H. Wu, Perspectives on the soy–breast cancer relation, *American Journal of Clinical Nutrition* 89 (2009): 1673S–1679S.
25. X. O. Shu and coauthors, Soy food intake and breast cancer survival, *Journal of the American Medical Association* 302 (2009): 2437–2443.
26. S. J. Nechuta and coauthors, Soy food intake after diagnosis of breast cancer and survival; an in-depth analysis of combined evidence from cohort studies of U.S. and Chinese women, *American Journal of Clinical Nutrition* 96 (2012): 123–132; S. A. Khan and coauthors, Soy isoflavones supplementation for breast cancer risk reduction: A randomized phase II trial, *Cancer Prevention Research* 5 (2012): 309–319.
27. S. Levis and coauthors, Soy isoflavones in the prevention of menopausal bone loss and menopausal symptoms, *Archives of Internal Medicine* 171 (2011): 1363–1369; Collaborators, The role of soy isoflavones in menopausal health: report of the North American Menopause Society, *Menopause* 18 (2011): 732–753.
28. D. L. Alekel and coauthors, The Soy Isoflavones for Reducing Bone Loss (SIRBL) Study: a 3-y randomized controlled trial in postmenopausal women, *American Journal of Clinical Nutrition* 91 (2010): 218–230.

29. R. Bosviel and coauthors, Can soy phytoestrogens decrease DNA methylation in BRCA1 and BRCA2 oncosuppressor genes in breast cancer? *Omics* 16 (2012): 235–244; T. T. Rajah and coauthors, Physiological concentrations of genistein and 17β-estradiol inhibit MDA-MB231 breast cancer cell growth by increasing BAX/BCL-2 and reducing pERK1/2, *Anticancer Research* 32 (2012): 1181–1191.
30. C. K. Taylor and coauthors, The effect of genistein aglycone on cancer and cancer risk: A review of in vitro, preclinical, and clinical studies, *Nutrition Reviews* 67 (2009): 398–415.
31. H. Kang and coauthors, Study on soy isoflavones consumption and risk of breast cancer and survival, *Asian Pacific Journal of Cancer Prevention* 13 (2012), epub, doi:http://dx.doi.org/10.7314/APJCP.2012.13.3.995; Shu and coauthors, Soy food intake and breast cancer survival, 2009.
32. American Cancer Society, Find Support and Treatment: Soybean, May 13, 2010, available at www.cancer.org.
33. J. R. Mein, F. Lian, and X-D. Wang, Biological activity of lycopene metabolites: Implications for cancer prevention, *Nutrition Reviews* 66 (2008): 667–668; N. Khan, F. Afaq, and H. Mukhtar, Cancer chemoprevention through dietary antioxidants: Progress and promise, *Antioxidants and Redox Signaling* 10 (2008): 475–510.
34. R. L. Roberts, J. Green, and B. Lewis, Lutein and zeaxanthin in eye and skin health, *Clinical Dermatology* 27 (2009): 195–201.
35. J. A. Satia and coauthors, Long-term use of beta-carotene, retinol, lycopene, and lutein supplements and lung cancer risk: Results from the VITamins and Lifestyle (VITAL) Study, *American Journal of Epidemiology* 169 (2009): 815–828.
36. P. Bogdanski and coauthors, Green tea extract reduces blood pressure, inflammatory biomarkers, and oxidative stress and improves parameters associated with insulin resistance in obese, hypertensive patients, *Nutrition Research* 32 (2012): 421–427; X. X. Zheng and coauthors, Green tea intake lowers fasting serum total and LDL cholesterol in adults: A meta-analysis of 14 randomized controlled trials, *American Journal of Clinical Nutrition* 94 (2011): 601–610.
37. C. H. Ruxton and P. Mason, Is black tea consumption associated with a lower risk of cardiovascular disease and type 2 diabetes? *Nutrition Bulletin* 37 (2012): 4–15; Z. Wang and coauthors, Black and green tea consumption and the risk of coronary artery disease: a meta-analysis, *American Journal of Clinical Nutrition* 93 (2011): 506–515; S. M. Henning, P. Wang, and D. Heber, Chemopreventive effects of tea in prostate cancer: Green tea versus black tea, *Molecular Nutrition and Food Research* 55 (2011): 905–920; O. K. Chun and coauthors, Estimation of antioxidant intakes from diet and supplements in U.S. adults, *Journal of Nutrition* 140 (2010): 317–324.
38. Zheng and coauthors, Green tea intake lowers fasting serum total and LDL cholesterol in adults: A meta-analysis of 14 randomized controlled trials, 2011.
39. K. Boehm and coauthors, Green tea (*Camellia sinensis*) for the prevention of cancer, *The Cochrane Database of Systematic Reviews* 3 (2009), epub, doi:10.1002/14651858.CD005004.pub2.
40. A. H. Schönthal, Adverse effects of concentrated green tea extracts, *Molecular Nutrition and Food Research* 55 (2011): 874–885; D. N. Sarma and coauthors, Safety of green tea extracts: a systematic review by the U.S. Pharmacopeia, *Drug Safety* 31 (2008): 469–484.
41. S. Egert and G. Rimbach, Which sources of flavonoids: Complex diets or dietary supplements? *Advances in Nutrition* 2 (2011): 8–14.
42. J. Sun and coauthors, A non-targeted approach to chemical discrimination between green tea dietary supplements and green tea leaves by HPLC/MS, *Journal of AOAC International* 94 (2011): 487–497.
43. L. M. Vislocky and M. L. Fernandez, Grapes and grape products: their role in health, *Nutrition Today* (2012), epub ahead of print, doi:10.1097/NT0b013e31823db374.
44. Q. Xu and L-Y. Si, Resveratrol role in cardiovascular and metabolic health and potential mechanisms of action, *Nutrition Research* 32 (2012): 648–658; J. M. Smoliga, J. A. Baur, and H. A.Hausenblas, Resveratrol and health—A comprehensive review of human clinical trials, *Molecular Nutrition and Food Research* 55 (2011): 1129–1141.
45. L. M. Chu and coauthors, Resveratrol in the prevention and treatment of coronary artery disease, *Current Atheroscerosis Reports* 13 (2011), 439–446; S. Hamed and coauthors, Red wine consumption improves in vitro migration of endothelial progenitor cells in young, healthy individuals, *American Journal of Clinical Nutrition* 92 (2010): 161–169.
46. A. J. Papoutsis and coauthors, Resveratrol prevents epigenetic silencing of BRCA-1 by the aromatic hydrocarbon receptor in human breast cancer cells, *Journal of Nutrition* 140 (2010): 1607–1614.
47. A. F. Fernández and M. F. Fraga, The effects of the dietary polyphenols resveratrol on human healthy aging and lifespan, *Epigenetics* 6 (2011): 870–874.
48. M. M. Dohadwala and J. A. Vita, Grapes and cardiovascular disease, *Journal of Nutrition* 139 (2009): 1788S–1793S.
49. Smoliga, Baur, and Hausenblas, Resveratrol and health, 2011; Chu and coauthors, Resveratrol in the prevention and treatment of coronary artery disease, 2011; Forester and A. L. Waterhouse, Metabolites are key to understanding health effects of wine polyphenolics, *Journal of Nutrition* 139 (2009): 1824S–1831S.
50. E. Guillemard and coauthors, Effects of consumption of a fermented dairy product containing the probiotic Lactobacillus casei DN-114 001 on common respiratory and gastrointestinal infections in shift workers in a randomized controlled trial, *Journal of the American College of Nutrition* 29 (2010): 455–468; M. E. Sanders, How do we know when something called "probiotic" is really a probiotic? A guideline for consumers and health care professionals, *Functional Food Reviews* 1 (2009): 3–12; G. J. Leyer and coauthors, Probiotic effects on cold and influenza-like symptom incidence and duration in children, *Pediatrics* 124 (2009): e172–e179.
51. S. Hemple and coauthors, Probiotics for the prevention and treatment of antibiotic-associated diarrhea: a systematic review and meta-analysis, *Journal of the American Medical Association* 307 (2012): 1959–1969; S. J. Allen and coauthors, Probiotics for treating acute infectious diarrhea (review), *Cochrane Database of Systematic Reviews* 11 (2010), epub, doi: 10.1002/14651858.
52. K. Whelan and C. E. Myers, Safety of probiotics in patients receiving nutritional support: A systematic review of case reports, randomized controlled trials, and nonrandomized trials, *American Journal of Clinical Nutrition* 91 (2010): 687–703; L. E. Morrow, Prebiotics in the intensive care unit, *Current Opinion in Critical Care* 15 (2009): 144–148.
53. N. Upadhyay and V. Moudgal, Probiotics: A review, *Journal of Clinical Outcomes Management* 19 (2012): 76–84.
54. C. Hughes and coauthors, Galactooligosaccharide supplementation reduces stress-induced gastrointestinal dysfunction and days of cold or flu: a randomized, double-blind, controlled trial in healthy university students, *American Journal of Clinical Nutrition* 93 (2011): 1305–1311; X. Tzounis and coauthors, Prebiotic evaluation of cocoa-derived flavanols in health humans by using a randomized, controlled, double-blind, crossover intervention study, *American Journal of Clinical Nutrition* 93 (2011): 62–72.
55. A. M. Brownawell and coauthors, Prebiotics and the health benefits of fiber: current regulatory status future research, and goals, *Journal of Nutrition* 142 (2012): 962–974; M. A. Conlon and coauthors, Resistant starches protect against colonic DNA damage and alter microbiota and gene expression in rats fed a Western diet, *Journal of Nutrition* 142 (2012): 832–840.
56. D. R. Jacobs, M. D. Gross, and L. C. Tapsell, Food synergy: An operational concept for understanding nutrition, *American Journal of Clinical Nutrition* 89 (2009): 1543S–1548S.
57. G. Yang and coauthors, Isothiocyanate exposure, glutathione S-transferase polymorphisms, and colorectal cancer risk, *American Journal of Clinical Nutrition* 91 (2010): 704–711; R. H. Dashwood and E. Ho, Dietary agents as histone deacetylase inhibitors: Sulforaphane and structurally related isothiocyanates, *Nutrition Reviews* 66 (2009): S36–S38.
58. A. J. Vargas and R. Burd, Hormesis and synergy: Pathways and mechanisms of quercetin in cancer prevention and management, *Nutrition Reviews* 68 (2010): 418–428; A. M. Knab and coauthors, Influence of quercetin supplementation on disease risk factors in community-dwelling adults, *Journal of the American Dietetic Association* 111 (2011): 542–549.
59. J. I. Boullata and L. M. Hudson, Drug-nutrient interactions: a broad view with implications for practice, *Journal of the Academy of Nutrition and Dietetics* 112 (2012): 506–517.

60. Egert and Rimback, Which sources of flavonoids: Complex diets or dietary supplements?, 2011.
61. R. E. Patterson, (editorial) Flaxseed and breast cancer: what should we tell our patients?, *Journal of Oncology* 29 (2011): 3723–3726; Hollman and coauthors, The biological relevance of direct antioxidant effects of polyphenols for cardiovascular health in humans is not established, 2011.
62. Egert and Rimback, Which sources of flavonoids, 2011.
63. C. Wang and coauthors, Cranberry-containing products for prevention of urinary tract infections in susceptible populations, *Archives of Internal Medicine* 172 (2012): 988–996; H. Shmuely and coauthors, Cranberry components for the therapy of infectious disease, *Current Opinion in Biotechnology* 23 (2012): 148–152; D. R. Guay, Cranberry and urinary tract infections, *Drugs* 69 (2009): 775–807.
64. Position of the American Dietetic Association, Functional foods, *Journal of the American Dietetic Association* 109 (2009): 735–746.

Chapter 3

1. I. R. Stienstra and coauthors, The inflammasome puts obesity in the danger zone, *Cell Metabolism* 15 (2012): 10-8; P. C. Calder and coauthors, Dietary factors and low-grade inflammation in relation to overweight and obesity, *British Journal of Nutrition* 106 (2011): S5–S78.
2. N. Chaudhari and S. D. Roper, The cell biology of taste, *Journal of Cell Biology* 190 (2010): 285–296; K. Kurihara, Glutamate: From discovery as a food flavor to role as a basic taste (umami), *American Journal of Clinical Nutrition* 90 (2009): 719S–722S.
3. K. D. Gifford, S. Baer-Sinnot, and L. N. Heverling, Managing and understanding sweetness: Common-sense solutions based on the science of sugars, sugar substitutes, and sweetness, *Nutrition Today* 44 (2009): 211–217.
4. A. Drewnowski and coauthors, Sweetness and food preference, *Journal of Nutrition* 142 (2012): 1142S–1148S.
5. J. A. Mennella and coauthors, The timing and duration of a sensitive period in human flavor learning: a randomized trial, *American Journal of Clinical Nutrition* 93 (2011): 1019–1024; R. Krebs, The gourmet ape: Evolution and human food preferences, *American Journal of Clinical Nutrition* 90 (2009): 707S–711S.
6. W. W. L. Hsiao and coauthors, The microbes of the intestine: An introduction to their metabolic and signaling capabilities, *Endocrinology and Metabolism Clinics of North America* 37 (2008): 857–871.
7. G. T. Macfarland and S. Macfarlane, Bacteria, colonic fermentation, and gastrointestinal health, *Journal of AOAC International* 95 (2012): 50–60; F. Shanahan and E. Murphy, The hybrid science of diet, microbes, and metabolic health, *American Journal of Clinical Nutrition* 94 (2011): 1–2; S. C. Bischoff and M. Zeitz, Scientific evidence for the medical use of probiotics, *Annals of Nutrition and Metabolism* 57 (2010): S1–S5.
8. N. Vakil, Acid inhibition and infections outside the gastrointestinal tract, *American Journal of Gastroenterology* 104 (2009): S17–S20.
9. R. M. Pluta, G. D. Perazza, and R. M. Golub, Gastroesophageal reflux disease, JAMA Patient's Page, *Journal of the American Medical Association* 305 (2011): 2024.
10. G. Sachs, D. R. Scott, and R. Wen, Gastric infection by *Helicobacter pylori*, *Gastroenterology Reports* 13 (2011): 540–546.
11. N. J. Talley, K. L. Lasch, and C. Baum, A gap in our understanding: Chronic constipation and its comorbid conditions, *Clinical Gastroenterology and Hepatology* 7 (2009): 9–19.
12. D. Keszthelyi, F. J. Troost, and A. A. Masclee, Irritable bowel syndrome: methods, mechanisms, and pathophysiology.
Methods to assess visceral hypersensitivity in irritable bowel syndrome, *American Journal of Physiology: Gastrointestinal and Liver Physiology* 303 (2012): G141–G154.
13. C. M. Surawicz, Mechanisms of diarrhea, *Current Gastroenterology Reports* 12 (2010): 236–241.
14. W. D. Heizer, S. Southern, and S. McGovern, The role of diet in symptoms of irritable bowel syndrome in adults: A narrative review, *Journal of the American Dietetic Association* 109 (2009): 1204–1214.

Controversy 3

1. Centers for Disease Control and Prevention, *Quick Stats: Binge Drinking*, 2009, available at www.cdc.gov/alcohol/quickstats/binge_drinking.htm.
2. K. J. Mukamal, A 42-year-old man considering whether to drink alcohol for his health, *Journal of the American Medical Association* 303 (2010): 2065–2073.
3. S. J. Nielsen and coauthors, Calories consumed from alcoholic beverages by U.S. adults, 2007–2010, NCHS Data Brief 110, November 2012, available at www.cdc.gov/nchs/data/databriefs/db110.htm.
4. R. W. Hingson, Z. Wenxing, and E. R. Weitzman, Magnitude of and trends in alcohol-related mortality and morbidity among US college students ages 18–24, 1998–2005, *Journal of Studies on Alcohol and Drugs* Supplement No. 16 (2009): 12–20.
5. Centers for Disease Control and Prevention, Vital signs: Binge drinking prevalence, frequency, and intensity among adults—United States, 2010, *Morbidity and Mortality Weekly Report* 61 (2012): 14–19.
6. E. Nova and coauthors, Potential health benefits of moderate alcohol consumption: current perspectives in research, *Proceedings of the Nutrition Society* 71 (2012): 307–315; J. W. Belens and coauthors, Alcohol consumption and risk of recurrent cardiovascular events and mortality in patients with clinically manifest vascular disease and diabetes mellitus: The Second Manifestations of ARTerial (SMART) disease study, *Atherosclerosis* 212 (2010): 281–286; S. E. Chiuve and coauthors, Light-to-moderate alcohol consumption and risk of sudden cardiac death in women, *Heart Rhythm* 7 (2010): 1374–1380; D. O. Ballunas and coauthors, Alcohol as a risk factor for type 2 diabetes: A systematic review and meta-analysis, *Diabetes Care* 32 (2009): 2123–2132.
7. S. E. Brien and coauthors, Effect of alcohol consumption on biological markers associated with risk of coronary heart disease: systematic review and meta-analysis of interventional studies, *British Medical Journal* 342 (2011) doi:10.1136/bmj.d636.
8. P. E. Ronksley and coauthors, Association of alcohol consumption with selected cardiovascular disease outcomes: A systematic review and meta-analysis, *British Medical Journal* 342 (2011) doi:10.1136/bmj.d671.
9. Q. Sun and coauthors, Alcohol consumption at midlife and successful ageing in women: A prospective cohort analysis in the Nurses' Health Study, *PLoS Medicine*, September 2011, doi:10.1371/journal.pmed.1001090; E. Lobo and coauthors, Is there an association between low-to-moderate alcohol consumption and risk of cognitive decline? *American Journal of Epidemiology* 172 (2010): 708–716; C. Cooper and coauthors, Alcohol in moderation, premorbid intelligence and cognition in older adults: Results from the Psychiatric Morbidity Survey, *Journal of Neurology, Neurosurgery and Psychiatry* 80 (2009): 1236–1239.
10. K. J. Mukamal and coauthors, Alcohol consumption and cardiovascular mortality among U.S. adults, 1987–2002, *Journal of the American College of Cardiology* 55 (2010): 1328–1335.
11. Centers for Disease Control and Prevention, *Quick Stats: Binge Drinking*, 2009.
12. U. A. Hvidtfeldt and coauthors, Alcohol intake and risk of coronary heart disease in younger, middle-aged, and older adults, *Circulation* 212 (2010): 1589–1597.
13. A. J. Barnes and coauthors, Prevalence and correlates of at-risk drinking among older adults: The project SHARE study, *Journal of General Internal Medicine* 25 (2010): 840–846; P. Meier and H. K. Seitz, Age, alcohol metabolism and liver disease, *Current Opinion in Clinical Nutrition and Metabolic Care* 11 (2008): 21–26.
14. J. Martin and coauthors, Alcohol-attributable mortality in Ireland, *Alcohol and Alcoholism* 45 (2010): 379–386.
15. J. Ruidavets and coauthors, Patterns of alcohol consumption and ischaemic heart disease in culturally divergent countries: the Prospective Epidemiological Study of Myocardial Infarction (PRIME), *British Medical Journal* 341 (2010): doi:10.1136/bmj.c6077; A. C. Carlsson, H. Theobald, and P. E. Wändell, Health factors and longevity in men and women: A 25-year follow-up study, *European Journal of Epidemiology* 25 (2010): 547–551.
16. M. Roerecke and J. Rehm, Irregular heavy drinking occasions and risk of ischemic heart disease: A systematic review and meta-analysis, *American Journal of Epidemiology* 171 (2010): 633–644.
17. L. M. Vislocky and M. L. Fernandez, Grapes and grape products: their role in health,

Nutrition Today (2012), epub ahead of print, doi:10.1097/NT0b013e31823db374.
18. S. C. Forester and A. L. Waterhouse, Metabolites are key to understanding health effects of wine polyphenolics, *Journal of Nutrition* 139 (2009): 1824S–1831S.
19. V. Fedirko and coauthors, Alcohol drinking and colorectal cancer risk: an overall and dose-response meta-analysis of published studies, *Annals of Oncology* 22 (2011): 1958–1972; A. Benedetti, M. E. Parent, and J. Siemiatycki, Lifetime consumption of alcoholic beverages and risk of 13 types of cancer in men: Results from a case-control study in Montreal, *Cancer Detection and Prevention* 32 (2009): 352–362.
20. Y. Liang and coauthors, Alcohol consumption and the risk of incident atrial fibrillation among people with cardiovascular disease, *Canadian Medical Association Journal* 6 (2012): E857–E866; J. Somes and N. S. Donatelli, Syndromes of "holiday heart," *Journal of Emergency Nursing* 37 (2011): 577–579; C. E. Balbão, A. V. de Paola, and G. Fenelon, Effects of alcohol on atrial fibrillation: myths and truths, *Therapeutic Advances in Cardiovascular Disease* 31 (2009): 53–63.
21. *Dietary Guidelines for Americans 2010*, www.dietaryguidelines.gov.
22. S. Jarjour, L. Bai, and C. Gianoulakis, Effect of acute ethanol administration on the release of opioid peptides from the midbrain including the ventral tegmental area, *Alcoholism: Clinical and Experimental Research* 23 (2009), published online, doi:10.1111/j.1530–0277.2009.00924.x.
23. *Dietary Guidelines for Americans 2010*, www.dietaryguidelines.gov.
24. *Dietary Guidelines for Americans 2010*, www.dietaryguidelines.gov.
25. Centers for Disease Control and Prevention, *Quick Stats: Binge Drinking*, 2009.
26. Centers for Disease Control and Prevention, *Quick Stats: Binge Drinking*, 2009.
27. Centers for Disease Control and Prevention, *Quick Stats: Binge Drinking*, 2009.
28. World Health Organization, *Global Status Report on Alcohol and Health* (Le Mont-sur-Lausanne, Switzerland: World Health Organization, 2011).
29. A. J. Birley and coauthors, Association of the gastric alcohol dehydrogenase gene ADH7 with variation in ethanol metabolism, *Human Molecular Genetics* 17 (2008): 179–189.
30. Y. Liang and coauthors, Alcohol consumption and the risk of incident atrial fibrillation among people with cardiovascular disease, *Canadian Medical Association Journal* 16 (2012): E857–E866; E. Childs, S. O'Connor, and H. de Wit, Bidirectional interactions between acute psychosocial stress and acute intravenous alcohol in health men, *Alcoholism: Clinical and Experimental Research* 35 (2011): 1794–1803.
31. J. Haorah and coauthors, Mechanism of alcohol-induced oxidative stress and neuronal injury, *Free Radical Biology and Medicine* 45 (2008): 1542–1550.
32. Centers for Disease Control and Prevention, National Center for Injury Prevention and Control (NCIPC), www.cdc.gov.
33. L. E. Beane Freemean and coauthors, Mortality from lymphohematopoietic malignancies among workers in formaldehyde industries: The National Cancer Institute cohort, *Journal of the National Cancer Institute* 101 (2009): 751–761.
34. J. Rehm and coauthors, Global burden of disease and injury and economic cost attributable to alcohol use and alcohol-use disorders, *Lancet* 373 (2009): 2223–2233; A. Z. Fan and coauthors, Patterns of alcohol consumption and the metabolic syndrome, *Journal of Clinical Endocrinology and Metabolism* 93 (2008): 3833–3838; Haorah and coauthors, 2008; Centers for Disease Control and Prevention, 2008.
35. Benedetti, Parent, and Siemiatycki, Lifetime consumption of alcoholic beverages and risk of 13 types of cancer in men, 2009; J. Q. Lew and coauthors, Alcohol and risk of breast cancer by histologic type and hormone receptor status in postmenopausal women, *American Journal of Epidemiology* 170 (2009): 308–317; N. E. Allen and coauthors, Moderate alcohol intake and cancer incidence in women, *Journal of the National Cancer Institute* 101 (2009): 296–305.
36. H. K. Seitz and coauthors, Epidemiology and pathophysiology of alcohol and breast cancer: update 2012, *Alcohol and Alcoholism*, 29 March 2012, epub ahead of print, doi:10.1093/alcalc/ags011.
37. Benedetti, Parent, and Siemiatycki, Lifetime consumption of alcoholic beverages and risk of 13 types of cancer in men, 2009.
38. P. A. Newcomb, No difference between red wine or white wine consumption and breast cancer, *Cancer Epidemiology, Biomarkers and Prevention* 18 (2009): 1007–1010.
39. W. Y. Chen and coauthors, Moderate alcohol consumption during adult life, drinking patterns, and breast cancer risk, *Journal of the American Medical Association* 306 (2011): 1884–1890; S. F. Brennan and coauthors, Dietary patterns and breast cancer risk: a systematic review and meta-analysis, *American Journal of Clinical Nutrition* 91 (2010): 1294–1302.
40. D. M. Fergusson, J. M. Boden, and L. J. Horwood, Tests of causal links between alcohol abuse or dependence and major depression, *Archives of General Psychiatry* 66 (2009): 260–266.
41. M. R. Lucey, P. Mathurin, and T. R. Morgan, Alcoholic hepatitis, *New England Journal of Medicine* 360 (2009): 2758–2769.
42. Centers for Disease Control and Prevention, Cost Calculators, available at www.cdc.gov.
43. L. Wang and coauthors, Alcohol consumption, weight gain, and risk of becoming overweight in middle-aged and older women, *Archives of Internal Medicine* 170 (2010): 453–461.
44. C. Sayon-Orea and coauthors, Type of alcoholic beverage and incidence of overweight/obesity in a Mediterranean cohort: The SUN project, *Nutrition* 27 (2011): 802–808.
45. Standing Committee on the Scientific Evaluation of Dietary Reference Intakes, Food and Nutrition Board, Institute of Medicine, *Dietary Reference Intakes for Energy, Carbohydrate, Fiber, Fat, Fatty Acids, Cholesterol, Protein, and Amino Acids* (Washington, D.C.: National Academy Press, 2002/2005), p. 109.
46. S. Lourenco, A. Oliveira, and C. Lopes, The effect of current and lifetime alcohol consumption on overall and central obesity, *Epidemiology* 66 (2012): 813–818; Sayon-Orea and coauthors, Type of alcoholic beverage and incidence of overweight/obesity in a Mediterranean cohort, 2011; M. Schütze and coauthors, Beer consumption and the 'beer belly': Scientific basis or common belief? *European Journal of Clinical Nutrition* 63 (2009): 1143–1149.
47. R. A. Breslow and coauthors, Alcoholic beverage consumption, nutrient intakes, and diet quality in the US adult population, 1999–2006, *Journal of the American Dietetic Association* 110 (2010): 551–562.
48. Q. Sun and coauthors, Alcohol consumption at midlife and successful ageing in women: a prospective cohort analysis in the Nurses' Health Study, *PLoS Medicine* 8 (2011), doi:10.1371/journal.pmed.1001090; N. Bakalar, Aging: Health gains from a small drink a day, *New York Times*, September 19, 2011, available at www.nytimes.com.

Chapter 4

1. Artificial photosynthesis: Turning sunlight into liquid fuels moves a step closer, *ScienceDaily*, 12 March 2009, available at http://www.sciencedaily.com.
2. U.S. Department of Agriculture and U.S. Department of Health and Human Services, *Dietary Guidelines for Americans 2010*, available at www.dietaryguidelines.gov.
3. U.S. Department of Agriculture and U.S. Department of Health and Human Services, *Dietary Guidelines for Americans 2010*; A. T. Merchant and coauthors, Carbohydrate intake and overweight and obesity among healthy adults, *Journal of the American Dietetic Association* 109 (2009): 1165–1172.
4. U.S. Department of Agriculture and U.S. Department of Health and Human Services, *Dietary Guidelines for Americans, 2010*, available at www.dietaryguidelines.gov; R. E. Kavey, How sweet it is: Sugar-sweetened beverage consumption, obesity, and cardiovascular risk in childhood, *Journal of the American Dietetic Association* 110 (2010): 1456–1460.
5. U.S. Department of Agriculture and U.S. Department of Health and Human Services, *Dietary Guidelines for Americans 2010*; R. K. Johnson and coauthors, Dietary sugars intake and cardiovascular health: A Scientific Statement from the American Heart Association, *Circulation* 120 (2009): 1011–1020.
6. J. W. Anderson and coauthors, Health benefits of dietary fiber, *Nutrition Reviews* 67 (2009): 188–205.
7. R. A. Othman, M. H. Moghadasian, and P. J. Jones, Cholesterol-lowering effects of oat β-glucan, *Nutrition Reviews* 69 (2011): 299–309; L. Van Horn and coauthors, The evidence for dietary prevention and treatment of cardiovascular disease, *Journal of the American Dietetic Association* 108 (2008): 287–331; M. O. Weickert and A. F. Pfeiffer, Metabolic effects of dietary fiber

consumption and prevention of diabetes, *Journal of Nutrition* 138 (2008): 439–442.
8. S. Chuang and coauthors, Fiber intake and total and cause specific mortality in the European Prospective Investigation into Cancer and Nutrition cohort, *American Journal of Clinical Nutrition* 96 (2012): 164–174; R. J. van de Laar and coauthors, Lower lifetime dietary fiber intake is associated with carotid artery stiffness: the Amsterdam Growth and Health Longitudinal Study, *American Journal of Clinical Nutrition* 96 (2012): 14–23.
9. K. C. Maki and coauthors, Whole-grain ready-to-eat oat cereal, as part of a dietary program for weight loss, reduces low-density lipoprotein cholesterol in adults with overweight and obesity more than a dietary program including low-fiber control foods, *Journal of the American Dietetic Association* 110 (2010): 205–214.
10. K. E. Andersson and coauthors, Oats (*Avena sativa*) reduce atherogenesis in LDL-receptor-deficient mice, *Atherosclerosis* 212 (2010): 93–99.
11. J. A. Nettleton and coauthors, Interactions of dietary whole-grain intake with fasting glucose- and insulin-related genetic loci in individuals of European descent: A meta-analysis of 14 cohort studies, *Diabetes Care* 33 (2010): 2684–2691.
12. A. Giacosa and M. Rondanelli, The right fiber for the right disease: An update on the psyllium seed husk and the metabolic syndrome, *Journal of Clinical Gastroenterology* 44 (2010): S58–S60.
13. A. M. Brownawell and coauthors, Prebiotics and the health benefits of fiber: Current regulatory status, future research, and goals, *Journal of Nutrition* 142 (2012): 962–974; J. E. Ravikoff and J. R. Korzenik, The role of fiber in diverticular disease, *Journal of Clinical Gastroenterology* 45 (2011): S7–S11.
14. A. F. Peery and coauthors, A high-fiber diet does not protect against asymptomatic diverticulosis, *Gastroenterology* 142 (2012): 266–272; F. L. Crow and coauthors, Diet and risk of diverticular disease in Oxford cohort of European Prospective Investigation into Cancer and Nutrition (EPIC): Prospective study of British vegetarians and non-vegetarians, *British Medical Journal* 343 (2011), epub, doi:10.1136/bmj.d4131; S. Tarleton and J. K. Dibaise, Low-residue diet in diverticular disease: Putting an end to a myth, *Nutrition in Clinical Practice* 26 (2011): 137–142; A. Rocco and coauthors, Treatment options for uncomplicated diverticular disease of the colon, *Journal of Clinical Gastroenterology* 43 (2009): 803–808.
15. Department of Health and Human Services, Centers for Disease Control and Prevention, and National Cancer Institute, United States Cancer Statistics: 1999–2007 Incidence and Mortality Web-based Report, 2010, available at http://www.cdc.gov/uscs.
16. D. Aune and coauthors, Dietary fibre, whole grains, and risk of colorectal cancer: systematic review and dose-response meta-analysis of prospective studies, *British Medical Journal* 343 (2011), epub, doi:10.1136/bmj.d6617; L. B. Sansbury and coauthors, The effect of strict adherence to a high-fiber, high-fruit and -vegetable, and low-fat eating pattern on adenoma recurrence, *American Journal of Epidemiology* 170 (2009): 576–584; N. Slimani and B. Margetts, Nutrient intakes and patterns in the EPIC cohorts from ten European countries, *European Journal of Clinical Nutrition* 63 (2009): S1–S274.
17. C. C. Dahm and coauthors, Dietary fiber and colorectal cancer risk: A nested case-control study using food diaries, *Journal of the National Cancer Institute* 102 (2010): 614–626.
18. M. H. Pan and coauthors, Molecular mechanisms for chemoprevention of colorectal cancer by natural dietary compounds, *Molecular Nutrition and Food Research* 55 (2011): 32–45.
19. K. Wallace and coauthors, The association of lifestyle and dietary factors with the risk for serrated polyps of the colorectum, *Cancer Epidemiology, Biomarkers, and Prevention* 18 (2009): 2310–2317.
20. L. A. Tucker and K. S. Thomas, Increasing total fiber intake reduces risk of weight and fat gains in women, *Journal of Nutrition* 139 (2009): 567–581.
21. N. Schroeder and coauthors, Influence of whole grain barley, whole grain wheat, and refined rice-based foods on short-term satiety and energy intake, *Appetite* 53 (2009): 363–369; M. Lyly and coauthors, Fiber in beverages can enhance perceived satiety, *European Journal of Nutrition* 48 (2009): 251–258; K. R. Juvonen and coauthors, Viscosity of oat bran-enriched beverages influences gastrointestinal hormonal responses in healthy humans, *Journal of Nutrition* 139 (2009): 461–466.
22. P. Vitaglione and coauthors, β-Glucan-enriched bread reduces energy intake and modifies plasma ghrelin and peptide YY concentrations in the short term, *Appetite* 53 (2009): 338–344.
23. H. Du and coauthors, Dietary fiber and subsequent changes in body weight and waist circumference in European men and women, *American Journal of Clinical Nutrition* 91 (2010): 329–336.
24. S. S. Jonnalagadda and coauthors, Putting the whole grain puzzle together: health benefits associated with whole grains—summary of American Society for Nutrition 2010 Satellite Symposium, *Journal of Nutrition* 141 (2011): 1011S–1022S; K. A. Harris and P. M. Kris-Etherton, Effects of whole grains on coronary heart disease risk, *Current Atherosclerosis Reports* 12 (2010): 368–376.
25. What We Eat in America, NHANES, 2005–2006, www.ars.usda.gov/ba/bhnrc/fsrg, 2008; Position of the American Dietetic Association: Health implications of dietary fiber, *Journal of the American Dietetic Association* 108 (2008): 1716–1731.
26. Food and Drug Administration, *The Scoop on Whole Grains*, 06 May 2009, available at www.fda.gov/consumer.
27. U.S. Department of Agriculture and U.S. Department of Health and Human Services, *Dietary Guidelines for Americans 2010*, available at www.dietaryguidelines.gov; M. Kristensen and coauthors, Jonnalagadda and coauthors, Putting the whole grain puzzle together: Health benefits associated with whole grains—Summary of American Society for Nutrition 2010 Satellite Symposium, 2011.
28. M. Kristensen and coauthors, Whole grain compared with refined wheat decreases the percentage of body fat following a 12-week, energy-restricted dietary intervention in postmenopausal women, *Journal of Nutrition* 142 (2012): 710–716; S. S. Jonnalagadda and coauthors, Putting the whole grain puzzle together: health benefits associated with whole grains—Summary of American Society for Nutrition 2010 Satellite Symposium, *Journal of Nutrition* 141 (2011): 1011S–1022S.
29. P. G. Williams, Evaluation of the evidence between consumption of refined grains and health outcomes, *Nutrition Reviews* 70 (2012): 80–99.
30. A. R. Bird and coauthors, Resistant starch, large bowel fermentation and a broader perspective of prebiotics and probiotics, *Beneficial Microbes* 1 (2010): 423–431.
31. M. A. Conlon and coauthors, Resistant starches protect against colonic DNA damage and alter microbiota and gene expression in rats fed a Western diet, *Journal of Nutrition* 142 (2012): 832–840.
32. T. A. Nicklas, H. Qu, and S. O. Hughes, Prevalence of self-reported lactose intolerance in a multi-ethnic sample of adults, *Nutrition Today* 44 (2009): 186–187.
33. NIH Consensus Development Conference: Lactose intolerance and health, available at consensus.nih.gov/2010/lactosestatement.htm.
34. Committee on Dietary Reference Intakes, *Dietary Reference Intakes for Calcium and Vitamin D* (Washington, D.C.: National Academies Press, 2011), pp. 8–14.
35. M. Figolet, V. R. Barragán, and M. T. González, Low-carbohydrate diets: a matter of love or hate, *Annals of Nutrition and Metabolism* 58 (2011): 320–334.
36. S. Thammongkol and coauthors, Efficacy of the ketogenic diet: which epilepsies respond? *Epilepsia* 53 (2012): e55–e59; N. E. Payne and coauthors, The ketogenic and related diets in adolescents and adults—A review, *Epilepsia* 52 (2011): 1941–1948; A. Patel and coauthors, Long-term outcomes of children treated with ketogenic diet in the past, *Epilepsia* 51 (2010): 1277–1282.
37. Standing Committee on the Scientific Evaluation of Dietary Reference Intakes, 2002/2005, p. 265.
38. F. Magkos, X. Wang, and B. Mittendorfer, Metabolic actions of insulin in men and women, *Nutrition* 28 (2010): 686–691.
39. W. J. Whelan and coauthors, The glycemic response is a personal attribute, *International Union of Biochemistry and Molecular Biology* 62 (2010): 637–641.
40. J. M. Jones, Glycemic Index, *Nutrition Today* 47 (2012): 207–213; H. Dodd and coauthors, Calculating meal glycemic index by using measured and published food values compared with directly measured meal glycemic index, *American Journal of Clinical Nutrition* 94 (2011): 992–996.
41. M. L. Wheeler and coauthors, Macronutrients, food groups, and eating patterns in the

management of diabetes: A systematic review of the literature, 2010, *Diabetes Care* 35 (2012): 434–445.
42. G. Livesey and H. Tagami, Interventions to lower the glycemic response to carbohydrate foods with a low-viscosity fiber (resistant maltodextrin): Meta-analysis of randomized controlled trials, *American Journal of Clinical Nutrition* 89 (2009): 114–125.
43. Centers for Disease Control and Prevention, Increasing prevalence of diagnosed diabetes, *Morbidity and Mortality Weekly Report* 61 (2012): 918–921.
44. Centers for Disease Control and Prevention, Diabetes: Successes and opportunities for population-based prevention and control, *At A Glance 2011*, available at www.cdc.gov/chronicdisease/resources/publications/AAG/ddt.htm; A. Whaley-Connell and coauthors, Diabetes mellitus and CKD awareness: The Kidney Early Evaluation Program (KEEP) and National Health and Nutrition Examination Survey (NHANES), *American Journal of Kidney Diseases* 53 (2009): S11–S12.
45. Centers for Disease Control and Prevention, *National Diabetes Fact Sheet, 2011.*
46. H. Zong, M. Ward, and A. W. Sitt, AGEs, RAGE, and diabetic retinopathy, *Current Diabetes Reports* 11 (2011): 244–252; R. C. Stanton, Oxidative stress and diabetic kidney disease, *Current Diabetes Reports* 11 (2011): 330–336.
47. National Institutes of Health, *Diabetic Neuropathies: The Nerve Damage of Diabetes* (2011), NIH publication 09–3185, available at www.diabetes.niddk.hih.gov.
48. L. Perreault and coauthors, Effect of regression from prediabetes to normal glucose regulation on long-term reduction in diabetes risk: results from the Diabetes Prevention Program Outcomes Study, *Lancet* 379 (2012): 2243–2251; A. G. Tabák and coauthors, Prediabetes: A high-risk state for diabetes development, *Lancet* 379 (2012): 2279–2290.
49. X. Zhuo and coauthors, Alternative HbA1c cutoffs to identify high-risk adults for diabetes prevention, *American Journal of Preventive Medicine* 42 (2012): 374–381; Executive summary: Standards of medical care in diabetes—2011, *Diabetes Care* 34 (2011): S4–S10.
50. S. E. Inzucchi and coauthors, Management of hyperglycemia in Type 2 diabetes: A patient-centered approach, *Diabetes Care*, 35 (2012): 1364–1379.
51. P. Concannon , S. S. Rich, and G. T. Nepom, Genetics of type 1a diabetes, *New England Journal of Medicine* 360 (2009): 1646–1654.
52. Diabetes Control and Complications Trial/Epidemiology of Diabetes Interventions and Complications Research Group, Modern-day clinical course of type 1 diabetes mellitus after 30 years' duration, *Archives of Internal Medicine* 169 (2009): 1307–1316.
53. A. Shapiro, State of the art of clinical islet transplantation and novel protocols of immunosuppression, *Current Diabetes Reports* 11 (2011): 345–354; G. N. Jofra and coauthors, Antigen-specific dependence of Tr1-cell therapy in preclinical models of islet transplantation, *Diabetes* 59 (2010): 433–439.
54. J. N. Cooke and coauthors, Genetic risks assessment of Type 2 diabetes-associated polymorphisms in African Americans, *Diabetes Care* 35 (2012): 287–292.
55. B. M. Filippi, P. I. Mighin, and T. K. T. Lam, Is insulin action in the brain clinically relevant? *Diabetes* 61 (2012): 773–775; E. Egeciouglu and coauthors, Hedonic and incentive signals for body weight control, *Reviews in Endocrine and Metabolic Disorders* 12 (2011): 141–151; S. C. Woods and D. S. Ramsay, Food intake, metabolism and homeostasis, *Physiology and Behavior* 104 (2011): 4–7.
56. P. W. Franks, Gene x environment interactions in type 2 diabetes, *Current Diabetes Reports* 11 (2011): 552–561.
57. E. Q. Ye and coauthors, Greater whole-grain intake is associated with lower risk of type 2 diabetes, cardiovascular disease, and weight gain, *Journal of Nutrition* 142 (2012): 1304–1313; A. J. Cooper and coauthors, A prospective study of the association between quantity and variety of fruit and vegetable intake and incident type 2 diabetes, *Diabetes Care* 35 (2012): 1293–1300; J. P. Reis and coauthors, Lifestyle factors and risk for new-onset diabetes, *Annals of Internal Medicine* 155 (2011): 292–299; D. Mozaffarian and coauthors, Lifestyle risk factors and new-onset diabetes mellitus in older adults: The Cardiovascular Health Study, *Archives of Internal Medicine* 169 (2009): 798–807.
58. M. C. Audelin and coauthors, Change of energy expenditure from physical activity is the most powerful determinant of improved insulin sensitivity in overweight patients with coronary artery disease participating in an intensive lifestyle modification program, *Metabolism* 61 (2012): 672–679; J. G. Karam and S. I. McFarlane, Update on the prevention of type 2 diabetes, *Current Diabetes Reports* 11 (2011): 56–63.
59. Centers for Disease Control and Prevention, Diabetes: Successes and opportunities for population-based prevention and control, *At A Glance*, 2011.
60. D. E. Arterburn and coauthors, A multisite study of long-term remission and relapse of type 2 diabetes mellitus following gastric bypass, *Obesity Surgery* (2012), epub, doi:10.1007/s11695-012-0802-1; R. V. Cohen and coauthors, Effects of gastric bypass surgery in patients with type 2 diabetes and only mild obesity, *Diabetes Care* 35 (2012): 1420–1428; P. R. Schauer and coauthors, Bariatric surgery versus intensive medical therapy in obese patients with diabetes, *New England Journal of Medicine* 366 (2012): 1567–1576.
61. S. E. Inzucchi and coauthors, Management of hyperglycemia in type 2 diabetes: a patient-centered approach, *Diabetes Care* 35 (2012): 1364–1379.
62. V. M. Montori and M. Fernandez-Balsells, Glycemic control in type 2 diabetes: Time for an evidence-based about-face? *Annals of Internal Medicine* 150 (2009): 803–808.
63. American Diabetes Association, Standards of medical care in diabetes—2011, *Diabetes Care* 34 (2011): S11–S61.
64. American Diabetes Association, Standards of medical care in diabetes—2011.
65. A. Grøntved and coauthors, A prospective study of weight training and risk of type 2 diabetes mellitus in men, *Archives of Internal Medicine*, August 6, 2012, epub ahead of print, doi:10.1001/archinternmed.2012.3138; S. R. Colberg and coauthors, Exercise and type 2 diabetes: The American College of Sports Medicine and the American Diabetes Association: Joint position statement, *Diabetes Care* 33 (2010): e147–e167.
66. M. A. McCrory and coauthors, Pulse consumption, satiety, and weight management, *Advances in Nutrition* 1 (2010): 17–30.
67. The *Dietary Guidelines for Americans 2010*, www.dietaryguidelines.gov.
68. Position of the Academy of Nutrition and Dietetics: Use of nutritive and nonnutritive sweeteners, *Journal of the Academy of Nutrition and Dietetics* 112 (2012): 739–758.
69. S. W. Rizkalla, Health implications of fructose consumption: A review of recent data, *Nutrition and Metabolism* 4 (2010): 82–98.

Consumer's Guide 4

1. U.S. Department of Agriculture and U.S. Department of Health and Human Services, *Dietary Guidelines for Americans 2010*, available at www.dietaryguidelines.gov.
2. P. G. Williams, Evaluation of the evidence between consumption of refined grains and health outcomes, *Nutrition Reviews* 70 (2012): 80–99.
3. Whole Grains Council, Whole grain statistics, May 2011, available at www.wholegraincouncil.org.
4. C. E. O'Neil and coauthors, Consumption of whole grains is associated with improved diet quality and nutrient intake in children and adolescents: The National Health and Nutrition Examination Survey 1999–2004, *Public Health Nutrition* 14 (2011): 347–355; C. E. O'Neil and coauthors, Whole-grain consumption is associated with diet quality and nutrient intake in adults: The National Health and Nutrition Examination Survey, 1999–2004, *Journal of the American Dietetic Association* 110 (2010): 1461–1468.

Controversy 4

1. U.S. Department of Agriculture and U.S. Department of Health and Human Services, *Dietary Guidelines for Americans 2010*, available at www.dietaryguidelines.gov.
2. U.S. Department of Agriculture, Agricultural Research Service, 2008, Nutrient intakes from food, available at www.ars.usda.gov/ba/bhnrc/fsrg.
3. U.S. Department of Agriculture and U.S. Department of Health and Human Services, *Dietary Guidelines for Americans 2010*, available at www.dietaryguidelines.gov.

4. B. A. Swinburn and coauthors, Estimating the changes in energy flux that characterize the rise in obesity prevalence, *American Journal of Clinical Nutrition* 89 (2009): 1723–1728.
5. Centers for Disease Control and Prevention, Trend in intake of kcalories and macronutrients: United States, 1971–2000, *Morbidity and Mortality Weekly Report* 53 (2004): 80–82; B. Swinburn, G. Sacks, and E. Ravussin, Increased food energy supply is more than sufficient to explain the U.S. epidemic of obesity, *American Journal of Clinical Nutrition* 90 (2009): 1453–1456.
6. S. B. Heymsfield, How large is the energy gap that accounts for the obesity epidemic? *American Journal of Clinical Nutrition* 89 (2009): 1717–1718; Centers for Disease Control and Prevention, Prevalence of regular physical activity among adults—United States, 2001 and 2005, *Morbidity and Mortality Weekly Report* 56 (2007): 1209–1212.
7. S. Haley, Sugar and sweeteners outlook: U.S. sugar, USDA Economic Research Service, *Electronic Outlook Report from the Economic Research Service*, May 2011, available at www.ers.usda.gov.
8. F. B. Hu and V. S. Malik, Sugar-sweetened beverages and risk of obesity and type 2 diabetes: Epidemiological evidence, *Physiology and Behavior* 100 (2010): 47–54; V. S. Malik and coauthors, Sugar-sweetened beverages, obesity, and type 2 diabetes mellitus, and cardiovascular disease risk, *Circulation* 121 (2010): 1356–1364; D. I. Jalal and coauthors, Increased fructose associates with elevated blood pressure, *Journal of the American Society of Nephrology* 21 (2010): 1543–1549; R. K. Johnson and B. A. Yon, Weighing in on added sugars and health, *Journal of the American Dietetic Association* 110 (2010): 1296–1299; V. S. Malik and coauthors, Sugar-sweetened beverages and risk of metabolic syndrome and type 2 diabetes: A meta-analysis, *Diabetes Care* 33 (2010): 2477–2483; L. Tappy and coauthors, Fructose and metabolic diseases: New findings, new questions, *Nutrition* 26 (2010): 1044–1049; T. T. Fung and coauthors, Sweetened beverage consumption and risk of coronary heart disease in women, *American Journal of Clinical Nutrition* 89 (2009): 1037–1042; L. M. Fiorito and coauthors, Beverage intake of girls at age 5 y predicts obesity and weight status in childhood and adolescence, *American Journal of Clinical Nutrition* 90 (2009): 935–942.
9. A. T. Merchant and coauthors, Carbohydrate intake and overweight and obesity among healthy adults, *Journal of the American Dietetic Association* 109 (2009): 1165–1172.
10. B. M. Popkin, L. S. Adair, and S. W. Ng, Global nutrition transition and the pandemic of obesity in developing countries, *Nutrition Reviews* 70 (2012): 3–21; A. Pan, V. Malik, and F. B. Hu, Exporting diabetes to Asia: the impact of Western-style fast food (editorial), *Circulation* 126 (2012): 163–165; Y. Wu and coauthors, The growing burden of overweight and obesity in contemporary China, *CVD Prevention and Control* 4 (2009): 19–26.
11. Hu and Malik, Sugar-sweetened beverages and risk of obesity and type 2 diabetes, 2010.
12. A. Mente and coauthors, A systematic review of the evidence supporting a causal link between dietary factors and coronary heart disease, *Archives of Internal Medicine* 169 (2009): 659–669.
13. J. P. Reis and coauthors, Lifestyle factors and risk for new-onset diabetes, *Annals of Internal Medicine* 155 (2011): 292–299; D. Mozaffarian and coauthors, Lifestyle risk factors and new-onset diabetes mellitus in older adults: The Cardiovascular Health Study, *Archives of Internal Medicine* 169 (2009): 798–807.
14. T. Adhavan, B. L. Luhovyy, and G. H. Anderson, Effect of drinking compared with eating sugars or whey protein on short-term appetite and food intake, *International Journal of Obesity (London)* 35 (2011): 562–569; D. V. Ranawana and C. J. Henry, Are caloric beverages compensated for in the short-term by young adults? An investigation with particular focus on gender differences, *Appetite* 55 (2010): 137–146.
15. B. A. Cassady, R. V. Considine, and R. D. Mattes, Beverage consumption, appetite, and energy intake: what did you expect? *American Journal of Clinical Nutrition* 95 (2012): 587–593.
16. R. P. LaComb and coauthors, *Beverage Choices of U.S. Adults: What We Eat in America, NHANES 2007–2008*, Data Brief No. 6, August 2011, available at http://ars.usda.gov.
17. D. F. Tate and coauthors, Replacing caloric beverages with water or diet beverages for weight loss in adults: main results of the Choose Healthy Options Consciously Everyday (CHOICE) randomized clinical trial, *American Journal of Clinical Nutrition* 95 (2012): 555–563.
18. K. L. Stanhope, Role of fructose-containing sugars in the epidemics of obesity and metabolic syndrome, *Annual Review of Medicine* 63 (2012): 19.1–19.15; L. de Koning and coauthors, Sugar-sweetened and artificially sweetened beverage consumption and risk of type 2 diabetes in men, *American Journal of Clinical Nutrition* 93 (2011): 1321–1327; K. J. Duffey and coauthors, Drinking caloric beverages increases the risk of adverse cardiometabolic outcomes in the Coronary Artery Risk Development in Young Adults (CARDIA) Study, *American Journal of Clinical Nutrition* 92 (2010): 954–959; V. S. Malik and coauthors, Sugar-sweetened beverages and risk of metabolic syndrome and type 2 diabetes, *Diabetes Care* 33 (2010): 2477–2483; Hu and Malik, Sugar-sweetened beverages and risk of obesity and type 2 diabetes, 2010; R. K. Johnson and coauthors, Dietary sugars intake and cardiovascular health. A scientific statement from the American Heart Association, *Circulation* 120 (2009): 1011–1020.
19. F. Magkos, X. Wang, and B. Mittendorfer, Metabolic actions of insulin in men and women, *Nutrition* 28 (2010): 686–691.
20. B. M. Filippi, P. I. Mighin, and T. K. T. Lam, Is insulin action in the brain clinically relevant? *Diabetes* 61 (2012): 773–775; E. Egecioglu and coauthors, Hedonic and incentive signals for body weight control, *Reviews in Endocrine and Metabolic Disorders* 12 (2011): 141–151; S. C. Woods and D. S. Ramsay, Food intake, metabolism and homeostasis, *Physiology and Behavior* 104 (2011): 4–7.
21. L. C. Dolan, S. M. Potter, and G. A. Burdock, Evidence-based review on the effect of normal dietary consumption of fructose on development of hyperlipidemia and obesity in healthy, normal weight individuals, *Critical Reviews in Food Science and Nutrition* 50 (2009): 53–84.
22. K. L. Teff and coauthors, Endocrine and metabolic effects of consuming fructose- and glucose-sweetened beverages with meals in obese men and women: Influence of insulin resistance on plasma triglyceride responses, *Journal of Clinical Endocrinology and Metabolism* 94 (2009): 1652–1659.
23. L. Tappy and coauthors, Fructose and metabolic diseases: New findings, new questions, *Nutrition* 26 (2010): 1044–1049.
24. M. D. Lane and S. H. Cha, Effect of glucose and fructose on food intake via malonyl-CoA signaling in the brain, *Biochemical and Biophysical Research Communications* 382 (2009): 1–5.
25. K. A. Page and coauthors, Circulating glucose levels modulate neural control of the desire for high-calorie foods in humans, *Journal of Clinical Investigation* 121 (2011): 4161–4169.
26. Tappy and coauthors, Fructose and metabolic diseases.
27. T. H. Moran, Fructose and satiety, *Journal of Nutrition* 139 (2009): 1253S–1256S.
28. J. L. Sievenpiper and coauthors, Effect of fructose on body weight in controlled feeding trials: a systematic review and meta-analysis, *Annals of Internal Medicine* 156 (2012): 291–304.
29. K. L. Stanhope and coauthors, Consuming fructose-sweetened, not glucose-sweetened, beverages increases visceral adiposity and lipids and decreases insulin sensitivity in overweight/obese humans, *Journal of Clinical Investigation* 119 (2009): 1322–1334.
30. N. K. Pollock and coauthors, Greater fructose consumption is associated with cardiometabolic risk markers and visceral adiposity in adolescents, *Journal of Nutrition* 142 (2012): 251–257.
31. L. de Koning and coauthors, Sweetened beverages consumption, incident coronary heart disease and biomarkers of risk in men, *Circulation* 125 (2012): 1735–1741; T. J. Angelopoulos and coauthors, The effect of high-fructose corn syrup consumption on triglycerides and uric acid, *Journal of Nutrition* 139 (2009): 1242S–1245S; Teff and coauthors, Endocrine and metabolic effects of consuming fructose- and glucose-sweetened beverages with meals in obese men and women.
32. M. J. Dekker and coauthors, Fructose: A highly lipogenic nutrient implicated in insulin resistance, hepatic steatosis, and the metabolic syndrome, *American Journal of Physiology, Endocrinology and Metabolism* 299 (2010): E685–E694.
33. K-A. Lê and coauthors, Fructose overconsumption causes dyslipidemia and ectopic lipid deposition in healthy subjects with and without a family history of type 2 diabetes, *American Journal of Clinical Nutrition* 89 (2009): 1760–1765.

34. J. A. Welsh and coauthors, Caloric sweetener consumption and dyslipidemia among US adults, *Journal of the American Medical Association* 303 (2010): 1490–1497.

35. K. J. Duffy and coauthors, Drinking caloric beverages increases the risk of adverse cardio-metabolic outcomes in the Coronary Artery Risk Development in Young Adults (CARDIA) Study, *American Journal of Clinical Nutrition* 92 (2010): 954–959.

36. I. Aeberli and coauthors, Low to moderate sugar-sweetened beverage consumption impairs glucose and lipid metabolism and promotes inflammation in healthy young men: a randomized controlled trial, *American Journal of Clinical Nutrition* 94 (2011): 479–485; K. L. Stanhope, Consumption of fructose and high fructose corn syrup increase postprandial triglycerides, LDL-cholesterol, and apolipoprotein-B in young men and women, *Journal of Clinical Endocrinology and Metabolism* 96 (2011): E1596–E1605; J. P. Bantle, Dietary fructose and metabolic syndrome and diabetes, *Journal of Nutrition* 139 (2009): 1263S–1268S.

37. J. C. Cohen, J. D. Horten, and H. H. Hobbs, Human fatty liver disease: Old questions and new insights, *Science* 332 (2011): 1519–1523; M. Krawczyk, L. Bonfrate, and P. Portincasa, Nonalcoholic fatty liver disease, *Best Practice and Research, Clinical Gastroenterology* 24 (2010): 695–708; G. Tarantino, S. Savastano, and A. Colao, Hepatic steatosis, low-grade chronic inflammation and hormone/growth factor/adipokine imbalance, *World Journal of Gastroenterology* 16 (2010): 4773–4783.

38. G. Lattuada, F. Ragogna, and G. Perseghin, Why does NAFLD predict type 2 diabetes? *Current Diabetes Reports* 11 (2011): 167–172; B. W. Smith and L. A. Adams, Nonalcoholic fatty liver disease and diabetes mellitus: pathogenesis and treatment, *Nature Reviews Endocrinology* 7 (2011): 456–465; L. Pacifico and coauthors, Pediatric nonalcoholic fatty liver disease, metabolic syndrome and cardiovascular risk, *World Journal of Gastroenterology* 17 (2011): 3082–3091.

39. K. F. Leavens and M. J. Birnbaum, Insulin signaling to hepatic lipid metabolism in health and disease, *Critical Reviews in Biochemistry and Molecular Biology* 46 (2011): 200–215.

40. Dolan, Potter, and Burdock, Evidence-based review on the effect of normal dietary consumption of fructose on development of hyperlipidemia and obesity in healthy, normal weight individuals.

41. L. C. Dolan, S. M. Potter, and G. A. Burdock, Evidence-based review on the effect of normal dietary consumption of fructose on blood lipids and body weight of overweight and obese individuals, *Critical Reviews in Food Science and Nutrition* 50 (2009): 889–918.

42. J. S. White, Misconceptions about high-fructose corn syrup: Is it uniquely responsible for obesity, reactive dicarbonyl compounds, and advanced glycation endproducts? *Journal of Nutrition* 139 (2009): 1219S–1227S.

43. M. T. Le and coauthors, Effects of high-fructose corn syrup and sucrose on the pharmacokinetics of fructose and acute metabolic and hemodynamic responses in healthy subjects, *Metabolism* 61 (2012): 641–651.

44. D. B. Allison and R. D. Mattes, Nutritively sweetened beverage consumption and obesity: The need for solid evidence on a fluid issue, *Journal of the American Medical Association* 301 (2009): 318–320.

Chapter 5

1. O. Quehenberger and E. A. Dennis, The human plasma lipidome, *New England Journal of Medicine* 365 (2011): 1812–1823; E. Fahy and coauthors, Update of the LIPID MAPS comprehensive classification system for lipids, *Journal of Lipid Research* 50 (2009): S9–S14.

2. N. Ouchi and coauthors, Adipokines in inflammation and metabolic disease, *Nature Reviews, Immunology* 11 (2011): 85–97; Y. Deng and P. E. Scherer, Adipokines as novel biomarkers and regulators of the metabolic syndrome, *Annals of the New York Academy of Sciences* 1212 (2010): E1–E19.

3. D. Schardt, Coconut oil, *Nutrition Action Healthletter,* June 2012, pp. 10–11; E. Cunningham, Is there science to support claims for coconut oil? *Journal of the American Dietetic Association* 111 (2011): 786.

4. P. M. Clifton, Palm oil and LDL cholesterol, *American Journal of Clinical Nutrition* 94 (2011): 1392–1393.

5. V. Remig and coauthors, *Trans* fats in America: A review of their use, consumption, health implications, and regulation, *Journal of the American Dietetic Association* 110 (2010): 585–592.

6. Quehenberger and Dennis, The human plasma lipidome, 2011.

7. D. A. Jenkins and coauthors, Effect of a dietary portfolio of cholesterol-lowering foods given at 2 levels of intensity of dietary advice on serum lipids in hyperlipidemia, *Journal of the American Medical Association* 306 (2011): 831–839; W. S. Harris and coauthors, Omega-6 fatty acids and risk for cardiovascular disease: A Science Advisory from the American Heart Association Nutrition Subcommittee of the Council of Nutrition, Physical Activity, and Metabolism; Council on Cardiovascular Nursing, and Council on Epidemiology and Prevention, *Circulation* 108 (2009): 902–907.

8. Jenkins and coauthors, Effect of a dietary portfolio of cholesterol-lowering foods, 2011; K. Zelman, The great fat debate: A closer look at the controversy—Questioning the validity of age-old dietary guidance, *Journal of the American Dietetic Association* 111 (2011): 655–658; A. Mente and coauthors, A systematic preview of the evidence supporting a causal link between dietary factors and coronary heart disease, *Archives of Internal Medicine* 169 (2009): 659–669.

9. A. H. Lichtenstein, The great fat debate: the importance of message translation, *Journal of the American Dietetic Association* 111 (2011): 667–670.

10. G. A. Bray, Is dietary fat important? *American Journal of Clinical Nutrition* 93 (2011): 481–482.

11. M. Miller and coauthors, Triglycerides and cardiovascular disease: a scientific statement from the American Heart Association, *Circulation* 123 (2011): 2292–2333.

12. U.S. Department of Agriculture and U.S. Department of Health and Human Services, *Dietary Guidelines for Americans 2010,* available at www.dietaryguidelines.gov.

13. P. R. Trumbo and T. Shimakawa, Tolerable upper intake levels for *trans* fat, saturated fat, and cholesterol, *Nutrition Reviews* 69 (2011): 270–278.

14. National Center for Health Statistics, Age-adjusted kilocalorie and macronutrient intake among adults aged ≥20 years, by sex—National Health and Nutrition Examination Survey, United States, 2007–2008, *Morbidity and Mortality Weekly Reports* 60 (2011): 252.

15. G. Kellner-Weibel and M. de la Llera-Moya, Update on HDL receptors and cellular cholesterol transport, *Current Atherosclerosis Reports* 13 (2011): 233–241; A. V. Khera and coauthors, Cholesterol efflux capacity, high-density lipoprotein function, and atherosclerosis, *New England Journal of Medicine* 364 (2011): 127–135.

16. J. Heineckem, HDL and cardiovascular-disease risk: time for a new approach? *New England Journal of Medicine* 364 (2011): 170–171.

17. A. K. Chhihatriwalla and coauthors, Low levels of low-density lipoprotein cholesterol and blood pressure and progression of coronary atherosclerosis, *Journal of the American College of Cardiology* 53 (2009): 1110–1115.

18. Centers for Disease Control and Prevention, Vital signs: Prevalence, treatment, and control of high levels of low-density lipoprotein cholesterol—United States, 1999–2002 and 2005–2008, *Journal of the American Medical Association* 305 (2011): 109–114; L. H. Kuller, The great fat debate: Reducing cholesterol, *Journal of the American Dietetic Association* 111 (2011): 663–664.

19. A. Astrup, The role of reducing intakes of saturated fats in the prevention of cardiovascular disease: Where does the evidence stand in 2010? *American Journal of Clinical Nutrition* 93 (2011): 684–688.

20. A. M. Brownawell and M. C. Falk, Cholesterol: Where science and public health policy intersect, *Nutrition Reviews* 68 (2010): 355–364; M. L. Fernandez and M. Calle, Revisiting dietary cholesterol recommendations: Does the evidence support a limit of 300 mg/d? *Current Atherosclerosis Reports* 12 (2010): 377–383.

21. U.S. Department of Agriculture and U.S. Department of Health and Human Services, *Dietary Guidelines for Americans 2010.*

22. J. M. Ordovas, Genetic influences on blood lipids and cardiovascular disease risk: Tools for primary prevention, *American Journal of Clinical Nutrition* 89 (2009): 1509S–1517S.

23. Fernandez and Calle, Revisiting dietary cholesterol recommendations, 2010.

24. J. D. Spence, D. J. Jenkins, and J. Davignon, Egg yolk consumption and carotid plaque, *Atherosclerosis* (2012), epub ahead of print, doi:10.1016/j.atherosclerosis2012.07.032; M. R. Flock and P. M. Kris-Etherton, *Dietary*

Guidelines for Americans 2010: Implications for cardiovascular disease, 2011.
25. M. D. Carroll and coauthors, Total and high-density lipoprotein cholesterol in adults: National Health and Nutrition Examination Survey, 2009–2010, *NCHS Data Brief* No. 92 (2012), available from www.cdc.gov/nchs/data/databriefs/db92.pdf.
26. P. W. Siri-Tarino and coauthors, Saturated fat, carbohydrate, and cardiovascular disease, *American Journal of Clinical Nutrition* 91 (2010): 502–509; W. S. Harris and coauthors, Omega-6 fatty acids and risk of cardiovascular disease: A science advisory from the American Heart Association Nutrition Subcommittee of the Council on Nutrition, Physical Activity, and Metabolism, 2009.
27. K. Zelman, The great fat debate, 2011.
28. J. P. Crosetti and coauthors, Cholesteryl ester transfer protein polymorphism (TaqIB) associates with risk in postinfarction patients with high c-reactive protein and high-density lipoprotein cholesterol levels, *Arteriosclerosis, Thrombosis, and Vascular Biology* 30 (2010): 1657–1664; K. M. Ali and coauthors, Cardiovascular disease risk reduction by raising HDL cholesterol—current therapies and future opportunities, *British Journal of Pharmacology,* 167 (2012): 1177–1194; AIM-HIGH Investigators, Niacin in patients with low HDL cholesterol level receiving intensive statin therapy, *New England Journal of Medicine* 365 (2011): 2255–2267.
29. F. Voight and coauthors, Plasma HDL cholesterol and risk of myocardial infarction: a mendelian randomization study, *Lancet* 380 (2012): 572–580.
30. K. M. Ali and coauthors, Cardiovascular disease risk reduction by raising HDL cholesterol—current therapies and future opportunities, 2012.
31. D. B. Jump, C. M. Depner, and S. Tripathy, Omega-3 fatty acid supplementation and cardiovascular disease, *Journal of Lipid Research* 53 (2012): 2525–2545.
32. Buczynski, Dumlao, and Dennis, An integrated omics analysis of eicosanoid biology, 2009.
33. Z. Mackhoul and coauthors, Associations of very high intakes of eicosapentaenoic and docosahexaenoic acids with biomarkers of chronic diseases among Yup'ik Eskimos, *American Journal of Clinical Nutrition* 91 (2010): 777–785; J. P. Middaugh, Cardiovascular deaths among Alaskan Natives, 1980–1986, *American Journal of Public Health* 80 (1990): 282–285; J. Dyerberg, Linolenate-derived polyunsaturated fatty acids and prevention of atherosclerosis, *Nutrition Reviews* 44 (1986): 125–134.
34. D. Mozaffarian and J. H Wu, Omega-3 fatty acids and cardiovascular disease; Effects on risk factors, molecular pathways, and clinical events, *Journal of the American College of Cardiology* 58 (2011): 2047–2067; M. R. Flock and P. M. Kris-Etherton, *Dietary Guidelines for Americans 2010*: Implications for cardiovascular disease, *Current Atherosclerosis Reports* 13 (2011): 499–507; R. Wall and coauthors, Fatty acids from fish: The anti-inflammatory potential of long-chain omega-3 fatty acids, *Nutrition Reviews* 68 (2010): 280–289.
35. R. Chowdhury and coauthors, Association between fish consumption, long chain omega 3 fatty acids, and risk of cerebrovascular disease: systematic review and meta-analysis, *British Medical Journal* (2012), epub, doi:10.1136/bmj.e6698; M. L. Chateau-Degat and coauthors, Obesity risks: Towards an emerging Inuit pattern, *International Journal of Circumpolar Health* 70 (2011): 166–177; M. L. Chateau-Degat and coauthors, Hypertension among the Inuit from Nunavik: Should we expect an increase because of obesity, *International Journal of Circumpolar Health* 69 (2010): 361–372.
36. D. Mozzaffarian and J. H. Y. Wu, (n-3) Fatty acids and cardiovascular health: are effects of EPA and DHA shared or complementary? *Journal of Nutrition* 142 (2012): 614S-625S; R. De Caterina, n-3 Fatty acids in cardiovascular disease, *New England Journal of Medicine* 364 (2011): 2439–2450; D. Mosaffarian and J. H. Wu, Omega-3 fatty acids and cardiovascular disease: Effects on risk factors, molecular pathways, and clinical events, *Journal of the American College of Cardiology* 58 (2011): 2047–2067; D. Mozaffarian and coauthors, Circulating long-chain ω-3 fatty acids and incidence of congestive heart failure in older adults: The Cardiovascular Health Study, *Annals of Internal Medicine* 155 (2011): 160–170.
37. J. Madden and coauthors, The impact of common gene variants on the response of biomarkers of cardiovascular disease (CVD) risk to increased fish oil fatty acid intakes, *Annual Review of Nutrition* 31 (2011): 203–234.
38. H. J. Murff and coauthors, Dietary intake of PUFAs and colorectal polyp risk, *American Journal of Clinical Nutrition* 95 (2012): 703–712; T. M. Brasky and coauthors, Specialty supplements and breast cancer risk in the VITamins And Lifestyle (VITAL) Cohort, *Cancer Epidemiology, Biomarkers and Prevention* 19 (2010): 1696–1708.
39. K. M. Szymanski, D. C. Wheeler, and L. A. Mucci, Fish consumption and prostate cancer risk: A review and meta-analysis, *American Journal of Clinical Nutrition* 92 (2010): 1223–1233.
40. T. M. Brasky and coauthors, Serum phospholipid fatty acids and prostate cancer risk: results from the Prostate Cancer Prevention Trial, *American Journal of Epidemiology* 173 (2011): 1429–1439.
41. A. P. Simopoulos, Evolutionary aspects of diet: The omega-6/omega-3 ratio and the brain, *Molecular Neurobiology* 44 (2011): 203–215; A. Lamaziere and coauthors, Differential distribution of DHA-phospholipids in rat brain after feeding: A lipidomic approach, *Prostaglandins, Leukotrienes and Essential Fatty Acids* 84 (2011): 7–11.
42. G. L. Bowman and coauthors, *Neurology* 78 (2012): 241–249; Z. S. Tan and coauthors, Red blood cell ω-3 fatty acid levels and markers of accelerated brain aging, *Neurology* 78 (2012): 658–664; J. Bradbury, Docosahexaenoic acid (DHA): An ancient nutrient for the modern human brain, *Nutrients* 3 (2011): 529–554; N. G. Bazan, M. F. Molina, and W. C. Gordon, Docosahexaenoic acid signalolipidomics in nutrition: Significance in aging, neuroinflammation, macular degeneration, Alzheimer's and other neurodegenerative diseases, *Annual Review of Nutrition* 31 (2011): 321–351; N. D. Riediger and coauthors, A systemic review of the roles of n-3 fatty acids in health and disease, *Journal of the American Dietetic Association* 109 (2009): 668–679.
43. A. Liu and coauthors, Long-chain and very long-chain polyunsaturated fatty acids in ocular aging and age-related macular degeneration, *Journal of Lipid Research* 51 (2010): 3217–3229.
44. R. S. Kuipers and coauthors, Fetal intrauterine whole body linoleic, arachidonic, and docosahexaenoic acid contents and accretion rates, *Prostaglandins, Leukotrienes, and Essential Fatty Acids* 86 (2012): 13–20; M. Guxens and coauthors, Breastfeeding, long-chain polyunsaturated fatty acids in colostrums, and infant mental development, *Pediatrics* 128 (2011): e880–e889; M. B. Imhoff-Kunsch and coauthors, Prenatal docosahexaenoic acid supplementation and infant morbidity: Randomized controlled trial, *Pediatrics* 128 (2011): e505–e515; E. E. Birch and coauthors, The DIAMOND (DHA Intake And Measurement Of Neural Development) Study: A double-masked, randomized controlled clinical trial of the maturation of infant visual acuity as a function of the dietary level of docosahexaenoic acid, *American Journal of Clinical Nutrition* 91 (2010): 848–859; R. K. McNamara and coauthors, Docosahexaenoic acid supplementation increases prefrontal cortex activation during sustained attention in healthy boys: A placebo-controlled, dose-ranging, functional magnetic resonance imaging study, *American Journal of Clinical Nutrition* 91 (2010): 1060–1067; S. E. Carlson, Early determinants of development: A lipid perspective, *American Journal of Clinical Nutrition* 89 (2009): 1523S–1529S.
45. U.S. Department of Agriculture and U.S. Department of Health and Human Services, *Dietary Guidelines for Americans 2010*, available at www.dietaryguidelines.gov.
46. C. Berr and coauthors, Increased selenium intake in elderly high-fish consumers may account for health benefits previously ascribed to omega-3 fatty acids, *Journal of Nutrition, Health and Aging* 13 (2009): 14–18.
47. A. L. Yaktine, eds., *Seafood Choices: Balancing Benefits and Risks* (Washington, D.C.: National Academies Press, 2007), p. 6.
48. U.S. Department of Agriculture and U.S. Department of Health and Human Services, *Dietary Guidelines for Americans 2010*, available at www.dietaryguidelines.gov.
49. U.S. Department of Agriculture and U.S. Department of Health and Human Services, *Dietary Guidelines for Americans 2010*, available at www.dietaryguidelines.gov.
50. P. C. Calder, Mechanisms of action of (n-3) fatty acids, *Journal of Nutrition* 142 (2012): 592S–599S; S. M. Kwak and coauthors, Efficacy of omega-3 fatty acid supplements (eicosapentaenoic acid and docosahexaenoic acid) in the

secondary prevention of cardiovascular disease: a meta-analysis of randomized, double-blind, placebo-controlled trials, *Archives of Internal Medicine* 172 (2012): 686–694; V. A. Andreeva and coauthors, B vitamin and/or ω-3 fatty acid supplementation and cancer, *Archives of Internal Medicine* 172 (2012): 540–547; Flock and Kris-Etherton, *Dietary Guidelines for Americans 2010*: Implications for cardiovascular disease, 2011; Brasky and coauthors, Serum phospholipid fatty acids and prostate cancer risk, 2011.

51. American Heart Association, Fish and omega-3 fatty acids, available at www.heart.org, updated September 2010.

52. E. A. Decker, C. C. Akoh, and R. S. Wilkes, Incorporation of (n-3) fatty acids in foods: challenges and opportunities, *Journal of Nutrition* 142 (2012): 610S–613S; M. I. Burgar and coauthors, MNR of microencapsulated fish oil samples during in vitro digestion, *Food Biophysics* 4 (2009): 32–41.

53. I. Fraeye and coauthors, Dietary enrichment of eggs with omega-3 fatty acids: a review, *Food Research International* 48 (2012): 961–969.

54. R. J. Deckelbaum and C. Torrejon, The omega-3 fatty acid nutritional landscape: health benefits and sources, *Journal of Nutrition* 142 (2012): 587S–591S; P. J. Gillies, W. S. Harris, and P. M. Kris-Etherton, Omega-3 fatty acids in food and pharma: The enabling role of biotechnology, *Current Atherosclerosis Reports* 13 (2011): 457–473.

55. I. A. Brouwer, A. J. Wanders, and M. B. Katan, Effect of animal and industrial *trans* fatty acids on HDL and LDL cholesterol levels in humans—A quantitative review, *PLoS One* 5 (2010): e9434; J. Y. Lee, L. Zhao, and D. H. Hwang, Modulation of pattern recognition receptor-mediated inflammation and risk of chronic diseases by dietary fatty acids, *Nutrition Reviews* 68 (2010): 38–61.

56. U.S. Department of Agriculture and U.S. Department of Health and Human Services, *Dietary Guidelines for Americans 2010*, available at www.dietaryguidelines.gov.

57. S. W. Ing and M. A. Belury, Impact of conjugated linoleic acid on bone physiology: Proposed mechanism involving inhibition of adipogenesis, *Nutrition Reviews* 69 (2011): 123–131; U.S. Department of Agriculture and U.S. Department of Health and Human Services, *Dietary Guidelines for Americans 2010*, available at www.dietaryguidelines.gov.

58. I. Rahkovsky, S. Martinez, and F. Kuchler, New food choices free of *trans* fats better align U.S. diets with health recommendations, April 2012, U.S. Department of Agriculture Economic Research Service, document EIB-95, available at www.ers.usda.gov/publications/eib-economic-information-bulletin/eib95.aspx.

59. W. H. Dietz and K. S. Scanlon, Eliminating the use of partially hydrogenated oil in food production and preparation, *Journal of the American Medical Association* 308 (2012): 143–144.

60. U.S. Department of Agriculture and U.S. Department of Health and Human Services, *Dietary Guidelines for Americans 2010*, available at www.dietaryguidelines.gov.

61. USDA Nutrient Data Laboratory, Release 24, available at www.USDA.gov.

62. U.S. Department of Agriculture and U.S. Department of Health and Human Services, *Dietary Guidelines for Americans 2010*, available at www.dietaryguidelines.gov.

63. U.S. Department of Agriculture and U.S. Department of Health and Human Services, *Dietary Guidelines for Americans 2010*, available at www.dietaryguidelines.gov.

64. S. R. Eussen and coauthors, Dose-dependent cholesterol-lowering effects of phytosterol/phytostanol-enriched margarine in statin users and statin non-users under free-living conditions, *Public Health Nutrition* 14 (2011): 1823–1832.

65. S. A. Doggrell, Lowering LDL cholesterol with margarine containing plant stanol/sterol esters: Is it still relevant in 2011?, *Complementary Therapies in Medicine* 19 (2011): 37–46; O. Weingärter and coauthors, Differential effects on inhibition of cholesterol absorption by plant stanol and plant sterol esters in apoE-/- mice, *Cardiovascular Research* 90 (2011): 484–492.

Consumer's Guide 5

1. A. M. Bernstein and coauthors, Major dietary protein sources and risk of coronary heart disease in women, *Circulation* 122 (2010): 876–883.

2. B. C. Scudder and coauthors, Mercury in fish, bed sediment, and water from streams across the United States, 1998–2005 (2009), U.S. Geological Survey Scientific Investigations Report 2009–5109, available at http://pubs.usgs.gov/sir/2009/5109/.

3. D. R. Laks, Assessment of chronic mercury exposure within the U.S. population, National Health and Nutrition Examination Survey, 1999–2006, *Biometals* 22 (2009): 1103–1114.

4. U.S. Department of Agriculture and U.S. Department of Health and Human Services, *Dietary Guidelines for Americans 2010*, available at www.dietaryguidelines.gov.

5. K. R. Mahaffey and coauthors, Balancing the benefits of n-3 polyunsaturated fatty acids and the risks of methylmercury exposure from fish consumption, *Nutrition Reviews* 69 (2011): 493–508.

6. D. B. Jump, C. M. Depner, and S. Tripathy, Omega-3 fatty acid supplementation and cardiovascular disease, *Journal of Lipid Research* 53 (2012): 2525–2545.

Controversy 5

1. U.S. Department of Agriculture and U.S. Department of Health and Human Services, *Dietary Guidelines for Americans 2010*, available at www.dietaryguidelines.gov; A. Mente and coauthors, A systematic review of the evidence supporting a causal link between dietary factors and coronary heart disease, *Archives of Internal Medicine* 169 (2009): 659–669; F. Sofi, The Mediterranean diet revisited: Evidence of its effectiveness grows, *Current Opinion in Cardiology* 24 (2009): 442–446.

2. *Third Report of the National Cholesterol Education Program (NCEP) Expert Panel on Detection, Evaluation, and Treatment of High Blood Cholesterol in Adults (Adult Treatment Panel III)*, 2002, National Institutes of Health publication no. 02–1205.

3. U.S. Department of Agriculture and U.S. Department of Health and Human Services, *Dietary Guidelines for Americans 2010*, available at www.dietaryguidelines.gov.

4. T. J. Angelopoulos and coauthors, The effect of high-fructose corn syrup consumption on triglycerides and uric acid, *Journal of Nutrition* 139 (2009): 1242S–1245S; K. L. Teff and coauthors, Endocrine and metabolic effects of consuming fructose- and glucose-sweetened beverages with meals in obese men and women: Influence of insulin resistance on plasma triglyceride responses, *Journal of Clinical Endocrinology and Metabolism* 94 (2009): 1562–1569.

5. A. Keys, *Seven Countries: A Multivariate Analysis of Death and Coronary Heart Disease* (Cambridge, Mass.: Harvard University Press, 1980).

6. H. Gardener and coauthors, Mediterranean-style diet and risk of ischemic stroke, myocardial infarction, and vascular death: the Northern Manhattan Study, *American Journal of Clinical Nutrition* 94 (2011): 1458–1464.

7. D. Kromhout and coauthors, The confusion about dietary fatty acids recommendations for CHD prevention, *British Journal of Nutrition* 106 (2011): 627–632.

8. D. J. A. Jenkins and coauthors, Effect of a dietary portfolio of cholesterol-lowering foods given at 2 levels of intensity of dietary advice on serum lipids in hyperlipidemia: A randomized controlled trial, *Journal of the American Medical Association* 306 (2011): 831–839.

9. U.S. Department of Agriculture and U.S. Department of Health and Human Services, *Dietary Guidelines for Americans 2010*, available at www.dietaryguidelines.gov.

10. T. Psaltopoulou and coauthors, Olive oil intake is inversely related to cancer prevalence: a systematic review and a meta-analysis of 13800 patients and 23340 controls in 19 observational studies, *Lipids in Health and Disease* 10 (2011) epub, doi:10.1186/1476–511X-10–127.

11. S. Cicerale, L. J. Lucas, and R. S. Keast, Antimicrobial, antioxidant, and anti-inflammatory phenolic activities in extra virgin olive oil, *Current Opinion in Biotechnology* 23 (2012): 129–135; C. Degirolamo and L. L. Rudel, Dietary monounsaturated fatty acids appear not to provide cardioprotection, *Current Atherosclerosis Reports* 12 (2010): 391–396; D. Raederstorff, Antioxidant activity of olive polyphenols in humans: A review, *International Journal of Vitamin Research* 79 (2009): 152–165.

12. S. Granados-Principal and coauthors, Hydoxytyrosol: From laboratory investigations to future clinical trials, *Nutrition Reviews* 68 (2010): 191–206.

13. L. Lucas, A. Russell, and R. Keast, Molecular mechanisms of inflammation. Anti-inflammatory benefits of virgin olive oil and the phenolic compound oleocanthal, *Current Pharmacological Design* 17 (2011): 754–768; Granados-Principal and coauthors, Hydoxytyrosol: From laboratory investigations to future clinical trials, 2010.

14. Degirolamo and Rudel, Dietary monounsaturated fatty acids appear not to provide cardioprotection, 2010.
15. A. Trichopoulou, C. Bamia, and D. Trichopoulos, Anatomy of health effects of Mediterranean diet: Greek EPIC prospective cohort study, *British Medical Journal* 338 (2009), epub, doi:10.1136/bmj/b2337; Mente and coauthors, A systematic review of the evidence supporting a causal link between dietary factors and coronary heart disease, 2009.
16. Psaltopoulou and coauthors, Olive oil intake is inversely related to cancer prevalence, 2011; G. Buckland and coauthors, Adherence to a Mediterranean diet and risk of gastric adenocarcinoma within the European Prospective Investigation into Cancer and Nutrition (EPIC) cohort study, *American Journal of Clinical Nutrition* 91 (2010): 381–390.
17. N. Scarmeas and coauthors, Mediterranean diet and mild cognitive impairment, *Archives of Neurology* 66 (2009): 216–225.
18. U.S. Department of Agriculture and U.S. Department of Health and Human Services, *Dietary Guidelines for Americans 2010*, available at www.dietaryguidelines.gov; S. Rajaram and coauthors, Walnuts and fatty fish influence different serum lipid fractions in normal to mildly hyperlipidemic individuals: A randomized controlled study, *American Journal of Clinical Nutrition* 89 (2009): 1657S–1663S.
19. P. M. Kris-Etherton and coauthors, The role of tree nuts and peanuts in the prevention of coronary heart disease: Multiple potential mechanisms, *Journal of Nutrition* 138 (2009): 1746S–1751S.
20. J. A. Novotny, S. K. Gebauer, and D. J. Baer, Discrepancy between the Atwater factor predicted and empirically measured energy values of almonds in human diets, *American Journal of Clinical Nutrition* 96 (2012): 296–301.
21. C. E. O'Neil and coauthors, Out-of-hand nut consumption is associated with improved nutrient intake and health risk markers in children and adults: National Health and Nutrition Examination Survey 1999–2004, *Nutrition Research* 32 (2012): 185–194; P. Casa-Agustench and coauthors, Cross-sectional association of nut intake with adiposity in a Mediterranean population, *Nutrition, Metabolism, and Cardiovascular Diseases* 21 (2011): 518–525.
22. Casa-Agustench and coauthors, Cross-sectional association of nut intake with adiposity in a Mediterranean population, 2011.
23. S. L. Tey and coauthors, Nuts improve diet quality compared to other energy-dense snacks while maintaining body weight, *Journal of Nutrition and Metabolism* 2011 (2011), epub, doi:10.1155/2011/357350.
24. M. Bes-Rastrollo and coauthors, Prospective study of nut consumption, long-term weight change, and obesity risk in women, *American Journal of Clinical Nutrition* 89 (2009): 1913–1919.
25. E. Ros, Health benefits of nut consumption, *Nutrients* 2 (2010): 653–682.
26. T. Y. Li and coauthors, Regular consumption of nuts is associated with a lower risk of cardiovascular disease in women with type 2 diabetes, *Journal of Nutrition* 139 (2009): 1333–1338.
27. Ros, Health benefits of nut consumption, 2010.
28. J. A. Vinson and Y. Cai, Nuts, especially walnuts, have both antioxidant quantity and efficacy and exhibit significant potential health benefits, *Food and Function* 3 (2012): 134–140.
29. J. Sabaté and Y. Ang, Nuts and health outcomes: New epidemiologic evidence, *American Journal of Clinical Nutrition* 89 (2009): 1643S–1648S.
30. Vinson and Cai, Nuts, especially walnuts, have both antioxidant quantity and efficacy and exhibit significant potential health benefits, 2012.
31. K. N. Aronis and coauthors, Short-term walnut consumption increases circulating total adiponectin and apolipoprotein A concentrations, but does not affect markers of inflammation or vascular injury in obese humans with the metabolic syndrome: Data from a double-blinded, randomized, placebo-controlled study, *Metabolism* 61 (2012): 577–582; E. Ros, Nuts and novel biomarkers of cardiovascular disease, *American Journal of Clinical Nutrition* 89 (2009):1649S–1656S.
32. D. K. Banel and F. B. Hu, Effects of walnut consumption on blood lipids and other cardiovascular risk factors: a meta-analysis and systematic review, *American Journal of Clinical Nutrition* 90 (2009): 56–63.
33. N. R. Damasceno and coauthors, Crossover study of diets enriched with virgin olive oil, walnuts or almonds: Effects on lipids and other cardiovascular risk markers, *Nutrition, Metabolism, and Cardiovascular Diseases* 21 (2011): S14–S20; O. J. Phung and coauthors, Almonds have a neutral effect on serum lipid profiles: A meta-analysis of randomized trials, *Journal of the American Dietetic Association* 109 (2009): 865–873.
34. Standing Committee on the Scientific Evaluation of Dietary Reference Intakes, Food and Nutrition Board, Institute of Medicine, *Dietary Reference Intakes for Energy, Carbohydrate, Fiber, Fat, Fatty Acids, Cholesterol, Protein, and Amino Acids* (Washington, D.C.: National Academy Press, 2002/2005), p. 835.
35. Standing Committee on the Scientific Evaluation of Dietary Reference Intakes, 2005, pp. 796–797.
36. J. A. Nettleton and coauthors, Dietary patterns and incident cardiovascular disease in the Multi-Ethnic Study of Atherosclerosis, *American Journal of Clinical Nutrition* 90 (2009): 647–654.
37. Jenkins and coauthors, Effect of a dietary portfolio of cholesterol-lowering foods given at 2 levels of intensity of dietary advice on serum lipids in hyperlipidemia: A randomized controlled trial, 2011.

Chapter 6

1. Standing Committee on the Scientific Evaluation of Dietary Reference Intakes, Food and Nutrition Board, Institute of Medicine, *Dietary Reference Intakes for Energy, Carbohydrate, Fiber, Fat, Fatty Acids, Cholesterol, Protein, and Amino Acids* (Washington, D.C.: National Academies Press, 2002/2005), pp. 589–768.
2. J.M. Brown, Managing the acutely ill adult with sickle cell disease, *British Journal of Nursing* 21 (2012): 90–96.
3. G. Wu, Amino acids: Metabolism, functions, and nutrition, *Amino Acids* 37 (2009): 1–17.
4. Position of the American Dietetic Association, Dietitians of Canada, and the American College of Sports Medicine, Nutrition and Athletic Performance, *Journal of the American Dietetic Association* 109 (2009): 509–527.
5. J. A. Gilbert and coauthors, Effect of proteins from different sources on body composition, *Nutrition, Metabolism & Cardiovascular Diseases* 21 (2011): B16–B31; D. K. Layman, Dietary Guidelines should reflect new understandings about adult protein needs, *Nutrition & Metabolism* 6 (2009), epub, doi:10.1186/1743-7075-6-12.
6. Position of the American Dietetic Association, Dietitians of Canada, and the American College of Sports Medicine, Nutrition and Athletic Performance, 2009.
7. G. Kreymann and coauthors, The ratio of energy expenditure to nitrogen loss in diverse patient groups—A systematic review, *Clinical Nutrition* 31 (2012): 168–175.
8. R. R. Wolfe and S. L. Miller, The Recommended Dietary Allowance of protein: A misunderstood concept, *Journal of the American Medical Association* 299 (2008): 2891–2893.
9. S. Tesseraud and coauthors, Role of sulfur amino acids in controlling nutrient metabolism and cell functions: Implications for nutrition, *British Journal of Nutrition* 101 (2009): 1132–1139; G. Wu, Amino acids: Metabolism, functions, and nutrition, *Amino Acids* 37 (2009): 1–17.
10. Position of the American Dietetic Association: Vegetarian Diets, *Journal of the American Dietetic Association* 109 (2009): 1266–1283.
11. Standing Committee on the Scientific Evaluation of Dietary Reference Intakes, *Dietary Reference Intakes for Energy, Carbohydrate, Fiber, Fat, Fatty Acids, Cholesterol, Protein, and Amino Acids*, 2002/2005.
12. J. D. Wright and C. Wang, Trends in intake of energy and macronutrients in adults from 1999–2000 through 2007–2008, NCHS Data Brief number 49, November 2010, available at www.cdc.gov/nchs/data/databriefs/db49.htm.
13. D. Wright and C. Wang, Trends in intake of energy and macronutrients in adults from 1999–2000 through 2007–2008, NCHS Data Brief number 49, November 2010, available at www.cdc.gov/nchs/data/databriefs/db49.htm.
14. U.S. Department of Agriculture, Agricultural Research Service, Beltsville Human Nutrition Research Center, Food Surveys Research Group (Beltsville, MD) and U.S. Department of Health and Human Services, Centers for Disease Control and Prevention, National Center for Health Statistics (Hyattsville, MD), *What We Eat in America*, NHANES 2007–2008, available at www.ars.usda.gov/ba/bhnrc/fsrg, 2010.
15. M. Journel and coauthors, Brain responses to high-protein diets, *Advances in Nutrition* 3

(2012): 322–329; A. D. Blatt and coauthors, Increasing the protein content of meals and its effect on daily energy intake, *Journal of the American Dietetic Association* 111 (2011): 290–294; S. Pombo-Rodrigues and coauthors, The effects of consuming eggs for lunch on satiety and subsequent food intake, *International Journal of Food Sciences and Nutrition* 62 (2011): 593–599; M. S. Westerterp-Plantenga and coauthors, Dietary protein, weight loss, and weight maintenance, *Annual Review of Nutrition* 29 (2009): 21–41.

16. A. Pan and coauthors, Red meat consumption and mortality, *Archives of Internal Medicine* 172 (2012): 555–563.

17. S. C. lrsson and A. Wolk, Red and processed meat consumption and risk of pancreatic cancer: meta-analysis of prospective studies, *British Journal of Cancer* 106 (2012): 603–607; A. M. Bernstein and coauthors, Dietary protein sources and the risk of stroke in men and women, *Stroke* 43 (2012): 637–644.

18. A. M. Fretts and coauthors, Associations of processed meat and unprocessed red meat intake with incident diabetes: The Strong Heart Family Study, *American Journal of Clinical Nutrition* 95 (2012): 752–758; Y. Wang and M. A. Beydoun, Meat consumption is associated with obesity and central obesity among U.S. adults, *International Journal of Obesity* 33 (2009): 621–628.

19. T. Huang and coauthors, Cardiovascular disease mortality and cancer incidence in vegetarians: a meta-analysis and systematic review, *Annals of Nutrition and Metabolism* 60 (2012): 233–240.

20. A. N. Friedman and coauthors, Comparative effects of low-carbohydrate high-protein versus low-fat diets on the kidney, *Clinical Journal of the American Society of Nephrology* 7 (2012): 1103–1111; H. Frank and coauthors, Effect of short-term high-protein compared with normal-protein diets on renal hemodynamics and associated variables in healthy young men, *American Journal of Clinical Nutrition* 90 (2009): 1509–1516.

21. V. Menon and coauthors, Effect of a very low-protein diet on outcomes: Long-term follow-up of the Modification of Diet in Renal Disease (MDRD) Study, *American Journal of Kidney Diseases* 53 (2009): 189–191.

22. J. Calvez and coauthors, Protein intake, calcium balance and health consequences, *European Journal of Clinical Nutrition* 66 (2012): 281–295.

23. M. P. Thorpe and E. M. Evans, Dietary protein and bone health: Harmonizing conflicting theories, *Nutrition Reviews* 9 (2011): 215–230; J. J. Cao, L. K. Johnson, and J. R. Hunt, A diet high in meat protein and potential renal acid load increases fractional calcium absorption and urinary calcium excretion without affecting markers of bone resorption or formation in postmenopausal women, *Journal of Nutrition* 141 (2011): 391–397; J. M. Beasley and coauthors, Is protein intake associated with bone mineral density in young women? *American Journal of Clinical Nutrition* 91 (2010): 1311–1316; A. L. Darling and coauthors, Dietary protein and bone health: A systematic review and meta-analysis, *American Journal of Clinical Nutrition* 90 (2009): 1674–1692; M. S. Westerterp-Plantenga and coauthors, Dietary protein, weight loss, and weight maintenance, *Annual Review of Nutrition* 29 (2009): 21–41.

24. J. J. Cao and F. H. Nielsen, Acid diet (high-meat protein) effects on calcium metabolism and bone health, *Current Opinion in Clinical Nutrition and Metabolic Care* 13 (2010): 698–702.

25. D. D. Alexander and coauthors, Meta-analysis of animal fat or animal protein intake and colorectal cancer, *American Journal of Clinical Nutrition* 89 (2009): 1402–1409.

26. A. Pan and coauthors, Red meat consumption and mortality, *Archives of Internal Medicine* 172 (2012): 555–563; A. T. Chan and E. L. Giovannucci, Primary prevention of colorectal cancer, *Gastroenterology* 138 (2010): 2029–2043; R. Sinha and coauthors, Meat intake and mortality: A prospective study of over half a million people, *Archives of Internal Medicine* 169 (2009): 543–545.

27. A. Sapone and coauthors, Spectrum of gluten-related disorders: consensus on new nomenclature and classification, *BMC Medicine*, February 7, 2012, epub ahead of print, doi:10.1186/1741-7015-10-13.

28. P. Fric, D. Gabrovska, and J. Nevoral, Celiac disease, gluten-free diet, and oats, *Nutrition Reviews* 69 (2011): 107–115.

29. A. D. Sabatino and G. R. Corazza, Non-celiac gluten sensitivity: Sense or sensibility? *Annals of Internal Medicine* 156 (2012): 309–311; M. Pietzak, Celiac disease, wheat allergy, and gluten sensitivity: When gluten free is not a fad, *Journal of Parenteral and Enteral Nutrition* 36 (2012): 68S–75S.

30. W. Marcason, Is there evidence to support the claim that a gluten-free diet should be used for weight loss? *Journal of the American Dietetic Association* 111 (2011): 1786.

31. A. Rubio-Tapia and coauthors, The prevalence of celiac disease in the United States, *American Journal of Gastroenterology*, July 2012, epub ahead of print, doi:10.1038/ajg.2012.219.

Consumer's Guide 6

1. A. D. Blatt and coauthors, Increasing the protein content of meals and its effect on daily energy intake, *Journal of the American Dietetic Association* 111 (2011): 290–294; S. Pombo-Rodrigues and coauthors, The effects of consuming eggs for lunch on satiety and subsequent food intake, *International Journal of Food Sciences and Nutrition* 62 (2011): 593–599.

2. D. K. Layman, Dietary Guidelines should reflect new understandings about adult protein needs, *Nutrition & Metabolism* 6 (2009): 12.

3. P. B. Pencharz, R. Elango, and R. O. Ball, Determination of the Tolerable Upper Intake Level of leucine in adult men, *Journal of Nutrition* 142 (2012): 2220S–2224S; M. Leenders and coauthors, Prolonged leucine supplementation does not augment muscle mass or affect glycemic control in elderly type 2 diabetic men, *Journal of Nutrition* 141 (2011): 1070–1076; S. M. Pasiakos and J. P. McClung, Supplemental dietary leucine and the skeletal muscle anabolic response to essential amino acids, *Nutrition Reviews* 69 (2011): 550–557.

4. M. W. Cashman and S. Brett Sloan, Nutrition and nail disease, *Clinics in Dermatology* 28 (2010): 420–425.

5. G. Wu, Amino acids: Metabolism, functions, and nutrition, *Amino Acids* 37 (2009): 1–17.

6. S. Métayer and coauthors, Mechanisms through which sulfur amino acids control protein metabolism and oxidative status, *Journal of Nutritional Biochemistry* 19 (2008): 207–215.

7. J. M. Prosser and coauthors, Adverse effects associated with arginine alpha-ketoglutarate containing supplements, *Human and Experimental Toxicology* 28 (2009): 259–262.

8. J. I. Boullata and L. M. Hudson, Drug-nutrient interactions: A broad view with implications for practice, *Journal of the Academy of Nutrition and Dietetics* 112 (2012): 506–507.

9. Standing Committee on the Scientific Evaluation of Dietary Reference Intakes, Food and Nutrition Board, Institute of Medicine, *Dietary Reference Intakes for Energy, Carbohydrate, Fiber, Fat, Fatty Acids, Cholesterol, Protein, and Amino Acids* (Washington, D.C.: National Academy Press, 2002/2005), pp. 589–768.

10. Standing Committee on the Scientific Evaluation of Dietary Reference Intakes, Food and Nutrition Board, Institute of Medicine, Dietary Reference Intakes—The Essential Guide to Nutrient Requirements (Washington, D.C.: National Academies Press, 2006), p. 152.

11. D. R. Jacobs, M. D. Gross, and L. C. Tapsell, Food synergy: An operational concept for understanding nutrition, *American Journal of Clinical Nutrition* 89 (2009): 1543S–1548S.

Controversy 6

1. A. Pan and coauthors, Red meat consumption and mortality: Results from 2 prospective cohort studies, *Archives of Internal Medicine* 172 (2012): 555–563; C. T. McEnvoy, N. Temple, and J. V. Woodside, Vegetarian diets, low-meat diets and health: a review, *Public Health Nutrition* 15 (2012): 2287–2294; W. J. Craig, Nutrition concerns and health effects of vegetarian diets, *Nutrition in Clinical Practice* 25 (2010): 613–620.

2. U.S. Department of Agriculture and U.S. Department of Health and Human Services, *Dietary Guidelines for Americans 2010*, available at www.dietaryguidelines.gov; Position of the American Dietetic Association: Vegetarian diets, *Journal of the American Dietetic Association* 109 (2009): 1266–1283.

3. B. Farmer and coauthors, A vegetarian dietary pattern as a nutrient-dense approach to weight management: an analysis of the National Health and Nutrition Examination Survey 1999–2004, *Journal of the American Dietetic Association* 111 (2011): 819–827.

4. Y. Wang and M. A. Beydoun, Meat consumption is associated with obesity and central obesity among U.S. adults, *International Journal of Obesity* 33 (2009): 621–628.

5. Pan and coauthors, Red meat consumption and mortality: Results from 2 prospective cohort studies, 2012; T. C. Su and coauthors, Arte-

rial function of carotid and brachial arteries in postmenopausal vegetarians, *Vascular Health Risk Management* 7 (2011): 517–523.
6. M. Messina, Insights gained from 20 years of soy research, *Journal of Nutrition* 140 (2011): 2289S–2295S; D. J. Jenkins and coauthors, Soy protein reduces serum cholesterol by both intrinsic and food displacement mechanisms, *Journal of Nutrition* 140 (2010): 23025–23115.
7. Jenkins and coauthors, Soy protein reduces serum cholesterol by both intrinsic and food displacement mechanisms, 2010.
8. Jenkins and coauthors, Soy protein reduces serum cholesterol by both intrinsic and food displacement mechanisms, 2010.
9. Messina, Insights gained from 20 years of soy research, 2010.
10. U.S. Department of Agriculture and U.S. Department of Health and Human Services, *Dietary Guidelines for Americans 2010*, available at www.dietaryguidelines.gov.
11. Pan and coauthors, Red meat consumption and mortality: results from 2 prospective cohort studies, 2012; B. Magalhaes and coauthors, Dietary patterns and colorectal cancer: systematic review and meta-analysis, *European Journal of Cancer Prevention* 21 (2012): 15–23; L. M. Ferrucci and coauthors, Meat consumption and the risk of incident distal colon and rectal adenoma, *British Journal of Cancer* 106 (2012): 608–616; E. H. Ruder and coauthors, Adolescent and mid-life diet: Risk of colorectal cancer in the NIH-AARP Diet and Health Study, *American Journal of Clinical Nutrition* 94 (2011): 1607–1619; D. S. M. Chan and coauthors, Red and processed meat and colorectal cancer incidence: Meta-analysis of prospective studies, *PLoS ONE* 6 (2011), epub, doi:10.1371/journal.pone.0020456.
12. R. Takachi and coauthors, Red meat intake may increase the risk of colon cancer in Japanese, a population with relatively low red meat consumption, *Asia Pacific Journal of Clinical Nutrition* 20 (2011): 603–612.
13. T. J. Key and coauthors, Cancer incidence in vegetarians: Results from the European Prospective Investigation into Cancer and Nutrition (EPIC-Oxford), *American Journal of Clinical Nutrition* 89 (2009): 1620S–1626S.
14. T. Huang and coauthors, Cardiovascular disease mortality and cancer incidence in vegetarians: a meta-analysis and systematic review, *Annals of Nutrition and Metabolism* 60 (2012): 233–240; C. T. McEvoy, N. Temple, and J. V. Woodside, Vegetarian diets, low-meat diets and health: a review, *Public Health Nutrition* 15 (2012): 2287–2294.
15. Pan and coauthors, Red meat consumption and mortality: Results from 2 prospective cohort studies, 2012.
16. N. E. Allen and coauthors, Moderate alcohol intake and cancer incidence in women, *Journal of the National Cancer Institute* 101 (2009): 296–305.
17. S. Tonstad and coauthors, Vegetarian diets and incidence of diabetes in the Adventist Health Study-2, *Nutrition, Metabolism, and Cardiovascular Diseases*, 2011, epub ahead of print, doi:10.1016/j.numecd.2011.07.004; B. J. Pettersen and coauthors, Vegetarian diets and blood pressure among white subjects: results from the Adventist Health Study-2 (AHS-2), *Public Health Nutrition* 15 (2012): 1909–1916; P. N. Appleby, N. E. Allen, and T. J. Key, Diet, vegetarianism, and cataract risk, *American Journal of Clinical Nutrition* 93 (2011): 1128–1135; F. S. Crowe and coauthors, Diet and risk of diverticular disease in Oxford cohort of European Prospective Investigation into Cancer and Nutrition (EPIC): Prospective study of British vegetarians and non-vegetarians, *British Medical Journal* 343 (2011), epub, doi:10.1136/bmj.d4131; W. J. Craig, Health effects of vegan diets, *American Journal of Clinical Nutrition* 89 (2009): 1627S–1633S; J. Sabaté and Y. Ang, Nuts and health outcomes: New epidemiologic evidence, *American Journal of Clinical Nutrition* 89 (2009): 1643S–1648S.
18. L. T. Ho-Pham and coauthors, Effect of vegetarian diets on bone mineral density: A Bayesian meta-analysis, *American Journal of Clinical Nutrition* 90 (2009): 943–950.
19. I. Elmadfa and I. Singer, Vitamin B-12 and homocysteine status among vegetarians: a global perspective, *American Journal of Clinical Nutrition* 89 (2009): 1693S–1698S.
20. M. R. Pepper and M. M. Black, B12 in fetal development, *Seminars in Cell and Developmental Biology* 22 (2011): 619–623.
21. N. F. Krebs and coauthors, Meat consumption is associated with less stunting among toddlers in four diverse low-income settings, *Food and Nutrition Bulletin* 32 (2011): 185–191.
22. M. Van Winckel and coauthors, Clinical practice: vegetarian infant and child nutrition, *European Journal of Pediatrics* 170 (2011): 1489–1494.
23. M. Amit, Vegetarian diets in children and adolescents, *Paediatrics & Child Health* 15 (2010): 303–314.
24. Position of the American Dietetic Association: Vegetarian diets, 2009.
25. Position of the American Dietetic Association: Vegetarian diets, 2009.
26. Position of the American Dietetic Association: Vegetarian diets, 2009.
27. Standing Committee on the Scientific Evaluation of Dietary Reference Intakes, Food and Nutrition Board, Institute of Medicine, *Dietary Reference Intakes for Vitamin A, Vitamin K, Arsenic, Boron, Chromium, Copper, Iodine, Iron, Manganese, Molybdenum, Nickel, Silicon, Vanadium, and Zinc* (Washington, D.C.: National Academies Press, 2001), p. 351.
28. R. Tupe and coauthors, Diet patterns of lactovegetarian adolescent girls: Need for devising recipes with high zinc bioavailability, *Nutrition* 26 (2010): 390–398.
29. J. B. Barnett, D. H. Hamer, and S. N. Meydani, Low zinc status: A new risk factor for pneumonia in the elderly? *Nutrition Reviews* 68 (2010): 30–37.
30. S. S. Genennt-Bonsmann and coauthors, Oxalic acid does not influence nonhaem iron absorption in humans: A comparison of kale and spinach meals, *European Journal of Clinical Nutrition* 62 (2008): 336–341.
31. A. M. Leung and coauthors, Iodine status and thyroid function of Boston-area vegetarians and vegans, *Journal of Clinical Endocrinology and Metabolism* 96 (2011): E1303–1307.
32. J. Chan, K. Jaceldo-Siegl, and Gary E. Fraser, Serum 25-hydroxyvitamin D status of vegetarians, partial vegetarians, and nonvegetarians: The Adventist Health Study, *American Journal of Clinical Nutrition* 89 (2009): 1686S–1692S.
33. I. Mangat, Do vegetarians have to eat fish for optimal cardiovascular protection? *American Journal of Clinical Nutrition* 89 (2009): 1597S–1601S; A. A. Welch and coauthors, Dietary intake and status of n-3 polyunsaturated fatty acids in a population of fish-eating and non-fish-eating meat-eaters, vegetarians, and vegans and the precursor-product ratio of α-linolenic acid to long-chain n-3 polyunsaturated fatty acids: Results from the EPIC-Norfolk cohort, *American Journal of Clinical Nutrition* 92 (2010): 1040–1051.
34. Mangat, Do vegetarians have to eat fish for optimal cardiovascular protection?, 2009.
35. W. S. Harris, Stearidonic acid-enhanced soybean oil: a plant-based source of (n-3) fatty acids for foods, *Journal of Nutrition* 142 (2012): 600S–604S; R. J. Deckelbaum and C. Torrejon, The omega-3 fatty acid nutritional landscape: Health benefits and sources, *Journal of Nutrition* 142 (2012): 587S–591S; A. M. Bernstein and coauthors, A meta-analysis shows that docosahexaenoic acid from algal oil reduces serum triglycerides and increases HDL-cholesterol and LDL cholesterol in persons without coronary heart disease, *Journal of Nutrition* 142 (2012): 99–104.

Chapter 7

1. A. Piro and coauthors, Casimir Funk: His discovery of the vitamins and their deficiency disorders, *Annals of Nutrition and Metabolism* 57 (2010): 85–88.
2. E. Reboul and P. Borel, Proteins involved in uptake, intracellular transport and basolateral secretion of fat-soluble vitamins and carotenoids by mammalian enterocytes, *Progress in Lipid Research* 50 (2011): 388–402.
3. J. von Lintig, Metabolism of carotenoids and retinoids related to vision, *Journal of Biological Chemistry* 287 (2012): 1627–1634; G. Tang, Bioconversion of dietary provitamin A carotenoids to vitamin A in humans, *American Journal of Clinical Nutrition* 91 (2010): 1468S–1473S.
4. J. von Lintig, Colors with functions: Elucidating the biochemical and molecular basis of carotenoid metabolism, *Annual Review of Nutrition* 30 (2010): 35–56; N. Noy, Between death and survival: Retinoic acid in regulation of apoptosis, *Annual Review of Nutrition* 30 (2010): 201–217.
5. M. Clagett-Dame and D. Knutson, Vitamin A in reproduction and development, *Nutrients* 3 (2011): 385–428.

6. A. Sommer, Vitamin A deficiency and clinical disease: An historical overview, *Journal of Nutrition* 138 (2008): 1835–1839.
7. J. C. Sherwin and coauthors, Epidemiology of vitamin A deficiency and xerophthalmia in at-risk populations, *Transactions of the Royal Society of Tropical Medicine and Hygiene* 106 (2012): 205–214.
8. E. Mayo-Wilson and coauthors, Vitamin A supplements for preventing mortality, illness, and blindness in children aged under 5: Systematic review and meta-analysis, *British Medical Journal* 343 (2011): d5094; H. Chen and coauthors, Vitamin A for preventing acute lower respiratory tract infections in children up to seven years of age, *Cochrane Database of Systematic Reviews*, 2009, epub, doi: 10.1002/14651858.CD006090.pub2.
9. S. Kato and R. Fujiki, Transcriptional controls by nuclear fat-soluble vitamin receptors through chromatin reorganization, *Bioscience, Biotechnology, and Biochemistry* 75 (2011): 410–413; A. C. Ross and R. Zolfaghari, Cytochrome P450s in the regulation of cellular retinoic acid metabolism, *Annual Review of Nutrition* 31 (2011): 65–87.
10. A. C. Ross, Vitamin A and retinoic acid in T cell-related immunity, *American Journal of Clinical Nutrition* 96 (2012): 1166S–1172S; Y. Yang and coauthors, Effects of vitamin A deficiency on mucosal immunity and response to intestinal infection in rats, *Nutrition* 27 (2011): 227–232.
11. World Health Organization, available at www.who.int/mediacentre/factsheets, f286, updated December 2009.
12. H. Ahmadieh and A. Arabi, Vitamins and bone health: Beyond calcium and vitamin D, *Nutrition Reviews* 69 (2011): 584–598; K. Tanaka and coauthors, Deficiency of vitamin A delays bone healing process in association with reduced BMP2 expression after drill-hole injury in mice, *Bone* 47 (2010): 1006–1012.
13. World Health Organization, *Global prevalence of vitamin A deficiency in populations at risk 1995–2005. WHOGlobal Database on Vitamin A Deficiency* (Geneva: World Health Organization, 2009).
14. CDC, Measles (Rubeola), in *The Yellow Book* (2012), available at www.cdc.gov; C. R. Sudfeld, A. M. Navar, and N. A. Halsey, Effectiveness of measles vaccine and vitamin A treatment, *International Journal of Epidemiology* 39 (2010): i48–i55.
15. Y. A. Shakur and coauthors, A comparison of micronutrient inadequacy and risk of high micronutrient intakes among vitamin and mineral supplement users and nonusers in Canada, *Journal of Nutrition* 142 (2012): 534–540; V. L. Fulgoni III and coauthors, Foods, fortificants, and supplements: Where do Americans get their nutrients? *Journal of Nutrition* 141 (2011): 1847–1854.
16. M. M. G. Ackermans and coauthors, Vitamin A and clefting: Putative biological mechanisms, *Nutrition Reviews* 69 (2011): 613–624.
17. A. Cyrulnik and coauthors, High-dose isotretinoin (Accutane) therapy: Positive results in nodulocystic acne, *Pharmacy and Therapeutics* 36 (2011): 294–296; F. Ghali and coauthors, The changing face of acne therapy, *Cutis* 83 (2009): 4–15.
18. W. Stahl and H. Sies, β-Carotene and other carotenoids in protection from sunlight, *American Journal of Clinical Nutrition* 96 (2012): 1179S–1184S; D. Aune and coauthors, Dietary compared with blood concentration of carotenoids and breast cancer risk: a systematic review and meta-analysis of prospective studies, *American Journal of Clinical Nutrition* 356 (2012): 356–373; K. Ried and P. Fakler, Protective effect of lycopene on serum cholesterol and blood pressure: Meta-analyses of intervention trials, *Maturitas* 68 (2011): 299–310; S. Carpentier, M. Knauss, and M. Suh, Associations between lutein, zeaxanthin, and age-related macular degeneration: An overview, *Critical Reviews in Food Science* 49 (2009): 313–326.
19. L. Ho and coauthors, Reducing the genetic risk of age-related macular degeneration with dietary antioxidants, zinc, and ω-3 fatty acids, *Archives of Ophthalmology* 129 (2011): 758–766; R. L. Roberts, J. Green, and B. Lewis, Lutein and zeaxanthin in eye and skin health, *Clinical Dermatology* 27 (2009): 195–201.
20. J. R. Evans and J. G. Lawrenson, Antioxidant vitamin and mineral supplements for preventing age-related macular degeneration, *Cochrane Database Systematic Reviews* (2012), epub, doi:10.1002/14651858.CD000253.pub3; J. H. Olson, J. C. Erie, and S. J. Bakri, Nutritional supplementation and age-related macular degeneration, *Seminars in Ophthalmology* 26 (2011): 131–136.
21. O. K. Chun and coauthors, Estimation of antioxidant intakes from diet and supplements in U.S. adults, *Journal of Nutrition* 140 (2009): 317–324.
22. R. D. Whitehead and coauthors, You are what you eat: Within-subject increases in fruit and vegetable consumption confer beneficial skin-color changes, *PLoS ONE* 7 (2012), epub, doi:10.1371/journal.pone.0032988.
23. V. Ganji, X. Zhang, and V. Tangpricha, Serum 25-hydroxyvitamin D concentrations and prevalence estimates of hypovitaminosis D in the U.S. population based on assay-adjusted data, *Journal of Nutrition* 142 (2012): 498–507; Committee on Dietary Reference Intakes, *Dietary Reference Intakes for Calcium and Vitamin D* (Washington, D.C.: National Academies Press, 2011), p. 496.
24. Committee on Dietary Reference Intakes, *Dietary Reference Intakes for Calcium and Vitamin D* (Washington, D.C.: National Academies Press, 2011), pp. 468–474.
25. Ganji, Zhang, and Tangpricha, Serum 25-hydroxyvitamin D concentrations and prevalence estimates of hypovitaminosis D in the U.S. population based on assay-adjusted data, 2012; A. C. Looker and coauthors, Vitamin D status: United States, 2001–2006, *NCHS Data Brief* 59 (2011): 1–8.
26. F. Bronner, Recent developments in intestinal calcium absorption, *Nutrition Reviews* 67 (2009): 109–113.
27. Committee on Dietary Reference Intakes, *Dietary Reference Intakes for Calcium and Vitamin D* (Washington, D.C.: National Academies Press, 2011), pp. 3–28.
28. G. J. Fung and coauthors, Vitamin D intake is inversely related to risk of developing metabolic syndrome in African American and white men and women over 20 y: The Coronary Artery Risk Development in Young Adults Study, *American Journal of Clinical Nutrition* 96 (2012): 24–29; I. Laaski, Vitamin D and respiratory infections in adults, *Proceedings of the Nutrition Society* 71 (2012): 90–97; Y. Liss and W. H. Frishman, Vitamin D: A cardioprotective agent? *Cardiology in Review* 20 (2012): 38–44; A. Zittermann and coauthors, Vitamin D deficiency and mortality risk in the general population: a meta-analysis of prospective cohort studies, *American Journal of Clinical Nutrition* 95 (2012): 91–100; C. D. Davis and J. A. Milner, Nutrigenomics, vitamin D and cancer prevention, *Journal of Nutrigenetics and Nutrigenomics* 4 (2011): 1–11; R. Jorde and G. Grimnes, Vitamin D and metabolic health with special reference to the effect of vitamin D on serum lipids, *Progress in Lipid Research* 50 (2011): 303–312; M. Hewison, Vitamin D and innate and adaptive immunity, *Vitamins & Hormones* 86 (2011): 23–62; E. M. Mowry, Vitamin D: Evidence for its role as a prognostic factor in multiple sclerosis, *Journal of Neurological Science* 311 (2011): 19–22; M. H. Hopkins and coauthors, Effects of supplemental vitamin D and calcium on biomarkers of inflammation in colorectal adenoma patients: A randomized, controlled clinical trial, *Cancer Prevention Research* 4 (2011): 1645–1654; K. Luong and L. T. Nguyen, Impact of vitamin D in the treatment of tuberculosis, *American Journal of Medical Sciences* 341 (2011): 493–498; A. E. Millen and coauthors, Vitamin D status and early age-related macular degeneration in postmenopausal women, *Archives of Ophthalmology* 129 (2011): 481–489; N. Parekh, Protective role of vitamin D against age-related macular degeneration: A hypothesis, *Topics in Clinical Nutrition* 25 (2010): 290–301; A. G. Pittas and coauthors, Systematic review: Vitamin D and cardiometabolic outcomes, *Annals of Internal Medicine* 152 (2010): 307–314; A. Zittermann, J. Gummert, and J Börgermann, Vitamin D deficiency and mortality, *Current Opinion in clinical Nutrition and Metabolic Care* 12 (2009): 634–639.
29. U.S. Preventive Services Task Force, Vitamin D and calcium supplementation to prevent cancer and osteoporotic fractures, Draft recommendations, August 2012, available at www.uspreventiveservicestaskforce.org/uspstf12/vitamind/vitdart.htm; M. Chung and coauthors, Vitamin D with or without calcium supplementation for prevention of cancer and fractures: An updated meta-analysis for the U.S. preventive services task force, *Annals of Internal Medicine* 155 (2011): 827–838; C. McGreevy and coauthors, New insights about vitamin D and cardiovascular disease, *Annals of Internal Medicine* 155 (2011): 820–826.

30. C. J. Rosen, Vitamin D insufficiency, *New England Journal of Medicine* 364 (2011): 248–254.
31. J. Kumar and coauthors, Prevalence and associations of 25-hydroxyvitamin D deficiency in US children: NHANES 2001–2004, *Pediatrics* 124 (2009): e362–e370.
32. S. Saintonge, H. Bang, and L. M. Gerber, Implications of a new definition of vitamin D deficiency in a multiracial U.S. adolescent population: The National Health and Nutrition Examination Survey III, *Pediatrics* 123 (2009): 797–803.
33. Committee on Dietary Reference Intakes, *Dietary Reference Intakes for Calcium and Vitamin D* (Washington, D.C.: National Academies Press, 2011), pp. 362–402.
34. K. Ukinc, Severe osteomalacia presenting the multiple vertebral fractures: a case report and review of the literature, *Endocrine* 36 (2009): 30–36.
35. U.S. Preventive Services Task Force, Vitamin D and calcium supplementation to prevent cancer and osteoporotic fractures, Recommendation Statement, 2012, available at www.uspreventiveservicestaskforce.org/uspstf12/vitamind/draftcvitdfig.htm; V. A. Moyer and the U.S. Preventive Services Task Force, Prevention of falls in community-dwelling older adults: U.S. Preventive Services Task Force Recommendation Statement, *Annals of Internal Medicine* 157 (2012): 197–204; P. Lips and coauthors, Once-weekly dose of 8400 IU vitamin D_3 compared with placebo: Effects on neuromuscular function and tolerability in older adults with vitamin D insufficiency, *American Journal of Clinical Nutrition* 91 (2010): 985–991.
36. Committee on Dietary Reference Intakes, *Dietary Reference Intakes for Calcium and Vitamin D* (Washington, D.C.: National Academies Press, 2011), pp. 490–491.
37. Committee on Dietary Reference Intakes, *Dietary Reference Intakes for Calcium and Vitamin D*, pp. 6–18.
38. Committee on Dietary Reference Intakes, *Dietary Reference Intakes for Calcium and Vitamin D*, pp. 125–344.
39. S. Sharma and coauthors, Vitamin D deficiency and disease risk among aboriginal Arctic populations, *Nutrition Reviews* 69 (2011): 468–478; L. M. Hall and coauthors, Vitamin D intake needed to maintain target serum 25-hydroxyvitamin D concentrations in participants with low sun exposure and dark skin pigmentation is substantially higher than current recommendations, *Journal of Nutrition* 140 (2010): 542–550; S. M. Smith and coauthors, Vitamin D supplementation during Antarctic winter, *American Journal of Clinical Nutrition* 89 (2009): 1092–1098.
40. Committee on Dietary Reference Intakes, *Dietary Reference Intakes for Calcium and Vitamin D*, p. 6.
41. Hall and coauthors, Vitamin D intake needed to maintain target serum 25-hydroxyvitamin D concentrations in participants with low sun exposure and dark skin pigmentation is substantially higher than current recommendations, 2010.
42. Committee on Dietary Reference Intakes, *Dietary Reference Intakes for Calcium and Vitamin D*, p. 9.
43. S. R. Koyyalamudi and coauthors, Vitamin D2 Formation and bioavailability from *Agaricus bisporus* button mushrooms treated with ultraviolet irradiation, *Journal of Agricultural and Food Chemistry* 57 (2009): 3351–3355.
44. V. Patel and coauthors, Oral tocotrienols are transported to human tissues and delay the progression of the model for end-stage liver disease score in patients, *Journal of Nutrition* 142 (2012): 513–519; B. B. Aggarwal and coauthors, Tocotrienols, the Vitamin E of the 21st century: its potential against cancer and other chronic diseases, *Biochemical Pharmacology* 80 (2010): 1613–1631; S. R. Wells, Alpha-, gamma- and delta-tocopherols reduce inflammatory angiogenesis in human microvascular endothelial cells, *Journal of Nutritional Biochemistry* 21 (2010): 589–597; C. S. Yang and coauthors, Inhibition of inflammation and carcinogenesis in the lung and colon by tocopherols, *Annals of the New York Academy of Sciences* 1203 (2010): 29–34; J. Ju and coauthors, Cancer-preventive activities of tocopherols and tocotrienols, *Carcinogenesis* 31 (2010): 533–542; J. M. Cook-Mills and C. A. McCary, Isoforms of vitamin E differentially regulate inflammation, *Endocrine, Metabolic & Immune Disorders—Drug Targets* 10 (2010): 348–366; S. R. Wells and coauthors, α-, γ-, and δ-tocopherols reduce inflammatory angiogenesis in human microvascular endothelial cells, *Journal of Nutritional Biochemistry* 21 (2009): 589–597.
45. A. Whaley-Connell, P. A. McCullough, and J. R. Sowers, The role of oxidative stress in the metabolic syndrome, *Reviews in Cardiovascular Medicine* 12 (2011): 21–29; M. Goodman and coauthors, Clinical trials of antioxidants as cancer prevention agents: Past, present, and future, *Free Radical Biology & Medicine* 51 (2011): 1068–1084; A. K. Smolarek and N. Suh, Chemopreventive activity of vitamin E in breast cancer: A focus on γ- and δ-Tocopherol, *Nutrients* 3 (2011): 962–986; L. K. Curtiss, Reversing atherosclerosis? *New England Journal of Medicine* 360 (2009): 1144–1146.
46. M. G. Traber and J. F. Stevens, Vitamins C and E: Beneficial effects from a mechanistic perspective, *Free Radical Biology & Medicine* 51 (2011): 1000–1013.
47. Ahmadieh and Arabi, Vitamins and bone health, 2011; Cook-Mills and McCary, Isoforms of vitamin E differentially regulate inflammation, *Endocrine, Metabolic & Immune Disorders—Drug Targets* 10 (2010): 348–366.
48. M. G. Traber, Vitamin E and K interactions—A 50-year-old problem, *Nutrition Reviews* 66 (2008): 624–629.
49. M. Schürks and coauthors, Effects of vitamin E on stroke subtypes: meta-analysis of randomized controlled trials, *British Medical Journal*, 2010, epub, doi:10.1136/bmj.c5702.
50. G. Bjelakovic and coauthors, Antioxidant supplements for prevention of mortality in healthy participants and patients with various diseases, *Cochrane Database of Systematic Reviews*, 2008, epub, doi:10.1002/14651858.CD007176.
51. S. Hercberg and coauthors, Incidence of cancers, ischemic cardiovascular disease and mortality during 5-year follow-up after stopping antioxidant vitamins and minerals supplements: a postintervention follow-up in the SU.VI.MAX study, *International Journal of Cancer* 127 (2010): 1875–1881; G. Pocobelli and coauthors, Use of supplements of multivitamins, vitamin C and vitamin E in relation to mortality, *American Journal of Epidemiology* 170 (2009): 472–483.
52. USDA Agricultural Research Service, Table 1, Nutrient Intakes from Food: Mean Amounts Consumed per Individual, One Day, 2005–2006 *What We Eat In America: NHANES*, available at www.ars.usda.gov.
53. U.S. Department of Agriculture and U.S. Department of Health and Human Services, *Dietary Guidelines for Americans 2010*, available at www.dietaryguidelines.gov.
54. Y. Lurie and coauthors, Warfarin and vitamin K intake in the era of pharmacogenetics, *British Journal of Clinical Pharmacology* 70 (2010): 164–170.
55. A. R. Miesner and T. S. Sullivan, Elevated international normalized ratio from vitamin K supplement discontinuation, *Annals of Pharmacotherapy* 45 (2011), epub, doi: 10.1345/aph.1P461.
56. R. P. Heaney, Diet, osteoporosis, and fracture prevention: The totality of the evidence, in *Preventive Nutrition* (New York: Humana Press, 2010), pp. 443–469.
57. T. Matsumoto, T. Miyakawa, and D. Yamamoto, Effects of vitamin K on the morphometric and material properties of bone in the tibiae of growing rats, *Metabolism Clinical and Experimental* 61 (2012): 407–414.
58. Ahmadieh and Arabi, Vitamins and bone health, 2011.
59. C. M. Gundberg, J. B. Lian, and S. L. Booth, Vitamin K-dependent carboxylation of osteocalcin: Friend or foe? *Advances in Nutrition* 3 (2012): 149–157; Heaney, Diet, osteoporosis, and fracture prevention: The totality of the evidence, 2010; K. D. Cashman and E. O'Connor, Does high vitamin K_1 intake protect against bone loss in later life? *Nutrition Reviews* 66 (2008): 532–538.
60. M. J. Shearer, X. Fu, and S. L. Booth, Vitamin K nutrition, metabolism, and requirements: current concepts and future research, *Advances in Nutrition* 3 (2012): 182–195; G. Lippi and M. Franchini, Vitamin K in neonates: Facts and myths, *Blood Transfusion* 9 (2011): 4–9.
61. M. N. Riaz, M. Asif, and R. Ali, Stability of vitamins during extrusion, *Critical Reviews in Food Science* 49 (2009): 361–368.
62. M. F. Garcia-Saura and coauthors, Nitroso-redox status and vascular function in marginal and severe ascorbate deficiency, *Antioxidants and Redox Signaling* 17 (2012): 937–950.
63. M. Levine, S. J. Padayatty, and M.G. Espey, Vitamin C: A concentration-function approach

yields pharmacology and therapeutic discoveries, *Advances in Nutrition* 2 (2011): 78–88.
64. A. Mathes and R. Bellanger, Herbs and other dietary supplements: Current regulations and recommendations for use to maintain health in the management of the common cold or other related infectious respiratory illnesses, *Journal of Pharmacy Practice* 23 (2010): 117–127.
65. N. W. Constantini and coauthors, The effect of vitamin C on upper respiratory infections in adolescent swimmers: a randomized trial, *European Journal of Pediatrics* 170 (2011): 59–63.
66. E. Fondell and coauthors, Dietary intake and supplement use of vitamins C and E and upper respiratory tract infection, *Journal of the American College of Nutrition* 30 (2011): 248–258.
67. H. Hemilä, The effect of vitamin C on the common cold, *Journal of Pharmacy Practice* 24 (2011): 241–242.
68. B. Barrett and coauthors, Placebo effects and the common cold: A randomized controlled trial, *Annals of Family Medicine* 9 (2011): 312–322.
69. R. L. Schleicher and coauthors, Serum vitamin C and the prevalence of vitamin C deficiency in the United States: 2003–2004 National Health and Nutrition Examination survey (NHANES), *American Journal of Clinical Nutrition* 90 (2009): 1252–1263.
70. A. D. Holley and coauthors, Scurvy: Historically a plague of the sailor that remains a consideration in the modern intensive care unit, *Internal Medicine Journal* 41 (2011): 283–285; Standing Committee on the Scientific Evaluation of Dietary Reference Intakes, Food and Nutrition Board, Institute of Medicine, *Dietary Reference Intakes*, p. 101.
71. R. L. Schleicher and coauthors, Serum vitamin C and the prevalence of vitamin C deficiency in the United States: 2003–2004 National Health and Nutrition Examination Survey (NHANES), *American Journal of Clinical Nutrition* 90 (2009): 1252–1263.
72. S. S. Jhala and A. S. Hazell, Modeling neurodegenerative disease pathophysiology in thiamine deficiency: Consequences of impaired oxidative metabolism, *Neurochemistry International* 58 (2011): 248–260.
73. Q. Long and coauthors, Riboflavin biosynthetic and regulatory factors as potential novel anti-infective drug targets, *Chemical Biology & Drug Design* 75 (2010): 339–347.
74. L. E. Torheim and coauthors, Women in resource-poor settings are at risk of inadequate intakes of multiple micronutrients, *Journal of Nutrition* 140 (2010): 2051S–2058S.
75. D. MacKay, J. Hathcock, and E. Guarneri, Niacin: Chemical forms, bioavailability, and health effects, *Nutrition Reviews* 70 (2012): 357–366.
76. J. E. Digby, N. Ruparelia, and R. P. Choudhury, Niacin in cardiovascular disease: Recent preclinical and clinical developments, *Arteriosclerosis, Thrombosis, and Vascular Biology* 32 (2012): 582–588.
77. AIM-HIGH Investigators and coauthors, Niacin in patients with low HDL cholesterol levels receiving intensive statin therapy, *New England Journal of Medicine* 365 (2011): 2255–2267.
78. R. P. Giugliano, Niacin at 56 years of age—time for an early retirement? *New England Journal of Medicine* 365 (2011): 2318–2320; A. M. Hill, J. A. Fleming, and P. M. Kris-Etherton, The role of diet and nutritional supplements in preventing and treating cardiovascular disease, *Current Opinion in Cardiology* 24 (2009): 433–441.
79. A. S. Tibbetts and D. R. Appling, Compartmentalization of mammalian folate-mediated one-carbon metabolism, *Annual Review of Nutrition* 30 (2010): 57–81.
80. T. M. Gibson and coauthors, Pre- and post-fortification intake of folate and risk of colorectal cancer in a large prospective cohort study in the United States, *American Journal of Clinical Nutrition* 94 (2011): 1053–1062; B. M. Oaks and coauthors, Folate intake, post-folic acid grain fortification, and pancreatic cancer risk in the Prostate, Lung, Colorectal, and Ovarian Cancer Screening Trial, *American Journal of Clinical Nutrition* 91 (2010): 449–455: World Cancer Research Fund and American Institute for Cancer Research, Summary, *Food, Nutrition, Physical Activity, and the Prevention of Cancer: A Global Perspective* (Washington, D.C.: AICR, 2008), pp. 6–7.
81. S. H. Blanton and coauthors, Folate pathway and nonsyndromic cleft lip and palate, *Birth Defects Research, Part A, Clinical and Molecular Teratology* 91 (2011): 50–60.
82. U.S. Department of Agriculture and U.S. Department of Health and Human Services, *Dietary Guidelines for Americans 2010,* available at www.dietaryguidelines.gov.
83. Centers for Disease Control and Prevention, *Second National Report on Biochemical Indicators of Diet and Nutrition in the U.S. Population 2012,* Executive Summary, available at www.cdc.gov/nutritionreport/.
84. R. L. Bailey and coauthors, Total folate and folic acid intake from foods and dietary supplements in the United States: 2003–2006, *American Journal of Clinical Nutrition* 91 (2010): 231–237; Q. Yang and coauthor, Folic acid source, usual intake, and folate and vitamin B-12 status in US adults: National Health and Nutrition Examination Survey (NHANES) 2003–2006, *American Journal of Clinical Nutrition* 91 (2010): 64–72.
85. J. B. Mason, Folate consumption and cancer risk: a confirmation and some reassurance, but we're not out of the woods quite yet, *American Journal of Clinical Nutrition* 94 (2011): 965–966; M. Ebbing and coauthors, Cancer incidence and mortality after treatment with folic acid and vitamin B_{12}, *Journal of the American Medical Association* 302 (2009): 2119–2126; U. C. Ericson and coauthors, Increased breast cancer risk at high plasma folate concentrations among women with the MTHFR 677T allele, *American Journal of Clinical Nutrition* 90 (2009): 1380–1389.
86. R. L. Bailey and coauthors, Total folate and folic acid intake from foods and dietary supplements in the United States: 2003–2006, *American Journal of Clinical Nutrition* 91 (2010): 231–237; L. H. Allen, How common is vitamin B-12 deficiency? *American Journal of Clinical Nutrition* 89 (2009): 693S–696S.
87. M. A. Caudill, Folate bioavailability: Implications for establishing dietary recommendations and optimizing status, *American Journal of Clinical Nutrition* 91 (2010): 1455S–1460S.
88. G. Varela-Moreiras, M. M. Murphy, and J. M. Scott, Cobalamin, folic acid, and homocysteine, *Nutrition Reviews* 67 (2009): S69–S72.
89. J. G. Walker and coauthors, Oral folic acid and vitamin B-12 supplementation to prevent cognitive decline in community-dwelling older adults with depressive symptoms—The Beyond Ageing Project: A randomized controlled trial, *American Journal of Clinical Nutrition* 95 (2012): 194–203; E. Moore and coauthors, Cognitive impairment and vitamin B12: A review, *International Psychogeriatrics* 24 (2012): 541–556; Y. Minn and coauthors, Sequential involvement of the nervous system in subacute combined degeneration, *Yonsei Medical Journal* 53 (2012): 276–278; L. Feng and coauthors, Vitamin B-12, apolipoprotein E genotype, and cognitive performance in community-living older adults: Evidence of a gene-micronutrient interaction, *American Journal of Clinical Nutrition* 89 (2009): 1263–1268.
90. E. A. Yetley, P. M. Coates, and F. L. Johnson, Overview of a roundtable on NHANES monitoring of biomarkers of folate and vitamin B-12 status: Measurement procedure issues, *American Journal of Clinical Nutrition* 94 (2011): 297S–302S.
91. Moore and coauthors, Cognitive impairment and vitamin B12, 2012.
92. L. Reinstatler and coauthors, Association of biochemical B_{12} deficiency with metformin therapy and vitamin B_{12} supplements, *Diabetes Care* 35 (2012): 327–333.
93. F. G. Bowling, Pyridoxine supply in human development, *Seminars in Cell & Developmental Biology* 22 (2011): 611–618.
94. S. C. Larsson, N. Orsini, and A. Wolk, Vitamin B_6 and risk of colorectal cancer: A meta-analysis of prospective studies, *Journal of the American Medical Association* 303 (2010): 1077–1083; J. Shen and coauthors, Association of vitamin B-6 status with inflammation, oxidative stress, and chronic inflammatory conditions: The Boston Puerto Rican Health Study, *American Journal of Clinical Nutrition* 91 (2010): 337–342; J. H. Page and coauthors, Plasma vitamin B(6) and risk of myocardial infarction in women, *Circulation* 120 (2009): 649–655.
95. S. H. Zeisel and K. A. da Costa, Choline: An essential nutrient for public health, *Nutrition Reviews* 67 (2009): 615–623.
96. D. N. Chester and coauthors, Dietary intakes of choline, Dietary Data Brief No. 9, October 2011, available at www.ars.usda.gov/ba/bhnrc/fsrg.
97. D. R. Jacobs, M. D. Gross, and L. C. Tapsell, Food synergy: An operational concept for understanding nutrition, *American Journal of Clinical Nutrition* 89 (2009): 1543S–1548S;

Position of the American Dietetic Association: Nutrient supplementation, *Journal of the American Dietetic Association* 109 (2009): 2073–2085.

Consumer's Guide 7

1. L. J. Black and coauthors, An updated systematic review and meta-analysis of the efficacy of vitamin D food fortification, *Journal of Nutrition* 142 (2012): 1102–1108.
2. J. I. Boullata, Vitamin D supplementation: A pharmacologic perspective, *Current Opinion in Clinical Nutrition & Metabolic Care* 15 (2012): 677–684.
3. Committee on Dietary Reference Intakes, *Dietary Reference Intakes for Calcium and Vitamin D* (Washington, D.C.: National Academies Press, 2011), pp. 88–89.
4. F. El Ghissassi and coauthors, A review of human carcinogens—Part D: Radiation, *The Lancet* 10 (2009): 751–752; B. A. Gilchrest, Sun exposure and vitamin D sufficiency, *American Journal of Clinical Nutrition* 88 (2008): 570S–577S.
5. E. Linos and coauthors, Sun protective behaviors and vitamin D levels in the US population: NHANES 2003–2006, *Cancer Causes and Control* 23 (2012): 133–140.

Controversy 7

1. J. Gahche and coauthors, Dietary supplement use among U.S. adults has increased since NHANES III (1988–1994), *National Center for Health Statistics: Data Brief* 61 (2011): 1–8; Position of the American Dietetic Association, Nutrient supplementation, *Journal of the American Dietetic Association* 109 (2009): 2073–2085.
2. M. J. Shearer, X. Fu, and S. L. Booth, Vitamin K nutrition, metabolism, and requirements: current concepts and future research, *Advances in Nutrition* 3 (2012): 182–195.
3. J. Visser, D. Labadarios, and R. Blaauw, Micronutrient supplementation for critically ill adults: A systematic review and meta-analysis, *Nutrition* 27 (2011): 745–258.
4. R. L. Bailey and coauthors, Examination of vitamin intakes among US adults by dietary supplement use, *Journal of the Academy of Nutrition and Dietetics* 112 (2012): 657–663; Y. A. Shakur and coauthors, A comparison of micronutrient inadequacy and risk of high micronutrient intakes among vitamin and mineral supplement users and nonusers in Canada, *Journal of Nutrition* 142 (2012): 534–540; R. L. Baily and coauthors, Dietary supplement use is associated with higher intakes of minerals from food sources, *American Journal of Clinical Nutrition* 94 (2011): 1376–1381; R. M. Bliss, Monitoring the populations food and supplement intakes, March 1, 2012, available at www.ars.usda.gov; D. B. McCormick, Vitamin/mineral supplements: of questionable benefit for the general population, *Nutrition Reviews* 68 (2010): 207–213.
5. D. B. McCormick, Vitamin/mineral supplements: Of questionable benefit for the general population, *Nutrition Reviews* 68 (2010): 207–213.
6. A. C. Bronstein and coauthors, 2007 Annual Report of the American Association of Poison Control Centers' national poisoning and exposure database, *Clinical Toxicology* 46 (2008): 927–1057.
7. FDA, Tainted products marketed as dietary supplements, *FDA Consumer Health Information*, December 2010, available at www.fda.gov/consumer.
8. W. R. Mindak and coauthors, Lead in women's and children's vitamins, *Journal of Agricultural and Food Chemistry* 56 (2008): 6892–6896.
9. ConsumberLab.com, Product review: Multivitamin and multimineral supplements review (2011), available at www.consumerlab.com.
10. P. A. Cohen, American roulette—Contaminated dietary supplements, *New England Journal of Medicine* 361 (2009): 1523–1525.
11. D. R. Jacobs, M. D. Gross, and L. C. Tapsell, Food synergy: An operational concept for understanding nutrition, *American Journal of Clinical Nutrition* 89 (2009): 1543S–1548S.
12. Committee on Dietary Reference Intakes, *Dietary Reference Intakes for Calcium and Vitamin D* (Washington, D.C.: National Academies Press, 2011), pp. 124–344.
13. U.S. Preventive Services Task Force, Vitamin D and calcium supplementation to prevent cancer and osteoporotic fractures, Draft recommendations, August 2012, available at www.uspreventiveservicestaskforce.org/uspstf12/vitamind/vitdart.htm.
14. M. E. Martinez and coauthors, Dietary supplements and cancer prevention: balancing potential benefits against proven harms, *Journal of the National Cancer Institute* 104 (2012): 732–739; M. G. O'Doherty and coauthors, Effect of supplementation with B vitamins and antioxidants on levels of asymmetric dimethylarginine (ADMA) and C-reactive protein (CRP): A double-blind, randomized, factorial design, placebo-controlled trial, *European Journal of Nutrition* 49 (2010): 483–492; G. J. Hankey and VITATOPS Trial Study Group, B vitamins in patients with recent transient ischaemic attack or stroke in the VITAmins TO Prevent Stroke (VITATOPS) trial: A randomized, double-blind, parallel, placebo-controlled trial, *The Lancet Neurology* 9 (2010): 855–865; S. Czernichow and coauthors, Effects of long-term antioxidant supplementation and association of serum antioxidant concentrations with risk of metabolic syndrome in adults, *American Journal of Clinical Nutrition* 90 (2009): 329–335; A. M. Hill, J. A. Fleming, and P. M. Kris-Etherton, The role of diet and nutritional supplements in preventing and treating cardiovascular disease, *Current Opinion in Cardiology* 24 (2009): 433–441; M. L. Neuhouser and coauthors, Multivitamin use and risk of cancer and cardiovascular disease in the Women's Health Initiative Cohorts, *Archives of Internal Medicine* 169 (2009): 294–304; G. Pocobelli and coauthors, Use of supplements of multivitamins, vitamin C, and vitamin E in relation to mortality, *American Journal of Epidemiology* 170 (2009): 472–483.
15. M. P. Rayman, Selenium and human health, *Lancet* 379 (2012): 1256–1268; G. Bjelakovic and C. Gluud, Vitamin and mineral supplement use in relation to all-cause mortality in the Iowa Women's Health Study, *Archives of Internal Medicine* 171 (2011): 1633–1634.
16. S. P. Juraschek and coauthors, Effects of vitamin C supplementation on blood pressure: A meta-analysis of randomized controlled trials, *American Journal of Clinical Nutrition* 95 (2012): 1079–1088; S. Rautiainen and coauthors, Vitamin C supplements and the risk of age-related cataract: A population-based prospective cohort study in women, *American Journal of Clinical Nutrition* 91 (2010): 487–493.
17. M. C. Mathew and coauthors, Antioxidant vitamin supplementation for preventing and slowing the progression of age-related cataract, *Chochrane Database Systematic Reviews* 6 (2012), epub, doi: 10.1002/14651858.CD004567.pub2.
18. G. Pocobelli and coauthors, Use of supplements of multivitamins, vitamin C and vitamin E in relation to mortality, *American Journal of Epidemiology* 170 (2009): 472–483.
19. A. M. Hill, J. A. Fleming, and P. M. Kris-Etherton, The role of diet and nutritional supplements in preventing and treating cardiovascular disease, *Current Opinion in Cardiology* 24 (2009): 433–441.
20. E. A. Klein and coauthors, Vitamin E and the risk of prostate cancer: The Selenium and Vitamin E Cancer Prevention Trial (SELECT), *Journal of the American Medical Association* 306 (2011): 1549–1556; S. Hercberg and coauthors, Incidence of cancers, ischemic cardiovascular disease and mortality during 5-year follow-up after stopping antioxidant vitamins and minerals supplements: A postintervention follow-up in the SU.VI.MAX study, *International Journal of Cancer* 127 (2010): 1875–1881.
21. M. C. Morris and C. C. Tangney, A potential design flaw of randomized trials of vitamin supplements, *Journal of the American Medical Association* 305 (2011): 1348–1349; J-M. Zingg, A. Azzi, and M. Meydani, Genetic polymorphisms as determinants for disease-preventive effects of vitamin E, *Nutrition Reviews* 66 (2008): 406–414.
22. M. C. Ledesma and coauthors, Selenium and vitamin E for prostate cancer: post-SELECT (Selenium and Vitamin E Cancer Prevention Trial) status, *Molecular Medicine* 17 (2011): 134–143; D. Farbstein, A. Kozak-Blickstein, and A. P. Levy, Antioxidant vitamins and their use in preventing cardiovascular disease, *Molecules* 15 (2010): 8098–8110.
23. C. S. Yang, N. Suh, and A. T. Kong, Does vitamin E prevent or promote cancer? *Cancer Prevention Research* 5 (2012): 701–705.
24. Y. G. J. van Helden and coauthors, β-Carotene metabolites enhance inflammation-induced oxidative DNA damage in lung epithelial cells, *Free Radical Biology and Medicine* 46 (2009): 299–304; J. A. Satia and coauthors, Long-term use of β-carotene, retinol, lycopene, and lutein supplements and lung cancer risk: Results from the VITamins And Lifestyle

(VITAL) Study, *American Journal of Epidemiology* 169 (2009): 815–828.

Chapter 8

1. E. E. Wilson, Three structural roles for water in bone observed by solid-state NMR, *Biophysical Journal* 90 (2006): 3722–3731.
2. B. M. Popkin, K. E. D'Anci, and I. H. Rosenberg, Water, hydration, and health, *Nutrition Reviews* 68 (2010): 439–458.
3. Standing Committee on the Scientific Evaluation of Dietary Reference Intakes, Food and Nutrition Board, Institute of Medicine, *Dietary Reference Intakes: Water, Potassium, Sodium, Chloride, and Sulfate* (Washington, D.C.: National Academies Press, 2004), pp. 269–423.
4. Popkin, D'Anci, and Rosenberg, Water, hydration, and health, 2010.
5. K. M. Kolasa, C. J. Lacky, and A. C. Grandjean, Hydration and health promotion, *Nutrition Today* 44 (2009): 190–201.
6. Standing Committee on the Scientific Evaluation of Dietary Reference Intakes, Food and Nutrition Board, Institute of Medicine, *Dietary Reference Intakes*, 2004.
7. A. K. Kant, B. I. Graubard, and E. A. Atchison, Intakes of plain water, moisture in foods and beverages, and total water in the adult U.S. population—nutritional, meal pattern, and body weight correlates: National Health and Nutrition Examination Surveys 1999–2006, *American Journal of Clinical Nutrition* 90 (2009): 655–663.
8. United Nations General Assembly, General Assembly adopts resolution recognizing access to clean water, sanitation as human right, 28 July 2010, available at http://www.un.org/News/Press/docs/2010/ga10967.doc.htm.
9. Actions You Can Take to Reduce Lead in Drinking Water, www.epa.gov/ogwdw/lead/lead1.html, updated April 2008.
10. S. D. Richardson and coauthors, Integrated disinfection by-products mixtures research: Comprehensive characterization of water concentrates prepared from chlorinated and ozonated/postchlorinated drinking water, *Journal of Toxicology and Environmental Health* 71 (2008): 1165–1186; L. D. Claxton and coauthors, Integrated disinfection by-products research: Salmonella mutagenicity of water concentrates disinfected by chlorination and ozonation/postchlorination, *Journal of Toxicology and Environmental Health* 71 (2008): 1187–1194.
11. United Nations Environment Program, World water day 2010 highlights solutions and calls for action to improve water quality worldwide, 22 March 2010, available at www.unep.org.
12. P. H. Gleick, U.S. water system needs better enforcement, smart investment to ensure quality, *The Washington Post*, 2010, available at www.washingtonpost.com.
13. U.S. Food and Drug Administration, Bottled water everywhere: Keeping it safe, 2010, available at www.fda.gov/ForConsumers/ConsumerUpdates.
14. Electronic Code of Federal Regulations (January 28, 2011), available at http://ecfr.gpoaccess.gov; Natural Resources Defense Council, *Bottled Water: Pure Drink or Pure Hype?*, available at www.nrdc.org/water/drinking/nbw.asp.
15. U. S. Food and Drug Administration, Bottled Water Regulation, 2009, available at http://www.fda.gov/Food/FoodSafety/Product-SpecificInformation/BottledWaterCarbonatedSoftDrinks/ucm077065.htm.
16. Consumer Recycling Institute, *Bottled Water*, February 2010, available at http://www.container-recycling.org/issues/bottledwater.htm.
17. Committee on Dietary Reference Intakes, *Dietary Reference Intakes for Calcium and Vitamin D* (Washington, D.C.: National Academies Press, 2011), pp. 463–465; A. Moshfegh and coauthors, *What We Eat in America, NHANES 2005–2006: Usual Nutrient Intakes from Food and Water compared to 1997 Dietary Reference Intakes for Vitamin D, Calcium, Phosphorus, and Magnesium* (Beltsville, Md.: USDA, 2009).
18. R. L. Bailey and coauthors, Estimation of total usual calcium and vitamin D intakes in the United States, *Journal of Nutrition* 140 (2010): 817–822.
19. Committee on Dietary Reference Intakes, *Dietary Reference Intakes for Calcium and Vitamin D*, 2011, p. 45.
20. J. Kaluza and coauthors, Dietary calcium and magnesium intake and mortality: A prospective study of men, *American Journal of Epidemiology* 171 (2010): 801–807; I. R. Reid and coauthors, Effects of calcium supplementation on lipids, blood pressure, and body composition in healthy older men: A randomized controlled trial, *American Journal of Clinical Nutrition* 91 (2010): 131–139; V. Centeno and coauthors, Molecular mechanisms triggered by low-calcium diets, *Nutrition Research Reviews* 22 (2009): 163–174.
21. M. Huncharek, J. Muscat, and B. Kupelnick, Colorectal cancer risk and dietary intake of calcium, vitamin D, and dairy products: A meta-analysis of 26,335 cases from 60 observational studies, *Nutrition and Cancer* 61 (2009): 47–69; Y. Park and coauthors, Dairy food, calcium, and risk of cancer in the NIH-AARP Diet and Health Study, *Archives of Internal Medicine* 169 (2009): 391–401.
22. J. A. Gilbert and coauthors, Milk supplementation facilitates appetite control in obese women during weight loss: A randomized, single-blind, placebo-controlled trial, *British Journal of Nutrition* 105 (2011): 133–143; D. R. Shahar and coauthors, Dairy calcium intake, serum vitamin D, and successful weight loss, *American Journal of Clinical Nutrition* 92 (2010): 1017–1022; R. P. Heaney and K. Rafferty, Preponderance of the evidence: An example from the issue of calcium intake and body composition, *Nutrition Reviews* 67 (2009): 32–39.
23. M. J. Soares, W. C. Ping-Delfos, M. H. Ghanbari, Calcium and vitamin D for obesity: A review of randomized controlled trials, *European Journal of Clinical Nutrition* 65 (2011): 994–1004.
24. Committee on Dietary Reference Intakes, *Dietary Reference Intakes for Calcium and Vitamin D*, pp. 2–4.
25. Khanal and Nemere, Regulation of intestinal calcium transport.
26. F. Bronner, Recent developments in intestinal calcium absorption, *Nutrition Reviews* 67 (2009): 109–113.
27. R. C. Schulman, A. J. Weiss, and J. I. Mechanick, Nutrition, bone, and aging: An integrative physiology approach, *Current Osteoporosis Reports* 9 (2011): 184–195.
28. M. P. Thorpe and E. M. Evans, Dietary protein and bone health: Harmonizing conflicting theories, *Nutrition Reviews* 69 (2011): 215–230.
29. S. A. Abrams, Setting Dietary Reference Intakes with the use of bioavailability data: Calcium, *American Journal of Clinical Nutrition* 91 (2010): 1474S–1477S.
30. U.S. Preventive Services Task Force, Vitamin D and calcium supplementation to prevent cancer and osteoporotic fractures, Draft recommendations, August 2012, available at www.uspreventiveservicestaskforce.org/uspstf12/vitamind/vitdart.htm; Committee on Dietary Reference Intakes, *Dietary Reference Intakes for Calcium and Vitamin D*, pp. 410–411.
31. M. S. Razzaque, Phosphate toxicity: New insights into an old problem, *Clinical Science* 120 (2011): 91–97.
32. Moshfegh and coauthors, *What We Eat in America*, 2009.
33. E. Takeda and coauthors, Dietary phosphorus in bone health and quality of life, *Nutrition Reviews* 70 (2012): 311–321.
34. A. M. Romani, Cellular magnesium homeostasis, *Archives of Biochemistry and Biophysics* 512 (2011): 1–23.
35. M. Shechter, Magnesium and cardiovascular system, *Magnesium Research* 23 (2010): 60–72.
36. S. C. Larsson, N. Orsini, and A. Wolk, Dietary magnesium intake and risk of stroke: a meta-analysis of prospective studies, *American Journal of Clinical Nutrition* 95 (2012): 362–366; S. E. Chiuve and coauthors, Plasma and dietary magnesium and risk of sudden cardiac death in women, *American Journal of Clinical Nutrition* 93 (2011): 253–260; F. H. Nielsen, Magnesium, inflammation, and obesity in chronic disease, *Nutrition Reviews* 68 (2010): 333–340.
37. R. K. Rude, F. R. Singer, and H. E. Gruber, Skeletal and hormonal effects of magnesium deficiency, *Journal of the American College of Nutrition* 28 (2009): 131–141.
38. Moshfegh and coauthors, *What We Eat in America*, 2009.
39. J. Otten, J. P. Hellwig, and L. D. Meyers, eds., *Dietary Reference Intakes: The Essential Guide to Nutrient Requirements* (Washington, D.C.: National Academies Press, 2006), p. 340.
40. J. Barsony, Y. Sugimura, and J. G. Verbalis, Osteoclast response to low extracellular sodium and the mechanism of hyponatremia-induced bone loss, *Journal of Biological Chemistry* 286 (2011): 10864–10875.
41. Committee on Dietary Reference Intakes, *Dietary Reference Intakes for Water, Potassium, Sodium, Chloride, and Sulfate* (Washington, D.C.: National Academies Press, 2005), p. 281.

42. Standing Committee on the Scientific Evaluation of Dietary Reference Intakes, Food and Nutrition Board, Institute of Medicine, *Dietary Reference Intakes*, 2006.
43. U.S. Department of Agriculture and U.S. Department of Health and Human Services, *Dietary Guidelines for Americans 2010*, available at www.dietaryguidelines.gov.
44. U.S. Department of Agriculture and U.S. Department of Health and Human Services, *Dietary Guidelines for Americans 2010*; K. Bibbins-Domingo and coauthors, Projected effect of dietary salt reductions on future cardiovascular disease, *New England Journal of Medicine* 362 (2010): 590–599.
45. P. K. Whelton and coauthors, Sodium, blood pressure, and cardiovascular disease: further evidence supporting the American Heart Association sodium reduction recommendations, *Circulation* 126 (2012): 2880–2889; H. Gardener and coauthors, Dietary sodium and risk of stroke in the Northern Manhattan Study, *Stroke* 43 (2012): 1200–1205; R. Takachi and coauthors, Consumption of sodium and salted foods in relation to cancer and cardiovascular disease: The Japan Public Health Center-based Prospective Study, *American Journal of Clinical Nutrition* 91 (2010): 456–464; F. J. He and G. A. MacGregor, A comprehensive review on salt and health and current experience of worldwide salt reduction programmes, *Journal of Human Hypertension* 23 (2009): 363–384.
46. J. P. Forman and coauthors, Association between sodium intake and change in uric acid, urine albumin excretion, and the risk of developing hypertension, *Circulation* 125 (2012): 3108–3116.
47. V. L. Roger and coauthors, Heart disease and stroke statistics—2012 update: A report from the American Heart Association, *Circulation* 125 (2012): e12–e230.
48. Forman and coauthors, Association between sodium intake and change in uric acid, urine albumin excretion, and the risk of developing hypertension, 2012; Standing Committee on the Scientific Evaluation of Dietary Reference Intakes, Food and Nutrition Board, Institute of Medicine, *Dietary Reference Intakes*, 2006.
49. Forman and coauthors, Association between sodium intake and change in uric acid, urine albumin excretion, and the risk of developing hypertension, 2012.
50. Y. Sun and coauthors, Role of the epithelial sodium channel in salt-sensitive hypertension, *Acta Pharmacologica Sinica* 32 (2011): 789–797.
51. J. P. Forman, M. J. Stampfer, and G. C. Curhan, Diet and lifestyle risk factors associated with incident hypertension in women, *Journal of the American Medical Association* 302 (2009): 401–411; E. B. Levitan, A. Wolk, and M. A. Mittleman, Consistency with the DASH Diet and incidence of heart failure, *Archives of Internal Medicine* 169 (2009): 851–857.
52. J. L. Bedford and S. I. Barr, Higher urinary sodium, a proxy for intake, is associated with increased calcium excretion and lower hip bone density in healthy young women with lower calcium intakes, *Nutrients* 3 (2011): 951–961.
53. L. D'Elia and coauthors, Habitual salt intake and risk of gastric cancer: A meta-analysis of prospective studies, *Clinical Nutrition* 31 (2012): 489–498; K. M. Fock and coauthors, Asia-Pacific consensus guidelines on gastric cancer prevention, *Journal of Gastroenterology and Hepatology* 23 (2009): 351–365.
54. M. E. Cogswell and coauthors, Sodium and potassium intakes among US adults: NHANES 2003–2008, *American Journal of Clinical Nutrition* 96 (2012): 647–657; A. Drewnowski, M. Maillot, and C. Rehm, Reducing the sodium-potassium ratio in the US diet: A challenge for public health, *American Journal of Clinical Nutrition* 96 (2012): 439–444.
55. M. Maillot and A. Drewnowski, A conflict between nutritionally adequate diets and meeting the 2010 *Dietary Guidelines* for sodium, *American Journal of Preventive Medicine* 42 (2012): 174–179.
56. Centers for Disease Control and Prevention, Where's the sodium? *VitalSigns*, February 2012, available at www.cdc.gov/VitalSigns/Sodium/index.html.
57. K. Bibbins-Domingo and coauthors, Projected effect of dietary salt reductions on future cardiovascular disease, *New England Journal of Medicine* 362 (2010): 590–599; Institute of Medicine (US) Committee on Strategies to Reduce Sodium Intake, *Strategies to Reduce Sodium Intake in the United States* (Washington, D.C.: National Academies Press, 2010).
58. Centers for Disease Control and Prevention, Vital signs: Food categories contributing the most to sodium consumption—United States, 2007–2008, *Morbidity and Mortality Weekly Report* 61 (2012): 92–98.
59. Q. Yang and coauthors, Sodium and potassium intake and mortality among US adults: Prospective data from the third National Health and Nutrition Examination Survey, *Archives of Internal Medicine* 171 (2011): 1183–1191; M. C. Houston, The importance of potassium in managing hypertension, *Current Hypertension Reports* 13 (2011): 309–317; C. J. Rodriguez and coauthors, Association of sodium and potassium intake with left ventricular mass: Coronary artery risk development in young adults, *Hypertension* 58 (2011): 410–416.
60. U.S. Department of Agriculture and U.S. Department of Health and Human Services, *Dietary Guidelines for Americans 2010*, available at www.dietaryguidelines.gov.
61. Yang and coauthors, Sodium and potassium intake and mortality among US adults, 2011; Standing Committee on the Scientific Evaluation of Dietary Reference Intakes, Food and Nutrition Board, Institute of Medicine, 2004, pp. 5–6.
62. Standing Committee on the Scientific Evaluation of Dietary Reference Intakes, Food and Nutrition Board, Institute of Medicine, 2004, pp. 5–35.
63. M. Andersson, V. Karumbunathan, and M. B. Zimmermann, Global iodine status in 2011 and trends over the past decade, *Journal of Nutrition* 142 (2012): 744–750.
64. M. B. Zimmermann, Iodine deficiency in pregnancy and the effects of maternal iodine supplementation on the offspring: A review, *American Journal of Clinical Nutrition* 89 (2009): 668S–672S.
65. R. C. Gordon and coauthors, Iodine supplementation improves cognition in mildly iodine-deficient children, *American Journal of Clinical Nutrition* 90 (2009): 1264–1271.
66. GAIN-UNICEF Universal Salt Iodization Partnership Program, 2011, available at www.gainhealth.gov/programs/usi.
67. Standing Committee on the Scientific Evaluation of Dietary Reference Intakes, Food and Nutrition Board, Institute of Medicine, *Dietary Reference Intakes for Vitamin A, Vitamin K, Arsenic, Boron, Chromium, Copper, Iodine, Iron, Manganese, Molybdenum, Nickel, Silicon, Vanadium, and Zinc* (Washington, D.C.: National Academies Press, 2001), p. 258.
68. Z. Sang and coauthors, Exploration of the safe upper level of iodine intake in euthyroid Chinese adults: A randomized double-blind trial, *American Journal of Clinical Nutrition* 95 (2012): 367–373.
69. C. G. Perrine and coauthors, Some subgroups of reproductive age women in the United States may be at risk for iodine deficiency, *Journal of Nutrition* 140 (2010): 1489–1494.
70. Centers for Disease Control and Prevention, *Second National Report on Biochemical Indicators of Diet and nutrition in the U.S. Population, 2012, Executive Summary*, p. 9, available at www.cdc.gov/nutritionreport.
71. J. R. Hunt, C. A. Zito and L. K. Johnson, Body iron excretion by healthy men and women, *American Journal of Clinical Nutrition* 89 (2009): 1792–1798.
72. Q. Liu and coauthors, Role of iron deficiency and overload in the pathogenesis of diabetes and diabetic complications, *Current Medicinal Chemistry* 16 (2009): 113–129.
73. G. J. Anderson and F. Wang, Essential but toxic: Controlling the flux of iron in the body, *Clinical and Experimental Pharmacology and Physiology* 39 (2012): 719–724.
74. M. Muñoz, J. A. García-Erce, and A. F. Remacha, Disorders of iron metabolism. Part 1: Molecular basis of iron homeostasis, *Journal of Clinical Pathology* 64 (2011): 281–286; M. D. Knutson, Iron-sensing proteins that regulate hepcidin and enteric iron absorption, *Annual Review of Nutrition* 30 (2010): 149–171.
75. Standing Committee on the Scientific Evaluation of Dietary Reference Intakes, Food and Nutrition Board, Institute of Medicine, *Dietary Reference Intakes for Vitamin A, Vitamin K, Arsenic, Boron, Chromium, Copper, Iodine, Iron, Manganese, Molybdenum, Nickel, Silicon, Vanadium, and Zinc*, pp. 9–14.
76. T. Ganz, Hepcidin and iron regulation: 10 years later, *Blood* 117 (2011): 4425–4433; M. Wessling-Resnick, Iron homeostasis and the inflammatory response, *Annual Review of Nutrition* 30 (2010): 105–122; D. M. Frazer and

G. J. Anderson, Hepcidin compared with prohepcidin: An absorbing story, *American Journal of Clinical Nutrition* 89 (2009): 475–476; M. D. Knutson, Into the matrix: Regulation of the iron regulatory hormone hepcidin by matriptase-2, *Nutrition Reviews* 67 (2009): 284–288; M. F. Young and coauthors, Serum hepcidin is significantly associated with iron absorption from food and supplemental sources in healthy young women, *American Journal of Clinical Nutrition* 89 (2009): 533–538.
77. Ganz, Hepcidin and iron regulation, 2011.
78. K. E. Finberg, Unraveling mechanisms regulating systemic iron homeostasis, *American Society of Hematology Education Program Book* 2011 (2011): 532–537.
79. E. C. Theil, Iron homeostasis and nutritional iron deficiency, *Journal of Nutrition* 141 (2011): 724S–728S.
80. L. E. Murray-Kolb, Iron status and neuropsychological consequences in women of reproductive age: What do we know and where are we headed? *Journal of Nutrition* 141 (2011): 747S–755S; K. Kordas, Iron, lead, and children's behavior and cognition, *Annual Review of Nutrition* 30 (2010): 123–148.
81. P. Vaucher and coauthors, Effect of iron supplementation on fatigue in nonanemic menstruating women with low ferritin: a randomized controlled trial, *Canadian Medical Association Journal* 184 (2012): 1247–1254; B. Lozoff, Early iron deficiency has brain and behavior effects consistent with dopaminergic dysfunction, *Journal of Nutrition* 141 (2011): 740S–746S.
82. L. M. De-Regil and coauthors, Intermittent iron supplementation for improving nutrition and development in children under 12 years of age (Review), *Cochrane Library* 12 (2011), epub, doi:10.1002/14651858.CD009085.pub2.
83. J. P. McClung and coauthors, Randomized, double-blind, placebo-controlled trial of iron supplementation in female soldiers during military training: Effects on iron status, physical performance, and mood, *American Journal of Clinical Nutrition* 90 (2009): 124–131.
84. S. L. Young, Pica in pregnancy: New ideas about an old condition, *Annual Review of Nutrition* 30 (2010): 403–422.
85. H. Njiru, U. Elchalal, and O. Paltiel, Geophagy during pregnancy in Africa: A literature review, *Obstetrics and Gynecological Survey* 66 (2011): 452–459.
86. S. R. Lynch, Why nutritional iron deficiency persists as a worldwide problem, *Journal of Nutrition* 141 (2011): 763S–768S.
87. L. M. Tussing-Humphreys and coauthors, Excess adiposity, inflammation, and iron-deficiency in female adolescents, *Journal of the American Dietetic Association* 109 (2009): 297–302.
88. A. C. Cepeda-Lopez and coauthors, Sharply higher rates of iron deficiency in obese Mexican women and children are predicted by obesity-related inflammation rather than by differences in dietary iron intake, *American Journal of Clinical Nutrition* 93 (2011): 975–983; L. Tussing-Humphreys and coauthors, Rethinking iron regulation and assessment in iron deficiency, anemia of chronic disease, and obesity: introducing hepcidin, *Journal of the Academy of Nutrition and Dietetics* 112 (2012): 391–400.
89. N. Milman, Anemia: Still a major health problem in many parts of the world, *Annals of Hematology* 90 (2011): 369–377.
90. Cepeda-Lopez and coauthors, Sharply higher rates of iron deficiency in obese Mexican women and children are predicted by obesity-related inflammation rather than by differences in dietary iron intake, 2011.
91. S. R. Lynch, Why nutritional iron deficiency persists as a worldwide problem, *Journal of Nutrition* 141 (2011): 763S–768S.
92. P. Brissot and coauthors, Molecular diagnosis of genetic iron-overload disorders, *Expert Review of Molecular Diagnostics* 10 (2010): 755–763; S. Lekawanvijit and N. Chattipakorn, Iron overload thalassemic cardiomyopathy: Iron status assessment and mechanisms of mechanical and electrical disturbance due to iron toxicity, *Canadian Journal of Cardiology* 25 (2009): 213–218.
93. Ganz, Hepcidin and iron regulation, 2011; C. Camaschella and E. Poggiali, Inherited disorders of iron metabolism, *Current Opinion in Pediatrics* 23 (2011): 14–20; P. Brissot and coauthors, Molecular diagnosis of genetic iron-overload disorders, *Expert Review of Molecular Diagnostics* 10 (2010): 755–763.
94. G. A. Ramm and R. G. Ruddell, Iron homeostasis, hepatocellular injury, and fibrogenesis in hemochromatosis: The role of inflammation in a noninflammatory liver disease, *Seminars in Liver Disease* 30 (2010): 271–287; S. Lekawanvijit and N. Chattipakorn, Iron overload thalassemic cardiomyopathy: Iron status assessment and mechanisms of mechanical and electrical disturbance due to iron toxicity, *Canadian Journal of Cardiology* 25 (2009): 213–218.
95. B. J. Cherayil, Iron and immunity: immunological consequences of iron deficiency and overload, *Archivum Immunologiae et Therapiae Experimentalis (Warsz)* 58 (2010): 407–415.
96. A. C. Bronstein and coauthors, 2009 Annual Report of the American Association of Poison Control Centers' National Poison Data System (NPDS): 27th Annual Report, *Clinical Toxicology* 28 (2010): 979–1178.
97. J. C. King, Zinc: An essential but elusive nutrient, *American Journal of Clinical Nutrition* 94 (2011): 679S–684S.
98. M. A. Ali and R. Schulz, Activation of MMP-2 as a key event in oxidative stress injury to the heart, *Frontiers in Bioscience* 14 (2009): 669–716; Y. Song and coauthors, Zinc deficiency affects DNA damage, oxidative stress, antioxidant defenses, and DNA repair in rats, *Journal of Nutrition* 139 (2009): 1626–1631.
99. S. D. Gower-Winter and C. W. Levenson, Zinc in the central nervous system: From molecules to behavior, *Biofactors* 38 (2012): 186–193; S. G. Bell and B. L. Vallee, The metallothionein/thionein system: An oxidoreductive metabolic zinc link, *ChemBioChem* 10 (2009): 55–62.
100. H. Haase and L. Rink, Functional significance of zinc-related signaling pathways in immune cells, *Annual Review of Nutrition* 29 (2009): 133–152.
101. C. L. Fischer Walker, M. Ezzati, and R. E. Black, Global and regional child mortality and burden of disease attributable to zinc deficiency, *European Journal of Clinical Nutrition* 63 (2009): 591–597.
102. J. B. Barnett, D. H. Hamer, and S. N. Meydani, Low zinc status: A new risk factor for pneumonia in the elderly? *Nutrition Reviews* 68 (2010): 30–37.
103. King, Zinc, 2011; U. Ramakrishnan, P. Nguyen, and R. Martorell, Effects of micronutrients on growth of children under 5 y of age: Meta-analyses of single and multiple nutrient interventions, *American Journal of Clinical Nutrition* 89 (2009): 191–201.
104. M. Science and coauthors, Zinc for the treatment of the common cold: a systematic review and meta-analysis of randomized controlled trials, *Canadian Medical Association* 184 (2012): E551–E561; FDA, Warnings on three Zicam intranasal zinc products, *For Consumers*, June 2009, available at www.fda.gov/forconsumers/ConsumerUpdates/ucm166931.htm.
105. W. J. Craig, Health effects of vegan diets, *American Journal of Clinical Nutrition* 89 (2009): 1627S–1633S; J. Sabaté and Y. Ang, Nuts and health outcomes: New epidemiologic evidence, *American Journal of Clinical Nutrition* 89 (2009): 1643S–1648S.
106. J. C. McCann and B. N. Ames, Adaptive dysfunction of selenoproteins from the perspective of the triage theory: Why modest selenium deficiency may increase risk of diseases of aging, *FASEB Journal* 25 (2011): 1793–1814.
107. F. P. Bellinger and coauthors, Regulation and function of selenoproteins in human disease, *Biochemical Journal* 422 (2009): 11–22.
108. D. L. St. Germain and coauthors, Minireview: Defining the roles of the iodothyronine deiodinases: Current concepts and challenges, *Endocrinology* 150 (2009): 1097–1107.
109. S. J. Fairweather-Tait and coauthors, Selenium in human health and disease, *Antioxidants and Redox Signaling* 14 (2011): 1337–1383.
110. R. Hurst and coauthors, Selenium and prostate cancer: Systematic review and meta-analysis, *American Journal of Clinical Nutrition* 96 (2012): 111–122; M. P. Rayman, Selenium and human health, *Lancet* 379 (2012): 1256–1268.
111. J. P. Richie and coauthors, Association of selenium status and blood glutathione concentrations in blacks and whites, *Nutrition and Cancer* 63 (2011): 367–375; B. K. Dunn and coauthors, A nutrient approach to prostate cancer prevention: The Selenium and Vitamin E Cancer Prevention Trial (SELECT), *Nutrition and Cancer* 62 (2010): 896–918; J. Brozmanová and coauthors, Selenium: A double-edged sword for defense and offence in cancer, *Archives of Toxicology* 84 (2010): 919–938; Position of the American Dietetic Association: Nutrient supplementation, *Journal of the American Dietetic Association* 109 (2009): 2073–2085; B. K. Dunn, A. Ryan, and L. G. Ford, Selenium and vitamin E cancer prevention trial: A nutrient approach to

prostate cancer prevention, *Recent Results in Cancer Research* 181 (2009): 183–193; N. Facompre and K. El-Bayoumy, Potential stages for prostate cancer prevention with selenium: Implications for cancer survivors, *Cancer Research* 69 (2009): 2699–2703.
112. C. Lei and coauthors, Is selenium deficiency really the cause of Keshan disease? *Environmental Geochemistry and Health* 33 (2011): 183–188; J. Yang and coauthors, Selenium level surveillance for the year 2007 of Keshan disease in endemic areas and analysis on surveillance results between 2003 and 2007, *Biological Trace Element Research* 138 (2010): 53–59.
113. S. Sun, Chronic exposure to cereal mycotoxin likely citreoviridin may be a trigger for Keshan disease mainly through oxidative stress mechanism, *Medical Hypotheses* 74 (2010): 841–842.
114. B. M. Aldosary and coauthors, Case series of selenium toxicity from a nutritional supplement, *Clinical Toxicology* 50 (2012): 57–64; Fairweather-Tait and coauthors, Selenium in human health and disease, 2011; J. K. MacFarquhar and coauthors, Acute selenium toxicity associated with a dietary supplement, *Archives of Internal Medicine* 170 (2010): 256–261.
115. Rayman, Selenium and human health, 2012; J. W. Finley, Selenium accumulation in plant foods, *Nutrition Reviews* 63 (2005): 196–202.
116. Position of the Academy of Nutrition and Dietetics, The impact of fluoride on health, *Journal of the Academy of Nutrition and Dietetics* 112 (2012): 1443–1453; J. Aaseth, G. Boivin, and O. Andersen, Osteoporosis and trace elements—An overview, *Journal of Trace Elements in Medicine and Biology* 26 (2012): 149–152.
117. Position of the Academy of Nutrition and Dietetics, The impact of fluoride on health, *Journal of the Academy of Nutrition and Dietetics*, 2012.
118. M. A. Buzalaf and coauthors, Mechanisms of action of fluoride for caries control, *Monographs in Oral Science* 22 (2011): 97–114.
119. E. D. Beltrán-Aguilar, L. Barker, and B. A. Dye, Prevalence and severity of dental fluorosis in the United States, *NCHS Data Brief* 53 (2010): 1–8.
120. S. B. Gopalakrishnan and G. Viswanathan, Assessment of fluoride-induced changes on physicochemical and structural properties of bone and the impact of calcium on its control in rabbits, *Journal of Bone Mineral Metabolism* 30 (2012): 154–163; E. T. Everett, Fluoride's effects on the formation of teeth and bones, and the influence of genetics, *Journal of Dental Research* 90 (2011): 552–560; D. Chachra and coauthors, The long-term effects of water fluoridation on the human skeleton, *Journal of Dental Research* 89 (2010): 1219–1223.
121. J. Aaseth, G. Boivin, and O. Andersen, Osteoporosis and trace elements—An overview, *Journal of Trace Elements in Medicine and Biology* 26 (2012): 149–152; S. Joshi and coauthors, Skeletal fluorosis due to excessive tea and toothpaste consumption, *Osteoporosis International* 22 (2011): 2557–2560; K. Izuora and coauthors, Skeletal fluorosis from brewed tea, *Journal of Clinical Endocrinology and Metabolism* 96 (2011): 2318–2324.
122. National Cancer Institute, Fact Sheet: Fluoridated water, February 21, 2012, available at www.cancer.gov.
123. A. Zhitkovich, Chromium in drinking water: sources, metabolism, and cancer risks, *Chemical Research in Toxicology* 24 (2011): 1617–1629.
124. Y. Hua and coauthors, Molecular mechanisms of chromium in alleviating insulin resistance, *Journal of Nutritional Biochemistry* 23 (2012): 313–319; Z. Q. Wang and W. T. Cefalu, Current concepts about chromium supplementation in type 2 diabetes and insulin resistance, *Current Diabetes Research* 10 (2010): 145–151.
125. D. L. de Romaña and coauthors, Risks and benefits of copper in light of new insights of copper homeostasis, *Journal of Trace Elements in Medicine and Biology* 25 (2011): 3–13.
126. Standing Committee on the Scientific Evaluation of Dietary Reference Intakes, Food and Nutrition Board, Institute of Medicine, *Dietary Reference Intakes for Vitamin A, Vitamin K, Arsenic, Boron, Chromium, Copper, Iodine, Iron, Manganese, Molybdenum, Nickel, Silicon, Vanadium, and Zinc*, p. 245.
127. Committee on Dietary Reference Intakes, *Dietary Reference Intakes for Calcium and Vitamin D*, 2011, pp. 463–464.
128. U.S. Department of Agriculture and U.S. Department of Health and Human Services, *Dietary Guidelines for Americans 2010*, available at www.dietaryguidelines.gov.

Consumer's Guide 8

1. M. D. Sorensen and coauthors, Impact of nutritional factors on incident kidney stone formation: A report from the WHI OS, *Journal of Urology* 187 (2012): 1645–1649; I. Tack, Effects of water consumption on kidney function and excretion, *Nutrition Today* 45 (2010): S37–S40.
2. D. F. Tate and coauthors, Replacing caloric beverages with water or diet beverages for weight loss in adults: Main results of the Choose Healthy Options Consciously Everyday (CHOICE) randomized clinical trial, *American Journal of Clinical Nutrition* 95 (2012): 555–563.
3. R. P. LaComb and coauthors, *Beverage Choices of U.S. Adults: What We Eat in America, NHANES 2007–2008*, Data Brief No. 6, August 2011, available at http://ars.usda.gov.
4. LaComb and coauthors, *Beverage Choices of U.S. Adults: What We Eat in America, NHANES 2007–2008*, 2011.
5. U.S. Department of Health and Human Services—Centers for Disease Control and Prevention, Rethink your drink, 2011, available at www.cdc.gov/healthyweight/healthy_eating/drinks.html.

Controversy 8

1. National Osteoporosis Foundation, available at www.nof.org, accessed July 2011; R. Nuti and coauthors, Bone fragility in men: Where are we? *Journal of Endocrinological Investigation* 33 (2010): 33–38.
2. A. Leboime and coauthors, Osteoporosis and mortality, *Joint Bone Spine* 77 (2010): S107–S112; C. A. Brauer and coauthors, Incidence and mortality of hip fractures in the United States, *Journal of the American Medical Association* 302 (2009): 1573–1579.
3. M. Agrawal and coauthors, Bone, inflammation, and inflammatory bowel disease, *Current Osteoporosis Reports* 9 (2011): 251–257; J. J. Cao, Effects of obesity on bone metabolism, *Journal of Orthopaedic Surgery and Research*, 2011, epub, doi:10.1186/1749-799X-6-30.
4. B. D. Mitchell and L. M. Yerges-Armstrong, The genetics of bone loss: Challenges and prospects, *Journal of Clinical Endocrinology and Metabolism* 96 (2011): 1258–1268.
5. H. F. Zheng, T. D. Spector, and J. B. Richards, Insights into the genetics of osteoporosis from recent genome-wide association studies, *Expert Reviews in Molecular Medicine* 13 (2011): e28.
6. C. Holroyd and coauthors, Epigenetic influences in the developmental origins of osteoporosis, *Osteoporosis International* 23 (2012): 401–410.
7. J. A. Cauley, Defining ethnic and racial differences in osteoporosis and fragility fractures, *Clinical Orthopaedics and Related Research* 469 (2011): 1891–1899.
8. J. C. Lo, S. A. Burnett-Bowie, and J. S. Finkelstein, Bone and the perimenopause, *Obstetrics & Gynecology Clinics of North America* 38 (2011): 503–517; B. Frenkel and coauthors, Regulation of adult bone turnover by sex steroids, *Journal of Cellular Physiology* 224 (2010): 305–310.
9. O. Svenjme and coauthors, Early menopause and risk of osteoporosis, fracture and mortality: A 34-year prospective observational study in 390 women, *BJOG An International Journal of Obstetrics and Gynaecology* 119 (2012): 810–816.
10. E. Gielen and coauthors, Osteoporosis in men, *Best Practice and Research: Clinical Endocrinology and Metabolism* 25 (2011): 321–335; N. Ducharme, Male osteoporosis, *Clinics in Geriatric Medicine* 26 (2010): 301–309; S. Khosla, Update in male osteoporosis, *Journal of Clinical Endocrinology and Metabolism* 95 (2010): 3–10.
11. F. Callewaert, S. Boonen, and D. Vanderschueren, Sex steroids and the male skeleton: a tale of two hormones, *Trends in Enfocrinology and Metabolism* 21 (2010): 89–95.
12. F. Oury, A crosstalk between bone and gonads, *Annals of the New York Academy of Sciences* 1260 (2012): 1–7.
13. J. J. Cao, Effects of obesity on bone metabolism, *Journal of Orthopaedic Surgery and Research* 6 (2011), epub, doi:10.1186/1749-799X-6-30; Y. Sheu and J. A. Cauley, The role of bone marrow and visceral fat on bone metabolism, *Current Osteoporosis Reports* 9 (2011): 67–75.
14. K. F. Janz and coauthors, Early physical activity provides sustained bone health benefits later in childhood, *Medicine and Science in Sports and Exercise* 42 (2010): 1072–1078; A. Guadalupe-Grau and coauthors, Exercise and

bone mass in adults, *Sports Medicine* 39 (2009): 439–468.
15. N. K. LeBrasseur and coauthors, Skeletal muscle mass is associated with bone geometry and microstructure and serum IGFBP-2 levels in adult women and men, *Journal of Bone and Mineral Research* (2012), epub ahead of print, doi:10.1002/jbmr.1666.
16. U.S. Department of Health and Human Services, *2008 Physical Activity Guidelines for Americans*, available at www.health.gov/paguidelines/default.aspx.
17. V. A. Moyer and the U.S. Preventive Services Task Force, Prevention of falls in community-dwelling older adults: U.S. Preventive Services Task Force Recommendation Statement, *Annals of Internal Medicine* 157 (2012): 197–204; DIPART (vitamin D Individual Patient Analysis of Randomized Trials) Group, Patient level pooled analysis of 68,500 patients from seven major vitamin D fracture trials in US and Europe, *British Medical Journal*, (2010), epub, doi:10.1136/bmj.b5463; H. A. Bischoff-Ferrari and coauthors, Fall prevention with supplemental and active forms of vitamin D: A meta-analysis of randomized controlled trials, *British Medical Journal* 339 (2009), epub, doi:10.1136/bmj.b362.
18. C. Yan, N. G. Avadhani, and J. Iqbal, The effects of smoke carcinogens on bone, *Current Osteoporosis Reports* 9 (2011): 202–209.
19. D. B. Maurel and coauthors, Alcohol and bone: Review of dose effects and mechanisms, *Osteoporosis International* 23 (2012): 1–16.
20. D. J. McLernon and coauthors, Do lifestyle choices explain the effect of alcohol on bone mineral density in women around menopause? *American Journal of Clinical Nutrition* 95 (2012): 1261–1269.
21. M. P. Thorpe and E. M. Evans, Dietary protein and bone health: Harmonizing conflicting theories, *Nutrition Reviews* 9 (2011): 215–230.
22. R. C. Schulman, A. J. Weiss, and J. I. Mechanick, Nutrition, bone, and aging: An integrative physiology approach, *Current Osteoporosis Reports* 9 (2011): 184–195.
23. J. E. Kerstetter, Dietary protein bone: A new approach to an old question, *American Journal of Clinical Nutrition* 90 (2009): 1451–1452.
24. J. Calvez and coauthors, Protein intake, calcium balance and health consequences, *European Journal of Clinical Nutrition* 66 (2012): 281–295; Schulman, Weiss, and Mechanick, Nutrition, bone, and aging; Thorpe and Evans, Dietary protein and bone health.
25. H. Ahmadieh and A. Arabi, Vitamins and bone health: beyond calcium and vitamin D, *Nutrition Reviews* 69 (2011): 584–598; R. P. Heaney, Diet, osteoporosis, and fracture prevention: The totality of the evidence, in *Preventive Nutrition* (New York: Humana Press, 2010), pp. 443–469; F. Gaucheron, Milk and dairy products: A unique micronutrient combination, *Journal of American College of Nutrition* 30 (2011): 400S–409S; A. J. Taylor and coauthors, Clinical and demographic factors associated with fractures among older Americans, *Osteoporosis International* 22 (2011): 1263–1274.
26. L. T. Ho-Pham and coauthors, Effect of vegetarian diets on bone mineral density: A Bayesian meta-analysis, *American Journal of Clinical Nutrition* 90 (2009): 943–950.
27. D. L. Alekel and coauthors, The Soy Isoflavones for Reducing Bone Loss (SIRBL) Study: a 3-y randomized controlled trial in postmenopausal women, *American Journal of Clinical Nutrition* 91 (2010): 218–230; A. M. Kenny and coauthors, Soy proteins and isoflavones affect bone mineral density in older women: A randomized controlled trial, *American Journal of Clinical Nutrition* 90 (2009): 234–242.
28. F. J. He and G. A. MacGregor, A comprehensive review on salt and health and current experience of worldwide salt reduction programmes, *Journal of Human Hypertension* 23 (2009): 363–384.
29. C. A. Nowson, A. Patchett, and N. Wattanapenpaiboon, The effects of a low-sodium base-producing diet including red meat compared with a high-carbohydrate, low-fat diet on bone turnover markers in women aged 45–75 years, *British Journal of Nutrition* 102 (2009): 1161–1170.
30. W. Pan and coauthors, The epithelial sodium/proton exchanger, NHE3, is necessary for renal and intestinal calcium (re)absorption, *American Journal of Physiology* 302 (2012): F943–F956.
31. Y. Zhou and coauthors, Caffeine inhibits the viability and osteogenic differentiation of rat bone marrow-derived mesenchymal stromal cells, *British Journal of Pharmacology* 161 (2010): 1542–1552.
32. E. Takeda and coauthors, Dietary phosphorus in bone health and quality of life, *Nutrition Reviews* 70 (2012): 311–321.
33. N. Emaus and coauthors, Vitamin K2 supplementation does not influence bone loss in early menopausal women: a randomized double-blind placebo-controlled trial, *Osteoporosis International* 21 (2010): 1731–1740; J-P. Bonjour and coauthors, Minerals and vitamins in bone health: The potential value of dietary enhancement, *British Journal of Nutrition* 101 (2009): 1581–1596.
34. R. K. Rude, F. R. Singer, and H. E. Gruber, Skeletal and hormonal effects of magnesium deficiency, *Journal of the American College of Nutrition* 28 (2009): 131–141.
35. J. E. Manson and coauthors, The VITamin D and OmegA-3 Trial (VITAL): Rationale and design of a large randomized controlled trial of vitamin D and marine omega-3 fatty acid supplements for the primary prevention of cancer and cardiovascular disease, *Contemporary Clinical Trials* 33 (2012): 159–171; E. K. Farina and coauthors, Protective effects of fish intake and interactive effects of long-chain polyunsaturated fatty acid intakes on hip bone mineral density in older adults: the Framingham Osteoporosis Study, *American Journal of Clinical Nutrition* 93 (2011): 1142–1151; E. K. Farina and coauthors, Dietary intakes of arachidonic acid and alpha-linolenic acid are associated with reduced risk of hip fracture in older adults, *Journal of Nutrition* 141 (2011): 1146–1153.
36. L. Langsetmo and coauthors, Dietary patterns and incident low-trauma fractures in postmenopausal women and men aged ≥50 y: A population-based cohort study, *American Journal of Clinical Nutrition* 93 (2011): 192–199.
37. R. Lorente-Ramos and coauthors, Dual-energy x-ray absorptiometry in the diagnosis of osteoporosis: A practical guide, *American Journal of Roentgenology* 196 (2011): 897–904.
38. T. D. Rachner, S. Khosla, and L. C.Hofbauer, Osteoporosis: Now and the future, *Lancet* 377 (2011): 1276–1287.
39. J. J. Body and coauthors, Extraskeletal benefits and risks of calcium, vitamin D and anti-osteoporosis medications, *Osteoporosis International* 23 (2012): S1–S23.
40. J. Aaseth, G. Boivin, and O. Andersen, Osteoporosis and trace elements—An overview, *Journal of Trace Elements in Medicine and Biology* 26 (2012): 149–152.
41. Committee on Dietary Reference Intakes, *Dietary Reference Intakes for Calcium and Vitamin D* (Washington, D.C.: National Academies Press, 2011), p. 463.
42. J. M. Quesada Gómez and coauthors, Calcium citrate and vitamin D in the treatment of osteoporosis, *Clinical Drug Investigation* 31 (2011): 285–298.
43. E. J. Samelson and coauthors, Calcium intake is not associated with increased coronary artery calcification: the Framingham Study, *American Journal of Clinical Nutrition* 96 (2012): 1274–1280; L. Kuanrong and coauthors, Associations of dietary calcium intake and calcium supplementation with myocardial infarction and stroke risk and overall cardiovascular mortality in the Heidelberg cohort of the European Prospective Investigation in to Cancer and nutrition study (EPIC-Heidelberg), *Heart* 98 (2012): 920–925; M. J. Bolland and coauthors, Calcium supplements with or without vitamin D and risk of cardiovascular events: Reanalysis of the Women's Health Initiative limited access dataset and meta-analysis, *British Medical Journal* 342 (2011): d2040.
44. U.S. Preventive Services Task Force, Vitamin D and calcium supplementation to prevent cancer and osteoporotic fractures, Draft recommendations, August 2012, available at www.uspreventiveservicestaskforce.org/uspstf12/vitamind/vitdart.htm.
45. Committee on Dietary Reference Intakes, *Dietary Reference Intakes for Calcium and Vitamin D*, 2011, p. 480.
46. E. Warensjö and coauthors, Dietary calcium intake an risk of fracture and osteoporosis: prospective longitudinal cohort study, *British Medical Journal* 342 (2011), epub, doi: 10.1136/bmj.d1473.
47. Committee on Dietary Reference Intakes, *Dietary Reference Intakes for Calcium and Vitamin D*, p. 45.
48. Committee on Dietary Reference Intakes, *Dietary Reference Intakes for Calcium and Vitamin D*, p. 38.

Chapter 9

1. G. Whitlock and coauthors, Body-mass index and cause-specific mortality in 900,000 adults: Collaborative analyses of 57 prospective studies, *Lancet* 373 (2009): 1083–1096.
2. N. R. Reyes and coauthors. Similarities and differences between weight loss maintainers and regainers: A qualitative analysis, *Journal of the Academy of Nutrition and Dietetics* 112 (2012): 499–505.
3. K. M. Flegal and coauthors, Prevalence and trends in obesity among US adults, 1999–2008, *Journal of the American Medical Association* 303 (2010): 235–241.
4. C. M. Apovian, The causes, prevalence, and treatment of obesity revisited in 2009: What have we learned so far? *American Journal of Clinical Nutrition* 91 (2010): 277S–279S.
5. C. L. Ogden and M. D. Carroll, Prevalence of obesity in the United States, 2009–2010, *NCHS Data Brief*, January 2012, available at www.cdc.gov/nchs/data/databriefs/db82.htm.
6. E. A. Finkelstein and coauthors, Obesity and severe obesity forecasts through 2030, *American Journal of Preventive Medicine* 42 (2012): 563–570; K. M. Flegal and coauthors, Prevalence and trends in obesity among U.S. adults, 1999–2008, *Journal of the American Medical Association* 303 (2010): 235–241.
7. Centers for Disease Control and Prevention, *Data and Statistics*, Obesity rates among all children in the United States 2011, available at http://www.cdc.gov/obesity/childhood/data.html.
8. B. M. Popkin, L. S. Adair, and S. W. Ng, Global nutrition transition and the pandemic of obesity in developing countries, *Nutrition Reviews* 70 (2012): 3–21; B. A. Swinburn and coauthors, The global obesity pandemic: Shaped by global drivers and local environments, *Lancet* 378 (2011): 804–814; B. M. Popkin, Recent dynamics suggest selected countries catching up to U.S. obesity, *American Journal of Clinical Nutrition* 91 (2010): 284S–288S.
9. C. D. Fryar and C. L. Odgen, Prevalence of underweight among adults: United States, 2003–2006, *NCHS Health E-Stats*, May 21, 2009, available at www.cdc.gov/nchs/data/hestat/underweight_children.htm.
10. W. Zheng and coauthors, Association between body-mass index and risk of death in more than 1 million Asians, *New England Journal of Medicine* 364 (2011): 719–729.
11. O. Bouillanne and coauthors, Fat mass protects hospitalized elderly persons against morbidity and mortality, *American Journal of Clinical Nutrition* 90 (2009): 505–510.
12. R. Gupta and coauthors, The effect of low body mass index on outcome in critically ill surgical patients, *Nutrition in Clinical Practice* 26 (2011): 593–597.
13. C. J. Lavie, R. V. Milani, and H. O. Ventura, Obesity and cardiovascular disease: Risk factor, paradox, and impact of weight loss, *Journal of the American College of Cardiology* 53 (2009): 1925–1932.
14. D. Lee and coauthors, Changes in fitness and fatness on the development of cardiovascular disease risk factors, *Journal of the American College of Cardiology* 59 (2012): 665–672; M. Patlak and S. J. Nass, *The Role of Obesity in Cancer Survival and Recurrence: Workshop Summary* (Bethesda: National Academies Press, 2012), pp. 4–46; Y. Wang and coauthors, Relationship between body adiposity measures and risk of primary knee and hip replacement for osteoarthritis: A prospective cohort study, *Arthritis Research and Therapy* 11 (2009), epub, doi:10.1186/ar2636.
15. E. A. Finkelstein and coauthors, Annual medical spending attributable to obesity: Payer- and service-specific estimates, *Health Affairs* 28 (2009): w822–w831.
16. E. A. Finkelstein and coauthors, The costs of obesity in the workplace, *Journal of Occupational and Environmental Medicine* 52 (2010): 971–976.
17. A. B. de Gonzolez and coauthors, Body-mass index and mortality among 1.46 million white adults, *New England Journal of Medicine* 363 (2010): 2211–2219.
18. H. Jia and E. I. Lubetkin, Trends in quality-adjusted life-years lost contributed by smoking and obesity: Does the burden of obesity overweight the burden of smoking? *American Journal of Preventive Medicine* 38 (2010): 138–144.
19. B. G. Nordestgaard and coauthors, The effect of elevated body mass index on ischemic heart disease risk: Causal estimates from a mendelian randomisation approach, *PLoS Medicine* 9 (2012) epub, doi:10.1371/journal.pmed.1001212; S. S. Dhaliwal and T. A. Welborn, Central obesity and multivariable cardiovascular risk as assessed by the Framingham prediction scores, *American Journal of Cardiology* 103 (2009): 1403–1407.
20. J. C. Cohen, J. D. Horton, and H. H. Hobbs, Human fatty liver disease old questions and new insights, *Science* 332 (2011): 1519–1523; M. Krawczyk, L. Bonfrate, and P. Portincasa, Nonalcoholic fatty liver disease, *Best Practice and Research, Clinical Gastroenterology* 24 (2010): 695–708; G. Tarantino, S. Savastano, and A. Colao, Hepatic steatosis, low-grade chronic inflammation and hormone/growth factor/adipokine imbalance, *World Journal of Gastroenterology* 16 (2010): 4773–4783.
21. M. L. Biggs and coauthors, Association between adiposity in midlife and older age and risk of diabetes in older adults, *Journal of the American Medical Association* 303 (2010): 2504–2512.
22. N. Ouchi and coauthors, Adipokines in inflammation and metabolic disease, *Nature Reviews Immunology* 11 (2011): 85–97; C. Stryjecki and D. M. Mutch, Fatty acid-gene interactions, adipokines and obesity, *European Journal of Clinical Nutrition* 65 (2011): 285–297.
23. F. P. de Heredia and coauthors, Obesity, inflammation and the immune system, *Proceedings of the Nutrition Society* 71 (2012): 332–338; W. V. Brown and coauthors, Obesity: Why be concerned? *American Journal of Medicine* 122 (2009): S4–S11; J. Korner, S. C. Woods, and K. A. Woodworth, Regulation of energy homeostasis and health consequences in obesity, *American Journal of Medicine* 122 (2009): S12–S18.
24. de Heredia and coauthors, Obesity, inflammation and the immune system, 2012.
25. I. Imayama and coauthors, Effects of a caloric restriction weight loss diet and exercise on inflammatory biomarkers in overweight/obese postmenopausal women: A randomized controlled trial, *Cancer Research* 72 (2012): 2314–2326.
26. E. M. McCarthy and M. E. Rinella, The role of diet and nutrient composition in nonalcoholic fatty liver disease, *Journal of the Academy of Nutrition and Dietetics* 112 (2012): 401–409; N. Kumashiro and coauthors, Cellular mechanism of insulin resistance in nonalcoholic fatty liver disease, *PNAS* 108 (2011): 16382–16385.
27. S. Havas, L. J. Aronne, and K. A. Woodworth, The obesity epidemic: Strategies in reducing cardiometabolic risk, *American Journal of Medicine* 122 (2009): S1–S3.
28. E. J. Jacobs and coauthors, Waist circumference and all-cause mortality in a large US cohort, *Archives of Internal Medicine* 170 (2010): 1293–1301; J. P. Reis and coauthors, Overall obesity and abdominal adiposity as predictors of mortality in U.S. white and black adults, *Annals of Epidemiology* 19 (2009): 134–142.
29. Korner, Woods, and Woodworth, Regulation of energy homeostasis and health consequences in obesity, 2009; Brown and coauthors, Obesity: Why be concerned?
30. D. Wormser and coauthors, Separate and combined associations of body-mass index and abdominal adiposity with cardiovascular disease: Collaborative analysis of 58 prospective studies, *The Lancet* 377 (2011): 1085–1095; K. M. Flegal and B. I. Graubard, Estimates of excess deaths associated with body mass index and other anthropometric variables, *American Journal of Clinical Nutrition* 89 (2009): 1213–1219.
31. P. Singh and coauthors, Effects of weight gain and weight loss on regional fat distribution, *American Journal of Clinical Nutrition* 96 (2012): 229–233.
32. Sayon-Orea and coauthors, Type of alcoholic beverage and incidence of overweight/obesity in a Mediterranean cohort: the SUN project, *Journal of Nutrition* 27 (2011): 802–808; M.Schütze and coauthors, Beer consumption and the "beer belly": Scientific basis or common belief? *European Journal of Clinical Nutrition* 63 (2009): 1143–1149; E. A. Molenaar and coauthors, Association of lifestyle factors with abdominal subcutaneous and visceral adiposity: the Framingham Heart Study, *Diabetes Care* 32 (2009): 505–510.
33. M. Hamer and E. Stamatakis, Metabolically healthy obesity and risk of all-cause and cardiovascular disease mortality, *Journal of Clinical Endocrinology and Metabolism* 97 (2012): 2482–2488; D. Lee and coauthors, Long-term effects of changes in cardiorespiratory fitness and body mass index on all-cause and cardiovascular disease mortality in men, *Circulation* 124 (2011): 2483–2490; D. E. Larson-Meyer and coauthors, Caloric restriction with or without exercise: The fitness versus fatness debate, *Medicine and Science in Sports and Exercise* 42 (2010): 152–159.

34. K. E. Giel and coauthors, Weight bias in work settings—A qualitative review, *Obesity Facts* 3 (2010): 33–40; M. S. Argugete, J. L. Edman, and A. Yates, Romantic interest in obese college students, *Eating Behaviors* 10 (2009): 143–145.
35. K. D. Hall and coauthors, Energy balance and its components: Implications for body weight regulation, *American Journal of Clinical Nutrition* 95 (2012): 989–994.
36. S. B. Hymsfield and coauthors, Energy content of weight loss: Kinetic features during voluntary caloric restriction, *Metabolism* 61 (2012): 937–943; Hall and coauthors, Energy balance and its components: Implications for body weight regulation, 2012; K. D. Hall and coauthors, Quantification of the effect of energy imbalance on bodyweight, *Lancet* 378 (2011): 826–837.
37. Standing Committee on the Scientific Evaluation of Dietary Reference Intakes, Food and Nutrition Board, Institute of Medicine, *Dietary Reference Intakes for Energy, Carbohydrate, Fiber, Fat, Fatty Acids, Cholesterol, Protein, and Amino Acids* (Washington, D.C.: National Academies Press, 2002/2005), p. 107.
38. G. Whitlock and coauthors, Prospective Studies Collaboration: Body-mass index and cause-specific mortality in 900,000 adults: collaborative analyses of 57 prospective studies, *Lancet* 373 (2009): 1083–1096.
39. N. R. Shah and E. R. Braverman, Measuring adiposity in patients: the utility of body mass index (BMI), percent body fat, and leptin, *PLoS One* 7 (2012), epub, doi:10.1371/journal.pone.0033308.
40. A. G. Dulloo and coauthors, Body composition phenotypes in pathways to obesity and the metabolic syndrome, *International Journal of Obesity* 34 (2010): S4–S17.
41. M. Heo and coauthors, Percentage of body fat cutoffs by sex, age, and race-ethnicity in the US adult population from NHANES 1999–2004, *American Journal of Clinical Nutrition* 95 (2012): 594–602; H. R. Hull and coauthors, Fat-free mass index: changes and race/ethnic differences in adulthood, *International Journal of Obesity (London)* 35 (2011): 121–127; N. Farah and coauthors, Comparison in maternal body composition between Caucasian Irish and Indian women, *Journal of Obstetrics and Gynecology* 31 (2011): 483–485; A. Liu and coauthors, Ethnic differences in the relationship between body mass index and percentage of body fat among Asian children from different backgrounds, *British Journal of Nutrition* 106 (2011): 1390–1397.
42. Shah and Braverman, Measuring adiposity in patients: the utility of body mass index (BMI), percent body fat, and leptin, 2012.
43. Dulloo and coauthors, Body composition phenotypes in pathways to obesity and the metabolic syndrome, 2010.
44. Position of the American Dietetic Association: Weight management, *Journal of the American Dietetic Association* 109 (2009): 330–346.
45. M. Heo and coauthors, Percentage of body fat cutoffs by sex, age, and race-ethnicity in the US adult population from NHANES 1999–2004, *American Journal of Clinical Nutrition* 95 (2012): 594–602.
46. X. Shan and G. S. Yeo, Central leptin and ghrelin signaling: comparing and contrasting their mechanisms of action in the brain, *Reviews in Endocrine and Metabolic Disorders* 12 (2011): 197–209.
47. S. C. Woods and D. S. Ramsay, Food intake, metabolism and homeostasis, *Physiology and Behavior* 104 (2011): 4–7.
48. C. Delporte, Recent advances in potential clinical application of ghrelin in obesity, *Journal of Obesity*, February 2012, epub, doi: 10.1155/2012/535624; T. R. Castañeda and coauthors, Ghrelin in the regulation of body weight and metabolism, *Frontiers in Neuroendocrinology* 31 (2010): 44–60.
49. M. P. St-Onge and coauthors, Sleep restriction leads to increased activation of brain regions sensitive to food stimuli, *American Journal of Clinical Nutrition* 95 (2012): 818–824; P. Lyytikäinen and coauthors, Sleep problems and major weight gain: A follow-up study, *International Journal of Obesity* 35 (2011): 109–114; L. Brondel and coauthors, Acute partial sleep deprivation increases food intake in healthy men, *American Journal of Clinical Nutrition* 91 (2010): 1550–1559; A. V. Nedeltcheva and coauthors, Sleep curtailment is accompanied by increased intake of calories from snacks, *American Journal of Clinical Nutrition* 89 (2009): 126–133.
50. J. A. Parker and J. R. Bloom, Hypothalamic neuropeptides and the regulation of appetite, *Neuropharmacology* (2012), epub, doi:org/10.1016/j.neuropharm.2012.02.004.
51. E. Egeciouglu and coauthors, Hedonic and incentive signals for body weight control, *Reviews in Endocrine and Metabolic Disorders* 12 (2011): 141–151.
52. N. Chaudhari and S. D. Roper, The cell biology of taste, *Journal of Cell Biology* 190 (2010): 285–296; K. Kurihara, Glutamate: From discovery as a food flavor to role as a basic taste (umami), *American Journal of Clinical Nutrition* 90 (2009): 719S–722S.
53. B. McFerran and coauthors, I'll have what she's having: Effects of social influence and body type on the food choices of others, *Journal of Consumer Research* 36 (2010): 915–929.
54. F. Bellisle and coauthors, Sweetness, satiation, and satiety, *Journal of Nutrition* 142 (2012): 1149S–1154S.
55. N. Zijlstra and coauthors, Effect of bite size and oral processing time of a semisolid food on satiation, *American Journal of Clinical Nutrition* 90 (2009): 269–275.
56. Y. Xu and Q. Tong, Expanding neurotransmitters in the hypothalamic neurocircuitry for energy balance regulation, *Protein and Cell* 2 (2011): 800–813.
57. J. M. Friedman, Leptin at 14 y of age: An ongoing story, *American Journal of Clinical Nutrition* 89 (2009): 973S–979S.
58. S. Nicholaidis, Metabolic and humoral mechanisms of feeding and genesis of the ATP/ADP/AMP concept, *Physiology and Behavior* 104 (2011): 8–14; E. Egecioglu and coauthors, Hedonic and incentive signals for body weight control, 2011.
59. G. Paz-Filho and coauthors, Changes in insulin sensitivity during leptin replacement therapy in leptin deficient patients, *American Journal of Physiology, Endocrinology, and Metabolism* 296 (2009): E1401–1408.
60. S. Nicholaidis, Metabolic and humoral mechanisms of feeding and genesis of the ATP/ADP/AMP concept, 2011.
61. M. Journel and coauthors, Brain responses to high-protein diets, *Advances in Nurition* 3 (2012): 322–329; J. M. Beasley and coauthors, Associations between macronutrient intake and self-reported appetite and fasting levels of appetite hormones: Results from the Optimal Macronutrient Intake Trial to Prevent Heart Disease, *American Journal of Epidemiology* 169 (2009): 893–900.
62. A. M. Johnstone, Safety and efficacy of high-protein diets for weight loss, *Proceedings of the Nutrition Society* 71 (2012): 339–349; J. A. Gilbert and coauthors, Milk supplementation facilitates appetite control in obese women during weight loss: A randomized, single-blind, placebo-controlled trial, *British Journal of Nutrition* 105 (2011): 133–143; E. R. Dove and coauthors, Skim milk compared with a fruit drink acutely reduces appetite and intake in overweight men and women, *American Journal of Clinical Nutrition* 90 (2009): 70–75.
63. M. S. Westerterp-Plantenga and coauthors, Dietary protein, weight loss, and weight maintenance, in *Annual Review of Nutrition* 29 (2009): 21–41.
64. K. A. Page and coauthors, Circulating glucose levels modulate neural control of desire for high-calorie foods in humans, *Journal of Conical Investigation* 121 (2011): 4161–4169.
65. M. D. Lane and S. H. Cha, Effect of glucose and fructose on food intake via malonyl-CoA signaling in the brain, *Biochemical and Biophysical Research Communications* 382 (2009): 1–5.
66. A. M. Brownawell and coauthors, Prebiotics and the health benefits of fiber: current regulatory status, future research, and goals, *Journal of Nutrition* 142 (2012): 962–974.
67. S. T. Tey and coauthors, Long-term consumption of high energy-dense snack foods on sensory-specific satiety and intake, *American Journal of Clinical Nutrition* 95 (2012): 1038–1047.
68. R. J. de Souza and coauthors, Effects of 4 weight-loss diets differing in fat, protein, and carbohydrate on fat mass, lean mass, visceral adipose tissue, and hepatic fat: Results from the POUNDS LOST trial, *American Journal of Clinical Nutrition* 95 (2012): 614–625.
69. Hall and coauthors, Energy balance and its components, 2012.
70. J. R. Speakman and coauthors, Set points, settling points and some alternative models: theoretical options to understand how genes and environments combine to regulate body adiposity, *Disease Models and Mechanisms* 4 (2011): 733–745; M. M. Farias, A. M. Cuevas, and F. Rodriguez, Set-point theory and obesity, *Meta-*

bolic Syndrome and Related Disorders 9 (2011): 85–89.
71. B. E. Levin, Developmental gene x environment interactions affecting systems regulating energy homeostasis and obesity, *Frontiers in Neuroendocrinology* 31 (2010): 270–283.
72. M. J. Müller, A. Bosy-Westphal, and S. B. Heymsfield, Is there evidence for a set point that regulates human body weight? *Medicine Reports* 2 (2010), epub, doi:10.3410/M2-59.
73. E. Ravussin and J. E. Galgani, The implication of brown adipose tissue for humans, *Annual Review of Nutrition* 31 (2011): 33–37; W. D. van Marken and coauthors, Cold-activated brown adipose tissue in healthy men, *New England Journal of Medicine* 360 (2009): 1500–1508.
74. P. Seale and M. A. Lazar, Brown fat in humans: Turning up the heat on obesity, *Diabetes* 58 (2009): 1482–1484.
75. J. Wu and coauthors, Beige adipocytes are a distinct type of thermogenic fat cell in mouse and human, *Cell* 150 (2012): 366–376.
76. K. L. Hoehn and coauthors, Acute or chronic upregulation of mitochondrial fatty acid oxidation has no net effect on whole-body energy expenditure or adiposity, *Cell Metabolism* 11 (2010): 70–76.
77. Li Shengxu and coauthors, Cumulative effects and predictive value of common obesity-susceptibility variants identified by genome-wide association studies, *American Journal of Clinical Nutrition* 91 (2010): 184–190.
78. A. Ichimura and coauthors, Dysfunction of lipid sensor GPR120 leads to obesity in both mouse and human, *Nature* 483 (2012): 350–354; F. Massiera and coauthors, A Western-like fat diet is sufficient to induce a gradual weight enhancement in fat mass over generations, *Journal of Lipid Research* 51 (2010): 2352–3261; P. Seale, S. Kajimura, and B. M. Spiegelman, Transcriptional control of brown adipocyte development and physiological function—of mice and men, *Genes and Development* 23 (2009): 788–797.
79. C. Lavebratt, M. Almgren, and T. J. Ekström, Epigenetic regulation in obesity, *International Journal of Obesity* 36 (2012): 757–765; M. Z. Alfaradhi and S. E. Ozanne, Developmental programming in response to maternal overnutrition, *Frontiers in Genetics*, June 2011, epub, doi:10.3389/fgene.2011.00027; H. Slomko, H. J. Heo, and F. H. Einstein, Minireview: Epigenetics of obesity and diabetes in humans, *Endocrinology* 153 (2012): 1025–1030; Y. Seki and coauthors, Minireview: Epigenetic programming of diabetes and obesity: animal models, *Endocrinology* 153 (2012): 1031–1038.
80. W. P. James, The fundamental drivers of the obesity epidemic, *Obesity Reviews* (supplement) 9 (2008): 6–13.
81. B. M. Popkin and K. J. Duffey, Does hunger and satiety drive eating anymore? Increasing eating occasions and decreasing time between eating occasions in the United States, *American Journal of Clinical Nutrition* 91 (2010): 1342–1347; K. J. Duffey and coauthors, Regular consumption from fast food establishments relative to other restaurants is differentially associated with metabolic outcomes in young adults, *Journal of Nutrition* 139 (2009): 2113–2118.
82. C. J. Moore and S. A. Cunningham, Social position, psychological stress, and obesity: a systematic review, *Journal of the Academy of Nutrition and Dietetics* 112 (2012): 518–526; J. P. Block and coauthors, Psychosocial stress and change in weight among U.S. adults, *American Journal of Epidemiology* 170 (2009): 181–192.
83. D. Ferriday and J. M. Brunstrom, 'I just can't help myself': Effects of food-cue exposure in overweight and lean individuals, *International Journal of Obesity* 35 (2011): 142–149; L. B. Shomaker and coauthors, Eating in the absence of hunger in adolescents: Intake after a large-array meal compared with that after a standardized meal, *American Journal of Clinical Nutrition* 92 (2010): 697–703.
84. B. Wansink and J. Kim, Bad popcorn in big buckets: Portion size can influence intakes as much as taste, *Journal of Nutrition Education and Behavior* 37 (2005): 242–245.
85. K. J. Duffey and B. M. Popkin, Energy density, portion size, and eating occasions: Contributions to increased energy intake in the United States, 1977–2006, *PLoS Medicine* 8 (2011): e1001050.
86. C. Davis and coauthors, Evidence that "food addiction" is a valid phenotype of obesity, *Appetite* 57 (2011): 711–717.
87. G. Wang and coauthors, Brain dopamine and obesity, *Lancet* 357 (2001): 354–357.
88. Y. M. Ulrich-Lai, M. M. Ostrander, and J. P. Herman, HPA axis dampening by limited sucrose intake: reward frequency vs. caloric consumption, *Physiology and Behavior* 103 (2011): 104–110; P. M. Johnson and P. J. Kenny, Addiction-like reward dysfunction and compulsive eating in obese rats: role for dopamine D2 receptors, *Nature Neuroscience* 13 (2010): 635–641.
89. L. Epstein and coauthors, Food reinforcement and obesity. Psychological moderators, *Appetite* 58 (2012): 157–162.
90. J. S. Schiller and coauthors, Summary health statistics for U.S. adults: National Health Interview Survey, *Vital and Health Statistics* 10 (2012): 1–217.
91. K. Wijndaele and coauthors, Increased cardiometabolic risk is associated with increased TV viewing time, *Medicine and Science in Sports and Exercise* 42 (2010): 1511–1518.
92. T. S. Church and coauthors, Trends over 5 decades in U.S. occupation-related physical activity and their association with obesity, *PLoS ONE* 6 (2011): e19657.
93. H. P. van der Ploeg and coauthors, Sitting time and all-cause mortality risk in 222,497 Australian adults, *Archives of Internal Medicine* 172 (2012): 494–500; P. T. Katzmarzyk and coauthors, Sitting time and mortality from all causes, cardiovascular disease, and cancer, *Medicine and Science in Sports and Exercise* 41 (2009): 998–1005.
94. J. F. Salis, Role of built environments in physical activity, obesity, and cardiovascular disease, *Circulation* 125 (2012): 729–737; D. Ding and coauthors, Neighborhood environment and physical activity among youth: A review, *American Journal of Preventive Medicine* 41 (2011): 442–455.
95. J. Beaulac, E. Kristjansson, and S. Cummins, A systematic review of food deserts, 1966–2007, *Preventing Chronic Disease: Public Health Research, Practice, and Policy*, 6 (2009): epub available at www.cdc.gov.
96. J. L. Blitstein, J. Snider, and W. D. Evans, Perceptions of the food shopping environment are associated with greater consumption of fruits and vegetables, *Public Health Nutrition* 21 (2012): 1–6; M. I. Larson, M. T. Story, and M. C. Nelson, Neighborhood environments: Disparities in access to healthy foods in the U.S., *American Journal of Preventive Medicine* 36 (2009): 74–81.
97. S. J. Marshall and coauthors, Translating physical activity recommendations into a pedometer-based step goal: 3000 steps in 30 minutes, *American Journal of Preventive Medicine* 36 (2009): 410–415; U.S. Department of Agriculture and U.S. Department of Health and Human Services, *2008 Physical Activity Guidelines for Americans*, available at www.health.gov/paguidelines/default.aspx.
98. E. Kirk, and coauthors, Minimal resistance training improves daily energy expenditure and fat oxidation, *Medicine and Science in Sports and Exercise* 41 (2009): 1122–1129.
99. American College of Sports Medicine, Position Stand: Appropriate physical activity intervention strategies for weight loss and prevention of weight regain for adults, *Medicine and Science in Sports and Exercise* 41 (2009): 459–471.
100. I. N. Bezerra, C. Curioni, and R. Sichieri, Association between eating out of home and body weight, *Nutrition Reviews* 70 (2012): 65–79; A. M. Fretts and coauthors, Associations of processed meat and unprocessed red meat intake with incident diabetes: The Strong Heart Family Study, *American Journal of Clinical Nutrition* 95 (2012): 752–758; H. J. Song and coauthors, Understanding a key feature of urban food stores to develop nutrition intervention, *Journal of Hunger & Environmental Nutrition* 7 (2012): 77–90.
101. Institute of Medicine (U.S.). Committee on Accelerating Progress in Obesity Prevention. *Accelerating Progress in Obesity Prevention: Solving the Weight of the Nation* (Washington, D.C.: National Academies Press, 2012), available at www.nap.edu.
102. C. M. Novak and C. K Gavini, Smokeless weight loss, *Diabetes* 61 (2012): 776–777.
103. J. W. Carbone, J. P. McClung, and S. M. Pasiakos, Skeletal muscle responses to negative energy balance: effects of dietary protein, *Advances in Nutrition* 3 (2012): 119–126.
104. R. J. de Souza and coauthors, Effects of 4 weight-loss diets differing in fat, protein, and carbohydrate on fat mass, lean mass, visceral adipose tissue, and hepatic fat: Results from the POUNDS LOST trial, *American Journal of Clinical Nutrition* 95 (2012): 614–625.

105. J. F. Trepanowski and coauthors, Impact of caloric and dietary restriction regimens on markers of health and longevity in humans and animals: A summary of available findings, *Nutrition Journal* 10 (2011), epub, doi:10.1186/1475–2891–10–107.
106. J. Karbowska and Z. Kochan, Intermittent fasting up-regulates *Fsp27/Cidec* gene expression in white adipose tissue, *Nutrition* 28 (2012): 294–299.
107. B. Wansink, A. Tal, and M. Shimizu, First foods most: After 18-hour fast, people drawn to starches first and vegetables last, *Archives of Internal Medicine* 172 (2012): 961–963.
108. K. D. Carr, Food scarcity, neuroadaptations, and the pathogenic potential of dieting in an unnatural ecology: Binge eating and drug abuse, *Physiology and Behavior* 104 (2011): 162–167; Trepanowski and coauthors, Impact of caloric and dietary restriction regimens on markers of health and longevity in humans and animals, 2011.
109. M. Schütze and coauthors, Beer consumption and the "beer belly," 2009.
110. E. W. Wamsteker and coauthors, Unrealistic weight-loss goals among obese patients are associated with age and causal attributions, *Journal of the American Dietetic Association* 109 (2009): 1903–1908.
111. D. R. Young and coauthors, Effects of the PREMIER interventions on health-related quality of life, *Annals of Behavioral Medicine* 40 (2010): 302–312.
112. L. Bacon and L. Aphramor, Weight science: evaluating the evidence for a paradigm shift, *Nutrition Journal* 10 (2011), epub, doi:10.1186/1475–2891–10–9.
113. Kirk and coauthors, Effective weight management practice: a review of the lifestyle intervention evidence, *International Journal of Obesity* 36 (2012): 178–185.
114. S. F. Kirk and coauthors, Effective weight management practice; D. Heber, An integrative view of obesity, *American Journal of Clinical Nutrition* 91 (2010): 280S–283S.
115. S. B. Heymsfield and coauthors, Energy content of weight loss: Kinetic features during voluntary caloric restriction, *Metabolism* 61 (2012): 937–943.
116. J. M. Nicklas and coauthors, Successful weight loss among obese U.S. adults, *American Journal of Preventive Medicine* 42 (2012): 481–485.
117. H. M. Seasle and coauthors, Position of the American Dietetic Association: Weight management, *Journal of the American Dietetic Association* 109 (2009): 330–346.
118. Hymsfield and coauthors, Energy content of weight loss: Kinetic features during voluntary caloric restriction, 2012.
119. Hymsfield and coauthors, Energy content of weight loss: Kinetic features during voluntary caloric restriction, 2012.
120. K. D. Hall and coauthors, Energy balance and its components: Implications for body weight regulation, *American Journal of Clinical Nutrition* 95 (2012): 989–994.
121. A. J. Tomiyama and coauthors, Low calorie dieting increases cortisol, *Psychosomatic Medicine* 72 (2010): 357–364.
122. C. M. Shay and coauthors, Food and nutrient intakes and their associations with lower BMI in middle-aged US adults: the International Study of Macro-/Micronutrients and Blood Pressure, *American Journal of Clinical Nutrition* 96 (2012): 483–491; M. Kristensen and coauthors, Whole grain compared with refined wheat decreases the percentage of body fat following a 12-week, energy restricted dietary intervention in postmenopausal women, *Journal of Nutrition* 142 (2012): 710–716.
123. M. Bes-Rastrollo and coauthors, Prospective study of nut consumption, long-term weight change, and obesity risk in women, *American Journal of Clinical Nutrition* 89 (2009): 1913–1919.
124. J. W. Carbone, J. P. McClung, and S. M. Pasiakos, Skeletal muscle responses to negative energy balance: Effects of dietary protein, *Advances in Nutrition* 3 (2012): 119–126.
125. Y. Wang and M. A. Beydoun, Meat consumption is associated with obesity and central obesity among U.S. adults, *International Journal of Obesity* 33 (2009): 621–628.
126. B. J. Rolls, L. S. Roe, and J. S. Meengs, Larger portion sizes lead to a sustained increase in energy intake over 2 days, *Journal of the American Dietetic Association* 106 (2006): 543–549.
127. M. A. Palmer, S. Capra, and S. K. Baines, Association between eating frequency, weight, and health, *Nutrition Reviews* 67 (2009): 379–390.
128. R. S. Sebastian, C. Wilkinson, and J. D. Goldman, Snacking patterns of U.S. adults: What We Eat in America, NHANES 2007–2008, Dietary Data Brief No. 4, June 2011, available at http://www.ars.usda.gov/Services/docs.htm?docid=19476.
129. P. Dshmukh-Taskar and coauthors, The relationship of breakfast skipping and type of breakfast consumed with overweight/obesity, abdominal obesity, other cardiometabolic risk factors and the metabolic syndrome in young adults, *Public Health Nutrition* (2012), epub ahead of print, doi.org/10.1017/S1368980012004296.
130. W. Milano and coauthors, Night eating syndrome: An overview, *Journal of Pharmacy and Pharmacology* 64 (2012): 2–10.
131. R. M. Morrison, L. Mancino, and J. N. Vaiyam, Will calorie labeling in restaurants make a difference? *Amber Waves* 9 (2011): 10–17.
132. B. Bruemmer and coauthors, Energy, saturated fat, and sodium were lower in entrées at chain restaurants at 18 months compared with 6 months following the implementation of mandatory menu labeling regulation in King County, Washington, *Journal of the Academy of Nutrition and Dietetics* 112 (2012): 1169–1176.
133. R. Pérez-Escamilla and coauthors, Dietary energy density and body weight in adults and children: a systematic review, *Journal of the Academy of Nutrition and Dietetics* 112 (2012): 671–684.
134. R. Pérez-Escamillia and coauthors, Dietary energy density and body weight in adults and children: a systematic review, *Journal of the Academy of Nutrition and Dietetics* 112 (2012): 671–684.
135. S. Phelan and coauthors, Use of artificial sweeteners and fat-modified foods in weight loss maintainers and always-normal weight individuals, *International Journal of Obesity* 33 (2009): 1183–1190.
136. D. F. Tate and coauthors, Replacing caloric beverages with water or diet beverages for weight loss in adults: Main results of the Choose Healthy Options Consciously Everyday (CHOICE) randomized clinical trial, *American Journal of Clinical Nutrition* 95 (2012): 555–563; U.S. Department of Agriculture and U.S. Department of Health and Human Services, *Dietary Guidelines for Americans, 2010*, available at www.dietaryguidelines.gov; R. D. Mattes and B. M. Popkin, Nonnutritive sweetener consumption in humans: Effects on appetite and food intake and their putative mechanisms, *American Journal of Clinical Nutrition* 89 (2009): 1–14.
137. E. Green and C. Murphy, Altered processing of sweet taste in the brain of diet soda drinkers, *Physiology and Behavior,* (2012), epub ahead of print, doi:10.1016/j.physbeh.2012.05.006; Bellisle and coauthors, Sweetness, satiation, and satiety, 2012.
138. M. Chen and coauthors, Effects of dairy intake on body weight and fat: a meta-analysis of randomized controlled trials, *American Journal of Clinical Nutrition* 96 (2012): 735–747.
139. C. L. Rock and coauthors, Effect of a free prepared meal and incentivized weight loss program on weight loss and weight loss maintenance in obese and overweight women: A randomized controlled trial, *Journal of the American Medical Association* 304 (2010): 1803–1810; Position of the American Dietetic Association: Weight management, *Journal of the American Dietetic Association* 109 (2009): 330–346.
140. Nicklas and coauthors, Successful weight loss among obese U.S. adults, 2012; B. H. Goodpaster and coauthors, Effects of diet and physical activity interventions on weight loss and cardiometabolic risk factors in severely obese adults: A randomized study, *Journal of the American Medical Association* 304 (2010): 1795–1802; A. L. Hankinson and coauthors, Maintaining a high physical activity level over 20 years and weight gain, *Journal of the American Medical Association* 304 (2010): 2603–2610.
141. P. S. MacLean and coauthors, Regular exercise attenuates the metabolic drive to regain weight after long-term weight loss, *American Journal of Physiology: Regulatory, Integrative, and Comparative Physiology* 297 (2009): R793–R802.
142. K. J. Guelfi and coauthors, Beneficial effects of 12 weeks of aerobic compared with resistance exercise training on perceived appetite in previously sedentary overweight and obese men, *Metabolism* 62 (2012): 235–243.
143. A. M. Knab and coauthors, A 45-minute vigorous exercise bout increases metabolic rate

for 14 hours, *Medicine and Science in Sports and Exercise* 43 (2011): 1642–1648.
144. A. R. Josse and coauthors, Diets higher in dairy foods and dietary protein support bone health during diet- and exercise-induced weight loss in overweight and obese premenopausal women, *Journal of Clinical Endocrinology and Metabolism* 97 (2012): 251–260.
145. G. M. Manzoni and coauthors, Can relaxation training reduce emotional eating in women with obesity? An exploratory study with 3 months of follow-up, *Journal of the American Dietetic Association* 109 (2009): 1427–1432.
146. U.S. Department of Agriculture and U.S. Department of Health and Human Services, *2008 Physical Activity Guidelines for Americans*.
147. P. C. Douris and coauthors, Comparison between Nintendo Wii fit aerobics and traditional aerobic exercise in sedentary young adults, *Journal of Strength and Conditioning Research* 26 (2012): 1052–1057; N. Mitre and coauthors, The energy expenditure of an activity-promoting video game compared to sedentary video games and TV watching, *Journal of Pediatric Endocrinology and Metabolism* 24 (2011): 689–695; A. Barnett, E. Cerin, and T. Baranowski, Active video games for youth: A systematic review, *Journal of Physical Activity and Health* 8 (2011): 724–737.
148. L. Lanningham-Foster and coauthors, Activity-promoting games and increased energy expenditure, *Journal of Pediatrics* 154 (2009): 819–823.
149. M. K. Robinson, Surgical treatment of obesity—Weighing the facts, *New England Journal of Medicine* 361 (2009): 520–521.
150. J. M. Nicklas and coauthors, Successful weight loss among obese U.S. adults, 2012.
151. U.S. Food and Drug Administration, Questions and answers about FDA's initiative against contaminated weight loss products, April 30, 2009, available at http://www.fda.gov/Drugs/ResourcesForYou/Consumers/QuestionsAnswers/ucm136187.htm.
152. Nicklas and coauthors, Successful weight loss among obese U.S. adults, 2012.
153. A. Nagle, Bariatric surgery—A surgeon's perspective, *Journal of the American Dietetic Association* 110 (2010): 520–523; G. L. Blackburn, S. Wollner, and S. B. Heymsfield, Lifestyle interventions for the treatment of class III obesity: A primary target for nutrition medicine in the obesity epidemic, *American Journal of Clinical Nutrition* 91 (2010): 289S–292S.
154. E. H. Livingston, Inadequacy of BMI as indicator for bariatric surgery, *Journal of the American Medical Association* 307 (2012): 88–89; M. L. Maciejewski and coauthors, Survival among high-risk patients after bariatric surgery, *Journal of the American Medical Association* 305 (2011): 2419–2426.
155. R. Padwal and coauthors, Bariatric surgery: A systematic review of the clinical and economic evidence, *Journal of General Internal Medicine* 26 (2011): 1183–1194.
156. M. Attiah and coauthors, Durability of Roux-en-Y gastric bypass surgery: A meta-regression study, *Annals of Surgery* 256 (2012): 251–254; L. Sjöström and coauthors, Bariatric surgery and long-term cardiovascular events, *Journal of the American Medical Association* 307 (2012): 56–65; P. R. Schauer and coauthors, Bariatric surgery versus intensive medical therapy in obese patients with diabetes, *New England Journal of Medicine* 366 (2012): 1567–1576; The Endocrine Society, Evaluating the benefits of treating type 2 diabetes with bariatric surgery, An Endocrine Society statement to providers on study findings related to medical versus surgical treatment of obese patients with type 2 diabetes, March 2012, available at www.endo-society.org/advocacy/index.cfm.
157. U.S. Food and Drug Administration, FDA targets gastric band weight-loss claims, *FDA Consumer Health Information*, December 2011, available at www.fda.gov/ForConsumers/ConsumerUpdates.
158. M. Ruz and coauthors, Iron absorption and iron status are reduced after Roux-en-Y gastric bypass, *American Journal of Clinical Nutrition* 90 (2009): 527–532.
159. S. Paganini and coauthors, Daily vitamin supplementation and hypovitaminosis after obesity surgery, *Nutrition* 28 (2012): 391–396; N. Gletsu-Miller and coauthors, Incidence and prevalence of copper deficiency following roux-en-y gastric bypass surgery, *International Journal of Obesity* 36 (2012): 328–335.
160. P. E. O'Brien and coauthors, Laparoscopic adjustable gastric banding in severely obese adolescents, *Journal of the American Medical Association* 303 (2010): 519–526.
161. Y. R. Krishna and coauthors, Acute liver failure caused by "fat burners" and dietary supplements: A case report and literature review, *Canadian Journal of Gastroenterology* 25 (2011): 157–160.
162. J. L. Kraschnewski and coauthors, Long-term weight loss maintenance in the United States, *International Journal of Obesity* 34 (2010): 1644–1654.
163. H. Shin and coauthors, Self-efficacy improves weight loss in overweight/obese postmenopausal women during a 6-month weight loss intervention, *Nutrition Research* 11 (2011): 822–828.
164. M. C. Daniels and B. M. Popkin, Impact of water intake on energy intake and weight status: A systematic review, *Nutrition Reviews* 68 (2010): 505–521; B. M. Davy and coauthors, Water consumption reduces energy intake at a breakfast meal in obese older adults, *Journal of the American Dietetic Association* 108 (2008): 1236–1239.
165. J. Stubbs and coauthors, Problems in identifying predictors and correlates of weight loss and maintenance: Implications for weight control therapies based on behaviour change, *Obesity Reviews* 12 (2011): 688–708.
166. J. P. Moreno and C. A. Johnston, Successful habits of weight losers, *American Journal of Lifestyle Medicine* 6 (2012): 113–115.
167. J. D. Akers and coauthors, Daily self-monitoring of body weight, step count, fruit/vegetable intake, and water consumption: A feasible and effective long-term weight loss maintenance approach, *Journal of the Academy of Nutrition and Dietetics* 112 (2012): 685–692; C. N. Sciamanna and coauthors, Practices associated with weight loss versus weight-loss maintenance: Results of a national survey, *American Journal of Preventive Medicine* 41 (2011): 159–166.
168. K. H. Pietiläinen and coauthors, Does dieting make you fat? A twin study, *International Journal of Obesity* 36 (2012): 456–464.
169. Nicklas and coauthors, Successful weight loss among obese U.S. adults, 2012; L. J. Appel and coauthors, Comparative effectiveness of weight-loss interventions in clinical practice, *New England Journal of Medicine* 365 (2011): 1959–1968.
170. S. Kodama and coauthors, Effect of web-based lifestyle modification on weight control: A meta-analysis, *International Journal of Obesity* 36 (2012): 675–685; A. G. Digenio and coauthors, Comparison of methods for delivering a lifestyle modification program for obese patients: A randomized trial, *Annals of Internal Medicine* 150 (2009): 255–262.
171. W. L. S. Falzon and coauthors, Interactive computer-based interventions for weight loss or weight maintenance in overweight or obese people, *Cochrane Database of Systematic Reviews* 8 (2012), epub, doi: 10.1002/14651858.cd007675.pub2.
172. L. K. Khan and coauthors, Recommended community strategies and measurements to prevent obesity in the United States, *MMWR Recommendations and Reports* 58 (2009): 1–26; A. M. Wolf and K. A. Woodworth, Obesity prevention: Recommended strategies and challenges, *The American Journal of Medicine* 122 (2009): S19–S23.
173. L. E. Burke, J. Wang, and M. A. Sevick, Self-monitoring in weight loss: A systematic review of the literature, *Journal of the American Dietetic Association* 111 (2011): 92–102.
174. N. R. Reyes and coauthors, Similarities and differences between weight loss maintainers and regainers: a qualitative analysis, *Journal of the Academy of Nutrition and Dietetics* 112 (2012): 499–505.

Consumer's Guide 9

1. J. M. Nicklas and coauthors, Successful weight loss among obese U.S. adults, *American Journal of Preventive Medicine* 42 (2012): 481–485.
2. W. S. Yancy and coauthors, A randomized trial of a low-carbohydrate diet vs orlistat plus a low-fat diet for weight loss, *Archives of Internal Medicine* 170 (2010): 136–145; F. M. Sacks and coauthors, Comparison of weight-loss diets with different compositions of fat, protein, and carbohydrates, *New England Journal of Medicine* 360 (2009): 859–873.
3. R. J. de Souza and coauthors, Effects of 4 weight-loss diets differing in fat, protein, and carbohydrate on fat mass, lean mass, visceral adipose tissue, and hepatic fat: Results from the POUNDS LOST trial, *American Journal of Clinical Nutrition* 95 (2012): 614–625.

4. C. B. Ebbeling and coauthors, Effects of dietary composition on energy expenditure during weight-loss maintenance, *Journal of the American Medical Association* 307 (2012): 2627–2634; P. Lagiou and coauthors, Low carbohydrate-high protein diet and incidence of cardiovascular diseases in Swedish women: Prospective cohort study, *British Medical Journal* 344 (2012), epub, doi:10.1136/bmj.e4026.
5. de Souza and coauthors, Effects of 4 weight-loss diets differing in fat, protein, and carbohydrate on fat mass, lean mass, visceral adipose tissue, and hepatic fat, 2012; Sacks and coauthors, Comparison of weight-loss diets, 2009.
6. The *Dietary Guidelines for Americans 2010*, available at www.dietaryguidelines.gov.
7. A. M. Johnstone, Safety and efficacy of high-protein diets for weight loss, *Proceedings of the Nutrition Society* 71 (2012): 339–349.
8. T. L. Hernandez and coauthors, Lack of suppression of circulating free fatty acids and hypercholesterolemia during weight loss on a high-fat, low-carbohydrate diet, *American Journal of Clinical Nutrition* 91 (2010): 578–585; S. Y. Foo and coauthors, Vascular effects of a low-carbohydrate high-protein diet, *Proceedings of the National Academies of Science* 106 (2009): 15418–15423; M. Miller and coauthors, Comparative effects of three popular diets on lipids, endothelial function, and C-reactive protein during weight management, *Journal of the American Dietetic Association* 109 (2009): 713–717.
9. T. D. Barnett, N. D. Barnard, and T. L Radak, Development of symptomatic cardiovascular disease after self-reported adherence to the Atkins diet, *Journal of the American Dietetic Association* 109 (2009): 1263–1265.

Controversy 9

1. D. S. Rosen and the Committee on Adolescence, Clinical report—Identification and management of eating disorders in children and adolescents, *Pediatrics* 126 (2010): 1240–1253.
2. N. I. Larson, D. Neumark-Sztainer, and M. Story, Weight control behaviors and dietary intake among adolescents and young adults: Longitudinal finding from project EAT, *Journal of the American Dietetic Association* 109 (2009): 1869–1877; K. C. Berg, P. Frazier, and L. Sherr, Change in eating disorder attitudes and behavior in college women: Prevalence and predictors, *Eating Behaviors* 10 (2009): 137–142.
3. D. Neumark-Sztainer and coauthors, Dieting and disordered eating behaviors from adolescence to young adulthood: Findings from a 10-year longitudinal study, *Journal of the American Dietetic Association* 111 (2011): 1004–1011.
4. T. K. Clarke, A. R. Weiss, and W. H. Berrettini, The genetics of anorexia nervosa, *Clinical Pharmacology & Therapeutics* 91 (2012): 181–188; R. Calati and coauthors, The 5-HTTLPR polymorphism and eating disorders: A meta-analysis, *International Journal of Eating Disorders* 44 (2011): 191–199; S. E. Mazzeo and C. M. Bulik, Environmental and genetic risk factors for eating disorders: What the clinician needs to know, *Child and Adolescent Psychiatric Clinics of North America* 18 (2009): 67–82; S. S. O'Sullivan, A. H. Evans, and A. J. Lees, Dopamine dysregulation syndrome: An overview of its epidemiology, mechanisms, and management, *CNS Drugs* 23 (2009): 157–170; T. D. Müller and coauthors, Leptin-mediated neuroendocrine alterations in anorexia nervosa: Somatic and behavioral implications, *Child and Adolescent Psychiatric Clinics of North America* 18 (2009): 117–129.
5. J. H. Slevec and M. Tiggemann, Predictors of body dissatisfaction and disordered eating in middle-aged women, *Clinical Psychology Review* 31 (2011): 515–524.
6. K. D. Carr, Food scarcity, neuroadaptations, and the pathogenic potential of dieting in an unnatural ecology: Binge eating and drug abuse, *Physiology and Behavior* 104 (2011): 162–167.
7. C. A. Hincapié and J. D. Cassidy, Disordered eating, menstrual disturbances, and low bone mineral density in dancers: A systematic review, *Archives of Physical Medicine and Rehabilitation* 91 (2010): 1777–1789.
8. T. G. Nazem and K. E. Ackerman, The female athlete triad, *Sports Health* 4 (2012): 302–311; Position of the American Dietetic Association, Dietitians of Canada, and the American College of Sports Medicine: Nutrition and Athletic Performance, *Journal of the American Dietetic Association* 109 (2009): 509–527.
9. C. J. Rosen and A. Klibanski, Bone, fat, and body composition: Evolving concepts in the pathogenesis of osteoporosis, *American Journal of Medicine* 122 (2009): 409–414.
10. Position of the American Dietetic Association, Dietitians of Canada, and the American College of Sports Medicine: Nutrition and Athletic Performance, 2009.
11. M. Misra and A. Klibanski, Bone metabolism in adolescents with anorexia nervosa, *Journal of Endocrinological Investigation* 34 (2011): 324–332.
12. G. G. Artioli and coauthors, Prevalence, magnitude, and methods of rapid weight loss among judo competitors, *Medicine and Science in Sports and Exercise* 42 (2010): 436–442.
13. R. D. Grave, Eating disorders: Progress and challenges, *European Journal of Internal Medicine* 22 (2011): 153–160.
14. D. L. G. Borzekowski and coauthors, e-Ana and e-Mia: A content analysis of pro-eating disorder Websites, *American Journal of Public Health* 100 (2010): 1526–1534.
15. Position of the American Dietetic Association, Nutrition intervention in the treatment of eating disorders, *Journal of the American Dietetic Association* (2011): 1236–1241.
16. *Diagnostic and Statistical Manual of Mental Disorders*, fifth edition (Washington, D.C.: American Psychiatric Association, 2013), prepublication, available at www.dsm5.org/ProposedRevision/Pages/FeedingandEatingDisorders.aspx.
17. A. P. Winston, The clinical biochemistry of anorexia nervosa, *Annals of Clinical Biochemistry* 49 (2012): 132–143.
18. Grave, Eating disorders, 2011.
19. Position of the American Dietetic Association, Nutrition intervention in the treatment of eating disorders, 2011.
20. ECRI Institute, *Bulimia Nervosa: Comparative Efficacy of Available Psychological and Pharmacological Treatments*, as cited in M. Mitka, Reports weighs options for bulimia nervosa treatment, *Journal of the American Medical Association* 305 (2011): 875.
21. E. Attia and B. T. Walsh, Behavioral management for anorexia nervosa, *New England Journal of Medicine* 360 (2009): 500–506.
22. Attia and Walsh, Behavioral management for anorexia nervosa, 2009.
23. Position of the American Dietetic Association, Nutrition intervention in the treatment of eating disorders, 2011; *Diagnostic and Statistical Manual of Mental Disorders*, fifth edition (Washington, D.C.: American Psychiatric Association, 2013), prepublication, available at www.dsm5.org/ProposedRevision/Pages/FeedingandEatingDisorders.aspx.
24. Position of the American Dietetic Association, Nutrition intervention in the treatment of eating disorders, 2011.
25. R. Rodgers and H. Chabrol, Parental attitudes, body image disturbance, and disordered eating amongst adolescents and young adults: A review, *European Eating Disorders Reviews* 17 (2009): 137–151.
26. Rodgers and Chabrol, Parental attitudes, body image disturbance, and disordered eating, 2009.
27. Rodgers and Chabrol, Parental attitudes, body image disturbance, and disordered eating, 2009.
28. M. J. Krantz and coauthors, Factors influencing QT prolongation in patients hospitalized with severe anorexia nervosa, *General Hospital Psychiatry* 34 (2012): 173–177.
29. P. S. Grigson and coauthors, Bilateral lesions of the thalamic trigeminal orosensory area dissociate natural from drug reward in contrast paradigms, *Behavioral Neuroscience* 126 (2012): 538–550.
30. K. D. Carr, Food scarcity, neuroadaptations, and the pathogenic potential of dieting in an unnatural ecology: Binge eating and drug abuse, *Physiology & Behavior* 104 (2011): 162–167.
31. D. Neumark-Sztainer, Preventing obesity and eating disorders in adolescents: What can health care providers do? *Journal of Adolescent Health* 44 (2009): 206–213.

Chapter 10

1. H. P. van der Ploeg and coauthors, Sitting time and all-cause mortality risk in 222,497 Australian adults, *Archives of Internal Medicine* 172 (2012): 494–500; A. Grøntved and F. B. Hu, Television viewing and risk of type 2 diabetes, cardiovascular disease, and all-cause mortality, *Journal of the American Medical Association* 305 (2011): 2448–2455; T. Y. Warren and coauthors, Sedentary behaviors increase risk of cardiovascular disease mortality in men, *Medicine and Science in Sports and Exercise* 42 (2010): 879–885; A. V. Patel and coauthors, Leisure time spent sitting

in relation to total mortality in a prospective cohort of adults, *American Journal of Epidemiology* 172 (2010): 419–429; P. T. Katzmarzyk and coauthors, Sitting time and mortality from all causes, cardiovascular disease, and cancer, *Medicine and Science in Sports and Exercise* 41 (2009): 998–1005.
2. I. Lee and coauthors, Effect of physical inactivity on major non-communicable diseases worldwide: An analysis of burden of disease and life expectancy, *Lancet* 380 (2012): 219–229.
3. D. C. Nieman and coauthors, Upper respiratory tract infection is reduced in physically fit and active adults. *British Journal of Sports Medicine* 45 (2011): 987–992.
4. J. S. Schiller and coauthors, Summary health statistics for U.S. adults: National Health Interview Survey, *Vital and Health Statistics* 10 (2012): 1–217.
5. C. A. Slentz and coauthors, The effects of aerobic versus resistance training on visceral and liver fat sources, liver enzymes and HOMA from STRRIDE AT/RT: A randomized trial, *American Journal of Physiology—Endocrinology and Metabolism* 301 (2011): E1033–E1039; T. S. Church and coauthors, Changes in weight, waist circumference and compensatory responses with different doses of exercise among sedentary, overweight postmenopausal women, *PLoS One* 4 (2009): e4515, epub, doi:101371/journal .pone.0004515.
6. J.C. Colado and coauthors, Effects of a short-term aquatic resistance program on strength and body composition in fit young men, *Journal of Strength and Conditioning* 23 (2009): 549–559.
7. R. S. Rector and coauthors, Lean body mass and weight-bearing activity in the prediction of bone mineral density in physically active men, *Journal of Strength and Conditioning Research* 23 (2009): 427–435. A. Guadalupe-Grau and coauthors, Exercise and bone mass in adults, *Sports Medicine* 39 (2009): 439–468.
8. J. Romeo and coauthors, Physical activity, immunity and infection, *Proceedings of the Nutrition Society* 69 (2010): 390–399.
9. C. Eheman and coauthors, Annual Report to the Nation on the status of cancer, 1975–2008, featuring cancers associated with excess weight and lack of sufficient physical activity, *Cancer* 118 (2012): 2338–2366; X. Sui and coauthors, Influence of cardiorespiratory fitness on lung cancer mortality, *Medicine and Science in Sports and Exercise* 42 (2010): 872–878; J. B. Peel and coauthors, A prospective study of cardiorespiratory fitness and breast cancer mortality, *Medicine and Science in Sports and Exercise* 41 (2009): 742–748; S. Y. Pan and M. DesMeules, Energy intake, physical activity energy balance, and cancer: Epidemiologic evidence, *Methods in Molecular Biology* 472 (2009): 191–215.
10. M. C. Peddie, N. J. Rehrer, and T. L. Perry, Physical activity and postprandial lipidemia: Are energy expenditure and lipoprotein lipase activity the real modulators of the positive effect? *Progress in Lipid Research* 51 (2012): 11–22.
11. M. C. Audelin and coauthors, Change of energy expenditure from physical activity is the most powerful determinant of improved insulin sensitivity in overweight patients with coronary artery disease participating in an intensive lifestyle modification program, *Metabolism* 61 (2012): 672–679; C. R. Mikus and coauthors, Lowering physical activity impairs glycemic control in healthy volunteers, *Medicine and Science in Sports and Exercise* 44 (2012): 225–231; A. Grøntved and coauthors, A prospective study of weight training and risk of type 2 diabetes mellitus in men, *Archives of Internal Medicine*, August 2012, epub ahead of print, doi:10.1001/archinternmed.2012.3138; J. Ralph and coauthors, Low-intensity exercise reduces the prevalence of hyperglycemia in type 2 diabetes, *Medicine and Science in Sports and Exercise* 42 (2010): 219–225.
12. P. J. Banim and coauthors, Physical activity reduces the risk of symptomatic gallstones: A prospective cohort study, *European Journal of Gastroenterology and Hepatology* 22 (2010): 983–988.
13. J. C. Sieverdes and coauthors, Association between leisure-time physical activity and depressive symptoms in men, *Medicine and Science in Sports and Exercise* 44 (2012): 260–265.
14. B. L. Willis and coauthors, Midlife fitness and the development of chronic conditions in later life, *Archives of Internal Medicine*, August 2012, epub ahead of print, doi:10.1001 /archinternmed.2012.3400; R. Liu and coauthors, Cardiorespiratory fitness as a predictor of dementia mortality in men and women, *Medicine and Science in Sports and Exercise* 44 (2012): 253–259; P. Wen and coauthors, Minimum amount of physical activity for reduced mortality and extended life expectancy: A prospective cohort study, *Lancet*, 378 (2011): 1244–1253.
15. P. Boström and coauthors, A PGC1-α-dependent myokine that drives brown-fat-like development of white fat and thermogenesis, *Nature* 481 (2012): 463–468; B. K. Pedersen, A muscular twist on the fate of fat, *New England Journal of Medicine* 366 (2012): 1544–1545.
16. U.S. Department of Health and Human Services, *Physical Activity Guidelines for Americans*.
17. Centers for Disease Control and Prevention, www.cdc.gov/physicalactivity/everyone; site updated February 16, 2011.
18. S. T. Boutcher, High-intensity intermittent exercise and fat loss, *Journal of Obesity* (2011), epub, doi:10.1155/2011/868305.
19. J. Sattelmair and coauthors, Dose response between physical activity and risk of coronary heart disease, *Circulation* 124 (2011): 789–795.
20. Nicklas and coauthors, Successful weight loss among obese U.S. adults, 2012; B. H. Goodpaster and coauthors, Effects of diet and physical activity interventions on weight loss and cardiometabolic risk factors in severely obese adults: A randomized study, *Journal of the American Medical Association* 304 (2010): 1795–1802; A. L. Hankinson and coauthors, Maintaining a high physical activity level over 20 years and weight gain, *Journal of the American Medical Association* 304 (2010): 2603–2610.
21. American College of Sports Medicine, Position Stand: Appropriate physical activity intervention strategies for weight loss and prevention of weight regain for adults, *Medicine and Science in Sports and Exercise* 41 (2009): 459–471.
22. American College of Sports Medicine, Position Stand: Quantity and quality of exercise for developing and maintaining cardiorespiratory, musculoskeletal, and neuromotor fitness in apparently healthy adults: Guidance for prescribing exercise, *Medicine and Science in Sports and Exercise* 43 (2011): 1334–1359.
23. P. J. Atherton and K. Smith, Muscle protein synthesis in response to nutrition and exercise, *Journal of Physiology* 590.5 (2012): 1049–1057.
24. B. E. Phillips, D. S. Hill, and P. J. Atherton, Regulation of muscle protein synthesis in humans, *Current Opinion in Clinical Nutrition and Metabolic Care* 15 (2012): 58–63.
25. Atherton and Smith, Muscle protein synthesis in response to nutrition and exercise, 2012.
26. Atherton and Smith, Muscle protein synthesis in response to nutrition and exercise, 2012.
27. T. Seene, P. Kaasik, and K. Alev, Muscle protein turnover in endurance training: a review, *International Journal of Sports Medicine* 32 (2011): 905–911.
28. American College of Sports Medicine, Position Stand: Progression models in resistance training for healthy adults, *Medicine and Science in Sports and Exercise* 41 (2009): 687–708.
29. F. Vega and R. Jackson, Dietary habits of bodybuilders and other regular exercisers, *Nutrition Research* 16 (1996): 3–10.
30. American College of Sports Medicine, Position Stand: Quantity and quality of exercise for developing and maintaining cardiorespiratory, musculoskeletal, and neuromotor fitness in apparently healthy adults, 2011.
31. Slentz and coauthors, The effects of aerobic versus resistance training on visceral and liver fat sources, liver enzymes and HOMA from STRRIDE AT/RT: A randomized trial, 2011.
32. E. Kirk and coauthors, Minimal resistance training improves daily energy expenditure and fat oxidation, *Medicine and Science in Sports and Exercise* 41 (2009): 1122–1129.
33. L. A. Bateman and coauthors, Comparison of aerobic versus resistance exercise training effects on metabolic syndrome (from the studies of a targeted risk reduction intervention through defined exercise—STRRIDE-AT/RT), *American Journal of Cardiology* 108 (2011): 838–844.
34. S. Marwood and coauthors, Faster pulmonary oxygen uptake kinetics in trained versus untrained male adolescents, *Medicine and Science in Sports and Exercise* 42 (2010): 127–134.
35. C. Bouchard and coauthors, Adverse metabolic response to regular exercise: Is it a rare or common occurrence? *PLoS ONE* 7 (2012): e37887.
36. American College of Sports Medicine, *ACSM's Guidelines for Exercise Testing and Prescription*, 8th ed. (Philadelphia: Lippincott, Williams, and Wilkins, 2010), pp. 18–39.
37. A. M. Knab and coauthors, A 45-minute vigorous exercise bout increases metabolic rate

for 14 hours, *Medicine and Science in Sports and Exercise* 43 (2011): 1642–1648.
38. E. L. Melanson, P. S. MacLean, and J. O. Hill, Exercise improves fat metabolism in muscle but does not increase 24-h fat oxidation, *Exercise and Sport Sciences Reviews* 37 (2009): 93–101.
39. M. Rosenkilde and coauthors, Body fat loss and compensatory mechanisms in response to different doses of aerobic exercise—a randomized controlled trial in overweight sedentary males, *American Journal of Physiology: Regulatory, Integrative and Comparative Physiology* 303 (2012): R571–R579.
40. T. E. Jensen and E. A. Richter, Regulation of glucose and glycogen metabolism during and after exercise, *Journal of Physiology* 590.5 (2012): 1069–1076; T. E. Graham and coauthors, The regulation of muscle glycogen: the granule and its proteins, *Acta Physiologica* 199 (2010): 489–498.
41. E. D. Berglund and coauthors, Glucagon and lipid interactions in the regulation of hepatic AMPK signaling and expression of PPARα and FGF21 transcripts in vivo, *American Journal of Physiology, Endocrinology, and Metabolism* 299 (2010): E607–E614.
42. J. Bergstrom and coauthors, Diet, muscle glycogen and physical performance, *Acta Physiologica Scandanavica* 71 (1967): 140–150.
43. Position of the American Dietetic Association, Dietitians of Canada, and the American College of Sports Medicine: Nutrition and athletic performance, *Journal of the American Dietetic Association* 109 (2009): 509–527.
44. Position of the American Dietetic Association, Dietitians of Canada, and the American College of Sports Medicine: Nutrition and athletic performance, 2009.
45. K. J. Stuempfle and coauthors, Race diet of finishers and non-finishers in a 100 mile (161 km) mountain footrace, *Journal of the American College of Nutrition* 30 (2011): 529–535.
46. Ali and Williams, Carbohydrate ingestion and soccer skill performance during prolonged intermittent exercise, 2009.
47. S. M. Phillips and coauthors, Carbohydrate gel ingestion significantly improves the intermittent endurance capacity, but not sprint performance, of adolescent team games players during a simulated team games protocol, *European Journal of Applied Physiology* 112 (2012): 1133–1141.
48. G. van Hall, Lactate kinetics in human tissues at rest and during exercise, *Acta Physiologica* 199 (2012): 499–508.
49. J. F. Moxnes and Ø. Sandbakk, The kinetics of lactate production and removal during whole-body exercise, *Theoretical Biology and Medical Modeling* 9 (2012), epub, doi: 10.1186/1742–4682-9-7.
50. K. Overgaard and coauthors, Effects of acidification and increased extracellular potassium on dynamic muscle contractions in isolated rat muscles, *Journal of Physiology* 588 (2010): 5065–5076; J. S. Baker, M. C. McCormick, and R. A. Robergs, Interaction among skeletal muscle metabolic energy systems during intense exercise, *Journal of Nutrition and Metabolism* 2010, epub, doi:10.1155/2010/905612; D. A. Jones, Changes in the force-velocity relationship of fatigued muscle: implications for power production and possible causes, *Journal of Physiology* 16 (2010): 2977–2986.
51. T. D. Noakes, Fatigue is a brain-derived emotion that regulates the exercise behavior to ensure the protection of whole body homeostasis, *Frontiers in Physiology*, April 2012, epub, doi:10.3389/phys.2012.00082.
52. B. I. Rapoport, Metabolic factors limiting performance in marathon runners, *PLoS Computational Biology* 6 (2010), epub, doi:10.1371/journal.pcbi.1000960.
53. Position of the American Dietetic Association, Dietitians of Canada, and the American College of Sports Medicine: Nutrition and athletic performance, *Journal of the American Dietetic Association* 109 (2009): 509–527.
54. Stuempfle and coauthors, Race diet of finishers and non-finishers in a 100 mile (161 km) mountain footrace, 2011.
55. J. Temesi and coauthors, Carbohydrate ingestion during endurance exercise improves performance in adults, *Journal of Nutrition* 141 (2011): 890–897; L. M. Burke, Fueling strategies to optimize performance: Training high or training low? *Scandinavian Journal of Medicine and Science in Sports* 20 (2010): 48–58; A. Ali and C. Williams, Carbohydrate ingestion and soccer skill performance during prolonged intermittent exercise, *Journal of Sport Science* 27 (2009): 1499–1508.
56. B. Pfeiffer and coauthors, Nutritional intake and gastrointestinal problems during competitive endurance events, *Medicine and Science in Sports and Exercise* 44 (2012): 344–351; B. Pfeiffer and coauthors, The effect of carbohydrate gels on gastrointestinal tolerance during a 16-km run, *International Journal of Sport Nutrition and Exercise Metabolism* 19 (2009): 485–503.
57. Position of the American Dietetic Association, Dietitians of Canada, and the American College of Sports Medicine: Nutrition and athletic performance, 2009.
58. K. N. Frayn, Fat as a fuel: emerging understanding of the adipose tissue-skeletal muscle axis, *Acta Physiologica* 199 (2010): 509–518.
59. Maughan and Shirreffs, Nutrition for sports performance: issues and opportunities, 2012.
60. Frayn, Fat as a fuel: emerging understanding of the adipose tissue-skeletal muscle axis, 2010.
61. L. H. Willis and coauthors, Effects of aerobic and/or resistance training on body mass and fat mass in overweight or obese adults, *Journal of Applied Physiology* 113 (2012): 1831–1837.
62. R. J. Maughan and S. M. Shirreffs, Nutrition for sports performance: issues and opportunities, *Proceedings of the Nutrition Society* 71 (2012): 112–119.
63. Position of the American Dietetic Association, Dietitians of Canada, and the American College of Sports Medicine, Nutrition and athletic performance, 2009.
64. Seene, Kaasik, and Alev, Muscle protein turnover in endurance training: a review, 2011.
65. T. A. Churchward-Venne, N. A. Burd, and S. M. Phillips, Nutrition regulation of muscle protein synthesis with resistance exercise: strategies to enhance anabolism, *Nutrition and Metabolism* 9 (2012), epub, doi:10.1186/1743-7075-9-40.
66. Atherton and Smith, Muscle protein synthesis in response to nutrition and exercise, 2012.
67. P. J. Atherton and coauthors, Muscle full effect after oral protein: time-dependent concordance and discordance between human muscle protein synthesis and mTORC1 signaling, *American Journal of Clinical Nutrition* 92 (2010): 1080–1088.
68. Churchward-Venne, Burd, and Phillips, Nutrition regulation of muscle protein synthesis with resistance exercise: strategies to enhance anabolism, 2012.
69. D. R. Moore and coauthors, Resistance exercises enhances mTOR and MARK signalling in human muscle over that seen at rest after bolus protein ingestion, *Acta Physiologica* 201 (2011): 365–372.
70. N. A. Burd and coauthors, Enhanced amino acid sensitivity of myofibrillar protein synthesis persists for up to 24 h after resistance exercise in young men, *Journal of Nutrition* 141 (2011): 568–573.
71. S. M. Pasiakos and J. P. McClung, Supplemental dietary leucine and the skeletal muscle anabolic response to essential amino acids, *Nutrition Reviews* 69 (2011): 550–557.
72. G. Wu, Amino acids: Metabolism, functions, and nutrition, *Amino Acids* 37 (2009): 1–17.
73. Maughan and Shirreffs, Nutrition for sports performance: issues and opportunities, 2012; S. S. Gropper, J. L. Smith, and J. L. Groff, Sports nutrition in integration and regulation of metabolism and the impact of exercise and sport, *Advanced Nutrition and Human Metabolism* (Belmont, Calif.: Wadsworth/Cengage Learning, 2009), pp. 265–275.
74. Position of the American Dietetic Association, Dietitians of Canada, and the American College of Sports Medicine, Nutrition and athletic performance, 2009; Committee on Dietary Reference Intakes, *Dietary Reference Intakes*.
75. Position of the American Dietetic Association, Dietitians of Canada, and the American College of Sports Medicine, Nutrition and athletic performance, 2009.
76. E. Coleman, Protein requirements for athletes, *Clinical Nutrition Insight* 38 (2012): 1–3.
77. Position of the American Dietetic Association, Dietitians of Canada, and the American College of Sports Medicine, Nutrition and athletic performance, 2009.
78. K. Janakiraman, S. Shenoy, and J. S. Sandhu, Firm insoles effectively reduce hemolysis in runners during long distance running—A comparative study, *Sports Medicine, Arthroscopy, Rehabilitation, Therapy and Technology* 3 (2011), epub, doi:10.1186/1758-255-3-12.
79. M. K. Newlin and coauthors, The effects of acute exercise bouts on hepcidin in women, *International Journal of Sport Nutrition and Exercise Metabolism* 22 (2012): 79–88.

80. American College of Sports Medicine, Position Stand, Exercise and fluid replacement, *Medicine and Science in Sports and Exercise* 39 (2007): 377–390.
81. C. A. Rosenbloom and E. J. Coleman, eds., *Sports Nutrition: A Practice Manual for Professionals* (Chicago: Academy of Nutrition and Dietetics, 2012), pp. 106–127; Position of the American Dietetic Association, Dietitians of Canada, and the American College of Sports Medicine: Nutrition and athletic performance, 2009.
82. R. M. Lopez and coauthors, Examining the influence of hydration status on physiological responses and running speed during trail running in the heat with controlled exercise intensity, *Journal of Strength and Conditioning Research* 25 (2011): 2944–2954; P. Carvalho and coauthors, Impact of fluid restriction and ad libitum water intake or an 8% carbohydrate-electrolyte beverage on skill performance of elite adolescent basketball players, *International Journal of Sport Nutrition and Exercise Metabolism* 21 (2011): 214–221.
83. Standing Committee on the Scientific Evaluation of Dietary Reference Intakes, Food and Nutrition Board, *Dietary Reference Intakes*.
84. C. A. Rosenbloom and E. J. Coleman, eds. *Sports Nutrition: A Practice Manual for Professionals* (Chicago: Academy of Nutrition and Dietetics, 2012), p. 116.
85. Position of the American Dietetic Association, Dietitians of Canada, and the American College of Sports Medicine: Nutrition and athletic performance, 2009.
86. A. Z. Jamurtas and coauthors, The effects of low and high glycemic index foods on exercise performance and beta-endorphin responses, *Journal of the International Society of Sports Nutrition* 8 (2011), epub, doi:10.1186/1550–2783-8-15.
87. C. Rosenbloom, Food and fluid guidelines before, during, and after exercise, *Nutrition Today* 47 (2012): 63–69.
88. W. R. Lunn and coauthors, Chocolate milk and endurance exercise recovery: protein balance, glycogen, and performance, *Medicine and Science in Sports and Exercise* 44 (2012): 682–691.
89. Hawley and coauthors, Nutritional modulation of training-induced skeletal muscle adaptations, 2011.

Consumer's Guide 10

1. American College of Sports Medicine, Position stand, exercise and fluid replacement, *Medicine and Science in Sports and Exercise* 39 (2007): 377–390.
2. D. C. Nieman and coauthors, Bananas as an energy sources during exercise: a metabolomics approach, *PLoS ONE* 7 (2012), epub, doi:10.1371/journal.pone.0037479.

Controversy 10

1. R. B. Kreider and coauthors, ISSN exercise & sport nutrition review: Research and recommendations, *Journal of the International Society of Sports Nutrition* 7 (2010), epub, doi:10.1186/1550–2783–7-7.
2. B. Östeman and coauthors, Coenzyme Q10 supplementation and exercise-induced oxidative stress in humans, *Nutrition* 28 (2012): 403–417; E. C. Gomes and coauthors, Effect of vitamin supplementation on lung injury and running performance in a hot, humid, and ozone-polluted environment, *Scandinavian Journal of Medicine and Science in Sports* 21 (2011): e452–460.
3. S. K. Powers, E. E. Talbert, and P. Adhihetty, Reactive oxygen and nitrogen species as intracellular signals in skeletal muscle, *Journal of Physicology*, 589 (2011): 2129–2138.
4. T. T. Peternelj and J. S. Coombes, Antioxidant supplementation during exercise training: beneficial or detrimental? *Sports Medicine* 41 (2011): 1043–1069; M. Ristow and coauthors, Antioxidants prevent health-promoting effects of physical exercise in humans, *Proceedings of the National Academy of Sciences* 106 (2009): 8665–8670.
5. T. A. Astorino and coauthors, Increases in cycling performance in response to caffeine ingestion are repeatable, *Nutrition Research* 32 (2012): 78–84; Kreider and coauthors, 2010; E. R. Goldstein and coauthors, International Society of Sports Nutrition position stand: Caffeine and performance, *Journal of the International Society of Sports Nutrition* 7 (2010), epub, doi:10.1186/1550-2783-7-5; G. L. Warren and coauthors, Effect of caffeine ingestion on muscular strength and endurance: A meta-analysis, *Medicine and Science in Sports and Exercise* 42 (2010): 1375–1387; J. K. Davis and J. M. Green, Caffeine and anaerobic performance: Ergogenic value and mechanisms of action, *Sports Medicine* 39 (2009): 813–832.
6. S. A. Conger and coauthors, Does caffeine added to carbohydrate provide additional ergogenic benefit for endurance? *International Journal of Sport Nutrition and Exercise Metabolism* 21 (2011): 71–84; M. J. Duncan and S. W. Oxford, The effect of caffeine ingestion on mood state and bench press performance to failure, *Journal of Strength and Conditioning Research* 25 (2011): 178–185; K. J. Pontifex and coauthors, Effects of caffeine on repeated sprint ability, reactive agility time, sleep and next day performance, *Journal of Sports Medicine and Physical Fitness* 50 (2010): 455–464.
7. Astorino and coauthors, Increases in cycling performance in response to caffeine ingestion are repeatable, 2012.
8. P. Jain and coauthors, A comparison of sports and energy drinks—physiochemical properties and enamel dissolution, *General Dentistry* 60 (2012): 190–197; W. Doyle and coauthors, The effects of energy beverages on cultured cells, *Food and Chemical Toxicology* 50 (2012): 3759–3768.
9. Standing Committee on the Scientific Evaluation of Dietary Reference Intakes, Food and Nutrition Board, Institute of Medicine, *Dietary Reference Intakes: Water, Potassium, Sodium, Chloride, and Sulfate* (Washington, D.C.: National Academies Press, 2004), pp. 269–423.
10. F. B. Stephens and coauthors, Vegetarians have a reduced skeletal muscle carnitine transport capacity, *American Journal of Clinical Nutrition* 94 (2011): 938–944.
11. Position of the American Dietetic Association, Dietitians of Canada, and the American College of Sports Medicine: Nutrition and athletic performance, *Journal of the American Dietetic Association* 109 (2009): 509–527.
12. R. Jäger and coauthors, Analysis of the efficacy, safety, and regulatory status of novel forms of creatine, *Amino Acids* 40 (2011): 1369–1383; M. A. Tarnopolsky, Caffeine and creatine use in sport, *Annals of Nutrition and Metabolism* 57 (2010): 1–8 (Supplement 2); D. H. Fukuda and coauthors, the effects of creatine loading and gender on anaerobic running capacity, *Journal of Strength and Conditioning Research* 24 (2010): 1826–1833; K. L. Kendall and coauthors, Effects of four weeks of high-intensity interval training and creatine supplementation on critical power and anaerobic working capacity in college-aged men, *Journal of Strength and Conditioning Research* 23 (2009): 1663–1669.
13. M. G. Bemben and coauthors, The effects of supplementation with creatine and protein on muscle strength following a traditional resistance training program in middle-aged and older men, *The Journal of Nutrition Health and Aging* 14 (2010): 155–159; M. Spillane and coauthors, The effects of creatine ethyl ester supplementation combined with heavy resistance training on body composition, muscle performance, and serum and muscle creatine levels, *Journal of the International Society of Sports Nutrition* 6 (2009), epub, doi:10.1186/1550–2783–6-6.
14. R. Harris, Creatine in health, medicine and sport, *Amino Acids* 40 (2011): 1267–1270.
15. R. M. Lopez and coauthors, Does creatine supplementation hinder exercise heat tolerance or hydration status? A systematic review with meta-analyses, *Journal of Athletic Training* 44 (2009): 215–223.
16. A. J. Carr, W. G. Hopkins, and C. J. Gore, Effects of acute alkalosis and acidosis on performance, *Sports Medicine* 41 (2011): 801–814.
17. J. Caruso and coauthors, Ergogenic effects of β-alanine and carnosine: proposed future research to quantify their efficacy, *Nutrients* 4 (2012): 585–601.
18. R. M. Hobson and coauthors, Effects of β-alanine supplementation on exercise performance: a meta-analysis, *Amino Acids* 43 (2012): 25–37; A. R. Jagim and coauthors, Effects of beta-alanine supplementation on sprint endurance, *Journal of Strength and Conditioning Research*, April 2012, epub ahead of print, doi:10.1519/JSC.0b013e318256bedc; A. E. Smith and coauthors, Exercise-induced oxidative stress: the effects of β-alanine supplementation in women, *Amino Acids* 43 (2012): 77–90; A. E. Smith-Ryan and coauthors, High-velocity intermittent running: effects of beta-alanine supplementation, *Journal of Strength and Conditioning Research*, July 2012, epub ahead of print, doi:10.1519/JSC.0b013e318267922b.
19. P. J. Atherton and K. Smith, Muscle protein synthesis in response to nutrition and exercise, *Journal of Physiology* 590.5 (2012): 1049–1057.

20. T. A. Churchward-Venne, N. A. Burd, and S. M. Phillips, Nutrition regulation of muscle protein synthesis with resistance exercise: strategies to enhance anabolism, *Nutrition and Metabolism* 9 (2012), epub, doi:10.1186/1743-7075-9-40; N. Gwacham and D. R. Wagner, Acute effects of a caffeine-taurine energy drink on repeated sprint performance of American college football players, *International Journal of Sport Nutrition and Exercise Metabolism* 22 (2012): 109–116; J. E. Tang and coauthors, Bolus arginine supplementation affects neither muscle blood flow nor muscle protein synthesis in young men at rest or after resistance exercise, *Journal of Nutrition* 141 (2011): 195–200.

21. L. E. Norton and coauthors, Leucine content of dietary proteins is a determinant of postprandial skeletal muscle protein synthesis in adult rats, *Nutrition and Metabolism* 9 (2012), epub, doi:10.1186/1743-7075-9-67.

22. K. D. Weisgarber, D. G. Candow, and E. S. Vogt, Whey protein before and during resistance exercise has no effect on muscle mass and strength in untrained young adults, *International Journal of Sport Nutrition and Exercise Metabolism* 22 (2012): 463–469; S. Graf, S. Egert, and M. Heer, Effects of whey protein supplements on metabolism: evidence from human intervention studies, *Current Opinion in Clinical Nutrition and Metabolic Care* 14 (2011): 569–580.

23. P. A. Cohen, DMAA as a dietary supplement ingredient, *Archives of Internal Medicine*, 172 (2012): 1038–1039; FDA challenges marketing of DMAA products for lack of safety evidence, *News and Events,* April 27, 2012, available at www.fda.gov/NewsEvents/Newsroom/PressAnnouncements/ucm302133.htm; P. Gee, S. Jackson, and J. Easton, Another bitter pill: A case of toxicity from DMAA party pills, *New Zealand Medical Journal* 123 (2012): 124–126.

24. P. Watson and coauthors, Urinary nandrolone metabolite detection after ingestion of a nandrolone precursor, *Medicine and Science in Sports and Exercise* 41 (2009): 766–772.

25. D. G. Finniss and coauthors, Biological, clinical, and ethical advances of placebo effects, *Lancet* 375 (2010): 686–695.

Chapter 11

1. Centers for Disease Control and Prevention, Plan to combat extensively drug-resistant tuberculosis: Recommendations of the Federal Tuberculosis Task Force, *Morbidity and Mortality Weekly Report* 58 (2009): 1–43.

2. N. J. Afacan, C. D. Fjell, and R. E. Hancock, A systems biology approach to nutritional immunology—Focus on innate immunity, *Molecular Aspects of Medicine* 33 (2012): 14–25; I. Laaksi, Vitamin D and respiratory infection in adults, *Proceedings of the Nutrition Society* 71 (2012): 90–97; J. M. Monk, T. Y. Hou, and R. S. Chapkin, Recent advances in the field of nutritional immunology, *Expert Review of Clinical Immunology* 7 (2011): 747–749.

3. S. L. Murphy and coauthors, Deaths: Preliminary data for 2010, *National Vital Statistics Reports* 60 (2012): 1–59.

4. Affacan, Fjell, and Hancock, A systems biology approach to nutritional immunology—Focus on innate immunity; Laaksi, Vitamin D and respiratory infection in adults, 2012; S. S. Percival, Nutrition and immunity: Balancing diet and immune function, *Nutrition Today* 46 (2011): 12–17; C. E. Taylor and C. A. Camargo Jr., Impact of micronutrients on respiratory infections, *Nutrition Reviews* 69 (2011): 259–269.

5. E. M. Gardner and coauthors, Energy intake and response to infection with influenza, *Annual Review of Nutrition* 31 (2011): 353–367.

6. Standing Committee on the Scientific Evaluation of Dietary Reference Intakes, Food and Nutrition Board, Institute of Medicine, *Dietary Reference Intakes for Vitamin A, Vitamin K, Arsenic, Boron, Chromium, Copper, Iodine, Iron, Manganese, Molybdenum, Nickel, Silicon, Vanadium, and Zinc* (Washington, D.C.: National Academies Press, 2001), pp. 442–501.

7. Position of the American Dietetic Association: Nutrition intervention and human immunodeficiency virus infection, *Journal of the American Dietetic Association* 110 (2010): 1105–1119.

8. B. K. Surmi and A. H. Hasty, the role of chemokines in recruitment of immune cells to the artery wall and adipose tissue, *Vascular Pharmacology* 52 (2010): 27–36.

9. J. D. Berry and coauthors, Lifetime risks of cardiovascular disease, *New England Journal of Medicine* 366 (2012): 321–329; S. E. Chiuve, and coauthors, Adherence to a low-risk, healthy lifestyle and risk of sudden cardiac death among women, *Journal of the American Medical Association* 306 (2011): 62–69; A. Astrup and coauthors, The role of reducing intakes of saturated fat in the prevention of cardiovascular disease: Where does the evidence stand in 2010? *American Journal of Clinical Nutriton* 93 (2011): 684–688; A. Mente and coauthors, A systematic review of the evidence supporting a causal link between dietary factors and coronary heart disease, *Archives of Internal Medicine* 169 (2009): 659–669; D. Li and coauthors, Body mass index and risk, age of onset, and survival in patients with pancreatic cancer, *Journal of the American Medical Association* 301 (2009): 2553–2562; A. Galimanis and coauthors, Lifestyle and stroke risk: A review, *Current Opinion in Neurology* 22 (2009): 60–68; M. U. Jakobsen and coauthors, Major types of dietary fat and risk of coronary heart disease: A pooled analysis of 11 cohort studies, *American Journal of Clinical Nutrition* 89 (2009): 1425–1432.

10. V. L. Roger and coauthors, Heart disease and stroke statistics—2012 update: A report from the American Heart Association, *Circulation* 125 (2012): e12–e230.

11. V. L. Roger and coauthors, Heart disease and stroke statistics—2012 update: A report from the American Heart Association, 2012.

12. C. E. Murry and R. T. Lee, Turnover after the fallout, *Science* 324 (2009): 47–48.

13. V. L. Roger and coauthors, Heart disease and stroke statistics—2012 update: A report from the American Heart Association, 2012; A. Towfighi, L. Zheng, and B. Ovbiagele, Sex-specific trends in midlife coronary heart disease risk and prevalence, *Archives of Internal Medicine* 169 (2009): 1762–1766.

14. L. Mosca and coauthors, Effectiveness-based guidelines for the prevention of cardiovascular disease in women—2011 update: A guideline from the American Heart Association, *Circulation* 123 (2011): 1243–1262.

15. S. E. Chiuve, and coauthors, Adherence to a low-risk, healthy lifestyle and risk of sudden cardiac death among women, *Journal of the American Medical Association* 306 (2011): 62–69; A. Astrup and coauthors, The role of reducing intakes of saturated fat in the prevention of cardiovascular disease: Where does the evidence stand in 2010? *American Journal of Clinical Nutriton* 93 (2011): 684–688; Mente and coauthors, A systematic review of the evidence supporting a causal link between dietary factors and coronary heart disease; S. S. Gidding, Implementing American Heart Association pediatric and adult nutrition guidelines: A scientific statement from the American Heart Association Nutrition Committee of the Council on Nutrition, Physical Activity and Metabolism, Council on Cardiovascular Disease in the Young, Council on Arteriosclerosis, Thrombosis and Vascular Biology, Council on Cardiovascular Nursing, Council on Epidemiology and Prevention, and Council for High Blood Pressure Research, *Circulation* 119 (2009): 1161–1175.

16. W. Insull, The pathology of atherosclerosis: Plaque development and plaque responses to medical treatment, *The American Journal of Medicine* 122 (2009): S3–S14.

17. Astrup and coauthors, The role of reducing intakes of saturated fat in the prevention of cardiovascular disease: Where does the evidence stand in 2010?, 2011; R. J. Berlin and coauthors, Diet quality and the risk of cardiovascular disease: The Women's Health Initiative (WHI), *American Journal of Clinical Nutrition* 94 (2011): 49–57; Jakobsen and coauthors, Major types of dietary fat and risk of coronary heart disease, 2009.

18. G. K. Hansson and A. Hermansson, The immune system in atherosclerosis, *Nature Immunology* 12 (2011): 204–212; M. Drechsler and coauthors, Neutrophilic granulocytes—promiscuous accelerators of atherosclerosis, *Thrombosis and Haemostasis* 106 (2011): 839–848; B. K. Surmi and A. H. Hasty, The role of chemokines in recruitment of immune cells to the artery wall and adipose tissue, *Vascular Pharmacology* 52 (2010): 27–36.

19. Insull, The pathology of atherosclerosis, 2009.

20. Insull, The pathology of atherosclerosis, 2009.

21. W. S. Harris and coauthors, Omega-6 fatty acids and risk for cardiovascular disease, A Science Advisory from the American Heart Association Nutrition Subcommittee of the Council on Nutrition, Physical Activity, and Metabolism; Council on Cardiovascular Nursing; and Council on Epidemiology and Prevention, *Circulation* 119 (2009): 902–907.

22. R. J. Belin and coauthors, Fish intake and the risk of incident heart failure: The Women's Health Initiative, *Circulation. Heart Failure* 4 (2011): 404–413; K. J. Newens and coauthors, DHA-rich fish oil reverses the detrimental effects of saturated fatty acids on postprandial vascular reactivity, *American Journal of Clinical Nutrition* 94 (2011): 742–748; P. C. Calder and P. Yagoob, Omega-3 (n-3) fatty acids, cardiovascular disease and stability of atherosclerotic plaques, *Cellular and Molecular Biology* 56 (2010): 28–37.

23. N. T. Artinian and coauthors, Interventions to promote physical activity and dietary lifestyle changes for cardiovascular risk factor reduction in adults: A Scientific Statement from the American Heart Association, *Circulation* 122 (2010): 406–441; C. J. Lavie, R. V. Milani, and H. O. Ventura, Obesity and cardiovascular disease: Risk factor, paradox, and impact of weight loss, *Journal of the American College of Cardiology* 53 (2009): 1925–1932.

24. Q. Yang and coauthors, Trends in cardiovascular health metrics and association with all-cause and CVD mortality, *Journal of the American Medical Association* 307 (2012): 1273–1283.

25. G. F. Tomaselli, Prevention of cardiovascular disease and stroke: Meeting the challenge, *Journal of the American Medical Association* 306 (2011): 2147–2148.

26. Insull, The pathology of atherosclerosis, 2009; Expert Panel on Detection, Evaluation, and Treatment of High Blood Cholesterol in Adults (Adult Treatment Panel III), Third Report of the National Cholesterol Education Program (NCEP), NIH publication no. 02–5215 (Bethesda, Md.: National Heart, Lung, and Blood Institute, 2002), p. II-18.

27. J. D. Berry and coauthors, Lifetime risks of cardiovascular disease, *New England Journal of Medicine* 366 (2012): 321–329.

28. V. L. Roger and coauthors, Heart disease and stroke statistics—2012 update: A report from the American Heart Association, 2012; L. Mosca and coauthors, Effectiveness-based guidelines for the prevention of cardiovascular disease in women—2011 update: A guideline from the American Heart Association, *Circulation* 123 (2011): 1243–1262.

29. V. L. Roger and coauthors, Heart disease and stroke statistics—2012 update: A report from the American Heart Association, 2012; Expert Panel on Detection, Evaluation, and Treatment of High Blood Cholesterol in Adults (Adult Treatment Panel III). Third Report of the National Cholesterol Education Program, 2002, p. II-19.

30. C. J. O'Donnell and E. G. Nabel, Genomics of cardiovascular disease, *New England Journal of Medicine* 365 (2011): 2098–2109; J. M. Ordovas, Genetic influences on blood lipids and cardiovascular disease risk: Tools for primary prevention, *American Journal of Clinical Nutrition* 89 (2009): 1509S–1517S.

31. M. H. Davidson, HDL and CETP inhibition: Will this DEFINE the future? *Current Treatment Options in Cardiovascular Medicine* 14 (2012): 384–390; A. K. Mahdy and coauthors, Cardiovascular disease risk reduction by raising HDL cholesterol—current therapies and future opportunities, *British Journal of Pharmacology* June 22, 2012, epub ahead of print, doi:10.1111/j.1476-5381.2012.02081.x.

32. A. K. Chhatriwalla and coauthors, Low levels of low-density lipoprotein cholesterol and blood pressure and progression of coronary atherosclerosis, *Journal of the American College of Cardiology* 53 (2009): 1110–1115.

33. Insull, The pathology of atherosclerosis, 2009.

34. P. J. Murray and T. A. Wynn, Protective and pathogenic functions of macrophage subsets, *Nature Reviews. Immunology* 11 (2011): 723–737; Surmi and Hasty, The role of chemokines in recruitment of immune cells to the artery wall and adipose tissue, 2010; Insull, The pathology of atherosclerosis, 2009.

35. E. S. Ford, Trends in the risk for coronary heart disease among adults with diagnosed diabetes in U.S.: Findings from the National Health and Nutrition Examination Survey, 1999–2008, *Diabetes Care* 34 (2011): 1337–1343; R. A. DeFronzo and M. Abdul-Ghani, Assessment and treatment of cardiovascular risk in prediabetes: Impaired glucose tolerance and impaired fasting glucose, *American Journal of Cardiology* 108 (2011): 3B–24B.

36. Expert Panel on Detection, Evaluation, and Treatment of High Blood Cholesterol in Adults (Adult Treatment Panel III), Third Report of the National Cholesterol Education Program, 2002, pp. II-16, II-50–II-53.

37. M. Hamer and coauthors, Physical activity and cardiovascular mortality risk: Possible protective mechanisms? *Medicine and Science in Sports and Exercise* 44 (2012): 84–88; J. Sattelmair and coauthors, Dose response between physical activity and risk of coronary heart disease: A meta-analysis, *Circulation* 124 (2011): 789–795; H. M. Ahmed and coauthors, Effects of physical activity on cardiovascular disease, *American Journal of Cardiology* 109 (2011): 288–295; A. Y. Arikawa and coauthors, Sixteen weeks of exercise reduces C-reactive protein levels in young women, *Medicine and Science in Sports and Exercise* 43 (2011): 1002–1009; R. J. F. Manders, J. W. M. Van Dujk, and L. J. C. Van Loon, Low-intensity exercise reduces the prevalence of hyperglycemia in type 2 diabetes, *Medicine and Science in Sports and Exercise* 42 (2010): 219–225; T. Y. Warren and coauthors, Sedentary behaviors increase risk of cardiovascular disease mortality in men, *Medicine and Science in Sports and Exercise* 42 (2010): 879–885.

38. Expert Panel on Detection, Evaluation, and Treatment of High Blood Cholesterol in Adults (Adult Treatment Panel III), Third Report of the National Cholesterol Education Program, 2002, p. II-16.

39. I. A. Brouwer, A. J. Wanders, and M. B. Katan, Effect of animal and industrial trans fatty acids on HDL and LDL cholesterol levels in humans—A quantitative review, *PLoS One* 5 (2010): e9434; D. Mozaffarian, A. Aro, and W.C. Willett, Health effects of trans-fatty acids: Experimental and observational evidence, *European Journal of Clinical Nutrition* 63 (2009): S5–S21.

40. Mente and coauthors, A systematic review of the evidence supporting a causal link between dietary factors and coronary heart disease, 2009.

41. D. Mozaffarian and J. H. Y. Wu, Omega-3 fatty acids and heart disease, *Journal of the American College of Cardiology* 58 (2011): 2047–2067; D. J. A. Jenkins and coauthors, Effect of a dietary portfolio of cholesterol-lowering foods given at 2 levels of intensity of dietary advice on serum lipids in hyperlipidemia, *Journal of the American Medical Association* 306 (2011): 831–839; R. J. Belin and coauthors, Diet quality and the risk of cardiovascular disease: The Women's Health Initiative (WHI), *American Journal of Clinical Nutrition* 94 (2011): 49–57; X. Zhang and coauthors, Cruciferous vegetable consumption is associated with a reduced risk of total and cardiovascular disease mortality, *American Journal of Clinical Nutrition* 94 (2011): 240–246; D. K. Banel and F.B. Hu, Effects of walnut consumption on blood lipids and other cardiovascular risk factors: A meta-analysis and systematic review, *American Journal of Clinical Nutrition* 90 (2009): 56–63.

42. R. Do and coauthors, The effect of chromosome 9p21 variants on cardiovascular disease may be modified by dietary intake: Evidence from a case/control and a prospective study, *PLoS Medicine* (2011): e1001106.

43. E. J. Gallagher, D. Leroith, and E. Karnieli, Insulin resistance in obesity as the underlying cause for the metabolic syndrome, *Mount Sinai Journal of Medicine* 77 (2010): 511–523.

44. S. M. Grundy, Pre-diabetes, metabolic syndrome, and cardiovascular risk, *Journal of the American College of Cardiology* 59 (2012): 635–643; Gallagher, Leroith, and Karnieli, Insulin resistance in obesity as the underlying cause for the metabolic syndrome, 2010; F. R. Jornayvaz, V. T. Samuel, and G. I. Shulman, The role of muscle insulin resistance in the pathogenesis of atherogenic dyslipidemia and nonalcoholic fatty liver disease associated with metabolic syndrome, *Annual Review of Nutrition* 30 (2010): 273–290.

45. P. Bhargava and C. H. Lee, Role and function of macrophages in metabolic syndrome, *Biochemical Journal* 442 (2012): 253–262; R. Lorenzet and coauthors, Thrombosis and obesity: Cellular bases, *Thrombosis Research* 129 (2012): 285–289; J. S. Chae and coauthors, Association of Lp-PLA(2) activity and LDL size with interleukin-6, an inflammatory cytokine and oxidized LDL, a marker of oxidative stress, in women with metabolic syndrome, *Atherosclerosis* 218 (2011): 499–506.

46. Roger and coauthors, Heart disease and stroke statistics—2012 update: A report from the American Heart Association, 2012; R. B. Ervin, Prevalence of metabolic syndrome among adults 20 years of age and over, by sex, age, race and ethnicity, and body mass index: United States, 2003–2006, *National Health Statistics Reports*, May 5, 2009.

47. T. Coutinho and coauthors, Central obesity and survival in subjects with coronary artery disease: A systematic review of the literature and collaborative analysis with individual subject data, *Journal of the American College of Cardiology* 57 (2011): 1877–1886; F. Schouten and coauthors, Increases in central fat mass and decreases in peripheral fat mass are associated with accelerated arterial stiffening in healthy adults: The Amsterdam Growth and Health Longitudinal Study, *American Journal of Clinical Nutrition* 94 (2011): 40–48; S. S. Satvinder and T. A. Welborn, Central obesity and multivariable cardiovascular risk as assessed by the Framingham prediction scores, *American Journal of Cardiology* 103 (2009): 1403–1407.
48. J. Sattelmair and coauthors, Dose response between physical activity and risk of coronary heart disease: A meta-analysis, *Circulation* 124 (2011): 789–795; H. M. Ahmed and coauthors, Effects of physical activity on cardiovascular disease, *American Journal of Cardiology* 109 (2011): 288–295; U. Ekelund and coauthors, Physical activity and gain in abdominal adiposity and body weight: Prospective cohort study in 288,498 men and women, *American Journal of Clinical Nutrition* 93 (2011): 826–835; A. K. Chomistek and coauthors, Vigorous physical activity, mediating biomarkers, and risk of myocardial infarction, *Medicine and Science in Sports and Exercise* 43 (2011): 1884–1890; T. Y. Warren and coauthors, Sedentary behaviors increase risk of cardiovascular disease mortality in men, *Medicine and Science in Sports and Exercise* 42 (2010): 879–885; M. Hamer and E. Stamatakis, Physical activity and risk of cardiovascular disease events: Inflammatory and metabolic mechanisms, *Medicine and Science in Sports and Exercise* 41 (2009): 1206–1211; F. Magkos and coauthors, Management of the metabolic syndrome and type 2 diabetes through lifestyle modification, *Annual Review of Nutrition* 29 (2009): 223–256.
49. M. Miller and coauthors, Triglycerides and cardiovascular disease: A Scientific Statement from the American Heart Association, *Circulation* 123 (2011): 2292–2333.
50. S. E. Chiuve and coauthors, Adherence to a low-risk, healthy lifestyle and risk of sudden cardiac death among women, *Journal of the American Medical Association* 306 (2011): 62–69; S. S. Gidding and coauthors, Implementing American Heart Association pediatric and adult nutrition guidelines: A Scientific Statement from the American Heart Association Nutrition Committee of the Council on Nutrition, Physical Activity and Metabolism, Council on Cardiovascular Disease in the Young, Council on Arteriosclerosis, Thrombosis and Vascular Biology, Council on Cardiovascular Nursing, Council on Epidemiology and Prevention, and Council for High Blood Pressure Research, *Circulation* 119 (2009): 1161–1175.
51. AHA Scientific Statement: Diet and lifestyle recommendations revision 2006, *Circulation* 114 (2006): 82–96; Expert Panel on Detection, Evaluation, and Treatment of High Blood Cholesterol in Adults (Adult Treatment Panel III), Third Report of the National Cholesterol Education Program, 2002, pp. V-1–V-28.
52. L. de Koning and coauthors, Sweetened beverage consumption, incident coronary heart disease and biomarkers of risk in men, *Circulation* 125 (2012): 1735–1741; P. W. Siri-Tarino and coauthors, Saturated fat, carbohydrate, and cardiovascular disease, *American Journal of Clinical Nutrition* 91 (2010): 502–509; Mente, A systematic review of the evidence supporting a causal link between dietary factors and coronary heart disease, 2009; G. Radulian and coauthors, Metabolic effects of low glycaemic index diets, *Nutrition Journal* 8 (2009), 5.
53. Miller and coauthors, Triglycerides and cardiovascular disease: A scientific statement from the American Heart Association, 2011; M. U. Jakobsen and coauthors, Intake of carbohydrates compared with intake of saturated fatty acids and risk of myocardial infarction: Importance of the glycemic index, *American Journal of Clinical Nutrition* 91 (2010): 1764–1768.
54. D. Mozaffarian and J. H. Y. Wu, Omega-3 fatty acids and cardiovascular disease, *Journal of the American College of Cardiology* 58 (2011): 2047–2067; R. J. Belin and coauthors, Fish intake and risk of incident heart failure: The women's Health Initiative, *Circulation: Heart Failure* 4 (2011): 404–413; Mente, A systematic review of the evidence supporting a causal link between dietary factors and coronary heart disease, 2009.
55. A. C. Skulas-Ray and coauthors, Dose-response effects of omega-3 fatty acids on triglycerides, inflammation, and endothelial function in healthy persons with moderate hypertriglyceridemia, *American Journal of Clinical Nutrition* 93 (2011): 243–252; J. V. Patel and coauthors, Omega-3 polyunsaturated fatty acids: A necessity for a comprehensive secondary prevention strategy, *Vascular Health and Risk Management* 5 (2009): 801–810.
56. S. M. Kwak and coauthors, Efficacy of omega-3 fatty acid supplements (eicosapentaenoic acid and docosahexaenoic acid) in the secondary prevention of cardiovascular disease, *Archives of Internal Medicine* 172 (2012): 686–694.
57. J. W. Anderson and H. M. Bush, Soy protein effects on serum lipoproteins: A quality assessment and meta-analysis of randomized, controlled studies, *Journal of the American College of Nutrition* 30 (2011): 79–91.
58. H. A. Tindle and coauthors, Optimism, cynical hostility, and incident coronary heart disease and mortality in the Women's Health Initiative, *Circulation* 120 (2009): 656–662; R. Lampert and coauthors, Anger-induced T-wave alternans predicts future ventricular arrhythmias in patients with implantable cardioverter-defibrillators, *Journal of the American College of Cardiology* 53 (2009): 774–778.
59. Chhatriwalla and coauthors, Low levels of low-density lipoprotein cholesterol and blood pressure, 2009.
60. Roger and coauthors, Heart disease and stroke statistics—2012 update: A report from the American Heart Association, 2012.
61. Roger and coauthors, Heart disease and stroke statistics—2012 update: A report from the American Heart Association, 2012.
62. L. J. Appel, American Society of Hypertension Writing Group, T. D. Giles and coauthors, ASH Position paper: Dietary approaches to lower blood pressure, *Journal of Clinical Hypertension* 11 (2009): 358–368.
63. Roger and coauthors, Heart disease and stroke statistics—2012 update: A report from the American Heart Association, 2012.
64. C. W. Mende, obesity and hypertension: A common coexistence, *Journal of Clinical* Hypertension 14 (2012): 137–138; J. Redon and coauthors, Mechanisms of hypertension in the cardiometabolic syndrome, *Journal of Hypertension* 27 (2009): 441–451; F. W. Visser and coauthors, Rise in extracellular fluid volume during high sodium depends on BMI in healthy men, *Obesity (Silver Spring)* 17 (2009): 1684–1688.
65. V. Savica, G. Bellinghieri, and J. D. Kopple, The effect of nutrition on blood pressure, *Annual Review of Nutrition* 30 (2010): 365–401; R. Takachi and coauthors, Consumption of sodium and salted foods in relation to cancer and cardiovascular disease: The Japan Public Health Center-based Prospective Study, *American Journal of Clinical Nutrition* 91 (2010): 456–464; F. J. He and G. A. MacGregor, A comprehensive review on salt and health and current experience of worldwide salt reduction programmes, *Journal of Human Hypertension* 23 (2009): 363–384.
66. Savica, Bellinghieri, and Kopple, The effect of nutrition on blood pressure, 2010.
67. A. J. Flint and coauthors, Whole grains and incident hypertension in men, *American Journal of Clinical Nutrition* 90 (2009): 493–498; J. P. Forman and coauthors, Diet and lifestyle risk factors associated with incident hypertension in women, *Journal of the American Medical Association* 302 (2009): 401–411.
68. L. Azadbakht and coauthors, The Dietary Approaches to Stop Hypertension Eating Plan affects C-reactive protein, coagulation abnormalities, and hepatic function tests among type 2 diabetic patients, *Journal of Nutrition* 141 (2011): 1083–1088; Y. Al-Solaiman and coauthors, DASH lowers blood pressure in obese hypertensives beyond potassium, magnesium and fibre, *Journal of Human Hypertension*, 24 (2010): 237–246.
69. L. J. Appel, On behalf of the American society of Hypertension Writing Group, ASH Position Paper: Dietary approaches to lower blood pressure, *Journal of Clinical Hypertension* 11 (2009): 358–368.
70. E. G. Ciolac and coauthors, Effects of high-intensity aerobic interval training vs. moderate exercise on hemodynamic, metabolic and neurohumoral abnormalities of young normotensive women at high familial risk for hypertension, *Hypertension Research* 33 (2010): 836–843; N. L. Chase and coauthors, The association of cardiorespiratory fitness and physical activity with incidence of hypertension in men, *American Journal of Hypertension* 22 (2009): 417–424.

71. S. Mohan and N. R. Campbell, Salt and high blood pressure, *Clinical Science (London)* 117 (2009): 1–11.
72. F. Dumier, Dietary sodium intake and arterial blood pressure, *Journal of Renal Nutrition* 19 (2009): 57–60; U.S. Department of Health and Human Services, National Institutes of Health, National Heart, Lung, and Blood Institute, *Your Guide to Lowering Your Blood Pressure with DASH* (NIH Publication No. 06–4082, 2006).
73. U.S. Department of Agriculture and U.S. Department of Health and Human Services, *Dietary Guidelines for Americans 2010*, available at www.dietaryguidelines.gov.
74. K. M. Dickinson, J. B. Keogh, and P. M. Clifton, Effects of a low-salt diet on flow-mediated dilation in humans, *American Journal of Clinical Nutrition* 89 (2009): 485–490.
75. Standing Committee on the Scientific Evaluation of Dietary Reference Intakes, Food and Nutrition Board, Institute of Medicine, *Dietary Reference Intakes for Water, Potassium, Sodium, Chloride, and Sulfate* (Washington, D.C.: National Academies Press, 2005): pp. 381–387.
76. *Dietary Guidelines for Americans 2010*, www.dietaryguidelines.gov; Centers for Disease Control and Prevention, Application of lower sodium intake recommendations to adults—United States, 1999–2006, *Morbidity and Mortality Weekly Report* 58 (2009): 281–283.
77. Savica, Bellinghieri, and Kopple, The effect of nutrition on blood pressure, 2010; X. Trudel and coauthors, Masked hypertension: Different blood pressure measurement methodology and risk factors in a working population, *Journal of Hypertension* 27 (2009): 1560–1567.
78. *Dietary Guidelines for Americans 2010*, www.dietaryguidelines.gov.
79. I. R. Reid and coauthors, Effects of calcium supplementation on lipids, blood pressure, and body composition in healthy older men: A randomized controlled trial, *American Journal of Clincial Nutrition* 91 (2010): 131–139; P. M. Kris-Etherton and coauthors, Milk products, dietary patterns and blood pressure management, *Journal of the American College of Nutrition* 28 (2009): 103S–119S; L. Wang and coauthors, Dietary intake of dairy products, calcium, and vitamin D and the risk of hypertension in middle-aged and older women, *Hypertension* 51 (2008): 1073–1079.
80. Savica, Bellinghieri, and Kopple, The effect of nutrition on blood pressure, 2010.
81. L. Cahill, P. N. Corey, and A. El-Sohemy, Vitamin C deficiency in a population of young Canadian adults, *American Journal of Epidemiology* 170 (2009): 464–471.
82. Savica, Bellinghieri, and Kopple, The effect of nutrition on blood pressure, 2010.
83. R. Siegel, D. Naishadham, and A. Jemal, Cancer statistics, 2012, *CA: Cancer Journal for Clinicians* 62 (2012): 10–29.
84. Siegel, Naishadham, and Jemal, Cancer statistics, 2012, 2012; N. J. Meropol and coauthors, American Society of Clinical Oncology Guidance Statement: The cost of cancer care, *Journal of Clinical Oncology* 27 (2009): 3868–3874.
85. D. Romaguera and coauthors, Is concordance with World Cancer Research Fund/American Institute for Cancer research guidelines for cancer prevention related to subsequent risk of cancer? Results from the EPIC study, *American Journal of Clinical Nutrition* 96 (2012): 150–163; L. H. Kushi and coauthors, American Cancer Society Guidelines on Nutrition and Physical Activity for Cancer Prevention, *CA: Cancer Journal for Clinicians* 62 (2012): 30–67; World Cancer Research Fund/American Institute for Cancer Research, *Policy and Action for Cancer Prevention—Food, Nutrition, and Physical Activity: A Global Perspective* (Washington, D.C.: AICR, 2009), pp. 12–28.
86. Kushi and coauthors, American Cancer Society Guidelines on Nutrition and Physical Activity for Cancer Prevention, 2012; L. D'Elia and coauthors, Habitual salt intake and risk of gastric cancer: A meta-analysis of prospective studies, *Clinical Nutrition* 31 (2012): 489–498; World Cancer Research Fund/American Institute for Cancer Research, Continuous Update Project, Colorectal Cancer Report 2010 Summary, *Food, Nutrition, Physical Activity, and the Prevention of Colorectal Cancer*, 2011.
87. C. Eheman and coauthors, Annual Report to the Nation on the status of cancer, 1975–2008, featuring cancers associated with excess weight and lack of sufficient physical activity, *Cancer* 118 (2012): 2338–2366; World Cancer Research Fund/American Institute for Cancer Research, Continuous Update Project, Colorectal Cancer Report 2010 Summary, *Food, Nutrition, Physical Activity, and the Prevention of Colorectal Cancer*, 2011.
88. K. B. Michels, The rise and fall of breast cancer rates, *British Medical Journal* 344 (2012), epub, doi:10.1136/bmj.d8003.
89. Kushi and coauthors, American Cancer Society Guidelines on Nutrition and Physical Activity for Cancer Prevention, 2012.
90. P. J. Murray and T. A. Wynn, Protective and pathogenic functions of macrophage subsets, *Nature Reviews/Immunology* 11 (2011): 723–737.
91. Kushi and coauthors, American Cancer Society Guidelines on Nutrition and Physical Activity for Cancer Prevention, 2012.
92. S. D. Hursting and coauthors, Calories and carcinogenesis: Lessons learned from 30 years of calorie restriction research, *Carcinogenesis* 31 (2010): 83–89.
93. S. Pendyala and coauthors, Diet-induced weight loss reduces colorectal inflammation: Implications for colorectal carcinogenesis, *American Journal of Clinical Nutrition* 93 (2011): 234–242; A. Vrieling and E. Kampman, The role of body mass index, physical activity, and diet in colorectal cancer recurrence and survival: A review of the literature, *American Journal of Clinical Nutrition* 92 (2010): 471–490; D. Li and coauthors, Body mass index and risk, age of onset, and survival in patients with pancreatic cancer, *Journal of the American medical Association* 301 (2009): 2553–2562; Y. Pan and M. DesMeules, Energy intake, physical activity, energy balance, and cancer: Epidemiologic evidence, *Methods in Molecular Biology* 472 (2009): 191–215.
94. N. H. Rod and coauthors, Low-risk factor profile, estrogen levels, and breast cancer risk among postmenopausal women, *International Journal of Cancer* 124 (2009): 1935–1940; World Cancer Research Fund/American Institute for Cancer Research, *Food, Nutrition, Physical Activity and the Prevention of Cancer*, 2007, pp. 30–46.
95. T. Boyle and coauthors, Timing and intensity of recreational physical activity and the risk of subsite-specific colorectal cancer, *Cancer Causes and Control* 22 (2011): 1647–1658; J. B. Peel and coauthors, A prospective study of cardiorespiratory fitness and breast cancer mortality, *Medicine and Science in Sports and Exercise* 41 (2009): 742–748; K. Y. Wolin and coauthors, Physical activity and colon cancer prevention: A meta-analysis, *British Journal of Cancer* 100 (2009): 611–616; J. B. Peel and coauthors, Cardiorespiratory fitness and digestive cancer mortality: Findings from the Aerobics Center Longitudinal Study, *Cancer Epidemiology Biomarkers Prevention*, 18 (2009): 1111–1117; S. Y. Pan and M. DesMeules, Energy intake, physical activity, energy balance, and cancer: Epidemiologic evidence, *Methods in Molecular Biology* 472 (2009): 191–215.
96. H. K. Na and S. Oliynyk, Effects of physical activity on cancer prevention, *Annals of the New York Academy of Sciences* 1229 (2011): 176–183; Peel and coauthors, A prospective study of cardiorespiratory fitness and breast cancer mortality, 2009; Wolin and coauthors, Physical activity and colon cancer prevention, 2009.
97. W. Y. Chen and coauthors, Moderate alcohol consumption during adult life, drinking patterns, and breast cancer risk, *Journal of the American Medical Association* 306 (2011): 1884–1890.
98. American Association for Cancer Research, High-fat/calorie diet accelerates development of pancreatic cancer, available at www.aacr.org/home/public—media/aacr-in-the-news.aspx?d=2816.
99. T. J. Key and coauthors, Dietary fat and breast cancer: Comparison of results from food diaries and food-frequency questionnaires in the UK Dietary Cohort Consortium, *American Journal of Clinical Nutrition* 94 (2011): 1043–1052; M. Gerber, Background review paper on total fat, fatty acid intake and cancers, *Annals of Nutrition and Metabolism* 55 (2009): 140–161.
100. M. Solanas and coauthors, Dietary olive oil and corn oil differentially affect experimental breast cancer through distinct modulation of the p21Ras signaling and the proliferation-apoptosis balance, *Carcinogenesis* 31 (2010): 871–879; J. Y. Lee, L. Zhao, and D. H. Hwang, Modulation of pattern recognition receptor-mediated inflammation and risk of chronic diseases by dietary fatty acids, *Nutrition Review* 68 (2009): 38–61.
101. T. M. Brasky and coauthors, Specialty supplements and breast cancer risk in the VITamins And Lifestyle (VITAL) Cohort, *Cancer Epidemiology, Biomarkers, and Prevention* 19 (2010): 1696–1708; Gerber, Background review paper on total fat, fatty acid intake and cancers, 2009.

102. D. S. Chan and coauthors, Red and processed meat and colorectal cancer incidence: Meta-analysis of prospective studies, *PLoS One* (2011), 6:e20456; A. Brevik and coauthors, Polymorphisms in base excision repair genes as colorectal cancer risk factors and modifiers of the effect of diets high in red meat, *Cancer, Epidemiology, Biomarkers, and Prevention* 19 (2010): 3167–3173; G. Randi and coauthors, Dietary patterns and the risk of colorectal cancer and adenomas, *Nutrition Reviews* 68 (2010): 389–408; S. Rohrmann, S. Hermann, and J. Linseisen, Herterocyclic aromatic amine intake increases colorectal adenoma risk: Findings from a prospective European cohort study, *American Journal of Clinical Nutrition* 89 (2009): 1418–1424; R. Sinha and coauthors, Meat intake and mortality: A prospective study of over half a million people, *Archives of Internal Medicine* 169 (2009): 562–571.
103. S. C. Larsson and A. Wolk, Red and processed meat consumption and risk of pancreatic cancer: Meta-analysis of prospective studies, *British Journal of Cancer* 106 (2012): 603–607; A. Wallin, N. Orsini, and A. Wolk, Red and processed meat consumption and risk of ovarian cancer: A dose-response meta-analysis of prospective studies, *British Journal of Cancer* 104 (2011): 1196–2011; D. D. Alexander and coauthors, A review and meta-analysis of red and processed meat consumption and breast cancer, *Nutrition Research and Review* 23 (2010): 349–365; N. Tasevska and coauthors, A prospective study of meat, cooking methods, meat mutagens, heme iron, and lung cancer risks, *American Journal of Clinical Nutrition* 89 (2009): 1884–1894; R. Sinha and coauthors, Meat and meat-related compounds and risk of prostate cancer in a large prospective cohort study in the United States, *American Journal of Epidemiology* 170 (2009): 1165–1177.
104. Y. Hua Loh and coauthors, N-nitroso compounds and cancer incidence: The European Prospective Investigation into Cancer and Nutrition (EPIC)—Norfolk Study, *American Journal of Clinical Nutrition* 93 (2011): 1053–1061; N. G. Hord, Y. Tang, and N. S. Bryan, Food sources of nitrates and nitrites: The physiologic context for potential health benefits, *American Journal of Clinical Nutrition* 90 (2009): 1–10.
105. S. Rohrmann, S. Hermann, and J. Linseisen, Heterocyclic aromatic amine intake increases colorectal adenoma risk: Findings from a prospective European cohort study, *American Journal of Clinical Nutrition* 89 (2009): 1418–1424.
106. Centers for Disease Control and Prevention, National Center for Environmental Health, *Second National Report on Biochemical Indicators of Diet and Nutrition in the U.S. Population 2012*, Executive Summary, available at www.cdc.gov/nutritionreport.
107. D. Aune and coauthors, Dietary fibre, whole grains, and risk of colorectal cancer: Systematic review and dose-response meta-analysis of prospective studies, *British Medical Journal* 343 (2011), epub, doi:10.1136/bmj.d6617; C. C Dahm and coauthors, Dietary fiber and colorectal cancer risk: A nested case-control study using food diaries, *Journal of the National Cancer Institute* 102 (2010): 614–626; L. B. Sansbury and coauthors, The effect of strict adherence to a high-fiber, high-fruit and -vegetable, and low-fat eating pattern on adenoma recurrence, *American Journal of Epidemiology* 170 (2009): 576–584.
108. T. M. Gibson and coauthors, Pre- and postfortification intake of folate and risk of colorectal cancer in a large prospective cohort study in the United States, *American Journal of Clinical Nutrition* 94 (2011): 1053–1062.
109. M. Huncharek, J. Muscat, and B. Kupelnick, Colorectal cancer risk and dietary intake of calcium, vitamin D, and dairy products: A meta-analysis of 26,335 cases from 60 observational studies, *Nutrition and Cancer* 61 (2009): 47–69.
110. Committee on Dietary Reference Intakes, *Dietary Reference Intakes for Calcium and Vitamin D* (Washington, D.C.: National Academies Press, 2011), pp. 134–147; Y. Ma and coauthors, Association between vitamin D and risk of colorectal cancer: A systematic review of prospective studies, *Journal of Clinical Oncology* 29 (2011): 3775–3782; C. F. Garland and coauthors, Vitamin D for cancer prevention: Global perspective, *Annals of Epidemiology* 19 (2009): 468–483.
111. World Cancer Research Fund/American Institute for Cancer Research, Continuous Update Project, Colorectal Cancer Report 2010 Summary, *Food, Nutrition, Physical Activity, and the Prevention of Colorectal Cancer*, 2011.
112. N. Annema and coauthors, fruit and vegetable consumption and the risk of proximal colon, distal colon, and rectal cancers in a case-control study in Western Australia, *Journal of the American Dietetic Association* 111 (2011): 1479–1490; T. J. Key, Fruit and vegetables and cancer risk, *British Journal of Cancer* 104 (2010): 6–11; F. J. B. van Duijnhoven and coauthors, Fruit, vegetables, and colorectal cancer risk: The European Prospective Investigation into Cancer and Nutrition, *American Journal of Clinical Nutrition* 89 (2009): 1441–1452.

Consumer's Guide 11

1. M. A. Alsawaf and A. Jatoi, Shopping for nutrition-based complementary and alternative medicine on the Internet: How much money might cancer patients be spending online? *Journal of Cancer Education* 22 (2007): 174–176.
2. M. Day, Mapping the alternative route, *British Medical Journal* 334 (2007): 929–931.
3. M. E. Wechsler and coauthors, Active albuterol or placebo, sham acupuncture, or no intervention in asthma, *New England Journal of Medicine* 365 (2011): 119–126.
4. D. Singh, R. Gupta, and S. A. Saraf, Herbs—Are they safe enough? An overview, *Critical Reviews in Food Science and Nutrition* 52 (2012): 876–898.
5. A. J. Vickers and coauthors, Acupunture for chronic pain, *Archives of Internal Medicine*, September 2012, epub, doi:10.1001/archinternmed.2012.3654; National Center for Complementary and Alternative Medicine, Acupuncture for pain, May 2009, available at http://nccam.nih.gov/health/acupuncture/acupuncture-for-pain.htm.
6. S. Barrett, Why NCCAAM should stop funding Reikki research, June 2009, available at http://www.nccamwatch.org.
7. S. Milazzo, S. Lejeune, and E. Ernst, Laetrile for cancer: A systematic review of the clinical evidence, *Supportive Care in Cancer* 15 (2007): 583–595.
8. D. M. Ribnicky and coauthors, Evaluation of botanicals for human health, *American Journal of Clinical Nutrition* 87 (2008): 472S–475S; M. Meadows, Cracking down on health fraud, *FDA Consumer*, November–December, 2006.
9. Meadows, Cracking down on health fraud, 2006.
10. R. B. Saper and coauthors, Lead, mercury, and arsenic in U.S.- and Indian-manufactured Ayurvedic medicines sold via the internet, *Journal of the American Medical Association* 300 (2008): 915–923.
11. U.S. Department of Health and Human Services, National Institutes of Health, National Center for Complementary and Alternative Medicine, *Herbs at a glance: A quick guide to herbal supplements*, January 2009, NIH Publication No. 096248; I. Meijerman, J. H. Beijnen, and J. H. M. Schellens, Herb-drug interactions in oncology: Focus on mechanisms of induction, *Oncologist* 11 (2006): 742–752.
12. National Institutes of Health, National Center for Complementary and Alternative Medicine, Herbs at a glance, Ginkgo, available at http://nccam.nih.gov/health/ginkgo/ataglance.htm; site updated November 2008; B. A. M. Messina, Herbal supplements: Facts and myths—Talking to your patients about herbal supplements, *Journal of PeriAnesthesia Nursing* 21 (2006): 268–278.
13. M. Tascilar and coauthors, Complementary and alternative medicine during cancer treatment: Beyond innocence, *Oncologist* 11 (2006): 732–741.

Controversy 11

1. M. Walsh and S. Kuhn, Developments in personalised nutrition, *Nutrition Bulletin* 37 (2012): 380–383; R. DeBusk, V. S. Sierpina, and M. J. Kreitzer, Applying functional nutrition for chronic disease prevention and management: bridging nutrition and functional medicine in 21st century healthcare, *Explore (NY)* 7 (2011): 55–57.
2. L. A. Afman and M. Müller, Human nutrigenomics of gene regulation by dietary fatty acids, *Progress in Lipid Research* 51 (2012): 63–70.
3. Centers for Disease Control and Prevention, Good laboratory practices for molecular genetic testing for heritable diseases and conditions, *Morbidity and Mortality Weekly Report* 58 (2009): 1–37.
4. L. R. Ferguson and M. Philpott, Nutrition and mutagenesis, *Annual Review of Nutrition* 28 (2008): 313–329.

5. C. Lavebratt, M. Almgren, and T. J. Eström, Epigenetic regulation in obesity, *International Journal of Obesity* 36 (2012): 757–765.
6. J. D. Clarke and coauthors, Differential effects of sulforaphane on histone deacetylases, cell cycle arrest and apoptosis in normal prostate cells versus hyperplastic and cancerous prostate cells, *Molecular Nutrition and Food Research* 55 (2011): 999–1009.
7. World Cancer Research Fund/American Institute for Cancer Research, *Food, Nutrition, Physical Activity and the Prevention of Cancer: A Global Perspective* (Washington, D.C.: AICR, 2007), pp. 75–115.
8. C. D. Davis and J. A. Milner, Nutrigenomics, vitamin D and cancer prevention, *Journal of Nutrigenetics and Nutrigenomics* 4 (2011): 1–11; E. Ho and coauthors, Dietary factors and epigenetic regulation for prostate cancer prevention, *Advances in Nutrition* 2 (2011): 497–510.
9. S. I. Khan and coauthors, Epigenetic events associated with breast cancer and their prevention by dietary components targeting the epigenome, *Chemical Research in Toxicology* 25 (2012): 61–73.
10. L. K. Park, S. Fiso, and S. Choi, Nutritional Influences on epigenetics and age-related disease, *Proceedings of the Nutrition Society* 71 (2012): 75–83.
11. S. Escott-Stump, A perspective on nutritional genomics, *Topics in Clinical Nutrition* 24 (2009): 92–113.
12. C. J. Field, Summary of a workshop, *American Journal of Clinical Nutrition* 89 (2009): 1533S–1539S.
13. Y. Zhang and coauthors, Global hypomethylation in hepatocellular carcinoma and its relationship to aflatoxin B exposure, *World Journal of Hepatology* 4 (2012): 169–175.
14. Z. A. Kaminsky and coauthors, DNA methylation profiles in monozygotic and dizygotic twins, *Nature Genetics* 41 (2009): 240–245; S. Escott-Stump, A perspective on nutritional genomics, *Topics in Clinical Nutrition* 24 (2009): 240–245.
15. J. P. Evans and coauthors, Deflating the genomic bubble, *Science* 331 (2011): 861–862.
16. S. Vakili and M. A. Caudill, Personalized nutrition: Nutritional genomics as a potential tool for targeted medical nutrition therapy, *Nutrition Reviews* 65 (2009): 301–315.
17. J. A. Satia and coauthors, Long-term use of β-carotene, retinol, lycopene, and lutein supplements and lung cancer risk: Results from the VITamins And Lifestyle (VITAL) Study, *American Journal of Epidemiology* 169 (2009): 815–828.
18. L. Kuanrong and coauthors, Associations of dietary calcium intake and calcium supplementation with myocardial infarction and stroke risk and overall cardiovascular mortality in the Heidelberg cohort of the European Prospective Investigation into Cancer and Nutrition study (EPIC-Heidelberg), *Heart* 98 (2012): 920–925.
19. DeBusk, Diet-related disease, nutritional genomics and food and nutrition professionals, 2009.

Chapter 12

1. Position of the American Dietetic Association: Food and water safety, *Journal of the American Dietetic Association* 109 (2009): 1449–1460.
2. M. T. Osterholm, Foodborne disease in 2011—The rest of the story, *New England Journal of Medicine* 364 (2011): 889–891; Centers for Disease Control and Prevention, *2011 Estimates of foodborne Illness in the United States*, available at www.cdc.gov/Features/dsFoodborneEstimates/.
3. S. Hoffman, U.S. food safety policy enters a new era, *Amber Waves*, December 2011, available at www.ers.usda.gov/amberwaves; K. Stewart and L. O. Gostin, Food and Drug Administration regulation of food safety, *Journal of the American Medical Association* 306 (2011): 88–89; U.S. Department of Health and Human Services, Food safety legislation key facts, 2011, available at http://www.fda.gov/Food/FoodSafety/FSMA/ucm237934.htm.
4. Centers for Disease Control and Prevention, Preliminary FoodNet Data on the Incidence of Infection with Pathogens Transmitted Commonly Through Food—10 States, 2008, *Morbidity and Mortality Weekly Report* 59 (2009): 333–337.
5. Centers for Disease Control and Prevention, Multistate Foodborne Outbreaks, 2012, available at http://www.cdc.gov/outbreaknet/outbreaks.html.
6. C. Cochran, USDA targeting six additional strains of *E. coli* in raw beef trim starting Monday, USDA News Release No. 0171.12, May 2012, available at www.usda.gov/wps/portal/usda/usdamediafb?contentid=2012/05/0171.xml&printable=true&contentidonly=true.
7. U.S. Food and Drug Administration, *Food Safety Facts for Consumers*, March 2007, available at www.cfsan.fda.gov/-dms/fs/-eggs.html.
8. E. L. Larson, B. Chen, and K. A. Baxter, Analysis of alcohol-based hand sanitizer delivery systems: Efficacy of foam, gel, and wipes against influenza A (H1N1) virus on hands, *American Journal of Infection Control* (February 2012), epub ahead of print, doi:10.1016/j.ajic.2011.10.016.
9. R. Montville and D. W. Schaffner, A meta-analysis of the published literature on the effectiveness of antimicrobial soaps, *Journal of Food Protection* 74 (2011): 1875–1882; E. H. Snyder, G. A. O'Connor, and D. C. McAvoy, Measured physicochemical characteristics and biosolids-borne concentrations of the antimicrobial Triclocarban (TCC), *Science of the Total Environment* 408 (2010): 2667–2673.
10. B. Sikorska and coauthors, B. Creutzfeldt-Jakob disease, *Advances in Experimental Medicine and Biology* 724 (2012): 76–90.
11. USDA, Fact Sheets: poultry preparation, September 2011, available at http://www.fsis.usda.gov/Factsheets/Turkey_Basics_Stuffing/index.asp.
12. S. J. Chai and coauthors, *Salmonella enterica* serotype Enteritidis: Increasing incidence of domestically acquired infections, *Clinical Infectious Diseases* 54 (2012): S488–S497.
13. FDA Improves Egg Safety, *FDA Consumer Updates*, July 7, 2009, available at www.fda.gov/ForConsumers/ConsumerUpdates/ucm170640.htm.
14. E. Alfano-Sobsey and coauthors, Norovirus outbreak associated with undercooked oysters and secondary household transmission, *Epidemiology and Infection* 140 (2012): 276–282; Centers for Disease Control and Prevention, Multistate outbreak of *Salmonella bareilly* and *Salmonella nchanga* infections associated with a raw scraped ground tuna product, May 2012, available at www.cdc.gov/salmonella/bareilly-04–12/advice-consumers.html.
15. Centers for Disease Control and Prevention, Majority of dairy-related disease outbreaks linked to raw milk, 2012, available at http://www.cdc.gov/media/releases/2012/p0221_raw_milk_outbreak.html.
16. I. B. Hanning, J. D. Nutt, and S. C. Ricke, Salmonellosis outbreaks in the United States due to fresh produce: Sources and potential intervention measures, *Foodborne Pathogens and Disease* 6 (2009): 645–648.
17. Centers for Disease Control and Prevention, Multistate outbreak of Listeriosis linked to whole cantaloupes from Jensen farms, Colorado: Update highlights, December 8, 2011, available at http://www.cdc.gov/listeria/outbreaks/cantaloupes-jensen-farms/index.html.
18. S. Reinberg, FDA: Dirty conditions likely to blame for Listeria outbreak at cantaloupe farm, 2011, available at www.consumer.healthday.com.
19. U.S. Food and Drug Administration, Supplement to the 2009 food code, September 29, 2011, available at www.fda.gov.
20. Food and Drug Administration, Acidic compounds with or without fatty acid surfactants, in *Safe Practices for Food Processes*, updated May 2009, available at www.fda.gov/Food/ScienceResearch/ResearchAreas/SafePracticesforFoodProcesses/default.htm.
21. U.S. Food and Drug Administration, Sprouts Safety Alliance, February 28, 2012, available at http://www.fda.gov/Food/FoodSafety/FSMA/ucm293429.htm.
22. U.S. Food and Drug Administration, Warning on raw alfalfa sprouts, *Consumer Update*, April 28, 2009, available at www.fda.gov.
23. E. R. Choffnes and coauthors, Improving food safety through a One Health approach: Workshop summary, September 2012, available at www.nap.edu/catalog.php?record_id=13423; U.S. Food and Drug Administration, Pathway to global product safety and quality, December 2011, available at http://www.fda.gov/AboutFDA/CentersOffices/OfficeofGlobalRegulatoryOperationsandPolicy/GlobalProductPathway/default.htm.
24. Centers for Disease Control and Prevention, CDC research shows outbreaks linked to imported foods increasing, March 14, 2012, available at http://www.cdc.gov/media/releases/2012/p0314_foodborne.html.
25. J. E. Riviere and G. J. Buckley, Eds., Institute of Medicine, *Ensuring safe foods and medical products through stronger regulatory systems abroad*

(Washington, D.C.: The National Academies Press, 2012); S. Daniells, FDA seeks 17% budget increase, with food safety and Chinese imports strategic targets, February 14, 2012, available at http://www.foodnavigator-usa.com.
26. Department of Agriculture, Mandatory Country of Origin Labeling of beef, pork, lamb, chicken, goat meat, wild and farm-raised fish and shellfish, perishable agricultural commodities, peanuts, pecans, ginseng, and macadamia nuts, *Federal Register* 74 (2009): 2658–2707.
27. U.S. Food and Drug Administration, On the road again: FDA's mobile laboratories, *FDA Consumer Health Information,* March 2009, available at www.fda.gov.
28. M. Wood, Oysters, clams, and mussels: keeping popular mollusks safe to eat, *Agricultural Research*, April 2011, pp. 16–17.
29. S. Ravishankar and coauthors, Edible apple film wraps containing plant antimicrobials inactivate foodborne pathogens on meat and poultry products, USDA Agricultural Research Service publication, June 2012, available at http://www.ars.usda.gov/research/publications/publications.htm?seq_no_115=232087.
30. M. Chu and T. F. Seltzer, Myxedema coma induced by ingestion of raw bok choy, *New England Journal of Medicine* 362 (2010): 1945–1946.
31. M. J. Tijhuis and coauthors, State of the art in benefit-risk analysis: Food and nutrition, *Food and Chemical Toxicology* 50 (2012): 5–25.
32. C. K. Winter and J. M. Katz, Dietary exposure to pesticide residues from commodities alleged to contain the highest contamination levels, *Journal of Toxicology* (2011), epub, doi:10.1155/2011/589674; B. M. Keikotlhaile, P. Spanoghe, and W. Steurbaut, Effects of food processing on pesticide residues in fruits and vegetables: A meta-analysis approach, *Food and Chemical Toxicology* 48 (2010): 1–6.
33. B. M. Keikotlhaile, P. Spanoghe, and W. Steurbaut, Effects of food processing on pesticide residues in fruits and vegetables: A meta-analysis approach, *Food and Chemical Toxicology* 48 (2010): 1–6.
34. C. K. Winter and J. M. Katz, Dietary exposure to pesticide residues from commodities alleged to contain the highest contamination levels, *Journal of Toxicology* 2011 (2011): 589–674.
35. J. Ross Fitzgerald, Human origin for livestock-associated methicillin-resistant *Staphylococcus aureus, mBio* 3 (2012): e00082–12.
36. L. B. Price and coauthors, *Staphylococcus aureus* CC398: Host adaptation and emergence of methicillin resistance in livestock, *mBio* 3 (2012): e00305–11.
37. U.S. Department of Health and Human Services, Guidance for industry the judicious use of medically important antimicrobial drugs in food-producing animals, April 2012, available at www.fda.gov/AnimalVeterinary/GuidanceComplianceEnforcement/GuidanceforIndustry/default.htm.
38. C. G. Gay, Strategies that work: alternatives to antibiotics in animal health, *Agricultural Research*, May/June 2012, pp. 4–7; R. M. Bliss, Keeping pathogens and chemical residues out of beef and poultry, *Agricultural Research*, April 2011, pp. 8–12.
39. F. W. Danby, Acne, dairy and cancer, *Dermato-Endocrinology* 1 (2009): 12–16.
40. K. E. Nachman and coauthors, Arsenic species in poultry feather meal, *Science of the Total Environment* 417–418 (2012): 183–188.
41. FDA looks for answers on arsenic in rice, *Consumer Updates*, September 2012, available at www.fda.gov/ForConsumers/ConsumerUpdates/ucm319827.htm.
42. Standing Committee on the Scientific Evaluation of Dietary Reference Intakes, *Dietary Reference Intakes for Vitamin A, Vitamin K, Arsenic, Boron, Chromium, Copper, Iodine, Iron, Manganese, Molybdenum, Nickel, Silicon, Vanadium, and Zinc* (Washington, D.C.: National Academy Press, 2001), pp. 503–510.
43. R. L. Newsome, Assessing food chemical risks, *Food Technology* 63 (2009): 36–40.
44. Environmental Protection Agency, Report on the environment: consumption of fish and shellfish, March 2011, available at http://cfpub.epa.gov/eroe/index.cfm?fuseaction=list.listBySubTopic&ch=47&s=287.
45. B. C. Scudder and coauthors, Mercury in fish, bed sediment, and water from streams across the United States, 1998–2005, U.S. Geological Survey Scientific Investigations report 2009–5109, available at www.usgs.gov/pubpord.
46. S. K. Sagiv and coauthors, Prenatal exposure to mercury and fish consumption during pregnancy and attention-deficit/hyperactivity disorder–related behavior in children, *Archives of Pediatrics and Adolescent Medicine* 166 (2012): 1123–1131; D. R. Laks, Assessment of chronic mercury exposure within the U.S. population, National Health and Nutrition examination survey, 1999–2006, *Biometals* 22 (2009): 1103–1114.
47. T. G. Neltner and coauthors, Navigating the U.S. food additive regulatory program, *Comprehensive Reviews in Food Science and Food Safety* 10 (2011): 342–368.
48. International Food Information Council (IFIC) and U.S. Food and Drug Administration, Food ingredients and colors, revised April 2010, available at http://www.fda.gov/food/foodingredientspackaging/ucm094211.htm.
49. R. Sinha and coauthors, Meat intake and mortality: A prospective study of over half a million people, *Archives of Internal Medicine* 169 (2009): 562–571.
50. Position of the Academy of Nutrition and Dietetics: Use of nutritive and nonnutritive sweeteners, *Journal of the Academy of Nutrition and Dietetics* 112 (2012): 739–758.
51. K. J. Duffey and coauthors, Dietary patterns matter: diet beverages and cardiometabolic risks in the longitudinal Coronary Artery Risk Development in Young Adults (CARDIA) study, *American Journal of Clinical Nutrition* 95 (2012): 909–915; C. Gardner and coauthors, Nonnutritive sweeteners: current use and health perspectives, *Diabetes Care* 35 (2012): 1798–1808.
52. E. S. Schernhammer and coauthors, Consumption of artificial sweetener- and sugar-containing soda and risk of lymphoma and leukemia in men and women, *American Journal of Clinical Nutrition* 96 (2012): 1419–1428; Position of the Academy of Nutrition and Dietetics: Use of nutritive and nonnutritive sweeteners, 2012.
53. E. T. Rolls, Functional neuroimaging of umami taste: What makes umami pleasant? *American Journal of Clinical Nutrition* 90 (2009): 804S–813S.
54. D. Melzer and coauthors, Urinary Bisphenol A concentration and risk of future coronary artery disease in apparently healthy men and women, *Circulation* 125 (2012): 1482–1490.
55. U.S. Food and Drug Administration, FDA continues to study BPA, *Consumer Health Information*, March 2012, available at www.fda.gov; K. Fiore, FDA rejects BPA ban, March 30, 2012, available at http://www.medpagetoday.com/PublicHealthPolicy/PublicHealth/31953.
56. D. Schardt, BPA: the saga continues, *Nutrition Action Healthletter*, June 2012, p. 9.
57. M. N. Riaz, M. Asif, and R. Ali, Stability of vitamins during extrusion, *Critical Reviews in Food Science and Nutrition* 49 (2009): 361–368.
58. Riaz, Asif, and Ali, Stability of vitamins during extrusion, 2009.

Consumer's Guide 12

1. M. Laux, Organic food trends profile, January 2012, available at www.agmrc.org; C. Dimitri and L. Oberholtzer, Marketing U.S. organic foods: Recent trends from farms to consumers, *USDA Economic Information Bulletin* 58 (2009), available at www.ers.usda.gov.
2. C. Smith-Spangler and coauthors, Are organic foods safer or healthier than conventional alternatives? A systematic review, *Annals of Internal Medicine* 157 (2012): 348–366; C. K. Winter and J. M. Katz, Dietary exposure to pesticide residues from commodities alleged to contain the highest contamination levels, *Journal of Toxicology* 2011 (2011): 589–674.
3. Environmental Working Group, *Shoppers Guide to Pesticides in Produce*, 2012, available at http://www.ewg.org/foodnews/summary/.
4. R. Reiss and coauthors, Estimation of cancer risks and benefits associated with a potential increased consumption of fruits and vegetables, *Food and Chemical Toxicology*, 2012, epub ahead of print, doi:10.1016/j.fct.2012.08.055.
5. Ø. Ueland and coauthors, State of the art in benefit-risk analysis: Consumer perception, *Food and Chemical Toxicology* 50 (2012): 67–76.
6. Reiss and coauthors, Estimation of cancer risks and benefits associated with a potential increased consumption of fruits and vegetables, 2012.
7. A. D. Dangour and coauthors, Nutritional quality of organic foods: A systematic review, *American Journal of Clinical Nutrition* 90 (2009): 680–685.
8. A. Vallverdú-Queralt and coauthors, Is there any difference between the phenolic content of organic and conventional tomato juices? *Food Chemistry* 130 (2012): 222–227.
9. A. D. Dangour and coauthors, Nutrition-related health effects of organic foods: A system-

atic review, *American Journal of Clinical Nutrition* 92 (2010): 203–210.
10. Dimitri and Oberholtzer, Marketing U.S. organic foods, 2009.
11. A. R. Sapkota and coauthors, Lower prevalence of antibiotic-resistant enterococci on U.S. conventional poultry farms that transitioned to organic practices, *Environmental Health Perspectives* 119 (2011): 1622–1628.
12. D. Zwerdling, In India: Bucking the "revolution" by going organic, Summary of the National Public Radio broadcast *Morning Edition*, June 1, 2009.

Controversy 12

1. Ø. Ueland and coauthors, State of the art in benefit-risk analysis: Consumer perception, *Food and Chemical Toxicology* 50 (2012): 67–76.
2. U. S. Department of Agriculture, Biotechnology: Glossary of agricultural biotechnology terms, available at www.usda.gov.
3. U.S. Food and Drug Administration, FDA issues final guidance on regulation of genetically engineered animals, *FDA Consumer Health Information*, January 15, 2009, available at www.fda.gov/consumer/updates/ge_animals011509.html.
4. Position of the American Dietetic Association: Agricultural and food biotechnology, *Journal of the American Dietetic Association* 106 (2006): 285–293, Reaffirmed, 2011.
5. G. Tang and coauthors, β-Carotene in Golden Rice is as good as β-carotene in oil at providing vitamin A to children, *American Journal of Clinical Nutrition* 98 (2012): 658–664; G. Tang and coauthors, Golden Rice is an effective source of vitamin A, *American Journal of Clinical Nutrition* 89 (2009): 1776–1783.
6. E. Leyva-Guerrero and coauthors, Iron and protein biofortification of cassava: Lessons learned, *Current Opinion in Biotechnology* 23 (2012): 257–264; K. D. Hirschi, Nutrient biofortification of food crops, *Annual Review of Nutrition* 29 (2009): 401–421; P. J. White and M. R. Broadley, Biofortification of crops with seven mineral elements often lacking in human diets—iron, zinc, copper, calcium, magnesium, selenium and iodine, *New Phytologist* 182 (2009): 49–84.
7. Leyva-Guerrero and coauthors, Iron and protein biofortification of cassava: Lessons learned, 2012.
8. J. C. Liao and J. Messing, Energy biotechnology, *Current Opinion in Biotechnology* 23 (2012): 287–289; C. S. Jones and S. P. Mayfield, Algae biofuels: Versatility for the future of bioenergy, *Current Opinion in Biotechnology* 23 (2012): 346–351; V. H. Work and coauthors, Improving photosynthesis and metabolic networks for the competitive production of phototroph-derived biofuels, *Current Opinion in Biotechnology* 23 (2012): 290–297.
9. U.S. Food and Drug Administration, FDA issues final guidance on regulation of genetically engineered animals, 2009.
10. USDA Natural Resources Conservation Service, USDA-Natural Resources Conservation Service (NRCS) expands project to control pigweed on cotton crops, 2012, available at www.ga.nrcs.usda.gov/news/Pigweed_Project_2012_Extended.html.
11. Union of Concerned Scientists, Protect our food: A campaign to take the harm out of pharma and industrial crops, available at www.ucsusa.org/food_and_environment/genetic_engineering/protect-our-food.html.
12. H. J. Atkinson, C. J. Lilley, and P.E. Urwin, Strategies for transgenic nematode control in developed and developing world crops, *Current Opinion in Biotechnology* 23 (2012): 251–256.
13. Just Label It Coalition, Archives: June 2012, available at http://justlabelit.org/2012/06/.
14. STOP the costly food labeling proposition—Fact sheet, available at www.stopcostlyfoodlabeling.com/page/fact-sheet2.

Chapter 13

1. J. Mendiola and coauthors, A low intake of antioxidant nutrients is associated with poor semen quality in patients attending fertility clinics, *Fertility and Sterility* 93 (2010) 1128–1133.
2. T. A. Simas and coauthors, Prepregnancy weight, gestational weight gain, and risk of growth affected neonates, *Journal of Women's Health* 21 (2012): 410–417.
3. J. L. Tarry-Adkins and S. E. Ozanne, Mechanisms of early life programming: Current knowledge and future directions, *American Journal of Clinical Nutrition* 94 (2011): 1765S–1771S; G. C. Burdge and K. A. Lillycrop, Nutrition, epigenetics, and developmental plasticity: Implications for understanding human disease, *Annual Review of Nutrition* 30 (2010): 315–339; C. Bouchard, Childhood obesity: Are genetic differences involved? *American Journal of Clinical Nutrition* 89 (2009): 1494S–1501S; M. E. Symonds, T. Stephenson, and H. Budge, Early determinants of cardiovascular disease: The role of early diet in later blood pressure control, *American Journal of Clinical Nutrition* 89 (2009): 1518S–1522S.
4. N. Wesglas-Kuperus and coauthors, Intelligence of very preterm or very low birthweight infants in young adulthood, *Archives of Disease in Childhood: Fetal and Neonatal Edition* 94 (2009): F196–200; D. S. Alam, Prevention of low birthweight, *Nestle Nutrition Workshop Series: Pediatric Program* 63 (2009): 209–225.
5. K. D. Kochanek and coauthors, Annual summary of vital statistics: 2009, *Pediatrics* 129 (2012): 338–348.
6. L. M. McCowan and coauthors, Spontaneous preterm birth and small for gestational age infants in women who stop smoking early in pregnancy: Prospective cohort study, *British Medical Journal* 338 (2009): b1081.
7. N. Kozuki, A. C. Lee, and J. Katz, Moderate to severe, but not mild, maternal anemia is associated with increased risk of small–for gestational–age outcomes, *Journal of Nutrition* 142 (2012): 358–362; Alam, Prevention of low birthweight; Position of the American Dietetic Association: Nutrition and lifestyle for a healthy pregnancy outcome, *Journal of the American Dietetic Association* 108 (2008): 553–561.
8. R. Retnakaran and coauthors, Effect of maternal weight, adipokines, glucose intolerance and lipids on infant birth weight among women without gestational diabetes mellitus, *Canadian Medical Association Journal* 184 (2012): 1353–1360.
9. A. A. Mamun and coauthors, Associations of maternal pre-pregnancy obesity and excess pregnancy weight gains with adverse pregnancy outcomes and length of hospital stay, *BMC Pregnancy and Childbirth* 11 (2011): 62.
10. Position of the American Dietetic Association and American Society for Nutrition: Obesity, reproduction, and pregnancy outcomes, *Journal of the American Dietetic Association* 109 (2009): 918–927.
11. Mamun and coauthors, Associations of maternal pre-pregnancy obesity and excess pregnancy weight gains with adverse pregnancy outcomes and length of hospital stay, 2011; Position of the American Dietetic Association and American Society for Nutrition: Obesity, reproduction, and pregnancy outcomes, 2009.
12. J. L. Mills and coauthors, Maternal obesity and congenital heart defects: A population–based study, *American Journal of Clinical Nutrition* 91 (2010): 1543–1549; K J. Stothard and coauthors, Maternal overweight and obesity and the risk of congenital anomalies: A systematic review and meta–analysis, *Journal of the American Medical Association* 301 (2009): 636–650.
13. M. Desforges and C. P. Sibley, Placental nutrient supply and fetal growth, *The International Journal of Developmental Biology* 54 (2010): 377–390.
14. A. F. M. van Abeelen and coauthors, Survival effects of prenatal famine exposure, *American Journal of Clinical Nutrition* 95 (2012): 179–183; Tarry-Adkins and Ozanne, Mechanisms of early life programming: Current knowledge and future directions, 2011; M. L. de Gusmão Correia and coauthors, Developmental origins of health and disease: Experimental and human evidence of fetal programming for metabolic syndrome, *Journal of Human Hypertension* June 23, 2011, pp. 1–15; L. C. Schulz, The Dutch hunger winter and the developmental origins of health and disease, *Proceedings of the National Academy of Sciences* 107 (2010): 16757–16758.
15. Standing Committee on the Scientific Evaluation of Dietary Reference Intakes, Food and Nutrition Board, Institute of Medicine, *Dietary Reference Intakes for Energy, Carbohydrate, Fiber, Fat, Fatty Acids, Cholesterol, Protein, and Amino Acids* (Washington, DC: National Academies Press, 2005), pp.185–194.
16. M. L. Blumfield and coauthors, Systematic review and meta-analysis of energy and macronutrient intakes during pregnancy in developed countries, *Nutrition Reviews* 70 (2012): 322–336.
17. P. Haggarty, Fatty acid supply to the human fetus, *Annual Review of Nutrition* 30 (2010): 237–255.
18. Centers for Disease Control and Prevention, Folic acid data and statistics, available at www

.cdc.gov/ncbddd/folicacid/data.html, updated July 7, 2010.
19. Centers for Disease Control and Prevention, Folic acid data and statistics, available at www.cdc.gov/ncbddd/folicacid/data.html, updated July 7, 2010.
20. S. H. Blanton and coauthors, Folate pathway and nonsyndromic cleft lip and palate, *Birth Defects Research. Part A, Clinical and Molecular Teratology* 91 (2011): 50–60.
21. I. Elmadfa and I. Singer, Vitamin B-12 and homocysteine status among vegetarians: A global perspective, *American Journal of Clinical Nutrition* 89 (2009): 1693S–1698S.
22. A. M. Molloy and coauthors, Maternal vitamin B12 status and risk of neural tube defects in a population with high neural tube defect prevalence and no folic acid fortification, *Pediatrics* 123 (2009): 917–923.
23. B. E. Young and coauthors, Maternal vitamin D status and calcium intake interact to affect fetal skeletal growth in utero in pregnant adolescents, *American Journal of Clinical Nutrition* 95 (2012): 1103–1112.
24. P. M. Brannon and M. F. Picciano, Vitamin D in Pregnancy and lactation in humans, *Annual Review of Nutrition* 31 (2011): 89–115; C. L. Wagner and F. R. Greer, and the Section on Breastfeeding and Committee on Nutrition, Prevention of rickets and vitamin D deficiency in infants, children, and adolescents, *Pediatrics* 122 (2008): 1142–1152.
25. Committee on Dietary Reference Intakes, *Dietary Reference Intakes for Calcium and Vitamin D* (Washington, DC: National Academies Press, 2011).
26. A. Merewood and coauthors, Widespread vitamin D deficiency in urban Massachusetts newborns and their mothers, *Pediatrics* 125 (2010): 640–647.
27. Committee on Dietary Reference Intakes, *Dietary Reference Intakes for Calcium and Vitamin D*, 2011, pp. 242–250.
28. A. N. Hacker, E. B. Fung, and J. C. King, Role of calcium during pregnancy: Maternal and fetal needs, *Nutrition Reviews* 70 (2012): 397–409.
29. K. J. Collard, Iron homeostasis in the neonate, *Pediatrics* 123 (2009): 1208–1216.
30. H. J. McArdle and coauthors, Role of the placenta in regulation of fetal iron status, *Nutrition Reviews* 69 (2011): S17–S22.
31. Scholl, Maternal iron status: Relation to fetal growth, length of gestation, and iron endowment of the neonate, 2011.
32. McArdle and coauthors, Role of the placenta in regulation of fetal iron status, 2011.
33. T. O. Scholl, Maternal iron status: Relation to fetal growth, length of gestation, and iron endowment of the neonate, *Nutrition Reviews* 69 (2011): S23–S29.
34. S. H. Zeisel, Is maternal diet supplementation beneficial? Optimal development of infant depends on mother's diet, *American Journal of Clinical Nutrition* 89 (2009): 685S–687S.
35. M. F. Picciano and M. K. McGuire, Use of dietary supplements by pregnant and lactating women in North America, *American Journal of Clinical Nutrition* 89 (2009): 663S–667S.
36. Position of the American Dietetic Association, Nutrition and lifestyle for a healthy pregnancy outcome, 2008.
37. J. M. Catov and coauthors, Periconceptional multivitamin use and risk of preterm or small-for-gestational-age births in the Danish National Birth Cohort, *American Journal of Clinical Nutrition* 94 (2011): 906–912.
38. WIC, The Special Supplemental Nutrition Program for Women, Infants, and Children, available at www.fns.usda.gov/fns, updated August 2011.
39. Position of the American Dietetic Association and American Society for Nutrition: Obesity, reproduction, and pregnancy outcomes, 2009; S. Y. Chu and coauthors, Gestational weight gain by body mass index among U.S. women delivering live births, 2004–2005: Fueling future obesity, *American Journal of Obstetrics and Gynecology* 200 (2009): 271, e1–e7.
40. Position of the American Dietetic Association and American Society for Nutrition: Obesity, reproduction, and pregnancy outcomes, 2009.
41. Institute of Medicine, *Weight Gain During Pregnancy: Reexamining the Guidelines* (Washington, DC: The National Academies Press, 2009).
42. J. H. Cohen and H. Kim, Sociodemographic and health characteristics associated with attempting weight loss during pregnancy, *Preventing Chronic Disease* 6 (2009): A07.
43. A. A. Mamun and coauthors, Association of excess weight gain during pregnancy with long-term maternal overweight and obesity: Evidence from 21 y postpartum follow-up, *American Journal of Clinical Nutrition* 91 (2010): 1336–1341; Position of the American Dietetic Association and American Society for Nutrition: Obesity, reproduction, and pregnancy outcomes, 2009; K. K. Vesco and coauthors, Excessive gestational weight gain and postpartum weight retention among obese women, *Obstetrics and Gynecology* 114 (2009): 1069–1075.
44. S. M. Ruchat and coauthors, Nutrition and exercise reduce excessive weight gain in normal-weight pregnant women, *Medicine and Science in Sports and Exercise* 44 (2012): 1419–1426; M. F. Mottola and coauthors, Nutrition and exercise prevent excess weight gain in overweight pregnant women, *Medicine and Science in Sports and Exercise* 42 (2010): 265–272.
45. B. E. Hamilton and S. J. Ventura, Birth rates for U. S. teenagers reach historic lows for all age and ethnic groups, *NCHS Data Brief*, no. 89 (Hyattsville, MD: National Center for Health Statistics, 2012).
46. P. N. Baker and coauthors, A prospective study of micronutrient status in adolescent pregnancy, *American Journal of Clinical Nutrition* 89 (2009): 1114–1124; S. Saintonge, H. Bang, and L. M. Gerber, Implications of a new definition of vitamin D deficiency in a multiracial U.S. adolescent population: The National Health and Nutrition Examination Survey III, *Pediatrics* 123 (2009): 797–803.
47. Trends in smoking before, during, and after pregnancy: Pregnancy Risk Assessment Monitoring system (PRAMS), United States, 31 sites, 2000–2005, *Morbidity and Mortality Weekly Report* 58 (2009): 1–29.
48. B. E. Gould Rothberg and coauthors, Gestational weight gain and subsequent postpartum weight loss among young, low-income, ethnic minority women, *American Journal of Obstetrics and Gynecology* 204 (2011): e1–e11; M. M. Thame and coauthors, Weight retention within the puerperium in adolescents: A risk factor for obesity? *Public Health Nutrition* 13 (2010): 283–288; E. P. Gunderson and coauthors, Longitudinal study of growth and adiposity in parous compared with nulligravid adolescents, *Archives of Pediatrics and Adolescent Medicine* 163 (2009): 349–356.
49. S. L. Young, Pica in pregnancy: New ideas about an old condition, *Annual Review of Nutrition* 30 (2010): 403–422.
50. J. R. Niebyl, Nausea and vomiting in pregnancy, *New England Journal of Medicine* 363 (2010): 1544–1550.
51. A. Matthews and coauthors, Interventions for nausea and vomiting in early pregnancy, *Cochrane Database of Systemic Reviews* 8 (2010): CD007575.
52. Trends in smoking before, during, and after pregnancy: Pregnancy Risk Assessment Monitoring system (PRAMS), United States, 31 sites, 2000–2005, 2009: 1–29.
53. E. Stéphan–Blanchard and coauthors, Perinatal nicotine/smoking exposure and carotid chemoreceptors during development, *Respiratory Physiology & Neurobiology* June 26, 2012, epub ahead of print, http://dx.doi.org/1016/j.resp.2012.06.023; J. M. Rogers, Tobacco and pregnancy: Overview of exposures and effects, *Birth Defects Research* (Part C) 84 (2008): 1–15.
54. C. C. Geerts and coauthors, Parental smoking and vascular damage in their 5–year–old children, *Pediatrics* 129 (2010): 45–54.
55. H. Burke and coauthors, Prenatal and passive smoke exposure and incidence of asthma and wheeze: Systematic review and meta–analysis, *Pediatrics* 129 (2012): 735–744; A. Bjerg and coauthors, A strong synergism of low birth weight and prenatal smoking on asthma in schoolchildren, *Pediatrics* 127 (2011): e905–912.
56. F. L. Trachtenberg and coauthors, Risk factor changes for sudden infant death syndrome after initiation of back-to-sleep campaign, *Pediatrics* 129 (2012): 630–638; H. C. Kinney and B. T. Thach, The sudden infant death syndrome, *New England Journal of Medicine* 361 (2009): 795–805.
57. A. Gunnerbeck and coauthors, Relationship of maternal snuff use and cigarette smoking with neonatal apnea, *Pediatrics* 128 (2011): 503–509.
58. Position of the American Dietetic Association, Nutrition and lifestyle for a healthy pregnancy outcome, 2008.
59. J. P. Ackerman, T. Riggings, and M. M. Black, A review of the effects of prenatal cocaine exposure among school-aged children, *Pediatrics* 125 (2010): 554–565.

60. M. O'Donnell and coauthors, Increasing prevalence of neonatal withdrawal syndrome: Population study of maternal factors and child protection involvement, *Pediatrics* 123 (2009): e614–e621.
61. A. Bloomingdale and coauthors, A qualitative study of fish consumption during pregnancy, *American Journal of Clinical Nutrition* 92 (2010): 1234–1240.
62. Position of the Academy of Nutrition and Dietetics: Use of nutritive and nonnutritive sweeteners, *Journal of the Academy of Nutrition and Dietetics* 112 (2012): 739–758; Position of the American Dietetic Association, Nutrition and lifestyle for a healthy pregnancy outcome, 2008.
63. M. Jarosz, R. Wierzejska, and M. Siuba, Maternal caffeine intake and its effect on pregnancy outcome, *European Journal of Obstetrics and Gynecology and Reproductive Biology* 160 (2012): 156–160; E. Maslova and coauthors, Caffeine consumption during pregnancy and risk of preterm birth: A meta-analysis, *American Journal of Clinical Nutrition* 92 (2010): 1120–1132.
64. R. Bakker and coauthors, Maternal caffeine intake, blood pressure, and the risk of hypertensive complications during pregnancy: The Generation R study, *American Journal of Hypertension* 24 (2011): 421–428; Maslova and coauthors, Caffeine consumption during pregnancy and risk of preterm birth: A meta-analysis, 2010; D. C. Greenwood and coauthors, Caffeine intake during pregnancy, late miscarriage and stillbirth, *European Journal of Epidemiology* 25 (2010): 275–280.
65. R. Bakker and coauthors, Maternal caffeine intake from coffee and tea, fetal growth, and the risks of adverse birth outcomes: The Generation R Study, *American Journal of Clinical Nutrition* 91 (2010): 1691–1698.
66. J. Gareri and coauthors, Potential role of the placenta in fetal alcohol spectrum disorder, *Paediatric Drugs* 11 (2009): 26–29; F. T. Crews and K. Nixon, Foetal alcohol spectrum disorders and alterations in brain and behaviour, *Alcohol and Alcoholism* 44 (2009): 108–114.
67. E. P. Riley, M. A. Infante, and K. R. Warren, Fetal alcohol spectrum disorders: An overview, *Neuropsychological Review* 21 (2011): 73–80; K. L. Jones and coauthors, Fetal alcohol spectrum disorders: Extending the range of structural defects, *American Journal of Medical Genetics* 152A (2010): 2731–2735.
68. Centers for Disease Control and Prevention, *Fetal Alcohol Spectrum Disorders*, www.cdc.gov/ncbddd/fasd/data.html, updated December 11, 2011; C. H. Denny and coauthors, Alcohol use among pregnant and nonpregnant women of childbearing age—United States, 1991–2005, *Morbidity and Mortality Weekly Report* 58 (2009): 529–532.
69. C. M. Marchetta and coauthors, Alcohol use and binge drinking among women of childbearing age—United States, 2006–2010, *Morbidity and Mortality Weekly Report* 61 (2012): 534–538; Denny and coauthors, Alcohol use among pregnant and nonpregnant women, 2009.
70. A. Ornoy and Z. Ergaz, Alcohol abuse in pregnant women: Effects on the fetus and newborn, mode of action and maternal treatment, *International Journal of Environmental Research and Public Health* 7 (2010): 364–379.
71. H. S. Feldman and coauthors, Prenatal alcohol exposure patterns and alcohol-related birth defects and growth deficiencies: A prospective study, *Alcoholism: Clinical and Experimental Research* 36 (2012): 670–676.
72. J. L. Kitzmiller and coauthors, Preconception care for women with diabetes and prevention of major congenital malformations, *Birth Defects Research Part A: Clinical and Molecular Teratology* 88 (2010): 791–803.
73. Position statement, Standards of medical care in diabetes—2011, *Diabetes Care* 34 (2011): S11–S61; Y. Yogev and G. H. Visser, Obesity, gestational diabetes and pregnancy outcome, *Seminars in Fetal and Neonatal Medicine* 14 (2009): 77–84.
74. Centers for Disease Control and Prevention, *Diabetes: At a glance 2009*, available at www.ced.gov/nccdphp/publications/aag/ddt.htm.
75. American Diabetes Association, Diagnosis and classification of diabetes mellitus, *Diabetes Care* 34 (2011): S62–S69.
76. P. E. Marik, Hypertensive disorders of pregnancy, *Postgraduate Medicine* 121 (2009): 69–76.
77. E. W. Seely and J. Ecker, Chronic hypertension in pregnancy, *New England Journal of Medicine* 365 (2011): 439–446.
78. H. Xu and coauthors, Role of nutrition in the risk of preeclampsia, *Nutrition Reviews* 67 (2009): 639–657; Position of the American Dietetic Association, Nutrition and lifestyle for a healthy pregnancy outcome, 2008.
79. R. Mustafa and coauthors, A comprehensive review of hypertension in pregnancy, *Journal of Pregnancy*, May 23, 2012, epub, doi:10.1155/2012/105918; Position of the American Dietetic Association, Nutrition and lifestyle for a healthy pregnancy outcome, 2008.
80. Xu and coauthors, Role of nutrition in the risk of preeclampsia, 2009.
81. A. Paxton and T. Wardlaw, Are we making progress in maternal mortality? *New England Journal of Medicine* 364 (2011): 1990–1993.
82. Committee on Dietary Reference Intakes, *Dietary Reference Intakes for Calcium and Vitamin D* (Washington, DC: National Academies Press, 2011), pp. 256–257.
83. J. A. Mennella and J. C. Trabulsi, complementary foods and flavor experiences: Setting the foundation, *Annals of Nutrition and Metabolism* 60 (2012): 40–50; G. Beauchanp and J. A. Mennella, Flavor perception in human infants: Development and functional significance, *Digestion* 83 (2011): 1–6.
84. F. R. Greer, S. H. Sicherer, A. Wesley Burks, and the Committee on Nutrition and Section on Allergy and Immunology, Effects of early nutritional interventions on the development of atopic disease in infants and children: The role of maternal dietary restriction, breastfeeding, timing of introduction of complementary foods, and hydrolyzed formulas, *Pediatrics* 121 (2008): 183–191.
85. A. J. Daley and coauthors, Maternal exercise and growth in breastfed infants: A meta-analysis of randomized controlled trials, *Pediatrics* 130 (2012): 108–114.
86. American Academy of Pediatrics, Policy statement: Breastfeeding and the use of human milk, *Pediatrics* 129 (2012): e827–e841, available at www.pediatrics.org/content/129/3/e827.full.
87. Trends in smoking before, during, and after pregnancy: Pregnancy Risk Assessment Monitoring system (PRAMS), United States, 31 sites, 2000–2005, 2009.
88. P. Bachour and coauthors, Effects of smoking mother's age, body mass index, and parity number on lipid, protein, and secretory immunoglobulin A concentration of human milk, *Breastfeeding Medicine* 7 (2012): 179–188.
89. G. Liebrechts–Akkerman and coauthors, Postnatal parental smoking: An important risk factor for SIDS, *European Journal of Pediatrics* 170 (2011): 1281–1291; G. Yilmaz and coauthors, Effect of passive smoking on growth and infection rates of breast–fed and non–breast–fed infants, *Pediatrics International* 51 (2009): 352–358.
90. American Academy of Pediatrics, Policy statement: Breastfeeding and the use of human milk, 2012.
91. N. Kapp, K. Curtis, and K. Nanda, Progestogen-only contraceptive use among breastfeeding women: A systematic review, *Contraception* 82 (2010): 17–37.
92. American Academy of Pediatrics, Policy statement: Breastfeeding and the use of human milk, 2012.
93. American Academy of Pediatrics, Policy statement: Breastfeeding and the use of human milk, 2012.
94. American Academy of Pediatrics, Policy statement: Breastfeeding and the use of human milk, 2012; P. L. Havens, L. M. Mofenson, and the Committee on Pediatric AIDS, Evaluation and management of the infant exposed to HIV-1 in the United States, *Pediatrics* 123 (2009): 175–187.
95. A. Koyanagi and coauthors, Effect of early exclusive breastfeeding on morbidity among infants born to HIV-negative mothers in Zimbabwe, *American Journal of Clinical Nutrition* 89 (2009): 1375–1382.
96. World Health Organization, *HIV and Infant Feeding*, available at http://www.who.int/child_adolescent_health/topics/prevention_care/child/nutrition/hivif/en/, site visited June 1, 2009; M. W. Kline, Early exclusive breastfeeding: Still the cornerstone of child survival, *American Journal of Clinical Nutrition* 89 (2009): 1281–1282.
97. Formula feeding of term infants, in *Pediatric Nutrition Handbook*, 6th ed., R. E. Kleinman, ed. (Elk Grove Village, IL: American Academy of Pediatrics, 2009), pp. 61–78.
98. Position of the American Dietetic Association: Promoting and supporting breastfeeding, *Journal of the American Dietetic Association* 109 (2009): 1926–1942.
99. American Academy of Pediatrics, Policy statement: Breastfeeding and the use of human

F

milk, 2012; Position of the American Dietetic Association, Promoting and supporting breastfeeding, 2009.
100. American Academy of Pediatrics, Policy statement: Breastfeeding and the use of human milk, 2012; Breastfeeding, in *Pediatric Nutrition Handbook*, 6th ed., pp. 29–59, 2009.
101. Breastfeeding, in *Pediatric Nutrition Handbook*, 6th ed., 2009.
102. S. M. Donovan, Human milk oligosaccharides—the plot thickens, *British Journal of Nutrition* 101 (2009): 1267–1269.
103. Fat and fatty acids, in *Pediatric Nutrition Handbook*, 6th ed., 2009, pp. 357–386; S. E. Carlson, Docosahexaenoic acid supplementation in pregnancy and lactation, *American Journal of Clinical Nutrition* 89 (2009): 678S–684S.
104. S. J. Meldrum and coauthors, Achieving definitive results in long-chain polyunsaturated fatty acid supplementation trials of term infants: Factors for consideration, *Nutrition Reviews* 69 (2011): 205– 214; L. G. Smithers, R. A. Gibson, and M. Makrides, Maternal supplementation with docosahexaenoic acid during pregnancy does not affect early visual development in the infant; A randomized controlled trial, *American Journal of Clinical Nutrition* 93 (2011): 1293–1299; E. E. Birch and coauthors, The DIAMOND (DHA Intake and Measurement of Neural Development) Study: A double–masked, randomized controlled clinical trial of the maturation of infant visual acuity as a function the dietary level of docosahexaenoic acid, *American Journal of Clinical Nutrition* 91 (2010): 848–859; S. E. Carlson, Early determinants of development: A lipid perspective, *American Journal of Clinical Nutrition* 89 (2009): 1523S–1529S.
105. Carlson, Early determinants of development: A lipid perspective, 2009.
106. M. Guxens and coauthors, Breastfeeding, long-chain polyunsaturated fatty acids in colostrums, and infant mental development, *Pediatrics* 128 (2011): e880–e889; Carlson, Early determinants of development: A lipid perspective, 2009.
107. S. A. Abrams, What are the risks and benefits to increasing dietary bone minerals and vitamin D intake in infants and small children? *Annual Review of Nutrition* 31 (2011): 285–297.
108. Fat–soluble vitamins, in *Pediatric Nutrition Handbook,* 6th ed., 2009, pp. 461–474.
109. C. L. Wagner, and F. R. Greer, and the Section on Breastfeeding and Committee on Nutrition, Prevention of rickets and vitamin D deficiency in infants, children, and adolescents, *Pediatrics* 122 (2008): 1142–1152.
110. C. G. Perrine and coauthors, Adherence to vitamin D recommendations among US infants, *Pediatrics* 125 (2010): 627–632.
111. American Academy of Pediatrics, Policy statement, 2012; Breastfeeding, in *Pediatric Nutrition Handbook,* 6th ed., 2009; Position of the American Dietetic Association, Promoting and supporting breastfeeding, 2009.
112. American Academy of Pediatrics, Policy statement: Breastfeeding and the use of human milk, 2012; A. Walker, Breast milk as the gold standard for protective nutrients, *Journal of Pediatrics* 156 (2010): 53–57.
113. American Academy of Pediatrics, Policy statement: Breastfeeding and the use of human milk, 2012; Breastfeeding, in *Pediatric Nutrition Handbook*, 6th ed., 2009; Position of the American Dietetic Association, Promoting and supporting breastfeeding, 2009.
114. American Academy of Pediatrics, Policy statement: Breastfeeding and the use of human milk, 2012.
115. American Academy of Pediatrics, Policy statement: Breastfeeding and the use of human milk, 2012.
116. F. R. Hauck, and coauthors, Breastfeeding and reduced risk of sudden infant death syndrome: A meta-analysis, *Pediatrics* 128 (2011): 103–110.
117. M. M. Vennemann and coauthors, Does breastfeeding reduce the risk of sudden infant death syndrome? *Pediatrics* 123 (2009): e406–e410.
118. K. Casazza, J. R. Fernandez, and D. B. Allison, Modest protective effects of breastfeeding on obesity, *Nutrition Today* 47 (2012): 33–38; A. Bererlein and R. von Kries, Breastfeeding and body composition in children: Will there ever be conclusive empirical evidence for a protective effect against overweight? *American Journal of Clinical Nutrition* 94 (2011): 1772S–1775S; L. Shields and coauthors, Breastfeeding and obesity at 21 years: A cohort study, *Journal of Clinical Nursing* 19 (2010): 1612–1617; L. Schack-Nielsen and coauthors, Late introduction of complementary feeding, rather than duration breastfeeding, may protect against adult overweight, *American Journal of Clinical Nutrition* 91 (2010): 619–627; L. Twells and L. A. Newhook, Can exclusive breastfeeding reduce the likelihood of childhood obesity in some regions of Canada? *Canadian Journal of Public Health* 101 (2010): 36–39; P. Chivers and coauthors, Body mass index, adiposity rebound and early feeding in a longitudinal cohort (Raine Study), *International Journal of Obesity* 34 (2010): 1169–1176; B. Koletzko and coauthors, Can infant feeding choices modulate later obesity risk? *American Journal of Clinical Nutrition* 89 (2009): 1502S–1508S.
119. Casazza, Fernandez, and Allison, Modest protective effects of breast-feeding on obesity, 2012; K. L. Whitaker and coauthors, Comparing maternal and paternal intergenerational transmission of obesity risk in a large population-based sample, *American Journal of Clinical Nutrition* 91 (2010): 1560–1567; R. Li, S. B. Fein, and L. M. Grummer-Strawn, Do infants fed from bottles lack self-regulation of milk intake compared with directly breastfed infants? *Pediatrics* 125 (2010): e1386–e1393.
120. W. Jedrychowski and coauthors, Effect of exclusive breastfeeding on the development of children's cognitive function in the Krakow prospective birth cohort study, *European Journal of Pediatrics* 171 (2012): 151–158; M. A. Quigley and coauthors, Breastfeeding is associated with improved child cognitive development: A population-based cohort study, *Journal of Pediatrics* 160 (2012): 25–32; C. McCrory and R. Layte, The effect of breastfeeding on children's educational test scores at nine years of age: Results of an Irish cohort study, *Social Science and Medicine* 72 (2011); 1515–1521.
121. M. Guxens and coauthors, Breastfeeding, long–chain polyunsaturated fatty acids in colostrum, and infant mental development *Pediatrics* 128 (2011): e880–e889.
122. Formula feeding of term infants, in *Pediatric Nutrition Handbook*, 6th ed., 2009.
123. Formula feeding of term infants, in *Pediatric Nutrition Handbook*, 6th ed., 2009.
124. Formula feeding of term infants, in *Pediatric Nutrition Handbook*, 6th ed., 2009.
125. Formula feeding of term infants, in *Pediatric Nutrition Handbook*, 6th ed., 2009.
126. E. E. Ziegler, Consumption of cow's milk as a cause of iron deficiency in infants and toddlers, *Nutrition Reviews* 69 (2011): S37–S42.
127. Expert Panel on Integrated Guidelines for Cardiovascular Health and Risk Reduction in children and Adolescents, Summary report, *Pediatrics* 128 (2011): S213–S256.
128. Complementary feeding, in *Pediatric Nutrition Handbook*, 6th ed., 2009, pp. 113–142.
129. H. Przyrembel, Timing of introduction of complementary food: Short- and long-term health consequences, *Annals of Nutrition and Metabolism* (supplement) 60 (2012): 8–20.
130. Iron, in *Pediatric Nutrition Handbook*, 6th ed., 2009, pp. 403–422.
131. Complementary feeding, in *Pediatric Nutrition Handbook*, 6th ed., 2009.
132. Feeding the child, in *Pediatric Nutrition Handbook*, 6th ed., 2009, pp. 145–174.
133. Feeding the child, in *Pediatric Nutrition Handbook*, 6th ed., 2009.
134. American Academy of Pediatrics Policy Statement—Prevention of choking among children, *Pediatrics* 125 (2010): 601–607.
135. Complementary feeding, in *Pediatric Nutrition Handbook*, 6th ed., 2009.

Consumer's Guide 13

1. A. Avery and coauthors, Confident commitment is a key factor for sustained breastfeeding, *Birth* 36 (2009): 141–148.
2. American Academy of Pediatrics, Breastfeeding and the use of human milk, *Pediatrics* 129 (2012): 598–601; Position of the American Dietetic Association, Promoting and supporting breastfeeding, *Journal of the American Dietetic Association* 109 (2009): 1926–1942; Centers for Disease Control and Prevention, April 19, 2010, Breastfeeding: Frequently asked questions, available at www.cdc.gov/breastfeeding.htm.
3. Centers for Disease Control and Prevention, Breastfeeding: Frequently asked questions, 2010.

Controversy 13

1. C. L. Ogden and coauthors, Prevalence of obesity and trends in body mass index among US children and adolescents, 1999–2010, *Journal of the American Medical Association* 307 (2012): 483–490; C. L. Ogden and coauthors, Preva-

lence of high body mass index in U.S. children and adolescents, *Journal of the American Medical Association* 303 (2010): 242–249.
2. F. M. Biro and M. Wien, Childhood obesity and adult morbidities, *American Journal of Clinical Nutrition* 91 (2010): 14995–15055; P. W. Franks and coauthors, Childhood obesity, other cardiovascular risk factors, and premature death, *New England Journal of Medicine* 362 (2010): 458–493.
3. M. de Onis and coauthors, Global prevalence and trends of overweight and obesity among preschool children, *American Journal of Clinical Nutrition* 92 (2010): 1257–1264; B. M. Popkin, Recent dynamics suggest selected countries catching up to US obesity, *American Journal of Clinical Nutrition* 91 (2010): 284S–288S; C. Bouchard, Childhood obesity: Are genetic differences involved? *American Journal of Clinical Nutrition* 89 (2009): 1494S–1501S.
4. C. L. Ogden and coauthors, Prevalence of obesity and trends in body mass index among US children and adolescents, 1999–2010.
5. M. D. Marcus and coauthors, Severe obesity and selected risk factors in a sixth grade multiracial cohort: the HEALTHY study, *Journal of Adolescent Health* 47 (2010): 604–607.
6. J. A. Mitchell and coauthors, Time spent in sedentary behavior and changes in childhood BMI: A longitudinal study from ages 9 to 15 years, *International Journal of Obesity* (2012), epub, doi:10.1038/ijo.2012.41; J. A. Mitchell and coauthors, Sedentary behavior and obesity in a large cohort of children, *Obesity* 8 (2009): 1596–1602.
7. C. L. Ogden and coauthors, Obesity and socioeconomic status in children: United States 1988–1994 and 2005–2008, *NCHS data brief no 51* (Hyattsville, MD: National Center for Health Statistics, 2010).
8. J. J. Reily. Assessment of obesity in children and adolescents: synthesis of recent systematic reviews and clinical guidelines, *Journal of Human Nutrition and Dietetics* 23 (2010): 205–211.
9. Centers for Disease Control and Prevention, Prevalence of abnormal lipid levels among youths—United States, 1999–2006, *Morbidity and Mortality Weekly Report* 59 (2010): 29–33; W. Insull, Jr., The pathology of atherosclerosis: Plaque development and plaque responses to medical treatment, *American Journal of Medicine* 122 (2009): S3–S14.
10. W. Tu and coauthors, Intensified effect of adiposity on blood pressure in overweight and obese children, *Hypertension* 58 (2011): 818–824.
11. M. Juonala and coauthors, Childhood adiposity, adult adiposity, and cardiovascular risk factors, *New England Journal of Medicine* 365 (2011): 1876–1885; D. S. Freedman and coauthors, Risk factors and adult body mass index among overweight children: The Bogalusa Heart Study, *Pediatrics* 123 (2009): 750–757.
12. C. M. Visness and coauthors, Association of childhood obesity with atopic and nonatopic asthma: Results From the National Health and Nutrition Examination Survey 1999–2006, *Journal of Asthma* 47 (2010): 822–829; M. Neovius, J. Sundström, and F. Rasmussen, Combined effects of overweight and smoking in late adolescence on subsequent mortality: Nationwide cohort study, *British Medical Journal* 338 (2009), epub, doi:10.1136/bmj.b496.
13. K. D. Graw-Panzer and coauthors, Effect of increasing body mass index on obstructive sleep apnea in children, *The Open Sleep Journal* 3 (2010): 19–23.
14. L. Pacifico and coauthors, Pediatric nonalcoholic fatty liver disease, metabolic syndrome and cardiovascular risk, *World Journal of Gastroenterology* 17 (2011): 3082–3091; U.S. Preventive Services Task Force, Screening for obesity in children and adolescents: U.S. Preventive Services Task Force Recommendation Statement, *Pediatrics* 125 (2010): 361–367.
15. R. L. Washington, Childhood obesity: Issues of weight bias, *Preventing Chronic Disease* 8 (2011): A94; J. C. Lumeng and coauthors, Weight status as a predictor of being bullied in third through sixth grades, *Pediatrics* 125 (2010): e1301–e1307; F. Wang and coauthors, The influence of childhood obesity on the development of self-esteem, *Health Reports* 20 (2009): 21–27.
16. R. M. Puhl and C. A. Heuer, Obesity stigma: Important considerations for public health, *American Journal of Public Health* 100 (2010): 1019–1028.
17. Centers for Disease Control and Prevention, Healthy weight—It's not a diet, it's a lifestyle!: About BMI for Children and Teens. 2011, available at www.cdc.gov.
18. C. L. Ogden and K. M. Flegal, Changes in terminology for childhood overweight and obesity, *National Health Statistics Reports* 2010.
19. Bouchard, Childhood obesity; J. M. Ordovas, Genetic influences on blood lipids and cardiovascular disease risk: Tools for primary prevention, *American Journal of Clinical Nutrition* 89 (2009): 1509S–1517S.
20. K. Uusi-Rasi and coauthors, Overweight in childhood and bone density and size in adulthood, *Osteoporosis International* 23 (2012): 1453–1461.
21. T. Rahman and coauthors, Contributions of built environment to childhood obesity, *Mount Sinai Journal of Medicine* 78 (2011): 49–57; K. Silventoinen and coauthors, The genetic and environmental influences on childhood obesity: A systematic review of twin and adoption studies, *International Journal of Obesity* 34 (2010): 29–40.
22. L. J. Benson and coauthors, Screening for obesity-related complications among obese children and adolescents: 1999–2008, *Obesity* 19 (2011): 1077–1082; S. B. Going and coauthors, Percent body fat and chronic disease risk factors in U.S. children and youth, *American Journal of Preventive Medicine* 41 (2011): S77–S86.
23. Centers for Disease Control and Prevention, Prevalence of abnormal lipid levels among youths—United States, 1999–2006, *Morbidity and Mortality Weekly Report* 59 (2010): 29–33.
24. L. J. Lloyd, S. C. Langley-Evans, and S. McMullen, Childhood obesity and adult cardiovascular disease risk: A systematic review, *International Journal of Obesity* 34 (2010): 18–28.
25. D. Jacobson and B. M. Melnyk, A primary care healthy choices intervention program for overweight and obese school-age children and their parents, *Journal of Pediatric Health Care* 26 (2012): 126–138; U.S. Preventive Services Task Force, Screening for obesity in children and adolescents: U.S. Preventive Services Task Force Recommendation Statement, 2010.
26. M. Barton, Childhood obesity: A lifelong health risk, *Acta Pharmacologica Sinica* 2 (2012): 189–193; J. Le and coauthors, "Vascular Age" is advanced in children with atherosclerosis-promoting risk factors, *Circulation: Cardiovascular Imaging* 3 (2010): 8–14.
27. S. Stabouli and coauthors, The role of obesity, salt and exercise on blood pressure in children and adolescents, *Expert Review of Cardiovascular Therapy* 9 (2011): 753–761.
28. K. E. Borradaile and coauthors, Snacking in children: The role of the urban corner stores, *Pediatrics* 124 (2009): 1292–1297.
29. S. B. Sisson and coauthors, Television, reading, and computer time: correlates of school-day leisure-time sedentary behavior and relationship with overweight in children in the U.S., *Journal of Physical Activity and Health Suppl* 2 (2011): S188–S197.
30. S. E. Anderson and R. C. Whitaker, Household routines and obesity in US preschool-aged children, *Pediatrics* 125 (2010): 420–428.
31. J. P. Chaput and coauthors, Video game playing increases food intake in adolescents: A randomized crossover study, *American Journal of Clinical Nutrition* 93 (2011): 1196–203; J. L. Harris, J. A. Bargh, and K. D. Brownell, Priming effects of television food advertising on eating behavior, *Health Psychology* 28 (2009): 404–413.
32. U. Ekelund and coauthors, Moderate to vigorous physical activity and sedentary time and cardiometabolic risk factors in children and adolescents, *Journal of the American Medical Association* 15 (2012): 704–712.
33. T. Hinkley and coauthors, Preschooler's physical activity, screen time, and compliance with recommendations, *Medicine & Science in Sports & Exercise* 44 (2012): 458–465; S. B. Sisson and coauthors, Profiles of sedentary behavior in children and adolescents: The US National Health and Nutrition Examination Survey, 2001–2006, *International Journal of Pediatric Obesity* 4 (2009): 353–359.
34. American Heart Association Advocacy Department, Obesity Prevention Fact Sheet, Unhealthy and unregulated: food advertising and marketing to children, 2012, available at http://www.heart.org/idc/groups/heart-public/@wcm/@adv/documents/downloadable/ucm_301781.pdf; L. M. Powell and coauthors, Trends in the nutritional content of television food advertisements seen by children in the United States analysis by age, food categories, and companies, *Archives of Pediatrics & Adolescent Medicine* 165 (2011): 1078–1086; R. Needlman, Food marketing to children and youth: Threat or

opportunity?, *Journal of Developmental Behavioral Pediatrics* 30 (2009): 183.
35. B. Kelly and coauthors, Television food advertising to children: a global perspective, *American Journal of Public Health* 100 (2010): 1730–1736; K. C. Montgomery and J. Chester, Interactive food and beverage marketing: Targeting adolescents in the digital age, *Journal of Adolescent Health* 45 (2009): S1–S98; A. E. Henry and M. Story, Food and beverage brands that market to children and adolescents on the internet: A content analysis of branded web sites, *Journal of Nutrition Education and Behavior* 41 (2009): 353–359.
36. L. Hebden and coauthors, Art of persuasion: An analysis of techniques used to market foods to children, *Journal of Paediatrics and Child Health* 47 (2011): 776–782.
37. H. K. M. Henry and D. L. G. Borzekowski, The nag factor, *Journal of Children and Media* 5 (2011): 298–317.
38. J. L. Harris and coauthors, US food company branded advergames on the Internet: Children's exposure and effects on snack consumption, *Journal of Children and Media* 6 (2012): 51–68; A. E. Henry and M. Story, Food and beverage brands that market to children and adolescents on the Internet: A content analysis of branded Web sites, *Journal of Nutrition Education and Behavior* 41 (2009): 353–359.
39. M. Ali and coauthors, Young children's ability to recognize advertisements in web page designs, *British Journal of Developmental Psychology* 27 (2009): 71–83.
40. J. Halliday, Industry is successfully self-regulating ads to kids, CFBAI, *Food Navigator-USA.com,* December 1, 2009, available at www.foodnavigator-sa.com; Center for Science in the Public Interest, Most food ads on Nickelodeon still for junk food, *CSPI Newsroom*, November 24, 2009, available at www.cspinet.org/new/200911241.html.
41. USDA Food and Nutrition Service, Core Nutrition Messages, 2012, available at www.fns.usda.gov; World Health Organization, Set of recommendations on the marketing of foods and non-alcoholic beverages to children, 2010, available at www.who.int.
42. White House Task Force on Childhood Obesity Report to the President, Solving The Problem of Childhood Obesity Within a Generation, 2010, available at www.letsmove.gov.
43. C. Danaher and coauthors, Early childhood feeding practices improved after short-term pilot intervention with pediatricians and parents, *Childhood Obesity* 7 (2011): 480–487.
44. American Heart Association, Overweight in children, 2012, available at http://www.heart.org/HEARTORG/GettingHealthy/Overweight-in-Children_UCM_304054_Article.jsp.
45. M. A. Kalarchian and coauthors, Family-based treatment of severe pediatric obesity: Randomized, controlled trial, *Pediatrics* 124 (2009): 1060–1068.
46. S. L. Anzman and coauthors, Parental influence on children's early eating environments and obesity risk: Implications for prevention, *International Journal of Obesity* 34 (2010): 1116–1124; K. J. Gruber and L. A. Haldeman, Using the family to combat childhood and adult obesity, *Preventing Chronic Disease* 6 (2009): A106.
47. T. V. E. Kral and E. M. Rauh, Eating behaviors of children in the context of their family environment, *Physiology and Behavior* 100 (2010): 567–573; M. A. Kalarchian and coauthors, Family-based treatment of severe pediatric obesity: Randomized, controlled trial, *Pediatrics* 124 (2009): 1060–1068.
48. K. A. Hinkle and coauthors, Parents may hold the keys to success in immersion treatment of adolescent obesity, *Child & Family Behavior Therapy* 33 (2011): 273–288; K. S. Geller and D. A. Dzewaltowski, Longitudinal and cross-sectional influences on youth fruit and vegetable consumption, *Nutrition Reviews* 67 (2009): 65–76.
49. A. L. Rogovik and R. D. Goldman, Pharmacologic treatment of pediatric obesity, *Canadian Family Physician* 57 (2011): 195–197.
50. T. H. Inge and coauthors, Reversal of type 2 diabetes mellitus and improvements in cardiovascular risk factors after surgical weight loss in adolescents, *Pediatrics* 123 (2009): 214–222.
51. Y. Latzer and coauthors, Managing childhood overweight: Behavior, family, pharmacology, and bariatric surgery interventions, *Obesity* 17 (2009): 411–423.
52. National Restaurant Association, Kids Live Well, Healthful Choices. Happy Kids, 2012, available at www.restaurant.org.
53. R. B. Ervin and coauthors, Consumption of added sugar among U.S. children and adolescents, 2005–2008, *NCHS Data Brief* 87 (2012).
54. C. B. Ebbeling and coauthors, A randomized trial of sugar-sweetened beverages and adolescent body weight, *New England Journal of Medicine* 367 (2012): 1407–1416; M. S. Vanselow and coauthors, Adolescent beverage habits and changes in weight over time: Findings from Project EAT, *American Journal of Clinical Nutrition* 90 (2009): 1489–1495.
55. Ervin and coauthors, Consumption of added sugar among U.S. children and adolescents, 2005–2008.
56. L. B. Anderson and coauthors, Physical activity and cardiovascular risk factors in children, *British Journal of Sports Medicine* 45 (2011): 871–876; P. K. Doyle-Baker and coauthors, Impact of a combined diet and progressive exercise intervention for overweight and obese children: The B.E.H.I.P. study, *Applied Physiology, Nutrition, and Metabolism* 36 (2011): 515–525.
57. R. Maddison and coauthors, Effects of active video games on body composition: A randomized controlled trial, *The American Journal of Clinical Nutrition* 94 (2011): 156–163; L. Lanningham-Foster and coauthors, Activity-promoting games and increased energy expenditure, *Journal of Pediatrics* 154 (2009): 819–823.
58. G. Antonogeorgos and coauthors, Breakfast consumption and meal frequency interaction with childhood obesity, *Pediatric Obesity* 7 (2012): 65–72.
59. A. Remington and coauthors, Increasing food acceptance in the home setting: A randomized controlled trial of parent-administered taste exposure with incentives, *The American Journal of Clinical Nutrition* 95 (2012): 72–77.

Chapter 14

1. S. A. Ramsay and coauthors, "Are you done?" child care providers' verbal communication at mealtimes that reinforce or hinder children's internal cues of hunger and satiation, *Journal of Nutrition Education and Behavior* 42 (2010): 265–270.
2. L. D. Ritchie, Less frequent eating predicts greater BMI and waist circumference in female adolescents, *American Journal of Clinical Nutrition* 95 (2012): 290–296; B. Koletzko and A. M. Toschke, Meal patterns and frequencies: Do they affect body weight in children and adolescents?, *Critical Reviews in Food Science and Nutrition* 50 (2010): 100–105.
3. Nutritional aspects of vegetarian diets, in *Pediatric Nutrition Handbook*, 6th ed., R. E. Kleinman, ed. (Elk Grove Village, IL: American Academy of Pediatrics, 2009), pp. 201–224.
4. USDA, Dietary Reference Intakes: Macronutrients, USDA National Agricultural Library, 2012, available at http://fnic.nal.usda.gov/dietary-guidance/dietary-reference-intakes/dri-tables.
5. Committee on Dietary Reference Intakes, *Dietary Reference Intakes for Energy, Carbohydrate, Fiber, Fat, Fatty Acids, Cholesterol, Protein, and Amino Acids* (Washington, DC: National Academies Press, 2005), Chapter 11.
6. U. Shaikh, R. S. Byrd, and P. Auinger, Vitamin and mineral supplement use by children and adolescents in the 1999–2004 National Health and Nutrition Examination Survey: Relationship with nutrition, food security, physical activity, and health care access, *Archives of Pediatrics and Adolescent Medicine* 163 (2009): 150–157.
7. Committee on Dietary Reference Intakes, *Dietary Reference Intakes for Calcium and Vitamin D* (Washington, DC: National Academies Press, 2011), pp. 5–35.
8. J. Kumar and coauthors, Prevalence and associations of 25-hydroxyvitamin D deficiency in US children: NHANES 2001–2004, *Pediatrics* 124 (2009): e362–e370.
9. R. D. Baker, F. R. Greer, and The Committee on Nutrition, Diagnosis and prevention of iron deficiency and iron-deficiency anemia in infants and young children (0–3 years of age), *Pediatrics* 126 (2010): 1040–1050.
10. H. Skouteris and coauthors, Parental influence and obesity prevention in pre-schoolers: A systematic review of interventions, *Obesity Reviews* 12 (2011): 315–328.
11. R. R. Briefel and coauthors, The Feeding Infants and Toddlers Study 2008: Study design and methods, *Journal of the Academy of Nutrition and Dietetics* 110 (2010): S16–S26.
12. A. M. Siega-Riz and coauthors, Food consumption patterns of infants and toddlers: where are we now?, *Journal of the Academy of Nutrition and Dietetics* 110 (2010): S38–S51.

13. M. Vadiveloo, L. Zhu, and P. A. Quatromoni, Diet and physical activity patterns of school-aged children, *Journal of the American Dietetic Association* 109 (2009): 145–151; Position of the American Dietetic Association, Nutrition guidance for healthy children ages 2 to 11 years, *Journal of the American Dietetic Association* 108 (2008): 1038–1047.
14. C. C. Tan and S. C. Holub, Maternal feeding practices associated with food neophobia, *Appetite* 59 (2012): 483–487.
15. S. Nicklaus, Development of food varity in children, *Appetite* 52 (2009): 253–255.
16. Cornell University Food and Brand Lab, Smarter lunchrooms: Consequences of belonging to the "Clean Plate Club," 2011 available at http://foodpsychology.cornell.edu/research/summary–consequences.html.
17. J. M. Brge and coauthors, Are parents of young children practicing healthy nutrition and physical activity behaviors?, *Pediatrics* 127 (2011): 881–887.
18. H. Coulthard and J. Blissett, Fruit and vegetable consumption in children and their mothers: Moderating effects of child sensory sensitivity, *Appetite* 52 (2009): 410–415.
19. N. Madan and coauthors, Developmental and neurophysiologic deficits in iron deficiency in children, *Indian Journal of Pediatrics* 78 (2011): 58–64.
20. A. F. Lulowski and coauthors, Iron deficiency in infancy and neurocognitive functioning at 19 years: evidence of long-term deficits in executive fundtion and recognition memory, *Journal of Nutritional Neuroscience* 13 (2010): 54–70.
21. P. Sant-Rayn, Should we screen for iron deficiency anemia? A review of the evidence and recent ecommendations, *Pathology* 44 (2012): 139–147; H. A. Eicher-Miller and coauthors, Food insecurity is associated with iron deficiency anemia in US adolescents, *American Journal of Clinical Nutrition* 90 (2009): 1358–1371.
22. Centers for Disease Control and Prevention, General lead information: Questions and answers, available at www.cdc.gov/nceh/lead/faq/about.htm.
23. A. Miodovnik and P. J. Landrigan, The U.S. Food and Drug Administration risk assessment on lead in women's and children's vitamins is based on outdated assumptions, *Environmental Health Perspectives* 117 (2009): 1021–1022.
24. J. Weuve and coauthors, Cumulative exposure to lead in relation to cognitive function in older women, *Environmental Health Perspectives* 117 (2009): 574–580.
25. J. J. Fadrowski and coauthors, Blood lead level and kidney function in U.S. adolescents, *Archives of Internal Medicine* 170 (2010): 75–82; K. Chandramouli and coauthors, Effects of early childhood lead exposure on academic performance and behavior of school-age children, *Archives of Disease in Childhood* 94 (2009): 844–848.
26. Chandramouli and coauthors, Effects of early childhood lead exposure on academic performance and behavior of school-age children; A. M. Wengroviz and M. J. Brown, Recommendations for blood lead screening of Medicaid-eligible childen aged 1–5 years: An updated approach to targeting a group at high risk, *Morbidity and Mortality Weekly Report* 58 (2009): 1–11.
27. J. Raloff, School-age lead exposures may do more harm than earlier exposures, *Science News*, June 6, 2009, p. 13.
28. Weuve, Cumulative exposure to lead in relation to cognitive function in older women, 2009.
29. L. Hubbs-Tait and coauthors, Main and interaction effects of iron, zinc, lead, and parenting on children's cognitive outcomes, *Developmental Neuropsychology* 34 (2009): 175–195.
30. R. S. Gupta and coauthors, The prevalence, severity, and distribution of childhood food allergy in the United States, *Pediatrics* 128 (2011): e9–17.
31. A. M. Hofmann and coauthors, Safety of a peanut oral immunotherapy protocol in children with peanut allergy, *Journal of Allergy and Clinical Immunology* 124 (2009): 286–291.
32. U.S. Food and Drug Administration, Food allergies: Reducing the risks, January 23, 2009, available at www.fda.gov/consumer/updates/foodallergies012209.htm.
33. Gupta, The prevalence, severity, and distribution of childhood food allergy in the United States, 2011; A. Cianferoni and J. M. Spergel, Food allergy: Review, classification, and diagnosis, *Allergology International* 58 (2009): 457–466.
34. M. C. Young, A. Muñoz-Furlong, and S. H. Sicherer, Management of food allergies in schools: A perspective for allergists, *Journal of Allergy and Clinical Immunology* 124 (2009): 175–182.
35. S. H. Sicherer and T. Mahr, Management of food allergy in the school setting, *Pediatrics* 126 (2010): 1232–1239.
36. H. N. Cho and coauthors, Nutritional status according to sensitized food allergens in children with atopic dermatitis, *Allergy, Asthma & Immunology Research* 3 (2010): 53–57.
37. USDHHS National Institute of Allergy and Infectious Diseases, Guidelines for the diagnosis and management of food allergy in the United States: Summary of the NIAID-Sponsored Expert Panel Report, 2010 NIH Publication No. 11-7700, available at http://www.niaid.nih.gov/topics/foodAllergy/clinical/Documents/FAGuidelinesExecSummary.pdf.
38. A. W. Burks and coauthors, Oral immunotherapy for treatment of egg allergy in children, *New England Journal of Medicine* 367 (2012): 233–243; A. M. Hoffman and coauthors, Safety of a peanut oral immunotherapy protocol in children with peanut allergy, *Journal of Allergy and Clinical Immunology* 124 (2009): 286–291.
39. M. M. Pieretti and coauthors, Audit of manufactured products: Use of allergen advisory labels and identification of labeling ambiguities, *Journal of Allergy and Clinical Immunology* 124 (2009): 337–341.
40. K. Stein, Are food allergies on the rise, or is it a misdiagnosis? *Journal of the American Dietetic Association* 109 (2009): 1832–1837.
41. Stein, Are food allergies on the rise, or is it a misdiagnosis?, 2009.
42. M. Gallo and R. Sayre, Removing allergens and reducing toxins from food crops, *Current Opinion in Biotechnology* 20 (2009): 191–196.
43. L. Batstra and A. Frances, DSM-5 further inflates attention deficit hyperactivity disorder, *The Journal of Nervous and Mental Disease* 200 (2012): 486–488; Centers for Disease Control and Prevention (CDC), Increasing prevalence of parent-reported attention-deficit/hyperactivity disorder among children–United States, 2003 and 2007, *Morbidity and Mortality Weekly Report* 59 (2010): 1439–1443.
44. T. E. Froehlich and coauthors, Update on environmental risk factors for Attention-Deficit/Hyperactivity Disorder, Current Psychiatry Reports 13 (2011): 333–344; J. Stevenson and coauthors, The role of histamine degradation gene polymorphisms in moderating the effects of food additives on children's ADHD symptoms, *American Journal of Psychiatry* 167 (2010): 1108–1115; M. J. Lidy and coauthors, A randomized controlled trial into the effects of food on ADHD, *European Child & Adolescent Psychiatry* 18 (2009): 12–19.
45. N. J. Wiles and coauthors, "Junk food" diet and childhood behavioural problems: Results from the ALSPAC cohort, *European Journal of Clinical Nutrition* 63 (2009): 491–498.
46. S. B. Sisson and coauthors, Television-viewing time and dietary quality among U.S. children and adults, *American Journal of Preventive Medicine* 43 (2012): 196–200.
47. D. C. Hernandez and A. Jacknowski, Transient, but not persistent, adult food insecurity influences toddler development, *Journal of Nutrition* 139 (2009): 1517–1524.
48. J. J. Rucklidge, J. Johnstone, and B. J. Kaplan, Nutrient supplementation approaches in the treatment of ADHD, *Expert Review of Neurotherapeutics* 9 (2009): 461–476.
49. R. Rader, L. McCauley, and E. C. Callen, Current strategies in the diagnosis and treatment of childhood attention-deficit/hyperactivity disorder, *American Family Physician* 79 (2009): 657–665.
50. B. A. Dye, O. Arevalo, and C. M. Vargas, Trends in paediatric dental caries by poverty status in the United States, 1988–1994 and 1999–2004, *International Journal of Paediatric Dentistry* 20 (2010): 132–143; R. A. Bagramian, F. Garcia-Godoy, and A. R. Volpe, The global increase in dental caries: A pending public health crisis, *American Journal of Dentistry* 22 (2009): 3–8.
51. Position of the Academy of Nutrition and Dietetics, The impact of fluoride on health, *Journal of the Academy of Nutrition and Dietetics* 112 (2012): 1443–1453.
52. FDA, About Dental Amalgam Fillings, 2009, available at www.fda.gov/MedicalDevices/ProductsandMedicalProcedures/DentalProducts/DentalAmalgam/ucm171094.htm.
53. V. E. Friedewald and coauthors, The *American Journal of Cardiology* and *Journal of Periodontology* editors' consensus: Periodontitis and

atherosclerotic cardiovascular disease, *Journal of Periodontology* 80 (2009): 1021–1032.
54. A. Borjian and coauthors, Pop-cola acids and tooth erosion: an *In Vitro*, *In Vivo*, electron-microscopic, and clinical report, *International Journal of Dentistry* 2010 (2010): 1–12.
55. T. Coppinger and coauthors, Body mass, frequency of eating and breakfast consumption in 9-13-year-olds, *Journal of Human Nutrition and Dietetics* 25 (2012): 43–49.
56. C. E. Basch, Breakfast and the achievement gap among urban minority youth, *Journal of School Health* 81 (2011): 635–640; S. B. Cooper, S. Bandelow, and M. E. Nevill, Breakfast consumption and cognitive function in adolescent schoolchildren, *Physiology & Behavior* 103 (2011): 431–439; P. R. Deshmukh-Taskar and coauthors, The relationship of breakfast skipping and type of breaskfast consumption with nutrient intake and weight status in children and adolescents: The National Health and Nutrition Examination Survay 1999–2006, *Journal of the American Dietetic Association* 110 (2010): 869–878.
57. M. D. Crepinsek and coauthors, Meals offered and served in U.S. public schools: Do they meet nutrient standards? *Journal of the American Dietetic Association* 109 (2009): S31–S43.
58. Position of the American Dietetic Association, Local support for nutrition integrity in schools, *Journal of the American Dietetic Association* 110 (2010): 1244–1254; National School Lunch Program: Participation and lunches served, USDA, Food and Nutrition Service, www.fns.usda.gov/pd/slsummar.htm, site accessed July 7, 2009.
59. M. Briggs and coauthors, Position of the American Dietetic Association, School Nutrition Association, and Society for Nutrition Education: comprehensive school nutrition services, *Journal of Nutrition Education and Behavior* 42 (2010): 360–371; D. H. Holben, Position of the American Dietetic Association: Food insecurity in the United States, *Journal of the American Dietetic Association* 110 (2010): 1368–1377.
60. M. Story, The Third School Nutrition Dietary Assessment Study: Findings and policy implications for improving the health of U.S. children, *Journal of the American Dietetic Association* 109 (2009): S7–S13.
61. Food and Nutrition Service (FNS), USDA, Nutrition standards in the National School Lunch and School Breakfast Programs. Final rule, *Federal Register* 77 (2012): 4088–4167.
62. Position of the American Dietetic Association: Local support for nutrition integrity in schools, 2010; M. K. Fox and coauthors, Availability and consumption of competitive foods in US public schools, *Journal of the American Dietetic Associaition* 109 (2009): S57–S66.
63. L. R. Turner and F. J. Chaloupka, Student access to competitive foods in elementary schools, trends over time and regional differences, *Archives of Pediatrics & Adolescent Medicine* 166 (2012): 164–169; Centers for Disease Control and Prevention, Availability of less nutritious snack foods and beverages in secondary schools—Selected states, 2002–2008, *Morbidity and Mortality Weekly Report* 58 (2009): 1102–1104.
64. Centers for Disease Control and Prevention, Availability of less nutritious snack foods and beverages in secondary schools—Selected states, 2002–2008, 2009, pp. 1102–1104; M. K. Fox and coauthors, Association between school food environment and practices and body mass index of U.S. public school children, *Journal of the American Dietetic Association* 109 (2009): S108–S117.
65. Food and Nutrition Service (FNS), USDA, Nutrition standards in the National School Lunch and School Breakfast Programs. Final rule, 2012.
66. Food and Nutrition Service (FNS), USDA, Nutrition standards in the National School Lunch and School Breakfast Programs. Final rule, 2012.
67. A. J. Hammons and B. H. Fiese, Is frequency of shared meals related to the nutritional health of children and adolescents?, *Pediatrics* 127 (2011): e1565–e1574; D. Neumark-Sztainer and coauthors, Family meals and adolescents: What have we learned from Project EAT (Eating Among Teens)?, *Public Health Nutrition* 13 (2010): 1113–1121.
68. Neumark-Sztainer and coauthors, Family meals and adolescents: What have we learned from Project EAT (Eating Among Teens)?, 2010.
69. A. S. Alberga and coauthors, Overweight and obese teenagers: Why is adolescence a critical period?, *Pediatric Obesity* 7 (2012): 261–273; R. R. Pate and coauthors, Age-related change in physical activity in adolescent girls, *Journal of Adolescent Health* 44 (2009): 275–282.
70. D. A. Cavallo and coauthors, Smoking expectancies, weight concerns, and dietary behaviors in adolescence, *Pediatrics* 126 (2010): e66–72.
71. N. Vendt and coauthors, Iron deficiency and *Helicobacter pylori* infection in children, *Acta Pediatrica* 100 (2011): 1239–1243; M. Mesías and coauthors, The beneficial effect of Mediterranean dietary patterns on dietary iron utilization in male adolescents aged 11–14 years, *International Journal of Food Sciences and Nutrition* 60 (2009): 335–368.
72. H. A. Eicher-Miller and coauthors, Food insecurity is associated with iron deficiency anemia in U.S. adolescents, *American Journal of Clinical Nutrition* 90 (2009): 1358–1371.
73. V. Matkovic, Calcium and peak bone mass, *Journal of Internal Medicine* 231 (2009): 151–160.
74. Committee on Dietary Reference Intakes, *Dietary Reference Intakes for Calcium and Vitamin D* (Washington, DC: National Academies Press, 2011), pp. 458–460; R. L. Bailey and coauthors, Estimation of total usual calcium and vitamin D intakes in the United States, *The Journal of Nutrition* 140 (2010): 817–822.
75. K. L. Keller and coauthors, Increased sweetened beverage intake is associated with reduced milk and calcium intake in 3- to 7-year-old children at multi-item laboratory lunches, *Journal of the American Dietetic Association* 109 (2009): 497–501.
76. L. Esterie and coauthors, Milk, rather than other foods, is associated with vertebral bone mass and circulating IGF-1 in female adolescents, *Osteoporosis International* 20 (2009): 567–575.
77. E. H. Spencer, H. R. Ferdowsian, and N. D. Barnard, Diet and acne: A review of the evidence, *International Journal of Dermatology* 48 (2009): 339–347.
78. Spencer, Ferdowsian, and Barnard, Diet and acne: A review of the evidence, 2009.
79. I. Y. Hur and M. Reicks, Relationship between whole-grain intake, chronic disease risk indicators, and weight status among adolescents in the National Health and Nutrition Examination Survey, 1999–2004, *Academy of Nutrition and Dietetics* 112 (2012): 46–55; T. L. Burgess-Champoux and coauthors, Longitudinal and secular trends in adolescent whole-grain consumption, 1999–2004, *American Journal of Clinical Nutrition* 91 (2009): 154–159.
80. C. M. McDonald and coauthors, Overweight is more prevalent than stunting and is associated with socioeconomic status, maternal obesity, and a snacking dietary pattern in school children from Bogota, Columbia, *Journal of Nutrition* 139 (2009): 370–376.
81. J. E. Holsten and coauthors, Children's food choice process in the home environment. A qualitative descriptive study, *Appetite* 58 (2012): 64–73; N. Pearson and coauthors, Predictors of change in adolescents' consumption of fruits, vegetables and energy-dense snacks, *The British Journal of Nutrition* 105 (2011): 795–803.
82. D. R. Keast, T. A. Nicklas, and C. E. O'Neil, Snacking is associated with reduced risk of overweight and reduced abdominal obesity in adolescents: National Health and Nutrition Examination Survey (NHANES) 1999–2004, *The American Journal of Clinical Nutrition* 92 (2010): 428–435.
83. B. R. Levy and coauthors, Age stereotypes held earlier in life predict cardiovascular events in later life, *Psychological Science* 20 (2009): 296–298.
84. T. Ahmed and N. Haboubi, Assessment and management of nutrition in older people and its importance to health, *Clinical Interventions in Aging* 5 (2010): 207–216.
85. S. Sabia and coauthors, Influence of individual and combined healthy behaviours on successful aging, *Canadian Medical Association Journal* (2012), epub, doi:10.1503/cmaj.121080.
86. National Center for Health Statistics, Life expectancy at age 65 years, by sex and race—United States, 2000–2006, *Morbidity and Mortality Weekly Report* 58 (2009): 473.
87. J. Xu, K. D. Kochanek, and B. Tehada-Vera, Deaths: Preliminary data for 2007, *National Vital Statistics Reports* 58, August 19, 2009, available at www.cdc.org.
88. A. M. Minino and coauthors, Deaths: Final data for 2008, *National Vital Statistics Reports* 59 (2011): 1–127; S. Harper, D. Rushani, and J. S. Kaufman, Trends in the black-white life expec-

tancy gap, 2003–2008, *Journal of the American Medical Association* 307 (2012): 2257–2259.
89. S. Harper, D. Rushani, and J. S. Kaufman, Trends in the black-white life expectancy gap, 2003–2008, *Journal of the American Medical Association* 307 (2012): 2257–2259.
90. B. M. Weon and J. H. Je, Theoretical estimation of maximum human lifespan, *Biogerontology* 10 (2009): 65–71.
91. I. M. Bensenor, R. D. Olmos, and P. A. Lotufo, Hypothyroidism in the elderly: diagnosis and management, *Clinical Interventions in Aging* 7 (2012): 97–111; M. P.St-Onge and D. Gallagher, Body composition changes with aging: The cause or the result of alterations in metabolic rate and macronutrient oxidation?, *Nutrition* 26 (2010): 152–155.
92. K. A. Intlekofer and C. W. Cotman, Exercise counteracts declining hippocampal function in aging and Alzheimer's disease, *Neurobiology of Disease* (2012), epub ahead of print, doi:10.1016/j.nbd.2012.06.011.
93. K. Anderson, C. Baraldi, and M. Supiano, Identifying failure to thrive in the long-term care setting, *Journal of the American Medical Directors Association* 13 (2012): 665.e15–665.e19.
94. E. Britton and J. T. McLaughlin, Ageing and the gut, *Proceedings of the Nutrition Society* 72 (2013): 173–177.
95. C. A. Zizza, D. D. Arsiwalla, and K. J. Ellison, Contribution of snacking to older adults' vitamin, carotenoid, and mineral intakes, *Journal of the American Dietetic Association* 110 (2010): 768–772.
96. R. Koopman and L. J. C. van Loon, Aging, exercise, and muscle protein metabolism, *Journal of Applied Physiology* 106 (2009): 2040–2048; M. D. Peterson and coauthors, Resistance exercise for muscular strength in older adults: A meta-analysis, *Ageing Research Reviews* 9 (2010): 226–237.
97. U.S. Department of Health and Human Services, Healthy People 2020 Physical Activity, 2012, available at http://www.healthypeople.gov/2020/topicsobjectives 2020/overview.aspx?topicid=33; B. Elsawy and K. E. Higgins, Physical activity guidelines for older adults, *American Family Physician* 81 (2010): 55–59.
98. M. G. Stineman and coauthors, All-cause 1-, 5-, and 10-year mortality in elderly people according to activities of daily living stage, *Journal of the American Geriatrics Society* 60 (2012): 485–492.
99. D. Paddon-Jones and B. B. Rasmussen, Dietary protein recommendations and the prevention of sarcopenia, *Current Opinion in Clinical Nutrition and Metabolic Care* 12 (2009): 86–90.
100. J. E. Morley and coauthors, Nutritional recommendations for the management of sarcopenia, *Journal of the American Medical Directors Association* 11 (2010): 391–396; D. K. Layman, Dietary guidelines should reflect new understandings about adult protein needs, *Nutrition & Metabolism* 6 (2009): 12; A. K. Surdykowski and coauthors, Optimizing bone health in older adults: The importance of dietary protein, *Aging Health* 6 (2010): 345–357.
101. American College of Sports Medicine, Position Stand: Exercise and physical activity for older adults, *Medicine and Science in Sports and Exercise* 41 (2009): 1510–1530.
102. J. Stessman and coauthors, Physical activity, function, and longevity among the very old, *Archives of Internal Medicine* 169 (2009): 1476–1483.
103. U. Raue and coauthors, Improvements in whole muscle and myocellular function are limited with high-intensity resistance training in octogenarian women, *Journal of Applied Physiology* 106 (2009): 1611–1617.
104. Paddon-Jones and Rassmussen, Dietary protein recommendations and the prevention of sarcopenia, 2009; R. Koopman and coauthors, Dietary protein digestion and absorption rates and the subsequent postprandial muscle protein synthetic response do not differ between young and elderly men, *Journal of Nutrition* 139 (2009): 1707–1713.
105. Q. L. Xue and coauthors, Patterns of 12-year change in physical activity levels in community-dwelling older women: Can modest levels of physical activity help older women live longer? *American Journal of Epidemiology* 176 (2012): 534–543; K. T. Morgan, Nutrition, resistance training, and sarcopenia: Their role in successful aging, *Topics in Clinical Nutrition* 27 (2012): 114–123.
106. D. E. King, A. G. Mainous 3rd, and C. A. Lambourne, Trends in dietary fiber intake in the United States, 1999–2008, *Journal of the Academy of Nutrition and Dietetics* 112 (2012): 642–648; J. W. Anderson and coauthors, Health benefits of dietary fiber, *Nutrition Reviews* 67 (2009): 188–205.
107. Centers for Disease Control and Prevention, Arthritis: Meeting the challenge, *At a Glance*, 2009, available at www.cdc.gov.
108. K. L. Bennell and R. S. Hinman, A review of the clinical evidence for exercise in osteoarthritis of the hip and knee, *Journal of Science and Medicine in Sport* 14 (2011): 4–9; N. J. Bosomworth, Exercise and knee osteoarthritis: benefit or hazard?, *Canadian Family Physician* 55 (2009): 871–878.
109. G. Smedslund and coauthors, Effectiveness and safety of dietary interventions for rheumatoid arthritis: a systematic review of randomized controlled trials, *Journal of the American Dietetic Association* 110 (2010): 727–735; C. Deighton and coauthors, Management of rheumatoid arthritis: summary of NICE guidance, *British Medical Journal* 338 (2009): b702.
110. S. Panicker and coauthors, Oral glucosamine modulates the response of the liver and lymphocytes of the mesenteric lymph nodes in a papain-induced model of joint damage and repair, *Osteoarthritis and Cartilage* 17 (2009): 1014–1021; G. Sobal, J. Menzel, and H. Sinzinger, Optimal (99m)Tc radiolabeling and uptake of glucosamine sulfate by cartilage. A potential tracer for scintigrapic detection of osteoarthritis, *Bioconjugate Chemistry* 20 (2009): 1547–1552.
111. Y. Zhang and coauthors, Purine-rich foods intake and recurrent gout attacks, *Annals of the Rheumatic Diseases* 71 (2012): 1448–1453; E. Krishnan, B. Lingala, and V. Bhalla, Low-level lead exposure and the prevalence of gout: An observational study, *Annals of Internal Medicine* 157 (2012): 233–241; T. Neogi, Gout, *The New England Journal of Medicine* 364 (2011): 443–452.
112. Committee on Dietary Reference Intakes, *Dietary Reference Intakes for Calcium and Vitamin D* (Washington, DC: National Academies Press, 2011), pp. 345–402.
113. Committee on Dietary Reference Intakes, *Dietary Reference Intakes for Calcium and Vitamin D* (Washington, DC: National Academies Press, 2011), pp. 345–402.
114. U.S. Preventive Services Task Force, Vitamin D and calcium supplementation to prevent cancer and osteoporotic fractures, Draft recommendations, August 2012, available at www.uspreventiveservicestaskforce.org/uspstf12/vitamind/vitdart.htm.
115. V. A. Moyer and the U.S. Preventive Services Task Force, Prevention of falls in community-dwelling older adults: U.S. Preventive Services Task Force Recommendation Statement, *Annals of Internal Medicine* 157 (2012): 197–204; DIPART (vitamin D Individual Patient Analysis of Randomized Trials) Group, Patient level pooled analysis of 68,500 patients from seven major vitamin D fracture trials in US and Europe, *British Medical Journal* (2010), epub, doi:10.1136/bmj.b5463; H. A. Bischoff-Ferrari and coauthors, Fall prevention with supplemental and active forms of vitamin D: A meta-analysis of randomized controlled trials, *British Medical Journal* 339 (2009), epub, doi:10.1136/bmj.b362.
116. L. H. Allen, How common is vitamin B-12 deficiency? *American Journal of Clinical Nutrition* 89 (2009): 693S–696S.
117. R. Green, Is it time for vitamin B-12 fortification? What are the questions? *American Journal of Clinical Nutrition* 89 (2009): 712S–716S.
118. J. W. Eichenbaum, Geriatric vision loss due to cataracts, macular degeneration, and glaucoma, *Mount Sinai Journal of Medicine: A Journal of Translational and Personalized Medicine* 79 (2012): 276–294; M. A. Alma and coauthors, Participation of the elderly after vision loss, *Disability and Rehabilitation* 33 (2011): 63–72.
119. L. A. Lott and coauthors, Non-standard vision measures predict mortality in elders: the Smith-Kettlewell Institute (SKI) Study, *Opthalmic Epidemiology* 17 (2010): 242–250.
120. S. Carpentier, M. Knauss, and M. Suh, Associations between lutein, zeaxanthin, and age-related macular degeneration: An overview, *Critical Reviews in Food Science* 49 (2009): 313–326.
121. J. R. Evans and J. G. Lawrenson, Antioxidant vitamin and mineral supplements for preventing age-related macular degeneration, *Cochrane Database Systematic Reviews* 6 (2012), epub, doi:10.1002/14651858.CD000253.pub3; N. Krishnadev, A. D. Meleth, and E. Y. Chew, Nutritional supplements for age-related macular degeneration, *Current Opinion in Opthamology* 21 (2010): 184–189.

122. M. Mvitu and coauthors, Regular, high, and moderate intake of vegetables rich in antioxidants may reduce cataract risk in Central African type 2 diabetics, *International Journal of General Medicine* 5 (2012): 489–493; J. A. Mares and coauthors, Healthy diets and the subsequent prevalence of nuclear cataract in women, *Archives of Opthamology* 128 (2010): 738–749.
123. S. Rautiainen and coauthors, Vitamin C supplements and the risk of age-related cataract: A population-based prospective cohort study in women, *American Journal of Clinical Nutrition* 91 (2010): 487–493.
124. D. Benton, Dehydration influences mood and cognition: a plausible hypothesis?, *Nutrients* 3 (2011): 555–573.
125. J. B. Barnett, D. H. Hamer, and S. N. Meydani, Low zinc status: A new risk factor for pneumonia in the elderly? *Nutrition Reviews* 68 (2010): 30–37.
126. V. W. Dolinsky and coauthors, Calorie restriction prevents hypertension and cardiac hypertrophy in the spontaneously hypertensive rat, *Hypertension* 56 (2010): 412–421; X. Liu and coauthors, Downregulation of Grb2 contributes to the insulin-sensitizing effect of calorie restriction, *American Journal of Physiology, Endocrinology, and Metabolism* 296 (2009): E1067–E1075.
127. R. J. Colman and coauthors, Caloric restriction delays disease onset of mortality in Rhesus monkeys, *Science* 325 (2009): 201–204.
128. J. A. Mattison and coauthors, Impact of caloric restriction on health and survival in rhesus monkeys from the NIA study, *Nature* 489 (2012): 318–322.
129. W. M. Teeuwisse and coauthors, Short-term caloric restriction normalizes hypothalamic neuronal responsiveness to glucose ingestion in patients with type 2 diabetes, *Diabetes* (2012); M. Lefevre and coauthors, Caloric restriction alone and with exercise improves CVD risk in healthy non-obese individuals, *Atherosclerosis* 203 (2009): 206–213; C. Cantó and J. Auwerx, Caloric restriction, SIRT1 and longevity, *Trends in Endocrinology and Metabolism* 20 (2009): 325–331.
130. K. Dorshkind, E. Montecino-Rodriguez, and R. A. Singer, The ageing immune system: Is it ever too old to become young again?, *Nature Reviews Immunology* 9 (2009): 57–62; T. Ahmed and coauthors, Calorie restriction enhances T-cell-mediated immune response in adult overweight men and women, *The Journals of Gerontology* 64A (2009): 1107–1113.
131. F. M. Cerqueira and A. J. Kowaltowski, Commonly adopted caloric restriction protocols often involve malnutrition, *Ageing Research Reviews* 9 (2010): 424–430.
132. J. P. de Magalhães and coauthors, Genome-environment interactions that modulate aging: powerful targets for drug discovery, *Pharmacological Reviews* 64 (2012): 88–101.
133. G. Candore and coauthors, Low grade inflammation as a common pathogenetic denominator in age-related diseases: novel drug targets for anti-ageing strategies and successful ageing achievement, *Current Pharmaceutical Design* 16 (2010): 584–596; A. Desai, A. Grolleau-Julius, and R. Yung, Leukocyte function in the aging immune system, *Journal of Leukocyte Biology* 87 (2010): 1001–1009.
134. M. A. Alaiti and coauthors, Kruppel-like factors and vascular inflammation: Implications for atherosclerosis, *Current Atherosclerosis Reports* (2012); H. Wood, Alzheimer disease: Prostaglandin E(2) signaling is implicated in inflammation early in the Alzheimer disease course, *Nature Review Neurology* (2012); V. Vachharajani and D. N. Granger, Adipose tissue: A motor for the inflammation associated with obesity, *International Union of Biochemistry and Molecular Biology Life* 61 (2009): 424–430.
135. P. Y. Perera and coauthors, The role of interleukin-15 in inflammation and immune responses to infection: Implications for its therapeutic use, *Microbes and Infection* 14 (2012): 247–261.
136. A. L. Kau and coauthors, Human nutrition, the gut microbiome and the immune system, *Nature* 474 (2011): 327–336; J. Romeo and coauthors, Physical activity, immunity and infection, *The Proceedings of the Nutrition Society* 69 (2010): 390–399.
137. A. B. Salmon, A. Richardson, and V. I. Pérez, Update on the oxidative stress theory of aging: does oxidative stress play a role in aging or healthy aging?, *Free Radical Biology and Medicine* 48 (2010): 642–655.
138. W. E. Sonntag and coauthors, Diverse roles of growth hormone and insulin-like growth factor-1 in mammalian aging: progress and controversies, *The Journals of Gerontology Series A, Biological Sciences and Medical Sciences* 67 (2012): 587–598; M. Srinivasan and coauthors, Effects on lipoprotein particles of long-term dehydroepiandrosterone in elderly men and women and testosterone in elderly men, *The Journal of Clinical Endocrinology and Metabolism* 95 (2010): 1617–1625.
139. J. C. Lambert and coauthors, Genome-wide association study identifies variants at CLU and CR1 associated with Alzheimer's disease, *Nature Genetics* 41 (2009): 1094–1099.
140. C. Féart and coauthors, Adherence to a Mediterranean diet, cognitive decline and risk of dementia, *Journal of the American Medical Association* 302 (2009): 638–648; N. Scarmeas and coauthors, Physical activity, diet and risk of Alzheimer Disease, *Journal of the American Medical Association* 302 (2009): 627–637; J. A. Luchsinger and D. R. Gustafon, Adiposity and Alzheimer's disease, *Current Opinion in Clinical Nutrition and Metabolic Care* 12 (2009): 15–21.
141. N. G. Bazan, M. F. Molina, and W. C. Gordon, Docosahexaenoic acid signalolipidomics in nutrition: Significance in aging, neuroinflammation, macular degeneration, Alzheimer's, and other neurodegenerative diseases, *Annual Review of Nutrition* 31 (2011): 321–351; E. Albanese and coauthors, Dietary fish and meat intake and dementia in Latin America, China, and India: A 10/66 Dementia Research Group population-based study, *American Journal of Clinical Nutrition* 90 (2009): 392–400; E. E. Devore and coauthors, Dietary intake of fish and omega-3 fatty acids in relation to long-term dementia risk, *American Journal of Clinical Nutrition* 90 (2009): 170–176.
142. C. Cooper and coauthors, Alcohol in moderation, premorbid intelligence and cognition in older adults: Results from the Psychiatric Morbidity Survey, *Journal of Neurology, Neurosurgery, and Psychiatry* 80 (2009): 1236–1239.
143. B. E. Snitz and coauthors, *Ginkgo biloba* for preventing cognitive decline in older adults, *Journal of the American Medical Association* 302 (2009): 2663–2670.
144. Position of the Academy of Nutrition and Dietetics: Food and nutrition for older adults: Promoting health and wellness, *Journal of the Academy of Nutrition and Dietetics* 112 (2012): 1255–1277; M. Dean and coauthors, Factors influencing eating a varied diet in old age, *Public Health Nutrition* 12 (2009): 2421–2427.
145. W. F. Nieuwenhuizen and coauthors, Older adults and patients in need of nutritional support: Review of current treatment options and factors influencing nutritional intake, *Clinical Nutrition* 29 (2010): 160–169.
146. Position of the Academy of Nutrition and Dietetics: Food and nutrition for older adults: promoting health and wellness, 2012.
147. C. M. Perissinotto, I. S. Cenzer, and K. E. Covinsky, Loneliness in older persons: A predictor of functional decline, *Archives of Internal Medicine* 172 (2012): 1078–1084.

Consumer's Guide 14

1. W. S. Biggs and R. H. Demuth, Premenstrual syndrome and premenstrual dysphoric disorder, *American Family Physician* 84 (2011): 918–924.
2. E. W. Freeman and coauthors, Core symptoms that discriminate premenstrual syndrome, *Journal of Women's Health* 20 (2011): 29–35.
3. J. Brown and coauthors, Selective serotonin reuptake inhibitors for premenstrual syndrome, *Cochrane Database of Systematic Reviews* 2 (2009), epub, doi:10.1002/14651858.CD001396.pub2.
4. M. A. McVay, A. L. Copeland, and P. J. Geiselman, Eating disorder pathology and menstrual cycle fluctuations in eating variables in oral contraceptive users and non-users, *Eating Behaviors* 12 (2011): 49–55.
5. E. R. Bertone-Johnson and coauthors, Timing of alcohol use and the incidence of premenstrual syndrome and probable premenstrual dysphoric disorder, *Journal of Women's Health* 18 (2009): 1945–1953.
6. Z. Ghanbari and coauthors, Effects of calcium supplement therapy in women with premenstrual syndrome, *Taiwanese Journal of Obstetrics and Gynecology* 48 (2009): 124–139.
7. E. R. Bertone-Johnson and coauthors, Dietary vitamin D intake, 25-hydroxyvitamin D3 levels and premenstrual syndrome in a college-aged population, *The Journal of Steroid Biochemistry and Molecular Biology* 121 (2010): 434–437.
8. A. Lasco, A. Catalano, and S. Benvenga, Improvement of primary dysmenorrhea caused by a single oral dose of vitamin D: Results of a ran-

domized, double-blind, placebo-controlled study, *Archives of Internal Medicine* 172 (2012): 366–367.
9. E. R. Bertone-Johnson and J. E. Manson, Vitamin D for menstrual and pain-related disorders in women, *Archives of Internal Medicine* 172 (2012): 367–368.
10. Lasco, Catalano, and Benvenga, Improvement of primary dysmenorrhea caused by a single oral dose of vitamin D: Results of a randomized, double-blind, placebo-controlled study, 2012; Bertone-Johnson and Manson, Vitamin D for menstrual and pain-related disorders in women, 2012.
11. I. Kwan and J. L. Onwude, Premenstrual syndrome, *Clinical Evidence* 12 (2009): 1–28.
12. A. Daley, Exercise and premenstrual symptomatology: A comprehensive review, *Journal of Women's Health* 18 (2009): 895–899.

Controversy 14

1. J. I. Boullata and L. M. Hudson, Drug-nutrient interactions: a broad view with implications for practice, *Journal of the Academy of Nutrition and Dietetics* 112 (2012): 506–517.
2. D. Singh, R. Gupta, and S. A. Saraf, Herbs—Are they safe enough? An overview, *Critical Reviews in Food Science and Nutrition* 52 (2012): 876–898.
3. D. Gnjidic and coauthors, Polypharmacy cutoff and outcomes: five or more medicines were used to identify community-dwelling older men at risk of different adverse outcomes, *Journal of Clinical Epidemiology* 65 (2012): 989–995.
4. K. A. Jaques and B. L. Erstad, Availability of information for dosing injectable medications in underweight or obese patients, *American Journal of Health-System Pharmacology* 15 (2010): 1948–1950.
5. B. S. Rich and coauthors, Cefepime dosing in the morbidly obese patient population, *Obesity Surgery* 22 (2012): 465–471.
6. S. L. Haber, K. A. Cauthon, and E. C. Raney, Cranberry and warfarin interaction: A case report and review of literature, *Consultant Pharmacist* 27 (2012): 58–65.
7. L. A. Kelly and coauthors, Genistein alters coagulation gene expression in ovariectomised rats treated with phytoestrogens, *Thrombosis and Haemostasis* 104 (2010): 1250–1257.
8. E. Perry and M. R. Howes, Medicinal plants and dementia therapy: herbal hopes for brain aging? *CNS Neuroscience and Therapeutics* 17 (2011): 683–698.
9. C. J. Derry, S. Derry, and R. A. Moore, Caffeine as an analgesic adjuvant for acute pain in adults, Cochrane Database Systematic Review March 14, 2012, epub, doi:10.1002/14651858.
10. E. S. Mitchell and coauthors, Differential contributions of theobromine and caffeine on mood, psychomotor performance and blood pressure, *Physiology and Behavior* 104 (2011): 816–822.
11. R. Sinha and coauthors, Caffeinated and decaffeinated coffee and tea intakes and risk of colorectal cancer in a large prospective study, *American Journal of Clinical Nutrition* 96 (2012): 374–381; R. A. Floegel and coauthors, Coffee consumption and risk of chronic disease in the European Prospective Investigation in to Cancer and Nutrition (EPIC)—Germany study, *American Journal of Clinical Nutrition* 95 (2012): 901–908; J. A. Nettleton, J. L. Follis, and M. B. Schabath, Coffee intake, smoking, and pulmonary function in the Atherosclerosis Risk in Communities Study, *American Journal of Epidemiology* 169 (2009): 1445–1453.
12. S. N. Bhupathiraju and coauthors, Caffeinated and caffeine-free beverages and risk of type 2 diabetes, *American Journal of Clinical Nutrition* 97 (2013): 155–166; F. Natella and C. Scaccini, Role of coffee in modulation of diabetes risk, *Nutrition Reviews* 70 (2012): 207–217; J. D. Lane and coauthors, Pilot study of caffeine abstinence for control of chronic glucose in type 2 diabetes, *Journal of Caffeine Research* 2 (2012): 45–47; W. Zhang and coauthors, Coffee consumption and risk of cardiovascular disease and all-cause mortality among men with type 2 diabetes, *Diabetes Care* 32 (2009): 1043–1045.

Chapter 15

1. A. Coleman-Jensen and coauthors, Economic Research Service Report Summary: Household food security in the United States in 2010, September 2011, available at www.ers.usda.gov.
2. Food and Agriculture Organization of the United Nations, *FAO Statistical Yearbook, 2012*, Hunger dimensions, available at www.fao.org.
3. C. P. Timmer, Preventing food crises using a food policy approach, *Journal of Nutrition* 140 (2010): 224S–228S.
4. A. Coleman-Jensen and coauthors, Household food security in the United States in 2011 *ERS Report Summary*, September 2012, available at www.ers.usda.gov.
5. K. Fiscella and H. Kitzman, Disparities in academic achievement and health: The intersection of child education and health policy, *Pediatrics* 123 (2009): 1073–1080.
6. B. Rutherford, Will U.S. food prices follow global trends? *Beef Magazine*, September 16, 2011, available at http://beefmagazine.com.
7. J. C. Eisenmann and coauthors, Is food insecurity related to overweight and obesity in children and adolescents? A summary of studies, 1995–2009, *Obesity Reviews* 12 (2011): e73–e83.
8. B. M. Popkin, L. S. Adair, and S. W. Ng, Global nutrition transition and the pandemic of obesity in developing countries, *Nutrition Reviews* 70 (2011): 3–21; A. Drewnowski, Obesity, diets, and social inequalities, *Nutrition Reviews* 67 (2009): S36–S39; B. J. Lohman and coauthors, Adolescent overweight and obesity: Links to food insecurity and individual, maternal, and family stressors, *Journal of Adolescent Health* 45 (2009): 230–237; E. Metallinos-Katsaras, B. Sherry, and J. Kallio, Food insecurity is associated with overweight in children younger than 5 years of age, *Journal of the American Dietetic Association* 109 (2009): 1790–1794.
9. World Health Organization, World Health Statistics 2012, available at http://www.who.int/gho/publications/world_health_statistics/2012/en/; H. K. Seligman and coauthors, Food insecurity and glycemic control among low-income patient with type 2 diabetes, *Diabetes Care* 35 (2012): 233–238; H. K. Seligman and D. Schillinger, Hunger and socioeconomic disparities in chronic disease, *New England Journal of Medicine* 363 (2010): 6–9.
10. A. M. Fretts and coauthors, Associations of processed meat and unprocessed red meat intake with incident diabetes: The Strong Heart Family Study, *American Journal of Clinical Nutrition* 95 (2012): 752–758; H. J. Song and coauthors, Understanding a key feature of urban food stores to develop nutrition intervention, *Journal of Hunger & Environmental Nutrition* 7 (2012): 77–90; Institute of Medicine and National Research Council, *The Public Health Effects of Food Deserts: Workshop Summary* (Washington, DC: The National Academies Press, 2009).
11. B. T. Izumi and coauthors, Associations between neighborhood availability and individual consumption of dark-green and orange vegetables among ethnically diverse adults in Detroit, *Journal of the American Dietetic Association* 111 (2011): 274–279.
12. A. Offer, R. Pechey, and S. Ulijaszek, Obesity under affluence varies by welfare regimes: The effect of fast food, insecurity, and inequality, *Economics and Human Biology* 8 (2010): 297–308.
13. A. Karnik and coauthors, Food insecurity and obesity in New York City primary care clinics, *Medical Care* 49 (2011): 658–661; N. I. Larson and M. T. Story, Food insecurity and weight status among U.S. children and families: A review of the literature, *American Journal of Preventative Medicine* 40 (2011): 166–173.
14. USDA Economic Research Service, Economic Bulletin 6–8, The food assistance landscape: 2010 annual report, available at http://www.ers.usda.gov/media/129642/eib6-8.pdf.
15. U.S. Census Bureau, The 2012 Statistical Abstract, available at http://www.census.gov/compendia/statab/cats/income_expenditures_poverty_wealth.html; Health and Human Services, 2012 HHS Poverty Guidelines, available at http://aspe.hhs.gov/poverty/12poverty.shtml.
16. Position of the American Dietetic Association, Food insecurity in the United States, *Journal of the American Dietetic Association* 110 (2010): 1368–1377.
17. L. Tiehen, D. Jolliffe, and C. Gundersen, Alleviating poverty in the United States: the critical role of SNAP benefits, ERR–132, U.S. Department of Agriculture, April 2012, available at http://www.ers.usda.gov.
18. Food Research Action Center, SNAP/Food Stamp participation 2012, available at http://frac.org.
19. Position of the American Dietetic Association: Food insecurity in the United States, *Journal of the American Dietetic Association* 110 (2010): 1368–1377.
20. Food and Agriculture Organization of the United Nations, *The State of Food Insecurity in the World, 2011*; M. W. Bloem, R. D. Semba, and K. Kraemer, Castel Gandolfo Workshop: An introduction to the impact of climate change, the economic crisis, and the increase in the food

process on malnutrition, *Journal of Nutrition* 140 (2010): 132S–135S.
21. Centers for Disease Control and Prevention, *Salmonella typhi* infections associated with contaminated water—Zimbabwe, October 2011–May 2012, *Morbidity and Mortality Weekly Report* 61 (2012): 435.
22. Food and Agriculture Organization of the United Nations, *The State of Food Insecurity in the World, 2011*.
23. Food and Agriculture Organization of the United Nations, *The State of Food Insecurity in the World, 2011*.
24. Food and Agricultural Organization of the United Nations, 925 million in chronic hunger worldwide, September 2010, available at http://www.fao.org/news/story/en/item/45210/icode/.
25. S. M. O'Neill and coauthors, Child mortality as predicted by nutritional status and recent weigh velocity in children under two in rural Africa, *Journal of Nutrition* 142 (2012): 520–525.
26. M. W. Bloem, R. D. Semba, and K. Kraemer, Castel Gandolfo workshop: An introduction to the impact of climate change, the economic crisis, and the increase in the food prices on malnutrition, *Journal of Nutrition* 140 (2010): 132S–135S.
27. Position of the American Dietetic Association, Food insecurity in the United States, 2010.
28. J. V. White and coauthors, Consensus statement of the Academy of Nutrition and Dietetics/American Society for Parenteral and Enteral Nutrition: Characteristics recommended for the identification and documentation of adult malnutrition (undernutrition), *Journal of the Academy of Nutrition and Dietetics* 112 (2012): 730–738.
29. Food and Agriculture Organization of the United Nations, *FAO Statistical Yearbook, 2012*.
30. E. Boy and coauthors, Achievements, challenges, and promising new approaches in vitamin and mineral deficiency control, *Nutrition Reviews* 67 (2009): S24–S30.
31. World Health Organization, *Global prevalence of vitamin A deficiency in populations at risk 1995–2005. WHO Global Database on Vitamin A Deficiency* (Geneva: World Health Organization, 2009); Standing Committee on the Scientific Evaluation of Dietary Reference Intakes, Food and Nutrition Board, Institute of Medicine, *Dietary Reference Intakes for Vitamin A, Vitamin K, Arsenic, Boron, Chromium, Copper, Iodine, Iron, Manganese, Molybdenum, Nickel, Silicon, Vanadium, and Zinc* (Washington, DC: National Academy Press, 2001), pp. 4-9–4-10.
32. World Health Organization, 10 Facts on Child Health, October 2012, available at http://www.who.int/features/factfiles/child_health2/en/index.html.
33. P. Svedberg, How many people are malnourished, *Annual Review of Nutrition* 31 (2011): 263–283.
34. N. F. Krebs and coauthors, Meat consumption is associated with less stunting among toddlers in four diverse low-income settings, *Food and Nutrition Bulletin* 32 (2011): 185–191.
35. T. E. Forrester and coauthors, Prenatal factors contribute to the emergence of kwashiorkor or marasmus in severe undernutrition: evidence for the predictive adaptation model, *PLoS ONE* (2012), epub, doi:10.1371/journal.pone.0035907; M. H. Golden, Evolution of nutritional management of acute malnutrition, *Indian Pediatrics* 47 (2010): 667–678.
36. World Health Organization/UNICEF, *WHO Child Growth Standards and the Identification of Severe Acute Malnutrition in Infants and Children* (Geneva: WHO Press, 2009), available at www.who.int/nutrition/publications/severemalnutrition/9789241598163_eng.pdf.
37. W. A. Petri Jr. and coauthors, Association of malnutrition with amebiasis, *Nutrition Reviews* 67 (2009): S207–S215; T. Ahmed and coauthors, Use of metagenomics to understand the genetic basis of malnutrition, *Nutrition Reviews* 67 (2009): S201–S206.
38. C. Best and coauthors, Can multi-micronutrient food fortification improve the micronutrient status, growth, health, and cognition of schoolchildren? A systematic review, *Nutrition Reviews* 69 (2011): 186–204; M. H. Golden, Proposed recommended nutrient densitites for moderately malnourished children, *Food and Nutrition Bulletin* 30 (2009): S267–S342.
39. D. R. Brewster, Inpatient management of severe malnutrition: Time for a change in protocol and practice, *Annals of Tropical Paediatrics* 31 (2011): 97–107.
40. S. van der Kam and coauthors, Ready-to-use therapeutic food for catch-up growth in children after an episode of Plasmodium falciparum malaria: An open randomised controlled trial, *PLoS One* 7 (2012), epub, doi: 10.1371/journal.pone.0035006.
41. K. M. Hendricks, Ready-to-use therapeutic food for prevention of childhood undernutrition, *Nutrition Reviews* 68 (2010): 429–435.
42. F. Dibari and coauthors, Low-cost, ready-to-use therapeutic foods can be designed using locally available commodities with the aid of linear programming, *Journal of Nutrition* 142 (2012): 955–961; V. O. Owino and coauthors, Development and acceptability of a novel milk-free soybean-maize-sorghum ready-to-use therapeutic food (SMS-RUTF) based on industrial extrusion cooking process, *Maternal and Child Nutrition* (2012), epub ahead of print, doi:10.1111/j.1740-8709.2012.00400.x.
43. U.S. Census Bureau, World Vital Events Per Time Unit: 2012, available at http://www.census.gov/population/international/data/idb/worldvitalevents.php.
44. U.S. Census Bureau, World Vital Events Per Time Unit: 2012.
45. International Energy Agency, Global carbon-dioxide emissions increase by 1.0 Gt in 2011 to record high, May 12, 2012, available at http://www.iea.org/newsroomandevents/news/2012/may/name,27216,en.html.
46. United States Court of Appeal for the District of Columbia circuit, Argued February 28 and 29, 2012 Decided June 26, 2012, No. 09-1322, available at http://www.cadc.uscourts.gov/internet/opinions.nsf/52AC9DC9471D374685257A290052ACF6/$file/09-1322-1380690.pdf; IOM (Institute of Medicine), *Global Environmental Health: Research Gaps and Barriers for Providing Sustainable Water, Sanitation, and Hygiene Services* (Washington, DC: The National Academies Press, 2009), p. 15.
47. National Research Council, Panel on Advancing the Science of Climate Change, *Advancing the Science of Climate Change; Report in Brief* (Washington, DC: National Academies Press, 2010), p. 1, available at http://www.nap.edu.
48. U.S. Environmental Protection Agency, *Inventory of U.S. Greenhouse Gas Emissions and Sinks, 1990–2007* (Washington, DC: Government Printing Office, 2009); International Energy Agency, Global carbon-dioxide emissions increase by 1.0 Gt in 2011 to record high, 2012.
49. National Research Council, Panel on Advancing the Science of Climate Change, *Advancing the Science of Climate Change; Report in Brief*, 2010, p. 2.
50. V. Paine, What causes ocean "dead zones"? *Scientific American*, September 25, 2012, available at http://www.scientificamerican.com/article.cfm?id=ocean-dead-zones; G. Eshel and P. A. Martin, Geophysics and nutritional science: Toward a novel, unified paradigm, *American Journal of Clinical Nutrition* 89 (2009): 1710S–1716S.
51. National Research Council, Panel on Advancing the Science of Climate Change, *Limiting the Magnitude of Climate Change; Report in Brief* (Washington, DC: National Academies Press, 2010), p. 1, available at http://www.nap.edu.
52. Food and Agricultural Organization of the United Nations, *State of the World Fisheries and Aquaculture, 2010*, available at www.fao.org/docrep/013/i1820e/i1820e00.htm.
53. Food and Agricultural Organization of the United Nations, *State of the World Fisheries and Aquaculture, 2010*.
54. Food and Agricultural Organization of the United Nations, *State of the World Fisheries and Aquaculture, 2010*.
55. E. Pikitch and coauthors, *Little Fish, Big Impact: Managing a Crucial Link in Ocean Food Webs* (Washington, DC: Lenfest Ocean Program, 2012), available at http://www.larecherche.fr/content/system/media/fish.pdf.
56. World Water Assessment Programme, *The United Nations World Water Development Report 3: Water in a Changing World* (Paris: UNESCO, and London: Earthscan, 2009).
57. United Nations, The Global Water Crisis, 2012, available at http://www.un.org/works/sub2.asp?lang=en&s=19.
58. D. Powell, Satellites show groundwater dropping globally, *Science News*, January 14, 2012, pp. 5–6.
59. Food and Agriculture Organization of the United States, *Global Food Losses and Food Waste: Extent, Causes and Prevention, 2011*, available at www.fao.org/ag/ags/ags-division/publications/en; K. D. Hall and coauthors, The progressive increase of food waste in America and its environmental impact, *PloSOne* 4 (2009): e7940.

60. Environmental Protection Agency, Basic Information about Food Waste, April 26, 2012, available at www.epa.gov.
61. Environmental Protection Agency, Generators of Food Waste, April 26, 2012, available at www.epa.gov.
62. National Research Council, Panel on Advancing the Science of Climate Change, *America's Climate Choices* (Washington, DC: National Academies Press, 2012), p. 3, available at http://www.nap.edu.
63. National Research Council, Panel on Advancing the Science of Climate Change, *America's Climate Choices*, 2012.

Consumer's Guide 15

1. Federal Trade Commission, FTC issues revised "Green Guides," October 2012, available at http://www.ftc.gov/opa/2012/10/greenguides.shtm.
2. A. S. Cohn and D. O'Rourke, Agricultural certification as a conservation tool in Latin America, *Journal of Sustainable Forestry* 30 (2011): 158–186.

Controversy 15

1. H. C. Godfray and coauthors, Food security: The challenge of feeding 9 billion people, *Science* 237 (2010): 812–818.
2. M. G. Paoletti, T. Gomiero, and D. Pimentel, Introduction to the special issue: Towards a more sustainable agriculture, *Critical Reviews in Plant Sciences* 30 (2011): 2–5.
3. R. O. Morawicki, *Handbook of Sustainability for the Food Sciences* (Oxford: Wiley-Blackwell, 2012), epub, doi: 10.1002/9780470963166.ch10.
4. Environmental Protection Agency, *Inventory of U.S. Greenhouse Gas Emissions and Sinks: 1990–2007* (Washington, DC: U.S. Government Printing Office, 2009), available at www.epa.gov.
5. A. D. Mohamed and coauthors, Urinary dialkyl phosphate levels before and after first season chlorpyrifos spraying amongst farm workers in the Western Cape, South Africa, *Journal of Environmental Science and Health, Part B: Pesticides, Food Contaminants, and Agricultural Wastes* 46 (2011): 163–172; G. E. Kisby and coauthors, Oxidative stress and DNA damage in agricultural workers, *Journal of Agromedicine* 14 (2009): 206–214; P. K. Mills, J. Dodge, and R. Yang, Cancer in migrant and seasonal hired farm workers, *Journal of Agromedicine* 14 (2009): 185–191.
6. Environmental Protection Agency, *Inventory of U.S. Greenhouse Gas Emissions and Sinks*, April 2012, available at http://www.epa.gov/climatechange/ghgemissions/usinventoryreport.html.
7. G. Eshel and P. A. Martin, Geophysics and nutritional science: Toward a novel, unified paradigm, *American Journal of Clinical Nutrition* 89 (2009): 1710S–1716S.
8. P. Ronald, Plant genetics, sustainable agriculture and global food security, *Genetics* 188 (2011): 11–20; P. Bagla, India: Hardy cotton-munching pests are latest blow to GM crops, *Science* 327 (2010): 1439.
9. H. J. Marlow and coauthors, Diet and the environment: Does what you eat matter? *American Journal of Clinical Nutrition* 89 (2009): 1699–1703.
10. D. S. Reay and coauthors, Global agriculture and nitrous oxide emissions, *Nature Climate Change* 2 (2012): 410–416; N. Gilbert, Palm-oil boom raises conservation concerns, *Nature* 487 (2012): 14–15.
11. J. C. Liao and J. Messing, Energy biotechnology, *Current Opinion in Biotechnology* 23 (2012): 287–289; C. S. Jones and S. P. Mayfield, Algae biofuels: Versatility for the future of bioenergy, *Current Opinion in Biotechnology* 23 (2012): 346–351; V. H. Work and coauthors, Improving photosynthesis and metabolic networks for the competitive production of phototroph-derived biofuels, *Current Opinion in Biotechnology* 23 (2012): 290–297; R. Ehrenberg, The biofuel future, *Science News*, August 1, 2009, pp. 24–29.
12. T. E. McKone and coauthors, Grand challenges for life-cycle assessment of biofuels, *Environmental Science and Technology* 45 (2011): 1751–1756.
13. D. Pimentel, Food for thought: A review of the role of energy in current and evolving agriculture, special issue: Towards a more sustainable agriculture, *Critical Reviews in Plant Sciences* 30 (2011): 35–44.
14. B. M. Popkin, Reducing meat consumption has multiple benefits for the world's health, *Archives of Internal Medicine* 169 (2009): 543–545.
15. The Global Partnership for Safe and Sustainable Agriculture, Global G.A.P. Annual Report 2011, available at http://www.globalgap.org.
16. Agroecology in Action (webpage), University of California, Berkeley, available at http://nature.berkeley.edu/~miguel-alt/index.html.
17. BIFS program overview, UC Sustainable Agriculture Research ad Education Program, available at www.sarep.ucdavis.edu.
18. USDA Natural Resources Conservation Service, Conservation Reserve program, updated June 23, 2009, available at www.nrcs.usda.gov/programs/CRP/.
19. P. L. Pingali, Green revolution: Impacts, limits, and the path ahead, Proceedings of the National Academy of Sciences of the United States of America (PNAS) 109 (July 31, 2012): 12302–12308; G. Brookes and P. Barfoot, The income and production effects of biotech crops globally 1996–2009, *International Journal of Biotechnology* 12 (2011): 1–49; R. Park and coauthors, The role of transgenic crops in sustainable development, *Plant Biotechnology Journal* 9 (2011): 2–21.
20. B. Knight, Regaining ground: A conservation reserve program right-sized for the times, a report by Strategic Conservation Solutions, June 2012, available at http://www.ngfa.org/files/SCSReGainingGroundResearchStudyforNGFF%286-8-2012%29.pdf.
21. N. Gilbert, Boost for conservation of plant gene assets, *Nature* online, 1 (June 2009), epub, doi:10.1038/news.
22. J. C. Quinn and coauthors, Current large-scale US biofuel potential from microalgae cultivated in photobioreactors, *BioEnergy Research* 5 (2012): 49-60; P. C. Hallenbeck (Ed.), *Microbial Technologies in Advanced Biofuels Production* (Springer Science+Business Media: New York, 2012).
23. E. Thompson and coauthors, Economic feasibility of converting cow manure to electricity: A case study of the CVPS Cow Power program in Vermont Agricultural and Applied Economics Association presented at the 2011 Annual Meeting, July 24–26, 2011, Pittsburgh, Pennsylvania, *AgEcon Search*, Waite Library, Dept. of Applied Economics, University of Minnesota.
24. Farms use organic waste to generate "Free" onsite power, *GE Energy*, April 29, 2009, available at www.waterandwastewater.com.
25. H. Blanco-Canqui, Crop residue removal for bioenergy reduces soil carbon pools: How can we offset carbon losses? BioEnergy Research, Online First, 30 May 2012.
26. Compost bin truckload sales event, 2009, available at http://www.talgov.com/you/solid/compost_event.cfm.
27. J. Skene, Methane moves from landfill to fuel tank, KQED Quest, January 2012, available at http://science.kqed.org/quest/2012/01/23/methane-moves-from-landfill-to-fuel-tank/.
28. A. Carlsson-Kanyama and A. D. González, Potential contributions of food consumption patterns to climate change, *American Journal of Clinical Nutrition* 89 (2009): 1704S–1709S.
29. E. A. Dodson and coauthors, Preventing childhood obesity through state policy: Qualitative assessment of facilitators and barriers, *Journal of Public Health Policy* Supplement 1 (2009): 161–176.
30. J. A. Foley, Can we feed the world & sustain the planet? *Scientific American* 305 (2011), epub, 18 October 2011 | doi:10.1038/scientificamerican1111-60; Pimentel, Food for thought: A review of the role of energy in current and evolving agriculture, 2011.
31. Marlow and coauthors, Diet and the environment.
32. Global Footprint Network, Ecological Footprint and Biocapacity, 2012, available at www.footprintwork.org.

Appendix G

Answers to Chapter Questions

Answers to Consumer's Guide Review and Self-Check Questions

CHAPTER 1

Consumer's Guide Review

1. d
2. b
3. b

Self-Check Questions

1. False. Heart disease and cancer are influenced by many factors with genetics and diet among them.
2. c
3. d
4. a
5. a
6. a
7. T
8. c
9. b
10. False. The choice of where, as well as what, to eat is often based more on taste and social considerations than on nutrition judgments.
11. b
12. a
13. T
14. F
15. b
16. a
17. d
18. False. In this nation, profiteers selling diplomas and certificates make it easy to obtain a bogus nutrition credential.

CHAPTER 2

Consumer's Guide Review

1. False. Restaurant portions are not held to standards and should not be used as a guide for choosing portion sizes.
2. True.
3. False. Most consumers overestimate both the calories and fat in restaurant foods.

Self-Check Questions

1. b
2. d
3. T
4. False. The DRI are estimates of the needs of healthy persons only. Medical problems alter nutrient needs.
5. c
6. T
7. c
8. d
9. False. People who choose to eat no meats or products taken from animals can still use the USDA Food Patterns to make their diets adequate.
10. a
11. False. A properly planned diet should include healthy snacks as part of the total daily food intake, if so desired.
12. c
13. False. By law, food labels must state as a percentage of the Daily Values the amounts of vitamins A, C, calcium, and iron present in a food.
14. T
15. T
16. d
17. T
18. False. Although they are natural constituents of foods, phytochemicals have not been proven safe to consume in large amounts.

CHAPTER 3

Self-Check Questions

1. a
2. False. Each gene is a blueprint that directs the production of one or more of the body's proteins, such as an enzyme.
3. c
4. a
5. b
6. T
7. T
8. d
9. c
10. d
11. False. Absorption of the majority of nutrients takes place across the specialized cells of the small intestine.

12. d
13. a
14. c
15. False. The kidneys straddle the cardiovascular system and filter the blood.
16. b
17. T
18. a
19. False. Alcohol is a natural toxin that can cause severe damage to the liver, brain, and other organs, and can be lethal in high enough doses.

CHAPTER 4

Consumer's Guide Review

1. b
2. b
3. a

Self-Check Questions

1. b
2. a
3. T
4. T
5. c
6. T
7. b
8. a
9. False. For people with diabetes, the risk of heart disease, stroke, and dying on any particular day is doubled.
10. False. Type I diabetes is most often controlled with insulin-injections or an insulin pump.
11. c
12. False. Regular physical activity can help by reducing the body's fatness and heightening tissue sensitivity to insulin.
13. d
14. False. Hypoglycemia as a true disease is rare.
15. d
16. T
17. T
18. a

CHAPTER 5

Consumer's Guide Review

1. False. Methylmercury is a highly toxic industrial pollutant found in highest concentrations in the flesh of large predatory species of fish.
2. False. Children and pregnant or lactating women should strictly follow recommendations set for them and choose fish species that are rich in omega-3 fatty acids *and* lower in mercury.
3. False. Cod provides little EPA and DHA.

Self-Check Questions

1. c
2. False. In addition to providing abundant fuel, fat cushions tissues, serves as insulation, forms cell membranes, and serves as raw material, among other functions.
3. b
4. False. Vegetable and fish oils are excellent sources of polyunsaturated fats.
5. c
6. T
7. b
8. d
9. T
10. T
11. False. Chylomicrons are produced in small intestinal cells.
12. False. Consuming large amounts of saturated fatty acids elevates serum LDL cholesterol and thus *raises* the risk of heart disease and heart attack.
13. d
14. False. Fish, not supplements, is the recommended source of fish oil.
15. T
16. b
17. b
18. d
19. T
20. T
21. d
22. c

CHAPTER 6

Consumer's Guide Review

1. False. Evidence does not support taking protein supplements such as commercial shakes and energy bars to lose weight.
2. True.
3. False. In high doses, tryptophan may induce nausea and skin disorders as unwanted side effects.

Self-Check Questions

1. b
2. c
3. a
4. a
5. b
6. T
7. T
8. d
9. a
10. T
11. T
12. d
13. False. Excess protein in the diet may have adverse effects, such as worsening kidney disease.
14. a
15. T
16. d
17. T

18. c
19. False. Fried banana or vegetable chips are often high in calories and saturated fat, and are best reserved for an occasional treat.

CHAPTER 7

Consumer's Guide Review

1. True.
2. True.
3. False. All things considered, the best and safest source of vitamin D for people in the United States is nutrient-dense foods and beverages.

Self-Check Questions

1. b
2. c
3. a
4. T
5. d
6. False. Vitamin A supplements have no effect on acne.
7. T
8. d
9. a
10. c
11. T
12. d
13. b
14. a
15. False. No study to date has conclusively demonstrated that vitamin C can prevent colds or reduce their severity.
16. d
17. c
18. T
19. a
20. b
21. b
22. False. FDA has little control over supplement sales.

CHAPTER 8

Consumer's Guide Review

1. a
2. d
3. d

Self-Check Questions

1. d
2. False. Water intoxication occurs when too much plain water floods the body's fluids and disturbs their normal composition.
3. c
4. b
5. T
6. a
7. d
8. b
9. d
10. False. After about age 30, the bones begin to lose density.
11. T
12. c
13. b
14. a
15. False. Calcium is the most abundant mineral in the body.
16. False. The Academy of Nutrition and Dietetics, among others, recommends the consumption of fluoridated water.
17. False. Butter, cream, and cream cheese contain negligible calcium, being almost pure fat. Some vegetables, such as broccoli, are good sources of available calcium.
18. T
19. T
20. b

CHAPTER 9

Consumer's Guide Review

1. False. A diet book that addresses eicosanoids and adipokines may or may not present accurate nutrition science or effective diet advice.
2. False. Calorie deficit is a key strategy for weight loss.
3. True.

Self-Check Questions

1. d
2. T
3. b
4. False. The BMI are unsuitable for use with athletes and adults over age 65.
5. False. The thermic effect of food is believed to have negligible effects on total energy expenditure.
6. c
7. d
8. d
9. a
10. b
11. False. Genomic researchers have identified multiple genes likely to play roles in obesity development but have not so far identified a single genetic cause of common obesity.
12. T
13. d
14. a
15. T
16. T
17. a
18. b
19. b
20. False. Disordered eating behaviors in early life set a pattern that likely continues into young adulthood.

CHAPTER 10

Consumer's Guide Review

1. a
2. a
3. b

Self-Check Questions

1. b
2. c
3. False. Weight-bearing exercise that improves muscle strength and endurance also helps maximize and maintain bone mass.
4. False. Muscle cells and tissues respond to a physical activity overload by altering the structures and metabolic equipment needed to perform the work.
5. c
6. c
7. a
8. T
9. a
10. d
11. T
12. T
13. a
14. False. Frequent nutritious between-meal snacks can help to provide some extra calories and help to maintain body weight.
15. T
16. d
17. a
18. b
19. d

CHAPTER 11

Consumer's Guide Review

1. True.
2. False. The National Center for Complementary and Alternative Medicine (NCCAM) does not promote laetrile therapy.
3. True.

Self-Check Questions

1. c
2. T
3. b
4. False. Chronic diseases have risk factors that show correlations with disease development but are not distinct causes.
5. d
6. b
7. False. Atherosclerosis is an accumulation of lipids within the artery wall, but it also involves a complex response of the artery to tissue damage and inflammation.
8. a
9. False. Men do have more heart attacks than women, but CVD kills more women than any other cause of death.
10. a
11. d
12. T
13. False. The prevalence of high blood pressure in African Americans is among the highest in the world.
14. T
15. d
16. T
17. False. Sufficient intakes of calcium-rich foods may help to prevent colon cancer.
18. False. The DASH diet is designed for helping people with hypertension to control the disease.
19. d
20. False. Currently, for the best chance of consuming adequate nutrients and staying healthy, people should eat a well-planned diet of whole foods, as described in Chapter 2.

CHAPTER 12

Consumer's Guide Review

1. d
2. b
3. a

Self-Check Questions

1. T
2. a
3. c
4. d
5. c
6. False. Today, the chance of getting a foodborne illness from eating produce is similar to the chance of becoming ill from eating meat, eggs, and seafood.
7. T
8. a
9. b
10. T
11. False. Nature has provided many plants used for food with natural poisons to fend off diseases, insects, and other predators.
12. False. The EPA and FDA warn of unacceptably high methylmercury levels in ocean fish and other seafood and advise all pregnant women not to eat certain types of fish.
13. T
14. c
15. c
16. d
17. T
18. b
19. b
20. T

G

CHAPTER 13

Consumer's Guide Review

1. False. Despite convincing advertising, no commercial formula can fully match the benefits of human milk.
2. False. Only about 23 percent of infants are still breastfeeding at 1 year of age.
3. False. Lactation consultants are employed by hospitals to help new mothers establish healthy breastfeeding relationships with their newborns and to help ensure successful long-term breastfeeding.

Self-Check Questions

1. c
2. T
3. b
4. d
5. d
6. T
7. b
8. False. The American Academy of Pediatrics urges all women to stop drinking as soon as they plan to become pregnant, and to abstain throughout the pregnancy.
9. a
10. T
11. d
12. T
13. a
14. a
15. d
16. d
17. False. There is no proof for the theory that "stuffing the baby" at bedtime will promote sleeping through the night.
18. T
19. False. In light of the developmental needs of one-year-olds, parents should discourage unacceptable behaviors, such as standing at the table or throwing food.
20. b

CHAPTER 14

Consumer's Guide Review

1. True.
2. False. Ongoing research suggests that taking multivitamins, magnesium, manganese, or diuretics is *not* useful.
3. False. During the two weeks *before* menstruation, women may experience a natural, hormone-governed increase of appetite.

Self-Check Questions

1. c
2. b
3. T
4. d
5. b
6. False. Research to date does not support the idea that food allergies or intolerances cause hyperactivity in children, but studies continue.
7. c
8. c
9. c
10. b
11. d
12. a
13. T
14. False. Vitamin A absorption appears to increase with aging.
15. False. To date, no proven benefits are available from herbs or other remedies.
16. b
17. a
18. False. However, people taking two or more drugs at the same time are more vulnerable to nutrient–drug interactions.
19. d

CHAPTER 15

Consumer's Guide Review

1. False. The term *green* is loosely regulated and is meaningless without scientific evidence to back it up.
2. True.
3. d

Self-Check Questions

1. b
2. a
3. c
4. d
5. T
6. a
7. c
8. False. Most children who die of malnutrition do not starve to death—they die because their health has been compromised by dehydration from infections that cause diarrhea.
9. a
10. c
11. d
12. c
13. T
14. T
15. T
16. d
17. False. The federal government, the states, local communities, big business and small companies, educators, and all individuals, including dietitians and food service managers, have many opportunities to make an impact in the fight against poverty, hunger, and environmental degradation.
18. T
19. c

Appendix H

Physical Activity Levels and Energy Requirements

Chapter 9 described how to calculate ranges of the estimated energy requirement (EER) for an adult by using an equation that accounts for age and gender alone. This appendix offers a way of establishing estimated calorie needs per day by age, gender, and physical activity level, as developed by the *Dietary Guidelines for Americans 2010*, and based on the equations of the *Committee on Dietary Reference Intakes*.

Table H–1 describes activity levels for three groups of people: sedentary, moderately active, and active. Once you have identified an activity level that approximates your own, find your daily calorie need in Table H–2.

TABLE H-1 Sedentary, Moderately Active, and Active People

Sedentary	A lifestyle that includes only the light physical activity associated with typical day-to-day life.
Moderately active	A lifestyle that includes physical activity equivalent to walking about 1.5 to 3 miles per day at 3 to 4 miles per hour in addition to the light physical activity associated with typical day-to-day life.
Active	A lifestyle that includes physical activity equivalent to walking more than 3 miles per day at 3 to 4 miles per hour in addition to the light physical activity associated with typical day-to-day life.

Source: U.S. Department of Agriculture and U.S. Department of Health and Human Services, Dietary Guidelines for Americans 2010, *available at www.dietaryguidelines.gov.*

TABLE H-2 Estimated Calorie Needs per Day by Age, Gender, and Physical Activity Level (Detailed)

Estimated amounts of calories needed to maintain calorie balance for various gender and age groups at three different levels of physical activity.[a] The estimates are rounded to the nearest 200 calories. An individual's calorie needs may be higher or lower than these average estimates.

Gender/ Activity level	Male/ Sedentary	Male/ Moderately Active	Male/Active	Female[b]/ Sedentary	Female[b]/ Moderately Active	Female[b]/Active
Age (years)						
2	1,000	1,000	1,000	1,000	1,000	1,000
3	1,200	1,400	1,400	1,000	1,200	1,400
4	1,200	1,400	1,600	1,200	1,400	1,400
5	1,200	1,400	1,600	1,200	1,400	1,600
6	1,400	1,600	1,800	1,200	1,400	1,600
7	1,400	1,600	1,800	1,200	1,600	1,800
8	1,400	1,600	2,000	1,400	1,600	1,800
9	1,600	1,800	2,000	1,400	1,600	1,800
10	1,600	1,800	2,200	1,400	1,800	2,000
11	1,800	2,000	2,200	1,600	1,800	2,000
12	1,800	2,200	2,400	1,600	2,000	2,200
13	2,000	2,200	2,600	1,600	2,000	2,200
14	2,000	2,400	2,800	1,800	2,000	2,400
15	2,200	2,600	3,000	1,800	2,000	2,400
16	2,400	2,800	3,200	1,800	2,000	2,400
17	2,400	2,800	3,200	1,800	2,000	2,400
18	2,400	2,800	3,200	1,800	2,000	2,400
19–20	2,600	2,800	3,000	2,000	2,200	2,400
21–25	2,400	2,800	3,000	2,000	2,200	2,400
26–30	2,400	2,600	3,000	1,800	2,000	2,400
31–35	2,400	2,600	3,000	1,800	2,000	2,200
36–40	2,400	2,600	2,800	1,800	2,000	2,200
41–45	2,200	2,600	2,800	1,800	2,000	2,200
46–50	2,200	2,400	2,800	1,800	2,000	2,200
51–55	2,200	2,400	2,800	1,600	1,800	2,200
56–60	2,200	2,400	2,600	1,600	1,800	2,200
61–65	2,000	2,400	2,600	1,600	1,800	2,000
66–70	2,000	2,200	2,600	1,600	1,800	2,000
71–75	2,000	2,200	2,600	1,600	1,800	2,000
76+	2,000	2,200	2,400	1,600	1,800	2,000

a. Based on Estimated Energy Requirements (EER) equations, using reference heights (average) and reference weights (healthy) for each age-gender group. For children and adolescents, reference height and weight vary. For adults, the reference man is 5 feet 10 inches tall and weighs 154 pounds. The reference woman is 5 feet 4 inches tall and weighs 126 pounds. EER equations are from the Institute of Medicine. Dietary Reference Intakes for Energy, Carbohydrate, Fiber, Fat, Fatty Acids, Cholesterol, Protein, and Amino Acids. Washington (DC): The National Academies Press; 2002.

b. Estimates for females do not include women who are pregnant or breastfeeding.

Source: U.S. Department of Agriculture and U.S. Department of Health and Human Services, Dietary Guidelines for Americans 2010, *available at www.dietaryguidelines.gov.*

Glossary

A

absorb to take in, as nutrients are taken into the intestinal cells after digestion; the main function of the digestive tract with respect to nutrients.

Academy of Nutrition and Dietetics (AND) the professional organization of dietitians in the United States (formerly the American Dietetic Association). The Canadian equivalent is the Dietitians of Canada (DC), which operates similarly.

acceptable daily intake (ADI) the estimated amount of a sweetener that can be consumed daily over a person's lifetime without any adverse effects.

Acceptable Macronutrient Distribution Ranges (AMDR) values for carbohydrate, fat, and protein expressed as percentages of total daily caloric intake; ranges of intakes set for the energy-yielding nutrients that are sufficient to provide adequate total energy and nutrients while minimizing the risk of chronic diseases.

accredited approved; in the case of medical centers or universities, certified by an agency recognized by the U.S. Department of Education.

acetaldehyde (ass-et-AL-deh-hide) a substance to which ethanol is metabolized on its way to becoming harmless waste products that can be excreted.

acid-base balance equilibrium between acid and base concentrations in the body fluids.

acidosis (acid-DOH-sis) the condition of excess acid in the blood, indicated by a below-normal pH (*osis* means "too much in the blood").

acid reducers prescription and over-the-counter drugs that reduce the acid output of the stomach; effective for treating severe, persistent forms of heartburn but not for neutralizing acid already present. Side effects are frequent and include diarrhea, other gastrointestinal complaints, and reduction of the stomach's capacity to destroy alcohol, thereby producing higher-than-expected blood alcohol levels from each drink. Also called *acid controllers*.

acids compounds that release hydrogens in a watery solution.

acne chronic inflammation of the skin's follicles and oil-producing glands, which leads to an accumulation of oils inside the ducts that surround hairs; usually associated with the maturation of young adults.

acupuncture (ak-you-punk-chur) a technique that involves piercing the skin with long, thin needles at specific anatomical points to relieve pain or illness. Acupuncture sometimes uses heat, pressure, friction, suction, or electromagnetic energy to stimulate the points.

added sugars sugars and syrups added to a food for any purpose, such as to add sweetness or bulk or to aid in browning (baked goods). Also called carbohydrate sweeteners, they include glucose, fructose, corn syrup, concentrated fruit juice, and other sweet carbohydrates.

additives substances that are added to foods but are not normally consumed by themselves as foods.

adequacy the dietary characteristic of providing all of the essential nutrients, fiber, and energy in amounts sufficient to maintain health and body weight.

Adequate Intakes (AI) nutrient intake goals for individuals; the recommended average daily nutrient intake level based on intakes of healthy people (observed or experimentally derived) in a particular life stage and gender group and assumed to be adequate. Set whenever scientific data are insufficient to allow establishment of an RDA value.

adipokines (AD-ih-poh-kynz) protein hormones made and released by adipose tissue (fat) cells.

adipose tissue the body's fat tissue, consisting of masses of fat-storing cells and blood vessels to nourish them.

adolescence the period from the beginning of puberty until maturity.

advertorials lengthy advertisements in newspapers and magazines that read like feature articles but are written for the purpose of touting the virtues of products and may or may not be accurate.

aerobic (air-ROH-bic) requiring oxygen. Aerobic activity strengthens the heart and lungs by requiring them to work harder than normal to deliver oxygen to the tissues.

aerobic activity physical activity that involves the body's large muscles working at light to moderate intensity for a sustained period of time. Brisk walking, running, swimming, and bicycling are examples. Also called *endurance activity*.

aflatoxin (af-lah-TOX-in) a toxin from a mold that grows on corn, grains, peanuts, and tree nuts stored in warm, humid conditions; a cause of liver cancer prevalent in tropical developing nations. (To prevent it, discard shriveled, discolored, or moldy foods.)

agave syrup a carbohydrate-rich sweetener made from a Mexican plant; a higher fructose content gives some agave syrups a greater sweetening power per calorie than sucrose.

agility nimbleness; the ability to quickly change directions.

agroecology a scientific discipline that combines biological, physical, and social sciences with ecological theory to develop methods for producing food sustainably.

AIDS acquired immune deficiency syndrome; caused by infection with human immunodeficiency virus (HIV), which is transmitted primarily by sexual contact, contact with infected blood, needles shared among drug users, or fluids transferred from an infected mother to her fetus or infant.

alcohol dehydrogenase (dee-high-DRAH-gen-ace) **(ADH)** an enzyme system that breaks down alcohol. The antidiuretic hormone listed below is also abbreviated ADH.

alcoholism a dependency on alcohol marked by compulsive uncontrollable drinking with negative effects on physical health, family relationships, and social health.

alcohol-related birth defects (ARBD) malformations in the skeletal and organ systems (heart, kidneys, eyes, ears) associated with prenatal alcohol exposure.

alcohol-related neurodevelopmental disorder (ARND) behavioral, cognitive, or central nervous system abnormalities associated with prenatal alcohol exposure.

alkalosis (al-kah-LOH-sis) the condition of excess base in the blood, indicated by an above-normal blood pH (*alka* means "base"; *osis* means "too much in the blood").

allergy an immune reaction to a foreign substance, such as a component of food. Also called *hypersensitivity* by researchers.

alpha-lactalbumin (lact-AL-byoo-min) the chief protein in human breast milk. The chief protein in cow's milk is *casein* (CAY-seen).

alternative (low-input, or **sustainable) agriculture** agriculture practiced on a small scale using individualized approaches that vary with local conditions so as to minimize technological, fuel, and chemical inputs.

amine (a-MEEN) **group** the nitrogen-containing portion of an amino acid.

amino acid chelates (KEY-lates) compounds of minerals (such as calcium) combined with amino acids in a form that favors their absorption. A chelating agent is a molecule that surrounds another molecule and can then either promote or prevent its movement from place to place (chele means "claw").

amino (a-MEEN-o) **acids** the building blocks of protein. Each has an amine group at one end, an acid group at the other, and a distinctive side chain.

amniotic (AM-nee-OTT-ic) **sac** the "bag of waters" in the uterus in which the fetus floats.

anabolic steroid hormones chemical messengers related to the male sex hormone testosterone that stimulate building up of body tissues (*anabolic* means "promoting growth"; *sterol* refers to compounds chemically related to cholesterol).

anaerobic (AN-air-ROH-bic) not requiring oxygen. Anaerobic activity is of high intensity and short duration.

anaphylactic (an-ah-feh-LACK-tick) **shock** a life-threatening whole-body allergic reaction to an offending substance.

androstenedione (AN-droh-STEEN-die-own) a precursor of testosterone that elevates both testosterone and estrogen in the blood of both males and females. Often called *andro*, it is sold with claims of producing increased muscle strength, but controlled studies disprove such claims.

anecdotal evidence information based on interesting and entertaining, but not scientific, personal accounts of events.

anemia the condition of inadequate or impaired red blood cells; a reduced number or volume of red blood cells along with too little hemoglobin in the blood. The red blood cells may be immature and, therefore, too large or too small to function properly. Anemia can result from blood loss, excessive red blood cell destruction, defective red blood cell formation, and many nutrient deficiencies. Anemia is not a disease, but a symptom of another problem; its name literally means "too little blood."

anencephaly (an-en-SEFF-ah-lee) an uncommon and always fatal neural tube defect in which the brain fails to form.

aneurysm (AN-you-rism) the ballooning out of an artery wall at a point that is weakened by deterioration.

anorexia nervosa an eating disorder characterized by a refusal to maintain a minimally normal body weight, self-starvation to the extreme, and a disturbed perception of body weight and shape; seen (usually) in teenage girls and young women (anorexia means "without appetite"; nervos means "of nervous origin").

antacids medications that react directly and immediately with the acid of the stomach, neutralizing it. Antacids are most suitable for treating occasional heartburn.

antibiotic-resistant bacteria bacterial strains that cause increasingly common and potentially fatal infectious diseases that do not respond to standard antibiotic therapy. An example is MRSA (pronounced MER-suh), a multi drug-resistant *Staphylococcus aureus* bacterium.

antibodies (AN-te-bod-ees) large proteins of the blood, produced by the immune system in response to an invasion of the body by foreign substances (antigens). Antibodies combine with and inactivate the antigens.

anticarcinogens compounds in foods that act in any of several ways to oppose the formation of cancer.

antidiuretic (AN-tee-dye-you-RET-ick) **hormone (ADH)** a hormone produced by the pituitary gland in response to dehydration (or a high sodium concentration in the blood). It stimulates the kidneys to reabsorb more water and so to excrete less. (This hormone should not be confused with the enzyme alcohol dehydrogenase, which is also abbreviated ADH.)

antigen a substance foreign to the body that elicits the formation of antibodies or an inflammation reaction from immune system cells.

antioxidant nutrients vitamins and minerals that oppose the effects of oxidants on human physical functions. The antioxidant vitamins are vitamin E, vitamin C, and beta-carotene. The mineral selenium also participates in antioxidant activities.

antioxidants (anti-OX-ih-dants) compounds that protect other compounds from damaging reactions involving oxygen by themselves reacting with oxygen (*anti* means "against"; *oxy* means "oxygen"). *Oxidation* is a potentially damaging effect of normal cell chemistry involving oxygen.

aorta (ay-OR-tuh) the large, primary artery that conducts blood from the heart to the body's smaller arteries.

appendicitis inflammation and/or infection of the appendix, a sac protruding from the intestine.

appetite the psychological desire to eat; a learned motivation and a positive sensation that accompanies the sight, smell, or thought of appealing foods.

appliance thermometer a thermometer that verifies the temperature of an appliance. An *oven thermometer* verifies that the oven is heating properly; a *refrigerator/freezer thermometer* tests for proper refrigerator (<40°F, or <4°C) or freezer temperature (0°F, or −17°C).

aquaculture the farming of aquatic organisms for food, generally fish, mollusks, or crustaceans, that involves such activities as feeding immature organisms, providing habitat, protecting them from predators, harvesting, and selling or consuming them.

aquifers underground rock formations containing water that can be drawn to the surface for use.

arachidonic (ah-RACK-ih-DON-ik) **acid** an omega-6 fatty acid derived from linoleic acid.

arsenic a poisonous metallic element. In trace amounts, arsenic is believed to be an essential nutrient in some animal species. Arsenic is often added to insecticides and weed killers and, in tiny amounts, to certain animal drugs.

arteries blood vessels that carry blood containing fresh oxygen supplies from the heart to the tissues.

artesian water water drawn from a well that taps a confined aquifer in which the water is under pressure.

arthritis a usually painful inflammation of joints caused by many conditions, including infections, metabolic disturbances, or injury; usually results in altered joint structure and loss of function.

artificial fats zero-energy fat replacers that are chemically synthesized to mimic the sensory and cooking qualities of naturally occurring fats but are totally or partially resistant to digestion.

ascorbic acid one of the active forms of vitamin C (the other is *dehydroascorbic* acid); an antioxidant nutrient.

-ase (ACE) a suffix meaning *enzyme*. Categories of digestive and other enzymes and individual enzyme names often contain this suffix.

atherosclerosis (ath-er-oh-scler-OH-sis) the most common form of cardiovascular disease; characterized by plaques along the inner walls of the arteries (*scleros* means "hard"; *osis* means "too much"). The term *arteriosclerosis* is often used to mean the same thing.

athlete a competitor in any sport, exercise, or game requiring physical skill; for the purpose of this book, anyone who trains at a high level of physical exertion, with or without competition. From the Greek *athlein*, meaning "to contend for a prize."

atrophy (AT-tro-fee) a decrease in size (for example, of a muscle) because of disuse.

autoimmune disorder a disease in which the body develops antibodies to its own proteins and then proceeds to destroy cells containing these proteins. Examples are type 1 diabetes and lupus.

B

baby water ordinary bottled water treated with ozone to make it safe but not sterile.

balance the dietary characteristic of providing foods of a number of types in proportion to each other, such that foods rich in some nutrients do not crowd out of the diet foods that are rich in other nutrients. Also called *proportionality*.

balance study a laboratory study in which a person is fed a controlled diet and the intake and excretion of a nutrient are measured. Balance studies are valid only for nutrients like calcium (chemical elements) that do not change while they are in the body.

basal metabolic rate (BMR) the rate at which the body uses energy to support its basal metabolism.

basal metabolism the sum total of all the involuntary activities that are necessary to sustain life, including circulation, respiration, temperature maintenance, hormone secretion, nerve activity, and new tissue synthesis, but excluding digestion and voluntary activities. Basal metabolism is the largest component of the average person's daily energy expenditure.

bases compounds that accept hydrogens from solutions.

B-cells lymphocytes that produce antibodies. *B* stands for bursa, an organ in the chicken where B-cells were first identified.

beer belly central-body fatness associated with alcohol consumption.

behavior modification alteration of behavior using methods based on the theory that actions can be controlled by manipulating the environmental factors that cue, or trigger, the actions.

beriberi (berry-berry) the thiamin-deficiency disease; characterized by loss of sensation in the hands and feet, muscular weakness, advancing paralysis, and abnormal heart action.

beta-carotene an orange pigment with antioxidant activity; a vitamin A precursor made by plants and stored in human fat tissue.

bicarbonate a common alkaline chemical; a secretion of the pancreas; also the active ingredient of baking soda.

bile a cholesterol-containing digestive fluid made by the liver, stored in the gallbladder, and released into the small intestine when needed. It emulsifies fats and oils to ready them for enzymatic digestion.

binge eating disorder an eating disorder whose criteria are similar to those of bulimia nervosa, excluding purging or other compensatory behaviors.

bioaccumulation the accumulation of a contaminant in the tissues of living things at higher and higher concentrations along the food chain.

bioactive having biological activity in the body.

bioactive food components compounds in foods, either nutrients or phytochemicals, that alter physiological processes.

biofilm a protective coating of proteins and carbohydrates exuded by certain bacteria; biofilm adheres bacteria to surfaces and can survive rinsing.

biofuels fuels made mostly of materials derived from recently harvested living organisms. Examples are *biogas*, *ethanol*, and *biodiesel*. Biofuels contribute less to the carbon dioxide burden of the atmosphere because plants capture carbon from the air as they grow and release it again when the fuel is burned; fossil fuels such as coal and oil contain carbon that was previously held underground for millions of years and is newly released into the atmosphere on burning.

biotechnology the science of manipulating biological systems or organisms to modify their products or components or create new products; biotechnology includes recombinant DNA technology and traditional and accelerated selective breeding techniques.

biotin (BY-o-tin) a B vitamin; a coenzyme necessary for fat synthesis and other metabolic reactions.

bladder the sac that holds urine until time for elimination.

blind experiment an experiment in which the subjects do not know whether they are members of the experimental group or the control group. In a *double-blind experiment*, neither the subjects nor the researchers know to which group the members belong until the end of the experiment.

blood the fluid of the cardiovascular system; composed of water, red and white blood cells, other formed particles, nutrients, oxygen, and other constituents.

body composition the proportions of muscle, bone, fat, and other tissue that make up a person's total body weight.

body mass index (BMI) an indicator of obesity or underweight, calculated by dividing the weight of a person by the square of the person's height.

body system a group of related organs that work together to perform a function. Examples are the circulatory system, respiratory system, and nervous system.

bone density a measure of bone strength, the degree of mineralization of the bone matrix.

bone meal or **powdered bone** crushed or ground bone preparations intended to supply calcium to the diet. Calcium from bone is not well absorbed and is often contaminated with toxic materials such as arsenic, mercury, lead, and cadmium.

botanical pertaining to or made from plants; any drug, medicinal preparation, dietary supplement, or similar substance obtained from a plant.

bottled water drinking water sold in bottles.

botulism an often fatal foodborne illness caused by botulinum toxin, a toxin produced by the *Clostridium botulinum* bacterium that grows without oxygen in nonacidic canned foods.

bovine spongiform encephalopathy (BOH-vine SPUNJ-ih-form en-SEH-fal-AH-path-ee) **(BSE)** an often fatal illness of the nerves and brain observed in cattle and wild game, and, rarely, in people who consume affected meats. Also called *mad cow disease*.

bran the protective fibrous coating around a grain; the chief fiber donator of a grain.

broccoli sprouts the sprouted seed of *Brassica italica*, or the common broccoli plant; believed to be a functional food by virtue of its high phytochemical content.

brown adipose tissue (BAT) a type of adipose tissue abundant in hibernating animals and human infants and recently identified in human adults. Abundant pigmented enzymes of energy metabolism give BAT a dark appearance under a microscope; the enzymes release heat from fuels without accomplishing other work. Also called *brown fat*.

brown bread bread containing ingredients such as molasses that lend a brown color; may be made with any kind of flour, including white flour.

brown sugar white sugar with molasses added, 95% pure sucrose.

buffers molecules that can help to keep the pH of a solution from changing by gathering or releasing H ions.

built environment the buildings, roads, utilities, homes, fixtures, parks, and all other man-made entities that form the physical characteristics of a community.

bulimia (byoo-LEEM-ee-uh) **nervosa** recurring episodes of binge eating combined with a morbid fear of becoming fat; usually followed by self-induced vomiting or purging.

C

caffeine a stimulant that can produce alertness and reduce reaction time when used in small doses but causes headaches, trembling, an abnormally fast heart rate, and other undesirable effects in high doses.

caffeine water bottled water with caffeine added.

calcium compounds the simplest forms of purified calcium. They include calcium carbonate, citrate, gluconate, hydroxide, lactate, malate, and phosphate. These supplements vary in the amount of calcium they contain, so read the labels carefully. A 500-milligram tablet of calcium gluconate may provide only 45 milligrams of calcium, for example.

caloric effect the drop in cancer incidence seen whenever intake of food energy (calories) is restricted.

calorie control control of energy intake; a feature of a sound diet plan.

calories units of energy. In nutrition science, the unit used to measure the energy in foods is a kilocalorie (also called *kcalorie* or *Calorie*): it is the amount of heat energy necessary to raise the temperature of a kilogram (a liter) of water 1 degree Celsius. This book follows the common practice of using the lowercase term *calorie* (abbreviated *cal*) to mean the same thing.

cancer a disease in which cells multiply out of control and disrupt normal functioning of one or more organs.

capillaries minute, weblike blood vessels that connect arteries to veins and permit transfer of materials between blood and tissues.

carbohydrase (car-boh-HIGH-drace) any of a number of enzymes that break the chemical bonds of carbohydrates.

carbohydrates compounds composed of single or multiple sugars. The name means "carbon and water," and a chemical shorthand for carbohydrate is CHO, signifying carbon (C), hydrogen (H), and oxygen (O).

carbonated water water that contains carbon dioxide gas, either naturally occurring or added, that causes bubbles to form in it; also called bubbling or sparkling water. Seltzer, soda, and tonic waters are legally soft drinks and are not regulated as water.

carcinogen (car-SIN-oh-jen) a cancer-causing substance (*carcin* means "cancer"; *gen* means "gives rise to").

carcinogenesis the origination or beginning of cancer.

cardiac output the volume of blood discharged by the heart each minute.

cardiorespiratory endurance the ability of the heart, lungs, and metabolism to sustain large-muscle exercise of moderate-to-high intensity for prolonged periods.

cardiovascular disease (CVD) a general term for all diseases of the heart and blood vessels. Atherosclerosis is the main cause of CVD. When the arteries that carry blood to the heart muscle become blocked, the heart suffers damage known as *coronary heart disease (CHD)*.

carnitine a nitrogen-containing compound, formed in the body from lysine and methionine, that helps transport fatty acids across the mitochondrial membrane. Carnitine is claimed to "burn" fat and spare glycogen during endurance events, but it does neither.

carotenoid (CARE-oh-ten-oyd) a member of a group of pigments in foods that range in color from light yellow to reddish orange and are chemical relatives of beta-carotene. Many have a degree of vitamin A activity in the body.

carrying capacity the total number of living organisms that a given environment can support without deteriorating in quality.

case studies studies of individuals. In clinical settings, researchers can observe treatments and their apparent effects. To prove that a treatment has produced an effect requires simultaneous observation of an untreated similar subject (a *case control*).

catalyst a substance that speeds the rate of a chemical reaction without itself being permanently altered in the process. All enzymes are catalysts.

cataracts (CAT-uh-racts) clouding of the lens of the eye that can lead to blindness. Cataracts can be caused by injury, viral infection, toxic substances, genetic disorders, and, possibly, some nutrient deficiencies or imbalances.

cathartic a strong laxative.

celiac (SEE-lee-ack) **disease** a disorder characterized by intestinal inflammation on exposure to the dietary protein gluten; also called *gluten-sensitive enteropathy* or *celiac sprue*.

cell differentiation (dih-fer-en-she-AY-shun) the process by which immature cells are stimulated to mature and gain the ability to perform functions characteristic of their cell type.

cells the smallest units in which independent life can exist. All living things are single cells or organisms made of cells.

cellulite a term popularly used to describe dimpled fat tissue on the thighs and buttocks; not recognized in science.

central obesity excess fat in the abdomen and around the trunk.

certified diabetes educator (CDE) a health-care professional who specializes in educating people with diabetes to help them manage their disease through medical and lifestyle means. Extensive training, work experience, and an examination are required to achieve CDE status.

certified lactation consultant a health-care provider, often a registered nurse or a registered dietitian, with specialized training and certification in breast and infant anatomy and physiology who teaches the mechanics of breastfeeding to new mothers.

cesarean (see-ZAIR-ee-un) **section** surgical childbirth, in which the infant is taken through an incision in the woman's abdomen.

chelating (KEE-late-ing) **agents** molecules that attract or bind with other molecules and are therefore useful in either preventing or promoting movement of substances from place to place.

chlorophyll the green pigment of plants that captures energy from sunlight for use in photosynthesis.

cholesterol (koh-LESS-ter-all) a member of the group of lipids known as sterols; a soft, waxy substance made in the body for a variety of purposes and also found in animal-derived foods.

choline (KOH-leen) a nonessential nutrient used to make the phospholipid lecithin and other molecules.

chromium picolinate a trace element supplement; falsely promoted to increase lean body mass, enhance energy, and burn fat.

chronic diseases degenerative conditions or illnesses that progress slowly, are long in duration, and that lack an immediate cure; chronic diseases limit functioning, productivity, and the quality and length of life. Examples include heart disease, cancer, and diabetes.

chronic hypertension in pregnant women, hypertension that is present and documented before pregnancy; in women whose prepregnancy blood pressure is unknown, the presence of sustained hypertension before 20 weeks of gestation.

chronic malnutrition malnutrition caused by long-term food deprivation; characterized in children by short height for age (stunting).

chylomicrons (KYE-low-MY-krons) lipoproteins formed when lipids from a meal cluster with carrier proteins in the cells of the intestinal lining. Chylomicrons transport food fats through the watery body fluids to the liver and other tissues.

chyme (KIME) the fluid resulting from the actions of the stomach upon a meal.

cirrhosis (seer-OH-sis) advanced liver disease, often associated with alcoholism, in which liver cells have died, hardened, turned an orange color, and permanently lost their function.

clone an individual created asexually from a single ancestor, such as a plant grown from a single stem cell; a group of genetically identical individuals descended from a single common ancestor, such as a colony of bacteria arising from a single bacterial cell; in genetics, a replica of a segment of DNA, such as a gene, produced by genetic engineering.

coenzyme (co-EN-zime) a small molecule that works with an enzyme to promote the enzyme's activity. Many coenzymes have B vitamins as part of their structure (*co* means "with").

cognitive skills as taught in behavior therapy, changes to conscious thoughts with the goal of improving adherence to lifestyle modifications; examples are problem-solving skills or the correction of false negative thoughts, termed *cognitive restructuring*.

cognitive therapy psychological therapy aimed at changing undesirable behaviors by changing underlying thought processes contributing to these behaviors; in anorexia, a goal is to replace false beliefs about body weight, eating, and self-worth with health-promoting beliefs.

collagen (COLL-a-jen) the chief protein of most connective tissues, including scars, ligaments, and tendons, and the underlying matrix on which bones and teeth are built.

colon the large intestine.

colostrum (co-LAHS-trum) a milklike secretion from the breasts during the first day or so after delivery before milk appears; rich in protective factors.

competitive foods unregulated meals, including fast foods, that compete side by side with USDA-regulated school lunches.

complementary and **alternative medicine (CAM)** a group of diverse medical and health-care systems, practices, and products that are not considered to be a part of conventional medicine. Examples include acupuncture, biofeedback, chiropractic, faith healing, and many others.

complementary proteins two or more proteins whose amino acid assortments complement each other in such a way that the essential amino acids missing from one are supplied by the other.

complex carbohydrates long chains of sugar units arranged to form starch or fiber; also called *polysaccharides*.

concentrated fruit juice sweetener a concentrated sugar syrup made from dehydrated, deflavored fruit juice, commonly grape juice; used to sweeten products that can then claim to be "all fruit."

conditionally essential amino acid an amino acid that is normally nonessential but must be supplied by the diet in special circumstances when the need for it exceeds the body's ability to produce it.

confectioner's sugar finely powdered sucrose, 99.9% pure.

congeners (CON-jen-ers) chemical substances other than alcohol that account for some of the physiological effects of alcoholic beverages, such as appetite, taste, and aftereffects.

constipation difficult, incomplete, or infrequent bowel movements associated with discomfort in passing dry, hardened feces from the body.

control group a group of individuals who are similar in all possible respects to the group being treated in an experiment but who receive a sham treatment instead of the real one. Also called *control subjects*. See also *experimental group* and *intervention studies*.

controlled clinical trial a research study design that often reveals effects of a treatment on human beings. Health outcomes are observed in a group of people who receive the treatment and are then compared with outcomes in a control group of similar people who received a placebo (an inert or sham treatment). Ideally, neither subjects nor researchers know who receives the treatment and who gets the placebo (a double-blind study).

cornea (KOR-nee-uh) the hard, transparent membrane covering the outside of the eye.

corn sweeteners corn syrup and sugar solutions derived from corn.

corn syrup a syrup, mostly glucose, partly maltose, produced by the action of enzymes on cornstarch. Includes corn syrup solids.

correlation the simultaneous change of two factors, such as the increase of weight with increasing height (a *direct* or *positive* correlation) or the decrease of cancer incidence with increasing fiber intake (an *inverse* or *negative* correlation). A correlation between two factors suggests that one may cause the other but does not rule out the possibility that both may be caused by chance or by a third factor.

cortex the outermost layer of something. The brain's cortex is the part of the brain where conscious thought takes place.

cortical bone the ivorylike outer bone layer that forms a shell surrounding trabecular bone and that comprises the shaft of a long bone.

country of origin label the required label stating the country of origination of many imported meats, chicken, fish and shellfish, other perishable foods, certain nuts, peanuts, and ginseng.

creatine a nitrogen-containing compound that combines with phosphate to burn a high-energy compound stored in muscle. Some studies suggest that creatine enhances energy and stimulates muscle growth but long-term studies are lacking; digestive side effects may occur.

cretinism (CREE-tin-ism) severe mental and physical retardation of an infant caused by the mother's iodine deficiency during pregnancy.

critical period a finite period during development in which certain events may occur that will have irreversible effects on later developmental stages. A critical period is usually a period of cell division in a body organ.

cross-contamination the contamination of a food through exposure to utensils, hands, or other surfaces that were previously in contact with a contaminated food.

cruciferous vegetables vegetables with cross-shaped blossoms—the cabbage family. Their intake is associated with low cancer rates in human populations. Examples are broccoli, brussels sprouts, cabbage, cauliflower, rutabagas, and turnips.

cuisines styles of cooking.

cultural competence having an awareness and acceptance of one's own and others' cultures and abilities leading to effective interactions with all kinds of people.

D

Daily Values nutrient standards that are printed on food labels and on grocery store and restaurant signs. Based on nutrient and energy recommendations for a general 2,000-calorie diet, they allow consumers to compare foods with regard to nutrients and calorie contents.

dead zones columns of oxygen-depleted ocean water in which marine life cannot survive; often caused by algae blooms that occur when agricultural fertilizers and waste runoff enter natural waterways.

dehydration loss of water. The symptoms progress rapidly, from thirst to weakness to exhaustion and delirium, and end in death.

denaturation the irreversible change in a protein's folded shape brought about by heat, acids, bases, alcohol, salts of heavy metals, or other agents.

dental caries decay of the teeth (*caries* means "rottenness"). Also called *cavities.*

dextrose, anhydrous dextrose forms of glucose.

DHEA (dehydroepiandrosterone) a hormone made in the adrenal glands that serves as a precursor to the male hormone testosterone; recently banned by the FDA because it poses the risk of life-threatening diseases, including cancer. Falsely promoted to burn fat, build muscle, and slow aging.

diabetes (dye-uh-BEET-eez) metabolic diseases characterized by elevated blood glucose and inadequate or ineffective insulin, which impair a person's ability to regulate blood glucose. The technical name is *diabetes mellitus* (*mellitus* means "honey-sweet" in Latin, referring to sugar in the urine).

dialysis (dye-AL-ih-sis) a medical treatment for failing kidneys in which a person's blood is circulated through a machine that filters out toxins and wastes and returns cleansed blood to the body. Also called *hemodialysis.*

diarrhea frequent, watery bowel movements usually caused by diet, stress, or irritation of the colon. Severe, prolonged diarrhea robs the body of fluid and certain minerals, causing dehydration and imbalances that can be dangerous if left untreated.

diastolic (dye-as-TOL-ik) **pressure** the second figure in a blood pressure reading (the "lubb" of the heartbeat is heard), which reflects the arterial pressure when the heart is between beats.

diet the foods (including beverages) a person usually eats and drinks.

dietary antioxidants compounds typically found in plant foods that significantly decrease the adverse effects of oxidation on living tissues. The major antioxidant vitamins are vitamin E, vitamin C, and beta-carotene. Many phytochemicals are also antioxidants.

dietary folate equivalent (DFE) a unit of measure expressing the amount of folate available to the body from naturally occurring sources. The measure mathematically equalizes the difference in absorption between less absorbable food folate and highly absorbable synthetic folate added to enriched foods and found in supplements.

Dietary Reference Intakes (DRI) a set of four lists of values for measuring the nutrient intakes of healthy people in the United States and Canada. The four lists are Estimated Average Requirements (EAR), Recommended Dietary Allowances (RDA), Adequate Intakes (AI), and Tolerable Upper Intake Levels (UL).

dietary supplements pills, liquids, or powders that contain purified nutrients or other ingredients.

dietetic technician a person who has completed a two-year academic degree from an accredited college or university and an approved dietetic technician program. A **dietetic technician, registered** (DTR) has also passed a national examination and maintains registration through continuing professional education.

dietitian a person trained in nutrition, food science, and diet planning. See also *registered dietitian.*

digest to break molecules into smaller molecules; a main function of the digestive tract with respect to food.

digestive system the body system composed of organs that break down complex food particles into smaller, absorbable products. The *digestive tract* and *alimentary canal* are names for the tubular organs that extend from the mouth to the anus. The whole system, including the pancreas, liver, and gallbladder, is sometimes called the *gastrointestinal*, or *GI*, system.

dipeptides (dye-PEP-tides) protein fragments that are two amino acids long (*di* means "two").

diploma mill an organization that awards meaningless degrees without requiring its students to meet educational standards. Diploma mills are not the same as diploma forgeries (fake diplomas and certificates bearing the names of real respected institutions). While virtually indistinguishable from authentic diplomas, forgeries can be unveiled by checking directly with the institution.

disaccharides pairs of single sugars linked together (*di* means "two").

distilled water water that has been vaporized and recondensed, leaving it free of dissolved minerals.

diuretic (dye-you-RET-ic) a compound, usually a medication, causing increased urinary water excretion; a "water pill."

diverticula (dye-ver-TIC-you-la) sacs or pouches that balloon out of the intestinal wall, caused by weakening of the muscle layers that encase the intestine. The painful inflammation of one or more of the diverticula is known as *diverticulitis.*

DNA an abbreviation for deoxyribonucleic (dee-OX-ee-RYE-bow-nu-CLAY-ick) acid, the thread-like molecule that encodes genetic information in its structure; DNA strands coil up densely to form the chromosomes.

DNA microarray technology research tools that analyze the expression of thousands of genes simultaneously and search for particular genes associated with a disease. DNA microarrays are also called *DNA chips.*

dolomite a compound of minerals (calcium magnesium carbonate) found in limestone and marble. Dolomite is powdered and is sold as a calcium-magnesium supplement but may be contaminated with toxic minerals, is not well absorbed, and interacts adversely with absorption of other essential minerals.

dopamine (DOH-pah-meen) a neurotransmitter with many important roles in the brain, including cognition, pleasure, motivation, mood, sleep, and others.

drink a dose of any alcoholic beverage that delivers half an ounce of pure ethanol.

drug any substance that when taken into a living organism may modify one or more of its functions.

dual-energy X-ray absorptiometry (ab-sorp-tee-OM-eh-tree) a noninvasive method of determining total body fat, fat distribution, and bone density by passing two low-dose X-ray beams through the body. Also used in evaluation of osteoporosis. Abbreviated DEXA.

E

eating disorder a disturbance in eating behavior that jeopardizes a person's physical or psychological health.

eating pattern habitual intake of foods and beverages over time; a person's usual diet.

eclampsia (eh-CLAMP-see-ah) a severe complication during pregnancy in which seizures occur.

edamame fresh green soybeans, a source of phytoestrogens.

edema (eh-DEEM-uh) swelling of body tissue caused by leakage of fluid from the blood vessels; seen in protein deficiency (among other conditions).

eicosanoids (eye-COSS-ah-noyds) biologically active compounds that regulate body functions.

electrolytes compounds that partly dissociate in water to form ions, such as the potassium ion (K^+) and the chloride ion (Cl^-).

elemental diets diets composed of purified ingredients of known chemical composition; intended to supply all essential nutrients to people who cannot eat foods.

embolism an embolus that causes sudden closure of a blood vessel.

embolus (EM-boh-luss) a thrombus that breaks loose and travels through the blood vessels (*embol* means "to insert").

embryo (EM-bree-oh) the stage of human gestation from the third to the eighth week after conception.

emergency kitchens programs that provide prepared meals to be eaten on-site; often called *soup kitchens.*

emetic (em-ETT-ic) an agent that causes vomiting.

emulsification the process of mixing lipid with water by adding an emulsifier.

emulsifier (ee-MULL-sih-fire) a compound with both water-soluble and fat-soluble portions that can attract fats and oils into water, combining them.

endosperm the bulk of the edible part of a grain, the starchy part.

energy the capacity to do work. The energy in food is chemical energy; it can be converted to mechanical, electrical, thermal, or other forms of energy in the body. Food energy is measured in calories.

energy density a measure of the energy provided by a food relative to its weight (calories per gram).

energy drinks and **energy shots** sugar-sweetened beverages in various concentrations with supposedly ergogenic ingredients, such as vitamins, amino acids, caffeine, guarana, carnitine, ginseng, and others. The drinks are not regulated by the FDA and are often high in caffeine or other stimulants.

energy-yielding nutrients the nutrients the body can use for energy—carbohydrate, fat, and protein. These also may supply building blocks for body structures.

enriched foods and **fortified foods** foods to which nutrients have been added. If the starting material is a whole, basic food such as milk or whole grain, the result may be highly nutritious. If the starting material is a concentrated form of sugar or fat, the result may be less nutritious.

enriched, fortified refers to the addition of nutrients to a refined food product. As defined by U.S. law, these terms mean that specified levels of thiamin, riboflavin, niacin, folate, and iron have been added to refined grains and grain products. The terms *enriched* and *fortified* can refer to the addition of more nutrients than just these five; read the label.

enterotoxins poisons that act upon mucous membranes, such as those of the digestive tract.

environmental tobacco smoke the combination of exhaled smoke (mainstream smoke) and smoke from lighted cigarettes, pipes, or cigars (sidestream smoke) that enters the air and may be inhaled by other people.

enzymes (EN-zimes) proteins that facilitate chemical reactions without being changed in the process; protein catalysts.

EPA, DHA eicosapentaenoic (EYE-cossa-PENTA-ee-NO-ick) acid, docosahexaenoic (DOE-cossa-HEXA-ee-NO-ick) acid; omega-3 fatty acids made from linolenic acid in the tissues of fish.

epidemiological studies studies of populations; often used in nutrition to search for correlations between dietary habits and disease incidence; a first step in seeking nutrition-related causes of diseases.

epigenetics (ep-ih-gen-EH-tics) the science of heritable changes in gene function that occur without a change in the DNA sequence.

epigenome (ep-ih-GEE-nohm) the proteins and other molecules associated with chromosomes that affect gene expression. The epigenome is modulated by bioactive food components and other factors in ways that can be inherited. *Epi* is a Greek prefix, meaning "above" or "on."

epinephrine (EP-ih-NEFF-rin) the major hormone that elicits the stress response. In allergy, an epinephrine drug that counteracts anaphylactic shock by opening the airways and maintaining heartbeat and blood pressure.

epiphyseal (eh-PIFF-ih-seal) **plate** a thick, cartilage-like layer that forms new cells that are eventually calcified, lengthening the bone (*epiphysis* means "growing" in Greek).

epithelial (ep-ith-THEE-lee-ull) **tissue** the layers of the body that serve as selective barriers to environmental factors. Examples are the cornea, the skin, the respiratory tract lining, and the lining of the digestive tract.

ergogenic (ER-go-JEN-ic) **aids** products that supposedly enhance performance, although few actually do so; the term *ergogenic* implies "energy giving" (*ergo* means "work"; *genic* means "give rise to").

erythrocyte (eh-REETH-ro-sight) **hemolysis** (HEE-moh-LIE-sis, hee-MOLL-ih-sis) rupture of the red blood cells that can be caused by vitamin E deficiency (*erythro* means "red"; *cyte* means "cell"; *hemo* means "blood"; *lysis* means "breaking"). The anemia produced by the condition is *hemolytic* (HEE-moh-LIT-ick) *anemia*.

essential amino acids amino acids that either cannot be synthesized at all by the body or cannot be synthesized in amounts sufficient to meet physiological need. Also called *indispensable amino acids*.

essential fatty acids fatty acids that the body needs but cannot make and so must be obtained from the diet.

essential nutrients the nutrients the body cannot make for itself (or cannot make fast enough) from other raw materials; nutrients that must be obtained from food to prevent deficiencies.

Estimated Average Requirements (EAR) the average daily nutrient intake estimated to meet the requirement of half of the healthy individuals in a particular life stage and gender group; used in nutrition research and policy making and is the basis upon which RDA values are set.

Estimated Energy Requirement (EER) the average dietary energy intake predicted to maintain energy balance in a healthy adult of a certain age, gender, weight, height, and level of physical activity consistent with good health.

ethanol the alcohol of alcoholic beverages, produced by the action of microorganisms on the carbohydrates of grape juice or other carbohydrate-containing fluids.

ethnic foods foods associated with particular cultural subgroups within a population.

euphoria (you-FOR-ee-uh) an inflated sense of well-being and pleasure brought on by a moderate dose of alcohol and by some other drugs.

evaporated cane juice raw sugar from which impurities have been removed.

excess postexercise oxygen consumption (EPOC) a measure of increased metabolism (energy expenditure) that continues for minutes or hours after cessation of exercise.

exchange system a diet-planning tool that organizes foods with respect to their nutrient content and calories. Foods on any single exchange list can be used interchangeably.

exclusive breastfeeding an infant's consumption of human milk with no supplementation of any type (no water, no juice, no nonhuman milk, and no foods) except for vitamins, minerals, and medications.

exercise planned, structured, and repetitive bodily movement that promotes or maintains physical fitness.

experimental group the people or animals participating in an experiment who receive the treatment under investigation. Also called *experimental subjects*. See also *control group* and *intervention studies*.

extracellular fluid fluid residing outside the cells that transports materials to and from the cells.

extreme obesity clinically severe overweight, presenting very high risks to health; the condition of having a BMI of 40 or above; also called *morbid obesity*.

extrusion processing techniques that transform whole or refined grains, legumes, and other foods into shaped, colored, and flavored snacks, breakfast cereals, and other products.

F

famine widespread and extreme scarcity of food that causes starvation and death in a large portion of the population in an area.

farm share an arrangement in which a farmer offers the public a "subscription" for an allotment of the farm's products throughout the season.

fast foods restaurant foods that are available within minutes after customers order them—traditionally, hamburgers, French fries, and milkshakes; more recently, salads and other vegetable dishes as well. These foods may or may not meet people's nutrient needs, depending on the selections made and on the energy allowances and nutrient needs of the eaters.

fasting plasma glucose test a blood test that measures current blood glucose in a person who has not eaten or consumed caloric beverages for at least 8 hours; the test can detect both diabetes and prediabetes. *Plasma* is the fluid part of whole blood.

fat cells cells that specialize in the storage of fat and form the fat tissue. Fat cells also produce fat-metabolizing enzymes; they also produce hormones involved in appetite and energy balance.

fat replacers ingredients that replace some or all of the functions of fat and may or may not provide energy.

fats lipids that are solid at room temperature (70°F or 21°C).

fatty acids organic acids composed of carbon chains of various lengths. Each fatty acid has an acid end and hydrogens attached to all of the carbon atoms of the chain.

fatty liver an early stage of liver deterioration seen in several diseases, including kwashiorkor and alcoholic liver disease, in which fat accumulates in the liver cells.

feces waste material remaining after digestion and absorption are complete; eventually discharged from the body.

female athlete triad a potentially fatal triad of medical problems seen in female athletes: disordered eating, amenorrhea, and osteoporosis.

fermentation the anaerobic (without oxygen) breakdown of carbohydrates by microorganisms that releases small organic compounds along with carbon dioxide and energy.

fertility the capacity of a woman to produce a normal ovum periodically and of a man to produce normal sperm; the ability to reproduce.

fetal alcohol spectrum disorders (FASD) a spectrum of physical, behavioral, and cognitive disabilities caused by prenatal alcohol exposure.

fetal alcohol syndrome (FAS) the cluster of symptoms including brain damage, growth restriction, mental retardation, and facial abnormalities seen in an infant or child whose mother consumed alcohol during her pregnancy.

fetus (FEET-us) the stage of human gestation from eight weeks after conception until the birth of an infant.

fibers the indigestible parts of plant foods, largely nonstarch polysaccharides that are not digested by human digestive enzymes, although some are digested by resident bacteria of the colon. Fibers include cellulose, hemicelluloses, pectins, gums, mucilages, and a few nonpolysaccharides such as lignin.

fibrosis (fye-BROH-sis) an intermediate stage of alcoholic liver deterioration. Liver cells lose their function and assume the characteristics of connective tissue cells (become fibrous).

fight-or-flight reaction the body's instinctive hormone- and nerve-mediated reaction to danger. Also known as the *stress response*.

filtered water water treated by filtration, usually through activated carbon filters that reduce the lead in tap water, or by reverse osmosis units that force pressurized water across a membrane, removing lead, arsenic, and some microorganisms from tap water.

fitness the characteristics that enable the body to perform physical activity; more broadly, the ability to meet routine physical demands with enough reserve energy to rise to a physical challenge; or the body's ability to withstand stress of all kinds.

fitness water lightly flavored bottled water enhanced with vitamins, supposedly to enhance athletic performance.

flavonoids (FLAY-von-oyds) a common and widespread group of phytochemicals, with over 6,000 identified members; physiologic effects may include antioxidant, antiviral, anticancer, and other activities. Flavonoids are yellow pigments in foods; *flavus* means "yellow."

flavored waters lightly flavored beverages with few or no calories, but often containing vitamins, minerals, herbs, or other unneeded substances. Not superior to plain water for athletic competition or training.

flaxseed small brown seed of the flax plant; used in baking, cereals, or other foods. Valued in nutrition as a source of fatty acids, lignans, and fiber.

flexibility the capacity of the joints to move through a full range of motion; the ability to bend and recover without injury.

fluid and electrolyte balance maintenance of the proper amounts and kinds of fluids and minerals in each compartment of the body.

fluid and electrolyte imbalance failure to maintain the proper amounts and kinds of fluids and minerals in every body compartment; a medical emergency.

fluorapatite (floor-APP-uh-tight) a crystal of bones and teeth, formed when fluoride displaces the "hydroxy" portion of hydroxyapatite. Fluorapatite resists being dissolved back into body fluid.

fluorosis (floor-OH-sis) discoloration of the teeth due to ingestion of too much fluoride during tooth development. *Skeletal fluorosis* is characterized by unusually dense but weak, fracture-prone, often malformed bones, caused by excess fluoride in bone crystals.

folate (FOH-late) a B vitamin that acts as part of a coenzyme important in the manufacture of new cells. The form added to foods and supplements is *folic acid*.

food medically, any substance that the body can take in and assimilate that will enable it to stay alive and to grow; the carrier of nourishment; socially, a more limited number of such substances defined as acceptable by each culture.

food aversion an intense dislike of a food, biological or psychological in nature, resulting from an illness or other negative experience associated with that food.

food banks facilities that collect and distribute food donations to authorized organizations feeding the hungry.

foodborne illness illness transmitted to human beings through food and water; caused by an infectious agent (*foodborne infection*) or a poisonous substance arising from microbial toxins, poisonous chemicals, or other harmful substances (*food intoxication*). Also commonly called *food poisoning*.

food contaminant any substance occurring in food by accident; any food constituent that is not normally present.

food crisis a steep decline in food availability with a proportional rise in hunger and malnutrition at the local, national, or global level.

food deserts urban and rural low-income areas with limited access to affordable and nutritious foods.

food group plan a diet-planning tool that sorts foods into groups based on their nutrient content and then specifies that people should eat certain minimum numbers of servings of foods from each group.

food intolerance an adverse reaction to a food or food additive not involving an immune response.

food neophobia (NEE-oh-FOE-bee-ah) the fear of trying new foods, common among toddlers.

food pantries community food collection programs that provide groceries to be prepared and eaten at home.

food poverty hunger occurring when enough food exists in an area but some of the people cannot obtain it because they lack money, are being deprived for political reasons, live in a country at war, or suffer from other problems such as lack of transportation.

food recovery collecting wholesome surplus food for distribution to low-income people who are hungry.

foodways the sum of a culture's habits, customs, beliefs, and preferences concerning food.

fork thermometer a utensil combining a meat fork and an instant-read food thermometer.

formaldehyde a substance to which methanol is metabolized on the way to being converted to harmless waste products that can be excreted.

fraud or **quackery** the promotion, for financial gain, of devices, treatments, services, plans, or products (including diets and supplements) claimed to improve health, well-being, or appearance without proof of safety or effectiveness. (The word *quackery* comes from the term *quacksalver*, meaning a person who quacks loudly about a miracle product—a lotion or a salve.)

free radicals atoms or molecules with one or more unpaired electrons that make the atom or molecule unstable and highly reactive.

fructose (FROOK-tose) a monosaccharide; sometimes known as fruit sugar (*fruct* means "fruit"; *ose* means "sugar").

fructose, galactose, glucose the monosaccharides.

fruitarian includes only raw or dried fruits, seeds, and nuts in the diet.

fufu a low-protein staple food that provides abundant starch energy to many of the world's people; fufu is made by pounding or grinding root vegetables or refined grains and cooking them to a smooth semisolid consistency.

functional foods whole or modified foods that contain bioactive food components believed to provide health benefits, such as reduced disease risks, beyond the benefits that their nutrients confer.

G

galactose (ga-LACK-tose) a monosaccharide; part of the disaccharide lactose (milk sugar).

gastric juice the digestive secretion of the stomach.

gastroesophageal (GAS-tro-eh-SOFF-ah-jeel) **reflux disease (GERD)** a severe and chronic splashing of stomach acid and enzymes into the esophagus, throat, mouth, or airway that causes injury to those organs. Untreated GERD may increase the risk of esophageal cancer; treatment may require surgery or management with medication.

gatekeeper with respect to nutrition, a key person who controls other people's access to foods and thereby affects their nutrition profoundly. Examples are the spouse who buys and cooks the food, the parent who feeds the children, and the caregiver in a day-care center.

generally recognized as safe (GRAS) list a list, established by the FDA, of food additives long in use and believed to be safe.

genes units of a cell's inheritance; sections of the larger genetic molecule DNA (deoxyribonucleic acid). Each gene directs the making of one or more of the body's proteins.

genetically engineered organism (GEO) an organism produced by genetic engineering; the term *genetically modified organism (GMO)* is often used to mean the same thing.

genetic engineering (GE) the direct, intentional manipulation of the genetic material of living things in order to obtain some desirable inheritable trait not present in the original organism. Also called *recombinant DNA technology*.

genetic profile the result of an analysis of genetic material that identifies unique characteristics of a person's DNA for forensic or diagnostic purposes.

genistein (GEN-ih-steen) a phytoestrogen found primarily in soybeans that both mimics and blocks the action of estrogen in the body.

genome (GEE-nohm) the full complement of genetic material in the chromosomes of a cell.

genomics the study of all the genes in an organism and their interactions with environmental factors.

germ the nutrient-rich inner part of a grain.

gestation the period of about 40 weeks (three trimesters) from conception to birth; the term of a pregnancy.

gestational diabetes abnormal glucose tolerance appearing during pregnancy.

gestational hypertension high blood pressure that develops in the second half of pregnancy and usually resolves after childbirth.

ghrelin (GREL-in) a hormone released by the stomach that signals the brain's hypothalamus and other regions to stimulate eating.

glucagon (GLOO-cah-gon) a hormone secreted by the pancreas that stimulates the liver to release glucose into the blood when blood glucose concentration dips.

glucose (GLOO-cose) a single sugar used in both plant and animal tissues for energy; sometimes known as blood sugar or *dextrose*.

glucose polymers compounds that supply glucose, not as single molecules, but linked in chains somewhat like starch. The objective is to attract less water from the body into the digestive tract.

gluten (GLOO-ten) a type of protein in certain grain foods that is toxic to the person with celiac disease.

glycemic index (GI) a ranking of foods according to their potential for raising blood glucose relative to a standard food such as glucose.

glycemic load (GL) a mathematical expression of both the glycemic index and the carbohydrate content of a food, meal, or diet.

glycerol (GLISS-er-all) an organic compound, three carbons long, of interest here because it serves as the backbone for triglycerides.

glycogen (GLY-co-gen) a highly branched polysaccharide that is made and stored by liver and muscle tissues of human beings and animals as a storage form of glucose. Glycogen is not a significant food source of carbohydrate and is not counted as one of the complex carbohydrates in foods.

goiter (GOY-ter) enlargement of the thyroid gland due to iodine deficiency is *simple goiter*; enlargement due to an iodine excess is *toxic goiter*.

gout (GOWT) a painful form of arthritis caused by the abnormal buildup of the waste product uric acid in the blood, with uric acid salt deposited as crystals in the joints.

grams units of weight. A gram (g) is the weight of a cubic centimeter (cc) or milliliter (ml) of water under defined conditions of temperature and pressure. About 28 grams equal an ounce.

granulated sugar common table sugar, crystalline sucrose, 99.9% pure.

granules small grains. Starch granules are packages of starch molecules. Various plant species make starch granules of varying shapes.

groundwater water that comes from underground aquifers.

growth hormone a hormone (somatotropin) that promotes growth and that is produced naturally in the pituitary gland of the brain.

growth spurt the marked rapid gain in physical size usually evident around the onset of adolescence.

H

hard water water with high calcium and magnesium concentrations.

hazard a state of danger; used to refer to any circumstance in which harm is possible under normal conditions of use.

Hazard Analysis Critical Control Point (HACCP) a systematic plan to identify and correct potential microbial hazards in the manufacturing, distribution, and commercial use of food products. *HACCP* may be pronounced "HASS-ip."

HbA_{1C} test a blood test that measures hemoglobin molecules with glucose attached to them (*Hb* stands for *hemoglobin*). The test reflects blood glucose control over the previous few months. Also called *glycosylated hemoglobin test*, or *A1C test*.

health claims claims linking food constituents with disease states; allowable on labels within the criteria established by the Food and Drug Administration.

heart attack the event in which the vessels that feed the heart muscle become closed off by an embolism, thrombus, or other cause with resulting sudden tissue death. A heart attack is also called a *myocardial infarction* (*myo* means "muscle"; *cardial* means "of the heart"; *infarct* means "tissue death").

heartburn a burning sensation in the chest (in the area of the heart) caused by backflow of stomach acid into the esophagus.

heat cramps painful cramps of the abdomen, arms, or legs, often occurring hours after exercise; associated with inadequate intake of fluid or electrolytes or heavy sweating.

heat stroke an acute and life-threatening reaction to heat buildup in the body.

heavy episodic drinking a drinking pattern that includes occasional or regular consumption of four or more alcoholic beverages in a short time period. Also called *binge drinking*.

heavy metal any of a number of mineral ions such as mercury and lead, so called because they are of relatively high atomic weight; many heavy metals are poisonous.

heme (HEEM) the iron-containing portion of the hemoglobin and myoglobin molecules.

hemoglobin (HEEM-oh-globe-in) the oxygen-carrying protein of the blood; found in the red blood cells (*hemo* means "blood"; *globin* means "spherical protein").

hemolytic-uremic (HEEM-oh-LIT-ic you-REEM-ick) **syndrome** a severe result of infection with Shiga toxin-producing *E. coli*, characterized by abnormal blood clotting with kidney failure, damage to the central nervous system and other organs, and death, especially among children.

hemorrhoids (HEM-or-oids) swollen, hardened (varicose) veins in the rectum, usually caused by the pressure resulting from constipation.

hepcidin (HEP-sid-in) a hormone secreted by the liver in response to elevated blood iron. Hepcidin reduces iron's absorption from the intestine and its release from storage.

herbal medicine a type of CAM that uses herbs and other natural substances to prevent or cure diseases or to relieve symptoms.

hernia a protrusion of an organ or part of an organ through the wall of the body chamber that normally contains the organ. An example is a *hiatal* (high-AY-tal) *hernia*, in which part of the stomach protrudes up through the diaphragm into the chest cavity, which contains the esophagus, heart, and lungs.

hiccups spasms of both the vocal cords and the diaphragm, causing periodic, audible, short, inhaled coughs. Can be caused by irritation of the diaphragm, indigestion, or other causes. Hiccups usually resolve in a few minutes but can have serious effects if prolonged. Breathing into a paper bag (inhaling carbon dioxide) or dissolving a teaspoon of sugar in the mouth may stop them.

high-carbohydrate energy drinks flavored commercial beverages used to restore muscle glycogen after exercise or as a pregame beverage.

high-carbohydrate gels semi-solid easy-to-swallow supplements of concentrated carbohydrate, commonly with potassium and sodium added; not a fluid source.

high-density lipoproteins (HDL) lipoproteins that return cholesterol from the tissues to the liver for dismantling and disposal; contain a large proportion of protein.

high food security a descriptor for households with no problems or anxiety about consistently accessing adequate food.

high-fructose corn syrup a commercial sweetener used in many foods, including soft drinks. Composed almost entirely of the monosaccharides fructose and glucose, its sweetness and caloric value are similar to sucrose.

high-quality proteins dietary proteins containing all the essential amino acids in relatively the same amounts that human beings require. They may also contain nonessential amino acids.

high-risk pregnancy a pregnancy characterized by risk factors that make it likely the birth will be surrounded by problems such as premature delivery, difficult birth, retarded growth, birth defects, and early infant death. A *low-risk pregnancy* has none of these factors.

histamine a substance that participates in causing inflammation; produced by cells of the immune system as part of a local immune reaction to an antigen.

histones proteins that lend structural support to the chromosome structure and that activate or silence gene expression.

homogenization a process by which milk fat is evenly dispersed within fluid milk; under high pressure, milk is passed through tiny nozzles to reduce the size of fat droplets and reduce their tendency to cluster and float to the top as cream.

honey a concentrated solution primarily composed of glucose and fructose, produced by enzymatic digestion of the sucrose in nectar by bees.

hormones chemical messengers secreted by a number of body organs in response to conditions that require regulation. Each hormone affects a specific organ or tissue and elicits a specific response.

hourly sweat rate the amount of weight lost plus fluid consumed during exercise per hour.

hunger the physiological need to eat, experienced as a drive for obtaining food; an unpleasant sensation that demands relief.

hunger an involuntary lack of sufficient quantity or quality of food; a consequence of food insecurity.

husk the outer, inedible part of a grain.

hydrochloric acid a strong corrosive acid of hydrogen and chloride atoms, produced by the stomach to assist in digestion.

hydrogenation (high-dro-gen-AY-shun) the process of adding hydrogen to unsaturated fatty acids to make fat more solid and resistant to the chemical change of oxidation.

hydroxyapatite (hi-DROX-ee-APP-uh-tight) the chief crystal of bone, formed from calcium and phosphorus.

hyperactivity (in children) a syndrome characterized by inattention, impulsiveness, and excess motor activity; usually diagnosed before age 7, lasts six months or more, and usually does not entail mental illness or mental retardation. Properly called *attention-deficit/hyperactivity disorder (ADHD)*.

hypertension higher than normal blood pressure.

hypertrophy (high-PURR-tro-fee) an increase in size (for example, of a muscle) in response to use.

hypoglycemia (HIGH-poh-gly-SEE-mee-ah) an abnormally low blood glucose concentration, often accompanied by symptoms such as anxiety, rapid heartbeat, and sweating.

hyponatremia (high-poh-nah-TREE-mee-ah) a decreased concentration of sodium in the blood.

hypothalamus (high-poh-THAL-uh-mus) a part of the brain that senses a variety of conditions in the blood, such as temperature, glucose content, salt content, and others. It signals other parts of the brain or body to adjust those conditions when necessary.

hypothermia a below-normal body temperature.

I

immune system a system of tissues and organs that defend the body against antigens, foreign materials that have penetrated the skin or body linings.

immunity protection from or resistance to a disease or infection by development of antibodies and by the actions of cells and tissues in response to a threat.

implantation the stage of development, during the first two weeks after conception, in which the fertilized egg (fertilized ovum or zygote) embeds itself in the wall of the uterus and begins to develop.

inborn error of metabolism a genetic variation present from birth that may result in disease.

incidental additives substances that can get into food not through intentional introduction, but as a result of contact with the food during growing, processing, packaging, storing, or some other stage before the food is consumed. Also called *accidental* or *indirect additives.*

infectious diseases diseases that are caused by bacteria, viruses, parasites, and other microbes and can be transmitted from one person to another through air, water, or food; by contact; or through vector organisms such as mosquitoes and fleas.

inflammation (in-flam-MAY-shun) part of the body's immune defense against injury, infection, or allergens, marked by increased blood flow, release of chemical toxins, and attraction of white blood cells to the affected area (from the Latin *inflammare*, meaning "to flame within"). Inflammation plays a role in many chronic diseases.

infomercials feature-length television commercials that follow the format of regular programs but are intended to convince viewers to buy products and not to educate or entertain them. The statements made may or may not be accurate.

initiation an event, probably occurring in a cell's genetic material, caused by radiation or by a chemical carcinogen that can give rise to cancer.

inositol (in-OSS-ih-tall) a nonessential nutrient found in cell membranes.

insoluble fibers the tough, fibrous structures of fruits, vegetables, and grains; indigestible food components that do not dissolve in water.

instant-read thermometer a thermometer that, when inserted into food, measures its temperature within seconds; designed to test temperature of food at intervals, and not to be left in food during cooking.

insulin a hormone secreted by the pancreas in response to a high blood glucose concentration. It assists cells in drawing glucose from the blood.

insulin resistance a condition in which a normal or high level of circulating insulin produces a less-than-normal response in muscle, liver, and adipose tissues; thought to be a metabolic consequence of obesity.

integrated pest management (IPM) management of pests using a combination of natural and biological controls and minimal or no application of pesticides.

intensity in exercise, the degree of effort required to perform a given physical activity.

Internet (the Net) a worldwide network of millions of computers linked together to share information.

intervention studies studies of populations in which observation is accompanied by experimental manipulation of some population members—for example, a study in which half of the subjects (the *experimental subjects*) follow diet advice to reduce fat intakes while the other half (the *control subjects*) do not, and both groups' heart health is monitored.

intestine the body's long, tubular organ of digestion and the site of nutrient absorption.

intracellular fluid fluid residing inside the cells that provides the medium for cellular reactions.

intrinsic factor a factor found inside a system. The intrinsic factor necessary to prevent pernicious anemia is now known to be a compound that helps in the absorption of vitamin B_{12}.

invert sugar a mixture of glucose and fructose formed by the splitting of sucrose in an industrial process. Sold only in liquid form and sweeter than sucrose, invert sugar forms during certain cooking procedures and works to prevent crystallization of sucrose in soft candies and sweets.

ions (EYE-ons) electrically charged particles, such as sodium (positively charged) or chloride (negatively charged).

iron deficiency the condition of having depleted iron stores, which, at the extreme, causes iron-deficiency anemia.

iron-deficiency anemia a form of anemia caused by a lack of iron and characterized by red blood cell shrinkage and color loss. Accompanying symptoms are weakness, apathy, headaches, pallor, intolerance to cold, and inability to pay attention. (For other anemias, see the index.)

iron overload the state of having more iron in the body than it needs or can handle, usually arising from a hereditary defect. Also called *hemochromatosis.*

irradiation the application of ionizing radiation to foods to reduce insect infestation or microbial contamination or to slow the ripening or sprouting process. Also called *cold pasteurization.*

irritable bowel syndrome (IBS) intermittent disturbance of bowel function, especially diarrhea or alternating diarrhea and constipation, often with abdominal cramping or bloating; managed with diet, physical activity, or relief from psychological stress. The cause is uncertain, but IBS does not permanently harm the intestines nor lead to serious diseases.

IU (international units) a measure of fat-soluble vitamin activity sometimes used in food composition tables and on supplement labels.

J

jaundice (JAWN-dis) yellowing of the skin due to spillover of the bile pigment bilirubin (bill-ee-ROO-bin) from the liver into the general circulation.

K

kefir (KEE-fur) a liquid form of yogurt, based on milk, probiotic microorganisms, and flavorings.

keratin (KERR-uh-tin) the normal protein of hair and nails.

keratinization accumulation of keratin in a tissue; a sign of vitamin A deficiency.

ketone (kee-tone) **bodies** acidic, water-soluble compounds that arise during the breakdown of fat when carbohydrate is not available.

ketosis (kee-TOE-sis) an undesirable high concentration of ketone bodies, such as acetone, in the blood or urine.

kidneys a pair of organs that filter wastes from the blood, make urine, and release it to the bladder for excretion from the body.

kwashiorkor (kwash-ee-OR-core, kwash-ee-or-CORE) severe malnutrition characterized by failure to grow and develop, edema, changes in the pigmentation of hair and skin, fatty liver, anemia, and apathy.

L

laboratory studies studies that are performed under tightly controlled conditions and are designed to pinpoint causes and effects. Such studies often use animals as subjects.

lactase the intestinal enzyme that splits the disaccharide lactose to monosaccharides during digestion.

lactate a compound produced during the breakdown of glucose in anaerobic metabolism.

lactation production and secretion of breast milk for the purpose of nourishing an infant.

lactoferrin (lack-toe-FERR-in) a factor in breast milk that binds iron and keeps it from supporting the growth of the infant's intestinal bacteria.

lacto-ovo vegetarian includes dairy products, eggs, vegetables, grains, legumes, fruits, and nuts; excludes flesh and seafood.

lactose a disaccharide composed of glucose and galactose; sometimes known as milk sugar (*lact* means "milk"; *ose* means "sugar").

lactose intolerance impaired ability to digest lactose due to reduced amounts of the enzyme lactase.

lactose, maltose, sucrose the disaccharides.

lacto-vegetarian includes dairy products, vegetables, grains, legumes, fruits, and nuts; excludes flesh, seafood, and eggs.

lapses periods of returning to old habits.

large intestine the portion of the intestine that completes the absorption process.

learning disability a condition resulting in an altered ability to learn basic cognitive skills such as reading, writing, and mathematics.

leavened (LEV-end) literally, "lightened" by yeast cells, which digest some carbohydrate components of the dough and leave behind bubbles of gas that make the bread rise.

lecithin (LESS-ih-thin) a phospholipid manufactured by the liver and also found in many foods; a major constituent of cell membranes.

legumes (leg-GOOMS, LEG-yooms) beans, peas, and lentils, valued as inexpensive sources of protein, vitamins, minerals, and fiber that contribute little fat to the diet.

leptin an appetite-suppressing hormone produced in the fat cells that conveys information about body fatness to the brain; believed to be involved in the maintenance of body composition (*leptos* means "slender").

leucine one of the essential amino acids; it is of current research interest for its role in stimulating muscle protein synthesis.

levulose an older name for fructose.

license to practice permission under state or federal law, granted on meeting specified criteria, to use a certain title (such as *dietitian*) and to offer certain services. Licensed dietitians may use the initials LD after their names.

life expectancy the average number of years lived by people in a given society.

life span the maximum number of years of life attainable by a member of a species.

lignans phytochemicals present in flaxseed, but not in flax oil, that are converted to phytoestrogens by intestinal bacteria and are under study as possible anticancer agents.

limiting amino acid an essential amino acid that is present in dietary protein in an insufficient amount, thereby limiting the body's ability to build protein.

linoleic (lin-oh-LAY-ic) **acid** an essential polyunsaturated fatty acid of the omega-6 family.

linolenic (lin-oh-LEN-ic) **acid** an essential polyunsaturated fatty acid of the omega-3 family. The full name of linolenic acid is *alpha-linolenic acid.*

lipase (LYE-pace) any of a number of enzymes that break the chemical bonds of fats (lipids).

lipid (LIP-id) a family of organic (carbon-containing) compounds soluble in organic solvents but not in water. Lipids include triglycerides (fats and oils), phospholipids, and sterols.

lipoic (lip-OH-ic) **acid** a nonessential nutrient.

lipoproteins (LYE-poh-PRO-teens, LIH-poh-PRO-teens) clusters of lipids associated with protein, which serve as transport vehicles for lipids in blood and lymph. The major lipoproteins include chylomicrons, VLDL, LDL, and HDL.

listeriosis a serious foodborne infection that can cause severe brain infection or death in a fetus or a newborn; caused by the bacterium *Listeria monocytogenes,* which is found in soil and water.

liver a large, lobed organ that lies just under the ribs. It filters the blood, removes and processes nutrients, manufactures materials for export to other parts of the body, and destroys toxins or stores them to keep them out of the circulatory system.

longevity long duration of life.

low birthweight a birthweight of less than $5^1/_2$ pounds (2,500 grams); used as a predictor of probable health problems in the newborn and as a probable indicator of poor nutrition status of the mother before and/or during pregnancy. Low-birthweight infants are of two different types. Some are *premature infants*; they are born early and are the right size for their gestational age. Other low-birthweight infants have suffered growth failure in the uterus; they are small for gestational age (small for date) and may or may not be premature.

low-density lipoproteins (LDL) lipoproteins that transport lipids from the liver to other tissues such as muscle and fat; contain a large proportion of cholesterol.

low food security a descriptor for households with reduced dietary quality, variety, and desirability but with adequate quantity of food and normal eating patterns. Example: a family whose diet centers on inexpensive, low-nutrient foods such as refined grains, inexpensive meats, sweets, and fats.

lungs the body's organs of gas exchange. Blood circulating through the lungs releases its carbon dioxide and picks up fresh oxygen to carry to the tissues.

lutein (LOO-teen) a plant pigment of yellow hue; a phytochemical believed to play roles in eye functioning and health.

lycopene (LYE-koh-peen) a pigment responsible for the red color of tomatoes and other red-hued vegetables; a phytochemical that may act as an antioxidant in the body.

lymph (LIMF) the fluid that moves from the bloodstream into tissue spaces and then travels in its own vessels, which eventually drain back into the bloodstream.

lymphocytes (LIM-foh-sites) white blood cells that participate in the immune response; B-cells and T-cells.

M

macrobiotic diet a vegan diet composed mostly of whole grains, beans, and certain vegetables; taken to extremes, macrobiotic diets can compromise nutrient status.

macrophages (MACK-roh-fah-jez) large scavenger cells of the immune system that engulf debris and remove it (*macro* means "large"; *phagein* means "to eat").

macular degeneration a common, progressive loss of function of the part of the retina that is most crucial to focused vision. This degeneration often leads to blindness.

major minerals essential mineral nutrients required in the adult diet in amounts greater than 100 milligrams per day. Also called *macrominerals.*

malnutrition any condition caused by excess or deficient food energy or nutrient intake or by an imbalance of nutrients. Nutrient or energy deficiencies are forms of undernutrition; nutrient or energy excesses are forms of overnutrition.

maltose a disaccharide composed of two glucose units; sometimes known as malt sugar.

malt syrup a sweetener made from sprouted barley.

maple syrup a concentrated solution of sucrose derived from the sap of the sugar maple tree. This sugar was once common but is now usually replaced by sucrose and artificial maple flavoring.

marasmus (ma-RAZ-mus) severe malnutrition characterized by poor growth, dramatic weight loss, loss of body fat and muscle, and apathy.

marginal food security a descriptor for households with problems or anxiety at times about accessing adequate food, but the quality, variety, or quantity of their food intake were not substantially reduced. Example: a parent worried that the food purchased would not last until the next paycheck.

margin of safety in reference to food additives, a zone between the concentration normally used and that at which a hazard exists. For common table salt, for example, the margin of safety is 1/5 (five times the amount normally used would be hazardous).

medical foods foods specially manufactured for use by people with medical disorders and administered on the advice of a physician.

medical nutrition therapy nutrition services used in the treatment of injury, illness, or other conditions; includes assessment of nutrition status and dietary intake and corrective applications of diet, counseling, and other nutrition services.

metabolic syndrome a combination of characteristic factors—high fasting blood glucose or insulin resistance, central obesity, hypertension, low blood HDL cholesterol, and elevated blood triglycerides—that greatly increase a person's risk of developing CVD. Also called *insulin resistance syndrome.*

metabolic water water generated in the tissues during the chemical breakdown of the energy-yielding nutrients in foods.

metabolism the sum of all physical and chemical changes taking place in living cells; includes all reactions by which the body obtains and spends the energy from food.

metastasis (meh-TASS-ta-sis) movement of cancer cells from one body part to another, usually by way of the body fluids.

methanol an alcohol produced in the body continually by all cells.

methyl groups small carbon-containing molecules that, among their activities, silence genes when applied to DNA strands by enzymes.

methylmercury any toxic compound of mercury to which a characteristic chemical structure, a methyl group, has been added, usually by bacteria in aquatic sediments. Methylmercury is readily absorbed from the intestine and causes nerve damage in people.

microbes a shortened name for *microorganisms*; minute organisms too small to observe without a microscope, including bacteria, viruses, and others.

microvilli (MY-croh-VILL-ee, MY-croh-VILL-eye) tiny, hairlike projections on each cell of every villus that greatly expand the surface area available to trap nutrient particles and absorb them into the cells (*singular*: microvillus).

milk anemia iron-deficiency anemia caused by drinking so much milk that iron-rich foods are displaced from the diet.

minerals naturally occurring, inorganic, homogeneous substances; chemical elements.

mineral water water from a spring or well that typically contains at least 250 parts per million (ppm) of naturally occurring minerals. Minerals give water a distinctive flavor. Many mineral waters are high in sodium.

miso fermented soybean paste used in Japanese cooking. Soy products are considered to be functional foods.

moderate drinkers people who do not drink excessively and do not behave inappropriately because of alcohol. A moderate drinker's health may or may not be harmed by alcohol over the long term.

moderation the dietary characteristic of providing constituents within set limits, not to excess.

modified atmosphere packaging (MAP) a technique used to extend the shelf life of perishable foods; the food is packaged in a gas-impermeable container from which air is removed or to which an oxygen-free gas mixture, such as carbon dioxide and nitrogen, is added.

molasses a thick brown syrup left over from the refining of sucrose from sugar cane. The major nutrient in molasses is iron, a contaminant from the machinery used in processing it.

monoglycerides (mon-oh-GLISS-er-ides) products of the digestion of lipids; a monoglyceride is a glycerol molecule with one fatty acid attached (*mono* means "one"; *glyceride* means "a compound of glycerol").

monosaccharides (mon-oh-SACK-ah-rides) single sugar units (*mono* means "one"; *saccharide* means "sugar unit").

monounsaturated fats triglycerides in which most of the fatty acids have one point of unsaturation (are monounsaturated).

monounsaturated fatty acid a fatty acid containing one point of unsaturation.

MSG symptom complex the acute, temporary, and self-limiting reactions, including burning sensations or flushing of the skin with pain and headache, experienced by sensitive people upon ingesting a large dose of MSG. Formerly called *Chinese restaurant syndrome.*

mucus (MYOO-cus) a slippery coating of the digestive tract lining (and other body linings) that protects the cells from exposure to digestive juices (and other destructive agents). The adjective form is *mucous* (same pronunciation). The digestive tract lining is a *mucous membrane.*

multi-grain a term used on food labels to indicate a food made with more than one kind of grain. Not an indicator of a whole-grain food.

muscle endurance the ability of a muscle to contract repeatedly within a given time without becoming exhausted. This muscle characteristic develops with increasing repetition rather than increasing workload and is associated with cardiorespiratory endurance.

muscle fatigue diminished force and power of muscle contractions despite consistent or increasing conscious effort to perform a physical activity; muscle fatigue may result from depleted glucose or oxygen supplies or other causes.

muscle power the efficiency of a muscle contraction, measured by force and time.

muscle strength the ability of muscles to overcome physical resistance. This muscle characteristic develops with increasing workload rather than repetition and is associated with muscle size.

mutation a permanent, heritable change in an organism's DNA.

myoglobin (MYE-oh-globe-in) the oxygen-holding protein of the muscles (*myo* means "muscle").

N

natural foods a term that has no legal definition but is often used to imply wholesomeness.

naturally occurring sugars sugars that are not added to a food but are present as its original constituents, such as the sugars of fruit or milk.

natural water water obtained from a spring or well that is certified to be safe and sanitary. The mineral content may not be changed, but the water may be treated in other ways such as with ozone or by filtration.

nectars concentrated peach nectar, pear nectar, or others.

nephrons (NEFF-rons) the working units in the kidneys, consisting of intermeshed blood vessels and tubules.

neural tube the embryonic tissue that later forms the brain and spinal cord.

neural tube defect (NTD) a group of abnormalities of the brain and spinal cord apparent at birth and caused by interruption of the normal early development of the neural tube.

neurotoxins poisons that act upon the cells of the nervous system.

neurotransmitters chemicals that are released at the end of a nerve cell when a nerve impulse arrives there. They diffuse across the gap to the next cell and alter the membrane of that second cell to either inhibit or excite it.

niacin a B vitamin needed in energy metabolism. Niacin can be eaten preformed or made in the body from tryptophan, one of the amino acids. Other forms of niacin are *nicotinic acid, niacinamide*, and *nicotinamide.*

niacin equivalents (NE) the amount of niacin present in food, including the niacin that can theoretically be made from its precursor tryptophan that is present in the food.

night blindness slow recovery of vision after exposure to flashes of bright light at night; an early symptom of vitamin A deficiency.

night eating syndrome a disturbance in the daily eating rhythm associated with obesity, characterized by no breakfast, more than half of the daily calories consumed after 7 p.m., frequent nighttime awakenings to eat, and often a greater total caloric intake than others.

nitrogen balance the amount of nitrogen consumed compared with the amount excreted in a given time period.

nonalcoholic a term used on beverage labels, such as wine or beer, indicating that the product contains less than 0.5% alcohol. The terms *dealcoholized* and *alcohol removed* mean the same thing. *Alcohol free* means that the product contains no detectable alcohol.

nonheme iron dietary iron not associated with hemoglobin; the iron of plants and other sources.

nonnutritive sweeteners sweet-tasting synthetic or natural food additives that offer sweet flavor but with negligible or no calories per serving; also called *artificial sweeteners, intense sweeteners, noncaloric sweeteners,* and *very low-calorie sweeteners.*

norepinephrine (NOR-EP-ih-NEFF-rin) a compound related to epinephrine that helps to elicit the stress response.

nori a type of seaweed popular in Asian, particularly Japanese, cooking.

nucleotide (NU-klee-oh-tied) one of the subunits from which DNA and RNA are composed.

nutraceutical a term that has no legal or scientific meaning but is sometimes used to refer to foods, nutrients, or dietary supplements believed to have medicinal effects. Often used to sell unnecessary or unproven supplements.

nutrient claims claims using approved wording to describe the nutrient values of foods, such as the claim that a food is "high" in a desirable constituent, or "low" in an undesirable one.

nutrient density a measure of nutrients provided per calorie of food. A *nutrient-dense food* provides vitamins, minerals, and other beneficial substances with relatively few calories.

nutrients components of food that are indispensable to the body's functioning. They provide energy, serve as building material, help maintain or repair body parts, and support growth. The nutrients include water, carbohydrate, fat, protein, vitamins, and minerals.

nutrition the study of the nutrients in foods and in the body; sometimes also the study of human behaviors related to food.

nutritional genomics the science of how food (and its components) interacts with the genome.

nutritionally enhanced beverages flavored beverages that contain any of a number of nutrients, including some carbohydrate, along with protein, vitamins, minerals, herbs, or other unneeded substances. Such "enhanced waters" may not contain useful amounts of carbohydrate or electrolytes to support athletic competition or training.

Nutrition Facts on a food label, the panel of nutrition information required to appear on almost every packaged food. Grocers may also provide the information for fresh produce, meats, poultry, and seafood.

nutritionist someone who studies nutrition. Some nutritionists are RDs, whereas others are self-described experts whose training is questionable and who are not qualified to give advice. In states with responsible legislation, the term applies only to people who have master of science (MS) or doctor of philosophy (PhD) degrees from properly accredited institutions.

O

obesity overfatness with adverse health effects, as determined by reliable measures and interpreted with good medical judgment. Obesity is officially defined as a body mass index of 30 or higher.

oils lipids that are liquid at room temperature (70°F or 21°C).

olestra a noncaloric artificial fat made from sucrose and fatty acids; formerly called *sucrose polyester*. A trade name is *Olean.*

omega-3 fatty acid a polyunsaturated fatty acid with its endmost double bond three carbons from the end of the carbon chain. Linolenic acid is an example.

omega-6 fatty acid a polyunsaturated fatty acid with its endmost double bond six carbons from the end of the carbon chain. Linoleic acid is an example.

omnivores people who eat foods of both plant and animal origin, including animal flesh.

oral rehydration therapy (ORT) oral fluid replacement for children with severe diarrhea caused by infectious disease. ORT enables parents to mix a simple solution for their child from substances that they have at home. A simple recipe for ORT: $^{1}/_{2}$ L boiled water, 4 tsp sugar, $^{1}/_{2}$ tsp salt.

organic carbon containing. Four of the six classes of nutrients are organic: carbohydrate, fat, protein, and vitamins. Organic compounds include only those made by living things and do not include compounds such as carbon dioxide, diamonds, and a few carbon salts.

organic foods foods meeting strict USDA production regulations for *organic*, including prohibition of synthetic pesticides, herbicides, fertilizers, drugs, and preservatives and produced without genetic engineering or irradiation.

organic gardens gardens grown with techniques of *sustainable agriculture*, such as using fertilizers made from composts and introducing predatory insects to control pests, in ways that have minimal impact on soil, water, and air quality.

organosulfur compounds a large group of phytochemicals containing the mineral sulfur. Organosulfur phytochemicals are responsible for the pungent flavors and aromas of foods belonging to the onion, leek, chive, shallot, and garlic family and are thought to stimulate cancer defenses in the body.

organs discrete structural units made of tissues that perform specific jobs. Examples are the heart, liver, and brain.

osteomalacia (OS-tee-o-mal-AY-shuh) the adult expression of vitamin D–deficiency disease, characterized by an overabundance of unmineralized bone protein (*osteo* means "bone"; *mal* means "bad"). Symptoms include bending of the spine and bowing of the legs.

osteoporosis (OSS-tee-oh-pore-OH-sis) a reduction of the bone mass of older persons in which the bones become porous and fragile (*osteo* means "bones"; *poros* means "porous"); also known as *adult bone loss*.

outbreak two or more cases of a disease arising from an identical organism acquired from a common food source within a limited time frame. Government agencies track and investigate outbreaks of foodborne illnesses, but tens of millions of individual cases go unreported each year.

outcrossing the unintended breeding of a domestic crop with a related wild species.

oven-safe thermometer a thermometer designed to remain in the food to give constant readings during cooking.

overload an extra physical demand placed on the body; an increase in the frequency, duration, or intensity of an activity. A principle of training is that for a body system to improve, it must be worked at frequencies, durations, or intensities that increase by increments.

overweight body weight above a healthy weight; BMI 25 to 29.9 (BMI is defined later).

ovo-vegetarian includes eggs, vegetables, grains, legumes, fruits, and nuts; excludes flesh, seafood, and milk products.

ovum the egg, produced by the mother, that unites with a sperm from the father to produce a new individual.

oxidants compounds (such as oxygen itself) that oxidize other compounds. Compounds that prevent oxidation are called antioxidants, whereas those that promote it are called prooxidants (*anti* means "against"; *pro* means "for").

oxidation interaction of a compound with oxygen; in this case, a damaging effect by a chemically reactive form of oxygen.

oxidative stress damage inflicted on living systems by free radicals.

oyster shell a product made from the powdered shells of oysters that is sold as a calcium supplement but is not well absorbed by the digestive system.

P

pancreas an organ with two main functions. One is an endocrine function—the making of hormones such as insulin, which it releases directly into the blood (*endo* means "into" the blood). The other is an exocrine function—the making of digestive enzymes, which it releases through a duct into the small intestine to assist in digestion (*exo* means "out" into a body cavity or onto the skin surface).

pancreatic juice fluid secreted by the pancreas that contains both enzymes to digest carbohydrates, fats, and proteins and sodium bicarbonate, a neutralizing agent.

pantothenic (PAN-to-THEN-ic) **acid** a B vitamin.

partial vegetarian a term sometimes used to mean an eating style that includes seafood, poultry, eggs, dairy products, vegetables, grains, legumes, fruits, and nuts; excludes or strictly limits certain meats, such as red meats. Also called *semi-vegetarian.*

pasteurization the treatment of milk, juices, or eggs with heat sufficient to kill certain pathogens (disease-causing microbes) commonly transmitted through these foods; not a sterilization process. Pasteurized products retain bacteria that cause spoilage.

PCBs stable oily synthetic chemicals once used in hundreds of U.S. industrial operations that persist today in underwater sediments and contaminate fish and shellfish. Now banned from use in the United States, PCBs circulate globally from areas where they are still in use. PCBs cause cancer, nervous system damage, immune dysfunction, and a number of other serious health effects.

peak bone mass the highest attainable bone density for an individual; developed during the first three decades of life.

pellagra (pell-AY-gra) the niacin-deficiency disease (*pellis* means "skin"; *agra* means "rough"). Symptoms include the "4 Ds": diarrhea, dermatitis, dementia, and, ultimately, death.

peptide bond a bond that connects one amino acid with another, forming a link in a protein chain.

performance nutrition an area of nutrition science that applies its principles to maintaining health and maximizing physical performance in athletes, firefighters, military personnel, and others who must perform at high levels of physical ability.

peripheral resistance the resistance to pumped blood in the small arterial branches (arterioles) that carry blood to tissues.

peristalsis (perri-STALL-sis) the wavelike muscular squeezing of the esophagus, stomach, and small intestine that pushes their contents along.

pernicious (per-NISH-us) **anemia** a vitamin B_{12}–deficiency disease, caused by lack of intrinsic factor and characterized by large, immature red blood cells and damage to the nervous system (*pernicious* means "highly injurious or destructive").

persistent of a stubborn or enduring nature; with respect to food contaminants, the quality of remaining unaltered and unexcreted in plant foods or in the bodies of animals and human beings.

pesticides chemicals used to control insects, diseases, weeds, fungi, and other pests on crops and around animals. Used broadly, the term includes *herbicides* (to kill weeds), *insecticides* (to kill insects), and *fungicides* (to kill fungi).

pH a measure of acidity on a point scale. A solution with a pH of 1 is a strong acid; a solution with a pH of 7 is neutral; a solution with a pH of 14 is a strong base.

phagocytes (FAG-oh-sites) white blood cells that can ingest and destroy antigens. The process by which phagocytes engulf materials is called *phagocytosis.* The Greek word *phagein* means "to eat."

phenylketonuria (PKU) an inborn error of metabolism that interferes with the body's handling of the amino acid phenylalanine, with potentially serious consequences to the brain and nervous system in infancy and childhood.

phospholipids (FOSS-foh-LIP-ids) one of the three main classes of dietary lipids. These lipids are similar to triglycerides, but each has a phosphorus-containing acid in place of one of the fatty acids. Phospholipids are present in all cell membranes.

photosynthesis the process by which green plants make carbohydrates from carbon dioxide and water using the green pigment chlorophyll to capture the sun's energy (*photo* means "light"; *synthesis* means "making").

physical activity bodily movement produced by muscle contractions that substantially increase energy expenditure.

phytates (FYE-tates) compounds present in plant foods (particularly whole grains) that bind iron and may prevent its absorption.

phytochemicals (FIGH-toe-CHEM-ih-cals) compounds in plants that confer color, taste, and other characteristics. Often, the bioactive food components of functional foods. *Phyto* means "plant."

phytoestrogens (FIGH-toe-ESS-troh-gens) phytochemicals structurally similar to the female sex hormone estrogen. Phytoestrogens weakly mimic estrogen or modulate hormone activity in the human body.

pica (PIE-ka) a craving and intentional consumption of nonfood substances. Also known as *geophagia* (gee-oh-FAY-gee-uh) when referring to clay eating and *pagophagia* (pag-oh-FAY-gee-uh) when referring to ice craving (*geo* means "earth"; *pago* means "frost"; *phagia* means "to eat").

placebo a sham treatment often used in scientific studies; an inert harmless medication. The *placebo effect* is the healing effect that the act of treatment, rather than the treatment itself, often has.

placenta (pla-SEN-tuh) the organ of pregnancy in which maternal and fetal blood circulate in close proximity and exchange nutrients and oxygen (flowing into the fetus) and wastes (picked up by the mother's blood).

plant pesticides substances produced within plant tissues that kill or repel attacking organisms.

plant sterols phytochemicals that resemble cholesterol in structure but that lower blood cholesterol by interfering with cholesterol absorption in the intestine. Plant sterols include sterol esters and stanol esters, formerly called *phytosterols.*

plaque (PLACK) a mass of microorganisms and their deposits on the surfaces of the teeth, a forerunner of dental caries and gum disease. The term *plaque* is also used in another connection—arterial plaque in atherosclerosis.

plaques (PLACKS) mounds of lipid material mixed with smooth muscle cells and calcium that develop in the artery walls in atherosclerosis (*placken* means "patch"). The same word is also used to describe the accumulation of a different kind of deposits on teeth, which promote dental caries.

plasma the cell-free fluid part of blood and lymph.

platelets tiny cell-like fragments in the blood, important in blood clot formation (*platelet* means "little plate").

point of unsaturation a site in a molecule where the bonding is such that additional hydrogen atoms can easily be attached.

polypeptide (POL-ee-PEP-tide) protein fragments of many (more than 10) amino acids bonded together (*poly* means "many"). A peptide is a strand of amino acids.

polysaccharides another term for complex carbohydrates; compounds composed of long strands of glucose units linked together (*poly* means "many"). Also called *complex carbohydrates*.

polyunsaturated fats triglycerides in which most of the fatty acids have two or more points of unsaturation (are polyunsaturated).

polyunsaturated fatty acid (PUFA) a fatty acid with two or more points of unsaturation.

pop-up thermometer a disposable timing device commonly used in turkeys. The center of the device contains a stainless steel spring that "pops up" when food reaches the right temperature.

prebiotic a substance that may not be digestible by the host, such as fiber, but that serves as food for probiotic bacteria and thus promotes their growth.

precursors compounds that can be converted into active vitamins. Also called *provitamins*.

prediabetes condition in which blood glucose levels are higher than normal but not high enough to be diagnosed as diabetes; a major risk factor for diabetes and cardiovascular diseases.

preeclampsia (PRE-ee-CLAMP-see-ah) a potentially dangerous condition during pregnancy characterized by hypertension and protein in the urine.

pregame meal a meal consumed in the hours before prolonged or repeated athletic training or competition to boost the glycogen stores of endurance athletes.

prehypertension borderline blood pressure between 120 over 80 and 139 over 89 millimeters of mercury, an indication that hypertension is likely to develop in the future.

premenstrual syndrome (PMS) a cluster of symptoms that some women experience prior to and during menstruation. They include, among others, abdominal cramps, back pain, swelling, headache, painful breasts, and mood changes.

prenatal (pree-NAY-tal) before birth.

prenatal supplements nutrient supplements specifically designed to provide the nutrients needed during pregnancy, particularly folate, iron, and calcium, without excesses or unneeded constituents.

pressure ulcers damage to the skin and underlying tissues as a result of unrelieved compression and poor circulation to the area; also called *bed sores*.

prion (PREE-on) a disease agent consisting of an unusually folded protein that disrupts normal cell functioning. Prions cannot be controlled or killed by cooking or disinfecting, nor can the disease they cause be treated; prevention is the only form of control.

probiotic a live microorganism which, when administered in adequate amounts, alters the bacterial colonies of the body in ways believed to confer a health benefit on the host.

problem drinkers or **alcohol abusers** people who suffer social, emotional, family, job-related, or other problems because of alcohol. A problem drinker is on the way to alcoholism.

processed foods foods subjected to any process, such as milling, alteration of texture, addition of additives, cooking, or others. Depending on the starting material and the process, a processed food may or may not be nutritious.

progressive weight training the gradual increase of a workload placed upon the body with the use of resistance.

promoters factors such as certain hormones that do not initiate cancer but speed up its development once initiation has taken place.

proof a statement of the percentage of alcohol in an alcoholic beverage. Liquor that is 100 proof is 50% alcohol, 90 proof is 45%, and so forth.

prooxidant a compound that triggers reactions involving oxygen.

protease (PRO-tee-ace) any of a number of enzymes that break the chemical bonds of proteins.

proteins compounds composed of carbon, hydrogen, oxygen, and nitrogen and arranged as strands of amino acids. Some amino acids also contain the element sulfur.

protein-sparing action the action of carbohydrate and fat in providing energy that allows protein to be used for purposes it alone can serve.

protein turnover the continuous breakdown and synthesis of body proteins involving the recycling of amino acids.

public health nutritionist a dietitian or other person with an advanced degree in nutrition who specializes in public health nutrition.

public water water from a municipal or county water system that has been treated and disinfected. Also called *tap water*.

purified water water that has been treated by distillation or other physical or chemical processes that remove dissolved solids. Because purified water contains no minerals or contaminants, it is useful for medical and research purposes.

pyloric (pye-LORE-ick) **valve** the circular muscle of the lower stomach that regulates the flow of partly digested food into the small intestine. Also called *pyloric sphincter*.

R

raw sugar the first crop of crystals harvested during sugar processing. Raw sugar cannot be sold in the United States because it contains too much filth (dirt, insect fragments, and the like). Sugar sold as "raw sugar" is actually evaporated cane juice.

reaction time the interval between stimulation and response.

ready-to-use therapeutic food (RUTF) highly caloric food products offering carbohydrate, lipid, protein, and micronutrients in a soft-textured paste used to promote rapid weight gain in malnourished people, particularly children.

recombinant bovine somatotropin (so-mat-oh-TROPE-in) **(rbST)** growth hormone of cattle, which can be produced for agricultural use by genetic engineering. Also called *bovine growth hormone (bGH)*.

recombinant DNA (rDNA) technology a technique of genetic modification whereby scientists directly manipulate the genes of living things; includes methods of removing genes, doubling genes, introducing foreign genes, and changing gene positions to influence the growth and development of organisms.

Recommended Dietary Allowances (RDA) nutrient intake goals for individuals; the average daily nutrient intake level that meets the needs of nearly all (97 percent to 98 percent) healthy people in a particular life stage and gender group. Derived from the Estimated Average Requirements (see below).

recovery drinks flavored beverages that contain protein, carbohydrate, and often other nutrients; intended to support postexercise recovery of energy fuels and muscle tissue. These can be convenient, but are not superior to ordinary foods and beverages, such as chocolate milk or a sandwich, to supply carbohydrate and protein after exercise. Not intended for hydration during athletic competition or training because their high carbohydrate and protein contents may slow water absorption.

reference dose an estimate of the intake of a substance over a lifetime that is considered to be without appreciable health risk; for pesticides, the maximum amount of a residue permitted in a food. Formerly called *tolerance limit*.

refined refers to the process by which the coarse parts of food products are removed. For example, the refining of wheat into white enriched flour involves removing three of the four parts of the kernel—the chaff, the bran, and the germ—leaving only the endosperm, composed mainly of starch and a little protein.

refined grains grains and grain products from which the bran, germ, or other edible parts of whole grains have been removed; not a whole grain. Many refined grains are low in fiber and are enriched with vitamins as required by U.S. regulations.

registered dietitian (RD) food and nutrition experts who have earned at least a bachelor's degree from an accredited college or university with a program approved by the Academy of Nutrition and Dietetics (or the Dietitians of Canada). The dietitian must also serve in an approved internship or coordinated program, pass the registration examination, and maintain professional competency through continuing education. Many states also require licensing of practicing dietitians.

registration listing with a professional organization that requires specific course work, experience, and passing of an examination.

requirement the amount of a nutrient that will just prevent the development of specific deficiency signs; distinguished from the DRI recommended intake value, which is a generous allowance with a margin of safety.

residues whatever remains; in the case of pesticides, those amounts that remain on or in foods when people buy and use them.

resistance training physical activity that develops muscle strength, power, endurance, and mass. Resistance can be provided by free weights, weight machines, other objects, or the person's own body weight. Also called *weight training* or *muscular strength exercises*.

resistant starch the fraction of starch in a food that is digested slowly, or not at all, by human enzymes.

resveratrol (rez-VER-ah-trol) a phytochemical of grapes under study for potential health benefits.

retina (RET-in-uh) the layer of light-sensitive nerve cells lining the back of the inside of the eye.

retinol one of the active forms of vitamin A made from beta-carotene in animal and human bodies; an antioxidant nutrient. Other active forms are *retinal* and *retinoic acid*.

retinol activity equivalents (RAE) a new measure of the vitamin A activity of beta-carotene and other vitamin A precursors that reflects the amount of retinol that the body will derive from a food containing vitamin A precursor compounds.

rhodopsin (roh-DOP-sin) the light-sensitive pigment of the cells in the retina; it contains vitamin A (*opsin* means "visual protein").

riboflavin (RIBE-o-flay-vin) a B vitamin active in the body's energy-releasing mechanisms.

rickets the vitamin D–deficiency disease in children; characterized by abnormal growth of bone and manifested in bowed legs or knock-knees, outward-bowed chest, and knobs on the ribs.

risk factors factors known to be related to (or correlated with) diseases but not proved to be causal.

S

safety the practical certainty that injury will not result from the use of a substance.

salts compounds composed of charged particles (ions). An example is potassium chloride (K^+Cl^-).

sarcopenia (SAR-koh-PEE-nee-ah) age-related loss of skeletal muscle mass, muscle strength, and muscle function.

satiation (SAY-she-AY-shun) the perception of fullness that builds throughout a meal, eventually reaching the degree of fullness and satisfaction that halts eating. Satiation generally determines how much food is consumed at one sitting.

satiety (sah-TIE-eh-tee) the perception of fullness that lingers in the hours after a meal and inhibits eating until the next mealtime. Satiety generally determines the length of time between meals.

saturated fats triglycerides in which most of the fatty acids are saturated.

saturated fatty acid a fatty acid carrying the maximum possible number of hydrogen atoms (having no points of unsaturation). A saturated fat is a triglyceride that contains three saturated fatty acids.

screen time sedentary time spent using an electronic device, such as a television, computer, or video game player.

scurvy the vitamin C–deficiency disease.

selective breeding a technique of genetic modification whereby organisms are chosen for reproduction based on their desirability for human purposes, such as high growth rate, high food yield, or disease resistance, with the intention of retaining or enhancing these characteristics in their offspring.

self-efficacy a person's belief in his or her ability to succeed in an undertaking.

senile dementia the loss of brain function beyond the normal loss of physical adeptness and memory that occurs with aging.

serotonin (SARE-oh-TONE-in) a compound related in structure to (and made from) the amino acid tryptophan. It serves as one of the brain's principal neurotransmitters.

set-point theory a theory stating that the body's regulatory controls tend to maintain a particular body weight (the set point) over time, opposing efforts to lose weight by dieting.

severe acute malnutrition (SAM) malnutrition caused by recent severe food restriction; characterized in children by underweight for height (wasting). *Moderate acute malnutrition* is a somewhat less severe form.

side chain the unique chemical structure attached to the backbone of each amino acid that differentiates one amino acid from another.

simple carbohydrates sugars, including both single sugar units and linked pairs of sugar units. The basic sugar unit is a molecule containing six carbon atoms, together with oxygen and hydrogen atoms.

single-use temperature indicator a type of instant-read thermometer that changes color to indicate that the food has reached the desired temperature. Discarded after one use, they are often used in commercial food establishments to eliminate cross-contamination.

skinfold test measurement of the thickness of a fold of skin and subcutaneous fat on the back of the arm (over the triceps muscle), below the shoulder blade (subscapular), or in other places, using a caliper; also called *fatfold test*.

small intestine the 20-foot length of small-diameter intestine, below the stomach and above the large intestine, that is the major site of digestion of food and absorption of nutrients.

smoking point the temperature at which fat gives off an acrid blue gas.

SNP a single misplaced nucleotide in a gene that causes formation of an altered protein. The letters SNP stand for *single nucleotide polymorphism*.

soft water water with a high sodium concentration.

solid fats fats that are high in saturated fatty acids and are usually solid at room temperature. Solid fats are found naturally in most animal foods but also can be made from vegetable oils through hydrogenation.

soluble fibers food components that readily dissolve in water and often impart gummy or gel-like characteristics to foods. An example is pectin from fruit, which is used to thicken jellies.

solvent a substance that dissolves another and holds it in solution.

soy milk a milklike beverage made from soybeans, claimed to be a functional food. Soy drinks should be fortified with vitamin A, vitamin D, riboflavin, and calcium to approach the nutritional equivalency of milk.

Special Supplemental Nutrition Program for Women, Infants, and Children (WIC) a USDA program offering low-income pregnant and lactating women and those with infants or preschool children coupons redeemable for specific foods that supply the nutrients deemed most necessary for growth and development. For more information, visit www.usda.gov/FoodandNutrition.

sphincter (SFINK-ter) a circular muscle surrounding, and able to close, a body opening.

spina bifida (SPY-na BIFF-ih-duh) one of the most common types of neural tube defects in which gaps occur in the bones of the spine. Often the spinal cord bulges and protrudes through the gaps, resulting in a number of motor and other impairments.

sports drinks flavored beverages designed to help athletes replace fluids and electrolytes and to provide carbohydrate before, during, and after physical activity, particularly endurance activities.

spring water water originating from an underground spring or well. It may be bubbly (carbonated) or "flat" or "still," meaning not carbonated. Brand names such as "Spring Pure" do not necessarily mean that the water comes from a spring.

staple foods foods used frequently or daily, for example, rice (in East and Southeast Asia) or potatoes (in Ireland). If well chosen, these foods are nutritious.

starch a plant polysaccharide composed of glucose. After cooking, starch is highly digestible by human beings; raw starch often resists digestion.

stem cell an undifferentiated cell that can mature into any of a number of specialized cell types. A stem cell of bone marrow may mature into one of many kinds of blood cells, for example.

sterols (STEER-alls) one of the three main classes of dietary lipids. Sterols have a structure similar to that of cholesterol.

stomach a muscular, elastic, pouchlike organ of the digestive tract that grinds and churns swallowed food and mixes it with acid and enzymes, forming chyme.

stone ground refers to a milling process using limestone to grind any grain, including refined grains, into flour.

stone-ground flour flour made by grinding kernels of grain between heavy wheels made of limestone, a kind of rock derived from the shells and bones of marine animals. As the stones scrape together, bits of the limestone mix with the flour, enriching it with calcium.

stroke the sudden shutting off of the blood flow to the brain by a thrombus, embolism, or the bursting of a vessel (hemorrhage).

stroke volume the volume of oxygenated blood ejected from the heart toward body tissues at each beat.

structure-function claim a legal but largely unregulated claim permitted on labels of foods and dietary supplements, often mistaken by consumers for a regulated health claim.

subclinical deficiency a nutrient deficiency that has no outward clinical symptoms. Also called *marginal deficiency*.

subcutaneous fat fat stored directly under the skin (*sub* means "beneath"; *cutaneous* refers to the skin).

sucrose (SOO-crose) a disaccharide composed of glucose and fructose; sometimes known as table, beet, or cane sugar and, often, as simply *sugar*.

sugar alcohols sugarlike compounds in the chemical family *alcohol* derived from fruits or manufactured from sugar dextrose or other carbohydrates; sugar alcohols are absorbed more slowly than sugars, are metabolized differently, and do not elevate the risk of dental caries. Also called *polyols*.

sugars simple carbohydrates; that is, molecules of either single sugar units or pairs of those sugar units bonded together. By common usage, *sugar* most often refers to sucrose.

surface water water that comes from lakes, rivers, and reservoirs.

sushi a Japanese dish that consists of vinegar-flavored rice, seafood, and colorful vegetables, typically wrapped in seaweed. Some sushi contains raw fish; other sushi contains only cooked ingredients.

sustainable able to continue indefinitely; the use of resources in ways that maintain both natural resources and human life into the future; the use of natural resources at a pace that allows the earth to replace them and does not cause pollution to accumulate.

systolic (sis-TOL-ik) **pressure** the first figure in a blood pressure reading (the "dupp" sound of the heartbeat's "lubb-dupp" beat is heard), which reflects arterial pressure caused by the contraction of the heart's left ventricle.

T

tannins compounds in tea (especially black tea) and coffee that bind iron. Tannins also denature proteins.

T-cells lymphocytes that attack antigens. *T* stands for the thymus gland of the neck, where the T-cells are stored and matured.

textured vegetable protein processed soybean protein used in products formulated to look and taste like meat, fish, or poultry.

thermic effect of food the body's speeded-up metabolism in response to having eaten a meal; also called *diet-induced thermogenesis*.

thermogenesis the generation and release of body heat associated with the breakdown of body fuels. *Adaptive thermogenesis* describes adjustments in energy expenditure related to changes in environment such as cold and to physiological events such as underfeeding or trauma.

thiamin (THIGH-uh-min) a B vitamin involved in the body's use of fuels.

thrombosis a thrombus that has grown enough to close off a blood vessel. A *coronary thrombosis* closes off a vessel that feeds the heart muscle. A *cerebral thrombosis* closes off a vessel that feeds the brain (*coronary* means "crowning" [the heart]; *thrombo* means "clot"; the cerebrum is part of the brain).

thrombus a stationary blood clot.

thyroxine (thigh-ROX-in) a principal peptide hormone of the thyroid gland that regulates the body's rate of energy use.

tissues systems of cells working together to perform specialized tasks. Examples are muscles, nerves, blood, and bone.

tocopherol (tuh-KOFF-er-all) a kind of alcohol. The active form of vitamin E is alpha-tocopherol.

tofu (TOE-foo) a curd made from soybeans that is rich in protein, often enriched with calcium, and variable in fat content; used in many Asian and vegetarian dishes in place of meat.

Tolerable Upper Intake Levels (UL) the highest average daily nutrient intake level that is likely to pose no risk of toxicity to almost all healthy individuals of a particular life stage and gender group. Usual intake above this level may place an individual at risk of illness from nutrient toxicity.

toxicity the ability of a substance to harm living organisms. All substances, even pure water or oxygen, can be toxic in high enough doses.

trabecular (tra-BECK-you-lar) **bone** the weblike structure composed of calcium-containing crystals inside a bone's solid outer shell. It provides strength and acts like a calcium storage bank.

trace minerals essential mineral nutrients required in the adult diet in amounts less than 100 milligrams per day. Also called *microminerals*.

training regular practice of an activity, which leads to physical adaptations of the body with improvement in flexibility, strength, or endurance.

***trans* fats** fats that contain any number of unusual fatty acids—*trans*-fatty acids—formed during processing.

***trans*-fatty acids** fatty acids with unusual shapes that can arise when hydrogens are added to the unsaturated fatty acids of polyunsaturated oils (a process known as *hydrogenation*).

transgenic organism an organism resulting from the growth of an embryonic, stem, or germ cell into which a new gene has been inserted.

triglycerides (try-GLISS-er-ides) one of the three main classes of dietary lipids and the chief form of fat in foods and in the human body. A triglyceride is made up of three units of fatty acids and one unit of glycerol (*fatty acids* and *glycerol* are defined later). In research, triglycerides are often called *triacylglycerols* (try-ay-seal-GLISS-er-ols).

trimester a period representing gestation. A trimester is about 13 to 14 weeks.

tripeptides (try-PEP-tides) protein fragments that are three amino acids long (*tri* means "three").

turbinado (ter-bih-NOD-oh) **sugar** raw sugar from which the filth has been washed; legal to sell in the United States.

type 1 diabetes the type of diabetes in which the pancreas produces no or very little insulin; often diagnosed in childhood, although some cases arise in adulthood. Formerly called *juvenile-onset* or *insulin-dependent diabetes*.

type 2 diabetes the type of diabetes in which the pancreas makes plenty of insulin but the body's cells resist insulin's action; often diagnosed in adulthood. Formerly called *adult-onset* or *non–insulin-dependent diabetes*.

U

ulcer an erosion in the topmost, and sometimes underlying, layers of cells that form a lining. Ulcers of the digestive tract commonly form in the esophagus, stomach, or upper small intestine.

ultra-high temperature a process of sterilizing food by exposing it for a short time to temperatures above those normally used in processing.

umbilical (um-BIL-ih-cul) **cord** the ropelike structure through which the fetus's veins and arteries reach the placenta; the route of nourishment and oxygen into the fetus and the route of waste disposal from the fetus.

unbleached flour a beige-colored refined endosperm flour with texture and nutritive qualities that approximate those of regular white flour.

underweight body weight below a healthy weight; BMI below 18.5.

unsaturated fatty acid a fatty acid that lacks some hydrogen atoms and has one or more points of unsaturation. An unsaturated fat is a triglyceride that contains one or more unsaturated fatty acids.

urban legends stories, usually false, that may travel rapidly throughout the world via the Internet gaining strength of conviction solely on the basis of repetition.

urea (yoo-REE-uh) the principal nitrogen-excretion product of protein metabolism; generated mostly by removal of amine groups from unneeded amino acids or from amino acids being sacrificed to a need for energy.

uterus (YOO-ter-us) the womb, the muscular organ within which the infant develops before birth.

V

variety the dietary characteristic of providing a wide selection of foods—the opposite of monotony.

vegan a vegetarian who eats only food from plant sources: vegetables, grains, legumes, fruits, seeds, and nuts; also called strict vegetarian.

vegetarians people who exclude from their diets animal flesh and possibly other animal products such as milk, cheese, and eggs.

veins blood vessels that carry blood, with the carbon dioxide it has collected, from the tissues back to the heart.

very-low-density lipoproteins (VLDL) lipoproteins that transport triglycerides and other lipids from the liver to various tissues in the body.

very low food security a descriptor for households that, at times during the year, experienced disrupted eating patterns or reduced food intake of one or more household members because of a lack of money or other resources for food. Example: a family in which one or more members went to bed hungry, lost weight, or didn't eat for a whole day because they did not have enough food.

villi (VILL-ee, VILL-eye) fingerlike projections of the sheets of cells lining the intestinal tract. The villi make the surface area much greater than it would otherwise be (*singular*: villus).

visceral fat fat stored within the abdominal cavity in association with the internal abdominal organs; also called *intra-abdominal fat*.

viscous (VISS-cuss) having a sticky, gummy, or gel-like consistency that flows relatively slowly.

vitamin B_6 a B vitamin needed in protein metabolism. Its three active forms are *pyridoxine*, *pyridoxal*, and *pyridoxamine*.

vitamin B_{12} a B vitamin that helps to convert folate to its active form and also helps maintain the sheath around nerve cells. Vitamin B_{12}'s scientific name, not often used, is *cyanocobalamin*.

vitamins organic compounds that are vital to life and indispensable to body functions but are needed only in minute amounts; noncaloric essential nutrients.

vitamin water bottled water with a few vitamins added; does not replace vitamins from a balanced diet and may worsen overload in people receiving vitamins from enriched food, supplements, and other enriched products such as "energy" bars.

$VO_{2\,max}$ the maximum rate of oxygen consumption by an individual (measured at sea level).

voluntary activities intentional activities (such as walking, sitting, or running) conducted by voluntary muscles.

W

waist circumference a measurement of abdominal girth that indicates visceral fatness.

wasting the progressive, relentless loss of the body's tissues that accompanies certain diseases and shortens survival time.

water balance the balance between water intake and water excretion, which keeps the body's water content constant.

water intoxication a dangerous dilution of the body's fluids resulting from excessive ingestion of plain water. Symptoms are headache, muscular weakness, lack of concentration, poor memory, and loss of appetite.

water stress a measure of the pressure placed on water resources by human activities such as municipal water supplies, industries, power plants, and agricultural irrigation.

wean to gradually replace breast milk with infant formula or other foods appropriate to an infant's diet.

websites Internet resources composed of text and graphic files, each with a unique URL (Uniform Resource Locator) that names the site (for example, www.usda.gov).

weight cycling repeated rounds of weight loss and subsequent regain, with reduced ability to lose weight with each attempt; also called *yo-yo dieting*.

well water water drawn from groundwater by tapping into an aquifer.

Wernicke-Korsakoff (VER-nik-ee KOR-sah-koff) **syndrome** a cluster of symptoms involving nerve damage arising from a deficiency of the vitamin thiamin in alcoholism. Characterized by mental confusion, disorientation, memory loss, jerky eye movements, and staggering gait.

wheat bread bread made with any wheat flour, including refined enriched white flour.

wheat flour any flour made from wheat, including refined white flour.

whey (way) the watery part of milk, a by-product of cheese production. Once discarded as waste, whey is now recognized as a high-quality protein source for human consumption.

white flour an endosperm flour that has been refined and bleached for maximum softness and whiteness.

white sugar granulated sucrose, produced by dissolving, concentrating, and recrystallizing raw sugar. Also called *table sugar*.

white wheat a wheat variety developed to be paler in color than common red wheat (most familiar flours are made from red wheat). White wheat is similar to red wheat in carbohydrate, protein, and other nutrients, but it lacks the dark and bitter, but potentially beneficial, phytochemicals of red wheat.

whole foods milk and milk products; meats and similar foods such as fish and poultry; vegetables, including dried beans and peas; fruits; and grains. These foods are generally considered to form the basis of a nutritious diet. Also called *basic foods*.

100% whole grain a label term for food in which the grain is entirely whole grain, with no added refined grains.

whole grains grains or foods made from them that contain all the essential parts and naturally occurring nutrients of the entire grain seed (except the inedible husk).

whole-wheat flour flour made from whole-wheat kernels; a whole-grain flour. Also called *graham flour*.

world food supply the quantity of food, including stores from previous harvests, available to the world's people at a given time.

World Health Organization (WHO) an agency of the United Nations charged with improving human health and preventing or controlling diseases in the world's people.

World Wide Web (the Web, commonly abbreviated **www**) a graphical subset of the Internet.

X

xerophthalmia (ZEER-ahf-THALL-me-uh) progressive hardening of the cornea of the eye in advanced vitamin A deficiency that can lead to blindness (*xero* means "dry"; *ophthalm* means "eye").

xerosis (zeer-OH-sis) drying of the cornea; a symptom of vitamin A deficiency.

Z

zygote (ZYE-goat) the product of the union of ovum and sperm; a fertilized ovum.

Index

The page letters A, B, and C that stand alone refer to the tables beginning on the inside front cover. Page letters Y and Z refer to the tables on the last two pages of the book. The page numbers preceded by A through H are appendix page numbers. The boldfaced page numbers indicate definitions. Terms are also defined in the glossary. Page numbers followed by *n* indicate footnotes. Page numbers followed by *t* indicate tables. Page numbers followed by *f* indicate figures.

A

D

E

O

X

Y

Z

Diet Analysis PLUS+ Student Guide

Table of Contents

Getting Started

Diet Analysis Plus is a health and nutrition management program that allows you to create personal profiles, estimate activity level, track diet and exercise, and create detailed reports. The Diet Analysis Plus online system (also referred to as DA+) makes it easy for you to evaluate the types and serving sizes of the foods you consume.

*Tip:*Diet Analysis Plus includes a**Tutorial** which includes useful step-by-step information, such as how to create your primary profile and other basic features. This tutorial is always available from the **Tutorial** link at the top of the page.

Accessing Helpful Information

When you're working in a specific report from the **Reports** page, you can click the Report Title to bring up an **Info box** which explains the functions on the screen.

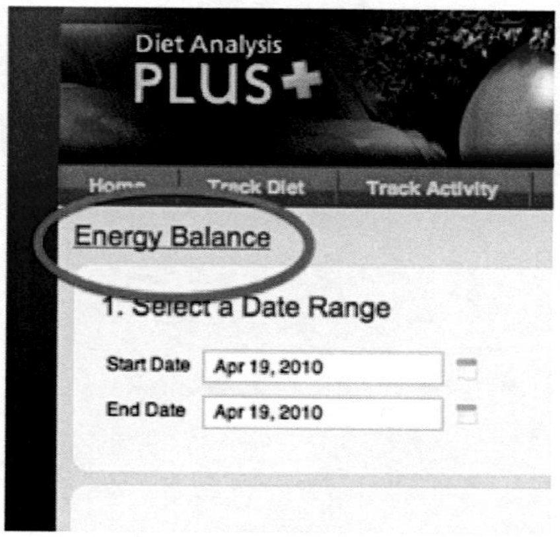

When you're working from the **Track Diet** page, you can click the Search Tips link to bring up a short **Help note** which contains an explanation of the report.

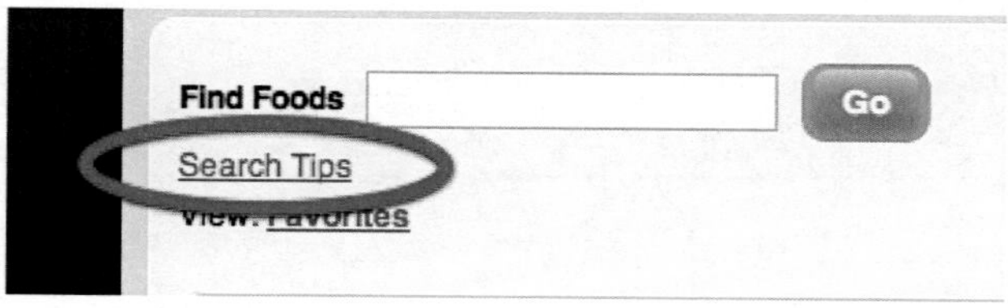

Entering a New Course Access Code

Your instructor can provide you with a **Course Access Code** that allows you to register for new or additional courses.

To register with a **Course Access Code:**.

Step 1. After you have received the **Course Access Code** from your instructor, you can log into DA+.

Step 2. Enter your Course Access Code in the **Course Access Code** text box.

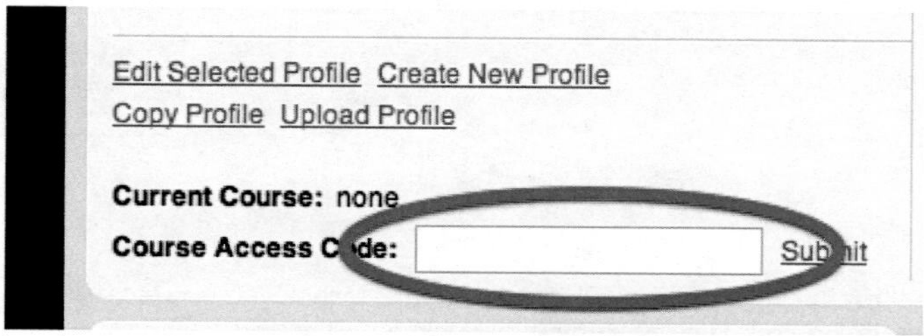

Step 3. Enter your new code exactly as it appears and click the Submit link. A confirmation text box will appear at the top of the page.

System Setup

To ensure the best experience with DA Plus and enjoy all of its features, please use the information in this section to optimize your computer system and browser settings. System requirements, browser configuration, and software conflicts are important issues when troubleshooting. Due to the constantly shifting nature of Web browsers, it is important to follow these recommendations, to take advantage of the most compatible configuration.

This section lists the basic hardware, software, and system settings you need to run Diet Analysis Plus. If your system meets the basic hardware needs, you can download any of the free software you need from the links in the following sections.

Windows System Requirements

- **Microsoft Windows** XP, Vista, or 7

- **Web browsers**: Firefox 3+, Chrome, and Internet Explorer 6, 7, or 8
- **Internet Connection**: Use of the Diet Analysis Plus web site requires an Internet connection speed of **56k** or higher
- To print tests, **Adobe Acrobat Reader** 4.05b or higher.
(Get Reader from get.adobe.com/reader/.)
- **Adobe Flash Player** 9 or higher is required.
(Get Player from http://get.adobe.com/flashplayer/.)

Macintosh System Requirements

- **Mac OS X** 10.4 or higher
- **Web browsers:** Firefox 3+ and Safari 3+
- **Adobe Flash Player** 9 or higher is required.
(Get Player from http://get.adobe.com/flashplayer/.)
- **Internet Connection**: Use of the Diet Analysis Plus web site requires an Internet connection speed of **56k** or higher

---

Linux System Requirements

- **Red Hat Linux 9.0** (or similar), X Windows System.
- **Web browser**: Firefox 3+
- **Adobe Acrobat Reader** 4.05b or higher is required to print tests.
(Get Reader from get.adobe.com/reader/.)
- **Adobe Flash Player** 9 or higher is required.
(Get Player from http://get.adobe.com/flashplayer/.)
- **Internet Connection**: Use of the Diet Analysis Plus web site requires an Internet connection speed of **56k** or higher

Setting Up Profiles

What is a Profile?

Nutritional requirements vary depending on an individual's height, weight, age, gender, and activity level. Your profile records this information and uses it to determine your Dietary Reference Intakes (DRIs) and to create custom reports. Before using Diet Analysis Plus, you must first create a profile.

Tip: Also see **Using Multiple Profiles** for more details on profile management.

Creating a Profile

Step 1. When you log into DA Plus for the first time, you will be asked to crate a primary profile. You can create multiple profiles for different analyses.

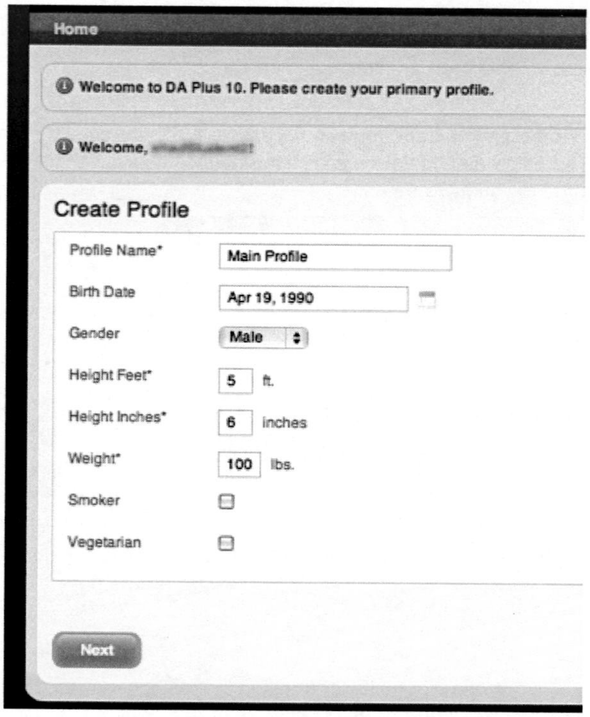

Step 2. Enter the information for your primary profile.

Step 3. When you are finished, click the **Next** button to see the **Activity Questionnaire** page.

Note: The**Long Activity Questionnaire** is the default when setting up a primary profile; you can use a shorter version when setting up additional profiles.

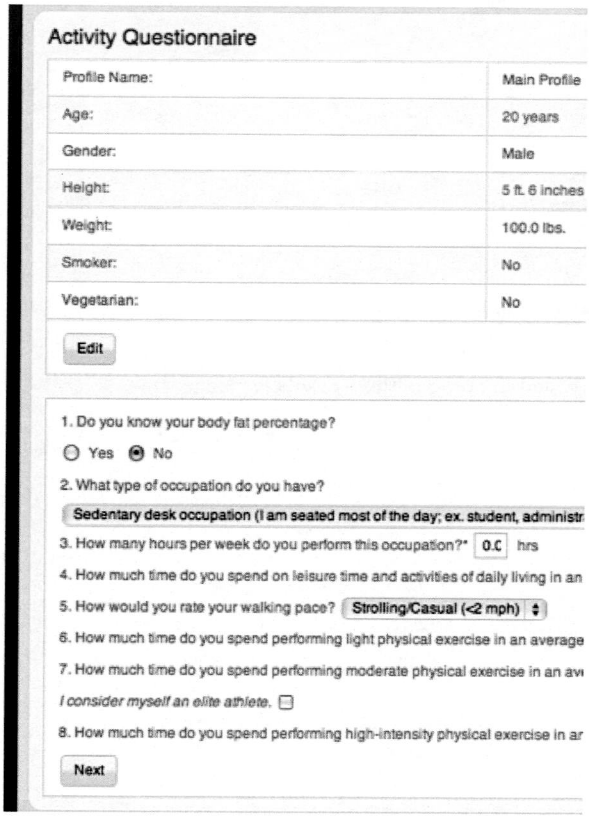

Step 4. Answer the questions on the questionnaire. The lifestyle related questions on this page will help the DA Plus system determine your activity level.

Step 5. When you are finished, click the **Next** button to see the **Confirm Profile** page.

Step 6. Click the **Edit** button(s) to change any of your information. When you are finished, click the **Save** button to save your profile.

After your primary profile is complete, you are directed to the **Home** page where you can access information on your profiles, food intakes, reports, and labs. To begin entering your dietary information, click the **Track Diet** tab on the main menu to find and list the foods you have eaten.

Viewing DRI Recommendations

After you have created a profile, Diet Analysis Plus determines recommendations for the **Dietary Reference Intakes** (DRI).

To see a profile's DRI recommendations:

Step 1. From the Home page, select the profile from the **Main Profile** menu in the **My Profile** pane.

Step 2. Click the **View Full DRI** button in the **Dietary Reference Intakes** pane to see the **Full DRI Page**.

Tip: You can edit the values in many of the fields on this page in order to customize the recommendations to your needs. For more information, see **Customizing DRI Recommendations**.

Managing an Existing Profile

Use the steps outlined below to view or edit an existing profile:

Step 1. From the Home page, select the profile from the **Main Profile** menu in the **My Profile** pane.

Step 2. Click the **Edit Selected Profile** link to makes changes to the profile and/or questionnaire.

Step 3. When you are finished, click **Save** to be returned to the **Home** page.

Importing a DA+ 9.0 Profile

To import a DA+ 9.0 profile:

Step 1. From the Home page, click the **Upload Profile** link to see the **Upload Profile** page.

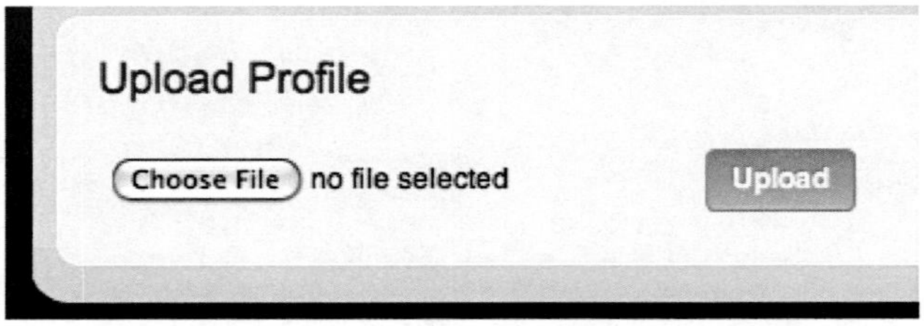

Step 4. Click the **Choose File** button and navigate to the **.dap** file you wish to upload.

Step 5. Once you selected the file, click the **Upload** button to import the profile. The newly uploaded profile will now appear in the **Active Profile** menu on the Home page.

Advanced Tips

+ Customizing DRI Recommendations

Diet Analysis Plus determines recommendations for each profile based on the Dietary Reference Intakes (DRI) published by the USDA. These recommendations are based on maintaining current weight. They do not take into account special needs based on athletic activity, health problems, weight loss goals, or other factors that may require variations for individuals.

Diet Analysis Plus is not a clinical tool, and will not create any recommendations that vary from the DRI. However, you may change many of the recommendations on the **View Full DRI** page.

To change the DRIs for a Profile:

Step 1. From the Home page, select the profile from the **Activity Profile** menu in the My Profile pang.

Step 2. Dietary Reference Intakes pane, click the **View Full DRI** button.

Step 3. Click the **Edit** button.

Step 4. Change the quantities in any editable field, then click **Save**. Your modified DRIs will now be used for the Recommendations reflected in all reports.

+ *Using Multiple Profiles*

A profile is two important things: a description of an individual, and the record of foods and activities recorded for that individual. You may need to use more than one more profile for yourself for several different reasons:

- Profile information changes (weight loss, pregnancy, age, etc.)
- Multiple class assignments with differing profile information for either yourself or a hypothetical profile
- Various profile settings and reports to show the effects of changes in diet and activity

---

+ *Deleting Profiles: CAUTION!*

A profile is not just the name and information in the **Profile** menu. A profile is also the record of all the foods recorded for that profile name in the **Track Diet** page, and all the activities recorded in the **Track Activities** page.

If you delete a profile, you will be deleting all the foods and activities as well as the name. Deleted profiles cannot be restored.

Tracking Your Diet

You can track your diet by recording everything you eat and drink during the day. Make sure you record everything you consume, including water, drinks, condiments, cooking oils, and alcoholic beverages.

Getting Started Tracking Your Diet

Step 1. On the main menu, click the **Track Diet** tab. The **Track Diet** page will open.

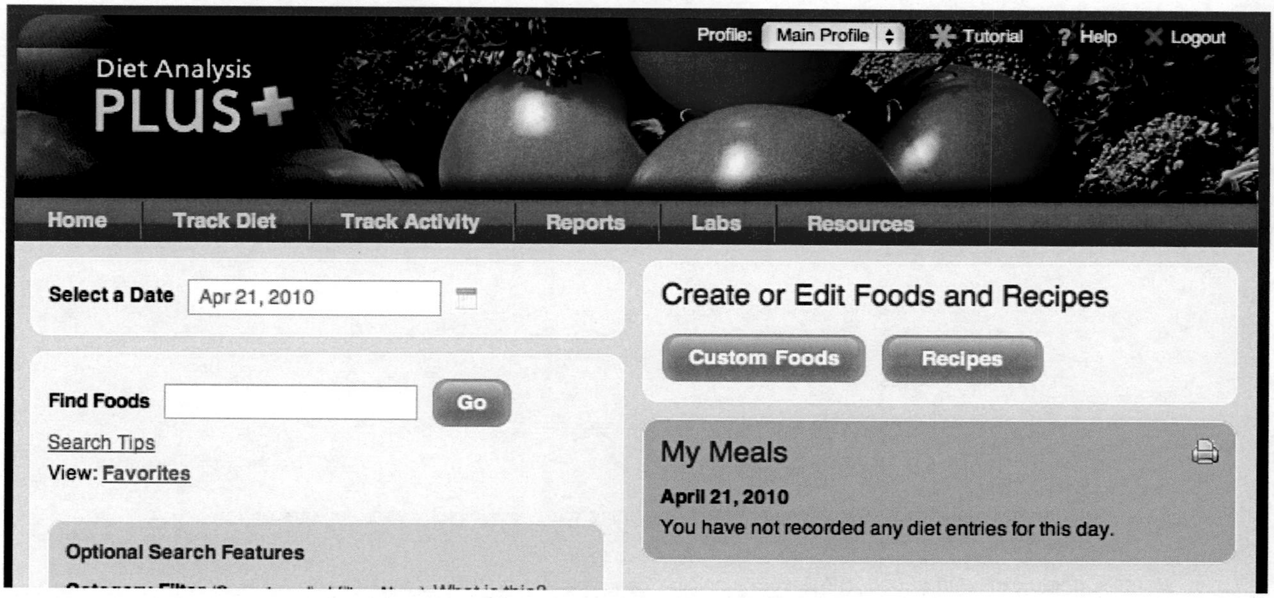

Step 2. Click the calendar icon next to the **Select a Date** field to open a drop-down **calendar**.

Step 3. Click a date to select it. The date you select will be highlighted.

Step 4. Use the arrows at the top to change the month or year.

Finding a Food

Step 1. In the **Find Foods** field, type the name of a food.

Step 2. Click the **Go** button.

A **Food List** will appear which provides a list of foods that contain the word or words you typed.

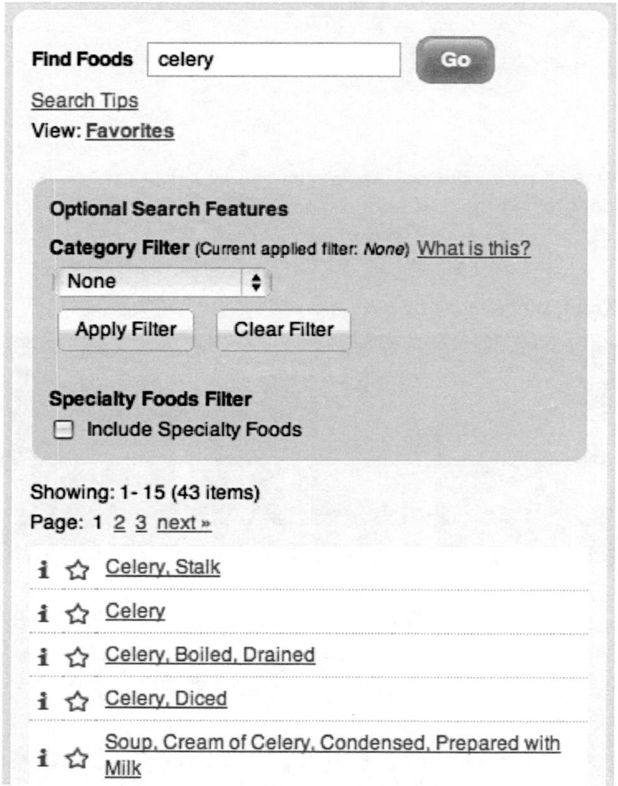

Search Tips:

• If you are looking for a brand-name food, try typing the name exactly as it appears on the packaging.

• If you can't find a food, check your spelling. If you are unsure how to spell or punctuate a food name, use just the first few letters of each word in the name. For example, "kell min whe" will find **"Kellogg's Mini-Wheats."**

• Ue the Category Filter and/or Specialty Foods Filter to fine-tune your search.

• If the list of foods is very long, try adding specifics. For example, "fried chicken" finds a shorter, more specific list than "chicken."

• Common foods come up first in a search, then generic foods, then brand names. To find your food, you may have to scroll through the list, using the page numbers and arrows that appear inside the **Food List** field.

• To scroll through a long list, use the page numbers and arrows that appear inside the Food List box.

• The **i** button to the left of each item opens a pop-up showing the nutrient information about the food. Sometimes this information can help you choose the right food.

Selecting Portions

Step 1. In the **Food List**, select the food name to open the **Serving Size** window for adding the food to a meal.

Step 2. Check the serving size of the food, the units of measurement, and the meal it was consumed.

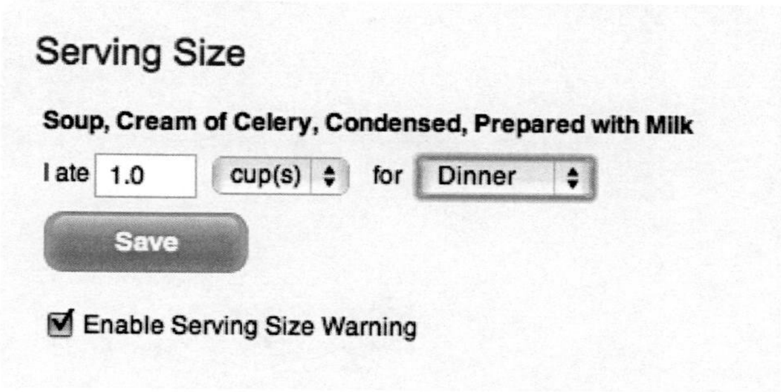

Careful! By default, all foods open showing the amount for one "serving," but servings are expressed in different units. Some food servings are "1 item." Others are "8 ounces" or "1 pint," or even "500 grams." If you are unsure what the serving size unit means, use the drop-down to change the unit of measurement. The quantity will automatically recalculate.

Step 3. Carefully estimate the amount of food you ate.

• **If you measured all the food in your meals** before you ate them and kept a record, this is an excellent practice. This is the most accurate way to track your diet.

• **If you did not measure all the food you ate**, estimate the amounts you ate. All the reports used in your assignments will be based on the information you enter about each food. Do your best to be realistic about your portion sizes. If you underestimate or overestimate, the reports you use for your assignments will be less reliable.

Step 4. Click the meal to which you want to add the food: **Breakfast**, **Lunch**, **Dinner**, or **Snack**.

Step 5. Click the **Save** button and you will see the food added to your **Food List**.

Step 6. Go to **Step 1**, and **Search** for your next food. Continue with these steps until you have listed everything you ate each day. Be as thorough as possible for best results.

Editing the My Meals Food List

Step 1. To **change** the amount, serving size, or meal of a food item, click the **Edit** button in the **My Meals** window.

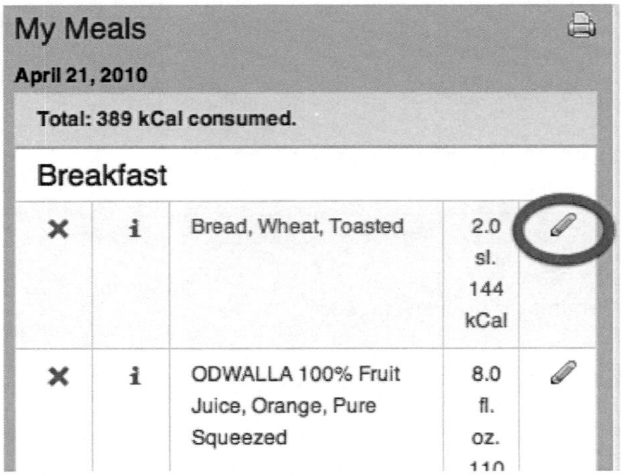

The **Serving Size** window will open.

Step 2. Make your changes as needed and click **Save**.

Step 3. To **delete** an item from the food list, click the ✖ **button** in the **My Meals** window.

Advanced Tips

Customize your list of foods by making a list of favorites, adding a food, or adding a recipe.

+ Making a Favorites List

You can create a "short list" of foods you eat frequently by adding them to your **Favorites** list.

Step 1. To **add** a food to your **Favorites** list, click the ☆ **button**.

Step 2. To **see** your **Favorites** list, click the **View: Favorites** link.

Tip: When you create a new food or recipe, it is automatically added to your **Favorites** list.

+ Creating Custom Foods

A **"Food"** is any food item for which nutritional information is available. To add a customized food to your account, you need to at least know the food's calories per serving -- and it is helpful to have nutrient information as well.

When adding a new food, it is helpful to have a Nutrition Facts label from a processed food. If you need to add an unprocessed food to your list of **Favorites**, you can research its nutritional information yourself.

To **add** a new food:

Step 1. On the **Track Diet** menu, click the **Custom Foods** button. The **Custom Food List** page will open.

Step 2. Click the **Create New Food** button. The **Nutrition Facts** page will open.

Step 3. Enter the information available for the food.

- For **packaged foods**, you can use the Daily Value for some nutrients on the **Nutrition Facts label**. You can also use grams or micrograms if the label lists nutrients in those units.
- For **supplements**, you can use grams, micrograms, or the Daily Value, if the label lists

nutrients in those units.

Step 4. When you are done, click the **Save** button at the bottom of the window. The **Custom Food List** will open with the new food listed.

Step 5. Click on any food name to do further edits or click **Close** to return to the **Track Diet** page.

Tip: The new Food is added to your **Favorites** list automatically.

+ *Creating Recipes*

A **"Recipe"** is a group of Foods. To create a Recipe, you don't need to know any nutrient information. The nutrient information comes from the Foods that are part of the Recipe.

Important Recipes that you create do not have My Pyramid or Exchange values. Those values cannot be calculated from the foods in the recipe.

To **add** a new recipe:

Step 1. On the Track Diet menu, click the **Recipes** button. The **Recipe List** page will open.

Step 2. Click the **Create New Recipe** button. Type the recipe name, and specify a number of portions. Then click the **Save** button to be returned to the Recipe List page.

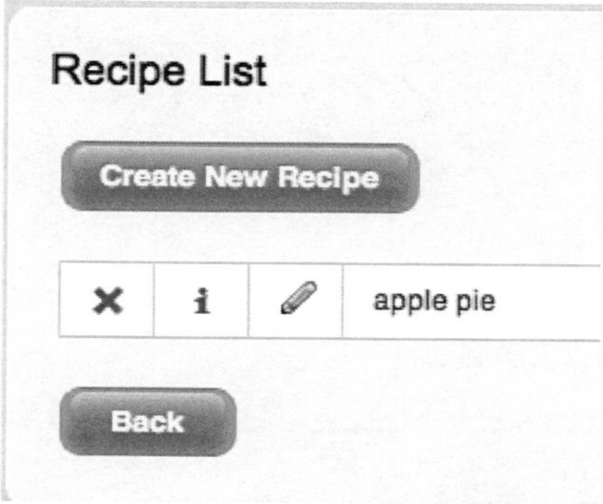

Tip: The **Number of Servings** is important. Diet Analysis Plus will divide the Recipe by this number, in order to calculate a serving size and nutrient content.

Step 3. In the **Recipe** box, click the **edit** button and then the **Add Ingredient** button to search for and choose a food to add to the recipe.

Step 4. Type the name of an ingredient and click the **Go** button to see a list of foods from which you can select.

Step 5. Click the **name** of the food and the **Add/Edit Ingredient** dialog will open.

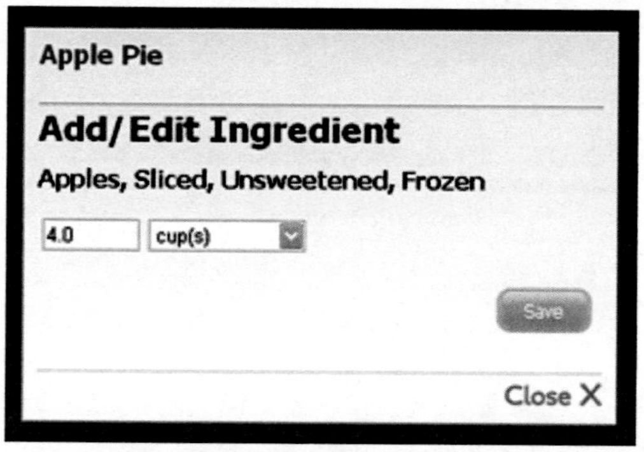

Step 6. Type in the amount to add to the recipe, and check the units of measurement in the drop-down menu. When you are finished, click the **Save** button to be returned to the **Recipe** page.

Step 7. Click the **Add Ingredient** button to add additional food items until the ingredient list for that recipe is complete.

Step 8. When you are finished adding food items to the recipe, click the **Back** button to be returned to the **Recipe List** page.

Step 8. Click the **Back** button to be returned to the **Track Diet** page.

Tip: The new Recipe is automatically added to your **Favorites** list.

Tracking Your Activities

You can track your activities by recording everything you do for 24 hours each day. Record everything, including activities that don't seem like "exercise," such as working on a computer or cooking. The calculation of calories burned will be more accurate if the activity list is as complete as possible.

Any time that is not accounted for in a 24 hour period will be calculated as "resting."

Getting Started Tracking Your Activities

Step 1. Click the **Track Activity** tab on the main menu.

Step 2. Click the calendar icon next to the **Select a Date** field to open a drop-down **calendar**.

Step 3. Click a date to select it. The date you select will be highlighted.

Step 4. Use the arrows at the top to change the month or year.

Adding an Activity

Step 1. In the **Find Activities** box, type the name of an activity and click the **Go** button.

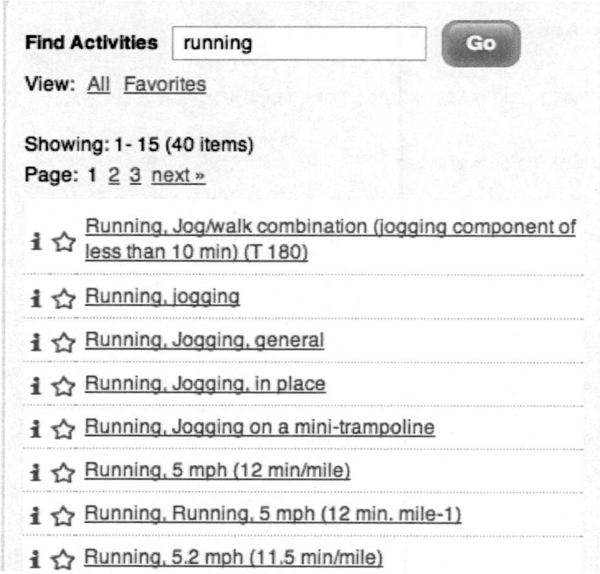

In the activity list that appears you will see a list of activities that contain the word or words you typed.

Search Tips:

- If you can't find an activity, try different or partial words. For example, instead of "bicycling," try "cycl."
- To browse the entire list of activities, click the **All** link.
- To scroll through a long list, use the page numbers.

Step 2. Select the activity that is the best match.

Tip: To the left of each item there is an **information** button. Click the **i** to see details about the energy burned by an activity, or MET value.

Step 3. Click a **name** to choose that activity. The **Activity Duration** window will open.

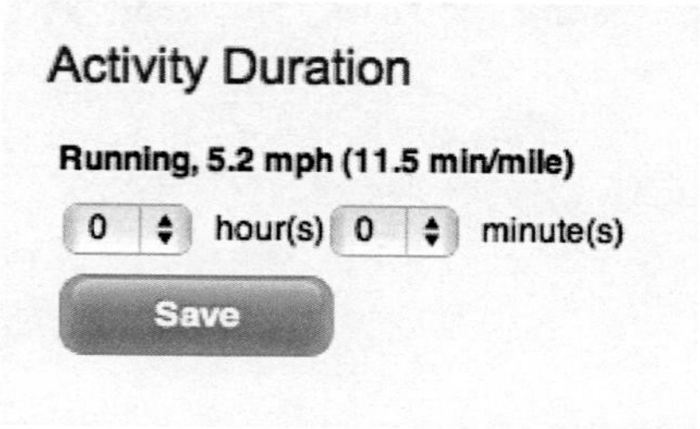

Step 4. Type in the amount of time you spent on the activity, then click **Save.**

Step 5. Go back to Step 1, and Search for the next activity. Continue with these steps until you have listed everything you did for 24 hours each day. Be as thorough as possible for best results.

Editing Activities

To edit the **Activity Duration**:

Step 1. Click the **Edit** icon to choose an activity in the My Activities window. The **Activity Duration** window will open.

Step 2. Make the necessary changes and click **Save**.

Step 3. Click the ✖ **button** to the left of the activity name to delete it.

Viewing Reports

Diet Analysis Plus allows you to generate a variety of reports showing analyses of the diet and activity information you have recorded for a profile. These reports can be part of an assignment and they are necessary when you are working with some of the Labs. You can easily print any of the reports for a profile in either PDF or RTF formats. These can then be printed, saved to disk, or emailed.

Use the **Reports** page to examine report information and to observe how reports are affected by changing variables such as dates viewed and meals included.

Getting Started Viewing Reports

Click the **Reports** tab on the main menu to open the **Reports** page. All available reports are shown under **Nutrients**, **Spreadsheets**, and **Advanced**.

Each report includes options for setting parameters that define the data to include. Below is an example of the **Macronutrient Ranges** report showing some of the report options.

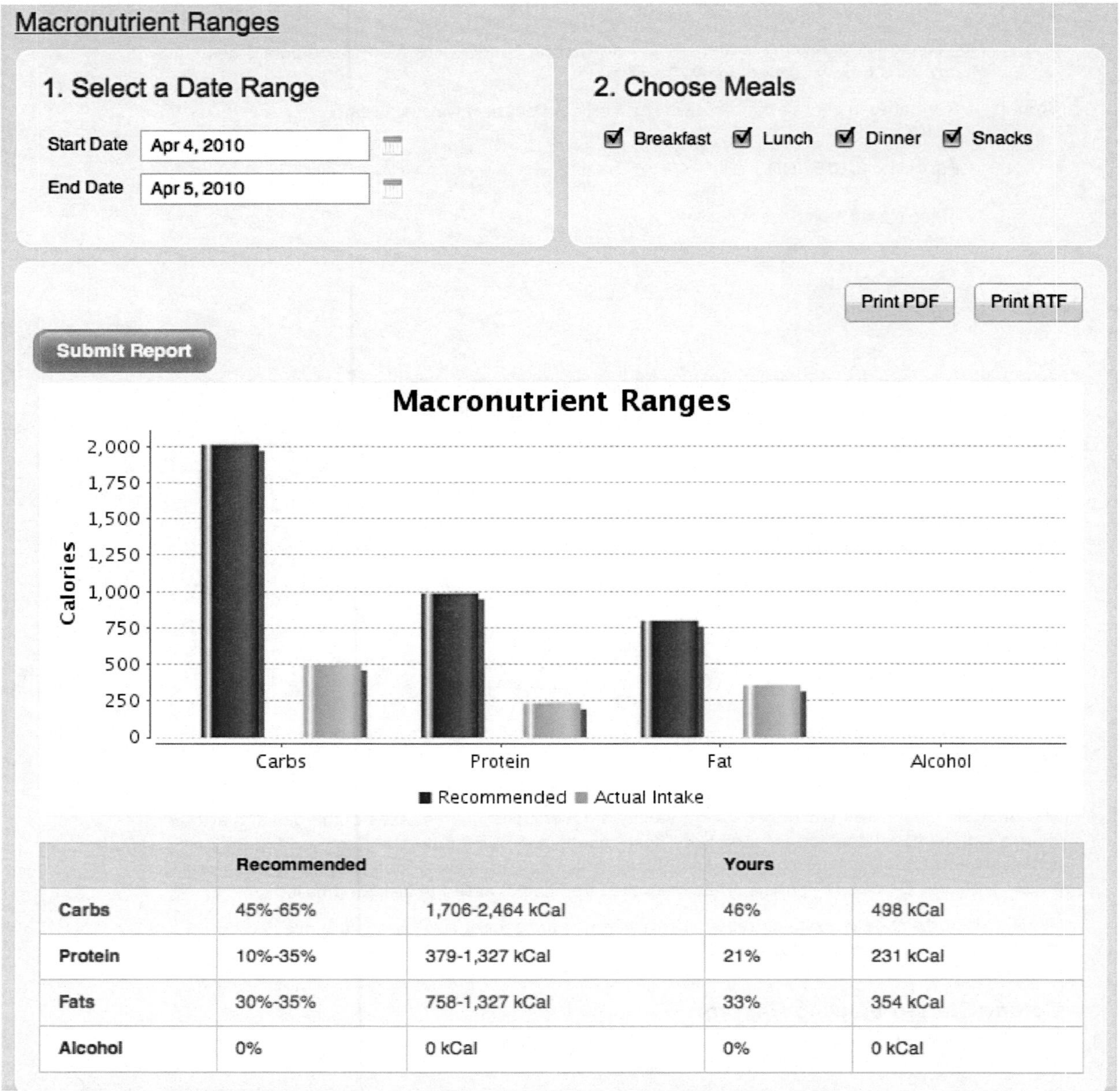

	Recommended		Yours	
Carbs	45%-65%	1,706-2,464 kCal	46%	498 kCal
Protein	10%-35%	379-1,327 kCal	21%	231 kCal
Fats	30%-35%	758-1,327 kCal	33%	354 kCal
Alcohol	0%	0 kCal	0%	0 kCal

Choosing Dates

To create a report, you must first choose the date range that the report should include.

Tip: Some reports use information from only one day at a time. Other reports let you choose a range of dates to include in a report.

Step 1. Click the calendar icon next to the **Select a Date** field to open a drop-down **calendar**.

If you have listed foods or activities for a day, it is highlighted.

Step 2. Click a date to select it.

Step 3. Use the arrows at the top to change the month or year.

Step 4. For many reports you will need to repeat this process for both the **Start Date** and **End Date**.

Tip: Generating reports may be slow if you select a range of dates that includes many days. If it is taking a very long time, try specifying fewer days.

Choosing Content

The **Choose Meals** pane allows you to choose the meals to include in a report.

Step 1. Click the check box to change the selections.

Step 2. In some reports, you may also be asked to choose a nutrient from a menu.

Using Reports

Diet Analysis Plus generates a variety of reports showing analyses of the diet and activity information recorded for a profile.

It is important to understand the data in the reports — what the scales measure, what conditions are being compared, and what the limitations are of the information provided.

Note: The **Custom Average** and **3-Day Average** reports are not generated on-screen and are only available as PDF's. See the section on <u>Printing Reports</u> for more details

---

IMPORTANT!: A Note about Nutrient Data

On the **Track Diet** page, you can search for foods that you ate. This search uses a database of thousands of foods, which includes information about the nutrients in the foods.

The Diet Analysis Plus database derives information from many sources. The data for brand name foods often comes directly from the companies that make them. Some of those foods have data for many nutrients. Others have data for only the nutrients the companies are required by regulation to report.

Sometimes you may see an asterisk (*) indicating that the nutrient content shown on the report may be lower than the actual nutrient content. This means that you may be getting more of a nutrient than the report shows.

Keep this in mind when reading reports, and consider the implications when drawing conclusions about actual dietary intake. Your instructor can provide more in-depth guidance on interpreting report data.

Profile DRI Goals Report

The **Profile DRI Goals** report shows the Dietary Reference Intakes (DRI) recommended for the profile. All the other reports use these DRI Goals as the baseline for comparisons with actual dietary intake.

The report shows the basic information from the profile that affects the recommendations, and then lists all the nutrients. Some nutrients have a dash next to them instead of a number. These nutrients don't have a recommendation.

You don't need to choose a date for this report, because it is based on the static personal information in the profile.

Intake vs. Goals Report

The **Intake vs. Goals** report compares actual dietary intake with the DRI goals for your profile.

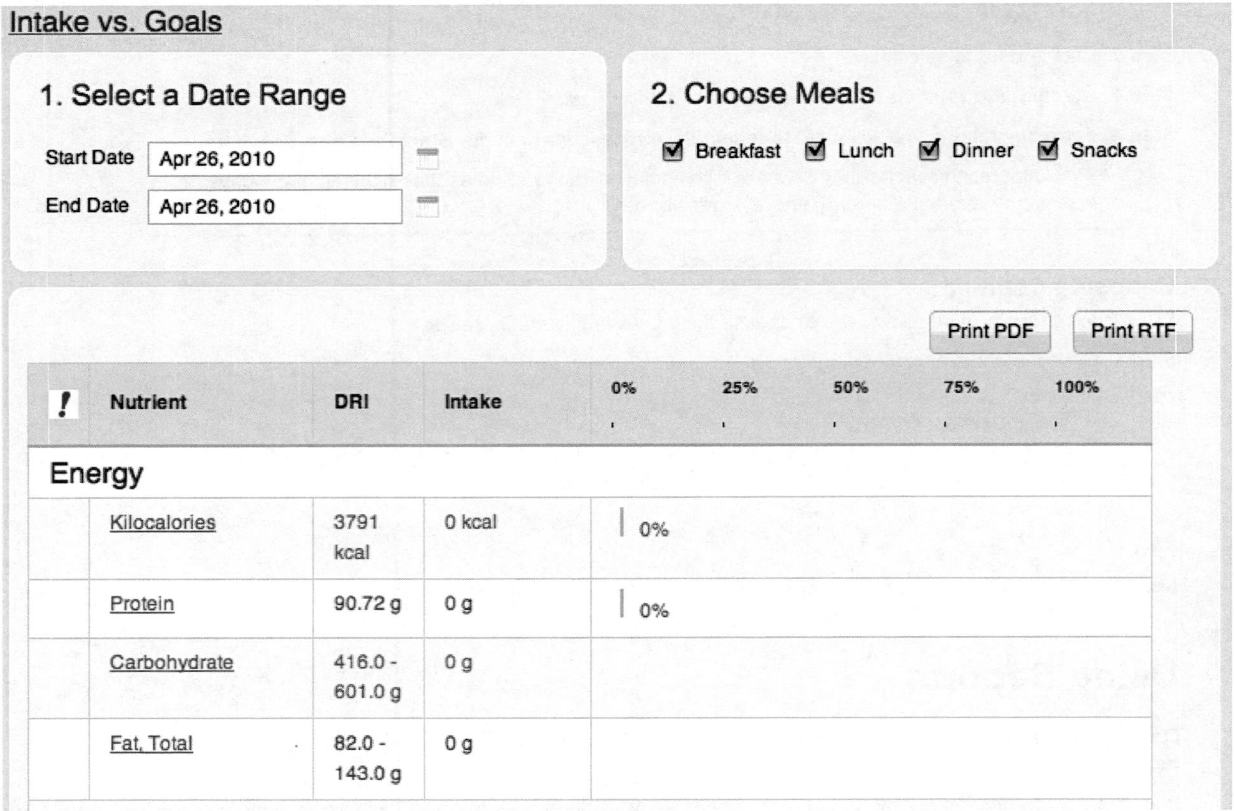

The **DRI** column shows the recommended amount of each nutrient you should have consumed based on the data in the profile. The **Intake** column shows the amount of each nutrient that you actually consumed. The graph shows the **percentage of the recommended DRI you consumed**of each nutrient.

To create an **Intake vs. Goals** report:

Step 1. Select a **Start Date** and **End Date** by using the **Calendar**.

Step 2. Use the check boxes under **Choose Meals** to select the meals to include. This limits the comparison by meal.

Step 3. When you are finished making selections, click **Print PDF** or **Print RTF**.

Macronutrient Ranges Report

The calories you eat come from the three **macronutrients: protein**, **carbohydrates**, and **fat**. (Calories can also come from alcohol, which is not a nutrient.)

The Dietary Reference Intakes (DRI) for these energy yielding nutrients are called the **Acceptable Macronutrient Distribution Ranges (AMDR)**. These ranges are the minimum and maximum percentages of each macronutrient that should comprise total caloric intake. Those percentages are:

- 45 to 65 percent from carbohydrate
- 10 to 35 percent from protein

- 20 to 35 percent from fat

The **Macronutrient Ranges** report compares the recommended percentage ranges with your actual dietary intake.

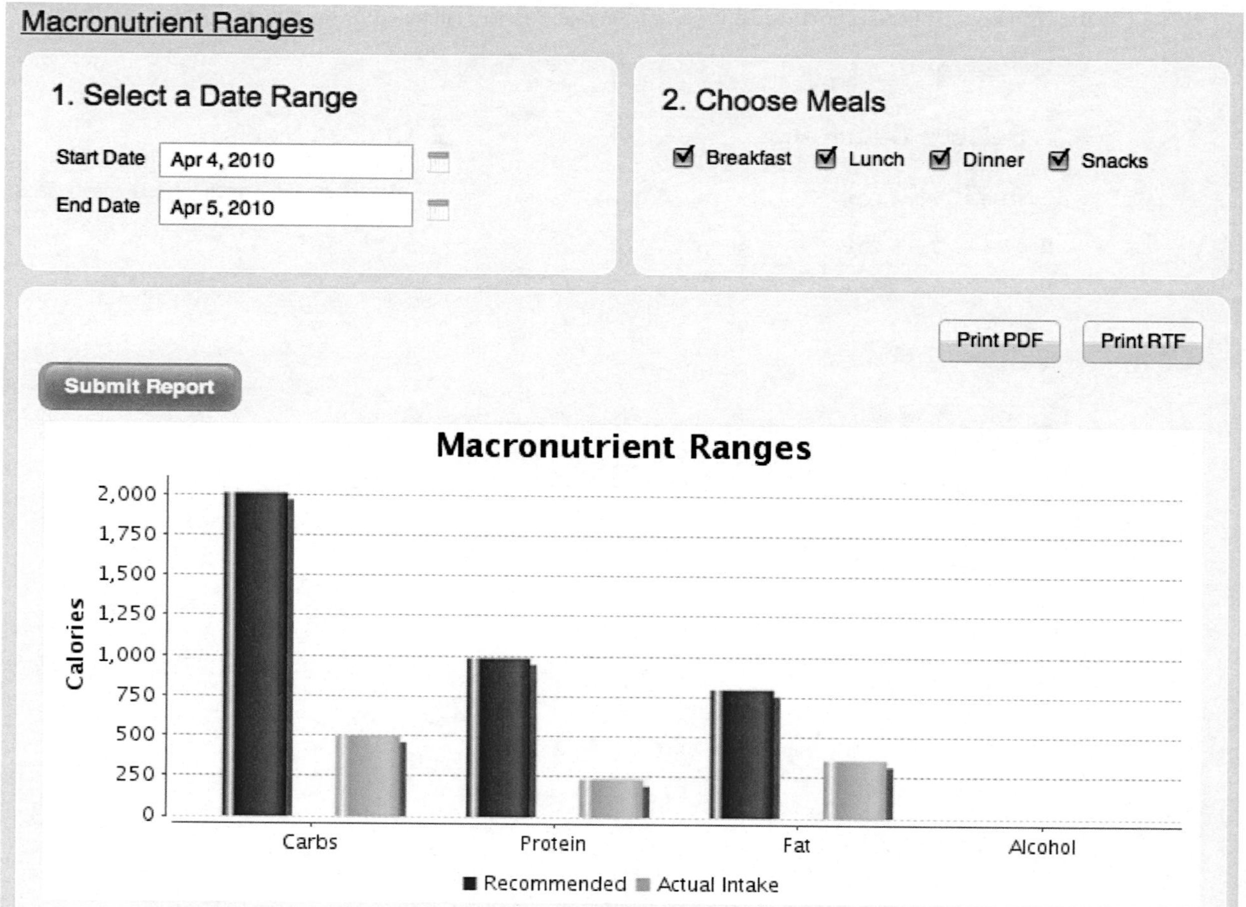

	Recommended		Yours	
Carbs	45%-65%	1,706-2,464 kCal	46%	498 kCal
Protein	10%-35%	379-1,327 kCal	21%	231 kCal
Fats	30%-35%	758-1,327 kCal	33%	354 kCal
Alcohol	0%	0 kCal	0%	0 kCal

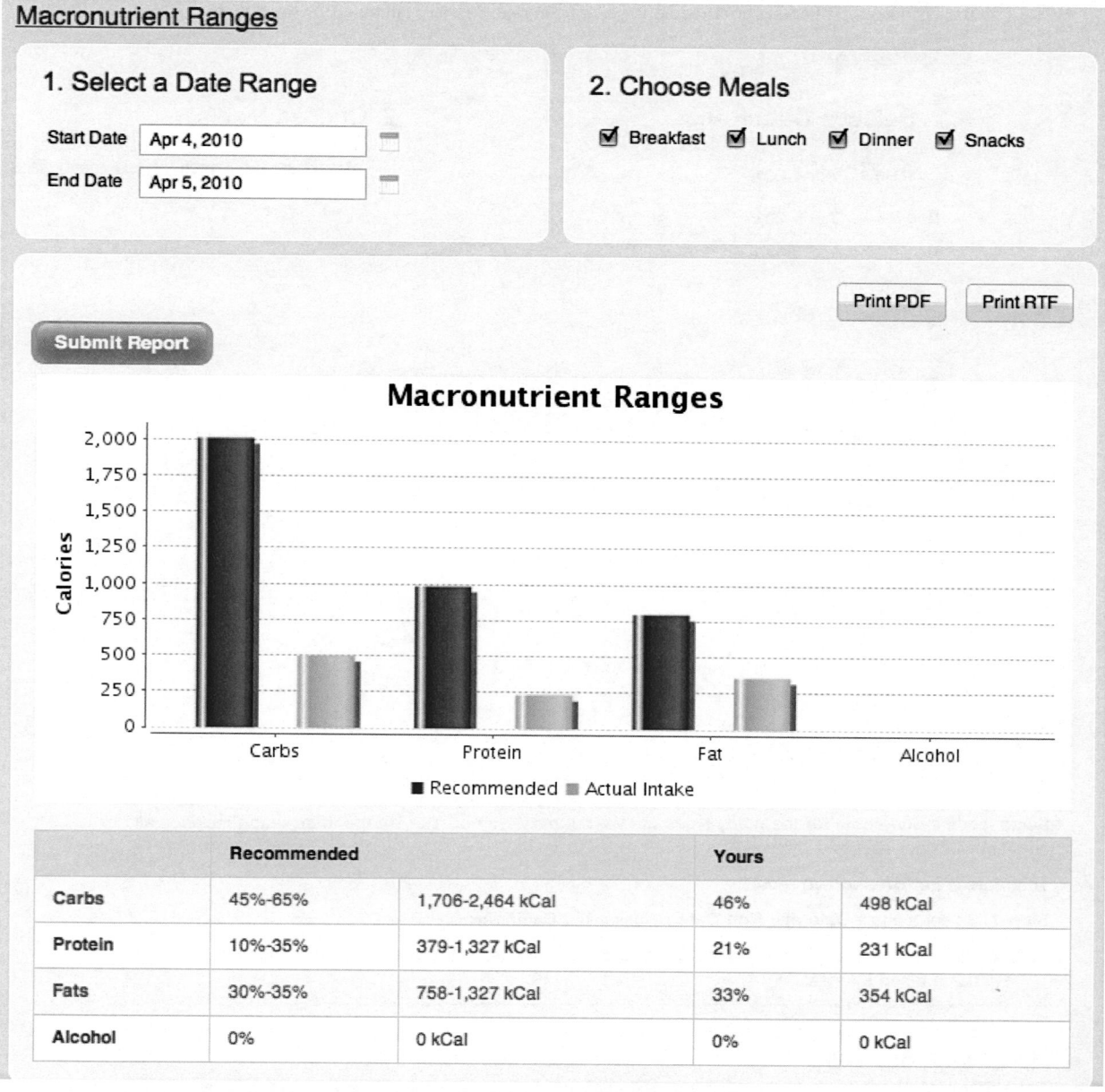

The first graph shows the **Recommended** percentages for each macronutrient. The second graph shows **Actual** caloric **(kcal)** intake. It also shows calories from **alcohol**.

The table shows the recommended percentages and calories or carbohydrates, protein, and fat, and the percentages and calories that you ate.

Note that it is important to consider the total number of calories consumed as well as the percentage. If you eat twice as many calories as is recommended, with 35% of those calories coming from fat, you will have eaten twice as much fat as the recommended maximum.

To create a **Macronutrient Ranges** report:

Step 1. Select a **Start Date** and **End Date** by using the **Calendar**.

Step 2. Use the check boxes under **Choose Meals** to select the meals to include. This limits the comparison by meal.

Fat Breakdown Report

There is no DRI recommendation for many fats, but nutritionists and other standards bodies often describe recommendations for different fats as percentages of total caloric intake.

The **Fat Breakdown** report shows what percentage of **total calories** is contributed by each type of fat.

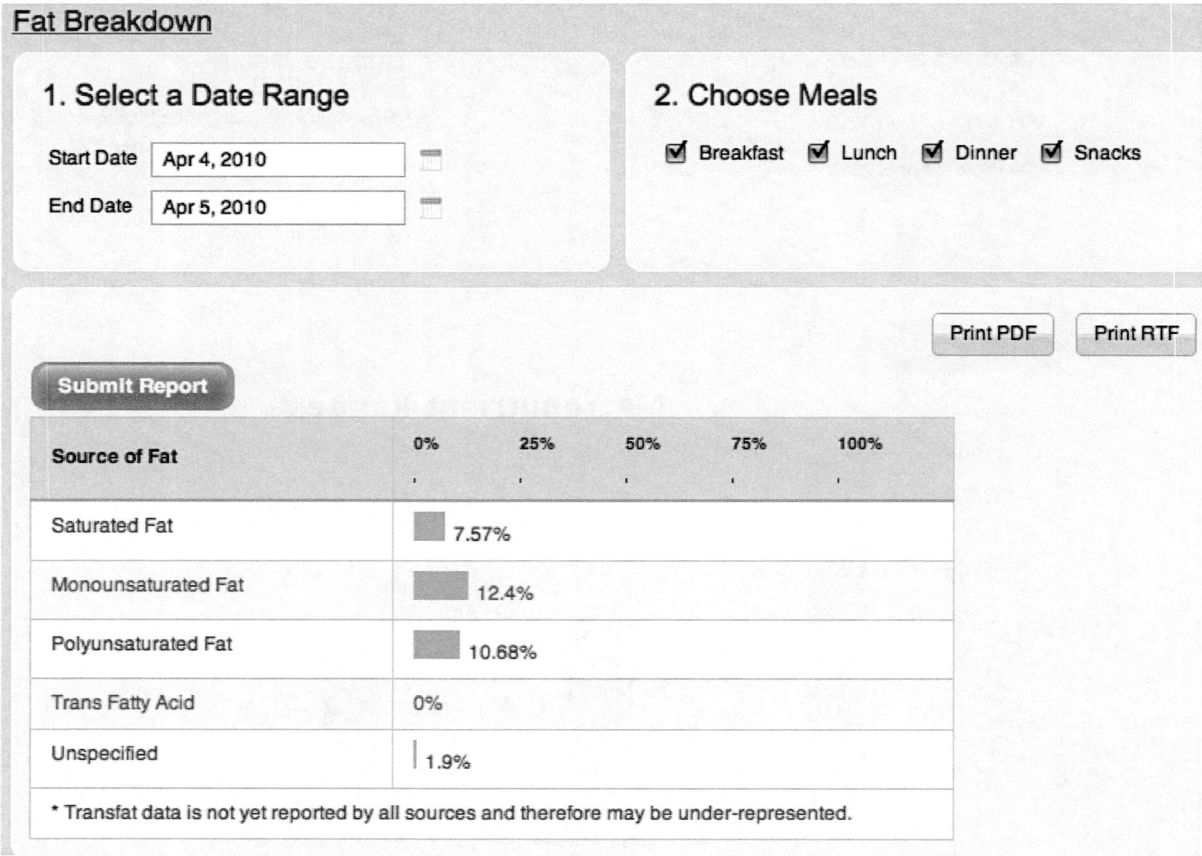

Some foods include data for the many types of fats, but most do not. The **Unspecified** graph includes all fats that were not specified by type.

To create a **Fat Breakdown** report:

Step 1. Select a **Start Date** and **End Date** by using the **Calendar**.

Step 2. Use the check boxes under **Choose Meals** to select the meals to include. This limits the comparison by meal.

MyPyramid Analysis

The USDA's MyPyramid shows a recommended number of daily servings for the food group categories of Grains, Vegetables, Fruits, Milk, Meats and Beans, and Discretionary Calories. It also includes recommendations for eating whole grains, various types of vegetables, and for oils.

For detailed explanations of the MyPyramid recommendations, click the links to specific items on the report screen, or click the MyPyramid logo to go to the www.mypyramid.gov home page.

The **MyPyramid Analysis** report graphically illustrates the comparison between the recommendations of MyPyramid and the number of servings from each food groups you ate.

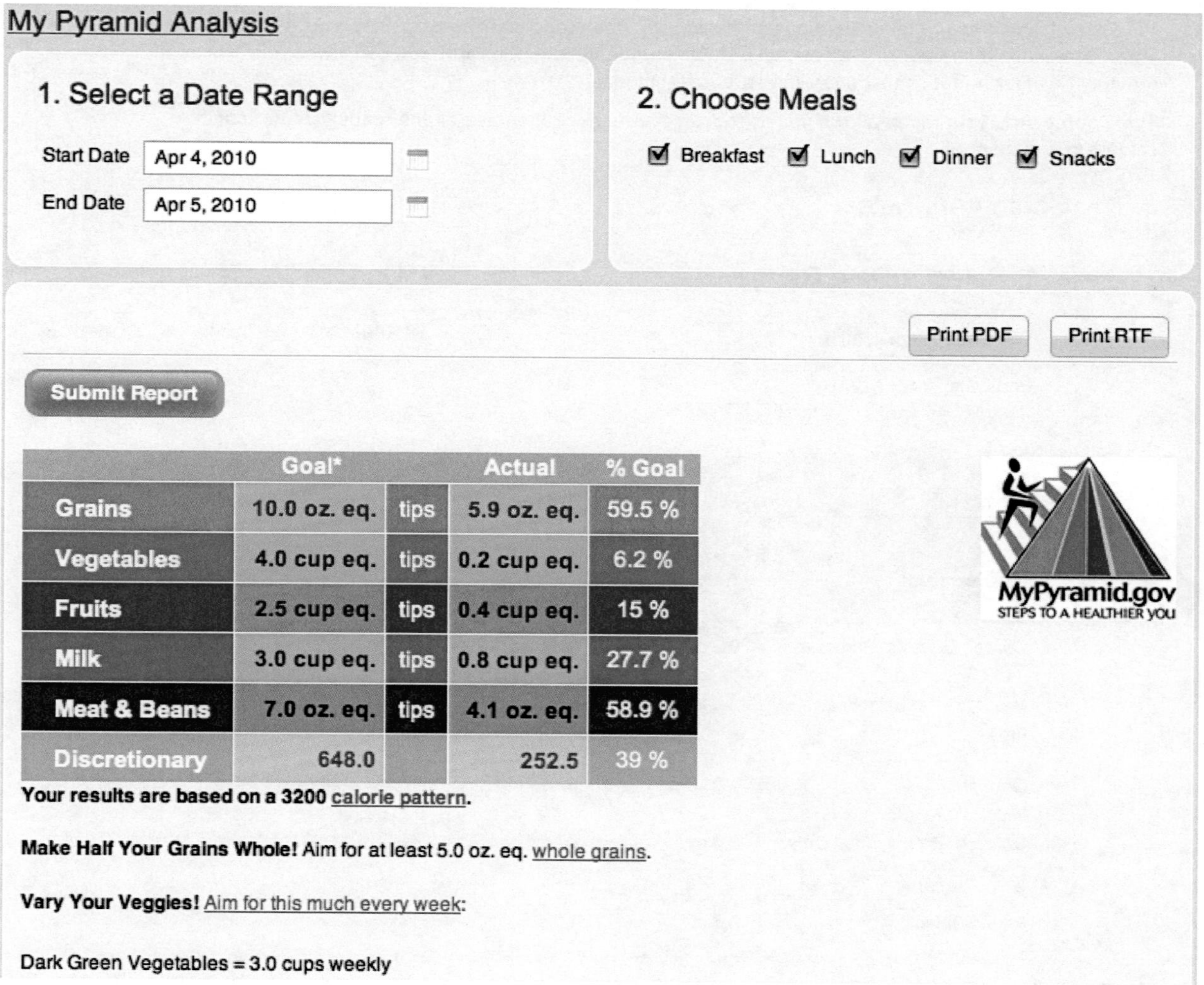

My Pyramid Analysis

1. Select a Date Range

Start Date Apr 4, 2010

End Date Apr 5, 2010

2. Choose Meals

☑ Breakfast ☑ Lunch ☑ Dinner ☑ Snacks

Print PDF | Print RTF

Submit Report

	Goal*		Actual	% Goal
Grains	10.0 oz. eq.	tips	5.9 oz. eq.	59.5 %
Vegetables	4.0 cup eq.	tips	0.2 cup eq.	6.2 %
Fruits	2.5 cup eq.	tips	0.4 cup eq.	15 %
Milk	3.0 cup eq.	tips	0.8 cup eq.	27.7 %
Meat & Beans	7.0 oz. eq.	tips	4.1 oz. eq.	58.9 %
Discretionary	648.0		252.5	39 %

Your results are based on a 3200 calorie pattern.

Make Half Your Grains Whole! Aim for at least 5.0 oz. eq. whole grains.

Vary Your Veggies! Aim for this much every week:

Dark Green Vegetables = 3.0 cups weekly

Important! Foods added using the **Create a Food** and **Create a Recipe** feature of Diet Analysis Plus do not have MyPyramid values, and are not included in the MyPyramid Analysis report.

To create a **Food Pyramid Analysis** report:

Step 1. Select a **Start Date** and **End Date** by using the **Calendar**.

Step 2. Use the check boxes under **Choose Meals** to select the meals to include. This limits the comparison by meal.

Intake Spreadsheet Report

The **Intake Spreadsheet** report shows all the foods eaten on the selected date, and the amount of each nutrient contained in each food.

This report can be saved to a file that can be imported into a spreadsheet program.

To create an **Intake Spreadsheet** report:

Step 1. Use the **Calendar** to select a date for the **Select a Date** field.

Step 2. Use the check boxes under **Choose Meals** to select the meals to include. This limits the comparison by meal.

(*Optional*) Click the **Export Excel File** button to export the report as a text file that can be opened in a spreadsheet program.

Source Analysis Report

You can influence your intake of various nutrients using the foods you choose to eat. You can raise your intake of a nutrient by choosing foods that are rich in that nutrient. Similarly, you can lower your intake of a nutrient by limiting foods that contain a lot of that nutrient.

The **Source Analysis** report allows you to choose one nutrient, then see all the foods you ate that contain that nutrient.

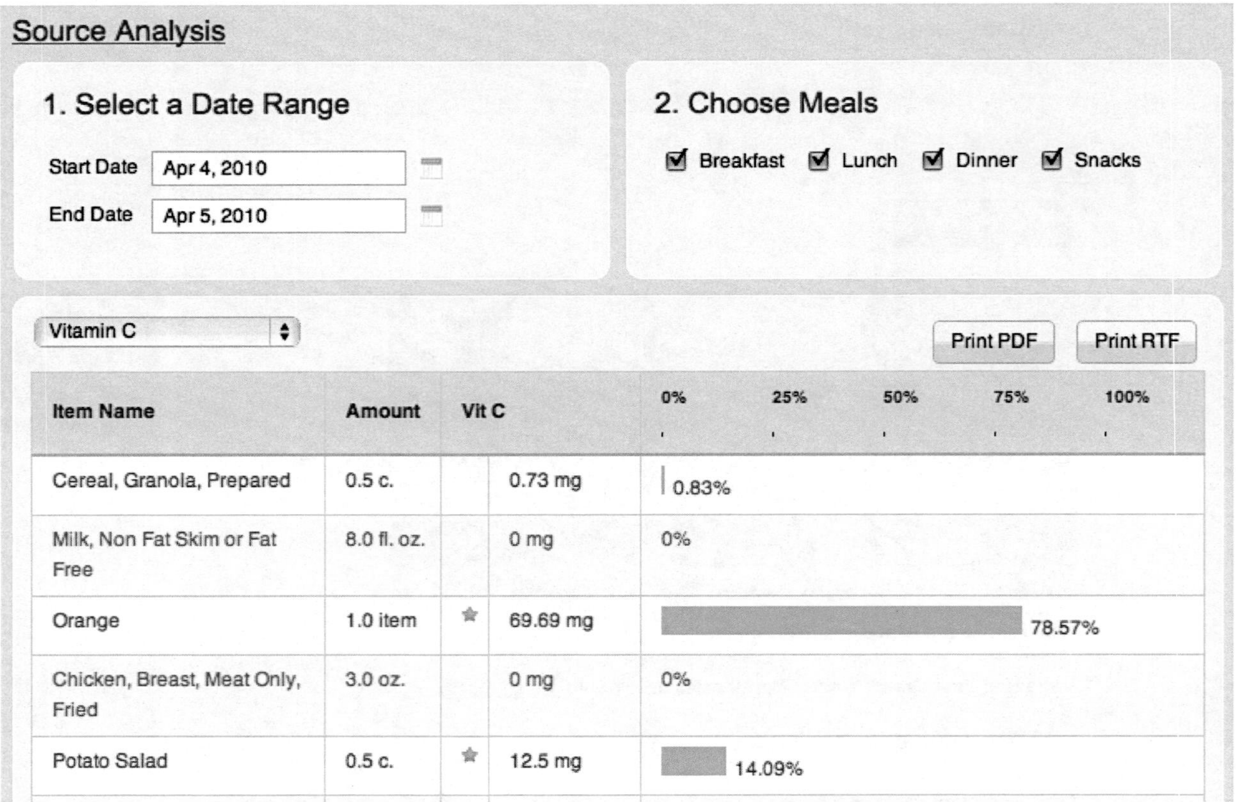

The **Amount** column shows how much you ate of each food.

To see the nutrient amounts in each food you ate, select the nutrient from the **Nutrient** drop-down list (in this example, "Vitamin C)."

The **graph** shows how each food contributes to the total intake of the nutrient selected. The graph does *not* show the percentage of the DRI, or the percentage of the Daily Value.

Note: "100%" on the graph is equal to "all that you actually ate." It does not express anything about how much you "should" have eaten.

You may see stars next to some of the foods on the list. A **orange star** marks foods containing more than 20% of the Daily Value for the nutrient. A **silver star** marks foods containing 10% to 20% of the Daily Value.

Tip: Daily Value is not the same as the DRI of a nutrient, which is based on a profile. Your instructor can provide guidance on the definition and uses of Daily Values.

To create a **Source Analysis** report:

Step 1. Select a day using the **Calendar**.

Step 2. Use the check boxes under **Choose Meals** to select the meals to include. This limits the comparison by meal.

Step 3. From the **Nutrient** drop-down list, choose the nutrient you want to analyse.

Energy Balance Report

Calories are different from nutrients in an important way. Their consumption is affected not just by food eaten, but also by energy expended in activity.

The **Energy Balance** report shows net caloric intake over time. It lets you see whether you retained more calories than you needed, fewer calories than you needed, or used about the same amount as you ate.

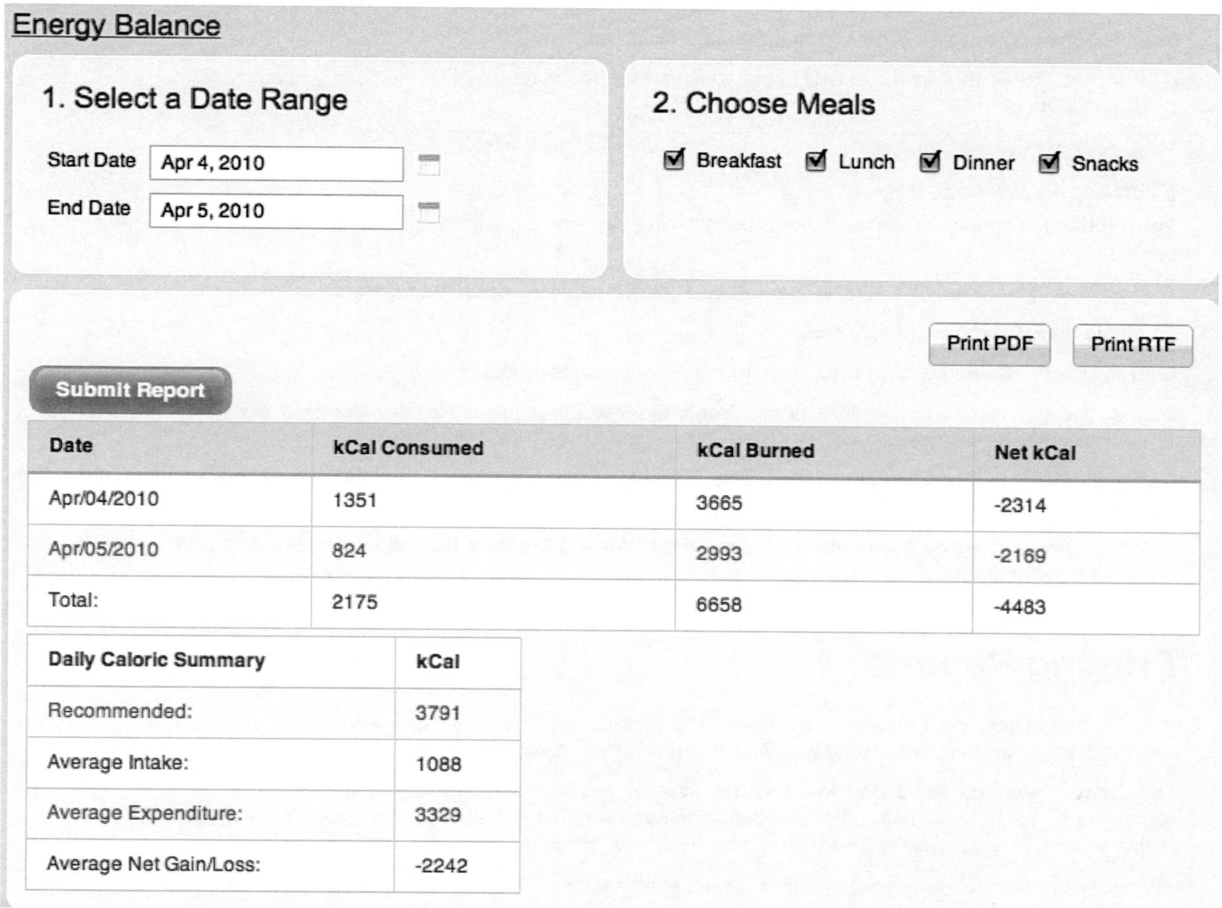

Date	kCal Consumed	kCal Burned	Net kCal
Apr/04/2010	1351	3665	-2314
Apr/05/2010	824	2993	-2169
Total:	2175	6658	-4483

Daily Caloric Summary	kCal
Recommended:	3791
Average Intake:	1088
Average Expenditure:	3329
Average Net Gain/Loss:	-2242

The **kcal Consumed** column shows how many calories you ate. The **kcal Burned** column shows how many calories you used performing activities and calories burned at rest. The calories burned at rest are calculated by multiplying the amount of time (in hours) left in the day not performing other activities by your profile weight in kilograms and again by 1.1(an estimated MET for resting). The **Net Calories** column shows whether you ate more than you burned, or burned more than you ate. The **Net kcal** column shows how many calories you needed per day according to the profile.

If the **total Net kcal** is a positive number, you ate more calories than you used. This might result in weight gain.

If the **total Net kcal** is a negative number, then you used more than you ate. This might result in weight loss.

Careful! Energy metabolism is complex and varies for each person. Your instructor can provide more detailed information about caloric expenditure and weight change.

To create an **Energy Balance Analysis** report:

Step 1. Select a **Start Date** and **End Date** by using the **Calendar**.

Step 2. Use the check boxes under **Choose Meals** to select the meals to include. This limits the comparison by meal.

Activities Spreadsheet

The **Activities Spreadsheet** report shows all the activities recorded for a day. It lists the rate at which calories were burned by the activity (in calories per kilogram per hour), the duration of the activity, and the total number of calories burned.

The report will account for all 24 hours in the day. If you did not record 24 hours of activity, Diet Analysis Plus will calculate unaccounted for time as having been spent resting.

This report can be saved to a text file that can be imported into a spreadsheet program.

To create an **Activities Spreadsheet** report:

Use the **Calendar** to select a date for the **Select a Date** field.

(*Optional*) Click the **Export Excel File** button to export the report as a text file that can be opened in a spreadsheet program.

Exchanges Spreadsheet

The **Exchanges** spreadsheet shows the diabetic exchanges, and the values contained in each food eaten.

This report can be saved to a text file that can be imported into a spreadsheet program.

To create an **Exchanges** Spreadsheet:

Step 1. Use the **Calendar** to select a date for the **Select a Date** field.

Step 2. Use the check boxes under **Choose Meals** to select the meals to include. This limits the comparison by meal.

(*Optional*) Click the **Export Excel File** button to export the report as a text file that can be opened in a spreadsheet program.

> *Important!* Foods added using the **Create a Food** and **Create a Recipe** feature of Diet Analysis Plus do not have exchange values.

Printing Reports

In addition to generating a browser-based version of reports, Diet Analysis Plus allows you to print all your reports by selecting either the **Print PDF** or **Print RTF** buttons.

The **Custom Average** and **3-Day Average Reports** are unique in that they are not generated electronically; the full reports are only available when you send an e-mail or print a copy. The instructions in this section describe how to print the reports which have a more specialized output process.

Tip: Diet Analysis Plus uses your system's default printer settings. You can't change things like page size or orientation from inside Diet Analysis Plus.

Printing the Intake Spreadsheet

The **Intake Spreadsheet** lists all the foods you have eaten and shows all the nutrient details for each food. For this type of report Diet Analysis Plus prints a spreadsheet for each day individually.

Step 1. Before generating your report, verify you have the correct information selected in the **Select a Date** and **Choose Meals** options. Make any necessary changes.

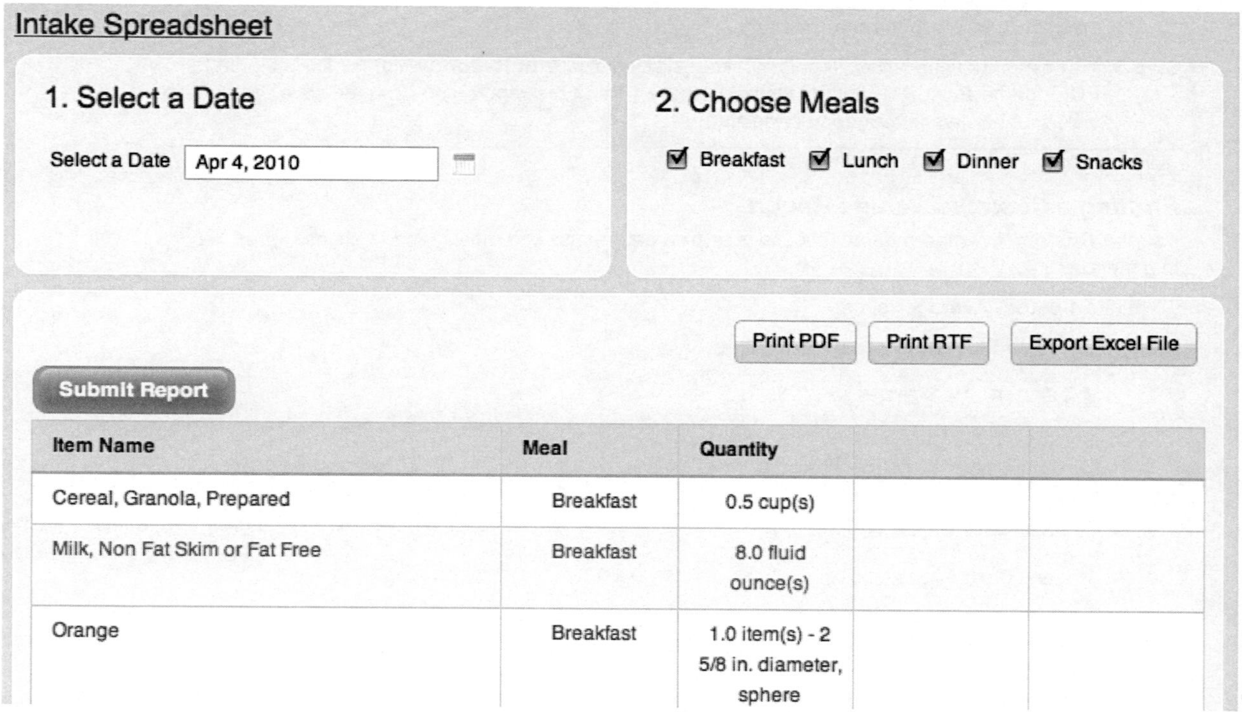

Step 2. Click **Print PDF** to open a PDF preview in a separate browser window. Click **Print RTF** to download an .rtf file to disk. Click **Export Excel File** to download a .csv file to disk.

Step 3. Once downloaded, you can email the file(s).

Printing the 3-Day Average Report

Most professors assign a 3-day Average assignment.

Diet Analysis Plus selects the first day on which you entered any food, and will average that day followed by the next two days with data. You can change the any day of the average by clicking on the Calendar next to the day you would like to change.

To print the 3-Day Average Report:

Step 1. Choose three dates from the **Calendar** icons to the right of the three date text boxes.

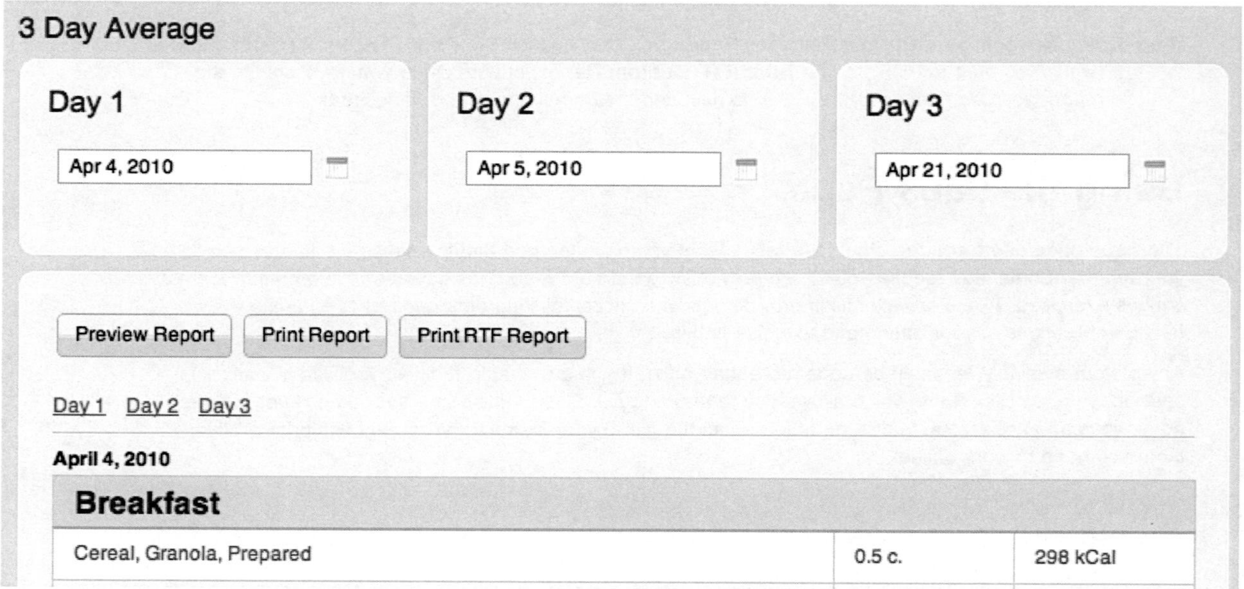

Step 2. Click the **Preview Report** button to preview your 3 Day Report. Press the **Back** button on your

browser to return to the main menu.

Step 3. You can print your 3 Day Report by either clicking the **Print Report** button (which will print to PDF), or the **Print RTF** button. Once you have printed the report, you can then save the report to disk and then email the report (if desired).

Printing a Custom Average Report

Use the **Custom Average** menu to choose your own date range and meals, and to choose which reports to generate.

To print a **Custom Average** report:

Step 1. Choose a **Start Date** and an **End Date**.

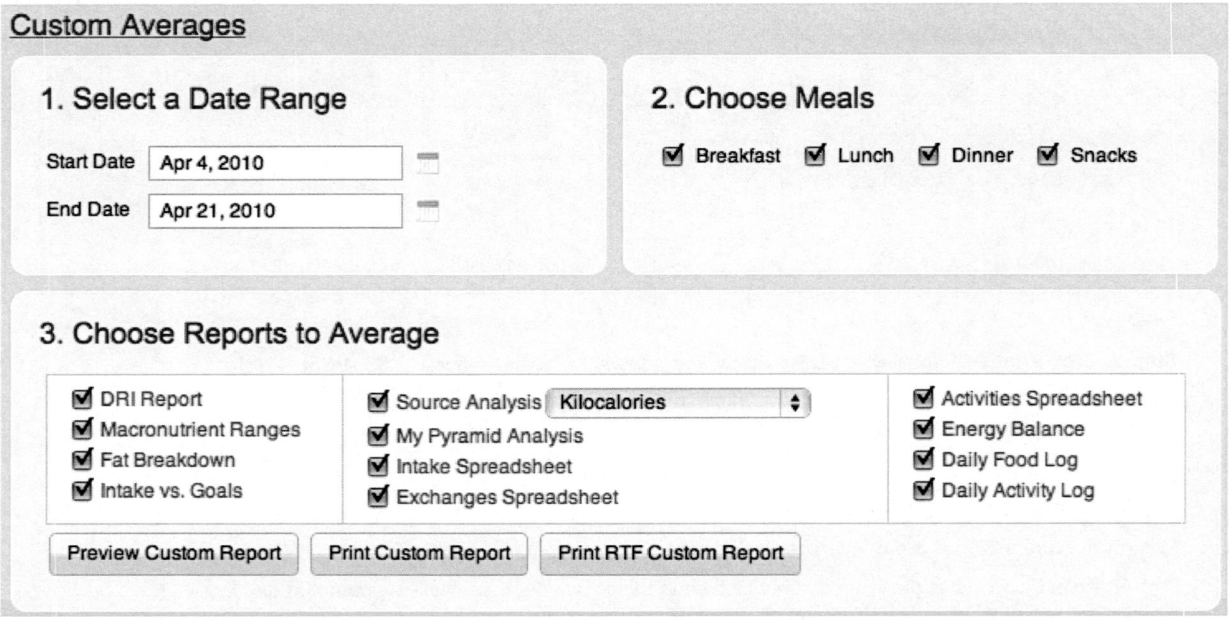

Step 2. Click the check boxes under **Choose Meals** to choose which meals to average.

Step 3. Click the check boxes under **Choose Reports to Average** to choose which reports to print.

Step 4. Click the **Preview Custom Report** button. A **Print Preview** window will open. Press the **Back** button on your browser to return to the main menu.

Step 5. You can print your Custom Averages Report by either clicking the **Print Custom Report** button (which will print to PDF), or the **Print RTF Custom Report** button. Once you have printed the report, you can then save the report to disk and then email the report (if desired).

Using the Labs Page

The **Labs** page offers you the chance to refine what you have learned and to assess the information in your reports. Some labs require you to use previously generated reports as sources of information for various exercises. These lab exercises provide you with critical thinking opportunities to evaluate the health implications of your eating and exercise habits.

Any of your available labs can be done more than once. If you are unable to finish your lab in one session, you can click **Save** which allows you to return to DA Plus to finish at a later time. After you are done, you can print a copy to disk and then email the file to your instructor. You can also print a blank lab if you need to work offline.

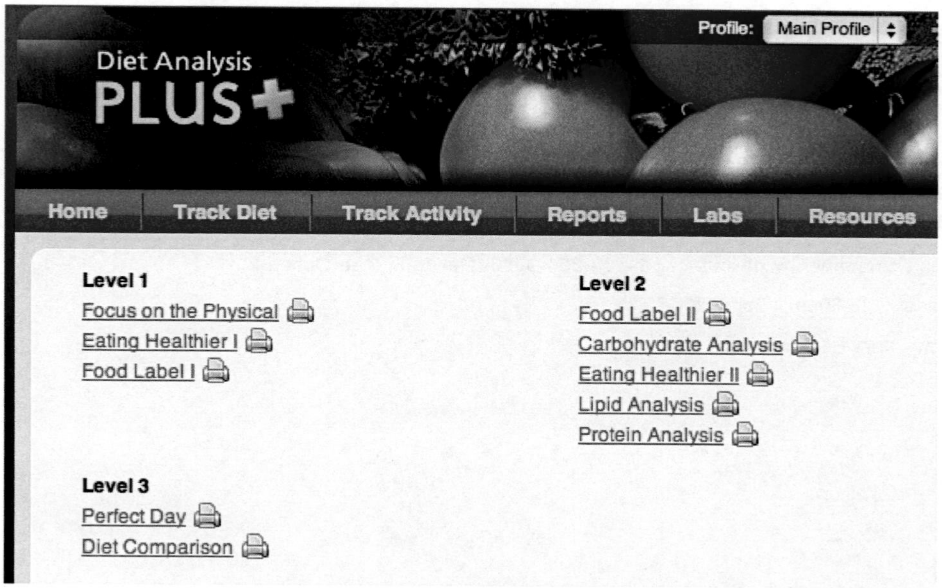

Lab Basics

Step 1. Click the **Labs** tab to see the labs that have been assigned to you.

Step 2. Click a lab **name** to open that activity.

Step 3. Review the **Objective** and **Lab Resources** before beginning.

Step 4. Enter your answers as required.

Step 5. Click **Next** to continue to the next page, or click **Save** to resume the lab at later time.

Step 6. When you have completed the lab, click the **Save and Submit** button to submit the lab results to your instructor.

Printing Labs

To print your lab:

Step 1. Click the **Labs** tab to see the labs that have been assigned to you.

Step 2. Click the print icon next to the lab you would like to print. A PDF version of your lab will open in a new window.

Step 3. Print or save the PDF from your browser toolbar.

E-mailing Labs

Step 1. Click the **Labs** tab to see the labs that have been assigned to you.

Step 2. Follow the steps in **Printing Labs** above to print a copy of your lab in PDF format. Use your favorite email program to attach this PDF file to your email.

Technical Support

If you have trouble using or accessing DA Plus, you can contact support to get further assistance. From the **Cengage Learning Technical Support** site you can get information on how to contact our experts by phone, e-mail form, or online chat.

To access **Online Chat/Customer Service Support** , you can either click the **Technical Support** link at the bottom of any page in DA Plus or direct your browser to www.cengage.com/support/.

Once you are at the Cengage Learning support site, you can follow these steps:

Step 1. Select **Diet Analysis Plus** from the **Student** drop-down menu.

Step 2. Click **Go** and the **Technical Support** page will open.

Step 3. Choose from any of the following technical support resources:

- Select **Chat Online** to chat with a technical support representative.
- Select **Submit your questions under Contact Us to access the online e-mail form.**

Note: Chat and e-mail support are available 24/7. When using an e-mail form, support requests are usually responded to within 48 hours.

To contact technical support personnel by **phone**, call **1-800-354-9706** (Option 5, then Option 2).

- Monday - Thursday: 8:30am - 9pm EST
- Friday: 8:30am - 6pm EST